Handbook on
Syntheses of Amino Acids

Handbook on
SYNTHESES of AMINO ACIDS

General Routes for the Syntheses of Amino Acids

MARK A. BLASKOVICH

OXFORD
UNIVERSITY PRESS

2010

OXFORD
UNIVERSITY PRESS

Oxford University Press, Inc., publishes works that further
Oxford University's objective of excellence
in research, scholarship, and education.

Oxford New York
Auckland   Cape Town   Dar es Salaam   Hong Kong   Karachi
Kuala Lumpur   Madrid   Melbourne   Mexico City   Nairobi
New Delhi   Shanghai   Taipei   Toronto

With offices in
Argentina   Austria   Brazil   Chile   Czech Republic   France   Greece
Guatemala   Hungary   Italy   Japan   Poland   Portugal   Singapore
South Korea   Switzerland   Thailand   Turkey   Ukraine   Vietnam

Copyright © 2010 by American Chemical Society

Developed and distributed in partnership by
the American Chemical Society and Oxford University Press

Published by Oxford University Press, Inc.
198 Madison Avenue, New York, New York 10016
www.oup.com

Oxford is a registered trademark of Oxford University Press

Library of Congress Cataloging-in-Publication Data
Blaskovich, Mark A.
Handbook on syntheses of amino acids : general routes to amino acids /
Mark A. Blaskovich.
    p. cm.
Includes bibliographical references and index.
ISBN 978-0-8412-7219-4   1. Amino acids—Synthesis.   I. Title.
QD431.25.S93B53 2010
572'.65—dc22   2009043469

9   8   7   6   5   4   3   2   1

Printed in the United States of America
on acid-free paper

Preface

The field of amino acid synthesis is vast, and any attempt to summarize it as in this book must of necessity skim the surface. Many of the topics contained in this monograph are worthy of a book in their own regard. The goal of this book is to present a comprehensive overview of the current state of amino acid synthesis, paying due respect to the many important historical amino acid syntheses of the past 150 years. Also included are overviews of closely related areas such as amino acid resolutions, *N*-protection and *N*-alkylation, and syntheses of β-, γ-, and δ-amino acids. It is hoped that this monograph will provide enough leading references that a topic can be examined in much greater depth if desired, while allowing a more casual practitioner of the art to gain an understanding of what has been achieved in the field.

Unfortunately, what this book is unable to provide is a description of the best method for synthesizing any particular amino acid of interest. Each of the general methods has its advantages and disadvantages, and some are more suited to certain types of amino acids than others. In many cases a simple racemic synthesis followed by enzymatic resolution may provide a faster, if less elegant, approach to the desired target. There are many additional methods that are specific for the preparation of either one class of amino acid or one individual amino acid. A second volume to this book is in progress, in which amino acids are grouped by general classes so as to allow more facile identification of relevant synthetic routes for specific amino acids. The need for comprehensive reviews for obtaining routes to specific amino acids is rapidly being superseded by the advent of electronic databases. However, it is hoped that readers of this monograph may be inspired to develop new and unique syntheses.

Acknowledgments

I must give my thanks to many people. First and foremost I am indebted to my wife, Celia, and children, Griffin, Quinlan, and Thea, who tolerated many years of "not now, I'm working on the book." My parents and sister provided a fertile environment to cultivate an inquisitive and tenacious (stubborn!) mind and have given great encouragement throughout my life. Dr. Gilles Lajoie introduced me to the world of amino acid synthesis, while Dr. M. Miller first suggested that I consider publishing what began as a naive attempt at a comprehensive literature review for the introduction of my thesis. A special thanks to those who spent many hours slaving over photocopiers in musty libraries to procure over 10,000 references—in particular Celia Lamond, Danielle Skropeta, and Julie Bowman—and to my colleagues at Molecumetics who reviewed much of the manuscript.

I am very grateful to the publications division of the American Chemical Society for an advance/grant which enabled this project to be initiated, to the Research School of Chemistry at the Australian National University for a visiting fellowship to write a significant proportion of this book, and to NSERC (Natural Sciences and Engineering Research Council of Canada) for financial support during most of my research education.

Last, but certainly not least, I give thanks to my many colleagues at various institutions over the years, who have provided great company, support, intellectual stimulation, and, most importantly, friendship.

Contents

Handbook on
Syntheses of Amino Acids

1 | Introduction

1.1 General Introduction

The class of compounds known as amino acids represents an enormous array of chiral molecules possessing considerable structural diversity and importance, both as biologically active substances and as substrates for the synthesis of more complex derivatives. Amino acids are ubiquitous in nature, forming the building blocks of which peptides and proteins are constructed. Their appearance in the "primordial soup" of prebiotic earth is often considered to be the origin of biological life (1) and they have been isolated from extraterrestrial sources (2–5). Whereas only 20 common α-amino acids are usually found in the peptides and proteins that form the structural basis for life, literally thousands of other structures have been isolated from a wide variety of natural sources, including many unusual amino acids from plants, sponges, bacteria, and fungi (6–10). The term "proteinogenic" is often applied to the 20 primary amino acids most commonly found in proteins, but this classification is misleading, as other amino acids, such as hydroxylysine and hydroxyproline, are also found in proteins. More accurately, the primary protein amino acids are coded for in the process of ribosomal translation of DNA via RNA, while the remaining "secondary" and "tertiary" proteinaceous amino acids arise from post-translational modifications of residues ("secondary"), including cross-linking of two amino acids ("tertiary") (11). These modifications can include acylation, phosphorylation, sulfation, glycosylation, hydroxylation, oxidation, nitration, methylation, and prenylation, as summarized in a 2005 review (12).

In contrast, the non-protein amino acids are "those amino acids which are not found in the protein main chains either for lack of a specific transfer RNA and codon triplet or because they do not arise from protein amino acids by post-translational modification" (13). Many of these non-protein amino acids are formed as secondary metabolites in bacteria, fungi, plants, or marine organisms. For example, a range of unusual α-, β-, and γ-amino acids is found in the lithistid sponges, as reviewed in 1998 (14). Various uncommon amino acids undoubtedly exist in natural sources in quantities too small to be detected; to demonstrate this possibility the nitrogenous fraction resulting from large-scale industrial processing of sugar beet was studied. N^ε-Acetyl-L-Lys, N^ε-acetyl-*allo*-δ-hydroxy-L-Lys, γ-N-lactyl-L-2,4-diaminobutyric acid, γ-N-acetyl-L-2,4-diaminobutyric acid, L-azetidine-2-carboxylate, and γ-L-glutamyl-GABA (GABA = γ-aminobutyric acid) were isolated (15).

However, even these distinctions are not clearcut. When the genetic code was unravelled in the 1960s, 61 of the 64 possible triplet codons (i.e. three nucleic acids coding for one amino acid) were divided among the 20 common amino acids, with the remaining three codons assigned as signals to terminate protein synthesis. Only Met and Trp are encoded by unique mRNA triplets, with degenerate codons available for the other amino acids. However, it has since been found that the UGA codon, originally assigned as a stop codon, can also act as a codon for the incorporation of selenocysteine (Sec), making Sec the 21st coded amino acid. A number of factors control how the codon is interpreted, including a structural element in the mRNA (16). The mechanism of Sec incorporation into proteins differs significantly from other amino acids, in that the Sec residue is biosynthesized on its unique Sec tRNA[Ser]Sec from Ser via a phosphoserine intermediate. The mammalian kinase responsible for phosphorylating the seryl moiety on the seryl-tRNA[Ser]Sec was identified by a comparative genomics approach in 2004 (17, 18). The labile selenium donor compound monoselenophosphate required to selenate the tRNA is synthesized from selenide and ATP by the enzyme selenophosphate synthetase (Sps). Two *Sps* genes from a human cancer cell have been cloned and expressed; *Sps1* appears to recycle L-Sec, while *Sps2* may be part of a selenite assimilation system (*19*). The expression of mammalian selenoproteins appears to be partly regulated by a novel mechanism, via methylation at the 2′-*O*-hydroxyl site of the ribosyl group at position 34 of the selenocysteine tRNA (*20*). Sec can also be incorporated into proteins non-specifically, via replacement of the sulfur atom during biosynthesis of Cys or Met. The biological incorporation of selenium was reviewed in 2002 (*16*), and new developments discussed in 2004 (*21, 22*).

In 2002, a pair of articles reported identification of the 22nd genetically coded amino acid, L-pyrrolysine (Pyl) (*23, 24*, summarized in *25–27*). The residue, a Lys with the ε-nitrogen acylated by a (4R,5R)-4-methyl-pyrroline-5-carboxylate group, was identified at the active site in a crystal structure of monomethylamine methyltransferase (MtmB) from the Archaeal microbe *Methanosarcina barkeri* (*24*). The identity of the methyl substituent at the 4-position could not be conclusively determined from the electon density map in the initial reports (proposed as either R = Me, OH or NH$_2$), but mass spectrometric observations reported in 2004 indicated the true structure (*28*). A chemical synthesis was reported the same year (*29*). The location of the pyrrolysine residue corresponds to a UAG stop codon in the MtmB gene. A nearby gene encodes an unusual tRNA with a CUA anticodon that can interact with the UAG stop codon, with another adjacent gene encoding an aminoacyl tRNA synthetase that charges the tRNA with a Lys residue (*23*). It thus appears possible that pyrrolysine is derived by modification of Lys-tRNA, just as selenocysteine is derived from Ser-tRNA. However, in 2004, synthetic L-pyrrolysine was attached to tRNA$_{CUA}$ by PylS, an Archaeal class II aminoacyl-tRNA synthetase that activated pyrrolysine with ATP and ligated it to the tRNA. When PylS and the corresponding tRNA$_{CUA}$ were expressed in *Escherichia coli*, pyrrolysine was incorporated into *Methanosarcina barkeri* methyltransferase at the normal position in response to the UAG stop codon, now acting as a sense codon (*28*). The fact that pyrrolysine was not incorporated into other proteins in the bacteria at the UAG stop codon (which would be expected to cause toxicity) indicates that only "special" UAG sequences are recognized, perhaps by RNA hairpin secondary structures similar to the selenocysteine insertion element (*30*). Another 2004 report demonstrated the possibility that two mechanisms are operative, with both PylS (a specialized aminoacyl-tRNA that charges tRNA^{Pyl} with pyrrolysine but not Lys) and a LysRS1:LysRS2 complex (that charges tRNA^{Pyl} with Lys but not pyrrolysine) acting on the tRNA (*31*). A 2005 report compared the decoding strategies for incorporation of pyrrolysine and selenocysteine. The organisms that utilize Pyl were found to have no genes that contained unambiguous UAG stop codons, a bias

not seen in non-Pyl-utilizing archaea or in Sec-utilizing organisms. Thus, UAG does not seem to be employed as a typical stop codon in microbes employing Pyl. Furthermore, Pyl was found to occur primarily in methylamine methyltransferase proteins (*32*).

Stop codons have been employed to insert unusual amino acids into genetically modified organisms (see Section 12.5.3d). For example, genes to synthesize 4′-aminophenylalanine and incorporate it into sperm whale myoglobin were inserted into *E. coli*, effectively generating a bacterium possessing a genetic code for 21 amino acids (*33*). Other methods have been developed to use 4-base codons to add unusual amino acids (*34*), or use some of the degenerate codons to encode non-natural residues (*35*).

Further complicating the codon–amino acid pairing relationship (traditionally viewed as sacrosanct) is the presence of additional or modified nucleobases in the third (degenerate) positions of tRNA anticodons, which provide the flexibility to recognize different bases at the third "wobble" position of degenerate codons. Many tRNAs cannot decode their cognate codons unless their anticodon loops are post-transcriptionally modified. Inosine, a guanine analog without the 2-amino group, can form base pairs with U, C or A. Crystal structures of the I–C and I–A pairings have been reported (*36, 37*). Crystal structures have also been obtained of tRNA^{Lys}$_{UUU}$ bound to both Lys codons AAA and AAG, showing the different modifications that have been made to the tRNA in order to enable it to bind and decode the degenerate codons (*37, 38*). Greater and greater complexity is being discovered in genetic coding. Bacteria contain shorter analogs of many aminoacyl-tRNA synthetases. One of these, YadB in *E. coli*, is a paralog of glutamyl-tRNA synthetase, but it acts to acylate tRNA^{Asp} with a Glu residue, instead of tRNA^{Glu} or tRNA^{Gln}. The tRNA^{Asp} is glutamylated at the hypermodified first anticodon nucleoside, queuosine (*39–41*).

The potential evolution of tRNA-supported direct pathways versus indirect pathways for amino acid incorporation was briefly reviewed in 2003 (*42*). The "late emerging" amino acids selenocysteine and pyrrolysine are only incorporated in contemporary organisms by indirect pathways via biosynthetic modifications of an amino acid attached to a tRNA scaffold. "Middle emerging" amino acids, Asn and Gln, are introduced by both direct and indirect methods, while "early emerging" amino acids (the remainder) are all only directly introduced through their corresponding tRNA synthetase/tRNA pairs (*42*). A mutagenesis study found that a single amino acid change in a specific aspartyl-tRNA synthetase converted it to a non-discriminating version that generated Asp-tRNA^{Asn} in addition to Asp-tRNA^{Asp}, suggesting a possible rapid evolution of this capability (*43*).

A "periodic table" of codons and amino acids has been designed to place the codons in regular locations, with AAA (Lys), UUU (Phe), GGG (Gly), and CCC (Pro) placed in the corners and connected by six axes. The resulting nucleic acid table showed perfect axial symmetry and, more interestingly, the corresponding amino acid table was periodic in terms of amino acid charge and hydrophobicity. The table predicts that central purine nucleotides determine the amino acid charge, while pyrimidine nucleotides determine polarity (*44*). Another strategy organized the 20 amino acids into a three-dimensional tetrahedral space based on the genetic codes; the hydrophilic and hydrophobic amino acids ended up distributed in distinct regions (*45*). The correlation of amino acid physicochemical properties with codons was reviewed in 1995 (*46*).

The usage of the complete repertoire of codons varies between genes and organisms. Highly expressed genes have a preference for abundant tRNA species, while organisms with a high GC or AT content (such as *Streptomyces* or *Campylobacter jejuni*, respectively) tend to use GC- or AT-rich codons, respectively. Amino acid usage within the prokaryote family Bacillaceae is affected by the genomic GC content, the optimal growth temperature, and usage of Cys (*47*). A formula has been derived to indicate the "effective number of codons" used by a gene, ranging from 20 for a gene relying on one codon per amino acid to 61 for a gene where all possible codons are used (*48*). This publication has led to some discussion over the best method of calculating the effective number of codons (*49*). The observation of certain regularities in the genetic code, namely that the first base of codon correlates with the biosynthetic pathway of the amino acids it encode (C derived from α-ketoglutarate, A from oxaloacetate, U from pyruvate), and the second base with the hydrophobicity of the encoded amino acid (U = most hydrophobic, A = most hydrophilic) has been used to propose an origin of the genetic code. If amino acids were initially synthesized by covalent complexes of catalytic dinucleotides with α-keto acids, it could create an association between the amino acids and the first two codon bases that was retained upon the evolution of translation (*50*). A series of publications has explored this same possibility of a linked prebiotic origin of RNA and coded peptides, using aminoacyl-RNA trimers as intermediates (*51–54*). An analysis of evolutionary changes in amino acid composition of proteins within taxa representing bacteria, Archaea, and Eukaryota found a univeral loss of Pro, Ala, Glu, and Gly, with a gain of Cys, Met, His, Ser, and Phe. The amino acids with declining frequencies are believed to be among the first incorporated into the genetic code; thus "newer" amino acids are replacing them (*55*). Cellular amino acid composition has been found to differ between archaeobacteria, eubacteria, protozoa, fungi, and mammalian cells, presumably due to evolutionary changes (*56–59*). The amino acid composition of proteins has also been employed to predict protein subcellular location (*60*).

A recent study examined whether the metabolic efficiency of individual amino acid synthesis might affect the composition of amino acids within the proteomes of *E.coli* and *Bacillus. subtilis* (*61*). The cost of amino acid synthesis varies between 12 and 74 high-energy phosphate bonds per molecule, with the overall energy cost of amino acid biosynthesis for a given codon in the genome a product of the cost per encoded amino acid and the number of times the codon is translated. In *E. coli*, individual protein concentrations can vary between several to 100,000 molecules per cell, so the potential energy savings by substituting less energy-expensive amino acids in highly expressed proteins are huge. Indeed, less costly amino acids such as Gly, Asp, and Glu were found to increase in abundance in proteins from highly expressed genes, while expensive amino acids like Trp, His, Phe, Cys, and Leu decreased, with an average reduction of 2.0–2.5 P bond energies per amino acid per protein synthesized. This suggests that natural selection for energetic efficiency does indeed play a role in the evolution of protein primary structure. Presumably the same factor would also be important in other organisms. A somewhat related study set out to see whether the full set of 20 naturally ocurring amino acids is required for protein function (*62*). Numerous cycles of mutagenesis were applied to *E. coli* orotate phosphoribosyltransferase (213 residues), substituting similar amino acid types (e.g., Arg for Lys and His) in order to restrict the amino acid alphabet. A catalytically viable enzyme with 73 substitutions, and only 13 amino acid types, was produced, demonstrating that the full amino acid repertoire is not necessarily essential for protein function.

Amino acids can also be incorporated into peptides without using the normal codon-based translation mechanism. This is accomplished by non-ribosomal peptide synthetases (NRPSs), which activate amino acids by formation of enzyme-bound thioesters, with amino acids loaded on a series of modules arranged in the order corresponding to the primary sequence of the final product. The activated amino acids are sequentially transferred to the *N*-terminus of the adjacent residue, with the final thioester linkage cleaved by hydrolysis or cyclization (*63*). The NRPSs assemble complex natural products containing many unusual amino acids, with the production of antibiotics particularly well researched (*64, 65*).

Additional layers of complexity in translating DNA to proteins have been proposed. Sense and complementary peptides, coded by reading the sense or antisense DNA strands, can interact specifically, suggesting a possible higher-order two-dimensional genetic code (*66*). The large tracts of non-coding DNA introns are now postulated to play an important role in the regulation of gene expression; some of these functions may include amino acid selection (*67–70*).

Another classification describing a set of amino acids are those termed the "essential" amino acids (Val, Ile, Leu, Lys, Met, Phe, Thr, Trp, and, for some species, His and Arg). These amino acids cannot be synthesized in sufficient quantities by mammals, so must be obtained from dietary sources, mainly plant protein (*71*). However, since the amino acid composition of plant proteins varies (with, for example, cereals deficient in Lys and soybeans in Met), diets reliant on only one source can result in nutritional deficiencies. The essential amino acid content of some edible Mexican insects was determined, and found to be higher than the adult requirements, making them a potentially useful protein source (72). The physiological and metabolic effects of 10 essential amino acids were reviewed in 1998 (73).

The presence of amino acids is ubiquitous, not only within biological organisms but also in the environment around us. Simple extraction of soil by shaking with demineralized water or 0.5 M ammonium acetate provided from 40 to 2000 ng (per g of dry soil) of each of 17 of the 20 common amino acids (Asn, Gln, and Trp were not tested) (74). Propanol has been identified as an effective solvent for the extraction of amino acids from soil samples, potentially useful for Martian in situ analysis since aqueous extractions would contaminate the search for water on Mars (75). Amino acids have been detected in interstellar objects, and are considered a potential sign of extraterrestrial life (see Section 2.1).

In recent years, interest in the synthesis and use of amino acids has increased exponentially, which can be attributed to two factors: their biological activity, and their relevance to chemical asymmetric synthesis. The importance of amino acids spans a diverse range of research fields, including organic chemistry, inorganic chemistry, physical chemistry, analytical chemistry, biochemistry, pharmacology, cell biology, nutrition, geology, archaeology, medicine, and physiology, as summarized in the inaugural issue of the journal *Amino Acids* (76).

1.1.1 Biological Activity

Amino acids are used to form structural proteins, enzymes, transport proteins, immune proteins, hormones, and neurotransmitters, and are the precursors for the biosynthesis of many other compounds. Amino acids themselves play a role in various signaling pathways, such as acting on excitatory amino acid neuroreceptors. It has recently been shown that *E. coli* appears to secrete glycine as a chemoattractant, inducing the bacterial cells to cluster at high density, potentially leading to biofilm formation (77). Humans and other mammals are not capable of producing all the required amino acids, and must obtain the "essential" amino acids (primarily Ile, Leu, Lys, Met, Phe, Thr, Trp, and Val), which originate in plants, from foodstuffs. Amino acids are often responsible for the aroma and taste of food. For example, Pro is important for the smell of bread crust and Cys is responsible for the formation of meat flavor (78). An amino acid taste receptor was identified and characterized in 2002; it responds to most of the 20 standard amino acids, but not their D-enantiomers (79). A report in 2006 found that direct injection of Leu into rat's brains was found to curb their appetite. Apparently the level of amino acids in the brain is monitored in order to judge how much food to eat (80, 81). Pharmaceutically, the common amino acids are used for infusion solutions and for diet supplements. However, they also have many other medical uses such

as the use of L-Trp as an antidepressant (for further examples, see 78).

Many non-proteinaceous amino acids, or the compounds in which they are contained, have been found to possess biological activity, most often antimicrobial properties. A number of uncommon amino acids are either themselves drugs, or critical components of drugs. Probably the best known examples are L-(2,4-dihydroxyphenyl)alanine (DOPA) and its analogs, used for the treatment of Parkinson's disease, and (4R)-4-[(E)-2-butenyl]-4,N-dimethyl-L-threonine (MeBmt). MeBmt is an essential constituent of the fungal undecapeptide immunosuppressant drug cyclosporin A (CsA), which is widely used in transplant patients and those with autoimmune diseases (82). The interest in synthetic routes to unusual amino acids arises because many are difficult to isolate and purify in significant quantities. By developing synthetic routes it is possible to provide useful quantities of novel amino acids, and to create new analogs for structure–activity relationship (SAR) studies. The growing importance of a wide variety of unusual amino acids is largely driven by the needs of the pharmaceutical drug discovery industry, and is reflected in the proliferation of commercially available unusual amino acids over the past decades.

Carefully designed amino acid analogs can also lead to valuable information on biochemical mechanisms when they are incorporated into peptides or proteins. By employing amino acids that are conformationally restricted or isotopically labeled or which act as suicide substrates, insight can be gained into mechanisms of enzymatic action or binding. These results can then be applied to design effective inhibitors. An overview of topographical design considerations for peptide and peptidomimetic ligands was published in 2001 (83). Isotopically labeled substrates or intermediates can also be used to study biosynthetic pathways. Recent advances in protein engineering allow the use of a codon to code for a transfer RNA containing an unusual amino acid, meaning that non-proteinaceous amino acids can be incorporated into biochemically prepared proteins (84). This protocol greatly simplifies the synthesis of large proteins containing unnatural amino acids, and provides even more impetus to develop improved methods for preparing non-proteinaceous amino acids. New enzymes have also been created by covalent modifications of existing proteins, such as by attachment of a new functional group to a Cys residue, as reviewed in 2001 (85).

D-Amino acids are also of interest. Ribosomally-directed protein biosynthesis requires L-amino acids, so the presence of D-amino acids has traditionally believed to be confined to prokaryotic microorganisms. Two examples of L-to-D conversions of amino acids in prokaryotic systems involve the post-translationally modified antimicrobial lantibiotic peptides lactocin S and lacticin 3147. With both peptides, the ribosomally encoded L-Ser residues are converted into D-Ala via enzymatic elimination to dehydroalanine followed by enzymatic reduction to L-Ala (86). In recent years, in contrast to the existing dogma, the presence of D-amino acids in eukaryotic systems has been found to be increasingly common. An analysis of free L- and D-amino acids in a variety of plants found significant quantities of D-amino acids, ranging from 0.2% to 8% of the corresponding L-amino acid. Most plants contained D-Asp, D-Asn, D-Glu, D-Gln, D-Ser, and D-Ala, while D-Pro, D-Val, D-Leu, and D-Lys were identified in at least some plants. Given that a variety of plant components (seeds, leaves, and fruits) in the major plant families were analyzed, it appears that low percentages of D-amino acids are normal constituents of plants (87). Significant quantities of D-amino acids are found in foodstuffs, such as milk, beer, vinegar, and green coffee (88). The quantities, relative amounts, and patterns of amino acid enantiomers have been proposed as a signature to identify and authenticate special beers (89). D-Amino acids in peptides and proteins can potentially be identified by the different patterns exhibited by the diastereomeric peptide/proteins when analyzed by tandem mass spectrometry (90).

Higher-order species also contain D-amino acids. A post-translational enzyme-catalyzed isomerization of amino acid residues in peptide chains has recently been observed in *Agelenopsis aperta* spider, with Ser, Cys, *O*-methyl-Ser, and Ala residues in the middle of peptide chains isomerized to the D-amino acid (91). A dehydro-alanine-containing inhibitor of the epimerase has been reported (92). A 46-amino acid peptide found in a group of conotoxins isolated from crude venom extracts of *Conus radiatus* contains a D-Phe residue at position 44 (93). This position, the third residue from the C-terminus, is in the same position as the D-amino acid in spider ω-agatoxin TK from *Agelenopsis aperta* spider, suggesting that there may be preferred positions within peptides for post-translational isomerizations to the D-amino acids. The D-Phe residue confers significant excitatory effects compared to the all-L peptide (93). Antibacterial and hemolytic peptides isolated from skin secretions of the frog *Bombinae varigiegata*, the bombinins H, contain either L-Ile or D-allo-Ile as the second amino acid from the *N*-terminus. An enzyme isolated from the same skin secretions catalyzes the epimerization of L-Ile in model peptides; proteins with similar sequences are present in other vertebrate species, including humans (94).

Other examples of D-amino acids in eukaryotic organisms have been reported. Antibodies highly specific for D-serine identified endogenous D-serine in rat brain localized astrocytes with release stimulated by agonists of non-NMDA (*N*-methyl-D-aspartate) glutamate receptors. These results indicate that D-Ser is the endogenous ligand for the glycine site of NMDA receptors (95, 96), and is involved in induction of long-term potentiation (97). A D-serine racemase was isolated from mammalian brain, giving further evidence that D-Ser in brains is not an artifact, but an important signaling molecule (98). The levels of D-Ser in the brains of mutant hyperkinetic mice were compared with normal mice to see if a physiological role could be determined, but no

differences were found (99). The levels of excreted D-amino acids in the urine of rats has been found to vary from the low percent level to over 40% of the total amino acid, depending on the specific amino acid. Levels in plasma were generally an order of magnitude lower. The quantities of excreted amino acids were found to depend upon the strain of rodent, its diet, age, the presence of advanced cancer, antibiotics, or pregnancy. Pregnant rats consistently possessed elevated levels of D-amino acids (100). Elevated levels of Ser are found in the urine of mutant rats lacking D-amino acid oxidase, with most of the serine present as D-Ser. The D-Ser thus appeared to be naturally biosynthesized, and since it is not metabolized in the mutant mice it accumulates and is excreted (101). A study of recombinant mouse serine racemase identified competitive dicarboxylic acid inhibitors such as malonate and L-*erythro*-3-hydroxy-Asp. The same enzyme induced β-elimination of the isomeric L-*threo*-3-hydroxy-Asp (102).

Inhibitors of human serine racemase (hSR) have been identified through a screening of combinatorial libraries, and may prove useful in identifying the biological and pharmacological role of hSR (103). The D-enantiomers of Phe, Tyr, and Trp have been detected in human urine at levels from 0.2 to 1.9% of the L-isomer; no D-Leu could be detected (104), but another report found D-Ser (up to 50%), D-Ala, D-Thr, D-Val, and D-Phe (105). The levels of D-amino acids in the urine of patients with chronic renal failure were elevated compared to normal amino acids, possibly due to depletion of D-amino acid oxidase (106). Partially racemized Tyr, with 79% ee (enantiomeric excess), has been detected in the insoluble protein fraction of aging human eye lenses (107), while racemized Asp has been detected in human skin (108). Analysis of the free amino acids in ventricular cerebrospinal fluid from Alzheimer and normal subjects found significantly higher total D-amino acids, particularly D-Ser and D-Asp, in the Alzheimer patients (109). A review of the distribution, function, origin, and analysis of D-amino acids in mammals was published in 2002 (110).

The presence of D-amino acids in peptides and proteins results in altered properties. The propensities of D-amino acids to destabilize helical secondary structures in peptides have been determined (111). Synthetic all-D-polypeptides are resistant to proteolytic degradation, giving them potential as long-lasting therapeutic agents. Even a few L-to-D substitutions confer proteolytic stability; a MUC2 peptide epitope used to elicit antibodies was prepared with up to three D-amino acids in the *N*- and *C*-terminal flanking regions. The partly D-peptides retained their epitope function, and also possessed high resistance to proteolytic degradation in human serum (112). A method for enzymatically ligating all-D-peptide fragments to form longer peptides (up to 34-mer) has been developed, using the cysteine protease clostripain (113). A cell-free protein expression system has been modified to allow for incorporation of D-amino acids, via mutagenesis of the peptidyltransferase and helix 89 regions of *E. coli* 23S ribosome (114).

Finally, the industrial importance (*78, 115*) of amino acid derivatives such as monosodium glutamate (seasoning agent), L-aspartyl-L-phenylalanine methyl ester (Nutrasweet or aspartame artificial sweetener), and D-phenylglycine (component of semisynthetic antibiotic ampicillin), and the increasing use of amino acids such as Lys, Met, and Thr as feed additives, gives an economic motive for improving synthetic methods. Alanine, glycine, and leucine have recently been evaluated as corrosion inhibitors for carbon steel, with 50–90% reduction in corrosion rates at concentrations of 0.01–0.1 M, and an obvious advantage over more conventional inhibitors that possess considerable toxicity (*116*).

1.1.2 Asymmetric Synthesis

The second factor governing the increased interest in amino acids is the dramatic growth in asymmetric organic synthesis in the past three decades. Amino acids have been an integral part of the development of this area, as substrates (*117–119*), targets (*120–122*), and, more recently, catalysts (*123, 124*). They provide a diverse range of economical chiral building blocks of high functionality available in both enantiomers, and can be used as chiral auxiliaries, reagents, and substrates.

1.2 A Brief Summary of Trends in Amino Acid Synthesis

Amino acid synthesis is one of the fundamental roots of organic chemistry, with variations of the first synthesis by Strecker in 1850 (*125*) still widely used today. Modified versions of other routes developed by pioneers such as Cahours (1858, (*126*)) and Sörensen (1903, (*127, 128*)) are also still in use. The evolution of amino acid synthesis has gone through several distinct phases, with initial developments taking place in the late 19th and early 20th centuries. The 1930s and 1940s saw a number of improvements in the alkylation of aminomalonate derivatives. A resurgence of interest in amino acid synthesis corresponded to the introduction of asymmetric synthetic techniques in the 1960s and 1970s. Amino acids have often been at the forefront of the development of asymmetric synthesis techniques, especially asymmetric hydrogenation. The closely dependent relationship between the development of asymmetric hydrogenation and the commerical synthesis of L-DOPA is nicely summarized in the 2001 Nobel lecture of William S. Knowles (*129*). The 1980s was a decade of rapid growth in general techniques for the asymmetric synthesis of amino acids. The ground-breaking Schöllkopf bislactim ether was soon followed by other cyclic templates, primarily the oxazolidinone and imidazolidinone templates of Seebach, the Evans oxazolidinone, and the Williams oxazinone (see Chapter 7). In recent years acyclic templates have become increasingly important for asymmetric syntheses, such as the camphor-sultam amide auxiliary of Oppolzer and the pseudoephedrine amide auxiliary of Myers (see Chapter 5). Continuing improvements in asymmetric hydrogenation of dehydroamino acids maintain its relevance as a useful synthetic method (see Chapter 4). Asymmetric alkylation of glycine Schiff base esters, catalyzed by various chiral phase-transfer catalysts, has produced extremely high enantioselectivities in recent years, and can be carried out relatively inexpensively on large scales.

The rapid development of new general methods for asymmetric α-amino acid synthesis appears to have peaked in the early 1990s, and much of the attention has been diverted to improving these methods and applying them to the synthesis of unusual amino acids. Many unusual amino acids are now commecially available. These new areas of interest have arisen in response to the realization that constrained peptides are potentially much more effective enzyme inhibitors/receptor ligands than flexible ones (*83,130*), and that analogs and isosteres often possess useful properties, such as resisting proteolytic degradation or having improved bioactivity and/or PK/ADMET (pharmacokinetic/absorption, distribution, metabolism, excretion, toxicology) properties. More recently, β-amino acids have attracted the same type of attention that α-amino acids did in previous years, driven by the presence of β-amino acids in important bioactive compounds such as the chemotherapeutic agent Taxol (paclitaxel).

Despite all these advances in asymmetric synthesis, the more traditional method of a simple racemic synthesis followed by enantiomer resolution is still widely employed, particularly on industrial-scale syntheses. This is highlighted in a 2004 review of industrial methods for the production of optically active intermediates, which includes a descriptions of the industrial production of amino acids (*131*). For this reason, a significant section of this book is devoted towards racemic methods of amino acid synthesis.

Due to the lengthy period of time covered by amino acid synthetic history, there exists a wealth of information that must not be ignored. An extremely important reference is the three-volume series *Chemistry of the Amino Acids*, by J.P. Greenstein and M. Winitz (1961, (*132*)). The first two volumes cover general properties of the amino acids, whereas the third volume describes the history, isolation, properties, methods of synthesis, and optical resolution of each common amino acid, as well as a number of others. *Chemistry and Biochemistry of the Amino Acids*, by G.C. Barrett (1985 (*6*)) is also a valuable resource for amino acid properties and historical syntheses. An informative and comprehensive review of the history of the discovery of the amino acids was published in 1931 (*133*). This paper contains many fascinating accounts of the efforts required to isolate unknown amino acids and identify their structure, such as the isolation of thyroxine from 3 tons of fresh thyroid gland! A chapter in the 1964 *Annual Review of Biochemistry* (*134*) describes the chemical synthesis and metabolism of amino acids isolated primarily from 1959 to 1963 (235 references, over 90 new amino acids), giving a good indication of the many new amino acids that were identified in a relatively brief period of time with comparatively primitive equipment. A 1965 two-volume treatise entitled *Biochemistry of Amino Acids* (*160*) includes chapters on the discovery of amino acids, the role of amino acids in nutrition, general biochemical properties, and metabolism of amino acids. The series *Amino Acids, Peptides and Proteins*, by the Royal Chemical Society, contains regular reviews of the past year's developments in amino acid chemistry, while the monograph series *Chemistry and Biochemistry of Amino Acids, Peptides and Proteins* (*135*) includes a number of chapters relevant to amino acid synthesis. The multi-volume monograph series *The Peptides: Analysis, Synthesis, Biology* covers a variety of facets of amino acid and peptide properties, and includes a chapter on syntheses of unusual amino acids with non-proteinogenic structures (*136–140*). A journal dedicated to amino acids, *Amino Acids*, has been published since 1991 by Springer-Verlag, although unfortunately few libraries appear to subscribe to it. However, the advent of online internet access has made it possible to more readily view issues. This journal focuses largely on the biological relevance of amino acids, but a number of synthetic articles and reviews have appeared, including entire issues in 1996 (*141*), 1999 (*142*), 2003 (*143*), and 2005 (*144*). The journal also includes abstracts from the International Congress on Amino Acids.

The only monograph to extensively discuss asymmetric synthesis of α-amino acids is R.M. Williams' comprehensive *Synthesis of Optically Active α-Amino Acids* (1989 (*145*)), which, since its publication, has been cited in nearly every paper of amino acid synthesis. A much shorter review of the same subject area had appeared in a journal several years earlier (*146*), while reviews of developments of both asymmetric and racemic syntheses in the early 1970s are also available (*9,10*). A 1994 review by R.O. Duthaler (*147*) covers many of the advances since Williams' book, as does a 1999 review (*148*). An overview of other developments was included in a 2002 article (*149*), while a summary of recent developments in catalytic asymmetric syntheses of α- and β-amino acids appeared in 2003 (*150*). A primer designed for introduction to the field of amino acid and peptide synthesis is part of the Oxford Chemistry Primer series, although it is more focused on derivatizations and peptide synthesis (*151*). Industrial syntheses were reviewed by Y. Izumi et al. (*152*) in 1978, with several more recent synopses available (*115,153*), including a comprehensive discussion in *Ullmann's Encyclopedia of Industrial Chemistry* (*78*), and a 2003 review of syntheses using *Corynebacterium* (*154*). A 2004 review of industrial methods for the production of optically active intermediates includes a significant section on the industrial production of amino acids (*131*). A monograph that describes methods of non-α-amino acid synthesis (primarily acyclic C_3–C_{10} ω-amino acids) was published in 1995 by M.B. Smith (*155*), while a monograph reviewing enantioselective synthesis of β-amino acids edited by E. Juaristi appeared in 1997 (*156*); developments in the synthesis of α-, β-, γ-, and

ω-amino acids from 1996 to 1997 were addressed in a 1998 review (*157*), a continuation of a series.

Methods for modifying amino acid amino groups, acid groups, and side chain groups are described in a 1999 monograph (*158*). The chromatography of amino acids, including enantiomeric resolution, is addressed in the *CRC Handbook of Chromatography of Amino Acids and Amines* (*159*), as well as a number of other monographs which are discussed later in this chapter. The analysis and separation of amino acids is thoroughly discussed in a 2005 monograph, *Quantitation of Amino Acids and Amines by Chromatography: Methods and Protocols* (*164*). Post-translational (in vivo) modifications of proteins, leading to unusual amino acids, were summarized in 1981 (*161*) and 2005 (*12*) reviews, with an issue of *Amino Acids* devoted to amino acid modifications by oxygen and nitrogen species (*162*). The occurrence and biosynthesis of D-amino acids was reviewed in a 1977 publication (*163*); traditionally most of these unusual amino acids have been identified in peptide antibiotics due to the screening methods used to isolate compounds of interest. More specific reviews of methods of synthesis of certain classes of amino acids have also been published, and are discussed in the appropriate chapter. Finally, the use of amino acids for the synthesis of other compounds has been addressed in the monograph by Coppola and Schuster (*120*), while two reviews of the use of amino acids in asymmetric catalysts appeared in 2002 (*123, 124*).

Amino acids are extensively used in medicinal chemistry and biotechnology, and a number of examples of their use are included in the chapters on specific classes of amino acids. For an introduction to some of the terms that are used, IUPAC (International Union of Pure and Applied Chemistry) has published a series of glossaries of terms used in medicinal chemistry (*165*) and biotechnology (*166*). Nomenclature for abbreviations for synthetic polypeptides were reported in 1973, referring primarily to polymer-like polypeptides rather than peptide sequences (*167*), while an earlier publication in 1966 gave preferred abbreviations for amino acids, polypeptides, and proteins, in addition to carbohydrates and nucleic acids (*168*).

A number of internet websites contain useful information related to amino acids. For example, the IUPAC recommendations for amino acid nomenclature are available online at http://www.chem.qmul.ac.uk/iupac/ Amino Acid/. Many properties of the amino acids (including mass, surface area, volume, side-chain pK_a, pI, solubility, density, hydrophobicity, and solvent accessibility) are tabulated at http://www.imb-jena.de/ IMAGE_AA.html and http://prowl.rockefeller.edu/aainfo/ contents.htm, while pK_1, pK_2 (and pK_3 if applicable) can be found at http://www.indstate.edu/thcme/mwking/ amino-acids.html, along with a general overview of amino acids and their properties. Other general information about each amino acid is contained at http://www. russell.embl-heidelberg.de/aas/aas.html and http://www. chemie.fu berlin.de/chemistry/bio/amino-acids_en.html. Names, structures, and limited properties of the amino acids are at http://www.chemie.fu-berlin.de/chemistry/bio/amino-acids_en.html. A convenient "Periodic Chart of the Amino Acids," and a "Periodic Chart of Unusual α-Amino Acids" are available in pdf format at www.bachem.com.

1.3 Industrial Synthesis

Most industrial processes focus on producing the L-amino acid enantiomers (78). Several amino acids (L-Cys$_2$, L-Tyr, L-Pro, L-Arg, L-His, L-Leu, L-Phe, L-Ser, L-Thr) are still produced by extraction from protein hydrolysates of natural raw materials (78); Cys is isolated from human hair (169). Microbiological fermentation using bacterial mutants is used to prepare many amino acids (L-Glu, L-Lys, L-Arg, L-Asp, L-Ile, L-Phe, L-Ser, L-Thr), while enzymatic processes are used for Asp and Ala (78). The Strecker or Bücherer synthesis is often used industrially to prepare racemic amino acids (especially DL-Met), which can then be resolved enzymatically (e.g., Ala, Val, Met, Phe, Trp) or by crystallization of diastereomers (78). Gly and Trp are the only common amino acids that are generally produced by chemical means (78). A potentially useful fermentation route toward unusual amino acids was reported in 2003. Metabolic engineering of the cysteine biosynthetic pathway in *E. coli* overproduced *O*-acetyl-L-Ser, which was then transformed into various β-substituted-Ala derivatives on a multiple g/L scale by *O*-acetylserine sulfhydrylase-catalyzed reaction with heteroatom nucleophiles (170). Significant quantities of Glu, Lys, Val, Ile, Thr, Asp, and Ala are produced by fermentation of *Corynebacterium*; this area of synthesis was reviewed in 2003 (154). Strains of *Rhizobium*, *Mesorhizobium*, and *Sinorhizobium* microorganisms were tested for their ability to produce amino acids from mannitol as the sole carbon source; high levels of Gly, Ala, and Ser were produced and released into the media, with Asp, Glu, His, Thr, Arg, Pro, Tyr, Val, Met, Lys, Ile, Leu, and Phe also detected (171). The potential for production of amino acids (Met, Lys, Arg, Trp, and Glu) by the nitrogen-fixing organisms *Azobacter* and *Azospirillum* were summarized in a 1995 minireview (172).

1.4 Amino Acid Properties

Many of the properties of amino acids are extensively discussed in the monographs by Barrett (6) and Greenstein and Winitz (132), and nicely tabulated in *Ullmann's Encyclopedia of Industrial Chemistry* (78), including melting (decomposition) point, specific rotation, pK_1, pK_2 (and pK_3 if applicable), pI, and solubility in H$_2$O. As mentioned earlier, many properties of the amino acids (including mass, surface area, volume, side-chain pK_a, pI, solubility, density, hydrophobicity, and solvent accessibility) are tabulated at

http://www.imb-jena.de/IMAGE_AA.html and http://prowl.rockefeller.edu/aainfo/contents.htm, while pK_1, pK_2 (and pK_3 if applicable) can be found at http://www.indstate.edu/thcme/mwking/amino-acids.html. In 1998 a multivariate characterization of 87 amino acids was published, employing 26 physicochemical descriptor variables, with the goal of being able to design biologically active peptides. These included 10 experimentally determined values—thin-layer chromatrography(TLC) R$_f$ values and nuclear magnetic resonance (NMR) chemical shift values—and 16 calculated values, such as log P, molecular weight, side-chain charge, hydrogen bond donor–acceptor properties, and side-chain van der Waals volume (173). The lipophilicities of the 20 common amino acids, as calculated by a number of different methods, were compared in 1994 (174). The hydrophobicity of amino acid side chains has been correlated with their dissolution enthalpies in aqueous solutions of urea (175). A review that evaluated methods for measuring amino acid hydrophobicities was published in 2003 (176). The pH, density, and water activity of solutions of Gly, Ala, Arg, and Pro (various concentrations in water, acid buffer, and basic buffer) were reported in 2001 (177). The ^{1}H NMR spectra of 20 amino acids in water were tabulated in a 2004 report (178). ^{13}C NMR chemical shifts of nine amino acids were reported in a 2004 paper, along with the observation that the addition of MgCl$_2$ resulted in spin-lattice relaxation so that the integral areas were proportional to the number of carbons (179).

Advances in X-ray crystallography have recently allowed for charge-density determinations of amino acids. A topological analysis of the electrostatic potential of L-Asn, DL-Glu, DL-Ser, and L-Thr has been reported (180), with further studies of Asn, Glu, Lys, Pro, Ser, and Val in 2002 (181). The adhesion of different amino acids to various inorganic surfaces, including metals, insulators, and semiconductors, was investigated in 2005. Significant differences in adhesion were observed between different amino acids, determined largely by the amino acid side-chain charge (182). A conformational flexibility scale for amino acids in peptides was developed, providing an absolute measure for the time scale of conformational changes in short peptides, depending on amino acid type. As might be expected, Gly is the most flexible and Pro the least, but the remainder of the ordering contains some interesting results (Gly > Ser > Asp, Asn, Ala > Thr, Leu > Phe, Glu, Gln > His, Arg > Lys > Val > Ile > Pro) (183). The propensity for amino acids to be contained within secondary structure types has, for many years, been predicted by rules developed by Chou and Fasman based on the few protein crystal structures available in the 1970s (184). A refined analysis was reported in 2006, based on over 2000 proteins in the protein data bank (185). The relative propensities of L-amino acids to induce various types of secondary structure have been extensively examined over the years; in 2000 a study on the helix-destabilizing properties of D-amino acids was published (111). Other attempts at correlating amino acids with protein structure have been reported (*186*). The amide bonds found in proteins and peptides have a strong preference for the *trans* conformation, with a considerable barrier to rotation that greatly influences protein structure. A detailed study of the kinetics and energetics of *cis/trans* isomerization of a Gly-Gly sequence was reported in 2001 (*187*).

Amino acids in solution at neutral pH exist as zwitterions, which are stabilized by solvent molecules and counterions. However, unsolvated zwitterions are much less stable, and gas phase glycine is not a zwitterion (*188*). Solid-state ^{13}C NMR analysis of L-His lyopholized from solutions at pHs varying from 1.8 to 10 allowed for observation of the three separate acid–base pairs resulting from successive deprotonations of the carboxylic acid, imidazolium cation, and α-ammonio group, and produced an estimation of pKs that corresponded well to classically measured values (*189*). The side-chain dissociation constants of amino acids within a peptide have been determined by two-dimensional NMR spectroscopy. The four Lys residues in one 15-mer peptide had pK_as varying from 10.95 to 11.14; six values (*N*-terminal α-amino group, Glu, His, Tyr, and two Cys side chains) were determined in a 13-mer peptide (*190*).

The rate of spontaneous decarboxylation of amino acids was assessed in 2000, with a half-life of 1.1 billion years estimated for Gly at 25 °C in neutral solution in the absence of catalyst (*191*). The ability of amino acid decarboxylases to accelerate this reaction rate, by approximately 7×10^{19}-fold for arginine decarboxylase, is truly remarkable.

1.4.1 Nomenclature

The 20 common amino acids, with their one- and three-letter codes, are presented in Table 1.1. A number of other commonly encountered amino acids are given in Table 1.2. Peptide chemistry makes use of a three-letter code for describing the amino acid nucleus of protected derivatives, with the *N*-protecting group as a prefix, the carboxyl protection as a suffix, and any side-chain protecting group in brackets. Thus, *N*-*t*-butyloxycarbonyl serine methyl ester *O*-benzyl ether becomes Boc-Ser(Bn)-OMe. The IUPAC rules of biochemical nomenclature for amino acid derivatives provide examples of these, and other, rules (*192, 193*). The nomenclature and symbolism for amino acids and peptides were extensively defined in a 1984 IUPAC publication, including definitions of three- and one-letter abbreviations, the formation of semisystematic names for amino acids and derivatives, the use of the prefixes "*homo*" and "*nor*," use of abbreviations for peptides and modified peptides, stereochemical configurations, and abbreviations for commonly used nitrogen, oxygen or sulfur substituents and protecting groups (*193, 194*). The prefix "*homo*" refers to amino acids that are similar to one of the common amino acids, but with one more methylene group in the carbon side chain. The "*nor*" prefix commonly refers to the straight-chain isomer of a

branched-chain compound, although its more rigorous definition denotes removal of a methylene group. Thus norvaline and norleucine describe straight-chain analogs (*n*-propyl and *n*-butyl side chains, respectively) of valine (isopropyl side chain) and leucine (isobutyl side chain), respectively. The IUPAC nomenclature is available online at http://www.chem.qmul.ac.uk/iupac/AminoAcid/.

Most amino acids possess a chiral center, with the exception of glycine and symmetrically disubstituted molecules. The historical literature designation to distinguish the two enantiomers originally relied on the rotation of polarized light of a neutral aqueous solution, giving the prefixes *d* (dextrorotatory) or *l* (levorotatory). However, confusion arose due to the realization that all protein-derived amino acids had the same configuration at the α-center even though their optical rotation sign differed (and could change depending on solvent). A modified nomenclature system was then devised based on an arbitrary *l* designation to all amino acids derived from proteins (except threonine, which was given a *d* prefix due to the configuration at the β-center!). This designation was used in combination with a (+) or (−) optical rotation sign. Unfortunately the confusion increased, as both systems were still employed (*195*). In 1947 a standard nomenclature was finally proposed, with the prefixes D- and L- used to designate the α-center configuration (derived from its relative configuration compared to the sugars, with serine interconverted with glyceraldehyde), and (+) and (−) or *dextro-* and *levo* designating the optical rotation. This naming system is still widely used, but is slowly being displaced by the Cahn–Prelog–Ingold *R/S* system (*196, 197*). With this unequivocal system the protein derived L-amino acids all have the *S* configuration, with the exception (there's always one!) of cysteine, which is an *R*-amino acid due to the side-chain sulfur atom changing the order of priority.

Further confusion arises when more than one chiral center exists. For example, the β-hydroxy amino acids conforming to the configuration of L-*allo*-threonine are interchangeably described as possessing an *erythro*, *anti*, or 2*S*,3*S* configuration, whereas the L-threonine analogs are *threo*, *syn*, or 2*S*,3*R* (see Scheme 1.1). Unfortunately there are a number of definitions of what constitutes a *threo* or *erythro* isomer, giving rise to an "almost Babylonic confusion" (*198*). The terms *syn* and *anti* introduced by Masamune are more rigorously defined (*199*), with adjacent groups on the same side of an extended (zig-zag) structure considered *syn*.

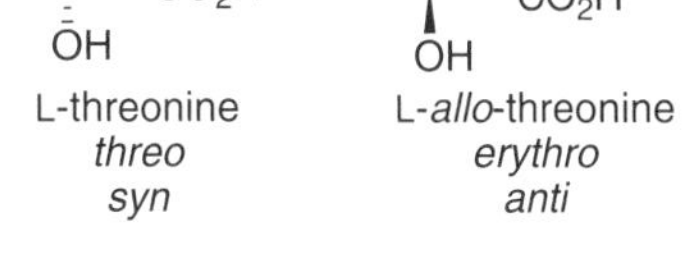

L-threonine
threo
syn

L-*allo*-threonine
erythro
anti

Scheme 1.1 Structures of Threonine and *allo*-Threonine.

Table 1.1 The Genetically Encoded Amino Acids.

Structure	Code		Name	Structure	Code		Name
H_3N^+–CO_2^-	G	Gly	Glycine	(–OH) H_3N^+–CO_2^-	S	Ser	L-Serine
H_3N^+–CO_2^-	A	Ala	L-Alanine	(–OH) H_3N^+–CO_2^-	T	Thr	L-Threonine
H_3N^+–CO_2^-	V	Val	L-Valine	(–SH) H_3N^+–CO_2^-	C	Cys	L-Cysteine
H_3N^+–CO_2^-	L	Leu	L-Leucine	(–S–) H_3N^+–CO_2^-	M	Met	L-Methionine
H_3N^+–CO_2^-	I	Ile	L-Isoleucine	(–C(=O)NH$_2$) H_3N^+–CO_2^-	N	Asn	L-Asparagine
N^+H_2 ring CO_2^-	P	Pro	L-Proline	(–C(=O)NH$_2$) H_3N^+–CO_2^-	Q	Gln	L-Glutamine
(benzyl) H_3N^+–CO_2^-	F	Phe	L-Phenylalanine	(–CO_2^-) H_3N^+–CO_2^-	D	Asp	L-Aspartic acid
(4-hydroxybenzyl, OH) H_3N^+–CO_2^-	Y	Tyr	L-Tyrosine	(–CO_2^-) H_3N^+–CO_2^-	E	Glu	L-Glutamic acid
(indolyl, NH) H_3N^+–CO_2^-	W	Trp	L-Tryptophan	(–NH_3^+) H_3N^+–CO_2^-	K	Lys	L-Lysine
(imidazolyl, N, N^+H_2) H_3N^+–CO_2^-	H	His	L-Histidine	(guanidinium: NH_2, NH_2^+) H_3N^+–CO_2^-	R	Arg	L-Arginine
(–SeH) H_3N^+–CO_2^-		Sec	L-Selenocysteine	(pyrroline amide) H_3N^+–CO_2^-		Pyl	L-Pyrrolysine

The nomenclature for stereochemical designations is defined in a 1984 IUPAC publication on biochemical nomenclature for amino acids, including definitions at both the α-carbon and other chiral centers (*193, 194*).

1.4.2 Melting Point

Amino acids generally do not have sharply defined melting points, but tend to decompose over a broad range (250–300 °C). Derivatives, such as the dinitrophenyl amino acid prepared by reaction with 2,4-dinitrofluorobenzene (Sanger's reagent), must be used for accurate measurements.

1.4.3 Solubility

Amino acids exist as zwitterions at neutral pH and generally have negligible solubility in aprotic organic

Table 1.2 Commonly Encountered Non-Coded Amino Acids.

Sarcosine Sar N-methyl glycine MeGly	Betaine	Hippuric acid	Creatine	L-2-Aminobutyric acid Abu	L-Norvaline Nva 2-aminovaleric acid Avl 2-aminopentanoic acid Ape	L-Norleucine Nle 2-aminohexanoic acid Ahx
L-Alloisoleucine alle	L-Cyclohexylalanine Cha	α-Aminoisobutyric acid Aib	β-Alanine β-Ala	Carnitine	γ-Aminobutyric acid GABA	
L-Pipecolic acid Pip	L-3-Hydroxyproline 3Hyp	L-4-Hydroxyproline 4Hyp	L-Kainic acid	L-Pyroglutamate Glp L-5-Oxoproline	L-5-Hydroxytryptophan 5-HO-Trp	
L-3,4-Dihydroxy-phenylalanine DOPA	L-Phenylserine Pse	L-Thyronine	L-Thyroxine Thx	L-allo-Threonine aThr	L-Homoserine Hse	
(4R)-4-[(E)-2-butenyl]-4,N-dimethyl-L-threonine MeBmt	L-3-Hydroxy glutamic acid	L-4-Carboxy glutamic acid Gla	L-2-Aminoadipic acid Aad 2-Aminohexane-dioic acid	L-2-Aminopimelic acid Apm 2-Aminoheptane-dioic acid	LL-2,6-Diaminopimelic acid Dpm, A²pm 2,6-Diaminoheptane-dioic acid	
L-Cystine Cys₂	LL-Lanthionine	LL-Cystathione	L-Cysteic acid Cya	L-Penicillamine	L-Homocysteine Hcy	L-Selenomethionine
L-2,3-Diamino-propionic acid Dpr, Dap, A₂pr	L-2,4-Diamino-butyric acid Dab, A₂bu	L-Ornithine Orn 2,5-Diamino-pentanoic acid	L-Citrulline Cit	L-5-Hydroxylysine 5Hyl	L-Allysine	

solvents, which can make transformations requiring aprotic solvents difficult. Strong acid salts are generally soluble, but strong base salts are not. Aqueous–organic solvent mixtures, such as 1,4-dioxane/water are often employed when basic conditions are required, such as for acylation with the Fmoc group (200, 201). Recently an effective organic solvent system for the dissolution of most amino acids (Asp, Asn, and Gln are the exceptions)

under basic conditions was described (202). The amino acid is dissolved in a solution of strong acid, such as 1M TFA (trifluoroacetic acid), in DMF (dimethylformamide), and an excess of tertiary base, preferably pyridine, is added. The basicity is sufficient to allow for acylation without racemization. A 19:1 pyridine–chlorotrimethylsilane solution has also been used to solubilize an unprotected amino acid (203).

The solubilities of the common amino acids in aqueous solution have recently been theoretically calculated as a function of temperature and pH, and compared to experimental values. Greatly increased solubility is observed at higher temperatures (204). Aqueous solubilities have been tabulated at http://www.imb-jena.de/IMAGE_AA.html.

One unusual property of many amino acid derivatives is their ability to form gels in aqueous and/or organic solutions. N-Benzoyl cystine and related derivatives have been found to be potent gelators that can rigidify water at concentrations as low as 0.25 mM (205). N^{ε}-Lauroyl-Lys ethyl esters with an N^{α}-[8-(1-pyridyl)octanoyl] positively charged substituent were able to gel water at 0.3 wt % (206, 207), while various bis(N-lauroyl)-Lys derivatives could gel both water and organic solvents (208). Glycosyl derivatives of Glu dialkyl esters showed a strong ability to gelate both water and organic solvents (209). Urea-bridged bis(Glu) long-chain diesters containing a methacrylate substituent were effective organogelators at concentrations as low as 1.5 nM; polymerization of the organogels by photo-irradiation significantly increased the gel stability (210). N-Cbz or -Aloc Phe, Phg, and Leu phthaloylhydrazide gelated apolar organic solvents such as toluene and carbon tetrachloride at concentrations as low as 0.2 wt % (211). Boc-Ala, N-nonanoyl Ala, and N-nonanoyl Ala gelated a variety of organic solvents (such as toluene, carbon tetrachloride, n-octane) at concentrations of 4–7 g/L (212). Bis(Leu-OH), bis(Val-OH), and bis(Phg-OH), linked as oxalylamides, were effective gelators of both water and organic solvents (213). Three Phe-OMe residues attached to a 1,3,5-tricarboxy-cyclohexane core induced hydrogelation at concentrations of 0.04 wt % as long as the compounds were heterochiral (LLD or DDL); the homochiral derivatives showed no gelation at up to 1 wt % (214). 11-Aminoundecanoic acid was used to form a number of derivatives that gelated both water and organic solvents at concentrations of 1–20 g/L (215, 216). When also incorporating a chiral α-amino acid unit, some racemic salts could gelate much greater quantities (up to 16 times more) of solvent than the pure enantiomers (216). A series of N-acyl ω-amino acids (ranging from 6-aminohexanoic acid to 13-aminotridecanoic acid) were found to be potent gelators in DMF at concentrations as low as 0.4 mg/mL (0.04 wt %), producing a thermoreversible gel that is stable and transparent for more than a year in a stoppered tube (217). Low molecular weight gelators were reviewed in 2005 (218).

Room temperature ionic liquids have been prepared from 1-ethyl-3-methylimidazolium salts of a number of amino acids. The amino acid ionic liquids are transparent colorless liquids that are miscible with organic solvents such as methanol, acetonitrile, and chloroform, but not ether (1236). A series of ionic liquids were also derived from various amino acid or amino ester chloride, nitrate, tetrafluoroborate, hexafluorophosphate, sulfate or trifluoroacetate salts (219). Alternatively, the amino group of amino esters has been embedded in a 1-methyl-imidazolium ring, with various anionic salt derivatives prepared (OTf, PF_6, NTf_2) (220).

1.4.4 Purification and Analysis

A number of chromatographic techniques have been employed to analyze and purify amino acids, including ion-exchange chromatography (221); liquid chromatography (222), and gas–liquid chromatography (223). Greenstein and Wintz cover more traditional methods, including partition, paper and ion-exchange chromatography (224). The analysis and separation of amino acids are thoroughly discussed in a 2005 monograph, Quantitation of Amino Acids and Amines by Chromatography: Methods and Protocols (164). The analysis of amino acids in physiological samples by high-performance liquid chromatography (HPLC) was reviewed in 1996, focusing primarily on derivatives formed with OPA (o-phthalaldehyde) or PITC (phenylisothiocyanate) (225). An earlier comprehensive review in 1986 examined the methods of analysis of amino acids in body fluids and tissues by HPLC, including OPA, dansyl chloride, dabsyl chloride, PTH (phenylthiohydantoin), PTC (phenylthiocarbamoyl), and 4-fluoro-7-nitrobenzo-2,1,3-oxadiazole derivatives (226). A 1993 review summarized analyses of amino acids in food and physiological samples, with ion-exchange chromatography of underivatized amino acids compared to HPLC separations of various derivatives (227). In 2000 the role of chromatography in the analysis of amino acids, sugars, and carboxylic acids in food was discussed, including a summary of HPLC, capillary (CE), and GC/elechrophoresis analysis of derivatized and underivatized amino acids (228). A 2003 review of HPLC quantitation of amino acids and amines in the same matrix included separations of dabsyl, Fmoc, OPA, fluorescein, PTC, furoylcarboxyl, carbazol, and underivatized amino acids (229). A 1999 review on fast (seconds to minutes) analytical scale separations by CE and HPLC included a number of examples of amino acid separations (230).

The identity of amino acids is usually established by derivatization with various reagents, followed by comparison of HPLC retention times with known standards. The overall composition of proteins is determined by protein hydrolysis followed by amino acid analysis, with the constituent amino acids traditionally separated on an ion-exchange resin then detected by postcolumn derivatization using ninhydrin, fluorescamine, or o-phthalaldehyde. Reviews of protein hydrolysis and determination of amino acid composition have been published in 1988 (231) and 1998 (232). A great number of new methods for the derivatization, separation, and analysis of amino acids have been developed, which will be discussed below. Chiral separations and analyses will be addressed in subsequent sections. The effect of the method of protein hydrolysis on the subsequent amino acid analysis (recovery, reliability, racemization) was examined in a 1998 report. Thermal solution hydrolysis with methanesulfonic acid was found to give the best results, followed by thermal HCl hydrolysis. Microwave-hydrolysis with

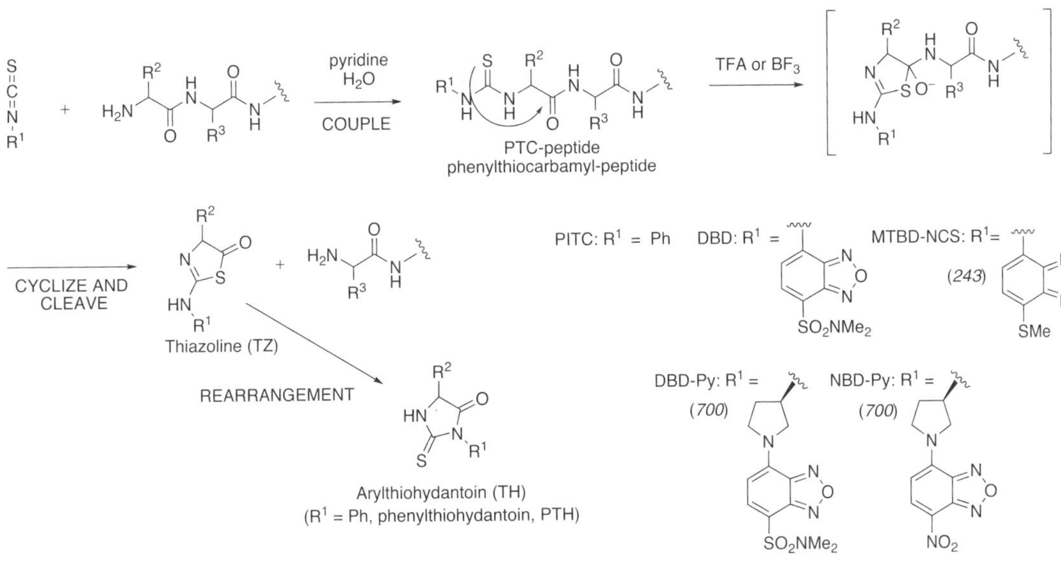

Scheme 1.2 Edman Sequencing.

either acid gave the next best results, although more racemization was observed (233).

The sequence of amino acids in peptides and proteins, as opposed to their overall composition, is generally established by the Edman degradation sequencing technique (see Scheme 1.2) (234). The N-terminal amino acid is reacted with an isothiocyanate (generally PITC) to form a phenylthiocarbamyl peptide intermediate. Under acidic conditions, this cyclizes to form a thiazoline, resulting in cleavage of the N-terminal residue and generation of a new free N-terminal amino group, ready for the next cycle of sequencing. The cleaved thiazoline undergoes rearrangement to a more stable thiohydantoin derivative (PTH amino acid), which is analyzed by chromatography, with picomolar sensitivity. The amino acid residue is identified by comparison of the thiohydantoin retention time with those of known amino acid derivatives (234–236). Both HPLC and TLC separations are possible (237), with a review of amino acid analysis using PTH derivatives in 1988 (231). Separation of PTH amino acids using an aqueous mobile phase is possible (and more environmentally friendly) by using a modified microparticulate silica gel that is polymer-grafted with poly(N-isopropylacrylamide-co-n-butyl methacrylate). Temperature can be used to modify the elution profiles on this support (238). A temperature-controlled separation of 19 amino acids on an ODS silica column has also been described, raising the temperature from 35 to 60 °C during the run to improve resolutions at the beginning and end (239). While PTH amino acids can be successfully separated on a variety of HPLC supports, such as ODS (normally employed) or carbon-coated zirconia, much better separation is achieved by using the two types of columns in tandem, with an ODS column at 30 °C and a C-ZrCO$_2$ column at 40 °C. Thirteen amino acid derivatives were separated

with baseline resolution (240). Micellar liquid chromatography conditions were optimized for the separation of nine PTH amino acids (241). Determination of the amino acid sequence can be combined with determination of D- or L-configuration by separating the PTH amino acids on a chiral column. A β-cyclodextrin-based chiral stationary phase provided enantiomeric separation of 18 PTH amino acids. The PTH derivatives are sensitive to racemization, so BF$_3$ and HCl–methanol replaced TFA as the cyclization/cleavage and conversion reagents to suppress racemization during the degradation process. Some racemization was still observed under these conditions, so the method is not suitable for determining enantiomeric purity, but is capable of distinguishing between antipodes (242). A highly fluorescent Edman reagent, 7-methylthio-4-(2,1,3-benzoxadiazolyl) isothiocyanate (MTBD-NCS), provides greater sensitivity than PITC and generates degradation byproducts that do not fluoresce, and therefore do not interfere with analysis. The adducts are also more stable, and enantiomeric amino acid derivatives can be separated by chiral columns during HPLC analysis (243). Analyses of the PTC intermediate derivatives obtained by reaction of amino acids with PITC (without further rearrangement to the thiohydantoin) are also possible; a 1999 study compared the effects of temperature, eluent flow rate, and different ODS columns on the separation of 27 PTC-amino acids. The method was used to determine amino acid profiles for three apple varieties (244).

Capillary electrophoresis has also been applied to the separation of PTH amino acids, with a gradient elution system resolving 12 PTHamino acids covering the polarity spectrum (Asp and Glu could not be analyzed) (245). The migration times of PTH or FTH (the thiohydantoins derived from FITC, fluorescein isothiocyanate) amino acids during CE can be standardized by the inclusion of two reference markers, allowing for more accurate reproducibility and identification (246). A mixed electrolyte with chiral selectors such as O-trimethyl β-cycodextrin, digitonin, and β-escin resolved 15 PTH-amino acid enantiomer pairs; His, Leu, Lys, and Pro could be resolved using individual selectors but not with the combination. The D-Ala residue in a Met-enkephalin analog was identified by sequencing using this method of analysis (247). A capillary electrochromatography (CEC) system separated 19 PTH amino acids on an ODS-silica support in one report (248), with 13 of 17 resolved in another (249). Both PTH- and NDA (naphthalene-2,3-dicarboxaldehyde)-derivatized amino acids were separated on a CE system using porous polymer monoliths as the stationary phase, with 19 of 21 PTH amino acids resolved (UV detection), and 14 of 15 NDA amino acids (LIF detection) (250). Unusual amino acids in peptides, such as S-ethyl-Cys, S-ethyl-3-methyl-Cys, γ-hydroxy-Lys, phospho-Tyr, 1-aminocyclohexane-1-carboxylic acid, and 4-aminotetrahydropyran-4-carboxylic acid, have been identifed via PTC derivatization followed by CE or HPLC separation, then matrix-assisted laser desorption ionization time- of- flight (MALDI-TOF) mass spectrometry (MS) analysis (251).

C-Terminal sequencing is also possible. One method activates the C-terminal amino acid with acetic anhydride, couples it with thiocyanate to form a thiohydantoin containing the amino acid, and then hydrolyzes the thiohydantoin derivative off the C-terminal end of the peptide (252).

More recently, peptide sequencing by mass spectrometry has become the definitive tool to study the primary structure of proteins, and is an integral component of the new field of proteomics. Mass spectrometry-based sequencing was reviewed in 2004 (253). Tandem MS is generally employed, coupled with computer programs to either match sequences to those in a database, or to provide de novo sequencing. One such de novo sequencing program is called Lutefish (254), while another, PepNovo, uses a probabilistic network modeling (255). An algorithm has been developed to provide de novo peptide sequencing on low-resolution quadrupole ion trap MS instruments (256), while a hidden Markov model NovoHMMM program provided superior results compared to other programs for de novo sequencing analysis (257). Automated de novo protein sequencing has been accomplished by using tandem MS data obtained from electron capture dissociation and collisionally activated dissociation of electrosprayed protein ions, with no additional chemistry required to help identify N- or C-terminal fragments. Proteins larger than approximately 10 kDa could be sequenced if first subjected to limited proteolysis (258). An on-column protein digestion has been combined with tandem MS sequencing, using information–dependent acquisition and an ion search database (259). Deuterium-labeled Lys, $[4,4,5,5-^2H_4]$-Lys, has been incorporated into proteins via in vitro cell culturing to aid in peptide de novo sequencing. The 4-Da mass tag at the C-terminal end of peptides resulting from proteolytic cleavage by

Lys-specific proteases provides a mass signature for all Lys-containing peptides, and can be used to distinguish between N- and C-terminal fragments during sequence assignment (260). N-Terminal amidination has been employed to aid in identification of N-terminal residues and fragments (261), as has N-terminal derivatization with a trityl cation-containing group, which generates a stable charge that survives collision-induced dissociation (262). Simultaneous peptide fragmentation and accurate mass measurement can be achieved using collision-induced dissociation in an electrospray ionization TOF MS instrument, giving superior mass measurements compared to standard tandem MS/MS instruments (263). New techniques are being developed, such as a C-terminal peptide sequencing via multistage MS/MS with quadrupole ion trap or ion cyclotron resonance mass spectrometers (264). A C-terminal sequencing of peptides is possible by using a mixture of carboxypeptidases to incompletely digest the protein, followed by MS analysis of the fragments. The utility of this method is increased if Lys and Cys residues in the peptide are prederivatized in order to distinguish the mass of Lys from that of Gln, and to create a Cys derivative that is a substrate for the carboxypeptidases (265). Peptides attached to a solid support have been sequenced via a partial Edman degradation using repeated treatments with a 9:1 mixture of PITC and PIC (phenyl isocyanate), N-hydroxysuccinimidyl benzoate (Bz-NHS) or N-hydroxysuccinimidyl nicotinooate (Nic-NHS), which provides a "peptide ladder" of sequence-related truncation products via a combination of degradation with PITC and capping with PTC/Bz-NHS/Nic-NHS. The mixture of peptide fragments is then analyzed by MALDI-MS (266, 267). This method is a modification of a procedure originally applied to proteins (268).

As mentioned earlier, considerable effort has been devoted towards developing new methods for the analysis of amino acids. These procedures are generally used to detect amino acids in physiological samples, but can also be applied to the analysis of hydrolyzed proteins or (depending on the derivatizing reagent) used for protein sequencing. One common pre- or post-column derivatization procedure forms fluorescent isoindole derivatives of amino acids using OPA and 2-mercaptoethanol (or other thiol, or sodium sulfite) (see Scheme 1.11 for asymmetric version). For example, immobilized proteins were quantified via hydrolysis, derivatization with OPA/2-mercaptoethanol, followed by HPLC analysis with fluorescence detection (269). A separation of 20 amino acids and 17 amines as their OPA/3-mercaptopropionic acid derivatives was achieved in 53 min on an ODS column with gradient elution; this method was used to analyze three different beers, three wines, and vinegar (270). This method has the sensitivity to detect picomole levels of total free amino acids in biological tissues (271). An isocratic RP-HPLC elution was employed for analysis of OBA/2-mercaptoethanol or OPA/Na_2SO_3 derivatives of amino acids in human serum, with quantitative determination of Ala, Ile, Leu,

Glu, Gln, Asn, Ser, His, Orn, Lys, Arg, Met, Phe, Trp, and taurine at concentrations as low as 0.5 pmol (*272*). Plasma levels of Cys, Hcy, Asp, Glu, Asn, Gln, and Ser were determined by prederivatization with OPA/2-mercaptoethanol followed by HPLC analysis with fluorometric detection (*273*). Post-column OPA derivatization was used to analyze the amino acids in effluents of landfills, composting plants, and fermentation plants, using ion-exchange chromatography for the separation of the underivatized amino acids (*274*). Amino acid fingerprinting of opium samples in order to identify their source was accomplished by post-column derivatization with OPA/*N*-acetyl Cys (*275*).

One problem with the OPA/thiol method is that secondary amines, such as Pro, do not react, while diamino acids such as Lys and Orn can give multiple products depending on how many amino groups react (*276*). A high-throughput method for the analysis of amino acids in foods, beverage, and feedstuff employed a pre-column derivatization of primary amines with OPA/3-mercaptopropionic acid, followed by secondary amine derivatization with Fmoc-Cl (9-fluorenylmethyl chloroformate). Data was obtained for a 25 amino acid standard mixture, along with analyses of soybean cattle cake feed (high in Glu), orange juice (high in Pro and Arg), and red wine (high in Pro) (*277*). Both OPA and Fmoc derivatization was used in a study of protein hydrolysis methods (*233*). The behavior of Orn and Lys with this two-step derivatization has been investigated (*276*). Despite the potential difficulties of multiple derivatives, OPA derivatization was employed for detection of the diaminodicarboxylic acids *meso*-lanthionine and *meso*-2,6-diaminopimelic acid in peptidoglycans from anaerobic bacteria (*278*), and for analysis of N^{ε}-monomethyl-, $N^{\varepsilon},N^{\varepsilon}$-dimethyl-, and $N^{\varepsilon},N^{\varepsilon},N^{\varepsilon}$-trimethyl-Lys (*279*). Diaminopimelic acid has been extracted from stromatolite rocks as a biomarker for the presence of eubacteria, with analysis by OPA//2-mercaptoethanol (*280*). His also undergoes additional reactions with OPA, which were analyzed in a 2004 report (*281*).

In 1999 the stability and characteristics of OPA/3-mercaptopropionic acid derivatives of 24 amino acids were examined (*282*). The derivatives of amino acids that have been reported to be unstable (such as Gly, β-Ala, Lys, and Orn) were found to be stable up to 50 min after formation. Amino acids that can also form double derivatives (Orn, Lys, etc.) required quantitation based on the total of all peaks. Extended reaction times of up to 30 min (with at least 7 min) were needed for quantitative derivatization, along with a 20-fold excess of OPA/thiol (with thiol added prior to amino acid). Impurities in the reagents and their stability towards storage were also tested (*282, 283*). A further study in 2000 optimized the separation and identification of 25 amino acids as their OPA/3-mercaptopropionic acid derivatives, testing different columns, flow rates, and temperature. The best results were provided by the highest possible flow rate at 50 °C. The amino acid content of five different apple types was then determined (*284*).

In the absence of a thiol additive, OPA and other *ortho*-dicarboxaldehydes such as naphthalene-2,3-dicarboxaldehyde (NDA) and anthracene-2,3-dicarboxaldehyde (ADA) form isoindolin-1-ones under acidic conditions. The derivatives are more stable than the isoindole fluorophore normally generated, and are resistant to racemization (*285*). The use of OPA amino acid derivatives was reviewed in a 2001 paper that focused on the different types of thiol additives (*286*) and a 2002 paper looking at the formation, stability, and characteristics of OPA/3-mercaptopropionic acid derivatives via HPLC-MS (*287*). Reactions of OPA with nucleophiles were reviewed in a 2004 *Chemical Reviews* article (*288*). By employing a chiral thiol, OPA derivatization can be used for the determination of enantiomeric purity (see Section 1.8.1, Scheme 1.11).

N-Fmoc (9-fluorenylmethoxycarbonyl) amino acids are highly fluorescent; Fmoc-Cl can be used as a precolumn derivatization reagent (*289, 290*). N^{ε}-methyl-Lys was analyzed in plasma samples using derivatization with Fmoc-Cl followed by HPLC analysis with ultraviolet (UV) and MS detection (*291*). Fmoc-OSu, *N*-(9-fluorenylmethoxycarbonyloxy)succinimide, has advantages over Fmoc-Cl, including reduced reactivity with side-chain functional groups, reduced dipeptide formation, and reduced need for extraction steps. Fmoc-OSu was used to derivatize 20 amino acids and four biogenic polymines, which were then separated within 32 min (other than Asn/Gln and Leu/Ile) using three short C18 columns and a dibutylamine phosphate elution buffer. Beer, wine, and cell extract samples were then analyzed (*292*). 2-(9-Anthryl)ethyl chloroformate (AEOC) gives a one order of magnitude better detection limit than the Fmoc group, and was demonstrated on a mixture of 17 amino acids and on hydrolysates of BSA, soybean, and sunflower seed (*293*). 6-Aminoquinolyl-*N*-hydroxysuccinimidyl carbamate (AQC) derivatizatization, combined with a quaternary eluent system, was able to resolve a mixture of 24 amino acids; collagen hydrolysate and cell culture media were then analyzed (*294*). AQC derivatives are suitable for fluorescence detection with detection limits of 3–7nM. Microdialysate samples from cerebrospinal fluid were analyzed for Asp, Glu, and GABA, with Gly, Pro, and taurine also detected (*295*). An unspecified alkyl choroformate is used in the Phenomenex (EZ) : Faast amino acid analysis kit; over 50 amino acids have been tested, with LC-MS standards shown for 31 amino acids (*296*).

Sulfonyl chloride derivatization has also been employed. A set of 22 dansyl amino acids could be separated within 40 min, with four polyamines also resolved in an additional 10 min. Detection limits of 1–20 pmol were obtained, and the method was used to analyze for amino acids in tissue cultures and foilage of trees (*297*). The separation of 11 dansyl amino acids by cationic micellar LC and reversed-phase LC provides quite different profiles; micelles also enhance the fluorescent intensity (*298*). Dabsyl amino acids have been detected by a sensitive laser-based absorbance method, with an injected 780 fmol detection limit for Gly (*299*). DABS

(4-dimethylaminoazobenzene-4'-sulphonyl chloride) derivatization of amino acids allows for the HPLC separation of 25 different amino acids on a C18 column, with femtomolar sensitivity (300).

Other reagents are used less frequently. N^G-methyl-Arg (NMMA), N^G, N^G-dimethyl-Arg (ADMA), and N^G, $N^{G'}$-dimethyl-Arg (SDMA) were assayed in plasma by derivatization with 4-fluoro-7-nitro-2,1,3-benzoxadiazole (NBD-F), followed by HPLC analysis with fluorescence detection. The method had detection limits of 12–20 nM (301). Highly fluorescent iminium ions have been generated from unprotected amino acids under aqueous conditions by Schiff base formation with a coumarin aldehyde (302). Another fluorescent tagging reagent, carbazole-N-(2-methyl)acetyl chloride (CMA-Cl) produced derivatives with good fluorescence and 10–65 fmol detection limits, with separation of an 18 amino acid mixture, and analysis of a collagen hydrolysate. Single peaks were observed despite the presence of a chiral center on the reagent (303). The Edman reagent, PITC, has also been used to derivatize amino acid mixtures from hydrolyzed whole cells, followed by HPLC analysis (58). The determination of the amino acid contents of plasma, serum, urine, thyroid gland, and algae was achieved by extraction, conversion to the PTC amino acids, and HPLC analysis. Nineteen amino acids were analyzed (304). The free amino acids in ripening cheese were also monitored by derivatization with PITC and reversed-phase (RP) HPLC analysis (305).

The ability to separate and analyze underivatized amino acids provides advantages over derivatization procedures. In 2002 a study compared detection methods for the analysis of underivatized amino acids. UV detection at 210 nm is suitable for some amino acids; urinary Met, Tyr, Phe, His, Trp, creatinine, and creatine were analyzed via separation on a cation-exchange column with phosphoric acid–methanol eluent (306). The advent of evaporative light scattering detection (ELSD) provides the opportunity to directly analyze underivatized amino acids, as the ELSD response does not require fluorophores or UV absorbance. Mixtures of amino acids can be readily separated by simple C_{18} chromatography (307). A porous graphitic carbon stationary phase was tested for ion-pair chromatography of 20 underivatized amino acids with ELSD detection: nonafluoropentanoic acid as ion-agent gave a good separation of all 20 amino acids in a 40-min run, with an elution order quite different from normal ODS packings, and faster equilibration times (308). Mass spectrometry also provides for analysis of underivatized amino acids. Ten polar underivatized amino acids (Asp, Asn, Ser, Gly, Gln, Glu, Cys, Thr, Ala, and Pro) were separated by ion-pair reversed-phase HPLC with pentadecafluorooctanoic acid as the ion-pairing agent, and ELSD detection (309); this method was extended to analysis of all 20 proteinaceous amino acids, with all 20 detected by ES-MS and 17 of 20 by ELSD (310). Underivatized amino acids can also be analyzed on a C_8 HPLC column by ion-pairing with sodium 1-heptanesulfonate (311). Twenty-two underivatized amino acids were separated on a C_{18} column in 36 min using perfluoroheptanoic acid and TFA as mobile-phase modifiers with an acetonitrile/water gradient. The amino acids were detected and quantified by MS/MS analysis; the method was applied to the analysis of amino acids in human serum, and gave better recovery and precision than the OPA-derivatization HPLC method (312). Another report used nonafluoropentanoic acid as the ion-pairing reagent, with 20 amino acids determined by ionspray MS-MS (313). A very rapid determination of amino acids has been achieved by using a teicoplanin chiral stationary phase coupled with electrospray tandem MS, with an analysis of 20 amino acids possible in less than 4 min. The rapid elution did not provide comprehensive separation of the amino acids, and no enantiomer discrimination. Quantification was achieved by using five deuterated amino acids as internal standards. The method had detection limits of 50–450 ng/mL, and was used to monitor fermentation profiles (314). An HPLC assay using hydrophilic interaction chromatography coupled to tandem MS was used to analyze underivatized Met and taurine in energy drinks (315). A 12-amino acid test mixture was separated using ion-pairing with pentadecafluorooctanoic acid on an ODS column, then analyzed using either a chemiluminescent nitrogen detector (CLND), ELSD, ES-MS, ES-MS-MS, UV, conductivity detector, refractive index, or NMR. The UV and RI detectors were unable to detect all of the test amino acids, while the CLND and ES-MS-MS detectors gave the best sensitivity (0.1–0.7 mg/L) and specificity; the ELSD was easy to use and had a reasonable limit of detection (1–10 mg/L) (316).

Another method of amino acid detection not requiring pre- or post-column derivatization employs integrated amperometry, based on amine oxidation at a gold electrode. The analysis of 17 amino acids gave concentrations that correlated well with those obtained by ninhydrin derivaization; Cys and Met required a preoxidation step (317). In another study underivatized amino acids and sugars were simultaneously measured by anion-exchange chromatography followed by integrated pulsed amperometric detection. Eighteen amino acids and three sugars (318), or 17 amino acids and nine carbohydrates (319), could be separated in a single run, with 0.5–5 pmol detection limits. Application to the analysis of green tea found that Glu and its side-chain ethyl amide derivative, theanine, were the major components (318). Factors affecting the separation and detection of amino acids by anion-exchange chromatography followed by pulsed amperometric detection were discussed in a 2003 paper; optimized conditions allowed for analysis of 19 amino acids in 60 min, with 0.15–4.5 pmol detection limits (320). A Ni-based composite electrode has also been applied to amperometric detection of underivatized amino acids, with the amino acid content of several cheeses determined by ion-exchange chromatography (321). Conductivity detection was employed for the analysis of carnitine in food supplement formulations, with ion-pair chromatography separation using octanesulfonate eluent (322).

Another analytical technique that has been employed for analysis of amino acids is gas chromatography–mass spectrometry (GC-MS) analysis, which requires formation of volatile amino acid derivatives (see Section 1.9 for derivatives used for chiral GC analysis). GC-MS analysis of N(O,S)-trifluoroacetyl amino acid isopropyl esters was applied to a mixture of 24 amino acids, and to adetermination of the amino acids contained in the pyoverdins (323). Trifluoroethyl ester ethylchloroformate-derivatized amino acids have also been analyzed (324). The amino acids in cycad seeds were analyzed as N-ethoxycarbonyl ethyl ester derivatives using GC-MS; reference samples of 52 protein and non-protein amino acids were first analyzed (325). Both MS and flame ionization detection (FID) were used during an analysis of 36 different amino acids that were derivatized with ethyl chloroformate (ECF) followed by N-methyl-N-(tert-butyldimethylsilyl)trifluoroacetamide (MTB-STFA). Carboxylic acids and keto acids were also detected, with limits of detection ranging from 0.01 to 55 ng/mL. The method was employed to analyze a urinary sample from a patient with phenylketonuria (326). Citrulline has been measured in watermelon by derivatization with MTB-STFA followed by GC-MS analysis, allowing for separation from Glu, which often coelutes in HPLC separations (327). Both MS and FID detection methods were also employed to analyze amino acids in honey, using a commercial derivatization kit (Phenomenex EZ:faast, an undefined alkyl chloroformate) to prepare the amino acids. A 32 amino acid standard was separated in under 8 min and limits of detection were determined for each amino acid for both detectors. Sample preparation and analysis was possible in less than 15 min (328). The EZ:faast alkyl chloroformate reagent has been tested with over 50 amino acids, with one GC analysis with FID detection separating 33 amino acids (296). Amino acid analysis on a femtomolar scale has been achieved by a GC-MS method. Hydrolyzed protein samples were mixed with a cocktail of all 20 common amino acids, in a stable isotope (^{13}C and ^{2}H) form. The mixture was acylated, benzylated, and silylated to form volatile derivatives; then analyzed by GC/MS. The retention times of each amino acid were used to provide identity, with the ratio of each amino acid to its stable isotope analog providing quantification of absolute amounts of each of the native amino acids in the original protein sample (329).

For detection of organic molecules on Mars, GC is still considered one of the best available space instruments. Three amino acid derivatization techniques were compared for compatibility with space travel—silylation with N,N-tert-butyl(dimethylsilyl)trifluoroacetamide, methylation with tetramethylammonium hydroxide, or acylation with methyl chloroformate. Standards of 18 amino acids were separated using each reagent. Silylation proved to be most suitable (330). Amino acids can be isolated from natural samples by a simple sublimation procedure that could be useful for extraterrestrial analysis, involving heating with 6 N HCl for 150 °C for 4 h, followed by vacuum drying, then a vacuum sublimation at 250 mTorr in a sealed tube at 450 °C for 30 s or 1100 °C for 5 s. Amino acids were sublimined with 80–100% recovery for Gly, Ala, and Val, but reduced recovery (10–50%) for Asp and Ser. No racemization was observed, and the isolated amino acids did not require desalting before derivatization/analysis (331). Pyrolysis of amino acids in the presence of hexamethyldisilazane followed by online GC analysis produces a characteristic distribution of products for each amino acid, with Gly, Ala, Val, Leu, Ile, Nle, Met, and Phe yielding predominantly the TMS ester of the parent amino acid, Trp, Tyr, and His releasing side-chain degradation products, Ser, Thr, and Asp undergoing reductive deamination, and Tyr, Cys, and Met producing decarboxylated amines (332).

Capillary electrophoresis (CE) and capillary electrochromatography (CEC) have become increasingly useful for amino acid analysis. Reviews of CE analyses of amino acids have been published in 2001 (333), 2003 (334), and 2006 (335), with pre-, post-, and on-column derivatization of amines for capillary electrophoresis reviewed in 1998 (336). Eight amino acids and four neuroactive amines were analyzed by CE using pre-column derivatization with OPA/t-butyl thiol, giving detection limits of 80–200 amol and allowing for analysis of neurotransmitters in microdialysis samples from the striatum of live rats, and tissue samples from butterfly brains (337). A microfluidic chip was employed to derivatize amino acids with OPA, then push the derivatives into a high-speed CE instrument with laser-induced fluorescence (LIF) detection. The probes could be implanted into the striatum of anesthetized rats with Glu, Asp, taurine, Gln, Ser, and Gly detected at 20 s intervals over 4 h (338). Levels of D-Ser (and L-Ser, Glu, Asp, and GABA) in rat striatum have been monitored by online microdialysis, derivatization with OPA/mercaptoethanol, CE separation on a hydroxypropyl-β-cyclodextrin separation buffer, and LIF detection. This was accomplished with a 10 s analysis time, allowing for continuous monitoring (339). Similarly, biological levels of Asp and Glu were detected by CE of derivatives formed online with OPA/β-mercaptoethanol and LIF detection, with analysis possible every 10 s (340). Amino acids have been derivatized with OPA in a zone in the middle of the capillary during analysis, with the OPA amino acids (Ala, Met, Glu, and Phe) then separated and detected (341). Post-column derivatization of separated amino acids with OPA/2-mercaptoethanol has also been employed, again using LIF detection (342). Sensitivity during analysis of amino acids by CE with OPA/tert-butyl thiol has been improved by using two scavenging reactions to remove excess derivatizing reagent. Sixteen amino acids were separated with detection limits of 90–350 pM (20-80 amol). The method was sensitive enough to analyze for neurotransmitter amino acids from probes in anesthetized rats, examining 200 μL samples collected every 10 s (343).

Naphthalene-2,3-dicarboxaldehyde (NDA) and cyanide were introduced into a living erythrocyte cell to derivatize the amino acids within the cell. The contents of a single cell were then analyzed by CE using amperometric detection, with six amino acids identified (344). Biological levels of GABA, Glu, and Ala can be monitored by derivatization with NDA, followed by separation by mixed micellar electrokinetic chromatography, with fluorescence detection. Levels of GABA in plant tissues were then determined (345). An analytical procedure to determine baclofen levels in human plasma (4-amino-3-p-chlorophenylbutyric acid, used clinically as a skeletal muscle relaxant) employed derivatization with NDA, CE separation, and LIF detection, with a detection limit of 10 ng/mL (346). Attomole detection of amino acids is possible using 3-(4-carboxybenzoyl)-2-quinoline-carboxaldehyde (CBQCA) for derivatization, with CE separation and LIF detection. An 18 amino acid mixture was separated, and the method was used to detect changes in amino acid concentration in media during bacterial growth (347). Amino acids in brain microdialysate samples were analyzed via pre-column derivatization with 5-furoylquinoline-3-carboxaldehyde (FQ) and LIF detection; a 13 amino acid standard was separated (348). An in-capillary derivatization with FQ separated a 5 amino acid standard and was used to analyze plasma samples, though peaks were broad (349). FQ was also used to derivatize amino acids in apple extracts in order to analyze for 1-aminocyclopropane-1-carboxylic acid; a 16 amino acid standard was separated (350).

Pre-column derivatization with FITC (fluorescein isothiocyanate) was employed for the analysis of 15 amino acids during wine aging, with micellar electrokinetic chromatography and LIF detection (351). FITC/LIF was also employed for an analysis of amino acids in plant cells; with SDS in the electrolyte, 14 of 21 amino acid standards were separated, but by adding α-cyclodextrin all 21 were separated (352). Amino acid amides have been analyzed by blocking the amino group by acetylation, converting the amide to an amine via a Hofmann rearrangement, and forming FITC derivatives for analysis by CE with LIF detection. Eleven amino acid amides were analyzed (353). 5-Carboxyfluorescein succinimidyl ester was tested as an amino acid derivatization reagent for CE analysis in 1998; it allowed for fluorescent detection of nanomolar levels of amino acids such as Lys and GABA, and had much better sensitivity than FITC (354). Post-column derivatization with fluorescamine gave up to twofold better sensitivity than OPA, depending on derivatization conditions (355).

Nineteen N-2,4-dinitrophenyl amino acids (DNP-AA) were separated by CE, with the addition of Mg^{2+} to the buffer found to enhance separation (356). Amino acids in serum were analyzed as DNP derivatives, with 2–8 µM detection limits for the 16 amino acids standardized. Cys, Trp, and Phe were found to be significantly elevated in patients with chronic renal failure, and levels of Ser, Ile, and Val were depressed (357). A pressurized gradient CE separation of 18 2,4-dinitrophenyl amino acids has also been described (358). The separation of DNP and dansyl amino acids by capillary isotachophoretic focusing can be improved by the addition of micelles (359). An organic polymer capillary formed from

poly(methylmethacrylate) was used to separate five OPA amino acids or 15 dansylated amino acids (360). A study compared the CE analysis of dansyl amino acids (11 examples) using direct detection to underivatized amino acids (five examples) using indirect UV detection (with dimethylbenzylamine as the electrolyte chromophore); the dansyl derivatives had more than an order of magnitude lower detection limit, and greater linear range (361). Acetonitrile gels of trans-(1S,2S)-1,2-bis(dodecylamido)cyclohexane provided superior CE separation of 14 dansyl amino acids compared to the corresponding buffer solution (362).

A micellar electrokinetic CE separation of 15 amino acids derivatized with 4-fluoro-7-nitro-2,1,3-benzoxadiazole (NBD-F) gave a rapid analysis (1 min derivatization with 22 min separation) with amol detection limits. This study also provided a comparison of the performance of CE separations of other derivatives (reaction time, derivative stability, analysis time, detection limits) (363). Five NBD-F derivatized amino acids were separated on a porous photopolymerized sol–gel monolithic column in 7 min, with the CE analysis applied to rat cerebrospinal fluid (364). 1,2-Naphthoquinone-4-sulfonate (NQS) derivatives have also been resolved by CE, though Phe/Ile/Trp were only partly resolved and His/Leu overlapped completely (365). A subsequent paper showed separation of a 19 amino acid mixture, using a continuous flow derivatization system. A feed sample hydrolysate and a pharmaceutical sample were also analyzed (366). An in-capillary derivatization strategy was also described, with a 12 amino acid mixture separated (367). An analysis of carnitine and acylcarnitines in urine samples derivatized the carboxyl group with 4'-bromophenacyl trifluoromethanesulfonate (368).

A number of methods have been used to analyze underivatized amino acids by CE. Direct UV detection at 214 nm is possible, with detection limits of 2–80 µg/mL for the amino acids tested; this method, with a fused silica capillary, was used to analyze the five amino acids contained in an enteral medicinal tablet (369). Underivatized Trp and Phe were analyzed using UV-pulsed laser-induced fluorescence detection, with 0.15 nM and 50 nM detection limits, respectively (370). Free amino acids in beverage samples were separated by CE in a strongly acidic octanesulfonic acid electrolyte, with UV detection. Different types of beer produced different amino acid profiles, with 10 different amino acids analyzed (371). A monolithic stationary phase formed from N,N-dimethylacrylamide-piperazine diacrylamide was used in a CEC separation of Phe, Trp, and His with UV detection at 214 nm, employing a combination of electrostatic and hydrophobic interactions (372). Another monolithic column was prepared by polymerization of polystyrene–divinylbenzene in the presence of copper–iminodiacetate, with Trp, Tyr, and Phe separated (373). Capillary zone electrophoresis and capillary isotachophoresis separations were employed for carnitine analysis in food supplements, with UV or conductivity detection, respectively (374). Underivatized amino acids have been analyzed by CE using indirect UV detection, in which the electrolyte contains a highly UV-absorbing species. Calixarenes have been employed as both indirect detection and selection agent. Seven amino acids were separated (375). Another group used a Rh(III) tetrakis(phenoxyphenyl)porphyrinate as a wall modifier during open tubular CEC of a 17-amino acid mixture; 15 amino acids were separated within 20 min (Glu and Asp were not), and limits of detection varied from 2 to 1350 µg/mL (376). Indirect fluorescence, with 2,6-pyridine dicarboxylic acid in the electrolyte, was also employed for a CE analysis that could simultaneously detect amino acids, organic acids, carbohydrates, and inorganic anions. The separations of 82 compounds, including 18 amino acids, were tested, with 43 compounds (14 amino acids) well separated in a single run. The amino acids in soy sauce and a nutrient tonic were also determined (377).

Twenty-one underivatized amino acids (the standard 20 protein amino acids plus the Arg analog canavanine) were simultaneously resolved and detected by capillary electrophoresis coupled with electrospray ionization mass spectrometry, providing a powerful analytical technique (CE/ESI-MS) (378, 379). Use of a chiral electrolyte, (+)-(18-crown-6)-2,3,11,12-tetracarboxylic acid, allowed for enantiomer resolution at the same time (378, 379). Refractometric detection was tested with a CE separation of seven underivatized amino acids (380).

Improvements to CE analyses include preconcentration of underivatized amino acids on-line using a positively charged sol–gel coating, providing 150,000-fold enrichment of an Ala sample. This method allows for injection of large volumes of samples onto a CE system (381). Very high resolution has been obtained by using a counterflow in the direction opposite that of the electrokinetic migration of the analyte, allowing for the resolution of underivatized Phe from isotopic d_8-Phe (382).

One significant advantage of CE and CEC instrumentation is that they lend themselves to miniaturization on microchips. A microchip CE analysis of seleno amino acids used OPA/ME derivatization to analyze for selenomethionine, selenoethionine, and selenocysteine, with amperometric detection (383). Amino acids prederivatized with NDA have been separated by electrochromatography in a microchip containing a lauryl acrylate polymer monolith stationary phase, with five amino acids (Arg, Ser, Gly, Phe, and Trp) separated in 2 min (384). The surface of a poly(methyl methacrylate) (PMMA) microchip was modified by dynamic coating with a quaternary ammonium starch derivative, allowing for separation of NBD-labeled amino acids (Asp, Glu, Gly, Val, and Leu) within 110 s (385). A PMMA microchip was used to analyze NBD-derivatized amino acids from Japanese green tea, showing high levels of theanine and Arg (386). Dansyl amino acids were analyzed in a microchip using chemiluminescence detection, giving a detection limit of 0.4 fmol for Lys. However, most amino acids coeluted with Gly (387). Three underivatized amino acids (Arg, His, Cys) were separated by microchip electrophoresis with pulsed ampherometric detection (388).

Indirect fluorescence can be used to detect underivatized amino acids separated on an electrophoretic microchip, using fluorescein in the buffer (389). Another PMMA microchip was used for an isotachophoresis electroseparation with conductivity detection, separating up to seven amino acids (390). A poly(dimethylsiloxane) (PDMS)/glass hybrid microchannel was employed for a CE analysis of four underivatized amino acids, using a single carbon fiber cyclinder working electrode for amperometric detection (391).

For thin layer chromatography (TLC), amino acids are traditionally detected using a ninhydrin solution, which produces a purple/violet color with most amino acids. A modified ninhydrin reagent was described in 2001. Addition of N-cyanoguanidine resulted in distinguishable color differences, depending on the amino acid (and before or after heating), ranging from a pale lilac for Gly and Cys, through a reddish pink for Ala, Val, Leu, Ile, Ser, Glu, and Gln, to a deep pink for Thr and Met, and a brownish yellow for Asn and hydroxy-Pro. The detection limit was similar to the traditional ninhydrin reagent, ranging from 0.03 to 0.1 mg after heating (392). 5-Sulphosalicylic acid has also been combined with ninhydrin as a spray reagent for the visualization of amino acids on TLC plates; a range of colors from orange to pink to grey was observed (393). Amino acids have been prederivatized with phenyl isothiocyanate (to introduce a sulfur atom), then developed by reaction with iodine-azide, which produces white spots on a violet-gray background. The method had significantly better sensitivity than iodine or UV detetion (394).

Simulated moving bed chromatography has been investigated as a possible industrial process of continuously separating amino acids, with Phe and Gly used as a test example (395).

Numerous efforts have been made at developing new methods to detect amino acids. Underivatized free amino acids in mammalian cell culture have been quantified by MALDI-TOF MS (matrix-assisted laser desorption ionization time-of-flight mass spectrometry) by monitoring the mass of each amino acid. Reference spectra were produced analyzing mixtures of 12 amino acids (396). Another MS-based method of analyzing for amino acids is based on electrospray ionization followed by ion mobility spectrometry, in which the mobilities of the different ionized amino acids are measured in various drift gases. The 20 common amino acids showed different drift times, and the amino acid components could be identified in mixtures of up to 10 different amino acids. The technique has potential as a simple, robust instrument for interstellar sampling (397, 398). The oxidation spectra of amino acids at a copper electrode provided a unique fingerprint for different amino acids when measured using sinusoidal voltammetry, though only Arg, Asn, and Gln were compared (399). Amino acids in beer were quantified by integration of their ^{1}H NMR spectra; reference spectra of 20 amino acids in water were tabulated. The method had a sensitivity of about 10 mg/L, and correlated well with HPLC or CE analyses (178).

A TOCSY (total correlation spectroscopy) NMR experiment was combined with statistical metabonomics analysis in order to test the feasibility of analyzing amino acids in honey (*400*).

A biosensor has been developed that uses three different enzymes to measure concentrations of L-Phe: L-Phe dehydrogenase, which converts Phe to phenylpyruvate and NAD⁺ to NADH; salicylate hydroxylase, which detects the NADH by using it and O_2 to convert salicylate to catechol; and tyrosinase, which converts catechol to *o*-quinone, which is detected by an electrode. The enzymes were immobilized on a carbon paste electode, and gave a linear detection range of 20–150 μM. This detection system could theoretically be readily extended to other amino acids by using the appropriate dehydrogenase (*401*). A Glu sensor with sub-micromolar sensitivity has been created by immobilization of Glu dehydrogenase on optical fiber probes, with detection by measurement of the fluorescence of NADH (*402*). Immobilized Glu oxidase/peroxidase and Lys oxidase/peroxidize have been packed sequentially in a tube and used to simultaneously detect L-Glu and L-Lys via generation of hydrogen peroxide followed by peroxidase-catalyzed luminol/hydrogen peroxide reaction, with chemiluminescent detection (*403*). A similar immobilized amino acid oxidase/horse radish peroxidase detector was attached to a CE instrument and used to detect six amino acids (Met, Leu, Tyr, Phe, His, and Trp) (*404*). These types of sensors can also be used for enantiomeric analysis of amino acids (see Section 1.13). The use of amperometric biosensors for analyzing amino acids was reviewed in 2004 (*405*). A Cys-selective electrode has been developed based on lead phthalocyanine incorporated into a plasticized poly(vinyl chloride) membrane, coated onto a graphite electrode. The sensor detected Cys over a wide linear range (1×10^{-6} to 5×10^{-2} M) and showed a much greater response to Cys than to other amino acids, including Met. It was able to measure Cys spiked in human serum samples (*406*).

A novel method that could potentially be used to prepare sensors for individual amino acids has been described. Cyclopeptides were immobilized on a glass support, with the presence or absence of an analyte binding to the peptide determined by measuring the change in optical film thickness. Good selectivity was observed for Arg over other amino acids with micromolar detection limits, although no enantiodiscrimination was observed. Libraries of cyclopeptides can be screened to identify selective receptors for other amino acids (*407*). A chemosensor array has been developed that allows for colorimetric identification of the 20 natural amino acids. It uses an indicator displacement assay (IDA), employing differential displacement of several indicators from a receptor. The IDA system consists of an organometallic Rh cyclopentadienyl complex as the receptor, with gallocyanine, xylenol orange, and calcein blue as the indicator dyes. The amino acids were tested at several different pH values (*408*). A palladium–azo complex formed from the azo dye methyl orange and tetrachloropalladate(II) produced solutions of different colors when reacted with aqueous solutions of amino acids with different side chains, ranging from mauve for Cys to various shades of blue for Ala, Lys, and Tyr (*409*). Half-sandwich complexes of rhodium(II) reversibly bound to an indicator dye selectively responded to aqueous solutions of peptides containing His and Met (*410*). A colorimetric sensing system, based on a guanylated 2,2′:6,2″ terpyridine-Zn(II) complex and pyrocatechol violet, provided some selectivity for Asp over other amino acids (*411*). A fluorescent receptor for zwitterionic amino acids was based on triaza-18-crown-6 ether with two pendant guanidinium groups and an anthryl fluorophore. Binding of Lys and GABA induced a very strong fluorescence, while Gly gave a small increase, and other amino acids (Ala, Phe, Val, Ser, Glu, Arg) showed no response (*412*). Another polyamine bicyclic macrocycle formed a dicopper(II) complex with L-Glu, displacing a quenched rhodamine indicator from the cage to give a fluorescent signal. The complex discriminated against related amino acids such as Gly, Ala, Asp, and GABA (*413*). A cholic acid-based fluorescent sensor binds dicarboxylic acids, including Glu and Asp, with a strong preference for Glu (*414*). A fluorescent detection system for His has also been reported, making use of a binuclear Cu(II) bisdien macrocycle which selectively binds the imidazole moiety (*415*), while a fluorescent sensor of DOPA was derived from lucifer yellow dye (*416*).

A fluorogenic aldehyde, based on 4-(carboxyethyl) benzaldehyde with an fluorophore amide substituent, formed a fluorescent thiazolidine with Cys. Only Cys, out of 10 amino acids, glutathione, and glucose, resulted in a significant fluorescence increase after 30 min (*418*). Homocysteine (Hcy) can be directly detected in the presence of structurally similar molecules such as Cys and glutathione via reaction with methyl viologen at gentle reflux, producing a blue solution via formation of a reducing α-aminoalkyl radical. Alternatively, reaction with fluorone black in the presence of triphenylphosphine at room temperature selectively produced a 510 nm-absorbing species (*419*). Another visual detection system for Cys and Hcy was reported in 2004, using an aldehyde-substituted yellow chromophore that reacts with the amino and thiol groups of Cys/Hcy to form a red-shifted derivative. Other common thiols (Met, glutathione) or amino acids did not react (*420*). An aqueous sensing mixture that showed high selectivity for Cys over other amino acids, including Hcy, consisted of *N,N,N′,N′*-tetra-(2-pyridylmethyl)-*p*-xylylenediamine, cadmium perchlorate, and pyrocatechol violet in a 1:2:1 ratio. The blue solution turned yellow in the presence of Cys, which could be quantified by the decrease in absorbance at 665 nm (*421*). Gold nanorods provided selective detection of Cys and glutathione over 20 other amino acids (including Met). The method measures the change in absorption of the Au nanorods at 900 nm in acetonitrile/water solution upon addition of the amino acids (*422*). A Phe-selective ionophore constructed from linked trifluoroacetophenone moieties allowed for potentiometric determination of Phe (*423*).

Crown ethers attached to heteroaromatic cations were found to give selectivity for Lys over other amino acids (His, Ser, Arg, Gly, Leu) (424), while binaphthyl-crown ether receptors with a pendant arylboronic acid or dinitrophenylurea showed discrimination for longer amino acids such as GABA (425). A crown ether attached to a xanthone unit showed differential extraction of amino acids from saturated water, with 98% of Gly extracted, 74% of Phe, and 25% of Ala (426). A ternary complex formed from a binaphthyl-xanthone-based macrocycle, 18-crown-6 ether, and unprotected amino acids showed even better extraction, with 100% of Phe, Phg, and Ala, but no extraction of more polar amino acids such as His or Asp. Chiral recognition was also demonstrated (427). A ferrocene-based receptor with two pendant arms demonstrated a strong 1:1 binding to unprotected α-amino acids; the structure potentially provides a template to build in enantiodiscrimination or selectivity for specific amino acid types (428). A cyclen macrocycle with four cholesterol appendages bound dansylated amino acid anions to give fluorescent complexes of varying intensity and λ_{max} (429). ^{1}H NMR titration was employed to test the binding of 11 amino acids to calix-[n]-arenes with various substituents; Lys and Arg bound strongly to all the derivatives, but the pendant groups altered the specificity for other amino acids (430). A macrocyclic dicopper(II) complex containing two 4,4'-di(aminomethyl)diphenylmethane moieties linked by dodecyloxy-substituted pyridine rings efficiently transported β-amino acids across a chloroform membrane, with a clear preference for β-amino acids (β-Phe) over α-amino acids (Phe) or γ-amino acids (GABA) (431). Substituted guanidine amides of pyrrole-2-carboxylic acids form strong complexes with amino acids, as determined by UV titration studies (432). Simple symmetrical acyclic thiourea receptors showed modest discrimination between N-protected amino acid enantiomers, in addition to discriminating against side chains (433). Modular building blocks with the potential to recognize amino acid side chains within peptides were disclosed in 2002, based on an aminopyrazole (to bind the peptide backbone) linked via Kemp's triacid to the side-chain recognition element (434).

Polyamine macrocycles containing 1H-pyrazole units act as artificial Glu receptors, complexing L-Glu in water at physiological pH values (435). A monocyclic polyamine binuclear Cu^{2+} complex selected L-Glu over L-Asp (436), while a macrobicyclic receptor with biaryl ring and thiourea linker bound Gln and Asn more strongly than Ala (437). Artificial Arg receptors have also been described, relying on phosphonate interactions with the guanidinium group (438, 439). A macrocyclic receptor with four strategically placed phosphonate anions strongly bound Lys in a 1:1 complex in water; His, Orn, and Arg associated less strongly in 2:1 complexes (440). A spirobisindane skeleton with two phosphonate substituents formed strong 1:1 complexes with Arg and Lys (441), while a water-soluble urea macrocycle formed from p-xylylenediamine and Asp showed

moderate affinity for Arg (442). A rigid planar heterocyclic system containing two pyridine-2-carboxylate rings and two fused pyridines was designed to strongly complex the Arg side-chain guanidinium group, and showed much stronger binding to Arg than Lys (443). Glycine forms a ternary complex with a Zn(II) complex of a monoxo-tetraamine macrocycle, mota; an X-ray crystal structure of the crystalline complex was obtained (444). Receptors with the ability to discriminate between amino acid enantiomers are also discussed in Sections 1.6 and 1.13.

High plasma total homocysteine (Hcy) concentration has recently become recognized as a risk factor for vascular diseases such as atherosclerosis and thromboembolism. The clinical importance of plasma Hcy levels has led to a fundamental patent dispute over the right to patent a natural phenomenon (445). A number of methods for the analysis of total plasma Hcy content have been published, including HPLC analysis after derivatization with SBD-F (ammonium-7-fluorobenzo-2-oxa-1,3-diazole-4-sulphonate) (446, 447), CMQT (2-chloro-1-methylquinolinium tetrafluoroborate) (448, 449), o-phthalaldehyde/2-mercaptoethanol (273), or monobromobimane (450), or CE analysis using fluorescein isothiocyanate (451) or 6-iodoacetamidofluorescein (452) derivatives and LIF detection, or in-capillary derivatization with 2,2'-dipyridyl disulfide followed by detection of the released 2-thiopyridone (453). A chiral separation of Hcy has been achieved via thiol derivatization with 4-aminosulfonyl-7-fluoro-2,1,3-benzoxadiazole followed by separation using CE with a γ-cyclodextrin electrolyte (454). A NIST MS-based method for certification of serum standard reference materials uses solid-phase anion extraction (SPE) followed by GC/MS, LC/MS, or LC/MS/MS analysis, or protein precipitation followed by LC/MS/MS analysis, using isotope dilution mass spectrometry with added $[3,3,3',3',4,4,4',4'-^2H_8]$ homocystine (455). A solid phase extraction of both Hcy and folate from plasma, with LC/tandem MS analysis, was reported in 2005 (456). Biological levels of N-acetyl Cys are also of interest; an analytical method for the determination of N-acetyl Cys in human plasma was developed for a phase 1 clinical trial, using analysis by LC/MS/MS. The detection limit was 50 ng/mL of plasma (457). The amount of N-acetyl Cys in a pharmaceutical formulation, along with Cys, cystine, N,N'-diacetylcystine, and N,S-diacetyl-Cys impurities, was determined by an LC-UV-MS method, where the UV detection quantified the large amounts of N-acetyl-Cys and the MS detection was able to analyze the low levels of impurities (458).

1.4.5 Racemization

An extremely effective asymmetric synthesis procedure is of little use if the amino acid product racemizes during subsequent reactions, such as protecting group removal. Conversely, intentional racemization of amino acids can be very useful for recycling an unwanted enantiomer, and is required for kinetic dynamic resolutions.

Racemization rates are also of considerable interest to geologists and archaeologists as they can be used to date geological samples via the increase in the ratio of D:L enantiomers after a living organism dies (459–464); total racemization occurs over about 5–10 million years (5). The age of fossil bone samples can be determined by D:L amino acid ratios, with His, Phe, Asp, Ala, Ile and Val contents useful for age determination of samples which are 2000–12,000, 3000–20,000, 5000–35,000, 10,000–80,000, >30,000, and >50,000 years old, respectively. The amino acid half-lives ranged from 5500 years for His to 180,000 years for Val (465). The ratio of D- and L-enantiomers of amino acids in fossils can also be compared to $\delta^{13}C$ and $\delta^{15}N$ values in order to determine the extent of amino acid contamination from external sources (466). Hair samples of varying ages, from a pre-dynastic man (3200 BC) to Egyptian mummies to 8–10th century graves, showed a correlation of age with extent of Pro racemization (467). Significant levels of D-amino acids are found in soil and litter samples, with up to 17% D-Asp, 22% D-Glu, 15% D-Ala, and 5% D-Phe, D-Tyr, and D-Leu. The high levels of D-Ala and D-Glu are probably due to their presence in bacterial cell walls, while Asp is prone to racemization. The fresh litter (<0.5 years old) contained less racemized amino acids than the decomposed litter (2 years old), and the amounts of D-amino acid in both litter samples were significantly less than in the soil samples (organic matter age 400–500 years) (468). Partially racemized Tyr, with 79% ee, was identified in the insoluble protein fraction of aging human eye lenses (107). D-Asp residues have been detected in sun-damaged skin in elderly people, and have been proposed to form through racemization of Asp residues in elastin. A kinetic study estimated the time for complete racemization of Asp residues within different elastin sequences to be 50–100 years at 37 °C (108). A good description of racemization is provided by J.L. Bada (469), while a monograph published in 2000 on perspectives in amino acid and protein geochemistry is predominantly concerned with amino acid racemization, kinetics, and practical examples of dating (470).

Amino acids can racemize under basic or acidic conditions, though for the proteinaceous amino acids elevated temperatures are required for significant racemization to occur. A number of methods have been developed to monitor racemization during hydrolysis by using deuterated or tritiated HCl (471–474). Strong base causes more rapid racemization than strong acid, with little racemization occurring during standard protein hydrolysis (6 N HCl for 1 h at 110 °C) (475). However, a study of 17 amino acids under acid-catalyzed conditions showed significant racemization after 24 h at 110 °C, especially for Cys (7.9%), Asp (5.0%), and His (3.4%) (476). Similar racemization rates were determined for Glu, Asp, Ala, and Phe in a 2001 report (477). Racemization even occurs slowly at neutral pH (459, 478), as evidenced by enolate formation of glycine methyl ester at pH 7.4 (479). Ionic strength, buffer concentration, and complexation by metal ions are all known

to affect racemization rates (459, 460, 475). A study of hydrolysis-induced racemization of 13 different amino acids from six different environmental sources (ranging from soil to fresh and decomposed litter to proteins) found that not only did the extent of racemization vary between amino acids but also it varied for each amino acid depending on the sample source. Anywhere between 0% (Glu) and 85% (allo-Ile) of the detected D-amino acids (which ranged from 0.7% for Thr to 17% for Asp) were due to hydrolysis (468).

However, if preparative-scale quantitative racemization is required, free amino acids are generally difficult to racemize. It was not until 2001 that it was reported that water appears to suppress the racemization of amino acids (480). Heating amino acids in polar organic solvents such as DMF, DMSO or ethylene glycol under alkaline conditions (KOH, K_2CO_3 or Et_3N) caused much greater racemization of Phg, Leu, Phe, and Trp than using water or water-containing solvents under the same conditions. For example, 2 h at 100 °C in DMF with 1.2 equivalents of K_2CO_3 produced Phg with 56% ee, while in water it was recovered with 94% ee. Similarly, Phg racemized extensively in AcOH, but this was suppressed in water or aqueous AcOH. Significant decomposition was also observed in the organic solvents (480). A comprehensive 1997 review of racemization (356 references) includes tables of thermal, redox, enzymatic, acid-catalyzed, and base-catalyzed racemization of amino acids or amino acid Schiff bases, describing types, conditions, and yields (481).

The mechanism of racemization involves rate-limiting loss of the α-methine proton, producing a resonance-stabilized planar anion (460, 469, 475). Supporting evidence for this mechanism includes the fact that the rates of racemization and of α-H isotope exchange are identical. Since a negative carboxylate group charge will inhibit α-C anion formation, esters and amides racemize more quickly than acids under basic conditions. A protonated amino group will increase racemization rates by stabilizing the incipient carbanion, so N-protected amino acids tend to racemize less quickly under acidic conditions. However, under basic conditions, an electronegative substituent such as an N-acyl group will accelerate the rate of racemization (460). The carbon acidities for $^+H_3NCH_2CO_2^-$, $^+H_3NCH_2CO_2Me$, $^+Me_3NCH_2CO_2^-$, and $^+Me_3NCH_2CO_2Me$ in aqueous solution were determined to be $pK_a = 28.9, 21.0, 27.3,$ and 18.0, respectively, confirming the beneficial effect of a positive charge on deprotonation, and the ability of a negative charge to reduce racemization (482). The paper also discussed the implications of these results on possible mechanisms of enzymatic racemizations. Internal amino acids within a protein generally racemize approximately 10 times faster than free amino acids, which indicates that a deprotonated carboxyl group is more effective at preventing racemization than a protonated amino group is at promoting it (469). Second-order rate constants have been determined in D_2O for deprotonations of the N-terminal α-amino carbon of Gly-Gly and Gly-Gly-Gly zwitterions, of Ac-Gly anion and Ac-Gly-NH$_2$, and of the central residue of a Gly-Gly-Gly anion, with pK_a values ranging from 23.9 to 30.8. The N-terminal residue of zwitterionic Gly-Gly-Gly ($pK_a = 25.1$) deprotonates approximately 130-fold faster than the central residue (483). Enzymatic deprotonations are known to be accelerated by Schiff base formation with pyridoxal phosphate; the increase in α-carbon acidity of Gly-OMe in the presence of acetone was found to be 7 pK units (from 21 to 14) by an NMR study (484). The racemase activity of pyridoxal enzyme models can be increased by the attachment of basic substituents (485).

Racemization rates are also very dependent on the amino acid side chain, with steric, inductive, resonance, electrostatic, and solvation effects, as well as intramolecular reaction, all playing a part (459, 460, 475). The α-hydrogen pK_a values of Ala, Val, Leu, Ile, Phe, and Phg were calculated to range from 14.9 to 17.0, based on their rates of racemization (486). A 1989 study reexamining the effect of the side chain of proteinaceous amino acids on racemization rates found that hydroxy amino acids are the most prone to racemization of the 13 examples studied under basic conditions (Ser > Ser(tbu) > Asp(OMe) > Asp >Thr > Met > Phe > Thr(tBu) > Ala > Glu(OMe) > Glu > α-Aaa > Pga > L) while alkyl groups retard the rate. Thus, Ser racemizes eight times faster than Ala at pH 8.0 and 140 °C (475). A similar relative stability was found in a previous report that estimated a racemization half-life for Ser of 4 days at 100 °C and pH 7.6, compared to about 300 days for Ile (469). Another study of racemization of amino acids in aqeuous solutions at 100 °C at pH 7.0 found Ser >> Asp, Met, Phe, Glu, Tyr, Lys, Pro > Ala, Val, Ile, Leu, while at pH 2.5 all rates were accelerated, with Asp > Ser > Glu, Ala > Met, Phe, Tyr, Lys > Leu, Pro > Val, Ile (487). A 2003 experimental analysis found L-Ser possessed only 52% ee after 30 h at pH 6 and 160 °C, while Thr racemized to 80% ee. [2,3,3-D₃]-Ser was employed to measure racemization rates by mass spectrometry analysis of the [3,3-D₂]-Ser product produced by loss of the α-deuterium (488). The high racemization reactivity of the hydroxy amino acids is proposed to result from inductive stabilization of the α-carbanion, combined with stabilization of the carboxylate anion, and solvation of the base participating in proton removal (475). Steric effects hindering attack of the base at the α-center were also found to be important, with Thr racemizing at approximately one-quarter the rate found for Ser (475). Bulky N-protecting groups have been used to help prevent racemization by sterically blocking proton removal (489).

Other amino acids prone to racemization include Phe, which racemizes significantly faster than Ala, Val, or Leu over a range of pH (490), and Asp acid (459, 475). Certain classes of non-proteinaceous amino acids are also inclined to racemize, with the β,γ-unsaturated amino acids probably the worst offenders due to isomerization to the α,β-unsaturated isomer. For example, (E)-3,4-didehydroornithine completely racemized in 15% AcOH at 95 °C in less than 10 min, and in 6 M HCl at 105 °C it racemized at least 550 times faster than Leu (491). Similarly, arylglycines racemize

from 40–500 times faster than Ala, depending on aromatic substitution (460). The effects of neighboring residues on racemization within dipeptides has been studied; a C-terminal Pro residue greatly increased the rate of epimerization of the N-terminal residue, while a more basic N-terminal residue increased racemization of the C-terminal group (492). Asp acid residues within peptides can racemize via cyclization to a succinimide species that is susceptible to CH bond fission; a racemization mechanism for Asn residues within peptides has also been identified (493).

Racemization during peptide coupling is obviously of great concern, and a number of reviews have been published (494–496). Studies of racemization during peptide coupling are accomplished by examining diastereomeric dipeptide products by NMR (497), HPLC (497, 498), or CE (499), or by enantiomeric amino acid analysis of peptide hydrolysates (500, 501). Even modern coupling methods are susceptible to racemization (502), as outlined in much greater detail in Section 12.5. His and Phe are recognized as being predisposed to racemization during peptide coupling, and are often used to measure the effectiveness of new coupling reagents to prevent racemization (e.g., see (503) and (504), respectively). The racemization of His has been reviewed (505). A report of cysteine racemization during solid-phase peptide synthesis has also been published (506). The intrinsic racemization tendencies of amino acids under conditions used for peptide synthesis were examined using urethane-protected N-carboxy anhydride (UNCA) derivatives, as the only method available for these compounds to racemize is via enolization (α-proton abstraction). In the presence of triethylamine (TEA) in toluene, Ser(Bn) had a racemization half-life of only 3 min, compared to 30 min for Phe, 11 min for Asp(OBn), and 18 h for Val. Diisopropylethylamine (DIEA) and TEA induced the most racemization of the tertiary amine bases examined, while the solvents dichloromethane and dimethylformamide resulted in faster racemization than toluene or THF (507). N-Trifluoroacetyl amino acids are also recognized as being especially prone to racemization during coupling (508). Silica gel caused extensive racemization of N-3,5-dinitrobenzoyl Leu during coupling to aminopropyl-functionalized Si gel, with EEDQ the only coupling reagent that minimized D-Leu formation (509).

It is sometimes desirable to racemize amino acids, such as during recycling of an unwanted enantiomer following resolution. As mentioned earlier, polar organic solvents increase racemization rates (480). Conditions have been reported for a practical preparative method to racemize optically active amino acids, by heating a solution of the amino acid in acetic acid for 1 h at 80–100 °C in the presence of 0.05 molar equivalents of an aromatic or aliphatic aldehyde (510). A protonated Schiff base intermediate is proposed to stabilize the α-C anion formation. Racemization of amino acid Schiff bases or other derivatives is used during dynamic resolutions of amino acids (see Section 12.1.7). Heating aqueous solutions of protein amino acids with glucose, fructose or saccharose induces partial racemization of the amino acids. This is proposed to occur during the Maillard non-enzymatic browning reaction, in which reducing sugars react with amino components to form flavor compounds and melanoidins (*487*). Enzymes which racemize *N*-acyl amino acids have been reported (*511–513*), including a serine racemase from a mammalian brain (*98*). The mechanism of racemization by glutamate racemase was reviewed in 2002 (*514*). A polarimetric assay for the measurement of α-amino acid racemase activity has been developed, using Glu racemase from *Lactobacillus fermentii*, expressed in *E. coli*. (*515*).

Another potentially useful racemization method involves grinding the amino acid in a ball mill. Leucine was completely racemized in a few minutes by steel balls in a steel drum (*516*). More recently, racemization has been catalyzed by transition metals. In the first report, 5 mol % of [Rh(cod)Cl]₂ and 50 mol % of P(*c*Hex)₃ induced nearly complete racemization of Ac-Phe-OH when heated together in acetonitrile at 60 °C for 48 h (*1233*). It was subsequently found that Pd(PPh₃)₄ was a more effective catalyst, with 2.5 mol % Pd(OAc)₂ and 12.5 mol % PPh₃ giving the same result for the same substrate (*517*). However, the extent of racemization with both catalysts was very dependent on the *N*-protecting group and type of side chain. *N*-Unprotected Phe-OMe was almost completely racemized (1% ee remaining) after heating in toluene at 110 °C for 12 h in the presence of a ruthenium cyclopentadienyl catalyst (*518*). The original conditions of 5 mol % [Rh(cod)Cl]₂ and 50 mol % P(*c*Hex)₃ were applied to Ac-Phe, Ac-Leu, Ac-Met, Ac-Asp, Ac-Phg, and Ac-Pro, producing products with 0–93% ee after 48 h at 60 °C (*519*).

Other methods of increasing the propensity of amino acids to racemize are discussed in Sections 12.1.6 and 12.1.7, in which racemization is employed for dynamic resolutions. The measurement of enantiomeric purity is obviously very important in assessing the extent of amino acid racemization, and is discussed in great detail in subsequent sections.

1.5 Absolute Configuration

The great majority of the general synthetic methods allow the absolute configuration of the resulting amino acid to be predetermined by the stereochemistry of the reaction sequence. However, in some cases when new amino acids are synthesized or new methodologies are developed it may be necessary to determine the absolute configuration of the product. There are two approaches—an unambiguous proof or a determination based on empirical rules. General reviews describing the determination of absolute configuration have been published (*520, 521*).

1.5.1 Absolute Determination

Historically, absolute configurations have been determined by rigorous chemical correlation, via conversion of an established compound into the amino acid of unknown configuration without breaking a bond at the asymmetric carbon, or by degradation of the amino acid of unknown configuration to a known compound (*522*). The other method of conclusive proof is via X-ray crystallographic analysis of a diastereomeric derivative formed with a chiral auxiliary of known configuration, or by a more sophisticated analysis of the crystal structure of the pure amino acid involving anomalous scattering. Obviously neither of these time-consuming methods is desirable for routine analysis.

1.5.2 Empirical Determination

All other methods of assigning absolute configuration rely upon the empirical creation of a set of rules describing some observable characteristic which is consistently found to differ between the (*R*)- and (*S*)-amino acids. Traditionally this has been done using biological or optical methods, but in the past 30 years these procedures have been largely superseded by NMR or chromatographic analysis of diastereomeric derivatives. Many of the methods used for determining enantiomeric composition (see Sections 1.6–1.13) can also be applied to assigning absolute configuration if a suitable parameter can be established, such as the order of elution for chromatographic separations. The danger with this empirical approach, especially when new unusual amino acids are being synthesized, is the possibility that there may be an exception to the rule. The more examples used to determine the empirical rule, the more assured one can be of a correct assignment of absolute configuration.

1.5.2a Biological Methods

Biological methods generally involve an enzyme-mediated transformation of an amino acid. The amino acid chirality is assigned based on whether or not it reacts with an enzyme whose enantiomeric preference has already been established (*523*). To be of use, the enzyme must have good optical specificity while tolerating a wide substrate specificity. A potentially useful development was described in 1998, when it was reported that antibodies can be raised against the amino acid chiral center. The antibodies show good L/D selectivity and broad side-chain tolerance (*524*).

1.5.2b Optical Methods

The first optical property used to establish absolute configuration was a positive change in optical rotation (measured at the sodium D-line) observed on acidification of L-amino acids (*522, 525*). However, a few exceptions have been noted (*525*). More consistent results are provided by the shape of a circular dichroism (CD) curve. Aqueous solutions of L-amino acids (55 examples) show a positive Cotton effect (CE) band at 200 nm in water and 208–210 nm in acid (*522*). A rule has also been developed relating the optical rotary dispersion (ORD) CE sign and amplitude to the conformation and absolute configuration of the amino acid (526). A CD detector can be used in combination with UV detection during the separation of amino acids by HPLC methods (527, 528). Amino acids have been N,N-dialkylated with 2-(bromomethyl)quinoline. Cu(II) complexes of these derivatives fix the geometrical relationship of the two chromophores, forming a propeller-like twist with axial chirality resulting from the amino acid absolute configuration. Analysis of 20 different amino acids demonstrated strong and predictive exciton-coupled circular dichroism spectra (529).

The optical properties of a number of other amino acid derivatives have been examined, in an attempt to shift the wavelength of interest to a frequency with fewer possible conflicting absorptions. Many of the early derivatives are based on compounds resulting from Edman stepwise degradation of peptides, potentially allowing for the identity and configuration of unknown amino acids to be determined in one step. N-Thiobenzoyl derivatives of L-amino acids were initially reported to have a positive CE at a useful 395 nm in etheral solutions (11 examples), but not in methanol (530). A number of exceptions were later found at this wavelength (531), but at 290 nm a solvent-independent positive CE was observed for all 22 amino acids examined, except for the cyclic imino acids (532). N-Dithiocarbamate derivatives (533, 534), with a positive CE at 330 nm, have been found to be superior to 3-phenyl-2-thiohydantoin and N-dithiocarbethoxy compounds (533). N-Methylthiocarbamoyl derivatives (535) have also been examined, but side reactions and racemization present potential problems. N-(2-pyridyl N-oxide) amino acids have a useful 330 nm ORD CE, but show opposite signs, depending on if a side chain chromophore absorbs between 215 and 330 nm (536). The absolute configuration of β-aryl-β-hydroxy-α-amino acid methyl esters has been determined by derivatization of the hydroxy and

amino groups with a 7-diethylaminocoumarin-3-carboxylate chromophore, followed by CD exciton measurement (537).

More recently the C-terminus of N-Boc amino acids has been derivatized with a chiral oxazolidin-2-selone (see Scheme 1.3) (538). All seven L-amino acids that were examined showed a positive CE at 265–270 nm. The absolute configuration was also indicated by TLC (L-amino acids eluted more slowly), UV (L-amino acids had a red shifted n→π* transition), and ^{77}Se chemical shift (L-amino acids were more shielded).

Metal complexes have been employed for optical determinations of absolute configuration. L-amino acid molybdenum complexes prepared from [Mo$_2$(OAc)$_4$] have negative 400 nm and positive 300 nm CE (539). Another complex, [Pd(dmba)(Aa)] is readily formed by reaction of the amino acid with [Pd(dmba)(acac)] (see Scheme 1.4) (540). A stable white solid is precipitated, which gives, for L-amino acid complexes, a characteristic negative CE band at 265–280 nm and a weaker positive CE band at 305–320 nm. This method also works for amino acids with an additional stereogenic center (16 examples, e.g., Thr, Ile), but not for those with an additional free carboxyl or amine group which can participate in coordination (e.g., Glu, Lys), although this group can be protected. The NMR shift reagent europium(III)-tris-(1,1,1,2,2,3,3)-heptafluoro-7,7-dimehyl-4,6-octanedione, [Eu(fod)$_3$], has been employed to form complexes of α-amino esters in chloroform. These complexes possessed characteristic CD spectra in the 250–350 nm region (541).

Amino acids have been amidated with 1,2-diaminoethane to give diamines. Complexation with a bis(zinc) porphyrin) "tweezer" allows for the determination of absolute configuration by circular dichroism (nine examples) (542). A variation of this approach acylated amino esters with N-(3-aminopropyl)-Gly, and complexed

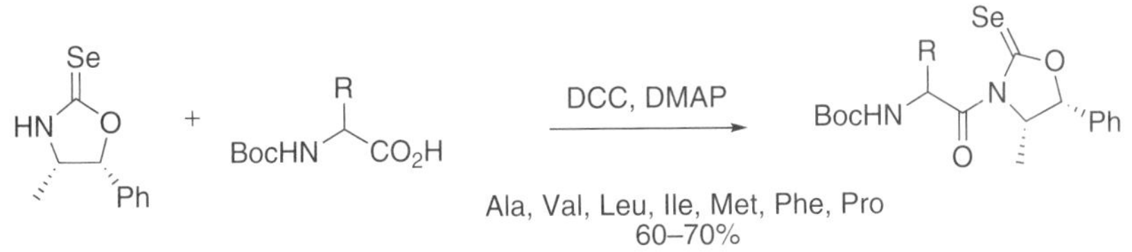

Scheme 1.3 Selone Chiral Derivatizing Reagent for Determination of Absolute Configuration (538).

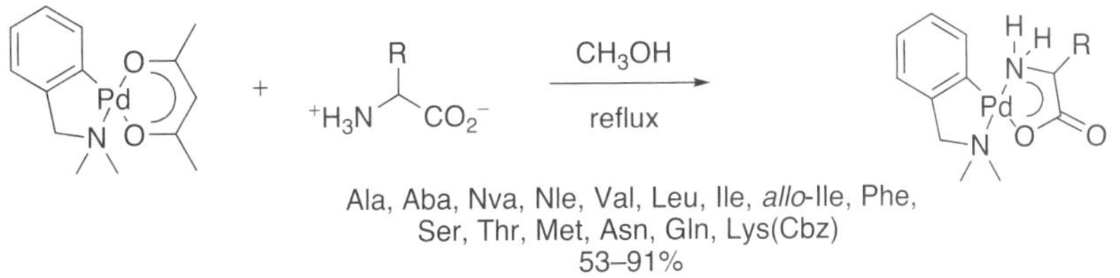

Scheme 1.4 Pd(dmba) Amino Acidato Complexes for Assignment of Absolute Configuration (540).

the resulting diamine bidentate carrier with a dimeric zinc porphyrin host. The optical rotation can be predicted based on steric and electronic factors (two examples, Ala-OMe and homoserine lactone) (543). A gadolinium(III) porphyrin complex of underivatized amino acids has been shown to give distinctive CD spectra for the enantiomers of 16 of 19 amino acids (544).

A novel method of detection of amino acid enantiomeric purity described in 2003 also appears to have great potential as a method for determining absolute configuration. A stereoregular poly(phenylacetylene) with a bulky crown ether attached to each aryl group (see Scheme 1.5) forms a one-handed helix in the presence of amino acids, creating a large induced circular dichroism (ICD) in the UV–visible region. The ICD effect is produced by both small quantities of amino acid (0.005 equivalents), and by very small enantiomeric excesses (as little as 0.005% ee for Ala). More importantly, all 19 L-amino acids tested produced a negative $\Delta\varepsilon_{2nd}$, while the one D-amino acid tested had a positive sign (545). Another crown ether derivative, attached to a carboxy-substituted ferrocene, showed distinctive CD spectra upon binding to hydrophobic D- or L-amino acids (546).

A review of the determination of absolute configuration by chiroptical means is included in Chapter 3 of Volume 4 of *The Peptides* series (139), and is also discussed by C. Toniolo in Barrett's monograph (547).

1.5.2c NMR Methods

The absolute configurations of many organic compounds have been determined by empirical rules derived from NMR studies of the different chemical shifts of diastereomeric derivatives. Many of the chiral auxiliaries used to determine enantiomeric purity have been applied to assign absolute configuration. A comprehensive review of the assignment of absolute configuration by NMR was published in 2004 (548), with a shorter "practical guide" appearing several years earlier (549). A number of amino acid examples were included. Thus the ^{1}H NMR spectra of Mosher's acid (methoxytrifluoromethylphenylacetic acid, MTPA) amide derivatives of amino acid methyl esters (nine examples, (550, 551)), O-methylmandelic acid (or methoxyphenylacetic acid,

MPA) amide derivatives of amino acid ethyl esters (three examples, (552)), other aryl (in particular 1-naphthyl and 9-anthryl) methoxyacetic acid amide derivatives of amino acid methyl esters (three examples, (553, 554)), (R)-O-aryllactic acid (ROAL) amide derivatives of amino acid methyl esters (three examples, (555)), and the ^{1}H and ^{31}P NMR spectra of diazacyclophosphamide derivatives of amino acid ethyl esters (one example, (556)) have all been described as useful for assigning absolute configurations of primary amines including amino acids (see Scheme 1.6). The assignment of absolute configuration by the Mosher method can be verified by the addition of the diamagnetic lanthanide shift reagent La(Hfaa)$_3$, which reverses the usual relative shift change and allows for assignment of substrates with only a few protons (557). Boc-Phenylglycine (Boc-Phe or BPG) proved to be better than MTPA or MPA at determining the absolute configuration of primary amines using ^{1}H NMR shifts, including Phe-OMe and Trp-OMe (558). α-Cyano-α-fluoro-p-tolylacetic acid (CFTA) has been used as a MPTA replacement, providing improved $\Delta\delta$ values in both ^{19}F and ^{1}H NMR spectra for several ester derivatives when compared to MTPA (559). In 2001 the use of CFTA ^{19}F NMR chemical shifts in determining the absolute configuration of amide derivatives of amino acid esters was described. A model consistent with the results was developed (560). A method for determining the absolute configuration of secondary alcohols derivatized as MPA esters, but using only one enantiomer of MPA, was described in 1998; the procedure should also be applicable to amino acids (561). A model for the determination of absolute configuration of primary amines derivatized with MPA or MTPA has been developed (562), but MPA amide derivatives of 4-oxo-α-amino acids give results that differ from model predictions (563). A "mix and shake" method of determining absolute configuration by NMR using MPA, MTPA or BPG has been developed. The chiral auxiliary is activated with an anhydride linkage to a polystyrene resin, then simply mixed with the amine and CDCl$_3$ in an NMR tube for 5 min. The resin floats to the top of the solution, allowing a ^{1}H NMR spectrum to be directly obtained of the amidated diastereomeric derivative. By using a 1:2 (R)/(S)-auxiliary resin mixture, a single spectrum allows determination of the absolute configuration, as the upfield/downfield shifts can be determined from the different ratios of the two diastereomeric amides (564). A very similar procedure, in which the resin was filtered off, was described several years previously (565). The ^{77}Se NMR spectra of oxazolidin-2-selone amide derivatives of amino acids (566, 567) (see Scheme 1.2) have also been applied to the assignment of chirality (seven examples, (538)).

Chiral lanthanoid shift reagents have found use in the determination of absolute configurations. 1,2-Propanediaminetetraacetatoeuropium(III) has been used to assign chirality to α-amino acids (18 examples, (568)) and α-methyl-α-amino acids (12 examples, (569)). A tris[3-((trifluoromethyl)hydroxymethylene)-(+-camphorato] samarium(III) complex, Sm(tfc)$_3$, gives good resolution

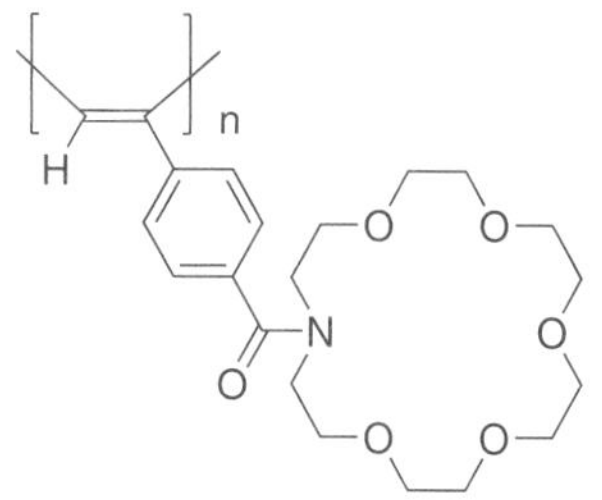

Scheme 1.5 Poly(phenylacetylene-crown ether) for Detection and Amplification of Enantiomeric Imbalances (545).

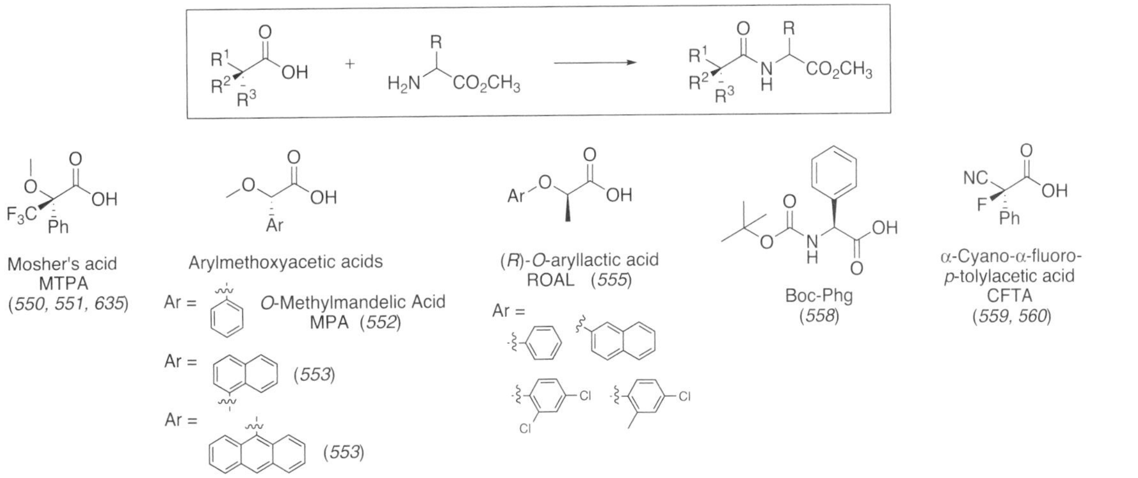

Scheme 1.6 Synthesis of Diastereomers for Determination of Absolute Configuration using NMR Analysis.

of amino acid esters with greatly reduced line broadening. Twelve amino acid methyl esters gave consistent correlations for their $\Delta\Delta\delta$ (570).

An achiral perdeuterated benzoyl derivatizing reagent allows for analysis of amino acid esters by deuterium NMR when used in combination with a chiral cholesteric liquid crystal solvent system; 15 amino acids showed consistent trends between enantiomers (571). The same methodology was carried out with amino acids doubly labeled on the amine and carboxylate with acetyl-d_3 chloride and methanol-d_4, allowing for two measurements in one sample (572). (R)-$(-)$-1-Phenyl-2,2,2-trifluoroethanol was employed as a chiral solvent to determine the absolute configuration of a 5,5-dimethyl-Pro-OMe derivative (573).

1.5.2d Chromatographic Methods

Many of the chromatographic methods for determination of optical purity have reported trends in retention times that could be useful for assigning absolute configuration. The application of chiral liquid chromatography to the determination of absolute configuration of enantiomers was reviewed in 2004 (574). The most common method for amino acid analysis employs Marfey's reagent (FDAA), with the L-amino acid diastereomer derivatives generally eluting before the D-isomers during HPLC analysis (575). However, a more recent study study using LC-MS and a racemization step to unambiguously assign chirality found an exception, Arg (576, 577). Further studies with a greater range (35) of amino acids and derivatives identified more exceptions (β-hydroxy-Asp, Orn, citrulline, N-methyl Asp) (578). A mechanism was proposed to account for the order of elution, based on the difference in hydrophobicity between the side chain and carboxyl group. The elution order can be elucidated from the average retention time of the L- and D-amino acid derivatives, with Ser/Asn situated at the turning point

for the elution order (578). Modified Marfey's reagents, FDLA (1-fluoro-2,4-dinitrophenyl-5-L-leucinamide) (579–581) and FDPEA (1-fluoro-2,4-dinitrophenyl-5-(R,s)-phenylethylamine) (582), have also been used to assign absolute configurations. Orn shows an unexpected reversal in elution order when derivatized with FDLA during determination of absolute configuration by Marfey's method; Lys and Dab do not show this abnormal behavior (583). FDLA gave better results if HPLC separation was combined with mass spectrometry analysis; the elution order of 28 amino acid derivatives was determined (581). Racemic FDLA could be employed instead of racemic analyte in order to compare elution times (581). Analysis with Marfey's reagent has become quite common for determining the absolute configuration of amino acids contained in natural products, including N-methylated amino acids (e.g. 584–591, 1237, 1238). The absolute configuration of thiazole-containing amino acids in peptides has been determined by derivatization with FDLA (592). Cys residues often decompose during acid hydrolysis of proteins. This can be prevented by protection with 4-vinylpyridine; after hydrolysis the liberated S-pyridyl-ethyl Cys can be derivatized with Marfey's reagent for chirality determination (593). The use of Marfey's reagent, and related derivatives, was reviewed in 2004 (594).

A variety of other derivatives have been tested. During the HLPC separation of o-phthalaldehyde/ N-isobutyryl-L-cysteine derivatives of amino acids, the L-amino acid isomer consistently elutes before the D-enantiomer (595). The elution order of amino acid α-methoxyphenylacetate derivatives varies, but a model has been developed to predict which isomer elutes first on a normal-phase column (596). Diastereomers of amino acids derivatized with α-methoxy-α-trifluoro-methylpropionic acid were separated on a polyethylene glycol column (597). GITC (2,3,4,6-tetra-O-acetyl-β-D-glucopyranosyl isothiocyanate) has been applied to the determination of absolute configuration of amino acids in natural products (589). D-Amino acid derivatives of (4S,5R)-4-methyl-5-phenyl-oxazolidin-2-selone migrate more rapidly than the L-enantiomers during TLC (538).

In 2004 the potential of four different derivatization reagents for chirality determination was evaluated with 17 amino acids, including unusual amino acids such as allo-Thr, allo-Ile, β-methoxy-Tyr, and Tle. FDAA, GITC, OPA/IBLC, and S-NIFE (N-(4-nitrophenoxy-carbonyl)-L-Phe methoxyethyl ester) were compared, with ES-MS detection. In most cases FDAA gave the best enantioselectivity, but lowest sensitivity. It was recommended that more than one CDA (chiral derivatizing agent) be used for structural assignment, and that authentic standards be used to confirm the results (598).

Absolute configurations have been assigned based on elution order of achiral amino acid derivatives on chiral stationary phases. Phenylthiocarbamoyl derivatives of amino acids (PTC-AAs) have been analyzed using a tandem column system, with a reversed-phase octyl silica providing an initial separation of the amino acids, followed by a column with a phenylcarbamoylated β-cyclodextrin chiral phase resolving the enantiomers. A mixture of 18 racemic amino acids and Gly gave 37 peaks with a 140 min run time, though all peaks were not completely resolved. Partial racemization of the PTC-amino acids was observed, but the method was suitable to show the presence of D-Thr in a Leu enkephalin analog (599). Determination of the amino acid sequence by Edman degradation can be combined with determination of D- or L-configuration by separating the PTH-amino acids on a chiral column. A β-cyclodextrin-based chiral stationary phase provided enantiomeric separation of 18 PTH-amino acids. The PTH derivatives are sensitive to racemization, so BF₃ and HCl/methanol replaced TFA as the cyclization/ cleavage and conversion reagents to suppress racemization during the degradation process. Some racemization was still observed under these conditions, so the method is not suitable for determining enantiomeric purity, but is capable of distinguishing between antipodes (242). PTH acid enantiomers could also be analyzed on (+)-tartaric acid-impregnated silica gel (600), and underivatized amino acids on berberine-impregnated silica TLC (601). Acetylated D-amino acid esters elute before the L-enantiomers during GC analysis on Chirasil-L-Val coated columns (602, 603).

1.5.2e Other Methods

Electrospray ionization of amino acids in the presence of a chiral selector, such as Boc-L-Phe, produces protonated trimers. Collision-induced dissociation spectra then generates protonated dimer products, with the ratio of trimer:dimer depending strongly on the chirality of the amino acid analyte. The chirality of all 19 common amino acids could be differentiated (604, 605).

A color-based method of determining absolute configuration utilizes a nematic liquid crystalline host, doped with a chiral auxiliary consisting of the imine formed from an amino acid ester and p-(p-methoxyphenyl)-benzaldehyde. Addition of the same imine derivative formed from an analyte results in a color shift, ranging from deep red for one enantiomer, to yellow for racemic sample, to violet for the other enantiomer. However, the method was only demonstrated with Phg as both auxiliary and analyte (606).

1.6 Determination of Enantiomeric Purity

One of the most important characterizations of amino acids is determination of their enantiomeric purity, especially with the development of increasingly powerful asymmetric synthetic techniques. For any asymmetric synthesis it is obviously critical that the optical purity of the product is established. It is also important to accurately determine the degree of asymmetry introduced by the asymmetry-inducing reaction, a value that is potentially masked by purification of the product before analysis. If some resolution occurs during purification the true degree of asymmetry caused by the reaction will be overstated. Older synthetic papers, in which many products are isolated by crystallization, are especially prone to these errors and should be examined carefully.

Assays for amino acid optical purity are also important in other areas of science. For example, the presence of homochirality in extraterrestrial amino acids is considered proof of extraterrestrial biological origin (5), and levels of racemization can be employed to determine the age of fossils (459–467). Very sensitive analytical techniques are required in order to determine optical purities of over 99% ee. The advent of new analytical methodologies has revealed limitations in what were once considered homochiral compounds — commercial sources of L-serine have been found to possess only 98–99% ee (607, 608), and virtually all other L- or D-amino acids are contaminated with their antipodes (104, 609, 610).

Optical rotation values have traditionally been employed to assign optical purity (e.g., see Greenstein and Wintz (582)). Unfortunately these measurements are notoriously unreliable as they are extremely dependent on solvent, concentration, and the presence of other ions, and often involve very small rotations. For example, alanine possesses $[\alpha]^{25}_D = +1.6°$ in H_2O, +13° in 5 N HCl, +2.7° in 1 N NaOH, and +29.4° in HOAc (c = 2, T = 25°) (611). Many older papers describe the enantiomeric purity of a product in terms of optical purity, which is usually determined by the formula: % op = $[\alpha]_{observed}/[\alpha]_{homochiral}$ × 100. The result effectively corresponds to the modern definition of enantiomeric excess, or % ee: % ee = (% R or S isomer) −(% S or R isomer); or % ee = 100 × ([R or S isomer]− [S or R isomer]) / ([R or S isomer] + [S or R isomer]). However, older papers should be closely examined, as some of them use the term optical purity to refer to the percentage of the major enantiomer (e.g., 85% optical purity in these papers describes an 85:15 ratio of enantiomers, corresponding to 70% ee). Given that many other methods are now available to accurately assess the extent of homochirality, optical rotation should not be relied upon as proof of enantiomeric purity. A 2006 article examines the relevance of the "% ee"concept, arguing that enantiomeric ratios provide a better description of stereoselectivity and stereocomposition, expressed as either a ratio with a denominator of 1 (e.g., for 96% ee $q = 98/2 = 49$), or as a percentage composition ratio (e.g., 98:2) (*612*).

A novel method of detection of amino acid enantiomeric purity described in 2003 appears to have great potential as a method for analyzing extremely small enantiomeric excesses, which could be very useful for applications such as explorations for evidence of extraterrestrial life (see Section 2.1). A stereoregular poly(phenylacetylene) with a bulky crown ether attached to each aryl group (see Scheme 1.5) forms a one-handed helix in the presence of amino acids, creating a large ICD in the UV–visible region. The ICD effect is produced by both small quantities of amino acid (0.005 equivalents), and by very small enantiomeric excesses (as little as 0.005% ee for Ala) via an amplification effect. The method is not of general use for enantiomeric determinations, as even 5% ee of Ala induced the same full ICD effect as pure L-Ala. However, a good linear relationship was found in the range of 0.1 to 0.005% ee for Ala. Nineteen other amino acids also gave good ICD effects (*613*). Another method for detecting trace amounts of enantiomeric impurities relies on stereoselective antibodies and an enzyme immunoassay. An antibody raised against *p*-amino-D-Phe could detect 0.002% ee D-Phe, a 0.1 μM concentration of D-Phe in the presence of 10 mM L-Phe (*614*).

All other methods commonly used to determine enantiomeric purity rely upon the formation of a diastereomeric derivative or complex. Quantitation can be provided by NMR spectroscopy (for review, see reference *615*) or by chromatographic separation using HPLC, GC or CE. A number of monographs address chromatographic separations (*616–621*), with several devoting significant space to amino acid analysis: reference *616* has a comprehensive discussion of HPLC chiral separations, including many applications to amino acids, although the information is fragmented throughout the book and the index is poor; reference *617* has sections on diastereomeric derivatives for GC and HPLC (pp 60–73), summaries of GC chiral stationary phases including applications to amino acids (pp 101–104), a section on amino acid analysis (pp 176–181), and a procedure for chiral TLC plate preparation (p 272); reference *618* includes a useful guide to the choice of chiral stationary phase for amino acid analysis (p 16), a brief discussion of HLPC analysis of complexes of dansyl amino acids with β-cyclodextrin (pp 61–64), and a section on analytical TLC with many examples of amino acid separations (pp 135–143); reference *619* has a table of diastereomers used in HPLC analysis (pp 97–107), a discussion of cyclodextrin-based TLC (pp 152–155), and

a chapter on chiral mobile phases for enantiomeric resolution of amino acids (pp 165–177); reference *620* has a table of HPLC diastereomer derivatives (pp 65–66) and, a table of the derivatizing agent structures (pp 41–49), and a section on enantiomeric resolution of amino acids by chiral ligand exchange chromatography (pp 89–95); and reference *621* has sections on GC separations of amino acids on chiral columns or as diastereomer derivatives (pp 76–77). A number of general reviews on chiral separations include numerous references on separations of amino acid enantiomers, such as biennial reviews in *Analytical Chemistry* (*474, 622, 623*) and a 2001 review of chiral GC separations (*624*). A 2006 summary of chiral recognition mechanisms provides a succinct overview of the types of chiral selectors employed for enantiomeric analyses (*625*).

In general, extreme care must be taken to ensure that the derivatizing reagent itself is enantiomerically pure, and that the deprotection and derivatization steps required to form the diastereomers do not themselves cause racemization of sensitive substrates, or discriminate against the formation of one diastereomer. However, it is not essential that the chiral probe be 100% enantiopure, as long as its enantiopurity is known. The optical purity of the derivatized substrate can then be calculated based on the optical purity of the derivatizing agent, as described in a 1998 paper (*626*). The enantiomeric purity of a number of commercial chiral resolving agents was reported in 1998 (*627*) and 1999 (*628*), with many possessing significantly less than 99% ee. Chiral auxiliaries have the great advantage that if both auxiliary antipodes are available, standards of both diastereomers can be prepared from only one enantiomer of the amino acid to be assayed. As mentioned previously, optical rotation measurements are unreliable for accurate determinations, and are useless if a pure standard is not available for comparison.

Various forms of cyclodextrins have been applied to chiral separations of amino acids by many of the methods that will be described in the following sections. Structural evidence of reasons for the chiral discrimination of amino acids by β-cyclodextrin has been provided by crystal structures of complexes formed by β-cyclodextrin with Ac-L-Phe (*629*), Ac-D-Phe (*629*), Ac-L-Phe-OMe (*630*), Ac-L-Phe-NH$_2$ (*630*) or *O*-Me Ac-L-Tyr-OMe (*631*). Titration microcalorimetry and ^{1}H NMR have also been employed to determine a correlation between complex conformation and chiral discrimination of amino acid derivatives by β- and γ-cyclodextrins (*632*). NMR investigations have been carried out on partially or exhaustively carbamoylated cyclodextrins in solution with analytes such as *N*-3,5-dinitrobenzoyl Ala, Val, and Phe (*633*), and on complexes of mixed methylated/carbamoylated cyclodextrins with *N*-3,5-dinitrobenzoyl Ala, Val, and Phe methyl esters (*634*).

The asymmetric complexation of amino acids by a wide range of other receptor-type molecules, which may have potential for use in asymmetric separation methodologies, is discussed further in Section 1.13.

1.7 Determination of Enantiomeric Purity: NMR Analysis

1.7.1 NMR Analysis: Diastereomeric Derivatives

One of the most widely used procedures for the determination of enantiomeric composition of organic compounds involves the derivatization of free hydroxyl or amine groups with Mosher's acid (2-methoxy-2-phenyl-3,3,3-trifluoropropionic acid, methoxytrifluoromethylphenyl-acetic acid, MTPA) (see Scheme 1.6), followed by assessment of the ratio of diastereomers by ^{1}H, ^{13}C, or ^{19}F NMR spectroscopy (635). However, care must be taken to ensure purity of the derivatizing reagent, as commercially available samples of Mosher's acid were recently reported to have variable purities of 97.9–99.8% ee (627, 636). A number of reports of amino acid or ester analysis using Mosher's acid and ^{19}F (637–639) or ^{1}H NMR (550, 551, 639) analysis have appeared, with one describing an improved synthesis of the reagent (637). The Mosher acid has been activated with an anhydride linkage to a polymer resin, then simply mixed with an amino ester in deuterated solvent. The resin can be filtered off before obtaining an NMR spectrum (565), or simply allowed to float to the top of the NMR tube (564). α-Cyano-α-fluoro-p-tolylacetic acid (CFTA) has been used as an MTPA replacement, providing improved Δδ values in both ^{19}F and ^{1}H NMR spectra for several ester derivatives when compared to MTPA. Amino acids were not examined in the initial report (559), but in 2001 the use of CFTA ^{19}F NMR chemical shifts in determining the absolute configuration of amide derivatives of amino acid esters was described (560).

O-Methylmandelic acid (methoxyphenylacetic acid, MPA) has also been applied to the assessment of amino ester optical purity, using ^{1}H NMR analysis (552). The aryl group was varied to improve the separation of NMR signals, with 1-naphthyl and 9-anthryl

derivatives doubling or tripling the resolution (553). The MPA auxiliary was employed to determine the optical purity of an α-alkylated δ-azidopentanoic acid (640). Closely related (R)-O-aryllactic acid (ROAL) amide derivatives (see Scheme 1.6) can also be used to quantitate amino ester purity by integration of the methyl ester peak in the ^{1}H NMR spectrum. The difference in chemical shifts is greater than with Mosher's acid, and the aryllactic acids are readily prepared from inexpensive lactic esters via Mitsunobu displacement with a phenol (641).

The amino group of amino acids can be used to form a Schiff base derivative with an aldehyde-substituted binaphthol-based auxiliary. The imine proton showed a good shift difference for 10 different amino acids (642). (S)-2-[(R)-fluoro(phenyl)methyl]oxirane was recently reported as a versatile general reagent for the derivatization and enantiomeric analysis of chiral amines (see Scheme 1.7). The diastereomeric derivatives were readily analyzed by ^{1}H, ^{19}F, or ^{13}C NMR. Phe-OMe and ethyl 3-aminobutyrate were among the amines tested (643).

A number of phosphorous-based chiral auxiliaries have been developed for use with ^{31}P NMR spectroscopy (see Scheme 1.7) (for a review see Reference 644). A cyclic phosphoric acid derived from butyraldehyde and benzaldehyde or other aromatic aldehydes forms phosphonic amides with amino acids (seven examined) or amino acid methyl esters (five examples), giving separated ^{31}P signals for the two diastereomers (645,646). A chiral phosphonate prepared from (S)-2-butanol and PCl$_3$ reacts readily with unprotected amino acids to form a phosphonic amide (647). Notably, the amino acid does not have to be protected in any way before analysis, and aqueous solutions of the amino acid can be analyzed. An analog prepared from 1-diethylamino-2-propanol, which can form a salt bridge to lock the conformation, unfortunately has limited stability (648). Signal resolution with the (S)-2-butanol-derived reagent

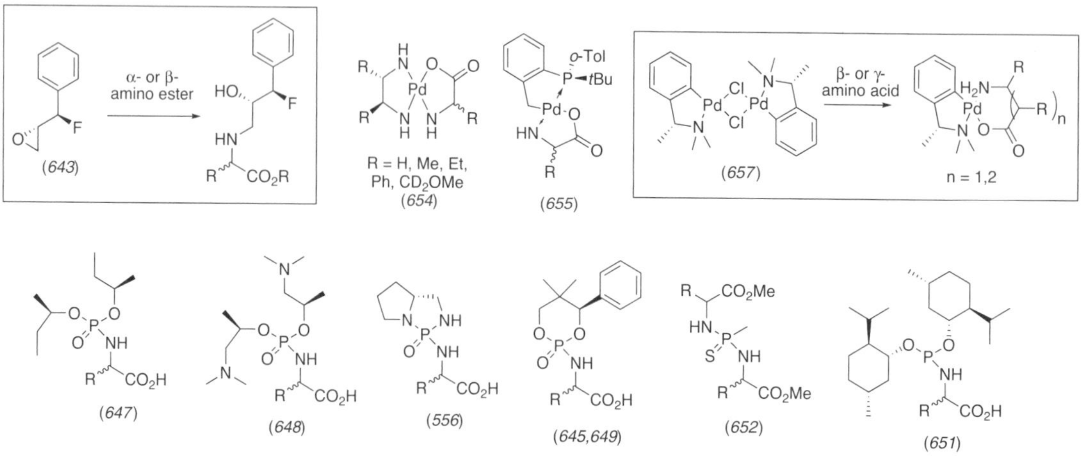

Scheme 1.7 Diastereomeric Derivatives for NMR Analysis of Amino Acid Enantiomeric Purity.

is sometimes insufficient, so an improved cyclic reagent has been prepared, (S)-2H-2-oxo-5,5-dimethyl-4(R)-1,3,2-dioxaphosphorinane (649). Amide derivatives are again readily prepared from unprotected amino acids in aqueous solution, and all 15 examples, including α-alkylated amino acids and amino esters and amides, gave satisfactory results by ^{31}P NMR. A similar cyclic phospholidine based on α-phenylethylamine reacted poorly with amino acids, although it gave accurate ee determinations (650). (2S,5R)-2-Chloro-3-phenyl-1,3,2-diazaphosphabicyclo[3.3.0]octane 2-oxide has been reacted with amines, including an amino ester, with chemical shift differences observed in both ^{1}P and ^{1}H NMR spectra (556). Dimenthyl chlorophosphite gave good chemical shift differences for Leu-OMe, Gly-Leu-OMe, and Gly-Gly-Leu-OMe in ^{13}P NMR (651).

A conceptually different approach to enantiomeric analysis involves derivatization of an achiral phosphorous reagent with two equivalents of amino acid, thus forming diastereomeric pairs (652). Methylphosphonothioic dichloride reacts with the amino group to give three possible distinguishable complexes, the enantiomeric RR/SS combinations, and two RS meso derivatives that are resolvable due to chirality at the phosphorus center. Baseline separation of the three derivatives was observed in the ^{31}P NMR spectrum for 13 of 15 amino acids examined.

All of the previous auxiliaries derivatize the amine group. A (4S,5R)-4-methyl-5-phenyl-oxazolidin-2-selone (566, 567) has been reacted with the carboxyl terminus of N-Boc amino acids (see Scheme 1.3), giving resolved signals by ^{77}Se NMR (538). Mosher has reported 2,2,2-trifluoro-1-phenylethylamine as a useful reagent to form amides with chiral acids (with ^{19}F NMR analysis), but amino acids were not addressed (653). The ^{1}H NMR methyl ester peak of Bz-Xaa-Lys(Bz)-OMe dipeptides has been used for estimating racemization (497).

A palladium complex formed from optically active C$_2$-chiral 1,2-diamines reacts with both amino and carboxy groups of unprotected amino acids in water to form a square-planar bicyclic compound. Separated signals were observed in both ^{13}C and ^{1}H NMR spectra for 10 amino acids, including a β-amino acid (654). Another Pd complex employs ^{31}P NMR to look at shifts of a chiral phosphine atom; very well resolved signals for unprotected Ala, Leu, Val, and Phg were obtained (655). ^{19}F NMR analysis of amino acid Pd complexes with N,N-dimethyl-(2,2,2-trifluoro-1-phenylethyl)amine showed excellent signal separation for complexes of Ala, Ile, Leu, Phe, Phg, Pro, and Val, allowing for integration of both resonance signals and determination of enantiomeric purity. A crystal structure of a complex was also reported (656). Spirocyclic Pd complexes formed from di-μ-chlorobis{[2-(1-dimethylamino-xN)ethyl]phenyl-xC}dipalladium and a range of β- or γ-amino acids provided distinct chemical shift differences in both ^{1}H and ^{13}C NMR spectra, allowing for the determination of enantiomeric purity (see Scheme 1.7) (657).

1.7.2 NMR Analysis: Diastereomeric Complexes

The advantage of forming non-covalent diastereomeric complexes is that no derivatization is needed, avoiding the concomitant problems of racemization or kinetic resolution during coupling. Complex formation with chiral lanthanide shift reagents is the most established procedure (see below), but several other methods have been reported. Perhaps the most intriguing of these is NMR analysis of a self-induced diastereomeric dimer of Ac-Val-OtBu, which associates via amide NH to ester C=O hydrogen bonding. Two amide NH doublets are observed in the ^{1}H NMR spectrum, with an intensity ratio corresponding to the enantiomeric ratio (658). Unfortunately the general usefulness of this procedure is limited.

Another approach is to use a chiral solvation agent (CSA). (L)-Chirasil-Val, the same polymer commonly employed as a chiral stationary phase in GC analyses of amino acids, causes a chemical shift non-equivalence of N-trifluoroacetyl-amino acid methyl esters in both ^{19}F and ^{1}H NMR. However only Leu and Val were examined (659). A naphthalimide derivative of 1,2-diaminocyclohexane allowed for discrimination of Phth-Ala, Phth-Leu, Ts-Ala, Ts-Val, and o- and p-Nbs-Ala (660), while a 1,3,5-triazine with 3 chiral α-methylbenzylamine substituents provided enantiodiscrimination of 3,5-dinitrobenzoyl derivatives of Ala-OMe, Phg-OMe, and Leu-OMe in CDCl$_3$ (see Scheme 1.8) (661, 662). Dihydroquinone derivatives with a carbamoyl-linked amino ester gave good separations of resonances of a dinitrobenzoyl Ala-OMe derivative in CDCl$_3$ (663). Attachment of the carbamoyl substituent at a different position also provided agents that could discriminate between Ala and Val derivatives (664). A spirobisindane skeleton with two phosphonate substituents formed strong 1:1 complexes with Arg and Lys, and provided discrimination between the enantiomers when the diastereomeric complexes were analyzed by NMR (441).

A cholesteric lyotropic liquid crystal medium prepared from poly(γ-benzyl L-glutamate) in dichloromethane provides a chiral environment to discriminate between amino acid benzophenone Schiff base trideuteromethyl esters. Proton-decoupled ^{2}H NMR analysis of seven amino acids showed resolved differences in the quadrupolar splitting of the methyl ester signal (665). This technique was also used in conjunction with a perdeuterated benzoyl derivatizing reagent, providing enantiomeric discrimination for 15 amino acids (571). The same methodology, when carried out with amino acids doubly labeled on the amine and carboxylate with acetyl-d$_3$ chloride and methanol-d$_4$, respectively, generates two measurements from one sample (572). An aqueous-based chiral liquid crystal system has been prepared from glucopon (a mixture of alkyl polyglucosides) in hexanol/buffered water. Deuterium NMR showed discrimination of α-deuterio-Ala and 3,3-dideutero-Phe (666). Preliminary experiments using stretched gelatin gels (with some derived from gummi bears!) as chiral

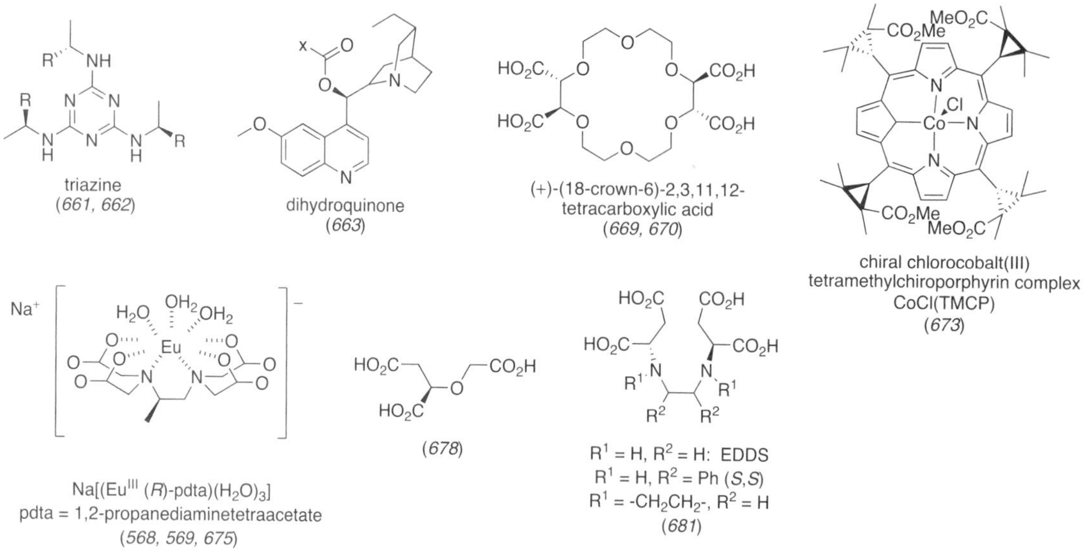

triazine
(661, 662)

dihydroquinone
(663)

(+)-(18-crown-6)-2,3,11,12-
tetracarboxylic acid
(669, 670)

chiral chlorocobalt(III)
tetramethylchiroporphyrin complex
CoCl(TMCP)
(673)

Na[(Eu^{III} (R)-pdta)(H$_2$O)$_3$]
pdta = 1,2-propanediaminetetraacetate
(568, 569, 675)

(678)

R^1 = H, R^2 = H: EDDS
R^1 = H, R^2 = Ph (S,S)
R^1 = -CH$_2$CH$_2$-, R^2 = H
(681)

Scheme 1.8 Chiral Shift Reagents for NMR Analysis of Amino Acid Enantiomeric Purity.

alignment media demonstrated an ability to discriminate between D- and L-Ala in a ^{1}H,^{13}C-BIRD-J-coupling spectrum (667).

Benzoylated cyclodextrins are potential chiral solvating agents for 3,5-dinitrophenyl derivatives of amines, but the enantiodiscriminating ability shown for Ala-OMe was poor (668). Complexes of underivatized amino acids or α-methyl amino acids with (+)-(18-crown-6)-2,3,11,12-tetracarboxylic acid show enantiomeric discrimination when analyzed by ^{1}H NMR; Ala, Val, Trp, Phg, α-methyl-Val, and α-methyl-Val-NH$_2$ were tested (669, 670). X-ray crystal structures and energy calculations have been made of complexes of 18-crown-6-tetracarboxylic acid with D- and L-Tyr, Ile, Met, and Phg in order to clarify the mechanism of chiral recognition (671). A chiral crown ether derived from methyl β-D-galactopyranoside provided enantiomeric discrimination of a number of amino esters in acetonitrile, as long as achiral lanthanide shift reagent was added (672). Two equivalents of amino acid methyl ester are bound by a chiral chlorocobalt(III) tetramethylchiroporphyrin complex in CDCl$_3$, giving simple and well-resolved spectra for 16 of the 20 common amino acids (more complex spectra are observed with Arg, Cys, His, and Lys), with up to 0.5 ppm diastereomeric dispersion observed for some protons (673).

Chiral lanthanide shift reagents are a well-established means of determining enantiomeric purity of organic compounds (615,674). Their major disadvantages are that it is difficult to predict which substrates and shift reagents will give useful results, and that high enantiomeric purities are often hard to ascertain. Typically, results are ±2% ee, except for purities >90% ee, when the error can be as large as ±10% (615). However, calibration plots have been used to measure optical purities as high as 99.7% ee by integration (675). Of the

lanthanide shift reagents, Eu(III) is most commonly used, since it causes the least line broadening (674). It generally causes shifts to lower fields. Pr(III) is more prone to causing line broadening, but shifts most signals to higher fields where there may be less interference. Tris[3-((trifluoromethyl)hydroxymethylene)-d-camphorato]europium(III) was used to determine the purity of enzymatically resolved Bz-Phg-OMe and Bz-Chg-OMe (676), and provided near baseline separation of the methyl ester group of β-(N-ethyl-carbazol-3-yl)-Ala-OMe (677).

Unfortunately, most underivatized amino acids are insoluble in the organic solvents generally applied to chiral shift analysis, and hydrolysis of the commonly used shift reagents leads to the formation of Eu$_2$O$_3$, which causes severe line broadening. For this reason special shift reagents have been developed for amino acid analysis (see Scheme 1.8). The Yb salt of (S)-carboxymethyloxysuccinic acid gave ^{1}H NMR spectral separation for the enantiomers of Ala and β-hydroxyPhe (678). 1,2-Propanediaminetetraacetato-europium(III) has been used to differentiate between amino acid enantiomers in D$_2$O (679) and assign absolute configuration to both α-amino acids (568) and α-methyl-α-amino acids (569). Spectral resolution of five amino acids in aqueous solutions was achieved by the chiral shift reagent formed from europium(III) and aspartic acid-derived (S,S)-ethylenediamine-N,N'-disuccinate (EDDS) (680). Modification of the ethylenediamine substituents led to three new reagents which were applied to 16 different α-amino acids and α-alkyl-α-amino acids (681); the diphenyl-substituted ligand gave the greatest shift differences. Samarium(II)-propylenediaminotetraacetate gave remarkably less line broadening than the corresponding europium complex during high-field NMR experiments, and provided clearly resolved spectra for 16 different underivatized amino acid enantiomeric pairs in D_2O. H^α signals were separated by as much as 0.146 ppm (*682*). A tris[3-((trifluoromethyl)hydroxymethylene)-(+-camphorato]samarium(III) complex, $Sm(tfc)_3$, gave good resolution of 12 amino acid methyl esters with minimal line broadening (*570*).

1.8 Determination of Enantiomeric Purity: HPLC Chromatographic Analysis

Three approaches are used for HPLC separation of enantiomeric compounds: formation of a diastereomeric derivative, elution using a chiral mobile phase, or elution using a chiral stationary phase. Despite the improvements that have been made to chiral stationary phases, diastereomeric derivatives are still predominantly used for routine HPLC analysis. Standards are generally required in order to identify which enantiomer or diastereomer corresponds to the eluted peaks, although a recent HPLC separation of underivatized amino acid racemates employed circular dichroism detection in order to identify the configuration of the resolved compounds (*527*). Amino acids are often derivatized by pre- or post-column modification in order to improve their separation, or their detectability. One commonly employed reagent is 2,4-dinitrofluorobenzene (Sanger's reagent), which generates N-(2,4-dinitrophenyl)(DNP)amino acids. N,N-Dimethyl-2,4-dinitro-5-fluorobenzylamine, was recently described as an alternative reagent, with the added advantage of improving detectability by positive-mode electrospray ionization mass spectrometry (*683*).

1.8.1 HPLC Analysis: Diastereomeric Derivatives

Many diastereomeric derivatives for chromatographic separation that are specific to α-amino acids have been reported (see Reference *620* for a table of HPLC diastereomer derivatives (pp 65–66) and a table of the derivatizing agent structures (pp 41–49)). One of the more successful methods involves derivatization of the amino acid (as either the free acid or the ethyl ester) with 2,3,4,6-tetra-O-acetyl-β-D-glucopyranosyl isothiocyanate (GITC, or AGIT), followed by separation on a standard C-18 reversed-phase column (see Scheme 1.9) (*684–686*); 17 amino acids can be resolved simultaneously (*686*). By changing to a porous graphite column, a more simple

gradient elution solvent system of 0.1% aqueous TFA–acetonitrile could be used, resolving 14 racemic amino acid pairs (*687*). The GITC derivatization method has been applied to β-methyl amino acids (*688*), conformationally constrained cyclic amino acids (tetrahydroisoquinoline or tetraline ring structures) (*689, 690*), arylalanines (*691, 692*), an axially chiral cyclic α,α-disubstituted amino acid (Bin) (*693*), and bicyclo[2.2.1]heptane and heptene-based β-amino acids (*694*). A study comparing GITC or FDAA (see below) derivatization for the separation of eight underivatized β-substituted β-amino acids found that FDAA was more efficient than GITC (*695*), but another report on various arylalanines found GITC derivatization tended to be more effective than FDAA (*692*). A modified reagent, 2,3,4,6-tetra-O-benzoyl-β-D-glucopyranosyl isothiocyanate (BGIT), has increased sensitivity. It resolved 17 natural and non-natural amino acids as the free acids, although not simultaneously (*696*). BGIT was used to determine the optical purity of a L-5,5-dimethyl-Pro sample (*573*). Another isothiocyanate, (R)-(+)-α-methylbenzyl isothiocyanate, was employed for determination of the optical purity of 4′-aminomethyl-L-Phe (*697*).

Chiral isothiocyanate reagents have not generally been successfully incorporated into the Edman sequencing of peptides (see Scheme 1.2), which would allow for enantiomeric analysis of each amino acid as it was cleaved from the peptide. Unfortunately, amino acid racemization occurs during cyclization and cleavage of the thiazoline (*698*). A recent examination of the process finds that the replacement proton is derived from the TFA used to induce cyclization, and that racemization can be greatly reduced (from 8–25% to <2.5%) by using a non-protic Lewis acid such as BF_3, to promote the cleavage (*699*). Two chiral isothiocyanate reagents have been developed specifically as replacements for phenyl isothiocyanate (PITC) during Edman sequencing. 4-(3-Isothiocyanatopyrrolidin-1-yl)-7-(N,N-dimethylaminosulfonyl)-2,1,3-benzoxadiazole (DBD-PyNCS) or 4-(3-isothiocyanatopyrroli-din-1-yl)-7-nitro-2,1,3-benzoxadiazole (NBD-PyNCS) react with amino acids under basic conditions The resulting derivatives are fairly stable, even under acidic conditions, with little conversion to the thiohydantoins. Fourteen derivatized amino acids were resolved on a reversed-phase column (*700*).

The most commonly used derivatizing reagent is now Marfey's reagent, 1-fluoro-2,4-dinitrophenyl-5-L-alanine-amide (FDAA). Derivatization of the amino acid

R¹ = Ac: AGIT/GITC (*684–687*)
R¹ = Bz: BGIT (*696*)

R³ = H, Et

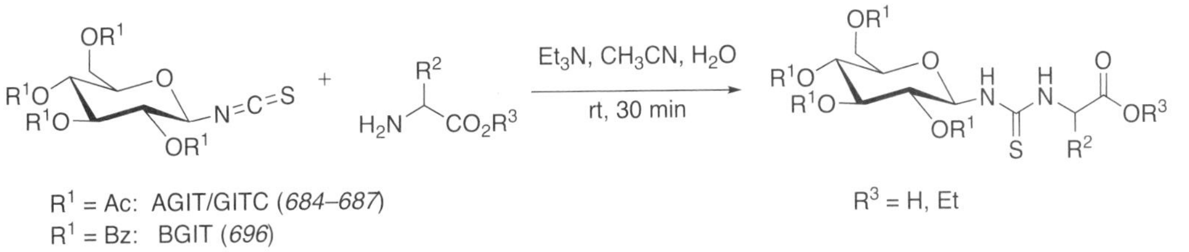

Scheme 1.9 AGIT, BGIT Diastereomeric Derivatization of Amino Acids for HPLC Analysis of Enantiomeric Purity.

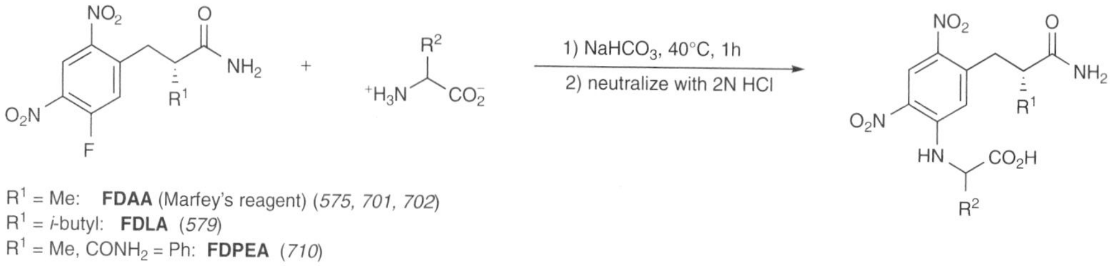

Scheme 1.10 Marfey's Reagent Diastereomeric Derivatization of Amino Acids for HPLC Analysis of Enantiomeric Purity.

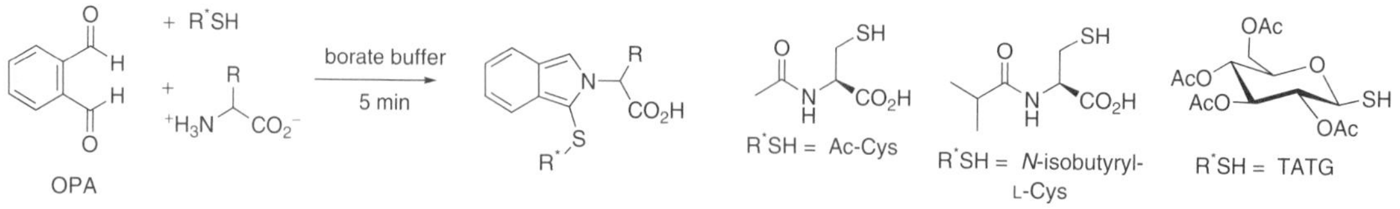

Scheme 1.11 OPA/Thiol Diastereomeric Derivatization of Amino Acids for HPLC Analysis of Enantiomeric Purity.

amine group via a nucleophilic aromatic substitution leads to a stable, highly absorbing chromophore which can be detected at 340 nm after separation on a standard reversed-phase column (see Scheme 1.10) (575,701–703). α-Alkylated amino acids (704) and β-leucine (705) were analyzed as Marfey reagent derivatives, while cyclic constrained amino acids (689, 690), β-methyl amino acids (706), eight underivatized β-substituted β-amino acids (695) and a number of unusual Phe, Tyr, and Trp analogs (691) were better resolved with FDAA than with GITC. However, another report on various arylalanines found GITC derivatization tended to be more effective than FDAA (692). Other amino acids to which Marfey's analysis has been applied include 5-hydroxynorvaline (707), an axially chiral binaphthyl-based β-amino acid (708), bicyclo[2.2.1]heptane and heptene-based β-amino acids (694), and 27 racemic α-alkyl-α-amino acids, which were more effectively resolved as the FDAA derivatives than as OPA/NAC or OPA/N-Boc-Cys diastereomers (709). Levels of free D-Ser in mammalian brain (99) or urine (101) were determined by Marfey's analysis. The Marfey derivatization has also been coupled with mass-spectral analysis to determine the enantiomeric purity of amino acids within synthesized peptides (472). The peptides were hydrolyzed using DCl/D$_2$O, derivatized and separated, and then analyzed by electrospray MS. This procedure distinguishes between racemization caused by coupling or by peptide hydrolysis, as amino acids affected by the latter will be deuterated.

The Marfey method has picomole sensitivity, but derivatization requires a reaction time of 90 min, a substantial reagent peak is produced, and chromatographic conditions vary greatly between amino acids (701). The method has been applied to the identification of unknown amino acids in a bacterial cyclic peptide by derivatization

of the hydrolysate and analysis with LC-MS. Racemization of half of the hydrolysate allowed the absolute configuration of the amino acid to be determined (576, 577). A modified method that also allows for absolute configuration to be established uses 1-fluoro-2,4-dinitrophenyl-5-L-leucine-amide (FDLA) (579, 580, 592). FDPEA, 1-fluoro-2,4-dinitrophenyl-5-(R,s)-phenylethylamine, is a similar derivatizing reagent useful for the determination of absolute configuration (710). A related reagent derived from trihalo-s-triazines and L-Ala amide has been reported, though the resolution for the five amino acids examined was not as good as with Marfey's reagent (711). The use of Marfey's reagent, and related derivatives, was reviewed in 2004, including a comparison with other derivatizing reagents (594).

N-Derivatization of amino acids with o-phthalaldehyde (OPA) and various N-acyl-L-cysteines (see Scheme 1.11) followed by chromatographic separation on a reversed-phase column and detection at 338 nm allows for 18 common amino acids and their enantiomers to be analyzed in a single run (Cys and Pro are the exceptions) (471, 595, 706, 712, 713). N-Isobutyryl-L-Cys (IBLC) was found to be the best acyl derivative; previously other chiral mercaptans such as Boc-Cys, Ac-Cys (NAC), and Ac-penicillamine had been used (471, 706, 712–714). N-IBLC-L-cysteine can be easily prepared by a modified procedure (608) of that originally reported (595). N-IBLC-L-cysteine (IBLC) and N-isobutyryl-D-Cys (IBDC) were both used for resolutions of a mixture of 23 amino acids, and for analysis of hydrolyzed seawater and fossil mollusks. Comparison of the two chromatograms allows for confirmation of enantiomer assignments and identification of impurity peaks (715). The OPA/IBLC derivatization provides resolution of both diastereomers of β-hydroxy amino acids, thus allowing quantification of all four stereoisomers in a single analysis (608). The OPA/IBLC procedure has been applied to the analysis of peptidic antibiotics, toxins, drugs, and pharmaceutically useful amino acids (716), the detection of D-amino acids in bacteria (717), fungi (717), plants (87, 717), invertebrates (717), vertebrates (717), human urine (105), and the Murchison meteorite (717), and the analysis of grape juice, beer, wine, vinegar, whey, cheese, fermented cabbage juice, yeast extract, and honey (718). The method was recently used (2001) for an analysis of the amino acid content of the Orgueil and Ivuna meteorites, employing UV fluorescence detection. Detection limits were better than 10 parts per billion, based on the approximately 100 mg of original sample that was extracted (719). It has also been used to look for enantiomeric excess in amino acids prepared under conditions mimicking primordial syntheses (720).

N-Acetyl-Cys has been employed for analysis of amino acids from hydrolyzed proteins during an evaluation of hydrolysis methods (233), for the separation of α-dialkylamino acids as found in geological samples (462), and for the analysis of other α-alkylated amino acids (704, 714). In all cases the L-amino acid isomer elutes before the D-enantiomer (104). In 1993, 27 racemic α-alkyl-α-amino acids were derivatized with either OPA/NAC or OPA/N-Boc-Cys, or with Marfey's reagent (FDAA). The Marfey derivatives were more effectively resolved (709).

The major disadvantage of the OPA/thiol procedure is that the diastereomeric derivative is not stable, so must be prepared shortly before each analysis. However, the derivatization is very rapid (5 min), so this is not a substantial drawback. Imino acids such as Pro cannot be analyzed. In 1999 the stability and characteristics of OPA/NAC derivatives of 24 amino acids were examined (282). For quantitative derivatization, extended reaction times of up to 30 min are required, with at least 7 min needed, and a 20-fold excess of OPA/thiol (with thiol added prior to amino acid) should be employed. The derivatives were found to be stable for at least 50 min after formation. Amino acids that can also form double derivatives (formed with Orn, Lys, etc.) need to be quantitated based on the total of all peaks. Impurities in the reagents and their stability towards storage were also tested (282, 283). A further study in 2000 optimized the separation and identification of 25 amino acids as their OPA/NAC derivatives, testing different columns, flow rates, and temperature, with the best results at the highest possible flow rate and 30 °C (284). His undergoes

irregular reactions with OPA, which were analyzed in a 2004 report (281). The use of OPA amino acid derivatives was reviewed in a 2001 paper that focused on the different types of thiol additives (286), and a 2002 paper looking at the formation, stability, and characteristics of OPA/NAC derivatives via HPLC-MS (287).

Several variations of the OPA/thiol method have been reported. An earlier report used a diastereomeric OPA/N-Ac-Cys derivative in combination with a mobile phase containing Cu(II) and L-Pro (721). A different chiral thiol, 2,3,4,5-tetra-O-acetyl-1-thio-β-glucopyranoside (TAGT) has also been tested; 14 of 16 amino acids (not Lys or Orn) were resolved and, unlike the other chiral thiol derivatives, the adducts were fairly stable (722). More recently, 1-thio-β-D-galactose has been used to resolve 15 of 16 racemic amino acids. However, a slightly different fluorescent response was noted between the D- and L-amino acid derivatives, which prevents accurate quantifications (722, 723). N-[(R)-2-hydroxy-2-phenylacetyl]-(S)-Cys has also been employed as a chiral thiol, though Phe was the only amino acid tested (724).

The final diastereomer derivatizing reagent of general use is the chiral Fmoc analog FLEC, or (+)-1-(9-fluorenyl)ethyl chloroformate (see Scheme 1.12). Amino acids are derivatized in 4 min in borate buffer, and 23 diastereomer pairs were resolved on a reversed-phase column (725). The disadvantages of this method include the requirement for a multistep ternary gradient of CH_3CN/THF/acetic acid buffer, different chromatographic conditions required for imino acids, and expense of the FLEC reagent.

Several groups have made use of dipeptide diastereomers to measure amino acid enantiomeric purity. One of the earliest analytical procedures involved formation of dipeptides by derivatization with the N-carboxy anhydride (NCA) of L-Leu or L-Glu, with subsequent separation by ion-exchange chromatography (726, 727). This method was used to assess racemization during peptide hydrolysis by tritiated HCl (473). Derivatization with the NCA of L-Leu or L-Phe followed by reversed-phase HPLC analysis has also been examined (728). Epimeric di-, tri-, and tetrapeptide derivatives were separated by Goodman et al. (498) and Benoiton et al. (729–731) in order to analyze the extent of racemization during peptide coupling. A standard reversed-phase column was used, with detection at 208 or 215 nm. Diastereomeric dipeptide derivatives of N-protected amino acids coupled to O-(4-nitrobenzyl)-tyrosine

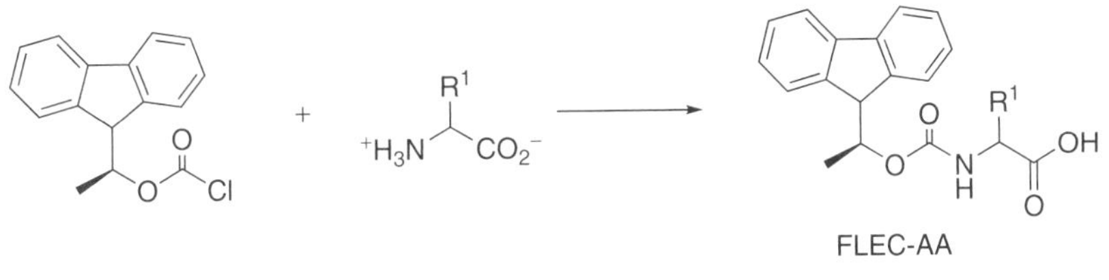

Scheme 1.12 FLEC Diastereomeric Derivatization of Amino Acids for HPLC Analysis of Enantiomeric Purity (725).

methyl ester were analyzed on a silica gel column, and gave good separation of the seven protected amino acid racemates examined (732). Amino acids have also been reacted with Boc-L-Ala-ONSu active ester, with the Boc group removed from the dipeptide before analysis on a standard amino acid analyzer, using an ion-exchange column with ninhydrin detection. The diastereomeric dipeptides formed from Ala, Val, Ile, and Glu were all resolved (733). A number of Fmoc amino acid N-carboxy anhydrides (NCAs) were examined as possible derivatizing reagents of amino acid methyl esters; Fmoc-L-Leu-NCA was selected as the best, then applied to the resolution of 25 amino esters, with only Pro not resolved (734). A new derivatizing reagent, N-(4-nitrophenoxycarbonyl)-L-Phe methoxyethyl ester (S-NIFE), has been used for the separation of a number of amino acids on a standard C_{18} column, with derivatives formed with β-substituted α-amino acids (such as β-methyl-Phe, -Tyr, Trp, separating all four amino acid diastereomers) (735), 19 different secondary amino acids (such as Pro, pipecolic acid, Tic) (736), and 18 β-amino acids (737). A new derivatizing reagent has been created by nucleophilic displacement of one chlorine in cyanuric chloride by an alkoxy/aryloxy group, the second chloride by L-Ala/Phe/ValNH₂ or L-Pro/Lys-OtBu, and the third chloride by the analyte amino group (see Scheme 1.13). The initial results were promising, with the methoxy/Val-NH₂ derivative giving some of the best results for seven underivatized amino acids (738).

Other diastereomers that have been used over the years include N-d-10-camphorsulphonyl p-nitrobenzoate amino acids (separated on silica column, eight examples (705, 739)), N-(-)-α-methoxy-α-methyl-1-naphthaleneacetyl amino acid methyl esters (separated on silica column, 20 examples (740, 741)), N-(+)-1-(1-naphthylethyl) carbamoyl derivatives of amino acids (14 of 17 examples resolved on reversed-phase column (742)), 1-(1-anthryl)- or 1-(2-anthryl)ethylamides of N-Ac amino acids (well resolved on a silica column, four examples (743)), or NSP-Cl, (S)-(−)-N-1-(2-naphthylsulfonyl)-2-pyrrolidinecarbonyl chloride (see Scheme 1.13) (13 of 19 amino acids resolved on a silica column (744)). Urea derivatives of amino acids formed with SIPEC (succinimido 1-phenylethylcarbamate) or SINEC (succinimido 1-(1-naphthyl)ethylcarbamate) (see Scheme 1.13), were resolved by reversed-phase HPLC (nine examples) (745). A fluorescent chiral tagging reagent, (R)-4-(3-isothiocyanatopyrrolidin-1-yl)-7-(N,N-dimethylaminosulfonyl)-2,1,3-benzoxadiazole (DBD-PyNCS), provided separation of 17 amino acid DL-amino acids. It was used to analyze for D-amino acids in yogurt (746). A reagent based on enantiomerically pure trans-1,2-diaminocyclohexane, (1R,2R)- or (1S,2S)-N-[(2-isothiocyanato)cyclohexyl]-6-methoxy-4-quinolinylamide (CDITC), resolved 19 amino acids with separation factors of α = 1.14–3.16, and good fluorescence (747). Another isothiocyanate, (1S,2S)- and (1R,2R)-1,3-diacetoxy-1-(4-nitrophenyl)-2-propylisothiocyanate (DANI), separated the enantiomers of 18 separate amino acids (748). The anhydride of (R)-(+)-1-methoxy-1-trifluoromethylphenylacetic acid (MTPA, Mosher's acid) was used to derivatize 4'-hydroxyphenylglycine methyl ester, allowing its resolution by reversed-phase HPLC (749). (S)-2-[(R)-fluoro(phenyl)methyl]oxirane (see Scheme 1.7) was recently reported as a versatile general reagent for the derivatization and enantiomeric analysis of chiral amines. The diastereomeric derivatives can be separated by HPLC, as well as analyzed by ¹H, ¹⁹F, or ¹³C NMR. Phe-OMe and ethyl 3-aminobutyrate were among the amines tested (643).

In 2004 the separations of 17 amino acids, including unusual amino acids such as allo-Thr, allo-Ile, β-methoxy-Tyr, and Tle, were compared using the

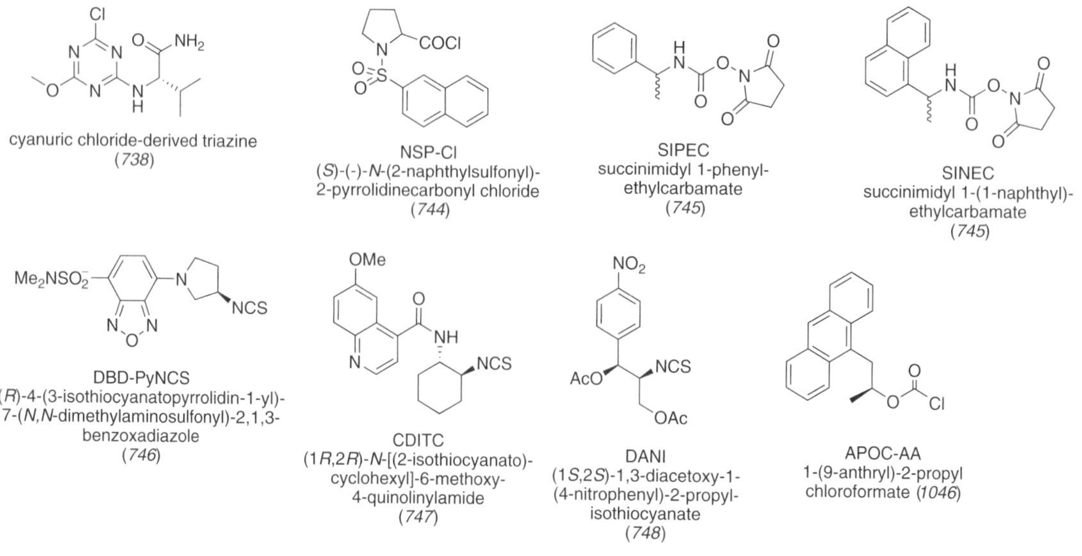

cyanuric chloride-derived triazine
(738)

NSP-Cl
(S)-(-)-N-(2-naphthylsulfonyl)-
2-pyrrolidinecarbonyl chloride
(744)

SIPEC
succinimidyl 1-phenyl-
ethylcarbamate
(745)

SINEC
succinimidyl 1-(1-naphthyl)-
ethylcarbamate
(745)

DBD-PyNCS
(R)-4-(3-isothiocyanatopyrrolidin-1-yl)-
7-(N,N-dimethylaminosulfonyl)-2,1,3-
benzoxadiazole
(746)

CDITC
(1R,2R)-N-[(2-isothiocyanato)-
cyclohexyl]-6-methoxy-
4-quinolinylamide
(747)

DANI
(1S,2S)-1,3-diacetoxy-1-
(4-nitrophenyl)-2-propyl-
isothiocyanate
(748)

APOC-AA
1-(9-anthryl)-2-propyl
chloroformate (1046)

Scheme 1.13 Other Derivatizing Reagents.

derivatizing reagents FDAA, GITC, OPA/IBLC, and S-NIFE, with ES-MS detection. In most cases FDAA gave the best enantioselectivity, but lowest sensitivity. For β-methoxy-Tyr, S-NIFE gave the best separation, but isomers such as D-*allo*-Ile and D-Ile were not completely resolved with any of the CDAs (*598*).

1.8.2 HPLC Analysis: Chiral Mobile Phase

A variety of chiral mobile phases have been used in combination with achiral stationary phases (usually silica normal-phase or ODS reversed-phase columns) to resolve amino acid enantiomers, with several different types of interactions used to effect the separation.

1.8.2a H-Bonding Complexes

The chiral additive *N*-acetyl-L-valine-*tert*-butyl amide was used to separate 12 of 13 *N*-(4-nitrobenzyl)amino acid isopropyl esters (not Pro) on a silica gel column (*750*). The mechanism of H-bonding responsible for the separation has been discussed (*751*).

1.8.2b Ligand Exchange: Chiral Mobile-Phase Mode

Ligand exchange chromatography involves a dynamic exchange of metal-based complexes formed between the analyte amino acid and a resolving amino acid, which can either be in the mobile phase (chiral mobile-phase mode) or attached to the stationary phase (chiral stationary-phase mode, see Section 1.8.3b). A summary of this area was published in 1989 (*752*), with more recent reviews in 2001 (*753*) and 2003 (*754*). Racemic dansylated amino acids were resolved by reversed-phase separation of the diastereomeric metal chelate complexes that are formed between the analyte amino acid and Cu(II) complexes of L-Pro, Arg, or His (*755*), His-OMe (10 of 13 amino acids separated in a single run (*756, 757*)), *N,N*-di-*n*-propyl-L-alanine (13 amino acids separated in single run, but not Pro (*758*)) or a Zn(II) complex of L-2-isopropyl-4-octyl-diethylenetriamine-Zn(II) (26 examples (*759, 760*)). However, care must be taken, as D- and L-Dns-amino acids have been found to give different fluorescence responses due to enantioselective fluorescence quenching by the copper(II) complexes (*761, 762*). Terdentate ligands based on *N*-aminoethyl Phe amide have been tested for separation of both dansyl and underivatized amino acids. Good resolutions were obtained with the dansyl derivatives (12 derivatives), but only three underivatized amino acids could be resolved (*763*). A number of L-amino acid amides have been examined as the chiral selectors for complex formation (13 amino acids separated (*764*)). A diastereomeric OPA/*N*-Ac-Cys derivative of 18 different amino acids was used in combination with Cu(II) and L-Pro in the mobile phase (*721, 757*). Discrimination is believed to result from metal chelate complexation with the Cys residue.

The underivatized imino acids Pro and pipecolic acid, which do not form Cu(II) complexes when *N*-derivatized, have been resolved using a Cu(II) complex of aspartame (*757*). Underivatized amino acids have also been resolved using a chiral mobile phase of L-Pro with Cu(II) acetate on an ion-exchange column (17 of 23 amino acids separated (*765*)) or a reversed-phase column (19 examples (*766*)). This ligand was also applied to α-methyl-α-amino acids and β-amino acids (*767*) and α-substituted Lys and Orn analogs (*768*). Racemic [1-^{11}C]-Val was preparatively separated by HPLC on a normal reversed-phase column by using a chiral mobile aqueous phase of L-Pro, cupric acetate, and sodium acetate (*769*). *N*-(*p*-toluenesulphonyl)-D-phenylglycine copper complexes were used to resolve 13 amino acids simultaneously on a reversed-phase column (*686*). An aqueous mobile phase containing the copper complex of *N,N*-dimethyl-L-Phe provided separations of eight amino acids, including the enantiomers of three of four diastereomers of 2-(2′-carboxy-3′-phenyl)cyclopropylglycine (*334*). The chirality of a 3-methylamino-2-aminobutyric acid hydrolysate obtained from an antibiotic was determined by comparison with standards analyzed on a Hypersil ODS column with a mobile phase containing *N,N*-di-*n*-propyl-L-Ala and cupric acetate (*770*). A tetradentate diaminodiamido-type ligand, derived by bisacylation of ethanediamine with Phe or *N*-Me-Phe, separated 14 different underivatized amino acids (*771*). A Leu-Leu-Leu tripeptide preloaded onto a C$_{18}$ column and then used in the eluent provided separation of Trp enantiomers; the same tripeptide bonded to stationary phases did not resolve the enantiomers (*772*).

Ligand exchange chromatography has been incorporated into a two-dimensional HPLC system for enantiomeric excess determination of underivatized amino acid mixtures. The amino acids were first separated on an ion-exchange column; as they eluted, they were switched to a reversed-phase column, resolved with a chiral Cu(II) complex of a phenylalanine derivative, and then derivatized and detected (*761*). This is an improvement over an earlier procedure, in which the amino acids were initially separated into three groups on a cation exchange resin, evaporated, and then resolved with a Cu(II) complex of *N,N*-di-*n*-propyl-L-alanine (*773*).

1.8.2c Ion Pairing (Chiral Counter Ions)

Several Cbz-, Boc-, or Dns-amino acids have been separated on a silica stationary phase by using quinine in the mobile phase as a chiral counter ion (*774,775*). Di- and tripeptides have been used to resolve Trp, Phe, and *p*-fluoro-Phe on a reversed-phase column (*776*).

1.8.2d Inclusion Complexes

Dansylated amino acids (19 of 20, not Trp) have been separated by micro-HPLC on a reversed-phase column by using β-cyclodextrin (β-CD) as a mobile-phase additive to form inclusion complexes with the aromatic dansyl group (777). The separation of Dns-DL-Glu and -Leu using β-cyclodextrin as a chiral mobile-phase additive has been compared to similar separations using CD as a chiral stationary phase in LC or as chiral mobile-phase additive in CE (778).

1.8.3 HPLC Analysis: Chiral Stationary Phases

The separation of enantiomeric amino acids on chiral stationary phase (CSP) columns has been the subject of much research. The number of commercially available columns is rapidly increasing, with some examples of commercially available chiral columns produced by one supplier contained in Scheme 1.14 (779). In the past CSP columns were often expensive, fragile, and of limited versatility, but they have become more affordable with greater separation potential. Research into CSPs can be divided into a number of categories, some of which use the same principles found with chiral mobile phases. A review of CSPs, with an emphasis on considerations of chiral recognition and Pirkle-type interactions, was published in 1989 (780), with a more recent review of low-molecular-mass CSPs in 2001 (781).

1.8.3a Imprinted Polymers

By polymerizing a monomer in the presence of a chiral imprinting molecule it is possible to obtain a support that is selective for one enantiomer of molecules similar to the template compound. This area of CSPs for HPLC enantioseparations was reviewed in 2001, with many examples of amino acid imprinting/separations (782). DL-Phe-OEt was resolved on paravinylbenzoic acid–divinylbenzene or ethylene glycol dimethylacrylate–acrylic acid copolymers formed with D- or L-Phe-OEt as template (783). Similarly, the polymer obtained from Cbz-Asp imprinted on polymethacrylic acid was selective for D- and L-Asp but not DL-Glu (and vice versa), while Cbz-Trp, Cbz-Trp-OMe, Boc-Phe, and

Boc-Pro-OSu imprinted polymers all resolved their respective print compounds (784). Aromatic amides of N-unprotected or N-alkyl-Phe, -Leu, and -Pro gave imprinted methacrylic acid supports that resolved their respective print compounds and, in some instances, other similar molecules (785,786). Better separations were obtained with a terpolymer support prepared from 2-vinylpyridine, methacrylic acid, and ethylene glycol dimethylacrylate. Boc-Trp, Boc-Phe, dansyl-Phe, and Cbz-Tyr were used as templates and analytes (787). A significant improvement in capacity, selectivity, and resolution was obtained by using pentaerythritol triacrylate and 2,2-bis(hydroxymethyl)butanol trimethylacrylate as cross-linkers with methacrylic acid. Cbz-Tyr, Cbz-Phe, Cbz-Glu, and some di- and tripeptides were imprinted and resolved (788). Ethylene glycol dimethylacrylate or trimethylolpropane trimethylacrylate have also been used as cross-linkers in combination with methacrylic acid or 4-vinylpyridine as monomer; enantioseparations of amino acids, peptides, and proteins were examined (789).

A polymer formed from methacrylic acid and ethylene glycol dimethylacrylate and imprinted with L-Phe anilide was used for a study on the thermodynamics and mass transfer kinetics for the resolution of Phe anilide (790); a similar study was carried out on Fmoc-L-Trp-imprinted 4-vinylpyridine/ethylene glycol dimethylacrylate polymer (791). The effect of pH on the behavior of the CSP has also been studied, with analytical separation best at pH 3.0 (good separation, short retention time) but preparative separation best at pH 5.8 (good compromise between separation and saturation capacity) (792). The effect of thermal annealing on a poly(methyacrylate-ethylene glycol dimethacrylate) polymer imprinted with L-Phe anilide were examined, and found to decrease the enantiomer resolution (793). Photo-induced and thermal-induced polymerization of acrylamide were compared for imprinted polymers of Boc-Trp and Boc-Tyr; both methods gave similar results (794). A polymer imprinted with L-Phe anilide was tested with supercritical fluid chromatography;

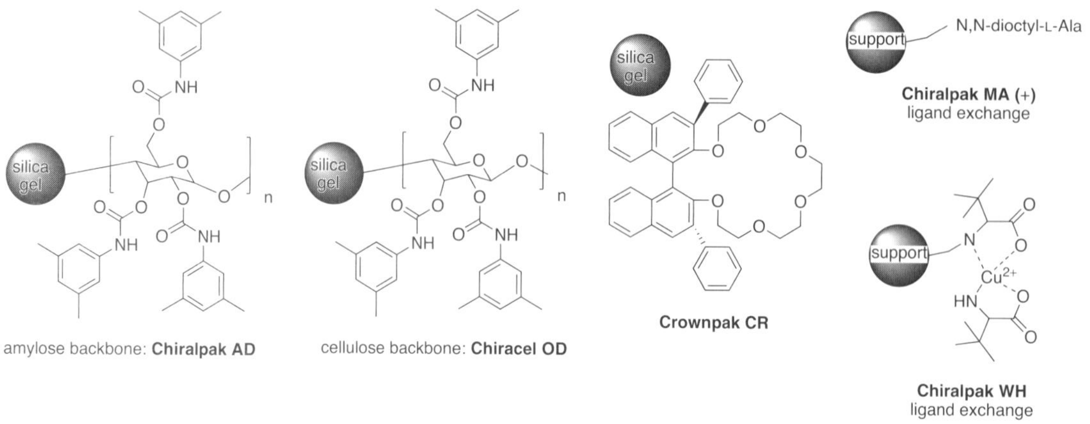

Scheme 1.14 Some Commercially Available Chiral HPLC Columns (779).

resolution of Phe anilide was observed, but peaks were very broad (795). Fmoc-L- and D-Trp were used as the templates for imprinting polymers formed from 4-vinylpyridine and ethylene glycol dimethylacrylate, produced as both traditional particulate and as monolithic columns. Thermodynamic studies showed different types of binding sites present; the monolithic column appears to have fewer non-selective binding sites, but the imprinted sites have lower association constants (796). Another monolithic column, imprinted with Fmoc- or Cbz-Trp using 4-vinylpyridine, methacrylic acid, and ethylene dimethylacrylate, was able to separate the corresponding analyte at high flow rates with both organic and aqueous solvents (797).

A urea-based polymer imprinted with Cbz-Glu separated Cbz-Glu from Cbz-Asp or Cbz-Gly, but failed to provide enantioseparation (798). N,O-Bis(methacryloyl)-Ser was employed as a functional cross-linker for the synthesis of non-covalent molecularly imprinted polymers; better selectivity was obtained for polymers imprinted with L-Trp-OMe than with D-Trp-OMe, indicating a diastereoselective effect (799). Another imprinted polymer has incorporated an acrylate-substituted 18-crown-6 moiety and 2-vinylpyridine as monomer components. The crown ether provides a receptor site for the primary ammonium group of amino acids, while the pyridyl group can interact with the carboxylate group. L-Phe BF_4 was employed for imprinting; good separation of D- and L-Phe was then observed (α = 1.28) (800). In 2004 it was reported that a single cross-linking monomer, N,O-bismethacryloyl ethanolamine (NOBE), provided a simple system for preparing imprinted polymers without the need for additional functional monomers. Imprinted polymers were prepared from Boc-Tyr, Cbz-Phe, dansyl-Phe, Cbz-Trp, and Cbz-Ser, and not only showed good enantioselectivity but also recognition of the imprinted amino acid over other similar residues (801).

Polymers containing an amide group, prepared from acrylamide, give good recognition when imprinted with N-protected Trp, Tyr or Phe. Polymers prepared with carboxyl groups as the H-bonding group showed comparatively weak recognition (802). L-Gln was imprinted in a nylon-6 polymer. The imprinted polymer effectively recognized L-Gln over D-Gln, and had little binding of either L- or D-Glu (803). An ultrathin multilayer of TiO_2 gel was imprinted with Cbz-L-Ala, and showed selectivity for Cbz-Ala/Cbz-Gly over other amino acids and high selectivity for amino acids over other acids. However, the selectivity for D-amino acids was not reported (804). Overoxidized polypyrrole films templated with L-Glu showed up to 30-times higher selectivity for L-Glu over D-Glu (805). Oxidized polypyrrole colloids have also been imprinted with L-lactate; these resolved Ala and other amino acids (806). A somewhat different approach was used by Fujii et al. (807), who imprinted a polymer with a N-benzyl-D-Val Co(III) Schiff base complex. The Co complex was covalently attached but the amino acid could be removed after polymerization. The resulting cavity was specific for N-benzyl-D-Val. Alternatively, a polymer was formed from Cu(II)-N-(4-vinylbenzyl)iminodiacetic acid and ethylene glycol dimethylacrylate, imprinted with the complex formed with L- or D-amino acids (Ala, Val, Leu, Ile, Phe, Trp). The resulting ligand-exchange CSPs (see following section) resolved predominantly only the amino acid they were imprinted with (417). A combination of imprinted polymers with chiral selectors has been achieved by copolymerizing a quinine tert-butyl-carbamate selector molecule with ethylene glycol dimethacrylate in the presence of analyte 3,5-dichlorobenzoyl (DCB) Leu. The resulting polymer showed improved chiral recognition of DCB-Leu compared to a silica-grafted reference CSP (808).

Imprinted polymers have also been employed for enantiomeric excess determination via UV measurement of the solution concentration. A copolymer of methacrylic acid and ethylene glycol dimethacrylate was imprinted with L-Phe anilide (PAA); 1.2 nM acetonitrile solutions of L- and D-PAA in varying ratios were then equilibrated with the polymer, resulting in concentration changes of up to 30% as the L-PAA was selectively absorbed by the polymer. A correlation curve could be established, and the ee of solutions determined without chromatography (809).

1.8.3b Chiral Ligand Exchange: Chiral Stationary-Phase Mode

Chiral ligand exchange chromatography was reviewed in 2001 (753) and 2003 (754). This method involves either covalently atttaching a Cu-ligand to the stationary phase, or using a dynamic coating of the ligand with an attached hydrophobic tail that embeds itself in a C_{18} column stationary phase. Ligand exchange HPLC with aqueous solutions of Cu(II) salts of analytes on a polystyrene resin containing L-proline (810), L-azetidine carboxylic acid (810) or L-allo-hydroxyproline (811) has been applied to the separation of up to 31 of 32 racemic underivatized amino acids. An L-Pro stationary phase has also been used to resolve α-alkyl-α-amino acids (704), although very broad peaks were observed. Similarly, a Cu(II) eluent with chemically bonded (1R,2S)-2-carboxymethylamino-1,2-diphenylethanol (15 amino acids (812)) or (−)-trans-1,2-cyclohexanediamine with a hydrocarbon spacer grafted on silica (10 amino acids (813)) has been used to separate underivatized amino acid racemates. Six unusual amino acids were resolved by ligand exchange on an unspecified support on a commercial column (714). Dansyl- (Dns) and dabsyl- (Dbs) amino acids were resolved by ligand exchange chromatography on silica gel modified with (S)- or (R)-phenylalaninamide; high enantioselectivity factors were observed for 11 Dns derivatives, with slightly lower selectivities for the Dbs derivatives (814). Silica-bonded amino acids (proline, hydroxyproline, or valine) were used with aqueous/acetonitrile solutions of copper sulfate or acetate to resolve a number of N-methyl amino acids and α-alkyl-α-amino acids (815). A ligand exchange CSP based on N-carboxymethyl, N-undecyl-(R)-phenylglycinol attached to silica was able to resolve all 20 α-amino acids and 15 β-amino acids tested (*816*). Silica-bound *N*-carboxymethyl, *N*-undecyl-(*S*)-leucinol was also prepared and tested with 24 amino acids, with Asn the only one not resolved. The covalently linked CSP was also compared to the same chiral selector coated on a C$_{18}$ column; the coated derivative gave better enantioseparations, but tended to have excessively long retention times (*817*). A chiral monolithic silica column was prepared by tethering the Cu(II) salts of Phe, Ala or Pro for use in micro-HPLC. Good separation factors were obtained for 12 dansyl amino acids with the Phe-derived support, but the Ala derivative only separated Dns-Ser, and the Pro support only three dansyl amino acids (*818*).

An ODS column coated with N^{τ}-*n*-decyl-L-His then loaded with Cu(II) ions separated a variety of underivatized racemic amino acids (20 amino acids tested) (*819*). Several *N*-substituted phenylglycinol derivatives were dynamically loaded onto a C$_{18}$ column, and then tested for separations of five amino acids with 1 mM Cu(II) eluent. An *n*-nonyl derivative tended to give the best results (*820*). *N,S*-Dioctyl-D-penicillamine embedded in reversed-phase C$_{18}$-silica provided separation of 1-amino-2-hydroxycyclohexane-1-carboxylic acid and 1-amino-2-methylcyclopentane-1-carboxylic acid, with a reversal in elution order observed depending on temperature. The pH and organic modifier content also had large effects, while the Cu(II) concentration had little impact (*821*). 2-Methyl-, 2-hydroxy-, and 2-amino-substituted 1-aminocyclohexane-1-carboxylic acid and 2-methyl-1-aminocyclopentane-1-carboxylic acid were resolved in an earlier report (*822*). Six *N*-substituted-L-Phe selectors have been coated onto a porous graphite carbon stationary phase, and then tested for the resolution of 18 underivatized amino acids with Cu(II) eluent. All derivatives worked for most amino acids, except those with basic side chains, but the order of enantiomer elution was reversed for CSPs based on *N*-alkyl Phe compared to *N*-aryl Phe (*823*). Similarly, *N*-alkyl Pro analogs were also coated onto porous graphite and tested with 18 underivatized amino acids. *N*-Naphthylmethyl Pro gave the best overall results (*824*). HPLC separations on a monolithic sol–gel column with Cu(II) salts of Pro, Phe, and Ala amides as chiral selectors were compared to capillary electrophoresis (CE) with the same selectors as electrolyte, and to capillary electrochromatography (CEC) with monolithic columns. Phe-NH$_2$ tended to give better resolutions of 12 dansyl amino acids by CEC and HPLC, while Pro gave much better results by CE. CEC gave the highest separation factors (*825*).

1.8.3c Synthetic Multiple Interaction Bonded Phases (Pirkle-Type Supports)

Much of the effort of CSP development, especially during the mid-1980s, has been devoted to Pirkle-type supports. These supports are designed to create the three interactions required for enantiomeric discrimination as originally proposed by Dalgliesh (*826*), generally

through two hydrogen bonds and one π-interaction (*751, 827, 828*).

A great many supports have been reported, with most based on amino acid derivatives. Valine is often employed. (*S*)-2-(4-chlorophenyl)isovaleroyl-L-Val amide attached to γ-aminopropyl silanized silica was used to resolve three *N*-3,5-dinitrobenzoyl amino acid methyl esters (*829*); by using urea derivatives of Val as the chiral molecule, six *N*-acetyl amino acid methyl esters were separated (*830, 831*). *N*-Acetyl amino acid methyl esters (eight examples) were resolved on a chiral phase prepared from *N*-acyl L-valylaminopropyl silanized silica (*832*), while *N*-acetyl amino acid *t*-butyl esters were separated on a similar (*N*-formyl-L-valylamino)propylsilica support with 2-propanol in hexane (16 of 18 examples, not Gln/Pro (*751, 833*)). The separation could be accomplished in less than 4 min by using supercritical CO$_2$ as mobile phase (11 of 12 examples, not Pro (*834*)). Polymers derived from *N*-acryloyl L-Val methyl amide and dimethyl amide linked to 3-mercaptopropyl silica gel provided resolutions of *N*-(3,5-dinitrobenzoyl) amino acid isopropyl esters (Ala, Val, and Leu) under aqueous conditions. The polymers were temperature sensitive, with enhanced retention and enantioselectivity at approximately 35 °C (*835*). Hydrophobic stationary phases composed of chiral cationic surfactants have been prepared, consisting of L-Val diamide moieties bound to silica gel supports, with a long alkyl quaternary ammonium amide substituent. *N*-3,5-Dinitrobenzoyl amino acid isopropyl esters of Ala, Val, Leu, and Phe were resolved (*836*).

Aromatic amino acids are also used in many CSPs. *N*-3,5-Dinitrobenzyl amino acid esters were separated on chiral stationary phases based on phenylglycyl-*R*/*S*-(α)-1-naphthylethylamide (10 of 11 amino acids separated, not Pro (*837*)), while *N*-3,5-dinitrobenzyl α- and β-amino acid esters and amides were resolved on (*S*)-(−)-11-siloxyundecanyl *N*-(2-naphthyl)alanine- and *N*-(2-naphthyl)valine-based CSPs (*838, 839*). CSPs using Boc- or Cbz-α-arylglycines on γ-aminopropyl-functionalized silica were applied to the separation of *N*-3,5-dinitrobenzoyl amino acid isopropyl esters (13 of 14 amino acids separated, not Pro (*840*)). Thirteen different chiral diamide-type stationary phases based on phenylglycine or valine with a number of different *N*-acyl groups were compared by examining the separation of 14 racemic dinitrobenzoyl amino acids (*841*). (*R*)-*N*-3,5-dinitrobenzoylphenylglycine covalently bonded to γ-aminopropylsilanized silica was able to resolve the Phg phosphonate analog *N*-phenyl diethyl α-phenyl aminomethyl phosphonate (*842*). Similarly, phenylglycine *n*-propylamide has been coupled to a silica matrix by its *N*-terminus via a 3-silyl propionyl linker; 12 of 13 racemic *N*-3,5-dinitrobenzoyl amino acid *n*-propylamides (not Pro) were separated (*843*). Porous graphitized carbon columns are a new alternative to traditional silica-based columns. Coating with *N*-(2-naphthlalenesulfonyl)-Phe allowed for separation of seven underivatized amino acid enantiomers (*844*).

Another aminopropylsilylsilica-based CSP has been prepared from L-Ala and pyrrolidinyl-disubstituted cyanuric chloride; 15 of 16 methyl esters of N-(3,5-dinitrobenzoyl) amino acids (not Pro) were resolved (845). Bis[carboxy(alkyl)methylamino]-6-chloro-s-triazine prepared from L-Ala, L-val, or L-Leu has been attached to 3-aminopropyl silica gel and used to resolve five N-dinitrobenzoyl amino acids (846). Macroporous polymer beads have been prepared by copolymerization of ethylene dimethacrylate with a chiral monomer containing a L-Pro-5-indananilide substituent. Separation factors of over 18 were obtained for N-3,5-dinitrobenzoyl Leu and Val dialkyl amides (847). Separations of seven N-3,5-dinitrobenzoyl amino acids on supports prepared from N-(3,5-dimethylbenzoyl) Leu-N(Ph)allyl or N-(3,5-dimethoxybenzoyl) Leu-N(Ph)allyl found that the dimethyl derivative gave better resolutions (848). N-(3,5-dinitrobenzoyl)-Leu-N(Ph)allyl has been attached to a silica support via the allyl group as a monomeric species, or copolymerized with a methacrylate-substituted silica gel to form a polymeric support. The monomeric CSP gave much sharper resolutions than the polymeric analog (849).

A combinatorial approach has been applied to the development of new CSPs. A library of 36 amino acid anilides was constructed from three L-amino acids and 12 anilines on functionalized macroporous beads, then screened for the resolution of N-(3,5-dinitrobenzoyl) amino acids/amides: L-Pro 1-indananilide was found to be the most effective CSP (850). Another library of DNB-dipeptides was constructed on a solid phase and then screened for its ability to resolve (2-naphthyl)-Ala-NEt₂, with a Dnb-Leu-Asp CSP providing excellent resolution (851). Chiral selectors for N-(1-naphthyl)-Leu were detected from a library by incubating the analyte with resin-bound selector, and then measuring the CD spectra of the supernatant, with the best selector most strongly removing one enantiomer from solution (852). A different procedure to screen for new selectors immobilizes both enantiomers of the target racemic analyte to silica gel, than elutes possible selectors. With N-(1-naphthyl) Leu ester as the racemic analyte, 3,5-dinitrobenzoyl-Leu-(4-aminobutyramide) was identified as the best selector from a 4 × 4 × 1 library (853). An earlier version of this procedure prepared enantiomeric CSP libraries, and looked for differences in their elution when passed over the immobilized analyte (854). Racemic libraries constructed via a Ugi four component condensation were screened using the target analyte, N-(3,5-dinitrobenzoyl)-L-Leu, attached to a solid support (855). A combinatorial library of 60 members was constructed with an amino acid, a structural turn element, and a large functionalized surface for binding interactions. Two dyes were employed in order to use a colorimetric binding assay to screen for enantiodiscrimination, with a blue dye attached to one enantiomer of Pro and a red dye to the other enantiomer (856). Another combinatorial approach prepared libraries of amidated amino acid residues on an optimized polymer support. Columns were prepared from mixtures of the resulting

resins and tested for selectivity in resolving DL-(3,5-dinitrobenzoyl)-Leu diallylamide. Deconvolution was then carried out to determine the most effective CSP, without needing to prepare a separate column for every combination of building blocks (857).

Chiral supports not based on amino acids have also been developed, with aromatic amines a common motif. N-(3,5-Dinitrobenzoyl)-amino acids (11 examples) were separated on a CSP derived from (R)-N-(11-triethoxysilylundecanoyl)cyclohexyl(6,7-dimethyl-1-naphthyl)methylamine, with methanol–water as eluent (858). Other naphthylmethylamine derivatives have been examined (827). N-3,5-Dinitrobenzoyl n-octyl esters of nine racemic β-substituted-β-alanines and nine α-substituted-β-alanines were resolved on CSPs derived from N-acetylated α-arylalkylamines on an undecanoyl triethoxysilyl linker (859). (R,R)-Tartaric acid mono amides of (R)-1-(α-naphthyl)ethylamine, (S)-valine-(S)-1-(α-naphthyl)ethylamide or (R)-valine-(R)-1-(α-naphthyl)ethylamide have been ionically bonded to silica gel or coated on a reversed-phase column as a copper complex and used to separate four dinitrobenzoyl amino acid, methyl esters or 18 underivatized amino acids, respectively (860). A polynaphthylethyl-urea CSP attached to aminopropyl silica was used to resolve five p-bromophenylcarbamyl amino acid enantiomers (861). A bifunctional CSP containing both an amino acid (N-3,5-dinitrobenzoyl D- or L-Val, Leu or -Phg) and a chiral amine ((R)- or (S)-1-(α-naphthyl) ethylamine) attached to a S-triazine scaffold provided good separation of a variety of enantiomers, including several N-3,5-dinitrobenzoyl amino esters (862–865).

A novel CSP has been prepared on 3-aminopropylsilyl silica using axially dissymmetric (aS)-1,1'-binaphthalene-2-2'dicarboxylic acid; three N-dinitrophenyl derivatized amino acid butyl esters were resolved (866). A similar 2'-substituted-1,1'-binaphthyl-2-carboxylic acid functionalized silica support was able to separate 5 N-(3,5-dinitrobenzoyl) amino acid butyl esters (867), while (aS)-1,1'-bianthracene-2-2'dicarboxylic acid resolved four of five amino acids (868). Polymeric (R,R)- or (S,S)-trans-1,2-diaminocyclohexane bis(acylamide) bonded to porous silica gel was effective at separating a wide range of enantiomers, including N-3,5-dinitrobenzoyl Leu, N,S-diacetyl Cys, and Fmoc-Phe (869). (S)-Biotin has been attached to silica gel and used to resolve five different N-3,5-dinitrobenzoyl amino acid diethyl amides (870). The helically chiral commercially available Chiralpak OP(+) column, consisting of (+)-poly(diphenyl-2-pyridylmethyl methacrylate) coated on silica gel, was able to resolve the Phg phosphonate analog N-phenyl diethyl α-phenyl aminomethyl phosphonate (842).

1.8.3d Immobilized Proteins/Peptides/Alkaloids

Protein-based CSPs for HPLC enantioseparations were reviewed in 2001, though with few examples of amino acid separations (871). This area of research developed from a report of tryptophan enantiomer resolution on a bovine serum albumin (BSA)-agarose column (*872*). The separation was later optimized and used to resolve Trp, 5-hydroxy-Trp, and other compounds (*873*). A review of optical resolution on immobilized BSA appeared in 1986 (*874*). A hypothesis for albumin enantioselectivity in LC or CE separations has been developed (*875*). BSA adsorbed on silica has been used to resolve Trp (*876*), N-(2-carboxybenzoyl)-Ala, N-phthalimido-Thr, N-benzenesulfonyl-Ser and -Ala (*877*), a number of N-aromatic acyl amino acids (five examples (*878*)), N-(2,4-dinitrophenyl) and N-dansyl protected DL-amino acids (eight examples of each (*879*)), and N-(chloroformyl) carbazole (CC, a highly fluorescent chromophore similar to an Fmoc group) and N-Fmoc amino acids (18 examples with separation factors ranging from $\alpha = 1.26$ to $\alpha = 5.8$ (*880*)). BSA has been bonded to silica gel via a triazine linker, with the CSP giving better resolution of Trp than a CSP prepared by immoblization using glutaric dialdehyde (*881*). BSA has also been immobilized on an ion-exchange stationary phase and used to resolve DL-Trp and N-benzoyl-DL-Ala (*882*). A polymeric flow-through support linked to BSA provided separations of eight dansyl amino acids and four 2,4-dinitrophenyl amino acids at high flow rates, allowing for baseline resolutions within 1–3 min (*883*). A BSA-coated hollow-fiber membrane provided analytical separation of underivatized Trp (*884*), while a highly stable BSA cross-linked with glutaraldehyde on porous membranes resolved Trp with a separation factor of 12 (*885*).

Human serum albumin (HSA) has also been tested as the chiral stationary phase, separating dansylated Nva, Val, Trp, and Phe (*886, 887*). A study on the effect of mobile-phase composition for the resolution of Trp on an HSA column found that optimizing the temperature, pH, ionic strength, and solvent polarity could improve separations (*888*). More recently, HSA was immobilized on a monolith column by several methods and tested for resolution of Trp; some differences in resolution were observed, depending on the attachment methods, but the separation was no better than on a traditional column (*889*). α-Chymotrypsin has also been applied to enantioseparations. After adsorption or covalent immobilization of the enzyme on silica, phosphate buffer was used to enantioseparate Cbz-Phe, Boc-Phe, Ac-Trp, and Trp (*890*).

Antibodies can be employed for resolutions. Monoclonal antibodies were raised against L- or D-p-amino-Phe conjugated to a protein. The antibodies were then immobilized on a synthetic high-flow-through support. A variety of aromatic amino acids in mild buffer were resolved, with very high separation factors for Tyr, p-amino-Phe, Phe, His, and DOPA, and reasonable separations of Trp, phospho-Tyr, cyclohexylalanine, and kynurenine (*891*). Mobile-phase conditions were examined for separations of Phe, Tyr, DOPA, Trp, and p-amino-Phe using immobilized monoclonal antibodies raised against p-amino-L- or D-Phe. Decreases in flow rate or temperature increased the retention of the retained

isomer; the unretained enantiomer eluted with the void volume (*892*).

A similar strategy has been employed with immobilized RNA aptamers that were selected for enantioselective binding to L-Tyr. Very high resolutions were obtained for Tyr and α-methyl Tyr, with good resolutions of DOPA, Trp, 1′-methyl-Trp, 5-hydroxy-Trp, Ac-Trp, β-(benzothien-3-yl)-Ala, β-(quinolin-2-yl)-Ala, and β-(2-naphthyl)-Ala, but no binding/resolution of O-methyl-Tyr, Tyr-OMe, Ac-Tyr, 3,5-diiodo-Tyr, Phe, p-amino-Phe, p-hydroxy-Phg, Trp-OMe, His, Val or Arg (*893, 894*). An anti-L-Arg D-RNA aptamer CSP was rapidly degraded by RNases under normal chromatographic conditions, but by using a CSP based on the enantiomeric L-RNA aptamer, a highly stable CSP was developed. The elution order of the Arg enantiomers was reversed (*895*).

One of the most promising approaches to amino acid enantiomeric analysis uses supports based on immobilized macrocyclic glycopeptide antibiotics, several of which are commercially available. Chiral separations with these CSPs were reviewed in 2001 (*896*). One of the most common, teicoplanin, was found to act as a very effective chiral selector for 54 underivatized amino acid and imino acid enantiomers, as well as some dipeptides. In most cases hydro-organic mobile phases with no buffer can be used, so the resolved amino acids can easily be isolated (*897*). A simultaneous analysis of 15 underivatized proteinogenic and non-proteinogenic amino acids was accomplished in under 25 min, with ionspray tandem mass spectrometry detection (*898*). The enantiomeric pairs of 14 N-(7-nitro-2,1,3-benzoxadiazol-4-yl) (NBD) amino acids were separated by a teicoplanin CSP using capillary LC with tandem MS analysis; under the same conditions DNB amino acids were not resolved (*899*). Six N-Boc amino acids were also separated on a teicoplanin CSP, with good resolutions at shorter separation times than on a hydroxypropyl-β-cyclodextrin-based column (*900*). Another study separated four Boc amino acids, comparing HPLC and CE resolutions with teicoplanin (*901*). Carnitine and O-acylcarnitine enantiomers have also been resolved using a teicoplanin CSP (*902*). A set of 31 unnatural amino acids, including arylalanines, pipecolic acid analogs, and cyclic and bicyclic β-amino acids were separated as underivatized amino acids on a commercial teicoplanin column (Chirobiotic T) using a hydro-organic mobile phase of predominantly water–methanol (*903*), with another six amino acids (2-methyl- and 2-hydroxy-substituted 1-aminocyclohexane-1-carboxylic acids, Cha and Tle) in another report (*904*). The Chirobiotic T column also separated ring-substituted phenylalanines and α-methyl-substituted Phe analogs, but not phenylalanine amides, which could be resolved on a Crownpak CR(+) column (chiral crown ether) (*692*). Trp enantiomer binding to a teicoplanin stationary phase was studied in 2003, looking at the role of sodium perchlorate in improving enantioselectivity (*905*), while the retention behavior of underivatized Phg on a Chirobiotic T column (*906*) and the effects of the mobile phase on the retention behavior of Phg and Trp (907), the effect of temperature on the resolution of β-methyl amino acids (both subambient and elevated temperatures allowed for improved enantiomer separation) (908) or the effect of eluent pH on the separation of five dansyl amino acids (909) were studied in other reports. A 2004 report studied the mechanism of chiral discrimination by teicoplanin-based CSPs by varying the mobile phase and examining resolutions of nine amino acids, seven N-acetyl amino acids, three amino esters, and four N-Boc amino acids. A deprotonated carboxyl group was found to be critical, and MeOH as organic modifier gave better resolutions than acetonitrile. For N-acetylated amino acids, the L-enantiomers appeared to be completely excluded by the CSP, while the D-enantiomers were retained (910).

Teicoplanin was compared to the teicoplanin aglycone, with the three attached carbohydrate moieties removed, in a 2000 study. Thirteen amino acids were tested, with the aglycone clearly responsible for the majority of enantioseparation of the amino acids; in fact, some amino acids were better separated on the aglycone CSP (911). The teicoplanin aglycone commercial column, Chirobiotic TAG, was used in an ion-exclusion chromatography mode to provide reasonable resolution for four of nine dansyl amino acids (912). Another study compared Chirobiotic T and Chirobiotic TAG (teicoplanin and teicoplanin aglycone) for their separation of 10 α-imino acids (e.g., underivatized Pro and pipecolic acid analogs). The aglycone tended to give better resolutions (913). The enantioseparation of 18 underivatized acyclic β³-amino acids was generally better on Chirobiotic T than Chirobiotic TAG columns (737). However, for 12 underivatized cyclic β-amino acids and 15 arylalanine α-amino acids, the aglycone TAG column usually provided better resolution (914). A teicoplanin analog, A-40,926, resolved unprotected β-amino acids, with semipreparative separations of 15–30 mg feasible (915). Teicoplanin stationary phase has also been employed for hydrophilic interaction chromatography, where polar compounds are separated using aqueous/organic mobile phases that establish a stagnant enriched water layer on the surface of the stationary phase. Seven of eight underivatized amino acids were resolved (Asp was unsuccessful), using evaporative light-scattering detection (916). Subcritical fluid-phase chromatography, with CO_2 as eluent, provided excellent resolution of Tyr and Trp on a Chirobiotic T column (917).

The macrocyclic antibiotic vancomycin was effective at resolving N-carbamyl-Phe, N-Bz Ala-OMe, -Val, -Phe, -Leu, -Trp, N-dansyl-Abu, -Nle, -Ser, -Val,- Glu, N-Cbz-Phe, -Ala, -Tyr, or -Trp, and N-Boc 4′-chloro-Phe (918). Vancomycin has been used for the enantiomeric separation of Trp and sulfur amino acids (Met, Cys_2) and their Se analogs, derivatized with AQC (6-aminoquinolyl-N-hydroxysuccinimidoyl carbamate) (919). Risocetin A is another glycopeptide antibiotic CSP. It provided resolutions for 28 unnatural amino acids, using either reversed-phase conditions (water–MeOH), or a polar–organic mobile phase (MeOH–HOAc–TEA) (920). In 2004 the

effects of temperature (5–45 °C) on the enantioseparation of 71 different chiral compounds were studied using four different commercial macrocyclic CSPs (teicoplanin: Chirobiotic T; its aglycone: Chirobiotic TAG; ristocetin A: Chirobiotic R; vancomycin: Chirobiotic V) and three separation modes (reversed-phase with MeOH/buffer, normal phase with hexane/ethanol, and polar ionic mode with 100% methanol/trace acid and base). In most cases enantioresolution decreased at increased temperatures. Asn was one of the analytes; it was noted that with teicoplanin the L-enantiomer eluted before the void volume, indicating repulsion, but the D-enantiomer was retained (921). A Crownpak CR(+) column (chiral crown ether) was compared against Chirobiotic T and R columns for the separation of eight underivatized β-substituted β-amino acids. The best chiral column was analyte-dependent (695). Chirobiotic T, V, and R stationary phases were used for enantiomer resolution during optimization of conditions for LC-ESI-MS analysis; the reversed-phase mode on the Chirobiotic T was found to give better results than the polar organic mode for amino acids. Ile, Leu, and Met were tested (335). Chirobiotic T, TAG, and R columns were tested for separations using supercritical (SFC) and subcritical fluid mobile phases; 111 chiral compound were tested, including 11 N-DNP amino acids, 9 N-Cbz amino acids, and 24 unprotected amino acids. The three columns enantioseparated all 24 underivatized amino acids and most of the N-blocked derivatives (Chirobiotic R was least successful), with baseline resolution in most cases. Conventional HPLC gave better resolutions than the SFC method, but the SFC technique provided separation in <5 min, compared to up to 20 min for HPLC (922).

A range of cinchona alkaloid derivatives have been tested as CSPs. OPA and other ortho-dicarboxaldehydes such as naphthalene-2,3-dicarboxaldehyde (NDA) and anthracene-2,3-dicarboxaldehyde (ADA) form isoindolin-1-ones when reacted with amino acids under acidic conditions in the absence of thiols, with the isoindolines more stable than the isoindole fluorophores normally generated, and resistant to racemization. The OPA derivatives of 23 amino acid enantiomer pairs were separated on a cinchona alkaloid-based CSP, with the separations of OPA, NDA, and ADA derivatives compared for a subset of nine amino acids. ADA tended to give the best results (285). A cinchona alkaloid has demonstrated extremely high chiral discriminatory powers for N-3,5-dinitrobenzoyl amino acids, with α = 32.6 for Dnb-Leu. The mechanism of chiral recognition has been studied (923). Quinine and carbamoylated quinine derivatives have been attached to 3-mercaptopropyl-modified silica gel. The carbamoylation significantly improved chiral recognition of DNB, Bz, Ac, Cbz, Fmoc, Boc, and dansyl-protected amino acids; six DNP amino acids were separated (924). A 2,6-di-t-butylphenylcarbamoyl quinine-derived chiral anion exchange stationary phase provided separations for nine different N-(2,4-dinitrophenyl) β-substituted-β-amino acids (925). Tert-butylcarbamoylquinine, attached to silica gel via a thiopropyl linker, was tested with five α- and β-amino acids as their N-Ac, Cbz, Bz, DNB, and 3,5-nitrobenzy-loxycarbonyl DNZ derivatives; the DNB derivatives gave significantly better resolution or all five amino acids. Enantioseparation increased at lower column temperatures (926). The same selector was chemically bonded to monolithic silica columns, with test separations using Cbz-, DNB-, DNZ-, Bz- and Ac- Leu and Phe. The monolith gave much faster separations compared to conventional particulate columns (927). CE separations of 24 Bz-, DNB-, and DNZ-amino acids using tert-butyl carbamoylated quinine under non-aqueous conditions were compared to HPLC separations using the same selector immobilized onto silica; a good correlation of enantioresolution was observed, meaning the CE technique could potentially be used as a fast screening method for the identification of new HPLC CSPs (928). 9-O-(Phenoxycarbonyl)quinine was bonded to carbon-clad zirconia to give a highly stable support that was tested for the resolution of 11 N-3,5-dinitrophenyl amino acids; resolutions tended to be higher than on quinine carbamate-bonded silica or quinine carbamate-coated zirconia columns (929). A quinine chiral selector has been combined with the imprinted polymers concept via copolymerization of a quinine tert-butylcarbamate selector molecule with ethylene glycol dimethacrylate in the presence of analyte 3,5-dichlorobenzoyl Leu. The resulting polymer showed improved chiral recognition of DCB-Leu compared to a silica-grafted quinine reference CSP (808).

Another natural product-derived CSP, based on the ergot alkaloid terguride, was applied to the enantioseparation of nine different N-dansyl-, 3,5-dinitrobenzoyl-, 2,4-dinitrophenyl-, benzoyl-, or β-naphthoyl-amino acids (930). A polyether antibiotic, lasalocid, was adsorbed onto porous graphitic carbon for use in capillary chromatography, with underivatized (1-naphthyl)Ala and (2-naphthyl)Ala both resolved (931). Substituted bile acids, derivatives of cholic and deoxycholic acid, were linked to silica gel to form CSPs. Several N-3,5-dinitrobenzoyl amino acid esters were among the test analytes, with resolution dependent on both side chain and type of ester (932). A commercial CSP based on an immobilized network polymer derived from L-tartaric acid, Kromasil CHI-DMB, gave good resolutions for seven underivatized aryl amino acids (933).

1.8.3e Inclusion Packings

CSPs that form inclusion complexes with analytes include cyclodextrin, crown ether, and calixarene derivatives. Cyclodextrin chiral phases were reviewed in 1986 (934) and further discussed in 1988 (935). The separation of 20 Fmoc-, Fmoc-glycyl-, or Fmoc-β-alanyl-amino acids on β- and γ-cyclodextrin columns has been examined (501). D-Ser can be quantified by derivatization with Fmoc-Cl or dansyl-Cl, followed by enantiomer separation using a γ-cyclodextrin column (936). The benz[f]isoindole derivatives resulting from reaction of amino acids with naphthalene-2,3-dicarboxaldehyde-cyanide reagent can also be resolved on β-cyclodextrin

(six amino acids or amino acid amides resolved (937)). The axially chiral binaphthyl-based α,α-disubstituted β-amino acid β-Bin was separated on β-cyclodextrin columns as the free amino acid, the ethyl ester, and the N-Boc derivative (708), while the corresponding cyclic α-amino acid (Bin) was separated as the free amino acid, N-Boc, and N-Cbz derivatives (693). The resolution of amino acid β-naphthylamides, β-naphthyl esters, and dansyl amino acids on β-cyclodextrin-bonded columns has been described (938). The separation of Dns-DL-Glu and Dns-DL-Leu using β-cyclodextrin as a CSP has been compared to separation using CD as chiral mobile-phase additive in LC or CE (778). More recently, seven aromatic N-derivatization labels were compared for the analysis of 19 amino acid racemates on an underivatized β-cyclodextrin stationary phase (939).

An attempt to combine multiple-interaction-type CSP properties with the inclusion characteristics of β–cyclodextrin introduced aromatic groups for π–π interactions, resulting in a stationary phase that resolved 10 Dns- or 2,4-dinitrophenyl amino acids, but actually had poorer selectivity than underivatized cyclodextrin for the Dns derivatives (940). A derivatized cyclodextrin bonded phase consisting of (R)- or (S)-(1-naphthylethyl) carbamoylated β-cyclodextrin was used to separate over 100 racemic compounds, including six amino acid esters and nine free amino acids. The results were used to develop an empirical prediction of separation factors based on the four substituents of a stereogenic center (941). Eight Fmoc imino acids have also been resolved, with sufficient sensitivity and precision to detect optical impurities below 0.01% (942). Pre-column derivatization of amino acids (26 natural and unnatural amino acids) with phenyl isocyanate allowed for resolution on naphthylethylcarbamate β-cyclodextrin bonded phases using acetonitrile-based mobile phase; N-benzoylated or N-Cbz amino acids were not resolved under the same conditions (943). N-2,4-Dinitrophenylated amino acids (obtained by nucleophilic displacement of 2,4-nitrofluorobenzene) were also resolved using acetonitrile-based mobile phase (25 amino acids tested) (944).

N-2,4-Dinitrophenyl Phe and Trp analogs were separated on native β-cyclodextrin and heptakis(3-O-methyl)-β-cyclodextrin, with the methylated CSP giving better results (945). Native γ-cyclodextrin and three methylated derivatives—octakis(2-O-methyl)-γ-cyclodextrin, octakis(3-O-methyl)-γ-cyclodextrin, and octakis(2,3-di-O-methyl)-γ-cyclodextrin—were compared for their ability to resolve 11 pairs of dansyl-amino acids. The octakis(3-O-methyl)-γ-cyclodextrin CSPs gave the best results for most amino acids; the 2-methylated derivatives showed no resolution for any of the analytes (946). Twenty-five pairs of racemic Boc-amino acids were separated with good resolution on hydroxypropyl derivatized β-cyclodextrin-bonded phase; the method was applied to determining the enantiomeric purity of starting materials (610). A cationic β-cyclodextrin, heptakis(6-hydroxyethylamino-6-deoxy-β-cyclodextrin), separated the enantiomers of dansyl-Asp and -Glu, though peak shape was poor (947).

Amino-functionalized β-cyclodextrin was tested with 42 different derivatized amino acids, N-Cbz, DNP, DNPyr, dansyl, Fmoc, Boc, Bz, phthaloyl, and OPA derivatives, which were all successfully resolved to some extent (948).

A systematic study of amino acid enantioseparation with four different N-protecting groups on six different β-cyclodextrin stationary phases found that 3,5-dinitro-2-pyridyl-, dabsyl- and 3,5-dinitrobenzoyl-amino acids were resolved best on (R)- or (S)-(1-naphthylethyl)carbamoylated β-cyclodextrin, while underivatized cyclodextrin was best for dansyl-amino acids. The paper also studied effects of organic modifiers, ionic strength, and pH (949). Similarly, 31 6-aminoquinolyl-N-hydroxysuccinimidyl carbamate (AQC) amino acids were examined on a number of cyclodextrin bonded stationary phases (α-, β-, γ-, 2-hydroxypropyl β-, acetylated β-, and napthylethylcarbamate β-cyclodextrins). The derivatives possessed a femtomole detection limit, giving the assay excellent sensitivity (detecting 0.05% D-Leu with 12% precision) (950).

As mentioned earlier, PTH (phenylthiohydantoin)-amino acids obtained during Edman sequencing of proteins can be analyzed for chirality, though racemization of the PTH derivative is problematic. BF_3 and HCl–methanol replaced TFA as the cyclization/cleavage and conversion reagents to suppress racemization during the degradation process. A β-cyclodextrin-based chiral stationary phase provided enantiomeric separation of 18 PTH-amino acids (242). The enantiomers of the Edman sequencing product arising from DBD-NCS (7-[(N,N-dimethylamino)sulfonyl]-4-(2,1,3-benzoxadiazolyl) isothiocyanate) derivatization can also be resolved on a β-cyclodextrin column (five examples (699)). Phenylthiocarbamoyl derivatives of amino acids (PTC-AAs) have been analyzed using a tandem column system, with a reversed-phase ODS column providing an initial separation of the amino acids, followed by a phenylcarbamoylated β-cyclodextrin CSP column resolving the enantiomers. A mixture of 18 racemic amino acids and Gly gave 37 peaks with a 140-min run time, though all peaks were not completely resolved. Partial racemization of the PTC-amino acids was observed (599).

A number of chiral crown ether CSPs have been developed. Binaphthol-derived chiral crown ethers attached to polystyrene resin were used to resolve 10 amino acid perchlorate salts (951), while a different binaphthol derivative separated 17 of 19 amino acids tested (not Pro or allo-Thr) (952). A chiral crown ether incorporating an α-D-mannoside unit successfully resolved α-phenylglycine (1234), while a pseudo-18-crown-6 ether containing a 1-phenyl-1,2-cyclohexanediol chiral unit successfully resolved 20 different amino acids and two amino esters (953). Another novel chiral crown ether was based on a diphenyl-substituted 1,1'-binaphthyl-based crown ether, covalently tethered to silica gel. Twenty-four proteinogenic and non-proteinogenic amino acids were tested, with all but Pro resolved, with separation factors from α = 1.48 (His) to α = 7.03 (4'-hydroxy-Phg) (954).

A commercially available chiral crown ether stationary-phase column, Crownpak CR(+), uses dynamic coating of a proprietary chiral crown ether onto an ODS stationary phase, limiting the column to temperatures below 50 °C and a mobile phase with <15% methanol. Hydrophobic compounds, such as N-(2-naphthyl)-Ala-NH$_2$, have long elution times (second peak at 204 min) with the standard mobile phase. Addition of β-cyclodextrin or potassium dihydrophosphate to the buffer greatly reduces elution times (61–68 min) while retaining enantioselectivity (955). Crownpak CR(+) was employed to separate a number of unusual Phe, Tyr, and Trp analogs (691), and bicyclic β-amino acids (694). The Crownpak CR(+) column separated ring-substituted phenylalanines and phenylalanine amides, but not α-methyl-substituted Phe analogs, which could be resolved on a teicoplanin Chirobiotic T column (692). A similar study compared a Crownpak CR(+) column (chiral crown ether) with Chirobiotic T and R columns for the separation of eight underivatized β-substituted β-amino acids. The best chiral column was analyte-dependent (695).

A novel CSP obtained by immobilizing (+)-18-crown-6 tetracarboxylic acid onto a 3-aminopropylsilanized silica gel provided resolutions of 13 out of 18 different underivatized amino acids in one study (with Asp, Asn, Ile, Val, and Thr not separated) (956), and 27 of 28 in another study (including those not separated in the other study, but not Pro) (957). Methylation of the amide nitrogens in this CSP resulted in a complementary system in which the elution orders of amino acids were sometimes opposite that of the original CSP, with 21 of 22 amino acids resolved (not Pro) (958). A comparison of the enantioseparation of α-substituted analogs of DOPA, β-substituted analogs of Tyr, and γ-substituted analogs of GABA using (+)-18-crown-6-tetracarboxylic acid as a stationary phase in LC or an additive in CE found similar separations for the aromatic amino acids, no separation for the α-substituted analogs, but much better resolution of the GABA analogs by LC (959). Adding triethylamine as a mobile-phase additive improved the resolutions of a number (15) of aryl amino acids on a covalently bonded (+)-18-crown-6-tetracarboxylic acid CSP (960). The (+)-18-crown-6-tetracarboxylic acid CSP also resolved 16 different N-(3,5-dinitrobenzoyl) amino acids, using an acetic acid/triethylamine/acetonitrile mobile phase (961). A commercial column, the Opticrown RCA, uses the 18-crown-6-tetracarboxylic acid chiral selector. It has been used in an ion-exclusion chromatography mode to provide reasonable resolutions for 12 of 19 underivatized amino acids, with a mobile phase of 10 mM TFA in 80:20 methanol:water (912). The complexes formed by amino acids and (+)-18-crown-6-tetracarboxylic acid have been studied by NMR (670), while X-ray crystal structures and energy calculations have been made of complexes formed with D- and L-Tyr, Ile, Met, and Phg in order to clarify the mechanism of chiral recognition (671).

Calixarenes derived from resorcinol were octaamidated with (R)-2-amino-1-butanol and immobilized on Amberlite or silica gel. Enantiomers of the sodium and potassium salts of Phg and Trp were readily discriminated (962). A calixarene has been linked by a urea linkage to a cinchona-type CSP to give a hybrid-type receptor, providing good enantioselectivities for six different amino acids as their DNB, DNZ, Bz, Fmoc, Cbz, or Boc derivatives (963).

1.8.3f Cellulose and Other Polysaccharides

A 1986 review of polysaccharide derivatives for HPLC enantioseparation contains several references to amino acid resolutions (964), but also contains the memorable statement "the behavior of polysaccharide derivatives summarized in this paper are as confusing as the case of the five blind men and the elephant" (unfortunately without any further elaboration as to what that case actually is!) (965). More recent reviews of polysaccharide derivatives for the separation of enantiomers by HPLC appeared in 1998 (966, 967) and 2001 (968), although again there are only a few references to resolutions of amino acids. DL-Trp, 5-hydroxy-Trp, and DOPA have been separated on a cellulose column (969, 970), and the resolution of 17 underivatized amino acids on a native cellulose column has also been described, although peaks were broad and elution times were up to 20 h (971). This resolution has been coupled with an amino acid analyzer (971). Cellulose derivatized with substituted benzoates and fixed to allylsilica gel provided resolution of N-3,5-dinitrobenzoyl-Phe-OMe (972).

Commercially available Chiracel OD/Chiralpak AD columns (779), based on carbohydrate supports, have become standard equipment for chiral analyses of a variety of substrates, among them amino acid derivatives. Chiralpak AD consists of tris(3,5-dimethylphenyl-carbamate)-derivatized amylose on a silica gel support, Chiracel OD is tris(3,5-dimethylphenylcarbamate)-derivatized cellulose coating on silica gel and Chiracel OJ is tris(4-methylbenzoate)-derivatized cellulose on silica gel. A solid-state NMR characterization of the Chiralpak AD support has been published (973). Amino acids derivatized with a highly fluorescent Edman reagent, 7-methylthio-4-(2,1,3-benzoxadiazolyl) isothiocyanate (MTBD-NCS) were resolved on Chiracel OJ-R or OD-RH columns during amino acid sequencing/enantiomer determination, with 22 amino acids tested (243). N-Cbz tert-Leu methyl ester was analyzed in 2002 (613), and series of N-acetyl β-methylarylalanine methyl esters (974) or α-aminophosphonates (975) in 2004. The enantiomers of ethyl nipecotate (β²-homoproline, piperidine-3-carboxylic acid) have low optical rotation, so enantiopurity was assessed by formation of the N-(2,4-dinitrophenyl) derivative and chromatography on a Chiracel-OD column (976).

In 2004 the separations of 30 analytes, including several amino esters, N-acetyl amino acids, and N-acetyl amino esters, were examined on the amylose-based

Chiralpak AD column with hexane/ethanol mobile phase containing various additives. Ethanesulfonic acid additive resulted in significantly better enantioresolution of amino acid esters, with a longer retention of the second eluting enantiomer (977). In a similar study the Chiralpak AD column fully resolved 20 of 25 underivatized amino acids, with a further four partially resolved (Arg the only exception) by using various sulfonic acid additives (e.g. ethanesulfonic acid) in place of TFA (978). The effects of both acidic (979) and amine (980) additives on improving enantioselectivity of 15 Phe analogs has also been studied. The enantiomers of various N-acyl α-hydroxy-Gly derivatives were resolved on a Chiralpak AD column (981). Chiralpak AD and AD-H columns have also been tested in subcritical fluid-phase chromatography, with CO_2 as eluent. Inclusion of amine mobile-phase additives, such as 1% cyclohexylamine, was essential for chiral resolution of amino esters in one study (982), though another report resolved Tyr-OMe, Leo-OBn, Phe-OMe, Phe, Pro, Tyr, and Phg using 0.1% ethanesulfonic acid and 20% ethanol (983). Another subcritical fluid-phase study resolved 18 N-trinitrophenyl amino acids with baseline separations on a Chiralpak AD column. Other N-substituents, and combinations with methyl esters, were also tested (917). Supercritical chromatography on Chiralpak AD or Chiracel OJ columns was able to resolve Cbz-Tyr, but not the underivatized amino acid (984).

The use of these columns has been extended to preparative separations of amino acids, though this method is still generally not applicable on a preparative scale due to the expense of chiral columns of sufficient size. For example, a 2004 synthesis of all diastereomers of β-methyl-Phe used chiral HPLC with a 10-undecanoate/3,5-dimethylphenylcarbamate-derivatized amylose stationary phase to resolve each set of erythro or threo enantiomers, but required 4–6 hs and 33–38 injections to separate 327–760 mg of racemic material (985). One of the drawbacks of polysaccharide-derived CSPs is their incompatibility with mobile phases other than hydrocarbons or alcohols, as the polysaccharide dissolves or swells. Mixed polysaccharide derivatives covalently bonded to an allylsilica gel matrix provide increased stability, and were used for a semipreparative resolution of cis- and trans-1-amino-2-phenylcyclobutane-1-carboxylic acid. Multiple (over 20) injections were still required for 700 mg to be purified (986).

Chitosan derivatives with 10-undecanoyl/phenylaminocarbonyl or benzoyl substituents have been immobilized on allylsilica gel. N-Acyl amino acids were among the analytes tested (987). The 10-undecanoyl/3,5-dimethylphenylcarbamates of amylose, chitosan, and cellulose bonded to silica gel were compared for the resolution of a number of enantiomers, including N-3,5-dimethylbenzoyl Ala-OMe: the cellulose derivative gave the best separation, with no resolution on the amylose derivative, and intermediate separation on the chitosan CSP (988).

1.9 Determination of Enantiomeric Purity: GC Chromatographic Analysis

1.9.1 GC Analysis: Diastereomeric Derivatives

The analysis of volatile derivatives by GC has been one of the standard assays for amino acid enantiomeric purity (*621, 989, 990*), though used less frequently in recent years. A number of diastereomeric derivatives have been developed for analysis on achiral columns (*991, 992*). For example, pentafluoropropionyl-amino acid (+)-3-methyl-2-butyl esters (*993*), and esters with other chiral 2-alkanols are resolved on a dimethylpolysiloxane-coated column; the procedure has also been applied to α-alkylated-α-amino acids (*704*) and *N*-methyl amino acid (seven examples (*994*)). Amino acid methyl esters *N*-derivatized with α-methoxy-α-trifluoromethylpropionic acid (MtPr, 12 examples) were resolved by GC, although some epimerization was noted during elution of phenylglycine. Heterochiral diastereomers eluted more quickly, so absolute configuration could be assigned (*597*). *N*-(+)- or (-)-menthyloxycarbonyl methyl or ethyl amino acid esters were readily prepared and analyzed, with 24 racemic amino acids separated; however, the Arg and His derivatives were not volatile (*995*). *N*-Methyl-alanine, pipecolic acid, isovaline, and the β-amino acids β-amino-*n*-butyric acid and β-amino-isobutyric acid were resolved using various *R*-(+)-2-alkanol esters (2-butanol, 2-pentanol, 2-hexanol, 2-heptanol, or 2-octanol) in combination with different *N*-protecting groups (*N*-trifluoroacetyl, *N*-pentafluoropropionyl, *N*-heptafluorobutyryl, or *N*-chlorodifluoroacetyl) (*996*).

1.9.2 GC Analysis: Chiral Stationary Phases

The separation of enantiomers by GC using chiral stationary phases was reviewed in 2001, with a number of amino acid examples (*624*). Most GC chiral stationary phases (CSPs) are, like HPLC CSPs, based on amino acid derivatives. Initially, carbonyl-bis(amino acid esters) and *N*-acyl-peptides esters were used as stationary-phase materials to analyze *N*-perfluoroacyl amino acid esters. The first publication of a successful GC chiral phase reported the separation of racemic *N*-TFA amino acid esters on a *N*-TFA-Ile-lauryl ester stationary phase (*997*). Soon after, *N*-TFA isopropyl esters of α-, β-, and γ-amino acids were resolved on *N*-dodecanoyl-L-valine *t*-butylamide, or 6-undecylamide liquid phases (*998*). Several diamide stationary phases (*tert*-butyl amide derivatives of *N*-lauroyl-Ala, -Leu, -Phg, or -Phe, or *N*-docosanoyl-Leu) have been examined for their ability to resolve *N*-TFA-α-methyl-α-amino acid isopropyl esters or *tert*-butyl amides (*999*). Racemization during solid-phase peptide synthesis was examined by separating *N*-TFA-amino acid isopropyl esters on an *N*-TFA-L-valyl-L-valine cyclohexyl ester stationary

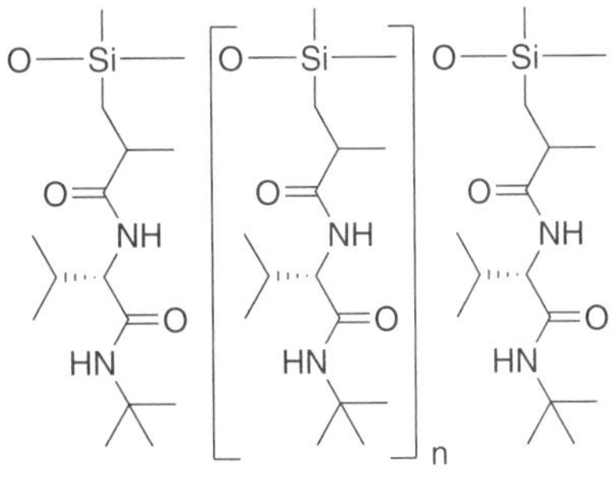

Chirasil-Val

Scheme 1.15 Chiral Stationary Phase for GC Analysis of Amino Acid Enantiomeric Purity (*603, 659, 1005–1007*).

phase (*1000*), as was the enantiomeric ratio of amino acids from a number of meteorites (*1001, 1002*). An *N*-TFA-L-Phe-L-Asp bis(cyclohexyl ester) stationary phase successfully separated 10 racemic amino acids derivatized with L-α-chloroisovaleryl chloride (*1003*).

However, these liquid phases were superseded by the development of other chiral phases (*989, 1004*), in particular polysiloxane-linked L-valine *tert*-butyl amide ("Chirasil-Val," see Scheme 1.15) (*603, 1005–1007*), which has become the standard for GC analysis of perfluoroacetylated amino acid esters (e.g., see *4, 463, 476*). *N*-Methyl, α-alkyl, β-, and γ- amino acids and amino phosphonic acids have all been analyzed by this method (*767, 768, 1008*). A study on the effects of the *N*-acyl group on the enantiomeric separation of six amino acid isopropyl esters on a Chirasil-Val-like stationary phase found that while *N*-trifluoroacetyl derivatives were the most volatile, the *N*-acetyl group gave the best separation; *N*-formyl and *N*-benzoyl groups were also examined (*1009*). The chiral resolution on a Chirasil-Val column has been coupled with mass-spectral analysis to determine the enantiomeric purity of amino acids within synthesized peptides (*500*), and the extent of racemization in amino acids from environmental samples (*468*). The peptides were hydrolyzed using DCl/D₂O, and the amino acids derivatized as *N,O,S*-trifluoroacetyl isobutyl esters, separated on a Chirasil-Val stationary phase, and then analyzed by electronionization (EI) or chemical ionization (CI) MS. This procedure distinguishes between racemization caused by coupling or by peptide hydrolysis, as amino acids affected by the latter will be deuterated. Chirasil-Val has also been employed for GC-MS analysis of enantiomeric excesses in amino acids isolated from meteorites (*1010*), and fossils (*466*), as well as a recent (2001) examination of the ability of calcite to selectively adsorb L- or D-Asp onto mirror-image crystal surfaces (*1011*). Racemization rates of amino acids were monitored by derivatization as *N*-trifluoroacetyl 1-propyl esters, followed by GC-FID analysis using a Chirasil-L-Val column (*487*), while *N*-trifluoroacetyl 2-propyl ester derivatives were separated for analysis of D-amino acids in human urine (105). A number of unusual Phe, Tyr, and Trp analogs were derivatized as their N-trifluoroacetyl isobutyl esters and separated on a Chirasil-L-Val column (691).

Other derivative combinations have been described. Diamino acids from the Murchison meteorite were derivatized as their N,N′-diethoxycarbonyl ethyl esters for analysis by GC-MS on a Chirasil-L-Val column (801). All stereoisomers of Thr were converted to N,O-bis(isobutoxycarbonyl) 2,2,2-trifluoroethyl esters and separated on a Chirasil-Val column, while 4-hydroxy-Pro required derivatization as the N,O-bis(isobutoxycarbonyl) isobutyl amide for complete separation (1012). Capillary GC analysis of amino acid N(O)-2,2,2-trifluoroethoxycarbonyl 2′,2′,2′-trifluoroethyl esters on a Chirasil-Val stationary phase gave almost complete separation of all enantiomeric pairs, except Pro, within 31 min (1013). This derivatization was applied to the analysis of amino acids produced by syntheses that mimicked primordial conditions (1014). N(O,S)-Ethoxycarbonyl heptafluorobutyl ester derivatization was employed for the GC quantification of amino acid enantiomers in food matrices. This combination was found to give the best compromise between short retention times, high yield responses, and good resolution for the 14 amino acids tested (88). N(O,S)-Trifluoroacetyl n-propyl esters and N(O,S)-pentafluoropropionyl isopropyl esters were both used for analysis of amino acids in beer on a Chirasil-L-Val column (89). A new method for derivatizing amino acids as their N(O)-pentafluoropropionyl 1-propyl esters was developed, based on using supercritical carbon dioxide as both the reaction medium, and as the agent used to extract the derivatives. A test mixture of 12 non-polar amino acids was derivatized and separated on a Chirasil-Val column (1015).

A similar polysiloxane L-Val-(R)-α-phenylethylamide stationary phase has been applied to the analysis of amino acid isopropyl esters (20 examples) and N-methyl amino acid isopropyl esters (eight examples) derivatized with tert-butyl isocyanate (1239). N-methyl amino acids (eight examples) or trifunctional amino acid isopropyl esters (six examples) were derivatized as oxazolidinediones by reaction with phosgene (1016, 1017) and analyzed on this support. The chiral stationary phase has also been used in a new peptide sequencing methodology, which cleaves the N-terminal amino acid as a tert-butylcarbamoyl amino acid isopropyl ester (1018). The advantage of this sequencing method is that it allows for the enantiomeric purity of each position to be determined. Conventional Edman sequencing produces a derivative which is not optically stable, as racemization occurs during cyclization/cleavage (698), while normal peptide hydrolysis followed by chiral analysis will give an averaged value for the optical purity of repeating residues.

Several cyclodextrin-based stationary phases have also been investigated for GC analysis. 2,6-Di-O-pentyl-3-O-butyryl-γ-cyclodextrin was extremely successful, resolving 35 of 36 enantiomer pairs of trifluoroacetyl amino acid methyl esters (1019). A 2,6-di-O-butyl-3-O-trifluoroacetylated-γ-cyclodextrin was unable to resolve Arg, Lys, Trp or Pro (1020). Fourteen of 15 N(O)-trifluoroacetyl methyl and isopropyl esters (not Trp) were enantioseparated on capillary columns coated with four types of cyclodextrin derivatives in a 1994 report (1021). Another study tested 2,6-di-O-pentyl-6-O-acyl-α-, -β-, and γ-cyclodextrins for the resolution of N-TFA,O-alkyl amino acids; the separation depended on both the cyclodextrin acyl substituent and amino acid ester group (1022). A review of cyclodextrin stationary phases for GC separation of enantiomers was published in 1990, but it contained no examples of amino acid separation (1023). A GC-based resolution of carnitine (γ-trimethylammonium-β-hydroxybutyrate) employed on-line conversion of carnitine into β-hydroxy-γ-butyrolactone, based on thermal intramolecular nucleophilic displacement of the trimethylammonium group by the carboxylate oxygen, followed by resolution on a β-cyclodextrin-based column (1024). Enantiomers of various derivatized nipecotic acids (piperidine-3-carboxylic acid) were separated by a commercial β-cyclodextrin-based column (CP Chirasil-Dex CB) during GC analysis (1025).

Another inclusion-type receptor CSP is formed by the chiral calixarenes, which favor much more nonpolar interactions. A resorcin[4]arene with four long ω-unsaturated alkyl chain substituents and pendant L-Val diamide groups (Chirasil-Calixval) was applied to the separation of the enantiomers of 18 N(O)-trifluoroacetyl amino acid ethyl esters (1026). A mixed chiral stationary phase consisting of permethylated β-cyclodextrin (Chirasil-Dex) and the resorcinarene Chirasil-Calixval retained the individual enantioselectivities of both components, allowing for the separation of both apolar and polar enantiomers. Resolutions were obtained of 16 N(O)-trifluoroacetyl amino acid ethyl esters in a single run, with some resolutions better than the individual CSPs, but some worse (1027). The molecular recognition of amino acids by a resorcin[4]arene calixarene-based receptor, with four pendant Val-OEt groups, was investigated by ion cyclotron resonance mass spectrometry (1028).

1.10 Determination of Enantiomeric Purity: Capillary Electrophoresis Analysis

While the development of HPLC chiral separations dominated the 1980s, capillary electrophoresis (CE) and capillary electrochromatography (CEC) separations have been the focus of the 1990s and early 2000s. These rapidly evolving analytical techniques have been extensively applied to the resolution of amino acids (over 150 papers by 1995 (1029)), adopting many of the same methods of chiral mobile-phase additives that were originally developed for chiral HPLC analysis. The importance of amino acids as an application for the development of new methods is indicated by their predominance as analytes in reviews of chiral CE separations published in 1994 (*1030, 1031*), 1997 (*1032*), 2000 (*1033, 1034*), 2001 (*1035*), 2005 (both CE (*1036*) and CEC (*1037*)), including specific reviews of chiral CE analyses of amino acids in 1995 (*1029*), 2000 (*1038*), 2001 (*333*), 2003 (*1039*), and 2006 (*1040*). Pre-, post-, and on-column derivatization of amines for capillary electrophoresis was reviewed in 1998 (*336*). The compact size and robust construction of CE devices provides the opportunity for unique applications. For example, a microdevice has been developed that could analyze for the presence of amino acids on Mars. The Mars Organic Analyzer (MOA) employs a microfabricated chip containing a portable CE instrument that is able to separate and detect amino acids present in soil extracts from the Atacama Desert in Chile, or in jarosite samples (a sulfate-rich mineral) from the Panoche Valley, California. The instrument possesses parts-per-trillion sensitivity, readily separates amino acids such as Val, Ala, Gly, Asp, and Glu, and potentially could employ chiral CE analysis to detect amino acid optical purity (*1041*).

1.10.1 CE Analysis: Diastereomeric Derivatives

Several diastereomeric derivatives, generally derived from HPLC applications, have been used for enantiomeric analysis. CE micellar electrokinetic separation of *o*-phthaldialdehyde (OPA)/*N*-acyl L-Cys (NAC) (see Scheme 1.11) derivatized Val diastereomers was successful, but the precision and detection limit was a factor of 3 times worse than when using HPLC (*1042*). CE analysis of other amino acids derivatized with OPA/NAC has been reported (*1235*), with online in-capillary derivatization separating 13 amino acid enantiomer pairs (*1043*). A 2006 report described a single-step method employing online sample preconcentration, in-capillary chemical derivatization with OPA/NAC, and diastereomer separation with UV detection. The procedure could analyze for D-amino acids in submicromolar solutions of 95% ee L-amino acids, and was applied to the detection of amino acids in bacterial cultures with minimal sample preparation (*1044*).

Amino acid derivatization with (+)- or (−)-1-(9-fluorenyl)ethyl chloroformate (FLEC) (16 amino acids separated) provided faster and better separation efficiencies than separation of *N*-Fmoc amino acids on a chiral support (*1045*). A related analog, (+) - and (−)-1-(9-anthryl)-2-propyl chloroformate (APOC), was tested with 17 racemic amino acids in one run using micellar electrokinetic chromatography with LIF detection (*1046*). APOC was also used to derivatize amino acids from hydrolyzed β-amyloid peptides and senile plaques in order to determine the levels of racemization of Asp, Glu, Ser, and Phe; a reference run separated the enantiomers of Glu, Asp, Ser, Ala, Thr, Gly, Val, Met, Tyr, Pro, Ile, Leu and Phe (*477*). A six amino acid mixture derivatized with NBD-PyNCS, (*R*)- or (*S*)-4-(3-isothio-cyanatopyrrolidin-1-yl)-7-nitro-2,1,3-benzoxadiazole,

was resolved by CE, with the method used to analyze the D- and L-amino acids in the hydrolyzed peptide gramicidin D (*1047*). A reagent based on enantiomerically pure *trans*-1,2-diaminocyclohexane, (1*R*,2*R*)- or (1*S*,2*S*)-*N*-[(2-isothiocyanato)cyclohexyl]-6-methoxy-4-quinolinylamide (CDITC), resolved 19 amino acids (*747*). (*S*)-1-(1-Naphthyl)ethyl isothiocyanate and (*S*)-1-phenylethyl isothiocyanate derivatives of Ala, Phe, and Val enantiomers were separated in a buffer containing a β-cyclodextrin derivative (*1048*). Isovaline (α-ethyl-Ala) has been analyzed for enantiomeric purity at picomolar concentrations via derivatization with a chiral fluorescein-Lys-based reagent, using CE separation and LIF detection (*1049*).

1.10.2 CE Analysis: Chiral Ligand Exchange

Dansyl amino acids (9 of 10) have been resolved by electrokinetic chromatography using a copper(II) histidine complex as the support electrolyte (*1050*), while 14 of 18 were separated using aspartame (aspartylphenylalanine methyl ester) (*1051*). Ligand exchange micellar electrokinetic capillary chromatography (LE-MEKC) using a copper(II) complex of L-4-hydroxy-Pro as the chiral selector along with sodium dodecyl sulfate separated the enantiomers of *o*-, *m*-, and *p*-fluoro-Phe and and *o*-, *m*-, and *p*-Tyr (*1052*). *N*-alkyl derivatives of 4-hydroxy-Pro showed improved enantioselectivity over 4-hydroxy-Pro itself for the resolution of 13 aromatic and six aliphatic underivatived amino acids (*1053*), while both Pro and Ile provided resolutions of eight unmodified amino acids (*1054*). Another study that compared Pro, Phe, and Ala amides as chiral selectors for CE (as electrolyte), CEC (on monolithic column), or micro-HPLC separations found Phe tended to give better resolutions of 12 dansyl amino acids by CEC and HPLC, while Pro gave much better results by CE. CEC gave the highest separation factors (*825*). Both underivatized (seven examples) and dansyl amino acids (11 examples) can be separated using Cu(II)-L-Val as a selector in micellar electrokinetic capillary chromatography (*1055*). A Cu(II) L-Lys complex resolved all four underivatized amino acids tested (Tyr, Phe, Trp, and phospho-Ser) (*1056*). Chiral ligand exchange chromatography was reviewed in 2001 (*753*) and 2003 (*754*).

1.10.3 CE Analysis: Chiral Surfactants/ Selectors

Phenylthiohydantoin (PTH) DL-amino acids have been analyzed by micellar electrokinetic chromatography using chiral surfactants such as sodium *N*-dodecanoyl-L-glutamate (*1057, 1058*), sodium *N*-dodecanoyl-L-valinate (*1059, 1060*), sodium *N*-dodecanoyl-L-serinate (*1061*) or digitonin-sodium taurodeoxycholate (*1058, 1060*). A number of amino acid racemates (Trp, Nle, Nva, Val, Ala, Abu, Met) were effectively separated by this method, although Ser was not (*1057–1062*). Separations have also been obtained with a chiral polymer, poly[sodium (10-undecenoyl)-L-valinate] (PSUV) (*1063*). Both PSUV and poly[sodium (10-undecenoyl)-L-threoninate] (PSUT) were compared for their interactions with five PTH amino acids; PSUV gave better resolutions (*1064*). Dipeptide micelle polymers formed from poly[sodium N-undecanoyl-L-leucylvalinate] (poly-L-SULV) were tested for the resolution of 75 diverse racemic compounds, with 58 analytes successfully separated. Among those resolved were PTH derivatives of Val, Nva, Tyr, Nle, Ile, Trp, Phe, His, and Arg and 2-aminooctanoic acid (*1065*). 3,5-Dinitrobenzylated amino acid isopropyl esters (8 of 9, not Pro) were resolved using micelles formed from sodium salts of N-dodecanoylated L-Val, Ala, or Thr (*1063*). (*S*)- and (*R*)- N-dodecanoylated Val was also used in combination with 6-aminoquinolyl-N-hydroxysuccinimidyl carbamate (AQC) derivatized amino acids; 12 examples were resolved, and AQC was found to give a much better signal-to-noise ratio than Boc, Cbz or PTH derivatives (*1066*). Six dansylated DL-amino acids were separated with bile salt micelles (*1067*). A giant vesicle forming surfactant, N-(4-dodecyloxybenzoyl)-L-valinate, has been used for chiral separations, though not yet of amino acids (*1068*).

Cyclic hexapeptides can be used for N-2,4-dinitrophenyl amino acid enantiomer separation. A combinatorial synthesis of libraries of cyclopeptides used amino acid resolution to screen for the most effective selectors (*1069*). Amphiphilic aminosaccharides, glucosamine with three fatty acid substituents, provided partial resolutions of five dansyl amino acids (*1070*). 1-S-Octyl-β-thioglucopyranoside resolved seven of eight dansyl amino acids as long as the pH was carefully controlled. Separation was enhanced by the addition of cyclodextrin (*1071*). The use of chiral glycosidic surfactants for enantiomer separation in CE was reviewed in 2000 (*1072*). The use of proteins as chiral selectors in CE was reviewed in 1997 (*1073*) and 2000 (*1074*), with a number of examples of amino acid resolutions. BSA has been incorporated into the electrophoresis gel and used to separate Trp enantiomers (*1075*) or, in combination with dextrin, Dns-Leu, and Dns-Nva (*1076*). A hypothesis for albumin enantioselectivity in LC or CE separations has been developed (*875*).

Macrocyclic antibiotics such as teicoplanin, vancomycin, and ristocetin A have been employed as chiral selectors. A summary of their applications to analytes, including many amino acid derivatives, was reported in 1998 (*1077*), with another review in 2001 (*896*). A 2000 study separated four Boc amino acids, comparing HPLC and CE resolutions with teicoplanin as the chiral selector (*901*). Vancomycin as chiral selector baseline resolved 12 Fmoc amino acids (*1078*). A mechanistic study of CE separations using macrocyclic antibiotics compared vancomycin and balhimycin, which contain the same aglycone core and almost identical sugar substituents, but with different attachment points to the aglycon. Balhimycin showed more than twofold greater

enantioselectivity for four dansyl amino acids, with a dimerization mechanism proposed to account for the increased resolution (*1079*). d-(+)-Tubocurarine chloride, a chiral antibiotic macrocycle, was employed as a chiral selector for organic carboxylates. N-Dansyl derivatives of Val, Nva, Nle, Phe, Glu, Met, and 2-aminoheptanedioic acid and N-(2,4-dinitrophenyl) derivatives of Abu, Nva, Leu, Leu, Phe, and ethionine were all resolved using 20% methanol and 80% 0.05 M phosphate buffer with 15 mM tubocurarine (*1080*).

A quinine carbamate chiral selector was used in packed CEC and non-aqueous CE for the resolution of N-derivatived (DNB, DNZ, DNP, Fmoc) amino acids. Leu, Pro, and Val derivatives were separated in the initial studies (*1081–1083*), with various derivatives of Phe, Tyr, Trp, Ala, Asn, Arg, Lys, and Ser also resolved in subsequent reports (*1084, 1085*). Other cinchona alkaloid derivatives were then assessed for their enantioseparation capabilities. Nine N-3,5-dinitrobenzoyl amino acids were best resolved with a cyclohexyl carbamoylated quinine selector on a silica support (*1086*), while non-aqueous CE conditions had the best results for nine DNB amino acids with adamantyl carbamoylated quinine (*1087*). CE separations of 24 Bz, DNB, and DNZ amino acids using *tert*-butyl carbamoylated quinine under non-aqueous conditions were compared to HPLC separations using the same selector immobilized onto silica; a good correlation of enantioresolution was observed, meaning the CE technique could potentially be used as a fast screening method for the identification of new HPLC CSPs (*928*). Dimeric quinine and quinidine chiral selectors were also synthesized and tested with 25 amino acid derivatives, with higher enantioselectivity values generally obtained than with the corresponding monomeric alkaloid. The best results were obtained with *trans*-1,4-cyclohexylene-bis(carbamoylated-11-dodecyl-thio-dihydroquinine (*1088*).

1.10.4 CE Analysis: Chiral Inclusion Complexes

A review of chiral CE separations via inclusion-complexation with cyclodextrins or crown ethers was published in 1997 (*1089*), with another review of enantioselective determination by CE with cyclodextrins as chiral selectors in 2000 (*1090*). A study examining the theory of enantiomer separation by CE employed cyclodextrin binding to demonstrate that enantioseparations are possible even if the analyte-chiral selector binding constants are the same for both enantiomers, as long as the resulting diastereomeric complexes have differing CE mobilities (*1091*).

Unmodified cyclodextrins have been employed for separations of a number of amino acid derivatives. Ten racemic dansylamino acids were resolved by using a solution of β-cyclodextrin in DMF (*1092*), eight with β-cyclodextrin in triethylammonium acetate aqueous buffer (*1093*) and 10 with γ-cyclodextrin in buffer solutions of different pH; the amino acids that are resolved change with pH (*1094*). Organic modifiers such as methanol improved the resolution of dansyl amino acids when added to a β-cyclodextrin solution in high pH phosphate buffer. Twelve amino acids were separated (1095). However, another study found organic additives reduced chiral selectivity (1096). A dramatic improvement in the enantioselectivity of 12 dansyl amino acids with unmodified β-cyclodextrin as chiral selector was obtained by adding urea (H-, methyl-, ethyl-, and 1,3-dimethyl-ureas) to the electrolyte (1097). Resolution was reduced with α- and γ- cyclodextrins and modified cyclodextrins. The reason for these changes is not known. A comparison of separations of Dns-DL-Glu and -Leu using β-cyclodextrin as chiral mobile-phase additive in CE and LC or as a CSP in LC found CE gave much better resolution for Glu, but the LC CSP gave equal or better separation for Leu (778). LIF detection has been employed to visualize the whole length of a CE capillary, rather than just a fixed point at the end of the capillary. The detector was used to examine the development of the chiral separation of dansylated amino acids using a β-cyclodextrin electrolyte (1098). An analysis for L- and D-Ala in hydrolyzed protein ferti-lisers used dansyl derivatization followed by analysis with a β-cyclodextrin selector; most fertilizers contained up to 40% D-Ala, and are thus likely to be less effective as plant nutrients (1099).

Both dansyl chloride and fluorescein isothiocyanate (FITC) were used to derivatize 15 amino acids for analysis by CE-MS, using β-cyclodextrin in the electrolyte. The MS sensitivity for both derivatives was similar, but FITC amino acids were more resolved. The method was used to analyze amino acids in commercial orange juice (1100). Combining β-cyclodextrin with sodium taurocholate provided superior resolution of 21 FITC amino acids compared to either selector alone (1101). Another derivatization reagent, 3-(4,6-dichloro-1,3,5-triazinylamino)-7-dimethylamino-2-methylphenazine (DTDP) was tested with three amino acids that were then resolved using β-cyclodextrin as the chiral selector (1102). Fifteen Fmoc amino acids were resolved with β-cyclodextrin in phosphate buffer (1045). Separation was slower and less efficient than by the FLEC amino acid diastereomer method. γ-Cyclodextrin has been applied to the separation of Fmoc amino acids to prove the enantiomeric purity of precursors during investigations of racemization during solid-phase peptide synthesis (three examples (499)), while Fmoc-protected β-amino acids (β-homoPhg and β-homoPhe) were examined using a buffer containing γ-cyclodextrin (1103). A chiral separation of homocysteine has been achieved via thiol derivatization with 4-aminosulfonyl-7-fluoro-2,1,3-benzoxadiazole, followed by separation using CE with a γ-cyclodextrin electrolyte (454). α-Cyclodextrin proved to be capable of separating underivatized Phe, Tyr, and Trp (1104), while the enantiomeric purity of β-heteroaryl-β-amino acids was successfully determined by CE with cyclodextrin (1105).

A range of derivatized cyclodextrins have been synthesized and tested for amino acid resolutions. Hydroxypropyl-substituted cyclodextrins in phosphate

buffer resolved 14 N-Boc amino acids (1106), while hydroxypropyl-γ-cyclodextrin provided the best resolution of several cyclodextrins for DOPA (1107). Hydroxypropyl β-cyclodextrin provided a "preparative-scale" (4 mg) enantiomer separation of dansyl-Phe; the same system was employed to examine some of the theoretical aspects of the separation (1108). The same resolving agent was used in a preparative-scale continuous free-flow isoelectric focusing separation of dansyl-Trp enantiomers (1109). Seven underivatized aromatic amino acids found in plants were resolved with a β-dimethylcyclodextrin chiral selector (1110). A mixed electrolyte with chiral selectors such as O-trimethyl β-cyclodextrin, digitonin, and β-escin resolved 15 PTH amino acids produced by protein sequencing; His, Leu, Lys, and Pro could be resolved using individual selectors but not with the combination. The D-Ala residue in a Met-enkephalin analog was identified by sequencing using this method (247,1111). The γ-amino acid carnitine was resolved via online derivatization with Fmoc-Cl, followed by separation using heptakis(2,6-di-O-methyl)-β-cyclodextrin (1112). The effect of selectively methylating β-cyclodextrin on the enantioseparation of 12 dansyl amino acids has been studied, with some unexpected results (1113).

An investigation of structure–resolution relationships examined N-terminal amino acid derivatives (dinitrobenzoyl, dinitrophenyl, dansyl, Cbz, Fmoc, and 6-aminoquinolyl-carbamoyl) with α-, β-, and γ-CDs, and CD derivatives. In general, the combination of nitro or dimethylamino groups on the protecting group, coupled with methylated or hydroxypropylated CDs, gave better resolutions, which were further improved by the addition of organic modifiers (1114). A subsequent report compared the separation of 19 N-(6-aminoquinolyl-carbamoyl) amino acids using native β-cyclodextrin, (2-hydroxypropyl)-β-cyclodextrin, heptakis(2,6-di-O-methyl)-β-cyclodextrin, heptakis(2,3,6-tri-O-methyl)-β-cyclodextrin, β-cyclodextrin polymer, and carboxy-methyl-β-cyclodextrin polymer. The optimum conditions/chiral selector for each amino acid were described (1115, 1116). The addition of 18-crown-6 ether improved resolutions of a number of underivatized aryl amino acids and esters (Phe, Trp, and Phg analogs), tested with α-, β-, and γ-cyclodextrin, hydroxypropyl-β-cyclodextrin, heptakis(2,6-di-O-methyl)-β-cyclodextrin, and heptakis (2,3,6-tri-O-methyl)-β-cyclodextrin (1104).

Allyl carbamoylated β-cyclodextrin copolymerized in polyacrylamide gel separated 9 of 12 dansyl amino acid enantiomers (1117) and 1-(1-naphthyl)ethylcar-bamoylated β-cyclodextrin was used as a chiral additive to separate three N-(3,5-dinitrobenzoyl) DL-amino acids (1118). β-Cyclodextrin capped with an L-Ala-O(CH$_2$CH$_2$O)$_3$-L-Ala linkage gave baseline resolution of N-dansyl DL-Glu or N-dansyl DL-Asp enantiomers (1119).

Cyclodextrins with charged substituents have been developed. Single isomer charged cyclodextrin, the sodium salt of heptakis(2,3-dimethyl-6-sulfato)-β-cyclodextrin, provided resolution of a wide variety of analytes, including underivatized or *N*-dansyl Trp (*1120*). Octa(6-*O*-sulfo)-γ-cyclodextrin resolved all seven dansyl amino acids tested, and one underivatized amino acid (Trp) out of two tested (*1121*). Octakis(2,3-diacetyl-6-sulfato)-γ-cyclodextrin has also been examined, with large separations of dansylated Asp, Met, Phe, Leu, Val, Ser, Abu, and Glu (*1122*). Heptakis(2,3-diacetyl-6-sulfato)-β-cyclodextrin gave the best resolution of DOPA out of a variety of cyclodextrin derivatives (*1123*). A highly sulfated β-cyclodextrin chiral selector provided resolution of eight different arylglycine amides (*1124*). Eight dansyl amino acids were resolved by a combination of sulfonated β-cyclodextrin along with a chiral surfactant, *N*-octoxycarbonyl-L-Leu (*1125*). A sulfated β-cyclodextrin was also used for electrokinetic capillary chromatography of Phe, Tyr and Trp, with temperature used to optimize resolution; 52 °C was optimal (*1126*). Three dansyl amino acids were separated with sulfobutyl ether γ-cyclodextrin (*1127*). A method to rapidly define experimental conditions for the separation of new compounds tested hydroxypropyl-, carboxymethyl- and sulfobutylether-β-cyclodextrin at various concentrations and several buffer pHs, along with different organic modifiers. Of 14 amino acid derivatives (six amino esters, six Boc amino acids, and two Cbz amino acids), resolution conditions were identified for 12 of them with less than nine experiments (*1128*).

Positively charged cyclodextrins include mono-6-ammonium-6-deoxy-β-cyclodextrin, which provided good separations of anionic analytes, including six dansyl amino acids (*1129*). A quaternary ammonium β-cyclodextrin provided resolutions of 11 dansyl amino acids and two Fmoc amino acids under non-aqueous conditions, with formamide the most effective solvent (*1130*). Mono-6^A-*N*-pentylammonium-6^A-deoxy-β-cyclodextrin chloride provided marginal resolutions of six dansyl and two benzoyl amino acids (*1131*), while an amphoteric β-cyclodextrin with both quaternary ammonium and carboxyl groups was used to resolve eight dansyl amino acids (*1132*). A copolymer of allylamine and 2-hydroxy-3-methacryloyl-β-cyclodextrin was much more effective than native β-cyclodextrin in the separation of seven *N*-(2,4-dinitrophenyl) amino acid enantiomers (*1133*).

Several instrumentation developments have been tested with amino acid resolutions using cyclodextrin selectors. A new technique has been developed to simultaneously concentrate and then separate enantiomers using a combination of temperature gradient focusing, an applied electric field, and a buffer with temperature-dependent ionic strength to move the analyte to an equilibrium point in the capillary. Addition of a chiral selector, γ-cyclodextrin, focuses the analyte enantiomers at different points in the capillary. While similar to CE, the method gives high-resolution separation in very short microchannels, and so is amenable to miniaturization. The method was demonstrated by resolving dansyl-Glu (*1134*). Enantiomeric separation of Ala, Val, Glu, and Asp within 4 min has been demonstrated on a microfabricated CE chip using fluorescein isothiocyanate to derivatize the

amino acids and γ-cyclodextrin as the chiral electrophoresis buffer. The test system has potential use in examining extraterrestrial environments for signs of life, and successfully analyzed extracts from the Murchison meteorite (*1135*). Hydroxypropyl-β-cyclodextrin was used to separate a mixture of five fluorescamine-labeled amino acids in another microfabricated CE device (*1136*). Levels of D-Ser (and L-Ser, Glu, Asp, and GABA) in rat striatum have been monitored by online microdialysis, derivatization with OPA/mercaptoethanol, CE separation on a hydroxypropyl-β-cyclodextrin separation buffer, and LIF analysis. This was accomplished with a 10 s analysis time, allowing for continuous monitoring (*339*). A rapid determinaton of Asp enantiomers in tissue samples employed microdialysis followed by online derivatization with *o*-phthalaldehyde/β-mercaptoethanol and then CE analysis using β-cyclodextrin chiral selector in the mobile phase. The enantiomers could be resolved with a run time of 3 s, with minimal sample preparation (*1137*). A similar resolution was reported 6 years later (*1138*). The D-Asp content of cells and subcellular domains has been determined via microvial sampling, derivatization with naphthalene-2,3-dicarboxaldehyde/KCN, CE separation and resolution using a cyclodextrin chiral selector, with LIF detection (*338*). Indirect fluorescence can be used to detect underivatized amino acids separated on a microchip, using fluorescein in the buffer (*389*).

Chiral crown ethers have also been used as chiral selectants. Eleven underivatized amino acids were simultaneously resolved and detected by capillary electrophoresis with 30 mM (+)-(18-crown-6)-2,3,11,12-tetracarboxylic acid (18C6H$_4$), using electrospray ionization mass spectrometry for detection (CE/ESI-MS), providing a powerful analytical technique (*378, 379*). This chiral selector was also used to determine the enantiomeric purity of Fmoc amino acids prior to their use in peptide synthesis (*502*), for measuring racemization during synthesis of a model tripeptide (*502*), and for resolution of dipeptide enantiomers and diastereomers (*1139*). A comparison of the enantioseparation of α-substituted analogs of DOPA, β-substituted analogs of Tyr, and γ-substituted analogs of GABA using (+)-18C6H$_4$ as a stationary phase in LC or an additive in CE found similar separations for the β-substituted analogs of Tyr, none for the α-substituted DOPA analogs, and better resolution of the GABA analogs by LC (*959*). Underivatized amino acids have also been resolved with 18C6H$_4$, then detected using contactless conductivity detection, providing an inexpensive and robust analysis system. Nine amino acids were tested (*1140*). X-ray crystal structures and energy calculations have been made of complexes of 18C6H$_4$ acid with D- and L-Tyr, Ile, Met, and Phg in order to clarify the mechanism of chiral recognition (*671*). An 18-crown-6 ether with two oxygens replaced by nitrogen, 1,4,10,13-tetraoxa-7,16-diazacyclooctadecane, has been alkylated with a chiral substituent. On its own, the diaza-crown ether did not provide enantiodifferentiation, but it enhanced the separation of of Trp and Tyr ester by cyclodextrin selectors (*1141*).

1.10.5 CE Analysis: Chiral Stationary Phase

Teicoplanin has been bonded to a silica support and used to resolved underivatized Trp and *N*-3,5-dinitrobenzoyl-Trp by CEC (*1142*). β-Cyclodextrin with a cyclam macrocycle substituent, bonded to a silica support, provided excellent resolution of Phe, the only amino acid tested (*1143*). A monolithic column chemically modified with the chiral selector L-Phe-NH$_2$ separated the enantiomers of 12 dansyl amino acids using a Cu(II)-containing mobile phase (*1144*). Another monolithic column, prepared by a copolymerization with a quinidine-functionalized chiral monomer, gave good separations of *N*-(3,5-dinitrobenzoyl)-Leu, Bz-Leu, Ac-Phe, Fmoc-Leu, Fmoc-Val, Cbz-Phe, *N*-(3,5-dinitro-Cbz)-Leu and -Phe, and *N*-(2,4-dinitrophenyl)-Ser, -Gln, -Leu, and -Val (*1145,1146*). The development and application of polymeric monolithic stationary phases for CEC was reviewed in 2004, including some examples of amino acid resolution (*1147*).

Molecular imprinting has been applied to CE analysis, just as it has been used for HPLC analysis. L-Phe anilide is often used as a test imprint species. A polypyrrole imprinted phase provided partial resolution of Pro, Phe, Lys, and Asp enantiomers via electrokinetic separation (*1148*). Reviews of imprinted stationary phases in CE/CEC analyses were published in 1998 (*1149*) and 2004 (*1150*).

Several sol–gel stationary phases have been reported using covalently bonded cyclodextrins, amino acid derivatives, or amino acid Cu(II) salts, as reviewed in 2004 (*1151*). CEC using a monolithic sol–gel column with Cu(II) salts of Pro, Phe, and Ala amides as chiral selectors was compared to CE with the same selectors as electrolyte, and with micro-HPLC separations on a monolithic column. Phe-NH$_2$ tended to give better resolutions of 12 dansyl amino acids by CEC and HPLC, while Pro gave much better results by CE. CEC gave the highest separation factors (*825*).

1.11 Determination of Enantiomeric Purity: TLC Chromatographic Analysis

The first enantioseparations of amino acids were observed during paper chromatography of aromatic amino acids (*826*). Cellulose has been used to resolve DL-Trp, His, Phe, Glu, Tyr, and 3-sulphonate-Tyr (*969, 1152*). Many of the same techniques employed for HPLC enantioseparation have also been applied to TLC analysis, as was reviewed in 1988 (*1153*). Enantiomers of underivatized amino acids can be separated on a commercially available (Chiralplate) chiral stationary phase using a methanol–water–acetonitrile eluent (33 natural and non-natural α-amino acids, eight halo α-amino acids, 14 α-alkyl-α-amino acids, and five *N*-Me-α-amino acids) (*601, 815, 1154–1157*). The stationary phase is obtained by treating C-18 modified silica with a copper acetate solution, followed by (2*S*,4*R*,2′*RS*)-4-hydroxy-1-(2-hydroxydodecyl)proline (*1158*). Similarly,

DL-dansyl amino acids (19 examples, all the proteinaceous amino acids except Pro) were separated on RP-TLC impregnated with Cu(II) complexes of amino acid-derived ligands, using water–acetonitrile for elution (*1159, 1160*); this procedure was later applied to a two-dimensional elution (*1161*).

Phenylthiohydantoin (PTH) amino acid enantiomers are very well resolved on (+)-tartaric acid-impregnated silica gel, with eight of nine examined being separated by an R$_f$ of >0.4 (*600*). The D-isomers eluted more quickly. Silica TLC plates impregnated with berberine (an optically active alkaloid) resolved 11 underivatized DL-amino acids (*601*), while a (−)-brucine coating separated nine different amino acids (*1162*), and borate-gelled galactomannan polysaccharide resolved 14 underivatized amino acids (*1163*). Both DL-dansyl amino acids (four examples) and amino acid β-napthyl amides (two examples) were resolved on a β-cyclodextrin-bonded silica gel (*1164*). Racemic aromatic amino acids have been separated on cellulose using mobile phases containing α- or β-cyclodextrin (*1165*), while DL-dansyl amino acids were separated on a reversed-phase support using β-cyclodextrin as a mobile-phase additive (13 of 15 separated, not imino acids) (*1166*). Cyclodextrin-based TLC separations were reviewed in 1986 (*934*) and 1988 (*1167*). Molecular imprinting has also been used in TLC to resolve DL-Phe anilide on a phase prepared from ground imprinted methacrylic acid polymer in a binder matrix (*1168*).

Diastereomers of amino acids derivatized with (*S*)-(−)-*N*-1-(2-naphthylsulphonyl)-2-pyrrolidinecarbonyl chloride could be resolved on silica (18 of 20 amino acids separated) (*744*).

1.12 Determination of Enantiomeric Purity: Mass Spectrometric Methods

Several techniques have been developed which allow for the use of mass-spectral analysis to determine enantiomeric purity. D-Amino acids in peptides and proteins can potentially be identified by the different patterns exhibited by the diastereomeric peptide/proteins when analyzed by tandem mass spectrometry (*90*). Another method reacts protonated complexes formed from amino acids and β-cyclodextrin with *n*-propyl amine for a fixed rate of time. The amine displaces the amino acid from the cyclodextrin, with the rate of guest exchange depending on the chirality of the amino acid. By construction of a calibration curve, enantiomeric purity can then be determined (*1169*). Another mass-spectrometry-based method relies on differential rates of reaction between the two amino acid enantiomers and pseudoenantiomeric chiral auxiliaries. The pseudoenantiomeric auxiliaries have different masses via additional substituents at sites remote from the site of bond formation, such as Bz-L-Pro and (4-Me-Bz)-D-Pro. Acylation of Phg-OMe or Val-OMe with these auxiliaries was examined, with good correlation between measured and actual enantiomeric excesses over a wide range (20–90%) (*1170*).

A related method employs electrospray ionization of complexes formed between amino acid derivatives, such as Boc-Pro-NH(3,4-Me$_2$-Ph), and a pseudoenantiomeric pair of Pirkle-type chiral selectors differing in molecular weight, such as *N*-(3,4-dinitrobenzoyl)-L-Leu-NH*n*Bu and *N*-(3,4-dinitrobenzoyl)-L-Leu-NH*n*Pent. The difference in intensity of the pseudoenantiomeric complex Li ion peaks (differing by 14 mass units) is proportional to the analyte enantiomeric purity (*1171*). A variation of this method attaches a tertiary amine to DNB-Leu, with the 2-(diethylamino)-ethylamine amide of DNB-L-Leu and the 3-(diethylamino)-propylamine amide of DNB-D-Leu used to analyze seven *N*-protected amino acid amides (*1172*).

The enantiomer-labeled (EL) **guest** method employs an oligosaccharide host complexed with amino acid guest molecules, consisting of a 1:1 mixture of unlabeled (*R*)-enantiomer and deuterium-labeled (*S*)-enantiomer. FAB-MS analysis of the mixture provides an estimate of the chiral recognition ability of the hosts, with enantioselectivity estimated by the relative peak intensity of the [complex + (*R*)-amino acid]/[complex + (*S*)-amino acid]. Permethylated cyclofructans showed small but significant discrimination of Trp, Phg, Tle, Met, Ser, and Pro isopropyl ester hydrochlorides (*1173*). A range of chiral crown ether hosts was also evaluated by this method (*1174*). In order to estimate enantiomeric purity of the guest molecules, underivatized D-mannose, D-galactose, and D-glucose were used as the polar hosts. The purity of Ala-OMe was then determined, using the enantiomeric methyl-d$_3$ ester as an internal standard, and the LSIMS (liquid secondary ion mass spectrometry) technique. A mixture of host, standard, and sample were loaded onto the target probe and analyzed, giving ions of the host + analyte discrimination complex. With an equimolar mixture of the deuterated standard and the enantiomeric sample, a chiral discrimination is observed in the peak heights of the two complex ions (deuterated and non-deuterated). A calibration plot can be prepared using different ratios of deuterated/undeuterated enantiomers (correlation = 0.99), with the calibration then applied to the analysis of enantiomeric excess. This method is particularly useful for very small (microgram) quantities of sample, such as produced by combinatorial syntheses (*1175*).

A variation of this technique is the enantiomer-labeled **host** method. Enantiomeric chiral 18-crown-6 hosts, in which the [*S,S,S,S*]-host is hexadeuterated, give high discriminations for (*R*)- and (*S*)-Phg, Asp, Asn, Phe, and Val methyl ester hydrochlorides. Enantiomeric excesses can again be determined by fast atom bambardment mass spectrometry (FAB-MS) analysis of complexes formed with amino acids of unknown enantiopurity, using calibration curves. Good linearity was observed with a calibration curve constructed from various ratios of (*R*)- and (*S*)-Phg-OMe (*1176*).

A different principle is utilized in another technique, which employs the kinetics of competitive fragmentation of trimeric Cu(II)-bound complexes, [Cu^{II}(A)(ref*)$_2$-H]$^+$ (where A = amino acid and ref* = chiral reference ligand, selected from the natural α-amino acids). The complex undergoes collision-induced dissociation in a quadrupole ion trap to generate [Cu^{II}(A)(ref*)-H]$^+$ and [Cu^{II}(ref*)$_2$-H]$^+$, with the relative abundance of the two fragments depending strongly on the chirality of the analyte amino acid. The preferred reference amino acid (Phe or Trp were generally best) was found to depend on the analyte amino acid, with large chiral distinctions obtained for all the natural amino acids except Cys and Arg. The method is quantitative, with correlation coefficients of 0.998 obtained from calibration curves of Pro or Phe, can measure ee values as small as 2%, and is essentially independent of concentration (*1177*). Pro has also been employed as the reference amino acid (*1178*). This method is called a single ratio (SR) method, with a two-point calibration line established using a racemic sample and sample of known ee, or using pure *R* and *S* enantiomers, or any two samples of known optical purity. A modified method has been developed that requires only one sample of known enantiomeric purity, called the quotient ratio (QR) method. In this case, the trimeric clusters generated contain two analyte molecules instead of two reference amino acids, [Cu^{II}(A)$_2$(ref*)-H]$^+$. The method was tested with Pro, Phe, Glu, Thr, Tyr, Asp, Leu, Met, Arg, and DOPA (*1179*). Other studies have shown that is not necessary to include the copper ion to form complexes. Electrospray ionization of amino acids in the presence of a chiral selector, such as Boc-L-Phe, produced protonated trimers. Collision-induced dissociation spectra then generated protonated dimer products, with the ratio of trimer:dimer again depending on the chirality of the amino acid analyte. The chirality of all 19 common amino acids could be differentiated (*604, 605*), and calibration curves could be constructed to determine enantiomeric excess (*1180*). A fundamentally similar method was outlined in an earlier report (*1181*).

1.13 Determination of Enantiomeric Purity: Other Methods

A method capable of detecting traces of enantiomeric purities, as well as assigning absolute configuration, has recently been described. It makes use of a technique called surface plasmon resonance (SPR), in which optical properties at the surface of a thin gold film on a glass support are used to monitor changes in molecular weight of binding partners immobilized on the sensor surface. For this assay, streptavidin was bound to the sensor surface, and D-phenylalanine was attached to the streptavidin. Polyclonal rabbit antibody sensitive to D-amino acids was then passed over the sensor, binding to the D-amino acid, causing an increase in molecular weight, and inducing a signal. In competitive assays, a fixed concentration of anti-D antibody was added to samples of varying concentration of D-amino acids and injected over the sensor surface. With more D-amino acid in solution, antibody was bound to the free amino acid and was therefore not available to bind to the sensor surface, reducing the signal observed. L-amino acids do not interact with the antibody, and so cause no inhibition of the antibody binding to the detector. The calibration curve obtained from various concentrations of D-amino acids could then be used to determine the amount of D-amino acid present in non-racemic mixtures. For DL-Trp, ratios of D:L as low as 1:2500 were readily determined, corresponding to accurate detection of 99.92% ee (1182). Enantioselective adsorption of Phe onto self-assembled monolayers of (R)- or (S)-1,1'-binaphthalene-2,2'-dithiol on a gold surface can be detected, perhaps leading to another type of sensor (1183).

A color-based method of determining enantiomeric purity utilized a nematic liquid crystalline host, doped with a chiral auxiliary consisting of the imine formed from an amino acid ester and p-(p-methoxyphenyl)-benzaldehyde. Addition of the same imine derivative formed from an amino acid analyte resulted in a color shift, ranging from deep red for one enantiomer, to yellow for racemic sample, to violet for the other enantiomer. However, the method was only demonstrated with Phg as both auxiliary and analyte (606). Near-infrared analysis of diastereomeric complexes formed from amino acids and carbohydrates can be used to determine enantiomeric excess values. The complex can be formed with either macrocyclic carbohydrates (such as cyclodextrins) or with simple inexpensive acyclic sugars, such as sucrose. Analysis of aqueous solutions with, for example, 2 mM Ala in 4.0 mM α-cyclodextrin, allowed for determination of ee values from −90 to +97%, and as low as 1.5%. Concentrations as low as micrograms gave results. Val, Leu, and Phe were also tested (1184).

An analytical method specific for D-amino acids was created by pre-column reaction with D-amino acid oxidase to form α-keto acids, followed by chemical derivatization with o-phenylenediamine and 2-mercaptoethanol, and then HPLC separation with fluorescence detection. The method was tested with D-Ala, -Leu, -Met, -Phe, and -Val, providing good separation and 0.2–1 mM detection limits. L-Amino acids were not detected (1185). A novel biocatalytic detector for the determination of L- and D-amino acids in foodstuffs has been developed. Amino acids are separated on a lithium cation-exchange HLPC column, and then detected by enzymatic conversion in keto acids and hydrogen peroxide by L- or D-amino acid oxidase. The hydrogen peroxide is detected amperometrically, allowing for specific detection of either L- or D-amino acids (1186).

Countercurrent chromatography (CCC) combines aspects of HPLC and enantioselective extraction. Enantioseparations in counter-current chromatography were reviewed in 2001; most use amino acid separations as examples (1187). Optimization of high-speed counter-current chromatography was reviewed in 2005 (1188). A number of Pro derivatives were evaluated as chiral selectors for counter-current separations of N-3,5-dinitrobenzoyl-Leu in a 2005 report (1189). Cinchona-derived anion-exchange-type chiral selectors were selected with the appropriate properties to maintain

chiral recognition but also partition correctly in the stationary organic phase (methyl isobutyl ketone) during the CCC run. The amino acid to be resolved was injected in an aqueous solution. Up to 900 mg of N-(3,5-dinitrobenzoyl)-Leu could be resolved in a single run in 100 min with 170 mL of stationary phase and 300 mL of mobile phase (1190).

A great deal of research has focused on developing chiral receptors for the specific recognition of amino acids, extending the use of cyclodextrins and other macrocycles already employed for chiral separation. These hosts may be useful for techniques such as HPLC or MS, as described in previous sections, or can be used to develop alternative methods for the analysis of optical purity or purification of enantiomers. For example, the use of molecularly imprinted polymers in biomimetic sensors was reviewed in 2000, with a number of examples of amino acids included (1191).

Binary complexes formed from two types of cyclo-dextrins and three different fluorophores (pyrene, xanthone, and anthraquinone) were tested as chiral selectors for three different amino acids via formation of ternary complexes. Significant enantioselectivity ratios were found for all three amino acids tested (Phe, Met, and His), as measured by changes in fluorescence spectra (1192). A β-cyclodextrin derivative with a dansylated Phe substituent showed different increases in fluorescence upon binding amino acid enantiomers in the presence of copper, with the difference particularly strong for Pro (1193, 1194). Arylseleno-modified β-cyclodextrins showed good L-enantioselectivity for aliphatic amino acids, particularly Leu (1195). Minor modifications to aminated cyclodextrins tailor selectivity towards either N-Cbz D-/L-Asp or N-Cbz D-/L-Glu pairs (1196).

A monoaza-15-crown-5 ether with chiral substituent showed chiral recognition of K+ or Na+ amino acid salts, with the best result shown with Trp (1197). Diaza-18-crown-6 ethers with arene sidearms demonstrated enantioselective transport of Phe, Phg, and Trp Na or K salts through a liquid membrane (1198, 1199). Another crown ether derivative, attached to a carboxy-substituted ferrocene, showed distinctive CD spectra upon binding to hydrophobic D- or L-amino acids (546). Primary organic ammonium salts were bound by a C2-symmetric 2,2'-bipyridine crown macrocycle, showing a preference for L-Phg-OMe over its enantiomer (1200). Another novel method employed an artificial host based upon a phenolphthalein skeleton with two appended phenolic 18-crown-6 moieties. The complex turned purple upon binding triamines of an appropriate spacing. A 1,6-diaminohexane amide of D-Ala formed a colored complex, whereas the L-Ala derivative solution was colorless (1201,1202).

Chiral calix[4]arenes have been synthesized with dansylated amino acid substituents. These fluorescent receptors showed much greater fluorescence quenching upon addition of Boc-L-Ala or Boc-L-Phe anions than upon addition of the enantiomeric anions, indicating enantiodiscrimination (1203). A crystal structure of p-sulfonatocalix[6]arene was obtained from a solution of racemic Leu, showing only bound L-Leu (*1204*). The molecular recognition of amino acids by a resorcin[4]-arene calixarene-based receptor, with four pendant Val-OEt groups, was investigated by ion cyclotron resonance mass spectrometry (*1028*). Several cyclophane receptors containing the side-chain imidazole rings of two His residues discriminated between a number of amino acid methyl esters (Ala, Val, Leu, Phe, Trp), and also showed some selectivity between amino acid types (*1205*).

Receptors based on a *cis-* or *trans*-tetrahydrobenzox-anthane skeleton with benzoxazole and amidopyridine substituents could complex *N*-triflate amino acids with enantiodifferentiation, and preferentially extract one enantiomer from water into chloroform (*1206, 1207*). This cleft-type tetrahydrobenzoxanthene receptor provided strong chiral discrimination between enantiomers of *N*-dinitrobenzoyl amino acids. The two different diastereomeric complexes possessed different colors, allowing for a potential colorimetric chiral determination. Furthermore, partition of the more strongly bound complex into organic solvent provides the possibility of large-scale enantioseparations (*1208,1209*). Preparative TLC plates impregnated with EtO₂C-Pro allowed for resolution of the racemic receptor (*1209*). A ternary complex formed from a binaphthyl-xanthone-based macrocycle, 18-crown-6 ether, and unprotected amino acids showed chiral selectivity for extraction of Phe, Phg, Ser, Trp, Val, and Ala from water into chloroform (*427*).

A chiral cobalt complex formed from β-(6-dimethylaminomethyl-2-pyridyl)-Ala bound amino acids with high stereospecificity, complexing one enantiomer of Ala, Phe or Trp more strongly (*1210*). A Ru-porphyrin picket-fence complex, bearing optically active substituents on both sides of the porphyrin plane, binds the amino group of amino esters and provides chiral discrimination (*1211*). Zinc porphyrin dimers linked by a binaphthyl moiety provided strong enantioselectivity towards Lys (*1212*). A pyridine bis(oxazoline) copper(II) complex, (*R*)-pybox-copper, showed moderate selectivity for binding a variety of D-amino acids over L-amino acids (Ala, Val, Phe, Asn, Gln), enough to allow for baseline resolution of (*R*)- and (*S*)-pybox by capillary electrophoresis using L-Lys as the chiral selector (*1213*).

A dibenzofuranyl/dibenzothiophene-based receptor containing a urethane substituent to bind the guest acid group and a chiral phosphonamide to hydrogen-bond the guest amino group provided moderate enantioselectivity, as determined by binding constants for L- and D-amino acids (*1214*). Molecular "tweezers" based on two chiral polyamine/amide macrocycles linked by a *m*-xylene bridge showed differential binding constants for a variety of amino acid methyl esters, with particularly strong recognition of L-Trp-OMe (*1215*). A steroidal guanidinium receptor, a cholic acid derivative with one guanidinium and two carbamate substituents, extracted five *N*-acetyl amino acids from aqueous to organic solutions (Phe, Val, Met, Ala, Trp) with >5:1 L:D

enantioselectivity, and 76–93% extraction efficiency; other amino acids were resolved less effectively (*1216, 1217*). Steroidal urea derivatives also gave good selectivities (*1218*). Guanidinium receptors with attached crown ethers or lasalocid A have also been reported (*1219*). A diphenylglycoluril-based molecular receptor showed significant preference in binding to D-Phe, D-Tyr, and D-Trp over their antipodes (*1220*). Chiral 4,4′-diamido-2,2′-bisimidazoles discriminated between Boc-L-Ser and Boc-D-Ser, interacting with the carboxylate group (*1221*). Lipophilic deoxyguanosine derivatives preferentially extracted one enantiomer of *N*-2,4-dinitrophenyl derivatives of Ala, Phe, Trp, Ile, and Pro, presumably via a self-assembled complex (*1222*). A tripodal imidazolium salt, consisting of a central benzyl core with three attached imidazolium groups, each substituted with a (−)-myrtanyl moiety, complexed with D-Ala but not L-Ala, based on NMR shifts (*1223*). The enzymes histidine ammonia lyase and phenylalanine ammonia lyase have been mutated to abolish their catalytic activity and act as artificial receptors. They were then immobilized in a membrane to facilitate enantioselective transport across the membrane (*1224*).

A 3,3′-aminomethyl substituted BINOL derivative showed a much greater fluorescence enhancement in the presence of Cbz-D-Phg compared to the L-enantiomer, showing potential as a sensor of chirality (*1225*). Fluorescent spectroscopy could also determine the enantiomeric excess of *N*-dansyl Phe in the presence of bovine α-acid glycoprotein as the receptor (*1226*). Chiral bisbinaphthyl-based macrocycles demonstrated increased fluorescence in the presence of one enantiomer of *N*-Cbz or *N*-Ac amino acids, but little enhancement in the presence of the other enantiomer (*1227*). A receptor showing enantioselectivity for L-Glu combined a binaphthyl core with pendant chromenone benzoxazole fragments (*1228*) while a macrocyclic bis(thiourea) bound Boc-L-Glu significantly more strongly than the D-enantiomer (and, to a lesser extent, Boc-L-Asp over its enantiomer) (*1229, 1230*).

A novel method with the potential to rapidly determine the enantiomeric purity of thousands of samples employs a pseudoenantiomeric pair of fluorescent reporter compounds, based on Cy3 or Cy5 fluorophores coupled to D- or L-Pro, respectively. A 1:1 mixture of the reporter molecules acylated the *N*-terminus of amino acids attached to a microarray solid support by their carboxylate group. Kinetic resolution during the amidation step converts the enantiomeric excess of the amino acid sample into differential incorporation of the fluorophores. Since the fluorophores fluoresce at different wavelengths, an image of the microarray can be generated with Cy3 fluorescence represented as green and Cy5 fluorescence as red. A racemic amino acid would thus produce a yellow signal, with graduations in color between the two pure enantiomers. Colorimetric discrimination proportional to their enantiomeric purity was demonstrated for Ala, Val, Leu, Pro, Ser, and Cys when the six amino acids were printed as mixtures of enantiomers ranging from 100% ee D to 100% ee L in steps of

10% ee. Experimentally determined ee values were within ±10% ee of 116 out of 126 samples. Even more impressively, a single >99% ee D-Pro and single >99% ee L-Pro sample were identified in <48 h from 15,552 samples, where the other 15,550 samples possessed between 0 and 20% ee (*1231, 1232*).

References

1. Brack, A. *Pure Appl. Chem.* **1993**, *65*, 1143–1151.
2. Lawless, J.G.; Kvenvolden, K.A.; Peterson, E.; Ponnamperuma, C.; Moore, C. *Science* **1971**, *173*, 626–627.
3. Kvenvolden, K.A.; Lawless, J.G.; Ponnamperuma, C. *Proc. Natl. Acad. Sci. U.S.A* **1971**, *68*, 486–490.
4. Engel, M.H.; Macko, S.A.; Silfer, J.A. *Nature* **1990**, *348*, 47–49.
5. Bade, J.L.; McDonald, G.D. *Anal. Chem.* **1996**, 668A–673A.
6. Barrett, G.C. *Chemistry and Biochemistry of the Amino Acids*; Chapman and Hall: New York, 1985, pp 55–138.
7. Wagner, I.; Musso, H. *Angew. Chem., Int. Ed. Engl.* **1983**, *22*, 816–828.
8. Musso, H. *Angew. Chem.* **1956**, *68*, 313–323.
9. Bell. E.A.; John, D.I. In *International Review of Science*, Organic Chemistry Series Two, Volume 6; Rydon, H.N. Ed.; Butterworths: Boston, Mass., 1976, pp 1–31.
10. Bell. E.A. In *International Review of Science*, Organic Chemistry Series One, Volume 6; Hey, D.H., John, D.I., Eds; University Park Press: Baltimore, Md. 1973, pp 1–16.
11. Hardy, P.M. In Barrett, G.C. *Chemistry and Biochemistry of the Amino Acids*; Chapman and Hall: New York, 1985, p 6.
12. Walsh, C.T.; Garneau-Tsodikova, S.; Gatto G.J., Jr., *Angew. Chem., Int. Ed.* **2005,** *44*, 7342–7372.
13. Hunt, S. In Barrett, G.C. *Chemistry and Biochemistry of the Amino Acids*; Chapman and Hall: New York, 1985; p 55.
14. Bewley, C.A.; Faulkner, D.J. *Angew. Chem., Int. Ed. Engl.* **1998**, *37*, 2162–2178.
15. Fowden, L. *Phytochemistry* **1972**, *11*, 2271–2276.
16. Hatfield, D.L.; Gladyshev, V.N. *Molec. Cell Biol.* **2002**, *22*, 3565–3576.
17. Carlson, B.A.; Xu, X.-M.; Kryukov, G.V.; Rao, M.; Berry, M.J.; Gladyshev, V.N.; Hatfield, D.L. *Proc. Natl. Acad. Sci. U.S.A.* **2004**, 101, 12848–12853.
18. Diamond, A.M. *Proc. Natl. Acad. Sci. U.S.A.* **2004**, 101, 13395–13396.
19. Tamura, T.; Yamamoto, S.; Takahata, M.; Sakaguchi, H.; Tanaka, H.; Stadtman, T.C.; Inagaki, K. *Proc. Natl. Acad. Sci. U.S.A.* **2004**, *101*, 16162–16167.
20. Carlson, B.A.; Xu, X.-M.; Gladyshev, V.N.; Hatfield, D.L. *J. Biol. Chem.* **2005**, *280*, 5542–5548.
21. Diamond, A.M. *Proc. Natl. Acad. Sci. U.S.A.* **2004**, *101*, 13395–13396.
22. Mehta, A.; Rebsch, C.M.; Kinzy, S.A.; Fletcher, J.E.; Copeland, P.R. *J. Biol. Chem.* **2004**, *279*, 37852–37859.
23. Srinivasan, G.; James, C.M.; Krzycki, J.A. *Science* **2002**, *296*, 1459–1462.
24. Hao, B.; Gong, W.; Ferguson, T.K.; James, M.; Krzycki, J.A.; Chan, M.K. *Science* **2002**, *296*, 1462–1466.
25. Atkins, J.F.; Gesteland, R. *Science* **2002**, *296*, 1409–1410.
26. Yarnell, A. *Chem. Eng. News* **2002**, *May 27*, 13.
27. Fenske, C.; Palm, G.J.; Hinrichs, W. *Angew. Chem. Int. Ed.* **2003**, *42*, 606–610.
28. Green-Church, K.B.; Chan, M.K.; Krzycki, J.A. *Nature* **2004**, *431*, 333–335.
29. Hao, B.; Zhao, G.; Kang, P.T.; Soares, J.A.; Ferguson, T.K.; Gallucci, J.; Krzycki, J.A.; Chan, M.K. *Chem. Biol.* **2004**, *11*, 1317–1324.
30. Schimmel, P.; Beebe, K. *Nature* **2004**, *431*, 257–258.
31. Polycarpo, C.; Ambrogelly, A.; Bérubé, A.; Winbush, S.M.; McCloskey, J.A.; Crain, P.F.; Wood, J.L.; Söll, D. *Proc. Natl. Acad. Sci. U.S.A.* **2004**, *101*, 12450–12454.
32. Zhang, Y.; Baranov, P.V.; Atkins, J.F.; Gladyshev, V.N. *J. Biol. Chem.* **2005**, *280*, 20740–20751.
33. Mehl, R.A.; Andersen, J.C.; Santoro, S.W.; Wang, L.; Martin,A.B.; King, D.S.; Horn, D.M.; Schultz, P.G. *J. Am. Chem. Soc.* **2003**, *125*, 935–939.
34. Hohsaka, T.; Ashizuka, Y.; Murakami, H.; Sisido, M. *J. Am. Chem. Soc.* **1996**, *118*, 9778–9779.
35. Kwon, I.; Kirshenbaum, K.; Tirrell, D.A. *J. Am. Chem. Soc.* **2003**, *125*, 7512–7513.
36. Murphy IV, F.V.; Ramakrishnan, V. *Nature Struct. Mol. Biol.* **2004**, *11*, 1251–1252.
37. Cochella, L.; Green, R. *Nature Struct. Mol. Biol.* **2004**, *11*, 1160–1162.
38. Murphy IV, F.V.; Ramakrishnan, V.; Malkiewicz, A.; Agris, P.F. *Nat. Struct. Mol. Biol.* **2004**, *11*, 1186–1191.
39. Ibba, M.; Francklyn, C. *Proc. Natl. Acad. Sci. U.S.A.* **2004**, *101*, 7493–7494.
40. Dubois, D.Y.; Blaise, M.; Becker, H.D.; Campanacci, V.; Keith, G.; Giegé, R.; Cambillau, C.; Lapointe, J.; Kern, D. *Proc. Natl. Acad. Sci. U.S.A.* **2004**, *101*, 7530–7535.
41. Salazar, J.C.; Ambrogelly, A.; Crain, P.F.; McCloskey, J.A.; Söll, D. *Proc. Natl. Acad. Sci. U.S.A.* **2004**, *101*, 7536–7541.
42. Francklyn, C. *Proc. Natl. Acad. Sci. U.S.A.* **2003**, *100*, 9650–9652.
43. Feng, L.; Tumbula-Hansen, D.; Toogood, H.; Söll, D. *Proc. Natl. Acad. Sci.* **2003**, *100*, 5676–5681.
44. Biro, J.C.; Benyó, B.; Sansom, C.; Szlávecz, A.; Fördös, G.; Micsik, T.; Benyó, Z. *Biochem. Biophys. Res. Commun.* **2003**, *306*, 408–415.
45. Zhang, R. *Amino Acids* **1997**, *12*, 167–177.

46. Siemion, I.Z. *Amino Acids* **1995**, *8*, 1–13.
47. Naya, H.; Zavala, A.; Romero, H.; Rodríguez-Maseda, H.; Musto, H. *Biochem. Biophys. Res. Commun.* **2004**, *325*, 1252–1257.
48. Fuglsang, A. *Bichem. Biophys. Res. Commun.* **2004**, *317*, 957–964.
49. Banerjee, T.; Gupta, S.K.; Ghosh, T.C. *Biochem. Biophys. Res. Commun.* **2005**, *330*, 1015–1018.
50. Copley, S.D.; Smith, E.; Morowitz, H.J. *Proc. Natl. Acad. Sci. U.S.A.* **2005**, *102*, 4442–4447.
51. Borsenberger, V.; Crowe, M.A.; Lehbauer, J.; Raftery, J.; Helliwell, M.; Bhutia, K.; Cox, T.; Sutherland, J.D. *Chemistry and Biodiversity* **2004**, *1*, 203–246.
52. Smith, J.M.; Borsenberger, V.; Raftery, J.; Sutherland, J.D. *Chemistry and Biodiversity* **2004**, *1*, 1418–1451.
53. Ace, K.; Sutherland, J.D. *Chemistry and Biodiversity* **2004**, *1*, 1678–1693.
54. Saewan, N.; Crowe, M.A.; Helliwell, M.; Raftery, J.; Chantrapromma, K.; Sutherland, J.D. *Chemistry and Biodiversity* **2005**, *2*, 66–83.
55. Jordan, I.K.; Kondrashov, F.A.; Adzhubei, I.A.; Wolf, Y.I.; Koonin, E.V.; Kondrashov, A.S.; Sunyaev, S. *Nature* **2005**, *433*, 633–638.
56. Sorimachi, K.; Itoh, T.; Kawarabayasi, Y.; Okayasu, T.; Akimoto, K.; Niwa, A. *Amino Acids* **2001**, *21*, 393–399.
57. Sorimachi, K. *Amino Acids* **2002**, *22*, 55–69.
58. Okayasu, T.; Ikeda, M.; Akimoto, K.; Sorimachi, K. *Amino Acids* **1997**, *13*, 379–391.
59. Sorimachi, K. *Amino Acids* **1999**, *17*, 207–226.
60. Gao, Y.; Shao, S.; Xiao, X.; Ding, Y.; Huang, Y.; Huang, Z.; Chou, K.-C. *Amino Acids* **2005**, *28*, 373–376.
61. Akashi, H.; Gojobori, T. *Proc. Natl. Acad. Sci. U.S.A.* **2002**, *99*, 3695–3700.
62. Akanuma, S.; Kigawa, T.; Yokoyama, S. *Proc. Natl. Acad. Sci. U.S.A.* **2002**, *99*, 13549–13553.
63. Bordusa, F. *ChemBioChem* **2001**, *2*, 405–409.
64. Walsh, C.T. *ChemBioChem* **2002**, *3*, 124–134.
65. Mootz, H.D.; Schwarzer, D.; Maahiel, M.A. *ChemBioChem* **2002**, *3*, 490–504.
66. Heal, J.R.; Roberts, G.W.; Raynes, J.G.; Bhakoo, A.; Miller, A.D. *ChemBioChem* **2002**, *3*, 136–151.
67. Mattick, J.S. *EMBO Rep.* **2001**, *2*, 986–991.
68. Mattick, J.S. *Nature Rev. Genetics* **2004**, *5*, 316–323.
69. Mattick, J.S. *BioEssays* **2003**, *25*, 930–939.
70. Mattick, J.S. *Science* **2005**, *309*, 1527–1528.
71. Hardy, P.M. In Barrett, G.C., Ed.; *Chemistry and Biochemistry of the Amino Acids*; Chapman and Hall: New York, 1985, pp 11–12.
72. Ladrón de Guevara, O.; Padilla, P.; García, L.; Pino, J.M.; Ramos-Elorduy, J. *Amino Acids* **1995**, *9*, 161–173.
73. Massy, K.A. Blaskeslee, C.H.; Pitkow, H.S. *Amino Acids* **1998**, *14*, 271–300.
74. Formánek, P.; Klejdus, B.; Vranova, V. *Amino Acids* **2005**, *28*, 427–429.
75. Buch, A.; Sternberg, R.; Meunier, D.; Rodier, C.; Laurent, C.; Raulin, F.; Vidal-Madjar, C. *J. Chromatogr., A* **2003**, *999*, 165–174.
76. Barrett, G.C.; Lubec, G. *Amino Acids* **1991**, *1*, 1–6.
77. Park, S.; Wolanin, P.M.; Yuzbashyan, E.A.; Silberzan, P.; Stock, J.B.; Austin, R.H. *Science* **2003**, *301*, 188.
78. Kleemann, A.; Leuchtenberger, W.; Hoppe, B.; Tanner, H. In *Ullmann's Encyclopedia of Industrial Chemistry*, Volume A2; VCH Publishers: Deerfield Beech, Fla., 1985, pp 57–97.
79. Nelson, G.; Chandrashekar, J.; Hoon, M.A.; Feng, L.; Zhao, G.; Ryba, N.J.P.; Zuker, C.S. *Nature* **2002**, *416*, 199–202.
80. Cota, D.; Proulx, K.; Blake Smith, K.A.; Kozma, S.C.; Thomas, G.; Woods, S.C.; Seeley, R.J. *Science* **2006**, *312*, 927–930.
81. Kahn, B.B.; Myers M.G., Jr., *Nat. Med.* **2006**, *12*, 615–617.
82. Schreiber, S.L. *Science* **1991**, *251*, 283–287.
83. Hruby, V.J. *Acc. Chem. Res.* **2001**, *34*, 389–397.
84. Hohsaka, T.; Ashizuka, Y.; Murakami, H.; Sisido, M. *J. Am. Chem. Soc.* **1996**, *118*, 9778–9779.
85. Qi, D.; Tann, C.-M.; Haring, D.; Distefano, M.D. *Chem. Rev.* **2001**, *101*, 3081–3111.
86. Cotter, P.D.; O'Connor, P.M.; Draper, L.A.; Lawton, E.M.; Deegan, L.H.; Hill, C.; Ross, R.P. *Proc. Natl. Acad. Sci. U.S.A.* **2005**, *102*, 18584–18589.
87. Brückner, H.; Westhauser, T. *Amino Acids* **2003**, *24*, 43–55.
88. Casal, S.; Oliveira, M.B.; Ferreira, M.A. *J. Chromatogr., A* **2000**, *866*, 221–230.
89. Erbe, T.; Brückner, H. *J. Chromatogr., A* **2000**, *881*, 81–91.
90. Adams, C.M.; Zubarev, R.A. *Anal. Chem.*, **2005**, *77*, 4571–4580.
91. Heck, S.D.; Faraci, W.S.; Kelbaugh, P.R.; Saccomano, N.A.; Thadeio, P.F.; Volkmann, R.A. *Proc. Natl. Acad. Sci. U.S.A.* **1996**, *93*, 4036–4039.
92. Murkin, A.S.; Tanner, M.E. *J. Org. Chem.* **2002**, *67*, 8389–8394.
93. Buczek, O.; Yoshikami, D.; Bulaj, G.; Jimenez, E.C.; Olivera, B.M. *J. Biol. Chem.* **2005**, *280*, 4247–4253.
94. Jilek, A.; Mollay, C.; Tippelt, C.; Grassi, J.; Mignogna, G.; Müllegger, J.; Sander, V.; Fehrer, C.; Bara, D.; Kreil, G. *Proc. Natl. Acad. Sci. U.S.A.* **2005**, *102*, 4235–4239.
95. Schell, M.J.; Molliver, M.E.; Snyder, S.H. *Proc. Natl. Acad. Sci. U.S.A.* **1995**, *92*, 3948–3952.
96. Hucho, F. *Angew. Chem., Int. Ed.* **2000, *39*, 2849–2850.
97. Yang, Y.; Ge, W.; Chen, Y.; Zhang, Z.; Shen, W.; Wu, C.; Poo, M.; Duan, S. *Proc. Nalt. Acad. Sci. U.S.A.* **2003**, *100*, 15194–15199.
98. Wolosker, H.; Blackshaw, S.; Snyder, S.H. *Proc. Natl. Acad. Sci. U.S.A.* **1999**, *96*, 13409–13414.
99. Nagata, Y.; Shoji, R.; Yonezawa, S.; Oda, S. *Amino Acids* **1997**, *12*, 95–100.

100. Armstrong, D.W.; Gasper, M.P.; Lee, S.H.; Ercal, N.; Zukowski, J. *Amino Acids* **1993**, *5*, 299–315.

101. Asakura, S.; Konno, R. *Amino Acids* **1997**, *12*, 213–223.

102. Strísovsky, K.; Jirásková, J.; Mikulová, A.; Rulísek, L.; Konvalinka, J. *Biochemistry.* **2005**, *44*, 13091–13100.

103. Dixon, S.M.; Li, P.; Liu, R.; Wolosker, H.; Lam, K.S.; Kurth, M.J.; Toney, M.D. *J. Med. Chem.* **2006**, *49*, 2388–2397.

104. Armstrong, D.W.; Duncan, J.D.; Lee, S.H. *Amino Acids* **1991**, *1*, 97–106.

105. Brückner, H.; Haasmann, S.; Friedrich, A. *Amino Acids* **1994**, *6*, 205–211.

106. Young, G.A.; Kendall, S.; Brownjohn, A.M. *Amino Acids* **1994**, *6*, 283–293.

107. Luthra, M.; Ranganathan, D.; Ranganathan, S.; Balasubramanian, D. *J. Biol. Chem.* **1994**, *269*, 22678–22682.

108. Kuge, K.; Fujii, N.; Miura, Y.; Tajima, S.; Saito, T. *Amino Acids* **2004**, *27*, 193–197.

109. Fisher, G.; Lorenzo, N.; Abe, H.; Fujita, E.; Frey, W.H.; Emory, C.; Di Fiore, M.M.; D'Aniello, A. *Amino Acids* **1998**, *15*, 263–269.

110. Hamase, K.; Morikawa, A.; Zaitsu, K. *J. Chromatog., B* **2002**, *781*, 73–91.

111. Krause, E.; Bienert, M.; Schmieder, P.; Wenschuh, H. *J. Am. Chem. Soc.* **2000**, *122*, 4865–4870.

112. Tugyi, R.; Uray, K.; Iván, D.; Fellinger, E.; Perkins, A.; Hudecz, F. *Proc. Natl. Acad. Sci. U.S.A.* **2005**, *102*, 413–418.

113. Wehofsky, N.; Thust, S.; Burmeister, J.; Klussmann, S.; Bordusa, F. *Angew. Chem., Int. Ed.* **2003**, *42*, 677–679.

114. Dedkova, L.M.; Fahmi, N.E.; Golovine, S.Y.; Hecht, S.M. *J. Am. Chem. Soc.* **2003**, *125*, 6616–6617.

115. Kamphuis, J.; Meijer, E.M.; Boesten, W.H.J.; Sonke, T.; van den Tweel, W.J.J.; Schoemaker, H.E. *Ann. N.Y. Acad. Sci.* **1992**, *672*, 510–527.

116. Lyon, S. *Nature* **2004**, *427*, 406–407.

117. ApSimon, J.W.; Seguin, R.P. *Tetrahedron* **1979**, *35*, 2797–2842

118. Mosher, H.S.; Morrison, J.D. *Science* **1983**, *221*, 1013–1019.

119. Kagan, H.B.; Fiaud, J.C. *Top. Stereochem.* **1987**, *10*, 175–285.

120. Coppola, G.M.; Schuster, H.F. *Asymmetric Synthesis. Construction of Chiral Molecules Using Amino Acids*; Wiley-Interscience: Toronto, Canada, 1987.

121. Ottenheijm, H.C.J. *Chimia* **1985**, *39*, 89–98.

122. Mertens, J. *Top. Curr. Chem.* **1984**, *125*, 165–246.

123. Jarvo, E.R.; Miller, S.J. *Tetrahedron* **2002**, *58*, 2481–2495.

124. List, B. *Tetrahedron* **2002**, *58*, 5573–5590.

125. Strecker, A. *Ann.* **1850**, *75*, 27–45.

126. Cahours, A. *Compt. rend.* **1858**, *46*, 1044.

127. Sörensen, S.P.L. *Compt. rend. trav. Lab. Carlsberg, Sér. chim.* **1903–1906**, *6*, 1.

128. Sörensen, S.P.L. *Z. Physiol. Chem.* **1905**, *44*, 448–460.

129. Knowles, W.S. *Angew. Chem., Int. Ed.* **2002**, *41*, 1998–2007.

130. Hruby, V.J.; Al-Obeidi, F.; Kazmierski, W. *Biochem. J.* **1990**, *268*, 249–262.

131. Breuer, M.; Ditrich, K.; Habicher, T.; Hauer, B.; KeBeler, M.; Stürmer, R.; Zelinski, T. *Angew. Chem., Int. Ed.* **2004**, *43*, 788–824.

132. Greenstein, J.P.; Winitz, M. *Chemistry of the Amino Acids, Volumes 1–3*; Wiley: New York, 1961.

133. Vickery, H.B.; Schmidt, C.L.A. *Chem. Rev.* **1931**, *9*, 169–318.

134. Fowden, L. *Ann. Rev. Biochem.* **1964**, *33*, 173–204.

135. *Chemistry and Biochemistry of Amino Acids, Peptides and Proteins. A Survey of Recent Developments*, Weinstein, B. Ed.; Marcel Dekker Inc.: New York (various years).

136. *The Peptides. Analysis, Synthesis, Biology. Volume 1 Major Methods of Peptide Bond Formation* Gross, E., Meienhofer, J., Eds.; Academic Press: New York, 1979.

137. *The Peptides. Analysis, Synthesis, Biology. Volume 2 Special Methods in Peptide Synthesis,* Gross, E, Meienhofer, J., Eds.; Academic Press: New York, 1980.

138. *The Peptides. Analysis, Synthesis, Biology. Volume 3 Protection of Functional Groups in Peptide Synthesis,* Gross, E, Meienhofer, J. Eds; Academic Press: New York, 1981.

139. *The Peptides. Analysis, Synthesis, Biology. Volume 4 Modern Techniques of Conformational, Structural and Configurational Analysis,* Gross, E.; Meienhofer, J. Eds; Academic Press: New York, 1981.

140. *The Peptides. Analysis, Synthesis, Biology. Volume 5 Special Method in Peptide Synthesis, Part B,* Gross, E, Meienhofer, J., Eds; Academic Press: New York, 1983.

141. *Amino Acids* **1996**, *11*, numbers 3–4, pp 257–434.

142. *Amino Acids* **1999**, *16*, numbers 3–4, pp 191–440.

143. *Amino Acids* **2003**, *24*, number 3, pp 229–333.

144. *Amino Acids* **2005**, *29*, number 2, pp 79–160.

145. Williams, R.M. *Synthesis of Optically Active α–Amino Acids*, Organic Chemistry Series, Volume 7; Pergamon Press: Toronto, Canada, 1989.

146. Kochetkov, K.A.; Belikov, V.M. *Russ. Chem. Rev.* **1987**, *56*, 1045–1067.

147. Duthaler, R.O. *Tetrahedron* **1994**, *50*, 1539–1650.

148. Calmes, M.; Daunis, J. *Amino Acids* **1999**, *16*, 215–250.

149. Nájera, C. *Synlett* **2002**, 1388–1403.

150. Ma, J.-A. *Angew. Chem., Int. Ed.* **2003**, *42*, 4290–4299.

151. Jones, J. *Amino Acid and Peptide Synthesis*; Oxford University Press: New York, 2000.

152. Izumi, Y.; Chibata, I.; Itoh, T. *Angew. Chem., Int. Ed. Engl.* **1978**, *17*, 176–183.

153. Scott, J.W. *Top. Stereochem.* **1989**, *19*, 209–226.

154. Hermann, T. *J. Biotech.* **2003**, *104*, 155–172.
155. Smith, M.B. *Methods of Non–α–Amino Acid Synthesis,* Marcel Dekker: New York, 1995.
156. *Enantioselective Synthesis of β–Amino Acids,* Juaristi, E., Ed.; VCH Publishers, John Wiley and Sons: New York, 1997.
157. Franklin, A.S. *J. Chem. Soc. Perkin Trans. 1* **1998**, 2451–2465.
158. *Amino Acid Derivatives: A Practical Approach,* Barrett, G.C., Ed; Oxford University Press: New York, 1999.
159. Blackburn, S. In *CRC Handbook of Chromatography, Volume 1, Amino Acids and Amines*; Zweig, G.; Sherma, J., Eds; CRC Press: Boca Raton, Fla, 1983.
160. Meister, A. *Biochemistry of the Amino Acids,* 2nd edn; Academic Press: New York, 1965.
161. Wold, F. *Ann. Rev. Biochem.* **1981**, *50*, 783–814.
162. Galli, F. *Amino Acids* **2003**, *25*, 205.
163. Davies, J.S. In *Chemistry and Biochemistry of Amino Acids, Peptides and Proteins. A Survey of Recent Developments, Volume 4,* Weinstein, B. Ed.; Marcel Dekker: New York, 1977, pp 1–27.
164. *Quantitation of Amino Acids and Amines by Chromatography: Methods and Protocols. Journal of* the Chromatography Library, Volume 70; Molnár-Perl, I. Ed.; Elsevier: Amsterdam, 2005.
165. Wermuth, C.G.; Ganellin, C.R.; Lindberg, P.; Mitscher, L.A. *Pure Appl. Chem.* **1998**, *70*, 1129–1143.
166. Nagel, B.; Dellweg, H.; Gierasch, L.M. *Pure Appl. Chem.* **1992**, *64*, 143–168.
167. IUPAC-IUB Commission on Biochemical Nomenclature *Pure Appl. Chem.* **1973**, *33*, 437–444.
168. IUPAC-IUB *Biochemistry.* **1966**, *5*, 1445–1453.
169. *Chem. Eng. News* **2004**, *April 26*, 16.
170. Maier, T.H.P. *Nat: Biotechno.* **2003**, *21*, 422–427.
171. Salmeron-Lopez, V.; Martinez-Toledo, M.V.; Salmeron-Miron, V.; Pozo, C.; Gonzalez-Lopez, J. *Amino Acids* **2004**, *27*, 169–174.
172. González-López, J.; Martínez-Toldeo, M.V.; Rodelas, B.; Pozo, C.; Salmerón, V. *Amino Acids* **1995**, *8*, 15–21.
173. Sandberg, M.; Eriksson, L.; Jonsson, J.; Sjöström, M.; Wold, S. *J. Med. Chem.* **1998**, *41*, 2481–2491.
174. van de Waterbeemd, H.; Karajiannis, H.; El Tayar, N. *Amino Acids* **1994**, *7*, 129–145.
175. Palecz, B. *J. Am. Chem. Soc.* **2005**, *127*, 17768–17771.
176. Biswas, K.M.; DeVido, D.R.; Dorsey, J.G. *J. Chromatogr., A* **2003**, *1000*, 637–655.
177. Ninni, L.; Meirelles, A.J.A. *Biotechno. Prog.* **2001**, *17*, 703–711.
178. Nord, L.I.; Vaag, P.; Duus, J.Ø. *Anal. Chem.* **2004**, *76*, 4790–4798.
179. Tian, J.; Yin, Y. *Amino Acids* **2004**, *26*, 175–181.
180. Flaig, R.; Koritsánzky, T.; Janczak, J.; Krane, H.-G.; Morgenroth, W.; Luger, P. *Angew. Chem., Int. Ed. Engl.* **1999**, *38*, 1397–1400.
181. Flaig, R.; Koritsanszky, T.; Dittrich, B.; Wagner, A.; Luger, P. *J. Am. Chem. Soc.* **2002**, *124*, 3407–3417.
182. Willett, R.L.; Baldwin, K.W.; West, K.W.; Pfeiffer, L.N. *Proc. Natl. Acad. Sci. U.S.A.* **2005**, *102*, 7817–7822.
183. Huang, F.; Nau, W.M. *Angew. Chem. Int. Ed.* **2003**, *42*, 2269–2272.
184. Chou, P.Y.; Fasman, G.D. *Biochemistry.* **1974**, *13*, 222–245.
185. Costantini, S.; Colonna, G.; Facchiano, A.M. *Biochem. Biophys. Res. Commun.* **2006**, *342*, 441–451.
186. Du, Q.; Wei, D.; Chou, K.-c. *Peptides* **2003**, *24*, 1863–1869.
187. Schiene-Fischer, C.; Fischer, G. *J. Am. Chem. Soc.* **2001**, *123*, 6227–6231.
188. Wyttenbach, T.; Witt, M.; Bowers, M.T. *J. Am. Chem. Soc.* **2000**, *122*, 3458–3464.
189. Henry, B.; Tekely, P.; Delpuech, J.-J. *J. Am. Chem. Soc.* **2002**, *124*, 2025–2034.
190. Rabenstein, D.L.; Hari, S.P.; Kaerner, A. *Anal. Chem.* **1997**, *69*, 4310–4316.
191. Snider, M.J.; Wolfenden, R. *J. Am. Chem. Soc.* **2000**, *122*, 11507–11508.
192. IUPAC-IUB Commission on Biochemical Nomenclature *J. Biol. Chem.* **1972**, *247*, 977–983.
193. IUPAC-IUB Joint Commission on Biochemical Nomenclature *Eur. J. Biochem.* **1984**, *138*, 9–37.
194. IUPAC-IUB Joint Commission on Biochemical Nomenclature *Pure Appl. Chem.* **1984**, *56*, 595–624.
195. Greenstein, J.P.; Winitz, M. *Chemistry of the Amino Acids, Volumes 1–3*; Wiley: New York, 1961, p 15.
196. IUPAC Commission on Nomenclature of Organic Chemistry: Section E: Stereochemistry. *Pure Appl. Chem.* **1976**, *45*, 13–30.
197. Cahn, R.S.; Ingold, C.; Prelog, V. *Angew. Chem., Int. Ed. Engl.* **1966**, *5*, 385–415.
198. Seebach, D.; Prelog, V. *Angew. Chem., Int. Ed. Engl.* **1982**, *21*, 654–660.
199. Masamune, S.; Ali, Sk. A.; Snitman, D.L.; Garvey, D.S. *Angew. Chem., Int. Ed. Engl.* **1980**, *19*, 557–558.
200. Sigler, G.F.; Fuller, W.D.; Chaturvedi, N.C.; Goodman, M.; Verlander, M. *Biopolymers* **1983**, *22*, 2157–2162.
201. Chang, C.-d.; Felix, A.M.; Jimenez, M.H.; Meienhofer, J. *Int. J. Pept. Prot. Res.* **1980**, *15*, 485–494.
202. Mitin, Y.V. *Int. J. Pept. Prot. Chem.* **1996**, *48*, 374–376.
203. Heeb, N.V.; Aberle, A.M.; Manbiar, K.P. *Tetrahedron Lett.* **1994**, *35*, 2287–2290.
204. Amend, J.P.; Helgeson, H.C. *Pure Appl. Chem.* **1997**, *69*, 935–942.
205. Menger, F.M.; Caran, K.L. *J. Am. Chem. Soc.* **2000**, *122*, 11679–11691.

206. Suzuki, M.; Yumoto, M.; Kimura, M.; Shirai, H.; Hanabusa, K. *Chem. Commun.* **2002**, 884–885.
207. Suzuki, M.; Yumoto, M.; Kimura, M.; Shirai, H.; Hanabusa, K. *Chem. Eur. J.* **2003**, *9*, 348–354.
208. Suzuki, M.; Nanbu, M.; Yumoto, M.; Shirai, H.; Hanabusa, K. *New J. Chem.* **2005**, *29*, 1439–1444.
209. Kiyonaka, S.; Shinkai, S.; Hamachi, I. *Chem.— Eur. J.* **2003**, *9*, 976–983.
210. Wang, G.; Hamilton, A.D. *Chem. Eur. J.* **2002**, *8*, 1954–1961.
211. Brosse, N.; Barth, D.; Jamart-Grégoire, B. *Tetrahedron Lett.* **2004**, *45*, 9521–9524.
212. Luo, X.; Liu, B.; Liang, Y. *Chem. Commun.* **2001**, 1556–1557.
213. Makarevic, J.; Jokic, M.; Peric, B.; Tomisic, V.; Kojic-Podic, B.; Zinic, M. *Chem.— Eur. J.* **2001**, *7*, 3328–3341.
214. Friggeri, A.; van der Pol, C.; van Bommel, K.J.C.; Heeres, A.; Stuart, M.C.A.; Feringa, B.L.; van Esch, J. *Chem.— Eur. J.* **2005**, *11*, 5353–5361.
215. D'Aléo, A.; Pozzo, J.-L.; Fages, F.; Schmutz, M.; Mieden-Gundert, G.; Vögtle, F.; Caplar, V.; Zinic, M. *Chem. Commun.* **2004**, 190–191.
216. Caplar, V.; Zinic, M.; Pozzo, J.-L.; Fages, F.; Mieden-Gundert, G.; Vögtle, F. *Eur. J. Org. Chem.* **2004**, 4048–4059.
217. Mieden-Gundert, G.; Klein, L.; Fischer, M.; Vögtle, F.; Heuzé, K.; Pozzo, J.-L.; Vallier, M.; Fages, F. *Angew. Chem., Int. Ed.* **2001**, *40*, 3164–3166.
218. de Loos, M.; Feringa, B.L.; van Esch, J.H. *Eur. J. Org. Chem.* **2005**, 3615–3631.
219. Tao, G.-h.; He, L.; Sun, N.; Kou, Y. *Chem. Commun.* **2005**, 3562–3564.
220. Jodry, J.J.; Mikami, K. *Tetrahedron Lett.* **2004**, *45*, 4429–4431.
221. Hare, P.E.; St. John, P.A.; Engel, M.H. In, Ed.; Barrett, G.C., Ed.; *Chemistry and Biochemistry of the Amino Acids*; Ed.; Chapman and Hall: New York, 1985, pp 415–425.
222. Perrett, D. In, Ed.; Barrett, G.C., Ed.; *Chemistry and Biochemistry of the Amino Acids*; Ed.; Chapman and Hall: New York, 1985, pp 426–461.
223. Engel, M.H.; Hare P.E. In, Ed.; Barrett, G.C., Ed.; *Chemistry and Biochemistry of the Amino Acids*; Ed.; Chapman and Hall: New York, 1985, pp 462–479.
224. Greenstein, J.P.; Winitz, M. *Chemistry of the Amino Acids, Volumes 2*; Wiley: New York, 1961, pp 1366–1511.
225. Fekkes, D. *J. Chromatog., B* **1996**, *682*, 3–22.
226. Deyl, Z.; Hyanek, J.; Horakova, M. *J. Chromatog., B* **1986**, *379*, 177–250.
227. Sarwar, G.; Botting, H.G. *J. Chromatog., B* **1993**, *615*, 1–22.
228. Molnár-Perl, I. *J. Chromatogr., A* **2000**, *891*, 1–32.
229. Molnár-Perl, I. *J. Chromatogr., A* **2003**, *987*, 291–309.
230. Kennedy, R.T.; German, I.; Thompson, J.E.; Witowski, S.R. *Chem. Rev.* **1999**, *99*, 3081–3131.
231. Cohen, S.A.; Strydom, D.J. *Anal. Biochem.* **1988**, *174*, 1–16.
232. Fountoulakis, M.; Lahm, H.-W. *J. Chromatogr., A* **1998**, *826*, 109–134.
233. Weiss, M.; Manneberg, M.; Juranville, J.-F.; Lahm, H.-W.; Fountoulous, M. *J. Chromatogr., A* **1998**, *795*, 263–275.
234. Edman, P. *Acta Chem. Scand.* **1950**, *4*, 283–293.
235. Edman, P. *Acta Chem. Scand.* **1950**, *4*, 277–282.
236. Sanger, F. *J. Polym. Sci.* **1961**, *49*, 3–29.
237. Bhushan, R.; Agarwal, R. *Biomed. Chromatogr.* **1998**, *12*, 322–325.
238. Kanazawa, H.; Sunamoto, T.; Matsushima, Y.; Kikuchi, A.; Okano, T. *Anal. Chem.* **2000**, *72*, 5961–5966.
239. Hayakawa, K.; Hirano, M.; Yoshikawa, K.; Katsumata, N.; Tanaka, T. *J. Chromatogr., A* **1999**, *846*, 73–82.
240. Mao, Y.; Carr, P.W. *Anal. Chem.* **2001**, *73*, 1821–1830.
241. Safa, F.; Hadjmohammadi, M.R. *J. Chromatogr., A* **2005**, *1078*, 42–50.
242. Iida, T.; Matsunaga, H.; Santa, T.; Fukushima, T.; Homma, H.; Imai, K. *J. Chromatogr., A* **1998**, *813*, 267–275.
243. Toriba, A.; Adzuma, K.; Santa, T.; Imai, K. *Anal. Chem.* **2000**, *72*, 732–739.
244. Vasanits, A.; Molnár-Perl, I. *J. Chromatogr., A* **1999**, *832*, 109–122.
245. Huber, C.G.; Choudhary, G.; Horváth, C. *Anal. Chem.* **1997**, *69*, 4429–4436.
246. Li, X.-f.; Ren, H.; Le, X.; Qi, M.; Ireland, I.D.; Dovichi, N.J. *J. Chromatogr., A* **2000**, *869*, 375–384.
247. Kurosu, Y.; Murayama, K.; Shindo, N.; Shisa, Y.; Satou, Y.; Senda, M.; Ishioka, N. *J. Chromatogr., A* **1998**, *802*, 129–134.
248. Seifar, R.M.; Kraak, J.C.; Poppe, H.; Th. Kok, W. *J. Chromatogr., A* **1999**, *832*, 133–140.
249. Qi, M.; Li, X.-F.; Stathakis, C.; Dovichi, N.J. *J. Chromatogr., A* **1999**, *853*, 131–140.
250. Shediac, R.; Ngola, S.M.; Throckmorton, D.J.; Anex, D.S.; Shepodd, T.J.; Singh, A.K. *J. Chromatogr., A* **2001**, *925*, 251–263.
251. Ivanov, A.R.; Nazimov, I.V. *J. Chromatogr., A* **2000**, *870*, 255–269.
252. Goto, M.; Kohara, N.; Yamashita, S. *Amino Acids* **1992**, *2*, 289–296.
253. Steen, H.; Mann, M. *Nat. Rev. Mol. Cell Biol.* **2004**, *5*, 699–711.
254. Taylor, J.A.; Johnson, R.S. *Anal. Chem.* **2001**, *73*, 2594–2604.
255. Frank, A.; Pevzner, P. *Anal. Chem.* **2005**, *77*, 954–973.
256. Zhang, Z. *Anal. Chem.* **2004**, *76*, 6374–6383.
257. Fischer, B.; Roth, V.; Roos, F.; Grossman, J.; Baginsky, S; Widmayer, P.; Gruissem, W.; Buhmann, J.M. *Anal. Chem.* **2005**, *77*, 7265–7273.

258. Horn, D.M.; Zubarev, R.A.; McLafferty, F.W. *Proc. Natl. Acad. Sci. U.S.A.* **2000**, *97*, 10313–10317.

259. Hedström, M.; Andersson, M.; Galaev, I.; Mattiasson, B. *J. Chromatogr., A* **2005**, *1080*, 117–123.

260. Gu, S.; Pan, S.; Bradbury, E.M.; Chen, X. *Anal. Chem.* **2002**, *74*, 5774–5785.

261. Fischer, B.; Roth, V.; Roos, F.; Grossman, J.; Baginsky, S; Widmayer, P.; Gruissem, W.; Buhmann, J.M. *Anal. Chem.* **2005**, *77*, 6300–6309.

262. Chagit, D.; Rabkin, E.; Tsoglin, A. *Org. Biomol. Chem.* **2005**, *3*, 2503–2504.

263. Williams, J.D.; Falanagan, M.; Lopez, L.; Fischer, S.; Miller, L.A.D. *J. Chromatogr., A* **2003**, *1020*, 11–26.

264. Lin, T.; Glish, G.L. *Anal. Chem.* **1998**, *70*, 5162–5165.

265. Bonetto, V.; Bergman, A.-C.; Jörnvall, H.; Sillard, R. *Anal. Chem.* **1997**, *69*, 1315–1319.

266. Sweeney, M.C.; Pei, D. *J. Comb. Chem.* **2003**, *5*, 218–222.

267. Wang, P.; Arabaci, G.; Pei, D. *J. Comb. Chem.* **2001**, *3*, 251–254.

268. Chait, B.T.; Wang, R.; Beavis, R.C.; Kent, S.B.H. *Science* **1993**, *262*, 89–92.

269. Salchert, K.; Pompe, T.; Sperling, C.; Werner, C. *J. Chromatogr., A* **2003**, *1005*, 113–122.

270. Kutlán, D.; Molár-Perl, I. *J. Chromatogr., A* **2003**, *987*, 311–322.

271. Fisher, G.H.; Arias, I.; Quesada, I.; D'Aniello, S.; Errico, F.; Di Fiore, M.M.; D'Aniello, A. *Amino Acids* **2001**, *20*, 163–173.

272. Tcherkas, Y.V.; Kartsova, L.A.; Krasnova, I.N. *J. Chromatogr., A* **2001**, *913*, 303–308.

273. Tcherkas, Y.V.; Denisenko, A.D. *J. Chromatogr., A* **2001**, *913*, 309–313.

274. Fischer, K.; Chodura, A.; Kotalik, J.; Bieniek, D.; Kettrup, A. *J. Chromatogr., A* **1997**, *770*, 229–241.

275. Reddy, M.M.K.; Ghosh, P.; Rasool, S.N.; Sarin, R.K.; Sashidhar, R.B. *J. Chromatogr., A* **2005**, *1088*, 158–168.

276. Hanczko, R.; Körös, Á.; Tóth, F.; Molnár-Perl, I. *J. Chromatogr., A* **2005**, *1087*, 210–222.

277. Heems, D.; Luck, G.; Fraudeau, C.; Vérette, E. *J. Chromatogr., A* **1998**, *798*, 9–17.

278. Satyanarayana, S.; Grossert, J.S.; Lee, S.F.; White, R.L. *Amino Acids* **2001**, *21*, 221–235.

279. Tsiboli, P.; Konstantinidis, G.; Skendros, Y.; Katsani, A.; Choli-Papadopoulou, T. *Amino Acids* **1997**, *13*, 13–23.

280. Borruat, G.; Roten, C.-A.H.; Marchant, R.; Fay, L.-B.; Karamata, D. *J. Chromatogr., A* **2001**, *922*, 219–224.

281. Csampai, A.; Kutlán, D.; Tóth, F.; Molnár-Perl, I. *J. Chromatogr., A* **2004**, *1031*, 67–78.

282. Molnár-Perl, I.; Vasanits, A. *J. Chromatogr., A* **1999**, *835*, 73–91.

283. Molnár-Perl, I.; Bozor, I. *J. Chromatogr., A* **1998**, *798*, 37–46.

284. Vasanits, A.; Kutlán, D.; Sass, P.; Molnár-Perl, I. *J. Chromatogr., A* **2000**, *870*, 271–287.

285. Gyimesi-Forrás, K.; Leitner, A.; Akasaka, K.; Lindner, W. *J. Chromatogr., A* **2005**, *1083*, 80–88.

286. Molnár-Perl, I. *J. Chromatogr., A* **2001**, *913*, 283–302.

287. Mengerink, Y.; Kutlán, D.; Tóth, F.; Csámpai, A.; Molnár-Perl, I. *J. Chromatogr., A* **2002**, *949*, 99–124.

288. Zuman, P. *Chem. Rev.* **2004**, *104*, 3217–3238.

289. Shangguan, D.; Zhao, Y.; Han, H.; Zhao, R.; Liu, G. *Anal. Chem.* **2001**, *73*, 2054–2057.

290. Yan, J.X.; Kett, W.C.; Herbert, B.R.; Gooley, A.A.; Packer, N.H.; Williams, K.L. *J. Chromatogr., A* **1998**, *813*, 187–200.

291. Kalász, H.; Szücs, Z.; Tihanyi, M.; Szilágyi, Á.; Lengyel, J. *J. Chromatogr., A* **2005**, *1079*, 208–212.

292. Lozanov, V.; Petrov, S.; Mitev, V. *J. Chromatogr., A* **2004**, *1025*, 201–208.

293. Björklund, J.; Einarsson, S.; Engström, A.; Grzegorczyk, A.; Becker, H.-D.; Josefsson, B. *J. Chromatogr., A* **1998**, *798*, 1–8.

294. van Wandelen, C.; Cohen, S.A. *J. Chromatogr., A* **1997**, *763*, 11–22.

295. Liu, H.; Sañuda-Peña, M.C.; Harvey-White, J.D.; Kalra, S.; Cohen, S.A. *J. Chromatogr., A* **1998**, *828*, 383–395.

296. Phenomenex EZ:faast amino acid analysis technical brochure; www.phenomenex.com/PHEN/Products/Brands/EZFaast.htm

297. Minocha, R.; Long, S. *J. Chromatogr., A* **2004**, *1035*, 63–73.

298. Takeuchi, T. *J. Chromatogr., A* **1997**, *780*, 219–228.

299. Wu, Z.; Tong, W.G. *J. Chromatogr., A* **1998**, *805*, 63–69.

300. Stocchi, V.; Palma, F.; Piccoli, G.; Biagiarelli, B.; Magnani, M.; Masat, L.; Cucchiarini, L. *Amino Acids* **1992**, *3*, 303–309.

301. Nonaka, S.; Tsunoda, M.; Imai, K.; Funatsu, T. *J. Chromatogr., A* **2005**, *1066*, 41–45.

302. Feuster, E.K.; Glass, T.E. *J. Am. Chem. Soc.* **2003**, *125*, 16174–16175.

303. You, J.; Lao, W.; Ou, Q.; Sun, X. *J. Chromatogr., A* **1999**, *848*, 117–130.

304. Campanella, L.; Crescentini, G.; Avino, P. *J. Chromatogr., A* **1999**, *833*, 137–145.

305. Izco, J.M.; Irigoyen, A.; Torre, P.; Barcina, Y. *J. Chromatogr., A* **2000**, *881*, 69–79.

306. Yokoyama, Y.; Tsuji, S.; Sato, H. *J. Chromatogr., A* **2005**, *1085*, 110–116.

307. *Underivatized Amino Acids*, Application Note 0040E, **1999**, Alltech Associates, Deerfield, IL.

308. Chaimbault, P.; Petritis, Elfakir, C.; Creux, M. *J. Chromatogr., A* **2000**, *870*, 245–254.

309. Petritis, K.N.; Chaimbault, P.; Elfakir, C.; Dreux, M. *J. Chromatogr., A* **1999**, *833*, 147–155.

310. Chaimbault, P.; Petritis, K.; Elfakir, C.; Dreux, M. *J. Chromatogr., A* **1999**, *855*, 191–202.

311. Frey, J.; Chamson, A.; Raby, N. *Amino Acids* **1993**, *4*, 45–51.

312. Qu, J.; Wang, Y.; Luo, G.; Wu, Z.; Yang, C. *Anal. Chem.* **2002**, *74*, 2034–2040.

313. Petritis, K.; Chaimbault, P.; Elfakir, C.; Dreux, M. *J. Chromatogr., A* **2000**, *896*, 253–263.

314. Dalluge, J.J.; Smith, S.; Sanchez-Riera, F.; McGuire, C.; Hobson, R. *J. Chromatogr., A* **2004**, *1043*, 3–7.

315. de Person, M.; Hazotte, A.; Elfakir, C.; Lafosse, M. *J. Chromatogr., A* **2005**, *1081*, 174–181.

316. Petritis, K.; Elfakir, C.; Dreux, M. *J. Chromatogr., A* **2002**, *961*, 9–21.

317. Clarke, A.P.; Jandik, P.; Rocklin, R.D.; Liu, Y.; Avdalovic, N. *Anal. Chem.* **1999**, *71*, 2774–2781.

318. Ding, Y.; Yu, H.; Mou, S. *J. Chromatogr., A* **2002**, *982*, 237–244.

319. Yu, H.; Ding, Y.-S.; Mou, S.-F.; Jandik, P.; Cheng, J. *J. Chromatogr., A* **2002**, *966*, 89–97.

320. Yu, H.; Ding, Y.-S.; Mou, S.-F. *J. Chromatogr., A* **2003**, *997*, 145–153.

321. Casella, I.G.; Gatta, M.; Cataldi, T.R.I. *J. Chromatogr., A* **2000**, *878*, 57–67.

322. Kakou, A.; Megoulas, N.C.; Koupparis, M.A. *J. Chromatogr., A* **2005**, *1069*, 209–215.

323. Dallakian, P.; Budzikiewicz, H. *J. Chromatogr., A* **1997**, *787*, 195–203.

324. Cao, P.; Moini, M. *J. Chromatogr., A* **1997**, *759*, 111–117.

325. Pan, M.; Mabry, T.J.; Cao, P.; Moini, M. *J. Chromatogr., A* **1997**, *787*, 288–294.

326. Paik, M.-J.; Kim, K.-R. *J. Chromatogr., A* **2004**, *1034*, 13–23.

327. Rimando, A.M.; Perkins-Veazie, P.M. *J. Chromatogr., A* **2005**, *1078*, 196–200.

328. Nozal, M.J.; Bernal, J.L.; Toribio, M.L.; Diego, J.C.; Ruiz, A. *J. Chromatogr., A* **2004**, *1051*, 137–146.

329. Duncan, M.W.; Poljak, A. *Anal. Chem.* **1998**, *70*, 890–896.

330. Rodier, C.; Sternberg, R.; Raulin, F.; Vidal-Madjar, C. *J. Chromatogr., A* **2001**, *915*, 199–207.

331. Glavin, D.P.; Bada, J.L. *Anal. Chem.* **1998**, *70*, 3119–3122.

332. Chiavari, G.; Fabbri, D.; Prati, S. *J. Chromatogr., A* **2001**, *922*, 235–241.

333. Pata, C.; Bonnafous, P.; Fraysse, N.; Treilhou, M.; Poinsot, V.; Couderc, F. *Electrophoresis* **2001**, *22*, 4129–4138.

334. Natalini, B.; Marinozzi, M.; Sardella, R.; Macchiarulo, A.; Pellicciari, R. *J. Chromatogr., A* **2004**, *1033*, 363–367.

335. Desai, M.J.; Armstrong, D.W. *J. Chromatogr., A* **2004**, *1035*, 203–210.

336. Bardelmeijer, H.A.; Lingeman, H.; de Ruiter, C.; Underberg, W.J.M. *J. Chromatogr., A* **1998**, *807*, 3–26.

337. McKenzie, J.A.M.; Watson, C.J.; Rostand, R.D.; German, I.; Witowski, S.R.; Kennedy, R.T. *J. Chromatogr., A* **2002**, *962*, 105–115.

338. Cellar, N.A.; Burns, S.T.; Meiners, J.-C.; Chen, H.; Kennedy, R.T. *Anal. Chem.* **2005**, *77*, 7067–7073.

339. Ciriacks, C.M.; Bowser, M.T. *Anal. Chem.* **2003**, *75*, 6582–6587.

340. Lada, M.W.; Vickroy, T.W.; Kennedy, R.T. *Anal. Chem.* **1997**, *69*, 4560–4565.

341. Taga, A.; Sugimura, M.; Honda, S. *J. Chromatogr., A* **1998**, *802*, 243–248.

342. Coble, P.G.; Timperman, A.T. *J. Chromatogr., A* **1998**, *829*, 309–315.

343. Boyd, B.W.; Witowski, S.R.; Kennedy, R.T. *Anal. Chem.* **2000**, *72*, 865–871.

344. Dong, Q.; Wang, X.; Zhu, L.; Jin, W. *J. Chromatogr., A* **2002**, *959*, 269–279.

345. Zhang, L.-Y.; Sun, M.-X. *J. Chromatogr., A* **2005**, *1095*, 185–188.

346. Chiang, M.-T.; Chang, S.Y.; Whang, C.-W. *J. Chromatogr., A* **2000**, *877*, 233–237.

347. Ummadi, M.; Weimer, B.C. *J. Chromatogr., A* **2002**, *964*, 243–253.

348. Chen, Z.; Wu, J.; Baker, G.B.; Parent, M.; Dovichi, N.J. *J. Chromatogr., A* **2001**, *914*, 293–298.

349. Veledo, M.T.; de Frutos, M.; Diez-Masa, J.C. *J. Chromatogr., A* **2005**, *1079*, 335–343.

350. Liu, X.; Li, D.-F.; Wang, Y.; Lu, Y.-T. *J. Chromatogr., A* **2004**, *1061*, 99–104.

351. Nousdje, G.; Siméon, N.; Dedieu, F.; Nertz, M.; Puig, Ph.; Couderc, F. *J. Chromatogr., A* **1997**, *765*, 337–343.

352. Arlt, K.; Brandt, S.; Kehr, *J. Chromatogr., A* **2001**, *926*, 319–3259.

353. Feng, L.; Johnson, M.E. *J. Chromatogr., A* **1999**, *832*, 211–224.

354. Lau, S.K.; Zaccardo, F.; Little, M.; Banks, P. *J. Chromatogr., A* **1998**, *809*, 203–210.

355. Zhu, R.; Th. Kok, W. *J. Chromatogr., A* **1998**, *814*, 213–221.

356. Pietrzyk, D.J.; Chen, S.; Chanthaway, B. *J. Chromatogr., A* **1997**, *775*, 327–338.

357. Shen, Z.; Sun, Z.; Wu, L.; Wu, K.; Sun, S.; Huang, Z. *J. Chromatogr., A* **2002**, *979*, 227–232.

358. Ru, Q.-H.; Yao, J.; Luo, G.-A.; Zhang, Y.-X.; Yan, C. *J. Chromatogr., A* **2000**, *894*, 337–343.

359. St'astná, M.; Slais, K. *J. Chromatogr., A* **1999**, *832*, 265–271.

360. Schneider, P.J.; Engelhardt, H. *J. Chromatogr., A* **1998**, *802*, 17–22.

361. Salimi-Moosavi, H.; Cassidy, R.M. *J. Chromatogr., A* **1997**, *790*, 185–193.

362. Mizrahi, S.; Gun, J.; Kipervaser, Z.G.; Lev, O. *Anal. Chem.* **2004**, *76*, 5399–5404.

363. Hu, S.; Li, P.C.H. *J. Chromatogr., A* **2000**, *876*, 183–191.

364. Kato, M.; Jin, H.; Sakai-Kato, K.; Toyo'oka, T.; Dulay, M.T.; Zare, R.N. *J. Chromatogr., A* **2003**, *1004*, 209–215.

365. Latorre, R.M.; Saurina, J.; Hernandez-Cassou, S. *J. Chromatogr., A* **2000**, *871*, 331–340.

366. Latorre, R.M.; Saurina, J.; Hernández-Cassou, S. *J. Chromatogr., A* **2002**, *976*, 55–64.

367. Latorre, R.M.; Hernández-Cassou, S.; Saurina, J. *J. Chromatogr., A* **2001**, *934*, 104–112.

368. Vernez, L.; Thormann, W.; Krähenbühl, S. *J. Chromatogr., A* **2000**, *895*, 309–316.

369. Jaworska, M.; Szulinska, Z.; Wilk, M. *J. Chromatogr., A* **2003**, *993*, 165–172.

370. Bayle, C.; Siri, N.; Poinsot, V.; Treilhou, M.; Caussé, E.; Couderc, F. *J. Chromatogr., A* **2003**, *1013*, 123–130.

371. Klampfl, C.W.; Buchbeger, W.; Turner, M.; Fritz, J.S. *J. Chromatogr., A* **1998**, *804*, 349–355.

372. Hoegger, D.; Freitag, R. *J. Chromatogr., A* **2003**, *1004*, 195–208.

373. Chuang, S.-C.; Chang, C.-Y.; Liu, C.-Y. *J. Chromatogr., A* **2004**, *1044*, 229–236.

374. Prokorátová, V.; Kvasnicka, F.; Sevcík, R.; Voldrich, M. *J. Chromatogr., A* **2005**, *1081*, 60–64.

375. Arce, L; Segura Carretero, A.; Ríos, A.; Cruces, C.; Fernández, A.; Valcárcel, M. *J. Chromatogr., A* **1998**, *816*, 243–249.

376. Charvátová, J.; Deyl, Z.; Kasicka, V.; Král, V. *J. Chromatogr., A* **2003**, *990*, 159–167.

377. Soga, T.; Ross, G.A. *J. Chromatogr., A* **1999**, *8372*, 231–239.

378. Schultz, C.L.; Moini, M. *Anal. Chem.* **2003**, *75*, 1508–1513.

379. Moini, M.; Schultz, C.L.; Mahmood, H. *Anal. Chem.* **2003**, *75*, 6282–6287.

380. Ivanov, A.R.; Nazimov, I.V.; Lobazov, A.P.; Popkovich, G.B. *J. Chromatogr., A* **2000**, *894*, 253–57.

381. Li, W.; Fries, D.; Alli, A.; Malik, A. *Anal. Chem.* **2004**, *76*, 218–227.

382. Henley, W.H.; Wilburn, R.T.; Crouch, A.M.; Jorgenson, J.W. *Anal. Chem.* **2005**, *77*, 7024–7031.

383. Wang, J.; Mannino, S.; Camera, C.; Chatrathi, M.; Scampicchio, M.; Zima, J. *J. Chromatogr., A* **2005**, *1091*, 177–182.

384. Throckmorton, D.J.; Shepodd, T.J.; Singh, A.K. *Anal. Chem.* **2002**, *74*, 784–789.

385. Kato, M.; Gyoten, Y.; Sakai-Kato, K.; Nakajima, T.; Toyo'oka, T. *Anal. Chem.* **2004**, *76*, 6792–6796.

386. Kato, M.; Gyoten, Y.; Sakai-Kato, K.; Toyo'oka, T. *J. Chromatogr., A* **2003**, *1013*, 183–189.

387. Hashimoto, M.; Tsukagoshi, K.; Nakajima, R.; Kondo, K.; Arai, A. *J. Chromatogr., A* **2000**, *867*, 271–279.

388. García, C.D.; Henry, C.S. *Anal. Chem.* **2003**, *75*, 4778–4783.

389. Munro, N.J.; Huang, Z.; Fnegold, D.N.; Landers, J.P. *Anal. Chem.* **2000**, *72*, 2765–2773.

390. Prest, J.E.; Baldock, S.J.; Fielden, P.R.; Goddard, N.J.; Brown, B.J.T. *J. Chromatogr., A* **2004**, *1051*, 221–226.

391. Xu, J.-J.; Peng, Y.; Bao, N.; Xia, X.-H.; Chen, H.-Y. *J. Chromatogr., A* **2005**, *1095*, 193–196.

392. Laskar, S.; Sinhababu, A.; Hazra, K.M. *Amino Acids* **2001**, *21*, 201–204.

393. Basak, B.; Bhattacharyya, U.K.; Lakar, S. *Amino Acids* **1993**, *4*, 193–196.

394. Kazmierczak, D. Ciesielski, W.; Zakrzewski, R.; Zuber, M. *J. Chromatogr., A* **2004**, *1059*, 171–174.

395. Molnár, Z.; Nagy, M.; Aranyi, A.; Hanák, L.; Argyelán, J.; Pencz, I.; Szánya, T. *J. Chromatogr., A* **2005**, *1075*, 77–86.

396. Dally, J.E.; Gorniak, J.; Bowie, R.; Bentzley, C.M. *Anal. Chem.* **2003**, *75*, 5046–5053.

397. Beegle, L.W.; Kanik, I.; Matz, L.; Hill. H.H., Jr., *Anal. Chem.* **2001**, *73*, 3028–3934.

398. Asbury, G.R.; Hill, H.H., Jr., *J. Chromatogr., A* **2000**, *902*, 433–437.

399. Brazill, S.A.; Singhal, P.; Kuhr, W.G. *Anal. Chem.* **2000**, *72*, 5542–5548.

400. Sanduskey, P.; Raftery, D. *Anal. Chem.* **2005**, *77*, 2455–2463.

401. Huang, T; Warsinke, A.; Kuwana, T.; Scheller, F.W. *Anal. Chem.* **1998**, *70*, 991–997.

402. Cordek, J.; Wang, X.; Tan, W. *Anal. Chem.* **1999**, *71*, 1529–1533.

403. Kiba, N.; Miwa, T.; Tachibana, M.; Tani, K.; Koizumi, H. *Anal. Chem.* **2002**, *74*, 1269–1274.

404. Tomer, S.; Dorsey, J.G.; Berthod, A. *J. Chromatogr., A* **2001**, *923*, 7–16.

405. Alaejos, M.S.; Montelongo, F.J.G. *Chem. Rev.* **2004**, *104*, 3239–3265.

406. Shahrokhian, S. *Anal. Chem.* **2001**, *73*, 5972–5978.

407. Leipert, D.; Nopper, D.; Bauser, M.; Gauglitz, G.; Jung, G. *Angew. Chem., Int. Ed.* **1998**, *37*, 3308–3311.

408. Buryak, A.; Severin, K. *J. Am. Chem. Soc.* **2005**, *127*, 3700–3701.

409. Hua, S.-H.; Yu, C.-W.; Xu, J.-G. *Chem. Commun.* **2005**, 450–452.

410. Buryak, A.; Severin, K. *Angew. Chem., Int. Ed.* **2004**, *43*, 4771–4774.

411. Aït-Haddou, H.; Wiskur, S.L.; Lynch, V.M.; Anslyn, E.V. *J. Am. Chem. Soc.* **2001**, *123*, 11296–11297.

412. Sasaki, S.-i.; Hashizume, A.; Citterio, D.; Fujii, E.; Suzuki, K. *Tetrahedron Lett.* **2002**, *43*, 7243–7245.

413. Bonizzoni, M.; Fabbrizzi, L.; Piovani, G.; Taglietti, A. *Tetrahedron* **2004**, *60*, 11159–11162.

414. Li, S.-Y.; Fang, L.; He, Y.-B.; Chan, W.-H.; Yeung, K.-T.; Cheng, Y.-K.; Yang, R.H. *Org. Lett.* **2005**, *7*, 5825–5828.

415. Hortalá, M.A.; Fabbrizzi, L.; Marcotte, N.; Stomeo, F.; Taglietti, A. *J. Am. Chem. Soc.* **2003**, *125*, 20–21.

416. Coskun, A.; Akkaya, E.U. *Org. Lett.* **2004**, *6*, 3107–3109.

417. Vidyasankar, S.; Ru, M.; Arnold, F.H. *J. Chromatogr., A* **1997**, *775*, 51–63.

418. Tanaka, F.; Mase,N.; Barbas III, C.F. *Chem. Commun.* **2004**, 1762–1763.

419. Wang, W.; Escobedo, J.O.; Lawrence, C.M.; Strongin, R.M. *J. Am. Chem. Soc.* **2004**, *126*, 3400–3401.

420. Rusin, O.; St. Luce, N.N.; Agbaria, R.A.; Escobedo, J.O.; Jiang, S.; Warner, I.M.; Dawan, F.B.; Lian, K.; Strongin, R.M. *J. Am. Chem. Soc.* **2004**, *126*, 438–439.

421. Han, M.S.; Kim, D.H. *Tetrahedron* **2004**, *60*, 11251–11257.

422. Sudeep, P.K.; Joseph, S.T.S.; Thomas, K.G. *J. Am. Chem. Soc.* **2005**, *127*, 6516–6517.

423. Sasaki, S.-i.; Hashizume, A.; Citterio, D.; Fuji, E.; Suzuki, K. *Angew. Chem., Int. Ed.* **2002**, *41*, 3005–3007.

424. Moghimi, A.; Rastegar, M.F.; Ghandi, M.; Taghizadeh, M.; Yari, A.; Shamsipur, M.; Yap, G.P.A; Rahbarnoohi, H. *J. Org. Chem.* **2002**, *67*, 2065–2074.

425. Tsubaki, K.; Tanaka, H.; Morikawa, H.; Fuji, K. *Tetrahedron* **2003**, *59*, 3195–3199.

426. Hernández, J.V.; Muñiz, F.M.; Oliva, A.I.; Simón, L.; Pérez, E.; Morán, J.R. *Tetrahedron Lett.* **2003**, *44*, 6983–6985.

427. Hernández, J.V.; Oliva, A.I.; Simón, L.; Muñiz, F.M.; Grande, M.; Morán, J.R. *Tetrahedron Lett.* **2004**, *45*, 4831–4833.

428. Debroy, P.; Banerjee, M.; Prasad, M.; Moulik, S.P.; Roy, S. *Org. Lett.* **2005**, *7*, 403–406.

429. Shinoda, S.; Okazaki, T.; Player, T.N.; Misaki, H.; Hori, K.; Tsukube, H. *J. Org. Chem.* **2005**, *70*, 1835–1843.

430. Da Silva, E.; Coleman, A.W. *Tetrahedron* **2003**, *59*, 7357–7364.

431. Pichler, U.; Scrimin, P.; Tecilla, P.; Tonellato, U.; Veronese, A.; Verzini, M. *Tetrahedron Lett.* **2004**, *45*, 1643–1646.

432. Schmuck, C.; Bickert, V. *Org. Lett.* **2003**, *5*, 4579–4581.

433. Kyne, G.M.; Light, M.E.; Hursthouse, M.B.; de Mendoza, J.; Kilburn, J.D. *J. Chem. Soc., Perkin Trans. 1* **2001**, 1258–1263.

434. Wehner, M.; Schrader, T. *Angew. Chem., Int. Ed.* **2002**, *41*, 1751–1754.

435. Miranda, C.; Escarti, F.; Lamarque, L.; Yunta, M.J.R.; Navarro, P.; García-España, E.; Jimeno, M.L. *J. Am. Chem. Soc.* **2004**, *126*, 823–833.

436. Verdejo, B.; Aguilar, J.; Doménech, A.; Miranda, C.; Navarro, P.; Jiménez, H.R.; Soriano, C.; García-España, E. *Chem. Commun.* **2005**, 3086–3088.

437. Jullian, V.; Shepherd, E.; Gelbrich, T.; Hursthouse, M.B.; Kilburn, J.D. *Tetrahedron Lett.* **2000**, *41*, 3963–3966.

438. Gschwind, R.M.; Armbrüster, M.; Zubrzycki, I.Z. *J. Am. Chem. Soc.* **2004**, *126*, 10228–10229.

439. Rensing, S.; Arendt, M.; Springer, A.; Grawe, T.; Schrader, T. *J. Org. Chem.* **2001**, *66*, 5814–5821.

440. Grawe, T.; Schrader, T.; Finocchiaro, P.; Consiglio, G.; Failla, S. *Org. Lett.* **2001**, *3*, 1597–1600.

441. Wehner, M.; Schrader, T.; Finocchiaro, P.; Failla, S.; Consiglio, G. *Org. Lett.* **2000**, *2*, 605–608.

442. Bhattacharyya, T.; Sundin, A.; Nilsson, U.J. *Tetrahedron* **2003**, *59*, 7921–7928.

443. Bell, T.W. Khasanov, A.B.; Drew, M.G.B.; Filikov, A.; James, T.L. *Angew. Chem., Int. Ed.* **1999**, *38*, 2543–2547.

444. Gao, J.; Martell, A.E.; Reibenspies, J. *Helv. Chim. Acta* **2003**, *86*, 196–203.

445. Kintisch, E. *Science* **2006**, *311*, 946–947.

446. Accinni, R.; Campolo, J.; Bartesaghi, S.; De Leo, G.; Lucarelli, C.; Cursano, C.F.; Parodi, O. *J. Chromatogr., A* **1998**, *828*, 397–400.

447. Minniti, G.; Piana, A.; Armani, U.; Cerone, R. *J. Chromatogr., A* **1998**, *828*, 401–405.

448. Chwatko, G.; Bald, E. *J. Chromatogr., A* **2002**, *949*, 141–151.

449. Bald, E.; Chwatko, G.; Glowacki, R.; Kusmierek, K. *J. Chromatogr., A* **2004**, *1032*, 109–115.

450. Ivanov, A.R.; Nazimov, I.V.; Baratova, L.; Lobazov, A.P.; Popkovich, G.B. *J. Chromatogr., A* **2001**, *913*, 315–318.

451. Caussé, E.; Terrier, R.; Champagne, S.; Nertz, M.; Valdiguié, P.; Salvayre, R.; Couderc, F. *J. Chromatogr., A* **1998**, *817*, 181–185.

452. Bayle, C.; Isaac, C.; Salvayre, R.; Couderc, F.; Caussé, E. *J. Chromatogr., A* **2002**, *979*, 255–260.

453. Sevcíková, P.; Glatz, Z.; Tomandl, J. *J. Chromatogr., A* **2003**, *999*, 197–204.

454. Kim, I.-J.; Park, S.-J.; Kim, H.-J. *J. Chromatogr., A* **2000**, *877*, 217–223.

455. Satterfield, M.B.; Sniegoski, L.T.; Welch, M.J.; Nelson, B.C.; Pfeiffer, C.M. *Anal. Chem.* **2003**, *75*, 4631–4638.

456. Nelson, B.C.; Satterfield, M.B.; Sniegoski, L.T.; Welch, M.J. *Anal. Chem.* **2005**, *77*, 3586–3593.

457. Celma, C.; Allué, J.A.; Pruñonosa, J.; Peraire, C.; Obach, R. *J. Chromatogr., A* **2000**, *870*, 13–22.

458. Toussaint, B.; Pitti, Ch.; Streel, B.; Ceccato, A.; Hubert, Ph.; Crommen, J. *J. Chromatogr., A* **2000**, *896*, 191–199.

459. Smith, G.G.; Williams, K.M.; Wonnacott, D.M. *J. Org. Chem.* **1978**, *43*, 1–5.

460. Smith, G.G.; Sivakua, T. *J. Org. Chem.* **1983**, *48*, 627–634.

461. Kvenvolden, K.A.; Peterson, E.; Brown, F.S. *Science* **1970**, *169*, 1079–1082.

462. Zhao, M.; Bada, J.L. *J. Chromatogr., A* **1995**, *690*, 55–63.

463. Kunnas, A.V.; Jauhiainen, T.-P. *J. Chromatogr.*, **1993**, *628*, 269–273.

464. Elster, H.; Gil-Av, E.; Weiner, S. *J. Archaeol. Sci.* **1991**, *18*, 605–617.

465. Csapó, J.; Csapó-Kiss, Z.; Némethy, S; Folestad, S.; Tivesten, A.; Martin, T.G. *Amino Acids* **1994**, *7*, 317–325.

466. Engel, M.H.; Goodfriend, G.A.; Qian, Y.; Macko, S.A. *Proc. Natl. Acad. Sci. U.S.A.* **1994**, *91*, 10475–10478.

467. Lubec, G.; Lubec, B. *Amino Acids* **1993**, *4*, 1–3.
468. Amelung, W.; Brodowski, S. *Anal. Chem.* **2002**, *74*, 3239–3246.
469. Bada, J.L. In *Chemistry and Biochemistry of the Amino Acids*; Barrett, G.C. Ed.; Chapman and Hall: New York, 1985, pp 399–414.
470. *Perspectives in Amino Acid and Protein Geochemistry,* Goodfriend, G.A.; Collins, M.J.; Fogel, M.L.; Macko, S.A.; Wehmiller, J.F., Eds; Oxford University Press : New York, 2000.
471. Buck, R.H.; Krummen, K. *J. Chromatogr.* **1987**, *387*, 255–265.
472. Goodlett, D.R.; Abuaf, P.A.; Savage, P.A.; Kowalski, K.A.; Mukherjee, T.K.; Tolan, J.W.; Corkum, N.; Goldstein, G.; Crowther, J.B. *J. Chromatogr., A* **1995**, *707*, 233–244.
473. Manning, J.M. *J. Biol. Chem.* **1971**, *246*, 2926–2929.
474. Ward, T.J. *Anal. Chem.* **2000**, *72*, 4521–4528.
475. Smith, G.G.; Reddy, G.V. *J. Org. Chem.* **1989**, *54*, 4529–4535.
476. Frank, H.; Woiwode, W.; Nicholson, G.; Bayer, E. *Liebigs Ann. Chem.* **1981**, 354–365.
477. Thorsén, G.; Bergquist, J.; Westlind-Danielsson, A.; Josefsson, B. *Anal. Chem.* **2001**, *73*, 2625–2631.
478. Bada, J.L. *J. Am. Chem. Soc.* **1972**, *94*, 1371–1373.
479. Rios, A.; Richard, J.P. *J. Am. Chem. Soc.* **1997**, *119*, 8375–8376.
480. Yokoyama, Y.; Hikawa, H.; Murakami, Y. *J. Chem. Soc., Perkin Trans. 1* **2001**, 1431–1434.
481. Ebbers, E.J.; Ariaans, G.J.A.; Houbiers, J.P.M.; Bruggink, A.; Zwanenburg, B. *Tetrahedron* **1997**, *53*, 9417–9476.
482. Rios, A.; Amyes, T.L.; Richard, J.P. *J. Am. Chem. Soc.* **2000**, *122*, 9373–9385.
483. Rios, A.; Richard, J.P.; Amyes, T.L. *J. Am. Chem. Soc.* **2002**, *124*, 8251–8259.
484. Rios, A.; Crugeiras, J.; Amyes, T.L.; Richard, J.P. *J. Am. Chem. Soc.* **2001**, *123*, 7949–7950.
485. Liu, L.; Breslow, R. *Tetrahedron Lett.* **2001**, *42*, 2775–2777.
486. Stroud, E.D.; Fife, D.J.; Smith, G.G. *J. Org. Chem.* **1983**, *48*, 5368–5369.
487. Brückner, H.; Justus, J.; Kirschbaum, J. *Amino Acids* **2001**, *21*, 429–433.
488. Takats, Z.; Nanita, S.C.; Cooks, R.G. *Angew. Chem., Int. Ed.* **2003,** *42*, 3521–3523.
489. Lubell, W.D.; Rapoport, H. *J. Am. Chem. Soc.* **1987**, *109*, 236–239.
490. Baum, R.; Smith, G.G. *J. Am. Chem. Soc.* **1986**, *108*, 7325–7327.
491. Havlícek, L.; Hanus, J.; Nemecek, J. *Collect. Czech. Chem. Commun.* **1989**, *54*, 3381–3386.
492. Smith, G.G.; Evans, R.C.; Baum, R. *J. Am. Chem. Soc.* **1986**, *108*, 7327–7332.
493. Li, B.; Borchardt, R.T.; Topp, E.M.; VanderVelde, D.; Schowen, R.L. *J. Am. Chem. Soc.* **2003**, *125*, 11486–11487.
494. Lloyd-Williams, P.; Albericio, F.; Giralt, E. *Chemical Approaches to the Synthesis of Peptides and Proteins;* CRC Press: Boca Raton, Fla., 1997; pp 114–121.
495. Kemp, D.S. In *The Peptides. Analysis, Synthesis, Biology. Volume 1 Major Methods of Peptide Bond Formation* Gross, E., Meienhofer, J., Eds; Academic Press: New York, 1979, pp 315–383.
496. Benoiton, N.L. In *The Peptides. Analysis, Synthesis, Biology. Volume 5 Special Method in Peptide Synthesis, Part B,* Gross, E, Meienhofer, J., Eds; Academic Press, New York, 1983, pp 217–285.
497. Benoiton, N.L.; Kuroda, K.; Chen, F.M.F. *Int. J. Pept. Prot. Chem.* **1979**, *13*, 403–408.
498. Goodman, M.; Keogh, P.; Anderson, H. *Bioorg. Chem.* **1977**, *6*, 239–247.
499. Riester, D.; Wiesmüller, K.-H.; Stoll. D.; Kuhn, R. *Anal. Chem.* **1996**, *68*, 2361–2365.
500. Celma, C.; Giralt, E. *J. Chromatogr.* **1991**, *562*, 447–458.
501. Tang, Y.; Zukowski, J.; Armstrong, D.W. *J. Chromatogr., A* **1996**, *743*, 261–271.
502. Riester, D.; Wiesmüller, K.-H.; Stoll, D.; Kuhn, R. *Anal. Chem.* **1996**, *68*, 2361–2365.
503. Seyer, R.; Aumelas, A.; Caraty, A.; Rivaille, P.; Castro, B. *Int. J. Pept. Prot. Chem.* **1990**, *35*, 465–472.
504. Gibson, F.S.; Rapoport, H. *J. Org. Chem.* **1995**, *60*, 2615–2617.
505. Rzeszotarska, B.; Masiukiewicz, E. *Org. Prep. Proc. Int.* **1989**, *21*, 393–450.
506. Han, Y.; Albericio, F.; Barany, G. *J. Org. Chem.* **1997**, *62*, 4307–4312.
507. Romoff, T.T.; Goodman, M. *J. Peptide Res.* **1997**, *49*, 281–292.
508. Benouargha, A.; Verducci, J.; Jacquier, R. *Bull. Soc. Chim. Fr.* **1995**, *132*, 824–828.
509. Yang, A.; Gehring, A.P.; Li, T. *J. Chromatogr., A* **2000**, *878*, 165–170.
510. Yamada, S.; Hongo, C.; Yoshioka, R.; Chibata, I. *J. Org. Chem.* **1983**, *48*, 843–846.
511. Grigg, R.; Gunaratne, H.Q.N. *Tetrahedron Lett.* **1983**, *24*, 4457–4460.
512. Tokuyama, S.; Miya, H.; Hatano, K.; Takahashi, T. *Appl. Microbiol. Biotechnol.* **1994**, *40*, 835–840.
513. Tokuyama, S.; Hatano, K.; Takahashi, T. *Biosci. Biotechno. Biochem.* **1994**, *58*, 24–27.
514. Tanner, M.E. *Acc. Chem. Res.* **2002**, *35*, 237–246.
515. Schönfeld, D.L.; Bornscheuer, U.T. *Anal. Chem.* **2004**, *76*, 1184–1188.
516. Zaikova, T.O.; Lomovskii, O.I.; Rukavishnikov, A.V. *Russ. J. Gen. Chem.* **1996**, *66*, 643–647.
517. Hateley, M.J.; Schichl, D.A.; Fischer, C.; Beller, M. *Synlett* **2001**, 25–28.
518. Pàmies, O.; Éll, A.H.; Samec, J.S.M.; Hermanns, N.; Bäckvakk, J.-E. *Tetrahedron Lett.* **2002**, *43*, 4699–4702.

519. Heteley, M.J.; Schichl, D.A.; Kreuzfeld, H.-J.; Beller, M. *Tetrahedron Lett.* **2000**, *41*, 3821–3824.

520. *Chromatographic Enantioseparation: Methods and Applications*, 2nd edn., Allenmark, S. Ed.; Ellis Horwood: New York, 1991, pp 41–48.

521. Greenstein, J.P.; Winitz, M. In. pp 46–244.

522. Fowden, L.; Scopes, P.M.; Thomas, R.N. *J. Chem Soc. (C)* **1971**, 833–840.

523. Greenstein, J.P.; Winitz, M. In pp 130–152.

524. Hofstetter, O.; Hofstetter, H.; Schurig, V.; Wilchek, M.; Green, B.S. *J. Am. Chem. Soc.* **1998**, *120*, 3251–3252.

525. Greenstein, J.P.; Winitz, M. In. pp 84–88.

526. Jorgensen, E.C. *Tetrahedron Lett.* **1971**, 863–866.

527. Takatori, K.; Toyama, S.; Fujii, S.; Kajiwara, M. *Chem. Pharm. Bull.* **1995**, *43*, 1797–1799.

528. *Applications in Chromatography,* July 2004, C–4; www.LaboratoryEquipment.com.

529. Zahn, S.; Canary, J.W. *Org. Lett.* **1999**, *1*, 861–864.

530. Barrett, G.C. *J. Chem. Soc.* **1965** Part 2, 2825–2830.

531. Barrett, G.C. *J. Chem. Soc. (C)* **1966**, 1771–1775.

532. Barrett, G.C. *J. Chem. Soc. (C)* **1967**, 1–5.

533. Djerassi, C.; Undheim, K.; Sheppard, R.C.; Terry, W.G.; Sjöberg, B. *Acta Chem. Scand.* **1961**, *15*, 903–912.

534. Sjöberg, B.; Fredga, A.; Djerassi, C. *J. Am. Chem. Soc.* **1959**, *81*, 5002–5003.

535. Toniolo, C. *Tetrahedron* **1970**, *26*, 5479–5488.

536. Tortorella, V.; Bettoni, G. *J. Chem. Soc., Chem. Commun.* **1967**, 321–323.

537. Lo, L.-C.; Yang, C.-T.; Tsai, C.-S. *J. Org. Chem.* **2002**, *67*, 1368–1371.

538. Peng, J.; Odom, J.D.; Dunlap, R.P.; Silks, L.A; III. *Tetrahedron: Asymmetry* **1994**, *5*, 1627–1630.

539. Frelek, J.; Majer, Z.; Perkowska, A.; Snatzke, G. *Pure Appl. Chem.* **1985**, *57*, 441–451.

540. Cantín, O.; Cativiela, C.; Díaz-de-Villegas, M.; Navarro, R.; Urriolabeitia, E.P. *Tetrahedron: Asymmetry* **1996**, *7*, 2695–2702.

541. Toome, V.; Wegrzynski, B. *Amino Acids* **1992**, *3*, 195–203.

542. Huang, X.; Rickman, B.H.; Borhan, B.; Berova, N.; Nakanishi, K. *J. Am. Chem. Soc.* **1998**, *120*, 6185–6186.

543. Kurtán, T.; Nesnas, N.; Li, Y-.-Q.; Huang, X.; Nakanishi, K.; Berova, N. *J. Am. Chem. Soc.* **2001**, *123*, 5962–5973.

544. Tmiaki, H.; Matsumoto, N.; Tsukube, H. *Tetrahedron Lett.* **1997**, *38*, 4239–4242.

545. Nonokawa, R.; Yashima, E. *J. Am. Chem. Soc.* **2003**, *125*, 1278–1283.

546. Tsukube, H.; Fukui, H.; Shinoda, S. *Tetrahedron Lett.* **2001**, *42*, 7583–7585.

547. Toniolo, C. In Barrett, G.C., Ed.; *Chemistry and Biochemistry of the Amino Acids*; Chapman and Hall: New York, 1985, pp 545–572.

548. Seco, J.M.; Quiñoá, E.; Riguera, R. *Chem. Rev.* **2004**, *104*, 17–117.

549. Seco, J.M.; Quiñoá, E.; Riguera, R. *Tetrahedron: Asymmetry* **2001**, *12*, 2915–2925.

550. Kusumi, T.; Fukushima, T.; Ohtani, I.; Kakisawa, H. *Tetrahedron Lett.* **1991**, *32*, 2939–2942.

551. Dale, J.A.; Mosher, H.S. *J. Am. Chem. Soc.* **1973**, *95*, 512–519.

552. Trost, B.M.; Bunt, R.C.; Pulley, S.R. *J. Org. Chem.* **1994**, *59*, 4202–4205.

553. Seco, J.M.; Latypov, Sh.; Quiñoá, E.; Riguera, R. *Tetrahedron Lett.* **1994**, *35*, 2921–2924.

554. Latypov, Sh.; Seco, J.M.; Quiñoá, E.; Riguera, R. *J. Org. Chem.* **1995**, *60*, 1538–1545.

555. Chinchilla, R.; Falvello, L.R.; Nájera, C. *J. Org. Chem.* **1996**, *61*, 7285–7290.

556. Oshikawa, T.; Yamashita, M.; Kumagai, S.; Seo, K.; Kobayashi, J. *J. Chem. Soc., Chem. Commun.* **1995**, 435–436.

557. Omata, K.; Fujiwara, T.; Kabuto, K. *Tetrahedron: Asymmetry* **2002**, *13*, 1655–1662.

558. Seco, J.M.N.; Quiñoá, E.; Riguera, R. *J. Org. Chem.* **1999**, *64*, 4669–4675.

559. Takeuchi, Y.; Konishi, M.; Hori, H.; Takahashi, T.; Kometani, T.; Kirk, K.L. *Chem. Commun.* **1998**, 365–366.

560. Fujiwara, T.; Omata, K.; Kabuto, K.; Kabuto, C.; Takahashi, T.; Segawa, M.; Takeuchi, Y. *Chem. Commun.* **2001**, 2694–2695.

561. Latypov, S.K.; Seco, J.M.; Quiñoá, E.; Riguera, R. *J. Am. Chem. Soc.* **1998**, *120*, 877–882.

562. Latypov, S.K.; Galiullina, N.F.; Aganov, A.V.; Kataev, V.E.; Riguera, R. *Tetrahedron* **2001**, *57*, 2231–2236.

563. Earle, M.A.; Hultin, P.G. *Tetrahedron Lett.* **2000**, *41*, 7855–7858.

564. Porto, S.; Durán, J.; Seco, J.M.; Quiñoá, E.; Riguera, R. *Org. Lett.* **2003**, *5*, 2979–2982.

565. Arnauld, T.; Barrett, A.G.M.; Hopkins, B.T.; Zécri, F.J. *Tetrahedron Lett.* **2001**, *42*, 8215–8217.

566. Peng, J.; Barr, M.E.; Ashburn, D.A.; Odom, J.D.; Dunlap, R.B.; Silks III, L.A. *J. Org. Chem.* **1994**, *59*, 4977–4987.

567. Silks III, L.A.; Peng, J.; Odom, J.D.; Dunlap, R.B. *J. Org. Chem.* **1991**, *56*, 6733–6736.

568. Kabuto, K.; Sasaki, Y. *J. Chem. Soc., Chem. Commun.* **1987**, 670–671.

569. Kabuto, K.; Sasaki, Y. *Tetrahedron Lett.* **1990**, *31*, 1031–1034.

570. Omata, K.; Aoyagi, S.; Kabuto, K. *Tetrahedron: Asymmetry* **2004**, *15*, 2351–2356.

571. Meddour, A.; Loewenstein, A.; Péchiné, J.-M.; Courtieu, J. *Tetrahedron: Asymmetry* **1997**, *8*, 485–494.

572. Chalard, P.; Bertrand, M.; Canet, I.; Théry, V.; Remuson, R.; Jeminet, G. *Org. Lett.* **2000**, *2*, 2431–2434.

573. An, S.S.A.; Lester, C.C.; Peng, J.-L.; Li, Y.-J.; Rothwarf, D.M.; Welker, E.; Thannhauser, T.W.;

Zhang, L.S.; Tam, J.P.; Scheraga, H.A. *J. Am. Chem. Soc.* **1999**, *121*, 11558–11566.

574. Roussel, C.; Delio, A.; Pierrot-Saunders, J.; Piras, P.; Vanthuyne, N. *J. Chromatogr., A* **2004**, *1037*, 311–328.

575. Marfey, P. *Carlsberg Res. Commun.* **1984**, *49*, 591–596.

576. Harada, K.-I.; Fujii, K.; Mayumi, T.; Hibino, Y.; Suzuki, M.; Ikai, Y.; Oka, H. *Tetrahedron Lett.* **1995**, *36*, 1515–1518.

577. Harada, K.-I.; Fujii, K.; Shimada, T.; Suzuki, M.; Sano, H.; Adachi, K.; Carmichael, W.W. *Tetrahedron Lett.* **1995**, *36*, 1511–1514.

578. Fujii, K.; Ikai, Y.; Mayumi, T.; Oka, H.; Suzuki, M.; Harada, K.-i. *Anal. Chem.* **1997**, *69*, 3346–3352.

579. Harada, K.-I.; Fujii, K.; Hayashi, K.; Suzuki, M. *Tetrahedron Lett.* **1996**, *37*, 3001–3004.

580. Fujii, K.; Shimoya, T.; Ikai, Y.; Oka, H.; Harada, K.-i. *Tetrahedron Lett.* **1998**, *39*, 2579–2582.

581. Fujii, K.; Ikai, Y.; Oka, H.; Suzuki, M.; Harada, K.-i. *Anal. Chem.* **1997**, *69*, 5146–5151.

582. Greenstein, J.P.; Winitz, M. *Chemistry of the Amino Acids, Volume 2*; Wiley: New York, 1961, pp 1724–1733; 1734–1749.

583. Harada, K.-i.; Matsui, A.; Shimizu, Y.; Ikemoto, R.; Fujii, K. *J. Chromatogr., A* **2001**, *921*, 187–195.

584. Itou, Y.; Ishida, K.; Shin, H.J.; Murakami, M. *Tetrahedron* **1999**, *55*, 6871–6882.

585. Randazzo, A.; Bifulco, G.; Giannini, C.; Bucci, M.; Debitus, C.; Cirino, G.; Gomez-Paloma, L. *J. Am. Chem. Soc.* **2001**, *123*, 10870–10876.

586. MacMillan, J.B.; Ernst-Russell, M.A.; de Ropp, J.S.; Molinski, T.F. *J. Org. Chem.* **2002**, *67*, 8210–8215.

587. Komatsu, K.; Shigemori, H.; Kobayashi, J. *J. Org. Chem.* **2001**, *66*, 6189–6192.

588. Tan, L.T.; Cheng, X.C.; Jensen, P.R.; Fenical, W. *J. Org. Chem.* **2003**, *68*, 8767–8773.

589. Bringmann, G.; Lang, G.; Steffens, S.; Schaumann, K. *J. Nat. Prod.* **2004**, *67*, 311–315.

590. Grach-Pogrebinsky, O.; Sedmak, B.; Carmeli, S. *J. Nat. Prod.* **2004**, *67*, 337–342.

591. Ishida, K.; Murakami, M. *J. Org. Chem.* **2000**, *65*, 5898–5900.

592. Fujii, K.; Yahashi, Y.; Nakano, T.; Imanishi, S.; Baldia, S.F.; Harada, K.-i. *Tetrahedron* **2002**, *58*, 6873–6879.

593. Jacobsen, P.G.; Sambandan, T.G.; Morgan, B. *J. Chromatogr., A* **1998**, *816*, 59–64.

594. Bhushan, R.; Brückner, H. *Amino Acids* **2004**, *27*, 231–247.

595. Brückner, H.; Wittner, R. Godel, H. *J. Chromatogr.* **1989**, *476*, 73–82.

596. Husain, P.A.; Debnath, J.; May, S.W. *Anal. Chem.* **1993**, *65*, 1456–1461.

597. Yasuhara, F.; Yamaguchi, S.; Takeda, M.; Ochiai, Y.; Miyano, S. *J. Chromatogr., A* **1995**, *694*, 227–236.

598. Hess, S.; Gustafson, K.R.; Milanowski, D.J.; Alvira, E.; Lipton, M.A.; Pannell, L.K. *J. Chromatogr., A* **2004**, *1035*, 211–219.

599. Iida, T.; Matsunaga, H.; Fukushima, T.; Santa, T.; Homma, H.; Imai, K. *Anal. Chem.* **1997**, *69*, 4463–4468.

600. Bhushan, R.; Ali, I. *J. Chromatogr.* **1987**, *392*, 460–463.

601. Günther, K.; Schickedanz, M.; Drauz, K.; Martens, J.; Fresenius Z. *Anal. Chem.* **1986**, *325*, 297–298.

602. König, W.A.; Hüthig, A. *The Practice of Enantiomer Separation by Capillary Gas Chromatography*; Huthig Heidelberg Verlag: New York, 1987; p 25.

603. Bayer, E. *Z. Naturforsch* **1983**, *38b*, 1281–1291.

604. Yao, Z.-P.; Wan, T.S.M.; Kwong, K.P.; Che, C.-T. *Anal. Chem.* **2000**, *72*, 5383–5393.

605. Yao, Z.-P.; Wan, T.S.M.; Kwong, K.P.; Che, C.-T. *Chem. Commun.* **1999**, 2119–2120.

606. van Delden, R.A.; Feringa, B.L. *Chem. Commun.* **2002**, 174–175.

607. Garner, P.; Park, J.M. *J. Org. Chem.* **1987**, *52*, 2361–2364.

608. Blaskovich, M.A.; Lajoie, G.A *J. Am. Chem. Soc.* **1993**, *115*, 5021–5030.

609. Armstrong, D.W.; Duncan, J.D.; Lee, S.H. *Amino Acids*, **1991**, *1*, 97.

610. Chang, S.C.; Wang, L.R.; Armstrong, D.W. *J. Liq. Chromatogr.* **1992**, *15*, 1411–1429.

611. Greenstein, J.P.; Winitz, M. In. p 1819.

612. Gawley, R.E. *J. Org. Chem.* **2006**, *71*, 2411–2416.

613. Chen, Y.K.; Lurain, A.E.; Walsh, P.J. *J. Am. Chem. Soc.* **2002**, *124*, 12225–12231.

614. Hofstetter, O.; Hofstetter, H.; Wilchek, M.; Schurig, V.; Green, B.S. *Chem. Comm.* **2000**, 1581–1582.

615. Parker, D. *Chem. Rev.* **1991**, *91*, 1441–1457.

616. *A Practical Approach to Chiral Separations by Liquid Chromatography*; Subramanian, A. Ed.; VCH: New York, 1994.

617. *Chromatographic Enantioseparation: Methods and Applications*, 2nd edn, Allenmark, S. Ed.; Ellis Horwood Limited: New York, 1991.

618. *Chiral Separations: Proceedings of the Chromatographic Society International Symposium on Chiral Separations;* Stevenson, D. Wilson, I.D., Eds.; Plenum Press: New York, 1988.

619. *Chromatographic Chiral Separations*, Zief, M., Crane, L.J., Eds.; Marcel Dekker, Inc.: New York, 1988.

620. *Chiral Liquid Chromatography* Lough, W.J., Ed.; Blackie: London, 1989.

621. König, W.A.; Hüthig, A. *The Practice of Enantiomer Separation by Capillary Gas Chromatography*; Verlag: New York, 1987.

622. Ward, T.J. *Anal. Chem.* **2002**, *74*, 2863–2872.

623. Ward, T.J.; Hamburg, D.-M. *Anal. Chem.* **2004**, *76*, 4635–4644.

624. Schurig, V. *J. Chromatogr., A* **2001**, *906*, 275–299.

625. Berthod, A. *Anal. Chem.* **2006**, *78*, 2093–2099.

626. Cawley, A.; Duxbury, J.P.; Kee, T.P. *Tetrahedron: Asymmetry* **1998**, *9*, 1947–1949.

627. Armstrong, D.W.; Lee, J.T.; Chang, L.W. *Tetrahedron: Asymmetry* **1998**, *9*, 2043–2064.

628. Armstrong, D.W.; He, L.; Yu, T.; Lee, J.T.; Liu, Y.-s. *Tetrahedron: Asymmetry* **1999**, *10*, 37–60.

629. Alexander, J.M.; Clark, J.L.; Brett, T.J.; Stezowski, J.J. *Proc. Natl. Acad. Sci. U.S.A.* **2002**, *99*, 5115–5120.

630. Clark, J.L.; Stezowski, J.J. *J. Am. Chem. Soc.* **2001**, *123*, 9880–9888.

631. Clark, J.L.; Booth, B.R.; Stezowski, J.J. *J. Am. Chem. Soc.* **2001**, *123*, 9889–9895.

632. Hembury, G.; Rekharsky, M.; Nakamura, A.; Inoue, Y. *Org. Lett.* **2000**, *2*, 3257–3260.

633. Uccello-Barretta, G.; Ferri, L.; Balzano, F.; Salvadori, P. *Eur. J. Org. Chem.* **2003**, 1741–1748.

634. Uccello-Barretta, G.; Balzano, F.; Sicoli, G.; Scarselli, A.; Salvadori, P. *Eur. J. Org. Chem.* **2005**, 5349–5355.

635. Dale, J.A.; Dull, D.L.; Mosher, H.S. *J. Org. Chem.* **1969**, *34*, 2543–2549.

636. König, W.A.; Nippe, K.-S.; Mischnick, P. *Tetrahedron Lett.* **1990**, *31*, 6867–6868.

637. Hull, W.E.; Seeholzer, K.; Baumeister, M.; Ugi, I. *Tetrahedron* **1986**, *42*, 547–552.

638. Breuer, W.; Ugi, I. *J. Chem. Res. (S)* **1982**, 271; *J. Chem. Res. (M)* **1982**, 2901–2945.

639. Dondoni, A.; Massi, A.; Minghini, E.; Sabbatini, S.; Bertolasi, V. *J. Org. Chem.* **2003**, *68*, 6172–6183.

640. Davey, A.E.; Horwell, D.C. *Biorg. Med. Chem.* **1993**, *1*, 45–58.

641. Chinchilla, R.; Falvello, L.R.; Nájera, C. *J. Org. Chem.* **1996**, *61*, 7285–7290.

642. Chin, J.; Kim, D.C.; Kim, H.-J.; Panosyan, F.B.; Kim, K.M. *Org. Lett.* **2004**, *6*, 2591–2593.

643. Rodríguez-Escrich, S.; Popa, D.; Jimeno, C.; Vidal-Ferran, A.; Pericàs, M.A. *Org. Lett.* **2005**, *7*, 3829–3832.

644. Hulst, R.; Kellogg, R.M.; Feringa, B.L. *Recl. Trav. Chim. Pays-Bas* **1995**, *114*, 115–138.

645. Hulst, R.; Zijlstra, R.W.J.; Feringa, B.L.; de Vries, N.K.; ten Hoeve, W.; Wynberg, H. *Tetrahedron Lett.* **1993**, *34*, 1339–1342.

646. ten Hoeve, W.; Wynberg, H. *J. Org. Chem.* **1985**, *50*, 4508–4514.

647. Hulst, R.; de Vries, N.K.; Feringa, B.L. *Angew. Chem., Int. Ed.* **1992**, *31*, 1092–1093.

648. Hulst, R.; de Vries, N.K.; Feringa, B.L. *Tetrahedron* **1994**, *50*, 11721–11728.

649. Hulst, R.; Zijlstra, R.W.J.; Koen de Vries, N.; Feringa, B.L. *Tetrahedron: Asymmetry* **1994**, *5*, 1701–1710.

650. Hulst, R.; de Vries, N.K.; Feringa, B.L. *Tetrahedron: Asymmetry* **1994**, *5*, 699–708.

651. Kolodiazhnyi, O.I.; Demchuk, O.M.; Gerschkovich, A.A. *Tetrahedron: Asymmetry* **1999**, *10*, 1729–1732.

652. Feringa, B.L.; Strijtveen, B.; Kellogg, R.M. *J. Org. Chem.* **1986**, *51*, 5484–5486.

653. Wang, Y.; Mosher, H.S. *Tetrahedron Lett.* **1991**, *32*, 987–990.

654. Staubach, B.; Buddrus, J. *Angew. Chem., Int. Ed. Engl.* **1996**, *35*, 1344–1346.

655. Dunina, V.V.; Gorunova, O.N.; Livantsov, M.V.; Grishin, Y.K. *Tetrahedron: Asymmetry* **2000**, *11*, 2907–2916.

656. Levrat, F.; Stoeckli-Evans, H.; Engel, N. *Tetrahedron: Asymmetry* **2002**, *13*, 2335–2344.

657. Böhm, A.; Seebach, D. *Helv. Chim. Acta* **2000**, *83*, 3262–3278.

658. Dobashi, A.; Saito, N.; Motoyama, Y.; Hara, S. *J. Am. Chem. Soc.* **1986**, *108*, 307–308.

659. Koppenhoefer, B.; Hummel, M. *Z. Naturforsch., B* **1992**, 1034–1036.

660. Yang, X.; Wang, G.; Zhong, C.; Wu, X.; Fu, E. *Tetrahedron: Asymmetry* **2006**, *17*, 916–921.

661. Uccello-Barretta, G.; Samaritani, S.; Menicagli, R.; Salvadori, P. *Tetrahedron: Asymmetry* **2000**, *11*, 3901–3912.

662. Iuliano, A.; Uccello-Barretta, G.; Salvadori, P. *Tetrahedron: Asymmetry* **2000**, *11*, 1555–1563.

663. Uccello-Barretta, G.; Bardoni, S.; Balzano, F.; Salvadori, P. *Tetrahedron: Asymmetry* **2001**, *12*, 2019–2023.

664. Uccello-Baretta, G.; Mirabella, F.; Balzano, F.; Salvadori, P. *Tetrahedron: Asymmetry* **2003**, *14*, 1511–1516.

665. Canet, I.; Meddour, A.; Courtieu, J.; Canet, J.L.; Salaün, J. *J. Am. Chem. Soc.* **1994**, *116*, 2155–2156.

666. Solgadi, A.; Meddour, A.; Courtieu, J. *Tetrahedron: Asymmetry* **2004**, *15*, 1315–1318.

667. Kobzar, K.; Kessler, H.; Luy, B. *Angew. Chem., Int. Ed.* **2005**, *44*, 3145–3147.

668. Uccello-Barretta, G.; Cuzzola, A.; Balzano, F.; Menicagli, R.; Iuliano, A.; Salvadori, P. *J. Org. Chem.* **1997**, *62*, 827–835.

669. Wenzel, T.J.; Thurston, J.E. *Tetrahedron Lett.* **2000**, *41*, 3769–3772.

670. Machida, Y.; Nishi, H.; Nakamura, K. *J. Chromatogr., A* **1998**, *810*, 33–41.

671. Nagata, H.; Nishi, H.; Kamiguchi, M.; Ishida, T. *Org. Biomol. Chem.* **2004**, *2*, 3470–3475.

672. Wenzel, T.J.; Thurston, J.E.; Sek, D.C.; Joly, J.-P. *Tetrahedron: Asymmetry* **2001**, *12*, 1125–1130.

673. Claeys-Bruno, M.; Toronto, D.; Pécaut, J.; Bardet, M.; Marchon, J.-C. *J. Am. Chem. Soc.* **2001**, *123*, 11067–11068.

674. Sullivan, G.R. *Top. Stereochem.* **1978**, *10*, 287–329.

675. Gupta, A.; Kazlauskas, R.J. *Tetrahedron: Asymmetry* **1992**, *3*, 243–246.

676. Bautista, F.M.; Campelo, J.M.; García, A.; Luna, D.; Marinase, J.M. *Amino Acids* **1992**, *2*, 87–95.

677. Taku, K.; Sasaki, H.; Kimura, S.; Imanishi, Y. *Amino Acids* **1994**, *7*, 311–316.

678. Peters, J.A.; Vijverberg, C.A.M.; Kieboom, A.P.G.; van Bekkum, H. *Tetrahedron Lett.* **1983**, *24*, 3141–3144.

679. Kabuto, K.; Sasaki, Y. *J. Chem. Soc., Chem. Commun.* **1984**, 316–318.

680. Kido, J.; Okamoto, Y.; Brittain, H.G. *J. Org. Chem.* **1991**, *56*, 1412–1415.

681. Hulst, R.; de Vries, N.K.; Feringa, B.L. *J. Org. Chem.* **1994**, *59*, 7453–7458.

682. Inamoto, A.; Ogasawara, K.; Omata, K.; Kabuto, K.; Sasaki, Y. *Org. Lett.* **2000**, *2*, 3543–3545.

683. Liu, Z.; Sayre, L.M. *Tetrahedron* **2004**, *60*, 1601–1610.

684. Nimura, N.; Ogura, H.; Kinoshita, T. *J. Chromatogr.* **1980**, *202*, 375–379.

685. Kinoshita, T.; Kasahara, Y.; Nimura, N. *J. Chromatogr.* **1980**, *210*, 77–81.

686. Nimura, N.; Toyama, A.; Kinoshita, T. *J. Chromatogr.* **1984**, *316*, 547–552.

687. Chan, W.C.; Micklewright, R.; Barrett, D.A. J. Chromatogr., A **1995**, *697*, 213–217.

688. Péter, A.; Tóth, G.; Török, T.; Tourwé, D. *J. Chromatogr., A* **1996**, *728*, 455–465.

689. Péter, A.; Tóth, G.; Olajos, E.; Fülöp, F.; Tourwé, D. *J. Chromatogr., A* **1995**, *705*, 257–265.

690. Péter, A.; Tóth, G.; Tourwé, D. *J. Chromatogr., A* **1994**, *668*, 331–335.

691. Péter, A.; Török, G.; Tóth, G.; Van Den Nest, W.; Laus, G.; Tourwé, D. *J. Chromatogr., A* **1998**, *797*, 165–176.

692. Péter, A.; Olajos, E.; Casimir, R.; Tourwé, D. Broxterman, Q.B.; Kaptein, B.; Armstrong, D.W. *J. Chromatogr., A* **2000**, *871*, 105–113.

693. Péter, A.; Török, G.; Mazaleyrat, J.-P.; Wakselman, M. *J. Chromatogr., A* **1997**, *790*, 41–46.

694. Török, G.; Péter, A.; Csomós, P.; Kanerva, L.T.; Fülöp, F. *J. Chromatogr., A* **1998**, *797*, 177–186.

695. Péter, A.; Lázár, L.; Fülöp, F.; Armstrong, D.W. *J. Chromatogr., A* **2001**, *926*, 229–238.

696. Lobell, M.; Schneider, M.P. *J. Chromatogr.,* **1993**, *633*, 287–294.

697. Stokker, G.E.; Hoffman, W.F.; Homnick, C.F. *J. Org. Chem.* **1993**, *58*, 5015–5016.

698. Davies, J.S.; Mohammed, A.K.A. *J. Chem. Soc., Perkin Trans. 2,* **1984**, 1723–1727.

699. Matsunaga, H.; Santa, T.; Iida, T.; Fukushima, T.; Homma, H.; Imai, K. *Anal. Chem.* **1996**, *68*, 2850–2856.

700. Toyo'oka, T.; Liu, Y.-M. *J. Chromatogr., A* **1995**, *689*, 23–30.

701. Adamson, J.G.; Hoang, T.; Crivici, A.; Lajoie, G.A. *Anal. Biochem.* **1992**, *202*, 210–214.

702. Carpino, L.A.; Ismail, M.; Truran, G.A.; Mansour, E.M.E.; Iguchi, S.; Ionescu, D.; El-Faham, A.; Riemer, C.; Warrass, R. *J. Org. Chem.* **1999**, *64*, 4324–4338.

703. Scaloni, A.; Simmaco, M.; Bossa, F. *Amino Acids* **1995**, *5*, 305–313.

704. Brückner, H.; Kühne, S.; Zivny, S.; Langer, M.; Kaminski, Z.J.; Leplawy, M.T. *Second Forum on Peptides*; Aubry, A., Marruad, M., Vitoux, B., Eds.; John Libbey and Company.: London, 1989, *174*, 291–295.

705. Aberhart, D.J.; Cotting, J.-A.; Lin, H.-J. *Anal. Biochem.* **1985**, *151*, 88–91.

706. Nimura, N.; Kinoshita, T. *J. Chromatogr.* **1986**, *352*, 169–177.

707. García, M.; Serra, A.; Rubiralta, M.; Diez, A.; Segarra, V.; Lozoya, E.; Ryder, H.; Palacios, J.M. *Tetrahedron: Asymmetry* **2000**, *11*, 991–994.

708. Török, G.; Péter, A.; Gaucher, A.; Wakselman, M.; Mazaleyrat, J.-P.; Armstrong, D.W. *J. Chromatogr., A* **1999**, *846*, 83–91.

709. Brückner, H.; Zivny, S. *Amino Acids* **1993**, *4*, 157–167.

710. Harada, K.-i.; Shimizu, Y.; Fujii, K. *Tetrahedron Lett.* **1998**, *39*, 6245–6248.

711. Brückner, H.; Strecker, B. *J. Chromatogr.* **1992**, *627*, 97–105.

712. Buck, R.H.; Krummen, K. *J. Chromatogr.* **1984**, *315*, 279–285.

713. Aswad, D.W. *Anal. Biochem.* **1984**, *137*, 405–409.

714. Florance, J.; Galdes, A.; Konteatis, Z.; Kosarych, Z.; Langer, K.; Martucci, C. *J. Chromatogr.* **1987**, *414*, 313–322.

715. Fitznar, H.P.; Lobbes, J.M.; Kattner, G. *J. Chromatogr., A* **1999**, *832*, 123–132.

716. Brückner, H.; Westhauser, T.; Godel, H. *J. Chromatogr. A*, **1995**, *711*, 201–215.

717. Brückner, H.; Haasmann, S.; Langer, M.; Westhauser, T.; Wittner, R.; Godel, H. *J. Chromatogr., A* **1994**, *666*, 259–273.

718. Brückner, H.; Langer, M.; Lüpke, M.; Westhauser, T.; Godel, H. *J. Chromatogr., A* **1995**, *697*, 229–245.

719. Ehrenfreund, P.; Glavin, D.P.; Botta, O.; Cooper, G.; Bada, J.L. *Proc. Natl. Acad. Sci. U.S.A.* **2001**, *98*, 2138–2141.

720. Bernstein, M.P.; Dworkin, J.P.; Sandford, S.A.; Cooper, G.W.; Allamandola, L.J. *Nature* **2002**, *416*, 401–403.

721. Lam, S. *J. Chromatogr.* **1986**, *355*, 157–164.

722. Einarsson, S.; Folestad, S.; Josefsson, B. *J. Liq. Chromatogr.* **1987**, *10*, 1589–1601.

723. Jegorov, A.; Tríska, J.; Trnka, T. *J. Chromatogr., A* **1994**, *673*, 286–290.

724. Guranda, D.T.; Kudryavtsev, P.A.; Khimiuk, A.Y.; Svedas, V.K. *J. Chromatogr., A* **2005**, *1095*, 89–93.

725. Einarsson, S.; Josefsson, B.; Möller, P.; Sanchez, D. *Anal. Chem.* **1987**, *59*, 1191–1195.

726. Manning, J.M.; Moore, S. *J. Biol. Chem.* **1968**, *243*, 5591–5597.

727. Mitchell, A.R.; Kent, S.B.H.; Chu, I.C.; Merrifield, R.B. *Anal. Chem.* **1978**, *50*, 637–640.

728. Takaya, Y.; Kishida, Y.; Sakakibara, S. *J. Chromatogr.* **1981**, *215*, 279–287.

729. Benoiton, N.L.; Lee, Y.; Liberek, B.; Steinauer, R.; Chen, F.M.F. *Int. J. Pept. Prot. Chem.* **1988**, *31*, 581–586.

730. Benoiton, N.L.; Lee, Y.; Chen, F.M.F. *Int. J. Pept. Prot. Chem.* **1988**, *31*, 443–446.

731. Steinauer, R.; Chen, F.M.F.; Benoiton, N.L. *J. Chromatogr.* **1985**, *325*, 111–126.

732. Görög, S.; Herényi, B.; Lów, M. *J. Chromatogr.* **1986**, *353*, 417–424.

733. Csapó, J.; Tóth-Pósfai, I.; Csapó-Kiss, ZS. *Amino Acids* **1991**, *1*, 331–337.

734. Pugniere, M.; Mattras, H.; Castro, B.; Previero, A. *J. Chromatogr., A* **1997**, *767*, 69–75.

735. Vékes, E.; Török, G.; Péter, A.; Sápi, J.; Tourwé, D. *J. Chromatogr., A* **2002**, *949*, 125–129.

736. Péter, A.; Vekes, E.; Tóth, G.; Toursé, D.; Borremans, F. *J. Chromatogr., A* **2002**, *948*, 283–294.

737. Péter, A.; Árki, A.; Vékes, E.; Tourwé, D.; Lázár, L.; Fülöp, F.; Armstrong, D.W. *J. Chromatogr., A* **2004**, *1031*, 171–178.

738. Brückner, H.; Wachsmann, M. *J. Chromatogr., A* **2003**, *998*, 73–82.

739. Furukawa, H.; Mori, Y.; Takeuchi, Y.; Ito, K. *J. Chromatogr.* **1977**, *136*, 428–431.

740. Goto, J.; Hasegawa, M.; Nakamura, S.; Kazutake, S.; Nambara, T. *J. Chromatogr.* **1978**, *152*, 413–419.

741. Goto, J.; Hasegawa, M.; Nakamura, S.; Shimada, K.; Nambara, T. *Chem. Pharm. Bull.* **1977**, *25*, 847–849.

742. Dunlop, D.S.; Neidle, A. *Anal. Biochem.* **1987**, *165*, 38–44.

743. Goto, J.; Ito, M.; Katsuki, S.; Saito, N.; Nambara,T. *J. Liq. Chromatogr.* **1986**, *9*, 683–694.

744. Nishi, H.; Ishii, K.; Taku, K.; Shimizu, R.; Tsumagari, N. *Chromatographia* **1989**, *27*, 301–305.

745. Iwaki, K.; Yoshida, S.; Nimura, N.; Kinoshita, T.; Takeda, K.; Ogura, H. *Chromatographia* **1987**, *23*, 899–902.

746. Jin, D.; Nagakura, K.; Murofushi, S.; Miyahara, T.; Toyo'oka, T. *J. Chromatogr., A* **1998**, *822*, 215–224.

747. Kleidernigg, O.P.; Lindner, W. *J. Chromatogr., A* **1998**, *795*, 251–261.

748. Péter, M.; Péter, A.; Fölöp, F. *J. Chromatogr., A* **2000**, *871*, 115–126.

749. Coleman, M.W. *Chromatographia* **1983**, *17*, 23–26.

750. Dobashi, A.; Hara, S. *Anal. Chem.* **1983**, *55*, 1805–1806.

751. Dobashi, A.; Dobashi, Y.; Hara, S. *J. Liq. Chromatogr.* **1986**, *9*, 243–267.

752. *Chiral Liquid Chromatography* Lough, W.J., Ed.; Blackie: London, 1989, pp 83–101.

753. Kurganov, A. *J. Chromatogr., A* **2001**, *906*, 51–71.

754. Davankov, V.A. *J. Chromatogr., A* **2003**, *1000*, 891–915.

755. Lam, S.; Chow, F.; Karmen, A. *J. Chromatogr.* **1980**, *199*, 295–305.

756. Lam, S.; Karmen, A. *J. Chromatogr.* **1984**, *289*, 339–345.

757. Lam, S.; Karmen, A. *J. Liq. Chromatogr.* **1986**, *9*, 291–311.

758. Weinstein, S.; Weiner, S. *J. Chromatogr.* **1984**, *303*, 244–250.

759. Lindner, W.; LePage, J.N.; Davies, G.; Seitz, D.E.; Karger, B.L. *J. Chromatogr.* **1979**, *185*, 323–344.

760. LePage, J.N.; Lindner, W.; Davies, G.; Seitz, D.E.; Karger, B.L. *Anal. Chem.* **1979**, *51*, 433–435.

761. Dossena, A.; Galaverna, G.; Corradini, R.; Marchelli, R. *J. Chromatogr., A* **1993**, *653*, 229–234.

762. Corradini, R.; Sartor, G.; Marchelli, R.; Dossena, A.; Spisni, A. *J. Chem. Soc., Perkin Trans. 2*, **1992**, *22*, 1979–1983.

763. Galaverna, G.; Corradini, R.; Dossena, A.; Chiavaro, E.; Marchelli, R.; Dallavalle, F.; Folesani, G. *J. Chromatogr., A* **1998**, *829*, 101–113.

764. Galaverna, G.; Corradini, R.; de Munari, E.; Dossena, A.; Marchelli, R. *J. Chromatogr., A* **1993**, *657*, 43–54.

765. Hare, P.E.; Gil-Av, E. *Science* **1979**, *204*, 1226–1228.

766. Gil-Av, E.; Tishbee, A.; Hare, P.E. *J. Am. Chem. Soc.* **1980**, *102*, 5115–5117.

767. Wagner, J.; Wolf, E.; Heintzelmann, B.; Gaget, C. *J. Chromatogr.* **1987**, *392*, 211–224.

768. Wagner, J.; Gaget, C.; Heintzelmann, B.; Wolf, E. *Anal. Biochem.* **1987**, *164*, 102–116.

769. Washburn, L.C.; Sun, T.T.; Byrd, B.L.; Callahan, A.P. *J. Nucl. Med.* **1982**, *23*, 29–33.

770. Boojamra, C.G.; Lemoine, R.C.; Lee, J.C.; Léger, R.; Stein, K.A.; Vernier, N.G.; Magon, A.; Lomovskaya, O.; Martin, P.K.; Chamberland, S.; Lee, M.D.; Hecker, S.J.; Lee, V.J. *J. Am. Chem. Soc.* **2001**, *123*, 870–874.

771. Galaverna, G.; Corradini, R.; Dallavalle, F.; Folesani, G.; Dossena, A.; Marchelli, R. *J. Chromatogr., A* **2001**, *922*, 151–163.

772. Kaufman, D.B.; Hayes, T.; Buettner, J.; Hammond, D.J.; Carbonell, R.G. *J. Chromatogr., A* **2000**, *874*, 21–26.

773. Weinstein, S.; Engel, M.H.; Hare, P.E. *Anal. Biochem.* **1982**, *121*, 370–377.

774. Pettersson, C.; No, K. *J. Chromatogr.* **1983**, *282*, 671–684.

775. Pettersson, C. *J. Chromatogr.* **1984**, *316*, 553–567.

776. Ravichandran, K.; Rogers, L.B. *J. Chromatogr.* **1987**, *402*, 49–54.

777. Takeuchi, T.; Asai, H.; Ishii, D. *J. Chromatogr.* **1986**, *357*, 409–415.

778. Penn, S.G.; Liu, G.; Bergström, E.T.; Goodall, D.M.; Loran, J.S. *J. Chromatogr., A* **1994**, *680*, 147–155.

779. Daicel Chemical Industries Ltd, www.chiraltech.com.

780. Pirkle, W.H.; Pochapsky, T.C. *Chem. Rev.* **1989**, *89*, 347–362.

781. Gasparrini, F.; Misiti, D.; Villani, C. *J. Chromatogr., A* **2001**, *906*, 35–50.

782. Sellergren, B. *J. Chromatogr., A* **2001**, *906*, 227–252.

783. Andersson, L.; Sellergren, B.; Mosbach, K. *Tetrahedron Lett.* **1984**, *25*, 5211–5214.

784. Andersson, L.I.; Mosbach, K. *J. Chromatogr.* **1990**, *516*, 313–322.

785. Andersson, L.I.; O'Shannessy, D.J.; Mosbach, K. *J. Chromatogr.* **1990**, *513*, 167–179.

786. Sellergren, B.; Lepistö, M.; Mosbach, K. *J. Am. Chem. Soc.* **1988**, *110*, 5853–5860.

787. Ramström, O.; Andersson, L.I.; Mosbach, K. *J. Org. Chem.* **1993**, *58*, 7562–7564.

788. Kempe, M. *Anal. Chem.* **1996**, *68*, 1948–1953.

789. Kempe, M.; Mosbach, K. *J. Chromatogr., A* **1995**, *691*, 317–323.

790. Sajonz, P.; Kele, M.; Zhong, G.; Sellergren, B.; Guiochon, G. *J. Chromatogr., A* **1998**, *810*, 1–17.

791. Kim, H.; Guiochon, G. *J. Chromatogr., A* **2005**, *1097*, 84–97.

792. Chen, Y.; Kele, M.; Quiñones, I.; Sellergren, B.; Guiochon, G. *J. Chromatogr., A* **2001**, *927*, 1–17.

793. Chen, Y.; Kele, M.; Sajonz, P.; Sellergren, B.; Guiochon, G. *Anal. Chem.* **1999**, *71*, 928–938.

794. Sun, R.; Yu, H.; Luo, H.; Shen, Z. *J. Chromatogr., A* **2004**, *1055*, 1–9.

795. Ellwanger, A.; Owens, P.K.; Karlsson, L.; Bayoudh, S.; Cormack, P.; Sherrington, D.; Sellergren, B. *J. Chromatogr., A* **2000**, *897*, 317–327.

796. Kim, H.; Guiochon, G. *Anal. Chem.* **2005**, *77*, 93–102.

797. Huang, X.; Zou, H.; Chen, X.; Luo, Q.; Kong, L. *J. Chromatogr., A* **2003**, *984*, 273–282.

798. Hall, A.J.; Manesiotis, P.; Emgenbroich, M.; Quaglia, M.; De Lorenzi, E.; Sellergren, B. *J. Org. Chem.* **2005**, *70*, 1732–1736.

799. Sibrian-Vazquez, M.; Spivak, D.A. *J. Org. Chem.* **2003**, *68*, 9604–9611.

800. Kim, H.; Spivak, D.A. *Org. Lett.* **2003**, *5*, 3415–3418.

801. Sibrian-Vazquez, M.; Spivak, D.A. *J. Am. Chem. Soc.* **2004**, *126*, 7827–7833.

802. Yu, C.; Mosbach, K. *J. Org. Chem.* **1997**, *62*, 4057–4064.

803. Reddy, P.S.; Kobayashi, T.; Fujii, N. *Chem. Lett.* **1999**, 293–294.

804. Lee, S.-W.; Ichinose, I.; Kunitake, T. *Chem. Lett.* **1998**, 1193–1194.

805. Deore, B.; Chen, Z.; Nagaoka, T. *Anal. Chem.* **2000**, *72*, 3989–3994.

806. Okuno, H.; Kitano, T.; Yakabe, H.; Kishimoto, M.; Deore, B.A.; Siigi, H.; Nagaoka, T. *Anal. Chem.* **2002**, *74*, 4184–4190.

807. Fujii, Y.; Matsutani, K.; Kikuchi, K. *J. Chem. Soc., Chem. Commun.* **1985**, 415–416.

808. Gavioli, E.; Maier, N.M.; Haupt, K.; Mosbach, K.; Lindner, W. *Anal. Chem.* **2005**, *77*, 5009–5018.

809. Chen, Y.; Shimizu, K.D. *Org. Lett.* **2002**, *4*, 2937–2940.

810. Davankov, V.A.; Zolotarev, Yu, A. *J. Chromatogr.* **1978**, *155*, 295–302.

811. Davankov, V.A.; Zolotarev, Yu, A. *J. Chromatogr.* **1978**, *155*, 303–310.

812. Yuki, Y.; Saigo, K.; Tachibana, K.; Hasegawa, M. *Chem. Lett.* **1986**, 1347–1350.

813. Carunchio, V.; Messina, A.; Sinibaldi, M.; Fanali, S. *J. High Res. Chromatog. Chromatog. Commun.* **1988**, 401–404.

814. Galli, B.; Gasparrini, F.; Misiti, D.; Villani, C.; Corradini, R.; Dossena, A.; Marchelli, R. *J. Chromatogr. A*, **1994**, *666*, 77–89.

815. Brückner, H. *Chromatographia* **1989**, *27*, 725–738.

816. Hyun, M.H.; Han, S.C.; Whangbo, S.H. *J. Chromatogr., A* **2003**, *992*, 47–56.

817. Hyun, M.H.; Han, S.C.; Lee, C.W.; Lee, Y.K. *J. Chromatogr., A* **2002**, *950*, 55–63.

818. Chen, Z.; Uchiyama, K.; Hobo, T. *J. Chromatogr., A* **2002**, *942*, 83–91.

819. Remelli, M.; Fornasari, P.; Pulidori, F. *J. Chromatogr., A* **1997**, *761*, 79–89.

820. Sliwka, M.; Slebioda, M.; Kolodziejczyk, A.M. *J. Chromatogr., A* **1998**, *824*, 7–14.

821. Schlauch, M.; Frahm, A.W. *Anal. Chem.* **2001**, *73*, 262–266.

822. Schlauch, M.; Volk, F.-J.; Fondekar, K.P.; Wede, J.; Frahm, A.W. *J. Chromatogr., A* **2000**, *897*, 145–152.

823. Wan, Q.-H.; Shaw, P.N.; Davies, M.C.; Barrett, D.A. *J. Chromatogr., A* **1997**, *765*, 187–200.

824. Wan, Q.-H.; Shaw, P.N.; Davies, M.C.; Barrett, D.A. *J. Chromatogr., A* **1997**, *786*, 249–257.

825. Chen, Z.; Niitsuma, M.; Uchiyama, K.; Hobo, T. *J. Chromatogr., A* **2003**, *990*, 75–82.

826. Dalgliesh, C.E. *J. Chem. Soc.* **1952**, 3940–3942.

827. Pirkle, W.H.; Hyun, M.H.; Bank, B. *J. Chromatogr.* **1984**, *316*, 585–604.

828. Pirkle, W.H.; Pochapsky, T.C In *Chiral Separations: Proceedings of the Chromatographic Society International Symposium on Chiral Separations;* Stevenson, D., Wilson, I.D., Eds; Plenum Press: New York, 1988, pp 23–35.

829. Ôi, N.; Nagase, M.; Inda, Y.; Doi, T. *J. Chromatogr.* **1983**, *265*, 111–116.

830. Ôi, N.; Kithara, H. *J. Chromatogr.* **1984**, *285*, 198–202.

831. Ôi, N.; Kithara, H. *J. Chromatogr.* **1984**, *285*, 198–202.

832. Hara, S.; Dobashi, A. *J. Chromatogr.* **1979**, *186*, 543–552.

833. Dobashi, A.; Oka, K.; Hara, S. *J. Am. Chem. Soc.* **1980**, *102*, 7123–7125.

834. Hara, S.; Dobashi, A.; Kinoshita, K.; Hondo, T.; Saito, M.; Senda, M. *J. Chromatogr.* **1986**, *371*, 53–158.

835. Kurata, K.; Shimoyama, T.; Dobashi, A. *J. Chromatogr., A* **2003**, *1012*, 47–56.

836. Kurata, K.; Ono, J.; Dobashi, A. *J. Chromatogr., A* **2005**, *1080*, 140–147.

837. Lloyd, M.J.B. *J. Chromatogr.* **1986**, *351*, 219–229.

838. Pirkle, W.H.; Pochapsky, T.C.; Mahler, G.S.; Corey, D.E.; Reno, D.S.; Alessi, D.M. *J. Org. Chem.* **1986**, *51*, 4991–5000.

839. Pirkle, W.H.; Pochapsky, T.C. *J. Am. Chem. Soc.* **1986**, *108*, 352–354.

840. Krüger, G.; Grötzinger, J.; Berndt, H. *J. Chromatogr.* **1987**, *397*, 223–232.

841. Sato, K.; Nakano, H.; Hobo, T. *J. Chromatogr., A* **1994**, *666*, 463–470.

842. Caccamese, S.; Failla, S.; Finocchiaro, P.; Hägele, G.; Principato, G. *J. Chem. Res (S)* **1992**, 242–243.

843. Kuropka, R.; Müller, B.; Höcker, H.; Berndt, H. *J. Chromatogr.* **1989**, *481*, 380–386.

844. Forgács, E. *J. Chromatogr., A* **2002**, *975*, 229–243.

845. Lin, C.-E.; Lin, C.-H. *J. Chromatogr., A* **1994**, *676*, 303–309.

846. Lin, J.-Y.; Yang, M.-H. *J. Chromatogr.* **1993**, *644*, 277–283.

847. Xu, M.; Barhmachary, E.; Janco, M.; Ling, F.H.; Svec, F.; Fréchet, J.M.J. *J. Chromatogr., A* **2001**, *928*, 25–40.

848. Hyun, M.H.; Choi, S.Y.; Um, B.H.; Han, S.C. *J. Chromatogr., A* **2001**, *922*, 119–125.

849. Lee, K.-P.; Choi, S.-H.; Kim, S.-Y.; Kim, T.-H.; Ryoo, J.J.; Ohta, K.; Jin, J.-Y.; Takeuchi, T.; Fujimoto, C. *J. Chromatogr., A* **2003**, *987*, 111–118.

850. Murer, P.; Lewandowski, K.; Svec, F.; Fréchet, J.M.J. *Anal. Chem.* **1999**, *71*, 1278–1284.

851. Welch, C.J.; Bhat, G.; Protopova, M.N. *J. Comb. Chem.* **1999**, *1*, 364–367.

852. Wang, Y.; Li, T. *Anal. Chem.* **1999**, *71*, 4178–4182.

853. Bluhm, L.H.; Wang, Y.; Li, T. *Anal. Chem.* **2000**, *72*, 5201–5205.

854. Wu, Y.; Wang, Y.; Ynag, A.; Li, T. *Anal. Chem.* **1999**, *71*, 1688–1691.

855. Brachmachary, E.; Ling, F.H.; Svec, F.; Fréchet, J.M.J. *J. Comb. Chem.* **2003**, *5*, 441–450.

856. Weingarten, M.D.; Sekanina, K.; Still, W.C. *J. Am. Chem. Soc.* **1998**, *120*, 9112–9113.

857. Murer, P.; Lewandowski, K.; Svec, F.; Fréchet, J.M.J. *Chem. Commun.* **1998**, 2559–2560.

858. Pirkle, W.H.; Hyun, M.H. *J. Chromatogr.* **1985**, *322*, 287–293.

859. Griffith, O.W.; Campbell, E.B.; Pirkle, W.H.; Tsipouras, A.; Hyun, M.H. *J. Chromatogr.* **1986**, *362*, 345–352.

860. Ôi, N.; Kitahara, H.; Aoki, F. *J. Chromatogr., A* **1994**, *666*, 457–462.

861. Iwaki, K.; Yamazaki, M.; Nimura, N.; Kinoshita, T. *J. Chromatogr.* **1992**, *625*, 353–356.

862. Iuliano, A.; Attolino, E.; Salvadori, P. *Eur. J. Org. Chem.* **2001**, 3523–3529.

863. Iuliano, A.; Attolino, E.; Salvadori, P. *Tetrahedron: Asymmetry* **2002**, *13*, 1805–1815.

864. Iuliano, A.; Lecci, C.; Salvadori, P. *Tetrahedron: Asymmetry* **2003**, *14*, 1345–1353.

865. Iuliano, A.; Pieroni, E.; Salvadori, P. *J. Chromatogr., A* **1997**, *786*, 355–360.

866. Yamashita, J.; Numakura, T.; Kita, H.; Suzuki, T.; Oi, S.; Miyano, S.; Hashimoto, H.; Takai, N. *J. Chromatogr.* **1987**, *403*, 275–279.

867. Oi, S.; Shijo, M.; Tanaka, H.; Miyano, S.; Yamashita, J. *J. Chromatogr.* **1993**, *645*, 17–28.

868. Oi, S.; Ono, H.; Tanaka, H.; Shijo, M.; Miyano, S. *J. Chromatogr., A* **1994**, *679*, 35–46.

869. Zhong, Q.; Han, X.; He, L.; Beesley, T.E.; Trahanovsky, W.S.; Armstrong, D.W. *J. Chromatogr., A* **2005**, *1066*, 55–70.

870. Kurata, K.; Sakamoto, S.; Dobashi, A. *J. Chromatogr., A* **2005**, *1068*, 335–337.

871. Haginaka, J. *J. Chromatogr., A* **2001**, *906*, 253–273.

872. Stewart, K.; Doherty, R.F. *Proc. Natl. Acad. Sci. U.S.A.* **1973**, *70*, 2850.

873. Allenmark, S.; Bomgren, B.; Borén, H. *J. Chromatogr.* **1982**, *237*, 473–477.

874. Allenmark, S. *J. Liq. Chromatogr.* **1986**, *9*, 425–442.

875. Simek, Z.; Vespalec, R. *J. Chromatogr., A* **1994**, *685*, 7–14.

876. Erlandsson, P.; Hansson, L.; Isaksson, R. *J. Chromatogr.* **1986**, *370*, 475–483.

877. Bomgren, B.; Allenmark, S. *J. Liq. Chromatogr.* **1986**, *9*, 667–672.

878. Allenmark, S.; Bomgren, B.; Borén, H. *J. Chromatogr.* **1984**, *316*, 617–624.

879. Allenmark, S.; Andersson, S. *J. Chromatogr.* **1986**, *351*, 231–238.

880. Allenmark, S.; Andersson, S. *Chromatographia* **1991**, *31*, 429–433.

881. Zhang, Q. Zou, H.; Wang, H.; Ni, J. *J. Chromatogr., A* **2000**, *866*, 173–181.

882. Jacobson, S.C.; Guiochon, G. *Anal. Chem.* **1992**, *64*, 1496–1498.

883. Hofstetter, H.; Hofstetter, O.; Schurig, V. *J. Chromatogr., A* **1997**, *764*, 35–41.

884. Nakamura, M.; Kiyohara, S.; Saito, K.; Sugita, K.; Sugo, T. *Anal. Chem.* **1999**, *71*, 1323–1325.

885. Nakamura, M.; Kiyohara, S.; Saito, K.; Sugita, K.; Sugo, T. *J. Chromatogr., A* **1998**, *822*, 53–58.

886. Peyrin, E.; Guillaume, Y.C.; Guinchard, C. *Anal. Chem.* **1997**, *69*, 4979–4984.

887. Peyrin, E.; Guillaume, Y.C.; Morin, N.; Guinchard, C. *J. Chromatogr., A* **1998**, *8085*, 113–120.

888. Yang, J.; Hage, D.S. *J. Chromatogr., A* **1997**, *766*, 15–25.

889. Mallik, R.; Jiang, T.; Hage, D.S. *Anal. Chem.* **2004**, *76*, 7013–7022.

890. Marle, I.; Karlsson, A.; Pettersson, C. *J. Chromatogr.* **1992**, *604*, 185–196.

891. Hofstetter, O.; Lindstrom, H.; Hofstetter, H. *Anal. Chem.* **2002**, *74*, 2119–2125.

892. Hofstetter, O.; Lindstrom, H.; Hofstetter, H. *J. Chromatogr., A* **2004**, *1049*, 85–95.

893. Ravelet, C.; Boulkedid, R.; Ravel, A.; Grosset, C.; Villet, A.; Fize, J.; Peyrin, E. *J. Chromatogr., A* **2005**, *1076*, 62–70.

894. Michaud, M.; Jourdan, E.; Ravelet, C.; Villet, A.; Ravel, A.; Grosset, C.; Peyrin, E. *Anal. Chem.* **2004**, *76*, 1015–1020.

895. Brumbt, A.; Ravelet, C.; Grosset, C.; Ravel, A.; Villet, A.; Peyrin, E. *Anal. Chem.* **2005**, *77*, 1933–1998.

896. Ward, T.J.; Farris III, A.B. *J. Chromatogr., A* **2001**, *906*, 73–89.

897. Berthod, A.; Liu, Y.; Bagwill, C.; Armstrong, D.W. *J. Chromatogr., A* **1996**, *731*, 123–137.

898. Petritis, K.; Valleix, A. Elfakir, C.; Dreux, M. *J. Chromatogr., A* **2001**, *913*, 331–340.

899. Song, Y.; Shenwu, M.; Zhao, S.; Hou, D.; Liu, Y.-M. *J. Chromatogr., A* **2005**, *1091*, 102–109.

900. Tesarová, E.; Bosáková, Z.; Pacáková, V. *J. Chromatogr., A* **1999**, *838*, 121–129.

901. Tesarová, E.; Bosáková, Z.; Zusková, I. *J. Chromatogr., A* **2000**, *879*, 147–156.

902. D'Acquarica, I.; Gasparrini, F.; Misiti, D.; Villani, C.; Carotti, A.; Cellamare, S.; Muck, S. *J. Chromatogr., A* **1999**, *857*, 145–155.

903. Péter, A.; Török, G.; Armstrong, D.W. *J. Chromatogr., A* **1998**, *793*, 283–296.

904. Schlauch, M.; Frahm, A.W. *J. Chromatogr., A* **2000**, *868*, 197–207.

905. Loukili, B.; Dufresne, C.; Jourdan, E.; Grosset, C.; Ravel, A.; Villet, A.; Peyrin, E. *J. Chromatogr., A* **2003**, *986*, 45–53.

906. Jandera, P.; Backovská, V.; Felinger, A. *J. Chromatogr., A* **2001**, *919*, 67–77.

907. Jandera, P.; Skavrada, M.; Klemmová, K.; Backovská, V.; Giuochon, G. *J. Chromatogr., A* **2001**, *917*, 123–133.

908. Péter, A.; Török, G.; Armstrong, D.W.; Tóth, G.; Tourwé, D. *J. Chromatogr., A* **1998**, *828*, 177–190.

909. Peyrin, E.; Ravelet, C.; Nicolle, E.; Villet, A.; Grosset, C.; Ravel, A.; Alary, J. *J. Chromatogr., A* **2001**, *923*, 37–43.

910. Cavazzini, A.; Nadalini, G.; Dondi, F.; Gasparrini, F.; Ciogli, A.; Villani, C. *J. Chromatogr., A* **2004**, *1031*, 143–158.

911. Berthod, A.; Chen, X.; Kullman, J.P.; Armstrong, D.W.; Gasparrini, F.; D'Acquarica, I.; Villani, C.; Carotti, A. *Anal. Chem.* **2000**, *72*, 1767–1780.

912. Steffeck, R.J.; Zelechonok, Y. *J. Chromatogr., A* **2003**, *983*, 91–100.

913. Péter, A.; Török, R.; Armstrong, D.W. *J. Chromatogr., A* **2004**, *1057*, 229–235.

914. Péter, A.; Árki, A.; Vékes, E.; Tourwé, D.; Forró, E.; Fülöp, F.; Armstrong, D.W. *J. Chromatogr., A* **2004**, *1031*, 159–170.

915. D'Acquarica, I.; Gasparrini, F.; Misiti, D.; Zappia, G.; Cimarelli, C.; Palmieri, G.; Carotti, A.; Cellamare, S.; Villani, C. *Tetrahedron: Asymmetry* **2000**, *11*, 42375–2385.

916. Risley, D.S.; Strege, M.A. *Anal. Chem.* **2000**, *72*, 1736–1739.

917. Medvedovici, A.; Sandra, P.; Toribio, L.; David, F. *J. Chromatogr., A* **1997**, *785*, 159–171.

918. Armstrong, D.W.; Tang, Y.; Chen, S.; Zhou, Y.; Bagwill, C.; Chen, J.-R. *Anal. Chem.* **1994**, *66*, 1473–1484.

919. Vespalec, R.; Corstjens, H.; Billiet, H.A.H.; Frank, J.; Luyben, K.Ch.A.M. *Anal. Chem.* **1995**, *67*, 3223–3228.

920. Péter, A.; Török, G.; Armstrong, D.W.; Tóth, G.; Tourwé, D. *J. Chromatogr., A* **2000**, *904*, 1–15.

921. Berthod, A.; He, B.L.; Beesley, T.E. *J. Chromatogr., A* **2004**, *1060*, 205–214.

922. Liu, Y.; Berthod, A.; Mitchell, C.R.; Xiao, T.L.; Zhang, B.; Armstrong, D.W. *J. Chromatogr., A* **2002**, *978*, 185–204.

923. Maier, N.M.; Schefzick, S.; Lombardo, G.M.; Feliz, M.; Rissanen, K.; Lindner, W.; Lipkowitz, K.B. *J. Am. Chem. Soc.* **2002**, *124*, 8611–8629.

924. Mandl, A.; Nicoletti, L.; Lámmerhofer, M.; Lindner, W. *J. Chromatogr., A* **1999**, *858*, 1–11.

925. Péter, A. *J. Chromatogr., A* **2002**, *955*, 141–150.

926. Oberleitner, W.R.; Maier, N.M.; Lindner, W. *J. Chromatogr., A* **2002**, *960*, 97–108.

927. Lubda, D.; Lindner, W. *J. Chromatogr., A* **2004**, *1036*, 135–143.

928. Piette, V.; Lämmehofer, M.; Lindner, W.; Crommen, J. *J. Chromatogr., A* **2003**, *987*, 421–427.

929. Park, J.H.; Lee, J.W.; Kwon, S.H.; Cha, J.S.; Carr, P.W.; McNeff, C.V. *J. Chromatogr., A* **2004**, *1050*, 151–157.

930. Messina, A.; Girelli, A.M.; Flieger, M.; Sinibaldi, M.; Sedmera, P.; Cvak, L. *Anal. Chem.* **1996**, *68*, 1191–1196.

931. Sandberg, A.; Markides, K.E.; Heldin, E. *J. Chromatogr., A* **1998**, *828*, 149–156.

932. Iuliano, A.; Félix, G. *J. Chromatogr., A* **2004**, *1031*, 187–195.

933. Weng, W.; Wang, Q.H.; Yao, B.X.; Zeng, Q.L. *J. Chromatogr., A* **2004**, *1042*, 81–87.

934. Ward, T.J.; Armstrong, D.W. *J. Liq. Chromatogr.* **1986**, *9*, 407–423.

935. Ward, T.J.; Armstrong, D.W. In *Chromatographic Chiral Separations*, Zief, M. Crane, L.J., Eds; Marcel Dekker: New York, 1988; pp 131–163.

936. Kim, T.-Y.; Kim, H.-J. *J. Chromatogr., A* **2001**, *933*, 99–106.

937. Duchateau, A.L.L.; Heemels, G.M.P.; Maesen, L.W.; de Vries, N.K. *J. Chromatogr.* **1992**, *603*, 151–156.

938. Armstrong, D.W.; DeMond, W. *J. Chromatogr. Sc.* **1984**, *22*, 411–415.

939. Rizzi, A.M.; Cladrowa-Runge, S.; Jonsson, H.; Osla, S. *J. Chromatogr. A*, **1995**, *710*, 287–295.

940. Li, S.; Purdy, W.C. *J. Chromatogr.* **1992**, *625*, 109–120.

941. Berthod, A.; Chang, S.-C.; Armstrong, D.W. *Anal. Chem.* **1992**, *64*, 395–404.

942. Zukowski, J.; Pawlowska, M.; Armstrong, D.W. *J. Chromatogr.* **1992**, *623*, 33–41.

943. Chen, S. *Amino Acids* **2004**, *26*, 291–298.

944. Chen, S. *Amino Acids* **2004**, *27*, 277–284.

945. Ryu, J.W.; Kim, D.W.; Lee, K.-P.; Pyo, D.; Park, J.H. *J. Chromatogr., A* **1998**, *814*, 247–252.

946. Araki, T.; Kashiwamoto, Y.; Tsunoi, S.; Tanaka, M. *J. Chromatogr., A* **1999**, *845*, 455–462.

947. O'Keeffe, F.; Shamsi, S.A.; Darcy, R.; Schwinté, P.; Warner, I.M. *Anal. Chem.* **1997**, *69*, 4773–4782.

948. Shpigun, O.A.; Shapovalova, E.N.; Ananieva, I.A.; Pirogov, A.V. *J. Chromatogr., A* **2002**, *979*, 191–199.

949. Lee, S.H.; Berthod, A.; Armstrong, D.W. *J. Chromatogr.* **1992**, *603*, 83–93.

950. Pawlowska, M.; Chen, S.; Armstrong, D.W. *J. Chromatogr.* **1993**, *641*, 257–265.

951. Sogah, G.D.Y.; Cram, D.J. *J. Am. Chem. Soc.* **1979**, 3035–3042.

952. Shinbo, T.; Yamaguchi, T.; Nishimura, K.; Sugiura, M. *J. Chromatogr.* **1987**, *405*, 145–153.

953. Hirose, K.; Yongzhu, J.; Nakamura, T.; Nishioka, R.; Ueshige, T.; Tobe, Y. *J. Chromatogr., A* **2005**, *1078*, 35–41.

954. Hyun, M.H.; Han, S.C.; Lipshutz, B.H.; Shin, Y.-S.; Welch, C.J. *J. Chromatogr., A* **2001**, *910*, 359–365.

955. Machida, Y.; Nishi, H.; Nakamura, K. *J. Chromatogr., A* **1999**, *830*, 311–320.

956. Machida, Y. Nishi, H.; Nakamura, K.; Nakai, H.; Sato, T. *J. Chromatogr., A* **1998**, *805*, 85–92.

957. Hyun, M.H.; Jin, J.S.; Lee, W. *J. Chromatogr., A* **1998**, *822*, 155–161.

958. Hyun, M.H.; Cho, Y.J.; Kim, J.A.; Jin, J.S. *J. Chromatogr., A* **2003**, *984*, 163–171.

959. Walbroehl, Y.; Wagner, J. *J. Chromatogr., A* **1994**, *685*, 321–329.

960. Jin, J.S.; Stalcup, A.M.; Hyun, M.H. *J. Chromatogr., A* **2001**, *933*, 83–90.

961. Hyun, M.H.; Tan, G.; Xue, J.Y. *J. Chromatogr., A* **2005**, *1097*, 188–191.

962. Seyhan, S.; Özbayrak, Ö.; Demirel, N.; Merdivan, M.; Pirinccioglu, N. *Tetrahedron: Asymmetry* **2005**, *16*, 3735–3738.

963. Krawinkler, K.H.; Maier, N.M.; Sajovic, E.; Lindner, W. *J. Chromatogr., A* **2004**, *1053*, 119–131.

964. Shibata, T.; Okamoto, I.; Ishii, K. *J. Liq. Chromatogr.* **1986**, *9*, 313–340.

965. Fortunately the story can now be found on the internet, with an entry in Wikipedia under "Blind Men and an Elephant," and a version of the story at http://www.robinwood.com/LivingtreeGrove/Stories/StoryPages/Elephant.html

966. Yashima, E.; Yamamoto, C.; Okamoto, Y. *Synlett* **1998**, 344–360.

967. Okamoto, Y.; Yashima, E. *Angew. Chem., Int. Ed. Engl.* **1998**, *37*, 1020–1043.

968. Tachibana, K.; Ohnishi, A. *J. Chromatogr., A* **2001**, *906*, 127–154.

969. Yuasa, S.; Shimada, A.; Kameyama, K.; Yasui, M.; Adzuma, K. *J. Chromatogr. Sci.* **1980**, *18*, 311–314.

970. Gübitz, G.; Jellenz, W.; Schönleber, D. *J. High Res. Chromatog. Chromatog. Commun.* **1980**, *3*, 31–32.

971. Yuasa, S.; Itoh, M.; Shimada, A. *J. Chromatogr. Sci.* **1984**, *22*, 288–292.

972. Garcés, J.; Franco, P.; Oliveros, L.; Minguillón, C. *Tetrahedron: Asymmetry* **2003**, *14*, 1179–1185.

973. Wenslow, R.M., Jr. Jn Wang, T. *Anal. Chem.* **2001**, *73*, 4190–4195.

974. Roff, G.J.; Lloyd, R.C.; Turner, N.J. *J. Am. Chem. Soc.* **2004**, *126*, 4098–4099.

975. Joly, G.D.; Jacobsen, E.N. *J. Am. Chem. Soc.* **2004**, *126*, 4102–4103.

976. Abele, S.; Vögtli, K.; Seebach, D. *Helv. Chim. Acta* **1999**, *82*, 1539–1558.

977. Ye, Y.K.; Stringham, R.W.; Wirth, M.J. *J. Chromatogr., A* **2004**, *1057*, 75–82.

978. Ye, Y.K.; Lord, B.S.; Yin, L.; Stringham, R.W. *J. Chromatogr., A* **2002**, *945*, 147–159.

979. Ye, Y.K.; Stringham, R. *J. Chromatogr., A* **2001**, *927*, 47–52.

980. Ye, Y.K.; Stringham, R. *J. Chromatogr., A* **2001**, *927*, 53–60.

981. McIninch, J.K. Geiser, F.; Prickett, K.B.; May, S.W. *J. Chromatogr., A* **1998**, *828*, 191–198.

982. Ye, Y.K.; Lynam, K.G.; Stringham, R.W. *J. Chromatogr., A* **2004**, *1041*, 211–217.

983. Stringham, R.W. *J. Chromatogr., A* **2005**, *1070*, 163–170.

984. Kraml, C.M.; Zhou, D.; Byrne, N.; McConnell, O. *J. Chromatogr., A* **2005**, *1100*, 108–115.

985. Alías, M.; López, M.P.; Cativiela, C. *Tetrahedron* **2004**, *60*, 885–891.

986. Lasa, M.; López, P.; Cativiela, C. *Tetrahedron: Asymmetry* **2005**, *16*, 4022–4033.

987. Senso, A.; Oliveros, L.; Minguillón, C. *J. Chromatogr., A* **1999**, *839*, 15–21.

988. Franco, P.; Senso, A.; Minguillón, C.; Oliveros, L. *J. Chromatogr., A* **1998**, *796*, 265–272.

989. *Chromatographic Enantioseparation: Methods and Applications*, 2nd edn; Allenmark, S. Ed.; Ellis Horwood: New York, 1991, pp 88–106.

990. Schurig, V. *Asymmetric Synthesis* **1983**, *1*, 59–86.

991. *Chromatographic Enantioseparation: Methods and Applications*, 2nd edn; Allenmark, S. Ed.; Ellis Horwood Limited: New York, 1991, pp 62–66.

992. König, W.A.; Hüthig, A. *The Practice of Enantiomer Separation by Capillary Gas Chromatography*; Verlag: New York, 1987, pp 76–77.

993. König, W.A.; Rahn, W.; Eyem, J. *J. Chromatogr.* **1977**, *133*, 141–146.

994. König, W.A.; Benecke, I.; Schulze, J. *J. Chromatogr.* **1982**, *238*, 237–240.

995. Domergue, N.; Pugniere, M.; Previero, A. *Anal. Chem.* **1993**, *214*, 420–425.

996. Pollock, G.E. *Anal. Chem.* **1972**, *44*, 2368–2372.

997. Gil-Av, E.; Feibush, B.; Charles-Sigler, R. *Tetrahedron Lett.* **1966**, 1009–1015.

998. Feibush, B.; Balan, A.; Altman, B.; Gil-Av, E. *J. Chem. Soc., Perkin Trans. 2* **1979**, 1230–1236.

999. Chang, S.-C.; Charles, R.; Gil-Av, E. *J. Chromatogr.* **1982**, *238*, 29–39.

1000. Bayer, E.; Gil-Av, E.; König, W.A.; Nakaparksin, S.; Oró, J.; Parr, W. *J. Am. Chem. Soc.* **1970**, *92*, 1738–1740.

1001. Oró, J.; Nakaparksin, S.; Lichtenstein, H.; Gil-Av, E. *Nature* **1971**, *230*, 107–108.

1002. Oró, J.; Gibert, J.; Lichtenstein, H.; Wikstrom, S.; Flory, D.A. *Nature* **1971**, *230*, 105–106.

1003. Koenig, W.A.; Stoelting, K.; Kruse, K. *Chromatographia* **1977**, *10*, 444–448.

1004. König, W.A.; Hüthig, A. *The Practice of Enantiomer Separation by Capillary Gas Chromatography*; Verlag: New York, **1987**, pp 25–29.

1005. Frank, H.; Nicholson, G.J.; Bayer, E. *Angew. Chem., Int. Ed. Engl.* **1978**, *17*, 363–365.

1006. Abe, I.; Izumi, K.; Kuramoto, S.; Musha, S. *J. High Res. Chromatog. Chromatog. Commun.* **1981**, 549–552.

1007. Frank, H.; Nicholson, G.J.; Bayer, E. *J. Chromatogr. Sci.* **1977**, *15*, 174–176.

1008. König, W.A.; Hüthig, A. *The Practice of Enantiomer Separation by Capillary Gas Chromatography*; Verlag: New York, 1987, pp 27–30.

1009. Lou, X.; Liu, X.; Zhu, D.; Wang, Q.; Zhou, L. *J. Chromatogr.* **1992**, *626*, 231–238.

1010. Cronin, J.R.; Pizzarello, S. *Science*, **1997**, *275*, 951–955.

1011. Hazen, R.M.; Filley, T.R.; Goodfriend, G.A. *Proc. Natl. Acad. Sci. U.S.A.* **2001**, *98*, 5487–5490.

1012. Fransson, B.; Ragnarsson, U. *Amino Acids* **1999**, *17*, 293–300.

1013. Abe, I.; Fujimoto, N.; Nishiyama, T.; Terada, K.; Nakahara, T. *J. Chromatogr., A* **1996**, *722*, 221–227.

1014. Muñoz Caro, G.M.; Meierhenrich, U.J.; Schutte, W.A.; Barbier, B.; Arcones Segovia, A.; Rosenbauer, H.; Thiemann, W.H.-P.; Brack, A.; Greenberg, J.M. *Nature* **2002**, *416*, 403–406.

1015. del Mar Caja López, M.; Blanch, G.P.; Herraiz, M. *Anal. Chem.* **2004**, *76*, 736–741.

1016. König, W.A.; Steinbach, E.; Ernst, K. *Angew. Chem., Int. Ed. Engl.* **1984**, *23*, 527–528.

1017. König, W.A.; Steinbach, E.; Ernst, K. *J. Chromatogr.* **1984**, *301*, 129–135.

1018. Bolte, T.; Yu, D.; Stüwe, H.-T.; König, W.A. *Angew. Chem., Int. Ed. Engl.* **1987**, *26*, 331–332.

1019. König, W.A.; Krebber, R.; Mischnick, P. *J. High Res.Chromatog.* **1989**, *12*, 732–738.

1020. Wan, H.; Zhou, X.; Ou, Q. *J. Chromatogr., A* **1994**, *673*, 107–111.

1021. Abe, I.; Fujimoto, N.; Nakahara, T. *J. Chromatogr., A* **1994**, *676*, 469–473.

1022. Spanik, I.; Krupcik, J.; Skacáni, I.; Sandra, P.; Armstrong, D.W. *J. Chromatogr., A* **2005**, *1071*, 59–66.

1023. Schurig, V.; Nowotny, H.-P. *Angew. Chem., Int. Ed. Engl.* **1990**, *29*, 939–957.

1024. Di Tullio, A.; D'Acquarica, I.; Gasparrini, F.; Desiderio, P.; Giannessi, F.; Muck, S.; Piccirilli, F.; Tinti, M.O.; Villani, C. *Chem. Commun.* **2002**, 474–475.

1025. McGachy, N.T.; Grinberg, N.; Variankaval, N. *J. Chromatogr., A* **2005**, *1064*, 193–204.

1026. Pfeiffer, J.; Schurig, V. *J. Chromatogr., A* **1999**, *840*, 145–150.

1027. Ruderisch, A.; Pfeiffer, J.; Schurig, V. *J. Chromatogr., A* **2003**, *994*, 127–135.

1028. Tafi, A.; Botta, B.; Botta, M.; Monache, G.D.; Filippi, A.; Speranza, M. *Chem. Eur. J.* **2004**, *10*, 4126–4135.

1029. Issaq, H.J.; Chan, K.C. *Electrophoresis* **1995**, *16*, 467–480.

1030. Ward, T.J. *Anal. Chem.* **1994**, *66*, 633A–640A.

1031. Novotny, M.; Soini, H.; Stefansson, M. *Anal. Chem.* **1994**, *66*, 646A–655A.

1032. Gübitz, G.; Schmid, M. *J. Chromatogr., A* **1997**, *792*, 179–225.

1033. Otsuka, K.; Terabe, S. *J. Chromatogr., A* **2000**, *875*, 163–178.

1034. Vespalec, R.; Bocek, P. *Chem. Rev.* **2000**, *100*, 3715–3753.

1035. Chankvetadze, B.; Blaschke, G. *J. Chromatogr., A* **2001**, *906*, 309–363.

1036. Lämmerhofer, M. *J. Chromatogr., A* **2005**, *1068*, 3–30.

1037. Lämmerhofer, M. *J. Chromatogr., A* **2005**, *1068*, 31–57.

1038. Wan, H.; Blomberg, L.G. *J. Chromatogr., A* **2000**, *875*, 43–88.

1039. Poinsot,V.; Bayle,C.; Couderc,F.*Electrophoresis* **2003**, *24*, 4047–4062.

1040. Poinsot, V.; Lacroix, M.; Maury, D.; Chataigne, G.; Feurer, B.; Couderc, F. *Electrophoresis* **2006**, *27*, 176–194.

1041. Skelley, A.M.; Scherer, J.R.; Aubrey, A.D.; Grover, W.H.; Ivester, R.H.C.; Ehrenfreund, P.; Grunthaner, F.J.; Bada, J.L.; Mathies, R.A. *Proc. Natl. Acad. Sci. U.S.A.* **2005**, *102*, 1041–1046.

1042. Houben, R.J.H.; Gielen, H.; van der Wal, Sj. *J. Chromatogr.* **1993**, *634*, 317–322.

1043. Oguri, S.;Yokoi, K.;Motohase,Y.*J. Chromatogr., A* **1997**, *787*, 253–260.

1044. Ptolemy, A.S.; Tran, L.; Britz-McKibbin, P. *Anal. Biochem.* **2006**, *354*, 192–204.

1045. Wan, H.; Andersson, P.E.; Engström, A.; Blomberg, L.G. *J. Chromatogr., A* **1995**, *704*, 179–193.

1046. Thorsén, G.; Engström, A.; Josefsson, B. *J. Chromatogr., A* **1997**, *786*, 347–354.
1047. Liu, Y.-M.; Schneider, M.; Sticha, C.M.; Toyooka, T.; Sweedler, J.V. *J. Chromatogr., A* **1998**, *800*, 345–354.
1048. Bonfichi, R.; Dallanoce, C.; Lociuro, S.; Spada, A. *J. Chromatogr., A* **1995**, *707*, 355–365.
1049. Vandenabeele-Trambouze, O.; Albert, M.; Bayle, C.; Couderc, F.; Commeyras, A.; Despois, D.; Dobrijevic, M.; Grenier Loustalot, M.-F. *J. Chromatogr., A* **2000**, *894*, 259–266.
1050. Gassman, E.; Kuo, J.E.; Zare, R.N. *Science* **1985**, *230*, 813–814.
1051. Gozel, P.; Gassmann, E.; Michelsen, H.; Zare, R.N. *Anal. Chem.* **1987**, *59*, 44–49.
1052. Chen, Z.; Lin, J.-M.; Uchiyama, K.; Hobo, T. *J. Chromatogr., A* **1998**, *813*, 369–378.
1053. Schmid, M.G.; Rinaldi, R.; Dreveny, D.; Gübitz, G. *J. Chromatogr., A* **1999**, *846*, 157–163.
1054. Karbaum, A.; Jira, T. *J. Chromatogr., A* **2000**, *874*, 285–292.
1055. Zheng, Z.-X.; Lin, J.-M.; Qu, F. *J. Chromatogr., A* **2003**, *1007*, 189–196.
1056. Lu, X.; Chen, Y.; Guo, L.; Yang, Y. *J. Chromatogr., A* **2002**, *945*, 249–255.
1057. Otsuka, K.; Kawakami, H.; Tamaki, W.; Terabe, S. *J. Chromatogr., A* **1995**, *716*, 319–322.
1058. Otsuka, K.; Kashihara, M.; Kawaguchi, Y.; Koike, R.; Hisamitsu, T.; Terabe, S. *J. Chromatogr., A* **1993**, *652*, 253–257.
1059. Otsuka, K.; Kawahara, J.; Tatekawa, K.; Terabe, S. *J. Chromatogr.* **1991**, *559*, 209–214.
1060. Otsuka, K.; Terabe, S. *J. Chromatogr.* **1990**, *515*, 221–226.
1061. Otsuka, K.; Karuhaka, K.; Higashimori, M.; Terabe, S. *J. Chromatogr., A* **1994**, *680*, 317–320.
1062. Otsuka, K.; Kawahara, J.; Tatekawa, K.; Terabe, S. *J. Chromatogr.* **1991**, *559*, 209–214.
1063. Dobashi, A.; Hamada, M.; Dobashi, Y.; Yamaguchi, J. *Anal. Chem.* **1995**, *67*, 3011–3017.
1064. Yarabe, H.H.; Shamsi, S.A.; Warner, I.M. *Anal. Chem.* **1999**, *71*, 3992–3999.
1065. Shamsi, S.A.; Valle, B.C.; Billiot, F.; Warner, I.M. *Anal. Chem.* **2003**, *75*, 379–387.
1066. Swartz, M.E.; Mazzeo, J.R.; Grover, E.R.; Brown, P.R. *Anal. Biochem.* **1995**, *231*, 65–71.
1067. Terabe, S.; Shibata, M.; Miyashita, Y. *J. Chromatogr.* **1989**, *480*, 403–411.
1068. Mohanty, A.; Dey, J. *Chem. Commun.* **2003**, 1384–1385.
1069. Chiari, M.; Desperati, V.; Manera, E.; Longhi, R. *Anal. Chem.* **1998**, *70*, 4967–4973.
1070. Horimai, T.; Arai, T.; Sato, Y. *J. Chromatogr., A* **1999**, *848*, 295–305.
1071. Tran, C.D.; Kang, J. *J. Chromatogr., A* **2002**, *978*, 221–230.
1072. El Rassi, Z. *J. Chromatogr., A* **2000**, *875*, 207–233.
1073. Lloyd, D.K.; Aubry, A.-F.; De Lorenzi, E. *J. Chromatogr., A* **1997**, *792*, 349–369.
1074. Haginaka, J. *J. Chromatogr., A* **2000**, *875*, 235–254.
1075. Birnbaum, S.; Nilsson, S. *Anal. Chem.* **1992**, *64*, 2872–2874.
1076. Sun, P.; Wu, N.; Barker, G.; Hartwick, R.A. *J. Chromatogr.* **1993**, *648*, 475–480.
1077. Desiderio, C.; Fanali, S. *J. Chromatogr., A* **1998**, *807*, 37–56.
1078. Kang, J.-W.; Yang, Y.-T.; You, J.-M.; Ou, Q.-Y. *J. Chromatogr., A* **1998**, *825*, 81–87.
1079. Kang, J.; Bischoff, D.; Jiang, Z.; Bister, B.; Süssmuth, R.D.; Schurig, V. *Anal. Chem.* **2004**, *76*, 2387–2392.
1080. Nair, U.B.; Armstrong, D.W.; Hinze, W.L. *Anal. Chem.* **1998**, *70*, 1059–1065.
1081. Lämmerhofer, M.; Lindner, W. *J. Chromatogr., A* **1999**, *839*, 167–182.
1082. Lámmerhofer, M.; Lindner, W. *J. Chromatogr., A* **1998**, *829*, 115–125.
1083. Lämmerhofer, M.; Zarbl, E.; Lindner, W. *J. Chromatogr., A* **2000**, *892*, 509–521.
1084. Tobler, E.; Lämmerhofer, M.; Lindner, W. *J. Chromatogr., A* **2000**, *875*, 341–352.
1085. Lämmerhofer, M.; Tobler, E.; Lindner, W. *J. Chromatogr., A* **2000**, *887*, 421–437.
1086. Piette, V.; Fillet, M.; Lindner, W.; Crommen, J. *J. Chromatogr., A* **2000**, *875*, 353–360.
1087. Piette, V.; Lindner, W.; Crommen, J. *J. Chromatogr., A* **2000**, *894*, 63–71.
1088. Piette, V.; Lindner, W.; Crommen, J. *J. Chromatogr., A* **2002**, *948*, 295–302.
1089. Fanali, S. *J. Chromatogr., A* **1997**, *792*, 227–267.
1090. Fanali, S. *J. Chromatogr., A* **2000**, *875*, 89–122.
1091. Chankvetadze, B.; Lindner, W.; Scriba, G.K.E. *Anal. Chem.* **2004**, *76*, 4256–4260.
1092. Valkó, I.E.; Sirén, H.; Riekkola, M.-L. *J. Chromatogr., A* **1996**, *737*, 263–272.
1093. Li, S.; Lloyd, D.K. *J. Chromatogr., A* **1994**, *666*, 321–335.
1094. Werner, A.; Nassauer, T.; Kiechle, P.; Erni, F. *J. Chromatogr., A* **1994**, *666*, 374–379.
1095. Ward, T.J.; Nichols, M.; Sturdivant, L.; King, C.C. *Amino Acids* **1995**, *8*, 337–344.
1096. Tang, L.; Silverman, C.E.; Blackwell, J.A. *J. Chromatogr., A* **1998**, *829*, 301–307.
1097. Yoshinaga, M.; Tanaka, M. *J. Chromatogr., A* **1995**, *710*, 331–337.
1098. Johansson, T.; Petersson, M.; Johansson, J.; Nilsson, S. *Anal. Chem.* **1999**, *71*, 4190–4197.
1099. Cavani, L.; Ciavatta, C.; Gessa, C. *J. Chromatogr., A* **2003**, *985*, 463–469.
1100. Simó, C.; Rizzi, A.; Barbas, C.; Cifuentes, A. *Electrophoresis* **2005**, *26*, 1432–1441.
1101. Lu, X.; Chen, Y. *J. Chromatogr., A* **2002**, *955*, 133–140.
1102. Ma, H.-M.; Wang, Z.-H.; Su, M.-H. *J. Chromatogr., A* **2002**, *955*, 125–131.

1103. Müller, A.; Vogt, C.; Sewald, N. *Synthesis* **1998**, 837–841.

1104. Dzygiel, P.; Wieczorek, P.; Jönsson, J.Å. *J. Chromatogr., A* **1998**, *793*, 414–418.

1105. Raatz, D.; Innertsberger, C.; Reiser, O. *Synlett* **1999**, 1907–1910.

1106. Yowell, G.G.; Fazio, S.D.; Vivilecchia, R.V. *J. Chromatogr., A* **1996**, *745*, 73–79.

1107. Shen, J.; Zhao, S. *J. Chromatogr., A* **2004**, *1059*, 209–214.

1108. Glukhovskiy, P.; Vigh, G. *Anal. Chem.* **1999**, *71*, 3814–3820.

1109. Spanik, I.; Vigh, G. *J. Chromatogr., A* **2002**, *979*, 123–129.

1110. Bjergegaard, C.; Hansen, L.P.; Møller, P.; Sørensen, H.; Sørensen, S. *J. Chromatogr., A* **1999**, *836*, 137–146.

1111. Kurosu, Y.; Murayama, K.; Shindo, N.; Shisa, Y.; Satou, Y.; Ishioka, N. *J. Chromatogr., A* **1997**, *771*, 311–317.

1112. Mardones, C.; Ríos, A.; Valcárcel, M.; Cicciarelli, R. *J. Chromatogr., A* **1999**, *849*, 609–616.

1113. Yoshinaga, M.; Tanaka, M. *J. Chromatogr., A* **1994**, *679*, 359–365.

1114. Lindner, W.; Böhs, B.; Seidel, V. *J. Chromatogr., A* **1995**, *697*, 549–560.

1115. Cladrowa-Runge, S.; Rizzi, A. *J. Chromatogr., A* **1997**, *759*, 157–165.

1116. Cladrowa-Runge, S.; Rizzi, A. *J. Chromatogr., A* **1997**, *759*, 167–175.

1117. Cruzado, I.D.; Vigh, G. *J. Chromatogr.* **1992**, *608*, 421–425.

1118. Gahm, K.-H.; Stalcup, A.M. *Anal. Chem.* **1995**, *67*, 19–25.

1119. Corradini, R.; Buccella, G.; Galaverna, G.; Dossena, A.; Marchelli, R. *Tetrahedron Lett.* **1999**, *40*, 3025–3028.

1120. Cai, H.; Nguyen, T.V.; Vigh, G. *Anal. Chem.* **1998**, *70*, 580–589.

1121. Zhu, W.; Vigh, G. *J. Chromatogr., A* **2003**, *987*, 459–466.

1122. Zhu, W.; Vigh, G. *Anal. Chem.* **2000**, *72*, 310–317.

1123. Sarac, S.; Chankvetadze, B.; Blaschke, G. *J. Chromatogr., A* **2000**, *875*, 379–387.

1124. Guo, L.; Lin, S.J.; Yang, Y.F.; Qi, L.; Wang, M.X.; Chen, Y. *J. Chromatogr., A* **2003**, *998*, 221–228.

1125. Ding, W.; Fritz, J.S. *J. Chromatogr., A* **1999**, *831*, 311–320.

1126. Zakaria, P.; Macka, M.; Haddad, P.R. *J. Chromatogr., A* **2004**, *1031*, 179–186.

1127. Francotte, E.; Brandel, L.; Jung, M. *J. Chromatogr., A* **1997**, *792*, 379–384.

1128. Perrin, C.; Vargas, M.G.; Heyden, Y.V.; Maftouh, M.; Massart, D.L. *J. Chromatogr., A* **2000**, *883*, 249–265.

1129. Tang, W.; Muderawan, I.W.; Ng, S.C.; Chan, H.S.O. *J. Chromatogr., A* **2005**, *1094*, 187–191.

1130. Wang, F.; Khaledi, M.G. *J. Chromatogr., A* **1998**, *817*, 121–128.

1131. Tang, W.; Muderawan, W.; Ong, T.-T.; Ng, S.-C. *J. Chromatogr., A* **2005**, *1091*, 152–157.

1132. Tanaka, Y.; Terabe, S. *J. Chromatogr., A* **1997**, *781*, 151–160.

1133. Chiari, M.; Cretich, M.; Crini, G.; Janus, L.; Morcellet, M. *J. Chromatogr., A* **2000**, *894*, 95–103.

1134. Baiss, K.M.; Vreeland, W.N.; Phinney, K.W.; Ross, D. *Anal. Chem.* **2004**, *76*, 7243–7249.

1135. Hutt, L.D.; Glavin, D.P.; Bada, J.L.; Mathies, R.A. *Anal. Chem.* **1999**, *71*, 4000–4006.

1136. Skelley, A.M.; Mathies, R.A. *J. Chromatogr., A* **2003**, *1021*, 191–199.

1137. Thompson, J.E.; Vickroy, T.W.; Kennedy, R.T. *Anal. Chem.* **1999**, *71*, 2379–2384.

1138. Cheng, Y.; Fan, L.; Chen, H.; Chen, X.; Hu, Z. *J. Chromatogr., A* **2005**, *1072*, 259–265.

1139. Schmid, M.G.; Gübitz, G. *J. Chromatogr. A* **1995**, *709*, 81–88.

1140. Gong, X.Y.; Kubán, P.; Tanyanyiwa, J.; Hauser, P.C. *J. Chromatogr., A* **2005**, *1082*, 230–234.

1141. Iványi, T.; Pál, K.; Lázár, I.; Massart, D.L.; Vander Heyden, Y. *J. Chromatogr., A* **2004**, *1028*, 325–332.

1142. Carter-Finch, A.S.; Smith, N.W. *J. Chromatogr., A* **1999**, *848*, 375–385.

1143. Gong, Y.; Lee, H.K. *Anal. Chem.* **2003**, *75*, 1348–1354.

1144. Chen, Z.; Hobo, T. *Anal. Chem.* **2001**, *73*, 3348–3357.

1145. Lämmerhofer, M.; Peters, E.C.; Yu, C.; Svec, F.; Fréchet, J.M.J.; Lindner, W. *Anal. Chem.* **2000**, *72*, 4614–4622.

1146. Lämmerhofer, M.; Svec, F.; Fréchet, J.M.J.; Lindner, W. *Anal. Chem.* **2000**, *72*, 4623–4628.

1147. Hilder, E.F.; Svec, F.; Fréchet, J.M.J. *J. Chromatogr., A* **2004**, *1044*, 3–22.

1148. Lee, H.S.; Hong, J. *J. Chromatogr., A* **2000**, *868*, 189–196.

1149. Schweitz, L.; Andersson, L.I.; Nilsson, S. *J. Chromatogr., A* **1998**, *817*, 5–13.

1150. Quaglia, M.; Sellergren, B.; De Lorenzi, E. . *J. Chromatogr., A* **2004**, *1044*, 53–66.

1151. Li, W.; Fries, D.P.; Malik, A. *J. Chromatogr., A* **2004**, *1044*, 23–52.

1152. Kotake, M.; Sakan, T.; Nakamura, N.; Senoh, S. *J. Am. Chem. Soc.* **1951**, *73*, 2973–2974.

1153. Wilson, I.D.; Ruane, R.J. In *Chiral Separations: Proceedings of the Chromatographic Society International Symposium on Chiral Separations;* Stevenson, D.; Wilson, I.D., Eds.; Plenum Press: New York, 1988, pp 135–143.

1154. Günther, K. *J. Chromatogr.* **1988**, *448*, 11–30.

1155. Günther, K.; Martens, J.; Schickedanz, M. *Angew. Chem., Int. Ed. Engl.* **1984**, *23*, 506.

1156. Günther, K.; Martens, J.; Schickedanz, M. *Fresenius Z. Anal. Chem.* **1985**, *322*, 512–513.

1157. Brinkman, U.A. Th.; Kamminga, D. *J. Chromatogr.* **1985**, *330*, 375–378.

1158. *Chromatographic Enantioseparation: Methods and Applications*, 2nd edn. Allenmark, S. Ed.; Ellis Horwood: New York, 1991, p 272.

1159. Marchelli, R.; Virgili, R.; Armani, E.; Dossena, A. *J. Chromatogr.* **1986**, *355*, 354–357.

1160. Weinstein, S. *Tetrahedron Lett.* **1984**, *25*, 985–986.

1161. Grinberg, N.; Weinstein, S. *J. Chromatogr.* **1984**, *303*, 251–255.

1162. Bhushan, R.; Ali, I. *Chromatographia* **1987**, *23*, 141–142.

1163. Mathur, V.; Kanoogo, N.; Mathur, R.; Narang, C.K.; Mathur, N.K. *J. Chromatogr., A* **1994**, *685*, 360–364.

1164. Alak, A.; Armstrong, D.W. *Anal. Chem.* **1986**, *58*, 582–584.

1165. Huang, M.-B.; Li, H.-K.; Li, G.-L.; Yan, C.-T.; Wang, L.-P. *J. Chromatogr. A*, **1996**, *742*, 289–294.

1166. LeFevre, J.W. *J. Chromatogr., A* **1993**, *653*, 293–302.

1167. *Chromatographic Chiral Separations*, Zief, M., Crane, L.J., Eds.; Marcel Dekker: New York, 1988, pp 152–155.

1168. Kriz, D.; Berggren Kriz, C.; Andersson, L.I.; Mosbach, K. *Anal. Chem.* **1994**, *66*, 2636–2639.

1169. Grigorean, G.; Ramirez, J.; Ahn, S.H.; Lebrilla, C.B. *Anal. Chem.* **2000**, *72*, 4275–4281.

1170. Gua, J.; Wu, J.; Siuzdak, G.; Finn, M.G. *Angew. Chem., Int. Ed. Engl.* **1999**, *38*, 1755–1758.

1171. Koscho, M.E.; Zu, C.; Brewer, B.N. *Tetrahedron: Asymmetry* **2005**, *16*, 801–807.

1172. Zu, C.; Brewer, B.N.; Wang, B.; Koscho, M.E. *Anal. Chem.* **2005**, *77*, 5059–5027.

1173. Sawada, M.; Shizuma, M.; Takai, Y.; Adachi, H.; Takeda, T.; Uchiyama, T. *Chem. Commun.* **1998**, 1453–1454.

1174. Sawada, M.; Takai, Y.; Yamada, H.; Hirayama, S.; Kaneda, T.; Tanaka, T.; Kamada, K.; Mizooku, T.; Takeuchi, S.; Ueno, K.; Hirose, K.; Tobe, Y.; Naemura, K. *J. Am. Chem. Soc.* **1995**, *117*, 7726–7736.

1175. Krishna, P.; Prabhakar, S.; Manoharan, M.; Jemmis, E.D.; Vairamani, M. *Chem. Commun.* **1999**, 1215–1216.

1176. Sawada, M.; Yamaoka, H.; Takai, Y.; Kawai, Y.; Yamada, H.; Azuma, T.; Fujioka, T.; Tanaka, T. *Chem. Commun.* **1998**, 1569–1570.

1177. Tao, W.A.; Zhang, D.; Nikolaev, E.N.; Cooks, R.G. *J. Am. Chem. Soc.* **2000**, *122*, 10598–10609.

1178. Tao, W.A.; Zhang, D.; Wang, F.; Thomas, P.D.; Cooks, R.G. . *Anal. Chem.* **1999**, *71*, 4427–4429.

1179. Tao, W.A.; Clark, R.L.; Cooks, R.G. *Anal. Chem.* **2002**, *74*, 3783–3789.

1180. Yao, Z.-P.; Wan, T.S.M.; Kwong, K.P.; Che, C.-T. *Anal. Chem.* **2000**, *72*, 5394–5401.

1181. Vékey, K.; Czira, G. *Anal. Chem.* **1997**, *69*, 1700–1705.

1182. Hofstetter, O.; Hofstetter, H.; Wilchek, M.; Schurig, V.; Green, B.S. *Nat. Biotechnol.* **1999**, *17*, 371–374.

1183. Nakanishi, T.; Yamakawa, N.; Asahi, T.; Osaka, T.; Ohtani, B.; Uosaki, K. *J. Am. Chem. Soc.* **2002**, *124*, 740–741.

1184. Tran, C.D.; Grishko, V.I.; Oliveira, D. *Anal. Chem.* **2003**, *75*, 6455–6462; correction **2004**, *76*, 2157.

1185. Oguri, S.; Nomura, M.; Fujita, Y. *J. Chromatogr., A* **2005**, *1078*, 51–58.

1186. Voss, K.; Galensa, R. *Amino Acids* **2000**, *18*, 339–352.

1187. Foucault, A.P. *J. Chromatogr., A* **2001**, *906*, 365–378.

1188. Ito, Y. *J. Chromatogr., A* **2005**, *1065*, 145–168.

1189. Delgado, B.; Pérez, E.; Santano, M.C.; Minguillón, C. *J. Chromatogr., A* **2005**, *1092*, 36–42.

1190. Franco, P.; Blanc, J.; Oberleitner, W.R.; Maier, N.M.; Lindner, W.; Minguillón, C. *Anal. Chem.* **2002**, *74*, 4175–4183.

1191. Haupt, K.; Mosbach, K. *Chem. Rev.* **2000**, *100*, 2495–2504.

1192. D'Anna, F.; Riela, S.; Gruttadauria, M.; Lo Meo, P.; Noto, R. *Tetrahedron* **2005**, *61*, 4577–4583.

1193. Pagliari, S.; Corradini, R.; Galaverna, G.; Sforza, S.; Dossena, A.; Marchelli, R. *Tetrahedron Lett.* **2000**, *41*, 3691–3695.

1194. Pagliari, S.; Corradini, R.; Galaverna, G.; Sforza, S.; Dossena, A.; Montalti, M.; Prodi, L.; Zaccheroni, N.; Marchelli, R. *Chem. Eur. J.* **2004**, *10*, 2749–2758.

1195. Liu, Y.; You, C.-C.; Zhang, H.-Y.; Zhao, Y.-L. *Eur. J. Org. Chem.* **2004**, 1411–1422.

1196. Rekharsky, M.; Yamamura, H.; Kawai, M.; Inoue, Y. *J. Am. Chem. Soc.* **2001**, *123*, 5360–5361.

1197. Togrul, M.; Askin, M.; Hosgoren, H. *Tetrahedron: Asymmetry* **2005**, *16*, 2771–2777.

1198. Demirel, N.; Bulut, Y.; Hosgören, H. *Tetrahedron: Asymmetry* **2004**, *15*, 2045–2049.

1199. Demirel, N.; Bulut, Y. *Tetrahedron: Asymmetry* **2003**, *14*, 2633–2637; **2004**, *15*, 2097.

1200. Lee, C.-S.; Teng, P.-F.; Wong, W.-L.; Kwong, H.-L.; Chan, A.S.C. *Tetrahedron* **2005**, *61*, 7924–7930.

1201. Tsubaki, K.; Nuruzzaman, M.; Kusumoto, T.; Hayashi, N.; Bin-Gui, W.; Fuji, K. *Org. Lett.* **2001**, *3*, 4071–4073.

1202. Tsubaki, K.; Tanima, D.; Nuruzzaman, M.; Kusumoto, T.; Fuji, K.; Kawabata, T. *J. Org. Chem.* **2005**, *70*, 4609–4616.

1203. Liu, S.-y.; He, Y.-b.; Qing, G.-y.; Xu, K.-x.; Qin, H.-j. *Tetrahedron: Asymmetry* **2005**, *16*, 1527–1534.

1204. Atwood, J.L.; Dalgarno, S.J.; Hardie, M.J.; Raston, C.L. *Chem. Commun.* **2005**, 337–339.

1205. You, J.-S.; Yu, X.-Q.; Zhang, G.-L.; Xiang, Q.-X.; Lan, J.-B.; Xie, R.-G. *Chem. Commun.* **2001**, 1816–1817.

1206. Oliva, A.I.; Simón, L.; Muñiz, F.M.; Sanz, F.; Morán, J.R. *Org. Lett.* **2004**, *6*, 1155–1157.

1207. Oliva, A.I.; Simón, L.; Muñiz, F.M.; Sanz, F.; Morán, J.R. *Chem. Commun.* **2004**, 426–427.

1208. Oliva, A.I.; Simón, L.; Muñiz, F.M.; Sanz, F.; Ruiz-Valero, C.; Morán, J.R. *J. Org. Chem.* **2004**, *69*, 6883–6885.

1209. Pérez, E.M.; Oliva, A.I.; Hernández, J.V.; Simón, L.; Morán, J.R.; Sanz, F. *Tetrahedron Lett.* **2001**, *42*, 5853–5856.

1210. Chin, J.; Lee, C.S.; Lee, K.J.; Park, S.; Kim, D.H. *Nature* **1999**, *401*, 254–257.

1211. Galardon, E.; Le Maux, P.; Bondon, A.; Simonneaux, G. *Tetrahedron: Asymmetry* **1999**, *10*, 4203–4210.

1212. Hayashi, T.; Aya, T.; Nonoguchi, M.; Mizutani, T.; Hisaeda, Y.; Kitagawa, S.; Ogoshi, H. *Tetrahedron* **2002**, *58*, 2803–2811.

1213. Kim, H.-J.; Asif, R.; Chung, D.S.; Hong, J.-I. *Tetrahedron Lett.* **2003**, *44*, 4335–4338.

1214. Tye, H.; Eldred, C.; Wills, M. *J. Chem. Soc., Perkin Trans. 1* **1998**, 457–465.

1215. Du, C.-p.; You, J.-s.; Yu, X.-q.; Liu, C.-l.; Lan, J.-b.; Xie, R.-g. *Tetrahedron: Asymmetry* **2003**, *14*, 3651–3656.

1216. Davis, A.P.; Lawless, L.J. *Chem. Commun.* **1999**, 9–10.

1217. Lawless, L.J.; Blackburn, A.G.; Ayling, A.J.; Pérez-Payán, M.N.; Davis, A.P. *J. Chem. Soc., Perkin Trans. 1* **2001**, 1329–1341.

1218. Siracusa, L.; Hurley, F.M.; Dresen, S.; Lawless, L.J.; Pérez-Payán, M.N.; Davis, A.P. *Org. Lett.* **2002**, *4*, 4639–4642.

1219. Breccia, P.; Van Gool, M.; Pérez-Fernández, R.; Martin-Santamaría, S.; Gago, F.; Prados, P.; de Mendoza, J. *J. Am. Chem. Soc.* **2003**, *125*, 8270–8284.

1220. Escuder, B.; Rowan, A.E.; Feiters, M.C.; Nolte, R.J.M. *Tetrahedron* **2004**, *60*, 291–300.

1221. Barnhill, D.K.; Sargent, A.L.; Allen, W.E. *Tetrahedron* **2005**, *61*, 8366–8371.

1222. Andrisano, V.; Gottarelli, G.; Masiero, S.; Heijne, E.H.; Pieraccini, S.; Spada, G.P. *Angew. Chem., Int. Ed.* **1999**, *38*, 2386–2388.

1223. Howarth, J.; Al-Hashimy, N.A. *Tetrahedron Lett.* **2001**, *42*, 5777–5779.

1224. Skolau, A.; Rétey, J. *Angew. Chem., Int. Ed.* **2002**, *41*, 2960–2962.

1225. Lin, J.; Rajaram, A.R.; Pu, L. *Tetrahedron* **2004**, *60*, 11277–11281.

1226. Yan, Y.; Myrick, M.L. *Anal. Chem.* **1999**, *71*, 1958–1962.

1227. Lin, J.; Li, Z.-b.; Zhang, H.-c.; Pu, L. *Tetrahedron Lett.* **2004**, *45*, 103–106.

1228. Hernández, J.V.; Oliva, A.I.; Simón, L.; Muñiz, F.M.; Mateos, A.A.; Morán, J.R. *J. Org. Chem.* **2003**, *68*, 7513–7516.

1229. Rossi, S.; Kyne, G.M.; Turner, D.L.; Wells, N.J.; Kilburn, J.D. *Angew. Chem., Int. Ed.* **2002**, *41*, 4233–4236.

1230. Ragusa, A.; Rossi, S.; Hayes, J.M.; Stein, M.; Kilburn, J.D. *Chem. Eur. J.* **2005**, *11*, 5674–5688.

1231. Korbel, G.A.; Lalic, G.; Shair, M.D. *J. Am. Chem. Soc.* **2001**, *123*, 361–362.

1232. Borman, S. *Chem. Eng. News* **2001**, *January 15*, 9.

1233. Hateley, M.J.; Schichl, D.A.; Kreuzfeld, H.J.; Beller, M. *Tetrahedron Lett.* **2000**, 3821–3824.

1234. Zukowski, J.; Pawlowska, M.; Pietraszkiewicz, M. *Chromatographia* **1991**, *32*, 82–84.

1235. Kang, L.; Buck, R.H. *Amino Acids* **1992**, *2*, 103–109.

1236. Fukumoto, K.; Yoshizawa, M.; Ohno, H. *J. Am. Chem. Soc.* **2005**, *127*, 2398–2399.

1237. Fusetani, N.; Warabi, K; Nogata, Y.; Nakao, Y.; Matsunaga, S.; van Soest, R.R.M. *Tetrahedron Lett.* **1999**, *40*, 4687–4690.

1238. Goetz, G.; Yoshida, W.Y.; Scheur, P.J. *Tetrahedron* **1999**, *55*, 7739–7746.

1239. König, W.A.; Benecke, I.; Lucht, N.; Schmidt, E.; Schulze, J.; Sievers, S. *J. Chromatogr.* **1983**, *279*, 555–564.

2 | Synthesis of Racemic α-Amino Acids: Amination and Carboxylation

2.1 Primordial Amino Acids

Discussions on the genesis of life on Earth invariably focus on the chemistry leading to today's biological keystones, proteins and RNA/DNA (*1, 2*). The hypothesis that both amino acids and nucleic acids led to a protein/RNA self-replicating system as the origin of life, with the proteins providing catalysis and the nucleic acids providing information storage, has since been challenged by a purely RNA-based theory (*3–5*). There have been some interesting efforts at catalyzing peptide bond formation via RNA catalysis, which has implications for the presence of a prebiotic "RNA world" before the development of proteins. Some of these developments were highlighted in 1998 (*6*), with further evidence in 2001 (*7*). Other evidence that nucleic acids can template a variety of organic reactions is provided in a 2004 review (*8*). Regardless of whether their presence was necessary for life to begin, amino acids are a requirement for the development of biological life as we know it. Considerable effort has been devoted to proving that amino acids could be synthesized in the "primordial soup" of primitive Earth, although the discovery of amino acids in meteorites (*9–11*) suggests the possibility of an extraterrestrial source. While an early study of amino acids found in the Orgueil, Mokoia, and Murray meteorites appeared to indicate terrestrial contamination due to the presence of mainly L-amino acids (*12*), this was not the case with the Murchison meteorite, which contained predominantly racemic amino acids (*13*). The possible presence of an excess of L-alanine in the Murchison meteorite (*14*) appears to be due to the presence of another amino acid interfering with the enantiomeric excess (ee) analysis (*15*). At least 12 amino acids in the Murchison meteorite (which fell in 1969) (*9*) and seven in the Murray meteorite (a 1950 strike) (*10*) are

non-proteinaceous. Furthermore, the amino acids are enriched in ^{13}C, ^{15}N, and ^2H, indicating extraterrestrial origin (*16, 17*). In 2004 racemic 2,3-diaminopropanoic acid, 2,3- and 2,4-diaminobutyric acid, and 3,3'-diaminoisobutanoic acid and 4,4'-diaminoisopentanoic acid were also identified in the Murchison meteorite (*18*). A 2001 analysis of amino acids in the Orgueil (an 1864 meteorite) and Ivuna (1938) meteorites found primarily Gly, β-Ala, and γ-aminobutyric acid (GABA), along with traces of Ala (racemic), Aib, and isovaline. The composition differs greatly from the Murchison/Murray samples, suggesting a different type of parent body, possibly comets instead of asteroids. Again, enrichment in ^{13}C supports extraterrestrial origin (*19*). A technique for the analysis of stable nitrogen isotope composition of each of a pair of amino acid enantiomers has been developed, which will allow for the assessment of the origins of the organic nitrogen of each enantiomer (*20*). A summary of the possible role of carbonaceous meteorites in the origin of life was published in 2006 (*21*).

The discovery of amino acids outside of Earth would provide further support to extraterrestrial origin theories, though it has recently been suggested that dust particles ejected from earth (or other life-containing planets) could then be propelled from the solar system by radiation pressure, allowing microorganisms to be transferred to protoplanetary systems and "seed" the whole galaxy within a few billion years (*22*). There has been significant debate over conclusions that the spectra of glycine are observed in star-forming interstellar clouds (*23*). A microdevice has been developed to analyze for the presence of amino acids on Mars. The Mars Organic Analyzer (MOA) employs a microfabricated chip containing a portable capillary electrophoresis (CE) instrument that was able to separate and detect amino acids present in soil extracts from the Atacama Desert in

Chile, or in jarosite samples (a sulfate-rich mineral) from the Panoche Valley, California. The instrument possesses parts-per-trillion sensitivity, readily separates amino acids such as Val, Ala, Gly, Asp, and Glu, and potentially could employ chiral CE analysis to detect amino acid optical purity (24). Enantiomeric separation of Ala, Val, Glu, and Asp within 4 min has been demonstrated on a microfabricated CE chip using fluorescein isothiocyanate to derivatize the amino acids and γ-cyclodextrin as the chiral electrophoresis buffer. The test system successfully analyzed extracts from the Murchison meteorite (25). Similarly, hydroxypropyl-β-cyclodextrin was used to separate a mixture of five fluorescamine-labeled amino acids in a microfabricated CE device (26). Indirect fluorescence can be used to detect underivatized amino acids separated on a microchip, using fluorescein in the buffer (27). A new method of analyzing for amino acids, based on electrospray ionization followed by ion mobility spectrometry (the mobilities of the different ionized amino acids in various drift gases), has potential as a simple, robust instrument for interstellar sampling (28). Extraterrestrial amino acid analysis also requires a method to isolate the amino acid from its environment and prepare it for analysis. Simple extraction of terrestrial soil, by shaking with demineralized water or 0.5 M ammonium acetate provided from 40–2000 ng (per g of dry soil) of each of 17 common amino acids (Asn, Gln, and Trp were not tested), demonstrated both the potential for contamination when looking for traces of amino acids and the stability of these ubiquitous building blocks (29). Propanol has been identified as an effective solvent for the extraction of amino acids from soil samples, potentially useful for Martian in situ analysis, since aqueous extractions would contaminate the search for water on Mars (30). Amino acids can also be isolated from natural samples by a simple procedure involving heating with 6 N HCl for 150 °C for 4 h, followed by vacuum drying, then a vacuum sublimation at 250 mTorr in a sealed tube at 450 °C for 30 s or 1100 °C for 5 s. Amino acids were sublimed with 80–100% recovery for Gly, Ala, and Val, but reduced recovery (10–50%) for Asp and Ser. No racemization was observed, and the isolated amino acids do not require desalting before derivatization/analysis (31).

The first proof that terrestrial amino acids could arise from a "primordial soup" came in 1953, when Miller circulated CH_4, NH_3, H_2O, and H_2 past an electric discharge (32, 33). After a week, Gly, Asp, Ala, β-Ala, Sar, Abu, and Aib were identified in the aqueous solution by paper chromatography. Met was also produced if H_2S was added to the mixture (34). Subsequently, a number of other energy sources have been used for similar experiments. A review of early experiments and discussion of energy sources was published in 1959 (35). In 2004 a simulated neutral prebiotic atmosphere of carbon dioxide, nitrogen, and water at 80 °C was subjected to electric discharges from copper electrodes for 2 weeks. Gly and Ala were identified as products (36). Contact glow discharge electrolysis of elemental carbon in the presence of aqueous ammonia produced mainly Gly, but

also Ala, Asp, Glu, Ser, and Thr (37). Asn, which is potentially significant for prebiotic peptide formation as it is the only amino acid known to polymerize without catalysts in aqueous media to give a water-soluble peptide (38), was synthesized by contact glow discharge electrolysis of Ala and formamide (39).

Thermal energy was used to produce at least 13 different amino acids from methane in an ammoniacal solution that was passed through silica gel, quartz sand, volcanic sand, or alumina heated to approximately 1000 °C (40, 41). A later study identified only six predominant amino acids (42). Asp, Thr, Ser, Glu, Gly, Ala, Val, Ile, and Leu were obtained by heating NH_3 and CO with zeolite catalysts at 300 °C for 120 h (43). The reaction of nitriles (malonitrile and succinonitrile) with HCN in aqueous ammonia solution produced Asp and Glu in potentially useful yields (44). Ultraviolet (UV) light, potentially the greatest energy source available on primitive earth (45), was used to irradiate a mixture of CH_4, C_2H_6, NH_3, and H_2O, with H_2S as the initial photon acceptor. Six amino acids, including cystine, were isolated (46, 47). Milder irradiation, using a Xenon lamp or solar radiation, of powdered suspensions of platinized TiO_2 in aqueous ammoniacal solutions saturated with methane produced Gly, Ala, Ser, Asp, and Glu. The TiO_2 absorbs near-UV light, but natural sources of TiO_2 were ineffective (48).

A number of "shock" energy sources have also been proposed. It has been estimated that heavy bombardment of the Earth's early atmosphere by asteroids and comets could not only inject large amounts of organic matter but also provide energy for the synthesis of amino acids (49). High-temperature shock waves have been simulated in an atmosphere of HCN, NH_3, and methane/ethane/ethylene/or isobutane, creating amino acids in surprisingly high yields via a Strecker-type reaction (50–52). Gly, Ala, β-Ala, Val, Leu, Abu, Asp, and Sar were identified (50, 52). Ala and Gly have been synthesized in aqueous solutions of formaldehyde and hydroxylamine using ultrasound waves, an energy source similar to cavitation in seas induced by earthquakes or submarine volcanoes (53). Regardless of energy source, a reducing atmosphere is an absolute requirement for the synthesis of organic compounds. Under oxidative conditions only traces are observed (35).

Studies have also examined other facets of the proposed development of life. Phe, Tyr, Phg, Ala, Val, Ile, Leu, Gly, and Glu were synthesized from the corresponding α-keto acid by simply adding ammonium carbonate and FeS in water at 100 °C (54). Much greater yields were obtained with CO_2 or Na_2CO_3 as catalyst. This process mimics modern amino acid biosynthesis, with the oxidative formation of pyrite from FeS providing a primitive energy source. However, direct reaction of CO_2, H_2S, NH_3, H_2O, and FeS at room temperature or 100 °C failed to produce any amino acids, even if butyrate was added as an alternate carbon source (55). It was noted that ion-exchange resins were the source of significant amino acid contamination. A 2003 study found that the state of FeS was critical to the successful synthesis of amino acids; with freshly precipitated FeS or $Fe(OH)_2$ and NH_3, $MeNH_2$ or Me_2NH and α-keto acids, good yields of Ala, Glu, Phe, and Tyr were produced at 50–75 °C. The reactions were also very sensitive to pH (best yields near the pK_a of the amino source), ionic strength, and large amounts of air (56). Some of these "cold soup" experiments were summarized in a 2000 article (57).

Another theory of the origin of amino acids is that they were synthesized in space, then delivered to earth. A potential interstellar-type synthesis of amino acids has been mimicked by condensing carbon vapor from a carbon arc with ammonia at –196 °C, followed by hydrolysis. Gly, Ala, β-Ala, Sar, Asp, and Ser were isolated (58). UV photolysis of an interstellar ice analog, a 20:2:1:1 frozen mixture of H_2O:MeOH:NH_3:HCN generated by vapour-deposition of gases onto a substrate at 15 K and 10^{-8} torr, produced significant quantitities of Ser, Gly, and Ala. Up to 0.5% carbon yield of Gly was produced (59). A very similar study with a 2:1:1:1:1 H_2O:MeOH: NH_3:CO:CO_2 composition produced 16 identified amino acids, with Gly again the most abundant (60).

Another primordial reaction of interest is the formation of peptides from amino acids. A 1998 study found that dipeptides were created under anaerobic, aqueous, conditions that modeled volcanic or hydrothermal settings, using Phe, Tyr or Gly in the presence of coprecipitated (Ni,Fe)S and CO in conjunction with H_2S as a catalyst at 100 °C and pH 7–10. Unlike previous experiments, dipeptides were formed preferentially over diketopiperazines (61). Further study has demonstrated that the same conditions also cause peptide degradation, suggesting the possibility of a primordial metabolic cycle that can recycle non-functional peptides (62). In 2004 the volcanic gas carbonyl sulfide (COS) was demonstrated to produce peptides from amino acids under mild conditions in aqueous solutions, resulting in di- and tripeptides in yields of up to 75%, depending on the presence of other additives (63). Activated alumina can act as an energy source for Ala dimerization reactions (64), while ferrihydrite (iron oxide hydroxide) has been shown to adsorb many amino acids and promote oligomerization (65). N-carboxyanhydrides (NCAs) have been postulated to be an intermediate for primordial polypeptide formation, and have been shown to activate inorganic phosphate in aqueous solution to form amino acid phosphate mixed anhydrides (66). Evidence for an initial RNA-dependent peptide synthesis mechanism includes the formation of di- and tripeptides from simple aminoacyl phosphate oligonucleotides and an RNA sequence (67).

An alternative theory of macromolecule evolution is that polypeptides are directly synthesized, without the need for an initial stepwise synthesis of amino acids. One study suggests that α-aminonitrile intermediates are unlikely to be converted to amino acids under less than strongly acidic conditions, instead releasing HCN which then polymerizes to give peptides (68). Exposure of aqueous solutions of NaCN, NH_4CN, HCN, acetonitrile or propionitrile to ionizing radiation from a cobalt-60 source resulted in oligomers, which released nine amino acids after hydrolysis. Predominantly Ala, Abu, and Asp were observed (69–71). Similarly, heating diaminomaleonitrile in water at 100 °C for 24 h gave free glycine and a peptide-like material. After hydrolysis 13 amino acids were identified (72). Cyanamide and potassium nitrate were reacted for 13 months at room temperature; few free amino acids were obtained until after hydrolysis of the product, when Asp, Thr, Ser, Glu, Pro, Gly, Ala, Val, Ile, Leu, Tyr, Phe, Lys, His, and Arg were detected (73). However, the possibility of microbial contamination does not seem to have been addressed in this study.

One factor that has yet to be adequately addressed is how the homochirality of biological amino acids arose. Indeed, homochirality is often considered a marker for biological life in the search for extraterrestrial life (74), although a report on enantiomeric excesses of α-methyl-α-amino acids isolated from the Murchison meteorite suggests that the asymmetry indicates a non-biogenic asymmetric influence, rather than indicating proof of biological life (75). Obviously, all of the amino acids produced during models of prebiotic syntheses are racemic. Discussions of current hypotheses have been published (76–78), including a comprehensive review of the origin, control, and amplification of chirality in 1999 (79), highlights of new developments in the origins of homochirality and chiral autocatalysis in 2000 (80, 81), a minireview of asymmetric photochemistry and photochirogenesis in 2002 (82), and a summary of the possible role of carbonaceous meteorites in the origin of homochirality in 2006 (21). Homochirality is proposed to have arisen from three processes: chiral selection (the most difficult step to explain, in which symmetry breaking allows one enantiomer to become predominant); chiral accumulation/amplification; and chiral transmission (in which homochirality in one molecule is transferred to a different type of molecule, e.g., from an amino acid to a different amino acid, or to a sugar) (83). The definition of symmetry breaking was discussed in 2004 (84), while a review of parity violation and how this could lead to cosmic homochirality appeared in 2000 (85). A parity-violating low-temperature difference in phase transition has been observed for crystals of L- and D-Ala (86).

One possible scenario for chiral selection is that neutron star remnants of supernova explosions provide an intense synchrotron radiation source of circularly polarized light (CPL) which could, by selective enantiomer destruction, introduce asymmetry into racemic organic compounds found on dust grains in interstellar clouds (76, 87). Strong infrared CPL has been observed in star formation regions, although UV-energy radiation is required to destroy amino acids (88). The ability of CPL to cause asymmetric photolysis of amino acids has been demonstrated using circularly polarized synchrotron radiation (89–91). Most of these experiments were carried out using solutions of amino acids, but in 2005 solid DL-Leu was exposed to vacuum UV left- or right-circularly polarized synchrotron radiation, with photodecomposition resulting in D-Leu with up to 2.6% ee from *r*-CPSR and L-Leu with 0.88% ee from *l*-CPSR (though, curiously, both *l*- and *r*-CPSR gave enhancement of D-Leu in one matched pair of experiments) (*92*). Though the levels of enantioenrichment are normally very low, CPL has induced high levels of asymmetry in other reactions when combined with asymmetric autocatalysis. Addition of *i*Pr$_2$Zn to pyrimidine-5-carbaldehyde was mediated by a chiral alkene of low ee produced by CPL irradiation, resulting in an alkanol addition product with up to 97% ee (*93*).

Enantioenriched molecules produced by CPL could then be delivered to Earth by accretion as the solar system passes through the gas clouds (*1, 76*), or by impacts of asteroids, comets, and meteorites (*49*). This theory is supported by the presence of non-racemic alanine in the Murchison meteorite; the possibility of terrestrial contamination of the sample has been addressed by the finding that both enantiomers are enriched in ^{13}C (*17*). Enantiomeric excesses of α-methyl-α-amino acids that do not occur naturally in the biosphere (α-Me-Ile, α-Me-Nva, and α-Me-Iva) have also been reported in the Murchison meteorite (*75*). The amino acids were enriched in ^{15}N, again indicating non-terrestrial origin, as the ^{15}N levels were similar to those in interstellar gas clouds (*14*). A potential mechanism for the transfer of chirality from enantiomerically enriched stellar α-methylated amino acids to terrestrial amino acids was demonstrated by the use of (*R*)-α-methylvaline to catalyze the formation of Ala and Phe with up to 10% ee from pyruvate and phenylpyruvate under prebiotic conditions (*94*). Another hypothesis employed homochiral α-methylated amino acid oxazolones to form polypeptides with proteinogenic α-amino acids, examining the extent of chirality transfer to the subsequent peptide formed (*95*). A quantum chemistry mechanism for the evolution of biomolecular homochirality from a possible precursor molecule, aziridine-2-carbonitrile, has been explored (*96*).

A new theory for the emergence of biochemical homochirality is based on selective adsorption of L- and D-amino acids onto calcite (*97*). Calcite crystal surfaces were probably widely abundant in prebiotic environments. The crystals possess surface structures related by mirror symmetry and are known to bind amino acids. Indeed, when calcite crystals were immersed in a solution of racemic Asp for 24 h, minor but consistent alterations were observed in the D:L ratios of the Asp subsequently washed from mirror image surfaces (up to 10% chiral excess). However, an explanation of how this would lead to the biological predominance of L-amino acids is still lacking. In a reversal of the direction of transfer of chiral information, addition of D- or L-Glu has been shown to influence formation of right- or left-handed inorganic helical crystals (*98*). In another study, scanning tunneling microscopy of racemic Cys adsorbed to a gold surface (the thiol group of Cys binds to gold with high affinity) showed molecular dimers were formed with LL and DD pairing, with no DL structures (*99*). Leu covalently attached to a self-assembled monolayer induces enantioselective crystal growth of Leu that is

dependent on the chirality of the attached enantiomer (*100*). One difficulty with theories based on adsorption of amino acids to minerals is that minerals more readily adsorb amino acids with charged side chains, while the majority of amino acids have uncharged side chains. Mechanisms involving lipids may play a role, as reviewed in 2004 (*101*). Recrystallization of DL-Asn results in individual crystals with ee values ranging from 60 to 89% ee. When cocrystallized with other amino acids, the Asn also induced the other amino acid to crystallize with an ee value similar to that of the Asn (*102*).

Mechanisms of chiral amplification, by which slight enantiomeric excesses are converted into near-homochiral molecules, have been proposed. Analogs of nylon polymers can be induced to form populations of either exclusively left- or right-handed helices by employing a chiral pendant group containing as little as 12% ee (*103*). In another study, it was found that a 32-residue peptide replicator template (designed to form intermolecular α-helix complexes) produced homochiral products from a racemic mixture of nucleophilic and electrophilic peptide fragments (15-mer and 17-mer, respectively) through a chiroselective autocatalytic cycle. D-Amino acid mutations in the template still produced homochiral products, demonstrating an error-correction mechanism. The paper demonstrates how an initially small enantiomeric excess can be amplified into a much larger monochiral pool (*104*). A chiroselective polymerization of pyranosyl RNA analogs demonstrates another possible route to homochiral polymers from initially low enantiomeric excesses. By polymerizing large molecules with large constitutional diversity it is possible that the number of theoretical combinations greatly exceeds the number of actually formed constitutions. The probability that the same sequences would occur in both D- and L-libraries would therefore be unlikely to be the same; hence, the combined libraries would no longer be racemic (*105*).

Another proposal is that two-dimensional (2D) crystalline self-assemblies can lead to chiral amplification of oligopeptides. Racemic Lys residues were side-chain acylated with a fatty acid and activated as thioethyl esters, and then spread on a water surface. Grazing incidence X-ray diffraction showed the presence of crystalline structures on the water surface. Polycondensation was initiated by adding I$_2$/KI or AgNO$_3$, with the racemic monomers showing a clear trend towards enhanced formation of homochiral sequences within diastereomeric peptides. With a non-racemic mixture of monomer, there was a distinct enhancement of the homochiral fraction of oligopeptide (*106–110*). γ-Stearyl-Glu and $N^ε$-stearoyl-Lys also showed homochiral oligopeptide formation under these conditions; chiral amplification was observed in polymerization of γ-stearyl-Glu *N*-carboxyanhydride (NCA) (*111*). Racemic mixtures of oligopeptides with homochiral sequences were generated by polymerization of Phe (NCA), again induced by a 2D crystallite monolayer at an air–water interface (*112*). Similarly, racemic or moderately enriched (20% ee) NCAs or of Leu or Glu oligomerized in the presence of quartz or hydroxylapatite to give oligopeptides with moderate to high homochirality (*113, 114*). Chiral amplification was also seen during Trp NCA polymerization in the presence of liposomes (*115*). A heterogeneous polymerization of racemic crystals of Phe NCA in hexane produced a series of oligopeptides, with oligopeptides longer than pentamers consisting primarily of homochiral sequences (*116*). Peptide formation from amino acids, induced by high salt concentrations (SIPF, salt-induced peptide formation), has been shown to preferentially produce L,L-dipeptides over the D,D-enantiomer (*117*). In contrast, competitive activated couplings of N-acyl derivatives of Gly, Ala, Val, Pro or Phe with mixtures of esters and amides of the same amino acids led to a significant preference for heterochiral outcomes, e.g., L,D-dipeptides (*118*).

Serine has been proposed to be a key player in chiral accumulation and chiral transmission. Small clusters of serine molecules undergo chiral selectivity (*119*), and can form homochiral octamer clusters when ionized (generally by electrospray ionization) (*120–122*) that react enantioselectively with racemic glyceraldehyde to form L-Ser:D-sugar clusters (*83, 123*). Two different structures of the Ser octamer appear to exist, with the more fragile form responsible for the strong chiral effects induced by these clusters (*124*). Chiral transmission from Ser to other amino acids substituted within the octamer complex has also been observed; a solution of 0.01 M L-Ser and L-Cys showed formation of octamer clusters containing one or two Cys substitutions, while a solution of 0.01 M D-Ser and L-Cys showed negligible substitutions (*125*). A sonic spray ionization source has been developed to generate amino acid clusters, with homochiral serine octamer clusters again observed. Thr and Cys could be substituted for more then two Ser molecules within the homochiral cluster (*126*). Vigorous evaporation of Ser solutions or pyrolysis of L-Ser crystals in a corona discharge also generates Ser octamer ions, a result of relevance for prebiotic scenarios as this method provides a more natural route for ion generation (*122*). The formation and reactions of serine octamer clusters, and their implications for biomolecule homochirality, were reviewed in 2006 (*127*). Protonated serine dimers have also been generated and studied by mass spectrometry and ab initio quantum chemical calculations (*128*).

Once non-racemic amino acids have been created, they are then capable of catalyzing asymmetric syntheses of other biomolecules, providing another mechanism of chiral transmission and amplification. Amino acids can catalyze the asymmetric formation of carbohydrates (*129, 130*), such as threose and erythrose from glycolaldehyde and formaldehyde (*131*). Pro has been found to catalyze an asymmetric synthesis of polyketide sugars from simple alkyl aldehydes (*132*), while acyclic amino acids can catalyze intermolecular aldol reactions with high stereoselectivity (*133*). The possibility of chiral amplification during autocatalysis was demonstrated in a chemical reaction in 1995, in the autocatalytic alkylation of pyrimidyl aldehydes with dialkylzincs (*134*).

The alcohol product accelerates the reaction rate, with a small amount of very low enantiomeric excess catalyst yielding catalyst with very high ee as product. Potential mechanisms for how this amplification occurs, and its implications for the origin of homochirality, were discussed in 2004 (*135, 136*), as was a theoretical discussion on symmetry-breaking via reversible chemical reactions (*137*). Developments in asymmetric autocatalysis were highlighted in 2005 (*138*). A proline-mediated α-aminoxylation of aldehydes also shows an amplification effect, with reaction rate and product enantiomeric excess both increasing over the course of the reaction (*139*). Poly-Leu made from Leu of varying enantiomeric purity (ranging from 0 to 100% ee) catalyzed an asymmetric epoxidation of 1,3-diphenylprop-2-enone with an amplification of chirality, with peptide made from 4.8% ee Leu giving epoxide with 26% ee, and peptide of 20% ee giving 73% ee epoxide (*140*).

Theories of how the relationship between the coding of nucleotide bases and the amino acids evolved have also been developed (*141*). A series of publications have explored the possibility of a linked prebiotic origin of RNA and coded peptides, using aminoacyl-RNA trimers as intermediates (*142–145*). A similar hypothesis suggested that if amino acids were initially synthesized by covalent complexes of catalytic dinucleotides with α-keto acids, it could create an association between the amino acids and the first two codon bases that was retained upon the evolution of translation (*146*).

A 1999 review of the origins of life addresses many issues related to prebiotic synthesis, in addition to other fundamental questions such as the development of self-replication and primitive catalysis (including the role of His) (*147*). A review of non-traditional pathways of extraterrestrial formation of organic compounds was published in 1997 (*148*), as was a review of prebiotic chemistry from a bioorganic perspective (*149*). An earlier (1974) review looks at syntheses of amino acids and peptides under possible prebiotic conditions (*150*). In 2000 a brief overview of the possible methods of formation of organic compounds on Earth was published (*151*), along with a monograph on perspectives in amino acid and protein geochemistry, which contains a section on extraterrestrial amino acids and origins of life (*152*). A 2001 article provides a brief review of state-of-the-art instruments for detecting extraterrestrial life, emphasizing the importance of detecting both amino acids and amino acid homochirality, and describes systems designed for the Mars Organic Detector instrument package (*153*). A review in 2004 provides an overview of the synthesis of amino acids and incorporation into homochiral peptides in a primordial atmosphere (*154*).

2.2 Addition of the Amino and Carboxy Groups to the Side Chain (Aminocarboxylation Reactions)

Aminocarboxylation reactions potentially provide a rapid method of amino acid synthesis from inexpensive components. The side chain and central carbon are usually supplied by an aldehyde component, while various sources can be employed for the amino and carboxyl groups. A brief review of aminocarboxylations was published in 1997 (155).

2.2.1 Strecker Synthesis

The first reported synthesis of an amino acid was in 1850 when Strecker, attempting to prepare lactic acid from acetaldehyde, instead produced alanine by treatment of acetaldehyde with ammonia followed by HCN (156). The intermediate α-aminonitrile can be hydrolyzed under basic or acidic conditions, although basic conditions are preferred. Thus both the amino and carboxy functions are introduced to the side chain provided by the aldehyde (see Scheme 2.1, upper path). A review of the synthesis and properties of α-aminonitriles was published in 1989 (157), with the biology, preparations, and synthetic applications of cyanohydrins extensively reviewed in 1999 (158). The Strecker synthesis is proposed as a route for the generation of amino acids found in extraterrestrial sources (19). The Strecker method was used for the first reported syntheses of a number of amino acids, such as Leu (159), Phe (160), and Ser (160). Improvements to the original procedure, still using HCN and NH₃, were reported in 1931 and applied to the synthesis of Ala (72% yield), Abu (61%), Val (68%), Aib (73%), α-Et-Ala (74%), and α-Me-Asp (51%) (161). The gaseous reagents have also been used in syntheses of β-(1-pyrazolyl)alanine (162), 2,5-diamino-3,4-dihydroxyadipic acid (12% yield) (163), β,γ-dihydroxy-α-aminobutyric acid (30% yield) (164), and Met and analogs (165).

In 1880 Tiemann suggested that the order of reagent addition could be reversed, so that a cyanohydrin is aminated (Scheme 2.1, lower path) instead of adding cyanide to a hydroxy amine or imine (166). A number of cyanohydrins were isolated from the reaction of aldehydes and anhydrous HCN (167) and one was reacted with aniline to give an α-aminonitrile. This method was used to prepare some allenic amino acids in low yields (168), in addition to Met (169), Aib and diethylglycine (170), and Tyr(Me) and α-Me-Phg (171).The original Strecker preparation required gaseous HCN and NH₃. However, a number of modified procedures have substituted more easily handled reagents. KCN and NH₃ were used for the first syntheses of Ile (172) and isovaline

(Iva, α-ethyl-Ala) (173, 174), and, more recently, 1-aminocyclopropanecarboxylic acid (175) and phosphinothricin, 2-amino-4-(p-methyl-phosphinyl)butyric acid (50–75% yield) (176). Isotopically labeled K¹³CN and NH₄OH were used to prepare [1-¹³C]-Val (3 g scale, 83% yield) (177) and [1-¹⁴C]-Leu (178). HCN was employed for addition to an imine generated intramolecularly during conversion of 1-aminomethyl-1-(2-ethanal)-cyclohexane into 4,4-spirocyclohexylproline (179).

Both reagents were used as alkali salts (KCN and NH₄Cl) for the synthesis of Ala, Abu, Aib, Phe, and a number of α,α-cyclic amino acids (180, 181), as well as the first synthesis of Met (182, 183). The alkali salt reagents are commonly used for more recent Strecker syntheses, such as an *Organic Syntheses* preparation of Ala (184) and preparations of 2-aminooctanoic acid (185), *trans*-α-(carboxycyclopropyl)Gly (186), Phg (98), 3′-carboxy-Phg (25% yield) (187), Ser (188), 4-hydroxy-Tle (189), vinylglycine (only 1.1% yield) (190), 3,4-didehydro-Ile (191, 192), and willardiine, a β-uracil-Ala derivative (193). Ketones have been used for α,α-disubstituted amino acids such as isovaline (α-methyl-Abu) (30% yield) (194, 195), α-methyl-Nva (39%) (196), α-methyldopa (>80%) (197), α-methyl-Glu (198), α-methyl-Ser and α-methyl-Ser homologs (199), α-methyl-β-fluoro-Ala (200), α-methyl-cyclopropylglycine (199), an *Organic Syntheses* preparation of α-methyl-Phg (201), α-methyl-Ala, -Nle, -Leu, and -α-aminooctanoic acid (202), diethylglycine (203), and a number of 4-aminopiperidine-4-carboxylic acid derivatives (204). A variety of cyclic and bicyclic amino acids (205–209) have been synthesized by both the Strecker and Bucherer syntheses, with the two routes producing different diastereomers (see end of Section 2.2.2 for details). Isotopically labeled amino acids, in particular ¹¹C/¹³C/¹⁴C carboxyl-labeled amino acids, are readily synthesized by using labeled cyanide. The first synthesis of [1-¹⁴C]-Ala used this method (210), as did a 1990 synthesis of [1-¹¹C]-Ala (211). Labeled amino acids such as [1-¹³C]-Glu (85% yield) and [1-¹³C]-Asp (45%) (212), various [¹³C]-labeled Leu derivatives (23–51%) (213), [2-¹³C,¹⁵N]-Asp (214, 215), and [2-¹³C,¹⁵N]-4′-HO-Phg (216) have also been prepared by the alkali salt modification of the Strecker procedure. Both [¹⁵N]-Aib and [¹⁵N]-isovaline were prepared by Strecker syntheses using [¹⁵N]NH₄Cl and NaCN. The Iva was obtained in good yield with high isotope incorporation, but initial attempts at the Aib synthesis found low levels

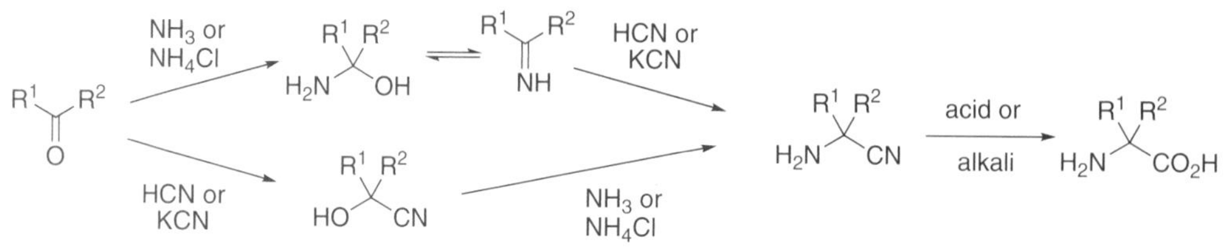

Scheme 2.1 Strecker Synthesis.

of aminonitrile intermediate (20%) with >75% of this unlabeled. The cyanohydrin intermediate appeared to be undergoing hydrolysis, releasing unlabeled ammonia in the process. By adding dichloromethane to the reaction, good yields of α-aminoisobutyronitrile with quantitative isotope incorporation were obtained (217).

A number of variations to reaction conditions have been developed to improve yields. Alcoholic ammonium cyanide was found to produce the α-aminonitrile intermediate when a number of other conditions failed (218). Alumina and ultrasound improved reactivity during a Strecker synthesis (KCN and NH₄Cl) of an amino acid with a side chain containing a phosphonate-substituted cyclopropyl ring (219), for a biphenylglycine derivative (220), and for a synthesis of 2-amino-[3.1.1] bicycloheptane-2,5-dicarboxylic acid (221). Ultrasound with alumina also facilitated the synthesis of α-aminonitriles using KCN in organic solvents, giving very high yields (82–100%) for a range of alkyl and aryl aldehydes and ketones (222). Ultrasound increased the yield of aminonitrile from 60–79% to 89–100% for the synthesis of a cyclic amino acid containing a benzoquinolizin-2-one nucleus, at the same time decreasing the reaction time from 12 days to 20–35 h (223). When aldehydes have low reactivity, sodium bisulfite can be used to prepare an α-hydroxy-sulfite intermediate; the sulfhite is then displaced with cyanide and the hydroxy with ammonia to give the usual α-aminonitrile intermediate (see Scheme 2.2). Gaudry compared this method with the original Strecker synthesis, the alkali salt method, and the Bücherer synthesis (see below) for the preparation of Val, with yields of 53%, 65%, 58%, and 65%, respectively) (224). The bisulfite method was used to prepare S-benzyl-Cys in 38% yield when the Bücherer synthesis failed (225), and has also been used to prepare 2,3-diaminopropionic acid (226), 3′-(phosphonomethyl)-Phg (227), 2′-(phosphonomethyl)-homoPhe (227), 2′-(phosphonomethyl)-bis(homo)Phe (227), [1-¹³C]-Leu (228), [1-¹³C]-Ile (41% yield) (228), [1-¹³C]-Val (40–60% yield) (228) (and, by use of ¹⁵NH₄OH, the corresponding isotopic N analogs) (228), S-benzyl-[1-¹³C]-Cys (41% yield) (229), and [1,1′-¹³C₂]-Cys₂ (31% yield) (229).

Substituted amines have also been incorporated into Strecker-type syntheses, allowing the preparation of N-substituted amino acids (see Scheme 2.3). Sarcosine and N-methylsarcosine were prepared in 1894 by addition of methylamine or dimethylamine to formaldehyde-derived cyanohydrin, followed by hydrolysis (230), while N-methyl-Phg was prepared in 1881 (231) and again in 1964 (232), and N-methyl-Aib in 1973 (204). Similarly, a number of cyclic N-arylaminonitriles were prepared from cyclohexanone, KCN, and aromatic amines (233). More recently, N-alkyl (N-alkyl = Me, Et, nPr) amino acids (side chain = Me, Et, nPr, iPr, nPr, iBu) were synthesized in 23–55% by this route (234). The unsubstituted amino acid can be obtained if the N-substitutent is subsequently removed, generally by hydrogenolysis of an N-benzyl substituent. Benzylamine was used in the synthesis of 3-aminooxetane-3-carboxylic acid (27% yield), 3-aminoazetidine-3-carboxylic acid (43%), and 1-amino-3-phosphonocyclobutyl-1-carboxylic acid (84%) from the corresponding cyclic ketones (235, 236), while dimethoxybenzylamine was used to prepare styrylglycine (45%) (237). Benzylamine has also been used in a modified Strecker reaction in which the α-benzylaminonitrile intermediate was prepared under non-aqueous conditions, facilitating the conversion of the normally insoluble organic carbonyl compounds. High yields (76–93%) were obtained for a number of examples (238).

5,5-Dimethyl-Pro was constructed by intramolecular imine formation from 1-amino-1,1,-dimethylbutanal, followed by reaction with KCN, then hydrolysis (239). A substituted amine was found to be an absolute requirement for a Strecker synthesis of 1-aminocyclopropanecarboxylic acid from cyclopropanone hemiacetal. Only polymeric material was obtained with ammonium chloride as the amine source, while methylbenzylamine gave an overall yield of 74% when used with NaCN and sonication (240). Iminodicarboxylic acids, N-(CH₂CO₂Et)-Ala (241), N-(CHMeCO₂Et)-Abu (242), and N-(CHMeCO₂Et)-Ala (243), were synthesized in 1907, using the Ala precursor α-aminopropionitrile or Abu as the amine component. A one-step synthesis of aminonitriles from amines, aldehydes/ketones, and KCN

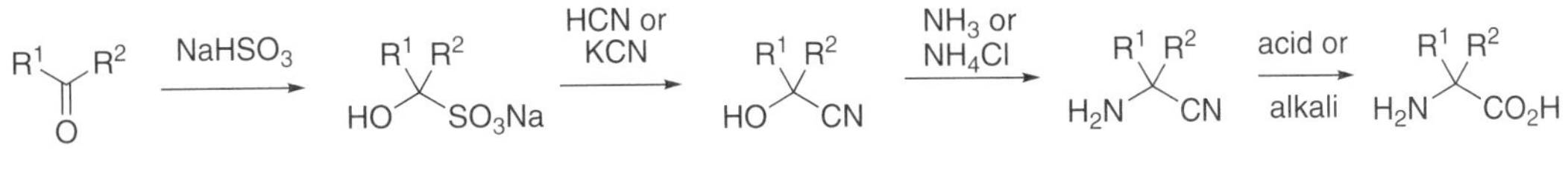

Scheme 2.2 Bisulphite Modification of Strecker Synthesis.

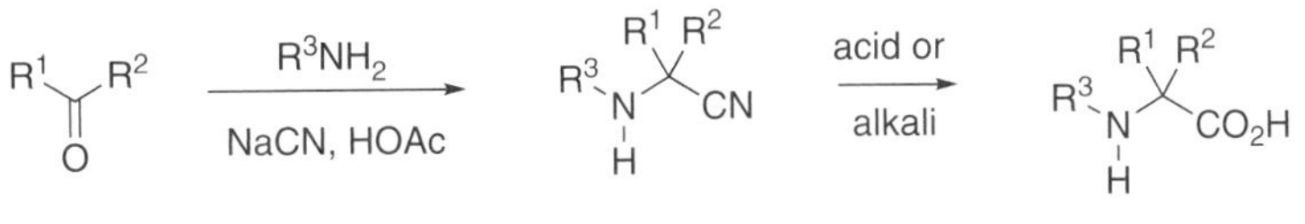

Scheme 2.3 Strecker Synthesis of N-substituted Amino Acids.

was catalyzed by indium trichloride in 2002, with 47–93% yields for 22 examples (*244*).

A more recent variation of the Strecker synthesis uses cyanotrimethylsilane (TMSCN) as the cyano ion source. Two synthetic routes to the α-aminonitrile precursors of amino acids are possible: addition of TMSCN to aldehydes followed by reaction with an amine, or addition of TMSCN to an imine prepared from an aldehyde and amine (Scheme 2.4).

The first route was used to prepare a number of α-aminonitriles in good yield (15–99%, generally >90%). Aldehydes were treated with TMSCN in the presence of catalytic ZnI to give silyloxynitrile intermediates, which were then reacted with various amines in methanol. Eighteen different aldehydes were combined with ammonia, benzylamine, dimethylamine or tetrahydropyrrole; ketones (three examples) reacted in much lower yield (*245*). No catalyst was needed for a similar reaction in which a primary amine was added to a mixture of aldehyde and TMSCN (*246*). The reactants were used in stoichiometric quantities and the self-catalyzed reaction proceeded at room temperature. Both aryl and alkyl amines were used. 3′,5′-Dihydroxy-4′-methoxy-Phg was prepared in good overall yield by this route (*247*).

The second route has been used much more frequently, although it suffers from an instability of the imine intermediates. Schiff bases and oximes were converted to α-aminonitriles in 66–77% yield (11 examples) by addition of TMSCN in the presence of a Lewis acid such as AlCl₃, ZnI₂, or Al(OR)₃ (*248*). A dimethoxybenzhydrylimine intermediate was used to prepare a number of β,γ-unsaturated amino acids, with yields of 7–68%

(*249, 250*). Many catalysts have been described for the addition of TMSCN to imines; the examples usually prepare α-aminonitriles but not amino acids. A lanthanide triflate, Yb(OTf)₃, effectively catalyzed the reaction of TMSCN with imines (94–100%), and also generated α-aminonitriles in excellent yield via a three-component reaction of aldehydes, amines, and TMSCN (84–92% for R¹ = Ph, α-Nap, c-C₆H₁₁, nPr, iPr, iBu; amine R² = MeOPh or Ph₂CH) (*251*). Cu(OTf)₂ was reported to be more effective than Yb(OTf)₃ at catalyzing the addition of TMSCN to various aldehydes and ketones, including aryl aldehydes for which the Yb catalyst was ineffective (*252*). In 2005 vanadyl triflate was described as an efficient catalyst of aminonitrile formation, with 74–95% yields of aminonitriles from TMSCN, aldehydes, and amines after 7–14 h at room temperature (10 different aldehydes, eight different amines). The catalyst could be recovered by simple aqueous extraction (*253*). A polymer-supported scandium catalyst allowed for a three-component reaction with TMSCN, giving α-aminonitriles in 83–99% yield (see Scheme 2.5) (*254*). Other silylated nucleophiles could be used, leading to β-amino esters (73–89% yield).

A 1998 report disclosed that lithium perchlorate/diethyl ether effectively catalyzed the reaction of aldehydes, amines, and TMSCN, giving both alkyl and aryl aminonitriles in 88–95% yield in 5 min (*255*). A solvent-free condensation of aliphatic or aromatic aldehydes with an amine and TMSCN was mediated by solid lithium perchlorate, producing α-aminonitriles in 74–94% yield after only 20 min reaction (14 examples) (*256*). In 2005 montmorillonite KSF clay was reported to catalyze the condensation of aldehydes, imines, and

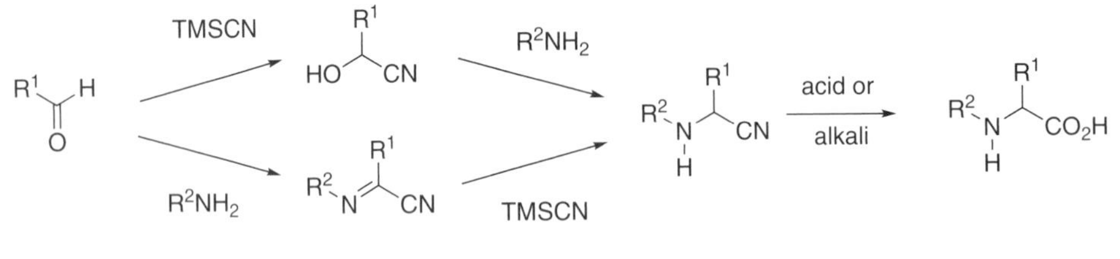

Scheme 2.4 Strecker Synthesis with TMSCN.

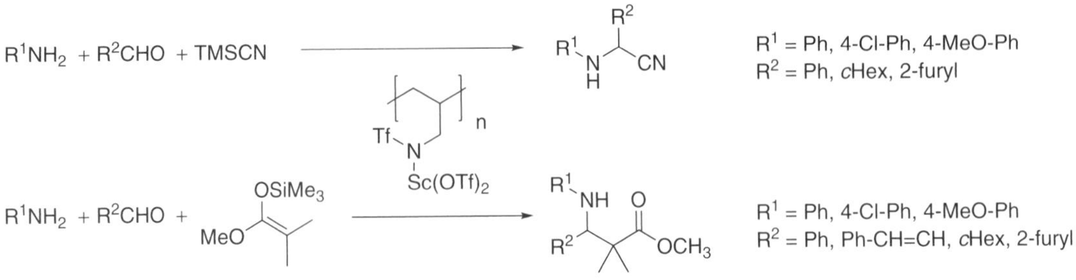

Scheme 2.5 Scandium Catalysed Three Component Reaction (*254*).

trimethylsilyl cyanide, producing 17 different α-aminonitriles in 85–94% yield in 2.5–5.5 h (257). Praseodymium trifluoromethylsulfonate gave a better yield than Sc(OTf)$_3$, InCl$_3$, or KSF-clay for one representative condensation with TMSCN, and produced 14 other aminonitriles in 74–91% yield. The catalyst could be recovered by simple aqueous extraction (258). RuCl$_3$ has also been tested as a catalyst with TMSCN, with 69–87% aminonitrile yields after 12–20 h (11 examples) (259). Similarly, bismuth trichloride gave 81–91% yields for 11 aminonitriles using TMSCN in acetonitrile for 5–10 h (260). Iodine was identified as another effective catalyst in 2005, giving much better yields (68%) of aminonitrile from acetophenone, trimethylaniline, and TMSCN than a number of Lewis acid catalysts (5–15%). A range of other aldehydes/ketones and amines also gave high yields (261). A {[2,6-bis (N-cyclohexyl)imino]phenyl}aquaoplatinum(II) complex catalyzed the room temperature addition of TMSCN to aromatic amines (262). In 2004, the addition of TMSCN to N-tosyl imines was reported, using LiClO$_4$ or BF$_3$.Et$_2$O as catalyst (887).

Ketones generally have poor reactivity unless catalysts are used. However, a series of studies starting in 2002 found that high pressure (0.6 GPa) gave good yields (0–99%) of aminonitriles from aniline, TMSCN, and a variety of 13 alkyl and aryl ketones (263–265). Although most Strecker reactions now employ TMSCN in the presence of a Lewis acid catalyst, a report in 2005 found that racemic Strecker reactions with TMSCN proceeded efficiently in acetonitrile solvent without any catalyst, cleanly forming the aminonitrile intermediate from alkyl and aryl aldehydes in 70 to >99% yields (266).

An alternative HCN substitute is tetrachlorosilane-KCN, which generates trichlorosilyl cyanide (TCSC) in situ. Reaction with 10 different N-aryl imines produced aminonitrile intermediates within 3 h in 76–96% yield (267). Cyanogen bromide has also been used as the cyanide source, with dehydrobromination and hydrolysis generating a number of α-methyl amino acids in 70–80% yield (R^1 = Me, R^2 = Me, Et, iBu, Ph, 4-Br-Ph) (268). Diethyl phosphorocyanidate provided α-aminonitrile intermediates for substrates such as 4-cholester-3-one, for which the classical Strecker synthesis cyanide source gave no product (269). Tributyltin cyanide has also been employed as the cyanide source, reacting with aldehydes

and benzhydrylamine in acetonitrile–toluene or water in the presence of scandium triflate catalyst to give products in 79–94% yield. The tin byproducts can be recovered and recycled (270). A polymer-supported scandium catalyst showed high activity in water, and could be readily recovered. Condensation of benzaldehyde, aniline, and tributyltin cyanide in water gave the aminonitrile in 77% yield (271). The use of toxic cyanide salts has been avoided in a procedure which reacted aryl aldehydes with LiHMDS and acetone cyanohydrin, producing the aryl aminonitriles in 80–92% yield (15 examples) (272). A ketone (dibenzosuberenone) for which standard Strecker methodology was ineffective was successfully converted to the α-aminonitrile by treatment of an imine intermediate with acetone cyanohydrin in ethanol and catalytic NaCN. The imine was obtained using NH$_3$/TiCl$_4$ (273).

The α-aminonitrile intermediate of the Strecker reaction can also be produced by oxidation of a secondary amine or a tertiary amine to an imine or iminium ion, respectively, followed by in situ trapping with cyanide (274). This reaction can be useful for ketone substrates or reactions with hindered amines. For example, treatment of a tertiary amine with sodium cyanide in the presence of molecular oxygen and catalytic ruthenium trichloride produced α-aminonitriles in 76–94% yield, leading to N,N-dialkyl products (275).

The Strecker synthesis has several advantages that should be apparent from the preceding examples, such as simple reaction conditions and the ability to prepare α,α-disubstituted amino acids in high yields by using a ketone as the starting material. Despite (or, more accurately, because of!) its longevity, variations of the Strecker synthesis are still widely used, especially in the industrial production of amino acids. Many of these syntheses incorporate some form of resolution after preparing the amino acid in order to obtain optically active product. Much effort has been expended at developing asymmetric versions of the Strecker reaction (see Section 5.2.1).

2.2.2 Bücherer Synthesis

The Bücherer synthesis is a modified Strecker synthesis that proceeds via a hydantoin intermediate (see Scheme 2.6). The overall approach was first described in

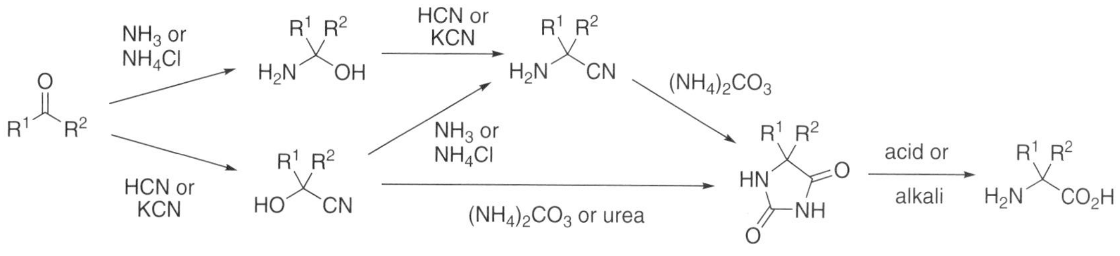

Scheme 2.6 Bücherer Synthesis.

1889 by Pinner and Spilker, who converted cyanohydrins into hydantoins by reaction with urea. The hydantoin derived from valeraldehyde was hydrolyzed under basic conditions to give norleucine (276). Urech had previously prepared Aib via hydrolysis of an acetone-derived hydantoin intermediate in 1872 (277). Wheeler and Hoffman (278) used hydantoin as a glycine anion equivalent, condensing it with aldehydes, and then reducing and hydrolyzing the product to give Phe and Tyr (see Section 4.2.2). The Strecker-like reaction did not gain widespread usage until after 1934, when Bucherer et al. reported that hydantoins were readily synthesized from α-oxy- or α-aminonitriles (279), and could therefore be prepared from aldehydes and ketones or their bisulfite adduct by reaction with KCN or HCN and $(NH_4)_2CO_3$ (280). Eleven mainly disubstituted hydantoins were synthesized and hydrolyzed to the corresponding amino acids (280). The synthesis is also often referred to as the Bücherer–Bergs synthesis, as Bergs published a patent in 1929 in which hydantoins were prepared from aldehydes or ketones by treating them with KCN, $(NH_4)_2CO_3$ and CO_2 under several atmospheres of pressure (281). Several α,α-disubstituted amino acids, including α-methyl-Asp and α-hydroxymethyl-Asp, were prepared around the same time (282–285).

Hydrolysis of the hydantoin intermediate generally involves treatment with aqueous barium hydroxide, although aqueous ammonium sulfide (286) and acid hydrolysis (287) have also been used. A 1950 *Chemical Reviews* review of hydantoin chemistry, including the Bücherer preparation and conversion into amino acids, listed 41 amino acids prepared via hydantions, (287). 5,5-Disubstituted-2,4-dithiohydantoins can be prepared from ketones, CS_2 and NH_4CN (288); they can then be converted into mono-2- or 4-thiohydantoins, or hydantoin (289).

Gaudry helped to popularize the method by publishing a comparative synthesis of Val using Strecker conditions of HCN/NH_3 (59% yield), KCN/NH_4Cl (65%), or $NaHSO_3/KCN/NH_3$ (54%), or via the Bücherer hydantoin (65%) (224). He and others then synthesized Ala (80% yield) (290), Phe (40%) (290), Met (69%) (291), and Lys (65%) (292) by the hydantoin method. Met was synthesized in 1948, with 67% yield compared to 38% for a Strecker synthesis (293). The procedure has become widely used, in part because the hydantoin intermediate is stable and easily isolated. In fact hydantoins have been used as derivatives for the identification of carbonyl compounds; over 100 were converted and 39 described in 1942 (294). For aromatic ketones it was found that reaction in a pressure vessel with fused acetamide as solvent gave much better yields than Bücherer's original conditions of diluted alcohol at 50 °C (295, 296). Much milder hydantoin hydrolysis conditions were developed in 1994, with di-Boc protected hydantoin cleaved by aqueous LiOH at room temperature (297). The procedure has been applied to the synthesis of many types of amino acids, such as 3,5-dihydroxy-phenylglycine (298), β-chloro-Ala (35%) (299), S-methyl and S-benzyl Cys (21% and 52%) (299), Met

(279) and Met analogs (300), 6,6,6-trifluoro-Met (301), γ-hydroxyvaleric acid (55%) (302), 6-hydroxy-Nle (303, 304), benz[f]tryptophan (a bathochromic Trp analog) (305), α-(cyclohexen-4-yl)glycine (306), and 2-amino-4-(3-hydroxy-5-methyl-4-isoxazolyl)butyric acid (28% hydantoin, 11% hydrolysis) (307). The hydantoin intermediate derived from 4-bromo-2-butenal was phosphorylated; hydrolysis provided 5-phosphono-3,4-dehydro-Nva (308).

As with the Strecker synthesis, the Bücherer synthesis has been used to prepare many labeled amino acids, such as $[1-^{11}C]$-Ala (309), $[1-^{11}C]$-Val (310), $[1-^{11}C]$-Leu (310, 311), $[1-^{11}C]$-Trp (310), $[1-^{11}C]$-α-methylVal (310), substituted $[carboxy-^{11}C]$-1-aminocyclopentane and $[carboxy-^{11}C]$-1-aminocyclohexane carboxylic acids (8–70%) (310), $[1-^{14}C]$-Lys (312–314), $[1-^{14}C]$-4′-hydroxy-Phg (315), $[1-^{14}C]$-2-aminopimelic acid (312, 314), and $[^{18}F]$-1-amino-3-fluorocyclopentane/hexane-1-carboxylic acids (316). Microwave heating was used to rapidly prepare $[1-^{11}C]$ analogs of Phe (40–60%), DOPA (10%), Met (40–60%), Tyr (40–60%), and Leu (70–80%) (317). The Bücherer synthesis has also been applied to the synthesis of $[2-^{11}C]$-amino acids (Leu, Ile, Val, Aib, 1-aminocyclopentane carboxylic acid, Phe, Phg, 4-methyl-Tyr) in 15–60% yield by incorporating $[1-^{11}C]$-aldehydes (318–321).

α,α-Disubstituted amino acids, including a large number of cyclic amino acids, are readily synthesized by using a ketone substrate. Twenty-three different racemic disubstituted amino acids were prepared by this method in 13–66% yield in 1960; the authors preferred the Bücherer method to the Strecker synthesis due to reduced decomposition and easier purification of the intermediates (322). Twenty-seven different α,α-dialkyl glycines were described in a 1972 report (323). A number of α-methyldopa analogs have been prepared in 67–81% yield (324–326). Other α-methyl-disubstituted amino acids include α-methyl-β-(4-methoxy-1-naphthyl)alanine (327), α-methyl-4-carboxyphenyl-glycine (a potent competitive antagonist of metabotropic glutamate receptors, 90–99% yield) (328, 329), α-(2-furyl)-Ala (330), α-methyl-Pro (331), α-methyl-Met (332), α-methyl-homocysteines with S-phosphono-methyl substituents (304), α-methyl-Orn (331, 333, 334), α-methyl-Arg (333, 334), α-methyl-Glu (332), and α,α′-dimethyl-diaminopimelic acid (332). Other Bücherer syntheses include α-fluoromethyl-glutamic acid (an irreversible inhibitor of glutamate decarboxylase, 23% yield) (335), diethylglycine (336), α-ethyl-Ser (330), α-hydroxymethyl-homoserine (337), O-substituted α-benzyl-Ser (338, 339) and related compounds (340), and a number of α-H and α-substituted γ-phospho analogs of Glu (341). An array of over 70 different α-substituted-α-(2-carboxycycloprop-1-yl)glycines, which were evaluated as antagonists of metabotropic glutamate receptors, were reported in 1998 (342–344).

Cyclic amino acids prepared by Bücherer synthesis include a series of 1-aminocycloalkane-1-carboxylic acids with 4-, 5-, 6-, 7-, and 8-membered rings described in a 1984 report (336). 1-Aminocyclooctane-1-carboxylic acid was also reported in 1964 (*232*), a cyclobutane amino acid derivative in 1995 (13% yield) (*345*), protected 3-hydroxymethyl-1-aminocyclobutane-1-carboxylic acid in 2002 (*346*), 3-(*p*-boronophenylalkyl)-1-amino-cyclobutane-1-carboxylic acids in 1999 (*347*), and car-boranyl-substituted cyclobutane derivatives (69%) in 1997 (*348–350*). In 1960 (*351*) and 1962 (*352*) a range of 1-aminocycloalkyl-1-carboxylic acids were prepared, including 2- and 3-carboxy and 2,3-benzo-1-aminocy-clopentane/hexane-1-carboxylic acids. However, the Bücherer reaction conditions gave lower yields compared to a Strecker synthesis of 1-aminocyclobutane-1,3-dicarboxylic acids (*209*). More recently, carboxylate, phosphonate or tetrazole substituted cyclopentyl-, cyclo-hexyl, and cycloheptyl-amino-1-carboxylic acids were synthesized in 9–58% yield (*353*), following earlier syn-theses of 1-amino-1,3-dicarboxy-cyclopentane (42%) (*354*), -cyclohexane (40%) (*355*), and -cyclohexene (29%) (*356*). Cyclohexyl cyclic analogs of DOPA were synthesized in 31–48% yield (*324, 357*), as were con-strained cyclic Phe analogs (*358, 359*) and a cyclic Trp analog with a methylene group between the α-carbon and indole C2-position (*360*). Heteroatom cyclic amino acids were derived from tetrahydrofuran-3-one and tetrahydropyran-4-one, as were the corresponding thia analogs (*361*). 3-Aminothietane-3-carboxylic acid (24%) (*236*), 4-aminopiperidine-4-carboxylic acid (*362, 363*), 3-amino-1,2,3,4-tetrahydrocarbazole-3-carboxylic acid (*364*), and 1,4-diaminocyclohexane-1-carboxylic acid (*364*) have also been synthesized. A range of differ-ent cyclic (71–81% yield) (*888*) and acyclic α,α-disubstituted amino acids, including 2-(aminomethyl)-Leu and -Ala (83–84%) (*366*) were prepared via the hydantoin synthesis; these were then resolved via for-mation of a diastereomer with L-Phe-cyclohexylamide (*365, 366*). A Bücherer synthesis produced a constrained Trp analog, 1-amino-2-(indol-3-yl)-cyclohexane-1-car-boxylic acid, in 2002 (*367*), while another group pro-duced a constrained Lys analog based on a octahydroanthracene skeleton, 1,8-diamino-1,2,3,4,5,6, 7,8-octahydroanthracene-1-carboxylate (*368*).

Several cyclic and bicyclic amino acids have been pre-pared via spirohydantoin Bücherer intermediates or via Strecker reactions. Different diastereomers are obtained from the two procedures, with the α-isomer produced by the Bücherer synthesis (kinetic product), and the β-isomer by the Strecker synthesis (thermodynamic product) (*253*). Thus *cis* 1-amino-4-*t*-butylcyclohexane-1-carboxylic acid was prepared from 4-*t*-butylcyclo-hexanone in 47% by the Strecker reaction, and the *trans* isomer in 31% by the Bücherer route (*369*). In a similar fashion 3-aminobicyclo[3.2.1]octane-3-carboxylic acid was prepared by both Strecker (β-isomer, 13%) and Bücherer–Bergs (α-isomer, 80%) syntheses (*205*), as was 2-aminonorbornane-2-carboxylic acid from nor-camphor (18–26% via Strecker (*208, 370*), 32–60% via Bücherer) (*370, 371*). *Cis* or *trans* 1,4-diaminocy-clohexane-1-carboxylic acids were obtained preferen-tially by Strecker or Bücherer–Bergs reactions of 4-aminocyclohexanone (*204*), as were the isomers of

3-aminotropane-3-carboxylic acid from tropinone (*372*), and both *cis* and *trans* 2-, 3-, and 4-methyl-1-aminocy-clohexane-1-carbocyclic acids (*373*). Both hydantoin intermediate isomers of tropinone, *cis*-3,4-dimethyl-cyclopentanone and *cis*-bicyclo[3.3.0]octan-3-one, have also been prepared (*207*). A number of other bicyclic amino acid isomers have been reported (27–80%) (*374, 375*), including a series of aminobicyclo[2.2.1] heptanedicarboxylic acid diastereomers (*376*), a bicyclic Met analog (*377*), and benzo-fused 2-aminobicyclo [2.2.2]octane-2-carboxylic acids (*378*). The tricyclic 2-aminoadamantane-2-carboxylic acid was also derived via a Bücherer reaction (*379*).

2.2.3 Ugi Synthesis

The Ugi three- or four-component condensation (Ugi 3CC or 4CC) is essentially a one-pot Strecker reaction in which either three components (acid, isocyanide, and imine) or four components (acid, isocyanide, amine, and aldehyde or ketone) are combined in one reaction, pro-ducing an amino acid carboxamide (see Scheme 2.7). The imine component (or amine and carbonyl compo-nents) contributes the side chain, α-*C* and α-*N* atoms, and any *N*-alkyl group. The isocyanide contributes the amino acid carboxyl group as a carboxamide, while the acid component forms an *N*-acyl group on either the α-*N* or carboxamide-*N* (depending on the number of α-*N* substituents). Peptides centered around a newly formed amino acid residue can be synthesized by using peptide-acid, imine, and cyano-peptide components. In many cases di- or tripeptide derivatives have been generated; diastereoselectivity at the newly generated amino acid center is often poor (<20% de) or unreported. These nonstereoselective reactions are included in this section, while reactions with greater stereoselectivity are described in Section 5.2.2. The Ugi reaction was first reported in 1959 (*380, 381*) and a number of reviews have since appeared (*382–384*), including a 2000 review by Dömling and Ugi of multicomponent reactions with isocyanides (*385*) and a 2006 *Chemical Reviews* treatise of recent developments in isocyanide-based multicom-ponent reactions (*386*). An aminomalonate derivative byproduct has been noted in some reactions (*387*). In 2004 a microwave-assisted Ugi condensation of levulinic acid with two different isonitriles and four different amines was optimized for equivalents of amine, imine pre-formation time, reaction time, tem-perature, and concentration using a Design of Experiment (DoE) approach. Yields of 17–90% in only 30 min were obtained, compared to conventional 48 h reaction times (*388*).

Tripeptides containing α,α-diisopropylglycine or α,α-diphenylglycine (which are normally very difficult to couple) as the middle residue were synthesized in 61% and 63% yield by a 4CC reaction at high pressure (0.9 GPa) (*389*), while an α,α-diisopropylglycine-containing dipeptide was prepared in 44% yield at atmo-spheric pressure (*390*). The isocyanides can be prepared from *N*-monosubstituted formamides (e.g., *N*-formyl

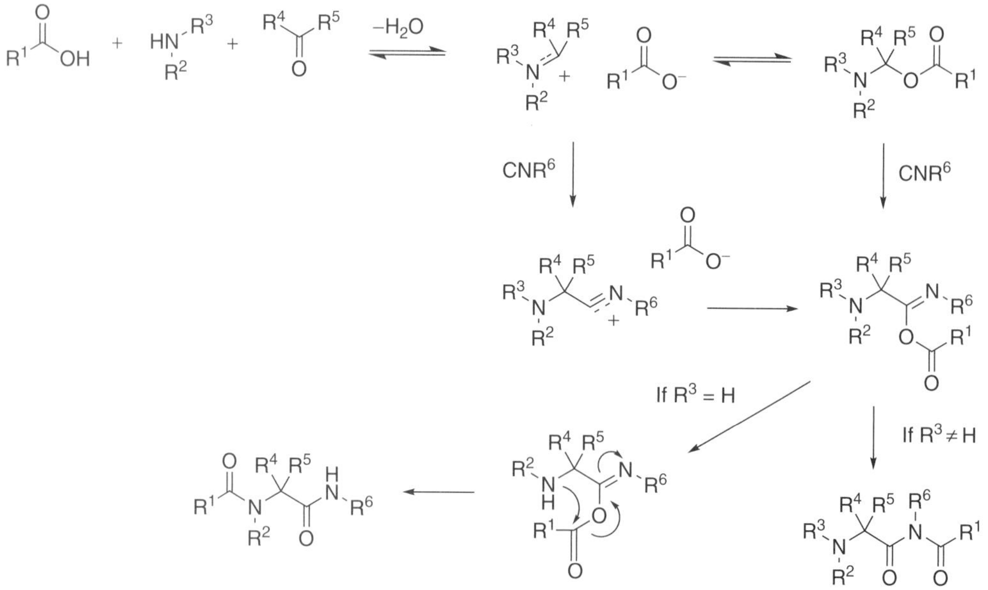

Scheme 2.7 Ugi 4CC Synthesis.

amino acids for peptide synthesis) by dehydration with diphosgene (perchloro(methyl)formate, which is easier to handle and gives better yields than phosgene) (*391*), or by more traditional methods (*392*). Other sterically congested tripeptides containing α,α-diphenylglycine as the central residue were prepared from diphenyl-methanimine, *N*-Cbz amino acids, and isocyanides derived from amino acids (*393*). A 50-fold rate accelera-tion in the Ugi reaction has been induced by carrying out the reaction in water; a tripeptide containing α,α-diisopropylglycine was synthesized, along with a 48-compound library with Abu, Val, and Leu (*394*). Dipeptide amides with an *N*-(arylalkyl) α,α-disubstituted *C*-terminal residue were constructed in 42–66% yield via Ugi condensations (*395*). A library of di- and tripeptides derivatives containing a *C*-terminal Asp Weinreb amide was constructed as potential caspase-3 inhibitors by condensing an isocyanide derivative of Asp(OBn)-Weinreb amide with phenylacetaldehyde, amines and acids (9–30% yield) (*396*). A Boc-Gly-*N*-Bn-α-methyl-homoallylglycine-allylglycine-OEt trip-eptide was constructed in 95% yield from Boc-Gly, an allylglycine-derived isocyanide, and an *N*-benzyl imine of hex-5-en-2-one (*397*). 2-Cyano-2-isocyanoalkanoates have been employed as the isocyanide component, pro-ducing tripeptide-like products with a *C*-terminal α-cyano-α-alkyl amino ester (*398*). Bioconjugates have been prepared by using the acid or amino groups of proteins such as bovine serum albumin (BSA and horseradish peroxidase (HRP), in combination with iso-cyanides that include a glycosyl cyanide, acids that include glycoside, biotin or rhodanine derivatives, and amines that include amino-functionalized glycosides and biotin (*399*).

The Ugi reaction has not been widely used for the preparative synthesis of amino acids, probably due to the difficulty in removing the amide and *N*-acyl groups, but there are a number of examples. α-Benzylphenylalanine was synthesized by condensing dibenzylketone, ben-zylamine, phenyl isocyanide, and acetic acid; acid hydrolysis (Ac removal) and catalytic hydrogenation (Bn removal) of the intermediate gave the deprotected amino acid in 56% yield (*400*). Tripeptides containing this residue were also prepared. 2,5-Cyclohexadiene-1-alanine (1,4-dihydroPhe) was prepared by Ugi reaction of the aldehyde, methylisonitrile, and ammonium triflu-oroacetate (45% yield) followed by hydrolysis of the *N*-TFA methyl amide (82%) (*401*).

Derivatized amino acids have also been prepared; a library of arylglycine amide derivatives was synthesized for evaluation as potential chiral stationary phases for chromatography (*402*). *N*-Protected heterocyclic amino acid amides derived from furfural, thiophenecarboxal-dehyde, and a five-membered sugar were obtained in 65%, 68%, and 68% yield, respectively (*403*). A number of oxazine Pro analogs were prepared in 14–75% yield via a Ugi 3CC reaction, as either the *N*-acyl amides or as part of di- and tripeptides (*404*). Thiazolidine and oxazo-lidine Pro analogs were synthesized in a similar fashion (18–85%) (*405*). *N*-Substituted α-methylpyroglutamate amides, and their six- seven- and eight-membered ring homologs, were synthesized by an intramolecular Ugi reaction, with ketoacids representing two of the four components (*406*). Piperazine-2-carboxamides were derived from *N*-alkyl ethylenediamines and chloroacet-aldehyde (*407*). A nucleo amino acid containing a uri-dine side chain was prepared via a 4CC condensation of the 3-(2-formylethyl)uridine aldehyde, which proceeded in 54% yield (*408*). Di- and tripeptides containing the amino acid were prepared by a similar procedure as a mixture of diastereomers (*409*). Quinolines and isoquinolines can be employed as the imine component, reacting with an acid chloride followed by isocyanide to give 1,2-dehydroquinoline-2-carboxylic acids or 1,2-isoquinoline-1-carboxylic acids (*410*).

The 4CC reaction has been applied to solid-phase combinatorial synthesis of *N*-acyl-α-amino amides by combining a polymer-linked amine with 12 acids, eight aldehydes, and one isocyanide (*411*). A more recent report used two isocyanides, three aldehydes, and three acids with a polymer-supported amine, with microwave irradiation to enhance reaction rates (*412*). Similarly, *C*-glycoside amino acid amides or di- and tripeptides were obtained from carbohydrate aldehydes (five types), an acid or diacid (15 types), an isocyanide (two types) and a polymer-linked amine or amino acid (six types) (*413*). Another report employed the Rink amino resin as the amine component, reacting it with 12 acids, eight aldehydes, and an isocyanide derived from *N*-formyl glycine. *N*-Acyl dipeptides were produced in 55–83% isolated yield on a 0.1 mmol scale (*414*). The Rink resin was also combined with 18 aldehydes, six isonitriles, and an acid mimicking phosphotyrosine in order to prepare a combinatorial library of protein tyrosine phosphatase inhibitors (*415*). Another library, of α-methylated amino acids, was prepared by both solution synthesis using primary amines, or by solid-phase synthesis with the Rink resin as the amine source (*416*). A library of

putative β-turn mimetics based on a diketopiperazine template was constructed via an Ugi condensation of resin-bound 2,3-diaminopropionic ester with an aldehyde, isocyanide, and 2-bromoalkyl acid (*417*). Isocyanocarboxylic acids (amino acids with the amine within an isonitrile group) have been attached to a resin support and used as the isonitrile component in a Ugi 4CC reaction, resulting in resin-bound dipeptide amides (*418*).

A number of modifications have been made to the reaction conditions in order to more easily modify the initial Ugi adducts. If "convertible" 1-isocyanocyclohexene is used as the isocyanide component, the amino acid cyclohexenamide products can undergo post-condensation modification to give amino acids, amino esters, and amino thioesters (see Scheme 2.8) (*419, 420*). Methyl esters can be obtained without isolation of the cyclohexylamide, giving an overall five-component condensation.

Unfortunately, the convertible isocyanide is difficult to synthesize, and is unstable (*421*). A new class of isocyanides, (β-isocyanoethyl)alkyl carbonates, were reported in 1999 to be useful reagents for the generation of α-amino acids and esters (Scheme 2.9). After Ugi condensation, the amide is treated with KO*t*Bu to form an ester, or with KO*t*Bu/H₂O to form an acid (*421*). The R group of the carbonate linkage has been attached to a solid support to give a resin-bound convertible isonitrile (*422*). Another improved convertible isocyanide is *O*-(TBS)-2-(hydroxymethyl)phenyl isocyanide.

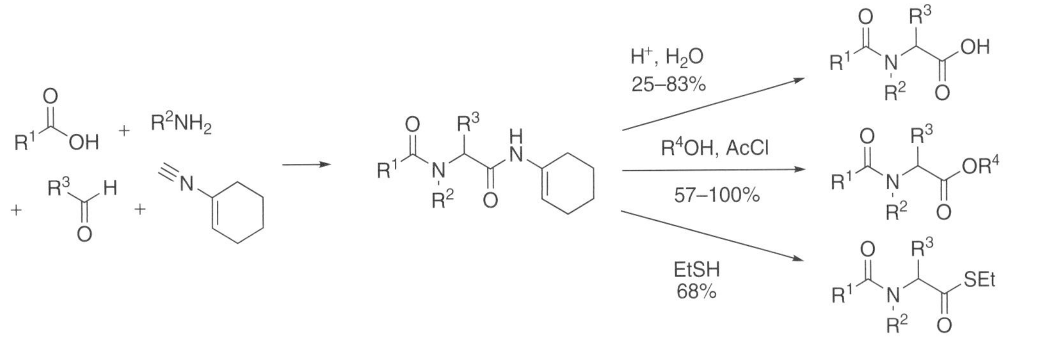

Scheme 2.8 Cyclohexenylisocyanide for Post-condensation Modification (*419, 420*).

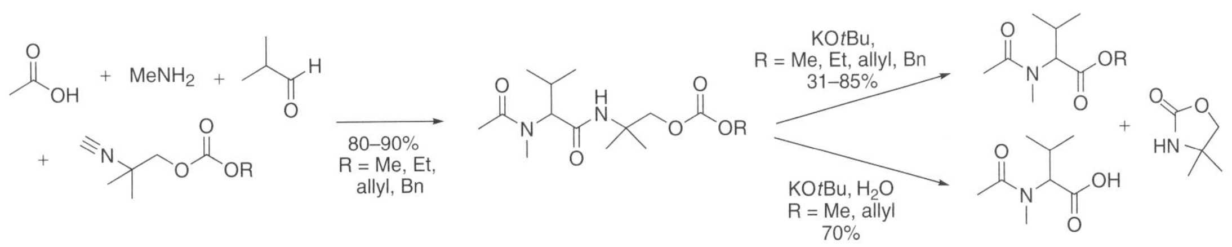

Scheme 2.9 (β-Isocyanoethyl) alkyl Carbonates for Post-condensation Modification (*421*).

Under acidic conditions the initial amide product undergoes intramolecular rearrangement to form an *o*-aminobenzyl ester (*423*). An alternative method of readily generating amino acids from the Ugi *N*-alkyl amino amide products makes use of the propensity of *N*,α,α-trialkyl glycine amides to undergo selective amide cleavage with trifluoroacetic acid (TFA). This is combined with cleavage of *N*-(4-methoxybenzyl) groups from the amine using boiling TFA, leading to *N*-acyl α,α-dialkyl-Gly products in 9–99% yield (*424, 425*).

A variety of different acid component analogs have been tested. Carbon dioxide was substituted for the acid component in one report; reaction of isobutyraldehyde, benzylamine, and an isocyanide in an autoclave at 50 atm pressure produced *N*-benzyl valine amides in 50–100% yield (*426*). A five-component reaction is also possible with an alcohol and CO_2 replacing the acid component, generating *N*-alkoxycarbonyl derivatives (*427*). By employing CS_2 or CS as the oxidized carbon source, *N*-H amino acid thioamides or *N*-alkoxythiocarbonyl amino acid thioamides were produced (*427*). A highly fluorinated silyated benzoic acid derivative, 4-[($C_{10}F_{21}CH_2CH_2)_3$Si]-$PhCO_2H$, can be employed as the acid component. This allows for purification of the product by extraction into a fluorous solvent. Treatment with TBAF (tetra *n*-butyl ammonium fluoride) removes the fluoroalkylsilyl substituent, leaving an *N*-benzoyl amino acid amide (*428*). The acid component can also be replaced by nitrophenols, leading to *N*-(nitrophenyl), *N*-substituted α-amino amides (*429*). A Petasis multicomponent reaction, combining glyoxylic acid with an amine and alkyl boronic acid, was used to generate the acid input for a Ugi 4CC reaction, resulting in dipeptide amides from six components, including a Rink resin amine component for the Ugi reaction step (*430, 431*).

Similarly, different amine components have been explored. As mentioned earlier, *N*-(4-methoxybenzyl) groups from 4-methoxybenzylamine can be removed by boiling TFA (*424, 425*). If 9-aminomethylfluorene is added as the amine component, the *N*-methylfluorene group can be easily removed by treatment with DBU. A Phg-containing dipeptide was prepared in 66% yield by this procedure (*432*). Alternatively, ammonia in methanol can be used as the amine component to give *N*-H *N*-acyl amino acids in good yield (*433*). Replacement of the Ugi 4CC amine component with an *N*-alkylated hydroxylamine (reacted with carbonyl, acid, and isocyanide) produced *O*-acylated *N*-hydroxy α-amino

amide products in variable (33–89%) yield. The carbonyl component could include alkyl aldehydes and ketones, but aryl aldehydes gave no product (*434*). Using *O*-benzyl hydroxyamine produced *N*-acyl, *N*-benzyloxy amino amides, which were hydrogenated to give *N*-hydroxy derivatives (*435*). Arylsulfonamides can also be employed as the amine input; 4-carboxybenzenesulfonamide bound to Rink resin was combined with aldehydes, isocyanides, and acetic acid at 60 °C for 24 h to give *N*-acetyl,*N*-arylsulfonyl amino amides. The acetyl group could be removed by treating the resin with 40% aq. $MeNH_2$, and the product then cleaved from the resin (*436*).

Iminoaziridines are highly reactive synthetic equivalents for three of the four Ugi components (amine, isocyanide, and aldehyde). They react with carboxylic acids, followed by a 1,3- or 1,4-migration of the acyl group (see Scheme 2.10). However, since the iminoaziridine must first be synthesized, the impressive flexibility of the Ugi reaction is lost (*437*). The imine component of the Ugi condensation can be replaced by an aziridine, yielding an acylated β-amino acid amide product (*438*).

The 4CC reaction has been used in conjunction with a glyoxal derivative as the aldehyde component, producing a symmetrical dicarboxamide product that is then oxidatively cleaved. This results in a net overall reaction of amide formation between the acid component and a primary amine (*439*).

2.2.4 Other Aminocarbonylation Reactions

Several other aminocarboxylation reactions reminiscent of Strecker reactions have been reported. α-Aminoarylacetic acids (phenylglycines) can be synthesized from bromoform, arylaldehydes, and ammonia under basic conditions, giving the crude products in 15–83% yield for 10 examples (*440*). The tribromomethyl carbanions derived from bromoform replace the cyanide in a conventional Strecker synthesis (see Scheme 2.11). The same reaction has been carried out with chloroform, aqueous ammonia, KOH, and LiCl under phase-transfer catalysis conditions (see Scheme 2.12) (*441*). With triethylbenzylammonium chloride as a phase-transfer catalyst, yields of 29–66% were obtained (*441*). Asymmetric versions of this reaction have been developed, including a useful version by Corey (see Section 5.3.2, Scheme 5.11). An earlier report used a

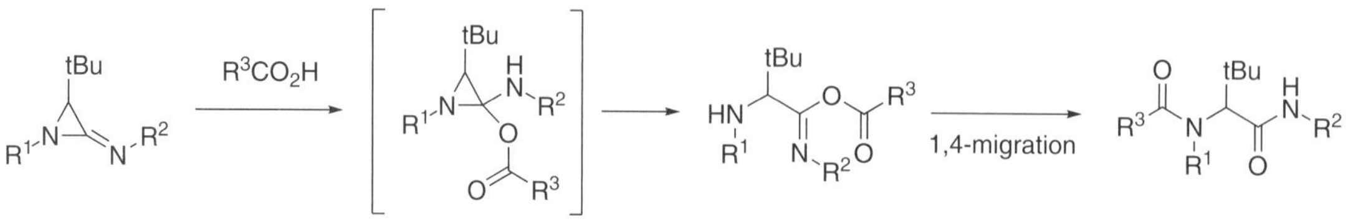

Scheme 2.10 Iminoaziridine as Ugi Synthon (*437*).

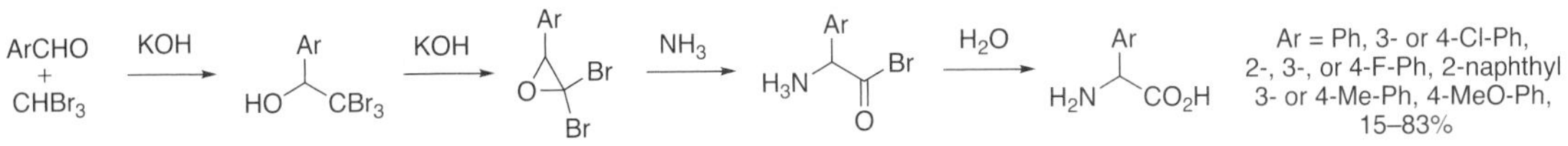

ArCHO + CHBr₃ →[KOH] HO—CBr₃/Ar →[KOH] Ar epoxide (O, Br, Br) →[NH₃] H₃N⁺—CH(Ar)—C(=O)Br →[H₂O] H₂N—CH(Ar)—CO₂H

Ar = Ph, 3- or 4-Cl-Ph, 2-, 3-, or 4-F-Ph, 2-naphthyl 3- or 4-Me-Ph, 4-MeO-Ph, 15–83%

Scheme 2.11 Haloform Aminocarboxylation (*440*).

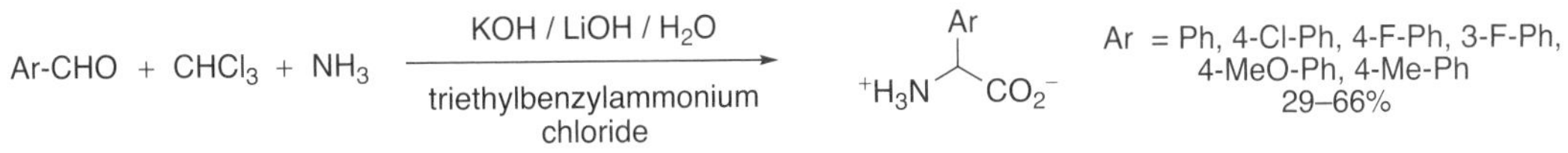

Ar-CHO + CHCl₃ + NH₃ →[KOH / LiOH / H₂O, triethylbenzylammonium chloride] ⁺H₃N—CH(Ar)—CO₂⁻

Ar = Ph, 4-Cl-Ph, 4-F-Ph, 3-F-Ph, 4-MeO-Ph, 4-Me-Ph 29–66%

Scheme 2.12 Phase-Transfer Catalyzed Haloform Reaction (*441*).

two-step procedure instead of the one-pot reaction, with trichloromethylcarbinols converted to amino acids (Val, Abu, Aib, Phg, 29–48% yield) by treatment with potassium amide in liquid ammonia. The substrate carbinols were prepared by several methods, but in poor yield (*442*). A similar reaction was employed to prepare a bicyclic Glu analog, α-[(4-[2.2.2]bicyclooctane-1-carboxylic acid]glycine, via addition of LiCCl₃ to the

formyl-substituted bicyclic acid, followed by reaction with DBU/NaN₃ and then azide hydrogenation (*443*).

Wakamatsu developed an aminocarboxylation synthesis of amino acids from aldehydes containing at least one α-proton, CO₂, H₂, and an amide (usually acetamide) in the presence of a cobalt catalyst (see Scheme 2.13). The developments of this reaction over the past 30 years were reviewed in 2000 (*444*). The original 1971 paper

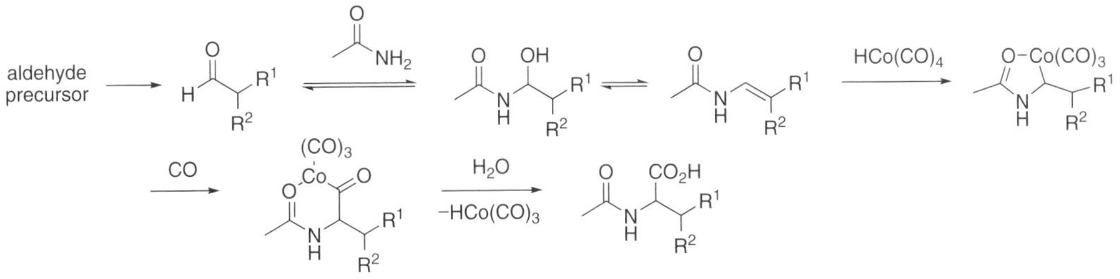

Substrate (Aldehyde or Aldehyde Precursor)	Product	Yield
R¹R²CHCHO: R¹ = R² = H, Me R¹ = H; R² = Me, Ph, CH₂CN, CH₂CO₂Me, CH₂SMe R¹ = Et, OAc; R² = Me	R¹R²CHCHO: R¹ = R² = H, Me R¹ = H; R² = Me, Ph, CH₂CN, CH₂CO₂Me, CH₂SMe R¹ = Et, OAc; R² = Me	26–80% (*445*)
R¹R²CHCHO: R¹ = H; R² = nPr, nHept, nDec R¹ = Et, OAc; R² = Me	R¹ = H; R² = nPr, nHept, nDec R¹ = Et, OAc; R² = Me	52–85% (*446*)
R¹R²CHCHO: R¹ = NPhth; R² = H, Bn, iBu R¹ = R² =(–CH₂CH₂CH₂NBn⁻)	R¹ = NPhth; R² = H, Bn, iBu R¹ = R² = (-CH₂CH₂CH₂NBn⁻)	30–50% (*451*)
R³CH₂OH: R³ = CH=CH₂, CH=CHMe, CH=CHMe₂, CH₂CH=CH₂, CH₂C(Me)=CH₂	R¹ = H; R² = Me, Et, iPr,	34–77% (*449, 453*)
R³ epoxide R³ = Me, Et, Ph, CH₂OPh	R¹ = H; R² = Me, Et, Ph, CH₂OPh	18–95% (*449*)
R³CH=CH₂: R³ = CF₃, CN, CH₂NPhth, (CH₂)₂NPhth, CH₂CH(CO₂Me)NPhth, CH₂CH(Me)NPhth, P(O)(Me)(OR)	R¹ = H; R² = CH₂CF₃, CH₂CN, (CH₂)₂NPhth, (CH₂)₃NPhth, (CH₂)₂CH(CO₂Me)NPhth, (CH₂)₂CH(Me)NPhth, P(O)(Me)(OR)	56–85% (*448, 449, 451, 452*)
RCHO: R = CH₂CH₂CF₃, CH(Me)CF₃	R¹ = H; R² = (CH₂)₂CF₃, CH(Me)CF₃	77–80% (*447*)
BnCl	R¹ = H; R² = Ph	82% (*450*)

Scheme 2.13 Co-Catalyzed Amidocarbonylation Reaction (*444*).

reported the synthesis of *N*-acetyl, *N*-benzoyl, and *N*-lauroyl Gly, Ala, Val, Abu, Phe, Glu, and Met in 26–70% yield (*445*), and this was extended to *N*-acetyl-Nva (*446*), Ile/*allo*-Ile (*446*), 2-aminotridecanoic acid (*446*), 2-aminopelargonic acid (*446*), 4,4,4-trifluorovaline (*447*), and 5,5,5-trifluoronorvaline (*447*) (78–86% yield). An attempt at asymmetric induction was unsuccessful (*446*). Olefins (*448*), epoxides (*449*), allylic alcohols (*449*), and benzylhalides (*450*) can also be used as substrates, as they are precursors capable of forming aldehydes under the amidocarbonylation conditions in the presence of other catalysts (see Scheme 2.13). A variety of α,β-, α,γ-, and α,δ-diamino acids were prepared from the corresponding aldehydes or olefins in 24–66% (*451*). Protected glufosinate (2-amino-4-[hydroxy)methylphosphinyl]butanoic acid) was synthesized in 85% yield from methylvinylphosphinate (*448*), and *N*-Ac-trifluoro-Val (82%) and -Nva (80%) were derived from trifluoropropene (*449, 452*). Allylic alcohols were used to prepare *N*-Ac-Abu (63%), Nva (77%), Val (49%), and Leu (66%) (*449, 453*), while homoallylic alcohols provided *N*-Ac-Nva (55%) and Leu (34%) (*449*). Oxiranes reacted to give *N*-Ac-Phe (95%), Abu (18%), Nva (27%), and *O*-Ph-Hse (36%) (*449*). Ac-Phe was prepared from benzyl chloride on a 5 g scale in 82% yield, with addition of NaHCO$_3$ greatly improving yields (*450*).

This reaction has more recently been carried out with a Pd co-catalyst (such as PdBr$_2$, [Pd(PPh$_3$)$_2$Br$_2$] or [Pd$_2$(dba)$_3$]), providing improvements such as higher catalyst activity and milder reaction conditions. The aldehydes leading to Gly, Leu, Phe, and Phg were reacted with acetamide and other amides, with 53–99% yields (*454*). Twelve different aryl aldehydes gave yields of 37–95% of the corresponding α-arylglycines (*455*), while Leu, Chg, Met, 4′-chloro-Phe, 4′-methoxy-Phg, 4′-chloro-Phg, and 4′-fluoro-Phg were synthesized in 72–99% yield on up to 10 g scale, then enzymatically resolved (*456*). Plain Pd/C has been identified as an even more effective catalyst, giving yields of up to 98%.

Alternative amine components have been developed. *N*-Acetyl amines were condensed with paraformaldehyde to give *N*-acetyl,*N*-substituted glycines, peptoid monomers (*457*). Nitriles have been reported as another amine source, being hydrolyzed in situ to form an amide which then reacts with CO and an aldehyde to give *N*-acyl amino acids in 44–92% yield (see Scheme 2.14) (*458*). *N*-Acyl imines and enamides can also be carboxylated under the same reaction conditions, indicating that they may be intermediates in the reaction mechanism (*459*). Alternatively, *N*-substituted ureas as the amine source lead to hydantoin products (see Scheme 2.15) (*460*).

Another metal-catalyzed aminocarboxylation condensation combines *N*-benzyl imines, BnN=CHR$_1$, with isocyanates, R$_2$-NCO, in the presence of TaCl$_5$/Zn to produce, after hydrolysis with 10% KOH, *N*-substituted amino amides, BnNHCH(R$_1$)CONHR$_2$ in 59–93% yield (11 examples, R$_1$ = Ph, 4-MeO-Ph, 4-Me-Ph, 4-Cl-Ph, 4-pyridyl, 2-furyl, *c*Hex) (*461*). A number of arylglycines were prepared by a four-component coupling approach in 2003, with benzaldehyde-derived imines condensed with acid chlorides and CO in the presence of a Pd catalyst to generate a cyclic Münchone intermediate; this was decomposed in situ by addition of methanol to form *N*-acylated,*N*-substituted arylglycine methyl esters in 31–91% yield (*462*). A one-pot domino-type reaction sequence used Pd catalysis to carry out two sequential carbonylations of aryl halides, followed by reductive amination of the intermediate α-ketoacid, to produce arylglycine amides in 60–88% yield (see Scheme 2.16) (*463*).

A three-component coupling used to prepare a wide variety of unnatural amino acids combined aldehydes, amides, and sodium *p*-toluene sulfinate to give intermediate α-amidoalkyl sulfones (see Scheme 2.17). The sulfone was then displaced by the sodium salt of nitromethane, with the nitromethane substituent converted to an acid group by oxidation with KMnO$_4$ (*464*). Alternatively, *N*-Boc α-amidosulfones derived from aldehydes, Boc carbamate, and sodium benzenesulfinate

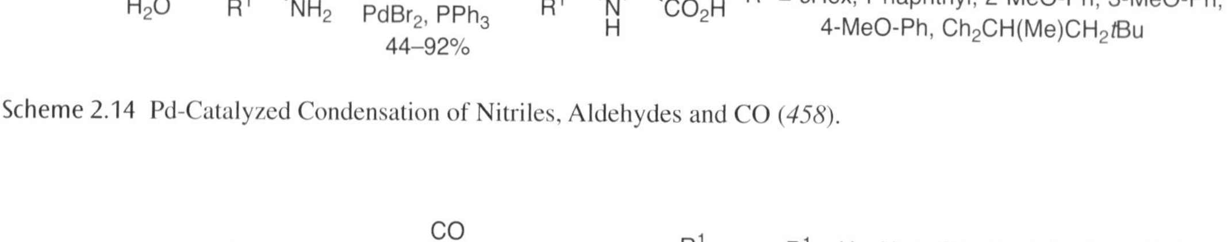

Scheme 2.14 Pd-Catalyzed Condensation of Nitriles, Aldehydes and CO (*458*).

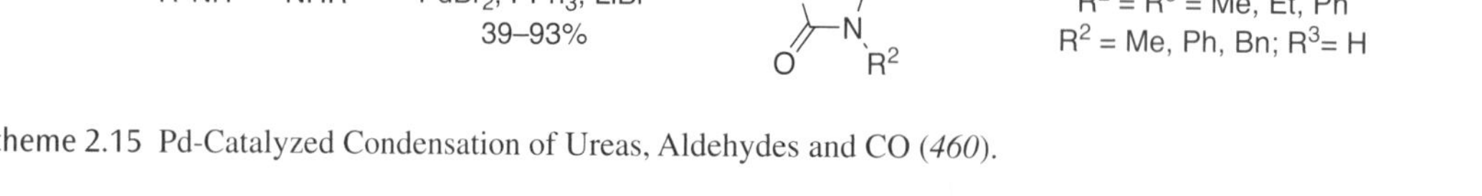

Scheme 2.15 Pd-Catalyzed Condensation of Ureas, Aldehydes and CO (*460*).

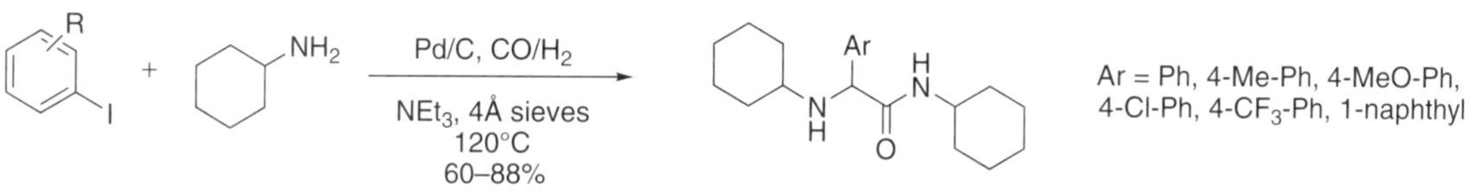

Scheme 2.16 Pd-Catalyzed Double Carbohydroamination of Aryl Halides (*463*).

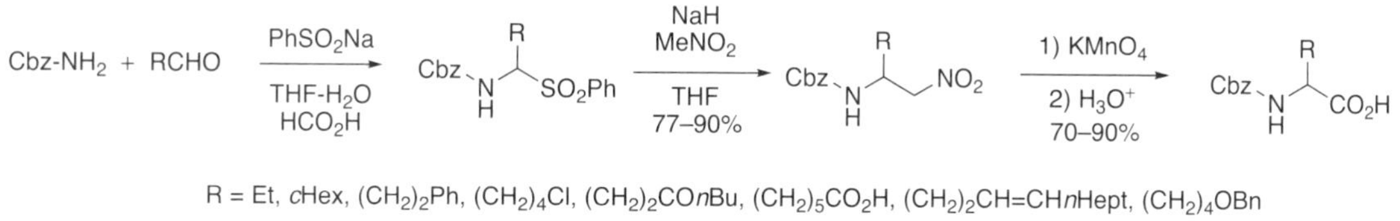

R = Et, cHex, (CH₂)₂Ph, (CH₂)₄Cl, (CH₂)₂COnBu, (CH₂)₅CO₂H, (CH₂)₂CH=CHnHept, (CH₂)₄OBn

Scheme 2.17 Condensation of Nitromethane with α-Amidoalkyl Sulfones (*464*).

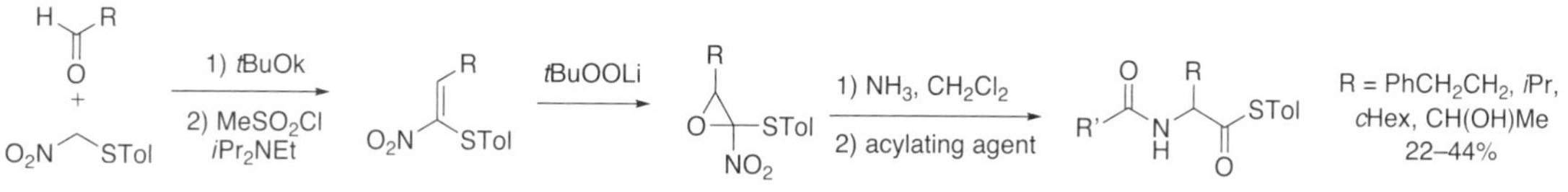

Scheme 2.18 Stereocontrolled Aminocarboxylation (*466*).

were treated with two equivalents of KCN in 2-propanol or CH₂Cl₂–H₂O under phase-transfer conditions, resulting in crystalline *N*-Boc α-aminonitriles in 36–88% yield (28 examples) (*465*).

A stereocontrolled aminocarboxylation reaction equivalent leading to racemic α-amino acid precursors condensed an aldehyde with [(4-methylphenyl)thio] nitromethane, followed by epoxidation to give a 2-thio-2-nitrooxirane (see Scheme 2.18). The oxirane was then regiospecifically ring-opened with aqueous ammonia to give an α-amino thioester in 24–44% overall yield (*466*). This method was used to prepare D-threonine and L-*allo*-threonine by starting with a chiral α-hydroxy aldehyde (*467*). α-Hydroxy-β-fluoronitriles have been obtained by opening 2-cyanoepoxides with HF. Amination with

ammonia and hydrolysis of the cyano group provided β-fluoro-α-amino acids (*468*).

A two-step reaction sequence introduces a masked carboxy group to aldehydes and ketones via a Wittig reaction (see Scheme 2.19). The α-oxygenated acrylonitrile is then aminated with sodium azide in the presence of ceric ammonium nitrate; sequential treatment with NaOAc and methanolic K₂CO₃ followed by hydrogenation gave the free amino acids in 29–43% overall yield (*469*).

A radical-based three-component condensation is possible. The oxime formed in situ from 2-hydroxy-2-methoxyacetic acid methyl ester reacts with alkyl radicals generated from alkyl iodides and Bu₃SnH/Et₃B as initiator. *N*-OBn amino esters were produced in 46–86%

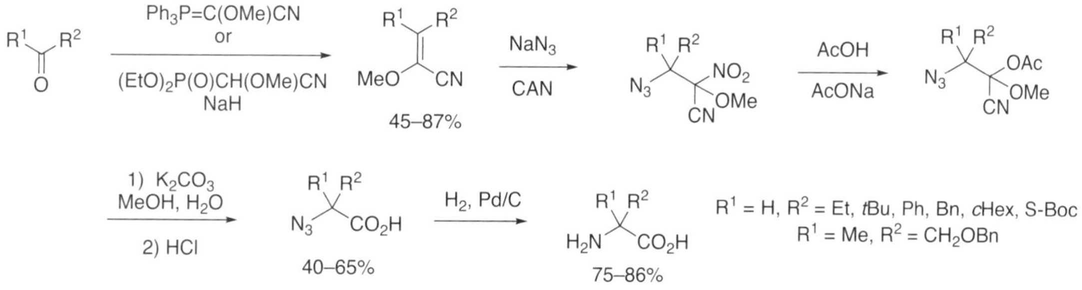

Scheme 2.19 Aminocarboxylation (*469*).

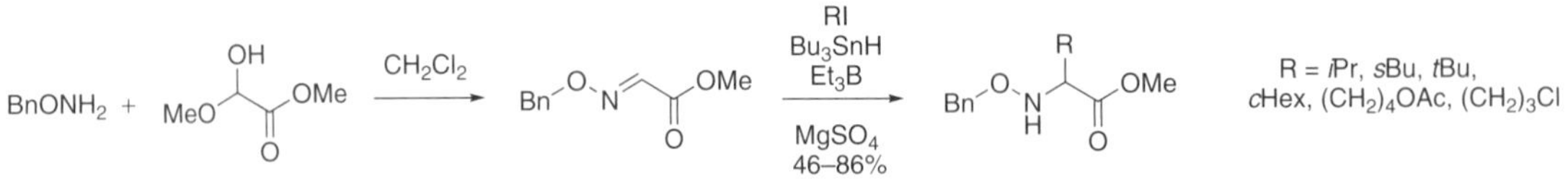

Scheme 2.20 Radical Based Aminocarboxylation (*470*).

yield (see Scheme 2.20) (*470*). Similar reactions are discussed in Section 3.10.

Another sequential aminocarboxylation reaction was used to synthesize 9-amino-4,5-diazafluorene-9-carboxylic acid, a rigid bipyridine amino acid that can potentially bind to transition metals. The ketone contained in 4,5-diazafluoren-9-one was converted to an imine by reaction with benzylamine, deprotonated with NaHMDS, and carboxylated with methyl choroformate (*471, 472*). Aldehyde-derived *N*-benzyl imines were reacted with a carbamoylsilane, Me$_3$SiCON(Me)$_2$, in the presence of BF$_3$·Et$_2$O to produce amino acid dimethylamides in 31–76% yield. Abu, Val, 3,3,3-trifluoro-Ala, Phg, α-(2-furyl)-Gly, α-(2-thienyl)-Gly, α-(2-pyridyl)-Gly, and styrylglycine were synthesized (*473*).

2.3 Addition of the Amino Group (Amination Reactions)

2.3.1 Cahours and Gabriel-Type Syntheses

Most amination reactions are based on the Cahours synthesis, in which a nitrogen nucleophile is used to displace the halide from an α-halo acid (see Scheme 2.21). The original Cahours reaction in 1858 prepared glycine from α-chloroacetic acid and ethanolic ammonia (*474, 475*), and was soon followed by Kolbe's synthesis of alanine in 1860 from α-chloropropionic acid (23% yield) (*476*). Increased yields for the preparation of glycine have been reported (50% recrystallized yield (*161, 477*), 54% yield (*478*), 60% yield (*479*)), mainly due to improvements in isolation. A diverse range of

amino acids have been prepared by this method (see Table 2.1), including a number of unusual amino acids in recent years. The substrates are usually prepared via α-bromination or α-chlorination of acids and then displaced with alcoholic ammonia or ammonium hydroxide. *Organic Syntheses* preparations of Ile (49% from 2-bromo-3-methylpentanoic acid and ammonia) (*480*), Leu 43–45%) (*481*), Phe (62%) (*482*), Ser·(quantitative) (*483*), and Thr (25%, can also separate and isolate *allo*-Thr with this procedure) (*484*) have been described. Pro was obtained via dialkylation of ammonia with ethyl α,δ-dibromopropylmalonate in 75% yield (*485*).

In a number of cases substituted malonic esters and acids have also been α-brominated and then aminated by displacement with ammonia. The extra carboxyl group is then removed by heating in HCl. For example, this method was applied to the synthesis of [3,4–^{2}H$_2$]-Met and -homocysteine (*486*).

The Gabriel reaction, published in 1887 (*487*), replaced ammonia with potassium phthalimide as the aminating reagent for the synthesis of primary amines, resulting in fewer byproducts. This reaction, which was reviewed in 1968 (*488*), was soon applied to the preparation of amino acids. In general, Gabriel rarely used solvents, but dimethylformamide (DMF) is now widely employed for the reaction due to the moderate solubility of potassium phthalimide in this solvent (*488–490*). The Gabriel synthesis has been used in recent years to displace the halide from 2-bromo-4-butyrolactone, giving a common intermediate that is then elaborated to a number of different amino acids (*491, 492*). [1-^{13}C]-Gly and [2-^{13}C]-Gly were obtained from ethyl [1-^{13}C]-bromoacetate or ethyl [2-^{13}C]-bromoacetate in 1989 via amination

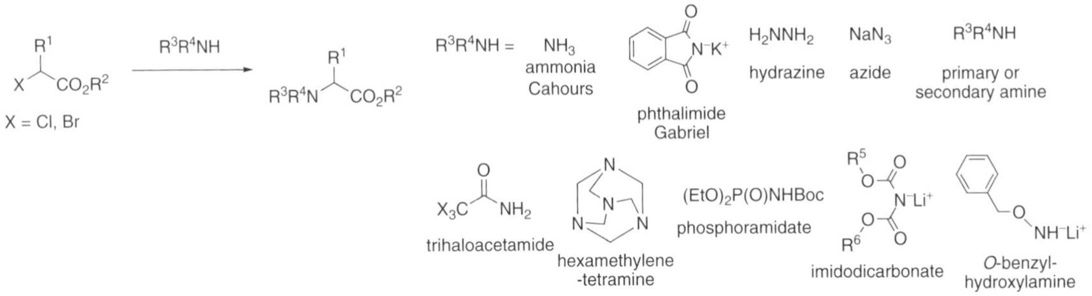

Scheme 2.21 Nucleophilic Amination of an α-Halo Acid.

Table 2.1 Amination of α-Halo Acids with Ammonia

Side Chain	Yield	Reference
H	— from α-Cl	1858 (474)
		1859 (475)
Me	23% from α-Cl	1860 (476)
CH$_2$NH$_2$	40–50% from dibromopropionic acid	1894 (813)
(CH$_2$)$_3$−(α-N)	75% from dibromo	1900 (485)
(CH$_2$)$_4$NH$_2$	42% amination	1901 (814)
(CH$_2$)$_2$NH$_2$	50% from α-Br-γ-NH$_2$-butyric acid	1901 (815)
nPr	60–62% amination	1902 (816)
iPr		
(CH$_2$)$_4$CH(CO$_2$H)NH$_2$	30–72% from dibromo	1905 (817)
(CH$_2$)$_6$CH(CO$_2$H)NH$_2$		
CH(NH$_2$)Me	20–24% from α-Br	1906 (818)
(CH$_2$)$_5$CH(CO$_2$H)NH$_2$		
CH$_2$CH$_2$OH	72% from α-Br	1907 (819)
Et (with α-Et)	— from α-Br	1909 (820)
(CH$_2$)$_4$NH$_2$	80% from α-Br-ε-NHBz-hexanoic acid	1909 (821)
(CH$_2$)$_3$NH$_2$	64% from dihalovaleric acid	1909 (822)
		1909 (823)
(CH$_2$)$_3$−(α-N)	65% from dihalovaleric acid	1909 (822)
		1909 (823)
CH$_2$(imidazol-3-yl)	42% amination	1911 (824)
CH$_2$(imidazol-3-yl)	38% from α-Cl	1911 (825)
CH(OH)CO$_2$H	65% amination	1921 (826)
		1955 (827)
		1963 (828)
C(OMe)(Me)$_2$	— from α-Br	1922 (829)
C(OH)(Me)$_2$		
CH(OMe)Ph		
CH(OMe)(4-MeO-Ph)		
H	50% amination	1927 (477)
		1931 (161)
H	54% amination	1930 (478)
CH$_2$CH$_2$SMe	24% bromination and amination	1930 (830)
CH(OH)Me	— from α-Cl	1930 (831)
CH(OH)Me with α-Me		
CH(OH)CO$_2$H		
CH$_2$OMe	— from α-Br	1933 (832)
CH$_2$OH		
CH(OMe)Me	— from α-Br	1934 (833)
Me	65–70% amination	1937 (834)
		1941 (835)
CH(OMe)Me	— from α-Br	1937 (836)
CH(OH)Me		
CHDCHDSMe	— from α-Br-malonic acid	1938 (486)
CHDCHDSH		
CH$_2$OH	from α-Br	1940 (837)
CH$_2$OEt		
H	57–61%	1941 (479)
Me		
Et		
nPr		
iPr		
nBu		
nBu	62–67%	1941 (838)
(CH$_2$)$_3$NH$_2$	— from α-Br-δ-NH$_2$-valeric acid	1941 (839)
(CH$_2$)$_4$N(Me)Bz	30% α-bromination and amination	1944 (840)
(CH$_2$)$_4$NH$_2$	60–81% amination	1947 (841)
		1948 (842)
		1949 (843)

Table 2.1 Amination of α-Halo Acids with Ammonia (continued)

Side Chain	Yield	Reference
(CH$_2$)$_3$-(α-N)	10–45% from dihalovaleric acid	1949 (*844*)
(CH$_2$)$_4$NH$_2$	62% from α-Br-ε-Br-hexanoic acid	1950 (*845*)
(CH$_2$)$_3$CO$_2$H	86% from α-Br	1950 (*846*)
CH$_2$(4-piperidyl)	65% from α-Cl	1950 (*847*)
CH(OMe)*i*Pr	quantitative from α-Br	1951 (*848*)
CH(OH)*i*Pr		
C(OH)Me$_2$	—	1952 (*849*)
*c*Hex	70% from α-Br	1952 (*850*)
CH$_2$*c*Hex		
(CH$_2$)$_4$NH$_2$ with α-[^{15}N]	73% from 2-Cl-6-NHBz-hexanoic acid	1952 (*851*)
CH(OH)Et	24% bromination and amination	1953 (*852*)
CH$_2$OH	from ethyl α-Br-β-MeO-propionate or α-Br-β-MeO-propionitrile	1954 (*853*)
(CH$_2$)$_5$CH(CO$_2$H)NH$_2$	—	1954 (*495*)
(CH$_2$)$_8$CH(CO$_2$H)NH$_2$		
*s*Bu	49% from α-Br	1955 (*480*)
*i*Bu	43–45% from α-Br	1955 (*481*)
Bn	62% from α-Br	1955 (*482*)
CH$_2$OH	quantitative from α-Br	1955 (*483*)
CH(OH)Me	25% from α-Br	1955 (*484*)
CH$_2$NH$_2$	44–55% from dibromopropionic acid	1955 (*854*)
CH(OH)*n*Pr	35–60% from α-Br	1955 (*855*)
CH(OH)*i*Pr		
*i*Pr	48% from α-Br	1955 (*856*)
*n*Bu	30–50% amination	1955 (*857*)
CH(OH)*i*Pr		
CH(OMe)Ph	66–74% from α-Br	1957 (*858*)
CH$_2$CH$_2$SBn	50–58% from α-Br	1957 (*859*)
CH$_2$OBn	67% bromination and amination	1957 (*860*)
*c*Hex	15% amination	1958 (*861*)
CH$_2$*c*Hex		
(CH$_2$)$_6$NH$_2$	90–100% chlorination	1959 (*686*)
(CH$_2$)$_7$NH$_2$	16–74% amination	
(CH$_2$)$_4$NHMe	47% from α-Cl	1959 (*552*)
4-Me-Bn	44–50% from α-Cl	1964 (*862*)
4-Et-Bn		
CH(NO$_2$) (CH$_2$)$_3$-(α-C)	from α-iodo	1966 (*863*)
CH$_2$F	17–81% bromination	1967 (*864*)
CH$_2$CH$_2$F	23–70% amination	
CH(OH)CH$_2$F		
CH(CH$_2$F)$_2$		
CH$_2$CH(CH$_2$F)$_2$		
(CH$_2$)$_3$NHCONHC(=NH)NH$_2$	84% from α-Cl	1969 (*865*)
2-Me-Bn	27–92% amination	1969 (*866*)
3-Me-Bn		
4-Me-Bn		
2-MeO-Bn		
3-MeO-Bn		
4-MeO-Bn		
3-NO$_2$-Ph		
3-CO$_2$H-Bn		
4-CO$_2$H-Bn		
4-CN-Bn		
4-MeCO-Bn		
4-H$_2$NCO-Bn		
CH$_2$(2-furyl)	from α,β-unsaturated acid	1973 (*867*)

Table 2.1 Amination of α-Halo Acids with Ammonia (continued)

Side Chain	Yield	Reference
CH$_2$F	7–32% amination	1973 (868)
CH(F)Me		
CF=C(Me)$_2$		
CH(F)Et		
CH(F)nPr		
CH(F)nBu		
CH(Me) (^{14}CH$_3$)	56% amination	1973 (869)
CH=CH$_2$	7% amination	1974 (190)
C^{2}H$_2$OH	55% bromination and amination	1976 (870)
CH$_2$(1-Br-2-naphthyl)	62–73% amination	1976 (871)
CH$_2$(1-Cl-2-naphthyl)		
CH=CH$_2$	29% bromination and amination	1977 (872)
CH$_2$F	44% amination	1978 (873)
CH(Me)CFMe$_2$		
Ph	—	1978 (874)
C(Ph)=CH$_2$	65% from dibromo	1979 (875)
CH=C(Me)$_2$	19–52% bromination	1979 (876)
cyclohexen-1-yl	5–16% amination	
cyclopenten-1-yl		
Z-CH=CHD	4% bromination and amination	1980 (877)
E-CD=CHD		
Z-CH=CHD	80% bromination	1981 (878)
E-CD=CHD	20% amination	
E- CH=CHMe	27–30% bromination and amination	1981 (879)
Z-CH=CHMe		
CH$_2$CH$_2$PO$_3$H$_2$	48–66% amination	1981 (880)
CH$_2$CH$_2$P(=O) (Me)OH		
3-(HO$_2$CCH$_2$O)-5-CF$_3$-isoxazol-4-yl	11% bromination and amination	1993 (881)
CH(Me)CH$_2$CF$_3$	— bromination and amination	2003 (534)

with potassium phthalimide (493), while 4-fluoro-Orn and 5-fluoro-Lys were prepared by KNPhth displacement of the α-bromoesters in 1976 (494). Both amino groups of 2,6-diaminopimelic acid were introduced by phthalimide displacement of the dibromo precursor (495, 496). 2,5-Diaminoadipic acid (490, 495), 2,4-diaminobutyric acid (497), and 4-hydroxy-Orn (498, 499) were also synthesized by the Gabriel procedure. Potassium [^{15}N]phthalimide can be used to synthesize labeled amino acids, such as [^{15}N]-Lys (886).

The phthalimide protecting group is difficult to remove, with cleavage requiring strong acidolysis, a two-step alkaline–acid hydrolysis, aminolysis, or hydrazinolysis (488); therefore, other nitrogen nucleophiles have been developed. Darapsky used hydrazine to prepare some aromatic amino acids, but cleavage of the N−N bond required a nitroso-hydrazine intermediate (500), although other cleavage methods have since been reported (501–503).

Azide is also commonly used to displace halides (especially in asymmetric syntheses), but again requires a subsequent reduction step to generate the amine. Hydrogenation is generally used, as illustrated in relatively recent preparations of racemic serine (504, 505),

phenylserine (506), 4,4,4-trifluoro-Abu (507), 4,4,4-trifluoro-Val (508), 5,5,5,5′,5′,5′-hexafluoro-Leu (509), and β-hydroxy-Orn (from benzyl 2-bromo-3-methoxy-5-phthalimidopentanoate) (510). Raney-Ni reduction was employed during a synthesis of Lys (511), while electrolytic cathodic reduction has been applied in other syntheses (512, 513). Triphenylphosphine can also be used (514), although triethylphosphine has been reported to be more effective than either triphenylphosphine or tributylphosphine (515). Stannous chloride in methanol is another useful reagent (516, 517). Reduction methods were summarized in a 1998 review of azide chemistry (518). Recent methods include reduction with zinc and ammonium chloride (519), In/NH$_4$Cl (520), N,N-dimethylhydrazine and catalytic ferric chloride (521), iodotrimethylsilane (522), Cu(NO$_3$)$_2$ (523), Zn-NiCl$_2$·6H$_2$O-THF (524), and Sm-NiCl$_2$·6H$_2$O-THF (525). Azides have been directly converted into Boc-protected amines using trimethylphosphine and Boc-ON (526), H$_2$/Pd-C/Boc$_2$O (527), or by using Et$_3$SiH and Boc$_2$O in the presence of a Pd catalyst (528), while reduction with trimethylphosphine in the presence of alkyl chloroformates led to N-Cbz, N-Aloc, N-Meoc, N-Troc, and N-CO$_2$Et protected amines in high yield (529). Amide bonds can be formed by reacting amino acid thio acids (generated from a trimethoxybenzyl thioester) with azides. The coupling appears to proceed via a concerted reduction/amidation process, and no racemization of the thioester component was observed. However, only simple alkyl azide partners were examined (e.g., no azido esters) (530). Alternatively, a Ph_2Se_2-Bu_3P system was employed as a direct amidation reagent via in situ generation of the amine component by reduction of α-azido esters using the benzeneselenol generated during carboxyl activation (531). A similar in situ reduction employed trialkylphosphines in combination with 3,5-dinitrobenzoyl mixed anhydride activation of the carboxyl component (532). A peptide with a C-terminal phosphinothioester (Xaa-SCH_2PPh_3) reacted with N-terminal azido-peptides to directly form an amide linkage, ligating the two peptide segments (533).

Azide displacement was used in the synthesis of the Ile analog 2-amino-3-trifluoromethylpentanoic acid, which was prepared by α-bromination of the fluoro ester, followed by displacement with azide, and then hydrogenation and ester hydrolysis (534). 2-Amino-4-methyl-5-hexenoic acid was produced by azide displacement of the 2-bromo or tosylated 2-hydroxy precursor, with hydrogenation over Raney Ni reducing the azide but not the alkene (535). The bromination/azide displacement method has been used to prepare the β-(heteroaryl)alanine compounds stizolobic acid and acromelobic acid (536). Sterically hindered azido acids corresponding to Aib and Dpg (diphenylglycine) were synthesized by azide displacement of the halides obtained by radical halogenation of α-branched acids. Azide reduction was carried out following SPPS coupling of the α-azido acid residues, using DTT (537).

Hexamethylenetetramine ($C_6H_{12}N_4$) has also been used as an amine source (538), reacting with a number of α-bromoacids to give, after hydrolysis with HCl, Gly (94%), Ala (93%), Val (91%), Nle (82%), Leu (71%), and Phe (79%). Diethyl N-(Boc)-phosphoramidate, $(EtO)_2P(O)NHBoc$, was used to prepare Gly (80% yield for amination and deprotection), Ala (36% for two steps), Abu (36% for amination), and Nva (38% for amination) (539). The free amine was generated by treatment with HCl in benzene.

The imidodicarbonates (or iminodicarboxylates), $(ROCO)_2NH$, are attractive nucleophilic aminating reagents as they produce products with the same carbamate protecting groups used in peptide synthesis, which are readily removed. Selective removal of one carbamate group after amination provides a monoprotected amino acid ready for coupling. Carpino first prepared the Boc_2NH reagent (540), but did not prepare any amino acids. Mixed alkyl imidodicarbonates, including Cbz_2NH, Cbz(Fmoc)NH, Cbz(Boc)NH, Cbz(Aloc)NH, Cbz(Troc)NH, and Cbz(Adoc)NH, can be prepared via reaction of Cbz-isocyanate with alcohols (541, 542). A methyl/t-butyl mixed iminodicarboxylate has been reacted with ethyl bromoacetate and ethyl 2-bromopropionate to give protected Gly (80% yield) and Ala (83%), respectively; the ethyl ester and N-methoxycarbonyl group were cleaved with NaOH to give the Boc-amino acids (543). N,N-Cbz_2-Gly-OtBu (94% crude yield) (541) and N,N-Boc_2-Gly-OBn (93% crude yield) (544) were prepared from the sodium salt of Cbz_2NH and t-butyl bromoacetate or benzyl bromoacetate, respectively. Similarly, the potassium salts of a number of imidodicarbonates were reacted with ethyl 2-bromopropionate to give N-Cbz,N-Boc- (78%), N-Cbz,N-Aloc- (83%), N-Cbz,N-Troc- (59%), N,N-Cbz_2- (80%), N-Cbz,N-Cbz(NO_2)- (62%), or N,N-Boc_2-Ala-OEt (78%) (545). Boc_2NK was also used in the synthesis of some γ-amino acid analogs by bromide displacement (55–69% amination yield) (546).

Trifluoroacetamide (547) and or the less expensive trichloroacetamide (548), have also been used as aminating reagents. Under phase-transfer catalysis conditions, they reacted with 2-bromocarboxylic esters in generally good yield (see Scheme 2.22). However, for R_1 = Bn elimination was the major reaction, while R_1 = iPr was not reactive. The R_1 = $(CH_2)_nBr$ reactants underwent an intramolecular diamination to give Pro and pipecolic acid. Basic hydrolysis of the N-(trihaloacetyl)-amino acid esters generated the free amino acids in 46–90% overall yield (547).

Nucleophilic displacements with substituted amines can be used to produce N-alkyl amino acids. Five N-alkyl glycines were prepared by amine displacement of ethyl bromoacetate in 1996 (549); hexadecylamine provided N-hexadecylglycine (550, 551). N^α,N^α-dimethyl-Lys was obtained in 51% yield by displacement of the α-bromo-ε-aminocaproic acid with dimethylamine (552); dimethylamine was also used to aminate α,α'-dibromosebacic acid and 1,10-dibromodecane-1,10-dicarboxylic acid

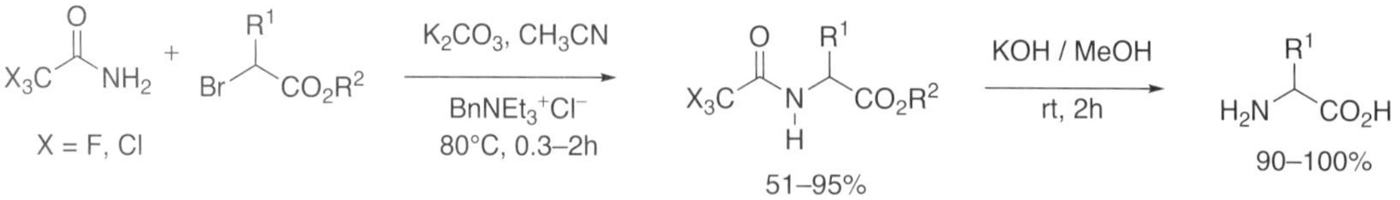

X = F, R^1 = H, Me, n-decane, Ph, 2-Me-Ph, 3-MeO-Ph, 4-F-Ph, 4-Cl-Ph, 4-Br-Ph
X = Cl, R^1 = H, Me, nHex, n-$C_{14}H_{29}$, iBu, Ph, Br$(CH_2)_3$, Br$(CH_2)_4$

Scheme 2.22 Nucleophilic Amination with Trihaloacetamide (547, 548).

(49–85% yield) (*495*). Piperidine was used as the nucleophile in the synthesis of 16 *N*,*N*-dialkyl α- and ω-amino acids in 1934 (*553*). The *N*-alkylglycines correspond to peptoid monomers, peptide analogs possessing the side-chain substituent on the amide nitrogen rather than the α-carbon. The peptoid polymers can be prepared by a submonomer approach in which alkyl amines displace the halide of a previously coupled bromoacetic acid moiety (*554*). A submonomer synthesis was applied to the preparation of peptoid oligomers with α-chiral aliphatic side chains (*555*), and has been combined with standard Fmoc-monomer synthesis to prepare mixed peptide–peptoid conjugates that mimic the cytostatic depsipeptide dolastatin 15 (*556*). The submonomer approach was employed to prepare a 5000-member combinatorial library of dimers and trimers (*557*), a library of tetrapeptoids containing a phosphotyrosine mimetic (*558*), a library of over 10,000 trimeric peptoids using 22 different primary amines (*559*), and very large libraries of 78,125 octamers, 100,000 pentamers, and 531,441 hexamers (*560*). A library of mixed peptide/peptoid derivatives was constructed on a continuous cellulose membrane surface, using the submonomer approach for peptoid synthesis and standard coupling techniques for introducing the amino acids (*561*).

A benzyl substituent can be used as a removable amino protecting group. Benzylamine was reacted with methyl 2,4-dibromopentanoate to give, after hydrogenolysis, 4-methylazetidine-2-carboxylate (*562*). Benzylamine displacement of α-bromo-β-hydroxypropionate (85% yield) led to serine after hydrogenation (90% yield) (*563*). *N*-methyl,*N*-triethoxysilylmethyl amino acids were prepared by amination with the corresponding amine (*564*). Alkyl diamines were used to prepare *N*-(ω-aminoalkyl) derivatives of Gly, Ala, and Leu (36–72% yield), providing building blocks for *N*-backbone cyclic peptides (*565*). *N*,*N*,*N*-trialkyl glycine derivatives, analogs of glycine betaine, have been prepared by amination of chloroacetic acid, esterified to a solid support resin, with a number of tertiary amines (*566*). *O*-benzylhydroxylamine has been used to prepare *N*-hydroxy amino acids; reaction with bromosuccinic acid esters led to *N*-benzyloxy-Asp ester in up to 93% yield (*567*). Another possible amination route is to react α-bromoacids with NaNO₂, resulting in an α-oxime acid that can be reduced to the amine with Zn/HCl (*568*).

Several reactions make use of compounds with masked carboxyl groups as the electrophile instead of α-halo acids. *N*-protected allylamines can be synthesized by palladium-catalyzed allylic substitution of allyl acetates with a variety of nitrogen nucleophiles in 41–78% yield. Oxidative cleavage of the alkene generated the *N*-protected amino acid in 49–95% yield (see Scheme 2.23) (*569*). Gly, Ala, Phg, and homo-Glu were prepared by this method. Homoallylic alcohols derivatized as an acylaminomethyl ether undergo an intramolecular cyclization in the presence of mercuric nitrate; oxidative demercuration and Jones oxidation introduces the carboxyl group to give γ-hydroxy-α-amino acids (*570*).

Ammonia can be added to α,β-unsaturated diesters to give Asp derivatives. For example, addition to diethyl fumarate followed by hydrolysis gave Asp in 76% yield (*571, 572*). A variety of alkyl amines (R = *n*Bu, *i*Bu, *c*Hex), imidazoles, and pyrazoles were used to aminate unsymmetrical fumaric esters. 1,4-Conjugate addition directly provided protected Asp or Asn derivatives in 20–80% yield (*573, 574*). Ammonia adds to 3-methoxy-isoxazol-5-yl-substituted propenoic acid in 25% yield in the presence of tin(IV) chloride, giving the α-amino acid homoibotenic acid in preference to the β-isomer (*575, 576*). A ketoaryl-substituted acrylate was aminated in 90% yield in a synthesis of 3′-hydroxykynurenine, (3′-hydroxy-2′-amino)-3-oxo-Phe (*577*). Dimethyl 2-phenylseleno fumarate is a good Michael acceptor of amines, including sterically hindered amines such as *t*BuNH₂, with 90–98% addition yields (*578*). β-Amino acids are accessible via nucleophilic Michael additions of amines to α,β-unsaturated acids or esters (see Section 11.2.2a), as reviewed in 1996 (*579*).

A regioselective and diastereoselective aminochlorination of cinnamic esters was described in 1999, using TsNCl₂ and catalytic CuOTf to produce *anti*-β-chloroarylalanines in 66–91% yield (*580*). Aminohydroxylation of fumaric or maleic acid produced racemic β-hydroxy-Asp in 88–98% yield using low levels (0.1–1.0 mol %) of osmium catalyst and stoichiometric chloramine T (*581*). Many asymmetric versions of these reactions have been developed (see Section 5.3.7).

2.3.2. Mitsunobu and Triflate Displacement Reactions

Nucleophilic displacement of the hydroxyl group in α-hydroxy ester derivatives is widely used for asymmetric syntheses (see Section 5.3.1) but is less common in racemic reactions as the α-halo ester is more readily obtained, and its susceptibility to racemization is not a concern. Nonetheless, Cbz-NHOBn and Troc-NHOBn have been used for Mitsunobu displacements of

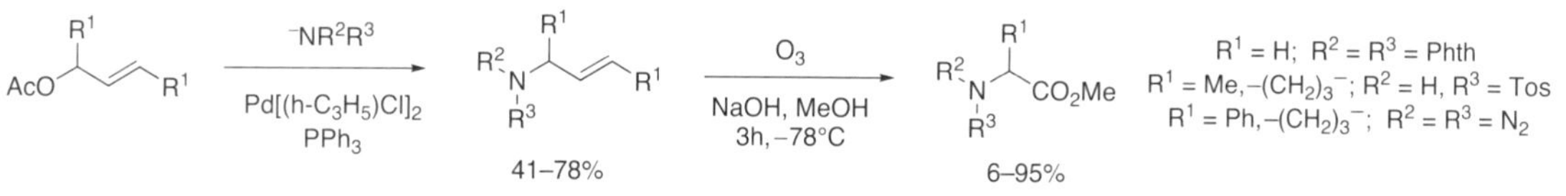

Scheme 2.23 Amination of Allylic Acetate (*569*).

α-hydroxy esters, providing *N*-hydroxy α-amino acids (*582*). Mitsunobu displacement of a hydroxyl group with phthalimide, which proceeded in 40–45% yield, was used in the synthesis of a fluoroallenyl γ-amino acid (*583*). *N*-Boc ethyl oxamate, EtO$_2$CC(=O)NHBoc, was recently reported to be a useful amine source for Mitsunobu reactions, leading to *N*-Boc amines after mild LiOH/THF/H$_2$O deprotection (*584*). A potential source of racemic α-hydroxy acids is alkyl or aryl halides, which can be converted to the desired substrate by introduction of both the α-CHOH and the carboxyl groups via a one-step palladium-catalyzed double-carbonylation reaction (73–81% yield, R = Ph, 4-Me-Ph, 1-naphthyl, 2-thiophene) (*585*). Racemic α-hydroxyarylacetic acids have been prepared from an aryl aldehyde, bromoform, KOH, and LiCl (*586*).

2.3.3 Electrophilic Amination

A number of electrophilic aminating reagents, which introduce an amine group to a carbanion, have been developed (see Scheme 2.24). A comprehensive review was published in 1989 (*587*), with a review of electrophilic α-amination of carbonyl compounds in 2004 (*588*). Many of these reagents have been used in the synthesis of amino acids, including *N*-haloamines, *O*-alkylhydroxylamines, *O*-phosphinylhydroxylamines, oximes, arenediazonium salts and dialkyl azodicarboxylates (*587*). The first electrophilic aminating reagent, *O*-(2,4-dinitrophenyl)hydroxylamine, was used to prepare 9-amino-9-fluorenecarboxylic acid (50% yield) and α-amino-α-phenylmalonate (53% yield) (*590, 591*), but more basic ester enolates were poorly aminated (*592*). One early report found *O*-methylhydroxylamine was the best electrophile of a number of *O*-alkyl and *O*-aryl hydroxylamines (*593*); reaction with α-lithiated acids gave Val (34%), Leu (11%), Met (9%), Phe (7%), and Phg (56%) in one step (*593, 594*). *O*-Mesitylenesulfonyloxyamine, which may explode when dry, was used to prepare methyl α-aminodiethylphosphonoacetate (47% yield) (*595*). The same reagent successfully reacted with a lithiated acid, leading to α-amino-3-hydroxy-4-methylisoxazol-5-ylacetic acid (20% yield), under conditions where methoxylamine gave no product (*575*). When *O*-(diphenylphosphinyl)hydroxylamine was introduced as a new electrophilic aminating reagent, a number of sodium and lithium carbanions were converted to diethylaminomalonate, 9-amino-9-fluorenecarboxylic acid, phenylglycine, and several α-aminonitriles in 31–96% yield (*596, 597*).

Chloramine is another reagent used to aminate lithium enolates generated from carboxylic acids or esters; this route has been used to prepare a number of unsaturated Glu analogs in 1–27% (*598*), β-methylene-Asn (68% yield for amination) (*599*), and β-methylene-Asp (65% yield for amination) (*600*). The trianions of several unsaturated dicarboxylic acids were also aminated, but yields were low (4–26%) (*601*). Tosylazide (*602*) (*p*-toluenesulfonyl azide, prepared from *p*-toluenesulfonylchloride (*603*)) or trisyl azide (2,4,6-triisopropylbenzenesulfonyl azide) (*604–608*) both introduce an azide group to enolates. Azide reduction is then needed to generate the amine, as discussed in Section 2.3.1. DPPA (diphenyl phosphorazidate) has been used as an electrophilic aminating reagent, reacting with the lithium enolates of *N*-methyl-*N*-phenylcarboxamides to give, after reduction and quenching with Boc$_2$O, the Boc amino acid amides (70–80% yield, R = Me, Et, Ph, 2-thiophene) (*609*).

Azodicarboxylate esters were popularized as general electrophilic aminating reagents by Evans et al. in 1988 (*610*), although there were a few previous reports of their use for this purpose. The most effective of these is di-*tert*-butyl azodicarboxylate (DBAD), as it is a commercially available, stable, crystalline solid. The Boc protecting groups are easily removed under mild conditions, while the sterically bulky *tert*-butyl substituents provide for high stereoselectivity (*611*). However a reductive cleavage of the hydrazine N−N bond is required after amination in order to generate the free amine. Evans et al. have found trisyl azide to be more generally useful than DBAD as an aminating reagent (*604*). In an attempt to avoid the two-step deprotection/reduction procedure needed for azides or hydrazides, lithium *tert*-butyl-*N*-tosyloxycarbamate (LiBTOC) was investigated as an electophilic aminating reagent. While the NHBoc moiety is directly introduced, ester enolates reacted in only 35% yield (to prepare Phe); an amide (oxazolidinone) enolate gave 55% yield in the presence of Cu(I) (*612, 613*).

N-Boc or *N*-Fmoc 3-aryloxaziridines have also been reported as electrophilic amination reagents (*614*); *N*-Boc 3-(4-cyanophenyl)oxaziridine was introduced as an amination reagent in 1993 (*615*). β-Silyl ketene acetals have been aminated with (ethoxycarbonyl)nitrene, generated by photolysis of N$_3$CO$_2$Et. *N*-Ethoxycarbonyl β-silylated-α-amino esters were produced in 36–44% yield (*616*). A number of 3-aryl-*N*-carboxamido oxaziridines were also tested for electrophilic amination, with the

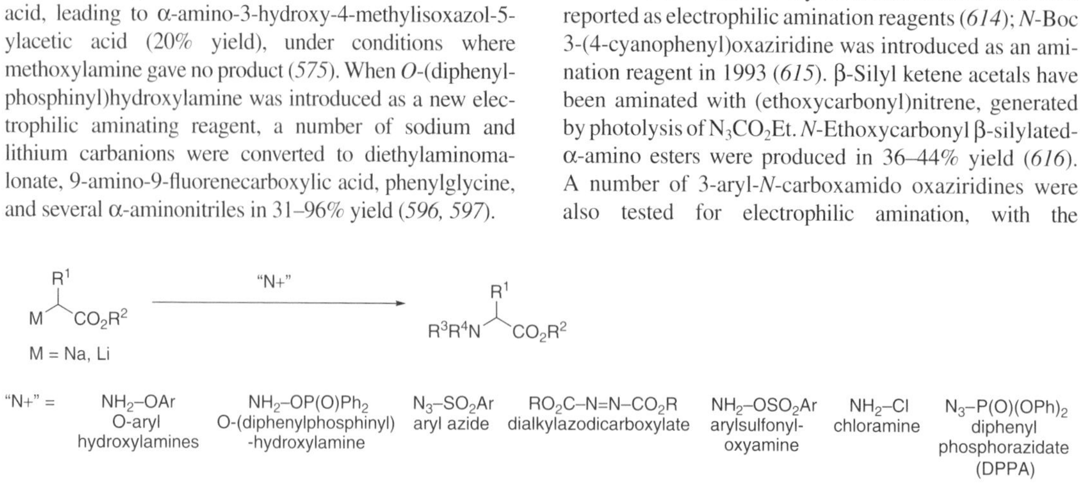

Scheme 2.24 Electrophilic Amination of an α-Metallated Acid/Ester.

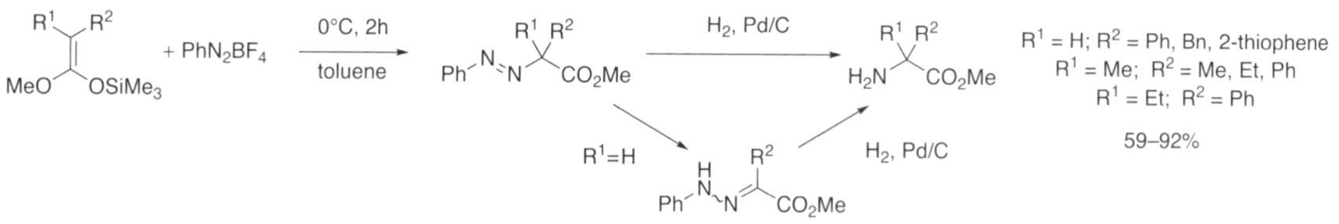

Scheme 2.25 Electrophilic Amination with Arenediazonium Tetrafluoroborates (617, 618).

3-(2-cyanophenyl) derivative giving the highest yields. N-(Et$_2$NCO) Gly-NMe$_2$, Ala-OtBu, and Aib-OEt were prepared in 51–55% yield (589).

Diazonium salts, specifically arenediazonium tetrafluoroborates, react with ketene silyl ketals to give α-azo- or α-hydrazono methyl esters in 59–92% yield (see Scheme 2.25). Hydrogenation gave quantitative yields of the amino acid methyl ester (617, 618). This reaction is an updated version of the Japp–Klingemann reaction, in which active methylenes bearing two electron-withdrawing groups, such as alkyl-substituted acetoacetic esters or malonic esters, are converted into hydrazono compounds or oximes by treatment with diazonium salts, sodium nitrite or ethyl nitrite. Amino acids such as Ala (619), Val (619), Nva (619), Leu (619), Nle (619), Phe (619), Met (619), Thr (620), phenylserine (621), β-hydroxy-Glu (622, 623), 4-hydroxy-5-chloro-2-aminopentanoic acid (624), Orn (625), [6-^{14}C]-Lys (626, 627), Glu (624), β-phenyl-Glu (628), and Trp (629) have been prepared via the Japp–Klingemann reaction (619, 620, 625). Nine amino acids (Ala, Abu, Nva, Nle, Ile, Asp, Glu, Phe, and O-methyl-Tyr) were prepared in 1942 by reaction of the substituted acetoacetates with butyl nitrite, producing α-oximino esters in 70–91% yield. Catalytic hydrogenation gave the amino esters (630). β-Hydroxy amino acids have also been prepared via oximes formed by treatment of esters with NaNO$_2$ or alkyl nitrites (631, 632).

2.3.4 Amination via Rearrangements

2.3.4a Curtius Rearrangement

The Curtius rearrangement converts carboxylic acid-derived acyl hydrazides or azides to amines or carbamates via a thermal rearrangement (see Scheme 2.26). The reaction was thoroughly reviewed in 1946, including applications to amino acid synthesis and general experimental conditions (633), and more recently in 1988 (518). Darapsky and Hillers first suggested the possibility of amino acid synthesis via a Curtius reaction in 1915, using ethyl cyanoacetate as substrate (634). The procedure has been used to prepare Gly (<1% yield) (635), Ala (635), Nva (31%) (635, 637), Val (635), Leu (20% from alkyl cyanoacetate) (636), Trp (637), Met (73%) (638), and 5-methyl-2-aminohexanoic acid (22%) (635, 636), with some of the monosubstituted cyanoacetic esters prepared by alkylation of cyanoacetic acid. The Darapsky method was extended by Gagnon et al. to the synthesis of Val (60%) (639, 640), Phe (50%) (639, 640), Tyr (11%) (639, 640), a number of substituted phenoxy alkyl amino acids (side chain R=(CH$_2$)$_n$ OAr) (641, 642), and a number of amino acids with alkyl side chains (638, 641–645) or aromatic side chains (642–645). Monosubstituted cyanoacetic esters were converted to the azides by reaction with hydrazine followed by diazotization with sodium nitrate or nitrous acid. The azides then underwent thermal rearrangement, and the resulting N-urethane α-amino-cyanoacetates were hydrolyzed. Attempted syntheses of Lys, Orn, and Asp failed (639). The hydrazine/sodium nitrate conditions were recently used to prepare a fluorinated analog of α-aminocyclopropane-1-carboxylic acid (646).

Malonic acid half-esters are another possible substrate (633), and were used for the synthesis of 1-amino-2-[N-(ethyl phosphonate)amino]cyclopropyl-1-carboxylic acid (647). The azide precursors can also be obtained by azide displacement of an acyl chloride, prepared from the acid by conventional means (PCl$_5$, SOCl$_2$, alkyl chloroformate, etc.). Reaction of mixed anhydrides

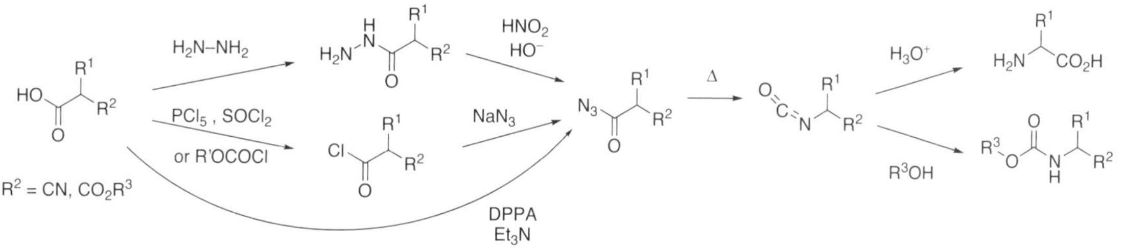

Scheme 2.26 Curtius-Darapsky Rearrangement.

with sodium azide avoids the need for isolation of pure acid chlorides (*648*). For example, ethyl chloroformate, followed by NaN₃, was employed during a Curtius synthesis of α,β-didehydrophenylalanine (26% yield from acid) from diethyl benzalmalonate (*649*). α-(Benzylthio)-Gly was prepared from the monomethyl thiomalonate via formation of the acid chloride with PCl₅ (*650*). A series of α-chlorofluoromethyl α-amino acids were obtained via alkylation of monomethyl mono-*tert*-butyl malonates, *tert*-butyl ester hydrolysis, acid chloride formation, reaction with NaN₃, and then rearrangement (*651*).

The traditional Curtius–Darapsky procedure is rather lengthy, with poor overall yields, but a significant improvement has been obtained via a modified Curtius reaction using diphenylphosphorylazide (DPPA, (PhO)₂PON₃), a reagent introduced by Yamada and co-workers (*652*). The modified reaction is much more simple than the conventional Curtius reaction and does not require the strong acid of the Schmidt reaction (see 2.3.4b) or the strong alkali of the Hofmann reaction (see 2.3.4c). Treatment of mono- or disubstituted malonic acid half-esters with DPPA/Et₃N, followed by refluxing with an alcohol, directly provided the *N*-alkyl/aryl-oxycarbonyl amino acid esters in 27–80% yield. Gly, Val, Phe, Trp, Phg, DOPA, and α-methyl-DOPA were prepared (*653, 654*). If the alcohol is present during the addition of DPPA, acid esterification takes place instead of rearrangement. The Curtius rearrangement with DPPA has been carried out on diacids attached to a polystyrene resin via an amide linkage to one of the acid groups (*655*).

More recently, both the Curtius rearrangement (DPPA/Et₃N, 81–94%) and the Hofmann rearrangement (Pb(OAc)₄/*t*BuOH, 43–7%) were used in two different routes to 2-substituted-3-hydroxymethyl-1-aminocyclopropane-1-carboxylic acids (*656, 657*). By carrying out the rearrangement in *tert*-BuOH the Boc-protected amino acid was obtained. A variety of other cyclic α-amino acids have been prepared by modified Curtius reactions, such as 1-aminocyclopropane-1-carboxylic acid (*658*), 1-aminocycloprop-2-ene-1-carboxylic acid (*659*), 2-hydroxymethyl-1-aminocyclopropane-1-carboxylic acid (*660*), 1-amino-2-methylenecyclopropane-1-carboxylic acid and a methylene-tritiated analog (*660, 661*), 1-amino-2-(but-3-enyl)cyclopropane-1-carboxylic acid (*662*), four diastereomers of 1-aminocyclopropyl-1-carboxylic acid with a 2-carboxycyclopropyl ring attached via a spirocyclic linkage at the 2-position (*663*), *cis*-2-aminocyclopropane-1-carboxylic acid (*664*), a series of *cis*-1-amino-2-substituted cyclopropane-1-carboxylic acids (*665*), a series of 3-substituted-1-aminocyclobutane-1-carboxylic acids (*666*), a series of 2-and 3-aryl-1-aminocyclobutane-1-carboxylic acids (*667*), *trans*-1-amino-2-phenyl-cyclobutane-1-carboxylic acid (*668*), both *cis*- and *trans*-3-(hydroxymethyl)-1-aminocyclobutane-1-carboxylic acid (*669*), and the lactone of 2,3-methanohomoserine (*670*). Other α,α-disubstituted amino acids prepared via Curtius rearrangements include α-(chlorofluoromethyl)-Orn and α-(chlorofluoromethyl)-*m*-Tyr (*671*).

2.3.4b Schmidt Rearrangement

The Schmidt rearrangement is closely related to the Curtius rearrangement, making use of HN₃ under acidic conditions to convert carboxylic acids to *N*-acyl amines in one step. A thorough review of the reaction, including its application to amino acid syntheses, appeared in 1946 (*672*), as did a brief comparison with the Curtius and Hofmann reactions (*633*). A short summary is also found in a more recent review of azides (*518*). Malonic acids directly yield α-amino acid products that do not react further with hydrazoic acid (*673*) as the newly generated α-amino group has an inhibitory effect on the reactivity of the carboxyl group, preventing reaction of the second acid moiety (*672*). A variety of alkyl amino acids (R = H, Me, *n*Pr, *i*Pr, *n*Bu, *i*Bu, *s*Bu, *n*Pent, *i*Pent, *n*Hex) were prepared in 38–89% yield in one step from the appropriate alkylmalonic acid by treatment with HN₃/H₂SO₄ in chloroform (*674*), although benzylmalonic acid gave Phe in only 16% yield (*675*). The ω-carboxyl group of α-aminodicarboxylic acids can be regioselectively converted to an amine, with Glu, α-aminoadipic acid, and α-aminopimelic acid transformed into 2,4-diaminobutyric acid, Orn, and Lys, respectively (41–75%) (*673, 676*). Homolysine was obtained by conversion of two acid groups of 1,1,6-hexanetricarboxylic acid (*677*).

β-Keto esters are also substrates for the Schmidt rearrangement, with the ketone carbonyl more reactive towards hydrazoic acid than the ester group (*672, 678*). For example, *N*-acetyl-α-benzyl-phe was prepared from ethyl dibenzylacetoacetate in trichloroacetic acid in 42–72% yield (*679, 680*), Tle-Leu from ethyl *tert*-butylacetoacetate in approximately 50% crude yield (*681*), Trp from ethyl α-aceto-β-(3-indolyl)propionate in 83% yield (*637, 682*), homoserine from 2-(2-alkoxyethyl)acetoacetate in 82–90% yield (*683*), and Met in 49% yield from the β-keto ester (*684*). A Schmidt rearrangement of ethyl 2-acetyl-5-fluoro-3-methylvalerate produced a mixture of *N*-acetyl 5-fluoroisoleucine and 5-fluoroisoleucine; the products were separated by crystallization (*685*).

The Schmidt rearrangement has also been used to synthesize a number of ω-amino acids (NH₂(CH₂)ₙCO₂H, n = 1, 2, 4, 7, 8; 58–83% yield) (*686*) and cyclic amino acids (*687*), although ester/acyl hydrolysis was problematic for the latter compounds. The rearrangement was recently used for a synthesis of a biaryl-containing cyclic amino acid (*688*) from a methyl/ethyl malonate precursor using methanesulfonic acid and NaN₃ in CHCl₃ (64% yield), and for several cyclic, bicyclic, and tricyclic α,α-disubstituted amino acids from both dicarboxyl and β-ketoester substrates (*689*).

2.3.4c Hofmann Rearrangement

The Hofmann rearrangement also accomplishes a net conversion of a carboxylic acid to an amine, but relies on treatment of an amide intermediate with hypohalide or lead tetraacetate. A review of the Hofmann reaction in

1946 included a number of examples of its application to amino acid synthesis, including experimental procedures for preparing isoserine (690). A brief comparison with the Curtius and Schmidt reactions appeared the same year (633). Amino acids that have been prepared via this procedure include isoserine (51% yield) (691), β-alanine (from succinimide in 65% yield) (692), methyl isoaminocamphonanate (70% yield) (693), aminodihydrocampholytic acid (aminolauronic acid, 65% yield) (694), and isoaminodihydrocampholytic acid (695). It has been used less frequently in recent years, although in 1983 it was employed to prepare dipropylglycine (696), which had previously been prepared by this method in the 1940s, along with a number of other α, α-dialkylglycines (697–701). A series of trans-1-amino-2-substituted-cyclopropane-1-carboxylic acids were prepared in 1985 (665), with an earlier synthesis of 1-aminocyclopropane-1-carboxylic acid and 1-amino-cyclobutane-1-carboxylic acid in 1922 (702), while in 1993 the Hofmann rearrangement was compared with the Curtius rearrangement for the synthesis of 2-substituted-3-hydroxymethyl-1-aminocyclopropane-1-carboxylic acids, using Pb(OAc)$_4$ and BuOH (43–77%) (656).

Mild electrochemically induced Hofmann rearrangement conditions have recently been reported, with a synthesis of a β-amino acid included among the examples presented (703). Alternatively, NBS/DBU induced the rearrangement to give methyl carbamate derivatives in high yield (704). The amide group of Asn can be converted into an amino group using mild conditions of iodosobenzene diacetate; thus, this reagent may also prove to be useful for rearrangements of α-carboxamides (705). The Hofmann rearrangement has also been used for a number of chiral substrates (see Section 5.3.8a).

2.3.4d Overman Rearrangement

Allyl alcohols functionalized with trichloroacetimidate undergo a [3,3] aza-Claisen rearrangement to introduce an acylated amine group (706). The alkene acts as a masked carboxyl group, which is revealed by oxidative cleavage and oxidation (see Scheme 2.27). Gly, Ala, Ser(Bn), Phe, Phg, Tle-Leu, and 1-amino-cyclopropane-1-carboxylic acid were prepared by this route, with 50–75% yield for formation of the trichloroacetimidate and rearrangement, and 60–84% yield for the oxidative cleavage (707). A complex α,α-disubstituted amino acid component of the neurotrophic factor lactacystin was

prepared from glucose, with the Overman rearrangement used for the amination step (708). Vinylglycine has been synthesized from 1,4-dihydroxy-2-butene by a modified version of this route. After rearrangement, the alkene was retained and the alcohol oxidized to generate the acid group (709). Asymmetric versions of this reaction have been developed (see Section 5.3.8b). A similar rearrangement of allylic N-benzoylbenzimidates has also been described, but not yet applied to an amino acid synthesis (708).

2.3.4e Neber Rearrangement

The Neber rearrangement introduces the amine group via an N-chloroimidate, which is derived from a nitrile in two steps (see Scheme 2.28). Treatment of the N-chloroimidate with alkoxide leads to amino acid ortho esters, which are easily hydrolyzed to the corresponding ester and then to the free acid. Gly, Ala, Abu, Ile, Phg, Phe, Tyr(Me), α-naphthylglycine, Ser(Me), and Lys (710, 711), Phe (712), and isotopically labeled β,γ-unsaturated-Val (713) ortho esters were prepared in 32–98% yield. More recently it was discovered that the rearrangement is induced by aqueous alkali, directly leading to the free amino acid. Vinylglycine was prepared by this route in 50% overall yield (714). The aqueous alkali conditions had previously been applied to a Neber-type rearrangement of allyl- or benzyl-glucosinolate, leading to vinylglycine (>100% yield!) or phenylglycine (39%) (715). Due to the glucose moiety, some asymmetric induction was observed. [1-^{13}C]-Homophenylalanine (Hfe) and [1-^{15}N]-Hfe have also been synthesized via the Neber rearrangement, with the N-chloroimidate derived from the reaction of labeled cyanide with α-chloro-3-phenylpropane (716).

2.3.4f Other Rearrangements

N-acylhydroxylamine-O-carbamates have been used to electrophilically aminate carboxylic acid derivatives via a hetero[3,3]sigmatropic shift (see Scheme 2.29) (717). Either LDA or KHMDS could be used to generate the di-enolate intermediate.

A number of racemic β,γ-unsaturated amino acids (both α-mono- and α,α-di-substituted) have been prepared by allylic oxidative rearrangement of α,β-unsaturated-β-seleno esters (12–87% yield), with BocNH$_2$, CbzNH$_2$ or TosNH$_2$ as the aminating nucleophile (718–720).

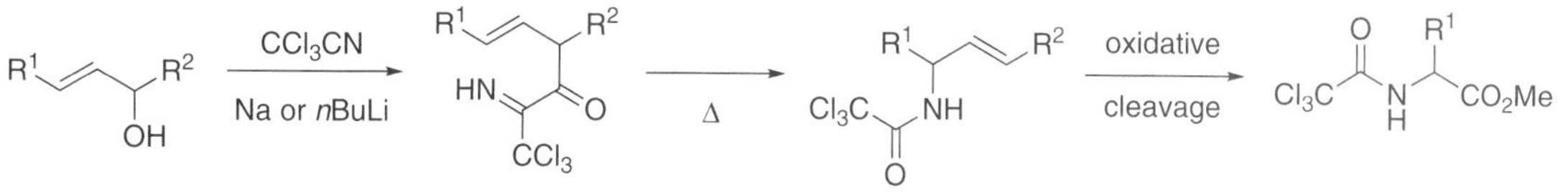

Scheme 2.27 Amination via Overman Rearrangement.

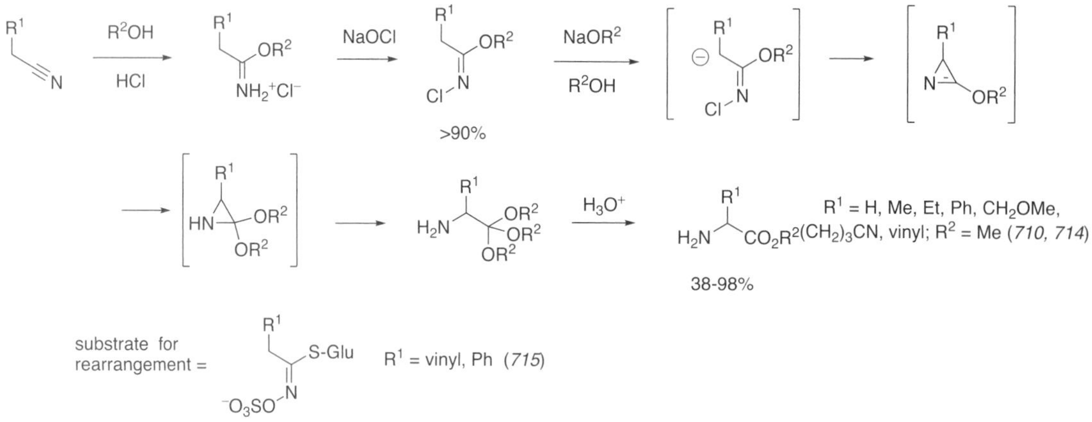

Scheme 2.28 Amination via Neber Rearrangement.

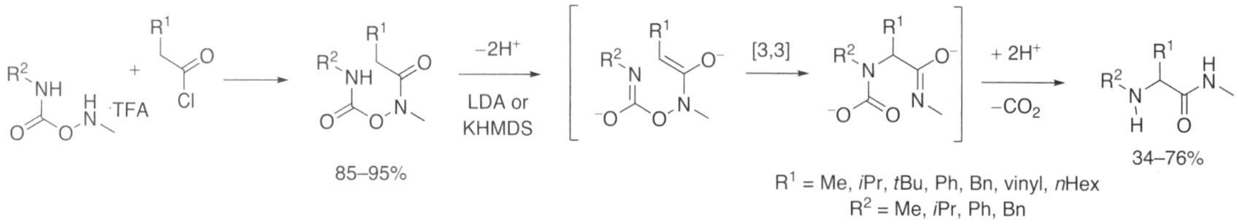

Scheme 2.29 Amination by Rearrangement (717).

2.3.5 Amination via Carbene Insertion into an N–H Bond

A carbene prepared from the tosyl hydrazone of the α-keto acid pyruvate was used in a marginally asymmetric synthesis of Ala (12–26% ee) by insertion into the N–H bond of chiral phenylethylamine (approximately 50% yield) (721); phenylglyoxalic acid was converted into racemic or slightly optically active Phg by a similar process (64–83% yield) (722). N-Substituted-trifluoroalanines were synthesized from 3,3,3-trifluoro-2-diazopropionate by insertion into the N–H bond of various N-mono- and N-di-substituted amines (68–85% yield) (723). With N-trisubstituted benzyldimethylamine, insertion occurred between the Bn–N bond, leading to α-trifluoromethyl-Phe. α-Trifluoromethyl-allylglycine was prepared in a similar fashion (65–72% yield) (723). N-Protected α-amino phosphonates have been obtained from diazophosphonates in 13–96% yield (722), while a

variety of N-substituted 2-amino-2-diethoxyphosphory-lacetates were synthesized by insertion of the carbene derived from ethyl 2-diazo-2-diethoxyphosphorylace-tate (724). An intramolecular carbenoid insertion reaction, employing an aromatic tether as a protecting group, led to α-amino acids and α,α-disubstituted amino acids (see Scheme 2.30). The α-allyl and N-unsaturated substituents could be cyclized via ruthenium-catalyzed metathesis to give α-substituted imino acids, with deprotected amino acids produced by hydrolysis and hydrogenation (725).

2.3.6 Amination via Reductive Amination of an α-Keto Ester

The biological synthesis of many amino acids proceeds by a transamination reaction of an α-keto ester involving pyridoxamine. Not surprisingly, many synthetic attempts

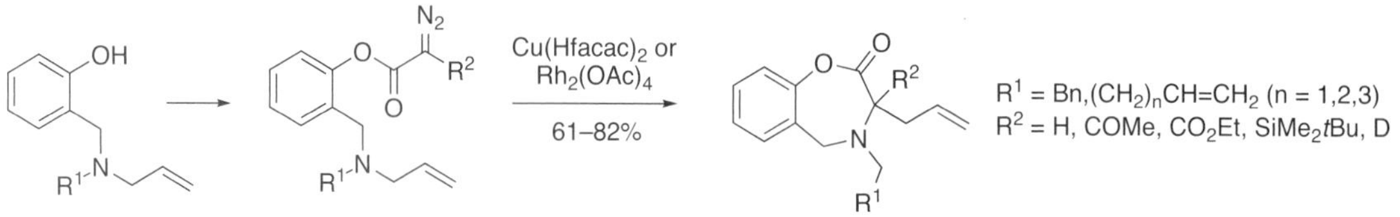

Scheme 2.30 Amination by Carbene Insertion (725).

have been made to mimic this reaction, but most attempts are stymied by poor reduction of the C=N bond. α-Keto acids are accessible by methods such as hydrolysis of acyl cyanides, which in turn can be obtained by cyanation of acyl chlorides with KCN (726), TMSCN (727) or tributyltin cyanide (728). α-Keto acids can also be prepared by the dehydrogenation of α-hydroxy esters and α-hydroxynitriles by t-butyl hydroperoxide in the presence of a ruthenium catalyst (729).

2.3.6a Imine/Oxime/Hydrazone Reduction via Hydride Addition or Hydrogenation

A number of simple reductive amination conditions have been described (see Scheme 2.31, Table 2.2). For example, Gly and Phg were prepared in 1885 via reductive amination of the α-keto acid using phenylhydrazine to form a hydrazone, followed by hydrazone reduction using sodium amalgam (730). The same method was used to prepare 2,3-diamino-1,4-butanedioic acid in 1893 (731). In 1936 Darapsky prepared a number of aromatic amino acids by reaction of the corresponding α-keto acids with hydrazine, reduction of the hydrazone with sodium amalgam, and then conversion of the hydrazino amino acid to an amino acid via a nitrosohydrazine intermediate (500). Gly, Ala, Val, Nle, Phe, and O-methyl-Tyr were prepared via the phenylhydrazones of the corresponding α-keto acids. Reductive cleavage of the phenylhydrazone by hydrogenation gave the amino acid in 85–98% yield (503). Ser has also been synthesized via a hydrazone (732). 2-Aminopimelic acid was obtained by reduction of the phenylhydrazone of α-ketopimelic acid with Zn/AcOH (733). In 1978 3′,5′-di-tert-butyl-4′-hydroxy-Phg was synthesized via hydrazone formation, with Zn/H⁺ reduction (734). Reductive amination of α-keto acids via hydrazone formation/reduction has been used to identify the α-keto acids contained in plants (735, 736).

Glu was prepared in 1890 via formation of an oxime and reduction with Zn/AcOH (737), while a 1914 synthesis prepared Tle via reduction of the oxime with Zn/AcOH or aluminum amalgam (738). The oximes of several heteroaryl ibotenic ester analogs were reduced using aluminum amalgam, with simultaneous oxime reduction and N–O bond cleavage (739, 740). Sodium amalgam was used in a synthesis of 3′-methoxy-Tyr (741). 4-Hydroxy-4-methylglutamic acid has been prepared by hydrogenation of the oxime of the lactone of

α-keto-γ-carboxy-pentanoic acid in unspecified yield (742). Oximes of α-keto acids can be reduced to N-hydroxy-α-amino acids with cyanoborohydride (743, 744) or by catalytic hydrogenation over Raney Ni (745), but oximes of α-keto esters require pyridine-borane (746) or $TiCl_3/NaBH_4$ (747). The latter conditions were employed in 2001 for a synthesis of hexafluoroleucine; L-tartaric acid was also present during the reduction, but it is unclear whether any asymmetry was induced (748). Oxime reduction with stannous chloride was employed in a synthesis of β-(4-pyridyl)alanine (749), several β-(diazophenyl)alanines (750), and both β-(2,4-dihydroxypyrimidin-6-yl)Ala and β-(4,6-dihydroxypyrimidin-2-yl)Ala (751); H_2/Pd-C was also employed for the β-(diazophenyl)alanines (750) and H_2/Raney Ni for the β-(dihydroxypyrimidinyl)Ala (751). β-Alkoxy-α-amino acids have been obtained in 66–84% yields from oximes via reduction with $NaBH_4/TiCl_3$ (752). A β-bromoalanine equivalent was prepared by reacting methoxyamine hydrochloride with ethyl bromopyruvate; the methoxyimine was maintained while the β-bromo group was converted to β-triphenylphosphine or β-trimethylphosphonate and then by a Wittig reaction to β,γ-unsaturated amino acids (753). The alkoxyimine was finally reduced with Zn/formic acid. An alkoxyimine has also been reduced with Red-Al, sodium bis(2-methoxyethoxy)aluminum hydride, in 80% yield (754). Oximes can also be produced by the reactions of esters with $NaNO_2$ or alkyl nitrites (see Section 2.3.3).

Imines can be generated from ammonia and α-keto acids/esters and then reduced by borohydride reagents (see Table 2.2). For example, erythro 3-fluorophenylalanine was prepared by reductive amination of 3-fluorophenylpyruvic acids using aqueous ammonia and sodium borohydride (755, 756); $NH_3/NaBH_4$ (25–60% yield) was found to be superior to $NH_3/NaBH_3CN$ (755). 3′-Bromo-4′-methoxy-5′-hydroxy-Phe (757) and the 3′-bromo-4′-hydroxy-5′-methoxy-Phe regioisomer (758) were prepared using $NH_4OH/NaBH_4$, as was β-fluoro-Ile (759). Deuterated phenylpyruvate was reductively aminated with $NH_4Br/NaBH_3CN$ in methanol-d_4 to give racemic [3,3-²H₂]-Phe (760); the same conditions were used to prepare [3-³H]-Val and [3-²H]-Val from isotopically labeled sodium α-ketoisovalerate (733). Reductive amination of phosphonopyruvates with $NH_3/NaBH_3CN$ produced α-amino-β-phosphono acids (761). β,β-Difluoro-α-keto esters can be reductively aminated in one step with aqueous $NH_3/NaBH_4$ or in two steps with H_2NOMe followed by Zn/HCO_2H; reaction with $NH_4Br/NaBH_3CN$ was unsuccessful (762). [2-²H]-Glycine was

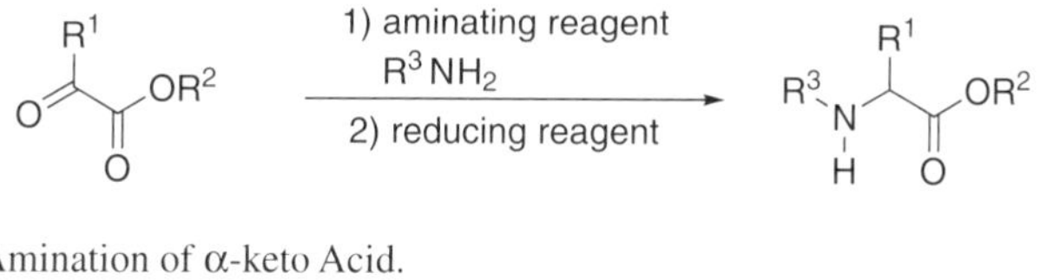

Scheme 2.31 Reductive Amination of α-keto Acid.

Table 2.2 Amino Acid Synthesis via Reductive Amination

Aminating Reagent	Reducing Reagent	R¹	R²	R³	Yield	Reference
PhNHNH₂	Na-Hg	H Ph	H	H	—	1885 (730)
PhNHNH₂	Na-Hg	CH(NH₂)CO₂H	H	H	—	1893 (731)
PhNHNH₂	Zn, AcOH	3,5-(tBu)₂-4-HO-Ph	H	H	85%	1978 (734)
[2,4-NO₂)₂-Ph]HNNH₂	Al-Hg	CH₂OH	H	H	24%	1946 (732)
PhNHNH₂ or PhN₂⁺Cl⁻	H₂, Pd-C	H Me iPr nBu Bn 4-MeO-Bn	H	H	85–95%	1973 (503)
H₂NNH₂	Na-Hg	Bn 4-Me-Bn (CH₂)₂Ph CH=CHPh (CH₂)₄Ph naphthyl	H	NH₂	21–52%	1936 (500)
NH₂OH	SnCl₂, HCl or H₂/Raney Ni		H	H	76–81% oxime 51–66% reduction	1965 (751)
NH₂OH	Na-Hg	CH₂CH₂CO₂H	H	H	—	1890 (737)
NH₂OH	Zn/AcOH or Al-Hg	tBu	H	H	40–80%	1914 (738)
NH₂OH	Na-Hg, D₂O	D	D	H	85%	1972 (882)
NH₂OH	Na-Hg	CH₂(2-oxo-2,3-dihydroindo-3-yl)	H	H	68%	1951 (883)
NH₂OH	Na-Hg	3-MeO-4-HO-Bn	H	H	48%	1950 (741)
NH₂OH	Al-Hg		Et	H	74%	1980 (739)
NH₂OH	Al-Hg		Me	H	56%	1978 (740)
NH₂OH	H₂, Raney-Ni	tBu	H	H	78%	1977 (745)
NH₂OH	H₂, Raney-Ni	CO(2-NO₂-3-HO-Ph)	H	H	64%	1957 (577)
NH₂OH	H₂, Pd-C or SnCl₂, HCl		H	H	58–89% oxime 10–80% reduction	1965 (750)
NH₂OH	SnCl₂, HCl	CH₂(4-pyridyl-1-oxide)	Et	H	50%	1958 (749)
NH₂OH	NaBH₄, TiCl₃	Me Et tBu Ph (CH₂)₂CO₂Me	Me, Et	H	63–82%	1989 (747)

Table 2.2 Amino Acid Synthesis via Reductive Amination (continued)

Aminating Reagent	Reducing Reagent	R^1	R^2	R^3	Yield	Reference
NH$_2$OH	NaBH$_4$, TiCl$_3$	CH$_2$CH(CF$_3$)$_2$	Et	H	12% overall from hexafluoro-acetone	2001 (748)
NH$_2$OH	NaBH$_4$, TiCl$_3$ or pyridine.BH$_3$	CH(Ph)OMe CH(4-Me-Ph)OMe CH(4-Cl-Ph)OMe CH(Ph)OEt CH(4-Me-Ph)OEt CH(4-Cl-Ph)OEt	Me, Et	H	66–84%	1999 (752)
NH$_2$OH	NaBH$_3$CN or LiBH$_3$CN	H Me Et nPr iPr (CH$_2$)$_2$CO$_2$H Ph Bn	H	OH	38–53%	1974 (744)
NH$_2$OH	NaBH$_3$CN or LiBH$_3$CN	4-OH-Bn	H	OH	74%	1977 (743)
NH$_2$OH	Red-Al, Bz-Cl	CH(Me)CH(OH)(4-MeO-Ph)	CO$_2$R$_2$= CONHtBu	Bz	78%	1991 (754)
NH$_2$OH	H$_2$, Pt	CH$_2$C(Me)(CO$_2$H)−R2	—	H	—	1967 (742)
NH$_2$OH, NH$_2$OBn	pyridine.BH$_3$	H Me Ph Bn	Et	OH, OBn	47–95%	1978 (746)
NH$_2$OMe	Zn, AcOH, H$_2$O	CF$_2$CH$_2$CH=CH$_2$	H	H	70–80%	1995 (762)
NH$_2$OMe	Zn, AcOH, H$_2$O	CH$_2$Br convert into CH$_2$PPh$_3$, CH=CHR	Et	OMe	—	1988 (753)
NH$_3$	Raney Ni	(CH$_2$)$_3$OH	H	H	48%	1950 (765)
NH$_3$, H$_2$O	NaBD$_4$	D	H	H	79%	1983 (763)
NH$_3$, H$_2$O	NaBH$_4$	3-Br-4-HO-5-MeO-Bn	H	H	69%	1969 (758)
NH$_3$, H$_2$O	NaBH$_4$	3-Br-4-MeO-5-HO-Bn	H	H	71%	1974 (757)
NH$_3$, H$_2$O	NaBH$_4$	CH$_2$C(OH)(Me)CO$_2$H	H	H	—	1967 (742)
NH$_3$, H$_2$O	NaBH$_4$	CF$_2$CH$_2$CH=CH$_2$ CF$_2$CH$_2$Ph	H	H	43–56%	1995 (762)
NH$_3$, H$_2$O	NaBH$_4$	CH$_2$F	H	H	60–80%	1978 (764)
NH$_3$, H$_2$O	NaBH$_4$	CHFPh CHF(4-Cl-Ph) CHF(4-NO$_2$-Ph)	H	H	25–60%	1984 (755)
NH$_3$, H$_2$O	NaBH$_4$	CF(Me)Et	H	H	—	1982 (759)
NH$_4$Br	NaBH$_3$CN	C^{2}H(Me)$_2$ C^{3}H(Me)$_2$	H	H	21–33%	1981 (733)
NH$_4$OAc	NaBH$_3$CN	CH$_2$PO$_3$H$_2$ CH(Me)PO$_3$H$_2$ CH(Et)PO$_3$H$_2$ C(Me)$_2$PO$_3$H$_2$	Me, Et	H	13–64%	1979 (761)
HCO$_2$NH$_4$	HCO$_2$H	tBu	Na	H	73%	1997 (767)

(Continued)

Table 2.2 Amino Acid Synthesis via Reductive Amination (continued)

Aminating Reagent	Reducing Reagent	R^1	R^2	R^3	Yield	Reference
HCO_2NH_4	Ir-H complex	Me Me with $[2\text{-}^2H,^{15}N]$ iPr iBu sBu Bn 4-HO-Bn 4-HO-Bn with $[2\text{-}^2H,^{15}N]$ $(CH_2)_2CO_2H$ $(CH_2)_2CO_2H$ with $[2\text{-}^2H,^{15}N]$	H	H	78–97%	2004 (768)
$ArNH_2$	H_2, Pd-C	Me Ph	Me	Ph $3\text{-}CF_3\text{-}Ph$ C_6Cl_5 $2\text{-}CO_2H\text{-}Ph$ $2,6\text{-}Me_2\text{-}Ph$	60–85%	1996 (769)
$H_2N(CH_2)_nCl$	$NaBH_3CN$, then LDA	$(CH_2)_n\text{-}R^3$	H	—	45–76%	1994 (770)
$H_2N(CH_2)_nSBn$ n = 2,3,4 $H_2N(CH_2)_3NHBoc$	$NaBH_3CN$, MeOH	H	H	$(CH_2)_nSBn$ n = 2,3,4 $(CH_2)_3NHBoc$	30–57%	1997 (771)
Pro-OBn, Lys(Cbz)-OEt	$NaBH_3CN$, MeOH	H	$CO_2R^2 =$ CO-Pro-OH	$CH(CO_2Et)$ $(CH_2)_4NHCbz$ $CH(CO_2Bn)$ $(CH_2)_3$- $(\alpha\text{-}N)$	—	1980 (772)
$BnNH_2$	$NaBH_3CN$	CHRSePh R = Me, Et, nPr, iPr, Bn	H	Bn	46–54%	2000 (884)
$H_2NCH_2PO_3H$	H_2, Pd-C	H	H	CH_2PO_3H	94%	1997 (773)
R^3NH_2	H_2, Pd-C	H	H	Et $(CH_2)_3NHC(=NH)$ NHPmc $(CH_2)_2CO_2t$Bu $(CH_2)_3CO_2t$Bu $(CH_2)_3$(imidazol-4-yl) $(CH_2)_3Ot$Bu $(CH_2)_3$(indol-3-yl) $(CH_2)_3$(4-HO-Ph) $(CH_2)_3$(1-adamantyl) $(CH_2)_2CHPh_2$ $(CH_2)_3CHPh_2$	62% (1 example)	1992 (774) 1992 (775)

prepared from glyoxylic acid in 79% crystallized yield by reduction amination with $NH_3/NaBD_4$; production of iminodiacetic acid byproduct was suppressed by reaction at increased temperature and reduced concentration (763). 3-Fluoro-$[2\text{-}^2H]$-alanine was prepared in a similar manner (764). 5-Hydroxy-2-aminopentanoic acid was prepared via Raney-Ni catalyzed reduction of an imine formed with ammonia (765). Imines of α-keto acids have also been generated within a cobalt complex and subsequently reduced with $NaBH_4$ or used as electrophiles to give amino acid cobalt complexes (766). A different procedure was used to obtain Tle; sodium trimethylpyruvate was reacted with ammonium formate

in formic acid at reflux. A second equivalent of the ke acid reacted to produce N-pivaloyl-*tert*-Leu (767). 2004 an acid-stable iridium hydride complex w employed to reductively aminate α-keto acids with amm nium formate. Ala, Val, Leu, Ile, Phe, Tyr, Glu, a $[2\text{-}^2H,^{15}N]$-Ala, -Tyr, and -Glu were produced wi 78–97% yield (768).

The reductive amination protocol provides the oppo tunity for preparing N-substituted amino acids. F example, N-aryl-Ala and -Phe methyl esters were p pared in 60–83% yield in a one-pot reaction by mixi the α-keto methyl ester and various substituted anilin (or nitrobenzene) along with Pd/C under H_2 (76

A number of Pro analogs were prepared by reductively aminating glyoxalic acid with an ω-chloroalkylamine; the N-(ω-chloroalkane) glycine was then deprotonated and cyclized to the imino acid by an intramolecular alkylation (770). In a similar manner (ω-thioalkyl) glycines and N-(ω-aminoalkyl)glycines were obtained from glyoxylic acid using NaBH$_3$CN for the reduction (771). N-Substituted Ala-Pro dipeptides were prepared by reductive amination of N-pyruvolyl-L-Pro with the α-amino group of Lys(Cbz)-OEt or the imino group of Pro-OBn, using NaBH$_3$CN for the reduction (772). The broad-spectrum herbicide glyphosate, N-(phosphono-methyl) glycine, has been prepared by hydrogenation of the imine formed from glyoxylic acid and (amino-methyl)phosphonic acid (773). Reductive amination is the preferred method for synthesis of peptoid monomers (N-alkyl glycines) (774, 775).

In recent years it has been found that the initial imine adduct of fluoroalkyl α-keto acids can be isomerized by a [1,3]-proton shift reaction to the corresponding Schiff base, without need for a reducing agent. The most simple trifluoro α-keto acid, trifluoropyruvate, does not readily form an imine with N-benzylamine, but it does with N-(1-phenylethylamine) (see Scheme 2.32). Triethyl-amine solution induced isomerization to the Schiff base of trifluoroalanine, with 1 N HCl hydrolysis producing the free amino acid (776). The same strategy can be used for the synthesis of fluorinated β-amino acids from β-keto esters (777–781).

A number of attempts have been made to construct transaminase enzyme mimics based on the enzymatic pyridoxamine cofactor; most of these artificial enzymes attempt to induce asymmetry during the reductive ami-nation and so are discussed in Section 5.3.9. Polyethyleneimine (PEI) has been partially N-alkylated with lauryl iodide and partially linked to pyridoxamine to produce a water-soluble polymeric pyridoxamine reagent. Treatment of pyruvic acid resulted in formation of Ala and pyridoxal, with a rate increase of 6700–8300-fold over pyridoxamine in solution at pH 5.0 (782). It is proposed that the PEI backbone provides nitrogens to act as general acid and base catalysts, while the alkyl chains form a hydrophobic cavity. Phenylpyruvic acid is aminated by the laurylated PEI catalyst 190-fold faster than by a methylated PEI analog (783). Lower-molecular-weight polyethylenimines were also effective (885). Accelerations have also been obtained using sep-arate pyridoxamine cofactors and polyethylenimine polymers, both with hydrophobic substituents to increase

binding (784). Pyridoxamine can be attached to den-drimers in order to increase its macromolecular, enzyme-like characteristics. Up to 1000-fold accelerations in the formation of racemic Ala and Phe were observed (785). The racemase activity of pyridoxal enzyme models can be increased relative to their transaminase activity by the attachment of basic substituents (786).

2.3.6b Imine/Oxime Reduction via Organometallic Addition or Cycloaddition

An alternative method of imine/oxime reduction is via addition of organometallic compounds. Regioselectivity is potentially a problem as the organometallic can add to either end of the C=N bond or to the carbonyl of the acid group. The electrophilic imine/oxime substrates have been derived from a variety of sources. If α-keto acids other than glyoxylate are used to form the imine/oxime C=N bond, organometallic addition leads to α,α-disubstituted amino acids. Many researchers have alky-lated imines prepared from α-halo- or α-acetoxyglycine derivatives, but the imines are usually a non-isolated (and often non-acknowledged) intermediate. The α-hydroxy/methoxy/acetoxy-glycine substrates are often prepared from glyoxylic acid by condensation with amines, amides or urethanes. All of these reactions will be discussed in the section on electrophilic glycine equivalents, as they involve introducing the side chain in addition to the amino group (see Section 3.9).

2.4 Addition of the Carboxyl Group to the Side Chain (Carboxylation Reactions)

Two general approaches have been used to synthesize amino acids by carboxylation of an amino compound: electrophilic carboxylation with an α-amino anion or nucleophilic carboxylation by addition of a $^-CO_2H$ syn-thon to imines.

2.4.1 Electrophilic Carboxylation

A number of α-amino acids have been synthesized by α-carboxylation of isocyano compounds, which are easily prepared by dehydration of formamides. Reaction with diethyl carbonate in the presence of NaH gave ethyl α-isocyanoacetate derivatives in 35–63% yield

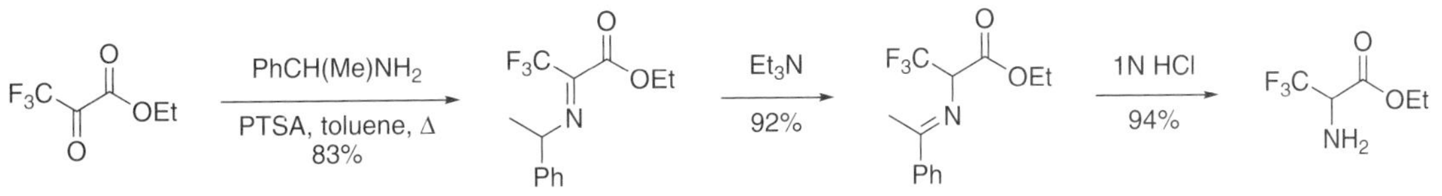

Scheme 2.32 Isomerization of Imines of Fluoroalkyl α-keto Acids (776).

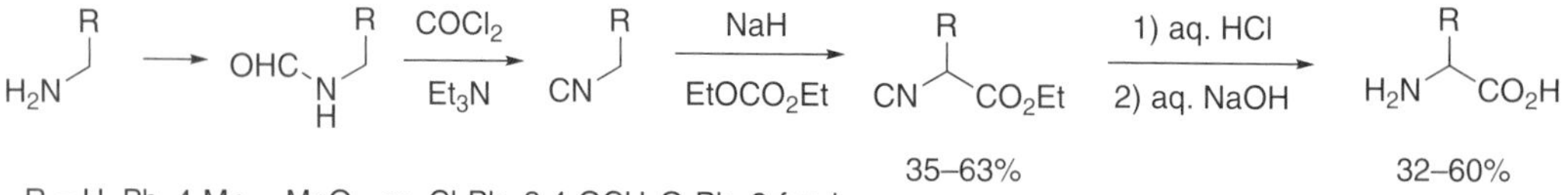

R = H, Ph, 4-Me-, -MeO-, or -Cl-Ph, 3,4-OCH$_2$O-Ph, 2-furyl

Scheme 2.33 Carboxylation of Isocyanides (*787*).

(see Scheme 2.33); a two-step hydrolysis produced the free amino acids in 32–60% yield (R = H, Ph, 4-Me-, MeO- or -Cl-Ph, 3,4-OCH$_2$O-Ph, 2-furyl) (*787*). Phenylglycine has been prepared by carboxylation of an α-lithiobenzylisocyanide anion or benzylamine Schiff base anion with CO$_2$. The desired amino acid was obtained in 81% and 52% yield, respectively (*788, 789*). A benzylimine anion was acylated with methyl chloroformate during two syntheses of α-vinyl-DOPA (*790, 791*), while α-lithiated isocyanides were carboxylated with ^{11}CO$_2$ to prepare [1-^{11}C]-tyrosine (*792*) and [1-^{11}C]-methionine (*793*). Alkyl chloroformates were used as the carboxyl group source in a synthesis of 2-amino-4-alkoxy-3-butenoic acid derivatives from lithiated isocyanopropanes (*794*).

α-Aminoorganolithiums have been carboxylated with CO$_2$ to give the N-protected amino acid in good yield (89–92%) (*795*). The α-aminoorganolithiums are accessible from aldehydes by addition of tributylstannane, Mitsunobu amination, and Sn–Li exchange (see Scheme 2.34). Esters can be generated by acylation with methyl chloroformate as electrophile (*796*). [1-^{14}C]-Sarcosine was synthesized by a similar procedure, using ^{14}CO$_2$

(52% yield for carboxylation) (*797*). Didehydroalanin (with a chiral amine auxiliary) was prepared via carbox lation of an organostannane; reaction of the organ lithium derivative was problematic (*798*).

A zirconocene aziridine complex has been used prepare N-aryl phenylglycine methyl ester in 50–72 yield by insertion of a CO$_2$ equivalent (ethylene carbo ate) into the Zr–C bond (*799*). Isocyanates also insert give phenylglycine amides (50–56% yield). An asym metric version of this reaction has produced other amir acids (see Section 5.4.1). Insertion is also involved in th reaction of an α-carbene ester with tertiary amine A transient ammonium ylide is generated, which the undergoes a Stevens [1,2]-shift to form a number amino acid esters in 26–97% yield, effectively insertin a methylene carboxy group between the amino grou and the side chain (*800*) (see Scheme 2.35).

A seven-membered cyclic analog of the excitato amino acid receptor agonist AMPA, 3-hydroxy-5,6,7, tetrahydro-4H-isoazolo[5,4-c]azepine-8-carboxyl acid, was prepared by carboxylation of an N-nitros derivative via treatment with n-butyllithium and meth chloroformate (*801*).

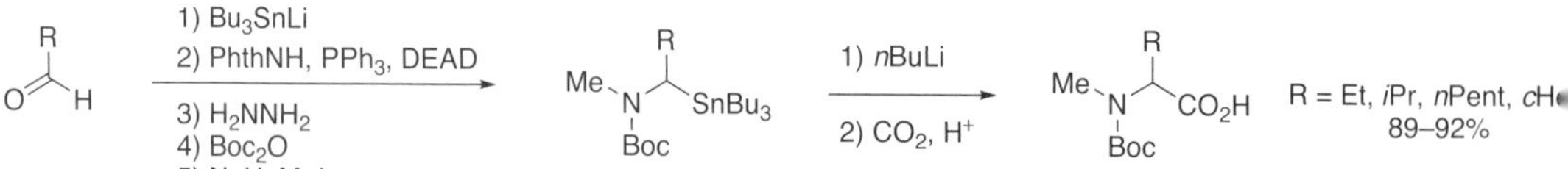

Scheme 2.34 Carboxylation of α-Aminoorganolithium (*795*).

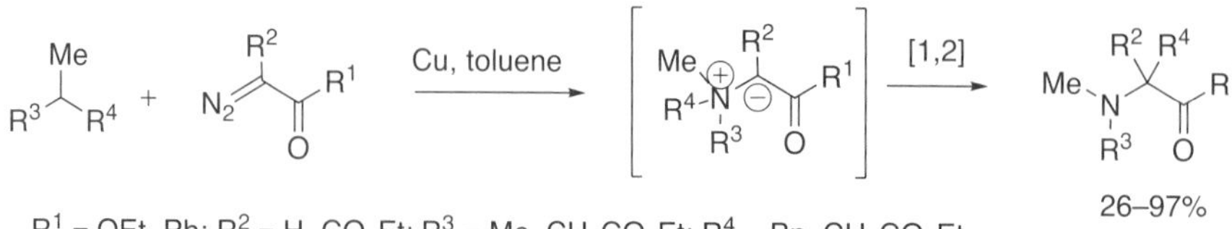

R^1 = OEt, Ph; R^2 = H, CO$_2$Et; R^3 = Me, CH$_2$CO$_2$Et; R^4 = Bn, CH$_2$CO$_2$Et

Scheme 2.35 Carboxylation via Carbene Insertion (*800*).

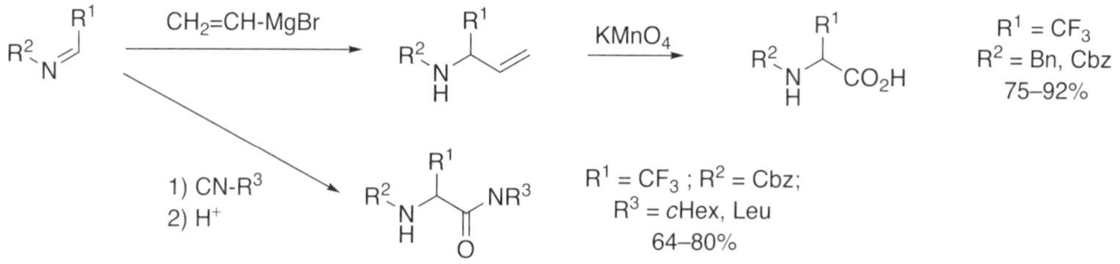

Scheme 2.36 Nucleophilic Carboxylation (*804*).

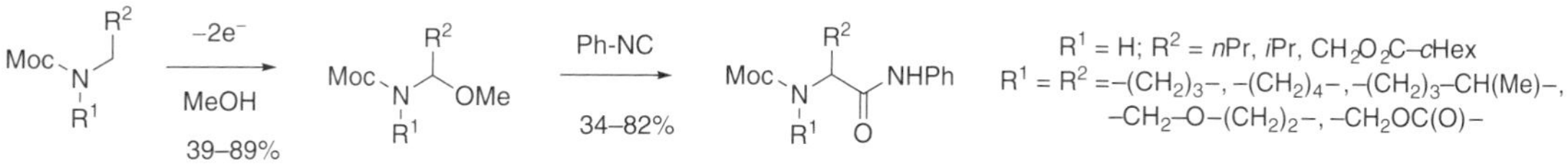

Scheme 2.37 Carboxylation via Electrolytic α-Methoxylation of Amine (*810, 811*).

2.4.2 Nucleophilic Carboxylation

α-Amino acids or β-amino acids can be synthesized by nucleophilic addition of masked carboxyl groups to imine equivalents prepared from aldehydes. Nucleophilic carboxylation is often accomplished by cyanide addition to an imine, essentially the last step of a Strecker synthesis (see Section 2.2). Proline derivatives have been obtained by the addition of cyanide to pyrrolines (*802*); 5,5-dimethylproline was also prepared via addition of cyanide to the corresponding nitrone, 5,5-dimethylpyrroline-1-oxide (*803*). Other carboxyl equivalents can be used instead of cyanide, such as vinylmagnesium bromide. Addition of this Grignard reagent to an imine followed by oxidative cleavage of the alkene generates the carboxyl group. This approach was employed in a synthesis of 3,3,3-trifluoroalanine, giving the product in 75% yield (see Scheme 2.36) (*804*). Addition of an isonitrile led to the trifluoroalanine amino acid amide instead (*804*). Direct conversion of alkenes to methyl esters using ozonolysis in methanolic NaOH has been reported (*805*). The addition of other C_2 units containing a masked carboxyl group can be used to synthesize β-amino acids (see Section 11.2.2c). For example, lithium enolates and ketene silyl acetals were added to an in situ generated cyclopropanone Schiff base to prepare a number of cyclopropane-containing β-amino acids (*806*).

N-Sulfonyl aldimines, generated in situ by reaction of aldehydes and N-sulfinyl sulfonamides, react much more cleanly than imines with organolithium and Grignard reagents. Vinyl and allyl reagents have been added in 66–90% yield to give tosyl-protected amines and, while not reported, the alkene groups could easily be oxidatively cleaved to generate the α- or β-amino acids (*807*). Similarly, diarylidenesulfamides are efficient Michael

acceptors for allyl reagents, with 83–92% yield after amine deprotection (*808*). α-Hydroxy-N-trimethylsilylimines, prepared in situ from α-hydroxy aldehydes by treatment with LiHMDS, also react well with allyl organometallic reagents (65–96% yield) (*809*).

Electrochemistry has been used to prepare α-alkoxy urethanes in 39–89% yield by anodic α-methoxylation (see Scheme 2.37). The carboxy group was then introduced by methoxy displacement with phenylisocyanide in the presence of a Lewis acid. Hydrolysis gives the N-acyl amino acid phenylamide in 34–82% (*810, 811*).

$[1-{}^{13}C]$-Gly was synthesized by displacement of N-(chloromethyl)phthalimide with Na^{13}CN, followed by hydrolysis (*812*).

References

1. Brack, A. *Pure Appl. Chem.* **1993**, *65*, 1143–1151.
2. Eschenmoser, A.; Kisakürek, M.V. *Helv. Chim. Acta* **1996**, *79*, 1249–1259.
3. Hager, A.J.; Pollard J.D.; Jr., Szostak, J.W. *Chem. Biol.* **1996**, *3*, 717–725.
4. Gilbert, W. *Nature* **1986**, *319*, 618.
5. Hughes, R.A.; Robertson, M.P.; Ellington, A.D.; Levy, M. *Curr. Opin. Chem. Biol.* **2004**, *8*, 629–633.
6. Frauendorf, C.; Jäschke, A. *Angew. Chem., Int. Ed. Engl.* **1998**, *37*, 1378–1381.
7. Kumar, R.K.; Yarus, M. *Biochemistry.* **2001**, *40*, 6998–7004.
8. Li, X.; Liu, D.R. *Angew. Chem., Int. Ed.* **2004**, *43*, 4848–4870.
9. Kvenvolden, K.A.; Lawless, J.G.; Ponnamperuma, C. *Proc. Natl. Acad. Sci. U.S.A.* **1971**, *68*, 486–490.

10. Lawless, J.G.; Kvenvolden, K.A.; Peterson, E.; Ponnamperuma, C.; Moore, C. *Science* **1971**, *173*, 626–627.

11. Kvenvolden, K.; Lawless, J.; Pering, K.; Peterson, E.; Flores, J.; Ponnamperuma, C.; Kaplan, I.R.; Moore, C. *Nature* **1970**, *228*, 923–926.

12. Oró, J.; Nakaparksin, S.; Lichtenstein, H.; Gil-Av, E. *Nature* **1971**, *230*, 107–108.

13. Oró, J.; Gibert, J.; Lichtenstein, H.; Wikstrom, S.; Flory, D.A. *Nature* **1971**, *230*, 105–106.

14. Engel, M.H.; Macko, S.A. *Nature*, **1997**, *389*, 265–268.

15. Pizzarello, S.; Cronin, J.R. *Nature* **1998**, *394*, 236.

16. Epstein, S.; Krishnamurthy, R.V.; Cronin, J.R.; Pizzarello, S.; Yuen, G.U. *Nature* **1987**, *326*, 477–479.

17. Engel, M.H.; Macko, S.A.; Silfer, J.A. *Nature* **1990**, *348*, 47–49.

18. Meierhenrich, U.J.; Muñoz Caro, G.M.; Bredehöft, J.H.; Jessberger, E.K.; Thiemann, W.H.-P. *Proc. Natl. Acad. Sci. U.S.A.* **2004**, *101*, 9182–9186.

19. Ehrenfreund, P.; Glavin, D.P.; Botta, O.; Cooper, G.; Bada, J.L. *Proc. Natl. Acad. Sci. U.S.A.* **2001**, *98*, 2138–2141.

20. Macko, S.A.; Uhle, M.E.; Engel, M.H.; Andrusevich, V. *Anal. Chem.* **1997**, *69*, 926–929.

21. Pizzarello, S. *Acc. Chem. Res.* **2006**, *39*, 231–237.

22. Napier, W.N. *Mon. Noti. R. Astron. Soc.* **2004**, *348*, 46.

23. Wilson, E. *Chem. Eng. News* **2004**, *February 14*, 44.

24. Skelley, A.M.; Scherer, J.R.; Aubrey, A.D.; Grover, W.H.; Ivester, R.H.C.; Ehrenfreund, P.; Grunthaner, F.J.; Bada, J.L.; Mathies, R.A. *Proc. Natl. Acad. Sci. U.S.A.* **2005**, *102*, 1041–1046.

25. Hutt, L.D.; Glavin, D.P.; Bada, J.L.; Mathies, R.A. *Anal. Chem.* **1999**, *71*, 4000–4006.

26. Skelley, A.M.; Mathies, R.A. *J. Chromatogr., A* **2003**, *1021*, 191–199.

27. Munro, N.J.; Huang, Z.; Fnegold, D.N.; Landers, J.P. *Anal. Chem.* **2000**, *72*, 2765–2773.

28. Beegle, L.W.; Kanik, I.; Matz, L.; Hill H.H., Jr. *Anal. Chem.* **2001**, *73*, 3028–3934.

29. Formánek, P.; Klejdus, B.; Vranova, V. *Amino Acids* **2005**, *28*, 427–429.

30. Buch, A.; Sternberg, R.; Meunier, D.; Rodier, C.; Laurent, C.; Raulin, F.; Vidal-Madjar, C. *J. Chromatogr., A* **2003**, *999*, 165–174.

31. Glavin, D.P.; Bada, J.L. *Anal. Chem.* **1998**, *70*, 3119–3122.

32. Miller, S.J. *Science* **1953**, *117*, 528–529.

33. Miller, S.L. *J. Am. Chem. Soc.* **1955**, *77*, 2351–2361.

34. Van Trump, J.E.; Miller, S.L. *Science* **1972**, *175*, 859–860.

35. Miller, S.L.; Urey, H.C. *Science* **1959**, *130*, 245–251.

36. Plankensteiner, K.; Reiner, H.; Schranz, B.; Rode, B.M. *Angew. Chem., Int. Ed.* **2004**, *43*, 1886–1888.

37. Harada, K.; Suzuki, S. *Nature* **1977**, *266*, 275–276.

38. Kovács, J.; Nagy, H. *Nature* **1961**, *190*, 531–532.

39. Munegumi, T.; Shimoyama, A.; Harada, K. *Chem. Lett.* **1997**, 393–394.

40. Harada, K.; Fox, S.W. *Nature* **1964**, *201*, 335–336.

41. Taube, M.; Zdrojewski, St.Z.; Samochocka, K.; Jezierska, K. *Angew. Chem., Int. Ed. Engl.* **1967**, *6*, 247.

42. Lawless, J.G.; Boynton, C.D. *Nature* **1973**, *243*, 405–407.

43. Poncelet, G.; Van Assche, A.T.; Fripiat, J.J. *Origins of Life* **1975**, *6*, 401–406.

44. Wolman, Y.; Miller, S.L. *Tetrahedron Lett.* **1972**, *13*, 1119–1200.

45. Ferris, J.P.; Chen, C.T. *J. Am. Chem. Soc.* **1975**, *97*, 2962–2967.

46. Khare, B.N.; Sagan, C. *Nature* **1971**, *232*, 577–579.

47. Khare, B.N.; Sagan, C. *Science* **1971**, *173*, 417–420.

48. Dunn, W.W.; Aikawa, Y.; Bard, A.J. *J. Am. Chem. Soc.* **1981**, *103*, 6893–6897.

49. Chyba, C.; Sagan, C. *Nature* **1992**, *355*, 125–132.

50. Barak, I.; Bar-Nun-A. *Origins of Life* **1975**, *6*, 483–506.

51. Bar-Nun, A. *Origins of Life* **1975**, *6*, 109–115.

52. Bar-Nun, A.; Bauer, S.H.; Sagan, C. *Science* **1970**, *162*, 470–473.

53. Sokolskaya, A. *Origins of Life* **1976**, *7*, 183–185.

54. Hafenbradl, D.; Keller, M.; Wächtershäuser, G. Stetter, K.O. *Tetrahedron Lett.* **1995**, *36*, 5179–5182.

55. Keefe, A.D.; Miller, S.L.; McDonald, G.; Bada, J. *Proc. Natl. Acad. Sci. U.S.A.* **1995**, *92*, 11904–11906.

56. Huber, C.; Wächtershäuser, G. *Tetrahedron Lett.* **2003**, *44*, 1695–1697.

57. Severin, K. *Angew. Chem., Int. Ed.* **2000**, *39*, 3589–3590.

58. Shevlin, P.B.; McPherson, D.W.; Melius, P. *J. Am. Chem. Soc.* **1981**, *103*, 7007–7009.

59. Bernstein, M.P.; Dworkin, J.P.; Sandford, S.A.; Cooper, G.W.; Allamandola, L.J. *Nature* **2002**, *416*, 401–403.

60. Muñoz Caro, G.M.; Meierhenrich, U.J.; Schutte W.A.; Barbier, B.; Arcones Segovia, A.; Rosenbauer H.; Thiemann, W.H.-P.; Brack, A.; Greenberg, J.M. *Nature* **2002**, *416*, 403–406.

61. Huber, C.; Wächtershäuser, G. *Science* **1998**, *281*, 670–672.

62. Huber, C.; Eisenreich, W.; Hecht, S.; Wächtershäuser, G. *Science* **2003**, *301*, 938–940.

63. Leman, L.; Orgel, L.; Ghadiri, M.R. *Science* **2004**, *306*, 283–286.

64. Bujdák, J.; Rode, B.M. *Amino Acids* **2001**, *2*, 281–291.

65. Matrajt, G.; Blanot, D. *Amino Acids* **2004**, *26*, 153–158.

66. Biron, J.-P.; Pascal, R. *J. Am. Chem. Soc.* **2004**, *126*, 9198–9199.

67. Tamura, K.; Schimmel, P. *Proc. Natl. Acad. Sci. U.S.A.* **2003**, *100*, 8666–8669.

68. Moser, R.E.; Matthews, C.N. *Experientia* **1968**, *24*, 658–659.

69. Draganic, I.; Draganic, Z.; Shimoyama, A.; Ponnamperuma, C. *Origins of Life* **1977**, *8*, 377–382.

70. Draganic, I.; Draganic, Z.; Shimoyama, A.; Ponnamperuma, C. *Origins of Life* **1977**, *8*, 371–376.

71. Sweeney, M.A.; Toste, A.P.; Ponnamperuma, C. *Origins of Life* **1976**, *7*, 187–189.

72. Moser, M.E.; Claggett, A.R.; Matthews, C.N. *Tetrahedron Lett.* **1968**, 1599–1603.

73. Wollin, G.; Ryan, W.B.F. *Biochim. Biophys. Acta* **1979**, *584*, 493–506.

74. Bade, J.L.*; McDonald, G.D. Anal. Chem.* **1996**, 668A–673A.

75. Cronin, J.R.; Pizzarello, S. *Science,* **1997**, *275*, 951–955.

76. Bonner, W.A.; Rubenstein, E. *BioSystems* **1987**, *20*, 99–111.

77. Bada, J.L. *Nature* **1995**, *374*, 594–595.

78. Podlech, J. *Angew. Chem., Int. Ed. Engl.* **1999**, *38*, 477–478.

79. Feringa, B.L.; van Delden, R.A. *Angew. Chem., Int. Ed.* **1999**, *38*, 3418–3438.

80. Buschmann, H.; Thede, R.; Heller, D. *Angew. Chem., Int. Ed.* **2000**, *39*, 4033–4036.

81. Avalos, M.; Babiano, R.; Cintas, P.; Jiménez, J.L.; Palacios, J.C. *Chem. Commun.* **2000**, 887–892.

82. Griesbeck, A.G.; Meierhenrich, U.J. *Angew. Chem., Int. Ed.* **2002**, *41*, 3147–3154.

83. Jacoby, M. *Chem. Eng. News* **2003**, *August 11*, 5.

84. Ávalos, M.; Babiano, R.; Cintas, P.; Jiménez, J.L.; Palacios, J.C. *Tetrahedron: Asymmetry* **2004**, *15*, 3171–3175.

85. Avalos, M.; Babiano, R.; Cintas, P.; Jiménez, J.L.; Palacios, J.C. *Tetrahedron: Asymmetry* **2000**, *11*, 2845–2874.

86. Wang, W.; Min, W.; Bai, F.; Sun, L.; Yi, F.; Wang, Z.; Yan, C.; Ni, Y.; Zhao, Z. *Tetrahedron: Asymmetry* **2002**, *13*, 2427–2432.

87. Rubenstein, E.; Bonner, W.A.; Noyes, H.P.; Brown, G.S. *Nature* **1983**, *306*, 118.

88. Bailey, J.; Chrysostomou, A.; Hough, J.H.; Gledhill, T.M.; McCall, A.; Clark, S.; Ménard, F.; Tamura, M. *Science* **1998**, *281*, 672–674.

89. Flores, J.J.; Bonner, W.A.; Massey, G.A. *J. Am. Chem. Soc.* **1977**, *99*, 3622–3624.

90. Nishino, H.; Kosaka, A.; Hembury, G.A.; Aoki, F.; Miyauchi, K.; Shitomi, H.; Onuki, H.; Inoue, Y. *J. Am. Chem. Soc.* **2002**, *124*, 11618–11627.

91. Nishino, H.; Kosaka, A.; Hembury, G.A.; Shitomi, H.; Onuki, H.; Inoue, Y. *Org. Lett.* **2001**, *3*, 921–924.

92. Meierhenrich, U.J.; Nahon, L.; Alcaraz, C.; Bredehöft, J.H.; Hoffmann, S.V.; Barbier, B.; Brack, A. *Angew. Chem., Int. Ed.* **2005**, *44*, 5630–5634.

93. Sato, I.; Sugie, R.; Matsueda, Y.; Furumura, Y.; Soai, K. *Angew. Chem., Int. Ed.* **2004**, *43*, 4490–4492.

94. Breslow, R.; Levine, M.S. *Tetrahedron Lett.* **2006**, *47*, 1809–1812.

95. Crisma, M.; Moretto, A.; Formaggio, F.; Kaptein, B.; Broxterman, Q.B.; Toniolo, C. *Angew. Chem., Int. Ed.* **2004**, *43*, 6695–6699.

96. Berger, R.; Quack, M.; Tschumper, G.S. *Helv. Chim. Acta* **2000**, *83*, 1919–1950.

97. Hazen, R.M.; Filley, T.R.; Goodfriend, G.A. *Proc. Natl. Acad. Sci. U.S.A.* **2001**, *98*, 5487–5490.

98. Oaki, Y.; Imai, H. *J. Am. Chem. Soc.* **2004**, *126*, 9271–9275.

99. Kühnle, A.; Linderoth, T.R.; Hammer, B.; Besenbacher, F. *Nature* **2002**, *415*, 891–893.

100. Banno, N.; Nakanishi, T.; Matsunaga, M.; Asahi, T.; Osaka, T. *J. Am. Chem. Soc.* **2004**, *126*, 428–429.

101. Zaia, D.A.M. *Amino Acids* **2004**, *27*, 113–18.

102. Kojo, S.; Uchino, H.; Yoshimura, M.; Tanaka, K. *Chem. Commun.* **2004**, 2146–2147.

103. Green, M.M.; Selinger, J.V. *Science* **1998**, *282*, 880–881.

104. Saghatelian, A.; Yokobayashi, Y.; Soltani, K.; Ghadiri, M.R. *Nature* **2001**, *409*, 797–801.

105. Bolli, M.; Micura, R.; Eschenmoser, A. *Chem. Biol.* **1997**, *4*, 309–320.

106. Zepik, H.; Shavit, E.; Tang, M.; Jensen, T.R.; Kjaer, K.; Bolbach, G.; Leiserowitz, L.; Weissbuch, I.; Lahav, M. *Science* **2002**, *295*, 1266–1269.

107. Weissbuch, I.; Bolbach, G.; Zepik, H.; Shavit, E.; Tang, M.; Frey, J.; Jensen, T.R.; Kjaer, K.; Leiserowitz, L.; Lahav, M. *J. Am. Chem. Soc.* **2002**, *124*, 9093–9104.

108. Rubinstein, I.; Bolbach, G.; Weygand, M.J.; Kjaer, K.; Weissbuch, I.; Lahav, M. *Helv. Chim. Acta* **2003**, *86*, 3851–3866.

109. Rubinstein, I.; Weissbuch, I.; Weygand, M.J.; Kjaer, K.; Leiserowitz, L.; Lahav, M. . *Chim. Acta* **2003**, *86*, 3867–3874.

110. Rubinstein, I.; Kjaer, K.; Weissbuch, I.; Lahav, M. *Chem. Commun.* **2005**, 5432–5435.

111. Weissbuch, I.; Zepik, H.; Bolbach, G.; Shavit, E.; Tang, M.; Jensen, T.R.; Kjaer, K.; Leiseroiwtz, L.; Lahav, M. *Chem. Eur. J.* **2003**, *9*, 1782–1794.

112. Nery, J.G.; Bolbach, G.; Weissbuch, I.; Lahav, M. *Angew. Chem., Int. Ed.* **2003**, *42*, 2157–2161.

113. Hitz, T.; Luisi, P.L. *Helv. Chim. Acta* **2002**, *85*, 3975–3983.

114. Hitz, T.; Luisi, P.L. *Helv. Chim. Acta* **2003**, *86*, 1423–1434.

115. Hitz, T.; Luisi, P.L. *Helv. Chim. Acta* **2001**, *84*, 842–848.

116. Neery, J.G.; Bolbach, G.; Weissbuch, I.; Lahav, M. *Chem. Eur. J.* **2005**, *11*, 3039–3048.

117. Plankensteiner, K.; Reiner, H.; Rode, B.M. *Peptides* **2005**, *26*, 535–541.

118. Hill, R.R.; Birch, D.; Jeffs, G.E.; North, M. *Org. Biomol. Chem.* **2003**, *1*, 965–972.

119. Julian, R.R.; Myung, S.; Clemmer, D.E. *J. Am. Chem. Soc.* **2004**, *126*, 4110–4111.

120. Cooks, R.G.; Zhang, D.; Koch, K.J.; Gozzo, F.C.; Eberlin, M.N. *Anal. Chem.* **2001**, *73*, 3646–3655.

121. Koch, K.J.; Gozzo, F.C.; Zhang, D.; Eberlin, M.N.; Cooks, R.G. *Chem. Commun.* **2001**, 1854–1855.

122. Gronert, S.; O'Hair, R.A.J.; Fagin, A.E. *Chem. Commun.* **2004**, 1944–1945.

123. Takats, Z.; Nanita, S.C.; Cooks, R.G. *Angew. Chem., Int. Ed.* **2003**, *42*, 3521–3523.

124. Takats, Z.; Nanita, S.C.; Schlosser, G.; Vekey, K.; Cooks, R.G. *Anal. Chem.* **2003**, *75*, 6147–6154.

125. Koch, K.J.; Gozzo, F.C.; Nanita, S.C.; Takats, Z.; Eberlin, M.N.; Cooks, R.G. *Angew. Chem. Int. Ed.* **2002**, *41*, 1721–1724.

126. Takats, Z.; Nanita, S.C.; Cooks, R.G.; Schlosser, G.; Vekey, K. *Anal. Chem.* **2003**, *75*, 1514–1523.

127. Nanita, S.C.; Cooks, R.G. *Angew. Chem., Int. Ed.* **2006**, *45*, 554–569.

128. Pollreisz, F.; Gömöry, Á.; Schlosser, G.; Vékey, K.; Solt, I.; Császár, A.G. *Chem. Eur. J.* **2005**, *11*, 5908–5916.

129. Córdova, A.; Engqvist, M.; Ibrahem, I.; Casas, J.; Sundén, H. *Chem. Commun.* **2005**, 2047–2049.

130. Córdova, A.; Ibrahem, I.; Casas, J.; Sundén, H.; Engqvist, M.; Reyes, E. *Chem. Eur. J.* **2005**, *11*, 4772–4784.

131. Pizzarello, S.; Weber, A.L. *Science* **2004**, *303*, 1151.

132. Casas, J.; Engqvist, M.; Ibrahem, I.; Kaynak, B.; Córdova, A. *Angew. Chem., Int. Ed.* **2005**, *44*, 1343–1345.

133. Bassan, A.; Zou, W.; Reyes, E.; Himo, F.; Córdova, A. *Angew. Chem., Int. Ed.* **2005**, *44*, 7028–7032.

134. Soai, K.; Shibata, T.; Morioka, H.; Choji, K. *Nature* **1995**, *378*, 767–768.

135. Blackmond, D.G. *Proc. Natl. Acad. Sci. U.S.A.* **2004**, *101*, 5732–5736.

136. Gridnev, I.D.; Brown, J.M. *Proc. Natl. Acad. Sci. U.S.A.* **2004**, *101*, 5727–5731.

137. Plasson, R.; Bersini, H.; Commeyras, A. *Proc. Natl. Acad. Sci. U.S.A.* **2004**, *101*, 16733–16738.

138. Podlech, J.; Gehring, T. *Angew. Chem., Int. Ed.* **2005**, *44*, 5776–5777.

139. Mathew, S.P.; Iwamura, H.; Blackmond, D.G. *Angew. Chem. Int. Ed.* **2004**, *43*, 3317–3312.

140. Kelly, D.R.; Meek, A.; Roberts, S.M. *Chem. Commun.* **2004**, 2021–2022.

141. Egami, F. *Origins of Life* **1981**, *11*, 197–202.

142. Borsenberger, V.; Crowe, M.A.; Lehbauer, J.; Raftery, J.; Helliwell, M.; Bhutia, K.; Cox, T.;

Sutherland, J.D. *Chem. and Biodiversity* **2004**, *1*, 203–246.

143. Smith, J.M.; Borsenberger, V.; Raftery, J.; Sutherland, J.D. *Chem. Biodiversity* **2004**, *1*, 1418–1451.

144. Ace, K.; Sutherland, J.D. *Chem. and Biodiversity* **2004**, *1*, 1678–1693.

145. Saewan, N.; Crowe, M.A.; Helliwell, M.; Raftery, J.; Chantrapromma, K.; Sutherland, J.D. *Chem. Biodiversity* **2005**, *2*, 66–83.

146. Copley, S.D.; Smith, E.; Morowitz, H.J. *Proc. Natl. Acad. Sci. U.S.A.* **2005**, *102*, 4442–4447.

147. Maurel, M.-C.; Décout, J.-L. *Tetrahedron* **1999** *55*, 3141–3182.

148. Gol'danskii, V.I. *Russ. Chem. Bull.* **1997**, *46*, 389–397.

149. Sutherland, J.D.; Whitfield, J.N. *Tetrahedron* **1997**, *53*, 11493–11527.

150. Harada, K. In *Chemistry and Biochemistry of Amino Acids, Peptides and Proteins. A Survey of Recent Developments, Volume 2,* Weinstein, B. Ed. Marcel Dekker.: New York, 1974, pp 297–351.

151. Severin, K. *Angew. Chem., Int. Ed.* **2000**, *39* 3589–3590.

152. *Perspectives in Amino Acid and Protein Geochemistry,* Goodfriend, G.A.; Collins, M.J. Fogel, M.L.; Macko, S.A.; Wehmiller, J.F. Eds. Oxford University Press: New York, 2000.

153. Bada, J.L. *Proc. Natl. Acad. Sci. U.S.A.* **2001**, *98*, 797–800.

154. Plankensteiner, K.; Reiner, H.; Rode, B.M. *Chem. Biodiversity* **2004**, *1*, 1308–1315.

155. Dyker, G. *Angew. Chem., Int. Ed. Engl.* **1997**, *36*, 1700–1702.

156. Strecker, A. *Ann.* **1850**, *75*, 27–45.

157. Shafran, Y.M.; Bakulev, V.A.; Mokrushin, V.S. *Russ. Chem. Rev.* **1989**, *58*, 148–162.

158. Gregory, R.J.H. *Chem. Rev.* **1999**, *99*, 3649–3682.

159. Limpricht, H. *Ann.* **1855**, *94*, 243–246.

160. Fischer, E.; Leuchs, H. *Ber.* **1902**, *35*, 3787–3805.

161. Cocker, W.; Lapworth, A. *J. Chem. Soc.* **1931** 1391–1403.

162. Sugimoto, N.; Watanabe, H.; Ide, A. *Tetrahedron* **1960**, *11*, 231–233.

163. Fischer, H.O.L.; Feldmann, L. *Helv. Chim. Acta* **1936**, *19*, 538–543.

164. Fischer, H.O.L.; Feldmann, L. *Helv. Chim. Acta* **1936**, *19*, 532–537.

165. Catch, J.R.; Cook, A.H.; Graham, A.R.; Heilbron, *J. Chem. Soc.* **1947**, 1609–1613.

166. Tiemann, F. *Ber.* **1880**, *13*, 381–385.

167. Ultee, A.J. *Recl. Trav. Chim.* **1909**, *2*, 248–256.

168. Black, D.K.; Landor, S.R. *J. Chem. Soc. (C)* **1968** 281–283.

169. Fiaud, J.C.; Horeau, A. *Tetrahedron Lett.* **197** *25*, 2565–2568.

170. Tiemann, F.; Friedländer, L. *Ber.* **1881**, *14*, 1967–1976.

171. Tiemann, F.; Köhler, K. *Ber.* **1881**, *14*, 1976–1982.

172. Ehrlich, F. *Ber.* **1907**, *40*, 2538–2562.

173. Slimmer, M.D. *Ber.* **1902**, *35*, 400–410.

174. Ehrlich, F. *Biochem. Z.*, **1908**, *8*, 438–466.

175. Salaün, J.; Marguerite, J.; Karkour, B. *J. Org. Chem.* **1990**, *55*, 4276–4281.

176. Gruszecka, E.; Soroka, M.; Mastalerz, P. *Polish J. Chem.* **1979**, *53*, 937–939.

177. Polach, K.J.; Shah, S.A.; LaIuppa, J.C.; LeMaster, D.M. *J. Lab. Cmpds. Radiopharm.* **1993**, *33*, 809–815.

178. Borsook, H.; Deasy, C.L.; Haagen-Smit, A.J.; Keighley, G.; Lowy, P.H. *J. Biol. Chem.* **1950**, *184*, 529–543.

179. Teetz, V.; Gaul, H. *Tetrahedron Lett.* **1984**, *25*, 4483–4486.

180. Zelinsky, N.; Stadnikoff, G. *Ber.* **1908**, *41*, 2061–2063.

181. Zelinsky, N.; Stadnikoff, G. *Ber.* **1906**, *39*, 1722–1732.

182. Barger, G.; Coyne, F.P. *Biochem. J.* **1928**, *22*, 1417–1425.

183. Holland, D.O.; Nayler, J.H.C. *J. Chem. Soc.* **1952**, 3403–3409.

184. Kendall, E.C.; McKenzie, B.F.; Marvel, C.S.; Moyer, W.W. *Org. Synth.* **1941**, *Coll. Vol. 1*, 21–23.

185. Marvel, C.S.; Noyes, W.A. *J. Am. Chem. Soc.*, **1920**, *42*, 2259–2278.

186. Landor, S.R.; Landor, P.D.; Kalli, M. *J. Chem. Soc., Perkin Trans. I* **1983**, 2921–2925.

187. Irreverre, F.; Kny, H.; Asen, S.; Thompson, J.F.; Morris, C.J. *J. Biol. Chem.* **1961**, *236*, 1093–1094.

188. Redemann, C.E.; Icke, R.N. *J. Org. Chem.* **1943**, *8*, 159–161.

189. Ackermann, W.W.; Shive, W. *J. Biol. Chem.* **1948**, *175*, 867–870.

190. Friis, P.; Helboe, P.; Larsen, P.O. *Acta Chem. Scand. B* **1974**, *28*, 317–324.

191. Cahill, R.; Crout, D.H.G.; Mitchell, M.B.; Müller, U.S. *J. Chem. Soc., Chem. Commun.* **1980**, 419–421.

192. Cahill, R.; Crout, D.H.G.; Gregorio, M.V.M.; Mitchell, M.B.; Muller, U.S. *J. Chem. Soc. Perkin Trans. I* **1983**, 173–180.

193. Dewar, J.H.; Shaw, G. *J. Chem. Soc.* **1962**, 583–585.

194. Levene, P.A.; Steiger, R.E. *J. Biol. Chem.* **1928**, *76*, 299–318.

195. Kurono, K. *Biochem. Z.* **1923**, *134*, 424–433.

196. Kurono, K. *Biochem. Z.* **1923**, *134*, 434–436.

197. Reinhold, D.F.; Firestone, D.F.; Gaines, W.A.; Chemerda, J.M.; Sletzinger, M. *J. Org. Chem.* **1968**, *33*, 1209–1213.

198. Gal, A.E.; Avakian, S.; Martin, G.J. *J. Am. Chem. Soc.* **1954**, *76*, 4181–4182.

199. Zelinsky, N.D.; Dengin, E.F. *Ber.* **1922**, *55*, 3354–3361.

200. McConathy, J.; Martarello, L.; Malveaux, E.J.; Camp, V.M.; Simpson, N.E.; Simpson, C.P.; Bowers, G.D.; Olson, J.J.; Goodman, M.M. *J. Med. Chem.* **2002**, *45*, 2240–2249.

201. Steiger, R.E. *Org. Syn.* **1944**, *24*, 9–12.

202. Gulewitsch, W.; Wasmus, T. *Ber.* **1906**, *39*, 1181–1194.

203. Rosenmund, K.W. *Ber.* **1909**, *42*, 4470–4481.

204. Christensen, H.N.; Cullen, A.M. *Biochim. Biophys. Acta* **1973**, *298*, 932–950.

205. Christensen, H.N.; Handlogten, M.E.; Vadgama, J.V.; de la Cuesta, E.; Ballesteros, P.; Trigo, G.C.; Avendaño, C. *J. Med. Chem.* **1983**, *26*, 1374–1378.

206. Edward, J.T.; Jitrangsri, C. *Can. J. Chem.* **1975**, *53*, 3339–3350.

207. Trigo, G.C.; Avendaño, C.; Santos, E.; Edward, J.T.; Wong, S.C. *Can. J. Chem.* **1979**, *57*, 1456–1461.

208. Tager, H.S.; Christensen, H.N. *J. Am. Chem. Soc.* **1972**, *94*, 968–972.

209. Allan, R.D.; Hanrahan, J.R.; Hambley, T.W.; Johnston, G.A.R.; Mewett, K.N.; Mitrovic, A.D. *J. Med. Chem.* **1990**, *33*, 2905–2915.

210. Frantz I.D., Jr.; Loftfield, R.B.; Miller, W.W. *Science* **1947**, *106*, 54–545.

211. Bjurling, P.; Antoni, G.; Watanabe, Y.; Långström, B. *Acta Chem. Scand. B* **1990**, *44*, 178–188.

212. Fotadar, U.; Cowburn, D. *J. Lab. Cmpds. Radiopharm.* **1983**, *20*, 1003–1009.

213. Aberhart, D.J.; Weiller, B.H. *J. Lab. Cmpds. Radiopharm.* **1983**, *20*, 663–668.

214. Baxter, R.L.; Abbot, E.M.; Greenwood, S.L.; McFarlane, I.J. *J. Chem. Soc., Chem. Commun.* **1985**, 564–566.

215. Baxter, R.L.; Abbot, E.M. *J. Lab. Cmpds. Radiopharm.* **1985**, *22*, 1211–1216.

216. Townsend, C.A.; Salituro, G.M. *J. Chem. Soc., Chem. Commun.* **1984**, 1631–1632.

217. Atherton, J.H.; Blacker, J. Crampton, M.R.; Grosjean, C. *Org. Biomol. Chem.* **2004**, *2*, 2567–2571.

218. Read, W.T. *J. Am. Chem. Soc.* **1922**, *44*, 1746–1755.

219. Dappen, M.S.; Pellicciari, R.; Natalini, B.; Monahan, J.B.; Chiorri, C.; Cordi, A.A. *J. Med. Chem.* **1991**, *34*, 161–168.

220. Baker, S.R.; Goldsworthy, J. *Synth. Commun.* **1994**, *24*, 1947–1957.

221. Baker, S.R.; Hancox, T.C. *Tetrahedron Lett.* **1999**, *40*, 781–784.

222. Hanafua, T.; Ichihara, J.; Ashida, T. *Chem. Lett.* **1987**, 687–690.

223. Menéndez, J.C.; Trigo, G.G.; Söllhuber, M.M. *Tetrahedron Lett.* **1986**, *27*, 3285–3288.

224. Gaudry, R. *Can. J. Res. B.* **1946**, *24*, 301–307.

225. Gawron, O.; Glaid III, A.J. *J. Am. Chem. Soc.* **1949**, *71*, 3232–3233.

226. Atkinson, R.O.; Poppelsdorf, F. *J. Chem. Soc.* **1952**, 2448.

227. Bigge, C.F.; Drummond, J.T.; Johnson, G.; Malone, T.; Probert Jr., A.W.; Marcoux, F.W.; Coughenour, L.L.; Brahce, L.J. *J. Med. Chem.* **1989**, *32*, 1580–1590.

228. Yuan, S.-S. *J. Lab. Cmpds. Radiopharm.* **1983**, *20*, 173–178.

229. Uchida, K.; Kainosho, M. *J. Lab. Cmpds. Radiopharm.* **1991**, *29*, 867–874.

230. Eschweiler, W. *Ann.* **1894**, *279*, 39–44.

231. Tiemann, F.; Piest, R. *Ber.* **1881**, *14*, 1982–1984.

232. Dvonch, W.; Fletcher III, H.; Alburn, H.E. *J. Org. Chem.* **1964**, *29*, 2764–2766.

233. Bucherer, H. Th.; Fischbeck, H. *J. Prakt. Chem.* **1934**, *140*, 69–89.

234. Groeger, U.; Drauz, K.; Klenk, H. *Angew. Chem., Int. Ed. Engl.* **1992**, *31*, 195–197.

235. Hanrahan, J.R.; Taylor, P.C.; Errington, W. *J. Chem. Soc., Perkin Trans. 1* **1997**, 493–502.

236. Kozikowski, A.P.; Fauq, A.H. *Synlett*, **1991**, 783–784.

237. Hines J.W., Jr.; Breitholle, E.G.; Sato, M.; Stammer, C.H. *J. Org. Chem.* **1976**, *41*, 1466–1469.

238. Georgiadis, M.P.; Haroutounian, S.A. *Synthesis* **1989**, 616–618.

239. Arnold, U.; Hinderaker, M.P.; Köditz, J.; Golbik, R.; Ulbrich-Hofmann, R.; Raines, R.T. *J. Am. Chem. Soc.* **2003**, *125*, 7500–7501.

240. Fadel, A. *Tetrahedron* **1991**, *47*, 6265–6274.

241. Stadnikoff, G. *Ber.* **1907**, *40*, 4350–4353.

242. Stadnikoff, G. *Ber.* **1907**, *40*, 4353–4356.

243. Stadnikoff, G. *Ber.* **1907**, *40*, 1014–1019.

244. Ranu, B.; Dey, S.S.; Harjra, A. *Tetrahedron* **2002**, *58*, 2529–2532.

245. Mai, K.; Patil, G. *Tetrahedron Lett.* **1984**, *25*, 4583–4586.

246. Leblanc, J.-P.; Gibson, H.W. *Tetrahedron Lett.* **1992**, *33*, 6295–6298.

247. Bois-Choussy, M.; Zhu, J. *J. Org. Chem.* **1998**, *63*, 5662–5665.

248. Ojima, I.; Inaba, S.-I.; Nakatsugawa, K. *Chem. Lett.* **1975**, 331–334.

249. Thornberry, N.A.; Bull, H.G.; Taub, D.; Greenlee, W.J.; Patchett, A.A.; Cordes, E.H. *J. Am. Chem. Soc.* **1987**, *109*, 7543–7544.

250. Greenlee, W.J. *J. Org. Chem.* **1984**, *49*, 2632–2634.

251. Kobayashi, S.; Ishitani, H.; Ueno, M. *Synlett* **1997**, 115–116.

252. Saravanan, P.; Anand, R.V.; Singh, V.K. *Tetrahedron Lett.* **1998**, *39*, 3823–3824.

253. De, S.K. *Synth. Commun.* **2005**, *35*, 1577–1582.

254. Kobayashi, S.; Nagayama, S.; Busujima, T. *Tetrahedron Lett.* **1996**, *37*, 9221–9224.

255. Heydari, A.; Fatemi, P.; Alizadeh, A.-A. *Tetrahedron Lett.* **1998**, *39*, 3049–3050.

256. Azizi, N.; Saidi, M.R. *Synth. Commun.* **2004**, *34*, 1207–1214.

257. Yadav, J.S.; Reddy, B.V.S.; Eeshwaraiah, B.; Srinivas, M. *Tetrahedron* **2004**, *60*, 1767–1771.

258. De, S.K.; Gibbs, R.A. *Synth. Commun.* **2005**, *35*, 961–966.

259. De, S.K. *Synth. Commun.* **2005**, *35*, 653–656.

260. De, S.K.; Gibbs, R.A. *Tetrahedron Lett.* **2004**, *45*, 7407–7408.

261. Royer, L.; De, S.K.; Gibbs, R.A. . *Tetrahedron Lett.* **2005**, *46*, 4595–4597.

262. Fossey, J.S.; Richards, C.J. *Tetrahedron Lett.* **2003**, *44*, 8773–8776.

263. Matsumoto, K.; Kim, J.C.; Iida, H.; Hamana, H.; Kumamoto, K.; Kotsuki, H.; Jenner, G. *Helv. Chim. Acta* **2005**, *88*, 1734–1753.

264. Jenner, G.; Salem, R.B.; Kim, J.C.; Matsumoto, K. *Tetrahedron Lett.* **2003**, *44*, 447–449.

265. Matsumoto, K.; Kim, J.C.; Hayashi, N.; Jenner, G. *Tetrahedron Lett.* **2002**, *43*, 9167–9169.

266. Martínez, R.; Ramón, D.J.; Yus, M. *Tetrahedron Lett.* **2005**, *46*, 8471–8474.

267. El-Ahl, A.-A.S. *Synth. Commun.* **2003**, *33*, 989–998.

268. Rai, M.; Singh, A.; Kalsi, P.S. *Ind. J. Chem.* **1979**, *18B*, 273–274.

269. Hrusawa, S.; Hamada Y.; Shioiri, T. *Tetrahedron Lett.* **1979**, *48*, 4663–4666.

270. Kobayashi, S.; Busujima, T.; Nagayama, S. *Chem. Commun.* **1998**, 981–982.

271. Nagayama, S.; Kobayashi, S. *Angew. Chem., Int. Ed.* **2000**, *39*, 5670–569.

272. Chu, G.-H.; Gu, M.; Gerard, B.; Dolle, R.E. *Synth. Commun.* **2004**, *34*, 4583–4590.

273. Gonzalez, J.; Carroll, F.I. *Tetrahedron Lett.* **1996**, *37*, 865–8658.

274. North, M. *Angew. Chem., Int. Ed.* **2004**, *43*, 4126–4128.

275. Murahashi, S.-I.; Komiya, N.; Terai, H.; Nakae, T. *J. Am. Chem. Soc.,* **2003**, *125*, 15312–15313.

276. Pinner, A.; Spilker, A. *Ber.* **1889**, *22*, 685–698.

277. Urech, F. *Ann.* **1872**, *164*, 255–279.

278. Wheeler, H.L.; Hoffman, C. *Am. Chem. J.* **1911** *45*, 568–583.

279. Bucherer, H. Th.; Steiner, W. *J. Prakt. Chem* **1934**, *140*, 291–316.

280. Bucherer, H. Th.; Lieb, V.A. *J. Prakt. Chem.* **1934** *141*, 5–43.

281. Bergs, H. German Patent 566,094, (May 26 1929); *Chem. Abstr.* **1933**, *27*, 1001.

282. Pfeiffer, P.; Hoyer, H. *J. Prak. Chem.* **1933**, *138*, 69–80.

283. Pfeiffer, P.; Heinrich, E. *J. Prak. Chem.* **1940** *156*, 241–259.

284. Pfeiffer, P.; Heinrich, E. *J. Prak. Chem.* **1936**, *147*, 93–98.

285. Pfeiffer, P.; Heinrich, E. *J. Prak. Chem.* **1936**, *147*, 105–112.

286. Boyd, W.J.; Robson, W. *Biochem. J.* **1935**, *29*, 546–554.

287. Ware, E. *Chem. Rev.* **1950**, *46*, 403–470.

288. Carrington, H.C. *J. Chem. Soc.* **1947**, 681–683.

289. Carrington, H.C. *J. Chem. Soc.* **1947**, 684–686.

290. Gaudry, R. *Can. J. Res. B* **1948**, *26*, 773–776.

291. Gaudry, R.; Nadeau, G. *Can. J. Res. B* **1948**, *26*, 226–229.

292. Rogers, A.O.; Emmick, R.D.; Tyran, L.W.; Phillips, L.B.; Levine, A.A.; Scott, N.D. *J. Am. Chem. Soc.* **1949**, *71*, 1837–1839.

293. Pierson, E.; Giella, M.; Tishler, M. *J. Am. Chem. Soc.* **1948**, *70*, 1450–1451.

294. Henze, H,R.; Speer, R.J. *J. Am. Chem. Soc.* **1942**, *64*, 522–523.

295. Henze, H.R.; Isbell, A.F. *J. Am. Chem. Soc.* **1954**, *76*, 4152–4156.

296. Henze, H.R.; Long, L.M. *J. Am. Chem. Soc.* **1941**, *63*, 1941–1943.

297. Kubik, S.; Meissner, R.S.; Rebek J., Jr., *Tetrahedron Lett.* **1994**, *35*, 6635–6638.

298. Baker, S.R.; Goldsworthy, J.; Harden, R.C.; Salhoff, C.R.; Schoepp, D.D. *Bioorg. Med. Chem. Lett.* **1995**, *5*, 223–228.

299. Nadeau, G.; Gaudry, R. *Can. J. Res. B* **1949**, *27*, 421–427.

300. Reisner, D.B. *J. Am. Chem. Soc.* **1956**, *78*, 2132–2135.

301. Dannley, R.L.; Taborsky, R.G. *J. Org. Chem.* **1957**, *22*, 1275–1276.

302. Gaudry, R. *Can. J. Chem.* **1951**, *29*, 544–551.

303. Culvenor, C.C.J.; Foster, M.C.; Hegarty, M.P. *Aust. J. Chem.* **1971**, *24*, 371–375.

304. Dreyfuss, P. *J. Med. Chem.* **1974**, *17*, 252–255.

305. Yokum, T.S.; Tungaturthi, P.K.; McLaughlin, M.L. *Tetrahedron Lett.* **1997**, *38*, 5111–5114.

306. Edelson, J.; Fissekis, J.D.; Skinner, C.G.; Shive, W. *J. Am. Chem. Soc.* **1958**, *80*, 2698–2700.

307. Madsen, U.; Frølund, B.; Lund, T.M.; Ebert, B.; Krogsgaard-Larsen, P. *Eur. J. Med. Chem.* **1993**, *28*, 791–800.

308. Natchev, I.A. *Bull. Chem. Soc. Jpn.* **1988**, *61*, 3711–3715.

309. Ropchan, J.R.; Barrio, J.R. *J. Nucl. Med.* **1984**, *25*, 887–892.

310. Hayes, R.L.; Washburn, L.C.; Wieland, B.W.; Sun, T.T.; Anon, J.B.; Butler, T.A.; Callahan, A.P. *Int. J. Appl. Rad. Isot.* **1978**, *29*, 186–187.

311. Barrio, J.R.; Keen, R.E.; Ropchan, J.R.; MacDonald, N.S.; Baumgartner, F.J.; Padgett, H.C.; Phelps, M.E. *J. Nucl. Med.* **1983**, *24*, 515–521.

312. Rothstein, M. *J. Am. Chem. Soc.* **1957**, *79*, 2009–2011.

313. Barry, J.M. *J. Biol. Chem.* **1952**, *195*, 795–803.

314. Rothstein, M.; Leak, J. *Biochem. Prep.* **1961**, *8*, 80–84.

315. Townsend, C.A.; Brown, A.M. *J. Am. Chem. Soc.* **1983**, *105*, 913–918.

316. Shoup, T.M.; Goodman, M.M. *J. Lab. Cmpds. Radiopharm.* **1997**, *40*, 46–47.

317. Giron, C.; Luurtsema, G.; Vos, M.G.; Elsinga, P.H.; Visser, G.M.; Vaalburg, W. *J. Lab. Cmpds. Radiopharm.* **1995**, *37*, 752–754.

318. Guddat, T.; Herdering, W.; Knöchel, A.; Zwernemann, O. *J. Lab. Cmpds. Radiopharm.*, **1989**, *26*, 79.

319. Fissekis, J.D.; Nielsen, C.; Dahl, J.R. *J. Lab. Cmpds. Radiopharm.* **1985**, *23*, 1083–1084.

320. Halldin, C.; Bergson, G.; Malmborg, P.; Långström, B. *J. Lab. Cmpds. Radiopharm.* **1983**, *21*, 1178–1179.

321. Haddin, C.; Långström, B. *J. Lab. Cmpds. Radiopharm.* **1985**, *22*, 631–640.

322. Goodson, L.H.; Honigberg, I.L.; Lehman, J.J.; Burton, W.H. *J. Org. Chem.* **1960**, *25*, 1920–1924.

323. Abshire, C.J.; Planet, G. *J. Med. Chem.* **1972**, *15*, 226–229.

324. Rastogi, S.N.; Bindra, J.S.; Anand, N. *Ind. J. Chem.* **1971**, *9*, 1175–1182.

325. Stein, G.A.; Bronner, H.A.; Pfister K., 3rd. *J. Am. Chem. Soc.* **1955**, *77*, 700–703.

326. Winn, M.; Rasmussen, R.; Minard, F.; Kyncl, J.; Plotnikoff, N. *J. Med. Chem.* **1975**, *18*, 434–437.

327. Ablewhite, A.J.; Wooldridge, K.R.H. *J. Chem. Soc. C* **1967**, 2488–2491.

328. Ndzié, E.; Cardinael, P.; Schoofs, A.-R.; Coquerel, G. *Tetrahedron: Asymmetry* **1997**, *8*, 2913–2920.

329. Coudert, E.; Acher, F.; Azerad, R. *Tetrahedron: Asymmetryetry* **1996**, *7*, 2963–2970.

330. Abshire, C.J.; Berlinguet, L. *Can. J. Chem.* **1965**, *43*, 1232–1234.

331. Ellington, J.J.; Honigberg, I.L. *J. Org. Chem.* **1974**, *39*, 104–106.

332. Pfister K.; 3rd, Leanza, W.J.; Conbere, J.P.; Becker, H.J.; Matzuk, A.R.; Rogers, E.F. *J. Am. Chem. Soc.* **1955**, *77*, 697–700.

333. Maehr, H.; Yarmchuk, L.; Leach, M. *J. Antibiotics* **1976**, *29*, 221–226.

334. Abdel-Monem, M.M.; Newton, N.E.; Weeks, C.E. *J. Med. Chem.* **1974**, *17*, 447–451.

335. Kuo, D.; Rando, R.R. *Biochemistry* **1981**, *20*, 506–511.

336. Tsang, J.W.; Schmied, B.; Nyfeler, R.; Goodman, M. *J. Med. Chem.* **1984**, *27*, 1663–1668.

337. Cappi, M.W.; Moree, W.J.; Qiao, L.; Marron, T.G.; Weitz-Schmidt, G.; Wong, C.-H. *Bioorg. Med. Chem.* **1997**, *5*, 283–296.

338. Pfeiffer, P.; Simons, H. *J. Prak. Chem.* **1942**, *160*, 83–94.

339. Pfeiffer, P.; Diebold, A. *J. Prak. Chem.* **1937**, *148*, 24–34.

340. Pfeiffer, P.; Epler, H. *Ann.* **1940**, *54*, 263–286.

341. Logusch, E.W.; Walker, D.M.; McDonald, J.F.; Leo, G.C.; Franz, J.E. *J. Org. Chem.* **1988**, *53*, 4069–4074.

342. Ornstein, P.L.; Bleisch, T.J.; Arnold, M.B.; Wright, R.A.; Johnson, B.G.; Schoepp, D.D. *J. Med. Chem.* **1998**, *41*, 346–357.

343. Ornstein, P.L.; Bleisch, T.J.; Arnold, M.B.; Kennedy, J.H.; Wright, R.A.; Johnson, B.G.; Tizzano, J.P.; Helton, D.R.; Kallman, M.J.; Schoepp, D.D.; Hérin, M. *J. Med. Chem.* **1998**, *41*, 358–378.

344. Collado, Ezquerra, J.; Mazón, A.; Pedregal, C.; Yruretagoyena, B.; Kingston, A.E.; Tomlinson, R.; Wright, R.A.; Johnson, B.G.; Schoepp, D.D. *Biorg. Med. Chem. Lett.* **1998**, *8*, 2849–2854.

345. Allan, R.D.; Apostopoulos, C.; Hambley, T.W. *Aust. J. Chem.* **1995**, *48*, 919–928.

346. Martarello, L.; McConathy, J.; Camp, V.M.; Malveaux, E.J., Simpson, N.E.; Simpson, C.P.; Olson, J.J.; Bowers, G.D.; Goodman, M.M. *J. Med. Chem.* **2002**, *45*, 2250–2259.

347. Srivastava, R.R.; Singhaus, R.R.; Kabalka, G.W. *J. Org. Chem.* **1999**, *64*, 8495–8500.

348. Srivastava, R.R.; Singhaus, R.R.; Kabalka, G.W. *J. Org. Chem.* **1997**, *62*, 4476–478.

349. Srivastava, R.R.; Kabalka, G.W. *J. Org. Chem.* **1997**, *62*, 8730–8734.

350. Srivastava, R.R.; Kbalka, G.W.; Longford, C.P.D. *J. Lab. Cmpds. Radiopharm.* **1997**, *40*, 395–396.

351. Conners, T.A.; Ross, W.C.J. *J. Chem. Soc.* **1960**, 2119–2132.

352. Mauger, A.B.; Ross, W.C.J. *Biochem. Pharmacol.* **1962**, *11*, 847–858.

353. Alonso, F.; Micó, I.; Nájera, C.; Sansano, J.M.; Yus, M.; Ezquerra, J.; Yruretagoyena, B.; Gracia, I. *Tetrahedron* **1995**, *51*, 10259–10280.

354. Stephani, R.A.; Rowe, W.B.; Gass, J.D.; Meister, A. *Biochemistry* **1972**, *11*, 4094–4100.

355. Gass, J.D.; Meister, A. *Biochemistry* **1970**, *9*, 842–846.

356. Trigalo, F.; Acher, F.; Azerad, R. *Tetrahedron* **1990**, *46*, 5203–5212.

357. Cannon, J.G.; O'Donnell, J.P.; Rosazza, J.P.; Hoppin, C.R. *J. Med. Chem.* **1974**, *17*, 565–568.

358. Schiller, P.W.; Weltrowska, G.; Nguyen, T.M.D.; Lemieux, C.; Chung, N.N.; Marsden, B.J.; Wilkes, B.C. *J. Med. Chem.* **1991**, *34*, 3125–3132.

359. Connors, T.A.; Ross, W.C.J.; Wilson, J.G. *J. Chem. Soc.* **1960**, 2994–3007.

360. Franceschetti, L.; Garzon-Aburbeh, A.; Mahmoud, M.R.; Natalini, B.; Pellicciari, R. *Tetrahedron Lett.* **1993**, *34*, 3185–3188.

361. Lavrador, K.; Guillerm, D.; Guillerm, G. *Biorg. Med. Chem. Lett.* **1998**, *8*, 1629–1634.

362. Wysong, C.L.; Yokum, T.S.; Morales, G.A.; Gundry, R.L.; McLaughlin, M.L.; Hammer, R.P. *J. Org. Chem.* **1996**, *61*, 7650–7651.

363. Maki, Y.; Masugi, T.; Hiramitsu, T.; Ogiso, T.; *Chem. Pharm. Bull.* **1973**, *21*, 2460–2465.

364. Yokum, T.S.; Bursavich, M.G.; Piha-Paul, S.A.; Hall, D.A.; McLaughlin, M.L. *Tetrahedron Lett.* **1997**, *38*, 4013–4016.

365. Obrecht, D.; Spiegler, C.; Schönholzer, P.; Müller, K. Heimgartner, H.; Stierli, F. *Helv. Chim. Acta* **1992**, *75*, 1666–1696.

366. Obrecht, D.; Karajiannis, H.; Lehmann, C.; Schönholzer, P.; Spiegler, C.; Müler, K. *Helv. Chim. Acta* **1995**, *78*, 703–714.

367. Liu, B.; Thalji, R.K.; Adams, P.D.; Fronczek, F.R.; McLaughlin, M.L.; Barkley, M.D. *J. Am. Chem. Soc.* **2002**, *124*, 13329–13338.

368. Stalker, R.A., Munsch, T.E.; Tan, J.D.; Nie, X.; Warmuth, R.; Beatty, A.; Aakeröy, C.B. *Tetrahedron* **2002**, *58*, 4837–4849.

369. Edward, J.T.; Jitrangsri, C. *Can. J. Chem.* **1975**, *53*, 3339–3350.

370. Christensen, H.N.; Handlogten, M.E.; Lam, I.; Tager, H.S.; Zand, R. *J. Biol. Chem.* **1969**, *244*, 1510–1520.

371. Cremlyn, R.J.W.; Chisholm, M. *J. Chem. Soc. C* **1967**, 1762–1764.

372. Trigo, G.G.; Avendaño, C.; Santos, E.; Christensen, H.N.; Handlogten, M.E. *Can. J. Chem.* **1980**, *58*, 2295–299.

373. Cremlyn, R.J.W. *J. Chem. Soc.* **1962**, 3977–3980.

374. Tellier, F.; Acher, F.; Brabet, I.; Pin, J.-P.; Bockaert J.; Azerad, R. *Bioorg. Med. Chem. Lett.* **1995**, *5*, 2627–2632.

375. Ma, D.; Tian, H.; Sun, H.; Kozikowski, A.P.; Pshenichkin, S.; Wroblewski, J.T. *Bioorg. Med. Chem. Lett.* **1997**, *7*, 1195–1198.

376. Tellier, F.; Acher, F.; Brabet, I.; Pin, J.-P.; Azerad R. *Biorg. Med. Chem.* **1998**, *6*, 195–208.

377. Fantin, G.; Fogagnolo, M.; Guerrini, R.; Marastoni, M.; Medici, A.; Pedrini, P. *Tetrahedron* **1994**, *50*, 12973–12978.

378. Grunewald, G.L.; Kuttab, S.H.; Pleiss, M.A.; Mangold, J.B. *J. Med. Chem.* **1980**, *23*, 754–758.

379. Nagasawa, H.T.; Elberling, J.A.; Shirota, F.N. *J. Med. Chem.* **1973**, *16*, 823–826.

380. Ugi, I.; Meyr, R.; Fetzer, U.; Steinbrückner, C. *Angew. Chem.* **1959**, *71*, 386.

381. Ugi, I.; Steinbrückner, C. *Angew. Chem.* **1960**, *72*, 267–268.

382. Ugi, I. *Angew. Chem., Int. Ed. Engl.* **1962**, *1*, 8–21.

383. Ugi, I.; Lohberger, S.; Karl, R. In *Comprehensive Organic Synthesis. Selectivity, Strategy and Efficiency in Modern Organic Chemistry;* Trost, B.M.; Fleming, I. Eds.; Pergamon Press: New York, **1991**, Vol 2.2, pp 1083–1109.

384. Ugi, I.; Marquarding, D.; Urban, R. In *Chemistry and Biochemistry of Amino Acids, Peptides and Proteins. A Survey of Recent Developments,*

Volume 6, Weinstein, B. Ed.; Marcel Dekker: New York, **1982**, pp 245–289.

385. Dömling, A.; Ugi, I. *Angew. Chem., Int. Ed.* **2000**, *39*, 3168–3210.

386. Dömling, A. *Chem. Rev.* **2006**, *106*, 17–89.

387. Gieren, A.; Dederer, B. *Tetrahedron Lett.* **1977**, *18*, 1503–1506.

388. Tye, H.; Whittaker, M. *Org. Biomol. Chem.* **2004**, *2*, 813–815.

389. Yamada, T.; Yanagi, T.; Omote, Y.; Miyazawa, T.; Kuwata, S.; Sugiura, M.; Matsumoto, K. *J. Chem. Soc., Chem. Commun.* **1990**, 1640–1641.

390. Hardy, P.M.; Lingham, I.N. *Int. J. Pept. Prot. Chem.* **1983**, *21*, 406–418.

391. Skorna, G.; Ugi, I. *Angew. Chem., Int. Ed. Engl.* **1977**, *16*, 259–260.

392. Saunders, J.H.; Slocombe, R.J. *Chem. Rev.* **1948**, *48*, 203–218.

393. Yamada, T.; Omote, Y.; Yamanaka, Y.; Miyazawa, T.; Kuwata, S. *Synthesis* **1998**, 991–998.

394. Pirrung, M.C.; Sarma, K.D. *J. Am. Chem. Soc.* **2004**, *126*, 444–445.

395. Shibata, N.; Das, B.K.; Takeuchi, Y. *J. Chem. Soc., Perkin Trans. 1* **2000**, 4234–4236.

396. Zhang, X.; Zou, X.; Xu, P. *Synth. Commun.* **2005**, *35*, 1881–1888.

397. Dietrich, S.A.; Banfi, L.; Basso, A.; Damonte, G.; Guanti, G. Riva, R. *Org. Biomol. Chem.* **2005**, *3*, 97–106.

398. Müller, S.; Neidlein, R. *Helv. Chim. Acta* **2002**, *85*, 222–2231.

399. Ziegler, T.; Gerling, S.; Lang, M. *Angew. Chem., Int. Ed.* **2000**, *39*, 2109–2112.

400. Maia, H.L.; Ridge, B.; Rydon, H.N. *J. Chem. Soc., Perkin Trans. 1* **1973**, 98–101.

401. Scholz, D.; Schmidt, U. *Chem. Ber.* **1974**, *107*, 2295–2298.

402. Brachmachary, E.; Ling, F.H.; Svec, F.; Fréchet, J.M.J. *J. Comb. Chem.* **2003**, *5*, 441–450.

403. Divanfard, H.R.; Lysenko, Z.; Wang, P.-C.; Joullié, M.M. *Synth. Commun.* **1978**, *8*, 269–273.

404. Gröger, H.; Hatam, M.; Martens, J. *Tetrahedron,* **1995**, *51*, 7173–7180.

405. Hatam, M.; Tehranfar, D.; Martens, J. *Synthesis* **1994**, 619–623.

406. Harriman, G.C.B. *Tetrahedron Lett.* **1997**, *38*, 5591–5594.

407. Rossen, K.; Sager, J.; DiMichele, L.M. *Tetrahedron Lett.* **1997**, *38*, 3183–3186.

408. Tsuchida, K.; Mizuno, Y.; Ikeda, K. *Heterocycles* **1981**, *15*, 883–887.

409. Boehm, J.C.; Kingsbury, W.D. *J. Org. Chem.* **1986**, *51*, 2307–2314.

410. Díaz, J.L.; Miguel, M.; Lavilla, R. *J. Org. Chem.* **2004**, *69*, 3550–3553.

411. Tempest, P.A.; Brown, S.D.; Armstrong, R.W. *Angew. Chem., Int. Ed. Engl.* **1996**, *35*, 640–642.

412. Hoel, A.M.L.; Nielsen, J. *Tetrahedron Lett.* **1999**, *40*, 3941–3944.

413. Sutherlin, D.P.; Stark, T.M.; Hughes, R.; Armstrong, R.W. *J. Org. Chem.* **1996**, *61*, 8350–8354.

414. Kim, S.W.; Bauer, S.M.; Armstrong, R.W. *Tetrahedron Lett.* **1998**, *39*, 6993–6996.

415. Li, Z.; Yeo, S.L.; Pallen, C.J.; Ganesan, A. *Biorg. Med. Chem. Lett.* **1998**, *8*, 2443–2446.

416. Kim, S.W.; Shin, Y.S.; Ro. S. *Biorg. Med. Chem. Lett.* **1998**, *8*, 165–1668.

417. Golebiowski, A.; Jozwik, J.; Klopfenstein, S.R.; Colson, A.-O.; Grieb, A.L.; Russell, A.F.; Rastogi, V.L.; Diven, C.F.; Portlock, D.E.; Chen, J.J. *J. Comb. Chem.* **2002**, *4*, 584–590.

418. Henkel, B.; Sax, M.; Dömling, A. *Tetrahedron Lett.* **2003**, *44*, 7015–7018.

419. Keating, T.A.; Armstrong, R.W. *J. Am. Chem. Soc.* **1996**, *118*, 2574–2583.

420. Keating, T.A.; Armstrong, R.W. *J. Am. Chem. Soc.* **1995**, *117*, 7842–7842.

421. Lindhorst, T.; Bock, H.; Ugi, I. *Tetrahedron* **1999**, *55*, 7411–7420.

422. Kennedy, A.L.; Fryer, A.M.; Josey, J.A. *Org. Lett.* **2002**, *4*, 1167–1170.

423. Linderman, RT.J.; Binet, S.; Petrich, S.R. *J. Org. Chem.* **1999**, *64*, 336–37; **1999**, *64*, 8058.

424. Jiang, W.-Q.; Costa, S.P.G.; Maia, H.L.S. *Org. Biomol. Chem.* **2003**, *1*, 3804–3810.

425. Costa, S.P.G.; Maia, H.L.S.; Pereira-Lima, S.M.M.A. *Org. Biomol. Chem.* **2003**, *1*, 1475–1479.

426. Haslinger, E. *Monatsh. Chem.* **1978**, *109*, 749–750.

427. Keating, T.A.; Armstrong, R.W. *J. Org. Chem.* **1998**, *63*, 867–871.

428. Studer, A.; Jeger, P.; Wipf, P.; Curran, D.P. *J. Org. Chem.* **1997**, *62*, 2917–2924.

429. Kaïm, L.; Grimaud, L.; Oble, J. *Angew. Chem., Int. Ed.* **2005**, *44*, 7961–7964.

430. Portlock, D.E.; Ostaszewski, R.; Naskar, D.; West, L. *Tetrahedron Lett.* **2003**, *44*, 603–605.

431. Portlock, D.E.; Naskar, D.; West, L.; Ostaszewski, R.; Chen, J.J. *Tetrahedron Lett.* **2003**, *44*, 5121–5124.

432. Hoyng, C.F.; Patel, A.D. *J. Chem. Soc., Chem., Commun.* **1981**, 491–492.

433. Floyd, C.D.; Harnett, L.A.; Miller, A.; Patel, S.; Saroglou, L.; Whittaker, M. *Synlett* **1998**, 637–639.

434. Basso, A.; Banfi, L.; Guanti, G.; Riva, R. *Tetrahedron Lett.* **2005**, *46*, 8003–8006.

435. Basso, A.; Banfi, L.; Guanti, G.; Riva, R.; Riu, A. *Tetrahedron Lett.* **2004**, *45*, 6109–6111.

436. Campian, E; Lou, B.; Saneii, H. *Tetrahedron Lett.* **2002**, *43*, 8467–8470.

437. Quast, H.; Aldenkortt, S. *Chem. Eur. J.* **1996**, *2*, 462–469.

438. Kern, O.T.; Motherwell, W.B. *Chem. Commun.* **2003**, 2988–2989.

439. König, S.; Klösel, R.; Karl, R.; Ugi, I. *Z. Naturforsch.* **1994**, 1586–1595.

440. Compere E.L., Jr., Weinstein, D.A. *Synthesis* **1977**, 852–853.
441. Landini, D.; Montanari, F.; Rolla, F. *Synthesis* **1979**, 26–27.
442. Reeve, W.; Fine, L.W. *J. Org. Chem.* **1964**, *29*, 1148–1150.
443. Baker, S.R.; Hancox, T.C. *Tetrahedron Lett.* **1999**, *40*, 781–784.
444. Beller, M.; Eckert, M. *Angew. Chem., Int. Ed.* **2000**, *39*, 1010–1027.
445. Wakamatsu, H.; Uda, J.; Yamakami, N. *J. Chem. Soc., Chem. Commun.* **1971**, 1540.
446. Parnaud, J.-J.; Campari, G.; Pino, P. *J. Mol. Catal.* **1979**, *6*, 341–350.
447. Ojima, I.; Kato, K.; Nakahashi, K.; Fuchikami, T.; Fujita, M. *J. Org. Chem.* **1989**, *54*, 4511–4522.
448. Sakakura, T.; Huang, X.-Y.; Tanaka, M. *Bull. Chem. Soc. Jpn.* **1991**, *64*, 1707–1709.
449. Ojima, I.; Hirai, K.; Fujita, M.; Fuchikami, T. *J. Organomet. Chem.* **1985**, *279*, 203–214.
450. de Vries, J.G.; de Boer, R.P.; Hogeweg, M.; Gielens, E.E.C.G. *J. Org. Chem.* **1996**, *61*, 1842–1846.
451. Amino, Y.; Izawa, K. *Bull. Chem. Soc. Jpn.* **1991**, *64*, 613–619.
452. Ojima, I.; Okabe, M.; Kato, K.; Kwon, H.B.; Horváth, I.T. *J. Am. Chem. Soc.* **1988**, *110*, 150–157.
453. Hirai, K.; Takahashi, Y.; Ojima, I. *Tetrahedron Lett.* **1982**, *23*, 2491–2494.
454. Beller, M.; Eckert, M.; Vollmüller, F.; Bogdanovic, S.; Geissler, H. *Angew. Chem., Int. Ed. Engl.* **1997**, *36*, 1494–1496.
455. Beller, M.; Eckert, M.; Holla, E.W. *J. Org. Chem.* **1998**, *63*, 5658–5661.
456. Beller, M.; Eckert, M.; Geissler, H.; Napierski, B.; Rebenstock, H.-P.; Holla, E.W. *Chem. Eur. J.* **1998**, *4*, 935–941.
457. Beller, M.; Moradi, W.A.; Eckert, M.; Neumann, H. *Tetrahedron Lett.* **1999**, *40*, 4523–4526.
458. Beller, M.; Eckert, M.; Moradi, W.A. *Synlett* **1999**, 108–110.
459. Freed, D.A.; Kozlowski, M.C. *Tetrahedron Lett.* **2001**, *42*, 3403–3406.
460. Beller, M.; Eckert, M.; Moradi, W.A.; Neumann, H. *Angew. Chem., Int. Ed.* **1999**, *38*, 1454–1457.
461. Shimizu, H.; Kobayashi, S. *Tetrahedron Lett.* **2005**, *46*, 7593–7595.
462. Dhawan, R.; Dghaym, R.D.; Arndtsen, B.A. *J. Am. Chem. Soc.* **2003**, *125*, 1474–1475.
463. Lin, Y.-S.; Alper, H. *Angew. Chem., Int. Ed.* **2001**, *40*, 779–781.
464. Ballini, R.; Petrini, M. *Tetrahedron Lett.* **1999**, *40*, 4449–4452.
465. Banphavichit, V.; Chaleawlertumpon, S.; Bhanthumnavin, W.; Vilaivan, T. *Synth. Commun.* **2004**, *34*, 3147–3160.
466. Jackson, R.F.W.; Palmer, N.J.; Wythes, M.J.; Clegg, W.; Elsegood, M.R.J. *J. Org. Chem.* **1995**, *60*, 6431–6440.
467. Jackson, R.F.W.; Palmer, N.J.; Wythes, M.J. *Tetrahedron* Lett. **1994**, *35*, 743–7434.
468. Ayi, A.I.; Guedj, R. *J. Fluorine Chem.* **1984**, *24*, 137–151.
469. Clive, D.L.J.; Etkin, N. *Tetrahedron Lett.* **1994**, 35, 2459–2462.
470. Miyabe, H.; Yoshioka, N.; Ueda, M.; Naito, T. *J. Chem. Soc., Perkin Trans. 1* **1998**, 3659–3660.
471. Mazaletrat, J.-P.; Wakselman, M.; Formaggio, F.; Crisma, M.; Toniolo, C. *Tetrahedron Lett.* **1999**, *40*, 6245–6248.
472. Mazaleyrat, J.-P.; Wright, K.; Wakselman, M.; Formaggio, F.; Crisma, M.; Toniolo, C. *Eur. J. Org. Chem.* **2001**, 1821–1829.
473. Chen, J.; Cunico, R.F. *Tetrahedron Lett.* **2003**, *44*, 8025–8027.
474. Cahours, A. *Compt. rend.*, **1858**, *46*, 1044.
475. Cahours, A. *Ann.* **1859**, *109*, 10–34.
476. Kolbe, H. *Ann.* **1860**, *113*, 220–223.
477. Robertson, G.R. *J. Am. Chem. Soc.* **1927**, *49*, 2889–2894.
478. Boutwell, P.W.; Kuick, L.F. *J. Am. Chem. Soc.* **1930**, *52*, 4166–4167.
479. Cheronis, N.D.; Spitzmueller, K.H. *J. Org. Chem.* **1941**, *6*, 349–375.
480. Marvel, C.S.; Bachmann, W.E.; Holmes, D.W. *Org. Syn. Coll. Vol. 3* **1955**, 495–498.
481. Marvel, C.S.; Adkins, H.; Gander, R. *Org. Syn. Coll. Vol. 3* **1955**, 523–525.
482. Marvel, C.S.; Smith, L.I.; Arnold, R.T.; Howard, K.L. *Org. Syn. Coll. Vol. 3* **1955**, 705–709.
483. Carter, H.E.; West, H.D.; Drake, N.L.; Stanton, W.A. *Org. Syn. Coll. Vol. 3* **1955**, 774–778.
484. Carter, H.E.; West, H.D.; Drake, N.L.; Stanton, W.A. *Org. Syn. Coll. Vol. 3* **1955**, 813–817.
485. Willstätter, R. *Ber. Deutsch. Chem. Ges.* **1900**, *33*, 1160–1666.
486. Patterson, W.I.; Du Vigneaud, V. *J. Biol. Chem.* **1938**, *123*, 327–334.
487. Gabriel, S. *Ber. Deutsch. Chem. Ges.* **1887**, *20*, 2224–2236.
488. Gibson, M.S.; Bradshaw, R.W. *Angew. Chem., Int. Ed. Engl.* **1968**, *7*, 919–930.
489. Sheehan, J.C.; Bolhofer, W.A. *J. Am. Chem. Soc.* **1950**, *72*, 2786–2788.
490. See Reference 489.
491. LaIuppa, J.C.; LeMaster, D.M. *J. Lab. Cmpds. Radiopharm.* **1993**, *33*, 913–919.
492. Logusch, E.W. *Tetrahedron Lett.* **1986**, *27*, 5935–5938.
493. Kurumaya, K.; Okazaki, T.; Seido, N.; Akasaka, Y.; Kawajiri, Y.; Kajiwara, M. ; Kondo, M. *J. Lab. Compds. Radiopharm.* **1989**, *27*, 217–235.
494. Tolman, V.; Benes, J. *J. Fluorine Chem.* **1976**, *7*, 397–407.
495. Simmonds, D.H. *Biochem. J.* **1954**, *58*, 520–523.
496. Work, E.; Birnbaum, S.M.; Winitz, M.; Greenstein, J.P. *J. Am. Chem. Soc.* **1955**,*77*, 1916–1918.

497. Talbot, G.; Gaudry, R.; Berlinguet, L. *Can. J. Chem.* **1958**, *36*, 593–596.

498. Talbot, G.; Gaudry, R.; Berlinguet, L. *Can. J. Chem.* **1956**, *34*, 911–914.

499. Tomita, M.; Nakashima, M. *Z. Physiol. Chem.* **1935**, *231*, 199–201.

500. Darapsky, A. *J. Pr. Chem.* **1936**, *146*, 268–306.

501. Mellor, J.M.; Smith, N.M. *J. Chem. Soc., Perkin Trans. 1* **1984**, 2927–2931.

502. Robinson, F.P.; Brown, R.K. *Can. J. Chem.* **1961**, *39*, 1171–1173.

503. Khan, N.H.; Kidwai, A.R. *J. Org. Chem.* **1973**, *38*, 822–825.

504. Effenberger, F.; Zoller, G. *Tetrahedron* **1988**, *44*, 5573–5582.

505. Bretschneider, H.; Karpitschka, N.; Piekarski, G. *Mon.* **1953**, *84*, 1084–1090.

506. Hönig, H.; Seufer-Wasserthal, P.; Weber, H. *Tetrahedron* **1990**, *46*, 3841–3850.

507. Walborsky, H.M.; Baum, M.E. *J. Org. Chem.* **1956**, *21*, 538–539.

508. Loncrini, D.F.; Walborsky, H.M. *J. Med. Chem.* **1964**, *7*, 369–370.

509. Lazar, J.; Sheppard, W.A. *J. Med. Chem.* **1968**, *11*, 138–140.

510. Wakamiya, T.; Teshima, T.; Kubota, I.; Shiba, T.; Kaneko, T. *Bull. Chem. Soc. Jpn.* **1974**, *47*, 2292–2296.

511. Brenner, M.; Rickenbacher, H.R. *Helv. Chim. Acta* **1958**, *41*, 181–188.

512. Knittel, D. *Monatsh. Chem.* **1985**, *116*, 1133–1140.

513. Knittel, D. *Monatsh. Chem.* **1984**, *115*, 1335–1343.

514. Hoffman, R.V.; Kim, H.-O. *Tetrahedron* **1992**, *48*, 3007–3020.

515. Urpí, F.; Vilarrasa, J. *Tetrahedron Lett.* **1986**, *27*, 4623–4624.

516. Hendry, D.; Hough, L.; Richardson, A.C. *Tetrahedron Lett.* **1987**, *28*, 4597–4600.

517. Maiti, S.N.; Singh, M.P.; Micetich, R.G. *Tetrahedron Lett.* **1986**, *27*, 1423–1424.

518. Scriven, E.F.V.; Turnbull, K. *Chem. Rev.* **1988**, *88*, 297–392.

519. Lin, W.; Zhang, X.; He, Z.; Jin, Y.; Gong, L.; Mi, A. *Synth. Commun.* **2002**, *32*, 3279–3284.

520. Reddy, G.V.; Rao, G.V.; Iyengar, D.S. *Tetrahedron Lett.* **1999**, *40*, 3937–3938.

521. Kamal, A.; Reddy, B.S.N. *Chem. Lett.* **1998**, 593–594.

522. Kamal, A.; Rao, N.V.; Laxman, E. *Tetrahedron Lett.* **1997**, *38*, 6945–6948.

523. Fringuelli, F.; Pizzo, F.; Rucci, M.; Vaccaro, L. *J. Org. Chem.* **2003**, *68*, 7041–7045.

524. Boruah, A.; Baruah, M.; Prajapati, D.; Sandhu, J.S. *Synlett* **1997**, 1253–1254.

525. Wu, H.; Chen, R.; Zhang, Y. *Synth. Commun.* **2002**, *32*, 189–193.

526. Ariza, X.; Urpí, F.; Viladomat, C.; Vilarrasa, J. *Tetrahedron Lett.* **1998**, *39*, 9101–9102.

527. Saito, S.; Nakajima, H.; Inaba, M.; Moriwake, T. *Tetrahedron Lett.* **1989**, *30*, 837–838.

528. Kotsuki, H.; Ohishi, T.; Araki, T. *Tetrahedron Lett.* **1997**, *38*, 2129–2132.

529. Ariza, X, Urpí, F.; Vilarrasa, J. *Tetrahedron Lett.* **1999**, *40*, 7515–7517.

530. Shangguan, N.; Katukojvala, S.; Greenberg, R.; Williams, L.J. *J. Am. Chem. Soc.* **2003**, *125*, 7754–7755.

531. Ghosh, S.K.; Verma, R.; Ghosh, U.; Mamdapur, V.R. *Bull. Chem. Soc. Jpn.* **1996**, *69*, 1705–1711.

532. Yokum, T.S.; Elzer, P.H.; McLaughlin, M.L. *J. Med. Chem.* **1996**, *39*, 3603–3605.

533. Nilsson, B.L.; Hondal, R.J.; Soellner, M.B.; Raines, R.T. *J. Am. Chem. Soc.* **2003**, *125*, 5268–5269.

534. Wang, P.; Tang, Y.; Tirrell, D.A. *J. Am. Chem. Soc.* **2003**, *125*, 6900–6906.

535. Snider, B.B.; Duncia, J.V. *J. Org. Chem.* **1981**, *46*, 3223–3226.

536. Baldwin, J.E.; Spyvee, M.R.; Whitehead, R.C. *Tetrahedron Lett.* **1994**, *35*, 6575–6576.

537. Meldal, M.; Juliano, M.A.; Jansson, A.M. *Tetrahedron Lett.* **1997**, *38*, 2531–2534.

538. Hillman, G.; Hillman, A. *Z. Physiol. Chem.* **1948**, *283*, 71–73.

539. Zwierzak, A.; Pilichowska, S. *Synthesis* **1982**, 922–924.

540. Carpino, L.A. *J. Org. Chem.* **1964**, *29*, 2820–2824.

541. Grehn, L.; Lurdes, M.; Almeida, S.; Ragnarsson, U. *Synthesis* **1988**, 992–994.

542. Grehn, L.; Ragnarsson, U. *Collect. Czech. Chem. Commun.* **1988**, *53*, 2778–2786.

543. Clarke, C.T.; Elliott, J.D.; Jones, J.H. *J. Chem. Soc., Perkin Trans. 1* **1978**, 1088–1090.

544. Grehn, L.; Ragnarsson, U. *Synthesis* **1987**, 275–276.

545. Degerbeck, F.; Fransson, B.; Grehn, L.; Ragnarsson, U. *J. Chem. Soc., Perkin Trans. 1* **1992**, 245–253.

546. Allan, R.D.; Johnston, G.A.R.; Kazlauskas, R.; Tran, H.W. *J. Chem. Soc., Perkin Trans. 1* **1983**, 2983–2985.

547. Landini, D.; Penso, M. *J. Org. Chem.* **1991**, *56*, 420–423.

548. Albanese, D.; Landini, D.; Penso, M. *J. Org. Chem.* **1992**, *57*, 1603–1605.

549. Goodfellow, V.S.; Marathe, M.V.; Kuhlman, K.G.; Fitzpatrick, T.D.; Cuadrado, D.; Hanson, W.; Zuzack, J.S.; Ross, S.E.; Wieczorek, M.; Burkard, M.; Whalley, E.T. *J. Med. Chem.* **1996**, *39*, 1472–1484.

550. Koppitz, M.; Huengs, M.; Gratias, R.; Kessler, H.; Goodman, S.L.; Jonczyk, A. *Helv. Chim. Acta.* **1997**, *80*, 1280–1300.

551. Stewart, F.H.C. *Aust. J. Chem.* **1961**, *14*, 654–656.

552. Poduska, K. *Collect. Czech. Chem. Commun.* **1959**, *24*, 1025–1028.

553. Drake, W.V.; McElvain, S.M. *J. Am. Chem. Soc.* **1934**, *56*, 697–700.

554. Zuckermann, R.N.; Kerr, J.M.; Kent, S.B.H.; Moos, W.H. *J. Am. Chem. Soc.* **1992**, *114*, 10646–10647.

555. Wu, C.W.; Kirshenbaum, K.; Sanborn, T.J.; Patch, J.A.; Huang, K.; Dill, K.A.; Zuckermann, R.N.; Barron, A.E. *J. Am. Chem. Soc.* **2003**, *125*, 13525–13530.

556. Schmitt, J.; Bernd, M.; Kutscher, B.; Kessler, H. *Biorg. Med. Chem. Lett.* **1998**, *8*, 385–388.

557. Zuckermann, R.N.; Martin, E.J.; Spellmeyer, D.C.; Stauber, G.B.; Shoemaker, K.R.; Kerr, J.M.; Figliozzi, G.M.; Goff, D.A.; Siani, M.A.; Simon, R.J.; Banville, S.C.; Brown, E.G.; Wang, L.; Richter, L.S.; Moos, W.H. *J. Med. Chem.* **1994**, *37*, 2678–2685.

558. Révész, L.; Bonne, F.; Manning, U.; Zuber, J.-F. *Biorg. Med. Chem. Lett.* **1998**, *8*, 405–408.

559. Humet, M.; Carbonell, T.; Masip, I.; Sánchez-Baeza, F.; Mora, P.; Cantón, E.; Gobernado, M.; Abad, C.; Pérez-Payá, E.; Messeguer, A. *J. Comb. Chem.* **2003**, *5*, 597–605.

560. Alluri, P.G.; Reddy, M.M.; Bachhawat-Sikder, K.; Olivos, H.J.; Kodadek, T. *J. Am. Chem. Soc.* **2003**, *125*, 13995–14004.

561. Ast, T.; Heine, N.; Germeroth, L.; Schneider-Mergener, J.; Wenschuh, H. *Tetrahedron Lett.* **1999**, *40*, 4317–4318.

562. Soriano, D.S.; Podraza, K.F.; Cromwell, N.H. *J. Heterocycl. Chem.* **1980**, *17*, 623–624.

563. Mattocks, A.M.; Hartung, W.H. *J. Biol. Chem.* **1946**, *165*, 501–503.

564. Lazareva, N.F.; Baryshok, V.P.; Voronkov, M.G. *Russ. Chem. Bull.* **1995**, *44*, 333–335.

565. Byk, G.; Gilon, C. *J. Org. Chem.* **1992**, *57*, 5687–5692.

566. Cosquer, A.; Pichereau, V.; Le Mée, D.; Le Roch, M.; Renault, J.; Carboni, B.; Uriac, P.; Bernard, T. *Biorg. Med. Chem. Lett.* **1999**, *9*, 49–54.

567. Kolasa, T. *Can. J. Chem.* **1985**, *63*, 2139–2142.

568. Treibs, W.; Reinheckel, H. *Chem. Ber.* **1956**, *89*, 51–57.

569. Jummah, R.; Willaims, J.M.J.; Williams, A.C. *Tetrahedron Lett.* **1993**, *34*, 6619–6622.

570. Harding, K.E.; Marman, T.H.; Nam, D.-H. *Tetrahedron* **1988**, *44*, 5605–5614.

571. Dunn, M.S.; Fox, S.W. *J. Biol. Chem.* **1933**, *101*, 493–497.

572. Greenstein, J.P.; Winitz, N. *Chemistry of the Amino Acids, Volume 3*; Wiley: New York, **1961**, pp 1856–1878.

573. Zaderenko, P.; López, M.C.; Ballesteros, P. *J. Org. Chem.* **1996**, *61*, 6825–6828.

574. Frankel, M.; Liwschitz, Y.; Amiel, Y. *J. Am. Chem. Soc.* **1953**, *75*, 330–332.

575. Hansen, J.J.; Krogsgaard-Larsen, P. *J. Chem. Soc., Perkin Trans. 1* **1980**, 1826–1833.

576. Hansen, J.J.; Krogsgaard-Larsen, P. *J. Chem. Soc. Chem. Commun.* **1979**, 87–88.

577. Butenandt, A.; Hallman, G.; Beckmann, R. *Chem. Ber.* **1957**, *90*, 1120–1124.

578. Bella, M.; D'Onofrio, F.; Margarita, R.; Parlanti, L.; Piancatelli, G.; Mangoni, A. *Tetrahedron Lett.* **1997**, *38*, 7917–7918.

579. Romnova, N.N.; Gravis, A.G.; Bundel, Y.G. *Russ. Chem. Rev.* **1996**, *65*, 1083–1092.

580. Li, G.; Wei, H.-X.; Kim, S.H.; Neighbors, M. *Org. Lett.* **1999**, *1*, 395–397.

581. Fokin, V.V.; Sharpless, K.B. *Angew. Chem., Int. Ed.* **2001**, *40*, 3455–3457.

582. Kolasa, T.; Miller, M.J. *J. Org. Chem.* **1987**, *52*, 4978–4984.

583. Castelhano, A.L.; Krantz, A. *J. Am. Chem. Soc.* **1987**, *109*, 3491–3493.

584. Berrée, F.; Michelot, G.; Le Corre, M. *Tetrahedron Lett.* **1998**, *39*, 8275–8276.

585. Kobayashi, T.; Sakakura, T.; Tanaka, M. *Tetrahedron Lett.* **1987**, *28*, 2721–2722.

586. Compere E.L. Jr.; *J. Org. Chem.* **1968**, *33*, 2565–2566.

587. Erdik, E.; Ay, M. *Chem. Rev.* **1989**, *89*, 1947–1980.

588. Erdik, E. *Tetrahedron* **2004**, *60*, 8747–8782.

589. Armstrong, A.; Atkin, M.A.; Swallow, S. *Tetrahedron Lett.* **2000**, *41*, 2247–2251.

590. Sheradsky, T.; Nir, Z. *Tetrahedron Lett.* **1969**, 77–78.

591. Sheradsky, T.; Salemnick, G.; Nir, Z. *Tetrahedron* **1972**, *28*, 3833–3843.

592. Radhakrishna, A.S.; Loudon, G.M.; Miller, M.J. *J. Org. Chem.* **1979**, *44*, 4836–4841.

593. Yamada, S.-I.; Oguri, T.; Shioiri, T. *J. Chem. Soc., Chem. Commun.* **1972**, 623.

594. Oguri, T.; Shioiri, T.; Yamada, S.-I. *Chem. Pharm. Bull.* **1975**, *23*, 167–172.

595. Scopes, D.I.C.; Kluge, A.F.; Edwards, J.A. *J. Org. Chem.* **1977**, *42*, 376–377.

596. Colvin, E.W.; Kirby, G.W.; Wilson, A.C. *Tetrahedron Lett.* **1982**, *23*, 3835–3836.

597. Boche, G.; Bernheim, M.; Schrott, W. *Tetrahedron Lett.* **1982**, *23*, 5399–5402.

598. Allan, R.D.; Duke, R.K.; Hambley, T.W.; Johnston, G.A.R.; Mewett, K.N.; Quickert, N.; Tran, H.W. *Aust. J. Chem.* **1996**, *49*, 785–791.

599. Dowd, P.; Kaufman, C.; Kaufman, P. *J. Org. Chem.* **1985**, *50*, 882–885.

600. Dowd, P.; Kaufman, C. *J. Org. Chem.* **1979**, *44*, 3956–3957.

601. Allan, R.D.; Duke, R.K.; Hambley, T.W.; Johnston, G.A.R.; Mewett, K.N.; Quickert, N.; Tran, H.W. *Aust. J. Chem.* **1996**, *49*, 785–791.

602. Frank, J.; Stoll, G.; Musso, H. *Liebigs Ann. Chem.* **1986**, 1990–1996.

603. Curtius, T. *J. Prak. Chem.* **1930**, *125*, 303–424.

604. Evans, D.A.; Britton, T.C.; Ellman, J.A.; Dorow, R.L. *J. Am. Chem. Soc.* **1990**, *112*, 4011–4030.

605. Evans, D.A.; Britton, T.C. *J. Am. Chem. Soc.* **1987**, *109*, 6881–6883.
606. Stone, M.J.; Maplestone, R.A.; Rahman, S.K.; Williams, D.H. *Tetrahedron Lett.* **1991**, *32*, 2663–2666.
607. Harmon, R.E.; Wellman, G.; Gupta, S.K. *J. Org. Chem.* **1973**, *38*, 11–16.
608. Leffler, J.E.; Tsuno, Y. *J. Org. Chem.* **1963**, *28*, 902–906.
609. Villalgordo, J.M.; Linden, A.; Heimgartner, H. *Helv. Chim. Acta* **1996**, *79*, 213–219.
610. Evans, D.E.; Britton, T.C.; Dorow, R.L.; Dellaria J.F., Jr., *Tetrahedron* **1988**, *44*, 5525–5540.
611. Trimble, L.A.; Vederas, J.C. *J. Am. Chem. Soc.* **1986**, *108*, 6397–6399.
612. Genet, J.P.; Mallart, S.; Greck, C.; Piveteau, E. *Tetrahedron Lett.* **1991**, *32*, 2359–2362.
613. Zheng, N.; Armstrong J.D., III; McWilliams, J.C.; Volante, R.P. *Tetrahedron Lett.* **1997**, *38*, 2817–2820.
614. Vidal, J.; Damestoy, S.; Collet, A. *Tetrahedron Lett.* **1995**, *36*, 1439–1442.
615. Vidal, J.; Guy, L.; Stérin, S.; Collet, A. *J. Org. Chem.* **1993**, *58*, 4791–4793.
616. Loreto, M.A.; Tardella, P.A.; Tedeschi, L.; Tofani, D. *Tetrahedron Lett.* **1997**, *38*, 5717–5718.
617. Sakakura, T.; Hara, M.; Tanaka, M. *J. Chem. Soc., Perkin Trans. 1*, **1994**, 289–292.
618. Sakakura, T.; Tanaka, M. *J. Chem. Soc., Chem. Commun.* **1985**, 1309–1310.
619. Phillips, R.R. *Org. React.* **1959**, *10*, 143–178.
620. Adkins, H.; Reeve, E.W. *J. Am. Chem. Soc.* **1938**, *60*, 1328–1331.
621. Bolhofer, W.A. *J. Am. Chem. Soc.* **1952**, *74*, 5459–5461.
622. Leanza, W.J.; Pfister K., 3rd, *J. Biol. Chem.* **1953**, *201*, 377–383.
623. Harington, C.R.; Randall, S.S. *Biochem. J.* **1931**, *25*, 1917–1925.
624. McIlwain, H.; Richardson, G.M. *Biochem. J.* **1939**, *33*, 44–46.
625. Shapiro, D.; Abramovitch, R.A. *J. Am. Chem. Soc.* **1955**, *77*, 6690–6691.
626. Olynyk, P.; Camp D.P.; Griffith, A.M.; Woislowski, S.; Helmkamp, R.W. *J. Org. Chem.* **1948**, *13*, 465–470.
627. Borsook, H.; Deasy, C.L.; Haagen-Smit, A.J.; Keighley, G.; Lowy, P.H. *J. Biol. Chem.* **1948**, *176*, 1383–1393.
628. Harington, C.R. *J. Biol. Chem.* **1925**, *64*, 29–39.
629. Holland, D.O.; Nayler, J.H.C. *J. Chem. Soc.* **1953**, 280–285.
630. Hamlin K.E., Jr., Hartung, W.H. *J. Biol. Chem.* **1942**, *145*, 349–357.
631. Scolastico, C.; Conca, E.; Prati, L.; Guanti, G.; Banfi, L.; Berti, A.; Farina, P.; Valcavi, U. *Synthesis* **1985**, 850–855.
632. Vidal-Cros, A.; Gaudry, M.; Marquet, A. *J. Org. Chem.* **1985**, 50, 3163–3167.
633. Smith, P.A.S. *Org. React.* **1946**, *3*, 337–449.
634. Darapsky, A.; Hillers, D. *J. Prak. Chem.* **1915**, *92*, 297–341.
635. Curtius, T. *J. Prak. Chem.* **1930**, *125*, 211–302.
636. Darapsky, A. *J. Prak. Chem.* **1936**, *146*, 250–267.
637. Holland, D.O.; Nayler, J.H.C. *J. Chem. Soc.* **1953**, 280–285.
638. Gagnon, P.E.; Savard, K.; Gaudry, R.; Richardson, E.M. *Can. J. Res. B* **1947**, *25*, 28–36.
639. Gagnon, P.E.; Gaudry, R.; King, F.E. *J. Chem. Soc.* **1944**, 13–15.
640. Gagnon, P.E.; Nadeau, G.; Côté, R. *Can. J. Chem.* **1952**, *30*, 592–597.
641. Gagnon, P.E.; Boivin, J.L.; Giguere, J. *Can. J. Res. B* **1950**, *28*, 352–357.
642. Gagnon, P.E.; Boivin, J.L. *Can. J. Res. B.* **1948**, *26*, 503–510.
643. Gagnon, P.E.; Boivin, J.L.; Biovin, P.A. *Can. J. Res. B.* **1950**, *28*, 207–212.
644. Gagnon, P.E.; Nolin, B. *Can. J. Res. B* **1949**, *27*, 742–748.
645. Gagnon, P.E.; Biovin, P.A.; Craig, H.M. *Can. J. Chem.* **1951**, *29*, 70–75.
646. Sloan, M.J.; Kirk, K.L. *Tetrahedron Lett.* **1997**, *38*, 1677–1680.
647. Dappen, M.S.; Pellicciari, R.; Natalini, B.; Monahan, J.B.; Chiorri, C.; Cordi, A.A. *J. Med. Chem.* **1991**, *34*, 161–168.
648. Weinstock, J. *J. Org. Chem.* **1961**, *26*, 3511.
649. Nitz, T.J.; Holt, E.M.; Rubin, B.; Stammer, C.H. *J. Org. Chem.* **1981**, *46*, 2667–2671.
650. Petrzilka, T.; Fehr, C. *Helv. Chim. Acta* **1973**, *56*, 1218–1224.
651. Bey, P.; Ducep, J.B.; Schirlin, D. *Tetrahedron Lett.* **1984**, *25*, 5657–5660.
652. Ninomiya, K.; Shioiri, T.; Yamada, S. *Tetrahedron* **1974**, *30*, 2151–2157.
653. Yamada, S.-I.; Ninomiya, K.; Shioiri, T. *Tetrahedron Lett.* **1973**, *26*, 2343–2346.
654. Ninomiya, K.; Shioiri, T.; Yamada, S-I. *Chem. Pharm. Bull.* **1974**, *22*, 1398–1404.
655. Richter, L.S.; Andersen, S. *Tetrahedron Lett.* **1998**, *39*, 8747–8750.
656. Koskinen, A.M.P.; Muñoz, L. *J. Org. Chem.* **1993**, *58*, 879–86.
657. Allan, R.D.; Hanrahan, J.R.; Hambley, T.W.; Johnston, G.A.R.; Mewett, K.N.; Mitrovic, A.D. *J. Med. Chem.* **1990**, *33*, 2905–2915.
658. Wheeler, T.N.; Ray, J.A. *Synth. Commun.* **1988**, *18*, 141–149.
659. Wheeler, T.N.; Ray, J. *J. Org. Chem.* **1987**, *52*, 4875–4877.
660. Zhao, Z.; Liu, H.-w. *J. Org. Chem.* **2002**, *67*, 2509–2514.
661. Zhao, Z.; Chen, H.; Li, K.; Du, W.; He, S.; Liu, H.-w. *Biochemistry* **2003**, *42*, 2089–2103.
662. Goudreau, N.; Brochu, C.; Cameron, D.R.; Duceppe, J.-S.; Faucher, A.-M.; Ferland, J.-M.;

Grand-Maître, C.; Poirer, M.; Simoneau, B.; Tsantrizos, Y.S. *J. Org. Chem.* **2004**, *69*, 6185–6201.

663. Pellicciari, R.; Marinozzi, M.; Camaioni, E.; del Carmen Nùnez, M.; Costantino, G.; GaspariniF.; Giorgi, G.; Macchiarulo, A.; Subramanian, N. *J. Org. Chem.* **2002**, *67*, 5497–5507.

664. Cannon, J.G.; Garst, J.E. *J. Org. Chem.* **1975**, *40*, 182–184.

665. Izquierdo, M.L.; Arenal, I.; Bernabé, M.; Alvarez, E.F. *Tetrahedron* **1985**, *41*, 215–220.

666. Fleet, G.W.J.; Seijas, J.A.; Tato, M.P.V. *Tetrahedron* **1988**, *44*, 2081–2086.

667. Burger, A.; Coyne, W.E. *J. Org. Chem.* **1964**, *29*, 3079–3082.

668. Lasa, M.; López, P.; Cativiela, C. *Tetrahedron: Asymmetry* **2005**, *16*, 4022–4033.

669. Fleet, G.W.J.; Seijas, J.A.; Tato, M.P.V. *Tetrahedron* **1988**, *44*, 2077–2080.

670. Koskinen, A.M.P.; Muñoz, L. *J. Chem. Soc., Chem. Commun.* **1990**, 1373–1374.

671. Schirlin, D.; Ducep, J.B.; Baltzer, S.; Bey, P.; Piriou, F.; Wagner, J.; Hornsperger, J.M.; Heydt, J.G.; Jung, M.J.; Danzin, C.; Weiss, R.; Fischer, J.; Mitschler, A.; De Cian, A. *J. Chem. Soc., Perkin Trans. 1* **1992**, 1053–1064.

672. Wolff, H. *Org. React.* **1946**, *3*, 307–336.

673. Adamson, D.W. *J. Chem. Soc.* **1939**, 1564–1568.

674. Takagi, S.; Hayashi, K. *Chem. Pharm. Bull.* **1959**, *7*, 96–98.

675. Briggs, L.H.; De Ath, G.C.; Ellis, S.R. *J. Chem. Soc.* **1942**, 61–63.

676. Rothchild, S.; Fields, M. *J. Org. Chem.* **1951**, *16*, 1080–1081.

677. Takagi, S.; Hayashi, K. *Chem. Pharm. Bull.* **1959**, *7*, 183–186.

678. Schmidt, K.F. *Ber.* **1924**, *57*, 704–706.

679. Barrett, G.C.; Hardy, P.M.; Harrow, T.A.; Rydon, H.N. *J. Chem. Soc., Perkin Trans.1* **1972**, 2634–2638.

680. Felkin, H. *Bull. Chim. Soc. Fr.* **1959**, 20–32.

681. Barrett, G.C.; Cousins, P.R. *J. Chem. Soc., Perkin Trans. 1* **1975**, 2313–2315.

682. Holland, D.O.; Nayler, J.H.C. *J. Chem. Soc.* **1953**, 280–285.

683. Hayashi, K. *Chem. Pharm. Bull.* **1959**, *7*, 187–191.

684. Riemschneider, R.; Kluge, A. *Mon.* **1953**, *84*, 522–526.

685. Hudlicky, M.; Jelínek, V.; Elsler, K.; Rudinger, J. *Collect. Czech. Chem. Commun.* **1970**, *35*, 498–503.

686. Takagi, S.; Hayashi, K. *Chem. Pharm. Bull.* **1959**, *7*, 99–102.

687. Moreno-Mañas, M.; Pleixats, R.; Roglans, A. *Liebigs Ann. Chem.* **1995**, 1807–1814.

688. Ridvan, L.; Abdallah, N.; Holakovsly, R.; Tichy, M.; Závada, J. *Tetrahedron: Asymmetryetry* **1996**, *7*, 231–236.

689. Moreno-Mañas, M.; Pleixats, R.; Roglans, A. *Liebigs Ann.* **1995**, 1807–1814.

690. Wallis, E.S.; Lane, J.F. *Org. React.* **1946**, *3*, 267–306.

691. Freudenberg, K. *Ber.* **1914**, *47*, 2027–2037.

692. Hale, W.J.; Honan, E.M. *J. Am. Chem. Soc.* **1919**, *41*, 770–776.

693. Noyes, W.A.; Skinner, G.S. *J. Am. Chem. Soc.* **1917**, *39*, 2692–2718.

694. Weir, J. *J. Chem. Soc.* **1911**, *99*, 1270–1277.

695. Noyes, W.A.; Nickell, L.F. *J. Am. Chem. Soc.* **1914**, *36*, 118–127.

696. Hardy, P.M.; Lingham, I.N. *Int. J. Pept. Prot. Res.* **1983**, *21*, 392–405.

697. Huang, Y.-t.; Lin, K.-h.; Li, L. *J. Chin. Chem. Soc.* **1941**, *8*, 81–91.

698. Huang, Y.-t.; Lin, K.-h.; Li, L.; Lu, M.-c. *J. Chin. Chem. Soc.* **1941**, *8*, 210–217.

699. Li, L.; Lin, K.-h.; Huang, Y.-t.; Kang, S.-a. *J. Chin. Chem. Soc.* **1942**, *9*, 1–13.

700. Li, L.; Lin, K.-h.; Huang, Y.-t.; Huang, A.Y.L. *J. Chin. Chem. Soc.* **1942**, *9*, 14–30.

701. Huang, Y.-t.; Lin, K-h.; Li, L.; Lu, M.-c. *J. Chin. Chem. Soc.* **1942**, *9*, 31–40.

702. Ingold, C.K.; Sako, S.; Thorpe, J.F. *J. Chem. Soc.* **1922**, *121*, 117–198.

703. Matsumura, Y.; Maki, T.; Satoh, Y. *Tetrahedron Lett.* **1997**, *38*, 8879–8882.

704. Huang, X.; Seid, M.; Keillor, J.W. *J. Org. Chem.* **1997**, *62*, 7495–7496.

705. Zhang, L.-h.; Kauffman, G.S.; Pesti, J.A.; Yin, J. *J. Org. Chem.* **1997**, *62*, 6918–6920.

706. Overman, L.E. *J. Am. Chem. Soc.* **1976**, *98*, 2901–2910.

707. Takano, S.; Akiyama, M.; Ogasawara, K. *J. Chem. Soc., Chem. Commun.* **1984**, 770–771.

708. Chida, N.; Takeoka, J.; Ando, K.; Tsutsumi, N.; Ogawa, S. *Tetrahedron* **1997**, *53*, 16287–16298.

709. Vyas, D.M.; Chiang, Y.; Doyle, T.W. *J. Org. Chem.* **1984**, *49*, 2037–2039.

710. Graham, W.H. *Tetrahedron Lett.* **1969**, 2223–2225.

711. Baumgarten, H.E.; Dirks, J.E.; Petersen, J.M.; Zey, R.L. *J. Org. Chem.* **1966**, *31*, 3708–3711.

712. Zemlicka, J.; Murata, M. *J. Org. Chem.* **1976**, *41*, 3317–3321.

713. Crout, D.H.G.; Lutstorf, M.; Morgan, P.J. *Tetrahedron* **1983**, *39*, 3457–3469.

714. Hallinan, K.O.; Crout, D.H.G.; Errington, W. *J. Chem. Soc., Perkin Trans. 1* **1994**, 3537–3543.

715. Friis, P.; Larsen, P.O.; Olsen, C.E. *J. Chem. Soc., Perkin Trans. 1* **1977**, 661–665.

716. Oldfield, M.F.; Botting, N.P. *J. Lab. Cmpds. Radiopharm.* **1998**, *41*, 29–36.

717. Endo, Y.; Hizatate, S.; Shudo, K. *Synlett*, **1991**, 649–650.

718. Fitzner, J.N.; Pratt, D.V.; Hopkins, P.B. *Tetrahedron Lett.* **1985**, *26*, 1959–1962.

719. Shea, R.G.; Fitzner, J.N.; Fankhauser, J.E.; Spaltenstein, A.; Carpino, P.A.; Peevey, R.M.; Pratt, D.V.; Tenge, B.J.; Hopkins, P.B. *J. Org. Chem.* **1986**, *51*, 5243–5252.

720. Boivin, S.; Outurquin, F.; Paulmier, C. *Tetrahedron* **1997**, *53*, 16767–16782.

721. Nicoud, J.-F.; Kagan, H.B. *Tetrahedron Lett.* **1971**, *23*, 2065–2068.

722. Aller, E.; Buck, R.T.; Drysdale, M.J.; Ferris, L.; Haigh, D.; Moody, C.J.; Pearson, N.D.; Sanghera, J.B. *J. Chem. Soc., Perkin Trans. 1*, **1996**, 2879–2884.

723. Osipov, S.N.; Sewald, N.; Kolomiets, A.F.; Fokin, A.V.; Burger, K. *Tetrahedron Lett.* **1996**, *37*, 615–618.

724. Ferris, L.; Haigh, D.; Moody, C.J. *J. Chem. Soc., Perkin Trans. 1* **1996**, 2885–2888.

725. Clark, J.S.; Middleton, M.D. *Org. Lett.* **2002**, *4*, 765–768.

726. Tanaka, M.; Koyanagi, M. *Synthesis* **1981**, 973–974.

727. Herrmann, K.; Simchen, G. *Synthesis* **1979**, 204–205.

728. Tanaka, M. *Tetrahedron Lett.* **1980**, *21*, 2959–2962.

729. Tanaka, M.; Kobayashi, T.-A.; Sakakura, T. *Angew. Chem., Int. Ed. Engl.* **1984**, *23*, 518.

730. Elbers, A. *Ann.* **1885**, *227*, 340–357.

731. Farchy, J.M.; Tafel, J. *Ber.* **1893**, *26*, 1980–1990.

732. Sprinson, D.B.; Chargaff, E. *J. Biol. Chem.* **1946**, *164*, 417–432.

733. Baldwin, J.E.; Wan, T.S. *Tetrahedron* **1981**, *37*, 1589–1595.

734. Teuber, H-J.; Krause, H.; Berariu, V. *Liebigs Ann. Chem.* **1978**, 757–770.

735. Towers, G.H.N.; Thompson, J.F.; Steward, F.C. *J. Am. Chem. Soc.* **1954**, *76*, 2392–2396.

736. See Reference 735.

737. Wolff, L. *Ann.* **1890**, *260*, 79–136.

738. Knoop, F.; Landmann, G. *A. Physiol. Chem.* **1914**, *89*, 157–159.

739. Honoré, T.; Lauridsen, J. *Acta. Chem. Scand. B* **1980**, *34*, 235–240.

740. Christensen, S.B.; Krogsgaard-Larsen, P. *Acta. Chem. Scand. B* **1978**, *32*, 27–30.

741. Clemo, G.R.; Duxbury, F.K. *J. Chem. Soc.* **1950**, 1795–1800.

742. Jadot, J.; Casimir, J.; Loffet, A. *Biochim. Biophys. Acta* **1967**, *136*, 79–88.

743. Møller, B.L.; McFarlane, I.J.; Conn, E.E. *Acta Chem. Scand. B* **1977**, *31*, 343–344.

744. Ahmad, A. *Bull. Chem. Soc. Jpn.* **1974**, *47*, 1819–1820.

745. Pospísek, J.; Bláha, K. *Coll. Czech. Chem. Commun.* **1977**, *42*, 1069–1076.

746. Hercheid, J.D.M.; Ottenheijm, H.C.J. *Tetrahedron Lett.* **1978**, *51*, 5143–5144.

747. Hoffman, C.; Tanke, R.S.; Miller, M.J. *J. Org. Chem.* **1989**, *54*, 3750–3751.

748. Tang, Y.; Tirrell, D.A. *J. Am. Chem. Soc.* **2001**, *123*, 11089–1090.

749. Bixler, R.L.; Niemann, C. *J. Org. Chem.* **1958**, *23*, 575–584.

750. Haggerty W.J., Jr.; Springer, R.H.; Cheng, C.C. *J. Heterocyl. Chem.* **1965**, *2*, 1–6.

751. Springer, R.H.; Haggerty Jr., W.J.; Cheng, C.C. *J. Heterocycl. Chem.* **1965**, *2*, 49–52.

752. Boukhris, S.; Souizi, A. *Tetrahedron Lett.* **1999**, *40*, 1669–1672.

753. Bicknell, A.J.; Burton, G.; Elder, J.S. *Tetrahedron Lett.* **1988**, *29*, 3361–3364.

754. Barrett, A.G.M.; Dhanak, D.; Lebold, S.A.; Russell, M.A. *J. Org. Chem.* **1991**, *56*, 1894–1901.

755. Tsushima, T.; Kawada, K.; Nishikawa, J.; Sato, T.; Tori, K.; Tsuji, T.; Misaki, S. *J. Org. Chem.* **1984**, *49*, 1163–1169.

756. Tsushima, T.; Sato, T.; Tsuji, T. *Tetrahedron Lett.* **1980**, *21*, 3593–3594.

757. Anhoury, M.L.; Crooy, P.; De Neys, R.; Eliaers, J. *Bull. Soc. Chim. Belg.* **1974**, *83*, 117–130.

758. Crooij, P.; Eliaers, J. *J. Chem. Soc. C* **1969**, 559–563.

759. Remli, M.; Avi, A.I.; Guedj, R. *J. Fluorine Chem.* **1982**, *20*, 677–682.

760. Hermes, J.D.; Weiss, P.M.; Cleland, W.W. *Biochemistry* **1985**, *24*, 2959–2967.

761. Varlet, J.-M.; Collignon, N.; Savignac, P. *Can. J. Chem.* **1979**, *57*, 3216–3220.

762. Shi, G.-q.; Cai, W.-l. *J. Org. Chem.* **1995**, *60*, 6289–6295.

763. White, R.H. *J. Lab. Cmpds. Radiopharm.* **1983**, *20*, 787–790.

764. Dolling, U.-H.; Douglas, A.W.; Grabowski, E.J.J.; Schoenewaldt, E.F.; Sohar, P.; Sletzinger, M. *J. Org. Chem.* **1978**, *43*, 1634–1640.

765. Plieninger, H. *Chem. Ber.* **1950**, *83*, 271–272.

766. Harrowfield, J.M.; Sargeson, A.M. *J. Am. Chem. Soc.* **1979**, *101*, 1514–1520.

767. Adger, B.M.; Dyer, U.C.; Lennon, I.C.; Tiffin, P.D.; Ward, S.E. *Tetrahedron Lett.* **1997**, *38*, 2153–2154.

768. Ogo, S.; Uehara, K.; Abura, T.; Fukuzumi, S. *J. Am. Chem. Soc.* **2004**, *126*, 3020–3021.

769. Fache, F.; Valot, F.; Milenkovic, A.; Lemaire, M. *Tetrahedron* **1996**, *52*, 9772–9784.

770. De Nicola, A.; Einhorn, C.; Einhorn, J.; Luche, J.L. *J. Chem. Soc., Chem. Commun.* **1994**, 879–880.

771. Bitan, G.; Muller, D.; Kasher, R.; Gluhov, E.V.; Gilon, C. *J. Chem. Soc., Perkin Trans. 1* **1997**, 1501–1510.

772. Patchett, A.A.; Harris, E.; Tristram, E.W.; Wyvratt, M.J.; Wu, M.T.; Taub, D.; Peterson, E.R.; Ikeler, T.J.; ten Broeke, J.; Payne, L.G.; Ondeyka, D.L.; Thorsett, E.D.; Greenlee, W.J.; Lohr, N.S.;

Hoffsommer, R.D.; Joshua, H.; Ruyle, W.V.; Rothrock, J.W.; Aster, S.D.; Maycock, A.L.; Robinson, F.M.; Hirschmann, R.; Sweet, C.S.; Ulm, E.H.; Gross, D.M.; Vassil, T..C.; Stone, C.A. *Nature* **1980**, *288*, 280–283.

773. Gavagan, J.E.; Fager, S.K.; Seip, J.E.; Clark, D.S.; Payne, M.S.; Anton, D.L.; DiCosimo, R. *J. Org. Chem.* **1997**, *62*, 5419–5427.

774. Simon, R.J.; Kania, R.S.; Zuckermann, R.N.; Huebner, V.D.; Jewell, D.A.; Banville, S.; Ng, S.; Wang, L.; Rosenberg, S.; Marlowe, C.K.; Spellmeyer, D.C.; Tan, R.; Frankel, A.D.; Santi, D.V.; Cohen, F.E.; Bartlett, P.A. *Proc. Natl. Acad. Sci. U.S.A.* **1992**, *89*, 9367–9371.

775. Simon, R.J.; Kania, R.S.; Zuckermann, R.N.; Huebner, V.D.; Jewell, D.A.; Banville, S.; Ng, S.; Wang, L.; Rosenberg, S.; Marlowe, C.K.; Spellmeyer, D.C.; Tan, R.; Frankel, A.D.; Santi, D.V.; Cohen, F.E.; Bartlett, P.A. *Proc. Natl. Acad. Sci. U.S.A.* **1992**, *89*, 9367–9371.

776. Soloshonok, V.A.; Kukhar, V.P. *Tetrahedron* **1997**, *53*, 8307–8314.

777. Soloshonok, V.A.; Ono, T.; Soloshonok, I.V. *J. Org. Chem.* **1997**, *62*, 7538–7539.

778. Soloshonok, V.A.; Kirilenko, A.G.; Galushko, S.V.; Kukhar, V.P. *Tetrahedron Lett.* **1994**, *35*, 5063–5064.

779. Soloshonok, V.A.; Kirilenko, A.G.; Kukhar, V.P. *Tetrahedron Lett.* **1993**, *35*, 3621–3624.

780. Soloshonok, V.A.; Kirilenko, A.G.; Fokina, N.A.; Shishkina, I.P.; Galushko, S.V.; Kukhar, V.P.; Svedas, V.K.; Kozlova, E.V. *Tetrahedron: Asymmetry* **1994**, *5*, 1119–1126.

781. Soloshonok, V.A.; Ono, T. *Tetrahedron* **1996**, *52*, 14701–14712.

782. Liu, L.; Rozenman, M.; Breslow, R. *J. Am. Chem. Soc.* **2002**, *124*, 4978–4979.

783. Liu, L.; Rozenman, M.; Breslow, R. *J. Am. Chem. Soc.* **2002**, *124*, 12660–12661.

784. Liu, L.; Zhou, W.; Chruma, J.; Breslow, R. *J. Am. Chem. Soc.* **2004**, *126*, 8136–8137.

785. Liu, L.; Breslow, R. *J. Am. Chem. Soc.* **2003**, *125*, 12110–12111.

786. Liu, L.; Breslow, R. *Tetrahedron Lett.* **2001**, *42*, 2775–2777.

787. Matsumoto, K.; Suzuki, M.; Miyoshi, M. *J. Org. Chem.* **1973**, *38*, 2094–2096.

788. Oguri, T.; Shioiri, T.; Yamada, S.-I. *Chem. Pharm. Bull.* **1975**, *23*, 173–177.

789. Vaalburg, W.; Strating, J.; Woldring, M.G.; Wynberg, H. *Synth. Commun.* **1972**, *2*, 423–425.

790. Taub, D.; Patchett, A.A. *Tetrahedron Lett.* **1977**, 2745–2748.

791. Metcalf, B.W.; Jund, K. *Tetrahedron Lett.* **1977**, *41*, 3689–3692.

792. Vaalburg, V; Bolster, J.M.; Paans, A.M.J.; Zijlstra, J.B.; van Djik, Th.; Smid, M.; Woldring, H.G. *J. Lab. Cmpds. Radiopharm.* **1983**, *21*, 1177.

793. Bolster, J.M.; Vaalburg, W.; Elsinga, Ph. H.; Ishiwata, K.; Vissering, H.; Woldring, M.G. *J. Lab. Cmpds. Radiopharm.* **1985**, *23*, 1081–1082.

794. Kobayashi, K.; Akamatsu, H.; Irisawa, S.; Takahashi, M.; Morikawa, O.; Konishi, H. *Chem. Lett.* **1997**, 503–504.

795. Chong, J.M.; Park, S.B. *J. Org. Chem.* **1992**, 2220–2222.

796. Burchat, A.F.; Ching, M.J.; Park, S.B. *Tetrahedron Lett.* **1993**, *34*, 51–54.

797. Ekhato, I.V.; Huang, C.C. *J. Lab. Cmpds. Radiopharm.* **1994**, *34*, 107–115.

798. Lander, P.A.; Hegedus, L.S. *J. Am. Chem. Soc.* **1994**, *116*, 8126–8132.

799. Gately, D.A.; Norton, J.R.; Goodson, P.A. *J. Am. Chem. Soc.* **1995**, *177*, 986–996.

800. West, F.G.; Glaeske, K.W.; Naidu, B.N. *Synthesis* **1993**, 977–980.

801. Madsen, U.; Frølund, B.; Lund, T.M.; Ebert, B.; Krogsgaard-Larsen, P. *Eur. J. Med. Chem.* **1993**, *28*, 791–800.

802. Bonnett, R.; Clark, V.M.; Giddey, A.; Todd, A. *J. Chem. Soc.* **1959**, 2087–2093.

803. Magaard, V.W.; Sanchez, R.M.; Bean, J.W.; Moore, M.L. *Tetrahedron Lett.* **1993**, *34*, 381–384.

804. Weygand, F.; Steglich, W.; Oettmeier, W.; Maierhofer, A.; Loy, R.S. *Angew. Chem., Int. Ed. Engl.* **1966**, *5*, 600–601.

805. Marshall, J.A.; Garofalo, A.W. *J. Org. Chem.* **1993**, *58*, 3675–3680.

806. Mertin, A.; Thiemann, T.; Hanss, I.; de Meijere, A. *Synlett* **1991**, 87–89.

807. Sisko, J.; Weinreb, S.M. *J. Org. Chem.* **1990**, *55*, 393–395.

808. Davis, F.A.; Giangiordano, M.A.; Starner, W.E. *Tetrahedron Lett.* **1986**, *27*, 3957–3960.

809. Cainelli, G.; Giacomini, D.; Mezzina, E.; Panunzio, M.; Zarantonello, P. *Tetrahedron Lett.* **1991**, *32*, 2967–2970.

810. Shono, T. *Tetrahedron* **1984**, *40*, 811–850.

811. Shono, T.; Matsumura, Y.; Tsubata, K. *Tetrahedron Lett.* **1981**, *22*, 2411–2412.

812. Sakami, W.; Evans, W.E.; Gurin, S. *J. Am. Chem. Soc.* **1947**, *69*, 1110–1112.

813. Klebs, E. *Z. Physiol. Chem.* **1894**, *19*, 301–338.

814. Fischer, E. *Ber. Deutsch. Chem. Ges.* **1901**, *34*, 454–464.

815. Fischer, E. *Ber.* **1901**, *34*, 2900–2906.

816. Slimmer, M.D. *Ber.* **1902**, *35*, 400–410.

817. Neuberg, C. *Z. Physiol. Chem.* **1905**, *45*, 92–109.

818. Neuberg, C. *Biochem. Z.* **1906**, *1*, 282–298.

819. Fischer, E.; Blumenthal, H. *Ber.* **1907**, *40*, 106–113.

820. Rosenmund, K.W. *Ber.* **1909**, *42*, 4470–4481.

821. von Braun, J. *Ber.* **1909**, *42*, 839–846.

822. Fischer, E.; Zemplén, G. *Ber.* **1909**, *42*, 1022–1026.

823. Fischer, E.; Zemplén, G. *Ber.* **1909**, *42*, 2989–2997.
824. Pyman, F.L. *J. Chem. Soc.* **1911**, *99*, 1386–1401.
825. See Reference 824.
826. Dakin, H.D. *J. Biol. Chem.* **1921**, *48*, 273–291.
827. Hauptmann, H.; Berl, H. *J. Am. Chem. Soc.* **1955**, *77*, 704–707.
828. Hedgcoth, C.; Skinner, C.G.; Lindstedt, G.; Lindahl, G. *Biochem. Prep.* **1963**, *10*, 67–72.
829. Schrauth, W.; Geller, H. *Ber.* **1922**, *55*, 2783–2796.
830. Windus, W.; Marvel, C.S. *J. Am. Chem. Soc.* **1930**, *52*, 2575–2578.
831. Burch, W.J.N. *J. Chem. Soc.* **1930**, 310–312.
832. Schiltz, L.R.; Carter, H.E. *J. Biol. Chem.* **1933**, *116*, 793–797.
833. Abderhalden, E.; Heyns, K. *Ber.* **1934**, *67*, 530–547.
834. Tobie, W.C.; Ayres, G.B. *J. Am. Chem. Soc.* **1937**, *59*, 950.
835. Tobie, W.C.; Ayres, G.B.; Johnson, J.R.; Hasbrouck, R.B. *Org. Synth.* **1941**, *Coll. Vol. 1*, 23–25.
836. West, H.D.; Carter, H.E. *J. Biol. Chem.* **1937**, *119*, 103–108.
837. Wood, J.L.; Du Vigneaud, V. *J. Biol. Chem.* **1940**, *134*, 413–416.
838. Marvel, C.S.; Du Vigneaud, V.; Clarke, H.T.; Taylor, E.R. *Org. Synth.* **1941**, *Coll. Vol. 1*, 48.
839. Fox, S.W.; Dunn, M.S.; Stoddard, M.P. *J. Org. Chem.* **1941**, *6*, 410–416.
840. Neuberger, A.; Sanger, F. *Biochem J.* **1944**, *38*, 125–129.
841. Galat, A. *J. Am. Chem. Soc.* **1947**, *69*, 86.
842. Eck, J.C.; Marvel, C.S.; Noller, C.R.; Munich, W. *Org. Synth.* **1948**, *Coll. Vol. 2*, 374–376.
843. Sayles, D.C.; Degering, E.F. *J. Am. Chem. Soc.* **1949**, *71*, 3161–3164.
844. Gaudry, R.; Berlinguet, L. *Can. J. Res. B.* **1949**, *27*, 282–290.
845. Degering, E.F.; Boatright, L.G. *J. Am. Chem. Soc.* **1950**, *72*, 5137–5139.
846. Waalkes, T.P.; Fones, W.S.; White, J. *J. Am. Chem. Soc.* **1950**, *72*, 5760.
847. Harris, J.I.; Work, T.S. *Biochem. J.* **1950**, *46*, 190–195.
848. Adams, R.T.; Niemann, C. *J. Am. Chem. Soc.* **1951**, *73*, 4260–4263.
849. Rüfenacht, K. *Helv. Chim. Acta* **1952**, *35*, 762–764.
850. Rudman, D.; Meister, A.; Greenstein, J.P. *J. Am. Chem. Soc.* **1952**, *74*, 551.
851. Arnstein, H.R.V.; Hunter, G.D.; Muir, H.M.; Neuberger, A. *J. Chem. Soc.* **1952**, 1329–1334.
852. Buston, H.W.; Churchman, J.; Bishop J. *J. Biol. Chem.* **1953**, *204*, 665–668.
853. Brockmann, H.; Musso, H. *Chem. Ber.* **1954**, *87*, 581–592.
854. Peduska, K.; Rudinger, J.; Sorm, F. *Coll. Czech. Chem. Commun.* **1955**, *20*, 1174–1182.
855. Buston, H.W.; Bishop, J. *J. Biol. Chem.* **1955**, *215*, 217–20.
856. Marvel, C.S.; Allen, C.F.H.; Van Allan, J. *Org. Synth. Coll. Vol.* **1955**, *3*, 848–850.
857. See Reference 855.
858. Suami, T.; Umezawa, S. *Bull. Chem. Soc. Jpn.* **1957**, *30*, 537–542.
859. Du Vigneaud, V.; Brown, G.B.; Sealock, R.R.; Blaney, D.J.; Law, J.H.; Carter, H.E. *Biochem. Prep.* **1957**, *5*, 84–91.
860. Okawa, K. *Bull. Chem. Soc. Jpn.* **1957**, *30*, 110.
861. Edelson, J.; Fissekis, J.D.; Skinner, C.G.; Shive, W. *J. Am. Chem. Soc.* **1958**, *80*, 2698–2700.
862. Zhuze, A.L.; Jost, K.; Kasafírek, E.; Rudinger, J. *Collect. Czech. Chem. Commun.* **1964**, *29*, 2648–2662.
863. Burrows, B.F.; Turner, W.B. *J. Chem. Soc. C* **1966**, 255–260.
864. Lettré, H.; Wölcke, U. *Liebigs Ann. Chem.* **1967**, *708*, 75–85.
865. Ito, K.; Hashimoto, Y. *Agr. Biol. Chem.* **1969**, *33*, 237–241.
866. Cleland, G.H. *J. Org. Chem.* **1969**, *34*, 744–747.
867. Ichihara, A.; Hasegawa, H.; Sato, H.; Koyama, M.; Sakamura, S. *Tetrahedron Lett.* **1973**, *1*, 37–38.
868. Gershon, H.; McNeil, M.W.; Bergmann, E.D. *J. Med. Chem.* **1973**, *16*, 1407–1409.
869. Baldwin, J.E.; Löliger, J.; Rastetter, W.; Nuess, N.; Huckstep, L.L.; De La Higuera, N. *J. Am. Chem. Soc.* **1973**, *95*, 3796–3797.
870. Cheung, Y.-f.; Walsh, C. *J. Am. Chem. Soc.* **1976**, *98*, 3397–3398.
871. McCord, T.J.; Watson, R.N.; DuBose, C.E.; Hulme, K.L.; Davis, A.L. *J. Med. Chem.* **1976**, *19*, 429–430.
872. Baldwin, J.E.; Haber, S.B.; Hoskins, C.; Kruse, L.I. *J. Org. Chem.* **1977**, *42*, 1239–1241.
873. Gershon, H.; Shanks, L.; Clarke, D.D. *J. Pharm. Sci.* **1978**, *67*, 715–717.
874. Stöckenius, O. *Chem. Ber.* **1978**, *11*, 2002–2004.
875. Chari, R.V.J.; Wemple, J. *Tetrahedron Lett.* **1979**, 111–114.
876. Allan, R.D. *Aust. J. Chem.* **1979**, *32*, 2507–2516.
877. Chang, M.N.T.; Walsh, C. *J. Am. Chem. Soc.* **1980**, *102*, 7368–7370.
878. Chang, M.N.T.; Walsh, C.T. *J. Am. Chem. Soc.* **1981**, *103*, 4921–4927.
879. Johnston, M.; Raines, R.; Chang, M.; Esaki, N.; Soda, K.; Walsh, C. *Biochemistry* **1981**, *20*, 4325–4333.
880. Wasielewski, C.; Antczak, K. *Synthesis* **1981**, 540–541.
881. Madsen, U.; Andresen, L.; Poulsen, G.A.; Rasmussen, T.B.; Ebert, B.; Krogsgaard-Larsen, P.;

Brehm, L. *Bioorg. Med. Chem. Lett.* **1993**, *3*, 1649–1654.

882. Behr, J.P.; Lehn, J.M. *J. Chem. Soc., Perkin Trans. 1* **1972**, 1488–1492.

883. Cornforth, J.W.; Cornforth, R.H.; Dalgliesh, C.E.; Neuberger, A. *Biochem. J.* **1951**, *48*, 591–597.

884. Boivin, S.; Outurquin, F.; Paulmier, C. *Tetrahedron Lett.* **2000**, *41*, 663–666.

885. Zhou, W.; Liu, L.; Breslow, R. *Helv. Chim. Acta* **2003**, *86*, 3560–3567.

886. Weissman, N.; Schoenheimer, R. *J. Biol. Chem.* **1941**, *140*, 779–795.

887. Prasad, B.A.; Bisai, A.; Singh, V.K. *Tetrahedron Lett.* **2004**, *45*, 9565–9567.

3 | Synthesis of Racemic α-Amino Acids: Introduction of the Side Chain

3.1 Introduction

The method most commonly employed for the synthesis of racemic amino acids is the alkylation of a nucleophilic glycine synthon with an electrophile. It is an attractive method as the side-chain diversity is theoretically limited only by the availability of an appropriate electrophile, although in reality some reactions, such as those with secondary halides, proceed in poor yield. A number of approaches have been used to reduce the acidity of the glycine α-proton to aid anion generation; the carbon acidities for $^+H_3NCH_2CO_2^-$ and $^+H_3NCH_2CO_2Me$ in aqueous solution were determined to be $pK_a = 28.9$, and 21.0, respectively (1). Second-order rate constants have been determined in D_2O for deprotonations of the zwitterion N-terminal α-amino carbon of Gly-Gly and Gly-Gly-Gly, of anionic Ac-Gly, of the central residue of a Gly-Gly-Gly, and of Ac-Gly-NH$_2$, with pK_a values ranging from 23.9 to 30.8. The N-terminal residue of zwitterionic Gly-Gly-Gly ($pK_a = 25.1$) deprotonates approximately 130-fold faster than the central residue (2). The α-hydrogen pK_a values of Ala, Val, Leu, Ile, Phe, and Phg were calculated to range from 14.9 to 17.0, based on their rates of racemization (3). N-Substituted aminomalonic esters, aminocyanoacetates, isocyanoacetates, and glycine Schiff bases are the four most prevalent glycine nucleophiles with increased carbon acidities. Enzymatic deprotonations are known to be accelerated by Schiff base formation with pyridoxal phosphate; the increase in α-carbon acidity of Gly-OMe in the presence of acetone was found to be 7 pK units (from 21 to 14) by an NMR study (4). One advantage of the aminomalonic acid and aminocyanoacetate synthons is that byproducts resulting from dialkylation are not possible.

3.2 Alkylation of Aminomalonic Acids

The strategy of amino acid synthesis via alkylation of a glycine equivalent was introduced by Sörensen in the early 1900s, using diethyl N-phthalimidomalonate as the glycine synthon (obtained from diethyl malonate by bromination followed by Gabriel amination with potassium phthalimide) (5–7). The sodium salt was condensed with an alkyl chloride or bromide, the phthalimide group removed by reaction with hydrazine, and the free amino acid obtained after de-esterification and decarboxylation by acid hydrolysis (which can also be used to remove the phthalimide group). A variety of aminomalonate derivatives have been used for amino acid synthesis (see Scheme 3.1). Initial efforts at modifying the Sörensen synthesis were directed towards using an unmasked amino group (e.g., for a synthesis of proline (8)) as the deprotected amino acid could be obtained under milder hydrolysis conditions (9), but products were contaminated by N-alkylated byproducts (10). Greater success was obtained with aminomalonic esters protected with N-acetyl (first prepared in 1931 (11) and alkylated in 1943–1946 (12–18)), N-benzoyl (prepared and alkylated in 1931 ($18,19$), applied for general use in 1939–40 ($10,20$), or N-formyl (introduced in 1947 (21)) groups. More recently, dialkyl N-Cbz aminomalonates (e.g., 22–24) and N-Boc aminomalonates (e.g., 23) have been employed for amino acid syntheses. Diethyl nitromalonate has also been applied (e.g., 25), while aminomalonate residues contained within a dipeptide or tripeptide were alkylated with 67–93% yields for a variety of electrophiles (26–28).

All of these glycine equivalents, including the original phthalimidomalonate of Sörensen, have found application in recent years (see Tables 3.1–3.5). Diethyl

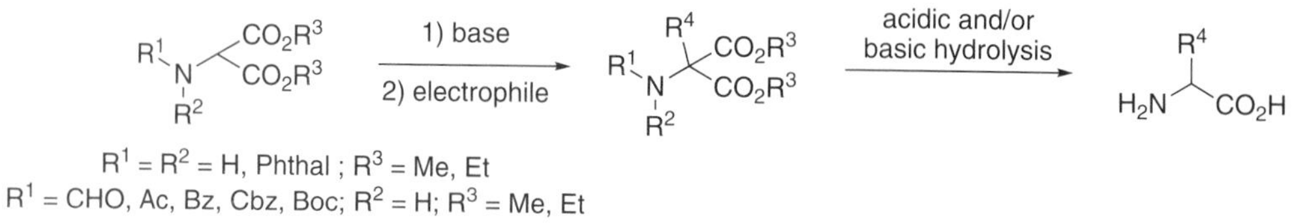

R^1, R^2, N, CO_2R^3, CO_2R^3 — 1) base 2) electrophile → R^1, N, R^2, R^4, CO_2R^3, CO_2R^3 — acidic and/or basic hydrolysis → R^4, H_2N, CO_2H

$R^1 = R^2 = H$, Phthal ; R^3 = Me, Et
R^1 = CHO, Ac, Bz, Cbz, Boc; R^2 = H; R^3 = Me, Et

Scheme 3.1 Alkylation of Dialkyl Acylaminomalonates.

Table 3.1 Alkylation of Diethyl Phthalimidomalonate

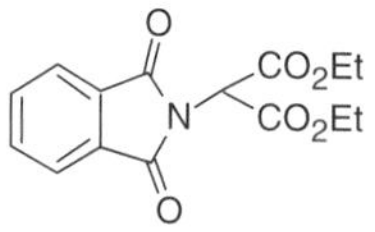

Phthalimido–N(CO_2Et)(CO_2Et) structure

Electrophile	Deprotonation Conditions	Decarboxylation Conditions	Reference
$Cl(CH_2)_2SMe$	NaOEt, EtOH, reflux 76–80%	1) 5N NaOH, reflux 2) conc. HCl, reflux 80–84%	1948 (416)
$ClCH_2CO_2Et$	Na, reflux 95–99%	conc. HCl, AcOH, reflux 42–43%	1950 (417) 1963 (418)
BnCl $Cl(CH_2)_3CN$ $ICH_2CH{=}CH_2$ $Br(CH_2)_3NPhth$ $Br(CH_2)_3Br$	NaOEt, EtOH 50–90%	aq. NaOH or HCl, reflux 70–90%	1905 (6,7,419)
$Br(CH_2)_3Br$ $Br(CH_2)_4Br$ converted into $-(CH_2)_nC(CO_2H)NH_2$ or $-(CH_2)_nOH$ depending on conditions	Na, Δ	1) NaOH, Δ 2) HCl, Δ	1908 (420)
$ICH_2CH{=}CH_2$	Na 80%	acid 85%	1908 (5)
4-MeO-BnBr 3,4-$(MeO)_2$-BnBr	K, EtOH then xylene	$Ba(OH)_2$ then conc. HCl	1914 (421)
ClCH(OEt)Et	Na, benzene	10N NaOH, conc. HCl 21% overall	1927 (422)
$ClCH_2OMe$	Na, Et_2O 60%	HBr or NaOH, HCl	1930 (423)
$ClCH_2CO_2Et$	Na, reflux 92%	conc. HCl, Δ 33%	1930 (424)
$Cl(CH_2)_2SMe$	NaOEt, EtOH, reflux 74%	1) aq. NaOH, reflux 2) HCl, reflux 81%	1931 (425)
ClCH(OEt)Et $Br(CH_2)_3Br$ converted into $-(CH_2)_3SH$	Na, EtOH 1) Na, Δ 2) NaSH	conc. HCl, reflux 1) 5N NaOH, reflux 2) conc. HCl, reflux 30% overall	1934 (426) 1934 (427)
$ClCH_2$(2-pyridyl) $HC{=}CHCO_2Me$ $ClCH_2CO_2Et$	Na, 150°C NaOEt, EtOH	35% HCl, 170°C 6N HCl 17–75% for 2 steps	1936 (428) 1938 (429)
$ClCH_2CH_2{}^{35}SBn$	Na, Δ	1) 5N NaOH, Δ 2) HCl, Δ	1939 (430)
$ClCH_2CH_2SBn$	Na, toluene, Δ 70%	1) NaOH, Δ 2) HCl, Δ 64%	1939 (431)
$ClCH_2SCH_2Cl$	Na, xylene 81%	conc. HCl 43%	1943 (432)

Table 3.1 Alkylation of Diethyl Phthalimidomalonate (continued)

Electrophile	Deprotonation Conditions	Decarboxylation Conditions	Reference
ClCH$_2$CH$_2$SEt	Na, xylene	1) NaOH 2) conc. HCl 65%	1943 (*433*)
BrCH$_2$CO(2-NO$_2$-Ph) converted into −CH$_2$CO(2-NH$_2$-Ph)	Na, EtOH, Δ 10%	conc. HCl, reflux 86% nitro reduction 75%	1943 (*434*)
Br(CH$_2$)$_4$Br	NaOEt, EtOH 93%	HCl, reflux 67%	1944 (*435*)
Cl^{13}CH$_2$^{13}CH$_2$^{34}SBn	Na, toluene, 170 °C	1) 5N NaOH, reflux 2) conc. HCl 70% for 3 steps	1944 (*436*)
ClCH$_2$(2-thiophene)	Na, toluene, Δ 93%	Ba(OH)$_2$, reflux 51%	1946 (*437*)
H$_2$C=CHCHO	NaOEt, EtOH, rt	not decarboxylated	1948 (*438*)
ClCH$_2$SeBn	Na, toluene	5N NaOH 59% for 2 steps	1948 (*439*)
Br(CH$_2$)$_4$Br converted into -(CH$_2$)$_4$CN, -(CH$_2$)$_4$CO$_2$H	1) Na, Δ 2)KCN, Δ	1) NaOH, Δ 2) HCl, Δ 12% overall	1949 (*440*)
ClCH$_2$CH$_2$^{35}SBn	Na, toluene, Δ 70%	1) NaOH, Δ 2) HCl, Δ 64%	1950 (*441*)
I(CH$_2$)$_4$NPhth	Na, EtOH	conc. HCl, reflux	1952 (*442*)
4-F-BnCl	Na, 160 °C 72%	1) 5N NaOH, reflux 2) 5N HCl, reflux 70%	1952 (*443*)
ClCH$_2$OCH$_2$Cl	Na, xylene 92%	aq. HCl 37%	1955 (*444*)
BrCD$_2$CO$_2$Et	Na, DMF 94%	SOCl$_2$, D$_2$O, reflux 87%	1966 (*445*)
BrCD$_2$CO$_2$Et	NaOEt 94%	DCl, reflux 95%	1966 (*446*)
Cl(CH$_2$)$_2$O(CH$_2$)$_2$NPhth	NaH, DMF	1) 5N NaOH, reflux 2) 6N HCl, reflux 62% for 3 steps	1976 (*447*)
ClCH$_2$^{14}CH$_2$O(CH$_2$)$_2$NPhth	Na, DMF 36%	1) EtOH, H$_2$NNH$_2$, reflux 2) 6N HCL, reflux 65%	1978 (*448*)
H$_2$C=CHCHO converted into -(CH$_2$)$_4$NH$_2$ or −(CH$_2$)$_3$^{13}CH$_2$^{15}NH$_2$ or −(CH$_2$)$_2$CDH^{13}CH$_2$^{15}NH$_2$	1) NaOEt 2) NaBH$_4$/NaBD$_4$ 3) MsCl, then NaCN or Na^{13}C^{15}N 4) H$_2$, PtO$_2$ 60% addition/reduction 45–60% mesylation/ displacement	HCl 25–39%	1989 (*449*)
I(CH$_2$)$_3$C(CN) (CH$_2$F)NPhth	KO*t*Bu, DMF 71%	conc. HCl, reflux 71%	1990 (*450*)
BrCH$_2$(4-C$_2$B$_{10}$H$_{11}$-Ph)	K$_2$CO$_3$, DMF, rt to 160°C 84%	6N HCl, reflux	1993 (*451*)
ClCH$_2$OCH$_2$PO$_{33}$Et$_2$	NaH, DMF, 80–120°C 52%	6N HCl, reflux 56%	1994 (*452*)

Table 3.2 Alkylation of Diethyl Acetamidomalonate

Electrophile	Deprotonation Conditions	Decarboxylation Conditions	Reference
Cl(CH$_2$)$_2$S(CH$_2$)$_2$S(CH$_2$)$_2$Cl	NaOEt, EtOH, reflux 25%	1) 10% NaOH, reflux 2) conc. HCl, reflux 96%	1943 (*14*)
	NaOMe, dioxane, reflux	HCl, reflux 50% for 2 steps	1944 (*17*)
IMe$_3$NCH$_2$(indol-3-yl)	Na, dioxane, reflux 63%	1) aq. NaOH, reflux 2) H$_2$O, reflux 3) aq. NaOH, reflux 81%	1944 (*15, 16*)
CH$_2$=CHCN	NaOEt, EtOH, rt 95%	conc. HCl, reflux 66%	1945 (*453*)
Br*n*Pr Br*i*Pr Br*n*Bu BnCl CH$_2$=CHCO$_2$Me ClCH$_2$C(Me)=CH$_2$	NaOEt, EtOH, reflux 46–82%	6N HCl, reflux or 1) 20% NaOH, reflux 2) conc. HCl, reflux 64–86%	1945 (*454*)
BnBr ClCH$_2$(imidazol-2-yl)	NaOEt, EtOH, reflux 67–90%	conc. HCl or HBr or aq. NaOH, reflux 48–87%	1945 (*13*)
Me$_2$NCH$_2$(indol-3-yl)	NaOH, toluene, reflux 90%	conc. HCl 73%	1945 (*455*)
Me$_2$NCH$_2$(indol-3-yl)	NaOEt, EtOH, EtI, reflux 73%	not decarboxylated	1945 (*456*)
ClCH$_2$CH$_2$SMe	Na, *t*BuOH	20% HCl, reflux 60% for 2 steps	1946 (*457*)
ICH$_2$CH$_2$CH(Me)Cl ClCH$_2$CH$_2$CH(Me)Cl converted into −CH$_2$CH$_2$CH(Me)-(α-N)	Na, dioxane	HCl, reflux	1946 (*458*)
ClCH$_2$(2-thiophene)	Na, EtOH 88%	48% HBr 78%	1946 (*437*)
IMe BrEt Br(CH$_2$)$_n$Me; n = 2,4,5,6,7,8 I*n*Bu I*i*Bu BrCH$_2$CH=CH$_2$ ClCH$_2$C(Me)=CH$_2$	NaOEt, EtOH, reflux	48% HBr, reflux or 37% HCl, reflux 32–85% for 2 steps	1946 (*12*)
BrCH$_2$CH$_2$OPh	NaOEt	1) 10% NaOH, reflux 2) conc. HCl, reflux 65% overall	1947 (*459*)
CH$_2$O	1N NaOH 99%	1) 1N NaOH, AcOH, Δ 2) conc. HCl 65% overall	1947 (*460*)
ClCH$_2$(1-naphthyl) ClCH$_2$(2-benzofuranyl)	Na, EtOH 73–92%	conc. HCl or KOH 79–80%	1947 (*461*)
HCOCH=CH$_2$ Cl(CH$_2$)$_2$CHO	NaOEt, EtOH, reflux or NaOMe, benzene 87%	not decarboxylated	1948 (*438*)
BrCH$_2$CH$_2$OPh	NaOEt	— 76% overall	1948 (*462*)
ClCH$_2$(1-naphthyl) BrCH$_2$(2-naphthyl)	Na, EtOH 66–100%	48% HBr, reflux 74–100%	1948 (*463*)

Table 3.2 Alkylation of Diethyl Acetamidomalonate (continued)

Electrophile	Deprotonation Conditions	Decarboxylation Conditions	Reference
ClCH$_2$CH=CHMe	NaOEt 80%	aq. NaOH 30%	1948 (52)
4-CN-BnBr converted into 4-CH$_2$NH$_2$-Bn	Na, EtOH 94% H$_2$, Pd-C: 70%	4N HCl, reflux 71%	1948 (464)
4-(4-NO$_2$-PhSO$_2$)-BnBr 4-NO$_2$-BnBr converted into 4-(4-NH$_2$- PhSO$_2$)-Bn, 4-NH$_2$-Bn, 4-[4-NH$_2$C(=NH) NH-PhSO$_2$]-Bn, and 4-NH$_2$C(=NH)NH-Bn	Na, EtOH 74–100%	3N HCl or HBr, reflux	1949 (465)
Me$_2$N^{14}CH$_2$(indol-3-yl)	NaOEt, EtOH, Me$_2$SO$_4$ 82%	1) NaOH, Δ 2) H$_2$SO$_4$ 76%	1949 (466)
ClCH$_2$(1-Bn-imidazol-2-yl)	NaOEt	12N HCl, Δ 38% for 2 steps	1949 (467)
ClCH$_2$(1-Bn-pyrazol-4-yl) ClCH$_2$(pyrazol-3-yl)	NaOEt	conc. HCl, Δ 22–84% for 2 steps	1949 (468)
2-, 3-, or 4-F-BnCl	Na, EtOH 68–89%	1) 2.5N NaOH, reflux 2) acidify, reflux 63–92%	1950 (469)
2-F-4-MeO-BnCl	NaOEt, EtOH 85%	1) 2.5N NaOH, reflux 2) 5N HCl, reflux 62%	1950 (470)
Cl(CH$_2$)$_5$NHBz	NaOEt 60%	1) KOH, Δ 2) HCl, Δ	1950 (471)
2- or 4-Cl-BnBr 3,4-Cl$_2$-BnBr 4-NO$_2$-BnBr 2-EtO-5-NO$_2$-BnBr 2-HO-5-NO$_2$-BnBr 4-NH$_2$-BnBr	NaOEt, EtOH, reflux 20–7%	40% HBr, reflux 68--5%	1951 (472)
2,4-Cl$_2$-BnCl	Na, EtOH 80%	48% HBr, reflux 100%	1951 (473)
HCOCH=CH$_2$ converted to (CH$_2$)$_2$CH(OH)CH$_2$NO$_2$	NaOEt, + MeNO$_2$ 56%	20% HCl, reflux 47%	1951 (474)
Me$_2$NCH$_2$(2-SR-4-Me- imidazol-5-yl) R = Me, Bn	Na, EtOH	H$_2$SO$_4$, reflux 65% for 2 steps	1951 (475)
IMe$_3$NCH$_2$(5-EtO-indol-3-yl) 	Na, EtOH 67–98%	1) NaOH, reflux 2) Δ 3) Ba(OH)$_2$ 25–78%	1951 (476)
3,4-(MeO)$_2$-BnCl	Na, EtOH 87%	conc. HCl, then HBr	1952 (477)
BrCH$_2$COR R = Me, Ph, 2-NO$_2$-Ph	NaOEt 19–71%	HCl, reflux 85–94%	1952 (478)
H$_2$C=C(Me)CO$_2$Et	NaOEt, EtOH, rt 71%	not decarboxylated	1952 (479)
H$_2$C=CHCHO converted into —(CH$_2$)$_2$CH(OH)^{14}CH$_2$NH$_2$	NaOEt, benzene	-	1953 (480)
2,5-(MeO)$_2$BnBr	NaOEt, EtOH 75%	HI, reflux 65%	1953 (481)

Table 3.2   Alkylation of Diethyl Acetamidomalonate (continued)

Electrophile	Deprotonation Conditions	Decarboxylation Conditions	Reference
CH_2O + Me_2NH + MeI forms $-CH_2NMe_3^+I^-$, converted into $-CH_2SBn$, $-CH_2SCH_2CH(NH_2)CO_2H$	70% initial condensation	HCl reflux 57–69% for displacement, hydrolysis	1953 (48)
BnCl IMe_3NCH_2(indol-3-yl)	NaOEt, EtOH 28%	1) 10% NaOH 2) 3N HCl 83–92%	1954 (482)
$Br(CH_2)_nBr$; n = 4,5,10	NaOEt 42–52%	48% HBr 85–92%	1954 (483)
$Cl(CH_2)_4Br$ converted into $-(CH_2)_4NH_2$	Na, EtOH 88–92%	HCl, reflux 89%	1954 (484)
Cl(cyclopenten-3-yl)	NaOEt, EtOH 28%	10% HCl, reflux 28%	1955 (485)
$ClCH_2CH(OAc)CO_2Et$	NaOEt, EtOH	20% HCl, reflux 35% for 2 steps	1955 (486)
$MeCH=CHCO_2Et$	Na, EtOH 60%	conc. HCl, reflux 74%	1955 (487)
$BrCH_2COMe$ converted into $-CH_2COCH_2NH_2$, $-CH_2$(2-SH-imidazol-4-yl)	Na, benzene 33%	LiOH, reflux 43%	1955 (488)
$Cl^{14}CH_2Ph$	NaOEt, EtOH 67%	HCl, reflux 70%	1955 (489)
Et_2NCH_2(4-Cl-indol-3-yl) Et_2NCH_2(5-Cl-indol-3-yl) Et_2NCH_2(6-Cl-indol-3-yl)	DMSO 48–49%	—	1955 (490)
IMe_3NCH_2(4-Cl-indol-3-yl) IMe_3NCH_2(6-Cl-indol-3-yl) or $-CH_2$(7-Cl-indol-3-yl) from 2-Cl-Ph-NHNH$_2$ + acraldehyde; $-CH_2$(5-Cl-indol-3-yl) from 4-Cl-Ph-NHNH$_2$ + acraldehyde	NaOEt or NaOH, toluene 40–70%	NaOH, reflux 40–85%	1955 (491)
$BrCH_2C(=CH_2)CO_2Et$	Na, EtOH 90%	conc. HCl, reflux 85%	1956 (492)
$ClCH(Ph)CH_2SMe$ $ClCH_2CH(Ph)SMe$	NaOEt, EtOH	20% HCl, reflux 49%	1956 (493)
$ClCH_2$(2-Me-imidazol-4-yl)	NaOEt 50%	HCl, reflux	1956 (494)
$IMe_3NCH_2C(NHAc)(CO_2Et)_2$ converted into $=C(NHAc)CO_2Et$	Na, EtOH	Ba(OH)$_2$ 68% overall	1956 (495)
$ClCH_2CH_2OCH_2Cl$ converted into $-CH_2OCH_2CH_2NH_2$	Na, EtOH then NH$_3$	6N HCl, reflux	1957 (496)
4-EtS-BnCl 4-MeS-BnCl	Na, EtOH 64–84%	conc. HCl, reflux 60–95%	1957 (497)
2-, 3- or 4-Me-BnBr 2,3-, 2,4-, 2,5-, 2,6-, 3,4-, or 3,5-Me$_2$-BnBr 2,4,6-Me$_3$-BnBr	NaH, toluene or Na, EtOH 47–88%	conc. HCl 85–100%	1957 (498)
2-R-3,4-(MeO)$_2$-BnCl 2-R-3,4-(HO)$_2$-BnCl R = F, Br, Et, NO$_2$	NaOEt 40–86%	HCl, reflux 50–81%	1957 (499)
Et_2NCH_2NHBz	NaOH 87%	conc. HBr 94%	1957 (500)
4-NO$_2$-BnBr converted into 4-NH$_2$-Bn, 4-F-Bn	1) NaOEt, 83% 2) H$_2$, Raney-Ni 3) HNO$_2$, HBF$_4$	acid, Δ 79%	1957 (501)
$BrCH_2$(4-pyridyl)	NaH, EtOH, benzene, reflux 70%	48% HBr, reflux 76%	1958 (502)

Table 3.2   Alkylation of Diethyl Acetamidomalonate (continued)

Electrophile	Deprotonation Conditions	Decarboxylation Conditions	Reference
BrCH$_2$COiPr	Na, benzene, reflux	conc. HCl, reflux 26% for 2 steps	1958 (503)
4-[B(OH)$_2$]-BnBr	NaOEt, EtOH 68%	1) NaOH 2) HCl 3) NaOH 64%	1958 (504)
ClCH$_2$SiMe$_2$ ICH$_2$Si(Me)$_2$CH$_2$SiMe$_3$	NaH, DMF 49%	40% HBr 1–2% overall	1958 (505)
BrCH$_2$Br	NaOEt	hydrolysis 23% for 2 steps	1958 (506)
BrCH$_2$$c$Hex	NaOEt 30%	6N HCl, Δ 100%	1958 (507)
ClCH$_2$CH$_2$OCH$_2$Cl converted into -CH$_2$OCH$_2$CH$_2$NH$_2$	Na, DMF	4N HCl, reflux 25% overall	1959 (508)
BrCH$_2$CH(Me)Et	NaOEt, EtOH	conc. HCl, reflux 25%	1959 (509)
ClCH$_2$NPhth	NaH	KOH 60% overall	1959 (510)
ClCH$_2$CH(OH)CH$_2$NPhth	Na	conc. HCl, Δ 8% for 2 steps	1958 (511)
BrCH$_2$CH=CHCO$_2$Et 2-NO$_2$-BnCl 2-[N(CH$_2$CH$_2$Cl)$_2$]-BnCl	NaOEt, EtOH NaOEt	conc. HCl, Δ conc. HCl, reflux	1960 (512) 1960 (513)
(Z)-ClCH$_2$CH=CHCH$_2$Cl converted into -CH$_2$CH=CHCH$_2$- (α-N)	NaOEt (N,C-dialkylation)	2.5N NaOH, reflux 29% overall	1960 (49)
BrCOMe	NaOEt 50%	10% HCl, Δ 78%	1960 (514)
ClCH$_2$(pyrazol-1-yl)	NaOEt	conc. HCl, Δ 60% for 2 steps	1960 (515)
BrCH(Me)CO$_2$Et	Na, EtOH	conc. HCl, reflux 45% for 2 steps	1961 (516)
ClCH$_2$CH=CHCl ClCH$_2$CH=C(Cl)Me	NaOEt 71–80%	NaOH 56–63%	1961 (517)
ClCH$_2$(3-MeO-4-Cl-pyrid-6-yl)	NaOEt 62%	48% HBr, reflux 100%	1961 (518)
H$_2$C=CHCOMe H$_2$C=CHCOPh H$_2$C=C(Me)COMe	NaOEt 75–94%	1) NaOH: 54–90% 2) H$_2$O, reflux: 77–95%	1961 (40)
H$_2$C=CFCO$_2$Et	NaOEt 59%	conc. HCl, Δ 61%	1961 (519)
3-CN-BnBr	NaOEt	6N HCl, Δ 50% for 2 steps	1961 (520)
ClCH$_2$CH(OAc)CO$_2$Et	Na, EtOH	20% HCl, reflux 35–55% for 2 steps	1962 (521)
BrCH$_2$^{14}CO(2-NO$_2$-3-MeO-Ph)	NaH, DMF	57% HI, P 35%	1962 (522)
ClCH$_2$C(=CH$_2$)CH$_2$Cl forms -CH$_2$C(=CH$_2$)CH$_2$-(α-N)	NaOEt 65%	NaOH, Δ 55%	1962 (523)
BrCH$_2$CH(Br)CO$_2$Et	NaOEt 35%	6N HCl, Δ 65%	1962 (524)
H$_2$C=CFCO$_2$Me	NaOEt	HCl 31% for 2 steps	1962 (525)
3,4-(OCH$_2$O)-BnCl	NaOEt, EtOH, reflux or NaH, benzene 64%	aq. NaOH, reflux or HCl, reflux 87%	1962 (526)
ClCH$_2$(2-thiophene)	Na, EtOH 54–61%	1) Na$_2$CO$_3$ 2) conc. HCl 77–85%	1963 (527)

Table 3.2 Alkylation of Diethyl Acetamidomalonate (continued)

Electrophile	Deprotonation Conditions	Decarboxylation Conditions	Reference
BrCH$_2$CH(CH$_2$Br)CO$_2$Et converted into $-$CH$_2$C($=$CH$_2$)CO$_2$Et	NaOEt	6N HCl, Δ 15% overall	1963 (528)
Br(CH$_2$)$_2$PO$_3$Et$_2$	NaOEt	6N HCl 46% for 2 steps	1964 (529)
BrCH$_2$COEt	Na, benzene, Δ 61%	6N HCl, Δ 79%	1964 (530)
HCOCH$=$CHPh similar adducts converted into $-$CH(Ph)CH$_2$CH$_2$-(α-N)	Na, EtOH 85%	—	1964 (45)
Br(cPent)	NaH, DMF	1) Na$_2$CO$_3$, reflux 2) 6N HCl 55% for 2 steps	1965 (531)
BrCH$_2$(4-fluorenyl)	Na, EtOH 90%	HCl, AcOH, reflux 66%	1965 (532)
3-CO$_2$H-4-MeO-BnCl 3-CO$_2$Et-4-HO-BnCl	NaOEt 50–85%	6N HCl, reflux 73–91%	1965 (533)
ClCH$_2$COAr Ar = 4-F-Ph, 4-Br-Ph, 4-Cl-Ph, 4-Me-Ph, 4-MeO-Ph, 4-PhO-Ph, 4-Ph-Ph, 2,4-F$_2$-Ph, 2,4-Cl$_2$-Ph, 4-AcNH-Ph converted to CH$_2$CH$_2$CH(Ar)-(α-N)	NaOEt 33–67%	NaOH, hydrolysis: 56–99% decarboxylation: 20–75% or acid hydrolysis: 52–97% imine formation/hydrogenation: 70–96%	1965 (534)
2,3,5,6-F$_4$-4-MeO-BnBr 2,3,4,5,6-F$_5$-BnBr	Na, DMF 81–88%	20% HCl, reflux 81–90%	1965 (535) 1969 (536)
RCH$=$CHCO$_2$Et R = Me, Et, nPr, nBu, iBu, nPent	NaOEt 67–88%	49% HBr, Δ 53–79%	1965 (537)
2,3,4,5,6-F$_5$-BnBr	Na, DMF 44%	20% HCl 73%	1965 (538)
![pyrimidine structure]	NaOMe, MeOH 59%	conc. HCl, Δ 64%	1965 (539)
BrCD$_2$CD$_2$CO$_2$Et	Na, EtOD 90%	SOCl$_2$, D$_2$O, reflux	1966 (445)
BrCD$_2$CD$_2$CO$_2$Et	NaOEt 90%	DCl, reflux 95%	1966 (446)
Br(CH$_2$)$_3$OH with [2-^{14}C] BrCH$_2$CH$=$CH$_2$ with [2-^{14}C] converted into (CH$_2$)$_3$SMe	NaOEt NaOEt 77%	4N HCl, Δ 6N HCl, Δ 47% or hv, MeSH then 6N HCl 77%	1966 (540) 1966 (540)
ClCH$_2$(4-MeO-1-naphthyl) ClCH$_2$(3-Cl-4-MeO-1-naphthyl)	NaH, DMF 51–77%	conc. HCl or KOH, then 2N HCl 70–89%	1967 (541)
BrCH$_2$CH$=$CFCO$_2$Et	Na, DMF 94%	HCl, AcOH 46%	1967 (542)
4-AcOCH$_2$-BnBr	NaOEt 81%	Na$_2$CO$_3$, Δ 79%	1967 (543)
2,3,4,5,6-(Me)$_3$-BnCl	NaOEt 68%	HBr 55%	1968 (544)
Cl(CH$_2$)$_2$N(Bn)Me	NaOEt 14%	6N HCl 65%	1968 (545)
ClCH$_2$(2-pyridyl-1-oxide) ClCH$_2$(3-pyridyl-1-oxide) ClCH$_2$(4-pyridyl-1-oxide)	NaOEt 30–45%	6N HCl 50–55%	1968 (546)
3,4-I$_2$-BnBr	NaOEt 72%	conc. HCl, AcOH 70%	1968 (547)
3-NO$_2$-4-F-BnCl	NaOEt 62%	conc. HCl 63%	1968 (548)

Table 3.2   Alkylation of Diethyl Acetamidomalonate (continued)

Electrophile	Deprotonation Conditions	Decarboxylation Conditions	Reference
ClCH$_2$(pyrazol-1-yl) ClCH$_2$(pyrazol-3-yl)	NaOEt	conc. HCl 30–95% for 2 steps	1968 (549)
Br(CH$_2$)$_5$CO$_2$Et	NaOEt	conc. HCl 77% for 2 steps	1968 (550)
ClCH$_2$(pyrid-2-yl)	NaOEt 91%	6N HCl 70%	1968 (551)
HC(Me)=CHCO$_2$Et	KOtBu, tBuOH 61%	1N NaOH, reflux 74%	1969 (552)
BrCH$_2$CH=CHCH$_2$Br converted into CH$_2$C(-SCH$_2$CH$_2$S-)	Na, NaOEt 100% then ozonolysis, (HSCH$_2$)$_2$ 37%	Na$_2$CO$_3$, reflux 78%	1969 (553)
3-Br-4-HO-5-MeO-BnCl	NaOEt 89%	HCl, AcOH, SO$_2$, 70°C 73%	1969 (554)
 R = iPr, sBu; R' = H, Me	Na, EtOH 75–90%	50% H$_2$SO$_4$ or HNO$_3$ 40–92%	1970 (555)
ClCH$_2$(3-MeO-4-Cl-pyridin-5-yl-1-oxide)	NaOEt 79%	25% HCl, 160°C 20–54%	1970 (556)
Cl(CH$_2$)$_6$NHBz	NaOEt	6N HCl, Δ	1970 (557)
BrCH$_2$CH=CH$_2$	NaOEt 78%	NaOH, Δ 84%	1971 (558)
BrCH$_2$(2-F-pyrid-4-yl) BrCH$_2$(2-F-pyrid-3-yl) BrCH$_2$(2-F-pyrid-6-yl)	Na, EtOH 14–42%	6N HCl 40–76%	1971 (559)
BrCH$_2$(2-X-pyrid-4-yl) BrCH$_2$(2-X-pyrid-3-yl) BrCH$_2$(2-X-pyrid-6-yl) X = Cl, Br	Na, EtOH 13–45%	6N HCl 31–68%	1971 (560)
ClCH$_2$CH=CH$_2$Cl converted into −CH$_2$CH=CH$_2$NPhth	Na, then KNPhth 35%	20% HCl 92%	1972 (561)
ClCH$_2$(2-Me-imidazol-4-yl) ClCH$_2$(5-Me-imidazol-4-yl) ClCH$_2$(5-NO$_2$-imidazol-4-yl)	Na, EtOH	conc. HCl 20–49% for 2 steps	1972 (562)
Br(CH$_2$)$_3$N(Ts)OBn	NaOEt 70%	HCl, AcOH, Δ 90%	1972 (563)
2-Me-BnBr BrCH$_2$(2-naphthyl) ClCH$_2$(6-quinolinyl)	Na, EtOH 75–84%	1) NaOH, EtOH 2) Δ 67–92%	1973 (22)
ClCH$_2$(4-F-5-imidazolyl)	NaOEt, EtOH 26%	6N HCl 82%	1973 (564)
ClCH$_2$CH=CHCH$_2$NPhth	NaOEt 76%	conc. HCl 87%	1973 (565)
BrCH$_2$CO(2-furyl)	Na, EtOH	5N HCl, reflux	1973 (566)
2,3,4,5,6-Me$_5$-BnBr	NaOEt 70%	1) NaOH 2) 6N HCl 96%	1973 (567)
BrCH(Ph)CO$_2$Et	NaOEt	9N HCl, reflux 82% for 2 steps	1973 (568)
3,4-(MeO)$_2$-5-NO$_2$-BnBr converted into 3,4-(MeO)$_2$-5-NH$_2$-Bn, 3,4-(MeO)$_2$-5-F-Bn, 3,4-(MeO)$_2$-5-^{18}F-Bn	NaOEt 60%	47% HBr 47%	1973 (569)

(Continued)

Table 3.2 Alkylation of Diethyl Acetamidomalonate (continued)

Electrophile	Deprotonation Conditions	Decarboxylation Conditions	Reference
BrCH=CH$_2$ converted to -CH(Br)CH$_2$Br, then -CH(OH)CH$_2$OH	1) NaOEt 2) Br$_2$ 67%	1) HBr, Δ 2) Ba(OH)$_2$ 10%	1973 (570)
3,4,5-(MeO)$_3$-BnCl	NaOEt	HCl or NaOH 5–34%	1973 (571)
AcOCH$_2$(N-Ts-imidazol-4-yl)	NaH, DMF 80%	conc. HCl, reflux 80%	1974 (572)
[structure: Cl~()$_n$-oxazolidinone] n = 1,2 converted into - (CH$_2$)$_{n+2}$NH(CH$_2$)$_2$Br (CH$_2$)$_{n+2}$NH(CH$_2$)$_2$SSO$_3$H	Na, DMA 45–53% conversion 63-65% Br, 77–96% sulfate	6N HCl then Ba(OAc)$_2$ 70–97%	1974 (573)
2-Br-4-HO-5-MeO-BnCl 2-Br-4,5-(MeO)$_2$-BnCl 3-Br-4-HO-5-MeO-BnCl 2,3-Br$_2$-4-HO-5-MeO-BnCl 2-Br-4-HO-5-MeO-BnCl 3-Br-4,5-(MeO)$_2$BnCl	NaOEt 44–92%	6N HCl, Δ 41–60%	1974 (574)
ClCH$_2$C(Cl)=CH$_2$	NaOEt	NaOH, Δ	1974 (575)
Cl(CH$_2$)$_2$(imidazol-4-yl)	NaOEt, EtOH 62%	6N HCl 93%	1975 (576)
BrCH$_2$CO[3,4-(MeO)$_2$-Ph]	NaH, benzene 84%	HCl 65%	1975 (577)
Br(CH$_2$)$_n$iPr n = 2,3,4	NaOEt	47% HBr, Δ 25–35% for 2 steps	1975 (578)
ClCH$_2$(2-Cl-4-BzO-pyrid-5-yl)	NaOEt	47% HBr, Δ	1975 (579)
BrCH$_2$CCH	NaOEt, EtOH 77%	1) aq. KOH 2) dioxane, reflux 3) aq. KOH 64%	1976 (580)
BrCH(Me)Ph	NaOEt, EtOH 73%	10% NaOH, reflux 88%	1976 (581)
ClCH$_2$SBn AcOCD$_2$NMe$_2$	NaH, DMF: 53% NMe$_2$: 70%	DCl, D$_2$O, reflux 65%	1976 (582)
ClCH$_2$CH$_2$OP(=O) (Me)CH=CH$_2$	Na, EtOH 95%	6N HCl, reflux 78%	1976 (583)
4-Ph-BnCl	NaOEt 93%	HCl, Δ 47%	1976 (584)
Br(CH$_2$)$_3$Ph Br(CH$_2$)$_3$(4-MeO-Ph)	NaOEt 64–77%	1) 2N NaOH 2) Δ 81-90%	1976 (585)
[structure: Cl-CH$_2$ pyridinone, HN, OBn]	NaOEt 88%	1) conc. NH$_4$OH, Δ 2) conc. HCl 85%	1976 (586)
[structure: Cl-CH$_2$ pyrimidine, Cl, OBn]	NaH, DMF 74%	conc. HCl 92%	1976 (587)
BnBr 4-MeO-BnBr BrCD$_2$(4-MeO-Ph)	NaOEt, EtOH 83–86%	DBr, D$_2$O 94–96%	1977 (588)
3-CN-BnBr 3-CO$_2$H-4-NO$_2$-BnBr	Na, EtOH	10N HCl, reflux 29–58% for 2 steps	1977 (589)

Table 3.2 Alkylation of Diethyl Acetamidomalonate (continued)

Electrophile	Deprotonation Conditions	Decarboxylation Conditions	Reference
ClCH(R) (imidazol-4-yl) R = Me, Et, nHex	NaOMe, EtOH 55–56%	12N HCl, Δ 58–70%	1977 (590)
ClCH₂(5-BnO-pyrid-2-yl)	NaOEt 97%	conc. HCl, Δ 80%	1977 (591)
BrCH(Me)COMe	NaH, DMF	6N HCl, Δ 30% overall	1977 (592)
Cl(CH₂)₂O(CH₂)₂Cl	NaH, DMF, 0-60°C 84%	H₂O, DMSO, NaCl, 170°C 77%	1978 (593)
4-MeO-BnBr	Na, EtOH 90%	48% HBr, 110°C 86%	1978 (594)
ClCH₂CH=CMe₂ converted into CH₂CH₂C(Me)₂SH	NaOEt, then BF₃·Et₂O- catalyzed addition of BnSH 63–74%	NaOH, then HCl 77%	1978 (595)
4-Bn-BnCl	NaOEt 16%	1) KOH, Δ 2) HCl, Δ 30%	1978 (596)
[3,5-¹³C₂]-BnBr	NaOEt, EtOH 87%	48% HBr, reflux 94%	1979 (597)
Br-CH₂-isoxazole-OMe	NaOEt 65%	48% HBr 55%	1979 (598)
BrCH₂CCH	Na, EtOH 75%	HCl 75%	1979 (599)
BrCH₂(benzofuran-3-yl)	NaOMe, DMF 72%		1979 (600)
Cl-CH₂-(HO-tropone-Me)	NaH, DMF 52%	6N HCl, Δ 27%	1979 (601)
Cl(5-NO₂-2-pyridyl)	NaH, DMF 51%	aq. NaOH, EtOH 93%	1979 (602)
Br-CH₂-isoxazole (OMe, Cl-CH₂-isoxazole OMe)	NaOEt, EtOH, reflux 45–65%	1N HCl or aq. HBr, reflux 80–85%	1980 (603)
Br-CH₂-isoxazole (OMe), Br-CH₂-isoxazole (OMe)	NaOEt, EtOH 6–31%	48% HBr, reflux 57–76%	1980 (604, 605)
BrCH₂CO(4-MeO-Ph)	Na, EtOH 92%	conc. HCl, reflux 93%	1980 (606)
ClCH₂CCCH₂NHBoc	Na, EtOH 75%	1) KOH 2) Δ 80%	1980 (607)
ClCH₂(selenothiophen-2-yl) ClCH₂(selenothiophen-3-yl)	Na, DMF 44–59%	1) aq. KOH 2) HCl, Δ 50–61%	1980 (608)
Br(CH₂)ₙCN convert into −(CH₂)ₙC(=NH)NH₂ n = 2,3,4,5	NaOEt	1) NaOH 2) Δ	1980 (609)
2,5,6-F₄-3,4-(MeO)₂-BnBr	Na, DMF 71%	47% HI 46%	1981 (610)

(Continued)

Table 3.2 Alkylation of Diethyl Acetamidomalonate (continued)

Electrophile	Deprotonation Conditions	Decarboxylation Conditions	Reference
BrCH$_2$CH(Ph)$_2$ 3-CF$_3$-BnBr 4-CF$_3$-BnBr 3,4,5-(MeO)$_3$-BnBr 2,4,6-Me$_3$-BnBr BrCH$_2$(1-naphthyl) BrCH$_2$(2-naphthyl) BrCH$_2$(1-Br-2-naphthyl) BrCH$_2$(9-fluorenyl) BrCH$_2$(9-anthryl)	NaOEt	acid hydrolysis 61–86% for 2 steps	1982 (*611*)
Br^{13}CH$_2$CH$_2$CO$_2$$n$Pr	NaOEt, EtOH	aq. HCl 47% for 2 steps	1983 (*612*)
H$_2$C=CHP(=O) (OMe)R R = Bn, 3,4-Cl$_2$-Bn, 3,5-Me$_2$-Bn, 4-Br-Bn, CH$_2$CO$_2$tBu, CH$_2$OMe	NaOEt	conc. HCl 29–71% overall	1983 (*613*) 1983 (*614*)
HCCPO$_3$Et$_2$	NaH, benzene 46%	6N HCl 20% for 2 steps	1983 (*615*)
2-CO$_2$Me-BnBr	NaOEt 19–51%	48% HBr, reflux 14–43%	1984 (*616*)
(cycloheptafuran/isoxazole structures: OMe and MeO substituted, Br–CH$_2$)			
Br(CH$_2$)$_2$PO$_3$Et$_2$ Br(CH$_2$)$_3$PO$_3$Et$_2$ BrCH$_2$(2-PO$_3$Et$_2$-Ph) Br(CH$_2$)$_2$(2-PO$_3$Et$_2$-Ph)	NaOEt	HCl, Δ 12–35% overall	1984 (*617*)
BrsO^{13}CH$_2$$i$Pr BrsO^{14}CH$_2$$i$Pr	NaH, 1,3- dimethylimidazolidinone 90%	6N HCl, reflux 92%	1984 (*618*)
(isoxazole ring structures, Br–(CH$_2$)$_n$, OMe, Br)	Na, EtOH 37–63%	48% HBr, reflux 54–77%	1985 (*619*)
I(CH$_2$)$_n$(3-MeO-5-furyl) n = 2,3			
Br(CH$_2$)$_5$NPhth Br(CH$_2$)$_6$NPhth	NaH, DMF 74–78%	6N HCl, reflux 91–94%	1985 (*620*)
2,6-F$_2$-4-MeO-BnBr 2,6-Cl$_2$-4-MeO-BnBr	NaOEt 57–71%	48% HBr 47–67%	1985 (*621*)
Me$_3$Si-CC-SO$_2$Ph	KOtBu, THF, 10°C 75%	6N HCl, reflux 80%	1986 (*622*)
H$_2$C=C=CHCO$_2$Et	NaOEt, EtOH 94%	20% HCl 67%	1986 (*623*)
ClCH$_2$(9-anthryl)	NaOEt, Δ	conc. HCl, Δ 49% overall	1986 (*624*)
BrCH=CHCN	NaOEt, EtOH 70%	2N HCl, reflux	1986 (*625*)
2-, 3-, or 4-F-BnBr 2-, 3-, or 4-Cl-BnBr 2,4- or 3,4-Cl$_2$-BnBr 4-Br-BnBr 2-, 3-, or 4-Me-BnBr 2,4,6-Me$_3$-BnBr 2- or 4-MeO-Bn 2,5-(MeO)$_2$-BnBr 4-SMe-BnBr	NaOEt 60–96%	48% HBr reflux 55–97%	1987 (*626*)

Table 3.2 Alkylation of Diethyl Acetamidomalonate (continued)

Electrophile	Deprotonation Conditions	Decarboxylation Conditions	Reference
3,4-(NO$_2$)$_2$-BnBr 4-CF$_3$-BnBr BrCH$_2$(1-naphthyl) BrCH$_2$(2-naphthyl) BrCH$_2$(5-R-indol-3-yl) R = Br, F, Cl, Me, OMe			
BnCl ClCH$_2$(3-benzothiophene)	NaOEt 50–83%	1) NaOH 2) dioxane, Δ 64–67%	1987 (627)
TfO(CH$_2$)$_n$CF$_3$ n = 1,2,3	KOtBu, THF 46–57%	conc. HCl, reflux 78–86%	1988 (628)
H$_2$C=C(R)CO$_2$Et R = Me, NHAc, OtBu, F	NaOEt, EtOH 63–97%	conc. HCl, reflux	1988 (628)
4-CN-BnBr converted to 4-CH$_2$NH$_2$-Bn 4-CH$_2$OH-Bn 4-CH$_2$Cl-Bn 4-CH$_2$PO$_3$Me$_2$-Bn 4-(SO$_3$Na)-Bn	Na, EtOH 87%	1) NaOH, Δ 2) HCl or MeSO$_3$H, Δ 46-55%	1988 (629)
2-Br-BnBr	NaOEt	6N HCl, Δ 77% for 2 steps	1988 (630)
Br(CH$_2$)$_2$CN Br(CD$_2$)$_2$CN converted into -(CH$_2$)$_3$NH$_2$, -(CD$_2$)$_3$NH$_2$	NaOEt then H$_2$, PtO$_2$	6N HCl	1989 (631)
H$_2$C=C=CH$_2$ + RX generates -CH$_2$C(R)=CH$_2$ R = Ph, CH=CH$_2$	NaH, DMSO, Pd(dba)$_2$	1) KOH, EtOH 2) 1N HCl 30%	1989 (46)
3- or 4-CH$_2$Cl-BnCl 3- or 4-CH$_2$Br-BnBr converted into 2-, 3- or 4-CH$_2$PO$_3$Et$_3$-Bn	NaOEt 19–91% NaPO$_3$Et$_3$ 30–98%	6N HCl, reflux 10–95%	1989 (632)
4-Cl-Bn	K$_2$CO$_3$, DMF, 18-crown-6 89%	6N HCl 90%	1989 (32)
Br(CH$_2$)$_4$CF$_2$PO$_3$Et$_2$	NaOEt, EtOH 50%	6N HCl, reflux 100%	1989 (633)
Br(CH$_2$)$_n$Me: n = 7,9,11,13,17	NaOEt, EtOH, reflux	conc. HCl. reflux 85–95% for 2 steps	1990 (634)
RCH=CHCHO R = nPr, Ph converted into CH(R)CH$_2$CH$_2$-(α-N)	NaOEt then Et$_3$SiH	NaOH or 6N HCl	1990 (41)
4-CH$_2$OAc-BnBr 4-Br-BnBr	K$_2$CO$_3$, TEBAC, MeCN or NaH, DMF 54–79%	Na$_2$CO$_3$, H$_2$O, reflux or 6N HCl 64–71%	1990 (33)
ClCH$_2$(4-thiazolyl)	NaOEt 95%	NaOH, Δ 89%	1990 (635)
ClCH$_2$(2-furyl)	—	1) KOH 2) dioxane, Δ 3) KOH 35%	1990 (636)
H$_2$C=CCO$_2$Me	TMG, DBU or DBN, CH$_2$Cl$_2$ 100%	6N HCl, reflux	1990 (34)
BrCH$_2$CCH	NaH, DMF, 60°C 95%	aq. NaOH, reflux 90%	1991 (637)
2-Me-BnBr	Na, EtOH	6N HCl, reflux 60%	1991 (638)

(Continued)

Table 3.2 Alkylation of Diethyl Acetamidomalonate (continued)

Electrophile	Deprotonation Conditions	Decarboxylation Conditions	Reference
BrCH₂(1-naphthyl) BrCH₂(2-naphthyl)	NaOEt, EtOH 90–92%	1) NaOH 2) Δ 78–82%	1991 (639)
BrCH₂CCH	NaH, DMF, 80°C 93%	LiCl, H₂O, DMF, 150°C 58%	1992 (640)
TsOCH₂CH₂Br TsOCD₂CD₂Br converted into −CH₂CH₂CN, −(CH₂)₃NH₂, (CH₂)₃NHOH	NaOEt 76%	6N HCl, reflux 61%	1992 (641)
2-, 3-, or 4-CN-BnBr CN converted into CH₂NH₂, CH₂OH, CH₂Cl, CH₂PO₃Et₂ or CH₂SO₃H	Na, EtOH 91–95%	9N HCl 61–91%	1992 (642)
2- or 4-Br-BnBr	NaOEt 53–74%	6N HCl, reflux 89%	1992 (643)
2-(BrCH₂)-BnBr	NaOMe 75–90%	KOH, H₂O, Δ 78–85%	1992 (644)
Br(CH₂)₂CH=CH₂	NaH, DMF 70%	NaOH, EtOH, Δ 84%	1992 (645)
BrCH₂COMe converted into −CH₂COCH₂Br, −CH₂COCH₂¹⁵NH₂	NaOEt 62%	37% HCl, Δ 78%	1992 (646)
H₂C=C(F)CO₂Et	NaH, EtOH, rt 88%	6N HCl, reflux	1993 (647)

<!-- Chemical structures in Electrophile column -->
$H_2C=C(F)CO_2Et$ region continued with quinoxaline-bromomethyl phosphonic acid structures:

Electrophile	Deprotonation Conditions	Decarboxylation Conditions	Reference
Quinoxaline structure: Br–CH₂– attached to quinoxaline with H₂O₃P–, R1, R2 substituents; H₂O₃P–CH=CH– quinoxaline R1 = H; R2 = 6-Cl, 7-Cl R1 = R2 = 6,7=Cl₂, 5,8-Cl₂, 6,7-F₂, 6,7-(MeO)₂	NaOEt 35–100%	6N HCl, reflux 60–93%	1993 (648)
Isoxazolone structure: Br–CH₂– on ring with O, N–CH₂–O– (methoxymethyl)	NaH, DMF 76%	TFA, H₂O, reflux 60%	1993 (37)
BrCH₂(Br)=CH₂	NaH, DMF	LiBr, H₂O, DMF 75% for 2 steps	1993 (649)
ClC(Me)CCH	NaOEt 18%	10% Na₂CO₃, Δ 70%	1993 (650)
Isoxazole structure: Cl–CH₂– on ring with Et, EtO, N, O	NaOMe, MeOH 51%	48% HBr, Δ 46%	1993 (651)
BrCH₂(6-Me-2-naphthyl) ClCH₂(6-Cl-2-naphthyl) BrCH₂(3-benzo[b]thiophene) BrCH₂(5,6,7,8-tetrahydro-2- naphthyl) ClCH₂(2,3-dihydro-1,4- benzodioxin-6-yl)	NaOEt 54–63%	NaOH, Δ 61–93%	1994 (652)
4CN-BnBr	NaOEt, EtOH, reflux	3N NaOH, 95°C 90% for 2 steps	1995 (653)
ClCH₂(2-F-pyrid-5-yl)	—	NaOH, EtOH, reflux 95%	1995 (654)

Table 3.2 Alkylation of Diethyl Acetamidomalonate (continued)

Electrophile	Deprotonation Conditions	Decarboxylation Conditions	Reference
4-Br-BnBr 4-BrCH$_2$-BnBr converted into 4-PO$_3$Et$_2$-Bn, 4-CH$_2$PO$_3$Et$_2$-Bn	NaOEt 62–83% 96–97% phosphonylation	6N HCl, reflux 77–94%	1995 (655)
Br(CH$_2$)$_5$Ph Br(CH$_2$)$_8$Ph	Na, EtOH 53%	1) KOH 2) HCl 90%	1995 (656)
4-CF$_3$-BnBr 4-CO$_2$Et-BnBr	Na, EtOH 85–91%	1) NaOH 2) 12N HCl, reflux 89–92%	1995 (657)
Br(CH$_2$)$_3$PO$_3$Et$_2$	Na, THF	HCl 50% overall	1996 (658)
BrCH$_2$cPr	NaH, DMF, 80°C 55%	1) 2N NaOH, reflux 2) acidify, reflux	1996 (659)
BrCH$_2$C(Br)=CH$_2$	NaH, EtOH 96%	-	1996 (172)
Br(CH$_2$)$_5$CO$_2$Et	NaOEt, EtOH	6N HCl, reflux 95% for 2 steps	1996 (660)
4-CN-BnBr	NaOEt 91%	1) NaOH 2) HCl 48%	1996 (661)
 R = H, Me	NaOEt, EtOH 67–69%	6N HCl, reflux 93–100%	1996 (662)
3-F-4-NO$_2$-BnBr 3-NO$_2$-4-F-BnbBr	NaH, DMF, rt	conc. HCl, reflux 85% overall	1997 (663)
BrCH$_2$(3-HO-6-pyridyl)	Na, EtOH/THF 59%	6N NaOH 81%	1996 (664)
IMe$_3$NCH$_2$(ferrocene)	NaOEt, EtOH	1) N NaOH, reflux 2) acidify, reflux 82%	1997 (665)
Br(CH$_2$)$_n$Me n = 7,13	NaOEt, EtOH	6N HCl, reflux 51–65% overall	1997 (666)
3-NO$_2$-BnBr	KOtBu, EtOH 74%	6N HCl, reflux 24–90%	1997 (667)
IEt I[^{13}C$_2$]Et I[^{2}H$_5$]Et	NaOEt, EtOH, reflux 75%	conc. HCl, reflux 76%	1997 (668)
3-MeO-4-NO$_2$-BnBr 2-NO$_2$-4-MeO-BnBr	NaOEt, EtOH, rt 71%	48% HBr 91%	1997 (669)
 R = Me, tBu	NaH 50–71%	48% HBr 25–31%	1997 (670)
	KOtBu 54–70%	47% HBr 58–82%	1997 (671)

Table 3.2 Alkylation of Diethyl Acetamidomalonate (continued)

Electrophile	Deprotonation Conditions	Decarboxylation Conditions	Reference
ClCH$_2$(2-anthraquinolyl)	NaOEt 95%	1) 1N NaOH, Δ 2) HCl 84%	1997 (*672*)
R = 2-, 3-, or 4-pyridyl, 6-Me-2-pyridyl, 2-quinolinyl	NaH 36–76%	48% HBr 32–81%	1998 (*673*)
R = 2-, 3-, or 4-F-Ph, *c*Hex, 1-cyclohexenyl	NaH, DMF 48–99%	HBr, reflux or TFA 19–65%	1998 (*674*)
BrCH$_2$CH$_2$CH=CH$_2$	NaH, DMF 73%	1) NaOH, H$_2$O-EtOH 2) 2N HCl 82%	1998 (*286*)
Br(CH$_2$)$_3$[4-C(CF$_3$) (-N=N-)-Ph]	NaH, DMF 75%	1) 3N NaOH 2) dioxane, 80°C 95%	1998 (*675*)
4-Cl-BnCl	NaOEt, EtOH 79%	20% HCl, 95%	1998 (*676*)
	—	50% overall	1998 (*677*)
BrCH$_2$(2-CN-thiophen-5-yl)	NaOEt, KI	NaOH, H$_2$O, reflux 67% for 2 steps	1998 (*678*)
4-*t*Bu-Bn	NaOEt 87%	1) NaOH, Δ 2) HCl, Δ 59%	1998 (*50*)
HCOCH=CHPh converted into –CH(Ph)CH$_2$CH$_2$-(α-N)	1) NaOEt, EtOH 2) Et$_3$SiH, TFA 78%	1) 6N HCl 2) H$_2$SO$_4$ 67%	1999 (*44*)
BrCH$_2$(4-pyridyl) converted into -CH$_2$(4-piperidyl)	—	1) KOH, EtOH 2) DME, reflux 67% 3) H$_2$, Pd/C	1999 (*679*)
	NaOEt 83%	1) NaOEt 2) 6N HCl 94%	1999 (*680*)
HOCH(CF$_3$) (indol-3-yl)	NaOEt 55%	1) 2N NaOH 2) conc. HCl 91%	1993 (*39*)
2,6-(BrCH$_2$)$_2$(pyridyl) 2,3-, 2,4-, 2,5-,2,6-, 3,4- or 3,5-F$_2$-BnBr	NaH, THF Na, EtOH 90%	6N HCl, reflux 1) 2N NaOH, EtOH, 40°C, 3h 2) 6N HCl 92%	1999 (*681*) 2000 (*682*)
TsOCH$_2$CH$_2$CH=CH$_2$ TsOCH$_2$CH$_2$CCH *E*- or *Z*- TsOCH$_2$CCH=CHMe TsO(CH$_2$)$_2$CF$_2$ TsO*n*Pent	KO*t*Bu, THF 44%	1) 10% NaOH, reflux 2) 1M HCl, reflux 63%	2000 (*35*)

Table 3.2 Alkylation of Diethyl Acetamidomalonate (continued)

Electrophile	Deprotonation Conditions	Decarboxylation Conditions	Reference
BrCH$_2$(5-benzofuryl)	NaOEt, EtOH 90%	LiCl, H$_2$O, DMF, 140°C 79%	2000 (835)
ICH$_2$(3-PhSO$_2$O-pyrid-6-yl) BrCH$_2$(2-F-pyrid-5-yl) (converted into 2-OH on hydrolysis)	NaOEt, EtOH 54–72%	6N HCl, Δ: 71% or 1) NaOH, H$_2$O, dioxane: 84% 2) 2N HCl, Δ: 78%	2001 (683)
H$_2$C=CHP(=O)(OEt)CH$_2$- CH(CO$_2$Bn)CH$_2$CH$_2$CO$_2$Bn	NaOEt, EtOH	1) H$_2$, Pd-C 2) 6N HCl, reflux 88% for alkylation and hydrolysis	2001 (684)
4-COMe-BnBr	Na, EtOH 95%	4N HCl 64% overall	2003 (685)
Br(CH$_2$)$_4$Br Br(CH$_2$)$_5$Br Br(CH$_2$)$_6$Br Br(CH$_2$)$_7$Br Br(CH$_2$)$_{10}$Br	NaOEt, EtOH 80-87%	1) NaOH, aq. EOH: 70–79% 2) PhH, reflux: 96–100% 3) NaOH, aq. EtOH: 70–90%	2004 (686)
Br(CH$_2$)$_4$Br	NaOEt, EtOH 73%	1) NaOH, EtOH, 0°C: 85% 2) PhH, reflux: 97% 3) NaOH, EtOH, 0°C: 81%	2005 (687)

Table 3.3 Alkylation of Other Dialkyl Acylaminomalonates

$$R^1{-}\underset{H}{N}{-}CH\begin{smallmatrix}CO_2R^2\\CO_2R^3\end{smallmatrix}$$

R^1	R^2,R^3	Electrophile	Deprotonation Conditions	Decarboxylation Conditions	Reference
R^{1}NH = NO$_2$ Ac	Et Bn	H$_2$C=CHCHO iPrNHCH(Me) (indol-3-yl)	Et$_3$N NaOMe, toluene 97%	NaOEt, AcOH, H$_2$ 1) H$_2$, Pd 2) pyridine, H$_2$O, reflux 34%	1957 (25) 1957 (688)
Cbz	Et	HCOC(R^1)=CH(R^2) converted into —CH(R^2)CH(R^1)CH$_2$-(α-N) R$_1$ = H; R$_2$ = Me, Ph R$_1$ = Me; R$_2$ = H	1) Na, EtOH 2) AcOH 45–80%	1) H$_2$, Pd 2) 6N HCl, Δ 71–80%	1964 (45)
Cbz	Et	Br-CH$_2$-(pyrimidin-5-yl)	Na, MeOH	KOH, conc. HCl 62% for 2 steps	1965 (689)
Cbz	Et	(Z)−ClCH=CHCO$_2$Et	KOtBu, tBuOH 51%	not decarboxylated	1969 (552)
Cbz	Bn	2,4,5-(BnO)$_3$-BnBr	NaH, toluene	1) H$_2$, Pd-C 2) Δ	1969 (24)
Cbz	Me	BnCl	NaOEt	6N NaOH 71% for 2 steps	1973 (22)
Cbz	Me, Et	4-Me-BnBr 3- or 4-F-BnBr 2,3,4,5,6-F$_5$-BnBr	NaOEt 65–82%	1) KOH, EtOH 78–91% 2) 1N H$_2$SO$_4$ 57–92%	1973 (23)
Me, Me$_2$, Me+Bn, Et+Bn, 3-Me-4-Cl-Ph, −(CH$_2$)$_2$O(CH$_2$)$_2$−, −(CH$_2$)$_2$CH(Me)(CH$_2$)$_2$ −, −(CH$_2$)$_n$−: n = 4,5,6	Et	H$_2$C=CPO$_3$Me$_2$	Na, EtOH	conc. HCl 48–62% for 2 steps	1983 (613) 1983 (614)

(Continued)

Table 3.3 Alkylation of Other Dialkyl Acylaminomalonates (continued)

R^1	R^2, R^3	Electrophile	Deprotonation Conditions	Decarboxylation Conditions	Reference
Boc	Et	$H_2C=C=CHCN$ forms $-C(=CH_2)CH_2CN$	NaOEt, EtOH 89%	20% HCl 23%	1986 (623)
Ac	Me	[Br-substituted pyranone with CO_2Me]	NaH, DMF 84%	—	1986 (690)
Ac	Me	$HCCCO_2tBu$	KF, 18-crown-6, MeCN, rt 89%	1) aq. NaOH, rt 2) aq. HCl, 50°C 69%	1988 (31)
Cbz	Et	$Et_2NI(Me)CH_2$(2-pyrrolyl)	NaOEt 54%	1) NaOH 2) dioxane, Δ 3) NaOH 62%	1990 (636)
Ac	Me	[isoxazole structure, OMe] R = nBu, nOct, $(CH_2)_2OMe$	Na, THF 15–50%	48% HBr, reflux 47–71%	1992 (691)
Ts-Val	Et	2-($TBSOCH_2$)-BnBr	NaOEt, EtOH 80%	1) KOH, MeOH 2) AcOH 75%	1993 (28)
Cbz Boc	$R^2 = H$ $R^3 = $ Me, Et	ClCOR R = nPent, Ph, Bn, $(CH_2)_2Ph$	nBuLi 70–85%	aq. NH_4Cl	1993 (39)
Boc, Troc	Me	$BrCH_2CH=CH_2$	Na, EtOH 80–85%	NaOH 81–83%	1993 (692)
Boc-Leu	$R^2 = $ Me, Bn $R^3 = $ Leu-OMe	IMe $BrCH_2CH=CH_2$ BnBr $H_2C=CHCO_2Me$ $H_2C=CHCN$ $NO_2CH=CHPh$ Br(cyclohexen-3-yl)	NaOMe or KOtBu 67–93%	for $R^2 = $ Bn, partiallly with H_2, Pd/C	1994 (26)
Boc	Et	$BrCH_2C(Br)=CH_2$ $MeO_2COCH_2C(OEt)=CH_2$	NaH, EtOH or Pd(0) 80–87%	—	1996 (172)
Ac	Me	[isothiazolone / isothiazole structures, OMe] R = Me, tBu	Na, DMF 50–71%	1M aq. TFA, reflux 28–31% or 4N HCl 20–25%	1997 (693)
Boc	Et	[isoxazole structure, OH] R = 2-furyl, 5-Br-2-furyl	NaH 67%	1) NaOH, EtOH 2) 4N HCl 46–81%	1998 (673)
Boc	Et	$BrCH_2$(3-PO_3Et_2-5-furyl)	NaH, EtOH 80%	1) NaOH, EtOH 2) dioxane, reflux 76%	1999 (694)
Cbz	Et	2-AcO-4-CO_2tBu-BnBr	NaH, DMF 95%	1) NaOH, EtOH 2) Δ 85%	1999 (695)

Table 3.3 Alkylation of Other Dialkyl Acylaminomalonates (continued)

R¹	R²,R³	Electrophile	Deprotonation Conditions	Decarboxylation Conditions	Reference
Boc	Et	BrCH₂(1-NH₂-isoquinolin-6-yl)	NaOEt, EtOH 56–68%	HCl, AcOH, 100°C 99%	1999 (696)
Boc	Et	Br(CH₂)₄Br	NaOEt, EtOH 78%	1) NaOH, EtOH 2) PhH, reflux 3) subtilisin 4) HCl	2004 (686)
Boc	Et	Br(CH₂)₄Br Br(CH₂)₃CH(Me)Br	NaOEt, EtOH 65–92%	1) NaOH, EtOH, 0°C: 80% 2) PhH, reflux: 85% 3) subtilisin	2005 (687)

Table 3.4 Alkylation of Diethyl Benzamidomalonate

Electrophile	Deprotonation Conditions	Decarboxylation Conditions	Reference
BrCH₂CH₂CO₂Et	NaOEt, EtOH, reflux	conc. HCl, reflux 52% for 2 steps	1931 (18)
I*i*Pr I*i*Bu BnCl ClCH₂CO₂Et	NaOEt, EtOH 66–88%	48% HBr, reflux 62–78%	1939 (10)
Br*n*Bu Br(CH₂)₂OH Br(CH₂)₂OPh	NaOEt, EtOH	conc. HCl	1940 (20)
BrCH(Me)CO₂Et	Na, EtOH	conc. HCl 46% overall	1941 (697)
BrCH₂(4-pyridyl) N₂CH₂(2-pyridyl)	Na, EtOH	49% HBr, reflux 3–17% overall	1942 (698)
IMe₃NCH₂(indol-3-y)	—	aq. NaOH, reflux 35% for 2 steps	1944 (15)
Me₂NCH₂(indol-3-y)	NaOEt, EtOH 50%	1) 10% NaOH, reflux 2) 2N H₂SO₄, reflux 88%	1945 (456)
BrCH₂CH₂OPh	NaOEt, EtOH	6N HCl, reflux 65% overall	1947 (459)
(Z)-ClCH=CHCO₂Et	KOtBu, *t*BuOH 69%	aq. NaOH, rt 76%	1969 (552)
2-AcO-4-CO₂Me-BnBr 2-AcO-4,5-(-CO₂CH₂-)-BnBr	NaH, DMF 72–75%	1) NaOH, EtOH 2) Δ 83–85%	1999 (695)
2-AcO-3-CO₂*t*Bu-BnBr	NaH, THF, DMF	1) 20% NaOH, EtOH 2) H⁺, EtOAc 3) TFA 4) 80°C	1999 (699)

Table 3.5 Alkylation of Diethyl Formamidomalonate

Electrophile	Deprotonation Conditions	Decarboxylation Conditions	Reference
BrCH$_2$CH=CH$_2$ converted into (CH$_2$)$_2$SMe	Na, EtOH: 96% MeSH addition: 91–96%	1) 1N KOH in MeOH 2) 6N HCl 85–90%	1955 (*700*)
ClCH$_2$CO$_2$Et H$_2$C=CHCN	NaOEt, EtOH, reflux	conc. HCl, reflux 55–60% for 2 steps	1947 (*21*)
ClCH$_2$(2,3-dihydro-3-benzothiophene)	Na, EtOH 83%	conc. HCl, reflux 75%	1948 (*701*)
R$_2$NH+HCHO forms CH$_2$NR$_2$ R = -(CH$_2$)$_5$-, -(CH$_2$)$_2$O(CH$_2$)$_2$-, iBu	62–93% condensation	HCl 69% for piperidine adduct	1949 (*702*)
Me$_2$NCH$_2$(indol-3-yl)	NaOH, toluene 98%	NaOH, reflux 96%	1949 (*703*)
BrCH$_2$CCH	NaOEt, EtOH 90%	8% NaOH, reflux 89%	1949 (*704*)
BrCH$_2$CH$_2$CH=CH$_2$ converted into -CH$_2$CH$_2$CH(-O-CH$_2$, -CH$_2$CH$_2$CH(OH)CH$_2$NH$_2$	Na, EtOH 55%	conc. HCl, reflux	1950 (*705*)
4-BrBnCl 4-NO$_2$-BnBr	NaOEt, EtOH	HCl 61–62% for 2 steps	1951 (*706*)
	Na, EtOH 34%	1) Na$_2$S$_2$O$_3$: 60% 2) 20% HCl: 75%	1952 (*707*)
HCHO, Me$_2$NH converted into CH$_2$CO$_2$H	73% for condensation then 1) MeI; 2) NaCN 71%	—	1952 (*47*)
BrCH$_2$CCH BrCH$_2$C(Cl)=CH$_2$ BrCH$_2$C(Br)=CH$_2$ BrCH$_2$(3-thienyl) BrCH$_2$(2-naphthyl)	NaH, toluene, benzene or DMF 80–96%	not decarboxylated	1953 (*29*)
BnCl IMe$_3$NCH$_2$(indol-3-yl)	NaOH 98%	1) 10% NaOH 2) 3N HCl 81%	1954 (*482*)
4-NO$_2$-BnCl	NaOEt 75%	—	1954 (*708*)
Me$_2$NCH$_2$(5-Bn-indol-3-yl) Me$_2$NCH$_2$(7-BnO-indol-3-yl)	NaOH, toluene 78–98%	1) NaOH, reflux 2) HCl, reflux 94–96%	1954 (*709*)
Cl(cyclopenten-3-yl) BrCH$_2$cPr	NaOEt, EtOH, rt NaOEt, EtOH 53%	conc. HCl, reflux 10% HBr, reflux 50%	1955 (*485*) 1955 (*710*)
Et$_2$NCH$_2$ (4-Cl-indol-3-yl) Et$_2$NCH$_2$ (6-Cl-indol-3-yl)	NaOH, toluene 69–88%	NaOH, reflux 61–85%	1955 (*490*)
Me$_2$NCH$_2$(4-Cl-indol-3-yl) Me$_2$NCH$_2$(6-Cl-indol-3-yl)	NaOH, toluene 69–88%	NaOH, reflux 61–85%	1955 (*491*)
4-SO$_2$NH$_2$-BnBr	Na, EtOH 63%	conc. HCl, reflux 76%	1957 (*497*)
ClCH$_2$(ferrocenyl)	Na, benzene, reflux 45%	NaOH, then HCl 75%	1957 (*711*)

Table 3.5 Alkylation of Diethyl Formamidomalonate (continued)

Electrophile	Deprotonation Conditions	Decarboxylation Conditions	Reference
TsOCH$_2$[1-Me-2-(=CH$_2$)-1-cyclopropyl]	NaH, DMF	1) NaOH 2) HOAc	1958 (712)
BrCH$_2$CCH BrCH$_2$CCPh BrCH$_2$CCCH$_2$Br BrCH$_2$CH=CHCH=CHCH$_2$Br converted into -nPr, CH$_2$CH=CH$_2$, diaminodicarboxylic acids linked by −CH$_2$CH=CHCH=CHCH$_2$−, CH$_2$CH=CHCH$_2$−, −CH$_2$CCCCCH$_2$−, −(CH$_2$)$_4$−, −(CH$_2$)$_6$−	Na, EtOH 40–95% reduction and oxidative couplings	6N HCl 44–84%	1958 (713)
IMe$_3$NCH$_2$(ferrocenyl)	Na, EtOH 73%	1) NaOH 2) 6N HCl 38%	1958 (714)
BrCOMe	Na, EtOH 75%	NaOH, then Δ	1960 (514)
Me$_2$NCH$_2$(5-BnO-indol-3-yl)	NaOH, toluene 90–92%	1) 2.5N NaOH, reflux 2) 2N HCl, reflux 84–88%	1962 (715)
4-SiMe$_3$-BnBr	Na, EtOH 93%	1) NaOEt, H$_2$O 2) 1.5N HCl, MeOH 73%	1963 (716)
HCOCH=CHR similar adducts converted into −CH(R)CH$_2$CH$_2$-(α-N) R = Ph, 3-NO$_2$-Ph	Na, EtOH 69–81%	—	1964 (45)
4-[SiMe$_2$(4-Me-Ph)]-BnBr 4-[SiMe$_2$(4-CH$_2$Br-Ph)-BnBr	NaOEt, EtOH 82–90%	1) NaOH 2) HCl, MeOH 53–68%	1967 (717)
3-SiMe$_3$-BnBr	Na, EtOH 87%	1) NaOEt, H$_2$O 2) HCl, MeOH 71%	1967 (718)
BrCH$_2$CH=C=CH$_2$, Br(CH$_2$)$_2$CH=C=CH$_2$, BrCH$_2$CH=CH=CHMe, BrCH(Me)−CH$_2$CH=C=CH$_2$, BrCH$_2$CH=CH-CCH, BrCH$_2$-CCH	NaH, benzene, reflux 41–86%	1) aq. NaOH, reflux 2) HCl, relfux 16–62%	1968 (719)
$\begin{array}{c} H \diagdown \\ Br \diagup \end{array} C=C \begin{array}{c} \diagup R^1 \\ \diagdown R^2 \end{array}$ R^1 = R^2 = tBu R^1 = H; R^2 = nPr R^1 = Me; R^2 = Et			
4-COMe-BnI	NaOEt 39%	conc. HCl 63%	1968 (548)
$\overset{+}{\bigcirc}$ BF4$^-$	Na, EtOH 54%	1) NaOH: 30% 2) 3N HCl: 70%	1969 (720)
Cl(CH$_2$)$_4$SMe	Na, EtOH	2.4N HCl, reflux 34% for 2 steps	1970 (721)
BrCH$_2$(3-benzothiophene)	Na, EtOH 100%	10% NaOH, reflux 85%	1970 (838)

(Continued)

Table 3.5 Alkylation of Diethyl Formamidomalonate (continued)

Electrophile	Deprotonation Conditions	Decarboxylation Conditions	Reference
IMe$_2$N [thieno-pyrrole structure] NH	Na, THF 76%	1) NaOH 2) 1N HCl 57%	1972 (722)
BrCH$_2$CCMe	NaOEt, reflux 100%	NaOH, reflux 25%	1972 (723)
I(CH$_2$)$_3$N(Bn)Ph	Na, EtOH 40%	6N HCl 78%	1972 (724)
3,4,5-(MeO)$_3$-BnCl	NaOEt 48%	NaOH, Δ 6%	1973 (571)
3,5-tBu$_2$-4-HO-BnCl	K$_2$CO$_3$, KI, acetone 81%	1) NaOH: 68% 2) reflux: 87% 3) aniline: 83%	1978 (725)
ClCH$_2$(3-benzothiophene)	NaOMe, DMF 81%	4N HCl, reflux 79%	1979 (600)
(N-Me-piperidine)CH$_2$(4,5,6,7-F$_4$-3-benzothiophene)	NaH, DMF 95%	4N HCl, reflux 72%	1979 (600)
Cl [imidazole-selenium structure] NHAc Se	NaOEt 16%	conc. HCl, Δ 80%	1981 (726)
IMe$_3$NCH$_2$[PhFe(CO)$_3$]	NaOEt 94%	1) NaOH, Δ 2) 2N HCl, Δ 75%	1981 (727)
BrCH$_2$$c$Pr	NaOEt, EtOH 53%	1) 1N KOH 2) H$_2$O, Δ 50%	1984 (728)
IMe$_3$NCH$_2$(ruthenocenyl)	NaOEt 80%	KOH, EtOH, H$_2$O, reflux 40%	1983 (729)
Br [benzyl-carborane structure] B$_{10}$H$_{10}$	NaOEt, EtOH, rt to reflux 87%	6N HCl, reflux 89%	1993 (451)
I(H$_2$C)$_3$-C≡CH [carborane structure] B$_{10}$H$_{10}$	NaH, DMF, reflux	conc. HCl, AcOH, reflux 100%	1995 (839)
BrCH$_2$(2-furyl)	NaOEt 91%	1) 10% NaOH, Δ 2) 50% AcOH, Δ 60%	1997 (730)
3,5-tBu$_2$-4-HO-BnBr	NaOEt 92%	hydrolysis 92%	1998 (731)

acetamidomalonate is by far the most commonly used synthon. The acetyl group can be difficult to remove chemically, but can be very useful if an acylase is used for a subsequent enzymatic resolution. Sodium ethoxide (or potassium t-butoxide) is generally employed for deprotonation, but sodium hydride in DMF, benzene or toluene has also been applied (29). In some reports the sodium salt of the dialkyl acylaminomalonate is isolated before alkylation (e.g., 29). Deprotonation has also been achieved with a system of KF and 18-crown-6 in acetonitrile (30), a potentially useful reagent combination when reacting the synthon with sensitive electrophiles (31). K$_2$CO$_3$/18-crown-6 in DMF at 100 °C was used for an alkylation of diethyl acetamidomalonate (32), as has phase-transfer catalysis with K$_2$CO$_3$ in the presence of

benzyltriethylammonium chloride in acetonitrile (33). For a Michael addition of diethylacetamidomalonate to methyl acrylate, anion generation using NaH gave only a 60% yield, while the amine bases tetramethylguanidine (TMG), 1,5-diazobicyclo[4.3.0]non-5-ene (DBN), and 1,5-diazobicyclo[5.4.0]undec-5-ene (DBU) catalyzed the reaction in quantitative yield (34). Alkyl halides are normally employed as electrophiles, though a synthesis in 2000 prepared a number of saturated and unsaturated amino acids from alkyl tosylates (35).

Strongly acidic or basic hydrolysis is commonly used to achieve deprotection and decarboxylation. The Krapcho decarboalkoxylation procedure avoids the use of acidic or basic conditions for sensitive substrates, relying on LiCl in refluxing H$_2$O/DMF to effect the desired transformation (*36*). TFA was used to induce deprotection/decarboxylation for a sensitive substrate prepared using diethyl acetamidomalonate (*37*). With a dibenzyl *N*-Cbz aminomalonate synthon complete deprotection can be achived via hydrogenation, allowing for the preparation of sensitive compounds (such a 6′-hydrox[8;-DOPA) that are unstable towards conditions commonly employed for aminomalonate hydrolysis (*24*). The diethyl aminomalonate synthon has been modified to make deprotection/decarboxylation possible under milder conditions by replacing one of the ester ethyl groups with a trimethylsilyethyl moiety. The nucleophile was generated via the *N*-TMS ketene silyl acetal and added to organoiron salts (see Scheme 3.2). After removal of the *N*-TMS group by mild acidolysis, desilylative-dealkylative-decarboxylation to the amino ester was achieved by refluxing with TBAF (tetra-*n*-butylammonium fluoride) in THF (*38*).

The electrophiles used for alkylating the aminomalonate anion are usually alkyl or aryl halides, carbonyl compounds, or Michael acceptors (see Tables 3.1–3.5), but several unusual systems have also been employed. Acyl chlorides provide β-keto products (*39*). Michael additions of malonates to α,β-unsaturated aldehydes or ketones can be followed by an intramolecular reductive alkylation of the amino group by the side-chain carbonyl to give 3-, 4-, and/or 5-substituted prolines (*40–43*). For example, cinnamaldehyde led to 3-phenylproline (*44, 45*) while crotonaldehyde provided 3-methylproline (*45*). Dienic or styryl side chains can be introduced by a catalytic carbopalladation of allenes followed by trapping of the intermediate π-allyl complex with the glycine anion (*46*). Side chains amenable to further reaction can be introduced. An electrophilic aminomethyl side chain was prepared by forming a Mannich base via reaction of diethyl acetamidomalonate or formamidomalonate with formaldehyde and dimethylamine, followed by methyl iodide. The trimethylammonium iodide group was displaced with NaCN, leading to Asp (*47*), or with thiol nucleophiles, leading to Cys derivatives (*48*). Other electrophilic side chains have been prepared by alkylation of the aminomalonates with dihalo electrophiles (see Tables 3.1–3.5). The second electrophile is displaced with nucleophiles to introduce diversity on

the side chain, or used for intramolecular alkylation of the α-amino group to form proline derivatives. For example, 4,5-dehydropipecolic acid was synthesized by *N,C*-dialkylation of diethyl acetamidomalonate with *cis*-1,4-dichloro-2-butene (*49*), while 1,2,3,4-tetrahydro-isoquinoline-3-carboxylic acid was obtained from 1,2-bis(halomethyl)benzene (*50*).

3.3 Alkylation of Aminocyanoacetic Acids

The use of aminocyanoacetic ester derivatives as nucleophilic glycine synthons was introduced by Albertson et al. in 1945 (*12, 51*) who prepared Val, Leu, Met, Phe, His, and Trp in 40–70% yield in only two steps (*51*) (see Table 3.6). A direct comparison between ethyl acetamidocyanoacetate and diethyl acetamidomalonate for alkylation by a range of alkyl halides showed little difference between the two synthons, although the reported yields were not assigned to either method (*12*). Hydrolysis and decarboxylation of the condensation product is possible under either acidic or basic conditions to yield the amino acid, and in some instances gives better yields than hydrolysis of the corresponding alkylated ethyl acetamidomalonate (e.g., see (*52*)). The Krapcho decarbalkoxylation procedure (refluxing DMSO/H_2O/LiCl) can also be employed (*36*). Recently, chemically sensitive β-silyl-Ala derivatives were synthesized using ethyl acetamidocyanoacetate and silyl alkyl iodides, with NaOH in DMSO for deprotonation (*53*). Aminocyanoacetate has also been incorporated into peptides, deprotonated and alkylated (*27*).

3.4 Alkylation of Schiff Bases

The discovery that glycine Schiff bases or related analogs could be deprotonated and alkylated was not achieved until 1975, when Hoppe reported that *N*-[bis(methylthio)methylene]glycine ethyl ester could be reacted with alkyl iodides or benzyl bromides in the presence of KO*t*Bu to give the monoalkylated products in 48–83% yield (*54, 55*) (see Scheme 3.3, Table 3.7). Little dialkylation was observed unless a second equivalent

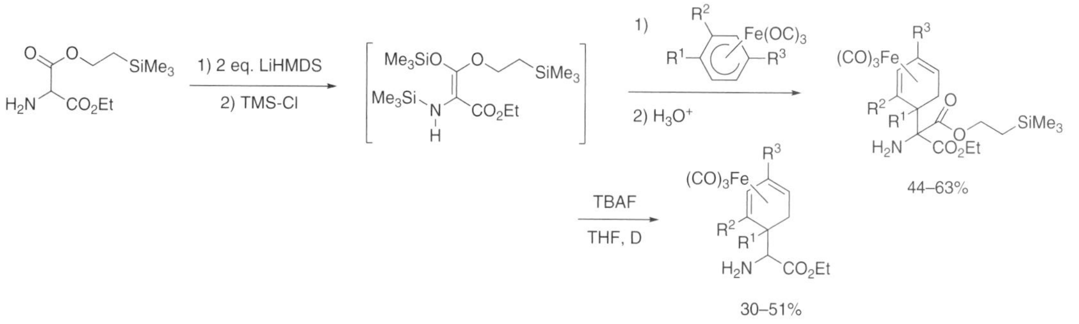

Scheme 3.2 Modified Ester Group for Decarboxylation (*38*).

Table 3.6 Alkylation of Ethyl Acetamidocyanoacetate

Electrophile	Deprotonation Conditions	Decarboxylation Conditions	Reference
BriPr BriBu BnCl Cl(CH$_2$)$_2$SMe Me$_2$NCH$_2$(indol-3-yl) ClCH$_2$(imidazol-4-yl)	NaOEt, EtOH 60–98%	aq. NaOH, H$_2$SO$_4$ or HCl, reflux 61–80%	1945 (51)
IMe BrEt Br(CH$_2$)$_n$Me: n = 2,4,5,6,7,8 ClCH$_2$C(Me)=CH$_2$	NaOEt, EtOH, reflux	37% HCl or 48% HBr or 10% NaOH, reflux 32–85% for 2 steps	1946 (12)
H$_2$C=CHCHO	NaOEt, EtOH, rt or NaOMe, benzene 82%	not decarboxylated	1948 (438)
BriPr IiBu BnCl 4-MeO-BnCl Cl(CH$_2$)$_2$SMe	NaOEt, EtOH 70%	HBr 84%	1948 (732)
ClCH$_2$(1-naphthyl)	Na, EtOH 98%	NaOH, reflux 78%	1948 (463)
Me$_2$NCH$_2$(2-pyrrole)	NaOEt, EtOH, Me$_2$SO$_2$ 90%	1) NaOH 2) HCl	1948 (733)
ClCH$_2$CH=CH$_2$	NaOEt, EtOH 76%	10% NaOH 50%	1948 (52)
Br(CH$_2$)$_2$CH=CHMe	NaOEt, EtOH 92%	10% NaOH 75%	1948 (734)
3-F-BnCl 4-F-BnCl	Na, EtOH 70%	1) 2.5N NaOH, reflux 2) acidify, reflux 40–42% overall	1950 (469)
4-NO$_2$-BnBr 4-NH$_2$-BnBr BrCH(Ph)$_2$	NaOEt, EtOH, reflux	30–92% conc. HCl. reflux 70–94%	1951 (472)
3-NO$_2$-4-Me-BnCl	NaOEt, EtOH	HCl, reflux 34% for 2 steps	1951 (473)
I(CH$_2$)$_3$NPhth I(CH$_2$)$_4$NPhth 4-MeO-BnBr with [2-^{14}C]	Na, EtOH 75–96%	conc. HCl, reflux 79–98%	1951 (735)
H$_2$C=C(Me)CO$_2$Me	NaOEt, EtOH	conc. HCl or H$_2$SO$_4$ 12% for 2 steps	1952 (479)
BrCH$_2$(2-thienyl)	NaH, toluene	20% KOH, reflux 50% for 2 steps	1953 (29)
Br(CH$_2$)$_3$NPhth	NaOEt 80%	HCl, H$_2$O 86%	1953 (736)
Me$_2$NCH$_2$(4-CN-indol-3-yl)	Na, EtOH 83%	50% KOH, reflux 83%	1955 (737)
BrCH(CO$_2$Et)CH$_2$CO$_2$Et	Na, EtOH	conc. HCl	1955 (738)
Br(cyclohexen-3-yl) BrCH$_2$(cyclohexen-2-yl)	Na, EtOH 54–64%	10% NaOH, reflux 33–68%	1957 (739)
IMe$_3$NCH$_2$(ferrocenyl)	NaOEt	NaOH 67% for 2 steps	1957 (740)
ClCH$_2$CH(OAc)CO$_2$Et	NaOEt, EtOH	20% HCl 50–60% for 2 steps	1957 (741)
ICH$_2$CH$_2$iPr BrCH$_2$C(Me)=CHMe	NaOEt, EtOH 70–80%	1N NaOH, reflux 2–50%	1959 (509)

Table 3.6 Alkylation of Ethyl Acetamidocyanoacetate (continued)

Electrophile	Deprotonation Conditions	Decarboxylation Conditions	Reference
Br*c*Pent	NaOEt, EtOH 80%	10% NaOH 75%	1963 (742)
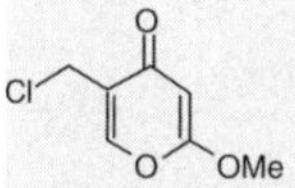 converted by ammonia into −CH₂[3,4-(OH)₂-pyrid-6-yl]	NaH, DMF 52%	NH₄OH then Ba(OH)₂ 50%	1967 (743)
BrCH₂CH=C(Me)₂	NaOEt, EtOH, reflux 78%	NaOH 15%	1968 (744)
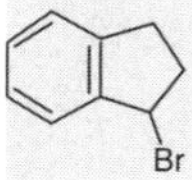 BrCH₂SiMe₃	NaOEt, DMSO 32–55%	10% NaOH, reflux 25–36%	1968 (745)
(Z)-ClCH=CHCO₂Et	KO*t*Bu, *t*BuOH 5%	not decarboxylated	1969 (552)
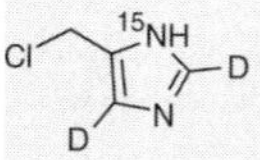	NaOEt 80%	3N H₂SO₄, Δ 59%	1977 (746)
I(CH₂)₃NPhth Cl(CH₂)₃N(Ts) (CH₂)₃NHAc	NaH, DMF 45–62%	conc. HCl 23%	1978 (747)
Br(CH₂)₃CH=CH₂	Na, EtOH	10% NaOH, reflux 88% for 2 steps	1978 (748)
Br*i*Bu	NaH, HMPT 70%	6N HCl, reflux 79%	1979 (749)
BrCH₂(cyclopenten-3-yl)	NaOEt	5N NaOH, reflux 43% for 2 steps	1980 (750)
4-COPh-BnCl	K₂CO₃, KI, acetone 69%	8N HCl, 100°C 92%	1986 (616)
MsOCHD(CH₂)₂NPhth	Na, DMSO 63–66%	6N HCl 87–95%	1988 (751)
Br(CH₂)₂CH=CHMe BrCH₂(2-Me-1-naphthyl) 2,3-Br₂-naphthyl	NaOEt, EtOH LDA, −78°C to reflux 50–60%	10% NaOH 12N HCl 65%	1994 (752) 1995 (753)
ICH₂SiMe₂Ph ICH₂SiMe[-(CH₂)₄] ICH₂SiMe[-(CH₂)₅]	NaOH, DMSO 44–49%	NaOH, H₂O 70–86%	1997 (53)
	K₂CO₃, acetone 47%	8N HCl 75%	1997 (754)

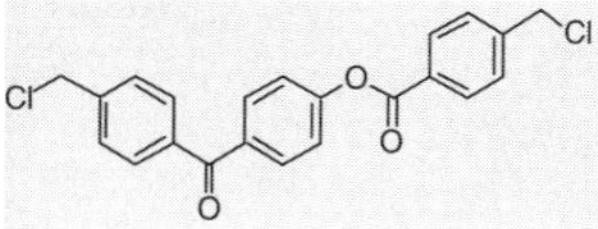

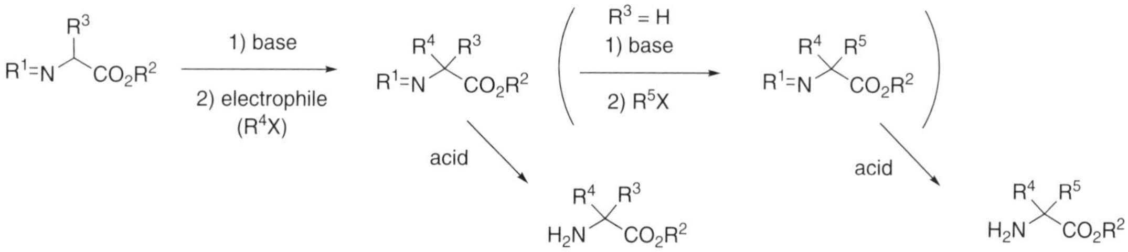

Scheme 3.3 Alkylation of Amino Acid Schiff Bases Under Anhydrous Conditions.

Table 3.7 Alkylation of Amino Acid Schiff Bases: Anhydrous Conditions with Strong Metal Bases (see Scheme 3.3)

R^1	R^2	R^3	Electrophile	Deprotonation Conditions	Hydrolysis Conditions	Reference
(4-NO₂-Ph)CH	COR² = R³ = CONH(CH₂)₃-	(CH₂)₃NH-(α-CO)	IMe	PhLi	2N HCl 67%	1974 (56)
PhCH	COR² = R³ = CONH(CH₂)₃-	(CH₂)₃NH-(α-CO)	IMe IEt InPr InBu InHex InOct BnBr	NaH, MeOH	2N HCl	1975 (57)
(MeS)₂C	Et	H	ICD₃ IMe IEt IiPr BnBr for R⁴ = 4-Br-Bn; R⁵X = IEt for R⁴ = iBu; R⁵X = IMe	KOtBu, THF 48–84%	HCO₂H, H₂O₂ 60% (1 example)	1975 (54) 1979 (755)
(MeS)₂C (−CH₂S)₂C	Et	H	R⁴X = R⁵X = IMe	KOtBu, THF 70–73%	—	1975 (55)
PhCH	Et	H	InOct IiPr BrCH₂CO₂Et Br(CH₂)₅CO₂Et	LDA, THF, HMPA 62–90%	6N HCl 90–100%	1976 (58)
PhCH	Et	H	InOct for R⁴ = nBu; R⁵X = InOct	KOtBu, THF 78–81%	6N HCl 90–100%	1976 (58)
PhCH	Et	H	cyclohex-2-enone RCH=CHCO₂Et R = H, Me, CO₂Et	NaOEt, EtOH 80–90%	6N HCl 90–100%	1976 (58)
PhCH	Et	Me	BrnBu	KOtBu, THF 77%	6N HCl 90–100%	1976 (58)
PhCH	Me	Me	IMe IiPr I(CH₂)₃NPhth BrCH₂CH=CH₂ ClCH₂CN BrCH₂CO₂Me ClCH₂OMe ClCH₂SMe BnBr 3,4-(MeO)₂-BnBr	NaH or LDA, THF 76–95%	1N HCl 77% (1 example)	1977 (60)
PhCH	tBu	H	BrnBu BnCl BnBr	LDA, THF	H₂NNHCONH₂ 25–76% for 2 steps (+3–11% dialkylation)	1977 (61)
Me₂NCH	Me	H	BrCH₂CH=CH₂ H₂C=CHPh	KOtBu or LDA, THF 65–90%	—	1977 (756)
Me₂NCH	Me	Ph, Bn, (CH₂)₂SMe	IMe InPr 4-Cl-BnCl H₂C=CHCO₂Me H₂C(-O-)CH₂	KOtBu or LDA, THF 60–88%	—	1977 (756)
PhCH	Me	Me	I(CH₂)ₙNPhth n = 2,3,4	LDA, THF	6N HCl, reflux	1978 (148)

Table 3.7 Alkylation of Amino Acid Schiff Bases: Anhydrous Conditions with Strong Metal Bases (see Scheme 3.3) (continued)

R^1	R^2	R^3	Electrophile	Deprotonation Conditions	Hydrolysis Conditions	Reference
PhCH	Me	Me	ClCH$_2$Br ClCHF$_2$ Cl$_2$CH$_2$ Br$_2$CF$_2$ ClCH$_2$F BrCF$_3$	LDA, THF	1N HCl	1978 (757)
Ph$_2$C	Et	H	IMe BrEt IiPr IsBu BnBr	LDA, THF, HMPA 65–93%	—	1978 (63)
PhCH	Me	Me	(E)-ClCH=CHSO$_2$Ph	LDA, THF	dilute HCl 83% for 2 steps	1978 (758)
PhCH, Me	Me	Me, (CH$_2$)$_3$NH$_2$, (CH$_2$)$_5$NH$_2$, (CH$_2$)$_2$SMe, Bn, 4-HO-Bn, 3,4-(MeO)$_2$-Bn, CH$_2$(imidazol-4-yl)	BrCH$_2$Cl ClCHF$_2$ ClCH$_2$F	LDA, THF 64–97%	2N HCl	1979 (759)
PhCH	Me	Me	(Z)- or (E)-BrCH=CHR	LDA, THF 86–100%	—	1980 (760)
PhCH	Et	Me, iPr, Et, Bn, Ph	CH$_2$O	LDA, THF 70–90%	6N HCl, reflux 70–85%	1981 (761)
PhCH	Me	H	R^{4}X = R^{5}X = BrCH$_2$CH=CHCH$_2$Br forms -CH$_2$CH(CH=CH$_2$)- (α-C)	LDA, THF	5N HCl 42% for 2 steps	1981 (762)
PhCH	Et	H	R^{4}X = R^{5}X = BrCD$_2$CD$_2$Br	LDA, THF, HMPA	6N HCl, Δ 28% overall	1982 (763)
(4-MeO-Ph) CH	COR^2 = R^3 = CONH(CH$_2$)$_3$-	—	IMe BrCH$_2$CO$_2$Et	LDA, THF 92–97%	1N HCl 85–90%	1983 (764)
PhCH	Me	H	Br(CH$_2$)$_5$CO$_2$Bn	KOtBu	KHSO$_4$, H$_2$O 72% for 2 steps	1983 (765)
PhCH	Et	H	R^{4}X = R^{5}X = BrCHDCHDBr	LDA, THF, HMPA	6N HCl, Δ 28% overall	1983 (766)
PhCH	Me, Et	Me, Bn, (CH$_2$)$_2$CH=CH$_2$, (CH$_2$)$_3$N=CHPh	Cl$_2$CHF	Na, K	1N HCl 20–32% for 2 steps	1984 (767)
Ph$_2$C PhCH	Me	H, Me	BrCH$_2$CH=CHCO$_2$Me BrCH(Me)CH= CHCO$_2$Me forms -CH(R) CH=CHCO$_2$Me or -CH(R)CH(CO$_2$Me)- (α-C) depending on alkylation conditions	LDA, THF or LDA, THF-HMPA 67–92% alkylation or addition/elimination	—	1986 (768)
PhCH	Me	H, Me, Bn	R^{4}X = R^{5}X = Cl(CH$_2$)$_n$Cl n = 2,3,4,5,6	LDA, Li$_2$CuCl$_4$ 40–90% for 1st alkylation 66–90% for 2nd alkylation	—	1986 (99)
Ph$_2$C	Et	H	2-F-4,5-(OMe)$_2$-BnBr	LDA	H$_2$, HI/P, Δ 55% for 2 steps	1986 (769)

(Continued)

Table 3.7 Alkylation of Amino Acid Schiff Bases: Anhydrous Conditions with Strong Metal Bases (see Scheme 3.3) (continued)

R^1	R^2	R^3	Electrophile	Deprotonation Conditions	Hydrolysis Conditions	Reference
PhC	tBu	(CH$_2$)$_3$CH(N=CHPh)CO$_2$tBu	ClCHF$_2$	LDA	HCl, AcOH 19% for 2 steps	1986 (770)
PhCH	Et	H	H$_2$C=CHP(O)(Me)OMe	KOH, EtOH	6N HCl 69% for 2 steps	1987 (771)
Ph$_2$C	Bn	H	HCO(CH$_2$)$_2$NHCbz	LiHMDS, THF	2N HCl 95% for 2 steps	1988 (772)
PhCH	Me	H	R^{4}X = BrCH$_2$CH$_2$CH=CH$_2$ R^{5}X = ClCHF$_2$	NaH, THF 27%, 69%	1N HCl 70%	1988 (773)
Ph$_2$C	Me, Et, tBu	H	H$_2$C=N$^+$Me$_2$I$^-$	CH$_2$Cl$_2$ 63–68%	—	1988 (774)
Ph$_2$C	tBu	H	BrCH$_2$CH$_2$Br	NaH, DMSO/Et$_2$O 72%	1N HCl 43%	1989 (775)
Ph$_2$C	tBu	H	I^{11}CH$_3$	KOH, DMF/DMSO	6N HCl	1989 (776)
Ph$_2$C	tBu	H	I^{11}CH$_3$	KOH, DMSO	6N HCl	1990 (777)
Ph$_2$C	Et, Bn	H	HCOCH$_2$CH$_2$N$_3$	LiHMDS, THF	2N HCl 60–74% for 2 steps 54–66% de thre	1990 (778)
Ph$_2$C	Et	H	RCHO R = Et, iPr, cHex, tBu, Ph, 4-NO$_2$-Ph, 2-thienyl, 2-furyl	1) LDA, TMS-Cl 2) ZnCl$_2$ 40–90% yield with 10–92% de	—	1991 (65)
PhCH	Et	H	HCOCF$_3$	LDA, THF 64%	—	1991 (779)
PhCH	Me	(CH$_2$)$_2$CH=CH$_2$ converted into (CH$_2$)$_2$CO$_2$H	Cl$_2$CHF	NaH, THF	—	1992 (780)
PhCH	Et	H	R^{4}X = R^{5}X = [structure, naphthalene-1,8-diyl bis(bromomethyl), Br ... Br]	NaHMDS, THF	1N HCl 92% for 2 steps	1992 (100)
PhCH	Et	H	R^{4}X = R^{5}X = Br(CH$_2$)$_4$Br Br(CH$_2$)$_5$Br 2-(BrCH$_2$)-BnBr 2-(BrCH$_2$)-3,4,5,6-Br$_4$-BnBr 2-(BrCH$_2$)-3,6-(MeO)$_2$-BnBr 2-(BrCH$_2$)-4,5-R$_2$-BnBr R = Me, OMe, Ph	LiHMDS or NaHMDS, THF	1N HCl 30–60% for 2 steps	1992 (101)
Ph$_2$C	Me	H	RCHO R = ribose, galactose	LDA, THF 55–65%	—	1993 (192)
PhCH	Me, Et, Bn	H	[structure] PO$_3$Me$_2$ Br ; [structure] R, R′ PO$_3$Me$_2$ Br R = R′ = H, Me	KOtBu, THF 24–98%	6N HCl 59–79%	1993 (648)
Ph$_2$C	Et	H	BrCH$_2$C(OEt)=CHPO$_3$Me$_2$	LiHMDS, THF	1N HCl, reflux	1993 (781)
PhCH	Et	Me				
Ph$_2$C	Et	H	Br(CH$_2$)$_n$CN n = 3,4,5,6	NaHMDS, THF 52–83%	—	1993 (782)

Table 3.7 Alkylation of Amino Acid Schiff Bases: Anhydrous Conditions with Strong Metal Bases (see Scheme 3.3) (continued)

R¹	R²	R³	Electrophile	Deprotonation Conditions	Hydrolysis Conditions	Reference
Ph$_2$C	Me	H	RCOCl R = iPr, tBu, CH(Et)$_2$, cPent, cHex, cHept, adamantyl, CH$_2$SMe, Ph, Bn, 1-naphthyl, 2-naphthyl, 3-pyridyl	KOtBu, THF, −78 °C	aq. HCl 41–95% for 2 steps	1993 (66)
PhCH	Me	Me	I^{11}CH$_3$	LDA, THF	1N HCl, 100°C 55% radiochemical yield	1994 (783)
Ph$_2$C	Me	H	BrCH$_2$CCPO$_3$Et$_2$	LiHMDS, THF	TMS-Br, pyridine 67% for 2 steps	1994 (452)
Ph$_2$C	H, Et, tBu	H	H$_2$C=CHPO$_3$R$_2$	KHMDS >90%	1N HCl	1994 (784)
Ph$_2$C	Me	H	Br—⟨indanyl⟩	nBuLi 33%	1N HCl 72%	1994 (785)
Ph$_2$C	tBu	H	BrCH$_2$CH(R)=CHR′ for R = H; R′ = CN, PO$_3$H$_2$ forms −CH$_2$CH=CHR′ for R = H; R′ = NO$_2$, CO$_2$Me, COMe, SO$_2$Ph or R = Me, R′ = NO$_2$ forms CH(R) (-CH$_2$CHR′-)	nBuLi, hexane-THF 56–91%	1N HCl 87–96%	1994 (786)
Ph$_2$C	Me	Me, Ph	IMe IEt	KH, THF, 18-crown-6 65–66%	HCl, reflux 63–100%	1995 (787)
Ph$_2$C	Et	H	IMe I(CH$_2$)$_2$Ph I(CH$_2$)$_3$Ph IiBu then R^{5}X = HCCCO$_2$Et	KOtBu, THF 20–88% for 1st alkylation 72–80% for 2nd alkylation	3N HCl 36–61%	1995 (788)
(4-Cl-Ph)CH	Me	nBu, Bn, (CH$_2$)$_4$NHCbz	BrCH$_2$NPhth	LDA	HCl 58–89% for 2 steps	1995 (789)
Ph$_2$C	Et	H	ClC(CF$_3$)=CHCO$_2$Et	LDA, THF 56%	12N HCl, Pd/C, H$_2$	1996 (790)
tBuCH	Et	Me	4-F-BnBr	KOtBu, toluene	1N HCl 90% for 2 steps	1996 (791)
PhCH	Me, Et, SiMe$_3$	H	ICH$_2$CH$_2$C(R)CC[2-O (2-NO$_2$-Bn)-5-CO$_2$R′-Ph) R = H, adenosine R′ = Me, 2-NO$_2$-Bn	LDA, HMPA, THF	80% HCO$_2$H or HOAc	1996 (792)
Ph$_2$C	Et	H	4-(BO$_2$C$_2$Me$_4$)-BnBr converted via reaction with RX, Pd(0) into 4-R-Bn R = 4-F-Ph, 3-MeO-Ph, 3,5-(MeO)$_2$-Ph, 2-pyridyl, 4-pyridyl, 2-CO$_2$Bn-3-thienyl, 2-CHO-4-thienyl, 2-pyrimidyl, 5-pyrimidyl	NaHMDS 20–75%	—	1997 (793)
Ph$_2$C	Et	H	2-NO$_2$-BnBr	LDA, THF, HMPA	1N HCl 85% for 2 steps	1997 (667)
Ph$_2$C	Et	H	4-(CH$_2$PO$_3$Bn$_2$)-BnBr	KHMDS, THF 58%	1N HCl, 66%	1997 (794)

(Continued)

Table 3.7 Alkylation of Amino Acid Schiff Bases: Anhydrous Conditions with Strong Metal Bases (see Scheme 3.3) (continued)

R¹	R²	R³	Electrophile	Deprotonation Conditions	Hydrolysis Conditions	Reference
Ph₂C	tBu	H	BrCH₂CH₂F BrCH₂CHFMe BrCH₂CHFEt BrCH₂CHnPr BrCH₂CHFiPr BrCH₂CHFnBu BrCH₂CHFnPent	LDA, THF, DMPU 21–39%	15% citric acid then 6N HCl 70–78%	1997 (795)
Ph₂C	Bn	H	4-(OCF₃CO₂tBu)-BnBr	LDA, DMP, THF 60%	H₂, Pd-C 81%	1998 (796)
(4-Cl-Ph)CH	tBu	Me	BrCH₂CH₂F BrCH₂CF(Me)₂ BrCH₂CHF(iPr) BrCH₂CHF(nPr) BrCH₂CF=CH₂	KOtBu, DMSO 20–88%	6N HCl 19–62%	1998 (797)
(4-Cl-Ph)CH	tBu	Me	BrCH₂CF=CH₂	LDA, THF 92%	1N HCl 73%	1998 (798)
Ph₂C	tBu	H	BrCH₂CF=CH₂	LDA, THF 95%	TFA 76%	1998 (798)
Ph₂C	tBu	H	ClCH₂B (-OCMe₂CMe₂O-)	LiHMDS	0.25 N HCl 86% for 2 steps	1998 (799)
Ph₂C	Et	H	BrCH₂CCH	NaH, DMF/THF	aq. HCl 75% for 2 steps	1999 (800)

of base was added, in which case a second alkyl group (different if desired) could be added. The imine was hydrolyzed by treatment with formic acid and H₂O₂. A previous report in 1974 had demonstrated that a 4-nitrobenzaldimine Schiff base derivative of Orn lactam could be α-methylated (56), with further alkylations of the benzaldimine analog by a number of electrophiles in 1975 (57). In 1976 the benzaldehyde-derived aldimine of glycine ethyl ester was deprotonated with LDA or KOtBu and reacted with alkyl halides or Michael acceptors in 62–90% yield (58). A chiral ketimine derived from 2-hydroxypinan-3-one and Gly was reported the same year (see Section 6.4.3a) (59), while two other groups published papers describing alkylations of benzaldehyde Schiff bases in 1977 (60, 61). The aldimine moiety is reasonably stable, surviving in the freezer for several months, but the perceived imine instability and the possibility of dialkylation restricted use of the method for many years. Glycine Schiff base derivatives have also been used as electrophilic glycine equivalents, as is discussed in Section 3.9.

A significant advancement in glycine Schiff base alkylation was achieved by O'Donnell et al. in 1978 (briefly reviewed in (62)), who reported that the benzophenone-derived ketimine of glycine ethyl ester was quite stable at room temperature for several months and could be successfully alkylated after deprotonation with LDA (63). The Schiff base, now commercially available, was prepared in 82% yield by refluxing benzophenone with glycine ethyl ester in xylene in the presence of BF₃.Et₂O (63), but can be more conveniently generated by transamination with benzophenone imine

(64). The LDA-deprotonated glycine ketimine can be trapped as the silyl enol ether, with the enolate subsequently regenerated with ZnCl₂ and added to aldehydes in 40–90% yield (65). A variety of amino acids have been prepared via anhydrous alkylation of glycine Schiff bases (see Scheme 3.3, Table 3.7), with alkyl halides the preferred electrophiles. Reaction with acyl chlorides leads to β-keto-α-aminoesters (e.g. 66). Michael additions to α,β-unsaturated esters are also possible, such as reactions of the pivalaldehyde Schiff base of Gly-OMe or Ala-OMe with methyl acrylate, methyl crotonate, dimethyl fumarate, dimethyl maleate, 3-penten-2-one, 4-phenyl-3-buten-2-one, and 1,5-diphenyl-2,4-pentadien-1-one in the presence of DBU/LiBr (67). A non-ionic strong base, proazaphosphatrane, effectively catalyzed Michael additions of the pivalaldehyde Schiff base of Gly-OMe to α,β-unsaturated esters in isobutyronitrile without the need for a chelating metal ion (68).

Another finding, of possibly greater importance, was that Schiff base deprotonation is also possible under phase-transfer catalysis (PTC) conditions using 10% NaOH and tetrabutylammonium hydrogen sulfate in dichloromethane (63) (see Scheme 3.4, Table 3.8). Although yields were slightly lower than the anhydrous anion generation conditions (59–89% vs 65–93%), the PTC method uses simple, mild, non-anhydrous conditions and inexpensive reagents, and can easily be scaled up. A subsequent publication reported improved PTC conditions (potassium carbonate in refluxing acetonitrile with tetrabutylammonium bromide), giving yields of 38–80% for alkylation and imine hydrolysis (69). Very similar PTC conditions have been used to saponify and

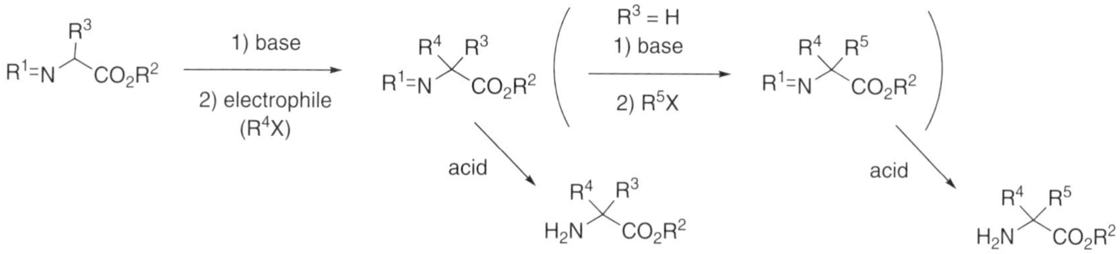

Scheme 3.4 Alkylation of Amino Acid Schiff Bases Under Phase-Transfer Catalysis Conditions.

O-alkylate benzophenone Schiff base esters of amino acids, leading to new ester derivatives (*70*). Aldimines derived from 4-chlorobenzaldehyde (which, unlike benzaldehyde, forms a crystalline derivative) and Gly or Ala were also successfully alkylated under these conditions (*71, 72*), as was the ketimine of aminoacetonitrile (*73, 74*). The latter imine has the advantage that only acidic hydrolysis, rather than basic ester hydrolysis, is required to generate the free carboxyl group (although dialkylation is potentially a problem). Phase-transfer alkylation conditions have found widespread application since they were first reported (see Scheme 3.4, Table 3.8), and are especially important for asymmetric syntheses (see Section 6.4.4).

O'Donnell found that aldimines could undergo dialkylation under PTC conditions (*75*), but that ketimines only underwent monoalkylation due to the decreased acidity of the monoalkylated ketimine preventing the second ionization step required for dialkylation (*76*). These results are supported by the relative acidities of various aldimines and ketimines measured in DMSO (*75*) (see Scheme 3.5), as the pK_a of the monoalkylated aldimine is essentially the same as that of the parent aldimine (19.2, 19.0 or 17.2 vs 18.8), whereas the monoalkylated ketimine is substantially less acidic (23.2, 22.8 or 21.2 vs 18.7). Steric crowding in the α-substituted benzophenone ketimines is proposed to reduce resonance stabilization, accounting for decreased acidity. The rates of alkylation of unsubstituted or mono-alkylated anions were comparable, indicating that their relative pK_a's are the dominant factor determining relative rates of mono- and dialkylation. The aldimine derived from 4-chlorobenzaldehyde was found to give the best results when introducing a second alkyl group during the synthesis of α,α-dialkyl amino acids (*77–79*), although the benzaldehyde aldimine has also been used. Amide derivatives (instead of esters) of *N*-benzylidene amino acids have also been alkylated (*80*), although reactions involving isopropyl groups as either the existing or introduced side-chain substituent were not successful.

The effect of different bases (KOH, K$_2$CO$_3$, and Na$_2$CO$_3$) on the alkylation of *N*-benzylidene Schiff bases of glycine or α-substituted amino acid esters has been investigated, and a reaction mechanism proposed (*81*). Mono-alkylated product can be obtained from glycine and active halides using K$_2$CO$_3$ or Na$_2$CO$_3$, but less active halides require KOH and K$_2$CO$_3$. A strong base (KOH) is required for introducing a second unreactive alkyl substituent. Michael additions are catalyzed by K$_2$CO$_3$ for either type of substrate. Aldol reactions are also possible, but give diastereomer mixtures as well as oxazolidines. More recently, and in contrast to O'Donnell's publications, a Spanish group has reported that by varying the PTC conditions, the ketimine ethyl *N*-(diphenylmethylene)glycinate can undergo monoalkylation, one-pot dialkylation with 2 equivalents of the same organic halide, or sequential dialkylation with two different active halides (*82–84*). Monoalkylations were achieved with powdered potassium carbonate as base in refluxing acetonitrile, without any phase-transfer catalyst, giving 33–92% yield after imine removal. However, a non-activated electrophile, phenyl ethylbromide, required KO*t*Bu in THF. Potassium hydroxide and tetrabutylammonium bromide allowed for dialkylation with activated halides (19–72% yield), while KOH or NaOEt with tetrabutylammonium bromide were suitable for introduction of a second electrophile to a monoalkylated ketimine. Michael additions to ethyl acrylate or ethyl propiolate were also catalyzed by NaOEt.

Schiff base alkylation under PTC conditions has been applied to a variety of syntheses. The simplicity and rapidity of the PTC process makes it a useful procedure to prepare ^{11}C-labeled amino acids. Due to the short half-life of this isotope a speedy synthesis is essential; the alkylations, aided by ultrasonication, were completed in 5–15 min (*85*). A Ru-based coordination complex has been used as a phase-transfer catalyst to promote faster alkylations of the benzophenone imine of Gly-O*t*Bu than achieved with quaternary ammonium salts (*86*). An iminium salt, *N,N*-dimethyl bisnaphthylmethylideneammonium iodide, acted as an effective PTC for alkylations of the benzophenone Schiff base of Gly-O*t*Bu with KOH as base (*87*).

O'Donnell et al. have applied the ketimine conditions to a new method of solid-phase peptide synthesis, in which new residues are constructed on the resin rather than coupled as complete moieties (see Scheme 3.6) (*88, 89*). Glycine was coupled to the resin-linked growing peptide chain, converted to the benzophenone ketimine, alkylated using the non-ionic phosphazene base BEMP conditions, then deprotected. Yields of 37–54% of deprotected cleaved mono-, di-, and tripeptides

Table 3.8 Alkylation of Amino Acid Schiff Bases: Aqueous Conditions or Non-Ionic Strong Bases (see Scheme 3.4)

R^1	R^2	R^3	Electrophile	Deprotonation Conditions	Hydrolysis Conditions	Reference
Ph_2C	$CO_2R^2 = CN$	H	Me_2SO_4 BrEt BriPr	50% NaOH BnEt$_3$NCl, toluene 75–95%	1) 1N HCl 2) 6N HCl	1978 (73)
Ph_2C	H	H	IMe IEt IiPr IsBu BnBr	10% NaOH 59–89%	90% (1 example)	1978 (63)
$(4\text{-Cl-Ph})CH$	Et	Me	4-BnO-BnCl 3,4-(BnO)$_2$-BnCl	KOH	1N HCl 55–93% for 2 steps	1982 (71)
$(4\text{-Cl-Ph})CH$	Et	Bn, 4-BnO-Bn, 3-BnO-Bn, 3,4-(BnO)$_2$-Bn	IMe	KOH	1N HCl 55–93% for 2 steps	1982 (71)
$(4\text{-Cl-Ph})CH$	Et	H	IEt InHex BrCH$_2$CH=CH$_2$	10% NaOH 59–89%	1N HCl 18–94% for 2 steps	1982 (72)
Ph_2C	Et	H	InHex BrnOct 4-Cl-BnBr	KOH or 50% NaOH BnEt$_3$NCl, CH$_2$Cl$_2$	1N HCl 47–86% for 2 steps	1982 (72)
Ph_2C $(4\text{-Cl-Ph})CH$	Et Me, Et	H, Me Me	BrnBu BrCH$_2$CH=CH$_2$ BrCH$_2$CCH 4-NO$_2$-BnCl 4-Cl-BnCl	K$_2$CO$_3$ 59–89%	1N HCl 38–80% for 2 steps	1984 (69)
Ph_2C	$CO_2R^2 = CN$	H	$R^4X = Br(CH_2)_3Br$ $R^4X = R^5X =$ $Br(CH_2)_nBr$; n = 2,4	50% NaOH BnEt$_3$NCl, toluene	1) 1N HCl 2) 6N HCl, reflux 70–87% for 2 steps	1984 (74)
Ph_2C	Bn	H	BrCHFPh	10% NaOH	1) TMSI, CHCl$_3$ 2) 1N HCl 70 for 2 steps	1985 (801)
Ph_2C	tBu	H	$I^{11}CH_3$ $I^{11}CH_2Me$ $I^{11}CH_2Et$ $I^{11}CH_2nPr$ $I^{11}CH_2iPr$ $I^{11}CH_2Ph$	10–20% NaOH sonication 59–89%	6N HCl 10–55% radiochemical yield for 2 steps	1987 (85)

PhCH	Et	H, Me, Ph	BriPr BriBu BrCH₂CH=CH₂ BrCH₂OBn BrCH₂CO₂Et BnCl Me₂-BnCl (MeO)₂-BnBr 4-MeO-BnCl	KOH or K₂CO₃ or K₂CO₃/KOH, BnEt₃NCl, MeCN 41–97%	—	1988 (81)
PhCH	Me	H, Me, (CH₂)₂CN, (CH₂)₂CO₂Me	RCH=CHCO₂Me RCH=CHCN RC(Ph)=CHCO₂Me RC(4-Cl-Ph)=CHCO₂Me	K₂CO₃, BnEt₃NCl, MeCN 63–94%	—	1988 (81)
PhCH	Et	H, Me	RCHO R = H, Ph, 4-Cl-Ph, 4-NO₂-Ph	K₂CO₃, BnEt₃NCl, MeCN or K₂CO₃, MeOH 64–95%	—	1988 (81)
Ph₂C	Me, Et	Me, CH₂(imidazol-4-yl)	IMe, AcOCH₂(imidazol-4-yl)	KOH 47–70%	1) 1N HCl 2) 6N HCl 58%	1989 (77)
(4-Cl-Ph)CH	Et	H	R⁴X = R⁵X = Ph₃BiCO₃	DMF, reflux	2N HCl, Et₂O 26% for 3 steps	1989 (76)
Ph₂C	Et	H	Ph₃BiCO₃	DMF, reflux	2N HCl, Et₂O 60% for 2 steps	1989 (76)
(4-Cl-Ph)CH	Me, Et	Me, Et, nOct, Ph, Bn, 4-Cl-Bn, CH₂CH=CH₂, CH₂CO₂Me, (CH₂)₂CO₂Et	Ph₃BiCO₃	DMF, reflux	1) 2N HCl, Et₂O 2) 6N HCl, reflux 21–54% for 3 steps	1989 (78)
Ph₂C	tBu	H	I¹¹CH₃	KOH	6N HCl, reflux 70% radiochemical yield	1990 (802)
tBuCH	Me	H, Me	RCH=CHCO₂Me R = H, Me, Ph, tBu, CO₂Me, RCH=CHCO₂Ph R = Me, CH=CHPh	DBU, LiBr 44–97%	—	1990 (67)
Ph₂C, (4-Cl-Ph)CH	Me	H, Me, iPr, iBu	Cr(CO)₃(F-Ph)	K₂CO₃, KOH	1) 1N HCl 2) K₂CO₃ 29–37% for 2 steps	1991 (803)
Ph₂C	Bn	H	4-(CO₂tBu)-BnBr 4-(CH₂CO₂tBu)-BnBr 4-(CF₂CO₂tBu)-BnBr 4-(2-tBu-tetrazol-5-yl)-BnBr 4-[CH₂(2-tBu-tetrazol-5-yl)]-BnBr	NaOH, (nBu₄N)₂SO₄	1) TsOH 2) H₂, Pd-C 67–81% overall	1991 (804)

(Continued)

167

Table 3.8 Alkylation of Amino Acid Schiff Bases: Aqueous Conditions or Non-Ionic Strong Bases (see Scheme 3.4) (continued)

R^1	R^2	R^3	Electrophile	Deprotonation Conditions	Hydrolysis Conditions	Reference
Ph_2C	Me, Et	H	H_2C=CHCOR, R = Me, Et, CH_2OAc, MeCH=CHCOMe (cyclize to Pro derivatives)	Cs_2CO_3, 60-78%	—	1991 (131)
Ph_2C	CO_2R^2 = $CONH_2$	Me, iBu, Bn, $(CH_2)_2$SMe, Ph	IMe, BniPr, BnBr, BrCH$_2$CH=CH$_2$, BrCH$_2$CH=CHPh	10N NaOH	aq. HCl 17–86% for 2 steps	1992 (80)
Ph_2C	CO_2R^2 = CN	H	3-Ar-5-CH$_2$PO$_3$Et$_2$-BnBr, Ar = Ph, 4-Cl-Ph, 4-Ph-Ph	NaOH, BnBu$_3$NCl, toluene	HCl, H$_2$O 16–68% overall	1992 (805)
Ph_2C	Et	H	H$_2$C=C(NO$_2$)	BnMe$_3$NOH, 13%	—	1992 (806)
Ph_2C	Et	H	4-(CH$_2$CO$_2$Me)-BnBr	BnMe$_3$NOH, KI	1N HCl 40% for 2 steps	1992 (807)
Ph_2C	Et	H	3-(CH$_2$CO$_2$R)-BnBr	BnMe$_3$NOH, KI	1N HCl then 6N HCl 25% for 2 steps	1992 (642)
Ph_2C	CO_2R^2 = CN	H	BrCH$_2$CCH	K$_2$CO$_3$ 59–89%	1) 6N HCl 2) 70% H$_2$SO$_4$ >70% for 2 steps	1992 (808)
Ph_2C	Me	H	BrCH$_2$(cyclooctatetraene)	NaOH	5% HCl 86%	1993 (809)
Ph_2C	Me	H	3-CF$_3$-BnBr, 3-Me-BnBr, 3-MeO-BnBr, 3-Cl-BnBr	KOH	—	1993 (810)
Ph_2C	Et	H	BrCH$_2$CH=CH$_2$, BrCH$_2$CCH, BrCH$_2$CH=CHPh, BrCH$_2$C(Br)=CH$_2$, BrCH$_2$CO$_2$Et, BnBr, 4-NO$_2$-BnBr	K$_2$CO$_3$, MeCN, reflux	1N HCl, Et$_2$O 33–92% for 2 steps	1993 (82) 1995 (83) 1996 (84)
Ph_2C	Et	H	I(CH$_2$)$_2$Ph	KOtBu, THF, rt	1N HCl, Et$_2$O 80% for 2 steps	1993 (82) 1995 (83) 1996 (84)

Ph$_2$C	Et	R^{4}X = R^{5}X = BrCH$_2$CH=CHPh; BrCH$_2$C(Br)=CH$_2$; BnBr; 4-NO$_2$-BnBr	KOH, MeCN, 0°C	1N HCl, Et$_2$O; 19–72% for 2 steps	1993 (82); 1995 (83); 1996 (84)
Ph$_2$C	Et	R^{4}X = BrCH$_2$C(Br)=CH$_2$; R^{5}X = BnBr, BrCH$_2$CH=CH$_2$, BrCH$_2$CCH, BrCH$_2$CH=CHPh; R^{4}X = BrCH$_2$CCH; R^{5}X = BrCH$_2$CH=CH$_2$; R^{4}X = BrCH$_2$CO$_2$Et; R^{5}X = BrCH$_2$CH=CHPh	KOH, MeCN for 1st alkylation; NaOEt, nBu$_4$NBr, MeCN for 2nd alkylation	1N HCl, Et$_2$O; 23–65% for 2 steps	1993 (82); 1995 (83); 1996 (84)
Ph$_2$C	Et	H$_2$C=CHCO$_2$Et	K$_2$CO$_3$, MeCN, rt	1N HCl, Et$_2$O; 93% for 2 steps	1993 (82); 1995 (83); 1996 (84)
Ph$_2$C	(CH$_2$)$_2$Ph	BrCH$_2$CCH	NaOEt, MeCN, rt	1N HCl, Et$_2$O; 14% for 2 steps	1993 (82); 1995 (83); 1996 (84)
PhCH	Et	R^{4}X = BrCH$_2$CCH; R^{5}X = BrCH$_2$CH=CH$_2$	K$_2$CO$_3$, MeCN for 1st alkylation; KOH, nBu$_4$NBr	15% citric acid; 67% for 3 steps	1995 (83)
Ph$_2$C	tBu	TfO^{11}CH$_3$	KOH, CH$_2$Cl$_2$	6N HCl; 70–90% radioactive yield for 2 steps	1997 (811)
Ph$_2$C	Et	(2-bromomethylquinoxaline structure)	NaOH	6N HCl; 54% for 2 steps	1995 (812)
Ph$_2$C	CO$_2$R^2 = CN	R^{4}X = R^{5}X = (binaphthyl bis(bromomethyl) and biphenyl bis(bromomethyl) structures)	50% NaOH	HCl, dioxane; 83% for 2 steps	1996 (105)
Ph$_2$C	Et	4-CH$_2$PO$_3$Et$_2$-BnBr; 4-CH$_2$Br-BnBr; converted into 4-CH$_2$SO$_3$Na-Bn	BnMe$_3$NOH, KI; 35–80%	1N HCl; 60–100%	1992 (813)

(Continued)

Table 3.8 Alkylation of Amino Acid Schiff Bases: Aqueous Conditions or Non-Ionic Strong Bases (see Scheme 3.4) (continued)

R^1	R^2	R^3	Electrophile	Deprotonation Conditions	Hydrolysis Conditions	Reference
Ph_2C	$CO_2R^2 = CO_2$ resin, CONH-Leu-Oresin, NH-Phe-Leu-Oresin	H	BnBr, BrCH(Ph)$_2$, BrCH$_2$(2-naphthyl), BrCH$_2$CH=CH$_2$, BrCH$_2$CCH	BEMP	aq. HCl, THF 37–55% after resin cleavage	1996 (88)
(2-Me-3-EtOCH$_2$CH$_2$O-4-BnOCH$_2$-4-pyridyl)CH	Bn	Me, Bn	IMe, BnBr, BrCH$_2$CH=CH$_2$, BrCH$_2$CCH, BrCH$_2$CO$_2$Et	LiOH	5% HCl 54–84% for 2 steps	1996 (137)
(4-Cl-Ph)CH	tBu	H	$R^4X = R^5X =$ (biphenyl bis-benzyl bromide structure)	K$_2$CO$_3$, KOH	NH$_2$OH.HCl, NaOAc 75–81% for 2 steps	1996 (102)
Ph_2C	Et	H	4-(CH$_2$PO$_3$Me$_2$)-BnBr, 4-(CHMePO$_3$Me$_2$)-BnBr	BnMe$_3$NOH, KI, dioxane	0.25N HCl 65–71% for 2 steps	1996 (814)
PhCH	Me	Me	4-(CH$_2$PO$_3$Me$_2$)-BnBr, 4-(CHMePO$_3$Me$_2$)-BnBr	K$_2$CO$_3$, KOH BnMe$_3$NCl, CH$_2$Cl$_2$	0.25N HCl 65–71% for 2 steps	1996 (814)
(4-Cl-Ph)CH	tBu	H	$R^4X = R^5X =$ (biphenyl bis-benzyl bromide structure)	K$_2$CO$_3$, KOH nBu$_4$NBr	SiO$_2$, CH$_2$Cl$_2$, MeOH 64% for 2 steps	1997 (815)
ArCH, Ar = Ph, 2-Cl-Ph, 4-Cl-Ph, 2,5-, 3,4-, or 3,5-Cl$_2$-Ph, 4-MeO-Ph, 4-NO$_2$-Ph, (2-naphthyl)	$CO_2R^2 = CO_2$ resin, CONH-Lys-Glu-Phe-resin	Me, (CH$_2$)$_2$$tBu, iPr, iBu, CH_2$OH, Bn, 4-HO-Bn, (CH$_2$)$_4NH_2$, (CH$_2$)$_3$NHC(=NH)NH$_2$, CH$_2$(imidazol-4-yl), CH$_2$(indol-3-yl)	BnBr, BrCH$_2$(2-naphthyl), BrCH$_2$CH=CH$_2$	BEMP	—	1997 (93)
PhCH	Et	Ph	BrCH$_2$CH=CH$_2$	NaOH, nBu$_4$NHSO$_4$ 90%	1N HCl 75%	1997 (816)
Ph_2C	tBu	H	BrCH$_2$CH$_2$F, BrCH$_2$CHFMe, BrCH$_2$CHFEt, BrCH$_2$CHnPr, BrCH$_2$CHFiPr, BrCH$_2$CHFnBu	50% NaOH. CH$_2$Cl$_2$ 18–42%	15% citric acid then 6N HCl 70–78%	1997 (795)

Ph$_2$C						
Ph$_2$C	H	resin	IMe I*i*Pr I*i*Bu I*n*Oct I*c*Hex BnBr BrCH(Me)Ph Br(CH$_2$)$_2$Ph Br(CH$_2$)$_2$OTBS	BEMP	NH$_2$OH.HCl	1997 (89)
Ph$_2$C	Et		Ar^{1}CH=NAr^2 Ar^1 = Ph; Ar^2 = Ph, 2-Cl-Ph, 4-Cl-Ph, 4-Me-Ph, 4-MeO-Ph Ar^1 = 2-Cl-Ph, 4-Cl-Ph, 4-Me-Ph, 4-MeO-Ph; Ar^2 = Ph	NaOH, TEBA 19–66%	—	1997 (817)
Ph$_2$C	H	Et	BrCH$_2$C=CH$_2$ BrCH$_2$CCH	K$_2$CO$_3$, *n*Bu$_4$NBr or KOH	1H HCl 42% overall	1998 (146)
Ph$_2$C	H	resin	MeCH=CHCN MeCH=CHCO$_2$Et MeCH=C(CO$_2$Me)$_2$ HC(Ph)=CHCN H$_2$C=C(Me)CN H$_2$C=C(Me)CO$_2$Et MeC(Me)=C(CO$_2$Et)$_2$ HC(4-NO$_2$-Ph)=CHCO$_2$Et HC(CO$_2$Me)=CHCO$_2$Me H$_2$C=CHR R = COMe, CO$_2$H, CN, SO$_2$Ph, SOPh	BEMP	NH$_2$OH.HCl, then TFA 61–88% for 2 steps	1998 (90)
Ph$_2$C	H	PEG-2000 or PEG-3400	Br*n*Bu Cl*n*Bu I*n*Bu I*i*Bu I*c*Hex BrCH$_2$CH$_2$Ph BnBr BnCl BrCH$_2$CH=CH$_2$ BrCH$_2$CCH BrCH$_2$C(=CH$_2$)CO$_2$Me BrCH$_2$C(Me)CO$_2$Me	K$_2$CO$_3$ or Cs$_2$CO$_3$ 75–97%	KCN/MeOH and TFA	1998 (96) 2000 (97)

(Continued)

Table 3.8 Alkylation of Amino Acid Schiff Bases: Aqueous Conditions or Non-Ionic Strong Bases (see Scheme 3.4) (continued)

R^1	R^2	R^3	Electrophile	Deprotonation Conditions	Hydrolysis Conditions	Reference
(4-Cl-Ph)CH	tBu	H	$R^4X = R^5X =$	KOH, K$_2$CO$_3$ 68%	TFA	1998 (8/8)
Ph$_2$C	tBu	H	IMe IEt BriPr InHex BrnOct BrCH$_2$CH=CH$_2$ BrCH$_2$cPr BnBr H$_2$C=CHCO$_2$Me H$_2$C=CHCOMe	CsOH Ru complex PTC catalyst 60–83%	—	1999 (86)
Ph$_2$C	PEG3400	H	BrnBu iBu BrCH$_2$CH=CH$_2$ BrCH$_2$CH=CPh BnBr BrCH$_2$CH(Me)CO$_2$Me H$_2$C=CHCO$_2$Me H$_2$C=C(Me)CO$_2$Me	Cs$_2$CO$_3$ 75–98%	microwave heating 30–60 min no solvent	2000 (98)
Ph$_2$C	tBu	H	IMe BrEt BrnBu BrCH$_2$CH=CH$_2$ BnBr	KOH, napthyl ketiminium salt as PTC catakyst	40–85%	2000 (87)
Ph$_2$C	NH Rink resin	H	EtI 4-Me-BnBr	BEMP	NH$_2$OH.HCl, then TFA	2001 (94)

(3,4-Cl$_2$-Ph)CH	WANG resin	Me iBu Bn 4-OH-Bn CH$_2$(indol-3-yl) CH$_2$OH (CH$_2$)$_4$NH$_2$ CH$_2$CO$_2$H	(CH$_2$)$_n$Cl; n = 2–5	BTPP, then mild hydrolysis, then DIEA to cyclize to give imino acids, TFA cleavage from resin	7–77% purified	2003 (107)
(3,4-Cl$_2$-Ph)CH	WANG resin	Me Bn	(CH$_2$)$_n$Cl; n = 2–5 then halide displacement with CN, OAc, SAc, S(CH$_2$)$_2$CO$_2$, SBn, NHBn, NHPh, pyrrolidine, N$_2$	BTPP, then halide displacement by nucleophiles, then mild hydrolysis, TFA cleavage from resin	30–74% purified yield	2002 (108)
Ph$_2$C	WANG resin	H	(CH$_2$)$_n$Cl/Br; n = 2–5 ClCH$_2$(1,2-Ph)CH$_2$Cl BrCH$_2$CH(Me)CH$_2$Cl then hydrolysis or halide displacement with CN, SBn, or cyclization with BTPP to Pro 1-amino-1-alkanoic acid analogs or N-deprotection and cyclization to Pro analogs	BTPP, then halide displacement by nucleophiles, then mild hydrolysis, TFA cleavage from resin	23–79% purified yield	2003 (109)
tBuCH	Me	H	H$_2$C=CHCO$_2$Me MeCH=CHCO$_2$Me MeO$_2$CCH=CHCO$_2$Me PhCH=CHCOMe MeCH=CHCOMe (Me)$_2$C=CHCOMe cyclohex-2-enone	proaza-phosphatrane base, isobutyronitrile	72–97%	2002 (68)
Ph$_2$C	Met	H	3,4-[-COCH=CHC(Me)$_2$-]-BnBr 3,4-[-COCH=CHO-]-BnBr 3,4-[-COCH=CHS-]-BnBr 3,4-[-SO$_2$CH=CHO-]-BnBr	NaOH, nBu$_4$NHSO$_4$ 42–60%	–1N HCl 68–100%	2003 (819)

(Continued)

Table 3.8 Alkylation of Amino Acid Schiff Bases: Aqueous Conditions or Non-Ionic Strong Bases (see Scheme 3.4) (continued)

R¹	R²	R³	Electrophile	Deprotonation Conditions	Hydrolysis Conditions	Reference
Ni (II) complex formed with (2-pyridyl)CONH-(2-COPh-Ph)	Ni complex	H	EtI iPrI BnBr BrCH₂CH=CH₂ ClCH(Ph)₂	NaOH, nBu₄NX CH₂Cl₂ 83 to >95%	aq. HCl	2003 (132)
Ni (II) complex formed with (2-pyridyl)CONH-(2-COPh-Ph)	Ni complex	Bn CH₂CH=CH₂ Et	BnBr BrCH₂CH=CH₂ BnBr	NaOH, nBu₄NX CH₂Cl₂ or NaOH/DMF or NaH/DMF 32 to >95%	aq. HCl	2003 (132)
Ph₂C (4-MeO-Ph)CH	tBu	H	PhBr or Cl 2-Me-PhBr or Cl 2-MeO-PhBr 4-MeO-PhBr or Cl 3,4-(OCH₂O)-PhBr or Cl 4-F-PhBr or Cl 4-CN-PhBr or Cl 4-CO₂Me-PhBr 4-CF₃-PhBr or Cl 4-Ph-PhBr or Cl 4-PhO-PhBr 1-Br-naphthyl 2-Br-naphthyl 3-Br-pyridine 3-Cl-pyridine	arylation of benzophenone or p-MeO-benzaldehyde Schiff base of Gly-OtBu with Ar X and Pd(dba)₂, P(tBu)₃, K₃PO₄	67–92% arylation	2001 (117)

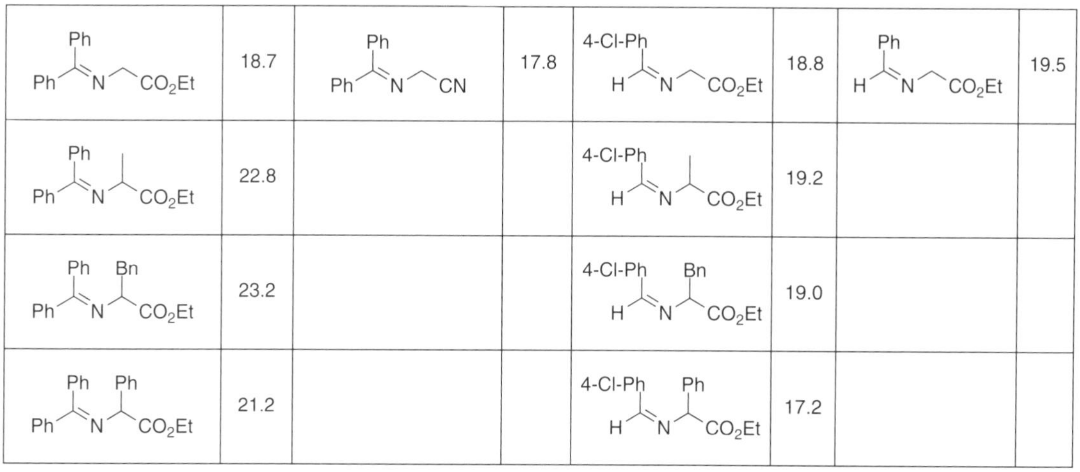

Structure	pK_a	Structure	pK_a	Structure	pK_a	Structure	pK_a
Ph–C(Ph)=N–CH₂–CO₂Et	18.7	Ph–C(Ph)=N–CH₂–CN	17.8	4-Cl-Ph–CH=N–CH₂–CO₂Et	18.8	Ph–CH=N–CH₂–CO₂Et	19.5
Ph–C(Ph)=N–CH(CH₃)–CO₂Et	22.8			4-Cl-Ph–CH=N–CH(CH₃)–CO₂Et	19.2		
Ph–C(Ph)=N–CH(Bn)–CO₂Et	23.2			4-Cl-Ph–CH=N–CH(Bn)–CO₂Et	19.0		
Ph–C(Ph)=N–CH(Ph)–CO₂Et	21.2			4-Cl-Ph–CH=N–CH(Ph)–CO₂Et	17.2		

Scheme 3.5 pK_a Values of Ketimines and Aldimines in DMSO (75).

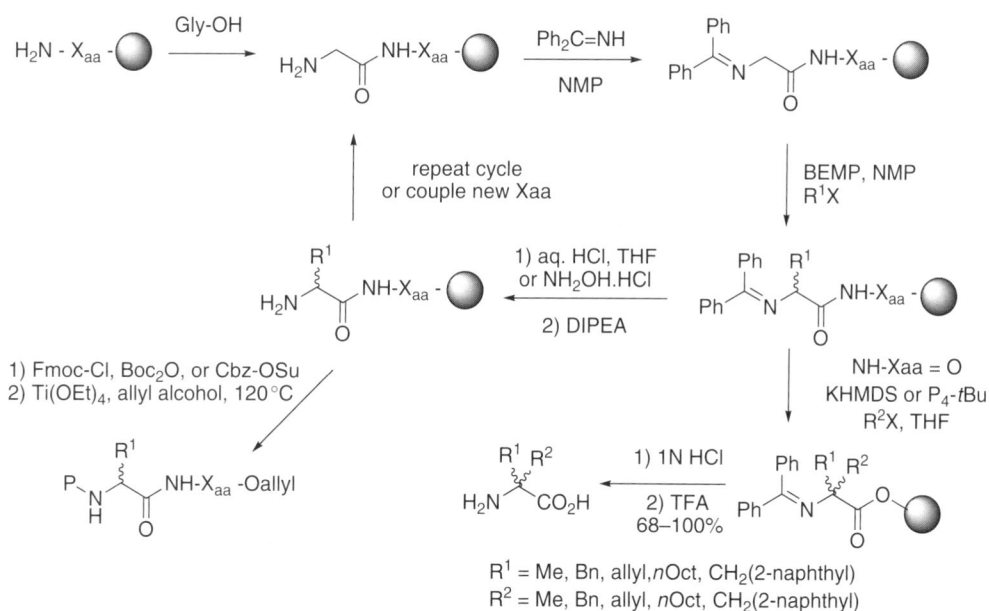

R^1 = Me, Bn, allyl, nOct, CH₂(2-naphthyl)
R^2 = Me, Bn, allyl, nOct, CH₂(2-naphthyl)

Scheme 3.6 Solid Phase Schiff Base Alkylation for Peptide Synthesis (88, 91).

were obtained. No diastereoselectivity was reported for alkylations where optically active amino acids were present. The BEMP deprotonated resin-linked benzophenone Schiff base of Gly has also been added to typical Michael acceptors (acrylates, acrylonitrile, methyl vinyl ketone) in good yield (90). The benzophenone ketimine of Gly esterified to a solid support can be consecutively deprotonated and alkylated to give α,α-disubstituted amino acids, using BEMP for the first deprotonation/alkylation. The second deprotonation/alkylation requires a stronger base (KHMDS or P₄-tBu) under anhydrous conditions (91, 92). α,α-Disubstituted

amino acids have also been synthesized by alkylation of various aryl aldimine derivatives of amino acids attached to a solid support, using BEMP as base (93). Benzophenone imine derivatives of Gly residues attached to the Rink resin via an amide linkage can be alkylated, leading to amino amide products after resin cleavage (94). The attachment point of the resin linkage has been reversed, using an imine formed from a benzaldehyde functionalized resin and Gly-OtBu instead of a C-terminal ester/amide linkage. Alkylations with a variety of benzyl and alkyl halides under PTC conditions, followed by resin cleavage with 1 N HCl and

N-benzoylation, gave Bz-Xaa-O*t*Bu products with 50–92% yield (*95*).

Soluble polymer supports have been employed, with the benzophenone Schiff base of Gly esterified to a PEG-2000 or PEG-3400 resin. The PEG acted as a phase transfer catalyst, allowing for deprotonation with K_2CO_3 or Cs_2CO_3 in acetonitrile. Alkylations with non-activated alkyl halides such as *n*-butyl chloride or isobutyl iodide proceeded in good (81–82%) yield with only 1.5 equivalents of electrophile and 3 equivalents of base. The PEG ester linkage was cleaved from the amino acid with KCN in MeOH (*96, 97*). It is possible to use the PEG as both solvent and polymer support, simply mixing the PEGylated Gly Schiff base with Cs_2CO_3 and electrophile with microwave heating. Yields of 75–98% were obtained with 30–60 min reaction times (*98*).

The original benzaldehyde aldimine of Stork has found favor in recent years for intramolecular dialkylation by dihalides, generating either α,α-disubstituted cyclic amino acids from *C*-dialkylation (e.g., *99–103*), or imino acids via sequential *C*-alkylation followed by *N*-alkylation (e.g., *99*). Dimeric bis(α-amino acids) are produced by reaction of two equivalents of the benzophenone imine of Gly-OEt with dihalides (*104*). Glycine benzophenone ketimine can also be used for forming cyclic amino acids, using PTC deprotonation conditions (*105*). Similarly, the benzophenone Schiff base of aminoacetonitrile was used for a tandem allylation/$S_{N'}$ cyclization with 1,4-dichloro-2-butene to give a 1-amino-cyclopropanecarboxylic acid derivative (*106*). Resin-bound amino acids have been converted into aldimines of 3,4-dichlorobenzaldehyde, deprotonated with the non-ionic base BTPP and alkylated with dihaloalkanes, $Br(CH_2)_nX$ (X = Cl, Br; n = 2–5). The imine intermediates were then hydrolyzed to a free amino group while still attached to the resin, with *N*-alkylation/cyclization induced by treatment with a 10% solution of DIEA. The resulting imino acids were *N*-protected with an Fmoc group, then cleaved from the resin in 26–73% purified yields (*107*). A library of 48 α-substituted prolines was prepared.

Resin-bound non-cyclized α-(ω-haloalkyl)-Ala and -Phe were also *N*-Fmoc protected and cleaved from the resin (*108*). The side-chain halide was then displaced using tetrabutylammonium cyanide, acetate, thioacetate, ethyl 3-mercaptopropanoate, benzyl mercaptide, azide, benzylamine, aniline, or pyrrolidine. Schiff base hydrolysis, *N*-acylation, and cleavage from the resin gave a wide range of α-substituted-Ala and -Phe analogs

with diverse functionality placed at 2–5-carbon chain atoms from the α-center (*108*). The same reaction sequences (alkylation with α,ω-dihaloakanes, followed by deprotection, halide displacement with thiols or cyanide, *N*-cyclization to Pro, or *C*-cyclization to 1-aminocycloalkanoic acids) has been carried out on benzophenone Gly Schiff base derivatives, using BTPP as base (*109*). Other side-chain derivatizations are possible, such as cyclization of the side chains of diallylated Schiff base derivatives to generate bicyclic and tricyclic α,α-disubstituted α-amino acids (*110*).

A variety of allylic electrophiles have been reacted with Gly and Ala ketimines or aldimines formed from benzophenone or 4-chlorobenzaldehyde (*111, 112*). The reaction could be carried out under neutral conditions by using allylic carbonates with Pd(dppe)$_2$, with 65–95% yields (see Scheme 3.7, Table 3.9) (*111, 112*). Palladium or molybdenum catalysts have been used in combination with basic conditions (LDA, NaH, or BSA) to add allyl acetates in 20–92% yield (*111, 113*). Allylation by alkylation with substituted allyl acetates is possible using *N,O*-bis(trimethylsilylacetamide) (BSA) and CsOAc as base, proceeding with moderate diastereoselectivity at the β-center (22–68% de) (*114*). Allyl esters of Schiff bases derived from amino acids undergo an intramolecular palladium-catalyzed α-allylation via an Ireland–Claisen-like rearrangement in the presence of base, resulting in allylglycine derivatives. Rearrangement yields of 72–90% were obtained for the 4-chlorobenzaldehyde imines of Gly, Ala, Val, Phg, and Phe with KO*t*Bu as base (*115*). Allenes have been catalytically carbopalladated with a palladium catalyst and a vinyl or aryl halide, and the intermediate π-allyl complex trapped by the lithium anion of glycine ketimine in 35–56% yield. Phosphonates of γ-hydroxyallenes were also used for Pd-catalyzed alkylations of the benzophenone Schiff base of Gly-OMe (*116*).

In 2001, 14 different α-arylglycines were produced by a Pd-catalyzed reaction of aryl halides with the benzophenone or *p*-methoxybenzaldehyde imines of Gly-O*t*Bu, with 67–92% yields (*117*). *N,N*-Dimethyl Gly-O*t*Bu also reacted (one example). A palladium-catalyzed reaction was used to introduce a styryl or 1,3-diene side chain to the benzophenone Schiff base of methyl glycinate (*46*).

Proline derivatives can be produced by a number of methods. Schiff bases of amino acids form 1,3-dipolar species in the presence of Lewis acid catalysts and triethylamine, or by heating. These can be trapped via

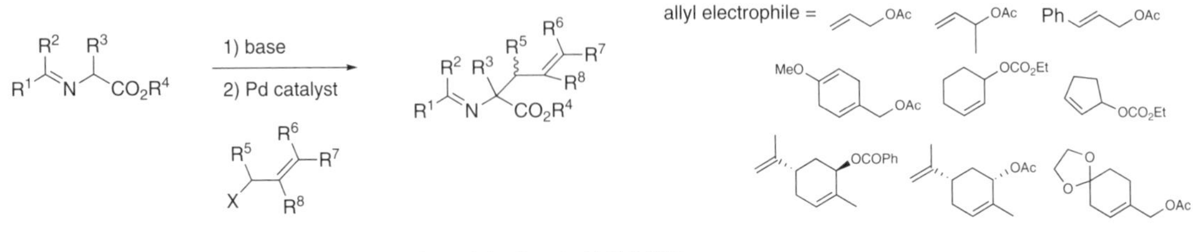

Scheme 3.7 Palladium Catalyzed Allylation of Amino Acid Schiff Bases.

Table 3.9 Palladium-Catalyzed Allylation of Amino Acid Schiff Bases (see Scheme 3.7)w

Schiff Base Substituents				Electrophile Substituents					Base, Catalyst	Reference
R^1	R^2	R^3	R^4	X	R^5	R^6	R^7	R^8		
(4-Cl-Ph)CH	H	H	Et	OAc	H	H	H	H Me	NaH, THF Pd(dppe)$_2$	1986 (*112*)
	Ph	H	Me	OCO$_2$Et	H	H	H	H Me	THF Pd(dppe)$_2$	1986 (*112*)
	Ph	H	CO$_2$R^4 = CN	OCO$_2$Et	H	H	CH$_2$OAc	H	THF Pd(dppe)$_2$	1986 (*112*)
(4-Cl-Ph)CH	H	H, Me, iPr, Ph, Bn	CH$_2$CH=CH$_2$	intramolecular from ester	H	H	H	H	KO,Bu Pd(dppe)$_2$	1988 (*115*)
	Ph	H, Me	Me	OCO$_2$Et	H	H	H	H	THF Pd(dppe)$_2$	1988 (*111*)
				OCO$_2$Et	H	Me	H	H		
				OCO$_2$Et	Me	H	H	H		
				OCO$_2$Et	H	H	H	Me		
				OCO$_2$Et	H	CH$_2$OAc	H	H		
				OCO$_2$Et	H	H	Ph	H		
				⌬—OCO$_2$Et (cyclopentenyl carbonate)						
(4-Cl-Ph)CH	H	H, Me	Me	Cl	H	H	H	Cl	LDA, NaH or BSA THF Pd(dba)$_2$, Pd(dppe)$_2$, or Pd(PPh$_3$)$_4$	1988 (*111*)
				OAc	H	H	Ph	H		
				OCO$_2$Et	Me	H	H	H		
				OAc	H	H	H	H		
				MeO—C$_6$H$_3$(OPh)—CH$_2$OAc (structure)						
				cyclohexenyl OPh (structure)						
				dioxolane-fused CH$_2$OAc / OCOPh (structure)						
				terpenyl OAc (structure)						

(*Continued*)

Table 3.9 Palladium-Catalyzed Allylation of Amino Acid Schiff Bases (see Scheme 3.7) (continued)

Schiff Base Substituents				Electrophile Substituents					Base, Catalyst	Reference
R^1	R^2	R^3	R^4	X	R^5	R^6	R^7	R^8		
Ph	Ph	H	Me	OPO_3Et_2	H	$R^6 = R^7 = =CH_2$		H	nBuLi $Pd(dba)_2$, PPh_3	1988 (*116*)
				OPO_3Et_2	H	$R^6 = R^7 = =CHMe$		H		
				OPO_3Et_2	Me	$R^6 = R^7 = =CH_2$		H		
				OPO_3Et_2	H	$R^6 = R^7 = =C(Me)_2$		H		
Ph	Ph	H	$CO_2R^4 = CN$	Cl	H	H	CH_2Cl	H	NaH, THF $Pd(dba)_2$, PPh_3	1995 (*106*)
Ph	Ph	H	Et	OAc	Me	H	Me	H	BSA, KOAc $[Pd(allyl)Cl]_3$	1995 (*114*)
Ph	Ph	H	Et	Br	H	H	H	CH_2X	LiHMDS, THF 5% $Pd(PPh_3)_4$	1995 (*104*)
				Br	H	H	H	$C(=CH_2)CH_2X$		
				Br	H	H	CH_2X	H		
Ph	Ph	H	Et	OAc	H	H	PO_3Me_2	H	BSA $Pd(dba)_2$	1997 (*113*)
				OAc	H	H	PO_3Et_2	H		
				OAc	H	H	PO_3iPr_2	H		
				OAc	H	Me	PO_3Me_2	H		
				OAc	Me	H	PO_3iPr_2	H		

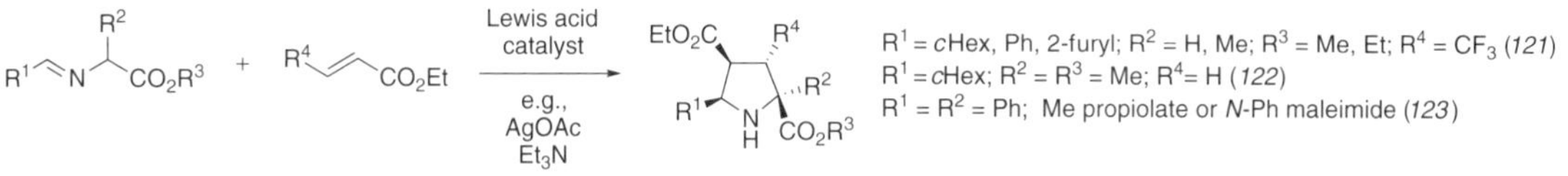

Scheme 3.8 Dipolar Cycloadditions of Amino Acid Schiff Bases.

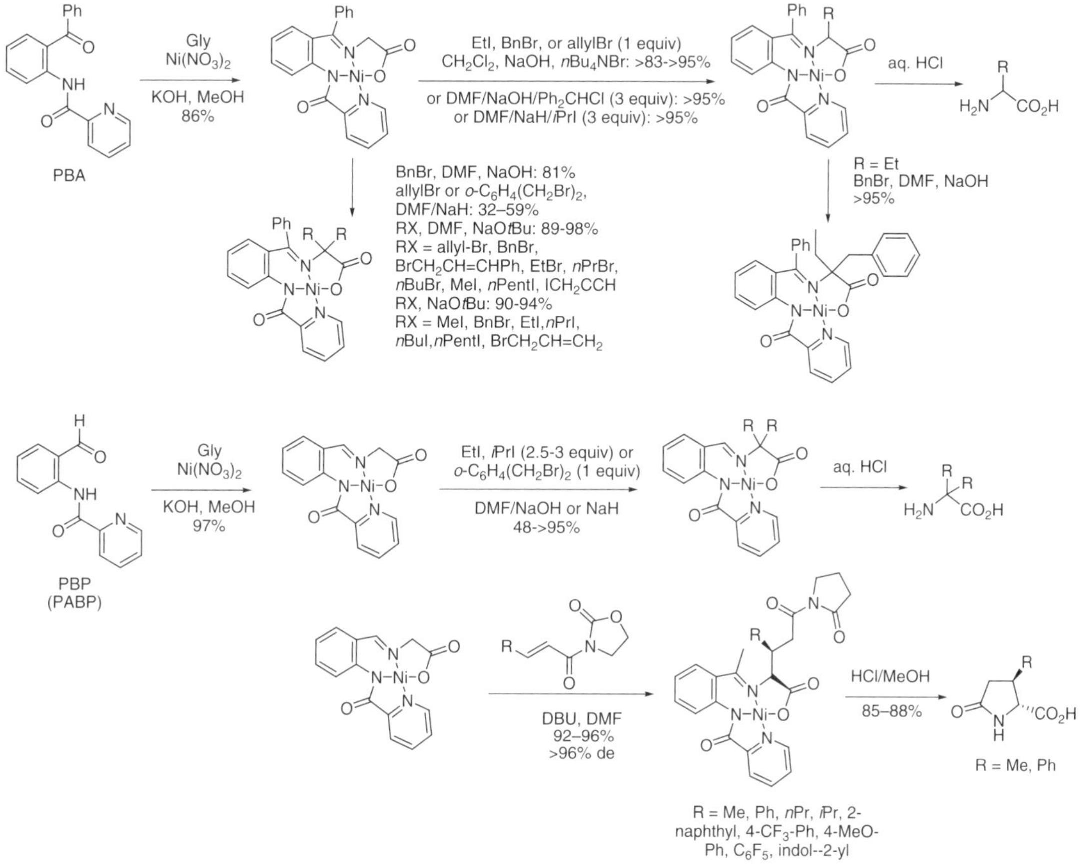

Scheme 3.9 Alkylation of Ni(II) Schiff Base Complex (132–136).

cycloadditions with dipolarophiles such as N-phenyl-maleimide, giving multi-substituted Pro derivatives (see Scheme 3.8) (118–125). The reaction rate and stereo/regioselectivity is influenced by the type of metal salt used for catalysis (126–128). Closely related reactions include the synthesis of α-substituted proline derivatives via Michael addition of amino acid Schiff bases to α,β-unsaturated ketones/aldehydes, followed by a 5-endo-trig cyclization (129, 130). Michael addition of the benzo-phenone Schiff base of Gly-OMe to α,β-unsaturated ketones, followed by N-deprotection, cyclization/imine formation, and imine reduction by hydrogenation pro-duced 3-alkyl- or 3,5-dialkyl prolines (131).

Belokon et al. reported PTC alkylations of two Schiff base Ni complexes in 2003. One or two alkyl groups (identical or different) were added to a Ni(II) complex

generated from the ketimine Schiff base formed from Gly and the 2-benzoyl-phenyl amide of pyridine-2-carboxy-lic acid (PBP, or 2-[N-(α-picolyl)amino]benzophenone, PABP) (see Scheme 3.9). Alkylations were achieved in CH_2Cl_2 with Bu_4NX or β-naphthol as the phase-transfer catalyst and NaOH as base, generally giving yields of >95% for monoalkyl products. Bis-alkylated product yields were generally lower, and required NaOH/DMF or NaH/DMF for deprotonation (132). The PBP com-plex reacted with one bromide of o-dibromoxylylene under PTC conditions in 97% yield. Cyclization with the second bromide was achieved in 93% yield by treat-ment with NaOtBu to give 2-aminoindane-2-carboxylic acid. The cyclic product was also obtained in one step using NaOH in DMF, but in only 67% yield (133). Achiral α,α-disubstituted amino acids were prepared in better yield from the less hindered aldimine complex formed from Gly and the 2-formyl-phenyl amide of pyridine-2-carboxylic acid (PBA), but monoalkylated products could not be prepared (*132*). However, anhydrous dialkylations of the PBP complex with NaOtBu as base proceeded in high yield for a variety of alkyl iodides. Sterically hindered isopropyl iodide and isobutyl iodide gave only monoalkylated products, as did attempted Michael additions with ethyl acrylate or acrylonitrile (*134, 135*). In contrast, a 2-acetyl analog of PBP formed a Ni(II) Gly complex that successfully added to a number of α,β-unsaturated acids derivatized as oxazolidin-2-one amides with high yield (92–96%) and very high diastereoselectivity (>96% de), using DBU as base. Hydrolysis produced β-substituted pyroglutamic acids (*136*). After alkylation, the Ni(II) complexes were decomposed by treatment with diluted methanolic HCl, accompanied by a solution color change from red to blue (*132, 134*). Asymmetric syntheses are possible by employing chiral PTCs (see Section 7.11).

A pyridoxal mimic has been used to form an aldimine with Ala-OBn and Phe-OBn, with the rationale that the pyridine ring should stabilize formation of the α-anion and the alkoxy substituent might form a chelate with the metal ion. Indeed, α-alkylation was possible with LiOH as base, and a clear preference was shown for Li over other cations (*137*). Schiff base-like amidines of amino acid methyl esters, Me₂NCH=N-CH(R)-CO₂Me, can be α-*C*-alkylated by deprotonation with LDA or KOtBu (*138*); *N*-alkylation is also possible by refluxing with a reactive electrophile in toluene (*139*).

Schiff base derivatives have other uses. Amino acid benzophenone ketimines were reductively *N*-alkylated by treatment with RCHO/NaBH₃CN, giving the *N*-alkyl,*N*-diphenylmethyl amino acid. The diphenylmethyl group was removed by hydrogenation (*140*). Schiff base formation has been used to catalyze the racemization of optically active amino acids. The amino acid is heated in acetic acid in the presence of 0.05 equivalent of an aromatic or aliphatic aldehyde (see Section 12.1.7 for further examples) (*141*).

3.5 Alkylation of Isocyanoacetates

The alkylation of isocyanoacetates is very similar to the alkylation of glycine Schiff bases, as the α-anion

stability is increased by resonance stabilization with the isocyano group. Dialkylation of the synthon is again possible (see Scheme 3.10, Table 3.10). The use of isocyanoacetates as glycine equivalents is a relatively recent development, with the first report of their alkylation in 1971 (*138*). Dialkylated isocyanoacetic esters were obtained from ethyl isocyanoacetate using 2 equivalents of electrophile or one equivalent of divalent electrophile, with KOtBu or NaH as base (*138, 142*). Monoalkylated products were obtained by using the *t*-butyl ester of isocyanoacetate instead of the ethyl ester (*138, 142*). PTC conditions have also been reported for alkylations (*143*), and have been used to prepare a number of cyclic amino acids, including *cis*-1-amino-2-phenyl-cyclobutane-1-carboxylic acid, by dialkylation using bis(bromomethyl)aryl electrophiles (*144*). PTC conditions were found to be effective for dialkylation with a dibromo substrate which failed to react under a range of other conditions (*145*). Cyclic amino acids have also been prepared via bis(allylation) or bis(homoallylation), followed by a ruthenium-catalyzed metathesis cyclization of the two alkene side chains (*146, 147*). Ethyl isocyanopropionate or other isocyanoalkanoates can also be alkylated, leading to α,α-disubstituted amino acids (see Scheme 3.11) (*142, 148–151, 397, 520, 525–527*). The isocyanoacetates can be converted into *N*-formylamino esters, *N*-formylamino acids, and amino acids. Reviews of the use of α-metallated isocyanides in organic synthesis, including the preparation of amino acids, appeared in 1974 (*152*) and 1977 (*153*).

Electrophiles other than organohalides have been used. Metallated isocyanocarboxylates, prepared by deprotonation with NaCN in EtOH (*154–156*), *n*-BuLi, NaH, or KOtBu in THF (*157*) or Cu₂O in benzene (*158*), react with aldehydes and ketones to give an oxazoline intermediate (see Scheme 3.12, Table 3.11). Hydrolysis of this intermediate gives *N*-formyl-β-hydroxy-α-amino esters (*157*), while treatment with base produces dide-hydroamino acids (*157*) (see Section 4.2.2) and hydrogenation of the aryl adducts leads to β-substituted amino acids (*159*). ZnCl₂ or CuCl/Et₃N also catalyze the aldol reaction (*159*), while Ag catalysis promotes high "*cis*" diastereoselectivity in aldol reactions with aryl fluoroalkyl ketones (*160*). The oxazoline intermediate can be treated with P₄S₁₀ to give thiazolines, with acid hydrolysis then producing β-mercapto amino acids (*161*). Similarly, α-cyanoacetamide was reacted with aldehydes in the presence of bases such as KOH

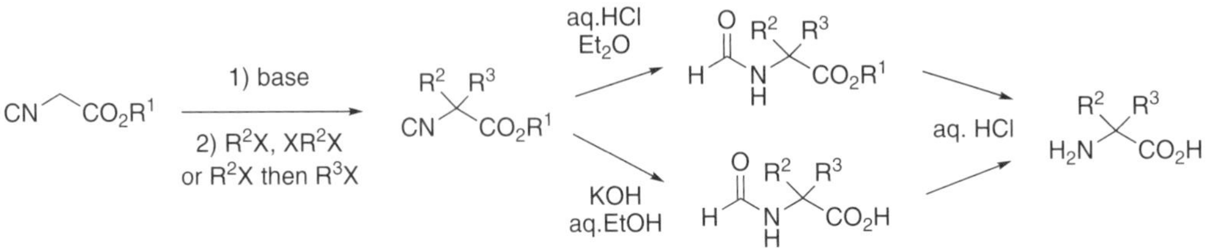

Scheme 3.10 Alkylation of Isocyanocarboxylates.

Table 3.10 Alkylation of Isocyanocarboxylates (see Scheme 3.10)

R¹	R², R³	Deprotonation Conditions	Reference
Et	$R^2 = R^3 = n$Bu, Bn, -(CH$_2$)$_2$-, -(CH$_2$)$_4$-, -(CH$_2$)O(CH$_2$)$_2$-	NaH, DMSO or KOtBu, THF 20–71%	1971 (138) 1975 (142)
tBu	$R^2 = R^3 = $ CH$_2$(indol-3yl) $R^3 = $ H; $R^2 = $ Me, iPr, iBu, CH$_2$C=CH$_2$, Bn	NaH, DMSO or KOtBu, THF 35–76%	1975 (142)
Et	$R^2 = R^3 = $ -CHDCHD-	36% dialkylation and hydrolysis	1983 (820)
Et	$R^2 = R^3 = $ -CH$_2$CH(cPr)- from Br-CH$_2$CH(cPr)Br	NaH 20–27%	1985 (821)
Et	$R^2 = R^3 = $ -CH(R⁴)C(R⁵) (R⁶)- from CH(R⁴) (-O-)C(R⁵) (R⁶) epoxides	1) nBuLi, BF$_3$.Et$_2$O: 54–80% 2) MsCl: 77–90% 3) KOtBu: 33–67%	1986 (822)
Et	$R^2 = R^3 = $ -CH(Me)CH$_2$-	Na, Et$_2$O 59%	1986 (823)
Et	$R^2 = R^3 = $ CH$_2$CCH	80% dialkylation	1997 (143)
Et	$R^2 = R^3 = $ 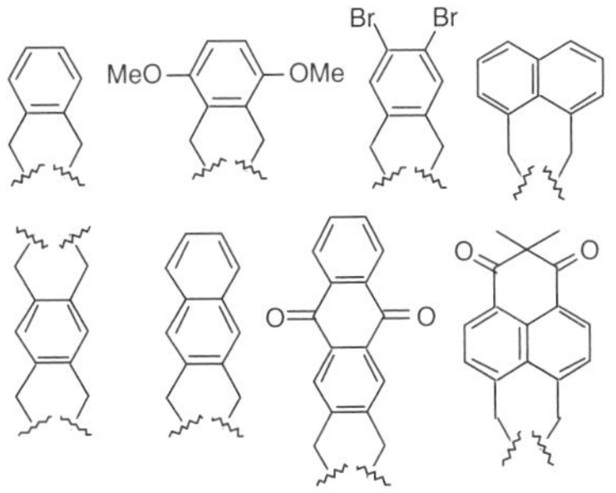 [chemical structures]	K$_2$CO$_3$, MeCN, PTC 40–89%	1997 (144)
Et	$R^2 = $ 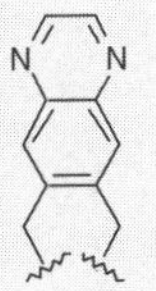 [chemical structure]	K$_2$CO$_3$, PTC 38%	1997 (145)
Et	$R^2 = R^3 = $ CH$_2$C=CH$_2$, (CH$_2$)$_2$C=CH$_2$	NaH or PTC conditions 62–88%	1998 (147)
Et	$R^2 = R^3 = $ -CH$_2$C(=CH$_2$)C(=CH$_2$)CH$_2$- -(CH$_2$)$_2$C(=CH$_2$)C(=CH$_2$) (CH$_2$)$_2$- from diiodo bis-electrophiles	NaH, DMSO 30–42%	1998 (151)
Et	$R^2 = $ CH$_2$CH=CH$_2$; $R^3 = $ CH$_2$CCH (then metathesis, then Diels-Alder)	42% alkylation	1998 (146)
Et	$R^2 = R^3 = $ CH(Ph)CH$_2$CH$_2$-	K$_2$CO$_3$, MeCN, PTC 50%	2005 (824)

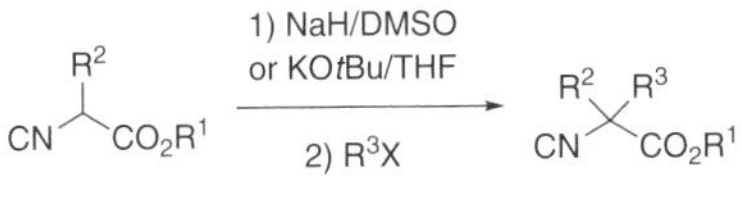

[Reaction scheme: starting material R^2-CH(CN)(CO$_2$R^1) → 1) NaH/DMSO or KOtBu/THF; 2) R³X → product R^2,R^3-C(CN)(CO$_2$R^1)]

R^1 =Et; R^2 =CH$_2$CH$_2$SMe; $R^3 = $ nBu, nHex, Bn, CH$_2$SPh, CH$_2$CO$_2$Et: 31–79% (149)
$R^1 = R^2 = $ Me; $R^3 = $ N-Ts-imidazoleCH$_2$OAc,indole-CH$_2$N⁺Me$_3$I⁻, 3,4-(MeO)$_2$-BnCl: 63–92% (149)
$R^1 = $ Me; $R^2 = $ 3,4-(MeO)$_2$-Bn; $R^3 = $ Me; iPr: 78% (149)
$R^1 = $ Me; $R^2 = $ Me, Ph, Bn; $R^3 = $ (CH$_2$)$_n$Br, n = 3,4,5 (150)
$R^1 = $ Bn; $R^2 = $ Me; $R^3 = $ (CH$_2$)$_3$NMeTs, (CH$_2$)$_3$NMe$_2$, (CH$_2$)$_3$N(Ts)(CH$_2$)$_3$NHAc: 30–74% (148)

Scheme 3.11 Alkylation of Isocyanoalkylcarboxylates.

and Na$_2$CO$_3$, with acid hydrolysis of the oxazoline intermediate giving the free *threo* β-hydroxyamino acids in good overall yield (68–90%) (162). A range of stereochemically defined β,β-disubstituted-β-hydroxy-α-amino acids were synthesized by aldol addition of methyl isocyanoacetate to prochiral ketones. The effects of metal catalyst and base on the reaction diastereoselectivity were systematically examined, as were the steric and electronic properties of the ketone (163). Ethyl isocyanoacetate attacks the carbonyl of α,β-unsaturated ketones in the presence of ZnCl$_2$ to give, after hydrolysis of the initial oxazoline adduct, β-hydroxy-allylglycine

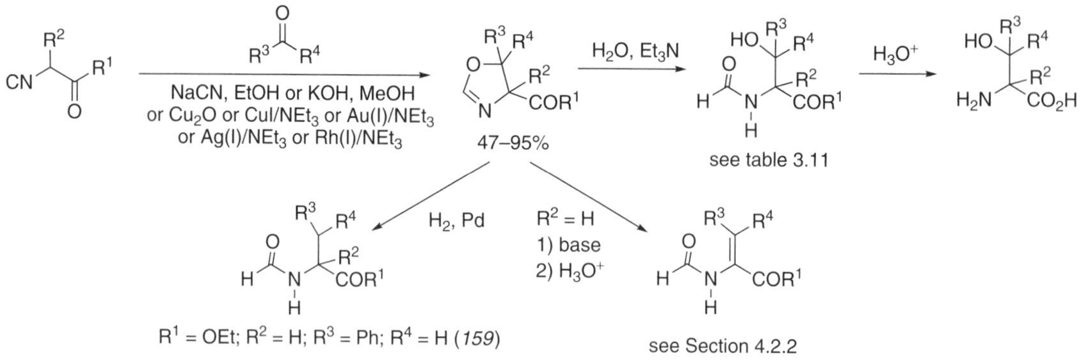

Scheme 3.12 reagents: NaCN, EtOH or KOH, MeOH or Cu₂O or CuI/NEt₃ or Au(I)/NEt₃ or Ag(I)/NEt₃ or Rh(I)/NEt₃. Product (oxazoline) 47–95%. Then H₂O, Et₃N; H₃O⁺ (see table 3.11). Branch: H₂, Pd. R² = H 1) base 2) H₃O⁺ (see Section 4.2.2).

R^1 = OEt; R^2 = H; R^3 = Ph; R^4 = H (*159*)

Scheme 3.12 Addition of Isocyanocarboxylates to Carbonyl Compounds.

Table 3.11 Aldol Reaction of Isocyanocarboxylates (see Scheme 3.12)

R^1	R^2	R^3	R^4	Reference
OEt	H	H	H, Me, Ph, iPr, CH₂CH=CH₂	1970 (*156*)
		Me	Me	
		(CH₂)₅-R⁴	(CH₂)₅-R³	
	Me	H	H, Bn	
		Me	Me	
OEt	H	H	H, Me, iPr, Ph, 4-Me-Ph, 4-MeO-Ph	1972 (*155*)
	H	Me	Me	
	H	(CH₂)₅-R⁴	(CH₂)₅-R³	
	Me	H	H, Ph, 3,4-(MeO)₂-Ph	
	(CH₂)₂SMe	H	Ph	
NH₂	H	H	H, Me, Et, iPr, CO₂H, Ph, 4-NO₂-Ph, 3-pyridyl	1979 (*162*)
OEt	H	CHCl₂	H	1981 (*165*)
		CH₂Cl	Me, Et, Ph, CH₂Cl	1987 (*166*)
		CH(Me)Cl	Me	
		CH(Et)Cl	H	
OEt	H	H	CH₂OMe	1982 (*154*)
OMe	H	Ph	CF₃, CClF₂, C₂F₅	1994 (*160*)
		C₆F₅, 4-MeO-Ph	CF₃	
OEt	H	H	CH=CH₂	1995 (*164*)
	H	H	CH=CHMe	
	H	H	CH=CHEt	
	H	H	CH=CHnPr	
	H	Me	CH=CH₂	
	H	H	C(Me)=CH₂	
OMe	H	CH₂Cl	CHCl₂	1997 (*163*)
		Ph	Me, CF₃, CClF₂, C₄F₉, CN	
		C₆F₅	Me, CF₃	
		Me	CH₂OMe, CH₂F, CH₂Cl, CHCl₂, CCl₃, CF₃, C₃F₇, Et, nPr, cHex	
		CF₃	Me, nHept, nOct, cHex, CCPh, CH₂CO₂Et, 4-CF₃-Ph, 4-MeO-Ph	

esters. These were rearranged to 5-hydroxyvinylglycine products via acetylation of the hydroxy group followed by treatment with Pd(II) in benzene, albeit in only 12–43% yield (*164*).

Ethyl isocyanoacetate was added to the carbonyl group of α-chloro carbonyl compounds in the presence of CuO, producing a chloro-substituted oxazoline

intermediate in 65–85% yield (see Scheme 3.13). Reductive elimination with zinc gave N-formyl-β, γ-unsaturated-α-amino acid derivatives, which were hydrolysed with 6 N HCl (*165*, *166*).

Acrylates undergo Michael addition of ethyl isocyanoacetate and ethyl isocyanopropionate in the presence of catalytic sodium ethoxide, giving glutamic acid

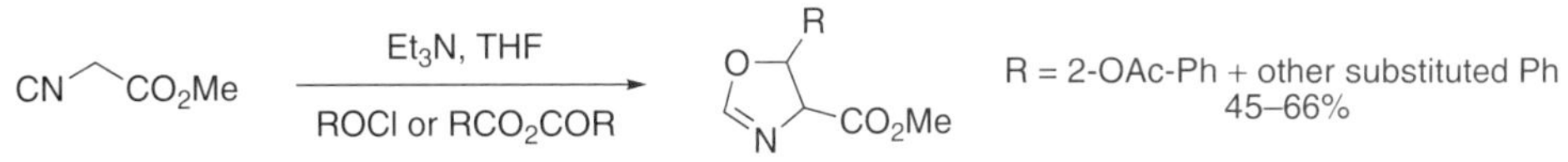

Scheme 3.13 Addition of Isocyanocarboxylates to Acyl Chlorides (*165*).

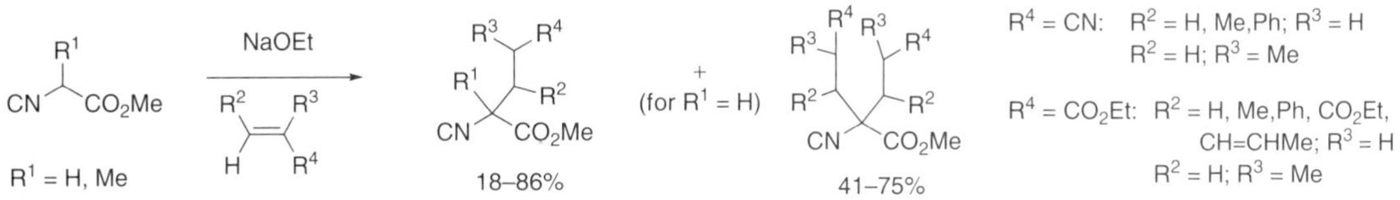

Scheme 3.14 Michael Reactions of Isocyanocarboxylates (*167,168*).

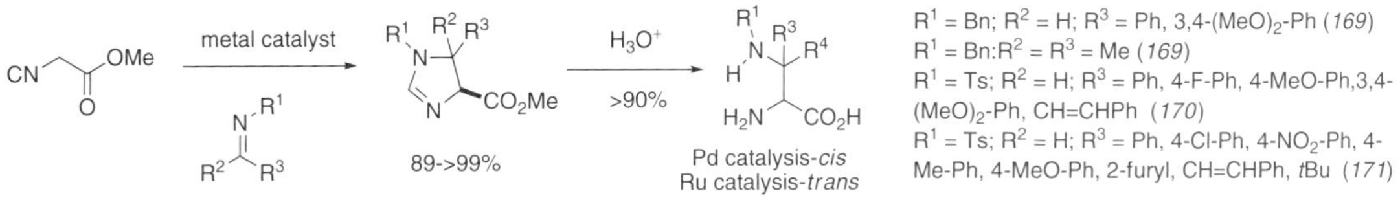

Scheme 3.15 Addition of Isocyanocarboxylates to Imine.

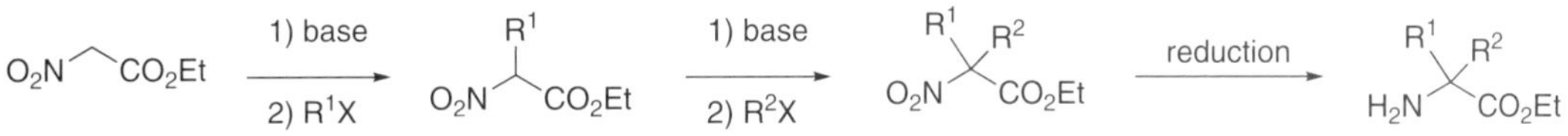

R^1X = CH$_2$=CHCH$_2$OAc; Pd(0) / alumina; 95% (*833*)
R^1X = CH$_2$=CHCH$_2$OR, CH$_2$=C(Me)CH$_2$OR, AcOCH$_2$CH=CHCH$_2$OR: R = Ac, Ph, CO$_2$Et,; Pd(0); 45–90% (*832*)
R^1X = R^2X = 1,1'-bis(halomethyl)ferrocene: 54% (*174*)
R^1X = ClCH$_2$[2,4,5-(BnO)$_3$-Ph]: 8%; H$_2$/PtO$_2$ for reduction: 20% (*834*)
R^1X = R^2X = ClCH$_2$C(=CH$_2$)Cl; H$_2$C=C(OEt)CH$_2$OCO$_2$Me; KH, Pd(0); 71–100% (*172*)
R^1X = R^2X = H$_2$C=C(Br)CH$_2$OCO$_2$Et, H$_2$C=CHCH$_2$OCO$_2$Et; 55–85% (*173*)
R^1X = R^2X = BnBr, 4-NO$_2$-BnBr, 4-CN-BnBr, 4-CO$_2$Me-BnBr, BrCH$_2$CO$_2$tBu, ICH$_2$CH=CH$_2$, CH$_2$=CHSO$_2$Ph,
CH$_2$=CHCO$_2$tBu, CH$_2$=CHCN; DIEA + cat. R$_4$N$^+$X$^-$, 45–89%, reduction by Zn/AcOH or H$_2$/Ni: 82–88% (*175*)
R^1X = R^2X = H$_2$C=C(Me)CH$_2$OCOMe, H$_2$C=CHCH$_2$OCOMe; DIEA, Pd(PPh$_3$)$_4$; 92–93% (*176*)

Scheme 3.16 Alkylations of Ethyl α-Nitroacetate.

derivatives after hydrolysis (see Scheme 3.14) (*167*). Acrylonitriles are also substrates for conjugate addition (*168*). In both cases dialkylation is a problem with the isocyanoacetate anion.

The isocyanocarboxylate anion can also be added to imines, producing imidazoline intermediates that are hydrolyzed to 2,3-diaminocarboxylic acids (see Scheme 3.15) (*169*). A similar addition to *N*-tosyldimines was catalyzed by a gold(I) complex, giving the imidazoline intermediates with high *cis* selectivity (*170*). The *trans* isomers could then be obtained by isomerization. Alternatively, Ru(II) catalysis gave the *trans* isomers directly (*171*). Hydrolysis of either isomer produced *erythro* and *threo* isomers of 2,3-diaminocarboxylic acids, respectively.

3.6 Alkylation of Other Activated Glycine Equivalents

Ethyl α-nitroacetate has been used as a glycine equivalent in several alkylations (see Scheme 3.16), although other glycine equivalents have generally been found to give better results (*172–174, 832–834*). Dialkylation with a variety of activated alkyl halides or Michael acceptors was achieved in the presence of 2 equivalents of DIEA as base and catalytic tetraalkylammonium salts; selective nitro reduction with Zn in acetic acid or hydrogenation over Raney Ni led to α,α-dibenzyl-Gly, α-carboxymethyl-Asp, and α-carboxyethyl-Glu (*175*). Pd-catalyzed dialkylation with allyl acetate or 2-methallyl acetate proceeded in 80–93% yield; reduction of the nitro and alkene groups of the methallyl adduct in a one-step hydrogenation produced α,α-diisobutylglycine (176). Alkyl azidoacetates were employed for aldol reactions with aryl aldehydes to give β-arylserines with threo selectivity (177). N-Boc,N-methyl α-tosylglycine has been coupled with allylic carbonates or vinyloxirane in the presence of a Pd catalyst. The tosyl group was removed by treatment with Mg, leaving a γ,δ-unsaturated N-methyl amino acid (178).

Alkylations of α-isocyanonitriles have also been reported (179), while the benzophenone Schiff base of aminoacetonitrile was alkylated with tosylated allyl alcohol and various deuterated isomers under PTC conditions. The products were hydrolyzed with 1 N HCl and then refluxed in conc. HCl to give allylglycine, [3-^{2}H$_1$]-allylglycine, Z-[5-^{2}H$_1$]-allylglycine, E-[4,5-^{2}H$_2$]-allylglycine, [3,3-^{2}H$_2$]-allylglycine, Z-[3,3,5-^{2}H$_3$]-allyl-glycine and E-[3,3,4,5-^{2}H$_4$]-allylglycine (180). The aminonitrile Schiff base has also been used for PTC-catalyzed benzylations (831) and other alkylations, as mentioned earlier (73, 74).

A benzodiazepine template containing a Gly residue, 7-chloro-1,3-dihydro-1-methyl-5-phenyl-2H-1,4-benzodiazepin-2-one, was employed for aldol reactions with α-methylcinnamaldehydes with either threo or erythro stereoselectivity, depending on reaction conditions. Hydrogenation of the alkene also proceeded stereoselectively to give racemic α-amino-β-hydroxy-γ-methyl acids (181). The nitrogen of glycine has been embedded within a triazine ring system in 2-(ethoxycarbonylmethyl)-1H-naphtho[1,8-de]-1,2,3-triazine; deprotonation with LDA and alkylation with a number of alkyl halides proceeded in 58–88% yield. Reductive cleavage (Al–Hg) generated the amino esters of Ala, Nle, allylglycine, 4'-Me-Phe, 4'-nitro-Phe, Glu and 1-aminocyclopentane-1-carboxylic acid (182).

3.7 Alkylation of Non-Activated Glycines

Nucleophilic glycine synthons in which the acidity of the α-proton has not been substantially reduced by the presence of activating groups are less common. Most non-activated glycine enolates have been used for aldol reactions to prepare β-hydroxy amino acids, though similar reactions also lead to dehydroamino acids (see Section 4.2.2). In 1894 hippuric acid (Bz-Gly) and benzaldehyde were combined to give phenylserine

(Pse) (183). Phenylserine ethyl ester (Pse-OEt) and Cbz-Pse-OEt were synthesized from Gly-OEt and Cbz-Gly-OEt, respectively, with the former compound obtained in 18% yield using Et$_3$N as base (184). Condensation of aldehydes and ketones with LDA-deprotonated Cbz-Gly-OEt gave threo isomers via 2-oxazolidone intermediates, which were obtained in 80–95% yield (185). Cbz-protected Ala, Val, Phe, and Abu t-butyl esters were deprotonated with LDA and reacted with aldehydes in the presence of metal salts, producing α-alkylated-β-hydroxy-α-amino acids with high diastereoselectivity (80–96% de of anti isomer) and good yields (60–85%) (186). Gly-OEt was used to prepare threo-p-nitrophenylserine in one report (187), and predominantly the erythro isomer in another (188). The α-protons of N-protonated glycine methyl ester have recently been determined to have pK_a = 21 (189).

The stereoselectivity of the aldol reaction of ethyl N-benzyl-N-methylglycinate can be somewhat controlled, with the LDA-derived lithium enolate giving the syn isomer (61–82% yield, 14–68% de) and its borane adduct, ethyl N-borane-N-benzyl-N-methylglycinate giving the anti product (48–67% yield, 58–88% de) (190). N,N-dibenzylglycine ethyl ester and LDA have also been used to prepare β-hydroxy amino acid diastereomer mixtures in 59–95% yield (191), including reaction with aldehydes of ribose and galactose (192). Addition of the anion to acetone produced β-hydroxyvaline in 72% overall yield after deprotection (193). The LDA-generated lithium enolate of N-Cbz- or N-(p-methoxyCbz)-sarcosine t-butyl ester is relatively stable, even at room temperature. It was reacted with aldehydes to give β-hydroxy amino acids in 71–83% yield, including MeBmt (194).

Silylated glycine ketene acetals are nucleophilic glycine synthons that avoid the need for a strong base during condensation. Both [N,N-bis(trimethylsilyl)amino]ketene bis(trimethylsilyl) acetal and its N-methyl-N-trimethylsilyl analog (see Scheme 3.17) have been prepared and condensed with carbonyl compounds to give, after acidic workup, deprotected β-hydroxy amino acids in 46–95% yield (195). The N-trifluoroacetyl analog has also been used, giving the β-silyloxy adducts in 76–93% yield (196). N,N-Bis(trimethylsilyl)glycine trimethylsilyl ester can be deprotonated with LDA and reacted with ketones and aldehydes in 50–90% yield, with the erythro isomers formed exclusively (185). The ethyl ester had previously been added to

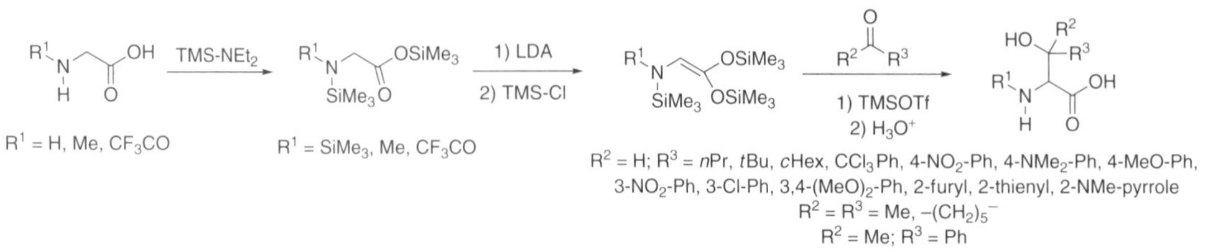

R^1 = H, Me, CF$_3$CO

R^1 = SiMe$_3$, Me, CF$_3$CO

R^2 = H; R^3 = nPr, tBu, cHex, CCl$_3$Ph, 4-NO$_2$-Ph, 4-NMe$_2$-Ph, 4-MeO-Ph, 3-NO$_2$-Ph, 3-Cl-Ph, 3,4-(MeO)$_2$-Ph, 2-furyl, 2-thienyl, 2-NMe-pyrrole
R^2 = R^3 = Me, –(CH$_2$)$_5$–
R^2 = Me; R^3 = Ph

Scheme 3.17 Aldol Reaction of a Glycine Ketene Silyl Acetal (195, 196).

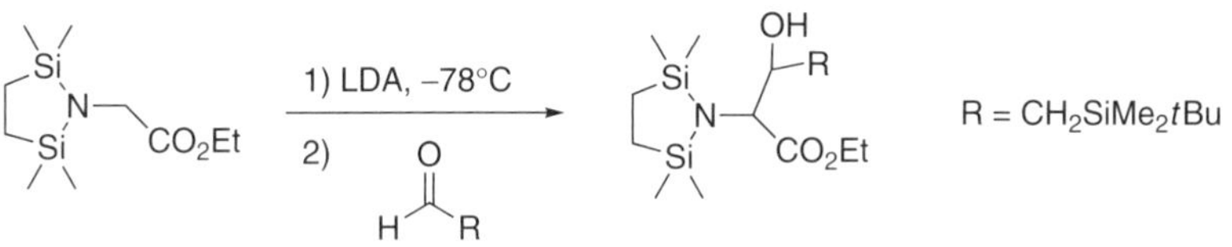

Scheme 3.18 Aldol Reaction of Stabase-protected Glycine.

benzaldehyde (*197*). The stabase adduct of ethyl glycinate (see Scheme 3.18) can be deprotonated with LDA and converted to a β-hydroxy derivative by reaction with an aldehyde in 85–93% yield (*198, 199*). An aldol reaction of the zinc enolate was used in a synthesis of an antifungal hydroxyamino acid, sphingofungin B (93% yield, 67% de) (*200*). The opposite diastereomer could be obtained by using the trimethysilyl enolate of the glycine derivative in the presence of BF₃·OEt₂ (73% de).

The condensation of unprotected glycine with benzaldehyde under alkaline conditions to give phenylserine (predominantly *threo*) has been known for many years (*201, 202*). More recently, unprotected glycine was reacted with benzaldehyde in aqueous medium at pH 7.0, using a bilayer membrane formed from a hydrophobic pyridoxal derivative and peptide lipid to mimic an aldolase enzyme. Phenylserine was obtained with significant rate enhancement (*203*). Unprotected glycine has also been reacted with isobutyraldehyde to give β-hydroxy-Leu (*204*), with 2-furaldehyde to give β-(2-furyl)serine (*205, 206*), with 2-thienaldehyde to give β-(2-thienyl)serine (*207, 208*), with 3,4-di(benzyloxy) benzaldehyde to give a mixture of *threo* and *erythro* β-hydroxy-DOPA (*209*), and with a great variety of other substituted benzaldehydes (*202, 207, 210–215*).

A copper complex of unprotected glycine has also been employed for many aldol reactions. Condensation of acetaldehyde with copper glycinate produced threonine in 64% yield as a 1.6:1 Thr:*allo*-Thr mixture (*216*), while reaction with furfural in 1 N KOH gave a mixture of *threo* (50%) or *erythro* (22%) β-furyl-Ser isomers that were separated by crystallization (*206*). Similarly, addition of copper glycinate to benzyloxyacetaldehyde gave a mixture of diastereomers of 2-amino-3-hydroxy-4-benzyloxybutyric acid which were separated by crystallization (*217*). Addition of the nucleophile to glyoxylate produced β-hydroxy-Asp in 64% yield (*218*), to pyruvic acid resulted in β-methyl-β-hydroxy-Asp in 60% yield (*219*), to *n*-butyraldehyde or tetrolaldehyde gave β-hydroxy-Nle or 2-amino-3-hydroxyhex-4-ynoic acid, respectively (*220*), and to 3-methylthiopropionaldehyde or methoxyacetaldehyde produced mixtures of diastereomers of β-hydroxyhomomethionine or β-hydroxy-γ-methoxy-Abu in 16–24% yield (*221*). Reaction of copper glycinate with formaldehyde in basic solution resulted in a 30% yield of Ser (*222*); by using excess formaldehyde a 75% yield of α-hydroxymethyl-Ser was produced instead (*223, 224*). Cu-catalyzed reaction of Ala or Abu with formaldehyde resulted in α-methyl-Ser and α-ethyl-Ser, respectively (*223*).

Alkylations of non-activated glycines with non-carbonyl electrophiles are less common. A trianion formed by LDA deprotonation of hippuric acid (*N*-benzoylglycine) reacted with MeI in 50–60% yield and BnBr in 40–50% yield, with less reactive electrophiles giving poor yields (*225*). Similar yields were obtained from the dianion of ethyl hippurate (RX, R = Me, Bn, *i*Pr; 17–61%) (*225*). Hoye et al. reported in 1985 that the monoanion of methyl hippurate or *N*-ethoxycarbonyl methyl glycinate could be alkylated with MeI, BnBr, or *n*BuBr in 52–61% (*226*). A more recent study examined the deprotonation and deuteration or alkylation of methyl hippurate, and found that the method of anion generation plays a significant role (*227*). In contrast to Hoye's report, one equivalent of LDA was not sufficient to cause *C*-alkylation. Efficient alkylation with MeI or BnBr was obtained with 2 equivalents of base (65–75% yield); 3 equivalents led to dialkylated products. HMPA as an additive gave slightly better yields than TMEDA (*227*). Proline derivatives have been synthesized by an intramolecular reaction via addition of a glycine zinc enolate to the alkene bond of an *N*-homoallyl group (96% yield) (*228*), or by condensation of *N*-carboxyethyl-Gly-OEt with diethyl fumarate in the presence of Na (*229*). The product of the latter reaction, 3-carboxyethyl-4-oxo-Pro, was then converted into 4-oxo-Pro or 4-hydroxy-Pro (*229*).

N,N-bis(trimethylsilyl) glycine ethyl or trimethylsilyl esters can be converted into the sodio derivatives using NaHMDS and alkylated with halides ((RX, R = Me, Et, Bn, CH₂-SiHMe₂, SiMe₃) in 52–77% yield (*197*). 1-Nitro-4-bromobutane was also employed as an electrophile (*230*). The ketene silyl acetal nucleophile reacted with α-glycosyl bromides to produce α-C-glycosylated amino acids in 75–91% yield (*231*). The lithiated stabase adduct of ethyl glycinate was alkylated with several halides in 80–91% yield (RX, R = Me, allyl, CC-SiMe₃) (*199*), while stabase-protected Ala-OEt was α-alkylated with [¹¹C]MeI (*232*). Chelated enolates of Cbz-, Boc-, Tfa- or Ts-Gly-O*t*Bu can be allylated in 59–82% yield by deprotonation with LHMDS/ZnCl₂, followed by Pd-catalyzed reaction with allyl carbonates. 1,3-Dimethyl allyl carbonate reacted with >96% de *anti* diastereoselectivity (*233, 234*).

A chelated enolate of Tfa-Gly-O*t*Bu formed with LHMDS/ZnCl₂ reacted with *O*-methoxycarbonyl (*E*)- or (*Z*)-4-hydroxybut-2-enoic acid methyl ester to form the side-chain lactone of β-hydroxymethyl-γ-carboxy-Glu in 80–91% yield via a Michael addition/ring-closure sequence. The corresponding lactone of β-(2-hydroxyethyl)-γ-carboxy-Glu was obtained from O-methoxycarbonyl (Z)-5-hydroxypent-2-enoic acid methyl ester. In contrast, Pd-catalyzed reaction with O-alkoxycarbonyl (Z)-4-hydroxy-4-methylbut-2-enoic acid methyl ester gave predominantly 2-amino-3-carboxy-hex-4-enoic acid, along with some (E)- and (Z)-2-amino-3-methyl-5-carboxy-pent-4-enoic acid (235). Michael addition of lithium enolates of N,N-dibenzyl Gly-OtBu or Ala-OtBu to β-substituted α,β-unsaturated esters produced *threo* adducts in 52–84% yield and >95% de (236). Ethylene oxide was used as a vinyl cation equivalent to introduce an α-vinyl group to N-benzoyl amino acid esters of Phe, DOPA, His, Lys, Orn, Val, Ala, and Hse that were deprotonated with 2 equivalents of LDA at −78°C; the monoanion generated from N-benzylidene-protected amino acid esters was not nucleophilic enough (237). Methyl phthalimidoacetate (N-phthalimide-glycine methyl ester) was deprotonated with sodium ethoxide and condensed with methyl formate to give a protected serine aldehyde derivative in 47% yield (238). The sodium salt of N-benzoyl-glycine reacted with acetic or propionic anhydride to give β-hydroxy-α,β-didehydro amino acid oxazolones, which were reduced to the β-hydroxy amino acid (239).

Possibly the most simple route of glycine alkylation was reported in 1992. Racemic Boc-protected amino acids were obtained in one step from Boc-Gly-OH by deprotonation with excess LDA, followed by trapping of the trianion with an electrophile. A variety of side-chain substituents (R = D, Me, Bn, allyl, nBu, iBu, iPr, CH$_2$OH, CH$_2$CONH$_2$, -(CH$_2$)$_3$-) were introduced in 40–80% yield (240); methallyl chloride is another electrophile that has been reacted (241). The method has also been used to prepare aryl and alkyl Arg analogs (242).

Aminomethylnitrile (see Scheme 3.19) is a glycine equivalent which possess higher reactivity towards dialkylation with 1,2-dielectrophiles than do α-amino esters. Dialkylation of N,N-dibenzyl aminomethylnitrile with epibromohydrin produced 2,3-methanohomoserine (243), while cyclic sulfate esters of 1,2-diols produced

substituted 1-aminocyclopropane-1-carboxylic acids (244). Aminonitrile precursors of 1-aminocyclopentane-1-carboxylic acid and 1-aminocyclohexane-1-carboxylic acid were obtained by dialkylation with dibromoalkanes (244). A C-terminal aminomethylnitrile residue within a dipeptide was converted into a 1-aminocyclopropane-1-carboxylic acid residue via deprotonation with 4 equivalents of LDA and 2 equivalents of n-BuLi followed by alkylation with ethylene sulfate (245).

Glycine residues within peptides have also been alkylated, but as asymmetric induction is possible in these reactions, they are discussed in Section 6.2.1d.

3.8 Alkylation of Glycine Equivalents via Other Methods

Allylic alcohol esters of amino acids undergo an Ireland–Claisen, Claisen ester enolate, or hetero-Cope rearrangement, which introduces an α-allyl substituent (see Scheme 3.20). The reaction has been used primarily for the synthesis of γ,δ-unsaturated acids, with the first report by Steglich in 1975 (246). This area was reviewed in 1996 (247) and 1997 (248). Both α,α-disubstituted (246, 249–254) and α-monosubstituted (249) allyl amino acids have been prepared. The rearrangement can be induced either by treatment with base (e.g., 251, 253–255), or by dehydration to an oxazole intermediate (e.g., 246, 250, 255). A very similar rearrangement has been carried out on amino acid Schiff bases, catalyzed by a Pd(0) complex (see Section 3.4) (115). *Syn* products are obtained from *trans*-substituted allyl esters and *anti* products from *cis*-substituted allyl esters, generally with high diastereoselectivity (>90% de), which is improved by use of a zinc chelate bridged enolate (256). β-Quaternary centers can be introduced by using trisubstituted allyl esters (253, 254). Baldwin et al. used a Claisen-enolate rearrangement of deuterated allyl esters of Boc-Gly to stereospecifically prepare the (2R,3R/2S,3S)- or (2R,3S/2S,3R)- diastereomers of [3-^{2}H$_1$]-allylglycine and [3,4-^{2}H$_2$]-allylglycine,

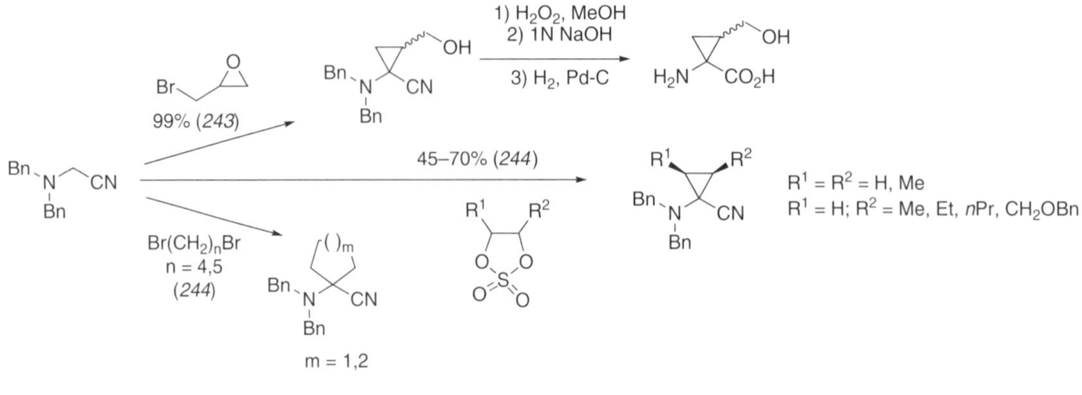

Scheme 3.19 Alkylation of Aminoacetonitrile.

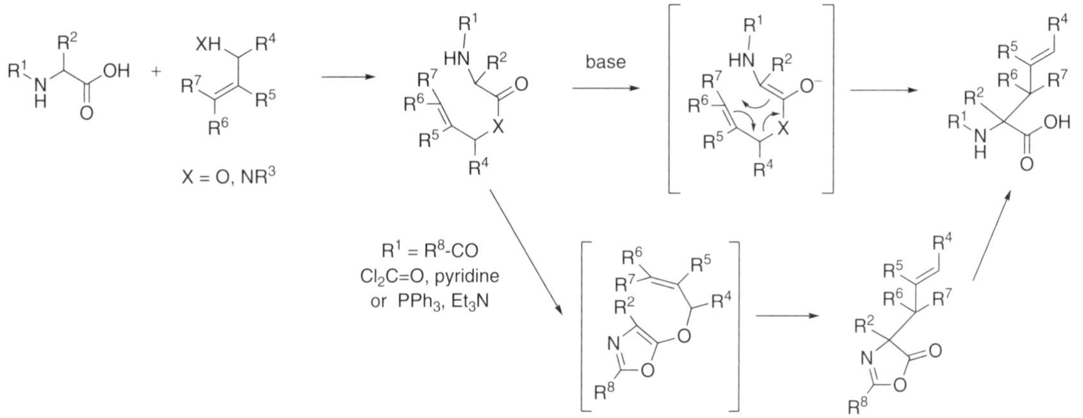

Scheme 3.20 Alkylation via Claisen Ester Enolate, Ireland-Claisen or Hetero-Cope Rearrangement.

respectively (*180*). β-Silyl-γ,δ-unsaturated acids were derived by rearrangements of Boc-Gly-OCH(R)CH=CHSiMe₃, with predominantly *syn* stereoselectivity (*257*). The allyl group can be converted to other functionalities.

The Claisen ester enolate rearrangement has also been employed for the synthesis of α-allyl-β-amino acids (*258*). The oxazolone intermediate variant works with propargyl esters, forming an α-allenic amino acid (*255, 259, 260*). The closely related aza-Claisen rearrangement uses an allyl amide instead of an allyl ester (X = NR*), which allows the use of a chiral auxiliary on the amide nitrogen for asymmetric induction (see Section 6.2.1c). 4,5-Dehydroisoleucine amide was prepared by this route (*261*). Alternatively, LDA-induced aza-[2,3]-Wittig sigmatropic rearrangements of N-allyl,N-substituted Gly derivatives produced substituted allylglycines (*262–264*). The same transformation was accomplished by using a phosphazene base and Lewis acid to rearrange N-allyl,N-benzyl Gly pyrrolidine amides (*265*).

3.9 Alkylation of an Electrophilic Glycine Equivalent

Electrophilic glycine equivalents contain a glycine framework to which nucleophiles can be added. The reactions described in this section include those that proceed, either explicitly or indirectly, via displacement of an α-substituted leaving group on the glycine equivalent. In reality most of these reactions actually proceed by an initial elimination of the α-leaving group, forming an imine intermediate. This intermediate is generally not isolated, and is often not acknowledged. The nucleophile then adds to the imine. This sequence thus requires 2 equivalents of basic nucleophile if the imine is not pre-generated. Much of the pioneering work in this area was carried out by the groups of Ben-Ishai, Steglich, and O'Donnell. Imines explicitly prepared from α-keto

acids, already described in Section 2.3.6, also qualify as electrophilic glycine synthons. Additions of side chains to the imines, as opposed to the simple imine reductions covered in Section 2.3.6, are thus included in this section. Some reactions of iminium ion electrophilic glycine equivalents for the synthesis of α-amino acids were reviewed in 2004 (*266*).

3.9.1 Synthesis of Electrophilic Synthons

Many reactions have been carried out on α-methoxy-, α-acetoxy- or α-haloglycine derivatives (see Table 3.12), with various combinations of the type of amine protection, acid protection, and leaving group. The electrophilic synthons are generally prepared from glyoxalic acid (see Scheme 3.21) or glycine (see Scheme 3.22). They can also be generated within peptides, but since the remaining peptide acts as a chiral auxiliary these reactions will be discussed in Section 6.5. The first electrophilic glycine derivatives were the α-hydroxy and α-methoxy compounds prepared by Ben-Ishai et al. by combining benzamide (*267*), acetamide (*268*), benzyl carbamate (*267*), methyl carbamate (*269*) or phenyl-acetamide (*269*) with glyoxalic acid monohydrate, with the methoxy derivatives obtained in acidic methanol (see Scheme 3.21). Other amides and carbamates (*270, 271*), including amino acid amides (*272*), have also been employed. Similarly, methyl N-Boc α-acetoxyglycine was prepared from methyl glyoxalate hemiacetal by amination with t-butylcarbamate and acetylation with acetic anhydride (*273*).

An alternative route for the synthesis of α-acetoxy-glycine is via oxidative cleavage of serine with lead tetra-acetate; the product can then be converted into α-(benzyloxy)glycine and α-hydroxyglycine (*274, 275*). A Wang resin-esterified benzophenone Schiff base of Gly was oxidized to resin-bound α-acetoxy-Gly via treatment with lead tetraacetate (*276*). α-Methoxyglycines can also be obtained from protected Gly-OMe via electrochemical oxidation in methanol (*277*), while Gly

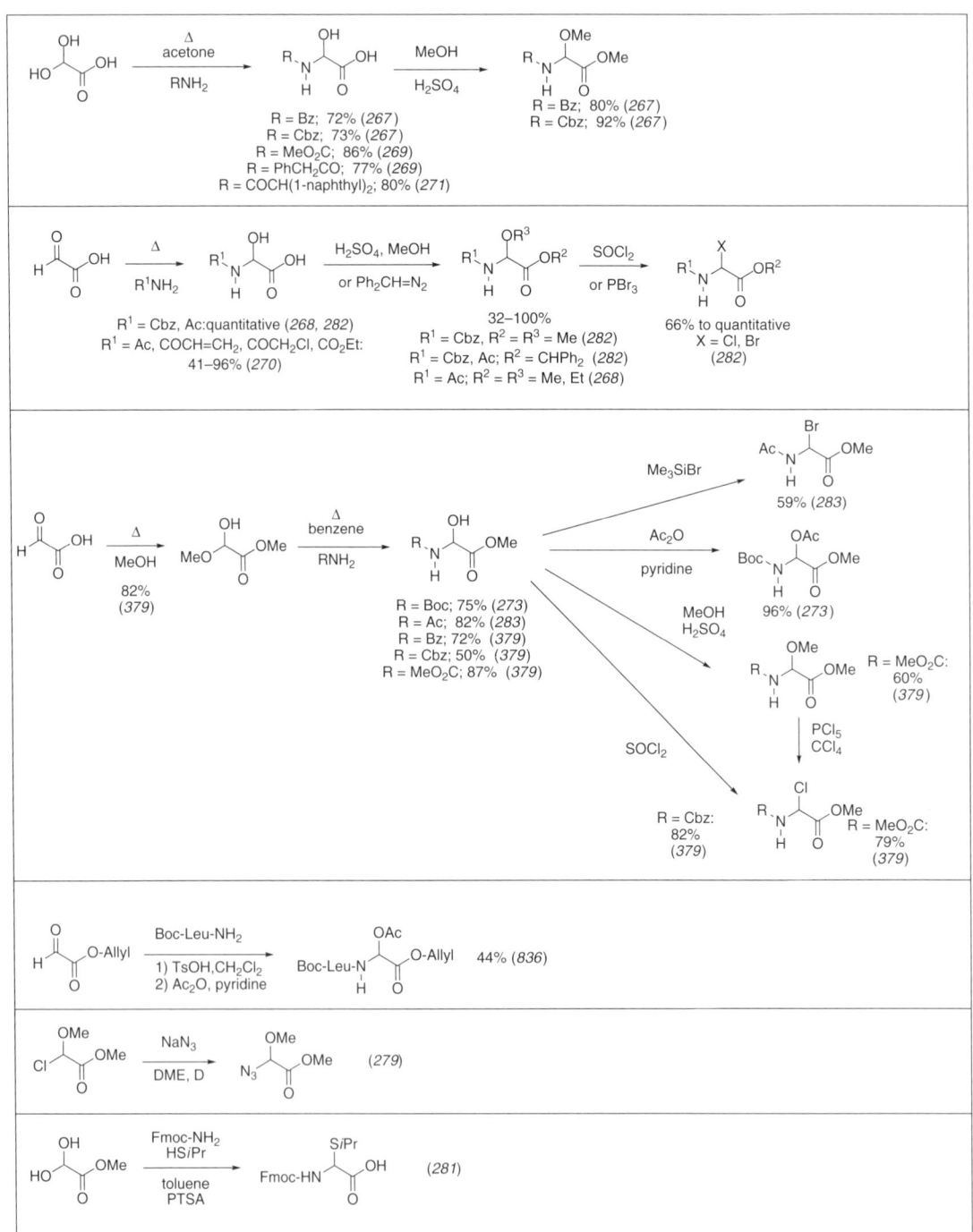

Scheme 3.21 Synthesis of Electrophilic Glycine Synthons from Glyoxalic Acid Derivatives.

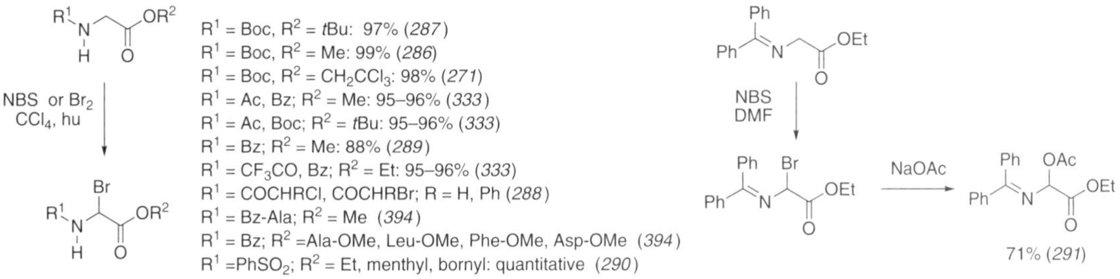

Scheme 3.22 Synthesis of Electrophilic Glycine Synthon from Glycine Derivatives.

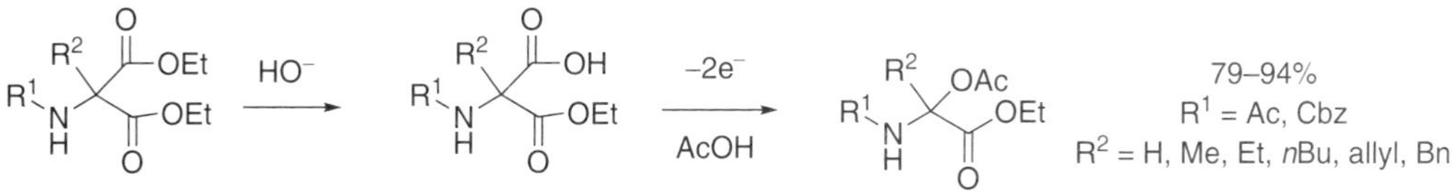

R¹ = Boc, R² = tBu: 97% (287)
R¹ = Boc, R² = Me: 99% (286)
R¹ = Boc, R² = CH₂CCl₃: 98% (271)
R¹ = Ac, Bz; R² = Me: 95–96% (333)
R¹ = Ac, Boc; R² = tBu: 95–96% (333)
R¹ = Bz; R² = Me: 88% (289)
R¹ = CF₃CO, Bz; R² = Et: 95–96% (333)
R¹ = COCHRCl, COCHRBr; R = H, Ph (288)
R¹ = Bz-Ala; R² = Me (394)
R¹ = Bz; R² =Ala-OMe, Leu-OMe, Phe-OMe, Asp-OMe (394)
R¹ =PhSO₂; R² = Et, menthyl, bornyl: quantitative (290)

71% (291)

Scheme 3.23 Synthesis of α-Substituted Electrophilic Synthons via Anodic Oxidation (293–295).

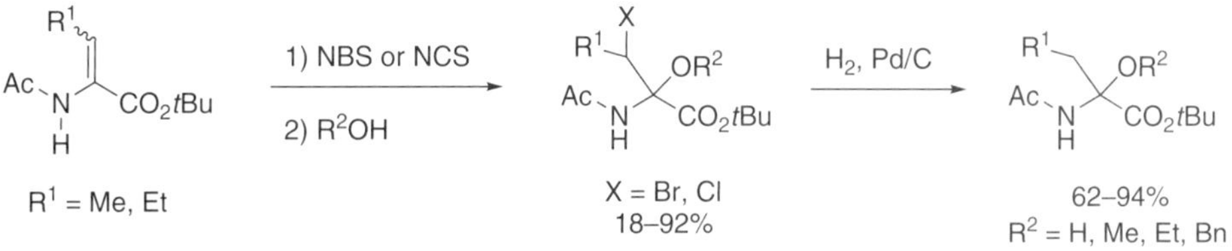

79–94%
R¹ = Ac, Cbz
R² = H, Me, Et, nBu, allyl, Bn

Scheme 3.24 Synthesis of α-Substituted Electrophilic Synthons via Halogenation of Didehydroamino Acids (297).

residues within cyclotetrapeptides were converted into α-hydroxy-Gly via heating an Ni(II) complex in air (278). An α-methoxy-α-azidocarboxylate was prepared from 2-chloro-2-methoxyacetate by nucleophilic displacement of the chloride with azide (279).

The electrophilic α-hydroxyglycine derivatives can be converted into α-haloglycine electrophilic synthons via an α-thiol intermediate (267). Alternatively, the α-thiol intermediate can be directly obtained by combining the thiol, a carbamate, and glyoxylic acid hydrate (280, 281). In recent years the methyl and benzhydryl esters of N-Cbz α-chloroglycine and the benzhydryl ester of N-Ac α-chloroglycine (282) were prepared by modified versions of this method, as was N-acetyl α-bromoglycine methyl ester (283). α-Chloroglycine was prepared within a peptide via incorporation and conversion of an α-(ethylthio)glycine residue (284), or from a serine residue via treatment with Pb(OAc)₄ followed by Et₃N/EtSH, then SO₂Cl₂ (285).

α-Haloacids can also be directly obtained by α-halogenation of glycine derivatives (see Scheme 3.22). Various ester derivatives of Boc-Gly were α-brominated by NBS or Br₂ in 97–98% yield (271,

286, 287), as were N-chloroacetyl (288), N-benzoyl (289), N-bromobenzoyl (288), N-bromophenylacetyl (288), and N-benzenesulfonyl (290) glycinates. The benzophenone Schiff base of Gly-OEt was α-brominated and then reacted with NaOAc to form a stable α-acetoxyglycine synthon in 71% yield (291). α-Fluoroglycinamides have been synthesized by reacting dihaloethanamides with amines (292).

α-Acetoxy amino acids other than glycine have also been prepared (see Schemes 3.23–3.26). The ethyl esters of N-Cbz- and N-Ac-protected α-acetoxy-Gly, Abu, Ala, Nle, Phe, and allylglycine were synthesized in 79–94% yield by anodic oxidation of the N-acylaminomalonic acid monoester derivatives (see Scheme 3.23) (293–296).

N-Acetyl-α-hydroxy- or α-alkoxy-Abu and -Nva ethyl esters have been prepared from the corresponding didehydroamino acids by a three-step procedure involving halogenation, reaction with water or alcohols, and hydrogenolytic reduction (see Scheme 3.24) (297).

α-Keto acids have been converted to α-methoxy-Gly, -Ala, -Abu, -Phe, and -MeCys by reductive amination of the 2-keto acid with O-benzylhydroxylamine (298), N-acetylation, and treatment with KOtBu in methanol (see Scheme 3.25) (299). The last step, which proceeds

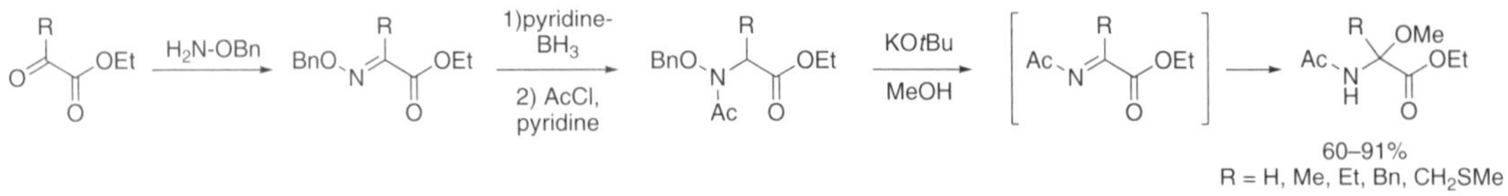

Scheme 3.25 Synthesis of α-Substituted Electrophilic Synthons via α-Keto Acid (*299*).

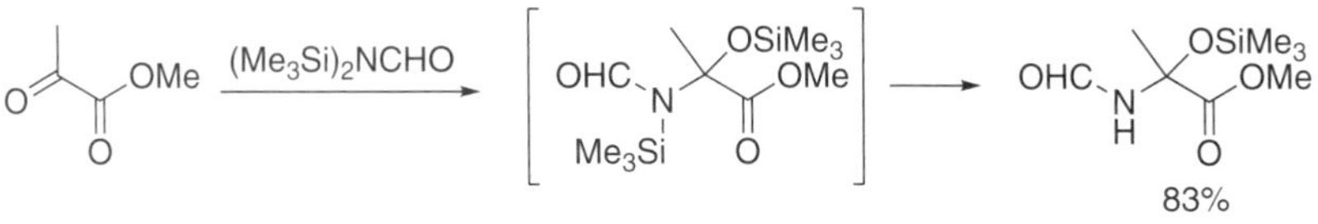

Scheme 3.26 Synthesis of α-Substituted Electrophilic Synthons via α-Keto Acid (*300*).

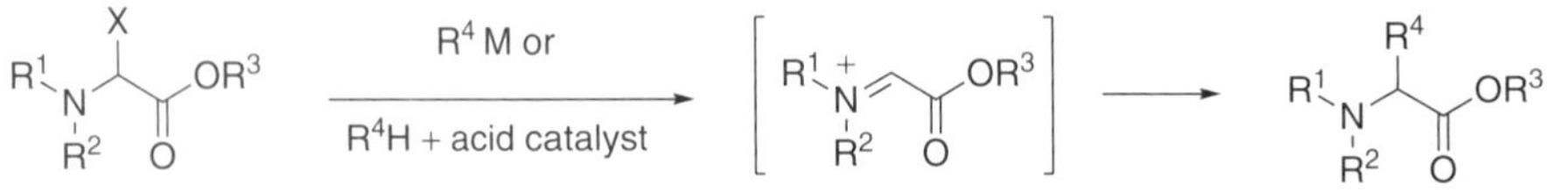

Scheme 3.27 Additions to Electrophilic Glycine Synthons (See Table 3.12).

via methanol addition to an imine intermediate, occurs quantitatively. *N*-Formyl-2-trimethylsiloxy-Ala was synthesized in 83% yield by condensation of *N,N*-bis(trimethylsilyl)formamide with the α-keto ester methyl pyruvate (see Scheme 3.26) (*300*).

Another electrophilic glycine equivalent, α-benzotriazolyl-glycine, was obtained by condensing glyoxylic acid, acylamines, and benzotriazole. The benzotriazole can be displaced with nucleophiles such as ammonia, leading to α-aminoglycine (*272, 301*). *N*-Fmoc-, -Cbz- or -Ac α-(isopropylthio)glycine were synthesized by condensing RCO-NH₂ with isopropylthiol and glyoxylic acid hydrate (*281, 302*).

The imines that are obtained as intermediates during reductive aminations of α-keto acids (or specifically prepared by other routes) can also be used as electrophilic glycine equivalents; these are discussed in Section 3.9.6.

3.9.2 Alkylation Reactions: α-Hydroxy and α-Alkoxy, and α-Acetoxy Derivatives (see Scheme 3.27, Table 3.12)

Methyl α-hydroxyhippurate and methyl α-methoxyhippurate react with thiols under acidic conditions to give α-thio-Gly derivatives in 68–92% yield, and the corresponding sulfoxides and sulfones can be obtained by oxidation of the thiol adducts (*267*). Displacement of the α-thiol intermediate with Boc-NH₂ (*281, 303*), Fmoc-NH₂ (*280*) or substituted amines (*271*) produces

N-substituted α-aminoglycines. α-Hydroxyglycine has also been directly converted into α-aminoglycine via displacement with azide using DPPA or Mitsunobu reaction conditions, followed by azide reduction (*304*). The latter two reactions have also been carried out on substrates within dipeptides. Amides and alkylcyanides also react with *N*-acyl α-hydroxyglycine under acidic conditions to produce α-aminoglycine derivatives (*270*). Cbz-Gly-α-acetoxy-Gly-OBn dipeptide was reacted with thiophenol, *p*-nitrophenol, and *N*-Boc 1-aminopropane-3-thiol in the presence of triethylamine to give 3-thia analogs of Phe, *p*-nitro-Phe, and Lys (*305*).

Both methyl α-hydroxyhippurate and methyl α-methoxyhippurate were combined with 1,3-diketones and β-keto esters in the presence of acid catalysts (sulfuric acid, trifluoroacetic acid, methanesulfonic acid, BF₃·Et₂O) to give γ,γ-diketo-α-amino acid derivatives in 30–89% yield (*306*). Arylglycines can be obtained from methyl α-hydroxyhippurate by reaction with a range of aromatic and heteroaromatic compounds in the presence of sulfuric acid (41–92% yield) (*269*). Similarly, a number of olefins were reacted with methyl α-methoxyhippurate in boiling benzene with a sulfonic acid catalyst to give *N*-benzoyl β,γ-unsaturated amino acid methyl esters (*307, 308*). Under slightly different conditions a 1,3-dipolar cycloaddition with a transient imine produced an oxazine which rearranged to a butyrolactone (see Section 3.9.6b).

N-Methoxycarbonyl-protected α-methoxyglycine was treated with allylsilanes under Lewis acid catalysis to

X	R¹	R²	R³	Nucleophile (R⁴H or R⁴M)	Catalyst or Base	Yield	Reference
OH	Bz Cbz	H	Me	MeSH nBuSH iPrSH PhSH	H_2SO_4	68–92%	1975 (267)
OH	Bz Cbz	H	H	MeCOCH₂COMe PhCOCH₂COPh MeCOCH₂COPh forms -CH(COR⁵)COR⁶ MeCOMe MeCOPh forms -CH₂COR⁵	H_2SO_4	33–71%	1975 (825)
OH	Bz Cbz MeO₂C BnCO	H	H	ArH Ar = Ph, 4-Me-Ph, 4-Cl-Ph, 4-AcNH-Ph, 4-MeO-Ph, 4-HO-Ph, naphthyl, anthryl, thienyl, furyl, 5-Me-furyl	H_2SO_4	41–92%	1975 (826) 1976 (269)
OH	Bz	H	H	3-CO₂H-4-HO-Ph	20% H_2SO_4/AcOH	32%	1977 (827)
OH	Cbz BnCO	H	Me	AcCl (gives R⁴ = Cl)	—	59%	1978 (828)
OH	Ac EtO₂C ClCHH₂CO	H	H	ArH Ar = Ph, 4-HO-Ph, 4-Me-Ph, 4-Cl-Ph, 4-F-Ph, 4-AcNH-Ph, 2.5-Me₂-Ph, 2,4-Me₂-Ph, 3-MeO-4-HO-Ph, 2-HO-5-AcO-Ph, 3-NO₂-4-HO-Ph, 2-HO-5-NO₂-Ph, 2-HO-5-F-Ph, 3-Cl-4-HO-Ph, 3,4-(HO)₂-Ph, 2,4-(HO)₂-Ph, 3,5-(MeO)₂-4-HO-Ph, 2,4-(MeO)₂-3-HO-Ph, 3,5-(tBu)₂-4-HO-Ph, 2-CO₂H-3,6-(HO)₂-Ph, 2,3,4-(HO)₃-6-CO₂nOct-Ph, 2-HO-1-naphthyl, 2-thienyl	H_2SO_4, HCl/AcOH or H_3PO_4	10–90%	1978 (270)
OH	Ac EtO₂C ClCHH₂CO	H	H	RCN (gives RCONH) R = Me, CH₂CH=CH₂, CH₂Cl	H_2SO_4, HCl/AcOH or H_3PO_4	45–70%	1978 (270)
OH	Bz Cbz	H	H	MeCOCH₂COMe PhCOCH₂COPh MeCOCH₂COPh MeCOCH₂CO₂Me MeCOCH₂CO₂Et forms -CH(COR⁵)COR⁶	H_2SO_4, TFA, BF₃.Et₂O or MeSO₃H	30–89%	1978 (306)
OH	Cbz	H	H	iPrSH then Boc-NH₂	H_2SO_4 then Hg²⁺	70%	1986 (303)

(Continued)

Table 3.12 Additions to Electrophilic Glycine Synthons (see Scheme 3.27) (continued)

X	R¹	R²	R³	Nucleophile (R⁴H or R⁴M)	Catalyst or Base	Yield	Reference
OH	Bz	H	H	Br(CH₂)ₙPh n = 0,1,2,3	MeSO₃H	43%	1989 (632)
OH	Cbz Ac Bz Boc	H	H Et	benzotriazole then H₂N, K₂CO₃	—	40–91% then 37–91%	1989 (301) 1990 (272)
OH	CO₂Me	(CH₂)₂CH=CH₂ (CH₂)₂CH=CHEt (CH₂)₂C(Me)=CH₂ CH₂CH(Me)CH=CH₂	Me	intramolecular from N-substituent	HCO₂H	70–87%	1991 (327)
OH	(1-naphthyl)₂-CHCO	H	H	MeSH EtOH	H₂SO₄	41–83%	1992 (271)
OH	Boc Fmoc	H	H	iPrSH then FmocNH₂, Boc-NH₂ or NBS	—	75–80%	1993 (280)
OH	Boc Cbz	H	Me Phe-OMe	N₃ (reduce to NHBoc, NHCbz)	DPPA or HN₃, PPh₃, DEAD	56–60%	1994 (304)
OMe	Bz Cbz	H	Me	MeSH nBuSH iPrSH PhSH	H₂SO₄	68–92%	1975 (267)
OMe	MeO₂C	H	Me	R⁵–⟍SiMe₃ → R⁵⟍R⁶ R⁶ R⁵ = H, nHex, cHex, Ph, (CH₂)₂OAc, CH=CHPh; R⁶ = H R⁵ = cHex; R⁶ = Me R⁵ = H; R⁶ = Me, Br	BF₃,Et₂O		1975 (307) 1977 (308)
OMe	Cbz	H	Me	MeCOCH₂COMe PhCOCH₂COPh MeCOCH₂COPh MeCOCH₂CO₂Me forms -CH(COR⁵)COR⁶	H₂SO₄, TFA, BF₃,Et₂O or MeSO₃H	40–83%	1978 (306)
OMe	Ac Et CO₂R³ = CONHBn	H	Me	ArH Ar = 2-furyl, 2-pyrrolyl, 3-pyrrolyl, 5-Me-2-furyl, 1-Me-2-pyrrolyl, 2-thienyl, 4-MeO-Ph, 2-HO-5-Me-Ph, 4-HO-Ph, 2-benzofuryl, 2-benzothienyl, 2-indolyl, PhSH	BF₃,Et₂O	4–94%	1988 (268)

OMe	Bz	H	Me	H$_2$C=CPh$_2$ H$_2$C=CH(Ph) H$_2$C=C(Me)(Ph)	naphthelane-sulphonic acid	84–66%	1989 (*309*)
OEt	Ac	H	CO$_2$R^3 = CONHBn	2-pyrrolyl	BF$_3$·Et$_2$O	9%	1990 (*314*)
OMe	Cbz Aloc MeO$_2$C	H	CO$_2$R^3 = CONHOMe	H$_2$C=CHCH$_2$SiMe$_3$ H$_2$C=C(Me)CH$_2$SiMe$_3$ 1-SiMe$_3$-cyclopen-2-enyl 1-SiMe$_3$-cyclohex-2-enyl	BF$_3$·Et$_2$O	41–90%	1992 (*311*)
OMe	Cbz	H	CO$_2$R^3 = CON(Me) OMe, CONMe$_2$	 R^5 = Me, Ph; R^6 = R^7 = H R^5 = R^7 = -(CH$_2$)$_4$-; R^6 = H R^5 = OMe; R^6 = R^7 = Me	—	32–71%	1992 (*312*)
OSiMe$_3$ with α-Me	CHO	H	Me	 R^5 = R^6 = R^7 = H R^5 = R^7 = -(CH$_2$)$_2$-, -(CH$_2$)$_3$-; R^6 = H R^5 = nHex, (CH$_2$)$_2$OAc; R^6 = R^7 = H	Me$_3$SiOTf	12–84%	1992 (*313*)
OAc	Cbz-Gly	H	Bn	PhSH 4-NO$_2$-PhSH BocNH(CH$_2$)$_3$SH	Et$_3$N	56–65%	1992 (*305*)
OAc with α-H, Me, Bn	Cbz	H	Et	PhSH BnSH *i*-amylSH MeOH EtOH PhOH 4-NO$_2$-PhOH *t*BuNH$_2$ 4-Me-PhNH$_2$ 4-morpholine	—	68–98%	1979 (*315*)
OAc with α-Me	Ac	H	Et	4-MeO-Ph	SnCl$_4$	58%	1979 (*315*)

(*Continued*)

Table 3.12 Additions to Electrophilic Glycine Synthons (see Scheme 3.27) (continued)

X	R¹	R²	R³	Nucleophile (R⁴H or R⁴M)	Catalyst or Base	Yield	Reference
OAc	Ac Cbz	H	Et	[heterocyclic structures; R = Cl, HS, NH₂; R = H, Me, F]	NaH or Et₃N	40–91%	1979 (*316*)
OAc	Bz Cbz	H	H	MeCOCH₂CO₂Et NCCH₂CO₂Et AcNHCHC(CO₂Et)₂ CbzNHCHC(CO₂Et)₂ HCONHCHC(CO₂Et)₂ forms -CH(R⁵) (R⁶)	Na	55–95%	1979 (*323*)
OAc	Ph₂C=	—	Et	MeOH EtOH *i*PrOH PhSH	—	65–95%	1985 (*291*)
OAc	Ph₂C=	—	Et	M*n*Bu M*t*Bu MPh M(1-naphthyl) M(2-thienyl)	M = Cu(CN)Li₂	46–71%	1985 (*317*)
OAc	Ph₂C=	—	Et	M*n*Oct M*t*Bu McHex MPh M(1-naphthyl) M(CH₂)₂(1-naphthyl) M(2-thienyl)	M = 9-BBN	47–92%	1985 (*319*)
OAc	Ph₂C=	—	Et	ArH Ar = 2-furyl, 4-MeO-Ph, 3-indolyl	TiCl₃	37–53%	1988 (*325*)
OAc	Ph₂C=	—	Et	MCH₂CH=CH₂ MCH₂C(Me)=CH₂ MOC(Ph)=CH₂	M = SiMe₃; + TiCl₄	16–46%	1988 (*325*)
OAc	Ph₂C=	—	Et	MCH(CO₂Me)₂ MCH(CO₂Bn)₂	M = Na + Pd(PPh₃)₄	36–79%	1990 (*321*) 1995 (*322*)
OAc	CO₂Me	(CH₂)₂CH=CH₂ (CH₂)₂CH=CHEt (CH₂)₂C(Me)=CH₂ CH₂CH(Me)CH=CH₂	Me	intramolecular from *N*-substituent	SnCl₄	33–80%	1991 (*326*)

OAc	Boc	H	Me	$BrCH_2CCH$, $BrCH_2CO_2Et$, $BrCH_2CH=CH_2$, $BrCH_2C(Me)=CH_2$, $BrCH_2Ph$, $BrCH_2(4\text{-MeO-Ph})$, $BrCH_2(4\text{-CN-Ph})$	Zn dust	37–99%	1994 (273)
OAc	$Ph_2C=$	—	Et	$MCH=C(Me)(nPr)$, $MCH=C(Me)(nHex)$, $MCH=C(Me)(tBu)$, $MCH=C(Me)(Ph)$, $MC(Et)=C(Me)(Et)$, $MC(Me)=C(Me)(nPr)$	$M = AlMe_2 + Pd(PPh_3)_4$	0–68%	1994 (320)
OAc	$Ph_2C=$	—	Et	McPr	$M = Cu(CN)Li_2$	75%	1997 (318)
OAc	$Ph_2C=$	—	Me	MAr (Ar = 2-Me-Ph, 4-MeO-Ph, 4-CN-Ph, 1-naphthyl)	$M = Zn$	65–86%	1997 (324)
OAc	$Ph_2C=$	—	Wang resin	MEt, MnBu, MnHex, MnOct, McPr, McPent, McHex, McOct, McHept, MCH_2cPent, MCH_2cHex, MPh, $M(CH_2)_2(2\text{-naphthyl})$	$M = BR_2$	51–99% with 81–99% purity	1999 (276)
Br	$Ph_2C=$	—	Et	MOCOPh	$M = Na$	72%	1985 (291)
Br	$Ph_2C=$	—	Et	Me_2NH_2	—	74%	1985 (291)
Br	Boc	H	tBu	$H_2C=C(OTMS)CH=CH_2$	$TiCl_4$	57%	1988 (829)
Br	$COCH_2Cl$, $COCH(Bn)Br$	H, Me	Et, Me	MeCOSH	K	62–71%	1980 (288)

(Continued)

Table 3.12 Additions to Electrophilic Glycine Synthons (see Scheme 3.27) (continued)

X	R¹	R²	R³	Nucleophile (R⁴H or R⁴M)	Catalyst or Base	Yield	Reference
Br	Boc	H	tBu	MEt MiPr MnBu McPent MCH=CH₂ M(CH₂)₂Ph MPh M(4-Me-Ph) M(4-MeO-Ph) M(1-naphthyl)	M = MgX (2 eq.) or 1 eq. MgX + DIEA	31–91%	1987 (287)
Br	Boc	H	tBu	[enamine structure] R⁵ = R⁶ = −(CH₂)₃−, −(CH₂)₄−, −(CH₂)₅−; R⁵ = Ph; R⁶ = H	enamine	52–95%	1987 (287)
Br	Bz Boc CF₃CO	H	Me Et tBu	MeCOCH₂COMe EtO₂CCH₂CO₂Et MeCOCH₂COPh MeCOCH₂CO₂Et 1,3-diketo-5,5-Me₂-cHex lactone of HOC(Me)=CHCOCH₂CO₂H forms −CH(COR⁵)COR⁶	Et₃N	48–92%	1988 (333)
Br	Ac Bz Boc	H	Et tBu	[silyl enol ether structure] R⁵ = Me, Ph; R⁶ = R⁷ = H; R⁵ = H, OMe; R⁶ = R⁷ = Me; R⁵ = R⁷ = −(CH₂)₄−; R⁶ = H	TiCl₄	25–89%	1988 (333)
Br	Ac Bz Boc	H	Me tBu	MnBu MtBu MCH=CH₂ MPh M(1-naphthyl)	M = Cu(CN)Li₂	30–83%	1988 (333)
Br	Bz-Val Bz	H	Me Val-OMe	CH₂(CO₂Et)₂	Na	18%	1988 (394)
Br	Bz	H	Me	MCH₂NO₂ MC(Me)₂NO₂ MC(Me)(Et)NO₂	Li	51–86%	1988 (339) 1995 (338)

Br	Ac	H	Me	MCH=CH(OMe)	Li	7%	1989 (283)
Br	Bz	H	Me	side chain of Bz-Lys-OMe, Bz-Ser-OMe, Bz-Cys-OMe	—	43–85%	1990 (329)
Br	Ac	H	Et	ArH, Ar = 2-furyl	ZnCl$_2$	87%	1990 (314)
Br	(1-naphthyl)$_2$CHCO	H	CH$_2$CCl$_3$	EtSH	NaH	77%	1992 (271)
Br	Bz	H	Me	BnSH, tBuSH, PhSH, PhCO$_2$H, MeOH	Et$_3$N	41–84%	1994 (328)
Br	Bz	H	Me	NH$_3$ (trialkylate)	—	64%	1995 (330)
Br	Bz, CF$_3$CO, Troc	H	Me	NaN$_3$		70–90%	1997 (331)
Br	Boc, Bz	H	Me, tBu	MCH$_2$CO(ferrocene), MCH$_2$CO[Ph-Fe(CO)(PPh$_3$)], [9-fluorenyl-Cr(CO)$_3$], MNH[Ph-Cr(CO)$_3$], CH(Ph)[Ph-Cr(CO)$_3$], MCH$_2$C(OMe)=Cr(CO)$_5$, MCH$_2$C(OMe)=W(CO)$_5$	M = Li, K	14–89%	1997 (340)
Br	Boc	H	Me	MCH=CH$_2$	M = MgBr	20%	1998 (286)
Br	Bz	H	Me	MCH$_2$Si(Me)$_2$Ph	M = MgCl	61%	2001 (289)
Br	Bz, Ac, EtOCO, C$_{11}$H$_{23}$CO	H	Et	R^6–O–SiMe$_3$, R^5, R^7 $\longrightarrow$ R^7 R^6 R^5 O; R^5 = Ph, 4-MeO-Ph, 4-Cl-Ph; R^6 = R^7 = H; R^5 = OMe; R^6 = R^7 = Me; R^5 = SEt; R^6 = H; R^7 = H,Me	M = SiMe$_3$ + polymer-SO$_3$Sc(OTf)$_2$	74–91%	2001 (342)
Br	Bz	H	Me	benzimidazole, pyrazole, imidazole, 1,2,4-triazole, imidazolidinone	DIEA	69–84%	2003 (332)

(Continued)

Table 3.12 Additions to Electrophilic Glycine Synthons (see Scheme 3.27) (continued)

X	R¹	R²	R³	Nucleophile (R⁴H or R⁴M)	Catalyst or Base	Yield	Reference
Br Cl	Cbz Ac	H	Me CH(Ph)₂	MCCMe MCCnPr MCCnBu MCCnHex MCCSiMe₃ MCCPh MCC(CH₂)₂OTBDMS	M = SnBu₃ + ZnCl₂	44–88%	1990 (282)
Cl	Cbz Bz BnCO	H	Me	MeOH EtOH PhSH (4-Cl-Ph)SH BnSH PhSSO₂H (4-Me-Ph)SO₂H AcNH(CH₂)₂SH N-OH-piperidine N-OH-succinimide	Et₃N	41–76%	1978 (828)
Cl	Bz	H	Me	MCH=C(Me)Et	Li	12%	1983 (830)
Cl	Cbz	H	Me	CH₂(CO₂R)₂ R = Et, Bn, tBu	Na	48–50%	1983 (336)
Cl	Cbz	H	Me	MEt MiPr McHex MCH=CH₂ MC(Me)=CH₂ MC(Me)=CHMe MC(Me)=C(Me)₂ MCH=CHEt MCH=CHPh MCF=CF₂ M(CH₂)₂CH=CH₂ M(4-MeO-Ph) MCCMe MCCnPr MCCnPent MCCPh	M = MgBr	10–85%	1986 (335) 1988 (259)
Cl	Cbz	H	Me	MCH=CHnBu MCH=CHPh MCH=CH(4-Me-Ph) MCH=CH(4-CF₃-Ph)	M = SiMe₃ + SnCl₄	64–95%	1987 (166)

Cl	MeO_2C	H	Me	[allylsilane diagram: R^5–SiMe$_3$ → R^5 R^6] R^5 = H, nHex, cHex, Ph, $(CH_2)_2OAc$, CH=CHPh; R^6 = H R^5 = cHex; R^6 = Me R^5 = H; R^6 = Me, Br	M = $SiMe_3$ + $SnCl_4$	58–99%	1989 (309)
Cl	MeO_2C	H	Me	[silyl enol ether diagram: R^6 O–SiMe$_3$ / R^7 → R^7 R^6] R^5 = Me, tBu; R^6 = R^7 = H R^5 = H, OMe; R^6 = R^7 = H R^5 = OMe; R^6 = CH=CH$_2$; R^7 = H R^5 = R^6 = $-(CH_2)_4-$; R^7 = H	M = $SiMe_3$ + $SnCl_4$	46–96%	1990 (341)
Cl	Cbz-Gly	H	Bn	iPrSH	Ag	34%	1992 (305)
F	CONHCO-R^2	CONHCO-R^1	CO_2R^3 = CONHBn	NaOEt EtSH piperidine	—	51–89%	1998 (292)
SMe	(1-naphthyl)$_2$CHCO	Sta-MBA	H	$EtNH_2$	$HgCl_2$	82%	1992 (271)
SiPr	Fmoc	H	H	NBS $Boc-NH_2$ Boc-NHMe	—	86%	1996 (281)
benzotriazol-1-yl	$-(CH_2)_5-R^2$ $-(CH_2)_2O(CH_2)_2-R^2$ $CH_2CH=CH_2$ Bn	$-(CH_2)_5-R^1$ $-(CH_2)_2O(CH_2)_2-R^1$ $CH_2CH=CH_2$ Bn	Et	ArH Ar = 2-HO-4-MeO-Ph, 2,4-(HO)$_2$-Ph, 2,4-(MeO)$_2$-Ph, 2-HO-4-Me-Ph, 4-NMe$_2$-Ph, 5-Me-2-furyl, 1-Me-2-pyrrolyl, 1-Me-3-indolyl, 2-HO-1-naphthyl	$AlCl_3$ $TiCl_4$	40–98%	1999 (343)
SiPr	Fmoc Cbz Ac	H	Bn tBu Bn	NBS + $Boc-NH_2$ iPrNH_2 CbzNHiPr MeNHEt $HNEt_2$ $BocNHCH_2CO_2tBu$ $BocNHOCO_2tBu$ $BocNH(CH_2)_2NHBoc$ $BocNH(CH_2)_3NHBoc$ $BocNH(CH_2)_4NHBoc$ BocNHPh $PhNH_2$ BocNH(4-BocO-Ph)	—	20–89%	2002 (302)

give γ,δ-unsaturated amino acids (34–66% yield), although the α-chlorinated derivative gave better yields (51–99%) (309, 310). Allyltrimethylsilane also added to N-Cbz α-methoxyglycine methyl ester in the presence of BF$_3$.Et$_2$O in 80% yield (259), while a methoxy amide derivative of N-Cbz α-methoxyglycine coupled with both allylsilanes (311) and silyl enol ethers (310, 312) under boron trifluoride etherate catalysis to produce allyl or γ-keto adducts. N-Formyl 2-trimethylsiloxy-Ala also reacted with allylsilanes in 12–84% yield with trimethylsilyl triflate as catalyst; dehydroalanine was a major byproduct (313). α-Methoxy-Phg has been allylated under Lewis acid catalysis (295). A range of aromatic and heteroaromatic substituents have been coupled to N-acetyl or N-benzoyl α-alkoxyglycine derivatives under BF$_3$.Et$_2$O catalysis to give arylglycines (268, 314); α-hydroxyglycine also reacts with aromatic compounds under acidic conditions (270).

N-Ac and N-Cbz α-acetoxy-Gly, -Ala, and -Phe derivatives, prepared by anodic oxidation, react in high yield with a range of heteroatom nucleophiles (thiols, alcohols, and amines) in the presence of triethylamine (68–98%) (315). Anisole and furan can be added in the presence of a Lewis acid (58% yield) (293, 315), while pyrimidine and purine bases react in the presence of Et$_3$N or NaH (316). The α-acetoxy-Gly Schiff base ethyl ester of O'Donnell et al. also reacts with neutral alcohols and thiols in good yields, but more basic amines and alkoxides attack the Schiff base instead (291). Some of these latter reagents can instead be successfully reacted with the α-bromo glycine Schiff base intermediate used to prepare the α-acetoxy derivative (291).

Carbon–carbon bonds can be formed in 46–71% yield by using higher-order mixed organocuprates to displace the α-acetoxy substituent, with successful reaction dependent on solvent, temperature, and mode of addition (317). Cyclopropylglycine has been prepared by this method (318). Alternatively, organoboranes react in 47–92% yield (319). This latter method allows the use of precursors not normally applied to amino acid synthesis, as the organoboranes can be prepared from alkenes, aryl halides, and organolithium compounds. Organoboranes have also been added to the benzophenone Schiff base of resin-bound α-acetoxyglycine, with high yields and purity in most cases (276).

Vinylalanes displace the α-acetoxy substrate in the presence of a palladium catalyst to give tri- and tetra-substituted β,γ-unsaturated amino acid derivatives (320). Allylsilanes and silyl enol ethers were combined with α-acetoxy-Ala, -Phe, and -allylglycine in the presence of Lewis acid catalysts, producing α-allyl-amino acids (295). A palladium catalyst was also used to aid the reaction of malonate anions with α-acetoxy-Gly, which led to β-carboxyAsp derivatives in 36–79% yield (321, 322). Diethyl aminomalonates reacted to form β-amino-Asp, while ethyl acetoacetate or ethyl cyanoacetate gave β-acetyl-Asp or β-cyano-Asp, respectively (323). Aryl groups can be introduced by using diarylzinc reagents (324). Neutral carbon nucleophiles can also be employed to displace the α-acetoxy group when reacted in the

presence of TiCl$_4$ although yields were poor, with furan, indole, and anisole adding in 37–53% yield and allylsilanes in 16–46% yield (325). Attempts to use α-thiophenol, α-dimethylamine or α-methoxy as the leaving group were unsuccessful. An intramolecular reaction of N-3-alkenyl-substituted-α-acetoxyglycines in the presence of SnCl$_4$ was used to prepare a number of pipecolic acid derivatives (326). Formic acid can also be used to induce the cyclization, giving formyl-substituted pipecolic acid derivatives instead of chloro- or hydroxy-substituted (327).

3.9.3 Alkylation Reactions: α-Halo Derivatives (see Scheme 3.27, Table 3.12)

The halides of α-haloglycines can be displaced with a number of heteroatoms. α-Alkoxy, α-methoxy and α-thioether glycines were prepared by displacement of N-Bz α-bromo-Gly-OMe with the corresponding heteroatom nucleophiles in the presence of triethylamine (328). Heteroatom-linked diaminodicarboxylic acids were produced by reaction of the same electrophile with the side chains of Cys, Ser, or Lys (329), while a nitrogen-linked glycine trimer resulted from trialkylation of ammonia (330). The fluoride of α-fluoroglycinamides has been displaced with NaOMe, NaSEt, or piperidine (292). Cbz-Gly-α-chloro-Gly-OBn dipeptide was reacted with 2-propanethiol in the presence of silver salt to give 3-thia-Leu (305). Bz-, TFA- or Troc-α-Br-Gly-OMe were reacted with sodium azide, and the resulting α-azido-Gly derivatives employed for cycloaddition reactions with alkynes to form α-triazol-1-yl-Gly products (331). Bz-α-Cl-Gly-OMe reacted with benzimidazole, pyrazole, imidazole, 1,2,4-triazole or imidazolidinone in 69–83% yield. Attempts to induce asymmetry by employing an N-galactopyranose substituent resulted in minimal stereoselectivity (25–41% de) and produced diastereomeric products that could not be resolved (332).

The reaction of organocuprates, silyl enol ethers, and β-dicarbonyl compounds with a variety of N-acyl (Ac, Bz, trifluoroacetyl, Boc) α-bromoglycine esters (Me, Et, tBu) has been investigated (333), with N-Boc α-bromoglycine t-butyl ester the synthon of choice due to ease of deprotection. However, the N-acetyl protecting group allows for enzymatic resolution by acylases. Carbon–carbon bonds are generally formed by "softer" organometallic reagents such as organocuprates and Grignard reagents, with organolithiums reacting in poor yield. For example, a vinyllithium reagent added to N-acetyl α-bromoglycine methyl ester in only 7% yield (283), but addition of a vinyllithium in the presence of CuI gave increased product (334). A number of Reformatsky or Barbier organozinc reagents prepared from the appropriate halide and zinc dust reacted with methyl N-Boc-2-acetoxyglycine in 37–99% yield (273), while Grignard reagents reacted with N-Cbz α-chloroglycine ethyl ester in 10–85% yield (259,335), and malonate added to give β-carboxy-Asp in 48–50% yield (336). N-Bz-protected α-bromoglycine methyl ester was treated with a Grignard reagent to give β-(dimethylphenylsilyl)-Ala in 61% yield (*289*). Both Grignard reagents and enamines were combined with *N*-Boc-protected α-bromoglycine *t*-butyl ester in good yield. The protecting groups were easily removed with 6 N HCl (*287*). Enamines were also added to an *N*-Bz α-bromoglycine-derived *N*-benzoyliminoacetate (*337*), with enantioselective synthesis using a chiral ester auxiliary also investigated (see Section 6.5.1). Organotin acetylides were combined with α-bromo- or α-chloroglycines to give β,γ-alkynylglycine derivatives in 44–86% yield (*282*), while alkyl nitronates gave β-nitro derivatives in 51–86% yield (*338, 339*). Amino acids with organometallic-containing side chains have been produced via addition of a variety of anionic organo-transition metal compounds to Boc- and Bz-α-bromo-Gly-OMe (*340*).

Allylsilanes can be added to *N*-methoxycarbonyl α-chloroglycine methyl ester under Lewis acid catalysis to give γ,δ-unsaturated amino acids (51–99% yield) (*309*), while vinylsilanes add to the *N*-Cbz analog in the presence of SnCl$_4$ to give β,γ-unsaturated derivatives (64–95% yield) (*166*). A hindered (*Z*)-silylstyrene did not add under the standard reaction conditions, but if a reactive acyliminium ion was first generated irreversibly with silver tetrafluoroborate the desired product was obtained in 83% yield (*166*). Silyl enol ethers add to *N*-methoxycarbonyl α-chloroglycine in the presence of SnCl$_4$, leading to γ-oxo-α-amino acid equivalents (*341*). A polymer-supported scandium catalyst was employed to aid the reaction of silyl enol ethers with various *N*-acyl α-bromo-Gly-OEt substrates, giving good yields of γ-ketohomophenylalanines and Asp derivatives, including 3-methyl-Asp and 3,3-dimethyl-Asp (*342*). Furan was coupled with *N*-Ac α-bromo-Gly-OEt under ZnCl$_2$ catalysis (*314*).

3.9.4 Alkylation Reactions: Other α-Substituted Glycines

The α-(benzotriazole)glycine derivative generated from secondary amines, benzotriazole and ethyl glyoxylate was treated with phenols, indoles, furans or pyrroles in the presence of Lewis acids (AlCl$_3$ or TiCl$_4$) to give α-(aryl)glycines in 40–98% yield (*343*). The benzotriazole was also displaced with nucleophiles such as ammonia, leading to α-aminoglycine (*272, 301*). In a similar manner, *N*-Fmoc α-(isopropylthiol)glycine was displaced with Boc carbamate or *N*-methyl Boc carbamate to give orthogonally protected α-aminoglycine.

The Boc group was removed and the amine acylated to introduce substituents corresponding to the common amino acid side chains (*281*). Other α-thiolglycines have also been reacted with other amines (*271, 280, 303*) or converted into α-haloglycines (*267, 282–285*). Fmoc- or Cbz-α-(isopropylthio)glycine was treated with various primary or secondary amines to give α-aminoglycine analogs of Ala, Leu, Ile, Glu, Ser, Lys, Ph, and Tyr, with aminoalkyl side chains (*302*).

3.9.5 Imines as Electrophilic Glycine Equivalents

The imines and oximes that are obtained as intermediates in reductive aminations of α-keto acids can also be used as electrophilic glycine equivalents. Instead of simply being reduced, they are used as substrates for the addition of organometallic compounds. Regioselectivity is potentially a problem as the organometallic can add to either end of the C=N bond, or the carbonyl of the acid group. The electrophilic imine/oxime substrates have been derived from a variety of sources. If α-keto acids other than glyoxylate are used to form the imine/oxime C=N bond, organometallic addition leads to α,α-disubstituted amino acids. Many researchers have alkylated imines prepared from α-halo- or α-acetoxyglycine derivatives, but the imines are usually non-isolated (and often non-acknowledged) intermediates. The α-hydroxy/methoxy/acetoxy-glycine substrates are often prepared from glyoxylic acid by condensation with amines, amides or urethanes (see Section 3.9.1). *N,N′*-Fmoc$_2$-α-aminoglycine was prepared in 59% yield from glyoxylic acid by refluxing with Fmoc-NH$_2$ in toluene/PTSA, and used as a masked glyoxylate during peptide synthesis (*344*).

3.9.5a Imine/Oxime Reduction via Organometallic Addition

Alkyllithium reagents (including hindered nucleophiles such as *t*-butyllithium) were found to add in good yield (63–80%) to *O*-benzoyl oximes of glyoxylate and pyruvate, giving *N*-hydroxy amino acids (see Scheme 3.28) (*345, 346*). The *N*-Boc or -Cbz imine of ethyl 3,3-difluoropyruvate was reacted with Grignard reagents in 61–80% yield to give protected α-difluoromethyl-α-amino acids, which were deprotected in 53–96% yield (*347*). Lithiated propargylamine also added in excellent yield, leading to α-substituted ornithine derivatives (*348*). A Grignard reagent, MgBr(CH$_2$)$_3$CH=CH$_2$, was

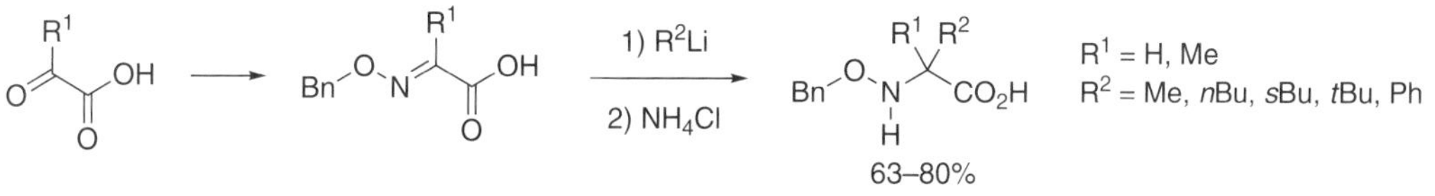

Scheme 3.28 Organometallic Addition to Oxime of α-Keto Acid (*345*).

added to an imine of trifluoropyruvate. The alkene was oxidized to provide α-(trifluoromethyl)-homoglutamic acid; Hofmann rearrangement of the side-chain acid gave α-(trifluoromethyl)-Orn, with guanylation resulting in α-(trifluoromethyl)-Arg (349). Nitroalkanes were added to imines formed from glyoxylate and ammonia or methylamine in aqueous KOH to form β-nitro-Nva, β-nitro-Val, β-nitro-Phe or β-nitro-cyclopentyl-Gly (350, 351). A resin-bound ester of the N-(p-methoxyphenyl)imine of glyoxylate was reacted with several silyl enol ethers in the presence of Sc(OTf)$_3$, with ester cleavage resulting in N-PMP derivatives of Asp, 3-methyl-Asp, 3,3-dimethyl-Asp, 3-benzyloxy-Asp, 4-oxo-homophenylalanine, and 3-methyl-4-oxo-homophenylalanine in 65–94% yield (352, 353). Another approach towards transforming glyoxylate imines to amino acids employs an imine-ene reaction, which involves Lewis acid-catalyzed addition of an alkene possessing an allylic hydrogen. A number of γ,δ-unsaturated amino acids were prepared from butyl N-(tosyl)-iminoacetate in 56–92% yield by this method (354).

Petasis and Zavialov have reductively aminated α-keto acids by adding an organoboronate to the imine generated by condensing the acid and an amine (see Scheme 6.34 for asymmetric version) (355). The organoboron compound was used to introduce an unsaturated side chain, providing vinylglycine derivatives from glyoxylic acid, and α-alkylvinylglycines from other α-keto acids. The alkene could be reduced. Arylglycines were obtained by using arylboronic acids (356). A closely related reaction employs a radical addition (see Scheme 3.40 in Section 3.10). The Petasis multicomponent synthesis of amino acids, combining glyoxylic acid with an amine and alkyl boronic acid, was used to generate the acid component input for a Ugi 4CC reaction, forming dipeptide amides from six components (357, 358).

Several four-component-type reactions have been described. A tertiary iminium salt was generated from ethyl glyoxylate, secondary amines, and acetyl chloride. It reacted with cyclohexanone or tetralone to give γ-oxo-α-amino acids (359). Another four-component reaction combined an amine and ethyl glyoxylate along with a cyclic enol ether in the presence of a Lewis acid. The oxocarbocation generated by addition of the

enol ether to the imine was then trapped by an alcohol. Amino acids with a 2-alkoxytetrahydropyran-3-yl side chain were produced (360). A three-component reaction between ethyl glyoxylate, anilines, and phenols in the presence of scandium triflate catalyst produced N-aryl 2'-hydroxyarylglycines in 30–86% yield (361).

Imine substrates can be obtained from sources other than α-keto acids, such as the imines generated in situ by treatment of α-haloglycines with base, as described previously. In a similar manner, Grignard reagents, alkyllithium enamines, and other C-anions were added to an acyliminomalonic ester (prepared by dehydrohalogenation of an α-bromo-α-aminomalonic ester) in 50–98% yield (see Scheme 3.29), resulting in the acylaminomalonic ester intermediate found in the Sörensen synthesis (see Section 3.2) (362, 363). Thioibotenic acid, α-(3-hydroxy-5-isothiazolyl)-Gly, was produced by this route (364). A series of lithiated 1-benzyloxy-5-substituted-pyrazoles, derived from the 4-iodopyrazoles, were also added to diethyl N-Boc iminomalonate in good yields (365).

Lipshutz et al. found that oxidative degradation of imidazoles (readily synthesized by a three-component cyclization) led to acylimines (see Scheme 3.30). If the carboxy group was derivatized as a hydroxamic acid, a variety of organolithium, organocuprate, and Grignard reagents could be added in 52–94% yield, resulting in α,α-disubstituted N-acyl amino acid hydroxamic acids (366). If the carboxy group was functionalized as an amide, no reaction occurred. Alternatively, the imine intermediate could be isomerized to an α,β-didehydroamino acid equivalent and catalytically hydrogenated, producing monosubstituted acyl amino acid amides (367).

The nitrile of O-allyl cyanohydrin, allyl-OCH$_2$CN, acts as a masked electrophilic glycine synthon, undergoing sequential addition of two Grignard reagents. Removal of the allyl protecting group and oxidation of the hydroxymethyl group produced α,α-disubstituted amino acids (368). An asymmetric version has been reported (369). A new electrophilic glycine equivalent has been prepared by NCS oxidation of an N-(2-nitrophenyl) sulfenamide (see Scheme 3.31) (370). This synthon (with a t-butyl ester) was a substrate for

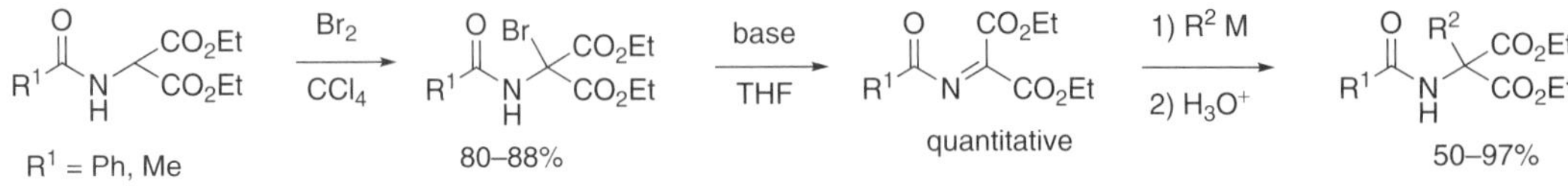

M = MgX, R^2 = Me, 1-naphthyl, CH=CH$_2$, CH=CH(Me), CC-Ph
M = Li, R^2 = nBu, 2-pyridyl
M = other base, R^2 = CH$_2$NO$_2$, 2-oxo-1-cyclopentyl,
CH(CN)-CO$_2$Me, CH(Me)-CONEt$_2$, Ph$_3$P=C(CO$_2$Et) (363)
M = Li, R^1 = tBuO; R^2 = 3-BnO-5-isothiazole (364)
M = Li, R^1 = tBuO; R^2 = 1-BnO-5-R-4-pyrazole; R = H, Cl, SMe, Me,
4-Me-Ph, 4-MeO-Ph, 2-MeO-Ph, 2-F-Ph, 2-thienyl, 4-NO$_2$-Ph (365)

Scheme 3.29 Alkylation of Electrophilic Glycine Equivalent via Imine.

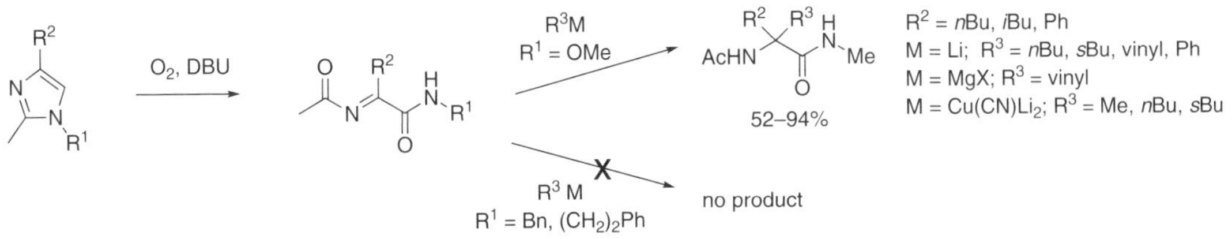

Scheme 3.30 Alkylation of Electrophilic Glycine Equivalent via Imine (*366*).

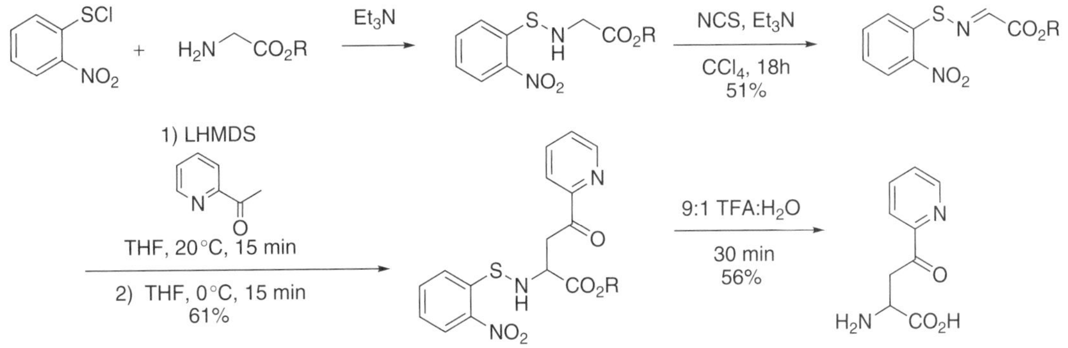

$R^2 = n$Bu, iBu, Ph
M = Li; $R^3 = n$Bu, sBu, vinyl, Ph
M = MgX; R^3 = vinyl
M = Cu(CN)Li$_2$; R^3 = Me, nBu, sBu

52–94%

R^3 M
R^1 = Bn, (CH$_2$)$_2$Ph
no product

Scheme 3.31 New Electrophilic Glycine Equivalent (*370*).

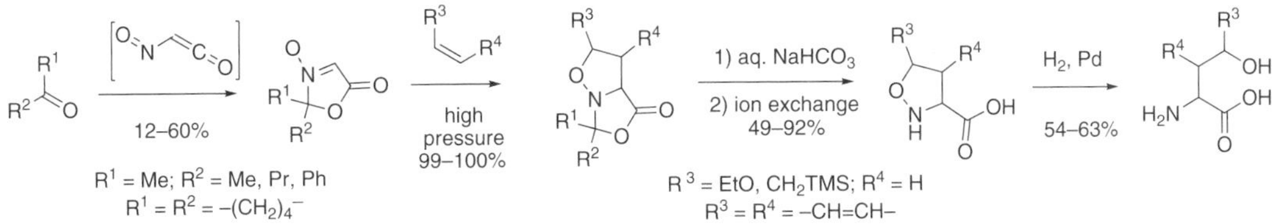

Et$_3$N

NCS, Et$_3$N
CCl$_4$, 18h
51%

1) LHMDS

THF, 20°C, 15 min

2) THF, 0°C, 15 min
61%

9:1 TFA:H$_2$O
30 min
56%

Scheme 3.32 Nitrone Synthesis and Cycloaddition (*371*).

12–60%
R^1 = Me; R^2 = Me, Pr, Ph
$R^1 = R^2 = -$(CH$_2$)$_4^-$

high pressure
99–100%

1) aq. NaHCO$_3$
2) ion exchange
49–92%

R^3 = EtO, CH$_2$TMS; R^4 = H
$R^3 = R^4 = -$CH=CH$-$

H$_2$, Pd
54–63%

addition of Grignard reagents, giving adducts in 35–65% yield. Lithium enolates were also added; a γ-oxo-2′-azo-homophenylalanine derivative was synthesized in 61% yield (*370*).

3.9.5b Imine/Oxime Reduction via Cycloaddition

A number of imines and imine-equivalents have been used as substrates for cycloaddition reactions. Nitrones, which are generally prepared by condensation of hydroxylamines with ketones, are potentially useful for amino acid synthesis as they undergo [3+2] dipolar cycloadditions with electron-rich olefins to give isoxazolidines, introducing a side-chain substituent. Cyclic nitrones have been prepared by a novel reaction involving a Meldrum's acid-derived nitrosoketene

intermediate (see Scheme 3.32). Cycloaddition, followed by hydrolysis and catalytic cleavage of the N–O bond, gave several amino acids (*371*).

Imines, generated in situ by acidic treatment of N-acyl α-hydroxy- or α-methoxyglycine precursors, predominantly act as dienes rather than as dienophiles in cycloadditions with dienes (see Scheme 3.33) (*372, 373*). Monoalkenes therefore also react in good yield (23–80%) (*374, 375*). The initial six-membered oxazine product rearranges to form a five-membered lactone under acidic conditions (*372*), but can be isolated if BF$_3$ is used to catalyze the cycloaddition (*374, 375*).

Reaction of the imine as a dienophile during the aza Diels–Alder reaction is favored by using an N-sulfonyl substituent as an electron-withdrawing group (see Scheme 3.34). N-Sulfonylimino acetates react with Danishefsky's diene, H$_2$C=C(OTMS)CH=CHOMe, to

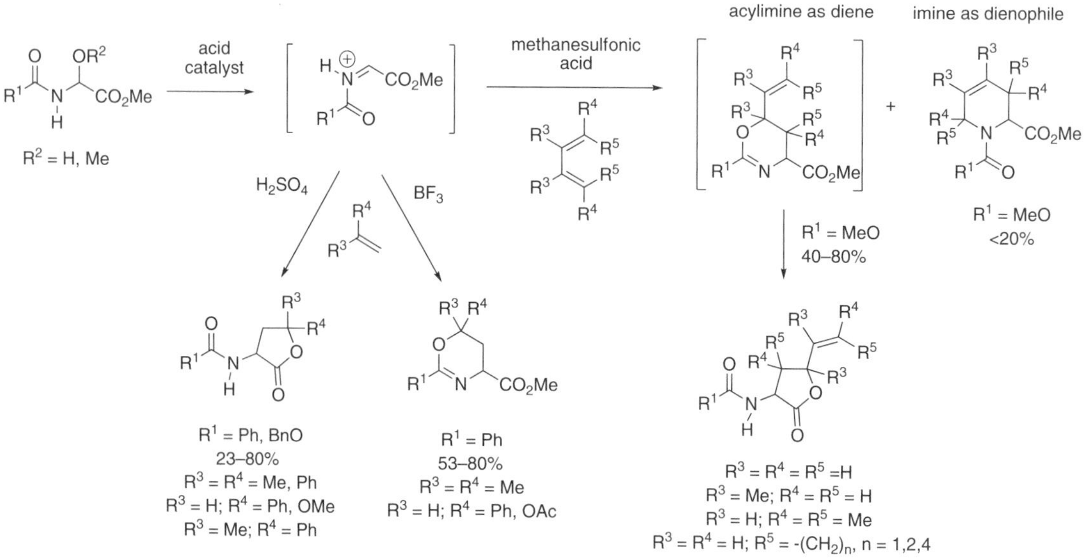

Scheme 3.33 Cycloaddition Reactions with Imines as Dienes (*372–375*).

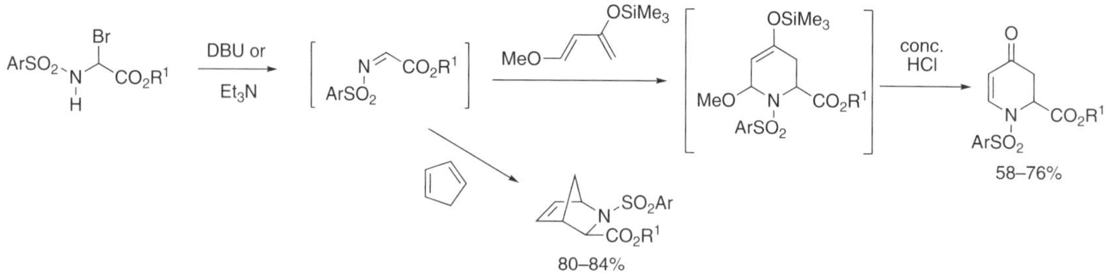

Scheme 3.34 Cycloaddition Reactions with Imines as Dienophiles (*376, 377*).

give pipecolic acid derivatives in 58–76% yield (*376*) and with cyclopentadiene to give bicyclic Pro analogs in 80–84% (*377*). The iminium ion derived from methyl glyoxylate and ammonium chloride acts as a dienophile with cyclopentadiene to give an azabicyclo[2.2.1]heptene Pro analog in over 70% yield (*378*). Similar reactions have been carried out using the resin-bound ester of *N*-aryl imines of glyoxylate. Reaction as the dienophile with Danishefsky's diene in the presence of Sc(OTf)$_3$ resulted in *N*-PMP methyl 4-oxo-5,6-dehydropipecolate in 69% yield after resin cleavage (*352*). Alternatively, when the imine acted as the diene component in aza Diels–Alder reactions with alkenes in the presence of Sc(OTf)$_3$, various 2-carboxy-tetrahydroquinolines were produced (*352, 353*).

Diazomethane is basic enough to generate imines from methyl *N*-acyl-α-chloroglycinate; a [3+2] cycloaddition provided aziridines in 14–34% yield (see Scheme 3.35) (*379*).

1,3-Dipolar cycloadditions of alkenes can be carried out by using the 1,3-dipole generated from methyl or ethyl chlorooximidoacetates after treatment with triethylamine (see Scheme 3.36) (*380, 381*). γ-Hydroxy amino acids, including the *N*-terminal amino acid component of the nikkomycins, have been prepared via this route (*382–386*).

3.10 Radical Alkylations

There are comparatively few examples of radical alkylations of glycine equivalents. The same substrates that are used for electrophilic glycine alkylation are often used for radical alkylation, and in some cases it is not clear which mechanism is operative (*310, 387*). Free radical reactions employed for the synthesis of amino acids were reviewed in 1997 (*388*). Glycine residues show a selective reactivity for radical generation at the α-center

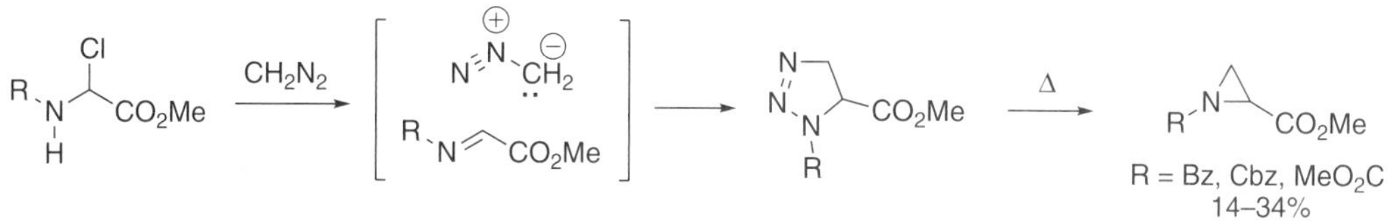

Scheme 3.35 [3+2] Cycloaddition of Diazomethane with in situ Generated Imine (*379*).

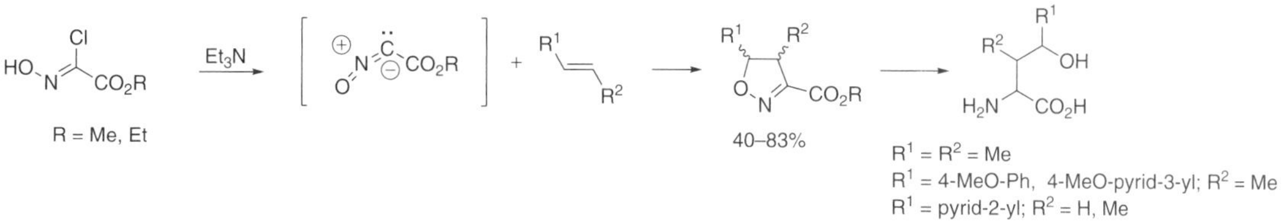

Scheme 3.36 1,3-Dipolar Cycloaddition of Alkenes with in situ Generated 1,3-Dipole (*380–386*).

Scheme 3.37 Radical Alkylation of Glycine (See Table 3.13).

compared to other amino acids, even though tertiary radicals are usually formed preferentially. This anomaly has been explained by a favorable geometry stabilizing the secondary glycine radical compared to tertiary radicals of other amino acids (*388, 389*). Thus glycine within a dipeptide can be selectively brominated with NBS and reduced with triphenyltin deuteride (*389*). Unsubstituted Gly residues within dipeptides have also been alkylated by radical reactions, with regioselectivity depending upon substituents (*388*). The prototypical 2-glycyl radical, $H_2N\text{-}CH\bullet\text{-}COOH$, has recently been shown to be a stable species in the gas phase (*390*). A theoretical study on the stability of α-carbon-centered peptide radicals correctly identified that, relative to Gly radicals, Ala and Val residues were destabilized (by approximately 9 and 18 kJ/mol, respectively). "Radical resistant" residues were predicted to be Tle-Leu and 3,3,3-trifluoro-Ala (36 and 41 kJ/mol destabilized) due to disruption of amide carbonyl planarity, and indeed these amino acids either reacted slowly with N-bromosuccinimide (Tle) or were inert (F$_3$-Ala) (*391*). The destabilization of glycyl radical formation in Gly derivatives has been correlated with their ability to inhibit peptidylglycine α-amidating monooxygenase, an enzyme that catalyzes the synthesis of peptide amides through a radical cleavage of a C-terminal Gly residue (*392*).

The first attempts at radical reactions used photochemical alkylations of glycine derivatives. Reasonable yields (10–50%) of Phe, Tyr, Nle or Asp could be obtained by irradiating a mixture of glycine-containing dipeptides with toluene, 4-methoxytoluene, 1-butene, or acetic acid, respectively, in the presence of a light-absorbing α-diketone and proton-abstracting peroxide (*393*). Subsequent attempts at alkylation of glycine synthons by radical reactions have generally relied upon α-haloglycinate derivatives. For example, N-benzoyl 2-bromoglycine methyl ester reacted with allylstannane reagents or Bu$_3$SnD in reasonable yields (*394–396*) (see Scheme 3.37; Table 3.13). The bromoglycine residue could also be located within a dipeptide, although yields were reduced (*394*). α-(Alkylthio)glycyl residues, derived from α-acetoxyglycine contained within a peptide, also react with tributyltin hydride or allyltributylstannane, though with no diastereoselectivity (*275*).

More recently, α-benzyloxy- and α-methoxyglycine derivatives, which are more stable than the corresponding halides, have been found to undergo analogous reactions with stannanes (see Scheme 3.37, Table 3.13) (*387*). The reactions are more consistent with a radical mechanism than cationic reactions as cross-linked glycine-dimer is obtained as a byproduct. Amino acids can be purposely cross-linked at the α-position by treatment

Table 3.13 Radical Alkylations of Glycine Synthons (see Scheme 3.37)

X	R¹	R²	Conditions	Products (R³)	Yield	Reference
Br	Bn	Me	Bu₃SnD	D	67%	1988 (395)
Br	Bn	Me	(PhCO₂)₂	OCOPh	64%	1988 (395)
Br	Bn	Me	R¹₃SnCH₂C(R²)=CH₂ AIBN, PhH, 80 °C R¹ = nBu, Ph	CH₂C(R²)=CH₂ R² = H, Cl, CN, CO₂Et	50–74%	1988 (395)
Br	Bz, Bz, Bz-Val	Me, Val-OMe, Me	Bu₃SnR	CH₂CH=CH₂	33–62%	1988 (394)
Br	Bz	Me	(structure with R⁶, R⁷, R⁴, R⁵, Bu₃Sn) AIBN, PhH, 80 °C	(structure) R⁴ = R⁵ = R⁶ = H; R⁷ = H, Me; R⁴ = Me; R⁵ = R⁶ = R⁷ = H; R⁴ = R⁵ = H; R⁶ = R⁷ = Me; R⁴ = R⁶ = −(CH₂)₂−; R⁵ = R⁷ = H	19–63%	1990 (396)
OMe, OBz	Bz	Me	Bu₃SnR³	H, D, CH₂CH=CH₂	34–92%	1992 (387)
OMe, OBz	Bz	Me	Bu₆Sn₂, PhH, reflux, then (tBuS)₂, (PhS)₂ or (Bz-Cys-OMe)₂	SPh, StBu, SCH₂CH(NHBz)CO₂Me	34–92%	1992 (387)
SEt	Cbz-Phe	Me, Val-OMe	Bu₃SnR	H, CH₂CH=CH₂	64–89%	1993 (275)
Br, OMe, OBz	Bz	Me	(RS)₂ (tBuO)₂ or nBu₆Sn₂ or Bu₃SnH or AIBN	SPh, SBn, StBu	5–50%	1994 (328)
S(2-pyridyl)	Bz	Phe-OMe, Val-OMe	SmI₂ with R¹COR² (generates enolate)	C(OH)[−(CH₂)₅−], C(OH)[−(CH₂)₄−], CH(OH)cHex, CH(OH)nHept, CH(OH)Ph, CH(OH)CH₂CH{O-CH(CH₂OBn)[CH(OBn)]₃−}	20–85%	2000 (415)
S	Bz or Bz-Xaa	OMe or Xaa-OMe	dilauroyl peroxide 1,2-DCE, reflux	(CH₂)₂CN, (CH₂)₃SiMe₃, CH₂CH=CH₂	14–57%	2001 (399)

with photochemically produced t-butoxy radicals in benzene, giving the dimer of methyl N-acetylglycinate or methyl pyroglutamate in 51% or 64% yield, respectively (397). α-Bromo-, α-benzoyloxy-, and α-methoxy-substituted glycine derivatives have been reacted with stannanes to give a glycyl radical, which was then trapped with dialkylsulfides or diarylsulfides, resulting in α-thioether-glycines (although often with quite poor yields) (328).

Although the additions of glycine radical equivalents to allylstannanes have been successful, additions to other alkenes have been problematic. Udding, Hiemstra, and Speckamp reported that a protected α-xanthate glycinate (with an α-SC(=S)OEt substituent, prepared from α-chloroglycine) coupled with alkenes using di-t-butyl peroxide as initiator (see Scheme 3.38) (398). The xanthate group was removed from several of the

adducts with Bu₃SnH. The α-xanthate-Gly residue has been incorporated within a number of dipeptides as either the N- or C-terminal residue. The radical was generated by addition of dilauroyl peroxide to a refluxing solution of xanthate in 1,2-dichloroethane, then trapped with acrylonitrile, allyltrimethylsilane, or allyl ethyl sulfone to produce 4-cyano-Abu, 5-trimethylsilyl-Nva, or allylglycine residues (with up to 2:1 reported diastereoselectivity) (399). Chai and Page successfully added the radical derived from bromoglycine contained within a piperazine-2,5-dione to methyl acrylate; other alkenes gave mainly reduced product (400).

Intramolecular radical reactions have also been investigated. α-Chloroglycine derivatives containing a 3-alkenyl substituent on the nitrogen were cyclized by a radical process, under copper(I) catalysis (310, 401, 402), leading to proline derivatives. Similar regio- and

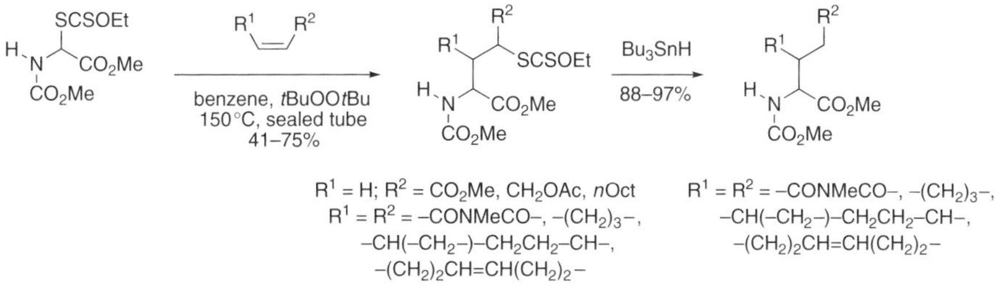
SCSOEt
CO_2Me
benzene, tBuOOtBu
150°C, sealed tube
41–75%
SCSOEt
Bu_3SnH
88–97%
R^1 = H; R^2 = CO_2Me, CH_2OAc, nOct
R^1 = R^2 = –CONMeCO–, $-(CH_2)_3-$,
$-CH(-CH_2-)-CH_2CH_2-CH-$,
$-(CH_2)_2CH=CH(CH_2)_2-$
R^1 = R^2 = –CONMeCO–, $-(CH_2)_3-$,
$-CH(-CH_2-)-CH_2CH_2-CH-$,
$-(CH_2)_2CH=CH(CH_2)_2-$

Scheme 3.38 Radical Reaction of Glycine Equivalent with Alkenes (*398*).

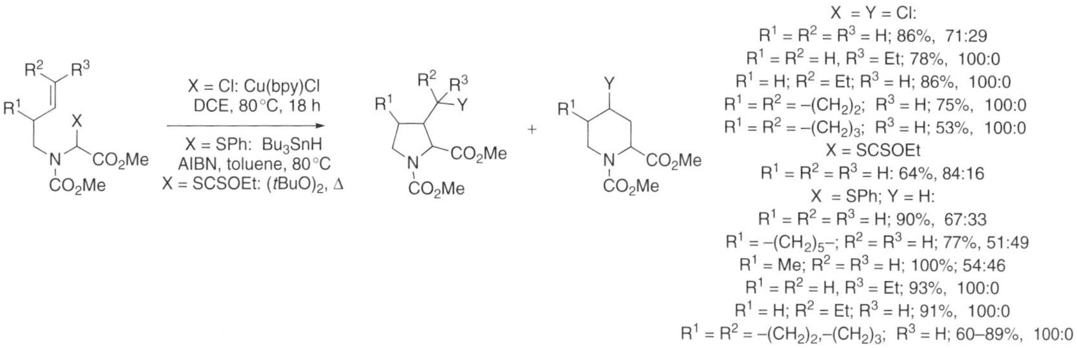
X = Cl: Cu(bpy)Cl
DCE, 80°C, 18 h
X = SPh: Bu_3SnH
AIBN, toluene, 80°C
X = SCSOEt: $(tBuO)_2$, Δ
X = Y = Cl:
$R^1 = R^2 = R^3$ = H; 86%, 71:29
$R^1 = R^2$ = H, R^3 = Et; 78%, 100:0
R^1 = H; R^2 = Et; R^3 = H; 86%, 100:0
$R^1 = R^2 = -(CH_2)_2$; R^3 = H; 75%, 100:0
$R^1 = R^2 = -(CH_2)_3$; R^3 = H; 53%, 100:0
X = SCSOEt
$R^1 = R^2 = R^3$ = H: 64%, 84:16
X = SPh; Y = H:
$R^1 = R^2 = R^3$ = H; 90%, 67:33
$R^1 = -(CH_2)_5-$; $R^2 = R^3$ = H; 77%, 51:49
R^1 = Me; $R^2 = R^3$ = H; 100%; 54:46
$R^1 = R^2$ = H, R^3 = Et; 93%, 100:0
R^1 = H; R^2 = Et; R^3 = H; 91%, 100:0
$R^1 = R^2 = -(CH_2)_2$,$-(CH_2)_3$; R^3 = H; 60–89%, 100:0

Scheme 3.39 Radical Cyclization of *N*-3-alkenyl-glycinate (*310, 401–403*).

stereoselectivity was obtained by the analogous Bu_3SnH-mediated cyclization of *N*-alkenyl α-pheny-lthio-glycinates (*310, 403, 404*) (see Scheme 3.39). The amino acid ester substituent was found to favour the radical transfer process leading to pyrrolidines over a cationic process, resulting in piperidines (*310, 405*). The cationic process could be preferentially selected by using an α-acetoxyglycine derivative under $SnCl_4$ catalysis (*310, 326*). α-Xanthate derivatives with *N*-(3-alkenyl) and *N*-(4-alkenyl) substituents have also been cyclized, using di-*t*-butyl peroxide as initiator. The conditions result in different ring systems than provided by the other methods (*406*). The xanthate group was reduced after the cyclization, providing conformationally restricted Met analogs. Proline derivatives have also been synthesized by photochemical conversion of *N*-acyl,*N*-(β-benzoylethyl)-glycinates (*407, 408*).

Radical reactions of glycine equivalent imine/oximes have also been reported. An *O*-benzyl oxime of glyoxylic acid, attached to a solid resin support, was employed as a substrate for addition of alkyl radicals derived from alkyl iodides, Bu_3SnH, and Et_3B (see Scheme 3.40 top) (*409*). The products were obtained in 24–78% yield following cleavage from the resin. The same reaction was conducted in solution with the oxime of methyl glyoxylate, producing the same products in 60–85% yield (*410*). A radical-based three-component condensation is possible. The oxime formed in situ from 2-hydroxy-2-methoxyacetic acid methyl ester and benzyloxyamine reacted with alkyl radicals generated from alkyl iodides and Bu_3SnH/Et_3B as initiator. *N*-OBn amino esters were produced in 46–86% yield (see Scheme 3.40 middle) (*411, 412*). A more extensive study of the reaction of the *O*-benzyl oxime of methyl glyoxylate with stabilized primary radicals generated from alkyl halides in the presence of triethylborane was reported in 2004. Reactivity varied greatly between different alkyl halides, with benzyl iodide, ethyl iodoacetate, isopropyl iodide, and α-bromobutyrolactone adding in 62–99% yield, but methyl iodide, allyl iodide, ethyl bromoacetate, and pentafluorobenzyl bromide all producing <10% yield of product. A variety of more complex secondary iodides gave good yields of unusual amino acids (see Scheme 3.40 bottom) (*413*). A glyoxylic hydrazone, $Ph_2NN{=}CHCO_2Me$, provided a good substrate for indium-mediated additions of radicals derived from alkyl halides; *N*-diphenylamino Val, Ile, Tle, and cyclopentylglycine were synthesized (*414*).

A radical reaction has been employed to selectively generate a Gly enolate within a peptide. Gly-containing

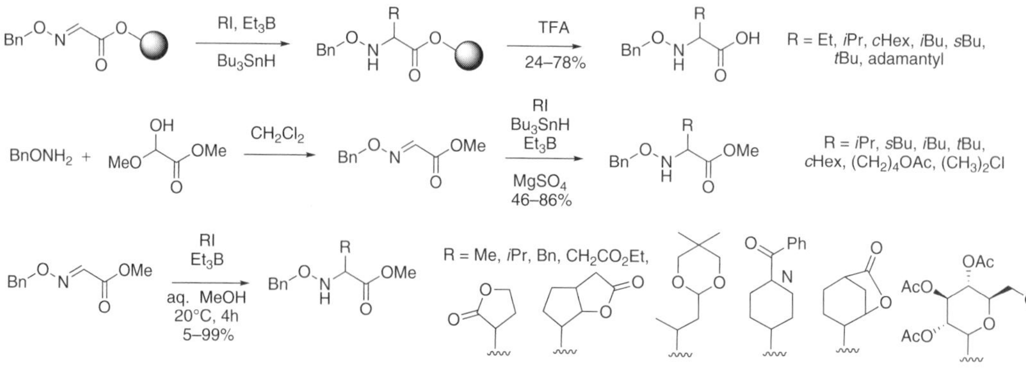

Scheme 3.40 Radical Addition to Oximes of Glyoxylic Acid (*409–413*).

di- or tripeptides were treated with NBS to generate α-bromo-Gly residues, which were converted to more stable α-(2-mercaptopyridyl)-Gly moieties. The glycyl sulfides were treated with samarium diiodide to form a glycyl radical that was further reduced to a samarium enolate; exposure to alkyl aldehydes and ketones produced β-hydroxy amino acid residues with no diastereoselectivity (*415*).

References

1. Rios, A.; Amyes, T.L.; Richard, J.P. *J. Am. Chem. Soc.* **2000**, *122*, 9373–9385.
2. Rios, A.; Richard, J.P.; Amyes, T.L. *J. Am. Chem. Soc.* **2002**, *124*, 8251–8259.
3. Stroud, E.D.; Fife, D.J.; Smith, G.G. *J. Org. Chem.* **1983**, *48*, 5368–5369.
4. Rios, A.; Crugeiras, J.; Amyes, T.L.; Richard, J.P. *J. Am. Chem. Soc.* **2001**, *123*, 7949–7950.
5. Sörensen, S.P.L. *Ber.* **1908**, *41*, 3387–3390.
6. Sœrensen, S.P.L. *Bull. Soc. Chim.* **1905**, *33*, 1042–1052.
7. Sörensen, S.P.L. *Z. Physiol. Chem.* **1905**, *44*, 448–460.
8. Putochin, N.J. *Ber.* **1923**, *56*, 2213–2216.
9. Locquin, R.; Cerchez, V. *Bull. Soc. Chim.* **1930**, *47*, 1386–1389.
10. Redemann, C.E.; Dunn, M.S. *J. Biol. Chem.* **1939**, *130*, 341–348.
11. Locquin, R.; Cerchez, V. *Bull. Soc. Chim.* **1931**, *49*, 42–57.
12. Albertson, N.F. *J. Am. Chem. Soc.* **1946**, *68*, 450–453.
13. Albertson, N.F.; Archer, S. *J. Am. Chem. Soc.* **1945**, *67*, 308–310.
14. Snyder, H.R.; Howe, E.E.; Cannon, G.W.; Nyman, M.A. *J. Am. Chem. Soc.* **1943**, *65*, 2211–2214.
15. Albertson, N.F.; Archer, S.; Suter, C.M. *J. Am. Chem. Soc.* **1944**, *66*, 500.
16. Snyder, H.R.; Smith, C.W. *J. Am. Chem. Soc.* **1944**, *66*, 350–251.
17. Dakin, H.D. *J. Biol. Chem.* **1944**, *154*, 549–555.
18. Dunn, M.S.; Smart, B.W.; Redemann, C.E.; Brown, K.E. *J. Biol. Chem.* **1931–2**, *94*, 599–609.
19. Locquin, R.; Cerchez, V. *Bull. Soc. Chim.* **1931**, *49*, 42–57.
20. Painter, E.P. *J. Am. Chem. Soc.* **1940**, *62*, 232–233.
21. Galat, A. *J. Am. Chem. Soc.* **1947**, *69*, 965.
22. Berger, A.; Smolarsky, M.; Kurn, N.; Bosshard, H.R. *J. Org. Chem.* **1973**, *38*, 457–460.
23. Bosshard, H.R.; Berger, A. *Helv. Chim. Acta* **1973**, *56*, 1838–1845.
24. Ong, H.H.; Creveling, C.R.; Daly, J.W. *J. Med. Chem.* **1969**, *12*, 458–461.
25. Okuda, T. *Bull. Chem. Soc. Jpn.* **1957**, *30*, 358–360.
26. Bossler, H.G.; Waldmeier, P.; Seebach, D. *Angew. Chem., Int. Ed. Engl.* **1994**, *33*, 439–440.
27. Matt, T.; Seebach, D. *Helv. Chim. Acta* **1998**, *81*, 1845–1895.
28. Kozikowski, A.P.; Ma, D.; Pang, Y.-P.; Shum, P.; Likic, V.; Mishra, P.K.; Macura, S.; Basu, A.; Lazo, J.S.; Ball, R.G. *J. Am. Chem. Soc.* **1993**, *115*, 3957–3965.
29. Shapira, J.; Shapira, R.; Dittmer, K. *J. Am. Chem. Soc.* **1953**, *75*, 3655–3657.
30. Belsky, I. *J. Chem. Soc., Chem. Commun.* **1977**, 237.
31. Tolman, V.; Sedmera, P. *Tetrahedron Lett.* **1988**, *29*, 6183–6184.
32. Lankiewicz, L.; Kasprzykowski, F.; Grzonka, Z.; Kettmann, U.; Hermann, P. *Bioorganic Chem.* **1989**, *17*, 275–280.
33. Bayle-Lacoste, M.; Moulines, J.; Collignon, N.; Boumekouez, A.; de Tinguy-Moreaud, E.; Neuzil, E. *Tetrahedron* **1990**, *46*, 7793–7802.
34. Potrzebowski, M.J.; Stolowich, N.J.; Scott, A.I. *J. Lab. Cmpds. Radiopharm.* **1990**, *28*, 355–361.
35. van Hest, J.C.M.; Kiick, K.L.; Tirrell, D.A. *J. Am. Chem. Soc.* **2000**, *122*, 1282–1288.
36. Krapcho, A.P.; Weimaster, J.F.; Eldridge, J.M.; Jahngen E.G.E., Jr.; Lovey, A.J.; Stephens, W.P. *J. Org. Chem.* **1978**, *43*, 138–147.
37. Begtrup, M.; Sløk, F.A. *Synthesis* **1993**, 861–863.
38. Hudson, R.D.A.; Osborne, S.A.; Stephenson, G.R. *Synlett*, **1996**, 845–846.

39. Schmidt, U.; Griesser, H.; Lieberknecht, A.; Schmidt, J.; Gräther, T. *Synthesis* **1993**, 765–766.
40. Gershon, H.; Scala, A. *J. Org. Chem.* **1961**, *26*, 2347–2350.
41. Chung, J.Y.L.; Wasicak, J.T.; Arnold, W.A.; May, C.S.; Nadan, A.M.; Holladay, M.W. *J. Org. Chem.* **1990**, *55*, 270–275.
42. Mauger, A.B.; Irreverre, F.; Witkop, B. *J. Am. Chem. Soc.* **1966**, *88*, 2019–2024.
43. Magerlein, B.J.; Birkenmeyer, R.D.; Herr, R.R.; Kagan, F. *J. Am. Chem. Soc.* **1967**, *89*, 2459–2464.
44. Damour, D.; Doeflinger, G.; Pantel, G.; Labaudinière, R.; Leconte, J.-P.; Sablé, S.; Vuilhorgne, M.; Mignani, S. *Synlett* **1999**, 189–192.
45. Cox, D.A.; Johnson, A.W.; Mauger, A.B. *J. Chem. Soc.* **1964**, 5024–5029.
46. Kopola, N.; Friess, B.; Cazes, B.; Gore, J. *Tetrahedron Lett.* **1989**, *30*, 3963–3966.
47. Atkinson, R.O. *J. Chem. Soc.* **1952**, 3317–3318.
48. Atkinson, R.O.; Poppelsdorf, F.; Williams, G. *J. Chem. Soc.* **1953**, 580–581.
49. Burgstahler, A.W.; Aiman, C.E. *J. Org. Chem.* **1960**, *25*, 489–492.
50. Jiang, J.; Miller, R.B.; Tolle, J.C. *Synth. Commun.* **1998**, *28*, 3015–3019.
51. Albertson, N.F.; Tullar, B.F. *J. Am. Chem. Soc.* **1945**, *67*, 502–503.
52. Goering, H.L.; Cristol, S.J.; Dittmer, K. *J. Am. Chem. Soc.* **1948**, *70*, 3310–3313.
53. Smith, R.J.; Bratovanov, S.; Bienz, S. *Tetrahedron* **1997**, *53*, 13695–13702.
54. Hoppe, D. *Angew. Chem., Int. Ed. Engl.* **1975**, *14*, 426–427.
55. Hoppe, D. *Angew. Chem., Int. Ed. Engl.* **1975**, *14*, 424–426.
56. Abdel-Monem, M.M.; Newton, N.E.; Weeks, C.E. *J. Med. Chem.* **1974**, *17*, 447–451.
57. Abdel-Monem, M.M.; Newton, N.E.; Ho, B.C.; Weeks, C.E. *J. Med. Chem.* **1975**, *18*, 600–604.
58. Stork, G.; Leong, A.Y.W.; Touzin, A.M. *J. Org. Chem.* **1976**, *41*, 3491–3493.
59. Yamada, S.-I.; Oguri, T.; Shioiri, T. *J. Chem. Soc., Chem. Commun.* **1976**, 136–137.
60. Bey, P.; Vevert, J.P. *Tetrahedron Lett.* **1977**, *7*, 1455–1458.
61. Oguri, T.; Shioiri, T.; Yamada, S.-i. *Chem. Pharm. Bull.* **1977**, *25*, 2287–2291.
62. O'Donnell, M.J.; Wu, S.; Huffman, J.C. *Tetrahedron* **1994**, *50*, 4507–4518.
63. O'Donnell, M.J.; Boniece, J.M.; Earp, S.E. *Tetrahedron Lett.* **1978**, *30*, 2641–2644.
64. O'Donnell, M.J.; Polt, R.L. *J. Org. Chem.* **1982**, *47*, 2663–2666.
65. van der Werf, A.W.; Kellogg, R.M.; van Bolhuis, F. *J. Chem. Soc., Chem. Commun.* **1991**, 682–683.
66. Singh, J.; Gordon, T.D.; Earley, W.G.; Morgan, B.A. *Tetrahedron Lett.* **1993**, *34*, 211–214.
67. Kanemasa, S.; Uchida, O.; Wada, E. *J. Org. Chem.* **1990**, *55*, 4411–4417.
68. Kisanga, P.B.; Ilankumaran, P.; Fetterly, B.M.; Verkade, J.G. *J. Org. Chem.* **2002**, *67*, 3555–3560.
69. O'Donnell, M.J.; Wojciechowski, K.; Ghosez, L.; Navarro, M.; Sainte, F.; Antoine, J.-P. *Synthesis* **1984**, 313–315.
70. O'Donnell, M.J.; Cook, G.K.; Rusterholz, D.B. *Synthesis* **1991**, 989–993.
71. O'Donnell, M.J.; LeClef, B.; Rusterholz, D.B.; Ghosez, L.; Antoine, J.-P.; Navarro, M. *Tetrahedron Lett.* **1982**, *23*, 4259–4262.
72. Ghosez, L.; Antoine, J.-P.; Deffense, E.; Navarro, M.; Libert, V.; O'Donnell, M.J.; Bruder, W.A.; Willey, K.; Wojciechowski, K. *Tetrahedron Lett.* **1982**, *23*, 4255–4258.
73. O'Donnell, M.J.; Eckrich, T.M. *Tetrahedron Lett.* **1978**, *47*, 4625–4628.
74. O'Donnell, M.J.; Bruder, W.A.; Eckrich, T.M.; Shullenberger, D.F.; Staten, G.S. *Synthesis* **1984**, 127–128.
75. O'Donnell, M.J.; Bennett, W.D.; Bruder, W.A.; Jacobsen, W.N.; Kruth, K.; LeClef, B.; Polt, R.L.; Bordwell, F.G.; Mrozack, S.R.; Cripe, T.A. *J. Am. Chem. Soc.* **1988**, *110*, 8520–8525.
76. O'Donnell, M.J.; Bennett, W.D.; Jacobsen, W.N.; Ma, Y.-a.; Huffman, J.C. *Tetrahedron Lett.* **1989**, *30*, 3909–3912.
77. O'Donnell, M.J.; Rusterholz, D.B. *Synth. Commun.* **1989**, *19*, 1157–1165.
78. O'Donnell, M.J.; Bennett, W.D.; Jacobsen, W.N.; Ma, Y.-a. *Tetrahedron Lett.* **1989**, *30*, 3913–3914.
79. O'Donnell, M.J.; Wu, S. *Tetrahedron: Asymmetry* **1992**, *3*, 59–594.
80. Kaptein, B.; Boesten, W.H.J.; Broxterman, Q.B.; Schoemaker, H.E.; Kamphuis, J. *Tetrahedron Lett.* **1992**, *33*, 6007–6010.
81. Yaozhong, J.; Changyou, Z.; Shengde, W.; Daimo, C.; Youan, M.; Guilan, L. *Tetrahedron* **1988**, *44*, 5343–5353.
82. Ezquerra, J.; Pedregal, C.; Moreno-Mañas, M.; Pleixats, R.; Roglans, A. *Tetrahedron Lett.* **1993**, *34*, 8535–8538.
83. Moreno-Mañas, M.; Pleixats, R.; Roglans, A. *Liebigs Ann. Chem.* **1995**, 1807–1814.
84. López, A.; Moreno-Mañas, M.; Pleixats, R.; Roglans, A.; Ezquerra, J.; Pedregal, C. *Tetrahedron* **1996**, *52*, 8365–8386.
85. Antoni, G.; Långstöm, B. *J. Lab. Cmpds. Radiopharm.* **1987**, *24*, 125–143.
86. Tzalis, D.; Knochel, P. *Tetrahedron Lett.* **1999**, *40*, 3685–3688.
87. Gmouh, S.; Jamal-Eddine, J.; Valnot, J.Y. *Tetrahedron* **2000**, *56*, 8361–8366.
88. O'Donnell, M.J.; Zhou, C.; Scott,W.L. *J. Am. Chem. Soc.* **1996**, *118*, 6070–6071.
89. O'Donnell, M.J.; Lugar, C.W.; Pottorf, R.S.; Zhou, C. *Tetrahedron Lett.* **1997**, *38*, 7163–7166.
90. Domiínguez, E.; O'Donnell, M.J.; Scott, W.L. *Tetrahedron Lett.* **1998**, *39*, 2167–2170.

91. Griffith, D.L.; O'Donnell, M.J.; Pottorf, R.S.; Scott, W.L.; Porco J.A., Jr., *Tetrahedron Lett.* **1997**, *38*, 8821–8824.

92. See Reference 91.

93. Scott, W.L.; Zhou, C.; Fang, Z.; O'Donnell, M.J. *Tetrahedron Lett.* **1997**, *38*, 3695–3698.

94. Scott, W.L.; Delgado, F.; Lobb, K.; Pottorf, R.S.; O'Donnell, M.J. *Tetrahedron Lett.* **2001**, *42*, 2073–2076.

95. Park, H.-g.; Kim, M.-J.; Park, M.-K.; Jung, H.-J.; Lee, J.; Lee, Y.-J.; Jeong, B.-S.; Lee, J.-H.; Yoo, M.-S.; Ku, J.-M.; Jew, S.-s. *Tetrahedron Lett.* **2005**, *46*, 93–95.

96. Sauvagnat, B.; Lamaty, F.; Lazaro, R.; Martinez, J. *Tetrahedron Lett.* **1998**, *39*, 821–824.

97. Sauvagnat, B.; Kulig, K.; Lamaty, F.; Lazaro, R.; Martinez, J. *J. Comb. Chem.* **2000**, *2*, 134–142.

98. Sauvagnat, B.; Lamaty, F.; Lazaro, R.; Martinez, J. *Tetrahedron Lett.* **2000**, *41*, 6371–6375.

99. Joucla, M.; El Goumzili, M. *Tetrahedron Lett.* **1986**, *27*, 1681–1684.

100. Kotha, S.; Anglos, D.; Kuki, A. *Tetrahedron Lett.* **1992**, *33*, 1569–1572.

101. Kotha, S.; Kuki, A. *Tetrahedron Lett.* **1992**, *33*, 1565–1568.

102. Mazaleyrat, J.-P.; Gaucher, A.; Wakselman, M.; Tchertanov, L.; Guilhem, J. *Tetrahedron Lett.* **1996**, *37*, 2971–2974.

103. Mazaletrat, J.-P.; Gaucher, A.; Savrda, J.; Wakselman, M. *Tetrahedron: Asymmetry* **1997**, *8*, 619–631.

104. Mazón, A.; Nájera, C. *Tetrahedron Lett.* **1995**, *36*, 7697–7700.

105. Ridvan, L.; Abdallah, N.; Holakovsly, R.; Tichy, M.; Závada, J. *Tetrahedron: Asymmetry* **1996**, *7*, 231–236.

106. Gaucher, A.; Dorizon, P.; Ollivier, J.; Salaün, J. *Tetrahedron Lett.* **1995**, *36*, 2979–2982.

107. Scott, W.L.; Alsina, J.; O'Donnell, M.J. *J. Comb. Chem.* **2003**, *5*, 684–692.

108. Scott, W.L.; O'Donnell, M.J.; Delgado, F.; Alsina, J. *J. Org. Chem.* **2002**, *67*, 2960–2969.

109. O'Donnell, M.J.; Alsina, J.; Scott, W.L. *Tetrahedron Lett.* **2003**, *44*, 8403–8406.

110. Meyer, L.; Poirer, J.-M.; Duhamel, P.; Duhamel, L. *J. Org. Chem.* **1998**, *63*, 8094–8095.

111. Genet, J.-P.; Juge, S.; Achi, S.; Mallart, S.; Ruiz Montes, J.; Levif, G. *Tetrahedron* **1988**, *44*, 5263–5275.

112. Ferroud, D.; Genet, J.P.; Kiolle, R. *Tetrahedron Lett.* **1986**, *27*, 23–26.

113. Attolini, M.; Maffei, M.; Principato, B.; Peiffer, G. *Synlett* **1997**, 384–386.

114. Baldwin, I.C.; Williams, J.M.J.; Beckett, R.P. *Tetrahedron: Asymmetry* **1995**, *6*, 1515–1518.

115. van der Werf, A.; Kellogg, R.M. *Tetrahedron Lett.* **1988**, *29*, 4981–4984.

116. Cazes, B.; Djahanbini, D.; Goré, J.; Genêt, J.-P.; Gaudin, J.-M. *Synthesis* **1988**, 983–985.

117. Lee, S.; Beare, N.A.; Hartwig, J.F. *J. Am. Chem. Soc.* **2001**, *123*, 8410–8411.

118. Grigg, R.; Gunaratne, H.Q.N. *Tetrahedron Lett.* **1983**, *24*, 4457–4460.

119. Grigg, R.; Kemp, J.; Sheldrick, G.; Trotter, J. *J. Chem. Soc., Chem. Commun.* **1978**, 109–111.

120. Grigg, R.; Kemp, J. *Tetrahedron Lett.* **1980**, 2461–2464.

121. Bonnet-Delpon, D.; Chennoufi, A.; Rock, M.H. *Bull. Soc. Chim. Fr.* **1995**, *132*, 402–405.

122. Grigg, R.; Montgomery, J.; Somasunderam, A. *Tetrahedron* **1992**, *48*, 10431–10442.

123. Tsuge, O.; Kanemasa, S.; Yoshioka, M. *J. Org. Chem.* **1988**, *53*, 1384–1391.

124. Grigg, R.; Gunaratne, H.Q.N. *J. Chem. Soc., Chem. Commun.* **1982**, 384–386.

125. Barr, D.A.; Grigg, R.; Gunaratne, H.Q.N.; Kemp, J.; McMeekin, P.; Sridharan, V. *Tetrahedron* **1988**, *44*, 557–570.

126. Amornraksa, K.; Barr, D.A.; Donegan, G.; Grigg, R.; Ratananukul, P.; Sridharan, V. *Tetrahedron* **1989**, *45*, 4649–4668.

127. Grigg, R.; Gunaratne, H.Q.N.; Sridharan, V. *Tetrahedron* **1987**, *43*, 5887–5898.

128. Barr, D.A.; Grigg, R.; Sridharan, V. *Tetrahedron Lett.* **1989**, *30*, 4727–4730.

129. Grigg, R.; Kemp, J.; Malone, J.; Tangthongkum, A. *J. Chem. Soc., Chem. Commun.* **1980**, 648–650.

130. Grigg, R.; Kemp, J.; Malone, J.F.; Rajviroongit, S.; Tangthongkum, A. *Tetrahedron* **1988**, *44*, 5361–5374.

131. van der Werf, A.; Kellogg, R.M. *Tetrahedron Lett.* **1991**, *32*, 3727–3730.

132. Belokon, Y.N.; Bespalova, N.B.; Churkina, T.D.; Císarová, I.; Ezernitskaya, M.G.; Harutyunyan, S.R.; Hrdina, R.; Kagan, H.B.; Kocovsky, P.; Kochetkov, K.A.; Larionov, O.V.; Lyssenko, K.A.; North, M.; Polásek, M.; Peregudov, A.S.; Prisyazhnyuk, V.V.; Vyskocil, S. *J. Am. Chem. Soc.* **2003**, *125*, 12860–12871.

133. Ellis, T.K.; Hochla, V.M.; Soloshonok, V.A. *J. Org. Chem.* **2003**, *68*, 4973–4976.

134. Ellis, T.K.; Martin, C.H.; Tsai, G.M.; Ueki, H.; Soloshonok, V.A. *J. Org. Chem.* **2003**, *68*, 6208–6214.

135. Ellis, .K.; Martin, C.H.; Ueki, H.; Soloshonok, V.A. *Tetrahedron Lett.* **2003**, *44*, 1063–1066.

136. Soloshonok, V.A.; Cai, C.; Hruby, V.J.; Van Meervelt, L.; Yamazaki, T. *J. Org. Chem.* **2000**, *65*, 6688–6696.

137. Miyashita, K.; Miyabe, H.; Kurozumi, C.; Tai, K.; Imanishi, T. *Tetrahedron* **1996**, *52*, 12125–12136.

138. Schöllkopf, U.; Hoppe, D.; Jentsch, R. *Angew. Chem., Int. Ed. Engl.* **1971**, *10*, 331–333.

139. O'Donnell, M.J.; Bruder, W.A.; Daugherty, B.W.; Liu, D.; Wojciechowski, K. *Tetrahedron Lett.* **1984**, *25*, 3651–3654.

140. Chruma, J.J.; Sames, D.; Polt, R. *Tetrahedron Lett.* **1997**, *38*, 5085–5086.

141. Yamada, S.; Hongo, C.; Yoshioka, R.; Chibata, I. *J. Org. Chem.* **1983**, *48*, 843–846.

142. Schöllkopf, U.; Hopppe, D.; Jentsch, R. *Chem. Ber.* **1975**, *108*, 1580–1592.

143. Kotha, S.; Brachmachary, E. *Tetrahedron Lett.* **1997**, *38*, 3561–3564.

144. Kotha, S.; Brahmachary, E. *Bioorg. Med. Chem. Lett.* **1997**, *7*, 2719–2722.

145. Kotha, S.; Brahmachary, E.; Kuki, A.; Lang, K.; Anglos, D.; Singaram, B.; Chrisman, W. *Tetrahedron Lett.* **1997**, *38*, 9031–9034.

146. Kotha, S.; Sreenivasachary, N.; Brahmachary, E. *Tetrahedron Lett.* **1998**, *39*, 2805–2808.

147. Kotha, S.; Sreenivasachary, N. *Biorg. Med. Chem. Lett.* **1998**, *8*, 257–260.

148. Bey, P.; Danzin, C.; Van Dorsselaer, V.; Mamont, P.; Jung, M.; Tardif, C. *J. Med. Chem.* **1978**, *21*, 50–55.

149. Suzuki, M.; Miyahara, T.; Yoshioka, R.; Miyoshi, M.; Matsumoto, K. *Agr. Biol. Chem.* **1974**, *38*, 1709–1715.

150. Yasuo, H.; Suzuki, M.; Yoneda, N. *Chem. Pharm. Bull.* **1979**, *27*, 1931–1934.

151. Kotha, S.; Brahmachary, E.; Sreenivasachary, N. *Tetrahedron Lett.* **1998**, *39*, 4095–4098.

152. Hoppe, D. *Angew. Chem., Int. Ed. Engl.* **1974**, *13*, 789–804.

153. Schöllkopf, U. *Angew. Chem., Int. Ed. Engl.* **1977**, *16*, 339–348.

154. Hoppe, I.; Schöllkopf, U. *Synthesis* **1982**, 129–131.

155. Hoppe, D.; Schöllkopf, U. *Liebigs Ann. Chem.* **1972**, *763*, 1–16.

156. Hoppe, D.; Schöllkopf, U. *Angew. Chem., Int. Ed. Engl.* **1970**, *9*, 300–301.

157. Schöllkopf, U.; Gerhart, F.; Schröder, R.; Hoppe, D. *Liebigs Ann. Chem.* **1972**, *766*, 116–129.

158. Saegusa, T.; Ito, Y.; Kinoshita, H.; Tomita, S. *J. Org. Chem.* **1971**, *36*, 3316–3323.

159. Schöllkopf, U.; Hoppe, D. *Angew. Chem., Int. Ed. Engl.* **1970**, *6*, 459.

160. Soloshonok, V.A.; Hayashi, T.; Ishikawa, K.; Nagashima, N. *Tetrahedron Lett.* **1994**, *35*, 1055–1058.

161. LeClercq, J.; Cossement, E.; Boydens, R.; Rodriguez, L.A.M.; Brouwers, L.; De Laveleye, F.; Libert, W. *J. Chem. Soc., Chem. Commun.* **1978**, 46–47.

162. Ozaki, Y.; Maeda, S.; Miyoshi, M.; Matsumoto, K. *Synthesis* **1979**, 216–217.

163. Soloshonok, V.A.; Kacharov, A.D.; Avilov, D.V.; Ishikawa, K.; Nagashima, N.; Hyashi, T. *J. Org. Chem.* **1997**, *62*, 3470–3479.

164. Kirihata, M.; Fukuari, M.; Izukawa, T.; Ichimoto, I. *Amino Acids* **1995**, *9*, 317–325.

165. Heinzer, F.; Bellus, D. *Helv. Chim. Acta* **1981**, *64*, 2279–2297.

166. Angst, C. *Pure Appl. Chem.* **1987**, *59*, 373–380.

167. Schöllkopf, U.; Hantke, K. *Liebigs Ann. Chem.* **1973**, 1571–1582.

168. Schöllkopf, U.; Porsch, P.-H. *Chem. Ber.* **1973**, *106*, 3382–3390.

169. Meyer, R.; Schöllkopf, U.; Böhme, P. *Liebigs Ann. Chem.* **1977**, 1183–1193.

170. Hayashi, T.; Kishi, E.; Soloshonok, V.A.; Uozumi, Y. *Tetrahedron Lett.* **1996**, *37*, 4969–4972.

171. Lin, Y.-R.; Zhou, X.-T.; Dai, L.-X.; Sun, J. *J. Org. Chem.* **1997**, *62*, 1799–1803.

172. David, K.; Cazes, B.; Goré, J. *J. Chem. Res. S* **1996**, 488–489.

173. Moreno-Mañas, M.; Pleixats, R.; Roglans, A. *Liebigs Ann. Chem.* **1995**, 1807–1814.

174. Carlström, A.-S.; Frejd, T. *J. Org. Chem.* **1990**, *55*, 4175–4180.

175. Fu, Y.; Hammarström, L.G.J.; Miller, T.J.; Fronczek, F.R.; McLaughlin, M.L.; Hammer, R.P. *J. Org. Chem.* **2001**, *66*, 7118–7124.

176. Fu, W.; Etienne, M.A.; Hammer, R.P. *J. Org. Chem.* **2003**, *68*, 9854–9857.

177. Geen, G.; Shaw, C.J.; Sweeney, J.B. *Synlett* **1999**, 1444–1446.

178. Alonsa, D.A.; Costa, A.; Nájera, C. *Tetrahedron Lett.* **1997**, *38*, 7943–7946.

179. Hantke, K.; Schöllkopf, U.; Hausberg, H.-H. *Liebigs Ann. Chem.* **1975**, 1531–1537.

180. Baldwin, J.E.; Bradley, M.; Turner, N.J.; Adlington, R.M.; Pitt, A.R.; Sheridan, H. *Tetrahedron* **1991**, *47*, 8203–8222.

181. Hamersak, Z.; Gaso, D.; Kovac, S.; Hergold-Brundié, A.; Vickovié, I.; Sunjié, V. *Helv. Chim. Acta* **2003**, *86*, 2247–2257.

182. Anikumar, R.; Chandrasekhar, S.; Sridhar, M. *Tetrahedron Lett.* **2000**, *41*, 6665–6668.

183. Erlenmeyer, E.; Früstück, E. *Ann.* **1894**, *284*, 36–49.

184. Hönig, H.; Seufer-Wasserthal, P.; Weber, H. *Tetrahedron* **1990**, *46*, 3841–3850.

185. Shanzer, A.; Somekh, L.; Butina, D. *J. Org. Chem.* **1979**, *44*, 3967–3969.

186. Grandel, R.; Kazmaier, U.; Nuber, B. *Liebigs Ann. Chem.* **1996**, 1143–1150.

187. Carrara, G.; Pace, E.; Cristiani, G. *J. Am. Chem. Soc.* **1952**, *74*, 4949–4950.

188. Holland, D.O.; Jenkins, P.A.; Nayler, J.H.C. *J. Chem. Soc.* **1953**, 273–280.

189. Rios, A.; Richard, J.P. *J. Am. Chem. Soc* **1997**, *119*, 8375–8376.

190. Ferey, V.; Le Gall, T.; Mioskowski, C. *J. Chem. Soc., Chem. Commun.* **1995**, 487–489.

191. Broxterman, H.J.G.; Liskamp, R.M.J. *Recl. Trav. Chim. Pays-Bas* **1991**, *110*, 46–52.

192. El Hadrami, M.; Lavergne, J.-P.; Viallefont, Ph.; Chiaroni, A.; Riche, C.; Hasnaoui, A. *Synth. Commun.* **1993**, *23*, 157–163.

193. Scott, A.I.; Wilkinson, T.J. *J. Lab. Compds. Radiopharm.* **1981**, *18*, 347–351.

194. Aebi, J.D.; Dhaon, M.K.; Rich, D.H. *J. Org. Chem.* **1987**, *52*, 2881–2886.

195. Hvidt, T.; Martin, O.R.; Szarek, W.A. *Tetrahedron Lett.* **1986**, *27*, 3807–3810.

196. Oesterle, T.; Simchen, G. *Synthesis* **1985**, 403–408.

197. Rühlmann, K.; Kuhrt, G. *Angew. Chem., Int. Ed. Engl.* **1968**, *7*, 809.
198. Hudrlik, P.F.; Kulkarni, A.K. *J. Am. Chem. Soc.* **1981**, *103*, 6251–6253.
199. Djuric, S.; Venit, J.; Magnus, P. *Tetrahedron Lett.* **1981**, *22*, 1787–1790.
200. Kobayashi, S.; Hayashi, T.; Iwamoto, S.; Furuta, T.; Matsumura, M. *Synlett*, **1996**, 672–674.
201. Shaw, K.N.F.; Fox, S.W. *J. Am. Chem. Soc.* **1953**, *75*, 3421–3424.
202. Erlenmeyer, E. *Ber.* **1892**, *25*, 3445–3447.
203. Kikuchi, J.-I.; Takashima, T.; Nakao, H.; Hie, K.-I.; Etoh, H.; Noguchi, Y.; Suehiro, K.; Murakami, Y. *Chem. Lett.* **1993**, 553–556.
204. Wieland, T.; Cords, H.; Keck, E. *Chem. Ber.* **1954**, *87*, 1312–1318.
205. Hayes, K.; Gever, G. *J. Org. Chem.* **1951**, *16*, 269–278.
206. Inui, T,; Ohta, Y.; Ujike, T.; Katsura, H.; Kaneko, T. *Bull. Chem. Soc. Jpn.* **1968**, *41*, 2148–2150.
207. Sola, R.; Frehel, D.; Mafrand, J.-P.; Brugidou, J. *Heterocycles* **1982**, *19*, 1797–1800.
208. Dullaghan, M.E.; Nord, F.F. *J. Am. Chem. Soc.* **1951**, *73*, 5455.
209. Hegedüs, B.; Krassó, A.F.; Noack, K.; Zeller, P. *Helv. Chim. Acta* **1975**, *58*, 147–162.
210. Carrara, G.; Weitnauer, G. *Gazz. Chim. Ital.* **1949**, *79*, 856–862.
211. Rosenmund, K.W.; Dornsaft, H. *Ber.* **1919**, *52*, 1734–1749.
212. Bolhofer, W.A. *J. Am. Chem. Soc.* **1954**, *76*, 1322–1326.
213. Bergmann, E.D.; Bendas, H.; Resnick, C. *J. Chem. Soc.* **1953**, 2564–2571.
214. Dalgliesh, C.E.; Mann, F.G. *J. Chem. Soc.* **1947**, 658–662.
215. Shaw, K.N.F.; Fox, S.W. *J. Am. Chem. Soc.* **1953**, *75*, 3417–3421.
216. Sato, M.; Okawa, K.; Akabori, S. *Bull. Chem. Soc. Jpn.* **1957**, *30*, 937–938.
217. Okawa, K.; Hori, K.; Hirose, K.; Nakagawa, Y. *Bull. Chem. Soc. Jpn.* **1969**, *42*, 2720–2722.
218. Kornguth, M.L.; Sallach, H.J. *Arch. Biochem. Biophys.* **1960**, *91*, 39–42.
219. Benoiton, L.; Winitz, M.; Colman, R.F.; Birnbaum, S.M.; Greenstein, J.P. *J. Am. Chem. Soc.* **1959**, *81*, 1726–1729.
220. Niimura, Y.; Hatanaka, S.-I. *Phytochemistry* **1974**, *13*, 175–178.
221. Otani, T.T.; Briley, M.R. *J. Pharm. Sci.* **1974**, *63*, 1253–1256.
222. Akabori, S.; Otani, T.T.; Marshall, R.; Winitz, M.; Grenstein, J.P. *Arch. Biochem. Biophys.* **1959**, *83*, 1–9.
223. Otani, T.T.; Winitz, M. *Arch. Biochem. Biophys.* **1960** *90*, 254–259.
224. Cappi, M.W.; Moree, W.J.; Qiao, L.; Marron, T.G.; Weitz-Schmidt, G.; Wong, C.-H. *Bioorg. Med. Chem.* **1997**, *5*, 283–296.

225. Krapcho, A.P.; Dundulis, E.A. *Tetrahedron Lett.* **1976**, 2205–2208.
226. Hoye, T.R.; Duff, S.R.; King, R.S. *Tetrahedron Lett.* **1985**, *26*, 3433–3436.
227. McIntosh, J.M.; Thangarasa, R.; Ager, D.J.; Zhi, B. *Tetrahedron* **1992**, *48*, 6219–6224.
228. Karoyan, P.; Chassaing, G. *Tetrahedron Lett.* **1997**, *38*, 85–88.
229. Kuhn, R.; Osswald, G. *Chem. Ber.* **1956**, *89*, 1423–1442.
230. Bayer, E.; Schmidt, K. *Tetrahedron Lett.* **1973**, *23*, 2051–2054.
231. Simchen, G.; Pürkner, E. *Synthesis* **1990**, 525–527.
232. Oberdorfer, F.; Zobeley, A.; Weber, K.; Prenant, C.; Haberkorn, U.; Maier-Borst, W. *J. Lab. Cmpds. Radiopharm.* **1993**, *33*, 345–353.
233. Kazmaier, U.; Zumpe, F.L. *Angew. Chem., Int. Ed. Engl.* **1999**, *38*, 1468–1470.
234. Zumpe, F.L.; Kazmaier, U. *Synthesis* **1999**, 1785–1791.
235. Pohlman, M.; Kazmaier, U.; Lindner, T. *J. Org. Chem.* **2004**, *69*, 6909–6912.
236. Yamaguchi, M.; Torisu, K.; Minami, T. *Chem. Lett.* **1990**, 377–380.
237. Pedersen, M.L.; Berkowitz, D.B. *J. Org. Chem.* **1993**, *58*, 6966–6875.
238. Sheehan, J.C.; Johnson, D.A. *J. Am. Chem. Soc.* **1954**, *76*, 158–160.
239. Attenburrow, J.; Elliott, D.F.; Penny, G.F. *J. Chem. Soc.* **1948**, 310–318.
240. De Nicola, A.; Einhorn, J.; Luche, J.-L. *Tetrahedron Lett.* **1992**, *33*, 6461–6464.
241. Papageorgiou, C.; Borer, X.; French, R.R. *Bioorg. Med. Chem. Lett.* **1994**, *4*, 267–272.
242. Vadon, S.; Custot, J.; Boucher, J.-L.; Mansuy, D. *J. Chem. Soc., Perkin Trans. 1* **1996**, 645–648.
243. Guilaume, D.; Aitken, D.J.; Husson, H.-P. *Synlett* **1991**, 747–749.
244. Guillaume, D.; Brum-Bousquet, M.; Aitken, D.J.; Husson, H.-P. *Bull. Soc. Chim. Fr.* **1994**, *131*, 391–396.
245. McMath, A.R.; Guillaume, D.; Aitken, D.J.; Husson, H.-P. *Bull. Soc. Chim. Fr.* **1997**, *134*, 105–110.
246. Kübel, B.; Höfle, G.; Steglich, W. *Angew. Chem., Int. Ed. Engl.* **1975**, *14*, 58–59.
247. Kazmaier, U. *Amino Acids* **1996**, *11*, 283–299.
248. Kazmaier, U. *Liebigs Ann./Recueil* **1997**, 285–295.
249. Bartlett, P.A.; Barstow, J.F. *J. Org. Chem.* **1982**, *47*, 3933–3941.
250. Holladay, M.W.; Nadzan, A.M. *J. Org. Chem.* **1991**, 3900–3905.
251. Kazmaier, U.; Maier, S. *Tetrahedron* **1996**, *52*, 941–954.
252. Kazmeier, U. *Tetrahedron Lett.* **1996**, *37*, 5351–5354.
253. Kazmaier, U. *Synlett*, **1995**, 1138–1140.
254. Kazmaier, U. *J. Org. Chem.* **1996**, *61*, 3694–3699.

255. Castelhano, A.L.; Pliura, D.H.; Taylor, G.J.; Hsieh, K.C.; Krantz, A. *J. Am. Chem. Soc.* **1984**, *106*, 2734–2735.

256. Kazmaier, U. *Angew. Chem., Int. Ed. Engl.* **1994**, *33*, 998–999.

257. Mohamed, M.; Brook, M.A. *Helv. Chim. Acta* **2002**, *85*, 4165–4181.

258. Dell, C.P.; Khan, K.M.; Knight, D.W. *J. Chem. Soc., Perkin Trans 1,* **1994**, 341–347.

259. Castelhano, A.L.; Horne, S.; Taylor, G.J.; Billedeau, R.; Krantz, A. *Tetrahedron* **1988**, *44*, 5451–5466.

260. See Reference 255.

261. Tsunoda, T.; Tatsuki, S.; Shiraishi, Y.; Akasaka, M.; Ito, S. *Tetrahedron Lett.* **1993**, *34*, 3297–3300.

262. Anderson, J.C.; Whiting, M. *J. Org. Chem.* **2003**, *68*, 6160–6163.

263. Coldham, I.; Middleton, M.L.; Taylor, P.L. *J. Chem. Soc., Perkin Trans. 1* **1998**, 2817–2821.

264. Honda, K.; Igarashi, D.; Asami, M.; Inoue, S. *Synlett* **1998**, 685–687.

265. Blid, J.; Brandt, P.; Somfai, P. *J. Org. Chem.* **2004**, *69*, 3043–3049.

266. Meester, W.J.N.; van Maarseveen, J.H.; Schoemaker, H.E.; Hiemstra, H.; Rutjes, F.P.J.T. *Eur. J. Org. Chem.* **2003**, 2519–2529.

267. Zoller, U.; Ben-Ishai, D. *Tetrahedron* **1975**, *31*, 863–866.

268. Legall, P.; Sawhney, K.N.; Conley, J.D.; Kohn, H. *Int. J. Pept. Prot. Res.* **1988**, *32*, 279–291.

269. Ben-Ishai, D.; Sataty, I.; Bernstein, Z. *Tetrahedron* **1976**, *32*, 1571–1573.

270. Schouteeten, A.; Christidis, Y.; Mattioda, G. *Bull. Soc. Chim. Fr.* **1978**, *11*, 248–254.

271. Repine, J.T.; Kalternbronn, J.S.; Doherty, A.M.; Hamby, J.M.; Himmelsbach, R.J.; Kornberg, B.E.; Taylor, M.D.; Lunney, E.A.; Humblet, C.; Rapundalo, S.T.; Batley, B.L.; Ryan, M.J.; Painchaud, C.A. *J. Med. Chem.* **1992**, *35*, 1032–1042.

272. Katritzky, A.R.; Urogdi, L.; Mayence, A. *J. Org. Chem.* **1990**, *55*, 2206–2214.

273. Abood, N.A.; Nosal, R. *Tetrahedron Lett.* **1994**, *35*, 3669–3672.

274. Bogenstätter, M.; Steglich, W. *Tetrahedron* **1997**, *53*, 7267–7274.

275. Apitz, G.; Jäger, M.; Jaroch, S.; Kratzel, M.; Schäffeler, L.; Steglich, W. *Tetrahedron* **1993**, *49*, 8223–8232.

276. O'Donnell, M.J.; Delgado, F.; Drew, M.D.; Pottorf, R.S.; Zhou, C.; Scott, W.L. *Tetrahedron Lett.* **1999**, *40*, 5831–5835.

277. Ginzel, K.-D.; Brungs, P.; Steckhan, E. *Tetrahedron* **1989**, *45*, 1691–1701.

278. Haas, K.; Dialer, H.; Piotrowski, H.; Schapp, J.; Beck, W. *Angew. Chem., Int. Ed.* **2002**, *41*, 1879–1881.

279. Ratcliffe, R.W.; Christensen, B.G. *Tetrahedron Lett.* **1973**, *46*, 4653–4656.

280. Qasmi, D.; René, L.; Badet, B. *Tetrahedron Lett.* **1993**, *34*, 3861–3862.

281. Rivier, J.E.; Jiang, G.; Koerber, S.C.; Porter, J.; Simon, L.; Craig, A.G.; Hoeger, C.A. *Proc. Natl. Acad. Sci. U.S.A.* **1996**, *93*, 2031–2036.

282. Williams, R.M.; Aldous, D.J.; Aldous, S.C. *J. Org. Chem.* **1990**, *55*, 4657–4663.

283. Alks, V.; Sufrin, J.R. *Synth. Commun.* **1989**, *19*, 1479–1486.

284. Paulitz, C.; Steglich, W. *J. Org. Chem.* **1997**, *62*, 8474–8478.

285. Jäger, M.; Polborn, K.; Steglich, W. *Tetrahedron Lett.* **1995**, *36*, 861–864.

286. Biagini, S.C.G.; Gibson, S.E.; Keen, S.P. *J. Chem. Soc., Perkin Trans. 1* **1998**, 2485–2499.

287. Münster, P.; Steglich, W. *Synthesis* **1987**, 223–225.

288. Lidert, Z.; Gronowitz, S. *Synthesis* **1980**, 322–324.

289. Trost, B.M.; Lee, C. *J. Am. Chem. Soc.* **2001**, *123*, 12191–12201.

290. Maggini, M.; Prato, M.; Scorrano, G. *Tetrahedron Lett.* **1990**, *31*, 6243–6246.

291. O'Donnell, M.J.; Bennett, W.D.; Polt, R.L. *Tetrahedron Lett.* **1985**, *26*, 695–698.

292. Bailey, P.D.; Baker, S.R.; Boa, A.N.; Clayson, J.; Rosair, G.M. *Tetrahedron Lett.* **1998**, *39*, 7755–7758.

293. Shono, T. *Tetrahedron* **1984**, *40*, 811–850.

294. Iwasaki, T.; Horikawa, H.; Matsumoto, K.; Miyoshi, M. *J. Org. Chem.* **1977**, *42*, 2419–2423.

295. Roos, E.C.; López, M.C.; Brook, M.A.; Hiemstra, H.; Speckamp, W.N.; Kaptein, B.; Kamphuis, J.; Schoemaker, H.E *J. Org. Chem.* **1993**, *58*, 3259–3268.

296. Iwasaki, T.; Horikawa, H.; Matsumoto, K.; Miyoshi, M. *Bull. Chem. Soc. Jpn.* **1979**, *52*, 826–830.

297. Shin, C-i.; Sato, Y.; Ohmatsu, H.; Yoshimura, J. *Bull. Chem. Soc. Jpn.* **1981**, *54*, 1137–1142.

298. Hercheid, J.D.M.; Ottenheijm, H.C.J. *Tetrahedron Lett.* **1978**, *51*, 5143–5144.

299. Herscheid, J.D.M.; Nivard, R.J.F.; Tijhuis, M.W.; Scholten, H.P.H.; Ottenheijm, H.C.J. *J. Org. Chem.* **1980**, *45*, 1880–1885.

300. Johnson, A.P.; Luke, R.W.A.; Steele, R.W. *J. Chem. Soc., Chem. Commun.* **1986**, 1658–1659.

301. Katritzky, A.R.; Urogdi, L.; Mayence, A. *J. Chem. Soc., Chem. Commun.* **1989**, 337–338.

302. Yaouancq, L.; René, L.; Huu Dau, M.-E.T.; Badet, B. *J. Org. Chem.* **2002**, *67*, 5408–5411.

303. Bock, M.G.; DiPardo, R.M.; Freidinger, R.M. *J. Org. Chem.* **1986**, *51*, 3718–3720.

304. Schmidt, U.; Stäbler, F.; Lieberknecht, A. *Synthesis* **1994**, 890–892.

305. Hermann, P.; Baumann, H.; Herrnstadt, Ch.; Glanz, D. *Amino Acids* **1992**, *3*, 105–118.

306. Ben-Ishai, D.; Altman, J.; Bernstein, Z.; Peled, N. *Tetrahedron* **1978**, *34*, 467–473.

307. Altman, J.; Moshberg, R.; Ben-Ishai, D. *Tetrahedron Lett.* **1975**, 3737–3740.

308. Ben-Ishai, D.; Moshenberg, R.; Altman, J. *Tetrahedron* **1977**, *33*, 1533–1542.

309. Mooiweer, H.H.; Hiemstra, H.; Speckamp, W.N. *Tetrahedron* **1989**, *45*, 4627–4636.

310. Speckamp, W.N. *J. Heterocycl. Chem.* **1992**, *29*, 653–674.

311. Roos, E.C.; Mooiweer, H.H.; Hiemstra, H.; Speckamp, W.N.; Kaptein, B.; Boesten, W.H.J.; Kamphuis, J.; *J. Org. Chem.* **1992**, *57*, 6769–6778.

312. Roos, E.C.; Hiemstra, H.; Speckamp, W.N.; Kaptein, B.; Kamphuis, J.; Schoemaker, H.E. *Recl. Trav. Chim. Pays-Bas* **1992**, *111,* 360–364.

313. Roos, E.C.; Hiemstra, H.; Speckamp, W.N.; Kaptein, B.; Kamphuis, J.; Schoemaker, H.E. *Synlett* **1992**, 451–452.

314. Kohn, H.; Sawhney, K.N.; LeGall, P.; Conley, J.D.; Robertson, D.W.; Leander, J.D. *J. Med. Chem.* **1990**, *33*, 919–926.

315. Ozaki, Y.; Iwasaki, T.; Horikawa, H.; Miyoshi, M.; Matsumoto, K. *J. Org. Chem.* **1979**, *44*, 391–395.

316. Nishitani, T.; Iwasaki, T.; Mushika, Y.; Miyoshi, M. *J. Org. Chem.* **1979**, *44*, 2019–2023.

317. O'Donnell, M.J.; Falmagne, J.-B. *Tetrahedron Lett.* **1985**, *26*, 699–702.

318. Kozhushkov, S.I.; Brandl, M.; Yufit, D.S.; Machinek, R.; de Meijere, A. *Leibigs Ann./Recueil* **1997**, 2197–2204.

319. O'Donnell, M.J.; Falmagne, J.-B. *J. Chem. Soc., Chem. Commun.* **1985**, 1168–1169.

320. O'Donnell, M.J.; Li, M.; Bennett, W.D.; Grote, T. *Tetrahedron Lett.* **1994**, *35*, 9383–9386.

321. O'Donnell, M.J.; Yang, X.; Li, M. *Tetrahedron Lett.* **1990**, *31*, 5135–5138.

322. O'Donnell, M.J.; Zhou, C.; Mi, A.; Chen, N.; Kyle, J.A.; Andersson, P.G. *Tetrahedron Lett.* **1995**, *36*, 4205–4208.

323. Ozaki, Y.; Iwasaki, T.; Miyoshi, M.; Matsumoto, K. *J. Org. Chem.* **1979**, *44*, 1714–1716.

324. Lamaty, F.; Lazaro, R.; Martinez, J. *Tetrahedron Lett.* **1997**, *38*, 3385–3386.

325. O'Donnell, M.J.; Bennett, W.D. *Tetrahedron* **1988**, *44*, 5389–5401.

326. Esch, P.M.; Boska, I.M.; Hiemstra, H.; de Boer, R.F.; Speckamp, W.N. *Tetrahedron* **1991**, *47*, 4039–4062.

327. Esch, P.M.; de Boer, R.F.; Hiemstra, H.; Boska, I.M.; Speckamp, W.N. *Tetrahedron* **1991**, *47*, 4063–4076.

328. Easton, C.J.; Peters, S.C. *Aust. J. Chem.* **1994**, *47*, 859–868.

329. Easton, C.J.; Peters, S.C. *Aust. J. Chem.* **1990**, *43*, 87–97.

330. Trojandt, G.; Polborn, K.; Steglich, W.; Schmidt, M.; Nöth, H. *Tetrahedron Lett.* **1995**, *36*, 857–860.

331. Achamlal, S.; Elachqar, A.; El Hallaoui, A.; El Hajji, S.; Roumestant, M.L.; Viallefont, Ph. *Amino Acids* **1997**, *12*, 257–263.

332. Bentama, A.; El Hadrami, E.M.; El Hallaoui, A.; Elachqar, A.; Lavergne, J.-P.; Roumestant, M.-L.; Viallefont, Ph. *Amino Acids* **2003**, *24*, 423–426.

333. Bretschneider, T.; Miltz, W.; Münster, P.; Steglich, W. *Tetrahedron* **1988**, *44*, 5403–5414.

334. Cahill, R.; Crout, D.H.G.; Mitchell, M.B.; Müller, U.S. *J. Chem. Soc., Chem. Commun.* **1980**, 419–421.

335. Castelhano, A.L.; Horne, S.; Billedeau, R.; Krantz, A. *Tetrahedron Lett.* **1986**, *27*, 2435–2438.

336. Rich, D.H.; Dhaon, M.K. *Tetrahedron Lett.* **1983**, *24*, 1671–1674.

337. Loh, T.-P.; Ho, D. S.-C.; Xu, K.-C.; Sim, K.-Y. *Tetrahedron Lett.* **1997**, *38*, 865–868.

338. Easton, C.J.; Roselt, P.D.; Tiekink, E.R.T. *Tetrahedron* **1995**, *51*, 7809–7822.

339. Burgess, V.A.; Easton, C.J. *Aust. J. Chem.* **1988**, *41*, 1063–1070.

340. Kayser, B.; Polborn, K.; Steglich, W.; Beck, W. *Chem. Ber./Recueil* **1997**, *130*, 171–177.

341. Mooiweer, H.H.; Ettema, K.W.A.; Hiemstra, H.; Speckamp, W.N. *Tetrahedron* **1990**, *46*, 2991–2998.

342. Kobayashi, S.; Kitagawa, H.; Matsubara, R. *J. Comb. Chem.* **2001**, *3*, 401–403.

343. Grumbach, H.-J.; Merla, B.; Risch, N. *Synthesis* **1999**, 1027–1033.

344. Far, S.; Melnyk, O. *Tetrahedron Lett.* **2004**, *45*, 1271–1273.

345. Kolasa, T.; Sharma, S.; Miller, M.J. *Tetrahedron Lett.* **1987**, *28*, 4973–4976.

346. Kolasa, T.; Sharma, S.K.; Miller, M.J. *Tetrahedron* **1988**, *44*, 5431–5440.

347. Osipov, S.N.; Golubev, A.S.; Sewald, N.; Michel, T.; Kolomiets, A.F.; Fokin, A.V.; Burger, K. *J. Org. Chem.* **1996**, *61*, 7521–7528.

348. Osipov, S.N.; Golubev, A.S.; Sewald, N.; Burger, K. *Tetrahedron Lett.* **1997**, *38*, 5965–5966.

349. Dal Pozzo, A.; Muzi, L.; Moroni, M.; Rondanin, R.; de Castiglione, R.; Bravo, P.; Zanda, M. *Tetrahedron* **1998**, *54*, 6019–6028.

350. Coghlan, P.A.; Easton, C.J. *ARKIVOC* **2004**, 101–108.

351. Coghlan, P.A; Easton, C.J. *J. Chem. Soc., Perkin Trans. 1* **1999**, 2659–2660.

352. Kobayashi, S.; Akiyama, R.; Kitagawa, H. *J. Comb. Chem.* **2000**, *2*, 438–440.

353. Kobayashi, S.; Akiyama, R.; Kitagawa, H. *J. Comb. Chem.* **2001**, *3*, 196–204.

354. Achmatowicz O. Jr.; Pietraszkiewicz, M. *J. Chem. Soc., Perkin Trans. 1* **1981**, 2680–2683.

355. Petasis, N.A.; Zavialov, I.A. *J. Am. Chem. Soc.,* **1997**, *119*, 445–446.

356. Petasis, N.A.; Goodman, A.; Zavialov, I.A. *Tetrahedron* **1997**, *53*, 16463–16470.

357. Portlock, D.E.; Ostaszewski, R.; Naskar, D.; West, L. *Tetrahedron Lett.* **2003**, *44*, 603–605.

358. Portlock, D.E.; Naskar, D.; West, L.; Ostaszewski, R.; Chen, J.J. *Tetrahedron Lett.* **2003**, *44*, 5121–5124.

359. Merla, B.; Grumbach, H.-J.; Risch, N. *Synthesis* **1998**, 1609–1614.

360. Jiménez, O.; de la Rosa, G.; Lavilla, R. *Angew. Chem., Int. Ed.* **2005**, *44*, 6521–6525.

361. Huang, T.; Li, C.-J. *Tetrahedron Lett.* **2000**, *41*, 6715–6719.

362. Kober, R.; Hammes, W.; Steglich, W. *Angew. Chem., Int. Ed. Engl.* **1982**, *21*, 203–204.

363. Havlícek, L.; Hanus, J.; Sedmera, P.; Nemecek, J. *Collect. Czech. Chem. Commun.* **1990**, *55*, 2074–2085.

364. Bunch, L.; Krogsgaard-Larsen, P.; Madsen, U. *J. Org. Chem.* **2002**, *67*, 2375–2377.

365. Calí, P.; Begtrup, M. *Tetrahedron* **2002**, *58*, 1595–1605.

366. Lipshutz, B.H.; Huff, B.; Vaccaro, W. *Tetrahedron Lett.* **1986**, *27*, 4241–4244.

367. Lipshutz, B.H.; Morey, M.C. *J. Am. Chem. Soc.* **1984**, *106*, 457–459.

368. Charette, A.B.; Gagnon, A.; Janes, M.; Mellon, C. *Tetrahedron Lett.* **1998**, *39*, 5147–5150.

369. Charette, A.B.; Mellon, C. *Tetrahedron* **1998**, *54*, 10525–10535.

370. Lovey, R.G.; Cooper, A.B. *Synlett* **1994**, 167–168.

371. Katagiri, N.; Sato, H.; Kurimoto, A.; Okada, M.; Yamada, A.; Kaneko, C. *J. Org. Chem.* **1994**, *59*, 8101–8106.

372. Ben-Ishai, D.; Hirsch, S. *Tetrahedron Lett.* **1983**, *24*, 955–958.

373. Ben-Ishai, D.; Hirsh, S. *Tetrahedron* **1988**, *44*, 5441–5449.

374. Altman, J.; Moshberg, R.; Ben-Ishai, D. *Tetrahedron Lett.* **1975**, 3737–3740.

375. Ben-Ishai, D.; Moshenberg, R.; Altman, J. *Tetrahedron* **1977**, *33*, 1533–1542.

376. McFarlane, A.K.; Thomas, G.; Whiting, A. *J. Chem. Soc., Perkin Trans. 1* **1995**, 2803–2808.

377. Maggini, M.; Prato, M.; Scorrano, G. *Tetrahedron Lett.* **1990**, *31*, 6243–6246.

378. Ward, S.E.; Holmes, A.B.; McCague, R. *Chem. Commun.* **1997**, 2085–2086.

379. Bernstein, Z.; Ben-Ishai, D. *Tetrahedron* **1977**, *33*, 881–883.

380. Kapeller, H.; Jary, W.G.; Hayden, W.; Grienl, H. *Tetrahedron: Asymmetry* **1997**, *8*, 245–251.

381. König, W.A.; Hass, W.; Dehler, W.; Fiedler, H.-P.; Zähner, H. *Liebigs Ann. Chem.* **1980**, 62.

382. Hass, W.; König, W.A. *Liebigs Ann. Chem.* **1982**, 1615–1622.

383. Jäger, V.; Grund, H.; Bub, V.; Schwab, W.; Müller, I.; Schohe, R.; Franz, R.; Ehrler, R. *Bull. Soc. Chim. Belg.* **1983**, *92*, 1039–1054.

384. Zimmermann, G.; Hass, W.; Faasch, H.; Schmalle, H.; König, W.A. *Liebigs Ann. Chem.* **1985**, 2165–2177.

385. König, W.A.; Hahn, H.; Rathmann, R.; Hass, W.; Keckeisen, A.; Hagenmaier, H.; Bormann, C.; Dehler, W.; Kurth, R.; Zähner, H. *Liebigs Ann. Chem.* **1986**, 407–421.

386. See Reference 384.

387. Easton, C.J.; Peters, S.C. *Tetrahedron Lett.* **1992**, *33*, 5581–5584.

388. Easton, C.J. *Chem. Rev.* **1997**, *97*, 53–82.

389. Easton, C.J.; Hay, M.P. *J. Chem. Soc., Chem. Commun.* **1986**, 55–57.

390. Turacek, F.; Carpenter, F.H.; Polce, M.J.; Wesdemiotis, C. *J. Am. Chem. Soc.* **1999**, *121*, 7955–7956.

391. Croft, A.K.; Easton, C.J.; Radom, L. *J. Am. Chem. Soc.* **2003**, *125*, 4119–4124.

392. Barratt, B.J.W.; Easton, C.J.; Henry, D.J.; Li, I.H.W.; Radom, L.; Simpson, J.S. *J. Am. Chem. Soc.* **2004**, *126*, 13306–13311.

393. Elad, D.; Schwarzberg, M.; Sperling, J. *J. Chem. Soc., Chem. Commun.* **1970**, 617–618.

394. Easton, C.J.; Scharfbillig, I.M.; Tan, E.W. *Tetrahedron Lett.* **1988**, *29*, 1565–1568.

395. Baldwin, J.E.; Adlington, R.M.; Lowe, C.; O'Neil, I.A.; Sanders, G.L.; Schofield, C.J.; Sweeney, J.B. *J. Chem. Soc., Chem. Commun.* **1988**, 1030–1031.

396. Easton, C.J.; Scharfbillig, I.M. *J. Org. Chem.* **1990**, *55*, 384–386.

397. Obata, N.; Niimura, K. *J. Chem. Soc., Chem. Commun.* **1977**, 238–239.

398. Udding, J.H.; Hiemstra, H.; Speckamp, W.N. *J. Org. Chem.* **1994**, *59*, 3721–3725.

399. Blakskjaer, P.; Pedersen, L.; Skrydstrup, T. *J. Chem. Soc.; Perkin Trans. 1* **2001**, 910–915.

400. Chai, C.L.L.; Page, D.M. *Tetrahedron Lett.* **1993**, *34*, 4373–4376.

401. Udding, J.H.; Tuijp, C.J.M.; Hiemstra, H.; Speckamp, W.N. *Tetrahedron* **1994**, *50*, 1907–1918.

402. Udding, J.H.; Hiemstra, H.; van Zanden, M.N.A.; Speckamp, W.N. *Tetrahedron Lett.* **1991**, *32*, 3123–3126.

403. Esch, P.M.; Hiemstra, H.; Speckamp, W.N. *Tetrahedron Lett.* **1990**, *31*, 759–762.

404. Esch, P.M.; Hiemstra, H.; de Boer, R.F.; Speckamp, W.N. *Tetrahedron* **1992**, *48*, 4659–4676.

405. Udding, J.H.; Tuijp, C.J.M.; Hiemstra, H.; Speckamp, W.N. *J. Chem. Soc., Perkin Trans. 2* **1992**, 857–858.

406. Udding, J.H.; Giesselink, J.P.M.; Hiemstra, H.; Speckamp, W.N. *Bull. Soc. Chim. Belg.* **1994**, *103*, 329–342.

407. Kernchen, F.; Henning, H.-G. *Monatsh. Chem.* **1989**, *120*, 253–267.

408. Hber, H.; Buchholz, H.; Sukale, R.; Henning, H.-G. *J. Prak. Chem.* **1985**, *327*, 51–62.

409. Miyabe, H.; Fujishima, Y.; Naito, T. *J. Org. Chem.* **1999**, *64*, 2174–2175.

410. Miyabe, H.; Ueda, M.; Yoshioka, N.; Naito, T. *Synlett* **1999**, 465–467.

411. Miyabe, H.; Yoshioka, N.; Ueda, M.; Naito, T. *J. Chem. Soc., Perkin Trans. 1* **1998**, 3659–3660.

412. Miyabe, H.; Ueda, M.; Yoshioka, N.; Yamakawa, K.; Naito, T. *Tetrahedron* **2000**, *56*, 2413–2420.

413. McNabb, S.B.; Ueda, M.; Naito, T. *Org. Lett.* **2004**, *6*, 1911–1914.

414. Miyabe, H.; Udea, M.; Nishimura, A.; Naito, T. *Org. Lett.* **2002**, *4*, 131–134.

415. Ricci, M.; Blakskjaer, P.; Skrydstrup, T. *J. Am. Chem. Soc.* **2000**, *122*, 12413–12421.

416. Barger, G.; Weichselbaum, T.E.; Clarke, H.T.; Gurin, S. *Org. Synth. Coll. Vol. 2*, **1948**, 384–386.

417. Dunn, M.S.; Smart, B.W.; Clarke, H.T.; Pearlman, W. *Org. Synth.* **1950**, *30*, 7–10.

418. Dunn, M.S.; Smart, B.W.; Clarke, H.T.; Pearlman, W. *Org. Syn. Coll. Vol. 4* **1963**, 55–58.

419. Sœrensen, S.P.L. *Bull. Soc. Chim.* **1905**, *33*, 1052–1055.

420. Sörensen, S.P.L.; Andersen, A.C. *Z. Physiol. Chem.* **1908**, *56*, 250–304.

421. Stephen, H.; Weizmann, C. *J. Chem. Soc.* **1914**, *105*, 1152–1155.

422. Osterberg, A.E. *J. Am. Chem. Soc.* **1927**, *49*, 538–540.

423. Mitra, S.K. *Ind. Chem. Soc. J.* **1930**, *7*, 799–802.

424. Dunn, M.S.; Smart, B.W. *J. Biol. Chem.* **1930**, *89*, 41–50.

425. Barger, G.; Weischelbaum, T.E. *Biochem. J.* **1931**, *25*, 997–1000.

426. Abderhalden, E.; Heyns, K. *Ber.* **1934**, *67*, 530–547.

427. Du Vigneaud, V.; Dyer, H.M.; Jones, C.B.; Patterson, W.I. *J. Biol. Chem.* **1934**, *106*, 401–407.

428. Overhoff, J.; Boeke, J.; Gorter, A. *Recl. Trav. Chim. Pays-Bas.* **1936**, *55*, 293–296.

429. Marvel, C.S.; Stoddard, M.P. *J. Org. Chem.* **1938**, *3*, 198–203.

430. Tarver, H.; Schmidt, C.L.A. *J. Biol. Chem.* **1939**, *130*, 67–80.

431. Wood, J.L.; Du Vigneaud, V. *J. Biol. Chem.* **1939**, *131*, 267–271.

432. Kuhn, R.; Quadbeck, G. *Chem. Ber.* **1943**, *76*, 527–528.

433. Kuhn, R.; Quadbeck, G. *Chem. Ber.* **1943**, *76*, 529–530.

434. Butenandt, A.; Weidel, W.; Weichert, R.; von Derjugin, W. *Z. Phys. Chem.* **1943**, *279*, 27–43.

435. Fink, R.M.; Enns, T.; Kimball, C.P.; Silberstein, H.E.; Bale, W.F.; Madden, S.C.; Whipple, G.H. *J. Exp. Med.* **1944**, *80*, 455–475.

436. Kilmer, G.W.; Du Vigneaud, V. *J. Biol. Chem.* **1944**, *154*, 247–253.

437. Dittmer, K.; Herz, W.; Chambers, J.S. *J. Biol. Chem.* **1946**, *166*, 541–543.

438. Moe, O.A.; Warner, D.T. *J. Am. Chem. Soc.* **1948**, *70*, 2763–2765.

439. Williams, L.R.; Ravve, A. *J. Am. Chem. Soc.* **1948**, *70*, 1244–1245.

440. Borek, E.; Waelsch, H. *J. Biol. Chem.* **1949**, *177*, 135–141.

441. Rachele, J.R.; Reed, L.J.; Kidwai, A.R.; Ferger, M.F.; Du Vigneaud, V. *J. Biol. Chem.* **1950**, *185*, 817–826.

442. Arnstein, H.R.V.; Hunter, G.D.; Muir, H.M.; Neuberger, A. *J. Chem. Soc.* **1952**, 1329–1334.

443. Bernhard, S.A. *J. Am. Chem. Soc.* **1952**, *74*, 4947–4948.

444. Zahn, H.; Dietrich, R.; Gerstner, W. *Chem. Ber.* **1955**, *88*, 1737–1746.

445. Blomquist, A.T.; Hiscock, B.F.; Harpp, D.N. *J. Org. Chem.* **1966**, *31*, 4121–4127.

446. See Reference 445.

447. Scannell, J.P.; Pruess, D.L.; Ax, H.A.; Jacoby, A.; Kellett, M.; Stempel, A. *J. Antibiotics.* **1976**, *29*, 38–43.

448. Liu, Y.-Y.; Thom, E.; Liebman, A.A. *Can. J. Chem.* **1978**, *56*, 2853–2855.

449. Reed, J.W.; Purvis, M.B.; Kingston, D.G.I.; Biot, A.; Gosselé, F. *J. Org. Chem.* **1989**, *54*, 1161–1165.

450. Ward, P.; Ewan, G.B.; Jordan, C.C.; Ireland, S.J.; Hagan, R.M.; Brown, J.R. *J. Med. Chem.* **1990**, *33*, 1848–1851.

451. Prashar, R.K.; Moore, D.E. *J. Chem. Soc., Perkin Trans. 1* **1993**, 1051–1053.

452. Harde, C.; Neff, K.-H.; Nordhoff, E.; Gerbling, K.-P.; Laber, B.; Pohlenz, H.-D. *Bioorg. Med. Chem. Lett.* **1994**, *4*, 273–278.

453. Albertson, N.F.; Archer, S. *J. Am. Chem. Soc.* **1945**, *67*, 2043–2044.

454. Snyder, H.R.; Shekleton, J.F.; Lewis, C.D. *J. Am. Chem. Soc.* **1945**, *67*, 310–312.

455. Howe, E.E.; Zambito, A.J.; Snyder, H.R.; Tishler, M. *J. Am. Chem. Soc.* **1945**, *67*, 38–39.

456. Albertson, N.F.; Archer, S.; Suter, C.M. *J. Am. Chem. Soc.* **1945**, *67*, 36–37.

457. Goldsmith, D.; Tishler, M. *J. Am. Chem. Soc.* **1946**, *68*, 144.

458. Dakin, H.D. *J. Biol. Chem.* **1946**, *164*, 615–620.

459. Painter, E.P. *J. Am. Chem. Soc.* **1947**, *69*, 232–234.

460. King, J.A. *J. Am. Chem. Soc.* **1947**, *69*, 2738–2741.

461. Erlenmeyer, H.; Grubenmann, W. *Helv. Chim. Acta.* **1947**, *30*, 297–304.

462. Armstrong, M.D. *J. Am. Chem. Soc.* **1948**, *70*, 1756–1759.

463. Dittmer, K.; Herz, W.; Cristol, S.J. *J. Biol. Chem.* **1948**, *173*, 323–326.

464. Elliott, D.F.; Fuller, A.T.; Harington, C.R. *J. Chem. Soc.* **1948**, 85–89.

465. Elliott, D.F.; Harington, C. *J. Chem. Soc.* **1949**, 1374–1378.

466. Heidelberger, C. *J. Biol. Chem.* **1949**, *179*, 139–142.

467. Jones, R.G. *J. Am. Chem. Soc.* **1949**, *71*, 383–386.

468. Jones, R.G. *J. Am. Chem. Soc.* **1949**, *71*, 3994–4000.

469. Bennett, E.L.; Nieman, C. *J. Am. Chem. Soc.* **1950**, *72*, 1800–1803.

470. Bennett, E.L.; Niemann, C. *J. Am. Chem. Soc.* **1950**, *72*, 1806–1807.

471. Harris, J.I.; Work, T.S. *Biochem. J.* **1950**, *46*, 190–195.

472. Burckhalter, J.H.; Stephens, V.C. *J. Am. Chem. Soc.* **1951**, *73*, 56–58.

473. Burckhalter, J.H.; Stephens, V.C. *J. Am. Chem. Soc.* **1951**, *73*, 3502–3503.

474. Van Zyl, G.; van Tamelen, E.E.; Zuidema, G.D. *J. Am. Chem. Soc.* **1951**, *73*, 1765–1767.

475. Heath, H.; Lawson, A.; Rimington, C. *J. Chem. Soc.* **1951**, 2220–2222.

476. Rydon, H.N.; Siddappa, S. *J. Chem. Soc.* **1951**, 2462–2467.

477. Bailey, A.S.; Bates, D.H.; Ing, H.R.; Warne, M.A. *J. Chem. Soc.* **1952**, 4534–4535.

478. Wiss, O.; Fuchs, H. *Helv. Chim. Acta* **1952**, *35*, 407–411.

479. Fillman, J.L.; Albertson, N.F. *J. Am. Chem. Soc.* **1952**, *74*, 4969–4970.

480. Lindstedt, S. *Acta Chem. Scand.* **1953**, *7*, 340–342.

481. Shulgin, A.T.; Gal, E.M. *J. Chem. Soc.* **1953**, 1316–1318.

482. Hellman, H.; Lingens, F. *Zeitschrift* **1954**, *297*, 283–287.

483. Simmonds, D.H. *Biochem. J.* **1954**, *58*, 520–523.

484. Servigne, M.; Szarvasi, E. *Bull. Ste. Chim. Biol.* **1954**, *36*, 1093–1100.

485. Dennis, R.L.; Plant, W.J.; Skinner, C.G.; Sutherland, G.L.; Shive, W. *J. Am. Chem. Soc.* **1955**, *77*, 2362–2364.

486. Benoiton, L.; Bouthillier, L.P. *Can. J. Chem.* **1955**, *33*, 1473–1476.

487. Morrison, D.C. *J. Am. Chem. Soc.* **1955**, *77*, 6072–6073.

488. Hegedüs, B. *Helv. Chim. Acta* **1955**, *38*, 22–27.

489. Bergel, F.; Burnop, V.C.E.; Stock, J.A. *J. Chem. Soc.* **1955**, 1223–1230.

490. Rydon, H.N.; Tweddle, J.C. *J. Chem. Soc.* **1955**, 3499–3503.

491. See Reference 490.

492. Hellmann, H.; Lingens, F. *Chem. Ber.* **1956**, *89*, 77–81.

493. Reisner, D.B. *J. Am. Chem. Soc.* **1956**, *78*, 2132–2135.

494. Ono, T.; Hirohata, R. *Z. Physiol. Chem.* **1956**, *304*, 77–81.

495. Hellmann, H.; Lingens, F.; Folz, E. *Ber.* **1956**, *89*, 2433–2436.

496. McCord, T.J.; Ravel, J.M.; Skinner, C.G.; Shive, W. *J. Am. Chem. Soc.* **1957**, *79*, 5693–5696.

497. Colescott, R.L.; Herr, R.R.; Dailey, J.P. *J. Am. Chem. Soc.* **1957**, *79*, 4232–4235.

498. Herr, R.R.; Enkoji, T.; Dailey, J.P. *J. Am. Chem. Soc.* **1957**, *79*, 4229–4232.

499. Kaiser, C.; Burger, A. *J. Am. Chem. Soc.* **1957**, *79*, 4365–4370.

500. Hellman, H.; Haas, G. *Ber.* **1957**, *90*, 1357–1363.

501. Okuda, T.; Tatsumi, S. *J. Biochem.* **1957**, *44*, 631–635.

502. Bixler, R.L.; Niemann, C. *J. Org. Chem.* **1958**, *23*, 575–584.

503. Renner, U.; Jöhl, A.; Stoll, W.G. *Helv. Chim. Acta* **1958**, *41*, 588–592.

504. Snyder, H.R.; Reedy, A.J.; Lennarz, W.J. *J. Am. Chem. Soc.* **1958**, *80*, 835–838.

505. Birkofer, L.; Ritter, A. *Liebigs Ann. Chem.* **1958**, *612*, 22–33.

506. Eberson, L. *Acta Chem. Scand.* **1958**, *12*, 314–323.

507. Edelson, J.; Fissekis, J.D.; Skinner, C.G.; Shive, W. *J. Am. Chem. Soc.* **1958**, *80*, 2698–2700.

508. Tesser, G.I.; Nefkens, G.H.L. *Rec. Trav. Chim. Pays-Bas* **1959**, *78*, 404–407.

509. Edelson, J.; Skinner, C.G.; Ravel, J.M.; Shive, W. *J. Am. Chem. Soc.* **1959**, *81*, 5150–5153.

510. Böhme, H.; Broese, R.; Eiden, F. *Ber.* **1959**, *92*, 1258–1262.

511. Fujita, Y. *Bull. Chem. Soc. Jpn.* **1958**, *31*, 439–442.

512. Conners, T.A.; Ross, W.C.J. *J. Chem. Soc.* **1960**, 2119–2132.

513. Connors, T.A.; Ross, W.C.J.; Wilson, J.G. *J. Chem. Soc.* **1960**, 2994–3007.

514. Miyake, A. *Chem. Pharm. Bull.* **1960**, *8*, 1074–1078.

515. Finar, I.L.; Utting, K. *J. Chem. Soc. C* **1960**, 5272–5273.

516. Barker, H.A.; Wawszkiewicz, E.J..; Winitz, M.; Birnbaum, S.M. *Biochem. Prep.* **1961**, *8*, 93–95.

517. Gershon, H.l.; Scala, A. *J. Org. Chem.* **1961**, *26*, 4517–4519.

518. Norton, S.J.; Skinner, C.G.; Shive, W. *J. Org. Chem.* **1961**, *26*, 1495–1498.

519. Hudlicky, M. *Collect. Czech. Chem. Commun.* **1961**, *26*, 1414–1421.

520. Thompson, J.F.; Morris, C.J.; Asen, S.; Irreverre, F. *J. Biol. Chem.* **1961**, *236*, 1183–1185.

521. Bouthillier, L.P.; Benoiton, L.; Dekker, E.E. *Biochem. Prep.* **1962**, *9*, 74–78.

522. Brown, R.R.; Hankes, L.V.; Kawahima, R.; Suhdolnik, R.J. *Biochem. Prep.* **1962**, *9*, 79–83.

523. Burgstahler, A.W.; Aiman, C.E. *Chem. Ind.* **1962**, 1430–1431.

524. Kaneko, T.; Lee, Y.K.; Hanafusa, T. *Bull. Chem. Soc. Jpn.* **1962**, *32*, 875–878.

525. Buchanan, R.L.; Dean, F.H.; Pattison, F.L.M. *Can. J. Chem.* **1962**, *40*, 1571–1575.

526. Yamada, S.-I.; Fuji, T.; Shioiri, T. *Chem. Pharm. Bull.* **1962**, *10*, 680–688.

527. Dunn, F.W.; Campaigne, E.; Neiss, E.S. *Biochem. Prep.* **1963**, *10*, 159–165.

528. Marcus, A.; Feeley, J.; Shannon, L.M. *Arch. Biochem. Biophys.* **1963**, *100*, 80–85.

529. Chambers, J.R.; Isbell, A.F. *J. Org. Chem.* **1964**, *29*, 832–836.

530. Barry, G.T.; Roark, E. *J. Biol. Chem.* **1964**, *239*, 1541–1544.

531. Hill, J.T.; Dunn, F.W. *J. Org. Chem.* **1965**, *30*, 1321–1322.

532. Morrison, D.C. *J. Chem. Soc.* **1965**, 2264–2265.

533. Leonard, F.; Wajngurt, A.; Tschannen, W.; Block, F.B. *J. Med. Chem.* **1965**, *8*, 812–815.

534. Gershon, H.; Scala, A.; Parmegiani, R. *J. Med. Chem.* **1965**, *8*, 877–881.

535. Filler, R.; Kang, H.H. *J. Chem. Soc., Chem. Commun.* **1965**, 626–627.

536. Filler, R.; Ayyangar, N.R.; Gustowski, W.; Kang, H.H. *J. Org. Chem.* **1969**, *34*, 534–538.

537. Kim, Y.C.; Cocolas, G.H. *J. Med. Chem.* **1965**, *8*, 509–512.

538. Filler, R.; Gustowski, W. *Nature* **1965**, *205*, 1105.

539. Springer, R.H.; Haggerty W.J., Jr.; Cheng, C.C. *J. Heterocycl. Chem.* **1965**, *2*, 49–52.

540. Chisholm, M.D.; Wetter, L.R. *Can. J. Biochem.* **1966**, *44*, 1625–1632.

541. Ablewhite, A.J.; Wooldridge, K.R.H. *J. Chem. Soc. C* **1967**, 2488–2491.

542. Tolman, V.; Veres, K. *Collect. Czech. Chem. Commun.* **1967**, *32*, 4460–4469.

543. Smith, S.C.; Sloane, N.H. *Biochim. Biophys. Acta* **1967**, *148*, 414–422.

544. Carrión, J.P.; Deranleau, D.A.; Donzel, B.; Esko, K.; Moser, P.; Schwyzer, R. *Helv. Chim. Acta* **1968**, *51*, 459–481.

545. McCord, T.J.; Booth, L.D.; Davis, A.L. *J. Med. Chem.* **1968**, *11*, 1077–1078.

546. Sullivan, P.T.; Kester, M.; Norton, S.J. *J. Med. Chem.* **1968**, *11*, 1172–1176.

547. Schatz, V.B.; O'Brien, B.C.; Sandusky, W.R. *J. Med. Chem.* **1968**, *11*, 140–141.

548. DeGraw, J.I.; Cory, M.; Skinner, W.A.; Theisen, M.C.; Mitoma, C. *J. Med. Chem.* **1968**, *11*, 225–227.

549. Tolman, V.; Hanus, J.; Veres, K. *J. Lab. Compds. Radiopharm.* **1968**, *4*, 243–247.

550. Hase, S.; Kiyoi, R.; Sakakibara, S. *Bull. Chem. Soc. Jpn.* **1968**, *41*, 1266–1267.

551. Watanabe, H.; Kuwata, S.; Naoe, K.; Nishida, Y. *Bull. Chem. Soc. Jpn.* **1968**, *41*, 1634–1638.

552. Kishida, Y.; Terada, A. *Chem. Pharm. Bull.* **1969**, 2417–2423.

553. Mertes, M.P.; Ramsey, A.A. *J. Med. Chem.* **1969**, *12*, 342–343.

554. Crooij, P.; Eliaers, J. *J. Chem. Soc. C* **1969**, 559–563.

555. Jorensen, E.C.; Wright, J. *J. Med. Chem.* **1970**, *13*, 367–370.

556. Norton, S.J.; Sullivan, P.T. *J. Heterocycl. Chem.* **1970**, *7*, 699–702.

557. Seely, J.H.; Benoiton, N.L. *Can. J. Biochem.* **1970**, *48*, 1122–1131.

558. Weinkam, R.J.; Jorgensen, E.C. *J. Am. Chem. Soc.* **1971**, *93*, 7028–7033.

559. Sullivan, P.T.; Sullivan, C.B.; Norton, S.J. *J. Med. Chem.* **1971**, *14*, 211–214.

560. Sullivan, P.T.; Norton, S.J. *J. Med. Chem.* **1971**, *14*, 557–558.

561. Hider, R.C.; John, D.I. *J. Chem. Soc., Perkin Trans. 1* **1972**, 1825–1830.

562. Trout, G.E. *J. Med. Chem.* **1972**, *15*, 1259–1261.

563. Isowa, Y.; Takashima, T.; Ohmori, M.; Kurita, H.; Sato, M.; Mori, K. *Bull. Chem. Soc. Jpn.* **1972**, *45*, 1461–1464.

564. Kirk, K.L.; Cohen, L.A. *J. Am. Chem. Soc.* **1973**, *95*, 4619–4624.

565. Davis, A.L.; Cavitt, M.B.; McCord, T.J.; Vickrey, P.E.; Shive, W. *J. Am. Chem. Soc.* **1973**, *95*, 6800–6802.

566. Couchman, R.; Eagles, J.; Hegarty, M.P.; Laird, W.M.; Self, R.; Synge, R.L.M. *Phytochemistry* **1973**, *12*, 707–718.

567. Tesser, G.I.; Slits, H.G.A.; van Nispen, J.W. *Int. J. Pept. Prot. Res.* **1973**, *5*, 119–122.

568. Chang, P.K.; Sciarini, L.J.; Handsschumacher, R.E. *J. Med. Chem.* **1973**, *16*, 1277–1280.

569. Firnau, G.; Nahmias, C.; Garnett, S. *J. Med. Chem.* **1973**, *16*, 416–418.

570. Lee, Y.K.; Kaneko, T. *Bull. Chem. Soc. Jpn.* **1973**, *46*, 2924–2926.

571. Sethi, M.L.; Rao, G.S.; Kapadia, G.J. *J. Pharm. Sci.* **1973**, *62*, 1802–1806.

572. Matsumoto, K.; Miyahara, T.; Suzuki, M.; Miyoshi, M. *Agr. Biol. Chem.* **1974**, *38*, 1097–1099.

573. Piper, J.R.; Rose, L.M.; Laseter, A.G.; Johnston, T.P.; Grenan, M.M. *J. Med. Chem.* **1974**, *17*, 1119–1120.

574. Anhoury, M.L.; Crooy, P.; De Neys, R.; Eliaers, J. *Bull. Soc. Chim. Belg.* **1974**, *83*, 117–130.

575. Hatanaka, S.-I.; Kaneko, S.; Niimura, Y.; Kinoshita, F.; Soma, G.-i. *Tetrahedron Lett.* **1974**, *45*, 3931–3932.

576. Bloemhoff, W.; Kerling, K.E.T. *Rec. Trav. Chim. Pays-Bas.* **1975**, *94*, 182–185.

577. Winn, M.; Rasmussen, R.; Minard, F.; Kyncl, J.; Plotnikoff, N. *J. Med. Chem.* **1975**, *18*, 434–437.

578. Shiba, T.; Mukunoki, Y.; Akiyama, H. *Bull. Chem. Soc. Jpn.* **1975**, *48*, 1902–1906.

579. Inouye, S.; Shomura, T.; Tsuruoka, T.; Ogawa, Y.; Watanabe, H.; Yoshida, J.; Niida, T. *Chem. Pharm. Bull.* **1975**, *23*, 2669–2677.

580. Leukart, O.; Caviezel, M.; Eberle, A.; Escher, E.; Tun-Kyi, A.; Schwyzer, R. *Helv. Chim. Acta* **1976**, *59*, 2181–2183.

581. Kataoka, Y.; Seto, Y.; Yamamoto, M.; Yamada, T.; Kuwata, S.; Watanabe, H. *Bull. Chem. Soc. Jpn.* **1976**, *49*, 1081–1084.

582. Upson, D.A.; Hruby, V.J. *J. Org. Chem.* **1976**, *41*, 1353–1358.

583. Gross, H.; Gnauk, T. *J. Prakt. Chem.* **1976**, *318*, 157–160.

584. Yabe, Y.; Miura, C.; Horikoshi, H.; Baba, Y. *Chem. Pharm. Bull.* **1976**, *24*, 3149–3157.

585. Shimohigashi, Y.; Lee, S.; Izumiya, N. *Bull. Chem. Soc. Jpn.* **1976**, *49*, 3280–3284.

586. Harris, R.L.N. *Aust. J. Chem.* **1976**, *29*, 1329–1334.

587. Harris, R.L.N. *Aust. J. Chem.* **1976**, *29*, 1335–1339.

588. Yamamoto, D.M.; Upson, D.A.; Linn, D.K.; Hruby, V.J. *J. Am. Chem. Soc.* **1977**, *99*, 1564–1570.

589. Larsen, P.O.; Wieczorkowska, E. *Acta Chem. Scand. B* **1977**, *31*, 109–113.

590. Kelley, J.L.; Miler, C.A.; McLean, E.W. *J. Med. Chem.* **1977**, *20*, 721–723.

591. Harris, R.L.N.; Teitei, T. *Aust. J. Chem.* **1977**, *30*, 649–655.

592. Perlman, D.; Perlman, K.L.; Bodanszky, M.; Bodanszky, A.; Foltz, R.L.; Matthews, H.W. *Bioorganic Chem.* **1977**, *6*, 263–271.

593. Keith, D.D.; Yang, R.; Tortora, J.A.; Weigele, M. *J. Org. Chem.* **1978**, *43*, 3713–3716.

594. Blumenstein, M.; Hruby, V.J.; Yamamoto, D.M. *Biochemistry* **1978**, *17*, 4971–4977.

595. Dilbeck, G.A.; Field, L.; Gallo, A.A.; Gargiulo, R.J. *J. Org. Chem.* **1978**, *43*, 4593–4596.

596. Podoscielny, W.; Podgórski, M.; Smulkowska, E. *Polish J. Chem.* **1978**, *52*, 2455–2459.

597. Viswanatha, V.; Hruby, V.J. *J. Org. Chem.* **1979**, *44*, 2892–2896.

598. Hansen, J.J.; Krogsgaard-Larsen, P. *J. Chem. Soc. Chem., Commun.* **1979**, 87–88.

599. Fauchère, J.-L.; Leukart, O.; Eberle, A.; Schwyzer, R. *Helv. Chim. Acta* **1979**, *62*, 1385–1395.

600. Rajh, H.M.; Uitzetter, J.H.; Westerhuis, L.W.; van den Dries, C.L.; Tesser, G.I. *Int. J. Pept. Prot. Res.* **1979**, *14*, 68–79.

601. Teitei, T. *Aust. J. Chem.* **1979**, *32*, 1631–1634,

602. Edgar, M.T.; Pettit, G.R.; Krupa, T.S. *J. Org. Chem.* **1979**, *44*, 396–400.

603. Hansen, J.J.; Krogsgaard-Larsen, P. *J. Chem. Soc., Perkin Trans. 1* **1980**, 1826–1833.

604. Honore, T.; Lauridsen, J. *Acta Chem. Scand. B* **1980**, *34*, 235–240.

605. See Reference 604.

606. Keller-Schierlein, W.; Joos, B. *Helv. Chim. Acta* **1980**, *63*, 250–254.

607. Sasaki, A.N.; Bricas, E. *Tetrahedron Lett.* **1980**, *21*, 4263–4264.

608. Frejd, T.; Davis, M.A.; Gronowitz, S.; Sadeh, T. *J. Heterocycl. Chem.* **1980**, *17*, 759–761.

609. Fujioka, T.; Satoh, T.; Tanizawa, K.; Kanaoka, Y. *J. Biochem.* **1980**, *87*, 1229–1234.

610. Filler, R.; Rickert, R.C. *J. Fluorine Chem.* **1981**, *18*, 483–495.

611. Nestor J.J., Jr.; Ho, T.L.; Simpson, R.A.; Horner, B.L.; Jones, G.H.; McRae, G.I.; Vickery, B.H. *J. Med. Chem.* **1982**, *25*, 795–801.

612. Fotadar, U.; Cowburn, D. *J. Lab. Cmpds. Radiopharm.* **1983**, *20*, 1003–1009.

613. Maier, L.; Lea, P.J. *Phosphorus Sulfur* **1983**, *17*, 1–19.

614. Maier, L.; Rist, G.; Lea, P.J. *Phosphorus Sulfur* **1983**, *18*, 349–352.

615. Monaghan, D.T.; McMills, M.C.; Chamberlin, A.R.; Cotman, C.W. *Brain Res.* **1983**, *278*, 137–144.

616. Krogsgaard-Larsen, P.; Nielsen, E.O.; Curtis, D.R. *J. Med. Chem.* **1984**, *27*, 585–591.

617. Matoba, K.; Yonemoto, H.; Fukui, M.; Yamazaki, T. *Chem. Pharm. Bull.* **1984**, *32*, 3918–3925.

618. Anastasis, P.; Overton, K. *J. Lab. Cmpds. Radiopharm.* **1984**, *21*, 393–400.

619. Lauridsen, J.; Honoré, T.; Krogsgaard-Larsen, P. *J. Med. Chem.* **1985**, *28*, 668–672.

620. Payne, L.S.; Boger, J. *Synth. Commun.* **1985**, *15*, 1277–1290.

621. Pascal R.A., Jr.; Chen, Y.-C. *J. Org. Chem.* **1985**, *50*, 408–410.

622. Sawada, S.; Nakayama, T.; Esaki, N.; Tanaka, H.; Soda, K.; Hill, R.K. *J. Org. Chem.* **1986**, *51*, 3384–3386.

623. Paik, Y.H.; Dowd, P. *J. Org. Chem.* **1986**, *51*, 2910–2913.

624. Egusa, S.; Sisido, M.; Imanishi, Y. *Bull. Chem. Soc. Jpn.* **1986**, *59*, 3175–3178.

625. Ego, D.; Beaucourt, J.P.; Pichat, L. *J. Lab. Cmpds. Radiopharm.* **1986**, *23*, 553–564.

626. Porter, J.; Dykert, J.; Rivieer, J. *Int. J. Pept. Prot. Res.* **1987**, *30*, 13–21.

627. Rao, P.N.; Burdett J.E., Jr.; Cessac, J.W.; DiNunno, C.M.; Peterson, D.M.; Kim, H.K. *Int. J. Pept. Prot. Res.* **1987**, *29*, 118–125.

628. Tsushima, T.; Kawada, K.; Ishihara, S.; Uchida, N.; Shiratori, O.; Higaki, J.; Hirata, M. *Tetrahedron* **1988**, *44*, 5375–5387.

629. Marseigne, I.; Roques, B.P. *J. Org. Chem.* **1988**, *53*, 3621–3624.

630. Koyama, M.; Saino, T. *J. Lab. Compds. Radiopharm.* **1988**, *25*, 1299–1306.

631. Gould, S.J.; Ju, S. *J. Am. Chem. Soc.* **1989**, *111*, 2329–2331.

632. Bigge, C.F.; Drummond, J.T.; Johnson, G.; Malone, T.; Probert A.W., Jr.; Marcoux, F.W.; Coughenour, L.L.; Brahce, L.J. *J. Med. Chem.* **1989**, *32*, 1580–1590.

633. Bigge, C.F.; Drummond, J.T.; Johnson, G. *Tetrahedron Lett.* **1989**, *30*, 7013–7016.

634. Gibbons, W.A.; Hughes, R.A.; Charalambous, M.; Christodoulou, M.; Szeto, A.; Aulabaugh, A.E.; Mascagni, P.; Toth, I. *Liebigs Ann. Chem.* **1990**, 1175–1183.

635. Hsiao, C.-N.; Leanna, M.R.; Bhagavatula, L.; De Lara, E.; Zydowsky, T.M.; Horrom, B.W.; Morton, H.E. *Synth. Commun.* **1990**, *20*, 3507–3517.

636. Bladon, C.M. *J. Chem. Soc., Perkin Trans. 1* **1990**, 1151–1158.

637. Baldwin, J.E., Bradley, M.; Abbott, S.D.; Adlington, R.M. *Tetrahedron* **1991**, *47*, 5309–5328.

638. Schiller, P.W.; Weltrowska, G.; Nguyen, T.M.D.; Lemieux, C.; Chung, N.N.; Marsden, B.J.; Wilkes, B.C. *J. Med. Chem.* **1991**, *34*, 3125–3132.

639. Rodriguez, M.; Bernad, N.; Galas, M.C.; Lignon, M.F.; Laur, J.; Aumelas, A.; Martinez, J. *Eur. J. Med. Chem.* **1991**, *26*, 245–253.

640. Crisp, G.T.; Glink, P.T. *Tetrahedron Lett.* **1992**, *33*, 4649–4652.

641. Gould, S.J.; Ju, S. *J. Am. Chem. Soc.* **1992**, *114*, 10166–10172.

642. Dorville, A.; McCort-Tranchepain, I.; Vichard, D.; Sather, W.; Maroun, R.; Ascher, P.; Roques, B.P. *J. Med. Chem.* **1992**, *35*, 2551–2562.

643. Ortwine, D.F.; Malone, T.C.; Bigge, C.F.; Drummond, J.T.; Humblet, C.; Johnson, G.; Pinter, G.W. *J. Med. Chem.* **1992**, *35*, 1345–1370.

644. Kammermeier, B.O.T.; Lerch, U.; Sommer, C. *Synthesis* **1992**, 1157–1160.

645. Baldwin, J.E.; Hulme, C.; Schofield, C.J. *J. Chem. Res. S* **1992**, 173.

646. Furuta, T.; Katayama, M.; Shibasaki, H.; Kasuya, Y. *J. Chem. Soc., Perkin Trans. 1* **1992**, 1643–1648.

647. Tolman, V. *J. Fluorine Chem.* **1993**, *60*, 179–183.

648. Emrey, T.A.; Simmonds, J.T.; Kowal, D.M.; Stein, R.P.; Tasse, R.P. *J. Med. Chem.* **1993**, *36*, 331–342.

649. Leanna, M.R.; Morton, H.E. *Tetrahedron Lett.* **1993**, *34*, 4485–4488.

650. Hasegawa, H.; Arai, S.; Shinohara, Y.; Baba, S. *J. Chem. Soc., Perkin Trans 1* **1993**, 489–494.

651. Madsen, U.; Frølund, B.; Lund, T.M.; Ebert, B.; Krogsgaard-Larsen, P. *Eur. J. Med. Chem.* **1993**, *28*, 791–800.

652. Hagiwara, D.; Miyake, H.; Igari, N.; Karino, M.; Maeda, Y.; Fujii, T.; Matsuo, M. *J. Med. Chem.* **1994**, *37*, 2090–2099.

653. Stüber, W.; Koschinsky, R.; Reers, M.; Hoffman, D.; Czech, J.; Dickneite, G. *Peptide Res.* **1995**, *8*, 78–85.

654. Andrews, D.M.; Gregoriou, M.; Page, T.C.M.; Peach, J.M.; Pratt, A.J. *J. Chem. Soc., Perkin Trans. 1* **1995**, 1335–1340.

655. Ruzza, P.; Deana, A.D.; Calderan, A.; Pavanetto, M.; Cesaro, L.; Pinna, L.A.; Borin, G. *Int. J. Pept. Prot. Res.* **1995**, *46*, 535–546.

656. Plummer, M.S.; Shahripour, A.; Kaltenbronn, J.S.; Lunney, E.A.; Steinbaugh, B.A.; Hamby, J.M.; Hamilton, H.W.; Sawyer, T.K.; Humblet, C.; Doherty, A.M.; Taylor, M.D.; Hingorani, G.; Batley, B.L.; Rapundalo, S.T. *J. Med. Chem.* **1995**, *38*, 2893–2905.

657. Matsoukas, J.M.; Agelis, G.; Wahhab, A.; Hondrelis, J.; Panagiotopoulos, D.; Yamdagni, R.; Wu, Q.; Mavromoustakos, T.; Maia, H.L.S.; Ganter, R.; Moore, G.J. *J. Med. Chem.* **1995**, *38*, 4660–4669.

658. Rozhko, L.F.; Ragulin, V.V.; Tsvetkov, E.N. *Russ. J. Gen. Chem.* **1996**, *66*, 1065–1067.

659. Hamon, C.; Rawlings, B.J. *Synth. Commun.* **1996**, *26*, 1109–1115.

660. Nishino, N.; Arai, T.; Ueno, Y.; Ohba, M. *Chem. Pharm. Bull.* **1996**, *44*, 212–214.

661. Pearson, D.A.; Lister-James, J.; McBride, W.J.; Wilson, D.M.; Martel, L.J.; Civitello, E.R.; Dean, R.T. *J. Med. Chem.* **1996**, *39*, 1372–1382.

662. Cheng, R.P.; Fisher, S.L.; Imperiali, B. *J. Am. Chem. Soc.* **1996**, *118*, 11349–11356.

663. Vergne, C.; Bois-Choussey, M.; Ouazzani, J.; Beugelmans, R.; Zhu, J. *Tetrahedron: Asymmetry* **1997**, *8*, 391–398.

664. Cooper, M.S.; Seton, A.W.; Stevens, M.F.G.; Westwell, A.D. *Bioorg. Med. Chem. Lett.* **1996**, *6*, 2613–2616.

665. Kira, M.; Matsubara, T.; Shinohara, H.; Sisido, M. *Chem. Lett.* **1997**, 89–90.

666. Koppitz, M.; Huengs, M.; Gratias, R.; Kessler, H.; Goodman, S.L.; Jonczyk, A. *Helv. Chim. Acta.* **1997**, *80*, 1280–1300.

667. Shearer, B.G.; Lee, S.; Franzmann, K.W.; White, H.A.R.; Sanders, D.C.J.; Kiff, R.J.; Garvey, E.P.; Furfine, E.S. *Bioorg. Med. Chem. Lett.* **1997**, *7*, 1763–1768.

668. Stirling, I.R.; Freer, I.K.A.; Robins, D.J. *J. Chem. Soc., Perkin Trans. 1* **1997**, 677–680.

669. Hill, R.A.; Wallace, L.J.; Miller, D.D.; Weinstein, D.M.; Shams, G.; Tai, H.; Layer, R.T.; Willins, D.; Uretsky, N.J.; Danthi, S.N. *J. Med. Chem.* **1997**, *40*, 3182–3191.

670. Matzen, L.; Engesgaard, A.; Ebert, B.; Didriksen, M.; Frøland, B.; Krogsgaard-Larsen, P.; Jaroszewski, J.W. *J. Med. Chem.* **1997**, *40*, 520–527.

671. Bang-Andersen, B.; Lenz, S.M.; Skjaerbaek, N.; Søby, K.K.; Hansen, H.O.; Ebert, B.; Bøgesø, K.P.; Krogsgaard-Larsen, P. *J. Med. Chem.* **1997**, *40*, 2831–2842.

672. Sisido, M. *Macromolecules* **1997**, *30*, 2651–2656.

673. Falch, E.; Brehm, L.; Mikkelsen, I.; Johansen, T.N.; Skjaerbaek, N.; Nielsen, B.; Stensbøl, T.B.; Ebert, B.; Krogsgaard-Larsen, P. *J. Med. Chem.* **1998**, *41*, 2513–2523.

674. Skjærbæk, N.; Brehm, L.; Johansen, T.N.; Hansen, L.M.; Nielsen, B.; Ebert, B.; Søby, K.K.; Stensbøl, T.B.; Falch, E.; Krogsgaard-Larsen, P. *Biorg. Med. Chem.* **1998**, *6*, 119–131.

675. Hashimoto, K.; Yoshioka, T.; Morita, C.; Sakai, M.; Okuno, T.; Shirahama, H. *Chem. Lett.* **1998**, 203–204.

676. Maryanoff, C.A.; Scott, L.; Shah, R.D.; Villani F.J., Jr. *Tetrahedron: Asymmetry.* **1998**, *9*, 3247–3250.

677. Lecointe, L.; Rolland-Fulcrand, V.; Roumestant, M.L.; Viallefont, P.; Martinez, J. *Tetrahedron: Asymmetry.* **1998**, *9*, 1753–1758.

678. Lee, K.; Hwang, S.Y.; Yun, M.; Kim, D.S. *Biorg. Med. Chem. Lett.* **1998**, *8*, 1683–1686.

679. Adang, A.E.P.; Peters, C.A.M.; Gerritsma, S.; de Zwart, E.; Veeneman, G. *Biorg. Med. Chem. Lett.* **1999**, *9*, 1227–1232.

680. Hohsaka, T.; Kajihara, D.; Ashizuka, Y.; Murakami, H.; Sisido, M. *J. Am. Chem. Soc.* **1999**, *121*, 34–40.

681. Chin, J.; Lee, C.S.; Lee, K.J.; Park, S.; Kim, D.H. *Nature* **1999**, *401*, 254–257.

682. Fujita, T.; Nose, T.; Matsushima, A.; Okada, K.; Asai, D.; Yamauchi, Y.; Shirasu, N.; Honda, T.; Shigehiro, D.; Shimohigashi, Y. *Tetrahedron Lett.* **2000**, *41*, 923–927.

683. Watkins, E.B.; Phillips, R.S. *Biochemistry.* **2002**, *40*, 14862–14868.

684. Valiaeva, N.; Bartley, D.; Konno, T.; Coward, J.K. *J. Org. Chem.* **2001**, *66*, 5146–5154.

685. Wang, L.; Zhang, Z.; Brock, A.; Schultz, P.G. *Proc. Natl. Acad. Sci. U.S.A.* **2003**, *100*, 56–61.

686. Watanabe, L.A.; Jose, B.; Kato, T.; Nishino, N.; Yoshida, M. *Tetrahedron Lett.* **2004**, *45*, 491–494.

687. Watanabe, L.A.; Haranaka, S.; Jose, B.; Yoshida, M.; Kato, T.; Moriguchi, M.; Soda, K.; Nishino, N. *Tetrahedron: Asymmetry.* **2005**, *16*, 903–908.

688. Snyder, H.R.; Matteson, D.S. *J. Am. Chem. Soc.* **1957**, *79*, 2217–2221.

689. Haggerty, W.J., Jr.; Springer, R.H.; Cheng, C.C. *J. Heterocycl. Chem.* **1965**, *2*, 1–6.

690. Stoll, G.; Frank, J.; Musso, H.; Henke, H.; Herrendorf, W. *Liebigs Ann. Chem.* **1986**, 1968–1989.

691. Christensen, I.T.; Ebert, B.; Madsen, U.; Nielsen, B.; Brehm, L.; Krogsgaard-Larsen, P. *J. Med. Chem.* **1992**, *35*, 3512–3519.

692. Schneider, H.; Sigmund, G.; Schricker, B.; Thirring, K.; Berner, H. *J. Org. Chem.* **1993**, *58*, 683–689.

693. Matzen, L.; Engesgaard, A.; Ebert, B.; Didriksen, M.; Frølund, B.; Krogsgaard-Larsen, P.; Jaroszewski, J.W. *J. Med. Chem.* **1997**, *40*, 520–527.

694. Schenkels, C.; Erni, B.; Reymond, J.-L. *Biorg. Med. Chem. Lett.* **1999**, *9*, 1443–1446.

695. Cabaret, D.; Adediran, S.A.; Garcia Gonzalez, M.J.; Pratt, R.F.; Wakselman, M. *J. Org. Chem.* **1999**, *64*, 713–720.

696. Rewinkel, J.B.M.; Lucas, H.; van Galen, P.J.M.; Noach, A.B.J.; van Dinther, T.G.; Rood, A.M.M.; Jenneboer, A.J.S.M.; van Boeckel, C.A.A. *Bioorg. Med. Chem. Lett.* **1999**, *9*, 685–690.

697. Dakin, H.D. *J. Biol. Chem.* **1941**, *141*, 945–950.

698. Niemann, C.; Lewis, R.N.; Hays, J.T. *J. Am. Chem. Soc.* **1942**, *64*, 1678–1682.

699. Adediran, S.A.; Cabaret, D.; Pratt, R.F.; Wakselman, M. *Biorg. Med. Chem. Lett.* **1999**, *9*, 341–346.

700. Kjaer, A.; Wagner, S. *Acta Chem. Scand.* **1955**, *9*, 721–726.

701. Avakian, S.; Moss, J.; Martin, G.J. *J. Am. Chem. Soc.* **1948**, *70*, 3075–3076.

702. Butenandt, A.; Hellmann, H. *Z. Phys. Chem.* **1949**, *284*, 168–175.

703. Hellman, H. *Z. Phys. Chem.* **1949**, *284*, 163–167.

704. Gershon, H.; Meek, J.S.; Dittmer, K. *J. Am. Chem. Soc.* **1949**, *71*, 3573–3574.

705. Weisiger, J.R. *J. Biol. Chem.* **1950**, *186*, 591–602.

706. Dornow, A.; Winter, G. *Chem. Ber.* **1951**, *84*, 307–313.

707. Behringer, H.; Weissauer, H. *Chem. Ber.* **1952**, *85*, 743–751.

708. Bergel, F.; Stock, J.A. *J. Chem. Soc.* **1954**, 2409–2417.

709. Ek, A.; Witkop, B. *J. Am. Chem. Soc.* **1954**, *76*, 5579–5588.

710. Meek, J.S.; Rowe, J.W. *J. Am. Chem. Soc.* **1955**, *77*, 6675–6677.

711. Schlögl, K. *Monatsh. Chem.* **1957**, *88*, 601–621.

712. Carbon, J.A.; Martin, W.B.; Swett, L.R. *J. Am. Chem. Soc.* **1958**, *80*, 1002.

713. Schlögl, K. *Monatsh Chem.* **1958**, *89*, 377–390.

714. Osgerby, J.M.; Pauson, P.L. *J. Chem. Soc.* **1958**, 656–660.

715. Shaw, K.N.F.; Morris, A.J.; Verbiscar, A.J. *Biochem. Prep.* **1962**, *9*, 92–99.

716. Frankel, M.; Gertner, D.; Shenhar, A.; Zilkha, A. *J. Chem. Soc.* **1963**, 5049–5051.

717. Rotman, A.; Gertner, D.; Zilkha, A. *Can. J. Chem.* **1967**, *45*, 2469–2471.

718. Zilkha, A.; Friedman, G.; Gertner, D. *Can. J. Chem.* **1967**, *45*, 2979–2985.

719. Black, D.K.; Landor, S.R. *J. Chem. Soc. C* **1968**, 283–287.

720. Hanessian, S.; Schütze, G. *J. Med. Chem.* **1969**, *12*, 347.

721. Lee, C.-J.; Serif, G.S. *Biochemistry.* **1970**, *9*, 2068–2070.

722. Humphries, A.J.; Keener, R.L.; Yano, K.; Skelton, F.S.; Freiter, E.; Snyder, H.R. *J. Org. Chem.* **1972**, *37*, 3626–3629.

723. Hatanaka, S.-I.; Niimura, Y.; Taniguchi, K. *Phytochemistry* **1972**, *11*, 3327–3329.

724. Skinner, W.A.; Johansson, J.G. *J. Med. Chem.* **1972**, *15*, 427–428.

725. Teuber, H-J.; Krause, H.; Berariu, V. *Liebigs Ann. Chem.* **1978**, 757–770.

726. Hanson, R.N.; Davis, M.A. *J. Heterocycl. Chem.* **1981**, *18*, 205–206.

727. Brunet, J.C.; Cuingnet, E.; Gras, H.; Marcincal, P.; Mocz, A.; Sergheraert, C.; Tartar, A. *J. Organo-metal. Chem.* **1981**, *216*, 73–77.

728. Muthukumaraswamy, N.; Day, A.R.; Pinon, D.; Liao, C.S.; Freer, R.J. *Int. J. Pept. Prot. Res.* **1983**, *22*, 305–312.

729. Soine, W.H.; Guyer, C.E.; Knapp F.F., Jr. *J. Med. Chem.* **1984**, *27*, 803–806.

730. Kitagawa, T.; Akiyama, N. *Chem. Pharm. Bull.* **1997**, *45*, 1865–1866.

731. Hutinec, A.; Ziogas, A.; El-Mobayed, M.; Rieker, A. *J. Chem. Soc., Perkin Trans. 1* **1998**, 2201–2208.

732. Ehrhart, G. *Chem. Ber.* **1949**, *82*, 60–63.

733. Herz, W.; Dittmer, K.; Cristol, S.J. *J. Am. Chem. Soc.* **1948**, *70*, 504–507.

734. Goering, H.L.; Cristol, S.J.; Dittmer, K. *J. Am. Chem. Soc.* **1948**, *70*, 3314–3316.

735. Fields, M.; Walz, D.E.; Rothchild, S. *J. Am. Chem. Soc.* **1951**, *73*, 1000–1002.

736. Gaudry, R. *Can. J. Chem.* **1953**, *31*, 1060–1062.

737. Uhle, F.C.; Robinson, S.H. *J. Am. Chem. Soc.* **1955**, *77*, 3544–3546.

738. Greenstein, J.P.; Izumiya, N.; Winitz, M.; Birnbaum, S.M. *J. Am. Chem. Soc.* **1955**, *77*, 707–716.

739. Edelson, J.; Pal, P.R.; Skinner, C.G.; Shive, W. *J. Am. Chem. Soc.* **1957**, *79*, 5209–5212.

740. Hauser, C.R.; Lindsay, J.K. *J. Org. Chem.* **1957**, *22*, 1246–1247.

741. Benoiton, L.; Winitz, M.; Birnbaum, S.M.; Greenstein, J.P. *J. Am. Chem. Soc.* **1957**, *79*, 6192–6198.

742. Govardham, L.K.; Skinner, C.G.; LeFlore, J.O.; Elliott, I.W. *Biochem. Prep.* **1963**, *10*, 40–42.

743. Norton, S.J.; Sanders, E. *J. Med. Chem.* **1967**, *10*, 961–963.

744. Dardenne, G.; Casimir, J.; Jadot, J. *Phytochemistry* **1968**, *7*, 1401–1406.

745. Porter, T.H.; Shive, W. *J. Med. Chem.* **1968**, *11*, 402–403.

746. SooHoo, C.; Lawson, J.A.; DeGraw, J.I. *J. Lab. Compds. Radiopharm.* **1977**, *13*, 97–102.

747. Bey, P.; Danzin, C.; Van Dorsselaer, V.; Mamont, P.; Jung, M.; Tardif, C. *J. Med. Chem.* **1978**, *21*, 50–55.

748. Karwoski, G.; Galione, M.; Starcher, B. *Biopolymers* **1978**, *17*, 1119–1127.

749. Viswanatha, V.; Larsen, B.; Hruby, V.J. *Tetrahedron* **1979**, *35*, 1575–1580.

750. Cramer, U.; Rehfeldt, A.G.; Spener, F. *Biochemistry* **1980**, *19*, 3074–3080.

751. Prabhakaran, P.C.; Woo, N.-T.; Yorgey, P.S.; Gould, S.J. *J. Am. Chem. Soc.* **1988**, *110*, 5785–5791.

752. Horwell, D.C.; Nichols, P.D.; Roberts, E. *Bioorg. Med. Chem. Lett.* **1994**, *4*, 2263–2266.

753. Wang, C.; Mosberg, H.I. *Tetrahedron Lett.* **1995**, *36*, 3623–3626.

754. Wilson, C.J.; Husain, S.S.; Stimson, E.R.; Dangott, L.J.; Miller, K.W.; Maggio, J.E. *Biochem.* **1997**, *36*, 4542–4551.

755. Hoppe, D.; Beckmann, L. *Liebigs AM. Chem.* **1979**, 2066–2075.

756. Fitt, J.J.; Gschwend, H.W. *J. Org. Chem.* **1977**, *42*, 2639–2641.

757. Bey, P.; Vevert, J.P. *Tetrahedron Let.* **1978**, *14*, 1215–1218.

758. Metcalf, B.W.; Bonilavri, E. *J. Chem. Soc., Chem. Commun.* **1978**, 914–915.

759. Bey, P.; Vevert, J.-P.; Van Dorsselaer, V.; Kolb, M. *J. Org. Chem.* **1979**, *44*, 2732–2742.

760. Bey, P.; Vevert, J.P. *J. Org. Chem.* **1980**, *45*, 3249–3253.

761. Calcagni, A.; Rossi, D.; Lucente, G. *Synthesis* **1981**, 445–447.

762. Walsh, C.; Pascal R.A., Jr.; Johnston, M.; Raines, R.; Dikshit, D.; Krantz, A.; Honma, M. *Biochemistry* **1981**, *20*, 7509–7519.

763. Adlington, R.M.; Aplin, R.T.; Baldwin, J.E.; Rawlings, B.J.; Osborne, D. *J. Chem. Soc., Chem. Commun.* **1982**, 1086–1087.

764. Unson, C.G.; Erickson, B.W. *Int. J. Pept. Prot. Res.* **1983**, *22*, 50–56.

765. Rich, D.H.; Singh, J.; Gardner, J.H. *J. Org. Chem.* **1983**, *48*, 432–434.

766. Adlington, R.M.; Baldwin, J.E.; Rawlings, B.J. *J. Chem. Soc., Chem. Commun.* **1983**, 290–292.

767. Bey, P.; Ducep, J.B.; Schirlin, D. *Tetrahedron Lett.* **1984**, *25*, 5657–5660.

768. Joucla, M.; Goumzili, E.E.; Fouchet, B. *Tetrahedron Lett.* **1986**, *27*, 1677–1680.

769. Grierson, J.R.; Adam, M.J. *J. Lab. Compds. Radiopharm.* **1986**, *23*, 1019–1018.

770. Kelland, J.G.; Arnold, L.D.; Palcic, M.M.; Pickard, M.A.; Vederas, J.C. *J. Biol. Chem.* **1986**, *261*, 13216–13223.

771. Minowa, N.; Hirayama, M.; Fukatsu, S. *Bull. Chem. Soc. Jpn.* **1987**, *60*, 1761–1766.

772. Baggaley, K.H.; Nicholson, N.H.; Sime, J.T. *J. Chem. Soc., Chem. Commun.* **1988**, 567–568.

773. Schirlin, D.; Gerhart, F.; Hornsperger, J.M.; Hamon, M.; Wagner, J.; Jung, M.J. *J. Med. Chem.* **1988**, *31*, 30–36.

774. Tarzia, G.; Balsamini, C.; Spadoni, G.; Duranti, E. *Synthesis* **1988**, 514–517.

775. Haddow, J.; Suckling, C.J.; Wood, H.C.S. *J. Chem. Soc., Perkin Trans. I* **1989**, 1297–1304.

776. Bjurling, P.; Watanabe, Y.; Tokushige, M.; Oda, T.; Långström, B. *J. Chem. Soc., Perkin Trans. I* **1989**, 1331–1334.

777. Bjurling, P.; Watanabe, Y.; Oka, S.; Nagasawa, T.; Yamada, H.; Långström, B. *Acta Chem. Scand. B* **1990**, *44*, 178–188.

778. Baggaley, K.H.; Elson, S.W.; Nicholson, N.H.; Sime, J.T. *J. Chem. Soc., Perkin Trans I* **1990**, 1521–1533.

779. Kitazume, T.; Lin, J.T.; Yamazaki, T. *Tetrahedron: Asymmetry* **1991**, *2*, 235–238.

780. Schirlin, D.; Ducep, J.B.; Baltzer, S.; Bey, P.; Piriou, F.; Wagner, J.; Hornsperger, J.M.; Heydt, J.G.; Jung, M.J.; Danzin, C.; Weiss, R.; Fischer, J.; Mitschler, A.; De Cian, A. *J. Chem. Soc., Perkin Trans. I* **1992**, 1053–1064.

781. Whitten, J.P.; Baron, B.M.; McDonald, I.A. *Bioorg. Med. Chem. Lett.* **1993**, *3*, 23–26.

782. Ornstein, P.L.; Arnold, M.B.; Evrard, D.; Leander, J.D.; Lodge, D.; Schoepp, D.D. *Bioorg. Med. Chem. Lett.* **1993**, *3*, 43–48.

783. Schmall, B.; Conti, P.S. *J. Lab. Cmpds. Radiopharm.* **1994**, *35*, 127–129.

784. Hamilton, R.; Shute, R.E.; Travers, J.; Walker, B.; Walker, B.J. *Tetrahedron Lett.* **1994**, *35*, 3597–3600.
785. Josien, H.; Lavielle, S.; Brunissen, A.; Saffroy, M.; Torrens, Y.; Beaujouan, J.-C.; Glowinski, J.; Chassaing, G. *J. Med. Chem.* **1994**, *37*, 1586–1601.
786. Zindel, J.; de Meijere, A. *Synthesis* **1994**, 190–194.
787. Shuman, R.T.; Rothenberger, R.B., Campbell, C.S.; Smith, G.F.; Gifford-Moore, D.S.; Paschal, J.W.; Gesellchen, P.D. *J. Med. Chem.* **1995**, *38*, 4446–4453.
788. Rubio, A.; Ezquerra, J. *Tetrahedron Lett.* **1995**, *36*, 5823–5826.
789. Gilbert, I.H.; Rees, D.C.; Crockett, A.K.; Jones, R.C.F. *Tetrahedron* **1995**, *51*, 6315–6336.
790. Bensadat, A.; Félix, C.; Laurent, A.; Laurent, E.; Faure, R.; Thomas, T. *Bull. Chem. Soc. Fr.* **1996**, *133*, 509–514.
791. Spero, D.M.; Kapadia, S.R. *J. Org. Chem.* **1996**, *61*, 7398–7401.
792. Burns, M.R.; Coward, J.K. *Bioorg. Med. Chem.* **1996**, *4*, 1455–1470.
793. Satoh, Y.; Gude, C.; Chan, K.; Firooznia, F. *Tetrahedron Lett.* **1997**, *38*, 7645–7648.
794. Arslan, T.; Mamaev, S.V.; Mamaeva, N.V.; Hecht, S.M. *J. Am. Chem. Soc.* **1997**, *119*, 10877–10887.
795. Kröger, S.; Haufe, G. *Amino Acids* **1997**, *12*, 363–372.
796. Fretz, H. *Tetrahedron* **1998**, *54*, 4849–4858.
797. Haufe, G.; Laue, K.W.; Triller, M.U.; Takeuchi, Y.; Shibata, N. *Tetrahedron* **1998**, *54*, 5929–5938.
798. Laue, K.W.; Haufe, G. *Synthesis* **1998**, 1453–1456.
799. Hsiao, G.K.; Hangauer, D.G. *Synthesis* **1998**, 1043–1046.
800. Park, K.-H.; Kurth, M.J. *J. Org. Chem.* **1999**, *64*, 9297–9300.
801. O'Donnell, M.J.; Barney, C.L.; McCarthy, J.R. *Tetrahedron Lett.* **1985**, *26*, 3067–3070.
802. Bjurling, P.; Långström, B. *J. Lab. Cmpds. Radiopharm.* **1990**, *28*, 427–432.
803. Chaari, M.; Jenhi, A.; Lavergne, J.-P.; Viallefont, P. *Tetrahedron* **1991**, *47*, 4619–4630.
804. Tilley, J.W.; Danho, W.; Lovey, K.; Wagner, R.; Swistok, J.; Makofske, R.; Michalewsky, J.; Triscari, J.; Nelson, D.; Weatherford, S. *J. Med. Chem.* **1991**, *34*, 1125–1136.
805. Müller, W.; Lowe, D.A.; Neijt, H.; Urwyler, S.; Herrling, P.L.; Blaser, D.; Seebach, D. *Helv. Chim. Acta* **1992**, *75*, 855–864.
806. Petrus, L.; BeMiller, J.N. *Carbohydrate Res.* **1992**, *230*, 197–200.
807. McCort-Tranchepain, I.; Ficheux, D.; Durieux, C.; Roques, B.P. *Int. J. Pept. Prot. Res.* **1992**, *39*, 48–57.
808. Wyzlic, I.M.; Soloway, A.H. *Tetrahedron Lett.* **1992**, *33*, 7489–7490.
809. Pirrung, M.C.; Krishnamurthy, N. *J. Org. Chem.* **1993**, *58*, 954–956.
810. Pirrung, M.C.; Krishnamurthy, N. *J. Org. Chem.* **1993**, *58*, 957–958.
811. Någren, K. *J. Lab. Cmpds. Radiopharm.* **1997**, *40*, 760–761.
812. Huang, X.; Long, E.C. *Bioorg. Med. Chem. Lett.* **1995**, *5*, 1937–1940.
813. Garbay-Jaureguiberry, C.; McCort-Tranchepain, I.; Barbe, B.; Ficheux, D.; Roques, B.P. *Tetrahedron: Asymmetry* **1992**, *3*, 637–650.
814. Liu, W.-Q.; Carreaux, F.; Meudal, H.; Roques, B.P.; Garbay-Jaureguiberry, C. *Tetrahedron*, **1996**, *52*, 4411–4422.
815. Mazaleyrat, J.-P.; Gaucher, A.; Goubard, Y.; Savrda, J.; Wakselman, M. *Tetrahedron Lett.* **1997**, *38*, 2091–2094.
816. Van Betsbrugge, J.; Tourwé, D.; Kaptein, B.; Kierkels, H.; Broxterman, R. *Tetrahedron* **1997**, *53*, 9233–9240.
817. Dryanska, V.; Pashkuleva, I.; Tasheva, D. *Synth. Commun.* **1997**, *27*, 1849–1856.
818. Mazaleyrat, J.-P.; Boutboul, A.; Lebars, Y.; Gaucher, A.; Wakselman, M. *Tetrahedron: Asymmetry.* **1998**, *9*, 2701–2713.
819. Yan, Z.; Kahn, M.; Qabar, M.; Urban, J.; Kim, H.-O.; Blaskovich, M.A. *Bioorg. Med. Chem. Lett.* **2003**, *13*, 2083–2085.
820. Pirrung, M.C. *J. Am. Chem. Soc.* **1983**, *105*, 7207–7209.
821. Pirrung, M.; McGeehan, G.M. *Angew. Chem., Int. Ed. Engl.* **1985**, *24*, 1044–1045.
822. Schöllkopf, U.; Hupfeld, B.; Gull, R. *Angew. Chem., Int. Ed. Engl.* **1986**, *25*, 754–755.
823. Pirrung, M.C.; McGeehan, G.M. *J. Org. Chem.* **1986**, *51*, 2103–2106.
824. Lasa, M.; López, P.; Cativiela, C. *Tetrahedron: Asymmetry.* **2005**, *16*, 4022–4033.
825. Ben-Ishai, D.; Berler, Z.; Altman, J. *J. Chem. Soc., Chem. Commun.* **1975**, 905–906.
826. Ben-Ishai, D.; Satati, I.; Berler, Z. *J. Chem. Soc., Chem. Commun.* **1975**, 349–350.
827. Larsen, P.O.; Wieczorkowska, E. *Acta Chem. Scand. B* **1977**, *31*, 109–113.
828. Matthies, D. *Synthesis* **1978**, 53–54.
829. Hartmann, P.; Obrecht, J.-P. *Synth. Commun.* **1988**, *18*, 553–557.
830. Cahill, R.; Crout, D.H.G.; Gregorio, M.V.M.; Mitchell, M.B.; Muller, U.S. *J. Chem. Soc. Perkin Trans. I* **1983**, 173–180.
831. Bolin, D.R.; Cottrell, J.; Garippa, R.; Michalewsky, J.; Rinaldi, N.; Simko, B.; O'Donnell, M. *Int. J. Pept. Prot. Res.* **1995**, *46*, 279–289.
832. Genet, J.P.; Ferroud, D. *Tetrahedron Lett.* **1984**, *25*, 3579–3582.
833. Ferroud, D.; Genet, J.P.; Muzart, J. *Tetrahedron Lett.* **1984**, *25*, 4379–4382.

834. Lee, F.G.H.; Dickson, D.E. *J. Med. Chem.* **1971**, *14*, 266–268.

835. Behrens, C.; Nielsen, J.N.; Fan, X.-J.; Doisy, X.; Kim, K.-H.; Praetorius-Ibba, M.; Nielsen, P.E.; Ibba, M. *Tetrahedron* **2000**, *56*, 9443–9449.

836. Niddam, V.; Camplo, M.; Le Nguyen, D.; Chermann, J.-C.; Kraus, J.-L. *Bioorg. Med. Chem. Lett.* 1996, *6*, 609–614.

837. Gong, Y.; Kato, K.; Kimoto, H. *Tetrahedron Lett.* **1999**, *40*, 5743–5744.

838. Campaigne, E.; Dinner, A. *J. Med. Chem.* **1970**, *13*, 1205–1208.

839. Sjöberg, S.; Hawthorne, M.F.; Wilmouth, S.; Lindström, P. *Chem. Eur. J.* **1995**, *1*, 430–435.

4 | Synthesis and Elaboration of Didehydroamino Acids

4.1 Introduction

Dehydroamino acids (didehydroamino acids, 2,3-didehydroamino acids, α,β-unsaturated amino acids, or α-aminoacrylates) are an important class of compounds, both as biologically interesting natural products, and as synthetic intermediates for the preparation of natural and unusual amino acids. They have become the subject of renewed interest in recent years for use in peptides, acting as constrained and enzymatically stable amino acid analogs. Dehydroamino acids are prochiral, and so form a bridge between the synthesis of racemic and optically active amino acids. Enantioselective reductions of these compounds provided some of the first attempts at asymmetric synthesis. Reviews of dehydroamino acids appeared in 1988 (*1*) and 2006 (*2*), with other discussions of the occurrence and synthesis of dehydroamino acids and their incorporation into peptides included in Volume 5 (1983) of *The Peptides* series (*3*) and Volume 6 (1982) of the *Chemistry and Biochemistry of Amino Acids, Peptides and Proteins* series (*4*). Dehydroamino acid derivatives of common amino acids (Xaa) are generally designated as ΔXaa, or sometimes Dhx.

There are a number of examples of naturally occurring compounds containing dehydroamino acids (see Scheme 4.1). Dehydroalanine (2,3-dehydro-Ala, ΔAla) is found in the phytotoxic cyclotetradepsipeptides AM-toxins I, II, and III, host-specific metabolites produced by *Alternaria mali* (the fungus causing leaf-spot disease of apple) that cause lesions on apple leaves (*5, 6*). The microcystins, toxic cyclic heptapeptides from a blue-green algae freshwater cyanobacterium, contain N-methyl-ΔAla (*7, 8*). The *Anabaena* hepatotoxic bacteria was assayed for its microcystin content and composition in 2002 (*9*). The thiazolyl peptide antibacterials

nocathiacins I–III have a ΔAla amide substituent, which can be selectively cleaved to give nocathiacin IV by treatment of nocathiacin I with iodomethane and hydroiodic acid (*10*). Nocathiacin analogs have been prepared via Michael addition of thiols and amines to the ΔAla group, although frozen water was required as solvent to avoid side reactions (*11*). Nisin, a 34 residue polypeptide containing both ΔAla and (Z)-2,3-dehydro-Abu (α-aminocrotonic acid, ΔAbu), has antibacterial and antimalarial properties and has been used as a food preservative (*12, 13*). The lantibiotics, a related group of polycyclic peptide antibiotics that include nisin and subtilin, were reviewed in 1991 (*14*) and again in 1995 (*15*). Mutacin 1140 is a new member of this family, on which NMR structural studies were carried out (*16*). Lacticin 481 is a three-ring lantibiotic from *Lactococcus lactis*, a bacterium used in cheese production. It possesses two lanthionine, one methyllanthionine, and a ΔAbu residue. The enzyme (LctM) responsible for lacticin 481 synthesis from a linear peptide precursor has been isolated; it dehydrates four Ser/Thr residues and then catalyzes the attack by Cys side chains on three of these dehydroamino acids to generate the lanthionine/methyllanthionine residues (*17*). A biomimetic synthesis of the A, B, and E rings of subtilin was reported in 2000, making use of an intramolecular formation of lanthionine residues via addition of the Cys side-chain thiol to ΔAla (*18*). Three ΔAla residues, including two in a consecutive sequence, are found in the powerful antibiotic thiostrepton. A total synthesis was described by Nicolaou et al. in 2004 (*19–21*), with other reports of advanced intermediates also in the literature (*22*).

(Z)-ΔAbu is contained in natural products such as Kahalalide F, a depsipeptide isolated from the marine mollusk *Elysia rufescens,* which is being tested in preclinical trials for treatment of lung and colon cancer (*23*),

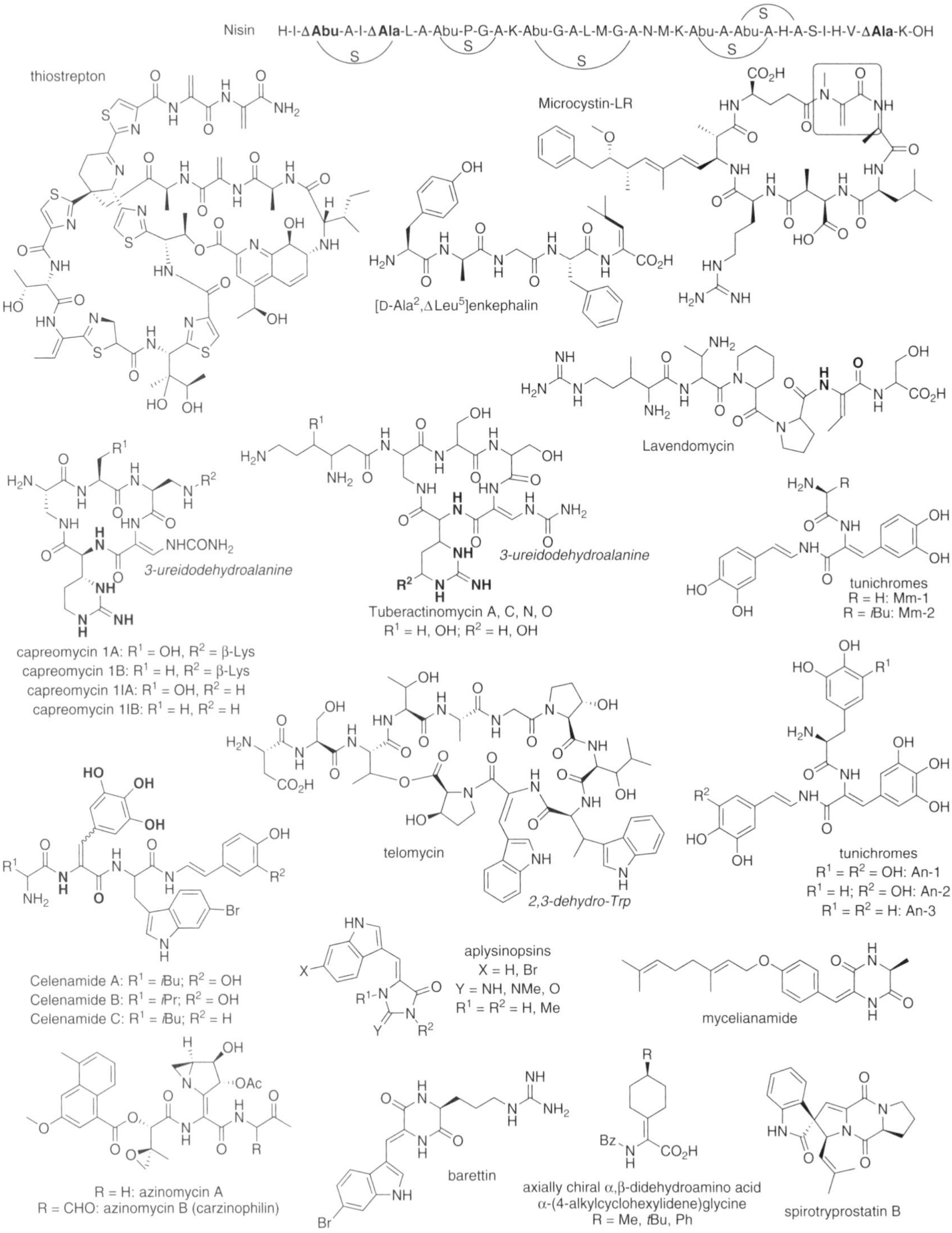

Nisin H-I-ΔAbu-A-I-ΔAla-L-A-Abu-P-G-A-K-Abu-G-A-L-M-G-A-N-M-K-Abu-A-Abu-A-H-A-S-I-H-V-ΔAla-K-OH

Scheme 4.1 α,β-Didehydroamino Acids in Natural Products and Unusual Dehdroamino Acids.

with a total synthesis and confirmation of structure reported in 2001 (24). The amino acid is also found in lavendomycin, an antibacterial peptide isolated from *Streptomyces lavendulae* (25, 26) and dolastatin 13, a cytostatic depsipeptide isolated from a marine mollusk (25 mg from 1600 kg of *Dolabella auricularia*!) (27).

The isomeric (E)-2,3-dehydro-2-aminobutyric acid is contained in the lipopeptide lobocyclamide A, isolated from the cyanobacteria *Lyngbya confervoides* (28), and in the cyclic depsipeptide phomalide, a host-selective phytotoxin produced by the fungus *Leptosphaeria mac-ulans* that causes blackleg disease of canola (*Brassica*) crops (29). Phomalide and analogs with a (Z)-2,3-dehydro-2-aminobutyric acid residue (isophomalide) or Abu replacement (dihydrophomalide) were evaluated for biological activity; only the (E)-isomer containing peptide caused lesions (29). ΔAbu was also identified in syringomycin, syringotoxin, and syringostatin (cyclic nonapeptide phytotoxins produced by *Pseudomonas syringae* pv. *syringae*, a phytopathogenic bacterium) (30), in motuporin (a cyclic pentapeptide isolated from the marine sponge *Theonella swinhoei* which inhibits protein phosphatase type I and is cytotoxic to human cancer cells) (31, 32), and in papuamides A–D (cyclic depsipeptides isolated from the sponges *Theonella mirabilis* and *Theonella swinhoei* which inhibit the infection of cells by HIV-1 and are cytotoxic against a number of human cancer cell lines) (33). Phomopsin A, a fungal metabolite that binds to microtubules, contains (E)-2,3-dehydroisoleucine (34). The same residue is found in cirratiomycin, an antibiotic heptapeptide from *Streptomyces cirratus* (35–37) and antrimycin A, a linear heptapeptide antibiotic isolated from *Streptomyces xanthocidicus* that has antitubercular activity (38, 39). The congeners antrimycins B, C, and D also contain 2,3-dehydroisoleucine, while antrimycins A$_V$, B$_V$, C$_V$, and D$_V$ contain 2,3-dehydrovaline (40).

N-Methyl-2,3-dehydro-Phe is found in the cyclic tetrapeptide tentoxin, a phytotoxic metabolite from the pathogenic fungus *Alternaria tenuis* Ness; a solid-phase synthesis of a library of analogs has been described (41). The iron-assimilating tunicate *Molgula manhattensis* contains unstable polyphenolic blood pigments (tunichromes Mm-1 and Mm-2) that are formed from a 2,3-dehydro-DOPA residue (42). 2,3-Dehydro-3′,5′-dihydroxy-Tyr is found in tunichromes An-1 to -3 (42), in tunichromes from the vanadium-assimilating tunicate *Ascidia nigra* (43), and in Celenamides A–D, linear peptides isolated from the sponge *Cliona celata* (44, 45). The antibiotic mycelianamide, from *Penicillium griseofulvum*, has a diketopiperazine containing O-geranyl 2,3-dehydrotyrosine (46). A 2,3-dehydro-2,3-diaminopropionic acid derivative, 3-ureidodehydroalanine, is contained in the tuberactinomycins, cyclic peptide antibiotics with antitubercular activity (47–51). Analogs of tuberactinomycin N have been prepared by semisynthesis in which a γ-hydroxy-β-lysine residue was removed and replaced by other α- and β-amino acids (52). 3-Ureidodehydroalanine is also found in viomycin, another antitubercular antibiotic (53), and in the closely related capreomycins isolated from *Streptomyces capreolus*; a total synthesis of capreomycin 1B was reported in 2003 (54). A complex aziridine-substituted dehydroamino acid with a 1-azabicyclo[3.1.0]hex-2-ylidene ring system is contained in the antitumor antibiotics azinomycins A and B (carzinophilin) isolated from *Streptomyces griseofuscus* (55, 56). Studies of the mechanism of action of azinomycin B, which acts via DNA cross-linking, were reported in 2002 (57) and 2004 (58), with several analogs prepared for the latter study. A total synthesis of azinomycin A was achieved in 2001 (59).

2,3-Dehydro-Trp is contained in the antibiotic telomycin (60); it is produced from Trp by a Trp side-chain oxidase and L-Trp 2′,3′-oxidase (61). It has also been identified in the cyclic antifungal hexapeptide microsclerodermin I from *Theonella* sponge (62). 2,3-Dehydro-Trp and 6′-Br-substituted derivatives are found in the aplysinopsins, isolated from marine sponges. A number of analogs have been synthesized (63). The structure of barettin, a pharmacologically active indole alkaloid isolated from the sponge *Geodia barretii* in 1986, was not correctly determined until 2002. It is a diketopiperazine formed from Arg and 6′-bromo-dehydro-Trp (64, 65). Spirotryprostatin B is also a diketopiperazine, formed from Pro and a Trp derivative cyclized to form a substituted 2,3-dehydroPro residue, isolated from *Aspergillus fumigatus*. A total synthesis was reported in 2000 (66). The virginiamycin class of antibiotics contain 2,3-dehydroproline; its formation was studied by using isotopically labeled prolines (67).

Dehydroamino acid residues are useful derivatives for restricting conformational mobility and inducing secondary structure. For example, in one study the conformations of three model dipeptides containing a C-terminal ΔAla residue were examined; the ΔAla appeared to induce a γ-turn structure (68). Homopeptides of ΔAla have been prepared and analyzed by FTIR, ^{1}H NMR, and X-ray crystallography. A flat, fully-extended conformation forming β-sheets predominates (69). ΔPhe residues induce β-bend and 3$_{10}$-helices in peptides, with three consecutive ΔPhe residues creating a left-handed helical conformation (70). A heptapeptide with four ΔPhe residues formed a 3$_{10}$ helix (71). Small tripeptides with a central ΔPhe residue adopting a 3$_{10}$ helical structure were converted into a β-turn structure via a metathesis cyclization of N-4-pentenoic/allyl amide substituents (72). An achiral helical H-(Aib-ΔPhe)$_4$-Aib-OMe peptide was induced to form a one-handed (chiral) helix by addition of various N-Boc amino acids, which interacted with the N-terminal amine (73). The crystal structure of a tripeptide containing ΔPhe has been reported (74), while the solution and X-ray structures of peptides containing a ΔPhe residue have been compared to a standard peptide and other peptide isosteres (75). Axially chiral 2,3-dehydroamino acids containing a substituted cyclohexyl ring, (4-alkylcyclohexylidene)glycines, have been incorporated into model dipeptides, with crystal structures showing a β-turn conformation (76, 77). A review of the influence of 2,3-dehydroamino acid residues on peptide conformational characteristics was published in 1996 (78); the size of the β-substituent appears to play a significant role in determining whether any secondary structure is induced. Another review examined crystal structures of dehydroamino acid-containing peptides (79).

The unusual properties of dehydroamino acids make them of interest for pharmaceutical research. For example, dehydroamino acid residues in enkephalin analogs (see Scheme 4.1) are completely resistant to carboxypeptidase cleavage at the amino–peptide linkage, and strongly resistant at the carboxy–peptide bond (80). A ΔAla residue is contained within a cyclic hexapeptide superpotent neurokinin A receptor antagonist. The ΔAla residue was accidentally obtained during attempted NaOMe-induced deprotection of an S-linked tetra-O-acetyl-β-D-Gal residue, with elimination giving the ΔAla residue (81). (E)- and (Z)-ΔPhe were used as conformationally restricted Phe replacements in peptides for structure–activity studies of the NK-1 receptor (82). A constrained analog of the chemotactic tripeptide N-formyl Met-Leu-Phe-OMe, containing (Z)-ΔPhe in addition to a cyclic Met analog, was synthesized and analyzed. X-ray and NMR analysis indicated a turn structure, but the peptide was almost inactive (83). A similar study used (Z)-ΔPhe in a chemotactic tetrapeptide that stimulates directed migration of human neutrophils; a β-turn structure was formed (84). Post-translational enzyme-catalyzed isomerization of amino acid residues in peptide chains has been observed in the spider *Agelenopsis aperta*, with Ser, Cys, O-methyl Ser, and Ala residues in the middle of peptide chains isomerized to the D-amino acid. A ΔAla-containing inhibitor of the epimerase was reported (85). Several β-(benzo[b]thiophene)-ΔPhe derivatives were found to possess significant activity against Gram-positive bacteria (86).

Dehydroamino acids can be difficult to couple into peptides, during both amidation of the carboxy group and acylation of the amino group. Coupling strategies were included in a 1997 review of coupling methods for non-coded amino acids (87). Papain can be used to enzymatically couple dehydroamino acids (88, 89). More commonly, coupling difficulties during the synthesis of peptides containing dehydroamino acids are avoided by generating the dehydroamino acid residue from a precursor residue already incorporated in the peptide (see Section 4.2.1d). For example, a heptapeptide containing four ΔPhe residues was synthesized by dehydration of a precursor peptide containing four β-hydroxy-Phe residues (71).

The tautomerism of dehydroamino acids between enamine/imine structures accounts for some of their unusual reactivity. The tautomerism of ΔVal, ΔPhe, and dehydropipecolic acid have been investigated (90). Under acidic conditions, the enamine tautomers are more stable than the imines, but under basic conditions the imines are more stable. Deamination by hydrolysis of the imine form of α,β-unsaturated α-amino acids to the α-keto acids (see Scheme 4.2) can occur in less than 5 min in neutral aqueous solution, a surprising lack of stability. Esters and sodium salts help to stabilize the enamine form (90). This deamination property can be useful: ΔAla has been employed as a C-terminal anchoring group for peptides. Treatment with HCl in glacial acetic acid deaminated the ΔAla residue to give a peptide amide, leaving pyruvic acid attached to the resin support (91). The ammonia lyase enzymes such as methylaspartase, phenylalanine ammonia-lyase, and histidine ammonia-lyase are believed to operate via the reversible conjugate addition of amino acid amino groups to the β-carbon of a ΔAla residue at the active site of the enzymes. Another possible mechanism is via amine addition to the α-carbon of the isomeric imine derivative. In order to determine if both of these mechanisms were plausible, the regioselectivity of propylamine addition to various N-alkoxycarbonyl-protected ΔAla esters and ethylamides was studied (92). Dehydroamino acids within peptides and proteins are potential substrates for Michael additions, though their electrophilicity compared to acrylamides is significantly reduced by the α-amino group. β-Fluorodehydroalanines have been synthesized to provide enhanced reactivity, and allow for irreversible additions via an addition–elimination pathway (93).

4.2 Synthesis of Dehydroamino Acids

α,β-Didehydroamino acids have been synthesized by two major processes: elimination reactions of amino acids containing an α- or β-leaving group substituent, or C–C bond-forming reactions between a glycine-type synthon and a side chain. Syntheses of dehydroamino acids often give mixtures of (E)- and (Z)-isomers. Treatment of the mixture with HCl in ether generates exclusively the (Z)-isomers in quantitative yield (94); the thermodynamically more stable (Z)-isomers can also be obtained by isomerization with $TiCl_4$ (95), by basic catalysis with triethylamine and charcoal (96), or by radical processes with thiophenol/AIBN/benzene (96).

4.2.1 Synthesis via Elimination Reactions

β-Elimination reactions are generally undesired side reactions when working with amino acids containing a β-leaving group. However, this reaction is synthetically useful when α,β-didehydroamino acids are desired (see Scheme 4.3, Table 4.1). These methods have an advantage over the C–C bond-forming Wittig-type syntheses (see Section 4.2.4) in that β,β-disubstituted-α-aminoacrylates can be prepared as easily as β-monosubstituted compounds. Unfortunately there are very subtle differences in the reaction conditions that can lead to either β-elimination, β-substitution, aziridine formation or oxazolone formation, as discussed more thoroughly in Section 10.3.1. Dehydroamino acids can also be prepared by elimination of α-leaving groups.

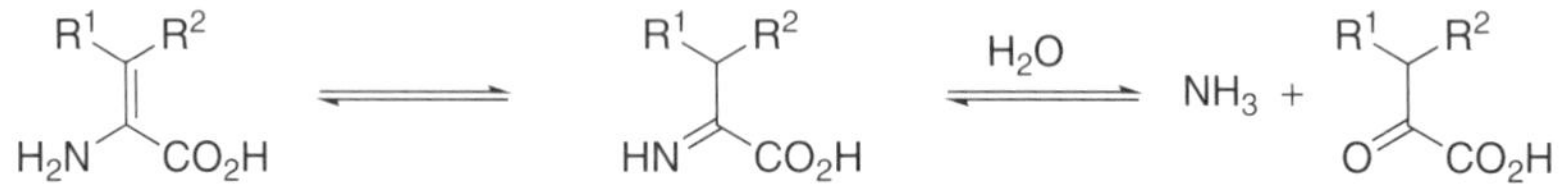

Scheme 4.2 Tautomerism/Deamination of Dehydroamino Acids.

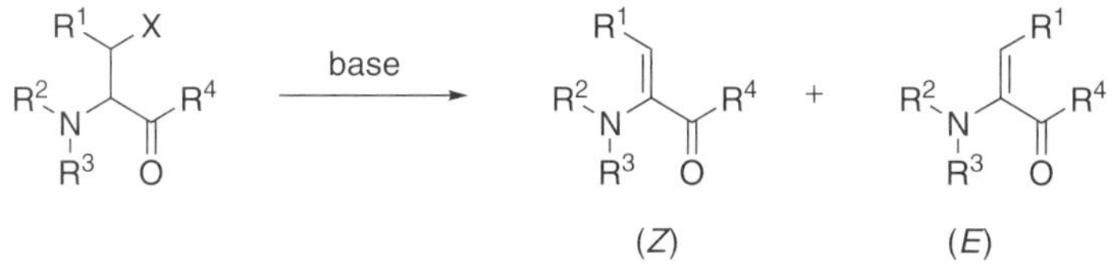

Scheme 4.3 Synthesis of Didehydroamino Acids via β-Elimination Reactions (see Table 4.1).

4.2.1a Dehydration

β-Hydroxy amino acids provide a common substrate source for dehydration reactions (see Scheme 4.3, Table 4.1), but, as mentioned above, side reactions can be a problem. Serine derivatives are often used to prepare ΔAla, with a variety of reaction conditions reported. A serine mesylate was treated with triethylamine to give the eliminated product in 84–96% yield (*93, 97, 98*) while the tosylate eliminates with diethylamine (*99*) or triethylamine (*100*) in 66–92% yield. Tosylation of Thr derivatives using tosyl chloride in pyridine at –5°C led to different outcomes depending on protecting groups; Fmoc-Thr-OBn gave a 100% yield of (Z)-ΔAbu, Fmoc-Thr-OMe gave a 30% yield, and Cbz-Thr-OMe gave a 10% yield, while Cbz-Thr-OMe, Boc-Thr-OBn, and Boc-Thr-OMe gave 100% yields of the O-tosyl Thr products (*101*). Elimination of O-tosyl Thr/*allo*-Thr derivatives with DABCO is stereospecific, with Thr giving (Z)-ΔAbu and *allo*-Thr the (E)-isomer (*102*). O-Tosyl derivatized *threo*- or *erythro*-β-hydroxyaspartic acid both eliminate to give the same fumaric acid product, and the reaction has been carried out on residues within dipeptides in 25–91% yield (*103*).

Several N-acyl serine methyl esters were O-acylated with acetyl chloride and then eliminated in 78–89% yield with DBU (*104*). N-Alkylation of Ser greatly increases the rate of elimination of acetylated (glycosylated) Ser derivatives, as formation of a protective aza-enolate from the amide bond is prevented (*105*). A chiral Ni(II) Schiff base complex of ΔAla was prepared from the Ni(II) Gly Schiff base complex by condensation with formaldehyde to form a Ser intermediate followed by dehydration using acetic anhydride/Na$_2$CO$_3$ (92% yield for elimination) (*106*). An improved procedure was reported in 2004 (*107*). The same procedure has been applied to the Schiff base complex of Thr (*108*). Thr-OMe was converted into N-acetyl (Z)-ΔAbu-OMe using Ac$_2$O/NaOAc and Et$_3$N (*109*). A variety of O-trimethylsilyl β-aryl-β-hydroxy amino acids were eliminated to the (Z)-didehydro derivatives in 83–95% yield by treatment with trifluoromethanesulfonic acid anhydride and 4-dimethylaminopyridine (*110*).

Treatment of N-Bz,N-CO$_2$Me Ser-OMe with HCl in MeOH followed by triethylamine induced dehydration in 82% yield (*111*). The N-acetyl methyl esters of a series of racemic *erythro* β-hydroxyamino acids were dehydrated to (Z)-didehydroamino acids by treatment

with dichloromethane presaturated with concentrated HCl (*112*). Both Ser and Thr derivatives can be dehydrated with PPh$_3$ and diethyl azodicarboxylate in 55–69% yield, with Thr producing a 1:1 mixture of (E)- and (Z)-isomers (*113*). In contrast, dehydration with disuccinimido carbonate(DSC)/triethylamine gave exclusively the (Z)-isomer from Thr, with yields of 70–90% from protected Ser and Thr (*114*). The dehydration of Thr with DSC was combined with concurrent peptide bond formation using the same reagent during a synthesis of the hexapeptide antibiotic lavendomycin (*115*). Stereospecific dehydration was also obtained with (diethylamino)sulfur trifluoride (DAST) and pyridine, with *threo* isomers again giving the (Z)-dehydroamino acid and *erythro* substrates giving the (E)-isomer in 65–90% yield (*116*). Good yields and minimal byproducts have been reported for dehydrations using Boc$_2$O/DMAP. Threonine and *threo*-β-phenylserine gave exclusively the Z-isomer product. β-Hydroxy amino acid residues within peptides have also been dehydrated (*117, 118*). Cbz-β-Hydroxy-Asp was converted into ΔAsp by dehydration, and then to ΔAsn via amidation of the β-carboxy group (*119*).

N,N'-Dicarbonylimidazole and triethylamine dehydrate a Schiff base of Ser-OMe in 62% yield (*120*), with the same conditions also working on Boc-Ser, Cbz-Ser, Cbz-Thr (with Thr giving the Z-alkene), and Ser contained in dipeptides (65–85% yield) (*121*). Cbz-Ser and -Thr methyl and benzyl esters were converted to the corresponding didehydroamino acids in 65–82% yield (without stereospecificity) by treatment with diisopropylcarbodiimide (DIC) and CuCl (*122*). However, under the same conditions the benzophenone Schiff base of the methyl esters of Ser, Thr, *allo*-Thr, and PhSer were stereoselectively dehydrated in 34–87% yield (*123*). The *threo* isomers gave the (E)-alkene isomer, which is difficult to prepare by other methods. The need to use the less common *erythro* hydroxy amino acids to prepare the (Z)-isomers was avoided by a simple thermal isomerization of the (E)-isomer (90–95% yield) (*123*). Boc-Ser-OMe was dehydrated with 1-(3-dimethylaminopropyl)-3-ethyl carbodiimide (EDC) and CuCl in 76% yield (*124*); the same conditions were used to dehydrate phenylserine in a resin-linked peptide (*41*) and Ser within a tripeptide (*18*). More recently, 100 equiv of EDC and 60 equiv of CuCl in DMF-CH$_2$Cl$_2$ induced sterespecific dehydration of a Thr residue within a peptide, giving exclusively the (Z)-isomer of Kahalalide F

Table 4.1 Elimination Reactions Leading to Dehydroamino Acids

Reaction scheme:

$$R^1\text{-CH}(X)\text{-C}(=O)R^4,\ \ \underset{\underset{R^3}{\overset{R^2}{N}}}{} \ \ \xrightarrow{\ base\ }\ \ (Z)\ \ +\ \ (E)$$

Side Chain R^1	Leaving Group X	R^2	R^3	R^4	Elimination Conditions	Yield	Reference
H	OH	Bz	CO$_2$Me	OH	MeOH, HCl(g), reflux then Et$_3$N	82%	1968 (111)
H	OH	Phth	—	OMe	PPh$_3$, DEAD	63–69%	1978 (113)
		Cbz	H				
		Boc	H				
H	OH	Cbz	H	OMe OBn	DIC, CuI	72–82%	1980 (122)
H	OH	Cbz	H	OMe	DSC, Et$_3$N	90%	1981 (114)
H	OH	Cbz	H	OMe	CDI, Et$_3$N	62–85%	1982 (121)
		Boc					
		Cbz-Gly					
		Cbz-Val					
H	OH	Cbz	H	OBz	DAST, pyridine	75%	1983 (116)
H	OH	=CHPh	—	OMe	CDI, Et$_3$N	62%	1984 (120)
H	OH	=CPh$_2$	—	OH Ni complex	Ac$_2$O, Na$_2$CO$_3$	92%	1988 (106)
H	OH	=CPh$_2$	—	OMe	DIC, CuI	87%	1990 (123)
H	OH	=C(SMe)$_2$	—	OMe	DIC, CuI	75%	1993 (95)
H	OH	Boc	H	OMe	EDAC, CuCl	76%	1998 (124)
H	OH	Boc-Leu	H	Ala-OH	EDC, CuCl	57%	2000 (18)
Me	OH	Boc-Phe	H	OMe	Martin's sulfurane [PhC(CF$_3$)$_2$O]$_2$SPh$_2$	52–96%	2002 (158)
Me$_2$		Boc-Gly		Ala-OMe			
Ph							
Ph	OH	Boc-Ala-Val	H	Gly-resin	EDAC, CuCl	—	2003 (41)
H	OH	Boc	H	OH	Boc$_2$O/DMAP	65–99%	1998 (117)
Me		Cbz		OMe			1999 (118)
Ph		NO$_2$-Cbz					
		Ts					
		Bz					
		Boc-Ala					
		Boc-Phe					
Me	OH	H	H	OMe	Ac$_2$O, NaOAc, Et$_3$N	—	2004 (109)
H	OCOCClY$_2$	Bn	H	OMe	DBU or Et$_3$N	78–89%	1996 (104)

R^1	X	Acyl	R^2		R^3 / O-terminal	Conditions	Yield	Year (Ref.)
H	OAc	Boc-Phe / Ac-Phe / Boc-MeLeu / Ac-Leu / Boc-Ala / Boc-Ala-Ala	H		Ala-OtBu / Val-OtBu-Leu-OMe	DBU, $LiClO_4$	69–91%	1993 (154)
H	OAc	Ac	H	$COCH_2NH$-R^4 / $COCH(Me)NH$-R^4	$NHCH_2CO$-R^3 / $NHCH(Me)CO$-R^3	Et_3N	81–89%	1995 (155)
H	OTs	Cbz	H		OMe	NaOH or $HNEt_2$	92%	1963 (99)
H	OTs	Cbz-Tyr(Bn)	H		Gly-Phe-OMe / Gly-Phe-Leu-OMe	Et_3N	35–45%	1982 (153)
H	OTs	Boc	$(CH_2)_n(2$-I-$Ph)$		OMe	Et_3N	66–78%	1997 (802) / 1997 (100)
H	OTs (from OH)	Boc	H		OMe	TsCl, pyridine then $HNEt_2$	73%	1998 (803)
Me	OTs	Fmoc	H		OBn	TsCl, pyridine, –5 °C	100%	2001 (101)
H	OMs	Phth	—		OMe	Et_3N	84%	1979 (98)
H	OMs	CO(pyrid-2-yl)	H		OMe	Et_3N	96%	1994 (97)
H	OMs	Boc-Ala	H		Ala-OMe	DBU	81%	2002 (21)
H	OMs (from OH)	Fmoc	H		ODpm	Et_3N	82%	2003 (93)
H	Cl	Cbz / CF_3CO / CF_3CO-Gly / Cbz-Xaa / Cbz-Xaa-Gly	H		OMe / Gly-OEt / Leu-OMe	Et_3N	40–91%	1977 (159)
H	Cl	Bz	H		OMe	KOtBu, THF	94%	1980 (126)
H	Cl	Cbz / Cbz-Phe / Cbz-Val / Cbz-Phe / Cbz-Ser / Cbz-Pro	H		OH / OMe / Ala-OMe / Val-OMe / Pro-OMe	DABCO, Et_3N	21–97%	1989 (160)
H	Cl (from OH)	$COCH_2Br$ / COCH(Me)Cl / COCH(Ph)Cl	Bn		OMe	PCl_5, then Et_3N	36–53%	1996 (104)
F	F	H-Ala	H		OH	dipeptide epimerase YcjG	—	2003 (93)
H	SO_2Me	Cbz	$OCH(tBu)$-R^4		$OCH(tBu)$-R^3	DBU	88–96%	1993 (141)

(Continued)

Table 4.1 Elimination Reactions Leading to Dehydroamino Acids (continued)

Side Chain R¹	Leaving Group X	R²	R³	R⁴	Elimination Conditions	Yield	Reference
H	SO₂Me (from SMe)	Boc	H	Pro-OMe Glu(OR)-OR	NaIO₄ then DBU or dioxane/Δ	—	1998 (162)
H	SO₂Me (from SMe)	Fmoc-Leu Boc-Lys(Boc)	H	Pro-OMe Leu-OMe	1) NaIO₄ 2) DBU	93–96%	2000 (18)
H	SO₂Me (from SMe)	Boc-Leu Boc-Lys(Boc) peptide	H	peptide	1) H₂O₂, Sc(OTf)₃ 2) DBU	—	2004 (144)
H	SO₂CH₂-resin	Cbz Boc (1-naphthyl)CO Cbz-Phe	H	OMe OBn NHBn Phe-OEt	DBU	31–86%	1998 (164)
H	SO₂Acm SO₂Bn	Ac Cbz Boc	H	O-Wang resin	DBU	—	1999 (163)
H	SO₂Bn (from SBn)	Fmoc	H	O-Wang resin	mCPBA then PhH, 110°C	—	2000 (142)
H	SMe₂⁺Br	Cbz	H	Gly	NaHCO₃	75%	1964 (139)
H	SCH(Ar)-R²	CH(Ar)S-X	H	OMe	Ag₂CO₃	77–88%	1979 (145)
H	S(DNP)	Cbz-Gly	H	OMe Gly	NaOMe	95–98%	1964 (139)
H	SCN or SOMe (from SH)	peptide	H	peptide	CDAP or MeI /NaIO₄ then 5% DIEA or 5% DBU	64–83%	2000 (143)
H	SePh	peptide	H	peptide	NaIO₄ in H₂O/MeOH	62–83%	2000 (165) 2001 (682)
H	Se(2-NHCO-resin-Ph)	peptide	H	peptide	tBuOOH in DCM/TFE	—	2001 (166)
H	Se(3-CO-NH-resin-Ph)	peptide	H	peptide	H₂O₂ in aq. THF o/n	—	2003 (167)
H	SePh	peptide	Me	peptide	NaIO₄ in H₂O/MeOH	76–83%	2002 (168)
H	SePh	peptide	H	peptide	H₂O₂ in H₂O/MeOH	—	2002 (85)
Me (threo) Me (erythro)	Se(3-CO₂tBu)-Ph	Fmoc	H	Oallyl	H₂O₂ in H₂O/THF	54–71%	2004 (146)
H	SePh	peptide	H	peptide	tBuOOH	68%	2004 (19, 20)
H	NO₂	Cbz-Ser Ac-Phe	H	OrBu	HNiPr₂	—	1999 (148)

R	X		N-protection	Ester	Conditions	Yield / selectivity	Year (Ref.)
H	NMe₂	=C(Ph)₂	—	OMe OEt OtBu	MeI, then K₂CO₃	92–94%	1988 (152)
H	NH₂	H		peptide	MeI, KHCO₃	18–24%	1981 (169)
H	NH₂	H		peptide	MeI, KHCO₃	22–74%	1999 (69)
H	CONH₂	H	Cbz-Ala	Phe-resin	1) PhI(Tfa)₂ 2) MeI. KHCO₃ 3) Et₃N	—	1994 (170)
Me (threo)	OTs or Cl (from OH)	H	Ac CF₃CO Bz BnCO Cbz Cbz-Gly	OMe	DABCO or DBU	46–90% Z from OTs/DABCO 3.3:1 to 1:6.7 E:Z from Cl/Dabco 19:1 to 5.7:1 E:Z from Cl/DBU	1977 (102)
Me (threo)	OH	H	Cbz	OMe	PPh₃, DEAD	55% 1:1 E:Z	1978 (113)
Me (threo)	OH	H	Cbz	OMe	DIC, CuI	65% 1:1 E:Z	1980 (122)
Me (threo)	OH	H	Cbz	OMe	DSC, Et₃N	97% Z	1981 (114)
Me (threo)	OH	H	Cbz Boc	OMe	CDI, Et₃N	71–76% Z	1982 (121)
iPr (threo)	OH	H	Cbz Boc	OMe OBn	DAST, pyridine	65–90% Z from threo E from erythro	1983 (116)
Ph (threo, erythro)	OH	=CHPh	—	OMe	DIC, CuI E converted to Z by Δ	34–84% E from threo Z from erythro	1990 (123)
Me (threo)	OH	H		within peptide	100 equiv. DIC, 60 equiv. CuI 6 days	—	2001 (24)
H Me Ph (threo)	OH	H		within peptide	100 equiv. DIC, 30 equiv. CuI 3 days	—	2001 (125)
Me (threo)	OH	H	Boc-Pro	OH	DSC, Et₃N	42%	1990 (115)
Me Ph (threo)	OH	=C(SMe)₂	—	OMe	DIC, CuI E converted to Z by TiCl₄	60–65% 90–95% isomerization to all Z	1993 (95)
Me (threo)	OH	H	Boc-Gln	OH	NaOAc, Ac₂O then aq. LiOH	78%	1994 (82)

(Continued)

Table 4.1 Elimination Reactions Leading to Dehydroamino Acids (continued)

Side Chain R^1	Leaving Group X	R^2	R^3	R^4	Elimination Conditions	Yield	Reference
Me	OH	Boc-Glu(OMe)	Me	OTse	Martin sulfarane	70%	1999 (*31*)
Me	OH	within peptide	H	within peptide	2N Ba(OH)$_2$	52%	2002 (*32*)
(*threo*)							
Et iPr iBu Ph	OH	Ac	H	Me	CH$_2$Cl$_2$ saturated with conc. HCl	75–83% all *Z*	2003 (*112*)
(*erythro*)							
Me	OAc	=CHPh	—	OH Ni complex	NaOAc	30% 1:1 *E*:*Z*	1990 (*713*)
(*threo*)							
Me	Cl	=CHPh	—	OMe	DBU	80% 1.5:1 *E*:*Z*	1978 (*129*)
(*threo*)							
Me	Cl OMs	Cbz-Dha	H	OMe	Et$_3$N	69–82% *E*	1982 (*161*)
(*threo*)							
Me with Me	SCH(Ar)-R^2	CH(Ar)S-X	H	OMe	Ag$_2$CO$_3$	77–88%	1979 (*145*)
F	S(4-MeO-Ph)	Fmoc Ac-Gly	H	ODpm	*m*CPBA then Δ	47%	2003 (*93*)
Ph 2,4-(MeO)$_2$-Ph	OTMS	CF$_3$CO	TMS	OMe	Tfa$_2$O, DMAP	83–95% *Z*	1986 (*110*)
Ph	OH	peptide	H	OH	NaOAc, Ac$_2$O, rt	80–92%	1997 (*71*)
Ph	OH	Boc-Ile	H	OH	NaOAc, Ac$_2$O	63%	1998 (*74*)
(CH$_2$)$_2$Cl (CH$_2$)$_2$F	NO$_2$	Bz	H	OMe	HN*i*Pr$_2$	—	1995 (*149*)
Me$_2$ CH$_2$CO$_2$ Me	NO$_2$	Cbz-Ser Ac-Phe	H	O*t*Bu	HN*i*Pr$_2$	—	1999 (*148*)
Et Me$_2$ –(CH$_2$)$_4$–	NO$_2$	Phe	H	Ser	HF/anisole	—	2004 (*151*)
CH$_2$SePh	Cl OMs	Cbz	H	OMe	DBU	45–95%	1990 (*130*)
CO$_2$H (*threo* or *erythro*)	OTs	Cbz Boc Cbz-Xaa	H	OMe	DABCO	73–91% *Z*	1982 (*103*)
CO$_2$H CONH$_2$	OAc	Cbz	H	OMe	—	—	1990 (*119*)

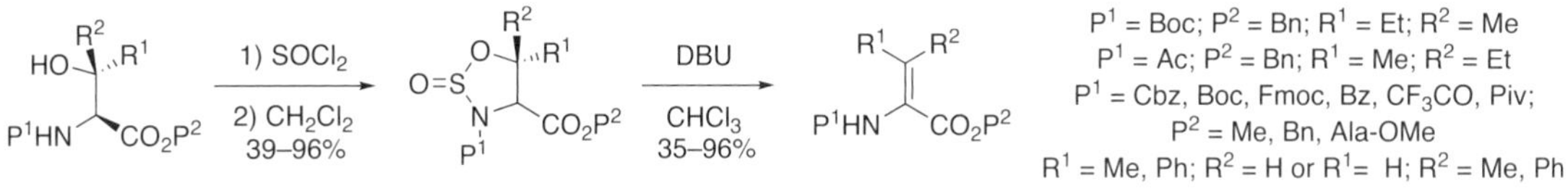

Scheme 4.4 Dehydration of β-Hydroxy-α-Amino Acids via Cyclic Sulfamidite (*34*).

(although the reaction took 6 days!) (*24*). Similar conditions induced dehydration of Ser, Thr, and phenylserine residues within peptides (*125*). Cativiela et al. have used the same dehydration conditions, but with N-[bis(methylthio)methylene] protection of the hydroxy amino acids (*95*). This protecting group is easily removed yet still gives stable derivatives. The same stereospecificity, (E)-alkenes from *threo* β-hydroxy amino acids, was observed. The thermodynamically more stable (Z)-isomers could be obtained by isomerization with TiCl₄ (*95*). Thr contained within a cyclic pentapeptide was converted into (Z)-ΔAbu by treatment of the peptide with 2 N Ba(OH)₂, as the final step in a total synthesis of motuporin (*32*).

A procedure has recently been reported for the stereospecific generation of dehydroamino acids, including the (E)-dehydroisoleucine residue found in phomopsin A. N-Protected β-hydroxy amino acids were treated with thionyl chloride to form a cyclic sulfamidite (see Scheme 4.4). Elimination was then induced by reaction with DBU. The reaction was successful for both secondary and tertiary alcohols (*34*). The elimination could also be carried out on β-hydroxy amino acid residues within peptides.

4.2.1b Dehydrohalogenation

β-Halo amino acids provide another route to dehydroamino acids. Serine has been dehydrated via conversion to a β-chloroalanine intermediate, which was then dehydrohalogenated to give ΔAla in 36–68% yield from Ser (see Scheme 4.5) (*104, 126, 127*). A number of N-acyl Thr methyl esters were also converted to ΔAbu via a β-chloro derivative, in 46–95% yield (*102*). Elimination with DABCO as base gave a mixture of (E)- and (Z)-isomers, while DBU gave predominantly the (E)-isomer. The *allo*-threonine-derived chloro intermediate led to (Z)-ΔAbu. As mentioned earlier, elimination of O-tosyl threonine derivatives with DABCO is

also stereospecific, but leads to the opposite isomers from the β-chloro procedure due to inversion at the β-center during the chlorination reaction (*102*). Attempted chlorination of β-arylserinates can lead to oxazoline formation instead of the desired β-(chloroaryl) alanine, depending on substituents and stereoisomer (*128*).

Dehydrohalogenation of β-chloro-α-aminobutyric acid in the presence of DBU proceeded in 80% yield (*129*), while similar treatment of a β-chloro-γ-seleno derivative obtained from vinylglycine gave the eliminated product in 45% yield for the two steps (*130*). 2,3-Dibromoalkanoates (R¹CHBr-CHBr-CO₂R², R¹= H, Me, Et, cHex; R² = Me, Et) react with three equivalents of sodium azide to give the 2-azido-2-alkenoates in 71–87% yield (mainly Z) (*131*). The ethyl 2-azidopropenoate undergoes conjugate addition of thiophenol and ethoxide, which is accompanied by loss of nitrogen and isomerization to give the (Z)-substituted 2-aminopropenoate in 59–73% yield (*132*). β-Aromatic and heteroaromatic-substituted α-azidoacrylic acids have been reduced to the corresponding α-aminodidehydro compounds by electrolytic cathodic reduction of the azido group in 44–86% yield (*133, 134*).

Alanine contained within an oxazolidinone or imidazolidinone template has been converted to ΔAla by bromination with NBS followed by treatment with NaI in acetone (56% yield (*135, 136*). Zimmermann and Seebach employed α-bromination of an alanine imidazole derivative with NBS followed by dehydrohalogenation with DBU to prepare a chiral ΔAla derivative in 67% yield (see Section 4.4) (*137*). Several other bromination/dehydrohalogenation procedures have been used to prepare chiral ΔAla-containing oxazolidinones (*108, 137, 138*).

Ala-β-fluoro-Ala and Ala-β,β-difluoro-Ala dipeptides have been converted into Ala-ΔAla and Ala-β-fluoro-ΔAla via fluoride elimination upon treatment with the dipeptide epimerase YcjG (*93*).

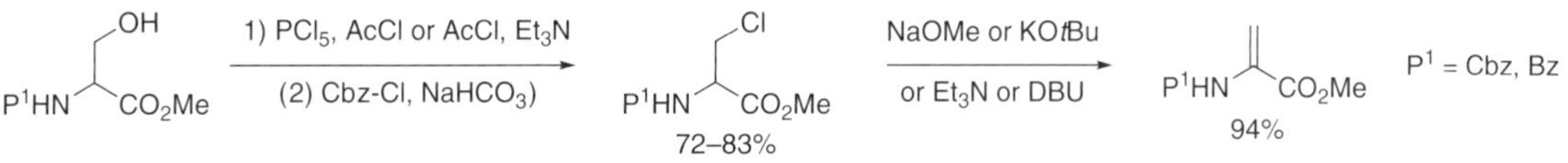

Scheme 4.5 Dehydration of Serine via β-Chloro-Alanine (*104, 126, 127*).

4.2.1c Other Eliminations

Cysteine derivatives are another potential source of β-elimination. The S-dinitrophenyl compound, prepared by reaction with dinitrofluorobenzene, eliminates under aqueous or non-aqueous basic conditions (139). Treatment of cysteine with $MeBr/HCO_2H$ and degradation of the sulfonium salt in aqueous $NaHCO_3$ also gives ΔAla (75% yield) (139, 140). Similarly, S-methylcysteine oxazolidinone has been oxidized to the sulfone, with elimination induced by treatment with DBU (141). Resin-bound Fmoc-Cys(Bn) was oxidized with mCPBA, and then eliminated by heating in toluene at 110 °C to generate Fmoc-ΔAla on a solid support (142). S-(4-Methoxyphenyl) Fmoc-β-fluoro-Cys-ODpm and dipeptides containing the same residue were oxidized to the sulfoxide with mCPBA, and then eliminated by heating to give a 1:1 Z:E mixture of β-fluoro-ΔAla product (93). Other S-alkyl Cys residues within peptides have been oxidized and then eliminated (see Section 4.2.1d) (18, 143, 144). Ac-Cys-OBn can be directly converted to ΔAla in 68% yield by reaction with diisopropylcarbodiimide and CuCl (122). 4-Thiazolidinecarboxylic esters, cyclic cysteine equivalents, have been converted to the aromatic Schiff bases of ΔVal and ΔAla by reaction with $AgNO_3$ (145).

Selenocysteine residues are often used to generate dehydroamino acids within peptides (see Section 4.2.1d). In one of the few examples of its application to isolated amino acids, Fmoc-Thr-Oallyl and Fmoc-allo-Thr-Oallyl were reacted with 3-t-butoxycarbonyl-phenylselenocyanide/PPh_3 to form the corresponding selenoether, with inversion at the β-center. Oxidation with H_2O_2 generated Z-ΔAbu from Thr, and E-ΔAbu from allo-Thr (146). A dehydroproline derivative has been prepared via elimination of an α-seleno compound, which was obtained by selenation of a proline α-anion (147).

β-Nitro-α-amino acids, easily accessible by addition of alkyl nitronates to α-bromoglycine derivatives,

eliminate under mildly basic conditions to give the didehydroamino acid (see Scheme 4.6) (148–150). The conditions are mild enough that ΔVal can be generated from a β-nitro-Val residue adjacent to a Ser residue within a peptide (148). Elimination is also induced under HF/anisole conditions used to cleave Boc-peptides from the resin (151).

ΔAla can be obtained in 53–57% yield from glycine via elimination of a β-dimethylamino intermediate (see Scheme 4.7) (152). N^α-protected 2,3-diaminopropionic acid can be converted into ΔAla by quaternization of the side-chain amine with $MeI/KHCO_3$, followed by elimination of the non-isolated N^β-trimethyl-Dap intermediate (69).

4.2.1d Dehydroamino Acid Generation within Peptides

Incorporation of dehydroamino acid residues into peptides by standard peptide synthesis techniques can be difficult, so several researchers have used β-elimination of a precursor already present in the peptides, generally employing conditions that have already been described above (see Scheme 4.2, Table 4.1). Serine has been tosylated and dehydrotosylated (153), or O-acylated and eliminated under mild conditions using DBU or DBN in the presence of large amounts of a Li salt (154). An O-acylated Ser residue in a cyclic diketopiperazine was converted to ΔAla by treatment with triethylamine (81–89% yield) (155). β-Phenylserine (Pse) was incorporated into a dipeptide and then eliminated with $NaOAc/Ac_2O$, followed by $LiOH/H_2O$, to give Z-ΔPhe (71, 82). Partial isomerization to the (E)-isomer (15% conversion) was achieved by photoirradiation with a UV lamp for 2 h (82). Application of the same dehydration conditions to Boc-Ile-Pse-OH generated an azlactone of Boc-Ile-ΔPhe, which was reacted with Trp-OMe to form a tripeptide (74). An oligo(ΔAla) peptide was

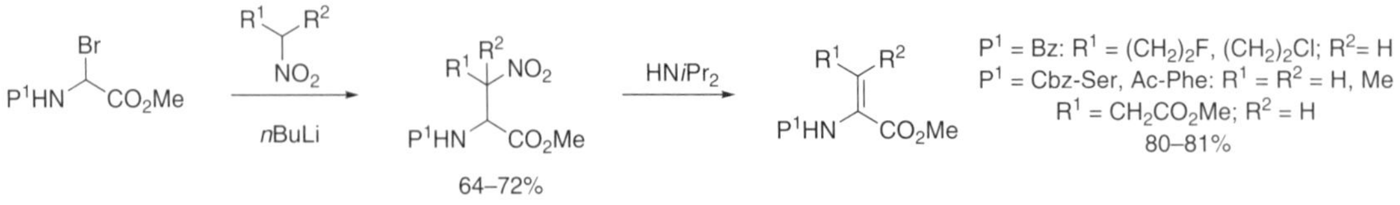

Scheme 4.6 Elimination of Nitro Group (148, 149).

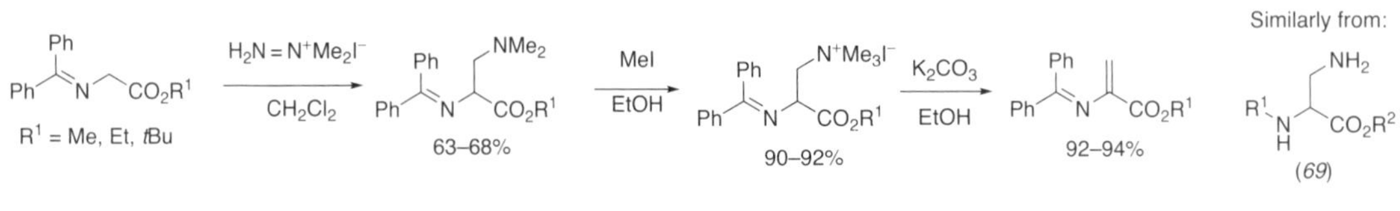

Scheme 4.7 Elimination of Trimethylamine (152).

synthesized by stepwise coupling of Ser(TBS), conversion to Ser(Ms), and elimination (97). This procedure was also employed to prepare a ΔAla residue within a tripeptide during a synthesis of part of the antibiotic thiostrepton, though DBU was employed for the elimination (21). DSC was used to convert Ser to ΔAla within a resin-linked peptide (156). Carbonyldiimidazole (CDI) and triethylamine also dehydrate Ser contained in dipeptides (121). As mentioned earlier, Ser, Thr, and phenylserine residues within peptides have been deydrated using EDC and CuCl over 3–6 days, giving the (Z)-isomers (18, 24, 125). Other β-hydroxyamino acid resides were also converted.

N-methyl-D-Thr was incorporated into a cyclic pentapeptide and induced to eliminate using aqueous Ba(OH)$_2$ (157). A solution of 15% Boc$_2$O/DMAP was used to dehydrate β-hydroxyamino acid residues in dipeptides, with 74–93% yields (117, 118). β-Hydroxyamino acids within peptides have been converted into cyclic sulfamidites by treatment with thionyl chloride, with elimination induced by DBU (34). Stereospecific dehydrations of β-hydroxy amino acids within dipeptides were achieved using Martin's sulfurane, [PhC(CF$_3$)$_2$O]$_2$SPh$_2$. Threo-amino acids (Thr, phenylserine) gave the (Z)-2,3-dehydroamino acids, while erythro isomers formed a 4,5-trans-oxazoline under the same conditions (158). Ser and β-hydroxy-Val also gave the dehydroamino acids (158). ΔAbu was introduced into a dipeptide during a synthesis of the cyclic tetrapeptide motuporin by dehydrating Thr with Martin's sulfurane as the reagent (31).

β-Chloroalanine residues within a dipeptide have been converted to ΔAla in 21–97% yield by treatment with Et$_3$N or DABCO (159, 160). The β-chloroalanine can be prepared in situ by treatment of serine-containing peptides with PCl$_5$ (37–81% yield) (159). Erythro 3-chloro-2-aminobutanoate, contained within a dipeptide, was stereoselectively dehydrohalogenated to the (E)-alkene in 69–82% yield (161).

Cys residues within peptides can be converted into ΔAla residues via thiol methylation/oxidation (MeI followed by NaIO$_4$) or cyanation (CDAP), followed by elimination induced with 5% DBU or 5% DIEA, respectively (143). This strategy was used to generate a ΔAla residue within a resin-bound peptide that was subsequently used for a Michael addition of a Cys thiol (144). S-Methylcysteine has been incorporated within peptides, then oxidized with NaIO$_4$ and eliminated with DBU or dioxane/reflux to generate ΔAla residues (18, 162). In a similar manner, resin-bound S-acetimidomethyl or S-benzyl Cys were oxidized with mCPBA and eliminated with DBU (163). Cysteine has also been linked to a Merrifield resin via a side-chain thioether linkage. The amino group was acylated and the carboxyl group amidated to form a peptide. Finally, the thioether was oxidized with mCPBA and then eliminated with DBU to release the ΔAla-containing peptide from the resin (164).

Peptides have prepared using Se-phenyl Fmoc-Sec (selenocysteine), with conversion to ΔAla achieved by treatment with H$_2$O$_2$ in H$_2$O/MeOH (85). The elimination was also induced using aqueous NaIO$_4$, conditions that are compatible with oxidation-sensitive Met and Trp residues (165). A total synthesis of the antibiotic thiostrepton employed three Se-Ph Sec residues to generate the desired dehydroalanines in the penultimate step using t-butylhydroperoxide oxidation/elimination (19, 20, 22). A selenocysteine derivative, Se-(2-aminophenyl) Fmoc-Sec-OAll, was linked to a polymer via the side-chain amino substituent, and then incorporated into a cyclic tetrapeptide. Oxidative cleavage from the resin with TBHP produced ΔAla-containing AM-toxin II (166). A similar derivative, Se-(3-carboxyphenyl) Fmoc-Sec-OAll, was coupled via the side-chain to an amino-methyl-substituted resin, elaborated in both directions to form an Ac-Arg-Gy-Asp-ΔAla-Phe-Ala-OMe peptide, and then cleaved with H$_2$O$_2$/aq. THF to generate the ΔAla residue (167). Se-Phenyl Fmoc β-methyl-selenocysteine has been incorporated into peptides and eliminated, generating ΔAbu (168).

Oligomers of ΔAla (n = 1–6) were prepared by peptide synthesis of the corresponding N^β-Cbz 2,3-diaminopropionic acid homopeptides, followed by removal of the side-chain Cbz groups and quaternization/elimination by treatment with MeI/K$_2$CO$_3$ (69). This method was also applied to a Dap residue within a cyclic peptide during a synthesis of AM toxin III, a host-specific phytotoxin (169). ΔAla can also be obtained from Asn in a resin-linked peptide by sequential application of a Hofmann-type degradation using bis(trifluoroacetoxy) iodobenzene to give a β-amino derivative, followed by a Hofmann elimination (MeI, KHCO$_3$, then 10% triethylamine) (170). β-Nitro-α-amino acids are very useful for dehydroamino acid generation within peptides as only a very mild base (diisopropylamine) is required to induce elimination; ΔVal and ΔGlu were formed adjacent to a Ser residue (148).

Intrapeptide 2-aziridine carboxylic acid and 3-methyl-2-aziridine carboxylic acid residues can be decomposed to the corresponding ΔAla and ΔAbu residues under basic conditions (Et$_3$N or DABCO, 77–100% yield) (171) or with NaI in acetone (71–92%) (172). ΔPhe residues in peptides were generated from 3-phenylaziridine-2-carboxylate in a tripeptide by treatment with TMSI/Et$_3$N (72).

4.2.2 Synthesis via Erlenmeyer Condensation

β-Hydroxy amino acids are non-isolated intermediates in the synthesis of N-acyl (Z)-dehydroamino acids from the condensation of N-acylglycine with aldehydes, which is known as the Erlenmeyer procedure and proceeds via an azlactone intermediate (see Scheme 4.8, Table 4.2). A comprehensive review of azlactones, including their preparation by the Erlenmeyer procedure and subsequent conversion into amino acids, was published in 1946 (173), while a more recent discussion of the synthesis of the intermediate alkylidene/arylidene-oxazolones emphasizes the preparation and interconversion of the

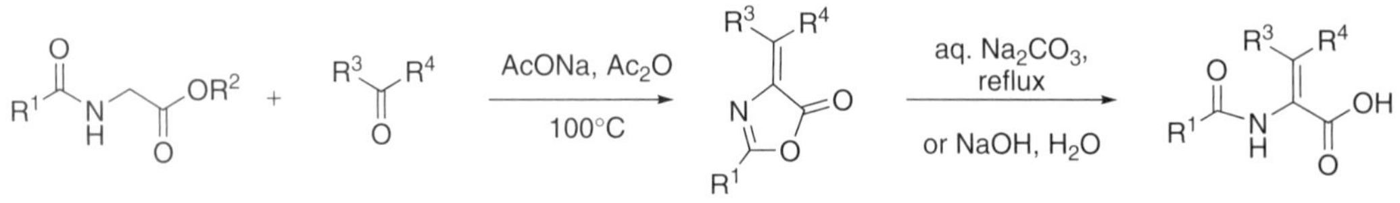

Scheme 4.8 Erlenmeyer Reaction.

Table 4.2 Erlenmeyer Condensations

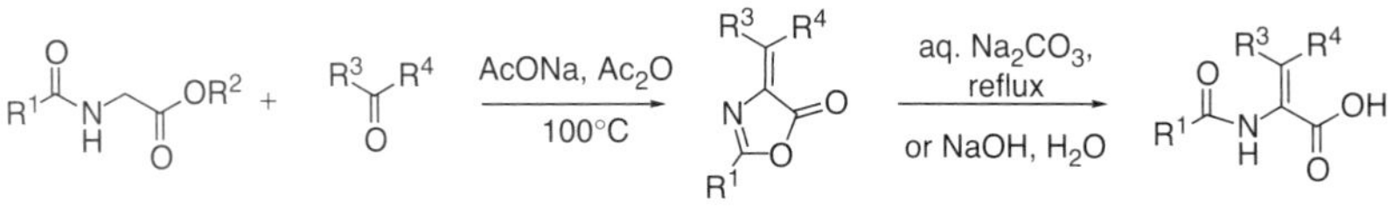

Glycine Equivalent	Aldehyde or Ketone	Yield	Reference
Bz-Gly-OH	PhCHO	—	1893 (*176*)
Bz-Gly-OH	(indol-3-yl)CHO	—	1907 (*804*)
Bz-Gly-OH	(4-Me-Ph)CHO	80%	1911 (*805*)
Bz-Gly-OH	(1-naphthyl)CHO	—	1911 (*806*)
	(2-naphthyl)CHO		
Bz-Gly-OH	(3,4-OCOO-Ph)CHO	74%	1911 (*807*)
	[3,4-HO)$_2$-Ph]CHO		
Bz-Gly-OH	(2-Me-indol-3-yl)CHO	—	1914 (*808*)
Bz-Gly-OH	(imidazol-2-yl)CHO	68–78%	1916 (*178*)
Bz-Gly-OH	(2-Me-indol-3-yl)CHO	—	1917 (*809*)
Bz-Gly-OH	[3,5-I$_2$-4-(4-MeO-PhO)-Ph]CHO	70–90%	1927 (*810*)
	(3-MeO-4-HO-Ph)CHO		
	[4-(4-MeO-PhO)-Ph]CHO		
Bz-Gly-OH	(2-F-Ph)CHO	34–75%	1932 (*811*)
	(3-F-Ph)CHO		
	(4-F-Ph)CHO		
Bz-Gly-OH	[2,4-(MeO)$_2$-Ph]CHO	44%	1936 (*812*)
Bz-Gly-OH	(2-thienyl)CHO	—	1938 (*813*)
Bz-Gly-OH	PhCHO	62–64%	1948 (*177*)
Bz-Gly-OH	[2,5-(MeO)$_2$-Ph]CHO	75%	1948 (*814*)
Bz-Gly-OH	(4-NMe$_2$-Ph)CHO	—	1948 (*815*)
Bz-Gly-OH	EtCHO	7–45%	1949 (*179*)
	*n*PrCHO		
	*i*PrCHO		
Bz-Gly-OH	(benzothiophen-3-yl)CHO	—	1949 (*816*)
Bz-Gly-OH	[2,3-(HO)$_2$-Ph]CHO	—	1950 (*817*)
Bz-Gly-OH	(3-HO-Ph)CHO	—	1951 (*818*)
Bz-Gly-OH	(4-NO$_2$-Ph)CHO	92%	1952 (*819*)
[1-^{13}C]-Bz-Gly-OH	(4-HO-Ph)CHO	—	1952 (*820*)
Bz-Gly-OH	[2,5-(HO)$_2$-Ph]CHO	47–51%	1953 (*821*)
Bz-Gly-OH	(3-MeO-4-HO-Ph)CHO	51%	1956 (*396*)
	(vanillin)		
Bz-Gly-OH	[6-Me-2,5-(HO)$_2$-Ph]CHO	45%	1958 (*822*)
Bz-Gly-OH	ferrocenyl-CHO	—	1957 (*823*)
Bz-Gly-OH	EtCOEt	—	1966 (*824*)
Bz-Gly-OH	(F$_5$-Ph)CHO	54%	1969 (*825*)
Bz-Gly-OH	[3,5-(*i*Pr)2-4-(3-*i*Pr-4-MeO-PhO)-Ph]CHO	25%	1969 (*826*)
Bz-Gly-OH	(8-quinolino-5-yl)CHO	59–80%	1969 (*827*)
hydantoin			
Bz-Gly-OH	[3,4-(OH)$_2$-6-(CH$_2$CH=CH$_2$)-Ph]CHO	53–72%	1971 (*828*)
Bz-Gly-OH	PhCDO	73%	1972 (*829*)
	PhCTO		

Table 4.2 Erlenmeyer Condensations (continued)

Glycine Equivalent	Aldehyde or Ketone	Yield	Reference
Bz-Gly-OH	PhCDO PhCTO (4-HO-Ph)CDO (4-HO-Ph)CTO	—	1973 (830)
Bz-Gly-OH	[3,4-(MeO)₂-Ph]CHO	96%	1975 (831)
Bz-Gly-OH	(CO)₃Mn-PhCHO	70%	1981 (832)
Bz-Gly-OH	(3-pyridyl)CHO	64%	1984 (833)
Bz-Gly-OEt	HCO₂Et	—	1942 (214) 1984 (215)
Bz-Gly-OH	(1-pyrenyl)CHO	60%	1985 (834)
Bz-Gly-OH	(CF₃)CH(Me)CHO CF₃(CH₂)₂CHO (C₆F₅) (CH₂)₂CHO (4,5,6,7-F₄-indol-3-yl)CHO	70–71%	1989 (183)
Bz-Gly-OH	(1-pyrenyl)CHO	—	1987 (835)
Bz-Gly-OH	(F₅-Ph)CHO	15%	1991 (836)
Bz-Gly-OH	(2-F-Ph)CHO (3-F-Ph)CHO (4-F-Ph)CHO (4-CF₃-Ph)CHO (F₅-Ph)CHO	42–80%	1992 (837)
Bz-Gly-OH	cyclohexanone PhCHO	19–84%	1992 (838)
Bz-Gly-OH	(2,2'-bipyridin-4-yl)CHO	75%	1993 (190)
Bz-Gly-OH	tBuCHO	81%	1994 (189)
Bz-Gly-OH	N-ethylcarbazolyl-3-CHO	—	1994 (839)
Bz-Gly-OH		—	1995 (193)
Bz-Gly-OH	[3-(4-CHO-PhO)-4-MeO-Ph]CHO	69%	1999 (195)
Bz-Gly-OH	MeCO(4-MeO-Ph)	—	1998 (191)
Bz-Gly-OH	(4-MeO-Ph)COMe	83%	1998 (191)
Bz-Gly-OH	2-F-PhCHO	71%	1999 (840) 2002 (841)
Bz-Gly-OH	4-Me-cyclohexanone 4-Ph-cyclohexanone 4-tBu-cyclohexanone	51–61%	1999 (76) 2004 (77)
Bz-Gly-OH	[3-(4-CHO-PhO)-4-MeO-Ph]CHO	—	1999 (195)
Bz-Gly-OH	cPrCHO	33% azlactone 85% Bz-dehydroAA- OBn	2003 (568)
Bz-Gly-OH	MeCOPh	38% pure (Z) isomerize to pure (E) with HBr/benzoyl peroxide 94%	2004 (192)
Ac-Gly-OH	[2,5-(HO)₂-Ph]CHO	60%	1948 (814)
Ac-Gly-OH	(3-HO-Ph)CHO	67–75%	1951 (818)
Ac-Gly-OH	[2,5-(MeO)₂-Ph]CHO [2,5-(MeO)₂-4-Me-Ph]CHO	55%	1974 (842)
Ac-Gly-OH Bz-Gly-OH	[3-(CH₂OH)-4-HO-Ph]CHO	74–81%	1974 (843)
Ac-Gly-OH	(2-naphthyl)CHO	33%	1984 (844)
Ac-Gly-OH Bz-Gly-OH	(ferrecene)CHO	11%	1993 (185)
Ac-Gly-OH Bz-Gly-OH	(thien-2-yl)CHO (thien-3-yl)CHO	—	1993 (188)

(Continued)

Table 4.2 Erlenmeyer Condensations (continued)

Glycine Equivalent	Aldehyde or Ketone	Yield	Reference
Ac-Gly-OH	(4-PhO-Ph)CHO	40%	1993 (845)
Ac-Gly-OH	iPrCHO	21–80%	1995 (184)
Bz-Gly-OH	PhCHO		
	(4-HO-Ph)CHO		
	(indol-3-yl)CHO		
	imidazol-4-yl)CHO		
Ac-Gly-OH	(pyrid-3-yl)CHO	—	1996 (187)
Bz-Gly-OH	(pyrid-4-yl)CHO		
Ac-Gly-OH	(3-NO₂-Ph)CHO	27%	1998 (186)
Ac-Gly-OH	(2-pyrenyl)CHO	30%	1999 (708)
Ac-Gly-OH	(1-naphthyl)CHO	40–50% Z	2005 (846)
Bz-Gy-OH			
azlactone of Ac-Gly	(2-naphthyl)¹⁴CHO	80%	1986 (472)
azlactone of Bz-Gly	(indol-3-yl)CHO	—	1944 (847)
azlactone of Bz-Gly	MeCHO	34–75%	1959 (180)
	EtCHO		
	nPrCHO		
	PhCHO		
	(4-MeO-Ph)CHO		
	(2-furyl)CHO		
	PhCH=CHCHO		
azlactone of Bz-Gly	glycosyl aldehydes	—	1976 (182)
			1978 (181)
azlactone of Bz-Gly	(anthracen-9-yl)CHO	59%	1982 (848)
hydantoin	PhCH=CHCHO	—	1915 (393)
hydantoin	(indol-3-yl)CHO	47%	1922 (849)
hydantoin	(5-Me-indol-3-yl)CHO	—	1924 (850)
hydantoin	(indolyl-3-yl)CHO	65%	1935 (851)
hydantoin	(thien-2-yl)CHO	—	1938 (813)
			1945 (852)
hydantoin	(6-MeO-indol-3-yl)CHO	65%	1938 (853)
hydantoin	(indol-3-yl)CHO	—	1944 (847)
hydantoin	(indol-3-yl)CHO	—	1944 (854)
[1-¹³C]-hydantoin	(indol-3-yl)CHO	95%	1948 (855)
hydantoin	(imidazol-4-yl)CHO	—	1949 (394)
[5-¹⁴C]-hydantoin	(2-furyl)CHO	56%	1968 (856)
hydantoin	[2,4,5-(MeO)₃-Ph]CHO	86%	1969 (857)
hydantoin	[3,4,6-(HO)₃-Ph]CHO	50%	1971 (858)
hydantoin	[3,4,5-(MeO)₃-Ph]CHO	66%	1973 (859)
2-thiohydantoin	PhCHO	94%	1912 (211)
2-thiohydantoin	(3-MeO-4-HO-Ph)CHO	—	1913 (209)
	[3,4-(MeO)₂-Ph]CHO		
2-thiohydantoin	(2-HO-Ph)CHO	—	1915 (207)
2-thiohydantoin	(2-HO-5-NH₂-Ph)CHO	—	1915 (208)
2-thiohydantoin	PhCH=CHCHO	—	1915 (393)
2-thiohydantoin	[3,4-(OCH₂O)-Ph]CHO	—	1932 (210)
2-thiohydantoin	(1-Me-imidazol-4-yl)CHO	95%	1944 (860)
2-thiohydantoin	(4-HO-1-naphthyl)CHO	40%	1966 (861)
2-mercaptothiazol-5-one	MeCHO	—	1949 (213)
	cHexCHO		
	MeCOMe		
	iPrCHO		
	nPrCHO		
	PhCHO		
	(4-HO-Ph)CHO		
Gly-Gly diketopiperazine	(2-furyl)CHO	—	1921 (221)
Gly-Gly diketopiperazine	(3-pyridyl)CHO	—	1942 (222)

Table 4.2 Erlenmeyer Condensations (continued)

Glycine Equivalent	Aldehyde or Ketone	Yield	Reference
Gly-Gly diketopiperazine	(3-Me-4-HO-Ph)CHO	—	1950 (817)
Val-Gly diketopiperazine	PhCHO	52–83%	2005 (223)
	iPrCHO		
creatinine	(4-NMe$_2$-Ph)CHO	—	1961 (224)
BnOC(=S)-Gly-OH	PhCHO	14–80%	1979 (226)
BnSC(=S)-Gly-OH	(4-HO-Ph)CHO		
	(4-MeO-Ph)CHO		
	(3-HO-Ph)CHO		
	(3-MeO-Ph)CHO		
	(3-Cl-Ph)-CHO		
	(4-Cl-Ph)CHO		
	(4-F-Ph)CHO		
	(4-Me-Ph)CHO		
	[3,4-(OCH$_2$O)-Ph]CHO		
	(3-HO-4-MeO-Ph)CHO		
	(3-MeO-4-HO-Ph)CHO		
N-Ph,N-Tfa-Gly-OMe	(4-Me-Ph)CHO	74–89% addition/	1988 (225)
TMS enolate	(4-MeO-Ph)CHO	dehydration	
	[3,4-(MeO)$_2$-Ph]CHO		
	[3,4,5-(MeO)$_3$-Ph]CHO		
	(2-furyl)CHO		
	(2-thienyl)CHO		
AcNHCH(CO$_2$Me)$_2$	HCHO	51%	1963 (216)
N$_3$CH$_2$CO$_2$Et	4-(CH$_2$PO$_3$tBu$_2$)-Bn	74%	1991 (228)
N$_3$CH$_2$CO$_2$Et	[4-(CH$_2$PO$_3$tBu$_2$)-Ph]CHO	57–68%	1993 (229)
	[4-(CF$_2$PO$_3$tBu$_2$)-Ph]CHO		1993 (227)
	[4-(CHF-PO$_3$tBu$_2$)-Ph]CHO		
	[4-(CHOH-PO$_3$tBu$_2$)-Ph]CHO		
	[4-(CO-PO$_3$tBu$_2$)-Ph]CHO		
N$_3$CH$_2$CO$_2$Et	[4-(CF$_2$PO$_3$Et$_2$)-Ph]CHO	17–76%	2000 (230)
	[4-[CH(OH)CF$_2$PO$_3$Et$_2$]-Ph]CHO		
NO$_2$CH$_2$CO$_2$Me	(4-MeO-Ph)CHO	46–88%	1998 (231)
	MeCOiPr		
	(2-furyl)CHO		
	(1-Me-indol-3-yl)CHO		
[1,2-^{13}C$_2$]-NO$_2$CH$_2$CO$_2$Me	iPrCHO	50%	1999 (232)
NO$_2$CH$_2$CO$_2$-Wang resin	PhCHO	—	2000 (233)
	(3-pyridyl)CHO		
	(2-imidazole)CHO		
	(2-furyl)CHO		
	iPrCHO		

geometric isomers (174). Azlactones have been identified as useful intermediates for dynamic kinetic resolutions, leading to enantiomerically enriched amino acids (e.g. 175) (see Section 12.1.7). In the original 1893 Erlenmeyer publication (176) hippuric acid (N-benzoyl-glycine) was condensed with benzaldehyde. An *Organic Syntheses* version of this preparation exists (177). Reports of reactions with other aldehydes such as 2-formyl-imidazole soon followed (68–78% yield of azlactone) (178), but aliphatic aldehydes generally gave poor yields (7–45% yield with RCHO, R = Et nPr, iPr) (179).

Yields for the Erlenmeyer reaction with aromatic (R = Ph, 4-MeO-Ph, furyl, PhCH=CH; 55–82%) and aliphatic aldehydes (R = Me, Et, nPr; 34–58%) and ketones (acetone, cyclohexanone; 35–74%) were improved by reacting the azlactone of hippuric acid instead of hippuric acid itself (180), aided by the use of lead acetate. The Bz-Gly azlactones have also been added to carbohydrate-derived aldehydes enroute to C-linked glycosyl amino acids (181, 182). A series of fluoroaliphatic amino acids (4,4,4-trifluoro-Leu, 5,5,5-trifluoro-Nle) were prepared via initial condensation of aliphatic fluoroaldehydes and hippuric acid in the presence of lead acetate and acetic anhydride, giving the azlactones in 70–71% yield. The lead acetate/acetic anhydride reagents were critical for obtaining good yields; under normal Erlenmeyer conditions no products were formed at all (*183*). The Erlenmeyer reaction was recently (1995) systematically examined with various *N*-acylglycines (*184*), and again found to give good yields for aromatic aldehydes (RCHO, R = Ph, 4-HO-Ph, 3-indole, 4-imidazole; 38–80%), but poor with aliphatic (R = *i*Pr, 20%). Other more recent examples of this venerable procedure include condensations of *N*-acyl glycines with ferrocenylcarbaldehyde (*185*), 3-nitrobenzaldehyde (*186*), pyridyl aldehydes (*187*), thienyl aldehydes (*188*), pivaldehyde (*189*), and 4-formyl-2,2′-bipyridine (*190*). Hippuric acid was condensed with 4′-methoxyacetophenone during a synthesis of β-methyl-tyrosine (*191*). It was also employed during a 2004 synthesis of all diastereomers of β-methyl-Phe. The initial (*Z*)-alkene adduct formed from hippuric acid and acetophenone was isomerized to the (*E*)-product by treatment with HBr and catalytic benzoyl peroxide (*192*). Condensation with D-glyceraldehyde acetonide produced a chiral didehydroamino acid azlactone intermediate in 75% yield, in a 80:20 *Z:E* ratio (*193, 194*). The Erlenmeyer condensation was recently used to prepare dehydroamino acids with an axis of chirality, via condensation with 4-alkylcyclohexanones (*76, 77*), and was employed during a synthesis of isodityrosine, via condensation with 3-(4-formyl-phenoxy)-4-methoxybenzaldehyde (*195*). Condensation of hippuric acid with dimethylformamide dimethylacetal produced Bz-β-dimethylamino-ΔAla-OMe in 75% yield, with ammonium acetate treatment leading to Bz-β-amino-ΔAla-OMe (*196*).

A variety of other glycine derivatives have been employed. The Wheeler and Hoffman method of amino acid synthesis (*197*) employed hydantoin (see Scheme 4.9) for condensation with aromatic (or aliphatic (*198*)) aldehydes using hot glacial acetic acid and anhydrous NaOAc (*197,199–203*), diethylamine (*204*) or piperidine (*205*), giving a didehydroamino acid hydantoin derivative enroute to aromatic or aliphatic amino acids. The didehydroamino acid hydantoin could be *N*-methylated (*206*). Thiohydantoins (see Scheme 4.9) have also been used for the condensation with aromatic aldehydes, giving better yields than hydantoins for substituted arylaldehydes (*207–212*). The closely related 2-mercaptothiazol-5-one (see Scheme 4.9) gave good yields for condensations with alkyl aldehydes; adducts leading to Abu, Chg, Val, Leu, Nle, Phe, and Tyr were prepared (*213*).

Ethyl hippurate was condensed with ethyl formate to produce dehydroserine, β-hydroxy-ΔAla (*214, 215*), while dimethyl acetamidomalonate reacted with formaldehyde to give *N*-Ac ΔAla in 51% yield on a 10 g scale (*216*). Glycine anhydride (*217–220*) and Gly-Gly diketopiperazine (*221, 222*) have been used as the nucleophile for Erlenmeyer-type condensations, with the latter resulting in a symmetrical didehydroamino acid diketopiperazine. *N,N*′-Bis(*p*-methoxybenzyl) Val-Gly diketopiperazines have also been employed for aldol condensations/eliminations (*223*). Creatinine was used for a condensation with *p*-dimethylaminobenzaldehyde; reaction with glycine anhydride was unsuccessful (*224*). More recently, the TMS-trapped enolate of *N*-phenyl-*N*-trifluoroacetyl glycine methyl ester was reacted with aromatic aldehydes (RCHO, R = 4-Me- or -MeO-Ph, 3,4-(MeO)$_2$-Ph, 3,4,5-(MeO)$_3$-Ph, 2-furyl, 2-thienyl) in the presence of zinc bromide to give TMS-protected β-hydroxy amino acids (74–89%). Treatment with pyridine/TFA$_2$O/DMAP gave the corresponding didehydroamino acids as a mixture of (*E*)- and (*Z*)-isomers (60–75% yield) (*225*). In 1979, 13 different dehydroamino acid thiazolones were obtained by condensing *N*-benzyloxythiocarbonylglycine or *N*-benzylthiothiocarbonylglycine with aryl aldehydes (*226*). Ethyl α-azidoacetate has also been reacted with aryl aldehydes, producing the adducts in 57–68% yield (*227–230*). The azide was reduced to an amine by hydrogenation, which also reduced the alkene bond. Methyl nitroacetate (*231*) and [1,2-^{13}C$_2$]nitroacetate (*232*) are other recent glycine equivalents used for Erlenmeyer-like condensations. Nitroacetate has been esterified to a Wang resin and then condensed with a variety of aldehydes to give resin-bound nitroalkenes (*233*).

α-Metallated methyl, ethyl, or *tert*-butyl isocyanoacetate have also be used as the nucleophile in an Erlenmeyer-type condensation, producing *N*-formyl didehydroamino acid esters after basic treatment of the 4-carboxyoxazoline intermediate (see Scheme 4.10) (*234–236*). Hydrolysis of the same intermediate led to β-hydroxy amino acids (*236*), while reduction gave alkyl amino acids (see Scheme 3.11) (*236*). Twelve aldehydes and ketones were condensed in the original report in 55–90% yield (*237*), while six more adducts were reported in 1996 (*238*). Reaction conditions were milder than the traditional Erlenmeyer synthesis, allowing for the use of more sensitive carbonyl electrophiles. The *N*-formyl group was converted back to an isocyano group by dehydration with COCl$_2$/Et$_3$N (*239*) or PPh$_3$/Et$_3$N (*240*), or changed to an *N*-acyl group by reaction of the dehydroamino acid derivative with an acyl chloride

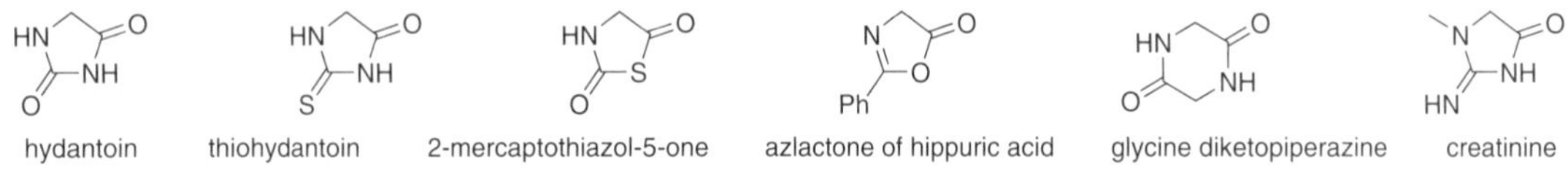

hydantoin   thiohydantoin   2-mercaptothiazol-5-one   azlactone of hippuric acid   glycine diketopiperazine   creatinine

Scheme 4.9 Other Glycine Equivalents for Condensations with Aldehydes.

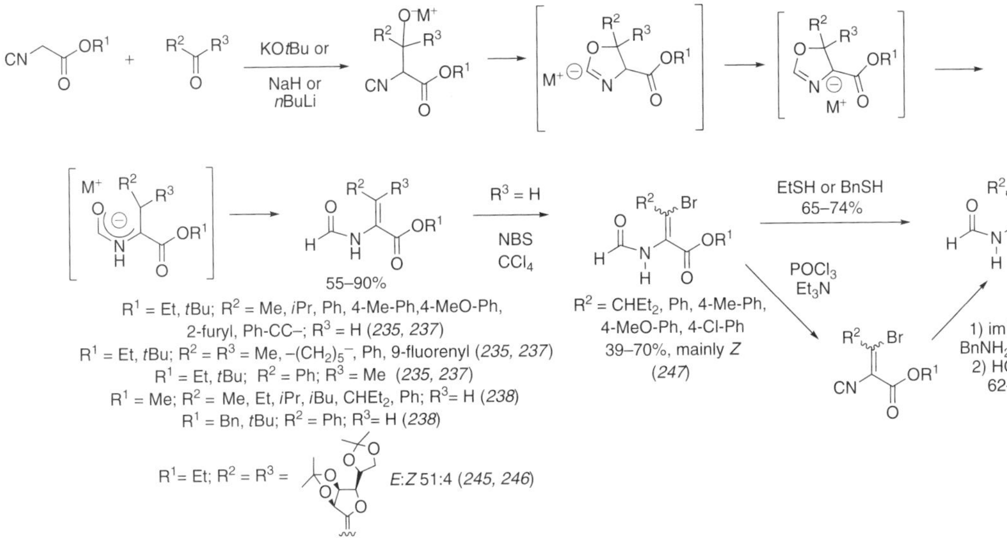

R^1 = Et, tBu; R^2 = Me, iPr, Ph, 4-Me-Ph,4-MeO-Ph,
2-furyl, Ph-CC–; R^3 = H (235, 237)
R^1 = Et, tBu; R^2 = R^3 = Me, –(CH$_2$)$_5$–, Ph, 9-fluorenyl (235, 237)
R^1 = Et, tBu; R^2 = Ph; R^3 = Me (235, 237)
R^1 = Me; R^2 = Me, Et, iPr, iBu, CHEt$_2$, Ph; R^3= H (238)
R^1 = Bn, tBu; R^2 = Ph; R^3= H (238)

E:Z 51:4 (245, 246)

55–90%

R^2 = CHEt$_2$, Ph, 4-Me-Ph,
4-MeO-Ph, 4-Cl-Ph
39–70%, mainly Z
(247)

EtSH or BnSH
65–74%

1) imidazole,
BnNH$_2$, or BnO
2) HOCO$_2$H
62–64%

Scheme 4.10 Modified Erlenmeyer Reaction.

R^1 = Ph; R^2 = H
R^1 = R^2 = Me, –(CH$_2$)$_4$–, –(CH$_2$)$_5$–
R^3 = Me, allyl, Bn
56–98%

R^1 = H; R^2 = Me, iPr, Ph
R^1 = Me; R^2 = Me, cPr
R^3 = OEt, OBn, Ph
45–98%

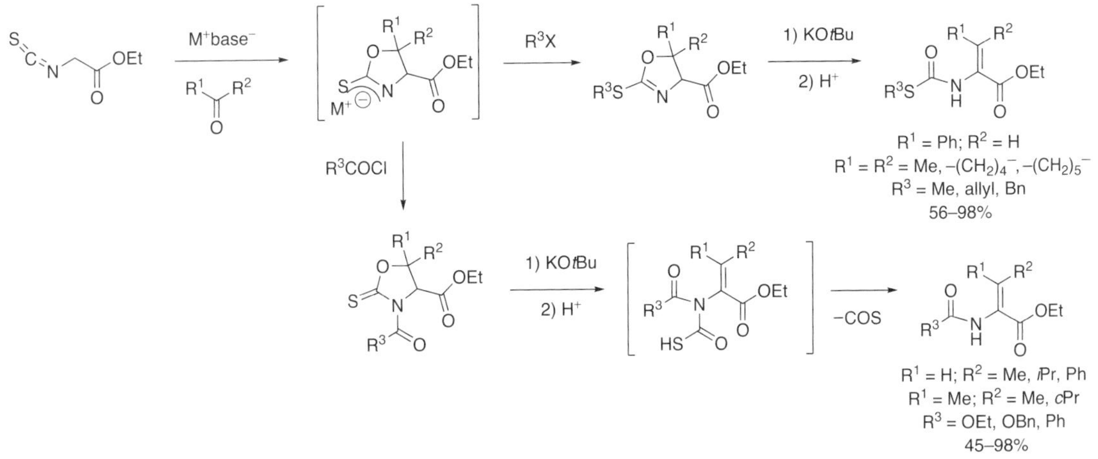

Scheme 4.11 Modified Erlenmeyer Reaction (248–250).

(Cbz, Bz, Ac) followed by deformylation (K$_2$CO$_3$/H$_2$O$_2$ or MeOH/HCl) (241). This modified Erlenmeyer procedure was recently used in the synthesis of a number of β-disubstituted didehydroamino acids (50–95% yield) (235), a dehydrohomoserine ether (45%) (242), and several cycloalkanone adducts (31–88% (243, 244). 2,3-Dehydro-α-glucosyl-glycines were produced via addition to a sugar-derived lactone (245, 246). The initial N-formyl dehydroamino acid has been β-brominated and the halogen displaced with thiols, but for other nucleophiles to be used for the displacement the N-formyl-amino function must first be dehydrated back to an isocyano group (see Scheme 4.10) (247).

A somewhat related synthesis reported by Hoppe in 1972 condensed ethyl α-isothiocyanatoacetates with carbonyl compounds (248), with acylation or alkylation of the intermediate 2-thioxo-oxazolidine-4-carboxylates (see Scheme 4.11). Mixtures of (E)- and (Z)-dehydroamino acid isomers were obtained in 45–98% yield after treatment with base (249, 250).

The chiral Boc-BMI imidazolidinone glycine template of Seebach has been condensed with acetaldehyde to give the dehydroamino acid derivative in low yield (24–54%) (251). Better results were obtained with the Schmidt-type procedure (see Section 4.2.4). Another chiral template containing Gly embedded within a 3,6-dihydro-2*H*-1,4-oxazin-2-one structure was condensed with aldehydes to give dehydroamino acids in 50–64% yield (*252*); a related 1,2,3,6-tetrahydropyrazin-2-one template was condensed with aldehydes and ketones in 47–88% yield (*253*).

4.2.3 Synthesis via Imine Rearrangements and Other Rearrangements

A number of syntheses of dehydroamino acids proceed via an initial *N*-chlorination of amino acid esters by *t*-butylhypochlorite. Treatment with a strong base results in elimination to the imine, followed by isomerization to the desired enamine promoted by HCl (see Scheme 4.12). The (*Z*)-didehydro derivatives of Val, Leu, Ile, and Abu methyl esters were prepared in 63–81% yield by this method, generally using DBU as the base (*254–256*). ΔLeu-OBn, later incorporated in an enkephalin analog, was synthesized in 91% yield (*80*), while Cbz-2,3-ΔLys was prepared in 63% yield and then used as a constrained Lys analog in the bioregulator tetrapeptide tuftsin (*257*). *N*-Cbz Δ Pro was produced in 79% yield by treatment of Pro-OMe with *t*BuOCl followed by Cbz-Cl (*119*).

Alternatively, treatment of the initial *N*-chloro derivative with methoxide gives an α-methoxy intermediate by addition of methanol to a transient imine; exposure to HCl then generates the dehydroamino acid (*256, 258*). Boc-ΔVal, ΔLeu, ΔPhe, and ΔAbu methyl esters were prepared in 66–93% yield (*258*). However, treatment of *N*-acyl Asp by this method led to *N*-acyl ΔAla as the imine rearrangement was accompanied by decarboxylation (*259*).

The chlorination/dechlorination procedure has been used to convert Phe to ΔPhe within a dipeptide in 28% yield using DABCO as the base (*260*). DABCO was also used to generate (*Z*)-β-(methoxymethyl)-α-aminoacrylic acid ester (60% yield) (*261*) and a related compound (*262*). However, the didehydro analog of the heterocyclic amino acid stizolobic acid was obtained in very poor yield (3%) (*263*). A much better result (24%) was provided by rearrangement of the same imine intermediate normally obtained by dehydrohalogenation, but generated in this instance by photolysis of an azido-malonate precursor (*264*).

A related synthesis involves treatment of *N*-hydroxy amino acids with tosyl chloride and triethylamine. The dehydroamino acid is isolated in a few minutes, again presumably via an imine intermediate. *N*-hydroxy-Ala, Asp, and Abu were dehydrated in this manner in 82–97% yield with, in contrast to other methods, the (*E*)-isomer predominating (*E*:*Z* 3:1–12:1) (*265*). Similarly, *O*-benzyl-*N*-(hydroxy) amino acids, readily obtained from α-keto acids by reductive amination with *O*-benzylhydroxylamine (*266, 267*), were converted to the dehydroamino acids by treatment with DBU.

α-Keto esters and acids and imine isomerization have also been used in another route to didehydroamino acids. The 2-oxoalkanoates were condensed with benzylcarbamate, *p*-toluenesulfonamide, chloroacetamide or bromoacetamide in the presence of an acid catalyst (sulfuric acid, *p*-toluenesulfonic acid or phosphorous oxychloride), directly giving the enamide in 24–92% yield as either predominantly (*Z*) (*268*) or (*E*/*Z*) (*261, 269–271*) isomer mixtures (see Scheme 4.13). *N*-Acetyl

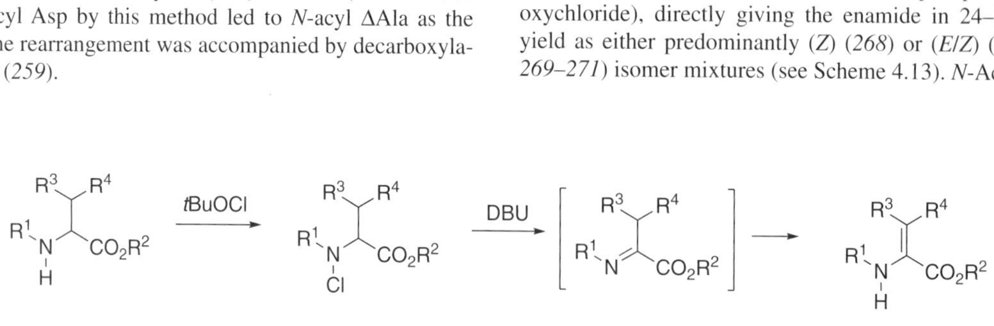

Scheme 4.12 *N*-Chlorination, Elimination and Isomerization of Amino Acids.

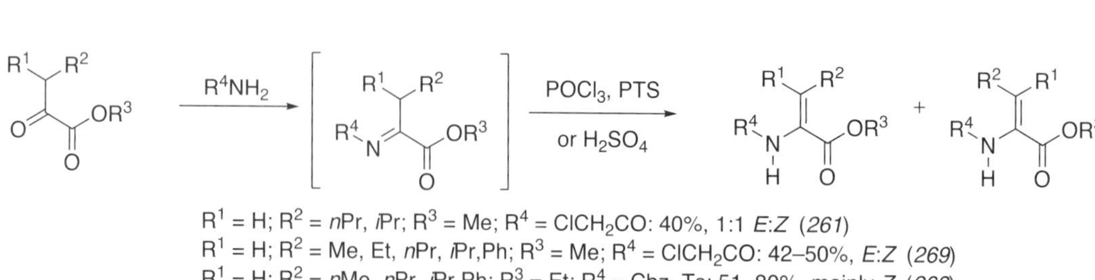

R^1 = H; R^2 = *n*Pr, *i*Pr; R^3 = Me; R^4 = ClCH$_2$CO: 40%, 1:1 *E*:*Z* (*261*)
R^1 = H; R^2 = Me, Et, *n*Pr, *i*Pr,Ph; R^3 = Me; R^4 = ClCH$_2$CO: 42–50%, *E*:*Z* (*269*)
R^1 = H; R^2 = *n*Me, *n*Pr, *i*Pr,Ph; R^3 = Et; R^4 = Cbz, Ts: 51–80%, mainly *Z* (*268*)
R^1 = H; R^2 = *n*Me, *n*Pr, *i*Pr,Ph;
R^1 = R^2 = H, Me; R^3 = Et; R^4 = ClCH$_2$CO: 24–58%, *E*:*Z* (*270, 271*)
R^1 = H; R^2 = H, me, Et, *i*Pr, Ph, (CH$_2$)$_2$CO$_2$*t*Bu; R^3 = Me; R^4 = Cbz
R^1 = Me; R^2 = Me, Et; R^3 = Me; R^4 = Cbz: 54–91% (*119*)
R^1 = CH$_2$CO$_2$Me.; R^2 = H; R^3 = Me; R^4 = Cbz: 47% (*274*)
R^1 = R^2 = R^3 = H; R^4 = Ac: no yield (*272*)

Scheme 4.13 Synthesis of Didehydro Amino Acids via Isomerization of Schiff Base.

ΔAla was obtained from pyruvic acid and acetamide (272), while N-Boc ΔAla-OMe was produced in 44–55% yield by reaction of Boc-NH₂ with methyl pyruvate in the presence of POCl₃ (273). The dehydroamino acid analogs of Ala, Abu, Nva, Val, Leu, Ile, Phe, and Glu were prepared by this method in 1990 using benzyl carbamate as the amine source; the ΔGlu was converted into ΔGln (119). In a separate report benzyl carbamate was condensed with dimethyl 2-oxoglutaric acid in refluxing benzene to give (Z)- ΔGlu-Glu dimethyl ester in 47% yield (274). In 1999 dehydrohomophenylalanine was obtained from the α-keto ester and methyl carbamate, followed by treatment with HCl/toluene (275). Imine isomerization has also been employed in the synthesis of dipeptides containing 2-amino-2-butenoic acid (Aba). 2-Oxobutanoic acid was condensed with Cbz-Val-NH₂ to give a separable 4:1 mixture of Cbz-Val-(Z)-Aba and Cbz-Val-(E)-Aba in 46% yield. After incorporation into the cyclic depsipeptide isophomalide, the (Z)-alkene was isomerized into the (E)-alkene found in phomalide via addition of PhSeLi followed by oxidative elimination with H₂O₂/acetone (29). (Z)-β-Fluoro-α,β-didehydroamino acids have been derived from α-ketoesters via enol silane formation and fluorination with Selectfluor, producing 3-fluoro-2-ketoester intermediates. Condensation with methyl carbamate provided the β-fluorinated didehydroamino acids. Unfortunately, attempts at reduction via catalytic hydrogenation or with NaBH₃CN gave very little of the desired amino acid product (276).

A number of α-keto acids were coupled with (S)-Pro and cyclized in the presence of ammonia and base; dehydration then gave a 2,5-diketopiperazine template containing the dehydroamino acid (see Scheme 4.14) (277).

Catalytic rhenium heptasulfide in acetic anhydride has been used to decompose α-azidocarboxylic acid

esters (94, 278) and amides (279) (obtained from α-halo esters via azide exchange under phase-transfer catalysis) with nitrogen loss creating an imine intermediate that then rearranged to the N-acetyl enamine in 33–96% yield (see Scheme 4.15). By carrying out the perrhenate catalyzed decomposition in the presence of acyl chlorides an N-acyl group was introduced along with the unsaturation with 32–92% yield as mixtures of (E)- and (Z)- isomers (94, 280). Treatment with HCl in ether generated exclusively the (Z)-isomers in quantitative yield (94). 2,3-ΔAla, -ΔNle, -ΔPhe, -ΔVal, and -S-Ph ΔCys were among those prepared. If an amino acid chloride was used as the acyl component dipeptides could be synthesized (281).

α,β-Dehydroamino acids can also be generated by the (usually undesired) isomerization of β,γ-unsaturated amino acids (e.g., see 282, 283). The reverse reaction is used to prepare vinylglycines.

N-(Substituted-vinyl) aziridine-2-carboxylates undergo a rearrangement when refluxed in dimethoxyethane to give the corresponding N-(substituted-vinyl) dehydroamino acids; for example aziridine-2-carboxylate formed ΔAla and 3-phenyl-aziridine-2-carboxylate formed ΔPhe. The aziridine E:Z stereochemistry did not transfer to the dehydroamino acid E:Z ratio (284). The aziridine decomposition can be carried out on aziridine-2-carboxylate residues contained within peptides (72, 171, 172).

4.2.4 Synthesis via Wittig-Type Olefination Reactions (Schmidt Procedure)

A large number of dehydroamino acids have been synthesized by what is generally described as the Schmidt procedure, in which aldehydes are reacted with an

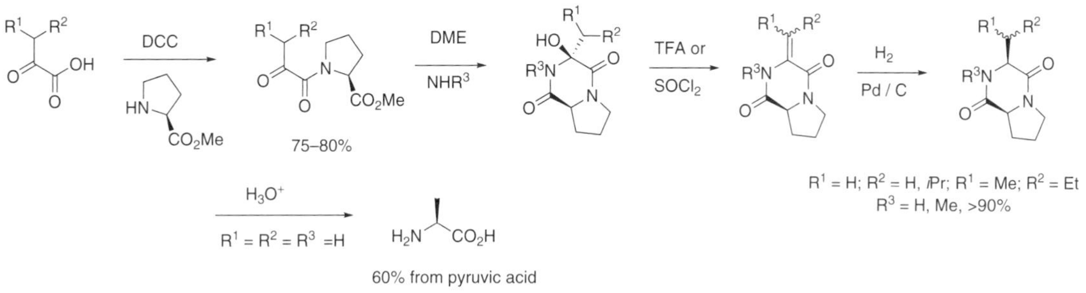

Scheme 4.14 Synthesis of Didehydro Amino Acids within Diketopiperazine (277).

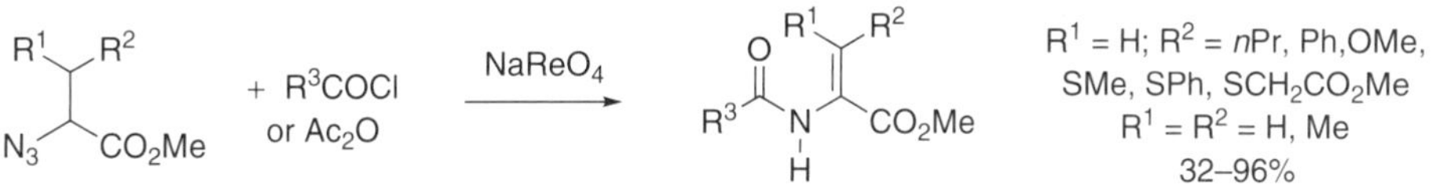

Scheme 4.15 Perrhenate Catalyzed Elimination (94, 278, 280).

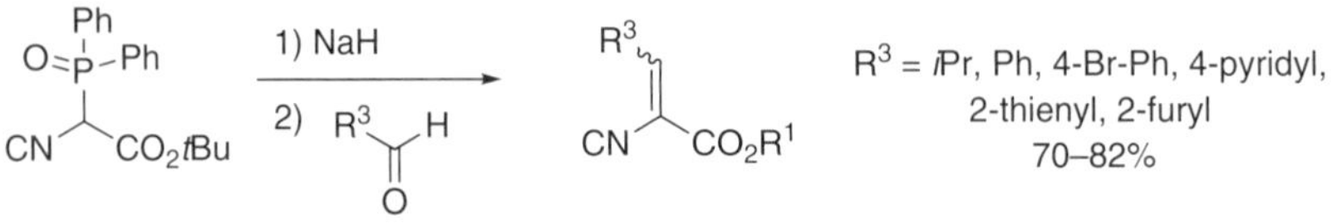

R³ = iPr, Ph, 4-Br-Ph, 4-pyridyl, 2-thienyl, 2-furyl
70–82%

Scheme 4.16 Synthesis of Didehydroamino Acids via Wittig-Type Reaction (288).

R¹ = Me; R² = Cbz, Boc, Ac, ClCH₂CO, HCO, Boc-(S)-Leu, Ac-(S)-Leu
R¹ = Et; R² = Cbz, Boc-(S)-Ile, Boc-(RS)-Trp
R¹ = H; R² = TMSE: 88% (297)
77–91%yield

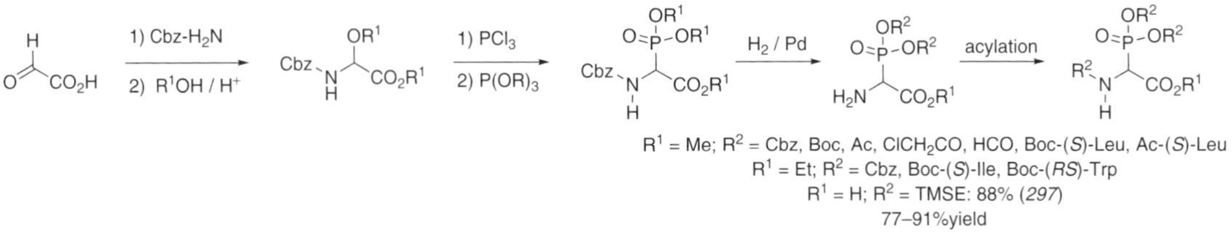

Scheme 4.17 Preparation of Glycine Phosphonate (295).

α-phosphonate-glycine derivative. However, the reaction concept predates Schmidt's report. In 1973 an α-thioformamido-diethylphosphonoacetate was prepared by carboxylation of the Schiff base of aminomethyl diethylphosphonate (285) and reacted with 1-chloro-2-propanone (83–100% yield) enroute to a synthesis of cephalothin (286). In 1977 methyl α-aminodiethylphosphonoacetate was prepared by electrophilic amination of methyl diethylphosphonoacetate, and the dithiocarbamate derivative reacted with chloroacetone to give a dehydroamino acid derivative in 81% yield (287). An analog of the glycine phosphonate, (diphenylphosphinyl) isocyanoacetic acid t-butyl ester, was reported in 1981 and reacted with several aromatic aldehydes in 71–82% yield (see Scheme 4.16) (288).

In 1982 Schmidt et al. reported a synthesis of the glycine phosphonate synthon by refluxing Cbz-α-chloroglycine ethyl ester with triethylphosphite (289, 290). Didehydroamino acids were then prepared by reaction with aldehydes and base, with 73–85% yield. A similar preparation of the phosphate derivative from N-acyl-α-bromoglycinate was published by a different research group the following year; the triphenylphosphonium salt was also synthesized by reaction of the bromide with triphenylphosphine (291). The phosphite was successfully reacted with several aldehydes in 47–64% yield after treatment with base, but the phosphonium salt self-condensed to form a fumarate derivative (291). The glycine phosphonate synthon has also been prepared within a dipeptide from an α-haloglycine residue, which itself was generated from Ser (292), and has been incorporated into a diketopiperazine formed with Pro (293). A hexafluoroacetone-derived oxazolidinone prepared from N-methylglycine was brominated (94% yield) and converted into α-phosphono-, α-phosphino-, and α-phosphinyl-sarcosine derivatives

(294), but no alkene-forming reactions were reported. An improved procedure for the large-scale synthesis of a variety of N-acyl phosphonoglycinates from glyoxalic acid in 77–90% yield was provided in 1984 (see Scheme 4.17) (295), and from methyl glyoxylate in a one-pot reaction in 1990 (67–80% yield) (296). The dimethyl phosphonate of Cbz-α-phosphono-Gly-OTMSE was prepared from α-hydroxy-Gly (297).

The deprotonated phosphonates react with a wide range of aldehydes in good yield (50–93%), but ketones are generally unreactive (see Scheme 4.18, Table 4.3) (295). In one study the best results were obtained with KOtBu in CH₂Cl₂ at –78 °C as base, giving predominantly the (Z)-isomer. (E/Z) mixtures were produced with LDA or NaH (295). Strong bases such as DBU (1,8-diazabicyclo[5.4.0]undec-7-ene) or TMG (tetramethylguanidine) also favored (Z)-isomer formation (298), with DBU giving >97% (Z) and 80–100% yield for 12 examples (96). However, it has been observed that the reaction temperature greatly influences the isomer distribution. In one study with KOtBu at –78 °C the (Z)-isomer was obtained preferentially with approximately a 90:10 Z:E ratio, whereas reaction at 25 °C gave a 15:85 ratio (299). Conflicting results were obtained

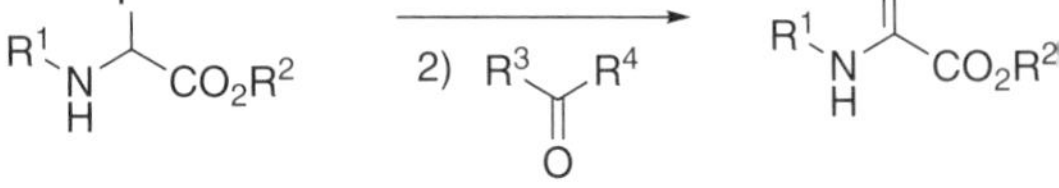

Scheme 4.18 Reactions of Glycine Phosphonate.

Table 4.3 Reactions of Glycine Phosphonate

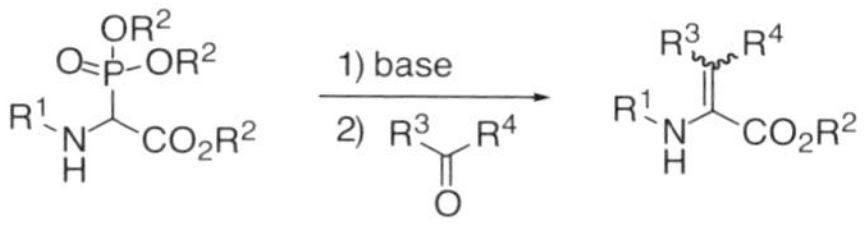

$$R^1{-}N(H){-}CH(PO(OR^2)_2){-}CO_2R^2 \xrightarrow[\text{2) } R^3C(O)R^4]{\text{1) base}} R^1{-}N(H){-}C({=}CR^3R^4){-}CO_2R^2$$

Protecting Groups		Carbonyl Compound		Base	Yield	Reference
R^1	R^2	R^3	R^4			
CH=S	Me	CH_2Cl	Me	K_2CO_3	83–100%	1973 (286)
	4-MeO-Bn		CH_2OAc			
Boc	Et	nPr	H	NaH or LDA	73–85%	1982 (289)
Cbz		iPr			mixture *E/Z*	1982 (290)
		Ph				
		2-MeO-Ph				
		2-Me-5-CO_2Bn-Ph				
Bz	Et	nPent	H	NaH	47–64%	1983 (291)
		cHex			mixture *E/Z*	
		4-Cl-Ph				
Ac	Me	Et	H	NaH or LDA or KOtBu	46–93%	1984 (295)
Boc	Et	nPr			mixture *E/Z* or	
Cbz		(CH_2)_4C(–OCH_2CH_2O–)(CH_2)_2Cl			mainly Z	
COCH_2Cl		(CH_2)_2CH=CH_2				
CHO		2-MeO-Ph				
Boc-Leu		4-MeO-Ph				
Ac-Leu		3,4,5-(BnO)_3-Ph				
Boc-Ile		3,4,5-(AcO)_3-Ph				
Boc-Trp		3,4,5-(MocO)_3-Ph				
COCH_2Cl	tBu	6-Br-indol-3-yl	H	KOtBu	<1:50 *E:Z*	1984 (862)
	Me					
Leu	Me	3,4,5-(OAc)_3-Ph	H	LDA	1:1 *E:Z*	1984 (862)
	CO_2R^2 = Trp					
Cbz	Me	(CH_2)_2NHCbz	H	KOtBu, CH_2Cl_2	85–90%	1987 (863)
	Et	(CH_2)_3NHCbz				

(Continued)

247

Table 4.3 Reactions of Glycine Phosphonate (continued)

Protecting Groups		Carbonyl Compound		Base	Yield	Reference
R^1	R^2	R^3	R^4			
Cbz	Et	nPr	H	20% NaOH, CH$_2$Cl$_2$	45–97% 1:1.4 to 1:42.8 E:Z	1988 (*300*)
		nHept				
		cHex				
		(E)-CH=CHPh				
		Ph				
		3-NO$_2$-Ph				
		4-NO$_2$-Ph				
		2-MeO-Ph				
		2-Cl-Ph				
		2-furyl				
Cbz	Et	4-MeO-Ph	H	NaOH or LiOH	59–92%	1988 (*864*)
		4-MeOCH$_2$-Ph				
		4-AcO-Ph				
Cbz	Me	3-AcO-4-MeO-Ph	H	KOtBu, CH$_2$Cl$_2$	73–78%	1988 (*317*)
	Bn	4-[2-MeO-4-CH$_2$CH(NHCbz)CO$_2$TMSE-PhO]-Ph				
Boc	Bn	3,4,5-(MeO)$_3$-Ph	H	NaH, THF	60% 1:3 E:Z	1989 (*43*)
	Me	3,4,5-(TBSO)$_3$-Ph				
Ac	Me	CH$_2$OMe	H	KOtBu	45–72% mainly Z	1989 (*795*)
Cbz		CH$_2$OiBu				
CHO						
Cbz	Me	1-Cbz-indol-3-yl	H	KOtBu	66–80%	1990 (*119*)
	Et	1-Boc-indol-3-yl				
		(CH$_2$)$_2$NHBoc				
		(CH$_2$)$_3$NHBoc				
Boc-Gly	Me	3,4-(TBSO)$_2$-Ph	H	LDA	64–70% 5:1 to 8:1 E:Z	1990 (*42*)
Boc-Leu	CO$_2$R^2 = CONHCH=CHAr					
CH(tBu)	CO$_2$R^2 = CON(Me)	H	H	DBU/LiBr	42–93% mainly E	1991 (*251*)
N(Me)-R^2	CH(Bu)-R^1	CH$_2$CF$_3$				
		nPr				
		iPr				
		nDec				
		tBu				
		CH(Et)$_2$				
		CH(Me)Et				
		CH(Me)Ph				

N-Acyl	R²	Substituent	R'	Base	Yield (E:Z)	Year (Ref.)
Cbz-Phe	Me	CH=CHMe; CH=CHPh; Ph; 2,4,6-(Me)₃-Ph; 4-Me-Ph	H	NaOMe	mixture *E/Z*	1991 (*292*)
Cbz	Me	CH₂PO₃Et₂	H	KOtBu, *n*BuLi or NaH	36–78%	1991 (*733*)
Cbz	Me	[structure: BnO-, dioxolane]	H	KOtBu	76–87% mainly Z	1991 (*439*)
Cbz	Me	[structure: dioxane, Ph]	H	KOtBu	81% Z	1991 (*865*)
L-Pro (diketo-piperazine)	Et	CO₂R² = L-Pro (diketo-piperazine)	H	KOtBu	75% 1:9 *E:Z*	1992 (*293*)
Boc	Me; TMS	[structure: dioxane, Ph]	H	DBU	95%	1992 (*866*)
Bz	Et	(p-MeO-BnO); Mmt–N; O(p-MeO-Bn)	H	LDA, TFA	72% 1:4 *E:Z*	1992 (*55*)
Ac; CHO	Me	CH₂OMe; CH₂OPh; CH₂OiBu; CH₂O(4-X-Ph) X = F, Cl, NO₂	H	KOtBu	47–81% mainly Z	1992 (*304*)
Boc-Leu	Me	[structure: dioxolane]	Me	DBU	75%	1992 (*331*) 1993 (*330*)
Boc; Cbz; Boc-Ala; Boc-Leu	Me; Et	Et; iPr; Ph; 2-MeO-Ph; 2,4-(BnO)₂-Ph; [structure: dioxane, Ph]	H	DBU	80–100% >97% Z	1992 (*96*)

(Continued)

Table 4.3 Reactions of Glycine Phosphonate (continued)

Protecting Groups		Carbonyl Compound		Base	Yield	Reference
R¹	R²	R³	R⁴			
Boc Cbz Boc-Ala Boc-Leu	Me Et	Me $(CH_2)_5$-R⁴ Et	Me $(CH_2)_5$-R³ Me	DBU	80–100% >97% Z	1992 (96)
Bz MeO₂C	Me	OR R = Bn, TBS	H	LDA or KOtBu or NaH	73–95% 1:1.2 Z:E to 100% Z	1993 (329)
Cbz Boc	Me	Ph	H	DBU	70–95%	1993 (867)
Bz	Me	$(CH_2)_2$NPhth	H	KOtBu	57% 100% Z	1993 (303)
Ac	Me	2-$(CH_2)_2PO_3Et_2$-cHex	H	NaH	15% mainly Z	1993 (307)
Cbz Boc	Me	2-furyl 3-furyl 2-thienyl 3-thienyl 1-H-pyrrol-2-yl 1-Boc-pyrrol-2-yl 1-Ts-pyrrol-2-yl	H	TMG	22–95% mainly Z	1994 (298)
Ac	Et	$(CH_2)_2CO_2Me$ $(CH_2)_3CO_2Me$ $(CH_2)_4CO_2Me$	H	KOtBu	55–64% 10:90 E:Z at −78°C 85:150 E:Z at −25°C	1994 (299)
Boc	Me		H	KHMDS	75%	1994 (868)
Boc	Me		H	LDA	65% 1:1 E:Z	1994 (869)

250

Cbz	Me	2-CO$_2$Me-cPr dialdehyde 2-CHO-cPr	H	LDA	61–65% mainly Z	1994 (*308*) 1995 (*309*)
Cbz	Me	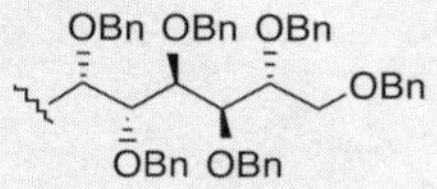 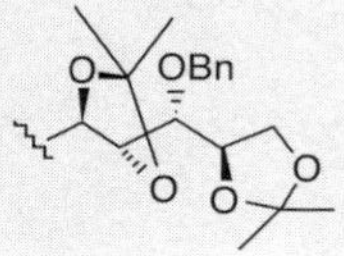 [structure: OBn OBn OBn OBn / OBn OBn / OBn isopropylidene dioxolane]	H	NaH	80% 1:1 to 2:3 *E:Z*	1994 (*321*)
Ac	Me	*i*Pr imidazol-4-yl	H	DBU	49–61% 100% Z	1995 (*184*)
Cbz Boc	Me	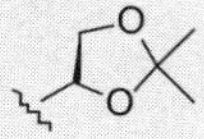 [structure: dimethyl dioxolane]	H	KO*t*Bu	80–85% mainly Z	1995 (*318*) 1994 (*319*)
Cbz	allyl	Me Et Bn	H	DBU	74–84% mainly Z	1996 (*689*)
Cbz	Me	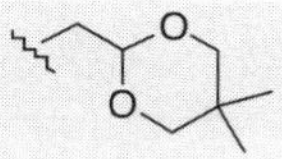 [structure: dimethyl dioxane]	H	TMG	90% mainly Z	1996 (*306*)
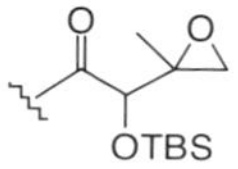 [structure: epoxide/ketone OTBS]	Et Bn CO$_2$R^2 = NHCH$_2$CH(OTBS) Me, Gly-OMe	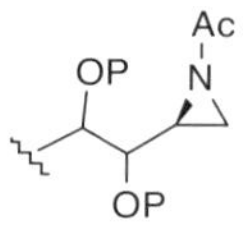 [structure: Ac-N aziridine OP OP]	H	LDA	18–75% mainly Z	1996 (*327*)
Cbz Boc	Me TMSE	dialdehyde 1,1'-(CHO)$_2$-ferrocenyl	H	TMG	61–69%	1997 (*314*) 2002 (*315*)
Boc	Me or TMSE	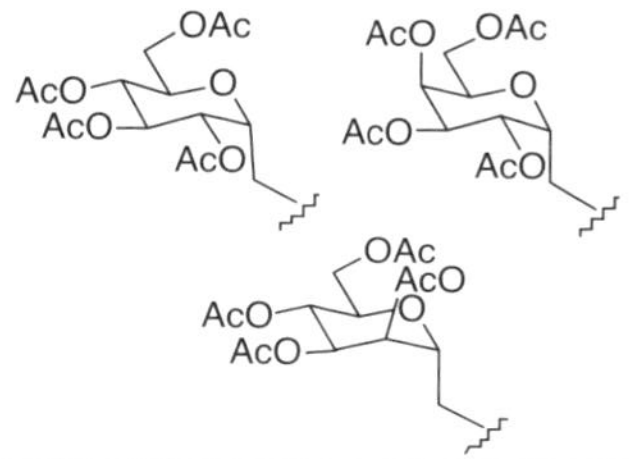 [structure: AcO sugar units]	H	TMG	65–88%	1997 (*323*) 2003 (*324*)

(*Continued*)

251

Table 4.3 Reactions of Glycine Phosphonate (continued)

Protecting Groups		Carbonyl Compound		Base	Yield	Reference
R^1	R^2	R^3	R^4			
Ac	Me	(E)-CH=CHCH$_2$OTBS (Z)-CH=CHCH$_2$OTBS CH=C(Me)CH$_2$CH$_2$CH=C(Me)$_2$ C(Et)=CH$_2$ CH=CHMe cyclohex-1-enyl	H	TMG, THF	17–77%	1998 (305)
Cbz	Me	TsO—[cyclohexyl]—OTs		DBU	76%	1998 (870)
Boc	Et	[sugar: OAc, AcO, AcO, O, NHAc]	H	LDA	73% 1:2 E:Z	1998 (322)
Ac	Me	[aziridine: Cbz, OAc, O(p-MeO-Bn)]	H	KOtBu	67% <1:4 E:Z	1998 (56)
Cbz	Me	trialdehyde 1,3,5-(CHO)$_3$-Ph	H	TMG	54%	1998 (313)
Ac Cbz Boc	Me	[cyclobutane dioxolane structure]	H	KOtBu	30–75%	1998 (871)
Cbz	Me	[cyclopropane]—CO$_2$Me	H	LDA or KOtBu	70–71% 25:75 or 7:93 E:Z respectively	1998 (310)
Ac Boc Cbz	Me	[cyclopropane]—CO$_2$Me [cyclopropane]—CO$_2$Me	H	LDA	58–79% 100% Z fro Ac, Cbz 1:4 E:Z for Boc	1998 (326)
Boc	Me	dialdehyde R^3 = R^4—(CH$_2$)$_2$N(Cbz)(CH$_2$)$_2$—	H	DBU	70%	1998 (302)

Cbz	Me	pyrid-2-yl 2-Me-pyrid-6-yl isoquinolin-2-yl quinoline-3-yl	H	—	45–66%	1999 (565)
Cbz	Me		H	DBU or TMG or NaH	85–95% 9:91, 6:94 or 48:52 E:Z respectively	1999 (872)
Cbz	Me	nPr nBu cHex	H	TMG	72–79%1:>20 E:Z	1999 (566)
Cbz Boc MeO$_2$C	Me	dialdehydes OHC-X-CHO X = -(CH$_2$)$_n$: n = 0,2,3,6,8 X = 1,4-Ph X = 1,3-Ph	H	DBU or TMG	40–80% mainly Z	1999 (311)
Boc	TMSE		H	TMG, THF	65–84%	1999 (297)
Cbz	ester tBu or Me phosphonate Me	OBn OBn OBn OTBS	H	DBU	54–83% Z 5–9% E	1999 (328)
Cbz Boc	Me	2-thienyl 3-thienyl 5-Me-2-furyl	H	TMG	68–88% Z	1999 (873)
Cbz Ac	Me	CH$_2$(2,2-Me$_2$-3-R-cyclobutane) R = CH=C(NHAc)CO$_2$Me, C(Me) (−OCH$_2$CH$_2$O−) (2,2-Me$_2$-3-R-cyclobutane) R = (CH$_2$)$_3$OBn, C(Me)(−OCH$_2$CH$_2$O−)	H	KOtBu, CH$_2$Cl$_2$	30–62% Z	2000 (874)

(Continued)

Table 4.3 Reactions of Glycine Phosphonate (continued)

Protecting Groups		Carbonyl Compound		Base	Yield	Reference
R¹	R²	R³	R⁴			
Cbz	Me	3-HO-Ph	H	DBU	90%	2000 (875)
Cbz Bz	Me	2-I-4-MeO-Ph	H	DBU	93–95%	2000 (876)
within diketo-piperazine with Pro	Me on phosphonate, Gly within diketopiperazine with Pro	C[=CHCH=C(Me)₂]-CONH(2-I-Ph)	H	KOtBu	61%	2000 (66)
Ac	Me	3-NO₂-4-F-Ph	H	TMG	99%	2001 (628)
Boc	CH₂CH₂TMS	(CH₂)₃O(carbohydrate) carbohydrate = α-Tn, Lewis^Y, fucosyl GM1, Globo-H (mono to hexasaccharides)	H	TMG	10–88%	2001 (235)
Bz	Me on phosphonate, CO₂R = CONHCH₂COMe on amino acid	(structure: Cbz, N, OAc, OSiEt₃)	H	KOtBu	40% Z major	2001 (59)
Cbz	Me	(structure: Boc, N, CO₂Et, Ph; Boc, N, CO₂Et, Ph)	H	DBU	90–100% Z	2002 (877)
Cbz	Me	3-(6-carboxy-2-oxo-4-pyridyl) 3-(6-carboxy-2-oxo-3-pyridyl)	H	TMG	76%	2002 (878)
Cbz	Me	2-Br-Ph 4-Br-Ph 3-BnO-Ph 5-Br-indol-3-yl	H	DBU	68–92%	2002 (879)

Ac, Cbz, Ar	CH₂-polystyrene resin	4-Br-Ph 4-HO-Ph 2-OMe-4-OH-Ph	H	DBU	—	2003 (334)
Ac	NH-Rink resin	H	H	K₂CO₃	—	2003 (335)
Ac	Me	(cyclobutane structure with CO₂Me)	H	KOtBu	31–38%	2003 (880)
Cbz	(cyclobutane structure: H···CO₂Me, H···CO₂Me)		H			2004 (881)
Cbz	Me	2-MeO-5-B(OR)₂-Ph 2-nBuO-5-B(OR)₂-Ph 2-MeO-4-B(OR)₂-Ph	H	DBU	99% Z:E = 89:11 to 96:4	
Boc-Arg(Boc)₂	Me	1-Boc-6-Br-indol-3-yl	H	DBU	55%	2004 (64)
Cbz	Me	(TBDMSO, OH dioxolane structure)	H	KOtBu	90%	2004 (882)
Boc-Val Boc-Ile-Val	Me	1-Boc-6-Br-indol-3-yl	H	DBU	83–90%	2004 (883)
Cbz	Et on phosphonate, Me ester	CH₂CH₂Ph CH₂CH₂(4-MeO-Ph) CH₂CH₂(5-Me-2-furyl) n-nonyl CH₂(N-Boc-piperidin-4-yl) Bn CH₂CH₂CH=CH₂ CH=CHMe Me iPr CH(Et)₂	H	KOtBu or TMG or DBU	56–94%	2005 (800)
Boc, Cbz	Me	4-[CH(CO₂H)]₂-Ph 4-[CH₂CH(CO₂H)]₂-Ph	H	TMG	75–80%	2005 (884)
Boc Cbz	Me	CH[CH₂CH₂SO₃CH₂tBu]NHCbz CHRNHBoc R = H, Bn, CH₂OtBu, CH₂CH₂CO₂tBu	H	DBU or nBui	24–41%	2005 (301)
Cbz	Et on phosphonate, Me ester	2-Ph-2-CO₂Me-cPr	H	DBU	87%	2005 (885)

255

when thienyl, furyl, and pyrrolylaldehydes were condensed with the glycine phosphonate using tetramethylguanidine as base (22–96% yield), as more (Z)-isomer was obtained at higher temperature (298). However in this case a bulky *tert*-butyl ester was used, compared to an ethyl ester in the previous study. A two-phase reaction system with 20% NaOH and CH$_2$Cl$_2$ has been reported, producing predominantly the (Z)-isomer (300). The glycine phosphonate method has found increasing popularity in the construction of a number of complex amino acids (see Table 4.3). For example, amino aldehydes derived from amino acids react to give γ-amino-α,β-didehydroamino acids (301). Despite their poor reactivity, some ketones have been successfully reacted; the unsaturated precursor of α-(4-piperidinyl)glycine was prepared from N-Boc piperidone, enroute to an Arg analog for a thrombin inhibitor (302).

A range of dehydroamino acids have been prepared by the Schmidt procedure (see Scheme 4.18, Table 4.3), such as (Z)-dehydroornithine (57% yield) (303), dehydrohomoserine ether derivatives (47–81% yield as predominantly the Z-isomer) (304), and six different α,γ-dienamide esters (17–77%) (305). The allyl ester of the glycine phosphonate also reacts well with alkyl and aryl aldehydes (74–81% yield) (292). The glycine phosphonate has been reacted with β,β-dialkoxypropionaldehyde (90% yield) (306), a cyclohexylphosphphosphonate aldehyde (15%) (307), and several cyclopropyl-ester aldehydes (61–65% yield (308–310). Reaction of two equivalents of phosphonate with alkyldialdehydes or bis(formyl)benzenes produced diaminodicarboxylic acids in 40–80% yields (311). Similarly, a 2,6-pyridine-linked bis-alanine was synthesized via condensation of 2,6-pyridinedicarbaldehyde with two equivalents of phosphonoglycine (312), while three equivalents reacted with 1,3,5-tris(formyl)benzene during a synthesis of phenyltrisalanine (313). The unsaturated precursor of 1,1′-ferrocenyl-bis(alanine) was also prepared by this method (314, 315), as were one (316) or both (317) halves of the aryl ether-linked bisamino acid isodityrosine.

Reaction of the glycine phosphonate with D-glyceraldehyde acetonide gave a chiral didehydroamino acid in 80–85% yield (318, 319) which was then used to induce diastereoselectivity in subsequent reactions. Other diol-acetonide aldehydes have also been reacted (76–87% yield) (320). The protected dehydroamino acid precursors of 4,5,6,7,8-pentahydroxy-2-aminooctanoic acid and 4,5,6,7,8,9-hexahydroxy-2-aminononanoic acid were obtained by Wittig reaction of N-Cbz phospho-Gly with D-mannose-derived aldehydes (321). Glucose, mannose, and galactose carbon-linked glycosyl amino acids were obtained via condensation of the glycosyl aldehydes (322–324), as were carba analogs of O-trisaccharide Ser (297). Glycosylated 6-hydroxy-2,3-dehydronorleucine derivatives were prepared by condensing complex carbohydrates (including hexasaccharides) with pentanal attached to the anomeric oxygen (325). Chiral cyclopropyl aldehydes have also been condensed (326).

The complex dehydroamino acid contained in the antitumor antibiotics azinomycins A and B, with a 1-azabicyclo[3.1.0]hex-2-ylidene ring system, has been prepared via a Wittig reaction of glycine phosphonate with protected homochiral 2,3-hydroxy-4,5-aziridylpentanal. The alkene was brominated, the aziridine nitrogen deprotected, and cyclization to the bicyclic system achieved by intramolecular Michael addition–elimination (55). This approach was employed during a total synthesis of azinomycin A, with the glycine phosphonate contained within the remainder of the peptidic natural product (59). Similar syntheses of complex didehydroamino acids related to those found in the azinomycins have been described (56, 327, 328). The bromination reaction has been employed on other γ-hydroxy-α,β-unsaturated amino esters. Bromination with NBS followed by treatment with DABCO provided (Z)-β-bromo-α,β-dehydroamino acids preferentially. However, by using other bases such as NaHMDS the (E)-isomer could be obtained with high stereoselectivity (329). Another complex natural product, (–)-spirotryprostatin B, was synthesized via a condensation of Gly-phosphonate-Pro diketopiperazine with an unsaturated aldehyde (66).

The glycine phosphonate has been incorporated into a dipeptide and used to form dehydroamino acids within peptides (330, 331). The glycine phosphonate dipeptide can be synthesized by a Rh-catalyzed carbenoid N–H insertion of an amino acid amide with triethyl diazophosphonoacetate (332). A tetrapeptide containing phosphonoglycine was treated with HCHO/K$_2$CO$_3$ to generate a ΔAla-containing peptide during a synthesis of analogs of the toxin microcystin LA (333). The glycine phosphonate has also been generated on a solid-phase resin and coupled with a range of benzaldehydes, using DBU as base (334). ΔAla was generated on peptides attached to a Rink resin via coupling of Ac-Gly phosphonate, followed by olefination with formaldehyde/K$_2$CO$_2$ (335).

The chiral Boc-BMI imidazolidinone glycine template of Seebach (see Section 7.9) has been brominated and converted into a chiral Schmidt-type phosphonate. Reaction with 15 aldehydes gave predominantly the (E)-isomer with >95:5 E:Z selectivity and 42–93% yield. In contrast with other phosphonates, more (E)-isomer was formed at lower temperature (251). In a similar fashion Williams' chiral oxazinone glycine template (see Scheme 4.19) has been brominated, converted to a phosphonate ester (67% yield), and reacted with a number of aldehydes (19–100% yield) (320, 336).

A Wittig-like condensation has been developed from N-Boc or N-Cbz α-tosyl-Gly-OEt, which were condensed with aldehydes and tributylphosphine in the presence of Na$_2$CO$_3$ and catalytic Bu$_4$NBr. Predominantly (Z)-isomers were generated in 60–96% yield for 11 different aldehydes, including the 2,3-dehydro analogs of Abu, Nva, Leu, Phe, Tyr, Arg, Trp, and His (337).

Little attention has been paid to the "reverse" Wittig reaction, using the phosphonate/phosphonium component to introduce the side chain. An α-keto glycine derivative was prepared from oxalyl chloride in 89%

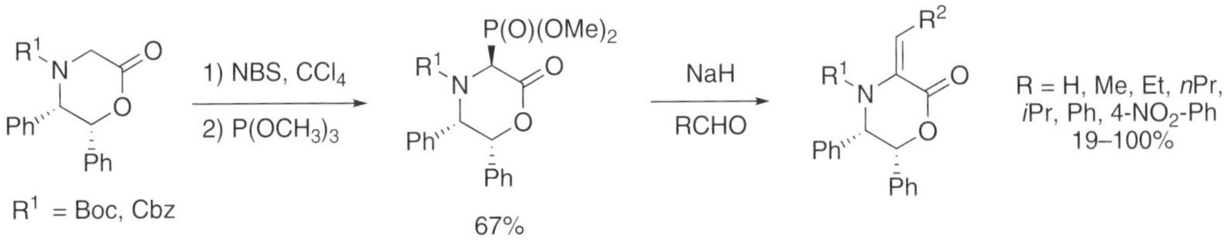

Scheme 4.19 Synthesis of Didehydroamino Acid within Oxazinone Template (*320, 336*).

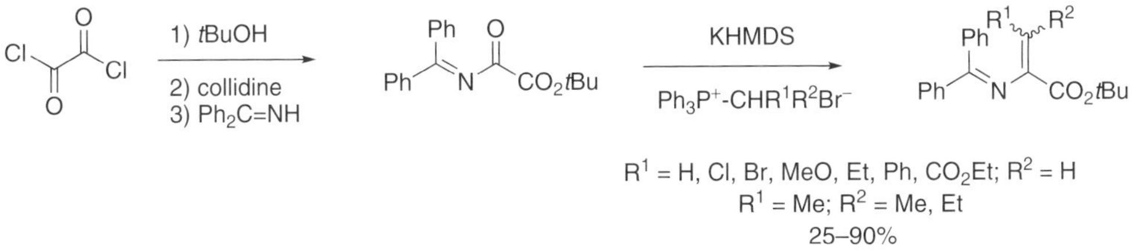

Scheme 4.20 Reverse Wittig Reaction (*338*).

yield (see Scheme 4.20). It reacted with a number of Wittig reagents in 25–90% yield, giving predominantly the (Z)-isomer (*338*).

4.2.5 Synthesis via Other Methods

A variety of other syntheses of dehydroamino acids have been described, including several amination methods. Nitrogen nucleophiles such as phthalimide or sulfonamides can be added to 2,3-alkynoates (see Scheme 4.21), with the desired α-addition promoted by phosphine catalysts (*339*). This method was employed to prepare α-phthalimidoacrylate in 85% yield from ethyl propiolate and phthalimide in 2001 (*340*). Trialkylamines were added to dimethyl acetylenedicarboxylate in the presence of SbCl$_3$, with the amine addition accompanied by dealkylation to give N,N-dialkyl ΔAsp (*341*). Amination of β-aryl-α,β-unsaturated esters can be achieved in 62–73% yield by a one-pot aminohalogenation with TsNCl$_2$/CuOTf, followed by elimination induced with DABCO, producing 2,3-dehydroarylalanines (*342*). ΔPhe has been prepared from ethyl benzalmalonate in 19%

yield, using a Curtius rearrangement to convert one of the ester groups into an amine (*343*).

β-Substituted-ΔAsp derivatives can be prepared from 3-substituted propiolic acids, via Curtius rearrangement conversion of the acid group to an isocyanate followed by Co-mediated dicarboxylation to generate the ΔAsp product as an anhydride (see Scheme 4.22) (*344*). Alkynoates have also been used as electrophiles for the addition of a Schiff base derivative of Gly-OEt, in the presence of a phosphazene base (see Scheme 4.23). The alkene of the initial adduct isomerizes to the α,β-unsaturated derivative, with the resulting side chain alkylated with alkyl halides. By using an 8-phenylmenthyl propiolate ester, asymmetry was induced in the alkylation step (*345*). Alternatively, the initial α,β-didehydroglutamate Schiff base adduct was deprotonated with LDA or KOtBu and used for aldol reactions with aryl aldehydes, as long as TMS-Cl was present as a trapping reagent (*346*). The addition of the benzophenone Schiff base of Gly-OtBu to ethyl propiolates has also been induced by KOtBu, producing 3-substituted-ΔGlu products. Deprotonation with NaOH and

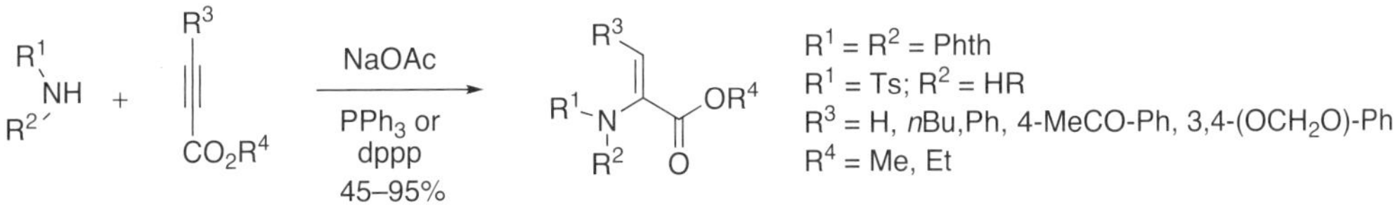

Scheme 4.21 Amine Addition to Alkynoates (*339*).

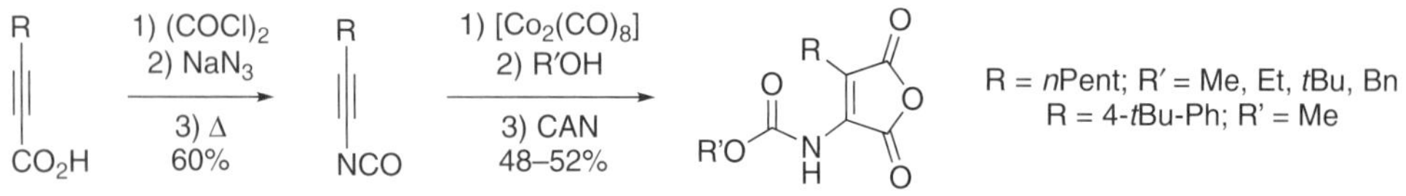

Scheme 4.22 β-Substituted Dehydro-Asp from 3-Substituted Propiolic Acids (*344*).

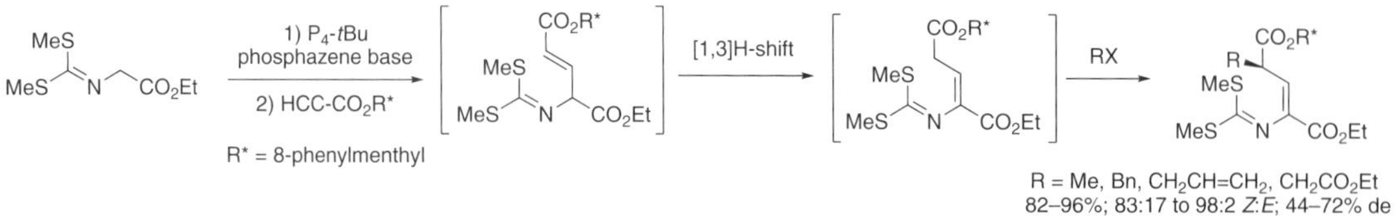

Scheme 4.23 Addition of Schiff Base to Propiolate (*345,346*).

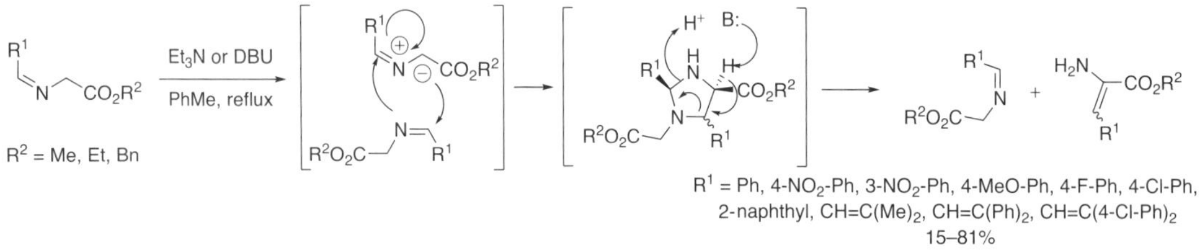

Scheme 4.24 Azomethine Formation from Glycine Schiff Bases (*348, 349*).

alkylation resulted in additional substituents at the 4-position (*347*).

Another synthesis of dehydroamino acids from glycine Schiff bases relies upon generation of an azomethine ylide by treatment of the aldimine with base (Et₃N or DBU). The ylide undergoes cycloaddition with a second equivalent of Schiff base to form an imidazolidine intermediate (see Scheme 4.24). Base-catalyzed ring-opening gave the desired dehydroamino acid product, containing the imine substituent in the side chain (*348, 349*). Backbone alkene-linked diaminodicarboxylic acids (3-amino-3-carboxy-ΔAla) were produced in good yield by dimerization of α-chloro-Gly in the presence of PPh₃ and NEt₃ (*350*).

N-Benzoyl dehydroamino acids with β-(pyrazol-4-yl) substituents were synthesized via reaction of

2H-pyran-2-ones with hydrazines (*351*). This reaction was extended to produce a range of β-(isoxazol-4-yl)- and β-(1-substituted-pyrazol-4-yl)-2,3-dehydroamino acids, using hydroxylamine or substituted hydrazines for the rearrangement (*352*). Trisubstituted imidazoles have been oxidatively cleaved and isomerized to give a didehydroamino acid (see Scheme 3.30) (*353*). Reaction of vinyldiazoacetates with imines in the presence of Rh catalysts produced 2,3-dehydro-Pro derivatives. However, in the presence of a Cu catalyst, the regioisomeric 2,5-dihydropyrrole-3-carboxylic acids were produced (*354*). 2,3-Dehydroproline and ring-expanded analogs with 6-, 7-, 8-, 9-, 13 -, and 16-membered ring sizes were synthesized by a Pd-catalyzed carboxylation of a ketene aminal phosphate derived from the corresponding lactam (see Scheme 4.25) (*355*).

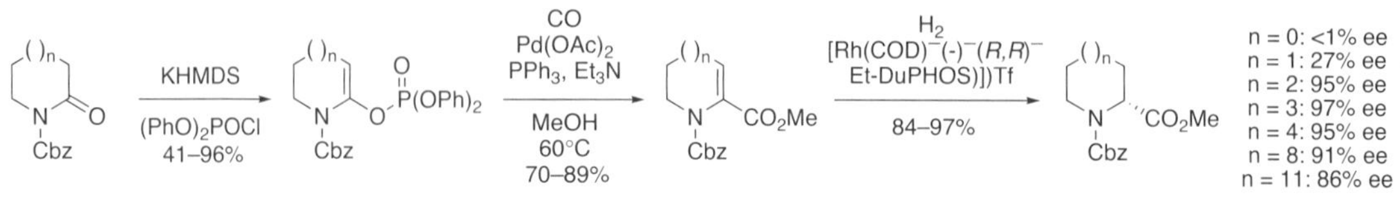

Scheme 4.25 Synthesis and Asymmetric Hydrogenation of 2,3-Dehydro Cyclic Imino Acids (*355*).

Some syntheses convert amino acids in the corresponding dehydroamino acids. Serine was transformed into O-tosyl ΔSer via oxidation of the hydroxymethyl group and tosylation of the resulting enol tautomer, using a one-pot reaction with a modified Moffat oxidation (see Scheme 4.26). The tosyl group was displaced by nucleophiles to give other dehydroamino acids (61, 356). The same α-formylglycine intermediate was incorporated into a peptide as the diethyl acetal during a total synthesis of capreomycin 1B; deprotection with 2 N HCl and reaction with urea generated 3-ureidodehydroalanine (54). Ser has also been transformed into β-fluoro-ΔAla via conversion to S-(4-methoxyphenyl) Cys followed by a fluoro-Pummerer rearrangement (oxidation with mCPBA, treatment with DAST/SbCl₃, and then oxidation with mCPBA followed by heating). The reaction could be carried out on a Ser residue within a dipeptide (357). (Z)-ΔTrp derivatives were obtained from Trp by oxidation with L-Trp 2′,3′-oxidase from *Chromobacterium violaceum* (358). The reaction has been carried out on Trp within peptides (359, 360).

More complex dehydroamino acids have been produced by elaboration of simple dehydroamino acids, in particular ΔAla. ΔAla was converted into (Z)-β-aryl didehydroamino acid derivatives in 12–80% yield by a Pd-catalyzed Heck coupling with aryl iodides (see Scheme 4.27) (361, 362); pyridyl bromides were also coupled in 60–64% yield (363, 364). The Heck coupling has been carried out using aryl iodides and ΔAla within di- and tripeptides (365), with aryl bromides and ΔAla derivatives attached to a solid support (334), and with aryl iodides and Rink resin-linked ΔAla (335). Consecutive Heck reactions of 3-bromoiodobenzene (366, 367) or 4-bromoiodobenzene (367) with two differentially protected ΔAla derivatives gave phenyl-bridged bis(ΔAla) compounds, while 1,3,5-triiodobenzene

reacted with three equivalents of Boc-ΔAla-OMe to give the unsaturated precursor of phenyltrisalanine (313). A single Heck coupling of ΔAla with 4′-iodo-Phe (367), or bis(coupling) of ΔAla with 4,4′-diiodobiphenyl (367) or 1,1′-diiodoferrocene (368), also led to bridged bis(alanines). An intramolecular Heck reaction of ΔAla with an N-(o-iodophenylalkyl) substituent was used to prepare conformationally constrained imino acid analogs of Phe with various ring sizes (100, 369). An intermolecular/intramolecular Heck reaction of 2 equivalents of N-[ω-(4-iodophenyl)alkyl]-ΔAla-OMe, with butyl, hepty, and decyl alkyl chains, produced ΔPhe-containing cyclophanes (370).

Attempts to functionalize ΔAla by Ru-catalyzed cross-metathesis reactions with other alkenes were generally unsuccessful (124, 371). However, a dipeptide formed from Ac-ΔAla and 5-allyl- or 5-homoallyl-Pro-OtBu underwent intramolecular metathesis to give azabicyclo[X.Y.0]derivatives containing the 2,3-dehydroamino acid moiety in 53–89% yield (372).

β-Bromodehydroamino acids can be obtained from N-formyl dehydroamino esters via bromination with NBS, initially producing alkyl 3-bromo-2-formylimino alkanoates. Treatment with an amine base (Et₃N or DABCO) induced isomerization to a mixture of (E)- and (Z)-β-bromo-2,3-dehydroamino esters. The (Z)-isomer formed preferentially, and in cases with bulky substituents, almost exclusively (238). This process was used to convert Boc-Ser-OMe to Boc-3-bromo-2,3-ΔAla-OMe, with MsCl/Et₃N inducing the initial dehydration (373). As mentioned earlier, γ-hydroxy-α, β-unsaturated amino acids were brominated with NBS then treated with DABCO to give (Z)-β-bromo-α, β-dehydroamino acids, or with NaHMDS to give the (E) isomer (329). α,γ-Dienamide esters were produced by a Suzuki cross-coupling of β-bromo-ΔAla with vinylboronates (see Scheme 4.28) (305).

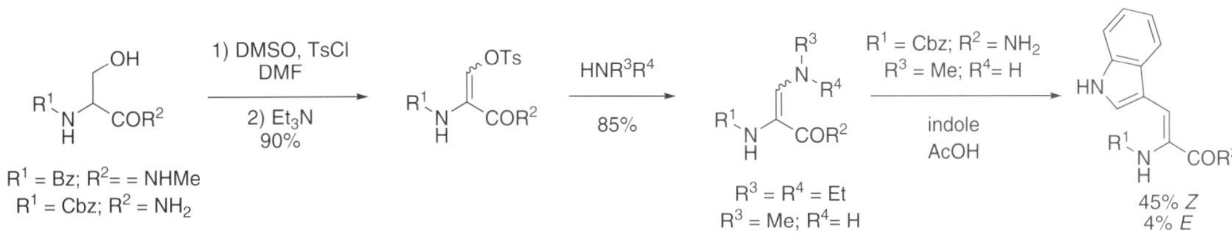

Scheme 4.26 Conversion of Serine to Dehydroserine and Displacement (61, 356).

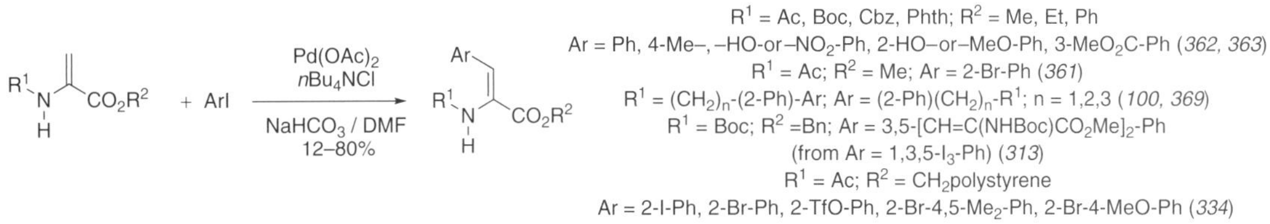

Scheme 4.27 Heck Reaction of Didehydroalanine.

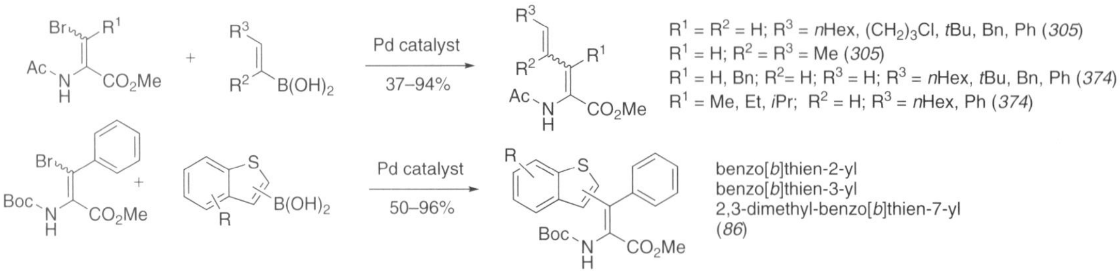

Scheme 4.28 Suzuki Coupling of β-Bromodehydroalanines.

Other β-bromodehydroamino acids have also been coupled (*374*). The configuration of the bromide (*E* or *Z*) determines the configuration of the product. A Suzuki coupling of indole-3-boronic acid with β-bromo-ΔAbu gave the unsaturated precursor of β-methyl-Trp (*375*). N-Boc 3-bromo-ΔAla-OMe was coupled with a Garner aldehyde-derived organoboron homoalanine equivalent, leading to (after asymmetric alkene hydrogenation and hydroxymethyl oxidation), *meso*-2,6-diaminopimelic acid (*373*).

N-acetyl(Z)-ΔAbu-OMe was iodinated with N-iodosuccinimide, giving predominantly the (Z)-isomer (5:1 ratio) when 2% TFA was added to the reaction mixture (1:1 ratio without). The (Z)-isomer was employed for a range of Suzuki couplings to produce (Z)-β-aryl-ΔAbu products (*109*). Boc-ΔPhe-OMe, obtained by TFA monodeprotection of the N,N-di-Boc precursor, was treated with NBS and then triethylamine to give a 2:1 Z:E mixture of β-bromo-2,3-ΔPhe in 89% yield. If the same sequence was carried out in one pot, a 6:1 ratio of isomers was generated. Suzuki cross-coupling with various benzo[b]thienylboronic acids produced the β-disubstituted β-(benzo[b]thienyl)-ΔPhe products (*86, 376*). In an earlier report β-bromo-ΔAla, -ΔAbu, and -ΔPhe were coupled to various benzo[b]thienylboronic acids in 40–61% yield (*377*). β,β-Dibromo-ΔAla was produced from N,N-di-Boc ΔAla-OMe in 52% yield by treatment with

TFA followed by 2.5 equiv of NBS (*378, 379*). The resulting Boc-β,β-dibromo-ΔAla-OMe was employed for Suzuki couplings with 2-, 3- or 7-benzo[b]thieneboronic acids, giving β,β-bis(benzo[b]thienyl)ΔAla derivatives in 80–90% yield (*378, 379*). Alternatively, both Boc-β-bromo-ΔAla-OMe and Boc-β,β-dibromo-ΔAla-OMe were employed for Pd/Cu-catalyzed Sonogashira cross-couplings with phenylacetylene, p-bromohenylacetylene or p-aminophenylacetylene to produce β-(2-phenylethynyl)- or β,β-bis(2-phenylethynyl)-ΔAla derivatives in 44–90% yield. Further derivatization was accomplished via Suzuki coupling of the p-bromoaryl group with thiophene-3-boronic acid, or cross-coupling of the p-aminoaryl group with 7-amino-2, 3-dimethylthiophene (*380*).

A range of heteroaryl-substituted ΔAla azlactones have been obtained from the azlactone of 3-chloro-ΔAla, via displacement with organometallic reagents or other nucleophiles, or by Friedel–Crafts acylation (see Scheme 4.29). Yields were generally better than those provided by the Erlenmeyer synthesis route (*381*). The β-chloro-ΔAla azlactone has also been used for Pd-catalyzed Stille coupling reactions with organostannanes, giving predominantly the (Z)-isomer (*382*).

Heteroaryl nucleophiles (1,2,4-triazole, 3-formylindole, imidazole, pyrazole, 3-formyl-2-methyl-5-nitroindole, thiophenol, methyl mercaptoacetate, octylthiol)

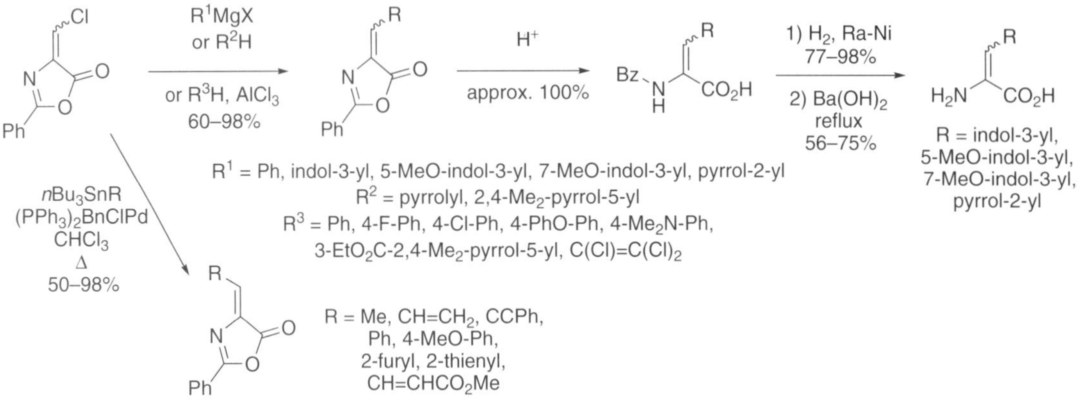

Scheme 4.29 Reaction of β-Chlorodehydroalanine Azlactone (*381, 382*).

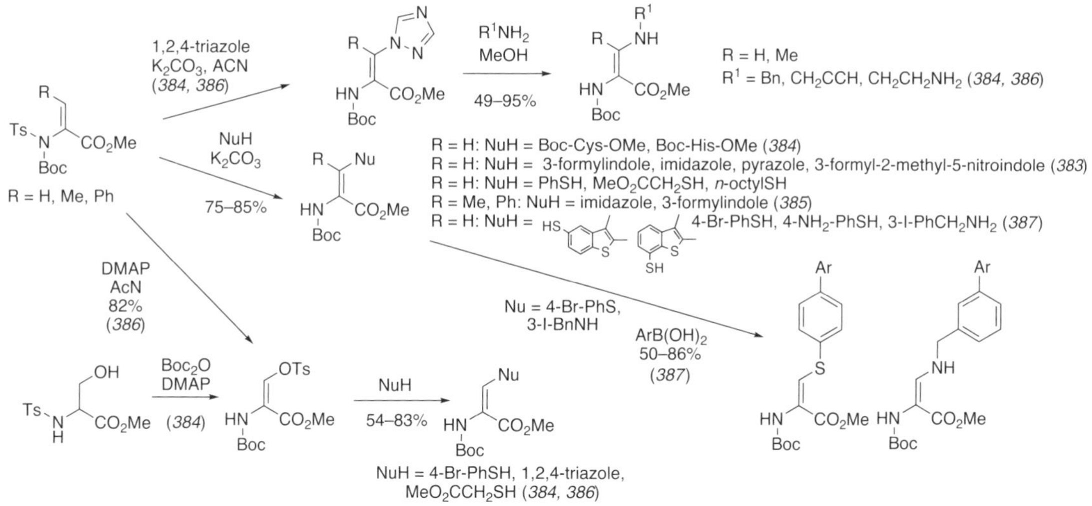

Scheme 4.30 Reactions of *N*-Boc,*N*-tosyl Dehydroalanines (*383*–*387*).

have also been added to *N*-Boc,*N*-tosyl-ΔAla-OMe, with subsequent addition of K₂CO₃, resulting in elimination/rearrangement of the *N*-tosyl group to generate β-heteroaryl *N*-Boc ΔAla derivatives (see Scheme 4.30) (*383*–*385*). β-Triazolyl-ΔAbu and -ΔPhe were also prepared by this method (*383*, *384*), with 3-formylindole and imidazole also successfully added (*385*). The triazole adduct could then be treated with amines (benzylamine, propargylamine) to give β-amino-ΔAla derivatives (*384*, *386*). The same reactions could be carried out with the corresponding 2,3-ΔAbu derivative (*384*, *386*). The adducts obtained by addition/elimination with 4-bromothiophenol or 3-iodobenzylamine were then employed for Suzuki couplings of the arylhalide substituent with arylboronates (*387*). Alternatively, treatment of *N*-Boc,*N*-tosyl ΔAla-OMe with DMAP induced rearrangement to Boc-β-tosyloxy-ΔAla-OMe, with the tosyloxy group then displaced by 4-bromothiophenol (*384*, *386*).

β-Dimethylamino-ΔAla was treated with a library of aniline hydrochlorides in a parallel solution-phase synthesis, producing 22 different (Z)-β-(arylamino)-ΔAla derivatives in 62–100% yield (*388*); exposure to ammonium acetate gave unsubstituted β-amino-ΔAla (*196*). Addition of a diacyloxyiodobenzene-derived radical based on Cbz-Glu-OMe to Trt-ΔAla-OBn resulted in a 5,6-dehydro-2,6-diaminopimelic acid derivative (*389*).

4.3 Reactions of Didehydroamino Acids

The alkene functionality of didehydroamino acids can be used to synthesize other α-amino acids by three major methods: reduction of the alkene, Michael addition to the alkene, or cycloaddition reactions of the alkene. Since didehydroamino acids are prochiral, asymmetry can be induced during these reactions by using a chiral auxiliary, a chiral reagent, or, most importantly, a chiral catalyst. Some derivatizations of dehydroamino acids contained within chiral templates were reviewed in 2002 (*390*).

4.3.1 Racemic Reductions

There are relatively few recent reports of dehydroamino acid reductions leading to racemic products, as most interest now resides in asymmetric reductions. However, most of the dehydroamino acids prepared during the first half of the 1900s were reduced using sodium amalgam, leading to amino acids such as phenylalanine (*176*), tyrosine (*391*), and histidine (*178*) (see Table 4.4). The condensation adducts formed from hydantoin and aromatic aldehydes that were obtained by Wheeler and Hofmann were reduced with refluxing hydroiodic acid and red phosphorus. These conditions also opened the hydantoin ring and hydrolyzed the urea group to give Phe and Tyr in 71% and 89% yield (*338*). Reduction with aluminum amalgam was unsuccessful, but aqueous ammonium sulfide gave good yields of the saturated hydantoins, which were then hydrolyzed to the amino acids (*199*). Refluxing HI alone has been used in hydantoin-based syntheses of 2'-chloro-Phe (*200*), 3',5'-dichloro-Phe (*392*), and 4'-methoxy-Phe (*392*). Dehydroamino acid hydantoins have also been reduced using SnCl₂ in HCl (*198*, *200*, *201*, *208*, *209*, *211*) or sodium amalgam (*202*, *203*, *207*, *209*, *210*, *393*, *394*); barium hydroxide is generally used for the subsequent hydrolysis.

The HI/P reduction method still has utility; it was employed during a 1989 synthesis of 5,5,5-trifluoroleucine and 6,6,6-trifluoronorleucine from the corresponding dehydroamino acid azlactones (*183*). The same paper used hydrogenation in the synthesis of 4,5,6,7-tetrafluoro-Trp.

Table 4.4　Racemic Reductions of Dehydroamino Acids

Product Side Chain	Glycine Equivalent	Reduction Method	Yield	Reference
4-HO-Bn	azlactone	Na-Hg	—	1879 (*391*)
Bn	azlactone	Na-Hg	—	1893 (*176*)
Bn	hydantoin	HI/P	71–89%	1911 (*197*)
4-HO-Bn				1911 (*205*)
				1912 (*206*)
4-Me-Bn	azlactone	Na-Hg	—	1911 (*805*)
CH$_2$(1-naphthyl)	azlactone	Na-Hg	—	1911 (*806*)
CH$_2$(2-naphthyl)				
3,4-(OH)$_2$-Bn	azlactone	Na-Hg	—	1911 (*807*)
3,5-Cl$_2$-Bn	hydantoin	HI	—	1911 (*392*)
4-MeO-Bn				
3-Br-4-HO-Bn	hydantoin	SnCl$_2$/HCl	75%	1912 (*201*)
Bn	thiohydantoin	SnCl$_2$/HCl	quant.	1912 (*211*)
3-MeO-4-HO-Bn	hydantoin	Na-Hg or SnCl$_2$/HCl	—	1913 (*209*)
3,4-(MeO)$_2$-Bn				
CH$_2$(2-Me-indol-3-yl)	azlactone	Na-Hg	—	1914 (*808*)
2-MeO-Bn	hydantoin	Na-Hg	—	1915 (*207*)
2-HO-Bn				
2-HO-5-NH$_2$-Bn	hydantoin	SnCl$_2$/HCl	—	1915 (*208*)
(CH$_2$)$_2$Ph	hydantoin	Na-Hg	85%	1915 (*393*)
CH$_2$(imidazol-4-yl)	azlactone	Na-Hg	42%	1916 (*178*)
CH$_2$(2-Me-indol-3-yl)	azlactone	Na-Hg	60%	1917 (*809*)
CH$_2$(1-naphthyl)	Gly anhydride	HI/P	87%	1921 (*218*)
CH$_2$(2-furyl)	diketopiperazine	Na-Hg	—	1921 (*221*)
2,3,4-(MeO)$_3$-Bn	hydantoin	Na-Hg	66%	1924 (*202*)
2,3,4-(HO)$_3$-Bn				
CH$_2$(5-Me-indol-3-yl)	hydantoin	Na-Hg	—	1924 (*850*)
2,4-(HO)$_2$-Bn	Gly anhydride	HI/P	47%	1926 (*220*)
2,5-(HO)$_2$-Bn	Gly anhydride	HI/P	—	1927 (*219*)
Bn	azlactone	HI-Ac$_2$O/P	50–88%	1927 (*810*)
4-HO-Bn				
3,5-I$_2$-4-(4-HO-PhO)-Bn				
3,4-(HO)$_2$-Bn				
4-(4-HO-PhO)-Bn				
2,4,5-(MeO)$_3$-Bn	hydantoin	Na-Hg	—	1931 (*203*)
3,4-(OCH$_2$O)-Ph	hydantoin	Na-Hg	85%	1932 (*210*)
2-F-Bn	azlactone	HI/P	—	1932 (*811*)
3-F-Bn				
4-F-Bn				
CH$_2$(indol-3-yl)	hydantoin	ammonium sulfide	70%	1935 (*851*)
2,4-(OH)$_2$-Bn	azlactone	Na-Hg	—	1936 (*812*)
CH$_2$(2-thienyl)	azlactone	Na-Hg	—	1938 (*813*)
CH$_2$(6-MeO-indol-3-yl)	hydantoin	H$_2$S	50%	1938 (*853*)
*n*Hept	hydantoin	SnCl$_2$/HCl	95%	1939 (*198*)
2-Cl-Bn	hydantoin	HI or SnCl$_2$/HCl	75%	1940 (*200*)
3-NH$_2$-Bn				
CH$_2$OH with [^{15}N]	Bz-Gly-OEt	Al-Hg	—	1942 (*214*)
CH$_2$(3-pyridyl)	diketopiperazine	HI/P	—	1942 (*222*)
CH$_2$(indol-3-yl)	azlactone	H$_2$/Ra-Ni	—	1944 (*847*)
	hydantoin			
CH$_2$(indol-3-yl)	hydantoin	H$_2$/Ra-Ni	100%	1944 (*854*)
CH$_2$(1-Me-imidazol-4-yl)	thiohydantoin	HI/P	83%	1944 (*860*)
CH$_2$(quinolin-4-yl)	hydantoin	HI	80%	1945 (*204*)
CH$_2$(2-thienyl)	hydantoin	ammonium sulfide	—	1945 (*852*)
Bn	azlactone	HI/P	—	1948 (*177*)
2,5-(HO)$_2$-Bn	azlactone	HI/P	50%	1948 (*814*)
2,5-(MeO)$_2$-Bn	azlactone	H$_2$/Ra-Ni	100%	1948 (*814*)
4-Me$_2$N-Bn	azlactone	HI/P		1948 (*815*)

Table 4.4 Racemic Reductions of Dehydroamino Acids (continued)

Product Side Chain	Glycine Equivalent	Reduction Method	Yield	Reference
CH$_2$(indol-3-yl) with [1-^{13}C]	hydantoin	H$_2$/Ra-Ni	71%	1948 (855)
CH$_2$(imidazol-4-yl)	thiohydantoin	H$_2$/Ra-Ni	43%	1949 (212)
Et	thiohydantoin	HI/P	—	1949 (213)
iPr				
iBu				
nBu				
cHex				
Bn				
4-HO-Bn				
CH$_2$(3-benzothienyl)	azlactone	HI/P	—	1949 (816)
CH$_2$(imidazoly-3-yl)	hydantoin	Na-Hg	70–71%	1949 (394)
	N-Me-hydantoin			
3-Me-4-HO-Bn	azlactone	HI/P	—	1950 (817)
2,3-(HO)$_2$-Bn				
3-HO-Bn	azlactone	HI/P	—	1951 (818)
4-NH$_2$-Bn	azlactone	HI/P	—	1952 (819)
2,5-(HO)$_2$-Bn	azlactone	HI/P	64%	1953 (821)
2,3-(OH)$_2$-Bn	azlactone	HI/P	—	1954 (886)
2,4-(OH)$_2$-Bn				
2,5-(OH)$_2$-Bn				
2,6-(OH)$_2$-Bn				
3,5-(OH)$_2$-Bn				
3,4-(OH)$_2$-Bn	azlactone	H$_2$/Ra-Ni	—	1956 (396)
CH$_2$(ferrocenyl)	azlactone	H$_2$/PtO$_2$	—	1957 (823)
2,4-(HO)$_2$-6-Me-Bn	azlactone	HI/P	38%	1958 (822)
4-NMe$_2$-Bn	creatinine	H$_2$/Pd-C	—	1961 (224)
CH$_2$(4-HO-1-naphthyl)	thiohydantoin	Na-Hg	20%	1966 (861)
CH(Et)$_2$	azlactone	HI/P	10% overall	1966 (824)
CH$_2$(pyrid-2-yl)	azlactone	HI/P	62–72%	1967 (887)
CH$_2$(pyrid-3yl)				
CH$_2$(pyrid-4-yl)				
CH$_2$(2-furyl) with [2-^{14}C]	hydantoin	Na-Hg	47%	1968 (856)
CH$_2$C$_6$F$_5$	azlactone	HI/P	60%	1969 (825)
3,5-iPr$_2$-4-(3-iPr-4-MeO-PhO)-Bn	azlactone	H$_2$/Pd-C	50%	1969 (826)
CH$_2$(8-HO-quinolinol-5-yl)	azlactone	HI/P	84–96%	1969 (827)
	hydantoin			
2,4,5-(MeO)$_3$-Bn	hydantoin	H$_2$/Pd-C	98%	1969 (857)
2,4,5-(HO)$_3$-Bn				
3,4,6-(HO)$_3$-Bn	hydantoin	Na-Hg	80%	1971 (858)
CHDPh	azlactone	H$_2$/Pd-C	90%	1972 (829)
CHTPh				
iPr	azlactone	H$_2$/Ni/MeOH-NH$_3$	72–100%	1972 (395)
nPr				
sBu				
nBu				
iBu				
(CH$_2$)$_3$Ph				
Bn				
4-HO-Bn				
4-MeO-Bn				
3,4-(MeO)$_2$-Bn				
3-MeO-4-HO-Bn				
CH$_2$(2-furyl)				
CHDPh	azlactone	H$_2$/Pd-C	—	1973 (830)
CHTPh				
CHD(4-HO-Ph)				
CHT(4-HO-Ph)				
3,4,5-(MeO)$_3$-Bn	hydantoin	H$_2$/PtO$_2$	70%	1973 (859)

(Continued)

Table 4.4 Racemic Reductions of Dehydroamino Acids (continued)

Product Side Chain	Glycine Equivalent	Reduction Method	Yield	Reference
2,5-(MeO)$_2$-Bn 2,5-(MeO)$_2$-4-Me-Bn	azlactone	H$_2$/Pd-C	quant.	1974 (842)
3-CH$_2$OH-4-HO-Bn	azlactone	H$_2$/Pd-C	—	1974 (843)
3,4-(MeO)$_2$-Bn	azlactone	H$_2$/Pd-C	96%	1975 (831)
3-CO$_2$H-4-HO-Bn	azlactone	HI/P	—	1977 (888)
CH$_2$[Ph-Mn(CO)$_3$]	azlactone	HI/P	90%	1981 (832)
CH(Me)(4-Me-Ph) CH(Me)(4-Cl-Ph) CH(Me)(4-NH$_2$-Ph)	hydrolyzed azlactone	H$_2$/Pd-C	82–92%	1981 (397)
CH$_2$(9,10-dihydro-9-anthryl)	azlactone	HI/P	71%	1982 (848)
CHT(2-naphthyl)	azlactone	T$_2$/PdRh	—	1984 (844)
CH$_2$(3-pyridyl)	azlactone	H$_2$/Pd-C	79%	1984 (833)
CH$_2$(1-pyrenyl)	azlactone	H$_2$/Pd-C	—	1985 (834)
CH$_2$(1-pyrenyl)	azlactone	H$_2$/Pd-C	—	1987 (835)
CH$_2$CH(Me)CF$_3$ (CH$_2$)$_3$CF$_3$	azlactone	HI/P	76%	1989 (183)
CH$_2$(4,5,6,7-F$_4$-indol-3-yl)	azlactone	H$_2$/Pd	—	1989 (183)
3,4,5-(HO)$_3$-Bn	glycine phosphonate	H$_2$/Pd-C	—	1989 (43)
cHex	hydrolyzed azlactone	H$_2$/Ni	100%	1992 (838)
CH$_2$(2,2′ bipyridin-4-yl)	azlactone	HI/P	77%	1993 (190)
4-(CH$_2$PO$_3$tBu$_2$)-Bn 4-(CF$_2$PO$_3$tBu$_2$)-Bn 4-(CHF-PO$_3$tBu$_2$)-Bn 4-(CHOH-PO$_3$tBu$_2$)-Bn 4-(CO-PO$_3$tBu$_2$)-Bn	azidoacetate adduct	HI	—	1993 (227)
4-PhO-Bn	azlactone	H$_2$/Pd-C	96%	1993 (845)
CH$_2$(N-Et-carbazol-3-yl)	hydrolyzed azlactone	H$_2$/Pd-C	—	1994 (839)
CHDiPr CHDPh CHD(4-HO-Ph) CHD(imidazol-4-yl) CHD(indol-3-yl)	glycine phosphonate	D$_2$/Pd-C	100%	1995 (184)
3-NH$_2$-Bn	azlactone	H$_2$/Pd-C		1998 (186)
CH(Me)(4-HO-Ph)	hydrolyzed azlactone	H$_2$/Pd-C	—	1998 (191)
CH$_2$(3-pyrrolyl)	glycine phosphonate	NaBH$_4$/NiCl$_2$	—	1998 (889)
iBu	nitroacetate adduct	NaBH$_4$ then H$_2$/Ra-Ni	—	1999 (232)
CH$_2$(1-pyrenyl)	azlactone	H$_2$/Pd-C	50%	1999 (708)
4-B(OH)$_3$-Bn	glycine phosphonate	H$_2$/Pd-C	—	1999 (872)
2-MeO-4-B(OH)$_3$-Bn 2-MeO-5-B(OH)$_3$-Bn 2-nBuO-5-B(OH)$_3$-Bn	glycine phosphonate	H$_2$/Pd-C	95–99%	2004 (881)
CH(Me)Ph	hydrolyzed azlactone	H$_2$/Pd-C	96–97%	2004 (192)
CH$_2$(2-Ph-2-CO$_2$H-cPr)	glycine phosphonate	H$_2$/Pd-C	33%	2005 (885)

Reduction of azlactones by catalytic hydrogenation did not become common until a 1972 report demonstrated the superiority of reduction by hydrogenation over reduction with HI/P. Thirteen different amino acids were prepared by hydrogenation catalyzed by Ni in alcoholic NH$_3$, followed by hydrolysis of the resulting amide (395). DOPA was prepared in 1956 by an Erlenmeyer condensation with vanillin; H$_2$/Ra-Ni was employed for hydrogenation, while azlactone and O-Me hydrolysis was achieved with HI (396). Azlactone reduction using HI/P lacks stereospecificity, but by using alkaline hydrolysis of pure (E)- or (Z)-oxazolone isomers, the resulting

2-benzoylamino-3-aryl-2-butenoic acid derivatives could be stereospecifically hydrogenated over a Pd-C catalyst (397). This method was used to selectively prepare threo- and erythro-β-methyl-Tyr (191), and threo- and erythro-β-methyl-Phe (192). Hydrogenation has also been used for reducing the condensation adducts of α-azidoacetate and aryl aldehydes, resulting in simultaneous azide and alkene reduction (227).

In the original report of Schmidt et al. on the preparation of didehydroamino acids by Horner–Wadsworth–Emmons olefination reactions, the alkene was reduced and Cbz-protecting group removed by catalytic hydrogenation over Pd-C. Alternatively, hydrogenation with rhodium on charcoal left the Cbz group intact (295). A number of β-aryl-didehydroamino acids were hydrogenated with either catalyst in 28–97% yield (362), while similar conditions were used to catalytically deuterate (Z)-N-acetyldehydroamino acids, giving predominantly the threo-[2,3-^{2}H$_2$]-isomer of Leu, Phe, Tyr, His, and Trp in quantitative yield (184). Significant amounts of the erythro isomer (up to 15%) were detected under certain conditions. A racemic β-cyclohexylphosphonate amino acid has also been prepared via a simple hydrogenation (307).

The Erlenmeyer-like condensation products of ethyl nitroacetate and aldehydes can be reduced by a conjugate reduction of the alkene with sodium borohydride, followed by hydrogenation of the nitro group with H$_2$/Ra-Ni (232).

4.3.2 Asymmetric Reductions

4.3.2a Chiral Auxiliaries

One of the first asymmetric syntheses of any kind ever reported (in 1956) involved a catalytic reduction of 2,3-dehydro-p-nitrophenylalanine, using menthyl and bornyl esters to induce diastereoselectivity. The menthyl ester provided an asymmetric induction of 33% de, and allowed for easy resolution by crystallization (398, 399). Protected ΔDOPA menthyl ester was also hydrogenated, but with only 2% de (400). Reductions of α-acetamidoacrylic acids with L-α-methylbenzylamine as a chiral amide auxiliary gave D-Val (39% ee) and D-Phe (6% ee) (401). Much better results were obtained from didehydrodipeptides containing a proline amide C-terminal residue (proline methylamide was best). Hydrogenation gave good diastereoselectivity for the (Z)-isomers (80–94% de for seven examples). The (R)-amino acid was obtained from (S)-proline, with the Pro auxiliary removed by hydrolysis (189). In a similar report, the Mg(II) or Ca(II) complex of an N-acetyl dipeptide containing ΔPhe or ΔVal was hydrogenated; the metal complex in combination with Pro as a chiral auxiliary induced 66–80% de, compared to 38–48% de for reduction of the dipeptide with no metal present (402, 403). In contrast, the N-benzoyl dehydroamino acid dipeptide showed little change in diastereoselectivity upon complexation, although in some cases the N-Bz derivative gave better selectivity than the N-Ac analog (404). The unsaturated precursor of 2,3-diaminosuccinic acid, derivatized as a bis(N,N'-Boc-Ala) tripeptide, was hydrogenated with a [Rh(COD)Cl]$_2$ catalyst to give a single diastereomer (405).

Improved stereoselectivities have been obtained with dehydroamino acid substrates derivatized with chiral auxiliaries within a cyclic template. A ΔAsp-containing chiral heterocycle formed between diphenyl-1,2-ethanolamine and acetylene dicarboxylate was hydrogenated with complete diastereoselectivity if the erythro amino-alcohol was used, although with only 50% de when the

threo isomer was employed (406). A number of dehydroamino acids contained within the chiral imidazolidinone template of Seebach (see Section 7.9) have been reduced with >95% ds using H$_2$ and Pd/C (137, 407). Another chiral template containing dehydroamino acids embedded within a 3,6-dihydro-2H-1,4-oxazin-2-one structure was reduced with H$_2$/PtO$_2$ to give 63–95% yields and 82–96% de (252).

Dehydroamino acids contained in a cyclic 2,5-diketopiperazine template with L-Pro (see Scheme 4.14) were reduced with >90% diastereoselectivity (277). Phe or Ala as the chiral auxiliary gave much poorer results. This 1975 report was based on a 1944 publication in which the diketopiperazine of Ac-ΔPhe-L-proline was reduced to give an L-Phe-L-Pro derivative (408). Diketopiperazine templates containing dehydroamino acids have been prepared by reaction of N,N'-Boc$_2$ Val-Gly diketopiperazine with aldehydes in the presence of KOtBu, with concomitant removal of the Boc group attached to the dehydroamino acid residue. Catalytic deuteration of the completely N-deprotected template produced the threo-[2,3-^{2}H$_2$] isomers of Phe, Tyr, and DOPA and the threo-[2,3,4-^{2}H$_3$] isomer of Leu with 88–94% de. The erythro isomer could be obtained with 80–94% de by reduction of the N,N-Boc$_2$-protected template (409, 410). An early version of this reaction prepared (2R,3R)-[2,3-^{2}H$_2$]-Phe via deuteration of ΔPhe within a diketopiperazine formed with D-Ala (411). In 2004 an N,N'-bis(PMB) diketopiperazine of L-Val-ΔPhe was reduced using SmI$_2$/THF/H$_2$O in 89–93%, with both the (Z)- and (E)-ΔPhe leading to L-Phe with 95–96% de. By using SmI$_2$ in D$_2$O, both isomers gave predominantly (2S,3R)-[2,3-^{2}H$_2$]-Phe with 84% ds of the major isomer. The protecting groups were removed with ceric ammonium nitrate and the template hydrolyzed with concentrated HCl. A similar synthesis provided (2S,3R)- or (2S,3S)-[2,3-^{2}H$_2$]-Leu with reduced stereoselectivity (58–70% ds), though after template hydrolysis the [2,3-^{2}H$_2$]-Leu and unlabeled Val products were not separated, but instead derivatized into dipeptides (223, 412). An efficient asymmetric synthesis of β-(1-pyrenyl)alanine (Pya) in 2001 employed the condensation of an N,N'-bisacetyl Ala-Gly diketopiperazine (1,4-diacetyl-3-methyl-piperazine-2,5-dione) with pyrene aldehyde in the presence of KOtBu in THF. Formation of the dehydroamino acid was accompanied by deacetylation of one acetyl group, with the second removed by hydrazine. Hydrogenation over Pd-C gave a 51% yield (after recrystallization) of pure c(L-Pya-L-Ala) diketopiperazine. Hydrolysis with 6 N HCl and N-Boc protection gave the desired amino acid after chromatography (413).

4.3.2b Chiral Substrates

Asymmetry can be induced at the α-center during hydrogenolytic reductions of dehydroamino acids that contain optically active side-chain substituents. This procedure is generally used with sugar-derived side chains, with diastereoselectivities varying from

0 to 88% de (*181, 182, 245, 246, 322*). A series of dehydroamino acids with optically active substituted cyclobutane side chains were reduced using Wilkinson's catalyst, with 50–60% de for a β-cyclobutyl-Ala substrate, but 0% de for a γ-cyclobutyl-Abu substrate. Employing a chiral catalyst improved stereoselectivity to 66–90% de for the β-substituted derivative, and >96% de for the γ-substituted (*414*).

4.3.2c Chiral Catalysts

Asymmetric catalytic hydrogenation (see Scheme 4.31) is one of the first asymmetric synthetic techniques developed, and dehydroamino acids have been substrates for asymmetric reduction since the first experiments began. α-Acylaminoacrylic acid substrates have consistently given some of the best enantioselectivities of all olefin substrates as they coordinate comparatively strongly with the chiral rhodium complexes via chelation, and also contain an electron-withdrawing carboxyl group that increases reactivity (*415, 416*). With most catalysts, higher H_2 pressures reduce optical yield (*416*).

The development of enantioselective hydrogenation catalysts was made possible by two developments in the 1960s: (1) the appearance of Wilkinson's homogeneous rhodium catalyst, [RhCl(PPh$_3$)$_3$], and (2) the discovery of methods to prepare optically active phosphines that could coordinate the catalytic metal. The evolution of the first synthetically useful asymmetric hydrogenations and their application to the commercial production of L-DOPA was outlined in the 2001 Nobel prize lecture of William S. Knowles (*417*). Most rhodium complexes containing chiral ligands are derivatives of the achiral 1,2-bis(diphenylphosphino)ethane (DIPHOS) ligand complex (see Scheme 4.31). A great number of catalysts have been developed (see Scheme 4.31, Table 4.5), but synthetically useful results for a range of substrates were not obtained until relatively recently with the introduction of the DuPHOS (see *418* for brief review) and BPE ligands. The optimum catalyst is often very dependent on the substrate substituents. It has been demonstrated that 5 mol % [Rh(cod)Cl]$_2$ and 50 mol % P(cHex)$_3$ can induce significant racemization of *N*-acetyl amino acids, so there is a possibility that hydrogenation catalysts may induce racemization of optically pure products over time (*419*).

The mechanism and stereoselectivity of early asymmetric catalysts was summarized in 1982 (*415*); more recent studies of catalytic mechanism include a multinuclear NMR examination of reductions by the Rh

complex of (*S,S*)-1,2-bis(*tert*-butylmethylphosphino)ethane (*420*), detection of a transient rhodium dihydride intermediate during hydrogenations with the PHANEPHOS ligand (*421*), and computational modeling of the reaction pathway and reasons for enantioselection with the DuPHOS ligand (*422*). A 2000 article provided a simple model to explain asymmetric hydrogenation of enamides by chiral diphosphine Rh(I) complexes (*423*), while a 2004 *Accounts of Chemical Research* summary reviewed the mechanism of hydrogenation stereoselection and described a general approach for predicting the sense of enantioselectivity provided by a variety of Rh-ligand catalysts (*424*). A 2005 article compared the mechanism for asymmetric hydrogenation of β-acylaminoacrylates using cationic Rh(I) complexes to that of α-acylaminoacrylates (*637*). Several more general reviews of enantioselective catalytic hydrogenation have appeared (*415, 416, 425, 426*), including a 2003 account of new catalyst/ligand developments (*427*), a 2003 *Chemical Reviews* summary of new chiral phosphorus ligands for asymmetric hydrogenation (*428*), a 2003 Chemical Reviews article on the use of combinatorial libraries of chiral ligands for enantioselective catalysis (*429*), two 2004 perspectives covering the design of chiral ligands for asymmetric catalysis (*430*) and mechanistic understanding for improving asymmetric hydrogenation (*431*), a 2004 review of chiral monodentate monophosphorus ligands in asymmetric hydrogenation (*432*), and a 2005 summary of new developments in biaryl-type phosphine ligands (*433*). A 2004 issue of *Tetrahedron: Asymmetry* was devoted to chiral phosphorus ligands (*434*).

The asymmetry of the catalyst phosphine ligands can reside on the phosphine atom, on the phosphine substituents, on a chiral backbone connecting two phosphine atoms, or on some combination of the three (see Scheme 4.31).

4.3.2ci Ligands with Chiral Phosphorus Center

The first ligands to contain chiral centers at the phosphorus atom were monodendate (*435, 436*). However, much better results were obtained with bidentate biphosphine ligands, of which the most successful was a C$_2$-symmetrical compound, (*R,R*)-1,2-ethanediylbis [(*o*-methoxyphenyl)phenylphosphine] (DIPAMP). The development of the DIPAMP ligand was outlined in the 2001 Nobel prize lecture of William S. Knowles (*417*). DIPAMP was used in a rhodium complex to hydrogenate several (*Z*)-dehydroamino acid derivatives with good selectivity (89–95% ee) in a 1975 report (*437*).

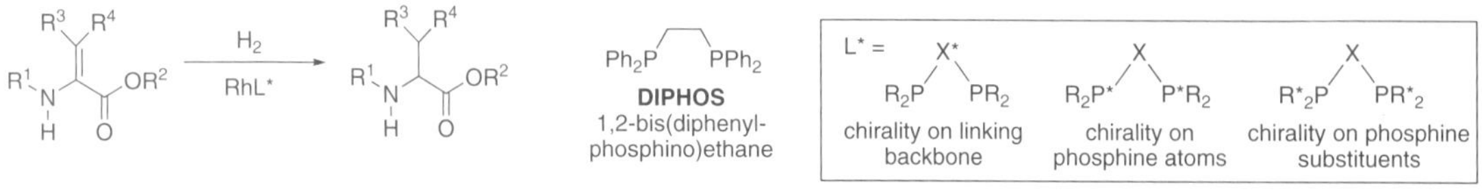

Scheme 4.31 Phosphine Ligands for Rhodium Complex Asymmetric Hydrogenation Catalysts.

Table 4.5 Asymmetric Reductions of Didehydroamino Acids with Chiral Catalysts

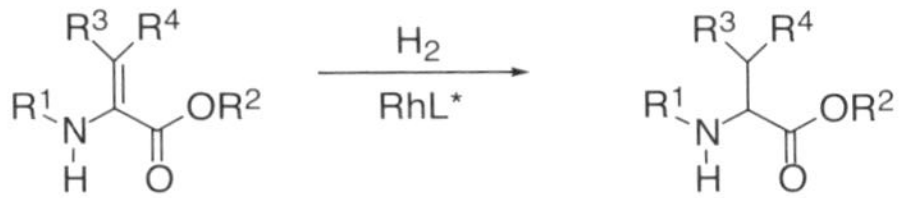

$$\underset{R^1-N}{\overset{R^3}{\underset{H}{\bigg|}}}\overset{R^4}{=}\overset{O}{\underset{OR^2}{\bigg|}} \xrightarrow[\text{RhL*}]{H_2} \underset{R^1-N}{\overset{R^3\ R^4}{\underset{H}{\bigg|}}}\overset{O}{\underset{OR^2}{\bigg|}}$$

Ligand	R¹	R²	R³	R⁴	Enantio-selectivity	Reference
Ligands Based on Chiral Phosphine						
MeO—(structure) P···Ph ; Ph—P··· —OMe **DIPAMP** 1,2-bis[2-(2-methoxyphenyl)phenyl]phos-phanyl]ethane	Ac	H	Ph	H	93.5–95.5% ee	1975 (437)
	Ac	H	3-MeO-4-AcO-Ph	H	89–95% ee	1977 (438)
	Bn	Me	1-Ac-indol-3-yl			
	CO-CH₂Cl	Et	H			
	CO₂Et		iPr			
			Ph			
	Ac	Me	nPr	H	65–96% ee	1981 (261)
			iPr			
			CH₂OMe			
			Ph			
	Cbz	tBu	3-MeO-4-AcO-Ph bis(amino acid) linked by 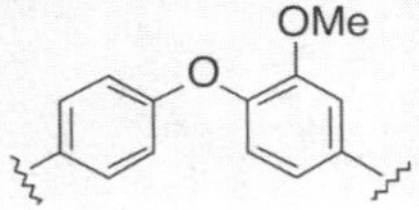 OMe (structure)	H	>98% ee	1988 (317)
	Ac	Me	CH₂OMe	H	95% ee	1989 (890)
	Ac	Me	3-pyridyl	H	—	2001 (364)
	Bz	Me	C₆F₅	H	80% ee	1991 (836)
	Boc	Me	(dioxolane structure)	H	94–>99% de	1991 (865)
	Cbz	Bn	BnO (dioxolane structure)	H		1991 (439)
	Ac	Me	3-pyridyl 4-pyridyl	H	86–>99% ee	1991 (363)
	Boc	TMSE	CH(OH)CH₂CH₂OTs	H	47% yield	1992 (866)

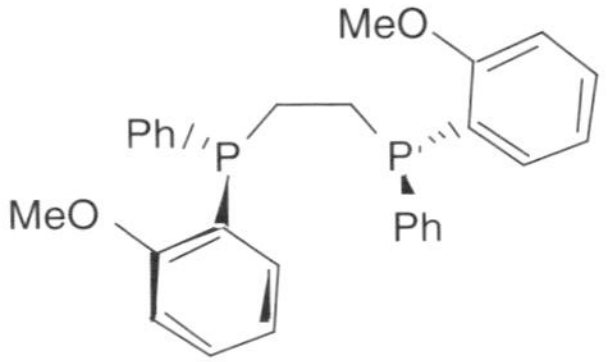

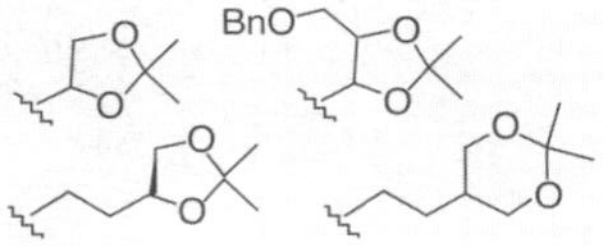

(Continued)

267

Table 4.5 Asymmetric Reductions of Didehydroamino Acids with Chiral Catalysts (continued)

Ligand	R¹	R²	R³	R⁴	Enantio-selectivity	Reference
	Cbz	Me	2-thienyl / 3-thienyl / 2-furyl / 3-furyl	H	85->99% ee	1994 (298)
	Xaa	Xaa	Ph	H	up to 88% ee	1994 (441)
	Ac	Me	CH₂(6,6-Me₂-1,5-dioxan-6-yl)	H	97% ee	1996 (306)
	Cbz	Me	bis(amino acid)	H	99% ee	1996 (312)
	Ac	OR² = Phe-OMe	indol-2-yl	H	82% ee	1997 (360)
	peptide	peptide	indol-3-yl	H	99% ee	1998 (359)
	peptide	peptide	indol-3-yl	H	99% ee if adjacent Nle / 0–26% ee if adjacent Met	
	Boc	Bn	bis(amino acid) linked by 1,3-Ph or 1,4-Ph or 4,4'-biphenyl	H	>98% ee	1998 (367)
	Ac	Me	H	H	92% ee	2004 (443)
DiPAMP on mesoporous aluminasilicate support	Ac	Me	H / Ph / 3-MeO-4-AcO-Ph	H	86–90% ee	1995 (445)
1,2-bis[2-(2-ethylphenyl)phenylphospha-nyl]ethane	Ac / Bz	H / Me	H / Ph / 3-MeO-4-AcO-Ph	H	91–99.9% ee	1998 (448)
R = tBu, CEt₃, 1-adamantyl, cPent, cHex BisP*	Ac / Bz	H / Me	Me / (CH₂)₅-R³	Me / (CH₂)₅-R⁴	91–99.9% ee	1998 (448)

Structure					Yield / ee	Year (Ref.)
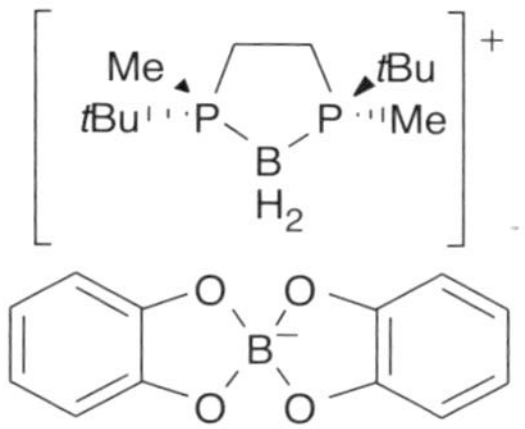 [Me, tBu, P⁺/P, B, H₂ catecholborane structure]	Ac	Me	Ph	H	14–99% yield; 83–94% ee	2004 (454)
BisP* with non-symmetrical R on P	Ac	Me	H; Ph; Me; (CH₂)₄-R⁴; (CH₂)₅-R⁴	H; H; Me; (CH₂)₄-R³; (CH₂)₅-R³	87.6–99.9% ee	2001 (451); 2002 (452)
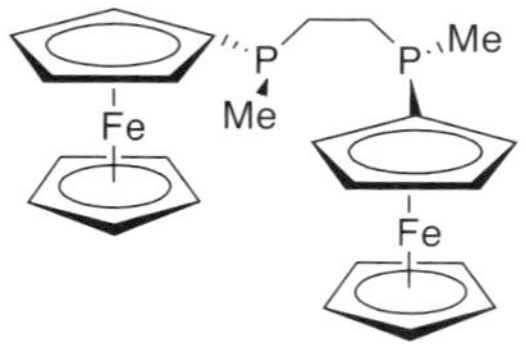 [ferrocene diphosphine structure]	Ac	H; Me	H; Ph; Me; (CH₂)₄-R⁴	H; H; Me; (CH₂)₄-R³; (CH₂)₅-R³	28–77% ee	2003 (449)
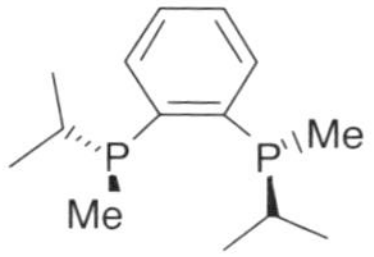 [o-phenylene diphosphine structure]	Ac	Me	H; Ph; 3-MeO-4-AcO-Ph	H	77–97% ee	1999 (446)
	Ac	Me	(CH₂)₄-R⁴; (CH₂)₅-R⁴	Me; (CH₂)₄-R³; (CH₂)₅-R³	77–97% ee	1999 (446)
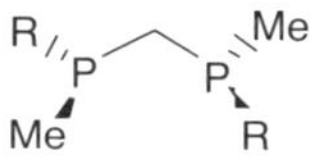 [R, P, Me structure]	Ac	H; Me	H; Ph; 3-MeO-4-AcO-Ph	H	85–>99.9% ee	1999 (453)
MiniPHOS R = tBu, cHex, Ph	Ac	H; Me	Me; (CH₂)₄-R⁴; (CH₂)₅-R⁴	Me; (CH₂)₄-R³; (CH₂)₅-R³	85–>99.9% ee	1999 (453)
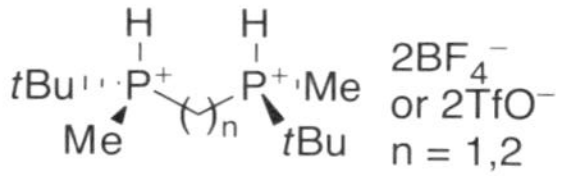 [tBu, Me, P⁺, 2BF₄⁻ or 2TfO⁻, n = 1,2 structure]	Ac	H	Ph	Me	>99% ee	2003 (455)

(Continued)

269

Table 4.5 Asymmetric Reductions of Didehydroamino Acids with Chiral Catalysts (continued)

Ligand	R¹	R²	R³	R⁴	Enantio-selectivity	Reference
	Ac	H	Ph 3,4-(OCH$_2$O)-Ph 4-AcO-Ph	H	91–96% ee	1987 (444)
	Ac	H	Ph	H	47% ee	1987 (444)
	Ac	H	Ph 3,4-(OCH$_2$O)-Ph 4-AcO-Ph 4-HO-Ph	H	79–90% ee	1987 (444)
 QuinoxP*	Ac	Me	Ph 3-MeO-4-AcO-Ph	H	99.6–99.9% ee	2005 (456)

Ligands Based on Chiral Backbone Connecting Phosphines

Ligand	R¹	R²	R³	R⁴	Enantio-selectivity	Reference
 DIOP 1-bis(diphenylphosphinomethyl)-2,2- dimethyl-1,3-dioxolane	Ac Bz	H	Me Ph	H	68–72% ee	1971 (458)
	Ac	H	H iPr Ph 4-HO-Ph 3,4-OCH$_2$O-Ph	H	22–80% ee	1972 (459)

Ac	Me	H iPr nPr CH$_2$OMe Ph	H	18–56% ee	1981 (*261*)
Ac	Me	3-pyridyl 4-pyridyl	H	70% ee	1991 (*363*)
Ac Bz	H Me	ferrocene	H	up to 85% ee	1993 (*185*)
CHO					
Ac Bz	H Me	2-thienyl 3-thienyl	H	20–77% ee	1993 (*188*)
Ac Bz Cbz Boc	Me	Ph	H	68% ee	1993 (*867*)
Ac Ac	OR^2 = Phe-OMe H	indol-3-yl Ph	H H	52% ee 52–86% ee	1997 (*360*) 1976 (*460*)
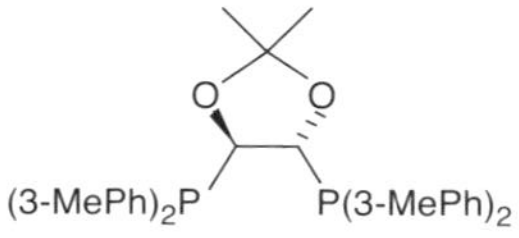 Ac	H	H nPr CH$_2$OMe	H	36–72% ee	1981 (*261*)
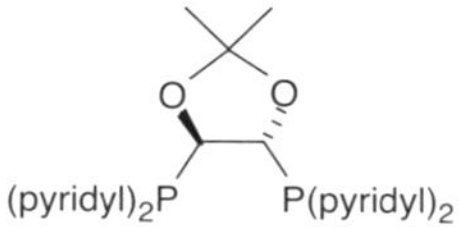 Ac	H Me	Ph 4-Me-Ph 4-Br-Ph 4-Cl-Ph 4-MeO-Ph 2-furyl	H	28–83% ee	1998 (*462*)
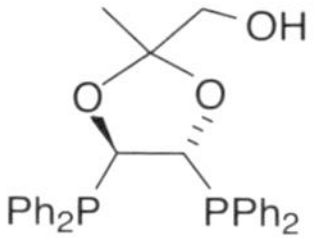 Ac	H Me	Ph	H	72–77% ee	1998 (*461*)
 Ac	H	H nPr iPr CH$_2$OMe	H	12–86% ee	1981 (*261*)

DIOP on polymer

(3-MePh)$_2$P / P(3-MePh)$_2$

3-CH$_3$DIOP

(pyridyl)$_2$P / P(pyridyl)$_2$

PyDIOP

OH

Ph$_2$P / PPh$_2$

HydroxyDIOP

Ph$_2$P / PPh$_2$

C$_4$DIOP

(Continued)

271

Table 4.5 Asymmetric Reductions of Didehydroamino Acids with Chiral Catalysts (continued)

Ligand	R¹	R²	R³	R⁴	Enantioselectivity	Reference
Ph₂P—PPh₂ **PROPHOS** 1,2-bis-(diphenyl-phosphino)-propane	Ac	Me	H nPr iPr CH_2OMe	H	53–88% ee	1981 (261)
	Boc	Et Bn	bis(amino acid) linked by 1,1'-ferrocene	H	84% ee 24% de	1990 (368)
	Ac Bz CHO	H Me	ferrocene	H	up to 88% ee	1993 (185)
	Ac	Me	$(CH_2)_2NPhth$	H	88% ee	1993 (303)
	Ac Bz	H	H iPr Ph 4-HO-Ph 3-MeO-4-AcO-Ph	H	83–>99% ee	1977 (467)
Ph₂P, PPh₂ **CHIRAPHOS** 2,3-bis(diphenylphosphino)butane	Ac Bz	H	H Ph 3-MeO-4-AcO-Ph	H	79–>99% ee	1980 (504)
	Ac	Me	H nPr iPr CH_2OMe	H	22–87% ee	1981 (261)
	Ac	Me	CH_2OMe	H	84% ee	1989 (890)
	Ac	H	$CH_2P(=O)(Me)OH$	H	91% ee	1991 (468)
	Boc	Me	(Cbz-oxazolidinone structure)	H	50% ee 80% yield	1994 (868)
	Ac	Me	(2,2-Me₂-3-R-cyclobutane) R = COMe, CO₂Me, CH₂OBn CH₃(2,2-Me₂-3-COMe-cyclobutane)	H	66–90% de	2001 (414)
	Ac	H Me	Ph	H	78–94.4% ee	2004 (469)
PPh₂ PPh₂ **BDPP (skewPHOS)**	Ac	Me	Ph	H	75–76% ee	2005 (470)

Ligand	Acyl	R¹	R²	ee	Year (Ref.)
(structure) PAr_2 PAr_2; Ar = 3,5-Me$_2$-Ph, 4-MeO-Ph, 3,5-Me$_2$-4-MeO-Ph	Ac	H	Ph	82–98% ee	2004 (469)
		Me	4-MeO-Ph		
(structure) R–PPh_2 PPh_2; R = Ph, cHex	Ac	Me	H	R = Ph: 20–25% ee	2003 (471)
			Ph	R = cHex: 53–60% ee	2005 (470)
(structure, norbornene) Ph_2P PPh_2 **NORPHOS**	Ac	Me	nPr	0–79% ee	1981 (261)
	Ac	H	$CH_2P(O)(Me)OH$	91% ee	1991 (468)
	Ac	H	ferrocene	up to 95% ee	1993 (185)
	Bz	Me			
	CHO				
2,3-bis(diphenylphosphino)-bicyclo[2.2.1]hex-5-ene (structure) Ph_2P PPh_2	Ac	H	H	8–78% ee	1981 (261)
			nPr		
			iPr		
			CH_2OMe		
BCOP (structures) Ph_2P ... PPh_2	Ac	H	Ph	7–90% ee	2002 (463)
		Me			

(Continued)

Table 4.5 Asymmetric Reductions of Didehydroamino Acids with Chiral Catalysts (continued)

Ligand	R¹	R²	R³	R⁴	Enantio-selectivity	Reference
Ph₂P⸍ ⟍N–R ⟍PPh₂	Ac	H	Ph	H	91–97.5% ee	1986 (474)
		Me				
R = H, Bn, Bz, CHO, Me, Boc, MEM						
PyrPHOS						
3,4-bis(diphenyl-phosphino)-pyrrolidine						
R = Bz						
BenzoylpyrPHOS	Ac	H	Me	Me	15–100% ee	1986 (474)
	Bz	Me	H	H		
			Ph	H		
			4-HO-Ph	H		
			4-MeO-Ph	H		
			3-MeO-4-HO-Ph	H		
			indol-3-yl	H		
BenzoylpyrPHOS on polymer	Ac	H	Ph	H	95% ee	1993 (188)
pyrPHOS on PEG polymer	Ac	H	H	H	90–96% ee	2001 (476)
pyrPHOS attached to dendrimer	Ac	H	Ph	H	88–93% ee	2002 (478)
pyrPHOS attached to dendrimer	Ac	H	Ph	H	98.7% ee	2004 (477)
	Ac	Et	CO₂Et	Me	58% ee	1978 (891)
	Ac	Me	H	H	37–95% ee	1981 (261)
			nPr			
			iPr			
			CH₂OMe			
			Ph			
BPPM	Ac	H	Ph	H	87–93% ee	1981 (473)
			4-AcO-Ph			
			4-HO-Ph			
			3-MeO-4-AcO-Ph			
(tBuO–C(=O)–N pyrrolidine, Ph₂P / PPh₂)	Ac	H	2-naphthyl	H	>96% ee / 83% yield	1986 (472)
	Ac	Me	Ph	H	82–93% ee	1993 (867)
	Bz					
	Cbz					
	Boc					

Ac	OR² = Phe-OMe	indol-3-yl	H	72% de	1997 (360)
Bz or Boc	OR² = D- or L-Phe-OMe	Ph, 4-F-Ph, 4-Me-Ph, 4-CF$_3$-Ph	H	59.6–91.8% de	1999 (892)
Bz or Boc	OR² = D- or L-Phe-OMe	2-F-Ph	H	96–98% de	1999 (840), 2002 (841)
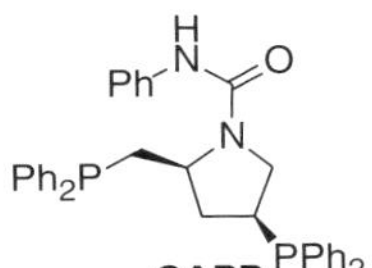 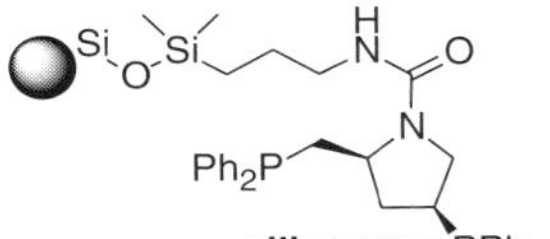 Ac	Me	Ph	H	90–94% ee	2004 (480)
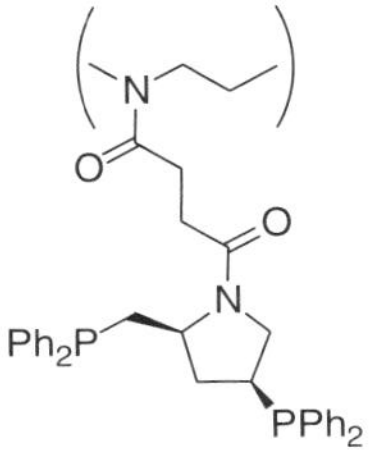 Ac	Me	Ph	H	88% ee	2003 (479)
H	H	nPr	 H	18% ee	1981 (261)
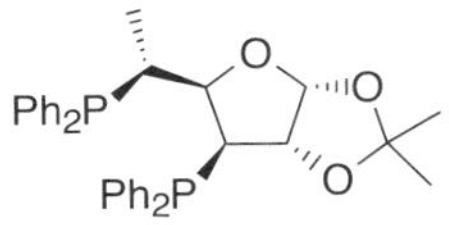 Ac	Me	H, Ph	H	98% ee	2000 (481)

CAPP

silica-ppm

NOPAPHOS

(Continued)

275

Table 4.5 Asymmetric Reductions of Didehydroamino Acids with Chiral Catalysts (continued)

Ligand	R^1	R^2	R^3	R^4	Enantio-selectivity	Reference
	Ac	H	nPr	H	42% ee	1981 (261)
PELLANPHOS	Ac	H Me	H Ph	H	11–57% ee	2002 (464)
	Ac	H	Ph	H	99.6% ee	2005 (465)
	Ac	Me	H Ph	H	79–84% ee	2004 (466)

Ligands Based on Chiral Backbone Connecting Phosphines on Heteroatoms

Ligand	R^1	R^2	R^3	R^4	Enantio-selectivity	Reference
DIMOP	Ac	H Me	Me Ph 2-Cl-Ph 3-Cl-Ph 4-Cl-Ph 2-MeO-Ph 3,4-(OCH$_2$O)-Ph 4-HO-Ph	H	90–97% ee	1999 (493)

276

Ligand					ee	Year
	Ac	Me	4-MeO-Ph 4-Me-Ph 4-NO₂-Ph 3-MeO-4-HO-Ph 4-Br-Ph 4-F-Ph	H	8–78% ee	2004 (494)
R = Ph, 4-F-Ph, 4-CF₃-Ph, 4-MeO-Ph, 3,5-Me₂-4-MeO-Ph, 2-thienyl, cHex	Ac	H	H Ph	H	68–91% ee	1999 (484)
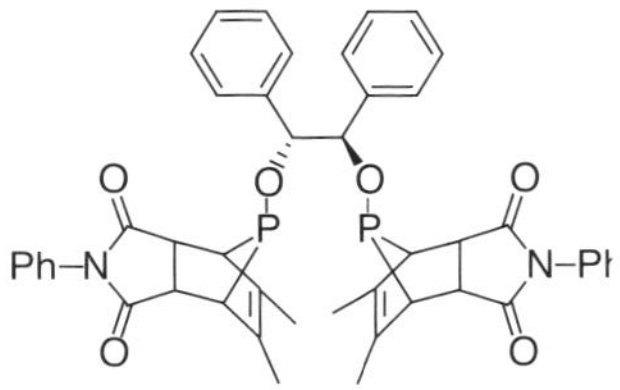	Ac	H	Ph	H	94% ee	2003 (488)
R = Me, –(CH₂)₄–, –(CH₂)₅–	Ac	H	H Ph 4-MeO-Ph	H	41–97% ee	2001 (490)
	Ac	H	H 4-MeO-Ph	H	68–98% ee	2001 (490)

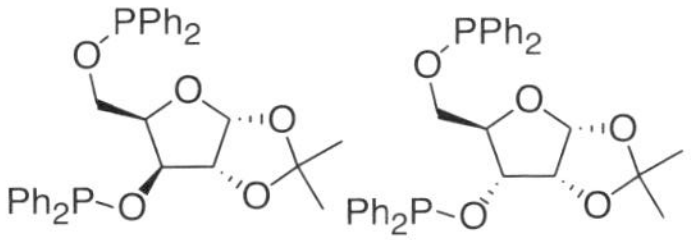

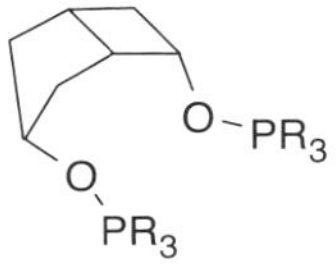

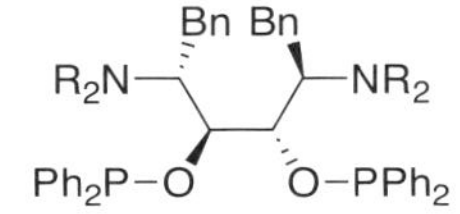

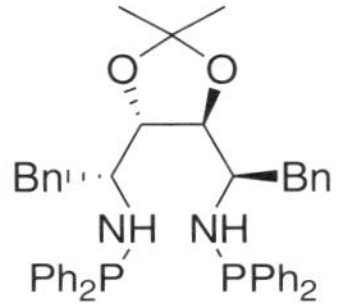

(Continued)

277

Table 4.5 Asymmetric Reductions of Didehydroamino Acids with Chiral Catalysts (continued)

Ligand	R¹	R²	R³	R⁴	Enantio-selectivity	Reference
DPAMPP	Ac Bz	H Me	Ph 4-Cl-Ph 4-F-Ph 4-Br-Ph 4-NO$_2$-Ph 4-Me-Ph 4-MeO-Ph 4-AcO-Ph 3-MeO-4-AcO-Ph	H	95.6–98.1% ee	1998 (485)
	Ac	Et	Bn	H	95.7 % ee	2000 (487)
	Ac Bz	Me Et	Ph 4-Cl-Ph 3-Cl-Ph 2-Cl-Ph 4-F-Ph 4-Br-Ph 4-NO$_2$-Ph 4-Me-Ph 4-MeO-Ph 4-AcO-Ph 2-OH-Ph 3-MeO-4-AcO-Ph 2-Cl-3-AcO-4-MeO-Ph 3,4-OCH$_2$O-Ph 2-furanyl H Bn	H	93.1–98.3 % ee	2000 (486)
PROPRAPHOS *N-iPr-O,N-bis(diphenylphosphino)-1-(1-naphthyl)-3-amino-2-butanol*	Bz	H Me	2-F-Ph 3-F-Ph 4-F-Ph 4-CF$_3$-Ph 2,3,4,5,6-F$_5$-Ph	H	86–91% ee	1992 (837)
	Ac Bz Cbz Boc	Me	Ph	H	86–93% ee	1993 (867)

Bz	H	2-Me-Ph	H	63–92% ee	1993 (482)
	Me	4-Cl-Ph			
		4-NO$_2$-Ph			
		4-CN-Ph			
		4-MeO-Ph			
		4-NMe$_2$-Ph			
		4-*i*Pr-Ph			
		4-Me-Ph			
		2,4-Me$_2$-Ph			
		2,4,6-Me$_3$-Ph			
		1-naphthyl			
		2-naphthyl			
		9-anthracenyl			
		9-phenanthryl			
Ac	H	2-thienyl	H	83–90% ee	1993 (188)
Bz	Me	3-thienyl			
Ac	H	3-pyridyl	H	70–90% ee	1996 (187)
Bz	Me	4-pyridyl			
BocCbz	H	2-thienyl	H	60–93% ee	1999 (873)
	Me	3-thienyl			
		5-Me-2-furyl			
Ac	H	2-thienyl	H	65–80% ee	1993 (188)
Bz	Me	3-thienyl			
Ac	H	3-pyridyl	H	74–86% ee	1996 (187)
Bz	Me	4-pyridyl			
Ac	Me	Ph	H	42–90% ee	2000 (483)
		2-Cl-Ph			
		3-Cl-Ph			
		4-Cl-Ph			
		4-F-Ph			
		4-AcO-Ph			
		3-Me-4-AcO-Ph			
		4-MeO-Ph			
		4-NO$_2$-Ph			
		4-Me-Ph			
		2-furyl			
		3,4-(OCH$_2$O)-Ph			

O,N-bis(diphenylphosphino)-2-*exo*-hydroxy-3-*endo*-methylamino-norbornane

(Continued)

Table 4.5 Asymmetric Reductions of Didehydroamino Acids with Chiral Catalysts (continued)

Ligand	R^1	R^2	R^3	R^4	Enantio-selectivity	Reference
Ph—O / O / OPh / (Ar)₂P (Ar)₂P	Ac, Cbz	H, Me	Ph, 3-MeO-Ph	H	97–99% ee	1994 (495)
	Ac, Cbz	H, Me	H, iPr, Ph, 3-MeO-Ph, 4-F-Ph, 4-Br-Ph, 2-F-Ph, 3-F-Ph, 3-Br-Ph, 3,5-(CF₃)₂-Ph, 2-naphthyl, 2-thienyl, 3-thienyl		91–99% ee	1997 (497)
Ar = Ph, 3,5-Me₂-Ph, 3,5-F₂-Ph, 3,5-(CF₃)₂-Ph, with Ar = 3,5-Me₂-Ph obtain L-amino acids						
OBz / BzO OMe / (Ar)₂P—O—(Ar)₂P	Ac	H, Me	3-Br-Ph, 3,5-F₂-Ph, 2-naphthyl, 2-thienyl	H	96–97% ee	1994 (495)
Ar = Ph, 3,5-Me₂-Ph, 3,5-F₂-Ph, 3,5-(CF₃)₂-Ph, with Ar = 3,5-Me₂-Ph obtain D-amino acids						
HO / OH / OH / OH / PPh₂ / Ph₂P—O (sugar diphosphine)	Ac	Me	Ph, 4-MeO-Ph, 2-naphthyl	H	80–98% ee	1999 (498)
	Ac	Me	Ph, 2-Cl-Ph, 3-Cl-Ph, 4-Cl-Ph, 4-NO₂-Ph, 4-MeO-Ph, 3,4-(MeO)₂-Ph, 1-naphthyl, 2-naphthyl	H	94->99% ee	1999 (499)

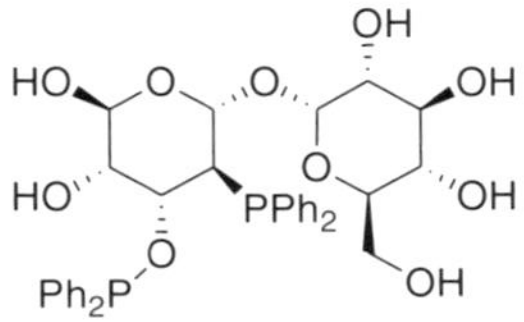

Ligand				ee	Year (Ref)
	Ac	H	H	10–73% ee	2002 (500)
		Me	Ph		
			4-MeO-Ph		
			3-Cl-Ph		
			2-naphthyl		
	Ac	H	H	46–96% ee	1998 (491)
	Bz		iPr		
			Ph		
			3-Br-Ph		
			2-Cl-Ph		
			4-F-Ph		
			4-MeO-Ph		
			3-MeO-4-AcO-Ph		
			2-naphthyl		
			2-thienyl		
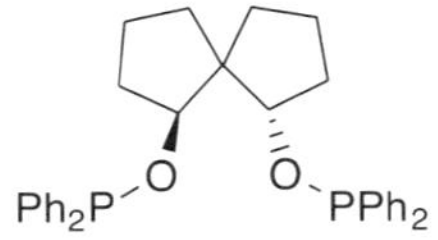 **spirOP**	Ac	H	Bn	96.2% ee	1999 (275)
	Ac	H	H	78->99% ee	2004 (492)
		Me	Ph		
			2-Cl-Ph		
			3-Cl-Ph		
			4-Cl-Ph		
			4-Br-Ph		
			4-F-Ph		
			4-MeO-Ph		
			4-Me-Ph		
			4-NO$_2$-Ph		
			3,4-(OCH$_2$O-)-Ph		
			2-furyl		
SpiroNP	Ac	Me	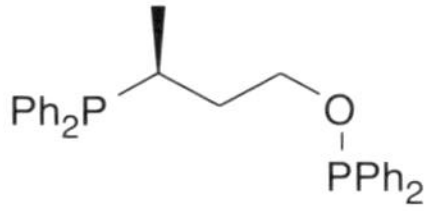 H	90.7–90.8% ee	2005 (489)
			Ph		

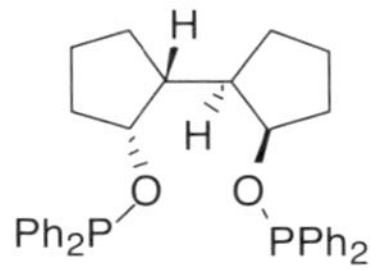

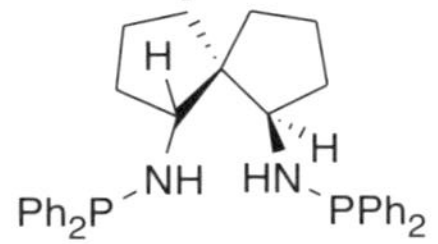

(Continued)

281

Table 4.5 Asymmetric Reductions of Didehydroamino Acids with Chiral Catalysts (continued)

Ligands Based on Biaryl, Paracyclophane or Ferrocene Chiral Backbones

Ligand	R^1	R^2	R^3	R^4	Enantio-selectivity	Reference
BINAP (Ph_2P, PPh_2; 2,2'-bis(diphenylphosphino)-1,1'-binaphthyl)	Ac	H	H	H	70–100% ee	1984 (505)
	Bz	Me	Ph			
			3-MeO-4-HO-Ph			
	Ac	Et	$(CH_2)_2CO_2Me$ $(CH_2)_3CO_2Me$ $(CH_2)_4CO_2Me$	H	70–96% ee	1994 (299)
	Cbz	Me	3-furyl	H	12% ee	1994 (298)
	$(CH_2)_2N(Boc)$-R^3 OR² = NHtBu		N(Boc) $(CH_2)_2$-R^1	H	99% ee	1995 (893)
	Cbz	Me	bis(amino acid) linked by 2,6-pyridine	H	30% ee	1996 (312)
diguanidinium- or PEG-BINAP	Ac	H	H	H	94–95% ee	2001 (510)
(Ar_2P, PAr_2; Ar = 4-($CH_2C_7F_{15}$)-Ph)	Ac	Me	Ph		15% ee	2003 (513)
(steroid-fused BINAP; Ph_2P, PPh_2)	Ac	H	H	H	87–97% ee	2000 (509)
			Ph			

282

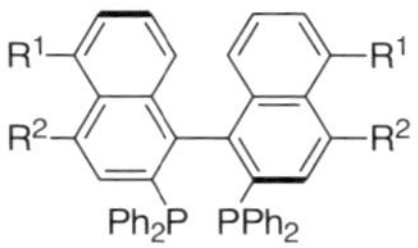 Ac	Me	H	H	74% ee	2004 (512)
Ac	H	H	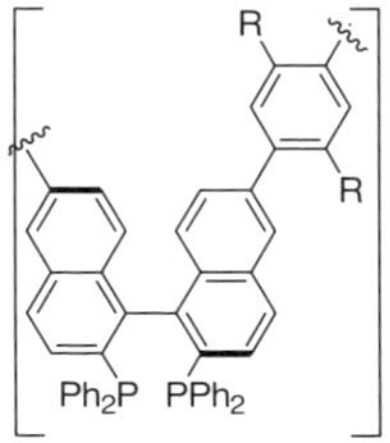 H	40–76% ee	2000 (511)
Bz	Me	Ph			
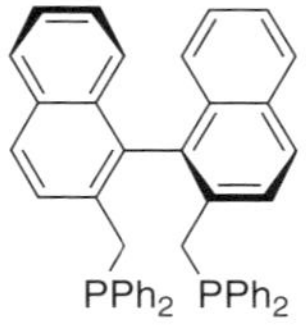 Ac	Me	H	H	54% ee	2002 (516)
Ac	Me	H	H	95.8–98.7% ee	2002 (516)
		Ph			2002 (515)
		4-F-Ph			
		4-MeO-Ph			
		3-Br-Ph			
		2-Cl-Ph			
		2-naphthyl			
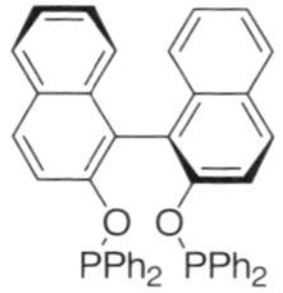 Ac	Me	Ph	H	64% ee	2002 (514)
Ac	Me	H	H	28–73% ee	2002 (516)
		Ph			

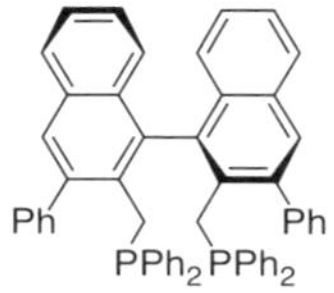

R¹, R²

R = O*n*Hex, *n*Hex

NAPHOS

Ph, Ph

Ph-*o*-NAPHOS

BINAPO

(Continued)

283

Table 4.5 Asymmetric Reductions of Didehydroamino Acids with Chiral Catalysts (continued)

Ligand	R^1	R^2	R^3	R^4	Enantio-selectivity	Reference
Ph-o-BINAPO	Ac	Me	H Ph 4-F-Ph 4-MeO-Ph 3-Br-Ph 2-Cl-Ph 2-naphthyl 4-CF$_3$-Ph 3-Me-Ph 4-Ph-Ph	H	81.5–99.9% ee	2002 (516)
H$_8$-BINAPO	Ac	Me	Me Ph 4-Me-Ph 4-Cl-Ph 2-Cl-Ph 2-furanyl	H	64–85% ee	2002 (514)
Xyl-H$_8$-BINAPO	Ac	Me	H Me Ph 4-Me-Ph 4-MeO-Ph 4-Cl-Ph 4-AcO-Ph 4-NO$_2$-Ph 2-Cl-Ph 2-furanyl	H	90–97% ee	2002 (514)
R = H: BIPNITE **R = Ph: o-BIPNITE**	Ac	H Me	Ph 4-F-Ph 3-Br-Ph 2-Cl-Ph 2-naphthyl 2-thienyl	H	95–>99% ee	2004 (520)

					ee	Year (Ref.)
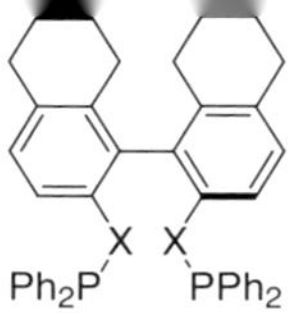	Ac Bz	Me	H Ph 2-Cl-Ph 3-Cl-Ph 4-Cl-Ph 4-Br-Ph 4-F-Ph 4-Me-Ph 4-MeO-Ph 4-NO$_2$-Ph 3,4-OCH$_2$O-Ph 2-furyl N-Ac-indol-3-yl CH=CHPh	H	90.2–97.1% ee	2001 (519)
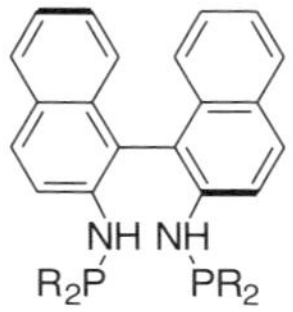	Ac	Me	H Ph 4-Me-Ph 4-MeO-Ph 4-NO$_2$-Ph 4-AcO-Ph 2-furanyl 3,4-(-OCH$_2$O)-Ph 2-Cl-Ph 3-Cl-Ph 4-Cl-Ph 4-F-Ph 4-Br-Ph	H	96–99% ee	2002 (517)
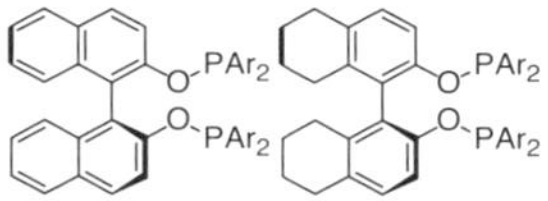	Ac	Me	Ph	H	29–100% 30.9–98.6% ee	2003 (518)
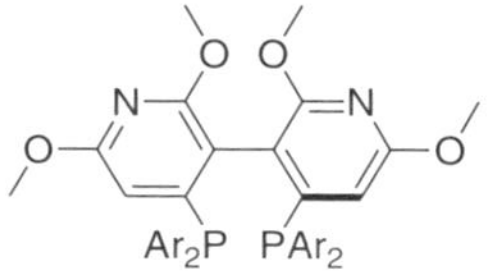	Ac	Me	Ph 2-Cl-Ph 3-Cl-Ph 4-Cl-Ph 4-Me-Ph 4-MeO-Ph	H	38–97% ee	2003 (525)

Ph$_2$P–X ··· X–PPh$_2$

X = O, NH

R$_2$P ··· NH NH ··· PR$_2$

R = Ph, cHex, 3,5-Me$_2$-Ph

O-PAr$_2$ / O-PAr$_2$

Ar = 4-Me-Ph, 4-MeO-Ph, 4-CF$_3$-Ph, 3,5-(CF$_3$)$_3$-Ph

Ar$_2$P PAr$_2$

Ar = Ph: **P-Phos**
Ar = 4-Me-Ph: **Tol-P-Phos**
Ar = 3,5-(Me$_2$ $^+$Ph): **Xyl-P-Phos**

(Continued)

285

Table 4.5 Asymmetric Reductions of Didehydroamino Acids with Chiral Catalysts (continued)

Ligand	R^1	R^2	R^3	R^4	Enantio-selectivity	Reference
	Ac	H Me	H Ph	H	21–45% ee	2004 (522)
MeO-BIPHEP						
	Ac	H Me	H Ph	H	98→99% ee	2004 (522)
	Ac	Me	Ph	H	98% ee	2004 (524)
R - *i*Pr, cHex						
	Ac	Me	Ph	H	29–45% ee	2004 (523)

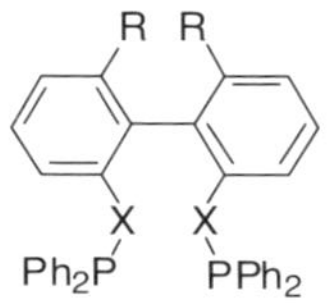

Ligand	Acyl	R	R	X	ee	Year (ref)
	Ac	H	H	H	86–95% ee (BDPAB)	1984 (526)
	Bz	Me	Ph			
	Ac	H	H	H	36–90% ee (Me-BDPAB)	1984 (526)
	Bz	Me	Ph			

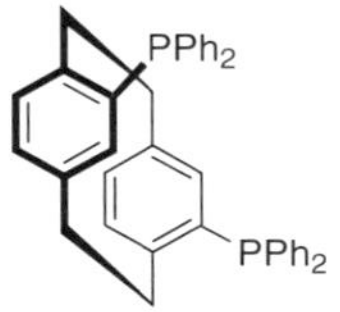

MABP R = Me; X = NH
NAPHOS R = H; X = CH$_2$
BDPAB R = H; X = NH
Me-BDPAB R = H; X = NMe

Ligand	Acyl	R	R	X	ee	Year (ref)
	Ac	Me	H	H	91–99.6% ee	1997 (527)
	Ac	Me	Ph	H		
	Ac	Me	Me	Me		
	Bz	Me	Me	H	except Val = 51% ee	
	Cbz	Me	Ph	H		
			H	H		

PhanePhos

Ligand	Acyl	R	R	X	ee	Year (ref)
	Ac	Me	H	H	99% ee	2001 (528)
			Ph			

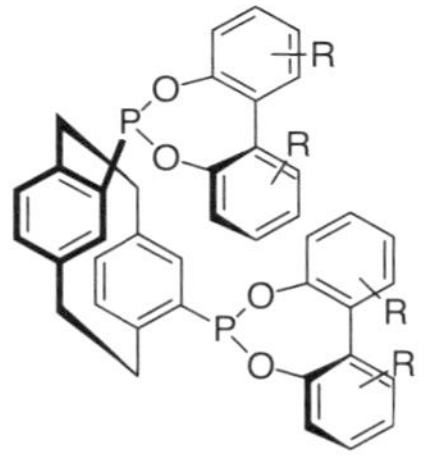

Ligand	Acyl	R	R	X	ee	Year (ref)
	Ac	H	Ph	H	77% ee	2000 (530)
		Me				

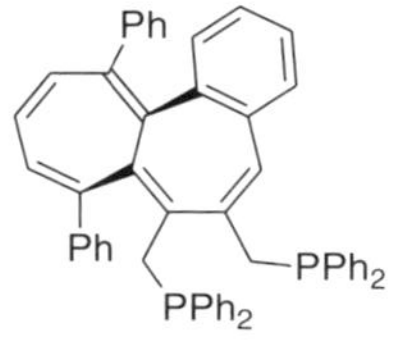

(Continued)

Table 4.5 Asymmetric Reductions of Didehydroamino Acids with Chiral Catalysts (continued)

Ligand	R¹	R²	R³	R⁴	Enantio-selectivity	Reference
Me͟PR₂ / R₂P͟Me, Fe / Fe (two ferrocene units) R = Me, Et, nBu, iBu **TRAP** 2,2'-bis[1-(dialkylphosphino)ethyl]-1,1'-biferrocene	Ac	Me	H Me iPr iBu Ph	H	57–96% ee	1995 (531)
	Ac	Me	Me tBu Ph 4-MeO-Ph 4-F-Ph Ph	Me Me Me Me Me Et	77–88% ee	1995 (531)
	Ac	Me	H	NHCbz	75–82% ee 90–100%	1998 (532)
	Ac	Me	OTBS	Me Et nPr CH₂OMe (CH₂)₃CO₂Et OCOtBu	92–97% ee 100% erythro 75–99% yield	1998 (533)
	Ac	Me	Me Et nBu Ph		73–97% ee 100% threo 12–99% yield	1998 (533)
	Boc Cbz Bz Ac	Me iBu OR² = NHtBu	(CH₂)₂-(α-N) (CH₂)₃-(α-N) (CH₂)₄-(α-N) O(CH₂)₂-(α-N) (1,2-Ph)CH₂-(α-N)	H	73–97% ee	1999 (535)
	Boc	OR² = NHtBu	N(Boc)(CH₂)₂-(α-N)	H	97% ee (S) with R = iBu 85% ee (R) with R = Me	1999 (534)
Fe with two Cp rings each bearing CH and PPh₂ (bisphosphine ferrocene ligand)	Ac	H	H Ph 2-naphthyl	H	95.7–98.7% ee	1998 (538)

Ac	H	H	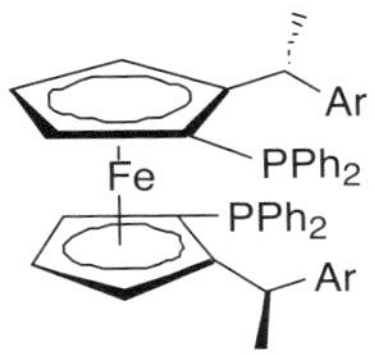 H	97.7–98.7% ee	1998 (536)
	Me	Ph			
		2-naphthyl			
Ac	Me	H	H	96.3–98.7% ee	1999 (537)
		Ph			
		4-F-Ph			
		4-Cl-Ph			
		2-naphthyl			
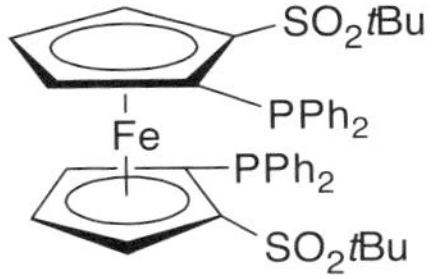 Ac	Me	H	H	93% ee	2005 (539)
Ac	Me	H	H	79–96% ee	1998 (544)
		Me			
		Et			
		iPr			
		Ph			
		4-MeO-Ph			
		4-Cl-Ph			
		4-NO2-Ph			
Ac	H	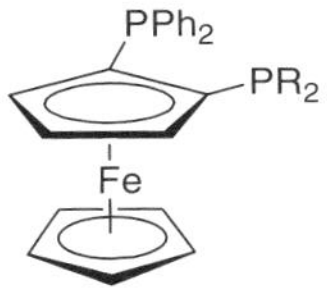 Ph	H	0–36% ee	2000 (546)
	Me				

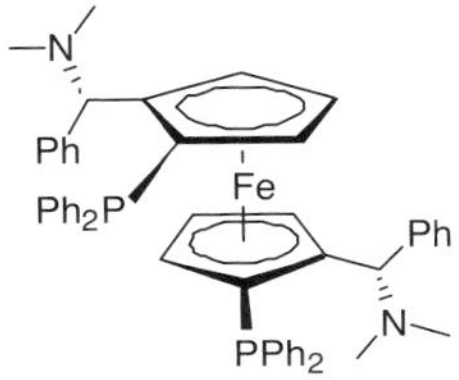

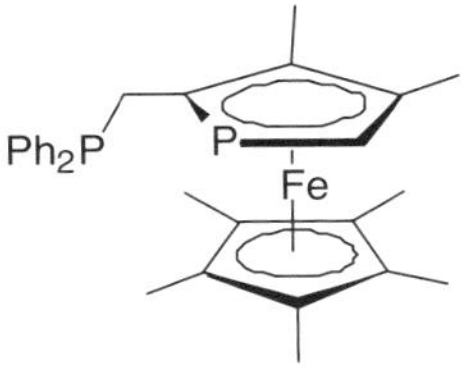

Ar = Ph, 2-Me-Ph, 2-naphthyl
FERRIPHOS

Diamino FERRIPHOS

R = Me, cHex

(Continued)

Table 4.5 Asymmetric Reductions of Didehydroamino Acids with Chiral Catalysts (continued)

Ligand	R^1	R^2	R^3	R^4	Enantio-selectivity	Reference
PR_2, PPh_2 ferrocene $R = c$-Pent, tBu, cHex, Ph	Ac	H Me	Ph	H	34–77% ee	2000 (545)
BoPhoz	Ac Bz Cbz Boc	Me H Bn	H cPr Bn Ph 4-MeO-Ph 3-MeO-Ph 2-MeO-Ph 4-Cl-Ph 4-F-Ph 4-NO$_2$-Ph 4-CN-Ph 1-naphthyl 2-naphthyl 2-furyl	H	97.2–99.5% ee	2005 (543)
Mandyphos R^1, $R^2 = c$Hex, Ph, 3,5-Me$_2$-Ph, 3,5-(CF$_3$)$_2$-Ph, 3,5-Me$_2$-4-MeO-Ph,	Ac	H	Ph	H	97.6–98.6% ee	2004 (540)
Taniaphos R^1, $R^2 = c$Hex, Ph, 3,5-Me$_2$-Ph, 3,5-(CF$_3$)$_2$-Ph, 3,5-Me$_2$-4-MeO-Ph,	Ac	H	Ph	H	96% ee	2004 (540)

  R = H, Me, *i*Pr R = Me, Et, *n*Pr, *n*Bu, *i*Bu	Ac	Me	Ph	H	95–96.6% ee	2002 (*542*)
	Ac	H	Ph	H	99% ee	2004 (*541*)
  R = Me, Et, Ph	Ac	H	Ph	H	50–94% ee	2004 (*541*)
	Ac	Me	H Ph 4-NO$_2$-Ph 4-Me-Ph 4-F-Ph 4-Br-Ph 4-Cl-Ph 3-Cl-Ph 4-CF$_3$-Ph 4-MeO-Ph	H	95.3–99.6% ee	2004 (*547*)
  R = H, *t*Bu R^1, R^2 = Ph, *c*Hex, *i*Pr	Ac	H Me	H Ph	H	31→95% ee	2004 (*548*)

(*Continued*)

Table 4.5 Asymmetric Reductions of Didehydroamino Acids with Chiral Catalysts (continued)

Ligand	R^1	R^2	R^3	R^4	Enantio-selectivity	Reference
chiral metal atom backbone	Ac	H Me Et	H Ph	H	82–93% ee	2001 (549)
streptavidin complex with	Ac	H	H Ph	H	93–94% ee	2004 (551) 2005 (552)

Ligands Based on Chiral Phosphine Substituents

Ligand	R^1	R^2	R^3	R^4	Enantio-selectivity	Reference
 R = Me, Et, nPr, iPr **DuPHOS** 1,2-bis(2,5-dialkylphosphalano)benzene	Ac Cbz	H Me	Me Et *n*Pr *i*Pr Ph thien-2-yl ferrocenyl 1-naphthyl 2-naphthyl 3-Br-Ph 4-Br-Ph 3-F-Ph 4-F-Ph 3-Me-Ph 4-Me-Ph 3-MeO-Ph 4-MeO-Ph 3-NO$_2$-Ph 3-BnO-Ph 2-F-Ph 3,5-Br$_2$-Ph 3,5-F$_2$-Ph 3,5-(MeO)$_2$-Ph 3,5-(CF$_3$)$_2$-Ph	H	>99% ee	1993 (559)

Ac	Me	2-Br-Ph 3-Br-Ph 4-Br-Ph 3,5-Br$_2$-Ph 4-Br-thien-2-yl 5-Br-thien-2-yl 5-Br-2-furyl 1-Br-2-naphthyl	H	>98.5% ee	1994 (570)
Cbz	Me tBu	thien-2-yl thien-3-yl 2-furyl 3-furyl 1-Boc-2-pyrrolyl 1-Ts-pyrrolyl	H	94–99.6% ee	1994 (298)
(CH$_2$)$_2$-N(Boc)-R^3 OR^2 = NHtBu	N(Boc)(CH$_2$)$_2$-R^1				1995 (893)
Ac	Me	Me Et iPr (CH$_2$)$_4$-R^4 (CH$_2$)$_5$-R^4 (CH$_2$)$_6$-R^4 (CH$_2$)$_2$O(CH$_2$)$_2$-R^4 (CH$_2$)$_2$S(CH$_2$)$_2$-R^4 (CH$_2$)$_2$CO(CH$_2$)$_2$-R^4	Me Et iPr (CH$_2$)$_4$-R^3 (CH$_2$)$_5$-R^3 (CH$_2$)$_6$-R^3 (CH$_2$)$_2$O(CH$_2$)$_2$-R^3 (CH$_2$)$_2$S(CH$_2$)$_2$-R^3 (CH$_2$)$_2$CO(CH$_2$)$_2$-R^3	70% ee 85–99.4% ee	1995 (564)
Ac	Me Me Me Me Ph	Me Me Me Me	Et iPr CH=CHPh CO$_2$Me CO$_2$Me	62–99.4% ee	1995 (564)
Boc	Me or TMSE	AcO OAc O AcO (sugar structures)	H	82–>95% de	1997 (323) 2003 (324)

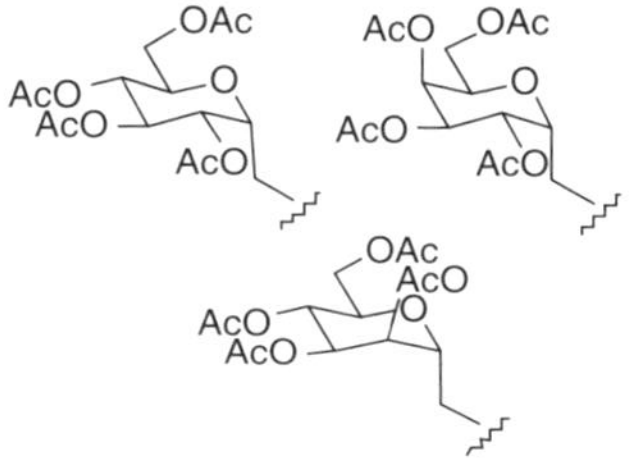

(Continued)

293

Table 4.5 Asymmetric Reductions of Didehydroamino Acids with Chiral Catalysts (continued)

Ligand	R¹	R²	R³	R⁴	Enantio-selectivity	Reference
	Boc	TMSE		H	88–98% yield >95% de	1999 (297)
	Ac	Me	2-Br-Ph	H	99% ee 90% yield	1997 (361)
	Ac	Me	(E)-CH=CHnHex (E)-CH=CHMe (E)-CH=CHnBu (E)-CH=CH(CH₂)₂Cl (E)-CH=CHPh (E)-CH=CHCH₂OTBS (Z)-CH=CHCH₂OTBS (E)-CH=CHBn (E)-C(Me)=CHMe (E)-CH=C(Me) (CH₂)₂CH=C(Me)₂ C(Et)=CH₂ cyclohex-1-enyl	H	86–99.5% ee	1998 (305)
	Boc Cbz	Me Bn	tris(amino acid) linked by 1,3,5-Ph	H	>98% ds	1998 (313)
	Cbz	Me	(CH₂)ₙ-(α-N) n = 2,3,4,5,6,10,13	H	0.4–97% ee 84–97% yield	1998 (355)
	Boc/Ac	Me/Bn	bis(amino acid) linked by 1,3-Ph	H	>98% ee 90% de 98% yield	1998 (366)
	Ac Me	H Me	1-TIPS-indol-3-yl	Me	97% ee 91% yield	1998 (375)
	Cbz	Me	2-Me-pyrid-6-yl quinolin-2-yl bis(amino acid) linked by	H	94–97% ee 78–100% yield	1999 (565)

Cbz	Me	nPr nPent cHex	H	98.7–99.6% ee	1999 (566)
Cbz	Me	bis(amino acid) linked by -(CH₂)ₙ- n = 0,2,3	H	98.5–99.5% ds (S,S)	1999 (311)
Ac	Me	bis(amino acid) linked by 3-CH₂CO₂Bn-1,5-Ph	H	>98% ee >98% de 99% yield	1999 (569)
Ac	Me	(E)-CH=CHnHex (E)-CH=CHPh (E)-CH=CHBn (E)-CH=CHtBu	Me Et	73–96% ee	1999 (563)
Cbz	Me	3-HO-Ph	H	98% ee	2000 (875)
Ac	Me	(2,2-Me₂-3-R-cyclobutane) R = COMe, CO₂Me, CH₂OBn CH₂(2,2-Me₂-3-COMe-cyclobutane)	H	60->96% de	2001 (414)
Bz	Me	NHCOMe NHCOPh NHBoc NHCbz NHFmoc NHAlloc	H	92–99% ee	2001 (796)
Bz	Et	NHCOMe	Me	>98% ee	2001 (571)
Boc	TSE	(CH₂)₃O(carbohydrate) carbohydrate = α-Tn, Lewisᵞ, fucosyl GM1, Globo-H (mono- to hexasaccharides)	H	>95% de	2001 (325)
Boc	Me	(CH₂)₂CH-N(Boc)C(Me₂)OCH₂-]CO₂H	H	>95% de	2001 (373)
Cbz	Me	3-(6-carboxy-2-oxo-4-pyridyl) 3-(6-carboxy-2-oxo-3-pyridyl)	H	89–93% >98% ee	2002 (878)
Cbz	Me	2-Br-Ph 4-Br-Ph 3-BnO-Ph 5-Br-indol-3-yl	H	95–100% >96% ee	2002 (879)
Trt	Bn	CH₂CH₂CH(CO₂Me)NHCbz	H	80% de	2002 (389)
Boc	Bn	cPr	H	100% 99.3% ee	2003 (568)

(Continued)

Table 4.5 Asymmetric Reductions of Didehydroamino Acids with Chiral Catalysts (continued)

Ligand	R^1	R^2	R^3	R^4	Enantio-selectivity	Reference
	Ac	NH-peptide-Rink resin	Ph 4-MeO-Ph 2-Me-Ph 3-Me-Ph 4-Me-Ph 1-naphthyl	H	86–98% de	2003 (335)
	Boc-Val Boc-Ile-Val	Me	1-Boc-6-Br-indol-3-yl	H	95–98% 89–>99% de	2004 (883)
	Ac	Me	Ph 4-F-Ph 4-CF$_3$-Ph 2-naphthyl 2,5-Me$_2$-Ph 4-MeO-Ph	Me	66–89% yield >99% ee	2004 (109)
	Boc Cbz Boc Cbz	Me Me Me Me	4-[CH(CO$_2t$Bu)]$_2$-Ph 4-[CH=C(CO$_2t$Bu)]$_2$-Ph	H	94–96% ee	2005 (884)
MeDuPHOS in polyvinyl alcohol film	Ac	Me	H	H	95.1–96.9% ee	2002 (572)
MeDuPHOS in polydimetylsiloxane film	Ac	Me	H	H	91.1–96.9% ee	2002 (574)
MeDuPHOS on mesoporous aluminasilicate support	Ac	Me	H	H	>98% ee	2004 (443)
MeDuPHOS in ionic liquid [bmim][PF$_6$]	Ac	Me	H Ph	H	94–96% ee over five cycles	2001 (575)
	Ac	Me	H iPr Ph	H	64–99% ee	1991 (560) 1993 (559)
BPE R = Me, Et, nPr, iPr 1.2-bis(2,5-dialkylphosphalano)ethane	Ac	Me	Me Et iPr (CH$_2$)$_4$-R^4 (CH$_2$)$_5$-R^4 (CH$_2$)$_6$-R^4 (CH$_2$)$_2$O(CH$_2$)$_2$-R^4 (CH$_2$)$_2$S(CH$_2$)$_2$-R^4 (CH$_2$)$_2$CO(CH$_2$)$_2$-R^4	Me Et iPr (CH$_2$)$_4$-R^3 (CH$_2$)$_5$-R^3 (CH$_2$)$_6$-R^3 (CH$_2$)$_2$O(CH$_2$)$_2$-R^3 (CH$_2$)$_2$S(CH$_2$)$_2$-R^3 (CH$_2$)$_2$CO(CH$_2$)$_2$-R^3	97.2–98.6% ee	1995 (564)

Ac	Me	Me	Et	80–99.0% ee	1995 (564)
			iPr		
			CH=CHPh		
			CO₂Me		
		Ph	CO₂Me		
Ac	Me	CH=CHPh	Me	78–96% ee	1999 (563)
		CH=CHBn	Me		
		CH=CHtBu	Me		
		CH=CHnhex	Me		
		Me	CH=CHPh		
		Me	CH=CHBn		
		Me	CH=CHtBu		
		Me	CH=CHnhex		
Ac	Me	Ph	H	89.5% ee	2004 (633)
Ac	Me	Ph	H	98% ee	1998 (578)
Ac	Me	Ph	H	98% ee	1998 (578)
Ac	H	H	H	98–>99% ee	1999 (579)
	Me	Ph			
		4-MeO-Ph			
		4-F-Ph			
		2-thienyl			

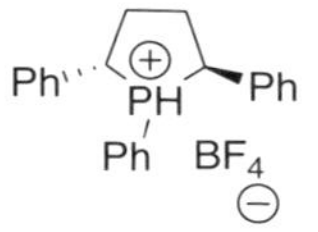

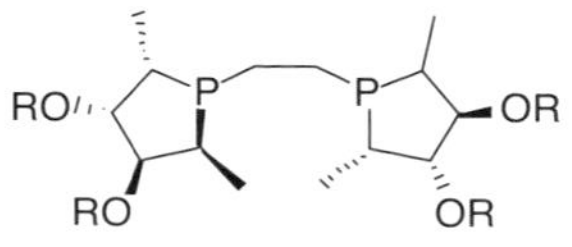

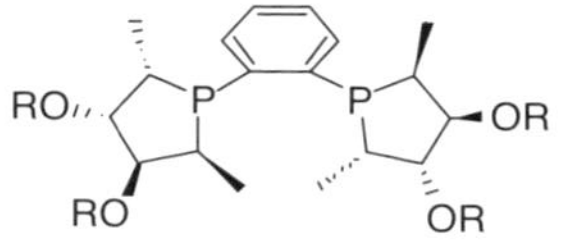

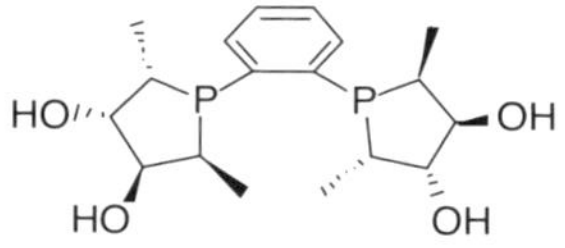

Ph—PH⁺—Ph / Ph BF₄⁻

R = Bn, tBu
RoPHOS

R = Bn

(Continued)

Table 4.5 Asymmetric Reductions of Didehydroamino Acids with Chiral Catalysts (continued)

Ligand	R^1	R^2	R^3	R^4	Enantio-selectivity	Reference
BASPHOS	Ac	H	H	H	99.6% ee	1999 (580)
MaiPHOS	Ac	Me	Ph	H	98.6% ee	2003 (577)
	Ac	H	H Ph	H	94.2–97.5% ee	2002 (581)
	Ac	Me	H	H	90% ee	1998 (590)
	Ac	H	Ph	H	18–48% ee	2003 (582)

R = H, Br, Me

Ligands Combining Two Types of Chirality

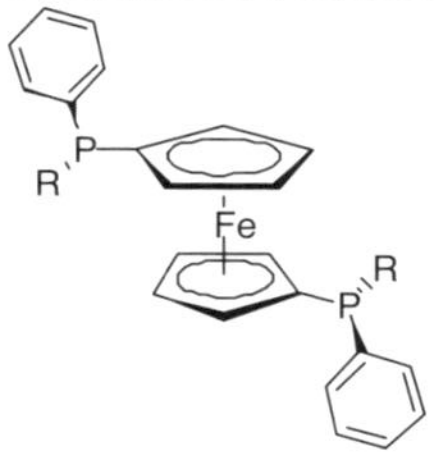 Ac	H	Ph	H	up to 98.7% ee	1999 (591)
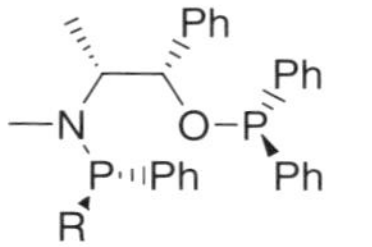 Ac	Me	Ph	H	99% ee with R = o-An	1999 (588)
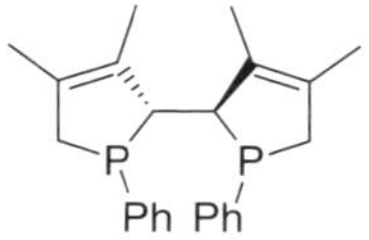 Ac	H	Ph	H-An	75% ee	1999 (587)
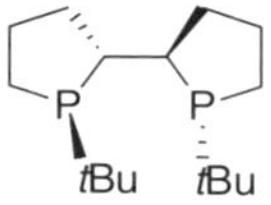 Ac	H	Ph 4-F-Ph 4-MeO-Ph 3-Br-Ph 2-Cl-Ph 2-thienyl 2-naphthyl	H	>99% ee	2002 (583)

R = 2-MeO-Ph, 2-Ph-Ph, 1-naphthyl, 2-naphthyl, 9-phenanthrene

R = Ph, Me, o-An, 1-naphthyl, 2-naphthyl, tBu

TangPhos

(*Continued*)

299

Table 4.5 Asymmetric Reductions of Didehydroamino Acids with Chiral Catalysts (continued)

Ligand	R¹	R²	R³	R⁴	Enantio-selectivity	Reference
	Ac	H	H Ph	H	>99% ee	2005 (584)
DuanPhos	Ac	H	Ph	H	96% ee	2004 (585)
	Ac	Me	H Ph	H	73–77% ee or 95–98% ee	2000 (586)
R = H: **BINAPHOS** R = Ph: **o-BINAPHOS**	Ac	H Me	Ph 4-F-Ph 3-Br-Ph 2-Cl-Ph 2-naphthyl 2-thienyl	H	95–>99% ee	2004 (520)
3-H³F⁶-BINAPHOS in supercritical CO₂/H₂O	Ac	H	H	H	97.3–99.1% ee over five cycles (recoverable catalyst by fluorous extraction)	2005 (521)

300

				ee	Year (Ref)
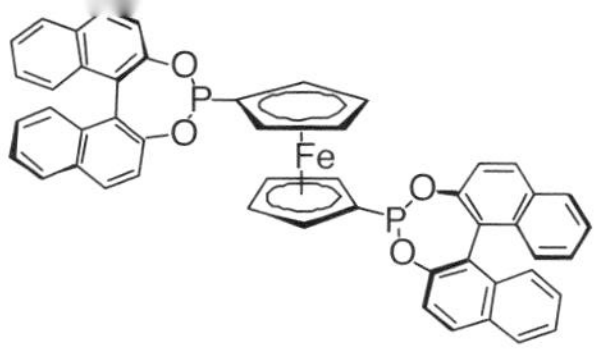 Ac	Me	H	H	99.5% ee	1998 (590)
Ac	Me	Ph	H	99.5–99.8% ee	2005 (593)
		2-Cl-Ph			
		4-Cl-Ph			
		2-MeO-Ph			
		4-MeO-Ph			
Ac	Me	Ph	H	81–99.9% ee	2005 (594)
Ac	H	Ph	H		
Ac	H	H	H		
Ac	Me	2-naphthyl	H		
Boc	Bn	cPr			
Ac	Me	Ph	H	97.4–99.6% ee	2004 (547)
		4-NO$_2$-Ph			
		4-Me-Ph			
		4-F-Ph			
		4-Br-Ph			
		4-Cl-Ph			
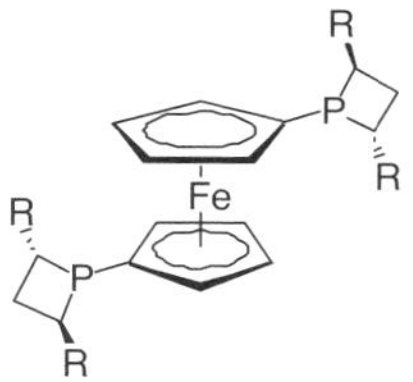 Ac	Me	H	H	R = Me 69–90% ee (S); R = iPr 83–94% ee (R)	1999 (589)
		Ph			
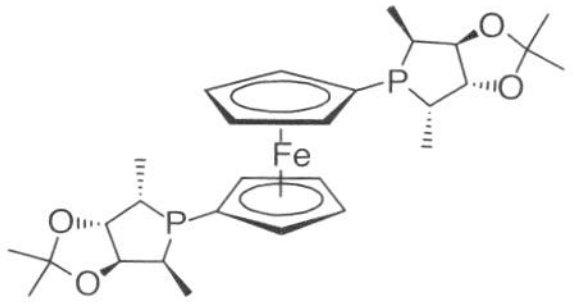 Ac	H, Me	H	H	98.4–>99.9 ee; 87.3% ee for Val	2002 (592)
		iPr	H		
		Ph	H		
		2-naphthyl	H		
		4-F-Ph	H		
		4-MeO-Ph	H		
		3-Br-Ph	H		
		2-Cl-Ph	H		
		Me	Me		

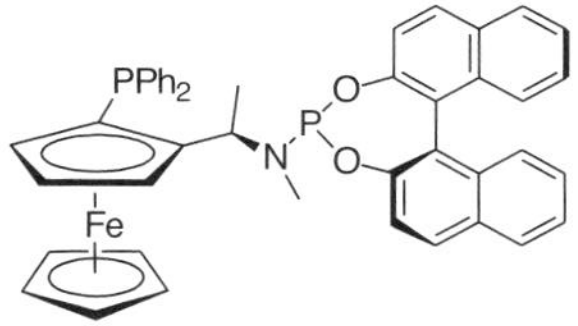

R = Me, iPr

(Continued)

301

Table 4.5 Asymmetric Reductions of Didehydroamino Acids with Chiral Catalysts (continued)

Ligand	R^1	R^2	R^3	R^4	Enantio-selectivity	Reference
	Ac	Me	H Ph	H	98.8–>99% ee	2000 (501)
	Ac	Me	H Ph	H	98.8–>99% ee	2001 (595)
	Ac	Me	H	H	37% ee 30% yield	2000 (596)
	Ac	Me	H Ph	H	98% ee	2001 (597)

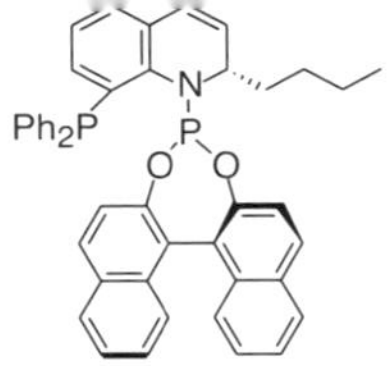 Ac	Me	H	H	97.8% ee >99% yield	2000 (423)

QUINAPHOS

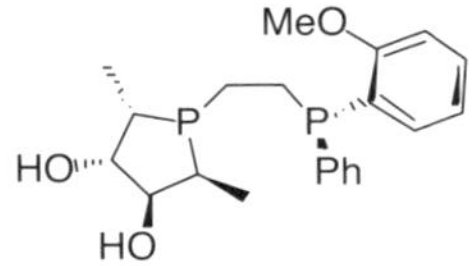 Ac Bz tBuCO	Me	Ph	H	88–92% ee	1999 (598)

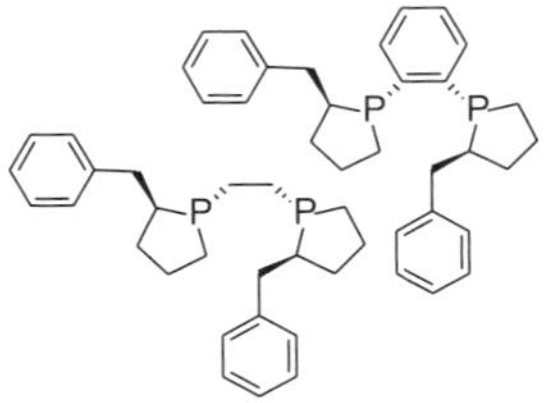 Ac	H Me	H Ph (CH$_2$)$_5$-R^4	H H (CH$_2$)$_5$-R^3	86–97% ee	2004 (599)

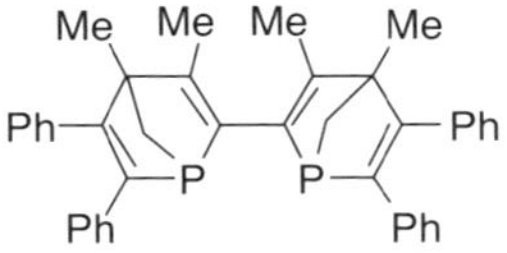 Ac	H	Ph	H	>98% ee	1997 (600)

BIPNOR

Ligands With no Phosphine

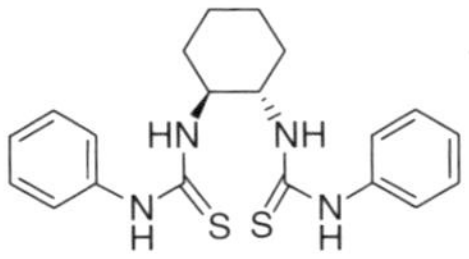 Ac	H	Me	Me	61–70% ee	2000 (558)

(Continued)

Table 4.5 Asymmetric Reductions of Didehydroamino Acids with Chiral Catalysts (continued)

Ligand	R¹	R²	R³	R⁴	Enantio-selectivity	Reference
Monodentate Ligands						
 R = Me: **MonoPhos**	Ac	H	H	H	94.4–99.8% ee	2000 (601)
		Me	Me			
			Ph			
			3-MeO-4-AcO-Ph			
	Ac	H	H	H	97–>99% ee	2003 (618)
		Me	Ph			
R = Et	Ac	Me	Ph	H	98.0–>99.9% ee	2003 (611)
	Bz		4-F-Ph			
			4-Cl-Ph			
			4-Br-Ph			
			3-Cl-Ph			
			2-Cl-Ph			
			4-Me-Ph			
			4-NO₂-Ph			
			4-MeO-Ph			
			4-AcO-Ph			
R = Et, -(CH₂)₅-	Ac	Me	Ph	H	94–95% ee with crude catalyst	2004 (612)
R = -(CH₂)₅-: **PipPhos**	Ac	Me	H	H	98–99% ee	2005 (613)
R = -(CH₂)₂O(CH₂)₂-: **MorfPhos**			Ph			
R = Me immobilized on aluminosilicate AlTUD-1	Ac	Me	H	H	83–97% ee little loss over three recycles	2004 (619)
	Ac	Me	H	H	>99.0% ee	2005 (614)

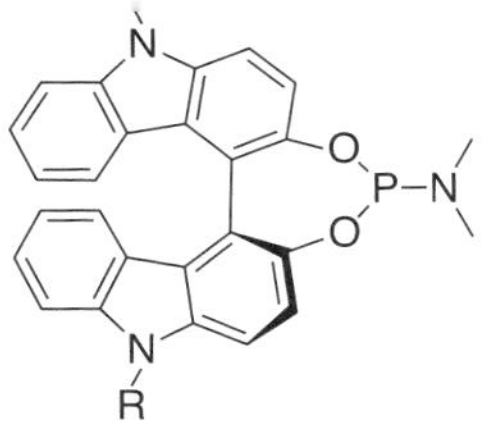

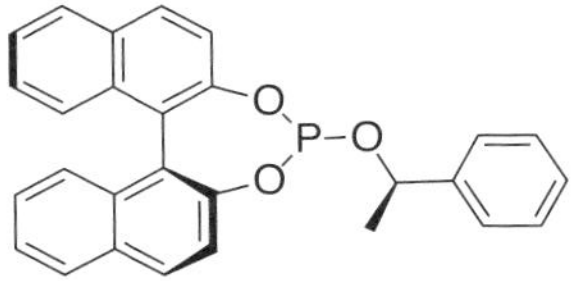 Ac	Me	H	H	94% ee 100% yield	2000 (604)
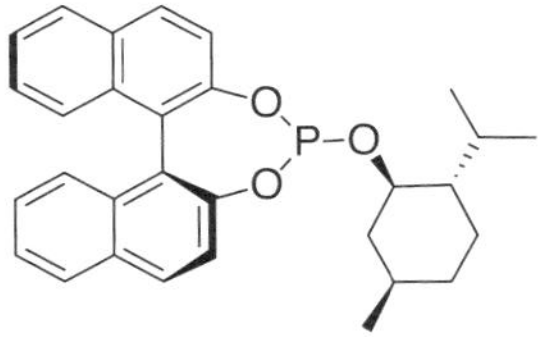 Ac	Me	H	H	85% ee 100% yield	2001 (606)
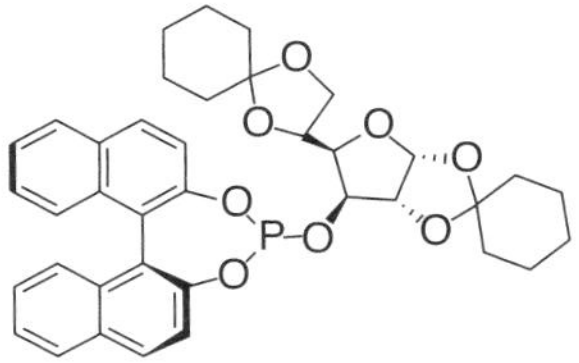 Ac	Me	H Ph 4-MeO-Ph 2-MeO-Ph 4-Cl-Ph 2-Cl-Ph	H	80–89% ee	2004 (607)
Ac	Me	Ph	H	7–50% ee	2004 (609)

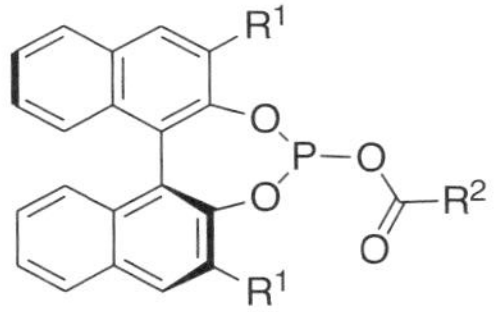

BICOL backbone

R = Me

R = CH$_2$CONH(CH$_2$)$_3$Si[(CH$_2$)$_3$Si-dendrimer]$_3$

R^1 = H, Br, Me, Ph, SiMe$_3$
R^2 = Me, Ph

(Continued)

Table 4.5 Asymmetric Reductions of Didehydroamino Acids with Chiral Catalysts (continued)

Ligand	R^1	R^2	R^3	R^4	Enantio-selectivity	Reference
 R = Me, Et, Ph, tBu	Ac	Me	H	H	95.5% ee 100% yield	2000 (602)
 R = Me, Ph, tBu, –CH₂–	Ac	Me	H Ph	H	10-92% ee	2000 (603)
 X = -, 1,4-Ph, 1,3-Ph	Ac	Me	H Me Ph	H	95.8–96.6% ee	2004 (616)
	Ac	Me	H	H	90% ee	2000 (617)

(Continued)

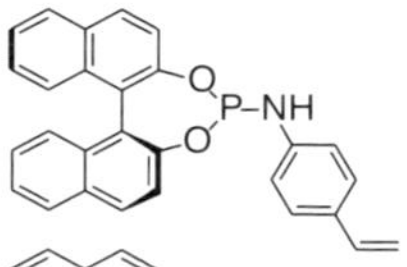 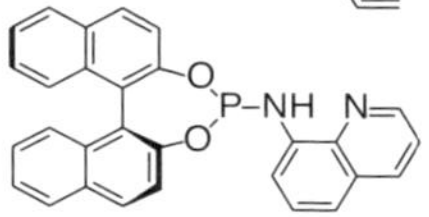 Ac	H Me	H Ph	H	33–84% ee	2003 (618)
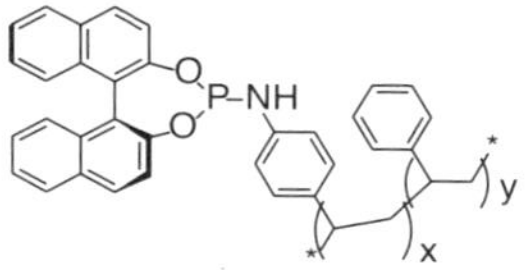 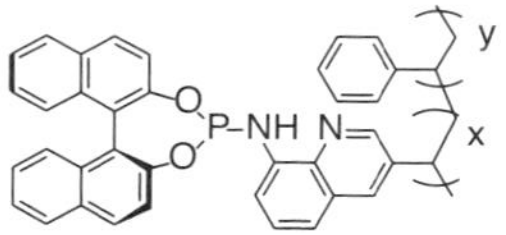 Ac	H Me	H Ph	H	49–80% ee	2003 (618)
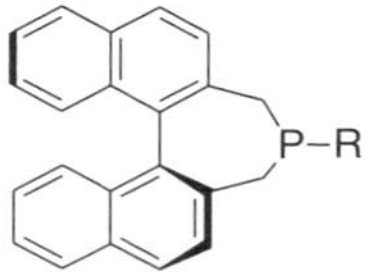 Ac	Me	Ph 4-Me-Ph 4-Br-Ph 4-NO₂-Ph	H	84–95% ee	2002 (630)
Ac	H Me	Ph	H	91% ee	2005 (631)

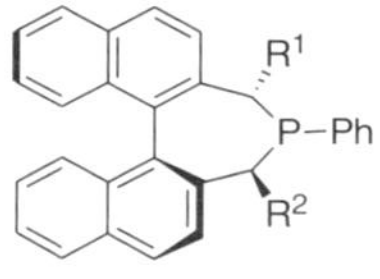

R = Ph (best results)

also tested R = Et, iPr, tBu, 4-MeO-Ph, 4-CF₃-Ph, 3,5-Me₂-Ph

$R^1 = R^2 = H, Me, Et, Bn$
$R^1 = Me, Et, Bn; R^2 = H$

Table 4.5 Asymmetric Reductions of Didehydroamino Acids with Chiral Catalysts (continued)

Ligand	R¹	R²	R³	R⁴	Enantio-selectivity	Reference
	Ac	Me	H	H	87% ee	2004 (620)
heterocombination of ligands						
	Ac	H Me	H Ph 4-Cl-Ph 2-Cl-Ph	H	up to 98% ee	2004 (621)
	Ac	Me	H	H	93% ee	2002 (623)
	Ac	Me	H 4-MeO-Ph 2-MeO-Ph 4-Cl-Ph 2-Cl-Ph	H	96.6–98.0% ee	2004 (608)
	Ac	Me	Ph 4-Me-Ph 4-MeO-Ph 2-Cl-Ph 3-Cl-Ph 4-Cl-Ph 2-NO₂-Ph 3-NO₂-Ph 4-NO₂-Ph	H	97–99% ee	2004 (627)

R = Me, nPr, iPr, tBu, Bn, CHPh₂, CH₂(1-naphthyl)

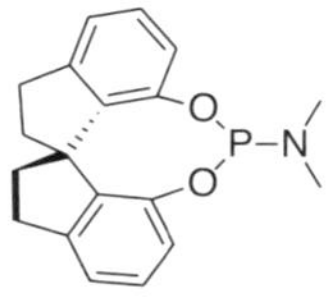

Ac	Me	H	H	95.6–99.3% ee	2002 (625)
		Me			
		Ph			
		4-MeO-Ph			
		2-Cl-Ph			
		4-Cl-Ph			
		3-NO$_2$-Ph			
		4-NO$_2$-Ph			
Ac	Me	Ph	H	94.5–99.1% ee	2003 (626)
		4-MeO-Ph			
		4-Cl-Ph			
		4-NO$_2$-Ph			
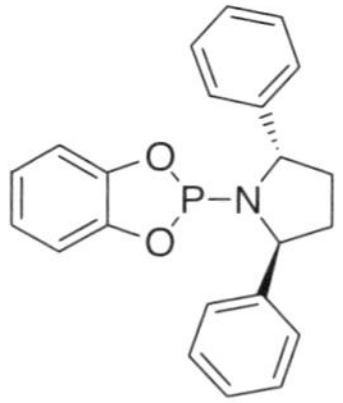 Ac	H	H	Me	66–92% ee	2004 (632)
		Ph			
Ac	H	H	H	15–85% ee	2004 (634)
Ac	Me	H	H		
Ac	H	Ph	H		
Ac	Me	Ph	H		
Ac	Me	Me	Me		
Cbz	Me	Me	Me		
Ac	Me	(CH$_2$)$_4$-	(CH$_2$)$_4$-		
Ac	Me	CH$_2$)$_5$-	CH$_2$)$_5$-		

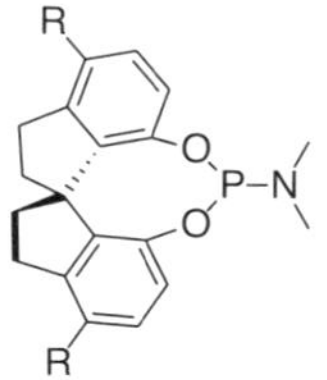

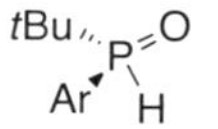

SIPHOS

R = H: SIPHOS
R = Br
R = Ph

tBu P=O
Ar H

Ar = Ph, 2-naphthyl

(Continued)

309

Table 4.5 Asymmetric Reductions of Didehydroamino Acids with Chiral Catalysts (continued)

Ligand	R^1	R^2	R^3	R^4	Enantio-selectivity	Reference
(phospholane: Ph, P–Ph, Ph, Ph)	Ac	Me	Ph	H	82% ee	1999 (894)
(menthyl OMe phosphine: OMe, P, Ph, Ph)	Ac	H	Ph	H	22% ee	1987 (444)
(Ph_2P–O, S–(3,5-dimethylphenyl), iPr)	Ac	Me	Me	H	89–97% ee	2003 (629)
	Ac	Me	Et	H		
	Ac	Me	iPr	H		
	Ac	Me	Ph	H		
	Ac	H	Ph	H		
	Bz	Me	Ph	H		
	Cbz	Me	Ph	H		
	Boc	Me	Ph	H		
	Ac	Me	4-MeO-Ph	H		
	Ac	Me	3-Br-Ph	H		
	Ac	Me	$3\text{-}NO_2\text{-}4\text{-}F\text{-}Ph$	H		
	Ac	Me	Me	$-(CH_2)_5-$		
	Ac	Me	$-(CH_2)_5-$			
	Ac	H	$3\text{-}NO_2\text{-}4\text{-}F\text{-}Ph$	H	94% ee	2001 (628)
(cyclohexyl: Ph_2P–O, S–tBu)	Ac	Me	Me	H	83–98% ee (except iPr; 36% ee)	2003 (629)
	Ac	Me	Et	H		
	Ac	Me	iPr	H		
	Ac	Me	Ph	H		
	Ac	H	Ph	H		
	Bz	Me	Ph	H		
	Cbz	Me	Ph	H		
	Boc	Me	Ph	H		
	Ac	Me	4-MeO-Ph	H		
	Ac	Me	3-Br-Ph	H		
	Ac	Me	$3\text{-}NO_2\text{-}4\text{-}F\text{-}Ph$	H		

The (E)-isomers were reduced much more slowly and with greatly reduced enantioselectivity (23–47% ee) (438). A 1981 comparison of DIPAMP with 10 other chiral ligand catalysts (DIOP, CH3DIOP, C4DIOP, BCOP, PROPHOS, CHIRAPHOS, NORPHOS, PELLANPHOS, NOPAPHOS, BPPM) for the reduction of nine E- and Z-2-acetamido-3-alkylacrylates, including a 3,3-disubstituted substrate, clearly demonstrated the superiority of DIPAMP (261). The Rh-DIPAMP catalyst has been applied to reductions of a γ-(1,3-dioxane)-substituted didehydroamino acid (97% ee) (306), phenyl-bridged bis(ΔAla) (367), 1,1′-ferrocenyl bis(ΔAla) (>95% ee, 60–80% de) (314, 315), and two pyridyl-substituted substrates (86–99% ee) (363, 364). DIOP gave only 70% ee for the latter compounds (363). Both halves of a pyridinyl-linked bis(ΔAla) were reduced with >99% ee, compared to 90% ee provided by BINAP (312). The DIPAMP rhodium catalyst has been used in conjunction with optically active didehydroamino acids derived from chiral diol aldehydes for a double asymmetric induction reduction, with 94–100% de for the matched pair and 57–100% de for the mismatched pair (439). For asymmetric reductions of dehydro-Trp within a dipeptide, DIPAMP gave 82% de, compared to 54% for DIOP and 72% for BPPM (360). The diastereoselectivities of DIPAMP asymmetric hydrogenations of ΔTrp contained within pentapeptides was found to be greatly influenced by the presence of the sulfur atom of an adjacent Met residue (0–26% de with Met, 99% de with Nle) (440). DIPAMP was also used to reduce ΔPhe contained within a Wang resin-anchored tripeptide with up to 88% ee, although Ph-CAPP gave up to 90% ee (441). DIPAMP is used industrially to synthesize L-DOPA (416, 442).

The DIPAMP catalyst has been non-covalently anchored on a mesoporous aluminosilicate support; Ac-ΔAla was reduced with the same enantioselectivity (92% ee) as with the regular catalyst. The catalyst could be filtered off and reused up to 3 times with no loss in enantioselectivity or activity (443). Analogs of DIPAMP include a homolog and both polydentate and monodentate derivatives (444), and a ligand in which the DIPAMP methoxy groups have been replaced by ethyl groups (445). The best results (comparable to DIPAMP) were obtained with the homolog (444). Rigidifying the DIPAMP backbone by replacing the ethyl linker with a phenyl group gave a ligand which could catalyze reductions of several dehydroamino acids, including β-unsubstituted (leading to Ala) and β,β-disubstituted derivatives, with 77–97% ee (446). A DIPAMP analog which also incorporates the chiral backbone of PYRPHOS gave improved enantioselectivity for the reduction of α-acetamidocinnamic acid (99% ee), depending on hydrogenation pressure (447).

A number of other P-chiral ligands have been developed. Bis(trialkylphosphine) ligands (BisP*), with an ethyl bridge between two chiral phosphine centers, reduced several dehydroamino acids with enantioselectivities of up to 90–99.9% ee. The chiral phosphines were substituted with a methyl group and a bulky alkyl substituent (tBu, CEt₃, 1-adamantyl, cPent or cHex). The most effective ligand varied with the substrate (448). A ferrocenyl substituent gave poor results (449). The Rh-BisP* catalyst reduced diethyl α-acetylaminoethenephosphonate to give the phosphonate analog of Ala with 90% ee (450). BisP* ligands with different substituents on the two phosphine centers reduced the β,β-disubstituted precursors leading to Val, cyclopentylglycine, and cyclohexylglycine with 88–98% ee (451), but ΔAla and ΔPhe were reduced with reduced values of 30–99% ee (452). A detailed study of the mechanism of asymmetric hydrogenation by the Rh complex of the t-Bu-substituted ligand, (S,S)-1,2-bis(t-butylmethylphosphino)ethane (BisP*), was reported in 2000, using multinuclear NMR (420). MiniPHOS is a closely related ligand with a methyl linker between the phosphines instead of an ethyl group. ΔAla was reduced with >99.9% ee, while the β,β-disubstituted enamide substrates leading to Val, cyclohexylglycine, and cyclopentylglycine were hydrogenated with 87–97% ee (453). One difficulty with P-chiral diphosphine ligands is that they are easily oxidized in air, and must be stored in a protected form. A phosphine–borane complex is often used, but requires complicated deprotection before use. However, it has been found that dihydroboronium catechol derivatives of BisP* ligands are convenient precursors of the ligand, being converted into the desired Rh-complex in situ (454). Trialkylphosphonium salt analogs of BisP* and MiniPHOS induced >99% ee for reduction of Ac-ΔPhe-OMe (455). An air-stable bisP*-like P-chiral ligand, QuinoxP*, was reported in 2005, using a quinoxaline backbone instead of the ethyl linker. Two dehydroarylalanines were reduced with 99.6–99.9% ee (456). However, BisP* analogs with a ferrocene bridge replacing the ethyl group gave poor (18–42% ee) enantioselectivity for reduction of dehydro-Phe (457).

4.3.2cii Ligands with Chiral Backbone

Many more chiral ligands have been prepared which rely on a chiral backbone connecting two phosphine atoms (normally diphenylphosphine) (see Scheme 4.31, Table 4.5). DIOP, 1-bis(diphenylphosphinomethyl)-2,2-dimethyl-1,3-dioxolane, was one of the first of these ligands and quickly gained popularity due to its easy preparation from tartaric acid (458). Although impressive at the time, Rh-DIOP catalysts give comparatively poor results with α-aminoacrylic acids (22–80% ee) and do not reduce β-disubstituted acrylates (458, 459). DIOP has been attached to a polymer to aid in product purification, with comparable enantioselectivity (460). A variety of DIOP analogs have been developed, including 3-CH₃DIOP and C4DIOP (261). A DIOP derivative with a hydroxy substituent gave poor enantioselectivities for reductions in aqueous solutions (1–34% ee), but when a micelle-forming amphiphile was added much better selectivity (72–76% ee) was obtained (461). Another DIOP analog, with pyridyl groups replacing the phosphine phenyls, gave similar enantioselectivites as DIOP

(28–84% ee), but the products had the opposite absolute configuration (462).

Bicyclic analogs of DIOP include BCOP, NORPHOS, NOPAPHOS, and PELLANPHOS (261). Several of these chiral catalyst complexes (NORPHOS, PROPHOS, DIOP) were examined for a reduction leading to ferrocenylalanine, with (+)- or (−)-NORPHOS giving the best results (90–95% ee, quantitative yield) (185). PROPHOS was used to prepare deuterated L-Orn (88% ee) (303). Diphosphines based on an oxygenated bicyclic skeleton were derived from camphor, giving from 3 to 90% ee for reductions of dehydroPhe derivatives (463). Camphor was also converted into a hydroxymethyl-substituted monocyclic backbone, but enantioselectivity was low (11–57% ee) for reductions of ΔAla and ΔPhe (464). Another diphosphine ligand based on a camphor backbone induced 99.6% ee for a hydrogenation leading to Phe (465). Several chiral diphosphine ligands based on a pinene core were reported in 2004, with the best giving 79–84% ee for reductions leading to Ala and Phe (466).

A more flexible chiral backbone ligand, 2,3-bis(diphenylphosphino)butane (CHIRAPHOS), was used to prepare Ala, Leu, Phe, Tyr, and DOPA with 83–100% ee (467). NORPHOS and CHIRAPHOS both gave 92% ee during hydrogenation of the substrate leading to phosphinothricin, 2-amino-4-methylphosphinobutyric acid (468). Another simple flexible derivative, 2,4-bis(diphenylphosphino)pentane (BDPP or skew-PHOS), reduced Ac-ΔPhe-OH with 94% ee and Ac-ΔPhe-OMe with 78% ee (469). Improved enantioselectivity (97–98% ee) was induced by analogs containing substituents on the phosphine phenyl groups (469). Other analogs have been prepared by replacing the backbone methyl substituents with phenyl or cyclohexyl rings. However, enantioselectivity for reduction leading to Phe was greatly reduced for the aryl analogs (2–20% ee) and only moderate for the cyclohexyl derivatives (53–75%), significantly lower than BDPP (75–76%). In contrast, for reductions leading to 3-aminobutyrate the analogs gave much better enantioselectivities than BDPP, with up to 98% ee (470, 471).

A variety of other chiral linker backbones have been developed. Pyrrolidine rings are used as the backbone in BPPM (261, 472), its diastereomer (473), and a number of N-substituted PyrPHOS ligands, of which BenzoylpyrPHOS was the most effective, reducing a number of dehydroamino acids with 15–100% ee (474). The latter ligand has been attached to the Merrifield resin and used to reduce α-(acetylamino)cinnamic acid with 95% ee (475). The nitrogen substituent of the PyrPHOS ligand has also been employed to attach the ligand to a PEG polymer (476), or to the focal point of an aryl-based dendrimer, with dendrimer analogs providing Phe with up to 98.7% ee (477). Another PyrPHOS-based dendrimer, based on bis(3-aminopropyl)amino monomers, gave 93% ee for the first generation dendrimer, reduced to 88% ee for the fourth generation (478). BPPM has also been linked to an amphiphilic polymer support; ΔPhe was reduced with 85% ee (479). A BPPM phenylurea analog, CAPP, was compared to another urea analog linked to a silica support. Both gave similar enantioselectivities for reduction of Ac-ΔPhe-OMe (90–94% ee), with some loss in ligand activity when the silica-supported reagent was recycled (480). A glucose-derived diphosphine based on a glucofuranoside backbone induced 98% de for reductions leading to Ala and Phe (481).

Heteroatoms have also been used to link the phosphine to the chiral backbone in several ligands, such as N-iPr-O,N-bis(diphenylphosphino)-1-(1-naphthyl)-3-amino-2-butanol (PROPRAPHOS). Twenty-one different (Z)-α-N-benzoyl-β-arylacrylic acids and methyl esters were reduced with a cationic rhodium complex of PROPRAPHOS with 65–92% ee (482). This catalyst was also used to reduce some β-pyridyl- (187) and β-thienyl- (188) didehydroamino acids, with 70–90% ee. A norbornane-based BCOP analog catalyst gave similar results (187, 188), while DIOP was not very successful (20–77% ee) (188). Another bicyclic ligand, derived from ketopinic acid, gave 42–90% ee for reductions leading to several arylalanines (483). A number of bisphosphinite ligands based on a bicyclic backbone gave modest enantioselectivities for reductions providing Phe (84–91% ee) and Ala (68–72% ee) (484). The DPAMPP ligand consists of a chiral aminoalcohol backbone, 1,2-diphenyl-2-(methylamino)-ethanol, with diphenylphosphines attached to the amine and oxygen. It induced >95% ee during reductions of aryl-substituted dehydroalanines, and >93% ee for Ala and homophenylalanine (485, 486). The reduction leading to Hfe, with 95.7% ee, was compared to other catalysts (BINAP, DIPAMP, BDPP, PPM), which only gave 14.4–69.4% ee (487). A 1,2-diphenylethane-1,2-diol backbone linking two 7-phosphanonorbornene-based phosphonite groups gave 94% ee for reduction of dehydroPhe (488). A simple phosphine/phosphinite hybrid ligand, O-(diphenylphosphino) 3-(diphenylphosphino) butanol, induced 91% ee for reductions leading to Ala and Phe (489). Several C2-symmetric bisphosphinite and a bisaminophosphine ligand based on the 1,6-diphenyl-2,5-diamino-3,4-dihydroxyhexane backbone gave up to 98% ee for reductions leading to Phe, and 68% ee for reductions leading to Ala (490).

A bisphosphinite ligand based on a bi(cyclopentanol) backbone gave enantioselectivities of 46–96% ee for alkyl- and aryl-substituted dehydroalanines (491); a bisphosphinite spirocyclic cyclopentane analog (spirOP) gave 96.2% ee for a reduction leading to homophenylalanine (275). The bisphosphinamidite analog of spirOP, SpiroNP, provided impressive enantioselectivities of 94->99% ee for reductions of 11 (Z)-2-acetamido-3-arylacrylic methyl esters, with slightly reduced 78->98% ee for the acid analogs (492). DIMOP is a ligand with a D-mannitol-derived backbone attached to two diphenylphosphines via ether linkages. Reductions leading to Ala and Phe proceeded with 90–97% ee (493). Two diphosphinite ligands based on a D-xylose backbone reduced ΔAla with up to 78% ee, but ΔPhe with only 35% ee (494). DIOP-like D-glycopyranoside-derived ligands have also been reported. Catalysts employing these ligands reduced ΔPhe with up to 99% ee depending on the substituents on the phosphinite aryl groups (495, 496). Either D- or L-amino acids can be produced using the same D-glucose backbone, depending on the juxtaposition of the diphosphinites (495). A range of other sugar-based ligands have been tested, consistently giving >95% ee for reductions of aryl-substituted dehydroamino acids (497). Hydrogenations of alkyl-substituted enamides leading to Ala and Leu gave lower optical purities (91–97% ee) (497). A water soluble chiral Rh bis(phosphonite) complex derived from β,β-trehalose gave 94–>99% ee for hydrogenations producing a range of β-arylalanines; the presence of sodium dodecyl sulfate was critical for maintaining high enantioselectivity, apparently due to the formation of micelles (498, 499). A very similar mixed phosphine-phosphite α,α-trehalose derivative gave reduced enantioselectivity (10–73% ee) (500). Another sugar backbone combined a diphenylphosphine substituent with several different biaryl phosphite substituents, with a tetra-t-butyl-substituted biaryl derivative producing 98.8–>99% ee for reductions producing Ala and Phe (501).

A number of bidentate ligands have been developed based on axially disymmetric 1,1-binaphthalene as the backbone, with syntheses of a number of these ligands included in reviews in 2003 (502, 503). The best recognized of these is the bisphosphine derivative BINAP (299, 504–506), the rhodium complex of which catalyzed the reduction of a number of derivatized aliphatic and aromatic (E)- and (Z)-didehydroamino acids with 70–100% ee. Reduction of the (Z)-isomer with (R)-BINAP led to the (R)-amino acid with 70–98% ee, while the (E)-isomer gave the (S)-amino acid with 70–96% ee (quantitative yield). However, reduction of an α-enamide with a 3-furyl substituent gave only 12% ee (298). The BINAP complex has broader utility than other catalysts (506) and has been used recently to prepare a number of α-aminodicarboxylates (299). The mechanism of asymmetric hydrogenation of α-(acylamino) acrylic esters by Ru(II)-(S)-BINAP was examined in 2002 using kinetic studies, deuterium labeling experiments, isotope effect measurements, and NMR/X-ray analysis; the sense of asymmetric induction is opposite to that provided by Rh(I)-(S)-BINAP catalysts (507). The development of the BINAP ligand was outlined in the 2001 Nobel prize lecture of Ryoji Noyori (508).

A BINAP analog with steroidal substituents gave 87–97% ee for reductions leading to Ala and Phe (509). Recyclable BINAP ligands with PEG or guanidine substituents to assist in catalyst recovery gave high enantioselectivity (94–95% ee) for reduction of Ac-ΔAla (510), while a polymeric BINAP derivative gave poor results (40–76% ee) for reductions of ΔAla and ΔPhe (511). A BINAP ligand with perfluoroalkyl substituents was employed for an asymmetric hydrogenation with supercritical CO_2 as solvent; methyl 2-acetamidoacrylate was reduced with 74% ee (512). Another fluorous BINAP derivative, with the perfluoroalkyl substituents on the phosphine aryl group, gave only 15% ee for reduction of ΔPhe (513).

Heteroatom-linked BINAP analogs have been developed. The oxygen-bridged phosphinoyl analog of BINAP, BINAPO (2,2′-bis(diphenylphosphinoyl)-1,1′-biphathyl), gave 64% ee for reduction of ΔPhe (514). Its octahydroanalog, H_8-BINAPO, gave improved enantioselectivity of 64–84% for a series of substrates (514). Replacing the phosphine phenyl groups with xylyl groups gave Xyl-H_8-BINAPO. Arylalanines, Ala, and Abu were produced with 90–97% ee (514). Introducing 3,3′-diphenyl substituents onto the BINAPO ligand gave Ph-O-BINAPO, which provided Ala with 99.9% ee and nine arylalanines with 81.5–97.3% ee, (515, 516). The corresponding carba analogs, NAPHOS and Ph-o-NAPHOS, gave 54% ee and 98.7% ee, respectively, for the reduction of ΔAla. Ph-o-NAPHOS was then used to produce six arylalanines, with 95.8–97.8% ee (516). A 1,1′-binaphthyl derivative, 2,2′-bis(dicyclohexyl-phosphinoamino) and the corresponding 2,2′-bis[bis (3,5-dimethylphenyl)phosphino-amino]-1,1′-binaphthyl ligand, both gave very high enantioselectivities (96–99% ee) for reductions leading to Ala and a range of 12 different substituted phenylalanines (517). 4-Methoxyphenyl substituents on the phosphine atom further improved enantioselectivity, giving 98.6% ee for a reduction of Ac-ΔPhe-OMe (518). BINAP analogs with oxygen- or nitrogen-linked diphenylphosphinite or diphenylphosphoramidite substituents on a 5,5′,6,6′, 7,7′,8,8′-octahydrobinaphthyl backbone ligands gave 90–97% ee for a range of substrates (519). Other BINAP analogs have been made in which one of the diphenylphosphine substituents is replaced with a phosphinite (BIPNITE) or chiral phosphite (BINAPHOS) substituent (520). For good chiral selectivity, an additional 3-phenyl substituent on one naphthyl ring was required, giving o-BIPNITE and o-BINAPHOS. These ligands induced 95–>99% ee for reductions of six different dehydroarylalanines (520). A fluorous version of BINAPHOS was used with a supercritical CO_2/aqueous biphasic solvent, allowing for efficient recycling of the catalyst. Ac-ΔAla-OMe was reduced with 97.3–99.1% ee over five cycles, with an average of 98.4% ee (521).

Biaryl-based analogs of BINAP have been developed. Biarylphosphine ligands based on 2,2′-bis(diphenylphosphine)-6,6′-dimethoxy-1,1′-biphenyl (BIPHEP) included a 6,6′-dimethoxy-3,3′-diphenyl analog, which gave much better enantioselectivities than BIPHEP for reductions leading to Ala and Phe (98–>99% ee, vs 21–46% ee (522). BIPHEP-type ligands were attached to a silica support, but both gave poor enantioselectivities (29–45% ee for reduction producing Phe) (523). A model for predicting the enantioselectivity of biaryl-type diphosphine ligands has been developed; some new biaryl ligands reduced ΔPhe with 98% ee (524). A bipyridyl backbone is employed in the ligand P-Phos, with other analogs made using different phosphine substituents. These ligands were examined for reductions leading to arylalanines, with variable results (38–97% ee) depending on ligand and substrate (525). Biaryl ligands with an oxygen or amine linker to the phosphine have been prepared, with the best giving giving 86–95% ee for reductions of ΔAla and ΔPhe (526).

A paracyclophane backbone was employed in the PhanePhos ligand, which gave 91–99.6% ee for reductions leading to Ala, Phe or Abu, but only 51% ee (with reversed enantioselectivity) for an Ac-ΔVal-OMe substrate (527). An analog of the PhanePhos ligand used phosphonites bearing biphenoxy or binapthoxy substituents instead of the phosphine substituents; reductions produced Ala and Phe with 99% ee (528). An iridium complex with a [2.2]paracyclophane skeleton containing a diphenylphosphine and imidazolylidene substituents gave moderate enantioselectivity (71% ee) for reduction of Ac-ΔPhe-OH (529). A chiral backbone based on an 8,12-diphenylbenzo[a]heptalene ring system produced a maximum of 77% ee for reduction leading to Phe, though a chiral amplification effect was observed with ligand of only 2% optical purity producing Phe with 25% ee (530).

A variety of ferrocene-based ligands have been developed. TRAP, a biferrocenyl-based ligand, was tested with a number of mono- and di-β-substituted α-enamides, in combination with various alkyl groups on the TRAP ligand phosphine atoms. Amino acids with 57–97% ee were obtained (531), including 2,3-diamino acids (532) and β-hydroxy amino acids (533). The latter products were obtained as the erythro diastereomers with 92–97% ee by reduction of (Z)-β-oxy-α-acetamidoacrylates, while the threo isomers were produced with 73–97% ee from the (E)-substrates (533). The TRAP ligands have been applied to the hydrogenation of 1,4,5,6-tetrahydropyrazine-2-carboxamide, producing 4-azapipecolic acid. The i-Bu-TRAP ligand resulted in the (S)-amino acid with 97% ee, while the Me-TRAP ligand gave the (R)-enantiomer with 85% ee (534). The Ph-TRAP ligand was also used to reduce 2,3-dehydro cyclic imino acids, producing Pro, Pip, homoPip, Tic, and 4-oxo-Tic with 73–97% ee (535). Another ferrocene-based ligand is the C2-symmetrical FERRIPHOS derivative, with chiral arylethyl substituents on the ferrocene backbone. Three didehydroamino acids were reduced with 97.7–98.7% ee (536). Dimethylamino substituents were added to this ligand to give diamino FERRIPHOS; Ala and arylalanines were produced with 96.3–98.7% ee (537). (S,S)-FerroPHOS is a cylindrically chiral ferrocenyldiphosphine ligand which produced Ala, Phe, and β-(naphthyl)alanine with 95.7–98.8% ee (538). An analog with bis(t-butyl)sulfonyl substituents provided 93% ee for reductions leading to Ala (539). A series of other chiral ligands based on a ferrocenyl nucleus, Mandyphos and Taniaphos, were screened for asymmetric reduction selectivity against a variety of substrates, including ΔPhe. A number of the derivatives induced >98% ee (540). Taniaphos analogs in which a dimethylamino substituent was replaced by a methoxy group gave 99% ee for reduction of ΔPhe; replacement with alkyl groups gave 50–94% ee (541). A variety of other Taniaphos-like ligands were reported in 2002, with up to 96.6% ee for reductions leading to Phe (542). A phosphinoferrocenylaminophosphine ligand, BoPhoz, induced high enantioselectivites (97.2–99.5% ee) for reductions of N-Ac, N-Bz, or N-Boc

substrates leading to Ala, β-cyclopropyl-Ala, β-furyl-Ala, Hfe, Phe, 4'-cyano-Phe, 4'-nitro-Phe, 4'-chloro-Phe, 4'-fluoro-Phe, 4'-methoxy-Phe, 3'-methoxy-Phe, 2'-methoxy-Phe, β-(1-naphthyl)-Ala and β-(2-naphthyl)-Ala (543).

A planar-chiral phosphaferrocene catalyst containing a phosphine within the ferrocene ring system provided reduction enantioselectivities of 79–96% ee for alkyl and aryl amino acids (544). Two other ferrocene-based bis(phosphine) ligands with only planar chirality have been reported: a 1,3-bis(phosphane) system gave 34–77% ee for reductions of ΔPhe derivatives (545), while a 1,2-bis(phosphane) system gave only 0–23% ee (546). A hybrid ligand combining a diaryl phosphite substituent with a ferrocenylphosphine gave enantioselectivities of 95.3–99.6% ee for eight arylalanines (547). Another planar chiral ligand employs an arene tricarbonylchromium(0) complex; with some combinations of substituents up to 95% ee for reduction of ΔPhe was possible (548). A diphosphine ligand based on a chiral rhenium complex, with chirality residing on the Re atom, was used in Rh complexes to give up to 92% ee for reduction of ΔAla, and 93% ee for reduction of ΔPhe (549).

A novel approach for induction of chirality involves attaching an achiral bis(diphenylphosphine)-Rh system to biotin, as reviewed in 2005 (550). The biotinylated ligand–catalyst was then complexed to a host protein, avidin or streptavidin, with the resulting artificial metalloenzyme inducing up to 96% ee during hydrogenation of acetamidoacrylic acid (551). Insertion of a Phe or Pro linker between the biotin and bis(diphenyl-phospinoethyl)amino moiety gave reduced enantioselectivity (552). A set of 360 artificial metallocatalysts based on this system was subsequently tested for catalytic activity and enantioselectivity for reductions of both ΔAla and ΔPhe (553). β-(Diphenylphosphino)-Ala and β-(Dicyclohexylphosphino)-Ala residues were incorporated into peptides designed to possess helical structure through the use of Aib and Ala residues. Rh complexes of these peptides induced up to 38% ee for reduction of Ac-ΔAla-OMe, albeit with only 2% conversion (554–556). An unusual class of ligands consists of helically chiral polymers formed from achiral monomers containing N and P atoms; the Rh complex of a phosphane-containing polyisocyanate reduced N-acetamidocinnamic acid to give Phe with 15% ee (557).

Very few ligands have been developed without a phosphorus atom. A chiral bis(thiourea) ligand based on a chiral 1,2-diaminocyclohexane backbone gave Ac-Ala-OMe with 61% ee and Ac-Phe-OMe with 70% ee (558).

4.3.2ciii Ligands with Chiral P-Substituents: DuPHOS Ligand

The third strategy for introducing chirality onto bis(phosphine)ligands based on DIPHOS is to incorporate the phosphine within a chiral ring system, rather than employing a chiral backbone linker or chiral phosphine atoms. The most effective of these ligands have been the C2-symmetrical 1,2-bis(2,5-dialkylphosphalano)benzene (DuPHOS) and 1,2-bis(2,5-dialkylphosphalano)ethane (BPE) ligands. As previously mentioned, the DuPHOS ligands provide some of the best results for hydrogenation of β-monosubstituted acetylaminoacrylic acids. Several different DuPHOS ligands have been prepared, varying in the substituents on the phosphalano ring systems. The *n*Pr-DuPHOS ligand appears to have the optimum steric requirements for the reduction of these substrates. Increasing the bulk of the ligand, such as with *i*Pr-DuPHOS, restricts substrate binding and reduces enantioselectivity (*425*). The DuPHOS catalyst ([(COD)Rh(DuPHOS)]⁺OTf⁻) has been applied to the reduction of a wide variety of α-acetamidoacrylate methyl esters, giving the *N*-Ac amino acid (25 examples) with >99% ee with both the Pr-DuPHOS and Et-DuPHOS catalyst (*559, 560*). Even substrates that commonly present problems, such as the sterically demanding *t*-Bu-substituted acrylate and the potentially catalyst poisoning methylthio-substituted-acrylate, are reduced with >95% ee. Both acrylate esters and acrylic acids are reduced. Similar enantioselectivities were encountered when supercritical CO_2 was used as solvent (*561*). An attempt to use ionic liquid phases for DuPHOS hydrogenations resulted in greatly reduced yields and enantioselectivity (*562*).

Perhaps most significantly, the DuPHOS ligand catalyzes the hydrogenation of both (*Z*)- and (*E*)-*N*-acyl-enamides with high enantiomeric excess (although the (*E*)-*N*-Cbz-enamides are reduced with somewhat lower selectivity), and both geometric isomers produce an amino acid with the same absolute configuration (*559*). Thus *E*/*Z* isomeric mixtures can be directly hydrogenated without need of an often difficult isomer separation. A computational modeling study in 2000 examined the reaction pathway and reasons for enantioselection (*422*). In addition to stereoselectivity, the DuPHOS reduction is very regioselective as the enamides are involved in chelation to the catalyst. Ketones, aldehydes, nitro groups, aryl and alkyl halides, and internal alkenes and alkynes, are not affected by the reduction conditions (30 psi H_2, 0.05 mol % catalyst) (*559*). α,γ-Dienamide esters were regioselectively reduced to γ,δ-unsaturated amino esters with 86–>99% ee (*305*) while β-substituted α,γ-dienamide esters were reduced by both DuPHOS and BPE-based catalysts to β-branched allylglycines with 82–96% ee (*563, 564*). The DuPHOS catalyst can also be used to hydrogenate *N*-Cbz-protected dehydroamino acid methyl esters without loss of the Cbz group, allowing for easy *N*-deprotection after synthesis (*141, 298, 559, 564–566*). The DuPHOS ligand is normally employed in a Rh complex with COD (cyclooctadiene). However, generation of the active catalyst requires prehydrogenation of the COD diene, which proceeds more slowly than generally assumed. In contrast, forming a precatalyst complex with norbornadiene (NBD) instead of COD gives a derivative which generates the active catalyst much more quickly, resulting in much faster conversion of Ac-ΔPhe-OMe substrate (4 min instead

of 15 min). Even when the substrate had been consumed at 15 min with the COD catalyst, only half of the COD had been reduced, meaning half of the catalyst was wasted (*567*).

The superiority of DuPHOS over other ligands such as BINAP and DIPAMP was confirmed in comparative reductions of thienyl-, furyl-, and pyrrolyl-substituted dehydroamino acids (*298*). Me-DuPHOS reduced all the (*Z*)-substrates with 94–99.5% ee, with *t*-butyl esters giving better enantioselectivities. The only poor result was with an (*E*)-*t*-butyl ester substrate (63% ee). Of the other ligands, BINAP gave very poor enantioselectivity (12% ee, one example) while DiPAMP was moderately successful (85–100% ee). In another study Et-DuPHOS gave the best results (99.3% ee) of a series of ligands for asymmetric reductions leading to Boc-β-cyclopropyl-Ala-OBn (Me- and Et-DuPHOS, 98.6 and 99.3% ee; BoPhoz, 98.6% ee; BINAP, 55% ee; DIOP, 40% ee; BPPFA, 34% ee; CHIRAPHOS, 0% yield; PROPHOS, 81.8% ee; PHANEPHOS, 5% ee; Josiphos, 86.8% ee) (*568*). The DuPHOS ligand reduced bis(dehydroamino acids) to give diaminodicarboxylic acids linked by various alkyl chain lengths as (*S*, *S*)- or (*R*,*R*)-enantiomeric diasteromers with 98.5–99.5% de and ee, considerably higher stereoselectivity than reductions catalyzed by either BINAP or CHIRAPHOS ligands (65–85% ds) and slightly higher than reductions with DIPAMP (97.5% ds) (*311*). Phenylbisalanine diaminodicarboxylic acids (*366, 569*) and the triaminotricarboxylic acid phenyltrisalanine (*313*) were also synthesized by reductions with this ligand.

A variety of other reductions using DuPHOS ligands have been reported (see Table 4.5), such as syntheses producing Cbz-Nle, Cbz-Cha, and Cbz-2-aminooctanoic acid (98.7–99.6% ee with the Et-DuPHOS-Rh catalyst) (*566*) and Bz-Dap(COR)-OMe (Dap = 2,3-diaminopropionic acid) with various acyl or urethane groups on the β-nitrogen (100% yield and 99% ee) (*196*). Glycosyl amino acids with glucosyl, galactosyl, and mannosyl side chains were synthesized with 82–>95% de at the α-center (*323, 324*), while both enantiomers of carba analogs of *O*-trisaccharide Ser were obtained with >98% de (*297*). Glycosylated 6-hydroxynorleucine derivatives were produced with >95% de, with no apparent asymmetric induction from the carbohydrate moiety (*325*). The DuPHOS ligand has been used in a tandem catalysis procedure developed for the synthesis of aryl-substituted Phe derivatives from a common dehydroamino acid intermediate. A variety of (bromoaryl)didehydroalanines were hydrogenated in the presence of Et-DuPHOS or Pr-DuPHOS with >98.5% ee (75–99% yield), and then subjected to a Suzuki cross-coupling to introduce an aryl substituent (*570*). A limitation of the DuPHOS ligand was observed in reductions of 2,3-dehydroproline and 2,3-dehydropipecolic acid, with the imino acids produced with only 0% and 26% ee, respectively. However, 2,3-dehydroimino acids of larger ring sizes (7-, 8-, 9-, 13-, and 16-membered rings) were reduced with 86–97% ee (*355*).

Given the tolerance of DuPHOS for both (*Z*)- and (*E*)-substrates, the hydrogenation of β-disubstituted

2,3-dehydroamino acids was investigated (*563, 564*). The least hindered Me-DuPHOS catalyst gave the best results (up to 99% ee) (*564*), which could be significantly improved by carrying out the hydrogenation in supercritical CO_2 (*561*). However, the bulkier tetrasubstituted acrylic acid substrates apparently cannot be readily accommodated within the rigid DuPHOS ligands. For these substrates the flexible and sterically less demanding Me-BPE ligand provided good enantioselectivity (*425*), producing a series of β-branched amino acids with >96% ee (*564*) and hydrogenating substrates that DuPHOS left unreduced. Both acrylate esters and free acids could be reduced. β-Branched amino acids containing disparate β-substituents could be selectively synthesized as either the *threo* or *erythro* diastereomers, depending on whether the (*Z*)- or (*E*)-substrate was reduced, although the (*E*)-isomer gave reduced selectivity. The discovery of the superiority of BPE ligands for sterically crowded substrates is somewhat ironic, as the BPE ligands were originally reported in the same papers as the DuPHOS ligands. However they did not give the high enantioselectivity desired for monosubstituted enamide reductions (from 64–99% ee, depending on ligand alkyl group), which led to the development of the more rigid DuPHOS ligands (*559, 560*). Despite the apparent limitations of the DuPHOS ligands for β-disubstituted substrates, asymmetric hydrogenation of *N*-acetyl (*Z*)-β-aryl-ΔAbu methyl esters with Rh(*R,R*-Et-DuPhos) or Rh(*S,S*-Et-DuPhos) catalyst produced (2*R*,3*S*)- or (2*S*,3*R*)-β-methylarylalanines, respectively, with 66–89% yield and >99% ee (*109*). Reductions of (*E*)- or (*Z*)-Bz-2,3-dehydro-Dab(Ac)-OMe (Dab = 2,3-diaminobutanoic acid) gave either (2*S*,3*R*)- or (2*S*,3*S*)-diastereomers with 96–>98% ee (*571*).

DuPHOS ligand derivatives include a recyclable heterogeneous catalytic system obtained by incorporating the Rh-MeDuPHOS catalyst into a polyvinyl alcohol film. Ac-ΔAla-OMe was reduced with up to 96% ee. The reaction could be carried out in water and the catalyst recycled without any subsequent loss in activity or enantioselectivity over two cycles (*572*). Rh-MeDuPHOS has also been occluded in a polydimethylsiloxane (PDMS) membrane, with Ac-ΔAla-OMe reduced with 91.1–96.9% ee, compared to 99.1% ee for the homogeneous catalyst (*573, 574*). Another heterogeneous system was created by using Rh-MeDuPHOS in an air-stable molten salt ionic liquid, [bmim][PF$_6$]. The catalyst was recycled 5 times, with enantioselectivity for reduction of ΔPhe changing from 96% ee to 94% ee (*575*). The Me-DuPHOS catalyst has been non-covalently anchored on a mesoporous aluminosilicate support; Ac-ΔAla was reduced with the same enantioselectivity (>98% ee) as with the regular catalyst. The catalyst could be filtered off and reused up to 3 times with no loss in enantioselectivity, though activity was significantly decreased (*443*).

Several similar types of ligands to DuPHOS, and other ligands based on chiral phosphine substituents, have been described. Degussa has commercialized a series of DuPHOS analogs in which the central linking phenyl ring has been exchanged for other four-, five-, or six-membered ring systems, allowing for fine-tuning of the P-Rh-P bond angle (*576*). A DuPHOS analog with a maleic anhydride backbone, MalPHOS, has a larger bite angle than DuPHOS, and gives similar enantioselectivity in reductions leading to Phe (*577*). ROPHOS is an ethyl-linked bis(phosphine) with the phosphine substituents derived from D-mannitol. Arylalanines were produced with up to 99% ee (*578, 579*). A ROPHOS analog with a phenyl linker backbone instead of the ethyl linker, and hydroxy substituents instead of alkoxy, also gave high enantioselectivity (98–>99% ee for reductions leading to Ala, Phe, and three other arylalanines) (*579*). Another analog (BASPHOS), with a different hydroxy-group substitution pattern, was water soluble and catalyzed the hydrogenation of AcΔAla-OH with 99.6% ee (*580*). Ethyl-linked phosphane-borohydrides, with a P-BH$_3$ substitution and the P contained in a chiral oxaphosphinane ring, produced Ala and Phe with 94.2–97.5% ee (*581*). Pyrophosphites based on two binaphthol phosphites bridged by an oxygen gave poor enantioselectivity (18–48% ee) for reduction giving Phe (*582*).

4.3.2civ Ligands with Multiple Chirality
There have been several other ligands combining two types of chirality. TangPhos, a ligand with both chiral backbone and chiral phosphine center contained in a five-membered ring, induced >99% ee for a series of arylalanine products (*583*). Aryl groups have been added to TangPhos to generate DuanPhos, which reduced ΔAla and ΔPhe with >99% ee (*584*). The ring-strained dibenzophosphetenyl 4-membered ring analog of DuanPhos reduced Ac-ΔPhe-OMe with 96% ee (*585*). The chiral phosphine substituents of the DuPHOS and BPR ligands have been combined with a chiral cyclopentyl backbone; depending on the backbone chirality they gave better (95–98% ee) or worse (73–77% ee) enantioselectivities than BPE (85–91% ee) for reductions leading to Ala and Phe (*586*). A 2,2'-bispholene ligand also combined backbone and phosphine-centered chirality, but gave only 75% ee for reduction of ΔPhe (*587*). A DIPAMPP pseudoephedrine-type backbone was combined with chiral phosphine centers to give a series of AMPP (aminophosphine phosphinite) ligands; ΔPhe was reduced with up to 99% ee depending on the phosphine substituent (*588*).

A ferrocenyl backbone was employed in combination with DuPHOS-type phosphalono ring systems. With methyl substituents on the phosphetane ring, the ligand induced 69–90% ee, producing (*S*)-amino acids. However, with isopropyl substituents, enantioselectivity was reversed and (*R*)-amino acids were produced with 83–94% ee (*589*). Another ligand employed a ferrocenyl backbone in conjunction with chiral substituents on two attached phosphonites, formed from homochiral binaphthol. This ligand catalyzed the hydrogenation of Ac-ΔAla-OMe to Ac-Ala-OMe with 99.5% ee (*590*). A ferrocenyl backbone ligand with chiral phosphine atoms containing a number of different substituents induced up to 98.7% ee in the hydrogenation of Ac-ΔPhe-OH (*591*). Combining a ferrocenyl backbone with D-mannitol-derived phosphine substituents produced a Rh catalyst ligand that induced 97.8–>99.9% ee for reductions leading to various Ala, Leu, and arylalanines. A β-disubstituted substrate was reduced to give Val with 87.3% ee (*592*). A hybrid ligand containing a BINOL-phosphoramidite and a ferrocenylphosphine was reported by a number of groups; reductions with the catalyst complex of this ligand produced six arylalanines with 99.5–99.8% ee (*593*), a different set of six arylalanines with 97.4–99.6% ee (*547*), or Ala, Phe, naphthylalanine, and cyclopropylalanine with 81–99.9% ee (*594*).

The QUINAPHOS ligand contains a chiral 2-alkyl-1,2-dihydroquinoline backbone, with a chiral binaphthol-substituted phosphoramidite at the quinoline nitrogen and a bisarylphosphine at the 8-position. ΔAla was reduced with 97.8% ee (*423*). A xylose-derived backbone was employed to link a diphenylphosphine substituent to a phosphite substituent with a chiral biphenol ligand. The resulting bidentate ligand induced 98.8–>99% ee for reductions leading to Ala and Phe (*595*). A similar ligand with two biphenol phosphite substituents only gave 37% ee for reduction of ΔAla (*596*), but a xylose-based ligand with chiral biphenol phosphite and biphenol phosphoramidite substituents gave 98% ee for reductions of Ac-ΔAla-OMe and Ac-ΔPhe-OMe (*597*). An ethyl-linked bis(phosphine) ligand joined one chiral phosphine atom and one phosphine contained within a chiral ring system. It produced Phe with up to 88–92% ee (*598*). An ethyl- or phenyl-bridged ligand containing two chiral phosphine substituents within benzyl-substituted 5-membered rings reduced ΔAla, ΔPhe, and 2,3-dehydrocyclopentylglycine with 86–98% ee (*599*). The BIPNOR ligand contains two chiral bridgehead phosphorus atoms which combine the properties of chiral phosphine, chiral phosphine substituent, and chiral backbone derivatives. Only one dehydroamino acid substrate was reduced, but N-Ac Phe was obtained with >98% ee (*600*).

4.3.2cv Monodentate Ligands

In 2000 a new ligand, MonoPhos, revisited an older strategy based on monodentate ligands. A phosphoramidite ligand prepared readily from (*S*)-1,1'-bi-2-naphthol and hexamethylphosphorus triamide induced enantioselectivities of 98.4–99.8% ee for Rh-catalyzed hydrogenations leading to Ala, Phe, and *O*-acetoxy-*p*-methoxy-Tyr (*601*). A very similar binaphthol-derived monophosphite ligand reported the same year also gave high enantioselectivity, with Ac-ΔAla-OMe reduced with up to 95.5% ee (*602*). The same binaphthol-derived and additional 9,9'-biphenanthrol monophosphite ligands were developed by a third group (up to 92% ee for Ala, 80% for Phe) (*603*), while a fourth group reported a series of binaphthol-derived monophosphonite ligands that reduced Ac-ΔAla-OMe with up to 94% ee (*604*). The binaphthol monophosphite has also been combined with a silsesquioxane silanol substituent

(only 64% ee for reductions leading to Phe) (*605*), a menthol substituent (85% ee for reduction of ΔAla) (*606*), a set of carbohydrate substituents (up to 80–89% ee for reductions leading to Ala and arylalanines) (*607*), and a D-mannitol-derived ligand (96.6–98.0% ee for hydrogenations leading to Ala and several arylalanines) (*608*). BINOL-derived monodentate acylphosphite ligands gave only 7–50% ee for reductions of Ac-ΔPhe-OMe (*609*).

A variety of MonoPhos phosphoramidite analogs have been described. A comparative rate study found that monodentate phosphoramidite catalysts could lead to higher rates and/or higher enantioselectivities than state-of-the-art bidentate catalysts (*610*). A MonoPhos analog with diethylamine replacing the dimethylamine P substituent induced 98.0–>99.9% ee for a series of 11 arylalanines (*611*). A parallel synthesis of 32 monodentate phosphoramidite binaphthol-derived ligands was combined with in situ screening for asymmetric hydrogenation of amino acid precursors. Different catalysts were optimal for reduction leading to Phe than for reduction leading to 3-aminobutyrate; one of the best was the analog with a diethylamine P substituent (*612*). Another library identified that a piperidine or morpholino phosphine substituent gave 98–99% ee for reductions leading to Ala or Phe (*613*). Adding a single *ortho* substituent to the MonoPhos binaphthol ring, which reduces the C_2 symmetry to C_1 and creates a stereogenic center at the P atom, allowed for hydrogenations of ΔAla with >99.0% ee (*614*). Other MonoPhos analogs are based on an octahydro H_8-BINOL template, with various substituents on the phosphoramidite amine, though no amino acids were tested as substrates (*615*).

In 2004 dimers of the MonoPhos ligand were created by cross-linking the binaphthyl aryl group. Formation of the Rh complex, which requires complexation by two separate dimers, results in a polymerization that produces a toluene-insoluble, yellow polymeric solid. The heterogeneous catalyst system produced Ala, Abu or Phe with 95.8–96.6% ee (*616*). Another dimeric MonoPhos-type ligand replaced the dimethylamine of two MonoPhos monomers with a bridging achiral azabicyclic[3.3.1] or [3.3.0] framework; Ac-ΔAla-OMe was reduced with 90% ee (*617*). Polymeric MonoPhos analogs have also been synthesized by co-polymerization of styrene with an analog in which the dimethylamine is replaced with *p*-vinylaniline or 3-vinyl-8-aminoquinoline. The monomeric ligands reduced ΔAla and ΔPhe with only 33–84% ee (compared to 97–>99% ee for MonoPhos), while the polymers induced 49–80% ee. One catalyst was reused 4 times with no loss in enantioselectivity (*618*). MonoPhos has been immobilized on an aluminosilicate support, AITUD-1. Depending on solvent, the catalyst could give as high enantioselectivity as the homogeneous catalyst (97% ee for reduction of Ac-ΔAla-OMe), could be used in water (95% ee), and could be recycled up to 3 times with minimal loss in activity and selectivity (*619*).

Monodentate ligands have been prepared based on a biphenol core, with both phosphite and phosphoramidite derivatives, and various substituents on the biphenol ring.

While individual ligands (16 tested) reduced Ac-ΔAla-OMe with only a maximum of 52% ee, screening binary combinations of ligands identified a pair that gave 87% ee (620). Another library screened 11 phosphite and eight phosphoramidite ligands based on a biphenol (TROPOS) unit with chiral alcohol or amine; Ala was obtained with up to 94% ee, Phe with up to 98% ee, and 3-aminobutyrate with up to 71% ee (621). The use of combinatorial libraries for chiral catalyst synthesis and screening was reviewed in 2004 (622). A biphenol phosphite derivative, with the biaryl ring tethered together by a 6,6'-oxyethyloxy linker, provided Ac-Ala-OMe with 93% ee (623). Another MonoPhos analog used a BICOL bicarbazole backbone instead of the binaphthol unit. N-Methylation of the carbazole nitrogen gave a ligand that reduced ΔPhe with 93% ee (vs 95% ee for MonoPhos), while attachment to a carbosilane dendritic wedge gave a catalyst that induced 95% ee (624). The SIPHOS dimethyl phosphoramidite ligand relies on a 1,1'-spirobiindane-7,7'-diol backbone. Ten N-acetyl dehydroamino acids were reduced with 97.1–99.3% ee (625). Three 4,4'-disubstituted SIPHOS ligands were also synthesized, with SIPHOS, 4,4'-dibromo-SIPHOS, and 4,4'-diphenyl-SIPHOS all giving similar enantioselectivities (94.5–99.1% ee) for reductions of dehydroarylalanines (626). Several chiral phosphonites were also derived from the 1,1'-spirobiindane-7,7'-diol backbone, giving 97–99% ee for reductions leading to nine different arylalanines (627).

The Rh(I) complex formed from a chiral S-trityl O-diphenylphosphine 2-hydroxycyclohexanethiol ligand gave 94% ee during reduction leading to 3'-nitro-4'-fluoro-Phe, notably allowing for chemoselective alkene reduction in the presence of the nitro group (628). A subsequent report greatly expanded the substrates tested with a S-t-butyl analog, with 83–98% ee for the production of a variety of arylalanines with different protecting groups, as well as Abu and Nva. Leu was synthesized with greatly reduced enantioselectivity (36% ee) (629). However, a modified acyclic ligand, O-diphenylphosphine 2-hydroxy-3-(3,5-dimethylphenylthio)-4-methylpentane, gave high enantioselectivity for reductions leading to Leu (89% ee) and other bulky alkyl amino acids, Val (93% ee) and cyclohexylglycine (95% ee) (629).

Other monodentate ligands include binaphthyl-derived phosphine ligands, which gave 84–95% ee for reductions of several dehydroarylalanines (630). Additional substituents, and thus stereocenters, have been added to the carbons between the phosphine and the binaphthyl unit in the monodentate phosphine ligand. The hydrogenation enantioselectivity (up to 91% ee for Phe) was substantially controlled by the α-C chirality, rather than the biaryl chirality (631). A catechol-based phosphoramidite with a chiral 2,5-diphenylpyrrolidine substituent (thus with chirality on the amine substituent, instead of the diol substituent) induced 66–92% ee for reductions leading to Ala and Phe (632). A monodentate ligand corresponding to one-half of the phospholane-type BPE ligand, 1,2,5-triphenylphospholane, was

protected from oxidation as the tetrafluoroborate salt, the ligand was mixed with Rh catalyst to reduce ΔPhe with up to 89.5% ee (633). Monodentate ligands based on chiral phosphine oxide atoms have also been reported; reductions of ΔAla and ΔPhe derivatives proceeded with only 25–53% ee, though sterically crowded β,β-disubstituted 2,3-dehydrocyclohexylglycine was reduced with up to 85% ee (634).

Mixtures of different monodentate ligands have been tested in a combinatorial fashion, with many of the heterocombinations providing higher enantioselectivities than the usual homocombinations (635). The revival of chiral monodentate monophosphorus ligands in asymmetric hydrogenation was highlighted in 2001 (636), with a more comprehensive review in 2004 (432).

4.3.2cvi Chiral catalysts for Reductions Producing β-Amino Acids

Asymmetric hydrogenation of β-(acylamino)acrylates, leading to β-amino acid products, is also possible (see also Section 11.2.3g). The mechanism for asymmetric hydrogenation of β-acylaminoacrylates using cationic Rh(I) complexes was compared to that of α-acylaminoacrylates in 2005. The reaction sequence for both hydrogenations was determined to be the same, but for the β-acylaminoacrylates the catalyst–substrate complex dominant in solution provided the major product of the reactions. In the case of α-acylaminoacrylates it is the minor solution complex that leads to product, due to the minor complex possessing much greater reactivity. This extreme difference in reactivity does not appear to apply in the case of hydrogenations leading to β-amino acids (637).

Two 2002 studies compared several phosphine ligands (Me- and Et-DuPHOS, Me-BPE, Me4-BASPHOS, DIOP, HO-DIOP, DIPAMP, Et-FerroTANE) for the reduction of (Z)- and (E)- methyl β-acetylamino butenoate; Me-DuPHOS gave the best results (88% 99% ee) for both substrates. Several other substrates were then reduced (638, 639). However, in another report Et-FerroTANE provided better enantioselectivity (98–>99% ee) than DuPHOS for reductions of β-alkyl and β-aryl (E)-isomers (640). Temperature was found to play a crucial role in the enantioselectivity for reductions of β-acetamido-β-alkylacrylates with Et-BPE or Et-DuPHOS catalysts, particularly the (Z)-isomer, with the best results around 30–40 °C (641). Rh-Me-DuPHOS and Rh-BICP catalysts both provided good enantioselectivity for reductions of (E)-β-substituted-β-(acylamino)acrylates, with DuPHOS tending to give slightly better results (97.6–99.6% ee) than BICP (90.9–97.0% ee). In contrast, for the (Z)-isomers, BICP (86.4–92.9% ee) was substantially better than DuPHOS (21.2–62.4% ee) (642). A subsequent report found that polar solvents greatly improved the enantioselectivity of DuPHOS reductions of the (Z)-isomers, with up to 87.8% ee for reduction leading to 3-aminobutanoic acid (643). For α-substituted-β-amidoacrylate substrates, leading to β²-amino acid products, enantioselectivities with DuPHOS and BPE only reached a maximun of 67% ee (*644*). The DuPHOS-Rh catalyst has also been applied to asymmetric hydrogenations of *N*-acetyl or *N*-Cbz enamido phosphonate substrates, producing the α-aminophosphonate analogs of Ala, Abu, Leu, and Phe with 71–95% ee (*645*). *N*-Protected pyrrolidine enamino acid methyl esters were reduced by asymmetric hydrogenation with a variety of catalysts. Me-DuPHOS produced pyrrolidine-2-acetic acid with >99% ee but only 37% conversion, while Me-BDPMI gave 100% conversion with 96% ee (*646*).

A variety of other catalysts have been tested in reductions leading to β-amino acids. Earlier attempts at reduction with a BPPM-Rh complex gave products with only 53–55% ee (*647*), while a BINAP-Ru(II) catalyst induced 90–96% ee (*648*). A biaryl diphosphine ligand produced β-methyl-, β-ethyl-, β-*n*-propyl-, β-*i*-propyl, or β-*t*-butyl-β-Ala with 97.7–99.8% ee (*649*), while a biaryl diphosphine ligand with a chiral bridge between the aryl groups produced β-methyl-, β-ethyl-, β-*n*-propyl-, β-*iso*-propyl, or β-*tert*-butyl-β-Ala with 97.7–99.8% ee (*649*). A Ru-bipyridyl complex induced 98.3–99.7% ee for reductions of (*E*)-β-substituted-β-acetamidoacrylates, but 68.3–82.3% ee for the (*Z*)-isomers (*650*). Analogs of the flexible BDDP ligand with phenyl or cyclohexyl rings, 1,3-bis(diphenylphosphino)-1,3-diphenyl- or -dicyclohexyl-propane, provided up to 98% ee for reductions leading to 3-aminobutyrate from the (*E*)-substrate, and 81% ee from the (*Z*)-substrate, despite very poor enantioselectivities for reductions leading to α-amino acids (*470, 471*). A "three-hindered quadrant" chiral ligand featuring three bulky *tert*-butyl groups attached to two phosphorus centers provided high enantioselectivity (98–99% ee) for reduction of a series of (*E*)-3-substituted-β-acetamidoacrylates. More importantly, the more common (*Z*)-isomers were also reduced with high stereoselectivity (92–98% ee) (*651*). A diphosphine ligand based on a camphor backbone induced 99.3% ee for a hydrogenation leading to 3-aminobutyric acid (*465*), while another ligand with an imidazolidin-2-one backbone, BDPMI, reduced a number of (*Z*)-substrates with equal or better enantioselectivity than the (*E*)-isomers, giving products with 75.6–97.4% ee (*652*).

The P-chirogenic *tert*-Bu-BisP and *tert*-Bu-Mini-PHOS catalysts reduced four different (*E*)-substrates with 97.2–99.7% ee (*653*). In 2005 a ligand with two chiral phosphine centers connected by a quinoxaline backbone was prepared, QuinoxP*. Unlike other P-chirogenic ligands, this ligand is an air-stable solid. Methyl 3-acetaminobutyrate was produced with 99.2% ee (*456*). An ethyl- or phenyl-bridged ligand containing two chiral phosphine substituents within benzyl-substituted 5-membered rings reduced methyl-3-acetamido-2-butenoate with up to 96% ee (*599*). P-Chirogenic trialkylphosphonium salts induced >99% ee for reduction of (*E*)-β-methyl-β-acetamidoacrylate, but only 64% ee for the (*Z*)-isomer (*455*). A new bisphospholane ligand, MalPHOS, was compared against DuPHOS for hydrogenations of both (*E*)- and (*Z*)-isomers, with similar results

for *E*-isomers (79–>99% ee) but improved enantioselectivity for the (*Z*)-substrates (58–90% ee vs 4–86% ee) (*577*). *N*-Unprotected β-substituted enamines were reduced with a Rh-ferrocenophosphine complex to give β-amino esters with 95–97% ee (*654*). A ferrocenyl-binaphthol phosphoramidite hybrid ligand gave 96–>99% ee for reductions of (*Z*)-β-aryl-β-(acylamino) acrylates, and 92–99% ee for (*E*)- or (*Z*)-β-alkyl substrates (*655*). TangPhos, a ligand with both chiral backbone and chiral phosphine center, induced 74–99.5% ee for reductions leading to a variey of β-substituted β-amino acids from mixtures of *E*/*Z* substrates (*656*). A related bis(binaphthophosphepine sulfide) catalyst provided 3-methylbutyric acid with 99.2% ee from the *Z*-isomer (but only 32.7% ee from the *E* isomer), and 96–>99% ee for a number of (*Z*)-β-aryl-β-(acetylamino)acrylate substrates (*657*). Aryl groups have been added to TangPhos to generate DuanPhos, which reduced several β-substituted substrates with 92–>99% ee (*584*).

Monodentate catalysts have also been applied to reductions leading to β-amino acids. A number of 2-substituted β-amino esters were produced with 98–99% ee using a Rh catalyst with a monodentate phosphoramidite ligand, an analog of binaphthol-derived MonoPhos (*658*). A parallel synthesis of 32 monodentate phosphoramidite binaphthol-derived ligands was combined with in situ screening for asymmetric hydrogenation of amino acid precursors. Different catalysts were optimal for reduction leading to 3-amino-butyrate than for reduction leading to Phe (*612*). A comparative rate study found that the monodentate phosphoramidite catalyst could lead to higher rates and/or higher enantioselectivities than state-of-the-art bidentate catalysts (*610*). Another library screened 11 phosphite and eight phosphoramidite ligands based on a biphenol (TROPOS) unit with chiral alcohol or amine substituent; 3-aminobutyrate was produced with up to 71% ee (*621*). A related catalyst replaced the MonoPhos dimethylamine with a D-mannitol-derived ligand, resulting in 98.7–99.9% ee for several hydrogenations (*608*). Other carbohydrate derivatives resulted in 93–98.4% ee (*607*). Several new monodentate chiral phosphonites were derived from 1,1′-spirobiindane-7,7′-diol, giving 85–98% ee for reductions leading to seven different β-aryl-β-amino acids (*627*).

Combining two different MonoPhos-type monodentate ligands provided much better enantioselectivity than either ligand alone. For example, reduction of ethyl β-(acetamino)-β-methacrylate with two different homogeneous ligands gave product with 54% and 80% ee, but using a mixture of the ligands gave product with 91% ee (*659*). A combinatorial approach toward identifying useful catalysts with mixtures of chiral monodentate P-ligands found that a mixture of BINOL-derived methyl phosphite with BINOL-derived *t*-butyl phosphonite consistently gave the best results for reductions leading to 3-substituted β-amino esters, providing 94–98.8% ee for five examples (*660*). A combination of BINOL-derived *t*-butyl phosphonite with achiral (configurationally

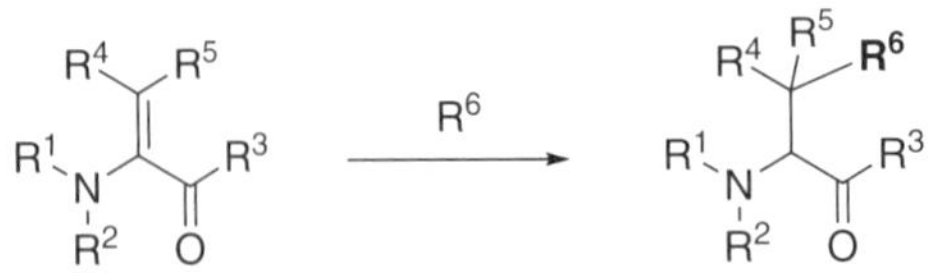

Scheme 4.32 Additions to Didehydroamino Acids (see Table 4.6).

fluxional) biphenyl-derived phosphite gave products with 84–97% ee (*661*).

4.3.3 Conjugate Additions

Didehydroamino acids can act as Michael acceptors for the addition of nucleophiles (see Scheme 4.32, Table 4.6). A number of synthetic routes to β-substituted amino acids designed to proceed via nucleophilic displacement of β-leaving groups from substrates such as β-chloroalanine actually proceed via elimination of the leaving group followed by conjugate addition of the nucleophile, as discussed further in Section 10.3. Various heteroatom nucleophiles have been explicitly reacted with dehydroamino acid intermediates (see Scheme 4.32, Table 4.6). One of the first additions reported was a synthesis of *O*-methyl Thr in 1939 via methoxide addition to the azlactone of ΔAbu (*662*), but there are few other reports of alkoxide reactions. Alcohols, water, and acetate were added to the α-position of ΔAbu and dehydronorvaline by a three-step process involving halogenation of the alkene, addition of the nucleophile, and reductive removal of the halogen (*336*).

Amines have often been employed as nucleophiles. A study of heteroatom additions to dehydroamino acids in 1948 found that amines (methylamine, benzylamine, and dimethylamine) added only to ΔAla, and not to ΔVal or ΔPhe (*663*). Dimethylamine was added to a Gly-ΔAla dipeptide to give 2-amino-3-dimethylaminopropionic acid in 50% yield in 1948 (*664*); in subsequent reports additions to Ac-ΔAla-OH of methylamine and ethylamine proceeded in 57–71% yield (*665, 666*), benzylamine in 32–64% yield (*273, 667*), isopropylamine, isobutylamine, *tert*-butylamine, and isopentylamine in 42–62% yield (*668*), an α-phosphonate-amine in 73% (*669*), and ethylamine, ethanolamine, *n*-propylamine, *n*-butylamine, amylamine, and diethylamine in 46–66% yield (*670*). *O,N*-Dimethylhydroxyamine was added in low (15%) yield (*671*), while the secondary imino group of 3-methoxy-4-pyridone was added to produce, after hydrolysis, leucenol in 26% yield (*672*). Ammonia and benzylamine were added to a number of *N*-formyl-didehydroamino esters, or their dehydrated isocyano equivalents, in 64–98% yield (*255*).

In 1995 ferric chloride was found to effectively catalyze the conjugate addition of nitrogen-based nucleophiles to Ac-ΔAla-OMe, with nine secondary amines added in 40–98% yield (*673*). Under these conditions primary amines reacted with two equivalents of the amino acid to give amine-bridged diaminodicarboxylic acids (*673*).

The regioselectivity of addition of propylamine to various *N*-alkoxycarbonyl-protected ΔAla esters and ethylamides has been studied. Boc- and Cbz-ΔAla-NHEt produced entirely the aminal resulting from α-addition, α-(aminopropyl)alanine, while the corresponding methyl esters and Ac-ΔAla-NHEt gave only the 1,2-diaminopropionic acid conjugate addition products (*92*). Benzylamine and cyclohexylamine were added to a variety of *N*-Boc, *N*-acylated ΔAla-OMe derivatives in 2001 (second acyl group = Boc, Cbz, NO$_2$-Cbz, Bz, NO$_2$-Bz) (*385*). Conjugate addition of 3-iodobenzylamine to *N,N*-Boc$_2$-ΔAla-OMe in the presence of K$_2$CO$_3$ produced the N^β-(3-iodobenzyl) Dap derivative in 90% yield. The aryl iodo group was then employed for Suzuki couplings with aryl boronates (*387*). In 2003 it was reported that water greatly accelerated the rate of addition of primary and secondary amines to Ac-ΔAla or Ac-ΔAbu amide derivatives (*674*). In 2004 thiols and amines were added to the ΔAla residue contained in the cyclic thiazolyl peptide antibiotic nocathiacin I. In order to avoid side reactions, unique conditions of employing frozen water as the solvent were identified. The same conditions allowed for addition of amines and thiols to Ac-ΔAla-NHCH$_2$CH$_2$ (4-HO-Ph) in 80–100% yield (*11*).

Thiols are particularly well suited for conjugate addition. A study of heteroatom additions to dehydroamino acids in 1948 found that thiophenol and benzyl mercaptan added to both ΔAla and ΔPhe, but not ΔVal (*663*). In an earlier report, benzylmercaptan was added to the azlactone derivative of ΔAla or to ΔAbu-OMe (*675*). *N,S*-Diacetyl Cys was synthesized in 1948 by addition of thiolacetic acid to Ac-ΔAla-OH (80–85% yield) (*272, 676*). Hydrolysis and oxidation produced Cys$_2$ in 72% yield (*272*). [^{34}S]Cysteine was prepared via addition of [^{34}S]thioacetic acid to α-acetamidoacrylic acid, followed by hydrolysis (66% yield) (*677*), while protected selenocysteine was produced in 95% yield from the addition of BnSeH to Ac-ΔAla-Ala-OMe (*216*). The addition of H$_2$S to ΔVal was used in a synthesis of penicillamine in 1953 (*678*). More recently, three trifluoromethyl-substituted thiophenol lithium salts were added to Ac-ΔAla-Ala-OMe in 28–40% yield (*679*). Seven other thiophenols added in 36–87% yield in piperidine/dioxane solvent if oxygen was excluded (*680*). In 2003 4-methoxythiophenol was added to Fmoc-ΔAla-ODpm in 70% yield (*93*). Ethyl mercaptan, benzyl mercaptan, and thiophenol added to a ΔAla benzophenone Schiff base in the presence of piperidine in 83–89% yield (*152*). An α-phosphonate-thiol has also been added (62% yield) (*669*). Thiophenol and methyl β-mercaptoacetate have been added to *N,N*-Boc$_2$-ΔAla-OMe in the presence of K$_2$CO$_3$ to give the *S*-substituted Cys derivatives in 81–93% yield (*681*). Similar additions to *N*-Boc,*N*-Cbz-, *N*-Boc,*N*-Cbz(NO$_2$)-, *N*-Boc,*N*-Bz-, and *N*-Boc,*N*-Bz(NO$_2$)-ΔAla-OMe and *N*-Boc,*N*-Bz-ΔAbu were also reported (*385*). Conjugate addition of 4-bromothiophenol to *N,N*-Boc$_2$-ΔAla-OMe in the presence of K$_2$CO$_3$ produced the *S*-aryl Cys derivative in 72% yield. The aryl bromo group was then employed for Suzuki couplings with aryl boronates (*387*). In 2003 it was

Table 4.6 Additions of Nucleophiles to Didehydroamino Acids

Reaction scheme (left: didehydroamino acid with substituents R⁴, R⁵, R³, R¹–N–R², C=O) → R⁶M → (right: product with R⁴, R⁵, R⁶, R³, R¹–N–R², C=O)

R¹	R²	R³	R⁴	R⁵	R⁶	M (reagent)	Yield	Reference
Heteroatom Additions								
=C(Ph)O-R³ —		OC(Ph)C=R¹	Me	H	OMe	Na	30–38%	1939 (662)
H	H	OMe	Me	H	SBn	H	70%	1941 (675)
=C(Ph)O-R³ —		OC(Ph)=R¹						
Gly		OH	H	H	NMe$_2$	H	50%	1948 (664)
Ac	H	OH	H	H	(4-methoxypyridin-2(1H)-one, N-linked)	H	26%	1949 (672)
Ac	H	OMe	H	H	S(2-CF$_3$-Ph), S(3-CF$_3$-Ph), S(4-CF$_3$-Ph)	Li	28–40%	1979 (679)
Ac	H	OH	H	H	SAc	H	80%	1947 (676)
Ac	H	OH	H	H	SCH$_2$CH(NH$_2$)CO$_2$H	H	84%	1948 (663)
Ac	H	OH	H	H	SAc	H	85%	1948 (272)
Ac	H	OMe	H	H	SPh, SBn, NHMe, NMe$_2$, HNBn	H	21–41%	1947 (910)
Ac	H	OMe	H	Ph	SPh, SBn	H	20–40%	1948 (663)
Ac	H	OH	H	H	SPh, S(4-Cl-Ph), S(4-Br-Ph), S(4-NO$_2$-Ph), ScHex	H	10–81%	1949 (895)
=C(Me)O-R³ —		OC(Me)=R¹	H	H	SH	H	75%	1953 (678)
Ac	H	OMe	H	H	NHBn	H	32%	1955 (667)
Alc	H	OMe	H	H	SeBn	H	95%	1963 (216)
Bz	H	OH	Me	Me	NHMe	H	59%	1967 (734)

(Continued)

Table 4.6 Additions of Nucleophiles to Didehydroamino Acids (continued)

R¹	R²	R³	R⁴	R⁵	R⁶	M (reagent)	Yield	Reference
Ac	H	OH	H	H	NHMe NHEt	H	57–60%	1968 (666)
Ac	H	OH	H	H	NHEt	H	71%	1968 (665)
Ac	H	OH	H	H	pyrazole	H	85%	1972 (896)
Ac	H	OH	H	H	NEt₂ NHEt NHnPr NHnBu NHnPent NH(CH₂)₂OH	H	46–66%	1972 (670)
Ac	H	OH	H	H	SCH₂CH(OH)CH₂OH	H	23%	1973 (898)
C≡	—	OEt	Me Ph 3,4-(MeO)₂-Ph	Me H H	NH₂ NHBn	H	64–98%	1977 (897)
C≡	—	OEt	Ph Ph Me Et (CH₂)₄-R⁵	H Me Me Et (CH₂)₄-R⁴	Et Ph	MgX	63–95%	1977 (240)
CHO	H	OEt	Me Ph 3,4-(MeO)₂-Ph	Me H H	NH₂ NHBn	H	—	1977 (897)
Ac	H	OH	H	H	NHiPr NHiBu NHtBu NHiPent	H	42–62%	1977 (668)
PhCH= H₂C= (2-HO-Ph) CH= (2-HO-4- NO₂-Ph) CH= Ac	H	OMe	Me	Me	SBn	H	72–82%	1978 (698)

CF_3CO	OMe	H	H		[purine ring: Cl, N, N, N, NH]	H	50%	1978 (899)
Cbz	H	Pro-NHMe	H		SMe	Na	76–100%	1978 (698)
Cbz-Pro		Pro-OEt					0–80% de	
		prolinol						
		Ala-NHMe						
		ValNHMe						
		NHMe						
=C(Ph)O-R³	—	OC(Ph)=R¹			NHOH	H	69%	1978 (900)
Bz	H	NH_2	H	Ph	NHOH	H	40–90%	1979 (901)
			$(CH_2)_5$-R⁴	$(CH_2)_5$-R⁵				
			H	4-NMe$_2$-Ph				
			H	4-HO-Ph				
			H	3-MeO-4-HO-Ph				
			H	3,4-OCH$_2$O-Ph				
			H	4-MeO-Ph				
			H	3,4-(MeO)$_2$-Ph				
			H	$(CH_2)_2$Ph				
			H	nPr				
			H	indol-3-yl				
Ph$_2$C=	—	OMe	H		SEt	H	23–90%	1988 (152)
		OEt			SPh	(piperidine)		
		OtBu			SBn			
Ac	H	OH	H		^{34}SAc	H	66%	1990 (677)
Ac	H	OH	H		S(2-Br-Ph)	(piperidine/dioxane)	36–87%	1990 (680)
					S(3-Br-Ph)			
					S(4-Br-Ph)			
					S(2,3-Br$_2$-Ph)			
					S(3,4-Br$_2$-Ph)			
					S[2,5-(OH)$_2$-Ph]			
					S[3-Br-2,5-(OH)$_2$-Ph]			
Ac	H	OH	H		P(OMe)$_3$	H	—	1990 (908)
peptide	H	peptide	H		SCH$_2$CH(NH-peptide)CO-peptide	H	—	1992 (156)
Ph$_2$C=	—	OMe	H		SPh	H, Na	59–73%	1992 (700)
		OEt			SCH$_2$CO$_2$Et			
					OEt			

(Continued)

Table 4.6 Additions of Nucleophiles to Didehydroamino Acids (continued)

R^1	R^2	R^3	R^4	R^5	R^6	M (reagent)	Yield	Reference
Boc	H	OMe	H	H	N(Me)OMe	H (NaHCO$_3$)	15%	1993 (671)
Boc Cbz Ac	H	OMe NHEt	H	H	NHnPr	H	—	1995 (92)
Ac	H	OMe	H	H	piperidine morpholine pyrrolidine imidazole 2-MeO$_2$C-piperidine 3-MeO$_2$C-piperidine 4-MeO$_2$C-piperidine pyrazole N(butyl)$_2$	H (FeCl$_3$)	40–98%	1995 (673)
Ac	H	OMe	H	H	SCH$_2$PO$_3$Et$_2$ NHCH$_2$PO$_3$Et$_2$	H (Et$_3$N)	62–73%	1996 (759)
Boc	H	OMe	H	H	NHCH$_2$(3,4-Cl$_2$-Ph)	H	41%	1998 (803)
Ac	H	O-Wang	H	H	triazole pyrazole	H (K$_2$CO$_3$)	54–78%	1998 (684)
Boc	H	OH	H	H	1-MeO-2-HO-indole	HOAc, Ac$_2$O	29%	1999 (703)
Boc	Boc	OMe	Me	Me	pyrazole 1,2,4-triazole imidazole 7-azaindole indole 3-CHO-indole 3-CO$_2$Et-carbazole	H (K$_2$CO$_3$)	49–99%	1999 (685)
Boc	Boc	OMe	H	H	SPh SCH$_2$CO$_2$Me	H	81–91%	2000 (681)

Boc	Boc	OMe	H	indole 3-formyl-indole 3-CHO-2-Me-5-NO₂-indole 2-Me-5-NO₂-indole 7-azaindole pyrazole imidazole 1,2,4-triazole 2-formyl-pyrrole 2-acetyl-pyrrole carbazole 3-CO₂Et-carbazole 3-F-carbazole 3-NO₂-carbazole	H	49–99%	2000 (681)
peptide	H	peptide	H	SCH₂CH(NH-peptide) CO-peptide	H	68–100%	2000 (18)
Boc	Boc Cbz NO₂-Cbz Bz NO₂-Bz Ts Ts-Gly Ts-Ala	OMe	H	SPh SCH₂CO₂Me NHBn NHcHex S(4-NH₂-Ph) CH(CO₂Et)₂ CH(COMe)₂CO₂Me CH(COMe)₂ CH(–COCH₂CH₂CH₂CO–) imidazole pyrazole 1,2,4-triazole 3-formylindole	H	33–98%	2001 (385)
Boc	Boc	OMe	Me	SPh SCH₂CO₂Me imidazole 1,2,4-triazole 3-formylindole	H	43–89%	2001 (385)

(Continued)

325

Table 4.6 Additions of Nucleophiles to Didehydroamino Acids (continued)

R^1	R^2	R^3	R^4	R^5	R^6	M (reagent)	Yield	Reference
peptide	H	peptide	H	H	SAc SCH_2CH=C(Me)_2 S(farnesyl) S(geranyl) S(geranylgeranyl) S(β-glucosyl) S(β-2-acetamido-3,4,6-triacetyl-2-deoxy-D-glucopyranose)	H or Ac NaOMe	62–76%	2001 (682)
peptide	H	peptide	Me	H	SCH_2CH(NH-peptide) CO-peptide	H	—	2002 (168)
Fmoc	H	ODpm	H	H	S(4-MeO-Ph)	E_3N	70%	2003 (93)
Boc	Boc	OMe	H	H	OMe OEt	Na	38–80%	2003 (384)
Boc	Boc	OMe	H	H	NH(3-I-Bn) S(4-Br-Ph)	H K_2CO_3	72–90%	2003 (387)
Ac	H	NHCH_2CH_2 (4-HO-Ph) N(Et)_2 NHCH_2CO_2H Bn	H	H	NHMe NHBn N(Me)_2 N(Me)CH_2CH_2OH N(Bn)NH_2NH_2OH morpholine N(OH)Bn NH_2 SiPr SCH_2CH_2NMe_2 SCH_2CH_2NH_2 SCH_2CH_2CO_2H SCH_2CH_2SO_3H	H MeOH/H_2O	62–100%	2003 (674)
Ac	H	NHCH_2CH_2 (4-HO-Ph)	Me	H	NHMe N(Me)_2 pyrrolidine SiPr SCH_2CH_2NMe_2	H MeOH/H_2O	63–100%	2003 (674)

Ac	H	NHCH$_2$CH$_2$(4-HO-Ph)	H	H		N(Me)$_2$, N(Me)CH$_2$CH$_2$OH, morpholine, SCH$_2$CH$_2$NMe$_2$, S(iPr)	H, frozen H$_2$O	80–100%	2004 (11)
peptide	H	peptide	Me	Me		SCH$_2$CH(NH-peptide) CO-peptide	H	—	2004 (144)
Carbon Additions									
=C(Ph)O-R^3	—	OC(Ph)O=R^1	Me	Me		Et	MgBr	30–89%	1955 (692)
=C(Ph)O-R^3	—	OC(Ph)O=R^1	(CH$_2$)$_5$-R^4	(CH$_2$)$_5$-R^5	Ph, 4-Cl-Ph, 4-NMe$_2$-Ph, 3,4-(MeO)$_2$-Ph	Et, nBu, iPr, Bn, (CH$_2$)$_2$Ph	MgBr	30–89%	1955 (692)
Ac	H	OMe	H	H		CF(CO$_2$Et)$_2$	Na	56%	1962 (687)
Ac	H	OEt	H	H		CF(CO$_2$Et)$_2$ (converted into CHFCO$_2$Et with conc. HCl)	Na	60%	1973 (903)
Phth	—	OMe	H	H		CH(CO$_2t$Bu)$_2$	Li	83%	1979 (98)
Ac	H	OMe	H	H		Et, nPr, iPr, nBu, tBu, Ph, 2-Me-Ph, 4-MeO-Ph, 3,4-OCH$_2$O-Ph, 1-naphthyl	MgBr (CuCl) α-anion quenched with H$_2$O, D$_2$O or MeI	40–85%	1985 (693)
(2-Ph-Ph)CH=	—	OMe	H	H		Me, nBu, sBu, tBu, Ph	CNCu-Li$_2$ or CuLi(LiI)	31–61%	1986 (697)
Ph$_2$C=	—	OMe, OEt, OtBu	H	H		indol-3-yl, CH(CO$_2$Et)$_2$, Ph	(LDA), Na, Li$_2$CuCN	23–90%	1991 (407)

(Continued)

327

Table 4.6 Additions of Nucleophiles to Didehydroamino Acids (continued)

R^1	R^2	R^3	R^4	R^5	R^6	M (reagent)	Yield	Reference
Ac	H	(+)-menthyl (−)-menthyl 8-Ph-menthyl bornyl 3-HO-isobornyl Di-O-isopropylidene-D-glucofuranosyl di-O-cyclohexylidene-D-glucofuranosyl	H	H	Ph	MgBr (CuI)	80–95% 0–44% de	1992 (902)
Bz	CH(tBu)O-R³	OCH(tBu)-R²	H	H			70–82%	1995 (691)
Boc	H	OEt	CH₂(Bn)-R⁵	C(Bn)CH₂-R⁴	CH(CO₂Et)₂ CH(COMe)CO₂Et CH(SO₂Ph)CO₂Me CH(SPh)CO₂Me	Na	40–70%	1996 (147)
Boc	H	OEt	CH₂C(Bn)-R⁵	C(Bn)CH₂-R⁴	Me Ph	CuLi	52–88%	1996 (147)
C≡	—	OEt	Me	Me	4-Me-Ph	MgBr	49%	1996 (235)
Boc	H	OMe	H	H	CH(CO₂Me)₂	Na	93%	1998 (686)
Boc	CH(tBu)N(Me)-R³	N(Me)CH(tBu)-R²	H Me nPr	H Me	Me Et nBu Ph	LiCu₂	23–86%	1991 (407)
Cbz	H	OMe	H	H	CH(NO₂)CH₂PO₃H₂ CH(NO₂)(CH₂)₃PO₃H₂	H (KFAl₂O₃)	37–64%	1992 (127)
Phth	—	OEt	H	H	Ph 3-Me-Ph 4-Me-Ph 4-tBu-Ph 4-Cl-Ph (1-naphthyl) (2-naphthyl)	ArSn-Me₃ with Rh catalyst	41–82%	2001 (340)

Phth	—	OEt	H	Me, nBu, nPent, nHex, iPr, tBu, CH$_2$CH(Me)Et, cPent, cHex, cHept	RI, Zn dust in aqueous NH$_4$Cl	31–97%	2002 (696)
Phth, Boc, Ac	—, H, H	OMe	H	Ph, 2-Me-Ph, 3-MeO-Ph, 3-Cl-Ph, (1-naphthyl), (2-naphthyl)	ArBF$_3$K	74–98%	2004 (695)

Lewis-Acid Catalyzed Additions

Ac	H	OH	H	indole, 7-Me-2,3-dihydro-indole, 2-Me-indole	H (AcOH)	19–58%	1955 (702)
Ph$_2$C=	—	OMe, OEt, OtBu	H	indol-3-yl, pyrrol-2-yl, 4-NHMe-Ph, 4-NMe$_2$-Ph	H (EtAlCl$_2$)	18–93%	1988 (152)
Ph$_2$C=	—	OMe, OEt	Me	indolyl, 5-MeO-indolyl, 6-MeO-indolyl, 6-Me-indolyl, 6-F-indolyl, 6,7-(MeO)$_2$-indolyl, 8-Me-indolyl	Me (EtAlCl$_2$)	52–79%	1992 (700)
Ac	H	OH	H	furan, anisole	H (Lewis acid)	95–100%	1992 (704)
Ph$_2$C=	—	OMe	H	2,5-disubstituted-indolyl	H (EtAlCl$_2$ or AlCl$_3$)	70–87%	1995 (701)

reported that water greatly accelerated the rate of addition of thiols to Ac-ΔAla or Ac-ΔAbu amides (674).

Unprotected Cys can be added to Ac-ΔAla to produce lanthionine in excellent (84%) yield (910); the addition has also been carried out on ΔAla (generated from Ser) within a peptide (156). A similar strategy employed oxidized Met as the source of the ΔAla residue; only a single diastereomer was isolated (144). (2S,3S,6R)-3-Methyllanthionine was produced by a diastereoselective cyclization via intramolecular Michael addition of L-Cys to ΔAbu within a peptide (168). A biomimetic synthesis of the A, B, and E rings of subtilin was reported in 2000, making use of an intramolecular formation of lanthionine residues via addition of the Cys side-chain thiol to a ΔAla residue three- or four-residues away within a linear peptide. A single stereoisomer of the B and E rings was obtained, corresponding to the natural meso D,L-lanthionine, while the A ring was formed with 50% de (18). ΔAla residues have been generated within peptides for the purpose of chemoselective ligations to generate glycopeptides, prenylated peptides, and lipopeptides. Farnesyl thiolacetate, geranylthiolacetate, geranylgeranylthiolacetate or thioglycosides successfully underwent conjugate addition to the residue, though with no diastereoselectivity observed (682).

An unusual reaction has been observed during addition of thiols to N,N-disubstituted 2-aminoalk-2-enals (see Scheme 4.33), with thioesters of α-amino acids the only observed products rather than the expected Michael adducts (683).

Heterocyclic nucleophiles can also be employed. Pyrazole and 1,2,4-triazole were successfully added to Ac-ΔAla attached to a Wang resin (684). Pyrazole, 1,2,4-triazole, imidazole, indole, 3-formyl-indole, 2-methyl-3-formyl-5-nitro-indole, 2-methyl-5-nitroindole, 7-azaindole, 2-formylpyrrole, 2-acetylpyrrole, carbazole, 3-ethoxycarbonyl-carbazole, 3-fluorocarbazole

and 3-nitro-carbazole all reacted with N,N-Boc₂-ΔAla-OMe in good to high yield (49–99%), using K₂CO₃ in MeCN (681, 685). Addition of 3-formylindole to N-Boc,N-Ts-ΔAla-OMe and N,N-Boc₂-ΔAbu-OMe was also reported (385). Pyrazole, imidazole, 1,2,4-triazole, and 3-formylindole were subsequently added to a variety of N-Boc,N-acylated ΔAla-OMe derivatives in 2001 (second acyl group = Boc, Cbz, NO₂-Cbz, Bz, NO₂-Bz), with imidazole and triazole also added to N,N-Boc₂-ΔAbu-OMe (385). However, lithiated indole added in poor yield (23% yield) to the benzophenone Schiff base of ΔAla-OMe (152).

Carbon–carbon bonds are formed by the addition of nucleophiles such as lithiated di-t-butyl malonate, which was added to Phth-ΔAla-OMe to give protected γ-carboxy-Glu in 83% yield (98). Dimethyl malonate/NaH reacted with N-Ac ΔAla in 90–93% yield (152, 686) and diethyl fluoromalonate in 56% yield (687), while diethyl acetamidomalonate added to Ac-ΔAla-OEt generated in situ from diethyl [(trimethyl)aminomethyl] acetamidomalonate to produce 2,4-diaminoglutaric acid in 68% yield (688). Diethyl malonate, methyl acetoacetate, pentane-1,4-dione, and cyclohexane-1,3-dione were added to a variety of N-Boc,N-acylated ΔAla-OMe derivatives in 2001 (second acyl group = Boc, Cbz, NO₂-Cbz, Bz, NO₂-Bz) (385). The Claisen ester enolate rearrangement of allyl esters (see Chapter 3, Scheme 3.20) also works with α,β-unsaturated amino acids, with the allyl group adding to the α-center and the alkene isomerizing to the β,γ-vinyl position (689). γ-Nitro amino acids were prepared by the conjugate addition of nitroalkenes to Phth-ΔAla-OMe, catalyzed by KF or nBu₄NF, or by additions to Cbz-ΔAla-OMe catalyzed by NaOMe or BnMe₃NOH (see Scheme 4.34). The nitro group was then converted into a chloro, hydroxy, oxo, or amino substituent (690). Nitrophosphonate derivatives (KF−Al₂O₃ catalysis, 37–64% yield) (127) and enamines

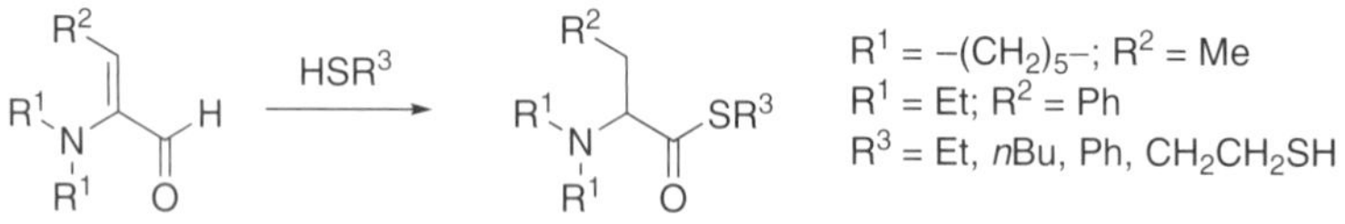

Scheme 4.33 Reaction of Thiols with N,N-Disubstitued 2-Aminoalk-2-Enals (683).

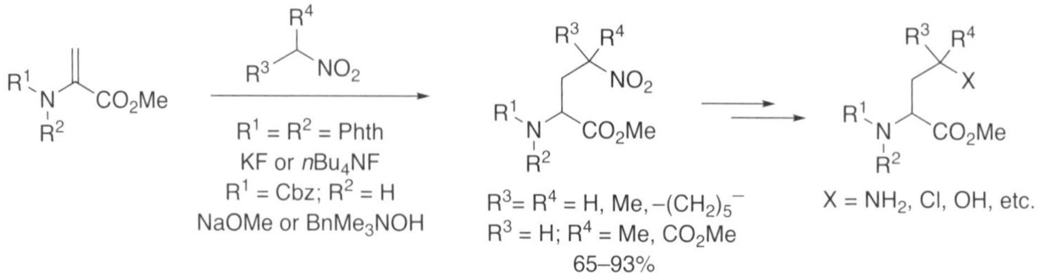

R¹ = –(CH₂)₅–; R² = Me
R¹ = Et; R² = Ph
R³ = Et, nBu, Ph, CH₂CH₂SH

R¹ = R² = Phth
KF or nBu₄NF
R¹ = Cbz; R² = H
NaOMe or BnMe₃NOH

R³= R⁴ = H, Me, –(CH₂)₅⁻
R³ = H; R⁴ = Me, CO₂Me
65–93%

X = NH₂, Cl, OH, etc.

Scheme 4.34 Addition of Nitroalkanes (690).

(70–82%, see Scheme 7.31 for asymmetric version) (*691*) have also been added to ΔAla derivatives in reasonable yield.

Organometallic reagents are potential nucleophiles for carbon–carbon bond formation. Five different Grignard reagents were added to six different dehydroamino acid azlactones in 1955, producing β-substituted Phe and other β-substituted amino acids in 30–89% yields (*692*). Grignard reagents also add to 2-isocyano-acrylic esters in good yields (63–95%), with subsequent hydrolysis to give the *N*-formyl amino acid (*240*). An aromatic Grignard reagent was added in 49% yield to a β-disubstituted didehydroamino acid analog (*235*), while a variety of Grignard reagents were added to ΔAla in the presence of CuI as catalyst in 45–80% yield (*693*). The reaction could be quenched with D_2O or MeI to introduce an electrophile at the α-position (*693*). Lithium diorganocuprates failed to add in this report, but a phenyl organocuprate has been added to ΔAla in 70% yield (*152*), and both enolates and organocuprates added in 40–88% yield to a ΔPro derivative (*147*). Organocuprates have been added to 2,3-dehydropipecolic acid contained within a morpholinone template, producing *cis*-3-substituted pipecolic acids (*694*). α-Nitroacrylates, dehydroamino acid equivalents produced by the condensation of methyl nitroacetate with aldehydes, are good substrates for the addition of organolithium or Grignard reagents, producing α-nitro precursors of β-branched amino acids. The nitro group can then be reduced by H_2/Raney-Ni (*231, 232*). A rhodium-catalyzed conjugate addition of aryltrialkyltins to ethyl α-phthalimidoacrylate produced arylalanines in 41–82% yields (*340*). In 2004 six different aryl trifluoroborate potassium salts were added to Ac-, Boc-, and Phth-ΔAla-OMe to give the β-arylalanines in good (70–98%) yield (*695*). In 2002 it was reported that simply mixing a suspension of Phth-ΔAla-OEt with alkyl halides in aqueous saturated NH_4Cl solution in the presence of zinc powder led to the alkyl-substituted amino acids in 31–94% yield (RX = MeI, *n*BuI, *n*PentI, *n*HexI, *i*PrI, 2-I-butane, *t*Bu-I, *c*PentI, *c*HexI, *c*HeptI) (*696*).

A pyridoxal Schiff base of ΔAla has been postulated as a reactive intermediate in the biosynthesis of some amino acids, such as Trp. An analog of the putative intermediate was synthesized (*120*) and found to be very reactive, undergoing a [4+2] cycloaddition with itself (*697*). A number of organocuprates added in a Michael

fashion in 31–61% yield, but Grignard and organolithium reagents failed to react (*697*). In a related report it was demonstrated that the rate of addition of phenylmethanethiol to a series of aromatic Schiff bases of dehydrovaline was greatly enhanced by an *o*-hydroxyl group on the Schiff base aromatic moiety, as found in pyridoxal (*698, 699*).

Lewis acid catalysis can also be used for C–C bond formation with dehydroamino acids (see Table 4.6). This approach was used in the synthesis of a number of substituted tryptophan derivatives, with the indole substituent adding in a 1,4-fashion to the benzophenone Schiff base of ΔAla-OMe in the presence of $AlCl_3$ or $EtAlCl_2$ in 52–87% yield (*700, 701*). $AcOH/Ac_2O$ was employed as catalyst in earlier syntheses (*702*), and for a synthesis of 1′-MeO-Trp using 1-methoxy-2-hydroxyindole (*703*). Indolyl, pyrrolyl, and *p*-aminophenyl groups were added in a similar manner in 18–93% yield (*152*). In contrast, anisole and furan underwent a 1,2-addition to Ac-ΔAla-OMe, giving the α-methyl-α-aryl disubstituted amino acid in varying yield, depending on Lewis acid catalyst ($BF_3·Et_2O$ was the best) (*704*). Reaction of the azlactone of ΔPhe with benzene in the presence of $AlCl_3$ gave β,β-diphenylalanine (*705*).

Most conjugate additions to dehydroamino acids lead to racemic products, but several asymmetric syntheses have been reported. An acyclic optically active β-enamino ester added to *N*-Ac ΔAla with >95% de at the α-center, resulting in a diaminodicarboxylic acid (*706*). Methanethiol was added to Cbz-ΔAla-L-Pro-amide to give a D-*S*-Me-Cys derivative in 77% yield, with the chiral auxiliary inducing >85% diastereoselectivity (*140*). A similar study with *N*-methyl-ΔAla-L-Pro-amide had previously found excellent yields (95%) and diastereoselectivity (80%) for thiol additions (*698*). However, little diastereoselectivity was observed if proline ester, prolinol, alanine amide or valine amide were used as carboxyl auxiliaries, or if Cbz-Pro was used as an amine auxiliary (*698*).

Lander and Hegedus (*707*) have prepared a ΔAla derivative possessing a chiral auxiliary on the amine (see Scheme 4.35). Organocuprates added with poor diastereoselectivity (approximately 40% de), but the copper-catalyzed addition of Grignard reagents proceeded with excellent yield (75–95%) and generally good stereoselectivity (72–92% de). Hydrogenation released the free amino acids. In contrast, addition of

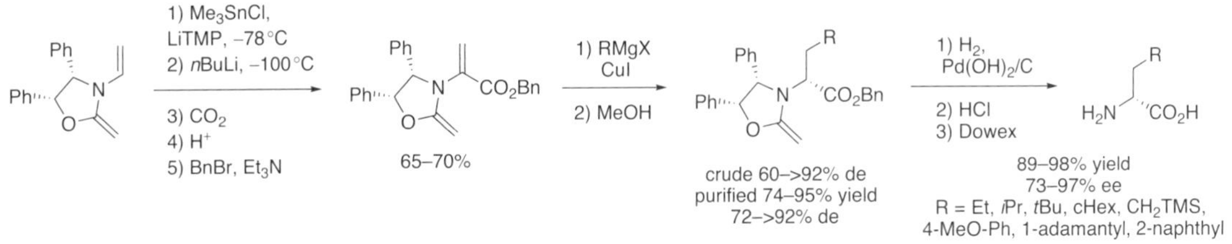

Scheme 4.35 Synthesis and Reaction of Dehydroamino Acid via Carboxylation of Organostannane (*707*).

PhMgBr in the presence of CuI to Ac-ΔAla protected with a variety of chiral ester groups gave a maximum diastereomeric excess of 44% (708).

Organocuprates have been added to dehydroamino acids contained within the imidazolidinone (BMI) template of Seebach (see Section 7.9) in 23–86% yield using BF₃·Et₂O catalysis, with complete diastereoselectivity (407). Ultrasonically induced conjugate addition of zinc–copper derivatives of alkyl iodides (nPentI, cHexI, InPrCO₂Me, InPrOH, IndodecylOH, Boc-β-iodo-Ala-OMe) to ΔAla within the Bz-BMI template under aqueous conditions produced the addition products in 38–95% yield with 44–90% de. The adducts were not deprotected (709). The imidazolidinone template has also been used to prepare an α,β-diaminodehydroamino acid intermediate. Organocuprate addition (Ph₂CuLi or Bu₂CuLi, 40–42% yield, 100% de) and hydrolysis led to β-substituted α,β-diamino acids (251). A chiral oxazolidinone derivative of ΔAla was used for stereoselective Michael additions of Schiff base enolates (65–81% yield), with a reversal in diastereoselectivity observed at higher temperature (see Scheme 7.31) (710). An N-Cbz oxazolidinone containing ΔAla was employed for aqueous reactions with alkyl iodides in the presence of Zn/CuI, producing protected Leu, 2-aminooctanoic acid, cyclohexylalanine, and 2-amino-1,7-pentanedioic acid with >96% de and 60–99% yields. Boc-β-iodo-Ala-OMe was also coupled to give 2,5-diaminoadipic acid (711).

A chiral Ni(II) Schiff complex of ΔAla prepared with (S)-o-N-(N-benzylprolyl)aminobenzophenone was used for conjugate additions of a variety of nucleophiles (thiols, amines, methanol, imidazole, diethyl malonate, and BnMgBr) in good yield (55–98%) and reasonable diastereoselectivity (50–96% de) (see Scheme 4.36) (106, 712). The free amino acids were readily obtained (75–93%) by acidic hydrolysis. The same substrate was employed for the addition of 3-amino-1,2,4-thiodiazole and 5-mercapto-1,2,4-triazole nucleophiles, producing β-heterocycle-substituted α-amino acids with >98.5% ee (107). The threonine-derived ΔAbu Ni(II) complex also underwent diastereoselective conjugate additions (713). More recently, an N',N'-bis(p-methoxybenzyl) diketopiperazine was converted into a cyclo[Val-ΔAla]

diketopiperazine template via deprotonation and reaction with paraformaldehyde. Addition of organocuprate reagents to the ΔAla residue gave cis-substituted bislactims with >95% de (R = Me, iPr, nBu, tBu, Ph, CN, CH=CH₂). Deprotection was achieved using CAN followed by 6 N HCl, giving amino acids with >99% ee in 79–84% yield (714, 715).

A series of arylalanines were constructed by a tandem 1,4-addition/enantioselective protonation procedure, using potassium trifluoro(aryl)borates to add to Ac-ΔAla-OMe followed by protonation with guaiacol and a Rh-complex with a (R)-BINAP ligand. The products were obtained with 81–89.5% ee (716).

4.3.4 Radical Reactions

ΔAla residues are also useful substrates for radical reactions, as reviewed in 1997 (717). Alkyl or aryl substituents have been introduced at the β-position using alkylmercury halides (RHgCl) with sodium borohydride. Yields of 30–85% were obtained for reactions with Tfa-ΔAla-OMe (RHgCl, R = iBu, cHex, cPr, iPr, CH₂tBu, tBu) (718). Tributyltin hydride and O-acyl-thiohydroxamate were reported to be unsuccessful radical sources for this reaction (160, 718). The reaction has also been successfully applied to Ac-ΔAla attached to a Wang resin support (719). A tBu radical was added to Phth-ΔAla in 41% yield (720). Vitamin-B₁₂-mediated photoelectrochemical 1,4-hydroaddition of alkyl halides (R = nBu, nPr-OAc) or acetic anhydride to Ac-ΔAla-OMe proceeded in 67–72% yield (718).

Minimal diastereoselectivity was observed when the alkylmercury halides were added to di- and tripeptides containing ΔAla, although yields were generally good (42–98% yield, 1–28% de) (160). Much better diastereoselectivity for the radical reaction was observed when the ΔAla residue was contained within a cyclic dipeptide piperazine-2,5-dione, rather than an acyclic peptide, although yields were reduced (see Scheme 4.37) (155). Only one diastereomer (the cis isomer) was observed when iPr and cHex groups were added (46–49% yield). Another template for radical addition is a chiral Ni(II)

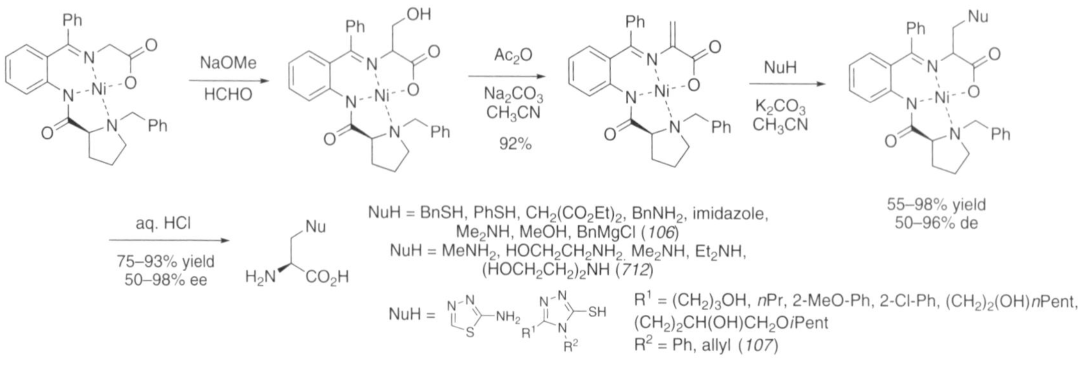

Scheme 4.36 Nucleophilic Addition to Chiral Ni(II) Schiff Base Complex of Dehydroalanine.

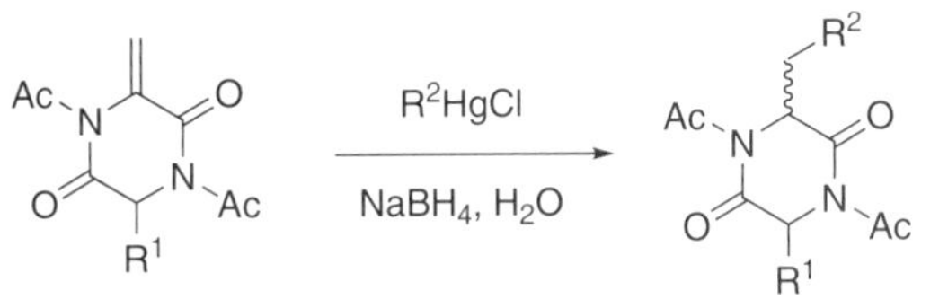

Scheme 4.37 Radical Reaction of Dehydroalanine in Diketopiperazine Template (155).

Schiff base complex of ΔAla. Reactions with carbon-centered radicals led to Nva, Leu, Hfe, and neopentylglycine with 40–92% de. The diastereoselectivity could be improved after reaction by NaOMe-catalyzed epimerization (721). A pantolactone ester of Phth-ΔAla provided 36–66% de for addition of stannyl radicals derived from *tert*-butyl iodide or cyclohexyl iodide, though with opposite stereoselectivity. (S)-γ-Methyl-Leu and (R)-cyclohexylalanine were obtained from (R)-pantolactone ester (722).

Racemic pyroglutamate derivatives can be obtained by an intramolecular radical cyclization of N-haloacyl ΔAla-OMe derivatives (see Scheme 4.38) (104, 723, 724). By using an 8-phenylmenthyl ester, diastereoselectivity of up to 70% was achieved, although yields were reduced (725). Proline-type compounds have been synthesized by intramolecular cyclizations of N-alkylhalo derivatives.

The ΔAla substrate has also been contained within a chiral oxazolidinone or imidazolidinone template (see Scheme 4.39, Table 4.7). The oxazolidinone compound can be prepared by radical bromination of an alanine oxazolidinone, followed by dehydrobromination with NaI/acetone (136). In contrast to previous studies, radical generation with tributylstannane and alkyl halides was found to give better addition yields than the alkylmercury hydride procedure, and a number of alkyl iodides (R = Me, Bn, cHex, tBu) were added in 60–73%

yield with 42–96% de (135). The free amino acid was obtained by hydrogenolysis of the protected adduct. The *trans* isomer was preferred in contrast to the *cis* isomer obtained during reactions of the piperazine-2,5-dione template. A further study found that the direction and degree of diastereoselectivity depended on the nature of both the addend radical and the N-acyl group (726). Reactions with tributylstannane as the hydrogen atom source gave a higher diastereoselectivity, but lower yield, than those with mercury hydrides. The best stereoselectivity was obtained with bulky alkyl radicals. Surprisingly, changing from N-benzoyl protection of the oxazolidinone to N-carbamate, N-acetyl or N-benzylacetyl protection resulted in a reversal of diastereoselectivity, with even higher asymmetric induction. An explanation for these results has not yet been put forward. Homoserine derivatives were produced by the addition of α-hydroxy or α-alkoxy radicals to ΔAla in an oxazolidinone template. Varying diasteroselectivity was observed in the addition, but the products were prone to racemization under the conditions required for deprotection (727). The ΔAla oxazolidinone derivative has also been used for addition of a difluoroglycosyl radical, providing a difluoromethylene-linked analog of a glycosylated serine residue (728).

Asymmetry has also been induced via addition of chiral radicals, such as those derived from glycosyl bromides, to ΔAla derivatives. The sugar moiety induced up

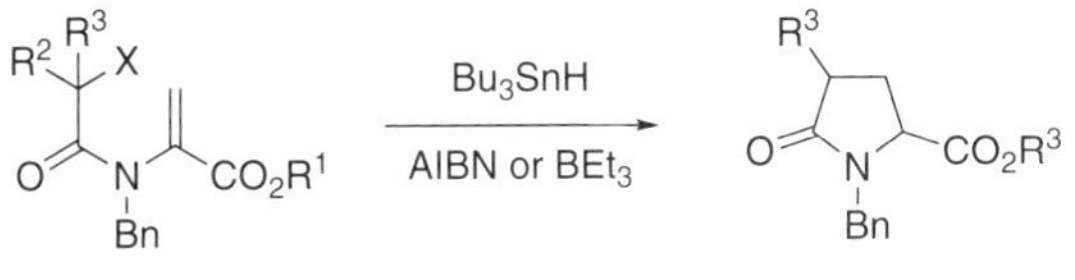

R¹ = Me; R² = H; R³ = SPh, Me, Ph; X = Cl: 46–56% (104, 723)
R¹ = Me; R² = R³ = H; X = Cl, Br, I: 38–52% (104)
R¹ = Me; R² = Cl; R³ = H; X = Cl: 70 (104)
R¹ = Me; R² = R³ = X = Cl: 84% (104)
R¹ = isopinocampheol, menthol, 8-phenylmenthol;
R² = R³ = H; X = Br: 16–62%, 0–70% de (725)
R¹ = Me; R² = Ph; R³ = H: 77% (724)

Scheme 4.38 Radical Cyclization of Dehydroalanine to Pyroglutamate Derivatives.

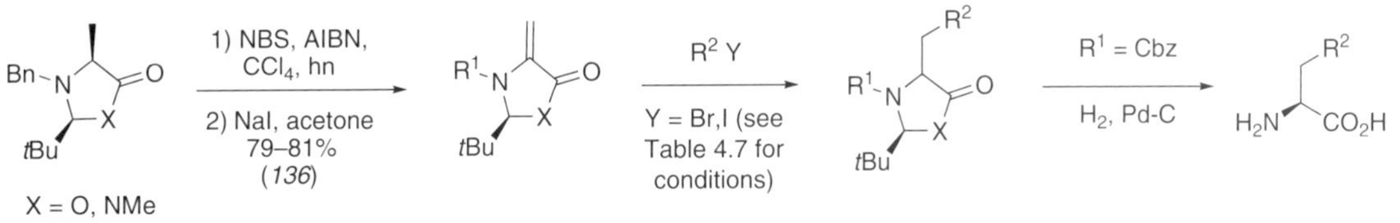

Scheme 4.39 Radical Reaction of Dehydroalanine in Imidazolidinone or Oxazolidinone Template (see Table 4.7).

Table 4.7 Radical Additions to Dehydroamino Acids in Template (see Scheme 4.39)

X	R¹	Conditions	R²	Yield	Reference
NMe	Bz	R²I, Bu₃SnH, AIBN, benzene	cHex	44% >75% de *trans*	1990 (*135*)
O	Bz	R²I, Bu₃SnH AIBN, benzene	Me iPr tBu cHex adamantyl (CF₂)₃CF₃ CH₂CO₂Me CH₂OMe Bn tetra-O-acetyl-glucosyl	24–73% 42–>96% de *trans*	1990 (*135*) 1995 (*726*)
O	Bz	R²HgCl, NaCNBH₄	cHex adamantyl	89–95% 68–86% de *trans*	1995 (*726*)
O	Ac Cbz CO₂Me CO₂Ph	R²HgCl, NaCNBH₄	cHex	47–86% 88–>96% de *cis*	1995 (*726*)
O	Bz CO(1-naphthyl)	R²HgCl, NaCNBH₄	cHex	77–89% 68% de *trans*	1995 (*726*)
O	Cbz	R²I, Bu₃SnH, NaCNBH₄	tetra-O-acetyl-glucosyl tetra-O-acetyl-galactosyl	73–88% 100% de *trans*	1995 (*726*)
O	Bz	R²SePh, AIBN, Bu₃SnH	CF₂(carbohydrate)	71–91% 60–72% de	1997 (*728*)

to 60% de at the α-center of the C-linked glycoamino acid products (*729*). The Williams oxazinone chiral synthon was used for addition of a chiral Gly radical to ΔAla. Deprotonation and treatment of the oxazinone template with PhSeBr produced an α-phenylselenyl-Gly radical precursor within the template. Radical reaction with Ac-ΔAla-OMe, induced by AIBN/n-Bu₃SnH, gave a 3:2 mixture of two of the possible four diastereomers of 4-aminoglutamic acid (*730*).

Chiral catalysis has also been employed. In 2001 a chiral Lewis acid bisoxazoline catalyst induced 27–85% ee for additions of various alkyl halide-derived radicals to (2-naphthoic acid)-ΔAla-OMe (R = Et, iPr, iBu, tBu, cHex, Ac, CH₂OMe) (*731*). α-Nitro-cinnamates have also been used as substrates for radical additions of alkyl halides in the presence of chiral Lewis acid catalysts. The *syn* products were generally preferred, with up to 84% ee. Nitro reduction produced β-isopropyl-Phe, O-methyl β-isopropyl-Tyr, and similar analogs (*732*).

4.3.5 Cycloaddition Reactions

Didehydroamino acids have been used for a number of cycloaddition reactions (see Schemes 4.40–4.44).

Conversion of the didehydroamino acid alkene to a cyclopropyl group is a direct route to conformationally constricted 2,3-methano saturated amino acids. Diazomethane has been added to a number of α,β-unsaturated acids to form an intermediate pyrazoline, which may or may not be isolated, and which decomposes to give the cyclopropane in 75–100% yield (see Scheme 4.40, Table 4.8) (*193, 226, 318, 319, 733–750*). Functionalized amino acids such as 2,3-methano-Cys (*741*) and 2,3-methano-Glu (*740*) have been produced, as was tricyclic 2,3-methano-tetrahydroquinoline-2-carboxylate (*751*). In 1979, 13 different 1-amino-2-arylcyclopropane-1-carboxylic acids were prepared via diazomethane cyclopropanation of the dehydroamino acid obtained by condensing N-benzyloxythiocarbonyl glycine or N-benzylthiothiocarbonylglycine with aryl aldehydes, (*226*). The azlactone intermediate obtained from an Erlenmeyer condensation of hippuric acid and D-glyceraldehyde acetonide has also been successfully reacted with diazomethane (*193, 194, 752*). Cyclopropanation of an oxazolone of 3-bromo-ΔAla was accompanied by methylene insertion to give predominantly 2-(bromomethyl)-1-aminocyclopropane-1-carboxylic acid (*753*). An asymmetric cyclopropanation with diazomethane has been carried out with >95% de on ΔNva contained within a diketopiperazine formed with Pro (*293*).

Substituted diazomethanes have also been employed (see Scheme 4.41, Table 4.9). The pyrazolines obtained from the p-nitrobenzyl ester of Boc-ΔAla and 2-diazopropane or 1-diazo-2-methylpropane were stable and required pyrolysis or thermolysis (>100 °C) for decomposition (*754*). Ethyl diazoacetate reacted with

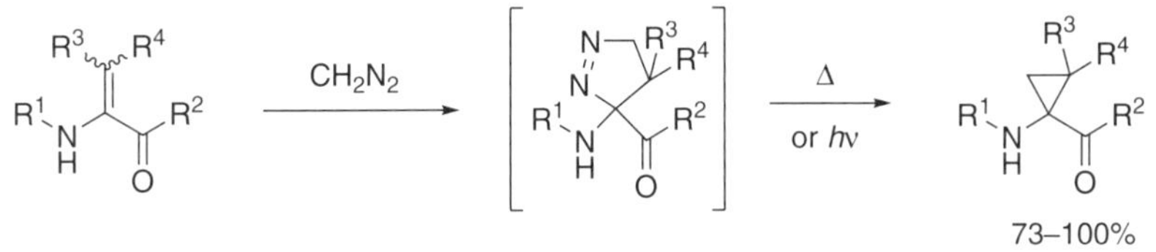

Scheme 4.40 Cycloaddition Reactions of Didehydroamino Acids with Diazomethane (see Table 4.8).

Table 4.8 Cyclopropanations of Didehydroamino Acids with Diazomethane (see Scheme 4.40)

R^1	R^2	R^3	R^4	Yield	Reference
=C(Ph)O-R^2	OC(Ph)=R^1	H	imidazol-4-yl	21%	1966 (745)
=C(Ph)O-R^2	OC(Ph)=R^1	H	3,5-I$_2$-4-AcO-Ph	29%	1967 (734)
			3,5-I$_2$-4-(4-AcO-Ph)-Ph		
=C(SBn)S-R^2	SC(SBn)=R^1	H	Ph	20–51%	1979 (226)
=C(OBn)-R^2	SC(OBn)=R^1		3-HO-Ph		
			4-HO-Ph		
			3-MeO-Ph		
			4-MeO-Ph		
			3-Cl-Ph		
			4-Cl-Ph		
			4-F-Ph		
			4-Me-Ph		
			3-MeO-4-HO-Ph		
			3-HO-4-MeO-Ph		
			3,4-OCH$_2$O-Ph		
			4-BnO-Ph		
=C(Ph)O-R^2	OC(Ph)=R^1	H	Ph	31–54%	1982 (743)
		Ph	H		
Ac	OEt	H	H	73–100%	1983 (735)
H					
=C(Ph)O-R^2	OC(Ph)=R^1	H	Me	40–75%	1983 (749)
=C(Me)O-R^2	OC(Me)=R^1	H	Ph		
		Me	H		
		Ph	H		
=C(Ph)O-R^2	OC(Ph)=R^1	H	Cl	89%	1984 (742)
		Cl	H		
=C(Ph)O-R^2	OC(Ph)=R^1	Me	H	25–60%	1985 (746)
		Ph			
		4-Me-Ph			
		4-MeO-Ph			
		3-MeO-Ph			
		4-Cl-Ph			
		2-Cl-Ph			
Cbz	OH	CH$_2$CO$_2$H	H	55%	1986 (738)
Bz	OMe	H	Ph	90–96%	1987 (747)
			4-Me-Ph		
			4-MeO-Ph		
			4-Cl-Ph		
			2-thienyl		
			3-thienyl		
=C(Ph)S-R^2	SC(Ph)=R^1	4-AcO-Ph	H	54%	1987 (750)
Cbz	OMe	CH$_2$CO$_2$Me	H	78%	1988 (274)
					1989 (740)
Cbz	OtBu	(CH$_2$)$_2$-(α-N)	H	66%	1989 (904)

(Continued)

Table 4.8 Cyclopropanations of Didehydroamino Acids with Diazomethane (see Scheme 4.40) (continued)

R^1	R^2	R^3	R^4	Yield	Reference
Pro diketopiperazine Bz	Pro diketopiperazine OCH(Ph)CH(Me) NHMe$_2$	Ph	H	57–63% up to >95% de	1989 (905)
Bz	OEt	CH$_2$CO$_2$Et	H	78%	1990 (744)
Cbz	OMe	CH$_2$PO$_3$Et$_2$	H	45–49%	1991 (733)
2-hydroxypinan -3-one Schiff base ester	2-hydroxypinan -3-one Schiff base ester	Me Et	H H	32–43% 100% de pyrazoline 40–60% de cyclopropyl	1991 (766)
Boc	Pro	Et	H	95% >95% de	1992 (293)
=C(SMe)$_2$	OMe	H	H	73–100%	1992 (736)
=C(Ph)O-R^2	OC(Ph)=R^1	(structure)	H	22–100%	1994 (194) 1995 (193) 1997 (752)
Boc Cbz	OMe	(structure)	H	73–100%	1994 (319) 1995 (318) 1996 (737) 1997 (739)
=C(Ph)O-R^2	OC(Ph)=R^1	(structure)	H	100% 66% de	1995 (748)
=C(Ph)O-R^2	OC(Ph)=R^1	SMe SPh SBn SCPh$_3$	H	54–73%	1999 (741)
CH(Ph)O-R^2	OCH(Ph)-R^1	CH$_2$CH$_2$(1,2- Ph)-(α-N)	H	62%	2002 (751)

ΔAla in much lower yield (12.5%) than diazomethane (95%) (735), although a more recent publication described conditions providing yields of up to 85% of 1-amino-cyclopropane-1,2-dicarboxylic acid by reaction of N,N-Boc$_2$ ΔAla with diazoacetate in the presence of copper complex catalysts (755). Attempts to induce asymmetry with chiral catalysts were unsuccessful. 2-Substituted-1-aminocyclopropane-1-carboxylic acids were produced by reaction of a Schiff base of ΔAla with diazoalkanes (756).

In 2003 a method of generating aryl and alkyl diazo compounds in situ from tosylhydrazone salts was applied to reactions with Boc-ΔAla-PNB. Non-metal-catalyzed reactions produced predominantly (E)-2,3-methanoamino acids in 36–50% yield, while reactions catalyzed by ClFeTPP gave the (Z)-substituted products in 44–82% yield. Coronamic acid, five different 2,3-methanophenylalanines, and 2,3-methano-5,5-diphenylallylglycine were prepared (757). A chiral sulfide component was employed with Rh catalyst and N-tosyl hydrazones to asymmetrically cyclopropanate Boc$_2$-ΔAla-OMe with PhCH=NNHTs or TMSCH=CHCH=NNTTs, providing predominantly cis-2,3-methano-Phe with 92% ee, or 2,3-methano-5-TMS-allylglycine with 75% ee (758).

2,3-Methanoamino acids can also be synthesized by addition of dimethylsulfoxonium methylide, which spontaneously cyclizes by elimination of DMSO (45–95% yield) (235, 759) (see Scheme 4.42). The reaction has also been carried out on β-substituted ethyl α-isocyanoacrylates (5–83% yield), with hydrolysis generating the 1-aminocyclopropane-1-carboxylates (239, 760). Schiff base derivatives of ΔAla-OMe have been cyclopropanated with diazoalkanes or oxosulfonium ylides to give 2-alkyl-1-aminocyclopentane-1-carboxylic acid derivatives; the oxosulfonium ylides provided much higher diastereoselectivity (756). Both diazomethane and Corey's sulfoxonium ylide gave poor diastereoselectivity when reacted with dehydroamino acids contained within William's oxazinone template. However, the ylide generated from [[(diethylamino) methyl]phenyl]oxosulfonium tetrafluoroborate by treatment with NaH gave the cyclopropyl adducts with >85% de and 79–97% yield (336). This reagent was also applied to dehydroamino acids in a chiral oxazolone substrate (761). Corey's dimethyl sulfoxonium methylide did give good diastereoselectivity and reasonable yields when reacted with ΔAbu or ΔNva within a 3,6-dihydro-2H-1,4-oxazin-2-one template (80% de, major isomer in 52–63% yield) (252, 762), while a chiral

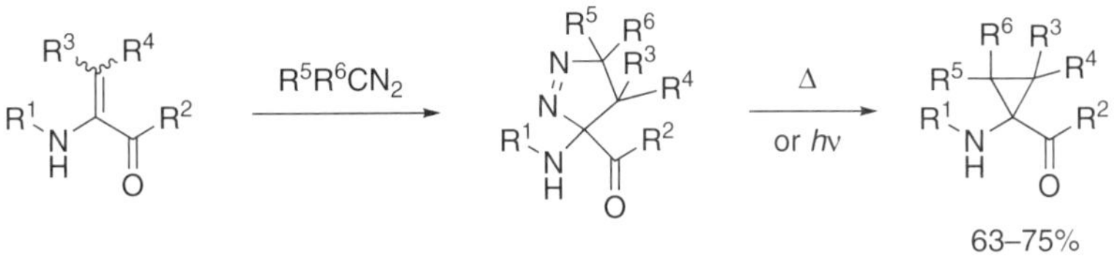

Scheme 4.41 Cycloaddition Reactions of Didehydroamino Acids with Diazoalkanes (see Table 4.9).

Table 4.9 Cyclopropanations of Dehydroamino Acids with Diazoalkanes (see Scheme 4.41)

R^1	R^2	R^3	R^4	R^5	R^6	Yield	Reference
Ac	OEt	H	H	CO_2Et	H	12.5%	1983 (735)
Boc	O(p-NO$_2$-Bn)	H	H	H	H	59–98% pyrazoline	1983 (906)
				Me		77–100% pyrolysis	
				Et			
				iPr			
				Ph			
Boc	O(p-NO$_2$-Bn)	H	H	Me	Me	88–91%	1989 (754)
				iPr	H		
=C(Ph)O–R^2	OC(Ph)=R^1	Ph	H	Ph	Ph	70%	1989 (907)
=CPh$_2$	OMe	H	H	H	H	100%	1994 (756)
=C(SMe)$_2$				Me		0–50% de	
				Ph			
Boc	OMe	H	H	CO_2Et	H	15–85%	1997 (755)
Boc$_2$							
Boc$_2$	OMe	H	H	Ph	H	65–72%	2000 (758)
				CH=CHTMS		70% de cis	
						75–92% ee with chiral	
						catalyst	
Boc	O(p-NO$_2$-Bn)	H	H	Ph	H	36–84%	2003 (757)
				4-MeO-Ph		32–90% de E with	
				4-Me-Ph		non-metal catalyst	
				4-F-Ph		62–84% de Z with Fe	
				3-TBSO-Ph		catalyst	
				CH=C(Ph)$_2$			
				CH=CH$_2$			

$R^1 = R^2 =$–CH(Ph)-CH(Ph)-O–
$R^3 =$ H, Me, nPr, iPr, Ph, 4-NO$_2$-Ph; $R^4 =$ H: 79–97% yield, >84% de (336)
R^1NH = CN; R^2 = OEt; $R^3 = R^4 =$ Me: 55% (760)
$R^1 = R^2 =$–CH(Ph)O–; $R^3 =$ [structure] ; $R^4 =$ H: 66% de (761)

R^1N = CN; R^2 = OEt; $R^3 =$ Me; $R^4 =$ 4-Me-Ph: 52% (235)
$R^3 =$ iPr, Ph; $R^4 =$ H; $R^3 = R^4 =$ H, Me, Et,–(CH$_2$)$_5$–, Ph: 5–83% (239)
$R^1 = R^2 =$ 2-hydroxypinan-3-one Schiff base ester;
$R^3 =$ Me, Et, iPr, iBu, Ph, 4-MeO-Ph, 3,4-(MeO)$_2$-Ph; $R^4 =$ H: 45–95% (759)
$R^1 = R^2 =$ =C(Ph)CH(iPr)O–; $R^3 =$ H; $R^4 =$ Me, Et:
80% de, 52–63% yield of major isomer (252, 762)
$R^1 = R^2 =$C(Ph)CH(iPr)N(Boc)–; $R^3 =$ Me, Et; $R^4 =$ H:
70–79% yield, 83–92% de (253, 763).

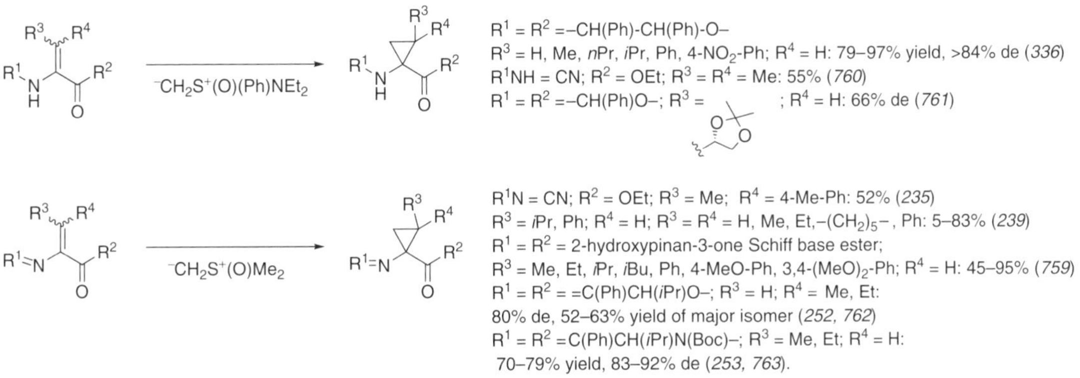

Scheme 4.42 Other Cyclopropanation Reactions of Didehydroamino Acids.

tetrahydropyrazin-2-one template induced 83–92% de for various dehydroamino acids (253, 763). Other cyclopropanations have been carried out on dehydroamino acid residues contained within Williams' oxazinone (320, 764), Seebach's imidazolidinone (407) or Seebach's oxazolidinone (765) templates, or a Schiff base/ester complex formed with 2-hydroxypinan-2-one (766). Δ within an N,N-bis(4-methoxybenzyl) Val-ΔAla diketopiperazine template was cyclopropanated with isopropylidenetriphenylphosphorane, ethylidenetriphenylphosphorane, phenyldiazomethane, or (CD₃)₂CD₂(Li)SO, resulting in 2,2-dimethyl-1-aminocyclopropane-1-carboxylate, 2-methyl-1-aminocyclopropane-1-carboxylate, 2-phenyl-1-aminocyclopropane-1-carboxylate, and [2,2-²H₂]-1-aminocyclopropane-1-carboxylate with >98% de at the α-center, and varying diastereoselectivity at the β-center (when applicable) (909).

Four-membered ring systems can also be produced from dehydroamino acids. 2,4-Methanoproline was obtained in 87% yield from an intramolecular [2+2] photoaddition of N-allyl ΔAla (126, 767, 768); N-acryloyl ΔAla gave 2,4-methano-pyroGlu (768), while N-Bz,N-(2-ethoxycarbonylallyl) ΔAla-OMe cyclized to 2,4-methano-4-carboxy-Pro in 35% yield (769). A thermal [2+2] cycloaddition between Ac-ΔAla-OMe and ketene diethyl acetal, H₂C=C(OEt)₂, produced 1-amino-2,2-diethoxy-cyclobutane-1-carboxylic acid in 64% yield. The adduct was elaborated to give either diastereomer of 1-amino-2-hydroxycyclobutane-1-carboxylic acid (770).

Heterocyclic five-membered cyclic amino acids can be synthesized by 1,3-dipolar cycloadditions (see Scheme 4.43). Benzonitrile oxide and C,N-diphenylnitrone were reacted with ΔAla in 75–85%

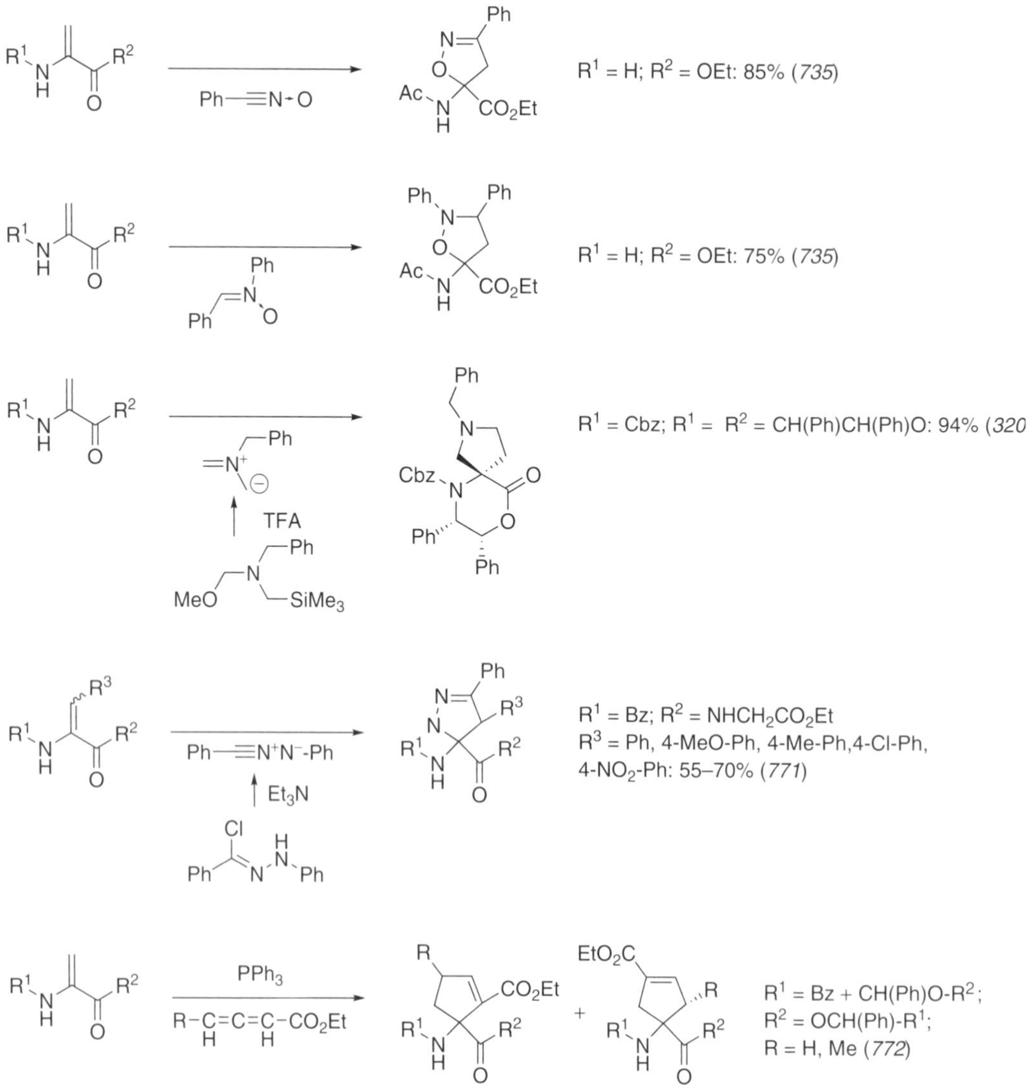

Scheme 4.43 Cycloaddition Reactions of Didehydroamino Acids to Form Five-Membered Rings.

yield (735). Benzonitrilium N-phenylimide added to N-terminal 2,3-didehydro aromatic amino acid residues in dipeptides in 55–70% yield (771). An azomethine ylide derived from N-benzyl-N-(methoxymethyl)[trimethylsilyl]methylamine reacted with ΔAla contained within a chiral oxazinone template in 94% yield, with complete diastereoselectivity (320). ΔAla within an oxazolidinone template has also been reacted with carboxy-substituted allenes to give carboxy-substituted 1-amino-cyclopentene-1-carboxylic acids (772).

Six-membered rings are formed by the classical Diels–Alder reaction (see Scheme 4.44). ΔAla and other 2,3-dehydro aminoacids undergo a Diels–Alder reaction with 1,3-butadiene to form 1-aminocyclohexane-1-carboxylic acid and 2-substituted analogs in 80–94% yield (773–778). Moderate diastereoselectivity can be induced by employing a chiral ester auxiliary (773), while almost complete diastereoselectivity can be induced by incorporating the ΔAla in a chiral oxazolidinone template (138, 141, 779, 780). ΔPhe has been cyclized with both 1,3-butadiene and 2,3-dimethyl-1,3-butadiene; the alkene in the product was hydrogenated to produce conformationally restricted Phe analogs (776). A resin-linked ΔPhe analog derived from condensation of nitroacetate (esterified to a Wang resin) with benzaldehyde was reacted with 2,3-dimethyl-1,3-butadiene under high pressure. The nitro group was reduced with $SnCl_2$ (233).

Cyclopentadiene and cyclohexa-1,3-diene react with dehydroamino acids to give bicyclic amino acids (25–100% yields) (141, 163, 781–786). Resin-bound

N-Fmoc ΔAla was condensed with cyclopentadiene, cyclohexa-1,3-diene, and O-TMS 1-hydroxy-1,3-butadiene with 48–56% yield; furan did not react (142). Asymmetry was introduced in Diels–Alder reactions of N-Ac ΔAla with cyclopentadiene by using chiral ester auxiliaries (787–789) or chiral Lewis acid catalysts (790), producing 2-aminonorbornene-2-carboxylic acids. Similarly, a menthyl ester of O-ethoxycarbonyl Bz-ΔSer underwent cycloaddition with cyclopentadiene with $Mg(ClO_4)_2$ catalysis to give 2-amino-3-hydroxynorbornene-2-carboxylic acid as a 77:23 mixture of exo:endo isomers, with 80% de for the exo isomer and 86% de for the endo isomer (791). Optically active dehydroamino acids with side chains derived from D-glyceraldehyde have also been reacted with cyclopentadiene to give optically active bicyclic amino acid derivatives (792, 793). ΔAla within a tetrahydropyrazin-2-one template was cyclized with both cyclopentadiene and cyclohexa-1,3-diene, producing single diastereomers in 42–95% yield. The deprotected amino acids were obtained in 61% and 38% yield (253, 763). In a similar manner, ΔAla within a 3,6-dihydro-2H-1,4-oxazin-2-one template was condensed with cyclopentadiene, cyclohexa-1,3-diene, and 1-methoxycyclohexa-1,3-diene (252, 794).

4.3.6 Other Reactions

Didehydroamino acid derivatives can be isomerized into the β,γ-unsaturated vinylglycine isomer by treatment

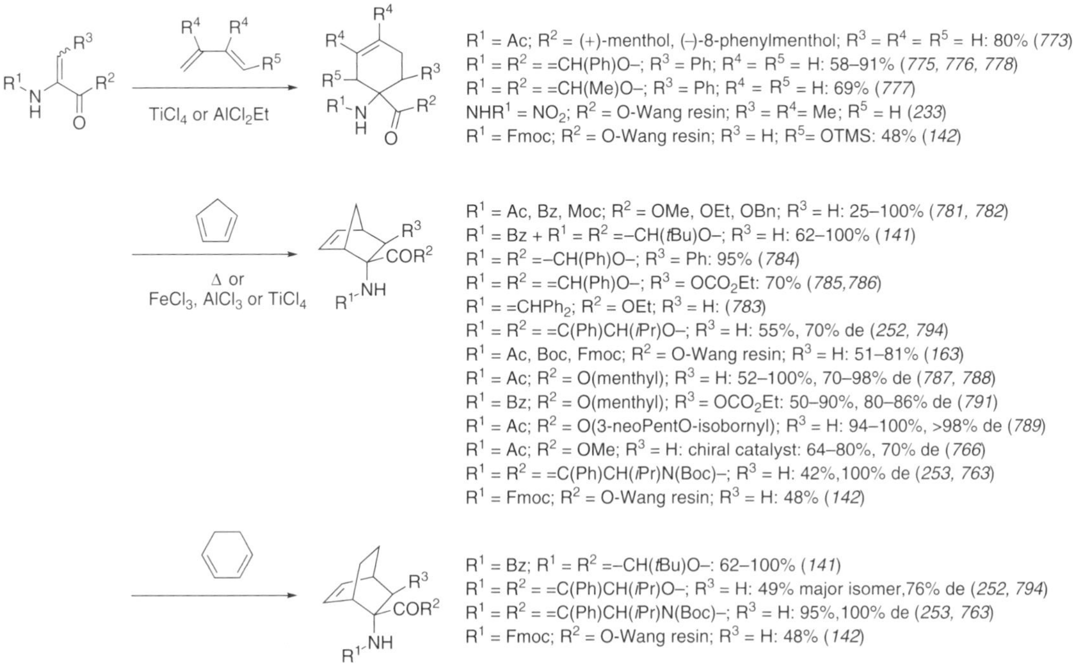

R^1 = Ac; R^2 = (+)-menthol, (–)-8-phenylmenthol; R^3 = R^4 = R^5 = H: 80% (773)
R^1 = R^2 = =CH(Ph)O–; R^3 = Ph; R^4 = R^5 = H: 58–91% (775, 776, 778)
R^1 = R^2 = =CH(Me)O–; R^3 = Ph; R^4 = R^5 = H: 69% (777)
NHR^1 = NO_2; R^2 = O-Wang resin; R^3 = R^4= Me; R^5 = H (233)
R^1 = Fmoc; R^2 = O-Wang resin; R^3 = H; R^5= OTMS: 48% (142)

R^1 = Ac, Bz, Moc; R^2 = OMe, OEt, OBn; R^3 = H: 25–100% (781, 782)
R^1 = Bz + R^1 = R^2 =–CH(tBu)O–; R^3 = H: 62–100% (141)
R^1 = R^2 =–CH(Ph)O–; R^3 = Ph: 95% (784)
R^1 = R^2 = =CH(Ph)O–; R^3 = OCO_2Et: 70% (785,786)
R^1 = =CHPh$_2$; R^2 = OEt; R^3 = H: (783)
R^1 = R^2 = =C(Ph)CH(iPr)O–; R^3 = H: 55%, 70% de (252, 794)
R^1 = Ac, Boc, Fmoc; R^2 = O-Wang resin; R^3 = H: 51–81% (163)
R^1 = Ac; R^2 = O(menthyl); R^3 = H: 52–100%, 70–98% de (787, 788)
R^1 = Bz; R^2 = O(menthyl); R^3 = OCO_2Et: 50–90%, 80–86% de (791)
R^1 = Ac; R^2 = O(3-neoPentO-isobornyl); R^3 = H: 94–100%, >98% de (789)
R^1 = Ac; R^2 = OMe; R^3 = H: chiral catalyst: 64–80%, 70% de (766)
R^1 = R^2 = =C(Ph)CH(iPr)N(Boc)–; R^3 = H: 42%,100% de (253, 763)
R^1 = Fmoc; R^2 = O-Wang resin; R^3 = H: 48% (142)

R^1 = Bz; R^1 = R^2 =–CH(tBu)O–: 62–100% (141)
R^1 = R^2 = =C(Ph)CH(iPr)O–; R^3 = H: 49% major isomer,76% de (252, 794)
R^1 = R^2 = =C(Ph)CH(iPr)N(Boc)–; R^3 = H: 95%,100% de (253, 763)
R^1 = Fmoc; R^2 = O-Wang resin; R^3 = H: 48% (142)

Scheme 4.44 Cycloaddition Reactions of Didehydroamino Acids to Form Six-Membered Rings.

with base (18–79% yield) followed by reprotonation (*129, 242–244, 262, 304, 795–799*). A range of eight *N*-Cbz dehydroamino acids, prepared via condensation of Cbz-Gly(phosphonate)-OMe with aliphatic aldehydes, were rearranged in 2005 using LiTMP/LiCl, producing (*E*)-4-substituted vinylglycines with 49–99% yields (*800*). A chiral acid has been used for the reprotonation step, giving vinylglycine with up to 36% ee (*801*). The α-centered anion can also be trapped with electrophiles (RX; R = Me, Et, *n*-Pr, *n*-Bu, *n*-Hex, allyl, CH₂CCH, CH₂CO₂ CH₃, CH₂-2-imidazole, Bn, 4-BnO-Bn, 3-BnO-Bn, 3,4-OCPh₂O-Ph) to generate α-substituted-vinylglycines (49–88% yield) (*129, 242, 244, 796, 797*). Deconjugation and electrophile trapping has also been applied to dehydroamino acids contained within the chiral imidazolidinone template of Seebach (77–92% yield) with >95% diastereoselectivity (*407*).

References

1. Schmidt, U.; Lieberknecht, A.; Wild, J. *Synthesis* **1988**, 159–172.
2. Bonauer, C.; Walenzyk, T.; König, B. *Synthesis* **2006**, 1–20.
3. *The Peptides. Analysis, Synthesis, Biology. Volume 5 Special Method in Peptide Synthesis, Part B*, Gross, E.; Meienhofer, J.; Eds.; Academic Press: New York, 1983, pp 285–339.
4. Stammer, C.H. In *Chemistry and Biochemistry of Amino Acids, Peptides and Proteins. A Survey of Recent Developments, Volume 6*, Weinstein, B., Ed.; Marcel Dekker: New York, 1982, pp 33–74.
5. Ueno, T.; Nakashima, T.; Hayashi, Y.; Fukami, H. *Agric. Biol. Chem.* **1975**, *39*, 1115–1122.
6. Ueno, T.; Nakashima, T.; Hayashi, Y.; Fukami, H. *Agric. Biol. Chem.* **1975**, *39*, 2081–2082.
7. Namikoshi, M.; Rinehart, K.L.; Sakai, R.; Stotts, R.R.; Dahlem, A.M.; Beasley, V.R.; Carmichael, W.W.; Evans, W.R. *J. Org. Chem.* **1992**, *57*, 866–872.
8. Sano, T.; Kaya, K. *Tetrahedron* **1998**, *54*, 463–470.
9. Fujii, K.; Sivonen, K.; Nakano, T.; Harada, K.-i. *Tetrahedron* **2002**, *58*, 6863–6871.
10. Regueiro-Ren, A.; Ueda, Y. *J. Org. Chem.* **2002**, *67*, 8699–8702.
11. Naidu, B.N.; Li, W.; Sorenson, M.E.; Connolly, T.P.; Wichtowski, J.A.; Zhang, Y.; Kim, O.K.; Matiskella, J.D.; Lam, K.S.; Bronson, J.J.; Ueda, Y. *Tetrahedron Lett.* **2004**, *45*, 1059–1063.
12. Wakamiya, T.; Shimbo, K.; Sano, A.; Fukase, K.; Shiba, T. *Bull. Chem. Soc. Jpn.* **1983**, *56*, 2044–2049.
13. Gross, E.; Morell, J.L. *J. Am. Chem. Soc.* **1971**, *93*, 4634–4635.
14. Jung, G. *Angew. Chem., Int. Ed. Engl.* **1991**, *30*, 1051–1068.
15. Sahl, H.-G.; Jack, R.W.; Bierbaum, G. *Europ. J. Biochem.* **1995**, *230*, 827–855.
16. Smith, L.; Zachariah, C.; Thirumoorthy, R.; Rocca, J.; Novák, J.; Hillman, J.D.; Edison, A.S. *Biochemistry* **2003**, *42*, 10372–10384.
17. Yarnell, A. *Chem. Eng. News* **2004**, *February 2*, 10.
18. Burrage, S.; Raynham, T.; Williams, G.; Essex, J.W.; Allen, C.; Cardno, M.; Swali, V.; Bradley, M. *Chem. Eur. J.* **2000**, *6*, 1455–1466.
19. Nicolaou, K.C.; Safina, B.S.; Zak, M.; Estrada, A.A.; Lee, S.H. *Angew. Chem., Int. Ed.* **2004**, *43*, 5087–5092.
20. Nicolaou, K.C.; Zak, M.; Safina, B.S.; Lee, S.H.; Estrada, A.A. *Angew. Chem., Int. Ed.* **2004**, *43*, 5092–5097.
21. Nicolaou, K.C.; Safina, B.S.; Funke, C.; Zak, M.; Zécri, F.J. *Angew. Chem., Int. Ed.* **2002**, *41*, 1937–1940.
22. Mori, T.; Tohmiya, H.; Satouchi, Y.; Higashibayashi, S.; Hashimoto, K.; Nakata, M. *Tetrahedron Lett.* **2005**, *46*, 6423–6427.
23. Goetz, G.; Yoshida, W.Y.; Scheur, P.J. *Tetrahedron* **1999**, *55*, 7739–7746.
24. López-Marcià, À.; Jiménez, J.C.; Royo, M.; Giralt, E.; Albericio, F. *J. Am. Chem. Soc.* **2001**, *123*, 11398–11401.
25. Komori, T.; Ezaki, M.; Kino, E.; Kohsaka, M.; Aoki, H.; Imanaka, H. *J. Antibiotics* **1985**, 691–698.
26. Uchida, I.; Shigematsu, N.; Ezaki, M.; Hashimoto, M. *Chem. Pharm. Bull.* **1985**, *33*, 3053–3056.
27. Pettit, G.R.; Kamano, Y.; Herald, C.L.; Fujii, Y.; Kizu, H.; Boyd, M.R.; Boettner, F.E.; Doubek, D.L.; Schmidt, J.M.; Chapuis, J.-C.; Michel, C. *Tetrahedron* **1993**, *49*, 9151–9170.
28. MacMillan, J.B.; Ernst-Russell, M.A.; de Ropp, J.S.; Molinski, T.F. *J. Org. Chem.* **2002**, *67*, 8210–8215.
29. Ward, D.E.; Vázquez, A.; Pedras, M.S.C. *J. Org. Chem.* **1999**, *64*, 1657–1666.
30. Fukuchi, N.; Isogai, A.; Nakayama, J.; Takayama, S.; Yamishita, S.; Suyama, K.; Takemoto, J.Y.; Suzuki, A. *J. Chem. Soc., Perkin Trans. 1* **1992**, 1149–1157.
31. Samy, R.; Kim, H.Y.; Brady, M.; Toogood, P.L. *J. Org. Chem.* **1999**, *64*, 2711–2728.
32. Hu, T.; Panek, J.S. *J. Am. Chem. Soc.* **2002**, *124*, 11368–11378.
33. Ford, P.W.; Gustafson, K.R.; McKee, T.C.; Shigematsu, N.; Maurizi, L.K.; Pannell, L.K.; Williams, D.E.; de Silva, E.D.; Lassota, P.; Allen, T.M.; Van Soest, R.; Andersen, R.J.; Boyd, M.R. *J. Am. Chem. Soc.* **1999**, *121*, 5899–5909.
34. Stohlmeyer, M.M.; Tanaka, H.; Wandless, T.J. *J. Am. Chem. Soc.* **1999**, *121*, 6100–6101.
35. Shiroza, T.; Ebisawa, N.; Kojima, A.; Furihata, K.; Shimazu, A.; Endo, T.; Seto, H.; Otake, N. *Agric. Biol. Chem.* **1982**, *46*, 1885–1890.

36. Shiroza, T.; Ebisawa, N.; Furihata, K.; Endo, T.; Seto, H.; Otake, N. *Agric. Biol. Chem.*, **1982**, *46*, 1891–1898.

37. Shiroza, T.; Ebisawa, N.; Furihata, K.; Endo, T.; Seto, H.; Otake, N. *Agric. Biol. Chem.*, **1982**, *46*, 865–867.

38. Shimada, N.; Morimoto, K.; Naganawa, H.; Takita, T.; Hamada, M.; Maeda, K.; Takeuchi, T.; Umezawa, H. *J. Antibiotics* **1981**, *34*, 1613–1614.

39. Morimoto, K.; Shimada, N.; Naganawa, H.; Takita, T.; Umezawa, H. *J. Antibiotics* **1981**, *34*, 1615–1618.

40. Morimoto, K.; Shimada, N.; Naganawa, H.; Takita, T.; Umezawa, H. *J. Antibiotics* **1982**, *35*, 378–380.

41. Jiménez, J.C.; Chavarría, B.; López-Macìa, À.; Royo, M.; Giralt, E.; Albericio, F. *Org. Lett.* **2003**, *5*, 2115–2118.

42. Kim, D.; Li, Y.; Horenstein, B.A.; Nakanishi, K. *Tetrahedron Lett.* **1990**, *31*, 7119–7122.

43. Horenstein, B.A.; Nakanishi, K. *J. Am. Chem. Soc.* **1989**, *111*, 6242–6246.

44. Stonard, R.J.; Andersen, R.J. *Can. J. Chem.* **1980**, *58*, 2121–2126.

45. Stonard, R.J.; Andersen, R.J. *J. Org. Chem.* **1980**, *45*, 3687–3691.

46. Gallina, C.; Romeo, A.; Tarzia, G.; Tortorella, V. *Gazz. Chim. Ital.* **1964**, *94*, 1301–1310.

47. Wakamiya, T.; Shiba, T. *J. Antibiotics* **1975**, *28*, 292–297.

48. Izumi, R.; Noda, T.; Ando, T.; Take, T.; Nagata, A. *J. Antibiotics* **1972**, *25*, 201–207.

49. Noda, T.; Take, T.; Nagata, A.; Wakamiya, T.; Shiba, T. *J. Antibiotics* **1972**, *25*, 427–428.

50. Nagata, A.; Ando, T.; Izumi, R.; Sakakibara, H.; Take, T.; Hayano, K.; Abe, J.-N. *J. Antibiotics* **1968**, *21*, 681–687.

51. Ando, T.; Matsuura, K.; Izumi, R.; Noda, T.; Take, T.; Nagata, A.; Abe, J.-N. *J. Antibiotics* **1971**, *24*, 680–686.

52. Wakamiya, T.; Teshima, T.; Sakakibara, H.; Fukukawa, K.; Shiba, T. *Bull. Chem. Soc., Jpn.* **1977**, *50*, 1984–1989.

53. Bycroft, B.W.; Cameron, D.; Croft, L.R.; Hassanali-Walji, A.; Webb, T. *J. Chem. Soc., Perkin Trans. I* **1972**, 827–834.

54. DeMong, D.E.; Williams, R.M. *J. Am. Chem. Soc.* **2003**, *125*, 8561–8565.

55. Armstrong, R.W.; Tellew, J.E.; Moran, E.J. *J. Org. Chem.* **1992**, *57*, 2208–2211.

56. Coleman, R.S.; Kong, J.-S. *J. Am. Chem. Soc.* **1998**, *120*, 3538–3539.

57. Coleman, R.S.; Perez, R.J.; Burk, C.H.; Navarro, A. *J. Am. Chem. Soc.* **2002**, *124*, 13008–13017.

58. Landreau, C.A.S.; LePla, R.C.; Shipman, M.; Slawin, A.M.Z.; Hartley, J.A. *Org. Lett.* **2004**, *6*, 3505–3507.

59. Coleman, R.S.; Li, J.; Navarro, A. *Angew. Chem., Int. Ed.* **2001**, *40*, 1736–1739.

60. Sheehan, J.C.; Mania, D.; Nakamura, S.; Stock, J.A.; Maeda, K. *J. Am. Chem. Soc.* **1968**, *90*, 462–470.

61. Nakazawa, T.; Ishii, M.; Musiol, H.-J.; Moroder, L. *Tetrahedron Lett.* **1998**, *39*, 1381–1384.

62. Qureshi, A.; Colin, P.L.; Faulkner, D.J. *Tetrahedron* **2000**, *56*, 3679–3685.

63. Selic, L.; Jakse, R.; Lampic, K.; Golic, L.; Golic-Grdadolnik, S.; Staznovnik, B. *Helv. Chim. Acta* **2000**, *83*, 2802–2811.

64. Johnson, A.-L.; Bergman, J.; Sjögren, M.; Bohlin, L. *Tetrahedron* **2004**, *60*, 961–965.

65. Sölter, S.; Dieckmann, R.; Blumenberg, M.; Francke, W. *Tetrahedron Lett.* **2002**, *43*, 3385–3386.

66. Overman, L.E.; Rosen, M.D. *Angew. Chem., Int. Ed.* **2000**, *39*, 4596–4599.

67. Purvis, M.B.; LeFevre, J.W.; Jones, V.L.; Lingston, D.G.I.; Biot, A.M.; Gosselé, F. *J. Am. Chem. Soc.* **1989**, *111*, 5931–5935.

68. Gupta, A.; Chauhan, V.S. *Biopolymers* **1990**, *30*, 395–403.

69. Crisma, M.; Formaggio, F.; Toniolo, C.; Yoshikawa, T.; Wakamiya, T. *J. Am. Chem. Soc.* **1999**, *121*, 3272–3278.

70. Jain, R.M.; Rajashankar, K.R.; Ramakumar, S.; Chauhan, V.S. *J. Am. Chem. Soc.* **1997**, *119*, 3205–3211.

71. Mitra, S.N.; Dey, S.; Karthkeyan, S.; Singh, T.P. *Biopolymers* **1997**, *41*, 97–105.

72. Sastry, T.V.R.S.; Banerji, B.; Kumar, S.K.; Kunwar, A.C.; Daas, J.; Nandy, J.P.; Iqbal, J. *Tetrahedron Lett.* **2002**, *43*, 7621–7625.

73. Inai, Y.; Tagawa, K.; Takasu, A.; Hirabayashi, T.; Oshikawa, T.; Yamashita, M. *J. Am. Chem. Soc.* **2000**, *122*, 11731–11732.

74. Vijayaraghavan, R.; Kumar, P.; Dey, S.; Singh, T.P. *J. Pept. Res.* **1998**, *52*, 98–94.

75. Aubry, A.; Marraud, M. *Biopolymers* **1989**, *28*, 109–122.

76. Cativiela, C.; Díaz-de-Villegas, M.D.; Gálvez, J.A. *Tetrahedron Lett.* **1999**, *40*, 1027–1030.

77. Cativiela, C.; Díaz-de-Villegas, M.D.; Gálvez, J.A.; Su, G. *Tetrahedron* **2004**, *60*, 11923–11932.

78. Jain, R.; Chauhan, V.S. *Biopolymers* **1996**, *40*, 105–119.

79. Benedetti, E. *Biopolymers* **1996**, *40*, 3–44.

80. Shimohiagashi, Y.; Stammer, C.H. *J. Chem. Soc., Perkin Trans. 1* **1983**, 803–808.

81. Lombardi, A.; D'Agostino, B.; Nastri, F.; D'Andrea, L.; Filippelli, A.; Falciani, M.; Rossi, F.; Pavone, V. *Biorg. Med. Chem. Lett.* **1998**, *8*, 1153–1156.

82. Josien, H.; Lavielle, S.; Brunissen, A.; Saffroy, M.; Torrens, Y.; Beaujouan, J.-C.; Glowinski, J.; Chassaing, G. *J. Med. Chem.* **1994**, *37*, 1586–1601.

83. Torrini, I.; Zecchini, G.P.; Paradisi, M.P.; Lucente, G.; Gavuzzo, E.; Mazza, F.; Pochetti, G.; Traniello, S.; Spisani, S.; Cerichelli, G. *Int. J. Pept. Prot. Res.* **1994**, *34*, 1291–1302.

84. Torrini, I.; Zecchini, G.P.; Paradisi, M.P.; Lucente, G.; Gavuzzo, E.; Mazza, F.; Pochetti, G.; Spisani, S. *Tetrahedron* **1993**, *49*, 489–496.

85. Murkin, A.S.; Tanner, M.E. *J. Org. Chem.* **2002**, *67*, 8389–8394.

86. Abreu, A.S.; Ferreira, P.M.T.; Monteiro, L.S.; Queiroz, M.-J.R.P.; Ferreira, I.C.F.R.; Calhelha, R.C.; Estevinho, L.M. *Tetrahedron* **2004**, *60*, 11821–11828.

87. Humphrey, J.M.; Chamberlin, A.R. *Chem. Rev.* **1997**, *97*, 2243–2266.

88. Sin, C.-G.; Kakusho, T.; Arai, K.; Seki, M. *Bull. Chem. Soc. Jpn.* **1995**, *68*, 3549–3555.

89. Shin, C.-G.; Arai, K.; Hotta, K.; Kakusho, T. *Bull. Chem. Soc. Jpn.* **1997**, *70*, 1427–1434.

90. Lu, S.-P.; Lewin, A.H. *Tetrahedron* **1998**, *54*, 15097–15104.

91. Noda, K.; Gazis, D.; Gross, E. *Int. J. Pept. Prot. Chem.* **1982**, *19*, 413–419.

92. Gulzar, M.S.; Morris, K.B.; Gani, D. *J. Chem. Soc., Chem. Commun.* **1995**, 1061–1062.

93. Zhou, H.; Schmidt, D.M.Z.; Gerlt, J.A.; van deer Donk, W.A. *ChemBioChem.* **2003**, *4*, 1206–1215.

94. Effenberger, F.; Beisswenger, T. *Angew. Chem., Int. Ed. Engl.* **1982**, *21*, 203.

95. Cativiela, C.; Diaz de Villegas, M.D. *Tetrahedron*, **1993**, *49*, 497–506.

96. Schmidt, U.; Griesser, H.; Leitenberger, V.; Lieberknecht, A.; Mangold, R.; Meyer, R.; Riedl, B. *Synthesis* **1992**, 487–490.

97. Shin, C.-G.; Okumura, K.; Ito, A.; Nakamura, Y. *Chem. Lett.* **1994**, 1301–1304.

98. Bory, S.; Gaudry, M.; Marquet, A.; Azerad, R. *Biochem. Biophys. Res. Commun.* **1979**, *87*, 85–91.

99. Photaki, I. *J. Am. Chem. Soc.* **1963**, *85*, 1123–1126.

100. Gibson, S.E.; Guillo, N.; Middleton, R.J.; Thuillie, A.; Tozer, M.J. *J. Chem. Soc., Perkin Trans. 1* **1997**, 447–455.

101. Somlai, C.; Lovas, S.; Forgó, P.; Murphy, R.F.; Penke, B. *Synth. Commun.* **2001**, *31*, 3633–3640.

102. Srinivassan, A.; Stephenson, R.W.; Olsen, R.K. *J. Org. Chem.* **1977**, *42*, 2256–2260.

103. Kolasa, T.; Gross, E. *Int. J. Pept. Prot. Chem.* **1982**, 259–266.

104. Goodall, K.; Parsons, A.F. *Tetrahedron* **1996**, *52*, 6739–6758.

105. Sjölin, P.; Kihlberg, J. *Tetrahedron Lett.* **2000**, *41*, 4435–4439.

106. Belekon, Y.N.; Sagyan, A.S.; Djamgaryan, S.M.; Bakhmutov, V.I.; Belikov, V.M. *Tetrahedron* **1988**, *44*, 5507–5514.

107. Saghiyan, A.S.; Geolchanyan, A.V.; Petrosyan, S.G.; Ghochikyan, T.V.; Haroutunyan, V.S.; Avetisyan, A.A.; Belokon, Y.N.; Fisher, K. *Tetrahedron: Asymmetry* **2004**, *15*, 705–711.

108. Crossley, M.J.; Tansey, C.W. *Aust. J. Chem.* **1992**, *45*, 479–481.

109. Roff, G.J.; Lloyd, R.C.; Turner, N.J. *J. Am. Chem. Soc.* **2004**, *126*, 4098–4099.

110. Seethaler, T.; Simchen, G. *Synthesis* **1986**, 390–392.

111. Hardegger, E.; Szabo, F.; Liechti, P.; Rostetter, C.; Zankowska-Jasinska, W. *Helv. Chim. Acta* **1968**, *51*, 78–85.

112. Griesbeck, A.G.; Bondock, S.; Lex, J. *J. Org. Chem.* **2003**, *68*, 9899–9906.

113. Wojciechowska, H.; Pawlowicz, R.; Andruszkiewicz, R.; Gryzbowska, J. *Tetrahedron Lett.* **1978**, *42*, 4063–4064.

114. Ogura, H.; Sato, O.; Takeda, K. *Tetrahedron Lett.* **1981**, *22*, 4817–4818.

115. Schmidt, U.; Mundinger, K.; Mangold, R.; Lieberknecht, A. *J. Chem. Soc., Chem. Commun.* **1990**, 1216–1218.

116. Somekh, L.; Shanzer, A. *J. Org. Chem.* **1983**, *48*, 907–908.

117. Ferreira, P.M.T.; Maia, H.L.S.; Monteiro, L.S. *Tetrahedron Lett.* **1998**, *39*, 9575–9578.

118. Ferreira, P.M.T.; Maia, H.L.S.; Monteiro, L.S.; Sacramento, J. *J. Chem. Soc., Perkin Trans. 1* **1999**, 3697–3703.

119. Shin, G.-G.; Takahashi, N.; Yonezawa, Y. *Chem. Pharm. Bull.* **1990**, *38*, 2020–2023.

120. Wulff, G.; Böhnke, H. *Angew. Chem., Int. Ed. Engl.* **1984**, *23*, 380–381.

121. Andruszkiewicz, R.; Czerwinski, A. *Synthesis* **1982**, 968–969.

122. Miller, M.J. *J. Org. Chem.* **1980**, *45*, 3131–3132.

123. Balsamini, C.; Duranti, E.; Mariani, L.; Salvatori, A.; Spadoni, G. *Synthesis* **1990**, 779–781.

124. Biagini, S.C.G.; Gibson, S.E.; Keen, S.P. *J. Chem. Soc., Perkin Trans. 1* **1998**, 2485–2499.

125. Royo, M.; Jiménez, J.C.; López-Macià, A.; Giralt, E.; Albericio, F. *Eur. J. Org. Chem.* **2001**, 45–48.

126. Hughes, P.; Martin, M.; Clardy, J. *Tetrahedron Lett.* **1980**, *21*, 4579–4580.

127. Bigge, C.F.; Wu, J.-P.; Drummond, J.T.; Coughenour, L.L.; Hanchin, C.M. *Bioorg. Med. Chem. Lett.* **1992**, *2*, 207–212.

128. Pines, S.H.; Kozlowski, M.A. *J. Org. Chem.* **1972**, *37*, 292–297.

129. Greenlee, W.J.; Taub, D.; Patchett, A.A. *Tetrahedron Lett.* **1978**, 3999–4002.

130. Meffre, P.; Vo-Quang, L.; Vo-Quang, Y.; Le Goffic, F. *Tetrahedron Lett.* **1990**, *31*, 2291–2294.

131. Kakimoto, M.; Kai, M.; Kondo, K. *Chem. Lett.* **1982**, 525–526.

132. Kakimoto, M.; Kai, M.; Kondo, K.; Hiyama, T. *Chem. Lett.* **1982**, 527–528.

133. Knittel, D. *Monatsh. Chem.* **1984**, *115*, 1335–1343.

134. Knittel, D. *Monatsh. Chem.* **1985**, *116*, 1133–1140.

135. Beckwith, A.L.J.; Chai, C.L.L. *J. Chem. Soc., Chem. Commun.* **1990**, 1087–1088.

136. Tozer, M.J.; White, A.H. *Aust. J. Chem.* **1993**, *46*, 1425–1430.

137. Zimmermann, J.; Seebach, D. *Helv. Chim. Acta* **1987**, *70*, 1104–1114.

138. Pyne, S.G.; Dikic, B.; Gordon, P.A.; Skelton, B.W.; White, A.H. *J. Chem. Soc., Chem. Commun.* **1991**, 1505–1506.

139. Sokolovsky, M.; Sadeh, T.; Patchornik, A. *J. Am. Chem. Soc.* **1964**, *86*, 1212–1217.

140. Schmidt, U.; Öhler, E. *Angew. Chem., Int. Ed. Engl.* **1976**, *15*, 42.

141. Pyne, S.G.; Dikic, B.; Gordon, P.A.; Skelton, B.W.; White, A.H. *Aust. J. Chem.* **1993**, *46*, 73–93.

142. Burkett, B.A.; Chain, C.L.L. *Tetrahedron Lett.* **2000**, *41*, 6661–6664.

143. Miao, Z.; Tam, J.P. *Org. Lett.* **2000**, *2*, 3711–3713.

144. Matteucci, M.; Bhalay, G.; Bradley, M. *Tetrahedron Lett.* **2004**, *45*, 1399–1401.

145. Öhler, E.; Schmidt, U. *Chem. Ber.* **1979**, *112*, 107–115.

146. Nakamura, K.; Isaka, T.; Toshima, H.; Kodaka, M. *Tetrahedron Lett.* **2004**, *45*, 7221–7224.

147. Ezquerra, J.; Escribano, A.; Rubio, A.; Remuiñán, M.J.; Vaquero, J.J. *Tetrahedron: Asymmetry* **1996**, *7*, 2613–2626.

148. Coghlan, P.A.; Easton, C.J. *Tetrahedron Lett.* **1999**, *40*, 4745–4748.

149. Easton, C.J.; Roselt, P.D.; Tiekink, E.R.T. *Tetrahedron* **1995**, *51*, 7809–7822.

150. Burgess, V.A.; Easton, C.J. *Aust. J. Chem.* **1988**, *41*, 1063–1070.

151. Coghlan, P.A.; Easton, C.J. *ARKIVOC* **2004**, 101–108.

152. Tarzia, G.; Balsamini, C.; Spadoni, G.; Duranti, E. *Synthesis* **1988**, 514–517.

153. Shimohigashi, Y.; Stammer, C.H. *Int. J. Pept. Prot. Chem.* **1982**, *19*, 54–62.

154. Sommerfeld, T.L.; Seebach, D. *Helv. Chim. Acta* **1993**, *76*, 1702–1714.

155. Chai, C.L.L.; King, A.R. *Tetrahedron Lett.* **1995**, *36*, 4295–4298.

156. Polinsky, A.; Cooney, M.G.; Toy-Palmer, A.; Ösapay, G.; Goodman, M. *J. Med. Chem.* **1992**, *35*, 4185–4194.

157. Valentekovich, R.J.; Schreiber, S.L. *J. Am. Chem. Soc.* **1995**, *117*, 9069–9070.

158. Yokokawa, F.; Shioiri, T. *Tetrahedron Lett.* **2002**, *43*, 8679–8682.

159. Srinivasan, A.; Stephenson, R.W.; Olsen, R.K. *J. Org. Chem.* **1977**, *42*, 2253–2256.

160. Crich, D.; Davies, J.W. *Tetrahedron* **1989**, *45*, 5641–5654.

161. Shin, C-G.; Yonezawa, Y.; Yamada, T.; Yoshimura, J. *Bull. Chem. Soc. Jpn.* **1982**, *55*, 2147–2152.

162. Burrage, S.A.; Raynham, T.; Bradley, M. *Tetrahedron Lett.* **1998**, *39*, 2831–2834.

163. Burkett, B.A.; Chai, C.L.L. *Tetrahedron Lett.* **1999**, *40*, 7035–7038.

164. Yamada, M.; Miyajima, T.; Horikawa, H. *Tetrahedron Lett.* **1998**, *39*, 289–292.

165. Okeley, N.M.; Zhu, Y.; van der Donk, W.A. *Org. Lett.* **2000**, *2*, 3603–3606.

166. Horikawa, E.; Kodaka, M.; Nakahara, Y.; Okuno, H.; Nakamura, K. *Tetrahedron Lett.* **2001**, *42*, 8337–8339.

167. Nakamura, K.; Ohnishi, Y.; Horiwaka, E.; Konakahara, T.; Kodaka, M.; Okuno, H. *Tetrahedron Lett.* **2003**, *44*, 5445–5448.

168. Zhou, H.; van der Donk, W.A. *Org. Lett.* **2002**, *4*, 1335–1338.

169. Izumiya, N.; Kato, T.; Waki, M. *Biopolymers* **1981**, *20*, 1785–1791.

170. Blettner, C.; Bradley, M. *Tetrahedron Lett.* **1994**, *35*, 467–470.

171. Nakajima, K.; Oda, H.; Okawa, K. *Bull. Chem. Soc. Jpn.* **1982**, *55*, 3232–3236.

172. Okawa, K.; Nakajima, K.; Tanaka, T.; Neya, M. *Bull. Chem. Soc. Jpn.* **1982**, *55*, 174–176.

173. Carter, H.E. *Org. React.* **1946**, *3*, 198–239.

174. Rao, Y.S.; Filler, R. *Synthesis* **1975**, 749–764.

175. Berkessel, A.; Cleeman, F.; Mukherjee, S.; Müller, T.N.; Lex, J. *Angew. Chem., Int. Ed.* **2005**, *44*, 807–811.

176. Erlenmeyer, E. *Ann.* **1893**, *275*, 1–20.

177. Gillespie, HB.; Snyder, H.R.; Hartman, W.W.; Dickey, J.B. *Org. Synth. Coll. Vol. 2*, **1948**, 489–494.

178. Pyman, F.L. *J. Chem. Soc.* **1916**, *109*, 186–202.

179. *J. Chem. Soc.* **1949**, 2726–2728 (misprinted journal, authors missing).

180. Crawford, M.; Little, W.T. *J. Chem. Soc.* **1959**, 729–731.

181. Rosenthal, A.; Dooley, K. *Carbohydrate Res.* **1978**, *60*, 193–199.

182. Rosenthal, A.; Brink, A.J. *Carbohydrate Res.* **1976**, *47*, 332–336.

183. Ojima, I.; Kato, K.; Nakahashi, K.; Fuchikami, T.; Fujita, M. *J. Org. Chem.* **1989**, *54*, 4511–4522.

184. Oba, M.; Ueno, R.; Fukuoka, M.; Kainosho, M.; Nishiyama, K. *J. Chem. Soc., Perkin Trans. 1*, **1995**, 1603–1609.

185. Brunner, H.; König, W.; Nuber, B. *Tetrahedron: Asymmetry* **1993**, *4*, 699–707.

186. Lowary, T.; Meldal, M; Helmboldt, A.; Vasella, A.; Bock, K. *J. Org. Chem.* **1998**, *63*, 9657–9668.

187. Döbler, C.; Kreuzfeld, H.-J.; Michalik, M.; Krause, H.W. *Tetrahedron: Asymmetry* **1996**, *7*, 117–125.

188. Döbler, C.; Kreuzfeld, H.-J.; Krause, H.W.; Michalik, M. *Tetrahedron: Asymmetry* **1993**, *4*, 1833–1842.

189. Schmidt, U.; Kumpf, S.; Neumann *J. Chem. Soc., Chem. Commun.* **1994**, 1915–1916.

190. Imperiali, B.; Prins, T.J.; Fisher, S.L. *J. Org. Chem.* **1993**, *58*, 1613–1616.

191. Tourwé, D.; Mannekens, E.; Diem, T.N.T.; Verheyden, P.; Jaspers, H.; Tóth, G.; Péter, A.; Kertész, I.; Török, G.; Chung, N.N.; Schiller, P.W. *J. Med. Chem.* **1998**, *41*, 5167–5176.

192. Alías, M.; López, M.P.; Cativiela, C. *Tetrahedron* **2004**, *60*, 885–891.

193. Cativiela, C.; Díaz-de-Villegas, M.; Jiménez, A.I. *Tetrahedron: Asymmetry* **1995**, *6*, 177–182.

194. Cativiela, C.; Díaz-de-Villegas, M.D.; Jiménez, A.I.; Lahoz, F. *Tetrahedron Lett.* **1994**, *35*, 617–620.

195. Bailey, K.L.; Molinski, T.F. *J. Org. Chem.* **1999**, *64*, 2500–2504.

196. Robinson, A.J.; Lim, C.Y.; He, L.; Ma, P.; Li, H.-Y. *J. Org. Chem.* **2001**, *66*, 4141–4147.

197. Wheeler, H.L.; Hoffman, C. *Am. Chem. J.* **1911**, *45*, 368–383.

198. Johnson, T.B. *J. Am. Chem. Soc.* **1939**, *61*, 2485–2487.

199. Boyd, W.J.; Robson, W. *Biochem. J.* **1935**, *29*, 542–545.

200. Henze, H.R.; Whitney, W.B.; Eppright, M.A. *J. Am. Chem. Soc.* **1940**, *62*, 565–568.

201. Johnson, T.B.; Bengis, R. *J. Am.Chem. Soc.* **1912**, *34*, 1061–1066.

202. Schaaf, F.; Labouchère, A. *Helv. Chim. Acta* **1924**, *7*, 357–363.

203. Jansen, M.P.J.M. *Recl. Trav. Chim. Pays-Bas* **1931**, *50*, 291–312.

204. Phillips, A.P. *J. Am. Chem. Soc.* **1945**, *67*, 744–748.

205. Wheeler, H.L.; Hoffman, C. *Am. Chem. J.* **1911**, *45*, 568–583.

206. Johnson, T.B.; Nicolet, B.H. *Am. Chem. J.* **1912**, *47*, 459–475.

207. Johnson, T.B.; Scott, W.M. *J. Am. Chem. Soc.* **1915**, *37*, 1846–1856.

208. Johnson, T.B.; Scott, W.M. *J. Am. Chem. Soc.* **1915**, *37*, 1856–1863.

209. Johnson, T.B.; Bengis, R. *J. Am. Chem. Soc.* **1913**, *35*, 1606–1617.

210. Deulofeu, V.; Mendive, J. *Z. Physiol. Chem.* **1932**, *211*, 1–4.

211. Johnson, T.B.; O'Brien, W.B. *J. Biol. Chem.* **1912**, *12*, 205–213.

212. Davis, A.C.; Levy, A.L. *J. Chem. Soc.* **1949**, 2179–2182.

213. Billimoria, J.D.; Cook, A.H. *J. Chem. Soc.* **1949**, 2323–2329.

214. Stetten D., Jr., *J. Biol. Chem.* **1942**, *144*, 501–506.

215. Bycroft, B.W.; Chhabra, S.R.; Grout, R.J.; Crowley, P.J. *J. Chem. Soc., Chem. Commun.* **1984**, 1156–1157.

216. Chu, S.-H.; Günther, W.H.H.; Mautner, H.G.; Redstone, M.O.; Herr E.B., Jr.; *Biochem. Prep.* **1963**, *10*, 153–158.

217. Hirai, K. *Biochem. Z.* **1921**, *114*, 67–70.

218. Sasaki, T.; Kinose, J. *Biochem. Z.* **1921**, *121*, 171–174.

219. Hirai, K. *Biochem. Z.* **1927**, *189*, 88–91.

220. Hirai, K. *Biochem. Z.* **1926**, *177*, 449–452.

221. Sasaki, T. *Ber.* **1921**, *54*, 2056–2059.

222. Niemann, C.; Lewis, R.N.; Hays, J.T. *J. Am. Chem. Soc.* **1942**, *64*, 1678–1682.

223. Davies, S.G.; Rodríguez-Solla, H; Tamayo, J.A.; Cowley, A.R.; Concellón, C.; Garner, A.C.; Parkes, A.L.; Smith, A.D. *Org. Biomol. Chem.* **2005**, *3*, 1435–1447.

224. Watanabe, K. *J. Antibiotics* **1961**, *14*, 1–13.

225. Jacobsen-Bauer, A.; Simchen, G. *Tetrahedron* **1988**, *44*, 5355–5360.

226. Bernabé, M.; Cuevas, O.; Fernandez-Alvarez, E. *Eur. J. Med. Chem.* **1979**, *14*, 33–45.

227. Burke T.R., Jr.; Smyth, M.S.; Nomizu, M.; Otaka, A.; Roller, P.P. *J. Org. Chem.* **1993**, *58*, 1336–1340.

228. Burke T.R., Jr.; Russ, P.; Lim, B. *Synthesis* **1991**, 1019–1020.

229. Burke T.R., Jr.; Smyth, M.S.; Otaka, A.; Roller, P.P. *Tetrahedron Lett.* **1993**, *34*, 4125–4128.

230. Chetyrkina, S.; Estieu-Gionnet, K.; Laïn, G.; Bayle, M.; Déléris, G. *Tetrahedron Lett.* **2000**, *41*, 1923–1926.

231. Fornicola, R.S.; Oblinger, E.; Montgomery, J. *J. Org. Chem.* **1998**, *63*, 3528–3529.

232. Fornicola, R.S.; Montgomery, J. *Tetrahedron Lett.* **1999**, *40*, 8337–8341.

233. Kuster, G.J.; Scheeren, H.W. *Tetrahedron Lett.* **2000**, *41*, 515–519.

234. Hoppe, D. *Angew. Chem., Int. Ed. Engl.* **1974**, *13*, 789–804.

235. Liu, W.-Q.; Carreaux, F.; Meudal, H.; Roques, B.P.; Garbay-Jaureguiberry, C. *Tetrahedron* **1996**, *52*, 4411–4422.

236. Hoppe, D.; Schöllkopf, U. *Liebigs Ann. Chem.* **1972**, *763*, 1–16.

237. Schöllkopf, U.; Gerhart, F.; Schröder, R.; Hoppe, D. *Liebigs Ann. Chem.* **1972**, *766*, 116–129.

238. Yamada, M.; Nakao, K.; Fukui, T.; Nunami, K.-i. *Tetrahedron* **1996**, *52*, 5751–5764. *erratum* **1998**, *54*, 10925.

239. Schöllkopf, U.; Harms, R.; Hoppe, D. *Liebigs Ann. Chem.* **1973**, 611–618.

240. Schöllkopf, U.; Meyer, R. *Liebigs Ann. Chem.* **1977**, 1174–1182.

241. Schöllkopf, U.; Meyer, R. *Liebigs Ann. Chem.* **1981**, 1469–1475.

242. Hoppe, I.; Schöllkopf, U. *Synthesis* **1982**, 129–131.

243. Nunami, K.-I.; Suzuki, M.; Yoneda, N. *J. Chem. Soc., Perkin Trans. 1* **1979**, 2224–2229.

244. Nunami, K.-I.; Suzuki, M.; Yoneda, N. *Chem. Pharm. Bull.* **1982**, *30*, 4015–4024.

245. Hall, R.H.; Bischofberger, K.; Eitelman, S.J.; Jordaan, A. *J. Chem. Soc., Perkin Trans. 1* **1977**, 743–753.

246. Bischofberger, K.; Hall, R.H.; Jordaan, A. *J. Chem. Soc., Chem. Commun.* **1975**, 806–807.

247. Nunami, K.-I.; Hiramatsu, K.; Hayashi, K.; Matsumoto, K. *Tetrahedron* **1988**, *44*, 5467–5478.

248. Hoppe, D. *Angew. Chem., Int. Ed. Engl.* **1972**, *11*, 933–934.

249. Hoppe, D. *Angew. Chem., Int. Ed. Engl.* **1973**, *12*, 656–658.

250. Hoppe, D. *Angew. Chem., Int. Ed. Engl.* **1973**, *12*, 658–659.

251. Schickli, C.P.; Seebach, D. *Liebigs Ann. Chem.* **1991**, 655–668.

252. Chinchilla, R.; Falvello, L.R.; Galindo, N.; Nájera, C. *J. Org. Chem.* **2000**, *65*, 3034–3041.

253. Abellán, T.; Nájera, C.; Sansano, J.M. *Tetrahedron: Asymmetry* **2000**, *11*, 1051–1055.

254. Poisel, H.; Schmidt, U. *Angew. Chem., Int. Ed. Engl.* **1976**, *15*, 294–295.

255. Schmidt, U.; Öhler, E. *Angew. Chem., Int. Ed. Engl.* **1977**, *16*, 327–328.

256. Poisel, H.; Schmidt, U. *Chem. Ber.* **1975**, *108*, 2547–2553.

257. Gupta, S.D.; Jain, R.; Chauhan, V.S. *Bioorg. Med. Chem. Lett.* **1994**, *4*, 941–944.

258. Poisel, H. *Chem. Ber.* **1977**, *110*, 942–947.

259. Seki, M.; Moriya, T.; Matsumoto, K. *Agric. Biol. Chem.* **1984**, *48*, 1251–1255.

260. Grim, M.D.; Chauhan, V.; Shimohigashi, Y.; Kolar, A.J.; Stammer, C.H. *J. Org. Chem.* **1981**, *46*, 2671–2673.

261. Scott, J.W.; Keith, D.D.; Nix, G., Jr.; Parrish, D.R.; Remington, S.; Roth, G.P.; Townsend, J.M.; Valentine D., Jr.; Yang, R. *J. Org. Chem.* **1981**, *46*, 5086–5093.

262. Keith, D.D.; Yang, R.; Tortora, J.A.; Weigele, M. *J. Org. Chem.* **1978**, *43*, 3713–3716.

263. Stoll, G.; Frank, J.; Musso, H.; Henke, H.; Herrendorf, W. *Liebigs Ann. Chem.* **1986**, 1968–1989.

264. Frank, J.; Stoll, G.; Musso, H. *Liebigs Ann. Chem.* **1986**, 1990–1996.

265. Kolasa, T. *Synthesis* **1983**, 539.

266. Herscheid, J.D.M.; Nivard, R.J.F.; Tijhuis, M.W.; Scholten, H.P.H.; Ottenheijm, H.C.J. *J. Org. Chem.* **1980**, *45*, 1880–1885.

267. Herscheid, J.D.M.; Ottenheijm, H.C.J. *Tetrahedron Lett.* **1978**, *51*, 5143–5144.

268. Yonezawa, Y.; Shin, C-g.; Ono, Y.; Yoshimura, J. *Bull. Chem. Soc. Jpn.* **1980**, *53*, 2905–2909.

269. Shin, C-g.; Hayakawa, M.; Suzuki, T.; Ohtsuka, A.; Yoshimura, J. *Bull. Chem. Soc. Jpn.* **1978**, *51*, 550–554

270. Shin, C.-g.; Fujii, M.; Yoshimura, J. *Tetrahedron Lett.* **1971**, *27*, 2499–2502.

271. Shin, C.-g.; Sato, K.-i.; Ohtsuka, A.; Mikami, K.; Yoshimura, J. *Bull. Chem. Soc. Jpn.* **1973**, *46*, 3876–3880.

272. Farlow, M.W. *J. Biol. Chem.* **1948**, *176*, 71–72.

273. Labia, R.; Morin, C. *J. Org. Chem.* **1986**, *51*, 249–251.

274. Elrod, L.F.; Holt, E.M.; Mapelli, C.; Stammer, C.H. *J. Chem. Soc., Chem. Commun.* **1988**, 252–253.

275. Li, X.; Yeung, C.-h.; Chan, A.S.C.; Lee, D.-S.; Yang, T.-K. *Tetrahedron: Asymmetry* **1999**, *10*, 3863–3867.

276. Okonya, J.F.; Johnson, M.C.; Hoffman, R.V. *J. Org. Chem.* **1998**, *63*, 6409–6413.

277. Bycroft, B.W.; Lee, G.R. *J. Chem. Soc., Chem. Commun.* **1975**, 988–989.

278. Effenberger, F.; Beisswenger, T. *Chem. Ber.* **1984**, *117*, 1497–1512.

279. Beisswenger, T.; Effenberger, F. *Chem. Ber.* **1984**, *117*, 1513–1522.

280. Effenberger, F.; Kühlwein, J.; Drauz, K. *Liebigs Ann. Chem.* **1993**, 1295–1301.

281. Effenberger, F.; Kühlwein, J.; Hopf, M.; Stelzer, U. *Liebigs Ann. Chem.* **1993**, 1303–1311.

282. Schöllkopf, U.; Schröder, J. *Liebigs Ann. Chem.* **1988**, 87–92.

283. Bory, S.; Gaudry, M.; Marquet, A. *New J. Chem.* **1986**, *10*, 709–713.

284. Gelas-Mialhe, Y.; Touraud, E.; Vessiere, R. *Can. J. Chem.* **1982**, *60*, 2830–2851.

285. Ratcliffe, R.W.; Christensen, B.G. *Tetrahedron Lett.* **1973**, *46*, 4645–4648.

286. Ratcliffe, R.W.; Christensen, B.G. *Tetrahedron Lett.* **1973**, *46*, 4653–4656.

287. Scopes, D.I.C.; Kluge, A.F.; Edwards, J.A. *J. Org. Chem.* **1977**, *42*, 376–377.

288. Rachon, J.; Schöllkopf, U. *Liebigs Ann. Chem.* **1981**, 99–102.

289. Schmidt, U.; Lieberknecht, A.; Schanbacher, U.; Beuttler, T.; Wild, J. *Angew. Chem. Suppl.* **1982**, 1682–1689.

290. Schmidt, U.; Lieberknecht, A.; Schanbacher, U.; Beuttler, T.; Wild, J. *Angew. Chem., Int. Ed. Engl.* **1982**, *21*, 776–777.

291. Kober, R.; Steglich, W. *Liebigs Ann. Chem.* **1983**, 599–609.

292. Apitz, G.; Steglich, W. *Tetrahedron Lett.* **1991**, *32*, 3163–3166.

293. Alcaraz, C.; Herrero, A.; Marco, J.L.; Fernández-Alvarez, E.; Bernabé, M. *Tetrahedron Lett.* **1992**, *33*, 5605–5608.

294. Burger, K.; Heistracher, E.; Simmerl, R.; Eggersdorfer, M. *Z. Naturforsch.*, **1992**, 425–433.

295. Schmidt, U.; Lieberknecht, A.; Wild, J. *Synthesis* **1984**, 53–60.

296. Daumas, M.; Vo-Quang, L.; Le Goffic, F. *Synth. Commun.* **1990**, *20*, 3395–3401.

297. Debenham, S.D.; Cossrow, J.; Toone, E.J. *J. Org. Chem.* **1999**, *64*, 9153–9163.

298. Masquelin, T.; Broger, E.; Müller, K.; Schmid, R.; Obrecht, D. *Helv. Chim. Acta* **1994**, *77*, 1395–1411.

299. Pham, T.; Lubell, W.D. *J. Org. Chem.* **1994**, *59*, 3676–3680.

300. Ciattini, P.G.; Morera, E.; Ortar, G. *Synthesis* **1988**, 140–142.

301. Inguimbert, N.; Dhôtel, H.; Coric, P.; Roques, B.P. *Tetrahedron Lett.* **2005**, *46*, 3517–3520.

302. Plummer, J.S.; Berryman, K.A.; Cai. C.; Cody, W.L.; DiMaio, J.; Doherty, A.M.; Edmunds, J.J.; He, J.X.; Holland, D.R.; Levesque, S.; Kent, D.R.; Narasimhan, L.S.; Rubin, J.R.; Rapundalo, S.T.; Siddiqui, M.A.; Susser, A.J.; St-Denis, Y.; Winocor, P.D. *Biorg. Med. Chem. Lett.* **1998**, *8*, 3409–3414.

303. Baldwin, J.E.; Merritt, K.D.; Schofield, C.J. *Tetrahedron Lett.* **1993**, *34*, 3919–3920.

304. Daumas, M.; Vo-Quang, L.; Le Goffic, F. *Tetrahedron* **1992**, *48*, 2373–2384.

305. Burk, M.J.; Allne, J.G.; Kiesman, W.F. *J. Am. Chem. Soc.* **1998**, *120*, 657–663.

306. Schmidt, U.; Braun, C.; Sutoris, H. *Synthesis* **1996**, 223–229.

307. Hamilton, G.S.; Huang, Z.; Patch, R.J.; Narayanan, B.A.; Ferkany, J.W. *Bioorg. Med. Chem. Lett.* **1993**, *3*, 27–32.

308. Hanafi, N.; Ortuño, R.M. *Tetrahedron: Asymmetry* **1994**, *5*, 1657–1660.

309. Le Corre, M.; Hercouet, A.; Bessieres, B. *Tetrahedron: Asymmetry* **1995**, *6*, 683–684.

310. Imogaï, H.; Bernardinelli, G.; Gränicher, C.; Moran, M.; Rossier, J.-C.; Müller, P. *Helv. Chim. Acta* **1998**, *81*, 1754–1764.

311. Hiebl, J.; Kollman, H.; Rovenszky, F.; Winkler, K. *J. Org. Chem.* **1999**, *64*, 1947–1952.

312. Basu, B.; Frejd, T. *Acta Chem. Scand.* **1996**, *50*, 316–322.

313. Ritzén, A.; Basu, B.; Wållberg, A.; Frejd, T. *Tetrahedron: Asymmetry* **1998**, *9*, 3491–3496.

314. Basu, B.; Chattopadhyay, S.K.; Ritzén, A.; Frejd, T. *Tetrahedron: Asymmetry* **1997**, *8*, 1841–1846.

315. Maricic, S.; Frejd, T. *J. Org. Chem.* **2002**, *67*, 7600–7606.

316. Rama Rao, A.V.; Chakraborty, T.K.; Reddy, K.L.; Rao, A.S. *Tetrahedron Lett.* **1992**, *33*, 4799–4802.

317. Schmidt, U.; Weller, D.; Holder, A.; Lieberknecht, A. *Tetrahedron Lett.* **1988**, *29*, 3227–3230.

318. Jiménez, J.M.; Rifé, J.; Ortuño, R.M. *Tetrahedron: Asymmetry* **1995**, *6*, 1849–1852.

319. Jiménez, J.M.; Casas, R.; Ortuño, R.M. *Tetrahedron Lett.* **1994**, *35*, 5945–5948.

320. Williams, R.M.; Fegley, G.J. *Tetrahedron Lett.* **1992**, *33*, 6755–6758.

321. Sato, K.-i.; Miyata, T.; Tanai, I.; Yonezawa, Y. *Chem. Lett.* **1994**, 129–132.

322. Fuchss, T.; Schmidt, R.R. *Synthesis* **1998**, 753–758.

323. Debenham, S.D.; Debenham, J.S.; Burk, M.J.; Toone, E.J. *J. Am. Chem. Soc.* **1997**, *119*, 9897–9898.

324. Debenham, S.D.; Snyder, P.W.; Toone, E.J. *J. Org. Chem.* **2003**, *68*, 5805–5811.

325. Allen, J.R.; Harris, C.R.; Danishefsky, S.J. *J. Am. Chem. Soc.* **2001**, *123*, 1890–1897.

326. Martín-Vilà, M.; Hanafi, N.; Jiménez, J.M.; Alvarez-Larena, A.; Piniella, J.F.; Branchadell, V.; Oliva, A.; Ortuño, R.M. *J. Org. Chem.* **1998**, *63*, 3581–3589.

327. Armstrong, R.W.; Tellew, J.E.; Moran, E.J. *Tetrahedron Lett.* **1996**, *37*, 447–450.

328. Konda, Y.; Sato, T.; Tsushima, K.; Dodo, M.; Kusunoki, A.; Sakayanagi, M.; Sato, N.; Takeda, K.; Harigaya, Y. *Tetrahedron* **1999**, *55*, 12723–12740.

329. Coleman, R.S.; Carpenter, A.J. *J. Org. Chem.* **1993**, *58*, 4452–4461.

330. Schmidt, U.; Riedl, B. *Synthesis* **1993**, 815–818.

331. Schmidt, U.; Riedl, B. *J. Chem. Soc., Chem. Commun.* **1992**, 1186–1187.

332. Moody, C.J.; Ferris, L.; Haigh, D.; Swann, E. *Chem. Commun.* 1997, 2391–2392.

333. Aggen, J.B.; Humphrey, J.M.; Gauss, C.-M.; Huang, H.-B.; Nairn, A.C.; Chamberlin, A.R. *Biorg. Med. Chem.* **1999**, *7*, 543–564.

334. Yamazaki, K.; Nakamura, Y.; Kondo, Y. *J. Org. Chem.* **2003**, *68*, 6011–6019.

335. Doi, T; Fujimoto, N.; Watanabe, J.; Takahashi, T. *Tetrahedron Lett.* **2003**, *44*, 2161–2165.

336. Williams, R.M.; Fegley, G.J. *J. Am. Chem. Soc.* **1991**, 113, 8796–8806.

337. Kimura, R.; Nagano, T.; Kinoshita, H. *Bull. Chem. Soc. Jpn.* **2002**, *75*, 2517–2525.

338. O'Donnell, M.J.; Arasappan, A.; Hornback W.J.; Huffman, J.C. *Tetrahedron Lett.* **1990**, *31* 157–160.

339. Trost, B.M.; Dake, G.R. *J. Am. Chem. Soc.* **1997** *119*, 7595–7596.

340. Huang, T.-S.; Li, C.-J. *Org. Lett.* **2001**, *3*, 2037–2039.

341. Cho, C.S. *Tetrahedron Lett.* **2005**, *46*, 1415–1417.

342. Chen, D.; Guo, L.; Liu, J.; Kirtane, S.; Cannon J.F.; Li, G. *Org. Lett.* **2005**, *7*, 921–924.

343. Nitz, T.J.; Holt, E.M.; Rubin, B.; Stammer, C.H. *J. Org. Chem.* **1981**, *46*, 2667–2671.

344. Schottelius, M.J.; Chen, P. *Helv. Chim. Acta* **1998** *81*, 2341–2347.

345. Alvarez-Ibarra, C.; Csákÿ, A.G.; Martin, M.E. Quiroga, M.L. *Tetrahedron* **1999**, *55*, 7319–7330.

346. Alvarez-Ibarra, C.; Csákÿ, A.G.; Murcia, M.C *J. Org. Chem.* **1998**, *63*, 8736–8740.

347. Alvarez-Ibarra, C.; Csákÿ, A.G.; de la Oliva, C.G *J. Org. Chem.* **2002**, *67*, 2789–2797.

348. Groundwater, P.W.; Sharif, T.; Arany, A.; Hibbs D.E.; Hursthouse, M.B.; Garnett, I.; Nyerges, M *J. Chem. Soc., Perkin Trans. 1* **1998**, 2837–2846.

349. Groundwater, P.W.; Sharif, T.; Arany, A.; Hibbs D.E.; Hursthouse, M.B.; Nyerges, M. *Tetrahedron Lett.* **1998**, *39*, 1433–1436.

350. Schumann, S.; Zeitler, K.; Jäger, M.; Polborn, K. Steglich, W. *Tetrahedron* **2000**, *56*, 4187–4195.

351. Vranicar, L.; Polanc, S.; Kocevar, M. *Tetrahedron* **1999**, *55*, 271–278.

352. Vranicar, L.; Meden, A.; Polanc, S.; Kocevar, M. *J. Chem. Soc., Perkin Trans. 1* 2002, 675–681.

353. Lipshutz, B.H.; Morey, M.C. *J. Am. Chem. Soc.* **1984**, *106*, 457–459.

354. Doyle, M.P.; Yan, M.; Hu, W.; Gronenberg, L.S. *J. Am. Chem. Soc.* **2003**, *125*, 4692–4693.

355. Nicolaou, K.C.; Shi, G.-Q.; Namoto, K.; Bernal, F. *Chem. Commun.* **1998**, 1757–1758.

356. Nakazawa, T.; Suzuki, T.; Ishii, M. *Tetrahedron Lett.* **199**7, *38*, 8951–8954.

357. Zhou, H.; van der Donk, W.A. *Org. Lett.* **2001**, *3*, 593–596.

358. Genet, R.; Bénetti, P.-H.; Hammadi, A.; Ménez, A. *J. Biol. Chem.* **1995**, *270*, 23540–23545.

359. Hammadi, A.; Meunier, G.; Ménez, A.; Genet, R. *Tetrahedron Lett.* **1998**, *39*, 2955–2958.

360. Hammadi, A.; Ménez, A.; Genet, R. *Tetrahedron* **1997**, *53*, 16115–16122.

361. Wagaw, S.; Rennels, R.A.; Buchwald, S.L. *J. Am. Chem. Soc.* **1997**, *119*, 8451–8458.

362. Carlström, A.-S.; Frejd, T. *Synthesis*, **1989**, 414–418.

363. Bozell, J.J.; Vogt, C.E.; Gozum, J. *J. Org. Chem.* **1991**, *56*, 2584–2587.

364. Melo, R.L.; Pozzo, R.C.B.; Pimenta, D.C.; Perissutti, E.; Caliendo, G.; Santagada, V.; Juliano, L.; Juliano, M.A. *Biochemistry* **2001**, *40*, 5226–5232.

365. Chattopadhyay, S.K.; Pal, B.K.; Biswas, S. *Synth. Commun.* **2005**, *35*, 1167–1175.

366. Ritzén, A.; Frejd, T. *J. Chem. Soc., Perkin Trans. 1* **1998**, 3419–3424.

367. Ritzén, A.; Basu, B.; Chattopadhyay, S.K.; Dossa, F.; Frejd, T. *Tetrahedron: Asymmetry* **1998**, *9*, 503–512.

368. Carlström, A.-S.; Frejd, T. *J. Org. Chem.* **1990**, *55*, 4175–4180.

369. Gibson, S.E.; Jones, J.O.; McCague, R.; Tozer, M.J.; Whitcombe, N.J. *Synlett* **1999**, 954–956.

370. Gibson, S.E.; Jones, J.O.; Kalindjian, S.B.; Knight, J.D.; Steed, J.W.; Tozer, M.J. *Chem. Commun.* **2002**, 1938–1939.

371. Schmidtmann, F.W.; Benedum, T.E.; McGarvey, G.J. *Tetrahedron Lett.* **2005**, *46*, 4677–4681.

372. Manzoni, L.; Colombo, M.; Scolastico, C. *Tetrahedron Lett.* **2004**, *45*, 2623–2625.

373. Collier, P.N.; Patel, I.; Taylor, R.J.K. *Tetrahedron Lett.* **2001**, *42*, 5953–5954.

374. Burk, M.J.; Allen, J.G.; Kiesman, W.F.; Stoffan, K.M. *Tetrahedron Lett.* **1997**, *38*, 1309–1312.

375. Hoerrner, R.S.; Askin, D.; Volante, R.P.; Reider, P.J. *Tetrahedron Lett.* **1998**, *39*, 3455–3458.

376. Abreu, A.S.; Ferreira, P.M.T.; Queiroz, M.-J.R.P.; Ferreira, I.C.F.R.; Calhelha, R.C.; Estevinho, L.M. *Eur. J. Org. Chem.* **2005**, 2951–2957.

377. Abreu, A.S.; Silva, N.O.; Ferreira, P.M.T.; Queiroz, M.-J.R.P. *Tetrahedron Lett.* **2003**, *44*, 6007–6009.

378. Abreu, A.S.; Silva, N.O.; Ferreira, P.M.T.; Queiroz, M.-J.R.P. *Tetrahedron Lett.* **2003**, *44*, 3377–3379.

379. Abreu, A.S.; Silva, N.O.; Ferreira, P.M.T.; Queiroz, M.-J.R.P.; Venanzi, M. *Eur. J. Org. Chem.* **2003**, 4792–4796.

380. Abreu, A.S.; Ferreira, P.M.T.; Queiroz, M.-J.R.P.; Gatto, E.; Venanzi, M. *Eur. J. Org. Chem.* **2004**, 3985–3991.

381. Behringer, H.; Taul, H. *Chem. Ber.* **1957**, *90*, 1398–1410.

382. Beccalli, E.M.; Clerica, F.; Gelmi, M.L. *Tetrahedron* **1999**, *55*, 781–786.

383. Ferreira, P.M.T.; Maia, H.L.S.; Monteiro, L.S.; Scaramento, J. *Tetrahedron Lett.* **2000**, *41*, 7437–7441.

384. Ferreira, P.M.T.; Maia, H.L.S.; Monteiro, L.S. *Eur. J. Org. Chem.* **2003**, 2635–2644.

385. Ferreira, P.M.T.; Maia, H.L.S.; Monteiro, L.S.; Sacramento, J. *J. Chem. Soc., Perkin Trans. 1* **2001**, 3167–3173.

386. Ferreira, P.M.T.; Maia, H.L.S.; Monteiro, L.S. *Tetrahedron Lett.* **2002**, *43*, 4495–4497.

387. Abreu, A.S.; Silva, N.O.; Ferreira, P.M.T.; Queiroz, M.-J.R.P. *Eur. J. Org. Chem.* **2003**, 1537–1544.

388. Cebasek, P.; Wagger, J.; Bevk, D.; Jakse, R.; Svete, J.; Stanovnik, B. *J. Comb. Chem.* **2004**, *6*, 356–362.

389. Sutherland, A.; Vederas, J.C. *Chem. Commun.* **2002**, 224–225.

390. Nájera, C. *Synlett* **2002**, 1388–1403.

391. Erlenmeyer, E.; Halsey, J.T. *Ann.* **1879**, *307*, 138–145.

392. Wheeler, H.L.; Hoffman, C.; Johnson, T.B. *J. Biol. Chem.* **1911**, *10*, 147–157.

393. Johnson, T.B.; Wrenshall, R. *J. Am. Chem. Soc.* **1915**, *37*, 2133–2144.

394. Deulofeu, V.; Mitta, A.E.A. *J. Org. Chem.* **1949**, *14*, 915–919.

395. Badshah, A.; Khan, N.H.; Kidwai, A.R. *J. Org. Chem.* **1972**, *37*, 2916–2918.

396. Ried, W.; Gebhardtsbauer, H.G. *Chem. Ber.* **1956**, *89*, 2933–2939.

397. Cativiela, C.; Melendez, E. *Synthesis* **1981**, 805–807.

398. Pedrazzoli, A. *Helv. Chim. Acta* **1957**, *40*, 80–85.

399. Pedrazzoli, A. *Chimia* **1956**, *11*, 260–261.

400. Yamada, S.-I.; Shioiri, T.; Fuji, T. *Chem. Pharm. Bull.* **1962**, *10*, 688–693.

401. Sheehan, J.C.; Chandler, R.E. *J. Am. Chem. Soc.* **1961**, *83*, 4795–4797.

402. Lisichkina, I.N.; Vinogradova, A.I.; Sukhorukova, N.B.; Tselyapina, E.V.; Saporovskaya, M.B.; Belikov, V.M. *Russ. Chem. Bull.* **1993**, *42*, 569–571.

403. Lisichkina, I.N.; Vasil'eva, T.Y.; Davtyan, D.A.; Belikov, V.M. *Russ. Chem. Bull.* **1997**, *46*, 331–333.

404. Lisichkina,, I.N.; Vinogradova, A.I.; Bachurina, I.B.; Kurkovskaya, L.N.; Belikov, V.M. *Russ. Chem. Bull.* **1994**, *43*, 828–831.

405. Zeitler, K.; Steglich, W. *J. Org. Chem.* **2004**, *69*, 6134–6136.

406. Vigneron, J.-P.; Kagan, H.; Horeau, A. *Bull. Soc. Chim. Fr.* **1972**, *10*, 3836–3841.

407. Seebach, D.; Bürger, H.M.; Schickli, C.P. *Liebigs Ann. Chem.* **1991**, 669–684.

408. Bergmann, M.; Tietzman, J.E. *J. Biol. Chem.* **1944**, *155*, 535–546.

409. Oba, M.; Terauchi, T.; Owari, Y.; Imai, Y; Motoyama, I.; Nishiyama, K. *J. Chem. Soc. Perkin Trans. 1* **1998**, 1275–1281.

410. Oba, M.; Nakajima, S.; Nishiyama, K. *Chem. Commun.* **1996**, 1875–1876.

411. Tanimura, K.; Kato, T.; Waki, M.; Lee, S.; Kodera, Y.; Izumiya, N. *Bull. Chem. Soc. Jpn.* **1984**, *57*, 2193–2197.

412. Davies, S.G.; Rodríguez-Solla, H.; Tamayo, J.A.; Garner, A.C.; Smith, A.D. *Chem. Commun.* **2004**, 2502–2503.

413. Szymanska, A.; Wiczk, W.; Lankiewicz, L. *Amino Acids* **2001**, *21*, 265–270.

414. Aguado, G.P.; Alvarez-Larena, A.; Illa, O.; Moglioni, A.G.; Ortuño, R.M. *Tetrahedron: Asymmetry* **2001**, *12*, 25–28.

415. Halpern, J. *Science* **1982**, *217*, 401–407.

416. Knowles, W.S. *Acc. Chem. Res.* **1983**, *16*, 106–112.

417. Knowles, W.S. *Angew. Chem., Int. Ed.* **2002**, *41*, 1998–2007.

418. Nugent, W.A.; RajanBabu, T.V.; Burk, M.J. *Science* **1993**, *259*, 479–483.

419. Heteley, M.J.; Schichl, D.A.; Kreuzfeld, H.-J.; Beller, M. *Tetrahedron Lett.* **2000**, *41*, 3821–3824.

420. Gridnev, I.D.; Higashi, N.; Asakura, K.; Imamoto, T. *J. Am. Chem. Soc.* **2000**, *122*, 7183–7194.

421. Giernoth, R.; Heinrich, H.; Adams, N.J.; Deeth, R.J.; Bargon, J.; Brown, J.M. *J. Am. Chem. Soc.* **2000**, *122*, 12381–12382.

422. Feldgus, S.; Landis, C.R. *J. Am. Chem. Soc.* **2000**, *122*, 12714–12727.

423. Franciò, G.; Faraone, F.; Leitner, W. *Angew. Chem., Int. Ed.* **2000**, *39*, 1428–1430.

424. Gridnev, I.D.; Imamoto, T. *Acc. Chem. Res.* **2004**, *37*, 633–644.

425. Albrecht, J.; Nagel, U. *Angew. Chem., Int. Ed. Engl.* 1996, 35, 407–409.

426. Kreuzfeld, H.J.; Döbler, Chr.; Schmidt, U.; Krause, H.W. *Amino Acids* **1996**, *11*, 269–282.

427. Genet, J.-P. *Acc. Chem. Res.* **2003**, *36*, 908–918.

428. Tang, W.; Zhang, X. *Chem. Rev.* **2003**, *103*, 3029–3069.

429. Gennari, C.; Pairulli, U. *Chem. Rev.* **2003**, *103*, 3071–3100.

430. Pfaltz, A.; Drury W.J. III, *Proc. Natl. Acad. Sci. U.S.A.* **2004**, *101*, 5723–5726.

431. Noyori, R.; Kitamura, M.; Ohkuma, T. *Proc. Natl. Acad. Sci. U.S.A.* **2004**, *101*, 5356–5362.

432. Jerphagnon, T.; Renaud, J.-L.; Bruneau, C. *Tetrahedron: Asymmetry.* **2004**, *15*, 2101–2111.

433. Shimizu, H.; Nagasaki, I.; Saito, T. *Tetrahedron* **2005**, *61*, 5405–5432.

434. Zhang, X. *Tetrahedron: Asymmetry* **2004**, *15*, 2099–2100.

435. Knowles, W.S.; Sabacky, M.J.; Vineyard, B.D. *J. Chem. Soc., Chem. Commun.* **1972**, 10–11.

436. Knowles, W.S.; Sabacky, M.J. *J. Chem. Soc., Chem. Commun.* **1968**, 1445–1446.

437. Knowles, W.S.; Sabacky, M.J.; Vineyard, B.D.; Weinkauff, D.J. *J. Am. Chem. Soc.* **1975**, *97*, 2567–2568.

438. Vineyard, B.D.; Knowles, W.S.; Sabacky, M.J.; Bachman, G.L.; Weinkauff, D.J. *J. Am. Chem. Soc.* **1977**, *99*, 5946–5952.

439. Schmidt, U.; Lieberknecht, A.; Kazmeier, U.; Griesser, H.; Jung, G.; Metzger, J. *Synthesis* **1991**, 49–55.

440. Hammadi, A.; Lam, H.; Gondry, M.; Ménez, A.; Genet, R. *Tetrahedron* **2000**, *56*, 4473–4477.

441. Ojima, I.; Tsai, C.-Y.; Zhang, Z. *Tetrahedron Lett.* **1994**, *35*, 5785–5788.

442. Scott, J.W. *Top Stereochem.* **1989**, *19*, 209–226.

443. Simons, C.; Hanefeld, U.; Arends, I.W.C.E.; Sheldon, R.A.; Maschmeyer, T. *Chem—Eur. J.* **2004**, *10*, 5829–5835.

444. Johnson, C.R.; Imamoto, T. *J. Org. Chem.* **1987**, *52*, 2170–2174.

445. Imamoto, T.; Tsuruta, H.; Wada, Y.; Masuda, H.; Yamaguchi, K. *Tetrahedron Lett.* **1995**, *36*, 8271–8274.

446. Miura, T.; Imamoto, T. *Tetrahedron Lett.* **1999**, *40*, 4833–4836.

447. Nagel, U.; Krink, T. *Angew. Chem., Int. Ed. Engl.* **1993**, *32*, 1052–1054.

448. Imamoto, T.; Watanabe, J.; Wada, Y.; Masuda, H.; Yamada, H.; Tsuruta, H.; Matsukawa, S.; Yamaguchi, K. *J. Am. Chem. Soc.* **1998**, *120* 1635–1636.

449. Oohara, N.; Katagiri, K.; Imamoto, T. *Tetrahedron Asymmetry* **2003**, *14*, 2171–2175.

450. Gridnev, I.D.; Yasutake, M.; Imamoto, T.; Beletskaya, I.P. *Proc. Natl. Acad. Sci. U.S.A.* **2004**, *101*, 5385–5390.

451. Ohashi, A.; Imamoto, T. *Org. Lett.* **2001**, *3*, 373–375.

452. Ohashi, A.; Kikuchi, S.-i.; Yasutake, M.; Imanoto, T. *Eur. J. Org. Chem.* **2002**, 2535–2546.

453. Yamanoi, Y.; Imamoto, T. *J. Org. Chem.* **1999**, *64* 2988–2989.

454. Miyazaki, T.; Sugawara, M.; Danjo, H.; Imamoto, T. *Tetrahedron Lett.* **2004**, *45*, 9341–9344.

455. Danjo, H.; Sasaki, W.; Miyazaki, T.; Imamoto, T. *Tetrahedron Lett.* **2003**, *44*, 3467–3469.

456. Imamoto, T.; Sugita, K.; Yoshida, K. *J. Am., Chem. Soc.* **2005**, *127*, 11934–11935.

457. Maienza, F.; Santoro, F.; Spindler, F.; Malan, C.; Mezzetti, A. *Tetrahedron: Asymmetry* **2002**, *13*, 1817–1824.

458. Dang, T.P.; Kagan, H.B. *J. Chem. Soc., Chem. Commun.* **1971**, 481.

459. Kagan, H.B.; Dang, T.-P. *J. Am. Chem. Soc.* **1972**, *94*, 6429–6433.

460. Takaishi, N.; Imai, H.; Bertelo, C.A.; Stille, J.K. *J. Am. Chem. Soc.* **1976**, *98*, 5400.

461. Selke, R.; Holz, J.; Riepe, A.; Börner, A. *Chem.— Eur. J.* 1998, 4, 769–771.

462. Hu, W.; Chen, C.-C.; Xue, G.; Chan, A.S.C. *Tetrahedron: Asymmetry* **1998**, *9*, 4183–4192.

463. Komarov, I.V.; Monsees, A.; Spannenberg, A.; Baumann, W.; Schmidt, U.; Fischer, C.; Börner, A. *Eur. J. Org. Chem.* **2003**, 138–150.

464. Komarov, I.V.; Monsees, A.; Kadyrov, R.; Fischer, C.; Schmidt, U.; Börner, A. *Tetrahedron: Asymmetry* **2002**, *13*, 1615–1620.

465. Kadyrov, R.; Ilaldinov, I.Z.; Almena, J.; Monsees, A.; Riermeier, T.H. *Tetrahedron Lett.* 200*5*, 46, 7397–7400.

466. Gavryushin, A.; Polborn, K.; Knochel, P. *Tetrahedron: Asymmetry* **2004**, *15*, 2279–2288.

467. Fryzuk, M.D.; Bosnich, B. *J. Am. Chem. Soc.* **1977**, *99*, 6262–6267.

468. Zeiss, H.-J. *J. Org. Chem.* **1991**, *56*, 1783–1788.

469. Herseczki, Z.; Gergely, I.; Hegedüs, C.; Szöllósy, A.; Bakos, J. *Tetrahedron: Asymmetry* **2004**, *15*, 1673–1676.

470. Dubrovina, N.V.; Tararov, V.I.; Monsees, A.; Spannenberg, A.; Kostas, I.D.; Börner, A. *Tetrahedron: Asymmetry* **2005**, *16*, 3640–3649.

471. Dubrovina, N.V.; Tararov, V.I.; Monsees, A.; Kadyrov, R.; Fischer, C.; Börner, A. *Tetrahedron: Asymmetry* **2003**, *14*, 2739–2745.

472. Parnes, H.; Shelton, E.J.; Huang, G.T. *Int. J. Pept. Prot. Res.* **1986**, *28*, 403–410.

473. Baker, G.L.; Fritschel, S.J.; Stille, J.R.; Stille, J.K. *J. Org. Chem.* **1981**, *46*, 2954–2960.

474. Nagel, U.; Kinzel, E.; Andrade, J.; Prescher, G. *Chem. Ber.* **1986**, *119*, 3326–3343.

475. Nagel, U. *Angew. Chem., Int. Ed. Engl.* **1984**, *23*, 435–436.

476. Fan, Q.-H.; Deng, G.-J.; Lin, C.-C.; Chan, A.S.C. *Tetrahedron: Asymmetry* **2001**, *12*, 1241–1247.

477. Yi, B.; Fan, Q.-H.; Deng, G.-J.; Li, Y.-M.; Qiu, L.-Q.; Chan, A.S.C. *Org. Lett.* **2004**, *6*, 1361–1364.

478. Engel, G.D.; Gade, L.H. *Chem.—Eur. J.* **2002**, *8*, 4319–4329.

479. Zarka, M.T.; Nuyken, O.; Weberskirch, R. *Chem.—Eur. J.* **2003**, *9*, 3228–3234.

480. Aoki, K.; Shimada, T.; Hayashi, T. *Tetrahedron: Asymmetry* **2004**, *15*, 1771–1777.

481. Diéguez, M.; Pàmies, O.; Ruiz, A.; Castillón, S.; Claver, C. *Tetrahedron: Asymmetry* **2000**, *11*, 4701–4708.

482. Taudien, S.; Schinkowski, K.; Krause, H.-W. *Tetrahedron: Asymmetry* **1993**, *4*, 73–84.

483. Li, X.; Lou, R.; Yeung, C.-H.; Chan, A.S.C.; Wong, W.K. *Tetrahedron: Asymmetry* **2000**, *11*, 2077–2082.

484. Derrien, N.; Dousson, C.B.; Roberts, S.M.; Berens, U.; Burk, M.J.; Ohff, M. *Tetrahedron: Asymmetry* **1999**, *10*, 3341–3352.

485. Mi, A.; Lou, R.; Jiang, Y.; Deng, J.; Qin, Y.; Fu, F.; Li, Z.; Hu, W.; Chan, A.S.C. *Synlett* **1998**, 847–848.

486. Lou, R.; Mi, A.; Jiang, Y.; Qin, Y.; Li, Z.; Fu, F.; Chan, A.S.C. *Tetrahedron* **2000**, *56*, 5857–5863.

487. Xie, Y.; Lou, R.; Li, Z.; Mi, A.; Jiang, Y. *Tetrahedron: Asymmetry* **2000**, *11*, 1487–1494.

488. Clochard, M.; Mattman, E.; Mercier, F.; Ricard, L.; Mathey, F. *Org. Lett.* **2003**, *5*, 3093–3094.

489. Boyer, N.; Léautey, M.; Jubault, P.; Pannecoucke, X.; Quirion, J.-C. *Tetrahedron: Asymmetry* **2005**, *16*, 2455–2458.

490. Zhang, A.; Jiang, B. *Tetrahedron Lett.* **2001**, *42*, 1761–1763.

491. Zhu, G.; Zhang, X. *J. Org. Chem.* **1998**, *63*, 3133–3136.

492. Lin, C.W.; Lin, C.-C.; Lam, L.F.-L.; Au-Yeung, T.T.-L.; Chan, A.S.C. *Tetrahedron: Asymmetry* **2004**, *45*, 7379–7381.

493. Chen, Y.; Li, X.; Tong, S.-k.; Choi, M.C.K.; Chan, A.S.C. *Tetrahedron Lett.* **1999**, *40*, 957–960.

494. Guimet, E.; Diéguez, M.; Ruiz, A.; Claver, C. *Tetrahedron: Asymmetry* **2004**, *15*, 2247–2251.

495. RajanBabu, T.V.; Ayers, T.A.; Casalnuovo, A.L. *J. Am. Chem. Soc.* **1994**, *116*, 4101–4102.

496. Selke, R.; Ohff, M.; Riepe, A. *Tetrahedron* **1996**, *52*, 15079–15102.

497. RajanBabu, T.V.; Ayers, T.A.; Halliday, G.A.; You, K.K.; Calabrese, J.C. *J. Org. Chem.* **1997**, *62*, 6012–6028.

498. Yonehara, K.; Hashizume, T.; Mori, K.; Ohe, K.; Uemura, S. *J. Org. Chem.* **1999**, *64*, 5593–5598.

499. Yonehara, K.; Ohe, K.; Uemura, S. *J. Org. Chem.* **1999**, *64*, 9381–9385.

500. Ohe, K.; Morioka, K.; Yonehara, K.; Uemura, S. *Tetrahedron: Asymmetry* **2002**, *13*, 2155–2160.

501. Pàmies, O.; Diéguez, M.; Net, G.; Ruiz, A. Claver, C. *Chem. Commun.* **2000**, 2383–2384.

502. Chen, Y.; Tekta, S.; Yudin, A.K. *Chem. Rev.* **2003**, *103*, 3155–3211.

503. Kocovsky, P.; Vyskocil, S.; Smrcina, M. *Chem. Rev.* 2003, 103, 3213–3245.

504. Miyashita, A.; Yasuda, A.; Takaya, H.; Toriumi, K.; Ito, T.; Souchi, T.; Noyori, R. *J. Am. Chem. Soc.* **1980**, *102*, 7932–7934.

505. Miyashita, A.; Takaya, H.; Souchi, T.; Noyori, R. *Tetrahedron* **1984**, *40*, 1245–1253.

506. Noyori, R.; Takaya, H. *Acc. Chem. Res.* **1990**, *23*, 345–350.

507. Kitamura, M.; Tsukamoto, M.; Bessho, Y.; Yoshimura, M.; Kobs, U.; Widhalm, M.; Noyori, R. *J. Am. Chem. Soc.* **2002**, *124*, 6649–6667.

508. Noyori, R. *Angew. Chem., Int. Ed.* **2002**, *41*, 2008–2022.

509. Enev, V.; Harre, M.; Nickisch, K.; Schneider, M.; Mohr, J.T. *Tetrahedron: Asymmetry* **2000**, *11*, 1767–1779.

510. Guerreiro, P.; Ratovelomanana-Vidal, V.; Genêt, J.-P.; Dellis, P. *Tetrahedron Lett.* **2001**, *42*, 3423–3426.

511. Yu, H.-B.; Hu, Q.-S.; Pu, L. *Tetrahedron Lett.* **2000**, *41*, 1681–1685.

512. Berthod, M.; Mignani, G.; Lemaire, M. *Tetrahedron: Asymmetry* **2004**, *15*, 1121–1126.

513. Bayardon, J.; Cavazzini, M.; Maillard, D.; Pozzi, G.; Quici, S.; Sinou, D. *Tetrahedron: Asymmetry* **2003**, *14*, 2215–2224.

514. Guo, R.; Au-Yeung, T.T.-L.; Wu, J.; Choi, M.C.K.; Chan, A.S.C. *Tetrahedron: Asymmetry* **2002**, *13*, 2519–2522.

515. Zhou, Y.-G.; Tang, W.; Wang, W.-B.; Li, W.; Zhang, X. *J. Am. Chem. Soc.* **2002**, *124*, 4952–4953.

516. Zhou, Y.-G.; Zhang, X. *Chem. Commun.* **2002**, 1124–1125.

517. Guo, R.; Li, X.; Wu, J.; Kwok, W.H.; Chen, J.; Choi, M.C.J.; Chan, A.S.C. *Tetrahedron Lett.* **2002**, *43*, 6803–6806.

518. Gergely, I.; Hegedüs, C.; Szöllösy, A.; Monsees, A.; Riermeier, T.; Bakos, J. *Tetrahedron Lett.* **2003**, *44*, 9025–9028.

519. Zhang, F.-Y.; Kwok, W.H.; Chan, A.S.C. *Tetrahedron: Asymmetry* **2001**, *12*, 2337–2342.

520. Yan, Y.; Chi, Y.; Zhang, X. *Tetrahedron: Asymmetry* **2004**, *15*, 2173–2175.

521. Burgemeister, K.; Franciò, G.; Hugl, H.; Leitner, W. *Chem. Commun.* **2005**, 6026–6027.

522. Wu, S.; He, M.; Zhang, X. *Tetrahedron: Asymmetry* **2004**, *15*, 2177–2180.

523. Steiner, I.; Aufdenblatten, R.; Togni, A.; Blaser, H.-U.; Pugin, B. *Tetrahedron: Asymmetry* **2004**, *15*, 2307–2311.

524. Shimizu, H.; Ishizaki, T.; Fujiwara, T.; Saito, T. *Tetrahedron: Asymmetry* **2004**, *15*, 2169–2172.

525. Wu, J.; Pai, C.C.; Kwok, W.H.; Guo, R.W.; Au-Yeung, T.T.L.; Yeung, C.H.; Chan, A.S.C. *Tetrahedron: Asymmetry* **2003**, *14*, 987–992.

526. Miyano, S.; Nawa, M.; Mori, A.; Hashimoto, H. *Bull. Chem. Soc. Jpn.* **1984**, 57, 2171–2176.

527. Pye, P.J.; Rossen, K.; Reamer, R.A.; Tsou, N.N.; Volante, R.P.; Reider, P.J. *J. Am. Chem. Soc.* **1997**, *119*, 6207–6208.

528. Zanotti-Gerosa, A.; Malan, C.; Herzberg, D. *Org. Lett.* **2001**, *3*, 3687–3690.

529. Focken, T.; Raabe, G.; Bolm, C. *Tetrahedron: Asymmetry* **2004**, *15*, 1693–1706.

530. Mohler, P.; Rippert, A.J.; Hansen, H.-J. *Helv. Chim. Acta* **2000**, *83*, 258–277.

531. Sawamura, M.; Kuwano, R.; Ito, Y. *J. Am. Chem. Soc.* **1995**, *117*, 9602–9603.

532. Kuwano, R.; Okuda, S.; Ito, Y. *Tetrahedron: Asymmetry* **1998**, *9*, 2773–2775.

533. Kuwano, R.; Okuda, S.; Ito, Y. *J. Org. Chem.* **1998**, *63*, 3499–3503.

534. Kuwano, R.; Ito, Y. *J. Org. Chem.* **1999**, *64*, 1232–1237.

535. Kuwano, R.; Karube, D.; Ito, Y. *Tetrahedron Lett.* **1999**, *40*, 9045–9049.

536. Perea, J.J.A.; Börner, A.; Knochel, P. *Tetrahedron Lett.* **1998**, *39*, 8073–8076.

537. Almena Perea, J.J.; Lotz, M.; Knochel, P. *Tetrahedron: Asymmetry* **1999**, *10*, 375–384.

538. Kang, J.; Lee, J.H.; Ahn, S.H.; Choi, J.S. *Tetrahedron Lett.* **1998**, *39*, 5523–5526.

539. Raghunath, M.; Gao, W.; Zhang, X. *Tetrahedron: Asymmetry* **2005**, *16*, 3676–3681.

540. Spindler, F.; Malan, C.; Lotz, M.; Kesselgruber, M.; Pittelkow, U.; Rivas-Nass, A.; Briel, O.; Blaser, H.-U. *Tetrahedron: Asymmetry* **2004**, *15*, 2299–2306.

541. Tappe, K.; Knochel, P. *Tetrahedron: Asymmetry* **2004**, *15*, 91–102.

542. Ireland, T.; Tappe, K.; Grossheimann, G.; Knochel, P. *Chem.—Eur. J.* **2002**, *8*, 843–852.

543. Boaz, N.W.; MacKenzie, E.B.; Debenham, S.D.; Large, S.E.; Ponasik J.A. Jr., *J. Am. Chem. Soc.* **2005**, *70*, 1872–1880.

544. Qiao, S.; Fu, G.C *J. Org. Chem.* **1998**, *63*, 4168–4169.

545. Argouarch, G.; Samuel, O.; Kagan, H.B. *Eur. J. Org. Chem.* **2000**, 2885–2891.

546. Argouarch, G.; Samuel, O.; Riant, O.; Daran, J.-C.; Kagan, H.B. *Eur. J. Org. Chem.* **2000**, 2893–2899.

547. Jia, X.; Li, X.; Lam, W.S.; Kok, S.H.L.; Xu, L.; Lu, G.; Yeung, C.-H.; Chan, A.S.C. *Tetrahedron: Asymmetry* **2004**, *15*, 2273–2278.

548. Gibson, S.E.; Ibrahim, H.; Pasquier, C.; Swamy, V.M. *Tetrahedron: Asymmetry* **2004**, *15*, 465–473.

549. Kromm, K.; Zwick, B.D.; Meyer, O.; Hampel, F.; Gladysz, J.A. *Chem.—Eur. J.* **2001**, *7*, 2015–2027.

550. Ward, T.R. *Chem.—Eur. J.* **2005**, *11*, 3798–3804.

551. Skander, M.; Humbert, N.; Collot, J.; Gradinaru, J.; Klein, G.; Loosli, A.; Sauser, J.; Zocchi, A.; Gilardoni, F.; Ward, T.R. *J. Am. Chem. Soc.* **2004**, *126*, 14411–14418.

552. Skander, M.; Malan, C.; Ivanova, A.; Ward, T.R. *Chem. Commun.* **2005**, 4815–4817.

553. Klein, G.; Humbert, N.; Gradinaru, J.; Ivanova, A.; Gilardoni, F.; Rusbandi, U.E.; Ward, T.R. *Angew. Chem., Int. Ed.* **2005**, *44*, 7764–7767.

554. Gilbertson, S.R.; Wang, X. *Tetrahedron* **1999**, *55*, 11609–11618.

555. Gilbertson, S.R.; Chen, G.; Kao, J.; Beatty, A.; Campana, C.F. *J. Org. Chem.* **1997**, *62*, 5557–5566.

556. Gilbertson, S.R.; Wang, X. *J. Org. Chem.* **1996**, *61*, 434–435.

557. Reggelin, M.; Doerr, S.; Klussmann, M.; Schultz, M.; Holbach, M. *Proc. Natl. Acad. Sci. U.S.A.* **2004**, *101*, 5461–5466.

558. Tommasino, M.L.; Casalta, M.; Breuzard, J.A.J.; Lemaire, M. *Tetrahedron Asymmetry* **2000**, *11*, 4835–4841.

559. Burk, M.J.; Feaster, J.E.; Nugent, W.A.; Harlow, R.L. *J. Am. Chem. Soc.* **1993**, *115*, 10125–10138.

560. Burk, M.J. *J. Am. Chem. Soc.* **1991**, *113*, 8518–8519.

561. Burk, M.J.; Feng, S.; Gross, M.F.; Tumas, W. *J. Am. Chem. Soc.* **1995**, *117*, 8277–8278.

562. Berger, A.; de Souza, R.F.; Delgado, M.R.; Dupont, J. *Tetrahedron: Asymmetry* **2001**, *12*, 1825–1828.

563. Burk, M.J.; Bedingfield, K.M.; Kiesman, W.F.; Allen, J.G. *Tetrahedron Lett.* **1999**, *40*, 3093–3096.

564. Burk, M.J.; Gross, M.F.; Martinez, J.P. *J. Am. Chem. Soc.* 1995, 117, 9375–9376.

565. Jones, S.W.; Palmer, C.F.; Paul, J.M.; Tiffin, P.D. *Tetrahedron Lett.* **1999**, *40*, 1211–1214.

566. Stammers, T.A.; Burk, M.J. *Tetrahedron Lett.* **1999**, *40*, 3325–3328.

567. Börner, A.; Heller, D. *Tetrahedron Lett.* **2001**, *42*, 223–225.

568. Boaz, N.W.; Debenham, S.D.; Large, S.E.; Moore, M.K. *Tetrahedron: Asymmetry* **2003**, *14*, 3575–3580.

569. Ritzén, A.; Frejd, T. *Chem. Commun.* **1999**, 207–208.

570. Burk, M.J.; Lee, J.R.; Martinez, J.P. *J. Am. Chem. Soc.* **1994**, *116*, 10947–10848.

571. Robinson, A.J.; Stanislawski, P.; Mulholland, D.; He, L.; Li, H.-Y. *J. Org. Chem.* **2001**, *66*, 4148–4152.

572. Wolfson, A.; Geresh, S.; Gottlieb, M.; Herskowitz, M. *Tetrahedron: Asymmetry* **2002**, *13*, 465–468.

573. Vankelecom, I.; Wolfson, A.; Geresh, S.; Landau, M.; Gottlieb, M.; Hershkovitz, M. *Chem. Commun.* **1999**, 2407–2408.

574. Wolfson, A.; Janssens, S.; Vankelecom, I.; Geresh, S.; Gottlieb, M.; Herskowitz, M. *Chem. Commun.* **2002**, 388–389.

575. Guernik, S.; Wolfson, A.; Herskowitz, M.; Greenspoon, N.; Geresh, S. *Chem. Commun.* **2001**, 2314–2315.

576. Rouhi, M. *Chem. Eng. News* **2004**, *February 16*, 20.

577. Holz, J.; Monsees, A.; Jiao, H.; You, J.; Komarov, I.V.; Fischer, C.; Frauz, K.; Börne, A. *J. Org. Chem.* **2003**, *68*, 1701–1707.

578. Holz, J.; Quirmbach, M.; Schmidt, U.; Heller, D.; Stürmer, R.; Börner, A. *J. Org. Chem.* **1998**, *63*, 8031–8034.

579. Li, W.; Zhang, Z.; Xiao, D.; Zhang, X. *Tetrahedron Lett.* **1999**, *40*, 6701–6704.

580. Holz, J.; Heller, D.; Stürmer, R.; Börner, A. *Tetrahedron Lett.* **1999**, *40*, 7059–7062.

581. Ostermeier, M.; Prieβ, J.; Helmchen, G. *Angew. Chem., Int. Ed.* **2002**, *41*, 612–614.

582. Korostylev, A.; Selent, D.; Monsees, A.; Borgmann, C.; Börner, A. *Tetrahedron: Asymmetry* **2003**, *14*, 1905–1909.

583. Tang, W.; Zhang, X. *Angew. Chem., Int. Ed.* **2002**, *41*, 1612–1614.

584. Liu, D.; Zhang, X. *Eur. J. Org. Chem.* **2005**, 646–649.

585. Imamoto, T.; Crépy, K.V.L.; Katagiri, K. *Tetrahedron: Asymmetry* **2004**, *15*, 2213–2218.

586. Fernandez, E.; Gillon, A.; Heslop, K.; Horwood, E.; Hyett, D.J.; Orpen, A.G.; Pringle, P.G. *Chem. Commun.* **2000**, 1663–1664.

587. Bienewald, F.; Ricard, L.; Mercier, F.; Mathey, F. *Tetrahedron: Asymmetry* **1999**, *10*, 4701–4707.

588. Moulin, D.; Darcel, C.; Jugé, S. *Tetrahedron: Asymmetry* **1999**, *10*, 4729–4743.

589. Marinetti, A.; Labrue, F.; Genêt, J.-P. *Synlett* **1999**, 1975–1977.

590. Reetz, M.T.; Gosberg, A.; Goddard, R.; Kyung, S.-H. *Chem. Commun.* **1998**, 2077–2078.

591. Nettekoven, U.; Kamer, P.C.J.; van Leeuwen, P.W.N.M.; Widhalm, M.; Spek, A.L.; Lutz, M. *J. Org. Chem.* **1999**, *64*, 3996–4004.

592. Liu, D.; Li, W.; Zhang, X. *Org. Lett.* **2002**, *4*, 4471–4474.

593. Zeng, Q.-H.; Hu, X.-P.; Duan, Z.-C; Liang, X.-M.; Zheng, Z. *Tetrahedron: Asymmetry* **2005**, *16*, 1233–1238.

594. Boaz, N.W.; Ponasik J.A.; Jr., Large, S.E. *Tetrahedron: Asymmetry* **2005**, *16*, 2063–2066.

595. Pàmies, O.; Diéguez, M.; Net, G.; Ruiz, A.; Claver, C. *J. Org. Chem.* **2001**, *66*, 8364–8369.

596. Pàmies, O.; Net, G.; Ruiz, A.; Claver, C. *Tetrahedron: Asymmetry* **2000**, *11*, 1097–1108.

597. Diéguez, M.; Ruiz, A.; Claver, C. *Chem. Commun.* **2001**, 2702–2703.

598. Carmichael, D.; Doucet, H.; Brown, J.M. *Chem. Commun.* **1999**, 261–262.

599. Hoge, G.; Samas, B. *Tetrahedron: Asymmetry* **2004**, *15*, 2155–1257.

600. Robin, F.; Mercier, F.; Ricard, L.; Mathey, F.; Spagnol, M. *Chem.—Eur. J.* **1997**, *3*, 1365–1369.

601. van den Berg, M.; Minnaard, A.J.; Schudde, E.P.; van Esch, J.; de Vries, A.H.M.; de Vries, J.G.; Feringa, B.L. *J. Am. Chem. Soc.* **2000**, *122*, 11539–11540.

602. Landis, C.R.; Feldgus, S. *Angew. Chem., Int. Ed.* **2000**, *39*, 2863–2866.

603. Claver, C.; Fernandez, E.; Gillon, A.; Heslop, K.; Hyett, D.J.; Martorell, A.; Orpen, A.G.; Pringle, P.G. *Chem. Commun.* **2000**, 961–962.

604. Reetz, M.T.; See, T. *Tetrahedron Lett.* **2000**, *41*, 6333–6336.

605. Ionescu, G.; van der Vlugy, J.I.; Abbenhuis, H.C.L.; Vogt, D. *Tetrahedron: Asymmetry* **2005**, *16*, 3970–3975.

606. Chen, W.; Xiao, J. *Tetrahedron Lett.* **2001**, *42*, 2897–2899.

607. Huang, H.; Liu, X.; Chen, S.; Chen, H.; Zheng, Z. *Tetrahedron: Asymmetry* **2004**, *15*, 2011–2019.

608. Huang, H.; Zheng, Z.; Luo, H.; Bai, C.; Hu, X.; Chen, H. *J. Org. Chem.* **2004**, *69*, 2355–2361.

609. Korostylev, A.; Monsees, A.; Fischer, C.; Börner, A. *Tetrahedron: Asymmetry* **2004**, *15*, 1001–1005.

610. Peña, D.; Minnaard, A.J.; de Vries, A.H.M.; de Vries, J.G.; Feringa, B.L. *Org. Lett.* **2003**, *5*, 475–478.

611. Jia, X.; Li, X.; Xu, L.; Shi, Q.; Yao, X.; Chan, A.S.C. *J. Org. Chem.* **2003**, *68*, 4539–4541.

612. Lefort, L.; Boogers, J.A.F.; de Vries, A.H.M.; de Vries, J.G. *Org. Lett.* **2004**, *6*, 1733–1735.

613. Bernsmann, H.; van den Berg, M.; Hoen, R.; Minnaard, A.J.; Mehler, G.; Reetz, M.T.; De Vries, J.G.; Feringa, B.L. *J. Org. Chem.* **2005**, *70*, 943–951.

614. Reetz, M.T.; Ma, J.-A.; Goddard, R. *Angew. Chem., Int. Ed.* **2005**, *44*, 412–415.

615. Li, X.; Jia, X.; Lu, G.; Au-Yeung, T.T.-L.; Lam, K.-H.; Lo, T.W.H.; Chan, A.S.C. *Tetrahedron: Asymmetry* **2003**, *14*, 2687–2691.

616. Wang, X.; Ding, K. *J. Am. Chem. Soc.* **2004**, *126*, 10524–10525.

617. Huttenloch, O.; Speiler, J.; Waldmann, H. *Chem.—Eur. J.* **2000**, *6*, 671–675.

618. Doherty, S.; Robins, E.G.; Pál, I.; Newman, C.R.; Hardacre, C.; Rooney, D.; Mooney, D.A. *Tetrahedron: Asymmetry* **2003**, *14*, 1517–1527.

619. Simons, C.; Hanefeld, U.; Arends, I.W.C.E.; Minnaard, A.J.; Maschmeyer, T.; Sheldon, R.A. *Chem. Commun.* **2004**, 2830–2831.

620. Monti, C.; Gennari, C.; Piarulli, U. *Tetrahedron: Asymmetry* **2004**, *45*, 6859–6862.

621. Monti, C.; Gennari, C.; Piarulli, U.; de Vries, J.G.; de Vries, A.H.M.; Lefort, L. *Chem.—Eur. J.* **2005**, *11*, 6701–6717.

622. Ding, K.; Du. H.; Yuan, Y.; Long, J. *Chem.—Eur. J.* **2004**, *10*, 2872–2884.

623. Hannen, P.; Militzer, H.-C.; Vogl, E.M.; Rampf, F.A. *Chem. Commun.* **2002**, 2210–2211.

624. Botman, P.N.M.; Amore, A.; van Heebeek, R.; Back, J.W.; Hiemstra, H.; Reek, J.N.H.; van Maarseveen, J.H. *Tetrahedron Lett.* **2004**, *45*, 5999–6002.

625. Fu, Y.; Xie, J.-H.; Hu, A.-G.; Zhou, H.; Wang, L.-X.; Zhou, Q.-L. *Chem. Commun.* **2002**, 480–481.

626. Zhu, S.-F.; Fu, Y.; Xie, J.-H.; Liu, B.; Xing, L.; Zhou, Q.L. *Tetrahedron: Asymmetry* **2003**, *14*, 3219–3224; corrigendum 2004, 15, 183.

627. Fu, Y.; Hou, G.-H.; Xie, J.-H.; Xing, L.; Wang L.-X.; Zhou, Q.-L. *J. Org. Chem.* **2004**, *69*, 8157–8160.

628. Evans, D.A.; Katz, J.L.; Peterson, G.S. Hintermann, T. *J. Am. Chem. Soc.* **2001**, *123*, 12411–12413.

629. Evans, D.A.; Michael, F.E.; Tedrow, J.S.; Campos K.R. *J. Am. Chem. Soc.* **2003**, *125*, 3534–3543.

630. Junge, K.; Oehme, G.; Monsees, A.; Riermeier, T.; Dingerdissen, U.; Beller, M. *Tetrahedron Lett* **2002**, *43*, 4977–4980.

631. Kasák, P.; Mereiter, K.; Widhalm, M. *Tetrahedron Asymmetry* **2005**, *16*, 3416–3426.

632. Hoen, R.; van den Berg, M.; Bernsmann, H. Minnaard, A.J.; de Vries, J.G.; Feringa, B.L. *Org Lett.* **2004**, *6*, 1433–1436.

633. Dobrota, C.; Toffano, M.; Fiaud, J.-C. *Tetrahedron Asymmetry* **2004**, *45*, 8153–8156.

634. Jiang, X.-b.; van den Berg, M.; Minnaard, A.J. Feringa, B.L.; de Vries, J.G. *Tetrahedron Asymmetry* **2004**, *15*, 2223–2229.

635. Reetz, M.T.; Sell, T.; Meiswinkel, A.; Mehler, G *Angew. Chem., Int. Ed.* **2003**, *42*, 790–792.

636. Komarov, I.V.; Börner, A. *Angew. Chem., Int. Ed* **2001**, *40*, 1197–1200.

637. Drexler, H.-J.; Baumann, W.; Schmidt, T.; Zhang S.; Sun, A.; Spannenberg, A.; Fischer, C. Buschmann, H.; Heller, D. *Angew. Chem., Int. Ed* **2005**, *44*, 1184–1188.

638. Heller, D.; Holz, J.; Komarov, I.; Drexler, H.-J You, J.; Drauz, K.; Börner, A. *Tetrahedron Asymmetry* **2002**, *13*, 2735–2741.

639. Heller, D.; Drexler, H.-J.; You, J.; Baumann, W Drauz, K.; Krimmer, H.-P.; Börner, A. *Chem.-Eur. J.* **2002**, *8*, 5196–5203.

640. You, J.; Drexler, H.-J.; Zhang, S.; Fischer, C Heller, D. *Angew. Chem., Int. Ed.* **2003**, *42*, 913–916.

641. Jerphagnon, T.; Renaud, J.-L.; Demonchaux, P Ferreira, A.; Bruneau, C. *Tetrahedron: Asymmetr* **2003**, *14*, 1973–1977.

642. Zhu, G.; Chen, Z.; Zhang, X. *J. Org. Chem.* **1999**, *64*, 6907–6910.

643. Heller, D.; Holz, J.; Drexler, H.-J.; Lang, J Drauz, K.; Krimmer, H.-P.; Börner, A. *J. Org Chem.* **2001**, *66*, 6816–6817.

644. Elaridi, J.; Thaqi, A.; Prosser, A.; Jackson, W.R Robinson, A.J. *Tetrahedron: Asymmetry* **2005**, *16*, 1309–1319.

645. Burk, M.J.; Stammers, T.A.; Straub, J.A. *Org Lett.* 1999, 1, 387–390.

646. Zhang, Y.J.; Park, J.H.; Lee, S.-g. *Tetrahedron Asymmetry* **2004**, *15*, 2209–2212.

647. Achiwa, K.; Soga, T. *Tetrahedron Lett.* **1978**, *1*, 1119–1120.

648. Lubell, W.D.; Kitamura, M.; Noyori, R Tetrahedron: Asymmetry **1991**, 2, 543–554.

649. Qiu, L.; Wu, J.; Chan, S.; Au-Yeung, T.T.-L.; J J.-X.; Guo, R.; Pai, C.-C.; Zhou, Z.; Li, X

Fan, Q.-h.; Chan, A.S.C. *Proc. Natl. Acad. Sci. U.S.A.* **2004**, *101*, 5815–5820.

650. Wu, J.; Chen, X.; Guo, R.; Yeung, C.-h.; Chan, A.S.C. *J. Org. Chem.* **2003**, *68*, 2490–2493.

651. Wu, H.-p.; Hoge, G. *Org. Lett.* **2004**, *6*, 3645–3647.

652. Lee, S.-g.; Zhang, Y.J. *Org. Lett.* **2002**, *4*, 2429–2431.

653. Yasutake, M.; Gridnev, I.D.; Higashi, N.; Imamoto, T. *Org. Lett.* **2001**, *3*, 1701–1704.

654. Hsiao, Y.; Rivera, N.R.; Rosner, T.; Krska, S.W.; Njolito, E.; Wang, F.; Sun, Y.; Armstrong J.D.; III, Grabowski, E.J.J.; Tillyer, R.D.; Spindler, F.; Malan, C. *J. Am. Chem. Soc.* **2004**, *126*, 9918–9919.

655. Hu, X.-P.; Zheng, Z. *Org. Lett.* **2005**, *7*, 419–422.

656. Tang, W.; Zhang, X. *Org. Lett.* **2002**, *4*, 4159–4161.

657. Tang, W.; Wang, W.; Chi, Y.; Zhang, X. *Angew. Chem., Int. Ed.* **2003**, *42*, 3509–3511.

658. Peña, D.; Minnaard, A.J.; de Vries, J.G.; Feringa, B.L. *J. Am. Chem. Soc.* **2002**, *124*, 14552–14553.

659. Peña, D.; Minnaard, A.J.; Boogers, J.A.F.; de Vries, A.H.M.; de Vries, J.G.; Feringa, B.L. *Org. Biomol. Chem.* **2003**, *1*, 1087–1089.

660. Reetz, M.T.; Li, X. *Tetrahedron* **2004**, *60*, 9709–9714.

661. Reetz, M.T.; Li, X. *Angew. Chem., Int. Ed.* **2005**, *44*, 2959–2962.

662. Carter, H.E.; Handler, P.; Melville, D.B. *J. Biol. Chem.* **1939**, *129*, 359–369.

663. Eiger, I.Z.; Greenstein, J.P. *Arch. Biochem.* **1948**, *19*, 467–473.

664. Price, V.E.; Greenstein, J.P. *J. Biol. Chem.* **1948**, *173*, 337–344.

665. Vega, A.; Bell, E.A.; Nunn, P.B. *Phytochemistry* **1968**, *7*, 1885–1887.

666. McCord, T.J.; Booth, L.D.; Davis, A.L. *J. Med. Chem.* **1968**, *11*, 1077–1078.

667. Fu, S.-C.J.; Greenstein, J.P. *J. Am. Chem. Soc.* **1955**, *77*, 4412–4413.

668. Asquith, R.S.; Yeung, K.W.; Otterburn, M.S. *Tetrahedron* **1977**, *33*, 1633–1635.

669. Harde, C.; Neff, K.-H.; Nordhoff, E.; Gerbling, K.-P.; Laber, B.; Pohlenz, H.-D. *Bioorg. Med. Chem. Lett.* **1994**, *4*, 273–278.

670. Asquith, R.S.; Carthew, P. *Tetrahedron* **1972**, *28*, 4769–4773.

671. Cohen, J.; Barlow, J.L.; Egan, D.A.; Tricarico, K.A.; Baker, W.R.; Kleinert, H.D. *J. Med. Chem.* **1993**, *36*, 449–459.

672. Adams, R.; Johnson, J.L. *J. Am. Chem. Soc.* **1949**, *71*, 705–708.

673. Pérez, M.; Pleixats, R. *Tetrahedron*, **1995**, *51*, 8355–8362.

674. Naidu, B.N.; Sorenson, M.E.; Connolly, T.P.; Ueda, Y. *J. Org. Chem.* **2003**, *68*, 10098–10102.

675. Carter, H.E.; Stevens, C.M.; Ney, L.F. *J. Biol. Chem.* **1941**, *139*, 247–254.

676. Schöberl, A.; Wagner, A. *Naturwiss* **1947**, *34*, 189.

677. Huynh-Ba, T.; Fay, L. *J. Lab. Cmpds. Radiopharm.* **1990**, *28*, 1185–1187.

678. Leach, B.E.; Hunter, J.H.; West, C.A.; Carter, H.E. *Biochem. Prep.* **1953**, *3*, 111–118.

679. Ondrus, T.A.; Christie, B.J.; Guy, R.W. *Aust. J. Chem.* **1979**, *32*, 2313–2316.

680. Hanzlik, R.P.; Weller, P.E.; Desai, J.; Zheng, J.; Hall, L.R.; Slaughter, D.E. *J. Org. Chem.* **1990**, *55*, 2736–2742.

681. Ferreira, P.M.T.; Maia, H.L.S.; Monteiro, L.S.; Sacramento, J.; Sebastião, J. *J. Chem. Soc., Perkin Trans. 1* **2000**, 3317–3324.

682. Zhu, Y.; van der Donk, W.A. *Org. Lett.* **2001**, *3*, 1189–1192.

683. Rulev, A.Y.; Larina, L.I.; Keiko, N.A.; Voronkov, M.G. *J. Chem. Soc., Perkin Trans. 1* **1999**, 1567–1569.

684. Barbaste, M.; Rolland-Fulcrand, V.; Roumestant, M.-L.; Viallefont, P.; Martinez, J. *Tetrahedron Lett.* **1998**, *39*, 6287–6290.

685. Ferreira, P.M.T.; Maia, H.L.S.; Monteiro, L.S. *Tetrahedron Lett.* **1999**, *40*, 4099–4102.

686. Dugave, C.; Cluzeau, J.; Ménez, A.; Gaudry, M.; Marquet, A. *Tetrahedron Lett.* **1998**, *39*, 5775–5778.

687. Buchanan, R.L.; Dean, F.H.; Pattison, F.L.M. *Can. J. Chem.* **1962**, *40*, 1571–1575.

688. Hellmann, H.; Lingens, F.; Folz, E. *Ber.* **1956**, *89*, 2433–2436.

689. Kazmeier, U. *Tetrahedron Lett.* **1996**, *37*, 5351–5354.

690. Crossley, M.J.; Fung, Y.M.; Potter, J.J.; Stamford, A.W. *J. Chem. Soc., Perkin Trans. 1* **1998**, 1113–1121.

691. Pyne, S.G.; Javidan, A.; Skelton, B.W.; White, A.H. *Tetrahedron* **1995**, *51*, 5157–5168.

692. Horner, L.; Schwahn, H. *Ann.* **1955**, *591*, 99–107.

693. Cardellicchio, C.; Fiandanese, V.; Marchese, G.; Naso, F.; Ronzini, L. *Tetrahedron Lett.* **1985**, *26*, 4387–4390.

694. Zaparucha, A.; Danjoux, M.; Chiaroni, A.; Royer, J.; Husson, H.-P. *Tetrahedron Lett.* **1999**, *40*, 3699–3700.

695. Navarre, L.; Darses, S.; Genet, J.-P. *Eur. J. Org. Chem.* **2004**, 69–73.

696. Huang, T.; Keh, C.C.K.; Li, C.-J. *Chem. Commun.* **2002**, 2440–2441.

697. Wulff, G.; Böhnke, H. *Angew. Chem., Int. Ed. Engl.* **1986**, *25*, 90–92.

698. Öhler, E.; Prantz, E.; Schmidt, U. *Chem. Ber.* **1978**, *111*, 1058–1076.

699. Schmidt, U.; Prantz, E. *Angew. Chem., Int. Ed. Engl.* **1977**, *16*, 328.

700. Spadoni, G.; Balsamini, C.; Bedini, A.; Duranti, E.; Tontini, A. *J. Heterocycl. Chem.* **1992**, *29*, 305–309.

701. Balsamini, C.; Diamantini, G.; Duranti, A.; Spadoni, G.; Tontini, A. *Synthesis* **1995**, 370–372.

702. Snyder, H.R.; MacDonald, J.A. *J. Am. Chem. Soc.* **1955**, *77*, 1257–1259.

703. Boger, D.L.; Keim, H.; Oberhauser, B.; Schreiner, E.P.; Foster, C.A. *J. Am. Chem. Soc.* **1999**, *121*, 6197–6205.

704. Cativiela, C.; López, M.P.; Mayoral, J.A. *Synlett* **1992**, 121–122.

705. Filler, R.; Rao, Y.S. *J. Org. Chem.* **1961**, *26*, 1685.

706. Cavé, C.; Le Porhiel-Castellon, Y.; Daley, V.; Riche, C.; Chiaroni, A.; d'Angelo, J. *Tetrahedron Lett.* **1997**, *38*, 8703–8706.

707. Lander, P.A.; Hegedus, L.S. *J. Am. Chem. Soc.* **1994**, *116*, 8126–8132.

708. Hohsaka, T.; Kajihara, D.; Ashizuka, Y.; Murakami, H.; Sisido, M. *J. Am. Chem. Soc.* **1999**, *121*, 34–40.

709. Suárez, R.M.; Sestelo, J.P.; Sarandeses, L.A. *Chem.—Eur. J.* **2003**, *9*, 4179–4187.

710. Javidan, A.; Schafer, K.; Pyne, S.G. *Synlett* **1997**, 100–102.

711. Suárez, R.M.; Sestelo, J.P.; Sarandeses, L.A. *Org. Biomol. Chem.* **2004**, *2*, 3584–3587.

712. Sagiyan, A.S.; Avetisyan, A.E.; Djamgaryan, S.M.; Djilavyan, L.R.; Gyulumyan, E.A.; Grigoryan, S.K.; Kuz'mina, N.A.; Orlova, S.A.; Ikonnikov, N.S.; Larichev, V.S.; Tararov, V.I.; Belekon, Y.N. *Russ. Chem. Bull.* **1997**, *46*, 483–486.

713. Belekon, Y.N.; Sagyan, A.S.; Djamgaryan, S.A.; Bakhmutov, V.I.; Vitt, S.V.; Batsanov, A.B.; Struchkov, Y.T.; Belikov, V.M. *J. Chem. Soc., Perkin Trans. 1* **1990**, 2301–2310.

714. Bull, S.D.; Davies, S.G.; O'Shea, M.D. *J. Chem. Soc., Perkin Trans. 1* **1998**, 3657–3658.

715. Bull, S.D.; Davies, S.G.; Garner, A.C.; O'Shea, M.D. *J. Chem. Soc., Perkin Trans. 1* **2001**, 3281–3287.

716. Navarre, L.; Darses, S.; Genet, J.-P. *Angew. Chem., Int. Ed.* **2004**, *43*, 719–723.

717. Easton, C.J. *Chem. Rev.* **1997**, *97*, 53–82.

718. Orlinki, R.; Stankiewicz, T. *Tetrahedron Lett.* **1988**, *29*, 1601–1602.

719. Yim, A.-M.; Vidal, Y.; Viallefont, P.; Martinez, J. *Tetrahedron Lett.* **1999**, *40*, 4535–4538.

720. Renaud, P.; Stojanovic, A. *Tetrahedron Lett.* **1996**, *37*, 2569–2572.

721. Gasanov, R.G.; Il'inskaya, L.V.; Misharin, M.A.; Maleev, V.I.; Raevski, N.I.; Ikonnikov, N.S.; Orlova, S.A.; Kuzmina, N.A.; Belokon, Y.N. *J. Chem. Soc., Perkin Trans. 1* **1994**, 3343–3348.

722. Yim, A.-M.; Vidal, Y.; Viallefont, P.; Martinez, J. *Tetrahedron: Asymmetry* **2002**, *13*, 503–610.

723. Goodall, K.; Parsons, A.F. *J. Chem. Soc., Perkin Trans. 1* **1994**, 3257–3259.

724. Baker, S.R.; Parsons, A.F.; Wilson, M. *Tetrahedron Lett.* **1998**, *39*, 2815–2818.

725. Goodall, K.; Parsons, A.F. *Tetrahedron Lett.* **199?**, *38*, 491–494.

726. Axon, J.R.; Beckwith, A.L.J. *J. Chem. Soc. Chem. Commun.* **1995**, 549–550.

727. Pyne, S.G.; Schafer, K. *Tetrahedron* **1998**, *5*, 5709–5720.

728. Herpin, T.F.; Motherwell, W.B.; Weibel, J.-M. *Chem. Commun.* **1997**, 923–924.

729. Kessler, H.; Wittmann, V.; Köck, M.; Kottenhahn, M. *Angew. Chem., Int. Ed. Engl.* **1992**, *31*, 902–904.

730. Kabat, M.M. *Tetrahedron Lett.* **2001**, *42*, 7521–7524.

731. Sibi, M.P.; Asano, Y.; Sausker, J.B. *Angew. Chem. Int. Ed.* **2001**, *40*, 1293–1296.

732. He, L.; Srikanth, G.S.C.; Castle, S.L. *J. Org. Chem.* **2005**, *70*, 8140–8147.

733. Kroona, H.B.; Peterson, N.L.; Koerner, J.F.; Johnson, R.L. *J. Med. Chem.* **1991**, *34*, 1692–1699.

734. McCord, T.J.; Foyt, D.C.; Kirkpatrick, J.L.; Davis, A.L. *J. Med. Chem.* **1967**, *10*, 353–355.

735. Horikawa, H.; Nishitani, T.; Iwasaki, T.; Inoue, I. *Tetrahedron Lett.* **1983**, *24*, 2193–2194.

736. Buñuel, E.; Cataviela, C.; Dias-de-Villegas, M.D.; Jimenez, A.I. *Synlett* **1992**, 579–581.

737. Jiménez, J.M.; Ortuño, R.M. *Tetrahedron: Asymmetry* **1996**, *7*, 3203–3208.

738. Wakamiya, T.; Oda, Y.; Fujita, H.; Shiba, T. *Tetrahedron Lett.* **1986**, *27*, 2143–2144.

739. Jiménez, J.M.; Bourdelande, J.L.; Ortuño, R.M. *Tetrahedron* **1997**, *53*, 3777–3786.

740. Mapelli, C.; Elrod, L.F.; Holt, E.M.; Stammer, C.H. *Tetrahedron* **1989**, *45*, 4377–4382.

741. Clerici, F.; Gelmi, M.L.; Pocar, D. *J. Org. Chem.* **1999**, *64*, 726–730.

742. Bland, J.M.; Stammer, C.H.; Varughese, K.I. *J. Org. Chem.* **1984**, *49*, 1634.

743. King, S.W.; Riordan, J.M.; Holt, E.M.; Stammer, C.H. *J. Org. Chem.* **1982**, *47*, 3270–3273.

744. Slama, J.T.; Satsangi, R.K.; Simmons, A.; Lynch, V.; Bolger, R.E.; Suttie, J. *J. Med. Chem.* **1990**, *33*, 824–832.

745. Pages, R.A.; Burger, A. *J. Med. Chem.* **1966**, *9*, 766–768.

746. Arenal, I.; Bernabé, M.; Fernández-Alvarez, E.; Penadés, S. *Synthesis* **1985**, 773–775.

747. Blasco, J.; Cativiela, C.; Diaz de Villegas, M.D. *Synth. Commun.* **1987**, *17*, 1549–1557.

748. Cativiela, C.; Díaz-de-Villegas, M.; Jiménez, A.I. *Tetrahedron: Asymmetry* **1995**, *6*, 177–182.

749. Arenal, I.; Bernabé, M.; Fernández-Alvarez, E.; Izquierdo, M.L.; Penadés, S. *J. Heterocycl. Chem.* **1983**, *20*, 607–613.

750. Suzuki, M.; Orr, G.F.; Stammer, C.H. *Bioorganic Chem.* **1987**, *15*, 43–49.

751. Szakonyi, Z.; Fülöp, F.; Tourwé, D.; De Kimpe, N. *J. Org. Chem.* **2002**, *67*, 2192–2196.

752. Cativiela, C.; Díaz-de-Villegas, M.D.; García, J.I.; Jiménez, A.I. *Tetrahedron* **1997**, *53*, 4479–4486.

753. Bland, J.; Shah, A.; Bortolussi, A.; Stammer, C.H. *J. Org. Chem.* **1988**, *53*, 992–995.

754. Srivastava, V.P.; Roberts, M.; Holmes, T.; Stammer, C.H. *J. Org. Chem.* **1989**, *54*, 5866–5870.

755. El Abdioui, K.; Martinez, J.; Viallefont, P.; Vidal, Y. *Bull. Soc. Chim. Belg.* **1997**, *106*, 425–431.

756. Cativiela, C.; Díaz-de-Villegas, M.D.; Jiménez, A.I. *Tetrahedron* **1994**, *50*, 9157–9166.

757. Adams, L.A.; Aggarwal, V.K.; Bonnert, R.V.; Bressel, B.; Cox, R.J.; Shepherd, J.; de Vicente, J.; Walter, M.; Whittingham, W.G.; Winn, C.L. *J. Org. Chem.* **2003**, *68*, 9433–9440.

758. Aggarwal, V.K.; Alonso, E.; Fang, G.; Ferrara, M.; Hynd, G.; Porcelloni, M. *Angew. Chem., Int. Ed.* **2001**, *40*, 1433–1436.

759. Calmes, M.; Daunis, J.; Escale, F. *Tetrahedron: Asymmetry* **1996**, *7*, 395–396.

760. Kirihata, M.; Sakamoto, A.; Ichimoto, I.; Udea, H.; Honma, M. *Agric. Biol. Chem.* **1990**, *54*, 1845–1846.

761. Cativiela, C.; Díaz-de-Villegas, M.D.; Jiménez, A.I. *Tetrahedron* **1995**, *51*, 3025–3032.

762. Chinchilla, R.; Falvello, L.R.; Galindo, N.; Nájera, C. *Tetrahedron: Asymmetry* **1998**, *9*, 2223–2227.

763. Abellán, T.; Mancheño, B.; Nájera, C.; Sansano, J.M. *Tetrahedron* **2001**, *57*, 6627–6640.

764. Williams, R.M.; Fegley, G.J.; Gallegos, R.; Schaefer, F.; Pruess, D.L. *Tetrahedron* **1996**, *52*, 1149–1164.

765. Chinchilla, R.; Nájera, C. *Tetrahedron Lett.* **1993**, *34*, 5799–5802.

766. Alami, A.; Clames, M.; Daunis, J.; Escale, F.; Jacquier, R.; Roumestant, M.-L.; Viallefont, P. *Tetrahedron: Asymmetry* **1991**, *2*, 175–178.

767. Pirrung, M.C. *Tetrahedron Lett.* **1980**, *21*, 4577–4578.

768. Hughes, P.; Clardy, J. *J. Org. Chem.* **1988**, *53*, 4793–4796.

769. Esslinger, C.S.; Koch, H.P.; Kavanaugh, M.P.; Philips, D.P.; Chamberlin, A.R.; Thompson, C.M.; Bridges, R.J. *Biorg. Med. Chem. Lett.* **1998**, *8*, 3101–3106.

770. Avenoza, A.; Busto, J.H.; Canal, N.; Peregrina, J.M. *J. Org. Chem.* **2005**, *70*, 330–333.

771. Abdallah, M.A.; Albar, H.A.; Shawali, A.S. *J. Chem. Res. S*, **1993**, *182–183*.

772. Pyne, S.G.; Schafer, K.; Skelton, B.W.; White, A.H. *Chem. Commun.* **1997**, 2267–2268; **1998**, 1607.

773. Avenoza, A.; Cativiela, C.; Fernández-Recio, M.A.; Peregrina, J.M. *Tetrahedron: Asymmetry* **1996**, *7*, 721–728.

774. Avenoza, A.; Cativiela, C.; Fernández-Recio, M.A.; Peregrina, J.M. *Synlett* **1995**, 891–892.

775. Avenoza, A.; Cativiela, C.; Peregrina, J.M. *Tetrahedron* **1994**, *50*, 10021–10028.

776. Cativiela, C.; Díaz-de-Villegas, M.D.; Avenoza, A.; Peregrina, J.M. *Tetrahedron* **1993**, *49*, 10987–10996.

777. Avenoza, A.; Campos, P.J.; Cativiela, C.; Peregrina, J.M.; Rodríguez, M.A. *Tetrahedron* **1999**, *55*, 1399–1406.

778. Avenoza, A.; Busto, J.H.; Cativiela, C.; Peregrina, J.M.; Rodríguez, F. *Tetrahedron* **1998**, *54*, 11659–11674.

779. Pyne, S.G.; Safaei-G, J.; Hockless, D.C.R.; Skelton, B.W.; Sobolev, A.N.; White, A.H. *Tetrahedron* **1994**, *50*, 941–956.

780. Pyne, S.; Safaei-G., J. *J. Chem. Res. S* **1996**, 160–161.

781. Horikawa, H.; Nishitani, T.; Iwasaki, T.; Mushika, Y.; Inoue, I.; Miyoshi, M. *Tetrahedron Lett.* **1980**, *21*, 4101–4104.

782. Bueno, M.P.; Cativiela, C.; Finol, C.; Mayoral, J.A.; Jaime, C. *Can. J. Chem.* **1986**, *65*, 2182–2186.

783. Cativiela, C.; Fraile, J.M.; Mayoral, J.A. *Bull. Chem. Soc. Jpn.* **1990**, *63*, 2456–2457.

784. Cativiela, C.; Mayoral, J.A.; Avenoza, A.; Gonzalez, M.; Roy, M.A. *Synthesis* **1990**, 1114–1116.

785. Clerici, F.; Gelmi, L.; Gambini, A. *J. Org. Chem.* **2001**, *66*, 4941–4944.

786. Clerici, F.; Gelmi, M.L.; Pellegrino, S.; Pilati, T. *J. Org. Chem.* **2003**, *68*, 5286–5291.

787. Cativiela, C.; López, P.; Mayoral, J.A. *Tetrahedron: Asymmetry* **1990**, *1*, 379–388.

788. Cativiela, C.; López, P.; Mayoral, J.A. *Tetrahedron: Asymmetry* **1990**, *1*, 61–64.

789. Cativiela, C.; López, P.; Mayoral, J.A. *Tetrahedron: Asymmetry* **1991**, *2*, 1295–1304.

790. Cativiela, C.; López, P.; Mayoral, J.A. *Tetrahedron: Asymmetry* **1991**, *2*, 441–450.

791. Abbiati, G.; Clerici, F.; Gelmi, M.L.; Gambini, A.; Pilati, T. *J. Org. Chem.* **2001**, *66*, 6299–6304.

792. Buñuel, E.; Cativiela, C.; Diaz-de-Villegas, M.D.; Garcîa, J.I. *Tetrahedron: Asymmetry* **1994**, *5*, 759–766.

793. Buñuel, E.; Cativiela, C.; Diaz-de-Villegas, M.D. *Tetrahedron: Asymmetry* **1994**, *5*, 157–160.

794. Chinchilla, R.; Falvello, L.R.; Galindo, N.; Nájera, C. *Tetrahedron: Asymmetry* **1999**, *10*, 821–825.

795. Daumas, M.; Vo-Quang, L.; Vo-Quang, Y.; Le Goffic, F. *Tetrahedron Lett.* **1989**, *30*, 5121–5124.

796. Nunami, K.-I.; Suzuki, M.; Yoneda, N. *Synthesis* **1978**, 840–841.

797. Hoppe, I.; Schöllkopf, U. *Synthesis* **1981**, 646–647.

798. Baldwin, J.E.; Haber, S.B.; Hoskins, C.; Kruse, L.I. *J. Org. Chem.* **1977**, *42*, 1239–1241.

799. Suzuki, M.; Nunami, K.-I.; Yoneda, N. *J. Chem. Soc., Chem. Commun.* **1978**, 270–271.

800. Alexander, P.A; Marsden, S.P.; Muñoz Subtil, D.M.; Reader, J.C. *Org. Lett.* **2005**, *7*, 5433–5436.

801. Duhamel, L.; Duhamel, P.; Fourquay, S.; Eddine, J.J.; Peschard, O.; Plaquevent, J.-C.; Ravard, A.; Solliard, R.; Valnot, J.-Y.; Vincens, H. *Tetrahedron* **1988**, *44*, 5495–5506.

802. Gibson, S.E.; Guillo, N.; Tozer, M.J. *Chem. Commun.* **1997**, 637–638.

803. Elliott, J.M.; Cascieri, M.A.; Chicchi, G.; Davies, S.; Kelleher, F.J.; Kurtz, M.; Ladduwahetty, T.; Lewis, R.T.; MacLeod, A.M.; Merchant, K.J.; Sadowski, S.; Stevenson, G.I. *Biorg. Med. Chem. Lett.* **1998**, *8*, 1845–1850.

804. Ellinger, A.; Flamand, C. *Ber.* **1907**, *40*, 3029–3033.

805. Dakin, H.D. *J. Biol. Chem.* **1911**, *9*, 151–160.

806. Tikkoji, T. *Biochem. Z.* **1911**, *35*, 57–87.

807. Funk, C. *J. Chem. Soc.* **1911**, 554–557.

808. Ellinger, A.; Matsuoka, Z. *Z. Phys. Chem.* **1914**, *91*, 45–57.

809. Barger, G.; Ewins, A.J. *Biochem. J.* **1917**, *11*, 58–63.

810. Harington, C.R.; McCartney, W. *Biochem. J.* **1927**, *21*, 852–856.

811. Schiemann, G.; Roselius, W. *Chem. Ber.* **1932**, *65*, 1439–1442.

812. Deulofeu, V. *Chem. Ber.* **1936**, *69*, 2456–2459.

813. Barger, G.; Easson, P.T. *J. Chem. Soc.* **1938**, 2100–2104.

814. Neuberger, A. *Biochem. J.* **1948**, *43*, 599–605.

815. Elliott, D.F.; Fuller, A.T.; Harington, C.R. *J. Chem. Soc.* **1948**, 85–89.

816. Elliott, D.F.; Harington, C. *J. Chem. Soc.* **1949**, 1374–1378.

817. Clemo, G.R.; Duxbury, F.K. *J. Chem. Soc.* **1950**, 1795–1800.

818. Sealock, R.R.; Speeter, M.E.; Schweet, R.S. *J. Am. Chem. Soc.* **1951**, *73*, 5386–5388.

819. Bernhard, S.A. *J. Am. Chem. Soc.* **1952**, *74*, 4947–4948.

820. Barry, J.M. *J. Biol. Chem.* **1952**, *195*, 795–803.

821. Gillespie, H.B.; Knox, W.E.; Aspen, A. *Biochem. Prep.* **1953**, *3*, 79–83.

822. Schneider, G. *Biochem. Z.* **1958**, *330*, 428–432.

823. Schlögl, K. *Monatsh. Clem.* **1957**, *88*, 601–621.

824. Eisler, K.; Rudinger, J.; Sorm, F. *Collect. Czech. Chem. Commun.* **1966**, *31*, 4563–4580.

825. Filler, R.; Ayyangar, N.R.; Gustowski, W.; Kang, H.H. *J. Org. Chem.* **1969**, *34*, 534–538.

826. Matsuura, T.; Nagamachi, T.; Nishinaga, A. *Chem. Pharm. Bull.* **1969**, *17*, 2176–2177.

827. Matsumura, K.; Kasai, T.; Tashiro, H. *Bull. Chem. Soc. Jpn.* **1969**, *42*, 1741–1743.

828. Morgenstern, A.P.; Schuijt, C.; Nauta, W.Th. *J. Chem. Soc. C* **1971**, 3706–3712.

829. Wightman, R.H.; Stauton, J.; Battersby, A.R.; Hanson, K.R. *J. Chem. Soc., Perkin Trans. I* **1972**, 2355–2364.

830. Kirby, G.W.; Michael, J. *J. Chem. Soc., Perkin Trans.* 1 1973, 115–120.

831. Saxena, A.K.; Jain, P.C.; Anand, N. *Ind. J. Chem.* **1975**, *13*, 230–237.

832. Brunet, J.C.; Cuingnet, E.; Gras, H.; Marcincal, P.; Mocz, A.; Sergheraert, C.; Tartar, A. *J. Organometal. Chem.* **1981**, *216*, 73–77.

833. Folkers, K.; Kubiak, T.; Stepinski, J. *Int. J. Pept. Prot. Res.* **1984**, *24*, 197–200.

834. Egusa, S.; Sisido, M.; Imanishi, Y. *Macromolecules* **1985**, *18*, 882–889.

835. López-Arbeloa, F.; Goedeweeck, R.; Ruttens, F.; De Schryver, F.C.; Sisido, M. *J. Am. Chem. Soc.* **1987**, *109*, 3068–3076.

836. Bovy, P.R.; Getman, D.P.; Matsoukas, J.M.; Moore, G.J. *Biochim. Biophys. Acta* **1991**, *1079*, 23–28.

837. Krause, H.-W.; Kreuzfeld, H.-J.; Döbler, C. *Tetrahedron: Asymmetry* **1992**, *3*, 555–566.

838. Bautista, F.M.; Campelo, J.M.; García, A.; Luna, D.; Marinase, J.M. *Amino Acids* **1992**, *2*, 87–95.

839. Taku, K.; Sasaki, H.; Kimura, S.; Imanishi, Y. *Amino Acids* **1994**, *7*, 311–316.

840. Döbler, Chr.; Kreuzfeld, H.-J.; Fischer, Chr.; Michalik, M. *Amino Acids* **1999**, *16*, 391–401.

841. Döbler, Chr.; Kreuzfeld, H.-J.; Fischer, Chr.; Michalik, M. *Amino Acids* **2002**, *22*, 325–331.

842. Coutts, R.T.; Malicky, J.L. *Can. J. Chem.* **1974**, *52*, 390–394.

843. Wang, T.S.T.; Vida, J.A. *J. Med. Chem.* **1974**, *17*, 1120–1122.

844. Parnes, H.; Shelton, E.J. *J. Lab. Compds. Radiopharm.* **1984**, *21*, 263–284.

845. Ljungqvist, A.; Bowers, C.Y.; Folkers, K. *Int. J. Pept. Prot. Res.* **1993**, *41*, 427–432.

846. Maekawa, K.; Kubo, K.; Igarashi, T.; Sakurai, T. *Tetrahedron* **2005**, *61*, 11211–11224.

847. Elks, J.; Elliott, D.F.; Hems, B.A. *J. Chem. Soc.* **1944**, 629–632.

848. Nestor J.J.; Jr., Ho, T.L.; Simpson, R.A.; Horner, B.L.; Jones, G.H.; McRae, G.I.; Vickery, B.H. *J. Med. Chem.* **1982**, *25*, 795–801.

849. Majima, R. *Ber.* **1922**, *55*, 3859–3865.

850. Robson, W. *J. Biol. Chem.* **1924–25**, *62*, 495–514.

851. Boyd, W.J.; Robson, W. *Biochem. J.* **1935**, *29*, 2256–2258.

852. Du Vigneaud, V.; McKennis H.; Jr., Simmonds, S.; Dittmer, K.; Brown, G.B. *J. Biol. Chem.* **1945**, *159*, 385–394.

853. Harvey, D.G.; Robson, W. *J. Chem. Soc.* **1938**, 97–101.

854. See Reference 847.

855. Bond, H.W. *J. Biol. Chem.* **1948**, *175*, 531–534.

856. Tolman, V.; Hanus, J.; Veres, K. *J. Lab. Compds. Radiopharm.* **1968**, *4*, 243–247.

857. Langemann, A.; Scheer, M. *Helv. Chim. Acta* **1969**, *52*, 1095–1097.

858. Lee, F.G.H.; Dickson, D.E. *J. Med. Chem.* **1971**, *14*, 266–268.

859. Sethi, M.L.; Rao, G.S.; Kapadia, G.J. *J. Pharm. Sci.* **1973**, *62*, 1802–1806.

860. Sakami, W.; Wilson, D.W. *J. Biol. Chem.* **1944**, *154*, 215–222.

861. Tsou, K.C.; Su, H.C.F.; Turner, R.B.; Mirachi, U. *J. Med. Chem.* **1966**, *9*, 57–60.

862. Schmidt, U.; Wild, J. *Angew. Chem., Int. Ed. Engl.* **1984**, *23*, 991–993.

863. Shin, C.-g.; Obara, T.; Segami, S.; Yonezawa, Y. *Tetrahedron Lett.* **1987**, *28*, 3827–3830.

864. Shin, C.-g.; Yonezawa, Y.; Obara, T.; Nishio, H. *Bull. Chem. Soc. Jpn.* **1988**, *61*, 885–891.

865. Schmidt, U.; Meyer, R.; Leitenberger, V.; Stäbler, F.; Lieberknecht, A. *Synthesis* **1991**, 409–413.

866. Schmidt, U.; Kleefeldt, A.; Mangold, R. *J. Chem. Soc., Chem. Commun.* **1992**, 1687–1689.

867. Kreuzfeld, H.-J.; Döbler, C.; Krause, H.W.; Facklam, C. *Tetrahedron: Asymmetry* **1993**, *4*, 2047–2051.

868. Holcomb, R.C.; Schow, S.; Ayral-Kaloustian, S.; Powell, D. *Tetrahedron Lett.* **1994**, *35*, 7005–7008.

869. Dumas, J.-P.; Germanas, J.P. *Tetrahedron Lett.* **1994**, *35*, 1493–1496.

870. Hansen, M.M.; Bertsch, C.F.; Harkness, A.R.; Huff, B.E.; Hutchison, D.R.; Khau, V.V.; LeTourneau, M.E.; Martinelli, M.J.; Misner, J.W.; Peterson, B.C.; Rieck, J.A.; Sullivan, K.A.; Wright, I.G. *J. Org. Chem.* **1998**, *63*, 775–785.

871. Moglioni, A.G.; García-Expósito, E.; Moltrasio, G.Y.; Ortuño, R.M. *Tetrahedron Lett.* **1998**, *39*, 3593–3596.

872. Park, K.C.; Yoshino, K.; Tomiyasu, H. *Synthesis* **1999**, 2041–2044.

873. Döbler, Chr.; Kreuzfeld, H.-J.; Michalik, M. *Amino Acids* **1999**, *16*, 21–27.

874. Moglioni, A.G.; García-Expósito, E.; Aguado, G.P.; Parella, T.; Branchadell, V.; Moltrasio, G.Y.; Ortuño, R.M. *J. Org. Chem.* **2000**, *65*, 3934–3940.

875. Nicolaou, K.C.; Murphy, F.; Barluenga, S.; Ohshima, T.; Wei, H.; Xu, J.; Gray, D.L.F.; Baudoin, O. *J. Am. Chem. Soc.* **2000**, *122*, 3830–3838.

876. Brown, J.A. *Tetrahedron Lett.* **2000**, *41*, 1623–1626.

877. Wang, W.; Yang, J.; Ying, J.; Xiong, C.; Zhang, J.; Cai, C.; Hruby, V.J. *J. Org. Chem.* **2002**, *67*, 6353–6360.

878. Adamczyk, M.; Akireddy, S.R.; Reddy, R.E. *Tetrahedron* **2002**, *58*, 6951–6963.

879. Wang, W.; Xiong, C.; Zhang, J.; Hruby, V.J. *Tetrahedron* **2002**, *58*, 3101–3110.

880. Aguado, G.P.; Moglioni, A.G.; Ortuño, R.M. *Tetrahedron: Asymmetry* **2003**, *14*, 217–223.

881. Wolan, A.; Laczynska, A.; Rafinski, Z.; Zaidlewicz, M. *Lett. Org. Chem.* **2004**, *1*, 238–245.

882. Krosigk, U.; Benner, S.A. *Helv. Chim. Acta* **2004**, *87*, 1299–1324.

883. Bentley, D.J.; Moody, C.J. *Org. Biomol. Chem.* **2004**, *2*, 3545–3547.

884. Chen, H.; Luzy, J.-P.; Garbay, C. *Tetrahedron Lett.* **2005**, *46*, 3319–3322.

885. Besong, G.E.; Bostock, J.M.; Stubbings, W.; Chopra, I.; Roper, D.I.; Lloyd, A.J.; Fishwick, C.W.G.; Johnson, A.P. *Angew. Chem., Int. Ed.* **2005**, *44*, 6403–6406.

886. Lambooy, J.P. *J. Am. Chem. Soc.* **1954**, *76*, 133–138.

887. Slater, G.; Somerville, A.W. *Tetrahedron* **1967**, *23*, 2823–2828.

888. Larsen, P.O.; Wieczorkowska, E. *Acta Chem. Scand. B* **1977**, *31*, 109–113.

889. Beecher, J.E.; Tirrell, D.A. *Tetrahedron Lett.* **1998**, *39*, 3927–3930.

890. Subramanian, P.K.; Kalvin, D.M.; Ramalingam, K.; Woodard, R.W. *J. Org. Chem.* **1989**, *54*, 270–276.

891. Achiwa, K. *Tetrahedron Lett.* **1978**, *29*, 2583–2584.

892. Kreuzfeld, H.-J.; Döbler, Chr.; Fischer, Chr.; Baumann, W. *Amino Acids* **1999**, *16*, 369–375.

893. Rossen, K.; Weissman, S.A.; Sager, J.; Reamer, R.A.; Askin, D.; Volante, R.P.; Reider, P.J. *Tetrahedron Lett.* **1995**, *36*, 6419–6422.

894. Guillen, F.; Fiaud, J.-C. *Tetrahedron Lett.* **1999**, *40*, 2939–2942.

895. Behringer, H.; Fackler, E. *Ann.* **1949**, *564*, 73–78.

896. Murakoshi, I.; Ohmiya, S.; Haginiwa, J. *Chem. Pharm. Bull.* **1972**, *20*, 609–611.

897. Meyer, R.; Schöllkopf, U.; Böhme, P. *Liebigs Ann. Chem.* **1977**, 1183–1193.

898. Hantke, K.; Braun, V. *Eur. J. Biochem.* **1973**, *34*, 284–296.

899. Duke, C.C.; MacLeod, J.K.; Summons, R.E.; Letham, D.S.; Parker, C.W. *Aust. J. Chem.* **1978**, *31*, 1291–1301.

900. Rakhshinda, M.A.; Khan, N.H. *Synth. Commun.* **1978**, *8*, 497–510.

901. Rakhshinda, M.A.; Khan, N.H. *Synth. Commun.* **1979**, *9*, 351–361

902. Cativiela, C.; Diaz-de-Villegas, M.D.; Galvez, J.A. *Can. J. Chem.* **1992**, *70*, 2325–2328.

903. Bergmann, E.D.; Chun-Hsu, L. *Synthesis* **1973**, 44–46.

904. Switzer, F.L.; Van Halbeeck, H.; Holt, E.M.; Stammer, C.H. *Tetrahedron* **1989**, *45*, 6091–6100.

905. Fernández, D.; de Frutos, P.; Marco, J.L.; Fernández-Alvarez, E.; Bernabé, M. *Tetrahedron Lett.* **1989**, *30*, 3101–3104.

906. Suzuki, M.; Gooch, E.E.; Stammer, C.H. *Tetrahedron Lett.* **1983**, *24*, 3839–3840.

907. Lalitha, K.; Iyengar, D.S.; Bhalerao, U.T. *J. Org. Chem.* **1989**, *54*, 1771–1773.

908. Smith, E.C.R.; McQuaid, L.A.; Paschal, J.W.; DeHoniesto, J. *J. Org. Chem.* **1990**, *55*, 4472–4474.

909. Buñuel, E.; Bull, S.D.; Davies, S.G.; Garner, A.C.; Savory, E.D.; Smith, A.D.; Vickers, R.J.; Watkin, D.J. *Org. Biomol. Chem.* **2003**, *1*, 2531–2542.

910. Schöberl, A. *Chem. Ber.* **1947**, *80*, 379–391.

5 | Synthesis of Optically Active α-Amino Acids: Extension of Achiral Methods—Amination and Carboxylation Reactions

5.1 Introduction

Many of the first methods developed for the asymmetric synthesis of amino acids were, not surprisingly, asymmetric versions of previously established methods of racemic synthesis. The results of early asymmetric syntheses must be interpreted carefully as the extent of asymmetric induction is often calculated by the optical rotation of a final product. Unintentional resolution during purification and crystallization can mask the extent of asymmetric induction during the actual asymmetry-inducing reaction, and the optical rotation measurement itself is quite prone to errors. The emergence of asymmetric amino acid syntheses corresponds with the evolution of asymmetric synthesis in general. For example, a modified Strecker synthesis of Ala was the first asymmetric non-enzymatic synthesis which did not involve catalytic hydrogenation in the asymmetry-inducing reaction (1, 2).

5.2 Asymmetric Aminocarboxylation Reactions

5.2.1 Asymmetric Strecker and Bücherer Reactions

Several different approaches have been taken in the development of asymmetric Strecker (see Section 2.2.1) and Bücherer (see Section 2.2.2) syntheses. Chiral amines, chiral aldehydes, and chiral catalysts have all been applied to these traditional routes, although most commonly with the Strecker conditions. An advantage of chiral auxiliaries or chiral substrates over chiral catalysts is that the intermediate diastereomer with

the desired configuration can often be purified from undesired stereoisomers, although the advent of chiral purification columns is making enantiomer purification more viable. A brief review of aminocarboxylation reactions was published in 1997 (3), with a highlight of recent developments in catalytic asymmetric Strecker-type reactions in 2001 (4). An entire issue of *Tetrahedron* in 2004 was devoted to the synthesis and applications of non-racemic cyanohydrins and α-aminonitriles, the intermediates in a Strecker synthesis (5), while a 2004 review discussed chemically catalyzed asymmetric cyanohydrin syntheses (6). The biology, preparations, and synthetic applications of cyanohydrins were extensively reviewed in 1999 (7), with the synthesis and applications of non-racemic cyanohydrins further reviewed in 2003 (8). Catalytic enantioselective cyanations of imines were included in a 1999 review of additions to imines (9), while a 2003 *Chemical Reviews* article extensively covered catalytic enantioselective Strecker reactions (10). Many Strecker reactions employ trimethylsilyl cyanide (TMSCN) in the presence of a Lewis acid catalyst; a report in 2005 found that racemic Strecker reactions with TMSCN proceeded efficiently in acetonitrile without any catalyst. It remains to be seen if this discovery will have any use in asymmetric reactions as well (11).

5.2.1a Chiral Amine Component

A number of chiral amines have been used to introduce asymmetry during the Strecker and Bücherer reactions (see Scheme 5.1). The intermediate α-aminonitrile is optically unstable and epimerizes over time (12), and some reactions make use of the thermodynamic equilibrated product while others transform the initial kinetic

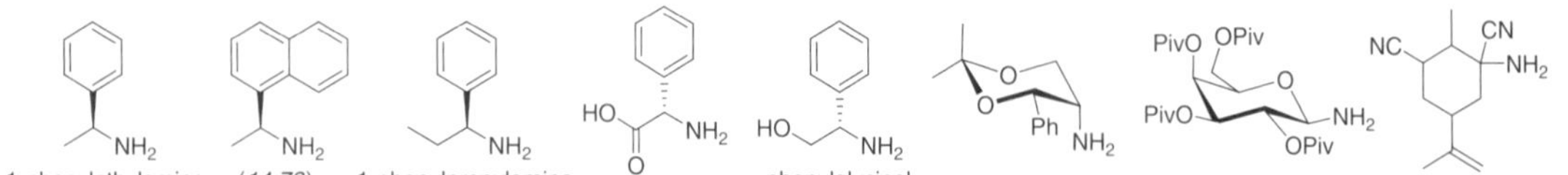

1-phenylethylamine
(1,2,14–16,19–39)

(14,78)

1-phenylpropylamine
(14)

phenylglycine
(14,60)

phenylglycinol
(40–56)

Weinges amine
(66–74,78)

Kunz galactosamine
(75–78)

ketone-derived
auxiliary/catalyst
(139,140)

Scheme 5.1 Chiral Amines Used as Auxiliaries in Asymmetric Strecker Synthesis.

adduct. This equilibration has been examined with α-alkylbenzylamines as the chiral auxiliary (12, 13). The first asymmetric Strecker synthesis treated acetaldehyde-derived cyanohydrin with D-(−)-α-methylbenzylamine (1-phenylethylamine). The chiral benzyl protecting group was removed by hydrogenolysis after acidic hydrolysis of the intermediate, producing Ala of approximately 75% ee, which improved to 86–99% ee after 1 recrystallization (1, 2). Other α-alkylbenzylamines, α-(1-naphthyl)ethylamine and phenylglycine, were examined for the synthesis of Ala, Val, Nva, and Leu, giving inductions of 22–52% de (14).

A report of the alternate Strecker route soon followed, with HCN added to optically active Schiff bases prepared from 1-phenylethylamine and aliphatic aldehydes (RCHO, R = nPr, nBu, iBu) (15). Impressively high optical purities (>98%) were reported, but a critical examination of the results (16) demonstrated the presence of some secondary source of enantioenrichment. If racemic 1-phenylethylamine was used, the crude product was found to possess only 60% de. A later study showed that the levels of diastereoselectivity for the addition of HCN to Schiff bases of several chiral amines (22–58% ee for the Ala, Abu, Val, and Leu products) (17, 18) were similar to the results for the addition of chiral amines to cyanohydrins. Significantly better diastereoselectivities were obtained from (R)-1-phenylethylamine by using a CN-modified hemincopolymer as the cyanide source. The deprotected amino acids (Ala, Abu, Phe, Val) were produced in 34–57% overall yield and 91–99% ee without any intermediate crystallizations (19). 1-Phenylethylamine has also been applied to a cyanogen bromide version of the Strecker synthesis, though with poor diastereoselectivity (21–50% de) and yield (20–64%) (20). Iodine was identified as an effective and inexpensive catalyst for Strecker reactions in 2005, giving much better yields (68%) of racemic aminonitrile from acetophenone, trimethylaniline, and TMSCN than a number of Lewis acid catalysts (5–15%). A range of other aldehydes/ketones and amines also gave high yields, including reactions with (R)- or (S)-1-phenylethylamine. Four aldehydes reacted with 36–68% de (21).

1-Phenylethylamine has been used in the preparation of neopentylglycine (33% de) (22), tert-Leu (46% yield of the free amino acid, with >98% ee) (23), L-adamantylalanine (24), and β-fluoro-α-amino acids (65–80% de for the aminonitrile intermediate) (25). γ-Carboxy-Glu was prepared from di-tert-butyl

(formylmethyl)malonate (26). (S)-1-Phenylethylamine was combined with TMSCN and catalytic LiClO₄ to give both alkyl and aryl aminonitrile intermediates in 93–97% yield in just 5 min, with 55–70% de (27). The first total synthesis of (2S,3R,4S)-4-hydroxyisoleucine, an insulinotropic amino acid isolated from the seeds of fenogreek (traditionally used in northern African countries to lower blood sugar), employed a chiral aldehyde substrate (THF-protected 2-methyl-4-hydroxybutanal) along with a chiral amine component (1-phenylethylamine) (64% de) (28). In 2004, imines formed from (S) 1-(4-methoxyphenyl)ethylamine were reacted with TMSCN using LiClO₄ or BF₃.Et₂O as catalyst, with the aminonitrile precursor of phenylglycine formed in 97% yield with 60% de. The diastereomers were separated, and the auxiliary removed in 91% yield by CAN or DDQ oxidation, with hydrolysis providing Phg in 88% yield with >99% ee (29).

The Strecker synthesis has also been applied to the transformation of methyl ketones into α-methyl-α-amino acids by reaction of the ketones (MeCOCH₂R; R = Ph, 3,4-MeO-Ph, Bn) with 1-phenylethylamine and NaCN. Enantiomerically pure products were produced in 53–65% yield after recrystallization (30). Complete diastereoselectivity was reported in the synthesis of (R)-α-methyl-Phe (>98% ee, 64% overall yield), prepared by reacting the hydrochloride salt of (S)-phenylethylamine with NaCN and phenylacetone at room temperature (31). Mixtures of diastereomers were observed under other conditions. 1-Amino-2-hydroxycyclohexane-1-carboxylic acid was produced from 2-methoxycyclohexanone with 36–52% de at the α-center, and from 48% de trans to 50% de cis stereoselectivity at the β-center, depending on the solvent used (32). (S)-Phenylethylamine was also used for a large-scale preparation of 1-amino-2-[1-(carboxy)cyclopentane]acetic acid (cyclopentyl-L-Asp) via reaction of the imine derivative with TMSCN in the presence of Et₂AlCl (93% de, 96% yield) (33). A constrained glutamic acid analog, 1-amino-1-carboxy-2-carboxymethylcyclohexane, was derived from racemic ethyl 2-(2-oxocyclohex-1-yl)ethanoate, (R)-1-phenylethylamine, and TMSCN (34).

Four different 2-substituted cyclopentanones were reacted with (S)-phenylethylamine followed by TMSCN/ZnCl₂ and then hydrolysis. For the α-methyl- and α-ethyl 1-aminocyclopentane-1-carboxylic acid products, all four possible stereoisomers could be isolated, with the product distribution depending on solvent and reaction temperature. However, for α-isopropyl and α-*tert*-butyl products, only the *trans*-substituted products were formed (*35*). Cyclobutanone, 2-phenyl-cyclobutanone, and 2-isopropyl-cyclobutanone were reacted with NaCN and several different chiral amines in several different solvents; with (*R*)-(+)-1-phenylethyl-amine in DMSO or MeOH the *cis* isomers were almost exclusively obtained (e.g., 56:42:1:<1 ratio for isopro-pyl substituent in DMSO) (*36*). Similarly, both *cis*- and *trans*-2,4-methanovaline were obtained from racemic 2-methylcyclobutanone, via Strecker reaction with (*R*)-(+)-1-phenylethylamine and TMSCN in the pres-ence of ZnCl$_2$. With methanol as solvent the two *trans* isomers were obtained as the major products (37*t*:36*t*:11*c*:6*c* ratio), while with hexane as solvent one of the *cis* isomers became a major product (41*t*:15*t*:33*c*:11*c*) (*37*). The four diastereomers of 2-methoxy- and 2-hydroxy-1-amino-cyclopentanecarboxylic acid (2,4-ethanothreonine) were obtained by this route, with 41:22:29:8 ratio obtained in methanol and 10:0:61:29 in hexane (*38*). In contrast, both solvents gave predominantly *trans* isomers with 2-benzoylaminocy-clohexanone as substrate (51:25:24:0 and 52:31:17:0 ratios), leading to both enantiomers of *trans*-1, 2-diaminocyclohexane-1-carboxylic acid after hydrolysis/auxiliary removal. The *cis* isomer was obtained as the 2-benzoylamino derivative, with the benzoyl protecting group remaining intact under hydrolysis conditions (*39*).

(*S*)- or (*R*)-Phenylglycinol has been used as an auxil-iary for several Strecker and Bücherer reactions, with the auxiliary removed by oxidative cleavage with lead tetraacetate under essentially neutral conditions. A vari-ety of alkyl and aromatic aldehydes (RCHO; R = Me, *i*Pr, *t*Bu, Bn, 4-MeOPh, 4-F-Ph, 1-naphthyl, 2-thienyl) were reacted with the amine, KCN, and NaHSO$_3$. The aminonitrile intermediates equilibrated over time, with 60–72% de after 3 h. Hydrolysis, recrystallization, and deprotection gave the free amino acids in 16–62% yield with 79–100% ee (*40*). The chiral amine has also been used to prepare imines from six different aldehydes (RCHO; R = Ph, 4-MePh, 4-MeOPh, Bn, *i*Pr, *t*Bu), which were then treated with TMSCN to give the amino-nitrile intermediates in 87–95% yield and 8–80% de (*41, 42*). A substituted phenylglycine present in vanco-mycin was prepared by this route (*43, 44*), as were a series of eight other arylglycines (82–88% yield of ami-nonitrile, 60–68% de) (*45*). Other amino acids synthe-sized using phenylglycinol as the auxiliary include 2-amino-4-methyloctanoic acid (*46*), 2′-carboxy-Phg (*47*), 3′,5′-dichloro-4′-methoxy-Phg (*48*), a biphenyl-linked bis(amino acid) component of vancomycin (*49*), 2′-methoxy-6′-(2,5-dimethylpyrrol-1-yl)-Phe (an inter-mediate in the synthesis of benzolactam-V8 protein kinase C inhibitor, 60% de) (*50*), all 16 stereoisomers of 2-(2′-carboxy-3′-phenylcyclopropyl)glycine (*51, 52*), the constrained Glu analog 2-(4′-carboxycubyl)glycine (*53*), and a series of α-substituted and α,γ-disubstituted Glu analogs (*54, 55*). Aryl methyl ketones were reacted with (*R*)-phenylglycinol, followed by treatment with TMSCN to provide α,α-disubstituted amino acids with

33–76% de (*56*). Both *O*-methyl phenylglycinol and 1-phenylethylamine were used in the synthesis of *allo*-coronamic and *allo*-norcoronamic acids, using 2-ethyl- and 2-methyl-1-methoxycyclopropan-1-ol as the ketone equivalents (*57*). An *N*,β,β-trimethyl-Trp residue found in the cytotoxic marine peptide (−)-hemiasterlin was constructed using an imine formed from (*R*)-2-phenylglycinol, which was reacted with Bu$_3$SnCN and scandium triflate catalyst. The aminonitrile was obtained in 94% yield with 78% de. Nitrile hydrolysis was unsuc-cessful; instead the nitrile was converted into an amide, with hydrogenation removing the chiral auxiliary (*58*).

Alternatively, phenylglycinol was condensed with aldehydes to give a chiral oxazolidine (RCHO, R = Me, Et, *i*Pr, *n*Bu, *i*Bu, PhCH$_2$CH$_2$). Ring opening with diethyl-aluminum cyanide followed by hydrolysis introduced the carboxyl group with 54–86% de (63–80% yield), which is comparable to the acyclic reactions (*59*). (*R*)-Phenylglycine amide (Phg-NH$_2$) induced high diastereoselectivity for the addition of NaCN to the imine formed from pivalaldehyde; the major diastere-omer selectively precipitated to provide a crystallization-induced asymmetric transformation, with up to 93% yields of aminonitrile with >98% de. Deprotection to (*S*)-Tle (>98% ee) proceeded in 73% yield. The amino-nitrile precursor of α-methyl-3′,4′-dimethoxy-Phe was also prepared with >98% de (*60*).

Other chiral amines have been employed. An intra-molecular chiral auxiliary/amine source was employed during a synthesis of α-methyl-Ser and α-methyl-Thr. The substrate α-hydroxyketones were esterified with the acid groups of the Val or Phe auxiliaries. Intramolecular Strecker reaction proceeded with high diastereoselectiv-ity (>98% de) at the α-center (*61, 62*). α-Benzylserine and α-carboxymethylserine were also prepared (*63*). A modification to this method oxidized the initial Strecker aminonitrile adduct with ozone, forming an imine that was readily hydrolyzed to release the free amino acid. A series of more complex amino aids were synthesized, including α-methyl-Ser, 2,4-diamino-2-methyl-5-hydroxypentanoic acid, 2-amino-2-(hydroxymethyl)-3-hydroxy-4-methylpentanoic acid, 1-amino-2-hydroxycyclopentane-1-carboxylic acid, and 1-amino-2-hydroxycyclohexane-1-carboxylic acid (*64*). Ephedrine was used as the amine source in a synthesis of *N*-methyl Phg, via the bisulfite-modified Strecker synthesis. One of the aminosulfonate intermediate diastereomers formed an oxazolidine derivative, allow-ing for easy resolution of the diastereomers (*65*).

Several chiral cyclic amines have been employed to induce asymmetry. Weinges et al. applied [4*S*,5*S*]-(+)-5-amino-2,2-dimethyl-4-phenyl-1,3-dioxane (*66*) to a number of Strecker syntheses. Reaction of this amine with methyl ketones and HCN gave α-aminonitrile pre-cursors of α-methyl-serine (*67*), α-methyl-hydroxyvaline (*67*), α-arylalanines (R = Ph, 4-MeO-Ph, 3,4-MeO-Ph (*67, 68*), α-alkylalanines (R = Me, Et, *n*Pr, *n*Bu, *n*Pent) (*69*), and α-cycloalkyl-alanines (*67*) with 38–83% yields and 22–100% de. Enantiomerically pure phenylglycine (44% yield (*70, 71*), anisylglycine (29% yield) (*70, 71*), and 2-(2-thienyl)- and 2-(3-thienyl)-glycines (72) were prepared by a similar route, although the diastereo-selectivity during the synthesis of the latter two compounds was only 40%. Both (2S,4S)- and (2S,4R)-5,5,5-trifluoroleucine were also prepared (73). The (S)-5-Phenylmorpholin-2-one provided aminonitriles with 38–88% de for a series of alkyl aldehydes when combined with CuCN; one example was elaborated into Val (74).

Some of the best asymmetric inductions during Strecker reactions have been obtained with a carbohydrate moiety acting as the stereodifferentiating amine auxiliary. Aldimines derived from O-pivaloylated D-galactosylamine and aromatic or aliphatic aldehydes (RCHO, R = iPr, tBu, Ph, 4-Me-Ph, 4-MeO-Ph, 4-Cl-Ph, 3-Cl-Ph, 4-F-Ph, 2-NO$_2$-Ph) reacted with TMSCN in the presence of Lewis acids to give the aminonitrile intermediates in >95% crude yield with 60–100% de; recrystallization followed by acid hydrolysis gave pure D-amino acids (75, 76). Direct reaction of the aldehyde, amine and NaCN gave reduced stereoselectivity. The stereoselectivity of the reaction could be reversed by using CHCl$_3$ as solvent instead of isopropanol or THF, giving the (S)-amino acids with high yield (>90%) but somewhat lower selectivity (50–80% de; R = iPr, tBu, PhCH$_2$CH$_2$, 4-Me-Ph, 4-F-Ph, 4-Cl-Ph, 3-Cl-Ph) (77).

Several of these chiral amine auxiliaries were compared for the preparation of 2,3-methanovaline using NaCN or TMSCN with sonication (78). Phenylethylamine gave a yield of 48% but with only 18% de, while the chiral galactosylamine of Kunz gave much better diastereoselectivity (>95:<5) but only 23% yield. The Weinges cyclic amine gave 68% yield and 74% de.

A significantly different approach was taken by Davis et al. (79, 80) by using a chiral sulfinimine. Reaction with Et$_2$AlCN in the presence of 2-propanol gave aminonitriles in 84–90% yield and 82–86% de (see Scheme 5.2), with diastereomerically pure products obtained after recrystallization. Hydrolysis with 6 N HCl simultaneously removed the chiral auxiliary and generated the carboxy group. This method was applied to the synthesis of (R)-4'-methoxy-3',5'-dihydroxy-Phg (the central amino acid of vancomycin), and related derivatives (81), and was also used to prepare β-fluoro-α-amino acids (82). Chiral sulfinimines derived from aliphatic aldehydes were successfully reacted with TMSCN in the presence of CsF to give the α-aminonitrile

intermediates, with the highest stereoselectivity (80->98% de) obtained with n-hexane as solvent. Only one example was hydrolyzed (with 6 N HCl) to give the amino acid (L-Leu) (83). 2-Aziridinesulfinimines also reacted with TMSCN, with the aziridine in the product opened by nucleophiles to give 2,3-diaminonitrile (83).

Sulfinimines derived from non-racemic α-hydroxyaldehydes led to β-hydroxy-α-amino acids (phenylserine and β-hydroxy-Leu), with the chirality of the sulfinyl group dominating the stereodifferentiation (74–87% de for mismatched pair, >96% de for matched pair) (84). A homochiral 3-methyl-4-benzyloxybutyraldehyde was combined with the chiral sulfinimine approach to produce (2S,4R)-5-hydroxy-Leu (85). Polyhydroxylated sulfinimines derived from protected 1,2-O-isopropylidene-L-threoses gave the α-aminonitriles with 66–82% de, with the stereoselectivity again controlled by the sulfinyl group rather than the aldehyde. Hydrolysis led to the lactones of 3,4,5-trihydroxy-2-aminopentanoic acid (86). Sulfinimines derived from ketones produced α-alkyl-α-aminonitriles in reasonable yields and 60–98% de from pure E-isomers, but with only 12–65% de if the sulfinimines were mixtures of (E)- and (Z)-isomers (87). The sulfinimine Strecker approach was summarized in a 2002 review of N-tert-butanesulfinyl imines (88), and in a 2004 review on the use of sulfinimines in asymmetric syntheses (89).

The (S)-1-amino-2-methoxymethylpyrrolidine (SAMP) hydrazine chiral auxiliary developed by Enders has been used to form hydrazones with alkyl and aryl aldehydes. Addition of TMSCN in the presence of TiCl$_4$ proceeded to give α-hydrazino nitriles in 75–93% yield with 88–91% de. The auxiliary was removed in 38–78% yield by N–N cleavage with magnesium monoperoxyphthalate and the cyano group hydrolyzed, giving amino acids with 94–97% ee (90).

5.2.1b Chiral Aldehyde Component

The second approach towards asymmetric Strecker syntheses requires the presence of a chiral center in the aldehyde component to induce stereoselectivity during formation of the α-center. Since the aldehyde chiral center remains in the product, each application is specific to the type of product being synthesized. This route was used to prepare 2-amino-3,4-dihydroxybutyric acid (91, 92), with the benzylamine Schiff base of D-glyceraldehyde giving the (2R,3S) product in

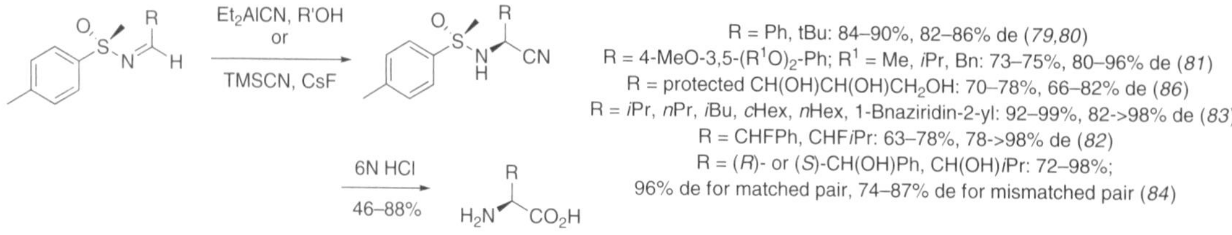

R = Ph, tBu: 84–90%, 82–86% de (79,80)
R = 4-MeO-3,5-(R^{1}O)$_2$-Ph; R^1 = Me, iPr, Bn: 73–75%, 80–96% de (81)
R = protected CH(OH)CH(OH)CH$_2$OH: 70–78%, 66–82% de (86)
R = iPr, nPr, iBu, cHex, nHex, 1-Bnaziridin-2-yl: 92–99%, 82->98% de (83)
R = CHFPh, CHF iPr: 63–78%, 78->98% de (82)
R = (R)- or (S)-CH(OH)Ph, CH(OH)iPr: 72–98%;
96% de for matched pair, 74–87% de for mismatched pair (84)

Scheme 5.2 Asymmetric Strecker Synthesis with Sulfinimine.

up to 80% de when treated with TMSCN. The (2S,3S) isomer was obtained in 70% de by reaction of D-glyceraldehyde with TMSCN, followed by displacement of the cyanohydrin hydroxyl with sodium azide. The two diastereomers were also obtained by using the aldehyde derived from lead tetraacetate cleavage of 1,2,5,6-diacetone mannitol (93). Another synthesis employed 2-O-benzyl-3-O-TBDMS-D-glyceraldehyde, benzylamine, and TMSCN, proceeding with 70% de. The TBDMS group of the adduct was removed, the N-benzyl amino group additionally protected with a methoxycarbonyl, and the hydroxy converted into a chloride or bromide. Treatment with base induced cyclization to give, after nitrile hydrolysis, 2,3-methanoserine (94).

Homochiral (2S)-[3,3,3-²H₃]isobutyraldehyde was used in a Strecker synthesis of labeled valine, but no stereoselectivity was observed at the α-center (95). (2R,3S)-Alloisoleucine was constructed by an asymmetric Strecker reaction of the chiral aldehyde derived from (S)-2-methyl-1-butanol, in combination with an enantiopure sulfinimine as the amine component. The aminonitrile was formed with 90% de (96). 5,5,5-Trifluoroleucine was obtained from a chiral aldehyde (97). Several syntheses of MeBmt have relied on the β-center chirality for a diastereoselective Strecker reaction, using either methylamine (98) or an intramolecular amino group (99) as the amine source. Modified polyoxins were obtained by reaction of uridine-5′-aldehyde with TMSCN and amines or amino acids (100). 3-Benzyl-4-formyl-1,3-thiazolidin-2-one (2-benzylamino-3-mercaptopropanal cyclized with a carbonyl group connecting the thiol and amine), was used for a Strecker reaction with benzylamine, NaHSO₃, and NaCN, producing a β-benzylaminohomocysteine derivative as an 11:1 mixture of isomers (101).

4-Oxoproline derivatives were employed as substrate for the preparation of 4-amino-4-carboxyprolines using the Bücherer procedure; the spirocyclic hydantoin intermediate was obtained with up to 92% de (74% yield) (102). A phosphonomethyl analog was synthesized in a similar fashion (103). Enzymatically resolved 3,4-dicarboxy-cyclopentanones were used as substrates for Strecker or Bücherer–Bergs reactions, resulting in 1-aminocyclopentane-1,3,4-tricarboxylic acids (104). In a similar manner, Bücherer aminocarboxylation of resolved 3-carboxycyclopentanone produced the *trans* product with 50% *trans* de, while the Strecker conditions gave a 1:1 mixture (105). (1S,2S)-1-Amino-2-methycyclopropanecarboxylic acid (norcoronamic acid) was prepared from (S)-2-methyl-3-hydroxypropanal; the initial acyclic aminonitrile adduct was obtained with only 10% de, but after conversion of the β-hydroxyl group to a chloro group, cyclization proceeded with 68% de (106, 107). Coronamic acid was also prepared (107). Chiral α-hydroxyketones induced significant diastereoselectivity in cyanosilylation reactions with TMSCN catalyzed by tetrabutylammonium cyanide, with 80–100% de for substituents larger than a methyl group (the adducts were not reacted further) (108).

5.2.1c Chiral Catalyst: Aminonitrile Synthesis

The use of chiral catalysts in enantioselective Strecker reactions was extensively reviewed in a 2003 *Chemical Reviews* article (10). Most catalysts involve a metal complex, and often only prepare the aminonitrile intermediate. A chiral Ti catalyst induced up to 59% ee in forming the aminonitrile precursor of α-Me-Phg from benzophenone and benzylamine (109). The complex formed from N-salicyl β-aminoalcohols and Ti(OiPr)₄ catalyzed aminonitrile formation from TMSCN and N-benzhydrylaldimines derived from aromatic and aliphatic aldehydes. Thirteen aryl aldehydes gave >96% conversion and 90–98% ee, while the aliphatic aldehydes cinnamaldehyde and pivalaldehyde reacted with 51–91% ee (110). With imines formed from benzylamine, the enantioselectivities were 39–81% ee (111). Another Ti-based catalyst, with a tripeptide Schiff base ligand, induced 85->99% ee for the addition of TMSCN to imines formed from aryl aldehydes or pivalaldehyde and benzhydrylamine (112). Parallel libraries were employed to identify tripeptide catalysts for Ti-induced addition of TMSCN to imines formed from benzhydrylamine and α,β-unsaturated aldehydes. The protected α-cyano amine precursors of a number of substituted vinylglycines were obtained with 76–97% ee. One example was converted into 3,4-dehydro-Val, with >99% ee (113). TMSCN was added to a benzhydrylamine imine of protected 3,4-dihydroxybenzaldehyde in the presence of Ti(OiPr)₄ catalyst and a Tle-Thr dipeptide derivative ligand, giving the cyano precursor of 3′-hydroxy-4′-methoxy-phenylglycine with 93% ee (796). The mechanism of peptide-based enantioselective Ti-catalyzed Strecker reactions was studied in 2001. The evidence presented supported a bifunctional ligand role, with the Ti coordinated to the Schiff base of the ligand, and the second amino acid residue associating with, and delivering, the HCN to the bound substrate (114).

Imines generated from aryl or alkyl aldehydes and 2-amino-3-methylphenol reacted with HCN in the presence of a chiral binaphthol Zr catalyst to give α-aminonitriles in 76–99% yield with 76–94% ee (14 examples). Three examples were converted into the amino acids D-Leu, D-homophenylalanine, and D-pipecolic acid (cyclized 2-amino-6-hydroxyhexanoic acid) (115). A Sc(BINOL)₂Li catalyst provided 55–95% ee for hydrocyanation of benzyl imines or aryl aldehydes and ketones by TMSCN. Aminonitrile precursors of Phg, 2-naphthyl-Gly, and α-methyl-Phg were prepared (116). An Al-based BINOL catalyst, with a TMSCN/HCN cyanide source and N-fluorenylimines, provided 70–95% ee for a number of aliphatic, α,β-unsaturated, aromatic and heteroaromatic aldehydes (117). The Al–BINOL type catalyst has been linked to a JandaJel resin to give an easily recovered catalyst, and still promoted Strecker-type reactions of aromatic imines with 83–87% ee (118). BINOL catalyst has also been employed with Et₂AlCN as the cyanide source, forming aminonitriles from N-benzhydryl imines of aryl aldehydes with 64–70% ee when 1.2 equiv of BINOL was employed. With 4.5 equiv of BINOL, a reversal in enantioselectivity was observed, giving the enantiomeric aminonitrile with 15% ee (*119*).

A chiral gadolinium-based catalyst with D-glucose-derived ligand was described in 2003; it induced 51–95% ee during reactions of ketone-derived N-diphenyl-phosphinoyl imines with TMSCN (11 examples). Two examples of the α-aminonitrile products were hydrolyzed to the α,α-disubstituted amino acid products (*120*). The new route was highlighted in a 2004 article (*121*). Improvements were described in 2004; addition of 2,6-dimethylphenol as a proton source greatly reduced reaction time and improved enantioselectivity. Seventeen aryl, heteroaryl, and alkyl ketimines were converted to the aminonitrile in 91–98% yield and with 69–99% ee. One product was elaborated to the cyclic amino acid found in the antidiabetic drug sorbinil, a 4-amino-6-fluorobenzopyran-1-carboxylic acid derivative (*122*). The Gd-based catalyst was also employed with a catalytic amount of TMSCN and stoichiometric amount of HCN, giving aminonitrile products from 6 N-diphenyl-phosphinoyl aryl ketimines in 97–99% yield with 90–99% ee, while using as little as 0.1% of metal catalyst (*123*).

Non-metal chiral catalysts have also been described. Library screening (*124, 125*) identified an extremely promising peptidic Schiff base derivative that acts as a metal-free catalyst for the reaction of HCN with benzyl-amine-derived imines of aldehydes and ketones. The N-benzyl α-aminonitrile precursors of Val, Leu, 2-amino-heptanoic acid, cyclopropylglycine, cyclohexylglycine, cyclooctylglycine, phenylglycine (Phg) and other aryl-glycines, *tert*-leucine (Tle), α-methyl-Phg (and other α-methyl-arylglycines), and α-methyl-Tle were produced with 86–99.3% ee (*125–128*). Unprotected Tle was obtained with >99% ee in 84% overall yield from the N-benzyl imine of pivalaldehyde (*125*). The peptidic catalyst could be attached to a solid-phase polystyrene resin (*125*). A rationally designed catalyst based on a cinchona alkaloid derivative (normally employed for asymmetric olefin dihydroxylation) successfully catalyzed the addition of HCN to imines formed from benzaldehydes and allylamine, giving 11 different aminonitriles with 79–>99% ee (*129*). Less successful catalysts include a bifunctional molecule bearing both thiourea and imidazole moieties on a chiral 1, 2-diamin-ocyclohexane scaffold which induced only a maximum of 68% ee during cyanation of benzaldehyde imine, and that was with only 17% conversion; at high conversion rates <25% ee was obtained (*130*). Other thiourea/imidazole or thiourea/pyridyl catalysts gave even worse results (*131*). Better enantioselectivities were obtained with an axially chiral N,N′dioxide biaryl pyridine Lewis base promoter, which catalyzed the addition of TMSCN to a series of aryl aldehyde N-benzhydryl imines with 49–95% ee (*132*). A chiral C_2-symmetric bicyclic guanidine catalyst provided enantioselectivities of 50–88% ee for reactions of HCN with imines formed from arylaldehydes and benzhydrylamine (10 examples);

benzylamine-derived imines reacted with only 0–25% ee (*133*).

In 1996 a cyclic dipeptide diketopiperazine prepared from (S)-Phe and (S)-norarginine was used as a catalyst for the Strecker reaction of preformed N-benzhydryl imines with HCN (*134, 135*). A theoretical study of the mechanism of catalysis has been published (*136*), as has a review of the reaction (*137*). Imines derived from aromatic aldehydes generally gave high enantioselectivities for the α-aminonitrile synthesis (80–>99% ee, 82–97% yield), but the results with aliphatic aldehydes were disappointing (<20% ee) (*134, 135*). However, in 2005 an attempt to repeat the cyclic peptide catalyst synthesis gave a compound with significantly different physical properties, with an X-ray crystal structure confirming the 2005 compound had the correct identify. Of greater concern was the fact that the diketopiperazine provided little catalytic rate enhancement for a Strecker synthesis with benzaldehyde, and induced no enantioselectivity at all (*138*).

A somewhat more convoluted approach has used a chiral cyclic ketone as a catalyst (*139, 140*). The ketone was converted to an aminonitrile with HCN, with the aminonitrile then acting as the chiral amine component (see Scheme 5.1), condensing with the substrate aldehyde (RCHO, R = IPr, Ph, Bn) to form an imine. Reaction with a second equivalent of HCN gave an aminodinitrile (2–67% yield, 24–79% de); hydrolysis resulted in an α-aminoamide and the recovered chiral ketone.

Another approach has been based upon a dynamic resolution of racemic aminonitriles derived from aldehydes RCHO, benzyl or isopropylamine, and HCN or TMSCN. Crystallization with (R)-mandelic acid initially produced 1:1 diasteromeric mixtures, but after stirring a suspension in ethanol for 12 h to 15 days, the optically labile aminonitriles epimerized to form preferentially one crystalline derivative of either the (S)- or (R)-aminonitrile. The pure aminonitrile and the chiral auxiliary acid were recovered after salt decomposition by treatment with cold aqueous $NaHCO_3$. The aminonitrile was then hydrolyzed to produce N-substituted amino acids. Hydrogenation of the N-benzyl products gave free amino acids, although some racemization was evident (72–99% ee, R = iPr, iBu, CH(Et)₂, tBu, sBu, Bn, CH(Me)Ph, Ph, 4-Me-Ph, 4-MeO-Ph, 4-F-Ph, 4-Cl-Ph) (*141*). The method is suitable for large-scale preparations (50 g scale).

Enzymatic approaches to enantioselective aminonitrile synthesis have also been taken. Nitrile hydratase from *Rhodococcus* sp. enantioselectively hydrolyzes α-aminonitriles to give D-amino acid amides and L-amino acids, with Phe, Chg, and a number of arylg-lycines produced with 50 to >99% ee (*142*). Several other bacterial whole cell isolates were tested for nitri-lase activity against phenylglycine nitrile, with both (R)- and (S)-selective nitrile hydratases identified that gave either enantiomer of Phg with >99% ee (*143*). Dynamic kinetic resolution conditions were established for N-acetyl phenylglycinonitrile and N-formyl 4-fluoro-phenylglycinonitrile, with spontaneous racemization at pH 8. The (R)-enantiomer was preferentially hydrolyzed by Nitrilase 5086 to give the phenylglycines in 87–95% yield and 91–99% ee on up to 1 g scales (144).

An asymmetric enzymatic version of the Bücherer reaction has been developed. Racemic hydantoin intermediates from the Bücherer synthesis were enantioselectively hydrolyzed by D-specific hydantoinases from thermophilic microorganisms. The hydantoins racemize under the reaction conditions, so conversions of up to 93% with >99% ee are possible (14 examples, 50–95%, 94 to >99% ee) (145).

5.2.1d Chiral Catalyst: Cyanohydrin Synthesis

The cyanohydrin intermediates formed by HCN addition to aldehydes during one version of the Strecker reaction have also been generated by asymmetric catalysis, and due to their usefulness as intermediates in other reactions, a large number of examples have appeared in recent years. As mentioned earlier, the biology, preparations, and synthetic applications of cyanohydrins were extensively reviewed in 1999 (7), with the synthesis and applications of non-racemic cyanohydrins further reviewed in 2003 (8). A BINOL-based lanthanide catalyst induced 48–73% ee for cyanohydrin formation from alkyl or aryl aldehydes and TMSCN (146), a Lewis acid catalyst formed from an AlCl complex of a binaphthol-based catalyst with Lewis base phosphine oxide substituents gave 86–100% yields and 90–95% ee for 10 different aliphatic and aryl aldehydes (147), and a monometallic aluminum complex with a 3,3′-bis(diethylaminomethyl)-binaphthol ligand (BINOLAM) provided 66 to >98% ee for nine different aldehydes (148). The bifunctional BINOLAM-AlCl catalyst was subsequently compared to the corresponding monofunctional BINOL–AlCl complex, and shown to be more efficient as well as easily recovered and recycled. Fifteen different aldehydes were converted in 55–99% yield with 20 to >98% ee (149). A BINOL–salen catalyst, in combination with Ti(O*i*Pr)₄, induced 89% ee for the addition of TMSCN to benzaldehyde (150).

A bimetallic bis[Ti(IV)salen] complex catalyzed the addition of TMSCN to aryl ketones with 32–72% ee, but additions to aldehydes proceeded with better enantioselectivity if a monomeric vanadium(IV)salen complex was employed (151). The vanadyl salen complex has been anchored to several solid supports in order to aid recovery of the catalyst. The highest enantiomeric excesses were obtained using a silica support (152). The bimetallic bis[Ti(IV)salen] complex was effective for the reaction of ethyl cyanoformate with aldehydes, producing O-ethoxycarbonyl cyanohydrins with 75–99% ee (12 aromatic and aliphatic examples). The catalyst has also been found to be effective at catalyzing the reaction of KCN, Ac₂O, and aldehydes, producing O-acetyl cyanohydrins with 62–93% ee (11 aryl and alkyl aldehydes). KCN is an inexpensive, non-volatile, and safe cyanide source compared to HCN or TMSCN (153, 154). The Ti–salen catalyst was also applied to the

addition of TMSCN to ketones in combination with an achiral tertiary amine N-oxide co-catalyst, producing the cyanohydrin trimethylsilyl ethers in 50–93% yield with 59–86% ee (13 examples) (155). In a similar manner, addition of 1 mol % achiral phenolic N-oxide to 10 mol % chiral salen-Ti(IV) complex improved yield and enantioselectivities for a series of 14 aromatic, aliphatic, and heterocyclic ketones to 58–96% yield and 56–82% ee (156). Alternatively, cross-linked or linear polymeric Ti(IV) and Va(V) salen complexes were tested for enantioselective O-acetyl cyanation of aldehydes (17 examples) with KCN and Ac₂O. Both catalysts provided good conversion and enantioselectivity (up to 94% ee with 99% conversion) and could be readily recovered and reused, with no loss in efficiency over six consecutive recyclings (157). Silylcyanation using TMSCN in the presence of an Mn(salen) chiral catalyst in combination with achiral POPh₃ cocatalyst proceeded with 44–62% ee for a series of aryl, vinyl, and alkyl aldehydes, with the POPh₃ improving ee and reducing reaction time (158). When combined with Al(salen), the POPh₃ procedure gave 72–86% ee for the same set of aldehydes (159).

Chiral phosphorus(V) reagents were used for the asymmetric addition of TMSCN to benzaldehydes catalyzed by samarium(II) chloride. The cyanohydrins were produced with up to 90% ee (160). A β-amino alcohol-titanium complex catalyzed TMSCN cyanosilylation of 18 different aromatic, conjugated, heteroaromatic, and aliphatic aldehydes in 90–99% yield with 57–94% ee (161). In a similar manner, β-hydroxyamide–titanium complexes allowed for cyanosilylation of nine aldehydes in 47–92% yield with 93 to >99% ee (162, 163). TADDOL, in combination with Zr(O*t*Bu)₄ and acetone cyanohydrin, induced 29–91% ee for 11 different aldehydes (164). A flow microreactor was used with a Lu(III)–pybox complex for TMSCN addition to benzaldehyde, giving similar enantioselectivity (73% ee) as a batch process (76% ee). However, Yb(III) gave reduced selectivity in the microreactor (53% ee vs 72% ee) (165). A tripeptide Schiff base with Al(O*i*Pr)₃ induced 80–91% ee for TMSCN addition to 14 aliphatic and aromatic ketones (166), while chiral Schiff base ligands derived from *tert*-butyl salicaldehydes and chiral amino alcohols induced up to 85% ee for Ti-catalyzed addition of TMSCN to benzaldehyde (167). The sodium salt of phenylglycine induced 94% ee (and 96% yield) for addition of TMSCN to acetophenone, and 55–97% ee for 12 other ketones (168).

Enzymatic approaches to cyanohydrin formation have also been taken. Oxynitrilases or hydroxynitrile lyases are enzymes found in a wide range of plants that catalyze the breakdown of a cyanohydrin into an aldehyde/ketone and HCN. However, the enzymes can also catalyze the reverse reaction, the enantioselective addition of HCN to aldehydes or ketones. The use of hydroxynitrile lyases in stereoselective catalysis was reviewed in 2000 (169). A hydroxynitrile lyase from *Hevea brasiliensis* produced (S)-cyanohydrins with 98–99% ee from aliphatic, unsaturated, aromatic and heteroaromatic aldehydes, methyl alkyl ketones, and methyl aryl ketones. A vigorously stirred two-phase system and HCN were employed (*170*). Cyanohydrin intermediates have also been produced by hydroxynitrile lyase-catalyzed addition of HCN to 2- and 3-substituted cyclohexanones (*171*), to 2-substituted cyclopentanones (*172*), and to heterocyclic saturated five- and six-membered ring ketones (*173*). The hydroxynitrile lyase from bitter almonds was found to be more effective than that from maniok for enantioselective HCN addition to *O*-allyl-protected α-hydroxyaldehydes, giving the cyanohydrins with >93% ee (*174*). The (*R*)-oxynitrilase contained in a suspension of 30 g of defatted almond meal was employed for cyanohydrin formation from 2-furaldehyde, giving 26 g of crude product with 98.6% ee. The cyanohydrin was converted into α-hydroxy-β-amino acids (*175*). (*R*)-Oxynitrilase from almond meal also produced cyanohydrins from fluorinated benzaldehydes and HCN under "micro-aqueous" conditions with 41–94% ee for mono- or difluorinated substrates; more fluorination or a trifluoromethyl substituent resulted in products with 0% ee (*176*).

A variety of other hydroxynitrile lyases have been tested. Hydroxynitrile lyases from *Prunus amygdalus* and *Hevea brasiliensis* were used with HCN in a two-phase system of aqueous buffer and various ionic liquids. Reaction rates were improved compared to organic solvents as the second phase, with similar enantioselectivities (*177*). Oxynitrilase from the seeds of *Pouteria sapota* (mamey seeds) was found to work best with diisopropyl ether as solvent in a biphasic system with 10% water and KCN/acid buffer as the HCN source. Benzaldehyde, 2-methyl-2-pentanal, *trans*-2-hexanal, cinnamaldehyde, and 1,4-hexadienal were converted with 79–95% ee (*178*). A hydroxynitrile lyase was isolated from the seed of the Japanese apricot, *Prunus mume*. Over 75 benzaldehydes, heteroaromatic aldehydes, aliphatic aldehydes, and aliphatic methyl ketones were tested as substrates. Thirty-six of these were then converted to cyanohydrins on a 5 g preparative scale, with 48–96% yields and 14–99% ee after 6–72 h (*179*). Guanabana (*Annona muricata*) seed meal was tested as another new source of (*S*)-oxynitrilase, giving 50–87% ee for aryl and heteroaryl aldehydes (but only 11–24% yield for aldehydes other than

furaldehyde), with no conversion of aliphatic aldehydes (*180*). The active site of almond hydroxynitrile lyase has been expanded by mutagenesis to accommodate cinnamaldehyde or 3-phenylpropanal as substrates; a V360I mutant gave 97.6% ee and 96.7% ee, respectively, at 97–98% conversion to the cyanohydrin (*181*). A Trp residue at the entrance to the active site of hydroxynitrile lyase from *Manihot esculenta* was also mutated, with a smaller Ala side chain allowing for HCN additions to bulky aryl or alkyl aldehyde substrates with higher enantioselectivity and faster reaction rates (*182*). When employed for HCN additions to racemic 2-phenylpropionaldehyde, the wild-type enzyme produced (2*S*,3*R*)-2-hydroxy-3-phenylbutyronitrile cyanohydrin with >96% de, while the Trp198Ala mutant reversed the stereoselectivity to give the (2*R*,3*S*)-cyanohydrin with 86% de (*183*).

Ethyl cyanoformate was employed as the cyanide source for a synthesis of an ethoxycarboxylated cyanohydrin of benzaldehyde, catalyzed by hydroxynitrile lyase from *Prunus amygdalus* (*184*). The hydroxynitrile lyase from *Hevea brasiliensis* was encapsulated in a sol–gel matrix, giving an aquagel that was successfully used to produce (*S*)-cyanohydrins from several aldehydes (RCHO, R = Ph, 2-furyl, hexyl, *m*PhO-Ph) with high enantioselectivity (>98% ee) (*185*). A cross-linked and polymer-entrapped (*R*)-oxynitrilase catalyzed the reaction of HCN and benzaldehyde to give the cyanohydrin with up to 99% ee. The entrapped enzyme could be readily recovered and reused up to 20 times with no loss of enantioselectivity (*186*). Another cross-linked aggregate of (*R*)-oxynitrilase was reported in 2005; four aryl and unsaturated aldehydes were hydrocyanated with 65–99% ee (*187*). A recombinant (*R*)-hydroxynitrile lyase was developed which retained activity at process chemistry scales and which could be used in emulsion systems at low pH values. Arylaldehydes were converted to cyanohydrins with 96% yield and 96.5% ee on a 20 g scale (*188*).

An oxynitrilase has been used to prepare optically pure cyanohydrins from aldehydes and acetone cyanohydrin. These were converted into aminonitriles by a Mitsunobu reaction with NHBoc(SES) as the amine nucleophile (see Scheme 5.3). Hydrolysis gave the amino acids with >95% ee, except for aryl side chains

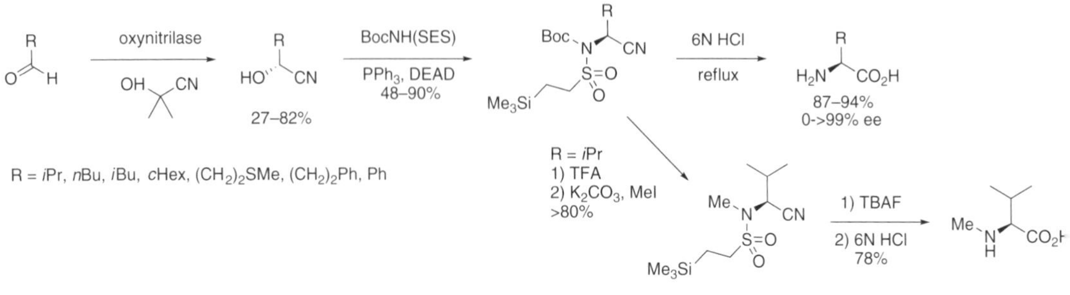

Scheme 5.3 Oxynitrilase-Catalyzed Addition of Cyanohydrin (*189*).

which gave lower enantiopurities. The intermediate ami-nonitriles could also be deprotected and *N*-methylated before hydrolysis (*189*).

Other enzymes have been applied to asymmetric cyanohydrin synthesis. Lipases can selectively acylate the cyanohydrin alcohol of ω-hydroxycyanohydrins protected at the primary alcohol. The effectiveness of the resolution was dependent on the alcohol protecting group, with a bulky trityl group giving the best results (*190*). A dynamic kinetic resolution using lipases on arylaldehyde-derived cyanohydrin substrates has also been described, with over 90% yields and 91->99% ee on gram-scale reactions (*191*). Lipase B from *Candida antarctica* was employed for a kinetic resolution of aryl aldehyde-derived racemic cyanohydrin acetates under mild conditions (*192*).

5.2.2 Chiral Ugi Reactions

Asymmetric reactions involving the Ugi synthesis (see Section 2.2.3) and its application to peptide synthesis have been reviewed (*193, 194*), and are also included in a 2000 review by Dömling and Ugi of multicomponent reactions with isocyanides (*195*), and in a 2006 *Chemical Reviews* treatise of recent developments in isocyanide-based multicomponent reactions (*196*). Chiral amine components are employed to induce asymmetry, often the same amines employed for asymmetric Strecker syn-theses (see Scheme 5.4). (*S*)-Phenylethylamine was used in a Ugi 4CC synthesis of *N*-benzoyl-*N*-(*S*)-α-phenylethyl(*S*)-valine *tert*-butyl amide, giving up to 58% de (*197*). The same amine was combined with *cis*- or *trans*-5-methyl-2,5-dihydro-2-furfural, benzoic acid, and *tert*-butyl isocyanide (*198, 201*), but no stereoselec-tivity was induced at the α-center so the amine essen-tially acted as a resolving agent. Hydrolysis gave the free amino acids, stereoisomers of the antibiotic furano-mycin, in 53–63% yield. Some stereoselectivity was observed in the synthesis of both diastereomers of 2-amino-2-(3-hydroxy-5-methyl-4-isoxazolyl)acetic acid (AMMA, 69% yield, 30% de). The diastereomers were easily separated, but unfortunately extensively racemized during deprotection (*202*). The same proce-dure was used to prepare 2-amino-4-(3-hydroxy-5-methyl-4-isoxazolyl)butyric acid (homo-AMPA), but racemization difficulties were again encountered (*203*).

(*S*)-1-(*m*-Methoxyphenyl)ethylamine provided diastere-omeric ratios of 75:25 to 95:5 during construction of tripeptide esters from Aloc-amino acids, allyl isocy-anoacetate, and aldehydes (*204*).

Amino acids can be used as either (or both) the amine or (and) carboxyl components of Ugi reactions, and pro-vide moderate diastereoselectivity at the newly formed center. Xaa-Val dipeptides were prepared in 37–84% yield by reacting a Cbz-amino acid, an amino acid methyl ester, isobutyraldehyde, and *p*-tosylmethyl isocyanide. The Val residue was formed with 10–64% de (*205*). *N*-Unprotected amino esters were combined with aldehydes R^1CHO, isocyanides R^2NC, and alco-hols R^3OH to produce 1,1'-iminodicarboxylic esters, R^2NC(O)CH(R^1)-NHCH(R)CO$_2R^3$, with 84–88% de. The new stereocenter possessed the same absolute con-figuration as the amino acid employed (*206–208*). Iminodiacetic acid chiral ligands for enantioselective transition metal catalysis were prepared from amino acids, 2-formylpyridine or 2-diphenylphosphinobenzal-dehyde, and several isocyanides. Diastereoselectivities of 41–84% de were observed (*209*). In a similar manner, Glu(OMe)-OH induced 66–82% de during reaction with nine different aryl aldehydes and *t*Bu-NC in the pres-ence of TiCl$_4$, while Val gave even greater induction for several of the aldehydes tested (*210*). L-Homoserine was employed as both the amino and acid components of a Ugi 4CC condensation, resulting in *N*-(3-butyrolactone) amino amides. Hindered aldehydes reacted with very high diastereoselectivity (94–98% de for pivalaldehyde and isobutyraldehyde) (*211*). The β-amino acid *N*-(2-hydroxy-1-phenylethyl) (*S*)-2,2-difluoro-3-phenyl-3-aminopropionic acid was used as the acid component in 4CC reactions with various amines, isonitriles, and aldehydes to form pseudopep-tides (with no diastereoselectivity) (*212*).

The first removable chiral auxiliary that provided synthetically useful stereoselectivities for Ugi 4CC syn-theses was a sterically bulky optically active ferrocenyl amine component. *N*-Benzoyl-(*S*)-valine *tert*-butyl amide was synthesized with up to 98% de (*213*), while a tetravaline peptide was prepared with 97% de by employing *N*-(2-isocyano-3-methylbutanoyl)-L-valine methyl ester as the isocyano component and *N*-formyl-L-valine as the acid (*214*). The ferrocenyl group was cleaved by acidolysis after the Ugi condensation.

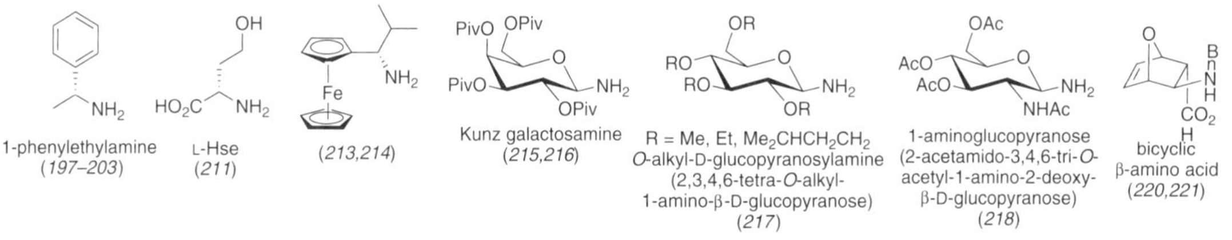

Scheme 5.4 Chiral Amines Used as Auxiliaries in Asymmetric Ugi Synthesis.

The best results for the Ugi 4CC reaction have been obtained with the same amino sugars used for the Strecker synthesis. The 1-amino-2,3,4,6-tetra-*O*-pivaloyl- and *O*-acetyl-β-D-galactopyranose derivatives of Kunz et al. produced *N*-formylated derivatives of D-amino acids in nearly quantitative yields and 82–94% de by reaction with various aldehydes (RCHO, R = *n*Pr, *i*Pr, *t*Bu, Ph, 2-furyl, 2-thienyl, 4-Cl–Ph, 4-NO₂-Ph, γ-cyanopropyl, CH=CHPh), *tert*-butyl isocyanide and formic acid in the presence of zinc chloride (*215, 216*). The products were obtained diastereomerically pure by recrystallization (75–93% yield) and could then be deprotected by a two-step acidic hydrolysis to give the free amino acids (side chain R = *t*Bu, Bn, Ph, 4-Cl-Ph, (CH₂)₃CO₂H) in 82–90% yield, enantiomerically pure by chiral TLC analysis. The chiral auxiliary could be recovered. Unlike the Strecker reaction, a change in solvent did not reverse the diastereoselectivity (*77*).

Ugi and co-workers modified the sugar auxiliary in an attempt to allow for easier cleavage of the sugar, so that peptide products are not hydrolyzed. *O*-alkylated-1-aminopyranoses gave reasonable stereoselectivity (aldehyde = *i*PrCHO, 59–98% de) and could be removed by mild acidolysis with various nucleophiles in the presence of a peptide bond, but yields for auxiliary removal were poor (7–50%) (*217*). Greater success was obtained with 2-acetamido-3,4,6-tri-*O*-acetyl-1-amino-2-deoxy-β-D-glucopyranose, readily prepared from *N*-acetyl glucosamine (*218*). Zinc catalysis was used to generate the imine, and the resulting rigid zinc complex reacted with high diastereoselectivity to generate *N*-acyl amino acid amides or peptides (89 to >99% de, R = Me, *i*Pr, Et, Bn, 4-AcNH-Ph). However, template removal was not optimized. A 1-amino-5-desoxy-5-thio-2,3,4-*O*-isobutanoyl-β-D-xylopyranose amine component provided stereoselectivity of 92% de with 92% yield when combined with isovaleraldehyde, *tert*-butylisocyanide, benzoic acid and zinc chloride. The *O*-acyl groups were removed using methylamine and the auxiliary with TFA/Hg(OAc)₂ followed by H₂S/H₂O, leaving Bz-Leu-NH*t*Bu in 58% yield (*219*).

A new bicyclic β-amino acid auxiliary, 3-benzyl-amino-7-oxabicyclo[2.2.1]hept5-ene-2-carboxylate, was reported in 2004, inducing nearly complete diastereoselectivity in reactions with alkyl and aryl aldehydes and benzyl isocyanide in methanol (see Scheme 5.5). The auxiliary was removed by a retro-Diels–Alder reaction followed by acidic hydrolysis; hydrogenolysis produced

benzylamides of Val, Leu, Abu, Phe, 4′-Cl-Phg or 3′-MeO-Phg with >95% ee (*220, 221*).

Anomeric glucosyl isonitriles have also been employed as chiral components in Ugi 4CC condensations. However, reaction with achiral amines, aldehydes, and acids resulted in minimal diastereoselectivity (0–20%) of the amino acid glycosyl amide products (*222, 223*). The modified Ugi 4CC reaction that uses an iminoaziridine as substrate (see Scheme 2.10) can also produce optically active products if a non-racemic iminoaziridine is employed (*224*).

5.2.3 Other Asymmetric Aminocarbonylation Reactions

Dondoni and co-workers have developed a two-step aminocarboxylation reaction which relies on chiral aldehydes (the eventual side chain of the amino acid products) to induce stereoselectivity at the α-center. The aldehydes are converted into nitrones, with a masked carboxyl group introduced by stereoselective addition of 2-lithiofuran or 2-lithiothiazole (see Scheme 5.6). The opposite diastereoselectivity can be obtained if a Lewis acid (Et₂AlCl) catalyst is employed. The reaction was originally used to prepare amino aldehydes (*225*). The furan can be converted into a carboxyl group by oxidative cleavage, while the thiazole is first converted into an aldehyde and then oxidized to the acid. These conditions have been used to prepare *N*-hydroxy amino acids (*226*), a number of carbon-linked α-glycosylglycines (*226*), the amino acid nucleoside polyoxin C (*227*) the polyhydroxy amino acid polyoxamic acid (*227*) other polyhydroxy α-amino acids (*228, 229, 818*), and the polyhydroxy ε-amino acid destomic acid (*225*). The use of thiazole as a carboxyl equivalent was reviewed in 1998 (*230*). A furyl group has also been employed as a masked carboxylate in a three-component condensation of optically active 3,3-difluorolactaldehyde, 2-furylboronic acid, and diallylamine. Pd-catalyzed de-allylation of the amino group and ozonolysis of the furyl group produced (2*S*,3*R*)-4,4-difluorothreonine (*231*). Acetylene is another carboxyl equivalent 3, 4-dihydroxy-2-aminobutyric acid was synthesized via addition of acetylide to the nitrone derivative of D-glyceraldehyde acetonide. Oxidative cleavage of the alkyne with RuCl₃/NaIO₄ generated the carboxyl group (*232*). Acetylide addition to the nitrone of the Garner aldehyde led to 2,3-diamino-3-hydroxybutyric acid (*233*).

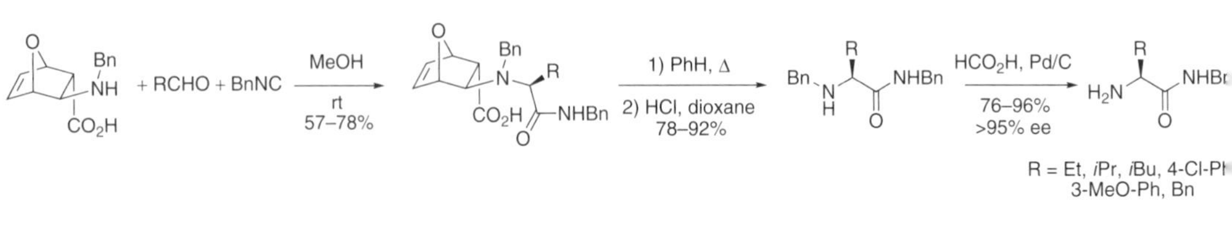

R = Et, *i*Pr, *i*Bu, 4-Cl-Ph 3-MeO-Ph, Bn

Scheme 5.5 Bicyclic β-amino Acid in Asymmetric Ugi Synthesis (*220, 221*).

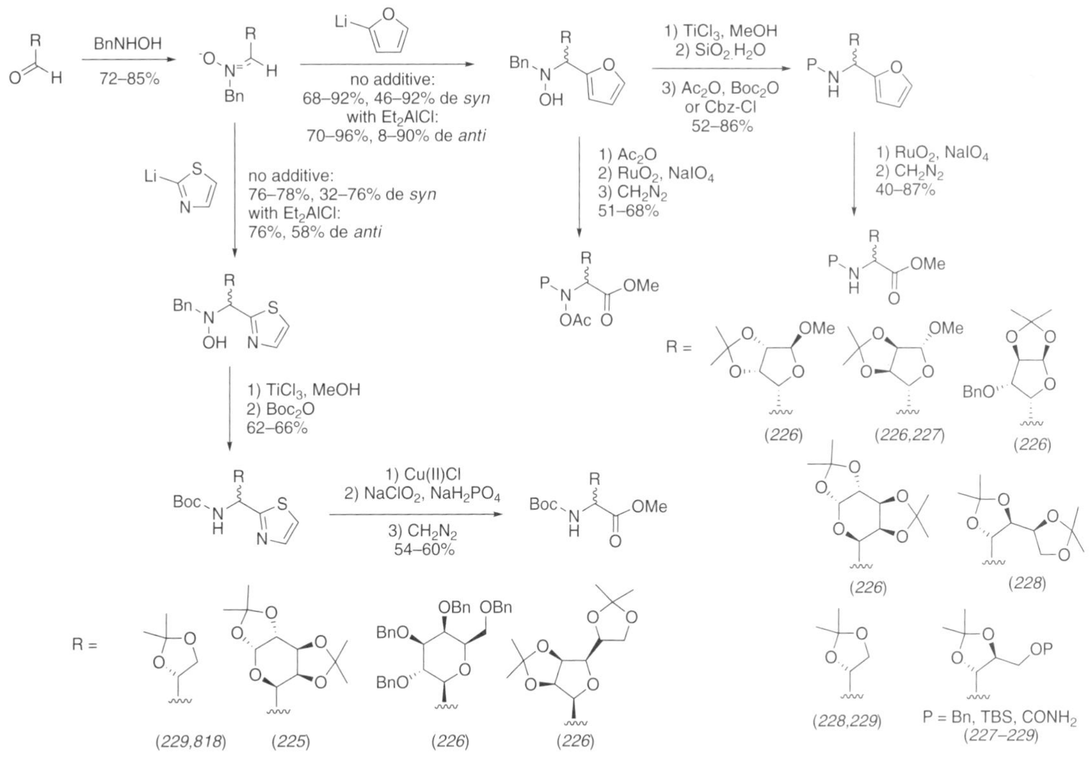

Scheme 5.6 Stereocontrolled Aminocarboxylation via Nitrone.

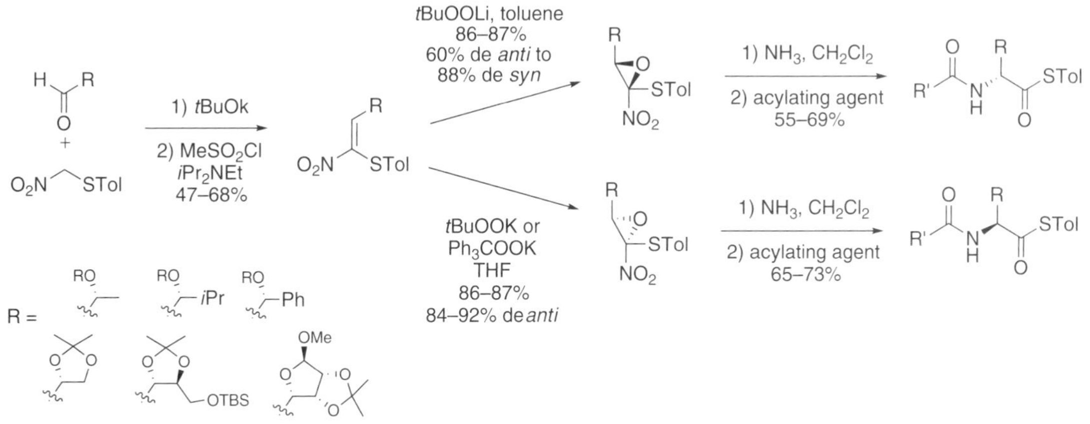

Scheme 5.7 Stereocontrolled Aminocarboxylation via Nitroalkene (234–238).

The "stereocontrolled Strecker reaction" of Jackson et al., in which an aldehyde is converted into a 1-(4-tolylthio)-1-nitroalkene (see Scheme 2.18), has been used for asymmetric syntheses by starting with chiral α-hydroxy aldehydes (234). Alkene epoxidation was stereoselectively controlled by the choice of nucleophilic epoxidizing reagent, with tBuOOLi giving the syn isomer and tBuOOK giving the anti isomer (235, 236). A number of β-hydroxy amino acids, including

polyoxamic acid derivatives (235, 237), γ-hydroxy-Thr (238), and D-Thr and L-allo-Thr (236), have been synthesized by this method (see Scheme 5.7).

Several other aminocarboxylation reactions rely upon chiral auxiliaries on the amino or carboxy components. 2-Lithiofuran was added to an imine formed from benzaldehyde and (S)-valinol. Removal of the chiral auxiliary and oxidation of the furyl group provided D-Phg (239). Organolithium carboxy synthons

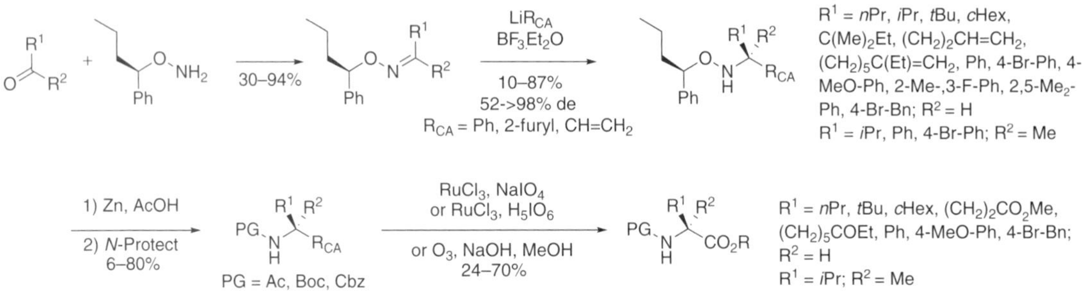
LiR$_{CA}$
BF$_3$Et$_2$O
30–94%
10–87%
52->98% de
R$_{CA}$ = Ph, 2-furyl, CH=CH$_2$
R^1 = *n*Pr, *i*Pr, *t*Bu, *c*Hex, C(Me)$_2$Et, (CH$_2$)$_2$CH=CH$_2$, (CH$_2$)$_5$C(Et)=CH$_2$, Ph, 4-Br-Ph, 4-MeO-Ph, 2-Me-,3-F-Ph, 2,5-Me$_2$-Ph, 4-Br-Bn; R^2 = H
R^1 = *i*Pr, Ph, 4-Br-Ph; R^2 = Me
1) Zn, AcOH
2) *N*-Protect
6–80%
PG = Ac, Boc, Cbz
RuCl$_3$, NaIO$_4$
or RuCl$_3$, H$_5$IO$_6$
or O$_3$, NaOH, MeOH
24–70%
R^1 = *n*Pr, *t*Bu, *c*Hex, (CH$_2$)$_2$CO$_2$Me, (CH$_2$)$_5$COEt, Ph, 4-MeO-Ph, 4-Br-Bn; R^2 = H
R^1 = *i*Pr; R^2 = Me

Scheme 5.8 Stereocontrolled Aminocarboxylation via ROPHy/SOPHy Oximes (*240*).

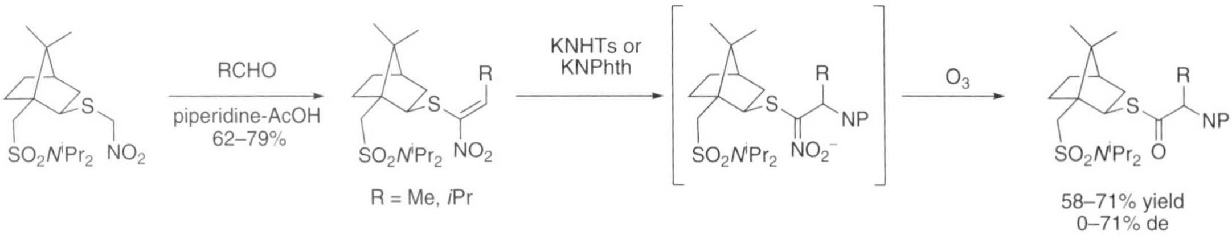
RCHO
piperidine-AcOH
62–79%
R = Me, *i*Pr
KNHTs or
KNPhth
O$_3$
58–71% yield
0–71% de

Scheme 5.9 Conjugate Additions to Nitro-Alkenes (*241*).

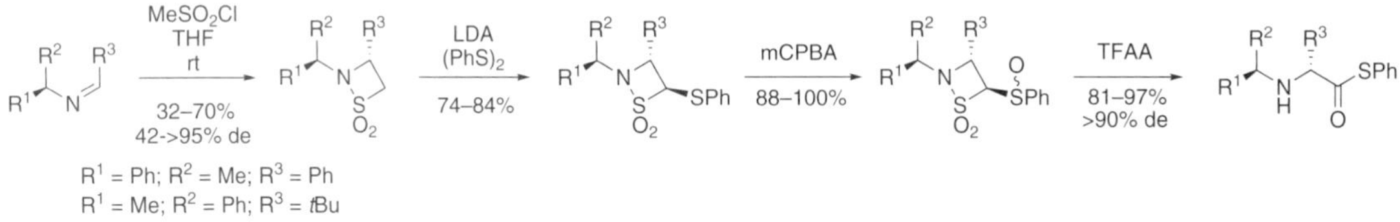
MeSO$_2$Cl
THF
rt
32–70%
42->95% de
LDA
(PhS)$_2$
74–84%
mCPBA
88–100%
TFAA
81–97%
>90% de
R^1 = Ph; R^2 = Me; R^3 = Ph
R^1 = Me; R^2 = Ph; R^3 = *t*Bu

Scheme 5.10 Pummerer Rearrangement of 3-Substituted-4-Sulfinyl-β-Sultams (*244*).

(furyl-, phenyl-, and vinyllithiums) have been added to chiral oximes formed from aldehydes or ketones and ROPHy/SOPHy, (*R*)- or (*S*)-*O*-(1-phenylbutyl)hydroxylamine (see Scheme 5.8). The best diastereoselectivities were obtained with vinyllithium addition (64–98% de for 10 examples). The N-O bond was then cleaved using Zn/acetic acid/ultrasound, the amine *N*-protected, and the vinyl group oxidatively cleaved by $RuCl_3$- or ozone-based methods (24–70% yield) (*240*).

(+)-Camphorsulfonic acid-derived nitroalkenes were formed from aldehydes and used as substrates for nucleophilic conjugate additions of potassium phthalimide or potassium tosylamide (see Scheme 5.9). Ozonolysis of the adducts produced amino thioesters, although diastereoselectivity varied (*241*).

α-Aminoamides can be formed from the reaction of aldehyde-derived imines, $R^1N{=}CHR^2$, with carbamoylsilanes, $TMSCONMeR^3$, in the presence of $BF_3.Et_2O$. If a chiral amide was employed to induce asymmetry (R^3 = 1-phenylethyl group), the diastereoselectivity induced was minimal, but an *N*-[1-(1-naphthyl)ethyl] or *N*-(1-phenylethyl) R^1 substituent on the imine component gave products with up to 50% de. Matched sets of chiral auxiliaries on both imine and amide resulted in up to 92% de for Abu, Val, and Phg (*242*). An Ugi-like reaction combined homochiral *N*-(1-arylethyl) imines, ArCH(Me)N=CHPh, with benzyl isocyanates, BnNCO, in the presence of $TaCl_5$/Zn to produce, after hydrolysis with 10% KOH, *N*-substituted phenylglycine benzylamide in 67–88% yield with 32–58% de (*243*).

A [2+2] cycloaddition of imines formed from a chiral amine and an aldehyde produced 3-substituted β-sultams (see Scheme 5.10). The sultams were converted into a 4-sulfinyl-β-sultam, with a Pummerer rearrangement providing α-amino acid thioesters (*244*).

Several aminocarboxylation reactions employing chiral catalysts have been developed. The phase-transfer catalyzed synthesis of α-aminoarylacetic acids from aromatic aldehydes, aqueous ammonia, and chloroform (see Scheme 2.12) has been converted into an asymmetric

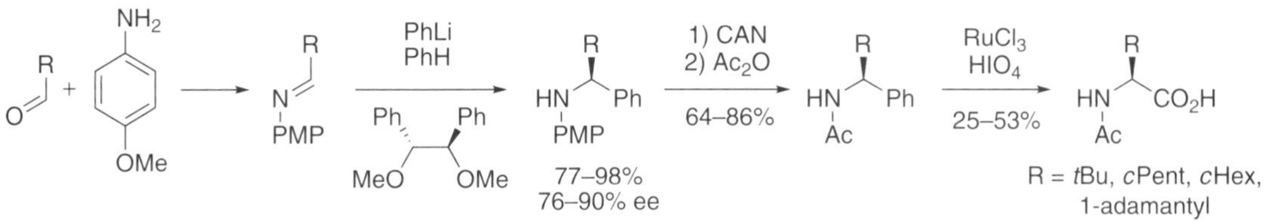

Scheme 5.11 Stepwise Aminocarboxylation Using PhLi as Masked Carboxyl Group (*248*).

reaction by using β-cyclodextrin as a catalyst (*245*), although enantioselectivities were marginal (3–28% ee, 19–84% yield, R = Ph, 3-Me-Ph, 4-Me-Ph, 4-MeO-Ph, 3-Br-Ph, 4-Br-Ph, 4-Cl-Ph, PhCH₂). A chiral surfactant has also been employed, again with limited enantio-enrichment (28% ee, 15–72% yield, R = Ph, 3-Me-Ph, 4-Me-Ph, 4-MeO-Ph, 3-Cl-Ph, 4-Cl-Ph, 4-Br-Ph) (*246*). Alternatively, this reaction has been applied to the synthesis of D-*allo*-isoleucine by using a chiral aldehyde substrate, (*S*)-2-methylbutanal, and azide. The azide was hydrogenated to give a 64:36 mixture of L-Ile and D-*allo*-Ile (*247*). An asymmetric amination by Corey et al. (see Scheme 5.16) can also be considered as an aminocarboxylation reaction, depending on the source of the trichloromethyl ketone substrate.

Another stepwise aminocarboxylation reaction employs addition of PhLi to imines formed from anisidine and aldehydes, with enantioselectivity (76–90% ee) induced by 1,2-dimethoxy-1,2-diphenylethane (see Scheme 5.11). The N-*p*-methoxyphenyl substituent was removed oxidatively with CAN, and the phenyl group converted to the carboxyl using RuCl₃/HIO₄. Four sterically hindered alkyl amino acids were prepared (*248*).

5.3 Asymmetric Amination Reactions

This section will discuss modified versions of the acyclic amination reactions introduced in Section 2.3. Four different approaches are used: amination of chiral substrates; amination of substrates containing chiral auxiliaries; amination in the presence of chiral catalysts; or amination with chiral aminating reagents. Several other asymmetric amination reactions have been developed, such as the nucleophilic opening of epoxides with amines, which will be discussed in later chapters. Since aminations with chiral ester or amide auxiliaries often employ the same auxiliaries for both nucleophilic and electrophilic aminations, they will be discussed together in Section 5.3.6. As mentioned in Chapter 2, α-azido esters are often obtained as intermediates in a number of these synthetic routes. While hydrogenation is generally employed for azide reduction, a number of other procedures have been reported. Recent methods include reduction with triphenylphosphine/H₂O (*249–251*), stannous chloride in methanol (*252–256*), N,N-dimethyl-hydrazine and catalytic ferric chloride (*257*), indium and ammonium chloride (*258*), iodotrimethylsilane

(*259*), Cu(NO₃)₂ (*260*), and Zn-NiCl₂.6H₂O-THF (*261*). Triethylphosphine is reportedly more effective for azide reduction than tributylphosphine, with both much more effective than triphenylphosphine (*262*). Direct conversion to Boc-protected amines has also been reported, using trimethylsilylphosphine and Boc-ON (*263*), H₂/Pd-C/Boc₂O (*264*), or triethylsilane and di-*tert*-butyl dicarbonate in the presence of a catalytic amount of 20% Pd(OH)₂/C (*265*). Trimethylphosphine and chloroformates were used to prepare N-Cbz, N-methoxycarbonyl, N-ethoxycarbonyl, N-Troc or N-Alloc derivatives (*266*).

5.3.1 Nucleophilic Aminations of Chiral α-Halo or α-Hydroxy Acid Substrates

For asymmetric syntheses via nucleophilic amination, chiral α-hydroxy acids are the substrates of choice. The conventional α-halo acids used in racemic syntheses are prone to racemization (*267*). Despite this, several asymmetric syntheses employ α-bromo acid intermediates. Nᵋ-Bz L-Lys can be converted into D-Lys via treatment with nitrosyl bromide to form the α-bromo derivative. Amination gave the product with equal but opposite optical rotation to the starting material, consistent with retention of configuration for the bromination step, and inversion for the amination (*268*). 3-Methyl-4-hydroxy-homotyrosine was prepared via azide displacement of a chiral bromo (*269*) or iodo (*270–273*) precursor. 3-Hydroxypyridine has been used to displace optically active α-bromoacids, with varying degrees of racemization, producing unusual amino acid analogs in which the amino group is a quaternary nitrogen within a pyridine nucleus (*274*).

The optical lability of the α-bromo derivatives has been utilized for dynamic kinetic resolutions, in which racemic α-bromoacids are esterified/amidated with homochiral alcohols/amines, and then displaced with amines (see Sections 5.3.6d and 12.1.7). α-Bromoesters have also been resolved by a dynamic kinetic resolution by using lipases for ester hydrolysis in the presence of bromide, under conditions in which the ester substrate racemizes more quickly than the acid product (*275*). Alternatively, the (*S*)-α-bromo-3-phenylpropionic acid derived from L-Phe was inverted into the (*R*)-enantiomer by simple crystallization with (*R*)-bornylamine in the presence of a bromide source (*276*). A rapid screening via parallel experimentation identified conditions suitable for crystallization-induced dynamic resolution of α-bromo-carboxylic acids using various chiral amines in the presence of a catalytic amount of tetrabutylammonium bromide (277). A direct organocatalytic asymmetric α-chlorination of aldehydes was reported in 2004, with eight aldehydes chlorinated with 81–97% ee. Four examples were subsequently oxidized to the α-chloro esters, which were obtained with 85–95% ee. The butyraldehyde-derived chloroester was then displaced with sodium azide to give, after azide reduction, 2-aminobutyric acid with 95% ee (278).

For α-hydroxyacids, the nitrogen nucleophile can be introduced via a Mitsunobu reaction of the hydroxy group (see Scheme 5.12), or by displacement of a triflate or other activated derivative (see Scheme 5.13). The optically active α-hydroxy acids can be derived from α-amino acids via reaction with nitrites, a conversion that proceeds with retention of configuration (e.g., see 279, 280). Other potential substrates are derived from optically active cyanohydrin intermediates, which can be prepared by a variety of methods (see Section 5.2.1c). The two steps of this conversion (cyanohydrin formation, amine displacement) correspond to a Strecker-like reaction, and are discussed in Section 5.2.

N-phenoxycarbonyl-O-Cbz and N-Cbz-O-Cbz hydroxylamines have been used as nucleophiles in the Mitsunobu reaction of chiral α-hydroxy esters (RCH(OH)CO$_2$P; R = Me, Bn, iPr, iBu, sBu), giving the protected N-hydroxy amino acids in 62–91% yield, with 86–96% ee (281). The N-hydroxy or N-unsubstituted amino acids could then be obtained by sequential deprotection. Ethyl (S)-lactate (R = Me) was reacted with a number of imidodicarbonate nucleophiles in a similar fashion to give N,N-diprotected ethyl (R)-alaninates with >95% ee, but yields varied from <5% (for NHBoc$_2$) to 93% (for TosNHBoc) depending on the electron-withdrawing properties of the imidodicarbonate alkyl groups (282). The Fmoc group was not suitable for these reactions. A much better yield (100%) of N,N-Boc$_2$ benzyl (R)-alaninate was obtained by displacement of the triflate of benzyl (S)-lactate with the lithium salt of HNBoc$_2$; deprotection to Boc-Ala proceeded in 83%

yield with 96% ee (282). This procedure was also used to prepare the ^{15}N-labeled amino acid (282). N-Boc, N-Cbz or N-Alloc 2-nitrobenzenesulfonamide were also used to aminate ethyl (S)-lactate via a Mitsunobu reaction. The N-nosyl group was removed using mercaptoacetic acid (283). N-(2-Thiazolyl) Ala was produced in 62% yield by a Mitsunobu reaction of (S)-(−) ethyl lactate with the trichloroethyl carbamate of 2-aminothiazole (284). N-Boc ethyl oxamate (BocNHCOCO$_2$Et) has been reported to be useful aminating reagent for Mitsunobu reactions, providing N-Boc amines after mild LiOH/THF/H$_2$O deprotection of the ethyl oxamate group. However, amino acids were not prepared (285).

An azide group was introduced under Mitsunobu conditions during syntheses of (R)-Ala (from ethyl (S)-lactate) (250), and of the 3-methyl-4-hydroxy-homoserine that forms the N-terminal amino acid of nikkomycin B (286). Several polyhydroxydiamino acids were prepared by a Mitsunobu displacement of a chiral α-hydroxy ester derived from the Garner aldehyde. The amination step, which employed DPPA as the azide source, proceeded in 50–80% yield (287). A Mitsunobu azidation with DEAD/DPPA was applied to the secondary alcohol of a chiral 1,2-dicarboxy-3-(1,2-dihydroxyethyl)cyclopropane substrate derived from (S)-glyceraldehyde acetonide; azide hydrogenation and hydroxymethyl oxidation produced (2S,1′R,2′R,3′R)-2-(2′,3′-dicarboxycyclopropyl)-Gly (288). In 2001 a series of α-hydrazino esters were prepared from optically active α-hydroxy esters via Mitsunobu reaction with N-Boc- or N-Cbz-aminophthalimides, producing N^{α}-Boc/Cbz,N^{β}-phthalimido triprotected products. The Boc/Cbz and phthalimido groups could be orthogonally removed to give either N^{α}- or N^{β}-protected products, while Raney nickel hydrogenation gave the amino acids with >97% ee. Analogs of Gly, Ala, Val, Leu, and Phe were synthesized (289).

Displacements of α-triflate esters are common. In one report a wide variety of alkyl and aryl amines (NR^{1}R^2: R^1 = H, R^2 = cHex, tBu, (CH$_2$)$_2$OH, allyl, Ph, 3-MeO-Ph, 3-HO-Ph, 4-Cl-Ph, 2-Cl-Ph, 3,4-Cl$_2$-Ph,

R^1 = Me iPr, Bu, (S)-sBu, Bn; R^2 = Me; R^3 = Cbz, PhOCO;
R^4 = OCbz, OBoc (281)
R^1 = Me; R^2 = Et; R^3 = Cbz; R^4 = Boc, Adoc, Aloc, TCBoc, Troc,
Cbz, Cbz(OMe), Cbz(Cl), Cbz(NO$_2$), Poc (282)
R^1 = Me; R^2 = Et; R^3 = Boc; R^4 = Boc (282)
R^1 = Me; R^2 = Et; R^3 = Tos; R^4 = Boc, Cbz, Cbz(NO$_2$) (282)
R^1 = Me; R^2 = Et; R^3 = Boc, Fmoc, Aloc; R^4 = (2-NO$_2$-Ph)SO$_2$ (283)

Scheme 5.12 Amination via Mitsunobu Displacement α-Hydroxy Acid.

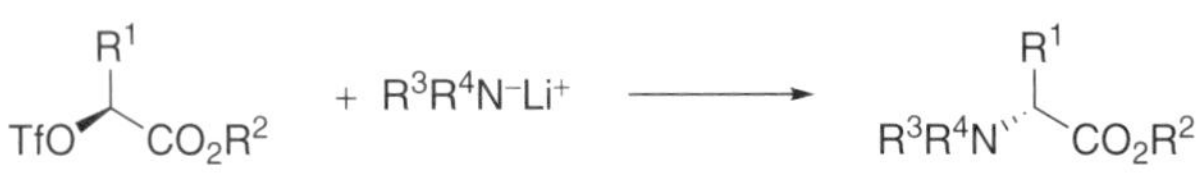

Scheme 5.13 Amination via Nucleophilic Displacement α-Triflate Acid.

2,6-Me$_2$-Ph, 1-naphthyl, 4-EtO$_2$C-Ph; R^1 = Me, R^2 = Ph, Bn; R^1 = Et, R^2 = Et; R^1 = R^2 = -(CH$_2$)$_4$-, R^1 = R^2 = -(CH$_2$)$_4$CH(Et)-) were used to displace triflates of chiral α-hydroxy carboxylic esters in 61–96% yield, producing N-alkylated Ala, Phe, and Asp products (267). The triflate was found to react much more quickly than mesylates, tosylates, bromides or chlorides. The diastereomer ratio observed after displacement of the chiral α-triflate, α-mesylate or α-bromide derivatives of ethyl propionate with (S)- or (R)-phenylethylamine was used to determine the extent of racemization during reaction. The α-bromo carboxylic ester led to extensive racemization (0–80% de) while the mesylate and triflate maintained most of their optical purity (92–96% de) (267). Lys was converted into the α-trifloxy derivative via nitrous deamination, and then reacted with a variety of amine nucleophiles to give N$^\alpha$-alkyl Lys products (280). As mentioned earlier, a much better yield (100%) of N,N-Boc$_2$ (R)-Ala-OBn was obtained by displacement of the triflate of benzyl (S)-lactate with the lithium salt of HNBoc$_2$ than by a Mitsunobu reaction with the α-hydroxy substrate (282).

Triflate displacement has also been used in the preparation of N-(ω-thioalkyl)- and N-(ω-amino)- and N-(ω-carboxyalkyl)- amino acids, avoiding the racemization observed when α-halo carboxylic acids were used as substrates. The nucleophile Bn-S-(CH$_2$)$_n$-NH$_2$ reacted in 19–89% yield (290), tBuO$_2$C-(CH$_2$)$_n$-NHBn in 72–100% yield (291), and BocNH-(CH$_2$)$_n$-NHBn in 51–100% yield (291). The chiral α-hydroxy carboxylic ester substrates were derived from amino acids. N-hydroxy-α-amino acids can be obtained from α-hydroxy ester triflates in 78–89% yield with 76–100% ee (R= Me, iBu, Bn, Ph, CH$_2$CO$_2$Me) by displacement with O-benzyl-hydroxylamine in the presence of lutidine (292, 293); interconversion of the (R)- and (S)-α-hydroxy esters was also described (293). A series of α,α′-iminodicarboxylic acids were prepared via N-alkylation of Ala-OEt, Phe-OMe, Tyr-OEt, Lys(Cbz)-OMe or Pro-OMe with triflate derivatives of α-hydroxy acids (294).

Azide triflate displacement was employed in the synthesis of a C-linked analog of β-D-glucopyranosyl serine, using DPPA in combination with tetrabutylammonium azide (295). Other C-linked glycosyl amino acids have also been prepared via azide displacement of triflates (296). A threose-derived α-triflyl ester was displaced with azide enroute to a synthesis of 3,4,5-trihydroxy-Nva (297). 4,5-Dihydroxy-L-norvaline was prepared in a similar fashion from D-ribonolactone (298), while a triflate of glucuronolactone was used in the synthesis of bulgecinine (a dihydroxyproline derivative) and di- and trihydroxypipecolic acid (299). The syn and anti isomers of β-methyl-Phe were prepared by azide displacement of the corresponding mesylates, which were obtained by organocuprate opening of aryl epoxides (300). erythro-β-Hydroxy-Asp was synthesized from a monotriflate derivative of tartaric acid (301, 302). Azide displacement of an L-Ile-derived α-hydroxy ester derivative was used to prepare D-allo-Ile (279). A series of D-amino acids (Ala, Val, Leu, Ile,

Phg, Phe, and Asp) were prepared from the L-amino acids via nitration/hydrolysis with NaNO$_2$/NH$_2$SO$_4$. Azide displacement of the nosylate derivatives gave the products with good enantiomeric purity (92–97% ee), except for Phg (35% ee) (249).

A number of β-hydroxy amino acids have been prepared by a route involving asymmetric dihydroxylation of an α,β-unsaturated ester, followed by activation of the α-hydroxyl group and azide displacement. Thus 3-hydroxy-5-oxo-Nva (303), (2R,3R)-3-hydroxy-Tyr (90% yield) (304), and (2R,3R)-m-chloro-3-hydroxyTyr (81%, 92% de) (305) were synthesized by azide displacement of a nosylate, (2R,3R)-3-hydroxy-Leu (306), (2S,3S)-3-hydroxy-Leu (307), (2S,3S)-3-hydroxy-Nle (307), 4,4,4-trifluoro-Thr (308), (2R,3R)-3-tert-butyl-Ser (296), and (2R,3R)-3-hydroxy-6-benzyloxy-Nle (296) via azide opening of a cyclic sulfate formed from the diol (80–92% yield for amination), and both enantiomers of allo-Thr and 3-hydroxy Val by azide opening of a cyclic sulfonate (90% yield) (309). A synthesis of L-DOPA began with an asymmetric dihydroxylation of ethyl (m,p-methylenedioxy)cinnamate. The diol was reacted with SOCl$_2$ to form a cyclic sulfite, which was regioselectively opened with sodium azide to give the 2-hydroxy-3-azido ester. This was converted into the correct regioisomer via azide reduction with PPh$_3$, which was accompanied by displacement of the α-hydroxy group to form an aziridine. Hydrogenolysis opened the aziridine, with ester/acetal hydrolysis giving DOPA with 85% ee (310). A very similar route was described by another group (see Scheme 8.5) (307), though in their synthesis azide opening of the sulfates gave β-hydroxy-α-azido esters. The use of cyclic sulfites and sulfates in organic synthesis was reviewed in 2000 (311), and is also discussed in Section 8.4.4. Asymmetric dihydroxylation has also been employed to generate other aziridine-2-carboxylates (see Section 8.2.2).

A complex aryl ether-linked bis(amino acid) contained in ustiloxin D consists of β-hydroxy-Ile linked to β-hydroxy-DOPA through an aryl ether linkage. The β-hydroxy-Ile amino acid center was constructed via asymmetric dihydroxylation of an O-3-methylpent-1-en-3yl β-hydroxy-DOPA derivative, followed by oxidation of the terminal hydroxy group to form the acid group, and azide displacement of the triflate to form the α-amino group (312). A synthesis of 3-hydroxy-Leu first converted the α-hydroxy group of a diol substrate to a bromide, and then displaced this with azide to give the (2R,3S)-isomer with net retention of configuration at the α-center (313). Conversion of one of the hydroxy groups of dihydroxylated products to an amino group with retention of configuration was also achieved via activation of the diol via a tin ketal, followed by reaction with benzoyl isothiocyanate, linking the diol by a cyclic N-benzoyl iminocarbonate (see Scheme 5.14). Treatment with bromide was followed by reflux-induced rearrangement to an N-benzoyloxazolidin-2-one, with the amino group regioselectively ending up adjacent to the substrate carboxyl group. Thr, β-hydroxy-Asp, and β-hydroxy-Phe were prepared, though only the

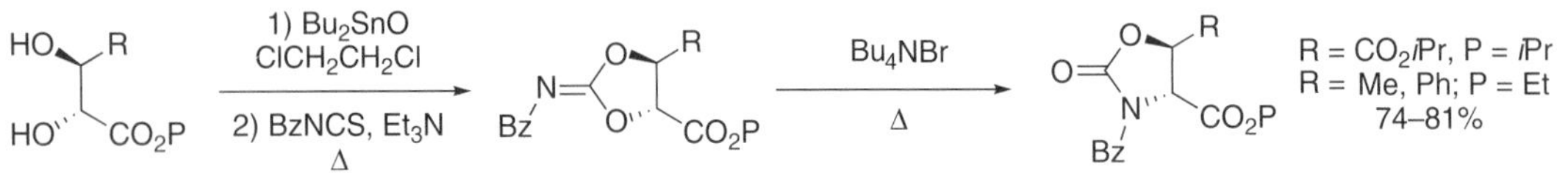

Scheme 5.14 Amination of Diols with Retention of Configuration (*314*).

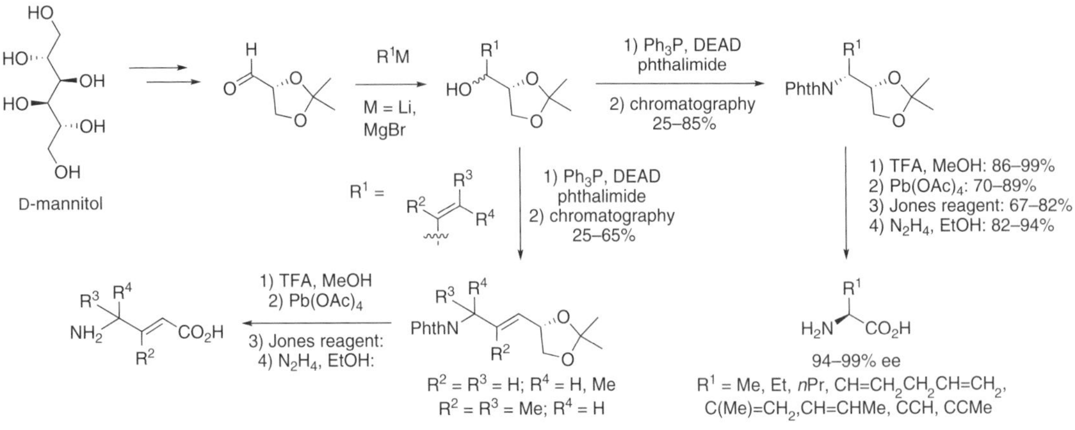

Scheme 5.15 Synthesis of Amino Acids from D-Mannitol (*315–318*).

Asp derivative was produced using non-racemic diol substrate (*314*).

5.3.2 Nucleophilic Aminations of Chiral Hydroxy Substrates with Masked Carboxyl Groups

A number of research groups have developed synthetic routes in which the carboxyl group is generated from an alcohol, diol, alkene, aryl group or trichloromethyl group after the amination step. Mulzer et al. have reported a general approach for amino acid synthesis which allows for easy preparation of the amination substrate. The key intermediate, (*R*)-2,3-isopropylidene-glyceraldehyde, was derived from D-mannitol and is an oxo-equivalent of the Garner aldehyde (see Section 10.6.1). Organometallic addition produced a triol derivative (*315–318*), which was aminated by a Mitsunobu displacement with phthalimide (see Scheme 5.15). After removal of the acetonide protecting group, the diol was oxidatively cleaved to generate the acid group (*315, 316*), producing amino acids with 94–99% ee. The glyceraldehyde synthon could also be used to prepare α,β-unsaturated-γ-amino acids, as the phthalimide anion can also attack allylic and propargylic alcohols by an S_N2-mechanism, depending on substituents (*315*). The same aldehyde synthon was used as a substrate for an enantioselective synthesis of 2,3-methano-L-glutamic acid (*319*).

Chiral arylmethyl alcohols were used as substrates for a phthalimide Mitsunobu reaction. The carboxyl group was generated from the aromatic moiety by oxidation with ozone or ruthenium tetraoxide, giving (*R*)- and (*S*)-[2-²H]-glycine on a large scale in 40% overall yield (*320*). A phenyl group was also used as a masked carboxyl group during a synthesis of L-*threo*-β-hydroxy-Asp from methyl cinnamate. The alkene was asymmetrically dihydroxylated, with the benzylic hydroxyl converted to a bromide and then displaced with azide (*321*). Several syntheses have relied on displacement of a glucose-derived triflate with azide. The carboxylate was then generated by PDC oxidation of the aldehyde in a synthesis of bulgecinine (a dihydroxyproline derivative) (*322*), or by lead tetraacetate oxidative diol cleavage during a synthesis of (2*S*,3*R*)-3-hydroxy-leucine (*323*).

Phenylglycines were derived from aryl-substituted styrenes via asymmetric dihydroxylation followed by a Mitsunobu amination of the secondary alcohol with DPPA as the azide source. The acid group was generated by oxidation of the primary hydroxylmethyl group. (*R*)-(4-Methoxy-3,5-dihydroxy)-Phg was obtained in 32% yield (*324*), as was (*R*)-3′,5′-dibromo-4′-diazo(pyrrolidin-1-yl)-Phg, which was prepared during a total synthesis of vancomycin (*325*). Azide displacement of a glycerol-derived triflate was used to introduce the amino group of a difluoromethylphosphonate analog of serine (*326*); oxidation of the hydroxymethy substituent revealed the masked carboxyl group. A hydroxymethyl group was also used as a masked carboxyl group during syntheses of (2*S*,3*S*)-4, 4,4-[²H₃]-Va (*327*) and 2-(carboxycyclopropyl)glycine (*328*), with the amino group introduced by azide displacement of a mesylate. A Mitsunobu displacement with phthalimide

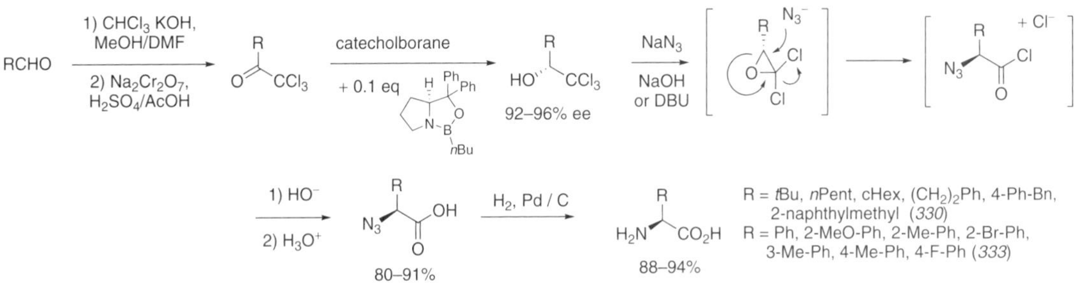

Scheme 5.16 Amination of Trichloromethyl Ketones.

was used on an optically active 1,2-diol substrate, with the carboxyl group generated by oxidation of the primary hydroxy group after a number of other steps (*329*).

Corey and Link have developed a modified version of α-hydroxy acid displacement, in which a trichloromethyl group is used both to activate the hydroxyl group for azide displacement and as a masked acid group (*330*). This synthetic route (see Scheme 5.16) begins with trichloromethyl ketones, which can be derived from reactions of aldehydes with nucleophilic trichloromethide reagents (making the overall sequence a stepwise aminocarboxylation reaction), or from trichloroacetyl chloride and nucleophilic carbon reagents. The trichloromethyl ketones undergo an asymmetric reduction by catecholborane in the presence of (S)-oxaborolidine catalyst with 92–98% ee (and 96% yield for the one example provided). Treatment of the (R)-(trichloromethyl) carbinols with sodium azide and base leads to (S)-α-azido acids; conventional hydrogenation gives the free amino acids in good overall yield. A modified Corey–Link procedure was applied to the synthesis of (1S,2S,5R,6S)-2-aminobicyclo[3.1.0]hexane-2,6-dicarboxylic acid, via addition of LiCHCl₃ to the protected homochiral ketone, followed by reaction with DBU/NaN₃ to generate the azido ester. The stereochemistry generated at the α-center was opposite to that provided by Strecker or Bücherer aminocarboxylation reactions (*331*). Another synthesis prepared (2S,4R,7R,9R)-7-amino-7-carboxy-octahydro-1H-indole-2-carboxylic acid (2-amino-7-aza-bicyclo[4.3.0]nonane-2,6-dicarboxylic acid) from 7-keto-octahydro-1H-indole-2-carboxylic acid (*332*). The Corey–Link procedure has also been adapted to produce arylglycines. The homochiral aryl trichloromethyl carbinols were obtained with >99% ee, but the original Corey–Link conditions for reaction with NaOH/NaN₃ resulted in racemic product. By carrying out the reaction with exactly one equivalent of DBU instead of NaOH, arylglycines with >99% ee were obtained (*333*).

An optically active (4-methylphenyl)(methyl)sulfinyl group was used as a masked carboxyl group during a synthesis of 3-fluoro-D-Ala. Amination was achieved via Mitsunobu displacement of a hydroxy group, with the carboxyl group revealed by a Pummerer rearrangement and oxidation (*334*).

5.3.3 Nucleophilic Aminations with Chiral Aminating Reagents

Chiral amines can be employed for nucleophilic amination reactions via conjugate addition reactions. A small amount of diastereoselectivity was obtained in an early asymmetric synthesis of Asp by addition of (S)- or (R)-α-methylbenzylamine to diethyl maleate (11–12% de) or diethylfumarate (6–8% de), with one amine producing the same diastereomer from both alkene isomers (*335*). Similarly, low stereoselectivities (8–23%) were obtained with an aminating reagent derived from ephedrine, which introduced a dimethylamino group (*336*). (S)-1-Phenylethylamine was added to (E)-4-keto-4-(methoxyphenyl)but-2-enoic acid with poor diastereoselectivity (10%). However, one isomer of the 4′-methoxy-homophenylalanine product precipitated with up to 96% de purity (*337*). Reaction at higher temperatures resulted in a retro-Michael equilibration that gave the product in 90% yield with 97% de (*337*). Chiral amides such as (S-α-methylbenzyl)benzylamide and (S-α-methylbenzyl)allylamide (*338–346*) have been used for the synthesis of β-amino acids, including the taxol side chain (*347*), via conjugate additions to acrylate derivatives (see Section 11.1). By incorporating a second amination, (2S,3S)- and (2R,3S)-2,3-diaminobutanoic acids were prepared (*348*). Enders has developed a version of the (S)-2-methoxymethyl-1-amino-pyrrolidine auxiliary which acts as a chiral ammonia equivalent. The TMS-SAMP underwent stereoselective conjugate additions to α,β-unsaturated enoates (93–98% de) enroute to β-amino acids (see Section 11.1) (*349–351*).

A homochiral oxazolidinone has been employed as a chiral aminating reagent, with Michael additions to nitroalkenes proceeding with complete diastereoselectivity (see Scheme 5.17). The nitromethyl substituent was oxidized to form the carboxyl group in 25–88% yield and the oxazolidinone ring cleaved using TMSI or Li/NH₃, producing D-amino acids with 95 to >96% ee (*352*).

Substituted acrylic acid amides of N-[(S)-1-(phenyl) ethyl]-N′-Cbz-diaminomethane were cyclized in the presence of Hg(TFA)₂ to form, after Hg cleavage, imidazolidin-4-one diastereomers (see Scheme 5.18). For R² = H the chiral amine had little effect on the

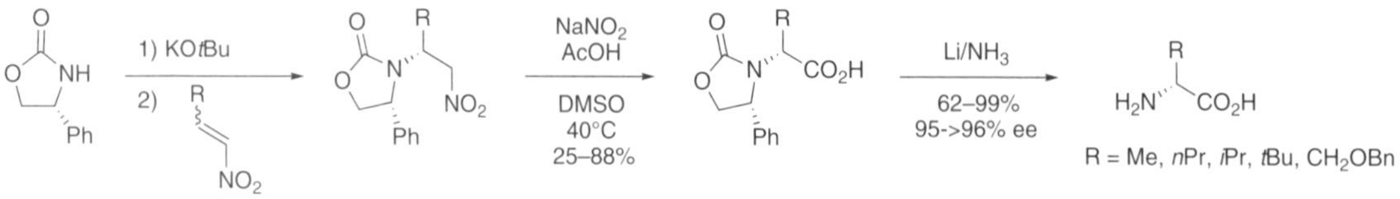

Scheme 5.17 Conjugate Additions of Oxazolidinones to Nitro-Alkenes (352).

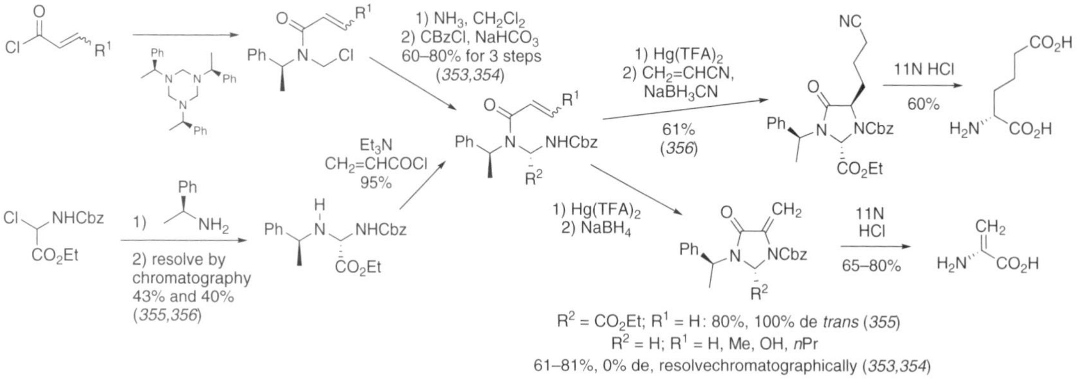

Scheme 5.18 Mercury-Catalyzed Amination.

stereoselectivity of the intramolecular amination and was effectively only used for diastereomer resolution, as 1:1 diastereomer mixtures were obtained (353, 354). However, for $R^2 = CO_2Et$ the diastereomers were resolved before cyclization and the imidazolidinone was obtained diastereomerically pure. Reductive removal of the mercury followed by hydrolysis led to enantiomerically pure D-Ala (355), while addition of the mercury-imidazolidinone to acrylonitrile provided a route to α-aminoadipic acid (356). Ala, Ser, Abu, and Nle were synthesized by using different acyl chlorides (353). A similar type of reaction has been applied to the synthesis of β-amino acids (357).

Aziridine-2-carboxylates are obtained by cyclization of alkyl 2, 3-dibromopropanoates with α-ethylbenzylamine or α-methylbenzylamine. Hydrolysis of the aziridine gave Ser with 25–38% ee (358).

5.3.4 Nucleophilic Aminations with Chiral Catalysts

The Gabriel synthesis has been adapted to asymmetric synthesis by using a chiral-phase transfer catalyst, (−)-N-Benzyl quininium chloride or (+)-N-benzyl cinchoninium chloride, to catalyze the amination of α-bromo esters with potassium phthalimide (359). Four deprotected amino acids (R = Me, Et, nPr, nBu) were synthesized in yields of 17–48% with 1.7–47% ee. The best results were obtained with a double asymmetric induction, in combination with a chiral ester on the 2-bromocarboxylate substrate.

Racemic 1,3-diphenylprop-2-enyl acetate has been asymmetrically aminated with phthalimide in the presence of a chiral palladium catalyst (see Scheme 5.19). The phenyl or alkene groups were used as masked carboxylates, with oxidation leading to D-Glu (31%, 96% ee) or L-Phg (33%, 96% ee) derivatives, respectively (360). Other disubstituted allylic acetates have been used (361), while benzylamine was used as the amine source in a synthesis of Ala, Nva, Val, and Phg (73–97% ee) (362).

Asp derivatives have been synthesized via amination of fumaric acid derivatives in the presence of aspartase enzymes (363–372).

5.3.5 Electrophilic Aminations

Electrophilic aminating reagents are discussed in Section 2.3.3; comprehensive reviews have been published in 1989 (373) and 2004 (374), along with a microreview of asymmetric electrophilic α-aminations of carbonyl groups in 2004 (375). Most asymmetric electrophilic aminations make use of substrates with chiral auxiliaries (see Section 5.3.6), but examples of aminations of chiral substrates or with chiral catalysts do exist, as discussed below. Recent advances in catalytic enantioselective α-aminations were reviewed in 2005 (376).

A number of β-hydroxy amino acids and hydrazino acids have been synthesized with excellent diastereoselectivity by electrophilic amination of the enolate derived from chiral β-hydroxy esters (see Scheme 5.20) (377–384). The chiral hydroxy esters are generally obtained

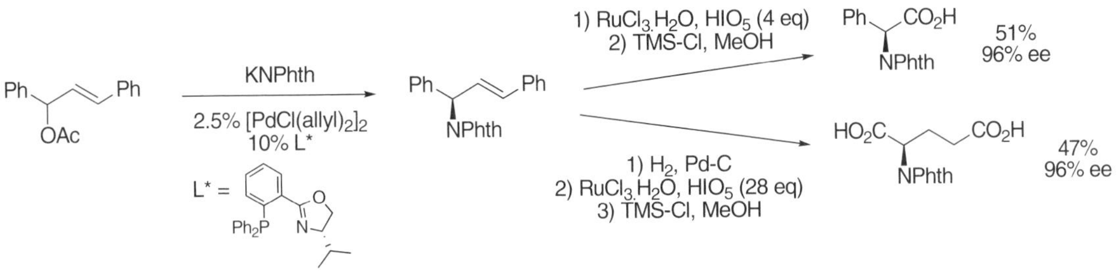

Scheme 5.19 Asymmetric Amination of Allyl Acetate with Chiral Catalyst (*360,361*).

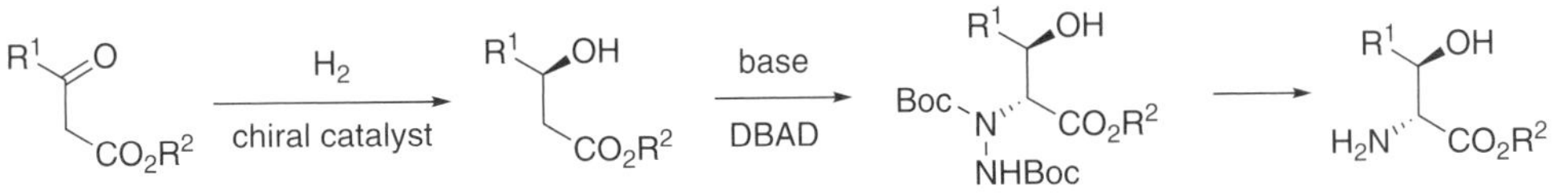

Scheme 5.20 Electrophilic Amination of Chiral Hydroxyesters (*377–386*).

by asymmetric reduction of β-keto esters. DBAD (di-*tert*-butyl azodicarboxylate) is the preferred electrophilic amine due to its greater stereoselectivity and easily removed Boc group (*385, 386*). The N–N bond of the hydrazino esters that are initially obtained are reductively cleaved to give the amino acids. Trisyl azide has been used to aminate an ester enolate enroute to some polyhydroxydiamino acids (65–72% yield, 70–100% de) (*287*), as well as the polyhydroxy amino acid component (polyoxin C) of the nikkomycins and polyoxins (*387*). Another synthesis of polyoxin C utilized TsN=Se=NTs for amination of an enolate (*388*). Methyl-3-hydroxy-5,5-dimethoxypentanoate was deprotonated and aminated with dibenzyl azodicarboxylate in 66% yield with >98% de (*389*). N-Boc 3-(4-cyanophenyl) oxaziridine was introduced as an electrophilic aminating reagent in 1993 (*390*). A review of new electrophilic amination reagents for the synthesis of hydroxy amino acids was published in 1997 (*391*).

Evans and Nelson reported an enantioselective electrophilic amination of achiral N-acyl oxazolidinones, using DBAD as the aminating reagent and a chiral magnesium bis(sulfonamide) complex for enolate generation. Arylglycine derivatives were produced with 80–99% ee, which improved to 96 to >99% ee upon recrystallization (*392*). α-Substituted-β-keto esters were aminated with dibenzylazodicarboxylate under the catalysis of [(S)-Ph-BOX-Cu(OTf)₂] to give α-aminated products with 97–98% ee (*393*). Cinchona alkaloids catalyzed the reaction of α-aryl-α-cyanoacetates, RCH(CN)CO₂Et, with dialkyl azodicarboxylates, producing α-cyano-arylglycine ester derivatives with 87–97% ee (*394*).

More recently, a potentially significant development in methods for amino acid synthesis was reported by two groups, employing L-Pro to catalyze an asymmetric

reaction between aldehydes and azodicarboxylates, without the need for preformed enolates. An oxidation step was still required to produce α-amino acids from the α-amino aldehyde adducts, but they were produced with >95% ee (side chain = Me, iPr, nPr, nBu, Bn) (*395–397*).

5.3.6 Amination of Acyclic Acids with Chiral Ester or Amide Auxiliaries

A number of research groups have developed synthetic routes involving asymmetric amination of an acyclic acid, in which a chiral ester or amide derivative is used to induce asymmetry. The best known of these are the Evans and Oppolzer procedures. Both nucleophilic and electrophilic aminations can be employed.

5.3.6a Evans' Oxazolidin-2-One Auxiliary

One of the most successful approaches to asymmetric amino acid synthesis makes use of an oxazolidinone auxiliary, prepared from chiral amino alcohols such as valinol (*398*), phenylalaninol (*399*) or norephedrine (*398*). Originally used to induce stereoselectivity in aldol condensations (*398*), the oxazolidinone has also been used for aldol reactions of a glycine enolate equivalent (see Section 6.4.2a) (*399, 400*). However, the more successful strategy involves nucleophilic or electrophilic amination of an acylated oxazolidinone auxiliary, as described below. In contrast to other synthetic routes involving cyclic templates such as oxazolidin-5-ones and imidazolidinones (see Chapter 7), the chiral center which is being generated lies outside the template ring. A summary of early results in this area has been published (*401*).

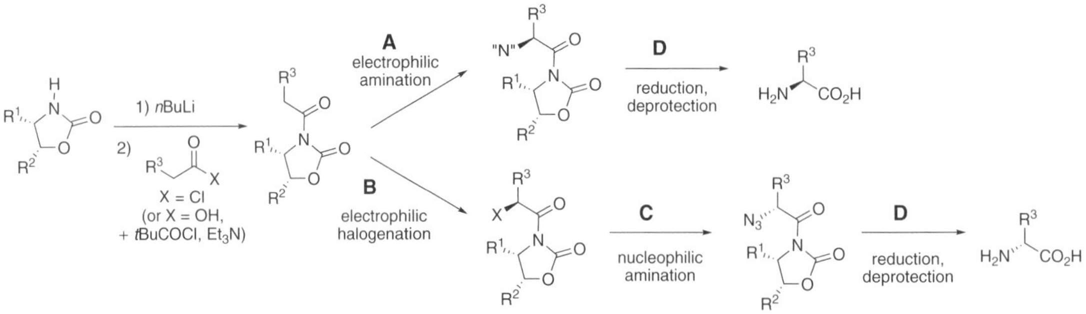

Scheme 5.21 Synthesis of Amino Acids via Amination with Oxazolidin-2-One Auxiliary (see Table 5.1).

The first reports of amino acid synthesis via amination of an acylated oxazolidinone appeared in 1986, when both Evans et al. (402) and Trimble and Vederas (403) published essentially the same procedure. A chiral oxazolidin-2-one template was acylated with the acid to be aminated, followed by generation of an enolate that was then electrophilically aminated with a dialkyl azodicarboxylate (see Scheme 5.21, Table 5.1). Full experimental details were later provided (385, 404). The best diastereoselectivities were achieved with di-tert-butyl or dibenzyl azodicarboxylate as an aminating reagent (403), while a benzyl substituent (402) on the oxazolidinone auxiliary resulted in better diastereoselectivity than an isopropyl group (403). Trisyl azide was subsequently reported to give equivalent amination results to the azodicarboxylates (404, 405), and the azido derivatives were found to be more versatile than the hydrazino equivalents (404). A significant advantage of the amination procedure over alkylations of a glycine equivalent is that sterically bulky side chains are accommodated without a decrease in yield. Indeed, stereoselectivity is increased.

The aminated diastereomers are generally readily resolved by chromatography (385). The deprotected amino acids are obtained with up to >99% ee after removal of the auxiliary and reduction of the hydrazide bond or azide group by hydrogenation. The azide group has also been reduced with PPh$_3$ (249, 251) and SnCl$_2$ (252–254); a variety of other methods for azide reduction have been developed in recent years, as outlined in the introduction of Section 5.3. The exocyclic imide carbonyl is quite reactive (on a par with a phenyl ester carbonyl (404)), allowing for cleavage of the auxiliary by saponification or transesterification under comparatively mild conditions. For sterically demanding substrates, the addition of hydrogen peroxide to the saponification solution prevents endocyclic imide carbonyl reaction (404). For a substrate where alkaline peroxidic hydrolysis was not compatible with the side chain, a transesterification using titanium tetraisopropoxide and excess benzyl alcohol proceeded smoothly (406). Another transesterification reaction using lanthanum(III) iodide and alcohols gives high yields of esters without racemization, but has not been applied to the synthesis

of amino esters (407). The azide group can also be reduced before removal of the auxiliary, although imidazolone formation can then be problematic (404).

A significant advantage of this route to amino acids is that it is possible to obtain the enantiomeric amino acid via a modified procedure, without needing to use the enantiomeric chiral auxiliary. The boron enolate of the acylated oxazolidinone is halogenated and the halide displaced with azide, resulting in a net inversion of amino acid stereochemistry (404, 408). Optimum stereoselectivity during halogenation was obtained by using bromide as the halogen, NBS as the halogen source, CH$_2$Cl$_2$ as solvent, and diisopropylethylamine as base (404). For displacement with azide, the best reagent was tetramethylguanidinium azide in CH$_2$Cl$_2$, which resulted in <1% epimerization (404). The aminated diastereomers were again readily resolved by chromatography, allowing for high optical purities of the final amino acid. A major explosion has been observed during solvent removal from a solution of tetrabutylammonium azide, and as this has been attributed to the dichloromethane solvent, the use of acetonitrile now is recommended for azide displacements (409, 410).

A wide variety of amino acids have been prepared using the above procedures (see Scheme 5.21, Table 5.1). Both electrophilic and nucleophilic amination were used in the synthesis of 3,5-dihydroxy-Phg derivatives, although in one example the amination with trisyl azide was not successful (411). For the synthesis of L-[1-^{14}C]-Phe the electrophilic hydrazide route was found to give better yields than the azide route, and hydrazide reduction was successfully accomplished at a lower H$_2$ pressure than Evans reported by using 2-propanol/water as solvent (412). 3,3-Diphenylalanine was prepared in four steps and 77% overall yield via electrophilic amination with trisyl azide (413); synthesis of the same compound by alkylation of a glycine Schiff base using Oppolzer's auxiliary gave a 46% overall yield (399). 5-Tritiated Orn (414), Lys homologs (415), 5-hydroxy-Leu (416), α-(1-carboxycyclopentyl) Gly (417), 3'-methoxy-4'-hydroxy Phe (418), (dicyclohexylphosphino) Ser (252), 2,3,4,5-tetrahydropyridazine-3-carboxylic acid (419), and piperazic acid (419–422) have also been prepared via the electrophilic amination route, with the

Table 5.1 Amination of Acids Derivatized with Oxazolidinone Auxiliary (see Scheme 5.21)

Auxiliary Substituents			A Electrophilic Amination		B,C Electrophilic Halogenation/ Nucleophilic Amination		D Deprotection		Reference
R^1	R^2	R^3	Conditions	Yield	Conditions	Yield	Conditions	Yield	
Bn	H	Me iPr tBu Ph Bn CH₂CH=CH₂	1) LDA 2) DBAD	91–96% 94–>98% de			1) LiOH, H₂O, or LiOBn 2) CH₂N₂ 3) TFA 4) H₂, Ra-Ni	1) 51–96% 98–>98% ee 2,3,4) 99% for R³ = Ph, 98% ee	1986 (402)
iPr	H	Me iPr Bn	1) LDA 2) (RO₂CN)₂ R = Me, Et, tBu, Bn	85–92% 38–94% de			1) LiOBn or LiSH 2) H₂ Pd-C or TFA 3) H₂, Ra-Ni	1) 74–97% 2+3) 81–98% 44–94% de	1986 (403)
Bn	H	Me iPr tBu Ph Bn CH₂CH=CH₂ (structure: NHBoc / CO₂Bn; O, OMe)	1) KHMDS 2) trisyl azide	74–91% 82–98% de			LiOH, H₂O₂	96–100%	1987 (405)
Bn Me	H Ph	H with α-Br or α-Cl reacted with 1) Bu₂BOTf, Et₃N 2) R⁴CHO to form CH(OH)R⁴ with α-Br R⁴ = Me, iPr, Ph, CH(Me) CH₂CH=CHMe 90–96% de pure isomer: 63–94%, >99% de			NaN₃, DMSO	90–95%	1) LiOH, H₂O 2) H₂, Pd/C	R = Me 82%	1987 (431)

(Continued)

Table 5.1 Amination of Acids Derivatized with Oxazolidinone Auxiliary (see Scheme 5.21) (continued)

Auxiliary Substituents			A Electrophilic Amination		B,C Electrophilic Halogenation/ Nucleophilic Amination		D Deprotection	Yield	Reference
R^1	R^2	R^3	Conditions	Yield	Conditions	Yield	Conditions		
Bn	H	H with α-Br reacted with 1) Bu$_2$BOTf, Et$_3$N 2) MeC(=CH$_2$)CHO to form CH(OH)C(Me)=Me$_2$ with α-Br			NaN$_3$, DMSO	82%	1) MeOMgX 2) (cHex)$_2$BH forms [structure: Me OH / CO$_2$Me / N-H pyrrolidine]	1) 87% 2) 72%	1987 (432)
Bn	H	iPr iBu Ph CH$_2$CH=CH$_2$			1) Bu$_2$BOTf, Et$_3$N 2) NBS, CH$_2$Cl$_2$ 3) TMG-N$_3$	56–90% de purified isomer: 67–86%, >99% de	1) H$_2$ Pd-C 2) LiOH, H$_2$O	95–100% 98–99% ee	1987 (408)
Bn	H	Me iPr tBu Ph Bn CH$_2$CH=CH$_2$ =CMe$_2$ (CH$_2$)$_2$CO$_2$Me	1) LDA 2) DBAD	51–96% 94–>98% de			1) LiOH, H$_2$O or LiOBn 2) CH$_2$N$_2$ 3) TFA 4) H$_2$, Ra-Ni	1) 51–96% 98–>98% ee 2+3+4) 83–89% 98–>99% ee	1988 (385)
Bn	H	Bn	1) KHMDS 2) trisyl azide	83–86%			1) Ti(OBn)$_4$, BnOH or TFA, thioanisole 2) H$_2$, Pd/C or Ra-Ni		1989 (427)
Bn		[structure: CO$_2$tBu, NHBoc, CO$_2$Bn, MeO, OMe, O]							
Bn	H	(R)- or (S)-CH(Me)(4-MeO-Ph)			1) Bu$_2$BOTf, DIEA 2) NBS, CH$_2$Cl$_2$ 3) TMG-N$_3$	65–70% 94–98% de	1) LiOH, H$_2$O$_2$ 2) H$_2$, Pd-C 3) 48% HBr	73%	1989 (444)
Ph	H	(R)- or (S)-CH(Me)Ph			1) Bu$_2$BOTf, DIEA 2) NBS, CH$_2$Cl$_2$ 3) TMG-N$_3$	88–99% >99% de	1) LiOH, H$_2$O$_2$ 2) H$_2$, Pd-C	64–73% 90–98% ee	1989 (440) 1992 (410)
Bn	H								1994 (441)

Bn	H	iPr iBu tBu Ph Bn $CH_2CH=CH_2$			1) Bu$_2$BOTf, DIEA 2) NBS, CH$_2$Cl$_2$ 3) TMG-N$_3$	56–90% de pure isomer: 67–86% >99% de	1) LiOH, H$_2$O$_2$ 2) H$_2$, Pd-C	88–97% 98→>99% ee	1990 (*404*)
Bn	H	iPr iBu tBu Ph Bn $CH_2CH=CH_2$	1) KHMDS 2) trisyl azide	82–98% 74–91% de			1) H$_2$ Pd-C 2) RCOCl, Et$_3$N 3) LiOH, H$_2$O	96–99% 98→>99% ee	1990 (*404*)

(structure: benzyl-protected compound bearing –CH$_2$– linked aromatic ring with O–, OMe substituents, and side chain –CH(NHBoc)CO$_2$Bn)

Bn	H	(CH$_2$)$_2$CH(OMe)$_2$	1) LDA 2) DBAD	93%			MeOMgI CH$_2$Cl$_2$, MeOH	96%	1991 (*419*)
Bn	H	(*R*)- or (*S*)-(CH$_2$)$_2$CHTN$_3$	1) KHMDS 2) trisyl azide	46–66%			1) LiOH, H$_2$O 2) H$_2$, Pd/C	1) 76% 2) 92%	1991 (*414*)
Bn	H	Bn	1) LDA 2) DBAD	46%			1) LiOH 2) TFA 3) H$_2$, Ra-Ni	97% >96% ee	1991 (*412*)
Bn	H	3,4-(MeO)$_2$-Ph	1) KHMDS 2) trisyl azide	82–98% 74–91% de	OR 1) Bu$_2$BOTf, DIEA 2) NBS, CH$_2$Cl$_2$ 3) TMG-N$_3$	54% 48% de	1) H$_2$, Pd-C 2) H$_3$O$^+$	>80% ee	1991 (*411*)
iPr	H	CH(OH)(oxazol-4-yl) CH(OH)(1-Boc-pyrrole-2-yl) with α-Br	NaN$_3$	92–95%			NaOMe	84–88%	1992 (*433*) 1997 (*436*)
Bn Me	H Ph	(CH$_2$)$_3$Br	1) LDA 2) DBAD 3) DMPU	55–76% 88–96% de	forms (Boc–N, N–Boc piperidine –COX* structure)		1) LiOH, H$_2$O 2) TFA	1) 89% 2) 94%	1992 (*420*) 1996 (*421*)
Me	Ph	3-HO-Ph	1) KHMDS 2) trisyl azide	30–68%			1) LiOH, H$_2$O$_2$ 2) H$_2$ Pd-C	1) 90% 2) 67%	1992 (*797*)
Me	Ph	4-Cl-Bn	1) KHMDS 2) trisyl azide	94% 100% de			1) LiOH, H$_2$O$_2$ 2) H$_2$ Pd-C	1) 99% 2) 77%	1992 (*413*)
Me	Ph	CH(Ph)$_2$							

(Continued)

Table 5.1 Amination of Acids Derivatized with Oxazolidinone Auxiliary (see Scheme 5.21) (continued)

Auxiliary Substituents			A Electrophilic Amination		B,C Electrophilic Halogenation/ Nucleophilic Amination		D Deprotection		Reference
R¹	R²	R³	Conditions	Yield	Conditions	Yield	Conditions	Yield	
Bn	H	H with α-Br reacted with 1) Bu₂BOTf, Et₃N 2) PhCHO to form CH(OH)Ph with α-Br 65%, 97% de			NaN₃, DMSO	82%	1) Mg(OMe)₂ 2) H₂, Pd/C	96%	1992 (437)
Bn	H	nPr							1992 (464)
Me	Ph	CH=CH₂; CH=CHMe; CH=CHPh coupled to oxazolidinone via acid chloride, CuCl₂							
Bn	H	Ph; 3,5-(MeO)₂-Ph; 3,5-(MeO)₂-4-Me-Ph [structures: SEMO₂C, NHBoc, OBn, BnO, Cl, OBn; and OBn, CN, O]	1) KHMDS 2) trisyl azide	60–82% 78–>90% de			1) SnCl₂, dioxane or H₂, Pd-C 2) LiOH, H₂O₂	94–96%	1992 (254)
Bn	H	[tetrahydrofuranyl structure]	1) KHMDS 2) trisyl azide	—			1) LiOH, NaHCO₃ 2) H₂, Pd-C	—	1993 (798)
Bn	H	[thiolane and 1,3-dioxolane structures]	1) NaHMDS 2) DBAD	67%			1) 1N LiOH 2) CH₂N₂	99%	1993 (799)
Bn	H	CH₂CF₂Me; CH₂CF₂CF₃; CH₂(CF₂)₂CF₃; (CH₂)₂CF₂CF₃; (CH₂)₂CF₂Me			1) Bu₂BOTf, DIEA 2) NBS, CH₂Cl₂ 3) TMG-N₃	1+2) 73–95% 40–76%de 3) 59–76%	1) LiOH, H₂O₂ 2) H₂, Pd-C	1) 70–94% 2) 77–99%	1993 (430)

Bn	H	4-Cl-Bn 3-BnO-4-MeO-Bn	1) KHMDS 2) trisyl azide	90–94% de 71–76% pure isomer			1) LiO₂H 2) H₂ Pd-C	1) 84–92% 2) 33–85%	1994 (800)
Bn	H	(R)- or (S)-CH(CHF₂) (2-BnO-Ph)	1) KHMDS 2) trisyl azide	50–56% >95% de			1) LiOH, H₂O₂ 2) H₂ Pd-C	27–30%	1994 (445) (446)
Bn	H	(CH₂)₂P(=S)(Ph)₂	1) KHMDS 2) trisyl azide	85%			1) LiOH, H₂O₂ 2) SnCl₂ MeOH	83%	1994 (801)
Ph	H	(R)- or (S)-CH(Me) (2,6-Me₂-Ph)			1) Bu₂BOTf, DIEA 2) NBS, CH₂Cl₂ 3) TMG-N₃		1) LiOH, H₂O₂ 2) H₂ Pd-C	51–70% overall	1995 (442)
H		(R)- or (S)-CH(Me)Ph			1) Bu₂BOTf, DIEA 2) NBS, CH₂Cl₂ 3) TMG-N₃	67–82% 76% de		54–69% overall	1995 (452)
Bn	H	CH₂SiMe₃ CH₂SiMe₂Ph CH₂SiMePh₂			1) Bu₂BOTf, DIEA 2) NBS, CH₂Cl₂ 3) TMG-N₃	65–87% >97% de	1) LiOH, H₂O₂ 2) PPh₃	1) 88–100% 2) 50–57%	1995 (251)
iPr	H	1-CO₂Bn-cPent	1) KHMDS 2) trisyl azide	78%			LiOH, H₂O₂	98% 88–94% ee	1995 (417)
Bn	H	(R)- or (S)-CH(CH=CH₂) (4-BnO-Ph)	1) KHMDS 2) trisyl azide	84–92%			1) LiOH, H₂O₂ 2) tBuOC(=NH)CCl₃, BF₃·Et₂O 3) (cHex)₂BH, CH₂Cl₂ forms 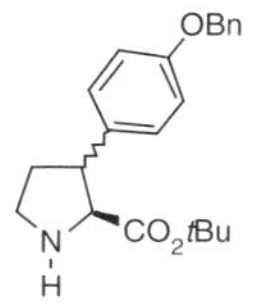	1) 72–75% 2) 78–84% 3) 48–60%	1996 (423)
Bn	H	CH₂P(=S)(cHex)₂	1) KHMDS 2) trisyl azide	83%			1) LiOH, THF/H₂O 2) SnCl₂, MeOH 3) Fmoc-OSucc or 1) SnCl₂, MeOH 2) Boc₂O, NaHCO₃	68% or 93%	1996 (802)

(Continued)

383

Table 5.1 Amination of Acids Derivatized with Oxazolidinone Auxiliary (see Scheme 5.21) (continued)

Auxiliary Substituents			A Electrophilic Amination		B,C Electrophilic Halogenation/ Nucleophilic Amination		D Deprotection	Yield	Reference
R^1	R^2	R^3	Conditions	Yield	Conditions	Yield	Conditions		
Bn	H	CH₂(o-carborane)			1) TiCl₄, DIEA, NBS 2) TMG-N₃	1) 65–83% 2) 82–89%	Ti(OiPr)₄, BnOH	89–99%	1996 (406)
Bn	H	CH(CH=CH₂)(4-BnO-Ph)	1) KHMDS 2) trisyl azide	84–92%					1996 (803)
Bn	H	CH₂P(=S)(cHex)₂	1) KHMDS 2) trisyl azide	83%			1) SnCl₂, MeOH 2) Boc₂O 3) LiOH, THF 4) TFA	1+2) 93% 3) 70%	1996 (252)
Bn	H	(arene structure)	1) KHMDS 2) trisyl azide	65–73% 60–95% de			1) SnCl₂, dioxane 2) Boc₂O 3) LiOH, THF 4) TFA	1+2) 56% 3+4) 95%	1996 (253)
Bn	H	3-TBSO-4-MeO-Bn	1) KHMDS 2) trisyl azide	75%			1) LiOH, H₂O₂ 2) H₂, Pd-C	1) 87% 2) 84%	1996 (418)
Ph	H	(R)- or (S)-CH(iPr)Ph			1) Bu₂BOTf, DIEA 2) NBS, CH₂Cl₂ 3) Amberlite-azide resin	64%	1) LiOH, H₂O₂ 2) H₂, Pd-C	1) 90% 2) 90%	1996 (443)
iPr	H	(structure with EtO, N, OEt, (CH₂)5)	1) KHMDS 2) trisyl azide	70%			1) LiOH, H₂O₂ 2) 0.25M HCl	75%	1997 (428) 2001 (429)
Bn	H	CH(CH=CH₂)CH₂OTBS	1) KHMDS 2) trisyl azide	80–86%			1) LiOH, H₂O₂ 2) tBuOC(=NH)CCl₃, BF₃.Et₂O 3) (cHex)₂BH, CH₂Cl₂ forms (structure)	1) 88–97% 2) 89–94% 3) 48–64%	1997 (804)

R^1	R^2	Substituent	Conditions (1, 2)	Yield (de)	Conditions	Yield	Year (Ref.)
Bn	H	(cyclopropyl)	1) KHMDS 2) trisyl azide	35%	1) SnCl₂, dioxane 2) LiOH, H₂O₂	1) 57–79% 2) 72–76%	1997 (426)
Bn	H	3,5-(BnO)₂-4-MeO-Ph	1) KHMDS 2) trisyl azide	82%	1) LiO₂H 2) H₂, Pd-C	1) 87% 2) 63%	1997 (425)
Bn	H	(CH₂)₄CH₂NHBoc; (CH₂)₄CH(Me)N(Me)Boc	1) KHMDS 2) trisyl azide	40–50%	1) LiOH, H₂O₂ 2) H₂, Pd-C	65–70%	1997 (415)
iPr	H	(sugar structure, OBn, BnO, ( )ₙ, n = 1,2)	1) LDA 2) DBAD	65% 98% de			1998 (805)
Bn	H	3,5-(iPrO)₂-4-MeO-Ph	1) KHMDS 2) trisyl azide	71% >85% de	1) 10% Pd/C, Boc₂O 2) LiOH, THF	81%	1998 (44)
Bn	H	(CH₂)₃CH=CH₂	1) KHMDS 2) trisyl azide	67%	1) SnCl₂, MeOH 2) Boc₂O 3) LiOH, THF	83%	1998 (806)
Bn	H	CH₂CH=CH₂	1) NaHMDS 2) DBAD	80%	LiO₂H	89%	1998 (807)
iPr	H	(cyclohexane structure)	1) LDA 2) DBAD	68–75% 72–98% de			1998 (808)
Bn	H	(sugar structure, OBn, BnO, OR, RO, R = Me, Bn)	1) KHMDS 2) trisyl azide	60–70%	1) LiOH, H₂O₂ 2) H₂, Pd-C	90–95%	1998 (424)
Bn	H	(CH₂)ₙBr, n = 3,4,5	1) KHMDS 2) trisyl azide	78%	1) LiOH, H₂O/THF 2) H₂, Pd-C	81%	1998 (809)
Bn	H	CH₂CH(Me)Et	1) KHMDS 2) trisyl azide	59%	1) H₂, Pd-BaSO₄ 2) LiOH, H₂O₂	77%	1999 (810)
Bn	H	3-BnO-Bn	1) KHMDS 2) trisyl azide	83%	1) MeOMgBr, MeOH 2) H₂, Pd-C	92%	1999 (811)
Bn	H	3-iPrO-4-MeO-Bn	1) KHMDS 2) trisyl azide	80%	1) LiOH, H₂O/THF 2) SnCl₂, Pd-C	76%	1999 (812)
Bn	H	CH(Me)CH₂SePh	1) KHMDS 2) trisyl azide				

(Continued)

Table 5.1 Amination of Acids Derivatized with Oxazolidinone Auxiliary (see Scheme 5.21) (continued)

Auxiliary substituents			A Electrophilic Amination		B,C Electrophilic Halogenation/ Nucleophilic Amination		D Deprotection		Reference
R^1	R^2	R^3	Conditions	Yield	Conditions	Yield	Conditions	Yield	
Me	Ph	3-Cl-4-Me-Ph 3-MeO-5-BnO-Ph	1) KHMDS 2) trisyl azide	66–72%			1) LiOH, H$_2$O/THF 2) H$_2$ Pd-C	50–62%	2000 (813)
Bn	H	CH$_2$(1-Boc-piperidin-4-yl)	1) KHMDS 2) trisyl azide	75%			1) LiOH, H$_2$O/THF 2) H$_2$ Pd-C	—	2001 (814)
Bn	H	CH(Me)CH$_2$(β-D-galactosyl)			1) Bu$_2$BOTf, DIEA 2) NBS, CH$_2$Cl$_2$ 3) TMG-N$_3$	57% Br 87% de 97% azide	1) LiOH, H$_2$O/THF	97%	2003 (438)
Ph	H	CH=CH(CH$_2$)$_2$-CO$_2$tBu			addition of 4-MeOBnMgBr, then NBS, then NaN$_3$	60%	1) LiOH, H$_2$O/THF 2) H$_2$ Pd-C	87%	2003 (815)
Bn	H	(CH$_2$)$_3$CH=CH$_2$ (CH$_2$)$_5$CH=CH$_2$	1) KHMDS 2) trisyl azide	67%			1) SnCl$_2$,MeOH 2) Boc$_2$O 3) LiOH, THF	42%	2004 (816)
Bn	H	CH$_2$CH(Me)CH$_2$-OBn	1) NaHMDS 2) trisyl azide	73%			1) H$_2$ Pd-C 2) LiOH, H$_2$O/THF	79%	2004 (416)
Bn	H	CH$_2$Si(Me)$_2$CH$_2$Cl converted to CH$_2$Si(Me)$_2$OH CH$_2$Si(Me)$_2$CH$_2$SMe			1) Bu$_2$BOTf, DIEA 2) NBS, CH$_2$Cl$_2$ 3) TMG-N$_3$	80% Br 99% azide	1) LiOH, H$_2$O/THF 2) H$_2$ Pd-C	89–99% hydrolysis 85–98% hydrogenation	2004 (439)

latter product obtained by an in situ intramolecular cyclization of the hydrazine adduct with a brominated side chain. A Pro derivative was synthesized by hydro-boration/cycloalkylation of an azido-olefin produced by electrophilic amination (*423*). The azido precursors of 5-bromo-Nva, 6-bromo-Nle, and 7-bromo-2-aminohep-tanoic acid were also prepared via azidation reactions of alkanoic acids derivatized with Evans' auxiliary (*424*).

A number of arylglycines related to those found in the peptide antibiotic vancomycin have been synthe-sized (*254, 425*). In some cases amination with trisyl azide gave a triazene intermediate that required decom-position to form the azide. Azide reduction was carried out with SnCl₂ for substrates with hydrogenation-sensitive protecting groups (*254*). A diaminodicarboxylic acid (phenyl-1,3- or -1,4-bisglycine) was synthesized by simultaneous reaction of both acyl groups of benzene-diacetic acid (*253*), as was a cyclopropane-bridged bis(glycine) (*426*), while the aryl-ether linked bis-amino acid isodityrosine was made by sequential elaboration (acylation with an oxazolidinone, electrophilic amina-tion, hydrolysis, and reduction) of either end of an aryl ether bis-carboxylic acid (*427*). A hexane-linked bis(amino acid) was prepared using Evans' auxiliary for amination to generate one amino acid center, with the other end supplied by alkylation of a glycine equivalent, Schöllkopf's bislactim ether template (*428, 429*).

The bromination/nucleophilic amination route is less popular but has been used to synthesize fluorinated ana-logs of norvaline and norleucine (*430*) and β-(trialkylsilyl) alanines (*251*). A variation of the nucleophilic amination procedure was used to synthesize *anti* β-hydroxy-α-amino acids (*431*). The halide substituent was present during aldol reactions of the enolate of the acylated oxazolidinone, and then was subsequently displaced with sodium azide. A MeBmt epimer (*431*), 3-hydroxy-4-methyl-Pro (*432*), β-hydroxy-His (*433–435*), and oxazole and pyrrole analogs of β-hydroxy-His (*436*) were prepared by this procedure, while β-phenyl-Cys was synthesized by displacement of the mesylate of a β-phenyl-Ser derivative with thiolacetic acid (*437*). A complex *C*-glycoside analog of β-D-galactosyl-Thr has also been synthesized (*438*), as have 4-dimethylsila analogs of homoserine and homomeThr (*439*).

The researchers of Hruby's group have made exten-sive use of oxazolidin-2-one auxiliaries in the synthesis of β-substituted aromatic amino acids. Initially the four isomers of β-methyl-Phe were prepared by acylating the L- or D-phenylalaninol-derived auxiliary with (S)- or (R)-3-phenylbutyric acid, followed by bromination and azide displacement (*410, 440*). Alternatively, racemic 3-phenylbutyric acid can be coupled and the two diastereomers resolved by crystallization or chro-matography (*441*). A similar procedure was used to prepare the four isomers of β-methyl-2',6'-dimethyl Phe (*442*), β-isopropyl Phe (*443*), and two isomers of β-methyl-Tyr (*444*), β-difluoromethyl-*m*-Tyr (*445, 446*) and 3-aryl-Pro (*423*). Improved bromination diastereo-selectivity was obtained with a phenyl substituent on the oxazolidinone, rather than a benzyl group (*442, 443*).

A more powerful methodology was developed to prepare the four isomers of β-methyl Trp, with the chiral auxiliary used to generate the stereochemistry of the amino acid β-center as well as the α-center (*409*). The oxazolidinone was acylated with an indole-substituted acrylic acid, followed by introduction of the methyl sub-stituent by a stereoselective 1,4-conjugate addition (70–80% de) (see Scheme 5.22, Table 5.2). The enolate present at the end of the Michael addition could be trapped with NBS or quenched with NH₄Cl. Nucleophilic displacement of the bromide with azide provided one C-2 diastereomer, while regeneration of the enolate with KHMDS and electrophilic amination generated the C-2 epimer. The remaining two diastereomers were obtained by starting with the enantiomeric chiral auxiliary. Minor improvements to this synthesis have since been reported (*447, 821*).

A study of the organocuprate conjugate addition step found that little diastereoselectivity (10% de) was obtained with a benzyl substituent on the oxazolidinone auxiliary, but with a phenyl group moderate to high stereoselectivity could be achieved (48–98% de) (*448*). Dibutylboron triflate has recently been found to promote conjugate addition reactions to similar *N*-acylated imi-dazolidinone derivatives with high diastereoselectivity (*449*). The tandem Michael addition/electrophilic bro-mination has also been investigated more thoroughly (*450*). The general synthetic route was applied to the preparation of the four isomers of β-methyl-Tyr (*451*), but in this report the *syn/anti* stereoisomers were obtained by reversing which β-substituent (e.g. methyl or aryl) was added by conjugate addition and which was already present on the acyl group, rather than by using both elec-trophilic and nucleophilic amination routes. Alternatively, the chiral auxiliary can be removed and replaced with its enantiomer after the conjugate addition, which is espe-cially useful if higher diastereoselectivity is observed with one of the two addition sequences (*451*). For a syn-thesis of β-methyl-Tyr, bromination was accomplished by a second enolate generation rather than by quenching the conjugate addition (*451*). The effect of the β-carbon chirality on bromination of an α-center boron enolate in the absence of a chiral center on the oxazolidinone has been studied, with up to 78% de observed (*452, 453*). Variations of the above procedures have been used to synthesize the four isomers of β-methyl-Phe (*452*), β-methyl-3-(2-naphthyl)-Ala (*454*), β-isopropyl-Phe (*455*), β-isopropyl-Tyr (*456*), *O*,2'-dimethyl-β-methyl-Tyr (*457*), *O*,2',6'-trimethyl-β-methyl-Tyr (*458*), β-aryl-*m*-Tyr (*459*), β-isopropyl-2',6'-dimethyl-Tyr (*460*), and β-methyl-His (*461*) (see Table 5.2). Dialkylaluminum chlorides also add to α,β-unsaturated *N*-acyl oxazolidi-nones, with the intermediate aluminum enolate trapped by *N*-halosuccinimide. The Evans' auxiliary resulted in mixtures of all four possible diastereomers, but oxazoli-dinone auxiliaries derived from glucosamine produced only one or two of the four possible isomers, even with aliphatic substituents (*462*).

A critical step of the Evans' amino acid synthesis route is the ability to acylate the oxazolidinone template.

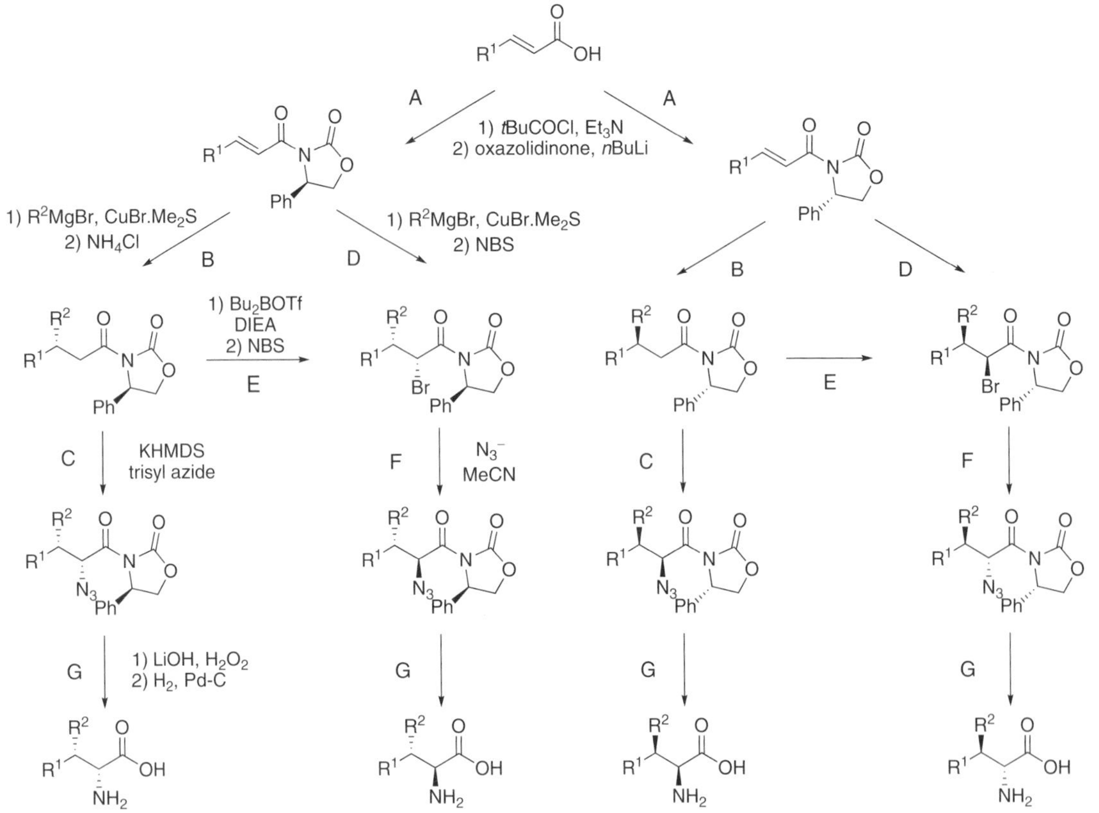

Scheme 5.22 Stereoselective Synthesis of β-Branched Amino Acid via Conjugate Addition (see Table 5.2).

Initially the lithiated oxazolidinone was reacted with an acyl chloride or a mixed pivalic acid anhydride (385, 402, 403, 463). However, the success of Evans' method for N-acylation depends on the nature of the acyl substituent. An improved acylation procedure has been published (464) which reacts an N-trimethylsilyl oxazolidinone intermediate with acyl chlorides in the presence of copper(II) chloride. For some substituents, such as acryloyls, better results can be obtained by reacting the acid anhydride or a mixed anhydride of the acid and pivalic acid in the presence of LiCl and Et₃N (465). However, neither procedure works with β,γ-unsaturated acyl compounds due to isomerization of the alkene bond. The desired acylation can be achieved using a mixed anhydride with N-methylmorpholine (466, 467). A one-pot method combining acid and oxazolidinone in the presence of pivaloyl chloride and triethylamine gave yields of 14–93% for nine different acids with three different oxazolidinones (468).

Another critical step of the reaction scheme is the ability to prepare the chiral oxazolidinone auxiliaries. An alternative procedure for the synthesis of oxazolidinones from α-amino acids has been described which avoids the potentially hazardous borane reduction of the acid group by using calcium borohydride reduction of the amino acid methyl ester. N-Carboxyethyl protection was employed to prevent difficulties in isolation of the amino alcohol intermediate. The oxazolidinones derived from Phe, Phg, Val, Ile, Met, and Trp were prepared in four steps and 69–87% overall yield (469). An even better approach employed NaBH₄ and LiI to achieve a one-pot reduction of N-alkoxycarbonyl amino acid esters, followed by cyclization to form the oxazolidinone. The method gave 80–100% yields of the oxazolidinone from the amino ester, while avoiding hazardous reduction and cyclization conditions (470). Xu and Sharpless have reported a new approach to enantiomerically enriched oxazolidinones via asymmetric dihydroxylation of conjugated dienes, followed by cyclization/displacement with p-tosylisocyanate (471). A number of years later, styrene and 2-vinylnaphthyl were converted into 4-phenyl- and 4-(2-naphthyl)-oxazolidinones by a route employing a Sharpless catalytic asymmetric aminohydroxylation reaction (472). Mild conditions for cyclization of 2-amino alcohols to oxazolidin-2-ones without racemization have been developed, via reaction of the amino alcohol with electrochemically generated tetraethylammonium peroxydicarbonate (NEt₄⁺O₂COCO₂O⁻NEt₄⁺) or tetraethylammonium carbonate (NEt₄⁺OCO₂⁻NEt₄⁺) (473). Another route to optically active oxazolidinones is via enzymatic resolution of racemic N-acyloxymethyl oxazolidinones via ester hydrolysis with lipases (474).

Table 5.2 Stereoselective Synthesis of β-Branched Amino Acids via Conjugate Additions (see Scheme 5.22)

R¹	A yield	R²	B yield	C yield	D yield	E yield	F Conditions	F Yield	G yield	Reference
N-Mes indol-3-yl	95–98%	Me	70–80% de; pure isomer: 57–71%; 100% de	59–62%; 88–92% de	72–82% de		TMG-N₃	78–79%	35–59%; >95% de	1992 (409); 1994 (821)
2-Me-4-MeO-Ph		Me			90–99% de		TMG-N₃		58–65% for D, F, G	1993 (457)
Me		2-Me-4-MeO-Ph			pure isomer: 74–80%; >99% de					
Me		Ph; 2-Me-Ph; 2,6-Me₂-Ph; 4-MeO-Ph; 2-Me-4-MeO-Ph	67–99% de; pure isomer: 52–89%; >99% de							1993 (450)
Me; 4-MeO-Ph	91–98%	4-MeO-Ph; Me	82–90%; 84–>98% de			92–>98% de		84–90%	70–76%	1993 (451)
Me; Ph; 4-MeO-Ph; 2-Me-4-MeO-Ph	97–98%	Me; Ph; 4-MeO-Ph; 2-Me-4-MeO-Ph	31–96%; 48–98% de							1993 (448)
2,6-Me₂-4-MeO-Ph; Me	75–94%	Me; 2,6-Me₂-4-MeO-Ph	85–95%; 80–>99% de	79%; 94% de		83–89%	Amberlite-azide	90–99%	100%	1995 (458)
iPr; 2,6-Me₂-4-MeO-Ph	75–87%	2,6-Me₂-4-MeO-Ph; iPr	90–94%; 50–>95% de	85%; >95% de		85–>95% de	TMG-N₃	96%; >95% de	97%; >99% de	1997 (460)
2-naphthyl	78%	Me	76%; >90% de	94%; >95% de		95%; >95% de	TMG-N₃	80%	73–80%	1997 (454)
iPr		Ph	80–84%; >90% de	73–85%		80–82%	TMG-N₃	78–80%	59–63%	1997 (455)
iPr	93%	4-MeO-Ph	81–86%	85%		85%	TMG-N₃	92%	87%	1997 (456)
3-MEMO-Ph		4-tBu-Ph; 4-Ph-Ph	90%	72%					75–92%	1998 (459)
N-Mts imidazol-4-yl	>80%	Me	100% de	>98% de					29%	2000 (461)
N-Boc indol-3-yl		Me			pure isomer: 50%; 100% de		NaN₃	>80%	87%	2001 (447)

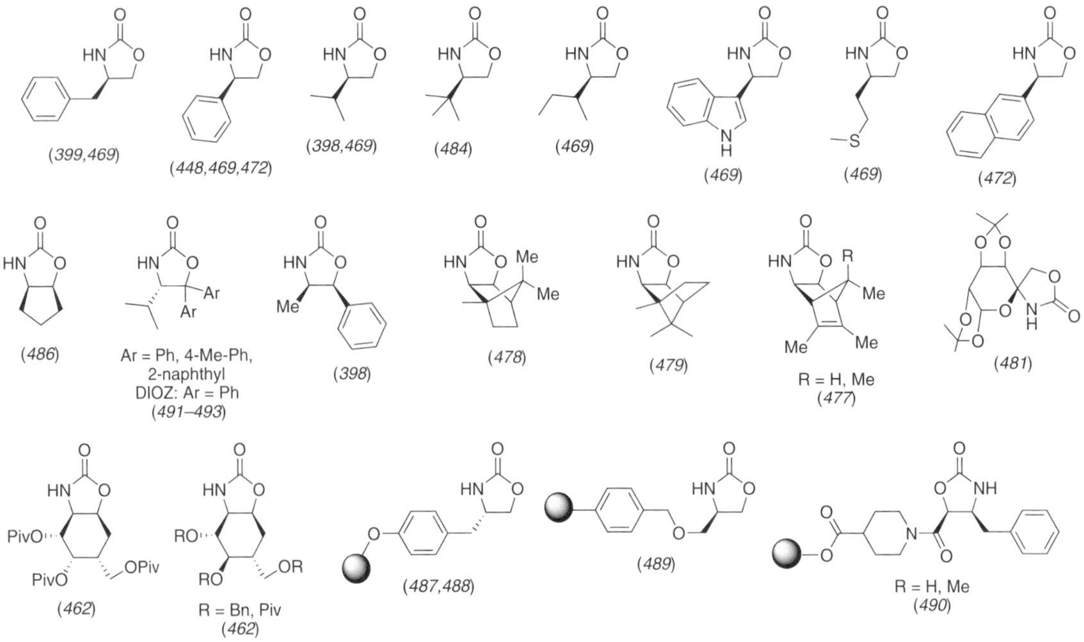

Scheme 5.23 Chiral Oxazolidin-2-one Auxiliaries.

The oxazolidinone auxiliaries that are commonly used are derived from valinol (398), phenylalaninol (399), phenylglycinol (448) or norephedrine (398). The enantiomeric purity of several commercially available oxazolidinone auxiliaries has been determined, with varying results (97.74-99.98% ee) (475, 476). A variety of other oxazolidinone auxiliaries have been described (see Scheme 5.23). Most of these have been used for aldol, alkylation, and Diels–Alder reactions, but have not yet been applied to amino acid synthesis. A sterically congested bicyclic oxazolidin-2-one was prepared via cycloaddition of hexa- or penta-methylcyclopentadienes with 2-oxazolone (477), while tricyclic derivatives were synthesized from 2-amino-3-borneol (478, 479) and D-xylose (480), and a spirocyclic oxazolidinone from D-galactose (481). Diphenylalaninol, derived from Ser by Grignard addition to the Garner aldehyde (482), was transformed into an oxazolidinone (483). A classical resolution of tert-leucinol by diastereomeric salt crystallization provided access to both enantiomers of the oxazolidinone with the bulky tert-butyl substituent (484). A review of the synthesis and use of tert-leucine (Tle) includes application of the Tle-based oxazolidinone to alkylations, Diels–Alder reactions, and radical reactions (485). cis-1-Amino-2-cyclopentanol was also converted into an oxazolidinone, which induced >99% de for alkylations and aldol reactions (486). An oxazolidinone auxiliary was constructed on a solid support via attachment of Boc-Tyr-OMe to the resin through the phenol group, reduction of the ester, and oxazolidinone formation (487). The same auxiliary has been prepared by a different route (488). Two other groups have prepared polymer-supported oxazolidin-2-ones, with alkylation of acyl derivatives proceeding in good yield (489, 490).

Galactosamine- and glucosamine-derived oxazolidinones have also been prepared, and induced much better diastereoselectivity for 1,4-additions of organocuprates to α,β-unsaturated N-acyl oxazolidinones than did the classic Evans' auxiliary. The 1,4-addition adducts were quenched with a halide source to give β-branched-α-halocarboxylic acid derivatives ready for amination (462). A 4-isopropyl-5,5-diaryloxazolidinone induced 85% de during electrophilic amination of 3-phenylpropionic acid, or 91% de during an electrophilic bromination/nucleophilic azide displacement sequence (491). The same year, Hintermann and Seebach reported the synthesis of a 4-isopropyl-5,5-diphenyloxazolidin-2-one auxiliary (DIOZ) from valine ester, PhMgBr, and ethyl chlorocarbonate (492). The oxazolidinone carbonyl is sterically protected from undesired nucleophilic attack by the phenyl and isopropyl groups; thus it is possible to employ nBuLi for deprotonation of the acylated oxazolidinone and remove the auxiliary with NaOH in MeOH/THF. The auxiliary was initially applied to the synthesis of β-amino acids. In 2002 Seebach and coworkers reported 32 crystal structures of derivatives of the DIOZ oxazolidinone auxiliary to support its superiority (493). The diphenyl groups buttress the isopropyl group, making it sterically equivalent to a tert-butyl group and providing for high stereoinduction. The DIOZ auxiliary has high crystallinity so that intermediates are readily purified, and the DIOZ can easily be recovered by precipitation from reaction mixtures (493).

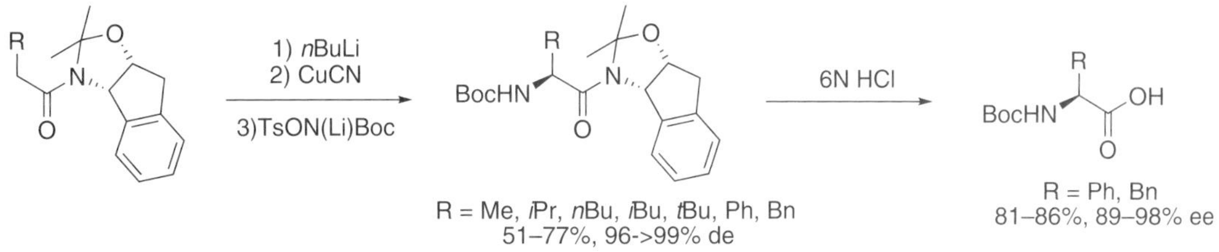

1) nBuLi
2) CuCN
3)TsON(Li)Boc
6N HCl
BocHN
R = Me, iPr, nBu, iBu, tBu, Ph, Bn
51–77%, 96->99% de
R = Ph, Bn
81–86%, 89–98% ee

Scheme 5.24 Amination of Acids with Aminoindanol-Derived Oxazolidinone Auxiliary (*494*).

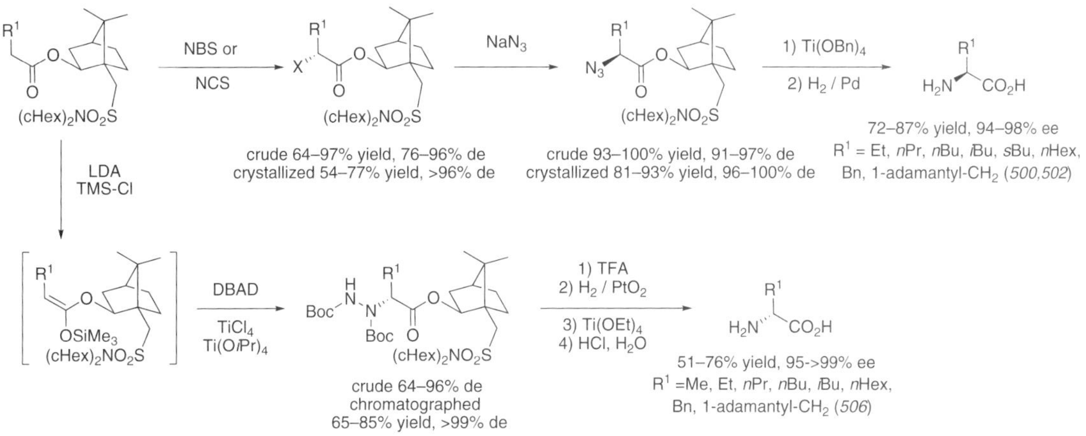

NBS or
NCS
NaN3
1) Ti(OBn)4
2) H2 / Pd
(cHex)2NO2S
crude 64–97% yield, 76–96% de
crystallized 54–77% yield, >96% de
crude 93–100% yield, 91–97% de
crystallized 81–93% yield, 96–100% de
72–87% yield, 94–98% ee
R1 = Et, nPr, nBu, iBu, sBu, nHex,
Bn, 1-adamantyl-CH2 (500,502)
LDA
TMS-Cl
DBAD
TiCl4
Ti(OiPr)4
1) TFA
2) H2 / PtO2
3) Ti(OEt)4
4) HCl, H2O
crude 64–96% de
chromatographed
65–85% yield, >99% de
51–76% yield, 95->99% ee
R1 =Me, Et, nPr, nBu, iBu, nHex,
Bn, 1-adamantyl-CH2 (506)

Scheme 5.25 Amination of Acids Esterified with Oppolzer Auxiliary.

Acids derivatized with the oxazolidinone prepared from (1*S*,2*R*)-*cis*-1-amino-2-indanol were aminated with a new $^{+}$NHBoc synthon, lithium *tert*-butyl-*N*-tosyloxycarbonate (LiBTOC), giving amino acid derivatives with good diastereoselectivity (see Scheme 5.24). However, some racemization was observed during hydrolysis of the auxiliary (*494*). Evans and Nelson have acylated an achiral oxazolidinone with aryl acids. A chiral catalyst induced 80–90% ee during the amination step (*392*).

A number of other amino acid syntheses make use of an oxazolidin-2-one template, but most of these are discussed in Section 7.8. γ-Amino acids (see Section 11.4) have been prepared by cyanomethylation of an acylated oxazolidinone (*495*), aldol addition of an acylated oxazolidinone to isoleucinal (*496*), or alkylation of an acylated oxazolidinone with benzyl bromoacetate. The ω-amino group was introduced by removal of the auxiliary and conversion of the newly exposed acid group to an amine (*497*). β-Amino acids were synthesized by alkylation with the aminomethylating reagents 1-(*N*-Cbz-aminomethyl)benzotriazole (*498*) or *N*-Cbz chloromethylamine (*492*) (see Sections 11.2.2c, 11.2.3j)

An unusual application of the Evans' auxiliary uses the auxiliary to introduce the α-carboxyl group as a masked hydroxymethyl substituent, and then removes the auxiliary and forms the α-amino moiety via a Hofmann rearrangement of the original acid group (*499*).

5.3.6b Oppolzer Camphor/Bornane-Derived Auxiliaries

Oppolzer and co-workers have developed a number of asymmetric syntheses of amino acids based on bornane-derived chiral auxiliaries. One of the great advantages of the Oppolzer auxiliary is that it tends to form diastereomerically pure crystalline derivatives, so that even if the induced diastereoselectivity during the reaction is poor the subsequent purification (both isomerically and chemically) is very easy. The initial incarnation of this synthesis employed a dicyclohexyl sulfonamido-isobornyl moiety (or its enantiomer) as an ester auxiliary for alkyl and aryl acids (Scheme 5.25, top route). Asymmetric halogenation with NBS or NCS generated the chiral α-halo esters with crude 76–96% de, which improved to >96% de after crystallization (*500, 501*). Nucleophilic amination by displacement with sodium azide gave the α-azido ester with >97% de after crystallization and 81–93% yield. The free amino acids were obtained in 42–57% overall yield with 94–98% ee by transesterification to the benzyl ester with $Ti(OBn)_4$ followed by hydrogenolysis (*502*). The auxiliary could be recovered.

The enolate of the ester auxiliary derivative has also been used for electrophilic amination (see Scheme 5.25, bottom route). The lithium, sodium or potassium enolate of 2-cyano-3-phenylpropanoate was trapped with O-(diphenylphosphinyl)-hydroxylamine with good yield (81–91%) but poor diastereoselectivity (44–60% de) (503). Alkylation of the same enolate led to β-amino acids (504, 505). Much better stereoselectivities were obtained by electrophilic amination of TMS-trapped kinetic enolates with di-tert-butyl azodicarboxylate in the presence of a Lewis acid, with >91% crude de obtained for all but one example (501, 506, 507). Amination has also been carried out by the photolysis of N₃CO₂Et in the presence of the silyl trapped enolate (508), which was used in the synthesis of N-(ethoxycarbonyl)-β-methylphenylalanine esters. The β-center chirality was assured by starting with (R)-3-phenylbutanoic acid; matched and mismatched α-center diastereoselectivities of 76% or 54% were obtained depending on the chirality of the Oppolzer auxiliary.

A number of α-aminomethyl-α-substituted amino acids were prepared from alkyl α-cyano esters, with electrophilic amination by O-(diphenylphosphinyl) hydroxylamine (see Scheme 5.26) (509). The intermediate 2-amino-2-cyanopropionates were obtained with 40–60% de and the purified diastereomers with 60–70% yield. Reduction of the cyano group and hydrolysis gave α-aminomethyl-Ala, -Nva, -Val, -Leu, and -Phe in 78–89% yield. The Oppolzer ester auxiliary has also been employed for amino acid synthesis by a route using a Curtius rearrangement for amination (see Section 5.3.8a) (510–515), and for β-amino acid synthesis by alkylation of a 2-cyanopropanoate ester followed by hydrolysis (see Section 11.2.3o) (516).

5.3.6c Oppolzer Camphor/Bornane-Derived Amide Auxiliary

Another route involving the bornane auxiliary proceeds via an electrophilic or nucleophilic amination of alkyl and aryl acids that have been amidated with the bornyl sultam (see Scheme 5.27). An Oppolzer bornyl amide derivative is also used as a chiral auxiliary during diastereoselective alkylations of a glycine Schiff base equivalent; this method is discussed in Section 6.4.2b. The use of the bornane[10,2]sultam as a chiral auxiliary for asymmetric synthesis, including the preparation of amino acids, has been reviewed (517).

N-Acylation of the bornane[10,2]sultam can be accomplished by reacting the lithium or sodium salt with acyl chlorides (518–520) or by using AlMe₃-mediated addition of acids (521), but a number of

difficulties have been reported (464, 522, 523). An improved procedure has been published (464) which reacts a N-trimethylsilyl bornane-2,10-sultam intermediate with acyl chlorides in the presence of copper(II) chloride (see Scheme 5.27). For some substituents, such as acryloyls, better results are obtained by reacting the acid anhydride or a mixed anhydride of the acid and pivalic acid in the presence of LiCl and Et₃N (465). For β,γ-unsaturated acyl compounds, Evans' auxiliary was acylated using a mixed anhydride with N-methylmorpholine as base; presumably the same procedure would be effective with the Oppolzer auxiliary (466, 467). Amide formation with DCC/DMAP has also been reported (524).

The amide enolate is generated with sodium hexamethyldisilazide and aminated with 1-chloro-1-nitrosocyclohexane, giving the N-hydroxy α-amino amides with complete diastereoselectivity (519, 525). In contrast to the route proceeding via alkylation of a glycine Schiff base derivative, (R)-amino acids are obtained from the (2R)-bornane sultam. It is also possible to generate the enolate by 1,4-addition of an organometallic nucleophile (EtMgBr, PhCu) to the amide derivative of an acrylic acid (526, 527), with asymmetric induction of 90% de at the C-3 chiral center (519). The amination adducts were deprotected and reduced to the free amino acids in 77–96% yield (>99% ee) by a two-step procedure (519, 525). N-Hydroxy amino acids can also be produced (519, 525), and further derivatized to N-alkyl amino acids (N-methyl, -ethyl, and iBu) by treatment with aldehydes in the presence of NaBH₃CN (528).

[α-¹⁵N]-Amino acids can be prepared by using 1-chloro-1-[¹⁵N]nitrosocyclohexene as the aminating reagent. L-[¹⁵N]-Ala, -Val, -Leu, -Phe, and [1-¹³C,¹⁵N]-Val were synthesized by this method (529). Some improvements in chemistry were reported during these syntheses. An excess of sultam (which is recovered) was used during acylation (88–96% yield) and the hydroxyamine reduction was carried out immediately, as the hydroxyamines were noticeably unstable (529). The intermediates were not purified, which resulted in slightly lower enantiomeric purities than methods involving isolation of the intermediates (products in 51–70% yield from the sultam with 97.2–99.5% ee).

The amination route has been used to synthesize both enantiomers of p-carboranylalanine and 2-methyl-o-carboranylalanine with >99% ee and 31–34% yield from the carboranylpropanoic acid (524); a synthesis of

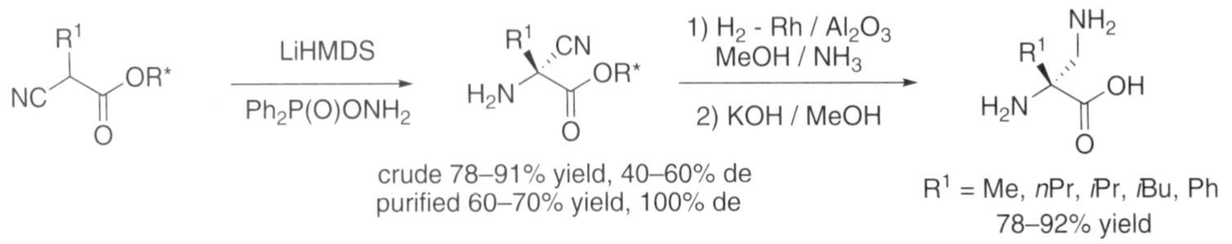

crude 78–91% yield, 40–60% de
purified 60–70% yield, 100% de

R¹ = Me, nPr, iPr, iBu, Ph
78–92% yield

Scheme 5.26 Synthesis of α-Aminomethyl-α-amino Acids (509).

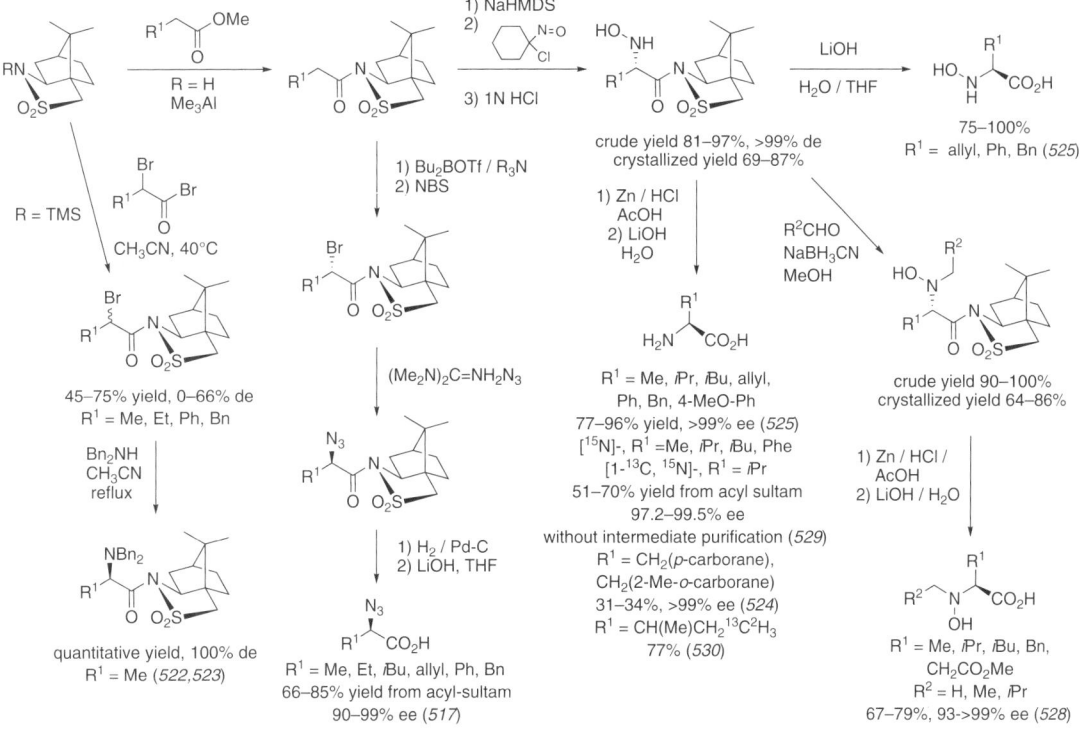

Scheme 5.27 Amination of Bornanesultam-Derived Amide Auxiliaries.

[5-^{13}C,5-^{2}H$_3$]-Ile was reported in 2001 (*530*). This auxiliary has also been used in a synthesis of substituted aziridine-2-carboxylic acids by aziridation of the amide of methacrylate (*531*). The Oppolzer auxiliary was also employed with acetoacetic acid and 3-oxostearic acid. The β-ketoamides underwent conjugate addition to dialkyl azodicarboxylates in the presence of an Ni(II) catalyst with 60–84% de; the Evans' oxazolidinone auxiliary gave slightly lower selectivity. Reduction of the ketone with BH$_3$.SMe$_{24}$ produced the (2S,3R)-β-hydroxy-α-hydrazino acids with up to 97% de. Curiously, the same result was obtained from both (2S)- and (2R)-substrates, apparently due to a dynamic kinetic resolution under the reduction conditions, resulting in epimerization of the α-center (*532*).

The Oppolzer bornane[10,2]sultam auxiliary has also been applied to nucleophilic aminations via an α-halo intermediate (see Scheme 5.27). Bromination of various acyl sultams followed by azide displacement, azide reduction, and sultam cleavage gave a number of amino acids in 66–85% yield from the acyl sultam, with 90–99% ee (*517*). The α-halo intermediate can be prepared by introducing the sultam auxiliary to an α-bromo acid halide (*522*, *523*). It was observed that the amide formation was diastereoselective, with up to 66% de obtained, indicating that epimerization was occurring at the α-center under the reaction conditions. Halide displacements with hard, unhindered nucleophiles such as

azide were found to proceed without epimerization, but a soft hindered nucleophile such as dibenzylamine allowed equilibration to occur, so that a bromide with minimal diastereomer enrichment could provide an amino acid derivative with 100% de (*522*, *523*). Unfortunately this procedure only worked for R^1 = Me.

A modified N-acetyl bornane auxiliary was treated with TiCl$_4$/DIEA, followed by Br$_2$/DIEA, followed by an aldehyde, resulting in α-bromo-β-hydroxyacid derivatives with >98% de. The amination step was not carried out (*533*).

5.3.6d Other Chiral Auxiliaries

Several other chiral auxiliaries have been applied to asymmetric aminations. An auxiliary similar to the Oppolzer sultam auxiliary resulted in >96% de during electrophilic aminations with trisyl azide (see Scheme 5.28) (*534*). The best results were provided by a *t*-Bu substituent on the auxiliary.

The E silyl ketene acetals derived from (1R,2S)-N-methylephedrine esters by LDA enolization and trapping with TMSCl were reacted with DBAD to give the α-hydrazino adducts with 78–91% de (see Scheme 5.29). Cleavage of the auxiliary and reduction of the hydrazine proceeded in good yield (*535*).

(S,S)-(+)-Pseudoephedrine amides of arylacetic acids have also been used to induce diastereoselectivity during

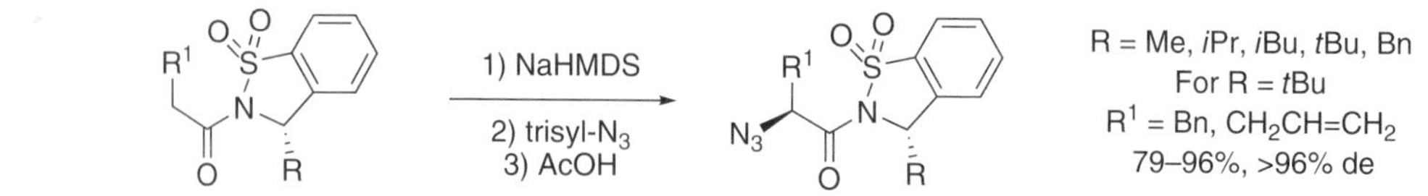

Scheme 5.28 Amination with Sultam Auxiliary (*534*).

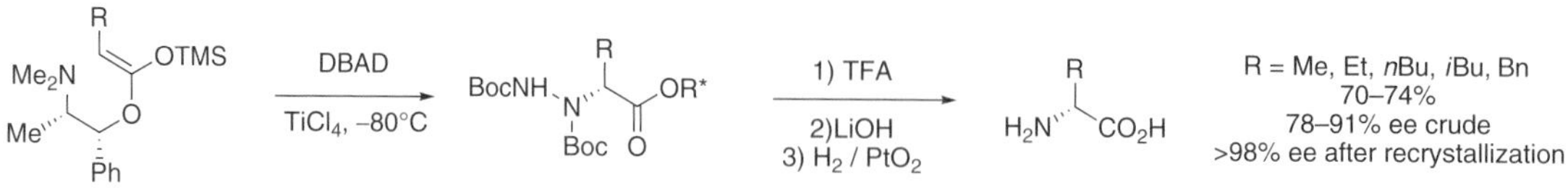

Scheme 5.29 Electrophilic Amination of Silyl Enolate with Pseudoephedrine Auxiliary (*535*).

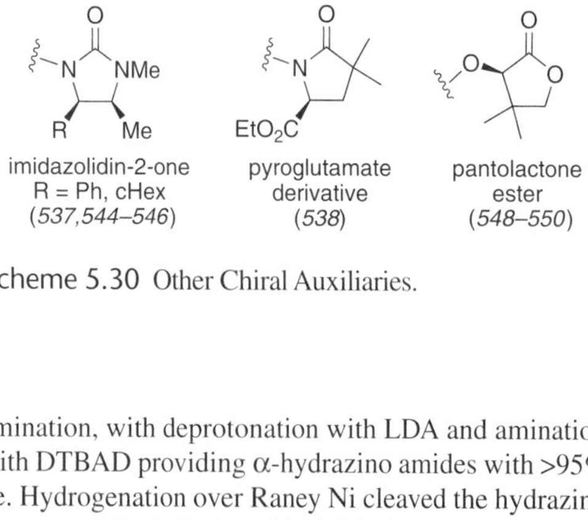

imidazolidin-2-one
R = Ph, cHex
(*537,544–546*)

pyroglutamate
derivative
(*538*)

pantolactone
ester
(*548–550*)

Scheme 5.30 Other Chiral Auxiliaries.

amination, with deprotonation with LDA and amination with DTBAD providing α-hydrazino amides with >95% de. Hydrogenation over Raney Ni cleaved the hydrazino group, and hydrolysis with 4 M H$_2$SO$_4$ removed the pseudoephedrine amide auxiliary to give arylglycines (*536*). An ephedrine-derived imidazolidin-2-one (see Scheme 5.30) was used in a similar fashion to Evans' oxazolidin-2-one for inducing diastereoselectivity in aldol reactions, but has not yet been applied to amino acid synthesis (*537*). Dibutylboron triflate promotes

conjugate addition of organocuprates to *N*-acylated. An oxazolidinone analog with the ring oxygen replaced with a quaternary carbon center (see Scheme 5.30) has been prepared from 4,4-dimethylpyroglutamate and used for diastereoselective aldol reactions, but not amino acid synthesis (*538*).

A symmetrical (2*R*,5*R*)-bis(methoxymethyl)pyrrolidine was used as an amide auxiliary for α-iodinations of 4-pentenamides, giving 78–96% de (*539*). The amine group was then introduced by azide displacement, with the amide auxiliary removed by iodolactonization followed by zinc reduction (see Scheme 5.31). Allylglycine, hydroxyproline, Leu, and Nle were prepared by this procedure.

β-Keto esters have been converted into β-enamino esters by reaction with optically active pyrrolidine. Electrophilic amination with NsONHCO$_2$Et resulted in β-keto-α-amino esters with up to 80% ee, with the pyrrolidine acting as a chiral auxiliary (*540*).

The ease of racemization of alkyl α-bromoalkanoates has been utilized for a dynamic kinetic resolution, in which one substrate diastereomer is depleted more

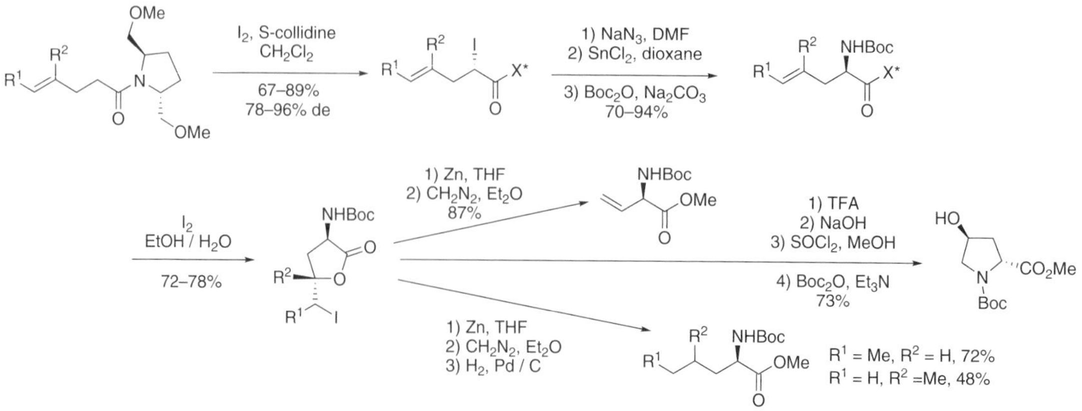

Scheme 5.31 Nucleophilic Amination (*539*).

quickly by preferential reaction with an aminating nucleophile, with the less reactive diastereomer rapidly epimerizing to the more reactive form. A chiral amide derivative of racemic 2-bromopropionate (with *tert*-butyl 1-methyl-2-oxoimidazolidine-4-carboxylate as the chiral auxiliary (541)) was treated with potassium phthalimide at room temperature to give the aminated adducts in yields of 74–90% with 90–94% de (see Scheme 5.32). N-Phthalimido-Ala was obtained in 68% overall yield after removal of the auxiliary (542). N-Benzyl-Ala was prepared in a similar fashion in 82% yield with 88% de for the intermediate adduct using Et₃N in HMPA to provide the basic conditions needed to epimerize the bromo substrate (543). An ephedrine-derived A 3,4-dimethyl-5-phenylimidazolidin-2-one auxiliary (see Scheme 5.30) gave up to 100% de for Phg derived from bromide displacements with piperidine or pyrrolidine, or 74% de with benzylamine, while N-benzyl Chg was produced in 97% yield with 86% de (537, 544, 545). Ala, Abu, and Nle were also produced, with up to 100% de (546). An imidazolidinone auxiliary enantiomer has also been used to selectively generate either enantiomer of the amino acid products, with one diastereomer produced by nucleophilic displacement with benzylamine under epimerizing conditions, and the other diastereomer obtained by bromide displacement with azide under non-epimerizing conditions (see Scheme 5.33) (547).

(R)-Pantolactone esters (see Scheme 5.30) of racemic α-bromo acids have also been used for kinetic dynamic resolutions by reaction with various amines, synthesizing N-benzyl-Pro (61% yield, 75% de), N hexyl-Pro (51% yield, 75% de), N-benzyl-pipecolic acid (66% yield, 82% de), N-benzyl-Phg (72% yield, 82% de), N,N-dibenzyl-Phg and 4′-bromo-Phg (70–76% yield, >98% de), N,N-dibenzyl-homophenylalanine (56% yield, 82% de), N,N-dibenzyl-γ-cyclohexyl-Nva (82% yield, 88% de), N,N-dibenzyl-Abu (66% yield, 86% de), and N-benzyl-Abu (70% yield, 75% de) (548–550). Amino acid esters have also been employed as an amide auxiliary, with α-bromophenylacetic acid-(S)-Leu-OBn reacting with dibenzylamine and tetrabutylammonium iodide to give (R)-Phg-(S)-Leu-OBn with 90% de (551, 552). Diacetone-D-glucose was employed as an ester chiral auxiliary for dynamic kinetic resolutions of α-chloroarylacetic acids; nucleophilic displacement with a variety of primary and secondary amines produced arylglycine derivatives with 86–94% de (553). Dynamic resolutions are also discussed in Section 12.1.7.

5.3.7 Aminohydroxylation Reactions

Catalytic asymmetric aminohydroxylation of olefins has recently emerged as a viable synthetic route to amino

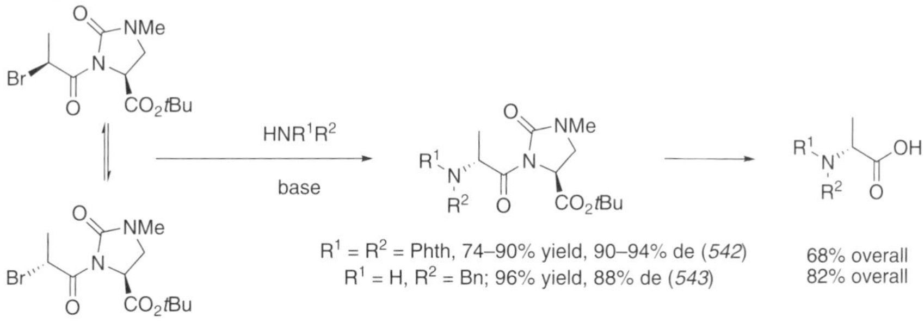

R¹ = R² = Phth, 74–90% yield, 90–94% de (542)
R¹ = H, R² = Bn; 96% yield, 88% de (543)

68% overall
82% overall

Scheme 5.32 Amination by Dynamic Kinetic Resolution with Imidazolidinone Auxiliary.

epimerizing conditions:
BnNH₂, Et₃N. *n*Bu₄NI

NaOMe, MeOH, rt
78–97%

77–98%
80–86% de

n = 1–5

non-epimerizing conditions:
TMGA, CH₂Cl₂, 0°C

LiOOH, MeOH/H₂O
75–99%

89–94%
95% de

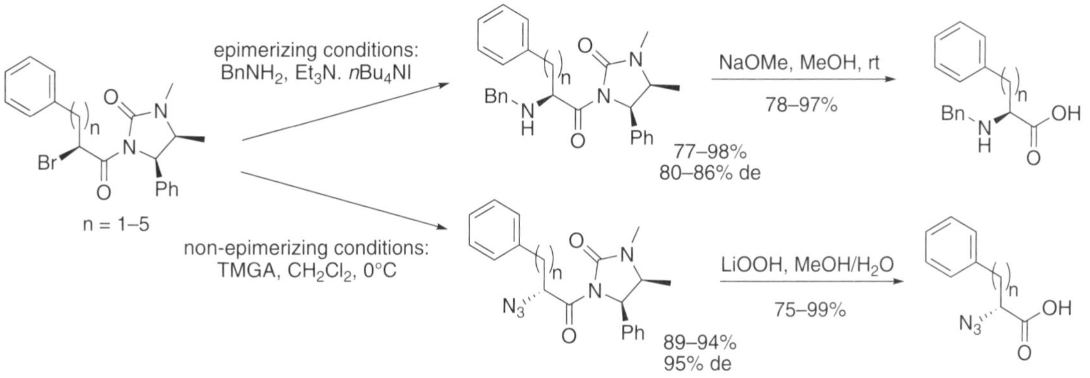

Scheme 5.33 Amination by Dynamic Kinetic Resolution or Inversion with Imidazolidinone Auxiliary (547).

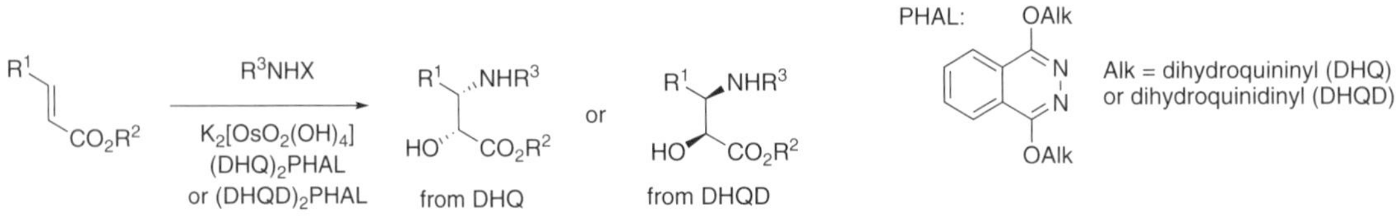

PHAL: OAlk

$$R^1 \underset{CO_2R^2}{\diagdown} \xrightarrow[\substack{K_2[OsO_2(OH)_4] \\ (DHQ)_2PHAL \\ or\ (DHQD)_2PHAL}]{R^3NHX} \quad R^1 \underset{HO \quad CO_2R^2}{\overset{\cdots NHR^3}{\diagdown}} \quad or \quad R^1 \underset{HO \quad CO_2R^2}{\overset{NHR^3}{\diagdown}}$$

from DHQ   from DHQD

Alk = dihydroquininyl (DHQ) or dihydroquinidinyl (DHQD)

R^1 = Ph; R^2 = Me; R^3 = Ts: 69%, 82% ee (*557*)
R^1 = Ph; R^2 = Me; R^3 = Ts
R^1 = Me; R^2 = Et; R^3 = Ts
R^1 = CO$_2$Me; R^2 = Me; R^3= Ts: 52–65%, 60–81% ee (*556*)
R^1 = Ph; R^2 = Me, *i*Pr; R^3= Ms
R^1 = Me; R^2 = *t*Bu; R^3= Ms
R^1 = CO$_2$Me; R^2 = Me; R^3=Ms: 49–76%, 80–95% ee (*565*)
R^1 =CH$_2$*c*Hex; R^2 = Et.; R^3 = Ts: 60%, 96% ee (*558*)
R^1 =CH$_2$*c*Hex; R^2 = Et.; R^3 = Ts: 65%, 89% ee (*559*)
R^1 = Ph; R^2 = *i*Pr; R^3 = Ac: 68%, >99% ee (*566*)

R^1 = (CH$_2$)$_3$OMOM; R^2 = Et; R^3= Ts: 70%, 85% ee (*561*)
R^1 = H, Ph, 2,6-Me$_2$-Ph, CO$_2$Me; R^2 = Me; R^3 = Cbz
R^1 = Ph, 4-AcO-Ph, 4-NO$_2$-Ph, 1-naphthyl; R^2 = Me; R^3 = Cbz:
      55–89%, 84–99% ee (*569*)
R^1 = Me; R^2 = *t*Bu; R^3 = Cbz: 62%, 90% ee (*579*)
R^1 = 2-furyl; R^2 = Et; R^3 = Cbz: 41%, >86% ee (*819*)
R^1 = Ph; R^2 = *i*Pr; R^3 = Bz: 46%, 97% ee (*567*)
R^1= Ph, 4-NO$_2$-Ph, 4-MeO-Ph; R^2 = Ac; R^3= Me: 30–81%, 92->99% ee
(*576*)
R^1 = 4-MeO-Ph; R^2 = Me; R^3 = Cbz: 71%, 99% ee (*588*)
R^1 = 3,5-*i*PrO$_2$-4-MeO-Ph; R^2 = Me; R^3= Boc: 70% (*571*)

Scheme 5.34 Aminohydroxylation Leading to β-Amino-α-Hydroxy Acids.

acids, particularly β-amino-α-hydroxy acids (see also Section 11.2.3e). The scope, limitations, and application to synthesis of the Sharpless asymmetric aminohydroxylation reaction were highlighted in a 1999 report (*554*). A much more extensive review appeared in 2002, including a number of examples of amino acid synthesis (*555*). In the initial report, methyl cinnamate and ethyl crotonate were reacted with chloramine T (TsNClNa.3H$_2$O, an oxidizing amine source) and a chiral osmium catalyst (formed from K$_2$OsO$_2$(OH)$_2$ and a phthalazine-based dihydroquininyl or dihydroquinidinyl-substituted ligand) to give the α-hydroxy-β-amino esters with 60–81% ee (see Scheme 5.34) (*556*). One important β-amino acid readily produced from cinnamate by this method is phenylisoserine, the side-chain component of taxol (*557*). Cyclohexylnorstatine has also been prepared (*558*, *559*). The *N*-tosyl products provided by chloramine-T can be *N*-deprotected by heating with 33% HBr in a sealed vessel (*557*), or by treatment with PPTS and 2,2-dimethoxypropane in dry toluene at 70 °C, forming an oxazolidine with the α-hydroxy group of (*559–561*). The latter procedure was employed during a synthesis of the γ-vinyl-γ-aminobutyric acid, (*S*)-vigabatrin (*561*).

The aminohydroxylation of α,β-unsaturated amides with chloramine-T proceeds efficiently without added ligands, but gives racemic products (*562*). A range of unsaturated acids were aminohydroxylated with 4-40× less osmium catalyst and 3× less chloramine-T than usually employed for the asymmetric version, without diol byproduct formation, and the reaction could be carried out in water (*563*). Acrylamide with a chiral amide auxiliary formed from 1-phenylethylamine was aminohydroxylated with 1.1 equivalents chloramine-T and 2 mol % K$_2$[OsO$_2$(OH)$_4$] to give α-hydroxy-β-aminopropionamide with 100% regioselectivity, 84% yield, and >99% de. The presence of α- or β-substituents greatly reduced the stereo- and regioselectivity (*564*).

Other aminating reagents were soon applied to the aminohydroxylation reaction. It was found that small substituents on the amine source resulted in higher enantioselectivities, with chloramine-M (MsNClNa) hydroxyaminating methyl cinnamate with 95% ee (*565*), compared to 82% ee for chloramine-T (*557*). *N*-Bromoacetamide was reported as a new nitrogen source, allowing for much easier removal of the *N*-substituent after aminohydroxylation and also providing higher enantioselectivities. Isopropyl cinnamate reacted with nearly complete regioselectivity to give either enantiomer of *syn*-phenylisoserine, with >99% ee. The procedure could be scaled up to reactions of 100 g, with 68% overall yield (*566*, *567*). This method was employed during a total synthesis of an NK1 antagonist that used (2*R*,3*S*)-phenylisoserine as an intermediate (*568*). Even better results were obtained with *N*-chlorocarbamates, with the added advantage of ease of removal of the *N*-substituent following synthesis. Methyl cinnamate reacted with 97% ee, under milder reaction conditions (*569*). Various β-heteroaryl-β-amino acids were obtained from 3-heteroarylacrylates and Cbz-NH$_2$/*t*Bu-hypochlorite (*570*). BocNNaCl was used for aminohydroxylation of 3,5-diisopropoxy-4-methoxy cinnamate, giving the α-hydroxy-β-amino acid in 70% yield and the α-amino-β-hydroxy ester in 11% yield. Three other byproducts were observed— benzaldehyde (7%), benzoic acid (2%), and 2, 2-diamino-3-oxo-3-arylpropionate (4%) derivatives—which may account for reduced chemical yields in other syntheses (*571*). *N*-Chloro-*N*-sodio-2-trimethylsilyl ethyl carbamate, TeocN(Cl)Na, was employed to prepare some *N*-Teoc arylisoserines (*572*). Amino-substituted heterocycles have also been reported as nitrogen sources, leading to *N*-heteroaryl β-amino acid products (*573*). *tert*-Butylsulfonamide, *t*BuSO$_2$N(Cl)Na, introduces an amine with an acid-labile protecting group (*574*).

A number of variations and applications of the aminohydroxylation procedure have been reported. Osmylated macroporous resins have recently been developed as a safe, readily recovered/reused source of the hazardous osmium catalyst. A series of alkenes were aminohydroxylated with AcNHBr, preparing several α-hydroxy-β-amino acids with >99% ee (575). A silica gel-supported bis(cinchona) alkaloid catalyst provided (2R,3S)-phenylisoserine and its β-methoxy analog with >99% ee, using AcNBrLi as the aminating oxidant. The catalyst could be recovered by filtration and reused (576). β-Substituted vinylphosphonates react to give β-amino-α-hydroxyphosphonates with 15–99% ee (577, 578).

Several groups have converted aminohydroxylation adducts into α-amino acids. Initially, only certain classes of products could be produced from special substrates such as fumarates or styrenes. Aminohydroxylation of dimethyl fumarate led to β-hydroxy-Asp (556). The β-amino-α-hydroxy aminohydroxylation product of tert-butyl crotonate was converted into both syn and anti diastereomers of 2,3-diaminobutanoic acid, with one isomer obtained by mesylation of the α-hydroxy group followed by azide displacement, and the other diastereomer provided by aziridine formation followed by aziridine ring opening with azide (579). Nineteen different aryl-substituted styrenes were converted into arylglycinols in 35–76% isolated yield and 74–99% ee using CbzNNaCl and BocNNaCl. Hydroxymethyl oxidation of eight of these products produced arylglycines (580). In another report haloamine salts of ethylcarbamate or tert-butyl carbamate were reacted to give the 1-aryl-2-hydroxyethylamine regioisomers as the major products with >87% ee, but oxidation to the arylglycines was not carried out (581, 582). Similarly, BocNNaCl was used for aminohydroxylation of 3-bromo-5-methoxystyrene and 3-benzyloxy-4-methoxystyrene, with TEMPO oxidation and hydrogenation giving 3′-bromo-5′-methoxy-Phg or 3′-hydroxy-4′-methoxy-Phg in 70–81% yields over three steps (583). In another report the amino alcohol precursors of 3′-fluoro-4′-nitro-Phg and 3′-hydroxy-5′-

methoxy-Phg were prepared by BocNNaCl aminohydroxylation. They were incorporated into peptides and cyclized/O-arylated before hydroxymethyl oxidation, leading to a total synthesis of the antibiotic teicoplanin aglycon (820). Both 3′- and 4′-amino-Phg were derived from the corresponding nitrostyrenes, then further functionalized by guanidation of the anilino nitrogen (584). A total synthesis of the ristocetin aglycon employed asymmetric aminohydroxylation to prepare 3′-fluoro-4′-nitro-Phg and 3′-hydroxy-4′-methyl-5′-methoxy-Phg residues, with the aminoalcohol intermediates coupled to the growing polycyclic ring system before oxidation (585). An aminohydroxylation of O-tert-butyl-4-carboxy-styrene, with hydrogenation of the aromatic ring before hydroxymethyl oxidation, produced cis-4′-carboxy-cyclohexylglycine. Treatment of the cyclohexyl amino alcohol intermediate with base induced epimerization to give the trans-substituted product (586). A total synthesis of the aglycon of ramoplanin A2, a lipoglycodepsipeptide, involved an asymmetric synthesis of orthogonally protected L-threo-β-hydroxy-Asn. Methyl p-methoxycinnamate was asymmetrically aminohydroxylated, and converted to the N-Boc,O-TBS amide of 2-hydroxy-3-amino-3-(p-methoxyphenyl)propanoic acid. Oxidative cleavage of the aryl ring with RuCl₃/NaIO₄ generated the α-carboxy group of the β-hydroxy-Asn product (587, 588).

More usefully, a reversal in the regioselectivity of aminohydroxylation has been reported, resulting in β-hydroxy-α-amino acids instead of β-amino-α-hydroxy acids, allowing for greater diversity in the substrates/products (see Scheme 5.35). The reversal is induced by employing cinchona ligands with an anthraquinone (AQN) core instead of a phthalazine (PHAL) core (589). This method was applied to the conversion of a cinnamate into (2S,3S)-β-hydroxy-Tyr, which was subsequently chlorinated to give (2S,3S)-3′-chloro-β-hydroxy-Tyr and used in a total synthesis of vancomycin (325). Another synthesis prepared β-hydroxy-Leu during a total synthesis of (+)-lactacystin (590), while the β-hydroxy-DOPA amino acid center of a complex aryl ether-linked bis(amino acid) was constructed during a

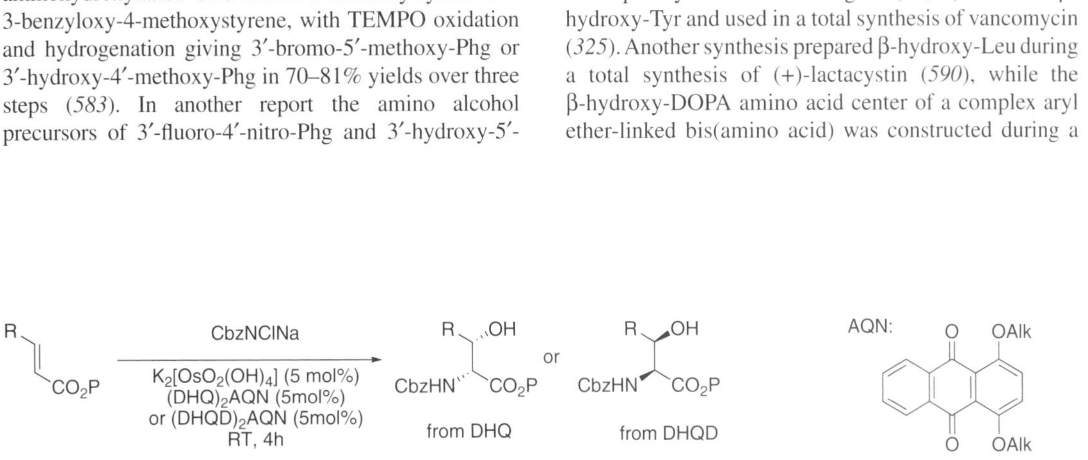

R = iPr; P = 4-Br-Ph: 60%, 7:1 α-amino:β-amino, 87% ee (590)
R = Ph, 4-F-Ph, 4-Cl-Ph, 4-Br-Ph, 4-Me-Ph, 4-MeO-Ph, 2,6-Me₂-Ph, 4-BnO-Ph;
P = Me: 40–68%, 66:34 to 82:18 α-amino:β-amino, 87–95% ee (589)
R = 4-BnO-Ph: 45%, 87% ee (325)
R = 3-[OC(Me)(Et)CH(NHBoc)CH₂OMOM]-4-BnO-Ph:
58%, 5:1 regioselectivity, 82% de (591)
R = N-[C(Me)₂CH=CH₂]-indol-3-yl; P = Et: 44%, 86% ee (590)

Scheme 5.35 Aminohydroxylation Leading to β-Hydroxy-α-Amino Acids.

total synthesis of ustiloxin D (*591*). The *N*-prenyl-β-hydroxy-Trp residue found in the cyclomarin heptapeptides was prepared via an aminohydroxylation of an indole α,β-unsaturated ester (*416*).

The regioselectivity can also be altered by changing the substrate steric and electronic parameters. For example, ethyl (*E*)-cinnamate and ethyl *O*-(*p*-methoxybenzoyl) 4-hydroxybut-2-enoate gave >20:1 regioselectivity for the β-amino acid. However, by using a bulky ester substituent the regioselectivity was directed towards the α-amino acid: (2-naphthyl)methyl *O*-(TBDPS) 4-hydroxybut-2-enoate gave a 17:1 ratio of the α-amino:β-amino products (*592*). Using Ts-Thr as the chiral additive instead of cinchona ligands led to phenylserine from methyl cinnamate with 2:1 regioselectivity, though with only 25% ee (*593*).

5.3.8 Amination by Rearrangement

5.3.8a Curtius, Hofmann, or Schmidt Rearrangements

Amino acids produced by amination rearrangement reactions require optically active substrate diacid equivalents. A Curtius rearrangement has been used in combination with the (1*S*,2*R*,4*R*)-10-dicyclohexylsulfamoylisobornyl ester auxiliary of Oppolzer to prepare a number of α,α-disubstituted amino acids (see Scheme 5.36). The asymmetry was induced during alkylation of a 2-cyano ester containing the auxiliary (20–70% de). After diastereomer purification, the ester auxiliary was cleaved and the acid group converted into an amine by Curtius rearrangement with DPPA. Hydrolysis of the cyano group completed the synthesis of (*R*)- and (*S*)-2-amino-2-methylbutanoic acid (*510*), α-methyl-Phe (*514*), α-methyl-Asp (*513*), α-methyl-Trp (*511*), and α-allyl-Ala, -Val and -Phe (*515*). For the synthesis of α-methyl-Val a slightly different approach was taken due to poor yields and diastereoselectivity during alkylation with the isopropyl group (*512*). The (*R*)-isomer was prepared in the conventional manner, although PCl$_5$/NaN$_3$ was used for the Curtius rearrangement (72% yield). However, the (*S*) isomer was prepared from the

(*R*)-cyano-acid intermediate by conversion of the cyano group to an amide, followed by a Hofmann rearrangement with Hg(OAc)$_2$ and NBS (59% yield). The Hofmann rearrangement has also been used to synthesize several amino acids from 2-substituted-2-(benzyloxymethyl)acetamides, with the benzyloxy group deprotected and oxidized to form the carboxylic acid after the aminative rearrangement. The chiral substrate was obtained by benzyloxymethylation of the enolate of 2-substituted acetic acids, derivatized with Evans' chiral auxiliary. Ala, Phe, 2-aminodecanoic acid, 2-aminooctadecanoic acid, and 2-amino-10-bromodecanoic acid were synthesized, though only the enantiomeric purity of Ala was measured with its optical rotation "matching" a commercial sample of Boc-D-Ala-OMe (*499*).

α,α-Dialkylglycines have been derived from optically active disubstituted malonic acid half-esters via Curtius rearrangement. The quaternary center of the substrate was stereoselectively generated by alkylation of monosubstituted half-esters with phenylmenthyl ester chiral auxiliaries, which induced 40–88% de (*594*). Enzymatic asymmetric hydrolysis of α,α-disubstituted malonic acids was also employed to generate asymmetric substrates for rearrangement, leading to (*R*)- and (*S*)-α-methyl-Phe (*595*) and (*R*)- and (*S*)-α-methyl-Cys (*596*). α,α-Bisalkylated β-keto esters, prepared via stereoselective alkylations with L-Val-O*t*Bu as a chiral auxiliary, were rearranged to amino acids without racemization by a Schmidt rearrangement (*597*). Stereoselectivity during alkylation leading to the β-keto ester substrates has also been induced using (*S*,*S*)-cyclohexane-1,2-diol, with Schmidt rearrangement again used to generate the α,α-disubstituted amino acids (*598*). This procedure was used to prepare α-ethyl-Leu (*599*).

A number of specific amino acid syntheses have made use of acid to amine rearrangements. Chiral 2-(ω-phosphonoalkyl)-3-hydroxypropanoic acids have been treated with DPPA and Et$_3$N to induce a Curtius rearrangement (42–71% yield). The hydroxymethyl group was later oxidized to generate the acid function of ω-phosphono-α-aminoalkanoic acids (*600*). The same

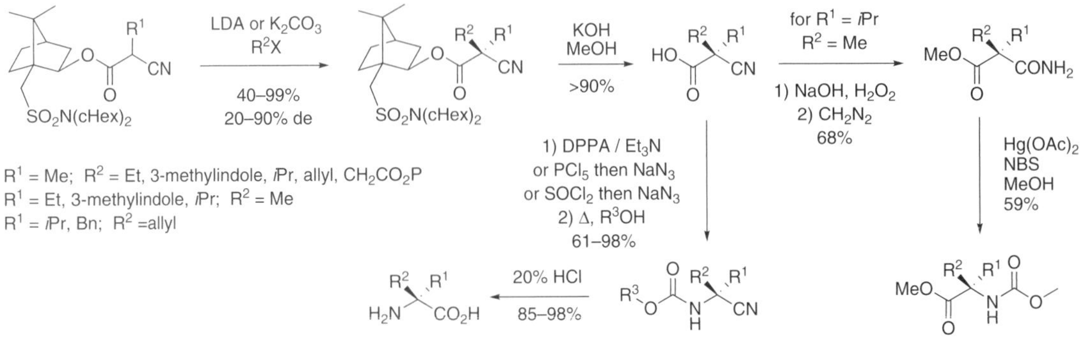

R^1 = Me; R^2 = Et, 3-methylindole, *i*Pr, allyl, CH$_2$CO$_2$P
R^1 = Et, 3-methylindole, *i*Pr; R^2 = Me
R^1 = *i*Pr, Bn; R^2 = allyl

Scheme 5.36 Amination via Curtius or Hofmann Rearrangement (*510–512, 515*).

reagents were employed for a Curtius rearrangement leading to all four 2,3-methanoLeu stereoisomers (601). By carrying out the reaction in tBuOH the Boc-protected amino acid is generated. Curtius rearrangements of homochiral monoesters of pyrrolidine-3,3-dicarboxylic acid or tetrahydrothiophene-3,3-dicarboxylic acid produced curcubitine and its thia analog (602). A variety of cyclic amino acids prepared via Curtius rearrangements of homochiral substrates include (R)- and (S)-[2,2-²H₂]-1-aminocyclopropane-1-carboxylic acid (603), (1R,2R)-1-amino-2-methylcyclopropane-1-carboxylic acid (604), (1R,2R)-2-phenyl-1-aminocyclopropane-1-carboxylic acid (2,3-methano-Phe) (605, 606), 2,3-diphenyl-1-aminocyclopropane-1-carboxylic acid (607), and 2,3-methano-Met (608). DPPA has also been used to induce a Curtius rearrangement in order to prepare a common intermediate that was then elaborated into 2,3-methano-Cys, -Glu, and -Arg (609). 1-Aminocyclopentane-1, 3-dicarboxylic acid was obtained by a Curtius rearrangement of homochiral 3-hydroxycyclopentane-1,1-dicarboxylate, with the 3-carboxy group introduced by cyanide displacement of the activated hydroxy group (610). Thermally induced Curtius rearrangement of the acylazide formed from the acid chloride of O-MOM α-methyl,α-hydroxymethylcyanoacetate provided α-methyl-Ser (611).

Protected derivatives of Orn- and Arg-2,3-methanologs were obtained from optically active substrates via a lead tetraacetate-induced Hofmann rearrangement (83% yield) (612, 613), as was α-methyl-Phe (614) and 2,3-methano-Met (608). A substituted 1-aminocyclohexane-1-carboxylic acid was synthesized in 92% yield by Hofmann rearrangement of a cyclohexyl-1-carboxymethyl-1-carboxamide substrate, using Hg(OAc)₂ in MeOH and NBS in DMF (615). Very mild conditions have recently been reported for the Hofmann rearrangement, with NBS/DBU giving excellent yields for a number of substrates. Amino acids were not prepared, but the method should be generally useful (616). Iodosobenzene diacetate has also been identified as an effective inducer of the Hofmann rearrangement. When applied to the side-chain amide group of Asn, 2,3-diaminopropionic acid was produced (617). Again,

the conditions could be useful for rearrangements of α-amides as well. A cyclic β-amino acid has been prepared via a new mild electrochemically induced Hofmann rearrangement (618).

Chiral isoserine, a β-amino acid intermediate used in the preparation of aziridines, and thence α-amino acids (see Section 8.2), was synthesized from (2S)-malic acid by either Hofmann or Curtius rearrangement conditions ([bis(trifluoroacetoxy)iodo]benzene treatment of amide, or thermal rearrangement of azide obtained via acid chloride) (619, 620). A cyclic β-amino acid, 2-amino-3-carboxy-norborn-5-ene was prepared via a Curtius rearrangement (isopropenyl chloroformate/NaN₃, 57%) using Pro as a chiral auxiliary (621). The amino group of 2-aminocyclopentane-1-carboxylic acid has also been introduced via a Curtius reaction (622). The DPPA-based Curtius rearrangement can also be applied to substrates with the amino acid carboxyl group masked as an alkene, leading to β-amino acids (361). Cyclic chiral γ-amino acids have been prepared from camphoric and isocamphoric acids via a Hofmann rearrangement (65–70% yield (623–625).

5.3.8b Overman Rearrangement

The Overman rearrangement involves the intramolecular amination of an allylic alcohol trichloroacetimidate. Optically active amino acids were prepared from enantiopure allylic alcohols, with the carboxyl group generated by oxidative cleavage of the alkene after the rearrangement. The [3,3] sigmatropic rearrangement proceeds with a high degree of transfer of chirality (626), although the thermal rearrangement was found to be accompanied by a decrease in enantiomeric excess (6% ee decrease) (627). However, palladium(II) catalysis provided complete transfer of chirality (627) as well as allowing for much milder reaction conditions by greatly increasing the rate of reaction (628).

An allyl derivative of D-glucose, prepared via an acetylenic intermediate, has been used as a chiral template for the reaction (see Scheme 5.37). Either L- or D- amino acids can be generated depending on if the alkene substituent is cis or trans. L- and D-Ala and

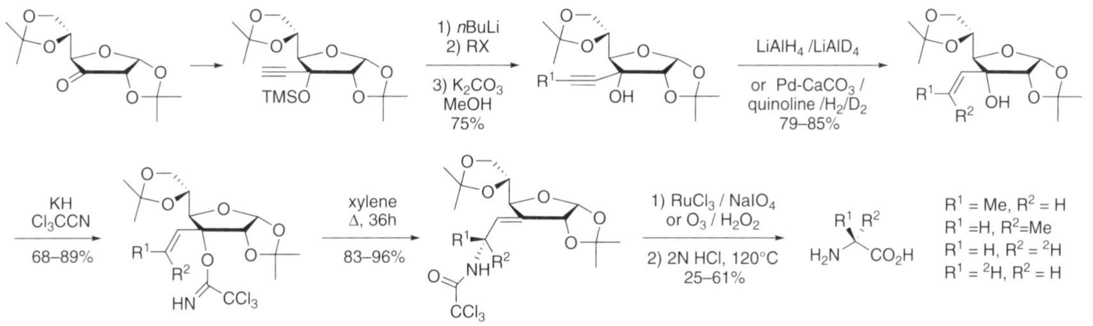

Scheme 5.37 Overman Rearrangement on a Diacetone-D-Glucose Template (629).

L- and D-[2-²H]glycine were prepared by this method (629). A D-glucose derivative has also been used to generate the quaternary carbon in a synthesis of the α-substituted pyroglutamate component of the neurotrophic factor lactacystin, with amination via Overman rearrangement (630, 631).

A number of α,α-disubstituted amino acids have been synthesized from chiral α-alkoxy aldehydes, which were reacted with vinylic organometallic compounds to generate the desired allylic alcohols (see Scheme 5.38). Rearrangement proceeded in 60–98% yield with only a single diastereomer observed; ozonolysis allowed for recovery of the starting α-alkoxyaldehyde as well as giving the α-amino aldehydes. The synthesis was completed by a nearly quantitative Jones oxidation of the aldehydes (632).

The [3,3] rearrangement of allylic chiral trifluoroacetamidates proceeds under milder conditions than the trichloroacetimidate analogs (633), and has been used to prepare D-valine (64% yield from trifluoroacetamide intermediate, 85% ee) (634, 635), polyoxamic acid (16% from trichloro (636), 37% from trifluoro, >96% de (633, 635)), and thymine polyoxin C (14%) (634, 637) from the appropriate chiral allylic alcohol (see Scheme 5.39). Rearrangements of allylic N-benzoylbenzimidates have also been reported, but not yet applied to the synthesis of amino acids (638).

Trichloroacetimidates of 2,3-epoxy alcohols, which can be prepared from allylic alcohols by a Sharpless epoxidation, also undergo an intramolecular amination

(639), but not via an Overman rearrangement (see Section 8.4.3). The Overman rearrangement has been incorporated into a general enantioselective route to α-amino acids, using the allyl group as a masked carboxylic acid (see Scheme 5.40). The route begins with a highly enantioselective and catalytic vinylation of benzaldehyde by alkynes. The resulting allylic alcohols are converted into allylic amines via the imidate rearrangement, with oxidative cleavage of the allyl group producing the α-amino acids in good overall yields and enantioselectivities. A series of bulky aliphatic amino acids were synthesized, including tert-Leu and adamantylglycine, along with functionalized analogs such as 6-chloro-2-aminohexanoic acid and 2,6-diaminopimelic acid. Obviously the side-chain substituent must be resistant to the oxidation conditions (640).

Another general route begins with an asymmetric allylation of α,β-unsaturated aldehydes, followed by Overman rearrangement and then oxidative alkene cleavage (see Scheme 5.41) (641).

5.3.8c Other Rearrangements

The oxidative rearrangement of optically active allylic selenides with Cbz carbamate as the aminating nucleophile (see Scheme 5.42) has been used to prepare several amino acids. The acid group is generated by oxidative cleavage of the alkene. Some racemization occurs during the synthesis as the final products have 78–84% ee (642).

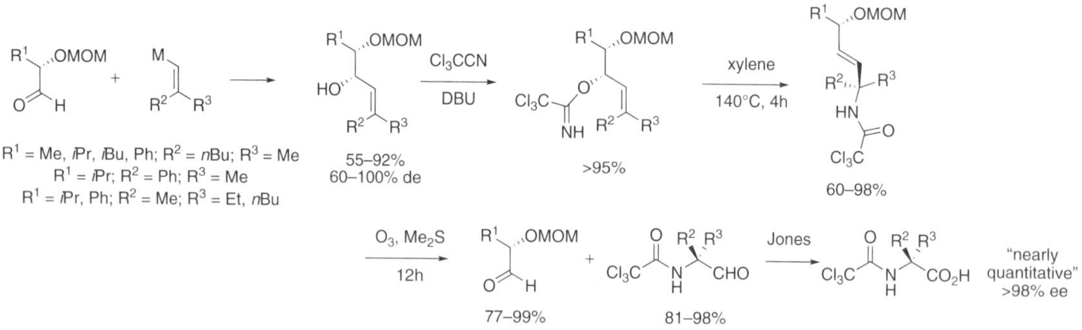

Scheme 5.38　Overman Rearrangement of Chiral Allylic Alcohols (632).

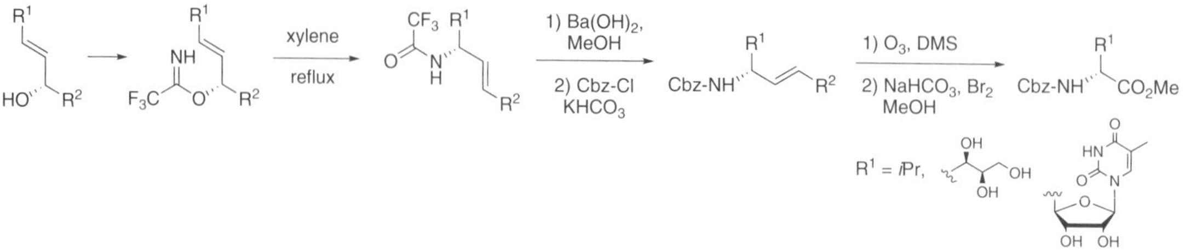

Scheme 5.39　Amination via Overman Rearrangement of Trifluoroacetamidates (633–637).

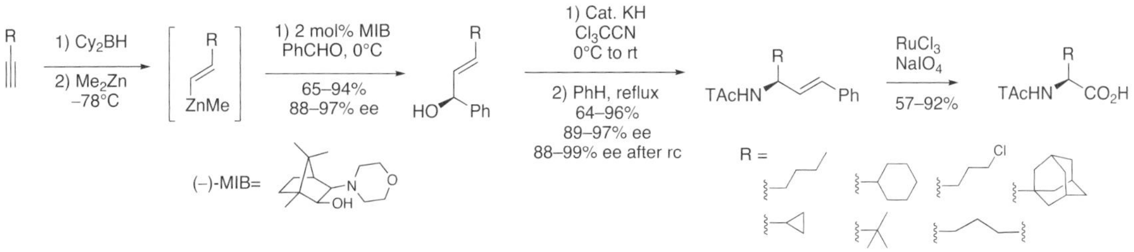

Scheme 5.40 General Synthesis of α-Amino Acids via Asymmetric Aldol Reaction and Overman Rearrangement (*640*).

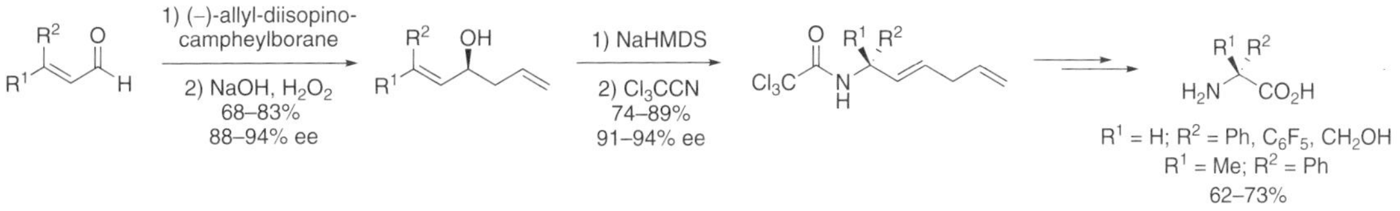

Scheme 5.41 General Synthesis of α-Amino Acids via Asymmetric Allylation Overman Rearrangement (*641*).

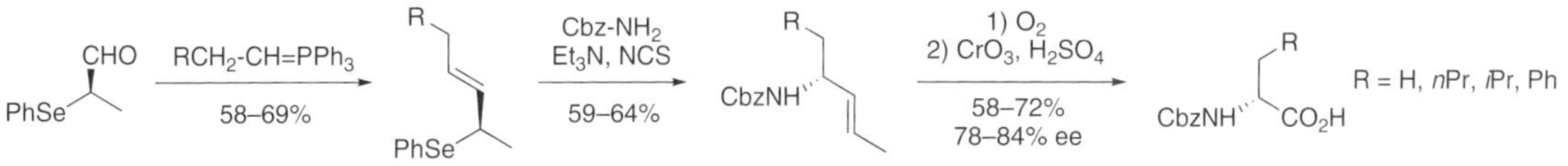

Scheme 5.42 Amination of Allylic Selenide (*642*).

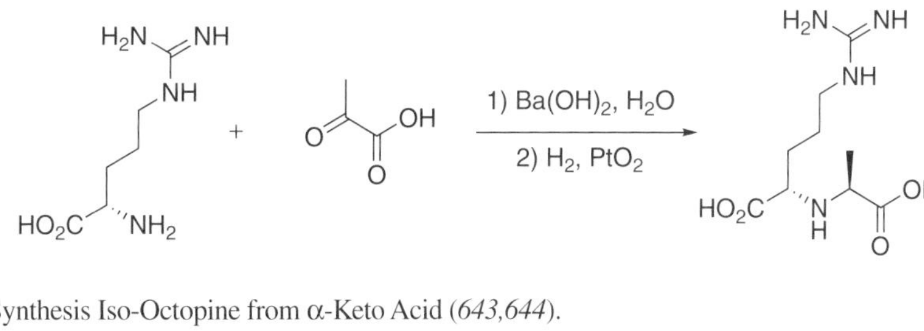

Scheme 5.43 Synthesis Iso-Octopine from α-Keto Acid (*643,644*).

5.3.9 Amination via Reductive Amination of an α-Keto Ester

The biological synthesis of many amino acids proceeds by a transamination reaction of an α-keto ester, with corresponding conversion of pyridoxamine to pyridoxal. Not surprisingly, many synthetic attempts have been made to mimic this reaction. An enantioselective synthesis requires an asymmetric reduction of the C=N bond, which can be accomplished by hydrogenation, hydride reduction, or nucleophilic addition. Several methods have been used to induce asymmetry during imine reduction, including chiral auxiliary groups on the amine and/or acid group, chiral reduction catalysts, or enzymatic methods. Nucleophilic addition of a substituent

corresponds to alkylation of an electrophilic glycine equivalent, and is discussed in Section 6.5.

5.3.9a Reductive Amination: Chiral Auxiliary on the Amine

Probably the first asymmetric synthesis involving reductive amination of an α-keto ester was the preparation of iso-octopine (*N*-(α-propionic acid) Arg) by Knoop and Martius in 1939 (*643*), although they erroneously thought they had prepared octopine (*644*). L-Arg was condensed with pyruvic acid and reduced with H_2/PtO$_2$ (see Scheme 5.43). Only one diastereomer was isolated. While in this case the "chiral auxiliary" remained part of the product, a very similar approach was taken 22 years later with a removable auxiliary. A Schiff base was formed between L- or D-α-methylbenzylamine and several α-keto-acids (645–648). Hydrogenation with a Pd catalyst reduced the C=N bond, while a second hydrogenation with Pd(OH)₂ cleaved the chiral auxiliary, giving Ala, Abu, Val or Phe in 16–85% yield with 8–70% ee (although some resolution may have occurred during crystallization). The Schiff bases formed from α-methylbenzylamine, α-(1-naphthyl)-ethylamine, α-ethylbenzylamine, and phenylglycinol or the alkyloximes of α-methylbenzylbenzoylformamide and α-ethylbenzylbenzoylformamide were also examined for the synthesis of Ala, Abu, Phg, Phe, and Glu (646, 649–652). α-Methylbenzylamine (13–77% ee) (649) and α-(1-naphthyl)ethylamine (61–80% ee) (652) gave the best results, while both oximes were poor auxiliaries. α-Methylbenzylamine has also been used to form imines with a number of α-ketocarboxylates in Pd(II) and Rh(III) complexes (653). Reduction by catalytic hydrogenation gave disappointing asymmetric induction (0–36% de), while reduction with NaBH₄ was slightly better (40% de). (S)-α-Methylbenzylamine was used for reductive amination of an α-keto ester that had a glycosyl side chain, but no diastereoselectivity was observed (654).

L- or D-Phg has also been tested as an auxiliary (655, 656), with the advantage that hydrogenation reduces the C=N bond and deprotects the amine in one step, giving Ala, Abu, Glu or Asp in 25–41% yield with 44–64% ee. The use of other amino acid tert-butyl esters as auxiliaries was examined in 1976 for the synthesis of Ala. After reduction of the Schiff base (H₂/Pd), the auxiliaries were oxidatively cleaved by hypochlorous acid derivatives (e.g., tBuOCl) to give Ala in 35–58% yield with 65–75% ee (657). The amino group of L-Ala-L-Pro was used in a reductive amination leading to L-homo-Phe, inducing 89% de. The resulting tripeptide analog was the desired product, the ACE inhibitor enalapril (658).

A very different approach was reported by Nicoud and Kagan in 1971 (659). Pyruvate was converted into the tosyl hydrazone derivative. Instead of undergoing reduction, the hydrazone group was transformed into a diazo group and used to generate a carbene, which

inserted into the NH bond of chiral phenylethylamine to give protected Ala with 12–26% ee (approximately 50% yield). Twenty-five years later phenylglyoxylic acid was converted into Phg by a very similar process (660), again with minimal diastereoselectivity (6–10% de, 64–83% yield) if a chiral amine was used. A chiral ester group did not aid stereoselectivity (up to 13% de, 42–71% yield).

5.3.9b Reductive Amination: Chiral Auxiliary on the α-Keto Acid

The menthyl esters of several α-keto amino acids were converted to their oximes or benzylamine Schiff bases and then catalytically hydrogenated. Ester hydrolysis gave D-Ala, D-Abu, and D-Phg with 0–25% ee in 25–50% yield (661). Ala and Val have been used as C-terminal amides of the benzylamine Schiff base of pyruvate, giving Ala-Ala or Ala-Val dipeptides with up to 64% de after hydrogenation (645, 469). The same procedure was used to prepare tert-Leu-Val-OMe with up to 56% de (48% yield), by hydrogenating the phenyl-hydrazone of 2-oxo-3,3-dimethylbutanoic acid L-Val-OMe amide (662).

5.3.9c Reductive Amination: Chiral Auxiliary on Both Amine and α-Keto Acid

The combination of chiral amine and ester auxiliaries has been examined by using the α-methylbenzylamine Schiff base of pyruvate menthyl ester. Much better results were obtained with the matched pair of auxiliaries (56–64% ee) than the mismatched pair (16–18% ee) (649, 661).

In 1970 Corey et al. published a reductive amination procedure which used chiral (S)-N-amino-2-hydroxy-methylindolines to form a cyclic hydrazone ester with an α-keto ester (see Scheme 5.44) (663, 664). Reduction of the hydrazone was difficult, requiring aluminum amalgam under carefully controlled conditions, but produced the hydrazino lactone with excellent diastereoselectivity. A single diastereomer was obtained by recrystallization at this stage. Hydrogenolysis and hydrolysis gave the free amino acid and the recovered

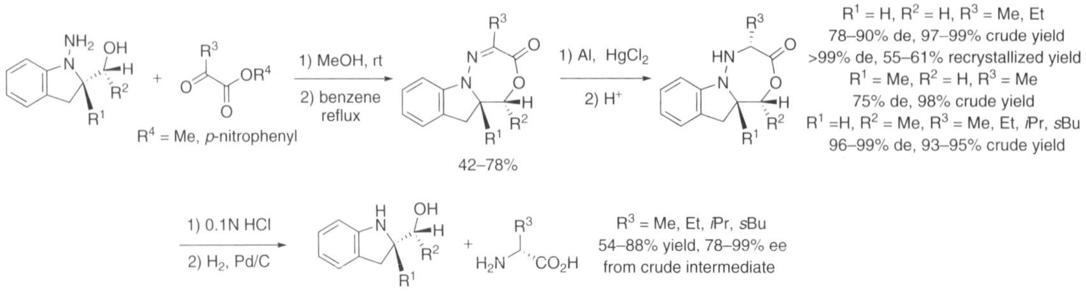

Scheme 5.44 Corey Amino Acid Synthesis from α-Keto Acid (663, 664).

deaminated auxiliary. D-Ala, D-Abu, D-Val, and D-Ile were synthesized with 98–99% ee if the intermediate was recrystallized, 96–99% ee if not.

A very similar approach was reported two years earlier by Vigneron et al., with phenylpyruvate cyclized as the Schiff base ester of a racemic amino alcohol (*erythro*-1,2-diphenyl-2-hydroxyethylamine) and hydrogenated with complete diastereoselectivity (*665*). This method was used for an asymmetric amino acid synthesis in 1992, with ethyl 4-diethoxyphosphonyl-2-oxo-butanoate cyclized with L-*erythro*-(+)-1,2-diphenyl-2-hydroxyethylamine. Reduction with Al/Hg, hydrogenation with H₂/Pd, and hydrolysis gave (S)-2-amino-4-phosphonobutanoic acid with 67% ee (*666*). Similarly, 2-oxobutanoic acid was first esterified with (R)- or (S)-phenylglycinol, and then cyclized to give the 3,5-dehydromorpholin-2-one Schiff base (*667*) (see Section 7.6, Scheme 7.27). Hydrogenation again found to be difficult, but PtO₂ catalyst in CH₂Cl₂ gave the reduced product as a single diastereomer in 53% yield after a single recrystallization (crude 87% de). Further hydrogenolysis gave Abu in 55–59% yield. Chiral Co(III) complexes of the imines of pyruvate and phenylpyruvate have been prepared, but showed little diastereoselectivity in reductions with NaBH₄ (*668*).

5.3.9d Reductive Amination: Chiral Catalyst

Catalytic asymmetric reduction of the C=N imine bond potentially provides a facile route to optically active amino acids from α-keto acids. Unfortunately the enantioselectivity and efficiency of imine-type reductions has generally been poor compared to the reduction of olefins and keto groups. A review of asymmetric reduction of carbon–nitrogen double bonds appeared in 1994 (*669*), with a brief summary of recent advances in 1993 (*670*). Many imine reductions have been reported, but only a few have been applied to amino acid synthesis.

One of the first attempts at a catalytic asymmetric reduction used palladium chloride adsorbed on silk fiber to reduce the imines, leading to Phe and Glu (*671*). The best results for both imine hydrogenation in general, and amino acid synthesis in particular, have been obtained with 1,2-bis(phospholano)-benzene (DuPHOS)-Rh complexes (see Scheme 5.45), which were also very successful for 2,3-dehydroamino acid reductions (see Section 4.3.2c). Optimization experiments on the catalytic reduction of N-acylhydrazones found that N-benzoyl acyl groups gave the best enantioselectivites. Solvent was found to play a critical role, with iPrOH giving 88% ee for one compound, whereas MeOH gave 72% ee and EtOAc gave 9% ee. Enantioselectivity also increased at lower temperatures. The optimized conditions were applied to a number of hydrazones derived from α-keto esters, giving the (S)-hydrazino esters from the (R,R)-Et-DuPHOS-Rh catalyst. The catalyst was quite chemoselective, with minimal reduction of ketones, aldehydes, esters, nitriles, nitro groups, alkenes and alkynes under the reaction conditions. The free amino acids were obtained by acidolytic cleavage of the benzoyl and ester groups, followed by Raney Ni hydrogenolysis of the hydrazine bond (*672–674*). In 2003 a series of phosphine ligands were screened for Rh-catalyzed hydrogenation of the imine formed from phenylpyruvic acid and benzylamine, with the Deguphos ligand producing Phe with 98% ee. Other α-keto acids reacted with lower enantioselectivity, giving Ala with 78% ee, homophenylalanine with 89% ee, Leu with 90% ee, β-*tert*-butyl-Ala with 86% ee, Glu with 60% ee, and Phg with only 19% ee (*675*).

3,3,3-Trifluoroalanine has been prepared from the corresponding aromatic imino ester by reduction with chiral borohydride reagents (*676, 677*). The best results (40–68% ee depending on ester and imine aryl group) were obtained with a catecholborane-chiral oxazaborolidine (see Scheme 5.46), which was also the only reagent that was completely regioselective for C=N reduction

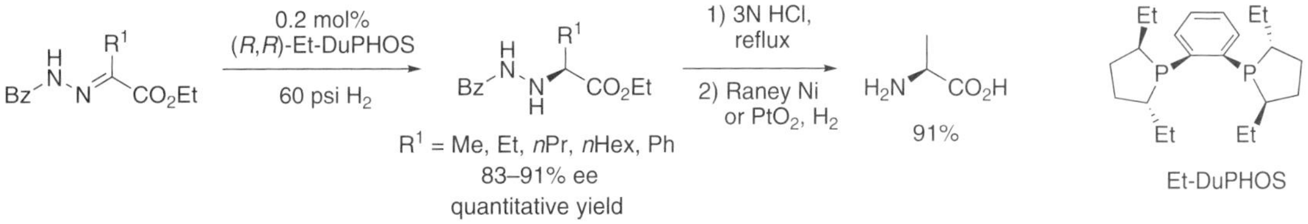

Scheme 5.45 Imine Reduction with DuPHOS Catalyst (*672*).

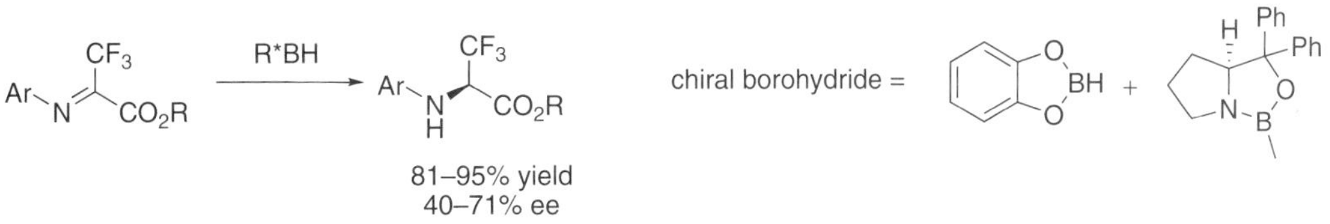

Scheme 5.46 Imine Reduction with Chiral Borohydride Reagent (*676, 677*).

over C=O reduction. A very slight improvement in stereoselectivity was obtained by double asymmetric induction with an L-menthyl ester (71% de, 85% yield).

A furyl group was used as a masked carboxyl group during enantioselective reductions of furylketone oxime ethers by chiral boron complexes formed from BH₃ and amino alcohols (see Scheme 5.47). The chirality of the aminofuryl product depended on the E- or Z-configuration of the oxime ether, which could be stereoselectively synthesized in either conformation. The acid group was revealed by ozonolysis of the furyl group, giving amino acids with 87–96% ee in 29–58% overall yield (678). This method was used to prepare α-(2-carboxycyclopropyl)-Gly (679).

An unusual attempt at asymmetric reduction used electrochemical reduction of an oxime on a poly-L-valine-coated graphite electrode (680). (S)-Ala and -Phe were obtained with very poor yields (10–18%) and enantioselectivity (0.4–6% ee).

5.3.9e Reductive Amination: Biomimetic Systems

Efforts to mimic the biological synthesis of amino acids offer an intriguing insight into the development of biomimetic chemistry, although the reactions are generally of little synthetic use due to poor yields, low enantioselectivity, and analytical-scale yields. Possibly the first system to mimic biological amino acid synthesis consisted of the transamination reaction between L-Ala or L-Phe and α-ketoglutarate in the presence of pyridoxal

and CuSO₄. L-Glutamate was produced in slight enantiomeric excess (up to 7% ee, 25% yield) (681).

Breslow and co-workers have developed a number of model systems incorporating a pyridoxamine equivalent in combination with a flexible amino group in a chiral environment, equivalent to the putative transaminase lysine residue that acts as a general acid–base catalyst to enantioselectively protonate the imine (see Scheme 5.48). Their efforts were summarized in 1988 (682). A chiral amine gave some enantioenrichment (39% ee for Nva) (683), while attachment of the amine to one face of a rigid template gave excellent enantioselectivity (86–92% ee, 68–89% conversion) for the preparation of Ala, Nva, and Trp (684). Cyclodextrin has been used to promote substrate binding and create a chiral environment for the reductive amination. Attachment of pyridoxamine resulted in a catalyst that produced amino acids with 0–66% ee (Phe, Trp), with a reversal of enantioselectivity depending on whether the pyridoxamine was attached on C-6 or C-3 (685, 686). Much better results were apparently obtained by regioselective attachment of pyridoxamine thiol to β-cyclodextrin along with ethylene diamine (687), creating an artificial B₆ transaminase enzyme. Trp, Phe, and Phg were reportedly synthesized with 90–96% ee and approximately 40% yield. However, a more recent reexamination of this work found significantly lower enantioselectivities (688). Better results were obtained by incorporation of an imidazole basic group, with either enantiomer of Phe produced with 60% ee L or 70% ee D, depending on the unambiguously defined isomeric catalyst that was employed (688).

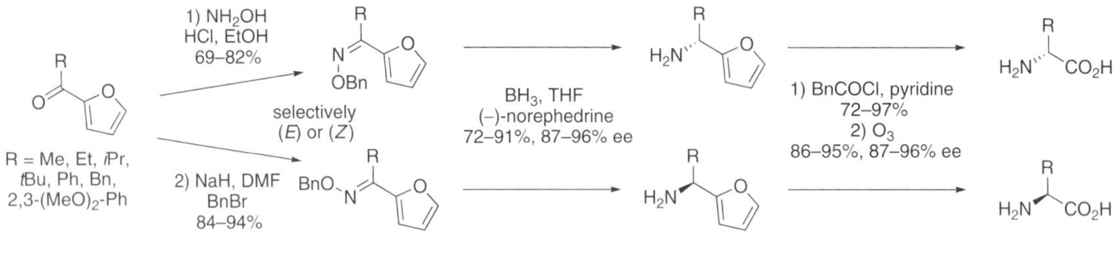

Scheme 5.47 Reductive Amination with Chiral Borane Complex (678).

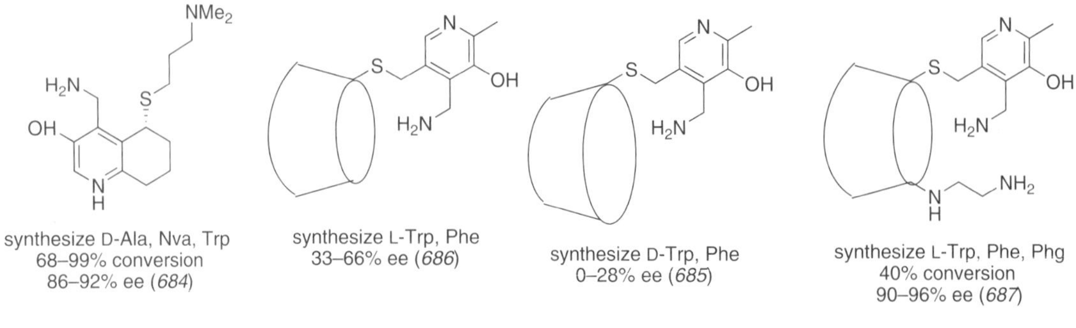

synthesize D-Ala, Nva, Trp
68–99% conversion
86–92% ee (684)

synthesize L-Trp, Phe
33–66% ee (686)

synthesize D-Trp, Phe
0–28% ee (685)

synthesize L-Trp, Phe, Phg
40% conversion
90–96% ee (687)

Scheme 5.48 Artificial Transaminase Enzymes.

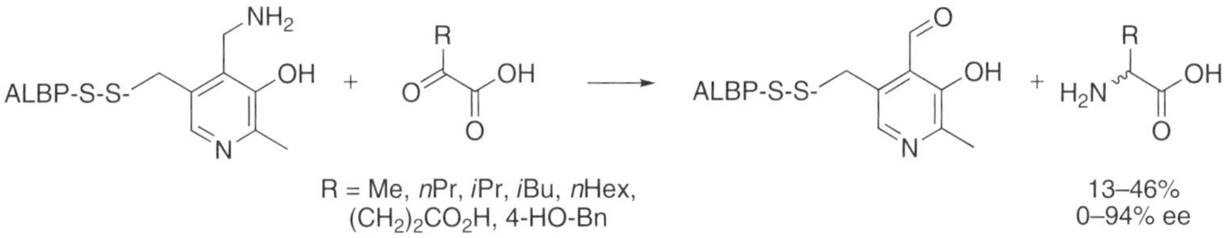

R = Me, *n*Pr, *i*Pr, *i*Bu, *n*Hex,
(CH₂)₂CO₂H, 4-HO-Bn

13–46%
0–94% ee

Scheme 5.49 Mimic of Enzymatic Transamination Reaction (*689*).

Pyridoxamine cofactor has been covalently attached to a thiol within the interior of adipocyte lipid-binding protein (ALBP), an 131-residue protein with a large cavity between two orthogonal planes of β-sheet secondary structure (see Scheme 5.49) (*689*). Seven different α-keto acids were added to a buffered solution of the enzyme complex, and the reaction monitored by HPLC analysis. The highest enantioselectivity (94%) was observed with a bulky β-branched amino acid (L-Val); no discrimination was observed for *n*-propyl or *n*-hexyl side chains. Curiously, Ala was produced with the opposite configuration of the other amino acids (D-Ala, 42% ee). Yields were quite low (13–46%). In a subsequent report, the effects of metal ions on the rates and enantioselectivities of the transamination reaction were examined (*690*). Pyridoxamine has also been attached to dendrimers in order to increase the macromolecular, enzyme-like characteristics. Up to 1000-fold accelerations in the formation of Ala and Phe were observed, but the products were racemic as no chiral centers were present (*691*). Accelerations have also been obtained using pyridoxamine cofactors and polyethylenimine polymers, both with hydrophobic substituents to increase binding (*692*).

A pyridoxamine analog with planar chirality, (*R*)- or (*S*)-15-aminomethyl-14-hydroxy-5,5-dimethyl-2, 8-dithia[9](2,5)pyridinophane (see Scheme 5.50) was used for Zn-catalyzed transamination reactions leading to [2-²H]-Val, [2-²H]-Leu, and [2-²H]-Phe, inducing 44 to >90% ee (*693*). A number of years later, 2′-, 3′-, and 4′-F-Phe and 2′-, 3′- and 4′-CF₃-Phe were prepared from the substituted phenylpyruvates using the same pyridoxamine analog, with 33–66% ee (*694*).

Another artificial transaminase was obtained by imprinting a polymer with a transition state analog

pyridoxamine

R = H, Me
(*693,694*)

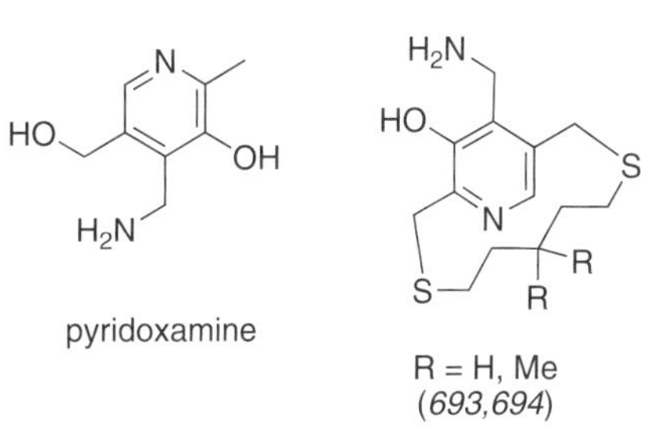

Scheme 5.50 Chiral Pyridoxamine Mimetic.

mimicking a pyridoxal imine of L-Phe-NH₂. The resulting polymer catalyzed the synthesis of Phe from phenylpyruvic acid with 32% ee and a 15-fold rate enhancement (*817*).

5.3.9f Reductive Aminations: Enzymatic Reactions

Given that amino acids are normally produced by enzymes, it is not surprising that a number of research groups have investigated enzymatic syntheses of amino acids. However it is quite remarkable that, compared to chemical syntheses, there has been comparatively little research into enzymatic routes, especially since enzymes are readily adaptable to industrial-scale syntheses. Most enzymes are simply used for resolutions of racemic amino acids produced by chemical means (see Section 12.1.4).

The most successful enzymatic synthetic route involves reductive amination of α-keto acids. A number of more specific enzymatic syntheses of amino acids via other mechanisms have been reported, but these are restricted to the preparation of certain classes of amino acids, such as the synthesis of tryptophan derivatives from serine using tryptophan synthase (see Section 10.3.4) (*368, 695–697*), of β-hydroxy amino acids by the aldolase-catalyzed condensation of glycine and aldehydes (see Section 6.2.2a) (*368, 698–704*), of aspartic acid derivatives via the amination of substituted fumaric acids by aspartases (see Section 5.3.4) (*363–372, 705–709*), and of Ala from Asp via L-aspartate-β-decarboxylase (*372, 710*).

The reductive amination of α-keto acids is catalyzed by amino acid dehydrogenase (see Scheme 5.51, Table 5.3) or transaminase (see Scheme 5.52, Table 5.4) enzymes, which generally have limited substrate tolerance. For example, glutamic oxaloacetic aminotransferase was used to catalyze the preparation of γ-hydroxy-L-glutamic acid (*711*), 4-ethyl-L-Glu (*712*), 4-propyl-L-Glu (*713*), and L-[4-¹¹C]-Asp (*714*). Glutamate dehydrogenase, which is primarily responsible for the biological fixation of ammonia and is commercially available, was employed to synthesize [¹⁵N]-labeled L-glutamate on a preparative scale in 80% yield by using ¹⁵NH₄Cl (*715*). Various ²H-labeled glutamates were produced from the appropriately labeled 2-oxoglutaric acid substrates (*716*). The corresponding labeled glutamines could be obtained by

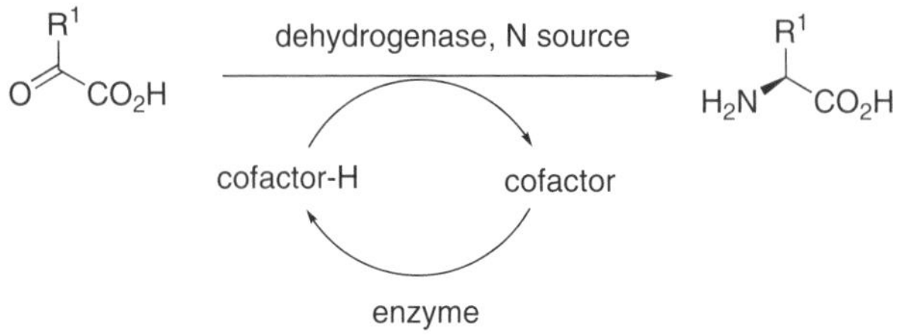

Scheme 5.51 Dehydrogenase-Catalyzed Reductive Amination of α-Keto Acids (see Table 5.3).

Table 5.3 Dehydrogenase-Catalyzed Reductive Amination of α-Keto Acids (see Scheme 5.51)

Enzyme	Nitrogen Source	Cofactors	R^1	Yield	Reference
alanine dehydrogenase	NH$_4$HCO$_2$	NADH formate dehydrogenase	CH$_2$Cl	90%, >99.9% ee (5 mmol scale)	1993 (722)
alanine dehydrogenase	^{15}NH$_4$Cl	NADH formate dehydrogenase	Me ^{13}CH$_3$ CH$_2$OH	66–93% (1 mmol scale)	1994 (723)
glutamate dehydrogenase	^{15}NH$_4$Cl	NADPH, glucose-6-phosphate glucose-6-phosphate dehydrogenase	(CH$_2$)$_2$CO$_2$H	80% (40 mmol scale)	1975 (715)
glutamate dehydrogenase	^{15}NH$_4$Cl	NADPH, glucose-6-phosphate glucose-6-phosphate dehydrogenase	(CH$_2$)$_2$CO$_2$H	41–68% (20 mmol scale)	1977 (721)
glutamate dehydrogenase	NH$_4$OH	NADH HL alcohol dehydrogenase	(CH$_2$)$_3$CO$_2$H	60–70% (50 mmol sclae)	1986 (719)
glutamate dehydrogenase	(NH$_4$)$_2$PO$_4$	NADH alcohol dehydrogenase	CHFCH$_2$CO$_2$H	96% (5 mmol scale)	1989 (718)
glutamate dehydrogenase	NH$_4$Cl or ^{15}NH$_4$Cl	NADH alcohol dehydrogenase	(CH$_2$)$_2$CO$_2$H [1-^{13}C], [2-^{13}C], or [1,2-^{13}C$_2$] substrate	75–86% (1 mmol scale)	1991 (717)
glutamate dehydrogenase	(^{15}NH$_4$)$_2$SO$_4$	NADPH, pyridoxal-5′-phosphate glucose dehydrogenase	(CH$_2$)$_2$CO$_2$H	—	1993 (720)
glutamate dehydrogenase	^{15}NH$_4$Cl	NADH or NAD^{2}H alcohol dehydrogenase	(CH$_2$)$_2$CO$_2$H CD$_2$CH$_2$CO$_2$H CH$_2$CCCO$_2$H (CH$_2$)$_2$^{13}CO$_2$H converted into amide by glutamine synthase and NH$_4$Cl or ^{15}NH$_4$Cl	70–88% (8 mmol scale)	1994 (716)
leucine dehydrogenase	NH$_4$Cl	NADH-PEG formate dehydrogenase	iBu	99.7% 42.5g/L/day continuous	1981 (729)
leucine dehydrogenase	NH$_4$HCO$_2$	NADH formate dehydrogenase	Me iBu (CH$_2$)$_2$SMe	>95%, >99% ee (0.1 mmol scale) keto acid prepared in situ from α-ketoglutarate, glutamate dehydrogenase	1990 (728)
leucine dehydrogenase	NH$_4$OH	NADH or PEG-NADH formate dehydrogenase or glucose dehydrogenase	CH(OH) (Me)$_2$	82% (8 mmol scale)	1990 (727)
leucine dehydrogenase	^{15}NH$_4$Cl	NADH formate dehydrogenase	iPr iBu	84–95% 92% (1 mmol scale)	1994 (723)
leucine dehydrogenase	NH$_4$HCO$_2$	NADH formate dehydrogenase	iBu CH$_2$CH(Me)CD$_3$	85% (1 mmol scale)	1995 (725)

Table 5.3 Dehydrogenase-Catalyzed Reductive Amination of α-Keto Acids (see Scheme 5.51) (continued)

Enzyme	Nitrogen Source	Cofactors	R^1	Yield	Reference
leucine dehydrogenase	NH₄HCO₂	NADH formate dehydrogenase	CH(Me)CD₃ CH(Et)¹³CH₃ CH(Et)CD₃	70–80%	1996 (724)
leucine dehydrogenase phenylalanine dehydrogenase	¹⁵NH₄HCO₂	NADH formate dehydrogenase	CH(OMe)Me CH(OMe)¹³CH₃ CH(OCH₂OMe)Me CH(OOCH₂OMe)¹³CH₃	61–93% (1 mmol scale)	1997 (726)
phenylalanine dehydrogenase	NH₄Cl	NADH	Bn 4-HO-Bn	(65–90 nmol scale)	1984 (732)
phenylalanine dehydrogenase	NH₄HCO₂	NADH formate dehydrogenase	nHept Bn 4-HO-Bn 4-F-Bn (CH₂)₂Ph (CH₂)₃Ph CH(Me)Ph	98–>99% (2–39 mmol scale, or produced continuously for 1 month, 10 g for R^1 = Bn)	1990 (733)
phenylalanine dehydrogenase	NH₄OH	NADH, NaHCO₂ formate dehydrogenase	(CH₂)₂Ph	63% (7.2 mmol scale)	1991 (734)
phenylalanine dehydrogenase	¹⁵NH₄Cl	NADH formate dehydrogenase	Bn	66–93% 84–95% 92% (1 mmol scale)	1994 (723)
mutant phenylalanine dehydrogenases	NH₄Cl	NADH	Bn 2-F-Bn 3-F-Bn 4-F-Bn 2-Cl-Bn 3-Cl-Bn 4-Cl-Bn 4-OMe-Bn 4-Me-Bn 4-CF₃-Bn CH₂(4-pyridyl) CH₂(cHex)	product detected	2004 (736)
N145A phenylalanine dehydrogenase on Celite	NH₄Cl	NADH	Bn 4-NO₂-Bn	89% (10 mg scale)	2005 (737)
bacterial culture	NH₄OH	—	Bn phenylpyruvate substrate obtained in situ from acetamidocinnamic acid	75% (4 mmol scale)	1984 (746)
tyrosine phenol lysase	NH₄Cl	pyridoxal-5′-phosphate	R^1 = Me converted into R^1 = CH₂Ar Ar = 4-HO-Ph, 2-F-4-HO-Ph, 3-F-4-HO-Ph, 2-Cl-4-HO-Ph, 3-Cl-4-HO-Ph, 2-Me-4-HO-Ph, 2-MeO-4-HO-Ph	54% (40 mmol scale)	1981 (739)
tyrosine phenol lysase	NH₄OAc	pyridoxal-5′-phosphate	R^1 = Me converted into R^1 = CH₂(2-N₃-4-HO-Ph)	54% (4.6 mmol scale)	1992 (740)

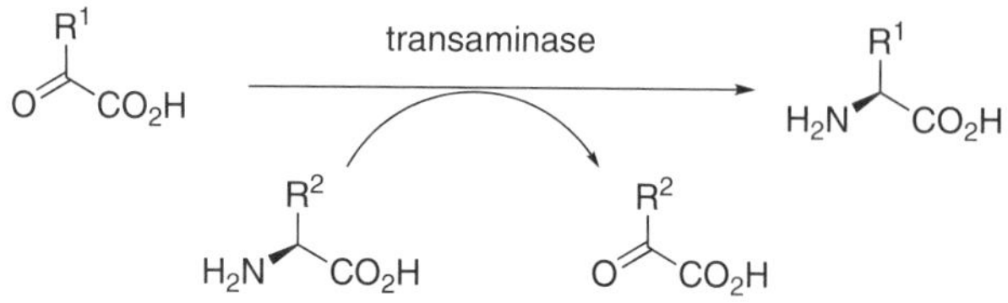

Scheme 5.52 Transaminase-Catalyzed Reductive Amination of α-Keto Acids (see Table 5.4).

Table 5.4 Transaminase-Catalyzed Reductive Amination of α-Keto Acids (see Scheme 5.52)

Enzyme	R^1	R^2	Yield	Reference
E. coli aspartate transaminase	Me Et iPr iBu (CH$_2$)$_2$SMe CH$_2$CO$_2$H (CH$_2$)$_2$CO$_2$H CH$_2$(indol-3-yl) CH$_2$(2-naphthyl) Bn 4-HO-Bn 2-Me-Bn	CH$_2$CO$_2$H (CH$_2$)$_2$CO$_2$H	35–84% >90% ee (1 mmol scale)	1987 (748)
E. coli aspartate transaminase	4-HO-Bn	CH$_2$CO$_2$H with ^{15}N	80% >95% ee	1987 (743)
glutamic oxaloacetic acid transaminase	CH$_2$CH(OH)CO$_2$H	CH$_2$CO$_2$H	86% >98% ee (150 mmol scale)	1987 (711)
glutamic oxaloacetic acid transaminase	4-HO-Bn	CH$_2$CO$_2$H with ^{13}N	57%	1985 (744)
E. coli branched-chain amino acid transaminase	iPr iBu (CH$_2$)$_2$SMe CDMe$_2$ CD$_2$iPr CD$_2$Ph	(CH$_2$)$_2$CO$_2$H with ^{15}N keto acid prepared in situ from α-ketoglutarate, glutamate dehydrogenase	65–81% (4–30 mmol scale)	1993 (720)
E. coli branched-chain amino acid transaminase	Et iPr iBu sBu (CH$_2$)$_2$SMe CH=CH$_2$ Bn (α-keto acid prepared from racemic amino acid via in situ conversion with D-amino acid oxidase)	(CH$_2$)$_2$CO$_2$H (excess)	61–96% 96.7–99.8% ee (10 mmol scale)	1994 (749)
E. coli branched-chain amino acid transaminase	iPr [1-^{13}C] substrate	(CH$_2$)$_2$CO$_2$H	75% 98.4% ee (25 mmol scale)	1993 (750)
E. coli whole cell	CH$_2$(2-thienyl) CH$_2$(3-thienyl) CH$_2$(3-Me-2-thienyl) CH$_2$(5-Me-2-thienyl) CH$_2$(4-Br-2-thienyl) CH$_2$(5-Br-2-thienyl)	CH$_2$CO$_2$H	3–61% >99.6% ee (up to 100 g scale)	1997 (747)
mutant E. coli aspartate aminotransferase	(CH$_2$)$_2$Ph	(CH$_2$)$_4$NH$_2$	97% >99.9% ee	2005 (751)

incubating the glutamate products with glutamine synthetase and $^{14/15}NH_4Cl$ (716). [3-^{13}C]-, [4-^{13}C]-, [5-^{13}C]-, and [3,4-^{13}C$_2$]-L-glutamate (717), 3-fluoro-Glu (718), and L-α-aminoadipic acid (719) were also synthesized with this enzyme.

Unfortunately glutamate dehydrogenase is very specific for α-ketoglutarate and closely related analogs, showing less than 1% of its activity with other α-keto acids (720). The amination can be forced to proceed by employing large quantities of the enzyme, as was used in syntheses of [^{15}N]-L-Met, [^{15}N]-L-Abu, and [^{15}N]-L-Val on a 1g scale in 41–68% yield (721). For other substrates, other dehydrogenases must be used. Alanine dehydrogenase was used for the synthesis of [^{15}N]-L-Ala, β-chloro-Ala (722), and [^{15}N]-L-Ser (723) while leucine dehydrogenase was used to prepare [^{15}N]-L-Val, [^{15}N]-L-Leu (723), [4,4,4-^{2}H$_3$]-L-Val, [^{15}N]-L-Ile, [4,4,4-^{2}H$_3$]-L-Ile, [4-^{13}C]-L-Ile, and [^{15}N]-L-*allo*-Ile (724), [5,5,5-^{2}H$_3$]-L-Leu (725), [^{15}N]-L-*allo*-Thr (726), [^{15}N,4-^{13}C]-L-*allo*-Thr (726), and β-hydroxy-Val (727). Leucine dehydrogenase has also been used to synthesize L-Met, L-Ala, and L-Leu from the corresponding racemic amino acid in a two-step process, with L-amino acid oxidase employed to generate the α-ketoacid from the racemic amino acid, and leucine dehydrogenase to reaminate it (728). Although the resolution proceeded with >95% yield and >99% ee, it was only carried out on a 100 μmol scale, and would only be applicable to aliphatic amino acids that are leucine hydrogenase substrates.

The potential industrial importance of a successful enzymatic synthesis is illustrated by a continuous enzymatic transformation in an enzyme membrane reactor, in which L-leucine dehydrogenase was used to reductively aminate 2-oxo-4-methylpentanoic acid, with formate dehydrogenase present to regenerate polyethylene glycol-linked NADH (729). L-Leucine was produced continuously for 48 days at a rate of 42.5 g/L/day, with a conversion of 99.7%. For industrial syntheses the use of immobilized enzymes (730) and cells (731) is potentially useful; alanine and glutamate dehydrogenases have been prepared in a cross-linked polyacrylamide gel (730).

Aromatic amino acids have been produced with several enzymes. A phenylalanine dehydrogenase was reported in 1984 that could be used to prepare Phe and Tyr, but not His (732). It has since been applied to the synthesis of [^{15}N]-L-Phe (723), Phe (462), β-methyl-Phe (733), Tyr (733), 4'-F-Phe (733), homophenylalanine (733, 734), [^{15}N]-L-Thr (726), [^{15}N,4-^{13}C]-L-Thr (726), and two alkyl amino acids derived from 3-substituted pyruvic acids with bulky substituents (733). L-Phe was produced continuously for a month using this enzyme in a dialysis tube (10 g overall yield) (733); the same dialysis method was used for a synthesis of the N-terminal amino acid component of the nikkomycins, 3-methyl-4-hydroxyhomotyrosine (735). A series of phenylalanine dehydrogenase mutants (N145A, N145L, and N145V) were tested for their substrate selectivity against 13 α-keto acids; the mutants were

more tolerant of non-natural substrates, especially phenylpyruvic acids with substituents at the 4-phenyl position (736). The N415A phenylalanine dehydrogenase was immobilized on Celite and successfully employed for reductive aminations of phenylpyruvic acid and p-nitrophenylpyruvic acid in non-polar organic solvents (737). N-Methyl-L-amino acid dehydrogenase from *Pseudomonas putida* produced N-methyl-L-Phe from phenylpyruvic acid and methylamine in 54% yield on a 1g scale, with >99% ee (738).

Tyrosine-phenol lyase simultaneously reductively aminates pyruvate and introduces a β-aryl substituent, and has been used to prepare Tyr and DOPA analogs (739) and 2'-azido-L-Tyr (740). By using α-ketobutyric acid as the substrate, β-methyl-Tyr was obtained (741). A modified version of this reaction utilized immobilized alanine racemase and D-amino acid oxidase on a column in order to convert [3-^{11}C]-DL-Ala into [3-^{11}C]-pyruvic acid. By also immobilizing β-tyrosinase on the same column, and adding phenol or pyrocatechol, [3-^{11}C]-L-Tyr and [3-^{11}C]-L-DOPA were produced in a single step (742). [2-^{13}C]-L-Phe and [^{15}N]-L-Phe have been prepared from (4-hydroxyphenyl)pyruvic acid by aspartate transaminase catalysis (743), while glutamate oxaloacetate transaminase was applied to the synthesis of [^{13}N]-L-Tyr (744). Glutamic/oxaloacetic acid transaminase immobilized on Sepharose was employed for the preparation of [3-^{11}C]-Phe from [3-^{11}C]phenylpyruvic acid and L-Glu (745). Intact bacteria have been used to transform acetamidocinnamic acid to L-Phe, with the reaction believed to proceed via conversion of the substrate to phenylpyruvic acid followed by transamination (746). A genetically engineered E. coli strain produced L-β-(thienyl)alanines on scales of up to 100 g (747).

The problem of substrate tolerance has been addressed by using E. coli transaminases, with Glu as the amine donor. A range of α-keto acids (RCOCO$_2$H; R = Me, Et, nPr, iBu, (CH$_2$)$_2$SMe, Bn, 4-HO-Bn, 2-Me-Bn, CH$_2$-3-indole) were converted into the corresponding amino acids by E. coli aspartate transaminase in 30–84% yield and >90% ee, although still only on a 1 mmol scale (748). Even better results were obtained with a coupled two-step procedure, which allowed for the synthesis of gram quantities of ^{15}N-labeled L-amino acids (720). In the first step [^{15}N]-L-glutamate was prepared in catalytic amounts from α-ketoglutarate and (^{15}NH$_4$)$_2$SO$_4$ in the presence of glutamate dehydrogenase. The labeled glutamate then served as an amine donor for reductive amination of the appropriate α-keto acid using E. coli branched-chain amino acid aminotransferase, which has a wide substrate tolerance. Then α- and β- deuteration could be introduced by deuterium exchange at the appropriate time. [^{15}N]- and [2,3-^{2}H$_2$, ^{15}N]-L-Val, [^{15}N]- and [2, 3,3-^{2}H$_3$, ^{15}N]-L-Leu, [^{15}N]-L-Met, and [2,3,3-^{2}H$_3$, ^{15}N]-L-Phe were prepared in 65–81% yield from the α-keto acid, with >99% ee.

The branched-chain amino acid aminotransferase has also been coupled with the D-amino acid oxidase system that was previously used in combination with leucine dehydrogenase as the aminating enzyme (728).

The oxidase provides the α-keto acid substrates from racemic amino acids. The amination reaction is driven by an excess of the glutamate amine donor. Racemic Ala, Val, Leu, Ile, Met, Phe, Abu, and allylglycine were converted to the L-amino acid in 61–96% yield, with 96.7–99.8% ee (*749*). [^{13}C]-L-Val has also been prepared by this method (*750*). Homophenylalanine was synthesized by using site-directed mutagenesis to change the substrate specificity of *E. coli* aspartate aminotransferase. A double mutant had a 13-fold increase in specific activity for the synthesis of L-Hfe from 2-oxo-4-phenylbutanoic acid, using Lys as the amino donor. The Lys transamination product (2-keto-6-aminocaproate) cyclized spontaneously to form 2,3-dehydropiperidine-2-carboxylic acid, helping to drive the reaction towards completion and giving a 97% yield of Hfe with >99.9% ee (*751*).

5.3.9g Additions to the C=N Bond

An alternative method of imine/oxime reduction is via addition of an organometallic compound. The electrophilic imine/oxime substrates have been derived from a variety of sources, not only α-keto acids, but also α-halo- or α-acetoxyglycine derivatives, although the latter imines are usually a non-isolated (and often non-acknowledged) intermediate. The α-hydroxy/methoxy/acetoxy-glycine substrates are often prepared from glyoxylic acid by condensation with amines, amides or urethanes. These reactions will be discussed in the section on acyclic electrophilic glycine equivalents (see Section 6.5).

5.4 Asymmetric Carboxylation Reactions

Two general approaches have been used in the synthesis of amino acids via introduction of a carboxyl group to a substrate containing the side chain and amino group: electrophilic carboxylation of an α-amino anion equivalent, or nucleophilic carboxylation by addition of a $^-CO_2H$ synthon to imines.

5.4.1 Electrophilic Carboxylation

5.4.1a Hegedus Chromium Carbene Complex

The most successful approach to carboxylation of amino compounds is via chromium aminocarbene complexes, as developed by Hegedus and co-workers and reviewed in 1995 (*752*) and 1997 (*753*). The required chromium carbene complex is prepared from tertiary amides (including *N*-substituted lactams) by reaction with Na$_2$Cr(CO)$_5$ and TMSCl (*754*), or by addition of the side chain to the chromium complex followed by addition of the amine (see Scheme 5.53) (*754–756*). An alternative approach to diversity of the chromium carbene intermediate involves derivatization of the alanine precursor by

deprotonation and alkylation with alkyl halides (*754*) or aldehydes (*757*).

Photolysis of alcohol solutions of the carbenes under an atmosphere of 60 psi carbon monoxide produced *N*-substituted α-amino esters via a ketene intermediate (see Scheme 5.53). Asymmetry was induced during photolysis by using an amide of a chiral oxazolidine in the complex, giving Ala, Glu, and homophenylalanine derivatives with >93% de (*754*), or γ-hydroxy amino acids (*757*). Chiral [2-^{2}H]Gly was prepared by photolysis of the unsubstituted aminocarbene in MeOD solution (*758*), with an oxazolidine auxiliary giving one diastereomer (86% de) and an oxazolidinone auxiliary giving the opposite diastereomer (74% de). [1,2-^{13}C$_2$]-Isotopically labeled amino acids were prepared by using a ^{13}C-enriched chromium hexacarbonyl complex (*759, 760*), with photolysis in MeOD allowing introduction of a [2-^{2}H] isotope as well (*760*).

Optically pure 1,2-diphenyl-2-aminoethanol has also been used as the amine source and chiral auxiliary, with the alcohol group replacing the solvent in forming an intramolecular ester of the amino acid product (*755, 756*). Oxazinones of Ala, cyclopropyl Ala (*755*), and various arylglycines (*756*) were prepared with 60–90% de and were readily separated and deprotected by mild reductive cleavage. The *syn* arylglycine oxazinones obtained are complementary to the *anti* isomers obtained by Williams' oxazinone synthesis (see Section 7.5).

The greatest advantage of the chromium carbene synthetic route is that the photolyzed carbene complex can be trapped by the amino group of an amino acid or peptide instead of an alcohol, resulting in direct incorporation of the newly synthesized amino acid into a peptide (see Scheme 5.53) (*759*). Dipeptides can be prepared from amino acid esters (*761*), including side-chain unprotected esters of Ser, Cys, Met, and Tyr, or diesters of Asp and Glu (*762*). Sterically hindered amino acids that are difficult to couple by conventional methods such as *N*-Me or α,α-disubstituted amino acids also couple efficiently, although in some cases diastereoselectivity was poor (*762*). This methodology has been applied to stepwise polypeptide synthesis on a Merrifield resin (*759*) but overall efficiencies were low, partly due to the multistep procedure required to remove the oxazolidine auxiliary (*759*). Somewhat better results were obtained with a soluble polyethylene glycol support (*763*).

The carbene complex can also be reacted with imines to give β-lactams (*764, 765*). These can be deprotonated, alkylated, and ring-opened, providing α-alkyl-α-amino acids with overall yields comparable to other approaches (see Scheme 5.54) (*764*).

5.4.1b Other Asymmetric Electrophilic Carboxylations

A chiral lithium amide base has been used to induce asymmetry via enantioselective kinetic deprotonation of Schiff bases in the presence of excess carboxylating reagent. Thus *N*-benzylidene benzylamine was

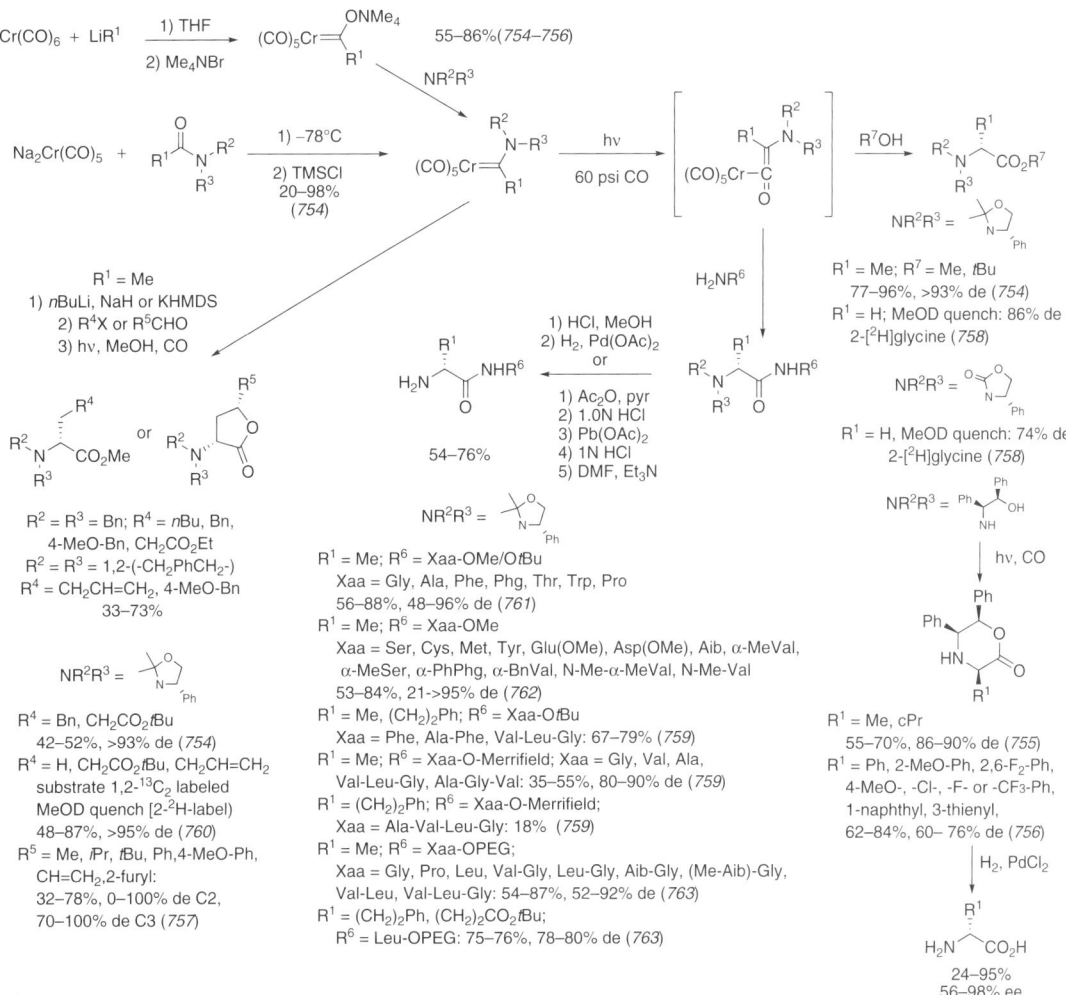

Scheme 5.53 Amino Acid Synthesis via Chromium Carbene Complex.

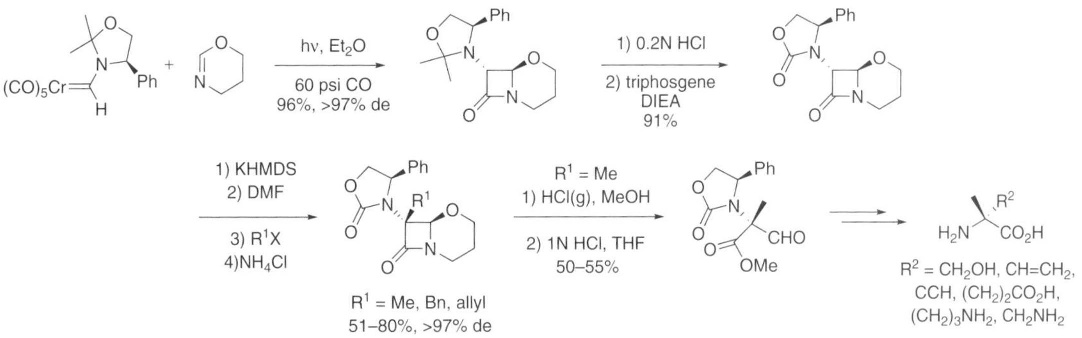

Scheme 5.54 Synthesis α-Alkyl Amino Acids via Chromium Carbene Complex (*764*).

α-deprotonated with lithium alkyl α-methylbenzylamide and carboxylated with various chloroformates to give phenylglycine with up to 40% ee and 60% yield (Scheme 5.55) (766). Enantioselective deprotonation of N-CO$_2$R,N-TMS benzylamines with sparteine/sBuLi, followed by carboxylation with CO$_2$ gave Phg and 4-substituted Phg with 27–96% ee in 10–86% yield (767–769). N-Boc,N-Me-Phg was prepared in the same manner, with up to 78% ee (770). By varying the electrophile (CO$_2$ or methyl chloroformate), it was possible to selectively produce either enantiomer of Phg in 83–95% yield and 86–92% ee (771). Other arylglycines were also prepared with >90% ee and >85% yield (aryl group = 1, 4-Me$_2$-Ph, 3-MeO-Ph, 4-F-Ph, 2-naphthyl, 3-thienyl) (771). This method can also be applied to the synthesis of β- and γ-amino acids by reacting the deprotonated amine with the appropriate electrophile, such as methyl bromoacetate for β-amino acids and acrolein for γ-amino acids (771).

Aziridine-containing chiral zirconium complexes have been prepared from arylamines (see Scheme 5.56). Regioselective insertion of ethylene carbonate (a CO$_2$ synthon) into the Zr–C bond resulted in an ortho ester; methanolysis generated the N-aryl amino acid methyl esters with 34–61% yield and 53–99% ee. The best results were obtained with aromatic substrates (772).

Amides can be obtained by insertion of isocyanate (31–62% yield, 80–99% ee). Higher enantioselectivities were obtained by employing a chiral cyclic carbonate as an optically active electrophilic carboxyl synthon. Phe and Phg were produced with 96–99% ee from an achiral zirconocene complex (773). This latter method was applied to the synthesis of several silyl amino acids (774).

Chiral α-aminoorganolithiums have been prepared from α-keto-organostannes by enantioselective reduction of the keto group, Mitsunobu amination, and Sn–Li exchange (see Scheme 5.57). The organolithium was carboxylated with CO$_2$ to give the N-protected amino acid in 75–93% yield and with 92% ee (775). N-Methyl Pro and N-methyl pipecolic acid were prepared from N-methyl 2-(tributylstannyl)-pyrrolidine or -piperidine, respectively, with 94–99% ee (776). Curiously, the α-aminoorganolithium intermediates of N-urea derivatives are not stable like the N-urethane anions. Instead, they undergo a rapid intramolecular 1,2-N-to-C migration of the -CONR$_2$ amine substituent to form α-aminocarboxamides with significant retention of chirality, depending on the urea substituent size (777). A much more convenient route to the α-aminoorgano-stannane intermediates was recently described, employing addition of Bu$_3$SnLi to tert-butanesulfinimines,

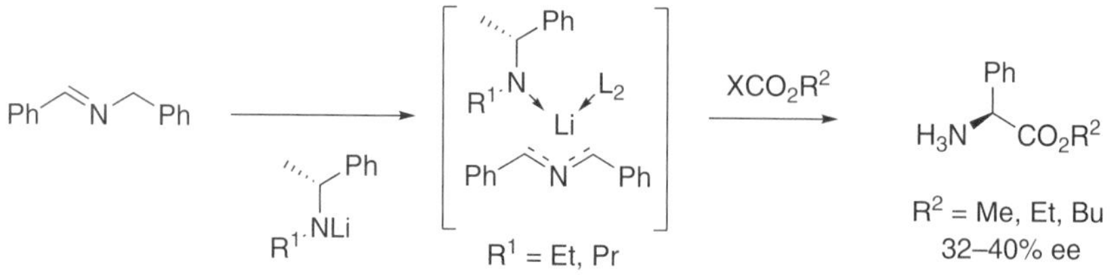

Scheme 5.55 Carboxylation by Chiral Deprotonation (766).

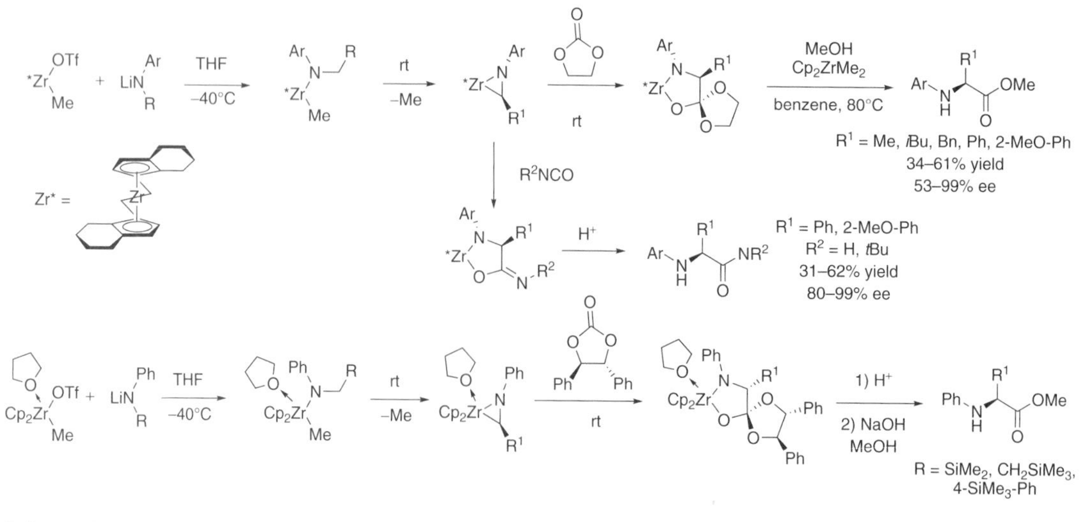

Scheme 5.56 Carboxylation of Zirconocene Aziridine Complex (772, 774).

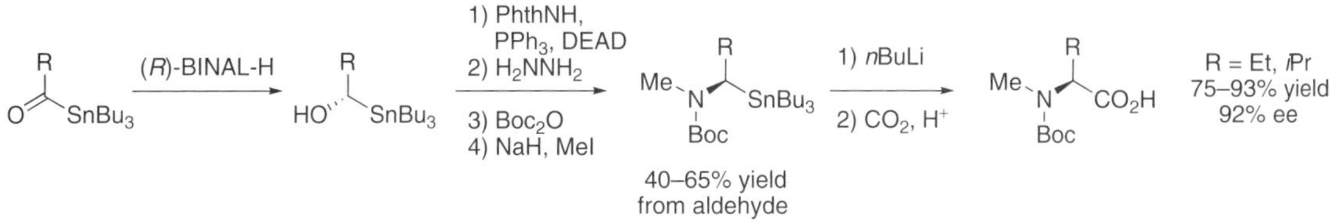

Scheme 5.57 Carboxylation of α-Aminoorganolithium (775).

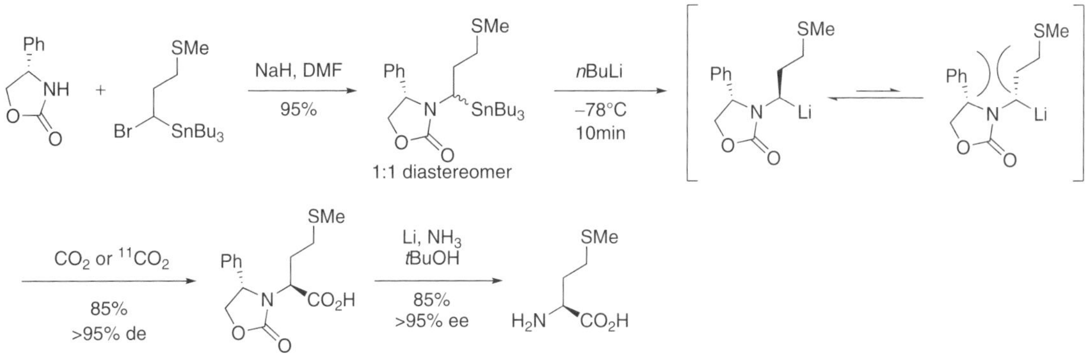

Scheme 5.58 Carboxylation of Chiral α-Aminoorganolithium (780, 781).

RCH=NS(O)tBu (with a chiral center on the sulfur). The additions proceeded with >99% de, potentially leading to a two-step aminocarboxylation of aldehydes procedure (778). A new synthesis of enantiopure tert-butanesulfinamide has been described (779).

This method was not successful for an attempted synthesis of Met. Instead, an organolithium derivative with the amino group contained within an optically active oxazolidinone was carboxylated, with 85% yield and >95% de (see Scheme 5.58). Deprotection was achieved by a Birch reduction. The total synthesis time was 35–40 min, making the route amenable for $^{11}CO_2$ incorporation (780–782). The method was subsequently expanded to include syntheses of homocysteine, Leu, and Ala, though overall yields were quite poor (782).

A dynamic resolution of 1-isobutyl-2-tributylstannane-pyrrolidine with nBuLi and various chiral amine ligands, followed by carboxamidation with PhNCO, produced N-isobutyl Pro-NHPh with up to 94% ee (783). The Seebach imidazolidinone Boc-BMI has been deoxygenated and the anion of the resulting imidazolidine carboxylated with CO_2 in 84% yield with >96% de. A variety of α,β-diamino acids were prepared (784).

5.4.2 Nucleophilic Carboxylation

A number of groups have prepared amino acids by addition of a $^-CO_2H$ synthon to chiral imines, though these reactions are essentially stereocontrolled aminocarboxylation reactions, as discussed in Section 5.2.3.

In an early version of this reaction, cyanide was added to chiral iminoboranes, RCH=N-B(Ipc)₂, derived from the addition of diisopinocampheylborane to alkylnitriles (RCN; R = Me, iPr). Hydrolysis gave (R)-Val and (R)-Ala (45% overall yield), though with marginal optical activity (12% ee) (785). More recently a number of reagents (KCN, TMSCN, (EtO)₂P(O)CN) were used to add cyanide to enantiomerically pure α-substituted-β'-sulfinylenamines. Diastereoselectivity was poor (10–40% de) but the isomers were easily purified before hydrolysis to give α-difluoromethyl-Ala and α-difluoromethyl-Ser (786). A chiral orthoacyl imine auxiliary derived from (R,R)-1, 2-phenylethylene-1,2-diol induced 92% de for addition of PhLi (a masked carboxyl equivalent) to an imine with a tert-butyl substituent (93% yield). Acidic hydrolysis removed the auxiliary to give α-tert-butyl benzylamine (70% yield). Protection of the amine with a trifluoroacetyl group, followed by oxidation of the phenyl group to form the α-carboxylate, gave tert-Leu in 68% yield (787).

There appears to be only one example of carboxylation with a chiral $^-CO_2H$ synthon. A chiral vinyl anion was added to a number of sulfonylimines with reasonable yield (20–63%) and good diastereoselectivity (92–96%) (see Scheme 5.59). Ozonolysis provided the N-protected amino acid methyl esters in 70–80% yield; one example was deprotected to give Phg in 78% yield and >98% ee (788).

Asymmetric carboxylations with a chiral catalyst are possible, via addition of cyanide to imines. However, as

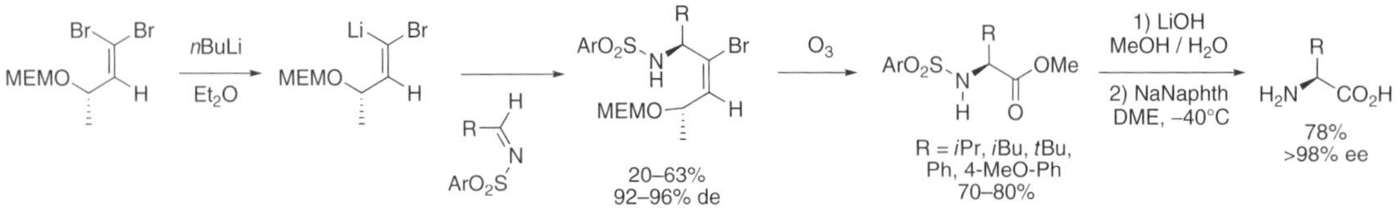

Scheme 5.59 Carboxylation with Chiral Carboxyl Synthon (*788*).

this corresponds to a Strecker-type reaction, it is discussed in Section 5.2.1c.

β-Amino acids can be synthesized by addition of a -CH₂CO₂H synthon to imines (see Section 11.2.3j). For example, this approach was used by Kunz and Schanzenbach (*789*), who added silyl ketene acetals to the chiral Schiff base formed from aldehydes and 2,3,4,6-tetra-*O*-pivaloyl-β-D-galactosylamine. Allylsilanes and allylstannanes have also been added to this imine, with subsequent oxidation to generate the acid (*790*). Both β- and γ-amino acids can be obtained by adding allylmagnesium bromide to chiral *N*-benzylidene-*p*-toluenesulfinamides (92–98% yield, 82–100% de), followed by ozonolysis/oxidation or hydroboration/oxidation of the alkene (*791*). Vinyl and allyl Grignard reagents were added in 50–87% yield and 20–80% de to nitrones bearing a chiral auxiliary (*792*), while a vinyl organocerium reagent added to a SAMP hydrazone in 82% yield and 92% de (*793*), and vinyllithium added to aryl imines in the presence of a catalytic chiral amino ether with 96–99% yield and 60–72% ee (*794*). 2-Lithiothiazole was added to 2-deoxy-2-azido-D-galactopyranolactone to give a precursor of a β-amino acid carbopeptoid (*795*).

References

1. Harada, K. *Nature* **1963**, *200*, 1201.
2. Harada, K.; Fox, S.W. *Naturwis.* **1964**, *51*, 106–107.
3. Dyker, G. *Angew. Chem., Int. Ed. Engl.* **1997**, *36*, 1700–1702.
4. Yet, L. *Angew. Chem., Int. Ed.* **2001**, *40*, 875–877.
5. *Tetrahedron Symposium-in-Print 109*, *Tetrahedron* **2004**, *60*, 10371–10568.
6. Brunel, J.-M.; Holmes, I.P. *Angew. Chem., Int. Ed.* **2004**, *43*, 2572–2778.
7. Gregory, R.J.H.; *Chem. Rev.* **1999**, *99*, 3649–3682.
8. North, M. *Tetrahedron: Asymmetry* **2003**, *14*, 147–176.
9. Kobayashi, S.; Ishitani, H. *Chem. Rev.* **1999**, *99*, 1069–1094.
10. Gröger, H. *Chem. Rev.* **2003**, *103*, 2795–2827.
11. Martínez, R.; Ramón, D.J.; Yus, M. *Tetrahedron Lett.* **2005**, *46*, 8471–8474.
12. Stout, D.M.; Black, L.A.; Matier, W.L. *J. Org. Chem.* **1983**, *48*, 5369–5373.
13. Inaba, T.; Fujita, M.; Ogura, K. *J. Org. Chem.* **1991**, 56, 1274–1279.
14. Harada, K.; Okawara, T.; Matsumoto, K. *Bull. Chem. Soc. Jpn.* **1973**, *46*, 1865–1868.
15. Patel, M.S.; Worsley, M. *Can. J. Chem.* **1970**, *48*, 1881–1884.
16. Fiaud, J.C.; Horeau, A. *Tetrahedron Lett.* **1972**, 25, 2565–2568.
17. Harada, K.; Okawara, T. *J. Org. Chem.* **1973**, *38*, 707–710.
18. Harada, K.; Okawara, T. *Bull. Chem. Soc. Jpn.* **1973**, *46*, 191–193.
19. Saito, K.; Harada, K. *Tetrahedron Lett.* **1989**, *30*, 4535–4538.
20. Phadtare, S.K.; Kamat, S.K.; Panse, G.T. *Ind. J. Chem.* **1985**, *24B*, 811–814.
21. Royer, L.; De, S.K.; Gibbs, R.A. *Tetrahedron Lett.* **2005**, *46*, 4595–4597.
22. Fauchère, J.-L.; Petermann, C. *Int. J. Pept. Prot. Res.* **1981**, *18*, 249–255.
23. Speelman, J.C.; Talma, A.G.; Kellogg, R.M.; Meetsma, A.; de Boer, J.L.; Beurskens, P.T.; Bosman, W.P. *J. Org. Chem.* **1989**, *54*, 1055–1062.
24. Do, K.D.; Thanei, P.; Caviezel, M.; Schwyzer, R. *Helv. Chim. Acta* **1979**, *62*, 956–964.
25. Ayi, A.I.; Guedj, R. *J. Fluorine Chem.* **1984**, *24*, 137–151
26. Oppliger, M.; Schwyzer, R. *Helv. Chim. Acta* **1977**, *60*, 43–47.
27. Heydari, A.; Fatemi, P.; Alizadeh, A.-A. *Tetrahedron Lett.* **1998**, *39*, 3049–3050.
28. Wang, Q.; Ouazzani, J.; Sasaki, N.A.; Potier, P. *Eur. J. Org. Chem.* **2002**, 834–839.
29. Prasad, B.A.; Bisai, A.; Singh, V.K. *Tetrahedron Lett.* **2004**, *45*, 9565–9567.
30. Weinges, K.; Gries, K.; Stemmle, B.; Schrank, W. *Chem. Ber.* **1977**, *110*, 2098–2105.
31. Subramanian, P.K.; Woodard, R.W. *Synth. Commun.* **1986**, *16*, 337–342.
32. Fondekar, K.P.; Volk, F.-J.; Frahm, A.W. *Tetrahedron: Asymmetry* **1999**, *10*, 727–735.
33. Moss, N.; Ferland, J.-M.; Goulet, S.; Guse, I.; Malenfant, E.; Plamondon, L.; Plante, R.; Déziel, R. *Synthesis* **1997**, 32–34.
34. Bisel, P.; Fondekar, K.P.; Volk, F.-J.; Frahm, A.W. *Tetrahedron* **2004**, *60*, 10541–10545.
35. Wede, J.; Volk, F.-J.; Frahm, A.W. *Tetrahedron: Asymmetry* **2000**, *11*, 3231–3252.

36. Truong, M.; Lecornué, F.; Fadel, A. *Tetrahedron: Asymmetry* **2003**, *14*, 1063–1072.

37. Volk, F.-J.; Wagner, M.; Frahm, A.W. *Tetrahedron: Asymmetry* **2003**, *14*, 497–502.

38. Meyer, U.; Breitling, E.; Bisel, P.; Frahm, A.W. *Tetrahedron: Asymmetry* **2004**, 15, 2029-2037.

39. Fondekar, K.P.P.; Volk, F.-J.; Khaliq-uz-Zaman, S.M.; Bisel, P. Frahm, A.W. *Tetrahedron: Asymmetry* **2002**, *13*, 2241–2249.

40. Inaba, T.; Kozono, I.; Fujita, M.; Ogura, K. *Bull. Chem. Soc. Jpn.* **1992**, *65*, 2359–2365.

41. Chakraborty, T.K.; Reddy, G.V.; Hussain, K.A. *Tetrahedron Lett.* **1991**, *32*, 7597–7600.

42. Chakraborty, T.K.; Hussain, K.A.; Reddy, G.V. *Tetrahedron* **1995**, *51*, 9179–9190.

43. Zhu, J.; Boillon, J.-P.; Singh, G.P.; Chastanet, J.; Beugelmans, R. *Tetrahedron Lett.* **1995**, *36*, 7081–7084.

44. Vergne, C.; Bouillon, J.-P.; Chastanet, J.; Bois-Choussy, M.; Zhu, J. *Tetrahedron: Asymmetry* **1998**, *9*, 3095–3103.

45. Dave, R.H.; Hosangadi, B.D. *Tetrahedron* **1999**, *55*, 11295–11308.

46. Boger, D.L.; Keim, H.; Oberhauser, B.; Schreiner, E.P.; Foster, C.A. *J. Am. Chem. Soc.* **1999**, *121*, 6197–6205.

47. Baek, D.-J.; Park, Y.-K.; Heo, H.I.; Lee, M.; Yang, Z.; Choi, M. *Biorg. Med. Chem. Lett.* **1998**, *8*, 3287–3290.

48. Roussi, G.; Zamora, E.G.; Carbonnelle, A.-C.; Beugelmans, R. *Tetrahedron Lett.* **1997**, *38*, 4401–4404.

49. Rao, A.V.R.; Chakraborty, T.K.; Joshi, S.P. *Tetrahedron Lett.* **1992**, *33*, 4045–4048.

50. Sakamuri, S.; Kozikowski, A.P. *Chem. Commun.* **2001**, 475–476.

51. Pellicciari, R.; Marinozzi, M.; Natalini, B.; Constantino, G.; Luneia, R.; Giorgi, G.; Moroni, F.; Thomsen, C. *J. Med. Chem.* **1996**, *39*, 2259–2269.

52. Marinozzi, M.; Natalini, B.; Costantino, G.; Tijskens, P.; Thomsen, C.; Pellicciari, R. *Bioorg. Med. Chem. Lett.* **1996**, *6*, 2243–2246.

53. Pellicciari, R.; Costantino, G.; Giovagnoni, E.; Mattoli, L.; Brabet, I.; Pin, J.-P. *Biorg. Med. Chem. Lett.* **1998**, *8*, 1569–1574.

54. Ma, D.; Tang, G.; Tian, H.; Zou, G. *Tetrahedron Lett.* **1999**, *40*, 5753–5756; *erratum* **1999**, *40*, 9385.

55. Tang, G.; Tian, H.; Ma, D. *Tetrahedron* **2004**, *60*, 10547–10552.

56. Ma, D.; Tian, H.; Zou, G. *J. Org. Chem.* **1999**, *64*, 120–125.

57. Fadel, A.; Khesrani, A. *Tetrahedron: Asymmetry* **1998**, *9*, 305–320.

58. Vedejs, E.; Kongkittingam, C. *J. Org. Chem.* **2001**, *66*, 7355–7364.

59. Andrés, C.; Maestro, A.; Pedrosa, R.; Perez-Encabo, A.; Vicente, M. *Synlett* **1992**, 45–47.

60. Boesten, W.H.; Seerden, J.-P.G.; de Lang, B.; Dielemans, H.J.A.; Elsenberg, H.L.M.; Kaptein, B.;

Moody, H.M.; Kellogg, R.M.; Broxterman, Q.B. *Org. Lett.* **2001**, *3*, 1121–1124.

61. Ohfune, Y.; Moon, S.-H.; Horikawa, M. *Pure Appl. Chem* **1996**, *68*, 645–648.

62. Moon, S.-H.; Ohfune, Y. *J. Am. Chem. Soc.* **1994**, *116*, 7405–7406.

63. Horikawa, M.; Nakajima, T.; Ohfune, Y. *Synlett* **1997**, 253–254.

64. Namba, K.; Kawasaki, M.; Takada, I.; Iwama, S.; Izumida, M.; Shinada, T.; Ohfune, Y. *Tetrahedron Lett.* **2001**, *42*, 3733–3736.

65. Neelakantan, L. *J. Org. Chem.* **1971**, *36*, 2253–2256.

66. Weinges, K.; Klotz, K.-P.; Droste, H. *Chem. Ber.* **1980**, *113*, 710–721.

67. Weinges, K.; Blackholm, H. *Chem. Ber.* **1980**, *113*, 3098–3102.

68. Weinges, K.; Graab, G.; Nagel, D.; Stemmle, B. *Chem. Ber.* **1971**, *104*, 3594–3606.

69. Weinges, K.; Stemmle, B. *Chem. Ber.* **1973**, 106, 2291–2297.

70. Weinges, K.; Brune, G.; Droste, H. *Liebigs Ann. Chem.* **1980**, 212–218.

71. See Reference 70.

72. Weinges, K.; Brachmann, H.; Stahnecker, P.; Rodewald, H.; Nixdorf, M.; Irngartinger, H. *Liebigs Ann. Chem.* **1985**, 566–578.

73. Weinges, K.; Kromm, E. *Liebigs Ann. Chem.* **1985**, 90–102.

74. Harwood, L.M.; Drew, M.G.B.; Hughes, D.J.; Vickers, R.J. *J. Chem. Soc., Perkin Trans. 1* **2001** 1581–1583.

75. Kunz, H.; Sager, W.; Schanzenbach, D.; Decker, M. *Liebigs Ann. Chem.* **1991**, 649–654.

76. Kunz, H.; Sager, W. *Angew. Chem., Int. Ed. Engl.* **1987**, *26*, 557–559.

77. Kunz, H.; Sager, W.; Pfrengle, W.; Schanzenbach, D. *Tetrahedron Lett.* **1988**, *29*, 4397–4400.

78. Fadel, A. *Synlett* **1993**, 503–505.

79. Davis, F.A.; Portonovo, P.S.; Reddy, R.E.; Chiu, Y.-H. *J. Org. Chem.* **1996**, *61*, 440–441.

80. Davis, F.A.; Reddy, R.E.; Portonova, P.S. *Tetrahedron Lett.* **1994**, *35*, 9351–9354.

81. Davis, F.A.; Fanelli, D.L. *J. Org. Chem.* **1998**, *63*, 1981–1985.

82. Davis, F.A.; Srirajan, V.; Titus, D.D. *J. Org. Chem.* **1999** *64*, 6931–6934.

83. Li, B.-F.; Yuan, K.; Zhang, M.-J.; Wu, H.; Dai, L.-X.; Wang, Q.R.; Hou, X.-L. *J. Org. Chem.* **2003**, *68*, 6264–6267.

84. Davis, F.A.; Srirajan, V.; Fanelli, D.L.; Portonovo, P. *J. Org. Chem.* **2000**, *65*, 7663–7666.

85. Hansen, D.B.; Starr, M.-L.; Tolstoy, N.; Joullié, M.M. *Tetrahedron: Asymmetry* **2005**, *16*, 3623–3627.

86. Davis, F.A.; Prasad, K.R.; Carroll, P.J. *J. Org. Chem.* **2002**, *67*, 7802–7806.

87. Davis, F.A.; Lee, S.; Zhang, H.; Fanelli, D.L. *J. Org. Chem.* **2000**, *65*, 8704–8708.

88. Ellman, J.A.; Owens, T.D.; Tang, T.P. *Acc. Chem. Res.* **2002**, *35*, 984–995.

89. Zhou, P.; Chen, B.-C.; Davis, F.A. *Tetrahedron* **2004**, *60*, 8003–8030.

90. Enders, D.; Mosere, M. *Tetrahedron Lett.* **2003**, *44*, 8479–8481.

91. Cativiela, C.; Díaz-de-Villegas, M.D.; Gálvez, J.A.; Garciá, J.I. *Tetrahedron* **1996**, *52*, 9563–9574.

92. Cativiela, C.; Díaz-de-Villegas, M.D.; Gálvez, J.A. *Tetrahedron Lett.* **1995**, *36*, 2859–2860.

93. Niemann, C.; Nichols P.L., Jr., *J. Biol. Chem.* **1942**, *143*, 191–202.

94. Badorrey, R; Cativiela, C.; Díaz-de-Villegas, M.D.; Gálvez, J.A. *Tetrahedron: Asymmetry.* **2000**, *11*, 1015–1025.

95. Aberhart, D.J.; Lin, L.J. *J. Am. Chem. Soc.* **1973**, *95*, 7859–7860.

96. Portonovo, P.; Liang, B.; Joullié, M.M. *Tetrahedron: Asymmetry.* **1999**, *10*, 1451–1455.

97. Taguchi, T.; Kawara, A.; Watanabe, S.; Oki, Y.; Fukushima, H.; Kobayashi, Y.; Okada, M.; Ohta, K.; Iitaka, Y. *Tetrahedron Lett.* **1986**, *27*, 5117–5120.

98. Wenger, R.M. *Helv. Chim. Acta* **1983**, *66*, 2308–2309.

99. Tung, R.D.; Rich, D.H. *Tetrahedron Lett.* **1987**, *28*, 1139–1142.

100. Fiandor, J.; Garcia-López, M.-T.; de las Heras, F.G.; Méndez-Castrilloón, P.P. *Synthesis* **1987**, 978–981.

101. Seki, M.; Hatsuda, M.; Yoshida, S.-I.. *Tetrahedron Lett.* **2004**, *45*, 6579–6581.

102. Tanaka, K.I.; Sawanishi, H. *Tetrahedron: Asymmetry* **1995**, *6*, 1641–1656.

103. Tanaka, K.-I.; Iwabuchi, H.; Sawanishi, H. *Tetrahedron: Asymmetry* **1995**, *6*, 2271–2279.

104. Acher, F.C.; Tellier, F.J.; Azerad, R.; Brabet, I.N.; Fagni, L.; Pin, J.-P.R. *J. Med. Chem.* **1997**, *40*, 3119–3129.

105. Curry, K.; Peet, M.J.; Magnuson, D.S.K.; McLennan, H. *J. Med. Chem.* **1988**, *31*, 864–867.

106. Gaucher, A.; Ollivier, J.; Salaün, J. *Synlett* **1991**, 151–153.

107. Gaucher, A.; Ollivier, J.; Marguerite, J.; Paugam, R.; Salaün, J. *Can. J. Chem.* **1994**, *72*, 1312–1327.

108. Amurrio, I.; Córdoba, R.; Csákÿ, A.G.; Plumet, J. *Tetrahedron* **2004**, *60*, 10521–10524.

109. Byrne, J.J.; Chavarot, M.; Chavant, P.-Y.; Vallée, Y. *Tetrahedron Lett.* **2000**, *41*, 873–876.

110. Banphavichit, V.; Mansawat, W.; Bhanthumnavin, W.; Vilaivan, T. *Tetrahedron* **2004**, *60*, 10559–10568.

111. Mansawat, W.; Bhanthumnavin, W.; Vilaivan, T. *Tetrahedron Lett.* **2003**, *44*, 3805–3808.

112. Krueger, C.A.; Kuntz, K.W.; Dzierba, C.D.; Wirschun, W.G.; Gleason, J.D.; Snapper, M.L.; Hoveyda, A.H. *J. Am. Chem. Soc.* **1999**, *121*, 4284–4285.

113. Porter, J.R.; Wirschun, W.G.; Kuntz, K.W.; Snapper, M.L.; Hoveyda, A.H. *J. Am. Chem. Soc.* **2000**, *122*, 2657–2658.

114. Josephsohn, N.S.; Kuntz, K.W.; Snapper, M.L.; Hoveyda, A.M. *J. Am. Chem. Soc.* **2001**, *123*, 11594–11599.

115. Ishitani, H.; Komiyama, S.; Hasegawa, Y.; Kobayashi, S. *J. Am. Chem. Soc.* **2000**, *122*, 762–766.

116. Chavarot, M.; Byrne, J.J.; Chavant, P.Y.; Vallée, Y. *Tetrahedron: Asymmetry* **2001**, *12*, 1147–1150.

117. Takamura, M.; Hamashima, Y.; Usada, H.; Kanai, M.; Shibasaki, M. *Angew. Chem., Int. Ed.* **2000**, *39*, 1650–1652.

118. Nogami, H.; Matsunaga, S.; Kanai, M.; Shibasaki, M. *Tetrahedron Lett.* **2001**, *42*, 279–283.

119. Nakamura, S.; Sato, N.; Sugimoto, M.; Toru, T. *Tetrahedron: Asymmetry* **2004**, *15*, 1513–1516.

120. Masumoto, S.; Usuda, H.; Suzuki, M.; Kanai, M.; Shibasaki, M. *J. Am. Chem. Soc.* **2003**, *125*, 5634–5635.

121. Spino, C. *Angew. Chem., Int. Ed.* **2004**, *43*, 1764–1766.

122. Kato, N.; Suzuki, M.; Kanai, M.; Shibasaki, M. *Tetrahedron Lett.* **2004**, *45*, 3147–3151.

123. Kato, N.; Suzuki, M.; Kanai, M.; Shibasaki, M. *Tetrahedron Lett.* **2004**, *45*, 3153–355.

124. Sigman, M.S.; Jacobsen, E.N. *J. Am. Chem. Soc.* **1998**, *120*, 4901–4902.

125. Sigman, M.S.; Vachal, P.; Jacobsen, E.N. *Angew. Chem., Int. Ed.* **2000**, *39*, 1279–1281.

126. Vachal, P.; Jacobsen, E.N. *J. Am. Chem. Soc.* **2002**, *124*, 10012–10014.

127. Vachal, P.; Jacobsen, E.N. *Org. Lett.* **2000**, *2*, 867–870.

128. See Reference 127.

129. Huang, J.; Corey, E.J. *Org. Lett.* **2004**, *6*, 5027–5029.

130. Tsogoeva, S.B.; Yalalov, D.A.; Hateley, M.J.; Weckbecker, C.; Huthmacher, K. *Eur. J. Org. Chem.* **2005**, 4995–5000.

131. Tsogoeva, S.B.; Hateley, M.J.; Yalalov, D.A.; Meindl, K.; Weckbecker, C.; Huthmacher, K. *Bioorg. Med. Chem.* **2005**, *13*, 5680–5685.

132. Jiao, Z.; Feng, X.; Liu, B.; Chen, F.; Zhang, G.; Jiang, Y. *Eur. J. Org. Chem.* **2003**, 3818–3826.

133. Corey, E.J.; Grogan, M.J. *Org. Lett.* **1999**, *1*, 157–160.

134. Kowalski, J.; Lipton, M.A. *Tetrahedron Lett.* **1996**, *37*, 5839–5840.

135. Iyer, M.S.; Gigstad, K.M.; Namdev, N.D.; Lipton, M *J. Am. Chem. Soc.* **1996**, *118*, 4910–4911.

136. Li, J.; Jiang, W.-Y.; Han, K.-L.; He, G.-Z.; Li, C. *J. Org. Chem.* **2003**, *68*, 8786–8789.

137. Iyer, M.S.; Gigstad, K.M.; Mandev, N.D.; Lipton, M. *Amino Acids* **1996**, *11*, 259–268.

138. Becker, C.; Hoben, C.; Schollmeyer, D.; Scherr, G.; Kunz, H. *Eur. J. Org. Chem.* **2005**, 1497–1499.

139. Bousquet, C.; Tadros, Z.; Tonnel, J.; Mion, L.; Taillades, J. *Bull. Soc. Chim. Fr.* **1993**, *130*, 513–520.

140. See Reference 139.

141. Hassan, N.A.; Bayer, E.; Jochims, J.C *J. Chem. Soc., Perkin Trans. 1* **1998**, 3747–3757.

142. Wang, M.-X.; Lin, S.-J. *J. Org. Chem.* **2002**, *67*, 6542–6545.

143. Hensel, M.; Lutz-Wahl, S.; Fischer, L. *Tetrahedron: Asymmetry* **2002**, *13*, 2629–2633.

144. Chaplin, J.A.; Levin, M.D.; Morgan, B.; Farid, N.; Li, J.; Zhu, Z.; McQuaid, J.; Nicholson, L.W.; Rand, C.A.; Burk, M.J. *Tetrahedron: Asymmetry* **2004**, *15*, 2793–2796.

145. Keil, O.; Schneider, M.P.; Rasor, J.P. *Tetrahedron: Asymmetry* **1995**, *6*, 1257–1260.

146. Qian, C.; Zhu, C.; Huang, T. *J. Chem. Soc., Perkin Trans. 1* **1998**, 2131–2132.

147. Hamashima, Y.; Sawada, D.; Nogami, H.; Kanai, M.; Shibasaki, M. *Tetrahedron* **2001**, *57*, 805–814.

148. Casas, J.; Nájera, C.; Sansano, J.M.; Saá, J.M. *Org. Lett.* **2002**, *4*, 2589–2592.

149. Casas, J.; Nájera, C.; Sansano, J.M.; Saá, J.M. *Tetrahedron* **2004**, *60*, 10487–10496.

150. Li, Z.-B.; Rajaram, A.R.; Decharin, N.; Qin, Y.-C.; Pu, L. *Tetrahedron Lett.* **2005**, *46*, 2223–2226.

151. Belekon, Y.N.; Green, B.; Ikonnikov, N.S.; North, M.; Parsons, T.; Tatarov, V.I. *Tetrahedron* **2001**, *57*, 771–779.

152. Baleizão, C.; Gigante, B.; García, H.; Corma, A. *Tetrahedron* **2004**, *60*, 10461–10468.

153. Belekon, Y.N.; Blacker, A.J.; Carta, P.; Clutterbuck, L.A.; North, M. *Tetrahedron* **2004**, *60*, 10433–10447.

154. Belekon, Y.N.; Gutnov, A.V.; Moskalenko, M.A.; Yashkina, L.V.; Lesovoy, D.E.; Ikonnikov, N.S.; Larichev, V.S.; North, M. *Chem. Commun.* **2002**, 244–245.

155. Chen, F.-X.; Qin, B.; Feng, X.; Zhang, G.; Jiang, Y. *Tetrahedron* **2004**, *60*, 10449–10460.

156. He, B.; Chen, F.-X.; Li, Y.; Feng, X.; Zhang, G. *Eur. J. Org. Chem.* **2004**, 4657–4666.

157. Huang, W.; Song, Y.; Wang, J.; Cao, G.; Zheng, Z. *Tetrahedron* **2004**, *60*, 10469–10477.

158. Kim, S.S.; Lee, S.H. *Synth. Commun.* **2005**, *35*, 751–759.

159. Kim, S.S.; Song, D.H. *Eur. J. Org. Chem.* **2005**, 1777–1780.

160. Yang, W.-B.; Fang, J.-M. *J. Org. Chem.* **1998**, *63*, 1356–1359.

161. Li, Y.; He, B.; Qin, B.; Feng, X.; Zhang, G. *J. Org. Chem.* **2004**, *69*, 7910–7913.

162. Uang, B.-J.; Fu, I.-P.; Hwang, C.-D.; Chang, C.-W.; Yang, C.-T.; Hwang, D.-R. *Tetrahedron* **2004**, *60*, 10479–10486.

163. Chang, C.-W.; Yang, C.-T.; Hwang, C.-D.; Uang, B.-J. *Chem. Commun.* **2002**, 54–55.

164. Ooi, T.; Miura, T.; Takaya, K.; Ichikawa, H.; Maruoka, K. *Tetrahedron* **2001**, *57*, 867–873.

165. Jönsson, C.; Lundgren, S.; Haswell, S.J.; Moberg, C. *Tetrahedron* **2004**, *60*, 10515–10520.

166. Deng, H.; Isler, M.P.; Snapper, M.L.; Hoveyda, A.H. *Angew. Chem., Int. Ed.* **2002**, *41*, 1009–1012.

167. Gama, A.; Flores-López, L.Z.; Aguirre, G.; Parra-Hake, M.; Somanathan, R.; Cole, T. *Tetrahedron: Asymmetry* **2005**, *16*, 1167–1174.

168. Liu, X.; Qin, B.; Zhou, X.; He, B.; Feng, X. *J. Am. Chem. Soc.* **2005**, *127*, 12224–12225.

169. Effenberger, F.; Förster, S.; Wajant, H. *Curr. Opin. Biotech.* **2000**, *11*, 532–539.

170. Griengl, H.; Klempier, N.; Pöchlauer, P.; Schmidt, M.; Shi, N.; Zabelinskaja-Mackova, A.A. *Tetrahedron* **1998**, *54*, 14477–14486.

171. Kobler, C.; Bohrer, A.; Effenberger, F. *Tetrahedron* **2004**, *60*, 10397–10410.

172. Kobler, C.; Effenberger, F. *Tetrahedron: Asymmetry* **2004**, *15*, 3731–3742.

173. Avi, M.; Fechter, M.H.; Gruber, K.; Belaj, F.; Pöchlauer, P.; Griengl, H. *Tetrahedron* **2004**, *60*, 10411–10418.

174. Roos, J.; Effenburger, F. *Tetrahedron: Asymmetry* **1999**, *10*, 2817–2828.

175. Tromp, R.A.; van der Hoeven, M.; Amore, A.; Brussee, J.; Overhand, M.; van der Marel, G.A.; van der Gen, A. *Tetrahedron: Asymmetry* **2003**, *14*, 1645–1652.

176. Han, S.; Chen, P.; Lin, G.; Huang, H.; Li, Z. *Tetrahedron: Asymmetry* **2001**, *12*, 843–846.

177. Gaisberger, R.P.; Fechter, M.H.; Griengl, H. *Tetrahedron: Asymmetry* **2004**, *15*, 2959–2963.

178. Solís, A.; Luna, H.; Manjarrez, N.; Pérez, H.I. *Tetrahedron* **2004**, *60*, 10427–10431

179. Nanda, S.; Kato, Y.; Asano, Y. *Tetrahedron* **2005**, *61*, 10908–10916.

180. Solís, A.; Luna, H.; Pérez, H.I.; Manjarrez, N. *Tetrahedron: Asymmetry* **2003**, *14*, 2351–2353.

181. Weis, R.; Gaisberger, R.; Skranc, W.; Gruber, K.; Glieder, A. *Angew. Chem., Int. Ed.* **2005**, *44*, 4700–4704.

182. Bühler, H.; Effenberger, F.; Förster, S.; Roos, J.; Wajant, H. *ChemBioChem* **2003**, *4*, 211–216.

183. Bühler, H.; Miehlich, B.; Effenberger, F. *ChemBioChem* **2005**, *6*, 711–717.

184. Purkarthofer, T.; Skranc, W.; Weber, H.; Griengl, H.; Wubbolts, M.; Scholz, G.; Pöchlauer, P. *Tetrahedron* **2004**, *60*, 735–739.

185. Veum, L.; Hanefeld, U.; Pierre, A. *Tetrahedron* **2004**, *60*, 10419–10425.

186. Gröger, H.; Capan, E.; Barthuber, A.; Vorlop, K.-D. *Org. Lett.* **2001**, *3*, 1969–1972.

187. van Langen, L.M.; Selassa, R.P.; van Rantwijk, F.; Sheldon, R.A. *Org. Lett.* **2005**, *7*, 327–329.

188. Glieder, A.; Weis, R.; Skranc, W.; Poechlauer, P.; Dreveny, I.; Majer, S.; Wubbolts, M.; Schwab, H.; Gruber, K. *Angew. Chem., Int. Ed.* **2003**, *42*, 4815–4818.

189. Decicco, C.P.; Grover, P. *Synlett* **1997**, 529–530.

190. de Gonzalo, G.; Lavandera, I.; Brieva, R.; Gotor, V. *Tetrahedron* **2004**, *60*, 10525–10532.

191. Paizs, C.; Tähtinen, P.; Tosa, M.; Majdik, C.; Irimie, F.-D.; Kanerva, L.T. *Tetrahedron* **2004**, *60*, 10533–10540.

192. Veum, L.; Kuster, M.; Telalovic, S.; Hanefeld, U.; Maschmeyer, T. *Eur. J. Org. Chem.* **2002**, 1516–1522.

193. Ugi, I.; Lohberger, S.; Karl, R. In *Comprehensive Organic Synthesis. Selectivity, Strategy and Efficiency in Modern Organic Chemistry;* Trost, B.M.; Fleming, I., Eds; Pergamon Press: New York, **1991**, Vol 2.2, pp 1083–1109.

194. Ugi, I.; Marquarding, D.; Urban, R. In *Chemistry and Biochemistry of Amino Acids, Peptides and Proteins. A Survey of Recent Developments, Volume 6*; Weinstein, B. Ed.; Marcel Dekker: New York, **1982**, pp 245–289.

195. Dömling, A.; Ugi, I. *Angew. Chem., Int. Ed.* **2000**, *39*, 3168–3210.

196. Dömling, A. *Chem. Rev.* **2006**, *106*, 17–89.

197. Ugi, I.; Offermann, K.; Herlinger, H.; Marquarding, D. *Liebigs Ann. Chem.* **1967**, 1–10.

198. Semple, J.E.; Wang, P.C.; Lysenko, Z.; Joullié, M.M. *J. Am. Chem. Soc.* **1980**, *102*, 7505–7510.

199. Joullié, M.M.; Wang, P.C.; Semple, J.E. *J. Am. Chem. Soc.* **1980**, *102*, 887–889.

200. Chen, S.-Y.; Joullié, M.M. *J. Org. Chem.* **1984**, *49*, 1769–1772.

201. See Reference 198.

202. Madsen, U.; Frydenvang, K.; Ebert, B.; Johansen, T.N.; Brehm, L.; Krogsgaard-Larsen, P. *J. Med. Chem.* **1996**, *39*, 183–190.

203. Ahmadian, H.; Nielsen, B.; Bräuner-Osborne, H.; Johansen, T.N.; Stensbøl, T.B.; Sløk, F.A.; Sekiyama, N.; Nakanishi, S.; Krogsgaard-Larsen, P.; Madsen, U. *J. Med. Chem.* **1997**, *40*, 3700–3705.

204. Hebach, C.; Kazmaier, U. *Chem. Commun.* **2003**, 596–597.

205. Yanada, T.; Motoyama, N.; Taniguchi, T.; Kazuta, Y.; Miyazawa, T.; Kuwata, S.; Matsumoto, K.; Sugiura, M. *Chem. Lett.* **1987**, 723–726.

206. Demharter, A.; Hörl, W.; Herdtweck, E.; Ugi, I. *Angew. Chem., Int. Ed. Engl.* **1996**, *35*, 173–175.

207. See Reference 206.

208. Ugi, I.; Demharter, A.; Hörl, W.; Schmid, T. *Tetrahedron* **1996**, *52*, 11657–11664.

209. Dyker, G.; Breitenstein, K.; Henkel, G. *Tetrahedron: Asymmetry* **2002**, *13*, 1929–1936.

210. Godet, T.; Bonvin, Y.; Vincent, G.; Merle, D.; Thozet, A.; Ciufolini, M.A. *Org. Lett.* **2004**, *6*, 3281–3284.

211. Park, S.J.; Keum, G.; Kang, S.B.; Koh, H.Y.; Kim, Y. *Tetrahedron Lett.* **1998**, *39*, 7109–7112.

212. Gouge, V.; Jubault, P.; Quirion, J.-C. *Tetrahedron Lett.* **2004**, *45*, 773–776.

213. Urban, R.; Ugi, I. *Angew. Chem., Int. Ed. Engl.* **1975**, *14*, 61–62.

214. Urban, R.; Eberle, G.; Marquarding, D.; Rehn, D.; Rehn, H.; Ugi, I. *Angew. Chem., Int. Ed. Engl.* **1976**, *10*, 627–628.

215. Kunz, H.; Pfrengle, W. *J. Am. Chem. Soc.* **1988**, *110*, 651–652.

216. Kunz, H.; Pfrengle, W. *Tetrahedron* **1988**, *44*, 5487–5494.

217. Goebel, M.; Ugi, I. *Synthesis* **1991**, 1095–1098.

218. Lehnhoff, S.; Goebel, M.; Karl, R.M.; Klösel, R.; Ugi, I. *Angew. Chem., Int. Ed. Engl.* **1995**, *34*, 1104–1107.

219. Ross, G.F.; Herdtweck, E.; Ugi, I. *Tetrahedron* **2002**, *58*, 6127–6133.

220. Basso, A.; Banfi, L.; Riva, R.; Guanti, G. *J. Org. Chem.* **2005**, *70*, 575–579.

221. Basso, A.; Banfi, L.; Riva, R.; Guanti, G. *Tetrahedron Lett.* **2004**, *45*, 587–590.

222. Ziegler, T.; Schlömer, R.; Koch, C. *Tetrahedron Lett.* **1998**, *39*, 5987–5960.

223. Ziegler, T.; Kaisers, H.-J.; Schlömer, R.; Koch, C. *Tetrahedron* **1999**, *55*, 8397–8408.

224. Quast, H.; Aldenkortt, S. *Chem. Eur. J.* **1996**, *2*, 462–469.

225. Dondoni, A.; Franco, S.; Junquera, F.; Merchán, F.L.; Merino, P.; Tejero, T.; Bertolasi, V. *Chem— Eur. J.* **1995**, *1*, 505–520.

226. Dondoni, A.; Junquera, F.; Merchán, F.L.; Merino, P.; Scherrmann, M.-C.; Tejero, T. *J. Org. Chem.* **1997**, *62*, 5484–5496.

227. Dondoni, A.; Franco, S.; Junquera, F.; Merchán, F.L.; Merino, P.; Tejero, T. *J. Org. Chem.* **1997**, *62*, 5497–5507.

228. Dondoni, A.; Junquera, F.; Merchán, F.L.; Merino, P.; Tejero, T.; *Synthesis* **1994**, 1450–1456.

229. Dondoni, A.; Franco, S.; Merchán, F.L.; Merino, P.; Tejero, T. *Tetrahedron Lett.* **1993**, 34, 5479–5382.

230. Dondoni, A. *Synthesis* **1998**, 1681–1706.

231. Prakash, G.K.S.; Mandal, M.; Schweizer, S.; Petasis, N.A.; Olah, G.A. *J. Org. Chem.* **2002**, *67*, 3718–3723.

232. Merino, P.; Franco, S.; Merchan, F.L.; Tejero, T. *Tetrahedron: Asymmetry* **1997**, *8*, 3489–3496.

233. Merino, P.; Frnaco, S.; Merchan, F.L.; Tejero, T. *J. Org. Chem.* **1998**, *63*, 5627–5630.

234. Adams, Z.M.; Jackson, R.F.W.; Palmer, N.J.; Rami, H.K.; Wythes, M.J. *J. Chem. Soc. Perkin Trans. 1* **1999**, 937–947.

235. Jackson, R.F.W.; Palmer, N.J.; Wythes, M.J.; Clegg, W.; Elsegood, M.R.J. *J. Org. Chem.* **1995**, *60*, 6431–6440.

236. Jackson, R.F.W.; Palmer, N.J.; Wythes, M.J. *Tetrahedron Lett.* **1994**, *35*, 7433–7434.

237. Jackson, R.F.W.; Palmer, N.J.; Wythes, M.J. *J. Chem. Soc., Chem. Commun.* **1994**, 95–96.

238. Jackson, R.F.W.; Kirk, J.M.; Palmer, N.J.; Waterson, D.; Wythes, M.J. *J. Chem. Soc., Chem. Commun.* **1993**, 889–890.

239. Alvaro, G.; Martelli, G.; Savoia, D.; Zoffoli, A. *Synthesis* **1998**, 1773–1777.

240. Cooper, T.S.; Laurent, P.; Moody, C.J.; Takle, A.K. *Org. Biomol. Chem.* **2004**, *2*, 265–276.

241. Barrett, A.G.M.; Braddock, D.C.; Christian, P.W.N.; Pilipauskas, D.; White, A.J.P.; Williams, D.J. *J. Org. Chem.* **1998**, *63*, 5818–5823.

242. Chen, J.; Pandey, R.K.; Cunico, R.F. *Tetrahedron: Asymmetry* **2005**, *16*, 941–947.

243. Shimizu, H.; Kobayashi, S. *Tetrahedron Lett.* **2005**, *46*, 7593–7595.

244. Iwama, T.Y.; Kataoka, T.; Muraoka, O.; Tanabe, G. *J. Org. Chem.* **1998**, *63*, 8355–8360.

245. Xu, W.; Zhang, Y. *Org. Prep. Proc. Int.* **1993**, *25*, 360–362.

246. Zhang, Y.; Li, W. *Synth. Commun.* **1988**, *18*, 1685–1689.

247. Lloyd-Williams, P.; Monerris, P.; Gonzalez, I.; Jou, G.; Giralt, E. *J. Chem. Soc., Perkin Trans. 1* **1994**, 1969–1974.

248. Hasegawa, M.; Taniyama, D.; Tomioka, K. *Tetrahedron* **2000**, *56*, 10153–10158.

249. Hoffman, R.V.; Kim, H.-O. *Tetrahedron* **1992**, *48*, 3007–3020.

250. Fabiano, E.; Golding, B.T.; Sadeghi, M.M. *Synthesis* **1987**, 190–192.

251. Walkup, R.D.; Cole, D.C.; Whittlesey, B.R. *J. Org. Chem.* **1995**, *60*, 2630–2634.

252. Gilbertson, S.R.; Wang, X. *J. Org. Chem.* **1996**, *61*, 434–435.

253. Falck-Pedersen, M.; Undheim, K. *Tetrahedron* **1996**, *52*, 7761–7770.

254. Evans, D.A.; Evrard, D.A.; Rychnovsky, S.D.; Früth, T.; Whittingham, W.G.; DeVries, K.M. *Tetrahedron Lett.* **1992**, *33*, 1189–1192.

255. Hendry, D.; Hough, L.; Richardson, A.C. *Tetrahedron Lett.* **1987**, *28*, 4597–4600.

256. Maiti, S.N.; Singh, M.P.; Micetich, R.G. *Tetrahedron Lett.* **1986**, *27*, 1423–1424.

257. Kamal, A.; Reddy, B.S.N. *Chem. Lett.* **1998**, 593–594.

258. Reddy, G.V.; Rao, G.V.; Iyengar, D.S. *Tetrahedron Lett.* **1999**, *40*, 3937–3938.

259. Kamal, A.; Rao, N.V.; Laxman, E. *Tetrahedron Lett.* **1997**, *38*, 6945–6948.

260. Fringuelli, F.; Pizzo, F.; Rucci, M.; Vaccaro, L. *J. Org. Chem.* **2003**, *68*, 7041–7045.

261. Boruah, A.; Baruah, M.; Prajapati, D.; Sandhu, J.S. *Synlett* **1997**, 1253–1254.

262. Urpí, F.; Vilarrasa, J. *Tetrahedron Lett.* **1986**, *27*, 4623–4624.

263. Ariza, X.; Urpí, F.; Viladomat, C.; Vilarrasa, J. *Tetrahedron Lett.* **1998**, *39*, 9101–9102.

264. Saito, S.; Nakajima, H.; Inaba, M.; Moriwake, T. *Tetrahedron Lett.* **1989**, *30*, 837–838.

265. Kotsuki, H.; Ohishi, T.; Araki, T. *Tetrahedron Lett.* **1997**, *38*, 2129–2132.

266. Ariza, X, Urpí, F.; Vilarrasa, J. *Tetrahedron Lett.* **1999**, *40*, 7515–7517.

267. Effenberger, F.; Burkard, U.; Willfahrt, J. *Liebigs Ann. Chem.* **1986**, 314–333.

268. Neuberger, A.; Sanger, F. *Biochem J.* **1944**, *38*, 125–129.

269. Mandville, G.; Ahmar, M.; Bloch, R. *J. Org. Chem.* **1996**, *61*, 1122–1124.

270. Barrett, A.G.M.; Lebold, S.A. *J. Org. Chem.* **1991**, *56*, 4875–4884.

271. Akita, H.; Chen, C.Y.; Uchida, K. *Tetrahedron: Asymmetry* **1995**, *6*, 2131–2134.

272. Barrett, A.G.M.; Lebold, S.A. *J. Org. Chem.* **1990**, *55*, 5818–5820.

273. Akita, H.; Chen, C.Y.; Kato, K. *Tetrahedron* **1998**, *54*, 11011–11026.

274. Undheim, K.; Grønneberg, T. *Acta Chem. Scand.* **1971**, *25*, 18–26.

275. Jones, M.M.; Williams, J.M.J. *Chem. Commun.* **1998**, 2519–2520.

276. Chen, J.G.; Zhu, J.; Skonezny, P.M.; Rosso, V.; Venit, J.J. *Org. Lett.* **2004**, *6*, 3233–3235.

277. Kiau, S.; Discordia, R.P.; Madding, G.; Okuniewicz, F.J.; Rosso, V.; Venit, J.J. *J. Org. Chem.* **2004**, *69*, 4256–4261.

278. Halland, N.; Braunton, A.; Bachmann, S.; Marigo, M.; Jørgensen, K.A. *J. Am. Chem. Soc.* **2004**, *126*, 4790–4791.

279. Schmidt, U.; Kroner, M.; Griesser, H. *Synthesis* **1989**, 832–835.

280. Weber, I.; Potier, P.; Thierry, J. *Tetrahedron Lett.* **1999**, *40*, 7083–7086.

281. Hanessian, S.; Yang, R.-Y. *Synlett* **1995**, 633–634.

282. Degerbeck, F.; Fransson, B.; Grehn, L.; Ragnarsson, U. *J. Chem. Soc., Perkin Trans. 1* **1992**, 245–253.

283. Fukuyama, T.; Cheung, M.; Kan, T. *Synlett* **1999**, 1301–1303.

284. Abarghaz, M.; Kerbal, A.; Bourguignon, J.-J. *Tetrahedron Lett.* **1995**, *36*, 6463–6466.

285. Berrée, G.; Michelot, G.; Le Corre, M. *Tetrahedron Lett.* **1998**, *39*, 8275–8276.

286. Mukai, C.; Miyakawa, M.; Hanaoka, M. *Synlett* **1994**, 164–166.

287. Hanessian, S.; Wang, W.; Gai, Y. *Tetrahedron Lett.* **1996**, *37*, 7477–7480.

288. Ma, D.; Cao, Y.; Yang, Y.; Cheng, D. *Org. Lett.* **1999**, *1*, 285–287.

289. Brosse, N.; Pinto, M.-F.; Bodiguel, J.; Jamart-Grégoire, B. *J. Org. Chem.* **2001**, *66*, 2869–2873.

290. Bitan, G.; Gilon, C. *Tetrahedron*, **1995**, *51*, 10513–10522.

291. Muller, D.; Zeltser, I.; Bitan, G.; Gilon, C. *J. Org. Chem.* **1997**, *62*, 411–416.

292. Feenstra, R.W.; Stokkingreef, E.H.M.; Nivard, R.J.F.; Otteneijm, H.C.J. *Tetrahedron Lett.* **1987**, *28*, 1215–1218.

293. Feenstra, R.W.; Stokkingreef, E.H.M.; Nivard, R.J.F.; Ottenheijm, H.C.J. *Tetrahedron* **1988**, *44*, 5583–5595.

294. Effenberger, F.; Burkard, U. *Liebigs Ann. Chem.* **1986**, 334–358.

295. Lay, L.; Meldal, M.; Nicotra, F.; Panza, L.; Russo, G. *Chem. Commun.* **1997**, 1469–1470.

296. Grison, C.; Coutrot, F. Coutrot, P. *Tetrahedron* **2002**, *58*, 2735–2741.

297. Uchida, K.; Kato, K.; Akita, H. *Synthesis* **1999**, 1678–1686.

298. Ariza, J.; Font, J.; Ortuño, R.M. *Tetrahedron Lett.* **1991**, *32*, 1979–1982.

299. Bashyal, B.P.; Chow, H.-F.; Fleet, G.W.J. *Tetrahedron Lett.* **1986**, *27*, 3205–3208.

300. Pastó, M.; Moyano, A.; Pericàs, M.A.; Riera, A. *J. Org. Chem.* **1997**, *62*, 8425–8431.

301. Hansson, T.G.; Kihlberg, J.O. *J. Org.Chem.* **1986**, *51*, 4490–4492.

302. Wagner, R.; Tilley, J.W.; Lovey, K. *Synthesis* **1990**, 785–786.

303. Boger, D.L.; Schüle, G. *J. Org. Chem.* **1998**, *63*, 6421–6424.

304. Nicolaou, K.C.; Jain, N.F.; Natarajan, S.; Hughes, R.; Solomon, M.E.; Li, H.; Ramanjulu, J.M.; Takayanagi, M.; Koumbis, A.E.; Bando, T. *Angew. Chem., Int. Ed. Engl.* **1998**, *37*, 2714–2716.

305. Rama Rao, A.V.; Chakraborty, T.K.; Laxman Reddy, K.; Srinivasa Rao, A. *Tetrahedron Lett.* **1994**, *35*, 5043–5046.

306. Hale, K.J.; Manaviazar, S.; Delisser, V.M. *Tetrahedron* **1994**, *50*, 9181–9188.

307. Xiong, C.; Wang, W.; Hruby, V.J. *J. Org. Chem.* **2002**, *67*, 3514–3517.

308. Jiang, Z.-X.; Qin, Y.-Y.; Qing, F.-L. *J. Org. Chem.* **2003**, *68*, 7544–7547.

309. Shao, H.; Goodman, M. *J. Org. Chem.* **1996**, *61*, 2582–2583.

310. Sayyed, I.A.; Sudalai, A. *Tetrahedron: Asymmetry* **2004**, *15*, 3111–3116.

311. Byun, H.-S.; He, L.; Bittman, R. *Tetrahedron* **2000**, *56*, 7051–7091.

312. Tanaka, H.; Sawayama, A.M.; Wandless, T.J. *J. Am. Chem. Soc.* **2003**, *125*, 6864–6865.

313. Soucy, F.; Grenier, L.; Behnke, M.L.; Destree, A.T.; McCormack, T.A.; Admas, J.; Plamondon, L. *J. Am. Chem. Soc.* **1999**, *121*, 9967–9976.

314. Cho, G.Y.; Ko, S.Y. *J. Org. Chem.* **1999**, *64*, 8745–8747.

315. Mulzer, J.; Funk, G. *Synthesis*, **1995**, 101–112.

316. Mulzer, J.; Angermann, A.; Schubert, B.; Seilz, C. *J. Org. Chem.* **1986**, *51*, 5294–5299.

317. Mulzer, J.; Greifenberg, S.; Beckstett, A.; Gottwald, M. *Liebigs Ann. Chem.* **1992**, 1131–1135.

318. Mead, K.; MacDonald, T.L. *J. Org. Chem.* **1985**, *50*, 422–424.

319. Jiménez, J.M.; Ortuño, R.M. *Tetrahedron: Asymmetry* **1996**, *7*, 3203–3208.

320. Ramalingam, K.; Nanjappan, P.; Kalvin, D.M.; Woodard, R.W. *Tetrahedron* **1988**, *44*, 5597–5604.

321. Deng, J.; Hamada, Y.; Shiori, T. *Synthesis* **1998**, 627–638.

322. Wakamiya, T.; Yamanoi, K.; Nishikawa, M.; Shiba, T. *Tetrahedron Lett.* **1985**, *26*, 4759–4760.

323. Yadav, J.S.; Chandrasekhar, S.; Ravindra Reddy, Y.; Rama Rao, A.V. *Tetrahedron* **1995**, *51*, 2749–2754.

324. Boger, D.L.; Borzilleri, R.M.; Nukui, S. *J. Org. Chem.* **1996**, *61*, 3561–3565.

325. Nicolaou, K.C.; Natarajan, S.; Li, H.; Jain, N.F.; Hughes, R.; Solomon, M.E.; Ramanjulu, J.M.; Boddy, N.C.; Takayanagi, M. *Angew. Chem., Int. Ed. Engl.* **1998**, *37*, 2708–2714.

326. Berkowitz, D.B.; Eggen, M.; Shen, Q.; Shoemaker, R.K. *J. Org. Chem.* **1996**, *61*, 4666–4675.

327. Shattuck, J.C.; Meinwald, J. *Tetrahedron Lett.* **1997**, *38*, 8461–8464.

328. Ma, D.; Ma, Z. *Tetrahedron Lett.* **1997**, *38*, 7599–7602.

329. Takano, S.; Moriya, M.; Iwabuchi, Y.; Ogasawara, K. *Tetrahedron Lett.* **1989**, *30*, 3805–3806.

330. Corey, E.J.; Link, J.O. *J. Am. Chem. Soc.* **1992**, *114*, 1906–1908.

331. Domínguez, C.; Ezquerra, J.; Baker, S.R.; Borrelly, S.; Prieto, L.; Espada, M.; Pedregal, C. *Tetrahedron Lett.* **1998**, *39*, 9305–9308.

332. Habay, S.A.; Schafmeister, C.E. *Org. Lett.* **2004**, *6*, 3369–3371.

333. Mellin-Morlière, C.; Aitken, D.J.; Bull, S.D.; Davies, S.G.; Husson, H.-P. *Tetrahedron: Asymmetry* **2001**, *12*, 149–155.

334. Bravo, P.; Cavicchio, G.; Crucianelli, M.; Poggiali, A.; Zanda, M. *Tetrahedron: Asymmetry.* **1997**, *8*, 2811–2815.

335. Harada, K.; Matsumoto, K. *J. Org. Chem.* **1966**, *31*, 2985–2991.

336. Boche, G.; Schrott, W. *Tetrahedron Lett.* **1982**, *23*, 5403–5406.

337. Yamada, M.; Nagashima, N.; Hasegawa, J.; Takahashi, S. *Tetrahedron Lett.* **1998**, *39*, 9019–9022.

338. Davies, S.G.; Fenwick, D.R. *J. Chem. Soc., Chem. Commun.* **1995**, 1109–1110.

339. Davies, S.G.; Ichihara, O.; Walters, I.A.S. *Synlett*, **1994**, 117–118.

340. Bunnage, M.E.; Davies, S.G.; Goodwin, C.J. *Synlett*, **1993**, 731–732.

341. Bunnage, M.E.; Burke, A.J.; Davies, S.G.; Goodwin, C.J. *Tetrahedron: Asymmetry* **1995**, *6*, 165–176.

342. Bunnage, M.E.; Burke, A.J.; Davies, S.G.; Goodwin, C.J. *Tetrahedron: Asymmetry* **1994**, *5*, 203–206.

343. Davies, S.G.; Ichihara, O.; Lenoir, I.; Walters, I.A.S. *J. Chem. Soc., Perkin Trans. 1*, **1994**, 1411–1415.

344. Davies, S.G.; Ichihara, O.; Walters, I.A.S. *J. Chem. Soc., Perkin Trans. 1*, **1994**, 1141–1147.

345. Davies, S.G.; Walters, I.A.S. *J. Chem. Soc., Perkin Trans. 1* **1994**, 1129–1139.

346. Bunnage, M.E.; Chernega, A.N.; Davies, S.G.; Goodwin, C.J. *J. Chem. Soc., Perkin Trans. 1*, **1994**, 2372–2384.

347. Bunnage, M.E.; Davies, S.G.; Goodwin, C.J. *J. Chem. Soc., Perkin Trans. 1* **1994**, 2385–2391.

348. Burke, A.J.; Davies, S.G.; Hedgecock, C.J.R. *Synlett* **1996**, 621–622.

349. Enders, D.; Wiedemann, J.; Bettray, W. *Synlett* **1995**, 369–371.

350. Enders, D.; Wahl, H.; Bettray, W. *Angew. Chem., Int. Ed. Engl.* **1995**, *34*, 455–457.

351. Enders, D.; Bettray, W.; Raabe, G.; Runsink, J. *Synthesis* **1994**, 1322–1326.

352. Sabelle, S.; Lucet, D.; Le Gall, T.; Mioskowski, C. *Tetrahedron Lett.* **1998**, *39*, 2111–2114.

353. Amoroso, R.; Cardillo, G.; Tomasini, C.; Tortoreto, P. *J. Org. Chem.* **1992**, *57*, 1082–1087.

354. Amoroso, R.; Cardillo, G.; Tomasini, C. *Tetrahedron Lett.* **1990**, *31*, 6413–6416.

355. Amoroso, R.; Cardillo, G.; Tomasini, C. *Tetrahedron Lett.* **1991**, *32*, 1971–1974.

356. Amoroso, R.; Cardillo, G.; Romero, M.S.; Tomasini, C. *Gaz. Chim. Ital.* **1993**, *123*, 75–78.

357. Amoroso, R.; Cardillo, G.; Tomasini, C. *Heterocycles* **1992**, *34*, 349–355.

358. Harada, K.; Nakamura, I. *J. Chem. Soc., Chem. Commun.* **1978**, 522–523.

359. Guifa, S.; Lingchong, Y. *Synth. Commun.* **1993**, *23*, 1229–1234.

360. Jumnah, R.; Williams, A.C.; Williams, J.M.J. *Synlett* **1995**, 821–822.

361. Bower, J.F.; Jumnah, R.; Williams, A.C.; Williams, J.M.J. *J. Chem. Soc., Perkin Trans. 1* **1997**, 1411–1420.

362. Hayashi, T.; Yamamoto, A.; Ito, Y.; Nishioka, E.; Miura, H.; Yanagi, K. *J. Am. Chem. Soc.* **1989**, *111*, 6301–6311.

363. Röhm, K.H.; Van Etten, R.L. *J. Lab. Cmpds. Radiopharm.* **1985**, *22*, 909–915.

364. Kluender, H.; Huang, F.-C.; Fritzberg, A.; Schnoes, H.; Sih, C.J.; Fawcett, P.; Abraham, E.P. *J. Am. Chem. Soc.* **1974**, *96*, 4054–4055.

365. Kluender, H.; Bradley, C.H.; Sih, C.J.; Fawcett, P.; Abraham, E.P. *J. Am. Chem. Soc.* **1973**, *95*, 6149–6150.

366. Gulzar, M.S.; Akhtar, M.; Gani, D. *J. Chem. Soc., Perkin Trans. 1* **1997**, 649–655.

367. Akhtar, M.; Botting, N.P.; Cohen, M.A.; Gani, D. *Tetrahedron* **1987**, *43*, 5899–5908.

368. Hamilton, B.K.; Hsiao, H.-Y.; Swann, W.E.; Anderson, D.M.; Delente, J.J. *Trends Biotech.* **1985**, *3*, 64–68.

369. Archer, C.H.; Thomas, N.R.; Gani, D. *Tetrahedron: Asymmetry* **1993**, *4*, 1141–1152.

370. Horwell, D.C.; Nichols, P.D.; Roberts, E. *Bioorg. Med. Chem. Lett.* **1994**, *4*, 2263–2266.

371. Akhtar, M.; Cohen, M.A.; Gani, D. *Tetrahedron Lett.* **1987**, *28*, 2413–2416.

372. Jandel, A.-S.; Hustedt, H.; Wandrey, C. *Eur. J. Appl. Microbiol. Biotech.* **1982**, *15*, 59–63.

373. Erdik, E.; Ay, M. *Chem. Rev.* **1989**, *89*, 1947–1980.

374. Erdik, E. *Tetrahedron* **2004**, *60*, 8747–8782.

375. Greck, C.; Drouillat, B.; Thomassigny, C. *Eur. J. Org. Chem.* **2004**, 1377–1385.

376. Janey, J.M. *Angew. Chem., Int. Ed.* **2005,** *44*, 4292–4300.

377. Genet, J.P. *Pure Appl. Chem.* **1996**, *68*, 593–596.

378. Greck, C.; Ferreira, F.; Genêt, J.P. *Tetrahedron Lett.* **1996**, *37*, 2031–2034.

379. Greck, C.; Bischoff, L.; Genêt, J.P. *Tetrahedron: Asymmetry* **1995**, *6*, 1989–1994.

380. Greck, C.; Bischoff, L.; Ferreira, F.; Pinel, C.; Piveteau, E.; Genêt, J.P. *Synlett*, **1993**, 475–477.

381. Genet, J.P.; Juge, S.; Mallart, S. *Tetrahedron Lett.* **1988**, *29*, 6765–6768.

382. Sting, A.R.; Seebach, D. *Tetrahedron* **1996**, *52*, 279–290.

383. Girard, A.; Greck, C.; Ferroud, D.; Genêt, J.P. *Tetrahedron Lett.* **1996**, *37*, 7967–7970.

384. Guanti, G.; Banfi, L.; Narisano, E. *Tetrahedron* **1988**, *44*, 5553–5562.

385. Evans, D.E.; Britton, T.C.; Dorow, R.L.; Dellaria J.F., Jr., *Tetrahedron* **1988**, *44*, 5525–5540.

386. Ferreira, F.; Greck, C.; Genêt, J.P. *Bull. Soc. Chim. Fr.* **1997**, *134*, 615–621.

387. Kapeller, H.; Griengl, H. *Tetrahedron* **1997**, *53*, 14635–14644.

388. Gethin, D.M.; Simpkins, N.S. *Tetrahedron* **1997**, *53*, 14417–14436.

389. Poupardin, O.; Greck, C.; Genêt, J.-P. *Synlett* **1998**, 1279–1281.

390. Vidal, J.; Guy, L.; Stérin, S.; Collet, A. *J. Org. Chem.* **1993**, *58*, 4791–4793.

391. Greck, C.; Genêt, J.P. *Synlett* **1997**, 741–748.

392. Evans, D.A.; Nelson, S.G. *J. Am. Chem. Soc.* **1997**, *119*, 6452–6453.

393. Marigo, M.; Juhl, K.; Jørgensen, K.A. *Angew. Chem. Int. Ed.* **2003**, *42*, 1367–1369.

394. Liu, X.; Li, H.; Deng, L. *Org. Lett.* **2005**, *7*, 167–169.

395. Borman, S. *Chem. Eng. News* **2002**, *June 3*, 47.

396. List, B. *J. Am. Chem. Soc.* **2002**, *124*, 5656–5657.

397. Bøgevig, A.; Juhl, K.; Kumaragurubaran, N.; Zhuang, W.; Jørgensen, K.A. *Angew. Chem., Int. Ed.* **2002**, *41*, 1790–1793.

398. Evans, D.A.; Bartroli, J.; Shih, T.L. *J. Am. Chem. Soc.* **1981**, *103*, 2127–2129.

399. Evans, D.A.; Weber, A.E. *J. Am. Chem. Soc.* **1986**, *108*, 6757–6761.

400. Rich, D.H.; Sun, C.-Q.; Guillaume, D.; Dunlap, B.; Evans, D.A.; Weber, A.E. *J. Med. Chem.* **1989**, *32*, 1982–1987.

401. Evans, D.A.; Weber, A.E.; Britton, T.C.; Ellman, J.A.; Sjogren, E.B. *Peptides:Chemistry and Biology, Proceedings of the 10th American Peptide Symposium*, Marshall, G.R., Ed.; ESCOM Science Publishers: AE Leiden, The Netherlands, 1988, pp 143–148.

402. Evans, D.A.; Britton, T.C.; Dorow, R.L.; Dellaria, J.F. *J. Am. Chem. Soc.* **1986**, *108*, 6395–6397.

403. Trimble, L.A.; Vederas, J.C. *J. Am. Chem. Soc.* **1986**, *108*, 6397–6399.

404. Evans, D.A.; Britton, T.C.; Ellman, J.A.; Dorow, R.L. *J. Am. Chem. Soc.* **1990**, *112*, 4011–4030.

405. Evans, D.A.; Britton, T.C. *J. Am. Chem. Soc.* **1987**, *109*, 6881–6883.

406. Radel, P.A.; Kahl, S.B. *J. Org. Chem.* **1996**, *61*, 4582–4588.

407. Fukuzawa, S.-I.; Hongo, Y. *Tetrahedron Lett.* **1998**, *39*, 3521–3524.

408. Evans, D.A.; Ellman, J.A.; Dorow, R.L. *Tetrahedron Lett.* **1987**, *28*, 1123–1126.

409. Boteju, L.W.; Wegner, K.; Hruby, V.J. *Tetrahedron Lett.* **1992**, *33*, 7491–7474.

410. Dharanipragada, R.; VanHulle, K.; Bannister, A.; Bear, S.; Kennedy, L.; Hruby, V.J. *Tetrahedron* **1992**, *48*, 4733–4748.

411. Stone, M.J.; Maplestone, R.A.; Rahman, S.K.; Williams, D.H. *Tetrahedron Lett.* **1991**, *32*, 2663–2666.

412. Lee, H.T.; Hicks, J.L.; Johnson, D.R. *J. Lab. Cmpds. Radiopharm.* **1991**, *29*, 1065–1072.

413. Chen, H.G.; Beylin, V.G.; Marlatt, M.; Leja, B.; Goel, O.P. *Tetrahedron Lett.* **1992**, *33*, 3293–3296.

414. Parry, R.J.; Ju, S.; Baker, B.J.*J. Lab. Cmpds. Radiopharm.* **1991**, *29*, 633–643.

415. Kennedy, K.J.; Lundquist IV, J.T.; Simandan, T.L.; Beeson, C.C.; Dix, T.A. *Bioorg. Med. Chem. Lett.* **1997**, *7*, 1937–1940.

416. Wen, S.-J.; Yao, Z.-J. *Org. Lett.* **2004,** *6*, 2721–2724.

417. Moss, N.; Beaulieu, P.; Duceppe, J.-S.; Ferland, J.-M.; Gauthier, J.; Ghiro, E.; Goulet, S.; Grenier, L.; Llinas-Brunet, M..; Plante, R.; Wernic, D.; Déziel, R. *J. Med. Chem.* **1995**, *38*, 3617–3623.

418. Pearson, A.J.; Zhang, P.; Lee, K. *J. Org. Chem.* **1996**, *61*, 6581–6586.

419. Nakamura, Y.; Shin, C.-G. *Chem. Lett.* **1991**, 1953–1956.

420. Hale, K.J.; Delisser, V.M.; Manaviazar, S. *Tetrahedron Lett.* **1992**, *33*, 7613–7616.

421. Hale, K.J.; Cai, J.; Delisser, V.; Manaviazar, S.; Peak, S.A.; Bhatia, G.S.; Collins, T.C.; Jogiya, N. *Tetrahedron* **1996**, *52*, 1047–1068.

422. Decicco, C.P.; Leathers, T. *Synlett* **1995**, 615–616.

423. Waid, P.P.; Flynn, G.A.; Huber, E.W.; Sabol, J.S. *Tetrahedron Lett.* **1996**, *37,* 4091–4094.

424. Lundquist J.T., IV, Dix, T.A. *Tetrahedron Lett.* **1998**, *39*, 775–778.

425. Pearson, A.J.; Chelliah, M.V.; Bignan, G.C. *Synthesis* **1997**, 536–540.

426. Neset, S.; Hope, H.; Undheim, K. *Tetrahedron* **1997**, *53*, 10459–10470.

427. Evans, D.E.; Ellman, J.A. *J. Am. Chem. Soc.* **1989**, *111*, 1063–1072.

428. Andrews, M.J.I.; Tabor, A.B. *Tetrahedron Lett.* **1997**, *38*, 3063–3066.

429. McNamara, L.M.A.; Andrews, M.J.I.; Mitzel, F.; Siligardi, G.; Tabor, A.B. *J. Org. Chem.* **2001**, *66*, 4585–4594.

430. Larsson, U.; Carlson, R.; Leroy, J. *Acta Chem. Scand.* **1993**, *47*, 380–390.

431. Evans, D.A.; Sjogren, E.B.; Weber, A.E.; Conn, R.E. *Tetrahedron Lett.* **1987**, *28*, 39–42.

432. Evans, D.A.; Weber, A.E. *J. Am. Chem. Soc.* **1987**, *109*, 7151–7157.

433. Boger, D.L.; Menezes, R.F. *J. Org. Chem.* **1992**, *57*, 4331–4333.

434. Boger, D.L.; Colletti, S.L.; Honda, T.; Menezes R.F. *J. Am. Chem. Soc.* **1994**, *116*, 5607–5618.

435. Owa, T.; Otsuka, M.; Ohno, M. *Chem. Lett.* **1988**, 1873–1874.

436. Boger, D.L.; Ramsey, T.M.; Cai, H. *Bioorg. Med. Chem.* **1997**, *5*, 195–207.

437. Lago, M.A. Samanen, J.; Elliot, J.D. *J. Org. Chem.* **1992**, *57*, 3493–3496.

438. Gustafsson, T.; Saxin, M.; Kihlberg, J. *J. Org. Chem.* **2003**, *68*, 2506–2509.

439. Smith, R.J.; Bienz, S. *Helv. Chim. Acta* **2004**, *87*, 1681–1696.

440. Dharanipragada, R.; Nicolas, E.; Toth, G.; Hruby, V.J. *Tetrahedron Lett.* **1989**, *30*, 6841–6844.

441. Li, G.; Patel, D.; Hruby, V.J. *J. Chem. Soc., Perkin Trans 1* **1994**, 3057–3059.

442. Xiang, L.; Wu, H.; Hruby, V.J. *Tetrahedron: Asymmetry* **1995**, *6*, 83–86.

443. Liao, S.; Hruby, V.J. *Tetrahedron Lett.* **1996**, *37*, 1563–1566.

444. Nicolas, E.; Dharanipragada, R.; Toth, G.; Hruby, V.J. *Tetrahedron Lett.* **1989**, *30*, 6845–6848.

445. Sabol, J.S.; McDonald, I.A. *Tetrahedron Lett.* **1994**, *35*, 1817–1820.

446. Sabol, J.S.; McDonald, I.A. *Tetrahedron Lett.* **1994**, *35*, 1821–1824.

447. Han, G.; Lewis, A.; Hruby, V.J. *Tetrahedron Lett.* **2001**, *42*, 4601–4603.

448. Nicolás, E.; Russell, K.C.; Hruby, V.J. *J. Org. Chem.* **1993**, *58*, 766–770.

449. van Heerden, P.S.; Bezuidenhoudt, B.C.B.; Ferreira, D. *Tetrahedron Lett.* **1997**, *38*, 1821–1824.

450. Li, G.; Russell, K.C.; Jarosinski, M.A.; Hruby, V.J. *Tetrahedron Lett.* **1993**, *34*, 2561–2564.

451. Nicolás, E.; Russell, K.C.; Knollenberg, J.; Hruby, V.J. *J. Org. Chem.* **1993**, *58*, 7565–7571.

452. Lung, F.-D.; Li, G.; Lou, B.-S.; Hruby, V.J. *Synth. Commun.* **1995**, *25*, 57–61.

453. Li, G.; Patel, D.; Hruby, V.J. *Tetrahedron Lett.* **1994**, *35*, 2301–2304.

454. Yuan, W.; Hruby, V. *Tetrahedron Lett.* **1997**, *38*, 3853–3856.

455. Liao, S.; Shenderovich, M.D.; Lin, J.; Hruby, V.J. *Tetrahedron* **1997**, *53*, 16645–16662.

456. Lin, J.; Liao, S.; Han, Y.; Qiu, W.; Hruby, V.J. *Tetrahedron: Asymmetry* **1997**, *8*, 3213–3221.

457. Li, G.; Russell, K.C.; Jarosinski, M.A.; Hruby, V.J. *Tetrahedron Lett.* **1993**, *34*, 2565–2568.

458. Quin, X.; Russell, K.C.; Boteju, L.W.; Hruby, V.J. *Tetrahedron* **1995**, *51*, 1033–1054.

459. Lin, J.; Liao, S.; Hruby, V.J. *Tetrahedron Lett.* **1998**, *39*, 3117–3120.

460. Han, Y.; Liao, S.; Qiu, W.; Cai, C.; Hruby, V.J. *Tetrahedron Lett.* **1997**, *38*, 5135–5138.

461. Wang, S.; Tang, X.; Hruby, V.J. *Tetrahedron Lett.* **2000**, *41*, 1307–1310.

462. Rück-Braun, K.; Stamm, A.; Engel, S.; Kunz, H. *J. Org. Chem.* **1997**, *62*, 967–975.

463. Evans, D.A.; Chapman, K.T.; Bisaha, J. *J. Am. Chem. Soc.* **1988**, *110*, 1238–1256.

464. Thom, C.; Kocienski, P. *Synthesis* **1992**, 582–586.

465. Ho, G.-J.; Mathre, D.J. *J. Org. Chem.* **1995**, *60*, 2271–2273.

466. Dobarro, A.; Velasco, D. *Tetrahedron* **1996**, *52*, 13525–13530.

467. Dobarro, A.; Velasco, D. *Tetrahedron* **1996**, *52*, 13733–13738.

468. Prashad, M.; Kim, H.-Y.; Har, D.; Repic, O.; Blacklock, T.J. *Tetrahedron Lett.* **1998**, *39*, 9369–9372.

469. Lewis, N.; McKillop, A.; Taylor, R.J.K.; Watson, R.J. *Synth. Commun.* **1995**, *25*, 561–568.

470. Sudharshan, M.; Hultin, P.G. *Synlett* **1997**, 171–172.

471. Xu, D.; Sharpless, K.B. *Tetrahedron Lett.* **1993**, *34*, 951–952.

472. Li, G.; Lenington, R.; Willis, S.; Kim, S.H. *J. Chem. Soc., Perkin Trans. 1* **1998**, 1753–1754.

473. Feroci, M.; Inesi, A.; Mucciante, V.; Rossi, L. *Tetrahedron Lett.* **1999**, *40*, 6059–6060.

474. Wakamatsu, H.; Terao, Y. *Chem. Pharm. Bull.* **1996**, *44*, 261–263.

475. Armstrong, D.W.; Lee, J.T.; Chang, L.W. *Tetrahedron: Asymmetry* **1998**, *9*, 2043–2064.

476. Armstrong, D.W.; He, L.; Yu, T.; Lee, J.T.; Liu, Y.-s. *Tetrahedron: Asymmetry* **1999**, *10*, 37–60.

477. Hashimoto, N.; Ishizuka, T.; Kunieda, T. *Tetrahedron Lett.* **1994**, *35*, 721–724.

478. Palomo, C.; Berrée, F.; Linden, A.; Villalgordo, J.M. *J. Chem. Soc., Chem. Commun.* **1994**, 1861–1862.

479. Tanaka, K.; Uno, H.; Osuga, H.; Suzuki, H. *Tetrahedron: Asymmetry* **1993**, *4*, 629–632.

480. Lützen, A.; Köll, P. *Tetrahedron: Asymmetry.* **1997**, *8*, 1193–1206.

481. Banks, M.R.; Blake, A.J.; Cadogan, J.I.G.; Dawson, I.M.; Gaur, S.; Gosney, I.; Gould, R.; Grant, K.J.; Hodgson, P.K.G. *J. Chem. Soc., Chem. Commun.* **1993**, 1147–1148.

482. Sibi, M.P.; Deshpande, P.K.; La Loggia, A.J.; Christensen, J.W. *Tetrahedron Lett.* **1995**, *36*, 8961–8964.

483. Sibi, M.P.; Deshpande, P.K.; Ji, J. *Tetrahedron Lett.* **1995**, *36*, 8965–8968.

484. Drauz, K.; Jahn, W.; Schwarm, M. *Chem.—Eur. J.* **1995**, *1*, 568–572.

485. Bommarius, A.S.; Schwarm, M.; Stingl, K.; Kottenhahn, M.; Huthmacher, K.; Drauz, K. *Tetrahedron: Asymmetry* **1995**, *6*, 2851–2888.

486. Ghosh, AK.; Cho, H.; Onishi, M. *Tetrahedron: Asymmetry* **1997**, *8*, 821–824.

487. Winkler, J.D.; McCoull, W. *Tetrahedron Lett.* **1998**, *39*, 4935–4936.

488. Desimoni, G.; Faita, G.; Galbiati, A.; Pasini, D.; Quadrelli, P.; Rancati, F. *Tetrahedron: Asymmetry* **2002**, *13*, 333–337.

489. Allin, S.M.; Shuttleworth, S.J. *Tetrahedron Lett.* **1996**, *37*, 8023–8026.

490. Kotake, T.; Hayashi, Y.; Rajesh, S.; Mukai, Y.; Takiguchi, Y.; Kimura, T.; Kiso, Y. *Tetrahedron* **2005**, *61*, 3819–3833.

491. Gibson, C.L.; Gillon, K.; Cook, S. *Tetrahedron Lett.* **1998**, *39*, 6733–6736.

492. Hintermann, T.; Seebach, D. *Helv. Chim. Acta* **1998**, *81*, 2093–2126.

493. Gaul, C.; Schweizer, B.W.; Seiler, P.; Seebach, D. *Helv. Chim. Acta* **2002**, *85*, 1546–1566.

494. Zheng, N.; Armstrong J.D., III; McWilliams, J.C.; Volante, R.P. *Tetrahedron Lett.* **1997**, *38*, 2817–2820.

495. Azam, S.; D'Souza, A.A.; Wyatt, P.B. *J. Chem. Soc., Perkin Trans. 1* **1996**, 621–627.

496. Pettit, G.R.; Burkett, D.D.; Williams, M.D. *J. Chem. Soc., Perkin Trans. 1* **1996**, 863–858.

497. Yuen, P.-O.; Kanter, G.D.; Taylor, C.P.; Vartanian, M.G. *Bioorg. Med. Chem. Lett.* **1994**, *4*, 823–826.

498. D'Souza, A.A.; Motevalli, M.; Robinson, A.J.; Wyatt, P.B. *J. Chem. Soc., Perkin Trans. 1,* **1995**, 1–2.

499. Chakraborty, T.K.; Ghosh, A. *Tetrahedron Lett.* **2002**, *43*, 9691–9693.

500. Oppolzer, W.; Dudfield, P. *Tetrahedron Lett.* **1985**, *26*, 5037–5040.

501. Oppolzer, W. *Tetrahedron* **1987**, *43*, 1969–2004, 4057.

502. Oppolzer, W.; Pedrosa, R.; Moretti, R. *Tetrahedron Lett.* **1986**, *27*, 831–834.

503. Cativiela, C.; Díaz-de-Villegas, M.D.; Gálvez, J.A. *Tetrahedron: Asymmetry* **1994**, *5*, 1465–1468; *6*, 2611.

504. Cativiela, C.; Díaz-de-Villegas, M.D.; Gálvez, J.A. *Tetrahedron: Asymmetry* **1993**, *4*, 229–238.

505. Cativiela, C.; Diaz-de-Villegas, M.D.; Galvez, J.A. *Tetrahedron: Asymmetry* **1992**, *3*, 1141–1144.

506. Oppolzer, W.; Moretti, R. *Helv. Chim. Acta* **1986**, *69*, 1923–1926.

507. Opplozer, W.; Moretti, R. *Tetrahedron* **1988**, *44*, 5541–5552.

508. Fioravanti, S.; Loreto, M.A.; Pellacani, L.; Sabbatini, F.; Tardella, P.A. *Tetrahedron: Asymmetry* **1994**, *5*, 473–478.

509. Badorrey, R.; Cativiela, C.; Diaz-de-Villegas, M.D.; Gálvez, J.A. *Tetrahedron: Asymmetry* **1995**, *6*, 2787–2796.

510. Cativiela, C.; Diaz-de-Villegas, M.D.; Galvez, J.A. *Tetrahedron: Asymmetry* **1993**, *4*, 1445–1448.

511. Cativiela, C.; Diaz de Villegas, M.D.; Gálvez, J.A. *Synlett* **1994**, 302–304.

512. Cativiela, C.; Díaz-de-Villegas, M.D.; Gálvez, J.A.; Lapeña, Y. *Tetrahedron* **1995**, *51*, 5921–5928.

513. Cativiela, C.; Díaz-de-Villigas, M.D.; Gálvez, J.A.; Lapeña, Y. *Tetrahedron* **1997**, *53*, 5891–5898.

514. Cativiela, C.; Diaz-de-Villegas, M.D.; Galvez, J.A. *Tetrahedron: Asymmetry* **1994**, *5*, 261–268.

515. Badorrey, R.; Cativiela, C.; Díaz-de-Villegas, M.D.; Gálvez, J.A.; Lapeña, Y. *Tetrahedron: Asymmetry.* **1997**, *8*, 311–317.

516. Cativiela, C.; Diaz-de-Villegas, M.D.; Gálvez, J.A. *Tetrahedron* **1996**, *52*, 687–694.

517. Oppolzer, W. *Pure Appl. Chem.* **1990**, *62*, 1241–1250.

518. Oppolzer, W.; Chapuis, C.; Bernardinelli, G. *Helv. Chim. Acta* **1984**, *67*, 1397–1401.

519. Oppolzer, W.; Tamura, O.; Deerberg, J. *Helv. Chim. Acta* **1992**, *75*, 1965–1978.

520. Oppolzer, W.; Moretti, R.; Thomi, S. *Tetrahedron Lett.* **1989**, *30*, 5603–5606.

521. Oppolzer, W.; Moretti, R.; Thomi, S. *Tetrahedron Lett.* **1989**, *30*, 6009–6010.

522. Ward, R.S.; Pelter, A.; Goubet, D.; Pritchard, M.C. *Tetrahedron: Asymmetry* **1995**, *6*, 93–96.

523. Ward, R.S.; Pelter, A.; Goubet, D.; Pritchard, M.C. *Tetrahedron: Asymmetry* **1995**, *6*, 469–498.

524. Malmquist, J.; Sjöberg, S. *Tetrahedron* **1996**, *52*, 9207–9218.

525. Oppolzer, W.; Tamura, O. *Tetrahedron Lett.* **1990**, *31*, 991–994.

526. Oppolzer, W. *Pure Appl. Chem.* **1988**, *60*, 39–48.

527. Oppolzer, W.; Poli, G.; Kingma, A.J.; Starkemann, C.; Bernardinelli, G. *Helv. Chim. Acta* **1987**, *70*, 2201–2214.

528. Oppolzer, W.; Cintas-Moreno, P.; Tamura, O.; Cardinaux, F. *Helv. Chim. Acta* **1993**, *76*, 187–196.

529. Lodwig, S.N.; Unkefer, C.J. *J. Lab. Cmpds. Radiopharm.* **1996**, *38*, 239–248.

530. Stocking, E.M.; Martinez, R.A.; Silks, L.A.; Sanz-Cervera, J.F.; Williams, R.M. *J. Am. Chem. Soc.* **2001**, *123*, 3391–3392.

531. Garner, P.; Dogan, O.; Pillai, S. *Tetrahedron Lett.* **1994**, *35*, 1653–1656.

532. Marchi, C.; Trepat, E.; Moreno-Mañas, M.; Vallribera, A.; Molins, E. *Tetrahedron* **2002**, *58*, 5699–5708.

533. Wang, Y.-C.; Su, D.-W.; Lin, C.-M.; Tseng, H.-L.; Li, C.-L.; Yan, T.-H. *Tetrahedron Lett.* **1999**, *40*, 3577–3580.

534. Ahn, K.H.; Kim, S.-K.; Ham, C. *Tetrahedron Lett.* **1998**, *39*, 6321–6322.

535. Gennari, C.; Colombo, L.; Bertolini, G. *J. Am. Chem. Soc.* **1986**, *108*, 6394–6395.

536. Vicario, J.L.; Badia, D.; Domínguez, E.; Crespo, A.; Carrillo, L.; Anakabe, E. *Tetrahedron Lett.* **1999**, *40*, 7123–7126.

537. Drewes, S.E.; Malissar, D.G.S.; Roos, G.H.P. *Chem. Ber.* **1993**, *126*, 2663–2673.

538. Ezquerra, J; Rubio, A.; Martín, J.; Navío, J.L.G. *Tetrahedron: Asymmetry* **1997**, *8*, 669–671.

539. Kitagawa, O.; Hanano, T.; Kikuchi, N.; Taguchi, T. *Tetrahedron Lett.* **1993**, *34*, 2165–2168.

540. Felice, E.; Fioravanti, S.; Pellacani, L.; Tardella, P.A. *Tetrahedron Lett.* **1999**, *40*, 4413–4416.

541. Kubota, H.; Kubo, A.; Nunami, K.-I. *Tetrahedron Lett.* **1994**, *35*, 3107–3110.

542. Kubo, A.; Kubota, H.; Takahashi, M.; Nunami, K.-I. *Tetrahedron Lett.* **1996**, *37*, 4957–4960.

543. Nunami, K.-I.; Kubota, H.; Kubo, A. *Tetrahedron Lett.* **1994**, *35*, 8639–8642.

544. Santos, A.G.; Candeias, S.X.; Afonso, C.A.M.; Jenkins, K.; Caddick, S.; Treweeke, N.R.; Pardoe, D. *Tetrahedron* **2001**, *57*, 6607–6614.

545. Caddick, S.; Afonso, C.A.M.; Candeias, S.X.; Hitchcock, P.B.; Jenkins, K.; Murtagh, L.; Pardoe, D.; Santos, A.G.; Treweeke, N.R.; Weaving, R. *Tetrahedron* **2001**, *57*, 6589–6605.

546. Caddick, S.; Jenkins, K.; Treweeke, N.; Candeias, S.X.; Afonso, C.A.M. *Tetrahedron Lett.* **1998**, *39*, 2203–2206.

547. Treweeke, N.R.; Hitchcock, P.B.; Pardoe, D.A.; Caddick, S. *Chem. Commun.* **2005**, 1868–1870.

548. Koh, K.; Ben, R.N.; Durst, T. *Tetrahedron Lett.* **1993**, *34*, 4473–4476.

549. Ben, R.N.; Durst, T. *J. Org. Chem.* **1999**, *64*, 7700–7706.

550. O'Meara, J.A.; Gardee, N.; Jung, M.; Ben, R.N.; Durst, T. *J. Org. Chem.* **1998**, *63*, 3117–3119.

551. Nam, J.; Chang, J.-y.; Shin, E.-k.; Kim, H.J.; Kim, Y.; Jang, S.; Park, Y.S. *Tetrahedron* **2004**, *60*, 6311–6318.

552. Nam, J.; Chang, J.-y.; Hahm, K.-S.; Park, Y.S. *Tetrahedron Lett.* **2003**, *44*, 7727–7730.

553. Kim, H.J.; Shin, E.-k.; Chang, J.-y.; Kim, Y.; Park, Y.S. *Tetrahedron Lett.* **2005**, *46*, 4115–4117.

554. O'Brien, P. *Angew. Chem., Int. Ed. Engl.* **1999**, *38*, 326–329.

555. Bodkin, J.A.; McLeod, M.D. *J. Chem. Soc., Perkin Trans. 1* **2002**, 2733–2746.

556. Sharpless, K.B. *Angew. Chem., Int. Ed. Engl.* **1996**, *35*, 451–454.

557. Li, G.; Sharpless, K.B. *Acta Chem. Scand.* **1996**, *50*, 649–651.

558. Upadhya, T.T.; Sudalai, A. *Tetrahedron: Asymmetry* **1997**, *8*, 3685–3689.

559. Chandrasekhar, S.; Mohapatra, S.; Yadav, J.S. *Tetrahedron* **1999**, *55*, 4763–4768.

560. Chandrasekhar, S.; Mohapatra, S. *Tetrahedron Lett.* **1998**, *39*, 695–698.

561. Chandrasekhar, S.; Mohapatra, S. *Tetrahedron Lett.* **1998**, *39*, 6415–6418.

562. Rubin, A.E.; Sharpless, K.B. *J. Am. Chem. Soc.* **1998**, *120*, 2637–2640.

563. Fokin, V.V.; Sharpless, K.B. *Angew. Chem., Int., Ed.* **2001**, *40*, 3455–3457.

564. Streuff, J.; Osterath, B.; Nieger, M.; Muñiz, K. *Tetrahedron: Asymmetry* **2005**, *16*, 3492–3496.

565. Rudolph, J.; Sennhenn, P.C.; Vlaar, C.P.; Sharpless, K.B. *Angew. Chem., Int. Ed. Engl.* **1996**, 35, 2810–2813.

566. Bruncko, M.; Schlingloff, G.; Sharpless, K.B. *Angew. Chem., Int. Ed. Engl.* **1997**, *36*, 1483–1486.

567. Song, C.E.; Oh, C.R.; Roh, E.J.; Lee, S.-G.; Choi, J.H. *Tetrahedron: Asymmetry* **1999**, *10*, 671–674.

568. Kandula, S.R.V.; Kumar, P. *Tetrahedron: Asymmetry* **2005**, *16*, 3579–3583.

569. Li, G.; Angert, H.H.; Sharpless, K.B. *Angew. Chem., Int. Ed. Engl.* **1996**, 35, 2813–2817.

570. Raatz, D.; Innertsberger, C.; Reiser, O. *Synlett* **1999**, 1907–1910.

571. Liu, Z.; Ma, N.; Jia, Y.; Bois-Choussy, M.; Malabarba, A.; Zhu, J. *J. Org. Chem.* **2005**, *70*, 2847–2850.

572. Reddy, K.L.; Dress, K.R.; Sharpless, K.B. *Tetrahedron Lett.* **1998**, *39*, 3667–3670.

573. Goossen, L.J.; Liu, H.; Dress, K.R.; Sharpless, K.B. *Angew. Chem., Int. Ed. Engl.* **1999**, *38*, 1080–1083.

574. Gontcharov, A.V.; Liu, H.; Sharpless, K.B. *Org. Lett.* **1999**, *1*, 783–786.

575. Jo, C.H.; Han, S.-H.; Yang, J.W.; Roh, E.J.; Shin, U.-S.; Song, C.E. *Chem. Commun.* **2003**, 1312–1313.

576. Song, C.E.; Oh, C.R.; Lee, S.W.; Lee, S.-G.; Canali, L.; Sherrington, D.C. *Chem. Commun.* **1998**, 2435–2436.

577. Cravotto, G.; Giovenzana, G.B.; Pagliarin, R.; Palmisano, G.; Sisti, M. *Tetrahedron: Asymmetry* **1998**, *9*, 745–748.

578. Thomas, A.A.; Sharpless, K.B. *J. Org. Chem.* **1999**, *64*, 8379–8385.

579. Han, H.; Yoon, J.; Janda, K.D. *J. Org. Chem.* **1998**, *63*, 2045–2048.

580. Reddy, K.L.; Sharpless, K.B. *J. Am. Chem. Soc.* **1998**, *120*, 1207–1217.

581. O'Brien, P.; Osborne, S.A.; Parker, D.D. *J. Chem. Soc., Perkin Trans. 1* **1998**, 2519–2526.

582. O'Brien, P.; Osborne, S.A.; Parker, D.D. *Tetrahedron Lett.* **1998**, *39*, 4099–4102.

583. Evans, D.A.; Katz, J.L.; Peterson, G.S.; Hintermann, T. *J. Am. Chem. Soc.* **2001**, *123*, 12411–12413.

584. Atkinson, R.N.; Moore, L.; Tobin, J.; King, S.B. *J. Org. Chem.* **1999**, *64*, 3467–3475.

585. Crowley, B.M.; Mori, Y.; McComas, C.C.; Tang, D.; Boger, D.L. *J. Am. Chem. Soc.* **2004**, *126*, 4310–4317.

586. Venkatraman, S.; Njoroge, F.G.; Girijavallabhan, V.; McPhail, A.T. *J. Org. Chem.* **2002**, *67*, 2686–2688.

587. Boger, D.L.; Lee, R.J.; Bounaud, P.-Y.; Meier, P. *J. Org. Chem.* **2000**, *65*, 6770–6772.

588. Jiang, W.; Wanner, J.; Lee, R.J.; Bounaud, P.-Y.; Boger, D.L. *J. Am. Chem. Soc.* **2003**, *125*, 1877–1887.

589. Tao, B.; Schlingloff, G.; Sharpless, K.B. *Tetrahedron Lett.* **1998**, *39*, 2507–2510.

590. Panek, J.S.; Masse, C.E. *Angew. Chem., Int. Ed. Engl.* **1999**, *38*, 1093–1095.

591. Cao, B.; Park, H.; Joullié, M.M. *J. Am. Chem. Soc.* **2002**, *124*, 520–521.

592. Han, H.; Cho, C.-W.; Janda, K.D. *Chem.— Eur. J.* **1999**, *5*, 1565–1569.

593. Andersson, M.A.; Epple, R.; Fokin, V.V.; Sharpless, K.B. *Angew. Chem., Int. Ed.* **2002**, *41*, 472–475.

594. Ihara, M.; Takahashi, M.; Taniguchi, N.; Yasui, K.; Niitsuma, H.; Fukumoto, K. *J. Chem. Soc. Perkin Trans 1* **1991**, 525–535.

595. Wu, Z.L.; Li, Z.Y. *J. Org. Chem.* **2003**, *68*, 2479–2482.

596. Kedrowski, B.L. *J. Org. Chem.* **2003**, *68*, 5403–5406.

597. Georg, G.I.; Guan, X.; Kant, J. *Tetrahedron Lett.* **1988**, *29*, 403–406.

598. Tanaka, M.; Oba, M.; Tamai, K.; Suemune, H. *J. Org. Chem.* **2001**, *66*, 2667–2673.

599. Oba, M.; Tanaka, M.; Kurihara, M.; Suemune, H. *Helv. Chim. Acta* **2002**, *85*, 3197–3218.

600. Yokomatsu, T.; Sato, M.; Shibuya, S. *Tetrahedron: Asymmetry* **1996**, *7*, 2743–2754.

601. Burgess, K.; Li, W. *Tetrahedron Lett.* **1995**, *36*, 2725–2728.

602. Morimoto, Y.; Achiwa, K. *Chem. Pharm. Bull.* **1987**, *35*, 3845–3849.

603. Hill, R.K.; Prakash, S.R.; Wiesendanger, R.; Angst, W.; Martinoni, B.; Arigoni, D.; Liu, H.-W.; Walsh, C.T. *J. Am. Chem. Soc.* **1984**, *106*, 795–796.

604. Hercouet, A.; Godbert, N.; Le Corre, M. *Tetrahedron: Asymmetry* **1998**, *9*, 2233–2234.

605. Davies, H.M.L.; Huby, N.J.S.; Cantrell W.R., Jr.; Olive, J.L. *J. Am. Chem. Soc.* **1993**, *115*, 9468–9479.

606. Davies, H.M.L.; Cantrell W.R; Jr. *Tetrahedron Lett.* **1991**, *32*, 6509–6512.

607. Moye-Sherman, D.; Jin, S.; Ham, I.; Lim, D.; Scholtz, J.M.; Burgess, K. *J. Am. Chem. Soc.* **1998**, *120*, 9435–9443.

608. Burgess, K.; Ho, K.-K. *J. Org. Chem.* **1992**, *57*, 5931–5936.

609. Lim, D.; Burgess, K. *J. Org. Chem.* **1997**, *62*, 9382–9384.

610. Ma, D.; Ma, J.; Dai, L. *Tetrahedron: Asymmetry* **1997**, *8*, 825–827.

611. Alías, M.; Cativiela, C.; Díaz-de-Villegas, M.D.; Gálvez, J.A.; Lapeña, Y. *Tetrahedron* **1998**, *54*, 14963–14974.

612. Burgess, K.; Ho, K.-K. *Tetrahedron Lett.* **1992**, *33*, 5677–5680.

613. Burgess, K.; Lim, D.; Ho, K.-K.; Ke, C.-Y. *J. Org. Chem.* **1994**, *59*, 2179–2185.

614. Jung, M.E.; D'Amico, D.C. *J. Am. Chem. Soc.* **1995**, *117*, 7379–7388.

615. Avenoza, A.; Cativiela, C.; París, M.; Peregrina, J.M.; Saenz-Torre, B. *Tetrahedron: Asymmetry* **1997**, *8*, 1123–1129.

616. Huang, X.; Seid, M.; Keillor, J.W. *J. Org. Chem.* **1997**, *62*, 7495–7496.

617. Zhang, L.-h.; Kauffman, G.S.; Pesti, J.A.; Yin, J. *J. Org. Chem.* **1997**, *62*, 6918–6920.

618. Matsumura, Y.; Maki, T.; Satoh, Y. *Tetrahedron Lett.* **1997**, *38*, 8879–8882.

619. Axelsson, B.S.; O'Toole, K.J.; Spencer, P.A.; Young, D.W. *J. Chem. Soc., Perkin Trans. 1*, **1994**, 807–815.

620. Milewska, M.J.; Polonski, T. *Synthesis* **1988**, 475.

621. Jones, I.G.; Jones, W.; North, M. *Synlett*, **1997**, 63–65.

622. Nöteberg, D.; Brånalt, J.; Kvarnström, I.; Classon, B.; Samuelsson, B.; Nillroth, U.; Danielson, U.H.; Karlén, A.; Hallberg, A. *Tetrahedron* **1997**, *53*, 7975–7984.

623. Noyes, W.A.; Skinner, G.S. *J. Am. Chem. Soc.* **1917**, *39*, 2692–2718.

624. Noyes, W.A.; Nickell, L.F. *J. Am. Chem. Soc.* **1914**, *36*, 118–127.

625. Weir, J. *J. Chem. Soc.* **1911**, *99*, 1270–1277.

626. Yamamoto, Y.; Shimoda, H.; Oda, J.; Inouye, Y. *Bull. Chem. Soc. Jpn.* **1976**, *49*, 3247–3249.

627. Mehmandoust, M.; Petit, Y.; Larchevêque, M. *Tetrahedron Lett.* **1992**, *33*, 4313–4316.

628. Metz, P.; Mues, C.; Schoop, A. *Tetrahedron* **1992**, *48*, 1071–1080.

629. Eguchi, T.; Koudate, T.; Kakinuma, K. *Tetrahedron* **1993**, *49*, 4527–4540.

630. Chida, N.; Takeoka, J.; Tsutsumi, N.; Ogawa, S. *J. Chem. Soc., Chem. Commun.* **1995**, 793–794.

631. Chida, N.; Takeoka, J.; Ando, K.; Tsutsumi, N.; Ogawa, S. *Tetrahedron* **1997**, *53*, 16287–16298.

632. Imogaï, H.; Petit, Y.; Larchevêque, M. *Tetrahedron Lett.* **1996**, *37*, 2573–2576.

633. Savage, I.; Thomas, E.J. *J. Chem. Soc., Chem. Commun.* **1989**, 717–719.

634. Chen, A.; Savage, I.; Thomas, E.J.; Wilson, P.D. *Tetrahedron Lett.* **1993**, *34*, 6769–6772.

635. Savage, I.; Thomas, E.J.; Wilson, P.D. *J. Chem. Soc., Perkin Trans. 1* **1999**, 3291–3303.

636. Saksena, A.K.; Lovey, R.G.; Girijavallabhan, V.M.; Ganguly, A.K.; McPhail, A.T. *J. Org. Chem.* **1986**, *51*, 5024–5028.

637. Chen, A.; Thomas, E.J.; Wilson, P.D. *J. Chem. Soc., Perkin Trans. 1* **1999**, 3305–3310.

638. Overman, L.E.; Zipp, G.G. *J. Org. Chem.* **1997**, *62*, 2288–2291.

639. Schmidt, U.; Respondek, M.; Lieberknecht, A.; Werner, J.; Fischer, P. *Synthesis* **1989**, 256–261.

640. Chen, Y.K.; Lurain, A.E.; Walsh, P.J. *J. Am. Chem. Soc.* **2002**, *124*, 12225–12231.

641. Ramachandran, P.V.; Burghardt, T.E.; Reddy, M.V.R. *J. Org. Chem.* **2005**, *70*, 2329–2331.

642. Shea, R.G.; Fitzner, J.N.; Fankhauser, J.E.; Spaltenstein, A.; Carpino, P.A.; Peevey, R.M.; Pratt, D.V.; Tenge, B.J.; Hopkins, P.B. *J. Org. Chem.* **1986**, *51*, 5243–5252.

643. Knoop, F.; Martius, C. *Z. Physiol. Chem.* **1939**, *258*, 238–242.

644. Herbst, R.M.; Swart, E.A. *J. Org. Chem.* **1946**, *11*, 368–377.

645. Hiskey, R.G.; Northrop, R.C. *J. Am. Chem. Soc.* **1965,** *8*, 1753–1757.

646. Matsumoto, K.; Harada, K. *J. Org. Chem.* **1968**, *33*, 4526–4528.

647. Hiskey, R.G.; Northrop, R.C. *J. Am. Chem. Soc.* **1961**, *83*, 4798–4800.

648. Tanaka, M.; Kobayashi, T.-A.; Sakakura, T. *Angew. Chem., Int. Ed. Engl.* **1984**, *23*, 518.

649. Harada, K.; Matsumoto, K. *J. Org. Chem.* **1967**, *32*, 1794–1800.

650. Harada, K.; Kataoka, Y. *Chem. Lett.* **1978**, 791–794.

651. Harada, K.; Tamura, M. *Bull. Chem. Soc. Jpn.* **1979**, *52*, 1227–1228.

652. Harada, K.; Matsumoto, K. *J. Org. Chem.* **1968**, *33*, 4467–4470.

653. Krämer, R.; Wnajek, H.; Polborn, K.; Beck, W. *Chem. Ber.* **1993**, *126*, 2421–2427.

654. Coutrot, P.; Grison, C.; Coutrot, F. *Synlett* **1998**, 393–395.

655. Harada, K. *Nature* **1966**, *212*, 1571–1572.

656. Harada, K. *J. Org. Chem.* **1967**, *32*, 1700–1793.

657. Yamada, S-I.; Hashimoto, S.-I. *Tetrahedron Lett.* **1976**, 997–1000.

658. Huffman, M.A.; Reider, P.J. *Tetrahedron Lett.* **1999**, *40*, 831–834.

659. Nicoud, J.-F.; Kagan, H.B. *Tetrahedron Lett.* **1971**, *23*, 2065–2068.

660. Aller, E.; Buck, R.T.; Drysdale, M.J.; Ferris, L.; Haigh, D.; Moody, C.J.; Pearson, N.D.; Sanghera, J.B. *J. Chem. Soc., Perkin Trans. 1* **1996**, 2879–2884.

661. Matsumoto, K.; Harada, K. *J. Org. Chem.* **1966**, *31*, 1956–1958.

662. Miyazawa, T.; Takashima, K.; Yamada, T.; Kuwata, S.; Watanabe, H. *Bull. Chem. Soc. Jpn.* **1982**, *55*, 341–342.

663. Corey, E.J.; McCaully, R.J.; Sachdev, H.S. *J. Am. Chem. Soc.* **1970**, *92*, 2476–2488.

664. Corey, E.J.; Sachdev, H.S.; Gougoutas, J.Z.; Saenger, W. *J. Am. Chem. Soc.* **1970**, *92*, 2488–2501.

665. Vigneron, J.P.; Kagan, H.; Horeau, A. *Tetrahedron Lett.* **1968**, 5681–5683.

666. Jiao, X.-Y.; Chen, W.-Y.; Hu, B.-F. *Synth. Commun.* **1992**, *22*, 1179–1186.

667. Cox, G.G.; Harwood, L.M. *Tetrahedron: Asymmetry* **1994**, *5*, 1669–1672.

668. Drok, K.J.; Harrowfield, J.M.; McNiven, S.J.; Sargeson, A.M.; Skelton, B.W.; White, A.H. *Aust. J. Chem.* **1993**, *46*, 1557–1593.

669. Zhu, Q.-C.; Hutchins, R.O. *Org. Prep. Proc. Int.* **1994**, *26*, 193–236.

670. Bolm, C. *Angew. Chem., Int. Ed. Engl.* **1993**, *32*, 232–233.

671. Akabori, S.; Sakurai, S.; Izumi, Y.; Fujii, Y. *Nature* **1956**, *178*, 323–324.

672. Burk, M.J.; Martinez, J.P.; Feaster, J.E.; Cosford, N. *Tetrahedron* **1994**, *50*, 4399–4428.

673. Nugent, W.A.; RajanBabu, T.V.; Burk, M.J. *Science* **1993**, *259*, 479–483.

674. Burk, M.J.; Feaster, J.E. *J. Am. Chem. Soc.* **1992**, *114*, 6266–6267.

675. Kadyrov, R.; Riermeier, T.H.; Dingerdissen, U.; Tararov, V.; Börner, A. *J. Org. Chem.* **2003**, *68*, 4067–4070.

676. Sakai, T.; Yan, F.; Kashino, S.; Uneyama, K. *Tetrahedron* **1996**, *52*, 233–244.

677. Sakai, T.; Yan, F.; Uneyama, K. *Synlett* **1995**, 753–754.

678. Demir, A.S. *Pure Appl. Chem.* **1997**, *69*, 105–108.

679. Demir, A.S.; Tanyeli, C.; Cagir, A.; Tahir, M.N.; Ulku, D. *Tetrahedron: Asymmetry* **1998**, *9*, 1035–1042.

680. Abe, S.; Fuchigami, T.; Nonaka, T. *Chem. Lett.* **1983**, 1033–1036.

681. Longenecker, J.B.; Snell, E.E. *Proc. Natl. Acad. Sci. U.S.A.* **1956**, *42*, 221–227.

682. Breslow, R.; Chmielewski, J.; Foly, D.; Johnson, B.; Kumabe, N.; Varney, M.; Mehra, R. *Tetrahedron* **1988**, *44*, 5515–5524.

683. Zimmerman, S.C.; Czarnik, A.W.; Breslow, R. *J. Am. Chem. Soc.* **1983**, *105*, 1694–1695.

684. Zimmerman, S.C.; Breslow, R. *J. Am. Chem. Soc.* **1984**, *106*, 1490–1491.

685. Breslow, R.; Czarnik, A.W. *J. Am. Chem. Soc.* **1983**, *105*, 1390–1391.

686. Breslow, R.; Hammond, M.; Lauer, M. *J. Am. Chem. Soc.* **1980**, *102*, 421–422.

687. Tabushi, I.; Kuroda, Y.; Yamada, M.; Higashimura, H.; Breslow, R. *J. Am. Chem. Soc.* **1985**, *107*, 5545–5546.

688. Fasella, E.; Dong, S.D.; Breslow, R. *Biorg. Med. Chem.* **1999**, *7*, 709–714.

689. Kuang, H.; Brown, M.L.; Davies, R.R.; Young, E.C.; Distefano, M.D. *J. Am. Chem. Soc.* **1996**, *118*, 10702–10706.

690. Qi, D.; Kuang, H.; Distefano, M.D. *Biorg. Med. Chem. Lett.* **1998**, *8*, 875–880.

691. Liu, L.; Breslow, R. *J. Am. Chem. Soc.* **2003**, *125*, 12110–12111.

692. Liu, L.; Zhou, W.; Chruma, J.; Breslow, R. *J. Am. Chem. Soc.* **2004**, *126*, 8136–8137.

693. Tachibana, Y.; Ando, M.; Kuzuhara, H. *Bull. Chem. Soc. Jpn.* **1983**, *56*, 3652–3656.

694. Ando, M.; Kuzuhara, H. *Bull. Chem. Soc. Jpn.* **1990**, *63*, 1925–1928.

695. Lee, M.; Phillips, R.S. *Bioorg. Med. Chem. Lett.* **1992**, *2*, 1563–1564.

696. Philips, R.S.; Cohen, L.A.; Annby, U.; Wensbo, D.; Gronowitz, S. *Bioorg. Med. Chem. Lett.* **1995**, *5*, 1133–1134.

697. Sloan, M.J.; Phillips, R.S. *Bioorg. Med. Chem. Lett.* **1992**, *2*, 1053–1056.

698. Vassilev, V.P.; Uchiyama, T.; Kajimoto, T.; Wong, C.-H. *Tetrahedron Lett.* **1995**, *36*, 5063–5064.

699. Herbert, R.B.; Wilkinson, B.; Ellames, G.J.; Kunec, E.K. *J. Chem. Soc., Chem. Commun.* **1993**, 205–206.

700. Vassilev, V.P.; Uchiyama, T.; Kajimoto, T.; Wong, C.-H. *Tetrahedron Lett.* **1995**, *36*, 4081–4084.

701. Chen, M.S.; Schirch, L.V. *J. Biol. Chem.* **1973**, *248*, 7979–7984.

702. Saeed, A.; Young, D.W. *Tetrahedron* **1992**, *48*, 2507–2514.

703. Svärd, H.; Antoni, G.; Jigerius, S.B.; Zdansky, G.; Långström, B. *J. Lab. Cmpds. Radiopharm.* **1983**, *21*, 1175–1176.

704. Kimura, T.; Vassilev, V.P.; Shen, G.-J.; Wong, C.-H. *J. Am. Chem. Soc.* **1997**, *119*, 11734–11742.

705. Akhtar, M.; Cohen, M.A.; Gani, D. *J. Chem. Soc., Chem. Commun.* **1986**, 1290–1291.

706. Barker, H.A.; Smyth, R.D.; Waszkiewicz, E.J.; Lee, M.N.; Wilson, R.M. *Arch. Biochem. Biophys.* **1958**, *78*, 468–476.

707. Barker, H.A.; Smyth, R.D.; Wilson, R.M.; Weissbach, H. *J. Biol. Chem.* **1959**, *234*, 320–328.

708. Barker, H.A.; Smyth, R.D.; Bright, H.J. *Biochem. Prep.* **1961**, *8*, 89–92.

709. Chibata, I.; Tosa, T.; Sato, T. *Methods Enzymol.* **1976**, *45*, 739–746.

710. Chibita, I.; Kakimoto, T.; Kato, J. *Appl. Microbiol.* **1965**, *13*, 638–645.

711. Passerat, N.; Bolte, J. *Tetrahedron Lett.* **1987**, *28*, 1277–1280.

712. Echalier, F.; Constant, O.; Bolte, J. *J. Org. Chem.* **1993**, *58*, 2747–2750.

713. Helaine, V.; Rossi, J.; Bolte, J. *Tetrahedron Lett.* **1999**, *40*, 6577–6580.

714. Barrio, J.R.; Egbeert, J.E.; Henze, E.; Schelbert, H.R.; Baumgartner, F.J. *J. Med. Chem.* **1982**, *25*, 93–96.

715. Greenaway, W.; Whatley, F.R.; *J. Lab. Cmpds.* **1975**, *11*, 395–400.

716. Ogrel, A.; Vasilenko, I.A.; Lugtenburg, J.; Raap, J. *Recl. Trav. Chim. Pays-Bas* **1994**, *113*, 369–375.

717. Cappon, J.J.; Baart, J.; van der Walle, G.A.M.; Raap, J.; Lugtenburg, J. *Recl. Trav. Chim. Pays-Bas* **1991**, *110*, 158–166.

718. Vidal-Cros, A.; Gaudry, M.; Marquet, A. *J. Org. Chem.* **1989**, *54*, 498–500.

719. Matos, J.R.; Wong, C.-H. *J. Org. Chem.* **1986**, *51*, 2388–2389.

720. Chanatry, J.A.; Schafer, P.H.; Kim, M.S.; LeMaster, D.M. *Analyt. Biochem.* **1993**, *213*, 147–151.

721. Greenaway, W.; Whatley, F.R. *FEBS Lett.* **1977**, *75*, 41–43.

722. Kato, Y.; Fukomoto, K.; Asano, Y. *Appl. Microbiol. Biotech.* **1993**, *39*, 301–304.

723. Kelly, N.M.; O'Neill, B.C.; Probert, J.; Reid, G.; Stephen, R.; Wang, T.; Willis, C.L.; Winton, P. *Tetrahedron Lett.* **1994**, *35*, 6533–6536.

724. Kelly, N.M.; Reid, R.G.; Willis, C.L.; Winton, P.L. *Tetrahedron Lett.* **1996**, *37*, 1517–1520.

725. Kelly, N.M.; Reid, R.G.; Willis, C.L.; Winton, P.L. *Tetrahedron Lett.* **1995**, *36*, 8315–8318.

726. Sutherland, A.; Willis, C.L. *Tetrahedron Lett.* **1997**, *38*, 1837–1840.

727. Hanson, R.L.; Singh, J.; Kissick, T.P.; Patel, R.N.; Szarka, L.J.; Mueller, R.H. *Bioorg. Chem.* **1990**, *18*, 116–130.

728. Nakajima, N.; Esaki, N.; Soda, K. *J. Chem. Soc., Chem. Commun.* **1990**, 947–948.

729. Wichmann, R.; Wandrey, C.; Bückmann, A.F.; Kula, M.-R. *Biotech. Bioeng.* **1981**, *23*, 2789–2802.

730. Pollak, A.; Blumenfeld, H.; Wax, M.; Baughn, R.L.; Whitesides, G.M. *J. Am. Chem. Soc.* **1980**, *102*, 6324–6336.

731. Chibata, I.; Tosa, T. *Ann. Rev. Biophys. Bioeng.* **1981**, *10*, 197–216.

732. Hummel, W.; Weiss, N.; Kula, M.-R. *Arch. Microbiol.* **1984**, *137*, 47–52.

733. Asano, Y.; Yamada, A.; Kato, Y.; Yamaguchi, K.; Hibino, Y.; Hirai, K.; Kondo, K. *J. Org. Chem.* **1990**, *55*, 5567–5571.

734. Bradshaw, C.W.; Wong, C.-H.; Hummel, W.; Kula, M.-R. *Bioorg. Chem.* **1991**, *19*, 29–39.

735. Henderson, D.P.; Shelton, M.C.; Cotterill, I.C.; Toone, E.J. *J. Org. Chem.* **1997**, *62*, 7910–7911.

736. Busca, P.; Paradisi, F.; Moynihan, E.; Maguire, A.R.; Engel, P.C. *Org. Biomol. Chem.* **2004**, *2*, 2684–2691.

737. Cainelli, G.; Engel, P.C.; Galletti, P.; Giacomini, D.; Gualandi, A.; Paradisi, F. *Org. Biomol. Chem.* **2005**, *3*, 4316–4320.

738. Muramatsu, H.; Mihara, H.; Kakutani, R.; Yasuda, M.; Ueda, M.; Kurihara, T.; Esaki, N. *Tetrahedron: Asymmetry* **2004**, *15*, 2841–2843.

739. Nagasawa, T.; Utagawa, T.; Goto, J.; Kim, C.-J.; Tani, Y.; Kumagai, H.; Yamada, H. *Eur. J. Biochem.* **1981**, *117*, 33–40.

740. Hebel, D.; Furlano, D.C.; Phillips, R.S.; Koushik, S.; Creveling, C.R.; Kirk, K.L. *Bioorg. Med. Chem. Lett.* **1992**, *2*, 41–44.

741. Kim, K.; Cole, P.A. *Biorg. Med. Chem. Lett.* **1999**, *9*, 1205–1210.

742. Sasaki, M.; Ikemoto, M.; Haradahira, T.; Tanaka, A.; Watanabe, Y.; Suzuki, K. *J. Lab. Cmpds. Radiopharm.* **1997**, *40*, 243–245.

743. Baldwin, J.E.; Ng, S.C.; Pratt, A.J.; Russell, M.A.; Dyer, R.L. *Tetrahedron Lett.* **1987**, *28*, 2303–2304.

744. Gelbard, A.S.; Nieves, E.; Filo-DeRicco, S.; Rosenspire, K.C. *J. Lab. Cmpds. Radiopharm.* **1985**, *23*, 1055.

745. Halldin, C.; Långström, B. *J. Lab. Compds. Radiopharm.* **1986**, *23*, 715–721.

746. Nakamichi, K.; Nabe, K.; Yamada, S.; Tosa, T.; Chibata, I. *Appl. Microbiol. Biotech.* **1984**, *19*, 100–105.

747. Meiwes, J.; Schudok, M.; Kretzschmar, G. *Tetrahedron: Asymmetry* **1997**, *8*, 527–536.

748. Baldwin, J.E.; Dyer, R.L.; Ng, S.C.; Pratt, A.J.; Russell, M.A. *Tetrahedron Lett.* **1987**, *28*, 3745–3746.

749. Shah, S.A.; Schafer, P.H.; Recchia, P.A.; Ploach, K.J.; LeMaster, D.M. *Tetrahedron Lett.* **1994**, *35*, 29–32.

750. Polach, K.J.; Shah, S.A.; LaIuppa, J.C.; LeMaster, D.M. *J. Lab. Cmpds. Radiopharm.* **1993**, *33*, 809–815.

751. Lo, H.-H.; Hsu, S.-K.; Lin, W.-D.; Chan, N.-L.; Hsu, W.-H. *Biotechnol. Prog.* **2005**, *21*, 411–415.

752. Hegedus, L.S. *Acc. Chem. Res.* **1995**, *28*, 299–305.

753. Hegedus, L.S. *Tetrahedron* **1997**, *53*, 4105–4128.

754. Hegedus, L.S.; Schwindt, M.A.; De Lombaert, S.; Imwinkelried, R. *J. Am. Chem. Soc.* **1990**, *112*, 2264–2273.

755. Hegedus, L.S.; de Weck, G.; D'Andrea, S. *J. Am. Chem. Soc.* **1988**, *110*, 2122–2126.

756. Vernier, J.-M.; Hegedus, L.S.; Miller, D.B. *J. Org. Chem.* **1992**, *57*, 6914–6920.

757. Schmeck, C.; Hegedus, L.S. *J. Am. Chem. Soc.* **1994**, *116*, 9927–9934.

758. Hegedus, L.S.; Lastra, E.; Narukawa, Y.; Snustad, D.C. *J. Am. Chem. Soc.* **1992**, *114*, 2991–2994.

759. Pulley, S.R.; Hegedus, L.S. *J. Am. Chem. Soc.* **1993**, *115*, 9037–9047.

760. Lastra, E.; Hegedus, L.S. *J. Am. Chem. Soc.* **1993**, *115*, 87–90.

761. Miller, J.R.; Pulley, S.R.; Hegedus, L.S.; DeLombaert, S. *J. Am. Chem. Soc.* **1992**, *114*, 5602–5607.

762. Dubuisson, C.; Fukumoto, Y.; Hegedus, L.S. *J. Am. Chem. Soc.* **1995**, *117*, 3697–3704.

763. Zhu, J.; Hegedus, L.S. *J. Org. Chem.* **1995**, *60*, 5831–5837.

764. Colson, P.-J.; Hegedus, L.S. *J. Org. Chem.* **1993**, *58*, 5918–5924.

765. Schmeck, C.; Hegedus, L.S. *J. Am. Chem. Soc.* **1991**, *113*, 5784–5791.

766. Duhamel, L.; Duhamel, P.; Fourquay, S.; Eddine, J.J.; Peschard, O.; Plaquevent, J.-C.; Ravard, A.; Solliard, R.; Valnot, J.-Y.; Vincens, H. *Tetrahedron* **1988**, *44*, 5495–5506.

767. Voyer, N.; Roby, J.; Chénard, S.; Barberis, C. *Tetrahedron Lett.* **1997**, *38*, 6505–6508.

768. Schlosser, M.; Limat, D. *J. Am. Chem. Soc.* **1995**, *117*, 12342–12343.

769. Barberis, C.; Voyer, N. *Synlett* **1999**, 1106–1108.

770. Voyer, N.; Roby, J. *Tetrahedron Lett.* **1995**, *36*, 6627–6630.

771. Park, Y.S.; Beak, P. *J. Org. Chem.* **1997**, *62*, 1574–1575.

772. Gately, D.A.; Norton, J.R. *J. Am. Chem. Soc.* **1996**, *118*, 3479–3489.

773. Tunge, J.A.; Gately, D.A.; Norton, J.R. *J. Am. Chem. Soc.* **1999**, *121*, 4520–4521.

774. Chen, J.-X.; Tunge, J.A.; Norton, J.R. *J. Org. Chem.* **2002**, *67*, 4366–4369.

775. Chong, J.M.; Park, S.B. *J. Org. Chem.* **1992**, *57*, 2220–2222.

776. Gawley, R.E.; Zhang, Q. *J. Am. Chem. Soc.* **1993**, *115*, 7515–7516.

777. Kells, K.W.; Ncube, A.; Chong, J.M. *Tetrahedron* **2004**, *60*, 2247–2257.

778. Keels, K.W.; Chong, J.M. *Org. Lett.* **2003**, *5*, 4215–4218.

779. Qin, Y.; Wang, C.; Huang, Z.; Xiao, X.; Jiang, Y. *J. Org. Chem.* **2004**, *69*, 8533–8536.

780. Jeanjean, F.; Pérol, N.; Goré, J.; Fournet, G. *Tetrahedron Lett.* **1997**, *38*, 7547–7550.

781. Jeanjean, F.; Fournet, G.; Le Bars, D.; Roidot, N.; Vichot, L.; Gore, J.; Comar, D. *J. Lab. Cmpds. Radiopharm.* **1997**, *40*, 722–724.

782. Jeanjean, F.; Fournet, G.; Le Bars, D.; Goré, J. *Eur. J. Org. Chem.* **2000**, 1297–1305.

783. Coldham, I.; Dufour, S.; Haxell, T.F.N.; Howard, S.; Vennall, G.P. *Angew. Chem., Int. Ed.* **2002,** *41*, 3887–3889.

784. Pfammatter, E.; Seebach, D. *Liebigs Ann. Chem.* **1991**, 1317–1322.

785. Diner, U.E.; Worsley, M.; Lown, J.W.; Forsythe, J.-A. *Tetrahedron Lett.* **1972**, 3145–3148.

786. Bravo, P.; Capelli, S.; Meille, S.V.; Seresini, P.; Volonterio, A.; Zanda, M. *Tetrahedron: Asymmetry* **1996**, *7*, 2321–2332.

787. Boezio, A.A.; Solberghe, G.; Lauzon, C.; Charette, A.B. *J. Org. Chem.* **2003**, *68*, 3241–3245.

788. Braun, M.; Opdenbusch, K. *Angew. Chem., Int. Ed. Engl.*, **1993**, *32*, 578–580.

789. Kunz, H.; Schanzenbach, D. *Angew. Chem., Int. Ed. Engl.* **1989**, *28*, 1068–1069.

790. Laschat, S.; Kunz, H. *Synlett* **1990**, 51–52.

791. Hua, D.H.; Miao, S.W.; Chen, J.S.; Iguchi, S. *J. Org. Chem.* **1991**, *56*, 4–6.

792. Chang, Z.-Y.; Coates, R.M. *J. Org. Chem.* **1990**, *55*, 3475–3483.

793. Denmark, S.E.; Weber, T.; Piotrowski, D.W. *J. Am. Chem. Soc.* **1987**, *109*, 2224–2225.

794. Tomioka, K.; Inoue, I.; Shindo, M.; Koga, K. *Tetrahedron Lett.* **1991**, *32*, 3095–3098.

795. Dondoni, A.; Scherrmann, M.-C. *J. Org. Chem.* **1994**, *59*, 6404–6412.

796. Deng, H.; Jung, J.-K.; Liu, T.; Kuntz, K.W.; Snapper, M.L.; Hoveyda, A.H. *J. Am. Chem. Soc.* **2003**, *125*, 9032–9034.

797. Pearson, A.J.; Park, J.G. *J. Org. Chem.* **1992**, *57*, 1744–1752.

798. Ghosh, A.K.; Thompson, W.J.; Holloway, M.K.; McKee, S.P.; Duong, T.T.; Lee, H.Y.; Munson, P.M.; Smith, A.M.; Wai, J.M.; Darke, P.L.; Zugay, J.A.; Emini, E.A.; Schleif, W.A.; Huff, J.R.; Anderson, P.S. *J. Med. Chem.* **1993**, *36*, 2300–2310.

799. Schmidt, U.; Riedl, B. *Synthesis* **1993**, 809–814.

800. Pearson, A.J.; Lee, K. *J. Org. Chem.* **1994**, *59*, 2304–2313.

801. Gilbertson, S.R.; Chen, G.; McLoughlin, M. *J. Am. Chem. Soc.* **1994**, *116*, 4481–4482.

802. Gilbertson, S.R.; Wang, X. *J. Org. Chem.* **1996**, *61*, 434–435.

803. Waid, P.P.; Flynn, G.A.; Huber, E.W.; Sabol, J.S. *Tetrahedron Lett.* **1996**, *37*, 4091–4094.

804. Sabol, J.S.; Flynn, G.A.; Friedrich, D.; Huber, E.W. *Tetrahedron Lett.* **1997**, *38*, 3687–3690.

805. Arya, P.; Ben, R.N.; Qin, H. *Tetrahedron Lett.* **1998**, *39*, 6131–6134.

806. Ripka, A.S.; Bohacek, R.S.; Rich, D.H. *Biorg. Med. Chem. Lett.* **1998**, *8*, 357–360.

807. Kamencka, T.M.; Danishefsky, S.J. *Angew. Chem., Int. Ed. Engl.* **1998**, *37*, 2995–2998.

808. Ben , R.N.; Orellana, A.; Arya, P. *J. Org. Chem.* **1998**, *63*, 4817–4820.

809. Nishida, A.; Fuwa, M.; Fujikawa, Y.; Nakahata, E.; Furuno, A.; Nakagawa, M. *Tetrahedron Lett.* **1998**, *39*, 5983–5986.

810. Bänteli, R.; Brun, I.; Hall, P.; Metternich, R. *Tetrahedron Lett.* **1999**, *40*, 2109–2112.

811. Bigot, A.; Dau, M.E.T.H.; Zhu, J. *J. Org. Chem.* **1999** *64*, 6283–6296.

812. Woiwode, T.F.; Wandless, T.J. *J. Org. Chem.* **1999**, *64*, 7670–7674.

813. Pearson, A.J.; Belmont, P.O. *Tetrahedron Lett.* **2000**, *41*, 1671–1675.

814. Melo, R.L.; Pozzo, R.C.B.; Pimenta, D.C.; Perissutti, E.; Caliendo, G.; Santagada, V.; Juliano, L.; Juliano, M.A. *Biochemistry* **2001**, *40*, 5226–5232.

815. Liu, F.; Zha, H.-Y.; Yao, Z.-J. *J. Org. Chem.* **2003**, *68*, 6679–6684.

816. Goudreau, N.; Brochu, C.; Cameron, D.R.; Duceppe, J.-S.; Faucher, A.-M.; Ferland, J.-M.; Grand-Maître, C.; Poirer, M.; Simoneau, B.; Tsantrizos, Y.S. *J. Org. Chem.* **2004**, *69*, 6185–6201.

817. Svenson, J.; Zhang, N.; Nicholls, I.A. *J. Am. Chem. Soc.* **2004**, *126*, 8554–8560.

818. Dondoni, A.; Franco, S.; Merchán, F.L.; Merino, P.; Tejero, T. *Tetrahedron Lett.* **1993**, *34*, 5475–5478.

819. Bushey, M.L.; Haukaas, M.H.; O'Doherty, G.A. *J. Org. Chem.* **1999**, *64*, 2984–2985.

820. Boger, D.L.; Kim, S.H.; Mori, Y.; Weng, J.-H.; Rogel, O.; Castle, S.L.; McAtee, J.J. *J. Am. Chem. Soc.* **2001**, *123*, 1862–1871.

821. Boteju, L.W.; Wegner, K.; Quin, X.; Hruby, V.J. *Tetrahedron* **1994**, *50*, 2391–2404.

6 Synthesis of Optically Active Amino Acids: Extension of Achiral Methods—Introduction of the Side Chain to Acyclic Systems

6.1 Introduction

A great deal of research effort has been expended in the development of asymmetric alkylations of glycine equivalents. Until recently the most successful of these have been based on cyclic templates, and will be discussed in Chapter 7. However, asymmetric alkylations of acyclic synthons have become increasingly important.

6.2 Asymmetric Alkylations of Glycine Enolates

Most asymmetric reactions of non-activated acyclic anionic glycine equivalents require reactive electrophiles, such as aldehydes, imines, or activated halides. The carbon acidities for $^+H_3NCH_2CO_2^-$ and $^+H_3NCH_2CO_2Me$ in aqueous solution were determined to be $pK_a = 28.9$ and 21.0, respectively (1). Second-order rate constants have been determined in D_2O for deprotonations of the N-terminal α-amino carbon of Gly-Gly and Gly-Gly-Gly zwitterions, of Ac-Gly anion and Ac-Gly-NH$_2$, and of the central residue of a Gly-Gly-Gly anion, with pK_a values ranging from 23.9 to 30.8. The N-terminal residue of zwitterionic Gly-Gly-Gly ($pK_a = 25.1$) deprotonates approximately 130-fold faster than the cental residue (2). Acidity is greatly increased by Schiff base formation; the change in α-carbon acidity of Gly-OMe in the presence of acetone was found to be 7 pK units (from 21 to 14) by an NMR study (3). Asymmetry can be induced by the use of chiral auxiliaries on the amino or acid group, by the use of chiral catalysts, or by the use of chiral substrates/ electrophiles.

6.2.1 Asymmetric Alkylations of Glycine Derivatives: Chiral Auxiliaries

6.2.1a C-Terminal Chiral Auxiliaries

Several ester and amide auxiliaries have been examined. The lithium enolate of N-stabase glycine (−)-menthyl ester reacted with imines to give chiral β-lactams (α,β-diamino acid precursor) in 38–70% yield with >99% ee (4). A (−)-menthyl ester was also used to induce diastereoselectivity in the alkylation of hippuric acid with allyl iodide. The auxiliary only induced 33% de, but the major isomer was readily separated and isolated in 53% yield (5).

A phenylglycinol amide of Boc-Sar was alkylated with methyl iodide or benzyl bromide in 58–84% yield with >95 to >99% de (6). Phenylglycinol has also induced high stereoselectivity for alkylations of a glycine residue contained within a 2-ketopiperazine template (7) (see Section 7.3). A preliminary report has described the use of (1S,2R)-1-aminoindan-2-ol as a chiral auxiliary. Amidation of hippuric acid with the acetonide of this auxiliary gave a glycine synthon that was alkylated with very high stereoselectivity (90–99% de) and moderate to good yields (37–92%), even by unreactive secondary alkyl iodides (see Scheme 6.1). Deprotection was not reported (8).

The Oppolzer sultam amide auxiliary (see Section 6.4.2b) has been used with Boc-Sar (instead of an amino acid Schiff base, as normally employed), giving very high diastereoselectivites (>98% de) for alkylations with benzyl bromides (9). [2,3]-Aza-Wittig rearrangements of N,N-dialkyl,N-allylic glycine ethyl ester ammonium salts, induced by treatment with NaH, have also been carried out on substrates with an Oppolzer camphorsultam amide chiral auxiliary. The allylglycine and β-substituted allylglycine products were obtained with

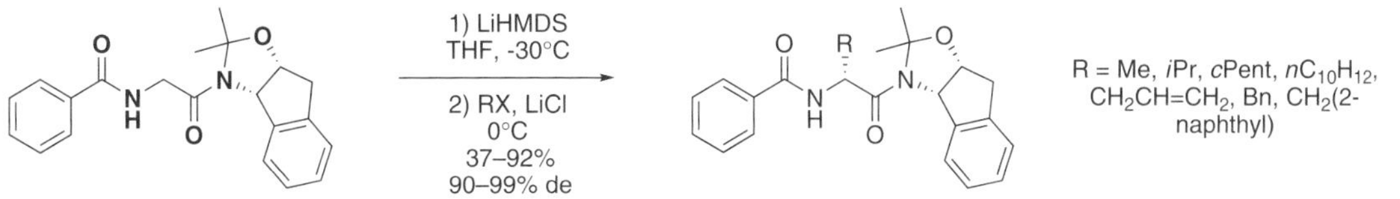

Scheme 6.1 Alkylation of Bz-Gly with Aminoindanol Auxiliary (8).

94 to >98% de (10). A 1,5-dimethyl-4-phenylimidazoli-din-2-one amide auxiliary, also employed for Schiff base alkylations (see Scheme 6.14), was applied to alkylations of N,N-dibenzyl Gly. Benzyl bromide, allyl bromide, and benzoyl chlorides reacted with >95% de, while aldol reactions proceeded with complete *syn* stereoselectivity, with up to 95% de. Auxiliary hydrolysis with NaOMe was successful for most products, but resulted in retro-aldol decomposition for those aldol adducts with β-aryl groups possessing electron-withdrawing substituents (11, 12).

A number of syntheses alkylate amino acids other than glycine to give α,α-disubstituted products. For example, the dianions of 8-phenylmenthol esters of Bz-Ala have been alkylated with nine alkyl halides (R = Et, *i*Bu, Bn, 3,4-(TBSO)$_2$-Bn, CH$_2$CO$_2$*t*Bu, CH$_2$OBn, CH$_2$CH=CHPh, CH$_2$CCH, CH$_2$CH=CH$_2$) with 67–86% yield and 78–88% de (13).

6.2.1b Myers Pseudoephedrine Amide Auxiliary

The most successful method for alkylation of an acyclic glycine residue with a chiral auxiliary, and one of the potentially most useful amino acid synthesis methodologies to emerge in recent years, is the pseudoephedrine glycinamide synthon of Myers et al. (see Scheme 6.2). Pseudoephedrine is a relatively inexpensive chiral auxiliary available in both enantiomeric forms, and many acids amidated with pseudoephedrine form crystalline derivatives. Pseudoephedrine can be acylated with glycine by using the mixed anhydride of Boc-Gly followed by acidic cleavage of the Boc group, or by direct condensation with Gly-OMe (14–16). For alkylation reactions the glycine nitrogen does not need to be protected, but the synthon must be scrupulously dried, according to initial reports. Deprotonation was achieved with 1.95 equiv of LDA or *n*BuLi, with the first equivalent deprotonating the hydroxyl group and the second generating the enolate (under thermodynamic conditions) (15). Excess base causes decomposition, so an accurate titer and anhydrous reagents are deemed essential for good yields. The best results were obtained by the dropwise addition of LDA in THF to a slurry of the synthon and LiCl (15). A wide range of alkyl halides reacted with 50–92% yield and 70–98% de (14, 15). A modified procedure was subsequently published, in which the synthon was prepared as a crystalline monohydrate by reaction of Gly-OMe.HCl with pseudoephedrine in

THF/LiO*t*Bu. The monohydrate was deprotonated with LiHMDS/LiCl or Li/*n*Hex-Cl/HN(TMS)$_2$ and then alkylated with good yields and stereoselectivity (17).

The N-Boc-protected pseudoephedrine glycinamide has also been alkylated, but the products were not crystalline. Proline-type derivatives can be produced by alkylation of the synthon with alkyl dihalides; heating in chloroform following alkylation induces an intramolecular alkylation of the unprotected nitrogen of the glycine adduct (14, 15). Improved alkylation diastereoselectivity was observed by using lower reaction temperatures, less reactive electrophiles, or iodide as the leaving group (15). Addition of LiCl to the enolate solution accelerated the rate of alkylation and improved stereoselectivity. Diastereomeric purities of >99% de were obtained for several examples by recrystallizing the product, or by chromatographic purification.

The free amino acids can be obtained by alkaline hydrolysis, typically by refluxing with 2 equiv of 0.5 N NaOH for 1.5–5 h (14, 15). Some enantiomeric enrichment by kinetic hydrolysis was observed. The products were generally N-acylated and then isolated as the N-Fmoc or N-Boc derivatives in good yield and high enantiomeric purity (15). Alternatively, the alkylation products were simply refluxed in pure water to give the hydrolyzed, salt-free amino acids (14, 15). In both cases the pseudoephedrine auxiliary was recovered by a simple extraction step. The products obtained by alkylation of the N-Boc glycinamide synthon can also be hydrolyzed under basic conditions to give the N-Boc products, although some epimerization was noted. An alternative route to the N-Fmoc amino acids introduces the Fmoc group before removal of the chiral auxiliary, with acidic hydrolysis generating the free carboxyl group.

The Myers procedure can be easily adapted to prepare N-methyl amino acids by substituting sarcosine for glycine in the initial pseudoephedrine acylation. The presence of 1 equiv of N-methylethanolamine during deprotonation was found to be necessary for good reproducibility, so 2.95 equiv of base were required (14, 15).

Due to its relatively recent development, the Myers synthetic route has not yet been applied to many amino acid syntheses. However, L-azatyrosine (18), L-prenyl-glycine (19), β-(8-hydroxyquinoline)-L-Ala derivatives (20), and a highly functionalized pyridoxamine coenzyme-amino acid chimera (21) have been prepared. The Myers pseudoephedrine route was used for a synthesis of Boc- and Cbz-allylglycine. With the chiral auxiliary still attached, the alkene was hydroborated and

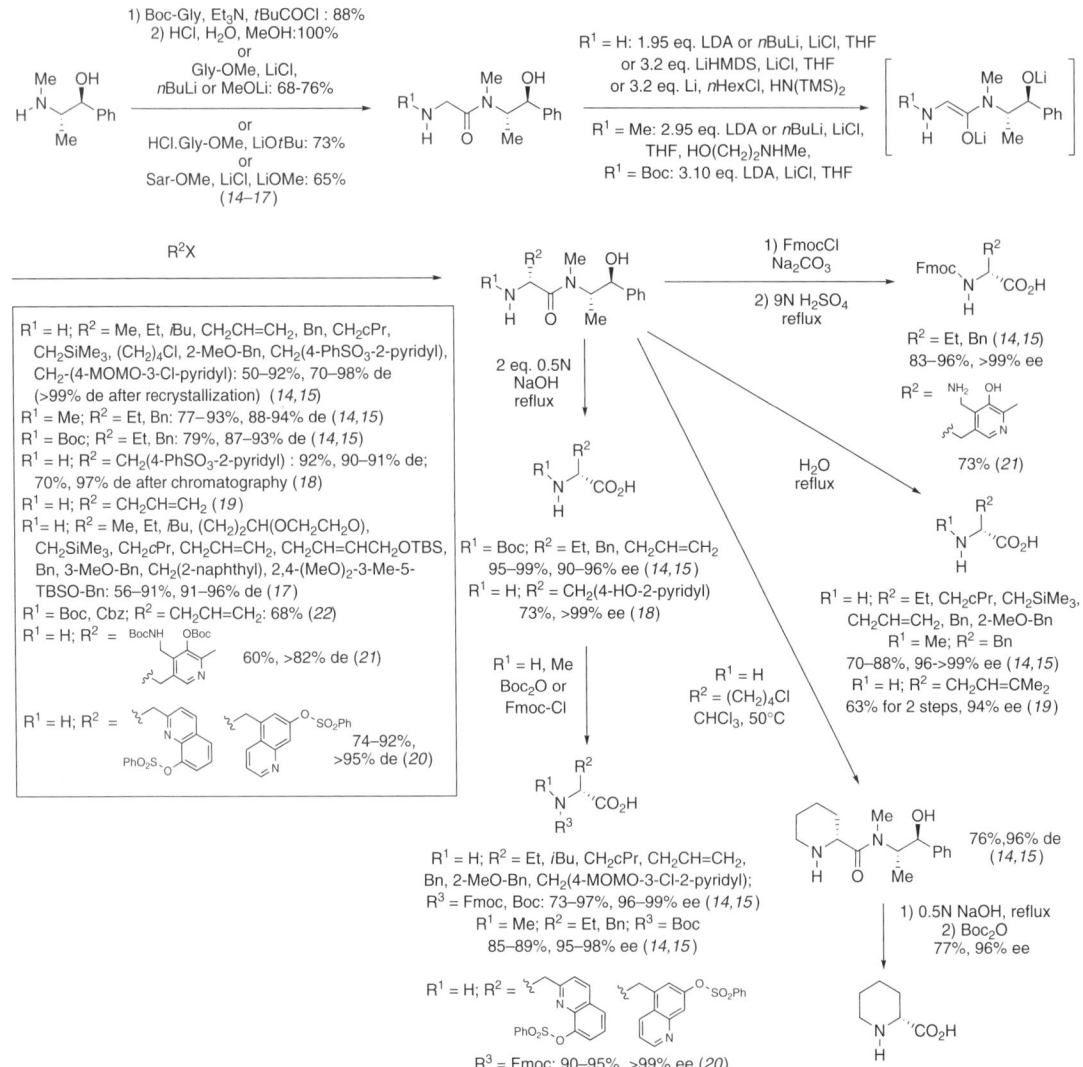

Scheme 6.2 Synthesis and Reaction of Pseudoephedrine Glycinamide.

coupled with heteroaryl bromides, producing five different 5-(heteroaryl)-2-aminopentanoic acid products (22). In another synthesis, N-Boc-α-amino ketones were obtained from N-Boc alkylated glycinamides by displacement of the pseudoephedrine with organolithium or Grignard reagents (23). The pseudoephedrine auxiliary has also been combined with β-alanine to prepare α-substituted-β-amino acids (see Section 11.2.3p) (24). A full report on the use of pseudoephedrine as a chiral auxiliary for asymmetric syntheses of various compounds via alkylations of acylated ephedrine was published in 1997, but no amino acid syntheses were included (25). The pseudoephedrine auxiliary has also been combined with Schiff base activation, with this synthon giving alkylation adducts in 40–60% yield and 68–96% de (26).

6.2.1c N-Terminal Chiral Auxiliaries

Alkylations of glycine equivalents with chiral auxiliaries on the amino group have also been reported. Glycine can be converted into proline derivatives via intramolecular cyclization of a Cu–Zn enolate of N-(4-but1-enyl)-Gly, using an N-phenylethyl auxiliary to induce stereoselectivity (27). α-Aminoacetonitrile, with the amino group contained within a chiral oxazolidinone ring, was asymmetrically dialkylated with the biselectrophiles epibromohydrin or glycidyl triflate, producing 2-hydroxymethyl-1-aminocyclopropane-1-carboxylic acid isomers (28, 29). Similarly, Gly-OtBu with the glycine nitrogen embedded within the same oxazolidine chiral auxiliary was employed for aldol reactions with aldehydes. The major anti-β-hydroxy-α-amino acid

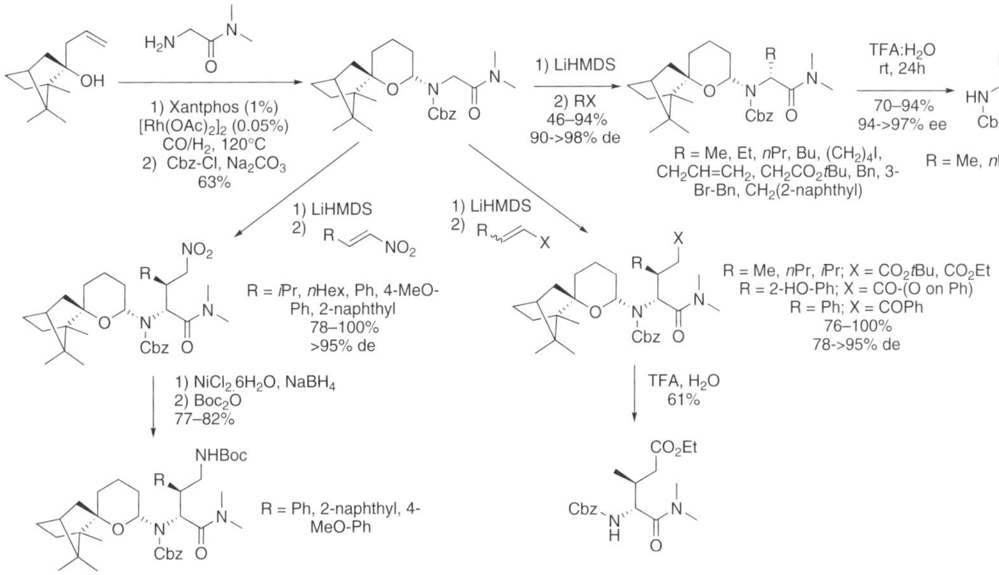

1) Xantphos (1%)
[Rh(OAc)₂]₂ (0.05%)
CO/H₂, 120°C
2) Cbz-Cl, Na₂CO₃
63%

1) LiHMDS
2) RX
46–94%
90->98% de

R = Me, Et, nPr, Bu, (CH₂)₄I,
CH₂CH=CH₂, CH₂CO₂tBu, Bn, 3-
Br-Bn, CH₂(2-naphthyl)

TFA:H₂O
rt, 24h
70–94%
94->97% ee

R = Me, nPr, iBu, Bn

1) LiHMDS
2) R⌁NO₂

R = iPr, nHex, Ph, 4-MeO-
Ph, 2-naphthyl
78–100%
>95% de

1) LiHMDS
2) R⌁X

R = Me, nPr, iPr; X = CO₂tBu, CO₂Et
R = 2-HO-Ph; X = CO-(O on Ph)
R = Ph; X = COPh
76–100%
78->95% de

1) NiCl₂·6H₂O, NaBH₄
2) Boc₂O
77–82%

NHBoc

R = Ph, 2-naphthyl, 4-
MeO-Ph

TFA, H₂O
61%

CO₂Et

Scheme 6.3 Synthesis and Reaction of Camphor-Derived δ-Lactol Auxiliary Gly Dimethylamide (*31, 32*).

isomer was formed with 84–98% de. The auxiliary was removed by treatment with formic acid followed by hydrogenolysis (*30*).

An *N*-Cbz,*N*-(camphor-derived δ-lactol) auxiliary (camTHP*) was prepared from Gly dimethylamide (see Scheme 6.3). Deprotonation with LiHMDS and alkylation with a variety of electrophiles gave the adducts in good yield (46–94%) with high diastereoselectivity (90 to >98% de). The chiral auxiliary was easily removed with TFA to give the *N*-Cbz dimethylamide products, but the amide was not hydrolyzed (*31*). The synthon could also be employed for Michael additions to nitroolefins and α,β-unsaturated ketones, esters, and lactones with >95% diastereoselectivity at both α- and β-centers. The nitro substituents were reduced to give β-substituted 2,4-diaminobutyric acid derivatives (*32*).

The Ireland–Claisen ester enolate rearrangement (see Section 3.8) can be carried out on a di- or tripeptide allyl ester, although the diastereoselectivity induced by the other amino acid residues was poor (<40%). Addition of a Pd(0) complex was found to improve yields (*33, 34*). Much better stereoselectivity for tripeptide allyl ester rearrangements was obtained by adding SnCl₂, with 84–88% de for the newly generated allylglycine or β-methylallylglycine residue (*35*). Manganese enolates of a number of dipeptide methallyl esters rearranged in 75–95% yield, with formation of only two of the four possible β-methylallylglycine diastereomers (2*S*,3*R*-, and 2*R*,3*S*-) with variable selectivity (*36*). Several chiral allyl ester derivatives were also rearranged (*36*). The ester enolate Claisen rearrangement was reviewed in 1996 (*37*) and 1997 (*38*). The closely related aza-Claisen rearrangement was used for an asymmetric synthesis of D-*allo*-isoleucine (*39*). By incorporating an

(*S*)-1-phenethyl chiral auxiliary on the amide nitrogen the rearrangement proceeded with 78% de at the α-center (89% yield). Alkene catalytic reduction and amide hydrolysis completed the synthesis.

N-Me,*N*-Bn Ala-OMe was reacted with borane to form a chiral derivative on the amino group. Deprotonation with LDA or KHMDS followed by alkylation with a range of electrophiles produced α-methyl amino acids in 50–82% yield with 36–82% ee (*40*). However, as described in greater detail below, *N*-Me,*N*-Boc-Phe-OMe could simply be deprotonated and alkylated with MeI with up to 82% ee. The enantioselectivity is possibly due to formation of a chiral complex of the enolate with unreacted starting material (*41*). Enantioselectivity has also been observed in an intramolecular alkylation of *N*-chloroacetyl amino acids (*42, 43*).

6.2.1d Glycine Residues within Peptides

A number of reports have examined the alkylation of glycine or *N*-methylglycine (sarcosine) residues within an unprotected peptide (for reviews see *44, 45*). Multiple equivalents of base are used to generate a polyanion, which undergoes alkylation at the glycine residue with surprising regio- and enantioselectivity. The peptide is solubilized in organic solvents by the addition of salts such as LiCl, which also increases alkylation yields.

Three different types of glycine enolates can be generated (see Scheme 6.4). Sarcosine enolates next to other *N*-alkyl residues within linear peptides (Type 1 enolate) gave 25–80% yields after alkylation with MeI (*46, 47*). D-Amino acids were usually obtained with

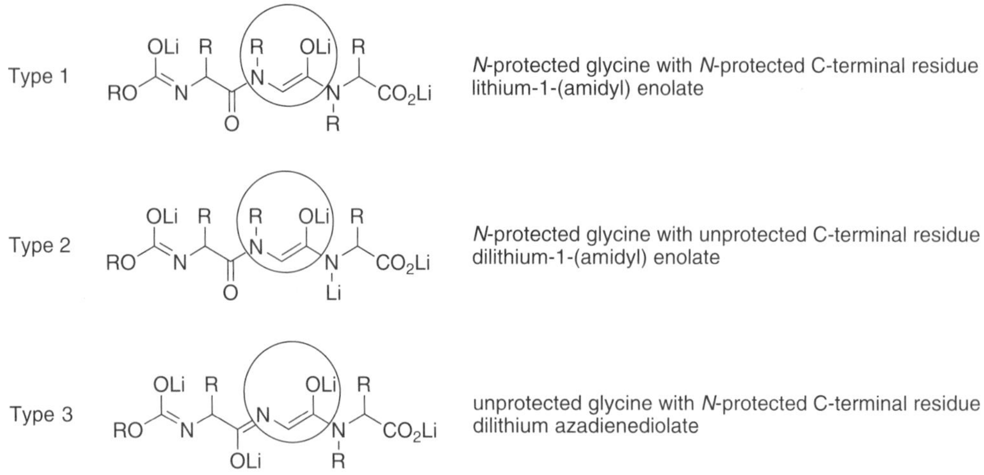

Type 1

OLi R   R   OLi   R
RO   N   N   N   CO₂Li
          O   R

N-protected glycine with N-protected C-terminal residue
lithium-1-(amidyl) enolate

Type 2

OLi R   R   OLi   R
RO   N   N   N   CO₂Li
          O   Li

N-protected glycine with unprotected C-terminal residue
dilithium-1-(amidyl) enolate

Type 3

OLi R   OLi   R
RO   N   N   N   CO₂Li
     OLi   R

unprotected glycine with N-protected C-terminal residue
dilithium azadienediolate

Scheme 6.4 Types of Glycine Enolates within Poly-Lithiated Peptides.

diastereoselectivities of around 60%, although warming of an LDA/BuLi-generated enolate to 0 °C before alkylation with MeI gave N-methyl L-Ala (47). Different bases were found to give different diastereoselectivities, while replacement of the sarcosine residue with N-benzylglycine resulted in reduced yields and benzylic alkylations (47). In cyclic peptides the configuration of the alkylated sarcosine residue depended on its location within the peptide and reaction conditions; sarcosine residues were alkylated in preference to glycine residues (48). The overall yield of the cyclization/alkylation sequence is competitive with conventional cyclic synthesis of the equivalent alkyated peptide, as the unsubstituted glycine peptide cyclizes with much better yield than one already α-substituted (48). Reactive electrophiles gave monoalkylated products with >90% de (49). This method has been used to prepare modified cyclosporin A compounds by alkylating the sarcosine residue contained within the cyclic undecapeptide with a range of electrophiles (13–92% yield) (50).

Sarcosine residues within peptides without neighboring N-alkyl groups (Type 2 enolate) give poor yields (47). The most reactive of the three types of enolates is the dilithium azadienediolate unit (Type 3), generated with tBuLi and LiBr from glycine residues within peptides with an adjacent C-terminal N-alkylated residue. Boc-protected poylithiated (up to Li₆) tri-, tetra-, and pentapeptides were alkylated by MeI, EtI, iPrI, allyl, and benzyl bromide, ethyl bromoacetate, CO₂, and CS₂ with 11–93% yield and 0–80% de (47). Diastereoselectivity was varied by manipulation of the reaction temperature. Proline could be used as the C-terminal N-protected residue, or N-benzyl protection could be used and then removed, although benzylic alkylation byproducts were present (47). N-Benzyl groups have been introduced to a cyclic tetrapeptide by benzylation with a metal-free, non-nucleophilic phosphazene P4 base and

benzylbromide (51); by using an excess of reagents, it was possible to also introduce a C-benzyl group (49). However, the N-benzyl groups could not be removed.

Alkylation of glycine-containing dipeptides has also been examined (52). Ts-Phe-Gly-OMe, after treatment with 3 equivalents of base, was alkylated with benzyl bromide with >70% yield and up to 58% de. A chiral menthyl ester reduced diastereoselectivity. Various Asp-Gly derivatives were also investigated to see if asymmetric induction could be increased by metal ion complexation of the Asp side chain, but side reactions predominated. A C-terminal sarcosine residue in dipeptides has been converted to the corresponding α-aminophosphonic acid residue and alkylated with MeI, BnBr, or allyl bromide (38–69% yield)(53). However, no diastereoselectivity was observed.

In the syntheses reported above, the glycine methylene proton is the least acidic of the protons removed. An alternative approach is to reverse the acidity, so that the position to be substituted is the most acidic. This has been done by using an aminomalonate residue. Aminomalonate residues contained in the center of a tripeptide were efficiently alkylated with a selection of electrophiles in 67–93% yield, using less than 2 equiv of NaOMe or KOtBu as base (54). The products could be decarboxylated after ester group cleavage. No diastereoselectivity was observed. This strategy failed when applied to a cyclotetrapeptide (49). A more comprehensive study of alkylations of peptides containing aminomalonate (Ama) or aminocyanoacetate (Aca) residues found that, while alkylations of Aca residues were nonstereoselective, conditions could be established for alkylation of the Ama residue which produced diastereoselectivity as high as 84% de for MeI as electrophile, 80% de for BnBr, and 90% de for $tert$-butyl acrylate (55). Unfortunately, racemization occurred upon decarboxylation.

6.2.2 Asymmetric Alkylations of Glycine: Chiral Catalysts

6.2.2a Enzymatic Catalysts

Enzymatic catalysis has been used in the synthesis of a number of β-hydroxy amino acids via condensation of glycine and an aldehyde (see Scheme 6.5, Table 6.1). Serine was produced in a bioreactor from glycine, formaldehyde, and serine hydroxymethyltransferase on a 400 g/L scale (56). Serine hydroxymethyltransferase was also used to prepare L-allo-threonine with 96% de from acetaldehyde (57), but other aldehydes gave L-amino acids in low yields with little stereoselectivity at the β-center (58). Similarly, L-threonine aldolase tolerated a range of substrates, but showed variable stereoselectivity at the β-center (0–100% de erythro) (59). An aldolase isolated from *Streptomyces* showed good *threo* selectivity for aromatic aldehydes (60). Purified recombinant L- and D-threonine aldolases have also been examined for use in the catalysis of the condensation of glycine with aldehydes; the *threo/erythro* stereoselectivity depended upon the substrate (61).

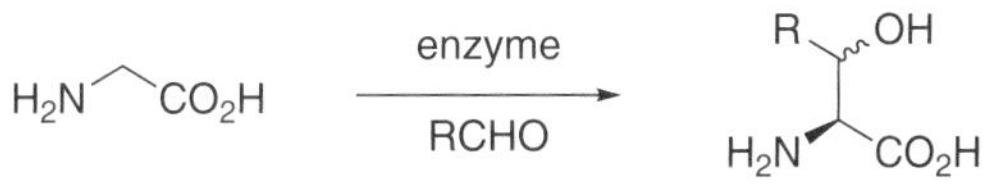

$$H_2N\frown CO_2H \xrightarrow[\text{RCHO}]{\text{enzyme}} \underset{H_2N \quad CO_2H}{R \frown OH}$$

Scheme 6.5 Enzyme-Catalyzed Aldol Reaction of Glycine (see Table 6.1).

6.2.2b Other Catalysts

An early (1959) attempt at an asymmetric synthesis of threonine reacted a chiral cobalt complex of glycine with acetaldehyde in aqueous sodium carbonate, giving an 80% yield of mainly Thr, though with only 8% ee (62). A pyridoxal mimetic consisting of various functional groups attached to a 5,6,7,8-tetrahydroquinoline ring system catalyzed the aldol reaction of Gly with acetaldehyde, producing mixtures of Thr/allo-Thr with up to 37% ee L-enantiomer at pH 9, or 62% ee D-enantiomer at pH 4 (63). Glycine, protected as the N-stabase ethyl ester, has been deprotonated with LDA and transmetallated with a chiral titanium–carbohydrate complex. Aldol reaction with a number of carbonyl compounds produced D-threo-β-hydroxy-α-amino acids in 43–70% yield, with >97% de and >96% ee (64). Reasonable diastereoselectivity (56–88% de anti) and enantioselectivity (85–97% ee) was achieved for aldol reactions of a silyl enolate of Tfa-Gly-OMe with aldehydes in the presence of a Zr binaphthol-based catalyst (11 examples) (65).

A chiral catalyst was also employed during alkylations of Phth-Gly-OEt with Me, nPrI, iPrBr or BnBr. Reverse micelles formed from chiral surfactants and apolar solvents (dichloromethane/n-hexanol) were used in combination with KOH, giving the products with enantioselectivities of 16–60% (66). Quinine and quinidine were used to catalyze an Ireland–Claisen ester enolate rearrangement (see Section 3.8) of N-Tfa glycine allyl esters, producing allylglycines with high

Table 6.1 Enzyme-Catalyzed Aldol Reactions of Glycine (see Scheme 6.5).

Enzyme	RCHO	Yield Diastereoselectivity	Reference
serine transhydroxymethylase	R = Me (with [1-^{14}C]-Gly)	96% de erythro	1973 (57)
serine transhydroxymethylase	R = H	>80%	1985 (56)
serine hydroxymethyltransferase	R = nHex, Ph, 2-furyl, 2-thienyl, 2-imidazolyl, 4-cyclohexenyl	11–55% on 1 mmol scale 28% de threo to 33% de erythro	1992 (58)
serine hydroxymethyl synthetase	RCHO = ^{11}CH$_2$O	10% radiochemical	1983 (470)
Streptomyces aldolase	R = Ph, 4-HO-Ph, 4-BnO-Ph, 4-Br-Ph, 4-NO$_2$-Ph, (CH$_2$)$_2$Ph	12–72% >94% de threo	1993 (60)
threonine aldolase	R = Me, nPr, CH$_2$N$_3$, (CH$_2$)$_2$SPh, (CH$_2$)$_2$OBn, (CH$_2$)$_8$CH=CHnBu, CH$_2$OBn, CH$_2$O(CH$_2$)$_3$OBn, CH$_2$NPhth, Ph, 4-HO-Ph, CH$_2$CH(Me)SPh, CH$_2$C(Me)$_2$SPh, CH$_2$CH(SPh)CO$_2$Et, CH$_2$O(CH$_2$)$_2$NPhth	10–>75% on 1 mmol scale 0–100% de erythro	1995 (59)
threonine aldolase	R = CH$_2$OBn	78% 84% de erythro	1995 (471)
threonine aldolase	R = Me, nPr, nPent, nHex, iPr, CH=CHMe, (CH$_2$)$_2$Ph, Ph, 4-NO$_2$-Ph, 2-NO$_2$-Ph, 3-HO-Ph, 4-HO-Ph, 4-Me-Ph, 4-F-3-NO$_2$-Ph, imidazol-2-yl, (R)- and (S)-CH[-CH$_2$OC(Me)$_2$O-]	0–98% 0–>98% de erythro	1997 (61)

diastereoselectivity (*42, 67–70*). Allylglycines with β-quaternary centers were obtained with enantiomeric excesses of 76–93%, although yields were often poor (12–70%) (*68*). The mono-β-substituted γ,δ-unsaturated amino acids were obtained with better enantioselectivity (79–90% ee) and yield (66–97%) (*67*).

N-Phthaloyl glycinal has been used as a glycine equivalent. Aldol reactions with α-branched aldehydes in the presence of L-Pro catalyst produced *anti*-β-hydroxy-α-amino aldehydes with high yield and stereoselectivity (66 to >99% de); direct oxidation of the aldol product with $NaClO_2$ produced the β-hydroxy-α-amino acids in 62–75% overall yield, with 86–98% ee (six examples, side chain = CH(OH)R with R = *i*Pr, CH(Et)$_2$, CH(*n*Bu)$_2$, CH(OMe)$_2$, *c*Pent, *c*Hex). Importantly, the procedure is readily scaled up, with (2*S*,3*S*)-β-hydroxy-Leu synthesized on a 1 g scale in 60% yield, including hydrazine deprotection of the phthaloyl group. α-Non-branched aldehydes did not form the desired adduct (*71*).

6.2.3 Asymmetric Alkylations of Glycine Derivatives: Chiral Electrophiles

Several syntheses rely upon chiral electrophiles. N,N-Bn$_2$-Gly-O*t*Bu was deprotonated and then added to the carbonyl group of protected α-D-ribohexofuranos-3-ulose, with 95% yield and >98% de at the α-center (*72*). Ts-Ala-O*t*Bu or Ts-Ala-OBn has been deprotonated with LDA, transmetallated with SnCl$_2$, and then added to chiral α-hydroxy aldehydes to generate α-methyl polyhydroxy amino acids. The products formed with predominantly 2,3-*syn*-3,4-*anti* configuration, with 66 to >99% ds for each set of isomers (*73*). Aldol reactions of N-Stabase-protected Gly-OEt with chiral aldehydes produced sphingofungin B. One of the four possible diastereomers was produced with up to 48% ds, depending on the Lewis acid employed (*74*). Both N-Stabase-protected Gly-OEt and N,N-dibenzyl Gly-OEt were added to chromium alkoxyalkenyl–carbene complexes with an (−)-8-phenylmenthyloxy auxiliary, producing *syn* diastereomers of β-phenyl-Glu. The Stabase derivative resulted in 100% diastereoselectivity but only 14% yield, with the dibenzyl nucleophile giving 86% yield and 80% de. The *anti* isomers were obtained if the benzophenone Schiff base of Gly-OEt was employed as the nucleophile. Oxidation to cleave the chromium complex followed by imine/ester deprotection was required to generate the free amino acid (*75*).

Tfa-Gly-O*t*Bu has been deprotonated with LHMDS/ZnCl$_2$ and then alkylated with chiral allylic acetates in the presence of a Pd catalyst with 1,3-chirality transfer to give 2-amino-3-methyl-5-phenylpent-4-enoic acid (>96% ee, >91% ds) and 2-amino-3-methyl-hex-4-enoic acid (96% ee, >96% ds) (*76, 77*). A number of examples of efficient 1,5-chirality transfer were reported several years later, with 2-amino-6-hydroxy-6-alkyl-pent-4-enoic acid products produced with 71–96% de (*78*). The Ireland–Claisen ester enolate rearrangement has been applied to the synthesis of enantioenriched amino acids

by using a chiral allylic alcohol to form the ester (*79, 80*). This area was reviewed in 1996 (*37*) and 1997 (*38*). A high transfer of chirality (90–98% ee at the α-center) was achieved by using a chelated ester enolate, with a number of polyhydroxylated γ,δ-unsaturated amino acids (*79*) or α-allyl-α-substituted amino acids (*80*) prepared. A C-glycosyl α-amino acid was obtained by rearrangement of a glycosyl ester containing an alkene in the allyl position (*81*). This method was also used to synthesize (2*R*,3*S*)-2-amino-3-methyl-4-butenoic acid, which was subsequently elaborated into β-methyl-Asp and the β-amino acid ADDA (*82*). Chiral cyclohexadienediol esters led to α-cyclohexenyl amino acids with 0–80% de (*83*).

6.2.4 Asymmetric Alkylations of Amino Acids: Memory Effects

A 2003 paper reported that intramolecular cyclizations via deprotonation of N-(bromoalkyl),N-Boc amino acid ethyl esters with KHMDS provided α-substituted proline or pipecolic acid derivatives with very high retention of optical purity (83–98% ee), with no chiral catalyst or auxiliary (*43*). Four- or seven-membered ring systems were also constructed (*43*), while α-substituted Asp lactams were obtained by cyclizations of N-chloroacetyl amino esters (*84*). The intramolecular cyclization occurs with an apparent "memory" of chirality, with choice of base (KHMDS), N-protecting group (Boc), and solvent (DMF) critical to success. A mechanism was proposed in which deprotonation of one conformer leading to an enolate is disfavored due to the Boc group blocking approach of the KHMDS (*43*). An earlier report described intermolecular alkylation of Phe derivatives using no external chiral sources (catalysts or auxiliaries) (*41*). The best results were obtained with N-Boc,N-Me Phe-O*t*Bu, which could be methylated with up to 82% ee (40% yield) and allylated with up to 88% ee (15% yield) when deprotonated with LTMP. Several possible complexes were proposed to account for the transfer of chirality, including formation of a chiral complex of the enolate with unreacted starting material. Another report described α-methylations of N-Boc,N-MOM Ile-OEt or *allo*-Ile-OEt using KHMDS or MeI. Both diastereomeric starting materials produced (2*S*)-products with 93% and 86% ds, respectively, indicating that memory of the C2 configuration was the major determinant of product stereochemistry, with the C3 chiral center having minimal effect (*85*).

6.3 Reactions of Isocyanocarboxylate Enolates

A number of papers have reported on the asymmetric addition of methyl isocyanocarboxylates to aldehydes in the presence of a chiral (aminoalkyl)ferrocenyl phosphine-gold(I) complex, which proceeds via an intermediate oxazoline (see Scheme 6.6, Table 6.2). Hydrolysis gives

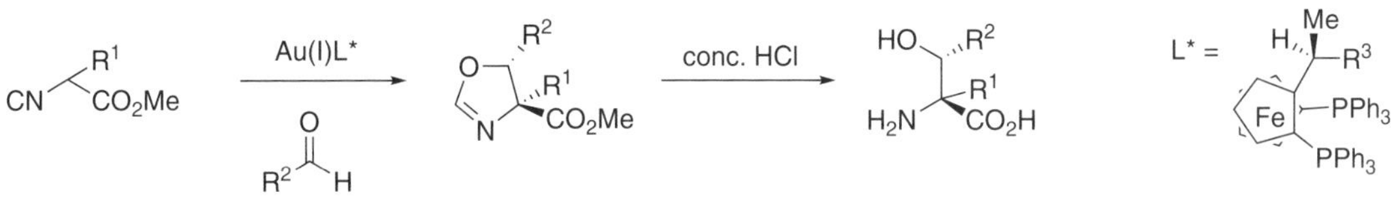

Scheme 6.6 Asymmetric Addition of Isocyanocarboxylates to Aldehydes Using Chiral Ferrocene Complex.

Table 6.2 Asymmetric Addition of Isocyanocarboxylates to Aldehydes Using Chiral Ferrocene Complex

Substrate	Ligand R^3	Aldehyde R^2	Yield Stereoselectivity	Reference
CN⌒CO$_2$Me	Me-N⌒NR^4; NR^4 = NMe$_2$, NEt$_2$	Me *i*Pr cHex *t*Bu Ph CH=CH*n*Pr C(Me)=CHMe	83–100% yield 60–100% de 72–97% ee	1986 (*86*)
CN⌒CO$_2$Me	Me-N⌒NR^4; NR^4 = N(*n*Bu)$_2$, N(*i*Pr)$_2$, pyrrolidino, piperidino, morpholino, 4-methylpiperazino, azepano	Me *i*Pr Ph CH=CH*n*Pr 3,4-OCH$_2$O-Ph	85–99% yield 74–90% de 89–96% ee	1987 (*88*)
CN–CH(R^1)–CO$_2$Me R^1 = H, Me, *i*Pr	Me-N⌒NR^4; NR^4 = NMe$_2$, piperidino, morpholino	Me Ph	86–100% yield 8–90% de 26–95% ee	1988 (*99*)
CN⌒C(O)NR$^5_2$ R^5 = Me, $-(CH_2)_5-$	Me-N⌒N(CH$_2$CH$_2$)$_2$X; X = O, CH$_2$	Me Et *i*Bu Ph 4-BnO-Bn	73–92% yield 82–92% de 94–99% ee	1988 (*94*)
CN⌒CO$_2$Me	Me-N⌒morpholino	CH=CH-*n*$C_{13}H_{27}$	100% yield 78% de 93% ee	1988 (*93*)
CN–CH(R^1)–CO$_2$Me R^1 = H, Me, Et, *i*Pr, Ph	Me-N⌒NR^4; NR^4 = NMe$_2$, piperidino	H	75–100% yield 52–81% ee	1988 (*98*)
CN⌒P(O)(OR)$_2$	Me-N⌒morpholino	*i*Pr Ph 3,4-OCH$_2$O-Ph	78–94% yield 100% de 88–96% ee	1989 (*100*)
CN⌒P(O)(OR)$_2$	Me-N⌒NMe$_2$	Ph	54% yield 96% de 85% ee	1989 (*101*)
CN⌒C(O)R^5 R^5 = OMe, NMe$_2$	Me-N⌒morpholino	R^6C(O)C(O)R^7 R^6 = OMe; R^7 = Me, *i*Bu, Ph R^6 = R^7 = Me	90–92% yield 2–76% de 42–90% ee	1989 (*95*)

(*Continued*)

Table 6.2 Asymmetric Addition of Isocyanocarboxylates to Aldehydes Using Chiral Ferrocene Complex (continued)

Substrate	Ligand R³	Aldehyde R²	Yield Stereoselectivity	Reference
CN–CO₂Me	Me–N–NMe₂	(CH₂=CH–CH₂–CH–CHO)	77% yield; 92% de (cis/trans); 88% de (α-center)	1989 (87)
CN–CO₂Me	Me–N–CO₂Me	Ph	78% de; 92% ee	1990 (89)
CN–CO₂Me	S–CH(Ph)–CH–N–Me	Ph	72–88% yield; 44–76% de; 13–89% ee	1990 (491)
CN–CO₂Me	Me–N–(CH₂)–N–morpholine; Ag(I) complex	iPr, tBu, Ph, C(Me)=CH₂	90–96% yield; 92–98% de; 80–90% ee	1991 (97)
CN–CO₂Me	Me–N–CH₂CH₂–N(Me)–CH₂CH₂–R⁴; Me–N–CH₂CH₂–N(Me)–CH₂CH₂–OR⁴, R⁵, R⁶	Ph	34–99% yield; 58–86% de; 88–97% ee	1991 (105)
CN–CO₂Me	R⁴–N–(R)–CH₂···CH(H)–C(=O)–NH–CR(R)(R)	Ph	33–96% yield; 28–80% de; 26–96% ee	1992 (90)
CN–CO₂Me	Me–N–CH₂CH₂–NR⁴; NR⁴ = NMe₂, NEt₂, N(nBu)₂, N(iPr)₂, N(nOct)₂, pyrrolidine, (various rings: N–S, N–O, N–NMe, N–piperidine, N–C(Me)₂, N–C(Me)₂, azepane, N-naphthyl, azabicyclic); Me₂-morpholine, iBu-morpholine, H-morpholine; **best results with** Me–N–CH₂CH₂–N–morpholine	Me; Me, iPr, iBu, tBu, CH=CHnPr, Ph, 3,4-OCH₂O-Ph, 3,4-BnO-Ph, 2-MeO-Ph, 4-Cl-Ph, 4-NO₂-Ph, ferrocenyl	67–100% yield; 40–88% de; 28–89% ee; 80–100% yield; 76–100% de; 86–96% ee; GEN	1992 (91)
CN–COR⁵; R⁵ = OMe, NMe₂	Me–N–CH₂CH₂–NR⁴₂; NR⁴₂ = piperidine, morpholine	Ph, 2-, 3-, or 4-F-Ph, 2,6-F₂-Ph, 2,4,6-F₃-Ph, 2,3,5,6-F₄-Ph, C₆F₅	74–99% yield; 6–90% de; 23–95% ee	1996 (92)

β-hydroxy-α-amino acids or β-hydroxy-α-alkyl-α-amino acids. The tertiary terminal amino group on the catalyst is believed to be important in abstracting one of the active hydrogens of the isocyanocarboxylate. In the original report (86), methyl isocyanoacetate was added to a number of aldehydes using a catalyst with an NMe$_2$ or NEt$_2$ substituent. The predominantly *trans* oxazoline was obtained with 60–100% de, 72–97% ee, and 83–100% yield. The method was applied to a synthesis of the critical amino acid component of cyclosporin, MeBmt (87). A variety of modified versions of the side-chain diamino moiety of the catalyst were then examined, primarily varying the terminal tertiary amine, with morpholino and piperidino groups found to give the best results (88–91). The morpholino-containing ligand was used to catalyze reaction with a range of aldehydes (91, 92) and was applied to the synthesis of sphingosine (93).

The stereoselectivity of reactions with less-hindered primary aldehydes could be improved by using isocyanoacetamides instead of methyl isocyanoacetate (94). Both isocyanoacetate and isocyanoacetamide were added to α-keto esters to give β-alkyl-β-hydroxyaspartic acid derivatives (95). The silver(I) complex of the same ferrocene ligand (96) produced the oxazoline intermediate with better diastereoselectivity, but poorer enantioselectivity (97). α-Substituted-β-hydroxy amino acids could also be prepared by starting with methyl isocyanocarboxylates, with a morpholino or piperidino substituent on the catalyst again giving the best results (98, 99). β-Hydroxy-α-aminophosphonic acids were prepared by using (isocyanomethyl)phosphonates as substrates (100, 101). An achiral oxazolidinone amide of isothiocyanoacetate added to aryl aldehydes in the presence of a bis(oxazoline) chiral catalyst and Mg(ClO$_4$)$_2$ to give protected β-aryl-β-hydroxy amino acids with *syn* selectivity (0–90% de, depending on aryl substituents) and 86–95% ee (102).

The diastereoselectivity of these reactions is determined by both the planar chirality of the ferrocene nucleus and the central chirality of the asymmetric carbon atom on the side-chain substituent (103). A large number of catalysts with various side chains containing one, two or three chiral centers were tested for the aldol reaction of isocyanoacetate with benzaldehyde (104, 105). A working model of the transition state that correctly predicts the stereochemical results has been developed (106), with a two-dimensional (2D)-NMR conformational study also reported (107).

This asymmetric aldol-type reaction has been extended to additions to *N*-sulfonyl imine electrophiles, producing *cis*-2-imidazoline intermediates that can be hydrolyzed to α,β-diamino acids. The *cis*-imidazoline intermediates can be isomerized to the *trans* epimers. The imidazoline precursors of a series of β-amino-β-arylalanines were produced in 76–91% yield with 46–88% ee using a Ru catalyst (108), or 22–56% yield and >97% ee after a single recrystallization using a Au catalyst (109).

The isocyanoacetate aldol reaction has also been carried out without chiral catalysts. Aryl aldehydes were complexed in a chiral chromium–tricarbonyl complex, with the addition of ethyl isocyanoacetate resulting in a single diastereomer with up to 94% yield (110). Deprotonation with LDA provided better results than using KCN. Ethyl isocyanoacetate has also been added to several chiral amino aldehydes with 62–74% diastereoselectivity and no racemization at the initial chiral center (111). A Cu(I)-catalyzed aldol reaction between methyl isocyanoacetate and a ketone containing a chiral tolylsulfinyl auxiliary was used in a synthesis of 3-(fluoromethyl)-threonine (112).

There are few reports of asymmetric alkylations of isocyanoalkanoates (as opposed to an aldol reactions). A bornyl or menthyl ester derivative of α-isocyanopropionate was used to prepare α-methyl-DOPA (113) and α-methyl-Orn (114). However, only 5–13% ee was obtained. A proline amide auxiliary induced diastereoselectivities of 11–55% de during sequential dialkylations of isocyanoacetates (115), or 10–45% de during Michael additions of the proline amide of isocyanopropionate (116). Isocyanoacetate has also been alkylated with a chiral electrophile, (S)-1,2-dibromopropane, producing (1R,2S)-2-methyl-1-aminocyclopropane-1-carboxylic acid (117).

Evans has reported a new glycine equivalent, consisting of a Gly imine embedded within an oxazole which reacts with aldehydes to generate the same *cis*-oxazoline adducts as formed from isocyanoacetates. The 2-aryl-5-methoxyoxazole reacted with aryl aldehydes in the presence of a chiral salen–aluminum complex to give *cis*-substituted 2-oxazoline-4-carboxylates in high (58–100%) yield with 46 to >98% de and 91 to >99% ee (see Scheme 6.7). Treatment with 5% DBU induced epimerization to give the *trans*-substituted isomer (118). A complex bis(amino acid) contained in ustiloxin D consists of β-hydroxy-Ile linked to β-hydroxy-DOPA through an aryl ether linkage. The β-hydroxy-DOPA half was constructed by an asymmetric aldol reaction between the Evans oxazole and 3-acetoxy-4-benzyloxy-benzaldehyde, catalyzed by the chiral Al complex. A *cis*-oxazoline adduct was produced with 99% yield and 98% ee (119).

6.4 Alkylations of Schiff Bases of Glycine or Other Amino Acids

Asymmetric syntheses based on alkylation of acyclic glycine Schiff bases utilize the same approaches as for other glycine derivatives: a chiral ester/amide auxiliary, a chiral Schiff base auxiliary, a chiral electrophile, or a chiral catalyst. Many of the chiral auxiliaries that have been examined possess a hydroxy substituent which aids stereoselectivity by forming a chelate with the enolate. Diastereoselectivity can often be varied by changing the cation of the base used for deprotonation. The Schiff base imine greatly increases the α-proton acidity: the increase in α-carbon acidity of Gly-OMe in the presence of acetone was found to be 7 pK units (from 21 to 14) by

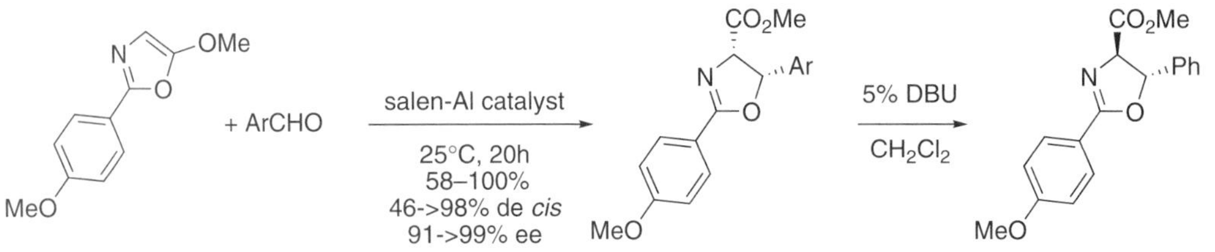

Ar = Ph, 4-F-Ph, 4-NO₂-Ph, 4-CN-Ph, 4-CO₂Me-Ph, 4-PhO-Ph, 4-AcO-Ph, 4-Me-Ph, 4-Ph-Ph, 3-Cl-Ph, 3-NO₂-Ph, 3-MeO-Ph, 3-Me-Ph, 2-Cl-Ph, 2-NO₂-Ph, 2-Br-6-OMe-Ph, 2-Me-Ph, 3-Cl-4-F-Ph, 3-Br-4-MeO-Ph, 3-NO₂-4-Cl-Ph, 2,5-(MeO)₂-3-NO₂-Ph, 1-naphthyl, 2-naphthyl, 2-furyl, CH=CH(4-NO₂)-Ph

Scheme 6.7 Chiral Salen-Al Catalysis of Aldol Reaction of 5-Alkoxyoxazoles (*118*).

an NMR study (*3*). Enzymatic deprotonations are known to be accelerated by Schiff base formation with pyridoxal phosphate. Due to the possibility of dialkylation, many of the methods described have been used to prepare α,α-disubstituted α-amino acids via dialkylation of glycine, or by alkylation of amino acids other than glycine. A thorough review of the preparation of optically active α-amino acids from the benzophenone imines of glycine derivatives was published in 2001 (*120*). Some developments in alkylations of glycine Schiff bases with chiral auxiliaries or with asymmetric catalysis were discussed in articles in 2000 (*121*) and 2002 (*122*), while the enantioselective synthesis of amino acids by phase-transfer catalysis with achiral Schiff base esters was reviewed in 2003 (*123*) and 2004 (*124, 125*). Asymmetric catalysis has also been extended to reactions of electrophilic glycine Schiff base derivatives (see Section 6.5.1).

6.4.1 Chiral Ester Auxiliaries

Binaphthol has been used as an ester auxiliary with the benzophenone Schiff base of glycine (*126*), with the enolate undergoing alkylation by a number of electrophiles in 62–77% yield and 69–86% de (see Scheme 6.8). Alkylating the binaphthol free hydroxyl group reduced diastereoselectivity.

(−)-8-Phenylmenthol has also been used to form an ester with the same glycine ketimine (see Scheme 6.9). Phase-transfer catalysis (PTC) alkylation with [¹¹C] methyl or [α-¹¹C]benzyl iodides followed by a two step hydrolysis was complete in less than 1 h, and gave the amino acids with 52–55% ee (15–40% radiochemical yield) (*127–129*).

The same chiral ester was used in combination with *N*-[bis(methylthio)methylene] protection (*130*). By careful selection of the deprotonation conditions, the *Z*- or

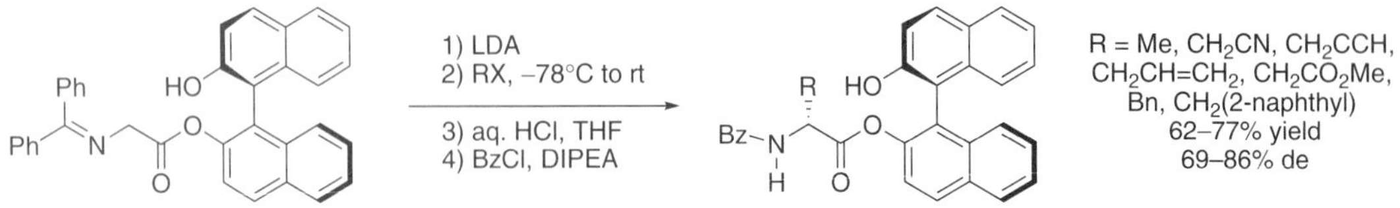

Scheme 6.8 Chiral Ester Auxiliary: Binaphthol (*126*).

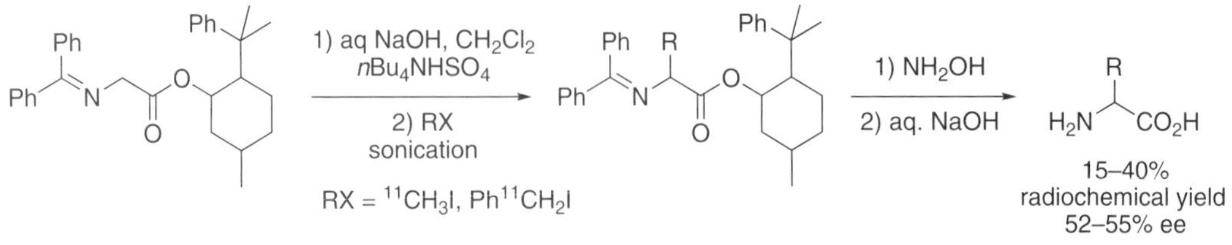

Scheme 6.9 Chiral Ester Auxiliary: 8-Phenylmenthol (*127–129*).

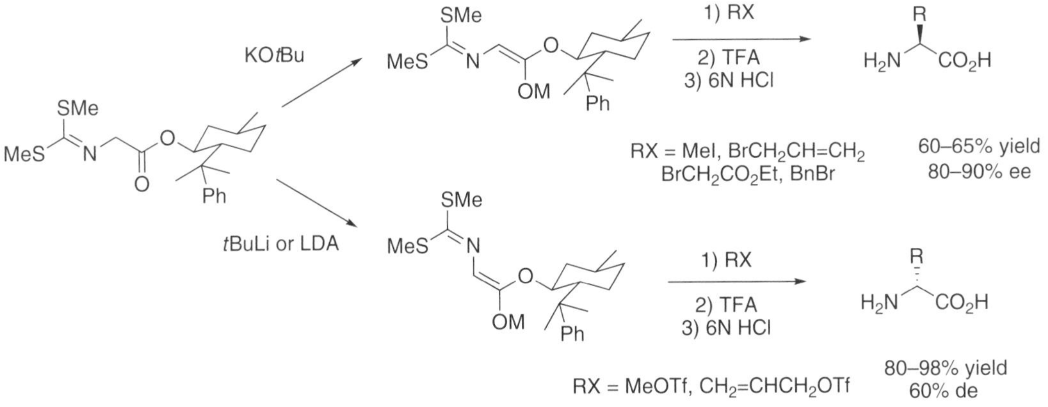

Scheme 6.10 Chiral Ester Auxiliary: 8-Phenylmenthol (*130*).

E-enolates were preferentially generated, and by then using specific alkylation conditions both *R*- and *S*-amino acids were produced from the same chiral auxiliary enantiomer (see Scheme 6.10). The study demonstrated that the diastereoselectivity was governed by the kinetics of the alkylation step as well as the enolate geometry. Deprotonation with KO*t*Bu and reaction with alkyl halides gave the 2*S* isomers with 80–90% de, while enolate generation with LDA or *t*BuLi and reaction with alkyl triflates gave the 2*R* products (60% de) (*130*).

6.4.2 Chiral Amide Auxiliaries

6.4.2a Oxazolidin-2-One, Pyrrolidine, and Imidazolidinone Auxiliaries

The oxazolidin-2-one chiral auxiliaries developed by Evans et al., prepared from chiral amino alcohols such as valinol (*131*), phenylalaninol (*132*) or norephedrine (*131*) and originally used to induce stereoselectivity in

aldol condensations (*131*), have also been used as amide auxiliaries with glycine Schiff base enolate equivalents. An isothiocyanate was used as the glycine synthon, with stannous triflate-mediated aldol additions to a number of aldehydes giving the *threo* adducts with 82–98% de (*132*) (see Scheme 6.11). Methanolysis or hydrolysis removed the amide auxiliary, giving the amino acid or ester. This method was used for the synthesis of Thr (*133, 134*), of MeBmt and related analogs (*132, 135*), and of several substituted phenylserines (*136–138*), as well for the preparation of 3-hydroxyhomotyrosine, a component of the antifungal cyclic hexapeptide echinocandin D (*139*). β-Phenylcysteine could be obtained from β-phenylserine prepared by this method, via displacement of the mesylate of β-phenylserine with thiolacetic acid (*140*). A complementary approach to the *erythro* diastereomers of both β-phenylcysteine (*140*) and β-hydroxyamino (*141*) acids has been published, but as it proceeds via an amination reaction it is discussed in Section 5.3.6a. The Evans auxiliary has also been used in the synthesis and alkylation of β-lactam

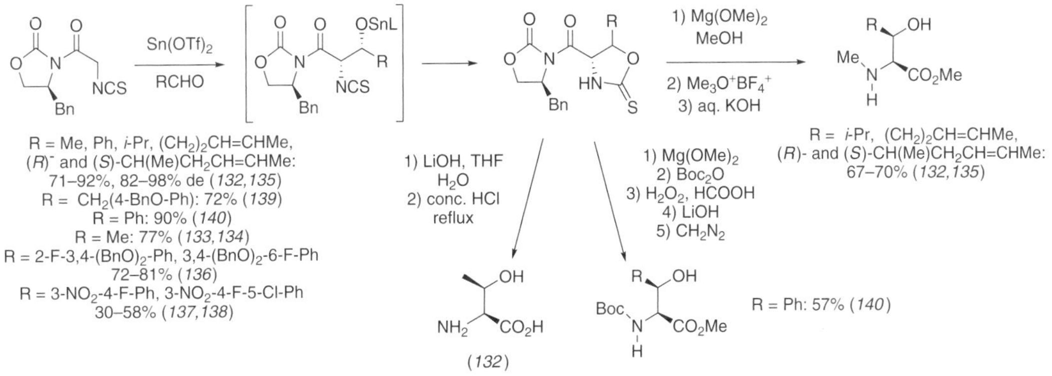

Scheme 6.11 Aldol Reaction of Glycine Enolate Equivalent with Evans' Oxazolidinone Auxiliary.

derivatives (see Section 8.5). Other oxazolidinone auxiliaries are summarized in Scheme 5.23.

Several 2,2-dimethyl-4-substituted oxazolidine amide auxiliaries (see Scheme 6.12) induced good *anti* diastereoselectivity and apparently complete stereoselectivity at the α-center during aldol reactions of the benzophenone Schiff base of Gly with alkyl aldehydes (*142*). The best results were provided by 2,2-dimethyl-4-benzyloxazolidine.

A chiral 2,5-disubstituted pyrrolidine was used as a chiral amide auxiliary of N-[bis(methylthio)methylene] glycine, inducing diastereoselectivities of 96–98% de during alkylation by aryl halides and alkyl triflates

(see Scheme 6.13) (*143, 144*). Aldol reactions gave poor yields and selectivity. The auxiliary was used for a synthesis of an unusual amino acid found in physiologically active peptides, 2-amino-8-oxo-9,10-epoxydecanoic acid (*144*).

1,5-Dimethyl-4-phenylimidazolidin-2-one has also been described as an amide auxiliary for the N-[bis(methylthio)methylene] Schiff base of Gly (see Scheme 6.14). Aprotic deprotonations and alkylations proceeded with 74–96% de, and the auxiliary was readily removed with LiO₂H (*145, 146*). The same imidazolidinone auxiliary was employed in conjunction with a benzophenone Schiff base. Deprotonation with DBU

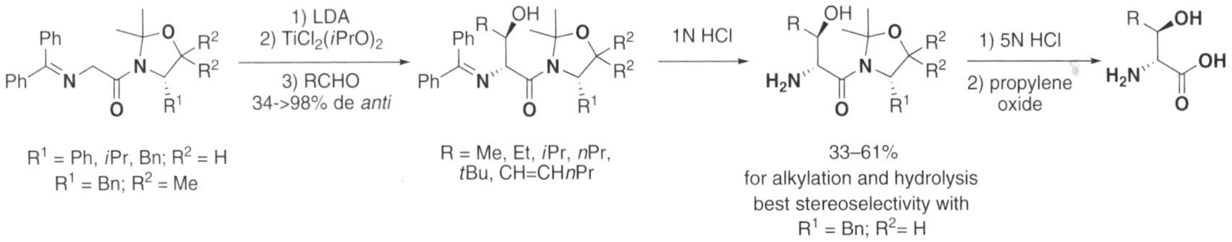

Scheme 6.12 2,2-Dimethyloxazolidine Auxiliaries (*142*).

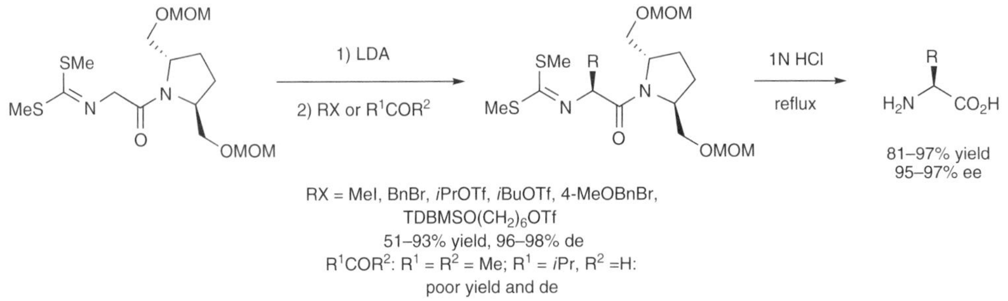

Scheme 6.13 Chiral Pyrrolidine Amide Auxiliary (*143, 144*).

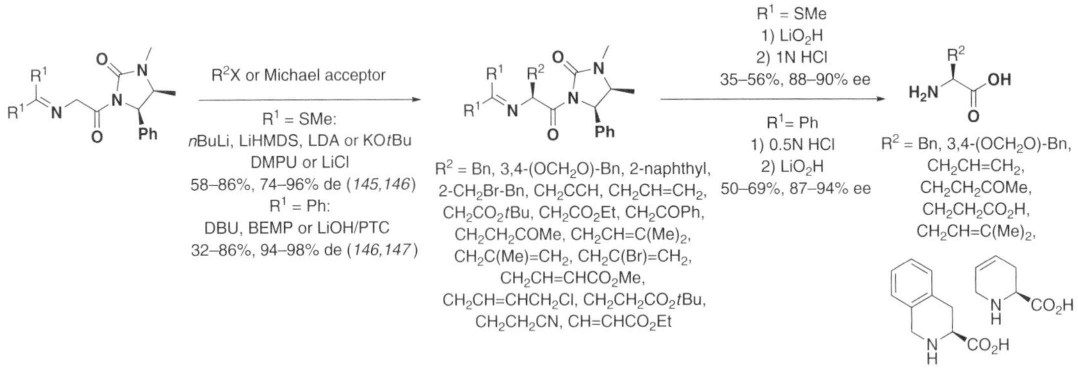

Scheme 6.14 Alkylation of Glycine Schiff Base with 1,5-Dimethyl-4-phenylimidazolidin-2-one Auxiliary.

under anhydrous conditions, or with LiOH under PTC conditions, allowed for alkylations and Michael additions with 32–86% yields and 94–98% de (*146, 147*).

6.4.2b Oppolzer Bornane[10,2]Sultam Auxiliary

The Oppolzer camphor-derived (*148*) bornane-sultam auxiliary has provided some of the best results for asymmetric alkylations of a Schiff base (see Scheme 6.15, Table 6.3). It is also used as an auxiliary for amination reactions (see Section 5.3.6c). The application of the bornane[10,2]sultam as a chiral auxiliary for asymmetric synthesis, including the preparation of amino acids, has been reviewed (*149*).

The desired synthon was prepared from *N*-[bis(methylthio)methylene]glycine or the benzophenone Gly Schiff base. Deprotonation and alkylation proceeded under either aprotic or phase-transfer catalysis (PTC) conditions to give the protected amino acid derivatives with good diastereoselectivity (crude product 84–98% de), which was readily improved to >99% de by purification (*150, 151*). PTC conditions required ultrasound assistance for good yields, and generally

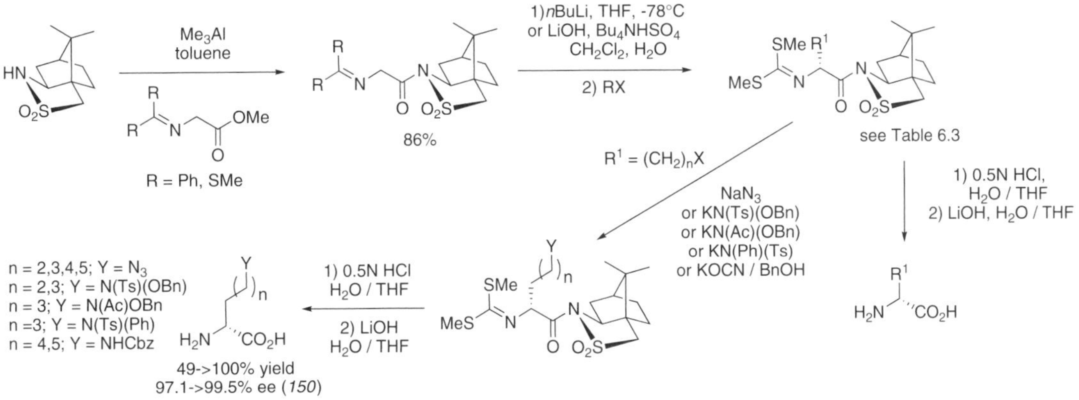

Scheme 6.15 Alkylation of Bornanesultam-Derived Amide Auxiliaries.

Table 6.3 Alkylation of Glycine Schiff Base with Oppolzer Auxiliary (see Scheme 6.15)

R	R^1	Alkylation Yield	Deprotection Yield	Reference
SMe	Me *i*Pr *n*Bu *i*Bu $CH_2CH=CH_2$ CH_2CO_2tBu Bn	87–100% crude 94.7–98.4% de 87–96% purified >99% de	75–>99% 99.5–>99.8% ee	1989 (*151*)
Ph	$CH(Ph)_2$ fluoren-9-yl	68–72% >95% de	65–77%	1991 (*160*) 1994 (*161*)
Ph	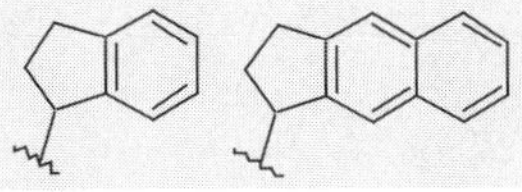	69–90%	38–52% >95% ee	1992 (*162*) 1994 (*161*)
SMe	Me *i*Pr *n*Bu *i*Bu CH_2Cl $(CH_2)_3Cl$	84–98.4% de crude 65–97% yield purified 98.3–>99.9% de purified	84–>99% 99.5–>99.8% ee	1994 (*150*)

(*Continued*)

Table 6.3 Alkylation of Glycine Schiff Base with Oppolzer Auxiliary (see Scheme 6.15) (continued)

R	R^1	Alkylation Yield	Deprotection Yield	Reference
	$(CH_2)_4I$			
	$(CH_2)_5I$			
	$(CH_2)_6I$			
	$CH_2CH=CH_2$			
	(Z)-$CH_2CH=CHCH_2OBn$			
	(Z)-$CH_2CH=CHCH_2OTBS$			
	CH_2CO_2tBu			
	Bn			
	4-Br-Bn			
	CH_2(1-naphthyl)			
SMe	Me	72–87%	59–89	1994 (152)
	CD_3		99.0–99.8% ee	
	Bn			
	CD_2Ph			
	4-BnO-Bn			
	with $[1,2-^{13}C_2,^{15}N]$			
Ph	4-$CH_2PO_3tBu_2$-Bn	—	46% overall	1995 (158)
			>97% ee	
SMe	$(CH_2)_3$[1,2-dicarba-closo-	90% crude	84–90%	1995 (156)
	dodecaboran(12)-1-yl]	60–71% purified	99.5–100% ee	
	$(CH_2)_3$[2-Me-1,2-dicarba-closo-	>99% de		
	dodecaboran(12)-1-yl]			
SMe	4-PO_3Me_2-Bn	84%	81%	1996 (155)
$=CR_2 = =OBn$	$CH_2CH=CH_2$	88–99%	—	1996 (393)
	$C(Me)_2CH=CH_2$	62–98% de		
	$CH_2C(Me)=CH_2$			
	$CH(Ph)CH=CH_2$			
SMe	$^{11}CH_3$	—	40–50% radiochemical	1997 (472)
			94% ee	
Ph	$2,3,5,6$-F_4-4-(CH_2PO_3MeH)-Bn	—	58% overall	1997 (159)
	4-(CF_2PO_3EtH)-Bn		92% ee	
SMe		—	80%	1997 (473)
			>99.5% ee	
Ph	CH_2CCH	74–90%	29–86%	1998 (163)
	$CH_2C(Br)=CH_2$			
	$CH_2(CO_2Et)=CH_2$			
	CH_2CH_2COMe			
	$CH_2CH_2CO_2Et$			
SMe	CH_2CCH	—	28% overall	2000 (157)
	converted into CH_2(o-carborane)		>99% ee	
Ph	2-F-3,4-$(MeO)_2$-Bn	88–98%	64–75%	2002 (164)
	5-F-3,4-$(MeO)_2$-Bn	>97% de	66–99% demethylation	
	6-F-3,4-$(MeO)_2$-Bn			
	$2,6$-F_2-3,4-$(MeO)_2$-Bn			
	2-F-3,4-$(HO)_2$-Bn			
	5-F-3,4-$(HO)_2$-Bn			
	6-F-3,4-$(HO)_2$-Bn			
	$2,6$-F_2-3,4-$(HO)_2$-Bn			
Ph	4-COEt-Bn (protected as 1,2-ethanediol	54% akylation	59% deprotection and	2002 (165)
	acetal during alkylation)	>95% de	N-Boc protection	

gave reduced diastereoselectivity compared to the apro-tic conditions. The stereoselectivity has been rational-ized as a C(α)-*Si*-face attack by the electrophile on a Li-chelated (Z)-enolate from the face opposite the sultam nitrogen lone pair (*150*). (*S*)-amino acids are obtained from the (2*R*)-bornanesultam, but since the chiral auxil-iary antipode is equally available both amino acid enantiomers are accessible. The free amino acids were generated with >99.5% ee by mild acid hydrolysis fol-lowed by LiOH saponification, a process that allows for recovery of the auxiliary. The protected adducts with haloalkane side chains could be converted into ω-azido-α-amino acids, ω-amino-α-amino acids or ω-*N*-hydroxylamine-α-amino acids by treatment with NaN₃ or the anions of TsNHPh, KOCN, *N*-tosyl-*O*-benzyl hydroxylamine or *O*-benzyl acetohydroxamate (*150*).

This method was directly compared with the Schöllkopf bislactim ether route (see Section 7.2) for the preparation of [1,2-¹³C₂,¹⁵N]-labeled Boc-Leu. Both procedures had comparable overall yields (34% vs 31%), but the Oppolzer method gave a product with 99.7% ee, compared to 97.2–97.4% ee for the Schöllkopf product (*152*). Labeled Ala, Leu, Phe, and Tyr were then prepared by the Oppolzer method starting with [1,2-¹³C₂,¹⁵N]-Gly-OEt (*119*), [1,2-¹³C₂, 2,2-²H₂,¹⁵N]-Gly-OEt (*153*), or [2,2-²H₂]-Gly-OMe (*154*). Another direct comparison of these two routes prepared 4′-dimethylphosphonomethyl-L-Phe in 38% yield via the Schöllkopf route, and 68% yield by the Oppolzer method (no enantiopurity reported) (*155*).

A similar direct comparison was made with Seebach's imidazolidinone template (see Section 7.9) for the syn-thesis of two carboranyl amino acids (*156*). The Seebach route provided the (*S*)-enantiomer of the first compound in 48% overall yield (99.6% ee) from the imidazolidi-none template, and the (*R*)-isomer in 28% yield (98.3% ee) by a slightly different route. The Oppolzer route gave a 64% yield of the (*S*)-isomer from the sultam glycine template with >99.9% ee. The second compound was prepared as the (*S*)-enantiomer in 67% and 57% yields, respectively, by the two methods, both with >99% ee. Another direct comparison of the same two methods for the synthesis of *o*-carboranylglycine found that the Seebach method produced the product with 91–96% ee after imidazolidinone hydrolysis, while the Oppolzer route gave the amino acid with >99% ee (*157*).

Other amino acids prepared using the Oppolzer auxiliary include di-*tert*-butyl-4-phosphonomethyl-L-phenylalanine, which was obtained in 46% overall yield and >97% ee by alkylation with di-*tert*-butyl-4-phosphonomethylbenzyl bromide. A benzophenone Schiff base was used instead of the *N*-[bis(methylthio) methylene] group, as it could be removed under much less acidic conditions (10% citric acid); milder conditions were also reported for removal of the auxiliary (0.5 N LiOH/THF, 4 °C, 3 h) (*158*). Other phosphotyrosine analogs have been synthesized (*159*). The benzophenone imine derivative was also used for the synthesis of L-diphenylalanine (*160, 161*), L-9-Fluorenylglycine (*160, 161*), L-1-indanylglycine (*161, 162*), and L-1-benz[f]

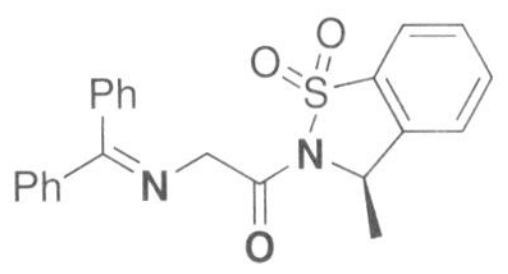

Scheme 6.16 Toluene-2,α-Sultam Auxiliary.

indanylglycine (*161, 162*), with >95% de at the α-center of all four adducts but <30% de at the β-center of the latter two compounds. Another report described alkyla-tion of the benzophenone imine under PTC conditions using K₂CO₃ as base, with propargyl bromide, 2,3-di-bromopropene, ethyl 2-bromomethylpropenoate, ethyl acrylate, and 3-buten-2-one as electrophiles (*163*). Alkylation of the benzophenone Schiff base derivative under PTC conditions was applied to syntheses of 2′-, 5′-, and 6′-fluoro-DOPA and 2′,6′-difluoro-DOPA, with the alkylated products obtained in 87–98% yield with >97% de (*164*). 4′-Propanoyl-Phe was synthesized as a potential photoactivatable Phe analog, with 95% de during alkylation (*165*).

The Oppolzer route has also been applied to the syn-thesis of α,α-disubstituted amino acids (*166*). Activated amino acids (Phe, Leu, and Met) were coupled with the sodium salt of the sultam auxiliary and converted into the 4-chlorobenzaldehyde Schiff bases. Deprotonation and alkylation with methyl iodide gave the protected crude α-methyl amino acid derivatives with 80 to >99% de, which improved to 94 to >99% de after crystalliza-tion (47–70% yield). As expected, the chirality of the starting amino acid had no effect on the reaction diaste-reoselectivity, but the topicity of the methylation was opposite that predicted based on the results of monoal-kylation of glycine derivatives.

As mentioned earlier, the Oppolzer sultam amide auxiliary has also been used with Boc-Sar instead of a Schiff base, giving very high diastereoselectivites (>98% de) for alkylations with benzyl bromides (*9*).

A toluene-2,α-sultam auxiliary (see Scheme 6.16) was described for alkylations of the benzophenone imine of glycine. Benzylation proceeded in 71% yield with >99% de (recrystallized product) (*167*).

6.4.2c Other Amide Auxiliaries

The pseudoephedrine amide auxiliary used for non-activated Gly alkylations (see Section 6.2.1b) has also been combined with Schiff base activation. This synthon produced alkylation adducts (Ala, Phe, Abu, Asp, alyl-glycine) in 40–60% yield with 68–96% de (*26*).

6.4.3 Chiral Schiff Base Auxiliaries

6.4.3a Hydroxypinanone

A number of chiral ketones have been used as chiral auxiliaries to form Schiff bases with glycine esters.

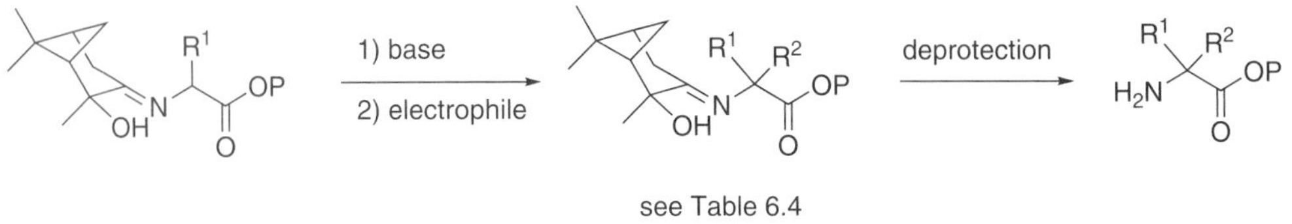

see Table 6.4

Scheme 6.17 Chiral Schiff Base Auxiliary: 2-hydroxypinan-3-one.

Many of these auxiliaries are bicyclic ketones, of which 2-hydroxypinan-3-one is the most studied and successful (see Scheme 6.17, Table 6.4). An added advantage of 2-hydroxypinan-3-one is that it generally provides for quantitative chromatographic resolution of the diastereomers, and has been used to resolve racemic amino acids (168, 169). (1S,2S,5S)-2-hydroxypinan-3-one was first used to form the Schiff base of glycine tert-butyl ester in 1976 (170). Deprotonation with LDA, alkylation, and hydrolysis gave D-Ala, D-Leu, D-Phe, and D-3,4-(MeO)$_2$Phe in 50–79% yield and a remarkable (for the time) 66–83% ee.

The Schiff base auxiliary was later applied to the α-alkylation of Ala and Nva with varying success (15–90% ee, 70–85% yield), with the same diastereomer obtained regardless of the order of introduction of the side chains (171). The chirality of the starting amino ester played a prominent role in the course of the alkylation (172), with (R)-amino acids readily alkylated but (S)-amino acids giving poor or no yields of product. The (S)-amino acid-derived products were the same configuration as those obtained from (R) starting material. A closer examination of reaction conditions found that a bulky ester group provided better diastereoselectivity, that diastereoselectivity was poor if the existing substituent and the electrophile were both of similar size, and that lithiated bases produced products with better diastereoselectivity than could be obtained with KOtBu, although of the opposite configuration (173). The results were rationalized on the basis of a folded-dimer model of the enolate, although this model does not appear to account for some of the results obtained in other papers (173).

A variety of electrophiles have been reacted (see Table 6.4). Conditions for alkylations with 4-iodo-1-butene were explored. The tert-butyl ester of Gly gave better results than the methyl ester, and LiHMDS gave the best diastereoselectivity of several bases, with >98% de and 90% yield (174). Alkylation with 1-iodo-2-propanol produced γ-hydroxynorvaline in 46% yield with complete stereoselectivity at the α-center (175). 2'-Methyl-Trp (169) and 2'-carboxy-Trp (176) were synthesized using this synthon. Several lipidic amino acids were prepared with 83–99% de by alkylation with bromides or iodides of dodecane, tetradecane, or hexadecane (177), while γ-fluoro-amino acids were derived from fluorinated alkyl or allyl electrophiles (178). Fluorobenzene tricarbonylchromium complexes were used to intoduce an α-phenyl group to Ala, Leu, and Val with 50–100% de (179). Alkylations with dihaloalkanes,

followed by N-deprotection and then cyclization, produced cyclic imino Pro derivatives (180, 181). Attempts to alkylate the chiral Gly synthon with bromoacetonitrile under the standard conditions gave disappointing results (20% de), but led to the discovery that by exchanging the lithium enolate with MgBr$_2$ asymmetric induction could be increased to 100% (182).

Aldol reactions of the derivatized Gly methyl ester were less successful, with poor yields (28–40%) and multiple diastereomers (183), although with ribose or galactose aldehyde derivatives one major diastereomer was isolated in 45% and 36% yield, respectively (184). However, aldol addition to 3-chloro-5-benzyloxy-benzaldehyde in the presence of Ti(OEt)$_3$Cl gave a single diastereomer in 55% isolated yield (185). A trans-metalated Ti enolate also gave good stereocontrol (94% de at C2 and 94% de anti at C3) for a synthesis of 3'-nitro-4'-fluoro-phenylserine (186). Sulfenylation of the hydroxypinanone Schiff base derivatives of Gly, Ala, Phe, and Phg with MeSSMe produced the α-methylthio amino acids in 55–76% yield and 85 to >98% de (187).

The Gly synthon has also been used for Michael additions, with addition to methyl acrylate leading to Glu, to diethyl vinylphosphonate providing the phosphonic analog of Glu, and to methyl vinylmethyl-phosphinate for a synthesis of phosphinothricin. Diastereoselectivities of up to 79% de were obtained (188, 189). Other Michael addition reactions have been reported (183, 189). Some surprising results were obtained by varying the concentration of starting enolate for additions to ethyl acrylate or acrylonitrile. With 0.1 equiv of base (NaH or phosphazene) one diastereomer was produced with 80–95% de. By employing one equivalent of DBU with MgBr$_2$ or LiBr, the other diastereomer was formed with 80–96% de (190). By employing MeMgBr and DBU for addition to ethyl crotonate, only two of four possible diastereomers were formed, in a 56:44 ratio. Diastereomer separation, then Schiff base/ester hydrolysis produced (2S,3S)- and (2S,3R)-3-methyl-Glu (191).

The 2-hydroxypinanone auxiliary has also been employed for deracemization of amino acids. The racemic amino acid Schiff base methyl ester was treated with KOtBu and then quenched with a saturated solution of NH$_4$Cl. The Schiff base amino esters of Nva, Val, Leu, Phe, and β-(2-naphthyl)-Ala were obtained in 92–96% yield with 79 to >98% de. Chromatographic purification followed by imine deprotection gave the amino esters in 72–94% yield with >98% ee (192).

Table 6.4 Alkylation 2-Hydroxypinan-3-one Schiff Base of Amino Esters (see Scheme 6.17)

P	R¹	Electrophile R²X	Base	Alkylation Yield	Deprotection Conditions	Deprotection Yield	Reference
tBu	H	RX: R = Me, iBu, Bn, 3,4-(MeO)₂-Bn	LDA	—	15% citric acid	50–79% for 2 steps 66–83% ee	1976 (170)
tBu	Me, nPr	RX: R = Me, nPr, CH₂CH=CH₂, CH₂CCH	LDA	—	15% citric acid	70–85% for 2 steps 15–90% ee	1983 (171)
tBu	H	¹¹CH₃I	LDA	—	1N HCl	80% for 2 steps radiochemical yield 80% ee	1983 (474)
Me	H	I(CH₂)₅CH=CH₂	LDA	40%	citric acid	85%	1984 (475)
Et	H	H₂C=CHCO₂Me, H₂C=CHP(=O)(Me)(OMeO), H₂C=CHPO₃Et₂	KOtBu	39–68% 35–79% de	—	—	1984 (189) 1987 (188)
Me	H, Me, nPr, Bn	I(CH₂)₃X, I(CH₂)₄X, I(CH₂)₅X, I(CH₂)₁₀X, X = Br, Cl	LDA	41–75% 5–>95% de	—	—	1986 (180) 1986 (181)
tBu	H	¹¹CH₃I	LiTMP	—	1) NH₂OAc 2) 6N HCl	12–18% for 2 steps radiochemical yield 87% ee	1986 (476)
Me	H	MeCHO, EtCHO, iPrCHO, tBuCHO, PhCHO	LDA	20–40% mixture of isomers	—	—	1988 (183)
Me	H, Me, nPr	H₂C=CHCO₂Me, MeCH=CHCO₂Et	LDA	52% 48–71% de	—	—	1988 (183)
Me	H, Me	HCCCO₂Et	LDA	40–64% >95% de	—	—	1988 (183)

(Continued)

Table 6.4 Alkylation 2-Hydroxypinan-3-one Schiff Base of Amino Esters (see Scheme 6.17) (continued)

P	R^1	Electrophile R^2X	Base	Alkylation Yield	Deprotection Conditions	Deprotection Yield	Reference
Me	Me	$H_2C=C=C(Me)CO_2Et$ $H_2C=C=CHCO_2$(menthyl)	LDA	35–74% >95% de	—	—	1988 (183)
lactone with 2-HO group of pinanone	H Me	MeI PhCHO	KOtBu	60–84% >95% de	—	—	1988 (183)
Me	Me nPr iPr iBu Ph	MeI	LDA	0–95% 81–95% de	—	—	1988 (172)
Et	H	RX: R = CH_2CN	LDA LDA then MgBr$_2$	100% 20% de 85% 100% de	—	—	1988 (182)
tBu	H Ph	RX: R = $^{11}CH_3$ $^{11}CH_2Ph$ $^{11}CH_2$(4-MeO-Ph)	nBuLi, TMP	—	NH_2OH	—	1989 (28)
tBu	Me nBu iPr Bn	RX: R = Me nBu iBu $(CH_2)_4I$ $CH_2CH=CH_2$ Bn	LDA	30–85% 5->98% de	—	—	1991 (173)
Me	Me iPr iBu	$Cr(CO)_3$(4-F-Ph)	LDA, HMPA	15–60% 50–100% de	—	—	1991 (179)
Me	H	ribose galactose	KOtBu, THF	55–65% 32–64% de	15% citric acid	75–80%	1993 (184)
Me	H Me Ph Bn	MeSSMe	LDA or KOtBu	55–76% 85->98% de	—	—	1994 (187)
tBu	H	$I(CH_2)_2CH=CH_2$	LiHMDS	90% >98% de	15% citric acid	89%	1996 (174)

R1	R2	Electrophile	Base	Alkylation result	Hydrolysis	Yield	Notes	Year (Ref)
tBu	H	ICH(Me)CH$_2$OTMS	LDA	46% >95% de C2 0% de C3	15% citric acid 3N HCl	80%		1997 (175)
Et	H	H$_2$C=CH-CO$_2$Et	0.1 eq. NaH or t-OctP2	90->98% 80-90% de (R)	—	—		1997 (190)
		H$_2$C=CHCN	1 eq. DBU, LiBr or MgBr$_2$	70->98% 60-98% de (S)				
Et	H	BrCH$_2$(2-Me-indol-3-yl)	LDA	65% >95% de	15% citric acid	85%		1998 (169)
Et	H	(3-Cl-4-BnO-Ph)CHO	NEt$_3$, Ti(OEt)$_3$Cl	55%	12N HCl	95% 100% de		1998 (185)
tBu	H	IMe$_3$NCH$_2$(2-CO$_2$Me-indol-3-yl)	LDA	—	15% citric acid	60% for 2 steps 87% ee		1999 (176)
tBu	H	CH$_2$CH$_2$F	LDA	36-90% de crude 13-73% yield purified 76->98% de	1) 15% citric acid 2) HCl	13-48% Schiff base 29-79% ester hydrolysis		2000 (178)
	H	CH$_2$CH(F)=CH$_2$						
	Me	CH$_2$CH$_2$F						
tBu	H	nC$_{12}$H$_{25}$I nC$_{12}$H$_{25}$Br nC$_{14}$H$_{29}$I nC$_{14}$H$_{29}$Br nC$_{16}$H$_{33}$I nC$_{16}$H$_{33}$Br	LDA	24-66% 83-99% de	NH$_2$OH·HCl, MeOH	88-91% Schiff base hydrolysis		2002 (177)
tBu	H	MeCH=CHCO$_2$Et	MeMgBr, DBU	75% 2 of 4 diastereomers 56:44	1) 15% citric acid 2) 6N HCl, reflux	48-64%		2003 (191)
lactone with 2-HO group of pinanone	H	C(OH)(Me)CO$_2$H C(OH)(CF$_3$)CO$_2$H C(OH)(iPr)CO$_2$H C(OH)(Ph)CO$_2$H C(OH)(CH$_2$Br)CO$_2$H C(-OCH$_2$-)CO$_2$H C(OH)(CH$_2$CH$_2$Ph)CO$_2$H	KHMDS	51-88% 2 of 4 diastereomers 4-86% de	1) anh. HF 2) 6N HCl, 90 °C	71-90%		2003 (193)
Et	H	(3-NO$_2$-4-F-Ph)CHO	NEt$_3$, Ti(OEt)$_3$Cl	75% 94% de C2 94% de anti C3	aq. HCl	100%		2004 (186)

A variation of the 2-hydroxypinanone auxiliary employed a rigid lactone derivative obtained by cyclizing the Gly carboxyl group with the 2-hydroxypinan-3-one hydroxyl group (*183*). Deprotonation was best achieved with KO*t*Bu, and MeI reacted with >95% de and 60% yield, although some dialkylated product was also obtained. In contrast with the acylic synthon, aldol reactions proceeded in much better yield but Michael additions were unsuccessful. This approach was used to prepare a number of β-substituted-β-hydroxy-Asp derivatives using KHMDS deprotonation, with only two of four diastereomers detected, and 8–86% de at the carbinol center (*193*).

6.4.3b Camphor-Related

Camphor can also be employed to form a ketimine with Gly esters (see Scheme 6.18), with the synthon initially prepared via a thione intermediate (*194*) but later by direct condensation with BF$_3$ catalysis (*195*). Deprotonation with LDA and alkylation with primary alkyl, allylic, and benzylic halides proceeded with 0–98% de (see Table 6.5). The best results were obtained from the *tert*-butyl ester derivative and alkylating agents containing a π-system (*194, 196*). Enolates generated with sodium, lithium, zinc, or zirconium showed little difference in diastereoselectivity, but potassium enolates gave inferior yields and selectivity (*197*). Racemic secondary halides with a π-system (1-phenethyl bromide or 3-bromocyclohexene) produced only two of four possible diastereomers (69–80% de), while 2-iodobutane gave all four diastereomers (*196, 198*). Michael additions were also possible (with poor diastereoselectivity), but aldol reactions were unsuccessful (*196*). Michael additions to α,β-unsaturated esters resulted in 3-substituted Glu products with (2*R*)-configuration and >90% de at the α-center and, if applicable, 50–76% de at C3 or 20% de at C4. Additions to 2-alkylidene malonates proceeded with 100% de (*199, 200*). The Michael adducts could be deprotected to form pyroglutamates, which were then reduced to form prolines (*201*). The camphor ketimine auxiliary has also been used to prepare several γ- and δ-fluoro-α-amino acids, although stereoselectivity at the α-center was only 32–42% de (*202*). A (*R*)-(+)-camphor imine of Ala-O*t*Bu was allylated with allyl bromide or 3-bromo-2-fluoropropene to give α-allyl-Ala or 2-amino-2-methyl-4-fluoropent-4-enoic acid in 52–64% yield and 49–56% ee; the corresponding Gly Schiff base gave 86% yields with 96% de (*203*).

A camphor imine has also been used in the asymmetric synthesis of α-aminophosphonates (*184*).

Attempts to form the corresponding imine of 3-hydroxycamphor were unsuccessful, while the imine derived from norcamphor gave no stereoselection and the ketimine from 10-hydroxymethylcamphor gave reduced stereoselectivity (see Scheme 6.19) (*195*). However, the latter imine did allow for an aldol condensation with benzaldehyde. A double asymmetric induction was attempted using camphor or 2-hydroxypinan-3-one to form a Schiff base in combination with a menthyl ester of glycine. The matched pair allowed for allylation, benzylation, or propylation with 94–96% ee (*204*). Alkylations of a glycine camphor derivative with a 10-sulfonamide substituent on the auxiliary have also been reported. Diastereoselectivities were variable, and were reversed with bulkier electrophiles (*205*). A synthesis of D-DOPA was subsequently described via alkylation with 3,4-*O*,*O*-methylenedihydroxybenzyl bromide, giving the final product in four steps and 96.4% ee (*206*).

The best diastereoselectivities for a camphor-based imine were provided by a tricyclic iminolactone derived from (1*R*)-camphor (see Scheme 6.20). Alkylations by a number of electrophiles proceeded with >98% de. However, hydrolysis required 8 N HCl at 87 °C and the chiral purity three deprotected amino acids (Ala, Phe, and allylglycine) varied from 94 to 98% ee (*207*).

6.4.3c Other Chiral Ketones

(+)-Ketopinic acid (see Scheme 6.21) was used as a Schiff base auxiliary for an aldol reaction of glycine with benzaldehyde, giving phenylserine with 3–60% ee, depending on the metal cation (*208*). Similarly, Zn(II) and Cu(II) complexes of the glycine imines of 3-hydroxymethylenebornan-2-one (see Section 7.11) reacted with benzaldehyde in the presence of KO*t*Bu, but with even poorer diastereoselectivity (9% ee) (*209*).

A new auxiliary derived from (*R*)-pulegone was reported in 1998. It induced excellent diastereoselectivities for alkylations of Gly-OMe, giving products with 99% ee after deprotection (see Scheme 6.22). The use of rare and expensive (*S*)-pulegone to obtain the enantiomeric amino acid products could be avoided by using an epimer produced as a byproduct during the synthesis of the (*R*)-pulegone aldehyde auxiliary. The epimeric auxiliary induced comparable, but opposite, diastereoselectivities during alkylations (98 to >99.5 de) (*210*).

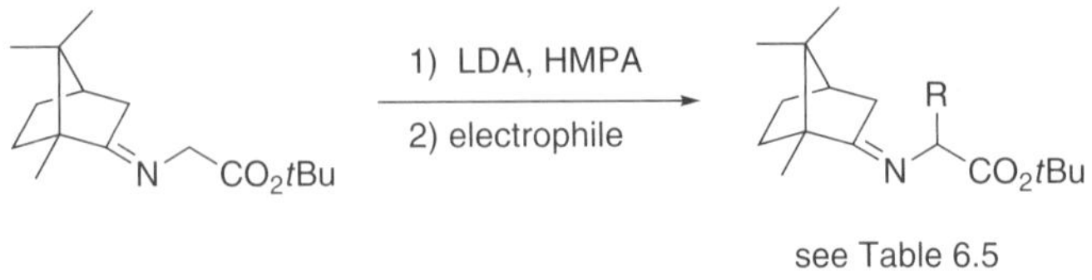

1) LDA, HMPA
2) electrophile

see Table 6.5

Scheme 6.18 Chiral Schiff Base: Camphor.

Table 6.5 Alkylation of Camphor Imine of Glycine (see Scheme 6.18)

Electrophile	Alkylation Yield	Reference
RX: R = Me, Et, *n*Bu, *i*Bu, CH_2CO_2Me, CH_2CO_2Et, CH_2CO_2tBu, CH_2SPh, $(CH_2)_2CH{=}CH_2$, $(CH_2)_2Ph$, $CH_2CH{=}CH_2$, $CH_2C(Me){=}CH_2$, $CH_2C(Me){=}C(Me)_2$, CH_2CCH, $CH_2CH{=}CHMe$, Bn, 2-Me-Bn, Me-Bn, 4-F-Bn, 4-*t*Bu-Bn, 4-CF_3-Bn, 4-CN-Bn, 4-MeO-Bn, 4-NO_2-Bn, CH_2(1-naphthyl)	31–89% 0–98% de	1986 (*194*) 1988 (*196*)
RX: R = *i*Pr, CH(Me)Et, 3-cyclohexenyl, *c*Hex, CH(Me)Ph, CH(Et)Ph, CH(Me)CCH	31–86% up to 80% de	1986 (*198*) 1988 (*196*)
$MeCH{=}CHCO_2Et$ $MeCH{=}C(CO_2Me)_2$ $MeCH{=}C(CO_2Et)_2$ $BrCH_2CO_2Me$	26–97%	1988 (*196*)
$H_2C{=}CHCO_2Me$ $H_2C{=}CHCO_2tBu$ $MeCH{=}CHCO_2Me$ $MeCH{=}CHCO_2tBu$ $H_2C{=}C(Me)CO_2Me$ $H_2C{=}C(Me)CO_2tBu$ $MeCH{=}C(CO_2Me)_2$ $EtCH{=}C(CO_2Me)_2$ $tBuCH{=}C(CO_2Me)_2$ $PhCH{=}C(CO_2Me)_2$ $PhCH{=}CHCH{=}C(CO_2Me)_2$ $MeCH{=}CHCH{=}C(CO_2Me)_2$ $MeOCH{=}CHCH{=}C(CO_2Me)_2$	49–100% 50–100% de	1989 (*199*) 1991 (*201*)
RX: R = *i*Pr, $CH_2CH{=}CH_2$, Bn	73% 85–96% de (with menthol ester)	1989 (*204*)
$HC(iPr){=}CHCO_2Et$ $HC[CH(-OCMe_2OCH_2-)]{=}CHCO_2Et$	73–74% 76–100% de	1993 (*200*)
$BrCH_2CH_2F$ $BrCH_2CH(F)Me$ $BrCH_2CH(F)Et$ $BrCH_2CH(F)nPr$ $BrCH_2CH(F)nBu$ $BrCH_2CH(F)iBu$ $BrCH_2CH(F)iPent$	54–73% 32–42% de at C2 24–46% de at C4	1997 (*202*)
$ICH_2CH{=}CHCH_2OBn$	87% 92% de	1999 (*477*)
$BrCH_2C(F){=}CH_2$ $BrCH_2CH{=}CH_2$	86% 96% de	1999 (*203*)

OH
O
3-hydroxycamphor
no imine formation
N
CO2tBu
norcamphor
no stereoselectivity
N
CO2tBu
OH
10-hydroxymethylcamphor
RX = BnBr: 80% de
(vs >98% for camphor)
PhCHO: 71% yield,
4 diastereomers
(161)
N
CO2P
SO2NHR
R = cHex, tBu
RX = MeI: >90% de
RX = BnBr: 33% de
opposite diastereomer
(170)

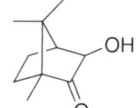

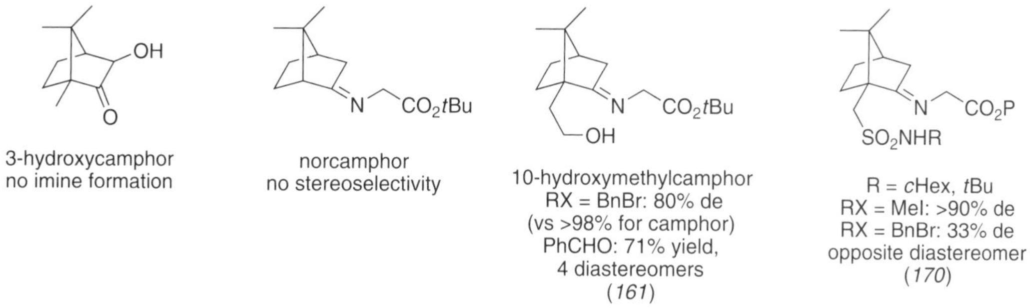

Scheme 6.19 Other Camphor-Related Schiff Base Auxiliaries.

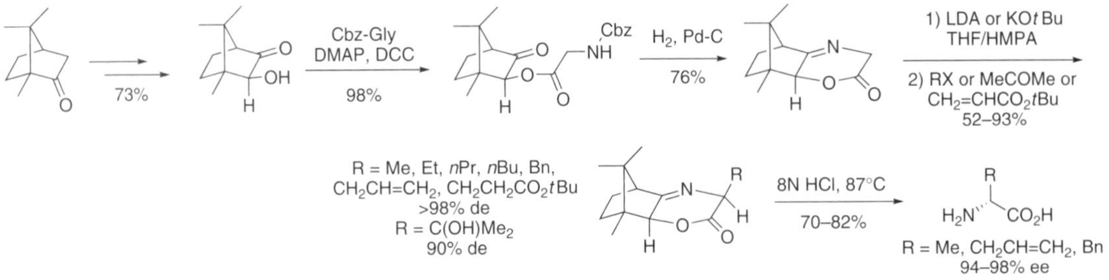

Scheme 6.20 Alkylation of Tricyclic Iminolactone (207).

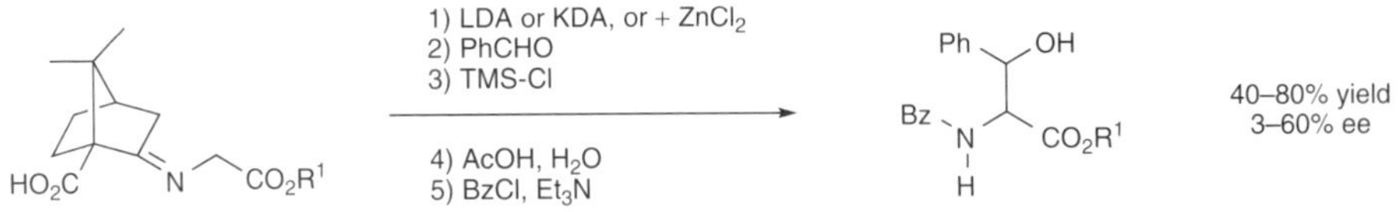

Scheme 6.21 Chiral Schiff Base: (+)-ketopinic Acid (208).

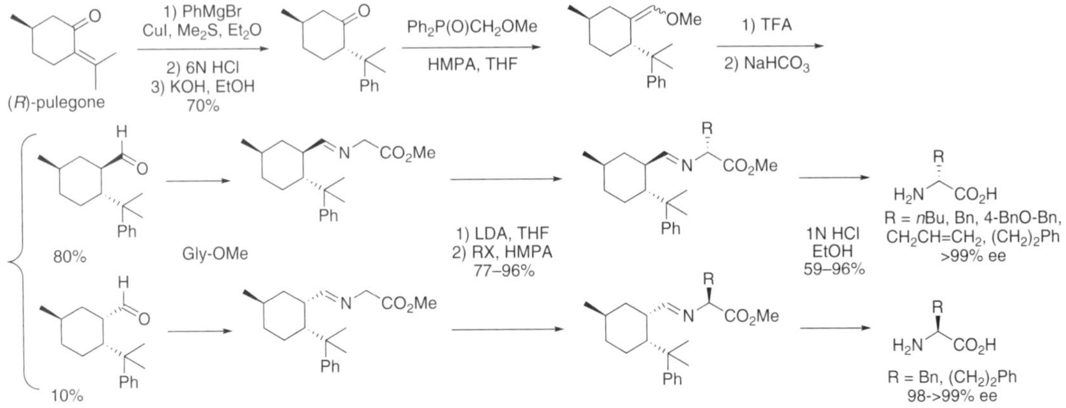

Scheme 6.22 (R)-Pulegone-Derived Imine Auxiliaries (210).

A much simpler chiral ketone is (3R)-3-phenyl-3-hydroxy-2-butanone, which induced diastereoselectivity of up to 64% when the Gly Schiff base was reacted with benzaldehyde (211). Menthone was examined for the α-benzylation of Ala, providing 21–24% de (212). Schöllkopf et al. have used protected D-galactodialdehyde (see Scheme 6.23) to form a Schiff base with a number of amino acid methyl esters (Val, Leu, Ile) (213). Deprotonation with nBuLi and alkylation with a number of electrophiles gave the α,α-disubstituted products with 23–95% de.

A derivatized paracyclophane was used as the chiral auxiliary for the aldol reaction of a Gly copper complex, and for anhydrous α-alkylations of other amino acids (see Scheme 6.24) (214).

A pyridoxal mimic possessing a chiral ionophore side chain and an additional chiral group has been used to form a Schiff base with Ala-OBn or Phe-OBn (see Scheme 6.25) (215, 216). Deprotonation with NaH produced significantly better results than LDA or KH as base, indicating that sodium ion may be recognized by the ionophore side chain. Ala was alkylated with BnBr with up to 96% ee, and Phe with MeI with up to 90% ee. With only the ionophore moiety to induce chirality, benzylation of A'a proceeded with up to 86% ee, and methylation of Phe with up to 82% ee (217). Another chiral pyridoxamine mimetic without the chiral ionophore was used to form an imine of Ala-Xaa dipeptides, with deprotonation/alkylation producing α-alkyl-Ala residues with 46–78% de (218, 219).

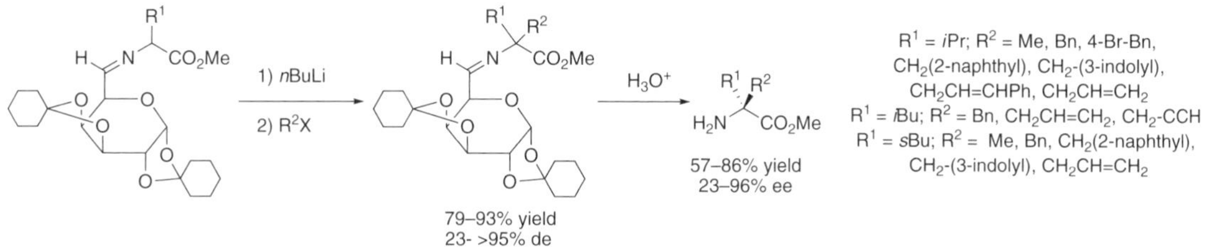
1) nBuLi
2) R2X
79–93% yield
23- >95% de
H3O+
57–86% yield
23–96% ee
R1 = iPr; R2 = Me, Bn, 4-Br-Bn, CH2(2-naphthyl), CH2-(3-indolyl), CH2CH=CHPh, CH2CH=CH2
R1 = iBu; R2 = Bn, CH2CH=CH2, CH2-CCH
R1 = sBu; R2 = Me, Bn, CH2(2-naphthyl), CH2-(3-indolyl), CH2CH=CH2

Scheme 6.23 Chiral Schiff Base: D-Galactodialdehyde (*213*).

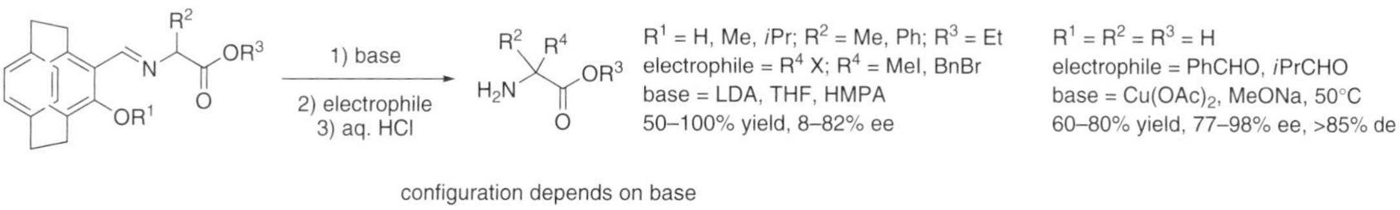
1) base
2) electrophile
3) aq. HCl
R1 = H, Me, iPr; R2 = Me, Ph; R3 = Et
electrophile = R4 X; R4 = MeI, BnBr
base = LDA, THF, HMPA
50–100% yield, 8–82% ee
R1 = R2 = R3 = H
electrophile = PhCHO, iPrCHO
base = Cu(OAc)2, MeONa, 50°C
60–80% yield, 77–98% ee, >85% de
configuration depends on base

Scheme 6.24 Chiral Schiff Base: 4-Formyl-5-hydroxy[2.2]paracyclophane (*214*).

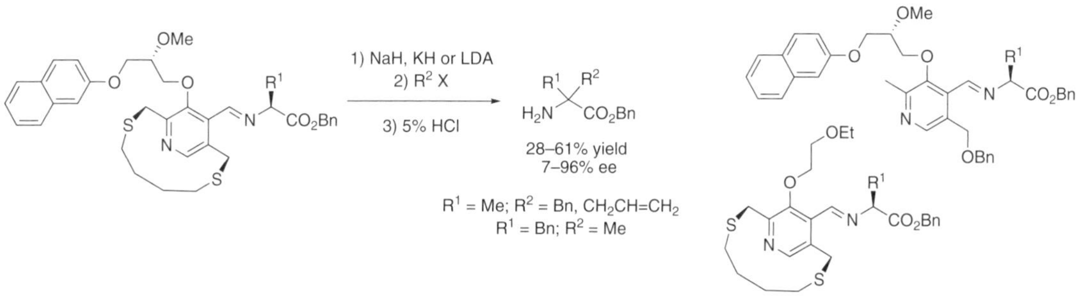
1) NaH, KH or LDA
2) R2 X
3) 5% HCl
28–61% yield
7–96% ee
R1 = Me; R2 = Bn, CH2CH=CH2
R1 = Bn; R2 = Me

Scheme 6.25 Chiral Schiff Base Auxiliary: Pyridoxal Mimic (*215–219*).

A chiral pyrrolidine-derived amidine was used to induce stereoselectivity during alkylations of amino acids (*220, 221*). The same auxiliary was used for alkylations of propargylamines, masked glycine equivalents, with the carboxyl group revealed by oxidative cleavage of the alkyne moiety (*220, 222*). In both cases the final amino acids were obtained with greatly varying enantiomeric purities of 15–84% ee.

Chiral Schiff base auxiliaries have been employed in conjunction with metal complexes for the alkylation of glycine and other amino acids, but this is discussed in Section 7.11.

6.4.4 Chiral Catalysts

Asymmetric alkylation of Schiff base derivatives using chiral phase-transfer catalysts has emerged as one of the most promising and convenient general synthetic routes to produce a range of unusual amino acids (see Scheme 6.26, Table 6.6). This procedure is an extension of O'Donnell's racemic phase-transfer catalysis (PTC) alkylation of ethyl glycinate benzophenone ketimine, introduced in 1989 (*223*) and briefly reviewed in 1994 (*224*). Some new developments in asymmetric PTC were briefly reviewed in 1999 (*225*) and 2002 (*122*), with more extensive reviews of the synthesis of amino acids by asymmetric PTC with achiral Schiff base esters published in 2003 (*123*) and 2004 (*124, 125*). The first phase-transfer catalysts were the pseudoenantiomeric *N*-benzyl quaternary salt derivatives of cinchonine or cinchonidine (see Scheme 6.27). The general catalytic mechanism is proposed to involve deprotonation at the interface between the two phases, extraction of the anion into the organic phase by ion-exchange with the cation of the chiral quaternary ammonium salt, and alkylation of the ion pair (*224*). There is evidence to suggest that the active catalyst may be the *N*-alkyl-*O*-alkyl cinchona salt formed in situ from the alkoxide of the catalyst (*224*). A theoretical study has been carried out to examine the possible molecular recognition processes between the catalyst and enolate involved in the asymmetric induction (*226*).

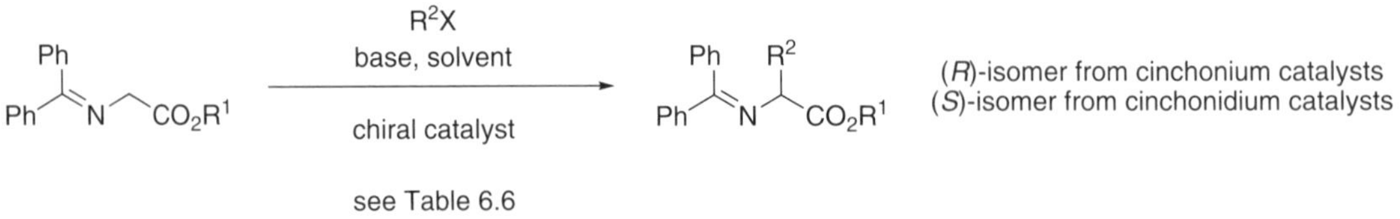

Scheme 6.26 Asymmetric Phase-Transfer Catalysis Alkylation of Schiff Bases.

Table 6.6 Asymmetric Alkylations of Glycine Imine with Chiral Phase-Transfer Catalyst (see Scheme 6.26)

R^1	Electrophile	Base, Catalyst	Alkylation Yield	Reference
tBu	R^2Br: $R^2 =$ Me nBu $CH_2CH=CH_2$ Bn 4-Cl-Bn CH_2(2-naphthyl)	50% aq. NaOH CH_2Cl_2 N-Bn cinchoninium chloride or N-Bn cinchonidinium chloride	60–82% 42–66% ee	1989 (223)
tBu	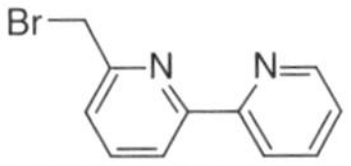 Br—[2,2'-bipyridine]	50% aq. NaOH CH_2Cl_2 N-Bn cinchonidinium chloride	83–85% 53% ee	1992 (478) 1993 (479)
4-Cl-PhCHO imine of Ala-OtBu	R^2Br: $R^2 =$ $CH_2CH=CH_2$ Bn 4-Cl-Bn 4-F-Bn 4-Br-Bn CH_2(2-naphthyl)	50% aq. NaOH or KOH:K_2CO_3 CH_2Cl_2 N-Bn cinchoninium chloride or N-Bn cinchonidinium chloride	78–87% 36–50% ee	1992 (227)
Me	R^2X: $R^2 =$ 3-CF_3-Bn 3-Me-Bn 3-MeO-Bn 3-Cl-Bn 2-Me-Bn 2-Cl-Bn 2-F-Bn	50% aq. NaOH CH_2Cl_2 N-Bn cinchoninium chloride	40–68% ee	1993 (480)
tBu	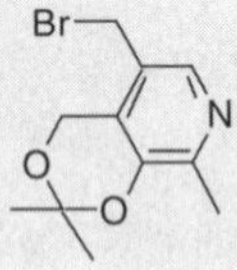 Br—[pyridine acetonide structure]	50% aq. NaOH CH_2Cl_2 N-Bn cinchonidinium chloride	68–74% 52% ee	1994 (481) 1995 (482)
tBu	3,4-$(OMe)_2$-BnBr 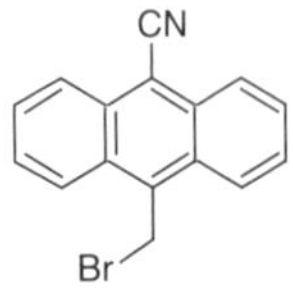 [9-cyano-10-bromomethyl anthracene structure, CN / Br]	50% aq. NaOH CH_2Cl_2 N-Bn cinchonidinium chloride	67–72% 50–53% ee	1996 (483)
tBu	MeI BnBr $BrCH_2CH=CH_2$ $BrCH_2C(Me)=CH_2$ nBuI $BrCH_2$(2-naphthyl) ICH_2CO_2tBu $BrCH_2COPh$	NaOH or KOH H_2O/PhMe or H_2O/CH_2Cl_2 N-(9-anthracenylmethyl) cinchoninium chloride N-(9-anthracenylmethyl) cinchoni- dinium chloride N-(9-anthracenylmethyl) dihydrocin- choninium chloride	40–86% 67–91% ee	1997 (231) 2001 (233)

Table 6.6 Asymmetric Alkylations of Glycine Imine with Chiral Phase-Transfer Catalyst (see Scheme 6.26) (continued)

R^1	Electrophile	Base, Catalyst	Alkylation Yield	Reference
		N-(9-anthracenylmethyl) dihydrocinchonidinium chloride N-(9-anthracenylmethyl) quininium chloride N-(9-anthracenylmethyl) quinidium chloride		
tBu	R^2Br: R^2 = Me Et nBu nOct iBu CH$_2$CH=CH$_2$ CH$_2$C(Me)=CH$_2$ Bn 4-NO$_2$-Bn 4-CN-Bn 4-CF$_3$-Bn 4-F-Bn 4-Ph-Bn 4-Me-Bn CH$_2$(2-naphthyl) CH$_2$CO$_2$tBu	BEMP or BTPP CH$_2$Cl$_2$ O-allyl,N-(9-anthracenylmethyl) cinchoninium bromide O-allyl,N-(9-anthracenylmethyl) cinchonidinium bromide	83–96% 84–97% ee (except CH$_2$CO$_2$tBu = 56% ee)	1998 (251)
tBu	(CH$_2$)$_4$Cl H$_2$C=CHCO$_2$Me H$_2$C=CHCOEt cyclohex-2-enone	CsOH·H$_2$O CH$_2$Cl$_2$ O-allyl,N-(9-anthracenylmethyl) cinchoninium bromide O-allyl,N-(9-anthracenylmethyl) cinchonidinium bromide	85–88% 95–99% ee	1998 (237)
tBu		50% NaOH CH$_2$Cl$_2$ N-Bn cinchonidinium chloride	83% 65% ee	1998 (484)
tBu (silyl ketene acetal)	RCHO: R = iPr iBu cHex nHex (CH$_2$)$_3$Cl (CH$_2$)$_2$Ph	silyl ketene acetal hexane-CH$_2$Cl$_2$ O-Bn,N-(9-anthracenylmethyl) cinchonidinium bifluoride	48–81% 0–86% de syn 70–95% ee	1999 (236)
tBu	BrCH$_2$CH=CHCH$_2$Br 2-(CH$_2$Br)-BnBr 3-(CH$_2$Br)-BnBr	50% aq. KOH PhMe N-(9-anthracenylmethyl) dihydrocinchonidinium bromide	48–52% 70–82% de >95% ee	1999 (234)
tBu		50% aq. KOH PhMe N-(9-anthracenylmethyl) cinchonidinium chloride	63–65% 80% de >95% ee	1999 (235)

(Continued)

Table 6.6 Asymmetric Alkylations of Glycine Imine with Chiral Phase-Transfer Catalyst (see Scheme 6.26) (continued)

R^1	Electrophile	Base, Catalyst	Alkylation Yield	Reference

resin	R^2Br: R^2 =
	 Me
	 Et
	 nOct
	 iBu
	 $CH_2CH=CH_2$
	 $CH_2C(Me)=CH_2$
	 Bn
	 4-Me-Bn
	 4-CN-Bn
	 4-CF_3-Bn
	 4-F-Bn
	 4-Br-Bn
	 4-NO_2-Bn
	 4-Ph-Bn
	 CH_2(2-naphthyl)
	 CH_2CO_2tBu
		BEMP or BTPP
		 CH_2Cl_2
		 O-allyl,N-(9-anthracenylmethyl)
		   cinchoninium bromide
		 O-allyl,N-(9-anthracenylmethyl)
		   cinchonidinium bromide
			72–100%
			 51–89% ee
				1999 (252)

tBu	MeI EtI $BrCH_2CH=CH_2$ $BrCH_2C(Me)=CH_2$ $BrCH_2CCH$ BnBr 4-F-BnBr 4-Me-BnBr $BrCH_2$(1-naphthyl)	50% aq. KOH PhMe binaphthyl-based quaternary ammonium bromide	41–95% 90–96% ee	1999 (270)
tBu	4-BnO-BnBr 3,4-$(BnO)_2$-BnBr	50% aq. KOH PhMe binaphthyl-based quaternary ammonium bromide	81–83% 98% ee	2000 (272)

tBu	$BrCH_2$(8-TBDMSO-
	   quinolin-2-yl)
	 $BrCH_2$(5-Ph-8-TBDMSO-
	   quinolin-2-yl)
		aq. CsOH
		 CH_2Cl_2
		 N-(9-anthracenylmethyl)
		   cinchonidinium bromide
			82% alkylation
			 91% ee
			 93% acidic
			   hydrolysis
				2001 (238)

tBu	MeI EtI $BrCH_2CH=CH_2$ $BrCH_2CH=CH(Ph)$ $BrCH_2CCH$ BnBr 4-F-BnBr 4-CF_3-BnBr $BrCH_2$(2-naphthyl)	1M KOH, Triton X-100 CH_2Cl_2 N-(9-anthracenylmethyl) cinchonidinium bromide	28–92% alkylation 64–85% ee	2001 (249)
tBu	MeI iPrI $BrCH_2CH=CH_2$ BnBr $BrCH_2$(1-naphthyl)	NaOH chiral Cu salen catalyst	12–>95% alkylation 7–81% ee	2001 (311)
tBu	$CH_2=CHCO_2Et$ $CH_2=CHCOMe$ $CH_2=CHCN$	chiral guanidine catalyst	79–98% yield 55–97% ee	2001 (305)

tBu	MeI
	 EtI
	 nHexI
	 $BrCH_2CH=CH_2$
	 $BrCH_2C(Me)=CH_2BrCH_2CH=$
	   $CCHPhBrCH_2CH_2CCH$
		50% aq. KOH
		 PhMe
		 bis(cinchona)-m-xylyl bridged
		   dimer catalyst
			50–98%
			 90–99% ee
				2001 (264)

Table 6.6 Asymmetric Alkylations of Glycine Imine with Chiral Phase-Transfer Catalyst (see Scheme 6.26) (continued)

R¹	Electrophile	Base, Catalyst	Alkylation Yield	Reference
	BnBr			
	4-CF₃-BnBr			
	4-CN-BnBr			
	4-*t*Bu-BnBr			
	BrCH₂(2-naphthyl)			
*t*Bu	MeI	50% aq. KOH	65–95%	2001 (*265*)
	BrCH₂CH=CH₂	PhMe	90–97% ee	
	BrCH₂CCH	trimeric phenyl-bridged		
	BnBr	cinchona catalyst		
	4-F-BnBr			
	4-Me-BnBr			
	4-CF₃-BnBr			
	3-I-BnBr			
	2-NO₂-BnBr			
	BrCH₂(2-naphthyl)			
*t*Bu	AcOCH₂CH=CH₂	Pd(0), (Pdo)₃P	67–94%	2002 (*245*)
	AcOCH₂CH=CHCH₂OAc	10% aq. KOH	85–96% ee	
	AcOCH(Ph)CH=CH₂	PhMe		
		N-(9-anthracenylmethyl)		
		cinchonidinium bromide		
*t*Bu	EtI	50% aq. KOH	65–95%	2002 (*261*)
	BrCH₂CH=CH₂	PhMe	32–90% ee	
	BrCH₂CCH	bis(cinchona)-anthracenyl bridged		
	BnBr	dimer catalyst		
	4-CN-BnBr			
	BrCH₂(1-naphthyl)			
*t*Bu	RCHO: R =	1% aq. NaOH	58–78%	2002 (*274*)
	Me	PhMe	90–98% ee	
	CH₂CH₂Ph	aryl-substituted binaphthyl-based	10–84% de *anti*	
	*n*Hex	quaternary ammonium bromide		
	*c*Hex			
	CH₂OSi(*i*Pr₃)			
	CH₂CH₂CH=CH₂			
*t*Bu	MeI	50% aq. KOH	61–91%	2002 (*283*)
	*i*PrI	PhMe	93–94% ee	
	BrCH₂CH=CH₂	aryl-substituted binaphthyl-		
	BnBr	biaryl-based quaternary		
	BrCH₂(1-naphthyl)	ammonium bromide		
*t*Bu	MeI	KOH (aq.)	61–85%	2002 (*306*)
	OctI	CH₂Cl₂	76–90% ee	
	BrCH₂CH=CH₂	chiral C₂-symmetric pentacyclic		
	BrCH₂CH=CHMe	guanidine catalyst		
	BrCH₂CCH			
	BnBr			
	4-NO₂-BnBr			
	BrCH₂(2-naphthyl)			
*t*Bu	CH₂=CHCO₂*t*Bu	CsOH	73%	2002 (*303*)
		Et₂O	77% ee	
		quaternary ammonium		
		spiropyrrolidine catalyst derived		
		from diethyl tartrate		
*t*Bu	BrCH₂CO(6-NMe₂-2-naphthyl)	40% KOH	83% alkylation	2002 (*485*)
		CH₂Cl₂	87% ee	
		O-allyl,*N*-(9-anthracenylmethyl)	96% hydrolysis	
		cinchonidinium bromide		

(*Continued*)

Table 6.6 Asymmetric Alkylations of Glycine Imine with Chiral Phase-Transfer Catalyst (see Scheme 6.26) (continued)

R^1	Electrophile	Base, Catalyst	Alkylation Yield	Reference
tBu	$(MeO)_2SO_2$ EtI nHexI $BrCH_2CH=CH_2$ $BrCH_2CH=CHPh$ $BrCH_2C(Me)=CH_2$ $BrCH_2CCH$ BnBr 4-F-BnBr 4-CN-BnBr 4-tBu-BnBr $3,4-(BnO)_2$-BnCl $ClCH_2$(anthracenyl)	50% aq. KOH PhMe dimeric naphthyl-bridged cinchona catalyst	70–95% 94–>99% ee	2002 (259)

tBu	(S)-$ICH_2CH_2CH(-O-)CH_2$	CsOH, O-allyl,N-
		    (9-anthracenylmethyl)
		    cinchonidinium bromide
			83%
			>96% ee
				2002 (486)

tBu	$BrCH_2CH=CH_2$
	$BrCH_2C(Me)=CH_2$
	$BrCH_2CH=CHCH_2Br$
	$BrCH_2CCD$
	BnBr
	4-Br-Bn
	$BrCH_2$(2-naphthyl)
		KOD
		D_2O, PhMe
		O-benzyl,N-(9-anthracenylmethyl)
		    cinchonidinium bromide
			70–92% for
			    alkylation and
			    hydrolysis
			86–95% ee
				2002 (242)

tBu	BnBr $BrCH_2C(Me)=CH_2$	KOH PhMe O-R^1,N-(R^2) dihydrocinchonidinium bromide	95% 93–98% ee	2002 (254)
iPr	nBuI BnBr 4-CN-BnBr $BrCH_2$(2-naphthyl)	50% aq. KOH or NaOH PhMe, N-[3,5-di(benzyloxy)benzyl] cinchonidinium bromide	78–98% 72% ee S to 40% ee R	2002 (258)
Et	$CH_2=CHCOMe$ $CH_2=CHCOEt$ $CH_2=CHCO_2Me$ $CH_2=CHCO_2Et$ $CH_2=CHCO_2tBu$ $CH_2=CHCN$	KOtBu CH_2Cl_2 chiral crown ether catalyst	65–80% 46–96% ee	2002 (307)

tBu	$BrCH_2C(Me)=CH_2$
	$BrCH_2CHCH_2$
	$BrCH_2CCH$
	$BrCH_2CH=CHPh$
	$CH_2=CCO_2Bn$
	BnBr
	4-Me-BnBr
	4-tBu-BnBr
	4-Br-BnBr
	4-F-BnBr
	$BrCH_2$(2-naphthyl)
	4-(TIPSO)-BnBr
	$BrCH_2$(4-oxo-cyclohex-1-en-1-yl)
	    (ketal protected)
		CsOH, PhH, CH_2Cl_2
		bis(quaternary ammonium)
		    1,3-dioxolane-linked catalyst
			76–93%
			88–94% ee
				2002 (301)
				2003 (299)
				2004 (300)

tBu	BnBr
	4-Cl-BnBr
	$BrCH_2$(2-naphthyl)
		KOH, N-(9-anthracenylmethyl)
		    cinchonidinium bromide
		    with clay or alumina support
			58–95%
			32–83% ee
				2003 (248)

tBu	nHexI $BrCH_2CH=CH_2$ $BrCH_2CCH$ BnBr 4-NO_2-BnBr $ClCH_2$(1-naphthyl)	50% aq. KOH PhMe dimeric F-Ph bridged cinchona catalyst	81–94% 97–>99% ee	2003 (263)

Table 6.6 Asymmetric Alkylations of Glycine Imine with Chiral Phase-Transfer Catalyst (see Scheme 6.26) (continued)

R¹	Electrophile	Base, Catalyst	Alkylation Yield	Reference
iPr	nBuI BnBr 4-CN-BnBr 3,5-(MeO)$_2$-BnBr BrCH$_2$(2-naphthyl)	50% aq. KOH PhMe dendritic cinchona catalyst	78–95% 44–72% ee	2003 (266)
tBu	BrCH$_2$CH=CH$_2$ BrCH$_2$C(Br)=CH$_2$ BrCH$_2$C(Me)=CH$_2$ BrCH$_2$CCH BnBr BrCH$_2$(2-naphthyl)	KOH, PhMe aryl-substituted biphenyl-based α-methyl-naphthylmethylamine quaternary ammonium bromide	71–100% 89–97% ee	2003 (285)
tBu	BnBr	CsOH CH$_2$Cl$_2$ chiral anthracenyl- cinchonidinium catalyst linked to PEG polymer	best result: 75% yield 64% ee	2003 (267)
tBu	BnBr HexI Ph$_2$CHBr	KOH PhMe chiral anthracenyl- cinchonidinium catalyst linked to PEG polymer	63–84% yield 20–81% ee	2003 (268)
tBu	TBDPSO / quinoline-sulfonamide structure with N, S(O$_2$)–N, and CH$_2$Br substituent (Br)	CsOH, O-allyl,N-(9- anthracenylmethyl) cinchonidinium bromide	70% alkylation, hydrolysis and Fmoc protection 96% ee	2003 (239)
tBu	BrCH$_2$CO(6-NMe$_2$-2-naphthyl)	CsOH, O-allyl,N-(9- anthracenylmethyl) cinchonidinium bromide	33%	2002 (240)
tBu	2-Br-BnBr 2-Br-5-MeO-BnBr 2-Br-5-Cl-BnBr BrCH$_2$(2-Br-pyrid-3-yl)	CsOH, O-allyl,N-(9- anthracenylmethyl) cinchonidinium or cinchonine bromide, intramolecular radical N-arylation to form indoline- 2-carboxylic acids	71–89% alkylation 86–96% ee 45–71% N-arylation	2003 (241)
tBu	BrCH$_2$(7-Br-indol-3-yl)	CsOH·H$_2$O CH$_2$Cl$_2$ O-allyl,N-(9-anthracenylmethyl) cinchonidinium bromide	75% 87% ee	2003 (487)
CO$_2$R¹ = CONEt$_2$	BnBr	CsOH PhMe-CH$_2$Cl$_2$ N-(9-anthracenylmethyl) cinchonidinium bromide	84% yield 80% ee	2003 (247)
tBu	BnBr	KOH, PhMe C3-symmetric tris(2-hydroxy- 3-methylbutane) quaternary ammonium salt	55% 58% ee	2003 (295)
tBu	ArCH=CHOPO$_3$Et$_2$ Ar = Ph, 4-F-Ph, 4-CF$_3$-Ph, 4-Me-Ph, 3-Cl-Ph, 2-naphthyl	alkylation of benzophenone Schiff base of Gly-OtBu or 4-chlorobenzaldehyde Schiff base of Ala-OtBu by allylic phosphonate with Ir catalyst and binaphthol-derived ligand	18–97% 30–98% ee 40–80% de	2003 (322) 2003 (488)
tBu	MeI EtI BrCH$_2$CH=CH$_2$ BrCH$_2$C(Me)=CH$_2$ BrCH$_2$CCH	50% aq. KOH PhMe C2-symmetric binaphthyl-based quaternary ammonium bromide	72–94% 96–99% ee	2003 (271)

(Continued)

Table 6.6 Asymmetric Alkylations of Glycine Imine with Chiral Phase-Transfer Catalyst (see Scheme 6.26) (continued)

R^1	Electrophile	Base, Catalyst	Alkylation Yield	Reference
	BnBr 4-F-BnBr 4-Me-BnBr 2,6-Cl$_2$-BnBr 2,6-Me$_2$-BnBr 4-Ph-BnBr 4-COPh-BnBr 2-Br-BnBr 3,4-(BnO)$_2$-BnBr BrCH$_2$(1-naphthyl)			

tBu	EtI
	BrCH$_2$CH=CH$_2$
	BrCH$_2$CCH
	BnBr
	4-F-BnBr
		50% aq. KOH
		PhMe
		C2-symmetric binaphthyl-based
		    quaternary ammonium bromide,
		    4,4′,6,6′-tetra(3,5-
		    diphenylphenyl)-binaphthyl-substi-
		    tuted ammonium bromide catalyst
			12–87%
			88–97% ee
				2003 (276)

tBu	EtI
	BrCH$_2$CH=CH$_2$
	BrCH$_2$C(Me)=CH$_2$
	BrCH$_2$CCH
	BnBr
	4-F-BnBr
	BrCH$_2$(1-naphthyl)
		50% aq. KOH
		PhMe
		4,4-diaryl-substituted
		    binaphthyl-based quaternary
		    ammonium bromide
			18–93%
			71–93% ee
				2003 (275)

CO$_2$P = CONHCHPh$_2$	BnBr BrCH$_2$CH=CH$_2$ nBuI iPrI cPentI cHexI cHeptI cHexCH$_2$I	50% aq. CsOH mesitylene C2-symmetric binaphthyl-based quaternary ammonium bromide	71–99% 82–98% ee	2003 (289)

Et	TsN=CHR
	R = Ph, 4-MeO-Ph, 2-Br-Ph,
	    3-Cl-Ph, 2-naphthyl, iPr,
	    nBu, cHex, 2-furyl, CO$_2$Et
		Et$_3$N, CuClO$_4$-chiral
		    phosphino-oxazoline
			61–99% addition
			8–>90% de syn
			60–97% ee
				2003 (323)

mono-, di- or
    tripeptide
	EtI
	BnBr
	BrCH$_2$CH=CH$_2$
	BrCH$_2$(2-naphthyl)
	BrCH$_2$(3-benzothiophene)
		50% KOH
		PhMe
		aryl-substituted binaphthyl-based
		    quaternary ammonium
		    bromide
			78–95% yield
			20–98% de
				2003 (279)
				2004 (278)

tBu	BnBr 4-Me-BnBr 2-Me-BnBr MeOCOCH=CH$_2$ EtOCOCH=CH$_2$	CsOH, PhH, CH$_2$Cl$_2$ bis(quaternary ammonium) 1,3-dioxolane-linked catalyst	13–100% 0–11% ee	2004 (302)
tBu	RCHO: R = CH$_2$OBn (CH$_2$)$_2$Ph (CH$_2$)$_2$(4-NO$_2$-Ph) (CH$_2$)$_2$(4-MeO-Ph) (CH$_2$)$_2$CH=CH$_2$ nHex	silyl ketene acetal hexane-CH$_2$Cl$_2$ O-allyl,N-(2,3,4-trifluorobenzyl) cinchonidinium	34–78% 0–9% de syn 6–83% ee	2004 (257)
tBu	RCHO: R = Me iPr CH$_2$CH$_2$Ph	1% aq. NaOH PhMe aryl-substituted binaphthyl-based quaternary ammonium bromide	39–83% 96–99% ee >92% de $anti$	2004 (277)

Table 6.6 Asymmetric Alkylations of Glycine Imine with Chiral Phase-Transfer Catalyst (see Scheme 6.26) (continued)

R^1	Electrophile	Base, Catalyst	Alkylation Yield	Reference
	nPent nHex cHex iBu CH$_2$OSiiPr$_3$ CH$_2$CH$_2$CH=CH$_2$			

tBu	BnBr	50% aq. KOH
		PhMe
		poly(binaphthyl-based quaternary
		    ammonium bromide) based on
		    polyamine backbone
			78%
			83% ee
				2004 (287)

Et	EtI
	BrCH$_2$CH=CH$_2$
	BrCH$_2$CCH
	BnBr
	4-PhCO-BnBr
	BrCH$_2$(2-naphthyl)
		50% aq. KOH
		PhMe
		3,4,5-triflorophenyl-substituted
		    binaphthyl-based quaternary
		    ammonium bromide
			72–99%
			93–98% ee
				2004 (280)

tBu	EtI
	BrCH$_2$CCH
	BnBr
	4-F-BnBr
	4-Me-BnBr
		50% aq. KOH
		PhMe
		fluorous-substituted
		    binaphthyl-based quaternary
		    ammonium bromide
			81–93%
			87–93% ee
				2004 (281)

tBu	nBuI
	BrCH$_2$CH=CH$_2$
	BrCH$_2$CO$_2$tBu
	BnBr
		50% aq. KOH
		PhMe/CHCl$_3$
		bis(cinchona)-anthracenyl
		    bridged dimer catalyst with
		    Br$^-$, BF$_4^-$ or PF$_6^-$ counterion
			54–98%
			44–80% ee with Br$^-$
			74–84% ee with BF$_4^-$
			84–90% ee with PF$_6^-$
				2004 (262)

tBu	3,4-(HO)$_2$-6-[^{18}F]-BnBr	CsOH
		PhMe, 0 °C
		and O-allyl,N-(9-anthracenylmethyl)
		    cinchonidinium bromide chiral
		    catalyst
			>90% radiochemical
			    yield
			96% ee
				2004 (243)

tBu	BnBr
	4-Cl-BnBr
	BrCH$_2$(2-naphthyl)
	BrCH$_2$CH=CH$_2$
	nBuI
	nHexI
		KOH preloaded on kaolin support
		    with 7% water
		PhMe/CHCl$_3$
		dimeric naphthyl-linked O-allyl
		    cinchonidinium bromide
			76–90%
			87–97% ee
				2005 (264)

CO$_2$R^1 = CONHBn,
    CONHCH(Ph)$_2$ or
    CON(OMe)Me
	EtI
	nBuI
	cPentI
	cHexI
	cHeptI
	iPrI
	CH(Me)Bn
	CH(Me)CH$_2$cHex
	CH(Me)CH$_2$(4-Me-Ph)
	CH(nBu)Bn
	BrCH$_2$CH=CH$_2$
	BrCH$_2$CCH
	BnBr
	BrCH$_2$(2-naphthyl)
	BrCH$_2$(1-naphthyl)
		50% KOH or sat.CsOH
		PhMe
		aryl-substituted binaphthyl-based
		    quaternary ammonium bromide
			71–99% yield
			89–98% ee
				2005 (288)

tBu	3-Cl-4-MeO-BnBr	50% aq. KOH PhMe dimeric naphthyl-bridged cinchona catalyst	87% 96% ee	2005 (260)

Table 6.6 Asymmetric Alkylations of Glycine Imine with Chiral Phase-Transfer Catalyst (see Scheme 6.26) (continued)

R^1	Electrophile	Base, Catalyst	Alkylation Yield	Reference

tBu	BocN=CHR
	R = Ph, 4-MeO-Ph, 4-Me-Ph,
	  3-Me-Ph, 2-Me-Ph, 4-F-Ph,
	  4-Cl-Ph, 2-naphthyl,
	  2-thienyl, CH-CHPh
		Cs$_2$CO$_3$, PhF
		(S,S)-TaDiAS
		  (tartrate-derived bis(ammonium)
		  catalyst)
			86–98% addition
			90–98% de syn
			60–82% ee
				2005 (296)

4-Cl-PhCHO imine
  of Ala or Phe
  NHCH(Ph)$_2$
  amide
	nBuI
	cPentI
	BrCH$_2$CH=CH$_2$
		sat.CsOH
		PhMe
		aryl-substituted binaphthyl-based
		  quaternary ammonium bromide
			81–93% yield
			82–93% ee
				2005 (288)

tBu
(also 4-Cl-PhCHO
  imine of
  Ala-OtBu
	BnBr
	BrCH$_2$(2-naphthyl)
	BrCH$_2$CH=CH$_2$
	BrCH$_2$CCH
	EtI
		50% aq. KOH
		PhMe, 0 °C
		aryl-substituted binaphthyl-based
		  quaternary ammonium bromide
		  with 2 alkyl substituents
			67–98% yield
			97–99% ee
				2005 (282)

resin-imine
  of Gly-OtBu
	nHexBr
	BrCH$_2$CH=CH$_2$
	BrCH$_2$C(Me)=CH$_2$
	BrCH$_2$CCH
	BnBr
	4-F-BnBr
	4-CN-BnBr
	4-Me-BnBr
	4-t-Bu-BnBr
	BrCH$_2$(2-naphthyl)
	BrCH$_2$(1-naphthyl)
		50% aq. CsOH
		binaphthyl-based quaternary
		  ammonium bromide or O-allyl,
		  N-(9-anthracenylmethyl)
		  cinchonidinium bromide
		  catalyst
			50–82% yield
			86–>99% ee
				2005 (253)

tBu	MeI
	BrCH$_2$CH=CH$_2$
	BrCH$_2$C(Me)=CH$_2$
	BrCH$_2$CCTMS
	BnBr
		50% aq. KOH
		PhH
		O-allyl,N-(1,8-difluoroanthracenyl-
		  10-methyl) cinchonidinium
		  bromide catalyst
			66–92% yield
			97–98% ee
				2005 (255)

tBu	RCH(OAc)C(=CH$_2$)CO$_2$Me
	  to give CH$_2$C(CO$_2$H)=CHR
	  side chain
	R = Ph, 4-MeO-Ph, 4-NO$_2$-Ph,
	  2-thienyl, 2-pyridyl,
	  2,6-F$_2$-Ph, nPr, tBu
		CsOH
		CH$_2$Cl$_2$
		O-allyl,N-(1,8-difluoroanthracenyl-
		  10-methyl) cinchonidinium
		  bromide catalyst
			63–92% yield
			80–97% ee
				2005 (246)

CHPh$_2$	CH$_2$=CHCOMe CH$_2$=CHCOEt CH$_2$=CHCOnPent	50 mol% Cs$_2$CO$_3$ iPr$_2$O biphenyl naphthylmethyl ammonium bromide catalyst	60–94% yield 91–94% ee	2005 (286)

CO$_2$R^1 = CONH
  (2,6-Me$_2$-Ph)
	Cl(CH$_2$)$_4$I	50% aq. CsOH
		binaphthyl-based quaternary
		  ammonium bromide
			85% yield
			96% ee
				2005 (290)

tBu	EtI
	nHexBr
	BrCH$_2$CH=CH$_2$
	BrCH$_2$CCH
	BnBr
	4-F-BnBr
	BrCH$_2$(3-pyridyl)
	BrCH$_2$(4-pyridyl)
	BrCH$_2$(1-naphthyl)
		50% aq. KOH
		O-allyl,N-(9-triarylcarbinol-
		  methyl) cinchonidinium
		  bromide catalyst
			68–93%
			89–94% ee
				2005 (256)

tBu	EtI
	BrCH$_2$CH=CH$_2$
	BrCH$_2$CCH
	BnBr
	4-Br-BnBr
		50% aq. KOH
		PhH
		4,4′,5,5′,6,6′-(MeO)$_6$-3,3′-
		  (3,4,5-F$_3$-Ph)$_2$-biphenyl
		  based quaternary ammonium
		  bromide
			80–99%
			94–98% ee
				2005 (284)

Table 6.6 Asymmetric Alkylations of Glycine Imine with Chiral Phase-Transfer Catalyst (see Scheme 6.26) (continued)

R¹	Electrophile	Base, Catalyst	Alkylation Yield	Reference
oxazoline derivative of Ser (α-CH₂OH substituent)	EtI BrCH₂CH=CH₂ BrCH₂C(Me)=CH₂ BrCH₂CCH BnBr 4-F-BnBr 4-CN-BnBr 4-MeO-BnBr 4-CF₃-BnBr 4-tBu-BnBr BrCH₂(2-naphthyl)	KOH PhH aryl-substituted binaphthyl-based quaternary ammonium bromide	48–99% yield 93–>99% ee	2004 (291)

oxazoline derivative of Ser (α-CH₂OH substituent)	BrCH₂CH=CH₂
	BnBr
	4-F-BnBr
	3-F-BnBr
	2-F-BnBr
	4-tBu-BnBr
	BrCH₂(2-naphthyl)
		CsOH
		CH₂Cl₂
		aryl-substituted binaphthyl-
		    based quaternary ammonium
		    bromide
			75–90% yield
			90–96% ee
				2005 (292)

tBu	EtI nHexI BrCH₂CH=CH₂ BnBr 4-F-BnBr BrCH₂(3-pyridyl) BrCH₂(1-naphthyl)	KOH PhH L-menthol-derived quaternary ammonium bromide	66–92% yield 62–72% ee	2005 (294)

tBu	MeI
	BrCH₂CH=CH₂
	BnBr
	2-Me-BnBr
	4-Me-BnBr
	BrCH₂(-naphthyl)
		CsOH
		CH₂Cl₂
		bis(quaternary ammonium salt)
		    based on 2,5-dimethylpyrroline
		    rings connected by tartaric
		    acid-derived backbone
			50–90% yield
			0–30% ee
				2005 (297)

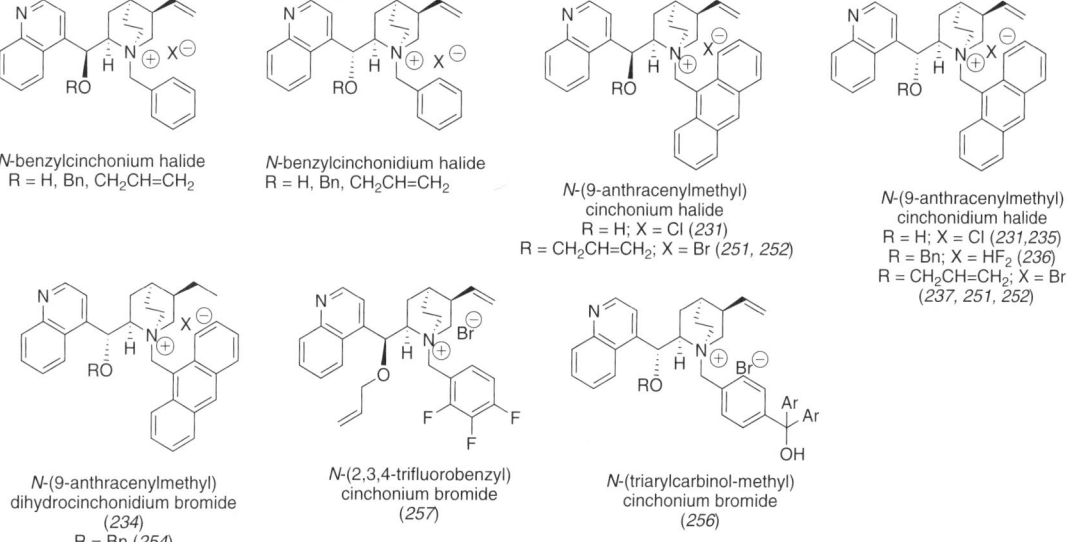

N-benzylcinchonium halide
R = H, Bn, CH₂CH=CH₂

N-benzylcinchonidium halide
R = H, Bn, CH₂CH=CH₂

N-(9-anthracenylmethyl)
cinchonium halide
R = H; X = Cl (231)
R = CH₂CH=CH₂; X = Br (251, 252)

N-(9-anthracenylmethyl)
cinchonidium halide
R = H; X = Cl (231,235)
R = Bn; X = HF₂ (236)
R = CH₂CH=CH₂; X = Br
(237, 251, 252)

N-(9-anthracenylmethyl)
dihydrocinchonidium bromide
(234)
R = Bn (254)

N-(2,3,4-trifluorobenzyl)
cinchonium bromide
(257)

N-(triarylcarbinol-methyl)
cinchonium bromide
(256)

Scheme 6.27 *Cinchona* Alkaloid Chiral Phase Transfer Catalysts.

For alkylation of the Schiff base of Gly, a *tert*-butyl ester gave the best results compared to other ester groups, resulting in monoalkylated products with 42–66% ee in 60–82% yield (*223*). α-Methyl amino acids were prepared with up to 87% ee by alkylation of aldimine Schiff bases of Ala-O*t*Bu with electrophiles other than methyl halide (*227, 228*). However, the incomplete enantioselectivities provided by this method mean that resolution by crystallization is required to achieve high enantiomeric excesses, since diastereomeric chiral auxiliaries are not available for chromatographic resolution. For those compounds that do not crystallize, an alternative resolution method must be found (normally enzymatic resolution), so that most of the advantages of an asymmetric alkylation are lost. Partially enantiomerically enriched amino acid Schiff base ester products give variable results during recrystallization attempts to obtain optically pure product. Fortunately, they can be readily converted into *N*-Fmoc *tert*-butyl ester derivatives, which have been identified as consistently providing higher enantiomeric purity upon recrystallization; a series of 10 compounds with initial 63–92% ee (average 76% ee) gained an average of 13.6% ee upon the first recrystallization (*229*).

Modified cinchona alkaloid catalysts have been developed. A 1999 report showed that variations in both the quinoline group sidearm and quinuclidine core greatly affected the enantioselectivity of alkylations (*230*). A new class of catalysts has been obtained by quaternization of the cinchona alkaloids with 9-chloromethylanthracene instead of benzyl chloride (see Scheme 6.27), either with or without *O*-alkylation. These gave significantly better enantioselectivities (67–91% ee) for alkylations of glycine ketimines compared to the *N*-benzyl catalysts (*231–233*); structure–selectivity studies on the catalysts showed that both the 1-quinolyl and anthracenylmethyl groups play a key role in enantioselectivity (*232*). The dihydrocinchonidium version of this catalyst was used in conjunction with bis(electrophiles) to construct aryl-bridged diaminodicarboxylic acids, giving products with 70–82% de and >95% ee (*234*). Dityrosine and isodityrosine were prepared using the appropriate biphenyl-bridged bis(benzyl bromide) electrophiles and the unsaturated catalyst (*235*). Similar anthracenyl catalysts were also applied to aldol reactions, producing predominantly *syn* (90–86% de) β-hydroxy amino acids with 72–95% ee (*236*). Michael additions were catalyzed with 91–99% ee (*237*).

The *O*-allyl-*N*-(9-anthracenylmethyl)cinchonidinium bromide catalyst has become a standard reagent (available from Aldrich) that, along with closely related analogs, has been applied to a number of amino acid syntheses (see Table 6.6). The anthracenyl catalyst was combined with CsOH as base to induce 96% ee in a synthesis of metal-binding fluorescent amino acids β-(8-hydroxyquinolin-2-yl)-Ala (*238*), β-(5-phenyl-8-hydroxyquinolin-2-yl)-Ala (*238*), and β-[8-hydroxy-5-(*N,N*-dimethylsulfonamido)-quinolin-2-yl]-Ala (*239*), and for a synthesis of the fluorescent amino acid 4-keto-4-(6-dimethylamino-6-naphthyl)-2-aminobutyric acid (*240*), with 91–96% ee. Another synthesis followed alkylation with *o*-bromobenzylbromides by an intramolecular radical aryl amination reaction to generate indoline-2-carboxylic acid derivatives (*241*). By using KOD in D$_2$O for alkylations of the benzophenone Schiff base of Gly-O*t*Bu with the *O*-benzyl,*N*-(9-anthracenylmethyl) cinchonidinium bromide catalyst, both the side chain and an α-deuterium atom were introduced. Seven [2-^{2}H] amino acids were prepared in 70–92% yield, with 86–96% ee (*242*). The rapidity and relatively mild conditions of the PTC Schiff base alkylation route were employed for a preparation of 6'-[^{18}F]fluoro-L-DOPA, an important radiopharmaceutical for positron emission tomography (PET). Speed is important, owing to the 109 min half-life of fluorine-18. With conditions optimized for a no-carrier added synthesis (only a few micrograms of alkylating reagent), the anthracenyl catalyst induced up to 96% ee using CsOH in toluene at 0 °C, with alkylation complete after 5 min (*243*). Care must be taken when selecting the electrophile, as in one reaction an allylic halide (bromomethylcyclooctatetraene) provided racemic product, possibly via electron transfer followed by addition via a radical process (*244*).

Pd-catalyzed allylations of the benzophenone Schiff base of Gly-O*t*Bu or the 4-chlorobenzaldehyde imine of Ala-O*t*Bu with allylacetates in the presence of the anthracenyl PTC catalyst gave allylglycines with 83–96% ee (*245*). The anthracenyl catalyst allowed for addition/elimination reactions with RCH(OAc)C(=CH$_2$) CO$_2$Me substrates, producing 4-alkylidenyl-Glu products with 80–97% ee in 63–92% yield (*246*). The anthacenyl cinchona catalyst also allowed for benzylations of various amide derivatives of Gly benzophenone Schiff base; the best enantioselectivity (80% ee) was obtained with a diethyl amide (*247*). Using the catalyst in the presence of alumina or clay provided alkylated products in 58–95% yield, but with only 34–83% ee (*248*). A micellar system with Triton X-100 and 1 M KOH has also been developed for use with the anthracenyl catalyst, giving products with 64–85% ee (*249*). An *O*-H anthracenyl catalyst was found to catalyze the alkylation of the benzophenone Schiff base of Gly-O*t*Bu in aqueous media without any organic solvent or with only 1% toluene, giving 81–97% yields with 82–92% ee. The reaction could be applied to multi-gram reactions (*250*).

An efficient homogeneous catalytic enantioselective system was described in 1998, using neutral non-ionic phosphazene bases (BEMP or BTPP) in CH$_2$Cl$_2$, the anthracenyl cinchona alkaloid chiral quaternary ammonium salts, and the benzophenone Schiff base of Gly-O*t*Bu. Alkyl, benzyl, and allyl bromides reacted in 83–95% yield with 83–97% ee, with the exception of *tert*-butyl bromoacetate (56% ee) (*251*). The anthracenyl PTC catalysts have also been combined with BEMP or BTPP bases for solid-phase syntheses of amino acids, using an ester linkage to the resin. However, enantioselectivities were not as impressive (51–89% ee) (*252*). Another group postulated that the resin–ester linkage was responsible for the reduced enantioselectivity compared to the normal *tert*-butyl ester. Linking Gly-OtBu to a resin via an aldimine linker provided a substrate that could be alkylated with 86 to >99% ee using 50% aqueous CsOH and the anthracenyl *Cinchona* catalyst (*253*).

Other catalysts based on cinchona alkaloids have been developed. A combinatorial approach towards optimizing the *O*- and *N*-substituents on *O*-alkyl, *N*-alkyl dihydrocinchonidium salts was described in 2002. Dihydrocinchonidine was first reacted with 1.2–1.5 equiv of the *N*-alkylating reagent (e.g., 9-bromomethylanthracene) followed by the *O*-alkylating reagent (in toluene at 60–75 °C for 5 h). In the same pot, the Gly Schiff base and alkylating electrophile were then added with 9M aq. KOH, with reaction at 25 °C for 18 h producing the alkylated product. Both *N*- and *O*-substituents were found to play a role in enantioselectivity, with an *O*-benzyl,*N*-anthracenylmethyl catalyst optimum for benzylation of the benzophenone Schiff base of Gly-OtBu, producing the Phe derivative obtained with 93% ee. The enantioselectivity was similar to that obtained with preformed catalyst, illustrating the validity of the approach. The optimized catalyst also induced 98% ee for an allylation reaction (*254*).

Substitution of the anthracenyl group with fluorine atoms in an *O*-allyl,*N*-(1,8-difluoroanthracenyl-10-methyl) cinchonidinium bromide catalyst allowed for PTC alkylations with methyl iodide, allyl, methallyl, or 3-TMS-propargyl bromide, or benzyl bromide, with 97–98% ee (*255*). A series of catalysts reported in 2005 replaced the anthracenyl group with several different triarylcarbinols; the best alkylated the benzophenone Schiff base of Gly-OtBu with a number of halides with 89–94% ee (*256*). Aldol reactions catalyzed by a trifluorobenzyl-substituted *Cinchona* alkaloid-based catalyst produced *syn* β-hydroxy-α-amino acids from alkyl aldehydes with 34–78% yields, 0–9% de *syn* stereoselectivity, but only 6–60% ee for *anti* products and 50–83% ee for *syn* products (*257*). An *N*-[3,5-di(benzyloxy)benzyl]cinchonidium catalyst had the unusual property of producing enantiomeric products depending on the base and electrophile employed: KOH gave the (*S*)-products with 44–66% ee, while NaOH gave (*R*)-Phe from benzyl bromide with 40% ee, but other products were (*S*)-enantiomers with 0–32% ee (*258*).

Instead of the anthracenyl unit, another cinchona-based catalyst employed a dimeric cinchona unit bridged with a 2,7-(dimethyl)-naphthyl moiety. Thirteen electrophiles alkylated Gly with 94–>99% ee (*259*). This catalyst was employed for a synthesis of *O*-methyl 3′-chloro-Tyr, with 96% ee (*260*). A dimethylanthracenyl-bridged cinchona catalyst gave 32–90% ee for a range of electrophiles (*261*). Altering the counterion from chloride or bromide to tetrafluoroborate or hexafluorophosphate improved the enantioselectivity of this dimeric catalyst significantly, with four electrophiles reacting with 44–80% ee with Br⁻, 74–84% ee with BF₄⁻, and 84–90% ee for PF₆⁻ (*262*). It would be interesting to see if similar improvements could be obtained with the monomeric catalysts. A dimeric catalyst bridged

by a fluoro-substituted phenyl ring induced 97 to >99% ee for a series of alkylations (*263*), while *o*-, *m*-, and *p*-xylyl-bridged catalysts induced 90–97% ee for 12 different alkyl, benzyl, and allyl halides (*264*). A trimeric cinchona catalyst linked by a phenyl ring has also been developed, with 90–97% ee for a series of alkylations (*265*). A cinchonidine catalyst was attached to first-, second-, and third-generation dendritic wedges; four electrophiles alkylated (diphenyliminomethylene)Gly-OiPr with 44–72% ee. Some preliminary results were presented with regards to recovering the catalyst by dialysis (*266*). Two groups have linked anthracenyl cinchona catalysts to poly(ethylene glycol) polymers by various attachment points. One group obtained only 58% ee for benzylations of the benzophenone Schiff base of Gly-OtBu (*267*), while the other group reached a maximum of 81% ee for benzylation but only 20–34% for hexyl iodide and diphenylmethylbromide (*268*). The use of KOH base preloaded on solid supports (kaolin, aluminum oxide, Montmorillonite K-10, Celite) was investigated in 2004, in combination with various catalyst types. The supported base, preferentially kaolin/KOH containing 7% water, allowed for alkylations in minimal volumes of toluene/chloroform with only 2 mol % catalyst, proceeding to completion in 5–15 min at 20 °C with good enantioselectivities (87–97% ee using a dimeric naphthyl-linked *O*-allyl cinchonidinium bromide). The solid support allowed for easy product isolation (*269*).

Other types of chiral phase-transfer catalysts have been employed (see Scheme 6.28). Several C₂-symmetric PTC catalysts were designed, based on binaphthyl quaternary ammonium salts. These induced enantioselectivities of 90–96% ee for a number of electrophiles (*270*), with improved results of 96–99% ee for 16 different electrophiles in a subsequent report (*271*). A synthesis of L-DOPA on a 5 g scale produced the *tert*-butyl ester with 98% ee in two steps with 75% overall yield (*272*). Sequential one-pot alkylations of the *p*-chlorobenzaldehyde imine of Gly-OtBu with these catalysts gave α,α-dialkyl-α-amino acids with 91–99% ee in 58–85% yield (*273*). Aldol reactions proceeded with 90–96% ee and *anti* diastereoselectivity (*274*). With 4,4′-aryl substituents on one binaphthyl group, enantioselectivities of 71–96% ee were obtained (*275*). A symmetrical 4,4′,6,6′-tetraarylbinaphthyl-substituted ammonium bromide catalyst (aryl group = 3,5-diphenylphenyl) gave improved enantioselectivities, with 93–97% ee for four electrophiles (*276*). Similarly, bulky aryl substituents on the binaphthyl catalyst induced 96–99% ee and >92% *anti* stereoselectivity during aldol reactions of the benzophenone imine of Gly-OtBu with a wide range of aldehydes, producing *anti*-β-hydroxy-α-amino esters (*277*). The same catalyst was employed to alkylate *N*-terminal glycine Schiff base residues in di-, tri- or tetrapeptides. The diastereoselectivity depended on both catalyst and peptide sequence. An Ala residue was also alkylated (*278, 279*). Another binaphthyl-based quaternary ammonium catalyst, with two trifluorophenyl substituents, was found to be effective at alkylating methyl or ethyl esters of the benzophenone Schiff base of Gly,

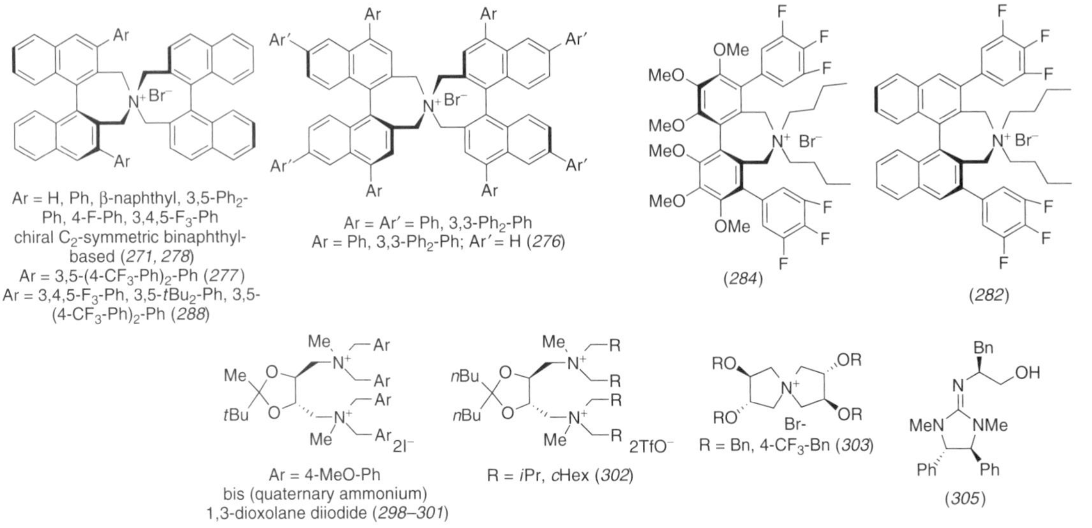

Ar = H, Ph, β-naphthyl, 3,5-Ph₂-
Ph, 4-F-Ph, 3,4,5-F₃-Ph
chiral C₂-symmetric binaphthyl-
based (271, 278)
Ar = 3,5-(4-CF₃-Ph)₂-Ph (277)
Ar = 3,4,5-F₃-Ph, 3,5-tBu₂-Ph, 3,5-
(4-CF₃-Ph)₂-Ph (288)

Ar = Ar' = Ph, 3,3-Ph₂-Ph
Ar = Ph, 3,3-Ph₂-Ph; Ar' = H (276)

(284)

(282)

Ar = 4-MeO-Ph
bis (quaternary ammonium)
1,3-dioxolane diiodide (298–301)

R = iPr, cHex (302)

R = Bn, 4-CF₃-Bn (303)

(305)

Scheme 6.28 Binaphthyl-Based Quaternary Ammonium and Other Chiral Phase Transfer Catalysts.

rather than the usual *tert*-butyl esters. Six alkyl, benzyl or allyl halides reacted with 93–98% ee. The *p*-chlorobenzaldehyde imine of Ala-OEt was also alkylated, with 82–98% ee (280).

A fluorous version of the binaphthyl catalyst has been developed, allowing for easy extractive recovery of the catalyst; enantioselectivity was retained over at least three cycles (281). Improved reactivity was obtained by replacing one side of the bis(binaphthyl) quaternary ammonium salt with more flexible alkyl substituents, giving a catalyst that still resulted in 97–99% ee for a number of alkylations of the benzophenone Schiff base of Gly-OtBu, or the *p*-chlorobenzaldehyde Schiff base of Ala-OtBu (282). A mixed biaryl–binaphthyl catalyst has also been described, giving 93–94% ee for a number of alkylations (283). A quaternary ammonium catalyst based on a 3,3′-(3,4,5-trifluorophenyl)-4,4′,5,5′,6,6′-hexamethoxybiphenyl chiral moiety induced 94–98% ee for alkylations of the benzophenone Schiff base of Gly-OtBu, using as little as 0.01 mol % of catalyst. Benzylation of the *p*-chlorobenzaldehyde Schiff base of DL-Ala-OtBu proceeded with 99% ee (284). A biphenyl naphthylmethyl quaternary ammonium catalyst provided 89–97% ee for allylations and benzylations (285), and 91–94% ee for Michael additions of the benzophenone Schiff base of Gly-OCHPh₂ to α,β-unsaturated ketones (286). The solvent (di-isopropylether) was critical for enantioselectivity, as was the diphenylmethyl ester group. *tert*-Butyl acrylate or *N,N*-dimethyl acrylamide gave <10% yield (286). The amino groups of several different linear and cyclic polyamines were reacted with 2,2′bis(bromomethyl)-binaphthalene to give poly(quaternary ammonium) catalysts; the best of these benzylated the benzophenone Schiff base of Gly-OtBu with 83% ee (287).

A 2005 report applied several of these binaphthyl catalysts to the alkylation of Gly benzophenone Schiff base amides (benzyl, diphenylmethyl, and Weinreb), with 92–98% ee for benzyl bromide, allyl bromide, or *n*-butyl iodide. The catalyst/substrate combination possessed high reactivity, allowing for alkylation with less reactive secondary alkyl halides (71–91% yield and 89–96% ee for isopropyl, cyclopentyl, cyclohexyl, and cycloheptyl iodides). Surprisingly, the catalyst also provided recognition of β-branched primary halides to give double stereodifferentiation. A matched pair of catalyst with 1-iodo-2-phenylpropane gave only the (2R,4R)-isomer in 89% yield. The mismatched pair reacted much more slowly and gave 31% of (2R,4S), 2.8% of (2S,4S), and 4.2% of (2R,4R) product. Kinetic dynamic resolutions using 1-iodo-2-(*p*-methyl-phenyl)propane, 1-iodo-2-phenylhexane, or 1-iodo-2-cyclohexylpropane to give predominantly the (2R,4S)-diastereomers were also reported. Alkylation of 4-chlorobenzaldehyde imines of Ala and Phe amides was also possible (288). The benzophenone imine of Gly diphenylmethylamide was alkylated in an earlier report, with 89–98% ee for a variety of primary and secondary alkyl iodides (289). Synthesis of a pipecolic acid amide via alkylation of the benzophenone Schiff base of Gly-NH(2,6-Me₂-Ph) with 4-chloro-iodobutane proceeded with only 64% ee using a cinchona catalyst, but with up to 96% ee with a binaphthyl catalyst (290). The binaphthyl catalysts also provided good enantioselectivity for alkylation of Gly-OtBu attached to a resin via an aldimine linker (253). The oxazoline derived from Ser-OtBu and benzaldehyde is a Ser Schiff base analog that can be alkylated under PTC conditions. Various chiral catalysts were tested, with a quaternary bis(binaphthyl)ammonium salt inducing 93 to >99% ee. Acidic hydrolysis provided α-alkylated-Ser products, with 11 different alkyl halides tested (*291*). Seven different α-substituted serines were prepared by this method in 2005, with 90–96% ee (*292*).

Other chiral quaternary ammonium salt catalysts include a different type of binaphthyl catalyst, 6,6′-di(triethylammoniummethyl)-2,2′-di(hexyloxy)-1,1′-binaphthalene, which showed catalytic activity but induced no enantioselectivity (*293*). An L-menthol-derived quaternary ammonium catalyst provided 62–72% ee for alkylations with seven different alkyl, allyl, and benzyl electrophiles (*294*). C_3-symmetric tris(2-hydroxy-3-methylbutane) quaternary ammonium salt induced 58% ee for benzylation (*295*). A tartrate-derived diammonium salt, TaDiAS, induced moderate enantioselectivity (58–82% ee) and high *syn* diastereoselectivity (90–98% de) for additions of the benzophenone imine of Gly-O*t*Bu to *N*-Boc imines of aryl aldehydes, producing 3-aryl-2,3-diamino acid products (*296*). However, another catalyst based on a chiral tartrate-derived backbone connecting two chiral 2,5-dimethyl-pyrroline quaternary ammonium groups produced only 0–30% ee (*297*). Another tartrate-derived catalyst with two quaternary ammonium groups connected by a chiral 1,2-dioxalane moiety (see Scheme 6.28) induced phase-transfer alkylations and Michael additions of the benzophenone imine of Gly-O*t*Bu or the 4-chlorobenzaldehyde imine of Ala-O*t*Bu with high enantioselectivity (88–94% ee) and yield (71–93%) (*298–301*). The catalyst was readily recovered by extraction, and could be reused with no loss in enantioselectivity. The reaction conditions were subsequently applied to the synthesis of three amino acids contained in Aeruginosin 298-A (*299, 300*). A similar catalyst prepared by another group induced only 0–11% de for a series of alkylations (*302*). A quaternary ammonium spiropyrrolidine tartrate-derived ligand, with 4-trifluoromethyl-benzyloxy substituents, induced up to 77% ee for conjugate additions to *tert*-butyl acrylate (*303*).

Non-quaternary ammonium catalysts include a chiral guanidine catalyst that was examined for induction of asymmetry during Michael additions to acrylates, but a maximum of 30% ee was obtained (*304*). A subsequent report identified a guanidine catalyst that induced up to 96% ee for addition of the benzophenone Schiff base of Gly-O*t*Bu to ethyl acrylate, 96% ee for addition to methyl vinyl ketone, and 55% ee for acrylonitrile (see Scheme 6.28) (*305*). A C_2-symmetric pentacyclic guanidine catalyst induced 76–90% ee for alkylations of the benzophenone Schiff base of Gly-O*t*Bu with nine different halides (*306*). A chiral crown ether derived from L-quebrachitol induced 46–96% ee for Michael additions of the benzophenone Schiff base of Gly-OEt to methyl vinyl ketone, ethyl vinyl ketone, acrylonitrile, and several acrylate derivatives (*307*). Enantiopure 2-hydroxy-2′-amino-1,1′-binaphthyl (NOBIN) induced up to 68% ee in PTC alkylations of aldimine Schiff bases of Ala esters with BnBr (*136, 308*), while tartaric acid-derived TADDOL provided up to 93% ee for benzyl bromide, 73% ee for allyl bromide, but only 12% ee for isopropyl bromide (*136, 309*). A number of catalysts based on rigid bicyclic natural products isomannide and isosorbide induced modest asymmetry (maximum of 48% ee) during benzylation of the benzophenone Schiff base of Gly-O*t*Bu (*310*).

A Cu–salen metal complex based on a bis(Schiff base) of 1,2-diaminocyclohexane induced up to 90% ee for allylation of the benzaldehyde Schiff base of Ala-O*i*Pr or 81% ee for allylation of the benzophenone Schiff base of Gly-O*t*Bu, but other electrophiles reacted with less selectivity (e.g., only 7–9% ee for alkylation of Gly with MeI or *i*PrI) (*311*). However, a series of α-substituted-Ala products were produced with 43–86% ee by alkylation of the benzaldehyde Schiff base of Ala-OMe (*312*). Alkylations of the 4-chlorobenzaldehyde Schiff bases of other amino acids had reduced enantioselectivity with increasing side-chain size (*313*). A mechanistic study on the alkylation of the 4-chlorobenzaldehyde Schiff base of Ala-OMe with various substituted benzyl halides in the presence of this Cu(II) salen catalyst indicated that the reaction proceeds via an S_N2 pathway with a build up of negative charge on the benzylic carbon (*314*).

Palladium complexes with a number of chiral phosphine ligands catalyzed the allylation of the benzophenone ketimine of *tert*-butyl or methyl glycinate, with moderate enantioselectivities (10–97% ee) (*315–318*). High diastereomeric excesses (90–99% ee in final product) were obtained by using the palladium chiral complex in combination with Oppolzer's *N,N*-cyclohexyl-sulfamoylisobornyl chiral ester auxiliary (*319*). Several attempted one-pot alkylation/cyclizations of various chiral or achiral imines of aminoacetonitrile with 1,4-dichlorobut-2-ene using achiral or chiral Pd catalysts resulted in 1-amino-2-cyclopropylcyclopropane-1-carboxylic acids with minimal enantio-/diastereoselectivity (2–32%) (*320*). Chiral Co(II) and Mn (II) catalysts induced diastereoselectivity during the cycloaddition of imines of Gly-OMe with methyl acrylate, producing Pro derivatives with up to 96% ee (*321*). An Ir-catalyzed allylic alkylation of the benzophenone Schiff base of Gly-O*t*Bu or 4-chlorobenzaldehyde Schiff base of Ala-O*t*Bu using a chiral bidentate binaphthol-derived ligand and allylic phosphates proceeded with up to 97% ee and 62% de for products with a β-substituent (*322*). Additions of the benzophenone Schiff base of Gly-OMe to *N*-tosyl imines, catalyzed by a chiral phosphino-oxazoline ligand and CuClO$_4$, produced 2,3-diamino acids with 8 to >90% *syn* diastereoselectivity and 60–97% ee (*323*).

Chiral catalysts have also been applied to alkylations of Gly Schiff base Ni(II) complexes (e.g., *298, 324–326*), but these are discussed in Section 7.11 on cyclic metal complexes.

The benzaldehyde Schiff base of L- or D-Trp-OMe can reportedly be deprotonated with LDA and alkylated with [^{11}C]MeI to give α-[^{11}C]methyl-L- or D-Trp with retention of configuration with no chiral auxiliary employed; a non-planar enolate intermediate is proposed to provide stereocontrol during the alkylation. Chiral HPLC analysis was provided to demonstrate enantiomeric purity of the product, but the reference cited to provide precedent for this asymmetric synthesis in fact only descibes racemic syntheses (*327*). However, other examples of asymmetry resulting from alkylations of deprotonated amino acids have been reported (see Section 6.2.4). Schiff bases of racemic amino acids have been employed for deracemization reactions via enantioselective protonation of enolates (see Section 12.1.7) (*328–332*).

6.4.5 Chiral Electrophiles

Several syntheses have relied on chiral electrophiles to induce asymmetry at the α-center during alkylations of glycine Schiff bases. The benzophenone Schiff base of Gly-O*t*Bu was alkylated with the triflate of 2-hydroxy-4-methylpentanoate, which was derived from D-Leu. Alkylation proceeded with 50% de *anti* stereoselectivity, producing β-isobutyl-Asp (*333*). The benzophenone Schiff bases of Gly-OEt or Ala-OEt were allylated with homochiral 2-acetoxy-4-aryl-3-butenes (allylic acetates) in the presence of $Pd_2(dba)_3$ to give 3-methyl-5-arylpent-4-enoic acids with 92 to >98% de at the α-center (*334*). Optically pure 2,2′-bis(bromomethyl)-1,1′-binaphthyl was used for a PTC bis-alkylation of the 4-chlorobenzaldehyde Schiff base of Gly-O*t*Bu, producing a cyclic 1,1′-binaphthyl-substituted α-aminoisobutyric acid (*335*); a crown-ether substituted binaphthyl analog was prepared by the same route (*489*).

Chiral auxiliaries can be employed on the electrophile. An electrophilic reagent possessing a variety of different chiral sulfonate leaving groups was used to alkylate a glycine or phenylalanine Schiff base (see Scheme 6.29). The best results, though generally poor, were obtained with sulfates derived from D-glucose (0–61% ee) (*328*). β-Substituted-Glu derivatives were obtained via addition of the benzophenone Schiff base of Gly-OR to chromium alkoxyalkenyl-carbene complexes with an (−)-8-phenylmenthyloxy auxiliary. *Anti* diastereomers of β-phenyl-Glu or β-(2- or 3-furyl)-Glu were produced in 65–89% yield with 92–94% de *anti*, and only two major isomers apparent. The *syn* isomers were obtained if *N,N*-dibenzyl Gly-OEt was employed as the nucleophile. Oxidation to cleave the chromium complex and imine/ester deprotection was required to generate the free amino acid (*75*). An achiral ketimine Ni(II) complex formed from Gly and the 2-acetyl-phenyl amide of pyridine-2-carboxylic acid added to acrylates derivatized with a chiral oxazolidinone auxiliary with nearly complete stereoselectivity at both centers (>94% de for 19 examples) (*336, 337*). The adducts were elaborated into β-substituted pyroglutamic acids and prolines. The reaction could readily be scaled up, with (2S,3S)-3-methyl-pyroglutamate prepared on a 10 g scale (*336*). A convenient procedure to acylate the oxazolidinones with enoyl chlorides was also reported (*338*).

More complex reactions have been reported. Methyl benzylideneglycinate was dialkylated with a chiral cyclic sulfate derived from (S)-1,2-propanediol in the presence of 2 equiv of sodium hydride to give a single diastereomer of the imine of allonorcoronamic acid (1-amino-2-methylcyclopropane-1-carboxylic acid) in nearly quantitative yield (*339*) (see Scheme 6.30). The benzophenone Schiff base of Gly-OEt underwent Michael addition to an menthol ester of 4-bromocrotonate, with a simultaneous cyclization producing a single diastereomer of *trans*-substituted 2-(2-carboxycyclopropyl)glycine (*340*). Similarly, Evan's oxazolidinone chiral auxiliary was used in the side chain of an allylic halide, which acted as a Michael acceptor for the methyl ester of the benzophenone Schiff base of GlyOMe; the adduct then cyclized to form a cyclopropane side chain with >95% de (*341*).

Menthyl acrylate has been used as a homochiral dipolarophile in a Ti(IV)-catalyzed cycloaddition reaction

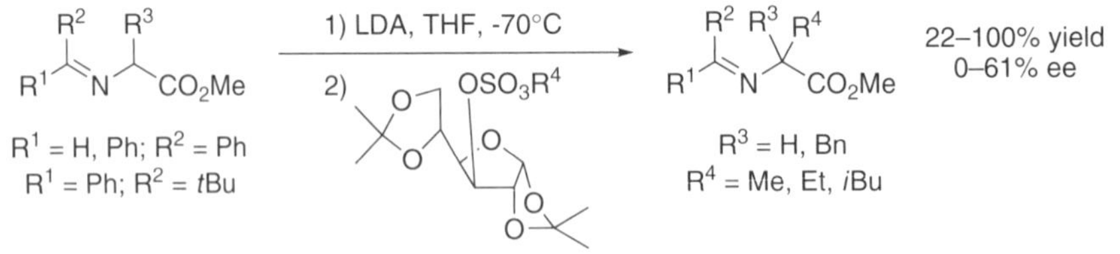

Scheme 6.29 Chiral Leaving Group on Alkylation Reagent (*328*).

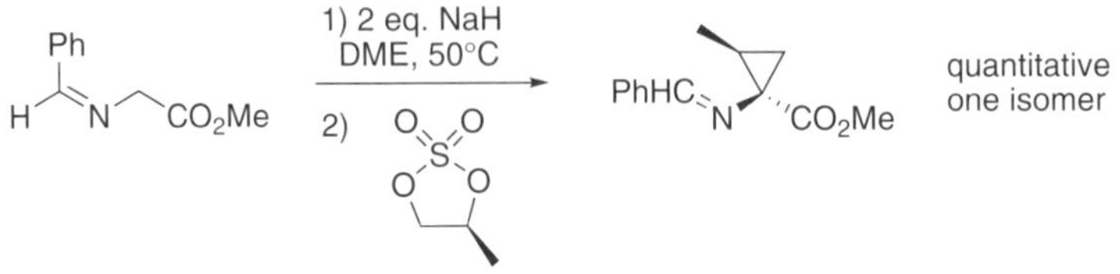

Scheme 6.30 Chiral Electrophile (*339*).

with α-amino ester imines, forming proline derivatives with complete asymmetric induction (*342*).

6.5 Asymmetric Alkylations of Electrophilic Glycine Equivalents

Most electrophilic Gly equivalents employed for asymmetric syntheses are based on one of the cyclic chiral templates that will be discussed in Chapter 8. However, several asymmetric reactions have been developed for acyclic electrophilic glycines. Many of these rely on amino acids as chiral auxiliaries by placing the electrophilic glycine residue within a peptide. Most alkylations of electrophilic Gly equivalents appear to proceed via displacement of an α-substituted leaving group on the Gly equivalent, as previously covered in Section 3.9. In reality most of these reaction actually proceed by an initial elimination of the α leaving group, forming an imine intermediate. This intermediate is generally not isolated, and is often not acknowledged. The nucleophile then adds to the imine. Other electrophilic Gly equivalents consist of the imines and oximes that are explicitly obtained as intermediates in reductive aminations of glyoxylic acid. Instead of simply being reduced, they are then used as substrates for the addition of organometallic compounds.

6.5.1 Displacements of α-Electrophiles on Acyclic Glycine Equivalents

The desired electrophilic Gly synthon has been generated from Ser and Thr residues within a chiral auxiliary peptide by treatment with lead tetraacetate, giving

α-acetoxyglycine. The acetoxy leaving group was converted into a more reactive α-chloro derivative by a two-step procedure involving chlorination of an ethanethiol intermediate (see Scheme 6.31) (*343–345*). Thiols, alcohols, amines, and organocuprates reacted with the resulting electrophile with minimal diastereoselectivity induced by the peptide, but chiral amino acids and enamines added with up to 90% de for the matched double stereoselection pair. Similarly, an α-chloroglycine residue within a cyclic pentapeptide was prepared from an α-(ethylthio)glycine residue. Thiol, amine, and alcohol nucleophiles displaced the chloride with 75 to >97% de (*346*). An α,α-dichloroglycyl dipeptide was prepared by chlorination of an α,α-di(alkylthio)glycine intermediate, which itself was obtained from the oxazolone of Bz-Gly (*347*). Thiol nucleophiles (e.g., propane-1,3-dithiol, 1,2-dithiophenol, Boc-Cys-OMe) replaced both chlorine atoms in a double elimination–addition reaction (72–91% yield).

Gly residues in dipeptides were α-brominated by photolysis with NBS to give unstable intermediates. Addition of methanol gave the α-methoxy derivative in 65–73% yield, while diethyl malonate anion added in 18% yield with minimal diastereoselectivity (*348*). A dipeptide containing α-acetoxy-Gly was also prepared by the addition of Boc-Leu-amide to allyl glyoxalate, followed by acetylation. Thiophenol added to give a 1:1 mixture of diastereomers (*349*).

Examples of electrophilic alkylations with non-peptide chiral auxiliaries include the addition of enamines containing a chiral amine to α-bromo-Gly-derived imines with a chiral menthyl or 8-phenylmenthyl ester. Excellent double asymmetric induction (>98% de) was observed, although this method could only be used for preparing α-amino-γ-oxo esters (*350*). Grignard reagents

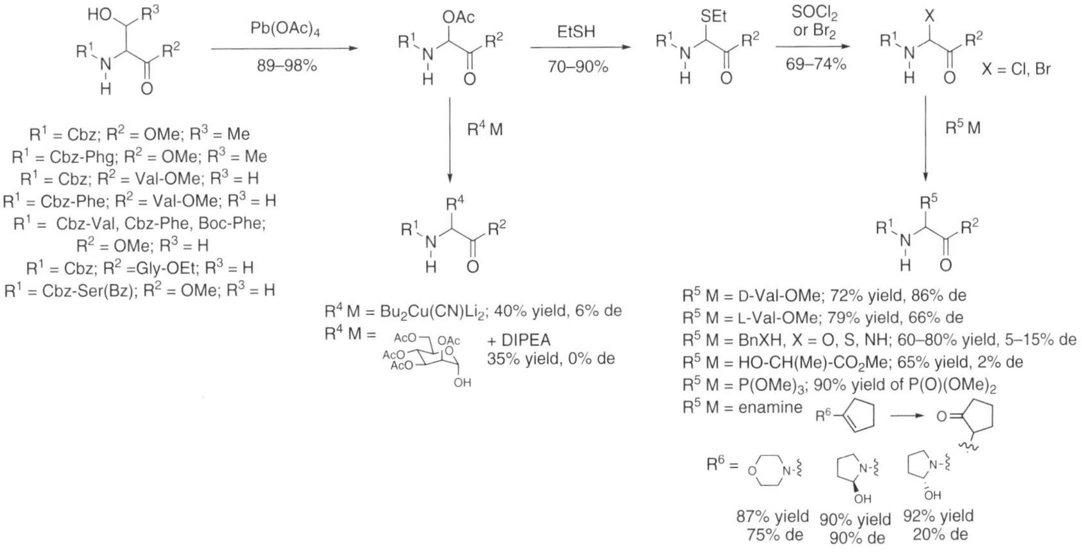

R^1 = Cbz; R^2 = OMe; R^3 = Me
R^1 = Cbz-Phg; R^2 = OMe; R^3 = Me
R^1 = Cbz; R^2 = Val-OMe; R^3 = H
R^1 = Cbz-Phe; R^2 = Val-OMe; R^3 = H
R^1 = Cbz-Val, Cbz-Phe, Boc-Phe;
    R^2 = OMe; R^3 = H
R^1 = Cbz; R^2 =Gly-OEt; R^3 = H
R^1 = Cbz-Ser(Bz); R^2 = OMe; R^3 = H

R^4 M = Bu$_2$Cu(CN)Li$_2$; 40% yield, 6% de
R^4 M =   AcO, AcO, AcO, OAc, O, OH   + DIPEA   35% yield, 0% de

R^5 M = D-Val-OMe; 72% yield, 86% de
R^5 M = L-Val-OMe; 79% yield, 66% de
R^5 M = BnXH, X = O, S, NH; 60–80% yield, 5–15% de
R^5 M = HO-CH(Me)-CO$_2$Me; 65% yield, 2% de
R^5 M = P(OMe)$_3$; 90% yield of P(O)(OMe)$_2$
R^5 M = enamine R^6

R^6 =

87% yield   90% yield   92% yield
75% de   90% de   20% de

Scheme 6.31 Generation and Reaction of an Electrophilic Glycine Synthon within a Peptide (*343–345*).

reacted with *N*-Boc α-bromo-Gly 8-phenylmenthyl ester with 54–78% yield and 82–92% de, but in order to remove the phenylmenthyl ester it was necessary to reduce the ester to an alcohol with LiAlH₄ and then reoxidize with RuCl₃/NaIO₄ (*351*). This reaction was subsequently reinvestigated with a number of other Grignard reagents. It was found that the phenylmenthyl ester could be successfully removed with 6 N HCl, without any racemization (*352*). Homochiral aluminum reagent complexes based on (*R*)-binaphthol delivered methyl, ethyl, and isopropoxide groups to racemic *N*-phenylsulfonyl-α-bromoglycinate ethyl ester in 55–97% yield, with 25–62% ee (*353*).

An α-hydroxy-Gly derivative, *N*-protected with a chiral carbamate derived from the Oppolzer *N*-dicyclohexylsulfamoylisobornol auxiliary, was arylated with anisole/BF₃ to give 4′-hydroxy-Phg with >92% de (*354*). A novel electrophilic synthon containing an α-alkoxy group attached intramolecularly to a chiral amide auxiliary was reported in 1996 (*355*) (see Scheme 6.32). Grignard reagents reacted in reasonable yield (68–76%) with moderate diastereoselectivity (47–67% de) while allylation with allyltrimethylsilane proceeded in the presence of a Lewis acid in 90% yield with 80% de.

Chiral catalysts have also been employed for electrophilic alkylations. O'Donnell and co-workers developed

an asymmetric coupling of malonate carbon nucleophiles with α-acetoxy glycine Schiff bases in the presence of a Pd catalyst containing chiral phosphine ligands. β-Carboxy-Asp acid derivatives were obtained with up to 85% ee, with a single recrystallization improving optical purity to 95% ee (*356–358*). The benzophenone Schiff base of α-acetoxy-Gly-O*t*Bu reacted with the benzophenone Schiff base of Gly-O*t*Bu in the presence of a Pd-BINAP catalyst and NaH to produce protected 3-amino-Asp as an approximately 1:1 mixture of *meso:dl* diastereomers, with up to 47% ee for the *dl* diastereomer (*359*). A Cu(II) diamine chiral Lewis acid catalyst formed from Cu(OTf)₂ and 1,2-bis(benzylamino)-1,2-diphenylethane mediated the addition of soft carbon nucleophiles (the TMS enol ether or morpholine enamine derivatives of acetophenone) to Cbz-α-acetoxy-Gly-OMe, producing protected 2-amino-4-oxo-4-phenylbutyric acid in 52–58% yield, with 73–85% ee (*360*).

A significant advancement in asymmetric electrophilic catalysis was reported in 2002, when the α-acetoxy derivative of the benzophenone imine of Gly-O*t*Bu was reacted with achiral organoborane reagents in the presence of a cinchona alkaloid and added lithium chloride (see Scheme 6.33) (*361, 362*). A β-substituted amino acid, α-cyclopentylglycine, was produced with up to 95% ee, and enantiomeric products could be obtained by employing either cinchonidine (*S*-product) or cinchonine

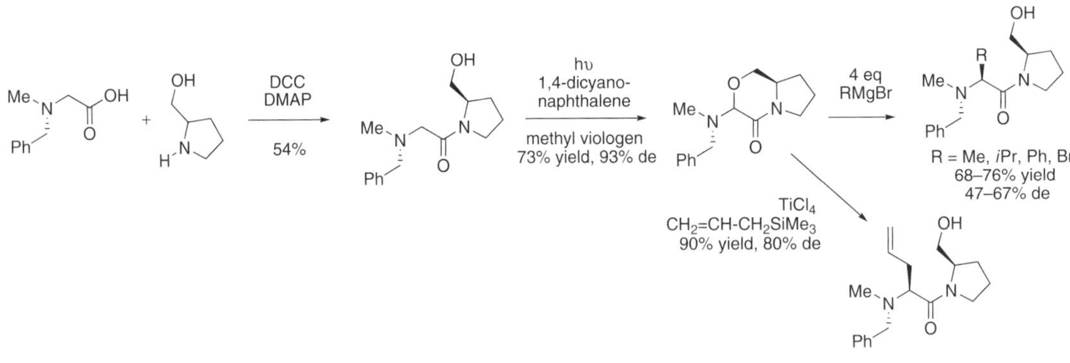

Scheme 6.32 Asymmetric Electrophilic Glycine Equivalent: Synthesis and Reaction (*355*).

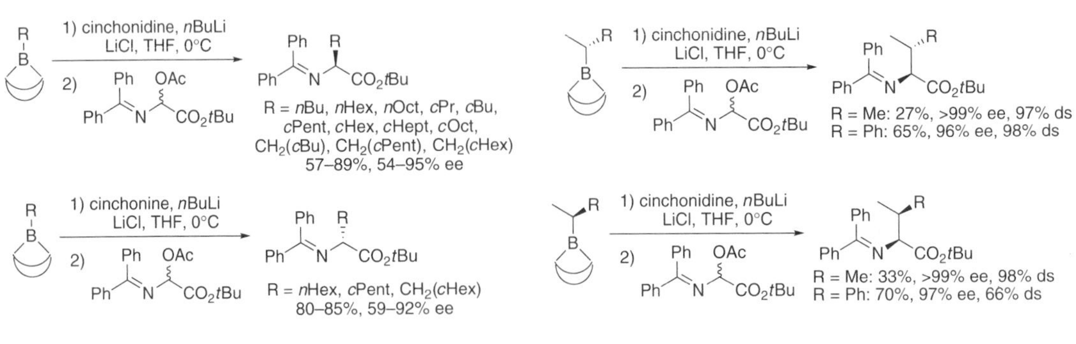

Scheme 6.33 Organoborane Addition to α-Acetoxy-Gly-O*t*Bu Benzophenone Schiff Base with Chiral Catalyst (*361, 362*).

(*R*-product) catalyst. By employing chiral, non-racemic β-substituted 9-BBNs as the nucleophile, the chirality of the β-center of β-substituted α-amino acids could also be controlled (*361*). With β-substitutents of similar size, such as the methyl and ethyl groups of Ile, all four diastereomers were stereoselectively prepared (>99% ee, 97–99% de). However, for mismatched substituents (e.g., a methyl and phenyl group in β-methyl-Phe), the *syn* product was preferred, so while the *syn* diastereomers could be obtained with 94–96% ee and 97–98% ds, the *anti* isomers were produced with a reduced 66% ds, although enantioselectivity was retained (94–96% ee for the major isomer).

6.5.2 Additions to the C=N Bond

The addition of nucleophiles to imines derived from α-keto acids is essentially the same reaction as nucleophile addition to electrophilic glycine equivalents (see Section 6.5.1), although for the latter reaction the imine is generally not isolated. Imines derived from glyoxalic acid lead to α-amino acids, while additions to imines prepared from higher α-keto acids provide α,α-disubstituted products. A comprehensive review of asymmetric syntheses via nucleophilic 1,2-addition of organometallic reagents to C=N bonds was published in 1997, including amino acid syntheses (*363*).

6.5.2a Chiral Nucleophiles

A number of different chiral nucleophiles have been added to imines of α-keto acids. Achiral imines derived from glyoxylate were treated with chiral 4- and 5-alkoxy-pent-2-enyl-1-(tributyl)stannanes in the presence of SnCl$_4$ to give γ,δ-unsaturated amino acids (which could then be further modified) (*364–366*). Surprisingly good 1,5-transfer of chirality was observed (50–98% de, 67–93% yield), presumably via a 4- or 5-alkoxy-pent-1-enyl-3-trichlorostannane transmetallated intermediate. The stereoselectivity was reversible, depending on the alkoxy substituent (*365*). The remote center exerted a greater control over the stereochemistry than did a chiral auxiliary on the imine component (*366*).

Oximes of α-ketoesters were enantioselectively allylated with allylzinc reagents containing a phenyl-substituted chiral bis(oxazoline) ligand. Protected allylglycines and α-allyl-Ala were obtained in 62–90% yield with 74–94% ee (*367*). *N*-Sulfonyl α-imino esters were employed for asymmetric additions of bis(allyl)titanium complexes with a chiral sulfonimidoyl substituent, S(O)(NMe)Ph, producing (*syn*,*E*)-β-alkyl-γ,δ-unsaturated-δ-sulfonimidoyl-α-amino acid derivatives (*368, 369*). The sulfonimidoyl group could be replaced by a phenyl group via a Ni-catalyzed cross-coupling reaction (*368*), or used for an intramolecular Pauson–Khand cyclization to give bicyclic pipecolic acid products (*369*).

Both enantiomers of α-trifluoromethyl-Ala were prepared from an *N*-alkoxycarbonyl imine of methyl 3,3,3-trifluoropyruvate by addition of the lithium anion of an optically active sulfoxide, (+)-(*R*)- or (−)-(*S*)-methyl-*p*-tolyl sulphoxide (*370, 371*). Diastereoselectivity was very poor (up to 10% de), but the isomers were easily separated. α-Trifluoromethyl-Phe was prepared by the same route (*371*), while α-trifluoromethyl-Ser was produced via a Pummerer rearrangement of the methyl sulfoxide adduct (*372*).

6.5.2b Chiral Substrates

Asymmetry can be induced during additions of nucleophiles to α-keto acid-derived imines by employing chiral auxiliaries on either the amine or α-keto acid components.

α-Imino esters derived from glyoxylate and the chiral amines (*S*)-α-methylbenzylamine or (−)-1-cyclohexylethylamine were *C*-allylated by allyl 9-borabicyclo[3.3.1]nonan-9-yl (9-BBN) with high yield (92–94%) and diastereoselectivity (92–96%). Yields and stereoselectivity were reduced for substituted allyl reagents (*373*). Phenylmenthyl esters as chiral auxiliaries gave much poorer diastereoselectivity (*374*). Benzylzinc bromide also added with the desired *C*-regioselectivity to the (*S*)-α-methylbenzylimine derivative (50% yield, 48% de), but the corresponding Al, Cu, Ti, and B reagents added exclusively to the nitrogen (*374*). In contrast, various metallated allyl reagents all added with the desired regioselectivity, but with generally poor diastereoselectivity (0–70%) (*374, 375*). The imine formed from (*S*)-α-methylbenzylimine and methyl glyoxylate reacted with a higher-order cyclohexyl zinc–copper reagent with 40% de (*376*). Additions of prop-2-enyl tin trichloride to the α-methylbenzylimine of *tert*-butyl glyoxylate resulted in the opposite diastereoselectivity (with 86% de) as additions of allyl boranes (*366*). Several organozinc reagents (RZnBr, R = *t*Bu, *c*Hex, *s*Bu, Bn) were added to the imine formed from ethyl glyoxylate and phenylglycinol methyl ether; precomplexation of the imine with ZnBr$_2$ provided diastereoselectivities of 78–92% (*377*). L-Val methyl ester has also been used as a chiral amine auxiliary to form an imine with glyoxalic acid. Allyl indium, preformed from allyl bromide and indium, added in 52% yield with a remarkable 98% de (*378*). Glyoxylic acid has been doubly derivatized with a menthyl ester and an imine formed from α-methylbenzylamine. Organocerium reagents added with complete regioselectivity but moderate diastereoselectivity to give Ala, Nva, Val, and Phe with 38–46% ee (*379*). Grignard reagents were also added (*379, 380*).

Petasis and Zavialov reductively aminated α-keto acids by condensing the acid, an amine, and a vinylboronate. The organoboron compound introduced an unsaturated side chain, providing vinylglycine derivatives from glyoxylic acid, and α-alkylvinylglycines from other α-keto acids (see Scheme 6.34). With phenylglycinol as the amine component, diastereoselectivity of

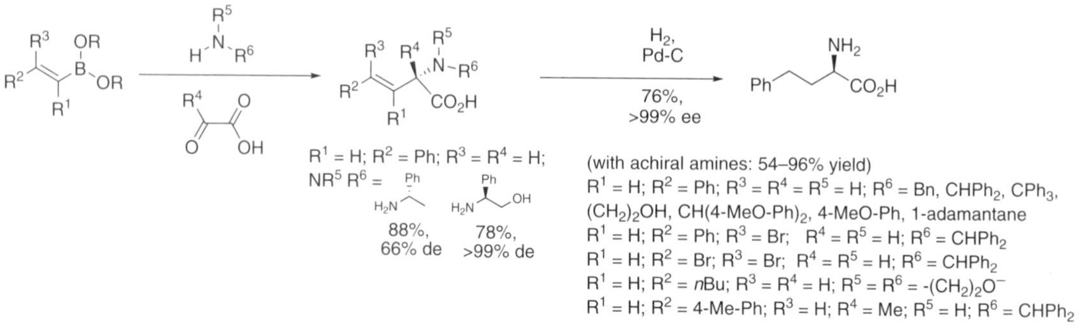

Scheme 6.34 α-Keto Acid Reductive Amination via Addition of Organoborane to Imine (*381*).

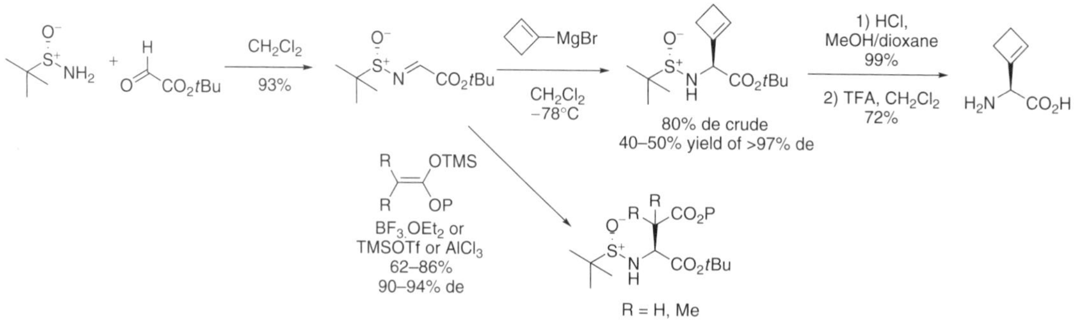

Scheme 6.35 Grignard or Ketene Acetal Addition to Chiral Sulfinimine (*383–385*).

>99% de was obtained. The alkene could be reduced (*381*). Attempts at inducing diastereoselectivity during the analogous reactions with aryl boronic acids gave only 28–35% de (*382*).

Chiral sulfinimines are another source of diastereoselectivity (see Scheme 6.35). Methyl, ethyl, benzyl, and phenyl Grignard/dialkylzinc reagents were added to the chiral imine formed from ethyl glyoxylate and (*S*)-*tert*-butanesulfinimide with 68 to >98% de (*383*). Similarly, L-α-(1-cyclobutenyl)-Gly was synthesized by diastereoselective addition of 1-cyclobutenylmagnesium bromide to the imine formed from *tert*-butyl glyoxylate and (*S*)-*tert*-butanesulfinimide, with the addition proceeding in 40–50% yield and 80% de. Deprotection of the *tert*-butylsulfinamide and *tert*-butyl ester was achieved in 71% yield by HCl in dioxane followed by TFA (*384*). Ketene silyl acetals added to produce Asp or 3,3-dimethyl-Asp with up to 94% de (*385*). A new synthesis of enantiopure *tert*-butanesulfinamide has been described (*386*). A chiral sulfinimine derivative of methyl 3,3,3-trifluoropyruvate reacted with Grignard reagents with 10–74% de, leading to α-trifluoromethyl amino acids (*387, 388*). The use of sulfinimines in asymmetric syntheses, including α-amino acids, was reviewed in 2004 (*389*).

Reactions of imines with auxiliaries attached to the acid group are less common. An imine formed from the

tert-butyl lactate ester of glyoxylate and 4-chloroaniline or 4-methoxyaniline reacted with allyltributyltin with 60% de. The adduct was hydrogenated to give Nva or ozonolyzed to Asp semialdehyde and further elaborated (*390*). Glyoxylate ester imines undergo an asymmetric imine-ene reaction to give γ-δ-unsaturated amino acid derivatives. (−)-8-Phenylmenthol ester auxiliaries provided high diastereofacial selectivity to give the γ-δ-unsaturated amino acids with 94–96% de. Chiral imine auxiliaries gave much less stereoselectivity than the ester auxiliary, in contrast to the better diastereoselectivity observed with imine auxiliaries for organometallic additions (*297, 391*).

Organolithium reagents add to oximes of α-keto amides, producing *N*-hydroxy α-amino acids with moderate diastereoselectivity when using chiral amide auxiliaries (0–47% de) or chiral *N*-hydroxy protecting groups (0–40% de) (*392*). An *O*-benzyl oxime of glyoxylic acid derivatized with Oppolzer's camphorsultam amide auxiliary was employed for the preparation of a number of substituted allylglycines, with the glycine equivalent treated with allylic bromides in the presence of powdered zinc in aqueous ammonium chloride. The addition generally took place at the γ-position of the allylic halide to give *N*-benzyloxy allylglycines with excellent yields (88–99%) and generally good diastereoselectivity (62–98% de). The N−O bond was cleaved with Mo(CO)₆ and the sultam auxiliary removed with aqueous LiOH, providing the free amino acids with enantiomeric purity corresponding to the diastereomeric purity (*393*). The allyl group could be ozonolyzed/reduced to give homoserine side chains, with activation/cyclization producing 3-substituted azetidin-2-carboxylates (*394*). Indium was also used to mediate the addition of allyl halides, with 90–96% yield and 59 to >98% de (*395*). Indium-mediated allyl addition to the closely-related benzylamine imine of glyoxylate derivatized with the Oppolzer sultam auxiliary proceeded with 80% yield and 90% de (*396*).

Several imine addition reactions have used imines contained with chiral cyclic complexes. Nucleophile addition to chiral Co(III) complexes of the imines of pyruvate and phenylpyruvate proceeded with some stereoselectivity (50–68% de for nitromethane), with reduction of the nitromethane/pyruvate adduct providing 2-(aminomethyl)Ala. N-Alkylation could be achieved via anion generation (*397*). An imine contained within a chiral morpholine template has also been used as an electrophile (see Scheme 7.27) (*398*). α-Trifluoromethyl-α-amino acids were synthesized from methyl 3,3,3-trifluoropyruvate using a chiral cyclic imine (see Scheme 6.36). The pyruvate was converted into a 2,5-diketopiperazine by reaction with Cbz-Ala-, -Leu-, or -Phe-amide. The acylimine was generated in situ by addition of an organometallic reagent, with a second equivalent of reagent then adding to the imine with 44–99% de (*399*). Acidolysis gave a dipeptide ester.

A number of syntheses employ masked carboxylic acid groups. Glyoxal was asymmetrically converted to α-aminoaldehydes (one oxidation removed from α-amino acids) by formation of a hydrazone with one aldehyde group and formation of a chiral aminal with the other (*400*) (see Scheme 6.37). Organolithium

reagents attacked the hydrazone with excellent diastereoselectivity (>99% de in THF) and good yield (52–74%). The hydrazone was hydrogenolyzed and the aminal hydrolyzed to give the amino aldehyde in 54% yield with no racemization (one example). Similarly, a glucose- or malic acid-derived acetal formed with glyoxal monohydrazone provided excellent diastereoselectivity for the addition of methyllithium and n-butyllithium, with 90–92% de and 72–80% yield (*401*). The amino acid was obtained by reductive cleavage of the hydrazine N–N bond, followed by perchloric acid cleavage/oxidation of the acetal. Unfortunately the amine must be temporarily protected as the phthalimide during the cleavage reaction.

The SAMP/RAMP ((S)- and (R)-1-amino-2-methoxymethylpyrrolidinone) derivatives of Enders et al. (*402*) have also been applied to amino acid synthesis (see Scheme 6.38). Organocerium reagents added to the hydrazone derived from diethoxy glyoxal with good yield (72–98%) and high diastereoselectivity (78–98%), giving a hydrazine that was easily purified to a single diastereomer by chromatography. Unfortunately, while the amino acetals were readily obtained, conversion to the amino acid required a tedious seven-step procedure.

D-Glyceraldehyde provides a glyoxalic acid equivalent with the carboxyl group masked as a chiral 1,2-diol. Grignard reagents added to the benzylamine Schiff base of hydroxyl-protected D-glyceraldehyde with complete diastereoselectivity. The opposite diastereomers were obtained by varying the hydroxyl protecting group. Amino acids (Ala and Phg) were synthesized following protecting group manipulation and diol oxidative cleavage (see Scheme 6.39) (*403*, *404*). Vinylglycine was synthesized via addition of vinylmagnesium bromide, with >98% de. Deprotection and oxidative cleavage was

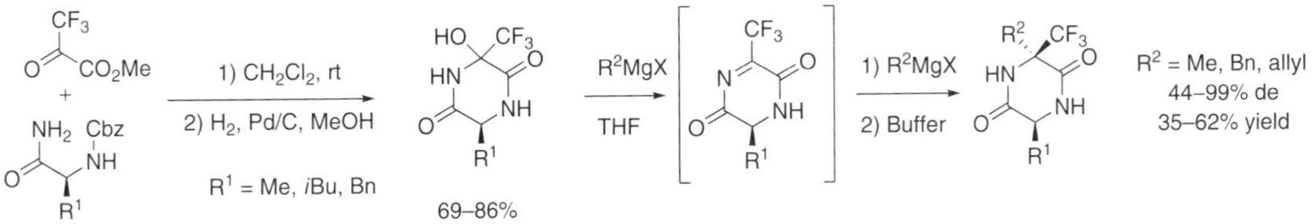

Scheme 6.36 Synthesis α-Trifluoromethyl Amino Acids via Addition to Imine (*399*).

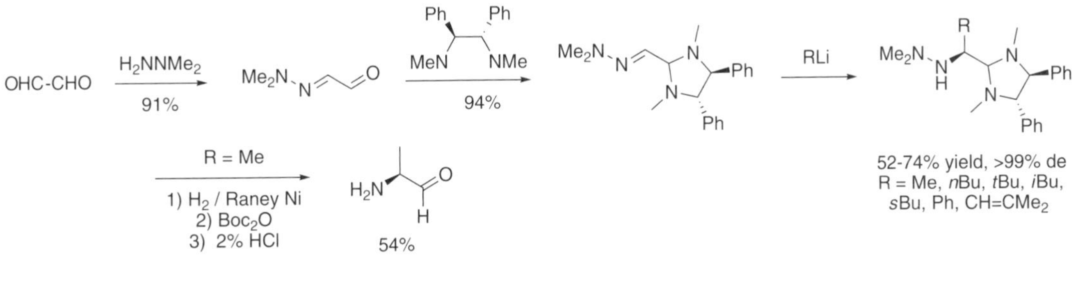

Scheme 6.37 Reduction of Imine to Chiral Aldehyde (*400*).

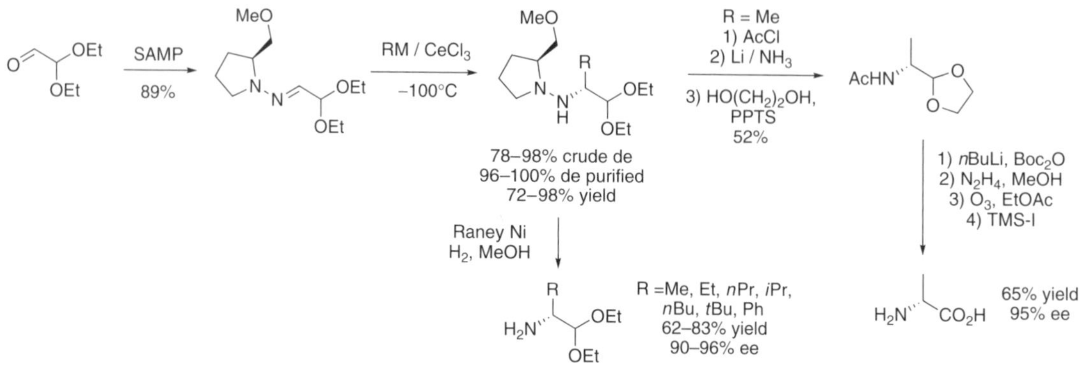

Scheme 6.38 Addition to SAMP Hydrazone (*402*).

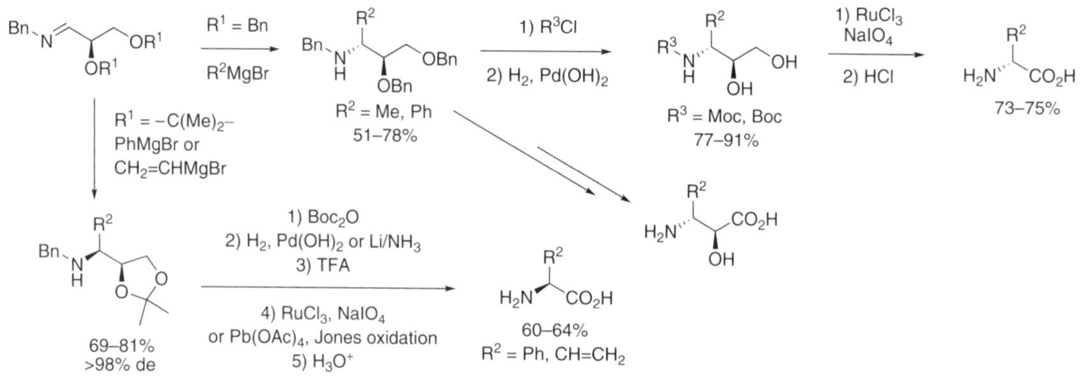

Scheme 6.39 Addition to a Chiral Imine (*403–405*).

accomplished with an overall yield of 45% (*405*). β-Amino acids were also accessible by selective oxidation of the terminal alcohol (*403*).

Similarly, a hydrazone derived from glyceraldehyde was used as a substrate for Grignard/CeCl₃ additions. Oxidative cleavage of the glyceraldehyde diol generated the carboxyl group of α-hydrazino acids (*406*). Organozinc reagents were added to an N-benzyl nitrone derivative of 2,3-O-isopropylidene-D-glyceraldehyde with 90–91% de. Oxidative diol cleavage and N–O reductive cleavage via hydrogenation led to D-Ala and D-Phg (*407*). Erythrulose has been used in a similar fashion; organolithium addition to an oxime ether derivative, followed by oxidative cleavage of the diol, provided α-substituted serines (*408*). A homochiral 2,3,4-trihydroxybutyronitrile derivative allowed for sequential addition of two nucleophiles to the nitrile to form a quaternary center. Oxidative cleavage of the triol substituent generated the acid group of α,α-disubstituted amino acids (*409*).

Good diastereoselectivity (71–93% de) was observed for organolithium additions to a chiral oxime derivative

of cinnamaldehyde formed with ROPHy, (*R*)-*O*-(1-phenylbutyl)hydroxylamine, although some bulky reagents (*i*PrMgBr, *t*BuMgBr) failed to react (see Scheme 6.40). The N–O bond was cleaved by Zn/AcOH, the amine protected with Cbz-Cl, and the carboxy group generated by oxidative cleavage of the alkene. Unfortunately, yields for the cleavage were poor (25–57%) (*410–412*).

An imine formed from 2-furaldehyde and (*S*)-valinol was reacted with organometallic reagents with >95% de. The auxiliary was cleaved with H₅IO₆, with the acid group revealed by oxidiation of the furyl moiety with NaIO₄/RuO₂ (*413*).

6.5.2c Chiral Catalysts

A variety of chiral catalysts have been employed for asymmetric alkylations of electrophilic glycines, with ketone/ester enolates the most common nucleophiles. These asymmetric Mannich-type reactions were reviewed in 2004 (*414*), while a number of catalytic enantioselective Mannich-type additions to imines were

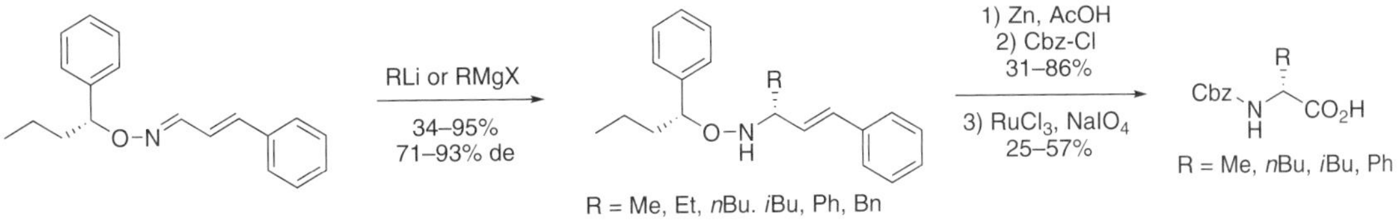

Scheme 6.40 Additions to Chiral Oxime of Cinnamaldehyde (*410–412*).

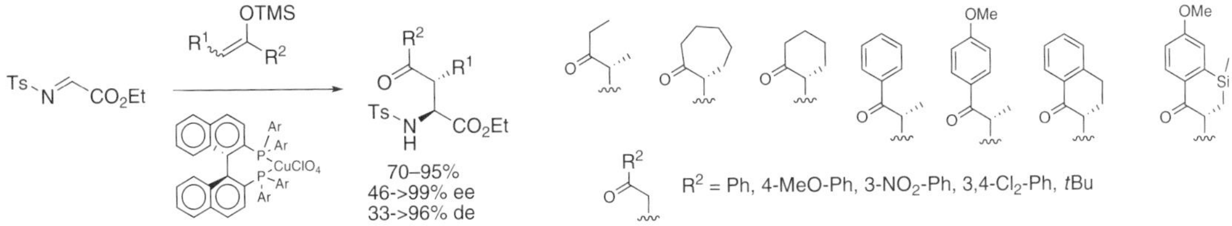

Scheme 6.41 Alkylation of α-Imino Esters with Enol Silanes Using Chiral Catalysts (*416, 417*).

included in a 1999 review of additions to imines (*415*). Enol silanes were added to the *N*-tosyl imine of ethyl glyoxylate in the presence of metal complexes with chiral phosphine ligands, adding with enantioselectivities of 65–95% ee. A single recrystallization gave the *N*-Ts 4-keto-2-amino ester products with 90 to >99% ee (see Scheme 6.41) (*416*). Enol silanes that gave products with a chiral β-center added with 46 to >99% ee and 33 to >96% de *anti* diastereoselectivity (*417*). These reactions were described in greater detail in a 2002 report, along with allylations with allylsilanes to give 4,5-unsaturated amino acids (*418*). Enol silanes were also added to imines formed from anilines and methyl or isopropyl glyoxylate with up to 90% ee (*419*). This approach was reviewed in 2003 (*420*).

The *N*-(*p*-methoxyphenyl) imine of ethyl glyoxylate was reacted with ketones in the presence of catalytic quantities of L-Pro, generating a transient chiral enamine that reacted to produce γ-keto-α-amino esters (seven examples) with excellent enantioselectivity (95 to >99% ee), and >95% de in examples when chiral β-centers were created (*421*). The same strategy was successfully applied to aldehydes to produce β-formyl amino acids

(seven examples); oxidation with NaClO$_2$ resulted in β-substituted Asp derivatives (*422*). Acetonide-protected dihydroxyacetone also reacted with the *N*-(*p*-methoxyphenyl) imine of ethyl glyoxylate using Pro catalyst, microwave irradiation, and TFE as solvent, producing a sugar-like amino acid derivative (protected 3,5-dihydroxy-4-keto-2-aminopentanoic acid) with up to 99% ee (*423*). Seven different aldehydes were also reacted with *N*-PMP (*p*-methoxy-phenyl) α-imino ethyl glyoxylate in the presence of 20 mol % SMP, (*S*)-2-methoxymethylpyrrolidine. The β-formyl amino acids were produced with 74–92% ee and generally high (>10:1) *anti* diastereoselectivity (see Scheme 6.42) (*424*). Addition of α-aryl-α-cyanoacetates to an *N*-Boc imine of ethyl glyoxylate in the presence of catalytic chiral amine catalyst produced β-aryl-β-cyano-Asp derivatives with 60–96% diastereoselectivity and 91–98% ee (*490*).

A catalytic amount of a Cu(OTf)$_2$-chiral diamine complex induced high enantioselectivities (67–96% ee) for the addition of a wide variety of ketone- or ester-derived silyl enol ethers to RCON=CHCO$_2$Et, producing γ-keto-α-amino acids or Asp analogs, respectively.

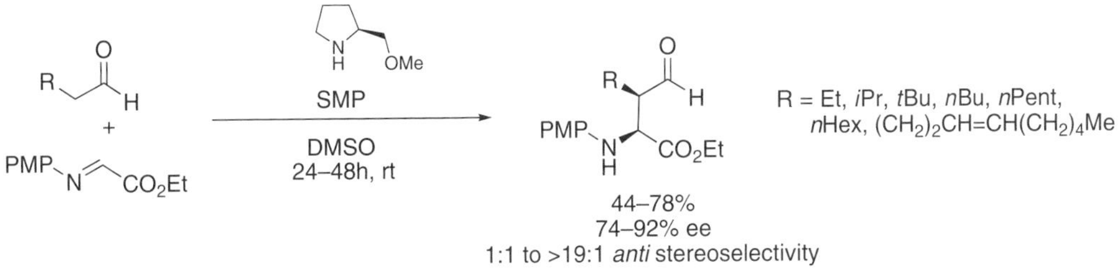

Scheme 6.42 SMP-Catalyzed Mannich Reactions of Unmodified Aldehydes (*424*).

Alkyl vinyl ethers also reacted. High *syn* diastereoselectivity (72–92% ds) was observed when asymmetric β-centers were generated (*425*). An earlier report produced β-substituted Asp derivatives by reactions of EtO₂CCH₂R with TsN=CHCO₂Et, catalyzed by a bis(oxazoline)cuprate catalyst, Ph-BOX-Cu(OTf)₂. Enantioselectivities of 78 to >98% ee and diastereomeric ratios of 3:1 to >10:1 were obtained (*426*). Chiral Cu(II) catalysis provided for 83–93% ee for the addition of enamides to *N*-acyl imines of ethyl or benzyl glyoxylate; acidic hydrolysis of the initial β-imino-α-amino ester products provided β-keto-α-amino esters (*427*).

An ene reaction of the *N*-tosyl imine of ethyl glyoxylate with alkenes in the presence of a chiral Cu catalyst produced substituted allylglycines, Trp, or β-(3-furyl)-Ala in 85–94% yield and 85–99% ee (see Scheme 6.43) (*418, 428, 429*). The same reaction was applied to a synthesis of 6′-bromo-Trp (*430*). Allylation reactions with allylsilanes (*418*) or allylstannanes (*431*) catalyzed by achiral Cu catalyst also produced substituted allylglycines. An enantioselective aza-Henry reaction proceeded via reaction of an *N*-(p-methoxyphenyl) imine of ethyl glyoxylate with trimethylsilyl nitronates, RCH=N⁺(O⁻)-OTMS, in the presence of various chiral bis(oxazoline) Cu catalysts. The products, β-nitro-α-amino esters, were obtained in good yield (67–94%) with good diastereose-

lectivity (5:1 to 39:1 *erythro*) and enantioselectivity (83 to >98% ee for *erythro* product). The nitro group could be reduced to give 2,3-diamino acids (*432*). Nitroalkanes could also be directly reacted, producing β-nitro-α-amino esters with 38–81% yield, 10–90% de *erythro* stereoselectivity, and 74–99% ee (*433*).

More recently, substituted furans, thiophenes, pyrroles, and aromatic compounds were added to *N*-alkoxycarbonyl imines of glyoxylate in the presence of a chiral BINAP-Cu(I) catalyst, producing α-(heteroaryl)glycines with up to 96% ee for furans, 93% ee for thiophenes, and 98% ee for aromatic compounds (see Scheme 6.44) (*434, 435*). In a similar manner, the *N*-tosyl imine of ethyl glyoxylate reacted with 5-substituted indoles, *N*-methylpyrrole, and 2-acetylpyrrole in the presence of CuPF₆ and Tol-BINAP to give α-(5-R-indol-3-yl)-Gly, α-(1-methyl-pyrrol-2-yl)-Gly, and α-(2-acetyl-pyrrol-4-yl)-Gly in 49–89% yield, with 78–97% ee (*436*).

Sparteine has been used to catalyze asymmetric additions of organolithium reagents to imines of cinnamaldehyde, with the CH=CHPh moiety acting as a masked carboxyl group. Enantioselectivities as high as 88% ee could be obtained with an *N*-(p-methoxyphenyl) group, but this substituent proved difficult to remove after the organometallic addition. More labile *N*-substituents,

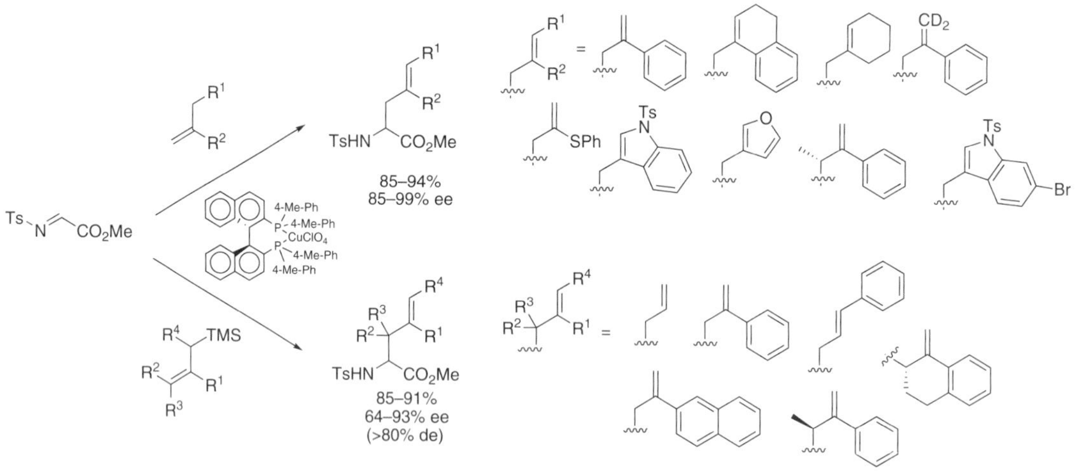

Scheme 6.43 Synthesis of Allylglycines via Asymmetric Ene Reaction or Allylation of α-Imino Esters (*418, 428, 429*).

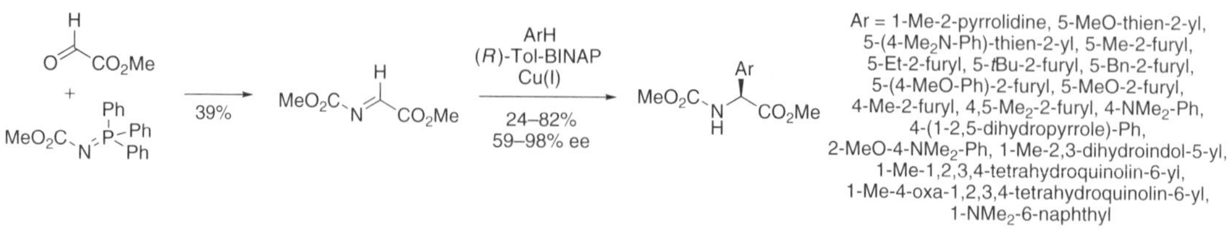

Scheme 6.44 Cu-BINAP Catalyzed Aryl Addition to Imine (*434, 435*).

such as the imine formed from *N*-trimethylsilylamine, resulted in poor enantioselectivities (*437*).

6.5.2d Cycloadditions of the C=N Bond

Several cycloaddition reactions of glyoxylate imines have been described. The imine formed from chiral α-methylbenzylamine and benzyl glyoxylate was employed for Diels–Alder reactions with cyclopentadiene or cyclohexadiene to form unsaturated bicyclic Pro analogs with 50% de. Purification and deprotection gave the free amino acid with >95% ee (*438, 439*). Selective hydrogenolysis resulted in cleavage of the allylic C−N bond, alkene reduction, and auxiliary removal to produce α-(cyclopentyl)Gly and α-(cyclohexyl)Gly (*439*). An improved synthesis of 2-azabicyclo[2.2.1]heptane-3-carboxylic acid, obtained by cycloaddition of the chiral imine with cyclopentadiene followed by alkene hydrogenation, was reported in 2002 (*440*). Diels–Alder reactions using chiral esters of in situ-generated *N*-sulfonyliminoacetates gave much poorer diastereoselectivity (6–20% de, 80–84% yield) (*441*). An imine formed from benzylamine and acyclic protected D-glyceraldehyde was used as a glycine equivalent to prepare pipecolic acid derivatives via cycloaddition reaction with Danishefsky's diene (*442*). An imine contained within a morpholine template has also been used for Diels–Alder reactions (see Section 7.6) (*443*).

The cinchona alkaloid catalyst benzoylquinine induced Staudinger lactam formation between ketenes derived from acid chlorides and the *N*-acylimine derived from Bz-α-Cl-Gly-OEt (see Scheme 6.45). The quinine catalyzed the dehydrohalogenations that generated the reactants, the cycloaddition itself, and also a subsequent methanolysis of the lactam product that generated β-substituted Asp derivatives. The products were produced with 94–96% ee and >10:1 diastereomeric ratio (*444*). Amidolysis could be employed to generate Asn analogs, and by using aryloxyacetic acid chlorides, β-hydroxy-Asp products were produced. The enantiomers were made with benzoylquinidine catalyst.

Glyoxylate can be converted into a nitrone derivative by condensation with a substituted hydroxylamine. The nitrone was used for a stereoselective 1,3-dipolar cycloaddition with the alkene group of acrylic acid derivatized with Oppolzer's bornane-sultam, leading to the synthesis of 4-hydroxy-L-Glu (*445*).

6.6 Asymmetric Radical Alkylations of Acyclic Glycine Equivalents

There are relatively few examples of asymmetric radical alkylations of Gly equivalents. Some of these involve cyclic piperazine-2,5-diones (*446–449*), and so will be discussed in Section 7.3. Other radical additions use didehydroalanine derivatives as substrates, and are discussed in Section 4.3.4. Asymmetric radical reactions of acyclic glycine equivalents were reviewed in 1997 (*450*). The radical can either be generated on the glycine equivalent or on the side chain to be added.

The 8-phenylmenthyl ester of *N*-Boc Gly can be α-brominated by NBS in quantitative yield. Reduction with tributyldeuterostannane proceeded at low temperature without any initiator to give chiral 2-deutero-Gly with up to 90% de (70% yield) (*449, 451*). Allyl transfer with allyltributylstannanes also proceeded with high diastereoselectivity (up to 92% de) (see Scheme 6.46), as did reaction with other organostannanes (*452, 453*). Gly radical equivalents within a peptide showed minimal asymmetric induction (*343, 348, 454*). There is some question over whether these reactions are radical or ionic processes (*452*). Similar reactions of tributyltin acetylides with the α-bromo derivative of Williams' oxazinone template (see Section 7.5) are believed to proceed via a cationic process (*455*).

A Gly derivative containing the amino group within an oxazolidinone auxiliary was α-brominated using NBS, and then allylated with allyl bromide, AIBN, and ZnCl₂, with up to 86% de (*456*). The menthyl ester of α-bromo-Gly can be reacted with Co(II) complexes of β-dicarbonyl amino acids, such as Co(Acac)₂ (see Scheme 6.47). The resulting δ-dioxo derivatives, obtained with 40% de and resolved by crystallization, were elaborated into heteroaryl-Gly derivatives (*457*).

Chiral imines of glyoxylate have also been used as substrates for the additions of alkyl radicals derived from alkyl iodides in the presence of triethylborane, leading to Abu, Nle, Chg or *t*Leu with 16–40% de for an acyclic substrate, and up to 100% de for a cyclic derivative (*458, 459*). A similar report described the synthesis of Abu, Chg, and Tle via radical reactions of alkyl iodides with chiral imines formed from methyl glyoxylate, mediated by Et₂Zn, with 23–90% de (*459, 460*). A number of conditions and amine auxiliaries were studied for additions of alkyl iodides (EtI, *i*PrI, *t*BuI, *c*HexI) to imines formed from ethyl glyoxylate; Et₂Zn tended to give slightly

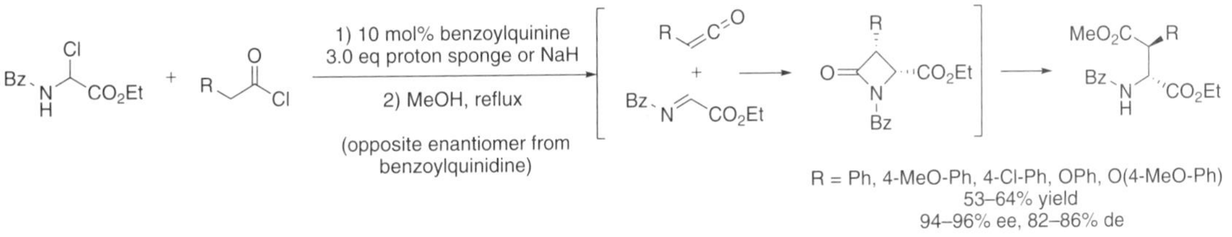

R = Ph, 4-MeO-Ph, 4-Cl-Ph, OPh, O(4-MeO-Ph)
53–64% yield
94–96% ee, 82–86% de

Scheme 6.45 Benzoylquinine-catalyzed Staudinger Reaction (*444*).

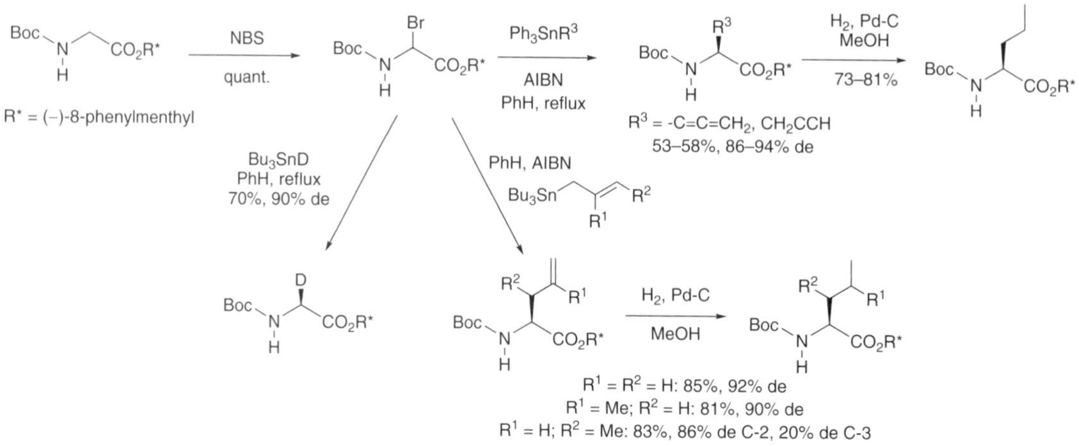

Scheme 6.46 Asymmetric Radical Alkylation of Glycine (449, 452, 453).

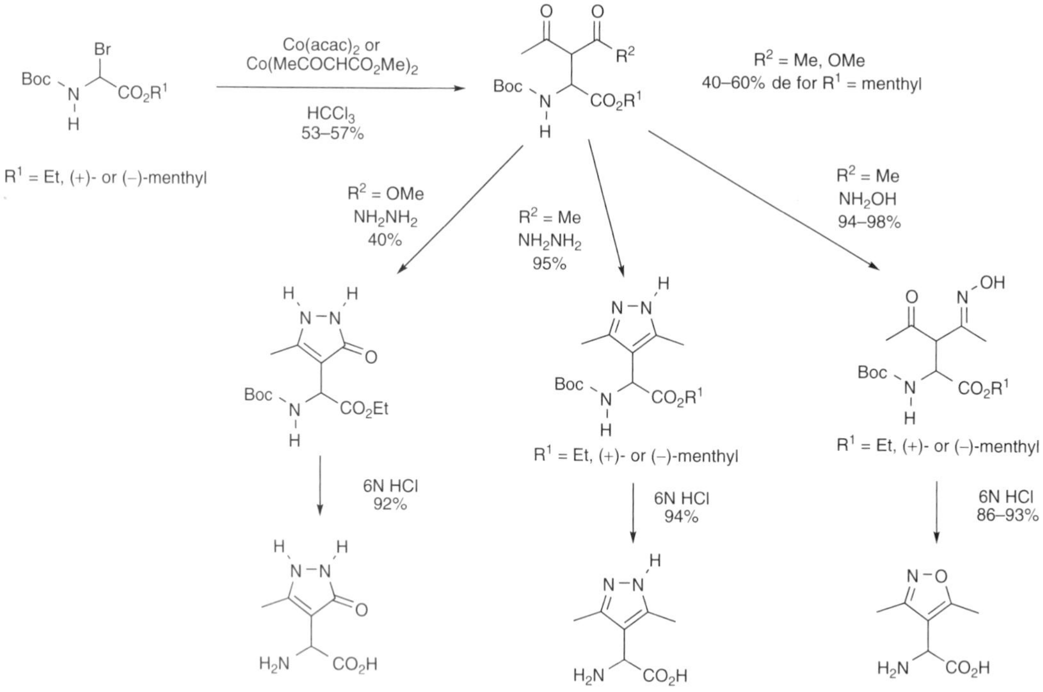

Scheme 6.47 Synthesis of Heteroarylglycines via Radical Alkylation of Bromoglycine (457).

higher diastereoselectivities than Et₃B, while 1-amino-1-phenyl-2,2-dimethylpropane as the amine auxiliary gave better stereoselectivity (72–94% de) but reduced yields (20–89%) compared to 1-amino-1-phenyl-2-methylpropane (62–94% de, 49–70% yield) (461).

The camphor sultam amide auxiliary of Oppolzer has been employed for induction of diastereoselectivity during radical additions to the oxime of glyoxylic acid (see Scheme 6.48). A variety of alkyl halides added with

90 to >96% de (461–463). A hydrazone derivative was also tested, instead of the oxime, with 58–84% yield and 86 to >95% de for additions of iPrI, cPentI, and tBuI (461). Indium has also been used to mediate the addition of allyl and alkyl halides, with 44–96% yield and 59 to >98% de (395).

β-, γ-, δ-, and ω-amino acids that already possess a chiral center were successfully C-allylated by first forming an α-selenophenyl or α-iodo derivative via enolate

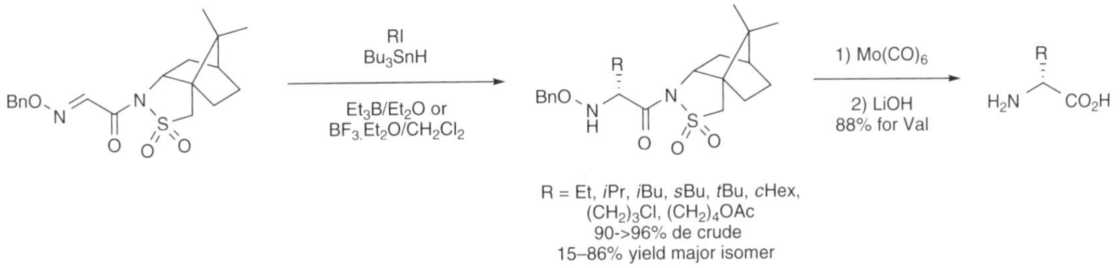

R = Et, iPr, iBu, sBu, tBu, cHex,
(CH₂)₃Cl, (CH₂)₄OAc
90->96% de crude
15–86% yield major isomer

Scheme 6.48 Radical Additions to Oxime of Glyoxylic Acid with Camphor Sultam Auxiliary (461–463).

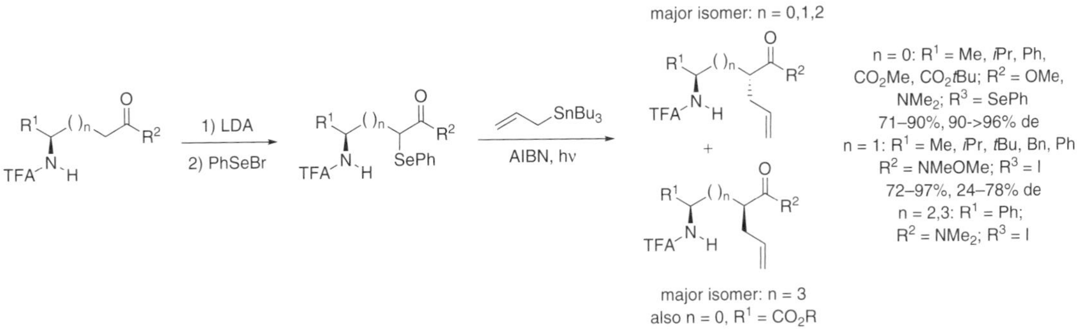

major isomer: n = 0,1,2

n = 0: R¹ = Me, iPr, Ph,
CO₂Me, CO₂tBu; R² = OMe,
NMe₂; R³ = SePh
71–90%, 90->96% de
n = 1: R¹ = Me, iPr, tBu, Bn, Ph
R² = NMeOMe; R³ = I
72–97%, 24–78% de
n = 2,3: R¹ = Ph;
R² = NMe₂; R³ = I

major isomer: n = 3
also n = 0, R¹ = CO₂R

Scheme 6.49 Radical Alkylation of ω-Amino Acids (464).

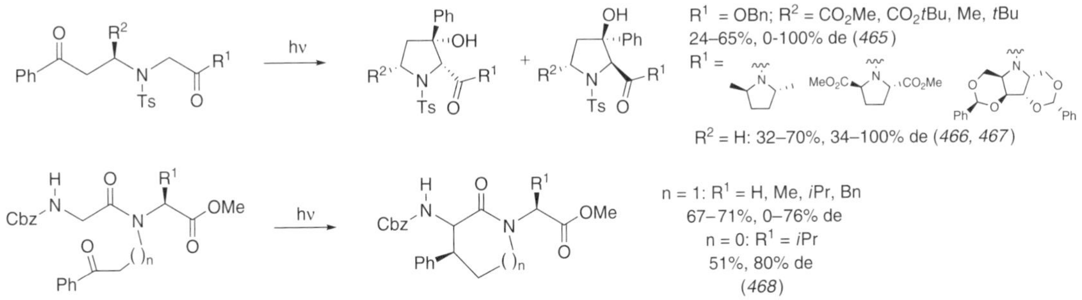

R¹ = OBn; R² = CO₂Me, CO₂tBu, Me, tBu
24–65%, 0-100% de (465)
R¹ =

R² = H: 32–70%, 34–100% de (466, 467)

n = 1: R¹ = H, Me, iPr, Bn
67–71%, 0–76% de
n = 0: R¹ = iPr
51%, 80% de
(468)

Scheme 6.50 Photocyclization of Amino Acids.

selenation, and then using allyltributylstannane and AIBN to introduce the allyl group (see Scheme 6.49). Surprisingly high diastereoselectivity was observed with even a remote γ-chiral center (464).

Photocyclization has been used to prepare cis-3-hydroxy-Pro esters from chiral N-(2-benzoylethyl)-N-tosyl Gly esters (see Scheme 6.50), with the chiral center on the N-alkyl group (465) or as an amide chiral auxiliary (466, 467). The cyclization can also be carried out on dipeptides with the benzoylmethyl or benzoylethyl group contained on the Gly amide bond, leading to dipeptide units containing a δ-lactam (468). Other Pro derivatives, kainic acid analogs, have been prepared by cobalt(I)-mediated cyclization of an N-allyl β-iodo amino acid derivative (469).

Most other radical reactions relating to amino acid synthesis involve radical reactions of the side chains of existing amino acids, and so will be discussed in Section 9.5.2a.

References

1. Rios, A.; Amyes, T.L.; Richard, J.P. J. Am. Chem. Soc. 2000, 122, 9373–9385.
2. Rios, A.; Richard, J.P.; Amyes, T.L. J. Am. Chem. Soc. 2002, 124, 8251–8259.
3. Rios, A.; Crugeiras, J.; Amyes, T.L.; Richard, J.P. J. Am. Chem. Soc. 2001, 123, 7949–7950.

4. Ojima, I.; Habus, I. *Tetrahedron Lett.* **1990**, *31*, 4289–4292.

5. Mehlführer, M.; Berner, H.; Thirring, K. *J. Chem. Soc., Chem. Commun.* **1994**, 1291.

6. Micouin, L.; Schanen, V.; Riche, C.; Chiaroni, A.; Quirion, J.-C.; Husson, H.-P. *Tetrahedron Lett.* **1994**, *35*, 7223–7226.

7. Schanen, V.; Riche, C.; Chiaroni, A.; Quirion, J.-C.; Husson, H.-P. *Tetrahedron Lett.* **1994**, *35*, 2533–2536.

8. Lee, J.; Choi, W.-B.; Lynch, J.E.; Volante, R.P.; Reider, P.J. *Tetrahedron Lett.* **1998**, *39*, 3679–3682.

9. Karoyan, P.; Sagan, S.; Clodic, G.; Lavielle, S.; Chassing, G. *Biorg. Med. Chem. Lett.* **1998**, *8*, 1369–1374.

10. Workman, J.A.; Garrido, N.P.; Sançon, J.; Roberts, E.; Wessel, H.P.; Sweeney, J.B. *J. Am. Chem. Soc.* **2005**, *127*, 1066–1067.

11. Caddick, S.; Parr, N.J.; Pritchard, M.C. *Tetrahedron* **2001**, *57*, 6615–6626.

12. Caddick, S.; Parr, N.J.; Pritchard, M.C. *Tetrahedron Lett.* **2000**, *41*, 5963–5966.

13. Berkowitz, D.B.; Smith, M.K. *J. Org. Chem.* **1995**, *60*, 1233–1238.

14. Myers, A.G.; Gleason, J.L.; Yoon, T. *J. Am. Chem. Soc.* **1995**, *117*, 8488–8489.

15. Myers, A.G.; Gleason, J.L.; Yoon, T.; Kung, D.W. *J. Am. Chem. Soc.* **1997**, *119*, 656–673.

16. Myers, A.G.; Yoon, T.; Gleason, J.L. *Tetrahedron Lett.* **1995**, *36*, 4555–4558.

17. Myers, A.G.; Schnider, P.; Kwon, S.; Kung, D.W. *J. Org. Chem.* **1999**, *64*, 3322–3327.

18. Myers, A.G.; Gleason, J.L. *J. Org. Chem.* **1996**, *61*, 813–815.

19. Smith A.B., III; Benowitz, A.B.; Favor, D.A.; Sprengeler, P.A.; Hirschmann, R. *Tetrahedron Lett.* **1997**, *38*, 3809–3812.

20. Walkup, G.K.; Imperiali, B. *J. Org. Chem.* **1998**, *63*, 6727–6731.

21. Roy, R.S.; Imperiali, B. *Tetrahedron Lett.* **1996**, *3*, 2129–2132.

22. Krebs, A.; Ludwig, V.; Pfizer, J.; Dürner, G.; Göbel, M.W. *Chem.—Eur. J.* **2004**, *10*, 544–553.

23. Myers, A.G.; Yoon, T. *Tetrahedron Lett.* **1995**, *36*, 9429–9432.

24. Nagula, G.; Huber, V.J.; Lum, C.; Goodman, B.A. *Org. Lett.* **2000**, *2*, 3527–3529.

25. Myers, A.G.; Yang, B.H.; Chen, H.; McKinstry, L.; Kopecky, D.J.; Gleason, J.L. *J. Am. Chem. Soc.* **1997**, *119*, 6496–6511.

26. Guillena, G.; Nájera, C. *Tetrahedron: Asymmetry* **2001**, *12*, 181–183.

27. Karoyan, P.; Chassaing, G. *Tetrahedron: Asymmetry* **1997**, *8*, 2025–2032.

28. Aitken, D.J.; Royer, J.; Husson, H.-P. *Tetrahedron Lett.* **1988**, *29*, 3315–3318.

29. Aitken, D.J.; Royer, J.; Husson, H.-P. *J. Org. Chem.* **1990**, *55*, 2814–2820.

30. Iwanowicz, E.J.; Blomgren, P.; Cheng, P.T.W.; Smith, K.; Lau, W.F.; Pan, Y.Y.; Gu, H.H.; Malley, M.F.; Gougoutas, Z. *Synlett* **1998**, 664–666.

31. Dixon, D.J.; Horan, R.A.J.; Monck, N.J.T. *Org. Lett.* **2004**, *6*, 4423–4426.

32. Dixon, D.J.; Horan, R.A.J.; Monck, N.J.T.; Berg, P. *Org. Lett.* **2004**, *6*, 4427–4429.

33. Kazmaier, U.; Maier, S. *Chem. Commun.* **1998**, 2535–2536.

34. Kazmaier, U. *J. Org. Chem.* **1994**, *59*, 6667–6670.

35. Kazmaier, U.; Maier, S. *J. Org. Chem.* **1999**, *64*, 4574–4575.

36. Maier, S.; Kazmaier, U. *Eur. J. Org. Chem.* **2000**, 1241–1251.

37. Kazmaier, U. *Amino Acids* **1996**, *11*, 283–299.

38. Kazmaier, U. *Liebigs Ann./Recueil* **1997**, 285–295.

39. Tsunoda, T.; Tatsuki, S.; Shiraishi, Y.; Akasaka, M.; Itô, S. *Tetrahedron Lett.* **1993**, *34*, 3297–3300.

40. Ferey, V.; Toupet, L.; Le Gall, T.; Mioskowski, C. *Angew. Chem., Int. Ed. Engl.* **1996**, *35*, 430–432.

41. Kawabata, T.; Wirth, T.; Yahiro, K.; Suzuki, H.; Fuji, K. *J. Am. Chem. Soc.* **1994**, *116*, 10809–10810.

42. Bakke, M.; Ohta, H.; Kazmaier, U.; Sugai, T. *Synthesis* **1999**, 1671–1677.

43. Kawabata, T.; Kawakami, S.; Majumdar, S. *J. Am. Chem. Soc.* **2003**, *125*, 13012–13013.

44. Seebach, D. *Angew. Chem., Int. Ed. Engl.* **1988**, *27*, 1624–1654.

45. Seebach, D. *Aldrichimica Acta*, **1992**, *25*, 59–66.

46. Seebach, D.; Bossler, H.; Gründler, H.; Shoda, S.-i.; Wenger, R. *Helv. Chim. Acta* **1991**, *74*, 197–224.

47. Bossler, H.G.; Seebach, D. *Helv. Chim. Acta* **1994**, *77*, 1124–1165.

48. Miller, S.A.; Griffiths, S.L.; Seebach, D. *Helv. Chim. Acta* **1993**, *76*, 563–595.

49. Seebach, D.; Bezençon, O.; Jaun, B.; Pietzonka, T.; Matthews, J.L.; Kühnle, F.N.M.; Schweizer, W.B. *Helv. Chim. Acta* **1996**, *79*, 588–608.

50. Seebach, D.; Beck, A.K.; Bossler, H.G.; Gerber, C.; Ko, S.Y.; Murtiashaw, C.W.; Naef, R.; Shoda, S.-I.; Thaler, A.; Krieger, M.; Wenger, R. *Helv. Chim. Acta* **1993**, *76*, 1564–1590.

51. Pietzonka, T.; Seebach, D. *Angew. Chem., Int. Ed. Engl.* **1992**, *31*, 1481–1482.

52. Ager, D.J.; Froen, D.E.; Klix, R.C.; Zhi, B.; McIntosh, J.M.; Thangarasa, R. *Tetrahedron*, **1994**, *50*, 1975–1982.

53. Gerber, C.; Seebach, D. *Helv. Chim. Acta* **1991**, *74*, 1373–1385.

54. Bossler, H.G.; Waldmeier, P.; Seebach, D. *Angew. Chem., Int. Ed. Engl.* **1994**, *33*, 439–440.

55. Matt, T.; Seebach, D. *Helv. Chim. Acta* **1998**, *81*, 1845–1895.

56. Hamilton, B.K.; Hsiao, H.-Y.; Swann, W.E.; Anderson, D.M.; Delente, J.J. *Trends Biotech.* **1985**, *3*, 64–68.

57. Chen, M.S.; Schirch, L.V. *J. Biol. Chem.* **1973**, *248*, 7979–7984.

58. Saeed, A.; Young, D.W. *Tetrahedron* **1992**, *48*, 2507–2514.

59. Vassilev, V.P.; Uchiyama, T.; Kajimoto, T.; Wong, C.-H. *Tetrahedron Lett.* **1995**, *36*, 4081–4084.

60. Herbert, R.B.; Wilkinson, B.; Ellames, G.J.; Kunec, E.K. *J. Chem. Soc., Chem. Commun.* **1993**, 205–206.

61. Kimura, T.; Vassilev, V.P.; Shen, G.-J.; Wong, C.-H. *J. Am. Chem. Soc.* **1997**, *119*, 11734–11742.

62. Murakami, M.; Takahashi, K. *Bull. Chem. Soc. Jpn.* **1959**, *32*, 308–309.

63. Liu, L.; Rozenman, M.; Breslow, R. *Bioorg. Med. Chem.* **2002**, *10*, 3973–3979.

64. Bold, G.; Duthaler, R.O.; Riediker, M. *Angew. Chem., Int. Ed. Engl.* **1989**, *28*, 497–498.

65. Kobayashi, J.; Nakamura, M.; Mori, Y.; Yamashita, Y.; Kobayashi, S. *J. Am. Chem. Soc.* **2004**, *126*, 9192–9193.

66. Wu, W.; Zhang, Y. *Tetrahedron: Asymmetry* **1998**, *9*, 1441–1444.

67. Kazmaier, U.; Krebs, A. *Angew. Chem., Int. Ed. Engl.* **1995**, *34*, 2012–2014.

68. Krebs, A.; Kazmaier, U. *Tetrahedron Lett.* **1996**, *37*, 7945–7946.

69. Kazmaier, U.; Krebs, A. *Tetrahedron Lett.* **1999**, *40*, 479–482.

70. Wen, S.-J.; Yao, Z.-J. *Org. Lett.* **2004**, *6*, 2721–2724.

71. Thayumanavan, R.; Tanaka, F.; Barbase C.F., III; *Org. Lett.* **2004**, *6*, 3541–3544.

72. Bouifraden, S.; Lavergne, J.-P.; Martinez, J.; Viallefont, P.; Riche, C. *Tetrahedron: Asymmetry* **1997**, *8*, 949–955.

73. Grandel, R.; Kazmaier, U.; Romonger, F. *J. Org. Chem.* **1998**, *63*, 4524–4528.

74. Gong, Y.; Kato, K.; Kimoto, H. *Tetrahedron Lett.* **1999**, *40*, 5743–5744.

75. Ezquerra, J.; Pedregal, C.; Merino, I.; Flórez, J.; Barluenga, J.; García–Granda, S.; Llorca, M.-A. *J. Org. Chem.* **1999**, *64*, 6554–6565.

76. Kazmaier, U.; Zumpe, F.L. *Angew. Chem., Int. Ed. Engl.* **1999**, *38*, 1468–1470.

77. Kazmaier, U.; Zumpe, F.L. *Eur. J. Org. Chem.* **2001**, 4067–4076.

78. Kazmaier, U.; Lindner, T. *Angew. Chem., Int. Ed.* **2005**, *44*, 3033–3306.

79. Kazmaier, U.; Schneider, C. *Synlett* **1996**, 975–977.

80. Kazmaier, U.; Schneider, C. *Synthesis* **1998**, 1321–1326.

81. Colombo, L.; Casiraghi, G.; Pittalis, A.; Rassu, G. *J. Org. Chem.* **1991**, *56*, 3897–3900.

82. Samy, R.; Kim, H.Y.; Brady, M.; Toogood, P.L. *J. Org. Chem.* **1999**, *64*, 2711–2728.

83. Gonzalez, D.; Schapiro, V.; Seoane, G.; Hudlicky, T.; Abboud, K. *J. Org. Chem.* **1997**, *62*, 1194–1195.

84. Gerona–Navarro, G.; Bonache, M.A.; Herranz, R.; García–López, M.T.; González–Muñiz, R. *J. Org. Chem.* **2001**, *66*, 3538–3547.

85. Kawabata, T.; Chen, J.; Suzuki, H.; Nagae, Y.; Kinoshita, T.; Chancharunee, S.; Fuji, K. *Org. Lett.* **2000**, *2*, 3883–3885.

86. Ito, Y.; Sawamura, M.; Hayashi, T. *J. Am. Chem. Soc.* **1986**, *108*, 6405–6406.

87. Togni, A.; Pastor, S.D.; Rihs, G. *Helv. Chim. Acta* **1989**, *72*, 1471–1478.

88. Ito, Y.; Sawamura, M.; Hayashi, T. *Tetrahedron Lett.* **1987**, *28*, 6215–6218.

89. Pastor, S.D.; Togni, A. *Tetrahedron Lett.* **1990**, *31*, 839–840.

90. Pastor, S.D.; Kesselring, R.; Togni, A. *J. Organomet. Chem.* **1992**, *429*, 415–420.

91. Hayashi, T.; Sawamura, M.; Ito, Y. *Tetrahedron* **1992**, *48*, 1999–2012.

92. Soloshonok, V.A.; Kacharov, A.D.; Hayashi, T. *Tetrahedron* **1996**, *52*, 245–254.

93. Ito, Y.; Sawamura, M.; Hayashi, T. *Tetrahedron Lett.* **1988**, *29*, 239–240.

94. Ito, Y.; Sawamura, M.; Kobayashi, M.; Hayashi, T. *Tetrahedron Lett.* **1988**, *29*, 6321–6324.

95. Ito, Y.; Sawamura, M.; Hamashima, H.; Emura, T.; Hayashi, T. *Tetrahedron Lett.* **1989**, *30*, 4681–4684.

96. Sawamura, M.; Hamashima, H.; Ito, Y. *J. Org. Chem.* **1990**, *55*, 5935–5936.

97. Hayashi, T.; Uozumi, Y.; Yamazaki, A.; Sawamura, M.; Hamashima, H.; Ito, Y. *Tetrahedron Lett.* **1991**, *32*, 2799–2802.

98. Ito, Y.; Sawamura, M.; Shirakawa, E.; Hayashizaki, K.; Hayashi, T. *Tetrahedron Lett.* **1988**, *29*, 235–238.

99. Ito, Y.; Sawamura, M.; Shirakawa, E.; Hayashizaki, K.; Hayashi, T. *Tetrahedron* **1988**, *44*, 5253–5262.

100. Sawamura, M.; Ito, Y.; Hayashi, T. *Tetrahedron Lett.* 1989, 30, 2247–2250.

101. Togni, A.; Pastor, S.D. *Tetrahedron Lett.* **1989**, *30*, 1071–1072.

102. Willis, M.C.; Cutting, G.A.; Piccio, V. J.-D.; Durbin, M.J.; John, M.P. *Angew. Chem., Int. Ed.* **2005**, *44*, 1543–1545.

103. Pastor, S.D.; Togni, A. *J. Am. Chem. Soc.* **1989**, *111*, 2333–2334.

104. Schöllkopf, U. *Angew. Chem., Int. Ed. Engl.* **1977**, *16*, 339–348.

105. Pastor, S.D.; Togni, A. *Helv. Chim. Acta* **1991**, *74*, 905–933.

106. Togni, A.; Pastor, S.D. *J. Org. Chem.* **1990**, *55*, 1649–1664.

107. Togni, A.; Blumer, R.E.; Pregosin, P.S. *Helv. Chim. Acta* **1991**, *74*, 1533–1543.

108. Zhou, X.-T.; Lin, Y.-R.; Dai, L.-X.; Sun, J.; Xia, L.-J.; Tang, M.-H. *J. Org. Chem.* **1999**, *64*, 1331–1334.

109. Zhou, X.-T.; Lin, Y.-R.; Dai, L.-X. *Tetrahedron: Asymmetry* **1999**, *10*, 855–862.

110. Colonna, S.; Manfredi, A.; Solladíe–Cavallo, A.; Quazzotti, S. *Tetrahedron Lett.* **1990**, *31*, 6185–6188.

111. Reetz, M.T.; Wünsch, T.; Harms, K. *Tetrahedron: Asymmetry* **1990**, *1*, 371–374.

112. Arnone, A.; Gestmann, D.; Meille, S.; Resnati, G.; Sidoti, G. *Chem. Commun.* **1996**, 2569–2570.

113. Suzuki, M.; Matsumoto, K.; Iwasaki, T.; Okumura, K. *Chem. Ind.* **1972**, 687–688.

114. Kirihata, M.; Mihara, S.; Ichimoto, I.; Ueda, H. *Agric. Biol. Chem.* **1978**, *42*, 185–186.

115. Yamamoto, Y.; Kirihata, M.; Ichimoto, I.; Ueda, H. *Agric. Biol. Chem.* **1985**, *49*, 2191–2193.

116. Yamamoto, Y.; Kirihata, M.; Ichimoto, I.; Ueda, H. *Agric. Biol. Chem.* **1985**, *49*, 1761–1765.

117. Pirrung, M.C.; McGeehan, G.M. *J. Org. Chem.* **1986**, *51*, 2103–2106.

118. Evans, D.A. Janey, J.M.; Magomedov, N.; Tedrow, J.S. *Angew. Chem., Int. Ed.* **2001**, *40*, 1884–1888.

119. Tanaka, H.; Sawayama, A.M.; Wandless, T.J. *J. Am. Chem. Soc.* **2003**, *125*, 6864–6865.

120. O'Donnell, M.J. *Aldrichimica Acta* **2001**, *34*, 3–15.

121. Abellán, T.; Chinchilla, R.; Galindo, N.; Guillena, G.; Nájera, C.; Sansano, J.M. *Eur. J. Org. Chem.* **2000**, 2689–2697.

122. Nájera, C. *Synlett* **2002**, 1388–1403.

123. Marouka, K.; Ooi, T. *Chem. Rev.* **2003**, *103*, 3013–3028.

124. O'Donnell, M.J. *Acc. Chem. Res.* **2004**, *37*, 506–517.

125. Lygo, B.; Andrews, B.I. *Acc. Chem. Res.* **2004**, *37*, 518–525.

126. Tanaka, K.; Ahn, M.; Watanabe, Y.; Fuji, K. *Tetrahedron: Asymmetry* **1996**, *7*, 1771–1782.

127. Fasth, K.-J.; Antoni, G.; Långström, B. *J. Chem. Soc., Perkin Trans. 1* **1988**, 3081–3084.

128. Fasth, K.J.; Antoni, G.; Malmborg, P.; Långström, B. *J. Lab. Cmpds. Radiopharm.* **1989**, *26*, 88–89.

129. Antoni, G.; Fasth, K.-J., Malmborg, P.; Långstrom, B. *J. Lab. Cmpds. Radiopharm.* **1985**, *23*, 1057–1058.

130. Alvarez-Ibarra, C.; Csákÿ, A.G.; Maroto, R.; Quiroga, M.L. *J. Org. Chem.* **1995**, *60*, 7934–7940.

131. Evans, D.A.; Bartroli, J.; Shih, T.L. *J. Am. Chem. Soc.* **1981**, *103*, 2127–2129.

132. Evans, D.A.; Weber, A.E. *J. Am. Chem. Soc.* **1986**, *108*, 6757–6761.

133. Boger, D.L.; Colletti, S.L.; Honda, T.; Menezes R.F. *J. Am. Chem. Soc.* **1994**, *116*, 5607–5618.

134. Boger, D.L.; Menezes, R.F. *J. Org. Chem.* **1992**, *57*, 4331–4333.

135. Rich, D.H.; Sun, C.-Q.; Guillaume, D.; Dunlap, B.; Evans, D.A.; Weber, A.E. *J. Med. Chem.* **1989**, *32*, 1982–1987.

136. Belekon, Y.N.; Kochetkov, K.A.; Churkina, T.D.; Ikonnikov, N.S.; Chesnokov, A.A.; Larionov, O.V.;

Singh, I.; Parmar, V.S.; Vyskocil, S.; Kagan, H.B. *J. Org. Chem.* **2000**, *65*, 7041–7048.

137. Zhu, J.; Boillon, J.-P.; Singh, G.P.; Chastanet, J.; Beugelmans, R. *Tetrahedron Lett.* **1995**, *36*, 7081–7084.

138. Evans, D.A.; Wood, M.R.; Trotter, W.B.; Richardson, T.I.; Barrow, J.C.; Katz, J.L. *Angew. Chem., Int. Ed. Engl.* **1998**, *37*, 2700–2704.

139. Evans, D.A.; Weber, A.E. *J. Am. Chem. Soc.* **1987**, *109*, 7151–7157.

140. Lago, M.A. Samanen, J.; Elliot, J.D. *J. Org. Chem.* **1992**, *57*, 3493–3496.

141. Evans, D.A.; Sjogren, E.B.; Weber, A.E.; Conn, R.E. *Tetrahedron Lett.* **1987**, *28*, 39–42.

142. Kanemasa, S.; Mori, T.; Tatsukawa, A. *Tetrahedron Lett.* **1993**, *34*, 8293–8296.

143. Ikegami, S.; Hayama, T.; Katsuki, T.; Yamaguchi, M. *Tetrahedron Lett.* **1986**, *27*, 3403–3406.

144. Ikegami, S.; Uchiyama, H.; Hayama, T.; Katsuki, T.; Yamaguchi, M. *Tetrahedron* **1988**, *44*, 5333–5342.

145. Guillena, G.; Nájera, C. *Tetrahedron: Asymmetry* **1998**, *9*, 1125–1129.

146. Guillena, G.; Nájera, C. *J. Org. Chem.* **2000**, *65*, 7310–7322.

147. Guillena, G.; Nájera, C. *Tetrahedron: Asymmetry* **1998**, *9*, 3935–3938.

148. Oppolzer, W.; Chapuis, C.; Bernardinelli, G. *Helv. Chim. Acta* **1984**, *67*, 1397–1401.

149. Oppolzer, W. *Pure Appl. Chem.* **1990**, *62*, 1241–1250.

150. Oppolzer, W.; Moretti, R.; Zhou, C. *Helv. Chim. Acta* **1994**, *77*, 2363–2380.

151. Oppolzer, W.; Moretti, R.; Thomi, S. *Tetrahedron Lett.* **1989**, *30*, 6009–6010.

152. Lankiewicz, L.; Nyassé, B.; Fransson, B.; Grehn, L.; Ragnarsson, U. *J. Chem. Soc., Perkin Trans. 1* **1994**, 2503–2510.

153. Elemes, Y.; Ragnarsson, U. *Chem. Commun.* **1996**, 935–936.

154. Elemes, Y.; Ragnarsson, U. *J. Chem. Soc., Perkin Trans. 1* **1996**, 537–540.

155. Larsson, E.; Lüning, B. *Acta Chem. Scand.* **1996**, *50*, 54–57.

156. Sjöberg, S.; Hawthorne, M.F.; Wilmouth, S.; Lindström, P. *Chem.—Eur. J.* **1995**, *1*, 430–435.

157. Lindström, P.; Naeslund, C.; Sjöberg, S. *Tetrahedron Lett.* **2000**, *41*, 751–754.

158. Liu, W.-Q.; Roques, B.P.; Garbay–Jaureguiberry, C. *Tetrahedron: Asymmetry* **1995**, *6*, 647–650.

159. Liu, W.-Q.; Roques, B.P.; Garbay, C. *Tetrahedron Lett.* **1997**, *38*, 1389–1392.

160. Josien, H.; Martin, A.; Chassaing, G. *Tetrahedron Lett.* **1991**, *32*, 6547–6550.

161. Josien, H.; Lavielle, S.; Brunissen, A.; Saffroy, M.; Torrens, Y.; Beaujouan, J.-C.; Glowinski, J.; Chassaing, G. *J. Med. Chem.* **1994**, *37*, 1586–1601.

162. Josien, H.; Chassaing, G. *Tetrahedron: Asymmetry* **1992**, *3*, 1351–1354.

163. López, A.; Pleixats, R. *Tetrahedron: Asymmetry* **1998**, *9*, 1967–1977.

164. Deng, W.-P.; Wong, K.A.; Kirk, K.L. *Tetrahedron: Asymmetry.* **2002**, *13*, 1135–1140.

165. Jullian, V.; Monjardet–Bas, V.; Fosse, C.; Lavielle, S.; Chassaing, G. *Eur. J. Org. Chem.* **2002**, 1677–1684.

166. Ayoub, M.; Chassaing, G.; Loffet, A.; Lavielle, S. *Tetrahedron Lett.* **1995**, *36*, 4069–4072.

167. Oppolzer, W.; Rodriguez, I.; Starkemann, C.; Walther, E. *Tetrahedron Lett.* **1990**, *31*, 5019–5022.

168. Bajgrowicz, J.A.; Cossec, B.; Pigière, Ch.; Jacquier, R.; Viallefont, Ph. *Tetrahedron Lett.* **1984**, *25*, 1789–1792.

169. Solladié–Cavallo, A.; Schwarz, J.; Mouza, C. *Tetrahedron Lett.* **1998**, *39*, 3861–3864.

170. Yamada, S.-I.; Oguri, T.; Shioiri, T. *J. Chem. Soc., Chem. Commun.* **1976**, 136–137.

171. Bajgrowicz, J.A.; Cossec, B.; Pigière, Ch.; Jacquier, R.; Viallefont, Ph. *Tetrahedron Lett.* **1983**, *24*, 3721–3724.

172. El Achqar, A.; Roumestant, M.-L.; Viallefont, P. *Tetrahedron Lett.* **1988**, *29*, 2441–2444.

173. Tabcheh, M.; El Achqar, A.; Pappalardo, L.; Roumestant, M.-L.; Viallefont, P. *Tetrahedron* **1991**, *47*, 4611–4618.

174. Hoarau, S.; Fauchère, J.L.; Pappalardo, L.; Roumestant, M.L.; Viallefont, P. *Tetrahedron: Asymmetry* **1996**, *7*, 2585–2593.

175. Jacob, M.; Roumestant, M.L.; Viallefont, P.; Martinez, J. *Synlett* **1997**, 691–692.

176. Sasaki, S.; Hamada, Y.; Shioiri, T. *Synlett* **1999**, 453–455.

177. Papini, A.M.; Nardi, E.; Nuti, F.; Uziel, J.; Ginanneschi, M.; Chelli, M.; Brandi, A. *Eur. J. Org. Chem.* **2002**, 2736–2741.

178. Laue, K.W.; Kröger, S.; Wegelius, E.; Haufe, G. *Eur. J. Org. Chem.* **2000**, 3737–3743.

179. Chaari, M.; Jenhi, A.; Lavergne, J.-P.; Viallefont, P. *Tetrahedron* **1991**, *47*, 4619–4630.

180. Bajgrowicz, J.; El Achquar, A.; Roumestant, M.-L.; Pigière, C.; Viallefont, P. *Tetrahedron Lett.* **1986**, *24*, 2165–2167.

181. See Reference *180*.

182. Solladié–Cavallo, A.; Simon, M.C. *Tetrahedron Lett.* **1989**, *30*, 6011–6014.

183. El Achqar, A.; Boumzebra, M.; Roumestant, M.-L.; Viallefont, P. *Tetrahedron* **1988**, *44*, 5319–5332.

184. El Hadrami, M.; Lavergne, J.-P.; Viallefont, Ph.; Chiaroni, A.; Riche, C.; Hasnaoui, A. *Synth. Commun.* **1993**, *23*, 157–163.

185. Solladié–Cavallo, A.; Nsenda, T. *Tetrahedron Lett.* **1998**, *39*, 2191–2194.

186. Crowley, B.M.; Mori, Y.; McComas, C.C.; Tang, D.; Boger, D.L. *J. Am. Chem. Soc.* **2004**, *126*, 4310–4317.

187. Bentama, A.; Hoarau, S.; Pappalardo, L.; Roumestant, M.L.; Viallefont, P. *Amino Acids* **1994**, *7*, 105–108.

188. Minowa, N.; Hirayama, M.; Fukatsu, S. *Bull. Chem. Soc. Jpn.* **1987**, *60*, 1761–1766.

189. Minowa, N.; Hirayama, M.; Fukatsu, S. *Tetrahedron Lett.* **1984**, *25*, 1147–1150.

190. Solladié–Cavallo, A.; Koesler, J.-L.; Isarno, T.; Roche, D.; Andriamiadanarivo, R. *Synlett* **1997**, 217–218.

191. Wehbe, J.; Rolland, V.; Roumestant, M.L.; Martinez, J. *Tetrahedron: Asymmetry.* **2003**, *14*, 1123–1126.

192. Tabcheh, M.; Guibourdenche, C.; Pappalardo, L.; Roumestant, M.-L.; Viallefont, P. *Tetrahedron: Asymmetry.* **1998**, *9*, 1493–1495.

193. Wehbe, J.; Kassem, T.; Rolland, V.; Rolland, M.; Tabcheh, M.; Roumestant, M.-L.; Martinez, J. *Org. Biomol. Chem.* **2003**, *1*, 1938–1942.

194. McIntosh, J.M.; Mishra, P. *Can. J. Chem.* **1986**, *64*, 726–731.

195. McIntosh, J.M.; Cassidy, K.C.; Matassa, L.C. *Tetrahedron* **1989**, *45*, 5449–5458.

196. McIntosh, J.M.; Leavitt, R.K.; Mishra, P.; Cassidy, K.C.; Drake, J.E.; Chadha, R. *J. Org. Chem.* **1988**, *53*, 1947–1952.

197. McIntosh, J.M.; Cassidy, K.C. *Can. J. Chem.* **1988**, *66*, 3116–3119.

198. McIntosh, J.M.; Leavitt, R.K. *Tetrahedron Lett.* **1986**, *27*, 3839–3842.

199. Kanemasa, S.; Tatsukawa, A.; Wada, E.; Tsuge, O. *Chem. Lett.* **1989**, 1301–1304.

200. Tatsukawa, A.; Dan, M.; Ohbatake, M.; Kawatake, K.; Fukata, T.; Wada, E.; Kanemasa, S.; Kakei, S. *J. Org. Chem.* **1993**, *58*, 4221–4227.

201. Kanemasa, S.; Tatsukawa, A.; Wada, E. *J. Org. Chem.* **1991**, *56*, 2875–2883.

202. Kröger, S.; Haufe, G. *Liebigs Ann./Recueil* **1997**, 1201–1206.

203. Laue, K.W.; Mück–Lichtenfeld, C.; Haufe, G. *Tetrahedron* **1999**, *55*, 10413–10424.

204. Yaozhong, J.; Changyou, Z.; Huri, P. *Synth. Commun.* **1989**, *19*, 881–888.

205. Yeh, T.-L.; Liao, C.-C.; Uang, B.-J. *Tetrahedron* **1997**, *53*, 11141–11152.

206. Chen, F.-Y.; Uang, B.-J. *J. Org. Chem.* **2001**, *66*, 3650–3652.

207. Xu, P.-F.; Chen, Y.-S.; Lin, S.-I.; Lu, T.-J. *J. Org. Chem.* **2002**, *67*, 2309–2314.

208. Casella, L.; Jommi, G.; Montanari, S.; Sisti, M. *Tetrahedron Lett.* **1988**, *29*, 2067–2068.

209. Casella, L.; Gullotti, M.; Jommi, G.; Pagliarin, R.; Sisti, M. *J. Chem. Soc., Perkin Trans. 1* **1990**, 771–775.

210. Meyer, L.; Poirier, J.-M.; Duhamel, P.; Duhamel, L. *J. Org. Chem.* **1998**, *63*, 8094–8095.

211. Nakatsuka, T.; Miwa, T.; Mukaiyama, T. *Chem. Lett.* **1981**, 279–282.

212. Oguri, T.; Shioiri, T.; Yamada, S.-i. *Chem. Pharm. Bull.* **1977**, *25*, 2287–2291.

213. Schöllkopf, U.; Tölle, R.; Egert, E.; Nieger, M. *Liebigs Ann. Chem.* **1987**, 399–405.

214. Antonov, D.Y.; Belokon, Y.N.; Ikonnikov, N.S.; Orlova, S.A.; Pisarevsky, A.P.; Raevski, N.I.; Rozenberg, V.I.; Sergeeva, E.V.; Struchkov, Y.T.; Tararov, V.I.; Vorontsov, E.V. *J. Chem. Soc., Perkin Trans. 1* **1995**, 1873–1879.

215. Miyashita, K.; Miyabe, H.; Tai, K.; Kurozumi, C.; Imanishi, T. *Chem Commun.* **1996**, 1073–1074.

216. Miyashita, K.; Miyabe, H.; Tai, K.; Iwaki, H.; Imanishi, T. *Tetrahedron* **2000**, *56*, 4691–4700.

217. Miyashita, K.; Miyabe, H.; Tai, K.; Kurozumi, C.; Iwaki, H.; Imanishi, T. *Tetrahedron* **1999**, *55*, 12109–12124.

218. Miyashita, K.; Iwaki, H.; Tai, K.; Murafuji, H.; Imanishi, T. *Chem. Commun.* **1998**, 1987–1988.

219. Miyashita, K.; Iwaki, H.; Tai, K.; Murafuji, H.; Sasaki, N.; Imanishi, T. *Tetrahedron* **2001**, *57*, 5773–5780.

220. Kolb, M.; Barth, J. *Liebigs Ann. Chem.* **1983**, 1668–1688.

221. Kolb, M.; Barth, J. *Tetrahedron Lett.* **1979**, *32*, 2999–3002.

222. Kolb, M.; Barth, J. *Angew. Chem., Int. Ed. Engl.* **1980**, *19*, 725–726.

223. O'Donnell, M.J.; Bennett, W.D.; Wu, S. *J. Am. Chem. Soc.* **1989**, *111*, 2353–2355.

224. O'Donnell, M.J.; Wu, S.; Huffman, J.C. *Tetrahedron* **1994**, *50*, 4507–4518.

225. Nelson, A. *Angew. Chem., Int. Ed. Engl.* **1999**, *38*, 1583–1585.

226. Lipkowitz, K.B.; Cavanaugh, M.W.; Baker, B.; O'Donnell, M.J. *J. Org. Chem.* **1991**, *56*, 5181–5192.

227. O'Donnell, M.J.; Wu, S. *Tetrahedron: Asymmetry* **1992**, *3*, 591–594.

228. Lygo, B.; Crosby, J.; Peterson, J.A. *Tetrahedron Lett.* **1999**, *40*, 8671–8674.

229. O'Donnell, M.J.; Delgado, F. *Tetrahedron* **2001**, *57*, 6641–6650.

230. Dehmlow, E.V.; Wagner, S.; Müller, A. *Tetrahedron* **1999**, *55*, 6335–6346.

231. Lygo, B.; Wainwright, P.G. *Tetrahedron Lett.* **1997**, *38*, 8595–8598.

232. Lygo, B.; Crosby, J.; Lowdon, T.; Wainwright, P.G. *Tetrahedron* **2001**, *57*, 2391–2402.

233. Lygo, B.; Crosby, J.; Lowdon, T.; Peterson, J.A.; Wainwright, P.G. *Tetrahedron* **2001**, *57*, 2403–2409.

234. Lygo, B.; Crosby, J.; Peterson, J.A. *Tetrahedron Lett.* **1999**, *40*, 1385–1388.

235. Lygo, B. *Tetrahedron Lett.* **1999**, *40*, 1389–1392.

236. Horikawa, M.; Busch–Petersen, J.; Corey, E.J. *Tetrahedron Lett.* **1999**, *40*, 3843–3846.

237. Corey, E.J.; Noe, M.C.; Xu, F. *Tetrahedron Lett.* **1998**, *39*, 5347–5350.

238. Jotterand, N.; Pearce, D.A.; Imperiali, B. *J. Org. Chem.* **2001**, *66*, 3224–3228.

239. Shults, M.D.; Pearce, D.A.; Imperiali, B. *J. Am. Chem. Soc.* **2003**, *125*, 10591–10597.

240. Cohen, B.E.; McAnaney, T.B.; Park, E.S.; Jan, Y.N.; Boxer, S.G.; Jan, L.Y. *Science* **2002**, *296*, 1700–1703.

241. Viswanathan, R.; Prabhakaran, E.N.; Plotkin, M.A.; Johnston, J.N. *J. Am. Chem. Soc.* **2003**, *125*, 163–168.

242. Lygo, B.; Humphreys, L.D. *Tetrahedron Lett.* **2002**, *43*, 6677–6679.

243. Lemaire, C.; Gillet, S.; Guillouet, S.; Plenevaux, A.; Aerts, J.; Luxen, A. *Eur. J. Org. Chem.* **2004**, 2899–2904.

244. Pirrung, M.C.; Krishnamurthy, N. *J. Org. Chem.* **1993**, *58*, 954–956.

245. Nakoji, M.; Kanayama, T.; Okino, T.; Takemoto, Y. *J. Org. Chem.* **2002**, *67*, 7418–7423.

246. Ramachandran, P.V.; Madhi, S.; Bland–Berry, L.; Ram Reddy, M.V.; O'Donnell, M.J. *J. Am. Chem. Soc.* **2005**, *127*, 13450–13451.

247. Kumar, S.; Ramachandran, U. *Tetrahedron: Asymmetry* **2003**, *14*, 2539–2545.

248. Yu, H.; Koshima, H. *Tetrahedron Lett.* **2003**, *44*, 9209–9211.

249. Okino, T.; Takemoto, Y. *Org. Lett.* **2001**, *3*, 1515–1517.

250. Mase, N.; Ohno, T.; Morimoto, H.; Nitta, F.; Yoda, H.; Takabe, K. *Tetrahedron Lett.* **2005**, *46*, 3213–3216.

251. O'Donnell, M.J.; Delgado, F.; Hostettler, C.; Schwesinger, R. *Tetrahedron Lett.* **1998**, *39*, 8775–8778.

252. O'Donnell, M.J.; Delgado, F.; Pottorf, R.S. *Tetrahedron* **1999**, *55*, 6347–6362.

253. Park, H.-g.; Kim, M.-J.; Park, M.-K.; Jung, H.-J.; Lee, J.; Choi, S.-h.; Lee, Y.-J.; Jeong, B.-S.; Lee, J.-H.; Yoo, M.-S.; Ku, J.M.; Jew, S.-s. *J. Org. Chem.* **2005**, *70*, 1904–1906.

254. Lygo, B.; Andrews, B.I.; Crosby, J.; Peterson, J.A. *Tetrahedron Lett.* **2002**, *43*, 8015–8018.

255. Andrus, M.B.; Ye, Z.; Zhang, J. *Tetrahedron Lett.* **2005**, *46*, 3839–3842.

256. Kumar, S.; Ramachandran, U. *Tetrahedron* **2005**, *61*, 7022–7028.

257. Mettath, S.; Srikanth, G.S.C.; Dangerfield, B.S.; Castle, S.L. *J. Org. Chem.* **2004**, *69*, 6489–6492.

258. Mazon, P.; Chinchilla, R.; Nájera, C.; Guillena, G.; Kreiter, R.; Kelin Gebbink, R.J.M.; van Koten, G. *Tetrahedron: Asymmetry* **2002**, *13*, 2181–2185.

259. Park, H.-G.; Jeong, B.-S.; Yoo, M.-S.; Lee, J.-H.; Park, M.-K.; Lee, Y.-J.; Kim, M.-J.; Jew, S.-S. *Angew. Chem., Int. Ed.* **2002**, *41*, 3036–3038.

260. Danner, P.; Bauer, M.; Phukan, P.; Maier, M.E. *Eur. J. Org. Chem.* **2005**, 317–325.

261. Chinchilla, R.; Mazon, P.; Nájera, C. *Tetrahedron: Asymmetry.* **2002**, *13*, 927–931.

262. Chinchilla, R.; Mazón, P.; Nájera, C.; Ortega, F.J. *Tetrahedron: Asymmetry* **2004**, *15*, 2603–2607.

263. Park, H.-g.; Jeong, B.-S.; Yoo, M.-S.; Lee, J.-H.; Park, B.-s.; Kim, M.G.; Jew, S.-s. *Tetrahedron Lett.* **2003**, *44*, 3497–3500.

264. Jew, S.-s.; Jeong, B.-S.; Yoo, M.-S.; Huh, H.; Park, H,-g. *Chem. Commun.* **2001**, 1244–1245.

265. Park, H.-g.; Jeong, B.-s.; Yoo, M.-s.; Park, M.-s.; Huh, H.; Jew, S.-s. *Tetrahedron Lett.* **2001**, *42*, 4645–4648.

266. Guillena, G.; Kreiter, R.; van de Coevering, R.; Klein Gebbink, R.J.M.; van Koten, G.; Mazón, P.; Chinchilla, R.; Nájera, C. *Tetrahedron: Asymmetry* **2003**, *14*, 3705–3712.

267. Danelli, T.; Annunziata, R.; Benaglia, M.; Cinquini, M.; Cozzi, F.; Tocco, G. *Tetrahedron: Asymmetry* **2003**, *14*, 461–467.

268. Thierry, B.; Plaquevent, J.-C.; Cahard, D. *Tetrahedron: Asymmetry* **2003**, *14*, 1671–1677.

269. Yu, H.; Takigawa, S.; Koshima, H. *Tetrahedron* **2004**, *60*, 8405–8410.

270. Ooi, T.; Kameda, M.; Maruoka, K. *J. Am. Chem. Soc.* **1999**, *121*, 6519–6520.

271. Ooi, T.; Kameda, M.; Marouka, K. *J. Am. Chem. Soc.* **2003**, *125*, 5139–5151.

272. Ooi, T.; Kameda, M.; Tannai, H.; Maruoka, K. *Tetrahedron Lett.* **2000**, *41*, 8339–8342.

273. Ooi, T.; Takeuchi, M.; Kameda, M.; Maruoka, K. *J. Am. Chem. Soc.* **2000**, *122*, 5228–5229.

274. Ooi, T.; Taniguchi, M.; Kameda, M.; Maruoka, K. *Angew. Chem., Int. Ed.* **2002**, *41*, 4542–4544.

275. Hashimoto, T.; Maruoka, K. *Tetrahedron Lett.* **2003**, *44*, 3313–3316.

276. Hashimoto, T.; Tanaka, Y.; Maruoka, K. *Tetrahedron: Asymmetry* **2003**, *14*, 1599–1602.

277. Ooi, T.; Kameda, M.; Taniguchi, M.; Maruoka, K. *J. Am. Chem. Soc.* **2004**, *126*, 9685–9694.

278. Maruoka, K.; Tayama, E.; Ooi, T. *Proc. Natl. Acad. Sci. U.S.A.* **2004**, *101*, 5824–5829.

279. Ooi, T.; Tayama, E.; Maruoka, K. *Angew. Chem., Int. Ed.* **2003**, *42*, 579–582.

280. Ooi, T.; Uematsu, Y.; Maruoka, K. *Tetrahedron Lett.* **2004**, *45*, 1675–1678.

281. Shirakawa, S.; Tanaka, Y.; Maruoka, K. *Org. Lett.* **2004**, *6*, 1429–1431.

282. Kitamura, M.; Shirakawa, S.; Maruoka, K. *Angew. Chem., Int. Ed.* **2005**, *44*, 1549–1551.

283. Ooi, T.; Uematsu, Y.; Kameda, M.; Maruoka, K. *Angew. Chem., Int. Ed.* **2002**, *41*, 1551–1554.

284. Han, Z.; Yamaguchi, Y.; Kitamura, M.; Maruoka, K. *Tetrahedron Lett.* **2005**, *46*, 8555–8558.

285. Lygo, B.; Allbutt, B.; James, S.R. *Tetrahedron Lett.* **2003**, *44*, 629–5632.

286. Lygo, B.; Allbutt, B.; Kirton, E.H.M. *Tetrahedron Lett.* **2005**, *46*, 4461–4464.

287. Kano, T.; Konishi, S.; Shirakawa, S.; Maruoka, K. *Tetrahedron: Asymmetry* **2004**, *15*, 1243–1245.

288. Ooi, T.; Takeuchi, M.; Kato, D.; Uematsu, Y.; Tayama, E.; Sakai, D.; Maruoka, K. *J. Am. Chem. Soc.* **2005**, *127*, 5073–5083.

289. Ooi, T.; Sakai, D.; Takeuchi, M.; Tayama, E.; Maruoka, K. *Angew. Chem., Int. Ed.* **2003**, *42*, 5868–5870.

290. Kumar, S.; Ramanchandran, U. *Tetrahedron Lett.* **2005**, *46*, 19–21.

291. Jew, S.-s.; Lee, Y.-J.; Lee, J.; Kang, M.K.; Jeong, B.-S.; Lee, J.-H.; Yoo, M.-S.; Kim, M.-J.; Choi, S.-h.; Ku, J.-M.; Park, H.-g. *Angew. Chem. Int. Ed.* **2004**, *43*, 2382–2385.

292. Lee, Y.-J.; Lee, J.; Kim, M.-J.; Kim, T.-S.; Park, H.-g.; Jew, S.-s. *Org. Lett.* **2005**, *7*, 1557–1560.

293. Kowtoniuk, W.E.; Rueffer, M.E.; MacFarland, D.K. *Tetrahedron: Asymmetry* **2004**, *15*, 151–154.

294. Kumar, S.; Sobhia, M.E.; Ramachandran, U. *Tetrahedron: Asymmetry* **2005**, *16*, 2599–2605.

295. Mase, N.; Ohno, T.; Hoshikawa, N.; Ohishi, K.; Morimoto, H.; Yoda, H.; Takabe, K. *Tetrahedron Lett.* **2003**, *44*, 4073–4075.

296. Okada, A.; Shibuguchi, T.; Ohshima, T.; Masu, H.; Yamaguchi, K.; Shibasaki, M. *Angew. Chem., Int. Ed.* **2005**, *44*, 4564–4567.

297. Kowtoniuk, W.E.; MacFarland, D.K.; Grover, G.N. *Tetrahedron Lett.* **2005**, *46*, 5703–5705.

298. Belokon, Y.N.; Bespalova, N.B.; Churkina, T.D.; Císarová, I.; Ezernitskaya, M.G.; Harutyunyan, S.R.; Hrdina, R.; Kagan, H.B.; Kocovsky, P.; Kochetkov, K.A.; Larionov, O.V.; Lyssenko, K.A.; North, M.; Polásek, M.; Peregudov, A.S.; Prisyazhnyuk, V.V.; Vyskocil, S. *J. Am. Chem. Soc.* **2003**, *125*, 12860–12871.

299. Ohshima, T.; Gnanadesikan, V.; Shibuguchi, T.; Fukuta, Y.; Nemoto, T.; Shibasaki, M. *J. Am. Chem. Soc.* **2003**, *125*, 11206–11207.

300. Fukuta, Y.; Ohshima, T.; Gnanadesikan, V.; Shibuguchi, T.; Nemoto, T.; Kisugi, T.; Okino, T.; Shibasaki, M. *Proc. Natl. Acad. Sci. U.S.A.* **2004**, *101*, 5433–5438.

301. Shibuguchi, T.; Fukuta, Y.; Akachi, Y.; Sekine, A.; Ohshima, T.; Shibasaki, M. *Tetrahedron Lett.* **2002**, *43*, 9539–9543.

302. Rueffer, M.E.; Fort, L.K.; MacFarland, D.K. *Tetrahedron: Asymmetry* **2004**, *15*, 3297–3300.

303. Arai, S.; Tsuji, R.; Nishida, A. *Tetrahedron Lett.* **2002**, *43*, 9535–9537.

304. Ma, D.; Cheng, K. *Tetrahedron: Asymmetry* **1999**, *10*, 713–719.

305. Ishikawa, T.; Araki, Y.; Kumamoto, T.; Seki, H.; Fukuda, K.; Isobe, T. *Chem. Commun.* **2001**, 245–246.

306. Kita, T.; Georgieva, A.; Hashimoto, Y.; Nakata, T.; Nagasawa, K. *Angew. Chem., Int. Ed.* **2002**, *41*, 2832–2834.

307. Akiyama, T.; Hara, M.; Fuchibe, K.; Sakamoto, S.; Yamaguchi, K. *Chem. Commun.* **2003**, 1734–1735.

308. Belokon, Y.N.; Kochetkov, K.A.; Churkina, T.D.; Ikonnikov, N.S.; Vyskocil, S.; Kagan, H.B. *Tetrahedron: Asymmetry* **1999**, *10*, 1723–1728.

309. Belekon, Y.N.; Kochetkov, K.A.; Churkina, T.D.; Ikonnikov, N.S.; Chesnokov, A.A.; Larionov, O.V.; Parmár, V.S.; Kumar, R.; Kagan, H.B. *Tetrahedron: Asymmetry* **1998**, *9*, 851–857.

310. Kumar, S.; Ramachandran, U. *Tetrahedron* **2005**, *61*, 4141–4148.

311. Belekon, Y.N.; North, M.; Churkina, T.D.; Ikonnikov, N.S.; Maleev, V.I. *Tetrahedron* **2001**, *57*, 2491–2498.

312. Belekon, Y.N.; Davies, R.G.; North, M. *Tetrahedron Lett.* **2000**, *41*, 7245–7248.

313. Belekon, Y.N.; Bhave, D.; D'Addario, D.; Groaz, E.; Maleev, V.; North, M.; Pertrosyan, A. *Tetrahedron Lett.* **2003**, *44*, 2045–2048.

314. Banti, D.; Belekon, Y.N.; Fu, W.-L.; Groaz, E.; North, M. *Chem. Commun.* **2005**, 2707–2709.

315. Baldwin, I.C.; Williams, J.M.J.; Beckett, R.P. *Tetrahedron: Asymmetry* **1995**, *6*, 1515–1518.

316. Genet, J.P.; Ferroud, D.; Juge, S.; Montes, J.R. *Tetrahedron Lett.* **1986**, *27*, 4573–4576.

317. Genêt, J.-P.; Jugé, S.; Montès, J.R.; Gaudin, J.-M. *J. Chem. Soc., Chem. Commun.* **1988**, 718–719.

318. Genet, J.-P.; Juge, S.; Achi, S.; Mallart, S.; Ruiz Montes, J.; Levif, G. *Tetrahedron* **1988**, *44*, 5263–5275.

319. Genet, J.P.; Kopola, N.; Juge, S.; Ruiz-Montes, J.; Antunes, O.A.C.; Tanier, S. *Tetrahedron Lett.* **1990**, *31*, 3133–3136.

320. Dorizon, P.; Su, G.; Ludvig, G.; Nikitina, L.; Paugam, R.; Ollivier, J.; Salaün, J. *J. Org. Chem.* **1999**, *64*, 4712–4724.

321. Allway, P.; Grigg, R. *Tetrahedron Lett.* **1991**, *32*, 5817–5820.

322. Kanayama, T.; Yoshida, K.; Miyabe, H.; Kimachi, T.; Takemoto, Y. *J. Org. Chem.* **2003**, *68*, 6197–6201.

323. Bernardi, L.; Gothelf, A.S.; Hazell, R.G.; Jørgensen, K.A. *J. Org. Chem.* **2003**, *68*, 2583–2591.

324. Belekon, Y.N.; Maleev, V.I.; Videnskaya, S.O.; Saporovskaya, M.B.; Tsyrypkin, V.A.; Belikov, V.M. *Bull. Acad. Sci. USSR Div. Chem. Sci.* **1991**, *40*, 110–118.

325. Belekon, Y.N.; Kochetkov, K.A.; Churkina, T.D.; Ikonnikov, N.S.; Larionov, O.V.; Harutyunyan, S.R.; Vyskocil, S.; North, M.; Kagan, H.B. *Angew. Chem., Int. Ed.* **2001**, *40*, 1948–1950.

326. Vyskocil, S.; Meca, L.; Tislerová, I.; Císarová, I.; Polásek, M.; Harutyunyan, S.R.; Belekon, Y.N.; Stead, R.M.H.; Farrugia, L.; Lockhart, S.C.; Mitchell, W.L.; Kocovsky, P. *Chem.—Eur. J.* **2002**, *8*, 4633–4648.

327. Chaly, T.; Diksic, M. *J. Nucl. Med.* **1988**, *29*, 370–374.

328. Duhamel, L.; Duhamel, P.; Fourquay, S.; Eddine, J.J.; Peschard, O.; Plaquevent, J.-C.; Ravard, A.; Solliard, R.; Valnot, J.-Y.; Vincens, H. *Tetrahedron* **1988**, *44*, 5495–5506

329. Duhamel, L.; Fourquay, S.; Plaquevent, J.-C. *Tetrahedron Lett.* **1986**, *27*, 4975–4978.

330. Duhamel, L.; Plaquevent, J.-C. *Tetrahedron Lett.* **1980**, *21*, 2521–2524.

331. Duhamel, L.; Plaquevent, J.-C. *J. Am. Chem. Soc.* **1978**, *100*, 7415–7416.

332. Duhamel, L.; Duhamel, P.; Plaqueven, J.-C. *Tetrahedron: Asymmetry* **2004**, *15*, 3653–3691.

333. Barlaam, B.; Koza, P.; Berriot, J. *Tetrahedron* **1999**, *55*, 7221–7232.

334. Ikeda, D.; Kawatsura, M.; Uenishi, J. *Tetrahedron Lett.* **2005**, *46*, 6663–6666.

335. Mazaletrat, J.-P.; Gaucher, A.; Savrda, J.; Wakselman, M. *Tetrahedron: Asymmetry* **1997**, *8*, 619–631.

336. Soloshonok, V.A.; Ueki, H.; Tiwari, R.; Cai, C.; Hruby, V.J. *J. Org. Chem.* **2004**, *69*, 4984–4990.

337. Soloshonok, V.A.; Cai, C.; Hruby, V.J. *Org. Lett.* **2000**, *2*, 747–750.

338. Soloshonok, V.A.; Ueki, H.; Jiang, C.; Cai, C.; Hruby, V.J. *Helv. Chim. Acta* **2002**, *85*, 3616–3623.

339. Hercouet, A.; Bessières, B.; Le Corre, M. *Tetrahedron: Asymmetry* **1996**, *7*, 283–284.

340. Chavan, S.P.; Sharma, P.; Sivappa, R.; Bhadbhade, M.M.; Gonnade, R.G.; Kalkote, U.R. *J. Org. Chem.* **2003**, *68*, 6817–6819.

341. Shibuya, A.; Kurishita, M.; Ago, C.; Taguchi, T. *Tetrahedron* **1996**, *52*, 271–278.

342. Barr, D.A.; Dorrity, M.J.; Grigg, R.; Malone, J.F.; Montgomery, J.; Rajviroongit, S.; Stevenson, P. *Tetrahedron Lett.* **1990**, *31*, 6569–6572.

343. Apitz, G.; Jäger, M.; Jaroch, S.; Kratzel, M.; Schäffeler, L.; Steglich, W. *Tetrahedron*, **1993**, *49*, 8223–8232.

344. Steglich, W.; Jäger, M.; Jaroch, S.; Zistler, P. *Pure Appl. Chem.* **1994**, *66*, 2167–2170.

345. Apitz, G.; Steglich, W. *Tetrahedron Lett.* **1991**, *32*, 3163–3166.

346. Paulitz, C.; Steglich, W. *J. Org. Chem.* **1997**, *62*, 8474–8478.

347. Jaroch, S.; Schwarz, T.; Steglich, W.; Zistler, P. *Angew. Chem., Int. Ed. Engl.* **1993**, *32*, 1771–1772.

348. Easton, C.J.; Scharfbillig, I.M.; Tan, E.W. *Tetrahedron Lett.* **1988**, *29*, 1565–1568.

349. Niddam, V.; Camplo, M.; Le Nguyen, D.; Chermann, J.-C.; Kraus, J.-L. *Bioorg. Med. Chem. Lett.* **1996**, *6*, 609–614.

350. Kober, R.; Papadopoulos, K.; Miltz, W.; Enders, D.; Steglich, W.; Reuter, H.; Puff, H. *Tetrahedron* **1985**, *41*, 1693–1701.

351. Ermert, P.; Meyer, J.; Stucki, C.; Schneebeli, J.; Obrecht, J.-P. *Tetrahedron Lett.* **1988**, *29*, 1265–1268.

352. Hamon, D.P.G.; Massy–Westropp, R.A.; Razzino, P. *Tetrahedron* **1992**, *48*, 5163–5178.

353. Morgan, P.E.; Whiting, A.; McCague, R. *Tetrahedron Lett.* **1996**, *37*, 4795–4796.

354. Harding, K.E.; Davis, C.S. *Tetrahedron Lett.* **1988**, *29*, 1891–1894.

355. Pandey, G.; Reddy, P.Y.; Das, P. *Tetrahedron Lett.* **1996**, *37*, 3175–3178.

356. O'Donnell, M.J.; Zhou, C.; Chen, N. *Tetrahedron: Asymmetry* **1996**, *7*, 621–624.

357. O'Donnell, M.J.; Chen, N.; Zhou, C.; Murray, A.; Kubiak, C.P.; Yang, F.; Stanley, G.G. *J. Org. Chem.* **1997**, *62*, 3962–3975.

358. O'Donnell, M.J.; Zhou, C.; Mi, A.; Chen, N.; Kyle, J.A.; Andersson, P.G. *Tetrahedron Lett.* **1995**, *36*, 4205–4208.

359. Chen, Y.; Yudin, A.K. *Tetrahedron Lett.* **2003**, *44*, 4865–4868.

360. Attrill, R.; Tye, H.; Cox, L.R. *Tetrahedron: Asymmetry* **2004**, *15*, 1681–1684.

361. O'Donnell, M.J.; Cooper, J.T.; Mader, M.M. *J. Am. Chem. Soc.* **2003**, *125*, 2370–2371.

362. O'Donnell, M.J.; Drew, M.D.; Cooper, J.T.; Delgado, F.; Zhou, C. *J. Am. Chem. Soc.* **2002**, *124*, 9348–9349.

363. Enders, D.; Reinhold, U. *Tetrahedron: Asymmetry* **1997**, *8*, 1895–1946.

364. Hallett, D.J.; Thomas, E.J. *Tetrahedron: Asymmetry* **1995**, *6*, 2575–2578.

365. Bradley, G.W.; Hallett, D.J.; Thomas, E.J. *Tetrahedron: Asymmetry* **1995**, *6*, 2579–2582.

366. Hallett, D.J.; Thomas, E.J. *J. Chem. Soc., Chem. Commun.* **1995**, 657–658.

367. Hanessian, S.; Yang, R.-Y. *Tetrahedron Lett.* **1996**, *37*, 8997–9000.

368. Schleusner, M.; Gais, H.-J.; Koep, S.; Raabe, G. *J. Am. Chem. Soc.* **2002**, *124*, 7789–7800.

369. Günter, M.; Gais, H.-J. *J. Org. Chem.* **2003**, *68*, 8037–8041.

370. Bravo, P.; Capelli, S.; Meille, S.V.; Viani, F.; Zanda, M.; Kukhar, V.P.; Soloshonok, V.A. *Tetrahedron: Asymmetry* **1994**, *5*, 2009–2018.

371. Bravo, P.; Viani, F.; Zanda, M.; Kukhar, V.P.; Soloshonok, V.A.; Fokina, N.; Sishkin, O.V.; Struchkov, Y.T. *Gazz. Chim. Ital.* **1996**, *126*, 645–652.

372. Bravo, P.; Viani, F.; Zanda, M.; Soloshonok, V. *Gazz. Chim. Ital.* **1995**, *125*, 149–150.

373. Yamamoto, Y.; Ito, W.; Maruyama, K. *J. Chem. Soc., Chem. Commun.* **1985**, 1131–1132.

374. Yamamoto, Y.; Ito, W. *Tetrahedron* **1988**, *44*, 5415–5423.

375. Yamamoto, Y.; Nishii, S.; Maruyama, K.; Komatsu, T.; Ito, W. *J. Am. Chem. Soc.* **1986**, *108*, 7778–7786.

376. Bandini, M.; Cozzi, P.G.; Umani–Ronchi, A.; Villa, M. *Tetrahedron* **1999**, *55*, 8103–8110.

377. Chiev, K.P.; Roland, S.; Mangeney, P. *Tetrahedron: Asymmetry* **2002**, *13*, 2205–2209.

378. Loh, T.-P.; Ho, D.S.-C.; Xu, K.-C.; Sim, K.-Y. *Tetrahedron Lett.* **1997**, *38*, 865–868.

379. Fiaud, J-C.; Kagan, H.B. *Tetrahedron Lett.* **1971**, *15*, 1019–1022.

380. Fiaud, J-C.; Kagan, H.B. *Tetrahedron Lett.* **1970**, *21*, 1813–1816.

381. Petasis, N.A.; Zavialov, I.A. *J. Am. Chem. Soc.*, **1997**, *119*, 445–446.

382. Petasis, N.A.; Goodman, A.; Zavialov, I.A. *Tetrahedron* **1997**, *53*, 16463–16470.

383. Davis, F.A.; McCoull, W. *J. Org. Chem.* **1999**, *64*, 3396–3397.

384. Jayathilaka, L.P.; Deb. M.; Standaert, R.F. *Org. Lett.* **2004**, *6*, 3659–3662.

385. Jacobsen, M.F.; Skrydstrup, T. *J. Org. Chem.* **2003**, *68*, 7112–7114.

386. Qin, Y.; Wang, C.; Huang, Z.; Xiao, X.; Jiang, Y. *J. Org. Chem.* **2004**, *69*, 8533–8536.

387. Bravo, P.; Crucianelli, M.; Vergani, B.; Zanda, M. *Tetrahedron Lett.* **1998**, *39*, 7771–7774.

388. Asensio, A.; Bravo, P.; Crucianelli, M.; Farina, A.; Fustero, S.; Soler, J.G.; Meille, S.V.; Panzeri, W.; Viani, F.; Volonterio, A.; Zanda, M. *Eur. J. Org. Chem.* **2001**, 1449–1458.

389. Zhou, P.; Chen, B.-C.; Davis, F.A. *Tetrahedron* **2004**, *60*, 8003–8030.

390. Di Fabio, R.; Alvaro, G.; Bertani, B.; Donati, D.; Giacobbe, S.; Marchioro, C.; Palma, C.; Lynn, S.M. *J. Org. Chem.* **2002**, *67*, 7319–7328.

391. Mikami, K.; Kaneko, M.; Yajima, T. *Tetrahedron Lett.* **1993**, *34*, 4841–4842.

392. Kolasa, T.; Sharma, S.K.; Miller, M.J. *Tetrahedron* **1988**, *44*, 5431–5440.

393. Hanessian, S.; Yang, R.-Y. *Tetrahedron Lett.* **1996**, *37*, 5273–5276.

394. Hanessian, S.; Bernstein, N.; Yang, R.-Y.; Maguire, R. *Biorg. Med. Chem. Lett.* **1999**, *9*, 1437–1442.

395. Miyabe, H.; Nishimura, A.; Ueda, M.; Naito, T. *Chem. Commun.* **2002**, 1454–1455.

396. Lee, J.G.; Choi, K.I.; Pae, A.N.; Koh, H.Y.; Kang, Y.; Cho, Y.S. *J. Chem. Soc., Perkin Trans. 1* **2002**, 1314–1317.

397. Drok, K.J.; Harrowfield, J.M.; McNiven, S.J.; Sargeson, A.M.; Skelton, B.W.; White, A.H. *Aust. J. Chem.* **1993**, *46*, 1557–1593.

398. Harwodd, L.M.; Vines, K.J.; Drew, M.G.B. *Synlett* **1996**, 1051–1053.

399. Seewald, N.; Seymour, L.C.; Burger, K.; Osipov, S.N.; Kolomiets, A.F.; Fokin, A.V. *Tetrahedron: Asymmetry* **1994**, *5*, 1051–1060.

400. Alexakis, A.; Lensen, N.; Mangeney, P. *Tetrahedron Lett.* **1991**, *32*, 1171–1174.

401. Thiam, M.; Chastrette, F. *Tetrahedron Lett.* **1990**, *31*, 1429–1432.

402. Enders, D.; Funk, R.; Klatt, M.; Raabe, G.; Hovestreydt, E.R. *Angew. Chem., Int. Ed. Engl.* **1993**, *32*, 418–421.

403. Cativiela, C.; Diaz–de–Villegas, M.D.; Gálvez, J.A. *Tetrahedron: Asymmetry* **1996**, *7*, 529–536.

404. Badorrey, R.; Cativiela, C.; Díaz–de–Villegas, M.; Gálvez, J.A. *Tetrahedron* **1997**, *53*, 1411–1416.

405. Badorrey, R.; Cativiela, C.; Díaz–de–Villegas, M.D.; Gálvez, J.A. *Synthesis* **1997**, 747–749.

406. Cativiela, C.; Díaz–de–Villegas, M.D.; Gálvez, J.A. *Tetrahedron: Asymmetry* **1997**, *8*, 1605–1610.

407. Merino, P.; Castillo, E.; Franco, S.; Merchan, F.L.; Tejero, T. *J. Org. Chem.* **1998**, *63*, 2371–2374.

408. Marco, J.A.; Carda, M.; Murga, J.; González, F.; Falomir, E. *Tetrahedron Lett.* **1997**, *38*, 1841–1844.

409. Charette, A.B.; Mellon, C. *Tetrahedron* **1998**, *54*, 10525–10535.

410. Moody, C.J.; Gallagher, P.T.; Lightfoot, A.P.; Slawin, A.M.Z. *J. Org. Chem.* **1999**, *64*, 4419–4425.

411. Moody, C.J.; Lightfoot, A.P.; Gallagher, P.T. *Synlett* **1997**, 659–660.

412. Moody, C.J. *Chem. Commun.* **2004**, 1341–1351.

413. Alvaro, G.; Martelli, G.; Savoia, D.; Zoffoli, A. *Synthesis* **1998**, 1773–1777.

414. Córdova, A. *Acc. Chem. Res.* **2004**, *37*, 102–112.

415. Kobayashi, S.; Ishitani, H. *Chem. Rev.* **1999**, *99*, 1069–1094.

416. Ferraris, D.; Young, B.; Dudding, T.; Lectka, T. *J. Am. Chem. Soc.* **1998**, *120*, 4548–4549.

417. Ferraris, D.; Young, B.; Cox, C.; Drury W.J., III; Dudding, T.; Lectka, T. *J. Org. Chem.* **1998**, *63*, 6090–6091.

418. Ferraris, D.; Young, B.; Cox, C.; Dudding, T.; Drury W.J., III; Ryzhkov, L.; Taggi, A.E.; Lectka, T. *J. Am. Chem. Soc.* **2002**, *124*, 67–77.

419. Hagiwara, E.; Fujii, A.; Sodeoka, M. *J. Am. Chem. Soc.* **1998**, *120*, 2474–2475.

420. Taggi, A.E.; Hafez, A.M.; Lectka, T. *Acc. Chem. Res.* **2003**, *36*, 10–19.

421. Córdova, A.; Notz, W.; Zhong, G.; Betancort, J.M.; Barbas C.F., III. *J. Am. Chem. Soc.* **2002**, *124*, 1842–1843.

422. Córdova, A.; Watanabe, S.–i.; Tanaka, F.; Notz, W.; Barbas C.F., III. *J. Am. Chem. Soc.* **2002**, *124*, 1866–1867.

423. Westermann, B.; Neuhaus, C. *Angew. Chem., Int. Ed.* **2005**, *44*, 4077–4079.

424. Córdova, A.; Barbas C.F., III. *Tetrahedron Lett.* **2002**, *43*, 7749–7752.

425. Kobayashi, S.; Matsubara, R.; Nakamura, Y.; Kitagawa, H.; Sugiura, M. *J. Am. Chem. Soc.* **2003**, *125*, 2507–2515.

426. Juhl, K.; Gathergood, N.; Jørgensen, K.A. *Angew. Chem., Int. Ed.* **2001**, *40*, 2995–2997.

427. Matsubara, R.; Nakamura, Y.; Kobayashi, S. *Angew. Chem., Int. Ed.* **2004**, *43*, 1679–1681.

428. Drury W.J., III; Ferraris, D.; Cox, C.; Young, B.; Lectka, T. *J. Am. Chem. Soc.* **1998**, *120*, 11006–11007.

429. Yao, S.; Fang, X.; Jørgensen, K.A. *Chem. Commun.* **1998**, 2547–2548.

430. Elder, A.M.; Rich, D.H. *Org. Lett.* **1999**, *1*, 1443–1446.

431. Fang, X.; Johannsen, M.; Yao, S.; Gathergood, N.; Hazell, R.G.; Jørgensen, K.A. *J. Org. Chem.* **1999**, *64*, 4844–4849.

432. Knudsen, K.R.; Risgaard, T.; Nishiwaki, N.; Gothelf, K.V.; Jørgersen, K.A. *J. Am. Chem. Soc.* **2001**, *123*, 5843–5844.

433. Nishiwaki, N.; Knudsen, K.R.; Gothelf, K.V.; Jørgensen, K.A. *Angew. Chem., Int. Ed.* **2001**, *40*, 2992–2995.

434. Saaby, S.; Bayón, P.; Aburel, P.S.; Jørgensen, K.A. *J. Org. Chem.* **2002**, *67*, 4352–4361.

435. Saaby, S.; Fang, X.; Gathergood, N.; Jørgensen, K.A. *Angew. Chem., Int. Ed.* **2000**, *39*, 4114–4116.

436. Johannsen, M. *Chem. Commun.* **1999**, 2233–2234.

437. Gittins (née Jones), C.A.; North, M. *Tetrahedron: Asymmetry* **1997**, *8*, 3789–3799.

438. Mellor, J.M.; Richards, N.G.J.; Sargood, K.J.; Anderson, D.W.; Chamberlin, S.G.; Davies, D.E. *Tetrahedron Lett.* **1995**, *36*, 6765–6768.

439. Alonso, D.A.; Bertilsson, S.K.; Johnsson, S.Y.; Nordin, S.J.M.; Södergren, M.J.; Andersson, P.G. *J. Org. Chem.* **1999**, *64*, 2276–2280.

440. Tararov, V.I.; Kadyrov, R.; Kadyrova, Z.; Dubrovina, N.; Börner, A. *Tetrahedron: Asymmetry* **2002**, *13*, 25–28.

441. Maggini, M.; Prato, M.; Scorrano, G. *Tetrahedron Lett.* **1990**, *31*, 6243–6246.

442. Badorrey, R.; Cativiela, C.; Díaz–de–Villegas, M.; Gálvez, J.A. *Tetrahedron Lett.* **1997**, *38*, 2547–2550.

443. Ager, D.; Cooper, N.; Cox, G.G.; Garro–Hélion, F.; Harwood, L.M. *Tetrahedron: Asymmetry* **1996**, *7*, 2563–2566.

444. Hafez, A.M.; Duddling, T.; Wagerle, T.R.; Shah, M.H.; Taggi, A.E.; Lectka, T. *J. Org. Chem.* **2003**, *68*, 5819–5825.

445. Gefflaut, T.; Bauer, U.; Airola, K.; Koskinen, A.M.P. *Tetrahedron: Asymmetry* **1996**, *7*, 3099–3102.

446. Chai, C.L.L.; Hay, D.B.; King, A.R. *Aust. J. Chem.* **1996**, *49*, 605–610.

447. Chai, C.L.L.; Page, D.M. *Tetrahedron Lett.* **1993**, *34*, 4373–4376.

448. Badran, T.W.; Eaaston, C.J.; Horn, E.; Kociuba, K.; May, B.L.; Schleibs, D.M.; Tiekink, E.R.T. *Tetrahedron: Asymmetry* **1993**, *4*, 197–200.

449. Hamon, D.P.; Massy–Westropp, R.A.; Razzino, P. *Tetrahedron* **1993**, *49*, 6419–6428.

450. Easton, C.J. *Chem. Rev.* **1997**, *97*, 53–82.

451. Hamon, D.P.G.; Razzino, P.; Massy–Westropp, R.A. *J. Chem. Soc., Chem. Commun.* **1991**, 332–333.

452. Hamon, D.P.G.; Massy-Westropp, R.A.; Razzino, P. *Tetrahedron* **1995**, *51*, 4183–4194.

453. Hamon, D.P.G.; Massy-Westropp, R.A.; Razzino, P. *J. Chem. Soc., Chem. Commun.* **1991**, 722–724.

454. Easton, C.J. *Pure Appl. Chem.* **1997**, *69*, 489–494.

455. Zhai, D.; Zhai, W.; Williams, R.M. *J. Am. Chem. Soc.* **1988**, *110*, 2501–2505.

456. Yamamoto, Y.; Onuki, S.; Yumoto, M.; Asao, N. *J. Am. Chem. Soc.* **1994**, *116*, 421–422.

457. Lloris, M.E.; Moreno-Mañas, M. *Tetrahedron Lett.* **1993**, *34*, 7119–7122.

458. Bertrand, M.P.; Feray, L.; Nouguier, R.; Stella, L. *Synlett* **1998**, 780–782.

459. Bertrand, M.P.; Coantic, S.; Feray, L.; Nouguier, R.; Perfetti, P. *Tetrahedron* **2000**, *56*, 3951–3961.

460. Bertrand, M.P.; Feray, L.; Nouguier, R.; Perfetti, P. *Synlett* **1999**, 1148–1150.

461. Singh, N.; Anand, R.D.; Trehan, S. *Tetrahedron Lett.* **2004**, *45*, 2911–2913.

462. Miyabe, H.; Ushiro, C.; Naito, T. *Chem. Commun.* **1997**, 1789–1790.

463. Miyabe, H.; Ushiro, C.; Ueda, M.; Yamakwa, K.; Naito, T. *J. Org. Chem.* **2000**, *65*, 176–185.

464. Hanessian, S.; Yang, H.; Schaum, R. *J. Am. Chem. Soc.* **1996**, *118*, 2507–2508.

465. Steiner, A.; Wessig, P.; Polborn, K. *Helv. Chim. Acta* **1996**, *79*, 1843–1862.

466. Wessig, P.; Wettstein, P.; Giese, B.; Neuburger, M.; Zehnder, M. *Helv. Chim. Acta* **1994**, *77*, 829–837.

467. Giese, B.; Müller, S.N.; Wyss, C.; Steiner, H. *Tetrahedron: Asymmetry* **1996**, *7*, 1261–1262.

468. Wyss, C.; Batra, R.; Lehmann, C.; Sauer, S.; Giese, B. *Angew. Chem., Int. Ed. Engl.* **1996**, *35*, 2529–2531.

469. Baldwin, J.E.; Li, C.-S. *J. Chem. Soc., Chem. Commun.* **1987**, 166–168.

470. Svärd, H.; Antoni, G.; Jigerius, S.B.; Zdansky, G.; Långström, B. *J. Lab. Cmpds. Radiopharm.* **1983**, *21*, 1175–1176.

471. Vassilev, V.P.; Uchiyama, T.; Kajimoto, T.; Wong, C.-H. *Tetrahedron Lett.* **1995**, *36*, 5063–5064.

472. Någren, K. *J. Lab. Cmpds. Radiopharm.* **1997**, *40*, 758–759.

473. Bennett, F.; Barlow, D.J.; Dodoo, A.N.O.; Hider, R.C.; Lansley, A.B.; Lawrence, M.J.; Marriott, C.; Bansai, S.S. *Tetrahedron Lett.* **1997**, *38*, 7449–7452.

474. See Reference *470*.

475. Jacquier, R.; Lazaro, R.; Ranririseheno, H.; Viallefont, P. *Tetrahedron Lett.* **1984**, *25*, 5525–5528.

476. Antoni, G.; Långström, B. *Acta Chem. Scand. B* **1986**, *40*, 152–156.

477. Sugiyama, H.; Yokokawa, F.; Shioiri, T.; Katagiri, N.; Oda, O.; Ogawa, H. *Tetrahedron Lett.* **1999**, *40*, 2569–2572.

478. Imperiali, B.; Fisher, S.L. *J. Org. Chem.* **1992**, *57*, 757–759.

479. Imperiali, B.; Prins, T.J.; Fisher, S.L. *J. Org. Chem.* **1993**, *58*, 1613–1616.

480. Pirrung, M.C.; Krishnamurthy, N. *J. Org. Chem.* **1993**, *58*, 957–958.

481. Imperiali, B.; Roy, R.S. *J. Am. Chem. Soc.* **1994**, *116*, 12083–12084.

482. Imperiali, B.; Roy, R.S. *J. Org. Chem.* **1995**, *60*, 1891–1894.

483. Torrado, A.; Imperiali, B. *J. Org. Chem.* **1996**, *61*, 8940–8948.

484. Kise K.J., Jr.; Bowler, B.E. *Tetrahedron: Asymmetry* **1998**, *9*, 3319–3324.

485. Nitz, M.; Mezo, A.R.; Ali, M.H.; Imperiali, B. *Chem. Commun.* **2002**, 1912–1913.

486. Boeckman R.K., Jr.; Clark, T.J.; Shook, B.C. *Helv. Chim. Acta* **2002**, *85*, 4532–4560.

487. Berthelot, A.; Piguel, S.; Le Dour, G.; Vidal, J. *J. Org. Chem.* **2003**, *68*, 9835–9838.

488. Kanayama, T.; Yoshida, K.; Miyabe, H.; Takemoto, Y. *Angew. Chem., Int. Ed.* **2003**, *42*, 2054–2056.

489. Mazaleyrat, J.-P.; Goubard, Y.; Azzinin, M.-V.; Wakselman, M.; Peggion, C.; Formaggio, F.; Toniolo, C. *Eur. J. Org. Chem.* **2002**, 1232–1247.

490. Poulsen, T.B.; Alemparte, C.; Saaby, S.; Bella, M.; Jørgensen, K.A. *Angew. Chem., Int. Ed.* **2005**, *44*, 2896–2899.

491. Togni, A.; Häusel, R. *Synlett* **1990**, 633–635.

7 | Synthesis of Optically Active α-Amino Acids: Alkylation of Cyclic Chiral Templates

7.1 Introduction

A diverse variety of cyclic chiral templates based on glycine equivalents have been developed in the past 25 years, and until recently these have provided some of the most useful methods for asymmetric syntheses of α-amino acids. Both nucleophilic and electrophilic glycine synthons have been employed, often with the same template. The cyclic template acts as a chiral auxiliary to induce diastereoselectivity in the newly formed α-center, and in some cases, the β-center as well. For simplicity in discussing diastereoselectivity in the following sections, the α- or C2 center and β- or C3 position refer to the numbering in the product amino acid, not the template. The cyclic template strategy is based on the belief that a rigid cyclic chiral glycine synthon can provide a higher degree of asymmetric induction than might be expected with the corresponding open-chain analog. More recent synthetic routes have shown that this is not necessarily true, with acyclic glycine equivalents based on chiral amide auxiliaries such as Myers pseudoephedrine, Evans' oxazolidinone (see Section 5.3.6a), or Oppolzer's bornane-sultam (see Section 6.4.2) giving equally high diastereoselectivity. More recently, chiral catalysts with acyclic substrates have provided equivalent stereoselection (see Section 6.4.4). In addition, the cyclic templates suffer from the same potential flaw as other chiral auxiliaries, in that even if the induction of stereoselectivity is 100%, the final product enantiopurity can only be as good as the optical purity of the template. A recent analysis of commercially available chiral synthons found, for example, that the Schöllkopf bislactim ether template possessed only 91–95% ee (1).

Many of the different cyclic templates have similar structures and names. For convenience a summary is presented (Scheme 7.1) in approximate order of descending

importance based on use. The summary also indicates the site of side-chain addition and whether an electrophilic (+) or nucleophilic (−) glycine equivalent is used for the reaction. Cyclic templates have also been employed for radical reactions, which were reviewed in 1997 (2). Some applications of cyclic templates were reviewed in 2002 (3).

7.2 Schöllkopf Bislactim Ether

The first cyclic template to be widely employed for amino acid synthesis was the bislactim ether developed by Schöllkopf et al., and this template has been applied to more amino acid syntheses than any other chiral synthon. Several summaries of this synthetic approach were published in 1983 (4–6). The synthon is derived from the cyclization of dipeptides into diketopiperazines, compounds that have been known since 1849 (7). An updated method for diketopiperazine synthesis was published in 1975 in an attempt to identify the bitter principle of cocoa, with a number of N-Cbz dipeptide methyl esters prepared, hydrogenated, and then cyclized (8). While diketopiperazines themselves have been used for amino acid synthesis (see following section), the Schöllkopf synthon removes the acidic amide protons by treatment with trimethyloxonium tetrafluoroborate, forming a dimethoxy lactim ether (84–88% yield) (9, 10). The derivatization also makes the two former amide bonds much more susceptible to hydrolysis, allowing for removal of the chiral auxiliary half of the template.

The first application of the synthon was reported in 1979, with the L-Ala-L-Ala derived template (optical purity only 93–95% ee) deprotonated with LDA or nBuLi and alkylated with a number of electrophiles (see Scheme 7.2, Table 7.1) (9). The electrophiles added

490

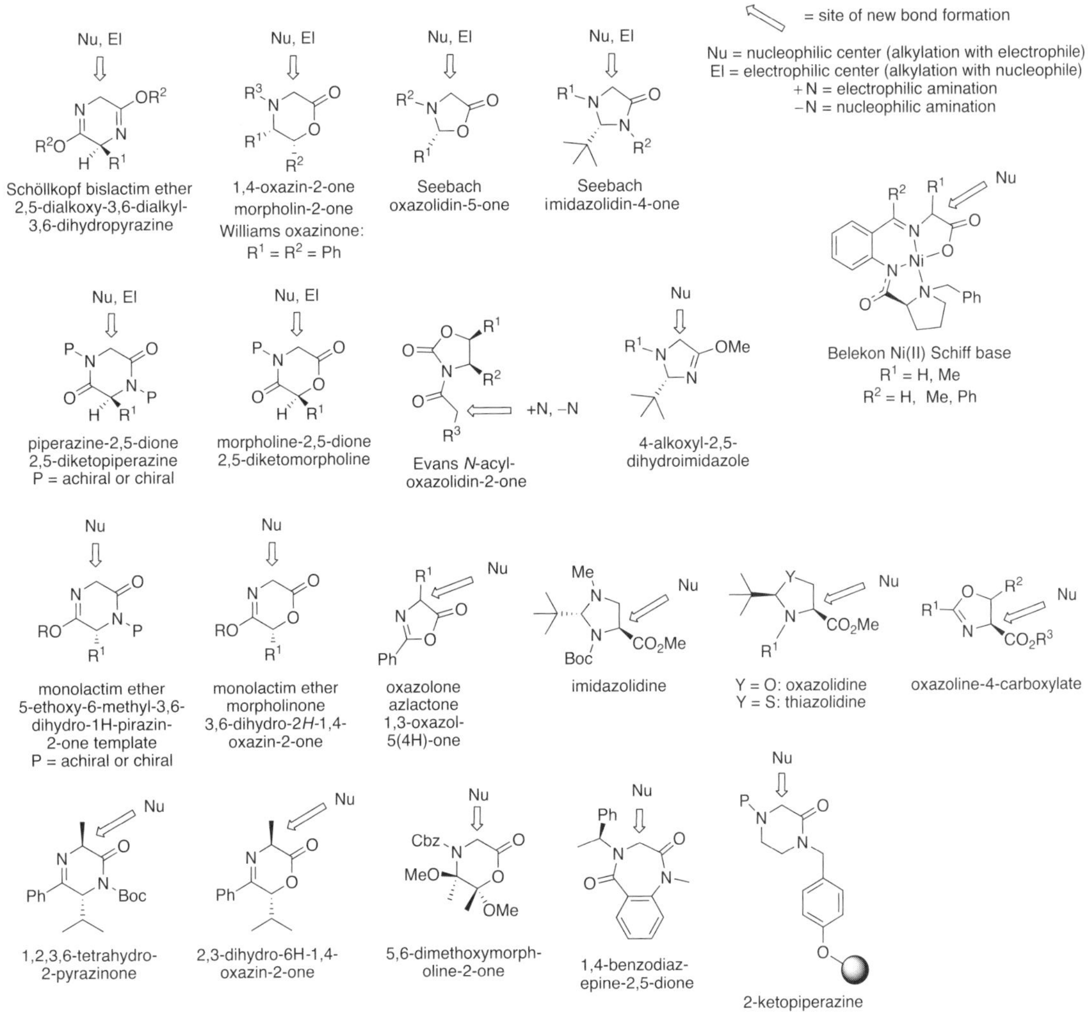

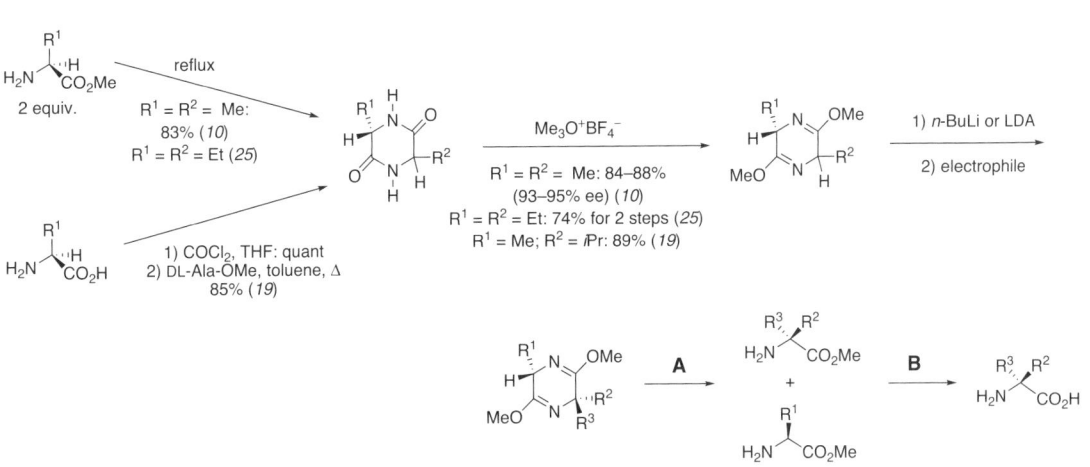

= site of new bond formation

Nu = nucleophilic center (alkylation with electrophile)
El = electrophilic center (alkylation with nucleophile)
   +N = electrophilic amination
   −N = nucleophilic amination

Schöllkopf bislactim ether
2,5-dialkoxy-3,6-dialkyl-3,6-dihydropyrazine

1,4-oxazin-2-one
morpholin-2-one
Williams oxazinone:
R^1 = R^2 = Ph

Seebach
oxazolidin-5-one

Seebach
imidazolidin-4-one

Belekon Ni(II) Schiff base
R^1 = H, Me
R^2 = H, Me, Ph

piperazine-2,5-dione
2,5-diketopiperazine
P = achiral or chiral

morpholine-2,5-dione
2,5-diketomorpholine

Evans N-acyl-oxazolidin-2-one

4-alkoxyl-2,5-dihydroimidazole

monolactim ether
5-ethoxy-6-methyl-3,6-dihydro-1H-pirazin-2-one template
P = achiral or chiral

monolactim ether
morpholinone
3,6-dihydro-2H-1,4-oxazin-2-one

oxazolone
azlactone
1,3-oxazol-5(4H)-one

imidazolidine

Y = O: oxazolidine
Y = S: thiazolidine

oxazoline-4-carboxylate

1,2,3,6-tetrahydro-2-pyrazinone

2,3-dihydro-6H-1,4-oxazin-2-one

5,6-dimethoxymorph-oline-2-one

1,4-benzodiaz-epine-2,5-dione

2-ketopiperazine

Scheme 7.1 Cyclic Templates for Syntheses of α-Amino Acids.

H$_2$N, R^1, H, CO$_2$Me
2 equiv.

reflux

R^1 = R^2 = Me:
83% (**10**)
R^1 = R^2 = Et (**25**)

H$_2$N, R^1, H, CO$_2$H

1) COCl$_2$, THF: quant
2) DL-Ala-OMe, toluene, Δ
   85% (**19**)

Me$_3$O$^+$BF$_4^-$

R^1 = R^2 = Me: 84–88%
(93–95% ee) (**10**)
R^1 = R^2 = Et: 74% for 2 steps (**25**)
R^1 = Me; R^2 = iPr: 89% (**19**)

1) n-BuLi or LDA
2) electrophile

A

H$_2$N, R^3, R^2, CO$_2$Me

+

H$_2$N, R^1, CO$_2$Me

B

H$_2$N, R^3, R^2, CO$_2$H

Scheme 7.2 Synthesis of α-Substituted-α-Amino Acids via Schöllkopf Bislactim Ether (see Table 7.1).

Table 7.1 Synthesis of α,α-Disubstituted-α-Amino Acids via Alkylation of the Schöllkopf Bislactim Ether Template (see Scheme 7.2)

R^1	R^2	Electrophile	Alkylation Yield	Derivatization of Side Chain	A	B	Reference
Me	Me	R^3X; $R^3 =$ Et iPr nOct $CH_2CH=CH_2$ $CH_2CH=CHPh$ Bn CH_2(2-naphthyl) CH_2(2-pyridyl) CH_2(2-quinolinyl)	75–90% 92–95% de		0.5N HCl 70–91% 80–93% ee	6N HCl, Δ $R^3 =$ Bn, CH_2(2-naphthyl), CH_2(2-pyridyl), CH_2(2-quinolinyl) 64–88%	1979 (9) 1981 (10)
Me	Me	H_2CO MeCOMe EtCOEt PhCOMe PhCHO (4-MeO-Ph)CHO	69–89% 73–95% de C2 21–48% de C3				1980 (11) 1981 (12)
iPr	Me	R^3X; $R^3 =$ Me nHept $CH_2CH=CH_2$ $CH_2CH=CPh$ $CH_2CH=CMe_2$ Bn 3,4-$(MeO)_2$-Bn CH_2CO_2tBu $(CH_2)_2OBn$ $(CH_2)_2SBn$	43–94% >95% de except $R^3 =$ Me, 85% de		0.25N HCl 47–90% >95% ee		1981 (19) 1983 (4)
iPr	Me	$MeCOR^4$ $R^4 =$ H, Me, Ph	85–94% >95% de C2	for $R^4 =$ H KOtBu, MeI to form $R^3 =$ CH(Me)OMe 92% for $R^4 =$ Me, Ph $SOCl_2$, MeOH to form $R^3 = C(=CH_2)R^4$ 81–92%	0.25N HCl 62–68%		1982 (15)
iPr	Me	$ClCH_2OBn$	91% >95% de		0.25 N HCl 81%	6N HCl 84%	1982 (22)
iPr	Me	CH_2Br_2	80% >95% de	R^4SH displacement $R^3 = CH_2Br$ to $R^3 = CH_2SR^4$	0.25N HCl 64–72%		1983 (21)

492

iPr	Me	CH$_2$CCH R^{3}X; R^3 = (CH$_2$)$_2$PO$_3$Et$_2$ (CH$_2$)$_3$N=CHPh (CH$_2$)$_2$ON=CMe$_2$ (CH$_2$)$_4$Br	69–81% >95% de	0.25N HCl 70–79%	4N HCl 90%	1986 (20)
iPr	Me	XC(Me)=CHCO$_2$Et XCH=CHCO$_2$Et XCH=C(Ph)CO$_2$Et 1-X-2-CO$_2$Et-cyclopent-1-ene	34–75% 95–>99% de	0.1 N HCl 38–55%	4N HCl, Δ 55% for R^3 = CH=CH(Ph) CO$_2$Et	1987 (13)
iPr	Me	CHOCH=CH$_2$ CHOC(Me)=CH$_2$ CHOC(Me)=CHMe	70–87% 94–97% de C2 92–98% de *threo* C3			1987 (13)
Me iPr	Me	MeCOCl PhCOCl PhCH=CHCHO	15–88% >95% de	0.25N HCl 49%	for R^3 = COMe, COPh NaH, Ph$_3$PMeBr to form R^3 = C(Me)=CH$_2$, C(Ph)=CH$_2$	1988 (16)
Me iPr	Me	H$_2$CO H$_2$C=CHCHO MeCH=CHCHO PhCH=CHCHO	77–95% 75–>95% de C2	0.25N HCl 49%	oxidize R^3 = CH(OH)R to form R^3 = COR for R^3 = CHO 1) CBr$_4$, PPh$_3$ 2) nBuLi to form R^3 = CCH	1988 (16)
Me	Me	R^{4}COCl + ZnCl$_2$, MnCl or Et$_2$Al; R^4 = CH$_2$Cl Ph 3,4-(AcO)$_2$-Ph	74–88% 80–>95% de		reduce R^3 = COR^4 to R^3 = CH(OH)R^4 39–82% >95% de *syn*	1993 (17)
Et	Et	R^{3}Br; R^3 = nPr nBu nHex nNon CH=CHMe	for R^3 = nBu 67%		6N HCl for R^3 = nBu 50% 92% ee	1995 (25)

(Continued)

Table 7.1 Synthesis of α,α-Disubstituted-α-Amino Acids via Alkylation of the Schöllkopf Bislactim Ether Template (see Scheme 7.2) (continued)

R^1	R^2	Electrophile	Alkylation Yield	Derivatization of Side Chain	A	B	Reference
iPr	CO_2Et	MeCH(OTBDPS)CHO with Sn(OTf)$_2$ or MgBr$_2$	70% 100% de	DIBAL reduction $R^2 = CO_2Et$ to $R^2 = CH_2OH$		12N HCl 51–68%	1995 (26)
Me	Me	C$_{13}$H$_{27}$CH=CHCH(OTBS)CH(OBn)CHO nHexCO(CH$_2$)$_6$CH=CHCH(OBn)CHO	83–92% 74–90% de C2 16–50% de C3			TsOH, EtOH 0.5N NaH, MeOH 55–58%	1997 (14)
iPr	CO_2Et	MeCHO iPrCHO nBuCHO CH=CMe$_2$CHO	70–86% 48–96% de C2 20–84% de C3	reduction $R^2 = CO_2Et$ to $R^2 = CH_2OH$			1998 (27)
iPr	CO_2Et	R^3X: R^3 = Me nHex CH$_2$CH=CH$_2$ Bn	58–84% 52–97% de	reduction $R^2 = CO_2Et$ to $R^2 = CH_2OH$ 56–84%	0.2 N HCl 16–58%		1998 (471)
iPr	Me	Br(CH$_2$)$_n$Br n = 4,5,6		displace R^3 = (CH$_2$)$_n$Br to form R^3 = (CH$_2$)$_n$N$_3$ dimerize to form —(CH$_2$)$_n$NH(CH$_2$)$_n$—			1998 (472)
iPr	Me	R^3X = Br(CH$_2$)$_n$Br n = 2,3,4	53–82% 88–90% de	react two equivalents of template	0.25N HCl 91–93%		1998 (95)
iPr	Me	H$_2$C=CH$_2$PO$_3$Et$_2$ PhCH=CH$_2$PO$_3$Et$_2$	75–86%			12N HCl, Δ 78–95%	1999 (473)
iPr	Me iPr	BrCH$_2$CH$_2$CH=CH$_2$	80–87% >95% de	alkene oxidation to CH$_2$CH$_2$COMe 65–74% formation diazoketone 42–45% Rh(II)-catalyzed insertion 30–40%		3M HCl, Boc$_2$O 53–58%	2004 (23)
iPr	(CH$_2$)$_3$CH=CH$_2$	MeBr nPrBr	78–80% >90% de	alkene oxidation to CH$_2$CH$_2$COMe 65–74% formation diazoketone 45–50% Rh(II)-catalyzed insertion 28–30%	0.5M HCl: dioxane 28–31%		2004 (24)
iPr	(CH$_2$)$_n$CH=CH$_2$ n = 1,2	propylene oxide MeCHO	62–75%	carbinol oxidation 71–81% alkene Wacker oxidation to methyl ketone 55–83%	0.1M TFA 47–59%		2004 (130)

trans to the methyl group of the other Ala residue, and diastereoselectivities of 90–93% de were obtained (*9, 10*). Hydrolysis produced α-methyl-α-amino methyl esters, which were first separated from the Ala-OMe byproduct and then further hydrolyzed to give the amino acids. Carbonyl compounds were also employed as electrophiles, producing α-methylserines with 73–95% de at the α-center (and 21–48% de at the β-center for asymmetrical ketone substrates) (*11, 12*). A titanium-exchanged anion added to the aldehyde group of α,β-unsaturated aldehydes with much better stereoselectivity (*13*). The lipidic amino acid sphingofungin F has been synthesized by this route, using a Sn(II) enolate of the *cyclo*[Ala-Ala] template (*14*).

Unfortunately the aldol adducts are susceptible to retro-aldol reactions during hydrolysis and purification. It is possible to dehydrate the aldol adducts with thionyl chloride/pyridine to give α-methyl-vinylglycine derivatives, and the acetaldehyde adduct has been *O*-methylated before deprotection (*15*). The aldol adducts can be oxidized to the corresponding ketones; the resulting β-oxo-derivatives can also be obtained by reaction of the lithiated template with acyl chlorides (*16*). Much better acylation yields were obtained from the cyclo [L-Ala-Ala] bislactim ether template by using aluminum, zinc, or manganese anions instead of the usual lithiated compound (*17*). The ketones could be reduced with hydride reagents, providing another route to β-hydroxy amino acids (*17*). The lithiated bislactim ether was also reacted with methyl acrylates containing a leaving group in the β-position, resulting in β,γ-didehydro-Glu derivatives (*18*).

The mixed bislactim ether prepared from L-Val-L-Ala has also been used as a substrate for alkylation (see Scheme 7.2, Table 7.1). Alkyl halides regiospecifically alkylate the Ala residue with high diastereoselectivity (>95% de) (*4, 19, 20*). α-Methyl-*S*-substituted-cysteines were prepared with this template via alkylation with dibromomethane followed by displacement of the second halide with potassium mercaptides (*21*), while α-methyl-Ser was prepared by alkylation with chloromethyl benzyl ether (*22*). The cyclo[L-Val-L-Ala] template was also reacted with acyl chlorides, and the

resulting β-keto amino acid derivatives further functionalized by Wittig reaction to provide α-alkenyl- or α-ethynyl-Ala derivatives (*16*). The cyclo[L-Val-L-Ala] and cyclo[L-Val-L-Val] templates were both alkylated with 4-bromo-1-butene, with the alkene bond chemoselectively oxidized to produce a 3-oxobutyl side chain. This was converted into a diazoketone, with a Rh(II)-catalyzed carbenoid cyclization to the lactim nitrogen and template hydrolysis producing α-substituted 4-keto-pipecolic acid products (*23*). In a similar manner, a 4-oxopentyl side chain was introduced and converted into the diazoketone, with Rh-catalyzed insertion now occurring at the side chain C3 carbon to give α-substituted-α-(cyclopentanon-3-yl)-Gly derivatives (*24*).

The cyclo[L-Abu-Abu] template has also been prepared and alkylated (*25*). After hydrolysis with 6 N HCl, the deprotected amino acids possessed approximately 90% ee. A mixed template containing Val as the chiral auxiliary and a CO₂Et substituent on the residue to be alkylated (i.e, an aminomalonate residue) was used for a synthesis of an α-substituted Ser; after an aldol-type alkylation the ester group was reduced to give the Ser hydroxymethyl side chain (*26*). Transmetallation with different metals was investigated to improve the aldol diastereoselectivity (*27*). The diastereoselective protonation of a number of cyclo[L-Val-Ala] and cyclo[L-Ala-Ala] bislactim ethers has been systematically examined using a range of bases and proton sources (*28*).

The Schöllkopf approach was soon applied to the preparation of α-monosubstituted amino acids by using a mixed cyclo[L-Val-Gly] bislactim ether prepared from L-Val (*29*) or Cbz-Val (*30*) and Gly-OMe (see Scheme 7.3 for template synthesis). One of the advantages of the Schöllkopf method is that the template is synthesized from inexpensive materials, with a 1992 report providing details for its preparation on a 50 g scale (*31*) and another article alluding to 500 g scale syntheses (*32*). An improved synthesis of the cyclo [L-Val-Gly] template on a 45 g scale was reported in 1998, with better methodology for the bis(*O*-methylation) step (*33*). In 2006 a microwave-assisted synthesis simplified the formation of the diketopiperazine on a

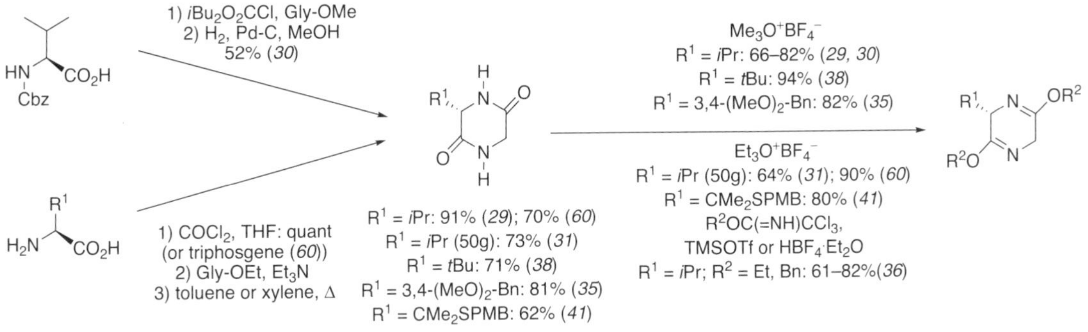

Scheme 7.3 Synthesis of Schöllkopf Bislactim Ether Template.

2–15 g scale by heating an aqueous solution of Boc-Val-Gly-OMe at 150 °C for 15 min; improvements to other steps of the sequence were also reported (*34*). However, a fundamental flaw of the Schöllkopf methodology, and all procedures relying on chiral auxiliaries, was revealed by a 2006 analysis of the enantiomeric purity of commercially available chiral synthons. The cyclo[Val-Gly] (*S*)-pyrazine template possessed only 95.26% ee, and the (*R*)-pyrazine template was even worse, with only 90.83% ee (*1*).

A number of electrophiles were reacted with the cyclo[L-Val-Gly] bislactim ether template in 62–92% yield to give, after hydrolysis, D-amino acid methyl esters in 52–89% yield with 75–95% ee (see Scheme 7.4, Table 7.2 for alkylation reactions) (*20, 29*). In an attempt to improve the diastereoselectivity of alkylation, the L-Val chiral auxiliary was replaced with *O,O'*-dimethyl-α-methyl-DOPA, but the results were mixed (*35*). The cyclo[L-Val-Gly] bislactim ether template has also been prepared as the bisbenzyl ether instead of the bismethyl analog (*36*), giving similar diastereoselectivity on alkylation with benzyl bromide (92% de). The benzyl ether derivative has the advantage that the product of hydrolysis is an amino acid benzyl ester, which can be deprotected by hydrogenolysis. Solid-state crystal structures have been obtained of two lithiated cyclo[L-Val-Gly] bislactim complexes, a racemic dimeric one with HMPA (hexamethylphosphoric triamide) and an enantiomerically pure monomeric one with PMDTA (pentamethyldiethylenetriamine) (*37*).

A number of amino acids have been tested as the chiral auxiliary. As previously mentioned, the Val- and α-methyl-DOPA-derived templates induced similar stereoselectivities upon alkylation. Much better results were obtained by using L-*tert*-leucine (Tle) as the chiral auxiliary, with alkyl halides reacting with 58–89% yield and >94% de (*38*) (except for MeI, with only 80% de) (*5*). However, aldol reactions with ketones proceeded with lower diastereoselectivity than obtained with the valine template (*39*). While Tle is not a naturally occurring amino acid, it is commercially available. A review of the synthesis and use of enantiomerically pure Tle was published in 1995 (*40*). Penicillamine (side chain R = CMe₂SH, thiol protected with 4-methoxybenzyl) has also been tested as the chiral auxiliary as it can potentially overcome the two major disadvantages of the Schöllkopf method: (1) moderate diastereoselectivity, and (2) difficulty in product purification due to the presence of the chiral auxiliary half of the template (*41*). The penicillamine bulky *tert*-butyl-like side chain provided improved diastereoselectivity for the one example

provided (alkylation with propargyl bromide: 92% de vs 60–65% de with the cyclo[L-Val-Gly] template). After hydrolysis the penicillamine ester was easily separated from ethyl propargylglycinate by a simple extraction. α-Methyl-Phe has also been used as an auxiliary. It allows for regioselective dialkylation of the Gly residue, as there is no competing α-proton on the auxiliary to abstract (*42*).

Diastereoselectivity can also be affected by the electrophile. A potentially useful effect of leaving group bulk on diastereoselectivity was reported in 1999. Alkylation of the cyclo[L-Val-Gly] template with electrophiles containing diphenylphosphonate leaving groups proceeded with considerably higher diastereoselectivity (95–98% de) compared to electrophiles with tosylate (67–92% de) or bromide (14–97% de) leaving groups, although yields were sometimes lower (*43*). In another paper, alkylation with 3-bromo-1-(TMS)-1-propyne proceeded with only 42% de. By replacing the bromide leaving group with an OPO(OEt)₂ group, the alkylation diastereoselectivity improved to 84% de, while with an OPO(OPh)₂ group 96% de (80% yield) was obtained (*44*). A synthesis of a sterically hindered γ-branched amino acid, 2-amino-4-isopropyl-5-[3-(3-methoxypropoxy)-4-methoxyphenyl]pentanoic acid, initially attempted to alkylate the cyclo[D-Val-Gly] or cyclo[L-Val-Gly] templates with (*R*)- or (*S*)-1-aryl-2-bromo-3-methylbutane. The matched set of enantiomers produced the desired adduct with >96% de at the newly generated α-center, but the mismatched pair resulted in 0–14% de. Instead, the electrophile was used to induce stereoselectivity with an achiral cyclo[Gly-Gly] bislactim ether template. Either electrophile enantiomer reacted with 90% de (*45*).

Both hydrogens of the Gly residue in the cyclo [L-Val-Gly] or cyclo[D-Val-Gly] bislactim ether templates have been exchanged with deuterium by refluxing in MeOD-D₂O containing 1 equiv of KOD (*30, 46*). This method avoids the need to use expensive deuterated Gly for the template synthesis. Conventional alkylation and work-up of the deuterated synthon provided [2-²H]-L-Ser, [2-²H]-L-Phe, [2-²H]-L-allylglycine, and [2-²H]-L-Asp acid (*30*). The Schöllkopf method has been compared with the Oppolzer alkylation method (see Section 6.4.2b) for the preparation of [1,2-¹³C₂,¹⁵N]-labeled Leu (*47*). Both methods gave similar yields, but the Schöllkopf product had 97.2–97.4% ee, while the Oppolzer product had 99.7% ee. A similar comparison for the preparation of 4'-dimethylphosphonomethyl-L-Phe found that the Oppolzer procedure gave a much better overall yield (68%) than the Schöllkopf method (38%) (*48*).

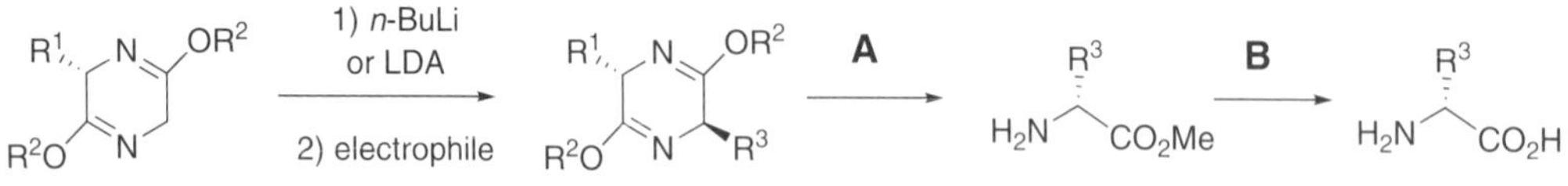

Scheme 7.4 Synthesis of α-Amino acids via Alkylation of Schöllkopf Bislactim Ether (see Table 7.2).

Table 7.2 Synthesis of α-Amino Acids via Alkylation of the Schöllkopf Bislactim Ether Template (see Scheme 7.4)

R¹	R²	Electrophile	Derivatization of Side Chain	Alkylation Yield	A	B	Reference
iPr	Me	R³Br; R³ = nHept CH₂CCH CH₂CCPh Bn 3,4-(MeO)₂-Bn CH₂(2-naphthyl)		62–92%	0.25N HCl 52–89% 75–>95% ee		1981 (29)
3,4-(MeO)₂-Bn	Me	R³X; R³ = nHept CH₂CH=CH₂ CH₂CH=CHMe CH₂CH=CHPh CH₂CH=C(Me)₂ C(Me)(CH=CH₂)₂ CH₂CCH C(Et)₂OMe C(Et)(Me)OMe C(nPr)₂OMe 1-MeO-cyclopent-1-yl 1-MeO-cyclohex-1-yl		21–95% 80–95% de	0.25N HCl 45–75% 70–95% ee	6N HCl 48–84%	1981 (35)
tBu	Me	R³X; R³ = Me nHept CH₂CH=CH₂ CH₂CCH CH₂CO₂tBu		59–89%	0.25N HCl 59–97% 94–>95% ee	6N HCl 73–90%	1982 (38) 1983 (5)
iPr 3,4-(MeO)₂-Bn	Me	BnOCH₂Cl		74–81% 93–95% de	0.25N HCl 56–62% 93–95% ee	6N HCl 76–82%	1982 (64)
iPr	Me	ICH₂F		80% 45% de			1983 (474)
iPr	Me	(CH₂)₂PO₃Et₂ (CH₂)₃PO₃Et₂ (CH₂)₄PO₃Et₂ (CH₂)₂ON=C(Me)₂ (CH₂)₃NBn₂		71–86% 58–96% de	0.25N HCl 64–80%	4N HCl, Δ	1986 (20)
tBu	Me	(CH₂)₂NBn₂ (CH₂)₃NBn₂		82–83% 88–93% de	0.25N HCl 64–80%	4N HCl, Δ	1986 (20)

(Continued)

Table 7.2 Synthesis of α-Amino Acids via Alkylation of the Schöllkopf Bislactim Ether Template (see Scheme 7.4) (continued)

R¹	R²	Electrophile	Alkylation Yield	Derivatization of Side Chain	A	B	Reference
iPr	Me	ClCH₂(1-Me-3-ArS-2-pyrrole)	84% / 66% de		0.25N HCl	6N hCl / 58% for 2 steps	1989 (54) / 1987 (55)
iPr	Me	ClCH₂CH₂P(O)(Me)OiBu	85% / 93% de		0.25N HCl / 96%	6N HCl / 87%	1987 (475)
iPr	Me	PhCOCl	90% -CH(OH)-Ph- bridged dimer / 88% COPh / 0% de				1988 (16)
iPr	Me	(structure: BnO, NHCbz, CO₂Me)			0.25N HCl / 57% for 2 steps	6N HCl / 100%	1989 (106) / 1989 (107) / 1990 (108)
iPr	Me	(structure: BnO, NHAc, CO₂Me, OMe)	43% / 62% de		0.25N HCl / 75%		1990 (105)
iPr	Me	(structure: (OC)₃Mn, BnO, NHAc, CO₂Me)	49–65% / 72% de		0.25N HCl / 56–60%		1991 (104)
iPr	Me	BrCH₂C(=CH₂) (2-TsNH-4-MeO-Ph)	93% / 100% de			2N HCl / 98%	1992 (31)
iPr	Me	4-CH₂PO₃Me₂-BnBr	72–83%		0.25N HCl / 71%	1N NaOH / 83%	1992 (476) / 1996 (48)
iPr	Me	(structure: Br...N-Boc)	71–80%		0.1N HCl / 85–89%	5N HCl / 59–83%	1992 (90) / 1992 (91)
iPr	Me	Br(CH₂)₃Br	71–80%	86% / 60% de (dimer)	0.1N HCl / 97%	5N HCl / 86%	1992 (90) / 1992 (91)
iPr / tBu	Me	ClCH₂CH=CHCH₂Cl	82–84%	1) intramolecular 1,3-S_N2-displacement via second deprotonation with nBuLi / R¹ = iPr: 34% de / R¹ = tBu: 66% de / 2) alkene reduction: 99%	0.25N HCl	6N HCl / 60–65% for 2 steps	1992 (116)
iPr	Me	BrCH₂CH=CHMe	60–80%		0.25N HCl	1) Boc₂O 2) Ba(OH)₂ / 80% for 3 steps	1992 (477)

498

iPr	Me	ClCH$_2$(2-Me-5-RS-imidazol-4-yl) R = H, Ph, 1-naphthyl, 4-MeO-Bn	75–93% 86–90% de	0.25N HCl 84–98%	6N HCl 83–96%	1992 (478) 1994 (56)
Bn	Bn	BnBr	73% 92% de	0.25N TFA 53% >95% de		1993 (36)
iPr	Me	(R)-BrCH(Me)CO$_2$Me	56% >80% de	0.07N HCl	1N NaOH 76% for 2 steps	1993 (51)
iPr	Me	iBuBr (with [1,2-^{13}C$_2$,^{15}N]-template)		0.01N HCl	aq. NaOH 45% for 3 steps 97.2–97.4% ee	1994 (47)
iPr	Me	BrCH$_2$C(Br)=CH$_2$	97% >98% de	1N HCl, MeOH 67%		1994 (52)
iPr	Me	ClCH$_2$(3-Bn-imidazol-4-yl) with [2′-^{13}C], [1′-^{15}N], [3′-^{15}N]	55–84%		1) 3N HCl, Δ 2) Pd-C, cyclohexene 72–94% 96–97% ee	1994 (53)
iPr	Me	BrCH$_2$CH=C(Me)$_2$ TfOCH$_2$CF(Me)$_2$	80%, 75% de 44%, >95% de	0.25N HCl 56–95%		1994 (479)
iPr	Me	[iodide template structure]	81% dimer 64:32:4 ratio			1994 (480)
iPr	Me	R^{3}Br; R^3 = CH$_2$OBn Bn CH$_2$CH=CH$_2$ CH$_2$CH=CH$_2$ CH$_2$OEt with [2,2-^{2}H$_2$]-template		0.2N HCl 31–82%	5N HCl, Δ 37–89% >95% ee	1995 (30)
iPr	Me	BrCH$_2$(1-SO$_2$Ph-5-MeO-indol-3-yl) BrCH$_2$(1-Boc-5-MeO-6-Cl-indol-3-yl)	90–96% 84–>95% de	2N HCl 93%	BBr$_3$ or NaOH, Δ 80–87%	1995 (57) 1997 (481)
iPr	Me	ClCH$_2$^{18}OBn Cl^{13}CH$_2$OBn		0.1N HCl 48–58% for 2 steps	1) H$_2$, Pd-C 2) Ba(OH)$_2$ or 6N HCl, Δ 83–89%	1995 (65)
iPr	Me	I(CH$_2$)$_3$CN	80%	6N HCl 49%	CN reduction with LiAlH$_4$	1995 (63)
iPr	Me	[isoxazole structure: Me, O–N ring, MeO, CH$_2$Br]	62% 86% de	0.25N HCl 70%	48% HBr 47% 97% ee	1996 (59)
CMe$_2$SPMB	Me	BrCH$_2$CCH	62% 92% de	TFA, MeCN 84% 92% ee		1996 (41)

(Continued)

Table 7.2 Synthesis of α-Amino Acids via Alkylation of the Schöllkopf Bislactim Ether Template (see Scheme 7.4) (continued)

R¹	R²	Electrophile	Alkylation Yield	Derivatization of Side Chain	A	B	Reference
iPr	Me	ClCH₂CCCH₂Cl	74% 78% de	1) displace Cl with NaN₃; 86% 2) reduce N₃ with PPh₃, protect with CCbz-Cl: 79% form R³ = CH₂CCCH₂NHCbz	1N HCl	NaOH 83% >97% ee	1996 (111)
iPr	Me	BrCH₂CH₂CH=CH₂		1) epoxidize alkene 2) open epoxide/dimerize with BnNH₂ 3) oxidize to give 4) aromatize			1997 (103)
iPr	Me	(E)-BrCH₂CH=CHCH₂Br (Z)-BrCH₂CH=CHCH₂Br (E)-BrCH₂C(Br)=CHCH₂Br (Z)-BrCH₂C(Br)=CHCH₂Br	30–71% of dimer	conversion –CH₂C(Br)=CHCH₂– with As(PPh₃)₄, Bu₃SnR to –CH₂C(R)=CHCH₂– R = CH=CH₂, CH=CHPh, CCPh, 2-thienyl 32–>95%	0.25N HCl 77–91%		1997 (96)
iPr	Me	BrCH₂CCH	60% >98% de	conversion R³ = CH₂CCH with RI/PdCl₂(PPh₃)₂/CuI to R³ = CH₂CCR R = Ph, 2-thienyl 75-81% conversion with IArI/PdCl₂(PPh₃)₂/CuI to –CH₂CC-Ar-CCCH₂– dimer Ar = 1,4-Ph, 2,5-thienyl 53–59% conversion with PdCl₂(PPh₃)₂/CuI to –CH₂CC-CCCH₂– dimer 86%	0.25N HCl 76–88%	0.25N LiOH >95%	1997 (101)

					0.2N HCl 77–98%	6N HCl, Δ or LiOH, H₂O/ dioxane 39–67%	1997 (94)
				94–>99% de			
*i*Pr	Me	4CN-BnBr		62%	0.25N HCl 67%		1997 (482)
*i*Pr	Me	Br(CH₂)₄Br	displace Br with (4-MeO-Bn)SH, KOtBu, form R³ = (CH₂)₄SAr		0.25N HCl	LiOH, H₂O/THF	1997 (110)
*i*Pr	Me	3-NO₂-4-F-BnBr		54%	TFA, MeCN 65%		1997 (483)
*i*Pr	Me	I(CH₂)₇-CO₂TBS	1) TBAF 2) *i*PrOCOCl, Evans oxazolidinone: 82% 3) KHMDS, trisyl azide: 70% form R³ = [structure]	90%			1997 (98) 2001 (99)
*i*Pr	Me	BrCH₂CO₂Et	convert to R³ = (CH₂)₂I diaplace I with second equiv. of template 50% de	88% de	6N HCl, Δ 80%		1998 (92)
*i*Pr	Me	Br(CH₂)₄Br	63% dialkylation (2 equiv. of template)		HCl, H₂O/dioxane 93%		1998 (93)
*i*Pr	Me	2-PMPS-3,5-(MeO)₂-BnCl ClCH₂(1-Me-4-SAr-imidazol-4-yl)		84–92% 26–86% de	0.25N HCl 97%		1998 (484)
*i*Pr	Me	[template structure]	Br displaced with NaN₃, *n*Bu₄NI 82% N₃ reduced with H₂, Pd/C form NH₂: 11% reductive coupled dimer: 56%	44% 97% de	0.1N HCl 85%	0.5N LiOH 59%	1998 (485)
*i*Pr	Me	2,6-Br₂-4-MeO-BnBr		74% 78% de	0.25N TFA 61%	BBr₃ 64%	1998 (486)
*i*Pr	Me	2-Cl-3-HO-4-MeO-BnBr		85% 96% de	0.5N HCl 88%	LiOH, THF 99%	1998 (487)
*i*Pr	Me	2-CN-BnBr 3-CN-BnBr 4-CN-BnBr		85–90%	0.25N HCl 46–58%	LiOH, MeOH quant.	1998 (488)
*i*Pr	Me	4-Br-BnBr	couple to give R³ = 4-(4-BnO-PhCO)-Bn 92%	88%	0.1N TFA 100%		1998 (489)

(Continued)

Table 7.2 Synthesis of α-Amino Acids via Alkylation of the Schöllkopf Bislactim Ether Template (see Scheme 7.4) (continued)

R^1	R^2	Electrophile	Alkylation Yield	Derivatization of Side Chain	A	B	Reference
iPr	Me	ClCH₂C(SPh)=CH₂	90%		TFA 79% >98% ee		1998 (490)
iPr	Me	ICH₂CH₂CH=CHCH₂CH₂I BrCH₂CH=CH₂	76–78% 84–89% de	displace I with malonate to give R³ = CH₂CH₂CH=CHCH₂CH₂C(CO₂Me)₂ 92%	0.25N HCl 68–89%	4N HCl	1998 (112)
iPr	Me	BrCH₂(1-Boc-indol-4-yl)	67%		0.1N TFA 100%	LiOH, H₂O/THF 62%	1998 (491)
iPr	Me	R³Br, R³OTs or R³OPO₃Ph₂; R³ = Et CH₂CH=CH₂ CH₂CCTMS CH₂CCTES 3,5-(MeO)₂-Bn	X = Br: 80–90% 14–97% de X = OTS: 80–89% 67–90% de X = OPO₃Ph₂ 24–90% 95–98% de				1999 (43)
iPr	Me	BrCH(R)CH=CHCO₂Et R = Me, Et, nPr	38–45% 0% de	forms 2-R-3-CO₂Et-cycloprop-1-yl	0.1N HCl 57–100%	6N HCl 54–90%	1999 (115)
iPr	Me	2-PMPS-3,5-(MeO)₂-BnCl	84%, 78% de 92%, 26% de		0.25N HCl 97%		1999 (109)
iPr	Me	2-Cl-3-BnO-4-MeO-BnBr	83% 90% de		0.5N HCl 88%	LiOH, THF 99%	1999 (492)
iPr	Me	BrCH₂CH₂CH=CH₂	84% 90% de		HCl 68%		1999 (493)
iPr	Et	4-[SiMe₂(allyl)]-BnBr	88% de 71% major isomer	link to bromopolystyrene resin by treatment with 9-BBN couple with 2-NH₂-nR-PhI, cyclize to give R³ = CH₂(nR-indol-2-yl) R = H, 5-NO₂, 5-F, 5,6-Me₂, 6-MeO 50–81%	1:1 MeOH: 1N HCl	5 eq. LiOH, THF/H₂O —	1999 (494)
iPr	Me	TMSCCCH₂OPO₃Ph₂	80% 96% de				1999 (44)

Me	iPr	TESCCCH$_2$OPO$_3$Ph$_2$	remove TES with TBAF: 96% Pd-catalyzed coupling with 2-I-aniline or 2-I-3-NH$_2$-naphthalene followed by cyclization to form indole: 64–75%	90% 95% de	1N HCl	1N NaOH 72–75%	2000 (61)
Me	iPr	TsOCH$_2$CH$_2$(1-TBDMS-indol-3-yl)	—	92% >95% de	1N HCl	1N NaOH 87%	2000 (61)
Me	iPr	ICH$_2$Si(Me$_2$)CH$_2$I	one step deprotection/cyclization to give Pro analog	67% 72% de	MeCN:H$_2$O: HCO$_2$H 70%		2000 (114)
Me	iPr	3-NO$_2$-4-F-BnBr	—	80% using nBuLi, CuCN	1) TFA: 87% 2) N protection	LiOH: 91–100%	2001 (49)
Me	iPr	ClCH$_2$CCCH$_2$Cl	1) second alkylation with BrCH$_2$CH=CH$_2$ (62%) 2) Ru-catalyzed metathesis cascade 92–97%	73% dimer	0.1M TFA 35%	—	2001 (127)
Me	iPr	ClCH$_2$CCCH$_2$Cl	1) second alkylation with ClCH$_2$CCTMS 2) Ru-catalyzed metathesis cascade 35%	71% dimer	0.1M TFA	—	2002 (126)
Me	iPr	BrCH$_2$CH=CH$_2$ BrCH$_2$C(Br)=CH$_2$	1) second alkylation with BrCH$_2$C(Br)=CH$_2$ or BrCH$_2$CH=CH$_2$ 60–71%, >95% de 2) Pd-catalyzed cyclization 52–55%	—			2003 (129)
Et	iPr	3-BnO-BnBr		62%	0.5N HCl	LiOH, THF/H$_2$O	2003 (495)
Me	iPr	BrCH$_2$(2-Cl-6-MeO-4-pyridinyl)	Stille ethoxycarboxylation of aromatic group 60%	84% 82% de	0.5N HCl 86%	1) LiOH, THF/H$_2$O 2) TMSI 71%	2003 (62)
Et	iPr	TMSCCCH$_2$OPO$_3$Ph$_2$	couple with 2-NH$_2$-nR-PhI, cyclize to give R^3 = CH$_2$(nR-indol-2-yl) R = 6-aza, 5- or 6-NO$_2$, 5-Cl, 5,6-Cl$_2$, 7-MeO 42–75%	90% >98% de	2N HCl/THF	1N NaOH 76–92% for 2 steps	2004 (60)
Et	iPr	BrCH$_2$CH(iPr)CH$_2$[3-(MeOPrO)-4-MeO-Ph]		68%	HCl, MeCN		2005 (496)
Me	iPr	3-Br-BnBr 4-Br-BnBr	conversion ArBr to organocuprate, alkylation with Bz-α-Br-Gly-OEt 56–60%	92–95%	A 0.1M TFA 75–84%		2005 (102)

The Schöllkopf methodology has been used by a number of other groups to prepare amino acids, relying primarily on the cyclo[D-Val-Gly] bislactim ether template to produce L-amino acids (see Schemes 7.3–7.6). 3′-Nitro-4′-fluoro-Phe was prepared as an intermediate in a total synthesis of the teicoplanin aglycon (49), while the side chain of a 3′-phosphono-L-Tyr derivative was introduced by an aldol-type reaction of a substituted benzaldehyde, followed by removal of the β-hydroxy group (50). 3-Methyl-Asp was prepared by alkylating the template with methyl (2R)-2-bromopropanoate (51). An analog of MeLeu, used in a modified cyclosporin derivative, was prepared by alkylation with bromoallyl bromide, followed by a Stille coupling with vinyltributyltin and subsequent hydrogenation of the resulting diene (52). [2′-¹³C]-, [1′-¹⁵N]-, and [3′-¹⁵N]-L-His were prepared via reaction of the corresponding 5-(chloromethyl) imidazoles (53); 5-arylthio-3-methyl-L-histidines have also been synthesized (54–56). Similarly, alkylation of the template with substituted 3-bromomethylindoles produced Trp derivatives (31, 32, 57, 58), with 4-bromo-methyl isooxazoles giving other heteroaromatic products (59). The Trp indole ring can also be prepared by reaction of propargyl-substituted bislactim ether template with o-iodoanilines (44, 60). This route was employed to prepare isotryptophan, β-(indol-2-yl)-Ala, and benzo[f]-Trp, β-(benzo[f]indol-3-yl)-Ala (61). A 2003 synthesis of (S)-(−)-acromelobic acid, 3-(6-car-boxy-2-oxo-4-pyridyl)-L-Ala, employed alkylation of the bislactim ether with 4-bromomethyl-2-chloro-6-methoxypyridine, followed by introduction of an ester on the aromatic ring by a Stille cross-coupling reaction (62). [2-¹³C]-Lys was produced by alkylation of a cyclo[D-Val-Gly] template, prepared using [2-¹³C]-; with 4-iodobutyronitrile. The nitrile was reduced with LiAlH₄ and the template hydrolyzed with 6 N HCl. [3,4-¹³C₂]- and [5,6-¹³C₂]-Lys were also synthesized, using the appropriately labeled butyronitrile (63).

Both the Val- and DOPA-derived templates were used to prepare (R)-Ser by alkylation with chloromethyl benzyl ether (64). The same procedure gave [3-¹⁸O]- and [3-¹³C]-L-Ser (65). Both templates have also been used

for the synthesis of β-hydroxy-Val by adding the lithiated templates to acetone, although deprotection was difficult (see Scheme 7.5, Table 7.3 for aldol reactions) (32). Low yields during hydrolysis of the bislactim ether template due to retro-aldol reaction can be minimized by use of 0.1 N trifluoroacetic acid (66), or by O-TBS protection before hydrolysis with aqueous HCl (67). The cyclo[L-Val-Gly] template has been reacted with a number of other aldehydes and ketones, giving adducts possessing 85 to >95% de at the α-center and 4–70% de at the carbinol center, with the threo isomers obtained preferentially (39). The diastereoselectivity could be slightly improved by using the O,O-diisopropyl bislactim ether instead of the dimethyl analog. A much more significant improvement in stereoselectvity was provided by exchanging the lithiated anion for a titanium derivative, with the aldol products possessing >95% de at the α-center and >94% de at the the β-center, still in favor of the threo isomer (6, 66, 68). The titanium derivative has been used to prepare β-hydroxy-Phe, β-hydroxy-Nle, β-hydroxy-Chg, β-hydroxy-4,5-dehydro-Nle, 2-amino-3-hydroxy-6-octenoic acid (66), 3′,4′-dihydroxy-6′-[¹⁸F] fluoro-phenylserine (69), and [3-¹⁸O]-, [3,4-¹³C₂]-, and [3-²H]-L-Thr (65). Double asymmetric inductions have also been investigated by reacting the lithium and titanium templates with protected (S)- or (R)-glyceraldehyde or (S)-lactaldehyde (70). The aldehyde chirality had little effect on stereoselectivity, as both glyceraldehyde enantiomers gave products with 96% de at C2 and >80% de at C3. An Sn enolate provided >90% de for addition to protected 2,3-O-isopropylidene-L-threose, compared to only 20% de for addition of the corresponding Li enolate (71). A transmetallated Zr anion was found to provide extremely high stereocontrol (>99% de at C2 and 66% de at C3) for a synthesis of 3′-nitro-4′-fluoro-phenylserine (67).

The titaniated species has also been added to the aldehyde group of α,β-unsaturated aldehydes, again with high stereoselectivity (13). The alkene of the adduct can be epoxidized (13, 72) or cyclopropanated (72), and the epoxide can be further reacted with nucleophiles (72). A major difficulty with all the aldol adducts is that

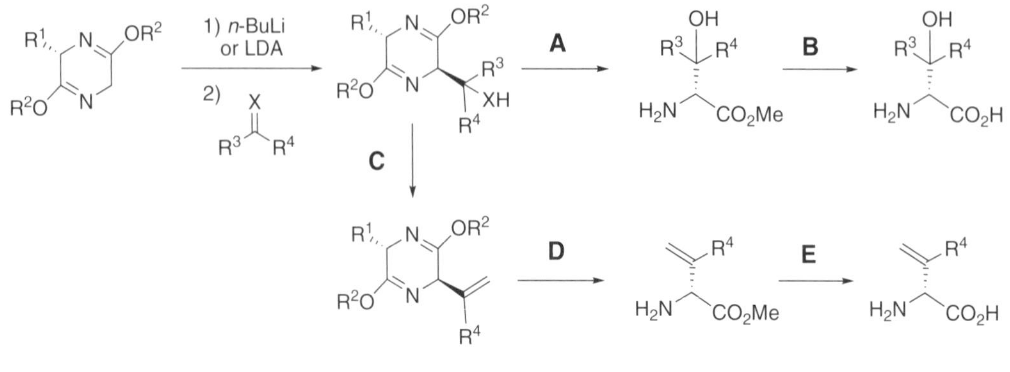

Scheme 7.5 Synthesis of α-Amino Acids via Aldol Reaction of Schöllkopf Bislactim Ether (see Table 7.3).

Table 7.3 Synthesis of α-Amino Acids via Aldol Reaction of the Schöllkopf Bislactim Ether Template (see Scheme 7.5)

R^1	R^2	Electrophile	Alkylation Yield	Derivatization of Side Chain	AD	BE	Reference
iPr 3,4-(MeO)₂-Bn	Me	MeCOMe MeCOPh	91%	C R^3 = Me; R^4 = Ph SOCl₂, 2,6-lutidine 88% with 25% α,β-unsaturated isomer	D 0.25N HCl 64% >95%ee	B 40% HBr 66–83%	1981 (88)
iPr 3,4-(MeO)₂-Bn	Me	MeCOMe	79–98%	MeI, *t*BuOK: 88% forms side chain = C(Me)₂OMe	A 0.1N HCl 59–81%	B	1982 (73)
iPr	Me	TBDMSCH₂CHO TBDMSC(Me)₂CHO TBDMSCH(Me)CHO	95–99% de C2 94–98% de C3 *threo*	(C) hydrolysis induces elimination to side chain = CH=CHMe CH=CH₂	A 0.25N HCl 53–58%	B 5N HCl 62–70% (induces elimination)	1983 (5) 1984 (89)
iPr	Me	EtCSEt *n*PrCS*n*Pr cyclothiapentanone MeCSEt	36–76% >95% de C2	C Ra-Ni 67–80% *E:Z* 1.5:1 to 2:1	D 0.25N HCl 64–76%	E 2N HCl 72–76%	1983 (5) 1984 (89)
iPr	Me	MeCHO *i*PrCHO PhCHO PhCOCH₂Cl	TiCl₄ 95–99% de C2 94–98% de C3 *threo*	for R^3 = Ph, R^4 = CH₂Cl treat with NaOH, form side chain 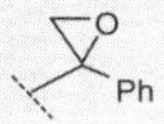 = Ph			1983 (6)
iPr	Me	H₂CO PhCHO MeCOMe MeCOPh		fluorinate with DAST	A	B 40–60% fluorination and hydrolysis	1983 (474)
iPr	Me	MeCOMe MeCOPh cyclopentanone MeCHO *i*PrCHO *t*BuCHO BnCHO PhCHO	77–98% 86–>95% de C2 4–70% de C3 *threo*	R^3 = Me; R^4 = Me, Ph SOCl₂, 2,6-lutidine 70–80% with 20-30% α,β-unsaturated isomer	A 0.25N HCl R^3 = R^4 = Me R^3 = H; R^4 = Me, *i*Pr, *t*Bu 48–59% >88–94% ee	B conc. HCl R^3 = H; R^4 = Me	1983 (39)
*t*Bu	Me	MeCOMe cyclopentanone	92–95% 85% de C2				1983 (39)

(Continued)

505

Table 7.3 Synthesis of α-Amino Acids via Aldol Reaction of the Schöllkopf Bislactim Ether Template (see Scheme 7.5) (continued)

R1	R2	Electrophile	Alkylation Yield	Derivatization of Side Chain	AD	BE	Reference
iPr	Me		69–79% 96–98% de C2 90–97% de C3		A 1) Ac$_2$O, Pyr. 82–92% 2) 0.2N HCl 63–80%		1985 (70)
iPr	Me	MeCOCH$_2$Cl PhCOCH$_2$Cl	91–94%	cyclize side chain with NaH to give side chain = 92%			1985 (77)
iPr	Me	CH$_2$=C(Me)CHO MeCH=C(Me)CHO CH$_2$=CHCHO MeCH=CHCHO PhCH=CHCHO EtO$_2$CCH=CHCHO	Ti anion 75–85% 96–>99% de C2 96–99% de C3 *threo*	cyclopropanate alkene: Et$_2$Zn, CH$_2$Cl$_2$ R^5 = H; R^6 = Me, Ph R^5 = Me; R^6 = Ph 61–76% epoxidize alkene: tBuOOH, Ti(OiPr)$_4$ R^5 = Me; R^6 = H, Me R^5 = H; R^6 = H, Me, Ph 55–83%, variable de open epoxide: Nu = BnSH, NaN$_3$ 40–80%	A 1) Ac$_2$O or MEM-Cl 82–95% 2) 0.1–0.2N HCl 51–82%		1987 (13) 1988 (72)

		Aldehyde	Notes	A	B	Year (ref)
iPr	Me	(4-Me-Ph)CHO (4-CN-Ph)CHO (4-Br-Ph)CHO (4-NO$_2$-Ph)CHO (2-BnO-Ph)CHO	Ti anion 68–83% >95% de C2 >95% de C3	A 0.5N HCl 44–66%		1989 (68)
iPr	Me		treat with DAST to form β-F 8–15%	36% 10–66% de	47–51%	1990 (75)
iPr	Me	PhCHO MeCH=CHCHO nPrCHO MeCH=CH(CH$_2$)$_2$CHO cyclopentanone		84–96%	A 0.1N TFA 63–73%		1991 (66)
iPr	Me	(S)-Boc$_2$NCH(CO$_2$tBu) CH$_2$CH$_2$CHO	Ti anion 71% 86% de	A 0.1N HCl 77%	B 5N HCl 75%	1992 (91)
iPr	Me	(3-PO$_3$Me$_2$-4-BnO-Ph)CHO	β-deoxygenation via 1) PhOC$_5$Cl 2) Bu$_3$SnH, AIBN 71% for 3 steps	A 0.25N HCl 77%	B 1N NaOH	1993 (50)
iPr	Me	^{13}CH$_3$^{13}CHO MeC^{2}HO	Ti anion 58%	A 0.25N HCl	B conc. HCl 88%	1995 (65)
iPr	Et			nBuLi, ZnCl$_2$ 97% 65% ds	A 42%	B 53%	1998 (74) 1998 (497)

(Continued)

Table 7.3 Synthesis of α-Amino Acids via Aldol Reaction of the Schöllkopf Bislactim Ether Template (see Scheme 7.5) (continued)

R¹	R²	Electrophile	Alkylation Yield	Derivatization of Side Chain	A D	B E	Reference
iPr	Et		Li enolate: 57%, 20% de Sn enolate: 70%, >90% de		A 65%		1999 (71)
iPr	Me	3-CH$_2$CH=C(Me)$_2$-4-MeO-PhCHO	74% 0% de C3		A 55–56%		1999 (498)
iPr	Me	HCOCH$_2$CH$_2$CH(NBoc$_2$)CO$_2$Me (= Glu semialdehyde)	73% >95% de C2 50% de C3	fluorination with DAST 52%		A+B 6N HCl 71%	1999 (76)
iPr	Me	3-NO$_2$-4-F-PhCHO	60% >98% de C2 66% de C3	protect β-OH with TBS	A aq. HCl 89%		2004 (67)

retro-aldol reactions often occur during hydrolysis, requiring "carefully controlled conditions" (5). Only a few amino acids, such as β-hydroxyVal (5, 6, 73) and Thr (39, 65), have been completely deprotected. Protection of the hydroxy group with an acetyl or MEM group allows for improved yields during hydrolysis (13), as does use of TFA instead of HCl for the template hydrolysis (66). Aldol reaction has also been employed for the synthesis of sphingofungins B and F (74).

The aldol adducts of ketones or aldehydes have been converted into β-fluoroamino acids, via fluorination of the aldol adduct with DAST. β-Fluoro-Val and β-fluoro-Phe were produced (20). Aldol addition of the cyclo[L-Val-Ala] template to an aldehyde derived by reduction of the side chain of a protected Glu derivative resulted in a precursor of 3-hydroxy-2,6-diaminopimelic acid. However, the hydroxyl group was reacted with DAST before template deprotection, resulting in either diastereomer of 3-fluoro-2,6-diaminopimelic acid (75). An aldol addition of the Schöllkopf bislactim ether template with N,N-Boc$_2$ Glu semialdehyde, followed by DAST fluorination of the hydroxy group, provided (2S,3R,6S)-3-fluoro-2,6-diaminopimelic acid in better yield (76). Addition of the template to 1-chloroacetone or chloroacetophenone, followed by treatment with NaOH, generated epoxide derivatives. The phenyl-substituted adduct was hydrolyzed to give the methyl ester of 2-amino-3,4-epoxy-3-phenylbutanoate, while the epoxide of the methyl-substituted analog was opened with azide and the product hydrolyzed to give methyl 3-amino-3-hydroxy-3-methyl-4-azidobutanoate (77).

Acylations of the cyclo[L-Val-Gly] template by acyl chlorides or related compounds are not as successful as with the cyclo[L-Val-Ala] template, with benzoyl chloride resulting in a bis-addition product dimer, and N-methoxymethyl-N-methylbenzamide giving a product with 0% de due to epimerization during work-up (16).

A number of electrophiles have been used for conjugate additions (see Scheme 7.6, Table 7.4). With α,β-unsaturated esters the lithiated [(L-Val-Gly-] template was found to add in a 1,4-Michael fashion, though with poor stereoselectivity at the β-center (0–86% de, with 50% de for E-methyl butenoate) (78). However, a more recent report found that the diastereoselectivity of the conjugate addition with E-methyl butenoate was >82%, with the C3 epimer obtained from Z-methyl butenoate with similar stereoselectivity (79). Deprotection yielded isomers of 3-methyl-Glu. The E/Z-isomers of methyl cinnamate also gave good diastereoselectivity (80).

In contrast, little C3 stereoselectivity was observed with additions to Z-4′-chlorocinnamate, although the E-isomer produced predominantly one diastereomer (81). Conjugate additions to (E)- or (Z)-vinyl sulfones also proceeded with high diastereoselectivity, and since the sulfonyl group can be reductively cleaved the method gave β-ethyl-branched amino acids of defined configuration (82). Alkylation with a bromoethylphosphonate ester resulted in 2-amino-4-phosphobutanoic acid (AP4), via addition to an in situ generated vinylphosphonate intermediate (83). 2-Amino-3-alkyl-4-phosphobutanoic acids were prepared by conjugate additions to the corresponding alkylvinylphosphonates, with >95% de at the product C2 center and >85% de at the C3 center (84–86). Additions of the cyclo[L-Ala-Gly] or cyclo[D-Val-DL-Ala] templates to a greatly expanded set of vinylphosphonates were reported in 2003, producing 3- or 4-monosubstituted and 2,3-, 2,4-, or 3,4-disubstituted AP4 derivatives in enantiomerically pure form (87).

Several routes to β,γ-unsaturated amino acid derivatives have been reported (see Scheme 7.5). The products from aldol additions can be converted into 3-substituted vinylglycine derivatives by dehydration with thionyl chloride/2,6-lutidine, although the isomeric dehydro-amino acid is a significant contaminant (39, 88). A potentially more useful synthesis proceeds via a Peterson-type elimination of the aldol adducts obtained with 2-($tert$-butyldimethylsilyl)aldehydes, but some of the adducts are unstable (5, 89). Alternatively, 3-substituted vinylglycines were produced by addition of the lithiated bislactim ether to thioketones, followed by methylthio-elimination induced by Raney-nickel (6,89). One of the disadvantages of the bislactim ether approach is apparent in this report, with several of the hydrolyzed products contaminated with significant quantities of Val-OMe ester (89). Attempts to eliminate dimethylsulfide from sulfonium salt derivatives of the thioketone adducts was less effective than the Raney-Ni-induced elimination (89). The diethylthioketone adduct has also been S-methylated and then hydrolyzed to give a Cys derivative (5, 89). The lithiated template can also be added to methyl acrylates containing a leaving group in the β-position, but the initial β,γ-didehydro amino acid derivatives isomerize to undesired α,β-unsaturated isomers (18).

2,6-Diaminoheptanedioic acids (2,6-diaminopimelic acids) were stereoselectively synthesized by reacting 2 equiv of the lithiated template with 1,3-dibromopropane, or by reacting 1 equiv of template with a

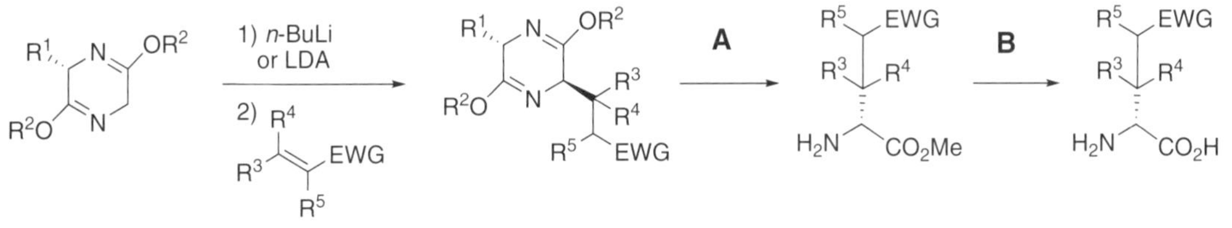

Scheme 7.6 Synthesis of α-Amino Acids via Michael Additions of Schöllkopf Bislactim Ether (see Table 7.4).

Table 7.4 Synthesis of α-Amino Acids via Michael Additions of Schöllkopf Bislactim Ether Template (see Scheme 7.6)

R^1	R^2	EWG	R^3	R^4	R^5	Alkylation Yield	Derivatization of Side Chain	A	B	Reference
iPr	Me	CO_2Me	H, Me, Ph, 4-MeO-Ph, 4-pyridyl	H	H	42–92% / 98% de C2 / 0–86% de C3		0.25N HCl / $R^3 = R^4 = R^5 = H$ / 65%	78%	1986 (78)
		CO_2Me	H	H	Me					
		CO_2Me	H	Ph	H					
		CO_2Me	Me	Me	H					
iPr	Me	CO_2Me	Ph	H	H	80–86% / 98–100% de C2 / 70–92% de C3				1987 (80)
		CO_2Me	H	Ph	H					
iPr	Me	$PO_3(allyl)_2$	H	H	H	80%		0.25N HCl 95%	chymotrypsin 76% >96% ee	1993 (83)
iPr	Me	PO_3R_2	Me	H	H	65–82% / >98% de C2 / 85–94% de C3		0.25N HCl 96%	lipase 55%	1994 (85) / 1994 (86)
		PO_3R_2 / R = allyl, Bn, 2-Br-Bn, 4-Ph-Bn	H	Me	H					
iPr	Me	CO_2Me	Me	H	H	55–65% / >82% de		0.25N HCl	6N HCl 53–62%	1994 (79)
		CO_2tBu	H	Me	H					
iPr	Me	SO_2Ph	Ph	H	H	82–84% / 100% de C2 / 80% de C3	reduce alkene with Na/NH_3 79–82%	0.5N HCl 90%	6N HCl 54%	1995 (82)
		SO_2Ph	indol-2-yl	H	H					
		SO_2Ph	H	Ph	H					
iPr	Me	CO_2Me	4-Cl-Ph	H	H	26–52% / 92–94% de C2 / 15–87% de C3		1N TFA	6N HCl 7–25%	1996 (81)
		CO_2Me	H	4-Cl-Ph	H					

iPr	Me	PO$_3$Et$_2$	iPr	H		61–95%	0.25N HCl 75–94%	12N HCl 88–92%	1996 (84)
		PO$_3$Et$_2$	Ph	H		>95% de C2			
		PO$_3$Et$_2$	CH$_2$OBn	H		>85% de C3			
		PO$_3$Et$_2$	CH$_2$NHCbz	H					
		PO$_3$Et$_2$	CH=CH$_2$	H					
iPr	Et	PO$_3$Et$_2$	H, iPr, iBu, Ph, Bn, 3-pyridyl, 2-thienyl, 2-furyl, CH$_2$OBn, CH$_2$NHCbz	H	H	39–92% >95% de C2 63–>95% de C3	0.25N HCl 72–93%	12N HCl 85–98% or 1) LiOH, H$_2$O 2) TMSBr 3) MeOH 63–98%	2003 (87)
		PO$_3$Et$_2$	iPr, iBu, Ph, 3-pyridyl, 2-thienyl, CH$_2$OBn	H	H				
		PO$_3$Et$_2$	H	H	Me, Ph, CO$_2$Et, PO$_3$Et$_2$, OPO$_3$Et$_2$, SiMe$_3$, SnPh$_3$				
		PO$_3$Et$_2$	Ph	H	CO$_2$Et				
		PO$_3$Et$_2$	Ph	H	SiMe$_3$				
		PO$_3$Et$_2$	Ph	H	SnPh$_3$				
		PO$_3$Et$_2$	H	Ph	SiMe$_3$				
		PO$_3$Et$_2$	H	Ph	SnPh$_3$				
		PO$_3$Et$_2$	(CH$_2$)$_3$-R^5	(CH$_2$)$_3$-R^4					

brominated Garner aldehyde derivative or Glu semialdehyde (90, 91). For 2,5-diaminohexanedioic acid, the
template was alkylated with ethyl bromoacetate, the
side chain converted into a 2-iodoethyl substituent, and
then the iodide displaced with a second equivalent
of template (92). 2,7-Diaminooctanedioic acid (2,7-
diaminosuberic acid) derivatives have also been
synthesized by reaction of two template equivalents
with 1,4-dihaloalkanes (93, 94). 1,2-Dibromoethane,
1,3-dibromopropane, and 1,4-dibromobutane were reacted
with 2 equiv of the cyclo[L-Val-Ala] template to give
α,α'-dimethyl diaminodicarboxylic acids (95). Alkenyl-
bridged diaminodicarboxylic acids have been synthesized by reaction of 2 equiv of template with
dibromoalkenes (96). Attempts to link two cyclo[L-Val-
Gly] bislactim ether templates or two allyl-substituted
templates with the biselectrophiles 2,3-dibromo-1,4-di-
chloro-2-butene or 1,2-dibromotetrafluoroethane were
unsuccessful, instead resulting in template dimerization
(homocoupling) at the α-centers. The 2,3-diaminosuc-
cinic acid derivatives were not deprotected (97). One
amino acid center of 2,9-diaminodecanedioic acid was
formed by alkylation of the cyclo[L-Val-Gly] bislactim
ether with 8-iodooctanoic acid TBS ester. The ester
group was then deprotected and acylated with Evans'

oxazolidinone auxiliary, with deprotonation/azidation/
reduction generating the second amino acid center
(98, 99).

A bis(amino acid) has also been obtained by
palladium-catalyzed coupling of a bromoalkene adduct
(see Scheme 7.7) (100). Alkyne-bridged diaminodicarboxylic acids were produced by Pd-catalyzed couplings
of propargylglycine derivatives with each other, or with
1,4-diiodobenzene (101). Alkylation of the bislactim
ether template with m- or p-bromobenzyl bromide, followed by conversion of the aryl bromide to an organocuprate and reaction with an electrophilic Gly derivative,
Bz-α-Br-Gly-OEt, produced phenyl-bridged diaminodicarboxylic acids (102). A tris(amino acid) based on a
central pyridine nucleus was also prepared by a route
beginning with alkylation of the bislactim ether template
(103). A Phg-Tyr bis(amino acid) linked by a biaryl
ether was prepared by arylation of the bislactim ether
template with an arene Mn(CO)₃ complex cation (104,
105). The biaryl ether-linked bis(amino acid) L,L-
isodityrosine was produced by introducing the second
amino acid center to a DOPA-derived biaryl ether amino
acid derivative (106–108). The Schöllkopf bislactim
ether template has been used to synthesize the complex
bis(amino acid) component of the starfish alkaloid

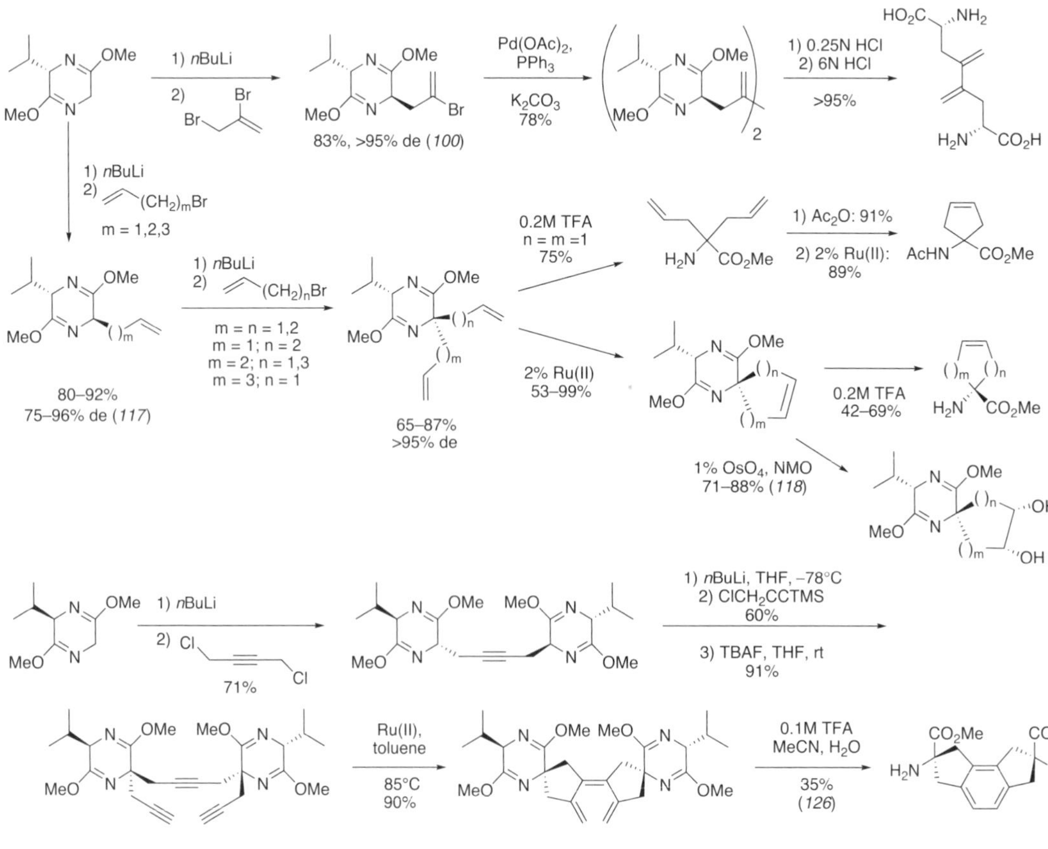

Scheme 7.7 Synthesis of Amino Acids via Alkene Coupling.

imbricatine, a benzyltetrahydroisoquinoline-3-carboxylic acid linked by a thioether to a His group. 2′-Thio-3′,5′-dimethoxy-Phe was first prepared and then cyclized to the benzyl-substituted tetrahydroisoquinoline moiety. The thiophenol was alkylated with an imidazole derivative, with the second amino acid center constructed by reaction of an imidazole chloromethyl substituent with a second equivalent of template (*109*).

A dibromoalkane electrophile has also been used to synthesize thiolysine, via alkylation of the template with 1,4-dibromobutane followed by displacement of the second bromide with the potassium salt of 4-methoxytoluenethiol (*110*). A similar methodology was used to prepare 4-alkynyl Lys using 1,4-dichloro-2-butyne for the initial alkylation followed by chloride displacement with NaN$_3$ (*111*). Alkylation of the template with 1,6-diiodo-3-hexene, followed by displacement of the second iodide by malonate, resulted in an unsaturated aminodicarboxylic acid derivative (*112*). An attempt at preparing unsaturated Pro derivatives via alkylation of the template with 1,3- or 1,4-dichloroalkenes, followed by cyclization with the α-amino group after template hydrolysis, was unsuccessful. The alkylation proceeded well, but only decomposition was observed upon template hydrolysis. The same approach was successful if the Seebach Boc-BMI template was employed (*113*). However, a 4,4-dimethyl-4-sila-Pro derivative was prepared via alkylation of the template with bis(iodomethyl)dimethylsilane, followed by a one-step template hydrolysis/ring closure induced with MeCN/H$_2$O/HCO$_2$H (*114*). 3,4-Methano-Glu derivatives, with substituents on the bridging methylene group, were obtained by an addition/elimination reaction of 4-alkyl-4-bromo-but-2-enoates (*115*).

α,α-Disubstituted cyclic amino acids can be produced by several routes. Alkylation of the template with *trans*-1,4-dichloro-2-butene, followed by a second deprotonation/cyclization step and reduction of the alkene bond, led to (1*R*,2*S*)-*allo*-coronamic acid (*cis*-1-amino-2-ethylcyclopropyl-1-carboxylic acid) (*116*). A variety of 1-aminocycloalkene-1-carboxylic acids were synthesized by an alternative route, involving introduction of two unsaturated substituents followed by a ring-forming metathesis reaction (see Scheme 7.7 for some simple examples) (*117*). An alkyl chain with a terminal vinyl group was introduced, followed by a second alkylation with another alkylvinyl compound of the same or different length. Ruthenium(II)-catalyzed ring closing metathesis was then employed to generate five-, six-, or seven-membered rings, with the position of the alkene bond (which provides asymmetry) determined by the relative lengths of the two alkenyl chains (and the order in which they were introduced) (*117*). The alkene bond could be dihydroxylated, giving predominantly one set of isomers (*118*). Rings containing a conjugated diene with one exocyclic double bond have also been prepared (*119*). A similar procedure was used to prepare cycloalkene amino acids with a β-hydroxy group. The second unsaturated substituent used for the metathesis was introduced by aldol reaction of the template with acrolein

(*120, 121*). The β-hydroxy group could be oxidized to a ketone before the cyclization metathesis step (*122*). Cycloalkenes with a β-hydroxy group were also prepared by alkylation of the template with an alkenyl halide, followed by a second alkylation with ethylene oxide. Oxidation of the hydroxyethyl substituent, Grignard addition of vinylmagnesium bromide to the aldehyde, and metathesis of the alkenes gave the cyclized products (*123*). The procedure was simplified by employing vinyloxirane for the second alkylation (*124*).

More complex versions of these metathesis reactions have been developed. Ru(II)-catalyzed metathesis of an α-(3-butenyl)-α-(1-oxoprop-2-enyl)-bislactim derivative led to 1-amino-1-carboxy-cyclohex-3-en-2-one; the α,β-unsaturated ketone could be used for conjugate addition of organocuprates to give 4-methyl- or 4-phenyl-1-amino-1-carboxy-cyclohexan-2-one (*125*). A very rigid bis(amino acid), consisting of two 1-amino-cyclopentane-1-carboxylic acid moieties fused to a phenyl ring, was constructed via an initial linkage of two templates by a C4-alkyne bridge formed by reaction with 1,4-dibromo-2-butyne. A second lithiation/alkynylation with 3-trimethylsilyl-propargyl chloride introduced a propargyl substituent to both templates. The cyclopentyl rings and linking fused phenyl ring were then constructed in one step by a Ru-catalyzed cascading metathesis reaction (see Scheme 7.7) (*126*). A similar sequence introduced allyl substituents instead of propargyl groups, with metathesis providing 3,3′-bis(1-aminocyclopent-3-ene-1-carboxylic acid). Diels–Alder reaction of the diene system with diethyl acetylenedicarboxylate led to a tricyclic bis(amino acid) (*127*).

A Pd-catalyzed Heck reaction has also been used to connect two unsaturated side chains, one with a bromide substituent (*128, 129*). Another synthesis introduced two *n*-oxo-alkane substituents, via alkylation with an ω-bromoalkene followed by a second alkylation with propylene oxide, or aldol reaction with acetaldehyde. The first ketone side chain was generated by oxidation of the carbinol, with the second ketone (or aldehyde, depending on substrate) formed by Wacker PdCl$_2$/CuCl oxidation of the alkene. Intramolecular aldol reaction induced by Cs$_2$CO$_3$ formed a range of methylketone-substituted 1-aminocyclopentene- or -hexene-1-carboxylates (*130*).

An anion is not the only reactive species that can be generated from the bislactim ether. Treatment of the anion with tosyl azide led to a diazo compound from which a bislactim carbene was generated (*131*) (see Scheme 7.8). Addition of the carbene to cyclohexene produced a bicyclic 1-amino-1-cyclopropanecarboxylic acid derivative.

The cyclo[L-Val-Gly] bislactim ether template has also been converted into an electrophilic species by chlorination of the lithiated template (*132*) (see Scheme 7.9). Surprisingly, the *cis* chloride is obtained. The chloride was displaced with inversion by "soft" nucleophiles such as sodium alkyl malonates (*132*) or used for Friedel–Crafts-like arylations with alkoxyarenes (*133*), although some of the latter reactions

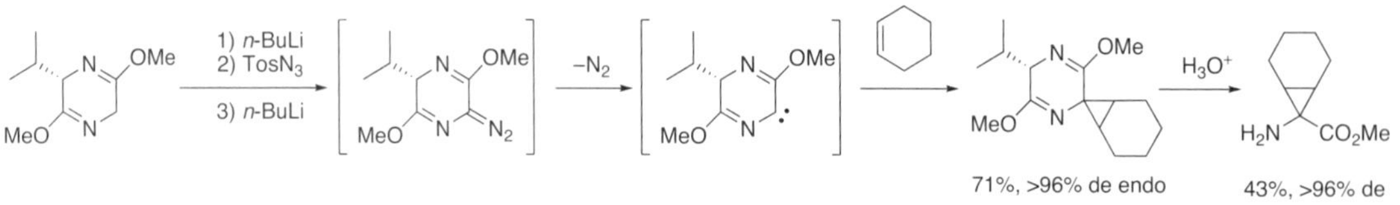

Scheme 7.8 Carbene Reaction of Schöllkopf Bislactim Ether (*131*).

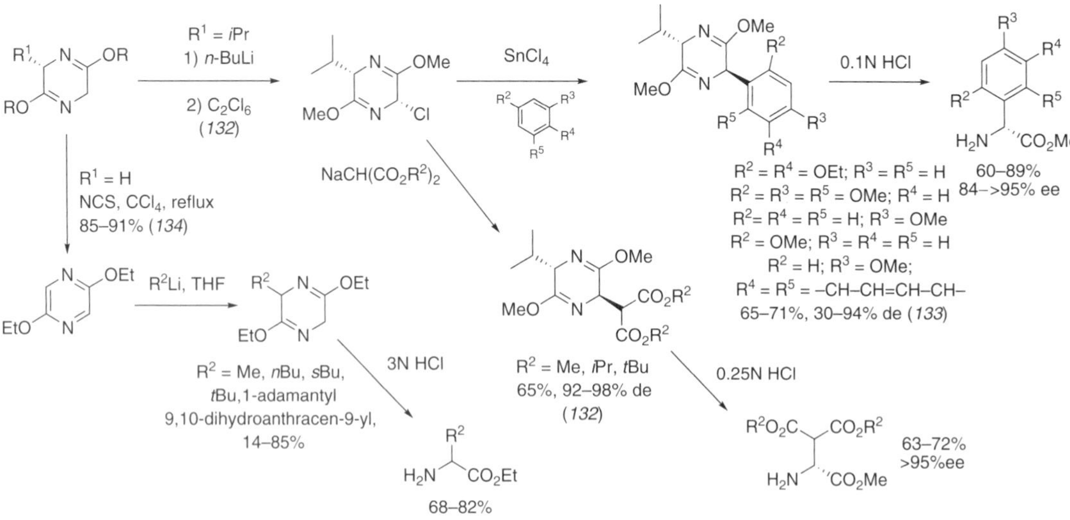

Scheme 7.9 Synthesis and Reaction of Cationic Schöllkopf Bislactim Ether.

had poor diastereoselectivity. The chloro-lactim ether tends to dehydrochlorinate and aromatize in the presence of organometallics. The resulting 2,5-alkoxypyrazine obtained from the cyclo[Gly-Gly] bislactim ether has also been used as an electrophilic species, with organolithium compounds adding in 14–85% yield (*134*). Racemic Tle was prepared by this method.

7.3 Piperazine-2,5-dione

The piperazine-2,5-dione (2,5-diketopiperazine) precursor of Schöllkopf's bislactim ether template has been applied to several asymmetric amino acid syntheses. Pro was used as the chiral auxiliary half of the template to induce asymmetric hydrogenation of dehydroamino acid derivatives derived from α-keto acids (see Section 4.3.2a) (*135*), while a cyclo[Val-dehydro-Phe] diketopiperazine template was deuterated to give [2,3-^{2}H$_2$]-Phe (*136*). Diketopiperazine templates containing dehydroamino acids have been prepared by reaction of *N,N'*-Boc$_2$ Val-Gly diketopiperazine with aldehydes in the presence of KO*t*Bu, with concomitant removal of the Boc group attached to the dehydroamino acid residue. Catalytic deuteration of the completely *N*-deprotected

template produced the *threo*-[2,3-^{2}H$_2$] isomers of Phe, Tyr, and DOPA and the *threo*-[2,3,4-^{2}H$_3$] isomer of Leu with 88–94% de. The *erythro* isomers were obtained with 80–94% de by reduction of the *N,'N*-Boc$_2$-protected template (*137*).

Alkylations of an *N,N'*-bis(4-methoxybenzyl) Val-Gly diketopiperazine have been described (see Scheme 7.10) and directly compared with the cyclo[L-Val-Gly] Schöllkopf bislactim ether template. Yields and diastereoselectivities were significantly increased (63–90% yield and 89–98% de vs 0–83% yield and 50–91% de) (*138, 139*). Both enantiomers of an amino acid could be obtained from the same template, with Phe used as an example. The initial benzylated template (>95% *trans* de) was re-deprotonated, and then quenched with a hindered proton source (di-*tert*-butylphenol), resulting in the *cis* diastereomer with 93% de (*140*). The same *cis* diastereoselectivity could be obtained by deprotonation of an initial 1:1 *cis:trans* mixture. Several of the alkylated templates were deprotected by *N*-deprotection with CAN, followed by hydrolysis with 6 N HCl, giving Ala (*138, 139*), both enantiomers of Phe (*140*), and homo-Tle (*141*). A dehydroalanine residue was generated within the template by deprotonation and reaction with paraformaldehyde (*141*). Organometallic additions to the alkene proceeded with >95% de, producing the *cis* diastereomer (*141*), while cyclopropanation with isopropylidenetriphenylphosphorane, ethylidenetriphenylphosphorane, phenyldiazomethane, or (CD$_3$)$_2$CD$_2$(Li) SO, resulted in 2,2-dimethyl-1-aminocyclopropane-1-carboxylate, 2-methyl-1-aminocyclopropane-1-carboxylate, 2-phenyl-1-aminocyclopropane-1-carboxylate, or [2,2-^{2}H$_2$]-1-aminocyclopropane-1-carboxylate with >98% de at the α-center, and varying diastereoselectivity at the β-center (when applicable) (*142*). The diketopiperazine template suffers from the same inherent disadvantage as the bislactim ether method, in that the auxiliary Val-OMe ester must be separated from the product methyl ester after template hydrolysis.

A cyclo[L-Ala-Gly] diketopiperazine which also contains chiral auxiliaries on the amide nitrogens was stereoselectively monoalkylated with a range of alkyl halides by Porzi and Sandri et al. (see Scheme 7.11, Table 7.5) (*143, 144*), giving the *cis* adduct with very high diastereoselectivity (>98% de) for the matched pair of chiral auxiliaries. This stereoselectivity is opposite to that of the Schöllkopf template, which gives the *trans* diastereomer. The template with the mismatched pair of auxiliaries still tended to give the product with the electrophile *cis* to the Ala methyl group, but diastereoselectivity was variable and dependent on the electrophile (*143, 144*). The template was also reacted with acyl chlorides (*145*) or aldehydes (*145, 146*), with the *cis/syn*

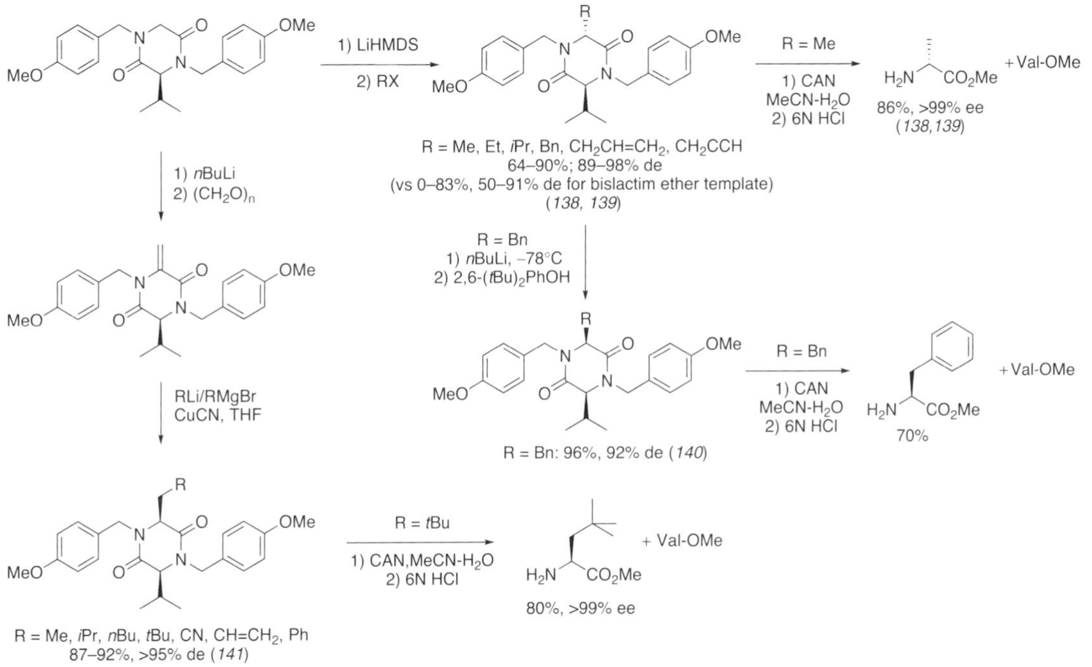

Scheme 7.10 Alkylation of *N,N*-bis(4-methoxybenzyl) Val-Gly Diketopiperazine.

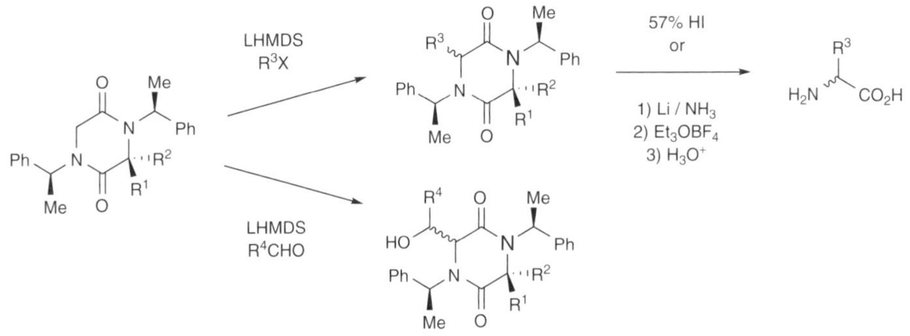

Scheme 7.11 Synthesis of Amino Acids via Piperazine-2,5-dione Template (see Table 7.5) (*143, 144*).

Table 7.5 Synthesis of α-Amino Acids via Piperazine-2,5-dione Template (see Table 7.11)

R^1	R^2	R^3	R^4	Alkylation Yield	Alkylation Diastero-Selectivity	Reference
Me	H	Me		92%	98% de *cis*	1992 (*144*)
H	Me	Me		90%	86% de *cis*	
Me	H	Bn $CH_2CH=CH_2$		92–96%	>98% de *cis*	1993 (*143*)
H	Me	$CH_2CH=CHMe$ $CH_2CH=CMe_2$ CH_2CCH		90–96%	10–50% de *cis* except $R^3 = CH_2CCH$, 98% de *anti*	
H	H	*n*Pr *i*Pr *i*Bu Bn $(CH_2)_2CH=CH_2$ $CH_2CH=CH_2$ $CH_2CH=CHMe$ $CH_2CH=CMe_2$ $CH_2CH=CHOBn$ CH_2CCH CH_2CO_2Et $CH_2CH=C(Me)$ $(CH_2)_2CH=CH_2$		90–95%	10–80% de	1994 (*147*)
H	*n*Pr Bn $CH_2CH=CH_2$ $CH_2CH=CMe_2$	*n*Pr Bn $CH_2CH=CH_2$ $CH_2CH=CMe_2$		90–96%	0–80% de	1994 (*147*)
*n*Pr Bn $CH_2CH=CH_2$ $CH_2CH=CMe_2$	H	*n*Pr Bn $CH_2CH=CH_2$ $CH_2CH=CMe_{21}$		90–96%	>98% de	1994 (*147*)
Bn	H		Me Ph $CH_2CH=CH_2$	30–75%	50–80% de C2 100% de C3	1995 (*146*)
H	Bn		Me Ph $CH_2CH=CH_2$	30–75%	33–88% de C2 100% de C3	
Me	H		Me Ph $CH_2CH=CH_2$		33–100% de C2 33–100% de C3	1995 (*145*)
H	Me		Me Ph $CH_2CH=CH_2$		33–60% de C2 100% de C3	
Me	H	MeCO PhCO		50–85%	100% de *cis*	1995 (*145*)
H	Me	MeCO PhCO		39–70%	0% de	
H	H	$(CH_2)_4-$ $CH_2(1,2\text{-Ph})\text{-}CH_2-$	—	80–90% yield	30–100% de	2000 (*148*) 2002 (*149*)
H	H	$CH_2C(=CH_2)$ CH_2-	—	90% yield	40% de	2003 (*151*)

diastereomers preferentially obtained in the latter reaction. In all cases heterocyclic ring cleavage to the amino acid ester required refluxing 57% HI or conversion to the bislactim ether before hydrolysis, a major drawback of this method.

The contribution of induction provided by the *N*-chiral auxiliaries was demonstrated by alkylation of the cyclo[Gly-Gly] derivative, which proceeded with poor to moderate diastereoselectivity (10–80% de) (*147*). The second Gly residue could react with a second equivalent of electrophile, with similar diastereoselectivity for matched (>98% de) or mismatched (0–80% de) chiral centers as was previously observed for the cyclo[Ala-Gly] diketopiperazine. However, a unique application of the cyclo-[Gly-Gly] derivative was the synthesis of the diaminodicarboxylic acids 2,6-diaminopimelic acid, 2,7-diaminosuberic acid, and o-phenylene-bis(Ala) via alkylation of both Gly residues in one equivalent of template with 1,3-diiodopropane, 1,4-diiodobutane, or α,α′-dibromo-o-xylene. Only the (2S,2′S)-diastereomer of the aryl-linked product was formed, while two of the four possible isomers of 2,7-diaminosubseric acid were generated in a 65:35 (2R,7R):(2S,7S) ratio (148, 149). The bicyclic bisalkylated templates could be further alkylated on one α-center, with deprotection providing α-alkyl (methyl, benzyl, allyl, methoxymethyl) 2,6-di-aminopimelic acids or o-phenylene-bis(alanine) (149, 150). A similar strategy was employed with 2-chorom-ethyl-3-chloropropene as the biselectrophile, leading to 2,6-diamino-4-methylene-1,7-heptanedioic acid, and α-substituted analogs (151).

A homolog of the diketopiperazine template, a 1,4-benzodiazepine-2,5-dione with an N-(α-methylbenzyl) substituent for asymmetric induction, was deprotonated with LDA and alkylated with alkyl halides in 55–91% yield, but only 54–82% de (see Scheme 7.12). Diastereoselectivity was reversed for alkylation with benzyl bromide compared to other alkyl halides (152). An achiral benzodiazepine template was employed to prepare racemic α-amino-β-hydroxy-γ-methyl acids (153). Another analog of the diketopiperazine template, a 2-oxopiperazine containing a chiral auxiliary on the lactam nitrogen, was alkylated with high diastereoselec-tivity (>90% de), but no amino acids were prepared from the adduct (154).

Porzi and Sandri have also reported a template corresponding to a monolactim ether, a 5-ethoxy-6-methyl-3,6-dihydro-1H-pirazin-2-one template, both with and without a chiral phenylethyl substituent on the amide nitrogen as a second auxiliary (see Scheme 7.13). Alkylations proceeded with unpredictable diaste-reoselectivity (0–99% de) (155). Consecutive alkyla-tions gave 66–96% de for the second alkylation, with N-deprotection and hydrolysis producing α,α-disubstituted amino acid Ala-OEt dipeptides (156). This template was employed to prepare Val-Xaa dipeptides, where the second amino acid was an α-alkylated imino acid derivative. The cyclo [Val-Gly] monolactim ether was deprotonated and N,C-bis(alkylated) with a number of dihalo derivatives (3-chloro-2-chloromethylpropene, 1-chloro-4-iodobutane, or α,α′-dibromo-o-xylene), with heating inducing cyclization with the amino group. The α-substituent could then be introduced by a second alkylation. Reversing the order of alkylation gave the opposite enantiomers (157). The resulting dipeptide products were investigated as possible conformationally constrained pseudopeptide scaffolds (158). In another synthesis, 2 equiv of template were condensed with 1 equiv of 2-iodomethyl-3-iodomethane, resulting in 2,6-diamino-4-methylene-1,7-heptanedioic acid with >96% trans stereoinduction. A second monoalkylation or dialkylation with alkyl iodides produced the corre-sponding α-dialkylated or α,α′-dialkylated analogs. Hydrolysis of the templates resulted in pseudopeptides containing the diaminodicarboxylic acid and two Val residues (159, 160).

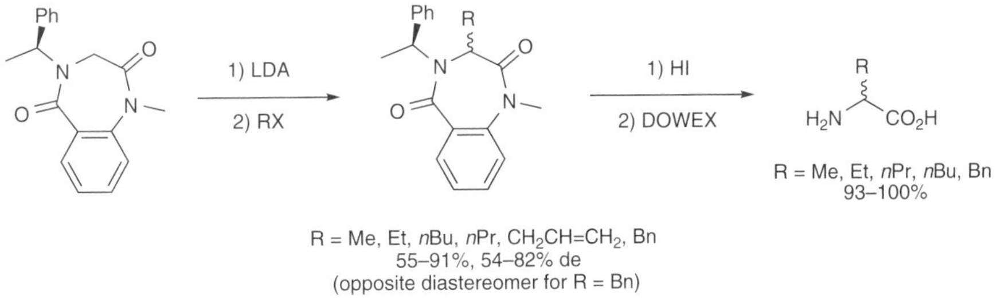

Scheme 7.12 Alkylation of 1,4-benzodiazepine-2,5-dione (152).

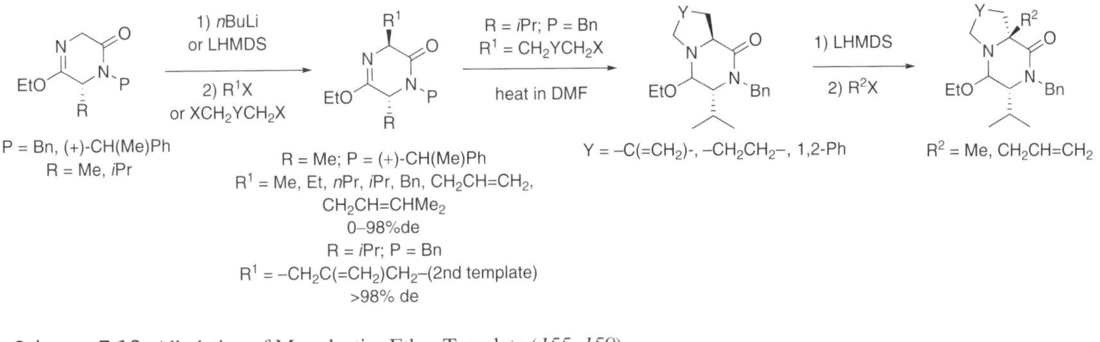

Scheme 7.13 Alkylation of Monolactim Ether Template (155–159).

An electrophilic 2,5-diketopiperazine template was prepared by mono- or dibromination of a cyclo[Gly-Gly] template (*161*). The bromide was displaced with 2-mercaptopyridine, and the sulfide exchanged with a silyl ketene acetal in the presence of silver(I) triflate. The piperazine ring was not cleaved, but used in a synthesis of the antibiotic bleomycin (*162*). Electrochemical methoxylative decarboxylation of a Pro-aminomalonate diketopiperazine produced a Pro-α-methoxy-Gly diketopiperazine (see Scheme 7.14). Reaction with allyl-trimethylsilane in the presence of TiCl₄ (99% de), followed by hydrolysis with 6 N HCl, gave allylglycine (*163*). The *N,N'*-bis(4-methoxybenzyl) cyclo[Val-Gly] diketopiperazine template described earlier (see Scheme 7.10) has also been converted into an electrophile, via formation of an α-fluoro- or α-chloro-Gly residue. However, attempted alkylations via Grignard additions gave variable results depending on the reagent, with significant addition to the lactam carbonyl (*164*). An *N,N'*-bis(4-methoxybenzyl) cyclo[L-Val-α-acetoxy-Gly] diketopiperazine template was allylated with allyltrim-ethylsilane/BF₃.Et₂O with 60% de; markedly lower stereoselectivity than provided by allylation of the corresponding Gly enolate (94% de) (*165*). Another electrophilic 2,5-diketopiperazine template was developed with a trifluoroacetate leaving group on the same carbon as a trifluoromethyl group, and Ala, Leu, or Phe as the chiral auxiliary. Organometallic reagents added with moderate stereoselectivity to give α-trifluoromethyl amino acid derivatives; hydrolysis with conc. HCl provided a dipeptide ester (*166*).

An electrophilic Gly template with a 2,4-dihydro-3,6-dioxo-(1H)-pyrazino[2,1-b]quinazolyl structure (see Scheme 7.15) has been reported. Arylations with electron-rich arenes in the presence of BF₃·Et₂O proceeded with >99% de, as did reactions with alkoxides. Free amino acids were not prepared (*167*).

Diketopiperazines have also been used for radical reactions (see Scheme 7.16). Preliminary tests of a cyclo[α-bromoGly-Gly] piperazine-2,5-dione with alkenes indicated that radical reactions were possible (*528*). Further studies with a cyclo[L-Ala-Gly] template showed that the Gly residue could be brominated and then deuterated with moderate diastereoselectivity (up to 56% de), depending on the *N*-substituents (*529*). The cyclo[L-Val-Gly] diketopiperazine gave a single α-bromo-Gly diastereomer (*trans*) on treatment with NBS (*530–532*). Allylation or deuteration also

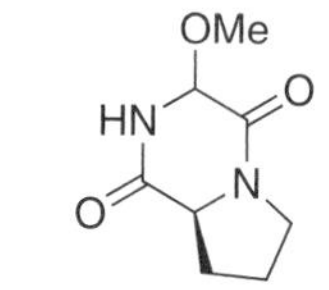

Scheme 7.14 Electrophilic Pro-α-Methoxy-Gly Diketo-piperazine Template (*163*).

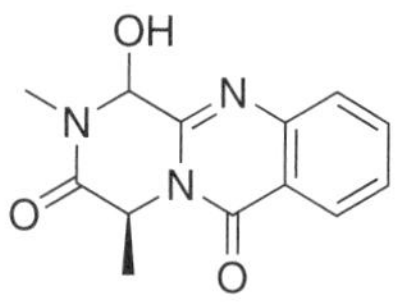

Scheme 7.15 2,4-dihydro-3,6-Dioxi-1-(1H)-pyrazino [2,1-b]quinazolyl Template (*167*).

gave the *trans* diastereomer preferentially (*530, 532*). Didehydroalanine contained within a piperazine-2, 5-dione template has also been employed for radical reactions (see Section 4.3.4) (*533*).

7.4 Morpholine-2,5-dione

The morpholine-2,5-dione (2,5-diketomorpholine) template is structurally very similar to the piperazine-2,5-dione template, with one of the amide groups replaced by an ester. Porzi and Sandri have extended their use of a chiral auxiliary on the amide nitrogens of diketopi-perazines to this template, in conjunction with an Ala auxiliary within the ring (see Scheme 7.17) (*168*). In contrast to the piperazine analog, alkylation took place *trans* to the ring methyl group with the *N*-chiral auxiliary having little effect, as >98% de was observed for both matched and mismatched chiral auxiliary pairs. Strong acid (57% refluxing HI) was still required to obtain the deprotected amino acid. This template has been applied to the synthesis of *cis*-4-hydroxyprolines and bulgecinines (*169*). α,α-Disubstituted amino acids were obtained via bis-alkylation (*170, 171*).

Several 2-substituted pipecolic acid derivatives have been prepared from a chiral 1,4-oxazin-2,5-dione

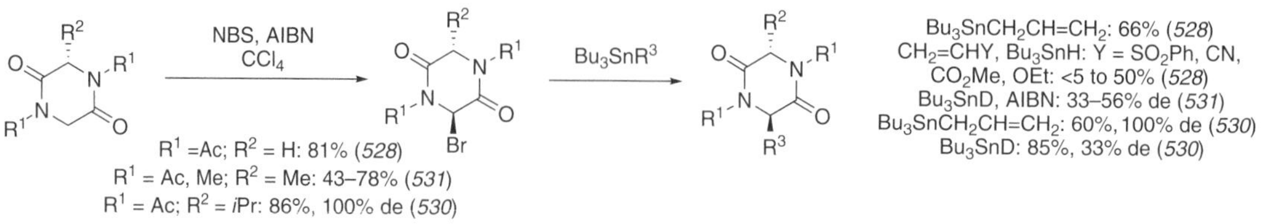

R¹ =Ac; R² = H: 81% (*528*)
R¹ = Ac, Me; R² = Me: 43–78% (*531*)
R¹ = Ac; R² = *i*Pr: 86%, 100% de (*530*)

Bu₃SnCH₂CH=CH₂: 66% (*528*)
CH₂=CHY, Bu₃SnH: Y = SO₂Ph, CN, CO₂Me, OEt: <5 to 50% (*528*)
Bu₃SnD, AIBN: 33–56% de (*531*)
Bu₃SnCH₂CH=CH₂: 60%, 100% de (*530*)
Bu₃SnD: 85%, 33% de (*530*)

Scheme 7.16 Radical Reaction of Piperazine-2,5-dione Template.

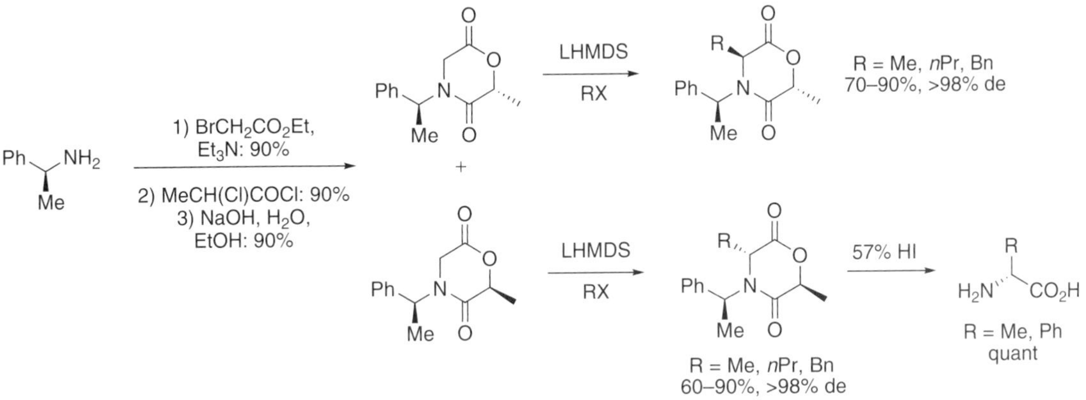

Scheme 7.17 Alkylation of Morpholine-2,5-dione Template (*168*).

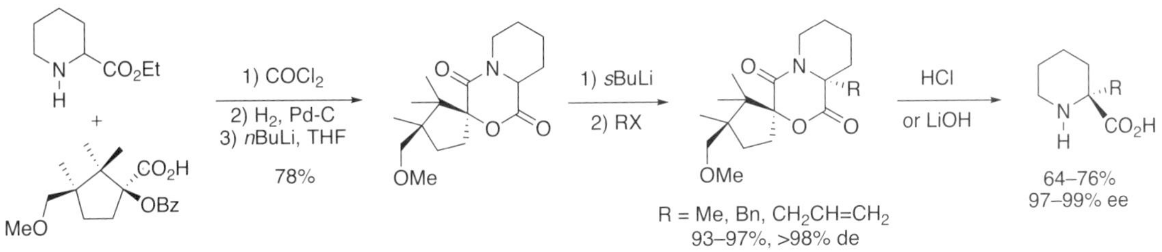

Scheme 7.18 Synthesis of Pipecolic Acids (*172*).

template incorporating pipecolic acid, with an α-hydroxy acid contained within the template providing the asymmetric induction (see Scheme 7.18) (*172*). The opposite diastereomer of the benzyl-substituted pipecolic acid could be prepared by alkylating the α-C and amino centers of a Phe-derived template with 1,4-dibromobutane.

An electrophilic 2,5-diketomorpholine template was obtained by electrochemical methoxylative decarboxylation of the template derived from chiral α-hydroxy acids and aminomalonate, producing an α-methoxy-Gly diketomorpholine template (see Scheme 7.19). Reaction with allyltrimethylsilane in the presence of TiCl$_4$ (69–90% de), followed by hydrolysis with 6 N HCl, gave allylglycine (*173*). The α-hydroxy acid auxiliary was readily separated and recovered after the template hydrolysis.

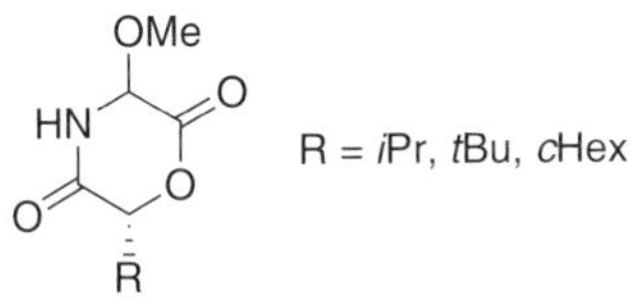

Scheme 7.19 Electrophilic α-Methoxy-Gly Morpholin-2, 5-dione Template (*173*).

7.5 Williams, Oxazinone (5,6-Diphenyl-1,4-Oxazin-2-one, 5,6-Diphenylmorpholin-2-One)

Most cyclic templates have been initially developed as chiral nucleophilic glycine equivalents. In contrast, the original synthon of Williams and co-workers was an electrophilic glycinate based on an oxazinone template (for a review of the use of oxazinones for amino acid synthesis, see (*174*)). Bromination (or chlorination) of a tetrahydrooxazinone derived from optically active α,β-diphenyl-β-hydroxyethylamine led to the desired synthon in 65% yield (see Scheme 7.20) (*175, 176*), with the bromide presumably *anti* to the phenyl substituents. The optically active diphenylaminoethanol was prepared by resolution of the racemic compound as a glutamic acid salt (*175–177*). A more recent report describes an alternative synthesis of the template via an enantiospecific enzymatic transformation of benzaldehyde, with an overall yield of 48% (compared to 13% from benzoin for the Williams' route) (*178*). An improved cyclization procedure was included in the new strategy.

The electrophilic oxazinone template is reacted with nucleophiles (see Scheme 7.21, Table 7.6). A significant advantage of this method is that the free amino acids are obtained in high yields and good optical purity (96–99% ee) by catalytic hydrogenation, which induces both ring opening and protecting group cleavage. Li/NH$_3$/EtOH

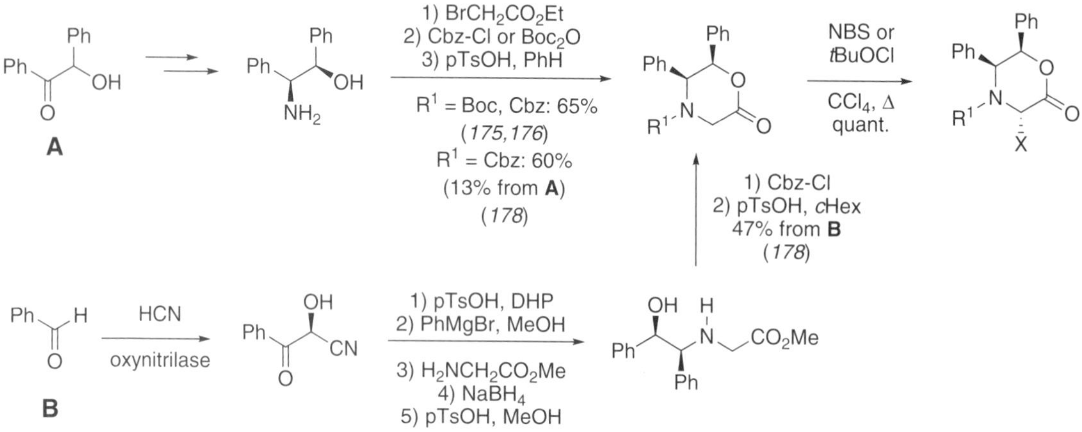

Scheme 7.20 Synthesis of Electrophilic Oxazinone Template.

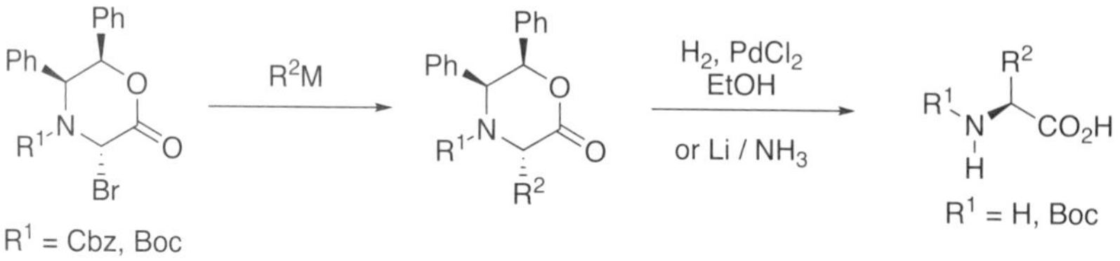

Scheme 7.21 Reaction of Electrophilic Oxazinone Template (see Table 7.6).

can be employed for molecules sensitive to hydrogenation (175). It was initially claimed that the zwitterionic free amino acids could be isolated without the need for an ion-exchange purification (176). However, a later review notes that some HCl accompanies the deprotected products, which must then be removed with an ion-exchange column (174).

A number of nucleophiles have been used to displace the bromide (see Scheme 7.21, Table 7.6). Neutral reagents, such as silyl enol ethers (175, 176, 179) or allylsilanes (175,176) in the presence of Lewis acids, give better yields (54–71%) than more basic organometallic reagents such as organocuprates (28–48%), which also cause reduction (175, 176). Most additions give the *anti* diastereomer adduct (corresponding to the L-amino acid deprotected product) with >96% ee, presumably through an S_N1-type reaction to an incipient carbocationic species from the less hindered ring face opposite the phenyl substituents. However, some nucleophiles (e.g., the TBDMS enol ether of ethyl acetate, or sodium phenylthiolate) provide the *syn* adduct with equally high stereoselectivity (>96% de), leading to the D-amino acid (175, 176). More powerful nucleophiles such as malonates appear to add via direct S_N2 displacement with inversion. However, an excess of a basic reagent will cause epimerization to the *anti* isomer (176). The *anti* adducts observed with organometallic reagents are explained by a possible radical mechanism (176).

The effect of varying the solvent and Lewis acid catalyst on the stereochemistry of addition of a number of silyl enol ethers or allylsilanes was examined (176). For some reagents the stereoselectivity could be reversed by altering the Lewis acid and solvent, while other nucleophiles were not affected by these factors.

Several classes of amino acids have been prepared using the oxazinone template. Tri-n-butyltin acetylides added in an *anti* fashion (180, 181), with the choice of solvent and metal cation critical for good yields. The resulting alkynes were reduced by hydrogenolysis during deprotection to give alkyl amino acids (180), or by dissolving metal reduction to give E-vinylglycines (181). Friedel–Crafts couplings produced arylglycines, but in order to prevent destruction of the product during removal of the chiral auxiliary a two-step procedure using TMSI followed by periodate oxidation was required (176, 182). Reduction of the brominated template with deuterium (183) or tritium (184, 185) gas in the presence of palladium chloride replaced the bromine (with retention of configuration), and simultaneously cleaved the template to give optically active isotopically-labeled glycine.

More recently, the enolate chemistry of the Williams' oxazinone template has been examined (see Scheme 7.22, Table 7.7) (186, 187). The LHMDS- or KHMDS-derived enolate (enolates prepared with other bases, such as LDA, decomposed) were reacted with activated

Table 7.6 Synthesis of α-Amino Acids via Alkylation of the Electrophilic Oxazinone Template (see Scheme 7.21)

R¹	R²M Nucleophile	Alkylation Yield	R² After Deprotection	Deprotection Conditions	Deprotection Yield	Reference
Cbz	$H_2C=C(OTBS)OEt$ AgOTf or ZnCl$_2$	74% 33–96% de syn	CH_2CO_2Et	H$_2$, PdCl$_2$ or Li / NH$_3$	52–94% 91–>99% ee	1986 (175) 1988 (176)
	$H_2C=C(OTMS)R$ R = Ph, 4-MeO-Ph, CO$_2$Bn AgOTf or ZnCl$_2$	72% 54% syn to 92% anti	CH_2CH_2Ar			
	Me$_3$SiCH$_2$CH=CH$_2$ 3-(Me$_3$Si)-cyclopentene AgOTf or ZnCl$_2$	66–82% >96% de anti	nPr or CH$_2$CH=CH$_2$ cPent or 3-cyclopentenyl			
	furan, ZnCl$_2$ 5-Me-furan, ZnCl$_2$	64–66%	2-tetrahydrofuryl 5-Me-2-furyl			
	MeZnCl Me$_2$CuCNLi$_2$ nBu$_2$CuCNLi$_2$	28–48%	Me nBu			
	PhSNa NaCH(CO$_2$Bn)$_2$	>96% de anti	—			
Boc	Me$_3$SiCH$_2$CH=CH$_2$ 3-(Me$_3$Si)-cyclopentene ZnCl$_2$	59–63%	CH$_2$CH=CH$_2$ cyclopenten-3-yl	Li / NH$_3$	70% >95% ee	1986 (175) 1988 (176)
Cbz	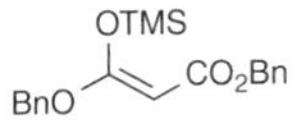 OTMS BnO⎯⎯CO$_2$Bn ZnCl$_2$	53% 70% de syn	$CH(CO_2Bn)_2$	H$_2$, PdCl$_2$	30% >98% ee	1988 (179)
Boc	R′$_3$SnCCR, ZnCl$_2$ R = Ph, nPr, nHex, (CH$_2$)$_2$OTBS	61–71% >95% de anti	CH=CHPh CH=CHnPr CH=CHnHex CH=CH(CH$_2$)$_2$OTBS	Na / NH$_3$ or Li / NH$_3$	71–80% 56–68% ee or 16–20% 65–>98% ee	1988 (181)
Cbz	R′$_3$SnCCR, ZnCl$_2$ R = Ph, nHex	53–55% >95% de anti	CH$_2$CH$_2$Ph nOct	H$_2$, Pd-C	57–68% 94–98% ee	1988 (180)
Boc	Ph$_2$CuLi (1-naphthyl)$_2$CuLi 1,3,5-(MeO)$_3$-Ph, ZnCl$_2$ furan, ZnCl$_2$ 5-Me-furan, ZnCl$_2$	39–83%	Ph 1-naphthyl 1,3,5-(MeO)$_3$-Ph 2-furyl 5-Me-2-furyl	1) TMSI 2) 10% HCl 3) NaIO$_4$, H$_2$O	26–73% 82–94% ee	1990 (182)
Cbz	furan, ZnCl$_2$ 5-Me-furan, ZnCl$_2$	64–66%	2-furyl 5-Me-2-furyl	H$_2$, Pd-C	57–82%	1990 (182)

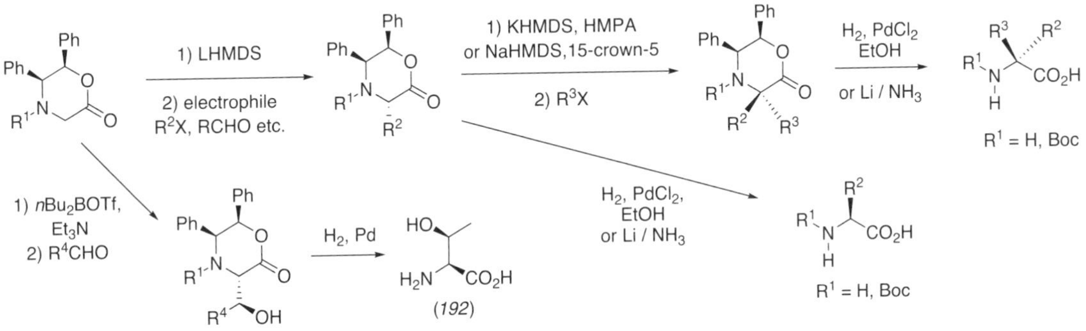

Scheme 7.22 Reaction of Nucleophilic Oxazinone Template (see Table 7.7).

Table 7.7 Synthesis of α-Amino Acids via Alkylation of Nucleophilic Oxazinone Template (see Scheme 7.22)

R^1	R^2 Electrophile (or R^2 for Second Alkylation)	R^3 Electrophile	Alkylation Yield	Further Derivatization of Side Chain	Deprotection Conditions	Deprotection Yield	Reference
Boc	R^2X: R^2 = Me Bn $CH_2CH=CH_2$ $CH_2CH=CMe_2$		48–91%		Li/NH_3 or 1) TFA 2) H_2, Pd/C	50–70% 97–100% ee 76% 98% ee	1988 (186)
Cbz	$BrCH_2CO_2Me$		61%		H_2, Pd/C	71% 96% ee	1988 (186)
Cbz	MeCHO nPrCHO iPrCHO		62–66% de (crude) 38–57% yield pure major isomer				1990 (192)
Boc	R^2X: R^2 = Me nPr Bn $CH_2CH=CH_2$ $CH_2CH=CMe_2$ CH_2CO_2Et $(CH_2)_3Cl$ $(CH_2)_3I$		64–91%		Li/NH_3	52–93% 96–>99% ee	1991 (187)
Cbz	R^2X: R^2 = Me nPr Bn $CH_2CH=CH_2$ $CH_2CH=CMe_2$ CH_2CO_2Et $(CH_2)_3Cl$ $(CH_2)_3I$		47–82%		R^2 = Bn H_2, Pd/C	93% 99% ee	1991 (187)

Protecting group	R2 / substrate	Reagent	Yield	Additional steps	Deprotection	Product yield / ee	Year (Ref)
Boc	Me Me nPr	$XCH_2CH=CH_2$ $XCH_2CH=CMe_2$ $XCH_2CH=CH_2$	80–90%		Li/NH_3	60–70% 100% ee	1991 (187)
Cbz	Me Me nPr $CH_2CH=CH_2$	BnBr $XCH_2CH=CHPh$ BnBr BnBr	80–85%		R^2 = Me; R^3 = Bn, $(CH_2)_2Ph$	93–95% 100% ee	1991 (187)
Cbz Boc	$X(CH_2)_2CH=CH_2$ $X(CH_2)_3CH=CH_2$ $(CH_2)_2CH=CH_2$	XCH_2OMe	47–88% 97%	1) O_3, MeOH 2) Me_2S convert $(CH_2)_nCH=CH_2$ into $(CH_2)_nCHO$ (A)	H_2, Pd/C		1991 (195) 1992 (193)
Cbz	aldol with **A** from above		55–65% 88–100% de	deoxygenate 1) PhOCSCl 2) nBu_3SnH, AIBN 30–83%	1) H_2, $PdCl_2$ 2) 48% HBr	72–99% >99% ee	1991 (195) 1992 (193)
Cbz	R^2X: R^2 = Me Bn Et nHept		64–83%				1992 (188)
Cbz	Me Me	BnBr $BrCH_2CH=CH_2$	90–94%				1992 (188)
Boc	R^2X: R^2 = Et iBu nHept		70–90%				1992 (188)
Boc	Me Me Et	nBuX iBuX nHeptX	38–76%				1992 (188)
Boc	$ClCH_2OSEM$ CH_2OSEM **A**	$Br(CH_2)_3Cl$	76% 70% 75%	NaI, acetone convert $(CH_2)_3Cl$ to $(CH_2)_3I$ (**A**)	1) TFA 2) H_2, Pd-C	83% [structure: HO_2C—OH—NH_2 / HO_2C—NH_2—CO_2H—NH_2]	1992 (188)
Boc	Bn $CH_2CH=CH_2$	$BrCH_2CH=CH_2$ BnBr	96% 60%		Li/NH_3		1992 (189)

(Continued)

Table 7.7 Synthesis of α-Amino Acids via Alkylation of Nucleophilic Oxazinone Template (see Scheme 7.22) (continued)

R^1	R^2 Electrophile (or R2 for Second Alkylation)	R^3 Electrophile	Alkylation Yield	Further Derivatization of Side Chain	Deprotection Conditions	Deprotection Yield	Reference
Cbz	X(CH₂)₃Cl **A**		52% 20%	NaI, acetone convert (CH₂)₃Cl to (CH₂)₃I (**A**)	H₂, Pd-C	99%	1992 (193)
Boc	R²X: R² = Bn CH₂CH=CH₂		93–96%		Li/NH₃		1992 (189)
Cbz	BrCH₂(5-SiPh₂tBu-pyrid-2-yl)		60% >95% de		1) TBAF 2) H₂, Pd(OH)₂ 3) H₂, Pd/C	51%	1994 (499)
Cbz	H₂C=C(CO₂tBu)₂		80%		H₂, Pd/C	74% >99% ee	1995 (177)
Boc	(CH₂)₂CH=CH₂		89%	1) O₃, MeOH 2) Me₂S convert (CH₂)₂CH=CH₂ into (CH₂)₂CHO 89%			1996 (203)
template = [structure] OR, RO, RO, Cbz, N	R²X: R² = Me Et CH₂CH=CH₂		57–74% (1:1 to 1:0 mono:di alkylation) >98% de				1996 (216)
Cbz	4-CF₂PO₃Et₂-BnBr		78%		H₂, PdCl₂	100%	1996 (197)
Cbz	3-BnO-BnBr		87%		H₂, PdCl₂	74% >96% ee	1997 (500)
Cbz	BrCH₂CO₂tBu CH₂CO₂tBu	MeI	81% 63%	CH₂CO₂tBu to CH₂CONH₂ 81–95%	H₂, PdCl₂		1998 (191)
Boc	ICH₂CH=CH₂		95%		Li/NH₃	60%	1998 (501)
Cbz	4-(OCF₂CO₂tBu)-BnBr		73%		H₂, 10% Pd/C	quant.	1998 (502)

				H2, Pd-C induces intramolecular reductive amination		
Cbz	CH2=CH(CH2)2CHO	69%	1) O3, MeOH 2) Me2S convert CH(OH)(CH2)2CH=CH2 into CH(OH)(CH2)2CHO			1998 (200)
Cbz	I(CH2)2CH(R)CH2N(Cbz)(CH2)4I R = H, OTBS	32–50%		1) TBAF 2) H2, Pd(OH)2	72–95%	1999 (198)
Cbz	4-CH2CO2tBu-BnBr 4-CF2CO2tBu-BnBr	42–56%		H2, Pd/C	95–99%	1999 (503)
Cbz Boc	ICH2CH(OH)CH2NHBoc ICH2CH(OH)CH2NHCbz protected as acetal	34–80%		1) H2, Pd/C 2) 1.5 N HCl	84%	2002 (504)
Cbz	3-AcO-4-CH2CO2tBu-BnBr	—	—	—	—	2000 (505)
Cbz	Me ICH2CH=CH2 Me I(CH2)2CH=CH2 Me I(CH2)3CH=CH2 Me I(CH2)4CH=CH2 Me I(CH2)6CH=CH2	— —	—	Na/NH3	93–95% —	2000 (190)
Cbz	BnN=CHCH2CH2OTBS	50–60% 50% de at C3	guanylation with BocN=(SMe) NHBoc, TBS deprotection, Mitsunobu cyclization	1) H2, Pd(OH)2 2) 0.5M HCl	95%	2003 (196)
Cbz	3-BocO-BnBr	60%		H2, PdCl2	92%	2002 (506)
Cbz	HC(OEt)3, TiCl4	85%	—	H2, Pd(OH)2	96%	2003 (196)
Cbz	3-(CO2H)-4-(CF2CO2H)-BnBr	57%	—	H2, Pd/C	86%	2004 (507)
Boc	I(CH2)2CH=CH2	60%	alkene epoxidation (89%), epoxide opening with azide (80%), azide reduction during deprotection	1) TFA-H2O 2) H2, Pd/C	78–88%	2004 (199)

halides to give the trans (*anti*) adduct with high diastereoselectivity (>99%) and reasonable yields (50–90%). Careful control of reaction conditions was essential to prevent dialkylation. Unactivated alkyl halides resulted in reduced yields, which could be improved by the addition of HMPA (*187*) or 15-crown-5 (*188*). α-Alkyl amino acids were prepared by repeating the sequence of enolate formation and reaction, although activated halides and KHMDS were required (*187, 189, 190*). Baldwin et al. found that second alkylations with less reactive electrophiles were possible by using NaHMDS in combination with 15-crown-5 (*188*). (*S*)-2-Methyl-Asn was prepared by this method (*191*).

The nucleophilic oxazinone template has also been used for aldol condensations, with the boron enolate reacting with aldehydes to give *anti* (*erythro*) β-hydroxy adducts in 38–57% yield for the major diastereomer (*192*). Diastereoselectivity of approximately 5:1 was observed at the carbinol center in the crude adduct. Aldol reactions were used in the diastereoselective synthesis of two isomers of 2,6-diaminopimelic acid (*193, 194*), 2,6-diamino-6-(hydroxymethyl)pimelic acid (*195*), and 2,7-diaminosuberic acid (*193, 194*), via reaction of the oxazinone boron enolate with an aldehyde substrate prepared from a second oxazinone template. The hydroxyl group was removed via radical reduction of a xanthate ester derivative. Baldwin et al. have also accomplished the synthesis of 2,6-diamino-6-(hydroxymethyl)pimelic acid by sequentially displacing both halides from 1-chloro-3-iodopropane with two different oxazinone templates (*188*). Williams et al. applied this procedure to the preparation of the unsubstituted 2,6-diaminopimelic acid, but the yield was lower than with their aldol reaction route (*193*).

Addition of the nucleophilic oxazinone tempate to an imine produced a β-amino product, as employed with the *N*-benzyl imine of *O*-TBS 3-hydroxypropanal during a synthesis of (2S,3R)-capreomycidine (*196*). Other amino acids prepared by alkylation of the diphenyloxazinone include 4-(phosphonodifluoromethyl)-L-Phe (*197*), the reduced collagen cross-links lysinonorleucine and 5-hydroxylysinonorleucine (*198*), as well as the collagen cross-link precursor 5-hydroxy-Lys (*199*). Michael addition of the template to di-*tert*-butyl methylenemalonate led to γ-carboxy-Glu (*177*), while aldol reaction with 4-pentenal, followed by ozonolysis of the alkene and cyclization with the α-amino group via reductive alkylation, provided (2R,3S)-β-hydroxypipecolic acid (*200*). An asymmetric synthesis of diethyl acetal-protected α-formylglycine was achieved by Ti-catalyzed alkylation of Williams' oxazinone template with triethylorthoformate; template deprotection gave (*R*)-α-formylglycine diethylacetal (*196*).

An alternative nucleophilic synthon, which leads to dehydroamino acids, was prepared by conversion of the bromide template to a phosphonate (*201*). Condensation with aldehydes provided didehydroamino acid derivatives, which could be cyclopropanated with reasonable diastereoselectivity by a number of reagents, providing 1-aminocyclopropane-1-carboxylic acid derivatives (see Scheme 7.23). The alkene was also used for a 1,3-dipolar cycloaddition with an azomethine ylide to give the cyclic amino acid (*S*)-cucurbitine (*202*). The phosphonate synthon has also been coupled with an another oxazinone template containing a propionaldehyde substituent. Cyclopropanation of the resulting alkene gave an 1-aminocyclopropane-1-carboxylic acid 2,6-diaminopimelic acid hybrid (*203*).

Williams' oxazinone template has also been used to prepare β-hydroxy-γ-amino acids, such as statine, and γ-hydroxy-δ-amino acids (*204, 205*). The deprotonated template was alkylated as normal, but instead of

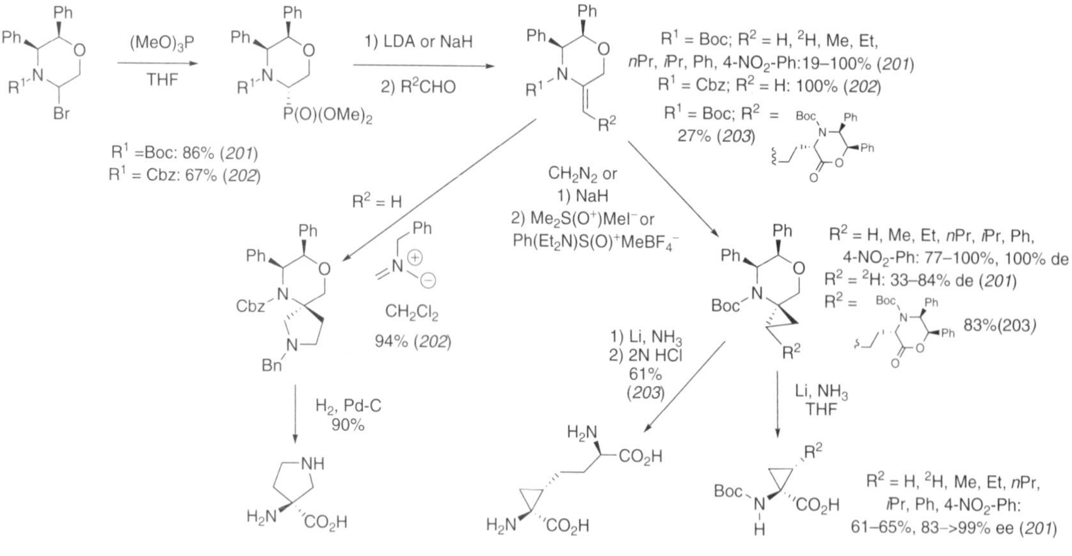

Scheme 7.23 Synthesis and Reaction of Dehydroamino Acids within Oxazinone Template.

template hydrolysis the lactone carbonyl was reduced and acetylated to give a hemiacetal. The silyl ketene acetal of methyl acetate was then condensed to reintroduce an acid group, with hydrolysis providing the γ-amino acid derivative (see Scheme 11.148). β-Amino acids have also been synthesized, by reaction of the acetal derivative with TMSCN (*393*).

The oxazinone synthon has also been applied to radical reactions. Deprotonation and treatment with PhSeBr produced an α-phenylselenyl-Gly residue within the template. Radical reaction with *N*-acetyl dehydroalanine methyl ester, induced by AIBN/*n*-Bu₃SnH, gave a 3:2 mixture of two of the four possible diastereomers of 4-amino-Glu acid (*206*).

7.6 Other Oxazinones (1,4-Oxazin-2-One, Morpholin-2-One)

In 1988, the same year that Williams et al. disclosed the application of their diphenyloxazinone template as an enolate, Dellaria and Santarsiero described an oxazinone enolate derived from phenylglycinol, containing only one of the two phenyl side chains present in Williams' synthon (see Scheme 7.24, Table 7.8) (*207, 208*). An improved one-pot synthesis of this template from phenylglycinol has since been reported (*209*), as has a synthesis proceeding via a SeO₂-promoted oxidative rearrangement of the phenylglycinol-derived oxazoline (*210*). A high level of *anti* diastereoselectivity was still observed during alkylation of the *N*-acyl template (>98% de), but the type of *N*-protecting group was found to play a critical role in addition stereoselectivity. If an *N*-alkyl group was used instead of an *N*-acyl group, *syn* addition occurred preferentially but with lower stereoselectivity (*207, 208*). The anion generation was also very sensitive to the base and solvent employed, with NaHMDS in DME/THF generally giving the best results (*207*). The deprotected amino ethyl esters were

obtained by a two-step procedure using refluxing ethanolic HCl followed by hydrogenolysis (*208*). The synthon has been used for enantioselective syntheses of [3-¹³C]-L-Phe (*211*), [3-¹³C]-L-Tyr (*211*), and 2-amino-5-phosphonopentanoic acid (*212*). A series of bifunctional electrophiles, 4-halobutyl-1-triflates, were reacted with 80–95% yields of diastereomerically pure halobutyl-substituted template; one example was deprotected by hydrogenation to give a 45% yield of pipecolic acid, resulting from *N*-cyclization, along with 12% of norleucine (*213*).

Baker et al. have modified the substituent in the monosubstituted oxazinone and examined the effect on alkylation of the enolate with allyl bromide (*214*). The *anti* adduct was obtained with diastereoselectivities ranging from 87% de (R = *i*Pr) to 100% de (R = Bn). Other electrophiles also reacted with the benzyl-substituted oxazinone with complete diastereoselectivity. However, addition to benzaldehyde resulted in no stereoselectivity at the carbinol center. The regioisomeric mono phenyl-substituted template (with the phenyl adjacent to the ring oxygen) was also prepared, using 1-amino-2-phenyl-2-hydroxyethane as the amino alcohol. α-Alkylated Phg derivatives were produced by starting with 2-bromophenylacetic acid as the other half of the template (*215*). An oxazinone template with the phenyl substituents replaced by a glucopyranose ring system was synthesized from D-glucopyranose (*216*). Alkylations proceeded with high diastereoselecivity (>98% de), but for electrophiles other than methyl iodide dialkylation was a significant side reaction. The phosphazene *tert*-butyl P4 base gave good results for some electrophiles, and might be useful with other oxazinone templates.

With all of the oxazinone templates described above, the comparatively expensive chiral auxiliary amino alcohol is destroyed during deprotection. An alternative approach makes use of Seebach's principle of self-reproduction of chirality (see Section 7.7) (*217*). Phe was

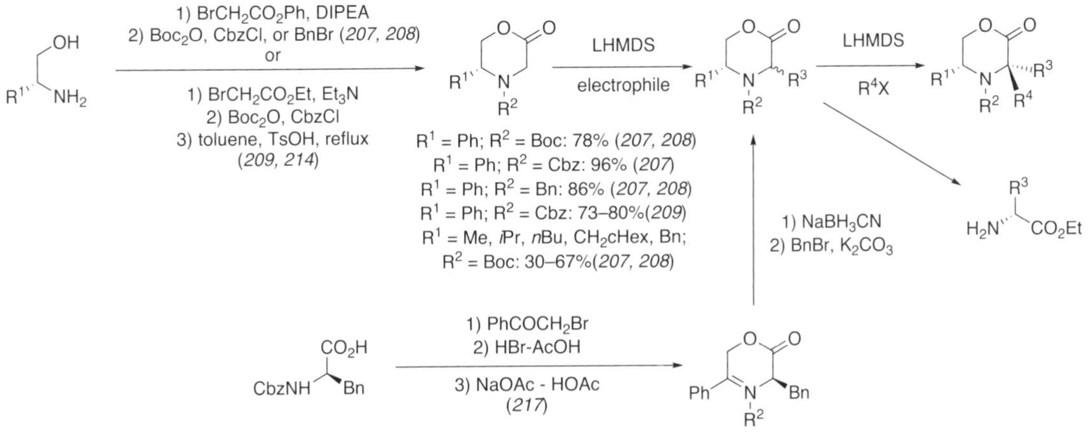

Scheme 7.24 Synthesis and Reaction of Oxazinone Template (see Table 7.8).

Table 7.8 Synthesis of α-Amino Acids via Alkylation of Nucleophilic Oxazinone Template (see Scheme 7.24)

R^1	R^2	R^3 Electrophile (or R^3 Substituent)	R^4 Electrophile	Alkylation Yield	Deprotection Conditions	Deprotection Yield	Reference
Ph	Boc	MeI nBuBr BnBr BrCH$_2$CH=CH$_2$ BrCH$_2$C(Me)=CH$_2$ BrCH$_2$CO$_2$Bn		59–90% 98–>99% de anti	1) HCl, EtOH 2) H$_2$, Pd/C	61–92% >99% ee	1988 (208) 1989 (207)
Ph	Cbz	MeI BnBr BrCH$_2$CO$_2$Bn		71–78% >98% de anti			1988 (208) 1989 (207)
Ph	Bn	BnBr		83% 88% de syn			1988 (208) 1989 (207)
Ph	Boc	BrCH$_2$CH=CHPO$_3$Et$_2$		60% >98% de	1) BrSiMe$_3$ 2) 6N HCl 3) H$_2$, Pd/C	75%	1989 (212)
Me iPr nBu Bn CH$_2$cHex	Boc	BrCH$_2$CH=CH$_2$		57–88% 87–100% de anti			1992 (214)
Bn	Boc	MeI nBuI BnBr Br(CH$_2$)$_2$Ph I(CH$_2$)$_2$Cl BrCH$_2$CH=CHMe		13–97% 100% de			1992 (214)
Bn	Boc	PhCHO		64% 100% de C2 18% de C3			1992 (214)
Ph	Bn	Bn	MeI BnBr BrCH$_2$CH=CH$_2$ BrCH$_2$CH=C(Me)$_2$ BrCH$_2$CH=CHPh	43–68% 100% de			1993 (217)
Ph	Boc	Br^{13}CH$_2$Ph Br^{13}CH$_2$(4-MOMO-Ph)		50–95% >98% de	1) HCl 2) H$_2$, Pd/C 3) 6N HCl	58–78%	1998 (211)
Ph	Cbz	TfO(CH$_2$)$_4$Br TfO(CH$_2$)$_4$Cl TfO(CH$_2$)$_2$CH=CH$_2$		80–95% >98% de	H$_2$, Pd/C	89% Nle from alkene 45% cyclized pipecolic acid	2003 (213)

incorporated into an oxazinone template, and used to induce asymmetry at a 5-phenyl substituent. A second substituent was then introduced at the Phe α-position with high *trans* diastereoselectivity, resulting in α-alkyl-Phe derivatives. The stereoselectivity was contrary to that observed by Dellaria et al. during monoalkylation of their analogous *N*-benzyl oxazinone (207, 208).

Pro derivatives can be obtained with high enantiomeric purity via reaction of 5-phenylmorpholin-2-one with aldehydes (218–228) (see Scheme 7.25, Table 7.9). An intermediate azomethine species is generated which undergoes diastereoselective [1,3]-dipolar cycloadditions with alkene or alkyne dipolarophiles under thermal conditions, or Lewis acid catalysis. Only one diastereomer was generally observed with the acyclic dipolarophiles dimethyl maleate or fumarate. By using an intramolecular cycloaddition, with the unsaturated dipolarophile attached to the aldehyde employed for the initial condensation, bicyclic proline derivatives were prepared (227, 228). In the absence of a dipolarophile, a cycloaddition takes place with a second equivalent of aldehyde, giving products that can be converted into *threo* β-hydroxyamino acids (223, 229). These adducts have also been observed as byproducts in the cycloaddition reactions, especially with unreactive dipolarophiles (224, 225). Exceptional diastereoselectivity was

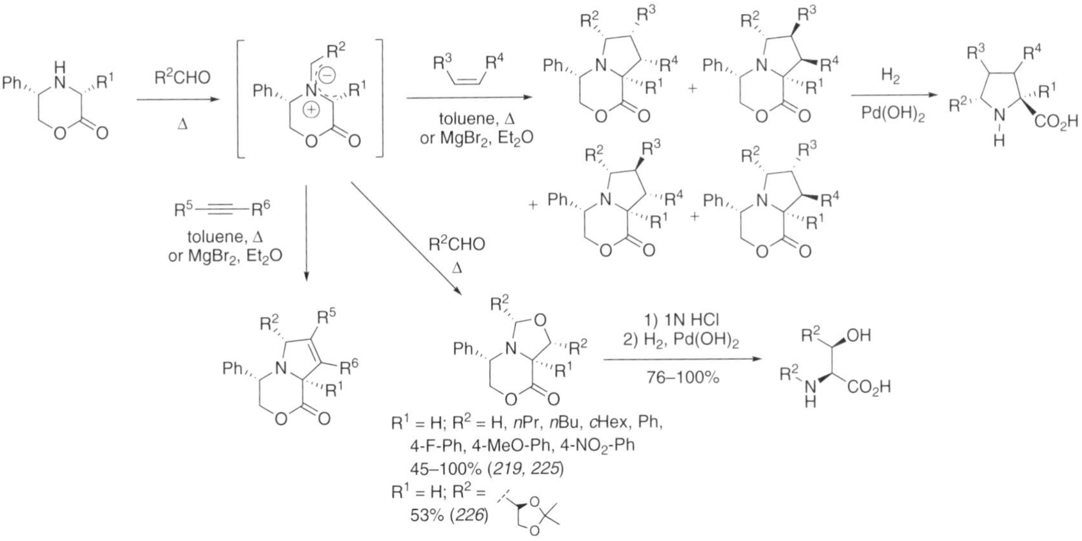

R² = H, nPr, nBu, cHex, Ph,
4-F-Ph, 4-MeO-Ph, 4-NO₂-Ph

1) 1N HCl
2) H₂, Pd(OH)₂
76–100%

R¹ = H; R² = H, nPr, nBu, cHex, Ph,
4-F-Ph, 4-MeO-Ph, 4-NO₂-Ph
45–100% (219, 225)
R¹ = H; R² =
53% (226)

Scheme 7.25 Synthesis of Proline Derivatives and β-hydroxy-α-amino Acids via 5-phenyl-morpholine-2-one Template (see Table 7.9).

observed in the matched double asymmetric induction product obtained by imine formation/cycloaddition with (S)-glyceraldehyde acetonide, leading to polyoxamic acid (230). A similar cycloaddition reaction occurs directly with imines, leading to threo 3-aryl-2,3-diamino-propionic acids (231).

The morpholin-2-one template used for these cycloadditions can be prepared using phenylglycinol as a chiral auxiliary for the source of chirality (220, 222), or with amino acids as both the original source of chirality and as part of the product (219, 221, 222, 226). The 5-phenyl auxiliary group has been replaced with a naphthyl group, with little effect on stereoselectivity but generally decreased yields (226). The morpholin-2-one template is removed by catalytic hydrogenolysis (218, 219, 221).

Williams' 5,6-diphenylmorpholin-2-one template has also been used for similar [1,3]-dipolar cycloadditions, reacting with a variety of aldehydes and dimethyl maleate to give the adducts in 32–71% yield but generally poor diastereoselectivity (232). A 5-phenylmorpholin-2-one derivative appeared in another Pro synthesis, but was used as an intermediate resulting from an aza-Cope process rather than a chiral template (233, 234). An alternative route to prolines is via Michael addition of the morpholin-2-one template to acrylates in the presence of CsF/Si(OMe)₄. Cyclization of the acrylate carboxy group with the morpholine amine, followed by reduction of the lactam carbonyl, produced α-phenyl-Pro derivatives (235). The 5-phenylmorpholin-2-one template has also been oxidized to a nitrone with dimethyl dioxirane, and then used for cycloaddition reactions with alkenes. The precursors of γ-hydroxy-α-amino acids were produced as single diastereomers in >90% yield (236).

An N-methyl 3-thiophenyl derivative of the 5-phenylmorpholin-2-one template, with the lactone ester reduced to a hemiketal, has been used as a cationic glycine equivalent (see Scheme 7.26) (237). The thiol group can be replaced with organocuprates (with inversion) or organozincates (with retention). Oxidation of the hemiketal back to a lactone followed by methanolysis provided N-methyl amino acid esters, with 70 to >99% ee.

An imine contained within the 5-phenyl-morpholine-2-one template can be prepared via esterification and cyclization of 2-oxoalkanoic acids with (R)- or (S)-phenylglycinol (238, 239) (see Scheme 7.27). The unsubstituted imine was prepared by bromination/dehydrobromination of 5-phenylmorpholin-2-one (240). An alternative route to the imine proceeds by an SeO₂-promoted oxidative rearrangement of 2-substituted oxazolines, which in turn are prepared from phenylglycinol or valinol and an acyl chloride or acetimidate (210). The imine can be used for the same types of reactions as acyclic α-keto acid imines (see Section 5.3.9, 6.5.2), and thus can be diastereoselectively reduced (leading to α-amino acids, 86–90% de) (210, 238), used as an electrophile for organometallic additions in the presence of a Lewis acid (leading to α-alkyl-α-amino acids if already substituted, > 95% de (239), or to α-amino acids if unsubstituted, >95% de (241)) or used as a dienophile in Diels–Alder reactions (leading to six-membered Pro analogs) (240). The free amino acids were obtained in one step by hydrogenolysis.

The regioisomeric 2,3-dihydro-6H-1,4-oxazin-2-one template (see Scheme 7.28), corresponding to a cyclic Schiff base imine of an Ala ester of a chiral aromatic α-hydroxyketone, has been used in the synthesis of α-methyl-α-amino acids (242–244). The closely related

Table 7.9 Synthesis of Proline Derivatives via 5-Phenyl-Morpholin-2-One Template (see Scheme 7.25)

R^1	R^2	Dipolarophile Substituents	Alkylation Yield	Hydrogenolysis Yield	Reference
H	Ph	$R^3 = R^4 = $ −CONHCO−, CO_2Me (Z)	8–72% 6–100% de		1991 (220) 1992 (223)
H	H	$R^3 = R^4 = $ −CONHCO−, −CONMeCO−, −CONPhCO−, −CO_2CO−, CO_2Me (Z)	49–68% 36–100% de		1991 (220) 1991 (223)
H	H	$R^5 = R^6 = CO_2Me$ $R^5 = H$; $R^6 = CO_2Me$	29–30% 100% de		1991 (220)
iPr	H	$R^3 = R^4 = $ −CONMeCO−, −CONPhCO−	41–46% 66–100% de	72–77%	1991 (221,222)
H	H Et iPr 2-furyl Ph 4-NO_2-Ph 4-MeO-Ph	$R^3 = R^4 = CO_2Me$ (Z)	32–71% 0–26% de	56–99%	1992 (232)
H	$CH_2SCH_2CH{=}CH_2$ $(CH_2)_4CH{=}CH_2$	intramolecular dipolarophile: product has $R^2 = R^3 =$ −CH_2SCH_2− −$(CH_2)_4$−	75–90% 100% de for derivative with S, treat with Raney Ni, acetone to give $R^2 = R^3 =$ Me		1993 (228)
H	$(CH_2)_3CCH$ $(CH_2)_4CCH$	intramolecular dipolarophile: product has $R^5 = R^6 =$ −$(CH_2)_3$− −$(CH_2)_4$−	61–63% 100% de	22%	1993 (227)
H	H	$R^3 = R^4 = $ −CONMeCO−, −CONPhCO−, CO_2Me (Z), CO_2Me (E)	56–80% 33–100% de		1993 (224)
H	H	$R^3 = H$; $R^4 = CN$, CO_2Me	62–70% 66–100% de		1993 (224)
H	H	$R^5 = R^6 = CO_2Me$	56% 100% de		1993 (224)
H	H	$R^3 = R^4 = $ −$CO_2CH_2CH_2$−, −CO_2CH_2−, −$CO_2C(-CH{=}CHCH{=}CH-){=}C-$, Ph (E), Ph (Z)	3–53% 68–100% de		1994 (225)
H	CO_2Et	$R^3 = R^4 = $ −CONMeCO−, −CONPhCO−, CO_2Me (Z), CO_2Me (E)	31–65% 18–100% de	75–90%	1995 (218)
H	CO_2Et	$R^5 = CO_2Me$; $R^6 = H$	21–40% 100% de		1995 (218)
Ph	H	$R^3 = R^4 = $ −CONMeCO−, −CONPhCO−, CO_2Me (Z), CO_2Me (E)	72–95% 15–100% de	70–78%	1995 (219)
Ph	H	$R^5 = CO_2Me$; $R^6 = H$, CO_2Me	77–79%		1995 (219)

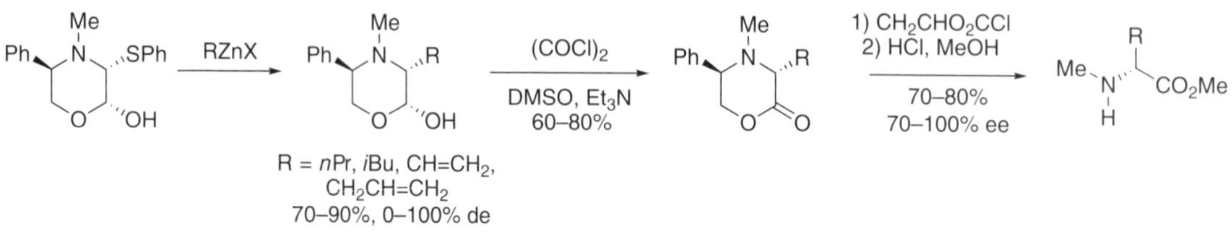

Scheme 7.26 Electrophilic Morpholine Template (237).

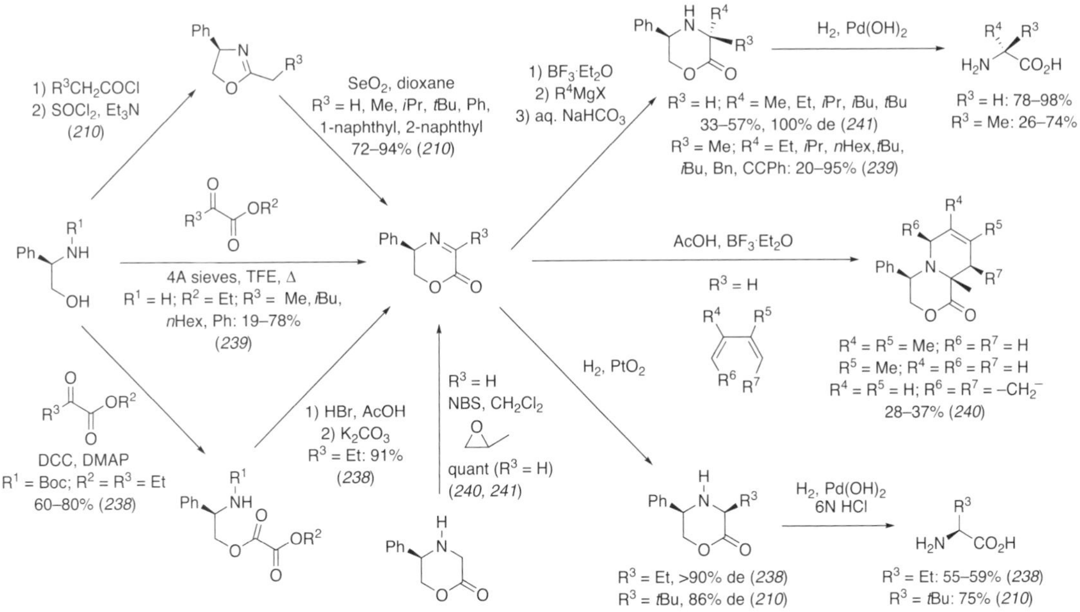

Scheme 7.27 Synthesis of Amino Acids via Imine within Morpholine Template.

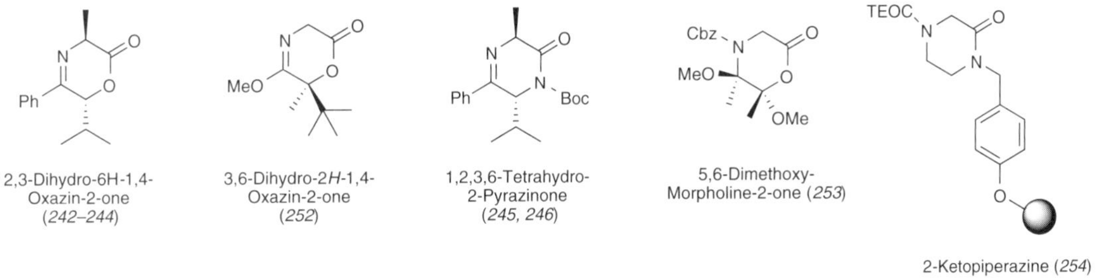

Scheme 7.28 2,3-dihydro-6H-1,4-Oxazin-2-one, 3,6-Dihydro-2H-1,4-Oxazin-2-one, 1,2,3,6-Tetrahydro-2-Pyrazinone, 5,6-Dimethoxymorpholin-2-one and 2-Ketopiperazine Templates.

amino analog 1,2,3,6-tetrahydro-2-pyrazinone template (see Scheme 7.28) was also used to prepare α-methyl-α-amino acids with very high diastereoselectivity under PTC alkylation conditions (245, 246). The Gly analog of the 2,3-dihydro-6H-1,4-oxazin-2-one template was condensed with aldehydes to provide dehydroamino acids within the template. Cyclopropanations gave 2-substituted-1-aminocyclopropane-1-carboxylic acids (247). Diels–Alder reactions with cyclopentadiene were also carried out (248). Reactions with these amino acid Schiff base analogs were reviewed in 2000 (249).

A monolactim ether of morpholine-2,5-dione was derived from DL-phenylglycine (Phg) and optically active 2-hydroxyalkanoic acids (250). The potassium derivative of the template was reacted with alkyl halides with 50–95% de, leading to α-alkyl-Phg derivatives. The same procedure was carried out with DL-2-(furyl) glycine (251). A similar chiral Gly equivalent template,

6-tert-butyl-5-methoxy-6-methyl-3,6-dihydro-2H-1,4-oxazin-2-one (see Scheme 7.28), was derived from 2,3,3-trimethyl-2-hydroxybutanoic acid via esterification with Cbz-Gly, lactam formation, and treatment with Meerwein's salt to form the monolactim ether (252). Monoalkylations with a number of halides proceeded with 85–99.4% de and 62–90% yield. A second alkylation was possible, with 60–86% yield and 88–99% de. Hydrolysis was achieved in 25–90% yield by a two-step procedure of TFA at 60 °C for 2 h (imidate cleavage) followed by aqueous NaOH (ester cleavage). Some loss of enantiopurity was observed, with 87–94% ee in the monosubstituted products (252). A 5,6-dimethoxy-5,6-dimethyl-morpholin-2-one template has also been prepared (see Scheme 7.28): alkylations with eight electrophiles proceeded in 67–92% yield with 33–91% de. A second alkylation was possible, with only one product isomer observed. Cleavage with aqueous TFA gave

N-Cbz protected amino acids in 63–74% yield with >99% ee (*253*).

The amino analog of a morpholin-2-one template, a 2-ketopiperazine, has been attached to a solid support and dialkylated. The preliminary account employed an achiral template (see Scheme 7.28) (*254*).

7.7 Seebach Oxazolidin-5-Ones

In 1981 Seebach et al. reported the synthesis of α-alkylated Pro derivatives via a new cyclic chiral template enolate, an oxazolidin-5-one derived from Pro and pivalaldehyde (see Scheme 7.29, Table 7.10) (*255*). This was the first demonstration of amino acid synthesis using an important concept pioneered by Seebach, the self-reproduction of chirality. This principle is based on using an optically active amino acid to induce a temporary second chiral center within a cyclic template. Asymmetry still exists at the second center when the α-center of the amino acid is racemized by enolate formation, with chirality transferred back to the original α-amino acid center upon reaction of the enolate with electrophiles. This method avoids the need for the preparation of expensive chiral auxiliaries that are later destroyed or which must be recovered. The development and application of this principle was thoroughly reviewed in 1996 (*256*).

A number of α-alkylated prolines were prepared with >98% ee at the α-center (incoming electrophile *cis* to the *tert*-butyl group), although less reactive alkyl halides gave poor yields (*255, 257*). Aldol-type reactions with carbonyl compounds gave from 51 to 100% diastereoselectivity at the carbinol center (*258*), while a Michael addition to a nitroolefin resulted in 85% de at the β-center (*259*). Removal of the *N,O*-acetal required strongly acidic hydrolysis (refluxing 48% HBr) (*258*). A number of years later 2-azetidinecarboxylic acid was α-alkylated in a similar fashion, but the only electrophiles that could be used were aldehydes, and even these gave low yields (*260*). The bicyclic β-lactam template was very susceptible to hydrolysis. The oxazolidinone of (2*S*,4*R*)-4-hydroxy Pro has also been α-alkylated (*261*).

Initial attempts at extending this reaction to acyclic amino acids were stymied by the inability to form the necessary oxazolidinone with pivalaldehyde, although imidazolidinone formation was successful (see Section 7.9) (*258*). However in 1984 another group successfully prepared oxazolidinones from aromatic aldehydes and Cbz-Ala or -Phe (*cis* major isomer) (*262*) and alkylated them with methyl iodide or benzyl bromide with complete stereoselectivity (incoming electrophile *trans* to the aromatic group). Subsequently, Seebach and Fadel established conditions that allowed for the synthesis of the pivalaldehyde oxazolidinones of Ala, Phe, Val, and Met, with the *cis* isomer preferentially formed (*263*). Alkylation proceeded with good stereoselectivity (incoming electrophile *trans* to the *tert*-butyl group), and the oxazolidinone could be cleaved under milder conditions (6 N HCl, reflux) than the corresponding imidazolidinone, since an amide bond did not need to be hydrolyzed (*263*).

Alkylation of the benzaldehyde-derived oxazolidinones of Ala, Val, Phe, and Met (*trans* oxazolidinone preferentially formed) with ethyl bromoacetate (addition *trans* to phenyl group) led to α-alkylated aspartic acids (*264*), while alkylations with epoxides produced α-alkyl homoserines (*265*). Isovaline was prepared by alkylation of the Ala template with ethyl iodide (*266*). Alkylations of Ala leading to α-methyl-Asp, -Glu, and -Lys (*267*) and α-methyl-Cys (*268, 269*) were also reported. α-Alkylated Met derivatives were oxidized to the sulfoxides and pyrolytically eliminated to give α-alkyl vinylglycine derivatives, with partially racemic (73–95% ee) products obtained after hydrolysis (*270*). The oxazolidinone prepared from *R*-norleucine and benzaldehyde (*trans* oxazolidinone preferentially formed) was alkylated with *N*-bromomethylphthalimide, leading to α-butyl-α,β,-diaminopropionic acid (*271, 272*). A Trp/pivalaldehyde-derived template (*cis*-oxazolidinone preferentially formed) was used to prepare α-methyl-Trp (*273*). A Phe-derived oxazolidinone was allylated and deprotected to give α-allyl-Phe. *N*-Allylation and Ru-catalyzed metathesis led to α-benzyl-3,4-dehydropipecolic acid (*261*). An extensive listing of α-alkylations of amino acids that have been carried out via an oxazolidinone template has been published (*256*).

An alternative synthesis of the Ala/benzaldehyde-derived oxazolidinone employed BF₃·Et₂O mediated cyclization of Cbz-Ala and benzaldehyde dimethyl

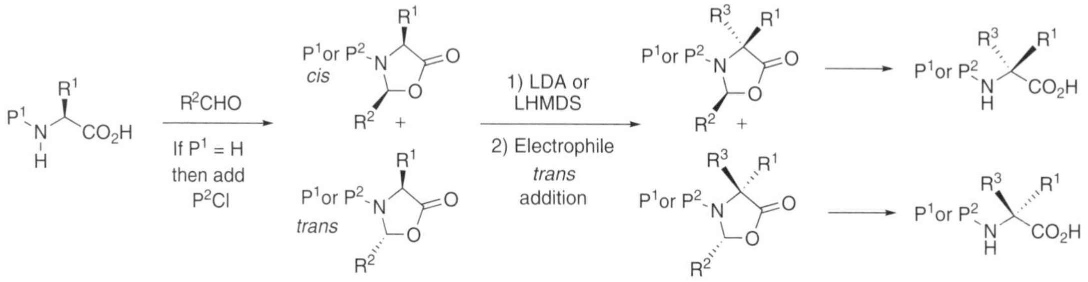

Scheme 7.29 Synthesis of α,α-Dialkyl Amino Acids from Seebach Oxazolidin-5-one (see Table 7.10).

Table 7.10 Synthesis of α,α-Disubstituted Amino Acids via the Seebach Oxazolidin-5-One Template (see Scheme 7.29)

P¹	P²	R¹	R² Oxazolidinone Formation	Electrophile	Alkylation Yield	Deprotection	Yield	Reference
-(CH₂)₃-R¹		-(CH₂)₃-P¹	tBu pentane, TFA, Δ 92% 100% de *trans*	R³X: R³ = CH₂CH=CH₂	92% >98% de			1981 (255)
-(CH₂)₃-R¹		-(CH₂)₃-P¹	tBu pentane, TFA, Δ 92% 100% de *trans*	R³X: R³ = D Me CH₂CH=CH₂ Bn CH₂CO₂Me CH₂CONMe₂ CH₂NMe₂ PhCr(CO)₃ CH₂SPh	30–93% >98% de	48% HBr R³ = Me, Bn		1983 (258)
-(CH₂)₃-R¹		-(CH₂)₃-P¹	tBu	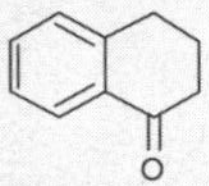 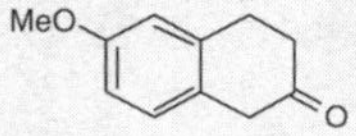 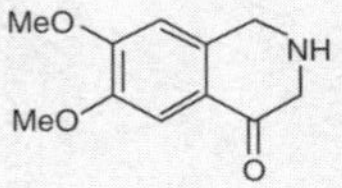 MeCHO tBuCHO PhCHO BnCHO MeCOMe MeCOCH₂CO₂Me NO₂CH₂CO[2,4-(MeO)₂-Ph] (chemical structures: MeO-substituted tetralone derivatives; 6,7-dimethoxy-tetrahydroisoquinolinone)	67–88% 100% de C2 51–100% de C3			1983 (258)

(Continued)

533

Table 7.10 Synthesis of α,α-Disubstituted Amino Acids via the Seebach Oxazolidin-5-One Template (see Scheme 7.29) (continued)

P¹	P²	R¹	R² / Oxazolidinone Formation	Electrophile	Alkylation Yield	Deprotection	Yield	Reference
Cbz		Me Bn	Ph 2,4-Cl₂-Ph pTSA, MeCCl₃ 40% 60–80% de *cis*	R³X; R³ = D	70–80% 100% de	1) 1N HCl 2) H₂, Pd/C		1984 (262)
-(CH₂)₃-R¹		-(CH₂)₃-P¹	*t*Bu	[3,4-(MeO)₂-Ph]CH=CHNO₂	40% 85% de			1985 (259)
H	Bz	Me *i*Pr Bn (CH₂)₂SMe	*t*Bu 1) *t*BuCHO, NaOH 2) BzCl 92–95% 60–100% de *cis*	MeI BnBr BnCHO	40–93% >95% de 96% >95% de	6N HCl, Δ R¹ = Me; R³ = Bn	95%	1985 (263)
	Bz	(CH₂)₂SMe	*t*Bu	H Me	convert side chain R¹	6N HCl, Δ	20–59% 73–95% ee	1986 (270)
H	Bz	Me *i*Pr Bn (CH₂)₂SMe	Ph 1) PhCHO, NaOH 2) BzCl, CH₂Cl₂ 88–94% 50–76% de *trans*	BrCH₂O₂Et	80–95% >95% de	40% HBr, Δ	94–95% 95% ee	1987 (264)
	Bz	Me	Ph	EtI	90% >95% de	conc. HCl, Δ	quant.	1988 (266)
	Bz	Me	*t*Bu	BrCH₂(indol-3-yl)	65% >95% de	6N HCl	87%	1988 (320)
salicaldehyde-derived		Me	salicaldehyde COCl₂, K₂CO₃	R³X; R³ = Me Et CH₂CH=CH₂	66–94% 67–94% de	LiOH H₂O, dioxane	64–95% 90–100% ee	1990 (280)

Substrate	PG	R	Conditions	Electrophile / Aldehyde	Yield, de	Cleavage	Yield	Year (ref)
-(CH$_2$)$_3$-R^1		tBu	1) Et$_2$NSiMe$_3$, 110 °C 2) tBuCHO, TMSOTf 89% >95% de *trans*	MeCHO EtCHO (4-Ph-Ph)CHO (4-NMe$_2$-Ph)CHO [3,4,5-(MeO)$_3$-Ph]CHO (2-pyridyl)CHO (3-pyridyl)CHO (4-pyridyl)CHO	17–39% >96% de except for MeCHO 15%, 60% de	oxazolidinone cleaved upon work up of aldol reaction		1990 (260)
-(CH$_2$)$_3$-R^1								1991 (257)
H	Cbz	Me	1) PhCHO, NaOH 2) Cbz-Cl crude: 42% de *trans* Pure: 30% yield, 100% de Pure: 30% yield, 100% de *trans*	Ph CH$_2$CH=CH$_2$ BrCH$_2$CO$_2$$t$Bu I(CH$_2$)$_4$N(Boc)$_2$ CH$_2$=CHCO$_2$$t$Bu	26–76%	6N HCl 2 eq. NaOH or LiOH 1:1 MeOH:H$_2$O, 45°C to form Cbz-Xaa-OH or 10:1 MeOH:H$_2$O, rt to form Cbz-Xaa-OMe	63–95% or 78–81%	1991 (267)
-CH$_2$CH(OAc)CH$_2$-R^1								
-CH$_2$CH-(OAc)CH$_2$-P^1		tBu tBuCHO, TFA 90%	MeI		55% >95% de	6N HCl	64%	1992 (508)
H	Bz Bz Ac	(CH$_2$)$_2$SMe tBu Ph	Ph tBu Ph 1) R^{2}CHO, NaOH 2) P^{2}COCl 23–49% 62% de *cis* to 88% de *trans*					1993 (277)
	Alloc	iPr iBu Bn	tBu 60–80%	BrCH$_2$CH=C(Me)$_2$	71–96% 90–98% de	1N NaOH, Δ	74–85%	1993 (509)
	Alloc	Me Bn iPr iBu	tBu	forms R = Me, CH$_2$OBn. CH(NHBoc)Bn	40–69% 89->95% de			1994 (265)

(Continued)

Table 7.10 Synthesis of α,α-Disubstituted Amino Acids via the Seebach Oxazolidin-5-One Template (see Scheme 7.29) (continued)

P¹	P²	R¹	R² Oxazolidinone Formation	Electrophile	Alkylation Yield	Deprotection	Yield	Reference
H	Bz	nBu	Ph 1) PhCHO, NaOH 2) BzCl, CH₂Cl₂ 76% 50% de trans	BrCH₂NPhth	89% >95% de	40% HBr, Δ	86%	1994 (271) 1995 (272)
H	CO₂Et	CH₂(indol-3-yl)	tBu 1) tBuCHO, NaOH 2) ClCO₂Et 67% 92% de trans	MeI	79% >98% de	6N HCl, Δ	87% 92% ee	1995 (273)
Cbz		Me	Ph BF₃·Et₂O, PhCH(OMe)₂ 96% 76% de trans	I(CH₂)₃CO₂tBu then treat with 1) TFA 2) DPPA, Et₃N 3) CuCl, tBuOH 70% to form (CH₂)₃NHBoc	88% 100% de	H₂, Pd(OH)₂	95%	1995 (274)
H	tBuCO	Me	ferrocene 1) ferrocene-CHO, NaOH 2) tBuCOCl 92% >98% de cis	R³X; R³ = Bn, 2'-Me-Bn, CH₂(2-naphthyl), CH₂(indol-3-yl), CH₂CN, CH₂CH=CH₂, CH₂CH=CHMe, CH₂CH=CHPh	71-95% 92->98% de	1) Amberlyst-15, acetone-H₂O 2) 2% NH₄OH	71-95% >98% ee	1994 (278) 1998 (279)
-COCH₂CH₂-R¹	-CH₂CH₂CO-P¹		tBu tBuCHO, THF, dioxane 40% 100% de	BnBr, 4-Br-BnBr, PhCHO, (4-Cl-Ph)CHO, (4-BnO-Ph)CHO, cyclohexanone	100% de			1995 (510)
H	Cbz	Me	Ph	BrCH₂SBn	60%	LiOH	quant.	1996 (269) 1997 (268)

H	Alloc	$CH_2CH=C(Me)_2$	tBu; tBuCHO, NaOH, EtOH, pentane, Δ; 64%; 34% de cis	BnBr; 4-BnO-BnBr; $ClCH_2OBn$; $I(CH_2)_4I$	70–88%	3N KOH, MeOH, Δ	58–91%	1997 (511)
	Bz	Me	tBu	3-BnO-BnBr	68%	1) 2N NaOH 2) 6N HCl	45%	1998 (276)
Bz	Bz	Ph; Bn	$BrCH_2CH=CH_2$		93% >95% de	NaOH, MeOH	>91%	1998 (276)
H	salicaldehyde-derived 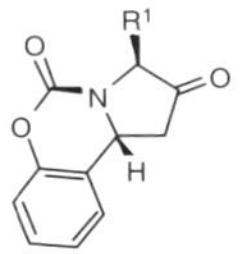	Me	salicaldehyde-derived 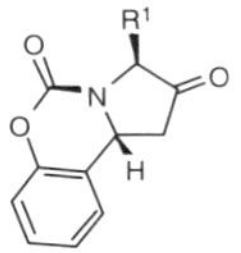 *(bicyclic lactone structure; labeled R^1, N, O, H)*	$BrCH_2$(indol-3-yl)	73% >85% de	LiOH	61%	1998 (281)
$-(CH_2)_3-R^1$	—	$-(CH_2)_3-P^1$	CCl_3; tBu; Cl_3CCHO, AcCN, rt or tBuCHO, TFA, pentane, Δ; 57%; 100% de	MeI; BnBr; $BrCH_2CH=CH_2$; $BrCH_2CO_2Et$	30–69% >95% de	NaOMe or 6N HCl or HCl/MeOH	47–90%	1999 (282)
Cbz	—	tBu	tBu; tBuCHO, pentane, Δ; 69%, 5.3:1	$BrCH_2CH=CH_2$; $BrCH_2CH=C(Me)_2$	90–94%	1N NaOH, MeOH, Δ	90–93%	1999 (512)
Cbz	—	Me	Ph; $PhCH(OMe)_2$, $SOCl_2$, $ZnCl_2$; 76%; 92% de cis	4-Br-BnBr	87%	1) HBr; AcOH 2) EtOH, HCl	87%	2001 (275)

(Salicaldehyde-derived bicyclic structures shown in the table are fused benzene–lactam–lactone scaffolds, labeled with R^1, N, O, and H.)

actetal, with 96% yield of product with 76% de *cis* ste-reoselectivity (*274*). The template was alkylated with *tert*-butyl-4-iodobutyrate and converted to α-methyl-D-Orn by a Curtius rearrangement of the side chain. In 2001 the conditions affecting the *cis:trans* ratio during oxazolidinone formation were examined, with cycliza-tion of Cbz-Ala-OH and the dimethylacetal of benzalde-hyde giving the *cis* isomer with anywhere from 70 to 92% de. The solvent and type of acid catalyst influenced the isomers formed, with the *cis* isomer the thermo-dynamic product. Alkylation with 4-bromobenzyl bromide gave the product in 75% yield if a mixture of the oxazolidinone and benzyl bromide were added to KHMDS in THF, but poor yields if the electrophile was added to the preformed enolate (*275*).

In most of the above examples the deprotected amino acid was obtained by hydrogenolysis and/or acid hydro-lysis. However, it is also possible to directly convert the oxazolidinone into an *N*-Cbz amino acid or methyl ester by base-catalyzed opening (*267*), as used for the prepa-ration of α-methyl-*meta*-Tyr (*276*). Note that the relative stereochemistry of the major diastereomer formed during oxazolidinone preparation varies with the alde-hyde, acylating reagent, and synthesis procedure (e.g., *263, 271, 274, 277*), but that alkylation almost always occurs *anti* to the chiral auxiliary substituent, except with the bicyclic (Pro and 2-azetidinecarboxylic acid) templates. The *cis* or *trans* isomers initially prepared must be well purified, as the presence of the opposite diastereomer will compromise the enantiomeric purity of the final α,α-disubstituted amino acid. Alonso and Davies have addressed this problem by using ferrocene carboxaldehyde to generate the oxazolidinone (*278, 279*). With Ala, only a single diastereomer (>98% de, *cis*) was obtained. Alkylation proceeded uneventfully with benzyl bromide and a range of other alkyl/benzyl bromides. The ferrocene template has the added advantage that hydrolysis is possible under very mild conditions (Amberlyst-15, acetone–H₂O); deprotected α-methyl-Phe was readily obtained. Similar advantages were described for the use of salicylaldehyde and phos-gene to form an oxazolidinone, as only one diastereomer of the resulting tricyclic oxazolidinone template was observed with Ala, Phe, or Leu (*280*). Alkylation pro-ceeded with 67–94% de, and the alkylated products could be deblocked under mild conditions using LiOH.

This template was employed for the synthesis of acid sensitive α-methyl-Trp (*281*).

Another Pro-derived oxazolidinone template has been formed from L-Pro and trichloroacetaldehyde, giving a single crystalline diastereomer. Deprotonation with LDA and alkylation gave 4-alkyl-2-trichloromethy-loxazolidin-5-ones in 30–69% yield as single *cis* iso-mers, which were also crystalline. A direct comparison with the Seebach *tert*-butyl-substituted oxazolidinone showed that the trichloromethyl analog was much less expensive to synthesize, could be prepared more quickly in higher yield, and had better physical properties (stability and crystallinity) (*282*). The alkylated oxazoli-dinones were converted into *N*-formyl α-alkyl-Pro methyl esters by treatment with NaOMe, into unpro-tected α-alkyl-Pro by hydrolysis with 6 N HCl at room temperature, or to the *N*-unprotected α-alkyl-Pro methyl esters by refluxing in acidic anhydrous methanol.

The oxazolidinone template can also be used to synthesize α-H amino acids (see Scheme 7.30, Table 7.11). However, the "self-reproduction of chirality" concept becomes invalid when Gly is the substrate, as there is no chiral center present to induce asymmetry. In order to synthesize the desired homochiral substrate, Seebach et al. prepared a variety of oxazolidinones derived from Gly and resolved them by HPLC on a chiral column (*283*). The initial report described some preliminary results at attempted amino acid synthesis: alkylation of the pivalaldehyde-derived oxazolidinone enolate with methyl iodide gave protected Ala, while condensation with benzaldehyde gave (*2R,3S*)-phenyl-serine (*threo* isomer) as the only observed diastereomer (see Scheme 7.30) (*283*). A Troc-protected *tert*-butyl-substituted oxazolidinone template was resolved on a Chiraspher chiral column on a 50 g scale (*284*).

In a subsequent publication (1991), the LiHMDS-derived enolate of the *N*-Cbz pivalaldehyde oxazolidi-none was added to a number of aldehydes (*285*). Attempts to generate the enolate with LDA, KHMDS, or NaHMDS led only to decomposition. Very high diastereoselectivities of the *threo* (*2S,3R*) adduct were observed (91 to >98% selectivity of the major isomer over the other three diastereomers), with high yields (70–99%) for aromatic aldehyde substrates. Yields were not as good with aliphatic aldehydes (38–81%), due in part to formation of bicyclic carbamates resulting from

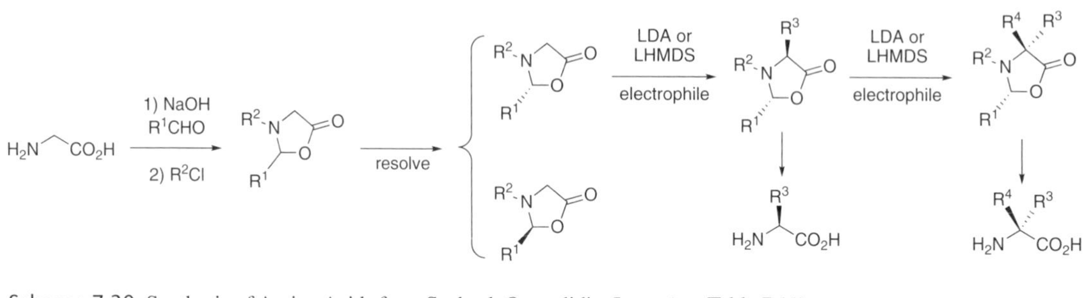

Scheme 7.30 Synthesis of Amino Acids from Seebach Oxazolidin-5-one (see Table 7.11).

Table 7.11 Synthesis of Amino Acids via the Seebach Oxazolidin-5-One Template (see Scheme 7.30)

R¹	R²	Electrophile R³	Electrophile R⁴	Alkylation Yield	Deprotection	Yield	Reference
tBu	Bz	MeI		63%	H$_2$, Pd/C	94%	1988 (283)
	Cbz	PhCHO		81%, >95% de			
tBu	CO(1-naphthyl)						1988 (283)
tBu	CO(2-naphthyl)						
iPr	CO(2-naphthyl)						
1-naphthyl	Bz						
1-naphthyl	CO$_2$Me						
Ph	Cbz						
Ph	tBuCO						
tBu	Cbz	EtCHO		alkylCHO:	1) H$_2$, Pd/C	>85%	1991 (285)
		iPrCHO		10–81%	2) H$_2$O		
		tBuCHO		>98% de			
		nDecCHO					
		nC$_{13}$H$_{27}$CHO		arylCHO:			
		PhCHO		46–95%			
		(2-F-Ph)CHO		91–>98% de			
		(4-CF$_3$-Ph)CHO					
		(4-Me$_2$N-Ph)CHO					
		(4-NO$_2$-Ph)CHO					
		(4-MeO-Ph)CHO					
		[3,4,5-(MeO)$_3$-Ph]CHO					
		(2-Cl-6-F-Ph)CHO					
		(2-naphthyl)CHO					
		(4-pyridyl)CHO					
		(2-furyl)CHO					
		(2-thiophene)CHO					
tBu	Cbz	(pent-3-enal / CH=CH–CH$_2$–CH$_2$–CH(–)CHO structure)		47–60%			1991 (286)
tBu	Bz	MeI		24–80%			1993 (287)
	Cbz			50–>96% de			
	tBuCHO						
	CO$_2$Me						
	4-NO$_2$-Cbz						
	4-MeO-Bz						
	4-(NMe$_2$)-Bz						
	4-NO$_2$-Bz						
	4-(CH$_2$NPr$_2$)-Bz						
tBu	Bz	R³X; R³ =		45–60%			1993 (287)
		nBu		88% de			
		Bn					
tBu	Cbz	R³X; R³ =		32–55%			1993 (287)
		CH$_2$CH=CH$_2$		>82% de			
		Bn					
tBu	4-(NMe$_2$)-Bz	R³X; R³ =		38–74%			1993 (287)
		Et		>96% de			
		CH$_2$CH=CH$_2$					
		Bn					
tBu	Bz		R⁴X; R⁴ =	40–73%			1993 (287)
		Me	Et	86–90% de			
			3-MeO-Bn				
			4-MeO-Bn				
		Bn	Et				
			3,4-(MeO)$_2$-Bn				
tBu	4-(NMe$_2$)-Bz	Me	R⁴X; R⁴ =	50%			1993 (287)
			CH$_2$CH=CH$_2$	>96% de			
tBu	Cbz	Me	R⁴X; R⁴ =	74–77%			1993 (287)
			Et	>94% de			
			CH$_2$CH=CH$_2$				
			Bn				

elimination of benzyl alcohol. The free amino acids were obtained by hydrogenolysis and neutral hydrolysis, generally in >90% yields. The oxazolidinone template was used in a synthesis of MeBmt via an aldol reaction with an appropriately functionalized chiral aldehyde (286). The addition produced predominantly the bicyclic carbamate, which ended up being a useful intermediate as it prevented retro-aldol reactions during N-methylation and oxazolidinone cleavage.

The oxazolidinone Gly template has also been monoalkylated with alkyl halides with reasonable yields (24–78%) and stereoselectivities (50 to >96% de) (287). A second consecutive alkylation provided α,α-disubstituted amino acids with better results (40–77% yield, 88 to >96% de). Hydrolysis to the free amino acids was much easier than with the corresponding imidazolidinone templates (Section 7.9). The effect of an N-acyl substituent on template conformation and alkylation stereoselectivity has been examined (288).

As with other chiral templates, dehydro-Ala was incorporated within the cyclic system and functionalized (see Scheme 7.31). An initial synthesis of the dehydro-Ala template from cyclized Ala via bromination/dehydrohalogenation (289) was later reported to be unreliable, so alternative preparations using modified bromination/dehydrohalogenation reaction conditions (290–293) or a different route from Met (277) have been described. The dehydro-Ala alkene was used for Diels–Alder reactions with cyclopentadiene or cyclohexa-1,3-diene, giving a single exo adduct with >94% de under thermal conditions (277, 292, 294, 295), although subsequent epimerization of the oxazolidinone aminoacetal carbon was possible in some cases (294). Reaction with a number of other dienes was also investigated (294), as was a PPh3-catalyzed cycloaddition with allenes (296).

Cyclopropanation of the dehydro-Ala template with isopropylidenetriphenylphosphorane resulted in a 1:1

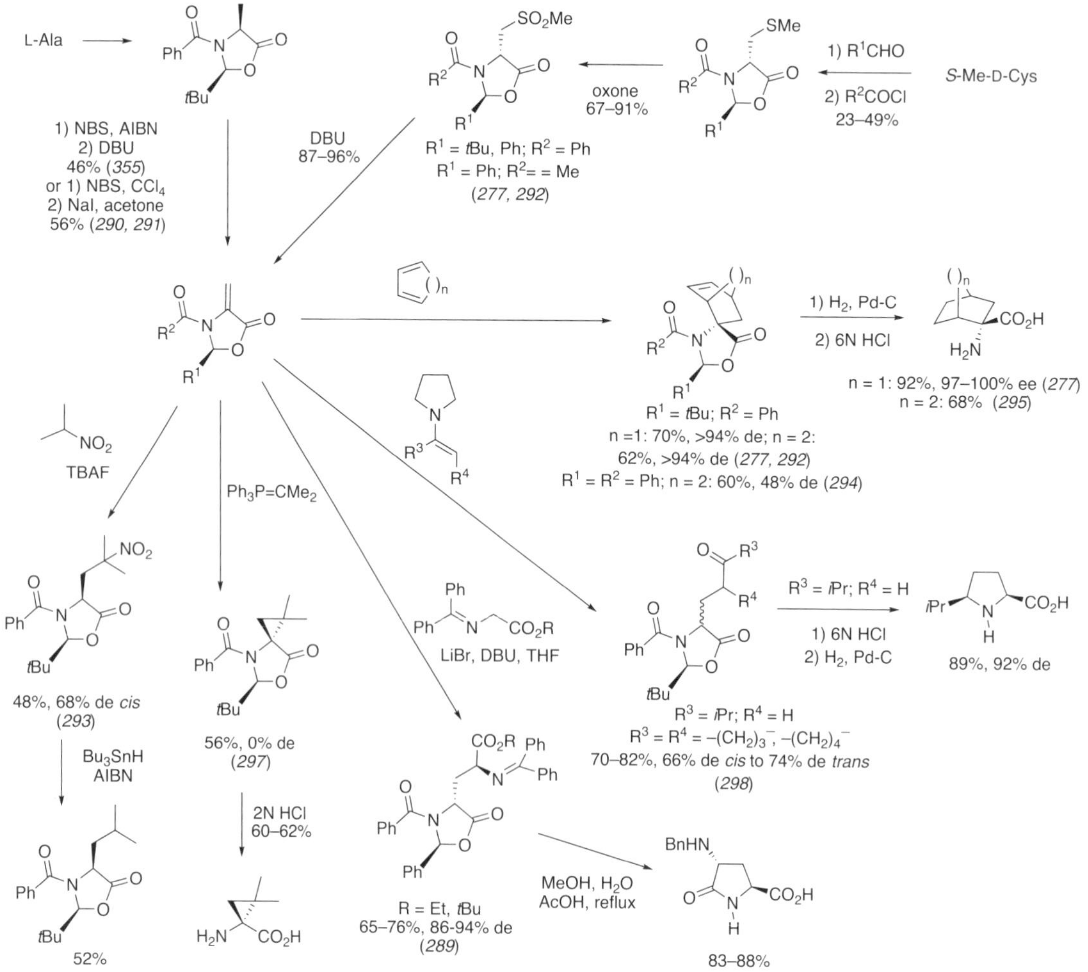

Scheme 7.31 Synthesis and Reaction of Dehydroalanine within Seebach Oxazolidin-5-one.

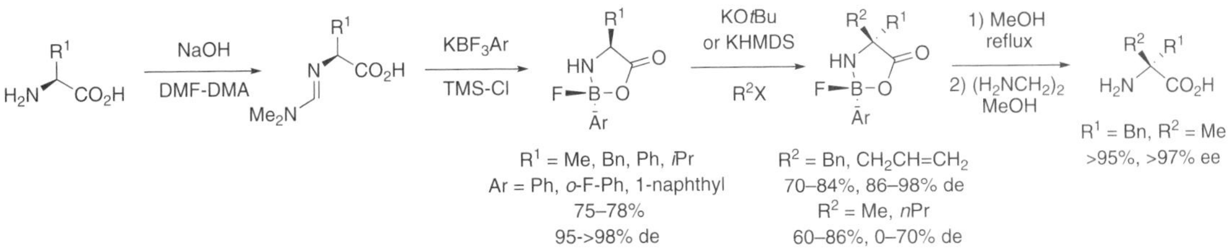

H_2N–CHR^1–CO$_2$H →(NaOH, DMF-DMA)→ Me$_2$N=N, R^1, CO$_2$H →(KBF$_3$Ar, TMS-Cl)→ HN, R^1, =O, F–B–O, Ar

R^1 = Me, Bn, Ph, iPr
Ar = Ph, o-F-Ph, 1-naphthyl
75–78%
95->98% de

→(KOtBu or KHMDS, R^{2}X)→ HN, R^2 R^1, =O, F–B–O, Ar

R^2 = Bn, CH$_2$CH=CH$_2$
70–84%, 86–98% de
R^2 = Me, nPr
60–86%, 0–70% de

→ 1) MeOH reflux 2) (H$_2$NCH$_2$)$_2$ MeOH → H_2N–CR^{2}R^1–CO$_2$H

R^1 = Bn, R^2 = Me
>95%, >97% ee

Scheme 7.32 Synthesis and Reaction of Oxazaborolidinone Template (*304, 307*).

mixture of diastereomers, which were separated and hydrolyzed to give both enantiomers of 2,3-methanovaline (*297*). Conjugate additions of enamines to the dehydro-Ala alkene provided γ-oxo derivatives which were cyclized to give substituted prolines (*298*), while additions of nitronates followed by radical denitration led to β-substituted alkyl amino acids (*293*). Reaction with azomethine ylides derived from the benzophenone Schiff base of Gly gave substituted pyroglutamates (*289*). Either *cis* or *trans* products were obtained as the major isomers from the Michael additions, depending on the substituents and nucleophile. Radical reactions of the alkene function have also been described (see Section 4.3.4) (*290, 299–302*). An N-Cbz oxazolidinone containing ΔAla was employed for aqueous reactions with alkyl iodides in the presence of Zn/CuI, producing protected Leu, 2-aminooctanoic acid, cyclohexylalanine, and 2-amino-1,7-pentanedioic acid with >96% de and 60–99% yields. Boc-β-iodo-Ala-OMe was also coupled to give 2,5-diaminoadipic acid (*303*).

A boron analog of Seebach's oxazolidinone template has been prepared from amino acids, with crystallization of an equilibrating mixture of diastereomers allowing for isolation of oxazaborolidinones with high diastereomeric purity (>95% de) (*304–307*). Enolate generation (KOtBu) and alkylation gave the α-substituted amino acid derivatives with high diastereoselectivity when benzyl bromide or allylbromide was used as alkylating reagent, but with poor to moderate selectivity with methyl iodide (see Scheme 7.32) (*304*). Deprotection was achieved by methanolysis. Care was required to avoid racemization resulting from reversible B–N bond cleavage.

glycine equivalent (e.g., *308, 309*) or amination of a derivative with the side chain and carboxyl group already present (e.g., *310–312*). These methods have already been discussed in Sections 6.4.2a and 5.3.6a, respectively, with a summary of similar oxazolidinone auxiliaries in Scheme 5.23.

A substantially different approach has been taken for the synthesis of 2-substituted-1-aminocyclopropane carboxylic acids and β,γ-cyclopropane-α,β-diamino acids (*313*). The oxazolidinone template was alkylated rather than acylated, and the nitrogen of the oxazolidinone incorporated into the final product as either the α- or β-amino group (see Scheme 7.33). A similar method was employed to prepare α,α-disubstituted amino aldehydes, which were then oxidized to form the corresponding amino acids (*314*).

An oxazolidinone auxiliary has also been used to induce asymmetry during conjugate additions to dehydro-Ala derivatives (see Section 4.3.3), with the nitrogen of the oxazolidinone again incorporated into the product α-amino acid (*315*). Homochiral ketenes generated with an oxazolidinone auxiliary were used for [2+2] cycloadditions with homochiral imines derived from amino acids; the resulting β-lactam derivatives (with the newly generated amino acid amino group within the oxazolidinone) were opened to give dipeptides (*316*).

An electrophilic oxazolidin-2-one template, with a methoxy leaving group, has also been reported (see Scheme 7.34). Addition of organocuprates in the presence of boron trifluoride provided amino alcohols with high diastereoselectvity. An oxidation step is required to produce amino acids (*317*).

7.8 Oxazolidin-2-Ones

A number of syntheses relying on oxazolidin-2-one chiral templates have been reported. The best-known methods employ the oxazolidinone auxiliaries of Evans, which are derived from chiral amino alcohols. However, in contrast to the other cyclic templates described in this section, the chiral center being generated lies outside the template ring, and the oxazolidinone therefore functions as a conventional chiral auxiliary. Two different strategies have been used: introduction of a side chain to a

7.9 Seebach Imidazolidin-4-One

The use of imidazolidinones as a template for the "self-reproduction of chirality" was pioneered and developed by Seebach and co-workers, using the same strategy described in Section 7.7 for the oxazolidin-5-one templates. A review was published in 1996 (*256*). Initially, a number of α-alkyl amino acids were synthesized from the N-methylimidazolidinones derived from pivalaldehyde and Ala, Phe, Met, Val, Phg (*318*), Asp, Glu (*319*), Orn, and Lys (*320*) (see Scheme 7.35, Table 7.12). The *cis* or *trans* imidazolidinones could be selectively

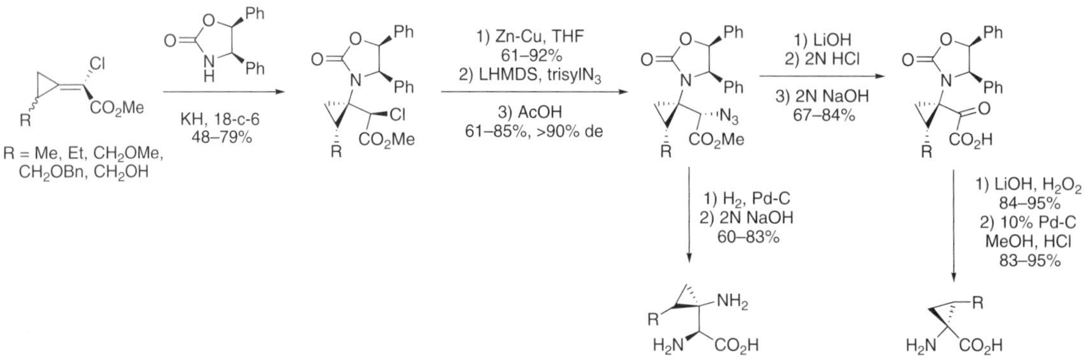

Scheme 7.33 Incorporation of Oxazolidinone Nitrogen in Product (*313*).

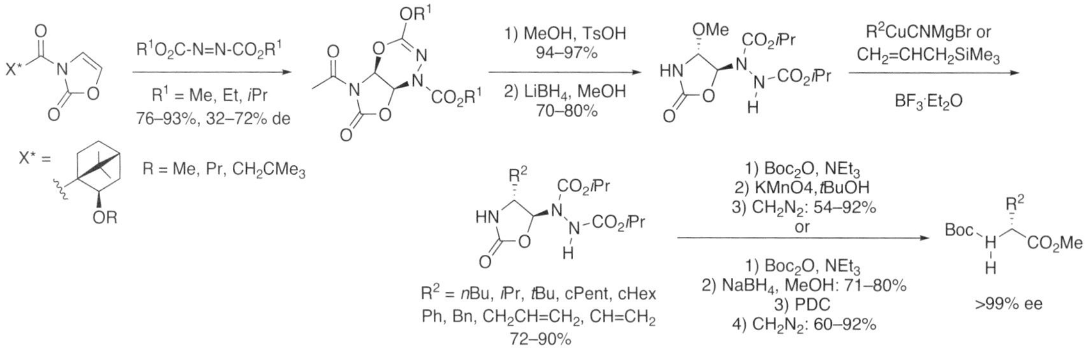

Scheme 7.34 Synthesis of Electrophilic Oxazolidin-2-one Template (*317*).

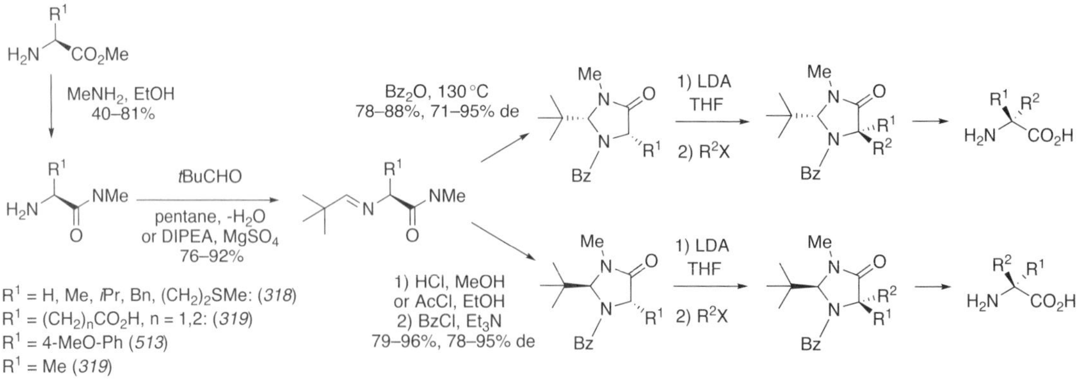

Scheme 7.35 Synthesis of α,α-Disubstituted Amino Acids with Imidazolidinone Template (see Table 7.12).

prepared, depending on cyclization conditions (*318*). Improved conditions for preparing the *trans* Ala-derived imidazolidinone have been described (*321*). Deprotonation of the imidazolidinone with LDA, reaction with alkyl halides, and deprotection by hydrolysis with refluxing 6 N HCl gave the amino acids with nearly

100% ee at the regenerated chiral center (*319, 322*). However, some adducts could not be hydrolyzed (*287*). Conditions to overcome this problem have since been described (*323*).

The imidazolidinone enolates are quite stable, and have been used for both aldol (*322, 324*) and Michael

Table 7.12 Synthesis of α,α-Disubstituted Amino Acids via the Seebach Imidazolidinone Template (see Scheme 7.35)

R¹	Electrophile	Alkylation Yield	Further Side Chain Derivatization	Deprotection	Deprotection Yield	Reference
Me	MeCH=CHNO₂	62% >90% de				1985 (259)
	R²X; R² =	41–95%		HBr, Δ	77–95%	1985 (319)
CH₂CO₂H	D					
CH₂CO₂H	Me					
CH₂CO₂H	Bn					
(CH₂)₂CO₂H	D					
(CH₂)₂CO₂H	Me					
(CH₂)₂CO₂H	Bn					
	R²X; R² =	73–90%		6N HCl, Δ	72–95%	1985 (322)
Me	Et			R¹ = iPr; R² = Me		
Me	Bn			R¹ = (CH₂)₂SMe;		
Me	3,4-(MeO)₂-Bn			R² = Me		
iPr	D					
iPr	Me					
iPr	Et					
iPr	CH₂CH=CH₂					
(CH₂)₂SMe	Me					
(CH₂)₂SMe	Et					
Ph	Et					
(CH₂)₂SMe	R²X; R² =	50–66% 93–95% de	H₂O₂ convert (CH₂)₂SMe to CH=CH₂	6N HCl, R¹ = CH=CH₂; R² = H, Et	37–63%	1986 (270)
	D					
	Me					
	Et					
	iPr					
CH=CH₂	R²X; R² =	62–77%		6N HCl, R¹ = CH=CH₂; R² = H, Et	37–63%	1986 (270)
	D					
	Me					
	Et					
	CH₂CH=CH₂					
	Bn					
(CH₂)₂SMe	MeCHO PhCHO Me₂CO	73–75% >95% de		6N HCl	forms	1987 (324)
CH=CH₂	MeCHO PhCHO Me₂CO	40–45% >90% de				1987 (324)
	R²X; R² =	34–63% 85–>95% de		HCl		1988 (320)
(CH₂)₃NH₂	Me					
(CH₂)₃NH₂	Bn					
(CH₂)₄NH₂	Me					
Me	CH₂(indol-3-yl)					
Me	AcCl		reduce with LiBHEt₃ to form 65%			1993 (325)
Me	BrCH(Me)Ph	3 eq. RX: 82%, 94% de 1 eq. RX: 63%, 50% de	37% aq. HCHO conc. HCl, Δ 71–78%	6N HCl	75–82%	1998 (231)

(Continued)

Table 7.12 Synthesis of α,α-Disubstituted Amino Acids via the Seebach Imidazolidinone Template (see Scheme 7.35) (continued)

R[1]	Electrophile	Alkylation Yield	Further Side chain Derivatization	Deprotection	Deprotection Yield	Reference
4-BnO-Ph	MeI	92%	1) Pd, H₂ 2) Tf₂O, 2,6-lutidine 3) CO, Pd(OAc)₂, dppp form 4-CO₂Et-Ph 89%	6N HCl	80%	1996 (513)
Me	R²X; R² = 4-^{18}F-Bn 2-^{18}F-Bn 6-^{18}F-3,4-(OH)₂-Bn 2-^{18}F-5-HO-Bn 2-^{18}F-4-HO-Bn	24–73%		6N HCl	7–13% radiochemical yield overall >97% ee	1997 (326)
Bn CH₂(indol-3-yl)	MeI	84%		10N HCl sealed tube		2000 (333)

additions (259). They also react with acyl chlorides, with reduction of the β-keto group leading to α-alkylated-β-hydroxy amino acids (325). The Ala-derived template has been used to prepare α-methyl [18F]-fluorinated aromatic amino acids, suitable for use as PET radiopharmaceuticals. The short reaction times of the synthetic route (approx. 120 min) are critical for radiopharmaceutical production (326). The Seebach imidazolidinone has been used to synthesize all four isomers of 2,3-dimethyl-Phe, starting with (R)- or (S)-Ala. Alkylation of one enantiomer with an excess of racemic (1-bromoethyl) benzene provided predominantly one diastereomer, out of the four possible (321). The diastereomer epimeric at the β-center was obtained by treating the enolate with only 1 equiv of the alkyl halide, providing a mixture of two diastereomers that could be separated. The products were deprotected to give 2,3-dimethyl-Phe, or cyclized under Pictet–Spenger conditions to give pipecolic acid derivatives.

The imidazolidinone template derived from Met has been used for the synthesis of α-alkylated vinylglycines. The alkene side chain was generated via oxidation of the Met side chain to a sulfoxide followed by pyrolytic elimination. This conversion could be carried out either before or after alkylation (270) or aldol reaction (324) of the template.

The chiral Gly-based imidazolidinone required for α-amino acid synthesis proved difficult to prepare; initial syntheses required the degradation of the Ser- (327) or Met- (270) derived imidazolidinones and gave poor yields (<30%) (see Scheme 7.36). Better results were obtained by recrystallizing the mandelate salt of the racemic 3-methyl-imidazolidinone (BMI) template prepared directly from Gly (328), or by resolving the ketogulonate salt of the 3-benzyl-imidazolidinone (BBI) template prepared from bromoacetyl bromide (329). Both enantiomers were isolated with >98% ee, and the cyclic aminals were acylated to give N-Bz-, N-Boc-, N-Cbz-, and N-formyl- derivatives (328, 329). In an

attempt to avoid the chiral resolution required for the Gly based oxazolidinone preparation, the N-Me group was replaced with an Ala residue before imidazolidinone formation. The cyclization resulted in a 2:1 preference for one diastereomer, but yields were low (30%) and the subsequent alkylation was troublesome since two possible enolates could be formed (330).

Deprotonation of the Gly template with LDA generated an enolate that could be reacted with electrophiles such as halides, nitroolefins, carbonyls, and epoxides (see Scheme 7.37, Table 7.13) (324, 329, 331). Products with one new stereocenter (the α-center in the deprotected amino acid) give a single diastereomer as judged by ¹H NMR analysis. Regeneration of the enolate followed by protonation provided the other amino acid enantiomer without needing the enantiomeric template, and also allowed for the construction of α-deuterated amino acids (331). Secondary halides or nitroalkenes gave reasonable selectivity at the β-center (77–85% ds) (329). Replacing the imidazolidinone tert-butyl auxiliary substituent with an isopropyl group reduced the stereoselectivity (324). The effect of the N-acyl substituent on the template conformation and alkylation diastereoselectivity has also been examined (288).

Two consecutive alkylations of the Gly imidazolidinone template provided α,α-disubstituted amino acids with much cleaner reactions than the corresponding oxazolidinones. Given that the Boc-BMI template is commercially available, α,α-disubstituted amino acids are produced much more quickly by this route than by alkylating the template prepared from the appropriate amino acid (287, 331, 332). However, an attempted synthesis of α-methyl-Trp via alkylation with bromo-methylindole followed by methyl iodide failed to give any product, while methylation of the cis-imidazolidinone template derived from Trp was successful (333). α,α-Cyclic amino acids (331), Pro analogs (113, 331), and dimeric diaminodicarboxylic acids (329, 331) have been obtained by alkylation with the appropriate

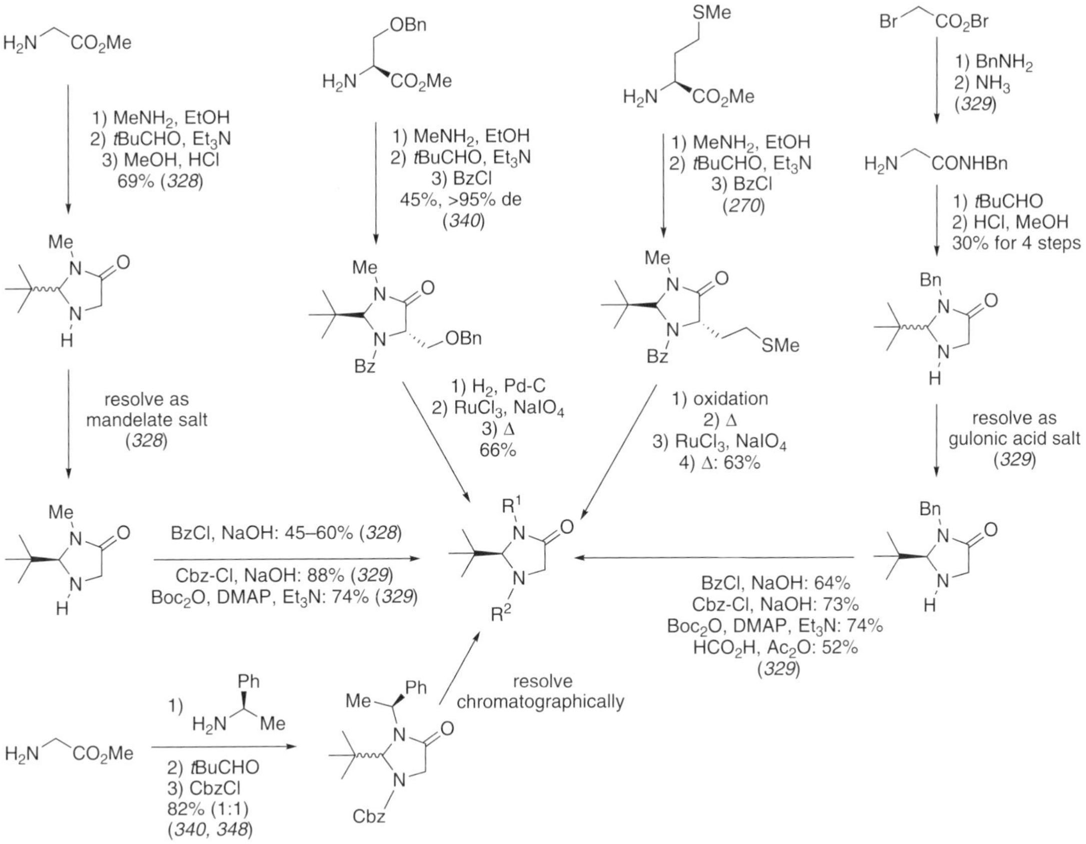

Scheme 7.36 Synthesis of Imidazolidinone Glycine Template.

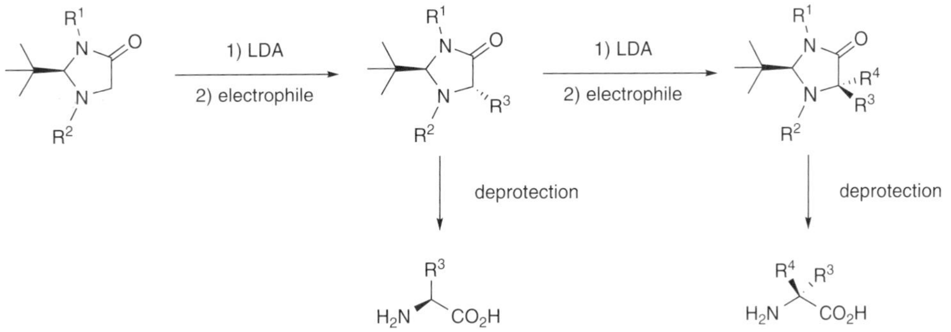

Scheme 7.37 Reactions of Imidazolidinone Template (see Table 7.13).

dihalo compounds. It must be noted that the enantiomeric purity of the commercial Boc-BMI templates have been reported to vary by several percent, so care must be taken if products with very high enantiomeric purity are required (334).

Enolate addition to aldehydes or ketones was found to be accompanied by transfer of the N-protecting group to the hydroxyl side chain, which fortuitously prevents retro-aldol reactions (see Scheme 7.38) (324, 329). Aldehydes gave 41–65% yields, with 63–96% diastereoselectivity in favor of the syn (threo) product (324). The side chain of MeBmt was introduced by this route, although complete deprotection was problematic (324). Greater threo selectivity was obtained by using a

Table 7.13 Synthesis of Amino Acids via the Seebach Imidazolidinone Template (see Scheme 7.37)

R¹	R²	Electrophile R³	Electrophile R⁴ or Side Chain Derivatization	Alkylation Yield	Deprotection	Yield	Reference
Me	Bz	R²X; R³ = Me iPr nBu Bu		27–90% >95% de			1985 (327) 1987 (234)
Me	Boc	R²X; R³ = Bn 4-D-Bn 4-F-Bn 4-Cl-Bn 4-Br-Bn 3-Cl-Ph CH₂(2-thienyl) CH₂(3-thienyl) CH₂(2-naphthyl) CH₂C₆D₅ CH₂CO₂Me	for R³ = CH₂CO₂Me, reduce with LiBEt₃D to give CH₂CD₂OH, activate and cyclize to give [D structure]	58–96%	1) TFA 2) 0.75N HCl Dowex 50W-X8	29–92%	1989 (331)
Me	Boc	(CH₂)₅Cl Me Me Et Et	H (CH₂)₅Cl (CH₂)₄Cl (CH₂)₄Cl (CH₂)₂Cl	68–96%	1) TFA 2) NaHCO₃ 3) Na₂CO₃, NaI, MeCN 4) H⁺ forms: [cyclic amino acid structure]	1+2) 68–88% 3) 79–89% 4) 34–90%	1989 (331)
Me	Boc	(CH₂)₄Cl	NaI, convert to (CH₂)₄I (A)				1989 (331)
Me	Cbz	A					
Me	Bz	H₂C=CHCF₃ (forms CH₂CH=CF₂)		15%			1989 (514)
Me	Bz	HCOCH₂CH₂CH-[N(Boc)₂] CO₂tBu		49% aldol and hydrolysis >95% de of C2 and C3 threo 50% fluorination			1990 (75)

546

Me	Boc	$^{11}CH_3I$ $I^{11}CH_2Ph$ $I(CH_2)_3{}^{11}CN$	9N HCl	20–75% radiochemical yield >98% ee	1991 (339) 1995 (334)
Me	Boc	3-Ar-4-$(CH_2PO_3Et_2)$-BnBr, Ar = Ph, 4-F-Ph, 4-Ph-Ph, 3-AcNH-Ph	1) TFA 2) HCl, Dowex 50W-X8	25–47% overall	1992 (338)
Me	Boc	RCH=CHCO$_2$(2,6-tBu$_2$-4-MeO-Ph), R = H, Me, Et, ipr, Ph, Bn	1) TFA 2) 0.75N HCl, sealed tube, 105°C	57–96% >95% de *threo*	1992 (337)
Me	Boc	R^{4}X; R^4 = Me Me Et 4-MeO-Bn Bn iPr Bn CH$_2$cHex Et iPr CH$_2$cHex Bn 3,4-(MeO)$_2$-Bn 3,4-(MeO)$_2$-Bn 3,4-(MeO)$_2$-Bn 3,4-(MeO)$_2$-Bn	6N HCl or 1) 6N HCl 2) Et$_3$N, BzCl 3) 4N HCl, dioxane 4) conc HCl, 100°C	50–86%	1993 (287) 1995 (323)
Me	Boc	MeI R^{2}X; R^2 = 4-^{18}F-Bn 2-^{18}F-Bn 6-^{18}F-3,4-(OH)$_2$-Bn 6-^{18}F-3-HO-Bn 2-^{18}F-4-HO-Bn 4-^{18}F-3-HO-Bn	6N HCl, Δ		1994 (515)
Me	Bz	Br(CH$_2$)$_3$Br displace side chain with [structure] hydrolysis gives (CH$_2$)$_3$CH(NH$_2$)PO$_3$H$_2$	12N HCl	57–66%	1994 (516)
Me	Bz	BrCH$_2$CH=CH$_2$ treat with I$_2$, THF-H$_2$O, form [structure]	2N HCl	90% >95% de	1994 (342)

(Continued)

Table 7.13 Synthesis of Amino Acids via the Seebach Imidazolidinone Template (see Scheme 7.37) (continued)

R¹	R²	Electrophile R³	Electrophile R⁴ or Side Chain derivatization	Alkylation Yield	Deprotection	Yield	Reference
Me	Boc	[2-(chloromethyl)quinoline]; [3-(chloromethyl)isoquinoline]; [chloromethyl-phenanthroline]		39–82% >95% de	1) TFA 2) HCl, Dowex 50W-X8	43–70% 98% ee	1994 (341)
Me	Bz	$Br(CH_2)_3PO_3Et_2$		59–69% >95% de	6N HCl, 140°C	33–44% ee 98%	1995 (340)
(+)-CH(Me)Ph	Bz	$Br(CH_2)_3PO_3Et_2$		63–69% >98% de	6N HCl, 115°C	>99% ee	1995 (345)
Me	Boc	$I(CH_2)_3CCH$	treat with $BH_{12}(NCMe_3)_2$ to form HC≡C–C(CH₂)₃ ($B_{10}H_{10}$) 53%	86%	1) TFA 2) HCl	60% 98.3% ee	1995 (345)
Me	Boc	TMS–C≡C–C ($B_{10}H_{10}$)		75%	1) TBAF 2) TFA 3) 3N HCl	86% 99.6% ee	1997 (113)
Me	Boc	$ClCH_2C(=CH_2)CH_2Cl$; (Z)-$ClCH_2CH=CHCH_2Cl$	cyclize to α-N after deprotection	48–70% 86–96% de	HCl, 110°C	33–35% 90–96% ee	1997 (332)
Me	Boc	MeI	EtI	95%	conc. HCl, Δ	quant.	1997 (343)
Me	Bz	$BrCH_2CH=CH_2$	elaborate into $CH_2CH(OH)CH_2CH(OH)CH(OH)CO_2H$				
(+)-CH(Me)Ph	Cbz	$Br(CH_2)_4PO_3Et_2$		61–62% >98% de	6N HCl	91–96% ee	1997 (349)
Me	Boc	[4-(bromomethyl)phenyl pinacol boronate]	Suzuki coupling with RX, Pd(O) to form 4-R-Bn R = Ph, 4-F-Ph, 2-thienyl, 1-cyclohexenyl, CH=CHPh, 3-quinolinyl 48–90%	36%			1997 (344)

		R³X; R³ =			
Me	Boc	CH₂cPr, CH₂CN, (CH₂)₃Cl, (CH₂)₄Cl, (CH₂)₂Ph, CH(Me)Ph, Bn, 4-Br-Bn, CH₂C₆F₅, CH₂CH=CH₂, CH₂CH=CHMe, CH₂C(=CH₂)CO₂Me	43–85%	1) TFA 2) 0.75N HCl	1998 (329)
Me	Boc	MeCH=CHNO₂ [structure] + BF₃·Et₂O [structure Br Br]	56–73%		1998 (329)
Me	Cbz	R³X; R³ = iPr, CH₂SiMe₃, CH₂CO₂Me, CH₂CO₂Et, (CH₂)₃Cl, (CH₂)₄Cl	for R³ = (CH₂)ₙCl, cyclize to [structure] 9–70% 41–63%	1) H₂, Pd-C 2) 0.75N HCl	1998 (329)
Me	Boc	[structures]	65–81%	1) TFA 2) 0.75N HCl	1998 (517)
Me	Boc	BrCH₂CCH	treat with B₁₀H₁₂/SEt₂ to form carborane 91–96% ee		2000 (346)
Me	Boc	MeI TsOCH₂CF=CH₂	TsOCH₂CF=CH₂ 84–89% 43–51%	1) 1N HCl, MeOH 2) Phth₂O 3) NaNO₂, Ac2O/AcOH	2005 (347)

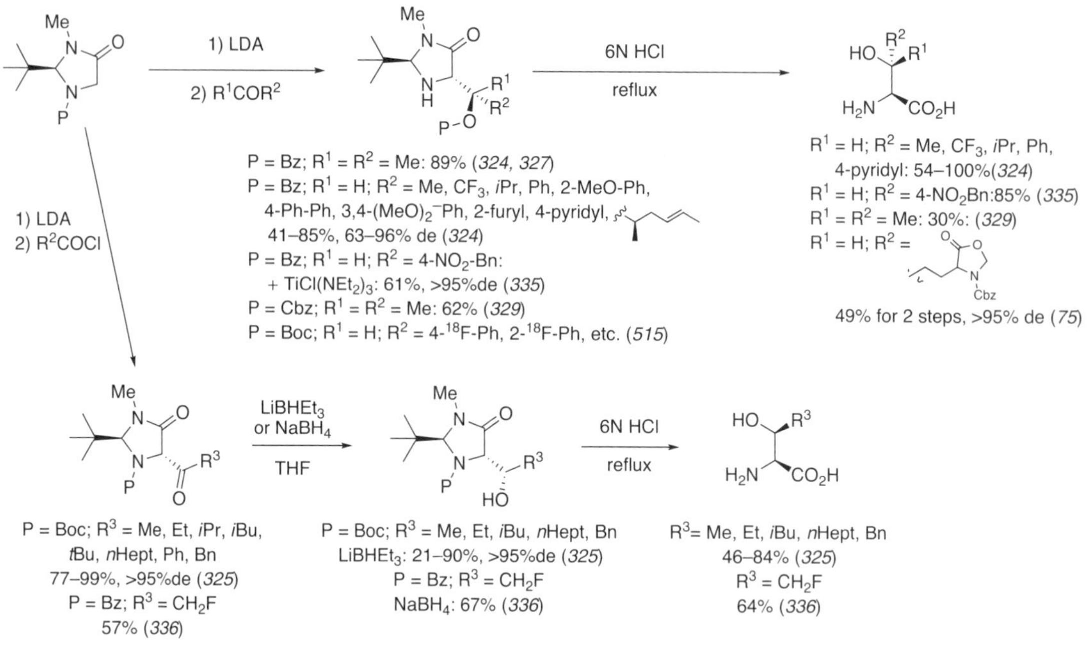

Scheme 7.38 Synthesis of β-Hydroxy Amino Acids with Imidazolidinone Template.

titanium enolate, though this aldol reaction was not accompanied by transacylation (*335*). An alternative route leading to the *allo* β-hydroxyamino acid isomers was achieved by acylation of the imidazolidinone enolate with acyl chlorides, followed by reduction of the carbonyl with LiBHEt$_3$ (*325*). (2*S*,3*S*)-4-Fluoro-Thr was prepared via this route (*336*). Michael additions of the template to α,β-unsaturated esters produced *threo* 3-substituted-Glu derivatives (*337*).

The Boc-BMI nucleophilic synthon has been applied to the synthesis of a variety of amino acids, including 3'-aryl-5'-phosphonomethyl-Phe derivatives (*338*), [3-^{11}C]-Ala, [3-^{11}C]-Phe, and [6-^{11}C]-2-aminoadipic acid (*334, 339*), 2-amino-5-phosphonopentanoic acid (*340*), and heteroaryl alanines (*341*). *trans*-4-Hydroxy-Pro was prepared by allylation of the imidazolidinone enolate, with cyclization to the nitrogen induced by iodine (*342*). The allyl adduct has also been elaborated into 2-amino-4,6-dihydroxypimelic acid (*343*). Aldol reaction of the template with Glu semialdehyde (derived from Glu by reduction of the side chain) gave *threo*-3-hydroxy-2,6-diaminopimelic acid in 49% yield with >95% de (*75*). The template has also been alkylated with 4-boronobenzyl bromide, with the aryl boronate then employed for Suzuki couplings with aryl halides and alkenyl halides. Deprotection gave biarylalanines and other 4'-substituted Phe derivatives without racemization (*344*). A carboranyl amino acid was synthesized using both the Seebach imidazolidinone and Oppolzer's camphor-derived sultam glycine Schiff base derivative (see Section 6.4.2b), with similar yields and enantiomeric purities (*345*). However, another comparison for

the preparation of *o*-carboranylAla found that the imidazolidinone route gave product with 91–96% ee, while the Oppolzer method resulted in >99% ee (*346*).

The Boc- and Cbz-BMI derivatives can be deprotected under milder conditions than the Bz-BMI template. After removal of the *N*-protection, the BMI aminal is heated with an acidic ion exchange resin to open the template (100 °C, 18–92 h), although not all derivatives were stable to this procedure (*329, 331*). Several of the β-hydroxy adducts were deprotected with 6 N HCl at 100 °C to give the free amino acids in 54 to >98% yield (*324*). A synthesis of 2-amino-4-fluoropent-4-enoic acid deprotected the Boc-BMI auxiliary in three steps, with 1 N HCl in MeOH/H$_2$O removing the Boc group and producing the amino acid *N*-methyl amide. Attempts to hydrolyze the amide with 2 N KOH resulted in nearly complete racemization, so a nitrosative deamination procedue was developed. The amine was protected as a phthalimide, followed by treatment with NaNO$_2$/Ac$_2$O/AcOH to produce the methyl ester (*347*).

An alternative imidazolidinone template has been prepared from Gly α-methylbenzylamide instead of Gly methylamide (*340, 348*). This template has two advantages: the chiral amine auxiliary allows the template diastereomers to be resolved, and hydrolysis of the template proceeds under milder conditions. Enantiomerically pure 2-amino-5-phosphonopentanoic acid (*340*) and 2-amino-6-phosphonohexanoic acid (*349*) were synthesized with this derivative, whereas with the BMI template extensive racemization occurred during deprotection (*340*). The methylamine has also been replaced with Gly-OEt, Ala-OMe, or Gly-Gly-OMe as

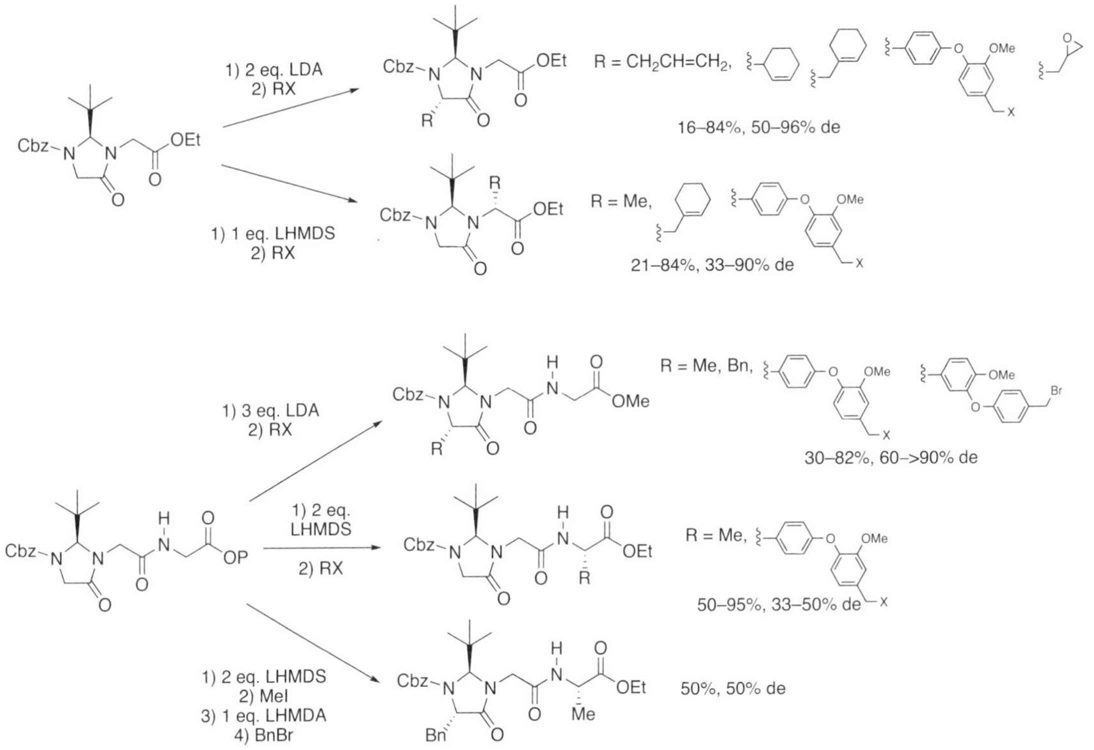

1) 2 eq. LDA
2) RX
Cbz
OEt
R = $CH_2CH=CH_2$,
16–84%, 50–96% de
1) 1 eq. LHMDS
2) RX
R = Me,
21–84%, 33–90% de
1) 3 eq. LDA
2) RX
OMe
R = Me, Bn,
30–82%, 60–>90% de
1) 2 eq. LHMDS
2) RX
OP
R = Me,
50–95%, 33–50% de
1) 2 eq. LHMDS
2) MeI
3) 1 eq. LHMDA
4) BnBr
50%, 50% de

Scheme 7.39 Reaction of Polyglycine Imidazolidinone (*350*).

the Gly amide (*350, 351*). The resulting di- or triGly derivatives were selectively alkylated via the mono-, di-, or trilithiated derivatives, with the Gly-OEt ester residue deprotonated first and the imidazolidinone enolate generated last (see Scheme 7.39) (*350*). Consecutive alkylations at different positions could be carried out with reasonable yields. The Ala residue was used to induce diastereoselectivity during imidazolidinone formation, with a variety of mono- and polyalkylations of the dipeptide derivative examined (*351*).

A number of didehydroamino acid derivatives have been prepared within the Boc-BMI template by dehydration of aldol adducts (*352*) or by Wittig–Horner olefination of an α-dimethylphosphonate derivative (*353*). The alkylidene derivatives were used as substrates for catalytic hydrogenation with H_2/D_2, or for addition of carbenes and organocuprates (see Section 4.3.2a) (*352*). In all cases there was complete selectivity of addition from the face opposite the *tert*-butyl group. However, few of the adducts were deprotected. An ultrasonically induced conjugate addition of alkyl iodide-derived zinc–copper species (*n*PentI, *c*HexI, I*n*PrCO$_2$Me, I*n*PrOH, I*n*dodecylOH, Boc-β-iodo-Ala-OMe) to dehydroAla within the Bz-BMI template under aqueous conditions produced the addition products in 38–95% yield with 44–90% de. The adducts were not deprotected (*354*). The dehydro-Ala template has also been used for radical additions (*290*).

As with other cyclic glycine templates, the imidazolidin-4-one template has been converted into an electrophilic synthon (see Scheme 7.40). Bromination of acylated BMI template (NBS, AIBN) gave the *trans* bromo BMI derivative, which was very reactive and could not be purified (*353, 355*). The brominated imidazolidinone was then used as a substrate for nucleophilic additions. Organocuprates gave poor yields, but malonates, silyl enol ethers, and hetereoatom nucleophiles gave reasonable results (*353*). Most nucleophiles added with inversion, resulting in a *cis*-substituted imidazolidinone with >25:1 diastereoselectivity, but a number of *O*-alkyl, *O*-acyl, and *N*-alkyl nucleophiles gave reduced or reversed selectivity. Deprotection of the adducts was not reported.

The racemic Boc-BMI template can be employed as a substrate for kinetic resolution via enantioselective deprotonation, using a homochiral lithium amide base to effect deprotonation and quenching the enolate with an excess of MeI (*356*). Ala was obtained with up to 80% ee at 75% conversion.

7.10 Other Imidazolidinones and Related Templates

The greatest disadvantage of the imidazolidinone template is the harsh conditions generally employed for

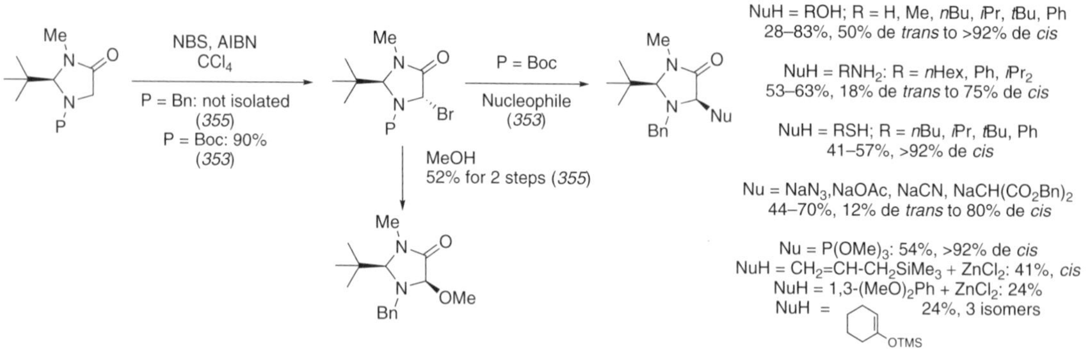

NuH = ROH; R = H, Me, nBu, iPr, tBu, Ph
28–83%, 50% de trans to >92% de cis

P = Bn: not isolated (355)
P = Boc: 90% (353)

NBS, AIBN CCl₄

P = Boc
Nucleophile (353)

MeOH
52% for 2 steps (355)

NuH = RNH₂: R = nHex, Ph, iPr₂
53–63%, 18% de trans to 75% de cis

NuH = RSH; R = nBu, iPr, tBu, Ph
41–57%, >92% de cis

Nu = NaN₃, NaOAc, NaCN, NaCH(CO₂Bn)₂
44–70%, 12% de trans to 80% de cis

Nu = P(OMe)₃: 54%, >92% de cis
NuH = CH₂=CH-CH₂SiMe₃ + ZnCl₂: 41%, cis
NuH = 1,3-(MeO)₂Ph + ZnCl₂: 24%
NuH =   24%, 3 isomers

Scheme 7.40 Synthesis and Reaction of Electrophilic Imidazolidinone.

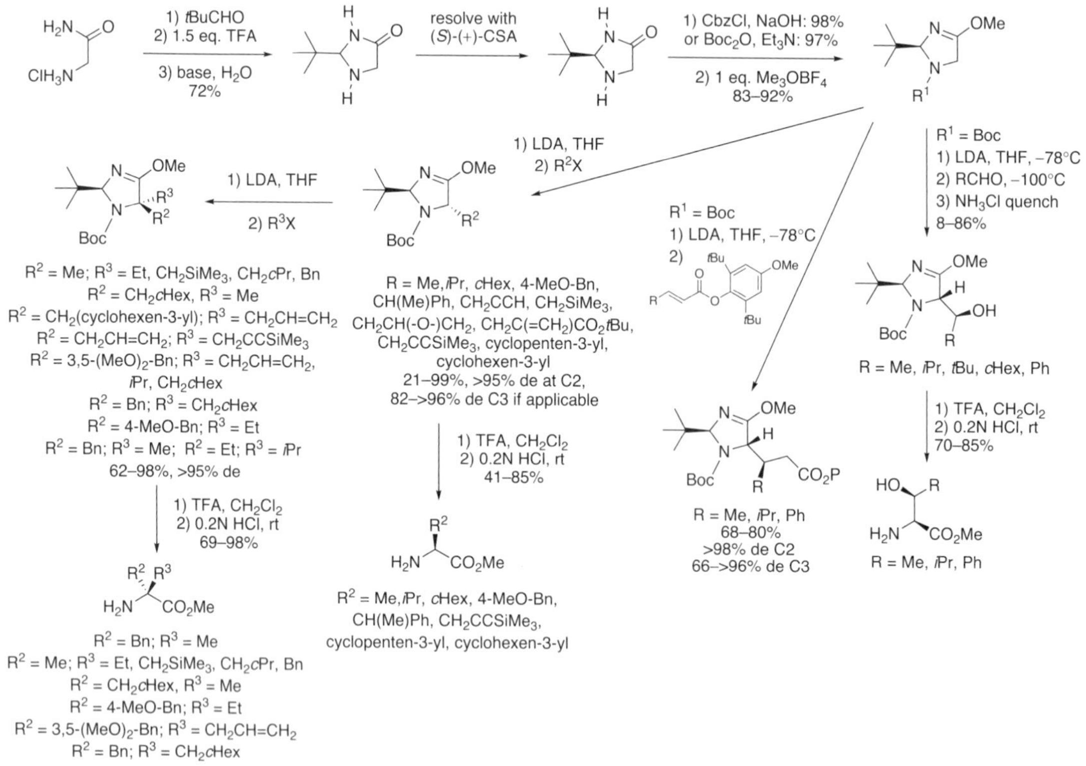

1) tBuCHO
2) 1.5 eq. TFA
3) base, H₂O
72%

resolve with (S)-(+)-CSA

1) CbzCl, NaOH: 98%
or Boc₂O, Et₃N: 97%
2) 1 eq. Me₃OBF₄
83–92%

R¹ = Boc
1) LDA, THF, –78°C
2) RCHO, –100°C
3) NH₄Cl quench
8–86%

1) LDA, THF
2) R²X

1) LDA, THF
2) R³X

R² = Me; R³ = Et, CH₂SiMe₃, CH₂cPr, Bn
R² = CH₂cHex, R³ = Me
R² = CH₂(cyclohexen-3-yl); R³ = CH₂CH=CH₂
R² = CH₂CH=CH₂; R³ = CH₂CCSiMe₃
R² = 3,5-(MeO)₂-Bn; R³ = CH₂CH=CH₂,
iPr, CH₂cHex
R² = Bn; R³ = CH₂cHex
R² = 4-MeO-Bn; R³ = Et
R² = Bn; R³ = Me; R² = Et; R³ = iPr
62–98%, >95% de

R = Me, iPr, cHex, 4-MeO-Bn,
CH(Me)Ph, CH₂CCH, CH₂SiMe₃,
CH₂CH(-O-)CH₂, CH₂C(=CH₂)CO₂tBu,
CH₂CCSiMe₃, cyclopenten-3-yl,
cyclohexen-3-yl,
21–99%, >95% de at C2,
82–>96% de C3 if applicable

R¹ = Boc
1) LDA, THF, –78°C
2) tBu OMe (acrylate ester)
R...

R = Me, iPr, tBu, cHex, Ph

1) TFA, CH₂Cl₂
2) 0.2N HCl, rt
70–85%

HO, R
H₂N   CO₂Me
R = Me, iPr, Ph

R¹ = Boc
1) TFA, CH₂Cl₂
2) 0.2N HCl, rt
41–85%

R = Me, iPr, Ph
68–80%
>98% de C2
66–>96% de C3

1) TFA, CH₂Cl₂
2) 0.2N HCl, rt
69–98%

R² R³
H₂N   CO₂Me
R² = Bn; R³ = Me
R² = Me; R³ = Et, CH₂SiMe₃, CH₂cPr, Bn
R² = CH₂cHex, R³ = Me
R² = 4-MeO-Bn; R³ = Et
R² = 3,5-(MeO)₂-Bn; R³ = CH₂CH=CH₂
R² = Bn; R³ = CH₂cHex

R²
H₂N   CO₂Me
R² = Me, iPr, cHex, 4-MeO-Bn,
CH(Me)Ph, CH₂CCSiMe₃,
cyclopenten-3-yl, cyclohexen-3-yl

Scheme 7.41 Synthesis and Reaction of Dihydroimidazole Template (357, 358).

hydrolysis; the milder deprotection conditions that have been developed for α,α-disubstituted products (323) require several steps. An imidazolidinone lactim ether derivative, 4-alkoxy-2-tert-butyl-2,5-dihydroimidazole-1-carboxylate (BDI), has been prepared from imidazolidinone in a similar fashion to Schöllkopf's bislactim ether and is easily hydrolyzed by 0.2 N HCl (see Scheme 7.41). Initial experiments with a racemic template indicated alkylations and aldol reactions

proceeded with high stereoselectivity, and preliminary reaction with a partially resolved template (87% ee) gave enantiomerically enriched L-valine (81% ee) (357). The Boc- and Cbz-protected O-methyl tert-butyl-substituted dihydroimidazole templates were resolved on Chiracel-OD or Chiralpak-AD chiral columns (284). A more comprehensive report prepared the (R)- or (S)-enantiomers on a multigram scale either by diastereomeric salt formation or by preparative HPLC with a

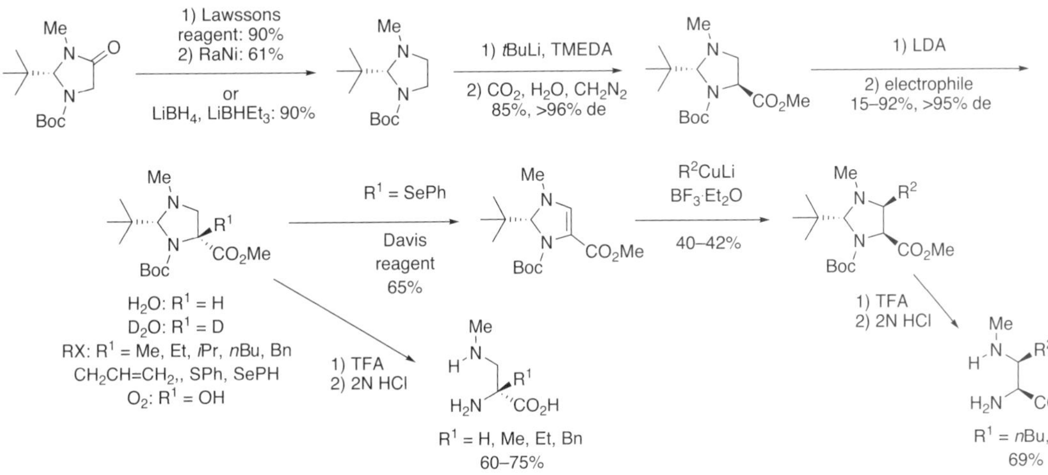

Scheme 7.42 Conversion of Imidazolidinone to Imidazoline (*359*).

chiral column. Deprotonation with LDA and alkylation with a number of primary, secondary, propargylic, allylic or benzylic halides gave diastereopure products, *trans* to the *tert*-butyl group. Products with a chiral β-center, such as those produced from *rac*-1-phenylethylbromide, 3-bromocyclopentene, and 3-bromocyclohexene, had high diastereoselectivity at the β-center (82->96% de), indicating a high degree of enantiomer-differentiating ability during the alkylation. The alkylated templates were converted into amino esters by removing the Boc group under aprotic conditions (trimethylsilyl triflate) and then cleaving the heterocycle with 0.1 N TFA in H₂O (*358*). A second alkylation of the template also proceeded with complete diastereoselectivity to give α,α-disubstituted amino acids, while Michael additions to β-substituted-α,β-unsaturated esters gave products with complete diastereoselectivity at the α-center, and with 66 to >96% de at the β-center. Aldol additions produced *erythro* β-hydroxy amino acids (analogs of *allo*-Thr), rather than the usual *threo* stereochemistry produced by other template aldol reactions (*358*).

The Boc-BMI template has been deoxygenated to give an imidazolidine synthon (see Scheme 7.42). The template was deprotonated and carboxylated, with hydrolysis of the product leading to α,β-diaminopropionic acid (*359*). A second deprotonation followed by alkylation provided α-substituted 2,3-diaminoalkanoic acid derivatives. Formation of an alkene within the imidazolidine template gave an electrophilic synthon, with nucleophilic additions leading to β-substituted derivatives. Unfortunately, racemization was observed during deprotection of the α-unsubstituted products.

Chiral 4-alkyl-2-imidazolin-5-ones, prepared by base-induced cyclization of (*S*)-*N*-phenylethyl-2-isocyano-alanamides, were used for the synthesis of α,α-dialkyl amino acids in 1978 and 1981. They have found little other application as poor diastereoselectivity was observed during alkylation with non-benzylic alkyl

halides (*360, 361*). An imidazolidinone was involved as an intermediate in a synthetic route which used a mercury-catalyzed cyclization to introduce the amino acid amine group to α,β-unsaturated acyl chlorides (see Scheme 5.18) (*362–365*). Another imidazolidinone has been used as a chiral auxiliary for a stereospecific amination using kinetic resolution (see Scheme 5.32) (*366*).

7.11 Cyclic Metal Complexes

Cyclic templates do not have to be purely organic: Belekon et al. have developed chiral Ni(II) Schiff base complexes of amino acids. The Schiff bases are formed with (*S*)-2-[*N*-(benzylprolyl)amino]acetophenone or (*S*)-2-[*N*-(benzylprolyl)amino]benzophenone (BPB) as chiral auxiliaries (*367*). Improved procedures for synthesizing both BPB and the subsequent Ni(II) Schiff base complexes of amino acids were reported in 1998 (*368*) and 2003 (*369*). Condensation of the Gly enolates with aldehydes and ketones followed by hydrolysis gave β-hydroxy amino acid products with 56–72% yield, 70–98% ee, and diastereoselectivity of 10:1 to >50:1 at the carbinol center in favor of the (2*R*,3*S*) (D-*threo*) isomer (see Scheme 7.43, Table 7.14) (*367*). The (2*S*) enantiomer was accessed by simply lowering the pH of the reaction, with the (2*S*)-isomer predominating with <0.01 N CH₃ONa, and the (2*R*)-isomer obtained with >0.1 N CH₃ONa. However, the *threo:erythro* selectivity was greatly reduced at low pH (2:1 for reaction with acetaldehyde). A further study of the stereocontrol of the aldol reaction with aliphatic aldehydes was subsequently published (*370*). The stereochemistry at the α-center was also found to be extremely dependent on reaction time, with the (*S*)-enantiomer obtained initially and the (*R*)-isomer produced after extended periods of time. In contrast to the reversal in selectivity previously obtained by varying the reaction pH, high

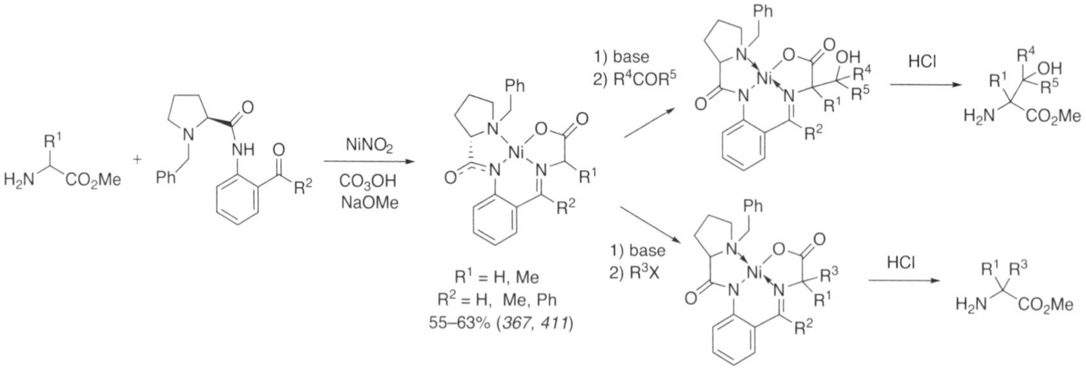

Scheme 7.43 Synthesis and Reaction of Ni(II) Complex of Amino Acid Schiff Base (see Table 7.14).

threo stereoselectivity (>90% de) was maintained with both enantiomers. The Schiff base metal complex also adds to imines, leading to β-substituted-2,3-diamino acids (*371*).

The Belekon procedure has been applied to large-scale syntheses (2–20 g) of enantiomerically pure β-hydroxy amino acids (*370*). Fluorinated phenylserines (*372, 373*) and fluoroalkyl β-hydroxy amino acids (*372*) have also been prepared by this method, but the fluoro-aliphatic aldehydes reacted with a reversal in α-center stereochemistry compared to aliphatic aldehydes (*372*). Trifluoromethyl ketones reacted with surprisingly high diastereoselectivity (>90% de) (*374, 375*). The aldol adducts of formaldehyde and acetaldehyde were dehydrated to give Ni(II) complexes of dehydro-Ala (*376*) or dehydroaminobutanoic acid (*377*), respectively. Nucleophiles were then added by conjugate addition with high diastereoselectivity (see Section 4.3.3), giving *allo*-Thr analogs in the case of dehydroaminobutanoic acid. For example, imidazole, methanol, and ethanol added to a Schiff base complex of dehydro-Abu formed with (*S*)-*N*-(2-benzoylphenyl)-1-(3,4-dichlorobenzyl) pyrrolidyl-2-carboxamide to give the β-substituted amino acids with >80% de at the α-center, and 11–98% de at the β-center (*378*).

The Belekon metal complex aldol addition chemistry was preceded by a similar synthesis of racemic β-hydroxy amino acids previously reported by Japanese workers, who reacted a Cu(II) complex of the pyruvate Schiff base of Gly with a number of aldehydes, obtaining the *threo* diastereomers (*379*). However, 4-form-ylimidazole gave exclusively *erythro* β-hydroxy-His, presumably due to coordination of the imidazole nitrogens (*380*). A pyruvate/Cu(II) complex of glycyl-D-Phe was later used to induce stereoselectivity at the α-center in the synthesis of β-hydroxy-His (50% de) (*381*). Much better enantioselectivity was obtained with an aromatic manganese complex as the chiral auxiliary. The Cu(II) complex of a Gly-Gly dipeptide Schiff base of this ligand reacted with acetaldehyde to give a Thr-Gly dipeptide, with >90% ee and a 2.4:1 preference for the *threo* isomer (*382*). The chiral auxiliary eventually used

for Belekon's Ni(II) complex was first examined as a Cu(II) complex; Gly was alkylated with acetaldehyde with 60–97% ee and 70–90% de *threo* (*383,384*).

The Belekon Ni(II) Gly complex has also been used for alkylations with alkyl halides, reacting in DMF with solid NaOH as base (*385*). Ala, Phe, Val, Trp, Ile, 3',4'-dimethoxy-Phe, and 2-aminohexanoic acid were obtained in 69–83% yield with 74 to >98% ee. The configuration of the substituted Gly within the complex can be deter-mined by NMR (*386*), and the free amino acids are read-ily obtained after acidic hydrolysis. This method was subsequently used to prepare Asp (*387*), [3-^{11}C]-amino acids (*339*), fluorine-containing phenylalanines (*388*), photoactivatable phenylalanine derivatives (*389, 390*), both enantiomers of β,β-diphenylalanine (*391*), 3-(1-pyrenyl)-L-Ala (*392*), and ω-phosphono-α-amino acids (*393*). Alkylation reactions were used to prepare the unsatu-rated amino acids allylglycine, propargylglycine, and 2-amino-tridec-12-enoic acid with 84–96% de. The alkenes were then hydroborated to produce ω-borono-α-amino acids (*394*). Alkylation with racemic *trans*-1-(iodomethyl)-2-nitrocyclopropane led to a mixture of both side-chain epimers of 3-(*trans*-2-nitrocyclopropyl)-Ala (*395*). Alkylations with acetone as solvent gave products with higher asymmetric induction than alkylations in acetonitrile or DMF (*396*). While dichlo-romethane is normally considered a "safe" solvent for alkylation reactions, dichloromethane reacted with 2 equiv of the Belekon Ni(II) chiral Schiff base complex formed from Gly and BPB (using a biphasic mixture with 30% aqueous NaOH in the presence of a phase-transfer catalyst, *n*Bu$_4$NBr for 24 h), resulting in diaste-reomerically pure (2*S*,4*S*)-4-amino-Glu after 24 h. The reaction was proposed to proceed via conjugate addition to an initial dehydro-Ala intermediate, with decomposi-tion of one diastereomer adduct (*397*).

Michael additions of the Gly–metal ion complex to a number of α,β-unsaturated ketones and aldehydes proceeded in the presence of Et$_3$N. Acid-catalyzed decomposition of the complexes was accompanied by cyclization to give dihydropyrrole-2-carboxylic acids, which were then reduced with NaBH$_3$CN to provide

Table 7.14 Synthesis of Amino Acids via Belekon Ni(II) Schiff Base (see Scheme 7.43)

R¹	R²	Base	Electrophile R³X or R⁴COR⁵	Alkylation Yield	Deprotection Yield (or Alkylation + Deprotection)	Reference
H	H Me	NaOMe, MeOH	MeCHO		60–97% ee 70–90% de *threo*	1982 (*383*) 1983 (*384*)
H	Me Ph	NaOMe, MeOH	HCHO MeCOMe PhCHO [3,4-MeO)₂-Ph]CHO		55–72% for 2 steps 72–98% ee (2*R*) >90% de *threo* 78% ee (2*S*) 33% de *threo*	1985 (*367*)
Me	H	*n*BuLi, THF or NaOH, TBA, CH₂Cl₂	R³X; R³ = Me Bn CH₂CH=CH₂	84–91% 12–50% de	75–77%	1985 (*411*)
H	Ph	Et₃N, MeOH	H₂C=CHCO₂Me PhCH=CHCO₂Me H₂C=C(Me)CO₂Me	69–83% 74–>98% de		1986 (*399*)
Me	H Ph	NaOH, DMF or McCN	R³X; R³ = Me Bn 4-BnO-Bn CH₂CH=CH₂		77–88%	1988 (*385*)
H	Ph	NaOH, DMF or McCN	R³X; R³ = Me *i*Pr *n*Bu *s*Bu Bn 3,4-(MeO)₂-Bn CH₂(indol-3-yl)		89–93% for 2 steps 82–87% ee	1988 (*385*)

(*Continued*)

Table 7.14 Synthesis of Amino Acids via Belekon Ni(II) Schiff Base (see Scheme 7.43) (continued)

R¹	R²	Base	Electrophile R³X or R³COR⁵	Alkylation Yield	Deprotection Yield (or Alkylation + Deprotection)	Reference
H	Ph	Et₃N, MeOH	H₂C=CHCHO MeCH=CHCHO PhCH=CHCHO H₂C=CHCOMe H₂C=C(Me)CHO	73–100% >90% de	69–74% >95% ee	1988 (398)
H	Ph	NaOMe, MeOH	HCHO MeCHO	30–92% aldol/elimination	75–93% 50–98% ee	1988 (376) 1990 (377)
			NuH = MeOH, EtOH, PhSH, BnSH, imidazole, Me₂NH, NaCH(CO₂Et)₂, BnMgCl	55–98% Nu: addition		
H	Ph	NaOH, acetone	R³X; R³ = ¹¹CH₃ ¹¹CH₂Ph ¹¹CH₂(4-MeO-Ph)		12–60% radiochemical yield for 2 steps 80–90% ee	1990 (396) 1991 (333)
Me	Ph	NaOMe, MeOH	HCHO	92% 40% de	75–76% >95% ee	1991 (410)
Me	H	NaOH, K₂CO₃ or NaOMe	iPrBr BrCH₂(indol-3-yl) H₂C=CHCO₂Me	75–98% 2–10% de	72–86%	1991 (412)
H	Ph	K₂CO₃, MeCN	BrCH₂CO₂Et	90–100% 25–80% de	71–90% 95–99% ee	1991 (387)
Me						

H	Ph	NaOMe or Et₃N, MeOH	PhCHO (2-F-Ph)CHO (4-F-Ph)CHO (2-CF₃-Ph)CHO (2-CHF₂O-Ph)CHO (4-CHF₂O-Ph)CHO (4-CF₃O-Ph)CHO (3-F-4-MeO-Ph)CHO [3,4,5-(MeO)₃-Ph]CHO C₆F₅CHO (4-NO₂-Ph)CHO (4-MeO-C₆F₄)CHO	70–82% 80–100% ee (2R) 100% de threo	77–95% 60–97% ee 70–90% de threo	1991 (373) 1991 (518) 1992 (519) 1992 (520) 1993 (372)
H	Ph	NaOMe or Et₃N, MeOH	CF_3CHO $H(CF_2)_4CHO$ $H(CF_2)_6CHO$ $F(CF_2)_6CHO$	70–82% 92% ee (2S) 100% de threo	77–95% 60–97% ee 70–90% de threo	1991 (373) 1991 (518) 1992 (519) 1992 (520) 1993 (372)
H	Ph	KOH, DMF	R³X; R³ = 2-F-Bn 3-F-Bn 4-F-Bn 3-CF₃-Bn 2-F-4-Br-Bn 2-F-6-Cl-Bn CH₂C₆F₅	69–84% >90% de	83–95%	1991 (521) 1993 (388)
Me	Ph	KOH, DMF	R³X; R³ = 2-F-Bn 3-F-Bn 4-F-Bn	69–84% >90% de	83–95%	1991 (521) 1993 (388)
H	Ph	KOH, TBAB, MeCN	BrCH₂PO₃Et₂ Br(CH₂)₂PO₃Et₂ Br(CH₂)₂P(O)(Me)OEt ICH₂PO₃iPr₂ H₂C=CHPPO₃Et₂ H₂C=CHP(O)(Me)OEt	30–74% 36–90% de	51–91%	1992 (522)
H	Ph	NaOH, MeCN		93% >95% de	95%	1994 (389)

(Continued)

Table 7.14 Synthesis of Amino Acids via Belekon Ni(II) Schiff Base (see Scheme 7.43) (continued)

R¹	R²	Base	Electrophile R³X or R⁴COR⁵	Alkylation Yield	Deprotection Yield (or Alkylation + Deprotection)	Reference
H	Ph	NaOMe, MeOH	MeCHO nHexCHO nOctCHO iPrCHO iBuCHO tBuCHO	75–99% 82–>96% ee (2S) after 0.5 min, >90% de threo 15–90% ee (2R) after 24h, >90% de threo	57–88%	1995 (370)
H	Ph	NaOMe, MeOH or DBU; MeCN	MeCOCF₃ nBuCOCF₃ nHeptCOCF₃ nOctCOCF₃ PhCH₂CH₂COCF₃ PhCCCOCF₃	56–87% 90–98% de (2S,3S)	87–94%	1996 (374) 1996 (375)
H	Ph	KOH, MeCN	Br(CH₂)₃PO₃Et₂ Br(CH₂)₄PO₃Et₂ Br(CH₂)₅PO₃Et₂ Br(CH₂)₆PO₃Et₂ Br(CH₂)₂O(CH₂)₂PO₃Et₂ 		64–87% for 2 steps	1996 (393)
H	Ph	NaOH, DMF	R³X; R³ = CHPh₂	98% 90% de	60% for 2 steps >99% ee	1997 (391)
H	Ph	Et₃N, LiCl, DMF	CF₃CH=N(4-MeO-Ph)	91% 98% de		1997 (371)
H	Ph	DBU, EtOH	CF₃CH=CHCO₂Et	64–98% 63–94% de		1997 (400)
H	Ph	NaOMe	BrCH₂CH=CH₂ BrCH₂CCH BrCH₂)₆CH=CH₂	76–80% 84–96% de	70–90%	1998 (394)
H	Ph	DBU, EtOH	MeCH=CHCO₂Et CF₃CH=CHCO₂Et	58–70% major (2S,3S)- isomer 3–11% (2S,3R)- isomer	85–88% as pyroglutamate	1999 (401)

H	Ph	MeC(CF$_3$)=CHCO$_2$Et CF$_3$CH=C(MeCO$_2$Et	DBU, EtOH	62–81% major (2S,3S)- isomer	93–95% as pyroglutamate		1999 (402)
H	Ph	RCH=CHCO(3-oxazolidin-2-one) R = Me, Et, nPr, iPr, Ph, 1-naphthyl, 2-naphthyl, 2-MeO-Ph, 3-MeO-Ph, 4-MeO-Ph, 2-CF$_3$-Ph, 3-CF$_3$-Ph, 4-CF$_3$-Ph, 3-indolyl, C$_6$F$_5$, 2,6-F$_2$-Ph, 2-F-Ph, 3,4-F$_2$-Ph, 4-MeO-C$_6$F$_4$, 3,4-Cl$_2$-Ph, 4-NO$_2$-Ph	DBU, EtOH	91–>98% de 100% de relative de C2–C3 2.1:1 to >26:1 (S,S):(R,R)	84–96% for 6 examples		1999 (405) 2001 (404)
H	Ph	RCH=CHCO(4-Ph-3-oxazolidin-2-one) R = Me, Et, nPr, iPr, Ph, Bn, 2-naphthyl, 4-CF$_3$-Ph, 3-CF$_3$-Ph, 2-CF$_3$-Ph, 4-NO$_2$-Ph, 4-MeO-Ph, 3-MeO-Ph, 2-MeO-Ph, 4-F-Ph, 3,5-F$_2$-Ph, 2,6-F$_2$-Ph, 4-Cl-Ph, 3,4-Cl$_2$-Ph, 4-Br-Ph, indol-3-yl	DBU, EtOH	>94% de use achiral Schiff base complex; chiral auxiliary on alkene electrophile	98–99% for 19 examples		2000 (406) 2000 (408) 2004 (407)
H	Ph	[structure: Br–CH$_2$–aryl with diazirine N=N–CF$_3$] R = OMe, CH$_2$CH$_2$NHBoc, (CH$_2$CH$_2$O)$_3$NHBoc	NaOH, MeCN	68–75% >95% de	65–72%		2002 (390)
H	Ph	ICH$_2$(2-NO$_2$-cPr) (racemic or chiral)	1) NaH, DMF/ MeCN 2) 60% aq. AcOH	84–85%	—		2003 (395)
H	Ph	CH$_2$CH(NH$_2$)CO$_2$H	30% NaOH, CH$_2$Cl$_2$, nBu$_4$NBr	>95% de of (S,S) product	—		2004 (397)
H	Ph	BrCH$_2$(1-pyrenyl)	KOH, DMF	89%	—		2004 (392)
H	Ph	RCH=CHCO(pyroglutamate) R = Me, Et, nPr, iPr, Ph, 4-CF$_3$-Ph, 4-MeO-Ph, 2,3,4,5,6-F$_5$-Ph	DBU, DMF	>88% de use achiral Schiff base complex; chiral auxiliary on alkene electrophile	—		2004 (409)

substituted prolines (*398*). Addition of the Schiff base complex to substituted methyl acrylates gave β- or γ-substituted Glu derivatives. The adducts were obtained with high (2*S*)-selectivity, while the (2*S*,3*R*)- or (2*S*,4*R*)-isomers were preferentially obtained with approximately 2:1 selectivity (*399*). 3-Trifluoromethylpyroglutamate was obtained via addition to 4,4,4-trifluorocrotonate; DBU/EtOH as base resulted in up to 94% de under kinetic conditions, if the reaction was quenched in 30 s (*400*). Subsequent reports prepared (2*S*,3*S*)-3-methyl-Glu, 3-trifluoromethyl-Glu, 3-methyl-3-trifluoromethyl-Glu, and (2*S*,3*S*,4*R*)-3-trifluoromethyl-4-methyl-Glu as the major isomers, isolated as the pyroglutamates (*401, 402*). An achiral version of this reaction with greatly improved diastereoselectivity has been reported, using oxazolidinone amides of α,β-unsaturated acids as the electrophile (*403*). The improved diastereoselectivity and reactivity of oxazolidinone amide electrophiles was then successfully transferred to the asymmetric reaction, producing only two of the four possible diastereomers, (2*S*,3*S*), and (2*R*,3*R*), in ratios of 2.4:1 to >26:1 (*404, 405*). By using a chiral oxazolidinone auxiliary as the acrylate amide in combination with an achiral ketimine Ni(II) complex formed from Gly and the 2-acetyl-phenyl amide of pyridine-2-carboxylic acid, much higher stereoselectivity of >94% de was obtained for19 examples (*406–408*). These were elaborated into β-substituted pyroglutamic acids and prolines (*407*). Much less expensive pyroglutamate could be employed as the chiral auxiliary instead of Evans' oxazolidinone, still providing high diastereoselectivity with an achiral Gly Schiff base (*409*). A chiral Ni(II) complex was also combined with the chiral oxazolidinone auxiliary on the acrylate electrophile (*408*).

The corresponding Ala Ni(II) Schiff base complex was also applied to aldol reactions (*410*) or alkylated with alkyl halides or methyl acrylate (*385, 387, 411, 412*), producing α-methyl amino acids with up to 86% ee. Fluorinated benzylic halides reacted with even better stereoselectivity (*388*). The steric bulk of the Schiff base chiral auxiliary must be reduced for alkylations with

bulky electrophiles (*412*). A Ni(II) complex was used in the synthesis of L-vinylglycine from racemic Met, with the critical step of asymmetric induction ocurring during enantioselective incorporation of the L-Met Schiff base into the complex accompanied by simultaneous racemization of the remaining D-enantiomer in solution. The diastereomerically pure complex was then converted to vinylglycine by oxidation of the sulfur followed by pyrolytic elimination and hydrolysis, but the free amino acid had only 58% ee (*377*).

A number of other chiral Schiff base auxiliaries have been used in combination with metal ion complexes (see Scheme 7.44). Zn(II) and Cu(II) complexes of (1*R*)-3-hydroxymethylenebornan-2-one glycine imine were used for diastereoselective aldol reactions with benzaldehyde, but gave phenylserine product with a maximum of only 27% ee (*413*). Other aldehydes were examined and also gave products with poor enantiomeric purity (*414*). A ketone possessing planar chirality, 4-formyl-5-hydroxy[2.2]paracyclophane, has also been employed as the chiral auxiliary. The Cu(II) complex of the Gly Schiff base reacted with isobutyraldehyde or benzaldehyde in the presence of NaOMe to give, after hydrolysis, β-hydroxy amino acids with 77–98% ee and 20:1 to 58:1 *threo* selectivity. The Schiff base enolate derived from Ala-OEt and LDA in THF could be *C*-alkylated with BnBr in 80% de; alkylation of the paracyclophane hydroxy group reversed (and lowered) the diastereoselectivity. Alkylation of Phe-OEt with MeI proceeded with up to 71% de (*415, 416*).

Achiral Ni(II) Gly Schiff base complexes can be alkylated under PTC conditions, using a chiral metal complex of a Pro derivative as the interfacial catalyst, but asymmetric induction was poor (3–21%) (*417*). Better results were reported by Belokon et al. in 2001 and 2003, with asymmetric PTC alkylations of two achiral Schiff base Ni complexes formed with Gly or Ala (*418, 419*). A number of catalysts were tested for benzylation of a ketimine Ni(II) complex formed from Gly and the 2-benzoyl-aniline amide of pyridine-2-carboxylic acid (PABP), with the best enantioselectivity

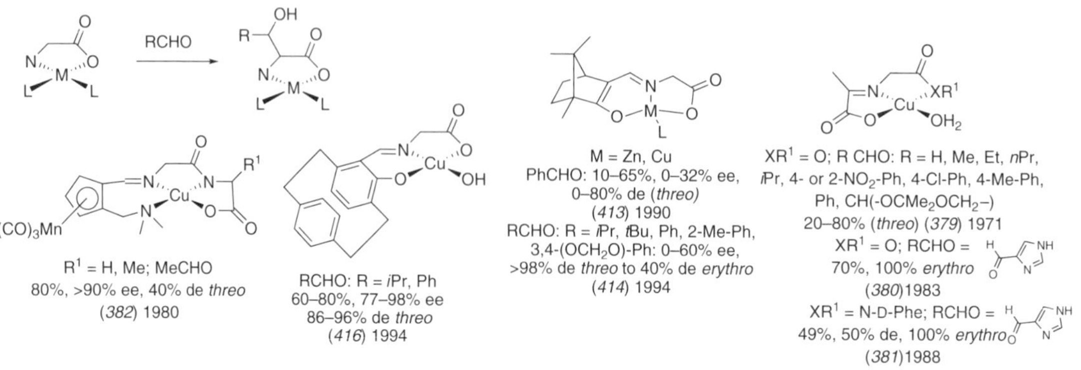

Scheme 7.44 Aldol Reactions of Other Glycine Schiff Base Metal Complexes.

(97% ee) provided by (R)- or (S)-NOBIN (the mono-amino analog of binaphthol) in combination with NaOH in CH$_2$Cl$_2$. A number of other benzyl and alkyl halides reacted with variable yields (10–92%) and excellent enantioselectivity (>90% ee, with the exception of *tert*-butyl bromoacetate) (see Scheme 7.45). Of note is the rapidity of the reaction: yields of up to 90% were obtained in under 10 min at room temperature for benzylic and allylic halides. Crystallization of the alkylated complex generally improved the product ee. The complexes were decomposed by treatment with diluted methanolic HCl, accompanied by a solution color change from red to blue (*418*). Benzylation of the analogous complex prepared from Ala gave the α-methyl-Phe product, but with only 40% ee. Asymmetric Michael additions were also tested, but NOBIN was not an effective catalyst. In contrast, acylated iso-NOBIN derivatives gave up to 96% ee for Gly additions to methyl acrylate, but only 39% ee for addition of the Ala complex (*418, 420*). An efficient large-scale synthesis of the Gly Ni(II) complexes of the 2-benzoylaniline or 2-acetylaniline amides of pyridine-2-carboxylic acid (PABP and PAAP) was reported in 2003 (*421*).

Chiral metal complexes of amino acid Schiff bases have been employed to resolve amino acids via preferential diastereomer formation, accompanied by simultaneous racemization of the less-favored enantiomer. The method was initially described by Belekon et al. (*384*) and used in a synthesis of allylglycine from racemic Met (*377*), but was more recently modified, with greater success (up to 99% ee with 83% yield), by De and Thomas (*422*).

Other complexes have been formed using electrophilic imines of α-ketocarboxylic acids, rather than nucleophilic Schiff base derivatives of amino acids. Palladium(II)- and rhodium(III)-complexes of imines and oximes of α-oxocarboxylates, prepared using chiral amine components, were catalytically hydrogenated or reduced with NaBH$_4$, but diastereoselectivity was poor (0–40%) (*423*). Chiral cobalt(III) complexes of Gly have been employed for amino acid synthesis, with the complex itself possessing chirality instead of a separate chiral auxiliary. An electrophilic Gly equivalent was prepared by oxidation of the Gly α-carbon to an imine; the equivalent sarcosine and Ala derivatives were also synthesized (*424*) (see Scheme 7.46). Dicarbonyl carbanionic reagents added with some stereoselectivity, which was greatly increased by a fortuitous precipitation of one diastereomer combined with epimerization of the remaining compound in the basic solution. Achiral cobalt complexes were previously used to form imines of α-keto acids, which were then reduced or used as electrophiles (*425*). However, chiral complexes of the imines of pyruvate and phenylpyruvate showed little diastereoselectivity in reductions with NaBH$_4$ (*426*). Nucleophiles such as nitromethane anion added with some stereoselectivity (50–68% de), and could be reduced to give α-(aminomethyl)-Ala.

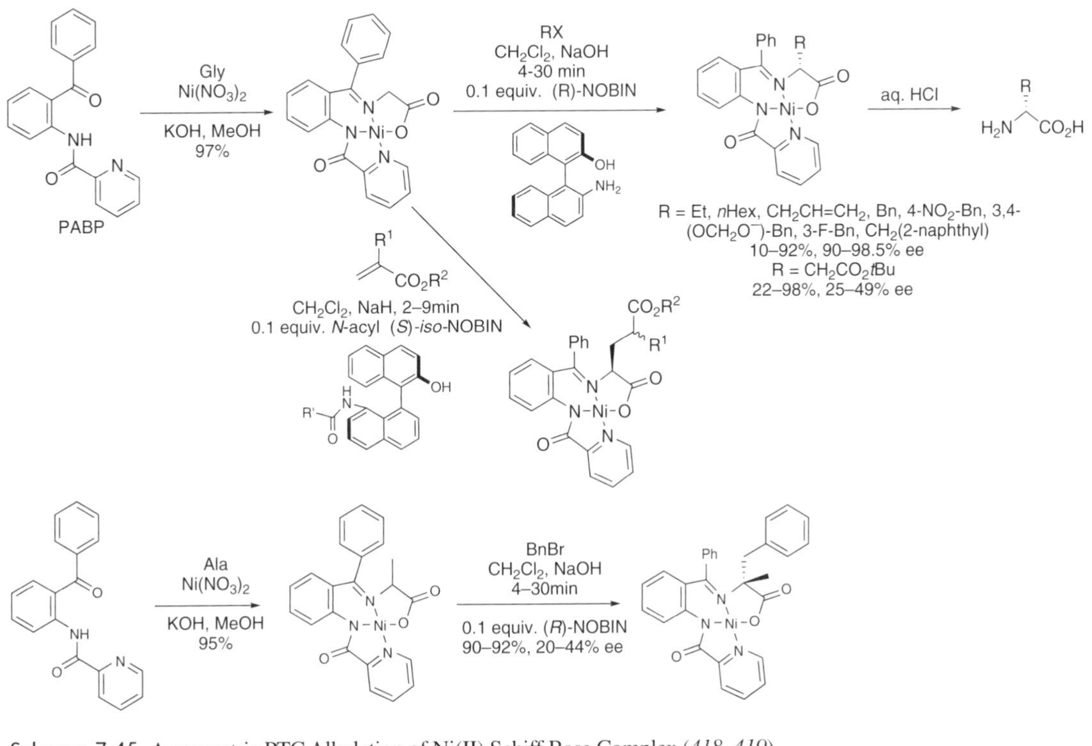

Scheme 7.45 Asymmetric PTC Alkylation of Ni(II) Schiff Base Complex (*418, 419*).

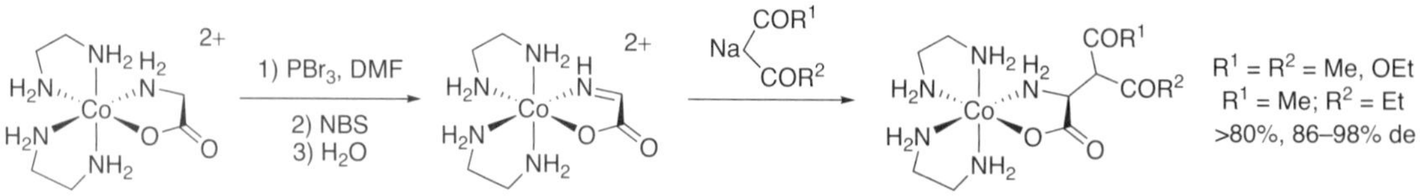

Scheme 7.46 Co(III) Complex of Glycine Imine (*424*).

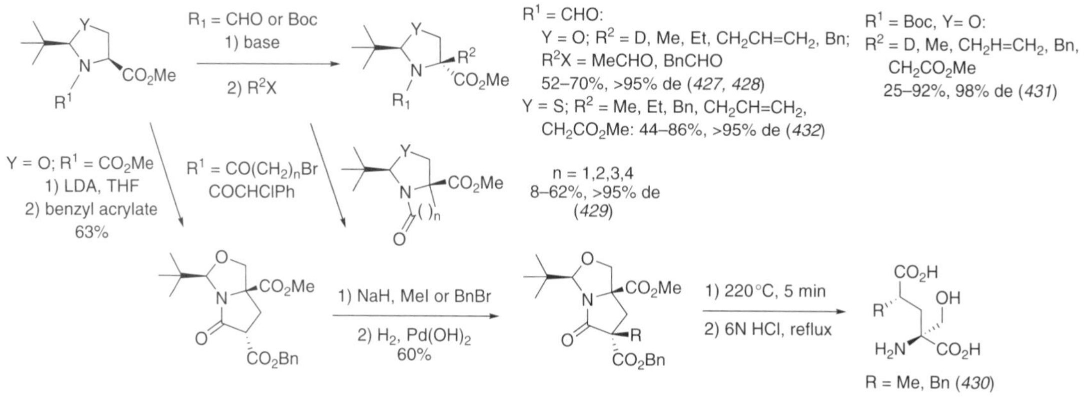

Scheme 7.47 Alkylation of Serine or Cysteine-Derived Oxazolidine/Thiazolidine.

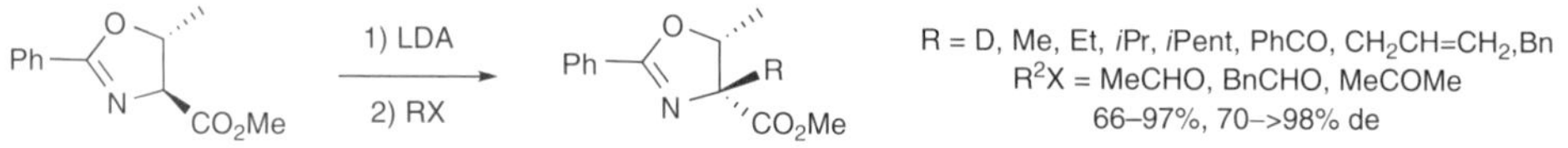

Scheme 7.48 Alkylation of Threonine-Derived Oxazoline (*427, 441*).

7.12 Other Cyclic Chiral Templates

A variety of other cyclic systems have been used for amino acid synthesis, mainly for preparing α-alkyl analogs. Seebach has used the "self-reproduction of chirality" principle to produce α-alkylated Ser from an *N*-formyl Ser oxazolidine (see Scheme 7.47) (*427, 428*), and this was applied to an intramolecular alkylation using an *N*-bromoacyl substituent (*429*). An *N*-methoxycarbonyl Ser oxazolidine underwent Michael addition to ethyl acrylate to give α-hydroxyethyl-Glu via a bicyclic intermediate; the Glu side chain could be γ-alkylated (*430*). Both *cis* and *trans* diastereomers of *N*-Boc Ser oxazolidine were alkylated with high diastereoselectivity in 2004 (*431*). Alkylation of the corresponding thiazolidine gave α-alkylated Cys (*432*). A template derived by cyclization of the α-amino and γ-amide nitrogens of Asn was used to prepare α-alkyl-Asn/Asp derivatives (*433*). A similar tetrahydropyrimidinone template was employed to prepare α-methyl-Asn (*434*).

Oxazoline-4-carboxylates can be obtained via cyclization of the side chain of β-hydroxy amino acids. A number of groups have used them to interconvert the configuration of the β-center, producing *allo*-The equivalents from Thr and vice-versa (*435–440*). However, the Bz-Thr-OMe-derived oxazoline has been deprotonated and alkylated at the α-center with a number of electrophiles, producing α-alkyl-threonines (see Scheme 7.48) (*427, 441*). Similarly, the oxazoline formed by treatment of *N*-4-methoxybenzoyl Thr-OMe with *p*-TsOH was deprotonated with LDA and alkylated with chloromethyl benzyl ether to give a 69% yield of protected α-hydroxymethyl-Thr (*442*).

Oxazolines have also been prepared from chiral β-amino alcohols (*443*) and used as protected chiral carboxyl groups, allowing for α-alkylation in the synthesis of ω-amino acids (see Section 11.4.3p) (*444*). Similar 2-substituted oxazolines can be converted into morpholin-2-one via a SeO$_2$-promoted oxidative rearrangement and imine reduction, leading to amino acids (see Section 7.6) (*210*). Homochiral oxazolines have been used as chiral carboxyl equivalents on a Gly synthon (see Scheme 7.49). Alkylation of *N*-benzyl,*N*-carbamate protected derivatives proceeded with reasonable asymmetric induction (40–92% de) compared to *N*,*N*-dimethyl or dibenzyl derivatives (0–18% de). The carboxyl group was not regenerated (*445*).

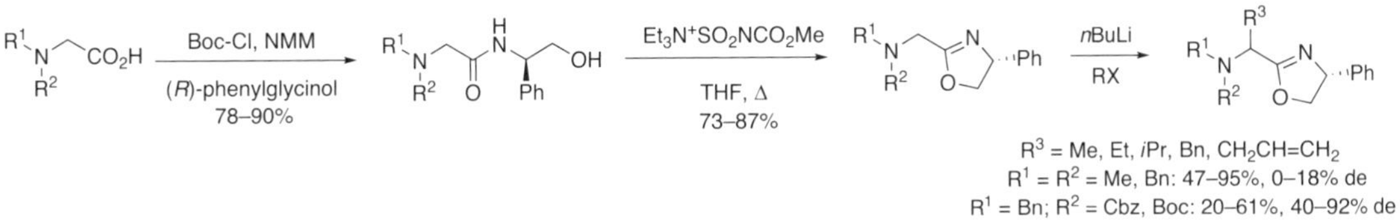
Boc-Cl, NMM
(R)-phenylglycinol
78–90%
Et3N+SO2NCO2Me
THF, Δ
73–87%
nBuLi
RX
R3 = Me, Et, iPr, Bn, CH2CH=CH2
R1 = R2 = Me, Bn: 47–95%, 0–18% de
R1 = Bn; R2 = Cbz, Boc: 20–61%, 40–92% de

Scheme 7.49 Chiral Carboxyl Equivalent (*445*).

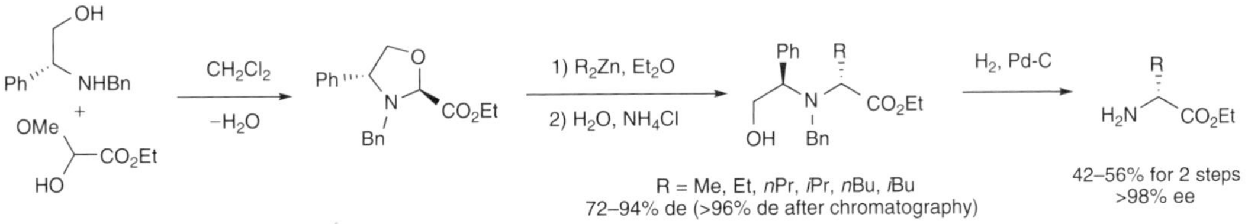
CH2Cl2
–H2O
1) R2Zn, Et2O
2) H2O, NH4Cl
H2, Pd-C
R = Me, Et, nPr, iPr, nBu, iBu
72–94% de (>96% de after chromatography)
42–56% for 2 steps
>98% ee

Scheme 7.50 Synthesis and Ring Opening of Oxazolidine (*449*).

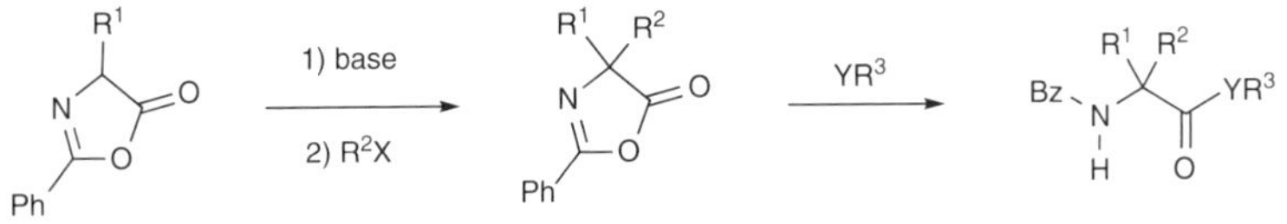
1) base
2) R2X
YR3

Scheme 7.51 Alkylation of Oxazolone (see Table 7.15).

Table 7.15 Synthesis of Amino Acids via Alkylation of Oxazolone (see Scheme 7.51)

Base	R^1	R^2	Alkylation Yield	R^3	Oxazolone Opening Yield	Reference
pyridine	$(CH_2)_3NHBoc$	CH_2OH		OH	70%	1978 (*523*)
$i\mathrm{Pr}_2\mathrm{NEt}$	*i*Pr	$CH_2CH{=}CH_2$	32–96%	OH	25–72%	1979 (*451*)
	*i*Pr	CH_2CCH		AcOH, HCl		
	*i*Pr	Bn				
	*i*Pr	2-NO_2-Bn				
	*i*Pr	4-NO_2-Bn				
	*i*Pr	CH_2COPh				
	Bn	Me				
	Bn	Et				
	Bn	$CH_2CH{=}CH_2$				
	Bn	CH_2CCH				
	Bn	CH_2CO_2Et				
	H	CPh_3				
Et_3N	Me	$CH{=}CH_2$		OMe	50–60% for	1980 (*453*)
	*i*Bu	from $PhSO_2CCH$		MeOH, Et_3N	3 steps	
	$(CH_2)_2CO_2H$					
	$(CH_2)_4NHBz$					
$i\mathrm{Pr}_2\mathrm{NEt}$	Ph	$(CH_2)_2OTHP$		$HNMe_2$	32–40% for	1985 (*452*)
		$(CH_2)_3OTHP$			2 steps	
		$(CH_2)_{12}OTHP$				
$i\mathrm{Pr}_2\mathrm{NEt}$	Me	*i*Pr	80–85%	Phe-NMe_2	64–80%	1992 (*454*)
	Me	Ph		Phe-NH*c*Hex	resolve	
	Me	Bn				
	$CH_2CH{=}CH_2$	Ph				

(Continued)

Table 7.15 Synthesis of Amino Acids via Alkylation of Oxazolone (see Scheme 7.51) (continued)

Base	R^1	R^2	Alkylation Yield	R^3	Oxazolone Opening Yield	Reference
iPr$_2$NEt	H converted into SiPr from iPrS-succinimide converted into Cl with SO$_2$Cl$_2$	SiPr from iPrS-succinimide Cl from SO$_2$Cl$_2$	71%	Gly-OMe Ala-OMe	69–72%	1993 (524)
pyridine	CO$_2$Et forms R^2 = H after deprotection	Ph 2-F-Ph 4-F-Ph 4-CF$_3$-Ph 2-MeO-Ph 4-MeO-Ph from ArPb(OAc)$_3$		OH NaOH, EtOH	75–94% for 2 steps (generates R^2 = H)	1993 (525)
NaH, DMF	(CH$_2$)$_2$CO$_2$$t$Bu (CH$_2$)$_2CO_2$$t$Bu CH$_2CO_2$$t$Bu iBu iBu Me Me Ph Ph iPr iPr iPr Bn Bn Bn Bn Bn Bn	Me CH$_2$CO$_2$Bn Me CH$_2$CO$_2$$t$Bu CH$_2$I 4-MeO-Bn CH$_2$I Me 4-MeO-Bn Me 4-MeO-Bn CH$_2$I Me 4-MeO-Bn CH$_2$CO$_2$Et (CH$_2$)$_2$Br CH$_2$CO$_2$$t$Bu CH$_2$I	30–88%			1994 (456)
Et$_3$N	Me Ph	CH$_2$CH$_2$PPPh$_3$Br treat with MeONa, (4-NO$_2$-Ph)CHO, Δ to form CH$_2$CH=CH(4-NO$_2$-Ph)	65%	OH or OMe H$_3$O$^+$Br- or MeOH	67–87%	1995 (458)
NaH, DMF	Me iPrBn CH$_2$CO$_2$$t$Bu	CH$_2$I, CH$_2$CO$_2$Et	35–80%	Phe-NHcHex	75–83% resolve	1996 (526)
NaH, DMF	Me iPr iBu Bn	CH$_2$I convert into CH$_2$OH	40–75%	Phe-NHcHex	94% resolve	1996 (527)
LDA	Me Me Me Me Me Bn Bn Bn (CH$_2$)$_2$SMe iBu CH$_2$CH=CH$_2$ iPr	CH(Ph)CH=CH$_2$ CH(2-thienyl)CH=CH$_2$ CH(2-furyl)CH=CH$_2$ CH[2,4-(MeO)$_2$-Ph] CH=CH$_2$ CH(2-Br-Ph)CH=CH$_2$ CH(Ph)CH=CH$_2$ CH(2-thienyl)CH=CH$_2$ CH[2,4-(MeO)$_2$-Ph] CH=CH$_2$ CH(Ph)CH=CH$_2$ CH(Ph)CH=CH$_2$ CH(Ph)CH=CH$_2$ CH(Ph)CH=CH$_2$	76–92% allylation 85–99% ee 92–>96% de	—	—	2002 (460)

Optically pure β-amino alcohol-derived oxazolidines have been used as chiral auxiliaries for the Hegedus chromium carbene complex (see Section 5.4.1a) (446), as well as substrates for organocerium (447) or organostannane (448) addition leading to β-amino acids (see Scheme 11.57). An electrophilic Gly synthon was prepared by the condensation of phenylglycinol with the methyl hemiacetal of ethyl glyoxylate. The 2-ethoxycarbonyl-1,3-oxazolidine was stereoselectively opened by dialkylzinc reagents with 72–94% de, providing α-amino acid ethyl esters in 42–56% overall yield and >98% ee after chromatography and hydrogenation (see Scheme 7.50) (449).

The alkylation of racemic oxazol-5-ones (azlactones), prepared from the corresponding N-benzoyl amino acids, was originally used by Kaminski et al. to prepare α-hydroxymethyl-α-amino acids in 1973 (450), and was again reported by Steglich in 1979 (451). Hydrolysis yielded α-alkyl-α-amino acids (451, 452), including α-alkyl-vinylglycine (see Scheme 7.51, Table 7.15) (453). In recent years the alkylation of oxazolones has been coupled with resolution via formation of a dipeptide with Phe amides (454, 455). More reliable alkylation conditions were reported in 1994 (456), using NaH instead of DIEA as base.

The improved alkylation and resolution conditions were used to prepare a number of α-substituted Ser and Asp derivatives (457). The oxazolones of Phe and Ala have been used for Michael additions to the vinyl group of triphenylvinylphosphonium bromide. The resulting α-substituted oxazolone derivatives contained a Wittig reagent on the side chain that was then condensed with aldehydes (458).

Azlactones (oxazolones) are also intermediates in the synthesis of N-acyl dehydroamino acids from the condensation of N-acylGly and aldehydes, which is known as the Erlenmeyer reaction (see Section 4.2.2). 2-Alkoxy-5(4H)-oxazolones are often obtained from N-urethane-protected amino acids under the activating conditions used for peptide bond formation (459). In 2002 a chiral Mo complex-catalyzed asymmetric allylic alkylation of oxazolones led to a number of α-substituted-β-aryl-allylglycines with 85–97% ee and 92 to >96% diastereoselectivity at the β-center (460).

A protected 2-hydroxypyrrole, prepared from pyrrole (461, 462), was used as an achiral glycine anion equivalent by addition to chiral aldehydes in the presence of a Lewis acid to give polyhydroxy amino acids (see Scheme 7.52) (463, 464). C2/C3 threo isomers were obtained by catalysis with $SnCl_4$, while

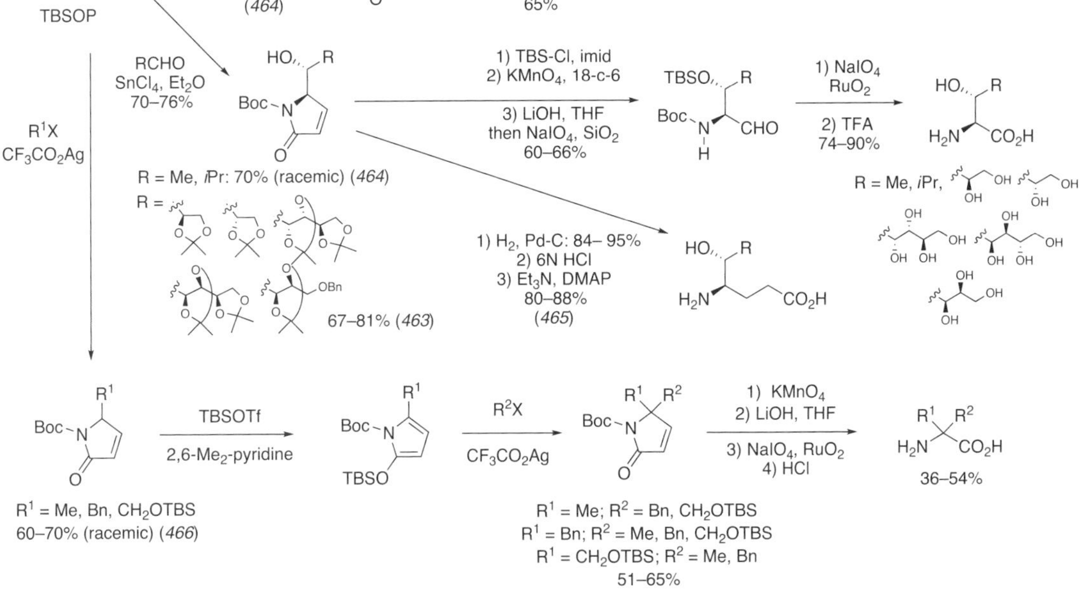

Scheme 7.52 Alkylation of Pyrrole (461–466).

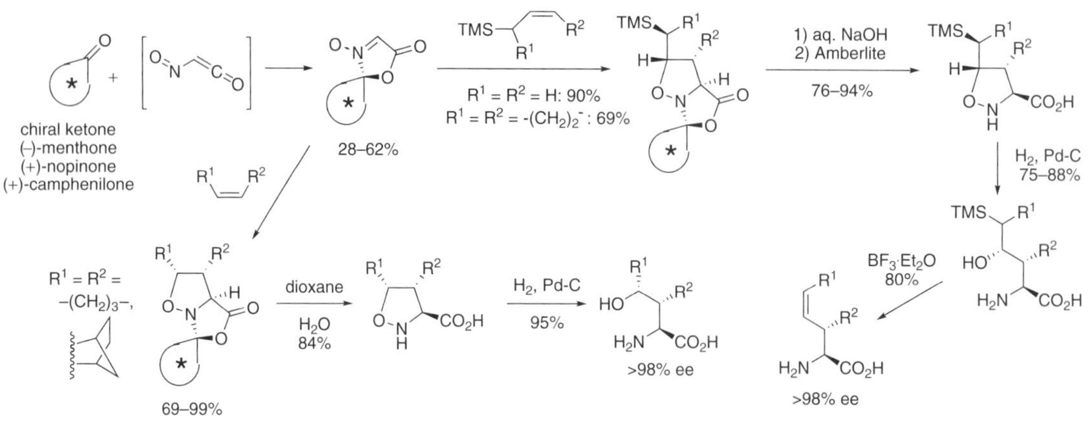

Scheme 7.53 Cycloaddition of Cylic Nitrone (*468–470*).

BF$_3$·Et$_2$O provided *erythro* diastereomers (which were also obtained by base-catalyzed epimerization of the initial *threo* adducts) (*463*). The pyrrole template has also been applied to the synthesis of polyhydroxylated γ-aminobutanoic acids (*465*). Consecutive alkylations of the siloxypyrrole substrate at the γ-position, with subsequent oxidative cleavage of the α,β-unsaturated lactam alkene, provided racemic α,α-disubstituted-α-amino acids (*466*).

Gly residues, with the Gly amine forming part of a chiral β-lactam, have been alkylated and sequentially dialkylated with very good diastereoselectivity. The β-lactam was cleaved to give a dipeptide product (see Section 8.5) (*467*).

A cyclic chiral nitrone corresponding to an imine of pyruvic acid was prepared by cycloaddition of chiral cyclic ketones with nitrosoketene. The nitrones were employed for 1,3-dipolar cycloadditions with electron-rich olefins to give oxazolidine derivatives, which were converted into optically pure unsaturated β-hydroxy amino acids (see Scheme 7.53) (*468–470*).

References

1. Huang, K.; Breitbach, Z.S.; Armstrong, D.W. *Tetrahedron: Asymmetry* **2006**, *17*, 2821–2832.
2. Easton, C.J. *Chem. Rev.* **1997**, *97*, 53–82.
3. Nájera, C. *Synlett* **2002**, 1388–1403.
4. Schöllkopf, U. *Top. Curr. Chem.* **1983**, *109*, 65–84.
5. Schöllkopf, U. *Tetrahedron* **1983**, *39*, 2085–2091.
6. Schöllkopf, U. *Pure Appl. Chem.* **1983**, *55*, 1799–1806.
7. Bopp, F. *Liebigs Ann. Chem.* **1849**, *69*, 16–37.
8. Pickenhagen, W.; Dietrich, P.; Keil, B.; Polonsky, J.; Nouaille, F.; Lederer, E. *Helv. Chim. Acta* **1975**, 1078–1086.
9. Schöllkopf, U.; Hartwig, W.; Groth, U. *Angew. Chem., Int. Ed. Engl.* **1979**, *18*, 863–864.
10. Schöllkopf, U.; Hartwig, W.; Groth, U.; Westphalen, K.-O. *Liebigs Ann. Chem.* **1981**, 696–708.
11. Schöllkopf, U.; Hartwig, W.; Groth, U. *Angew. Chem., Int. Ed. Engl.* **1980**, *19*, 212–213.
12. Schöllkopf, U.; Groth, U.; Hartwig, W. *Liebigs Ann. Chem.* **1981**, 2407–2418.
13. Schöllkopf, U.; Bardenhagen, J. *Liebigs Ann. Chem.* **1987**, 393–397.
14. Kobayashi, S.; Matsumura, M.; Furuta, T.; Hayashi, T.; Iwamoto, S. *Synlett* **1997**, 301–303.
15. Groth, U.; Schöllkopf, U.; Chiang, Y.-C. *Synthesis* **1982**, 864–865.
16. Schöllkopf, U.; Westphalen, K.-O.; Schröder, J.; Horn, K. *Liebigs Ann. Chem.* **1988**, 781–786.
17. Tolstikov, G.A.; Kresteleva, I.V.; Spivak, A.Y.; Fatykhov, A.A.; Sultanmuratova, V.R. *Russ. Chem. Bull.* **1993**, *42*, 557–563.
18. Schöllkopf, U.; Schröder, J. *Liebigs Ann. Chem.* **1988**, 87–92.
19. Schöllkopf, U.; Groth, U.; Westphalen, K.-O.; Deng, C. *Synthesis* **1981**, 969–971.
20. Schöllkopf, U.; Busse, U.; Lonsky, R.; Hinrichs, R. *Liebigs Ann. Chem.* **1986**, 2150–2163.
21. Schöllkopf, U.; Groth, U. *Synthesis* **1983**, 37–38.
22. Groth, U.; Chiang, Y.-C.; Schöllkopf, U. *Liebigs Ann. Chem.* **1982**, 1756–1757.
23. Andrei, M.; Römming, C.; Undheim, K. *Tetrahedron: Asymmetry* **2004**, *15*, 1359–1370.
24. Andrei, M.; Römming, C.; Undheim, K. *Tetrahedron: Asymmetry* **2004**, *15*, 2711–2717.
25. Liu, W.; Ray, P.; Benezra, S.A. *J. Chem. Soc., Perkin Trans. 1*, **1995**, 553–559.
26. Sano, S.; Liu, X.-K.; Takebayashi, M.; Kobayashi, Y.; Tabata, K.; Shiro, M.; Nagao, Y. *Tetrahedron Lett.* **1995**, *36*, 4101–4104.
27. Sano, S.; Miwa, T.; Liu, X.-K.; Ishii, T.; Takehisa, T.; Shiro, M.; Nagao, Y. *Tetrahedron: Asymmetry* **1998**, *9*, 3615–3618.
28. Hünig, S.; Klaunzer, N.; Wenner, H. *Chem. Ber.* **1994**, *127*, 165–172.

29. Schöllkopf, U.; Groth, U.; Deng, C. *Angew. Chem., Int. Ed. Engl.* **1981**, *20*, 798–799.

30. Rose, J.E.; Leeson, P.D.; Gani, D. *J. Chem. Soc., Perkin Trans. 1* **1995**, 157–165.

31. Allen, M.S.; Hamaker, L.K.; La Loggia, A.J.; Cook, J.M. *Synth. Commun.* **1992**, *22*, 2077–2102.

32. Zhang, P.; Liu, R.; Cook, J.M. *Tetrahedron Lett.* **1995**, *36*, 7411–7414.

33. Bull, S.D.; Davies, S.G.; Moss, W.O. *Tetrahedron: Asymmetry* **1998**, *9*, 321–327.

34. Carlsson, A.–C.; Jam, F.; Tullberg, M.; Pilotti, Å.; Ioannidis, P.; Luthman, K.; Grøtli, M. *Tetrahedron Lett.* **2006**, *47*, 5199–5201.

35. Schöllkopf, U.; Hartwig, W.; Pospischil, K.–H.; Kehne, H. *Synthesis* **1981**, 966–969.

36. Groth, U.; Schmeck, C.; Schöllkopf, U. *Liebigs Ann. Chem.* **1993**, 321–323.

37. Andrews, P.C.; Maguire, M.; Pombo–Villar, E. *Helv. Chim. Acta* **2002**, *85*, 3516– 3524.

38. Schöllkopf, U.; Neubauer, H.–J. *Synthesis* **1982**, 861–864.

39. Schöllkopf, U.; Groth, U.; Gull, M.–R.; Nozulak, J. *Liebigs Ann. Chem.* **1983**, 1133–1151.

40. Bommarius, A.S.; Schwarm, M.; Stingl, K.; Kottenhahn, M.; Huthmacher, K.; Drauz, K. *Tetrahedron: Asymmetry* **1995**, *6*, 2851–2888.

41. Richter, L.S.; Gadek, T.R. *Tetrahedron: Asymmetry* **1996**, *7*, 427–434.

42. Subramanian, P.K.; Woodard, R.W. *J. Org. Chem.* **1987**, *52*, 15–18.

43. Ma, C.; He, X.; Liu, X.; Yu, S.; Zhao, S.; Cook, J.M. *Tetrahedron Lett.* **1999**, *40*, 2917–2918.

44. Ma, C.; Liu, X.; Yu, S.; Zhao, S.; Cook, J.M. *Tetrahedron Lett.* **1999**, *40*, 657–660.

45. Göschke, R.; Stutz, S.; Heinzelmann, W.; Maibaum, J. *Helv. Chim. Acta* **2003**, *86*, 2848–2870.

46. Rose, J.E.; Leeson, P.D.; Gani, D. *J. Chem. Soc., Perkin Trans. 1* **1992**, 1563–1564.

47. Lankiewicz, L.; Nyassé, B.; Fransson, B.; Grehn, L.; Ragnarsson, U. *J. Chem. Soc., Perkin Trans. 1* **1994**, 2503–2510.

48. Larsson, E.; Lüning, B. *Acta Chem. Scand.* **1996**, *50*, 54–57.

49. Boger, D.L.; Kim, S.H.; Mori, Y.; Weng, J.–H.; Rogel, O.; Castle, S.L.; McAtee, J.J. *J. Am. Chem. Soc.* **2001**, *123*, 1862–1871.

50. Paladino, J.; Guyard, C.; Thurieau, C.; Fauchère, J.–L. *Helv. Chim. Acta* **1993**, *76*, 2465–2472.

51. Archer, C.H.; Thomas, N.R.; Gani, D. *Tetrahedron: Asymmetry* **1993**, *4*, 1141–1152.

52. Papageorgiou, C.; Florineth, A.; Mikol, V. *J. Med. Chem.* **1994**, *37*, 3674–3676.

53. Cappon, J.J.; Witters, K.D.; Baart, J.; Verdegem, P.J.E.; Hoek, A.C.; Luiten, R.J.H.; Raap, J.; Lugtenburg, J. *J. Recl. Trav. Chim. Pays–Bas* **1994**, *113*, 318–328.

54. Holler, T.P.; Ruan, F.; Spaltenstein, A.; Hopkins, P.B. *J. Org. Chem.* **1989**, *54*, 4570–4575.

55. Holler, T.P.; Spaltenstein, A.; Turner, E.; Klevit, R.E.; Shapiro, B.M.; Hopkins, P.B. *J. Org. Chem.* **1987**, *52*, 4421–4423.

56. Ohba, M.; Mukaihira, T.; Fuji, T. *Chem. Pharm. Bull.* **1994**, *42*, 1784–1790.

57. Zhang, P.; Cook, J.M. *Synth. Commun.* **1995**, *25*, 3883–3900.

58. Zhang, P.; Liu, R.; Cook, J.M. *Tetrahedron Lett.* **1995**, *36*, 9133–9136.

59. Amici, R.; Pevarello, P.; Colombo, M.; Varasi, M. *Synthesis* **1996**, 1177–1179.

60. Li, X.; Yin, W.; Srirama, P.V.V.; Zhou, H.; Ma, J.; Cook, J.M. *Tetrahedron Lett.* **2004**, *45*, 8569–8573.

61. Ma, C.; Yu, S.; He, X.; Liu, X.; Cook, J.M. *Tetrahedron Lett.* **2000**, *41*, 2781–2785.

62. Wild, N.; Groth, U. *Eur. J. Org. Chem.* **2003**, 4445–4449.

63. Raap, J.; Wolthuis, W.N.E.; Hehenkamp, J.J.J.; Lugtenburg, J. *Amino Acids* **1995**, *8*, 171–186.

64. Nozulak, J.; Schöllkopf, U. *Synthesis* **1982**, 866–868.

65. Karstens, W.F.J.; Berger, H.J.F.F.; van Haren, E.R.; Lugtenburg, J.; Raap. J. *J. Lab. Cmpds. Radiopharm.* **1995**, *36*, 1077–1096.

66. Beulshausen, T.; Groth, U.; Schöllkopf, U. *Liebigs Ann. Chem.* **1991**, 1207–1209.

67. Crowley, B.M.; Mori, Y.; McComas, C.C.; Tang, D.; Boger, D.L. *J. Am. Chem. Soc.* **2004**, *126*, 4310–4317.

68. Schöllkopf, U.; Beulshausen, T. *Liebigs Ann. Chem.* **1989**, 223–225.

69. Lambin, D.; Lemaire, C.; Plenevaux, A.; Damhaut, P.; Luxen, A. *J. Lab. Cmpds. Radiopharm.* **1997**, *40*, 17–19.

70. Grauert, M.; Schöllkopf, U. *Liebigs Ann. Chem.* **1985**, 1817–1824.

71. Ruiz, M.; Ruanova, T.M.; Ojea, V.; Quintela, J.M. *Tetrahedron Lett.* **1999**, *40*, 2021–2024.

72. Schöllkopf, U.; Tiller, T.; Bardenhagen, J. *Tetrahedron* **1988**, *44*, 5293–5305.

73. Schöllkopf, U.; Nozulak, J.; Groth, U. *Synthesis* **1982**, 868–870.

74. Kobayashi, S.; Furuta, T. *Tetrahedron* **1998**, *54*, 10275–10294.

75. Gelb, M.H.; Lin, Y.; Pickard, M.A.; Song, Y.; Vederas, J.C. *J. Am. Chem. Soc.* **1990**, *112*, 4932–4942.

76. Sutherland, A.; Vederas, J.C. *Chem. Commun.* **1999**, 1739–1740.

77. Neubauer, H.–J.; Baeza, J.; Freer, J.; Schöllkopf, U. *Liebigs Ann. Chem.* **1985**, 1508–1511.

78. Schöllkopf, U.; Pettig, D.; Buse, U. *Synthesis* **1986**, 737–740.

79. Hartzoulakis, B.; Gani, D. *J. Chem. Soc., Perkin Trans. 1* **1994**, 2525–2531.

80. Hartwig, W.; Born, L. *J. Org. Chem.* **1987**, *52*, 4352–4358.

81. Jane, D.E.; Chalmers, D.J.; Howard, J.A.K.; Kilpatrick, I.C.; Sunter, D.C.; Thompson, G.A.; Udvarhelyi, P.M.; Wilson, C.; Watkins, J.C. *J. Med. Chem.* **1996**, *39*, 4738–4743.

82. Shapiro, G.; Buechler, D.; Marzi, M.; Schmidt, K.; Gomez–Lor, B. *J. Org. Chem.* **1995**, *60*, 4978–4979.

83. Shapiro, G.; Buechler, D.; Ojea, V.; Pombo–Villar, E.; Ruiz, M.; Weber, H.–P. *Tetrahedron Lett.* **1993**, *39*, 6255–6258.

84. Ojea, V.; Fernández, M.C.; Ruiz, M.; Quintela, J.M. *Tetrahedron Lett.* **1996**, *37*, 5801–5804.

85. Ruiz, M.; Ojea, V.; Shapiro, G.; Weber, H.–P.; Pombo–Villar, E. *Tetrahedron Lett.* **1994**, *35*, 4551–4554.

86. Ojea, V.; Ruiz, M.; Shapiro, G.; Pombo–Villar, E. *Tetrahedron Lett.* **1994**, *35*, 3273–3276.

87. Ruiz, M.; Fernández, M.C.; Díaz, A.; Quintela, J.M.; Ojea, V. *J. Org. Chem.* **2003**, *68*, 7634–7645.

88. Schöllkopf, U.; Groth, U.; *Angew. Chem., Int. Ed. Engl.* **1981**, *20*, 977–978.

89. Schöllkopf, U.; Nozulak, J.; Groth, U. *Tetrahedron* **1984**, *40*, 1409–1417.

90. Jurgens, A.R. *Tetrahedron Lett.* **1992**, *33*, 4727–4730.

91. Bold, G.; Allmendinger, T.; Herold, P.; Moesch, L.; Schär, H.-P.; Duthaler, R.O. *Helv. Chim. Acta* **1992**, 865–882.

92. Bull, S.D.; Chernega, A.N.; Davies, S.G.; Moss, W.O.; Parkin, R.M. *Tetrahedron* **1998**, *54*, 10379–10388.

93. Lange, M.; Fischer, P.M. *Helv. Chim. Acta* **1998**, *81*, 2053–2061.

94. Kremminger, P.; Undheim, K. *Tetrahedron* **1997**, *53*, 6925–6936.

95. Lange, M.; Undheim, K. *Tetrahedron* **1998**, *54*, 5337–5344.

96. Efskind, J.; Benneche, T.; Undheim, K. *Acta Chem. Scand.* **1997**, *51*, 942–952.

97. Efskind, J.; Hope, H.; Undheim, K. *Eur. J. Org. Chem.* **2002**, 464–467.

98. Andrews, M.J.I.; Tabor, A.B. *Tetrahedron Lett.* **1997**, *38*, 3063–3066.

99. McNamara, L.M.A.; Andrews, M.J.I.; Mitzel, F.; Siligardi, G.; Tabor, A.B. *J. Org. Chem.* **2001**, *66*, 4585–4594.

100. Møller, B.S.; Benneche, T.; Undheim, K. *Tetrahedron* **1996**, *52*, 8807–8812.

101. Rødbotten, S.; Benneche, T.; Undheim, K. *Acta Chem. Scand.* **1997**, *51*, 873–880.

102. Furenes, E.B.; Luijendijk, J.; Efskind, J.; Undheim, K. *Synth. Commun.* **2005**, *35*, 193–200.

103. Waelchli, R.; Beerli, C.; Meigel, H.; Révész, L. *Bioorg. Med. Chem. Lett.* **1997**, *7*, 2831–2836.

104. Pearson, A.J.; Bruhn, P.R. *J. Org. Chem.* **1991**, *56*, 7092–7097.

105. Pearson, A.J.; Lee, S.–H.; Gouzoules, F. *J. Chem. Soc. Perkin Trans. 1* **1990**, 2251–2254.

106. Boger, D.L.; Yohannes, D. *J . Org. Chem.* **1989**, *54*, 2498–2502.

107. Boger, D.L.; Yohannes, D. *Tetrahedron Lett.* **1989**, *30*, 2053–2056.

108. Boger, D.L.; Yohannes, D. *J. Org. Chem.* **1990**, *55*, 6000–6017.

109. Ohba, M.; Nishimura, Y.; Kato, M.; Fujii, T. *Tetrahedron* **1999**, *55*, 4999–5016.

110. Phelan, J.C.; Skelton, N.J.; Braisted, A.C.; McDowell, R.S. *J. Am. Chem. Soc.* **1997**, *119*, 455–460.

111. Heerding, D.; Bhatnagar, P.; Hartmann, M.; Kremminger, P.; LoCastro, S. *Tetrahedron: Asymmetry* **1996**, *7*, 237–242.

112. Fink, B.E.; Kym, P.R.; Katzenellenbogen, J.A. *J. Am. Chem. Soc.* **1998**, *120*, 4334–4344.

113. Mazón, A.; Nájera, C. *Tetrahedron: Asymmetry* **1997**, *8*, 1855–1859.

114. Vivet, B.; Cavelier, F.; Martinez, J. *Eur. J. Org. Chem.* **2000**, 807–811.

115. Mazón, A.; Pedregal, C.; Prowse, W. *Tetrahedron* **1999**, *55*, 7057–7064.

116. Groth, U.; Halfbrodt, W.; Schöllkopf, U. *Liebigs Ann. Chem.* **1992**, 351–355.

117. Hammer, K.; Undheim, K. *Tetrahedron* **1997**, *53*, 2309–2322.

118. Hammer, K.; Wang, J.; Falck–Pedersen, M.L.; Römming, C.; Undheim, K. *J. Chem. Soc., Perkin Trans. 1* **2000**, 1691–1695.

119. Hammer, K.; Undheim, K. *Tetrahedron* **1997**, *53*, 10603–10614.

120. Hammer, K.; Undheim, K. *Tetrahedron* **1997**, *53*, 5891–5898.

121. Hammer, K; Undheim, K. *Tetrahedron* **1997**, *53*, 5925–5936.

122. Krikstolaitytè, S.; Hammer, K.; Undheim, K. *Tetrahedron Lett.* **1998**, *39*, 7595–7598.

123. Hammer, K.; Rømming, C.; Undheim, K. *Tetrahedron* **1998**, *54*, 10837–10850.

124. Hammer, K.; Undheim, K. *Tetrahedron: Asymmetry* **1998**, *9*, 2359–2368.

125. Krikstolaityté, S.; Hammer, K.; Rømming, C.; Undheim, K. *Synth. Commun.* **2002**, *32*, 571–580.

126. Hoven, G.B.; Efskind, J.; Rømming, C.; Undheim, K. *J. Org. Chem.* **2002**, *67*, 2459–2463.

127. Efskind, J.; Römming, C.; Undheim, K. *J. Chem. Soc., Perkin Trans. 1* **2001**, 2697–2703.

128. Møller, B.; Undheim, K. *Tetrahedron* **1998**, *54*, 5789–5804.

129. Møller, B.; Undheim, K. *Eur. J. Org. Chem.* **2003**, 332–336.

130. Andrei, M.; Undheim, K. *Tetrahedron: Asymmetry* **2004**, *15*, 53–63.

131. Schöllkppf, U.; Hauptreif, M.; Dippel, J.; Nieger, M.; Egert, E. *Angew. Chem., Int. Ed. Engl.* **1986**, *25*, 192–193.

132. Schöllkopf, U.; Neubauer, H.–J.; Hauptreif, M. *Angew. Chem., Int. Ed. Engl.* **1985**, *24*, 1066–1067.

133. Schöllkopf, U.; Grüttner, S.; Anderskewitz, R.; Egert, E.; Dyrbusch, M. *Angew. Chem., Int. Ed. Engl.* **1987**, *26*, 683–684.

134. Groth, U.; Huhn, T.; Porsch, B.; Schmeck, C.; Schöllkopf, U. *Liebigs Ann. Chem.* **1993**, 715–719.

135. Bycroft, B.W.; Lee, G.R. *J. Chem. Soc., Chem. Commun.* **1975**, 988–989.

136. Oba, M.; Nakajima, S.; Nishiyama, K. *Chem. Commun.* **1996**, 1875–1876.

137. Oba, M.; Terauchi, T.; Owari, Y.; Imai, Y; Motoyama, I.; Nishiyama, K. *J. Chem. Soc., Perkin Trans. 1* **1998**, 1275–1281.

138. Bull, S.D.; Davies, S.G.; Epstein, S.W.; Leech, M.A.; Ouzman, J.V.A. *J. Chem. Soc., Perkin Trans. 1* **1998**, 2321–2330.

139. Bull, S.D.; Davies, S.G.; Epstein, S.W.; Ouzman, J.V.A. *Chem. Commun.* **1998**, 659–660.

140. Bull, S.D.; Davies, S.G.; Epstein, S.W.; Ouzman, J.V.A. *Tetrahedron: Asymmetry* **1998**, *9*, 2795–2798.

141. Bull, S.D.; Davies, S.G.; O'Shea, M.D. *J. Chem. Soc., Perkin Trans. 1* **1998**, 3657–3658.

142. Buñuel, E.; Bull, S.D.; Davies, S.G.; Garner, A.C.; Savory, E.D.; Smith, A.D.; Vickers, R.J.; Watkin, D.J. *Org. Biomol. Chem.* **2003**, *1*, 2531–2542.

143. Orena, M.; Porzi, G.; Sandri, S. *J. Chem. Res. S* **1993**, 318–319; *J. Chem. Res. M* **1993**, 2125–2152.

144. Orena, M.; Porzi, G.; Sandri, S. *J. Org. Chem.* **1992**, *57*, 6532–6536.

145. D'Arrigo, M.C.; Porzi, G.; Rossetti, M.; Sandri, S. *J. Chem. Res. S* **1995**, 162–163; *M* **1995**, 1038–1050.

146. D'Arrigo, M.C.; Porzi, G.; Sandri, S. *J. Chem. Res. S* **1995**, 430–431; *M* **1995**, 2612–2621.

147. Porzi, G.; Sandri, S. *Tetrahedron: Asymmetry* **1994**, *5*, 453–464.

148. Paradisi, F.; Porzi, G.; Rinaldi, S.; Sandri, S. *Tetrahedron: Asymmetry* **2000**, *11*, 4617–4622.

149. Ferioli, F.; Piccinelli, F.; Porzi, G.; Sandri, S. *Tetrahedron: Asymmetry* **2002**, *13*, 1181–1187.

150. Paradisi, F.; Piccinelli, F.; Porzi, G.; Sandri, S. *Tetrahedron: Asymmetry* **2002**, *13*, 497–502.

151. Piccinelli, F.; Porzi, G.; Sandri, M.; Sandri, S. *Tetrahedron: Asymmetry* **2003**, *14*, 393–398.

152. Juaristi, E.; León–Romo, J.L.; Ramírez–Quirós, Y. *J. Org. Chem.* **1999**, *64*, 2914–2918.

153. Hamersak, Z.; Gaso, D.; Kovac, S.; Hergold–Brundié, A.; Vickovié, I.; Sunjié, V. *Helv. Chim. Acta* **2003**, *86*, 2247–2257.

154. Schanen, V.; Riche, C.; Chiaroni, A.; Quirion, J.–C.; Husson, H.–P. *Tetrahedron Lett.* **1994**, *35*, 2533–2536.

155. Favero, V.; Porzi, G.; Sandri, S. *Tetrahedron: Asymmetry* **1997**, *8*, 599–612.

156. Porzi, G.; Sandri, S.; Verrocchio, P. *Tetrahedron: Asymmetry* **1998**, *9*, 119–132.

157. Balducci, D.; Grandi, A.; Porzi, G.; Sandri, S. *Tetrahedron: Asymmetry* **2005**, *16*, 1453–1462.

158. Balducci, D.; Emer, E.; Piccinelli, F.; Porzi, G.; Recanatini, M.; Sandri, S. *Tetrahedron: Asymmetry* **2005**, *16*, 3785–3794.

159. Balducci, D.; Crupi, S.; Galeazzi, R.; Piccinelli, F.; Porzi, G.; Sandri, S. *Tetrahedron: Asymmetry* **2005**, *16*, 1103–1112.

160. Piccinelli, F.; Porzi, G.; Sandri, M.; Sandri, S. *Tetrahedron: Asymmetry* **2004**, *15*, 1085–1093.

161. Williams, R.M.; Armstrong, R.W.; Maruyama, L.K.; Dung, J.–S.; Anderson, O.P. *J. Am. Chem. Soc.* **1985**, *107*, 3246–3253.

162. Williams, R.M.; Armstrong, R.W.; Dung, J.–S.; *J. Am. Chem. Soc.* **1985**, *107*, 3253–3266.

163. Kardassis, G.; Brungs, P.; Steckhan, E. *Tetrahedron* **1998**, *54*, 3471–3478.

164. Bull, S.D.; Davies, S.G.; Garner, A.C.; Savory, E.D.; Snow, E.J.; Smith, A.D. *Tetrahedron: Asymmetry* **2004**, *15*, 3989–4001.

165. Bull, S.D.; Davies, S.G.; Garner, A.C.; O'Shea, M.D.; Savory, E.D.; Snow, E.J. *J. Chem. Soc., Perkin Trans. 1* **2002**, 2442–2448.

166. Seewald, N.; Seymour, L.C.; Burger, K.; Osipov, S.N.; Kolomiets, A.F.; Fokin, A.V. *Tetrahedron: Asymmetry* **1994**, *5*, 1051–1060.

167. Martín–Santamaría, S.; Espada, M.; Avendaño, C. *Tetrahedron* **1999**, *55*, 1755–1762.

168. Porzi, G.; Sandri, S. *Tetrahedron: Asymmetry* **1996**, *7*, 189–196.

169. Graziani, L.; Porzi, G.; Sandri, S. *Tetrahedron: Asymmetry* **1996**, *7*, 1341–1346.

170. Carloni, A.; Porzi, G.; Sandri, S. *Tetrahedron: Asymmetry* **1998**, *9*, 2987–2998.

171. Porzi, G.; Sandri, S. *Tetrahedron: Asymmetry* **1998**, *9*, 3411–3420.

172. Wanner, K.T.; Stamenitis, S. *Liebigs Ann. Chem.* **1993**, 477–484.

173. Kardassis, G.; Brungs, P.; Nothhelfer, C.; Steckhan, E. *Tetrahedron* **1998**, *54*, 3479–3488.

174. Williams, R.M. *Aldrichimica Acta* **1992**, *25*, 11–25.

175. Sinclair, P.J.; Zhai, D.; Reibenspies, J.; Williams, R.M. *J. Am. Chem. Soc.* **1986**, *108*, 1103–1104.

176. Williams, R.M.; Sinclair, P.J.; Zhai, D.; Chen, D. *J. Am. Chem. Soc.* **1988**, *110*, 1547–1557.

177. Schuerman, M.A.; Keberline, K.I.; Hiskey, R.G. *Tetrahedron Lett.* **1995**, *36*, 825–828.

178. van den Nieuwendijk, A.M.C.H.; Warmerdam, E.G.J.C.; Brussee, J.; van der Gen, A. *Tetrahedron: Asymmetry* **1995**, *6*, 801–806.

179. Williams, R.M.; Sinclair, P.J.; Zhai, W. *J. Am. Chem. Soc.* **1988**, *110*, 482–483.

180. Zhai, D.; Zhai, W.; Williams, R.M. *J. Am. Chem. Soc.* **1988**, *110*, 2501–2505.

181. Williams, R.M.; Zhai, W. *Tetrahedron* **1988**, *44*, 5425–5430.

182. Williams, R.M.; Hendrix, J.A. *J. Org. Chem.* **1990**, *55*, 3723–3728.

183. Williams, R.M.; Zhai, D.; Sinclair, P.J. *J. Org. Chem.* **1986**, *51*, 5021–5022.

184. Ramer, S.E.; Cheng, H.; Palcic, M.M.; Vederas, J.C. *J. Am. Chem. Soc.* **1988**, *110*, 8526–8532.

185. Ramer, S.E.; Cheng, H.; Vederas, J.C. *Pure Appl. Chem.* **1989**, *61*, 489–492.

186. Williams, R.M.; Im, M.–N. *Tetrahedron Lett.* **1988**, *29*, 6075–6078.

187. Williams, R.M.; Im, M.–N. *J. Am. Chem. Soc.* **1991**, *113*, 9276–9286.

188. Baldwin, J.E.; Lee, V.; Schofield, C.J. *Synlett* **1992**, 249–251.

189. Baldwin, J.E.; Lee, V.; Schofield, C.J. *Heterocycles* **1992**, *34*, 903–906.

190. Schafmeister, C.E.; Po, J.; Verdine, G.L. *J. Am. Chem. Soc.* **2000**, *122*, 5891–5892.

191. Aoyagi, Y.; Williams, R.M. *Synlett* **1998**, 1099–1101.

192. Reno, D.S.; Lotz, B.T.; Miller, M.J. *Tetrahedron Lett.* **1990**, *31*, 827–830.

193. Williams, R.M.; Yuan, C. *J. Org. Chem.* **1992**, *57*, 6519–6527.

194. Williams, R.M.; Yuan, C. *J. Org. Chem.* **1994**, *59*, 6190–6193.

195. Williams, R.M.; Im, M.–N.; Cao, J. *J. Am. Chem. Soc.* **1991**, *113*, 6976–6981.

196. DeMong, D.E.; Williams, R.M. *J. Am. Chem. Soc.* **2003**, *125*, 8561–8565.

197. Solas, D.; Hale, R.L.; Patel, D.V. *J. Org. Chem.* **1996**, *61*, 1537–1539.

198. van der Nieuwendijk, A.M.C.H.; Benningshof, J.C.J.; Wegmann, V.; Bank, R.A.; te Koppele, J.M.; Brussee, J.; van der Gen, A. *Biorg. Med. Chem. Lett.* **1999**, *9*, 1673–1676.

199. Allevi, P.; Anastasia, M. *Tetrahedron: Asymmetry* **2004**, *15*, 2091–2096.

200. Scott, J.D.; Tippie, T.N.; Williams, R.M. *Tetrahedron Lett.* **1998**, *39*, 3659–3662.

201. Williams, R.M.; Fegley, G.J. *J. Am. Chem. Soc.* **1991**, *113*, 8796–8806.

202. Williams, R.M.; Fegley, G.J. *Tetrahedron Lett.* **1992**, *33*, 6755–6758.

203. Williams, R.M.; Fegley, G.J.; Gallegos, R.; Schaefer, F.; Pruess, D.L. *Tetrahedron* **1996**, *52*, 1149–1164.

204. Williams, R.M.; Colson, P.–J.; Zhai, W. *Tetrahedron Lett.* **1994**, *35*, 9371–9374.

205. Aoyagi, Y.; Williams, R.M. *Tetrahedron* **1998**, *54*, 10419–10433.

206. Kabat, M.M. *Tetrahedron Lett.* **2001**, *42*, 7521–7524.

207. Dellaria J.F., Jr.; Santarsiero, B.D. *J. Org. Chem.* **1989**, *54*, 3916–3926.

208. Dellaria J.F., Jr.; Santarsiero, B.D. *Tetrahedron Lett.* **1988**, *29*, 6079–6082.

209. Dastlik, K.A.; Giles, R.G.F.; Roos, G.H.P. *Tetrahedron: Asymmetry* **1996**, *7*, 2525–2526.

210. Schafer, C.M.; Molinski, T.F. *J. Org. Chem.* **1996**, *61*, 2044–2050.

211. Takatori, K.; Nishihara, M.; Nishiyama, Y.; Kajiwara, M. *Tetrahedron* **1998**, *54*, 15861–15869.

212. Ornstein, P.L. *J. Org. Chem.* **1989**, *54*, 2251–2253.

213. Roos, G.H.P.; Dastlik, K.A. *Synth. Commun.* **2003**, *33*, 2197–2208.

214. Baker, W.R.; Condon, S.L.; Spanton, S. *Tetrahedron Lett.* **1992**, *33*, 1573–1576.

215. Remuzon, P.; Soumeillant, M.; Dussy, C.; Bouzard, D. *Tetrahedron* **1997**, *53*, 17711–17726.

216. Keynes, M.N.; Earle, M.A.; Sudharshan, M.; Hultin, P.G. *Tetrahedron* **1996**, *52*, 8685–8702.

217. Boa, A.N.; Guest, A.L.; Jenkins, P.R.; Fawcett, J.; Russell, D.R.; Waterson, D. *J. Chem. Soc., Perkin Trans. 1* **1993**, 477–481.

218. Harwood, L.M.; Lilley, I.A. *Tetrahedron: Asymmetry* **1995**, *6*, 1557–1560.

219. Anslow, A.S.; Harwood, L.M.; Lilley, I.A. *Tetrahedron: Asymmetry* **1995**, *6*, 2465–2468.

220. Anslow, A.S.; Harwood, L.M.; Phillips, H.; Watkin, D. *Tetrahedron: Asymmetry* **1991**, *2*, 169–172.

221. Anslow, A.S.; Harwood, L.M.; Phillips, H.; Watkin, D. *Tetrahedron: Asymmetry* **1991**, *2*, 997–1000.

222. Anslow, A.S.; Harwood, L.M.; Phillips, H.; Watkin, D.; Wong, L.F. *Tetrahedron: Asymmetry* **1991**, *2*, 1343–1358.

223. Harwood, L.M.; Macro, J.; Watkin, D.; Williams, C.E.; Wong, L.F. *Tetrahedron: Asymmetry* **1992**, *3*, 1127–1130.

224. Harwood, L.M.; Manage, A.C.; Robin, S.; Hopes, S.F.G.; Watkin, D.J.; Williams, C.E. *Synlett* **1993**, 777–780.

225. Baldwin, J.E.; MacKenzie Turner, S.C.; Moloney, M.G. *Synlett* **1994**, 925–928.

226. Anslow, A.S.; Cox, G.C.; Harwood, L.M. *Chem. Heterocycl. Cmpd.* **1995**, *31*, 1222–1230.

227. Harwood, L.M.; Kitchen, L.C. *Tetrahedron Lett.* **1993**, *34*, 6603–6606.

228. Harwood, L.M.; Lilley, I.A. *Tetrahedron Lett.* **1993**, *34*, 537–540.

229. Alker, D.; Hamblett, G.; Harwood, L.M.; Robertson, S.M.; Watkin, D.J.; Williams, C.E. *Tetrahedron* **1998**, *54*, 6089–6098.

230. Harwood, L.M.; Robertson, S.M. *Chem. Commun.* **1998**, 2641–2642.

231. Alker, D.; Harwood, L.M.; Williams, C.E. *Tetrahedron Lett.* **1998**, *39*, 475–478.

232. Williams, R.M.; Zhai, W.; Aldous, D.J.; Aldous, S.C. *J. Org. Chem.* **1992**, *57*, 6527–6532.

233. Agami, C.; Couty, F.; Poursoulis, M. *Synlett* **1992**, 847–848.

234. Agami, C.; Couty, F.; Lin, J.; Mikaeloff, A.; Poursoulis, M. *Tetrahedron* **1993**, *49*, 7239–7250.

235. Harwood, L.M.; Hamblett, G.; Jiménez–Díaz, A.I.; Watkin, D.J. *Synlett* **1997**, 935–938.

236. Baldwin, S.W.; Young, B.G.; McPhail, A.T. *Tetrahedron Lett.* **1998**, *39*, 6819–6822.

237. Agami, C.; Couty, F.; Daran, J.–C.; Prince, B.; Puchot, C. *Tetrahedron Lett.* **1990**, *31*, 2889–2892.

238. Cox, G.G.; Harwood, L.M. *Tetrahedron: Asymmetry* **1994**, *5*, 1669–1672.

239. Harwood, L.M.; Vines, K.J.; Drew, M.G.B. *Synlett* **1996**, 1051–1053.
240. Ager, D.; Cooper, N.; Cox, G.G.; Garro–Hélion, F.; Harwood, L.M. *Tetrahedron: Asymmetry* **1996**, *7*, 2563–2566.
241. Harwood, L.M.; Tyler, S.N.G.; Anslow, A.S.; MacGilp, I.D.; Drew, M.G.B. *Tetrahedron: Asymmetry* **1997**, *8*, 4007–4010.
242. Chinchilla, R.; Galindo, N.; Nájera, C. *Tetrahedron: Asymmetry* **1998**, *9*, 2769–2772.
243. Chinchilla, R.; Galindo, N.; Nájera, C. *Synthesis* **1999**, 704–717.
244. Chinchilla, R.; Falvello, L.R.; Galindo, N.; Nájera, C. *Ang. Chem., Int. Ed. Engl.* **1997**, *36*, 995–997.
245. Abellán, T.; Nájera, C.; Sansano, J.M. *Tetrahedron: Asymmetry* **1998**, *9*, 2211–2214.
246. Nájera, C.; Abellán, T.; Sansano, J.M. *Eur. J. Org. Chem.* **2000**, 2809–2820.
247. Chinchilla, R.; Falvello, L.R.; Galindo, N.; Nájera, C. *Tetrahedron: Asymmetry* **1998**, *9*, 2223–2227.
248. Chinchilla, R.; Falvello, L.R.; Galindo, N.; Nájera, C. *Tetrahedron: Asymmetry* **1999**, *10*, 821–825.
249. Abellán, T.; Chinchilla, R.; Galindo, N.; Guillena, G.; Nájera, C.; Sansano, J.M. *Eur. J. Org. Chem.* **2000**, 2689–2697.
250. Hartwig, W.; Schöllkopf, U. *Liebigs Ann. Chem.* **1982**, 1952–1970.
251. Schöllkopf, U.; Scheuer, R. *Liebigs Ann. Chem.* **1984**, 939–950.
252. Koch, C.–J.; Simonyiová, S.; Pabel, J.; Kärtner, A.; Polborn, K.; Wanner, K.T. *Eur. J. Org. Chem.* **2003**, 1244–1263.
253. Dixon, D.J.; Harding, C.I.; Ley, S.V.; Tilbrook, M.G. *Chem. Commun.* **2003**, 468–469.
254. Zhu, Z.; McKittrick, B. *Tetrahedron Lett.* **1998**, *39*, 7479–7482.
255. Seebach, D.; Naef, R. *Helv. Chim. Acta* **1981**, *64*, 2704–2708.
256. Seebach, D.; Sting, A.R.; Hoffmann, M. *Angew. Chem., Int. Ed. Engl.* **1996**, *35*, 2708–2748.
257. Hinds, M.G.; Welsh, J.H.; Brennand, D.M.; Fisher, J.; Glennie, M.J.; Richards, N.G.J.; Turner, D.L.; Robinson, J.A. *J. Med. Chem.* **1991**, *34*, 1777–1789.
258. Seebach, D.; Boes, M.; Naef, R.; Schweizer, W.B. *J. Am. Chem. Soc.* **1983**, *105*, 5390–5398.
259. Calderari, G.; Seebach, D. *Helv. Chim. Acta* **1985**, *68*, 1592–1604.
260. Seebach, D.; Vettiger, T.; Müller, H.–M.; Plattner, D.A.; Petter, W. *Liebigs Ann. Chem.* **1990**, 687–695.
261. Weber, T.; Seebach, D. *Helv. Chim. Acta* **1985**, *68*, 155–161.
262. Karady, S.; Amato, J.S.; Weinstock, L.M. *Tetrahedron Lett.* **1984**, *25*, 4337–4340.
263. Seebach, D.; Fadel, A. *Helv. Chim. Acta* **1985**, *68*, 1243–1250.
264. Fadel, A.; Salaün, J. *Tetrahedron Lett.* **1987**, *28*, 2243–2246.
265. Smith A.B., III; Pasternak, A.; Yokoyama, A.; Hirschmann, R. *Tetrahedron Lett.* **1994**, *35*, 8977–8980.
266. Nebel, K.; Mutter, M. *Tetrahedron* **1988**, *44*, 4793–4796.
267. Altmann, E.; Nebel, K.; Mutter, M. *Helv. Chim. Acta* **1991**, *74*, 800–806.
268. Kuriyama, N.; Akaji, K.; Kiso, Y. *Tetrahedron* **1997**, *53*, 8323–8334.
269. Akaji, K.; Kuriyama, N.; Kiso, Y. *J. Org. Chem.* **1996**, 61, 3350–3357.
270. Weber, T.; Aeschimann, R.; Maetzke, T.; Seebach, D. *Helv. Chim. Acta* **1986**, *69*, 1365–1377.
271. Jones, R.C.F.; Crockett, A.K.; Rees, D.C.; Gilbert, I.A. *Tetrahedron: Asymmetry* **1994**, *5*, 1661–1664.
272. Gilbert, I.H.; Rees, D.C.; Crockett, A.K.; Jones, R.C.F. *Tetrahedron* **1995**, *51*, 6315–6336.
273. Zhang, L.; Finn, J.M. *J. Org. Chem.* **1995**, *60*, 5719–5720.
274. Shrader, W.D.; Marlowe, C.K. *Bioorg. Med. Chem. Lett.* **1995**, *5*, 2207–2210.
275. Kapadia, S.R.; Spero, D.M.; Eriksson, M. *J. Org. Chem.* **2001**, *66*, 1903–1905.
276. Ma, D.; Ma, Z.; Kozikowski, A.P.; Pshenichkin, S.; Wroblewski, J.T. *Biorg. Med. Chem. Lett.* **1998**, *8*, 2447–2450.
277. Pyne, S.G.; Dikic, B.; Gordon, P.A.; Skelton, B.W.; White, A.H. *Aust. J. Chem.* **1993**, *46*, 73–93.
278. Alonso, F.; Davies, S.G. *Tetrahedron: Asymmetry* **1994**, *5*, 353–356.
279. Alonso, F.; Davies, S.G.; Elend, A.S.; Haggitt, J.L. *J. Chem. Soc., Perkin Trans. 1* **1998**, 257–264.
280. Zydowsky, T.M.; de Lara, E.; Spanton, S.G. *J. Org. Chem.* **1990**, *55*, 5437–5439.
281. Goodman, M.; Zhang, J.; Gantzel, P.; Benedetti, E. *Tetrahedron Lett.* **1998**, *39*, 9589–9592.
282. Wang, H.; Germanas, J.P. *Synlett* **1999**, 33–36.
283. Seebach, D.; Müller, S.G.; Gysel, U.; Zimmermann, J. *Helv. Chim. Acta* **1988**, *71*, 1303–1318.
284. Seebach, D.; Hoffman, M.; Sting, A.R.; Knkel, J.N.; Schulte, M.; Küsters, E. *J. Chromatog., A* **1998**, *796*, 299–307.
285. Blaser, D.; Seebach, D. *Liebigs Ann. Chem.* **1991**, 1067–1078.
286. Blaser, D.; Ko, S.Y.; Seebach, D. *J. Org. Chem.* **1991**, *56*, 6230–6233.
287. Seebach, D.; Gees, T.; Schuler, F. *Liebigs Ann. Chem.* **1993**, 785–799; erratum **1994**, 529.
288. Seebach, D.; Lamatsch, B.; Amstutz, R.; Beck, A.K.; Dobler, M.; Egli, M.; Fitzi, R.; Gautschi, M.; Herradón, B.; Hidber, P.C.; Irwin, J.J.; Locher, R.; Maestro, M.; Maetzke, T.; Mouriño, A.; Pfammatter, E.; Plattner, D.A.; Schickli, C.; Schweizer, W.B.; Seiler, P.; Stucky, G.; Petter, W.; Escalante, J.; Juaristi, E.; Quintana, D.; Miravitlles, C.; Molins, E. *Helv. Chim. Acta* **1992**, *75*, 913–934.

289. Javidan, A.; Schafer, K.; Pyne, S.G. *Synlett* **1997**, 100–102.

290. Beckwith, A.L.J.; Chai, C.L.L. *J. Chem. Soc., Chem. Commun.* **1990**, 1087–1088.

291. Beckwith, A.L.J.; Pyne, S.G.; Dikic, B.; Chai, C.L.L.; Gordon, P.A.; Skelton, B.W.; Tozer, M.J.; White, A.H. *Aust. J. Chem.* **1993**, *46*, 1425–1430.

292. Pyne, S.G.; Dikic, B.; Gordon, P.A.; Skelton, B.W.; White, A.H. *J. Chem. Soc., Chem. Commun.* **1991**, 1505–1506.

293. Crossley, M.J.; Tansey, C.W. *Aust. J. Chem.* **1992**, *45*, 479–481.

294. Pyne, S.G.; Safaei–G, J.; Hockless, D.C.R.; Skelton, B.W.; Sobolev, A.N.; White, A.H. *Tetrahedron* **1994**, *50*, 941–956.

295. Pyne, S.; Safaei–G., J. *J. Chem. Res. S* **1996**, 160–161.

296. Pyne, S.G.; Schafer, K.; Skelton, B.W.; White, A.H. *Chem. Commun.* **1997**, 2267–2268; **1998**, 1607.

297. Chinchilla, R.; Nájera, C. *Tetrahedron Lett.* **1993**, *34*, 5799–5802.

298. Pyne, S.G.; Javidan, A.; Skelton, B.W.; White, A.H. *Tetrahedron* **1995**, *51*, 5157–5168.

299. Axon, J.R.; Beckwith, A.L.J. *J. Chem. Soc., Chem. Commun.* **1995**, 549–550.

300. Beckwith, A.L.J.; Chai, C.L.L. *Tetrahedron* **1993**, *49*, 7871–7882.

301. Pyne, S.G.; Schafer, K. *Tetrahedron* **1998**, *54*, 5709–5720.

302. Herpin, T.F.; Motherwell, W.B.; Weibel, J.–M. *Chem. Commun.* **1997**, 923–924.

303. Suárez, R.M.; Sestelo, J.P.; Sarandeses, L.A. *Org. Biomol. Chem.* **2004**, *2*, 3584–3587.

304. Vedejs, E.; Fields, S.C.; Schrimpf, M.R. *J. Am. Chem. Soc.* **1993**, *115*, 11612–11613.

305. Vedejs, E.; Fields, S.C.; Lin, S.; Schrimpf, M.R. *J. Org. Chem.* **1995**, *60*, 3028–3034.

306. Vedejs, E.; Chapman, R.W.; Fields, S.C.; Lin, S.; Schrimpf, M.R. *J. Org. Chem.* **1995**, *60*, 3020–3027.

307. Vedejs, E.; Fields, S.C.; Hayashi, R.; Hitchcock, S.R.; Powell, D.R.; Schrimpf, M.R. *J. Am. Chem. Soc.* **1999**, *121*, 2460–2470.

308. Evans, D.A.; Weber, A.E. *J. Am. Chem. Soc.* **1986**, *108*, 6757–6761.

309. Evans, D.A.; Weber, A.E. *J. Am. Chem. Soc.* **1987**, *109*, 7151–7157.

310. Evans, D.A.; Britton, T.C. *J. Am. Chem. Soc.* **1987**, *109*, 6881–6883.

311. Evans, D.A.; Britton, T.C.; Dorow, R.L.; Dellaria, J.F. *J. Am. Chem. Soc.* **1986**, *108*, 6395–6397.

312. Trimble, L.A.; Vederas, J.C. *J. Am. Chem. Soc.* **1986**, *108*, 6397–6399.

313. Es–Sayed, M.; Gratkowski, C.; Krass, N.; Meyers, A.I.; de Meijere, A. *Synlett* **1992**, 962–964.

314. Wenglowsky, S.; Hegedus, L.S. *J. Am. Chem. Soc.* **1998**, *120*, 12468–12473.

315. Lander, P.A.; Hegedus, L.S. *J. Am. Chem. Soc.* **1994**, *116*, 8126–8132.

316. Ojima, I.; Chen, H.–J.C. *J. Chem. Soc., Chem. Commun.* **1987**, 625–626.

317. Matsunaga, H.; Ishizuka, T.; Kuneida, T. *Tetrahedron* **1997**, *53*, 1275–1294.

318. Naef, R.; Seebach, D. *Helv. Chim. Acta* **1985**, *68*, 135–143.

319. Aebi, J.D.; Seebach, D. *Helv. Chim. Acta* **1985**, *68*, 1507–1518.

320. Gander–Coquoz, M.; Seebach, D. *Helv. Chim. Acta* **1988**, *71*, 224–236.

321. Kazmierski, W.M.; Urbancyzk–Lipkowska, Z.; Hruby, V.J. *J. Org. Chem.* **1994**, *59*, 1789–1795.

322. Seebach, D.; Aebi, J.D.; Naef, R.; Weber, T. *Helv. Chim. Acta* **1985**, *68*, 144–154.

323. Studer, A.; Seebach, D. *Liebigs Ann. Chem.* **1995**, 217–222.

324. Seebach, D.; Juaristi, E.; Miller, D.D.; Schickli, C.; Weber, T. *Helv. Chim. Acta* **1987**, *70*, 237–261.

325. Blank, S.; Seebach, D. *Liebigs Ann. Chem.* **1993**, 889–896.

326. Damhaut, P.; Lemaire, C.; Plenevaux, A.; Brihaye, C.; Christiaens, L.; Comar, D. *Tetrahedron* **1997**, *53*, 5785–5796.

327. Seebach, D.; Miller, D.D.; Müller, S.; Weber, T. *Helv. Chim. Acta* **1985**, *68*, 949–952.

328. Fitzi, R.; Seebach, D. *Angew. Chem., Int. Ed. Engl.* **1986**, *25*, 345–346; *25*, 766.

329. Fitzi, R.; Seebach, D. *Tetrahedron* **1988**, *44*, 5277–5292.

330. Polt, R.; Seebach, D. *Helv. Chim. Acta* **1987**, *70*, 1930–1936.

331. Seebach, D.; Dziadulewicz, E.; Behrendt, L.; Cantoreggi, S.; Fitzi, R. *Liebigs Ann. Chem.* **1989**, 1215–1232.

332. Jaun, B.; Tanaka, M.; Seiler, P.; Kühnle, F.N.M.; Braun, C.; Seebach, D. *Liebigs Ann./Recueil* **1997**, 1697–1710.

333. Mzengeza, S.; Venkatachalam, T.K.; Diksic, M. *Amino Acids* **2000**, *18*, 81–88.

334. Fasth, K.–J.; Hörnfeldt, K.; Långstrom, B. *Acta Chem. Scand.* **1995**, *49*, 301–304.

335. Lowe, C.; Pu, Y.; Vederas, J.C. *J. Org. Chem.* **1992**, *57*, 10–11.

336. Amin, M.R.; Harper, D.B.; Moloney, J.M.; Murphy, C.D.; Howard, J.A.K.; O'Hagan, D. *Chem. Commun.* **1997**, 1471–1472.

337. Suzuki, K.; Seebach, D. *Liebigs Ann. Chem.* **1992**, 51–61.

338. Müller, W.; Lowe, D.A.; Neijt, H.; Urwyler, S.; Herrling, P.L.; Blaser, D.; Seebach, D. *Helv. Chim. Acta* **1992**, *75*, 855–864.

339. Fasth, K.J.; Malmborg, P.; Långström, B. *J. Lab. Cmpds. Radiopharm.* **1991**, *30*, 401.

340. García–Barradas, O.; Juaristi, E. *Tetrahedron* **1995**, *51*, 3423–3434.

341. Krippner, G.Y.; Harding, M.M. *Tetrahedron: Asymmetry* **1994**, *5*, 1793–1804.

342. Mehlführer, M.; Berner, H.; Thirring, K. *J. Chem. Soc., Chem. Commun.* **1994**, 1291.

343. Mehlführer, M.; Thirring, K.; Berner, H. *J. Org. Chem.* **1997**, *62*, 4078–4081.

344. Satoh, Y.; Gude, C.; Chan, K.; Firooznia, F. *Tetrahedron Lett.* **1997**, *38*, 7645–7648.

345. Sjöberg, S.; Hawthorne, M.F.; Wilmouth, S.; Lindström, P. *Chem.—Eur. J.* **1995**, *1*, 430–435.

346. Lindström, P.; Naeslund, C.; Sjöberg, S. *Tetrahedron Lett.* **2000**, *41*, 751–754.

347. Shendage, D.M.; Fröhlich, R.; Bergander, K.; Haufe, G. *Eur. J. Org. Chem.* **2005**, 719–727.

348. Juaristi, E.; Rizo, B.; Natal, V.; Escalante, J.; Regla, I. *Tetrahedron: Asymmetry* **1991**, *2*, 821–826.

349. García–Barradas, O.; Juaristi, E. *Tetrahedron: Asymmetry* **1997**, *8*, 1511–1514.

350. Bezençon, O.; Seebach, D. *Liebigs Ann. Chem.* **1996**, 1259–1276.

351. Polt, R.; Seebach, D. *J. Am. Chem. Soc.* **1989**, *111*, 2622–2632.

352. Seebach, D.; Bürger, H.M.; Schickli, C.P. *Liebigs Ann. Chem.* **1991**, 669–684.

353. Schickli, C.P.; Seebach, D. *Liebigs Ann. Chem.* **1991**, 655–668.

354. Suárez, R.M.; Sestelo, J.P.; Sarandeses, L.A. *Chem.—Eur. J.* **2003**, *9*, 4179–4187.

355. Zimmermann, J.; Seebach, D. *Helv. Chim. Acta* **1987**, *70*, 1104–1114.

356. Coggins, P.; Simpkins, N.S. *Synlett* **1991**, 515–516.

357. Blank, S.; Seebach, D. *Angew. Chem., Int. Ed. Engl.* **1993**, *32*, 1765–1766.

358. Seebach, D.; Hoffman, M. *Eur. J. Org. Chem.* **1998**, 1337–1351.

359. Pfammatter, E.; Seebach, D. *Liebigs Ann. Chem.* **1991**, 1317–1322.

360. Schöllkopf, U.; Hausberg, H.H.; Hoppe, I.; Segal, M.; Reiter, U. *Angew. Chem., Int. Ed. Engl.* **1978**, *17*, 117–119.

361. Schöllkopf, U.; Hausberg, H.–H.; Segal, M.; Reiter, U.; Hoppe, I.; Saenger, W.; Lindner, K. *Liebigs Ann. Chem.* **1981**, 439–458.

362. Amoroso, R.; Cardillo, G.; Tomasini, C.; Tortoreto, P. *J. Org. Chem.* **1992**, *57*, 1082–1087.

363. Amoroso, R.; Cardillo, G.; Tomasini, C. *Tetrahedron Lett.* **1991**, *32*, 1971–1974.

364. Amoroso, R.; Cardillo, G.; Tomasini, C. *Tetrahedron Lett.* **1990**, *31*, 6413–6416.

365. Amoroso, R.; Cardillo, G.; Romero, M.S.; Tomasini, C. *Gazz. Chim. Ital.* **1993**, *123*, 75–78.

366. Kubota, H.; Kubo, A.; Nunami, K.–I. *Tetrahedron Lett.* **1994**, *35*, 3107–3110.

367. Belokon, Y.N.; Bulychev, A.G.; Vitt, S.V.; Struchkov, Y.T.; Batsanov, A.S.; Timofeeva, T.V.; Tsyryapkin, V.A.; Ryzhov, M.G.; Lysova, L.A.; Bakhmutov, V.I.; Belikov, V.M. *J. Am. Chem. Soc.* **1985**, *107*, 4252–4259.

368. Belokon, Y.N.; Tararov, V.I.; Maleev, V.I.; Savel'eva, T.F.; Ryzhov, M.G. *Tetrahedron: Asymmetry* **1998**, *9*, 4249–4252.

369. Ueki, H.; Ellis, T.K.; Martin, C.H.; Boettiger, T.U.; Bolene, S.B.; Soloshonok, V.A. *J. Org. Chem.* **2003**, *68*, 7104–7107.

370. Soloshonok, V.A.; Avilov, D.V.; Kukhar, V.P.; Tararov, V.I.; Savel'eva, T.F.; Churkina, T.D.; Ikonnikov, N.S.; Kochetkov, K.A.; Orlova, S.A.; Pysarevsky, A.P.; Struchkov, Y.T.; Raevsky, N.I.; Belekon, Y.N. *Tetrahedron: Asymmetry* **1995**, *6*, 1741–1756.

371. Soloshonok, V.A.; Avilov, D.V.; Kukhar, V.P.; Van Meervelt, L.; Mischenko, N. *Tetrahedron Lett.* **1997**, *38*, 4671–4674.

372. Soloshonok, V.A.; Kukhar, V.P.; Galushko, S.V.; Svistunova, N.Y.; Avilov, D.V.; Kuz'mina, N.A.; Raevski, N.I.; Struchkov, Y.T.; Pysarevsky, A.P.; Belekon, Y.N. *J. Chem. Soc., Perkin Trans. 1* **1993**, 3143–3155.

373. Soloshonok, V.A.; Kukhar, V.P.; Galushko, S.V.; Kolycheva, M.T.; Rozhenko, A.B.; Belekon, Y.N. *Bull. Acad. Sci. USSR Div. Chem. Sci.* **1991**, *40*, 1046–1054.

374. Soloshonok, V.A.; Avilov, D.V.; Kukhar, V.P. *Tetrahedron: Asymmetry* **1996**, *7*, 1547–1550.

375. Soloshonook, V.A.; Avilov, D.V.; Kukhar, V.P. *Tetrahedron* **1996**, *52*, 12433–12442.

376. Belekon, Y.N.; Sagyan, A.S.; Djamgaryan, S.M.; Bakhmutov, V.I.; Belikov, V.M. *Tetrahedron* **1988**, *44*, 5507–5514.

377. Belekon, Y.N.; Sagyan, A.S.; Djamgaryan, S.A.; Bakhmutov, V.I.; Vitt, S.V.; Batsanov, A.B.; Struchkov, Y.T.; Belikov, V.M. *J. Chem. Soc., Perkin Trans. 1* **1990**, 2301–2310.

378. Saghiyan, A.S.; Hambardzumyan, H.H.; Manasyan, L.L.; Petrosyan, A.A.; Maleev, V.I.; Peregudov, A.S. *Synth. Commun.* **2005**, *35*, 449–459.

379. Ichikawa, T.; Maeda, S.; Okamoto, T.; Araki, Y.; Ishido, Y. *Bull. Chem. Soc. Jpn.* **1971**, *44*, 2779–2786.

380. Saito, S.–I.; Umezawa, Y.; Yoshioka, T.; Takita, T.; Umezawa, H. *J. Antibiotics* **1983**, *36*, 92–95.

381. Owa, T.; Otsuka, M.; Ohno, M. *Chem. Lett.* **1988**, 83–86.

382. Belekon, Y.N.; Zel'tzer, I.E.; Loim, N.M.; Tsiryapkin, V.A.; Aleksandrov, G.G.; Kursanov, D.N.; Parnes, Z.N.; Struchkov, Y.T.; Belikov, V.M. *Tetrahedron* **1980**, *36*, 1089–1097.

383. Belekon, Y.N.; Zel'tzer, I.E.; Ryzhov, M.G.; Saporovskaya, M.B.; Bakhmutov, V.I.; Belikov, V.M. *J. Chem. Soc., Chem. Commun.* **1982**, 180–181.

384. Belekon, Y.N.; Zel'tzer, I.E.; Bakhmutov, V.I.; Saporovskaya, M.B.; Ryzhov, M.G.; Yanovsky, A.I.; Struchkov, Y.T.; Belikov, V.M. *J. Am. Chem. Soc.* **1983**, *105*, 2010–2017.

385. Belekon, Y.N.; Bakhmutov, V.I.; Chernoglazova, N.I.; Kochetkkov, K.A.; Vitt, S.V.; Garbalinskaya, N.S.; Belikov, V.M. *J. Chem. Soc., Perkin Trans. 1* **1988**, 305–312.

386. Jirman, J.; Popkov, A. *Collect. Czech. Chem. Commun.* **1995**, *60*, 990–998.

387. Belekon, Y.N.; Tararov, V.I.; Maleev, V.I.; Motsishkite, S.M.; Vitt, S.V.; Chernoglazova, N.I.; Savel'eva, T.F.; Saporovskaya, M.B. *Bull. Acad. Sci. USSR Div. Chem. Sci.* **1991**, *40*, 1361–1365.

388. Kukhar, V.P.; Belekon, Y.N.; Soloshonok, V.A.; Svistunova, N.Y.; Rozhenko, A.B.; Kuz'mina, N.A. *Synthesis* **1993**, 117–120.

389. Fishwick, C.W.G.; Sanderson, J.M.; Findlay, J.B.C. *Tetrahedron Lett.* **1994**, *35*, 4611–4614.

390. Hashimoto, M.; Hatanaka, Y.; Sadakane, Y.; Nabeta, K. *Bioorg. Med. Chem. Lett.* **2002**, *12*, 2507–2510.

391. Tararov, V.I.; Savel'eva, T.F.; Kuznetsov, N.Y.; Ikonnikov, N.S.; Orlova, S.A.; Belekon, Y.N.; North, M. *Tetrahedron: Asymmetry* **1997**, *8*, 79–83.

392. Alves, I.; Cowell, S.; Lee, Y.S.; Tang, X.; Davis, P.; Porreca, F.; Hruby, V.J. *Biochem. Biophys. Res. Commun.* **2004**, *318*, 335–340.

393. Andronova, I.G.; Maleev, V.I.; Ragulin, V.V.; Il'in, M.M.; Tsvetkov, E.N.; Belokon, Y.N. *Russ. J. Gen. Chem.* **1996**, *66*, 1068–1071.

394. Collet, S.; Bauchat, P.; Danion–Bougot, R.; Danion, D. *Tetrahedron: Asymmetry* **1998**, *9*, 2121–2131.

395. Larionov, O.V.; Savel'eva, T.F.; Kochetkov, K.A.; Ikonnokov, N.S.; Kozhushkov, S.I.; Yufit, D.S.; Howard, J.A.K.; Khrustalev, V.N.; Belekon, Y.N.; de Meijere, A. *Eur. J. Org. Chem.* **2003**, 869–877.

396. Fasth, K.J.; Långström, B. *Acta Chem. Scand.* **1990**, *44*, 720–725.

397. Taylor, S.M.; Yamada, T.; Ueki, H.; Soloshonok, V.A. *Tetrahedron Lett.* **2004**, *45*, 9159–9162.

398. Belekon, Y.N.; Bulychev, A.G.; Pavlov, V.A.; Fedorova, E.B.; Tsyrypkin, V.A.; Bakhmutov, V.I.; Belikov, V.M. *J. Chem. Soc., Perkin Trans. 1* **1988**, 2075–2083.

399. Belekon, Y.N.; Bulychev, A.G.; Ryzhov, M.G.; Vitt, S.V.; Batsanov, Y.T.; Bakhmutov, V.I.; Belikov, V.M. *J. Chem. Soc., Perkin Trans* 1 **1986**, 1865–1872.

400. Soloshonok, V.A.; Avilov, D.V.; Kukhar, V.P.; Van Meervelt, L.; Mischenko, N. *Tetrahedron Lett.* **1997**, *38*, 4903–4904.

401. Soloshonok, V.A.; Cai, C.; Hruby, V.J.; Van Meervelt, L.; Mischenko, N. *Tetrahedron* **1999**, *55*, 12031–12044.

402. Soloshonok, V.A.; Cai, C.; Hruby, V.J.; Van Meervelt, L. *Tetrahedron* **1999**, *55*, 12045–12058.

403. Soloshonok, V.A.; Cai, C.; Hruby, V.J. *Tetrahedron Lett.* **2000**, *41*, 135–139.

404. Cai, C.; Soloshonok, V.A.; Hruby, V.J. *J. Org. Chem.* **2001**, *66*, 1339–1350.

405. Soloshonok, V.A.; Cai, C.; Hruby, V.J. *Tetrahedron: Asymmetry* **1999**, *10*, 4265–4269.

406. Soloshonok, V.A.; Cai, C.; Hruby, V.J. *Org. Lett.* **2000**, *2*, 747–750.

407. Soloshonok, V.A.; Ueki, H.; Tiwari, R.; Cai, C.; Hruby, V.J. *J. Org. Chem.* **2004**, *69*, 4984–4990.

408. Soloshonok, V.A.; Cai, C.; Hruby, V.J. *Tetrahedron Lett.* **2000**, *41*, 9645–9649.

409. Cai, C.; Yamada, T.; Tiwari, R.; Hruby, V.J.; Soloshonok, V.A. *Tetrahedron Lett.* **2004**, *45*, 6855–6858.

410. Belekon, Y.N.; Tararov, V.I.; Savel'eva, T.F. *Bull. Acad. Sci. USSR Div. Chem. Sci.* **1991**, *40*, 1054–1058.

411. Belekon, Y.N.; Chernoglazova, N.I.; Kochetkov, C.A.; Garbalinskaya, N.S.; Belikov, V.M. *J. Chem. Soc., Chem. Commun.* **1985**, 171–172.

412. Belekon, Y.N.; Motsishkite, S.M.; Tararov, V.I.; Maleev, V.I. *Bull. Acad. Sci. USSR Div. Chem. Sci.* **1991**, *40*, 1355–1360.

413. Casella, L.; Gullotti, M.; Jommi, G.; Pagliarin, R.; Sisti, M. *J. Chem. Soc., Perkin Trans. 1* **1990**, 771–775.

414. Jommi, G.; Laudi, A.; Pagliarin, R.; Sello, G.; Sisti, M. *Gazz. Chim. Ital.* **1994**, *124*, 299–300.

415. Antonov, D.Y.; Belokon, Y.N.; Ikonnikov, N.S.; Orlova, S.A.; Pisarevsky, A.P.; Raevski, N.I.; Rozenberg, V.I., Sergeeva, E.V.; Struchkov, Y.T.; Tararov, V.I.; Vorontsov, E.V. *J. Chem. Soc., Perkin Trans. 1* **1995**, 1873–1879.

416. Rozenberg, V.; Kharitonov, V.; Antonov, D.; Sergeeva, E.; Aleshkin, A.; Ikonnikov, N.; Orlova, S.; Belekon, Y. *Angew. Chem., Int., Ed. Engl.* **1994**, *33*, 91–92.

417. Belekon, Y.N.; Maleev, V.I.; Videnskaya, S.O.; Saporovskaya, M.B.; Tsyrypkin, V.A.; Belikov, V.M. *Bull. Acad. Sci. USSR Div. Chem. Sci.* **1991**, *40*, 110–118.

418. Belokon, Y.N.; Bespalova, N.B.; Churkina, T.D.; Císarová, I.; Ezernitskaya, M.G.; Harutyunyan, S.R.; Hrdina, R.; Kagan, H.B.; Kocovsky, P.; Kochetkov, K.A.; Larionov, O.V.; Lyssenko, K.A.; North, M.; Polásek, M.; Peregudov, A.S.; Prisyazhnyuk, V.V.; Vyskocil, S. *J. Am. Chem. Soc.* **2003**, *125*, 12860–12871.

419. Belekon, Y.N.; Kochetkov, K.A.; Churkina, T.D.; Ikonnikov, N.S.; Larionov, O.V.; Harutyunyan, S.R.; Vyskocil, S.; North, M.; Kagan, H.B. *Angew. Chem., Int. Ed.* **2001**, *40*, 1948–1950.

420. Vyskocil, S.; Meca, L.; Tislerová, I.; Císarová, I.; Polásek, M.; Harutyunyan, S.R.; Belekon, Y.N.; Stead, R.M.H.; Farrugia, L.; Lockhart, S.C.; Mitchell, W.L.; Kocovsky, P. *Chem.—Eur. J.* **2002**, *8*, 4633–4648.

421. Ueki, H.; Ellis, T.K.; Martin, C.H.; Soloshonok, V.A. *Eur. J. Org. Chem.* **2003**, 1954–1957.

422. De, B.B.; Thomas, N.R. *Tetrahedron: Asymmetry* **1997**, *8*, 2687–2691.

423. Krämer, R.; Wnajek, H.; Polborn, K.; Beck, W. *Chem. Ber.* **1993**, *126*, 2421–2427.

424. Bendahl, L.; Hammershøi, A.; Jensen, D.K.; Kaifer, E.; Sargeson, A.M.; Willis, A.C. *Chem. Commun.* **1996**, 1649–1650.

425. Harrowfield, J.M.; Sargeson, A.M. *J. Am. Chem. Soc.* **1979**, *101*, 1514–1520.

426. Drok, K.J.; Harrowfield, J.M.; McNiven, S.J.; Sargeson, A.M.; Skelton, B.W.; White, A.H. *Aust. J. Chem.* **1993**, *46*, 1557–1593.

427. Seebach, D.; Aebi, J.D.; Gander–Coquoz, M.; Naef, R. *Helv. Chim. Acta* **1987**, *70*, 1194–1216.

428. Seebach, D.; Aebi, J.D. *Tetrahedron Lett.* **1984**, *25*, 2545–2548.

429. Andrews, M.D.; Brewster, A.G.; Moloney, M.G.; Owen, K.L. *J. Chem. Soc., Perkin Trans.1* **1996**, 227–228.

430. Zhang, J.; Flippen–Anderson, J.L.; Kozikowski, A.P. *J. Org. Chem.* **2001**, *66*, 7555–7559.

431. Brunner, M.; Saarenketo, P.; Straub, T.; Rissanen, K.; Koskinen, A.M.P. *Eur. J. Org. Chem.* **2004**, 3879–3883.

432. Pattenden, G.; Thom, S.M.; Jones, M.F. *Tetrahedron* **1993**, *49*, 2131–2138.

433. Juaristi, E.; López–Ruiz, H.; Madrigal, D.; Ramírez–Quirós, Y.; Escalante, J. *J. Org. Chem.* **1998**, *63*, 4706–4710.

434. Hopkins, S.A.; Ritsema, T.A.; Konopelski, J.P. *J. Org. Chem.* **1999**, *64*, 7885–7889.

435. Wipf, P.; Miller, C.P. *J. Org. Chem.* **1993**, *58*, 1575–1578.

436. Fischer, P.M.; Sandosham, J. *Tetrahedron Lett.* **1995**, *36*, 5409–5412.

437. Fry, E.M. *J. Org. Chem.* **1949**, *14*, 887–894.

438. Attenburrow, J.; Elliott, D.F.; Penny, G.F. *J. Chem. Soc.* **1948**, 310–318.

439. Elliott, D.F. *J. Chem. Soc.* **1949**, 589–594.

440. Elliott, D.F. *J. Chem. Soc.* **1950**, 62–68.

441. Seebach, D.; Aebi, J.D. *Tetrahedron Lett.* **1983**, *24*, 3311–3314.

442. Reddy, L.R.; Saravanan, P.; Corey, E.J. *J. Am. Chem. Soc.* **2004**, *126*, 6230–6231.

443. Meyers, A.I.; Knaus, G.; Kamata, K.; Ford, M.E. *J. Am. Chem. Soc.* **1976**, *98*, 567–576.

444. Rottmann, A.; Liebscher, J. *Tetrahedron Lett.* **1996**, *37*, 359–362.

445. Le Bail, M.; Aitken, D.J.; Vergne, F.; Husson, H.–P. *J. Chem. Soc., Perkin Trans. 1* **1997**, 1681–1689.

446. Hegedus, L.S.; Schwindt, M.A.; De Lombaert, S.; Imwinkelried, R. *J. Am. Chem. Soc.* **1990**, *112*, 2264–2273.

447. Wu, M–J.; Pridgen, L.N. *Synlett* **1990**, 636–63.

448. Mokhallalati, M.K.; Wu, M.–J.; Pridgen, L.N. *Tetrahedron Lett.* **1993**, *34*, 47–50.

449. Andrés, C.; González, A.; Pedrosa, R.; Pérez–Encabo, A.; García–Granda, S.; Salvadó, M.A.; Gómez–Beltrán, F. *Tetrahedron Lett.* **1992**, 4743–4746.

450. Kaminski, Z.J.; Leplawy, M.T.; Zabrocki, J. *Synthesis* **1973**, 792–793.

451. Kübel, B.; Gruber, P.; Hurnaus, R.; Steglich, W. *Chem. Ber.* **1979**, *112*, 128–137.

452. Obrecht, D.; Heimgartner, H. *Tetrahedron Lett.* **1985**, *25*, 1717–1720.

453. Steglich, W.; Wegmann, H. *Synthesis* **1980**, 481–483.

454. Obrecht, D.; Spiegler, C.; Schönholzer, P.; Müller, K. Heimgartner, H.; Stierli, F. *Helv. Chim. Acta* **1992**, *75*, 1666–1696.

455. Obrecht, D.; Bohdal, U.; Daly, J.; Lehmann, C.; Schönholzer, P.; Müller, K. *Tetrahedron* **1995**, 10883–10900.

456. Obrecht, D.; Bohdal, U.; Ruffieux, R.; Müller, K. *Helv. Chim. Acta* **1994**, *77*, 1423–1429.

457. Obrecht, D.; Abrecht, C.; Altorfer, M.; Bohdal, U.; Grieder, A.; Kleber, M.; Pfyffer, P.; Müller, K. *Helv. Chim. Acta* **1996**, *79*, 1315–1337.

458. Clerici, F.; Gelmi, M.L.; Pocar, D.; Rondena, R. *Tetrahedron* **1995**, *51*, 9985–9994.

459. Crisma, M.; Valle, G.; Formaggio, F.; Toniolo, C.; Bago, A. *J. Am. Chem. Soc.* **1997**, *119*, 4136–4142.

460. Trost, B.M.; Dogra, K. *J. Am. Chem. Soc.* **2002**, *124*, 7256–7257.

461. Rassu, G.; Casiraghi, G.; Spanu, P.; Pinna, L.; Fava, G.G.; Ferrari, M.B.; Pelosi, G. *Tetrahedron: Asymmetry* **1992**, *3*, 1035–1048.

462. Casiraghi, G.; Rassu, G.; Spanu, P.; Pinna, L. *J. Org. Chem.* **1992**, *57*, 3760–3763.

463. Casiraghi, G.; Rassu, G.; Spanu, P.; Pinna, L. *Tetrahedron Lett.* **1994**, *35*, 2423–2426.

464. Rassu, G.; Zanardi, F.; Cornia, M.; Casiraghi, G. *J. Chem. Soc., Perkin Trans. 1* **1994**, 2431–2437.

465. Rassu, G.; Pinna, L.; Spanu, P.; Ulgheri, F.; Cornia, M.; Zanardi, F.; Casiraghi, G. *Tetrahedron* **1993**, *49*, 6489–6496.

466. Zanardi, F.; Battistini, L.; Rassu, G.; Cornia, M.; Casiraghi, G. *J. Chem. Soc., Perkin Trans. 1* **1995**, 2471–2475.

467. Ojima, I.; Komata, T.; Qiu, X. *J. Am. Chem. Soc.* **1990**, *112*, 770–774.

468. Katagiri, N.; Okada, M.; Morishita, Y.; Kaneko, C. *Tetrahedron* **1997**, *53*, 5725–5746.

469. Katagiri, N.; Okada, M.; Morishita, Y.; Kaneko, C. *Chem. Commun.* **1996**, 2137–2138.

470. Katagiri, N.; Okada, M.; Kaneko, C.; Furuya, T. *Tetrahedron Lett.* **1996**, *37*, 1801–1804.

471. Sano, S.; Takebayashi, M.; Miwa, T.; Ishii, T.; Nagao, Y. *Tetrahedron: Asymmetry* **1998**, *9*, 3611–3614.

472. Lange, M.; Pattersen, A.L.; Undheim, K. *Tetrahedron* **1998**, *54*, 5745–5752.

473. Ruiz, M.; Ojea, V.; Fernández, M.C.; Conde, S.; Díaz, A.; Quintela, J.M. *Synlett* **1999**, 1903–1906.

474. Groth, U.; Schöllkopf, U. *Synthesis* **1983**, 673–675.

475. Zeiss, H.–J. *Tetrahedron Lett.* **1987**, *28*, 1255–1258.

476. Cushman, M.; Lee, E.–S. *Tetrahedron Lett.* **1992**, *33*, 1193–1196.

477. Guillerm, D.; Guillerm, G. *Tetrahedron Lett.* **1992**, *33*, 5047–5050.

478. Ohba, M.; Mukaihira, T.; Fuji, T. *Heterocycles* **1992**, *33*, 21–26.

479. Papageorgiou, C.; Borer, X.; French, R.R. *Bioorg. Med. Chem. Lett.* **1994**, *4*, 267–272.

480. Mueller, R.; Revesz, L. *Tetrahedron Lett.* **1994**, *35*, 4091–4092.

481. Liu, R.; Zhang, P.; Gan, T.; Cook, J.M. *J. Org. Chem.* **1997**, *62*, 7447–7456.

482. Pecunioso, A.; Papini, D.; Tamburini, B.; Tinazzi, F. *Org. Proc. Prep. Int.* **1997**, *29*, 218–220.

483. Roussi, G.; Zamora, E.G.; Carbonnelle, A.–C.; Beugelmans, R. *Tetrahedron Lett.* **1997**, *38*, 4401–4404.

484. Ohba, M.; Nishimura, Y.; Imasho, M.; Fujii, T.; Kubanek, J.; Andersen, R.J. *Tetrahedron Lett.* **1998**, *39*, 5999–6002.

485. Kremminger, P.; Undheim, K. *Tetrahedron: Asymmetry* **1998**, *9*, 1183–1189.

486. Hasegawa, H.; Shinohara, Y. *J. Chem. Soc., Perkin Trans. 1* **1998**, 243–247.

487. Hinterding, K.; Hagenbuch, P.; Rétey, J.; Waldmann, H. *Angew. Chem., Int. Ed. Engl.* **1998**, *37*, 1236–1239.

488. Kent, D.R.; Cody, W.L.; Doherty, A.M. *J. Pept. Res.* **1998**, *52*, 201–207.

489. Fauq, A.H.; Ziani–Cherif, C.; Richelson, E. *Tetrahedron: Asymmetry* **1998**, *9*, 2333–2338.

490. Gao, Y.; Lane–Bell, P.; Vederas, J.C. *J. Org. Chem.* **1998**, *63*, 2133–2143.

491. Fauq, A.H.; Hong, F.; Cusack, B.; Tyler, B.M.; Ping–Pang, Y.; Richelson, E. *Tetrahedron: Asymmetry* **1998**, *9*, 4127–4134.

492. Hinterding, K.; Hagenbuch, P.; Rétey, J.; Waldmann, H. *Chem.—Eur. J.* **1999**, *5*, 227–236.

493. Löhr, B.; Orlich, S.; Kunz, H. *Synlett* **1999**, 1139–1141.

494. Lee, Y.; Silverman, R.B. *J. Am. Chem. Soc.* **1999**, *121*, 8407–8408.

495. Sedrani, R.; Kallen, J.; Cabrejas, L.M.M.; Papageorgiou, C.D.; Senia, F.; Rohrbach, S.; Wagner, D.; Thai, B.; Jutzi Eme, A.–M.; France, J.; Obere, L.; Rihs, G.; Zenke, G.; Wagner, J. *J. Am. Chem. Soc.* **2003**, *125*, 3849–3859.

496. Dong, H.; Zhang, Z.–L.; Huang, J.–H.; Ma, R.; Chen, S.–H.; Li, G. *Tetrahedron Lett.* **2005**, *46*, 6377–6340.

497. Kobayashi, S.; Furuta, T.; Hayashi, T.; Nishijima, M.; Hanada, K. *J. Am. Chem. Soc.* **1998**, *120*, 908–919.

498. Li, Y–Q.; Sugase, K.; Ishiguro, M. *Tetrahedron Lett.* **1999**, *40*, 9097–9100.

499. Schow, S.R.; DeJoy, S.Q.; Wick, M.M.; Kerwar, S.S. *J. Org. Chem.* **1994**, *59*, 6850–6852.

500. Bender, D.M.; Williams, R.M. *J. Org. Chem.* **1997**, *62*, 6690–6691.

501. Williams, R.M.; Liu, J. *J. Org. Chem.* **1998**, *63*, 2130–2132.

502. Fretz, H. *Tetrahedron* **1998**, *54*, 4849–4858.

503. Yao, Z.–J.; Gao, Y.; Voight, J.H.; Ford H., Jr.; Burke T.R., Jr. *Tetrahedron* **1999**, *55*, 2865–2874.

504. van den Nieuwendijk, A.M.C.H.; Kriek, N.M.A.J.; Brusse, J.; van Boom, J.H.; van der Gen, A. *Eur. J. Org. Chem.* **2002**, 3683–3691.

505. Gao, Y.; Wu, L.; Luo, J.H.; Guo, R.; Yang, D.; Zhang, Z.–Y.; Burke T.R., Jr. *Bioorg. Med. Chem. Lett.* **2000**, *10*, 923–927.

506. Paquette, L.A.; Duan, M.; Konetzki, I.; Kempmann, C. *J. Am. Chem. Soc.* **2002**, *124*, 4257–4270.

507. Kang, S.U.; Gao, Y.; Wu, L.; Zhang, Z.–Y.; Burke T.R., Jr. *Chem. Biodiversity* **2004**, *1*, 626–633.

508. Remuzon, P.; Massoudi, M.; Bouzard, D.; Jacquet, J.–P. *Heterocycles* **1992**, *33*, 679–684.

509. Smith A.B., III; Holcomb, R.C.; Guzman, M.C.; Keenan, T.P.; Sprengeler, P.A.; Hirschmann, R. *Tetrahedron Lett.* **1993**, *34*, 63–66.

510. Dikshit, D.K.; Maheshwari, A.; Panday, S.K. *Tetrahedron Lett.* **1995**, *36*, 6131–6134.

511. Smith A.B., III; Benowitz, A.B.; Favor, D.A.; Sprengeler, P.A.; Hirschmann, R. *Tetrahedron Lett.* **1997**, *38*, 3809–3812.

512. Smith A.B., III; Benowitz, A.B.; Sprengeler, P.A.; Barbosa, J.; Guzman, M.C.; Hirschmann, R.; Schweiger, E.J.; Bolin, D.R.; Nagy, Z.; Campbell, R.M.; Cox, D.C.; Olson, G.L. *J. Am. Chem. Soc.* **1999**, *121*, 9286–9298.

513. Ma, D.; Tian, H. *Tetrahedron: Asymmetry* **1996**, *7*, 1567–1570.

514. Kendrick, D.A.; Kolb, M. *J. Fluorine Chem.* **1989**, *45*, 265–272.

515. Damhaut, P.; Lemaire, C.; Plenevaux, A.; Christiaens, L.; Comar, D. *J. Lab. Compds. Radiopharm.* **1994**, *35*, 178–180.

516. Song, Y.; Niederer, D.; Lane–Bell, P.M.; Lam, L.K.P.; Crawley, S.; Palcic, M.M.; Pickard, M.A.; Pruess, D.L.; Vederas, J.C. *J. Org. Chem.* **1994**, *59*, 5784–5793.

517. Pajouhesh, H.; Curry, K. *Tetrahedron: Asymmetry* **1998**, *9*, 2757–2760.

518. Soloshonok, V.A.; Kukhar, V.P.; Batsanov, A.S.; Galakhov, M.A.; Belekon, Y.N.; Struchkov, Y.T. *Bull. Acad. Sci. USSR Div. Chem. Sci.* **1991**, *40*, 1366–1372.

519. Soloshonok, V.A.; Svistunova, N.Y.; Kukhar, V.P.; Kuz'mina, N.A.; Belekon, Y.N. *Bull. Acad. Sci. USSR Div. Chem. Sci.* **1992**, *42*, 540–544.

520. Soloshonok, V.A.; Kukhar, V.P.; Galushko, S.V.; Rozhenko, A.B.; Kuz'mina, N.A.; Kolycheva, M.T.; Belekon, Y.N. *Bull. Acad. Sci. USSR Div. Chem. Sci.* **1992**, *42*, 1692–1699.

521. Soloshonok, V.A.; Belekon, Y.N.; Kukhar, V.P.; Chernoglazova, N.I.; Saporovskaya, M.B.; Bkhmutov, V.I.; Kolycheva, M.T.; Belikov, V.M. *Bull. Acad. Sci. USSR Div. Chem. Sci.* **1991**, *40*, 1479–1485.

522. Soloshonok, V.A.; Belekon, Y.N.; Kuz'mina, N.A.; Maleev, V.I.; Svistunova, N.Y.; Solodenko, V.A.; Kukhar, V.P. *J. Chem. Soc., Perkin Trans. 1* **1992**, 1525–1529.

523. Bey, P.; Danzin, C.; Van Dorsselaer, V.; Mamont, P.; Jung, M.; Tardif, C. *J. Med. Chem.* **1978**, *21*, 50–55.

524. Jaroch, S.; Schwarz, T.; Steglich, W.; Zistler, P. *Angew. Chem., Int. Ed. Engl.* **1993**, *32*, 1771–1772.

525. Koen, M.J.; Morgan, J.; Pinhey, J.T. *J. Chem. Soc., Perkin Trans. 1* **1993**, 2383–2384.

526. Obrecht, D.; Abrecht, C.; Altorfer, M.; Bohdal, U.; Grieder, A.; Kleber, M.; Pfyffer, P.; Müller, K. *Helv. Chim. Acta* **1996**, *79*, 1315–1337.

527. Obrecht, D.; Altorfer, M.; Lehmann, C.; Schönholzer, P.; Müller, K. *J. Org. Chem.* **1996**, *61*, 4080–4086.

528. Chai, C.L.L.; Page, D.M. *Tetrahedron Lett.* **1993**, *34*, 4373–4376.

529. Chai, C.L.L.; Hay, D.B.; King, A.R. *Aust. J. Chem.* **1996**, *49*, 605–610.

530. Badran, T.W.; Easton, C.J.; Horn, E.; Kociuba, K.; May, B.L.; Schleibs, D.M.; Tiekink, E.R.T. *Tetrahedron: Asymmetry* **1993**, *4*, 197–200.

531. Chai, C.L.L.; Hockless, D.C.R.; King, A.R. *Aust. J. Chem.* **1996**, *49*, 1229–1233.

532. Easton, C.J. *Pure Appl. Chem.* **1997**, *69*, 489–494.

533. Chai, C.L.L.; King, A.R. *Tetrahedron Lett.* **1995**, *36*, 4295–4298.

8 | Synthesis of Optically Active α-Amino Acids: Opening of Small Ring Systems

8.1 Introduction

Cyclic ring systems have also been utilized for a different approach to amino acid synthesis than that discussed in the previous chapter, with the cyclic system opened during the bond-forming procedure. Five classes of small strained cyclic compounds have been used for most of these types of syntheses: aziridines, azirines, epoxides, β-lactams, and β-lactones (see Scheme 8.1). The β-lactone synthons are conventionally derived from Ser (or Thr), and so will be covered in Section 10.3.7.

8.2 Aziridines

Aziridine ring opening has provided a useful synthetic route to α- or β-amino acids, depending on the regioselectivity of the ring opening. Many of the simple aziridine-2-carboxylate synthons are derived from Ser, and so the reaction can be considered a net displacement of the Ser hydroxyl by a nucleophile. The synthesis and use of aziridines has been reviewed in 1993 (*1*), 1994 (*2*), and 1997 (*3*), with a general review of nucleophilic ring opening of aziridines (not focused on amino acids)

in 2004 (*4*). A 2006 monograph on aziridines and epoxides in organic synthesis includes aziridine-2-carboxylates (*5*).

8.2.1 Aziridine Synthesis from Hydroxy Amino Acids

Much of the pioneering work on Ser-derived aziridines was carried out by Okawa and co-workers, who reported the first synthesis of optically active L-aziridine-2-carboxylate (L-Azy) in 1978 via an intramolecular cyclization of an *N*-trityl-*O*-tosylated L-Ser intermediate (*6*) (see Scheme 8.2). Baldwin et al. used the same method in 1993, with the trityl group replaced by a tosyl group. The aziridine was formed in 43% overall yield (*7*). Larsson and Carlson prepared the *N*-trityl methyl ester of Azy-from the mesylate of Ser-OMe in 89% yield, with subsequent conversion to the *N*-Cbz derivative in 90% yield (*8, 9*). It is interesting to note that the reaction conditions for aziridine formation are almost identical to those used to tosylate Cbz-Ser to form Cbz-Ser(Ts) (see Section 10.3), except that *N*-trityl protection is used instead of the Cbz group, and the reaction is heated for aziridine formation. This is a good example

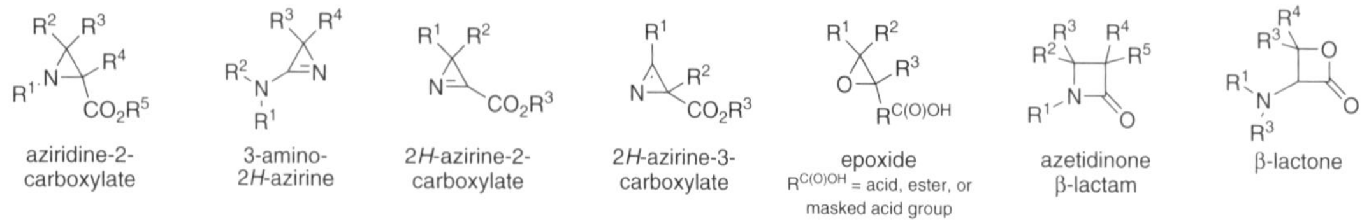

| aziridine-2-carboxylate | 3-amino-2*H*-azirine | 2*H*-azirine-2-carboxylate | 2*H*-azirine-3-carboxylate | epoxide R^C(O)OH = acid, ester, or masked acid group | azetidinone β-lactam | β-lactone |

Scheme 8.1 Small Cyclic Ring Systems.

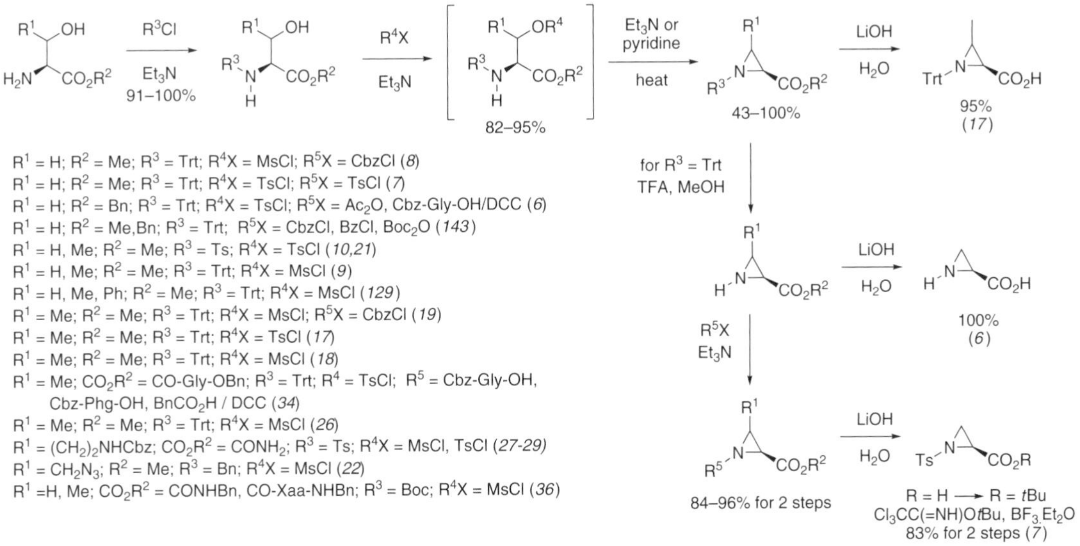
R¹ = H; R² = Me; R³ = Trt; R⁴X = MsCl; R⁵X = CbzCl (8)
R¹ = H; R² = Me; R³ = Trt; R⁴X = TsCl; R⁵X = TsCl (7)
R¹ = H; R² = Bn; R³ = Trt; R⁴X = TsCl; R⁵X = Ac₂O, Cbz-Gly-OH/DCC (6)
R¹ = H; R² = Me,Bn; R³ = Trt; R⁵X = CbzCl, BzCl, Boc₂O (143)
R¹ = H, Me; R² = Me; R³ = Ts; R⁴X = TsCl (10,21)
R¹ = H, Me; R² = Me; R³ = Trt; R⁴X = MsCl (9)
R¹ = H, Me, Ph; R² = Me; R³ = Trt; R⁴X = MsCl (129)
R¹ = Me; R² = Me; R³ = Trt; R⁴X = MsCl; R⁵X = CbzCl (19)
R¹ = Me; R² = Me; R³ = Trt; R⁴X = TsCl (17)
R¹ = Me; R² = Me; R³ = Trt; R⁴X = MsCl (18)
R¹ = Me; CO₂R² = CO-Gly-OBn; R³ = Trt; R⁴ = TsCl; R⁵ = Cbz-Gly-OH, Cbz-Phg-OH, BnCO₂H / DCC (34)
R¹ = Me; R² = Me; R³ = Trt; R⁴X = MsCl (26)
R¹ = (CH₂)₂NHCbz; CO₂R² = CONH₂; R³ = Ts; R⁴X = MsCl, TsCl (27-29)
R¹ = CH₂N₃; R² = Me; R³ = Bn; R⁴X = MsCl (22)
R¹ =H, Me; CO₂R² = CONHBn, CO-Xaa-NHBn; R³ = Boc; R⁴X = MsCl (36)
R³Cl
Et₃N
91–100%
R⁴X
Et₃N
82–95%
Et₃N or pyridine
heat
43–100%
LiOH
H₂O
95%
(17)
for R³ = Trt
TFA, MeOH
100%
(6)
R⁵X
Et₃N
84–96% for 2 steps
R = H ⟶ R = tBu
Cl₃CC(=NH)OtBu, BF₃.Et₂O
83% for 2 steps (7)

Scheme 8.2 Synthesis of Aziridine-2-Carboxylate from Serine or Threonine.

of the fine balance between various possible reactions of Ser derivatives, such as *O*-acylation, *O*-acylation displacement, aziridine formation, oxazoline formation, or elimination to generate dehydroalanine (see Scheme 10.3). *N*-Tosyl-*O*-mesyl Ser (Ts-Ser(Ms)-OMe) gave a mixture of aziridinyl and dehydroalanyl products when heated (*7*). However, it was reported in 1991 that while *N,O*-ditosyl-L-Ser-OEt gave dehydroalanine on treatment with Et_3N in THF, the desired aziridine could be prepared in 43% yield by using MeOH as solvent instead (*10*).

The balance between elimination/cyclization reactions is further illustrated by an attempted conversion of *N*-Boc Ser *tert*-butyl ester (Boc-Ser-O*t*Bu) to the aziridine using Mitsunobu conditions, which instead provided the dehydroalanine derivative in 96% yield (*11*). However, *N*-Boc Ser *tert*-butyl amide (Boc-Ser-NH*t*Bu) was successfully converted to the aziridine under the same conditions in 74% yield (*11*), as were several tertiary amides (Boc-Ser-NR_2, 57–84%) (*12*). The *p*-toluidide amide of Ser, Cbz-Ser-NH-*p*-MeC_6H_4, gave a β-lactam derivative via cyclization with the amide amine (*13*). Treatment of Ts-Ser-OMe with DEAD/PPh_3 gave the dehydroalanine elimination product, but Ts-Ser-O*t*Bu resulted in a 60% yield of aziridine (*14*). *N*-(2-nitrobenzenesulfonyl) Ser-O*t*Bu was also cyclized to Azy in 92% yield using DEAD/PPh_3 (*15*). In 2001, the acid group of *N*-(2-nitrobenzenesulfonyl)-Ser was protected as a 4-methyl-2,6,7-trioxabicyclo[2.2.2]octane (OBO) ortho ester. The OBO ester group reduces the acidity of the α-proton, reducing elimination and epimerization side reactions. Treatment with DEAD/PPh_3 gave the aziridine in 98% yield (*16*).

3-Methyl-aziridine-2-carboxylate (MeAzy) can be prepared from Thr by similar methods to those described above, such as via *N*-trityl-*O*-tosyl (*17*), *N*-trityl-*O*-mesyl (*9, 18–20*), or *N*-tosyl-*O*-tosyl (*10, 21*) intermediates, by treatment of Bn-Thr-OMe with PPh_3/CCl_4 (*22*), or by Mitsunobu reaction of various derivatives with DEAD/PPh_3 and DEAD (*13, 23*). The latter reaction, using β-hydroxy amino acid amides as substrates, also shows a balance between various possible reaction pathways. The reaction could be directed towards β-lactam formation (via displacement of the activated side chain with the amide amine) by using *N*-phthalimido protection, or by replacing the triphenylphosphine reagent with hexamethylphosphorus triamide (*13*). Dehydration products were obtained by substituting tributylphosphine or triethylphosphine for the triphenylphosphine (*13*). *N*-Tosyl Thr cyclohexyl ester was converted to the aziridine in 85% yield upon cyclization with DEAD/PPh_3. The *N*-tosyl group was removed with Mg in MeOH in 75% yield with no loss of enantiomeric purity (*24*).

An early synthesis of all four diastereomers of 2,3-diaminobutyric acid proceeded via reaction of ammonia with the *O*-tosylates of Ts-L-Thr-OMe and the corresponding diastereomers. Retention of configuration was observed, indicating double displacement via a non-isolated aziridine intermediate (*25*). The mesylate of *N*-trityl Thr formed the *cis*-aziridine upon treatment with nucleophiles, but the epimeric alloThr mesylate produced mixtures of products (*26*). α-*N*-Tosyl-β-hydroxy ornithine was also cyclized to an aziridine via an *O*-mesyl derivative (*27–29*), as was β-hydroxy-γ-azido-Abu (*22*). (2*S*,3*S*)-*N*-Phenylfluorenyl 3,4-dihydroxy-2-aminobutanoic acid methyl ester was converted into a dimesylate. Cyclization using DMF-Et_3N gave a 4:1 mixture of 3-hydroxymethyl-aziridine-2-carboxylate product and 3-hydroxy-azetidine-2-carboxylate, while dioxane-$LiClO_4$-s-collidine resulted exclusively in the aziridine. However, attempted cyclization of the (2S,3R)-diastereomer gave only the azetidine (30).

An improved synthesis of L-Azy from L-Ser benzyl ester was reported in 1989 (31). Treatment with diethoxytriphenylphosphorane (DTPP, Ph₃P(OEt)₂ (32)) gave the L-Azy benzyl ester in >92% yield and high optical purity. Baldwin et al. have also used this reagent to cyclize L-Ser tert-butyl ester, with yields of 25–69% for the cyclization and a subsequent introduction of an N-Boc group (11). The yield was noted to be greatly dependent on the batch of DTPP used (11). Thr could not be cyclized with DTPP (31), but a cyclic sulfamidate derivative of both Ser and Thr underwent an intramolecular ring opening to give the aziridine derivatives in 90–94% yield (see Scheme 8.3) (33). A bulky N-trityl substituent was essential for the desired reaction to proceed.

Aziridine residues can also be generated within peptides from Ser or Thr residues. N-Trityl Thr-Gly-OBn was converted to the corresponding Trt-MeAzy-Gly-OBn dipeptide in 88% yield via elimination of an O-tosyl intermediate (34); reaction of Cbz-Gly-Thr-Gly-OBn under the same conditions led to oxazoline formation instead (35). Azy or 3-methyl-Azy were generated from Boc-Ser or Boc-Thr within dipeptides via mesylation/ cyclization (36). There are conflicting reports on the products of the Mitsunobu reaction of Ser and Thr residues contained within peptides, with the differences likely due to variations in the position of the hydroxy amino acids in the peptides. The most recent publication

contends that Thr contained as the C-terminal residue in a dipeptide was successfully cyclized to the MeAzy residue under Mitsunobu reaction conditions in 56–84% yield, but that Ser and allo-Thr instead resulted in oxazoline formation (35). However, both Ser and Thr, when located as the Cbz-protected N-terminal residues of tripeptides, were previously reported to give the corresponding aziridine in 80% yield under the same conditions (37).

Acidic deprotection of N-trityl aziridines with formic acid or TFA can induce aziridine ring cleavage. By adding a reducing reagent (e.g., solutions of TFA/ Et₃SiH, methanesulfonic acid/Et₃SiH, or TFA/Me₃N-BH₃) much better yields were obtained (38).

8.2.2 Aziridine Syntheses from Other Sources

A number of other methods lead to aziridine-2-carboxylates. Optically active aziridine-2-carboxylates were produced with 85–95% ee from glycidic esters by treatment with sodium azide followed by a Staudinger reaction (39–41) (see Scheme 8.4). The glycidic esters were obtained from allylic alcohols via a Sharpless epoxidation (39–43) or from tartaric acid (43). This method allows for the introduction of substituents on the aziridine template. A similar procedure has been used to prepare aziridine-2-hydroxymethyl derivatives, with the hydroxymethyl group oxidized to the desired acid group after the aziridine ring-opening step (44–47).

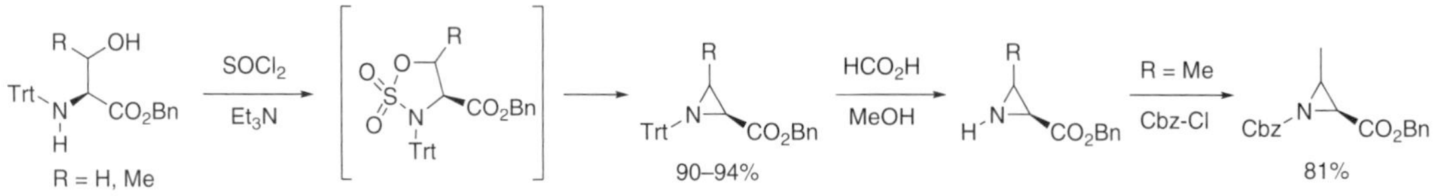

Scheme 8.3 Synthesis of Aziridine-2-Carboxylate from Serine and Threonine via Sulfamidate (33).

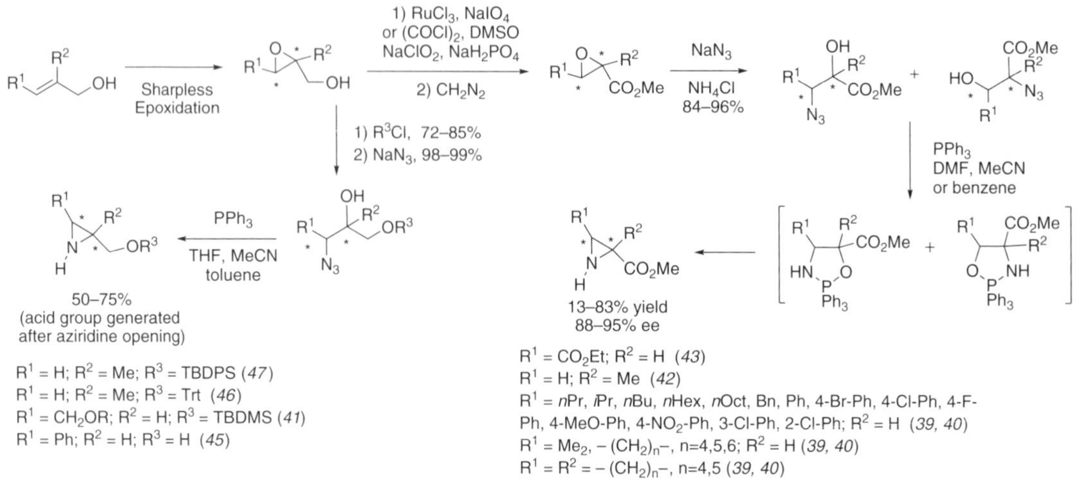

Scheme 8.4 Synthesis of Aziridine-2-Carboxylate from Epoxide.

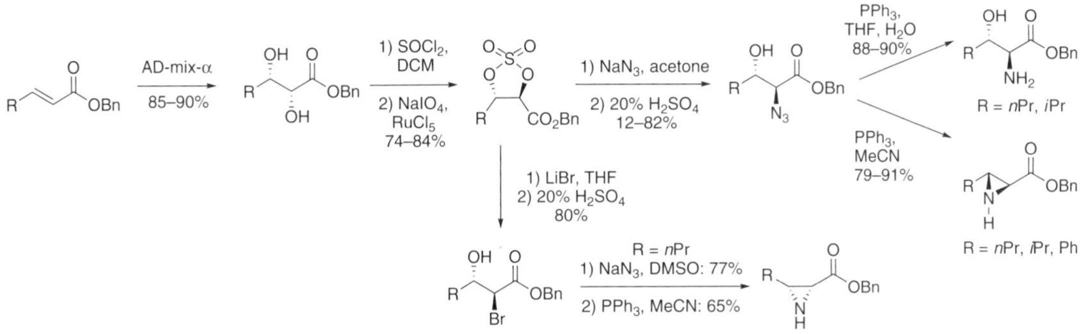

Scheme 8.5 Asymmetric Aziridine Synthesis via Asymmetric Dihydroxylation (48).

Asymmetric dihydroxylation of α,β-unsaturated benzyl esters, followed by formation of a cyclic sulfate with the diol, then azide opening of the sulfate, produced β-hydroxy-α-azidoesters. Treatment with triphenylphosphine in acetonitrile (with no water) generated the *trans*-3-substituted-aziridine-2-carboxylates (see Scheme 8.5). The *cis* isomers could be produced by bromide opening of the sulfate, azide displacement of the bromide, and aziridine formation (48). Azide reduction by hydrogenation or PPh$_3$/H$_2$O directly gave β-hydroxy-α-amino acids, with no aziridine formation (49, 50). A very similar synthesis was described by another group, starting with an asymmetric aminohydroxylation of ethyl (*m,p*-methylenedioxy)cinnamate. The diol was reacted with SOCl$_2$ to form a cyclic sulfite, which was regioselectively opened with sodium azide to give the 2-hydroxy-3-azido ester. This was converted into the correct regioisomer via azide reduction with PPh$_3$, which was accompanied by displacement of the α-hydroxy group to form an aziridine. Hydrogenolysis opened the aziridine, with ester/acetal hydrolysis giving DOPA with 85% ee (51). The use of cyclic sulfites and sulfates in organic synthesis was reviewed in 2000 (52).

Enantiopure aziridines have also been synthesized by a chemoenzymatic route from fumaric acid (or Asp) via an isoserine intermediate (see Scheme 8.6). Initially, the D-enantiomer was prepared (53, 54), but a subsequent modification allowed for the synthesis of the L-isomer via an intramolecular inversion of the isoserine intermediate (55). The advantage of this route is that stereospecifically deuterated aziridines can be prepared, and readily elaborated into amino acids. Another chemoenzymatic route produced optically active β-chloro-Ala by enzymatic reductive amination of the β-chloro-α-keto acid. A set of N-trityl ester derivatives were then cyclized to the aziridine in 90–98% yield using KHSO$_3$ in MeCN, with the choice of solvent and salt critical for good yields. Applying the same conditions to β-chloro-2-aminobutyric acid gave the 3-methyl-Azy product from the *threo* substrate in 98% yield, but the *erythro* substrate gave only eliminated product in 99% yield (56). Four diastereomers of N-acetyl 3-methylaziridine-2-carboxylate, as the 1-phenylethylamine amides, were prepared by cyclization of 2-chloro-3-amino-butyric acid amides, themselves obtained by halogenation and hydrolysis of a 6-methylperhydropyrimidin-4-one

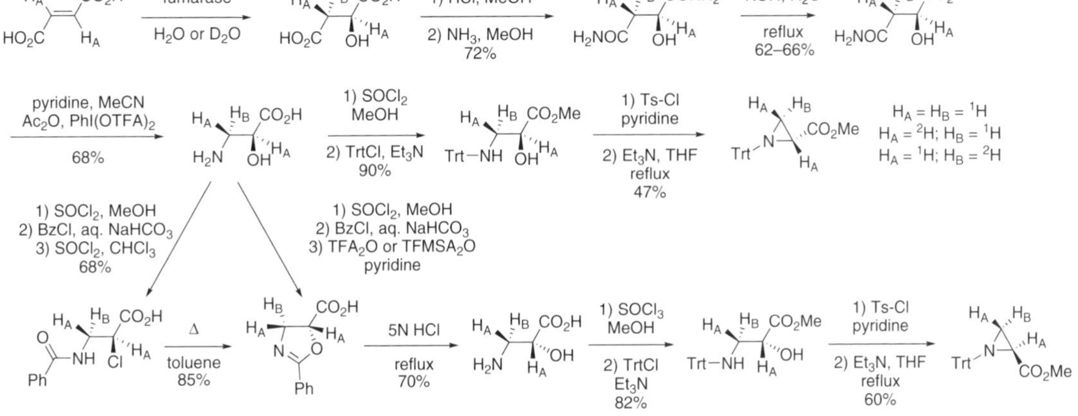

Scheme 8.6 Chemo-Enzymatic Synthesis of Aziridine-2-Carboxylate (53, 55, 65).

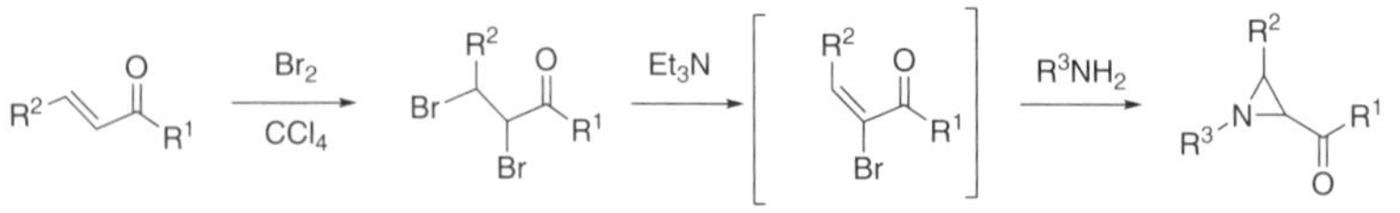

COR^1 = CN; R^2 = Ph, 4-Cl-Ph; R^3 = H, Me: 70–75% (*62*)
R^1 = OMe; R^2 = 4-Ph-Ph; R^3 = Et, Bn, *t*Bu: 31–100% (*63*)
R^1 = O*i*Pr; R^2 = Ph; R^3 = H: quantitative (*60*)
R^1 = O-(–)-menthyl; R^2 = Ph; R^3 = H: 48% yield, 24%de (*61, 376*)
COR^1 = CN; R^2 = Ph; R^3 = H (*59*)
R^1 = OEt or COR^1 = CN; R^2 = H, Me, Ph; R^3 = H: 25–79% (*64*)
R^1 = O*t*Bu, Et; R^2 = H; R^3 = (*R*)- or (*S*)-CH(Ph)CH$_2$OMe: 80–94%, 0% de (*67*)

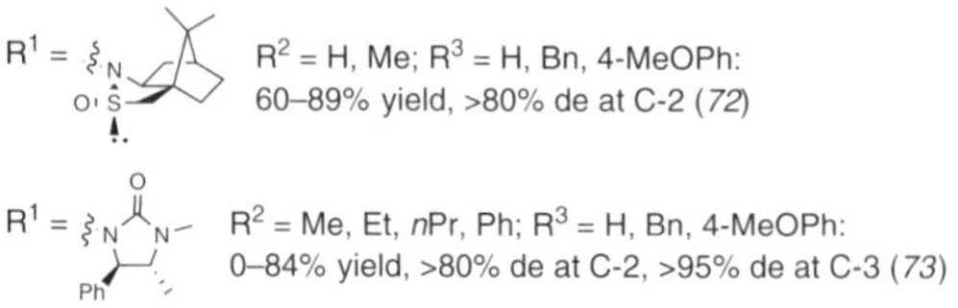

R^1 = [sultam structure]   R^2 = H, Me; R^3 = H, Bn, 4-MeOPh:
60–89% yield, >80% de at C-2 (*72*)

R^1 = [imidazolidinone structure]   R^2 = Me, Et, *n*Pr, Ph; R^3 = H, Bn, 4-MeOPh:
0–84% yield, >80% de at C-2, >95% de at C-3 (*73*)

Scheme 8.7 Synthesis of Aziridine-2-Carboxylate from α,β-Dibromocarboxylate.

templates (*57*). Racemic β-phenylseleno-α-amino esters were cyclized to give 3-alkyl-aziridine-2-carboxylates (*58*).

An alternative route to aziridine-2-carboxylates proceeds via reaction of ammonia with an α,β-dibromoester or cyanide intermediate (*59–64*) using conditions originally developed for α,β-dibromoketones (*65, 66*) (Scheme 8.7). Racemic aziridines were produced. Menthyl ester (*46*) or chiral amine (*67–69*) auxiliaries induced minimal diastereoselectivity, but allowed for resolution of the diastereomers by crystallization or chromatography. Racemic *tert*-butyl aziridine-2-carboxylates containing chiral substituents on the amine were kinetically resolved via hydrolysis of the ester with KO*t*Bu in THF, producing one diastereomer of the acid with >95% de (*70*). Racemic aziridines have also been resolved enzymatically (*71*).

The camphor-derived sultam of Oppolzer was used as an amide chiral auxiliary in a similar synthesis (*72*). In this case, aziridines were obtained from the addition of ammonia, benzylamine, or 4-methoxyaniline to brominated acrylate or methacrylate derivatives in 60–89% yield with 80–100% diastereoselectivity at the α-center but little stereocontrol at the β-center (*72*). In contrast, a chiral imidazolidinone auxiliary provided excellent *trans* selectivity (only traces of the *cis*-aziridine observed) for the reaction of ammonia with brominated 3-substituted acrylates, as well as providing >80% de at

the α-center (*73*). α-Bromocarboxylates containing a chiral diol acetonide substituent have also been converted into aziridine-2-carboxylates via reaction with ammonia. The diol substituent was subsequently converted into a 3-vinyl substituent (*74*).

A chiral imidazolidinone auxiliary was used to induce asymmetry during the addition of O-benzylhydroxylamine to α,β-unsaturated acids, initially generating β-amino acid derivatives (see Scheme 8.8). The stereochemistry at the newly formed chiral center could be almost completely controlled by the type of Lewis acid employed during the addition, allowing for a single chiral auxiliary enantiomer to be used for the selective preparation of either diastereomer of the Michael addition adduct. The same Lewis acids were then combined with triethylamine to induce cyclization of the β-amino amides to the aziridines, giving products with *trans* stereochemistry via intramolecular enolate displacement of the N-OBn group (*75–77*). N-Boc O-Bz hydroxylamine has also been added using NaH for a one-step conjugate addition/cyclization (*78*). The benzyl ester of the aziridine was obtained by treatment of the aziridine imidazolidinone with lithiated benzyl alcohol, a procedure that also allows for recovery of the chiral auxiliary. This reaction is an asymmetric version of earlier cyclizations in which the methoxy group was displaced from methoxylamine (*45, 63*). An optically active diaryl sulfinimide amine nucleophile also undergoes reaction with

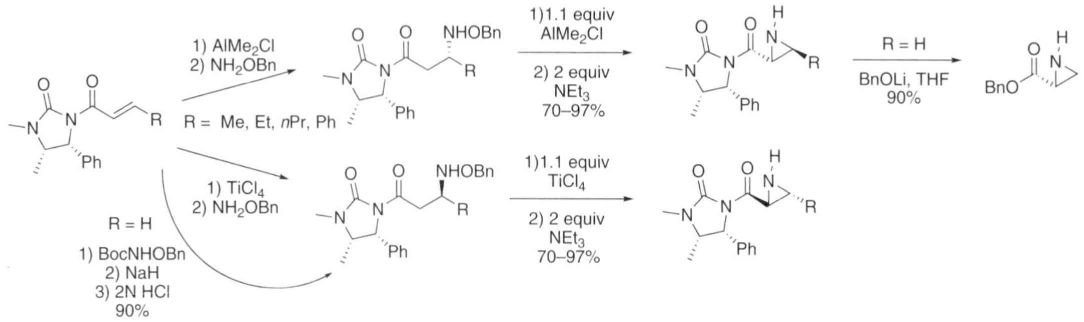

Scheme 8.8 Asymmetric Aziridine Synthesis via Intramolecular Cyclization (*75–78*).

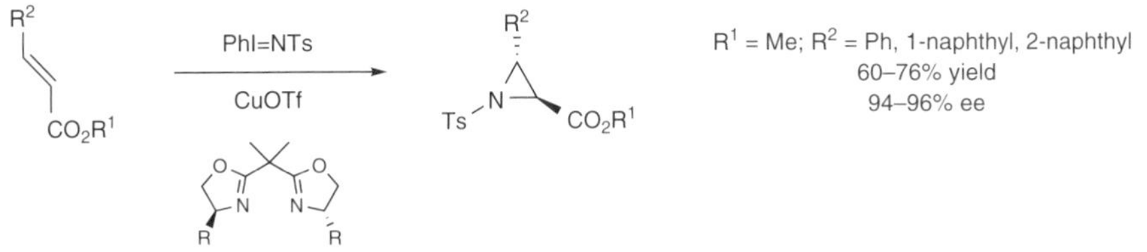

Scheme 8.9 Cu-catalyzed Aziridation of Olefins (81).

acrylates, with Michael addition followed by an intra-molecular displacement of the diaryl sulfide to give the aziridine with modest diastereoselectivity (79).

Copper catalysts have been employed for the aziridi-nation of olefins with PhI=NTs (see Scheme 8.9) (80–82). Chiral bis(oxazoline) copper complexes induced enantioselectivities of 94–96% ee for reaction with aryl-substituted acrylates (80). Acrylates can also be converted into aziridines using homochiral 3-ace-toxyamino-2-[(S)-1-hydroxy-2,2-dimethylpropyl]qui-nazolin-4-[3H]-one as an aziridating reagent. In the presence of Ti(OtBu)$_4$, methyl acrylate, tert-butyl acry-late, and methyl methacrylate reacted with 90% de, >98% de, and 70% de, respectively. Without Ti catalyst the opposite diastereomer was produced, with 28–50% de. However, the auxiliary was not removed from any of these products (83). A practical electrochemical olefin aziridation was reported in 2002, with methyl cinna-mate, dimethyl fumarate, and methyl 2-methylene-3-hydroxy-3-phenylpropanoate aziridinated in 73–92% yield with N-aminophthalimide as the amine source (84). The same aminating reagent was employed in combination with lead tetraacetate for aziridination of eight different acrylates, with asymmetry induced by a camphor amide auxiliary (85). Alternatively, a chiral ligand provided 67–95% ee for aziridations of achiral acrylate oxazolidinone amides (86).

Two new diaziridine-based chiral aziridation reagents have been applied to α,β-unsaturated amide substrates (see Scheme 8.10). An optically active 3-monosubsti-tuted diaziridine generated aziridines with high trans selectivity and up to 98% ee, while an optically active 3,3-dimethyl diaziridine gave cis products with >99% ee, albeit in low (4%) yield (87). Achiral versions of these reagents provide good yields of both cis and trans-aziridine-2-carboxamides.

Aziridines have been obtained from imines and diaz-oacetates (see Scheme 8.11). Initially, copper complexes were employed as catalyst, although yields were modest (88, 89). Diastereoselectivity (cis/trans) varied depend-ing on the substituents, while an attempt to induce asym-metry by using (−)-menthyl diazoacetate gave only 25% de (88). Chiral ligands on the copper catalyst also gave poor results, with a maximum of 67% ee (88, 89). Another report examined various catalysts (BF$_3$·Et$_2$O, Zn(OTf)$_2$, Yb(OTf)$_3$, TiCl$_2$(OiPr)$_2$, Cu(OTf)$_2$) for the reaction of ethyl diazoacetate with imines, preferentially producing cis-aziridines. Chiral ligands were again examined, but gave poor enantioselectivity (90). A race-mic synthesis of cis-3-CF$_3$-aziridine-2-carboxylates combined CF$_3$-imines with ethyl diazoacetate in the presence of BF$_3$·Et$_2$O catalyst, giving cis products with 95:5 cis:trans selectivity and 90–93% yield (91). Nine lanthanide triflates in protic media (EtOH) were

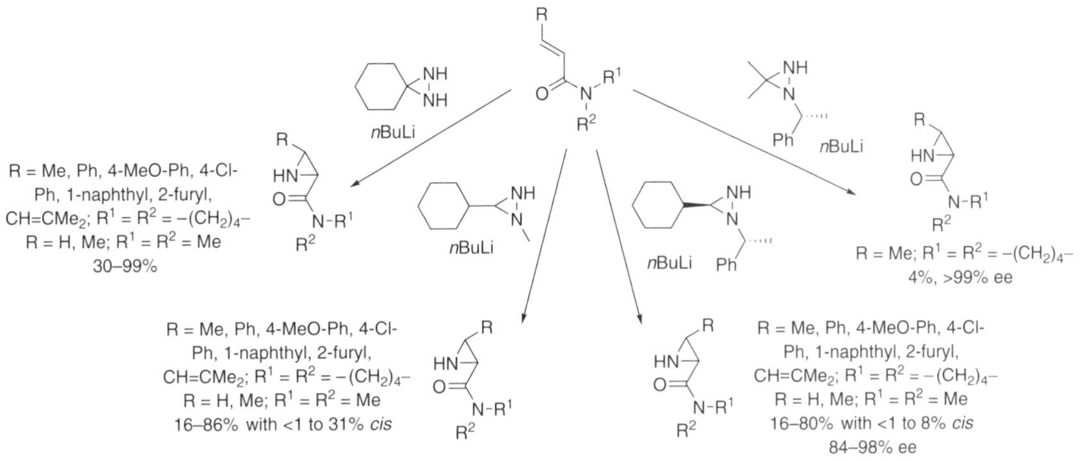

Scheme 8.10 Racemic or Asymmetric Aziridation of Olefins with Diaziridine (87).

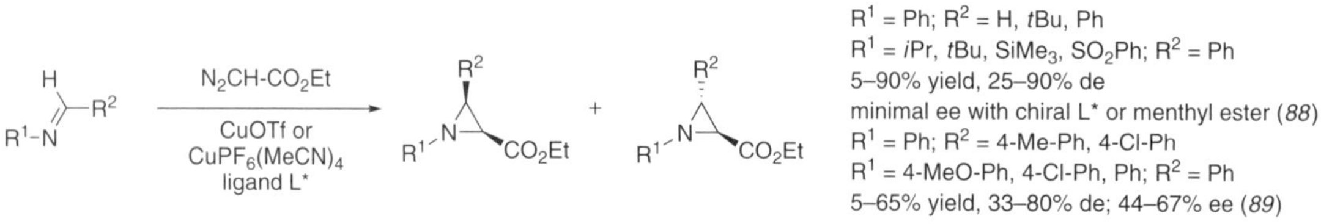
N2CH-CO2Et
CuOTf or
CuPF6(MeCN)4
ligand L*
CO2Et
CO2Et
R1 = Ph; R2 = H, tBu, Ph
R1 = iPr, tBu, SiMe3, SO2Ph; R2 = Ph
5–90% yield, 25–90% de
minimal ee with chiral L* or menthyl ester (88)
R1 = Ph; R2 = 4-Me-Ph, 4-Cl-Ph
R1 = 4-MeO-Ph, 4-Cl-Ph, Ph; R2 = Ph
5–65% yield, 33–80% de; 44–67% ee (89)

Scheme 8.11 Cu-catalyzed Aziridation of Imines.

compared for the aziridation of imines with ethyl diazoacetate, with $Nd(OTf)_3$ giving the best yields (*92*). $InCl_3$ catalyzed the reaction of ethyl diazoacetate with *N*-phenyl imines to give *cis*-3-aryl aziridine-2-carboxylates in 45–50% yield (*93*). *N*-Aryl imines (formed in situ) reacted with ethyl diazoacetate using $LiClO_4$ catalysis, producing *cis* isomers of *N*-aryl 3-substituted-aziridine-2-carboxylates in 75–91% yield with 64–100% de (depending on the C3-substituent) (*94*). Methyl diazoacetate has been added to aryl/alkyl imines with Rh(II)- and Mn(II)-exchanged Montmorillonite K10 clays as catalysts, giving the aziridine-2-carboxylates with 30–75% yields (*95*).

Hexahydro-1,3,5-triazines have been used as an alternative amine source for the reaction with alkyl diazoacetates, with $SnCl_4$ as catalyst (50–86% yield). A chiral triazine provided up to 52% de (*96*). The best asymmetry of these imine/diazoacetate reactions was induced in a 2000 report: ethyl diazoacetate was reacted with benzhydrylamine-derived imines in the presence of a (*S*)-VAPOL-triphenylborate catalyst, producing *cis*-substituted *N*-benzhydrylamine 3-substituted aziridine-2-carboxylates in 54–91% yield with 90–98% ee (3-substituent = aryl, *n*-propyl, *tert*-butyl, cyclohexyl) (*97*). A single diastereomer of a *cis*-aziridine product was obtained in 83% yield from ethyl diazoacetate and a chiral imine formed from benzhydrylamine and glyceraldehyde, using triflic acid as catalyst (*98*).

A variation of this procedure uses the ester group on the imine insead of the diazo group. Racemic Azy was prepared by the reaction of methyl *N*-Cbz-2-chloroglycinate with diazomethane, which proceeds via a [3+2] cycloaddition reaction with a transient acylimine intermediate (*99*) (see Section 3.9.6b). A similar synthesis reacted the benzophenone Schiff base of ethyl glycinate with difluorocarbene (*100*).

A more versatile route to enantiomerically pure aziridine-2-carboxylates, which allows for substitution at the C2 and C3 positions, was reported by Davis et al. (*101*). The lithium enolates of α-bromocarboxylates were added to enantiopure sulfinimines (see Scheme 8.12) in an aza-Darzens-type synthesis, producing *N*-sulfinylaziridines in 65–85% yield (*101–104, 343*). The *N*-sulfinyl auxiliary is easily removed under acid or base conditions, and can be converted to the *N*-tosyl group (which often provides for superior ring-opening selectivity) by a simple and quantitative oxidation with *m*-CPBA (*102*). The *N*-sulfinyl or *N*-tosyl aziridines were converted into azirines (see Section 8.3) by treatment with LDA or by other methods (*105–107*); Grignard addition then regenerated the aziridine, introducing a C3 substituent (*105, 107*). Chiral 2*H*-azirine-2-carboxylic esters also undergo diastereoselective addition of five-membered aromatic nitrogen heterocycles to give *trans*-3-(heteroaryl)-aziridine-2-carboxylates (*108*). A mild procedure for removing the tosyl group from an ester of Ts-Azy, using Mg in MeOH, has been described (*24*).

An aza-Darzens reaction of the chiral enolate derived from (2*S*)-bromoacetyl camphor sultam with C3-aryl

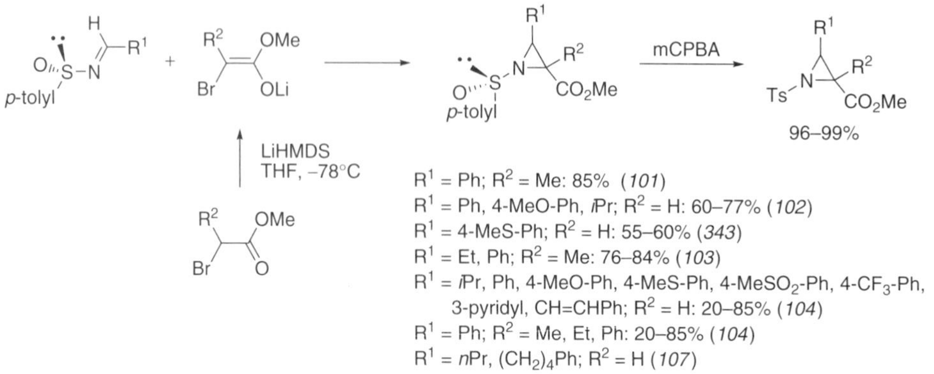
p-tolyl
OMe
OLi
Br
LiHMDS
THF, −78°C
OMe
mCPBA
CO2Me
p-tolyl
Ts
CO2Me
96–99%
R1 = Ph; R2 = Me: 85% (101)
R1 = Ph, 4-MeO-Ph, iPr; R2 = H: 60–77% (102)
R1 = 4-MeS-Ph; R2 = H: 55–60% (343)
R1 = Et, Ph; R2 = Me: 76–84% (103)
R1 = iPr, Ph, 4-MeO-Ph, 4-MeS-Ph, 4-MeSO2-Ph, 4-CF3-Ph,
3-pyridyl, CH=CHPh; R2 = H: 20–85% (104)
R1 = Ph; R2 = Me, Et, Ph: 20–85% (104)
R1 = nPr, (CH2)4Ph; R2 = H (107)

Scheme 8.12 Aziridine Synthesis via Bromoenolate Addition to Sulfinimine.

N-diphenylphosphinyl imines gave predominantly *cis*-substituted 3-aryl-aziridine-2-carboxylates when the aryl group was unsubstituted or 3- or 4-substituted, but *trans* products with sterically hindered 2-substituted aryl groups (*109*). *N*-Sulfonyl imines reacted with *N,N*-dialkylcarbamoylmethyl dimethylsulfonium bromides in the presence of solid KOH to produce aziridinyl carboxamides in 75–98% yield (see Scheme 8.13). A chiral sulfonium salt induced up to 70% ee (*110*).

A general route to 3-substituted aziridine-2-carboxylates from amino acids was reported in 1999. The amino acid was converted into an *E*-methoxy alkene via an amino aldehyde intermediate (see Scheme 8.14). Addition of PhSeCl followed by oxidation generated the aziridine (*111*).

Once formed, homochiral aziridine-2-carboxylates with a chiral *N*-substituent can be deprotonated and α-alkylated with high diastereoselectivity (*67*). For example, *N*-(1-phenyl-2-methoxyethyl) aziridine-2-carboxylates were deprotonated and α-alkylated. Special conditions were required to avoid decomposition or self-condensation, and the products were obtained with retention of chirality at the α-center, rather than being controlled by the *N*-auxiliary (*69*). 2,3-Disubstituted aziridine-2-carboxylates were prepared via alkylation of homochiral 3-substituted *N*-diphenylmethyl aziridine-2-carboxylate ethyl esters, using LDA for deprotonation. The electrophile added *trans* to the 3-substituent with complete C2 diastereoselectivity (*112*).

8.2.3 Ring-Opening Reactions: Heteroatom Nucleophiles

A variety of nucleophiles have been used to open aziridine-2-carboxylates, with several methods used to ensure that the desired regiospecificity (C3–N bond cleavage for α-amino acid synthesis) is achieved. Heteroatomic nucleophiles, Wittig reagents, and organometallic reagents have all been added (see Schemes 8.15, 8.16 and Tables 8.1, 8.2). The reactivity of *N*-unactivated aziridines toward nucleophiles is generally low, so an electron-withdrawing protecting group on the nitrogen is required (such as an acyl or sulfonyl group). Protonation or Lewis acid coordination of the aziridine further increases the reactivity.

Initial attempts at opening aziridine-2-carboxylates involved heteroatom nucleophiles. Alcohol or thiol nucleophiles were added in large excess or used as a co-solvent in combination with catalysis by boron trifluoride etherate (*113–115*). Alcohols can also open *N*-unactivated aziridines (e.g., with *N*-alkyl groups) under these conditions (*116*). In 1995 it was reported that good yields of product were obtained from *N*-acyl aziridine-2-carboxylates with only 3 equiv of alcohol nucleophile (rather than a large excess) when BF$_3$ catalysis was employed (*117*); a similar observation was made in 1991 when a reasonable yield (64%) was achieved with an almost equimolar ratio of reactants (*118*). A variety of alcohol and glycol amino acid side chains have since been introduced by this method (*8*).

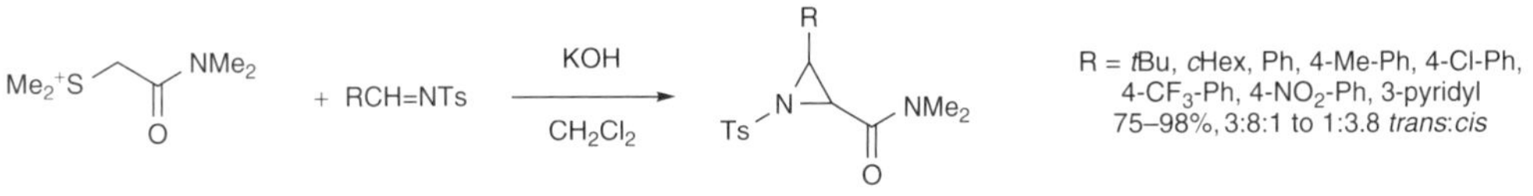

Scheme 8.13 Aziridination with *N*-Sulfonylimines (*110*).

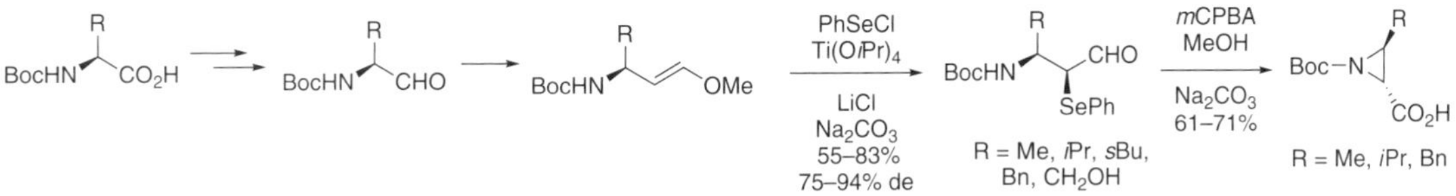

Scheme 8.14 Synthesis of Aziridine-2-Carboxylates from Amino Acids (*111*).

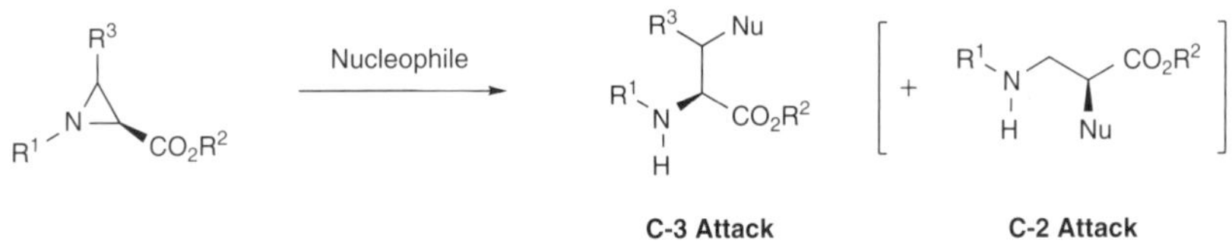

Scheme 8.15 Opening of Aziridine-2-Carboxylates with Nucleophiles (see Table 8.1).

Table 8.1 Opening of Aziridine-2-carboxylates with Nucleophiles (see Scheme 8.15)

R^1	R^2	R^3	Nucleophile (Nu)	Conditions	Yield	Reference
Ts	CO_2R^2 = CONHPh	Me	OAc Cl	AcOH, BF_3 HCl	88–100%	1968 (21)
H	Et	CO_2Et	OH	$HClO_4$	—	1971 (372)
CO_2Et	Et	Me,Me	OAc	AcOH, reflux	28%	1971 (373)
Ts	CO_2R^2 = $CONH_2$	$(CH_2)_2NH_2$	NH_2	NH_3	52–87%	1977 (29) 1978 (27) 1980 (28)
CH(Me)Ph	Me Et iPr tBu	H	OH	$HClO_4/H_2O$	—	1978 (68)
Cbz-Gly Cbz-Ala	CO_2R^2 = CO-Gly-OBn	H Me	OPO_3H_2	85% H_3PO_4	66–82%	1979 (137)
H	iPr	H Me	F	HF, pyridine	45–50%	1979 (60)
Cbz-Gly, Cbz-Phg	CO_2R^2 = CO-Gly-OBn	H, Me	OAc OCOR	Ac_2O RCO_2H RCO_2H = Cbz-Gly-OH, Cbz-Ser-OH, Cbz-MeVal-OH, Boc-Leu-OH, Boc-Met-OH, Boc-Pro-OH, Boc-Phe-PH, Boc-Leu-Leu-OH, Boc-Pro-Ser-OH	80–89% 74–97%	1979 (138)
Cbz-Gly	CO_2R^2 = CO-Gly-OBn	H	NHPh NEt_2	$PhNH_2$ Et_2NH	72%	1980 (139)
(2-NO_2-3-BnO-4-Me-Ph)CO	CO_2R^2 = CO-Val-Pro-OtBu	Me	OCOR	Cbz-Sar-MeVal-OH	54%	1980 (17)
Trt	Me	H Me Ph	F	HF, pyridine	48–51%	1980 (129)
H	iPr	Ph	F	HF, pyridine	70%	1980 (130)
H	iBu	H Me	F	HF, pyridine	60–95%	1980 (128)
Cbz-Gly Cbz-Ala Cbz-Phe	CO_2R^2 = CONHBn	Me	NHR OCOR	H_2NR H-Ala-OEt, H-Phe-OEt, H-Val-OEt RCO_2H Boc-Leu-OH, Cbz-Gly-OH, Boc-Leu-Leu-OH	85–100% 80–97%	1981 (18)
Cbz	Me	Me	SCOPh	PhCOSH, $BF_3·Et_2O$	55%	1982 (19)

R^1	R^2	OCOR	Reagent	Yield	Year (Ref.)
Cbz	CO_2R^2 = CO-Val-Pro-OtBu; H; Me	OCOR	Boc-Sar-MeVal-OH	80–81%	1982 (37)
Cbz	H; Me; Bn	OMe, OiPr, OsBu, OtBu, OcHex, OBn, OPh	ROH, $BF_3 \cdot Et_2O$	27–100%	1982 (114)
CO_2Et	Bn	$SCH_2CH(Me)CO_2Me$	RSH, $BF_3 \cdot Et_2O$	—	1983 (374)
H; Me; tBu; Bn	CO_2R^2 = CN, $CONH_2$; H; Me; Ph; 4-Cl-Ph	F	HF, pyridine	45–85%	1983 (62)
Cbz	H; Me	SH, SMe, SiPr, SsBu, StBu, ScHex, SBn, SPh, $SCH_2CH(NHCbz)CO_2Bn$	RSH, $BF_3 \cdot Et_2O$	52–92%	1983 (113); 1983 (115)
Ts; Cbz; p-NO_2-Cbz; p-NO_2-Bz	Me; H	$C(=PPh_3)CO_2Me$	$Ph_3P=CHCO_2Me$	30–66% C3 attack; 0–13% C2 attack	1987 (146)
Cbz	Bn	$O(CH_2)_2Cl$, $S(CH_2)_2Cl$	RXH, $BF_3 \cdot Et_2O$	77–88%	1987 (375)
H	Me	SPh, $SCH_2CH(NH_2)CO_2H$	RSH, $BF_3 \cdot Et_2O$	76–90%; 1:1 to 4.3:1 C2:C3 attack	1987 (135)
H	(−)-menthyl	S(4-MeO-Bn)	RSH, $BF_3 \cdot Et_2O$	60–70%	1988 (376)
Cbz	Ph; H; Me; Bn	5-Me-indol-3-yl, 5-MeO-indol-3-yl, 4-CbzNH-indol-3-yl, 4-MeO-indol-3-yl, 6-Cl-indol-3-yl, 4-NO_2-indol-3-yl, 4-(MeVal-OMe)-indol-3-yl	indole, $Zn(OTf)_2$	3–46%	1989 (143)

(Continued)

Table 8.1 Opening of Aziridine-2-carboxylates with Nucleophiles (continued)

R^1	R^2	R^3	Nucleophile (Nu)	Conditions	Yield	Reference
Ts	tBu	H	Me Et nPr iPr nBu	$RMgBr + CuBr \cdot SMe_2$	10–50% C3 attack 20–55% C2 attack	1989 (*147*) 1993 (*7*)
H	Me	Ph	SPh	$RSH, BF_3 \cdot Et_2O$	67%	1989 (*39*)
Cbz	Me	H	1-Me-indol-3-yl	indole, $BF_3 \cdot Et_2O$	44–69%	1990 (*144*)
Boc	Me	Me	5-MeO-1-Me-indol-3-yl 5-Me-1-Me-indol-3-yl 5-F-1-Me-indol-3-yl 5-Cl-1-Me-indol-3-yl			
Ts	Et	CO_2Et	Me nBu N_3 I Br	$LiMe_2Cu$ $LiBu_2Cu$ NaN_3 MgI_2 $MgBr_2$	54–81%	1990 (*43*)
Ts	Me	H Me	OCOR	H-His-OH. NaOH	45–46%	1991 (*10*)
Cbz	Me	H	OH Cl SBn	20% $HClO_4$ $TiCl_4$ $RSH, BF_3 \cdot Et_2O$	40–100%	1991 (*54*)
Trt		^{2}H				
Boc	Me	H	$O(CH_2)_2OBn$ $O(CH_2)_2OCH_2CH(NHBoc)CO_2Me$	$ROH, BF_3 \cdot Et_2O$	64%	1991 (*118*)
Cbz	Pac					
iPr	Me	H	OMe OiPr OtBu OiBu OCH_2CCH	$ROH, BF_3 \cdot Et_2O$	79–89%	1992 (*116*)
Bn						
H	Me	Ph	Cl SPh indol-3-yl	HCl $ROH, BF_3 \cdot Et_2O$ indole, $BF_3 \cdot Et_2O$	53–85%	1992 (*136*)
		4-MeO-Ph				
		4-NO_2-Ph				
Ts	H	H	$CCSiMe_3$	$LiCCSiMe_3$	30–79%	1994 (*148*)
Pmc		^{2}H				
Cbz	Me	H	$O(CH_2)_2OBn$ $O(CH_2CH_2O)_2Bn$ $O(CH_2CH_2O)_3Bn$ $O(CH_2CH_2O)_2N_3$ $OCH_2CH \cdot O_3N_3$	$ROH, BF_3 \cdot Et_2O$		1994 (*8*)

Ac	Me	H	NaN₃ → N₃	50% 1:1 C3:C3 attack	1995 (71)
Ts	H	H	R₂CuCNLi₂ → Me, nBu, tBu, Ph, CH=CH₂	54–68% C3 attack	1995 (150)
Ts	H	H	LiCCTMS → CCTMS	70% C3 attack	1995 (150)
Ts	H	H	R₂CuCNLi₂ → Me, nBu	30–34% mainly C2 attack	1995 (150)
Ac	Bn	H	ROH, BF₃·Et₂O → OBn, O(CH₂)₂Ph, OCH(Me)Ph, OCH₂CH(OMe)Ph	45–71%	1995 (117)
Ts	tBu	H	RNH₂ → NH₂, NHMe, NHEt, NHnPr	86–95% 86:14 to 89:11 C3:C2 attack	1995 (123)
Ac	Me	H	AcOH, AcNH₂, TsOH, HCl, MeOH BF₃·Et₂O, BnSH BF₃·Et₂O, CH₂CN BF₃·Et₂O, indole ZnOTf, NaN₃ → OAc, NHAc, OTs, Cl, OMe, SBn, CH₂CN, indol-3-yl, N₃	25–55% C3 attack 0–40% C2 attack	1995 (134)
Boc	tBu	H	pyrazole, 1,2,4-oxadiazolidin-3,5-dione → pyrazol-1-yl, 1,2,4-oxadiazolidin-3,5-dione	49–80%	1996 (377)
Cbz	Me	H	2-Me-indole, ZnOTf → 2-Me-indol-3-yl	68%	1996 (378)
SO(p-tolyl)	Me	CH=CHnC₁₃H₂₇	H₂O → OH	59–72% 100% C3 attack	1996 (379)
Boc	tBu	H	RMgX, CuBr·Me₂S → Me, nPr, iPr, nBu, iBu, nHex, Ph, CH=CH₂	50–85%	1996 (11)

(Continued)

Table 8.1 Opening of Aziridine-2-carboxylates with Nucleophiles (continued)

R^1	R^2	R^3	Nucleophile (Nu)	Conditions	Yield	Reference
Boc	$CO_2R^2 = CONHtBu$	H	Me nBu iPr Ph	$RMgX$, $CuBr\cdot Me_2S$	29–82%	1996 (11)
CO_2Et	Me Et	nPr iPr $nC_{12}H_{25}$ cHex Ph	I Br	MgI_2 or $MgBr_2$, Et_2O or NaI or NaBr, Amberlyst 15, acetone	quantitative >99:1 C3:C2 attack quantitative 1:1 to 1:99 C3:C2 attack	1996 (132)
H Ac CO_2Me SO_2Me	Me	CO_2Me	SBn Cl I N_3 OMe $COCF_3$ OAc	—	20–94%	1997 (161)
CO_2Et	Me	nC_7H_{15}	Br	NaBr, Amberlyst	70% 100% C3 attack	1997 (265)
Boc	iPr	H	^{11}CN convert into $^{11}CO_2H$ $^{11}CONH_2$ $^{11}CH_2NH_2$	$^{11}CN^-$	—	1997 (131)
Boc	Et	Ph	^{18}F	$^{18}F^-$	—	1997 (131)
Cbz	Bn	H	indol-3-yl 1,2-Me$_2$-indol-3-yl 1-Me-indol-3-yl 2-Me-indol-3-yl 2-Me-5-MeO-indol-3-yl 1-Bn-indol-3-yl 2-Me-5-NO$_2$-indol-3-yl 2-Me-4-OH-indol-3-yl	indole, $Sn(OTf)_3$	22–85%	1998 (145)
Pmc	H	H ^2H	CCTMS	LiCCTMS	30% C2 attack	1998 (149)

Ac	$CO_2R^2 = CONHCH(Me)Ph$	Me	OAc	Ac_2O, pyridine	80–85%	1998 (57)
Cbz	$CO_2R^2 = CONHMe$	H	HN(3-Me-Bn)	H_2N(3-Me-Bn), Δ	79%	1998 (124)
p-NO_2-Cbz	Me	H	$OC(Me)_2CH=CH_2$	$HOC(Me)_2CH=CH_2$, $BF_3 \cdot Et_2O$	75%	1999 (380)
Cbz	Me	H; Me	$OC(Me)_2CH=CH_2$	$HOC(Me)_2CH=CH_2$, $BF_3 \cdot Et_2O$	72–88%	1999 (381)
H	$CO_2R^2 = CO$(imidazolidinone)	Ph	OAc; OBz	Ac_2O; BzCl	90%	1999 (76)
Ts	Me; Et	H; Ph; 4-MeO-Ph; 3-PhO-Ph; 3,4-(OCH_2O)-Ph; $(CH_2)_2$Ph	H	polymethyl hydrosiloxane; Pd/C	80–95%; C3 attack for R^3 = aryl; C2 attack for R^3 = alkyl	1999 (133)
$CH(Ph)_2$	Et	3,4-$(AcO)_2$-Ph	H	HCO_2H/MeOH/Pd black	72%	2000 (97)
o-Ns	OBO ortho ester	H	CCTMS; CCPh; $CCCH_2OTHP$; $CCCH_2CH_2OTHP$; CC(6,7-dideoxy-α-D-glucopyranoside); CC(β-D-galacto-pyranosyl)	LiCCR; THF; 0°C to rt	82–98%	2001 (16)
Bn	c-CF_3	Et	Cl; OH; SPh; SBn; SEt	HCl; CF_3CO_2H; PhSH/CF_3SO_3H; BnSH/CF_3SO_3H; EtSH/CF_3SO_3H	50–98%; 100% C2 attack	2001 (91)
o-Ns	tBu	H	NHBn; $NHCH(Me)CO_2Bn$; $NHCH(CH_2Ph)CO_2Me$	Et_3N	88–96%; 3.4:1 to 17:1 C3:C2 attack	2001 (15)
H	tBu	nPr; nPr; $n$pr	Me; iPr; nBu	H_2/Raney Ni	71–86%; 100% C2 attack	2002 (107)
Cbz	Bn	nPr; iPr; Ph	S(4-MeO-Ph); OAc	HS(4-MeO-Ph), BF_3; Ac_2O	60–95%	2002 (48)
Cbz	Me	Me	SBn	BnSH, $BF_3 \cdot Et_2O$	57%	2002 (20)

(Continued)

Table 8.1 Opening of Aziridine-2-carboxylates with Nucleophiles (continued)

R^1	R^2	R^3	Nucleophile (Nu)	Conditions	Yield	Reference
cinnamyl-Pro	Et	Ph	OH	TsOH, H_2O	72–74%	2002 (382)
			MeOH, allyl-OH	ROH, TsOH		
Boc-Ala Boc-Val Boc-Thr Boc-PheAc-Gly- Gly-Phe- Ac–Gly-Gly-Phe Ac-Lys-Ser-Gly- Phe	CO_2R^2 = CO-Gly-OBn, Ala-OH, Asp-OH	H	SEt SₜBu GalNAcα1-S NeuNAcα2-6GalNAcα1-S	DBU on solid phase	17–93%	2004 (140)
Boc	CO_2R^2 = $CONR_2$ R = Me, Et, $(CH_2)_5$	H	nBu	RMgX, CuBr·Me_2S	19–43%	2004 (12)
H	Et	3,4-($-CH_2OCH_2-$)-Ph	H	HCO_2NH_4, 10% Pd/C, MeOH, reflux	92%	2004 (51)
CH(Me)Ph	Et	H	N_3	NaN_3, $AlCl_3$	86–95%	2005 (127)
Boc	CO_2R^2 = CONHBn, CO-Xaa-NHBn	H	PhSeH	Et_3N	62–93% 100:0 to 0:100 C2:C3 attack	2005 (36)

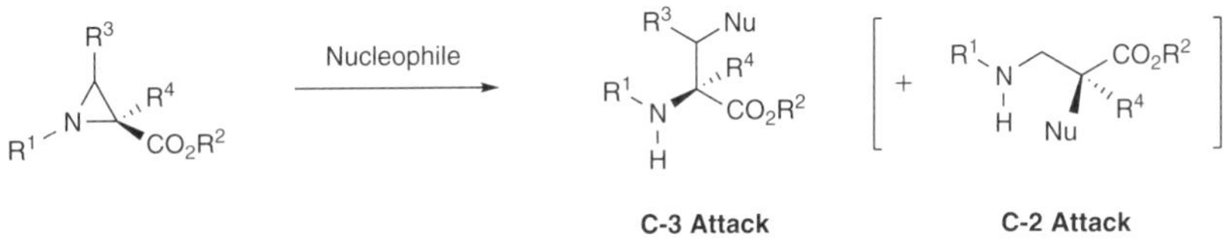

Scheme 8.16 Opening of 2-Substituted Aziridine-2-Carboxylates with Nucleophiles.

Table 8.2 Opening of 2-Substituted Aziridine-2-carboxylates with Nucleophiles (see Scheme 8.16)

R^1	R^2	R^3	R^4	Nucleophile (Nu)	Conditions	Yield	Reference
H	Bn	H	Me	S(4-MeO-Bn)	(4-MeO-Bn)SH, $BF_3 \cdot Et_2O$	78%	1995 (42)
Ts	Me	Ph	Me	H	Pd, HCO_2H	99%	1996 (101)
				OH	50% TFA, MeCN	73–75%	
Ts	tBu	H	Me	Cl	HCl	48–95%	1996 (47)
				OH	HCO_2H	100:0 to 1:4 C3:C2	
				NHBn	$BnNH_2$	attack	
				N_3	NaN_3		
				SPh	PhSH		
				indol-3-yl	indole, MeMgBr, $CuBr \cdot SMe_2$		
				iPr	$RMgX$, $CuBr \cdot SMe_2$		
				Ph			
$CH(Ph)_2$	Et	4-Br-Ph	Me	H	$BH_3 \cdot Me_3N/TFA$	87%	2005 (112)

Hydrolytic opening of unsubstituted aziridine-2-carboxylate (Azy) with $HClO_4/H_2O$ produced Ser (68). A *cis* N-sulfinyl 3-substituted-aziridine-2-carboxylate with an unsaturated substituent was regioselectively hydrolyzed using acetone/TFA/H_2O to give the *threo* β-hydroxy α-amino acid isomer exclusively. Surprisingly, by employing trifluoroacetic anhydride in CH_2Cl_2 the *erythro* isomer was obtained with 76% de, possibly via a Pummerer-type rearrangement (68). A regioselective intramolecular ring opening has been reported for N-acyl aziridine-2-carboxylate amides (see Scheme 8.17). In chloroform, the oxygen of the N-acyl carbonyl group spontaneously attacked the C3 position to form an oxazoline. Oxazoline hydrolysis provided β-hydroxy-α-amino acids (119). This rearrangement has been promoted by Lewis acid catalysis combined with microwave irradiation for N-Boc Azy-2-carboxylate amides and

esters, giving >99:1 regioselectivity for the Ser oxazoline over the isoserine β-amino acid (120). N-Acyl aziridines can also be isomerized into oxazolines by treatment with NaI in acetone (61, 121). MeAzy residues with an imidazolidinone amide group have been acylated with an amino acid, ring-expanded to an oxazoline by treatment with $BF_3 \cdot Et_2O$, and then opened with TsOH/MeOH-H_2O to give optically pure Thr-containing dipeptides (122).

For amine nucleophiles, a bulky *tert*-butyl ester has been used to direct the regioselectivity of attack to the C3 carbon of N-tosyl Azy-2-carboxylate. With no ester group present, attack occurred at both C2 and C3 (123). In this case, C2 attack still generates 2,3-diaminopropionic acid, but with reversed N-substituents and inverted C2 stereochemistry. N^β-(3-Me-Bn)-2,3-diaminopropionic acid was prepared by opening Cbz-Azy-NHMe

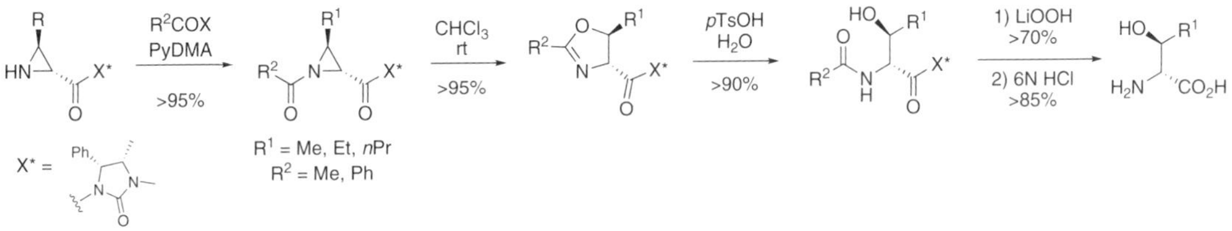

Scheme 8.17 Intramolecular Ring Opening (119).

with *m*-methylbenzylamine, with complete regioselectivity observed (*124*). Benzyl amine and amino esters were used to open *N*-(2-nitrobenzenesulfonyl) Azy-O*t*Bu in the presence of Et₃N, with 88–96% yields and 3.4:1 to 17:1 C3:C2 regioselective attack (*15*). Racemic Ts-Azy-OMe and Ts-α-methyl-Azy-OMe were both treated with (*R*)-(+)-α-methylbenzylisocyanate, forming both diastereomers of a five-membered 2-imidazolidinone system. The isomers could be separated by chromatography or crystallization. *N*-Tosyl removal with Mg-MeOH and imidazolidinone hydrolysis with 6 N HCl produced either enantiomer of 2,3-diaminopropionic acid, or 2-methyl-2,3-diaminopropionic acid (*125*). An aziridinium-2-carboxylate has been employed as a non-isolated intermediate for ring opening. Ethyl *trans*-2,3-epoxy-3-phenylpropionic acid was converted into 2-(morpholin-4-yl)-3-chloro-3-phenylpropionate; reaction with primary and secondary amines generated an aziridinium ion that was then regioselectively opened to give *N*-substituted 2,3-diamino-3-phenylpropionic acids (*126*).

Sodium azide demonstrated no regioselectivity in ring opening *N*-Ac-Azy-2-carboxylate, but the products were used to confirm that the reaction proceeds via an S_N2-like mechanism, with inversion at the attacked center (*71*). *N*-(1-Phenylethyl) Azy-2-carboxylate ethyl or (−)-menthol esters were opened with NaN₃ in the presence of catalytic AlCl₃ to give β-azido-Ala in 86–95% yield, even on several hundred gram scale. Catalytic hydrogenation in the presence of Boc₂O provided *N*,*N*′-Boc₂ 2,3-diaminopropionate, while Ph₃P/H₂O azide reduction and *N*-Fmoc protection before hydrogenation gave Boc-Dap(Fmoc)-OR in 48% overall yield (*127*).

The regioselective introduction of halides by aziridine opening has proven to be difficult, although fluoride (from HF/pyridine) opened *N*-unsubstituted or substituted aziridines with good C3 regioselectivity (*60, 62, 128–130*). Ytterbium(III) triflate was used as a catalyst for the ring opening of *N*-Boc-2-cyano-3-phenylaziridine with ¹⁸F-fluoride, with hydrolysis giving β-[¹⁸F]fluoro-Phe (*59*). The corresponding 3-phenylaziridine-2-carboxylate has also been opened with ¹⁸F⁻ (*131*). Magnesium halides gave C-3 attack with *N*-ethoxycarbonyl aziridine-2-carboxylates with >99:1 regioselectivity, while sodium halides with Amberlyst 15 in acetone produced the C2 opened β-amino acid with 1:1 to 99:1 selectivity (*132*).

Polymethylhydrosiloxane/Pd-C, a soluble hydride source, has been used to open 3-substituted aziridine-2-carboxylates. With 3-aryl groups, the hydride opened the aziridine to give α-amino esters, while with 3-alkyl substituents β-amino esters were formed (*133*). 2,3-Disubstituted aziridine-2-carboxylates, prepared via alkylation of homochiral 3-substituted *N*-diphenylmethyl Azy-OEt, were reductively opened with BH₃.Me₃N/TFA, followed by diphenylmethane removal with Et₃SiH, producing α-substituted amino acids (*112*). Ethyl *N*-benzhydryl 3-(3, 4-diacetoxyphenyl)-Azy was opened by transfer hydrogenation with HCO₂H/MeOH/Pd black to give the ethyl ester of *O*-acetoxy L-DOPA in 72%

yield (*97*). The same procedure led to DOPA from ethyl 3-(3,4-methylenedioxy)-Azy (*51*). *N*-Unsubstituted-3,3-dialkyl-aziridine-2-carboxylates undergo regioselective reductive opening by H₂/Raney Ni to give 3,3-disubstituted-β-amino acids. The chiral substrates were derived from 2*H*-azirine-2-carboxylates via Grignard addition of the second 3-alkyl substituent (*107*).

A study of the addition of nucleophiles to Ac-Azy-OMe found that methanol, *p*-toluenesulfonic acid, acetic acid or acetonitrile added exclusively to the C2 carbon, but benzyl thiol, sodium azide, hydrogen chloride, and indole were less regioselective (*134*). Thiol addition could be directed to C2 by reacting *N*-unsubstituted aziridines in aqueous solution (*135*). As previously noted for the opening of aziridines with hydride, 3-aryl substituents induced C3 opening of *N*-unsubstituted aziridines (*136*). In contrast, racemic *N*-benzyl ethyl *cis*-3-CF₃-aziridine-2-carboxylate was opened with a variety of nucleophiles (Cl, OH, SPh, SBn, SEt), with the CF₃ group directing the ring opening to give exclusively C2 attack and β-amino ester products (*91*).

Unsubstituted L-Azy residues can be incorporated into peptides without ring opening by standard coupling techniques (*18*), or generated within the peptides from Ser and Thr residues. New amino acids can then be created within the peptide by ring opening with nucleophiles (see Scheme 8.15, Table 8.1). For example, Azy- or 3-MeAzy-containing peptides were reacted with phosphoric acid to give phosphonopeptides (*137*), with carboxylic acids or the carboxylic acid group of amino acids to give depsipeptides (*18, 37, 138*), or with secondary amines (*139*) or the amino group of amino acids (*18*) to give diaminopropionic acid derivatives. In all cases the desired C3–N cleavage occurred preferentially. Boc-L-Azy-OH and Fmoc-L-Azy-OH were incorporated into tri-, penta- and hexa-peptides by solid-phase synthesis. On-resin treatment with thiols in the presence of DBU produced peptides with *S*-alkyl Cys residues, including thioglycopeptides (*140*). PhSeH/Et₃N opened Azy residues within peptides to give predominantly the β-phenylseleno product, but 3-methyl-Azy residues tended to give mainly the α-opened products, depending on the peptide sequence (*36*). Peptides containing Azy or MeAzy residues can also be converted into the corresponding dehydroamino acid peptides by treatment with NaI in acetone (71–92% yield) (*141*) or with tertiary amines (56–93% yield) (*142*).

8.2.4 Ring-Opening Reactions: Carbon Nucleophiles

The successful opening of aziridine-2-carboxylates with carbon nucleophiles has only been achieved relatively recently (see Scheme 8.15, Table 8.1), as previous attempts generally resulted in attack of the nucleophile at the ester group. An exception is ring opening with indole, which proceeded in low yields in the presence of zinc triflate (*143*) and moderate yields with BF₃·Et₂O (*136, 144*) or Sc(OTf)₃ (*145*) catalysis. Baldwin et al. found

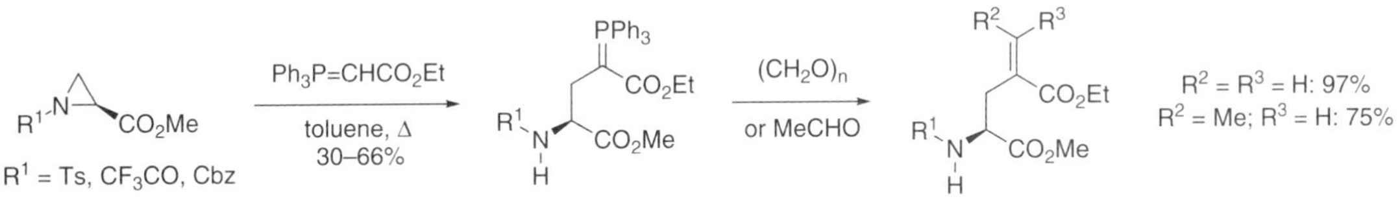

Scheme 8.18 Opening of Aziridine-2-Carboxylate with Wittig Reagents (*146*).

that *N*-tosyl or *N*-acyl aziridine-2-carboxylates could be opened with stabilized Wittig reagents. The resulting phosphorus ylide side chain was then reacted with carbonyl compounds (see Scheme 8.18) (*146*).

With *N*-sulfonamide activated aziridines, organolithium and Grignard reagents reacted preferentially with the ester carbonyl, but higher-order organocuprates gave mixtures resulting from both C2 and C3 attack, as well as reduced products from hydride attack (*7, 147*). Church and Young found that the undesired reaction of lithium trimethylsilylacetylide with the ester group of Ts-Azy-OMe was prevented by using the free acid of the aziridine (see Scheme 8.15, Table 8.1). Attack was directed entirely to C3 in 79% yield (*148, 149*). Removal of the tosyl group was difficult, so the more labile Pmc (2,2,5,7,8-pentamethylchroman-6-sulfonyl) protecting group was employed. These conditions were successfully extended to allow the reaction of a number of organocuprate reagents in 54–68% yield (*150*). Lithium acetylide, but not organocuprates, could also be used to open *N*-Ts-MeAzy (*150*). [2-^{2}H]- and [1,2-^{2}H$_2$]-D-propargylglycine were prepared via this route (*149*). The *N*-diphenylphosphinoyl group also activates aziridines towards ring opening and is much more easily cleaved than *N*-sulfonyl protecting groups, but when used to protect 2-carboxylate-substituted aziridines they underwent attack at the ester group instead of ring opening (*151*). It remains to be seen whether this problem can also be alleviated by using the unprotected acid. A mild method for removing the tosyl group without ring cleavage was recently reported, employing sodium naphthalenide (*152*).

In 2001, the acid group of *N*-(2-nitrobenzenesulfonyl)-Azy was protected as a bulky 4-methyl-2,6,7-trioxabicyclo[2.2.2]octane (OBO) ortho ester, preventing attack on the carbonyl. Ring opening with a variety of lithium acetylides proceeded in 82–98% yield, with the *N-o*-Ns group readily removed with K$_2$CO$_3$/PhSH (*16*). Baldwin et al. have instead used acid-labile *N*-Boc/*tert*-butyl ester protection and successfully reacted Grignard reagents in the presence of a Cu salt (*11*). Regioselectivity was much better than with the *N*-sulfonamide-activated aziridines (*7, 147*), although C2 attack was detectable at higher temperatures. No significant racemization was evident. The aziridine *tert*-butyl amide could also be opened (*11*). Several Boc-Azy dialkyl amides (derived from Boc-Ser) were opened with *n*-BuMgBr/CuBr in 19–43% yield (*12*).

Another carbon nucleophile is cyanide. Boc-Azy was opened with ^{11}CN$^-$, with the β-cyanoAla product converted into [4-^{11}C]-Asp, -Asn and -2,4-diaminobutyric acid. Unfortunately, complete racemization was reported (*131*).

8.2.5 Ring-Opening Reactions: Other Aziridine Substrates

Most amino acid syntheses from aziridines rely on aziridine-2-carboxylates as substrates, but it is also possible to use aziridine-2-hydroxymethyl derivatives, with the carboxylate group generated by oxidation of the hydroxymethyl group after ring opening. The aziridine-2-hydroxymethyl substrates can be obtained from 3-amino-1,2-diols, which in turn were obtained by regioselective opening of chiral epoxides derived from a Sharpless epoxidation (*153, 154*). This method was used in the enantioselective synthesis of α-methylserine (*46*) and a variety of β-arylalanines and β-methyl-β-arylalanines (*154*). The reaction of a variety of nucleophiles with 2-methyl-2-silyloxymethyl-substituted aziridine was compared with the corresponding *tert*-butyl 2-methylaziridine-2-carboxylate (*47*). NaN$_3$ and PhSH, which tended to attack at C2 of the carboxylate-substituted aziridine, gave exclusively C3 attack with the hydroxymethyl derivative, while HCl and HCO$_2$H gave C2 attack. Organocuprates were successfully added to both substrates, with ester addition no longer a competing side reaction (*47*).

Another route to 2-hydroxymethylaziridine is via reduction of the ester of Boc-L-Ser(TBS)-OMe, followed by cyclization of the serinol under Mitsunobu conditions to give *O*-TBS,*N*-Boc 2-hydroxymethylaziridine. Aziridine opening with organocuprates proceeded in 61–100% yields, with oxidation providing D-amino acids with high optical rotation values. Abu, Phe, Leu, β-cyclopentyl-Ala, and allylglycine were produced. The reaction sequence results in a net inversion of chirality, as the Ser side chain becomes the product carboxyl group (*155*). 2-Hydroxymethylaziridines can also be produced by LiAlH$_4$ reduction of aziridine-2-carboxylates. The reduction has been used in conjunction with a chelated hydride aziridine ring opening, allowing for regioselective opening to a γ-amino alcohol. Reoxidation of the hydroxymethyl group with NaIO$_4$/RuCl$_3$ produced β-amino acids (*103*).

Aziridine-2-carboxaldehydes are also useful. Organometallic addition to the aldehyde group of aziridine-2-carboxaldehyde, followed by aziridine ring opening with AcOH, produced 1-aryl-2-amino-1,3-propanediols. Cyclization to regenerate an aziridine and regioselective cleavage of the C–N bond by catalytic hydrogenation resulted in phenylalaninols, which were oxidized to phenylalanines (156).

D-Mannitol has been used to prepare chiral bis(aziridines) (157, 158). A range of nucleophiles, including organocuprates, halides, azide, and thiols, were employed to ring open both aziridines with the desired regioselectivity (158–160). Oxidative cleavage of the central diol then gave 2 equiv of the amino acid product, although only one example was carried through this procedure (158, 160) (see Scheme 8.19).

One strategy which ensures that poor regioselectivity during nucleophilic aziridine ring opening does not produce regioisomers is to use a C2-symmetric aziridine-2,3-dicarboxylate derived from tartaric acid (43). Organocuprates, halides, and azide anion reacted to give β-substituted Asp derivatives in 54–81% yield (see Scheme 8.20). Racemic trans-aziridine-2,3-dicarboxylates have also been opened with nucleophiles, giving exclusively erythro β-substituted Asp isomers (substituents = Cl, I, SBn, N₃, NH₂, OAc, OCOCF₃, OMe) (161).

D-Ribose has been converted into bicyclic analogs of aziridine-2-carboxylate (see Scheme 8.21) (162, 163). These bicyclic compounds show a different reactivity with nucleophiles compared to monocyclic Azy, as "soft" nucleophiles (e.g., thiols, acetate) attack the C2 position of the aziridine instead of C3 (44, 162, 164). C2 attack is also observed with N-methyl indole under

Lewis acid conditions (165). Hard nucleophiles (alcohols, benzylamine) tend to deacetylate the aziridine nitrogen and open the lactone ring (44). The desired C3 opening can be accomplished with alcohols as long as an appropriate N-protecting group is present, and is accompanied by lactone opening to give a polyhydroxy amino acid (44, 162). MNDO calculations have been used to attempt to explain the reactivity patterns of the rigid aziridine lactones compared to the aziridine carboxylates (44, 165).

As mentioned earlier, it is possible to convert N-sulfinylaziridine-2-carboxylates to optically active azirines by LDA-induced β-elimination of the sulfinyl group (see Section 8.3, Scheme 8.26) (166), with organometallic addition then regenerating an aziridine (105, 107). Aziridine carboxylates can also be converted to 3-halo-2-azetidinones by reaction with carboxylate activating reagents such as oxalyl chloride (167). Isomerization of N-acyl aziridine-2-carboxylates into oxazolines is possible under a variety of conditions (61, 119–121). Aziridines without a 2-carboxyl function have been used to prepare pipecolic acid derivatives by an LDA-induced aza-Wittig rearrangement-ring expansion (45) (see Scheme 8.22). Aziridines with an α,β-unsaturated ester substituent can be opened by reductive cleavage of the C–N bond or by organocuprates to give γ- or δ-amino acids, including (E)-alkene dipeptide isosteres (168–173).

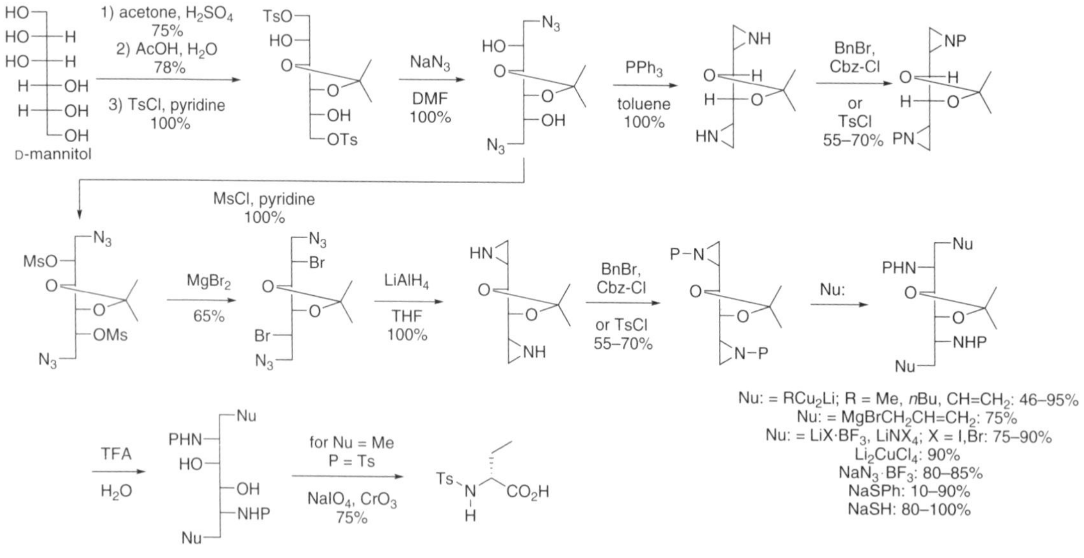

Scheme 8.19 Synthesis and Reaction of D-Mannitol-derived Bis-Aziridines (157–160).

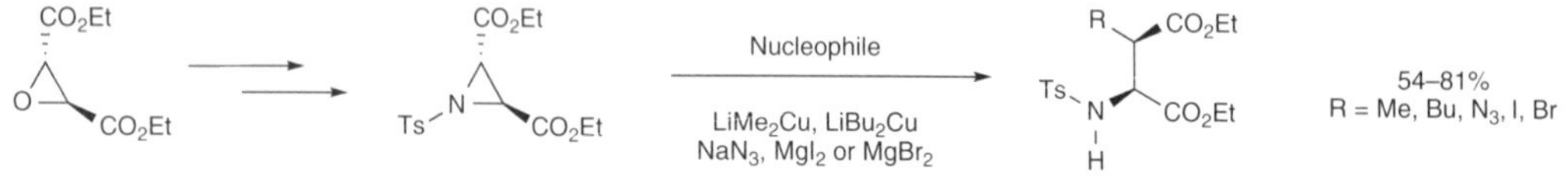

Scheme 8.20 Synthesis and Reaction of C2-Symmetric Aziridine-2,2-Dicarboxylate (43).

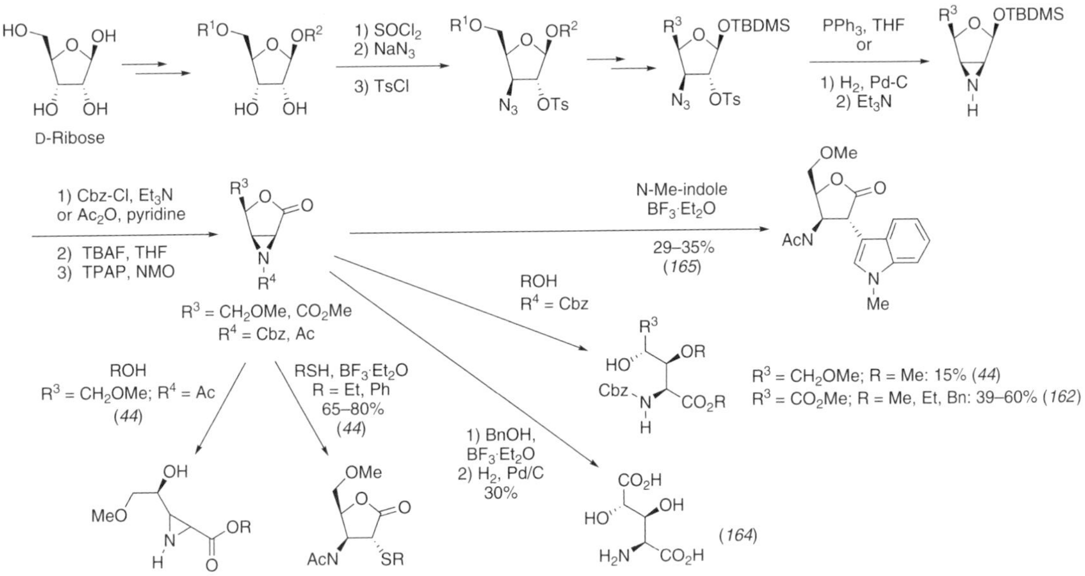

Scheme 8.21 Synthesis and Reaction of D-Ribose-derived Aziridine Lactone (44, 162–165).

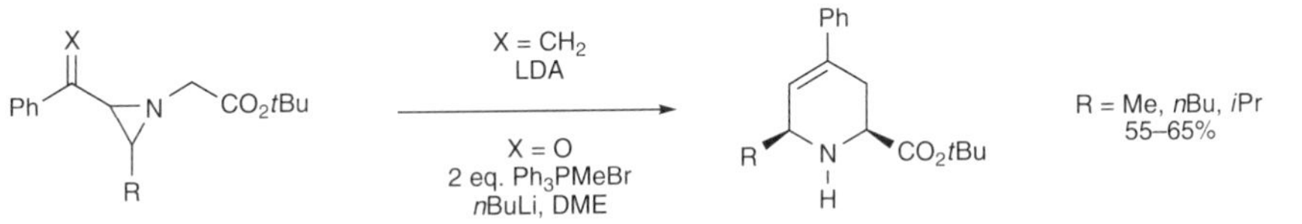

Scheme 8.22 Intramolecular Aziridine Ring-Opening (45).

8.3 Azirines

For amino acid synthesis, the most useful azirines are the cyclic amidines, 3-amino-2H-azirines, which were first synthesized in 1970 (174). A review of their synthesis and reactions was published in 1991 (175), with another overview in 2001 (176). A 2001 minireview of 2H-azirines includes some examples of azirine-2-carboxylate and 3-amino-azirine synthesis, and their use in amino acid synthesis (177). They are usually prepared from disubstituted carboxylic amides via a keteniminium salt (generated directly from the amide or via a thioamide intermediate), which reacts with sodium azide and eliminates N_2 to form the three-membered ring (see Scheme 8.23) (97, 174, 175, 178–183).

The 3-amino-2H-azirines are generally stable compounds which, under different reactions conditions, can be selectively ring opened by cleavage at either of the three ring bonds. Photolysis and thermolysis cleave the C–C bond, while strong protic acids break the C–N single bond. For amino acid synthesis the C=N double bond is hydrolyzed by nucleophilic attack of water at the C atom, leading to racemic α,α-disubstituted-α-amino acid amides or acids (Scheme 8.24) (174, 175).

Other nucleophiles can also be used, such as activated phenols (184) or enolizable cyclic diketones (185), which lead to N-substituted α,α-disubstituted-α-amino acid amides. Reaction with HF results in the addition of 2 equiv of fluoride to 2-substituted azirine-3-carboxylates, producing β,β-difluoro-β-substituted-α-amino-propionates (186, 187). Carboxylic acids add to give N-acyl derivatives (185, 188) via 1,2-addition of the acid followed by 1,2-ring cleavage and transfer of an acyl group. This reaction can be very useful for introducing an α,α-disubstituted-α-amino acid to the C-terminus of a peptide chain, a coupling which otherwise can be difficult to accomplish by conventional means due to steric hindrance (Scheme 8.25) (189–191). Hydrolysis of the terminal amide to liberate the free acid group can be achieved under quite mild conditions as long as the amide is disubstituted (192), with N-methylanilide removed particularly easily (189). No diastereoselectivity was observed during the ring opening of non-symmetrically substituted azirines (175), though the diastereomeric products (if applicable) can sometimes be separated.

Amino acids synthesized and incorporated into peptides by this method include Aib (189, 190, 193, 194),

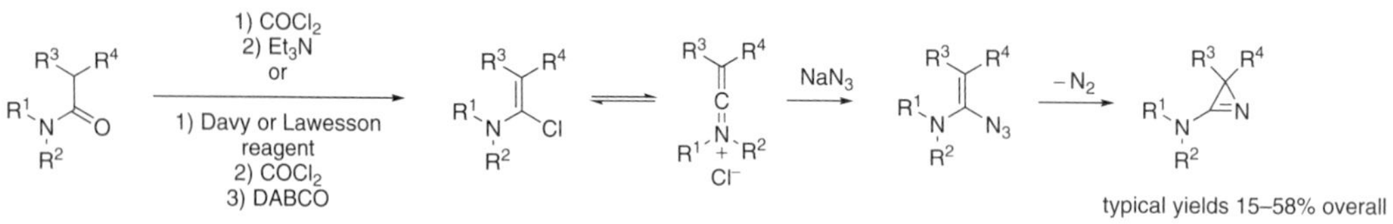

Scheme 8.23 Synthesis of 2-Aminoazirines.

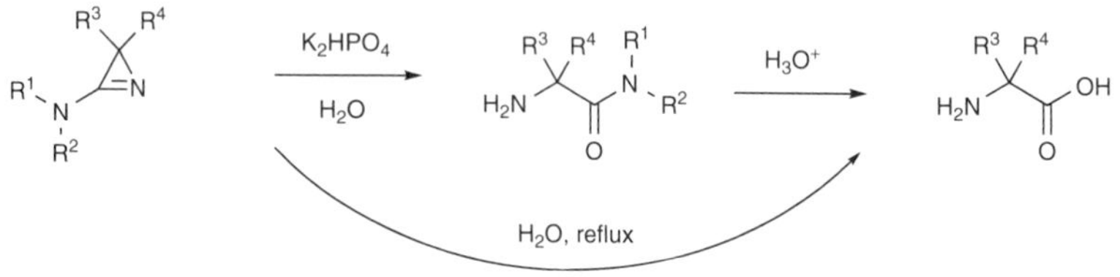

Scheme 8.24 Hydrolysis of Azirine (*174, 175*).

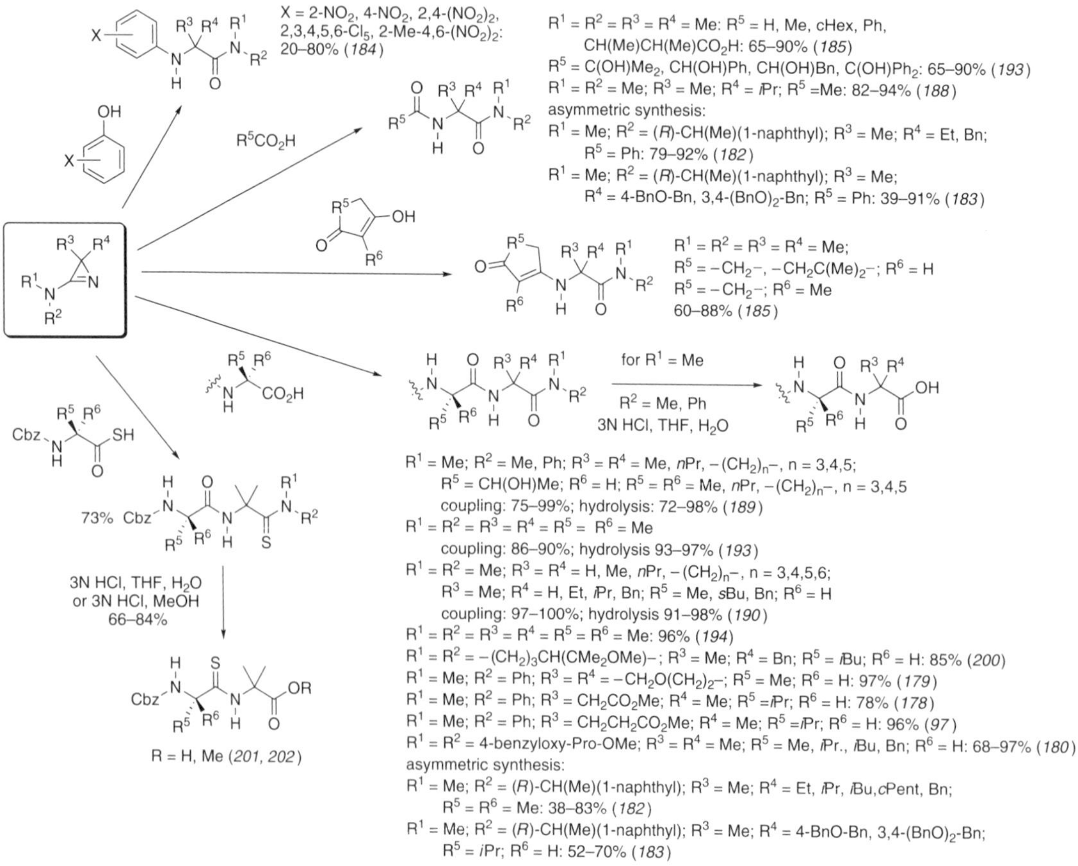

Scheme 8.25 Reactions of 2-Aminoazirines.

di(*n*-propyl)glycine (*189, 190*), α-methyl-Abu, α-methyl-Val, α-methyl-Phe, α-methyl-Asp (*178*), α-methyl-Glu (*391*), 1-amino-cyclobutane-1-carboxylic acid (*189, 190*), 1-amino-cyclopentane-1-carboxylic acid (*189, 190*), 1-amino-cyclohexane-1-carboxylic acid (*189, 190*), 1-amino-cycloheptane-1-carboxylic acid (*190*), and 3-amino-tetrahydrofuran-3-carboxylic acid (*179*). Peptides containing poly-α,α-disubstituted amino acid sequences have been assembled by this method (*189, 193, 194*). For example, azirines were employed to synthesize cyclic hexapeptides containing up to three Aib and one α-methyl-Phe residue (*195*), or cyclic hexadepsipeptides containing up to five Aib residues (*196*). The tetradecapeptide peptaibol antibiotic trichovirin I 1B was synthesized with four of five Aib residues introduced as azirines (*197*). Cyclic depsipeptides containing four α,α-disubstituted amino acids (a combination of Aib, α-methyl-Phe, and 1-aminocyclopentane-1-carboxylic acid) and one 3-hydroxypropionic acid derivative were constructed from azirine synthons (*198*). A solid-phase *N*- to *C*-terminus peptide synthesis procedure employing azirines to introduce Aib residues has been reported (*199*).

The azirines can be constructed upon the amino group of an amino acid, creating an azirine-containing dipeptide synthon. Azirine opening with the carboxylic acid of an *N*-protected amino acid produces a tripeptide with, for example, an Xaa-Aib-4-hydroxyproline sequence (*180*). A number of other Aib-Xaa dipeptide synthons were subsequently prepared (*181*). Non-symmetrically 2,2-disubstituted azirines have been prepared using a chiral amine auxiliary, *N*-methyl 1-(1-naphthyl)ethylamine, to form the initial acyl amide. The diastereomers were chromatographically separated to give optically pure amino acid synthons before azirine opening with carboxylic acids or thiocarboxylic acids or amino acids. The amino acids generated included α-methyl-Abu (*182*), α-methyl-Val (*182*), α-methyl-Leu (*182*), α-methyl-cyclopentylglycine (*182*), α-methyl-Phe (*182*), α-methyl-Tyr (*183*), and α-methyl-DOPA (*183*). The *C*-terminal chiral auxiliary amide could be removed by acid hydrolysis, though a methylamide byproduct was also generated (*183*). An azirine with the external 3-amino group contained within a chiral Pro derivative was employed for a synthesis of α-Me-Phe. The chiral substituent acts as a chiral auxiliary for chromatographic separation of the azirine diastereomers (*200*).

Reaction of amino acid thioamides with 2,2-dimethyl-3-dialkylamino-2*H*-azirines results in a dipeptide with a *C*-terminal Aib thioamide residue, which undergoes a rearrangement in 3 N HCl/MeOH to give an endothioamide linkage to Aib-OMe (*201, 202*). This reaction sequence was also used to form an endothiopeptide with a *C*-terminal Aib residue (*203, 204*). If the rearrangement is carried out with ZnCl$_2$, the Aib thioamide is retained as an amide instead of hydrolyzed to an ester (*202, 205*). The Xaa-Ψ[CSNH]Aib-Xaa sequence in Aib-containing thiopeptides can be converted into a 1,3-thiazol-5(4*H*)-imine upon treatment with coupling

reagents such as PyBOP (*205, 206*), or by reaction with camphor-10-sulfonic acid (*205*).

Racemic unsubstituted benzyl 2*H*-azirine-3-carboxylate has been prepared from benzyl acrylate via dibromination with Br$_2$ followed by treatment with NaN$_3$ to give benzyl 2-azido-acrylate, which was heated to induce formation of the azirine (*207*). Racemic 2-(nucleobase)-aziridine-2-carboxylates were obtained in low to moderate yields (13–50%) from this benzyl 2*H*-azirine-3-carboxylate via addition of thymine, uracil, adenine, and substituted guanines (*207*). Alkyl 2-aryl-2-*H*-azirine-3-carboxylates undergo Diels–Alder reactions with furans to form a variety of bicyclic and α-substituted aziridine-2-carboxylates (*208*). Benzyl 2*H*-azirine-3-carboxylate was also employed as a dienophile in a Diels–Alder reaction with cyclohexa-1,4-diene; halide opening of the resulting tricyclic aziridine product produced 1-aza-3-halobicyclo[3.2.1]oct-4-ene-3-carboxylate (*209*). Methyl 2-aryl-2*H*-azirine-3-carboxylates were reacted with thiols, propargyl alcohol, and enamines to give 2-substituted-3-arylaziridine-2-carboxylates (*210*). When esterified with Oppolzer's isobornyl-10-sulfonamide chiral auxiliary, thiol addition proceeded with complete diastereoselectivity and Diels–Alder reaction with cyclopentadiene proceeded with 30% de (*211*). Several thiols, benzimidazole, thymine, and PhMgBr were subsequently added to 2*H*-azirine-3-carboxylate esterified with the isobornyl-10-sulfonamide chiral auxiliary, with variable (0–100%) stereoselectivity (*199*). In a similar manner, 2*H*-azirine-3-carboxylates esterified with 8-phenylmenthol were used as substrates for the asymmetric addition of alkyl radicals generated from trialkylboranes, producing 2-substituted aziridine-2-carboxylates in 28–85% yield and 10–92% de. Oppolzer's bornane-sultam chiral auxiliary gave reduced stereoselectivity (*212*).

There are few reports of asymmetric syntheses of azirines. 2*H*-Azirine-2-carboxylic acids have been prepared from optically active *N*-sulfinylaziridine-2-carboxylates by β-elimination of the sulfinyl group induced by treatment with LDA (Scheme 8.26) (*105, 106, 166*). Elimination using Lewis acids such as TMSCl tended to give better yields (*106*). Alternatively, treatment with 2 equivalents of MeMgBr removed the *N*-sulfinyl group, with Swern oxidation forming the azirine (*107*). Organometallic addition generated a chiral center corresponding to the β-center of the product amino acid, regenerating an aziridine-2-carboxylate. Hydrogenation opened aziridines with a C3 phenyl group to give α-amino acids (*105*), while with other C3 substituents reductive opening produced β-amino acids (*107*). Chiral 2*H*-azirine-2-carboxylic esters undergo diastereoselective addition of five-membered aromatic nitrogen heterocycles to give *trans*-3-(heteroaryl)-aziridine-2-carboxylates (*108*).

An asymmetric synthesis of 2*H*-azirine-carboxylic esters based on a Neber rearrangement has been developed. Asymmetry was induced by carrying out the reaction in the presence of chiral alkaloid bases (*213*).

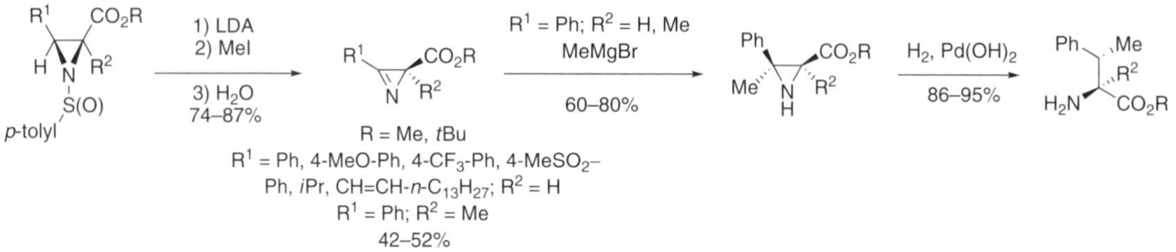

Scheme 8.26 Asymmetric Synthesis of 2*H*-Azirine-2-Carboxylate and Reaction with Grignard Reagents (*105, 106, 166*).

8.4 Epoxides

Given the close correlation between the development of asymmetric amino acid synthesis and advances in asymmetric synthesis in general, it is not surprising that the Sharpless epoxidation of allylic alcohols has been employed in several synthetic schemes leading to α-amino acids. The requisite homochiral epoxide intermediates can also be prepared by a catalytic asymmetric dihydroxylation of an alkene, followed by conversion of the diol to an epoxide group (reviewed in *214*). Ring openings of 2,3-epoxy acids generally lead to β-hydroxy-α-amino acids via C2 opening by an amine, and so have limited versatility. The greatest difficulty lies in obtaining the correct regioselectivity of attack, the same problem encountered with aziridine openings. C2 attack, generating an α-amino acid, is generally preferred, but the usefulness of C3 attack has been recognized in recent years due to the increasing importance of β-amino acids (see Sections 11.2.2a, 11.2.3c). A 2006 monograph on aziridines and epoxides in organic synthesis includes epoxy ester syntheses and ring openings (*5*).

8.4.1 Reaction of 2,3-Epoxy Acids

Chiral 2,3-epoxy acids are generally synthesized by a Sharpless epoxidation of an allylic alcohol followed by oxidation of the terminal alcohol (*215–219*). Epoxidation of a substrate with a chiral auxiliary has also been considered (*220*). A highly enantioselective epoxidation of α,β-unsaturated esters using a chiral dioxirane reagent was reported in 2002, producing 2,3-epoxy esters

with 82–96% ee for *E*-alkene substrates (*221*). Racemic epoxy esters can be prepared by the Darzens reaction, the cyclization of β-hydroxy esters with an α-leaving group, such as *syn*- or *anti*-2-chloro-3-hydroxy esters. The Darzens substrate can be synthesized by the condensation of an aldehyde or ketone with an α-halo ester (*222–226*). Optically active precursors can be prepared for this reaction by using a chiral auxiliary (*223*), a chiral bromoborane catalyst (*227*), or by microbial reduction of 2-chloro-3-oxoesters (*228*). Chemoenzymatic syntheses, via bromination of α-keto esters (*229*) or β-keto esters (*230*), bioreduction of the ketone, and cyclization to the epoxide, were reported in 2005. 2,3-Epoxyamides have been synthesized by an enantioselective Darzens reaction employing a camphor-derived sulfonium amide salt that was reacted with a number of aldehydes, giving *trans*-substituted epoxides with 63–97% ee (*231*). Alternatively, an asymmetric Darzens reaction of the chiral enolate derived from (1*R*)-2-bromoacetyl-isoborneol with various aldehydes produced epoxy ketone products with good stereoselectivity. The camphor auxiliary was removed by treatment with ceric ammonium nitrate to give *cis*-substituted 2,3-epoxy acids from alkyl aldehydes, and the *trans* isomers from aryl aldehydes (*232*).

A number of studies have examined the regioselectivity of attack of nitrogen-based nucleophiles on 2,3-epoxyacids, esters, and amides (see Scheme 8.27, Table 8.3) (*217, 233, 234*), including a theoretical study of the opening of glycidic acids by ammonia (*235*). Fortunately for those desiring to synthesize α-amino acids, ammonia (*223, 224, 236–238*), amines (*227, 237, 239–242*), and azides (*243, 244*) tend to attack at C2 of 2,3-epoxy acids, although 3-aryl substituted (*222, 231*,

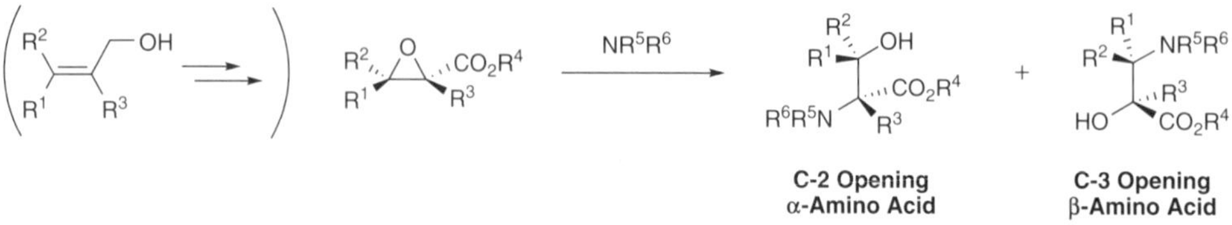

Scheme 8.27 Nucleophilic Opening of 2,3-Epoxy Acids, Esters, and Amides (see Table 8.3).

Table 8.3 Opening of 2,3-Epoxy Acids, Esters, and Amides with Amine Nucleophiles (see Scheme 8.27)

R^1	R^2	R^3	R^4	Amine Nucleophile NR^5R^6	Yield	C2:C3 Opening	Reference
Me	H	H	H	NH_2OH	90%	100:0	1949 (225)
CF_3	H	H	Et	NH_2OH	13%	100:0	1958 (224)
CO_2H	Me	H	H	$BnNH_2$	55–100%	100:0	1962 (237)
Me	CO_2H	H	H				
H	Me	H	H				
iPr	H	H	H				
H	Me	H	H	NH_2OH	54%	100:0	1962 (237)
CO_2H	H	H	H	$BnNH_2$	87–90%	100:0	1962 (242)
H	CO_2H	H	H				
CO_2H	H	H	H	NH_2OH	72%	—	1963 (255)
Me	H	H	Na	$BnNH_2$	68%	100:0	1966 (383)
Me	H	H	H	NH_4OH	63%	100:0	1982 (223)
⟨drawn epoxide structure⟩	H	H	Me	$Mg(N_3)_2$, $BnNH_2$ or NH_4OH	60–78%	11:1 to 1.5:1	1983 (217)
$(CH_2)_2Ph$	H	H	Me	$Mg(N_3)_2$ or $BnNH_2$	53–82%	1:1.6 to 1:2.3	
H	$(CH_2)_2Ph$	H	Bn	$Mg(N_3)_2$	82%	0:100	
nHept, cHex	H	H	CO_2R^4 = CO-Val-OMe	NaN_3, $MgSO_4$	41–95%	1:6 to 1:10	1983 (233)
$(CH_2)_2Ph$	H	H	CO_2R^4 = CONHBn, CON(Bn)$_2$	NaN_3, $MgSO_4$	78–82%	1.5:1 to 1:2.3	1983 (233)
$^{13}CH_3$, $^{13}CD_3$	Me	H	H	NH_4OH	—	1:1	1983 (384)
nHept	H	H	H	Et_2NH, H_2O	71–95%	>20:1 to 1.3:1	1985 (215)
cHex	H	H	H	Et_2NH, $Ti(OiPr)_4$	71–95%	1:10 to 1:>20	
CH_2OSiPh_2tBu	H	H	H	$(H_2C=CHCH_2)_2NH$	71–95%	1:2.4 to 2.4:1	
H	nHept	H	H	$(H_2C=CHCH_2)_2NH$, $Ti(OiPr)_4$	71–95%	1:>20	
H	CH_2OSiPh_2tBu	H	H	LiN_3, $Ti(OiPr)_4$	88–98%	1:>20	
nHept	H	H	CO_2R^4 = CONHBn	NaN_3, NH_4Cl	94%	1.5:1	1985 (234)
H	Me	H	Me	$Mg(N_3)_2$	71%	100:0	1985 (244)
H	CH_2OBn	H	H	NH_4OH	57%	90:10	1985 (385)
CO_2Et	H	H	Et	HN_3 in DMF	97%	100:0	1985 (253)
$CH(Me)CH_2OBn$	H	H	H	NH_4OH	70%	100:0	1986 (218)

(Continued)

Table 8.3 Opening of 2,3-Epoxy Acids, Esters, and Amides with Amine Nucleophiles (continued)

R^1	R^2	R^3	R^4	Amine Nucleophile NR^5R^6	Yield	C2:C3 Opening	Reference
Me	H	H	Et	$MeCN$, $AlPO_4$, Al_2O_3 (adds NHAc after hydrolysis)	40–85%	0:100	1986 (251)
Me	Me	H	Et				
$-(CH_2)_4-R^2$	$-(CH_2)_4-R^1$	H	Et				
(oxazolidine: Ph, Me, N–Ts)	H	H	H	NH_4OH	100%	>50:1	1988 (220)
Me	H	H	H	NH_4OH	40%	100:0	1990 (216)
H	H	H	H		33%	0:100	
iPr	H	H	H	$BnNH_2$, $NaOH$, H_2O	66%	100:0	1990 (239)
Ph	H	H	Et	NaN_3, NH_4Cl	60–65%	0:100	1990 (246)
H	Ph	H	Et	NH_3, $MeOH$	77%	0:100	
CH_2OBn	H	H	Me	HN_3, DIEA	74–92%	100:0	1991 (243)
CH_2OTBS	H	H	Me				
H	CH_2OTBS	H	Et		62%	100:0	
iPr	H	H	H	$BnNH_2$, $NaOH$, H_2O	60%	100:0	1992 (227)
$CH(Me)CH_2CH_2CH=CHMe$	H	H	H	$MeNH_2$, H_2O	70%	100:1	1992 (240)
CF_3	H	H	Et	Me_3SiN_3	69%	100:0	1992 (386)
$CH(Me)CH_2OBn$	H	H	H	NH_4OH	72%	100:0	1993 (236)
$CH(cHex)POPh_2$ / $CH(iPr)POPh_2$	H	H	H	$BnNH_2$, H_2O	40–77%	100:1	1993 (263)
H	$CH(Me)CH_2CCH$	H	H	$MeNH_2$, H_2O	60%	100:0	1994 (260)
H	Ph	H	Et	NH_3	54%	0:100	1995 (222)
2-furyl	H	H	Et	$BnNH_2$	20–50%	0:100	1995 (245)
2-thienyl				$TMS\text{-}N_3$	50–80%	12:1 to 0:100	
(dioxolane epoxide structure)	H	H	$CO_2R^4 = CONMe_2$	NH_3, NH_4Cl; $BnNH_2$; Me_2NH; NaN_3; $Mg(N_3)_2$	49–97%	100:0	1995 (247)
$CH(OPPh_2)iPr$ / $CH(OPPh_2)cHex$	H	H	H	$BnNH_2$	40–77%	100:0	1995 (262)
H ([C2,C3-$^{13}C_2$]-substrate)	CH_2OBn	H	H	$BnNH_2$	87%	3.5:1	1995 (387)

R	R	R	R	Conditions	Yield	Ratio	Year (Ref)
H			Me	NaN_3	35%	0:100	1995 (388)
				$PhCH(Me)NH_2$	86%	0:100	
nC_5F_{11}	H		Et	NaN_3, NH_4Cl	60%	75:25	1996 (261)
nC_7F_{15}	H		Et	NaN_3, Et_2NH	80%	90:10	
H	nC_5F_{11}		Et	$C_{12}H_{25}NH_2$, THF	60%	100:0	
	nC_7F_{15}		Et	Et_2NH	0%	0:0	
$CH(Me)$	H		H	$MeNH_2$, H_2O	70%	100:0	1996 (259)
$CH_2CH=CHMe$							
$CH(Me)CH_2CCH$							
iPr	H		H	NH_4OH	95%	100:0	1997 (389)
CO_2Et	H		Et	$TMS\text{-}N_3$	80%	100:0	1997 (254)
H	Ph		10-deacetyl-baccatin III	NaN_3, HCO_2Me	78%	0:100	1999 (248)
				NH_2OBn, $Yb(OTf)_3$	20%		
Me	Me			MeCN, $BF_3{\cdot}Et_2O$	60–92%	0:100	2000 (252)
Me	nPr						
Me	iPr						
H	iPr		Me				
H	H						
Me	Me		$CO_2R^4 = CONH$				
H	H						
Me	H		H	NaN_3, $InCl_3$	>99%	0:100<1:>99	2001 (249)
nPr	nPr		H				
Ph	Ph		H				
Et	Me	Et	H				
Ph	Me	Me	H				
$(CH_2)_4$-C2	$(CH_2)_4$-C3		H				
Me	H		H				
4-Me-Ph	H		$CO_2R^4 = CONE$	$TMS\text{-}N_3$, $Yb(OTf)_3$	79%	0:100	2002 (231)
Me	H			NaN_3, $Cu(NO_3)_2$	79–91%	0:100	2003 (250)
nPr	H						
Ph	H						
Et	$(CH_2)_4$-C3	H					
Me	Me						
Ph	Me						
Bn	H	H	Et	PhCN, $BF_3{\cdot}Et_2O$	78% as oxazoline / 83% hydrolysis	0:100	2005 (229)
Bn	H	Et		PhCN. or MeCN $BF_3{\cdot}Et_2O$	95–98% as oxazoline	0:100	2005 (230)
Ph	H	Et			95–98% hydrolysis		
iBu	H	Et			>99% ee		

245, 246) or 3-unsubstituted 2,3-epoxy acids (216) and 2,3-epoxy amides (217) preferentially undergo C3 opening. Amine nucleophiles can be directed to the C3 center by activation with transition metal alkoxides (Et$_2$NH, allyl$_2$NH, LiN$_3$; titanium tetraisopropoxide) (215). Behrens and Sharpless reported that 3-alkyl-2,3-epoxy amides reacted with amines or Mg(N$_3$)$_2$ to give exclusively the C3 adduct (233, 234), while NH$_4$N$_3$ opened the epoxide with azide with almost no regioselectivity (234). Valpuesta et al. reported that a number of nitrogen nucleophiles gave exclusively C2 attack on 4,5-dihydroxy-2,3-epoxy amides (247).

The preference for C3 opening has been used to prepare β-amino acids such as bestatin and 2-methylbestatin (217) and the phenylisoserine side chain of taxol (222, 248). Indium trichloride catalyzed the azidolysis of 2,3-epoxy acids by NaN$_3$ to give the α-hydroxy-β-azido products with >99:1 regioselectivity and >99% yield (249). Cu(NO$_3$)$_2$ not only catalyzed regioselective C3-azide opening of 2,3-epoxy acids, but also catalyzed the subsequent azide reduction step, giving a one-pot synthesis of α-hydroxy-β-amino acids (250). AlPO$_4$-Al$_2$O$_3$ catalyzed the addition of acetonitrile to give 3-acetamido adducts exclusively via an oxazoline intermediate, but with basic amines no reaction took place (251). cis-4-Phenyl-2,3-epoxy-butanoic acid was treated with benzonitrile and BF$_3$·Et$_2$O to induce C3 opening, resulting in a phenyl-substituted oxazoline of ethyl 2-hydroxy-3-amino-4-phenylbutanoate. Hydrolysis with 6 N HCl gave the free amino acid (229). The same approach was employed for isobutyl- and phenyl-substituted epoxides, using either benzonitrile or acetonitrile (230); an earlier synthesis prepared α-alkyl-α-hydroxy-β-amino acids using acetonitrile for the opening (252).

Regioselectivity problems have been avoided by using a C$_2$-symmetrical 2,3-epoxysuccinate derived from tartaric acid (253–256), an asymmetric version of a racemic synthesis reported 23 years previously (242, 257). This method was used to prepare β-hydroxy-L-Asp

(253, 255, 256), β-hydroxy-D-Asp (254), β-hydroxy-D-Asp β-hydroxamate (254), and β-hydroxymethyl-L-Ser (253), but its application to the synthesis of other amino acids is limited. The initial product of azide opening (azido precursor of β-hydroxy-Asp) has been treated with PPh$_3$/DMF/Δ, with azide reduction/cyclization producing (2S,3S)-aziridine-2,3-dicarboxylic acid (258).

Epoxide opening has been used in a number of syntheses of the difficult target MeBmt (240, 259, 260), as well as 3-hydroxy-4-methyl-L-Pro (236), 3-(fluoroalkyl)-3-hydroxy amino acids (261), and 3-hydroxy-Leu (239). A 2,3-epoxy acid with a 4-diphenylphosphinoyl substituent was opened with benzylamine to give a β-hydroxy-γ-(diphenylphosphinoyl)-α-amino acid. Elimination of the 3-hydroxy-4-phosphine oxide intermediate resulted in β,γ-unsaturated amino acids (262, 263).

Amino acids can also be obtained from epoxides by an indirect route, with initial epoxide opening using a halide followed by halide displacement with an amine (see Scheme 8.28). C2:C3 regioselectivity of 1.5:1 to 11.5:1 was achieved with NaI or NaBr and Amberlyst-15 in acetone (264). Azide displacement and hydrogenation give the syn β-hydroxy-α-amino acid, starting from the trans epoxide. The methodology is thus complementary to direct azide opening of trans epoxides, which produces anti β-hydroxy-α-amino acids. This is important, as the trans epoxide can be obtained with much greater enantiomeric purity than the cis isomer. A remarkable reversal in regioselectivity of epoxide opening with halides was achieved by using MgBr$_2$ etherate, which gave the β-halo ester with >97:3 selectivity (265, 266) and was used to prepare α-hydroxy-β-amino acids (265).

Another indirect route for conversion of 2,3-epoxy esters into α-amino acids proceeds via C3 ring opening with nucleophiles to produce α-hydroxy esters. The hydroxy group is then converted into a leaving group and displaced with an amine nucleophile. This route was used to prepare both diastereomers of β-methyl-Phe, via azide displacement of an α-mesylate (267).

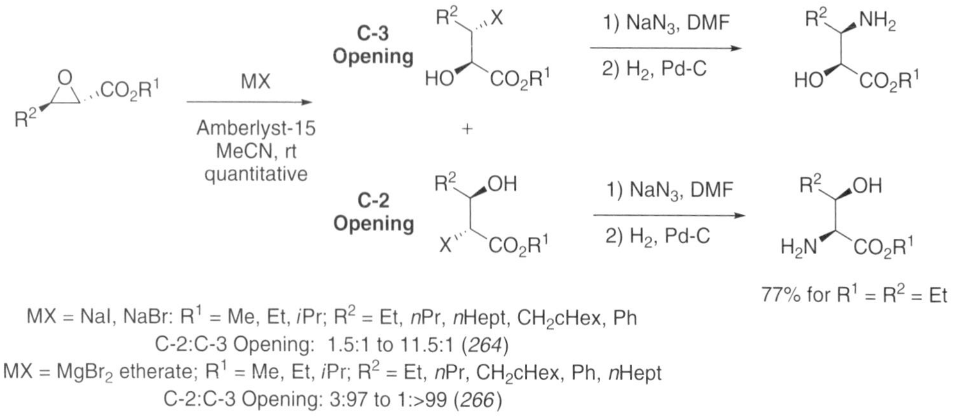

MX = NaI, NaBr: R^1 = Me, Et, iPr; R^2 = Et, nPr, nHept, CH$_2$cHex, Ph
C-2:C-3 Opening: 1.5:1 to 11.5:1 (264)
MX = MgBr$_2$ etherate; R^1 = Me, Et, iPr; R^2 = Et, nPr, CH$_2$cHex, Ph, nHept
C-2:C-3 Opening: 3:97 to 1:>99 (266)

Scheme 8.28 Regioselective Halide Opening of 2,3-Epoxy Esters (264–266).

8.4.2 Reaction of 2,3-Epoxy-1-Alkanols

A more versatile synthetic route to amino acids from epoxides is achieved using 2,3-epoxy-1-alkanols (see Scheme 8.29, Table 8.4). Regioselective C3 opening with amines gives 3-amino-1,2-diols, which can be converted into an α-amino acid by oxidative cleavage of the diol moiety. The same epoxide synthon can be used to prepare β-hydroxy-α-amino acids via C2 opening and oxidation of the terminal alcohol, β-amino acids by C3 opening and oxidation of the terminal hydroxyl group, or γ-amino acids by displacement of the terminal hydroxyl group with cyanide followed by hydrolysis.

A number of studies have examined the regioselectivity of nucleophilic attack on 2,3-epoxy-1-alkanols (233, 234), which depends collectively on steric, electronic,

and conformational effects (see Scheme 8.29, Table 8.4). Epoxide opening generally occurs by an S_N2 mechanism, but, depending on the nucleophile and substrate, a *syn* attack is also possible (245). A preference for C3 attack is observed with NaN_3/NH_4Cl or $NHEt_2/EtOH$ unless there is a very bulky C3 substituent, and an increase in C3 selectivity can be achieved by the addition of metal ions (Li^+, Mg^{2+}, Zn^{2+}), normally with $LiClO_4$ (268). This effect is probably due to a chelated structure, with the metal cation coordinated to the oxirane and alkoxide oxygens. Lithium trifluoromethanesulfonate in acetonitrile has recently been reported as an excellent substitute for the hazardous lithium perchlorate (269). Regioselective C3 addition of azide can also be achieved with $Ti(OiPr)_2(N_3)_2$, although this is not very effective for *cis*-substituted epoxides (270). Primary amines also

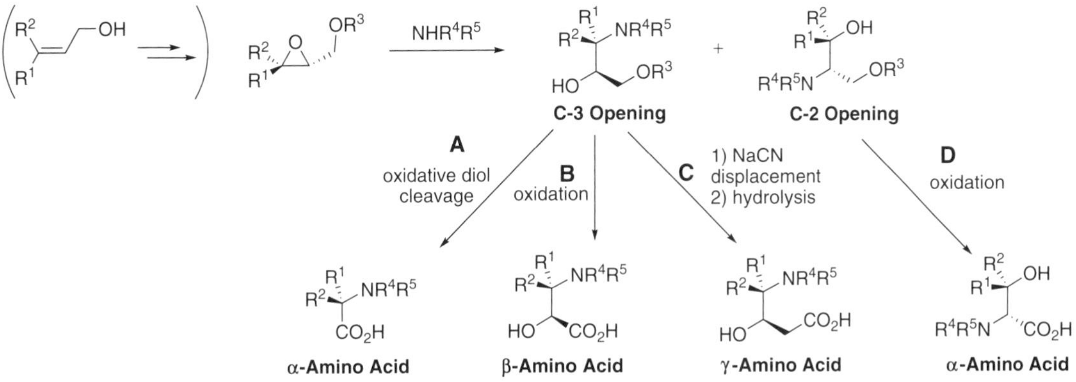

Scheme 8.29 Nucleophilic Opening of 2,3-Epoxy Alkanols (see Table 8.4).

Table 8.4 Nucleophilic Opening of 2,3-Epoxy Alkanols (see Scheme 8.29)

R^1	R^2	R^3	Amine Nucleophile NR^5R^6	Yield	C3:C2 Opening	Oxidation	Reference
nHept	H	H, Bn	NaN_3, NH_4Cl	88–98%	1.7:1 to 4.5:1		1983 (233)
cHex	H	H, Bn					1985 (234)
tBu	H	H		47%	0:100		
H	H	H, Bn					
				83–90%	1:7 to 1:15		
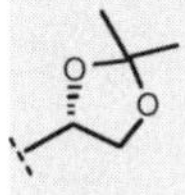	H	H, Bn					
nPr	H	H	NaN_3, NH_4Cl	71–100%	1:1 to 100:1	A	1988 (270)
cHex			$Ti(OiPr)_2(N_3)_2$	76–96%	20:1 to 100:1	R^1 = Ph, Bn	
Bn						$RuCl_3$, H_5IO_5	
Ph						85–98%	
$(CH_2)_2CH=C(I)Me$							
CH_2OBn							
$CH_2N(Bn)Ts$							

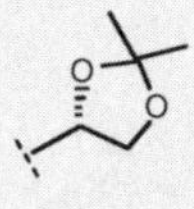

Table 8.4 Nucleophilic Opening of 2,3-Epoxy Alkanols (continued)

R¹	R²	R³	Amine Nucleophile NR⁵R⁶	Yield	C3:C2 Opening	Oxidation	Reference
H	nHept CH₂OBn CH₂N(Bn)Ts	H	Ti(OiPr)₂(N₃)₂	78–89%	1:1 to 6:1		1988 (270)
Me iPr Ph	H	H	RNH₂, Ti(OiPr)₄ R = nHex tBu Ph₂CH Bn 4-MeO-Bn	18–96%	88:12 to 100:0		1991 (271)
Me nPr Ph (CH₂)₂Ph	H	H	Ph₂CHNH₂, Ti(OiPr)₄			A RuCl₃, NaIO₄ 52–90%	1993 (390)
iPr	−C(OMe) Me₂-R³	−C(OMe) Me₂-R²	LiN₃, LiClO₄	100%	>20:1	C	1993 (277)
Et	H	H, Bn	NaN₃, NH₄Cl		7:93 to 87:13		1993 (268)
tBu	H	H, Bn	NaN₃, LiClO₄		40:60 to 96:4		
cHex	H	H, Bn	Et₂NH		73:27 to 87:13		
H	nPr	Bn	Et₂NH, LiClO₄		88:12 to >99:<1		
Ph	H	H	NaN₃, NH₄Cl	100%	100:0		1994 (280)
H	(CH₂)₂NH₂	H	intramolecular from R²	100%	100:0	A RuCl₃, NaIO₄ 52%	1995 (275)
Me	H	H	Ph₂CHNH₂, Ti(OiPr)₄	68–74%	100:0	C	1996 (278)
H	(CH₂)ₙMe n = 11–19	H	NaN₃, NH₄Cl	86%	>10:1	A RuCl₃, NaIO₄ >85%	1996 (273)
1-naphthyl	H	H	NaN₃, LiClO₄ Ph₂CHNH₂, Ti(OiPr)₄	80% 60%	100:0 6:1	A RuCl₃, NaIO₄ 68%	1997 (274)
H	H	H, Bn	EtAlN₃	38–89%	>25:<1		1998 (272)
H	Me	H, Bn					
Me	H	H, Bn					
Me	Me	H, Bn					
Ph	H	H					

substrate
   epoxides =

| | H | H | Ti(OiPr)₂(N₃)₂ | 96% | 3:1 | A RuCl₃, NaIO₄ 64% | 1999 (276) |

give good regioselectivity in the presence of Ti(IV) catalysts (*271*). 2,3-Epoxy alcohols have been opened with high C3 regioselectivity (>25:1) by employing diethylaluminum azide. The high regioselectivity is obtained with both *cis*- and *trans*-substituted epoxide substrates, and with epoxides containing bulky C3 substituents (*272*). One possible side reaction when unprotected 2,3-epoxy alcohols are treated with simple amines is a Payne rearrangement, in which the amine displaces the terminal hydroxy group, which intramolecularly opens the epoxide (*233*).

Lipidic α-amino acids (*273*) and (1-naphthyl)-Gly (*274*) have been synthesized from 2,3-epoxy-1-alkanols, while an intramolecular amine attack was used to prepare pipecolic acid (*275*). Polyoxamic acid was obtained via a Ti-catalyzed azide opening of an epoxide that was prepared via Sharpless epoxidation (*276*). The γ-amino acids statine (*277*) and cyclohexylstatine (*278*) were synthesized via route C in Scheme 8.29. Trost and Bunce (*279*) have used a chiral palladium complex to catalyze the addition of phthalimide to racemic 3,4-epoxy-1-butene, ring opening generating the vinylglycine precursor *N*-Phth-vinylglycinol with >75:1 regioselectivity, 99:1 enantioselectivity, and 99% yield.

A Japanese group has used an aryl group as a masked carboxyl moiety instead of the hydroxymethyl group, with conversion to the acid by oxidation with ruthenium tetraoxide (see Scheme 8.30) (*280*). This method was used to synthesize a number of polyhydroxy amino acids which act as phytosiderophores, transporting iron in plants: mugineic acid (*281, 282*), 3-epi-hydroxymugineic

acid, 2′-hydroxynicotianamine, and distichonic acid A (*283, 284*). The polyhydroxy amino acid component of the antifungal antibiotics polyoxins, polyoxamic acid, has also been prepared by this route (*271*).

8.4.3 Intramolecular Opening of 2,3-Epoxy-1-Alkanols

Nucleophilic C2 opening of terminal epoxides by an amine leads to a 2-amino-1-alkanol, which can then be converted into an α-amino acid by oxidation of the terminal alcohol group. The desired regioselectivity of attack at the more sterically hindered site can be forced by using an intramolecular attack from the nitrogen of a carbamate derivative of an adjacent C3 alcohol (see Scheme 8.31), although the end product is again a β-hydroxy-α-amino acid (*285*). The oxazolidinone formation is accompanied by a facile N-to-O migration of the benzoyl group (*286*). β-Hydroxy-Tyr (*285*), β-hydroxy-Phe, β-hydroxy-Leu, and Thr (*287*) have been prepared by this method.

A closely related synthesis uses a terminal alcohol to deliver the carbamate amine to an internal epoxide to form the oxazolidinone (see Scheme 8.32, Table 8.5). The carbamate is formed by treatment of the hydroxyl group with an isocyanate. The desired nucleophilic epoxide opening by the carbamate nitrogen is induced by treatment with strong base (*238, 288–290*); under Lewis acid conditions the urethane oxygen is the nucleophile (*289, 291*). The free secondary hydroxyl can be

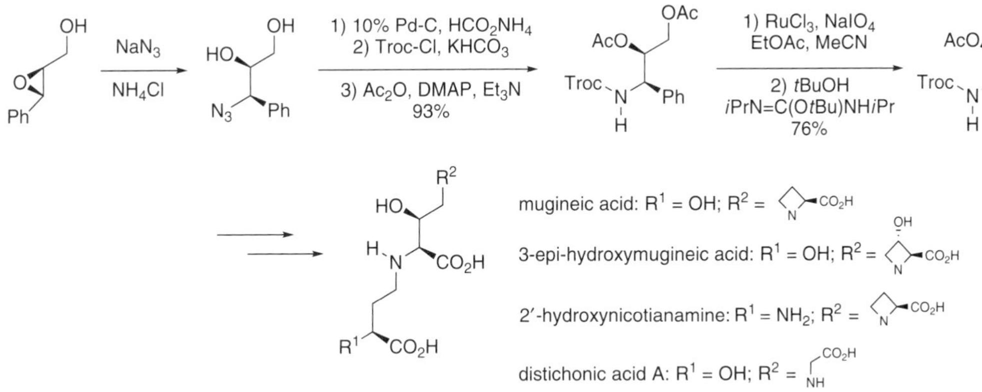

Scheme 8.30 Use of Phenyl Group as Masked Carboxyl Group (*280–284*).

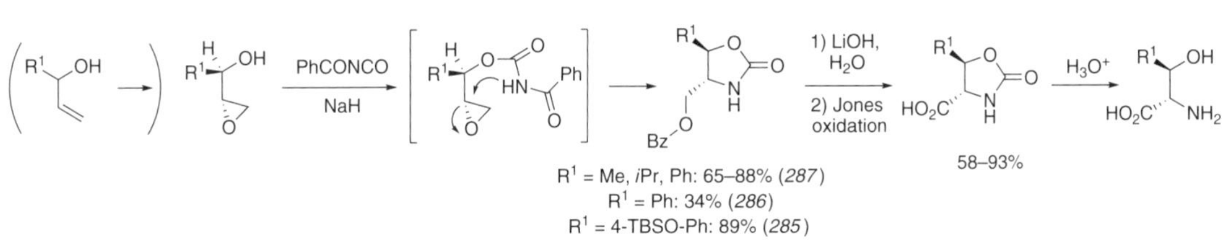

R¹ = Me, *i*Pr, Ph: 65–88% (*287*)
R¹ = Ph: 34% (*286*)
R¹ = 4-TBSO-Ph: 89% (*285*)

Scheme 8.31 Intramolecular Opening of Terminal Epoxides.

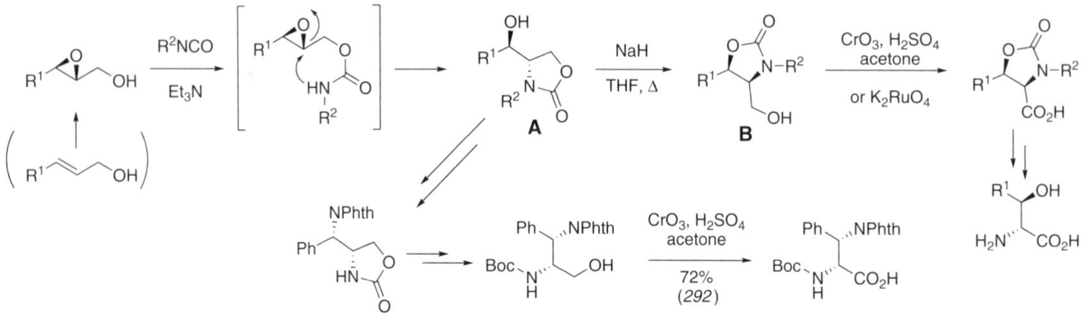

Scheme 8.32 Intramolecular Opening of Internal Epoxides (see Table 8.5).

Table 8.5 Intramolecular Opening of Internal Epoxides (see Scheme 8.32)

R^1	R^2	Epoxide Opening Yield	Oxidation Yield	Reference
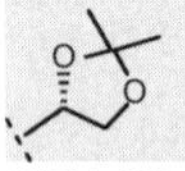 (structure)	Bn	81% A (using KOtBu/THF)	—	1982 (238)
$(CH_2)_2OBn$	Ph	92% B		1983 (289)
$nC_{15}H_{31}$	Bn	55–81% mixture of A and B		1985 (288)
cHex				
CH_2OBn				
$(CH_2)_3OTBS$				
$CH(Me)CH_2OBn$	Me	—	60–70%	1988 (294)
$CH(Me)CH_2CH=CHMe$	Me	29–37% B	68–75%	1988 (295) 1989 (298)
$(CH_2)_2CH=CHMe$	Me	80–87% B	50–70%	1988 (297)
$CH_2CH(Me)CH=CHMe$				
$(CH_2)_2CH=CHMe$	Me	87% B	67%	1990 (296)
iPr	Bn	75% B	100%	1993 (293)
Me, Me	Me	66–85% B		1999 (299) 1999 (300)
$CH_2CH=CH_2$	Bn	86% A (using NaHMDS)	65%	2005 (301)

converted to other groups by displacement reactions (Scheme 8.32, Path A). The terminal hydroxymethyl moiety is regenerated by hydrolysis of the oxazolidinone, followed by oxidation to form the acid group. 2,3-Diamino acids (292) have been prepared by this route. Alternatively, the oxazolidinone can rearrange to form an isomeric oxazolidinone with the secondary hydroxyl group, releasing the primary hydroxyl, which is then available for oxidation to the acid group (Scheme 8.32, Path B). Oxazolidinone rearrangement is promoted by NaH in refluxing THF. This method was used to prepare β-hydroxy-Leu (293), β-hydroxy-Orn (290), and MeBmt analogs (294–298). For synthesis of the latter compounds, methylisocyanate was used to form the oxazolidinone, with the advantage that the desired N-methyl group was introduced with this step. A similar route was employed to prepare N-methyl β-hydroxy-Val (299), which was subsequently employed in a total synthesis of luzopeptin

and quinoxapeptin cyclic depsidecapeptides (300). In 2005 the Path A route was employed in a synthesis of erythro-β-hydroxy-homoallylglycine enroute to erythro-β-hydroxy-Glu, using a vinyl group as the masked side-chain carboxylic acid. The use of NaHMDS as base in the carbamate cyclization minimized acyl migration (Path B) during oxazolidinone formation (301).

A similar synthetic route uses trichloroacetimidic esters of 2,3-epoxy alcohols to deliver the amino group (see Scheme 8.33) (302). The regioselectivity of addition is determined by the structure of the substrates and the catalyst. C2 attack, leading to five-membered oxazolines, occurs in the presence of methanesulfonic acid with epoxy alcohols containing one aliphatic C-3 substituent (302), or by using Et$_2$AlCl catalysis with C2-substituted epoxides (303). The oxazolines can be converted into α-amino acids by transformation into an oxazolidinone and oxidation of the free terminal

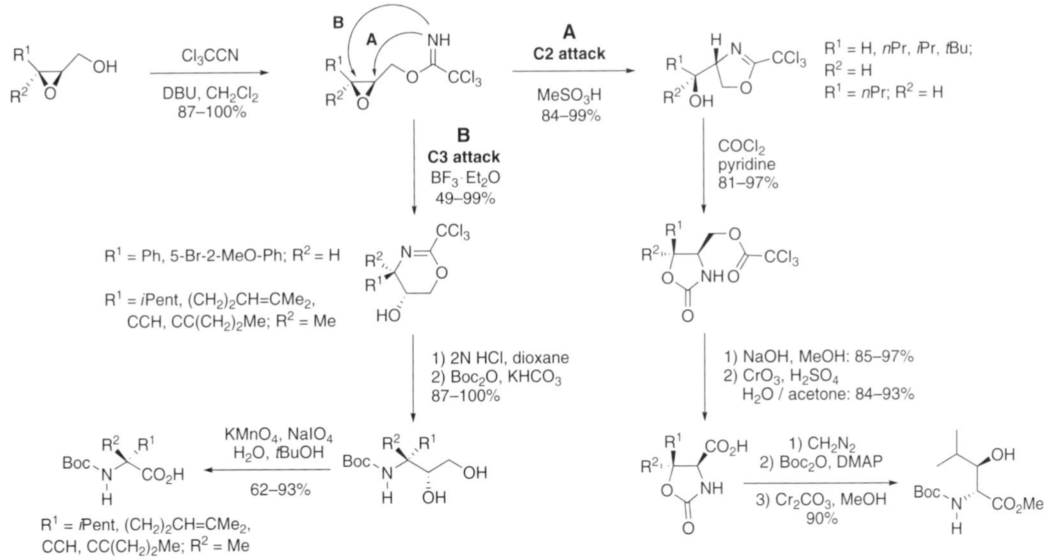

Scheme 8.33 Intramolecular Opening of 2,3-Epoxy-1-Alkanols (*302*).

hydroxymethyl group. Both enantiomers of α-methyl-Ser were prepared by this route (*303*), as was a β-hydroxy-Leu derivative found in a cyclopeptide alkaloid (*304*), and Sphingofungin F, 2-amino-2-methyl-3,45-trihydroxy-14-oxo-didec-6-enoic acid (*305*). The trichloroacetimidic esters can also be induced to undergo C3 attack, giving six-membered dihydrooxazines, via catalysis with boron trifluoride etherate of substrates with aromatic, acetylenic, or dialkyl C3 substituents (*302*). The dihydrooxazines, after hydrolysis, give a 1,2-diol that can be oxidatively cleaved to generate α-amino acids. The immunosuppressant α,α-disubstituted amino acid (−)-myriocin, with hydroxymethyl and substituted C_{18} side chains, has also been synthesized via an intramolecular trichloroacetimidate epoxide opening. However, the carboxyl group was generated from a vinyl substituent, rather than a hydroxymethyl group (*306*).

8.4.4. Other Reactions of Epoxides and Epoxide Equivalents

Another possible route to amino acids via epoxides is to open 1-amino-2,3-epoxides with a carboxylate equivalent.

C3 opening with cyanide, followed by hydrolysis, provided γ-amino acids (see Sections 11.4.3c) (*307*). Opening with an oxygen nucleophile, followed by hydroxymethyl oxidation of the 3-amino-1,2-diol product, resulted in α-hydroxy-β-amino acids (see Scheme 11.47) (*308*).

The opening of (arylthio)nitrooxiranes with ammonia has been used for polyhydroxy α-amino acid synthesis (*309*–*312*) (see Scheme 2.18 for racemic example). A bis-sulfoxide-substituted epoxide (see Scheme 8.34) was opened with benzhydrylamine to give a phenylglcyine amide product, with the carboxylic acid arising from one of the epoxide carbon atoms. The homochiral substrate was prepared from a 2-phosphonyl-1,3-dithiane via an asymmetric oxidation (*313*).

Homochiral diols, which can be produced by asymmetric dihydroxylation reactions of alkenes, have been converted into amino acids via an epoxide equivalent, a cyclic sulfate. The diol sulfate, which is more reactive than an epoxide, is formed by treatment of the diol with $SOCl_2$ followed by $NaIO_4/RuCl_3$ oxidation (see Scheme 8.5) (e.g., *48*, *214*, *314*, *315*). Ring opening with azide, followed by azide reduction, produces β-hydroxy-α-amino acids, while azide reduction with triphenylphosphine in

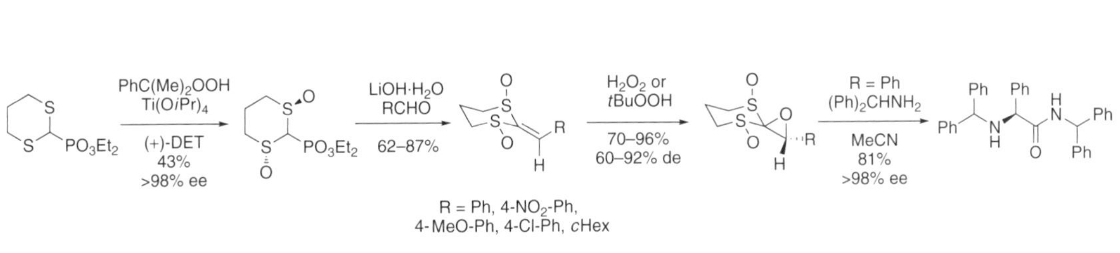

Scheme 8.34 Opening of Bis(sulphoxide)-Epoxide (*313*).

acetonitrile instead of THF/H$_2$O produces aziridine-2-carboxylates (48). The procedure was applied to a syntheses of 4,4,4-trifluoro-Thr(49),(2R,3R)-3-hydroxy-Leu (315), (2S,3S)-3-hydroxy-Leu (48), (2S,3S)-3-hydroxy-Nle (48), 3-tert-butyl-Ser (50), and 3-hydroxy-6-benzyloxy-Nle (50). The use of cyclic sulfites and sulfates in organic synthesis was reviewed in 2000 (52) and is also discussed in Section 5.3.1.

8.5 β-Lactams

The synthesis and reactions of β-lactams is an area which would require several monographs to thoroughly explore, but fortunately their application to α-amino acid synthesis is more limited. This review will restrict itself to preparations of β-lactams subsequently applied to α-amino acid synthesis. Since β-lactams contain a β-amino acid core, more general syntheses are described in Sections 11.2.2f and 11.2.3s. In order to obtain an α-amino acids from β-lactams, a second amino or carboxy group must be introduced in the appropriate position, or a rearrangement must be used to change the existing amino-carboxy group proximity. Some applications of β-lactams to α-amino acid synthesis were included in a 1988 review (316), with other syntheses summarized in 1999 (317), and the use of β-lactams as intermediates in the synthesis of α- and β-amino acids in 2001 (318).

8.5.1 β-Lactams with Additional Carboxy Group

Optically active 4-carboxy-2-azetidinone was first prepared from Asp acid in 1980 (319) using conditions previously reported for cyclizing other β-amino acids (320) (see Scheme 8.35). Baldwin et al. have since reported more experimental detail and an improved version of this synthesis (321), and introduced N-Boc (322) and N-TBDMS (321) protection. Other groups have employed this procedure (323). Another route to azetidin-2-one-4-carboxylates is via a Staudinger [2+2] cycloaddition of chiral imine derivatives of methyl glyoxylate with

substituted ketenes. The reaction proceeds in good yield, but with minimal diastereoselectivity (324). Racemic 3,4-dialkyl-4-carboxyazetidin-2-ones have also been synthesized by a [2+2] cycloaddition (325). An electron-deficient imino ester, TsN=CHCO$_2$Et, reacted with a number of acid chloride-derived ketenes in the presence of a chiral catalyst, benzoylquinine, to give 2-carboxy-3-substituted-azetidin-2-ones in 36–65% yields, with 95–99% ee and >98% de cis stereoselectivity (326, 327). Bifunctional catalysis, employing a chiral cinchona alkaloid in combination with an In(III) metal cocatalyst, provided a similar series of β-lactams in 92–98% yield with 96–98% ee and 10:1 to 60:1 dr (328).

The chiral N-Boc β-lactam benzyl ester can be opened without loss of chiral integrity with trimethylsulfoxonium ylide or sulfone stabilized carbon nucleophiles, giving a γ-keto intermediate amenable to furthur modification (see Scheme 8.36) (329). For the sulfones, attack at the ester carbonyl was avoided by using a bulky tert-butyl ester. A higher-order organocuprate was also successfully reacted, but in low (37%) yield. Stabilized Wittig reagents added to the carbonyl to give 2-exo-methylene derivatives in 20–98% yield (322). Cyclization of the intermediate produced 4-oxoprolines (330).

The dianion of Asp acid-derived 4-carboxy-2-azetidinone (321) can be alkylated or allylated (321, 331–333) at the C3 carbon with complete diastereoselectivity (see Scheme 8.37). Acidic hydrolysis opened the β-lactam to give β-substituted Asp. Hydrolysis in a phosphate buffer has been studied by NMR (334). Alternatively, the dianion can be quenched with a cyclic sulfate to introduce a hydroxyethyl substituent, with subsequent manipulations resulting in the synthesis of 3-carboxy-Pro and 3-amino-Pro (333). The benzyl ester of N-TBDMS 4-carboxy-azetidin-2-one was allylated, followed by reduction of the lactam carbonyl to give a four-membered Pro analog. Oxidative cleavage of the allyl group resulted in (−)-trans-2-carboxyazetidine-3-acetic acid, while hydroboration and oxidation of the alkene produced the homologous (−)-trans-2-carboxyazetidine-3-propionic acid (323).

3-Methyl-4-carboxyazetin-2-ones have been successfully ring opened with Grignard reagents to give β-methyl-γ-keto-α-amino acids (see Scheme 8.38).

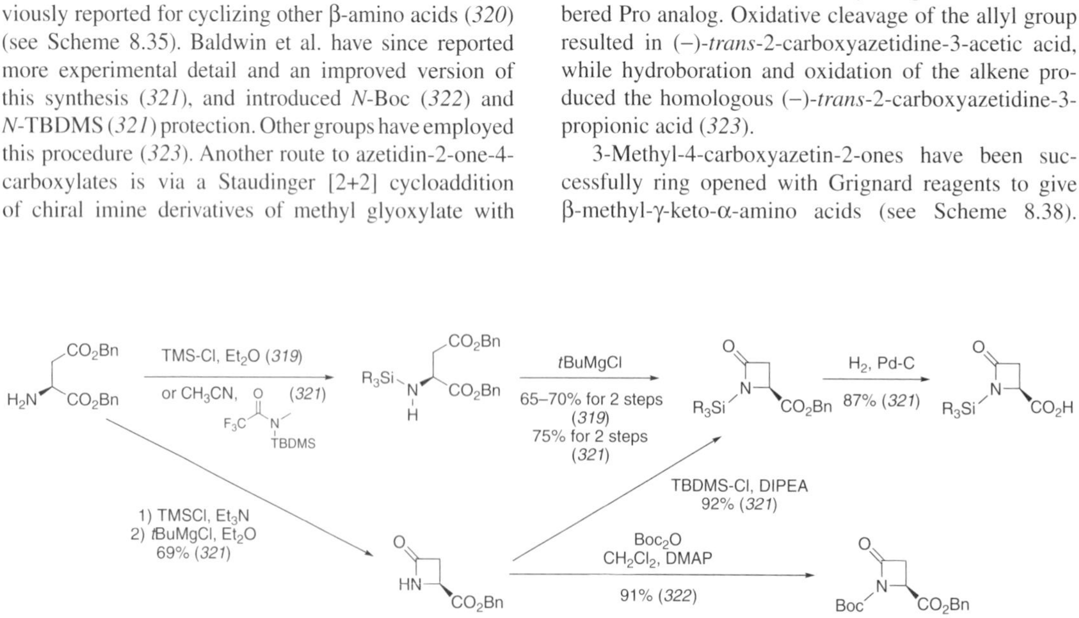

Scheme 8.35 Synthesis of 4-Carboxyl-2-azetidinone from Asp.

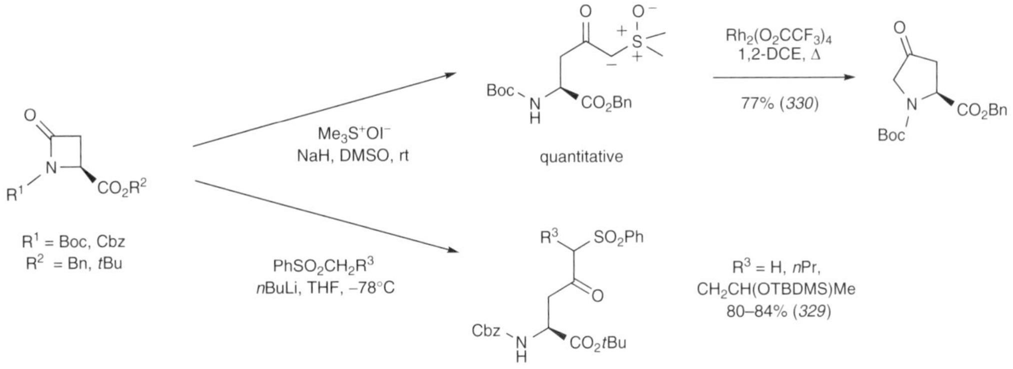

Scheme 8.36 Ring Opening of β-Lactam With Carbon Nucleophiles.

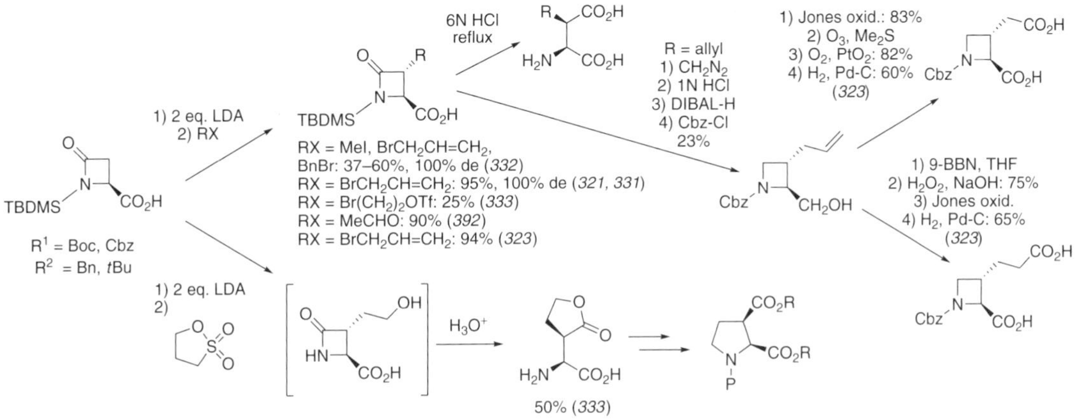

Scheme 8.37 Alkylation and Hydrolysis of 4-Carboxyl-2-Azetidinone.

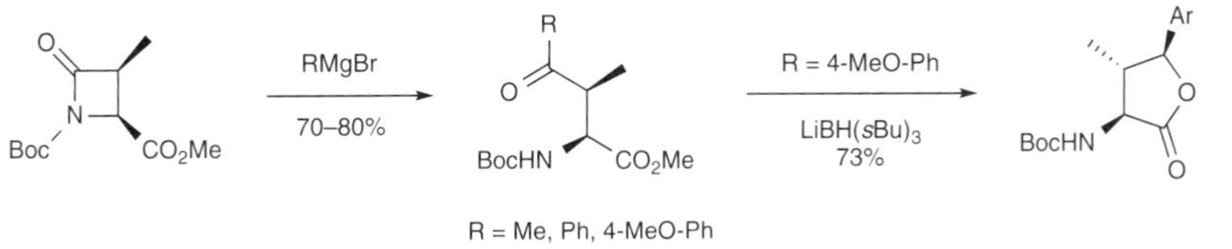

Scheme 8.38 Grignard Opening of 4-Carboxy-2-Azetidinones (*335*).

The carbonyl group was stereoselectively reduced to provide the corresponding γ-hydroxy derivatives (*335*).

The Ser-derived Garner aldehyde was converted into a 4-carboxy-2-azetidinone via a cycloaddition reaction of an imine derivative (see Scheme 8.39). The amino alcohol revealed by acetonide deprotection was oxidatively cleaved to form a *cis*-substituted carboxyl substituent, which could be epimerized to the *trans* diastereomer. Hydrolysis of the lactam produced protected β-hydroxy-Asp. The side-chain ester was also regioselectively reduced to give 3,4-dihydroxy-2-aminobutyric acid (*336*).

It is also possible to use a β-lactam with a masked carboxyl group, which is then revealed after azetidinone cleavage. This approach was used in a synthesis of polyoxamic acid. Treatment of a 3-hydroxyazetidinone with LiAlH₄ opened the N–CO bond to give a 1,2-diol, which was then converted to a carboxylic acid group with NaIO₄/RuCl₃ (*337*).

8.5.2 β-Lactams with Additional Amino Group

3-Amino-substituted β-lactams can be prepared via cyclization of β-hydroxy α-amino acid hydroxamic acids under Mitsunobu conditions (*338*). For example, L-Ser hydroxamide was cyclized to give an *N*-hydroxy-3-aminoazetidinone in good yield (see Scheme 8.40) (*339, 340*). N–O reduction and hydrolysis (*339*) or treatment with methanolic ammonia followed by hydrogenolysis (*341*) produced 2,3-diaminopropionic acid, while exposure to lithium ethanethiolate induced isomerization to an isooxazolidin-5-one, a precursor to a number of 2,3-diaminopropionic acid derivatives (*342, 343*). A similar strategy was applied to Boc-Thr-NHOBn, with cyclization to the azetidinone followed by ring opening with NaOH producing *N*-Boc,*O*-Bn 2-amino-3-(hydroxyamino)-butyric acid. The side-chain amino group was reductively methylated and the benzyloxy group removed by hydrogenation to give 2-amino-3-methylamino-butyric acid (*344*). Boc-Ser-NHOMe and Boc-Thr-NHOMe led to the corresponding β-methoxyamino-Ala and 2-amino-3-methoxyamino-butyric acids (*345*). The *p*-tolyl amides of *N*-protected β-hydroxy amino acids can also be cyclized to β-lactams via Mitsunobu reaction with triphenylphosphine and DEAD (*15*), although aziridine formation and dehydration are competing reactions. The reaction can be directed towards β-lactam formation by using *N*-phthalimido protection or by replacing triphenylphosphine with hexamethylphosphorus triamide (*13*).

Alkylation at the C3 carbon of homochiral 3-amino-4-aryl-substituted azetidones is also possible, again with high diastereoselectivities (>99.5% de) (*346, 347*). Alkylations of substrates with non-aryl C4 substituents have also been reported (*348*). The precursors were obtained by an asymmetric [2+2] cycloaddition between homochiral ketimines and achiral or chiral imines (*347–350*). The C4–N bond of the β-lactam was reductively cleaved to give β-aryl-α,α-disubstituted amino acid amides (*346, 347, 349, 351*) (see Scheme 8.41). This method has been applied to dipeptide synthesis by using a benzaldehyde Schiff base of an amino acid as the imine component of the cycloaddition (*347*). However, if alkylation is attempted with these substrates, the electrophile adds to the amino acid α-carbon instead of the β-lactam carbon, again with high diastereoselectivity (*347, 352*). These results have been summarized in short reviews (*353, 354*). More recently, dipeptides containing one or two α,α-disubstituted residues were prepared via dialkylations of a Gly residue that has the Gly nitrogen contained with a chiral 4-phenyl β-lactam template (see Scheme 8.42) (*355*). The β-lactam contains a 3-amino substituent contained within a chiral oxazolidinone, and can also be alkylated at the 3-position. The lactam was reductively opened at the benzylic position to give the dipeptide. These results have also been reviewed (*354, 356*).

Amino-substituted β-lactams have also been synthesized via a [2+2] cycloaddition between an optically active chromium carbene complex and an achiral imine (*357, 358*). Alkylation of the lactam provided a route to

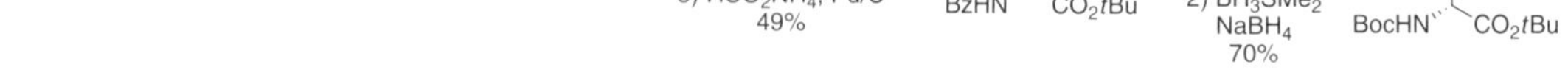

Scheme 8.39 Formation of 4-Carboxy-2-Azetidinone from Garner Aldehyde (*336*).

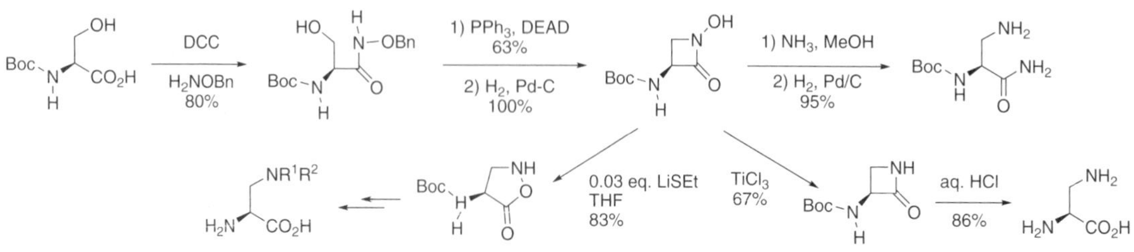

Scheme 8.40 Synthesis of 3-Amino-2-Azetidinone from Ser (*339–342*).

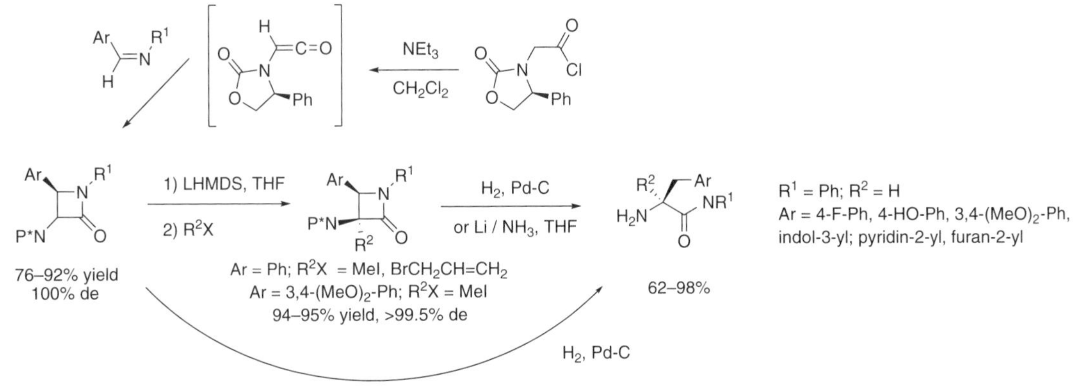

Scheme 8.41 Synthesis, Alkylation and Reductive Cleavage of 3-Amino-4-aryl-2-Azetidinone (*346–353*).

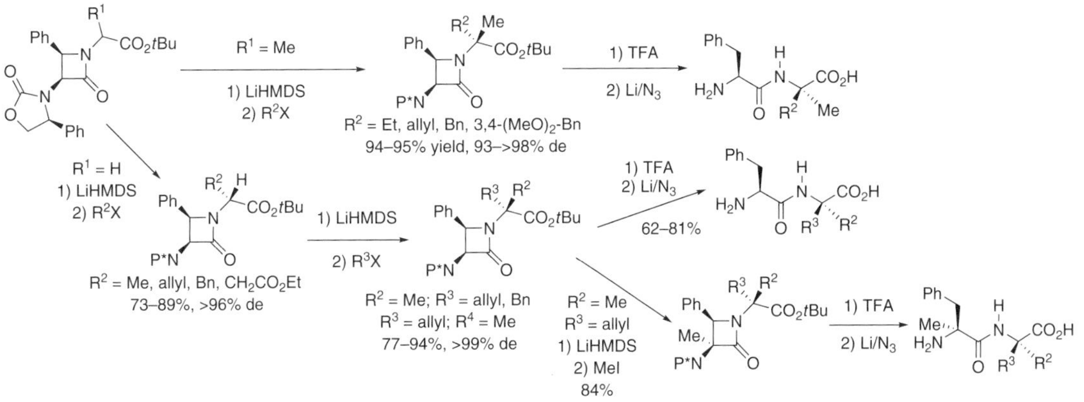

Scheme 8.42 Alkylation of Residues Attached to β-Lactam (*355*).

a number of α-methyl-α-amino acids (see Scheme 5.54) (*357*). Another route to 3-aminoazetidin-2-ones is via azide displacement of a 3-tosyl precursor. Hydrolysis of the lactam produced a 2,3-diamino acid analog of the phenylisoserine taxol side chain (*359*).

8.5.3 Rearrangement of β-Lactams

A Baeyer–Villiger ring expansion rearrangement has been used to convert optically active α-keto β-lactams into amino acid N-carboxy anhydrides (NCAs) (see Scheme 8.43), which are convenient amine-protected carboxyl-activated precursors for peptide synthesis (*360–362*). The β-lactam precursors were obtained by the cycloaddition reaction of achiral alkoxyketenes with chiral aldehyde-derived imines (*363*), or by the addition of Grignard reagents to chiral 3-(benzyloxy)-4-formyl-lazetidin-2-one (*360*), followed by oxidation of the

α-hydroxy group (see Scheme 8.43). Treatment with m-chloroperbenzoic acid at −40°C gave the NCA in almost quantitative yield. The crude product could either be converted to the amino acid methyl ester by refluxing in methanol, or reacted with another amino acid to give peptides. A series of polyhydroxy amino acids, including γ-hydroxy-Thr and polyoxamic acid, were prepared via this route (*364*), while the ketene-derived β-lactam was also applied to the synthesis of β- and γ-amino acids (*362*). A similar expansion of 3-hydroxyazetidine-2-ones was used in the synthesis of *tert*-leucine (*365*), polyoxamic acid (*366*), (2*S*,3*S*,4*S*)-3,4-dihydroxyhomo-tyrosine (*367*), α-methyl-α-amino acids (*368, 369*) and piperazine-2-carboxylic acids (*370*), activated as their N-carboxy anhydrides. Alternatively, the aldehyde of a 3-benzyloxy-4-formyl-2-azetidinone was olefinated via the Wittig–Horner reaction, and the alkene subjected to a Sharpless asymmetric dihydroxylation reaction

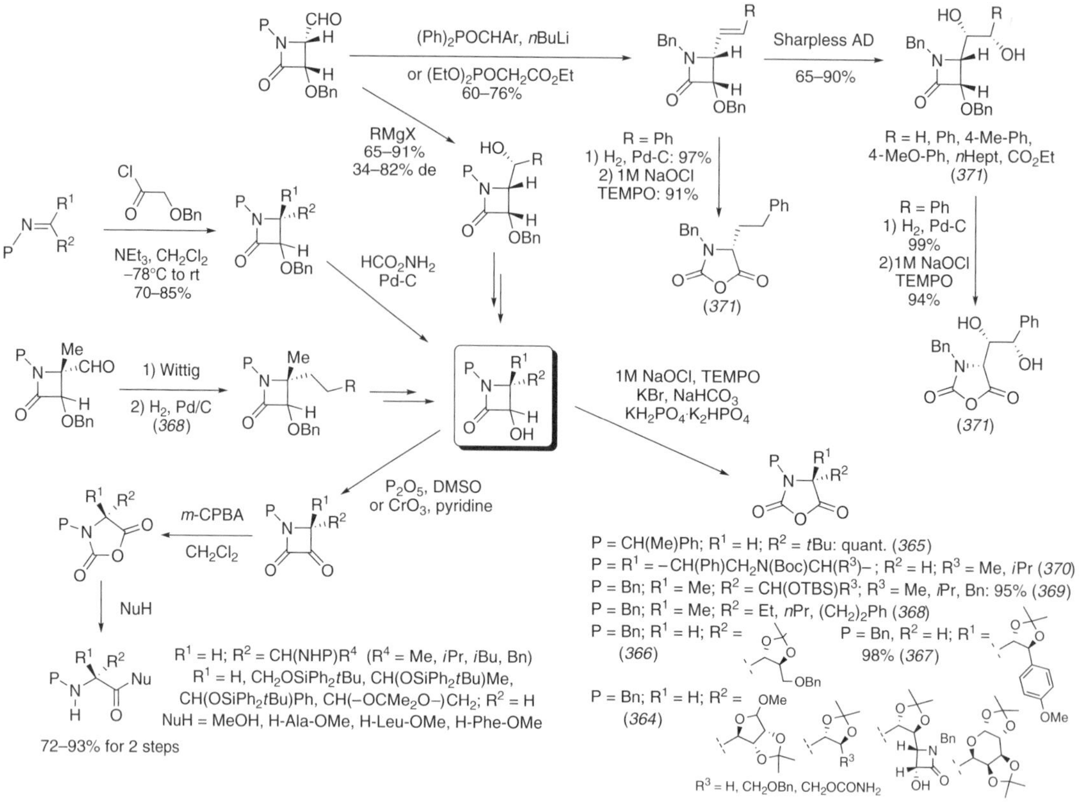

Scheme 8.43 Synthesis of β-Lactams and Baeyer-Villiger Ring Expansion to Amino Acid NCA (*360–371*) .

(see Scheme 8.43). Baeyer–Villiger oxidation produced the NCA of 2-amino-3,4-dihydroxy-4-phenylbutanoic acid; the β-lactam precursors of several analogs with other substituents replacing the phenyl group (R = H, 4-Me-Ph, 4-MeO-Ph, Bn, *n*Hept, CO_2Et) were also prepared. The alkene could be hydrogenated instead of dihydroxylated, with homophenylalanine synthesized as an example (see Scheme 8.43) (*371*).

References

1. Tanner, D. *Pure Appl. Chem.* **1993**, *65*, 1319–1328.
2. Tanner, D. *Angew. Chem., Int. Ed. Engl.* **1994**, *33*, 599–619.
3. Osborn, H.M.I.; Sweeney, J. *Tetrahedron: Asymmetry* **1997**, *8*, 1693–1715.
4. Hu, X.E. *Tetrahedron* **2004**, *60*, 2701–2743.
5. Yudin, A.K. Ed. *Aziridines and Epoxides in Organic Synthesis;* Wiley-VCH: Weinheim, **2006**.
6. Nakajima, K.; Takai, F.; Tanaka, T.; Okawa, K. *Bull. Chem. Soc. Jpn.* **1978**, *51*, 1577–1578.
7. Baldwin, J.E.; Spivey, A.C.; Schofield, C.J.; Sweeney, J.B. *Tetrahedron* **1993**, *49*, 6309–6330.
8. Larsson, U.; Carlson, R. *Acta Chem. Scand.* **1994**, *48*, 511–516.
9. Willems, J.G.H.; Hersmis, M.C.; de Gelder, R.; Smits, J.M.M.; Hammink, J.B.; Dommerholt, F.J.; Thijs, L.; Zwanenberg, B. *J. Chem. Soc., Perkin Trans. 1* **1997**, 963–967.
10. Imae, K.; Kamachi, H.; Yamashita, H.; Okita, T.; Okuyama, S.; Tsuno, T.; Yamasaki, T.; Sawada, Y.; Ohbayashi, M.; Naito, T.; Oki, T. *J. Antibiotics* **1991**, *44*, 76–85.
11. Baldwin, J.E.; Farthing, C.N.; Russell, A.T.; Schofield, C.J.; Spivey, A.C. *Tetrahedron Lett.* **1996**, *37*, 3761–3764.
12. Kells, K.W.; Ncube, A.; Chong, J.M. *Tetrahedron* **2004**, *60*, 2247–2257.
13. Bose, A.K.; Sahu, D.P.; Manhas, M.S. *J. Org. Chem.* **1981**, *46*, 1229–1230.
14. Solomon, M.E.; Lynch, C.L.; Rich, D.H. *Synth. Commun.* **1996**, *26*, 2723–2729.
15. Turner, J.J.; Sikkema, F.D.; Filippov, D.V.; van der Marel, G.A.; van Boom, J.H. *Synlett* **2001**, 1727–1730.
16. Turner, J.J.; Leeuwenburgh, M.A.; van der Marel, G.A.; van Boom, J.H. *Tetrahedron Lett.* **2001**, *42*, 8713–8716.
17. Tanaka, T.; Nakajima, K.; Okawa, K. *Bull. Chem. Soc. Jpn.* **1980**, *53*, 1352–1355.
18. Okawa, K.; Nakajima, K. *Biopolymers* **1981**, *20*, 1811–1821.

19. Wakamiya, T.; Shimbo, K.; Shiba, T.; Nakajima, K.; Neya, M.; Okawa, K. *Bull. Chem. Soc. Jpn.* **1982**, *55*, 3878–3881.

20. Zhou, H.; van der Donk, W.A. *Org. Lett.* **2002**, *4*, 1335–1338.

21. Okawa, K.; Kinutani, T.; Sakai, K. *Bull. Chem. Soc. Jpn.* **1968**, *41*, 1353–1355.

22. Shaw, K.J.; Luly, J.R.; Rapoport, H. *J. Org. Chem.* **1985**, *50*, 4515–4523.

23. Ibuka, T.; Nakai, K.; Habashita, H.; Hotta, Y.; Fujii, N.; Mimura, N.; Miwa, Y.; Taga, T.; Yamamoto, Y. *Angew. Chem., Int. Ed. Engl.* **1994**, *33*, 652–654.

24. Alonso, D.A.; Andersson, P.G. *J. Org. Chem.* **1998**, *63*, 9455–9461.

25. Atherton, E.; Meienhofer, J. *J. Antibiotics* **1972**, *25*, 539–540.

26. Dugave, C.; Ménez, A. *Tetrahedron: Asymmetry* **1997**, *8*, 1453–1465.

27. Wakamiya, T.; Mizuno, K.; Ukita, T.; Teshima, T.; Shiba, T. *Bull. Chem. Soc. Jpn.* **1978**, *51*, 850–854.

28. Teshima, T.; Konishi, K.; Shiba, T. *Bull. Chem. Soc. Jpn.* **1980**, *53*, 508–511.

29. Shiba, T.; Ukita, T.; Mizuno, K.; Teshima, T.; Wakamiya, T. *Tetrahedron Lett.* **1977**, *31*, 2681–2684.

30. Fernández-Megía, E.; Montaos, M.A.; Sardina, F.J. *J. Org. Chem.* **2000**, *65*, 6780–6783.

31. Kuyl-Yeheskiely, E.; Dreef-Tromp, C.M.; van der Marel, G.A.; van Boom, J.H. *Recl. Trav. Chim. Pays-Bas* **1989**, *108*, 314–316.

32. Kelly, J.W.; Eskew, N.L.; Evans S.A., Jr. *J. Org. Chem.* **1986**, *51*, 95–97.

33. Kuyl-Yeheskiely, E.; Lodder, M.; van der Marel, G.A.; van Boom, J.H. *Tetrahedron Lett.* **1992**, *33*, 3013–3016.

34. Okawa, K.; Nakajima, K.; Tanaka, T.; Kawana, Y. *Chem. Lett.* **1975**, 591–594.

35. Wipf, P.; Miller, C.P. *Tetrahedron Lett.* **1992**, *33*, 6267–6270.

36. Ide, N.D.; Galonic, D.P.; van der Donk, W.A.; Gin, D.Y. *Synlett* **2005**, 2011–2014.

37. Nakajima, K.; Tanaka, T.; Neya, M.; Okawa, K. *Bull. Chem. Soc. Jpn.* **1982**, *55*, 3237–3241.

38. Vedejs, E.; Klapars, A.; Warner, D.L.; Weiss, A.H. *J. Org. Chem.* **2001**, *66*, 7542–7546.

39. Legters, J.; Thijs, L.; Zwanenburg, B. *Tetrahedron Lett.* **1989**, *30*, 4881–4884.

40. Legters, J.; Thijs, L.; Zwanenburg, B. *Recl. Trav. Chim. Pays-Bas* **1992**, *111*, 1–15.

41. Tanner, D.; Somfai, P. *Tetrahedron Lett.* **1987**, *28*, 1211–1214.

42. Shao, H.; Zhu, Q.; Goodman, M. *J. Org. Chem.* **1995**, *60*, 790–791.

43. Tanner, D.; Birgersson, C.; Dhaliwal, H.K. *Tetrahedron Lett.* **1990**, *31*, 1903–1906.

44. Dauben, P.; Dubois, L.; Tran Huu Dau, M.E.; Dodd, R.H. *J. Org. Chem.* **1995**, *60*, 2035–2043.

45. Coldham, I.; Collis, A.J.; Mould, R.J.; Rathmell, R.E. *J. Chem. Soc., Perkin Trans. 1* **1995**, 2739–2745.

46. Wipf, P.; Venkatraman, S.; Miller, C.P. *Tetrahedron Lett.* **1995**, *36*, 3639–3642.

47. Burgaud, B.G.M.; Horwell, D.C.; Padova, A.; Pritchard, M.C. *Tetrahedron* **1996**, *52*, 13035–13050.

48. Xiong, C.; Wang, W.; Hruby, V.J. *J. Org. Chem.* **2002**, *67*, 3514–3517 (note that structures and *R/S* configurations in the text are inconsistent).

49. Jiang, Z.-X.; Qin, Y.-Y.; Qing, F.-L. *J. Org. Chem.* **2003**, *68*, 7544–7547.

50. Alonso, M.; Riera, A. *Tetrahedron: Asymmetry* **2005**, *16*, 3908–3912.

51. Sayyed, I.A.; Sudalai, A. *Tetrahedron: Asymmetry* **2004**, *15*, 3111–3116.

52. Byun, H.-S.; He, L.; Bittman, R. *Tetrahedron* **2000**, *56*, 7051–7091.

53. Axelsson, B.S.; O'Toole, K.J.; Spencer, P.A.; Young, D.W. *J. Chem. Soc., Perkin Trans. 1*, **1994**, 807–815.

54. Axelsson, B.S.; O'Toole, K.J.; Spencer, P.A.; Young, D.W. *J. Chem. Soc., Chem. Commun.* **1991**, 1085–1086.

55. Beresford, K.J.M.; Young, D.W. *Tetrahedron* **1996**, *52*, 9891–9900.

56. Kato, Y.; Fukumoto, K. *Chem. Commun.* **2000**, 245–246.

57. Cardillo, G.; Gentilucci, L.; Tolomelli, A.; Tomasini, C. *J. Org. Chem.* **1998**, *63*, 3458–3462.

58. Boivin, S.; Outurquin, F.; Paulmier, C. *Tetrahedron Lett.* **2000**, *41*, 663–666.

59. Gillings, N.M.; Venkachatalam, T.K.; Gee, A.D. *J. Lab. Cmpds. Radiopharm.* **1995**, *37*, 133–134.

60. Wade, T.N.; Gaymard, F.; Guedj, R. *Tetrahedron Lett.* **1979**, *29*, 2681–2682.

61. Lown, J.W.; Itoh, T.; Ono, N. *Can. J. Chem.* **1973**, *51*, 856–869.

62. Ayi, A.I.; Guedj, R. *J. Chem. Soc., Perkin Trans. 1* **1983**, 2045–2051.

63. Nagel, D.L.; Woller, P.B.; Cromwell, N.H. *J. Org. Chem.* **1971**, *36*, 3911–3917.

64. Gelas-Mialhe, Y.; Touraud, E.; Vessiere, R. *Can. J. Chem.* **1982**, *60*, 2830–2851.

65. Cromwell, N.H.; Hoeksema, H. *J. Am. Chem. Soc.* **1949**, *71*, 708–711.

66. Cromwell, N.H.; Mercer, G.D. *J. Am. Chem. Soc.* **1957**, *79*, 3819–3823.

67. Alezra, V.; Micouin, L.; Bonin, M.; Husson, H.-P. *Tetrahedron Lett.* **2000**, *41*, 651–654.

68. Harada, K.; Nakamura, I. *J. Chem. Soc., Chem. Commun.* **1978**, 522–523.

69. Alezra, V.; Bonin, M.; Micouin, L.; Policar, C.; Husson, H.-P. *Eur. J. Org. Chem.* **2001**, 2589–2594.

70. Alezra, V.; Bouchet, C.; Micouin, L.; Bonin, M.; Husson, H.-P. *Tetrahedron Lett.* **2000**, *41*, 655–658.

71. Davoli, P.; Forni, A.; Moretti, I.; Prati, F. *Tetrahedron: Asymmetry* **1995**, *6*, 2011–2016.

72. Garner, P.; Dogan, O.; Pillai, S. *Tetrahedron Lett.* **1994**, *35*, 1653–1656.

73. Cardillo, G.; Gentilucci, L.; Tomasini, C.; Castejon-Bordas, M.P.V. *Tetrahedron: Asymmetry* **1996**, *7*, 755–762.

74. Jähnisch, K. *Liebigs Ann./Recueil* **1997**, 757–760.

75. Cardillo, G.; Casolari, S.; Gentilucci, L.; Tomasini, C. *Angew. Chem., Int. Ed. Engl.* **1996**, *35*, 1848–1849.

76. Cardillo, G.; Gentilucci, L.; Tolomelli, A. *Tetrahedron Lett.* **1999**, *40*, 8261–8264.

77. Bongini, A.; Cardillo, G.; Gentilucci, L.; Tomasini, C. *J. Org. Chem.* **1997**, *62*, 9148–9153.

78. Cardillo, G.; Gentilucci, L.; Bastardas, I.R.; Tolomelli, A. *Tetrahedron* **1998**, *54*, 8217–8222.

79. Furukawa, N.; Yoshimura, T.; Ohtsu, M.; Akasaka, T.; Oae, S. *Tetrahedron* **1980**, *36*, 73–80.

80. Li, Z.; Conser, K.R.; Jacobsen, E.N. *J. Am. Chem. Soc.* **1993**, *115*, 5326–5327.

81. Evans, D.A.; Faul, M.M.; Bilodeau, M.T.; Anderson, B.A.; Barnes, D.M. *J. Am. Chem. Soc.* **1993**, *115*, 5328–5329.

82. Dauban, P.; Dodd, R.H. *Tetrahedron Lett.* **1998**, *39*, 5739–5742.

83. Atkinson, R.S.; Ayscough, A.P.; Gattrell, W.T.; Raynham, T.M. *J. Chem. Soc. Perkin Trans. 1* **1998**, 2783–2793.

84. Siu, T.; Yudin, A.K. *J. Am. Chem. Soc.* **2002**, *124*, 530–531.

85. Yang, K.-S.; Chen, K. *J. Org. Chem.* **2001**, *66*, 1676–1679.

86. Yang, K.-S.; Chen, K. *Org. Lett.* **2002**, *4*, 1107–1109.

87. Ishihara, H.; Hori, K.; Sugihara, H.; Ito, Y.N.; Katsuki, T. *Helv. Chim. Acta* **2002**, *85*, 4272–4286.

88. Rasmussen, K.G.; Jørgensen, K.A. *J. Chem. Soc., Chem. Commun.* **1995**, 1401–1402.

89. Hansen, K.B.; Finney, N.S.; Jacobsen, E.N. *Angew. Chem., Int. Ed. Engl.* **1995**, *34*, 676–678.

90. Rasmussen, K.G.; Jørgensen, K.A. *J. Chem. Soc., Perkin Trans. 1* **1997**, 1287–1291.

91. Crousse, B.; Narizuka, S.; Bonnet-Delpon, D.; Bégué, J.-P. *Synlett* **2001**, 679–681.

92. Xie, W.; Fang, J.; Li, J.; Wang, P.G. *Tetrahedron* **1999**, *55*, 12929–12938.

93. Sengupta, S.; Mondal, S. *Tetrahedron Lett.* **2000**, *41*, 6245–6248.

94. Yadav, J.S.; Reddy, B.V.S.; Rao, M.S.; Reddy, P.N. *Tetrahedron Lett.* **2003**, *44*, 5275–5278.

95. Mohan, J.M.; Uphade, B.S.; Choudhary, V.R.; Ravindranathan, T.; Sudalai, A. *Chem. Commun.* **1997**, 1429–1430.

96. Ha, H.-J.; Kang, K.-H.; Suh, J.-M.; Ahn, Y.-G. *Tetrahedron Lett.* **1996**, *37*, 7069–7070.

97. Antilla, J.C.; Wulff, W.D. *Angew. Chem., Int. Ed.* **2000, 39**, 4518–4521.

98. Williams, A.L.; Johnston, J.N. *J. Am. Chem. Soc.* **2004**, *126*, 1612–1613.

99. Bernstein, Z.; Ben-Ishai, D. *Tetrahedron* **1977**, *33*, 881–883.

100. McCarthy, J.R.; Barney, C.L.; O'Donnell, M.J.; Huffman, J.C. *J. Chem. Soc., Chem. Commun.* **1987**, 469–470.

101. Davis, F.A.; Liu, H.; Reddy, G.V. *Tetrahedron Lett.* **1996**, *37*, 5473–5476.

102. Davis, F.A.; Zhou, P.; Reddy, G.V. *J. Org. Chem.* **1994**, *59*, 3243–3245.

103. Davis, F.A.; Reddy, G.V.; Liang, C.-H. *Tetrahedron Lett.* **1997**, *38*, 5139–5142.

104. Davis, F.A.; Liu, H.; Zhou, P.; Fang, T.; Reddy, G.V.; Zhang, Y. *J. Org. Chem.* **1999**, *64*, 7559–7567.

105. Davis, F.A.; Liang, C.-H.; Liu, H. *J. Org. Chem.* **1997**, *62*, 3796–3797.

106. Davis, F.A.; Liu, H.; Liang, C.-H.; Reddy, G.V.; Zhang, Y.; Fang, T.; Titus, D.D. *J. Org. Chem.* **1999**, *64*, 8929–8935.

107. Davis, F.A.; Deng, J.; Zhang, Y.; Haltiwanger, R.C. *Tetrahedron* **2002**, *58*, 7135–7143.

108. Alves, M.J.; Fortes, A.G.; Gonçalves, L.F. *Tetrahedron Lett.* **2003**, *44*, 6277–6279.

109. McLaren, A.B.; Sweeney, J.B. *Org. Lett.* **1999**, *1*, 1339–1341.

110. Zhou, Y.-G.; Li, A.-H.; Hou, X.-L.; Dai, L.-X. *Tetrahedron Lett.* **1997**, *38*, 7225–7228.

111. Demarcus, M.; Filigheddu, S.N.; Mann, A.; Taddei, M. *Tetrahedron Lett.* **1999**, *40*, 4417–4420.

112. Patwardhan, A.P.; Pulgam, V.R.; Zhang, Y.; Wulff, W.D. *Angew. Chem., Int. Ed.* **2005, 44**, 6169–6172.

113. Nakajima, K.; Oda, H.; Okawa, K. *Bull. Chem. Soc. Jpn.* **1983**, *56*, 520–522.

114. Nakajima, K.; Neya, M.; Yamada, S.; Okawa, K. *Bull. Chem. Soc. Jpn.* **1982**, *55*, 3049–3050.

115. Nakajima, K.; Okawa, K. *Bull. Chem. Soc. Jpn.* **1983**, *56*, 1565–1566.

116. Bodenan, J.; Chanet-Ray, J.; Vessiere, R. *Synthesis* **1992**, 288–292.

117. Lee, T.; Jones, J.B. *Tetrahedron*, **1995**, *51*, 7331–7346.

118. Ho, M.; Wang, W.; Douvlos, M.; Pham, T.; Klock, T. *Tetrahedron Lett.* **1991**, *32*, 1283–1286.

119. Cardillo, G.; Gentilucci, L.; Tolomelli, A.; Tomasini, C. *Tetrahedron Lett.* **1997**, *38*, 6935–6956.

120. Cardillo, G.; Gentilucci, L.; Gianotti, M.; Tolomelli, A. *Synlett* **2000**, 1309–1311.

121. Heine, H.W.; Fetter, M.E.; Nicholson, E.M. *J. Am. Chem. Soc.* **1959**, *81*, 2202–2204.

122. Cardillo, G.; Gentilucci, L.; Tolomelli, A. *Chem. Commun.* **1999**, 167–168.

123. Solomon, M.E.; Lynch, C.L.; Rich, D.H. *Tetrahedron Lett.* **1995**, *36*, 4955–4958.

124. Harada, H.; Morie, T.; Suzuki, T.; Yoshida, T.; Kato, S. *Tetrahedron* **1998**, *54*, 10671–10676; corrigenda **1998**, *54*, 14635.

125. Nadir, U.K.; Krishna, R.V.; Singh, A. *Tetrahedron Lett.* **2005**, *46*, 479–482.

126. Chuang, T.-H.; Sharpless, K.B. *Org. Lett.* **1999**, *1*, 1435–1437.

127. Kim, Y.; Ha, H.-J.; Han, K.; Ko, S.W.; Yun, H.; Yoon, H.J.; Kim, M.S.; Lee, W.K. *Tetrahedron Lett.* **2005**, *46*, 4407–4409.

128. Wade, T.N.; Kheribet, R. *J. Chem. Res. S* **1980**, 210–211.

129. Bapama, A.; Condom, R.; Guedj, R. *J. Fluorine Chem.* **1980**, *16*, 183–187.

130. Wade, T.N. *J. Org. Chem.* **1980**, *45*, 5328–5333.

131. Gillings, N.M.; Gee, A.D. *J. Lab. Cmpds. Radiopharm.* **1997**, *40*, 764–765.

132. Righi, G.; D'Achille, R.; Bonini, C. *Tetrahedron Lett.* **1996**, *37*, 6893–6896.

133. Chandrasekhar, S.; Ahmed, M. *Tetrahedron Lett.* **1999**, *40*, 9325–9327.

134. Bucciarelli, M.; Forni, A.; Moretti, I.; Prati, F.; Torre, G. *Tetrahedron: Asymmetry* **1995**, *6*, 2073–2080.

135. Hata, Y.; Watanabe, M. *Tetrahedron* **1987**, *43*, 3881–3888.

136. Legters, J.; Thijs, L.; Zwanenburg, B. *Recl. Trav. Chim. Pays-Bas* **1992**, *111*, 16–21.

137. Okawa, K.; Yuki, M.; Tanaka, T. *Chem. Lett.* **1979**, 1085–1086.

138. Tanaka, T.; Nakajima, K.; Maeda, T.; Nakamura, A.; Hayashi, N.; Okawa, K. *Bull. Chem. Soc. Jpn.* **1979**, *52*, 3579–3581.

139. Nakajima, K.; Tanaka, T.; Morita, K.; Okawa, K. *Bull. Chem. Soc. Jpn.* **1980**, *53*, 283–284.

140. Galonic, D.P.; van der Donk, W.A.; Gin, D.Y. *J. Am. Chem. Soc.* **2004**, *126*, 12712–12713.

141. Okawa, K.; Nakajima, K.; Tanaka, T.; Neya, M. *Bull. Chem. Soc. Jpn.* **1982**, *55*, 174–176.

142. Nakajima, K.; Oda, H.; Okawa, K. *Bull. Chem. Soc. Jpn.* **1982**, *55*, 3232–3236.

143. Sato, K.; Kozikowski, A.P. *Tetrahedron Lett.* **1989**, *30*, 4073–4076.

144. Shima, I.; Shimazaki, N.; Imai, K.; Hemmi, K.; Hashimoto, M. *Chem. Pharm. Bull.* **1990**, *38*, 564–566.

145. Bennani, Y.L.; Zhu, G.-D.; Freeman, J.C. *Synlett* **1998**, 754–756.

146. Baldwin, J.E.; Adlington, R.M.; Robinson, N.G. *J. Chem. Soc., Chem. Commun.* **1987**, 153–155.

147. Baldwin, J.E.; Adlington, R.M.; O'Neil, I.A.; Schofield, C.; Spivey, A.C.; Sweeney, J.B. *J. Chem. Soc., Chem. Commun.* **1989**, 1852–1854.

148. Church, N.J.; Young, D.W. *J. Chem. Soc., Chem. Commun.* **1994**, 943–944.

149. Church, N.J.; Young, D.W. *J. Chem. Soc., Perkin Trans. 1* **1998**, 1475–1482.

150. Church, N.J.; Young, D.W. *Tetrahedron Lett.* **1995**, *36*, 151–154.

151. Osborn, H.M.I.; Sweeney, J.B. *Tetrahedron Lett.* **1994**, *35*, 2739–2742.

152. Bergmeier, S.C.; Seth, P.P. *Tetrahedron Lett.* **1999**, *40*, 6181–6184.

153. Poch, M.; Verdaguer, X.; Moyano, A.; Pericàs, M.A.; Riera, A. *Tetrahedron Lett.* **1991**, *32*, 6935–6938.

154. Medina, E.; Moyano, A.; Pericàs, M.A.; Riera, A. *J. Org. Chem.* **1998**, *63*, 8574–8578.

155. Travins, J.M.; Etzkorn, F.A. *Tetrahedron Lett.* **1998**, *39*, 9389–9392.

156. Chang, J.-W.; Bae, J.H.; Shin, S.-H.; Park, C.S.; Choi, D.; Lee, W.K. *Tetrahedron Lett.* **1998**, *39*, 9193–9196.

157. Le Merrer, Y.; Duréault, A.; Greck, C.; Micas-Languin, D.; Gravier, C.; Depezay, J.-C. *Heterocycles* **1987**, *25*, 541–548.

158. Duréault, A.; Greck, C.; Depezay, J.C. *Tetrahedron Lett.* **1986**, *27*, 4157–4160.

159. Duréault, A.; Tranchepain, I.; Greck, C.; Depezay, J.-C. *Tetrahedron Lett.* **1987**, *28*, 3341–3344.

160. Duréault, A.; Tranchepain, I.; Depezay, J.-C. *J. Org. Chem.* **1989**, *54*, 5324–5330.

161. Antolini, L.; Bucciarelli, M.; Caselli, E.; Davoli, P.; Forni, A.; Moretti, I.; Prati, F.; Torre, G. *J. Org. Chem.* **1997**, *62*, 8784–8789.

162. Dauban, P.; Chiaroni, A.; Riche, C.; Dodd, R.H. *J. Org. Chem.* **1996**, *61*, 2488–2496.

163. Dubois, L.; Dodd, R.H. *Tetrahedron* **1993**, *49*, 901–910.

164. Dauban, P.; de Saint-Fuscien, C.; Dodd, R.H. *Tetrahedron* **1999**, *55*, 7589–7600.

165. Dubois, L.; Mehta, A.; Tourette, E.; Dodd, R.H. *J. Org. Chem.* **1994**, *59*, 434–441.

166. Davis, F.A.; Reddy, G.V.; Liu, H. *J. Am. Chem. Soc.* **1995**, *117*, 3651–3652.

167. Deyrup, J.A.; Clough, S.C. *J. Org. Chem.* **1974**, *39*, 902–907.

168. Fujii, N.; Nakai, K.; Tamamura, H.; Otaka, A.; Mimura, N.; Miwa, Y.; Taga, T.; Yamamoto, Y.; Ibuka, T. *J. Chem. Soc., Perkin Trans 1*, **1995**, 1359–1371.

169. Satake, A.; Shimizu, I.; Yamamoto, A. *Synlett* **1995**, 64–68.

170. Ohno, H.; Mimura, N.; Otaka, A.; Tamamura, H.; Fujii, N.; Ibuka, T.; Shimizu, I.; Satake, A.; Yamamoto, Y. *Tetrahedron*, **1997**, *53*, 12933–12946.

171. Ibuka, T.; Mimura, N.; Ohno, H.; Nakai, K.; Akaji, M.; Habashita, H.; Tamamura, H.; Miwa, Y.; Taga, T.; Fujii, N.; Yamamoto, Y. *J. Org. Chem.* **1997**, *62*, 2982–2991.

172. Ibuka, T.; Mimura, N.; Aoyama, H.; Akaji, M.; Ohno, H.; Miwa, Y.; Taga, T.; Nakai, K.; Tamamura, H.; Fujii, N.; Yamamoto, Y. *J. Org. Chem.* **1997**, *62*, 999–1015.

173. Tamamura, H.; Yamashita, M.; Muramatsu, H.; Ohno, H.; Ibuka, T.; Otaka, A.; Fujii, N. *Chem. Commun.* **1997**, 2327–2328.

174. Rens, M.; Ghosez, L. *Tetrahedron Lett.* **1970**, 43, 3765–3768.

175. Heimgartner, H. *Angew. Chem., Int. Ed. Engl.* **1991**, *30*, 238–264.

176. Gilchrist, T.L. *Aldrichimica Acta* **2001**, *34*, 51–55.

177. Palacios, F.; Ochoa de Retana, A.M.; Martínez de Marigorta, E.; Manuel de los Santos, J. *Eur. J. Org. Chem.* **2001**, 2401–2414.

178. Brun, K.A.; Heimgartner, H. *Helv. Chim. Acta* **2005**, *88*, 2951–2959.

179. Stamm, S.; Linden, A.; Heimgartner, H. *Helv. Chim. Acta* **2003**, *86*, 1371–1396.

180. Breitenmoser, R.A.; Hirt, T.R.; Luykx, R.T.N.; Heimgartner, H. *Helv. Chim. Acta* **2001**, *84*, 972–979.

181. Breitenmoser, R.A.; Heimgartner, H. *Helv. Chim. Acta* **2002**, *85*, 885–912.

182. Brun, K.A.; Linden, A.; Heimgartner, H. *Helv. Chim. Acta* **2001**, *84*, 1756–1777.

183. Brun, K.A.; Linden, A.; Heimgartner, H. *Helv. Chim. Acta* **2002**, *85*, 3422–3443.

184. Chandrasekhar, B.P.; Heimgartner, H.; Schmid, H. *Helv. Chim. Acta* **1977**, *60*, 2270–2287.

185. Vittorelli, P.; Heimgartner, H.; Schmid, H.; Hoet, P.; Ghosez, L. *Tetrahedron* **1974**, *30*, 3737–3740.

186. Wade, T.N.; Guedj, R. *Tetrahedron Lett.* **1979**, *41*, 3953–3954.

187. Wade, T.N.; Khéribet, R. *J. Org. Chem.* **1980**, *45*, 5333–5335.

188. Obrecht, D.; Spiegler, C.; Schönholzer, P.; Müller, K.; Heimgartner, H.; Stierli, F. *Helv. Chim. Acta* **1992**, *75*, 1666–1696.

189. Sahebi, M.; Wipf, P.; Heimgartner, H. *Tetrahedron* **1989**, *45*, 2999–3000.

190. Wipf, P.; Heimgartner, H. *Helv. Chim. Acta* **1988**, *71*, 140–154.

191. Heimgartner, H. *Israel J. Chem.* **1986**, *27*, 3–15.

192. Obrecht, D.; Heimgartner, H. *Helv. Chim. Acta* **1981**, *64*, 482–487.

193. Obrecht, D.; Heimgartner, H. *Helv. Chim. Acta* **1987**, *70*, 102–115.

194. Basu, G.; Bagchi, K.; Kuki, A. *Biopolymers* **1991**, *31*, 1763–1774.

195. Jeremic, T.; Linden, A.; Heimgartner, H. *Chem. Biodivers.* **2004**, *1*, 1730–1761.

196. Koch, K.N.; Heimgartner, H. *Helv. Chim. Acta* **2000**, *83*, 1881–1900.

197. Luykx, R.T.N.; Linden, A.; Heimgartner, H. *Helv. Chim. Acta* **2003**, *86*, 4093–4110.

198. Lehmann, J.; Heimgartner, H. *Helv. Chim. Acta* **2000**, *83*, 233–257.

199. Álvares, Y.S.P.; Alves, M.J.; Azoia, N.G.; Bickley, J.F.; Gilchrist, T.L. *J. Chem. Soc., Perkin Trans. 1* **2001**, 1911–1919.

200. Bucher, C.B.; Linden, A.; Heimgartner, H. *Helv. Chim. Acta* **1995**, *78*, 935–946.

201. Lehmann, J.; Linden, A.; Heimgartner, H. *Helv. Chim. Acta* **1999**, *82*, 888–908.

202. Breitenmoser, R.A.; Heimgartner, H. *Helv. Chim. Acta* **2001**, *84*, 786–796.

203. Lehmann, J.; Linden, A.; Heimgartner, H. *Tetrahedron* **1998**, *54*, 8721–8736.

204. Lehmann, J.; Linden, A.; Heimgartner, H. *Tetrahedron* **1999**, *55*, 5359–5376.

205. Lehmann, J.; Heimgartner, H. *Helv. Chim. Acta* **1999**, *82*, 1899–1915.

206. Breitenmoser, R.A.; Linden, A.; Heimgartner, H. *Helv. Chim. Acta* **2002**, *85*, 990–1018.

207. Gilchrist, T.L.; Mendonça, R. *Synlett* **2000**, 1843–1845.

208. Alves, M.J.; Azoia, N.G.; Bickley, J.F.; Fortes, A.G.; Gilchrist, T.L.; Mendonca, R. *J. Chem. Soc., Perkin Trans. 1* **2001**, 2969–2976.

209. Timén, A.S.; Somfai, P. *J. Org. Chem.* **2003**, *68*, 9958–9963.

210. Alves, M.J.; Gilchrist, T.L.; Sousa, J.H. *J. Chem. Soc., Perkin Trans. 1* **1999**, 1305–1310.

211. Alves, M.J.; Bickley, J.F.; Gilchrist, T.L. *J. Chem. Soc., Perkin Trans. 1* **1999**, 1399–1401.

212. Risberg, E.; Fischer, A.; Somfai, P. *Chem. Commun.* **2004**, 2088–2089.

213. Verstappen, M.M.H.; Ariaans, G.J.A.; Zwanenburg, B. *J. Am. Chem. Soc.* **1996**, *118*, 8491–8492.

214. Kolb, H.C.; VanNieuwenhze, M.S.; Sharpless, K.B. *Chem. Rev.* **1994**, *94*, 2483–2547.

215. Chong, J.M.; Sharpless, K.B. *J. Org. Chem.* **1985**, *50*, 1560–1563.

216. Pons, D.; Savignac, M.; Genet, J.-P. *Tetrahedron Lett.* **1990**, *31*, 5023–5026.

217. Sharpless, K.B.; Behrens, C.H.; Katsuki, T.; Lee, A.W.M.; Martin, V.S.; Takatani, M.; Viti, S.M.; Walker, F.J.; Woodard, S.S. *Pure Appl. Chem.* **1983**, *55*, 589–604.

218. Kurokawa, N.; Ohfune, Y. *J. Am. Chem. Soc.* **1986**, *108*, 6041–6043.

219. Carlsen, P.H.J.; Katsuki, T.; Martin, V.S.; Sharpless, K.B. *J. Org. Chem.* **1981**, *46*, 3939–3940.

220. Cardani, S.; Bernardi, A.; Colombo, L.; Gennari, C.; Scolastico, C.; Venturini, I. *Tetrahedron* **1988**, *44*, 5563–5572.

221. Wu, X.-Y.; She, X.; Shi, Y. *J. Am. Chem. Soc.* **2002**, *124*, 8792–8793.

222. Cabon, O.; Buisson, D.; Larcheveque, M.; Azerad, R. *Tetrahedron: Asymmetry* **1995**, *6*, 2211–2218.

223. Murakami, M.; Mukaiyama, T. *Chem. Lett.* **1982**, 1271–1274.

224. Walborsky, H.M.; Baum, M.E. *J. Am. Chem. Soc.* **1958**, *80*, 187–192.

225. Carter, H.E.; Zirkle, C.L. *J. Biol. Chem.* **1949**, *178*, 709–714.

226. Newman, M.S.; Magerlein, B.J. *Org. React.* **1949**, *5*, 413–440.

227. Corey, E.J.; Lee, D.-H.; Choi, S. *Tetrahedron Lett.* **1992**, *33*, 6735–6738.

228. Cabon, O.; Buisson, D.; Larcheveque, M.; Azerad, R. *Tetrahedron: Asymmetry* **1995**, *6*, 2199–2210.

229. Feske, B.D.; Steart, J.D. *Tetrahedron: Asymmetry* **2005**, *16*, 3124–3127.

230. Rodrigues, J.A.R.; Milagre, H.M.S.; Milagre, C.D.F.; Moran, P.J.S. *Tetrahedron: Asymmetry* **2005**, *16*, 3099–3106.

231. Aggarwal, V.K.; Hynd, G.; Picoul, W.; Vasse, J.-L. *J. Am. Chem. Soc.* **2002**, *124*, 9964–9965.

232. Palomo, C.; Oiarbide, M.; Sharma, A.K.; González-Rego, M.C.; Linden, A.; García, J.M.; González, A. *J. Org. Chem.* **2000**, *65*, 9007–9012.

233. Behrens, C.H.; Sharpless, K.B. *Aldrichimica Acta* **1983**, *16*, 67–79.

234. Behrens, C.H.; Sharpless, K.B. *J. Org. Chem.* **1985**, *50*, 5696–5704.

235. Petit, Y.; Hutin, P. *Bull. Chem. Soc. Fr.* **1996**, *133*, 1081–1094.

236. Kurokawa, N.; Ohfune, Y. *Tetrahedron* **1993**, *49*, 6195–6222.

237. Liwschitz, Y.; Rabinsohn, Y.; Perera, D. *J. Chem. Soc.* **1962**, 1116–1119.

238. Minami, N.; Ko, S.S.; Kishi, Y. *J. Am. Chem. Soc.* **1982**, *104*, 1109–1111.

239. Caldwell, C.G.; Bondy, S.S. *Synthesis* **1990**, 34–36.

240. Genêt, J.P.; Durand, J.O.; Savignac, M.; Pons, D. *Tetrahedron Lett.* **1992**, *33*, 2497–2500.

241. Clayden, J.; Collington, E.W.; Warren, S. *Tetrahedron Lett.* **1993**, *34*, 1327–1330.

242. Liwschitz, Y.; Rabinsohn, Y.; Haber, A. *J. Chem. Soc.* **1962**, 3589–3591.

243. Saito, S.; Takahashi, N.; Ishikawa, T.; Moriake, T. *Tetrahedron Lett.* **1991**, *32*, 667–670.

244. Mukaiyama, T.; Yura, T.; Iwasawa, N. *Chem. Lett.* **1985**, 809–812.

245. Alcaide, B.; Biurrun, C.; Martínez, A.; Plumet, J. *Tetrahedron Lett.* **1995**, *36*, 5417–5420.

246. Hönig, H.; Seufer-Wasserthal, P.; Weber, H. *Tetrahedron* **1990**, *46*, 3841–3850.

247. Valpuesta, M.; Durante, P.; López-Herrera, F.J. *Tetrahedron Lett.* **1995**, *36*, 4681–4684.

248. Yamagichi, T.; Harada, N.; Ozaki, K.; Hayashi, M.; Arakawa, H.; Hashiyama, T. *Tetrahedron* **1999**, *55*, 1005–1016.

249. Fringuelli, F.; Pizzo, F; Vaccaro, L. *J. Org. Chem.* **2001**, *66*, 3554–3558.

250. Fringuelli, F.; Pizzo, F.; Rucci, M.; Vaccaro, L. *J. Org. Chem.* **2003**, *68*, 7041–7045.

251. Riego, J.; Costa, A.; Saa, J.M. *Chem. Lett.* **1986**, 1565–1568.

252. García Ruano, J.L.; García Paredes, C. *Tetrahedron Lett.* **2000**, *41*, 5357–5361.

253. Saito, S.; Bunya, N.; Inaba, M.; Moriwake, T.; Torii, S. *Tetrahedron Lett.* **1985**, *26*, 5309–5312.

254. Charvillon, F.B.; Amouroux, R. *Synth. Commun.* **1997**, *27*, 395–403.

255. Kaneko, T.; Katsura, H. *Bull. Chem. Soc. Jpn.* **1963**, *36*, 899–903.

256. Mattingly, P.G.; Miller, M.J. *J. Org. Chem.* **1983**, *48*, 3556–3559.

257. Hedgcoth, C.; Skinner, C.G.; Lindstedt, G.; Lindahl, G. *Biochem. Prep.* **1963**, *10*, 67–72.

258. Legters, J.; Thijs, L.; Zwanenberg, B. *Tetrahedron* **1991**, *47*, 5287–5294.

259. Genet, J.P. *Pure Appl. Chem.* **1996**, *68*, 593–596.

260. Savignac, M.; Durand, J.-O.; Genêt, J.-P. *Tetrahedron: Asymmetry* **1994**, *5*, 717–722.

261. Lanier, M.; Le Blanc, M.; Pastor, R. *Tetrahedron* **1996**, *52*, 14631–14640.

262. Clayden, J.; McElroy, A.B.; Warren, S. *J. Chem. Soc., Perkin Trans 1* **1995**, 1913–1934.

263. Clayden, J.; Collington, E.W.; Warren, S. *Tetrahedron Lett.* **1993**, *34*, 1327–1330.

264. Righi, G.; Rumboldt, G. *Tetrahedron* **1995**, *51*, 13401–13408.

265. Righi, G.; Chionne, A.; D'Achille, R.; Bonini, C. *Tetrahedron: Asymmetry* **1997**, *8*, 903–907.

266. Righi, G.; Rumboldt, G.; Bonini, C. *J. Org. Chem.* **1996**, *61*, 3557–3560.

267. Pastó, M.; Moyano, A.; Pericàs, M.A.; Riera, A. *J. Org. Chem.* **1997**, *62*, 8425–8431.

268. Chini, M.C.; Crotti, P.; Flippin, L.A.; Gardelli, C.; Giovani, E.; Macchia, F.; Pineschi, M. *J. Org. Chem.* **1993**, *58*, 1221–1227.

269. Augé, J.; Leroy, F. *Tetrahedron Lett.* **1996**, *37*, 7715–7716.

270. Caron, M.; Carlier, P.R.; Sharpless, K.B. *J. Org. Chem.* **1988**, *53*, 5185–5187.

271. Canas, M.; Poch, M.; Verdaguer, X.; Moyano, A.; Pericàs, M.A.; Riera, A. *Tetrahedron Lett.* **1991**, *32*, 6931–6934.

272. Benedatti, F.; Berti, F.; Norbedo, S. *Tetrahedron Lett.* **1998**, *39*, 7971–7974.

273. Kokotos, G.; Padrón, J.M.; Noula, C.; Gibbons, W.A.; Martin, V.S. *Tetrahedron: Asymmetry* **1996**, *7*, 857–866.

274. Medina, E.; Vidal-Ferran, A.; Moyano, A.; Pericàs, M.A.; Riera, A. *Tetrahedron: Asymmetry* **1997**, *8*, 1581–1586.

275. Fernández-García, C.; McKervey, M.A. *Tetrahedron: Asymmetry* **1995**, *6*, 2905–2906.

276. Ghosh, A.K.; Wang, Y. *J. Org. Chem.* **1999**, *64*, 2789–2795.

277. Bertelli, L.; Fiaschi, R.; Napolitano, E. *Gazz. Chim. Ital.* **1993**, *123*, 521–524.

278. Castejón, P.; Moyano, A.; Pericàs, M.A.; Riera, A. *Tetrahedron* **1996**, *52*, 7063–7086.

279. Trost, B.M.; Bunt, R.C. *Angew. Chem., Int. Ed. Engl.* **1996**, *35*, 99–102.

280. Shioiri, T.; Matsuura, F.; Hamada, Y. *Pure Appl. Chem.* **1994**, *66*, 2151–2154.

281. Matsuura, F.; Hamada, Y.; Shioiri, T. *Tetrahedron* **1993**, *49*, 8211–8222.

282. Matsuura, F.; Hamada, Y.; Shioiri, T. *Tetrahedron Lett.* **1992**, *33*, 7917–7920; errata **1993**, *34*, 2394.

283. Matsuura, F.; Hamada, Y.; Shioiri, T. *Tetrahedron Lett.* **1992**, *33*, 7921–7924; errata **1993**, *34*, 2394.

284. Matsuura, F.; Hamada, Y.; Shioiri, T. *Tetrahedron* **1994**, *50*, 265–274.

285. Jung, M.E.; Jung, Y.H. *Synlett* **1995**, 563–564.

286. Knapp, S.; Kukkola, P.J.; Sharma, S.; Dhar, T.G.M.; Naughton, A.B.J. *J. Org. Chem.* **1990**, *55*, 5700–5710.

287. Jung, M.E.; Jung, Y.H. *Tetrahedron Lett.* **1989**, *30*, 6637–6640.

288. Roush, W.R.; Adam, M.A. *J. Org. Chem.* **1985**, *50*, 3752–3757.

289. Roush, W.R.; Brown, R.J.; DiMare, M. *J. Org. Chem.* **1983**, *48*, 5083–5093.

290. Rao, B.V.; Krishna, U.M.; Gurjar, M.K. *Synth. Commun.* **1997**, *27*, 1335–1345.

291. Roush, W.R.; Brown, R.J. *J. Org. Chem.* **1982**, *47*, 1371–1373.

292. Rossi, F.M.; Powers, E.T.; Yoon, R.; Rosenberg, L.; Meinwald, J. *Tetrahedron* **1996**, *52*, 10279–10286.

293. Sunazuka, T.; Nagamitsu, T.; Tanaka, H.; Omura, S. *Tetrahedron Lett.* **1993**, *34*, 4447–4448.

294. Tung, R.D.; Sun, C.-Q.; Deyo, D.; Rich, D.E. *Peptides: Proc. 10th Ann Peptide Symp.* **1988**, 149–151.

295. Rama Rao, A.V.; Murali Dhar, T.G.; Chakraborty, T.K.; Gurjar, M.K.; *Tetrahedron Lett.* **1988**, *29*, 2069–2072.

296. Aebi, J.D.; Deyo, D.T.; Sun, C.Q.; Guillaume, D.; Dunlap, B.; Rich, D.E. *J. Med. Chem.* **1990**, *33*, 999–1009.

297. Sun, C.-Q.; Rich, D.H. *Tetrahedron Lett.* **1988**, *29*, 5205–5208.

298. Rama Rao, A.V.; Murali Dhar, T.G.; Subhas Bose, D.; Chakraborty, T.K.; Gurjar, M.K. *Tetrahedron* **1989**, *45*, 7361–7370.

299. Boger, D.L.; Ledeboer, M.W.; Kume, M. *J. Am. Chem. Soc.* **1999**, *121*, 1098–1099.

300. Boger, D.L.; Ledeboer, M.W.; Kume, M.; Searcey, M.; Jin, Q. *J. Am. Chem. Soc.* **1999**, *121*, 11375–11383

301. Ginestra, X.; Pericàs, M.A.; Riera, A. *Synth. Commun.* **2005**, *35*, 289–297.

302. Schmidt, U.; Respondek, M.; Lieberknecht, A.; Werner, J.; Fischer, P. *Synthesis* **1989**, 256–261.

303. Hatakeyema, S.; Matsumoto, H.; Fukuyama, H.; Mukugi, Y.; Irie, H. *J. Org. Chem.* **1997**, *62*, 2275–2279.

304. Schmidt, U.; Zäh, M.; Lieberknecht, A. *J. Chem. Soc., Chem. Commun.* **1991**, 1002–1004.

305. Liu, D.-G.; Wang, B.; Lin, G.-Q. *J. Org. Chem.* **2000**, *65*, 9114–9119.

306. Hatakeyama, S.; Yoshida, M.; Esumi, T.; Iwabuchi, Y.; Irie, H.; Kawamoto, T.; Yamada, H.; Nishizawa, M. *Tetrahedron Lett.* **1997**, *38*, 7887–7890.

307. Bessodes, M.; Saïah, M.; Antonakis, K. *J. Org. Chem.* **1992**, *57*, 4441–4444.

308. Pégorier, L.; Haddad, M.; Larchevêque, M. *Synlett* **1996**, 585–586.

309. Jackson, R.F.W.; Palmer, N.J.; Wythes, M.J.; Clegg, W.; Elsegood, M.R.J. *J. Org. Chem.* **1995**, *60*, 6431–6440.

310. Jackson, R.F.W.; Palmer, N.J.; Wythes, M.J. *Tetrahedron Lett.* **1994**, *35*, 7433–7434.

311. Jackson, R.F.W.; Kirk, J.M.; Palmer, N.J.; Waterson, D.; Wythes, M.J. *J. Chem. Soc., Chem. Commun.* **1993**, 889–890.

312. Jackson, R.F.W.; Palmer, N.J.; Wythes, M.J. *J. Chem. Soc., Chem. Commun.* **1994**, 95–96.

313. Aggarwal, V.K.; Barrell, J.K.; Worrall, J.M.; Alexander, R. *J. Org. Chem.* **1998**, *63*, 7128–7129.

314. Gao, Y.; Sharpless, K.B. *J. Am. Chem. Soc.* **1988**, *110*, 7538–7539.

315. Hale, K.J.; Manaviazar, S.; Delisser, V.M. *Tetrahedron* **1994**, *50*, 9181–9188.

316. Manhas, M.S.; Wagle, D.R.; Chiang, J.; Bose, A.K. *Heterocycles* **1988**, *27*, 1755–1802.

317. Paloma, C.; Aizpurua, J.M.; Ganboa, I.; Oiarbide, M. *Amino Acids* **1999**, *16*, 321–343.

318. Palomo, C.; Aizpurua, J.M.; Ganboa, I.; Oiarbide, M. *Synlett* **2001**, 1813–1826.

319. Salzmann, T.N.; Ratcliffe, R.W.; Christensen, B.G.; Bouffard, F.A. *J. Am. Chem. Soc.* **1980**, *102*, 6163–6165.

320. Birkofer, L.; Schramm, J. *Liebigs Ann. Chem.* **1975**, 2195–2200.

321. Baldwin, J.E.; Adlington, R.M.; Gollins, D.W.; Schofield, C.J. *Tetrahedron* **1990**, *46*, 4733–4748.

322. Baldwin, J.E.; Edwards, A.J.; Farthing, C.N.; Russell, A.T. *Synlett* **1993**, 49–50.

323. Kozikowski, A.P.; Liao, Y.; Tückmantel, W.; Wang, S.; Pshenichkin, S.; Surin, A.; Thomsen, C.; Wroblewski, J.T. *Bioorg. Med. Chem. Lett.* **1996**, *6*, 2559–2564.

324. Barreau, M.; Commercon, A.; Mignani, S.; Mouysset, D.; Perfetti, P.; Stella, L. *Tetrahedron* **1998**, *54*, 11501–11516.

325. Palomo, C.; Aizpurua, J.M.; Galarza, R.; Iturburu, M.; Legido, M. *Bioorg. Med. Chem. Lett.* **1993**, *3*, 2461–2466.

326. Taggi, A.E.; Hafez, A.M.; Wack, H.; Young, B.; Drury W.J., III; Lectka, T. *J. Am. Chem. Soc.* **2000**, *122*, 7831–7832.

327. Taggi, A.E.; Hafez, A.M.; Wack, H.; Young, B.; Ferraris, D.; Lectka, T. *J. Am. Chem. Soc.* **2002**, *124*, 6626–6635.

328. France, S.; Shah, M.H.; Waetherwax, A.; Wack, H.; Roth, J.P.; Lectka, T. *J. Am. Chem. Soc.* **2005**, *127*, 1206–1215.

329. Baldwin, J.E.; Adlington, R.M.; Godfrey, C.R.A.; Gollins, D.W.; Smith, M.L.; Russel, A.T. *Synlett*, **1993**, 51–53.

330. Baldwin, J.E.; Adlington, R.M.; Godfrey, C.R.A.; Gollins, D.W.; Vaughan, J.G. *J. Chem. Soc., Chem. Commun.* **1993**, 1434–1435.

331. Baldwin, J.E.; Adlington, R.M.; Gollins, D.W.; Schofield, C.J. *J. Chem. Soc., Chem. Commun.* **1990**, 720–721.

332. Hanessian, S.; Sumi, K.; Vanasse, B. *Synlett* **1992**, 33–34.

333. Baldwin, J.E.; Adlington, R.M.; Gollins, D.W.; Godfrey, C.R.A. *Tetrahedron*, **1995**, *51*, 5169–5180.

334. Westwood, N.J.; Schofield, C.J.; Claridge, T.D.W. *J. Chem. Soc., Perkin Trans. 1* **1997**, 2725–2729.

335. Palomo, C.; Aipurua, J.M.; García, J.M.; Iturburu, M.; Odriozola, J.M. *J. Org. Chem.* **1994**, *59*, 5184–5188.

336. Palomo, C.; Cabré, F.; Ontoria, J.M. *Tetrahedron Lett.* **1992**, *33*, 4819–4822.

337. Banik, B.K.; Manhas, M.S.; Bose, A.K. *J. Org. Chem.* **1993**, *58*, 307–309.

338. Miller, M.J.; Mattingly, P.G.; Morrison, M.A.; Kerwin J.F., Jr. *J. Am. Chem. Soc.* **1980**, *102*, 7026–7032.

339. Mattingly, P.G.; Miller, M.J. *J. Org. Chem.* **1980**, *45*, 410–415.

340. Miller, M.J. *Acc. Chem. Res.* **1986**, *19*, 49–56.

341. Arai, H.; Hagmann, W.K.; Suguna, H.; Hecht, S.M. *J. Am. Chem. Soc.* **1980**, *102*, 6631–6633.

342. Baldwin, J.E.; Adlington, R.M.; Birch, D.J. *J. Chem. Soc., Chem. Commun.* **1985**, 256–257.

343. Davis, F.; Zhou, P. *Tetrahedron Lett.* **1994**, 35, 7525–7528.

344. Boojamra, C.G.; Lemoine, R.C.; Lee, J.C.; Léger, R.; Stein, K.A.; Vernier, N.G.; Magon, A.; Lomovskaya, O.; Martin, P.K.; Chamberland, S.; Lee, M.D.; Hecker, S.J.; Lee, V.J. *J. Am. Chem. Soc.* **2001**, *123*, 870–874.

345. Carrasco, M.R.; Brown, R.T. *J. Org. Chem.* **2003**, *68*, 8853–8858.

346. Ojima, I.; Chen, H.-J.C.; Nakahashi, K. *J. Am. Chem. Soc.* **1988**, *110*, 278–281.

347. Ojima, I.; Chen, H.-J.C.; Qui, X. *Tetrahedron* **1988**, *44*, 5307–5318.

348. Palomo, C.; Aizpurua, J.M.; Benito, A.; Miranda, J.I.; Fratila, R.M.; Matute, C.; Domercq, M.; Gago, F.; Martin-Santamaria, S.; Linden, A. *J. Am. Chem. Soc.* **2003**, *125*, 16243–16160.

349. Ojima, I.; Chen, H.-J.C. *J. Chem. Soc., Chem. Commun.* **1987**, 625–626.

350. Palomo, C.; Aizpurua, J.M.; Legido, M.; Mielgo, A.; Galarza, R. *Chem.—Eur. J.* **1997**, *3*, 1432–1441.

351. Ojima, I.; Suga, S.; Abe, R. *Chem. Lett.* **1980**, 853–856.

352. Ojima, I.; Qiu, X. *J. Am. Chem. Soc.* **1987**, *109*, 6537–6538.

353. Ojima, I.; Shimizu, N.; Qiu, X.; Chen, H.-J.C.; Nakahashi, K. *Bull. Soc. Chim. Fr.* **1987**, 649–658.

354. Ojima, I.; Delaloge, F. *Chem. Soc. Rev.* **1997**, *26*, 377–386.

355. Ojima, I.; Komata, T.; Qiu, X. *J. Am. Chem. Soc.* **1990**, *112*, 770–74.

356. Ojima, I. *Acc. Chem. Res.* **1995**, *28*, 383–389.

357. Colson, P.-J.; Hegedus, L.S. *J. Org. Chem.* **1993**, *58*, 5918–5924.

358. Schmeck, C.; Hegedus, L.S. *J. Am. Chem. Soc.* **1991**, *113*, 5784–5791.

359. Moyna, G.; Williams, H.J.; Scott, A.I. *Synth. Commun.* **1997**, *27*, 1561–1567.

360. Palomo, C.; Aizpurua, J.M.; Ganboa, I.; Carreaux, F.; Cuevas, C.; Maneiro, E.; Ontoria, J.M. *J. Org. Chem.* **1994**, *59*, 3123–3130.

361. Cossío, F.P.; López, C.; Oiarbide, M.; Palomo, C.; Aparicio, D.; Rubiales, G. *Tetrahedron Lett.* **1988**, *29*, 3133–3136.

362. Palomo, C.; Aizpurua, J.M.; Ganboa, I. *Russ. Chem. Bull.* **1996**, *45*, 2463–2483.

363. Palomo, C.; Aizpura, J.M.; Cabré, F.; Garcia, J.M.; Odriozola, J.M. *Tetrahedron Lett.* **1994**, *35*, 2721–2724.

364. Palomo, C.; Oiarbide, M.; Esnal, A.; Landa, A.; Miranda, J.I.; Linden, A. *J. Org. Chem.* **1998**, *63*, 5836–5846.

365. Palomo, C.; Ganboa, I.; Odriozola, B.; Linden, A. *Tetrahedron Lett.* **1997**, *38*, 3093–3096.

366. Palomo, C.; Oiarbide, M.; Esnal, A. *Chem. Commun.* **1997**, 691–692.

367. Palomo, C.; Oiarbide, M.; Landa, A. *J. Org. Chem.* **2000**, *65*, 41–46.

368. Palomo, C.; Aizpurua, J.M.; Ganboa, I.; Odriozola, B.; Urchegui, R.; Görls, H. *Chem. Commun* **1996**, 1269–1270.

369. Palomo, C.; Aizpurua, J.M.; Urchegui, R.; Garcia, J.M. *J. Chem. Soc., Chem. Commun.* **1995**, 2327–2328.

370. Palomo, C.; Ganboa, I.; Cuevas, C.; Boschetti, C.; Linden, A. *Tetrahedron Lett.* **1997**, *38*, 4643–4646.

371. Palomo, C.; Oiarbide, M.; Landa, A.; Esnal, A.; Linden, A. *J. Org. Chem.* **2001**, *66*, 4180–4186.

372. Berlin, K.D.; Williams, L.G.; Dermer, O.C. *Tetrahedron Lett.* **1971**, *7*, 873–876.

373. Berse, C.; Bessette, P. *Can. J. Chem.* **1971**, *49*, 2610–2611.

374. Parry, R.J.; Naidu, M.V. *Tetrahedron Lett.* **1983**, *24*, 1133–1134.

375. Kogami, Y.; Okawa, K. *Bull. Chem. Soc. Jpn.* **1987**, *60*, 2963–2965.

376. Ploux, O.; Caruso, M.; Chassaing, G.; Marquet, A. *J. Org. Chem.* **1988**, 53, 3154–3158.

377. Farthing, C.N.; Baldwin, J.E.; Russell, A.T.; Schofield, C.J.; Spivey, A.C. *Tetrahedron Lett.* **1996**, *37*, 5225–5226.

378. Fukami, T.; Yamakawa, T.; Niiyama, K.; Kojima, H.; Amano, Y.; Kanda, F.; Ozaki, S.; Fukuroda, T.; Ihara, M.; Yano, M.; Ishikawa, K. *J. Med. Chem.* **1996**, 39, 2313–2330.

379. Davis, F.A.; Reddy, G.V. *Tetrahedron Lett.* **1996**, *37*, 4349–4352.

380. McKeever, B.; Pattenden, G. *Tetrahedron Lett.* **1999**, *40*, 9317–9320.

381. Wipf, P.; Uto, Y. *Tetrahedron Lett.* **1999**, *40*, 5165–5169.

382. Saha, B.; Nandy, J.P.; Shukla, S.; Siddiqui, I.; Iqbal, J. *J. Org. Chem.* **2002**, *67*, 7858–7860.

383. Oh-Hashi, J.; Harada, K. *Bull. Chem. Soc. Jpn.* **1966**, *39*, 2287–2289.
384. Aberhart, D.J. *J. Lab. Cmpds. Radiopharm.* **1983**, *20*, 605–620.
385. Scolastico, C.; Conca, E.; Prati, L.; Guanti, G.; Banfi, L.; Berti, A.; Farina, P.; Valcavi, U. *Synthesis* **1985**, 850–855.
386. von dem Bussche-Hünnefeld, C.; Seebach, D. *Chem. Ber.* **1992**, *125*, 1273–1281.
387. Wolf, E.; Spenser, I.D. *J. Org. Chem.* **1995**, *60*, 6937–6940.
388. Díaz, M.; Branchadell, V.; Oliva, A.; Ortuño, R.M. *Tetrahedron* **1995**, *51*, 11841–11854.
389. Fadnavis, N.W.; Sharfuddin, M.; Vadivel, S.K.; Bhalerao, U.T. *J. Chem. Soc., Perkin Trans. 1* **1997**, 3577–3578.
390. Poch, M.; Alcón, M.; Moyano, A.; Pericàs, M.A.; Riera, A. *Tetrahedron Lett.* **1993**, *34*, 7781–7784.
391. Hilty, F.M.; Brun, K.A.; Heimgartner, H. *Helv. Chim. Acta* **2004**, *87*, 2539–2548.
392. Reider, P.J.; Grabowski, E.J.J. *Tetrahedron Lett.* **1982**, *23*, 2293–2296.
393. Aoyagi, Y.; Jain, R.P.; Williams, R.M. *J. Am. Chem. Soc.* **2001**, *123*, 3472–3477.

Synthesis of Optically Active α-Amino Acids: Elaboration of Amino Acids Other Than Serine

9.1 Introduction

Amino acids provide a tempting starting synthon for the synthesis of more complex amino acids as several important features are already in place: namely, the amino and carboxyl groups and α-center of chirality. Moreover, the common amino acids are readily available as either enantiomer and are comparatively inexpensive, at least for the L-isomer. The amino acids that have proven to be particularly useful as general synthons for preparing a range of other amino acids (excluding achiral Gly) are Glu, Asp, and Ser. Glu and Asp possess both an activated methylene group and a side-chain carboxyl group that are amenable to further modification, while the β-hydroxyl group of Ser provides several possible options for elaboration. The use of Ser will be discussed separately in Chapter 10. Most other common amino acids have been used as substrates at one time or another, but usually for a specific synthesis of a single derivative or class of derivatives. Some syntheses of amino acids from Asp were reviewed in 1996 (1).

9.2 Synthesis of Optically Active Amino Acids from Aspartic Acid

9.2.1 Enolate Formation

Many amino acid syntheses originating from Asp rely on generation of a β-anion (see Scheme 9.1, Table 9.1). The desired enolate regioselectivity is obtained by two possible methods, which are often used together. The N-protecting group can be chosen to electronically discourage α-deprotonation by acidifying the adjacent nitrogen proton so that it is removed first, giving an N,

β-enolate dianion (2, 3). Alternatively, bulky N- and α-carboxy-protecting groups can be used to sterically prevent α-deprotonation (4).

The first report of general alkylation of an Asp β-anion appeared in 1981, with Seebach and Wasmuth's description of dideprotonation of N-formyl L-Asp(OtBu)-OtBu with LDA. Mixtures of β- and α-substituted products (7:2 ratios) were produced (5). In 1989 Baldwin et al. dideprotonated Cbz-Asp(OR)-OtBu with LDA or LHMDS and reacted the enolate with a number of electrophiles, giving adducts with moderate to low diastereoselectivity (6–8). No racemization at the α-center was observed with LHMDS as base, but LDA gave products with approximately 85% ee (6, 8). Reactive halides alkylated the side-chain in 50–60% yield (although MeI gave a mixture of C- and N-alkylated products), but less reactive alkylating agents (e.g., nBuLi) did not give product. The alkylated product could be decarboxylated by a radical Barton decarboxylation (8). Additions of the dianion to aldehydes and ketones gave β-carboxy-γ-hydroxy derivatives (7), which could be converted into β,γ-unsaturated amino acids by treatment with tribenzylphosphine/DEAD (7) (see Scheme 9.2). By carefully controlling the reaction conditions, Cbz-Asp(OMe)-OtBu was β-alkylated with allyl bromide or tert-butyl chloroacetate with >98% de anti diastereoselectivity (2S,3R). Under the same conditions, propargyl bromide and bromoacetonitrile gave only 33–60% de, while benzaldehyde resulted in a 1:1 mixture of diastereomers (presumably epimeric at the γ-center) (9). Styrylglycine has been prepared via an aldol reaction of Cbz-Asp(Alloc)-OtBu with benzaldehyde, followed by a decarboxylative elimination to generate the alkene (10). Boc-Asp(OMe)-OMe was β-alkylated with methyl iodide with 0% de, and with benzyl bromide or

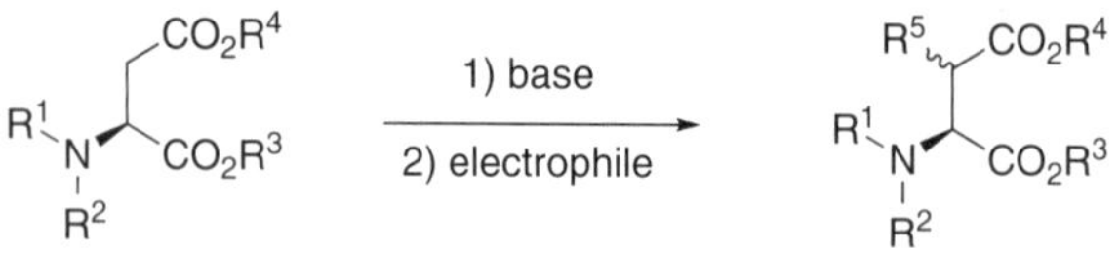

Scheme 9.1 Alkylation of Aspartate β-Enolate (see Table 9.1).

Table 9.1 Alkylation of Aspartate β-Enolate (see Scheme 9.1)

R^1	R^2	R^3	R^4	Base	Electrophile	Yield	Reference
PhFl	$(CH_2)_3I$	Bn	$CO_2R^4 = CN$	LDA	intramolecular then quench with H_2O or EtI	63–76% 0% de	1985 (292)
PhFl	$(CH_2)_3I$	tBu	Me	LDA	intramolecular then quench with I_2O or MeI	72–85% 0% de	1986 (14) 1993 (27)
PhFl	H	tBu Me	Me	KHMDS	MeI EtOTf iPrOTf	43–95% 20–66% de	1989 (13)
Cbz	H	tBu	Me $CH_2CH=CH_2$	LHMDS	PhCHO EtCHO BnBr $BrCH_2CH=CH_2$	45–60% 2 of 4 isomers for aldol, 0% de 50–66% de for alkylation	1989 (6) 1989 (7)
Cbz	H	tBu	Me	LDA (85% ee) LHMDS (100% ee)	BnBr $BrCH_2CH=CH_2$	50–60% 50–66% de	1989 (8)
PhFl	H Bn	Me tBu	Me Bn	KHMDS	BnBr BnI	75–80% 20–94% de	1990 (17)
Bn	Bn	tBu	Bn	KHMDS	BnBr BnI	75–80% 20–94% de	1990 (17)
PhFl	$(CH_2)_3I$	tBu Me	Me	LDA	intramolecular then quench with EtI	82% 96% de	1990 (28)
PhFl	H	Me tBu	Me	KHMDS	$Cl(CH_2)_3OTf$	76% 60–70% de	1990 (28)
Boc	H	Me	Me	LDA	HCO_2iPr RCO(1-imidazolyl) R = Me, Et, ipr	20–92%	1991 (12)
PhFl	H	H Me tBu	Me	LHMDS KHMDS LDA LTMP	MoOPH (R^5 = OH)	22–95% 84–90% de at C3	1992 (20) 1994 (19)
PhFl	H	H Me tBu	Me	LHMDS KHMDS LDA LTMP	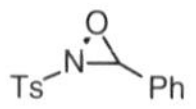 Ts-N—Ph (epoxide) (R^5 = OH)	80–98% 33–60% de	1992 (20) 1994 (19)
PhFl	H	H Me tBu	Me	LHMDS KHMDS LDA LTMP	trisyl azide or di-tert-butyl azodicarboxylate or dibenzyl azodicarboxylate (R^5 = NR)	75–85% 0–90% de	1992 (20) 1994 (19)
PhFl	H	Me	Me	KHMDS	EtOTf	87% 90% de	1993 (16)

(Continued)

Table 9.1 Alkylation of Aspartate β-Enolate (see Scheme 9.1) (continued)

R^1	R^2	R^3	R^4	Base	Electrophile	Yield	Reference
Cbz Ts	H	Me	Me	LHMDS	MoOPH $\begin{array}{c}\text{O}\\\text{Ts}^{-\text{N}}\diagdown\text{Ph}\end{array}$ (R^5 = OH)	40–70% 45:1 to 1:4 de	1993 (21)
		tBu	tBu				
PhFl	Bn	Me	Me	LHMDS	MeI EtI BnBr ICH₂CH=CH₂ ICH₂C(Me)=CH₂	75–98% 82–>96% de	1994 (18)
PhFl	Bn	Me	Me	KHMDS	ICH₂C(Me)=CH₂	0–99% 33–96% de (other diastereomer from above)	1994 (18)
Cbz	H	tBu Bn	Me	LHMDS	EtBr BrCH₂CH=CH₂ BrCH₂CH=CMe₂ H₂C=CHCHO	31–72% 50–60% de	1995 (24)
PhFl	Bn	Me	Me	LHMDS	¹³CH₃I	98% 100% de	1995 (34)
Alloc	H	CH₂CH=CH₂	4-MeO-Bn	LHMDS	2,4-(MeO)₂-BnSTs	63% 100% de	1996 (22) 1996 (23)
Cbz	H	tBu	CH₂CH=CH₂	LHMDS	PhCHO (then decarboxylatively dehydrate)	86% 33% de	1997 (10)
Cbz	H	Me TMSE	TMSE Me	LHMDS	BrCH₂CH=CH₂ BrCH₂CH=CHMe BrCH₂C(Me)=CH₂ BrCH₂CH=CHPh BrCH₂C(CO₂Me)=CH₂ BrCH₂(cyclohexen-3-yl)	58–87% >98% de anti except CH₂(cyclohexen-3-yl), 50% de	1998 (419)
Cbz	H	tBu	Me	LHMDS	ClCO₂tBu BrCH₂CH=CH₂ BrCH₂CN BrCH₂CCH	33–>98% de anti	1999 (9)
Boc	H	Me	Me	LHMDS	BnBr MeI BrCH₂CO₂Me	38–73% 0–75% de	2001 (11)

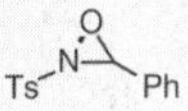

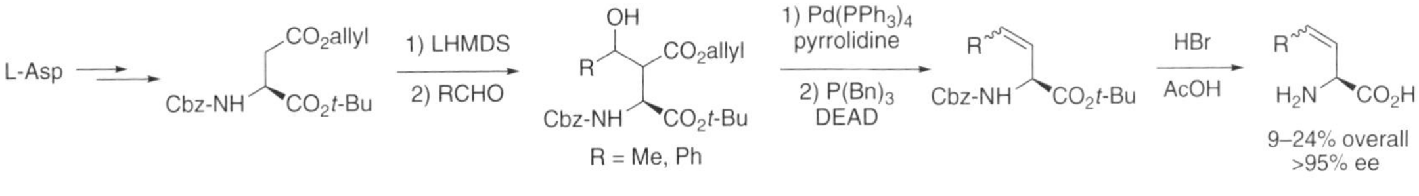

Scheme 9.2 Elaboration of Aspartic Acid (7).

methyl bromoacetate with 50–75% de (11); it has also been deprotonated with LDA and acylated with isopropyl formate or N-acyl imidazoles (12).

Rapoport et al. employed a steric approach to direct deprotonation to the β-position, using a bulky 9-phenyl-fluoren-9-yl (PhFl) group for N-protection in combination with an α-tert-butyl ester (13). A number of electrophiles were introduced at the β-position after deprotonation with KHMDS, although with minimal diastereoselectivity. The β-carboxyl ester could then be regioselectively reduced with DIBAL and oxidized to give an aldehyde. Reductive amination of the aldehyde with the amine group of Ala-OMe, followed by lactam formation with the α-carboxyl group, produced

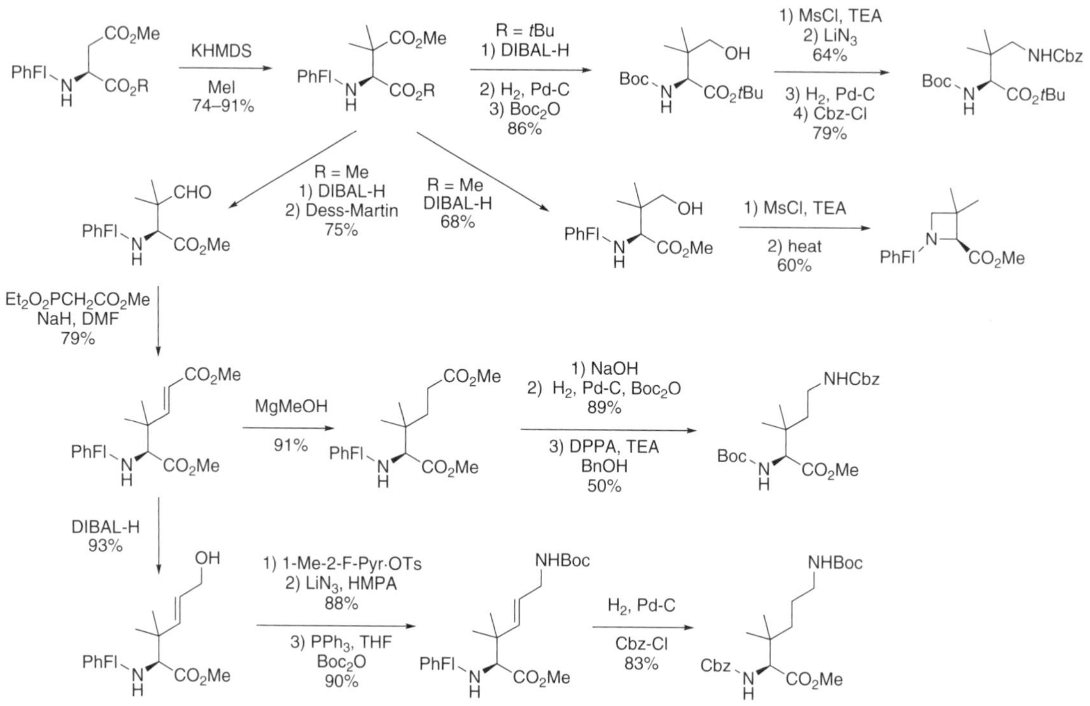

Scheme 9.3 Dimethylation and Further Derivatization of PhFl-Asp(OMe)-OR (15).

dipeptide isosteres containing a cyclic γ-lactam bridge between residues. The same protecting group combination was used in 1986 for a synthesis of a pipecolic acid derivative, using an intramolecular β-alkylation of a protected aspartate (14). PhFl-Asp-OtBu/OMe was dialkylated with excess KHMDS/MeI to produce β,β-dimethyl-Asp in 74–91% yield (see Scheme 9.3). The β-carboxyl group was reduced to provide β,β-dimethyl-Hse, with azide displacement of the mesylated hydroxyl leading to β,β-dimethyl-2,4-diaminobutyric acid, or cyclization of the mesylate providing 3,3-dimethylazetidine-2-carboxylate. The dimethylated Asp was also converted into Asp semialdehyde, and then further derivatized to give β,β-dimethyl-Orn, β,β-dimethyl-Lys, and β,β-dimethyl-homoGlu (15).

A 1993 report on the alkylation of PhFl-protected Asp with EtOTf found that the diastereoselectivity and yield could be significantly improved by modifying the reaction conditions (16). Similarly, modifications of electrophiles and protecting groups allowed a benzyl group to be introduced with diastereoselectivities of up to 94% de (17). After alkylation, the β-carboxyl group was regioselectively converted to an amine by a Curtius rearrangement, resulting in 2,3-diamino acid derivatives (17).

Better alkylation diastereoselectivity was also achieved using an N-Bn,N-PhFl diprotected Asp derivative (18). KHMDS deprotonation resulted in adducts with 33–96% de, while LHMDS gave even better stereoselectivity (82–>96% de), but the opposite diastereomer.

The allyl adduct could be oxidatively cleaved and cyclized to give 3-carboxy-Pro.

β-Hydroxylation and amination of the anion of PhFl-Asp(OMe)-OtBu, -OMe, and -OH has also been examined (19, 20). Diastereoselectivity was highly dependent on the reaction conditions, with base, solvent, cosolvent, and electrophilic reagent all affecting the stereochemical outcome. For hydroxylation with MoOPH, one diastereomer was obtained with 84% de using LHMDS in THF/HMPA, while the C3 epimer was prepared with 90% de using nBuLi/LHMDS in THF. Mechanistic arguments for the observed stereoselectivity of both sets of reactions have been presented (18, 19). A number of other protecting groups have been tested for the β-hydroxylation of Asp dianion, having little effect on yield but significantly altering the diastereomer ratio (21); the diastereomer ratio could also be reversed by using MoOPH or of oxaziridine reagents. The dianion of Alloc-Asp(O-4-methoxybenzyl)-OAllyl was sulfenylated with 2,4-dimethoxybenzylthio-4-methylphenylsulfonate, with a 63% yield of a single diastereomer of the mono-sulfenylated adduct (22, 23). Unfortunately N-Alloc removal resulted in epimerization at the β-center.

The β-enolate of Cbz-Asp(OMe)-OtBu has been used to prepare 3-carboxy-Pro, 5-substituted carboxy-prolines, and β-alkylated Asp derivatives (24). Reaction of the LHMDS-generated N,C-dianion with 1,2-bis-electrophiles to form the Pro template was unsuccessful, but the desired cyclization could be achieved via ozonolysis of a β-allyl adduct and intramolecular reductive amination (see Scheme 9.4) (24, 25). This method has also been used by other groups (26). The β-allyl adduct could also be transformed into other derivatives by reduction, epoxidation or further alkylation (24).

Ring closure to form imino acids has also been achieved by intramolecular alkylation of the β-carbon with an N-(3-iodopropane) (27, 28) or N-(3-chloropropane) (29) derivative, followed by trapping of the β-anion with other electrophiles, giving pipecolic acid derivatives (see Scheme 9.5). A reversal in alkylation order is also possible (28). Further elaboration, such as reaction of the β-carboxyl group with dimethyl lithium methylphosphonate (27), has been carried out. Asn can be used as a substrate for similar reaction sequences (30).

It is also possible to alkylate the dianion formed from L-Asp acid-derived 4-carboxy-2-azetidinone, but this is discussed in Section 8.5.1 (see Scheme 8.37) (30–33).

The β-carboxyl group can be elaborated into other groups after β-alkylation, usually via an initial regioselective reduction to an alcohol (18), as mentioned earlier

(see Schemes 9.6 and 9.3), and is discussed more extensively in the following section. This method was used in the synthesis of diastereoselectively labeled (3S)-[4-^{13}C]-L-Val and [3-^{13}CH$_3$]-L-Ile (34).

9.2.2 Modification of the β-Carboxyl Group

A number of methods have been used to modify the β-carboxyl group of Asp. A Curtius or Hofmann rearrangement of the side-chain acid group produces 2,3-diaminopropionic acid (see Section 9.4). Reduction of the β-carboxyl group of differentially protected Asp derivatives gives homoserine (Hse) (35–44). Side-chain oxidation then produces Asp semialdehyde, a useful intermediate that will be discussed in the following Section 9.2.3. Other reactions of the Hse intermediate are possible. In one synthesis, the side-chain of Boc-Asp-OFm was reduced via formation of a mixed anhydride with NMM/MeOCOCl, followed by reduction with NaBH$_4$. The homoserine was iodinated with

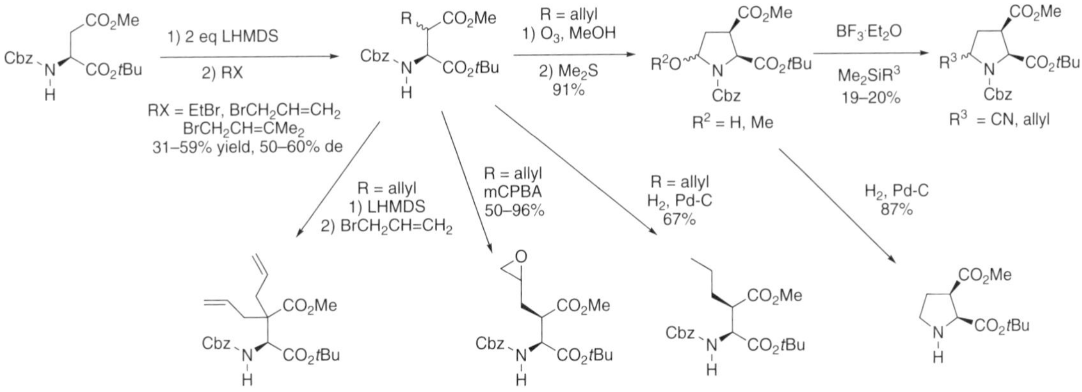

Scheme 9.4 Alkylation of Aspartate β-Enolate and Further Elaboration (24, 25).

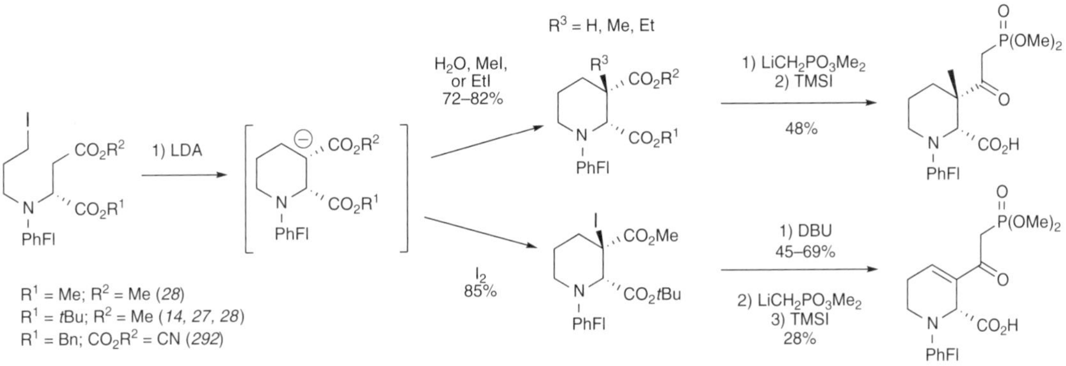

Scheme 9.5 Intramolecular Alkylation of Aspartate β-Enolate and Further Elaboration.

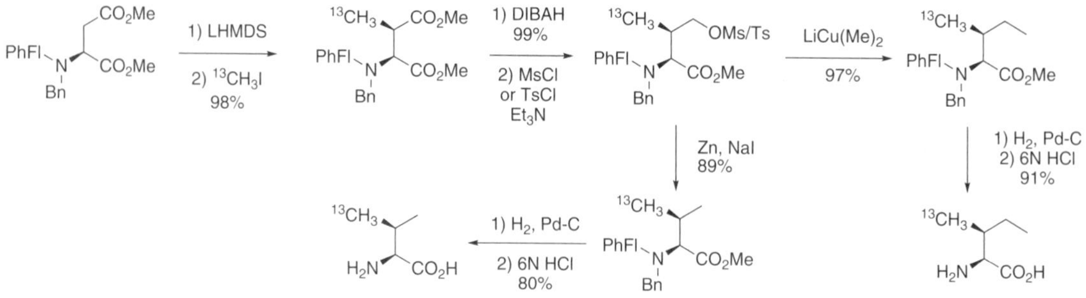

Scheme 9.6 Alkylation of Aspartate β-Anion and Further Elaboration (*34*).

I$_2$/PPh$_3$ to give γ-iodo-Abu, which was employed for alkylation of the side-chain of Cys (*44*). The iodide can also be used to form iodozinc reagents similar to the Jackson β-iodo-Ala- and δ-iodo-Nva-derived reagents (see below, Section 10.5.3) (*430*). Bromination of Hse with CBr$_4$/PPh$_3$ produced γ-bromo-Abu; treatment with NaH induced cyclization with the α-carbon to give 1-amino-cyclopropane-1-carboxylate (*45*). Side-chain reduction of *N*-phenylfluorenyl 3-hydroxy-Asp dimethyl ester with NaBH$_4$/BH$_3$.SMe$_2$ gave 3,4-dihydroxy-Abu (*46*).

Grignard addition to the Asp side-chain ester is possible if the α-ester group is masked. A synthesis of 4-fluoro-Leu in 2005 employed Boc-L-Asp(OBn)-OH as the starting material. The α-carboxyl group was reduced to a hydroxymethyl group and protected within an oxazolidinone (along with the α-amino group); 2 equiv of MeMgBr were added to the side-chain carboxyl (72%) and the resulting hydroxy group fluorinated with DAST (70%). Oxazolidinone hydrolysis and hydroxy-methyl oxidation (65%) completed the synthesis (*47*).

Several procedures derivatize the β-carboxyl group to provide γ-keto-α-amino acid products. One method of achieving this transformation is via a Stille palladium-catalyzed cross-coupling reaction of protected aspartyl chloride (*48*) with an aryl or vinyl organostannane (see Scheme 9.7) (*48–51*). The vinyl adducts can be elaborated to 4-oxopipecolic acid derivatives. The acyl chloride side-chain has also been reacted with trimethylsilyl cyanide for a synthesis of protected 4-oxo-L-Orn (*50*), and with cuprate-modified dimethyl alkylphosphonates to prepare 4-oxo-5-phosphono-Nva derivatives (*52, 53*). Treatment of the acid chloride with diazomethane provides 2-amino-4-oxo-5-diazopentanoic acid, a versatile intermediate that has been used to synthesize 5-hydroxy-4-oxo-L-Nva (*50, 54, 55*), β-(thiazol-4-yl)-Ala (*56*), and other 2-amino-4-oxohexanoic acid derivatives (*55*). Oxidation of the diazoketone with dimethyldioxirane provided 4,5-diketo-Nva (*57*), while displacement with HCl resulted in a chloroketone that was further reacted with potassium phthalimide to give 2,5-diamino-4-keto-pentanoic acid. Cyclization of this side chain resulted in His (*58*). A similar conversion of Glu to homo-His has been described (*59*). An intramolecular Rh-catalyzed

insertion of the diazoketone-derived carbene into the α-amino N—H bond produced Pro and 4-hydroxy-Pro (*60*). In a similar manner, treatment of the acid chloride with ethyl diazoacetate, followed by Rh-catalyzed carbene formation, generated an intermediate leading to the amino acid component of bulgecinine, 4-hydroxy-5-hydroxymethyl-Pro (*61*). The β-carboxyl group of 3-methyl-Asp has been homologated to give 3-methyl-Glu via conversion of the β-acid chloride to a diazo-ketone followed by rearrangement (Arndt–Eistert homologation) (*62*). The side chain of *N*-methyl Asp, protected as a hexafluoroacetone oxazolidinone, was activated as the acid chloride, reacted with diazomethane, converted to the bromomethyl ketone, and cyclized with thiobenzamide to give *N*-methyl β-(2-phenyl-1,3-thiazol-4-yl)-Ala (*432*).

Another activated Asp derivative, the Weinreb amide of Boc-Asp-O*t*Bu, was converted into an alkynyl ketone side chain via reaction with acetylide anions (see Scheme 9.8). Cyclocondensation with amidines resulted in a pyrimidin-4-yl substituent (*63, 64*), while other heterocyclic derivatives were produced by cyclocondensations with enamines, phenylhydrazine, hydroxylamine, and phenyl azide (*65, 66*). The homologs were similarly prepared from Glu.

The Asp β-carboxyl group can be converted into a γ-keto-δ-carboxyl functionality by activation with carbonyldiimidazole and displacement with a malonate enolate (see Scheme 9.9) (*67–70*). Reduction of the ketone gave a mixture of protected 2-amino-4-hydroxy-adipic acid diastereomers (*67*), while treatment with base (Na$_2$CO$_3$) and alkylation produced δ-substituted derivatives, which could then be decarboxylated (*68*). The β-keto ester has been condensed with a variety of phenols in the presence of methanesulfonic acid (von Pechman condensation) to produce amino acids with substituted coumarin side chains (*71*). The malonate adduct can also be cyclized to give a Pro derivative (*69*), while a 4-hydroxypipecolic acid product was obtained by a 1-carbon homologation of the malonate adduct, followed by cyclization (*72*). Alternatively, coupling of Boc-Asp-OBn with Meldrum's acid, followed by heating with an alcohol, produced the side-chain ester of 2-amino-4-oxo-adipic acid. This was further elaborated

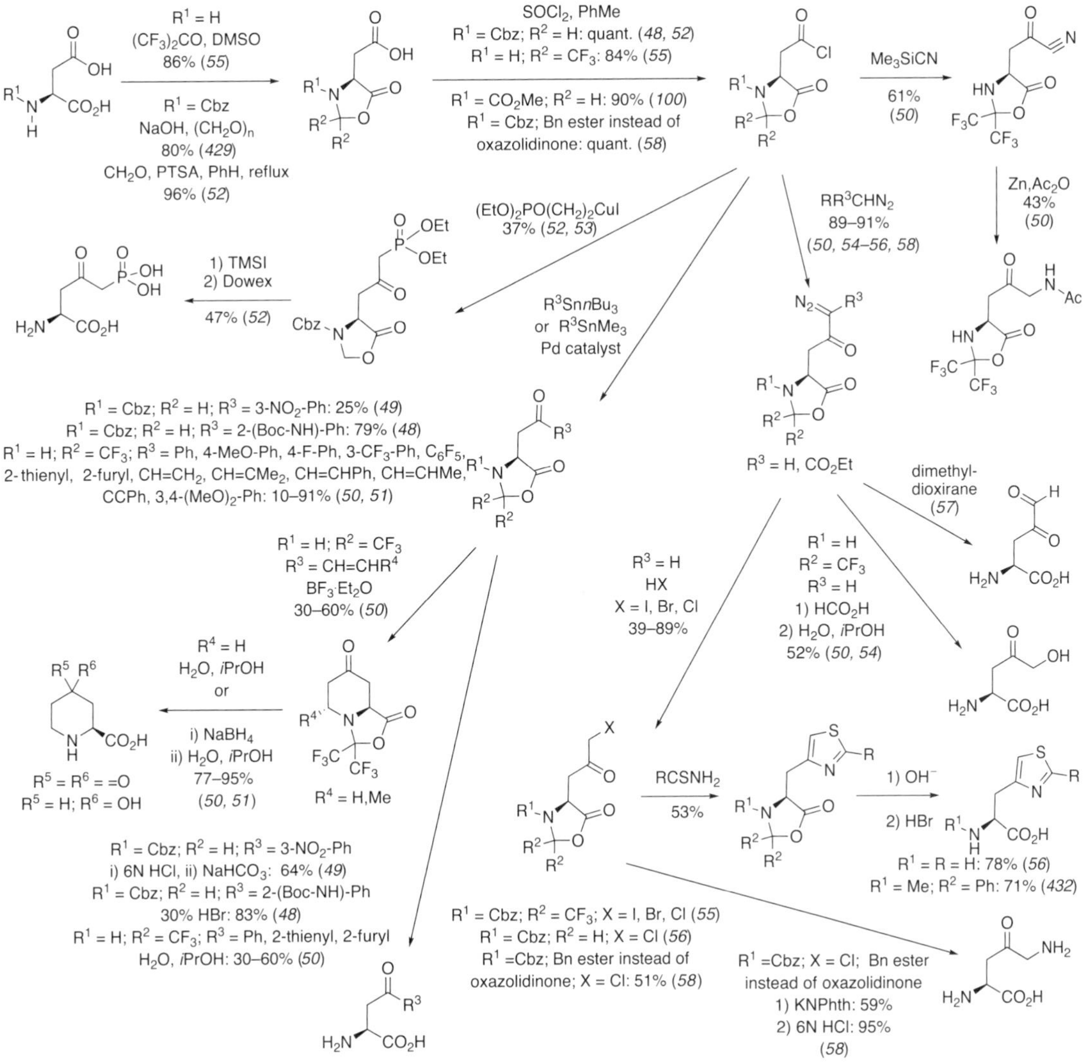

Scheme 9.7 Synthesis and Elaboration of Aspartyl Chloride.

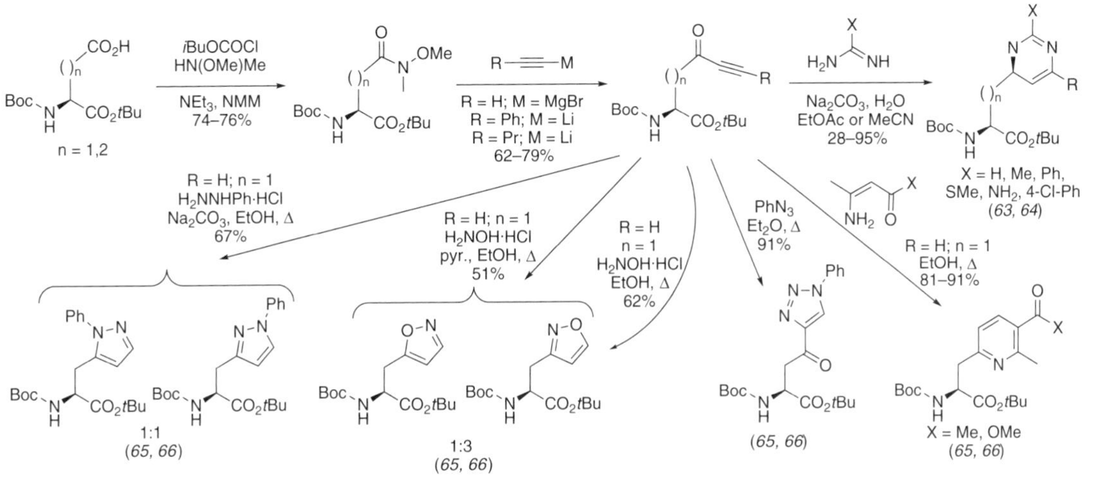

Scheme 9.8 Acetylide Addition to Weinreb Amide Derivatives.

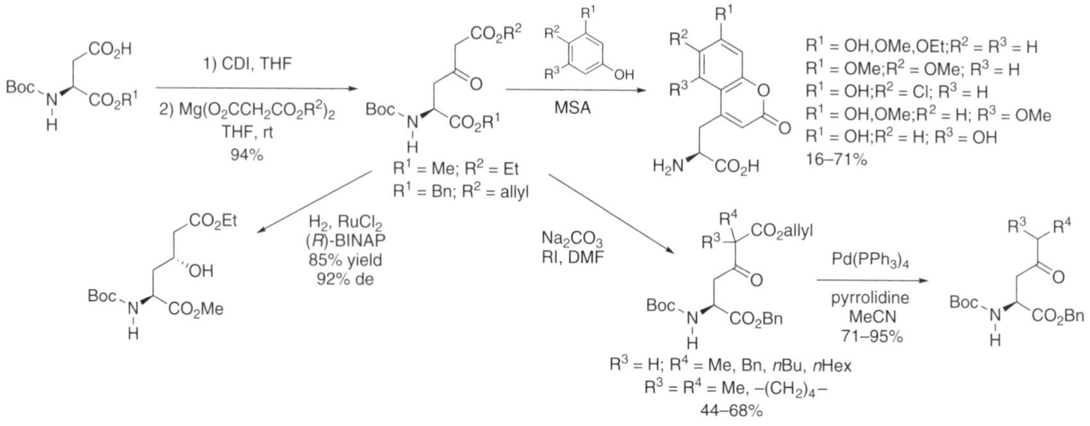

Scheme 9.9 Selective Reduction of Aspartate β-Carboxyl Group via Malonate Displacement (*67–71*).

to give a mimetic of an *O*-linked glycosylated amino acid (*73*). The same Meldrum's acid adduct was also converted into 2-amino-5,6-dihydroxyhexanoic acid, 2-amino-5-hydroxy-6-iodohexanoic acid, 2-amino-5,6-epoxyhexanoic acid, and 5-hydroxy-Lys (*74*).

The side-chain of Boc₂-Asp-OtBu has been activated with DCC/DMAP and then condensed with (*tert*-butoxycarbonylmethylene)triphenylphosphorane, with ozonolysis generating a vicinal tricarbonyl side-chain (CH₂COCOCO₂tBu). Reaction with ethylenediamine or *o*-phenylenediamine followed by oxidation produced β-(2-carboxy-pyrazine-3-yl)-Ala and β-(2-carboxy-quinoxalin-3-yl)-Ala, while condensation with *S*-methyl isothiosemicarbazide produced the regioisomers β-(3-methylthio-5-carboxytriazin-6-yl)-Ala and β-(3-methyl-thio-6-carboxytriazin-5-yl)-Ala (*75, 76*). The same reactions were carried out with Glu to make γ-(heteroaryl)-Abu homologs.

Regioselective modification of the β-carboxyl group can be achieved without differential ester protection, via an intramolecular anhydride of Trt-Asp (*77*) (see Scheme 9.10). Two equivalents of Grignard reagents were added to give *N*-trityl γ-disubstituted Hse, with the bulky Trt group discouraging α-attack. The side-chain was deoxygenated by cyclization to the lactone and cat-alytic hydrogenolysis. Wittig reagents also reacted regio-selectively with the γ-carbonyl function of the anhydride to form olefinic products. However, attempts at regio-selective reduction of the anhydride derivative gave a mixture of lactone products resulting from both α- and γ-carbonyl attack, with most reagent/protecting group combinations leading to the β-amino acids resulting from α-carbonyl reduction (see Section 9.2.4, Scheme 9.15). The desired Hse lactone was only obtained as the major product by using bulky reducing reagents. The anhydride can also be used for Friede–Crafts acylation

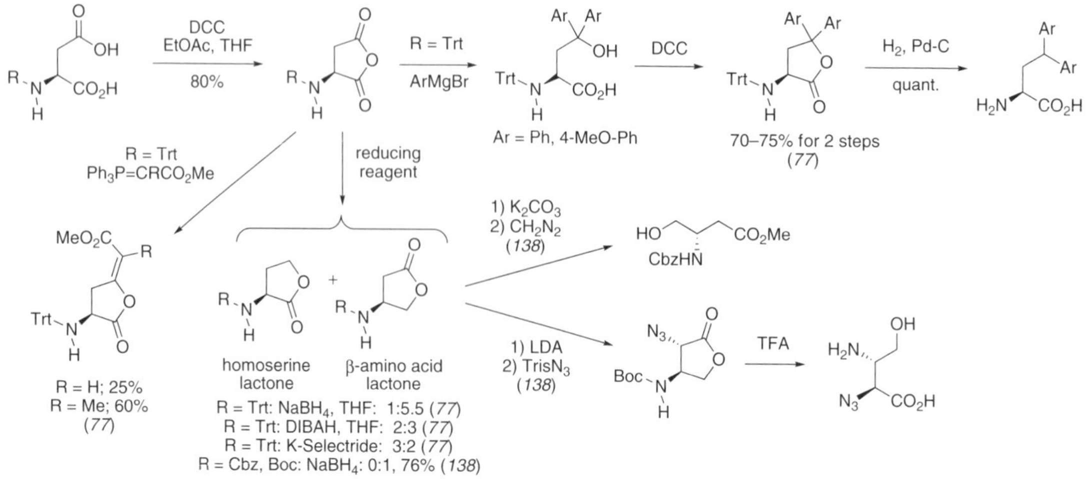

Scheme 9.10 Reactions of Aspartyl Anhydride (*77*).

of arenes, with the α:β amino acid product ratio dependent on the arene employed (78). The preference for α-attack has been used in the synthesis of β-amino acids (see Section 9.2.4, Scheme 9.15). The possibility of regioselective opening of the anhydride with amines to prepare Asn derivatives was also investigated, with mixtures of isomers obtained (77).

Other derivatizations of the side-chain carboxyl group include reaction of PhFl-Asp(OMe)-OtBu with LiCH$_2$PO(OR)$_2$, leading to a β-ketophosphonate, which was then employed for Horner–Wadsworth–Emmons olefinations (79, 80). Pro and pipecolic acid derivatives can be obtained via intramolecular cyclization of an enolate on an N-substituent with the β-carboxyl group (69, 72). A number of β-(1,2,4-oxadiazol-3-yl)-Ala derivatives, with substituents at the 5-position of the oxadiazole ring, were prepared by esterification of Fmoc-Asp(OH)-OtBu with amidoxamines, HON=C(R)NH$_2$. Treatment with a solution of sodium acetate in ethanol/water at 86 °C induced cyclization to the oxadiazoles in 20–56% overall yield (81). Tetrazole amino acids analogs of Asp and Glu can be prepared from

Boc-Asn-OMe or Boc-Gln-OMe via dehydration of the side-chain amide to a nitrile using benzenesulfonyl chloride in pyridine, followed by conversion of the nitrile to a tetrazole group using nBu$_3$SnN$_3$ (82, 83). An earlier version of this reaction employed NaN$_3$/NH$_4$Cl/DMF for the tetrazole formation (84), while an improved tetrazole-forming reaction reacted sodium azide in toluene in the presence of triethylamine hydrochloride (85).

The ω-carboxyl group of both Asp and Glu can be removed by a reductive radical decarboxylation, using photolysis of N-hydroxypyridine-2-thione esters (86–88) (see Scheme 9.11). By carrying out the reaction in the presence of radical trapping reagents, the carboxyl group was replaced by bromide, chloride, iodide, or other groups. This method was used to prepare L-seleno-Met and L-selenocystine (89). The γ-iodo-Abu product obtained by this procedure can be converted into an organozinc/copper reagent (see Scheme 9.23 below), in an analogous manner to the Ser-derived β-iodo-Ala reagents described in Section 10.5.3 (e.g., see 40, 90, 91).

The radical intermediate generated by photolysis can also be trapped by various olefins, with the resulting

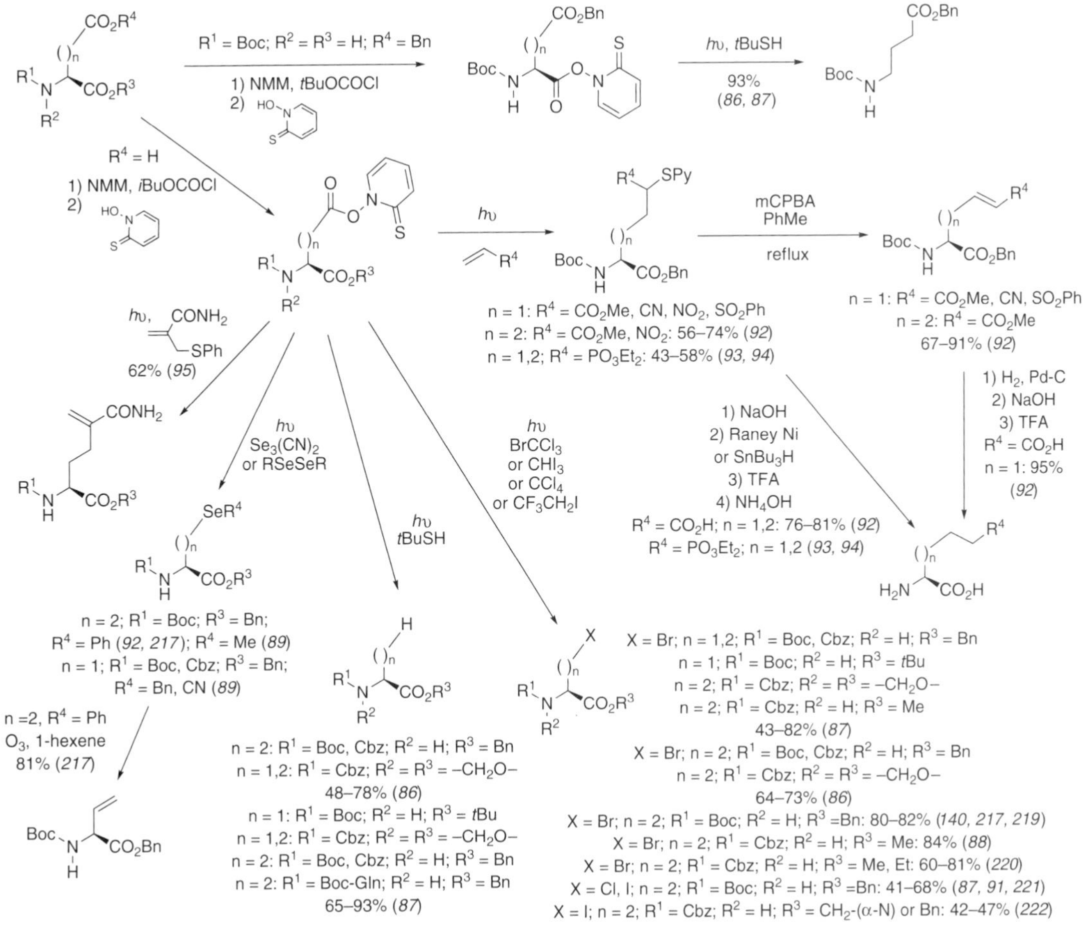

Scheme 9.11 Radical Decarboxylation and Elaboration of Aspartate and Glutamate Derivatives.

adducts further transformed into side-chain unsaturated amino acids (*92*), 2-aminoadipic and 2-aminopimelic acids (*92*) or their ω-phosphonate analogs (*93, 94*). Trapping with phenylthiomethyl acrylamide led to δ-methylene-homoglutamine (*95*), while intramolecular trapping with an *N*-cyclohexenyl alkene gave a perhydroindole-2-carboxylic acid derivative (*89*).

Decarboxylation has also been achieved via Kolbe electrolysis of Boc-Asp-OBn or Cbz-Asp-OMe, with the decarboxylated radicals combining to give protected 2,5-L,L-diaminoadipic acid in 17–25% yield (*96*).

9.2.3 Aspartate Semialdehyde

Asp semialdehyde is a homolog of Ser aldehyde (see Section 10.6), but the Asp derivative is much more stable as enolization of the aldehyde is not promoted by conjugation with the α-carboxyl group. The aldehyde is an intermediate in the biosynthesis of Lys, Thr, and Met. Asp semialdehyde has been prepared by ozonolysis of allylglycine (*97*), by reduction of Asp-β-acid chloride via hydrogenation (*98–100*) or with *n*Bu₃SnH/Pd(Ph₃)₄ (*101*), by reduction of Asp β-ethyl thioester (*102, 103*), by regioselective DIBAL-H reduction of *N,N*-Boc₂-Asp(OMe)-OMe (*104, 105*), PhFl-Asp(OMe)-O*t*Bu (*106*) or the Weinreb amide of Boc-Asp(OH)-O*t*Bu (*107*) (a reduction that can also be accomplished with the substrate attached to a solid support (*108*)), LiAlH(O*t*Bu)₃ reduction of the Weinreb amide of Boc-Asp(OH)-O*t*Bu (*109*), or by oxidation of a Hse intermediate (*35, 36, 43, 79, 105, 106, 110–118*) (see Scheme 9.12, Table 9.2). The Hse intermediate has been prepared from Asp (*35, 37, 43, 79, 111, 113, 114, 116*) or Met (*110*). A DIBAL-H reduction of *N,N*-Boc₂-Asp(OMe)-OMe gave a 68% yield of aldehyde and 22% yield of Hse, with the Hse oxidized to the aldehyde with

Dess–Martin periodinane (*105*). Deuterated Asp semialdehydes were stereospecifically synthesized from Asp or labeled-Asp derivatives via reduction to Hse followed by oxidation (*42*). Racemic Asp semialdehyde has been prepared from methyl α-methoxyhippurate via reaction with vinyl acetate (*119, 120*).

Asp semialdehyde is a substrate for a number of other reactions (see Scheme 9.13). Dimethyl trimethylsilyl phosphite was added to the aldehyde in good yield, with deoxygenation giving a 2-amino-4-phosphonobutanoic acid derivative (*35, 113, 121*). A similar reaction with a difluoromethylene phosphonate anion, followed by oxidation of the aldol adduct, produced 2-amino-4-oxo-5,5-difluoro-5-phosphonopentanoic acid (*105*). Aldol reaction of the aldehyde with methyl ketone enolates, followed by dehydration of the adduct, reduction of the alkene, and intramolecular reductive amination of the side-chain ketone, resulted in 6-*cis*-substituted pipecolic acids (*106*). Dimerization of Asp semialdehyde via a triazole-catalyzed acyloin condensation produced 4-oxo-5-hydroxy-2,7-diaminosuberic acid. *O*-Acetylation and SmI₂-mediated dehydroxylation gave 4-oxo-2,7-diaminosuberic acid, with selective *N*-protection and intramolecular reductive amination resulting in 5-(2-amino-2-carboxyethyl)-Pro (*122*). Conversion of the aldehyde into a dimethyl acetal followed by methanol elimination produced β,γ-unsaturated enol ethers, including the unusual amino acid rhizobitoxine (*111*).

Treatment of Asp semialdehyde with diazomethane produces the homologous 4-oxo-L-Nva (*115*). Fluorination or chlorination of Asp semialdehyde, with the α-amino and carboxyl groups protected as a hexafluoroacetone oxazolidinone, led to 4,4-difluoro-Abu or 4,4-dichloro-Abu after hydrolysis (*99*). Wittig reagents have been successfully coupled with the aldehyde to give a number of amino acids (*79, 101, 104, 123*), with methylenation producing allylglycine (*109*). The alkene

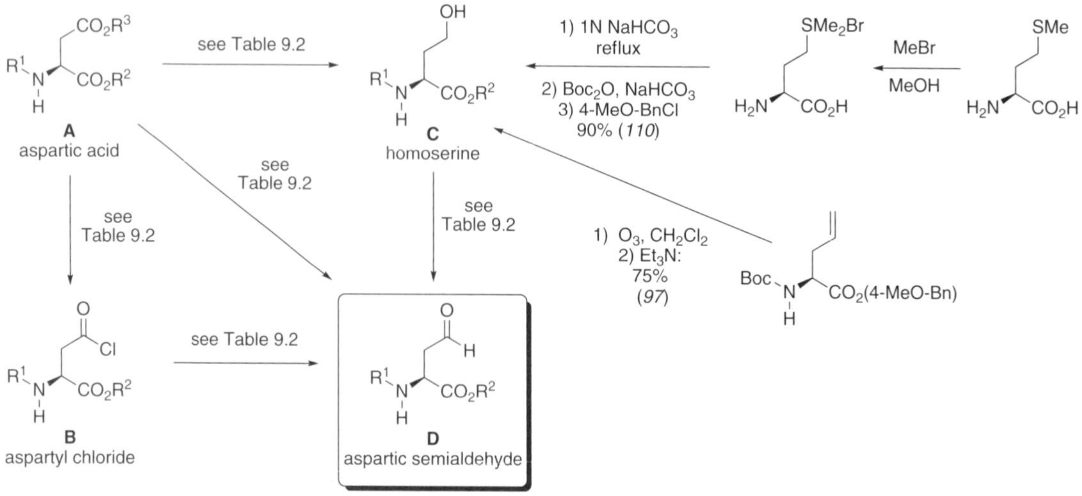

Scheme 9.12 Synthesis of Aspartate Semialdehyde (see Table 9.2).

Table 9.2 Synthesis of Aspartate Semialdehyde (see Scheme 9.12)

R^1	R^2	R^3	A to B conditions	Yield	B to D conditions	Yield	Reference
Cbz + CH_2-R^2	CH_2-R^1	H	$SOCl_2$	79%	H_2, Pd, $BaSO_4$	87%	1990 (*98*)
MeO_2C+ CH_2-R^2	CH_2-R^1	H	$SOCl_2$		H_2, Pd, $BaSO_4$		1992 (*100*)
Cbz	Me	H	$SOCl_2$		nBu_3SnH, $Pd(PPh_3)_4$	77%	1994 (*101*)
$C(CF_3)_2$-R^2	$C(CF_3)_2$-R^1	H	$SOCl_2$	85%	H_2, Pd, $BaSO_4$	64%	1996 (*99*)

R^1	R^2	R^3	A to C conditions	Yield	C to D conditions	Yield	Reference
Cbz	Bn				CrO_3, pyridine	68%	1975 (*111*)
Boc	tBu				DMSO, $(COCl)_2$, NEt_3	93%	1980 (*118*)
Boc	tBu	H	1) EtOCOCl, Et_3N 2) $NaBH_4$	73%	CrO_3, pyridine	74%	1982 (*36*)
Boc	Bn				PCC, NaOAc, CH_2Cl_2	44%	1987 (*110*)
Boc	tBu	H	1) EtOCOCl, Et_3N 2) $NaBH_4$ or $NaBD_4$	66–74%	Collins reagent	70–72%	1988 (*42*)
Cbz	tBu	H	1) EtOCOCl 2) $NaBH_4$	85%	CrO_3, pyridine	60%	1988 (*43*)
Cbz	Bn				PCC	70%	1988 (*112*)
Cbz	Bn				DMSO, $(COCl)_2$, NEt_3	quant.	1990 (*117*)
Tfa	Me	H	$NaBD_4$ or BH_3	92–96%			1990 (*41*)
H	H	H	BEt_3 then BH_3	82%			1990 (*41*)
Boc	tBu	H	1) iBuOCOF, NMM 2) $NaBH_4$	96%	NaOCl, TEMPO	82%	1990 (*113*) 1992 (*35*) 1992 (*121*) 1994 (*116*)
Boc Cbz	Bn		1) iBuOCOCl, NMM 2) $NaBH_4$	61–84%			1991 (*37*)
R^1NH = dibenzyl-triazone	tBu	H	1) iPrOCOCl 2) $NaBH_4$	84%	DMSO, $(COCl)_2$, NEt_3	92%	1992 (*114*)
PhFl	tBu	Me	DIBAL-H	96%	DMSO, $(COCl)_2$, NEt_3	91%	1998 (*79*) 1999 (*106*)
Boc_2	Me	Me	DIBAL-H	22% (byproduct of A to D reaction)	Dess-Martin periodinane	75%	1998 (*104*) 2000 (*105*)

R^1	R^2	R^3	A to D conditions	Yield	Reference
Boc	tBu	H	1) HN(Me)OMe, DCC 2) DIBAL-H	70% 98%	1989 (*107*)
Boc	tBu	H	1) HN(Me)OMe, DCC 2) $LiAlH(OtBu)_3$	88% 93%	2001 (*109*)
Boc	tBu	H	1) EtSH, DCC 2) Et_3SiH, 10% Pd-C	93% 83%	1993 (*102*)
Boc_2	Me	Me	1 equiv. DIBAL-H	88% 68%	1998 (*104*) 2001 (*105*)
Fmoc	tBu	H	1) EtSH, DCC 2) Et_3SiH, 10% Pd-C	96% 93%	1999 (*103*)
PhFl	tBu	Me	1 equiv. DIBAL-H	>95%	1999 (*106*)
Boc	tBu	resin	1) HN(Me)OMe, DCC 2) DIBAL-H	70% 98%	1999 (*108*)

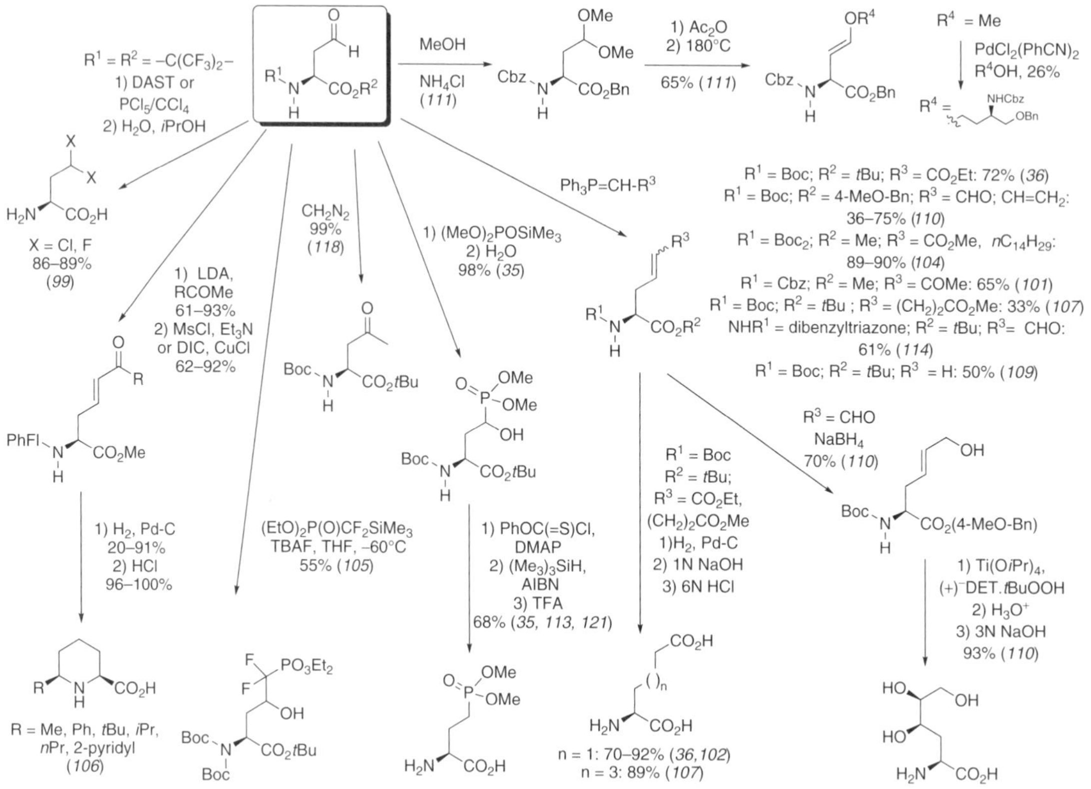

$R^1 = R^2 = -C(CF_3)_2-$
1) DAST or
PCl$_5$/CCl$_4$
2) H$_2$O, iPrOH

R^1-N(H)-CO$_2$R^2 with O=CH side chain

MeOH
NH$_4$Cl
(111)

OMe / OMe
Cbz-N(H)-CO$_2$Bn (111)

1) Ac$_2$O
2) 180°C
65% (111)

OR^4 R^4 = Me
Cbz-N(H)-CO$_2$Bn

PdCl$_2$(PhCN)$_2$
R^{4}OH, 26%

$R^4 = $ NHCbz / OBn

X
H$_2$N-CH(X)-CO$_2$H
X = Cl, F
86–89%
(99)

CH$_2$N$_2$
99%
(118)

1) (MeO)$_2$POSiMe$_3$
2) H$_2$O
98% (35)

Ph$_3$P=CH-R^3

R^1-N(H)-CO$_2$R^2 with R^3 side chain

R^1 = Boc; R^2 = tBu; R^3 = CO$_2$Et: 72% (36)
R^1 = Boc; R^2 = 4-MeO-Bn; R^3 = CHO; CH=CH$_2$: 36–75% (110)
R^1 = Boc$_2$; R^2 = Me; R^3 = CO$_2$Me, nC$_{14}$H$_{29}$: 89–90% (104)
R^1 = Cbz; R^2 = Me; R^3 = COMe: 65% (101)
R^1 = Boc; R^2 = tBu ; R^3 = (CH$_2$)$_2$CO$_2$Me: 33% (107)
NHR^1 = dibenzyltriazone; R^2 = tBu; R^3= CHO: 61% (114)
R^1 = Boc; R^2 = tBu; R^3 = H: 50% (109)

1) LDA,
RCOMe
61–93%
2) MsCl, Et$_3$N
or DIC, CuCl
62–92%

Boc-N(H)-CO$_2$tBu with O= ketone side chain

O=P(OMe)(OMe)OH

Boc-N(H)-CO$_2$tBu

R^1 = Boc
R^2 = tBu;
R^3 = CO$_2$Et,
(CH$_2$)$_2$CO$_2$Me
1)H$_2$, Pd-C
2) 1N NaOH
3) 6N HCl

R^3 = CHO
NaBH$_4$
70% (110)

Boc-N(H)-CO$_2$(4-MeO-Bn) with OH side chain

PhFl-N(H)-CO$_2$Me with R side chain

1) H$_2$, Pd-C
20–91%
2) HCl
96–100%

(EtO)$_2$P(O)CF$_2$SiMe$_3$
TBAF, THF, –60°C
55% (105)

1) PhOC(=S)Cl,
DMAP
2) (Me$_3$)$_3$SiH,
AIBN
3) TFA
68% (35, 113, 121)

CO$_2$H
( )$_n$
H$_2$N-CO$_2$H
n = 1: 70–92% (36, 102)
n = 3: 89% (107)

1) Ti(OiPr)$_4$,
(+)‾DET, tBuOOH
2) H$_3$O$^+$
3) 3N NaOH
93% (110)

R-N(H)-CO$_2$H (cyclic)
R = Me, Ph, tBu, iPr,
nPr, 2-pyridyl
(106)

F / PO$_3$Et$_2$
F / OH
Boc-N(Boc)-CO$_2$tBu

O=P(OMe)(OMe)
H$_2$N-CO$_2$H

HO, / OH
HO,
H$_2$N-CO$_2$H

Scheme 9.13 Reactions of Aspartate Semialdehyde.

bond can then be reduced (as in the synthesis of α-aminoadipic acid (36, 102) and 2-amino-6-oxohexanoic acid (114)), epoxidized to give polyhydroxy amino acids (110) or used for intramolecular Diels–Alder reactions (101). A two-carbon homolog of Asp semialdehyde was prepared via Wittig reaction of Asp semialdehyde with Ph$_3$P=CHCO$_2$Me followed by alkene reduction and side-chain ester reduction. A second Wittig reaction with Ph$_3$P=CHCO$_2$Me gave 2-amino-oct-6-ene-1,8-dioic acid, with regioselective ester reduction providing 2-amino-8-hydroxy-oct-6-enoic acid. Alkene epoxidation, azide opening, and diol oxidative cleavage led to 2,6-diaminopimelic acid (123).

Another Asp semialdehyde synthon is a homolog of Garner's aldehyde (see Section 10.6.1). The α-carboxy group of Boc-Asp(OBn)-OH was reduced and cyclized to an oxazolidine, followed by reduction of the side-chain ester to an aldehyde (see Scheme 9.14) (124). Dibromomethylenation of the aldehyde produced protected propargylglycinol, which could then be modified by carbonylation and further reaction. As with amino acid syntheses using Garner's aldehyde, the α-carboxy group must be regenerated by oxidation at the end of the synthesis. Propargylglycine, 4-methylene-L-Glu acid and two 4,4-disubstituted derivatives were prepared via this route.

9.2.4 Modification of the α-Carboxyl Group

Asp can also be used to synthesize β-amino acids by regioselectively reducing or modifying the α-carboxy group instead of the β-carboxy group. N-Boc, Fmoc, or Cbz protected Asp β-esters were regioselectively reduced by a two-step procedure (i-BuO$_2$CCl/NMM, then NaBH$_4$, 84–97% yield) to give protected 3-amino-4-hydroxybutanoic acids (37). The alcohol can then be converted into a leaving group and displaced by organocuprates (125, 126). Completely regioselective reduction of the α-carboxyl group has also been accomplished via NaBH$_4$ reduction of the N-tosyl (127–130), N-Tfa (41), or N-Cbz (131, 132) protected Asp anhydride (see Scheme 9.15). Under the same conditions the N-trityl protected anhydride gave a 5.5:1 mixture of α-:β- carbonyl reduction (77). The reduced N-trityl lactone has also been prepared in 69% yield from H-Asp(OMe)-OH (77) via regioselective reduction of the acyclic derivative using a protocol developed for the Glu equivalent (133). The N-tosyl lactone has been α-hydroxylated (127, 128) and both the hydroxylated and non-hydroxylated lactones opened with TMSI to give 3-amino-4-iodobutanoic acid derivatives. Displacement of the iodo group with organocuprates gave a number of enantiomerically pure protected β-amino acids in 42–96% yield (127–129,

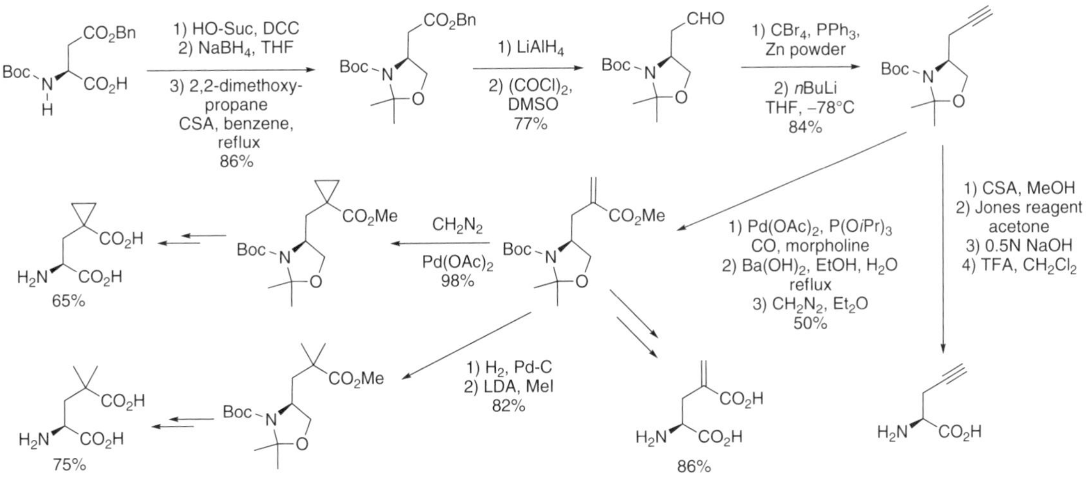

Scheme 9.14 Synthesis of Aspartate Semialdehyde and Further Elaboration (*124*).

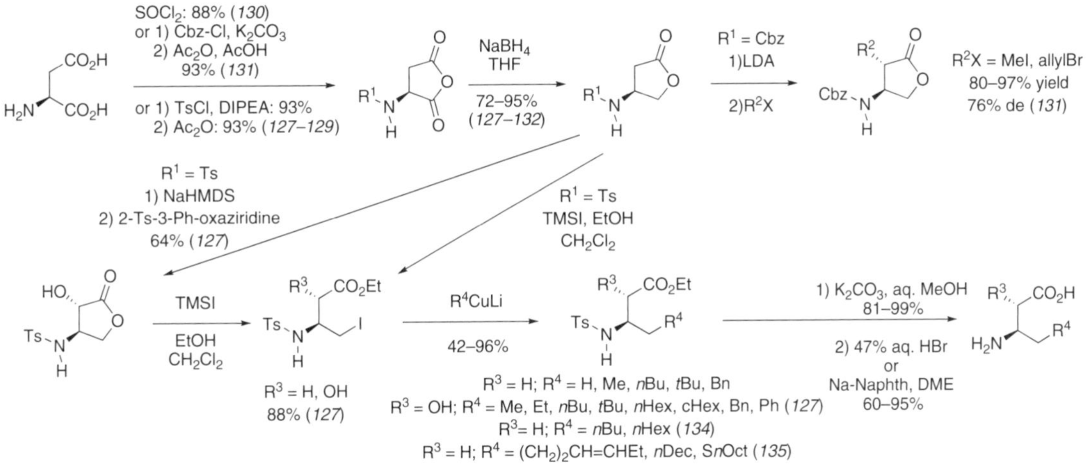

Scheme 9.15 Regioselective Reduction of Aspartate α-Carboxyl via Anhydride.

134, 135). The iodide was also displaced with octane-1-thiol (*135*). The N-Cbz lactone has been alkylated (*131*). Similarly, Boc-Asp(OBn)-OH was reduced with isobutyl chloroformate/NaBH₄, and the alcohol displaced by Mitsunobu reaction with 3-benzoyluracil or 3-benzoylthymine, or mesylated and displaced with 3-benzoylthymine or Cbz-cytosine, producing nucleoso-β-amino acids (*136*). Another synthesis reduced Cbz-Asp(OtBu)-OH via the isobutyl chloroformate/NaBH₄ method, and then converted the resulting amino alcohol into *tert*-butyl N-Cbz aziridine-2-acetate by a Mitsunobu reaction with PPh₃/DEAD. The aziridine could be opened with amine nucleophiles to form 3,4-diamino-butyrate products (*137*). Alternatively, 2,3-diamino-4-hydroxybutyric acid was prepared via NaBH₄ reduction of the α-carboxyl group of Boc- and Cbz-protected Asp

derivatives. Enolate generation and electrophilic azidation adjacent to the original β-carboxyl group generated the new α-amino acid center (*138*). The side-chain β-carboxy group of N,N-Bn₂-Asp-OBn has been protected as a 2,6,7-trioxabicyclo[2.2.2]octane (OBO) ester, with the α-carboxyl group converted into an aldehyde. The OBO ester is resistant to base and nucleophilic addition (*433*).

Homologation of the α-carboxyl group of Asp by a Wolff rearrangement of the corresponding α-diazoketone produced 3-amino-pentan-1,5-dioic acid. A γ-amino group was then introduced by a regioselective Curtius rearrangement of one of the carboxyl groups (*139*). Alternatively, regioselective reductive radical decarboxylation of the α-carboxyl group of Asp, using photolysis of N-hydroxypyridine-2-thione esters in the presence of *tert*-butyl thiol, provided β-amino acids (*86*). Photoreductive decarboxylation of the benzophenone oxime esters of amino acids has also been reported (*140*). More recently, easily prepared phenylselenoesters of amino acids were decarboxylated (*141*).

9.3 Synthesis of Optically Active Amino Acids from Glutamic Acid

9.3.1 Enolate Formation: Acyclic Glutamic Acid

The same strategies that were used to promote regioselective side-chain enolate formation for Asp (namely steric or electronic protection of the α-proton) have also been applied to Glu (see Scheme 9.16, Table 9.3). The steric approach was used in the first reports of Glu alkylation by Zee-Cheng and Olson in 1980, when Trt-Glu(OBn)-OBn was deprotonated with LDA and carboxylated (*4*). Baldwin et al. also used this approach, with Trt-Glu(OMe)-O*t*Bu capable of addition to carbonyl electrophiles when deprotonated with lithium isopropyl-cyclohexylamide (but not LDA) (*142, 143*). The hydroxyl acid group of the aldol adducts could be converted to an alkene by heating with dimethylformamide dimethylacetal, resulting in γ,δ-unsaturated-α-amino acid derivatives (*144*).

The steric approach was also employed by Rapoport et al., with the bulky 9-phenylfluoren-9-yl (PhFl) *N*-protecting group combined with an α-*tert*-butyl ester to promote selective formation of the γ-enolate. Alkylation with a number of electrophiles proceeded in reasonable yield, but poor diastereoselectivity (*145*). After alkylation, the 4-carboxyl ester was reduced to a 5-hydroxy group with LiAlH₄. Intramolecular displacement of the hydroxyl group via Mitsunobu activation resulted in 4-substituted prolines (*145*). Alkylation of the γ-enolate with PhSeCl, followed by oxidative elimination, allowed for a synthesis of optically active protected 3,4-didehydroglutamate (*146*), which is generally too unstable to isolate. The α,β-unsaturated ester system was then dihydroxylated enroute to polyoxamic acid, or used as a Michael addition substrate for reaction with organocuprates (*146*).

The Rapoport protection strategy was also used in an electrophilic amination of the γ-enolate that proceeded with 50% de (*147*). δ-Keto-α-amino esters were prepared via acylation of the lithium enolate with acid

chlorides, followed by γ-ester hydrolysis and decarboxylation (*148*). The 9-phenylfluoren-9-yl (PhFl) *N*-protecting group has also been used without α-carboxylate protection, by double deprotonation of the PhFl-Glu(OMe)-OH (*149*). The δ-keto-α-amino esters were converted to *cis*-5-substituted Pro derivatives by intramolecular reductive amination of the intermediate ketone, using catalytic hydrogenation to deprotect the amine and reduce the subsequent imine (*148, 149*) (see Scheme 9.17). The *trans* isomer could also be prepared, but by a more laborious procedure (*149*).

A more reactive γ-anion synthon was reported by North and co-workers in 1995, who replaced the original *N*-trityl protecting group of Baldwin et al. with an *N*-Cbz group, which provides electronic protection from α-deprotonation (*3, 150*). Dianion generation under the same conditions used for the analogous Asp equivalent (see Section 9.2.1) led to an undesired intramolecular cyclization to pyroglutamate. However, by keeping the reaction temperature below −40 °C the anion successfuly reacted with reactive halides and carbonyl compounds (see Scheme 9.18). The diastereomeric excess of the new chiral center that was formed ranged from 40 to 100%, with no racemization detected at the α-center. The adducts could be further elaborated into other derivatives. Cbz-Glu(OMe)-OMe was also γ-hydroxylated in good yield, with 80% de (*21*).

An *N-p*-nitrobenzoyl protecting group is also proposed to prevent racemization via electronic effects. Very high diastereoselectivity was observed during alkylation of LHMDS-deprotonated *N-p*-nitrobenzoyl protected Glu diester. Only the *threo* diastereomer was detected (*2*). The high diastereoselectivity was proposed to result from a chelated dianion transition state, with careful control of the reaction temperature at −78 °C perhaps accounting for the improved results compared to other Glu dianions (at higher temperatures the other diastereomer began to appear). Acidic removal of the protecting groups caused approximately 3% epimerization at the C4 position (*2*).

More recently, the dianions of Boc or Cbz-Glu(OMe)-OMe or -OTMSE were γ-alkylated with methyl iodide, benzyl bromide, or allylic halides with very high (>98%) de) *anti* stereoselectivity (*151*). The allylated adducts were cyclized to form 4-*cis*-carboxy pipecolic acid derivatives. A synthesis of the complex amino acid dysiherbaine began with an alkylation at the γ-carbon of Boc-Glu(OMe)-OMe with a protected 2-trifloxy-methyl-3,4,5-trihydroxypyran derivative derived from D-mannose (*152*).

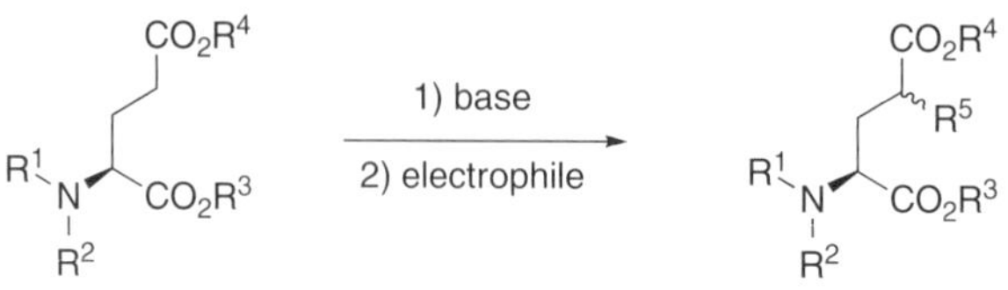

Scheme 9.16 Alkylation of Glutamate γ-Enolate (see Table 9.3).

Table 9.3 Alkylation of Glutamate γ-Enolate (see Scheme 9.16)

R¹	R²	R³	R⁴	Base	Electrophile	Yield	Reference
Trt	H	Bn	Bn	LDA	BnO₂CCl	>50%	1980 (*4*)
Trt	H	*t*Bu	Me	LICA	HCHO MeCHO EtCHO *i*PrCHO *n*PrCHO PhCHO (4-NO₂-Ph)CHO (4-MeO-Ph)CHO MeCOMe	30–95% mixture of diastereomers	1988 (*142*) 1989 (*143*)
PhFl	H	*t*Bu	Me	LTMP LHMDS KHMDS	MeI *n*PrOTf BrCH₂CN I₂	52–94% 0–58% de	1989 (*145*)
Cbz	H	Me	Me	LHMDS	(oxaziridine: Ts–N–O ring bearing Ph) (R⁵ = OH)	70% 80% de	1993 (*21*)
PhFl	H	*t*Bu	Me	LHMDS	ROCl R = Et, *i*Pr, *n*Bu, *t*Bu, Ph, Bn	58–90%	1993 (*148*)
PhFl	H	Me	Me	KHMDS	PhSeCl (followed by oxidative elimination with mCPBA)	84% for 2 steps	1993 (*146*)
PhFl	H	*t*Bu	Me	KHMDS	trisyl azide (R⁵ = N₃)	54% 50% de	1994 (*147*)
Cbz	H	*t*Bu	Me	LHMDS	MeI BrCH₂CH=CH₂ BnBr AcCl PhCHO	21–75% 40–100% de	1995 (*150*)
4-NO₂-Bz	H	Me, Et	Me, Et	LHMDS	MeI EtBr BrCH₂CH=CH₂ BnBr	33–65% 100% de	1995 (*2*)
PhFl	H	H	Me	LHMDS	*t*BuCOCl	75%	1996 (*149*)
Cbz	H	Me	Me	LHMDS	MeI BrCH₂CH=CH₂ BrCH₂CH=CHMe BrCH₂C(Me)=CH₂ BrCH₂CH=CHPh cyclohex-2-enyl bromide BrCH(CO₂Me)CH=CH₂ BnBr	65–93% >98% de *anti* except cylohexenyl Br, 60%, 50% de	1998 (*151*)
Boc	H	Me	Me	LHMDS	BrCH₂CH=CH₂ BrCH₂CH=CHPh BrCH(CO₂Me)CH=CH₂ BnBr	75–92% >98% de *anti*	1998 (*151*)
Boc	H	TMSE	Me	LHMDS	BrCH₂CH=CH₂ BrCH₂C(Me)=CH₂ BrCH₂CH=CHPh BrCH(CO₂Me)CH=CH₂	75–89% >98% de *anti*	1998 (*151*)
Boc	H	Me	Me	LHMDS	TfOCH₂CH[−CH(OR) CH(OR)−CH(OR)CH₂O−]	75% >95% de	2002 (*152*)

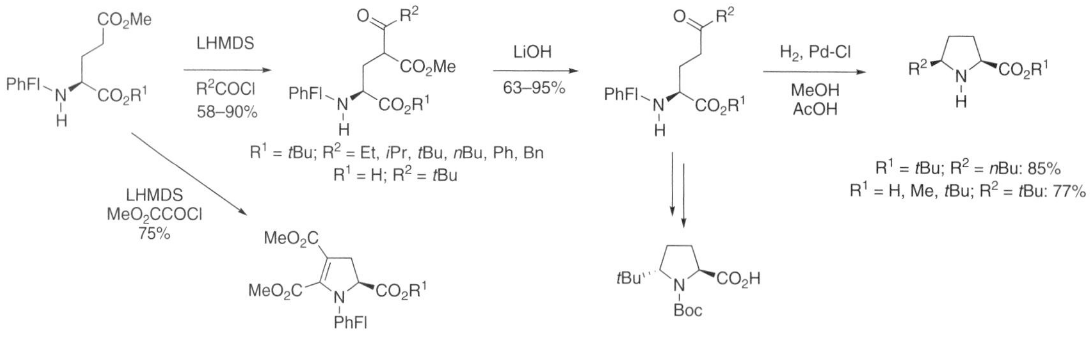

Scheme 9.17 Alkylations and Elaboration of Glutamate (*148, 149*).

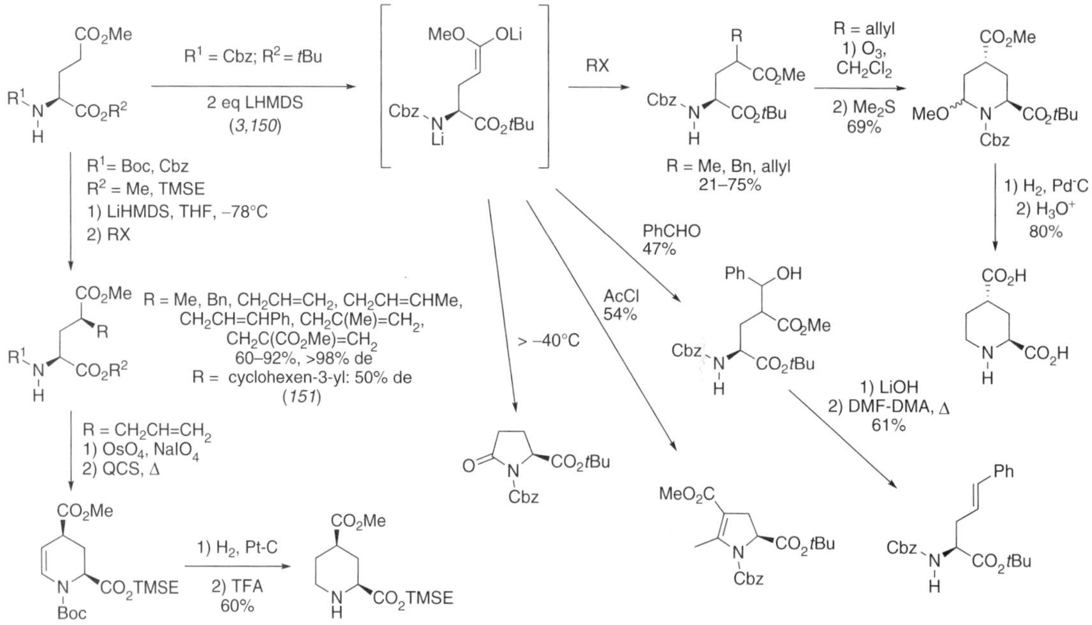

Scheme 9.18 Alkylation of Glutamate γ-Enolate and Further Elaboration.

An electrophilic synthon has also been prepared from Glu (and from 2-aminoadipic acid), via selective bromination α- to the side-chain carboxyl group using Br₂/PBr₅. The halogen was then displaced with potassium phthalimide (*153*).

9.3.2 Enolate Formation: Pyroglutamate

Pyroglutamate (see Section 9.3.5a for preparation) can also be regioselectively alkylated to give 4-substituted pyroglutamates, often with excellent stereoselectivity. A 1999 review summarized asymmetric syntheses using pyroglutamic acid, including ring alkylation or functionalization reactions, transformations of the lactam carbonyl group, and ring-opening reactions (*154*).

A number of syntheses have alkylated a pyroglutaminol derivative, in which the pyroglutamate α-carboxyl group has been reduced to a hydroxymethyl substituent and protected with bulky groups. This analog was initially believed to be necessary to prevent racemization and ensure 1,3-asymmetric induction (see Scheme 9.19, Table 9.4) (*155–158*). Alkylation generally produces the *trans* isomer preferentially. It can be converted into the *cis* diastereomer via a three-step process consisting of a second alkylation with PhSeCl, oxidative elimination to the 3,4-dehydroglutaminol, and reduction of the double bond by catalytic hydrogenation, with reduction of the alkene from the less hindered side (*156*). The pyroglutaminol has also been protected as a bicyclic derivative via hemiaminal formation. The diastereoselectivity

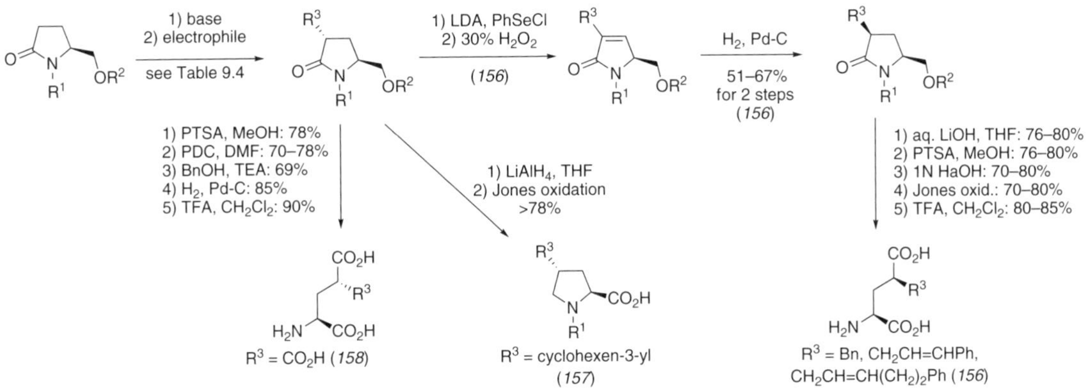

Scheme 9.19 Alkylation and Elaboration of Pyroglutaminol (see Table 9.4).

Table 9.4 Alkylation of Pyroglutaminol and Further Elaboration (see Scheme 9.19)

R^1	R^2	Base	Electrophile	Yield	Reference
—CH(Ph)-R$_2$	—CH(Ph)-R$_1$	LDA	3-Br-cyclohexene	100%	1986 (157)
Boc	TBS	LDA	Cbz-Im (R_3 = CO$_2$Bn)	65%	1986 (158)
Boc	TBS	LDA	BnBr BrCH$_2$CH=CHPh BrCH$_2$CH=CH(CH$_2$)$_2$Ph	64–66%	1990 (156)
Ts	TBDPS	NaHMDS	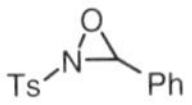 Ts–N⟨O⟩Ph (R^3 = OH)	72%	1991 (266)
—CH(Ph)-R^2	—CH(Ph)-R^1	LDA	BnBr (4-NO$_2$-Bn)Br BnO$_2$CCl PhCHO (4-NO$_2$-Ph)CHO MeO$_2$CCO$_2$Me PhSeCl	25–74% 28–99% de trans	1991 (267)
Boc	TBDPS	LDA	PhSeCl MoOPH	23–86% 57%	1991 (155)
—C(Ph)(Me)R^2	—C(Ph)(Me)R^1	LDA	BnBr 4-NO$_2$-BnBr MeI 3-(CH$_2$Br)-1-Boc-indole TsCl	52–82% 40–88% de cis	2000 (159)

then becomes dependent on the hemiaminal substituents, with a benzylidene derivative giving mostly the same *trans* stereochemistry as the monocyclic substrates, but an acetophenone derivative giving *cis*-substituted products (*159*).

4-Substituted glutamic acids can be generated from the pyroglutaminol derivatives by ring opening and oxidation of the primary alcohol back to a carboxyl group (*156, 158*), while substituted prolines can be obtained by reduction of the lactam carbonyl (*157*). Variations of this

procedure have been used to prepare other amino acids, such as polyoxamic acid (*160*), as well as a range of non-amino acid products such as deoxynojirimycin (*160*) and pilocarpine (an alkaloid used to treat glaucoma (*16*)). A bicyclic lactam derived from pyroglutaminol has also been alkylated (*161*).

More recently, the reduction of the pyroglutamate α-carboxyl group was found to be unecessary, and alkyl or aryl (*162–165*), acyl (*12, 166*), hydroxyl (*167, 168*), and alkylamine (*162, 169*) substituents were successfully

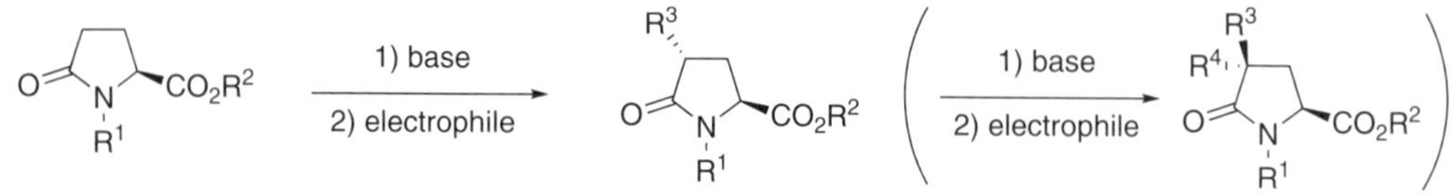

Scheme 9.20 Alkylation of Pyroglutamate (see Table 9.5).

introduced by reaction with the anion of protected pyro-glutamate (see Scheme 9.20, Table 9.5). Baldwin et al. had reported that reactive electrophiles other than benzyl bromide gave products resulting from multiple alkyla-tion and ring cleavage (164), but both Ezquerra et al. (163) and Dikshit and Panday (170) found that the desired products could be isolated in reasonable yield. The *trans* isomer is again generally preferred, although Ezquerra et al. (163) corrected an earlier report on alky-lation with alkyl bromoacetate (171), finding that the major adduct (considered *trans* in the original report) was in fact *cis* based on spectroscopic evidence. The ste-reoselectivity of C4 alkylation of Boc-pyroglutamate-O*t*Bu has recently been reinvestigated, and it was found that under certain conditions some electrophiles such as methyl triflate gave the thermodynamically less stable *cis* adduct. The *trans* product that was still obtained from electrophiles such as benzyl bromide could be con-verted to the *cis* isomer by adding more base after the initial alkylation, and then quenching with the hindered proton source di-*tert*-butylphenol (172, 173). N-Alkyl pyroglutamate acids can be *trans*-alkylated at the 4-position without loss of enantiomeric purity by form-ing the dilithio dianion (174). The regioselectivity of aldol reactions/alkylations of N-protected pyroglutamate-OMe is dependent on the protecting group, with N-benzyl protection resulting in α-alkylation, and N-Boc protec-tion leading to γ-alkylation (175).

Double alkylations with the same, or two different, electrophiles have been reported (165, 176). Additions to aldehydes and ketones proceeded with good yields (177–179), and the resulting adducts were dehydrated and then hydrogenated to give alkyl side chains (177), or oxi-dized to γ-acyl products (12). Titanium enolates reacted with carbonyl compounds to give exclusively the *trans* isomer (180). Pyroglutamate enolates have also been added to activated imines, with the *trans* isomer predom-inating (180). A 4-exo-methylene group was introduced via alkylation with Eschenmoser's salt followed by base-induced elimination of the methylated amine (169, 181), or by reaction with Bredereck's reagent, bis(dimethylamino)-*tert*-butoxymethane (182–185). The methylene group was subsequently cyclopropanated (169) or used for a Diels–Alder reaction (183). Reduction of the alkene gave 4-methyl-Glu, with both γ-epimers accessible via epimerization of the γ-center using KCN in DMF, followed by chromatographic resolution of the diastereomers (186). Alternatively, stereoselective reduc-tion and further derivatization led to (2S,4R)-[5,5,5-²H₃]-Leu (185). A series of 4-(arylalkyl)-pyroglutamates with varying length of alkyl linker (n = 1,2,3) were obtained from N-methoxycarbonyl pyroGlu-OMe via deprotona-tion and alkylation with benzyl bromides (n = 1, *trans* products), aldol reaction with phenylacetaldehyde fol-lowed by elimination and hydrogenation (n = 2, *cis* prod-uct), or alkylation with cinnamyl bromides (n = 3, *trans* products). The lactam of each derivative was activated by formation of the hemiaminal acetate (reduction with LiEt₃BH and then acetylation), followed by an intramo-lecular Friedel–Crafts cyclization to give tricyclic pro-line products, 1,3a,8,8a-tetrahydro-2*H*-3-azacyclopenta [a]indene-3-carboxylic acid, 2,3,3a,4,5,9b-hexahydro-benzo[g]-indole-2-carboxylic acid or 3,3a,4,5,6,10b-hexa-hydro-2*H*-1-azabenzo[e]azulene-1-carboxylic acid (187).

As an alternative route to direct alkylation of pyro-glutamate enolate, which can result in variable dia-stereoselectivity, (E)-4-alkylidenepyroglutamates were prepared by reaction of Grignard reagents with the inter-mediate enaminone obtained from pyroglutamate and Bredereck's reagent (see Scheme 9.21) (188, 425). Hydrolysis of these adducts produced 4-alkylidene-Glu products, but hydrogenation gave *cis*-4-alkyl-pyrogluta-mates as single diastereomers (189). The pyroglutamates were then either hydrolyzed to 4-alkyl-Glu products, or the lactam carbonyl reduced to give 4-alkylprolines (189). The enaminone has also been used to prepare het-eroarylalanines, with the heterocyclic ring formed with simultaneous opening of the pyroglutamate ring (190). Five-membered rings could be produced directly via hydrolysis of the enaminone to an aldehyde and cycliza-tion/ring opening, while six-membered rings were obtained after reaction of the aldehyde with diaz-omethane (190). The aldehyde was also reduced to a hydroxymethyl substituent, giving 4-(hydroxymethyl)-Glu after pyroglutamate hydrolysis; reduction of the side-chain acid then gave stereospecifically labeled 5,5'-dihydroxy-Leu (191).

Once alkylated, the pyroglutamate nucleus is ame-nable to further modification (see Section 9.3.5b), or can be hydrolyzed to produced 4-substituted-glutamic acid derivatives (169, 177, 184, 188, 189, 192, 425).

9.3.3 Modification of the γ-Carboxyl Group: Acyclic Glutamic Acid

A number of synthetic schemes rely upon a regioselec-tive reduction of the γ-carboxyl group of Glu. Two groups reported the first regioselective reduction of Boc-Glu-O*t*Bu in 1984 (with more reports in subsequent years),

Table 9.5 Alkylation of Pyroglutamate

R¹	R²	Base	Electrophile R³	Yield	Base	Electrophile R⁴	Yield	Reference
Cbz	Bn		Bredereck's reagent (Me₂N)₂CHOtBu R³ = R⁴ = = CHNMe₂	95%				1979 (182) 1981 (183)
Boc	Bn	LHMDS	Ts–N (structure) Ph (R⁵ = OH)	30–61% 100% de				1988 (167) 1992 (168)
Cbz	tBu	LDA	EtCHO	25–79%				1989 (164)
Boc	Bn	LHMDS	iPrCHO PhCHO BnCHO					
Boc	tBu	LDA	BnOCOCl	52%				1990 (166)
Boc	Me	LDA	EtCHO iPrCHO cPrCHO cPentCHO PhCHO MeCO(imidazol-1-yl) HCO₂tBu	29–99%				1991 (12)
Boc	Bn	LHMDS	PhCH=NTs (structure)	62–95% mixture of 2 of 4 diastereomers 42–60% de				1991 (162) 1997 (420)
Boc Cbz	tBu	LHMDS	BnBr PhCHO (4-MeO-Ph)CHO (4-F-Ph)CHO (4-Cl-Ph)CHO (3-MeO-Ph)CHO (2-furyl)CHO	54–64% mixture of 2 of 4 diastereomers				1992 (170)
Boc	tBu		Bredereck's reagent (Me₂N)₂CHOtBu R₃ = R⁴ = = CHNMe₂	91%				1992 (184)

(Continued)

Table 9.5 Alkylation of Pyroglutamate (continued)

R¹	R²	Base	Electrophile R³	Yield	Base	Electrophile R⁴	Yield	Reference
Boc	Et, tBu	LHMDS	BrCH₂CO₂Et BrCH₂CH=CHPh BrCH₂CH=CH₂ ICH₂CN (PhS)₂ PhSeCl PhCOCl BnBr 4-Me-BnBr 4-CN-BnBr 4-Br-BnBr 4-CF₃-BnBr BrCH₂(2-naphthyl)	36–75% 0–100% de				1993 (163)
CO₂Me	Bn	LHMDS	BrCH₂CO₂Me BrCH₂CO₂tBu	56–82% 14–54%				1993 (171)
Boc	Et	LHMDS	MeI BnBr BrCH₂CH=CHPh BrCH₂CO₂Et BrCH₂CH=CH₂	dialkylate		dialkylated R₃ = R₄	50–89%	1994 (165)
Boc	Et	LHMDS	BnBr		LHMDS	MeI 4-NO₂-BnBr BrCH₂CH=CHPh ICH₂CN BrCH₂CH=CH₂	40–67% 0–100% de%	1994 (165)
Boc	Et	LHMDS	PhSSPh		LHMDS	BnBr BrCH₂CH=CHPh	40–67% 0–100% de%	1994 (165)
Boc	Et	LHMDS	4-NO₂-BnBr BrCH₂CH=CHPh ICH₂CN BrCH₂CH=CH₂		LHMDS	BnBr	40–67% 0–100% de%	1994 (165)
Boc	Et	LHMDS	[CH₂=NMe₂]⁺I⁻ (R³ = CH₂NMe₂)	65% 100% de				1994 (169)
Boc	Bn	LHMDS	[CH₂=NMe₂]⁺I⁻ (R³ = CH₂NMe₂)			oxidize mCPBA, Δ (R³ = R⁴ = =CH₂)	75%	1994 (181)
Boc	Et	LHMDS + BF₃·Et₂O	MeCHO EtCHO Ph(CH₂)₃CHO	61–83%		dehydrate (R³ = R⁴ = = CR⁵R⁶)	36–84%	1995 (177)

N	R	Base	Conditions	Electrophile	Yield / de	Year (Ref)
				Ph₂CH(CH₂)₃CHO PhCHO (4-Me-Ph)CHO (3-MeO-Ph)CHO (4-CF₃-Ph)CHO MeCOMe 4,4-Ph₂-cyclohex-2-enone		1995 (178) 1995 (180)
Boc	tBu	TiCl₄		rBuCHO		1996 (280)
CO₂Et	Me	DIEA		PhCHO (4-Me-Ph)CHO (4-Cl-Ph)CHO PhCH=CHCHO cyclopentanone		1997 (176)
Boc	Et	LHMDS		BnBr		1998 (172)
Boc	Et	LHMDS (2 eq.)	MeI	MeI	70%	2001 (173)
Boc	tBu	LHMDS	LHMDS 2,6-di-tert-butyl-phenol (R⁴ = H)	MeI MeOTf BnBr BrCH₂CO₂tBu	63–75% 60–100% de cis	1998 (175)
Boc	Me	LHMDS		BrCH₂(1-Boc-2-Br-indol-3-yl) (1-Ts-indol-3-yl)CH₂CHO (1-Boc-indol-3-yl)CH=CHNO₂	55–59%	1998 (421)
Boc	tBu	LHMDS		ICH₂CN	65% 33% de	1998 (253)
Boc	Bn	LHMDS		BnBr	77% 100% de	1999 (174)
Me Bn 4-Me-Bn 4-MeO-Bn	H	LHMDS		BnBr 4-Cl-BnBr 4-Br-BnBr MeI BrCH₂CH=CH₂	8–60% 100% de	2002 (179) 2002 (422)
Boc	Me	LHMDS		MeCOMe	85%	2003 (187)
Boc	Et	LHMDS		BnBr	76% 100% de	
MeO₂C	Me	LHMDS		BnBr 3-Br-BnBr 3-MeO-BnBr PhCH₂CHO PhCH=CHCH₂Br (3-MeO-Ph)CH=CHCH₂Br	43–59%	

643

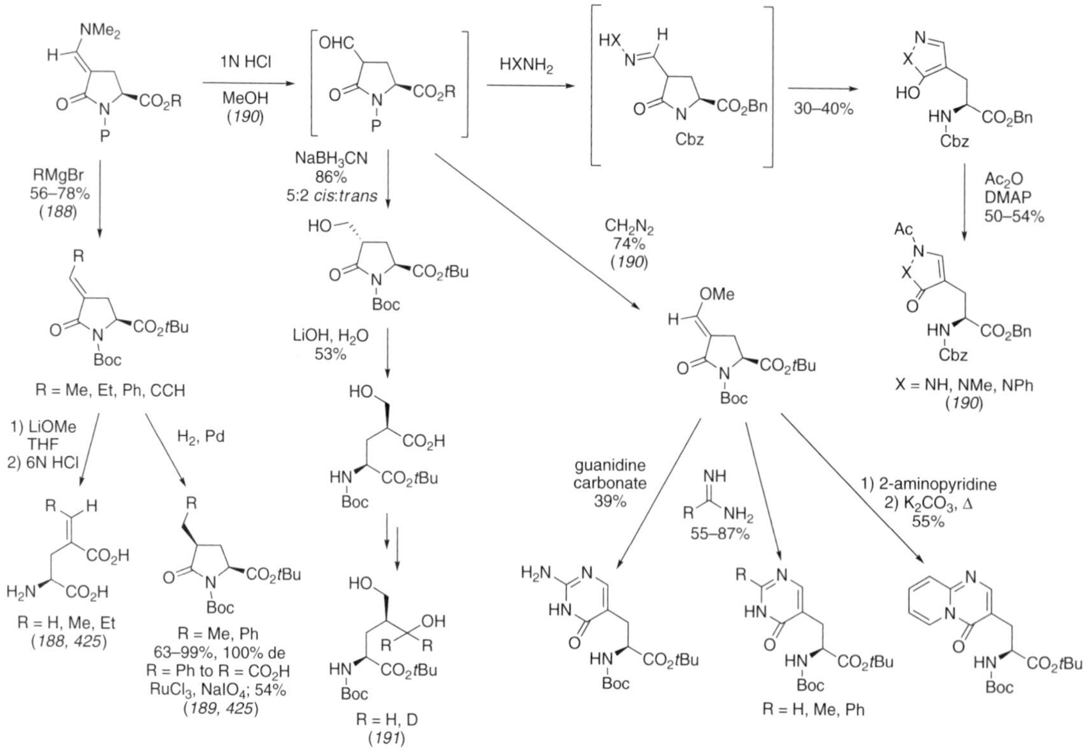

Scheme 9.21 Reactions of Pyroglutamate Enaminone.

activating the unprotected side-chain carboxyl group with ethyl chloroformate and then reducing the mixed anhydride with NaBH₄ to give protected 5-hydroxy-Nva (193–195) (see Scheme 9.22, Table 9.6). Cbz-Glu-OtBu (196), Cbz-Glu-OMe (195), and Fmoc-Glu-OBn (197) were also reduced under these conditions. A similar procedure was used more recently with isobutyl chloroformate (37) in conjunction with a number of protecting groups (45, 198), while the side chain of Boc-Glu-OFm was reduced in 80% yield using NMM/methyl chloroformate, followed by reduction with NaBH₄ (44), and Boc-Glu-OBn was activated with N-hydroxysuccinimide/ DCC and reduced with NaBH₄ (40). Attempted reduction of Cbz-Glu-OMe or Boc-Glu-OtBu via activation with oxalyl chloride/DMF followed by treatment with NaBH₄ instead gave the Boc- or Cbz-pyroglutamate esters in 66–81% yield (195). The side chain of Boc₂-Glu(OMe)-OtBu was regioselectively reduced with NaBH₄ in 64% yield (199).

Direct diborane reduction of Cbz-Glu-OMe resulted in the desired alcohol in 58% yield, along with 18% of Cbz-L-Pro-OMe byproduct (195) Under the same conditions, Boc-Glu-OtBu instead gave the Pro byproduct in 56% yield, with only traces of the alcohol (195). A convenient conversion of unprotected Glu to 5-hydroxy-Nva was reported in 2000. Treatment with triethylborane formed a boroxazolidone in 92% yield, protecting the α-amino and carboxyl groups. Side-chain

reduction with BH₃.THF, followed by hydrolysis of the non-isolated product with 1.5 N HCl to cleave the boroxazolidone, gave 5-hydroxy-Nva in 70% yield. Enantiomeric purity was confirmed by derivatization with Marfey's reagent (200). Regioselective borane reduction of the γ-carboxy group of Cbz-Glu, protected as an oxazolidinone formed with formaldehyde, has also been reported (38, 201). In 1987 either the α- and γ-carboxy groups of N-Trt-protected glutamate monomethyl esters were regioselectively reduced with LiAlH₄, giving 4-amino- or 2-amino-5-hydroxypentanoic acids in good yields (72–87%) (133). The γ-carboxy groups of 4-substituted Boc-Glu(OH)-OtBu derivatives have also been reduced to hydroxymethyl groups (184, 191).

The δ-hydroxy group that is produced by reduction of the γ-carboxyl can be displaced if first converted into a leaving group. The hydroxy group has been activated as an iodo substituent (5-iodo-Nva) via a mesylate intermediate (202) or by reaction with methyltriphenoxyphosphonium iodide (40, 184) or I₂/PPh₃ (44). The iodide in turn can be displaced with other nucleophiles, including deuteride (184) and the thiol side chain of Cys (44). The iodide displacement route was used to prepare L-indospicine, an isosteric analog of Arg found in the pasture legume *Indigofera spicata* (202). The iodide can be used to form iodozinc reagents similar to the Jackson β-iodo-Ala and γ-iodo-Abu derived reagents (see below, Section 10.5.3) (430).

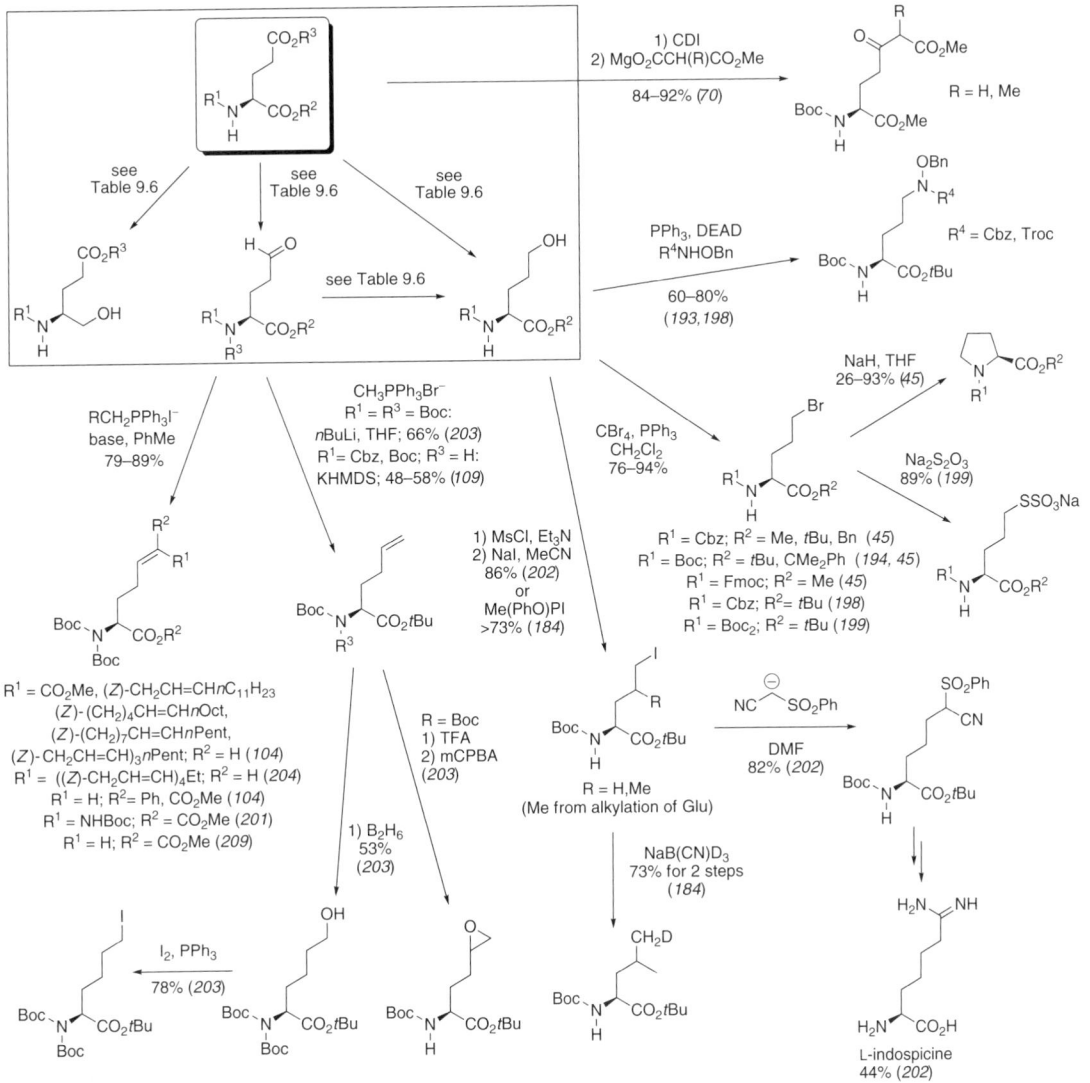

Scheme 9.22 Regioselective Reduction of Glutamate γ- or α-Carboxyl Group.

A similar procedure has been used to prepare the bromide, with direct conversion of the hydroxy group using CBr₄/PPh₃ (45, 198, 199). The bromide was then displaced with sodium thiosulfate enroute to sulfon-amide substituents (199), or cyclized by Na-induced reaction with the α-amino group to give protected pro-lines (45). Under the same conditions, the analogous Asp-derived γ-bromo-Nva gave 1-amino-cyclopropane-1-carboxylate resulting from α-alkylation, rather than the azetidine-2-carboxylate (45).

Activation of the δ-hydroxy Nva hydroxyl as a tosy-late, displacement with Na¹³CN, and nitrile reduction provided labeled Lys (195). The hydroxyl group has been converted into an amine via Mitsunobu reaction, leading to an Orn derivative (38, 193). Alternatively, an intramo-lecular Mitsunobu reaction provided Pro derivatives from

4-substituted PhFl-Glu(OMe)-OtBu precursors that were regioselectively reduced at the γ-carboxy ester with LiAlH₄ (145).

Several other regioselective modifications of the γ-carboxyl group have been reported. Glu acid semialde-hyde has been prepared by hydrogenation of the side-chain acid chloride (98), by regioselective reduction of N,N-Boc₂ Glu(OMe)-OtBu (203) or N,N-Boc₂ Glu(OMe)-OMe (104, 204) with DIBAL-H, or by reduction of the Weinreb amide of Boc₂-Glu(OH)-OtBu (205), Cbz-Glu(OH)-OtBu (109), or Boc-Glu(OH)-OtBu (109) with DIBAL-H (205) or LiAlH(OtBu)₃ (109). The side-chain Weinreb amide reduction could also be carried out on a solid support (108). Another route to Glu semialdehyde converted Phth-Glu-OMe to the γ-ethylthiol ester, fol-lowed by side-chain reduction with triethylsilane (206).

Table 9.6 Regioselective Reduction of Glutamate (see Scheme 9.22)

R^1	R^2	R^3	γ-Carboxyl Reduction Conditions (to CH_2OH)	Yield	Reference
Boc	tBu	H	1) EtOCOCl, NEt₃ 2) NaBH₄	75–93%	1984 (193) 1984 (194) 1996 (202) 2003 (316)
Trt	H	Me	LiAlH₄, THF	87%	1987 (133)
Cbz Alloc Ph	tBu	H	1) iBuOCOCl, NMM 2) NaBH₄	56–90%	1990 (198)
Cbz	tBu	H	1) EtOCOCl, NEt₃ 2) NaBH₄	68%	1991 (196)
Boc	tBu	H (with C4-Me)	1) iBuOCOCl, NEt₃ 2) NaBH₄	75%	1992 (184)
$-CH_2-R^2$	$-CH_2-R^1$	H	BH₃·SMe		1994 (201)
$-CH_2-R^2$	$-CH_2-R^1$	H	B₂H₆, THF	84%	1995 (38)
Cbz Boc Fmoc	Me tBu Bn CMe₂Ph	H	1) iBuOCOCl, NMM 2) NaBH₄	75–95%	1996 (45)
Boc	Bn	H	1) iBuOCOCl, NMM 2) NaBH₄	84%	1998 (197)
Boc₂	Me	Me	1) DIBAL-H 2) NaBH₄	78%	1999 (208)
$-B(Et)_2-R^2$	$-B(Et)_2-R^1$	H	BH₃·THF	70%	2000 (200)
Boc₂	tBu	Me	NaBH₄	64%	2003 (199)
Boc	Fm	H	1) MeOCOCl, NMM 2) NaBH₄	80%	2004 (44)
R^1	R^2	R^3	γ-Carboxyl Reduction Conditions (to CHO)	Yield	Reference
Phth	Me	H	1) EtSH, DAMP, EDAC 2) Et₃SiH, Pd-C, MeCN	74%	1994 (206)
Boc	Me	H	1) HN(OMe)Me, NEM 2) Boc₂O, DMAP 3) DIBAL-H	92% 100% 62%	1997 (205)
Boc₂	Me	Me	DIBAL-H	85%	1998 (204) 1998 (104)
Boc₂	tBu	Me	DIBAL-H	82%	1999 (203)
R^1	R^2	R^3	α-Carboxyl Reduction Conditions (to CH_2OH)	Yield	Reference
Boc	H	Et	BH₃, THF	57%	1982 (230)
Trt	Me	H	LiAlH₄, THF	72–77%	1987 (133)
Boc	OSu	Me	NaBH₄, THF	83%	1989 (229)
Boc	H	Me	1) DCC, HOSucc 2) NaBH₄	83%	1991 (268)
R^1	R^2	R^3	δ-Hydroxy Oxidation Conditions	Yield	Reference
allyl₂	tBu		Swern	—	1989 (423)
Cbz + Boc	tBu		PCC	—	1990 (198)
Cbz + $-CH_2-R^2$	$-CH_2-R^1$		PCC	51%	1994 (201)
Cbz	tBu		Swern	—	1998 (424)

Oxidation of the reduced 5-hydroxyl-Nva intermediate described previously also gives the aldehyde (*201*).

Glu semialdehyde was used in a synthesis of 2,6-diaminoheptanedioic acid (2,6-diaminopimelic acid) derivatives (*110*). An aldol addition of the Schöllkopf bislactim ether template to N,N-Boc$_2$ Glu semialdehyde, followed by DAST fluorination of the hydroxy group, provided (2S,3R,6S)-3-fluoro-2,6-diaminopimelic acid (*207*). The aldehyde has also been used as a substrate for the addition of a glycosyl dianion (*205*), and can be reduced with NaBH$_4$ to give 5-hydroxy-Nva (*208*). Wittig olefinations have been carried out on solid support (*108*) or in solution, and used to prepare α-aminoarachidonic acid (*204*), a series of related lipidic derivatives (*104*), and 2-aminohex-5-enoic acid (*203*). The alkene of the latter product was either epoxidized or hydroborated to give further derivatives (*203*). Methylenation of Boc- or Cbz-Glu semialdehyde-OtBu produced homoallylglycine in 48–48% yield (*109*). Another Wittig reaction employed N,N-Boc$_2$ Glu semialdehyde *tert*-butyl ester as the substrate and hydrogenated the (E)-2-aminohept-5-enedioic acid product to give 2-aminoheptanedioic acid; side-chain ester DIBAL reduction then produced 2-amino-7-oxoheptanoic acid (*209*). Alternatively, the (E)-2-aminohept-5-enedioic acid side-chain ester was first reduced to give 2-amino-7-hydroxyhept-5-enoic acid, and then oxidized to 2-amino-7-oxohept-5-enoic acid (*209*). The two-carbon homolog of Glu semialdehyde, prepared via Wittig reaction of Glu semialdehyde with Ph$_3$P=CHCO$_2$Me, alkene reduction and side-chain ester reduction, was employed for a second Wittig reaction with Ph$_3$P=CHCO$_2$Me to give 2-amino-non-7-ene-1,9-dioic acid. Regioselective ester reduction provided 2-amino-9-hydroxy-non-7-enoic acid. Alkene epoxidation, azide opening, and diol oxidative cleavage led to 2,6-diaminosuberic acid (*123*). A review of the synthesis and use of N,N-(Boc)$_2$ Glu semialdehyde was published in 2003 (*427*).

The γ-carboxyl group of Glu can be converted into a δ-keto-ε-carboxyl functionality by activation with carbonyldiimidazole and displacement with a malonate enolate, enroute to 2-amino-5-oxoheptanedioic acid (*68, 70*), as previously discussed for Asp (see Scheme 9.6). Treatment of the initial adduct with base (Na$_2$CO$_3$), followed by alkylation, produced ε-substituted derivatives, which were then decarboxylated (*68*). The β-keto ester could be condensed with a variety of phenols in the presence of methanesulfonic acid (von Pechman condensation) to produce substituted coumarin side chains (*71*). Similarly, formation of the Glu γ-carboxyl Weinreb amide and displacement with acetylide anions provided alkynyl ketones, which can be converted into heteroaromatic side chains (see Scheme 9.8) (*63*). The side-chain of Boc$_2$-Glu-OtBu has been converted into a vicinal tricarbonyl, with a –CH$_2$CH$_2$COCOCO$_2$$t$Bu side chain, via reaction with DCC/DMAP and (*tert*-butoxycarbonylmethylene) triphenylphosphorane (45%) followed by ozonolysis (84%). Reaction with ethylenediamine or o-phenylenediamine followed by oxidation produced γ-(2-carboxy-pyrazine-3-yl)-Abu and γ-(2-carboxy-quinoxalin-3-yl)-Abu, while

condensation with S-methyl isothiosemicarbazide produced regioisomeric γ-(3-methylthio-5-carboxytriazin-6-yl)-Abu and γ-(3-methylthio-6-carboxytriazin-5-yl)-Abu (*75, 76*). The same reactions were carried out with Asp to make β-(heteroaryl)-Ala homologs.

Reaction of a Glu derivative with a side-chain acid chloride with diazomethane or the anion of bis(TBDMS) succinate produced other δ-keto-α-amino acids (*210*). The acid chloride has also been treated with 2-phenyl-5-oxazolone enroute to a synthesis of racemic 2,6-diamino-3-hydroxypimelic acid (*211*). Cbz-Glu-OMe was activated with ethyl chloroformate and then converted into a diazoketone side-chain with diazomethane. Rh-catalyzed intramolecular insertion into the N−H bond of the α-amino group produced 5-oxopipecolic acid (*212*). The diazoketone side chain was also displaced with HCl to give a chloroketone that was further reacted with potassium phthalimide to form 2,6-diamino-5-ketopentanoic acid. Cyclization of this side chain resulted in homo-His (*59*). The side chain of N-methyl Glu, protected as a hexafluoroacetone oxazolidinone, was activated as the acid chloride, reacted with diazomethane, converted into the bromomethyl ketone and cyclized with thiobenzamide to give N-methyl γ-(2-phenyl-1,3-thiazol-4-yl)-Abu, or homologated to N-methyl 2-aminoadipic acid via the diazomethane intermediate in 55% yield (*432*). A variety of other transformations have been described for Asp-derived diazoketones (see Section 9.2.2).

Regioselective reaction of LiCH$_2$PO$_3$Me$_2$ with the side chain of PhFl-Glu(OMe)-OtBu led to a β-ketophosphonate that was then employed for Horner–Wadsworth–Emmons olefinations (*79*). N-Ethoxycarbonyl Glu(OMe)-OH was treated with excess PhMgBr; 2 equiv added to the ester, accompanied by elimination of the resulting carbinol to form 5,5-diphenylallylglycine. Esterification and ozonolysis produced Asp semialdehyde in 12% overall yield, along with its hydrate and dimethyl acetal (*213*).

Unprotected Glu undergoes a regioselective Schmidt rearrangement leading to 2,4-diaminobutyric acid (Dab), with the α-amino group inhibiting reaction of the α-carboxyl group (*214*). Curtius or Hoffmann rearrangements also provide Dab (e.g., *215*) (see Section 9.4).

The γ-carboxy group has been removed from Cbz-Glu-OMe by a decarboxylative elimination with lead tetraacetate, which is catalyzed by cupric acetate and leads to vinylglycine (*216*). A reductive radical decarboxylation is also possible, using photolysis of an N-hydroxypyridine-2-thione ester in the presence of *tert*-butyl thiol (*86, 87*) (see Scheme 9.11). N-Hydroxy-2-selenopyridine esters can be used instead of the thio analogs for the decarboxylative rearrangement (*217*), and the side chain decarboxylation has been carried out within a dipeptide (*87*). Phenylseleno esters have recently been found to be suitable for decarboxylations of amino acids, although not yet applied to side-chain carboxyl groups (*141*). Kolbe electrolysis of Boc- or Cbz- Glu-OBn/OMe produced the decarboxylated dimerized derivative, 2,7-L,L-diaminosuberic acid, in 9–27% yield (*96*). Cbz-Glu-OMe was converted into a diacyloxyiodobenzene derivative, which generates a radical under thermolysis conditions. Addition to Trt-dehydroAla-OBn resulted in a 5,6-dehydro-2,6-diamino-pimelic acid derivative, with DuPHOS hydrogenation providing 2,6-diaminopimelic acid (218).

The intermediate radical obtained during decarboxy-lation of Glu can be trapped by alkenes (50, 93) or diphenyl diselenide, with the latter compound giving a γ-seleno derivative that was readily transformed into vinylglycine by oxidative elimination (92, 217). The radical has also been trapped by halogens, with 4-bromo-2-aminobutanoic acid (γ-bromo-Abu) obtained by photolytic reaction of the substrate in the presence of BrCCl$_3$ (86, 87, 140, 217, 219, 220) and the γ-chloro or γ-iodo derivatives by thermolysis of the substrate with CCl$_4$ or CHI$_3$, respectively (87). Improved yields of the iodo compound (68% overall) were obtained by using

photolytic decomposition of the intermediate N-hydroxy-2-thiopyridone ester rather than thermolysis (221). A 42–47% yield was provided by using 2-iodo-1,1,1-trifluoroethane as the iodide donor (222). Nucleophilic displacements of the Glu-derived γ-halo-Abu have been reported with secondary amines (223), phenols (223), thiols (223), malonate (223), phenylselenide (217), nucleobases (thymine, adenine, cytosine, and 2-amino-6-chloropurine) (140, 219), or phosphites (220). An aza carbacyclic analog of S-adenosylmethionine (SAM), in which the Met sulfur is replaced with nitrogen and the ribose ring oxygen with carbon, was synthesized via a displacement reaction of Glu-derived γ-iodo-Abu with the N-methyl aza/carba adenosine analog (224), as were aza analogs with normal adenosine moieties (222).

Jackson et al. and others have prepared a γ-anion equivalent homolog of their Ser-derived β-iodozinc

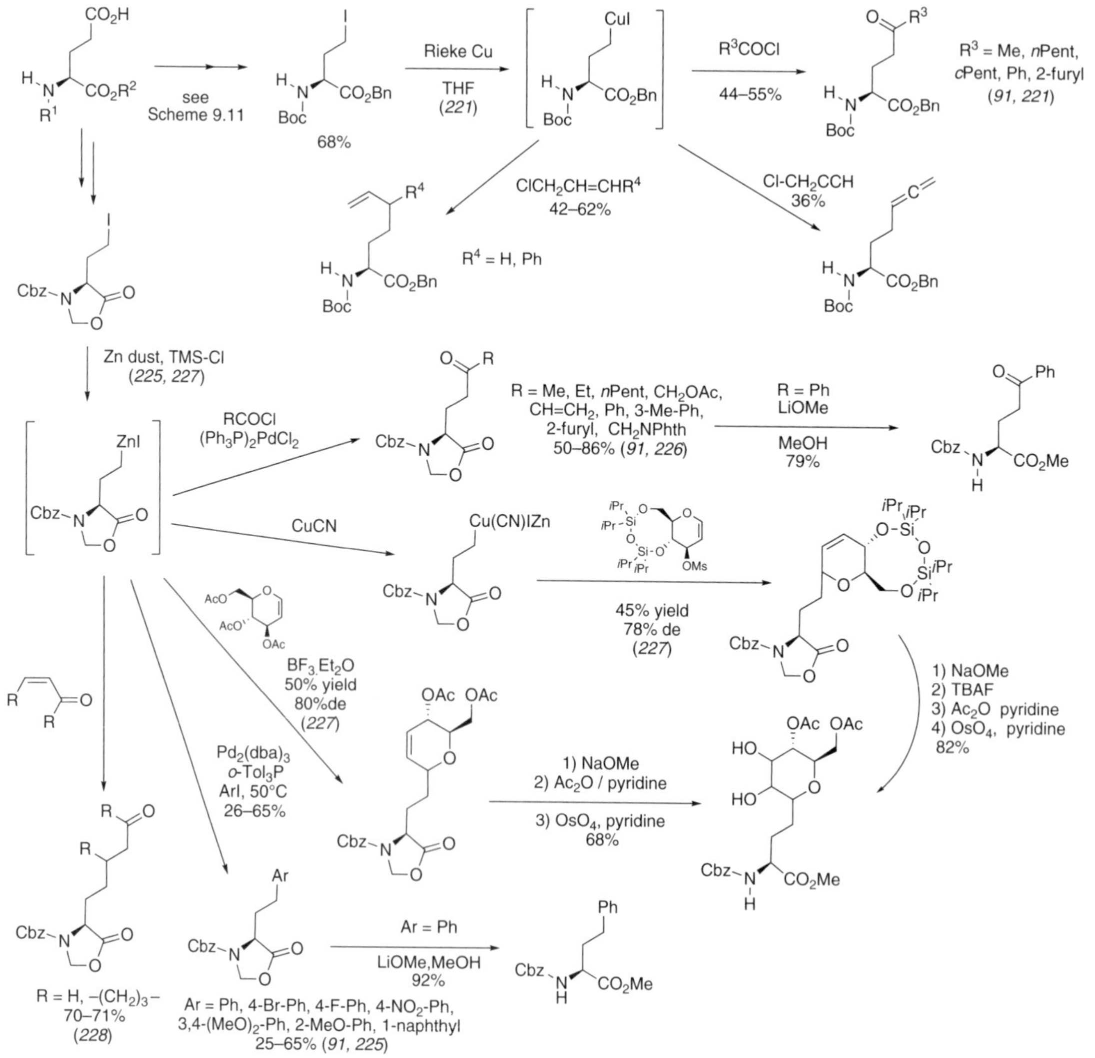

Scheme 9.23 γ-Anion Synthon Derived from Glutamate.

synthon (see Section 10.5.3), using Glu to synthesize the requisite γ-iodo intermediate. Initial attempts to prepare the iodozinc compound from 4-iodo-Abu were not completely successful, so an organocuprate reagent was prepared instead and reacted with a number of electrophiles (see Scheme 9.23) (*91, 221*). The zinc reagent was later successfully prepared and coupled with aryl iodides (*91, 225*) or acyl chlorides (*91, 226*) under palladium catalysis. The zinc/copper reagent has also been used to displace a mesylate of a glucal derivative, and the organozinc analog for a C1 selective attack on tri-*O*-acetyl-D-glucal, with both reactions providing *C*-linked glycosyl amino acid derivatives (*227*). The organozinc reagent underwent Cu(I)-catalyzed Michael additions to cyclohexenone and acrolein (*228*). The development and application of these amino acid-derived organometallics was reviewed in 2005 (*430*).

9.3.4 Modification of the α-Carboxyl Group: Acyclic Glutamic Acid

Several syntheses modify the α-carboxyl group of Glu, resulting in γ-amino acids. The α-carboxy group of *N*-Trt-protected Glu(OH)-OMe was regioselectively reduced with LiAlH$_4$, giving 4-amino-5-hydroxypentanoic acid in good yield (72–87%) (*133*). The α-carboxy group of Boc-Glu(OMe)-OH has also been reduced to a hydroxymethyl group (*229, 230*). The resulting hydroxyl was converted into a leaving group and displaced with organocuprates, producing substituted γ-amino acids (*231*).

The α-carboxyl group of Glu was removed by a reductive radical decarboxylation, using photolysis of *N*-hydroxypyridine-2-thione esters in the presence of *tert*-butyl thiol (*86, 87*) (see Scheme 9.11), leading to γ-amino acids. More recently, phenylseleno esters have been found to be suitable for decarboxylations of amino acids (*141*).

9.3.5 Modification of the γ-Carboxyl Group: Pyroglutamic Acid

9.3.5a Pyroglutamate Synthesis

One method of regioselectively differentiating between the two carboxyl groups of Glu is to use a cyclic

N-protected pyroglutamate ester, as the γ-lactam carbonyl becomes chemically distinct from the α-ester or α-acid carbonyl. Pyroglutamates are readily prepared from Glu by treatment with thionyl chloride in ethanol followed by neutralization, evaporation, and thermal cyclization (see Scheme 9.24, Table 9.7) (*118, 232*). Racemization during cyclization is avoided by isolating the diester intermediate with an aqueous extractive work-up before the cyclization step (*233*). α-Amino-adipic acid has also been cyclized under these conditions (*233*). Direct thermal cyclization of unprotected Glu at 200 °C also provides unprotected pyroglutamic acid (*234*), but this reaction is accompanied by racemization (*233–235*). Once prepared, L-pyroglutamic acid can be modified by protection of the acid and amide groups via esterification and urethane formation (*162, 166*) (see Scheme 9.24, Table 9.7). *N*-Protected pyroglutamate can also be obtained from *N*-protected Glu via formation of Glu anhydride, followed by a dicyclohexylamine-mediated rearrangement. The pyroglutamate is apparently formed with minimal racemization (*236–238*). A solid-phase synthesis of pyroglutamates from Fmoc-Glu(O*t*Bu)-NH-peptide resin proceeded via deprotection of the acid and amine groups, followed by reaction with DPPA/Et$_3$N (*239*).

Amino acids other than Glu can also provide pyroglutamate. Oxidation of *N*-protected Pro with ruthenium tetraoxide gives the corresponding *N*-protected pyroglutamate (see Scheme 9.24, Table 9.7) (*191, 240–243*). No racemization was observed. *N*-Alkyl-4-hydroxyprolines can be regioselectively oxidized with RuCl$_3$/NaIO$_4$ (*244*). Serine-derived dehydroalanine derivatives have been converted into racemic 4-substituted pyroglutamates in up to 84% yield by a radical cyclization process (*245, 246*); optically active products were obtained by using chiral ester auxiliaries (*245*) (see Scheme 4.38). Treatment of Tfa-Asp-OMe with diazomethane, followed by photolysis in methanol in the presence of sodium bicarbonate, produced pyroglutamate with retention of configuration at all centers, as determined by ^{2}H-labeling studies (*247*). α-Substituted pyroglutamates have been prepared via a cycloaddition reaction of Sm(II)-azomethine ylides with α,β-unsaturated esters (*248*).

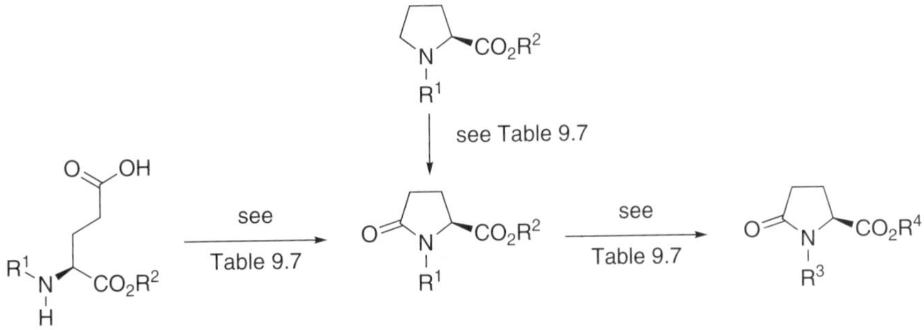

Scheme 9.24 Synthesis of Pyroglutamate from Glu or Pro.

Table 9.7 Synthesis of Pyroglutamate (see Scheme 9.24).

R^1	R^2	Cyclization of Glu			Yield	Reference
Cbz Ts CHO	H	1) DCC, THF 2) DCHA, THF			55–85%	1961 (236)
Boc 4-NO$_2$-Cbz	H	1) DCC, THF 2) DCHA, THF			62–74%	1964 (237)
H	H	1) SOCl$_2$, EtOH 2) Δ			72–84%	1980 (118) 1989 (233)
R^1	R^2	Oxidation of Pro			Yield	Reference
Boc Boc Boc Troc 4-NO$_2$-Cbz	Me tBu 4-NO$_2$-Bn Me Me	RuO$_4$			87–98%	1986 (240)
Boc	Et	RuO$_2$, NaIO$_4$			79%	1997 (241)
CH$_2$CH$_2$NH-Boc	Me	RuCl$_3$, NaIO$_4$			30–45%	2004 (244)
R^1	R^2	Protecting Group Conversion	R3	R4	Yield	Reference
Cbz	H	tBuOH, pyridine	Cbz	tBu	41%	1989 (164)
Cbz	H	1) K$_2$CO$_3$ 2) BnBr, DME	Cbz	Bn	51%	1989 (164)
Boc	H	tBuOH, DCC, DMAP	Boc	tBu	56%	1989 (164)
H	H	1) isobutene, H$_2$SO$_4$ 2) Boc$_2$O, DMAP	Boc	tBu	40%	1990 (166)
H	H	1) BnOH 2) Boc$_2$O, DMAP	Boc	Bn	70%	1991 (162)
H	Bn	MeO$_2$CCl, Et$_3$N, DMF	CO$_2$Me	Bn	80%	1993 (264)

9.3.5b Reactions of Pyroglutamate

As mentioned earlier, a 1999 review summarized asymmetric syntheses using pyroglutamic acid, including ring alkylation or functionalization reactions, transformations of the lactam carbonyl group, and ring-opening reactions (154). Ring opening of pyroglutamic acids with alcohols (158, 182, 249), amines, and thiols (249) results in Glu γ-esters, Glu γ-amides, and Glu γ-thioesters respectively (see Scheme 9.25, Table 9.8). Pyroaminoadipate esters are also substrates for openings with alcohols, which proceed under neutral (with KCN catalysis) or basic conditions to give asymmetrical 2-aminoadipate diesters (249, 250). No racemization was detected for any of the reactions.

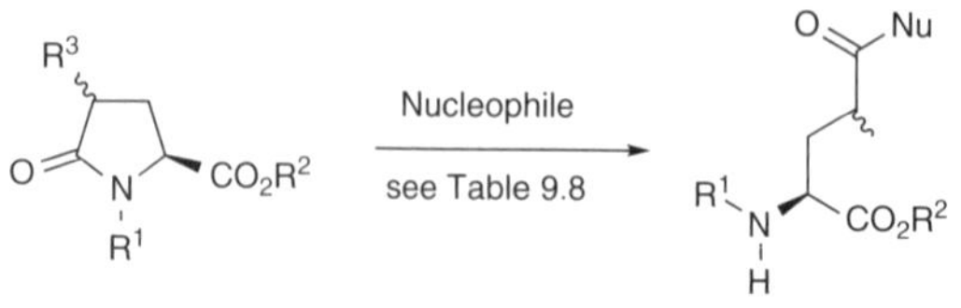

Scheme 9.25 Regioselective Opening of Pyroglutamate.

N-Boc or N-Cbz pyroglutamate were regioselectively ring-opened with Grignard reagents (167, 251, 252), lithium enolates, and other carbon nucleophiles (253–256, 428) to give δ-oxo-α-amino acids (see Scheme 9.25, Table 9.8). Reaction with lithium trimethylsilyldiazomethane produced 5-oxo-6-diazo-Nle (257). Once ring-opened, the resulting δ-ketones have been cyclized via intramolecular reductive alkylation of the α-amino group, producing Pro derivatives (253). When a 5-phenyl substituent was present, hydrogenolysis reopened the Pro pyrrolidine ring, resulting in net removal of the δ-ketone (see Scheme 9.26) (253). Treatment of pyroglutamate amides (258) or Cbz-pyroGlu-OtBu (198) with LiBH$_4$ or NaBH$_4$, respectively, reductively cleaved the CO–N bond, forming 5-hydroxy-Nva.

The α-carboxyl group of pyroglutamate can be regioselectively reduced (118, 155, 233), as can the lactam carbonyl (e.g., 165, 170, 171, 177, 181, 189, 241, 247, 259–262), leading to γ-amino acids or Pro derivatives, respectively (see Scheme 9.27, Table 9.9). For example, a chemoselective route for reduction of the lactam carbonyl was reported in 1994, using a two-step procedure in which the amide carbonyl was first reduced to a hemiaminal using lithium triethylborohydride, followed by

Table 9.8 Regioselective Ring Opening of Pyroglutamate (see Scheme 9.25)

R^1	R^2	R^3	Nucleophile	Conditions	Yield	Reference
Cbz	Bn	$CO_2CH_2CCl_3$	OBn	BnOH, Et_3N	62%	1979 (*182*)
H	H	(cyclohexenyl)–OH	OH	NaOH, EtOH	—	1981 (*183*)
Boc	Bn	CO_2Bn	OBn	BnOH, Et_3N	69%	1986 (*158*)
Boc	Bn	H	Me	RMgBr	52–92%	1987 (*251*)
Cbz	tBu		$CH{=}CH_2$			
			$CH_2CH{=}CH_2$			
			CCnBu			
Boc	Et	OBn	$CH{=}CH_2$	RMgBr	80%	1988 (*167*)
Boc	Bn	H	CH_2CO_2P	Li enolate	71–97%	1988 (*255*)
Cbz	Me		$CH(Me)CO_2P$			
			$CH(Et)CO_2P$			
			$CH(nPr)CO_2P$			
			$CH(Bn)CO_2P$			
Boc	tBu	CO_2Bn	OBn	NaOBn, BnOH	64%	1990 (*166*)
Boc	Me	H	Ph	$Li_2Cu(CN)Ph_2$	42%	1990 (*254*)
Boc	tBu	Me	OH	1N LiOH, THF	94%	1992 (*184*)
Boc	Et	H	$CH(-SCH_2CH_2CH_2S-)$	LDA, THF	60–75%	1992 (*428*)
			$CH_2PO_3Et_2$	$CH_2(-SCH_2CH_2CH_2S-)$		
			$CH{=}C(OH)Ph$	or $MePO_3Et_2$		
			$CH{=}C(OH)$ (3-MeO-Ph)	or MeCOPh		
			CH_2CO_2Et	or MeCO(3-MeO-Ph)		
				or $MeCO_2Et$		
Boc	Me	H	Et	RMgX	74–97%	1993 (*252*)
			Ph			
			4-Cl-Ph			
			4-F-Ph			
			4-Me-Ph			
Boc	Et	H	OBn	KCN, THF, ultrasound	32–97%	1993 (*249*)
			O(4-MeO-Bn)	4-MeO-BnOH or HOCH$_2$CCH		
			OCH_2CCH	or		
			$OCH_2CH{=}CH_2$	NaH, THF		
			NHBn	BnOH or HOCH$_2$CH=CH$_2$ or		
			$NHCH_2CH{=}CH_2$	BnNH$_2$ or H$_2$NCH$_2$CH=CH$_2$		
			SBn	or BnSH		
Boc	Et	H	$CH_2S(O)Tol$	$LiCH_2S(O)Tol$	70%	1993 (*256*)
Boc	Et	$={}CH_2$	OH	1) aq. LiOH, THF	80%	1994 (*169*)
				2) HCl, EtOAc		
Boc	Et	alkyl	OH	1N HCl, reflux	68–98%	1995 (*177*)
		aryl		or		
				1) 2.5N LiOH		
				2) HCl, EtOAc		
Boc	Bn	H	Ph	PhLi	65–77%	1998 (*253*)
		Bn				
Boc	Bn	H	CH_2N_2	$LiCH_2N_2$	61–75%	1998 (*257*)
Alloc	tBu					
Fmoc						

further reduction with triethylsilane/boron trifluoride etherate. No loss of chirality was observed. N-Boc-protected L-Pro, 4-*cis*-benzyl-L-Pro, 4-*trans*-allyl-L-Pro, 4-*cis*-carboxymethyl-L-Pro and 4-*trans*-cyanomethyl-L-Pro were synthesized from the corresponding pyroglutamates in 70–85% yield (*263*). The lactam carbonyl can also be converted to a carboxyl group by reduction with DIBAL-H, displacement with TMS-CN/SnCl₄, and

hydrolysis (*264*). Other methods of lactam reduction involve introduction of a substituent at the 5-position. For example, reduction of Boc-pyroGlu-OBn with Superhydride produced the hemiaminal intermediate, which was coupled with *tert*-butyl diethylphosphono-acetate to introduce a 5-carboxymethyl substituent with 7:1 *trans* stereochemistry (*265*).It is possible to introduce substituents to the 5-position of pyroglutamate via

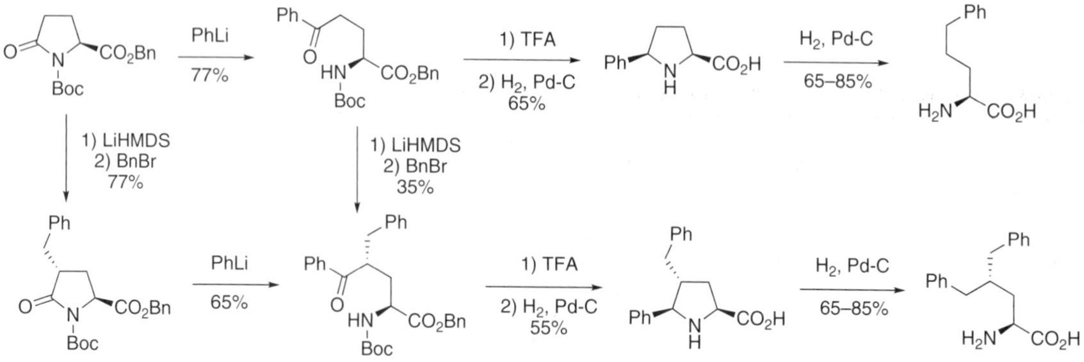

Scheme 9.26 Derivatization of Pyroglutamate (253).

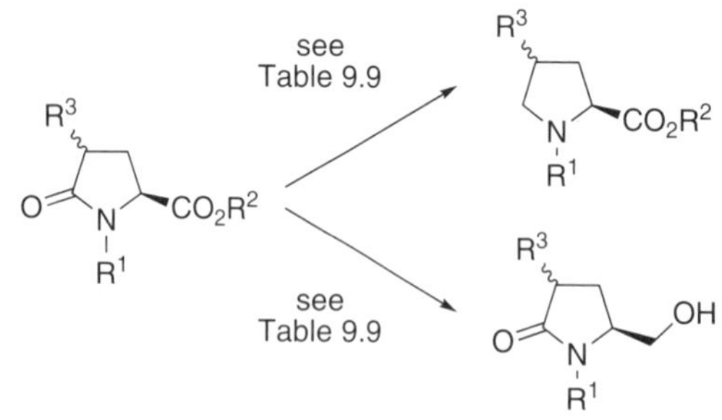

Scheme 9.27 Regioselective Reductions of Pyroglutamate.

organoaluminum addition to the lactam carbonyl. Removal of the hydroxyl group by hydrogenation resulted in a 5-*cis* substituent on a Pro nucleus (192).

Unsaturation can be introduced at the C3,C4 position within α-carboxyl-reduced pyroglutamates via deprotonation and selenation at the C4 position, followed by oxidative elimination (155, 259, 266–270) (see Scheme 9.28). α-Carboxyl group reduction is necessary to prevent C2 configurational instability. Deuteration of *N*-Boc 3,4-dehydropyroglutaminol over 10% Pd on carbon, followed by two-step lactam reduction with LiEt₃BH and then (TMS)₃SiD, provided (2S,3S,4R,5R)-[3,4,5-D₃]-Pro after oxidation of the hydroxymethyl group. Alternatively, if the lactam reduction was carried after the hydroxymethyl oxidation, using LiEt₃BH followed by TsOH/MeOH and then Bu₃SnD/BF₃·OEt₂, the C5-epimeric (2S,3S,4R,5S)-[3,4,5-D₃]-Pro product was obtained (271). A synthesis of (2S,3S,4S)-[5-¹³C;3,4,5′5′-D₅]-Leu began with di-deprotonation of *O*-TBDMS Boc-pyroglutaminol with NaHMDS and alkylation at the 4-position with ¹³CH₃I and PhSeCl. Oxidative elimination produced 3,4-dehydro-4-[¹³CH₃] pyroglutaminol, which was stereoselectively deuterated using D₂/PtO₂ (see Scheme 9.28). Lactam hydrolysis

was followed by conversion of the carboxyl group to the trideuteromethyl group (272).

The 3,4-dehydropyroglutaminol derivative is a substrate for Michael additions, giving products equivalent to a nucleophilic β-alkylation of Glu (see Scheme 9.29, Table 9.10). Grignard cuprates, Gilman cuprates (260, 266, 269), lithium benzyl phenyl sulfide (273, 274) or dialkyl sodium malonate (259, 267) added with complete diastereoselectivity to give the *trans* 3-substituted derivative (273). The *cis* isomer was prepared from the *trans* derivative by a second phenylselenylation and oxidative elimination, with reduction of the alkene from the less hindered side by catalytic hydrogenation (260). After a substituent has been added at the 3-position via Michael addition, it is also possible to introduce another substituent at the 4-position by deprotonation and alkylation (275). Conjugate addition of the Reformatsky reagent derived from *tert*-butyl bromoacetate to 3,4-dehydro-4-carboxypyroglutaminol introduced a 3-*trans*-acetic acid substituent. A 4-aryl group was then appended using ArPb(OAc)₃ in refluxing pyridine (276).

The alkene bond of 3,4-dehydropyroglutaminol can also be cyclopropanated (229), dihydroxylated (277, 278), epoxidized (279) or aziridinated (279). The epoxide can be nucleophilically opened to provide 3,4-disubstituted products (279). In all of the above reactions, the hydroxymethyl group must be oxidized to regenerate the α-carboxyl group.

A variation of the selenation–elimination reaction described above uses 4-alkylprolines derived from C4 alkylated pyroglutamates via removal of the lactam carbonyl group. Enolate generation, C2 selenation, and oxidative elimination gave 2,3-didehydroprolinates, with a chiral center maintained at the 4-position. Michael addition reactions at the 3-position gave homochiral 3,4-disubstituted prolines (280, 281).

Reduction of the α-carboxyl group is necessary during synthesis and reaction of 3,4-dehydropyrogluta-mates due to the configurational instability of the C2 center. However, the 2,6,7-trioxabicyclo[2.2.2]octane (OBO) ester-protecting group used to stabilize side-chain

Table 9.9 Regioselective Reductions of Pyroglutamate (see Scheme 9.27)

R^1	R^2	R^3	Lactam Carbonyl Reduction Conditions	Yield	Reference
H	tBu	Bn	1) Lawesson's reagent 2) NaBH$_4$, NiCl$_2$	34%	1992 (170)
CO$_2$Me	Bn	CH$_2$CO$_2$$t$Bu	BH$_3$·Me$_2$S or 1) DIBAL-H 2) TsOH, MeOH 3) NaBH$_3$CN, AcOH	61%	1993 (171)
Boc	Me	H	BH$_3$·Me$_2$S	—	1993 (247)
Boc	Et	alkyl aryl	1) LiEt$_3$BH, THF, −78°C 2) Et$_3$SiH, BF$_3$·Et$_2$O, −78°C 3) HCl, reflux	54–81%	1994 (165) 1995 (177) 1996 (280)
Boc	Bn	CH$_2$NMe$_2$	1) DIBAL-H 2) NaBH$_3$CN	—	1994 (181)
Boc	tBu	Me Et nPr	BH$_3$·Me$_2$S	53–66%	1994 (189)
Boc	Me Et	H Bn CH$_2$CH=CH$_2$ CH$_2$CO$_2$Et CN	1) LiEt$_3$BH, THF 2) Et$_3$SiH, BF$_3$·Et$_2$O	70–85%	1994 (263)
Boc	Et	H	1) LiEt$_3$BH 2) TsOH, MeOH 3) Et$_3$SiD, BF$_3$·Et$_2$O	81%	1997 (241)
Boc	Me	3,4-methano	BH$_3$·THF	31–64%	2004 (262)
R^1	R^2	R^3	α-Carboxyl Reduction Conditions (to CH$_2$OH)	Yield	Reference
H	Et	H	LiBH$_4$	88%	1980 (118) 1989 (233)
H	H	H	1) SOCl$_2$, MeOH 2) NaBH$_4$	98%	1986 (157) 1991 (155)

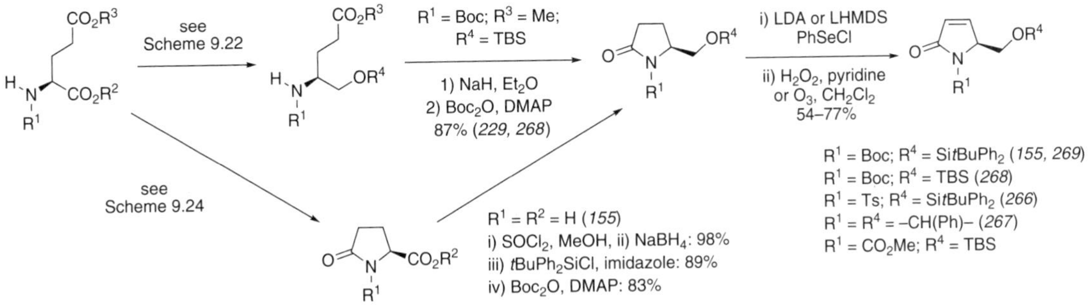

Scheme 9.28 Synthesis of Reduced, Unsaturated Pyroglutamate.

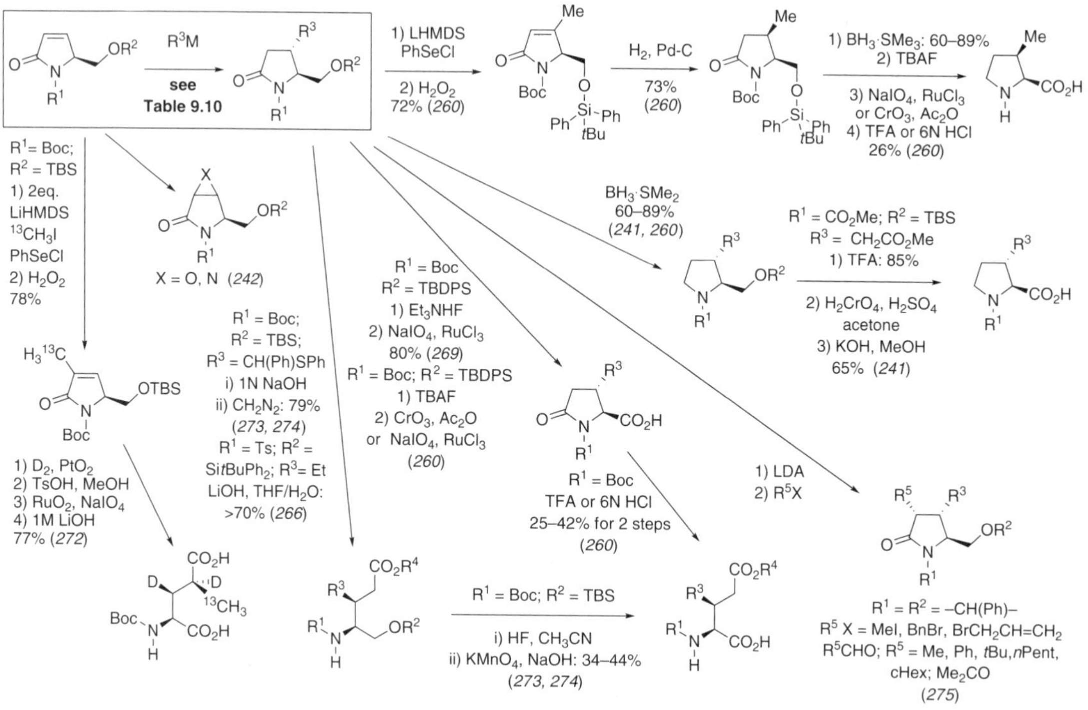

Scheme 9.29 Reactions of Reduced, Unsaturated Pyroglutamate.

Table 9.10 Alkylations of 3,4-Dehydropyroglutaminol (see Scheme 9.29)

R^1	R^2	R^3M	Yield	Reference
Boc	TBS	PhCH$_2$SPh, nBuLi	73%	1989 (274)
				1996 (273)
−CH(Ph)-R^2	−CH(Ph)-R^1	Me$_2$CuLi	84–90%	1990 (275)
		nBu$_2$CuLi		
CO$_2$Me	TBS	NaCH(CO$_2$Me)$_2$	85–91%	1991 (259)
		NaCH(CO$_2t$Bu)$_2$		
		LiCH(SPh)CO$_2t$Bu		
Ts	TBDPS	Et$_2$CuLi	75%	1991 (266)
−CH(Ph)-R^2	−CH(Ph)-R^1	NaCH(CO$_2$Me)$_2$	—	1991 (267)
Boc	TBDPS	H$_2$C=CHCuMgBr	60–85%	1992 (269)
		H$_2$C=CHCH$_2$CuMgBr		1994 (260)
		(4-Cl-Ph)CuMgBr		
		Me$_2$CuLi/TMS-Cl		
		nBuCuLi/TMS-Cl		
		PhCuLi/TMS-Cl		

serine aldehydes (see Section 10.7) has also been applied to pyroglutamate (see Scheme 9.30). Cbz-pyroGlu was converted into the OBO ester and then deprotonated, phenylselenated, and oxidatively eliminated to generate the 3,4-dehydro-pyroGlu intermediate. Michael addition of Grignard cuprates gave only the *trans* isomers, presumably due to the steric bulk of the OBO ester group. These could be epimerized to the *cis* isomers via a second selenation/oxidation/elimination sequence, followed by hydrogenation of the alkene. Deprotection with 6 N HCl or a TFA/NaOH/HCl sequence produced 3-substituted Glu products (282). A closely related synthesis employed an ABO ester, a 2,7,8-trioxabi-cyclo[3.2.1]octyl ring system, which was used during formation and cyclopropanation of 3,4-dehydropyroglu-tamate (262, 426). Michael addition of Me$_2$CuLi to the

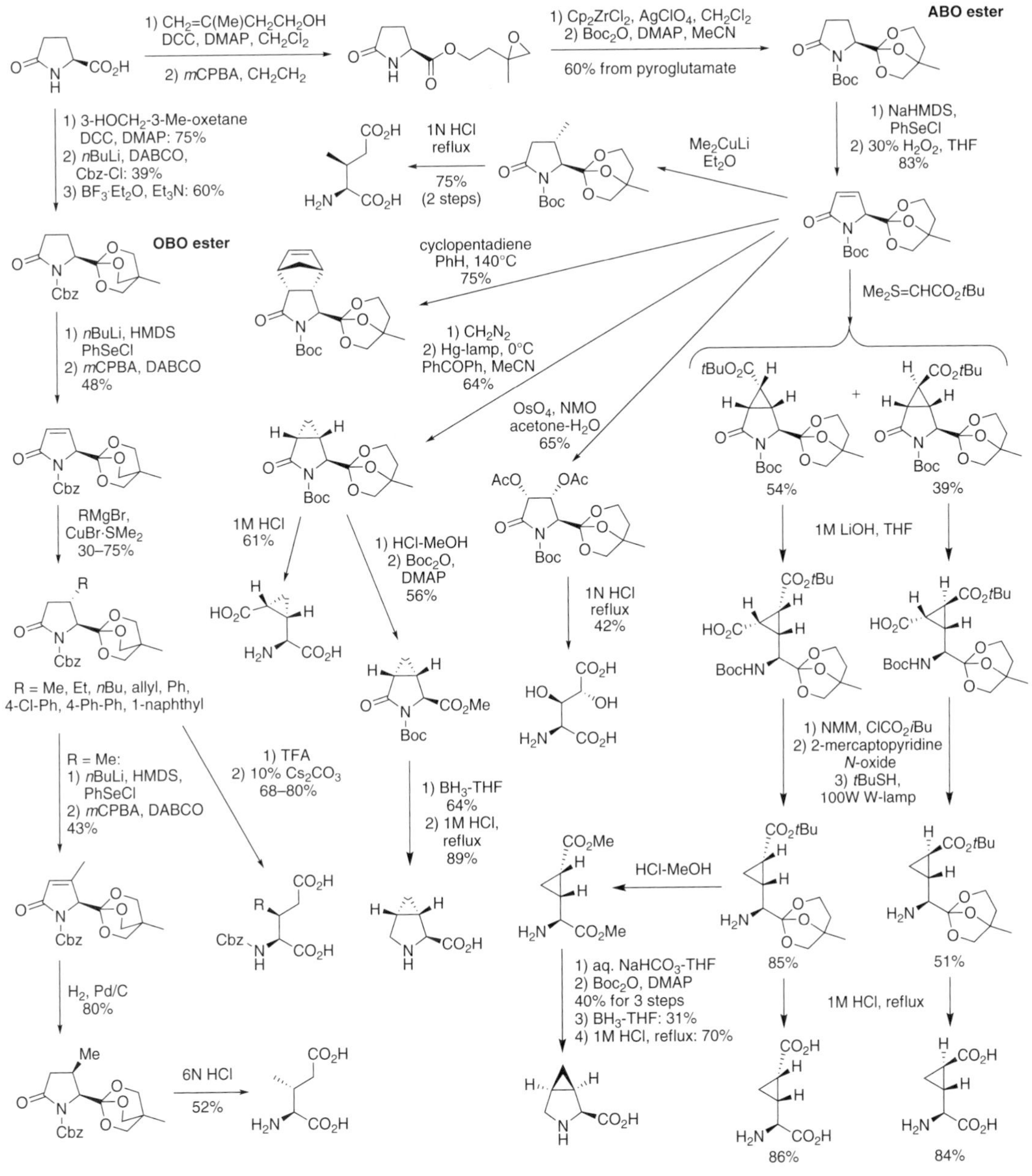

Scheme 9.30 Use of OBO or ABO Ester for Pyroglutamate Protection (*262, 282, 426*).

unsaturated lactam system led to (2*S*,3*S*)-3-methyl-Glu after hydrolysis, with >99% ee, while catalytic osmylation gave a single diastereomer of the *cis*-diol in 65% yield, and Diels–Alder reaction with cyclopentadiene gave a single isomer of the *endo* adduct in 75% yield (*426*). The ABO ester is reportedly easier to generate and gave a much better yield of the 3,4-dehydro-pyroglutamate intermediate than the OBO ester (see Scheme 9.30), though the presence of an additional chiral center in the protecting group may complicate

NMR assignments. Both *cis*- and *trans*-substituted 3,4-methano-Pro derivatives were also generated.

Other modifications of pyroglutamates have been described. Pyroglutamate was dehydrated to a pyrrole by treatment with thionyl chloride or oxalyl chloride (*283*), or to a 4,5-dichloro-pyrrole with phosphorus pentachloride (*284*). γ-Amino acids were prepared via reduction of the α-carboxyl group, oxidation to an aldehyde, conversion to an acetylene, and deprotection (see Scheme 9.31) (*285*).

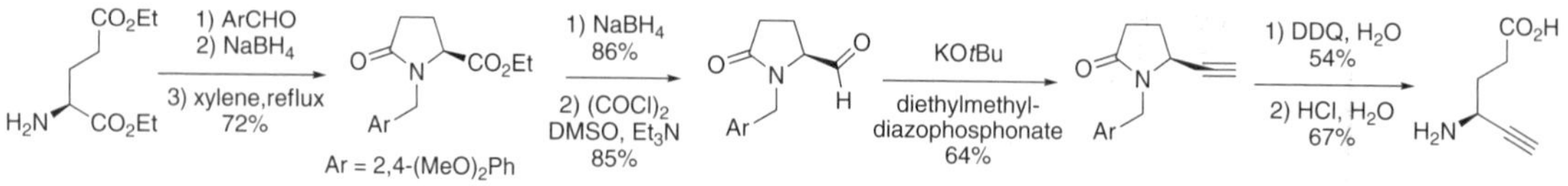

Scheme 9.31 Regioselective Reduction of Glutamate α-Carboxyl Group (285).

9.4 Synthesis of Optically Active Amino Acids from Asparagine/Glutamine

Both asparagine and glutamine have been applied to the synthesis of more complex amino acids, with the advantage over Glu and Asp that the ω-carboxyl group is already differentiated from the α-carboxyl group. Dehydration of the side-chain amide of Gln produces γ-cyano-Abu; this reaction has been carried out on Cbz-Gln-OMe with cyanuric chloride (286), on Boc-Gln-OH using Ac$_2$O/pyridine (287) or pyridine/DCC (288), on Cbz-Gln with Ac$_2$O/pyridine (289), on Gln with phosgene (290) or on Cbz-L-Gln-OMe with benzenesulfonyl chloride in pyridine (291). Reaction of the dehydrated ω-amide group with tributylstannane azide produced a tetrazole substituent (see Scheme 9.32) (82). Alternatively, the nitrile can be used to generate a C3 anion, with intramolecular C3/N bis-alkylation producing pipecolic acid derivatives (292) (see Scheme 9.5 for a similar cyclization).

Hofmann rearrangements of Asn and Gln lead directly to 2,3-diaminopropionic acid and 2,4-diaminobutyric acid. Careful protecting group selection is required to avoid imidazolidinone formation during the rearrangement of Asn, although the imidazolidinone can be hydrolyzed (293). Boc-L-Asn was rearranged with NaOBr in 12% yield with no racemization (294, 295), while Ts-Asn and Ts-Gln rearranged in 56–70% yield (296), even on kg scale reactions (297). Improved yields (60–84%) were obtained from the rearrangement of Fmoc-, Cbz-, and Boc-L-Asn with bis(trifluoroacetoxy)-iodobenzene (298–304); iodosobenzene diacetate has been found to be an even better reagent (305–308).

A Hofmann-type degradation and Hofmann elimination (309) have been used to convert Asn to dehydroalanine within peptides (see Section 4.2.1d). Asn can also be converted into α-substituted-β-amino acids via a

regioselective oxidative decarboxylation of the α-carboxyl group, alkylation adjacent to the amide, and hydrolysis (see Scheme 11.89) (310).

9.5 Syntheses from Other Amino Acids

9.5.1 Other Amino Acid Substrates

Most of the remaining common (proteinogenic) amino acids have been used at one time or another to prepare other amino acids. However, they are normally employed to synthesize structurally related compounds, such as substituted prolines from Pro or sulfur-containing amino acids from Cys. Nonetheless, there are a number of non-proteinogenic amino acids that have been elaborated to provide a range of other amino acids.

Unsaturated amino acids have been used for a variety of syntheses. The use of allylglycine, vinylglycine, and other unsaturated amino acids as substrates for radical additions was reviewed in 1997 (311). Halolactonization of allylglycine with NBS in THF introduced a γ-hydroxy group (312–314). Further elaboration gave a number of highly functionalized intermediates, and eventually led to 4-hydroxyPro and a protected version of 4,5-dihydroxy-Orn (see Scheme 9.33). Dihydroxylation of allylglycine produced 4,5-dihydroxy-Nva (315), hydroboration gave 5-hydroxy-Nva (316), ene reaction with ethylglyoxylate provided 2-amino-3,4-dehydro-6-hydroxypimelic acid (317), and ozonolysis led to the useful synthon Asp acid β-semialdehyde (see Section 9.2.3) (97). Conversion of the allylglycine alkene to a cyclopropene group has also been reported (318, 319), while cyclopropanations were achieved by reaction with electrophilic carbenes (320). Homomethionine analogs were prepared via radical additions to allylglycines (321).

n = 1,2

Scheme 9.32 Dehydration of Asparagine/Glutamine (82).

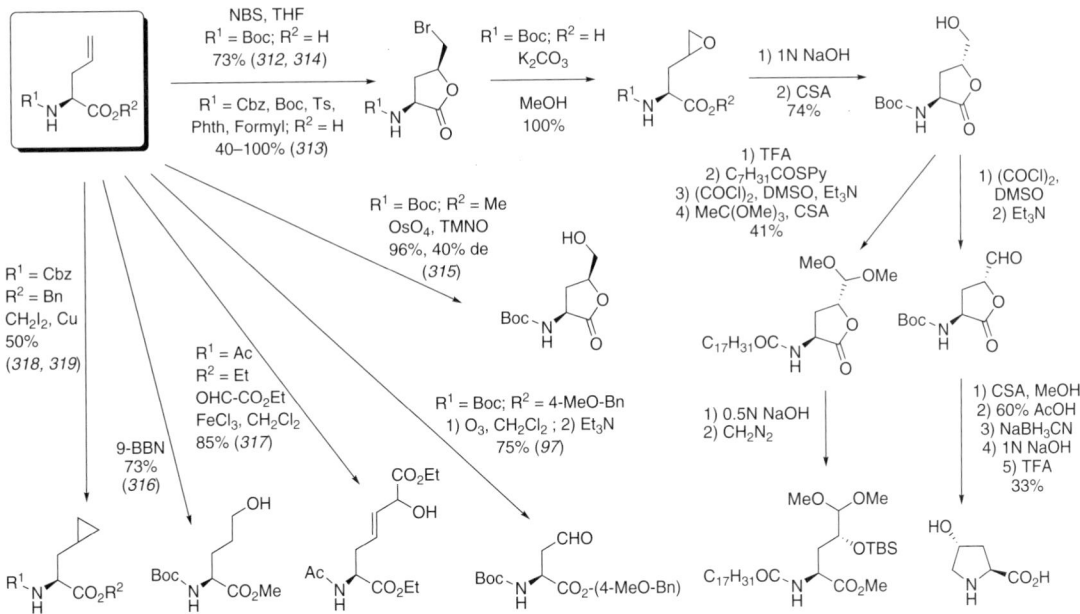

Scheme 9.33 Elaboration of Allylglycine.

Allylglycine and homoallylglycine have both been employed for Ru-catalyzed cross-metathesis reactions with alkenes. Under the same conditions, vinylglycine and dehydroalanine failed to react (*322*). Another study provided similar results: allylglycine gave good yields for a symmetrical metathesis with itself, or with allyl or vinyl-substituted *C*-glycosides. Dehydroalanine did not react, while vinylglycine gave <5% of the self-metathesis product and only reacted with the allyl-substituted *C*-glycoside (*323*). A tripeptide of Boc-Gly-*N*-Bn-α-methyl-homoallylglycine-allylglycine-OEt underwent intramolecular metathesis to give a cyclic dipeptide containing 2,8-diamino-2-methylnon-5-endioic acid (*324*).

Intramolecular metathesis of unsaturated amino acids with alkenes attached to the α-amino group produces cyclic imino acids, such as 4,5-dehydro-pipecolic acid from *N*-allyl allylglycine (*325–329*).

Hydroboration of protected allylglycines with 9-BBN produces an organoborane that is suitable for Suzuki cross-coupling with a variety of aryl and alkenyl halides (53–64% yield) (see Scheme 9.34). The adduct of 2-bromopyridine *N*-oxide was elaborated into the natural product pyrimine. Attempts to extend this hydroboration/Suzuki coupling methodology to vinylglycine were less successful, with only 32% yield for coupling with iodobenzene to give homophenylalanine (*330, 331*).

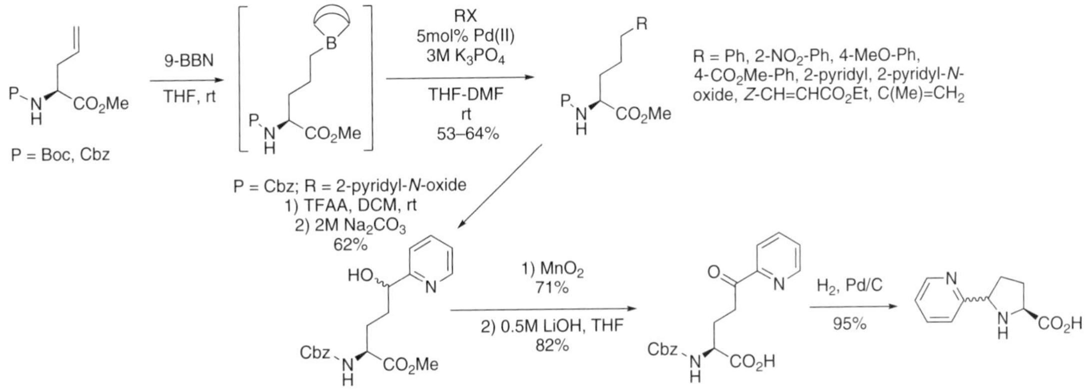

Scheme 9.34 Hydroboration and Suzuki Coupling of Allylglycine (*330, 331*).

However, another group successfully hydroborated and coupled N-Cbz,O-TBS vinylglycinol with aryl and heteroaryl bromides to produce six different γ-aryl-Abu precursors with 83–97% yields; PDC oxidation to generate the carboxyl group proceeded in 56–86% yield. Five homologous δ-aryl-Nva products were obtained by hydroboration/coupling of an amide derivative of Cbz-allylglycine (332). Homoallylglycine was hydroborated and oxidized to give 2-amino-6-hydroxyhexanoic acid, or used for Suzuki couplings to give 2-amino-6-arylhexanoic acids (431).

Boc-L-(2-bromoallyl)glycine is a fairly versatile intermediate that has been elaborated into a number of other amino acids, including β-thiazolyl-Ala and 4-oxo Nva (see Scheme 9.35) (333). Suzuki or Stille couplings gave other unsaturated amino acids.

Protected L-propargylglycine was carbonylated to give 4-methylene-L-Glu and two 4,4-disubstituted derivatives (see Scheme 9.14) (124). Propargylglycine has also been used in a synthesis of a boron analog of Phe, L-o-carboranylalanine, via reaction with decaborane (50–60% yield) (334). Racemic Boc-propargylglycine-OMe was brominated with NBS/AgNO₃ to give 5-bromopropargylglycine in 87% yield. This intermediate was employed for coupling with alkyl iodides (phenethyl iodide, phenpropyl iodide, cyclohexyl iodide, heptyl iodide, or (S)-Boc-iodoalanine-OMe) in the presence of Zn/CuCN/LiCl, giving the substituted propargylglycine adducts in 30–81% yield (335). The sequence with optically active propargylglycine was not reported, but should be viable.

Vinylglycine is another potentially useful synthon which has been used in several syntheses. Epoxidation gave a 3,4-epoxyamino acid intermediate which was regioselectively opened by sodium thiophenoxide or

4-chloroaniline (see Scheme 9.36) (336). Attempts to open the epoxide with oxygen nucleophiles led to an oxazine of the dehydroamino acid instead. Ring opening of the epoxide derived from carboxyl-reduced 2-amino-3-butenol with carbon nucleophiles was more successful, and was used to prepare 3-hydroxy-L-homotyrosine (314). The alkene bond of vinylglycine (or related derivatives (337, 338) has also been used for cycloaddition reactions to prepare amino acids with cyclic heteroatom side chains (337, 339–343), or for cyclopropanation reactions via treatment with carbenes (344). Reaction with N-benzylnitrone gave an isoxazolidinylglycine derivative, which was hydrogenolyzed to 3-hydroxy-Orn (345). Dipolar cycloaddition of Cbz-vinylglycine-OBn with the non-stabilized carbonyl ylide generated from bis(chloromethyl) ether produced a 57:43 mixture of diastereomers of L-α-(3-tetrahydrofuranyl)-Gly (epimeric at the β-center) in 65% yield (346). Vinylglycine has also been used to prepare wybutine, a fluorescent amino acid nucleoside, via a palladium-catalyzed Heck reaction (347). Hydrostannation of vinylglycine gave γ-triphenylstannyl-α-aminobutyric acid (348).

Homoserine (Hse), which is a naturally occurring amino acid, has already been described in Section 9.2.3 as it is often prepared from Asp as an intermediate in the synthesis of Asp semialdehyde (see Scheme 9.11). In turn, oxidation of Hse obtained from other sources generates Asp semialdehyde, which can then be homologated to 4-oxo-L-Nva by treatment with diazomethane (115). Hse contained within a peptide was transformed into 1,4-diaminobutyric acid or canaline (the 5-oxa analog of Orn) by a Mitsunobu displacement of the γ-hydroxy group with phthalimide or N-hydroxyphthalimide, respectively (40–58% yield) (349, 350) (see Scheme 9.37). Treatment of Hse with 2-nitrophenylselenocyanate

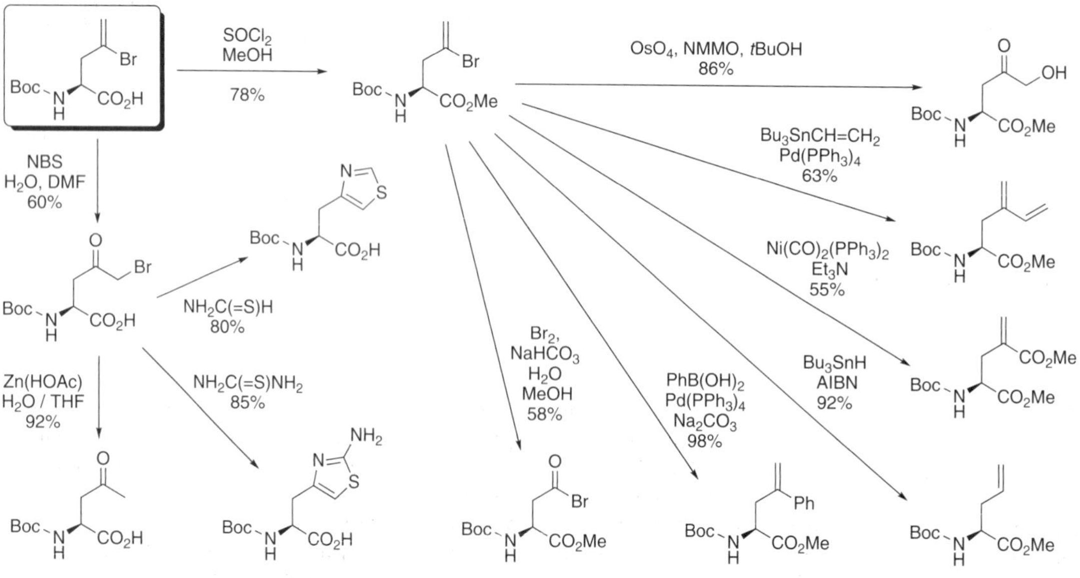

Scheme 9.35 Elaboration of 4-Bromo-Allylglycine (333).

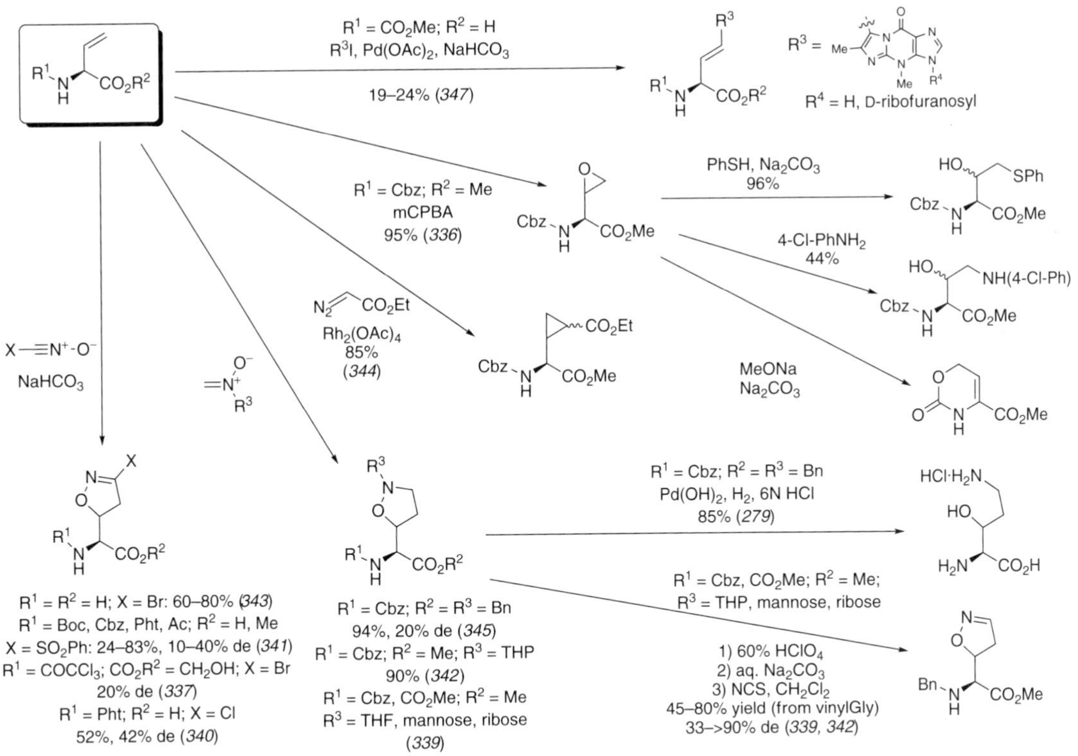

Scheme 9.36 Elaboration of Vinylglycine.

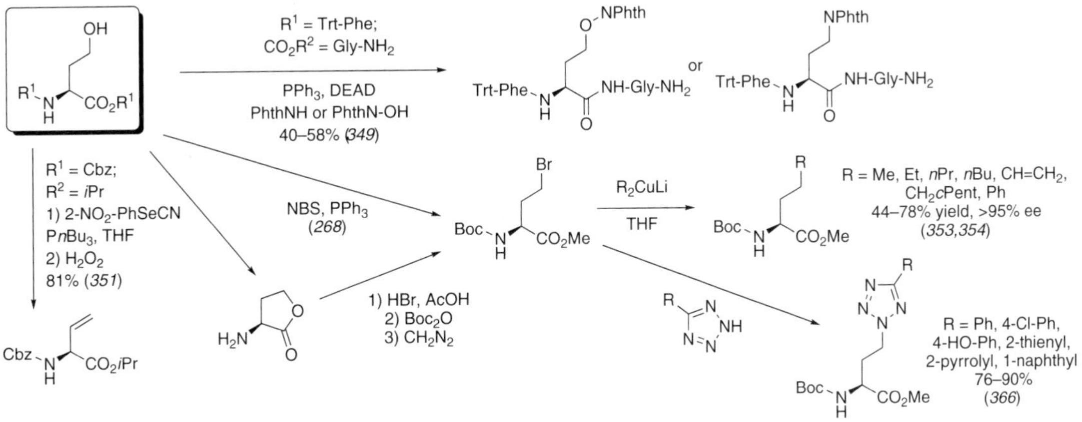

Scheme 9.37 Elaboration of Homoserine.

followed by in situ oxidative elimination with hydrogen peroxide provided vinylglycine (351); a photochemical route to vinylglycine has also been reported (352).

Homoserine-derived γ-butyrolactone was opened by HBr to give α-amino-γ-bromobutyric acid, which was then coupled with organocuprates in 44–78% yield (353–355). Met was obtained by reaction of homoserine lactone with sodium methyl mercaptide, either directly

(356) or via initial opening with bromide to give a γ-bromo-Abu hydantoin intermediate (though with a racemic substrate) (357). The γ-bromohydantoin was also used to synthesize canaline and canavanine, the 5-oxa analogs of Orn and Arg, respectively (358). Trp was obtained by opening the lactone with bromide, followed by bromide displacement with sodium ethyl acetoacetate and then formation of the indole ring (359).

In a similar manner, racemic *N*-Bz homoserine lactone was opened with HCl to give γ-chloro-Abu in 78% yield, with the chloride displaced by methyl mercaptide to produce Bz-Met in 72% yield (*360*).

Homoserine has been incorporated within a peptide, converted to γ-chloro-Abu by treatment with triphenylphosphine dichloride, and then cyclized with a Cys residue in the same peptide to form cystathionine (*361*). A diketopiperazine derivative of γ-chloro-Abu was reacted with cystine or homocysteine to provide the thioether-linked diaminodicarboxylic acids cystathionine and homolanthionine (*362*). The γ-hydroxy of Boc-Hse-OMe has also been converted into a bromide with NBS/PPh₃ (*363, 364*), while Boc-Hse-Oallyl was mesylated and reacted with LiBr to give the bromo amino acid (*365*). The bromide was displaced with *N*-methylhydroxylamine or methoxylamine (*365*), with 5-substituted tetrazoles (*366*), or with purine and pyrimidine nucleobases (*363, 364*). The corresponding α-amino-γ-iodobutyric acid has been prepared by NaI displacement of the tosylate of Trt-Hse-OBn (*350*), or by iodination of Boc-Hse-OBn with I₂/PPh₃ (*40*). Displacement of the iodide with NaONHBoc gave a canaline derivative (*350*). Homoserine can be cyclized with the α-amino group via either *O*-tosyl or γ-bromo derivatives, forming Pro (*258*). An enzymatic thiolation of homoserine with [³⁵S]H₂S by (*S*)-homoserine sulfhydrylase, followed by *S*-methylation, gave [³⁵S]-(*S*)-Met in 10–15% overall yield (*367*). *O*-Acetylhomoserine was converted into γ-[¹¹C]cyano-Abu using γ-cyano-2-aminobutyric acid synthase, with hydrolysis producing [5-¹¹C]-L-Glu (*368*).

A cyclic sulfamidate was prepared from *N*-[9-(9-phenylfluorenyl)] homoserine *tert*-butyl ester by treatment with thionyl chloride, followed by oxidation with sodium periodate (see Scheme 9.38). The synthon was opened with a variety of nucleophiles (azide, thiocyanate, imidazole, primary and secondary amines, phenols and thiophenols) in 65–95% yield. The piperidine and morpholine adducts were deprotected, with derivatization of the morpholino product establishing >97% ee optical purity (*369*).

Alkyl amino acids with halogenated side chains are useful substrates for preparing other amino acids, as illustrated by the halo amino acids prepared from Asp, Glu, and Hse in this chapter, and from Ser in Section 10.3. L-α-Amino-ω-bromoalkanoic acids with longer side chains are also potential synthons, as was demonstrated in 2004 for a series of bromo amino acids with side-chain lengths varying from 4 to 10 methylene units. The requisite bromo amino acids were obtained by alkylation of diethyl acetamidomalonate with dibromoalkanes followed by enzymatic resolution. Two examples were then converted into other amino acids, with *N*-Boc 2-amino-7-bromoheptanoic acid TMSE ester displaced with P(OMe)₃ (98% yield), KSAc (100% yield), or BnONHC(=O)H (90% yield) (*370*). Displacements with phthalimide and azide were also mentioned, but no details provided.

Cys has been converted into the thiol equivalent of the Ser-derived Garner aldehyde (see Section 10.6.1) and elaborated into β,γ-unsaturated amino acids (*371*), while cystine was used to prepare β-chloro-Ala (*372*). Met can be converted to allylglycine (*373*), vinylglycine (*352, 374–377*) or vinylglycinol (*332*) and homoserine (*110*).

The side-chain heteroatoms of Ser, Cys, and Lys have been employed for Michael additions to conjugated alkynes (alkynones, alkynoic esters, and alkynoic amides), giving heterosubstituted conjugated vinyl side chains (*378*). Various fluorescent labels and dyes were attached to amino acids by this method, with the dyes linked to propiolic acid as esters or amides (*379*).

Substituted Pro derivatives, which can be prepared from Glu (see Section 9.3) or from other Pro, have been converted into Glu derivatives via C5 oxidation of the proline to form pyroglutamate, followed by hydrolysis (see Scheme 9.39) (*158, 240, 242, 243, 380*).

The acid groups of a variety of α-amino acids have been homologated to give β-amino acids by Arndt–Eistert reaction, and γ-amino acids by two consecutive Arndt–Eistert reactions or other methods (see Chapter 11). Asn/Asp can also be applied to the synthesis of β-amino acids, and Gln/Glu to γ-amino acids (*381*).

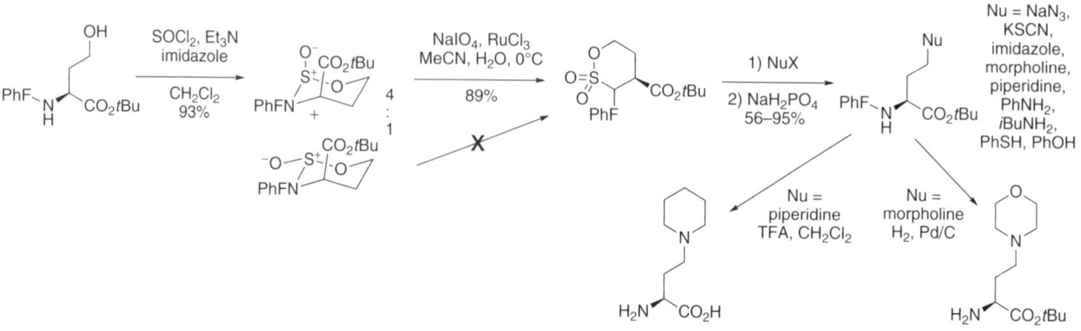

Scheme 9.38 Elaboration of Homoserine Cyclic Sulphamidite (*369*).

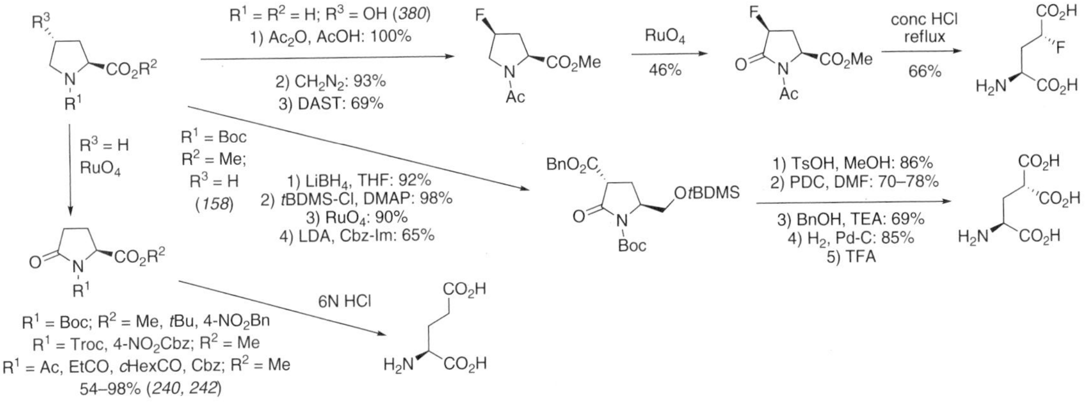

$R^1 = R^2 = H; R^3 = OH$ (380)
1) Ac$_2$O, AcOH: 100%
2) CH$_2$N$_2$: 93%
3) DAST: 69%

RuO$_4$
46%

conc HCl
reflux
66%

$R^3 = H$
RuO$_4$

$R^1 = Boc$
$R^2 = Me;$
$R^3 = H$
(158)
1) LiBH$_4$, THF: 92%
2) tBDMS-Cl, DMAP: 98%
3) RuO$_4$: 90%
4) LDA, Cbz-Im: 65%

BnO$_2$C
OtBDMS

1) TsOH, MeOH: 86%
2) PDC, DMF: 70–78%
3) BnOH, TEA: 69%
4) H$_2$, Pd-C: 85%
5) TFA

6N HCl

$R^1 = Boc; R^2 = Me, tBu, 4-NO_2Bn$
$R^1 = Troc, 4-NO_2Cbz; R^2 = Me$
$R^1 = Ac, EtCO, cHexCO, Cbz; R^2 = Me$
54–98% (240, 242)

Scheme 9.39 Elaboration of Prolines.

9.5.2 General Procedures for Elaborating Amino Acids

There are several synthetic procedures that are generally applicable to the derivatization of a number of different amino acids.

9.5.2a Radical Reactions

Free-radical halogenation and photochemical reactions have been applied to a number of amino acid substrates. A review of free-radical reactions in the synthesis of amino acids, including modifications of amino acid side chains, appeared in 1997 (311). The use of N-phthaloyl amino acid derivatives in the synthesis of side-chain-functionalized amino acid derivatives was reviewed in 1998 (382).

As mentioned previously, carboxyl groups of amino acids, including the side chains of Glu and Asp, can be removed by a radical decarboxylation of the Barton ester (see Sections 9.2.2, 9.3.3). An alternative procedure, using more readily prepared phenylseleno esters, was more recently reported for α-carboxy groups (141). The resulting 1-amidoalkyl radicals were trapped by methyl acrylate to give γ-amino acids (141). Photoreductive decarboxylation of amino acid benzophenone oxime esters has also been reported (383). Simple irradiation of N-phthaloyl amino acid methyl esters of Val, Nva, Ile, and tert-Leu gave a variety of products, with Val producing β,γ-dehydrovaline and tert-Leu forming a benzannelated pyrrolizidinone in synthetically useful yields (85% and 90%, respectively) (384). Ile, Abu, and Nva produced the β,γ-unsaturated amino acids in 15–75% yield (385). A minireview of photochemical activation of amino acids with phthalimido chromophores, summarizing the work of Griesbeck et al., appeared in 1996 (386). Amino acids can be cross-linked at the α-position by treatment with photochemically produced tert-butoxy radicals in benzene, with the dimers of methyl N-acetylglycinate or methyl pyroglutamate produced in 51% or 64% yield, respectively (387).

A number of research groups have made use of radical halogenation reactions to prepare, after hydrolysis, hydroxy amino acid derivatives. However, for most amino acid substrates, regioselectivity and/or stereoselectivity is a problem. Regioselectivity depends on both reagents and protecting groups. For example, free-radical chlorination and hydrolysis of unprotected L-Val gave a mixture of the two diastereomers of γ-hydroxyVal in 14% yield (388), while L-Glu and α-aminobutyric acid gave mixtures of β- and γ-chlorination (389). Abu, Val, Ile, Leu, Nle, and Tle have all been chlorinated and hydrolyzed to give γ-hydroxy amino acids as racemic mixtures of diastereomers, with yields of approximately 25% (390).

However, for some substrates, regioselective reaction is possible. Thus, γ-chloro-L-Lys was prepared from unprotected L-Lys under acidic conditions in 74% yield and hydrolyzed to the hydroxyl amino acid in 66% yield (389, 391), while γ-hydroxy-Orn was obtained from Orn in 45% yield (389), and β-chloro-Ala from L-Ala in 78% yield (389). L-Ile was selectively γ-chlorinated under acidic conditions and then cyclized under basic conditions to give 3-methyl-Pro (392). Photofluorination of unprotected L- or D-Ala by CF$_3$OF in HF gave the corresponding 3-fluoro-Ala without racemization in 54–59% yield (393). N-benzoyl Val-OMe and Ala-OMe were β-chlorinated with sulfuryl chloride (311, 394, 395), although the regioisomeric γ-chloro-Val was also produced (311, 391). In contrast, bromination of the N-benzoyl amino esters with NBS also caused α-bromination (395, 396). Reaction of NBS at the isopropyl methine proton of Val-OMe and Leu-OMe has been favoured by using a bulky N-phthaloyl protecting group to discourage α-bromination (311, 397–399). Further reaction of the β-bromo-Val and γ-bromo-Leu derivatives with silver nitrate in aqueous acetone gave the corresponding hydroxy amino acids without racemization (399, 400) (see Scheme 9.40). In order to selectively introduce a γ-hydroxy group to Val, a β,γ-dehydro-Val derivative was needed as substrate. This compound was prepared by dehydrohalogenation of β-bromo-Val.

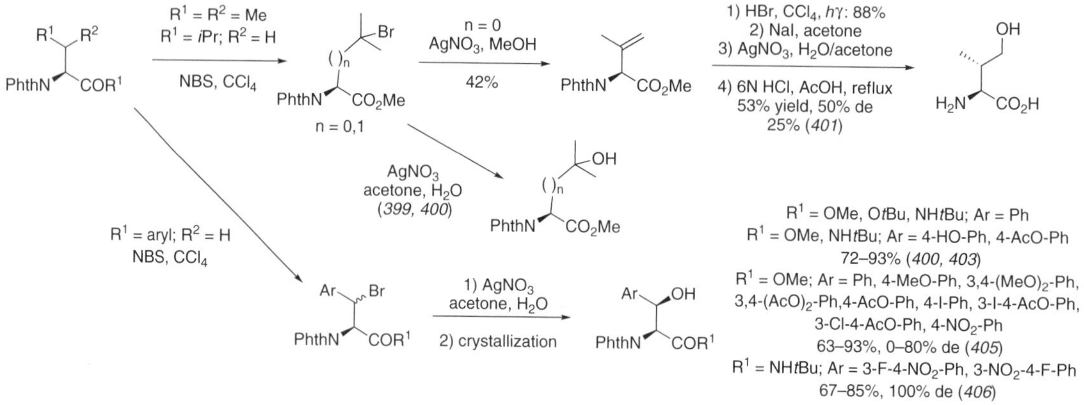

Scheme 9.40 Radical Bromination of Amino Acids and Conversion to Hydroxy Group.

Anti-Markovnikov hydrobromination followed by hydrolysis gave the desired γ-hydroxy-Val product with 50% de (401).

Phe and Tyr derivatives were regioselectively brominated at C3 by NBS with no diastereoselectivity (402). However, the resulting 1:1 diastereomer mixture was separated by fractional crystallization (400, 403). Elimination gave E- and Z-2,3-dehydrophenylalanines (404), whereas treatment of the bromides with silver nitrate in aqueous acetone afforded the corresponding β-hydroxy-α-amino acids without racemization (see Scheme 9.40) (400, 403). The hydrolysis reaction at the β-center was completely stereoselective for the (2S,3S)-bromide (S$_N$2 with inversion) but not for the (2S,3R)-bromide, giving the (2S,3R)-hydroxy diastereomer preferentially in both cases. The pure (2S,3R)-hydroxy product could therefore be isolated from the hydrolysis products of the crude bromide diastereomer mixture by crystallization. A tert-butyl ester gave improved diastereoreoselectivity for the hydroxylation reaction, while a tert-butyl amide produced the (2S,3R) hydroxy product with >99.9% de. A number of β-hydroxy-arylalanine derivatives have been prepared by this method (405), including β-hydroxy-4′-nitro-Phe (399), threo-β-hydroxy-3′-nitro-4′-fluoro-Phe (406), and threo-β-hydroxy-3′-fluoro-4′-nitro-Phe (406) (the latter two derivatives are useful intermediates for syntheses of vancomycin-type cyclic peptides). The tert-butyl amide of Phth-Phe was converted into β-methoxy-Phe via bromination, hydroxylation with silver nitrate in aqueous acetone, and then methylation with MeI (407).

The halogenated amino acids produced by radical reactions can be employed for syntheses other than conversion to hydroxy amino acids. Thus N-phthaloyl 4′-bromo-3,4-dehydrovaline methyl ester (methyl 2-phthalamido-3-bromomethylbut-3-enoate), obtained by radical bromination of 3,4-dehydrovaline in 77% yield, was deuterated by Bu$_3$SnD to give 4′-deutero-3,4-dehydrovaline in 57% yield, cyclized to 2-methylene-ACC in 20% yield by treatment with NaH, or displaced with dimethylmalonate to give trimethyl 2-phthalimido-3-methylene-5,5-pentanoate in 26% yield (408). Chlorination of the 3,4-dehydrovaline methyl ester with Cl$_2$ in CCl$_4$ produced 4′-chloro-3,4-dehydrovaline in 53% yield, but attempted bromination with Br$_2$ in CCl$_4$ instead gave a mixture of diastereomers of 3,4-dibromo-Val (408). Similarly, chlorination of methyl N-phthaloyl 2-amino-3-methylpent-3(E)-enoate or 2-amino-3-methylpent-3(Z)-enoate produced two diastereomers of 2-amino-3-methylene-4-chloropentenoate, while chlorination of 2-amino-3-ethylbut-3-enoate gave 2-amino-3-chloromethylpent-3-enoate, and chlorination or bromination of 2-amino-4-methylpent-4-enoate (4,5-dehydro-Leu) gave 2-amino-4-halomethylpent-4-enoate, in 28–52% yields (408). The bromide of β-bromo-Phe was exchanged with ^{2}H, with retention of configuration, by treatment with D$_2$/Pd-C (409). N-Phthaloyl γ-halo-Leu, γ-halo-homoPhe, and γ-halo-Nva were all cyclized to the 2,3-methano amino acids by treatment with NaH or DBU (410).

Side-chain radical reactions of amino acids have also been used to examine radical reaction mechanisms, without synthetic application (411). Ser-derived iodo-Ala is a substrate for radical reactions (e.g., 412), but this is discussed in Section 10.4.

A promising technology for converting racemic N-trifluoroacetyl α-bromo amino acid benzyl esters into amino acids with >99% ee has been developed by an Australian company, Chirogen (413). Treatment of the substrate with a chiral menthyl stannane at −78 °C in toluene in the presence of MgBr$_2$ produced either L- or D-tert-Leu.

9.5.2b Oxidation Reactions

Regioselective oxidation of a number of alkyl amino acids (Val, Nva, Leu, Ile) with catalytic K$_2$PtCl$_4$ in the presence of 7 equiv. CuCl$_2$ in water led to the γ-hydroxy amino acid lactones in 15–35%, along with some Pro derivatives resulting from γ-chlorination/cyclization.

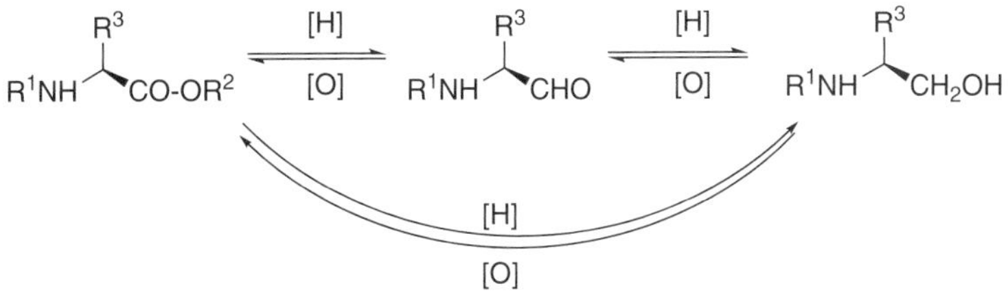

Scheme 9.41 Interconversion of α-Amino Acids, α-Amino Aldehydes, and β-Amino Alcohols.

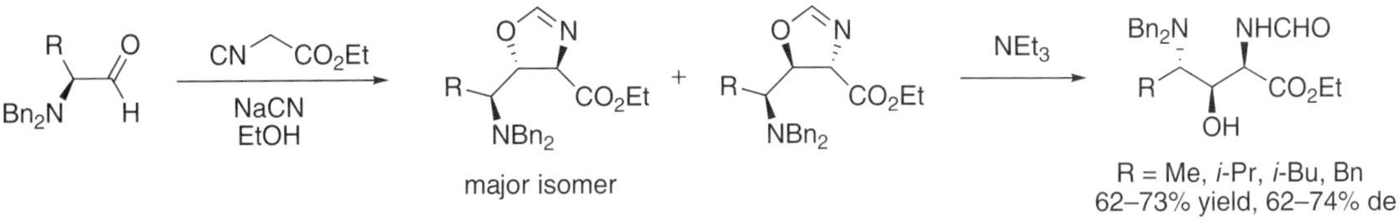

major isomer

R = Me, *i*-Pr, *i*-Bu, Bn
62–73% yield, 62–74% de

Scheme 9.42 Synthesis of β-Hydroxy-α,γ-diamino Acids from α-Amino Aldehydes (*418*).

Chelate formation was proposed to account for the regioselectivity (*414*).

3,3-Dimethyldioxirane oxidation was applied to a number of amino acids (Leu, Thr, Trp, His, and Pro) (*415*). The most useful result was obtained from Leu, which formed good yields of the lactone of γ-hydroxy-Leu; the procedure could also be applied to Leu residues within peptides. Thr formed β-keto-Abu, Pro produced 5-hydroxy-Pro as a mixture of diastereomers, His led to β-(2,3-dihydro-2-oxoimidazol-4-yl)-Ala, and Trp provided β-(2,3-dihydro-2-oxoindol-3-yl)-Ala (*415*).

9.5.2c Amino Acid Synthesis from α-Amino Aldehydes

Homochiral α-amino aldehydes are of considerable interest as substrates for the synthesis of more complex molecules. They can both be obtained from, and converted into, α-amino acids. Several reviews of this area have appeared (*416, 417*). Enantiomeric purity is a serious concern with α-amino aldehydes. Most α-amino aldehydes are obtained by reduction of an amino acid ester, or by oxidation of the corresponding N-protected β-amino alcohols (which in turn can also be obtained by reduction of amino acids) (see Scheme 9.41). A number of other routes to α-amino aldehydes are discussed in the above reviews. Any enantioselective synthesis of α-amino aldehydes is potentially a route to α-amino acids, via a simple oxidation of the aldehyde group.

N-protected α-amino aldehydes can undergo additions of organometallic reagents, participate in aldol condensations and related reactions, react in Diels–Alder cycloadditions, and form alkenes in Wittig-type additions. Once the carbonyl of an α-amino aldehyde has been functionalized, it is obviously no longer available for oxidation to regenerate an α-amino acid. However,

ω-amino acids can be synthesized if an acid group is introduced during elaboration of the aldehyde. The amino aldehydes are particularly useful for syntheses of α-hydroxy-β-amino acids, such as bestatin, or of β-hydroxy-γ-amino acids, such as statine, as discussed in Chapter 11. α-Amino acids can be obtained from amino aldehydes if another amino acid center is introduced. α,γ-Diamino-β-hydroxy amino acids were produced in 51–73% yield with 62–74% ee by addition of a Gly enolate equivalent to the amino aldehyde (see Scheme 9.42) (*418*). The *syn* diastereomer was obtained exclusively.

References

1. Pires, R.; Fehn, S.; Golubev, A.; Winkler, D.; Burger, K. *Amino Acids* **1996**, *11*, 301–312.
2. Gu, Z.-Q.; Hesson, D.P. *Tetrahedron: Asymmetry* **1995**, *6*, 2101–2104.
3. Johnstone, A.N.C.; Lopatriello, S.; North, M. *Tetrahedron Lett.* **1994**, *35*, 6335–6338.
4. Zee-Chang, R.K-Y.; Olson, R.E. *Biochem. Biophys. Res. Commun.* **1980**, *94*, 1128–1132.
5. Seebach, D.; Wasmuth, D. *Angew. Chem., Int. Ed. Engl.* **1981**, *20*, 971.
6. Baldwin, J.E.; Moloney, M.G.; North, M. *J. Chem. Soc., Perkin Trans. 1* **1989**, 833–834.
7. Baldwin, J.E.; Moloney, M.G.; North, M. *Tetrahedron* **1989**, *45*, 6319–6330.
8. Baldwin, J.E.; Moloney, M.G.; North, M. *Tetrahedron* **1989**, *45*, 6309–6318.
9. Parr, I.B.; Dribben, A.B.; Norris, S.R.; Hinds, M.G.; Richards, N.G.R. *J. Chem. Soc., Perkin Trans. 1* **1999**, 1029–1038.
10. Andrews, M.D.; O'Callaghan, K.A.; Vederas, J.C. *Tetrahedron* **1997**, *53*, 8295–8306.

11. Park, J.-i.; Tian, G.R.; Kim, D.H. *J. Org. Chem.* **2001**, *66*, 3696–3703.

12. Tamura, N.; Matsushita, Y.; Iwama, T.; Harada, S.; Kishimoto, S.; Itoh, K. *Chem. Pharm. Bull.* **1991**, *39*, 1199–1212.

13. Wolf, J.-P.; Rapoport, H. *J. Org. Chem.* **1989**, *54*, 3164–3173.

14. Feldman, P.L.; Rapoport, H. *J. Org. Chem.* **1986**, *51*, 3882–3890.

15. Kawahata, N.; Weisberg, M.; Goodman, M. *J. Org. Chem.* **1999**, *64*, 4362–4369.

16. Dener, J.M.; Zhang, L.-H.; Rapoport, H. *J. Org. Chem.* **1993**, *58*, 1159–1166.

17. Dunn, P.J.; Häner, R.; Rapoport, H. *J. Org. Chem.* **1990**, *55*, 5017–5025.

18. Humphrey, J.M.; Bridges, R.J.; Hart, J.A.; Chamberlin, A.R. *J. Org. Chem.* **1994**, *59*, 2467–2472.

19. Fernández-Megía, E.; Paz, M.M.; Sardina, F.J. *J. Org. Chem.* **1994**, *59*, 7643–7652.

20. Sardina, F.J.; Paz, M.M.; Fernández-Megía, E.; de Boer, R.F.; Alvarez, M.P. *Tetrahedron Lett.* **1992**, *33*, 4637–4640.

21. Hanessian, S.; Vanasse, B. *Can. J. Chem.* **1993**, *71*, 1401–1406.

22. Shibata, N.; Baldwin, J.E.; Jacobs, A.; Wood, M.E. *Synlett* **1996**, 519–520.

23. Shibata, N.; Baldwin, J.E.; Jacobs, A.; Wood, M.E. *Tetrahedron* **1996**, *52*, 12839–12852.

24. Cotton, R.; Johnstone, A.N.C.; North, M. *Tetrahedron*, **1995**, *51*, 8525–8544.

25. Cotton, R.; Johnstone, A.N.C.; North, M. *Tetrahedron Lett.* **1994**, *35*, 8859–8862.

26. Bridges, R.J.; Lovering, F.E.; Humphrey, J.M.; Stanley, M.S.; Blakely, T.N.; Cristofaro, M.F.; Chamberlin, A.R. *Bioorg. Med. Chem. Lett.* **1993**, *3*, 115–121.

27. Whitten, J.P.; Cube, R.V.; Baron, B.M.; McDonald, I.A. *Bioorg. Med. Chem. Lett.* **1993**, *3*, 19–22.

28. Gmeiner, P.; Feldman, P.L.; Chu-Moyer, M.Y.; Rapoport, H. *J. Org. Chem.* **1990**, *55*, 3068–3074.

29. Whitten, J.P.; Muench, D.; Cube, R.V.; Nyce, P.L.; Baron, B.M.; McDonald, I.A. *Bioorg. Med. Chem. Lett.* **1991**, *1*, 441–444.

30. Hanessian, S.; Sumi, K.; Vanasse, B. *Synlett* **1992**, 33–34.

31. Salzmann, T.N.; Ratcliffe, R.W.; Christensen, B.G.; Bouffard, F.A. *J. Am. Chem. Soc.* **1980**, *102*, 6163–6165.

32. Baldwin, J.E.; Adlington, R.M.; Gollins, D.W.; Schofield, C.J. *Tetrahedron* **1990**, *46*, 4733–4748.

33. Baldwin, J.E.; Adlington, R.M.; Gollins, D.W.; Schofield, C.J. *J. Chem. Soc., Chem. Commun.* **1990**, 720–721.

34. Humphrey, J.M.; Hart, J.A.; Chamberlin, A.R. *Bioorg. Med. Chem. Lett.* **1995**, *5*, 1315–1320.

35. Perich, J.W. *Synlett* **1992**, 595–596.

36. Ramsamy, K.; Olsen, R.K.; Emery, T. *Synthesis* **1982**, 42–43.

37. Rodriguez, M.; Llinares, M.; Doulut, S.; Heitz, A.; Martinez, J. *Tetrahedron Lett.* **1991**, *32*, 923–926.

38. Okayama, T.; Seki, S.; Ito, H.; Takeshima, T.; Hagiwara, M.; Morikawa, T. *Chem. Pharm. Bull.* **1995**, *43*, 1683–1691.

39. Barclay, F.; Chrystal, E.; Gani, D. *J. Chem. Soc., Perkin Trans. 1* **1996**, 683–689.

40. Jackson, R.F.W.; Moore, R.J.; Dexter, C.S.; Elliott, J.; Mowbray, C.E. *J. Org. Chem.* **1998**, *63*, 7875–7884.

41. Gong, B.; Lynn, D.G. *J. Org. Chem.* **1990**, *55*, 4763–4765.

42. Ramalingam, K.; Woodard, R.W. *J. Org. Chem.* **1988**, *53*, 1900–1903.

43. Valerio, R.M.; Alephosphowood, P.F.; Johns, R.B. *Synthesis* **1988**, 786–789.

44. Campiglia, P.; Gomez-Monterrey, I.; Longobardo, L.; Lama, T.; Novellino, E.; Grieco, P. *Tetrahedron Lett.* **2004**, *45*, 1453–1456.

45. Yamaguchi, J.-I.; Ueki, M. *Chem. Lett.* **1996**, 621–622.

46. Fernández-Megía, E.; Montaos, M.A.; Sardina, F.J. *J. Org. Chem.* **2000**, *65*, 6780–6783.

47. Truong, V.L.; Gauthier, J.Y.; Boyd, M.; Roy, B.; Scheigetz, J. *Synlett* **2005**, 1279–1280.

48. Salituro, F.G.; McDonald, I.A. *J. Org. Chem.* **1988**, *53*, 6138–6139.

49. Natalini, B.; Mattoli, L.; Pellicciari, Carpenedo, R.; Chiarugi, A.; Moroni, F. *Bioorg. Med. Chem. Lett.* **1995**, *5*, 1451–1454.

50. Golubev, A.S.; Sewald, N.; Burger, K. *Tetrahedron* **1996**, *52*, 14757–14776.

51. Golubev, A.; Sewald, N.; Burger, K. *Tetrahedron Lett.* **1995**, *36*, 2037–2040.

52. Whitten, J.P.; Baron, B.M.; Muench, D.; Miller, F.; White, S.; McDonald, I.A. *J. Med. Chem.* **1990**, *33*, 2961–2963.

53. Whitten, J.P.; Baron, B.M.; McDonald, I.A. *Bioorg. Med. Chem. Lett.* **1993**, *3*, 23–26.

54. Golubev, A.; Sewald, N.; Burger, K. *Tetrahedron Lett.* **1993**, *34*, 5879–5880.

55. Burger, K.; Rudolph, M.; Neuhauser, H.; Gold, M. *Synthesis* **1992**, 1150–1156.

56. Hsiao, C.-n.; Leanna, M.R.; Bhagavatula, L.; De Lara, E.; Zydowsky, T.M.; Horrom, B.W.; Morton, H.E. *Synth. Commun.* **1990**, *20*, 3507–3517.

57. Groarke, M.; McKervey, M.A.; Niewenhuyzen, M. *Tetrahedron Lett.* **2000**, *41*, 1275–1278.

58. Furuta, T.; Katayama, M.; Shibasaki, H.; Kasuya, Y. *J. Chem. Soc. Perkin Trans. 1* **1992**, 1643–1648.

59. Bloemhoff, W.; Kerling, K.E.T. *Recl. Trav. Chim. Pays-Bas* **1975**, *94*, 182–185.

60. Burger, K.; Rudolph, M.; Fehn, S. *Angew. Chem., Int. Ed. Engl.* **1993**, *32*, 285–287.

61. Fehn, S.; Burger, K. *Tetrahedron: Asymmetry* **1997**, *8*, 2001–2005.

62. Hartzoulakis, B.; Gani, D. *J. Chem. Soc., Perkin Trans. 1* **1994**, 2525–2531.

63. Adlington, R.M.; Baldwin, J.E.; Catterick, D.; Pritchard, G.J. *Chem. Commun.* **1997**, 1757–1758.

64. Adlington, R.M.; Baldwin, J.E.; Catterick, D.; Pritchard, G.J. *J. Chem. Soc., Perkin Trans. 1* **1999**, 855–866.

65. Adlington, R.M.; Baldwin, J.E.; Catterick, D.; Pritchard, G.J.; Tang, L.T. *J. Chem. Soc., Perkin Trans. 1* **2000**, 303–305.

66. Adlington, R.M.; Baldwin, J.E.; Catterick, D.; Pritchard, G.J.; Tang, L.T. *J. Chem. Soc., Perkin Trans. 1* **2000**, 2311–2316.

67. Tohdo, K.; Hamada, Y.; Shioiri, T. *Synlett* **1994**, 105–106.

68. Aubry, N.; Plante, R.; Déziel, R. *Tetrahedron Lett.* **1990**, *31*, 6311–6312.

69. Honma, T.; Tada, Y.; Adachi, I.; Igarashi, K. *J. Heterocycl. Chem.* **1989**, *26*, 629–634.

70. Tamura, N.; Iwama, T.; Itoh, K. *Chem. Pharm. Bull.* **1992**, *40*, 381–386.

71. Brun, M.-P.; Bischoff, L.; Garbay, C. *Angew. Chem., Int. Ed.* **2004**, *43*, 3432–3436.

72. Bousquet, Y.; Anderson, P.C.; Bogri, T.; Duceppe, J.-S.; Grenier, L.; Guse, I. *Tetrahedron* **1997**, *53*, 15671–15680.

73. Venturi, F.; Venturi, C.; Liguori, F.; Cacciarini, M.; Montalbano, M.; Nativi, C. *J. Org. Chem.* **2004**, *69*, 6153–6155.

74. Marin, J.; Didierjean, C.; Aubry, A.; Briand, J.-P.; Guichard, G. *J. Org. Chem.* **2002**, *67*, 8440–8449.

75. Adlington, R.M.; Baldwin, J.E.; Catterick, D.; Pritchard, G.J. *J. Chem. Soc., Perkin Trans. 1* **2000**, 299–302.

76. Adlington, R.M.; Baldwin, J.E.; Catterick, D.; Pritchard, G.J. *J. Chem. Soc., Perkin Trans. 1* **2001**, 668–679.

77. Athanassopoulos, C.; Tzavara, C.; Papaioannou, D.; Sindona, G.; Maia, H.L.S. *Tetrahedron* **1995**, *51*, 2679–2688.

78. Griesbeck, A.G.; Heckroth, H. *Synlett* **1997**, 1243–1244.

79. Gosselin, F.; Lubell, W.D. *J. Org. Chem.* **1998**, *63*, 7463–7471.

80. Werner, R.M.; Williams, L.M.; Davis, J.T. *Tetrahedron Lett.* **1998**, *39*, 9135–9138.

81. Hamzé, A.; Hernandez, J.-F.; Fulcrand, P.; Martinez, J. *J. Org. Chem.* **2003**, *68*, 7316–7321.

82. Ornstein, P.L.; Arnold, M.B.; Evrard, D.; Leander, J.D.; Lodge, D.; Schoepp, D.D. *Bioorg. Med. Chem. Lett.* **1993**, *3*, 43–48.

83. Trach Van, T.; Kojro, E.; Grzonka, Z. *Tetrahedron* **1977**, *33*, 2299–2302.

84. Morley, J.S. *J. Chem. Soc. C* **1969**, 809–813.

85. Koguro, K.; Oga, T.; Mitsui, S.; Orita, R. *Synthesis* **1998**, 910–914.

86. Barton, D.H.R.; Hervé, Y.; Potier, P.; Thierry, J. *J. Chem. Soc., Chem. Commun.* **1984**, 1298–1299.

87. Barton, D.H.R.; Hervé, Y.; Potier, P.; Thierry, J. *Tetrahedron* **1988**, *44*, 5479–5486.

88. Strazewski, P.; Tamm, C. *Synthesis* **1987**, 298–299.

89. Barton, D.H.R.; Bridon, D.; Hervé, Y.; Potier, P.; Thierry, J.; Zard, S.Z. *Tetrahedron* **1986**, *42*, 4983–4990.

90. Zeng, B.; Wong, K.K.; Pompliano, D.L.; Reddy, S.; Tanner, M.E. *J. Org. Chem.* **1998**, *63*, 10081–10086.

91. Jackson, R.F.W.; Fraser, J.L.; Wishart, N.; Porter, B.; Wythes, M.J. *J. Chem. Soc., Perkin Trans. 1* **1998**, 1903–1912.

92. Barton, D.H.R.; Hervé, Y.; Potier, P.; Thierry, J. *Tetrahedron* **1987**, *43*, 4297–4308.

93. Barton, D.H.R.; Géro, S.D.; Quiclet-Sire, B.; Samadi, M. *J. Chem. Soc., Chem. Commun.* **1989**, 1000–1001.

94. Barton, D.H.R.; Géro, S.D.; Quiclet-Sire, B.; Samadi, M. *Tetrahedron* **1992**, *48*, 1627–1636.

95. Barton, D.H.R.; Géro, S.D.; Quiclet-Sire, B.; Samadi, M. *J. Chem. Soc., Perkin Trans. 1* **1991**, 981–985.

96. Hiebl, J.; Blanka, M.; Guttman, A.; Kollman, H.; Leitner, K.; Mayrhofer, G.; Rovenszky, F.; Winkler, K. *Tetrahedron* **1998**, *54*, 2059–2074.

97. Tudor, D.W.; Lewis, T.; Robins, D.J. *Synthesis* **1993**, 1061–1062.

98. Bold, G.; Steiner, H.; Moesch, L.; Walliser, B. *Helv. Chim. Acta* **1990**, *73*, 405–410.

99. Winkler, D.; Burger, K. *Synthesis* **1996**, 1419–1421.

100. Hoffmann, M.G.; Zeiss, H.-J. *Tetrahedron Lett.* **1992**, *33*, 2669–2672.

101. Ornstein, P.L.; Melikian, A.; Martinelli, M.J. *Tetrahedron Lett.* **1994**, *35*, 5759–5762.

102. Bergmeier, S.C.; Cobás, A.A.; Rapoport, H. *J. Org. Chem.* **1993**, *58*, 2369–2376.

103. Han, Y.; Chorev, M. *J. Org. Chem.* **1999**, *64*, 1972–1978.

104. Padrón, J.M.; Kokotos, G.; Martín, T.; Markidis, T.; Gibbons, W.A.; Martín, V.S. *Tetrahedron: Asymmetry.* **1998**, *9*, 3381–3394.

105. Cox, R.J.; Hadfield, A.T.; Mayo-Martín, M.B. *Chem. Commun.* **2001**, 1710–1711.

106. Swarbrick, M.E.; Gosselin, F.; Lubell, W.D. *J. Org. Chem.* **1999**, *64*, 1993–2002.

107. Wernic, D.; DiMaio, J.; Adams, J. *J. Org. Chem.* **1989**, *54*, 4224–4228.

108. Paris, M.; Douat, C.; Heitz, A.; Gibbons, W.; Martinez, J.; Fehrentz, J.-A. *Tetrahedron Lett.* **1999**, *40*, 5179–5182.

109. Douat, C.; Heitz, A.; Martinez, J.; Fehrentz, J.-A. *Tetrahedron Lett.* **2001**, *42*, 3319–3321.

110. Baldwin, J.E.; Flinn, A. *Tetrahedron Lett.* **1987**, *28*, 3605–3608.

111. Keith, D.D.; Tortura, J.A.; Ineichen, K.; Leimgruber, W. *Tetrahedron* **1975**, *31*, 2633–2636.

112. Fushiya, S.; Maeda, K.; Funayama, T.; Nozoe, S. *J. Med. Chem.* **1988**, *31*, 480–483.

113. Tong, G.; Perich, J.W.; Johns, R.B. *Tetrahedron Lett.* **1990**, *31*, 3759–3762.

114. Knapp, S.; Hale, J.J.; Bastos, M.; Molina, A.; Chen, K.Y. *J. Org. Chem.* **1992**, *57*, 6239–6256.

115. Werner, R.M.; Shokek, O.; Davis, J.T. *J. Org. Chem.* **1997**, *62*, 8243–8246.

116. Perich, J.W. *Int. J. Pept. Prot. Res.* **1994**, *44*, 288–294.

117. Walker, D.M.; McDonald, J.F.; Franz, J.E.; Logusch, E.W. *J. Chem. Soc., Perkin Trans I* **1990**, 659–665.

118. Silverman, R.B.; Levy, M.A. *J. Org. Chem.* **1980**, *45*, 815–818.

119. Altman, J.; Moshberg, R.; Ben-Ishai, D. *Tetrahedron Lett.* **1975**, 3737–3740.

120. Ben-Ishai, D.; Moshenberg, R.; Altman, J. *Tetrahedron* **1977**, *33*, 1533–1542.

121. Tong, G.; Perich, J.W.; Johns, R.B. *Aust. J. Chem.* **1992**, *45*, 1225–1240.

122. Dietrich, E.; Lubell, W.D. *J. Org. Chem.* **2003**, *68*, 6988–6996.

123. Hernández, N.; Martin, V.S. *J. Org. Chem.* **2001**, *66*, 4934–4938.

124. Ouerfelli, O.; Ishida, M.; Shinozaki, H.; Nakanishi, K.; Ohfune, Y. *Synlett* **1993**, 409–410.

125. Bland, J.M. *Synth. Commun.* **1995**, *25*, 467–477.

126. El Marini, A.; Roumestant, M.L.; Viallefont, P.; Razafindramboa, D.; Bonato, M.; Follet, M. *Synthesis* **1992**, 1104–1108.

127. Jefford, C.W.; McNulty, J.; Lu, Z.-H.; Wang, J.B. *Helv. Chim. Acta* **1996**, *79*, 1203–1216.

128. Jefford, C.W.; Lu, Z.-H.; Wang, J.B. *Pure Appl. Chem.* **1994**, *66*, 2075–2078.

129. Jefford, C.W.; Wang, J. *Tetrahedron Lett.* **1993**, *34*, 1111–1114.

130. Bergmeier, S.C.; Lee, W.K.; Rapoport, H. *J. Org. Chem.* **1993**, *58*, 5019–5022.

131. McGarvey, G.J.; Hiner, R.N.; Matsubara, Y.; Oh, T. *Tetrahedron Lett.* **1983**, *24*, 2733–2736.

132. McGarvey, G.J.; Williams, J.M.; Hiner, R.N.; Matsubara, Y.; Oh, T. *J. Am. Chem. Soc.* **1986**, *108*, 4943–4952.

133. Barlos, K.; Mamos, P.; Papaioannou, D.; Patrianakou, S. *J. Chem. Soc., Chem. Commun.* **1987**, 1583–1584.

134. Ensch, C.; Hesse, M. *Helv. Chim. Acta* **2002**, *85*, 1659–1673.

135. Detterbeck, R.; Guggisberg, A.; Popaj, K.; Hesse, M. *Helv. Chim. Acta* **2002**, *85*, 1742–1758.

136. Brückner, A.M.; Schmidt, H.W.; Diederichsen, U. *Helv. Chim. Acta* **2002**, *85*, 3855–3866.

137. Thierry, J.; Servajean, V. *Tetrahedron Lett.* **2004**, *45*, 821–823.

138. Nitta, H.; Ueda, I.; Hatanaka, M. *J. Chem. Soc., Perkin Trans. 1* **1997**, 1793–1798.

139. Misiti, D.; Santaniello, M.; Zappia, G. *Bioorg. Med. Chem. Lett.* **1992**, *2*, 1029–1032.

140. Ciapetti, P.; Soccolini, F.; Taddei, M. *Tetrahedron* **1997**, *53*, 1167–1176.

141. Stojanovic, A.; Renaud, P. *Synlett* **1997**, 181–182.

142. Baldwin, J.E.; North, M.; Flinn, A.; Moloney, M.G. *J. Chem. Soc., Chem. Commun.* **1988**, 828–829.

143. Baldwin, J.E.; North, M.; Flinn, A.; Moloney, M.G. *Tetrahedron* **1989**, *45*, 1453–1464.

144. Baldwin, J.E.; North, M.; Flinn, A.; Moloney, M.G. *Tetrahedron* **1989**, *45*, 1465–1474.

145. Koskinen, A.M.P.; Rapoport, H. *J. Org. Chem.* **1989**, *54*, 1859–1866.

146. Paz, M.M.; Sardina, F.J. *J. Org. Chem.* **1993**, *58*, 6990–6995.

147. Mulzer, J.; Schröder, F.; Lobbia, A.; Buschmann, J.; Luger, P. *Angew. Chem,. Int. Ed. Engl.* **1994**, *33*, 1737–1739.

148. Ibrahim, H.H.; Lubell, W.D. *J. Org. Chem.* **1993**, *58*, 6438–6441.

149. Beausoleil, E.; L-Archevêque, B.; Bélec, L.; Atfani, M.; Lubell, W.D. *J. Org. Chem.* **1996**, *61*, 9447–9454.

150. Del Bosco, M.; Johnstone, A.N.C.; Bazza, G.; Lopatriello, S.; North, M. *Tetrahedron* **1995**, *51*, 8545–8554.

151. Hanessian, S.; Margarita, R. *Tetrahedron Lett.* **1998**, *39*, 5887–5890.

152. Phillips, D.; Chamberlin, A.R. *J. Org. Chem.* **2002**, *67*, 3194–3201.

153. Krasnov, V.P.; Zhdanova, E.A.; Korolynova, M.A.; Bukrina, I.M.; Kodess, M.I.; Kravtsov, V.K.; Biyushkin, V.N. *Russ. Chem. Bull.* **1997**, *46*, 319–323.

154. Nájera, C.; Yus, M. *Tetrahedron: Asymmetry* **1999**, *10*, 2245–2303.

155. Woo, K.-C.; Jones, K. *Tetrahedron Lett.* **1991**, *32*, 6949–6952.

156. Hon, Y.-s.; Chang, Y.-c.; Gong, M.-l. *Heterocycles* **1990**, *31*, 191–195.

157. Thottathil, J.K.; Moniot, J.L.; Mueller, R.H.; Wong, M.K.Y.; Kissick, T.P. *J. Org. Chem.* **1986**, *51*, 3140–3143.

158. Tanaka, K.-I.; Yoshifuji, S.; Nitta, Y. *Chem. Pharm. Bull.* **1986**, *34*, 3879–3884.

159. Bailey, J.H.; Byfield, A.T.J.; Davis, P.J.; Foster, A.C.; Leech, M.; Moloney, M.G.; Müller, M.; Prout, C.K. *J. Chem. Soc., Perkin Trans. 1* **2000**, 1977–1982.

160. Ikota, N. *Heterocycles* **1989**, *29*, 1469–1472.

161. Zhang, R.; Brownewell, F.; Madalengoitia, J.S. *Tetrahedron Lett.* **1999**, *40*, 2707–2710.

162. Bowler, A.N.; Doyle, P.M.; Hitchcock, P.B.; Young, D.W. *Tetrahedron Lett.* **1991**, *32*, 2679–2682.

163. Ezquerra, J.; Pedregal, C.; Rubio, A.; Yruretagoyena, B.; Escribano, A.; Sánchez-Ferrando, F. *Tetrahedron* **1993**, *49*, 8665–8678.

164. Baldwin, J.E.; Miranda, T.; Moloney, M.; Hokelek, T. *Tetrahedron* **1989**, *45*, 7459–7468.

165. Ezquerra, J.; Pedregal, C; Rubio, A.; Vaquero, J.J.; Matía, M.P.; Martín, J.; Diaz, A.; Navío, J.L.G.; Deeter, J.B. *J. Org. Chem.* **1994**, *59*, 4327–4331.

166. Attwood, M.R.; Carr, M.G.; Jordan, S. *Tetrahedron Lett.* **1990**, *31*, 283–284.

167. Ohta, T.; Hosoi, A.; Nozoe, S. *Tetrahedron Lett.* **1988**, *29*, 329–332.

168. Avent, A.G.; Bowler, A.N.; Doyle, P.M.; Marchand, C.M.; Young, D.W. *Tetrahedron Lett.* **1992**, *33*, 1509–1512.

169. Ezquerra, J.; Pedregal, C.; Micó, I.; Nájera, C. *Tetrahedron: Asymmetry* **1994**, *5*, 921–926.

170. Dikshit, D.K.; Panday, S.K. *J. Org. Chem.* **1992**, *57*, 1920–1924.

171. Langlois, N.; Rojas, A. *Tetrahedron Lett.* **1993**, *34*, 2477–2480.

172. Charrier, J.-D.; Duffy, J.E.S.; Hitchcock, P.B.; Young, D.W. *Tetrahedron Lett.* **1998**, *39*, 2199–2202.

173. Charrier, J.-D.; Duffy, J.E.S.; Hitchcock, P.B.; Young, D.W. *J. Chem. Soc., Perkin Trans. 1* **2001**, 2367–2371.

174. Dikshit, D.K.; Maheshwari, A. *Tetrahedron Lett.* **1999**, *40*, 44121–4412.

175. Braña, M.F.; Garranzo, M.; Pérez-Castells, J. *Tetrahedron Lett.* **1998**, *39*, 6569–6572.

176. Ezquerra, J; Rubio, A.; Martín, J.; Navío, J.L.G. *Tetrahedron: Asymmetry* **1997**, *8*, 669–671.

177. Ezquerra, J.; Pedregal, C.; Yruretagoyena, B.; Rubio, A; Carreño, M.C.; Escribano, A.; Ruano, J.L.G. *J. Org. Chem.* **1995**, *60*, 2925–2930.

178. Dikshit, D.K.; Bajpai, S.N. *Tetrahedron Lett.* **1995**, *36*, 3231–3232.

179. Hanessian, S.; Claridge, S.; Johnstone, S. *J. Org. Chem.* **2002**, *67*, 4261–4274.

180. See reference *178*.

181. Panday, S.K.; Griffart-Brunet, D.; Langlois, N. *Tetrahedron Lett.* **1994**, *35*, 6673–6676.

182. Danishefsky, S.; Berman, E.; Clizbe, L.A.; Hirama, M. *J. Am. Chem. Soc.* **1979**, *101*, 4385–4386.

183. Danishefsky, S.; Morris, J.; Clizbe, L.A. *J. Am. Chem. Soc.* **1981**, *103*, 1602–1604.

184. August, R.A; Khan, J.A.; Moody, C.M.; Young, D.W. *Tetrahedron Lett.* **1992**, *33*, 4617–4620.

185. August, R.A.; Khan, J.A.; Moody, C.M.; Young, D.W. *J. Chem. Soc., Perkin Trans. 1* **1996**, 507–514.

186. Coudert, E.; Acher, F.; Azerad, R. *Synthesis* **1997**, 863–865.

187. Hanessian, S.; Papeo, G.; Angiolini, M.; Fettis, K.; Beretta, M.; Munro, A. *J. Org. Chem.* **2003**, *68*, 7204–7218.

188. Moody, C.M.; Young, D.W. *Tetrahedron Lett.* **1993**, *34*, 4667–4670.

189. Moody, C.M.; Young, D.W. *Tetrahedron Lett.* **1994**, *35*, 7277–7280.

190. Bowler, A.N.; Dinsmore, A.; Doyle, P.M.; Young, D.W. *J. Chem. Soc., Perkin Trans. 1* **1997**, 1297–1306.

191. Durand, X.; Hudhomme, P.; Khan, J.A.; Young, D.W. *J. Chem. Soc., Perkin Trans. 1,* **1996**, 1131–1139.

192. Flynn, D.L.; Zelle, R.E.; Grieco, P.A. *J. Org. Chem.* **1983**, *48*, 2424–2427.

193. Lee, B.H.; Gerfen, G.J.; Miller, M.J. *J. Org. Chem.* **1984**, *49*, 2418–2423.

194. Olsen, R.K.; Ramasamy, K.; Emery, T. *J. Org. Chem.* **1984**, *49*, 3527–3534.

195. Sutherland, A.; Willis, C.L. *J. Lab. Cmpds. Radiopharm.* **1996**, *38*, 95–102.

196. Dolence, E.K.; Lin, C.-E.; Miller, M.J. *J. Med. Chem.* **1991**, *34*, 956–968.

197. Broddefalk, J.; Bergquist, K.-E.; Kihlberg, J. *Tetrahedron* **1998**, *54*, 12047–12070.

198. Kolasa, T.; Miller, M.J. *J. Org. Chem.* **1990**, *55*, 1711–1721.

199. Cama, E.; Shin, H.; Christianson, D.W. *J. Am. Chem. Soc.* **2003**, *125*, 13052–13057.

200. García, M.; Serra, A.; Rubiralta, M.; Diez, A.; Segarra, V.; Lozoya, E.; Ryder, H.; Palacios, J.M. *Tetrahedron: Asymmetry* **2000**, *11*, 991–994.

201. Holcomb, R.C.; Schow, S.; Ayral-Kaloustian, S.; Powell, D. *Tetrahedron Lett.* **1994**, *35*, 7005–7008.

202. Feldman, P.L.; Chi, S.; Sennequier, N.; Stuehr, D.J. *Bioorg. Med. Chem. Lett.* **1996**, *6*, 111–114.

203. Adamczyk, M.; Johnson, D.D.; Reddy, R.E. *Tetrahedron: Asymmetry* **1999**, *10*, 775–781.

204. Kokotos, G.; Padrón, J.M.; Martín, T.; Gibbons, W.A.; Martín, V.S. *J. Org. Chem.* **1998**, *63*, 3741–3744.

205. Burkhart, F.; Hoffman, M.; Kessler, H. *Angew. Chem., Int. Ed. Engl.* **1997**, *36*, 1191–1192.

206. Robl, J.A. *Tetrahedron Lett.* **1994**, *35*, 393–396.

207. Sutherland, A.; Vederas, J.C. *Chem. Commun.* **1999**, 1739–1740.

208. Clive, D.L.J.; Yeh, V.S.C. *Tetrahedron Lett.* **1999**, *40*, 8503–8507.

209. Shin, H.; Cama, E.; Christianson, D.W. *J. Am. Chem. Soc.* **2004**, *126*, 10278–10284.

210. Scholtz, J.M.; Bartlett, P.A. *Synthesis* **1989**, 542–544.

211. Stewart, J.M.; Woolley, D.W. *J. Am. Chem. Soc.* **1956**, *78*, 5336–5338.

212. Ko, K.-y.; Lee, K.I.; Kim, W.J. *Tetrahedron Lett.* **1992**, *33*, 6651–6652.

213. Lüsch, H.; Uzar, H.C. *Tetrahedron: Asymmetry* **2000**, *11*, 4965–4973.

214. Adamson, D.W. *J. Chem. Soc.* **1939**, 1564–1568.

215. Nouvet, A.; Binard, M.; Lamaty, F.; Martinez, J.; Lazaro, R. *Tetrahedron* **1999**, *55*, 4685–4698.

216. Hanessian, S.; Sahoo, S.P. *Tetrahedron Lett.* **1984**, *25*, 1425–1428.

217. Barton, D.H.R.; Crich, D.; Hervé, Y.; Potier, P.; Thierry, J. *Tetrahedron* **1985**, *41*, 4347–4357.

218. Sutherland, A.; Vederas, J.C. *Chem. Commun.* **2002**, 224–225.

219. Lenzi, A.; Reginato, G.; Taddei, M. *Tetrahedron Lett.* **1995**, *36*, 1713–1716.

220. Malachowski, W.P.; Coward, J.K. *J. Org. Chem.* **1994**, *59*, 7625–7634.

221. Jackson, R.F.W.; Wishart, N.; Wythes, M.J. *Synlett* **1993**, 219–220.
222. Thompson, M.J.; Mekhalfia, A.; Hornby, D.P.; Blackburn, G.M. *J. Org. Chem.* **1999**, *64*, 7467–7473.
223. Ciapetti, P.; Mann, A.; Schoenfelder, A.; Taddei, M. *Tetrahedron Lett.* **1998**, *39*, 3843–3846.
224. Yang, M.; Ye, W.; Schneller, S.W. *J. Org. Chem.* **2004**, *69*, 3993–3996.
225. Fraser, J.L.; Jackson, R.F.W.; Porter, B. *Synlett* **1994**, 379–380.
226. Fraser, J.L.; Jackson, R.F.W.; Porter, B. *Synlett* **1995**, 819–820.
227. Dorgan, B.J.; Jackson, R.F.W. *Synlett* **1996**, 859–861.
228. Tamaru, Y.; Tanigawa, H.; Yamamoto, T.; Yoshida, Z.-I. *Angew. Chem., Int. Ed. Engl.* **1989**, *28*, 351–353.
229. Shimamoto, K.; Ohfune, Y. *Tetrahedron Lett.* **1989**, *30*, 3803–3804.
230. Oppolzer, W.; Thirring, K. *J. Am. Chem. Soc.* **1982**, *104*, 4978–4979.
231. El Marini, A.; Roumestant, M.L.; Viallefont, P.; Razafindramboa, D.; Bonato, M.; Follet, M. *Synthesis* **1992**, 1104–1108.
232. Adkins, H.; Billica, H.R. *J. Am. Chem. Soc.* **1948**, 3121–3125.
233. Huang, S.-B.; Nelson, J.S.; Weller, D.D. *Syn. Commun.* **1989**, *19*, 3485–3496.
234. Schmidt, U.; Schölm, R. *Synthesis* **1978**, 752–753.
235. Hardegger, E.; Ott, H. *Helv. Chim. Acta* **1955**, *36*, 312–320.
236. Gibian, H.; Klieger, E. *Liebigs Ann. Chem.* **1961**, *640*, 145–156.
237. Schröder, E.; Klieger, E. *Liebigs Ann. Chem.* **1964**, *673*, 196–207.
238. See Reference *236*.
239. Alvarez-Gutierrez, J.M.; Nefzi, A.; Houghten, R.A. *Tetrahedron Lett.* **2000**, *41*, 851–854.
240. Yoshifuji, S.; Tanaka, K.-I.; Kawai, T.; Nitta, Y. *Chem. Pharm. Bull.* **1986**, *34*, 3873–3878.
241. Oba, M.; Terauchi, T.; Hashimoto, J.; Tanaka, T.; Nishiyama, K. *Tetrahedron Lett.* **1997**, *38*, 5515–5518.
242. Yoshifuji, S.; Matsumoto, H.; Tanaka, K.-I.; Nitta, Y. *Tetrahedron Lett.* **1980**, *21*, 2963–2964.
243. Yoshifuji, S.; Kaname, M. *Chem. Pharm. Bull.* **1995**, *43*, 1617–1620.
244. Sharma, N.K.; Ganesh, K.N. *Tetrahedron Lett.* **2004**, *45*, 1403–1406.
245. Goodall, K.; Parsons, A.F. *Tetrahedron Lett.* **1997**, *38*, 491–494.
246. Goodall, K.; Parsons, A.F. *J. Chem. Soc., Perkin Trans. 1* **1994**, 3257–3259.
247. Dieterich, P.; Young, D.W. *Tetrahedron Lett.* **1993**, 34, 5455–5458.
248. Alvarez-Ibarra, C.; Csákÿ, A.G.; de Silanes, I.L.; Quiroga, M.L. *J. Org. Chem.* **1997**, *62*, 479–484.
249. Molina, M.T.; del Valle, C.; Escribano, A.M.; Ezquerra, J.; Pedregal, C. *Tetrahedron* **1993**, *49*, 3801–3808.
250. Schoenfelder, A.; Mann, A. *Synth. Commun.* **1990**, *20*, 2585–2588.
251. Ohta, T.; Hosoi, A.; Kimura, T.; Nozoe, S. *Chem. Lett.* **1987**, 2091–2094.
252. Ezquerra, J.; Pedregal, C.; Rubio, A.; Valenciano, J.; Navio, J.L.G.; Alvarez-Builla, J.; Vaquero, J.J. *Tetrahedron Lett.* **1993**, *34*, 6317–6320.
253. Van Betsbrugge, J.; Van Den Nest, W.; Verheyden, P.; Tourwé, D. *Tetrahedron* **1998**, *54*, 1753–1762.
254. Ackermann, J.; Matthes, M.; Tamm, C. *Helv. Chim. Acta* **1990**, *73*, 122–132.
255. Ohta, T.; Kimura, T.; Sato, N.; Nozoe, S. *Tetrahedron Lett.* **1988**, *29*, 4303–4304.
256. Ezquerra, J.; Rubio, A.; Pedregal, C.; Sanz, G.; Rodriguez, J.H.; Ruano, J.L.G. *Tetrahedron Lett.* **1993**, *34*, 4989–4992.
257. Coutts, I.G.C.; Saint, R.E. *Tetrahedron Lett.* **1998**, *39*, 3243–3246.
258. Pravda, Z.; Rudinger, J. *Coll. Czech. Chem. Commun.* **1955**, *20*, 1–8.
259. Langlois, N.; Andrialmialisoa, R.Z. *Tetrahedron Lett. 1991, 32, 3057–3058.*
260. Herdeiss, C.; Hubmann, H.P.; Lotter, H. *Tetrahedron: Asymmetry* **1994**, *5*, 351–354.
261. Soloshonok, V.A.; Ueki, H.; Tiwari, R.; Cai, C.; Hruby, V.J. *J. Org. Chem.* **2004**, *69*, 4984–4990.
262. Oba, M.; Nishiyama, N.; Nishiyama, K. *Tetrahedron* **2004**, *60*, 8456–8464.
263. Pedregal, C.; Ezquerra, J.; Escribano, A.; Carreño, M.C.; Ruano, J.L.G. *Tetrahedron Lett.* **1994**, *35*, 2053–2056.
264. Langlois, N.; Rojas, A. *Tetrahedron* **1993**, *49*, 77–82.
265. Stapon, A.; Li, R.; Townsend, C.A. *J. Am. Chem. Soc.* **2003**, *125*, 8486–8493.
266. Somfai, P.; He, H.M.; Tanner, D. *Tetrahedron Lett.* **1991**, *32*, 283–286.
267. Baldwin, J.E.; Moloney, M.G.; Shim, S.B. *Tetrahedron Lett.* **1991**, *32*, 1379–1380.
268. Shimamoto, K.; Ishida, M.; Shinozaki, H.; Ohfune, Y. *J. Org. Chem.* **1991**, *56*, 4167–4176.
269. Herdeis, C.; Hubmann, H.P. *Tetrahedron: Asymmetry* **1992**, *3*, 1213–1221.
270. Ohfune, Y.; Tomita, M. *J. Am. Chem. Soc.* **1982**, *104*, 3511–3513.
271. Oba, M.; Miyakawa, A.; Nishiyama, K.; Terauchi, T.; Kainosho, M. *J. Org. Chem.* **1999**, *64*, 9275–9278.
272. Oba, M.; Kobayashi, M.; Oikawa, F.; Nishiyama, K.; Kainosho, M. *J. Org. Chem.* **2001**, *66*, 5919–5922.
273. Hashimoto, M.; Hashimoto, K.; Shirahama, H. *Tetrahedron* **1996**, *52*, 1931–1942.
274. Yanagida, M.; Hashimoto, K.; Ishida, M.; Shinozaki, H.; Shirahama, H. *Tetrahedron Lett.* **1989**, *30*, 3799–3802.

275. Hanessian, S.; Ratovelomanana, V. *Synlett* **1990**, 501–503.

276. Dyer, J.; Keeling, S.; Moloney, M.G. *Chem. Commun.* **1998**, 461–462.

277. Ikota, N. *Tetrahedron Lett.* **1992**, *33*, 2553–2556.

278. Ikota, N.; Hanaki, A. *Chem. Pharm. Bull.* **1989**, *37*, 1087–1089.

279. Herdeis, C.; Aschenbrenner, A.; Kirfel, A.; Schwabenländer, F. *Tetrahedron: Asymmetry* **1997**, *8*, 2421–2432.

280. Ezquerra, J.; Escribano, A.; Rubio, A.; Remuiñán, M.J.; Vaquero, J.J. *Tetrahedron: Asymmetry* **1996**, *7*, 2613–2626.

281. Ezquerra, J.; Escribano, A.; Rubio, A.; Remuiñán, M.J.; Vaquero, J.J. *Tetrahedron Lett.* **1995**, *36*, 6149–6152.

282. Herdeis, C.; Kelm, B. *Tetrahedron* **2003**, *59*, 217–229.

283. Müller, W.; Dorsch, W.; Effenberger, F. *Chem. Ber.* **1987**, *120*, 55–59.

284. Effenberger, F.; Müller, W.; Isak, H. *Chem. Ber.* **1987**, *120*, 45–54.

285. McAlonan, H.; Stevenson, P.J. *Tetrahedron: Asymmetry* **1995**, *6*, 239–244.

286. Boger, D.L.; Keim, H.; Oberhauser, B.; Schreiner, E.P.; Foster, C.A. *J. Am. Chem. Soc.* **1999**, *121*, 6197–6205.

287. Xue, C.-b.; DeGrado, W.F. *Tetrahedron Lett.* **1995**, *36*, 55–58.

288. Chen, Y.; Bilban, M.; Foster, C.A.; Boger, D.L. *J. Am. Chem. Soc.* **2002**, *124*, 5431–5440.

289. Yoneta, T.; Shibahara, S.; Fukatsu, S.; Seki, S. *Bull. Chem. Soc. Jpn.* **1978**, *51*, 3296–3297.

290. Wilchek, M.; Ariely, S.; Patchornik, A. *J. Org. Chem.* **1968**, *33*, 1258–1259.

291. Trach Van, T.; Kojro, E.; Grzonka, Z. *Tetrahedron* **1977**, *33*, 2299–2302.

292. Christie, B.D.; Rapoport, H. *J. Org. Chem.* **1985**, *50*, 1239–1246.

293. Karrer, P.; Schlosser, A. *Helv. Chim. Acta* **1923**, *6*, 411–418.

294. Otsuka, M.; Kittaka, A.; Iimori, T.; Yamashita, H.; Kobayashi, S.; Ohno, M. *Chem. Pharm. Bull.* **1985**, *33*, 509–514.

295. Umezawa, Y.; Morishima, H.; Saito, S.-i.; Takita, T.; Umezawa, H.; Kobayashi, S.; Otsuka, M.; Narita, M.; Ohno, M. *J. Am. Chem. Soc.* **1980**, *102*, 6630–6631.

296. Rudinger, J.; Poduska, K.; Zaoral, M. *Coll. Czech. Chem. Commun.* **1960**, *25*, 2022–2028.

297. Amato, J.S.; Bagner, C.; Cvetovich, R.J.; Gomolka, S.; Hartner F.W., Jr.; Reamer, R. *J. Org. Chem.* **1998**, *63*, 9533–9534.

298. Waki, M.; Kitajima, Y.; Izumiya, N. *Synthesis* **1981**, 266–268.

299. Al-Obeidi, F.; de L. Castrucci, A.M.; Hadley, M.E.; Hruby, V.J. *J. Med. Chem.* **1989**, *32*, 2555–2561.

300. Rew, Y.; Goodman, M. *J. Org. Chem.* **2002**, *67*, 8820–8826.

301. Nicolaou, K.C.; Trujillo, J.I.; Jandeleit, B.; Chibale, K.; Rosenfeld, M.; Diefenbach, B.; Cheresh, D.A.; Goodman, S.L. *Biorg. Med. Chem.* **1998**, *6*, 1185–1208.

302. Chhabra, S.R.; Mahajan, A.; Chan, W.C. *Tetrahedron Lett.* **1999**, *40*, 4905–4908.

303. Ruan, F.; Chen, Y.; Itoh, K.; Sasaki, T.; Hopkins, P.B. *J. Org. Chem.* **1991**, *56*, 4347–4354.

304. Stanfield, C.F.; Felix, A.M.; Danho, W. *Org. Prep. Proc. Int.* **1990**, *22*, 597–603.

305. Zhang, L.-h.; Kauffman, G.S.; Pesti, J.A.; Yin, J. *J. Org. Chem.* **1997**, *62*, 6918–6920.

306. Chhabra, S.R.; Mahajan, A.; Chan, W.C. *J. Org. Chem.* **2002**, *67*, 4017–4029.

307. Grieco, P.A.; Reilly, M. *Tetrahedron Lett.* **1998**, *39*, 8925–8928.

308. Andruskiewicz, R.; Rozkiewicz, D. *Synth. Commun.* **2004**, *34*, 1049–1056.

309. Blettner, C.; Bradley, M. *Tetrahedron Lett.* **1994**, *35*, 467–470.

310. Juaristi, E.; Quintana, D.; Balderas, M.; García-Pérez, E. *Tetrahedron: Asymmetry* **1996**, *7*, 2233–2246.

311. Easton, C.J. *Chem. Rev.* **1997**, *97*, 53–82.

312. Kurokawa, N.; Ohfune, Y. *J. Am. Chem. Soc.* **1986**, *108*, 6041–6043.

313. Ohfune, Y.; Hori, K.; Sakaitani, M. *Tetrahedron Lett.* **1986**, *27*, 6079–6082.

314. Kurokawa, N.; Ohfune, Y. *Tetrahedron* **1993**, *49*, 6195–6222.

315. Girard, A.; Greck, C.; Genêt, J.P. *Tetrahedron Lett.* **1998**, *39*, 4259–4260.

316. Ohshima, T.; Gnanadesikan, V.; Shibuguchi, T.; Fukuta, Y.; Nemoto, T.; Shibasaki, M. *J. Am. Chem. Soc.* **2003**, *125*, 11206–11207.

317. Agouridas, K.; Girodeau, J.M.; Pineau, R. *Tetrahedron Lett.* **1985**, 3115–3118.

318. Dappen, M.S.; Pellicciari, R.; Natalini, B.; Monahan, J.B.; Chiorri, C.; Cordi, A.A. *J. Med. Chem.* **1991**, *34*, 161–168.

319. Ohta, T.; Nakajima, S.; Sato, Z.; Aoki, T.; Hatanaka, S.-i.; Nozoe, S. *Chem. Lett.* **1986**, 511–512.

320. Zaragoza, F. *Tetrahedron* **1997**, *53*, 3425–3439.

321. Broxterman, Q.B.; Kaptein, B.; Kamphuis, J.; Schoemaker, H.E. *J. Org. Chem.* **1992**, *57*, 6286–6294.

322. Biagini, S.C.G.; Gibson, S.E.; Keen, S.P. *J. Chem. Soc., Perkin Trans. 1* **1998**, 2485–2499.

323. Schmidtmann, F.W.; Benedum, T.E.; McGarvey, G.J. *Tetrahedron Lett.* **2005**, *46*, 4677–4681.

324. Dietrich, S.A.; Banfi, L.; Basso, A.; Damonte, G.; Guanti, G. Riva, R. *Org. Biomol. Chem.* **2005**, *3*, 97–106.

325. Varray, S.; Gauzy, C.; Lamaty, F.; Lazaro, R.; Martinez, J. *J. Org. Chem.* **2000**, *65*, 6787–6790.

326. Rutjes, F.P.J.T.; Schoemaker, H.E. *Tetrahedron Lett.* **1997**, *38*, 677–680.

327. Miller, S.J.; Blackwell, H.E.; Grubbs, R.H. *J. Am. Chem. Soc.* **1996**, *118*, 9606–9614.

328. Zumpe, F.L.; Kazmaier, U. *Synthesis* **1999**, 1785–1791.
329. Kanayama, T.; Yoshida, K.; Miyabe, H.; Kimachi, T.; Takemoto, Y. *J. Org. Chem.* **2003**, *68*, 6197–6201.
330. Collier, P.N.; Campbell, A.D.; Patel, I.; Taylor, R.J.K. *Tetrahedron* **2002**, *58*, 6117–6125.
331. Collier, P.N.; Campbell, A.D.; Patel, I.; Taylor, R.J.K. *Tetrahedron Lett.* **2000**, *41*, 7115–7119.
332. Krebs, A.; Ludwig, V.; Pfizer, J.; Dürner, G.; Göbel, M.W. *Chem.— Eur. J.* **2004**, *10*, 544–553.
333. Leanna, M.R.; Morton, H.E. *Tetrahedron Lett.* **1993**, *34*, 4485–4488.
334. Leukart, O.; Caviezel, M.; Eberle, A.; Escher, E.; Tun-Kyi, A.; Schwyzer, R. *Helv. Chim. Acta* **1976**, *59*, 2184–2187.
335. IJsselstijn, M.; Kaiser, J.; van Delft, F.L.; Schoemaker, H.E.; Rutjes, F.P.J.T. *Amino Acids* **2003**, *24*, 263–266.
336. Meffre, P.; Vo-Quang, L.; Vo-Quang, Y.; Le Goffic, F. *Tetrahedron Lett.* **1990**, *31*, 2291–2294.
337. Vyas, D.M.; Chiang, Y.; Doyle, T.W. *Tetrahedron Lett.* **1984**, *25*, 487–490.
338. Baldwin, J.E.; Hoskins, C.; Kruse, L. *J. Chem. Soc., Chem. Commun.* **1976**, 795–796.
339. Mzengeza, S.; Yang, C.M.; Whitney, R.A. *J. Am. Chem. Soc.* **1987**, *109*, 276–27.
340. Wade, P.E.; Pillay, M.K.; Singh, S.M. *Tetrahedron Lett.* **1982**, *23*, 4563–4566.
341. Wade, P.A.; Singh, S.M.; Pillay, M.K. *Tetrahedron* **1984**, *40*, 601–611.
342. Mzengeza, S.; Yang, C.M.; Whitney, R.A. *J. Chem. Soc., Chem. Commun.* **1984**, 606–607.
343. Hagedorn A.A., III; Miller, B.J.; Nagy, J.O. *Tetrahedron Lett.* **1980**, *21*, 229–230.
344. Pellicciari, R.; Natalini, B.; Marinozzi, M.; Monahan, J.B.; Snyder, J.P. *Tetrahedron Lett.* **1990**, *31*, 139–142.
345. Krol, W.J.; Mao, S.-S.; Steele, D.L.; Townsend, C.A. *J. Org. Chem.* **1991**, *56*, 728–731.
346. Rajesh, S.; Ami, E.; Kotake, T.; Kimura, T.; Hayashi, Y.; Kiso, Y. *Bioorg. Med. Chem. Lett.* **2002**, *12*, 3615–3617.
347. Itaya, T.; Shimomichi, M.; Ozasa, M. *Tetrahedron Lett.* **1988**, *29*, 4129–4132.
348. Dölling, K.; Krug, A.; Hartung, H.; Weichmann, H. *Z. Naturforsch* **1997**, *52*, 9–16.
349. Barlos, K.; Mamos, P.; Papaioannou, D.; Sanida, C.; Antonopoulos, C. *J. Chem. Soc., Chem. Commun.* **1986**, 1258–1259.
350. Barlos, K.; Papaioannou, D.; Sanida, C. *Liebigs Ann. Chem.* **1986**, 287–291.
351. Pellicciari, R.; Natalini, B.; Marinozzi, M. *Synth. Commun.* **1988**, *18*, 1715–1721.
352. Griesbeck, A.G.; Hirt, J. *Liebigs Ann.* **1995**, 1957–1961.
353. Bajgrowicz, J.A.; El Hallaoui, A; Jacquier, R.; Pigière, Ch.; Viallefont, Ph. *Tetrahedron Lett.* **1984**, 2231–2234.
354. Bernardini, A.; El Hallaoui, A.; Jacquier, R.; Pigière, Ch.; Viallefont, Ph.; Bajgrowicz, J. *Tetrahedron Lett.* **1983**, *24*, 3717–3720.
355. Bajgrowicz, J.A.; Hallaoui, A.El.; Jacquier, R.; Pigière, Ch.; Viallefont, Ph. *Tetrahedron* **1985**, *41*, 1833–1843.
356. Plieninger, H. *Chem. Ber.* **1950**, *83*, 265–268.
357. Livak, J.E.; Britton, E.C.; VanderWeele, J.C.; Murray, M.F. *J. Am. Chem. Soc.* **1945**, *67*, 2218–2220.
358. Nyberg, D.D.; Christensen, B.E. *J. Am. Chem. Soc.* **1957**, *79*, 1222–1226.
359. Plieninger, H. *Chem. Ber.* **1950**, *83*, 268–271.
360. Hill, E.M.; Robson, W. *Biochem. J.* **1936**, *30*, 248–251.
361. Yu, L.; Lai, Y.; Wade, J.V.; Coutts, S.M. *Tetrahedron Lett.* **1998**, *39*, 6633–6636.
362. Weiss, S.; Stekol, J.A. *J. Am. Chem. Soc.* **1951**, *73*, 2497–2499.
363. Howarth, N.M.; Wakelin, L.P.G. *J. Org. Chem.* **1997**, *62*, 5441–5450.
364. Baddiley, J.; Jamieson, G.A. *Chem. & Ind.* **1954**, 375.
365. Carrasco, M.R.; Brown, R.T. *J. Org. Chem.* **2003**, *68*, 8853–8858.
366. Alami, A.; El Hallaoui, A.; Elachqar, A.; Roumestant, M.L.; Viallefont, P. *Bull. Soc. Chim. Belg.* **1996**, *105*, 769–772.
367. Boullais, C.; Riva, M.; Noel, J.-P. *J. Lab. Compds. Radiopharm.* **1997**, *39*, 621–624.
368. Antoni, G.; Omura, H.; Sundin, A.; Takalo, R.; Valind, S.; Watanabe, Y.; Långström, B. *J. Lab. Cmpds. Radiopharm.* **1997**, *40*, 807–809.
369. Atfani, M.; Wei, L.; Lubell, W.D. *Org. Lett.* **2001**, *3*, 2965–2968.
370. Watanabe, L.A.; Jose, B.; Kato, T.; Nishino, N.; Yoshida, M. *Tetrahedron Lett.* **2004**, *45*, 491–494.
371. Duthaler, R.O. *Angew. Chem., Int. Ed. Engl.* **1991**, *30*, 705–707.
372. Benoiton, L. *Can. J. Chem.* **1968**, *46*, 1549–1552.
373. Guo, Z.-X.; Schaeffer, M.J.; Taylor, R.J.K. *J. Chem. Soc., Chem. Commun.* **1993**, 874–875.
374. Rosegay, A.; Taub, D. *Synth. Commun.* **1989**, *19*, 1137–1145.
375. Carrasco, M.; Jones, R.J.; Kamel, S.; Rapoport, H.; Truong, T.; Grützmann, A.; Winterfeldt, E. *Org. Synth.* **1992** *70*, 29–34.
376. Afzali-Ardakani, A.; Rapoport, H. *J. Org. Chem.* **1980**, *45*, 4817–4820.
377. Meffre, P.; Vo-Quang, L.; Vo-Quang, Y.; Le Goffic, F. *Synth. Commun.* **1989**, *19*, 3457–3468.
378. Crisp, J.T.; Millan, M.J. *Tetrahedron* **1998**, *54*, 637–648.
379. Crisp, J.T.; Millan, M.J. *Tetrahedron* **1998**, *54*, 649–666.
380. Hulicky, M. *J. Fluorine Chem.* **1993**, *60*, 193–210.
381. Gmeiner, P.; Hummel, E.; Haubmann, C. *Liebigs Ann.* **1995**, 1987–1992.

382. Easton, C.J.; Hutton, C.A. *Synlett* **1998**, 457–466.

383. Hasebe, M.; Tsuchiya, T. *Tetrahedron Lett.* **1987**, *28*, 6207–6210.

384. Griesbeck, A.G.; Mauder, H. *Angew. Chem., Int. Ed. Engl.* **1992**, *31*, 73–75.

385. Griesbeck, A.G.; Mauder, H.; Müller, I. *Chem. Ber.* **1992**, *125*, 2467–2475.

386. Griesbeck, A.G. *Liebigs Ann.* **1996**, 1951–1958.

387. Obata, N.; Niimura, K. *J. Chem. Soc., Chem. Commun.* **1977**, 238–239.

388. Usher, J.J. *J. Chem. Res. S* **1980**, 30.

389. Kollonitsch, J.; Rosegay, A.; Doldouras, G. *J. Am. Chem. Soc.* **1964**, *86*, 1857–1858.

390. Faulstich, H.; Dölling, J.; Michl, K.; Wieland, T. *Liebigs Ann. Chem.* **1973**, 560–565.

391. Fujita, Y.; Kollonitsch, J.; Witkop, B. *J. Am. Chem. Soc.* **1965**, *87*, 2030–2033.

392. Kollonitsch, J.; Scott, A.N.; Doldouras, G.A. *J. Am. Chem. Soc.* **1966**, *88*, 3624–3626.

393. Kollonitsch, J.; Barash, L. *J. Am. Chem. Soc.* **1976**, *98*, 5591–5593.

394. Bowman, N.J.; Hay, M.P.; Love, S.G.; Easton, C.J. *J. Chem. Soc., Perkin Trans. 1* **1988**, 259–264.

395. Easton, C.J.; Hay, M.P. *J. Chem. Soc., Chem. Commun.* **1985**, 425–427.

396. Udding, J.H.; Tuijp, C.J.M.; Hiemstra, H.; Speckamp, W.N. *J. Chem. Soc., Perkin Trans 2* **1992**, 857–858.

397. Easton, C.J.; Hutton, C.A.; Rositano, G.; Tan, E.W. *J. Org. Chem.* **1991**, *56*, 5614–5618.

398. Easton, C.J.; Tan, E.W.; Hay, M.P. *J. Chem. Soc., Chem. Commun.* **1989**, 385–386.

399. Easton, C.J.; Hutton, C.A.; Merrett, M.C.; Tiekink, E.R.T. *Tetrahedron* **1996**, 52, 7025–7036.

400. Easton, C.J.; Hutton, C.A.; Tan, E.W.; Tiekink, E.R.T. *Tetrahedron Lett.* **1990**, *31*, 7059–7062.

401. Easton, C.J.; Merrett, M.C. *Tetrahedron* **1997**, *53*, 1151–1156.

402. Easton, C.J. *Pure Appl. Chem.* **1997**, *69*, 489–494.

403. Easton, C.J.; Hutton, C.A.; Roselt, P.D.; Tiekink, E.R.T. *Tetrahedron* **1994**, *50*, 7327–7240.

404. Easton, C.J.; Hutton, C.A.; Roselt, P.D.; Tiekink, E.R.T. *Aust. J. Chem.* **1991**, *44*, 687–694.

405. Hutton, C.A. *Tetrahedron Lett.* **1997**, *38*, 5899–5902.

406. Hutton, C.A. *Org. Lett.* **1999**, *1*, 295–297.

407. Wen, S.-J.; Yao, Z.-J. *Org. Lett.* **2004**, *6*, 2721–2724.

408. Easton, C.J.; Edwards, A.J.; McNabb, S.B.; Merrett, M.C.; O'Connell, J.L.; Simpson, G.W.; Simpson, J.S.; Willis, A.C. *Org. Biomol. Chem.* **2003**, *1*, 2492–2498.

409. Easton, C.J.; Hutton, C.A. *J. Chem. Soc., Perkin Trans 1* **1994**, 3545–3548.

410. Easton, C.J.; Tan, E.W.; Ward, C.A *Aust. J. Chem.* **1992**, *45*, 395–402.

411. Easton, C.J.; Merrett, M.C. *J. Am. Chem. Soc.* **1996**, *118*, 3035–3036.

412. Baldwin, J.E.; Fieldhouse, R.; Russell, A.T. *Tetrahedron Lett.* **1993**, *34*, 5491–5494.

413. Rouhi, A.M. *Chem. Eng. News* **2003**, *July 14*, 34–35.

414. Dangel, B.D.; Johnson, J.A.; Sames, D. *J. Am. Chem. Soc.* **2001**, *123*, 8149–8150.

415. Saladino, R.; Mezzetti, M.; Mincione, E.; Torrini, I.; Paradisi, M.P.; Mastropietro, G. *J. Org. Chem.* **1999**, *64*, 8468–8474.

416. Reetz, M.T. *Angew. Chem., Int. Ed. Engl.* **1991**, *30*, 1531–1546.

417. Jurczak, J.; Golebiowski, A. *Chem. Rev.* **1989**, *89*, 149–164.

418. Reetz, M.T.; Wünsch, T.; Harms, K. *Tetrahedron: Asymmetry* **1990**, *1*, 371–374.

419. Hanessian, S.; Margarita, R.; Hall, A.; Luo, X. *Tetrahedron Lett.* **1998**, *39*, 5883–5886.

420. Bowler, A.N.; Doyle, P.M.; Hitchcock, P.B.; Young, D.W. *Tetrahedron* **1997**, *53*, 10545–10554.

421. Murray, P.J.; Starkey, I.D.; Davies, J.E. *Tetrahedron Lett.* **1998**, *39*, 6721–6724.

422. Wang, W.; Yang, J.; Ying, J.; Xiong, C.; Zhang, J.; Cai, C.; Hruby, V.J. *J. Org. Chem.* **2002**, *67*, 6353–6360.

423. Park, S.B.; Meier, G.P. *Tetrahedron Lett.* **1989**, *30*, 4215–4218.

424. Cowart, M.; Kowaluk, E.A.; Daanen, J.F.; Kohlhaas, K.L.; Alexander, K.M.; Wagenaar, F.L.; Kerwin, J.F., Jr. *J. Med. Chem.* **1998**, *41*, 2636–2642.

425. Moody, C.M.; Young, D.W. *J. Chem. Soc., Perkin Trans. 1* **1997**, 3519–3530.

426. Oba, M.; Nishiyama, N.; Nishiyama, K. *Chem. Commun.* **2003**, 776–777.

427. Constantinou-Kokotou, V.; Magrioti, V. *Amino Acids* **2003**, *24*, 231–243.

428. Ezquerra, J.; de Mendoza, J.; Pedregal, C.; Ramírez, C. *Tetrahedron Lett.* **1992**, *33*, 5589–5590.

429. Itoh, M. *Chem. Pharm. Bull.* **1969**, *17*, 1679–1686.

430. Rilatt, I.; Caggiano, L.; Jackson, R.F.W. *Synlett* **2005**, 2701–2719.

431. Rodríguez, A.; Miller, D.D.; Jackson, R.F.W. *Org. Biomol. Chem.* **2003**, *1*, 973–977.

432. Burger, K.; Spengler, J. *Eur. J. Org. Chem.* **2000**, 199–204.

433. Andrés, J.M.; Muñoz, E.M.; Pedrosa, R.; Pérez-Encabo, A. *Eur. J. Org. Chem.* **2003**, 3387–3397.

10 | Synthesis of Optically Active α-Amino Acids: Elaboration of Serine

10.1 Introduction

Serine (Ser) is potentially an excellent starting material for amino acid synthesis. It is inexpensive and both enantiomers are readily available. The amino and carboxyl groups are already present, while the hydroxyl side chain provides an ideal site for further elaboration. Unfortunately, the advantages of Ser as a synthon also form its Achilles heel. The multiple functional groups require careful protection and, more importantly, efforts at modifying the side chain of a protected Ser derivative generally lead to unwanted side reactions. Attempts at nucleophilic displacement of an activated hydroxyl functionality often result in elimination, producing the stable dehydroalanine derivative, while oxidation of the Ser side chain to an aldehyde is usually accompanied by enolization, causing racemization at the α-center. Despite these obstacles, several useful synthetic routes have been cultivated, depending on either Ser side-chain displacement or side-chain oxidation. The susceptibility of Ser towards elimination or epimerization side reactions can be greatly reduced by protection of the carboxyl group as an *ortho* ester group. A third derivatization scheme, in which Ser is converted into a β-anion synthon, has also been developed in recent years. Some of the synthetic transformations applied to Ser have also been carried out on other β-hydroxy amino acids (particularly Thr), or on homologous hydroxy amino acids, such as homoserine. These will also be discussed in this chapter, although syntheses and derivatizations of homoserine are also included in Section 9.2.3.

10.2 Racemization of Serine

Unfortunately, Ser is one of the amino acids most prone to racemization. Racemization rates of amino acids depend on the type of side chain, temperature, pH, ionic strength, metal ion chelation, and a number of other effects (see Section 1.4.5). A recent re-examination of the racemization rates of amino acids found that the hydroxy amino acids racemize faster than amino acids with carboxy, alkoxy, carboalkoxy, alkyl, aryl, or thioether side chains (*1*). The racemization half-life for Ser at 100 °C and pH 7.6 is estimated at 4 days, compared to about 300 days for isoleucine (*2*). The optical instability of the hydroxy amino acids is proposed to result from inductive stabilization of the α-carbanion in combination with the side-chain hydroxyl stabilizing the carboxylate anion and helping to solvate the base participating in proton removal (see Scheme 10.1) (*1*). The importance of non-inductive effects is supported by the reduced racemization rates observed with the corresponding β-hydroxy ethers. Steric effects that hinder attack of base at the α-center were found to be as important as inductive effects at altering racemization rates. Thr racemizes at approximately one-quarter the rate found for Ser. It must be noted that commercial samples of L-Ser have been found to be contaminated with 1–2% of D-Ser, which is not removed by recrystallization (*3, 4*). Thus, great care must be taken when using Ser as a substrate for reactions where a high degree of enantiopurity is required in the product, and in assessing the extent of racemization caused by synthetic transformations, as the starting material may not be sufficiently optically pure.

10.3 Nucleophilic Displacement of an Activated Serine Hydroxyl Group

10.3.1 Possible Reaction Products

In 1963 Photaki noted that the previously reported displacement of *N*-Cbz-*O*-tosyl-L-Ser methyl ester with

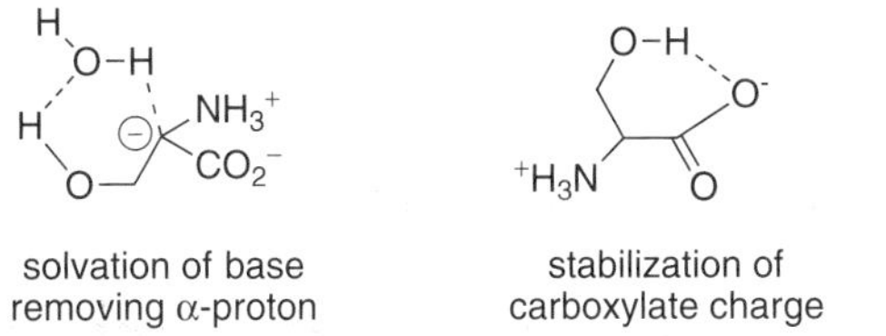

solvation of base
removing α-proton

stabilization of
carboxylate charge

Scheme 10.1 Intramolecular Participation of Serine Hydroxyl Group Assisting Racemization.

the sodium salt of tritylthiol led to racemic Cys (5). Closer examination revealed that the thiol was in fact adding to a 2,3-dehydroalanine intermediate (see Scheme 10.2), and that β-elimination of tosylate from O-tosyl Ser derivatives occurs readily under the action of bases (90% elimination in 15 min at room temperature with 0.01 N NaOH; instantly with 1 equiv of Et$_2$NH in EtOAc) (5). However, thiol displacement of N-Cbz-O-tosyl-L-Ser-Gly-OEt gave an optically active Cys derivative (6). A similar elimination was also observed during displacements of a Ser-derived β-chloro-Ala derivative (7).

This reaction is a good example of the fine balance between possible reaction paths of activated Ser derivatives. The desired β-displacement is generally not the preferred reaction. Instead, elimination, aziridine formation (nucleophilic attack by the α-amino group), or oxazoline formation (nucleophilic attack by the carbonyl oxygen of N-acyl protecting groups) tend to occur (see Scheme 10.3). β–lactone, oxazolidone or oxazolinone formation is also possible. N-Alkylation of

Ser greatly increases the rate of elimination, as formation of a protective aza-enolate from the amide bond is prevented (8). The conditions required for elimination have been discussed in Section 4.2.1, as Ser (and other β-hydroxy amino acids) are often precursors of didehydroamino acids. Similarly, the reaction environment required to generate aziridine-2-carboxylates has been covered in Section 8.2.1, as β-hydroxy amino acids are also a useful source of these synthons. Conditions for oxazoline and oxazolidone formation will be discussed below. These products are not as useful for amino acid synthesis, although Ser- and Thr-derived oxazolines and oxazolidones have been α-alkylated (see Section 7.12). Other reaction condition/protecting group combinations lead to β-lactams and β-lactones, which are discussed in Sections 8.5 and 10.3.7, respectively.

10.3.2 Heterocycle Byproducts

A number of heterocyclic byproducts have been produced during attempts to displace the hydroxyl group of β-hydroxy amino acids. N-Acyl-β-hydroxy amino acids are known to form oxazolines when treated with thionyl chloride, via conversion of the hydroxyl into a leaving group, which is then displaced by the carbonyl oxygen of the N-acyl group (see Scheme 10.4) (9–17). Cyclization under Mitsunobu conditions is also possible (18, 19). N-Acyl-aziridines are alternative products of these reaction conditions (see Section 8.2.1), and can be isomerized to oxazolines in the presence of NaI, acid, or heat (20–22). Aziridine/oxazoline interchange has also been observed in the Mitsunobu-mediated (Ph$_3$P/DIAD)

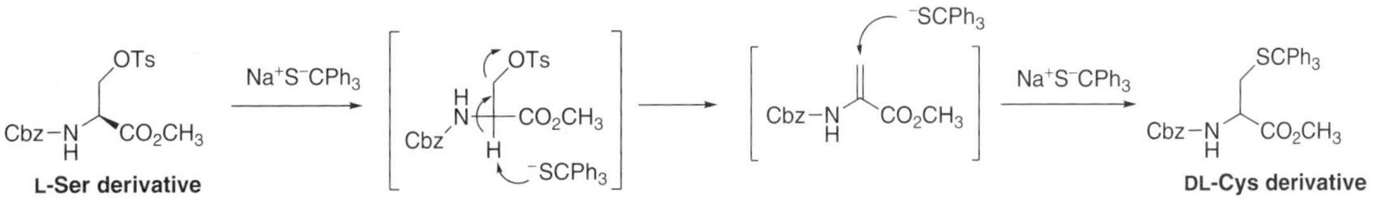

Scheme 10.2 Apparent Tosylate Displacement via Elimination / Addition (5).

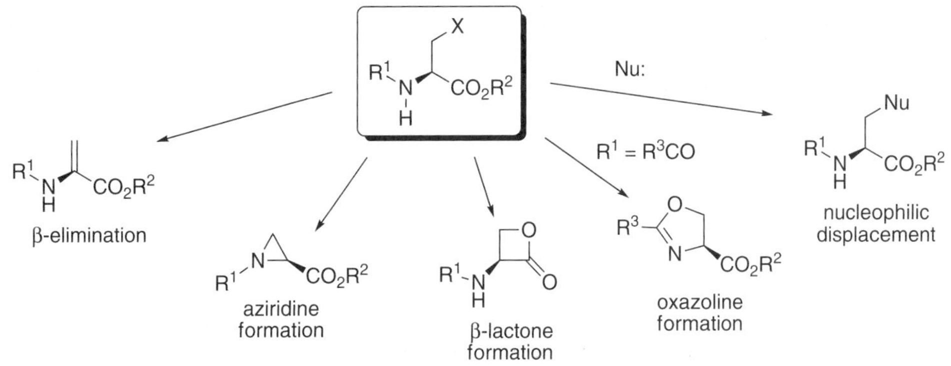

Scheme 10.3 Possible Reactions of Activated Serine.

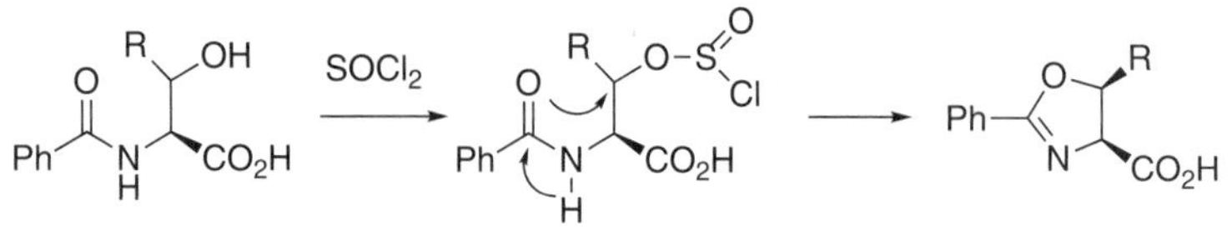

Scheme 10.4 Oxazoline Formation.

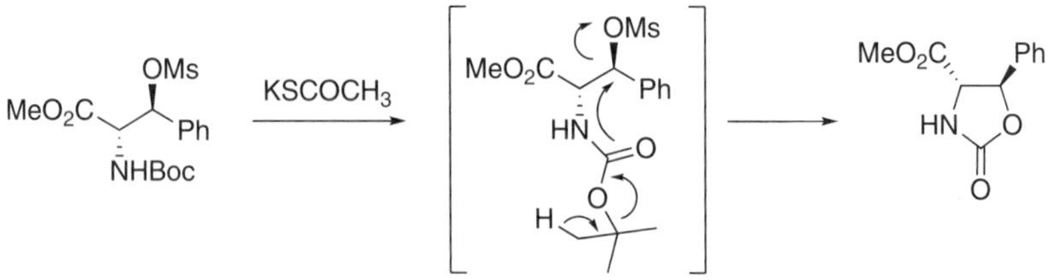

Scheme 10.5 Oxazolidinone Formation via Attack of Urethane Carbonyl on Activated Hydroxyl Group.

cyclizations of derivatives of Ser, Thr, and *allo*-Thr (*19*). Ser and Thr were found to form aziridines, but *allo*-Thr cyclized to the oxazoline under identical conditions (*19*). This is presumably due to stereochemical effects, but again demonstrates the subtle balance between different reaction pathways. When peptides incorporating an acyl-2-aziridine carboxylic acid residue were treated with NaI, the dehydroalanine derivative was obtained via elimination of the ring-opened iodoalanine intermediate (*23*). An oxazoline has also been obtained in 46% yield from attempted conversion of Ac-Ser-NHBn to β-fluoro-Ala with TMSF (*24*).

The aziridine/oxazoline structures are difficult to tell apart, as illustrated by a report of oxazoline formation (*18*), which was later shown to be an aziridine (*19*). The two structures can be distinguished by their NMR spectra. A theoretical study of 1-acylaziridine/oxazoline isomerization has been carried out (*25*). Oxazolines have been employed as a substrate for nucleophilic attack (*7, 26, 27*).

A similar intramolecular nucleophilic attack by an *N*-Boc urethane carbonyl was accompanied by elimination of the *tert*-butyl group, giving an oxazolidinone byproduct in 12% yield when a protected phenylserine mesylate was treated with thiol anions (*28*). The desired displacement product was produced in 69% yield (see Scheme 10.5).

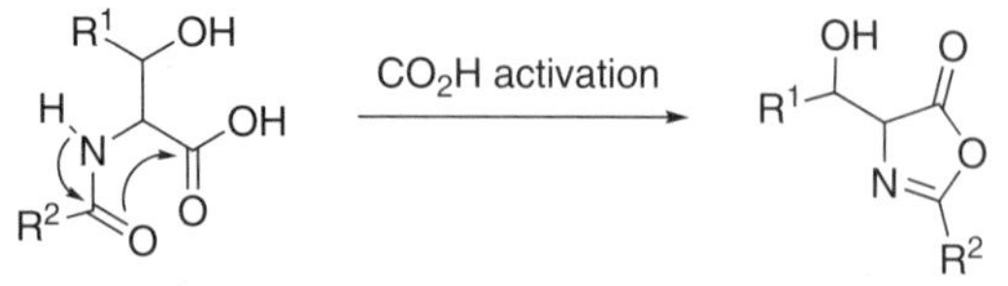

Scheme 10.6 Oxazolinone Formation via Intramolecular Attack of Urethane Oxygen on Carboxyl Group.

The urethane carbonyl can also attack the carboxyl group, giving oxazolinones. These heterocycles were obtained during attempts to synthesize β-lactones by the cyclization of *N*-protected Thr derivatives through activation of the carboxyl group (see Scheme 10.6) (*29, 30*).

An unprotected β-hydroxyl group can also act as an intramolecular nucleophile under basic conditions. Cbz-protected Ser and Thr form oxazolidones upon treatment with alkali; in this case the side-chain hydroxyl oxygen attacks the carbonyl of the urethane and displaces benzyl alcohol (see Scheme 10.7) (*31, 32*).

10.3.3 Synthesis of Activated Serines

The hydroxyl group of Ser has been converted into a number of different leaving groups (see Scheme 10.8,

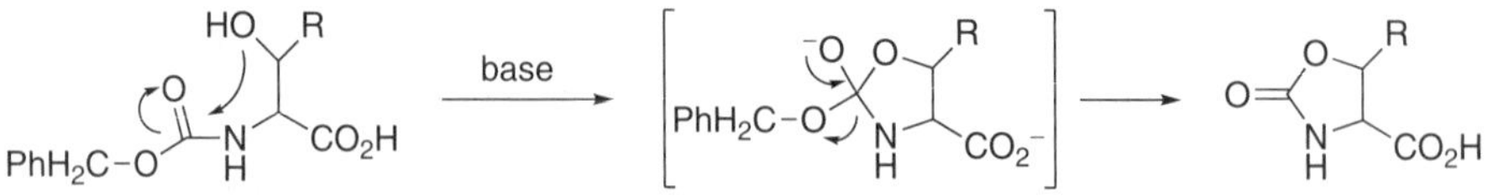

Scheme 10.7 Oxazolidone Formation via Intramolecular Attack of Hydroxyl Group on Urethane.

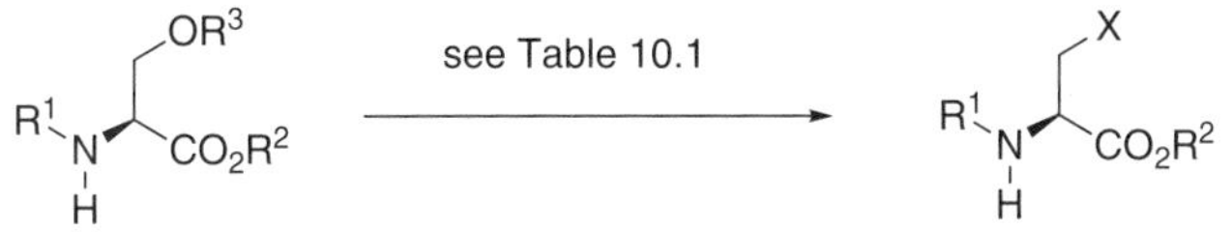

Scheme 10.8 Synthesis of Activated Serine Equivalent.

Table 10.1). Tosyl derivatives of Trt-Ser-OBn (33), Cbz-Ser-OMe (5), Cbz-Ser-OBn (34, 35), Boc-Ser-OBn (36), Boc-Ser-OpNB (34), and other protected serines (34) have been isolated, as has a mesylate of Boc-Ser-OMe (36). Other mesylates have been employed as non-isolated intermediates (36, 37–39). Sibi et al. have taken a different approach, with the carboxyl group of oxazolidine-protected Ser converted into a tosylated hydroxymethyl moiety (see Scheme 10.10 below) (40). The carboxyl group is later regenerated from the Ser side chain, similar to syntheses with the Garner aldehyde (see Section 10.6.1).

Halogens have also been used as the leaving group. Protected β-chloro-Ala can be prepared by treatment of Ser-OMe with phosphorus pentachloride in chloroform (41–46), modified versions of a synthesis first reported by Fischer and Raske in 1907 (47). These authors determined that the β-chloro-Ala product retained optical activity, as it could be reduced to give L-Ala (48). A more recent report described an improved isolation of the product, and also notes that the deprotected chloro-Ala is not significantly racemized (49). Other routes to β-chloro-Ala are possible: several N-acyl β-chloro-Ala esters were prepared by chlorination of L-cystine diester (50), while an enzymatic reductive amination of 3-chloropyruvate gave β-chloro-Ala with >99.9% ee and 90% yield on a 5 mmol scale (51).

A number of side-chain-tosylated protected serines have been converted to β-iodo-Ala by tosylate or mesylate displacement with NaI in acetone (35, 36, 52–54), while a synthesis of β-iodo-Ala via NaI displacement of the mesylate of Alloc-Ser-Oallyl gave the desired product in 78% yield, along with some eliminated dehydro-Ala product (38). However, a reported conversion of Trt-Ser-OR via mesylation and NaI displacement (55) was actually found to proceed via aziridine formation, giving a mixture of β-iodo-Ala/α-iodo-β-Ala regioisomers in which the undesired 2-iodo-3-aminopropionic acid predominated (56, 57). This side reaction was avoided by direct conversion of Trt-Ser-Oallyl into β-iodo-Ala in 72% yield via Mitsunobu reaction with MeI/PPh₃/DEAD (38). β-Iodo-Ala has also been prepared from Cbz-Ser-OMe by treatment with methyl triphenoxyphosphonium iodide (58); Cbz-Thr-OtBu was converted to one diastereomer of β-iodo-Abu in 55–60% yield using the same reagent; the other diastereomer (epimeric at the β-center) was then obtained as the major isomer by treatment with NaI (59).

Displacement of tosylated Ser with NaBr in acetone gave β-bromo-Ala (54). β-Bromo-Ala was also produced from Boc-D-Ser-OBn in 71% yield using CBr₄/PPh₃ (60), while β-bromo-Abu was obtained in 50% yield by treatment of Cbz-Thr-OtBu with CBr₄/PPh₃ (59). Formation of β-bromo-Ala from Ser within a peptide is also possible (48, 61). Trimethylsilyl halides are effective reagents for the conversion of serines with unprotected hydroxyl groups into β-haloalanine derivatives in one step, with 20–74% yields and no apparent racemization (24).

10.3.4 Displacement Reactions

Despite the pronounced tendency of activated Ser derivatives to undergo elimination reactions, a number of nucleophilic displacements have been successfully carried out without the accompanying racemization that would indicate an undetected dehydroalanine intermediate (see Scheme 10.9, Table 10.2). However, many of the older literature references are unclear as to the optical activity of starting material and/or product, and so should be viewed judiciously. Strong nucleophiles possessing poor basicity are required for successful displacements. These requirements were initially met by thiol or selenol nucleophiles. For example, sodium benzyl selenoate displaced the tosylate from O-tosyl N-Boc or N-Cbz Ser methyl or benzyl esters to give the enantiomerically pure selenocysteine adducts (34), as did thioacetate (6). Benzyl mercaptan (5, 6) and alkyl thiol anions (6) gave optically active Cys derivatives from tosylated Ser residues within peptides, but racemic products from Cbz-Ser(Ts)-OMe (6). Attempts at employing potassium phthalimide (50) or the enolate of malonic acid esters (62) for displacement of the tosylate from Cbz-Ser(Ts)-OMe also gave racemic products. Bz-Ser has been treated with PS₅ to directly give racemic Cys, with no side-chain activation (63). In 2002 Boc-Ser-OtBu was activated by reaction with benzyl sulfonyl chloride, followed by displacement with sodium azide and hydrogenation. The 2,3-diaminopropionic acid product had optical rotation equivalent to approximately 92% ee when compared to an older literature value (64). The carbonate of Ser(CO₂Me)-OBn was displaced by thiols in the presence of a Schiff-base-forming pyridoxal mimetic, but proceeded via a dehydroalanine intermediate and so gave racemic S-substituted Cys products (65). Attempted mesylation and azide displacement of Boc-, Fmoc- or Cbz-Ser methyl esters gave 77–82% yields of dehydroalanine. However, the Weinreb amide derivatives allowed for a mesylation/azide displacement sequence to produce β-azido-Ala in

Table 10.1 Synthesis of Activated Serine Equivalents

R¹	R²	R³	Leaving Group X	Conditions	Yield	Reference
H	Me	H	Cl	PCl_5	88–90%	1907 (47) 1907 (468)
H	Me	H	Cl	PCl_5	85–90%	1941 (43)
H	Me	H	Cl	PCl_5, $CHCl_3$	—	1949 (41)
H	Me	H	Cl	PCl_5, $CHCl_3$	82%	1968 (46)
CO_2Me	H	H	Cl	$SOCl_2$	28%	1968 (46)
H	Me	H	Cl	PCl_5, $CHCl_3$	66%	1971 (42)
H	Me	H	Cl	PCl_5	—	1980 (45)
H	Me	H	Cl	PCl_5, $CHCl_3$	81%	1987 (49)
Cbz	Me CO_2R^2 = CO-Gly-OEt	H	OTs	TsCl, pyridine	60%	1963 (5)
Cbz-Glu	Me CO_2R^2 = CO-Gly-OEt	H	OTs	TsCl, pyridine	80%	1965 (6)
Cbz Boc	Bn CHPh₂ 4-NO₂-Bn CO_2R^2 = CO-Gly-OBn	H	OTs	TsCl, pyridine	52–77%	1967 (34)
Trt	Bn	H	OTs	TsCl, pyridine	95%	1978 (33)
Boc	Me	H	OTs	TsCl, pyridine	90%	1987 (180)
Cbz	Bn	H	OMs I	1) MsCl, Et₃N 2) NaI, acetone	—	1985 (469)
CO_2Me Boc Cbz	Me Bn Bn	H H H	OTs I	1) TsCl, pyridine 2) NaI, acetone	81–98% 96–97%	1993 (52)
Trt	Me Bn	H	OMs I	1) MsCl, Et₃N 2) NaI , acetone	quant. 95–100%	1996 (37)
Cbz	Bn	H	OTs I	1) TsCl, pyridine 2) NaI, acetone	75% for 2 steps	1997 (53)
Boc	Me	H	OTs I Br	1) TsCl, pyridine 2) NaI or NaBr, acetone	71% 85%	1997 (53)
Boc	Bn	H	OTs I	1) TsCl, pyridine 2) NaI, acetone	85% 80%	1997 (53)
Boc	Bn	H	OTs I	1) TsCl, pyridine 2) NaI, acetone	86% 60–81%	1997 (36)
Boc	Me	H	OMs I	1) MsCl, DIPEA 2) NaI , acetone	94% 60–81%	1997 (36)
Alloc	allyl	H	OMs I	1) MsCl, DIPEA 2) NaI , acetone	92% 78%	2005 (38)
Boc	Me	H	Br	CBr₄, PPh₃	73%	1996 (176)
Boc	Bn	H	Br	CBr₄, PPh₃	71%	2003 (60)
Fmoc-Tyr(tBu)	CO_2R^2 = CO-Gly-Phe-Cys(Trt)-resin	H	Br	CBr₄, PPh₃	—	1998 (61)
Cbz	Me	H	I	MePI(OPh)₃, DMF	81%	1994 (58)
Ac	CO_2R^2 = CONHBn	H	Cl Br I	TMS-X, MeCN, reflux	20–74%	1995 (24)
Trt	allyl	H	I I	MeI/PPh₃/DEAD	72%	2005 (38)

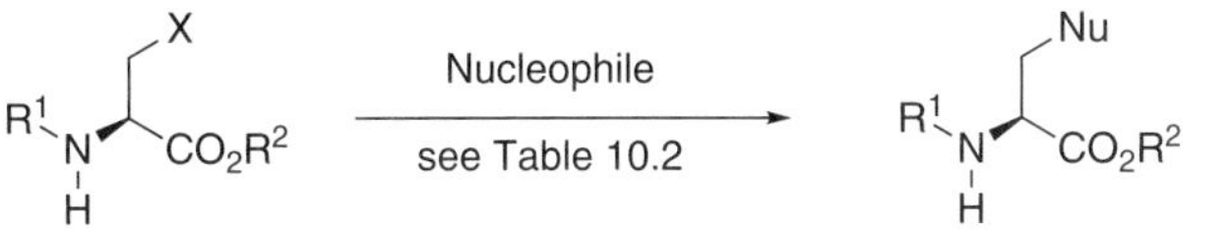

Scheme 10.9 Displacement Reactions of Activated Serine.

88–92% yield. The Weinreb amide could be hydrolyzed with LiOH in 88–90% yield (*39*).

In contrast to Ser, the tosylate of Boc-Thr-OMe or Boc-*allo*-Thr-OMe can be readily displaced with potassium thiolacetate without racemization or elimination (*66*). Azide displaced the mesylate of Thr in good yield if the carboxyl group was first reduced to a hydroxymethyl group; the acid group was then regenerated by Jones oxidation (*67*). The diastereomer epimeric at the β-center was accessed by first inverting the β-center via a Mitsunobu reaction with benzoic acid (*67*). Tosylation of Thr derivatives using tosyl chloride in pyridine at −5 °C led to different outcomes depending on protecting groups; Fmoc-Thr-OBn gave 100% yield of dehydro-Abu, Fmoc-Thr-OMe a 30% yield, and Cbz-Thr-OMe a 10% yield, while Cbz-Thr-OMe, Boc-Thr-OBn, and Boc-Thr-OMe gave 100% yields of the *O*-tosyl Thr products (*68*).

The Ser-derived β-haloalanines have been used for many displacements. Fischer and Raske converted Ser-derived β-chloro-Ala into optically active Cys by displacement with barium hydrogen sulfide in 1908 (*69*). Racemic [^{35}S]Cys was reported in 1947 via displacement of β-chloro-Ala with sodium benzyl [^{35}S]mercaptide (*70*). Initial syntheses of racemic selenocysteine employed benzyl or phenyl selenomercaptan to displace β-chloro-Ala-OMe or Bz-β-chloro-Ala-OMe (*71*). Dilithioselenide and dilithiotelluride have been reacted with β-halo-Ala methyl esters to give L-selenocystine and L-tellurocystine with >99% ee (*54*), an extension of earlier racemic syntheses (*72*). Boc-Ser-OMe was tosylated, displaced with NaBr, and then with Li$_2$Se$_2$ to give selenocystine (*73*). Better yields were provided by displacement of Fmoc-Ser(Ts)Oallyl with PMBSeH (*73*). Displacement of L-β-chloro-Ala with selenocysteamine (2-aminoethylselenol) gave 4-seleno-Lys with an optical rotation that indicated at least some optical purity was maintained (*74*); another report that claimed non-racemic product provided no evidence for optical purity (*75*). Several *S*-aryl-substituted selenocysteine derivatives have been prepared via reaction of D- or L-β-chloro-Ala with aryl selenides (*76*). Thr-derived (2*R*,3*S*)-3-methylselenocysteine was produced in 47–58% yield from Boc-Thr-OBn or Fmoc-Thr-ODpm via tosylation and displacement with PhSeH (*77*). The iodo group of Cbz-β-iodo-Ala-OBn can be displaced with a thioglucopyranoside (*35, 78*).

Reaction of optically active β-haloalanines with the side-chain thiols of Cys (*79–81*) or homocysteine (*43, 44, 82, 83*) gave the thioether-linked diaminodicarboxylic acids lanthionine or cystathionine, respectively.

Trt-β-iodo-Ala-OBn, prepared from Trt-Ser-OBn via a mesylate intermediate, was employed for displacement reactions with protected Cys. The authors noted the presence of additional ^{1}H NMR signals, which were attributed to rotamers (*55*). However, a subsequent attempt to prepare lanthionine via Trt-β-iodo-Ala-Oallyl under the same conditions found that the major signals were in fact due to the regioisomer α-iodo-β-Ala, produced via iodide opening of an aziridine generated from the mesylate (*56, 57*). Thus, in the original paper the major products likely contained norlanthionine analogs, with a β-Ser residue. The correct lanthionine regioisomer was obtained in good yield (88%) by displacement of Alloc-β-iodo-Ala-Oallyl with Fmoc-Cys-O*t*Bu. The carbamate protecting group resulted in no formation of the β-Ala regioisomer, but an led to an 85:15 ratio of diastereomers, due to partial formation of dehydroalanine followed by conjugate addition of the Cys residue (*38*). A single lanthionine diastereomer was produced in 90% yield by Cys displacement of a Trt-β-iodoAla-Oallyl substrate, prepared by a different route to ensure the correct regiochemistry (*38*). Alternatively, good yields of single isomers of differentially protected lanthionines were obtained by displacement of Cbz-β-bromo-L-Ala-OBn, Boc-β-bromo-D-Ala-OBn or Fmoc-β-bromo-L-Ala-O*t*Bu with Boc-L-Cys-OBn, or of Boc-β-bromo-D-Ala-OBn with Fmoc-L-Cys-Oallyl (*60*).

Racemic 2,3-diaminopropionic acid (Dap) was prepared in 1907 by treatment of β-chloro-Ala with ammonia (*47*). β-Bromo-Ala was generated from Ser within a Boc-Gly-Ser-OMe dipeptide via treatment with PPh$_3$/CBr$_4$; azide displacement and hydrogenation of the azide produced Boc-Gly-Dap-OMe in 63% yield (*48*). Fmoc-β-chloro-Ala-O*t*Bu was converted into β-iodo-Ala in 99% yield using NaI in acetone (*84*). The iodo group of Cbz-β-iodo-Ala-OBn was transformed into a dimethylphosphonate by a Michaelis–Arbusov reaction (*53*).

Carbon-bond forming reactions are obviously of greater synthetic use. Viallefont and co-workers have successfully used organocuprates for displacement reactions of protected *O*-tosyl Ser, β-chloro-Ala or β-iodo-Ala, giving the alkyl adduct (0–75% yield) along with varying amounts (0–90% yields) of the eliminated dehydroalanine byproduct (*85–87*). The best results were generally obtained from the β-iodo-Ala substrate. Some racemization was observed (products possessed 84 to >95% ee), especially for reactions when little dehydroalanine was formed (*85*). Boc-L-Ser(Ts)-OMe reacted with the anion of *tert*-butyl cyanoacetate to give racemic γ-cyano-Glu (*88*).

Table 10.2 Displacement Reactions of Activated Serine Equivalents

$$R^1\text{-NH-CH}(CH_2X)\text{-}CO_2R^2 \xrightarrow{\text{Nucleophile}} R^1\text{-NH-CH}(CH_2Nu)\text{-}CO_2R^2$$

R^1	R^2	Leaving Group X	Nucleophile Nu	Conditions	Yield	Reference
H	H	Cl	NH_2	NH_3	55% (racemic)	1907 (47)
H	H	Cl	SH	$Ba(SH)_2$	20–25%	1908 (69)
H	Me	Cl	$S(CH_2)_2CH(NH_2)CO_2H$	homoCys, KOH	18%	1941 (43)
H	Me	Cl	$SCH_2CH(NH_2)CO_2H$	Cys, KOH	—	1941 (79)
H	H	Cl	$SCH_2CH(NH_2)CO_2H$	L-Cys, Na, KOH, H_2O	34%	1941 (80)
H	H	Cl	$S(CH_2)_2CH(NH_2)CO_2H$	L-homoCys, KOH	36%	1942 (82)
H	H	Cl	$S(CH_2)_2CH(NH_2)CO_2H$	L- or D-homoCys, Na	12–37%	1946 (83)
H	Me	Cl	SPh SeBn	NaSPh or Na SeBn, EtOH	56–60%	1947 (71)
Bz	Et	Cl	^{35}SBn	$K^{35}SBn$, EtOH	80% (racemic)	1947 (70)
H	H	Cl	SeH	BaHSeH, H_2O, reflux	20% (racemic)	1948 (72)
Bz	Me	Cl	$^{35}S(CH_2)_2CH(NHBz)CONHPh$	$Na^{35}S(CH_2)_2CH(NHBz)CONHPh$	83%	1950 (44)
H	H	Cl	SCH=CHMe	HSCH=CHMe, Li, NH_3	65%	1975 (470)
H	H	Cl	$SeCH_2CH_2NH_2$	$HSeCH_2CH_2NH_2$, NaOH	40%	1975 (74)
H	H	Cl	$Se(CH_2)_2NHBz$	$HSe(CH_2)_2NHBz$	53%	1976 (75)
H	H	Cl	$SCH_2CH(NH_2)CO_2H$	Cys	70–71%	1986 (81)
H	H	Cl	SeAr	HSeAr, NaOH	25–61%	1996 (76)
Boc	tBu	Cl	I	NaI, acetone	99%	2001 (84)
Cbz Cbz-Glu	Me CO_2R^2 = CO-Gly-OEt	OTs	SCOMe STrt $S(CH_2)_2NH_2$ SCH_2CO_2H	HSR, NaOMe, DMF	56–86%	1965 (6)
Cbz Boc	Bn $CHPh_2$ $4\text{-}NO_2\text{-Bn}$ CO_2R^2 = CO-Gly-OBn	OTs	SeBn	HSeBn, NaOH, DMF	64–88%	1967 (34)
Cbz	$CHPh_2$	OTs	SeH $SeCH_2CH(NHCbz)CO_2CHPh_2$	NaHSe $NaSeCH_2CH(NHCbz)CO_2CHPh_2$	78–93%	1970 (471)

N-Protecting group	Ester	Leaving group	Nucleophile / Product	Conditions	Yield	Year (Ref)
Boc	Me	OTs	nPr, nBu	R_2CuLi, Et_2O, −60°C	72–75%	1983 (87)
Ts	Me	Br	nPr		74%	
Bz	Me	Cl	Me, nPr, nBu		0–29%	
Boc	Me	OTs	Me	R_2CuLi, Et_2O, −60°C	0–75%	1984 (85)
Ts	Me	I	Et, nPr, nBu, Ph, $CH=CH_2$		84–95% ee	1985 (86)
Boc	Me	OTs	SPh	NaSPh, DMF, 0°C	quant.	1987 (180)
Boc	Me	OTs	$CH(CN)CO_2tBu$	$CH_2(CN)CO_2tBu$, NaH, DMF	69%	1991 (88)
Boc	Me	OTs	Br	NaBr	41%	2001 (73)
Boc	tBu	OSO_2Bn	N_3	NaN_3, DMF	70%	2002 (64)
Fmoc, Boc, Cbz	$CO_2R^2 = CO\text{-}NH(Me)OMe$	OMs	N_3	NaN_3	88–92% for mesylation and azidation	2004 (39)
Boc	Me	Br, I	SeH, TeH	Li_2Se_2 or Li_2Te_2, THF	56–85%, >99% ee	1997 (54)
Fmoc-Tyr(tBu)	$CO_2R^2 = CO\text{-}Gly\text{-}Phe\text{-}Cys\text{-}resin$	Br	$SCH_2CH(CO_2\text{-resin})\text{-}NH\text{-peptide}$	intramolecular from Cys side chain	—	1998 (61)
Boc	Me	Br	$Se)_2$	Li_2Se_2	87%	2001 (73)
Cbz	Bn	I	(peracetylated sugar thiouronium: OAc, OAc, OAc, AcO, OAc, S, NH_2, $NH_2^+Br^-$)		72%	1977 (35)
Boc	Me	I	(peracetylated sugar: OAc, O, AcO, AcO, AcHN, SH)	NaHMDS, DMF	84%	2001 (78)
Trt	Me, Bn	I	$CH(CO_2Me)_2$, $CH(CO_2tBu)_2$, $CH(CN)CO_2Me$, $CH(CN)CO_2tBu$, $CH(CN)_2$, $CH(CN)SO_2Ph$, $CH(CN)N=CPh_2$, $CH(CO_2Et)N=CPh_2$	CHR^3R^4, NaH or LDA THF-HMPA	30–90%	1996 (37)
Cbz	Bn	I	$P(=O)(OMe)_2$	$P(OMe)_3$, 60°C	—	1997 (53)

(Continued)

Table 10.2 Displacement Reactions of Activated Serine Equivalents (continued)

R¹	R²	Leaving Group X	Nucleophile Nu	Conditions	Yield	Reference
Trt	Bn	I	$SCH_2CH(NHFmoc)CO_2tBu$ $SCH_2CH(NHBoc)CO_2Me$ $SCH(Me)CH(NHFmoc)CO_2tBu$ $SCH(Me)CH(NHBoc)CO_2Me$ $SC(Me)_2CH(NHBoc)CO_2Me$	Fmoc-Cys-OtBu, Boc-Cys-OMe, Fmoc-β-Me-Cys-OtBu, Boc-β-Me-Cys-OMe or Boc-Pen-OMe Cs_2CO_3 **major regioisomer of starting material likely α-iodo-β-Ala**	41–88%	1997 (55) 2002 (57) 2003 (56)
Cbz	Bn	Br	$SCH_2CH(NHBoc)CO_2Bn$	pH 8.5 solution of $NaHCO_3$,	84–98%	2003 (60)
Boc	Bn	Br	$SCH_2CH(NHBoc)CO_2Bn$	TBAHS, EtOAC		
Fmoc	tBu	Br	$SCH_2CH(NHBoc)CO_2Bn$	Boc-Cys-OBn, Fmoc-Cys-Oallyl		
Boc	Bn	Br	$SCH_2CH(NHFmoc)CO_2allyl$			
Trt	allyl	I	$SCH_2CH(NHFmoc)CO_2tBu$ (using correct β-iodo-Ala regioisomer)	Fmoc-Cys-OtBu (using correct β-iodo-Ala regioisomer)	90%	2005 (38)
Alloc	allyl	I	$SCH_2CH(NHFmoc)CO_2tBu$	Fmoc-Cys-OtBu	88% 85:15 mixture diaster-eomers	2005 (38)
Bz	H	OH	SH	PS_5	— (racemic)	1903 (63)
Cbz	Et	OH	SH	PS_5	—	1904 (472)
Cbz	Me	OH	N_3	HN_3, PPh_3, DEAD or DIAD	53%	1984 (99)
Boc	Me	OH	N_3	HN_3, PPh_3, DEAD	73%	1985 (94)
Cbz	H Me	OH	N_3	HN_3, PPh_3, DEAD	35–41%	1987 (100) 1992 (119)
H	H	OH	4-Cl-indol-3-yl 5-Cl-indol-3-yl 6-Cl-indol-3-yl 7-Cl-indol-3-yl 4-aza-indol-3-yl 5-aza-indol-3-yl 6-aza-indol-3-yl 7-aza-indol-3-yl	tryptophan synthase	35–88%	1975 (62) 1998 (61)
Boc	Me	OH	N_3	HN_3, DEAD, PPh_3	73% + 9% ΔAla	1993 (102)

			Substituent	Reagents	Yield	Year (ref)
H	H	OH	(thiophene / indole structures)	tryptophan synthase	47%	1995 (122)
Trt / PhFl	Me	OH	phthalimide / O2C(4-NO2-Ph) / O(4-MeO-Ph) / O(4-CO2Me-Ph)	phthalimide, (4-NO2-Ph)CO2H, 4-MeO-PhOH or 4-CO2Me-PhOH, PPh3, DEAD	40–95%	1996 (103)
Boc-Gly	Me	OH	N3	NaN3, CBr4, PPh3, DMF	70%	1997 (48)
Trt	Me	OH	N(Mts)[4-S(4-Me-Bn)-Ph]	MtsHN[4-S(4-Me-Bn)-Ph] PPh3, DEAD	78%	1997 (104)
Boc	OBO ester	OH	N3	HN3, PPh3, DEAD	55–96%	1999 (105)
Cbz	tBu	OH	6-Cl-purin-9-yl	DEAD, PPh3	50%	2000 (114)
Fmoc	allyl	OH	CNSe(2-NO2-Ph)	Bu3P, pyridine	83%	2001 (110)
H	H	OH	7-Br-indol-3-yl	tryptophan synthase	—	2002 (120)
Boc	CO2R2 = CO-Phe-NHOBn	OH	C-terminal - NHOBn	DIAD, PPh3	75%	2003 (107)
Trt	allyl	OH	Fmoc-Cys-OtBu	ADDP, Me3P, zinc tartrate	50%	2003 (56)
Fmoc	allyl	OH	CNSe(3-CO2tBu-Ph)	Bu3P	74%	2003 (111)
Fmoc / Boc / Cbz	CO2R2 = CO-NH(Me)OMe	OH	N3	HN3, PPh3, DEAD	88–90%	2004 (39)
Boc	tBu	OH	SH / SeH	DIAD/Ph3P/AcSH / Ph3P/Br2/imidazole/N2H4/Se	76% / 74%	2004 (113)
H	H	OAc	pyrazole / 3,4-(OH)2-pyridine / 3,5-dioxadiazolidine / 3-isoxazolin-5-one / zeatin / uracil / 3-HO-5-Me-isoxazole / 3-NH2-1,2,4-triazole / ascorbic acid / 6-BnNH-purine / histidine	Ga3+, pyridoxal-5′-phosphate pH 4.0-4.5, acetate buffer	0.1–45%	1986 (128)
H	H	OAc	11CN	β-cyanoalanine synthase	93%	1995 (126) / 1997 (127)

Steric protection has been applied to nucleophilic displacements of Ser-derived β-iodo-Ala by malonate anions and other carbon nucleophiles (*37*). The *N*-trityl protecting group was critical for successful reaction, preventing α-proton abstraction and α-ester group saponification. *N*-trityl protection has traditionally been used in the synthesis of aziridines from Ser, and aziridine formation was a major byproduct of the malonate displacements. Indeed, only aziridine was obtained when a mesylate leaving group was employed instead of the iodide. The protected 4-carboxy-Glu products retained high enantiomeric purity (97–98% ee), and the trityl group was easily removed and replaced by a Boc group (*37*). Iodoalanine has also been employed for the preparation of nucleophilic Ser-derived synthons (see Section 10.5).

Mitsunobu reactions of Ser derivatives have been used to synthesize dehydroalanine (see Section 4.2.1) (*89*), β-lactones (see Section 10.3.7) (*90*), β-lactams (*91–93*) (see Section 8.5), oxazolines (*18, 19*), and aziridines (*19, 92, 94–98*) (see Section 8.2.1). These competing reactions have prevented Mitsunobu conditions from being used for many intermolecular displacements. Until recently only an azide group had been introduced, with hydrazoic acid as the nucleophile (*99–101*). Reduction of the azide gave Dap. For example, Boc-Ser-OMe reacted with HN$_3$/PPh$_3$/DEAD to give β-azido-L-Ala in 73% yield, along with 9% of the dehydro-Ala byproduct (*102*), although much higher elimination yields were obtained in a 2004 report (see below) (*39*). By employing a bulky *N*-phenylfluorenyl or *N*-trityl protecting group, Mitsunobu displacement products were obtained in good yield under conditions where, with *N*-Cbz or *N*-Boc protection, dehydroalanine was the predominant product. Phthalimide, 4-nitrobenzoic acid, and 4-carboxymethylphenol reacted successfully, although with the latter nucleophile aziridine byproduct formation was still a problem (*103*). A Mitsunobu displacement of Trt-Ser-OMe with the amino group of *N*-Mts,*S*-(4-MeBn)-4-aminothiophenol led to aminothiotyrosine, *N*^{β}-(4-thiophenyl)-Dap (*104*).

An OBO *ortho* ester protecting group strategy, developed by Blaskovich and Lajoie for the synthesis and reaction of Ser aldehydes (see Section 10.7), was also successfully applied to the synthesis of α,β-diamino acids from β-hydroxy-α-amino acids (Ser, Thr, and phenylSer) via a Mitsunobu replacement of the hydroxy group with azide. The OBO *ortho* ester carboxyl protecting group reduces elimination side reactions and epimerization at the α-carbon via both steric and electronic effects. For substrates with β-alkyl substituents, the displacement occurred with inversion to give stereochemically pure product, while with a β-phenyl group (from (2*S*,3*R*)-phenylSer) a 78:12 mixture of (2*S*,3*S*):(2*S*,3*R*) diastereomers was obtained. Azide reduction and OBO ester hydrolysis provided the *N*^{α}-monoprotected diamino acids (*105*). The OBO ester protection has also been employed during conversion of Ser to an aziridine-2-carboxylate, followed by nucleophilic ring opening (*106*). Alternatively, a Weinreb

amide has been found to prevent the elimination normally associated with Mitsunobu azidation reactions. Attempted Mitsunobu reactions of Fmoc-, Cbz-, or Boc-Ser-OMe with DEAD/PPh$_3$/HN$_3$ gave only dehydroalanine derivatives in 88–92% yield, but by using Weinreb amide derivatives the desired β-azido-Ala products were obtained in 88–90% yield, with 92–97% ee. The Weinreb amides were hydrolyzed with LiOH to give 88–90% yields of the free acids, with 97–98% ee (*39*). An amide derivative also provided protection from elimination during cyclization of a Boc-Ser-Phe-NHOBn dipeptide hydroxamate, providing a side-chain acylated *N*-benzyloxy derivative of 2,3-Dap in up to 75% yield (*107*). Mitsunobu displacements of Thr are possible if Thr amides are used as substrates; (2*S*,3*S*)-2,3-diaminobutyric acid was prepared by using azide as the nucleophile (*108, 109*).

Alkyl thiols are normally not compatible with Mitsunobu conditions, but lanthionine was successfully synthesized from Trt-Ser-Oallyl and Fmoc-Cys-O*t*Bu using ADDP and Me$_3$P in the presence of catalytic zinc tartrate (*56*). A selenocysteine derivative, *Se*-2-nitrophenyl Fmoc-Sec-Oallyl, was obtained in 83% from Fmoc-Ser-Oallyl, *o*-nitrophenyl selenocyanide, and Bu$_3$P/pyridine (*110*). Similarly, *m*-(*tert*-butoxycarbonyl)phenyl selenocyanide and Bu$_3$P selenated Fmoc-Ser-Oallyl in 74% yield and Fmoc-D-*allo*-Thr-Oallyl in 55% yield. Under the same reaction conditions, Fmoc-L-Thr-Oallyl gave only eliminated 2,3-dehydro-Abu product (*111*). However, in a subsequent report, Fmoc-L-Thr-Oallyl and Fmoc-*allo*-L-Thr-Oallyl were reacted under similar conditions to form the corresponding selenoethers (with inversion at the β-center) in 29% and 73% yield, respectively (*112*). Another report formed unprotected Cys or Sec from Boc-Ser-O*t*Bu via Mitsunobu reaction with Ph$_3$P/DIAD/AcSH (76%) or Ph$_3$P/Br$_2$/imidazole/N$_2$H$_4$/Se (74%) followed by 1 M KOH/TFA or NaBH$_4$/TFA, respectively (*113*). Allyl-Ser-O*t*Bu was employed for a Mitsunobu reaction with 6-chloropurine to give the β-nucleo amino acid in 50% yield, though enantiomeric purity was not reported (*114*).

The electrophilic alaninol synthon of Sibi et al. (*40*) employs a tosylated hydroxymethyl side-chain that is actually derived from the Ser carboxyl group. It was displaced by a number of organocuprates or Grignard/CuX reagents in excellent yield (82–97%, see Scheme 10.10). The displaced adducts were not converted into amino acids (which requires ring opening and oxidation of the Ser side-chain hydroxymethyl group to an acid), but this has been carried out for a range of similar compounds produced from the Garner aldehyde (see Section 10.6.1). The tosylate was also converted into an iodo leaving group; reaction with silyl copper species led to β-silylalanines after ring opening/oxidation (*115*). Displacement of a mesylated analog with LiCF$_2$PO$_3$Et$_2$ provided a synthesis of 4,4-difluoro-4-(diethylphosphono)-Abu (*116*). A similar synthon was prepared from the amino alcohol precursor of the Garner aldehyde (see Section 10.6.1a); tosylation and displacement with LiPPh$_2$, followed by treatment with S$_8$, led to

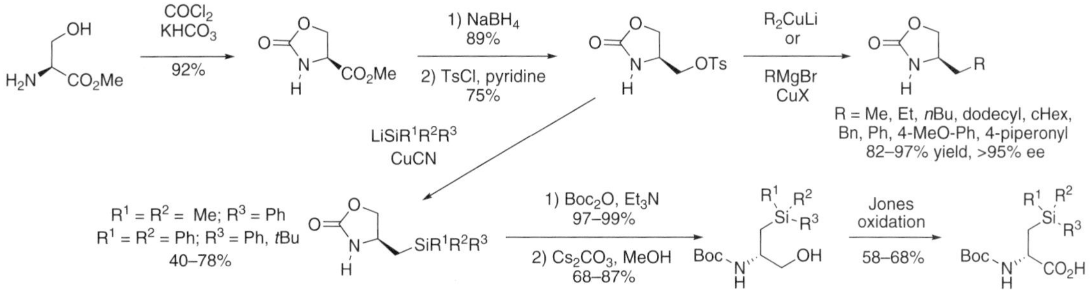

Scheme 10.10 Synthesis and Reaction of Electrophilic Alaninol Synthon (40).

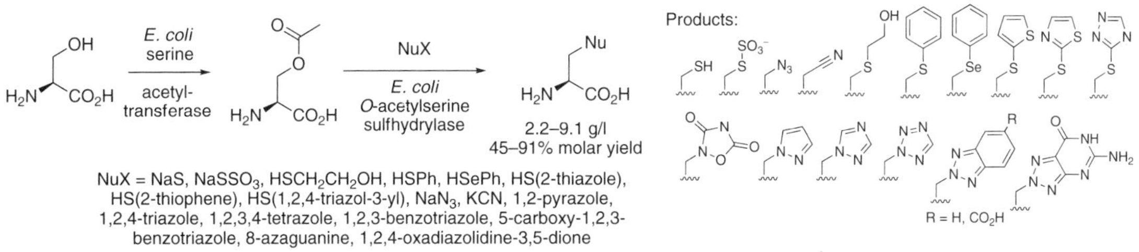

Scheme 10.11 Fermentation Synthesis of β-Substituted Alanines via O-Acetylserine Sulfhydrylases (129).

β-[P(=S)(Ph)$_2$]-Ala (the sulfide of diphenylphosphoryl-alanine) in 22% overall yield with >99% ee (117).

Several enzymatic amino acid syntheses employ Ser as the substrate. Tryptophan synthase enzymatically converted unprotected Ser to a number of optically pure analogs of L-Trp by reaction with various isotopically labeled indoles (118), substituted chloroindoles (119), 7-bromoindole (120), aza-indoles (121), or thienopyr-roles (122). Tyr analogs were produced from isotopically labeled or substituted phenols (123). The use of trypto-phan synthase in preparing Trp analogs was reviewed in 2004 (124). L-Mimosine synthase catalyzed the displacement of O-acetyl-L-Ser by 3,4-dihydroxypyri-dine, pyridine, or 3-amino-1,2,4-triazole (125), while β-cyanoalanine synthase and H^{11}CN provided isotopi-cally labeled β-cyano-Ala (126, 127). Hydrolysis, reduc-tion or percarbonate oxidation gave [4-^{11}C]-L-Asp, [4-^{11}C]-L-2,4-diaminobutyric acid or [4-^{11}C]-L-Asn in 30–85% radiochemical yield (126, 127). A biomimetic synthesis employed pyridoxal 5′-phosphate and metal ions to catalyze the displacement of O-acetyl-L-Ser by various heterocyclic compounds, leading to mimosine, quisqualic acid, lupinic acid, willardine, isowillardine, ascorbalamic acid, β-(6-benzylaminopurin-9-yl)-Ala, β-(3-amino-1,2,4-triazol-1-yl)-Ala, β-(3-isoxazolin-5-on-2-yl)-Ala, β-(5-methylisoxazolin-3-on-2-yl)-Ala, histi-dinio-Ala, and β-(pyrazolyl)-Ala, albeit in variable (0.1–45%) yields (128).

A potentially useful fermentation route for industri-al-scale production of unusual amino acids was reported in 2003. Metabolic engineering of the Cys biosynthetic

pathway in E. coli overproduced O-acetyl-L-Ser, which was then transformed into various β-substituted-Ala derivatives on a multiple g/L scale by O-acetylSer sulf-hydrylase-catalyzed reaction with heteroatom nucleo-philes (see Scheme 10.11). Thiols and heteroaryl thiols worked well, as did N-heterocycles as long as they were five-membered with at least two neighboring nitrogen atoms. Nucleophiles such as pyrrole, imidazole, or pyra-zine were unsuccessful. The unnatural amino acid prod-ucts were conveniently isolated from the culture broth, as they were efficiently secreted by the cells (129).

10.3.5 Aziridines and Azetidinones Derived from Serine

Regioselective C3 ring openings of aziridine-2-carboxy-lates derived from Ser correspond to a net displacement of the Ser hydroxyl group by a nucleophile, but as the aziridine substrates can also be prepared by other meth-ods these reactions are discussed in Section 8.2.1. Similarly, β-lactams can be derived from Ser and opened to give amino acid products, but these are covered in Section 8.5.

10.3.6 Isooxazolidinone, Sulfamidate, and Oxazoline Derivatives of Serine

Baldwin et al. have developed two other cyclic Ser derivatives. A β-lactam was prepared from Ser (see Section 8.5) and then catalytically isomerized to an

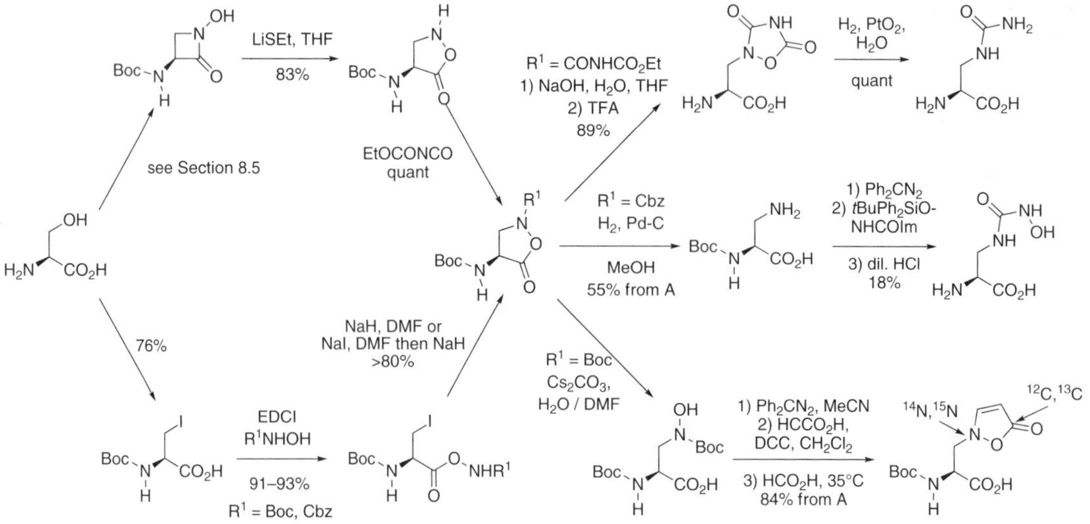

Scheme 10.12 Synthesis and Reactions of Isooxazolidinone (*130, 131*).

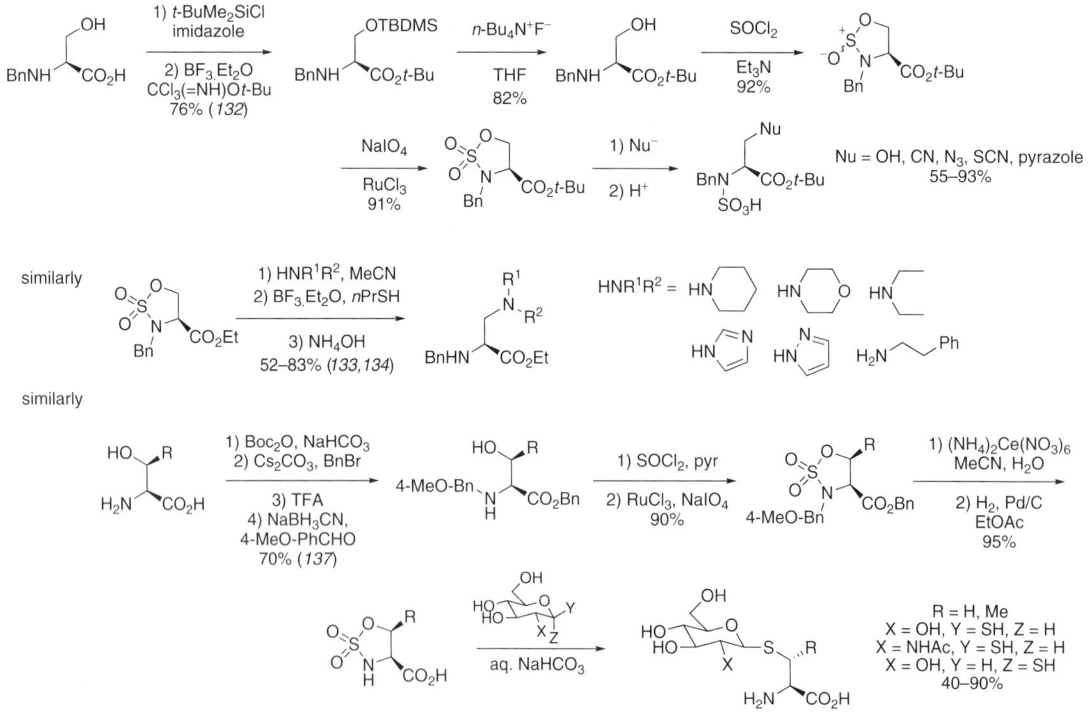

Scheme 10.13 Synthesis and Reaction of Cyclic Sulfamidate .

isooxazolidin-5-one (see Scheme 10.12). *N*-acylation and ring opening produced an α,β-diamino acid derivative (*130*). The isooxazolidin-5-one was more readily prepared by intramolecular displacement of Ser-derived β-iodo-Ala (see Scheme 10.12) (*131*). Hydrogenation gave protected 2,3-Dap acid in 55% yield from β-chloro-Ala. A number of other α,β-diamino acid derivatives were prepared by this method.

Ser can also be converted to a chiral sulfamidate in 52% yield (see Scheme 10.13). Ring opening under neutral or acidic conditions with a variety of nucleophiles gave β-substituted alanines in 55–91% yield (*132*).

Attempts at opening with more basic organometallic reagents were unsuccessful. Resynthesis of Ser gave a product with 91% ee by optical rotation; thus, some racemization occurred during the derivatization procedure. The N-benzyl ethyl ester sulfamidate derivative was opened with secondary amines, imidazole, and pyrazole (133, 134). Aziridine formation via intramolecular ring-opening was promoted by using a bulky N-protecting group and non-hindered ester group (135). An N-(9-phenylfluorenyl)-protected sulfamidate was opened with enolates of β-keto esters, β-keto ketones, and dimethyl malonate to give various γ-substituted-Glu or δ-keto amino acid derivatives. However, the products were obtained in racemic form due to NaH-induced formation of dehydroalanine from the cyclic sulfamidate, with the adducts predominantly produced by conjugate addition to this intermediate (136).

A sulfamidate derived from homoSer has also been reported (see Scheme 9.38) (478), as has one from allo-Thr (137). The sulfamidates derived from N-(p-methoxybenzyl) Ser or allo-Thr benzyl esters were deprotected by oxidative removal of the PMB-protecting group and hydrogenation of the benzyl ester. Reaction with unprotected 1-thio sugars in aqueous bicarbonate generated S-linked glycosyl Ser and Thr analogs with excellent yields for the Ser-derived products (85–90%) and moderate yields (40–60%) for the analogous β-methyl-substituted products (137). Alternatively, the benzyl ester could be selectively hydrogenated and the carboxylic acid amidated with an amino acid residue (solution or solid-phase) before N-deprotection; the sulfamidate was then opened with the glycosyl thiols (137). The use of cyclic sulfites and sulfates (including sulfamidates) in organic synthesis was reviewed in 2000 (138), with another review in 2003 discussing the synthesis and reactivity of cyclic sulfamidites and sulfamidates (139). Cyclic sulfamidates of 1,2-diols have been employed as epoxide equivalents (see Section 8.4.4).

Both Ser and Thr methyl esters have been cyclized to oxazolines by condensation with iminoether hydrochlorides (see Scheme 10.14). Ring opening/halogenation was accomplished by reaction with trimethylsilyl halides, alkyl chloroformates, or alkyl chloroformates in the presence of NaBr or NaI to give the β-chloro-, -bromo- or -iodo-α-amino acids in 30–89% yield (26). The oxazoline formed from Ser has also been opened with H₂S to give optically active Cys (7). The β-halo-Ala esters were converted into β-(triphenylphosphino)-Ala salts via reaction with PPh₃, giving enantiomerically pure products. However, attempts at ester hydrolysis led

to partial racemization. Instead, enantiomerically pure β-(triphenylphosphino)-Ala product was obtained by hydrolyzing the oxazoline ester before the ring-opening/phosphine-formation sequence. A preliminary Wittig coupling of the optically pure product with benzaldehyde gave a 30% yield of styrylglycine, unfortunately optically inactive (27).

10.3.7 β-Lactones Derived from Serine

One of the more successful methods for amino acid synthesis based on displacement of an activated Ser hydroxyl proceeds through a β-lactone intermediate, as developed by Vederas and co-workers (90). N-Protected Ser is cyclized under modified Mitsunobu conditions to give an α-amino-β-lactone in 60–72% yield without racemization (see Scheme 10.15). The cyclization has been shown to proceed via activation of the hydroxyl group rather than the carboxyl group (140). A slightly modified procedure helps to prevent undesired β-lactone oligomerization (141). Hints for improving yields of the lactone synthon when prepared on a large (kg) scale have been published (142).

The lactone ring opening can proceed by two pathways (see Scheme 10.15, Table 10.3). The desired alkyl–oxygen cleavage proceeds with "soft" nucleophiles such as acetate or imidazole, while "hard" nucleophiles such as ammonia or methoxide cause unwanted acyl–oxygen cleavage (90, 143). The correct regioselectivity for nitrogen nucleophiles can be achieved with trimethylsilyamines, which gave good yields of β-amino-Ala derivatives (144). Azide opening of the N-Boc lactone, followed by azide reduction and N-protection, produced differentially protected Dap (145). Opening of the N-Cbz lactone with a secondary amine gave an Nᵝ,Nᵝ,-disubstituted Dap derivative (146), while opening with N,N-Boc₂ 1,4,7-triazacyclononane gave an amino acid that can be employed as a ligand for transition metals (147). The procedure was employed to prepare Nᵝ-methyl and Nᵝ,Nᵝ-dimethyl Dap in 2003 (148). Opening with allylamine, followed by olefin oxidative cleavage and then intramolecular reductive amination, produced piperazine-2-carboxylic acid (149). No racemization was observed in any of these lactone openings.

A large-scale synthesis of N-Cbz,S-phenyl-L-Cys opened the N-Cbz lactone with thiophenol (142); 3,4-dichlorobenzylthiol has also been used to open the lactone (150), while the N-Boc lactone was also opened with thiophenols (151). Prenylated cysteines were prepared via lactone opening with prenyl thiolate (152),

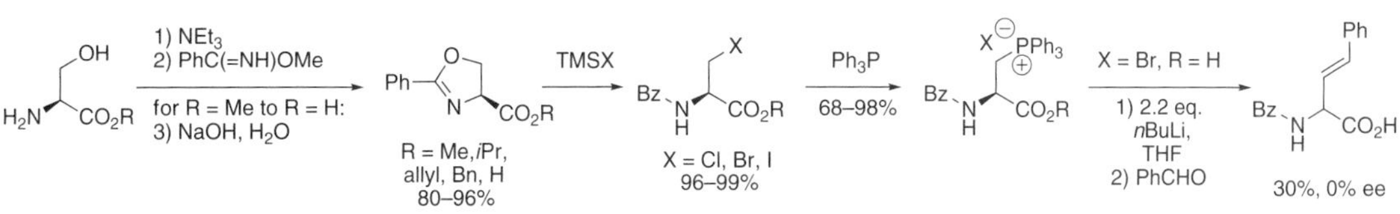

Scheme 10.14 Synthesis and Reactions of Oxazoline (26, 27).

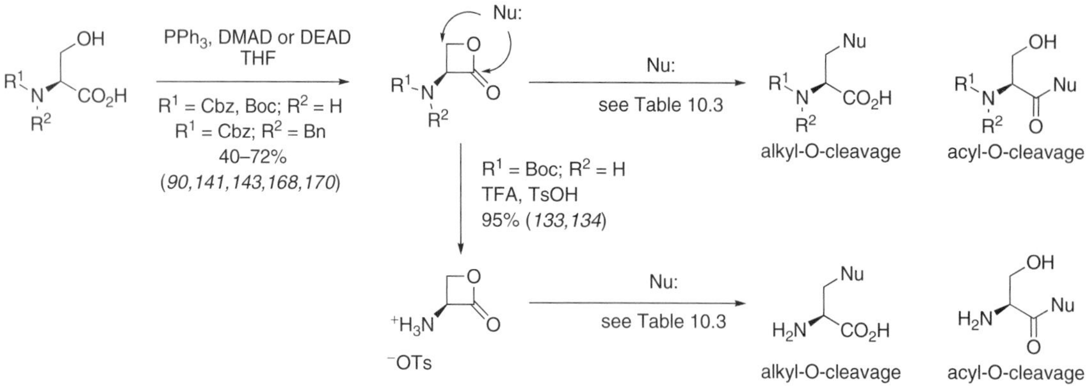

OH

PPh₃, DMAD or DEAD
THF

$R^1 = $ Cbz, Boc; $R^2 = $ H
$R^1 = $ Cbz; $R^2 = $ Bn
40–72%
(90, 141, 143, 168, 170)

Nu:

see Table 10.3

Nu
alkyl-O-cleavage

OH
Nu
acyl-O-cleavage

$R^1 = $ Boc; $R^2 = $ H
TFA, TsOH
95% (133, 134)

^+H_3N ^-OTs

Nu:

see Table 10.3

Nu
H_2N CO_2H
alkyl-O-cleavage

OH
H_2N Nu
acyl-O-cleavage

Scheme 10.15 Synthesis and Nucleophilic Opening of Serine-Derived β-Lactone.

while lanthionine or β,β-dimethyllanthionine resulted from opening the N-Cbz lactone with the side-chain thiol of Boc-Cys-OMe (153, 154) or Boc-penicillamine-OMe (153, 155). Selenocysteine could be obtained by opening with $(PhSe)_2$/NaHB(OMe)₃ (156) or PhSeH/ DMF (157). The susceptibility of the Cbz-β-lactone to thiol ring opening was employed to acylate the nucleophilic thiol of cysteine protease enzymes; the L-enantiomer irreversibly inhibited the hepatitis A virus 3C proteinase with 35 µM IC_{50}. The D-enantiomer possessed a 6 µM IC_{50} but was a competitive reversible inhibitor, presumably due to binding in a position incompatible with thiol attack at the β-carbon (158).

L-[4–¹³C]-Asp was prepared by opening the lactone with K¹³CN; the presence of dialkyl azodicarboxylate was found to prevent side reactions during the addition (141). The phosphonic acid analog, 2-amino-3-phosphonopropionic acid, was produced by reaction of the lactone with trialkyl phosphites (159, 160). Nucleic amino acids have been synthesized by opening the N-Boc lactone with thymine or adenine in the presence of DBU or K_2CO_3 (161, 162), with uracil or 6-chloropurine in the presence of NaH in anhydrous DMF (163), with Cbz-cytosine or 2-amino-6-chloropurine with DBU/DMSO (164), or with 2-amino-7-carba-6-chloropurine and 7-carba-6-chloropurine in the presence of DBU/DMSO (165). A series of substituted willardiine derivatives were prepared as kainate receptor antagonists via opening with 3-substituted-pyrimidin-2,4-diones and NaH/ DMF (166).

The N-deprotected lactone trifluoroacetate or tosylate salts are also substrates for nucleophilic opening, thus avoiding the need for amine deprotection (potentially harmful to some substituents) after ring opening (167, 168). The lactone tosylate salt is a stable solid. 4-Alkyl-prolines were produced by N-alkylation of the lactone, followed by lactone opening with sodium benzeneselenoate and pyrrolidine ring formation via an intramolecular radical cyclization (169).

Carbon–carbon bonds can be formed by reaction of the β-lactone with organometallic reagents (170).

Organocuprates were found to open the lactone with the desired regioselectivity, but Grignard or organolithium reagents attacked the carbonyl. Di-N-protected β-lactones were better substrates than mono-N-protected lactones for most additions of $R_2Cu(CN)Li_2$ and R_2CuLi reagents, with average yields of 70–80% for the diprotected substrate and 50% for the monoprotected. However, the diprotected synthons suffered from greater decreases in optical purity during the additions (10–14% decrease in ee for diprotected, <2% for monoprotected). These problems were avoided by using Cu(I)-catalyzed Grignard additions, which gave good yields and <1% racemization with both types of substrates. Opening the N-Boc β-lactone with n-pentylMgBr/CuBr.Me₂S gave 2-amino octanoic acid in 55–65% yield (139).

Attempts to prepare the corresponding Thr-derived lactone from N-acyl protected substrates failed (30). However, the N-phenylsulfonyl derivatives of Thr, allo-Thr, and 2-amino-3-hydroxypentanoic acid could be cyclized in 40–55% yield by treatment with 4-bromobenzenesulfonyl chloride in pyridine (see Scheme 10.16, Table 10.4). Unfortunately, only certain nucleophiles attacked (with inversion) the β-carbon to give the β-substituted products. Most useful reagents, such as organocuprates, attacked at the carbonyl instead. Apparently steric or electronic properties of the N-substituent alter the relative reactivity of the two positions. The phenylsulfonyl group was removed to give unprotected tosyl salts, which were then acylated (29). Nucleophilic ring opening of several of these derivatives by bromide ion gave pure 2-amino-3-bromobutanoic acids, but other nucleophiles were again ineffective (29).

The lactone derived from α-methyl-Ser, N,N-dibenzyl-α-methyl-Ser-β-lactone, has been prepared via activation/cyclization with HBTU. Ring opening with various Grignard-derived organocuprates gave a variety of α-methyl amino acids in 43–98% yield (171). Similarly, a series of other N-Boc protected α-substituted serines (methyl, isopropyl, isobutyl, benzyl) were cyclized to the lactones using DEAD/PPh₃ (92–98% yield). The lactones were N-deprotected, coupled with

Table 10.3 Nucleophilic Opening of Serine-Derived β-Lactone (see Scheme 10.15)

R¹	R²	Nucleophile Nu	Conditions	Alkyl-O Cleavage	Acyl-O Cleavage	Reference
Boc	H	NH_2	NH_3/THF	—	79%	1985 (90)
Cbz	H	NH_2	NH_3/THF		77–88%	1985 (90)
		OMe	NaOMe, MeOH-THF			
Cbz	H	$N(Me)_2$	$HN(Me)_2$	43–99%		1985 (90)
		$S(CH_2)_2NH_2$	$HS(CH_2)_2NH_3Cl$			1992 (143)
		OAc	NaOAc			
		SBn	NaSBn			
		Cl	$MgCl_2$-Et_2O			
		Br	$MgBr_2$-Et_2O			
		imidazol-1-yl	imidazole			
		$NHC(=S)NH_2$	$H_2NC(=S)NH_2$			
Cbz	H	Me	RLi, CuCN or CuBr.SMe_2	25–92%		1987 (170)
Cbz	Bn	iPr	R = Me, nBu, sBu, tBu, Ph, $CH=CH_2$	33–100% ee		
		nBu	RMgX, CuCN or CuBr.SMe_2			
		sBu	R = iPr, Ph, $CH=CH_2$			
		tBu				
		Ph				
		$CH=CH_2$				
Ts salt	H	$S(CH_2)_2NH_2$	$HS(CH_2)_2NH_3Cl$	77–96%		1988 (167)
		SH	LiSH			
		$SCH_2CH(NH_2)CO_2H$	L-Cys			
		S_2O_3H	$Na_2S_2O_3$			
		$COCF_3$	CF_3CO_2-			
		OPO_3H	K_2HPO_4			
		Cl	HCl			
		CN	nBu_4CN			
		N_3	NaN_3			
		imidazol-1-yl	imidazole			
Boc	H	$P(=O)(OMe)_2$	$P(OMe)_3$	82%		1990 (160)
Boc	$CH_2CH(R^1)=CHR^2$	SePh	NaSePh	75–90%		1991 (169)
Boc	H	^{13}CN	$K^{13}CN$, DMSO, DMAD	29%	—	1992 (141)
Cbz	H	$SCH_2CH(NHBoc)CO_2Me$	Boc-Cys-OMe, Cs_2CO_3	50%		1993 (154)

(Continued)

Table 10.3 Nucleophilic Opening of Serine-Derived β-Lactone (see Scheme 10.15) (continued)

R¹	R²	Nucleophile Nu	Conditions	Alkyl-O Cleavage	Acyl-O Cleavage	Reference
Boc	H	imidazol-1-yl pyrazol-1-yl	imidazole pyrazole	61% 32%		1993 (102)
Cbz	H	NH_2 NHMe NHBn $N(Me)_2$ $N(Et)_2$ $N[-(CH_2)_4-]$ $N[-CH=CH-N=CH-]$ $N[-(CH_2)_2O(CH_2)_2-]$	$R^3R^4NSiMe_3$, MeCN	40–88%		1994 (144)
Cbz	H	$SCH_2CH(NHBoc)CO_2Me$ $SC(Me)_2CH(NHBoc)CO_2Me$ $SC[-(CH_2)_5-]CH(NHBoc)CO_2Me$	(R)- or (S)- $HSC(R^1)$ $(R^2)CH(NHBoc)CO_2Me$ $R^1 = R^2 = H$, Me, $-(CH_2)_5-$ Cs_2CO_3, DMF	50–92%		1995 (153)
Cbz	H	$SC(Me)_2CH(NHBoc)CO_2Me$	$HSC(Me)_2CH(NHBoc)CO_2Me$, Cs_2CO_3	—		1996 (155)
Boc	H	thymine adenine	thymine, DBU adenine, K_2CO_3	22–58%		1996 (161)
Boc	H	$NHCH_2CH=CH_2$	$NH_2CH_2CH=CH_2$	52%	41%	1997 (149)
Boc	H	$SCH_2CH=C(Me)C_{11}H_{19}$ $SCH_2CH=C(Me)C_{18}H_{27}$	$NaSCH_2CH=C(Me)C_{11}H_{19}$ $NaSCH_2CH=C(Me)C_{18}H_{27}$	52–60%		1997 (152)
Cbz	H	$N(Me)CH_2CH_2NHCOC(=CHPh)NHBoc$	$HN(Me)CH_2CH_2NHCOC(=CHPh)NHBoc$, Cs_2CO_3	73%		1998 (146)
Cbz	H	[structure: Boc–N ⋯ N–Boc pyrrolidine ring]	[structure: Boc–N ⋯ NH, N–Boc] MeCN, 48h	80%		1998 (147)
Fmoc	H	$P(=O)(Oallyl)_2$	$P(OTMS)(Oallyl)_2$	42%		1998 (159)
Boc	H	adenine	adenine, K_2CO_3	32%		1998 (162)

Boc	H	uracil		56–67%	1998 (163)
		6-Cl-purine	6-Cl-purine		
Cbz	Bn	S(3,4-Cl₂-Bn)	HS(3,4-Cl₂-Bn)	100%	1998 (150)
Boc	H	N₃	NaN₃, DMF	85%	1999 (145)
		(to NH₂)	(then H₂, Pd/C)	98%	
Boc	H	Cbz-cytosine	Cbz-cytosine, DBU/DMSO	60–65%	1999 (164)
		2-NH₂-6-Cl-purine	2-NH₂-6-Cl-purine, DBU/DMSO		
Boc	H	PhSe	(PhSe)₂, NaHB(OMe)₃	93%	2000 (156)
Boc	H	6-chloro-7-carbapurin-9-yl	6-chloro-7-carbapurine, DBU, DMSO	15–53%	2003 (165)
		2-amino-6-chloro-7-carbapurin-9-yl	2-amino-6-chloro-7-carbapurine		
Boc	S	S(4-MeO-Ph)	NaH, HS(4-MeO-Ph)	—	2003 (151)
Cbz	H	NHMe	(Me)₃SiNHMe, MeCN	70–72%	2003 (148)
		NMe₂	(Me)₃SiNMe₂, MeCN		
Boc	H	nPent	nPentMgBr, CuBr.Me₂S	55–65%	2004 (473)
Boc	H	3-Me-pyrimidin-2,4-dion-1-yl	NaH, DMF	20–73%	2005 (166)
		3-CH₂CO₂H-pyrimidin-2,4-dion-1-yl	3-R-pyrimidin-2,4-dione or 1-R-pyrimidin-2,4-dione		
		1-CH₂CO₂H-pyrimidin-2,4-dion-3-yl			
		3-(CH₂)₂CO₂H-pyrimidin-2,4-dion-1-yl			
		3-(CH₂)₃CO₂H-pyrimidin-2,4-dion-1-yl			
		3-(2-CO₂H-Bn)-pyrimidin-2,4-dion-1-yl			
		3-(3-CO₂H-Bn)-pyrimidin-2,4-dion-1-yl			
		3-(4-CO₂H-Bn)-pyrimidin-2,4-dion-1-yl			
		3-(4-CO₂H-Bn)-5-1-pyrimidin-2,4-dion-1-yl			
Boc	H	PhSe	PhSeH, DMF	82%	2005 (157)

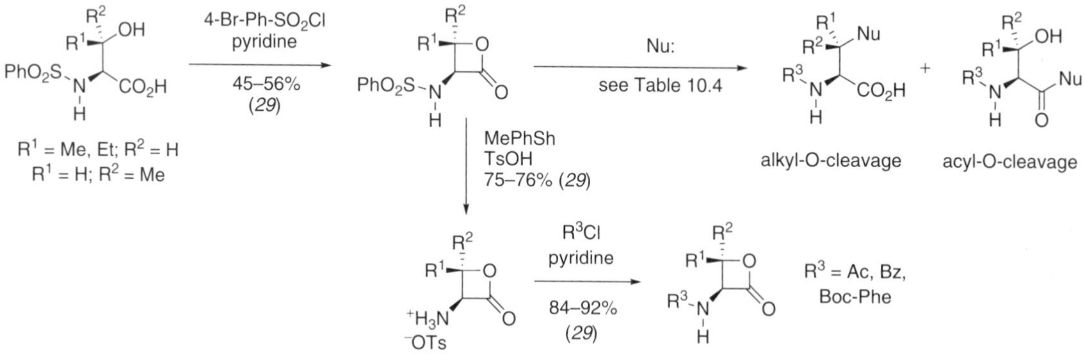

Scheme 10.16 Nucleophilic Opening of β-Hydroxy Amino Acid-Derived β-Lactone.

Table 10.4 Nucleophilic Opening of β-Hydroxy Amino Acid-Derived β-Lactone (see Scheme 10.16)

R^1	R^2	R^3	Nucleophile Nu	Conditions	Alkyl-O Cleavage	Acyl-O Cleavage	Reference
Me	H	SO$_2$Ph	SH	LiSH or NaSH		30–86%	1989 (30)
			pyrazol-1-yl	pyrazole			
			NHBn	BnNH$_2$			
			Et	EtMgCl + CuBr.SMe$_2$			
Me	H	SO$_2$Ph	NHC(=S)NH$_2$	N$_2$HC(=S)NH$_2$	51–99%		1989 (30)
			OAc	NaOAc			
			Cl	MgCl$_2$.Et$_2$O + nBu$_4$NCl			
			Br	MgBr$_2$.Et$_2$O			
			I	MgI$_2$.Et$_2$O			
Et	H	SO$_2$Ph	Br	MgBr$_2$	77–80%		1989 (30)
H	Me	SO$_2$Ph					
Me	H	SO$_2$Ph	Br	HBr	68–92%		1991 (29)
Me	H	Ts salt					
H	Me	SO$_2$Ph					

Boc-Asp(OBn)-OH, and then opened with 4-methoxy-benzylthiol to give α-substituted-Cys products (172).

10.4 Radical Reactions

Ser-derived β-haloalanines are also useful substrates for radical reactions. Photolytically (173) or thermally (174) initiated coupling of protected iodo-Ala with allylstannanes gave coupled products in 62–70% yield (see Scheme 10.17). One of the product alkenes was subsequently epoxidized and converted into a precursor of tabtoxinine β-lactam, the causative agent of wildfire disease in tobacco plants (173). A propargylstannane was also coupled, giving an allenic amino acid (175). In a similar type of reaction, Boc-3-bromo-Ala-OMe was added to ethyl acrylate in 46% yield using vitamin B$_{12}$ as radical catalyst (176), while the radical derived from Cbz-β-iodo-Ala-OBn by treatment with Bu$_3$SnH added to acrylic acid in 30% yield to give optically pure homo-Glu

(2-aminoadipic acid) (177). Alternatively, intramolecular reaction of the radical with an N-propargyl group produced 4-exomethylene-Pro (178). Boc-3-iodo-Ala-OMe was added to a glycosyl difluoroalkene in the presence of AIBN and SnBu$_3$H, resulting in a difluoromethylene-linked analog of O-glycosyl Ser (179). An oxazolidinone derivative of β-bromo-Ala gave improved yields (179). A serinol-derived iodo-oxazolidine (similar to Sibi's nucleophilic serinol, see Scheme 10.21 below, Section 10.5) was coupled with α,β-unsaturated acids under zinc/copper radical addition reaction conditions, but the products were not elaborated to amino acids (58).

10.5 Nucleophilic Alaninol Synthons Derived from Serine

Several groups have reported preparing a nucleophilic alaninol equivalent derived from Ala: i.e., a β-anionic synthon.

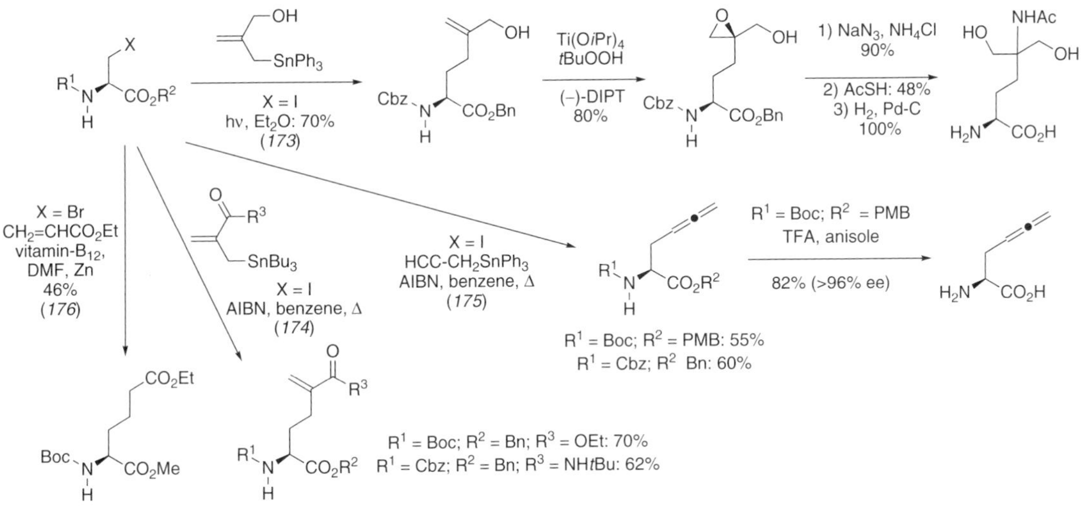

Scheme 10.17 Radical Reactions of 3-Haloalanine.

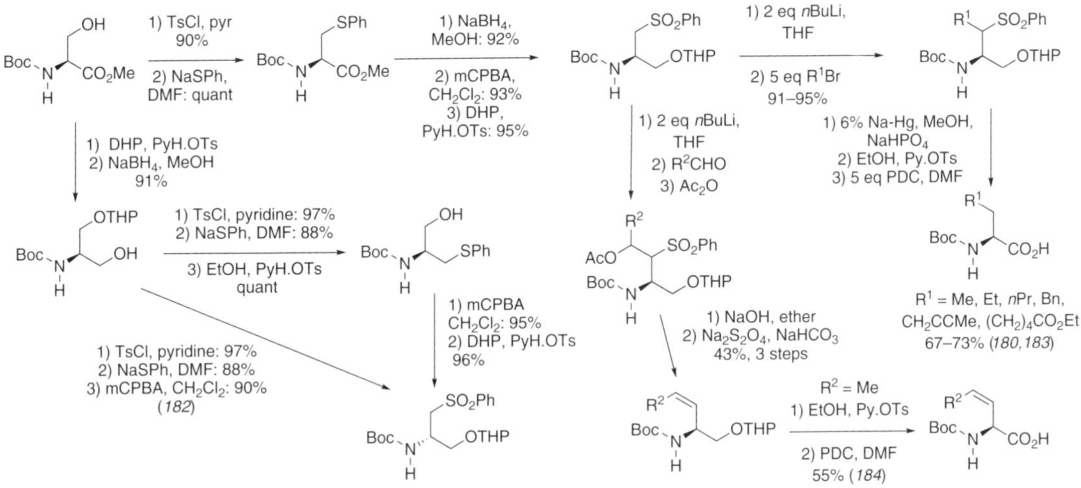

Scheme 10.18 Serine-Derived Alaninol Equivalent: Sasaki Aminosulfone.

10.5.1 Sasaki Aminosulfone

Sasaki et al. prepared a β-anion synthon from L-Ser via displacement of a tosylated intermediate with a thiol anion. The resulting alkyl thioether was oxidized to a sulfone with mCPBA (180) or, more recently and in better yield, with magnesium monoperoxyphthalate (181). The ester was then reduced to a protected alcohol (see Scheme 10.18). The enantiomeric synthon could also be synthesized from L-Ser by converting the carboxyl group to the methylene-sulfone group (180, 182). Deprotonation of either enantiomer with 2 equiv of nBuLi generated the β-anion, which was reacted with alkyl halides. Removal of the phenylsulfonyl group by

reductive cleavage, followed by deprotection and oxidation of the alcohol, gave Boc-amino acids with >98% ee (180, 182, 183). The dilithiated synthon was also added to aldehydes, with the aldol adduct subsequently dehydrated to give vinylglycine derivatives (184).

Two types of cyclic amino acids were prepared by using an α,ω-dihaloalkane for the initial alkylation, with the cyclization regioselectivity based on the different reactivities of the carbamate and sulfonyl anions (181) (see Scheme 10.19). Pro-like derivatives were obtained by cyclization with the monoanionic carbamate anion remaining after initial alkylation. Alternatively, by adding a second equivalent of base, the sulfonyl anion was regenerated and used for a second alkylation to give

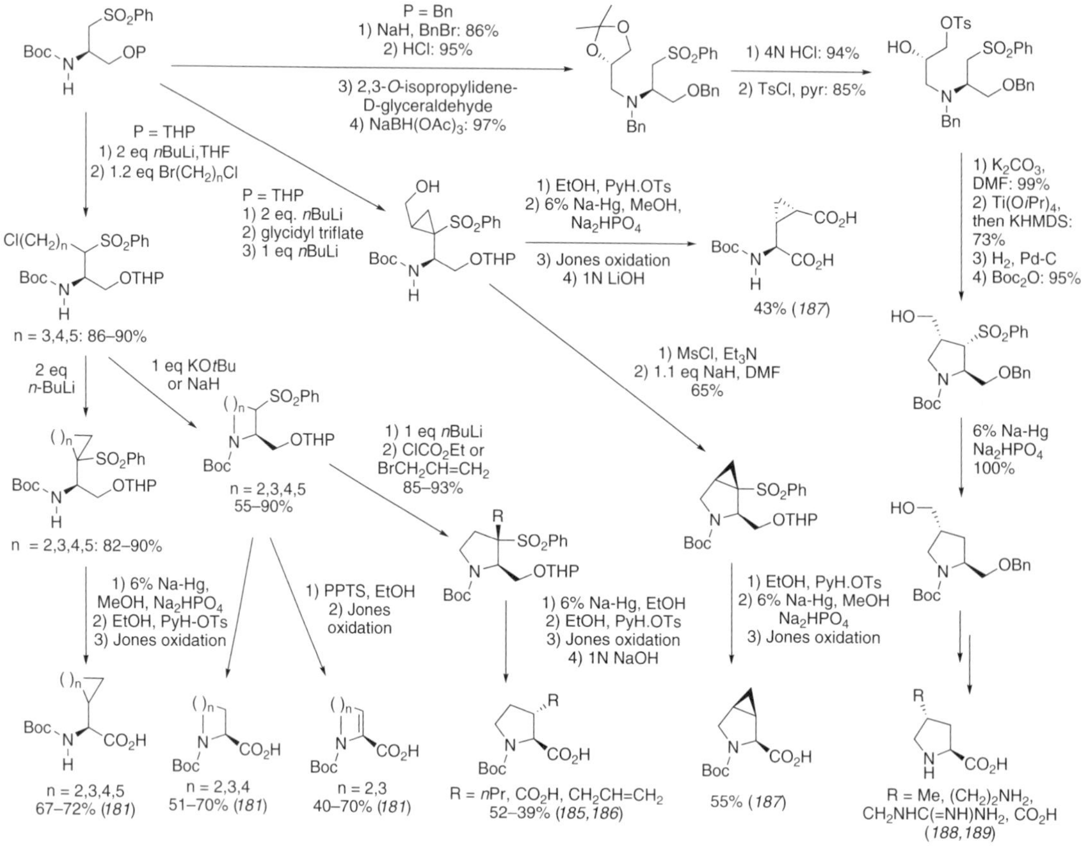

Scheme 10.19 Serine-Derived Alaninol Equivalent: Sasaki Aminosulfone.

α-(cycloalkyl)glycines. The initial Pro-like adducts could be further functionalized by another alkylation before removal of the phenylsulfone group (185), or unsaturation could be introduced (181). By reversing the order of alkylation it was possible to selectively produce either cis- or trans-3-substituted Pro derivatives (186). A similar divergent strategy stemmed from an initial alkylation with glycidyl triflate, giving 3,4-methano-Pro or 3,4-methano-Glu(187). 4-Substituted Pro derivatives have also been prepared from the Sasaki β-anion synthon, via reductive amination with (2R)-2,3-isopropylideneglyceraldehyde, isopropylidene deprotection, and cyclization via activation of the terminal hydroxy group and alkylation of the β-anion (188, 189).

10.5.2 Serine-Derived Wittig Reagents

A different alaninol synthon, based on a Wittig ylide, has been developed by Itaya et al. (190). Serine-derived β-iodo-Ala methyl or benzyl ester were treated with triphenylphosphine to give the phosphonium iodide (see Scheme 10.20). Attempts to generate an ylide from this substrate failed due to competing β-elimination. However, the ylide of the corresponding derivative without carboxyl group protection could be formed, due

to the decreased stability of an α-anion with a carboxyl anionic charge present. Reaction of this ylide with benzaldehyde gave the alkene of styrylglycine, with reduction providing homophenylalanine with 99% ee (190). A heterocycle was also coupled to give wybutine (190). Unfortunately, yields of these Wittig reactions were very poor (4–13%). Slightly improved yields with benzaldehyde (27–28%) were obtained by converting the phosphonium iodide to a chloride (191) or an internal salt (52). The chloride was also coupled to give wybutine (16% yield) (191). Varying the N-protecting group had little effect on reaction yields when tested with piperonal (38% yield for MeO₂C, 28% for Boc, 39% for Cbz, and 12% for Phth) (52). In all cases the adducts had high optical purity. A similar strategy in 2003 converted β-halo-Ala into β-(triphenylphosphino)-Ala salts via reaction with PPh₃, giving enantiomerically pure products. A preliminary Wittig coupling of the optically pure product with benzaldehyde gave a 30% yield of styrylglycine, but it was optically inactive in this report (27).

A racemic version of the phosphonium salt has been developed, prepared from the oxime of ethyl bromopyruvate (192). Reaction with aldehydes (RCHO; R = Et, iPr, Ph, 4-CF₃-Ph) proceeded in much better yield

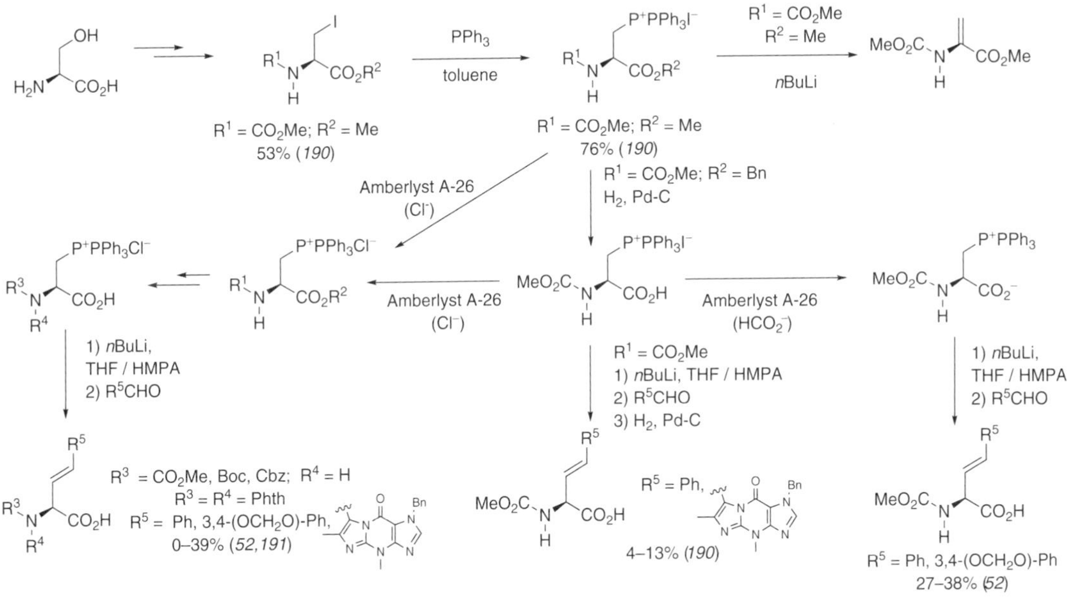

Scheme 10.20 Synthesis and Reactions of Acyclic Wittig Reagent.

(50–99%), with reduction of the oxime providing the racemic amino ester. For reaction with ketones the corresponding methyl phosphonate was prepared; yields were somewhat lower (24–54% for RCOR; R = Et, $-(CH_2)_5-$, $-(CH_2)_2S(CH_2)_2-$).

A more successful Ser-derived ylide has been prepared by Sibi and Renhowe (193). To avoid elimination problems the carboxy group was reduced to a hydroxymethyl group. The Ser side chain is then used as a masked carboxyl group protected within an oxazolidinone ring, while the carboxyl group becomes the side-chain alcohol that is transformed into the ylide (see Scheme 10.21). This inversion of Ser functionality is an approach that will be encountered again in the discussion on Ser aldehyde equivalents, and is used with the popular Garner aldehyde (see Section 10.6.1). Ylide generation and reaction with aromatic aldehydes gave alkenes in high yield with excellent E selectivity (>99:1), while aliphatic aldehydes tended to give a slight preference of the Z isomer (193, 194). Enantiomeric purities were >93–95%. The oxazolidinone-alkene can be opened to the amino alcohol, but only one example of deprotection was given in the initial reports. The final oxidation of the hydroxymethyl group that is required to generate β,γ-unsaturated α-amino acids was also not performed in the initial reports (although similar substrates have been oxidized by other groups). The alkenes can be hydroborated, but again the products were not converted into amino acids (195). However, the Sibi synthon has been applied to the synthesis of a carbon-linked glycosyl amino acid via condensation of the ylide with a galactose aldehyde, with Jones oxidation employed to regenerate the acid (196, 197). A full paper of the method was published

in 1999 (198). Another Wittig ylide dependent on a Ser-derived oxazolidine is described in Section 10.6.1d and Scheme 10.40.

10.5.3 Jackson β-Iodozinc Reagent

The most successful β-anion synthon has been developed by Jackson et al., comprising an organozinc reagent that is prepared from Ser-derived Boc-β-iodo-Ala-OBn by ultrasonication with zinc in benzene/dimethylacetamide (see Scheme 10.22) (199–201). An improved procedure allows the reagent to be prepared in THF (202, 203), and the corresponding methyl ester can also be prepared (202). The development and application of this amino acid-derived organometallic was thoroughly reviewed in 2005 (204). The β-anion reagent was initially coupled with aryl iodides (199, 205, 206), bromides (see 199, 205, 207), or triflates (with lower yields than aryl halides) (208) (see Scheme 10.22), and acyl chlorides (see Scheme 10.23) (199, 200). No racemization was detected in any of the products from these couplings (199). The homologous reagents derived from γ-iodo-Abu and δ-iodo-Nva have also been prepared and arylated (206), as have reagents derived from the conversion of the α-carboxyl groups of Asp and Glu, which were employed to prepare β- or γ-amino acid derivatives (e.g., see Sections 11.2.3n, 11.4.3m) (208–211). More recently, Fmoc-β-iodo-Ala-OtBu was converted into the iodozinc reagent and coupled with 12 aryl iodides (21–59%) and five acyl chlorides (42–47%) (84).

The aryl coupling reaction has been used to prepare p-(dimethylphosphonomethyl)-L-Phe (212),

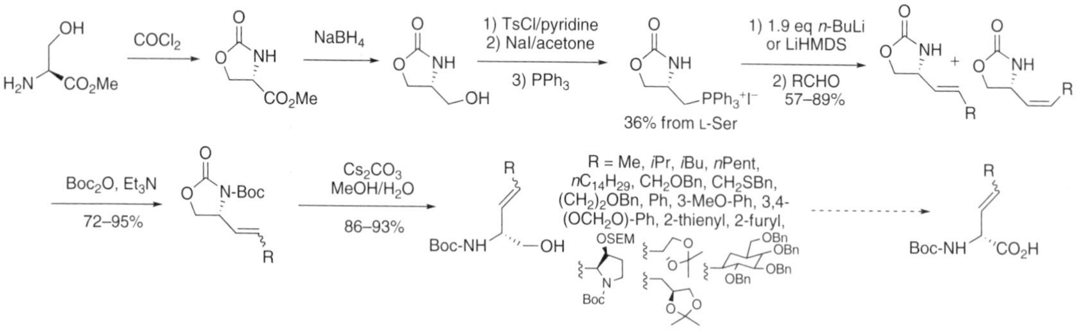

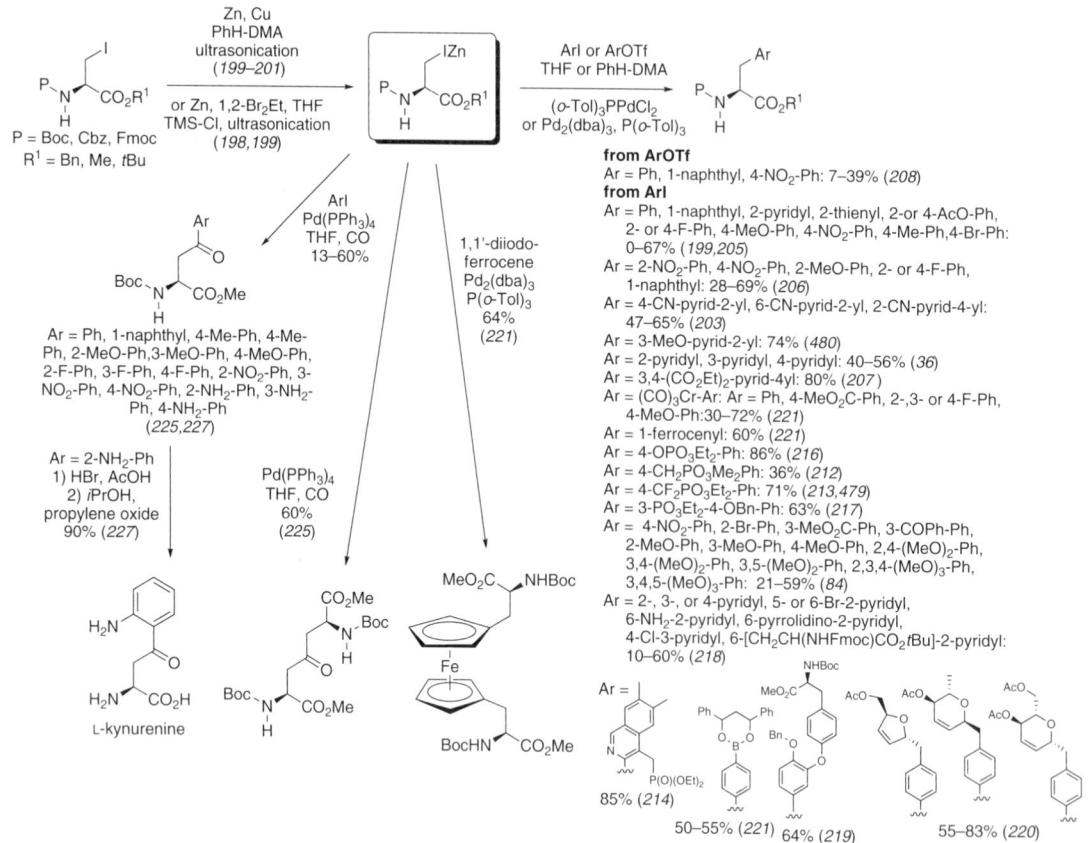

Scheme 10.21 Synthesis and Reaction of Alaninol Synthon (*193–198*).

Scheme 10.22 Synthesis of β-Iodozinc Synthon and Reaction with Aryl Iodides.

p-(diethylphosphonodifluoromethyl)-L-Phe (*213, 479*), heteroaryl-spaced phosphono α-amino acids (*214*), 4-borono-L-Phe (*215*), and 4-*O*-diethylphospho-L-azatyrosine (*216*), while 3′-phosphono-Tyr was synthesized and then subsequently alkylated to give 4′-carboxymethyloxy-3′-phosphono-Phe (*217*). The three regioisomers of β-(pyridyl)-Ala were prepared by coupling the corresponding bromopyridine regioisomers. Reaction yields were found to be dependent on the source

of Zn used for the reaction, while enantiomeric purity varied with the pyridine regiochemistry. The 2′- and 3′-aza-Phe products were enantiomerically pure, but the 4′-aza-Phe had only 40% ee (*36*). A series of 5′- or 6′-substituted β-(2-pyridyl)alanines were also synthesized, with 2 equiv of iodozinc reagent reacting with 2,6-dibromopyridine to give a 10% yield of the pyridyl-linked bis(amino acid) (*218*); β-(3′-methoxy-pyrid-2-yl)-Ala was also prepared (*480*). The first efficient synthesis of

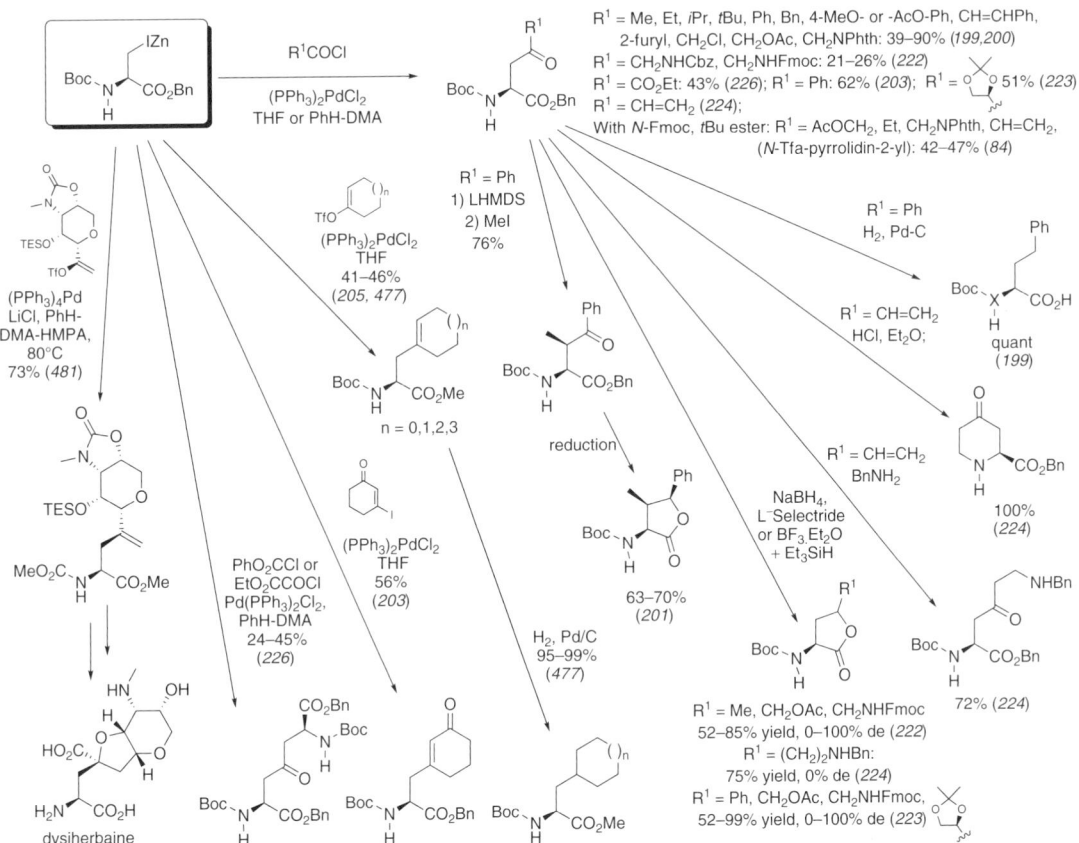

Scheme 10.23 Reaction of β-Iodozinc Synthon with Acyl Halides and Other Reagents.

isodityrosine, a biphenyl ether-linked diaminodicarboxy-lic acid, coupled the iodo-Ala organozinc synthon with an aryl iodide already substituted with the biaryl ether-linked second amino acid (219). Several C-glycosyl analogs of O-glycosyl Tyr were obtained by coupling the glycosyl aryl iodides (220). Metal-complexed aromatic amino acid derivatives were synthesized by coupling (haloarene) tricarbonylchromium complexes or 1-iodoferrocene (221). The chromium complexes were de-metallated by photolysis. A disubstituted ferrocene compound was prepared by reacting 1,1'-diiodoferrocene (221).

The 4-oxo-α-amino acids resulting from coupling with acyl chlorides can be diastereoselectively reduced to give 4-hydroxy-α-amino acids (as their lactones) with variable diastereoselectivity (0–100% de) dependent on reagent and substituent (see Scheme 10.23) (222). Ketone reduction with triethylsilane-BF₃·Et₂O produced the *cis*-substituted lactone, while L-Selectride in THF gave the *trans* epimer preferentially (223). The unlac-tonized hydroxy amino acids were produced in some instances (222, 223). The acyl chloride coupling reac-tion has been used to prepare L-4-oxopipecolic acid and L-4-oxo-Lys (224) as well as clavalanine, 4-hydroxy-Orn, and (+)-bulgecinine (223). The initial adduct of

coupling with benzoyl chloride was stereoselectively β-methylated by deprotonation/methylation. Carbonyl reduction provided 2-amino-3-methyl-4-hydroxy-4-phenylbutyrate (225). By using bifunctional carbonyl compounds (phenyl chloroformate or ethyloxalyl chlo-ride), modest yields of 2,6-diaminopimelic acid deriva-tives could be obtained (226).

The coupling of aryl iodides can be carried out under a carbon monoxide atmosphere, producing 4-aryl-4-oxo-α-amino acid derivatives such as L-kynurenine (see Scheme 10.22) (227). Strongly electron-withdraw-ing or *meta* substituents on the aryl ring resulted in low yields. By omitting the aryl iodide, a carbonyl-linked dimer (4-oxo-2,6-diaminopimelic acid) was pro-duced in 60% yield (225). In 2005 the iodozinc reagent was coupled with a series of cycloalken-1-yl triflates of varying ring size (five- to eight-membered rings). The β-(cycloalk-1-en-1-yl)-Ala adducts were hydrogenated to give β-(cycloalkyl)-Ala products; β-(1-methylcyclopent-1-yl)-Ala and β-(1-methylcyclohex-1-yl)-Ala were prepared by slightly more complex routes (477). Coupling with a complex vinyl triflate derivative gave an advanced intermediate for a total synthesis of the neuroexcitotoxin (−)-dysiherbaine (481).

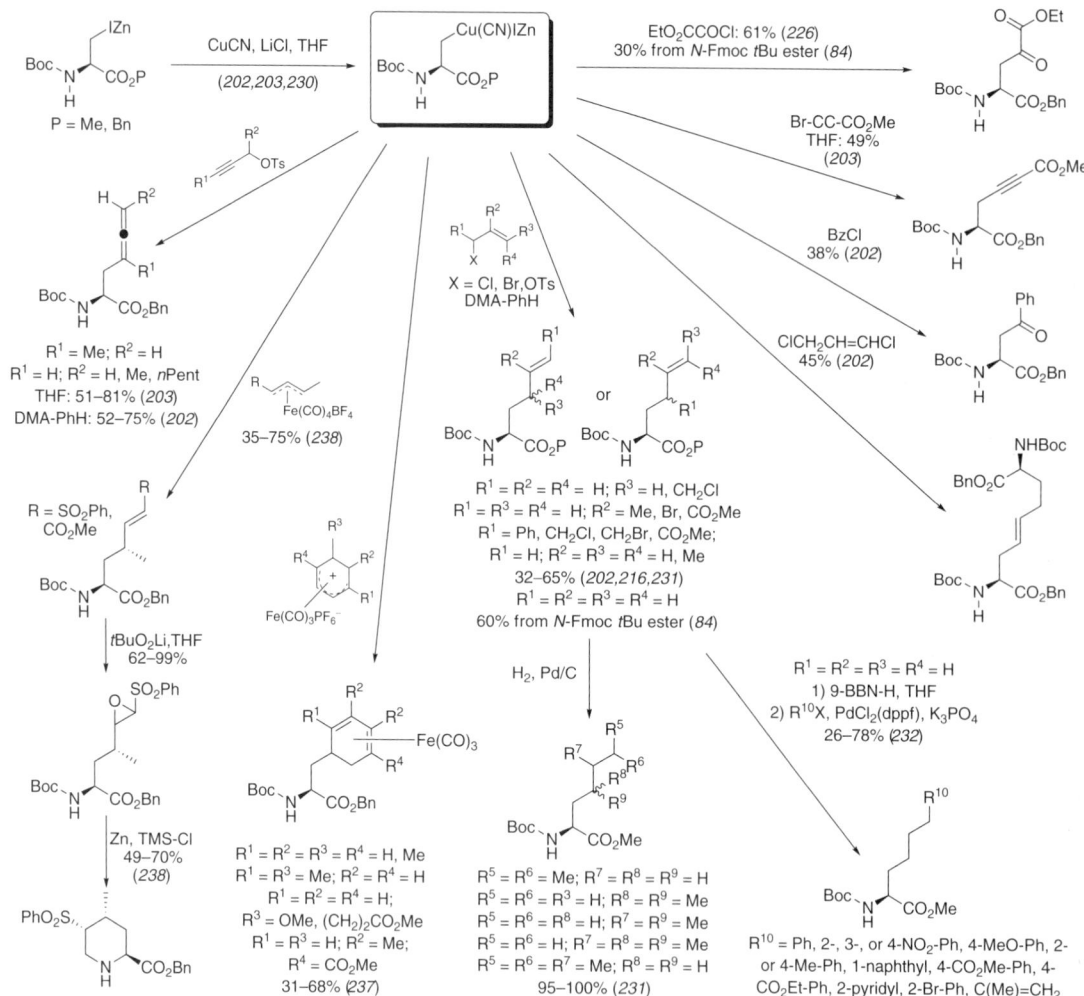

Scheme 10.24 Reaction of β-Iodocopper/zinc Synthon.

The organozinc reagent has also been generated from iodo-Ala contained within Boc-Ala-Ala(β-I)-OMe or Boc-Ala(β-I)-Ala-OMe dipeptides (228, 229). The C-terminal reagent was used for palladium-catalyzed reactions with acyl chlorides (RCOCl, R = Ph, 2-furyl, 4-Br-Ph: 25–51% yield) or aryl iodides (ArI, Ar = Ph, 4-NO₂-Ph, 4-Br-Ph, 1-naphthyl: 15–63%), but reaction of the N-terminal reagent appeared to be inhibited.

A slightly different reagent, an organozinc/copper derivative, has been synthesized by transmetallation of the organozinc benzyl ester synthon with CuCN (see Scheme 10.24). The transmetallation can be carried out in either the original solvent system (202, 230) or in THF (202, 203). The reagent is more reactive than the iodozinc analog, and does not require palladium catalysis for reaction. The methyl ester analog can also be prepared, which is useful for derivatives where hydrogenolytic deprotection is undesirable (202). The zinc/copper reagent has been applied to the synthesis of unsaturated α-amino acids where the site of unsaturation is remote

from the α-center. The reagent reacted with allylic halides and tosylates in either solvent system to give δ,ε-unsaturated-α-amino acids (202, 230, 231), and with propargylic tosylates to give substituted allenic amino acid derivatives (202, 203). The unsaturated side chains were then hydrogenated to provide 2-amino-6-methylheptanoic acid, 2-amino-4,4-dimethylhexanoic acid, 2-amino-5,6-dimethylheptanoic acid, and 4,4,5-trimethylhexanoic acid (231). An unsaturated 2,8-diaminoocta-1,9-dioic acid derivative (with >99.5% ee) was prepared from the bifunctional halide 1,3-dichloroprop-1-ene (202). The homoallylglycine adduct formed in 60% yield from coupling with allyl chloride was hydroborated and used for Suzuki couplings with aryl iodides to give 2-amino-6-arylhexanoic acids, or with 2-bromopropene to give 2-amino-7-methyloct-7-enoic acid (232).

Treatment of Boc-iodo-Ala-OMe with Zn/CuCN/ LiCl and iodoacetylenes produced 5-TMS-propargylglycine or 5-(n-pentyl)-propargylglycine in 52–62% yield (233). Another synthesis used the organocuprate reagent

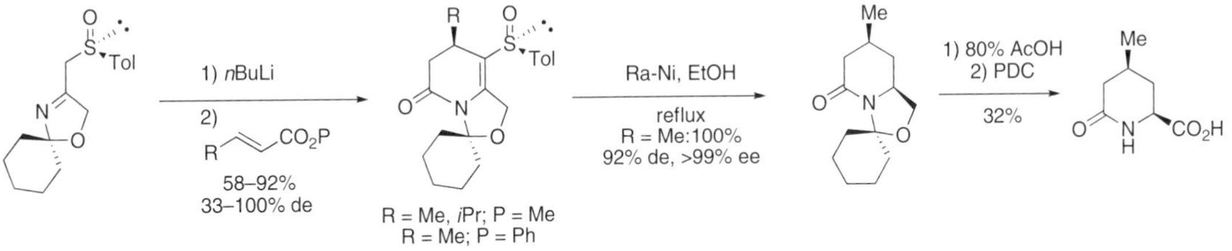

Scheme 10.25 Synthesis of 4-Substituted-2-aminoadipic Acid Lactoms (*246*).

for a Pd-catalyzed coupling reaction with protected 3-iodohept-2-ene-1,6-diol. The adduct was elaborated into lycoperdic acid (*234*). An ultrasonically induced aqueous conjugate addition of the zinc–copper derivative of Boc-β-iodo-Ala-OMe to dehydroAla contained within the Seebach Bz-BMI imidazolidinone template produced the diamino dicarboxylic acid addition product in 61% yield with 60% de. The adduct was not deprotected (*235*). The same reaction with dehydroalanine in an *N*-Cbz oxazolidinone template produced protected 2,5-diaminoadipic acid in 76% yield (*236*).

The zinc/copper reagent was also reacted with ethyloxalyl chloride without a palladium catalyst to give protected 4-oxo-Glu in 61% yield (*226*). Cyclohexadiene complexes were produced from tricarbonyl(η^5-cyclohexadienyl)iron salts in 31–68% yield (*237*), while (η^3-allyl) iron tetracarbonyl salts gave allylated derivatives (homoallylglycines) (*238*). The metals were then removed from the complexes by treatment with trimethylamine *N*-oxide (*237*) or ceric ammonium nitrate (*238*), respectively. The alkenes of the allylated substrates were epoxidized and the epoxides opened intramolecularly by the α-amino group to give pipecolic acid derivatives (*238*). Finally, the zinc/copper reagent was employed to displace a mesylate from a glucal derivative to give a C-linked glucosyl amino acid derivative (*239*). More recently, Fmoc-β-iodo-Ala-O*t*Bu was converted into the iodozinc/copper reagent, and reacted with ethyl oxalyl chloride (30%) or allyl chloride (60%) (*84*).

As mentioned earlier, homologs of the Ser-derived iodozinc reagent have been prepared from Asp and Glu, and employed for similar reactions (e.g., see Scheme 9.23) (*206*, *209–211*, *239–245*), as reviewed in 2005 (*204*). Analogous reagents derived from 3-amino-4-iodobutyrate have been used for the synthesis of β-amino acids (see Section 11.2.3n) and 4-amino-5-iodopentanoate for γ-amino acids (see Section 11.4.3m). Likewise, Cbz-Thr-O*t*Bu was converted into both diasteromers of β-iodo-Abu, and then into both the iodozinc and iodozinc/copper reagents. These were coupled with aryl iodides to give β-methyl-arylalanines in low yields (17–44%) as 1:1 mixtures of C3 epimers, or with allyl bromides to give 1:1 mixtures of C3 epimers in 34–38% yield (*59*).

10.5.4 Other β-Anions

C4 Substituted 2-aminoadipic acid lactams have been synthesized via stereoselective Michael addition of a

homochiral alanine β-anion equivalent derived from a β-iminosulfoxide (see Scheme 10.25) (*246*). Ser-derived β-iodo-Ala can be used to generate radicals for radical reactions (see Section 10.4). β-Iodo-Ala has also been employed for a Pd-catalyzed coupling with phosphomethyl-substituted 4-pyridone triflate (*247*).

A β-lithiated serinol derivative was added to ketones, acylated with Weinreb amides, and alkylated with TMSCl or CD₃OD (see Scheme 10.26). Oxidation to regenerate the α-carboxyl group produced lactones of γ-hydroxy amino acids, β-trimethylsilyl-Ala, β-borono-Ala, γ-keto-Hfe, or β-deutero-Ala (*248*). Quenching of the anion with triisopropylborate led to a boronic acid side-chain that was successfully used for Suzuki couplings with nine aryl halides. The hydroxymethyl groups of two examples were oxidized to the acid, giving Phe and 4′-nitro-Phe with optical rotations consistent with literature values (*249*).

10.6 Serine Aldehydes: Conversion of the Serine Carboxyl Group into an Aldehyde/Ketone

Two possible routes exist for forming a Ser aldehyde, which can then be elaborated into a more complex α-amino acid. Both analogs have been called serinal derivatives:

1. Direct oxidation of the side-chain hydroxyl to the aldehyde without altering the oxidation state of the carboxylic acid group (forming α-formylglycine).
2. Reduction of the acid group to the aldehyde (forming 2-amino-3-hydroxypropanal)). The aldehyde is generally functionalized before oxidization of the hydroxyl side-chain to regenerate a carboxylic acid group, effectively inverting the stereochemistry at the α-center.

The synthesis of α-amino acids from serinals is a relatively unexplored area compared to other approaches. This inattention is due primarily to the tendency of α-formylglycine serinals to enolize and racemize; stable serinals have only been reported since the mid-1980s and their use has been limited. Three major Ser aldehyde equivalents have been developed: those of Garner and Rapoport both rely on the indirect approach via

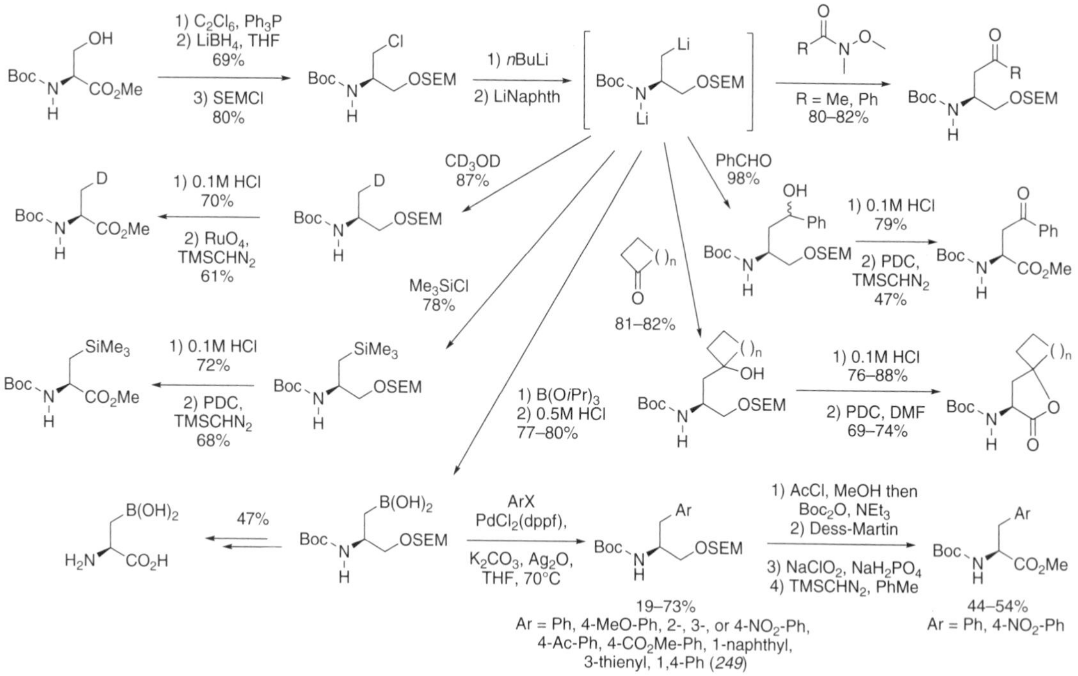

Scheme 10.26 Reactions of Lithiated Serinol Derivative (*248*).

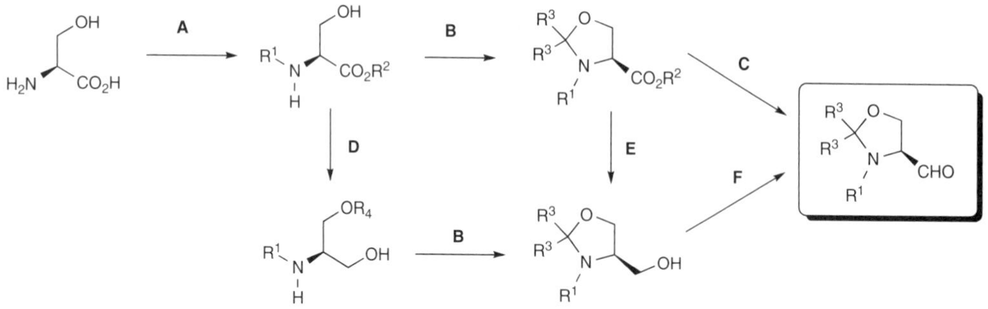

Scheme 10.27 Synthesis of the Garner Aldehyde and Related Analogs (see Table 10.5).

reduction of the acid group, whereas that of Blaskovich and Lajoie employs a true side-chain oxidized Ser aldehyde derivative. The Garner aldehyde has found widespread use as a chiral synthon in asymmetric syntheses, with a comprehensive review in 2001 including over 200 references (*250*). A number of other aldehyde equivalents have also been reported. The indirect approach will be discussed first.

10.6.1 The Garner Aldehyde

10.6.1a Garner Aldehyde Synthesis

The Garner aldehyde is a 2,2-dimethyloxazolidine carboxylate in which the side-chain hydroxyl and α-amino

groups of Ser or Thr have been protected within an oxazolidine. The heterocycle is prepared by reacting Boc-Ser-OMe with dimethoxypropane in the presence of a catalytic amount of *p*-toluenesulfonic acid. The aldehyde is then produced by reducing the ester with DIBAL-H (see Scheme 10.27, Table 10.5). The Ser-derived synthon was first reported in 1984 as an intermediate in the synthesis of *threo* β-hydroxy-Glu (*251*). In 1987 the preparation of both the Ser- and Thr-derived aldehydes was published as a separate paper with full details (*3*). It was noted that some racemization occurs during the synthesis, as the final product possessed 93–95% ee. This lack of enantiomeric purity is one of the drawbacks of the Garner approach. An optimized procedure has since appeared in *Organic Syntheses*,

Table 10.5 Synthesis of the Garner Aldehyde and Related Analogs (see Scheme 10.27)

R^1	R^2	R^3	R^4	A	B	C	D	E	F	Reference
Boc	Me	Me	—	—	DMP, TsOH benzene 60%	DIBAL, PhMe 80%	—	—	—	1984 (251)
Boc	Me	Me	—	1) Boc₂O, NaOH 2) MeI, K₂CO₃, DMF 80–90%	DMP, TsOH benzene 70–89%	DIBAL, PhMe 76–85%	—	—	—	1987 (3) 1992 (252)
Cbz	Me	Me	—	—	DMP, TsOH benzene 96%	DIBAL, PhMe 76%	—	—	—	1988 (264)
Boc	Me	Me	—	—	—	—	—	LiAlH₄, Et₂O	Swern oxidation	1989 (258)
Bn	Me	Me	—	PhCHO, NaBH₄	DMP, H⁺	—	—	LiAlH₄	DMSO, (COCl)₂, Et₃N	1990 (257)
Boc	$CO_2R^2 = CONMe(OMe)$	Me	—	(from Boc-Ser) 1) ClCO₂iBu, NMM 2) HNMe(OMe), THF 65%	DMP, PPTS benzene 71%	LiAlH₄, Et₂O 78%	—	—	—	1992 (261)
Boc	Me	Me	—	(from Boc-Ser) MeOH, PPh₃, DEAD, Et₂O 98% (up to 100g)	DMP, TsOH, PhMe 70%	—	—	—	—	1993 (254)
Boc	Me	Me	—	1) MeOH, HCl 2) Boc₂O, Et₃N 90%	DMP, acetone, BF₃·Et₂O 91%	DIBAL, PhMe 70%	—	—	—	1994 (253)
Boc Cbz	H	H	—	**B**, then **A**	1) HCHO, 2N NaOH 2) Boc₂O or Cbz-Cl 94%	DIBAL, PhMe 61%	—	—	—	1995 (266)
Boc	Me	—(CH₂)₅—	—	1) MeOH, SOCl₂ 2) Boc₂O, Et₃N 94%	cyclohexanone, TsOH 82%			NaBH₄, LiCl 92%	Dess-Martin periodinane 75% or TFAA, DMSO, Et₃N 75%	1995 (255) 1996 (256)

(Continued)

Table 10.5 Synthesis of the Garner Aldehyde and Related Analogs (see Scheme 10.27) (continued)

R^1	R^2	R^3	R^4	A	B	C	D	E	F	Reference
Cbz	Me	Me	—	1) Cbz₂O, Et₃N 2) MeOH, AcCl 87%	DMP, TsOH 86%	—	—	NaBH₄, THF/ MeOH 88%	—	1995 (405)
Boc	Me	Me	—	—	—	—	—	LiAlH₄, Et₂O 96%	DMSO, (COCl)₂, Et₃N 88%	1995 (259)
Boc	Me	Me	—	1) MeOH, SOCl₂ 2) Boc₂O, Et₃N	DMP, TsOH, benzene 71% for 3 steps	—	—	NaBH₄, LiCl 96%	Dess–Martin periodinane 93% or TFAA, DMSO, Et₃N 60%	1996 (256)
Boc	Me	Me	TBDPS or Bn	—	1) DMP, acetone, BF₃·Et₂O 94–98% 2) nBu₄NH, CH₂Cl₂ 98% or 3) H₂, Pd-C, AcOEt 95%	—	1) TBDPSCl, imidazole 2) LiBH₄ 90% or from Boc-Ser(Bn)-OMe): LiBH₄ 98%	—	DMSO, (COCl)₂, Et₃N 85% 70% ee or Dess–Martin periodinane 93%, >98% ee	1997 (263)
Boc	Me	Me	—	—	—	—	—	LiAlH₄, THF 93–96%	DMSO, (COCl)₂, DIPEA 99% 96–98% ee	1998 (260)
Boc	CO₂R² = CONMe(OMe)	Me	—	—	DMP, acetone, BF₃·Et₂O 88% from D-Ser	LiAlH₄, THF 100% 94% ee	—	—	—	1998 (262)

with the Boc-protected serinal obtained from L-Ser in 46–58% yield (252).

Several improvements to the synthon preparation have been reported (see Scheme 10.27, Table 10.5). A modified procedure which avoids using methyl iodide (for Ser esterification) and benzene (for oxazolidine formation) gave the synthon in 57–60% yield (253). A Mitsunobu reaction has also been used to prepare the methyl ester, and can be applied to large-scale (100 g) reactions (254). A number of groups have found the DIBAL ester reduction to be capricious, and so have generated the aldehyde by a two-step procedure involving LiAlH₄ reduction of the ester group to an alcohol, followed by a Dess–Martin (255, 256) or Swern (256–259, 484) oxidation. The two-step reduction/oxidation sequence is reportedly more reliable, especially on a large scale (256, 484). In one of two reports where enantiomeric purity was established for the two-step procedure, it was only 86–87% ee (259), although the DIBAL reduction route gave a similar result when used by these authors. In the other report, DIEA was employed as the base for the Swern oxidation, producing product with 96–98% ee (484). More recently, oxidation of the alcohol to the aldehyde with TEMPO has been described as giving product in 90% yield with no racemization (260).

It is also possible to directly reduce the acid to the aldehyde via LiAlH₄ reduction of a Weinreb amide intermediate. The initial report of this route gave an overall yield from Boc-Ser of 38% (261); modified conditions allowed for an overall yield of 88% on a 10 g scale, with enantiomeric purity of approximately 94% ee based on optical rotation (262). A modified version of the reduction/oxidation route is also possible, with the α-ester group of side-chain protected Boc-Ser-OMe reduced to give an amino diol prior to formation of the oxazolidine ring (263). Both benzyl and tert-butyldiphenylsilyl protecting groups were employed. The oxazolidinone was then formed with the hydroxymethyl group generated from the ester, followed by deprotection and oxidation of the original Ser hydroxymethyl side chain to generate the aldehyde in 75% overall yield (see Scheme 10.27, Table 10.5). Thus, inexpensive L-Ser can be used to produce both the D- and L-Garner aldehydes. It was observed that Swern oxidation conditions (with triethylamine as base) caused 15% racemization during the oxidation step, but that Dess–Martin periodinane oxidation proceeded without racemization (263).

Several related compounds have been prepared (see Table 10.5, Scheme 10.28). Syntheses of the N-Cbz (264) and N-benzyl (257) protected analogs have been published. A synthesis of the Garner aldehyde from L-Met has also been developed, with the Met side-chain converted into the aldehyde via a vinyglycinol-type intermediate. Ironically, this unsaturated intermediate is one of the products that has been prepared from the Garner aldehyde (265). A spirooxazolidine analog was obtained by using cyclohexanone to form the oxazolidine (255), while an unsubstituted oxazolidine was prepared from formaldehyde (266). The side chain of Thr was used to make a methyl ketone-substituted equivalent of the Garner aldehyde (267–269), while an α-methyl-Ser analog has also been prepared (see Scheme 10.28) (270). A conformational analysis of the N-Boc Garner aldehyde and its α-methyl-substituted analog was reported in 2003, showing a dynamic equilibrium between the two possible conformers of the carbamate group (271). A more functionalized analog of the Garner aldehyde (see Scheme 10.28) has been prepared from D-isoascorbic acid, and elaborated into both 2-hydroxy-3-aminopropionic acids and MeBmt (272). Duthaler has reported the Cys-derived analog of the Garner aldehyde (see Scheme 10.28) (273).

10.6.1b Garner Aldehyde: Organometallic Additions

The Garner aldehyde was quickly applied to the synthesis of chiral amino alcohol and amino sugars, based on organometallic additions to the aldehyde. In 1988 five reports appeared on the preparation of the amino alcohol sphingosine, based on anti diastereoselective additions of 2-trimethylsilylthiazole (274, 275) or lithium 1-pentadecyne (276) to the carbonyl. Additions of the lithium enolate of 2-acetylthiazole (277), of alkynyl lithium reagents (278, 279), or of alkyl/aryl lithium and Grigard reagents (280) led to other amino alcohols or amino sugars, also with the anti configuration.

The first conversion of one of these adducts to an amino acid was reported in 1988, when Garner and Park published a synthesis of 5-O-carbamoyl polyoxamic acid (3,4,5-trihydroxynorvaline), starting with addition of vinylMgBr and proceeding via a dihydroxylation reaction. After side-chain elaboration, the oxazolidine was opened and the original Ser side-chain hydroxyl oxidized to form the acid group (281). In 1989 the glycosyl α-amino acid thymine polyoxin C

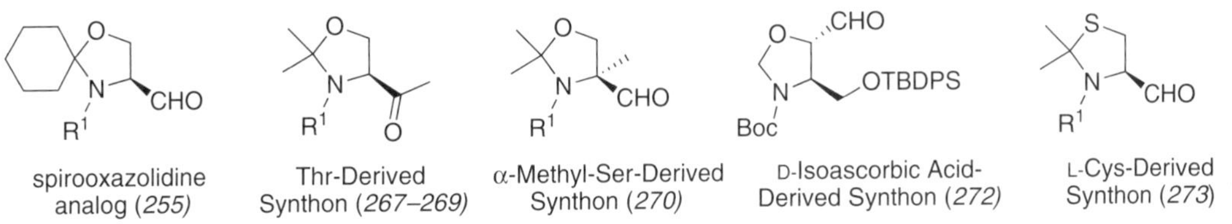

Scheme 10.28 Garner Aldehyde Analogs (272).

| spirooxazolidine analog (255) | Thr-Derived Synthon (267–269) | α-Methyl-Ser-Derived Synthon (270) | D-Isoascorbic Acid-Derived Synthon (272) | L-Cys-Derived Synthon (273) |

was prepared (*282*). An alkynyl lithium reagent was added in 78% yield with 86% de (*anti*), with the carboxyl group again regenerated after side-chain elaboration. Related compounds were prepared by a similar route (*283*). Another glycosyl amino acid was prepared via BF$_3$·Et$_2$O-catalyzed addition of 2-(trimethylsiloxy) furan, which added with complete diastereoselectivity in 86% yield (*284–286*). Addition of an acetylide anion to the Garner aldehyde led to (2*S*,3*S*)-β-hydroxy-Asp, via RuO$_4$ oxidation of the ethynyl moiety to generate the β-carboxy group (*287*).

A variety of carbanions have been added to the Garner aldehyde (see Scheme 10.29, Table 10.6), with many attempts to control the addition diastereoselectivity. The metal counterion is critical to stereocontrol, as was previously found for the isopropylidene glyceraldehyde oxo-analog of the Garner aldehyde (*288*). The first successful reversal of diastereoselectivity appears to be that reported by Herold (*289*), who found that lithiated ethynyl trimethylsilane in THF/HMPA gave the conventional *anti* adduct with 95% de, but the corresponding Grignard reagent with CuI or the lithium reagent with ZnBr$_2$ gave the *syn* isomer, also with up to 95% de. Similarly, vinylzinc chloride was found to give up to 72% de of the *syn* adduct when used in non-polar solvents, while vinylmagnesium bromide in THF gave 50% de *anti* product, and vinyllithium gave 66% de *anti* product (*290*). The *anti* products are believed to arise from a non-chelated attack of a Felkin–Ahn transition state, while the *syn* products are possibly due to a coordinated delivery of the nucleophile, rather than a chelated transition state (*290*). The vinyl adduct was converted into an allylic primary bromide by treatment with CBr$_4$/PPh$_3$, and the bromide used to alkylate a Schöllkopf bis-lactim ether enolate enroute to a synthesis of 2,6-diaminoheptandioic acid (*261*). Addition of trimethylsilylcyanocuprate to the vinyl adduct led to (*E*)-2-amino-5-trimethylsilylpent-3-enoic acid (*291*).

The Garner aldehyde was used to prepare both diastereomers of D-phenylserine via Grignard addition of phenylmagnesium chloride; the 3:2 mixture of diastereomers was separated by fractional crystallization (*292*). The hydroxyl group was protected as a TBS ether to prevent problems during deprotection and generation of the acid group. A much better diastereoselectivity (66% de, *anti* preferred) was observed with phenylmagnesium bromide, while isopropylmagnesium chloride surprisingly gave the opposite (*syn*) diastereomer with 72% de (*256*). The reversal in stereoselectivity

is possibly due to less reactive Grignard reagents adding via a chelation pathway (*256*). Conversion of the isopropyl adduct to 3-hydroxy-Leu by the conventional route of oxazolidinone opening followed by hydroxymethyl group oxidation was found to give poor yields. Instead, a bicyclic oxazolidine/oxazolidinone was prepared before oxazolidine ring cleavage by cyclizing the β-hydroxy/α-amino groups (see Table 10.6). The bicyclic ring formation was accompanied by inversion at the β-hydroxy center; thus, the *anti* isomer (2*S*,3*S*)-3-hydroxy-Leu was obtained as the final product (*255, 256*). The carbinol of the isopropyl adduct has also been *O*-arylated before oxidation to generate the carboxy group, leading to the *O*-aryl-β-hydroxy-Leu component of the 14-membered cyclopeptide alkaloids (*293*).

Beaulieu obtained 0% de for the addition of MeMgBr to the Cbz-protected Garner aldehyde (*294*). The poor diastereoselectivity observed with MeMgBr or MeLi was overcome by using a bulkier trichloromethane anion, which added in 82% yield to give a single (*syn*) diastereomer (*294*). The trichloromethyl group was then converted to a methyl group by hydrogenation (after the oxazolidine opening and hydroxyl oxidation steps), producing D-*allo*-Thr or the ³H-labeled isomer. Increasing the steric bulk of the oxazolidine ring by employing a spirooxazolidine analog of Garner's aldehyde improved the diastereoselectivity of addition for PhMgBr (78% de *anti*) and *i*PrMgCl (80% de *syn*) (*255, 256*). Addition of other Grignard reagents to the more hindered Garner aldehyde also resulted in the *syn* isomer, with the exception of MeMgBr, which gave the *anti* diastereomer with 34% de (*256*). Additions of aryl Grignard reagents to the conventional Garner aldehyde mediated by CuI-Me$_2$S proceeded with high (>90% de) *syn* selectivity (*295*). Organotitanium complexes formed from Ti(O*i*Pr)$_4$/*i*PrMgX and allyl bromides, acetate or phosphate added in 86–95% yield with 84–85% de *anti* stereoselectivity (*296*).

Reversals in the diastereoselectivity of addition have also been obtained with other reagents. The addition of *ortho*-metallated phenols can be stereocontrolled depending on the metal counterion, with Mg-based phenolates giving the *syn* aryl product with 92–98% de and Ti-based phenolates the *anti* epimer with similar stereocontrol (*297*). Lithiated trimethylsilylacetylene produced the *anti* adduct, while the *syn* isomer was obtained from the magnesium anion in the presence of CuI (*289*). Lithiated *O*-TBS propargyl alcohol added with >90% *anti* de when HMPA was employed as additive, but with >90% *syn* de with SnCl$_4$ as additive (*298*).

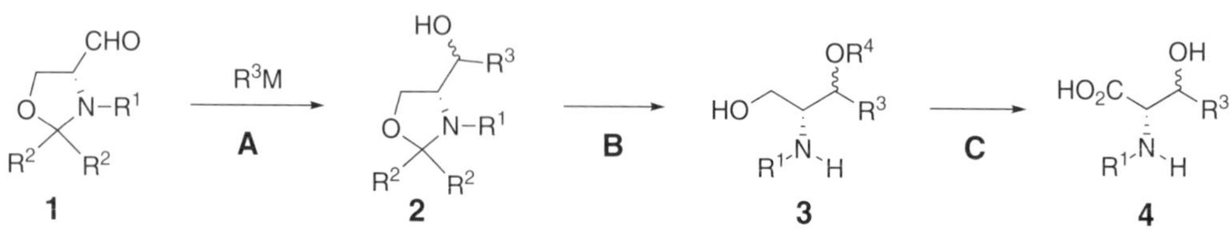

Scheme 10.29 Organometallic Addition to the Garner Aldehyde (see Table 10.6).

Table 10.6 Organometallic Addition to the Garner Aldehyde (see Scheme 10.29)

R¹	R²	Organometallic R³M	Yield and Stereoselectivity	Side-Chain Derivatization	Deprotection/Oxidation	Reference
Boc	Me	MgBrCH=CH₂	80% 70% de anti	convert side chain to $CH(OH)CH(OH)CH_2OCONH_2$		1988 (281)
Boc	Me	$\xrightarrow{nBu_4NF}$	85% 84% de			1988 (274) 1990 (275) 1998 (276)
Boc	Me	ArMgBr or ArTi(OiPr)₃ 2-HO-4-nC₁₅H₃₁-Ph 2,6-(HO)₂-4-nC₁₅H₃₁-Ph 2-HO-5-tBu-Ph 2-HO-4,5-(OCH₂O)-Ph	ArMgBr: 63–74% 92–98% de syn ArTi(OiPr)₃: 60–64% 91–98% de anti			1988 (297)
Boc	Me	LiCCnC₁₃H₂₇	74% 78% de anti			1988 (276)
Boc	Me	LiPh LinBu LiCCnC₅H₁₁ LiCCnC₁₃H₂₇ nBuMgBr PhMgBr	40–96% de anti			1988 (280)
Boc	Me	R³ = CCSiMe₃, CCnC₁₃H₂₇ M = Li, HMPA/THF M = MgBr +CuI, THF M = Li +ZnBr₂, Et₂O	Li: 71–86% 95% de anti MgBr: 85% 95% de syn Zn: 87–89% 91–95% de syn			1988 (289)
Boc	Me	LiCCnC₁₃H₂₇	90% 80% de anti			1988 (474)
Boc	Me	LiCCCO₂Et, THF, HMPA	78% 86% de anti			1989 (282)

(Continued)

Table 10.6 Organometallic Addition to the Garner Aldehyde (see Scheme 10.29) (continued)

R^1	R^2	Organometallic R^3M	Yield and Stereoselectivity	Side-Chain Derivatization	Deprotection/Oxidation	Reference
Boc	Me	$LiCC(CH_2)_2CH=CHnC_{13}H_{27}$	49% 92% de *anti*			1990 (278)
Boc	Me	MeCO(thiazol-2-yl) LiOtBu, THF	70% 60% de *anti*			1990 (277)
Boc	Me	$LiCCCH(OMe)_2$	87% 78% de *anti*			1990 (283)
Boc	Me	[furan–OSiMe₃] $\xrightarrow{BF_3 \cdot Et_2O}$ [butenolide]	86% 90% de *anti*	1) TMSCl, pyr 2) KMnO₄ 3) TMSCl, pyr 4) DIBAL 46% [TMSO/HO furanose OTMS structure]	1) 5% citric acid: 75% 2) Ac₂O, pyr: 90% 3) 70% AcOH: 70% 4) NaIO₄, RuO₂ 5) CH₂N₂ 88% [OAc pyranose AcO OAc structure]	1990 (285) 1991 (284) 1991 (286)
Boc	Me	MCCTMS		convert side chain to CH(OBn)CO₂Bn 1) RuCl₃, NaIO₄ 2) Bnbr, K₂CO₃	1) pTsOH, MeOH 73% 2) Na₂Cr₂O₇ 74%	1990 (287)
Cbz	Me	MeMgBr MeLi ᶜCCl₃ (generated from Al, PbCl₂, CCl₄ or from Cl₃CCO₂TMS, K₂CO₃, 18-c-6, 90 °C)	0% de 0% de 62–82%, 100% de *syn*		1) MeOH, Dowex 2) Jones oxidation 3) H₂ or T₂, Pd(OH)₂/C R^3 = Me or CT₃ 92%	1991 (294)
Boc	Me	$LiCCnC_{15}H_{31}$	76% 78% de *anti*			1991 (279)
Boc	Me	MCH=CH₂ M = Li, MgBr, CuLi, AlEt₂, ZnCl, MgBr + BF₃·Et₂O, MgBr + ZnCl₂, Li + TiCl₄, Li + Et₂AlCl, Li + ZnCl₂	70–90% 72% de *syn* to 66% de *anti*			1992 (290)

Boc	Me	MgBrCH=CH₂	79%	convert side chain to CH=CHCH₂Br by treatment with CBr₄, PPh₃ 50%	1992 (261)
Boc	Me	LiCCSiMe₃, HMPA MgBrCCSiMe₃, CuI	88%, anti 80%, syn		1994 (299)
Boc	Me	PhMgBr	92%, 66% de anti	1) BF₃·2AcOH: 80–100% 2) Jones oxidation: 74% 3) 5N HCl: 99% — For R² = Me, –(CH₂)₅–; R³ = iPr: Tf₂O, 2,6-tBu₂-4-Me-Pyr	1994 (299)
Boc	Me	iPrMgCl	85%, 72% de syn		
Boc	–(CH₂)₅–	PhMgBr	72%, 78% de anti	**2**	
Boc	–(CH₂)₅–	MeMgCl	85%, 34% de anti	81–83% yield inversion at β-center	
Boc	–(CH₂)₅–	RMgX R = Et, iPr, tBu, cHex	15–74% 80–88% de syn		
Boc	Me	LiCF₂PO₃Et₂	58% 64% de anti	deoxygenation via 1) PhOCSCl, DMAP 2) nBu₃SnH, AIBN 58% convert side chain to CH₂CF₂PO₃Et₂ — 1) HCl, EtOH 78% 2) RuCl₃, NaIO₄ 41%	1995 (475)
Boc	Me	PhMgBr	100% 20% de syn	1) TBSOTf, DIEA: 96% 2) 80% AcOH: 84% 3) PDC 4) LiOH, THF/H₂O 5) TFA 95%	1995 (292)
Boc	Me	MgBr(3-CH₂CH=CMe₂-4-MeO-Ph)	72% 33% de anti	convert syn adduct to anti via: 1) PDC, CH₂Cl₂; 88% 2) Zn(BH₄)₂, PhH: 75% — 1) Ac₂O, DMAP 2) 80% AcOH 80% 3) PDC, DMF: 70% 4) NaOMe, MeOH: 100%	1997 (302)
Boc	Me	LiCCCH₂OTBS + HMPA LiCCCH₂OTBS + SnCl₄	85%, >90% de anti 46%, >90% de syn		1997 (298)

(continued)

Table 10.6 Organometallic Addition to the Garner Aldehyde (see Scheme 10.29) (continued)

R¹	R²	Organometallic R³M	Yield and Stereoselectivity	Side-Chain Derivatization	Deprotection/Oxidation	Reference
Boc	Me	BrCH₂CH=CH₂ BrCH₂CH=CHMe CrCl₂, THF	80–98% 24–70% de *anti*			1998 (476)
Boc	Me	LiCCCH(Me)OTBS with ZnBr₂	86% 80% de *syn*	convert alkyl to 5-Me-1,5-dihydrofuran-2-yl	TsOH, MeOH: 95% 1) Dess–Martin 2) NaClO₂, NaH₂PO₄: 77%	2000 (319)
Boc or CO₂Me	Me	Li(Boc-5-allyl-Pro-OMe)	60–86%	convert allyl group to amino acid substituent	TsOH, MeOH: 85% Jones oxidation: 53%	2001 (314)
Boc	Me	ZnBrCF₂CO₂Et	81% 75% de	Barton deoxygenation (80%)	TFA/MeOH: 30% Jones oxidation or PDC: 50–60%	2001 (316)
Boc	Me	MgCl(CH₂CH=CH₂)	86% 26% de	reductive ozonolysis of alkene, displacement of both side chain OH by azide, reduction and cyclization with Br¹³CN to give guanidine	6N HCl KMnO₄ 10%	2002 (317)
Boc	Me	Li(4-BnO-Ph)	50–71% 50–60% de *anti*	oxidize carbinol center (100%) K-Selectride reduction (98%, 94% de *syn*)	TsOH, MeOH: 76% SO₃·Py, DMSO, Et₃N then NaClO₂, KH₂PO₄: 90%	2002 (303)
Boc	Me	LiCC(glycosyl)	45–88%	reduce alkyne, dehydroxylate 51–89%	Jones oxidation 75–95%	2002 (318)

Boc	Me	CuI, THF-Me$_2$S with PhMgBr 4-MeO-PhMgBr 3,4-(OCH$_2$CH$_2$O)-Ph	64–74% >90% de syn	2002 (295)
Boc	Me	Ti(OiPr)$_4$ / iPrMgCl with BrCH$_2$CH=CH$_2$ EtO$_2$COCH$_2$CH=CH$_2$ (EtO)$_2$P(O)OCH$_2$CH=CH$_2$nPrCCnPr nBuCCSiMe$_3$ TBSOCH$_2$CCSiMe$_3$ EtO$_2$COCH(Me)CH=CH$_2$ EtO$_2$COC(Me)$_2$CH=CH$_2$ EtO$_2$COCH$_2$CCH BrCH$_2$CCH EtO$_2$COC(Me)$_2$CCnHex EtO$_2$COCH$_2$CCSiMe$_3$	56–95% 75–98% de anti	2002 (296)
Boc	Me	InBrCH$_2$CH=CHOAc (results in diol with CH(OH)CH=CH$_2$ substituent)	95% 68% de anti	2003 (313) —
Boc	Me	LiCH$_2$CON(Me)CH(Me)CH(OH)Ph	79%, 96% de for matched pair 61%, 12% de for mismatched pair	2004 (315) —
Boc	Me	MgCl(CH=CH$_2$)	32% de syn	2006 (291)

1) AcCl, Et$_3$N
2) (TMS)$_2$CuLi.
CuCN

SiMe$_3$

1) TFA, MeOH
2) H$_5$IO$_6$/CrO$_3$,3) MeI, KHCO$_3$
43%

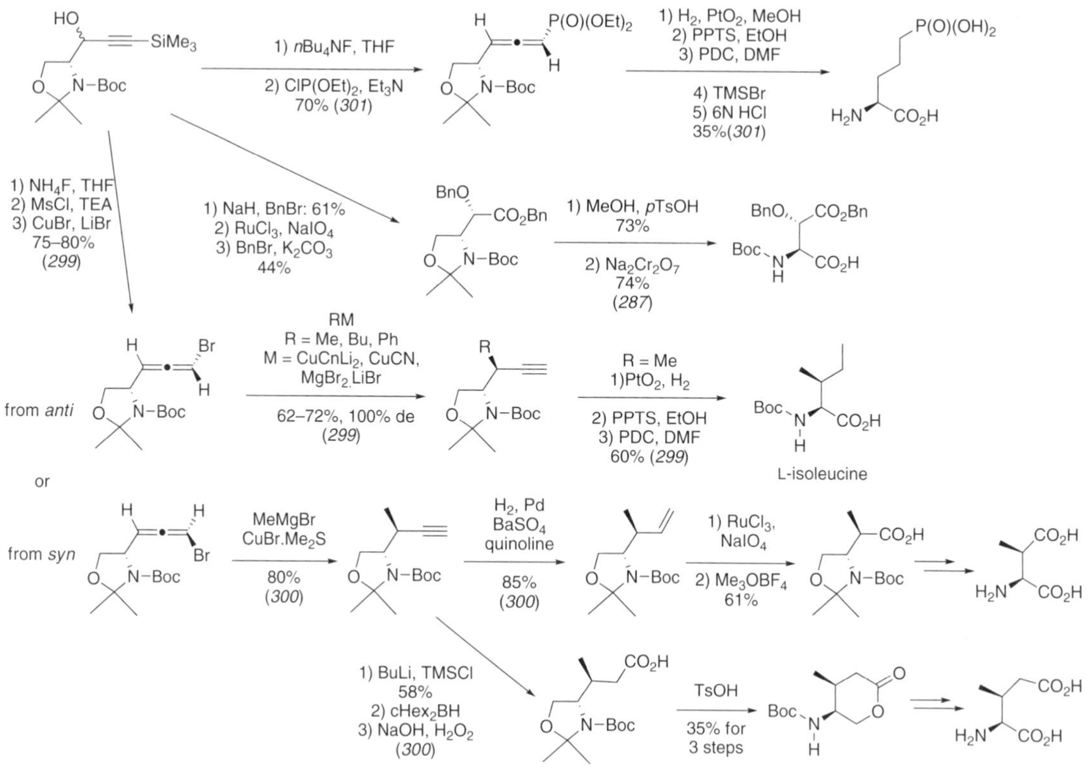

Scheme 10.30 Elaboration of Alkynol Adduct.

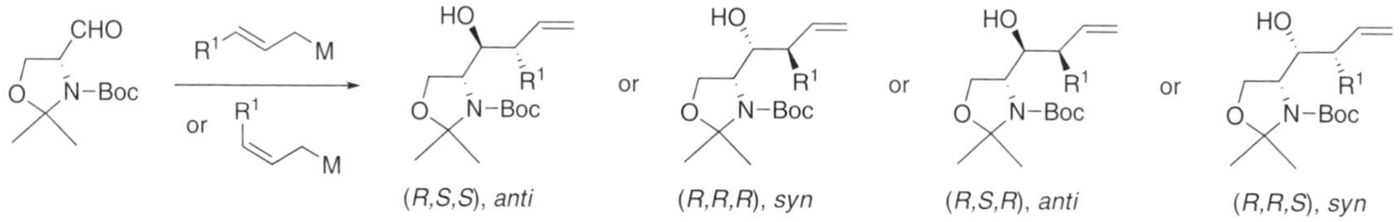

Scheme 10.31 Allyl Addition to the Garner Aldehyde (see Table 10.7).

The alkynol products are useful intermediates (see Scheme 10.30), and were converted into bromoallenes (299), which are substrates for completely diastereose-lective organocuprate additions. β-Branched amino acids, such as L-Ile or L-allo-Ile, were prepared by this method (299), while β-substituted-Glu derivatives were obtained by oxidizing the terminal position of the alkyne adduct resulting from organocuprate addition (300), and β-alkylated-Asp derivatives via reduction of the alkyne to an alkene followed by oxidative cleavage (300). The original alkynol adduct was also transformed into an allenic phosphonate, with reduction, deprotec-tion, and oxidation giving (2R)-2-amino-5-phosphono-pentanoic acid (301).

A syn adduct isomer has also been converted to the anti epimer by oxidation of the initial carbinol with PDC followed by reduction with Zn(BH₄)₂ (302). All four

diastereomers of β-methoxy-Tyr were derived from L- or D-Ser via additions of 4-benzyloxyphenyllithium to the Garner aldehyde. The crude product, with 50–60% de anti selectivity, was obtained diastereomerically pure after one recrystallization. Oxidation of the carbinol center followed by reduction with K-Selectride gave the syn isomer with 94% de. Either diastereomer was then O-methylated prior to oxazolidine hydrolysis and hydroxymethyl oxidation (303).

A number of reports have examined the stereoselec-tive allylation of the Garner aldehyde (see Scheme 10.31, Table 10.7), and a comprehensive review of the reactions of allylic metals includes several examples of additions to the Garner aldehyde (304). Allylsilane adds in the presence of a Lewis acid, giving high anti stereoselectivity (90% de) with 1 equiv of TiCl₄ and reasonable syn selectivity (78% de) with 0.5 equiv of the

Table 10.7 Allyl Addition to the Garner Aldehyde (see Scheme 10.31)

Reagent	Stereoselectivity	Further Derivatization	Reference
~~SiMe$_3$	0.5 eq. TiCl$_3$: 47%, 78% de *syn* 1.0 eq. TiCl$_3$: 84%, 90% de *anti*		1989 (*305*)
~~M	M = MgBr: 86%, 1-% de *syn* M = TiCp(O-2-Pr)$_2$: 89%, 26% de *anti* M = Ti(*R,R*-L*): 93%, 96% de *syn* M = Ti(*S,S*-L*): 95%, 99% de *anti*		1992 (*306*)
~~SnBu$_3$	(added to Thr-derived Garner aldehyde) BF$_3$.Et$_2$O: 84%, 76% de *anti* MgBr$_2$.Et$_2$O: 87%, 66% de *syn*		1994 (*307*)
~~B(O)(O)—CO$_2i$Pr / CO$_2i$Pr	PhMe, (*R,R*) reagent: 77%, 82% de *syn* Et$_2$O, (*S,S*) reagent: 84%, 92% de *anti*		1995 (*259*)
Ti(OiPr)$_4$ / *i*PrMgCl with allylBr allylOCO$_2$Et allylOPO$_3$Et$_2$	86–95% 84–85% de *anti*		2002 (*296*)
allylMgCl	86% 26% de	reductive ozonolysis of alkene, displacement of both side-chain OH by azide, reduction and cyclization with Br^{13}CN to give guanidine	2002 (*317*)
In/BrCH$_2$CH=CHOAc	88% 94% de (only observe 2 of 4 possible diastereomers under optimized conditions)	major diastereomer:	2004 (*312*)
~~B(O)(O)—CO$_2i$Pr / CO$_2i$Pr	PhMe, (*R,R*) reagent: 84%, 66% de *syn* Et$_2$O, (*S,S*) reagent: 78%, 64% de *anti*	1) BH$_3$-THF 2) 30% H$_2$O$_2$, 2M NaOH 77–90% 1) TsCl, Et$_3$N, DMAP 54–59% 2) *p*TsOH 3) PDC 4) 4.5N HCl 33–41% + 3 other diastereomers separately	2005 (*308*)

(Continued)

Table 10.7 Allyl Addition to the Garner Aldehyde (see Scheme 10.31) (continued)

Reagent	Stereoselectivity	Further Derivatization	Reference
⟋⟍⟋⟍SiMe₃	1.0 eq. TiCl₃: 66%, 90% de *anti* (*R,S,R*)	1) Pd-C, cHex: 94% 2) NaH, CS₂, imid 3) AIBN, H₃PO₄, Et₃N: 72% 4) PPTS,EtOH 5) PDC, DMF: 73% or 1) TBS-Cl, imid 2) BH₃, THF 3) NaOH, H₂O₂ 4) DMSO,COCl₂, Et₃N 85% 1) CBr₄, PPh₃, Et₃N 2) BuLi, Hex 3) MeI 74% 1) Na, NH₃: 96% 2) PPTS 3) PDC, DMF: 70% 4) TBAF: 81%	1996 (*311*)
⟋⟍⟋⟍M	M = Ti(*R,R*-L*): 93%, >98% de *syn* (*S,S,S*)		1992 (*306*)
*i*Pr–N(Si)(*i*Pr)...B(Ipc)₂	d-Ipc reagent: 57%, 96% de *anti* (*R,S,S*) l-Ipc reagent: 45%, 34% de *anti* (*R,S,S*)		1992 (*310*) 1994 (*309*)
⟍⟋B(O)(O)	46%, 50% de *anti* (*R,S,R*)		1994 (*309*)
TBSO...SnBu₃	MgBr₂.Et₂O: 93%, 100% de *syn*	1) TBAF 2) TsOH, acetone: 90% for 2 steps 3) PDC 4) CH₂N₂: 70% for 2 steps 5) O₃, MeOH, NaBH₄: 71%	1994 (*307*)

same catalyst (*305*). The reversal of stereocontrol is proposed to be due to the formation of different complexes. Organotitanium complexes formed from Ti(O*i*Pr)₄/ *i*PrMgX and allyl bromides, acetate or phosphate added in 86–95% yield with 84–85% de *anti* stereoselectivity (*296*). Almost complete diastereocontrol has been obtained from a chiral allyltitanium complex, with the matched enantiomer giving the *anti* adduct with 99% de, and the mismatched isomer producing the *syn* product with 96% de (*306*). Allylstannane can also produce either diastereomer upon addition to both Ser- and Thr-derived Garner aldehydes, with BF₃.Et₂O catalysis affording predominantly the *anti* isomer and MgBr₂.OEt₂ catalysis the *syn* diastereomer (*307*). A γ-oxygenated allylic stannane was added with 100% de at both the β- and γ-centers, leading to a protected polyhydroxy amino acid, polyoxamic acid (*307*). An enantiomeric pair of tartrate-derived chiral allylborane reagents gave slightly

lower selectivity, with 92% de *anti* from one reagent and 82% de *syn* from the enantiomer (*259*). A subsequent synthesis used this pair of reagents on the D- and L-Garner aldehydes to generate all four possible diastereomers. Hydroboration/oxidation of the alkene, followed by cyclization, generated a tetrahydrofuryl ring, with deprotection/oxidation producing the four diastereomers of 2-(tetrahydrofuran-2-yl)-Gly (*308*).

Crotyl-type organometallic reagents lead to γ-branched- β-hydroxy amino acids (see Scheme 10.31, Table 10.7), with three contiguous chiral centers. An achiral (*Z*)-crotylboronate gave the (*S,R,S*) (*anti* aldehyde addition) isomer with 50% de (*309*) while a silyl substituted chiral allylic (*E*)-B(Ipc)₂ reagent produced even better diastereoselectivity for the matched pair of isomers (*anti–anti*, with 96% de, 57% yield), although the mismatched pair had only 34% de (*309, 310*). Reaction of (*E*)-crotyltrimethylsilane with the (R)-Garner aldehyde in the presence of TiCl₄ resulted in the (R,S,R) adduct (anti-addition at aldehyde) in 66% yield with 90% de (311), while a chiral (E)-crotyltitanium reagent reacted with the (S)-aldehyde to give the (S,S,S) configuration (syn-addition at aldehyde) with complete diastereoselectivity (306). The side chain was further elaborated by extension to a homologous alkene or by reduction and hydroxyl removal, prior to conventional deprotection/oxidation to generate the amino acid (311).

Hydroxyallylation of the Garner aldehyde has been achieved under Barbier conditions with 3-bromoprope-nyl acetate (AcOCH=CHCH₂Br) in the presence of zinc or indium, resulting in a CH(OH)CH(OH)CH=CH₂ side-chain on the oxazolidine template. One of the four possible diastereomers with anti/anti stereochemistry was always obtained preferentially, and could be produced in 88% yield with 94% ds (312, 313).

A range of complex amino acids has been prepared via additions to the Garner aldehyde. A total synthesis of the trisaminotriscarboxylic acid kaitocephalin, with a 2,5-disubstituted Pro core, proceeded via an aldol reaction of Boc-5-allyl-Pro-OBn with the Garner aldehyde (314). The enolate of acetate, derivatized as an amide of pseudoephedrine, has also been added to the Garner aldehyde. The matched pair of enantiomers produced the aldol adduct with 96% de and the mis-matched pair with 12% de. Hydrolysis gave 3,5-dihy-droxy-4-aminopentanoic acid (315). The Reformatsky reagent derived from ethyl bromodifluoroacetate added to the Garner aldehyde in 81% yield, with dehydroxyla-tion, oxazolidine hydrolysis, and hydroxymethyl oxidation providing 4,4-difluoro-Glu. However, the oxazolidine hydrolysis took 4 days and proceeded in only 30% yield, while the oxidation step only gave 50–60% yield (316).

A synthesis of (2S,3R)-[guanidino-¹³C]capreomyci-dine employed an initial addition of allylMgCl to the Garner aldehyde. Reductive ozonolysis of the alkene and sequential azide displacements of both side-chain alcohols gave protected 3-amino-ornithinol. The amines were cyclized to form the capreomycidine guanidine by

treament with Br¹³CN, with the synthesis completed by oxazolidine deprotection and hydroxymethyl oxidation (317). C-Glycosyl acetylenes were deprotonated with nBuLi and added to the Garner aldehyde. Alkyne reduc-tion and Barton–McCombie deoxygenation produced 3-carbon-linked glycosyl amino acids, C-analogs of gly-cosylated Asn. Twelve different compounds (six pairs of α- and β-anomers) were synthesized (318). A synthesis of furanomycin, α-(5-methyl-1,5-dihydrofuran-2-yl)-Gly, began with addition of homochiral acetylide LiCCCH(Me)OTBS to the Garner aldehyde. Inclusion of ZnBr₂ in the reaction resulted in chelation control and non-Felkin–Ahn addition stereoselectivity (80% de). The dihydrofuran ring system was constructed via an allenic intermediate, followed by oxazolidine cleavage and hydroxymethyl oxidation (319).

The Ser-derived oxazolidine ester precursor of the Garner aldehyde has also been used as a substrate for organometallic additions (see Scheme 10.32). DiphenylSer was synthesized via addition of 2 equiv of phenyl Grignard reagent to the ester group (292); removal of the β-hydroxyl group by hydrogenation provided diphenyl-alanine (292). Similarly, the precursor of β-hydroxy-Val was prepared by reaction of 2 equiv of MeMgBr to the Garner aldehyde ester precursor (256). A β-hydroxy-Ile skeleton was obtained via reaction of the Weinreb amide precursor of the Garner aldehyde with MeLi, followed by Grignard reaction of the resulting ketone with EtMgBr. This was converted into the complex aryl ether-linked bis(amino acid) contained in ustiloxin, consisting of β-hydroxy-Ile linked to β-hydroxy-DOPA. The hydroxy group was alkylated with a 4-bromo-2-fluorobenzonitrile moiety to introduce the aryl ether linkage. The aromatic group was elaborated into a cinnamate derivative, with the second amino acid center (forming the β-hydroxy-DOPA residue) constructed by asymmetric aminohydroxylation of the acrylate. The synthesis was completed by oxidation of the original D-Ser side chain (320). The Weinreb amide was also employed for a synthesis of the Abbott neuramini-dase inhibitor A-315675, via reaction with allylMgBr, alkene hydrogenation, and addition of MeMgBr to

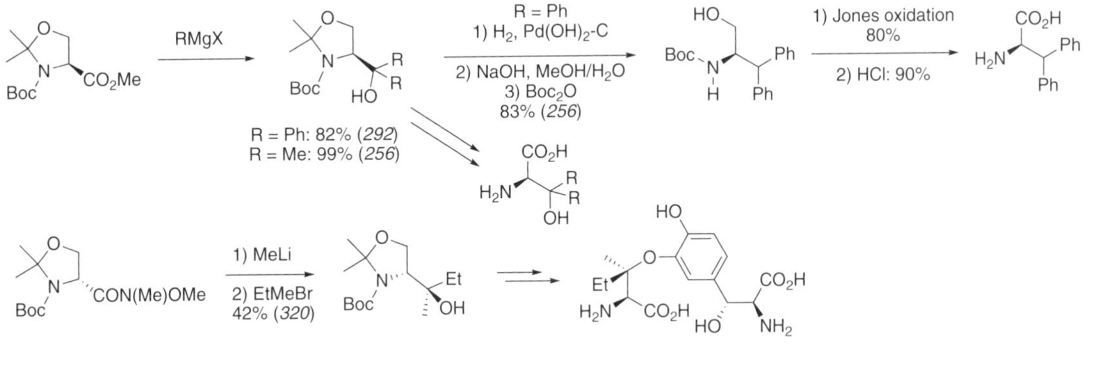

Scheme 10.32 Organometallic Additions to the Garner Aldehyde Precursor.

the ketone. Further elaboration produced the desired 4-(1Z-propenyl)-5-(1-acetamino-2-methoxy-2-methylpentane)-D-Pro product (*321*).

The alcohol intermediate that is oxidized to generate the Garner aldehyde has instead been converted into an electrophilic tosylated group for nucleophilic displacements (see Section 10.3.4) or a nucleophilic β-anionic Ala equivalent (see Section 10.5.2).

10.6.1c Garner Aldehyde: Olefination Reactions

The first report of alkene formation from the Boc-protected Garner aldehyde was in 1987 by Moriwake et al. (*322*). Attempts to react the aldehyde with the methylene ylide derived from Ph₃PMeBr and KH gave the desired alkene, but with complete racemization (see Scheme 10.33, Table 10.8). Methylenation under neutral conditions with AlMe₃-Zn-CH₂I₂ successfully led to optically pure vinylglycinol, but vinylglycine itself was not synthesized. A more recent report found that olefination with methyltriphenylphosphonium bromide and KHMDS under salt-free conditions gave the vinylglycinol derivative in 93% yield and high optical purity, and was much more convenient than the methylenation reaction for multigram syntheses (*253*). This method was employed in a 2003 synthesis of vinylglycinol (*323*). As mentioned earlier, this vinylglycinol compound has paradoxically been obtained from L-Met as an intermediate for the synthesis of the Garner aldehyde synthon (*265*). The alkene obtained by methylenation of the Garner aldehyde was hydroborated to give an organoborane that was then used for Pd-catalyzed Suzuki coupling with aryl halides. Oxazolidine cleavage and hydroxymethyl oxidation provided homophenylalanine derivatives (*324*). The organoboron homoalanine equivalent was also coupled with N-Boc 3-bromo-2,3-dehydro-Ala-OMe to give, after asymmetric alkene hydrogenation and hydroxymethyl oxidation, *meso*-2,6-diaminopimelic acid (*325*).

The (Z)-α,β-unsaturated ester adduct was employed to synthesize β-substituted glutamic acids as acyclic analogs of the kainoids, compounds that possess potent neurological effects (*258, 326*). The *cis* α,β-unsaturated ester was derived from Garner's Boc-protected aldehyde in 87% yield using the stabilized phosphonate anion of Still and Gennari (*327*) (see Scheme 10.34). A number of other olefinations of the Garner aldehyde have also relied on this reagent (*326, 328–332*). Rearrangement of the oxazolidine to an unsaturated lactone gave a substrate for Michael

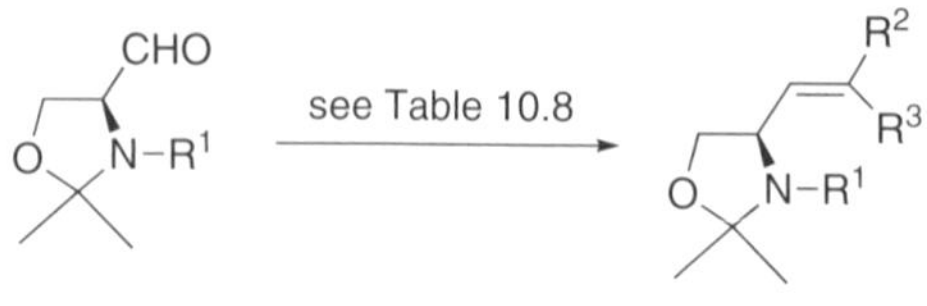

additions of lithiated benzylphenylsulfide (*258, 326*) or organocuprates (*326*). The additions proceeded with complete stereoselectivity. Treatment of the adducts with aqueous permanganate followed by TFA generated deprotected β-substituted glutamic acids. The alkene bond of the Z-α,β-unsaturated ester has also been *cis*-dihydroxylated with osmium tetraoxide (*330, 332*) and then converted to a polyhydroxy-γ-amino acid (*330*).

The corresponding *trans* α,β-unsaturated ester was preferentially obtained in a 60:40 E:Z ratio and 91% yield by reaction of the Garner aldehyde with the stabilized Ph₃P=CHCO₂CH₃ ylide (*333*) (see Scheme 10.35). Other syntheses of this isomer have been reported (*257, 328, 331, 334, 335*), with up to an 95:5 E:Z ratio obtained in 88% yield with (MeO)₂P(O)CH₂CO₂Me/K₂CO₃ (*331, 335*). The E-alkene is a substrate for a number of amino acid syntheses. A γ-substituted Glu derivative was prepared via reduction of the alkene bond of the α,β-unsaturated ester followed by alkylation of an enolate of the ester group (*334*). Dihydroxylation of the alkene gave a 1:1 mixture of two diastereomers (compared to the one diasteromer obtained from the analogous Z-alkene) (*332*). 3,4-Methano-Glu was produced by olefination with Ph₃P=CHCO₂CH₃, ester reduction and protection of the resulting alcohol, cyclopropanation, oxazolidinone deprotection, and oxidation of both hydroxymethyl groups (*336*). A synthesis of *trans*-4-methyl-Pro olefinated the Garner aldehyde with Ph₃P=C(Me)CO₂Me, and then reduced the acid group with DIBAL and the alkene with H₂/Raney Ni. The resulting 2-methyl-3-hydroxypropyl substituent was converted to a bromide before hydrolysis of the Garner hemiaminal, with cyclization to the prolinol induced with KHMDS. Oxidation completed the synthesis (*337*).

Michael additions of organocuprate reagents to the *trans* α,β-unsaturated ester adduct in the presence of TMS-Cl gave 70–97% yields with 60–96% de in favor of the *syn* isomer (*331, 335, 338, 339*). The methyl and benzyl adducts were converted into the corresponding β-substituted Glu derivatives (*335*). Alternatively, the methyl adduct was further elaborated by electrophilic amination or hydroxylation of a KHMDS-generated enolate (70% de), Mitsunobu amination of the hydroxylated adduct allowing for the stereoselective synthesis of either amine isomer. Oxazolidine opening gave 2,4-diamino-5-hydroxypentanoic acid, with no oxidation required, as the α-amino and α-carboxyl groups were introduced rather than originating from the Garner aldehyde nucleus (*339*). A Michael addition has also been carried out on the *trans* α,β-unsaturated ester amino alcohol obtained after oxazolidine cleavage, with a tandem Michael addition/cyclization of methyl vinyl ketone leading to allokainic acid (*257*). α-Kainic acid was produced by using 2-nitro-3-methyl-buta-1,3-diene for the cyclization (*340*).

Both the Z- and E-α,β-unsaturated esters can be reduced to the corresponding allyl alcohols (*328, 329, 341*) (see Scheme 10.36). These intermediates have been used in several syntheses of highly functionalized amino acids. Epoxidation and further elaboration of the

Scheme 10.33 Olefination of the Garner Aldehyde.

Table 10.8 Olefination of the Garner Aldehyde (see Scheme 10.33)

R^1	R^2	R^3	Olefination Reagent	Yield	Further Derivatization	Reference
Boc	H	H	Ph_3PMeBr, KH, PhH; Zn, CH_2I_2, $AlMe_3$	66%, 0% ee		1987 (322)
Cbz	H	H	$Ph_3PCHR^2R^3X$	75%, 100% ee	1) convert CN to CH_2NHBoc with $NaBH_4$	1988 (264)
Boc	H	Me	LHMDS or LDA	27–96%		1991 (346)
	H	Ph		100% E to 98% Z		
	H	nPent			1) TsOH, MeOH 2) Jones oxidation 87–92%	
	H	$(CH_2)_2Ph$				
	H	$(CH_2)_2CO_2H$			or	
	H	$(CH_2)_2CN$			1) H_3O^+: 75–98%	
	H	$(CH_2)_3CN$			2) Jones oxidation	
	Ph	H			3) CH_2N_2 0–92%, 82–>95% ee	
	CO_2Me	H				
	Et	Me			or	
					$R^3 = (CH_2)_nNHBoc$, n = 3,4; n-Pent	
					1) H_2, $Pd(OH)_2$ 2) Jones oxidation 3) CH_2N_2 70–91% R^4 = Et, NHBoc, CH_2NHBoc	
Boc	Ph	H	Ph_3PCH_2RBr	70%, 3:1 $E:Z$	1) $CHBr_3$, NaOH, $BnEt_3NCl$ 45–87%, 46–70% de	1991 (350)
	H	Et	nBuLi, hex	63%, 1:3 $E:Z$	2) Bu_3SnH: 80–97% 3) TsOH, MeOH 4) Jones oxidation 5) 6N HCl: 77–81% R = Et, Ph	1992 (349)

(Continued)

Table 10.8 Olefination of the Garner Aldehyde (see Scheme 10.33) (continued)

R¹	R²	R³	Olefination Reagent	Yield	Further Derivatization	Reference
Boc	H	H	Ph₃PMeBr, KHMDS	93%	–	1994 (253)
Boc	H	CH₂C[-O(CH₂)₃O-]	Zn, CH₂I₂, AlMe₃ / BrPh₃P(CH₂)₂C[-O(CH₂)₃O-] NaH, THF	70% / 80:20 *Z:E*	1) CHBr₃, KOH TEBA, benzene 73%, 14%de 2) *n*Bu₃SnH, hex: 85% 3) Jones oxidation 4) Dowex-50W: 75% (HO₂C cyclopropane H₂N⁺ CO₂H)	1994 (345)
Boc	CO₂Me	NHCbz	K(MeO)₂P(O)CH(NHCbz)CO₂Me	77%	1) H₂, *i*PrOH 97%, 88% de 2) CSA, MeOH 3) Jones oxidation 4) CH₂N₂ 66% for 3 steps 5) 10N HCl: 89% (HO₂C···NH₂···H₂N···CO₂H)	1996 (356)
Boc	H	H	BnO, OBn, OBn sugar CH₂PPh₃ (OBn)	62% mainly Z	1) H₂, Pd-C: 84% 2) Jones oxidation: 94%	1998 (351)
Boc	H	H	Ph₃P=CH₂	80%	BR₂ → Arl, Pd(Cl)₂(dppf) 68–76% → Ar. Ar = Ph, 2-CH=CH₂-Ph, 2-NHBoc-Ph, 4-MeO-Ph, 3-NO₂-Ph then Jones oxidation 30–59%	1999 (324)
Boc	H	H	Ph₃P=C(Br)CO₂Me	55%	formation of aziridine via reaction with BnNH₂	2000 (354)
Boc	H	H	Ph₃P=C(Me)CO₂Me	72% *E:Z* >96:4	1) DIBAL: 88% 2) H₂,Raney Ni: 95%,72% de 3) PPh₃, CBr₄: 90% 4) AcOH: 80% 5) PhCOCl, pyr 6) KHMDS: 71% 7) 1N HCl then Fmoc-Cl: 96% 8) Jones oxidation: 59% (pyrrolidine, Fmoc, CO₂H)	2001 (337)

714

Boc	H	H	Reagent	Conditions	Yield	Year (ref.)	Notes
Boc	H	H	Ph₃P=CHCO₂Me	1) DIBAL-H: 87% 2) TBDPSCl, im: 94% 3) Et₂Zn, CH₂I₂: 93% 4) deprotection/oxidation: 74%	92%	2001 (336)	
Boc	H	H	Ph₃P=CHCOCH₂OTBS	1) H₂, Pd/C: 80% 2) TMSCHN₂, BuLi: 69% 3) H₂, Pd/C 4) HF: 81% 5) NaIO₄, RuCl₃: 49%	87%	2002 (353)	
Boc	H	H	Ph₃P=C(Br)CO₂Me	1) FSO₂CF₂CO₂Me, CuI, DMF/HMPA 2) H₂, Raney Ni 3) LiAlH₄ 1) BnBr, NaH 2) 80% AcOH 3) TBDMSCl, im 4) H₂, 10% Pd/C 5) MsCl, Et₃N 6) KHMDS 7) TBAF 8) Jones	55%	2002 (355)	
Boc	H	H	Ph₃PMeBr, KHMDS		63%	2003 (323)	
Boc	H	H	Ph₃P=CHC₆F₅	–	92%	2004 (348)	
Boc	H	H	Ph₃P=CHMe Ph₃P=CHnPr Ph₃P=CHnOct Ph₃P=CHnUndecyl Ph₃P=CHiPr	1) H₂, Pd-C: 984% 2) Jones oxidation: 64%	48–81%	2004 (347)	E:Z 93:7 to 95:5 when quenched with MeOH at –78°C E:Z 6:94 when not E:Z 32:68 with quenching

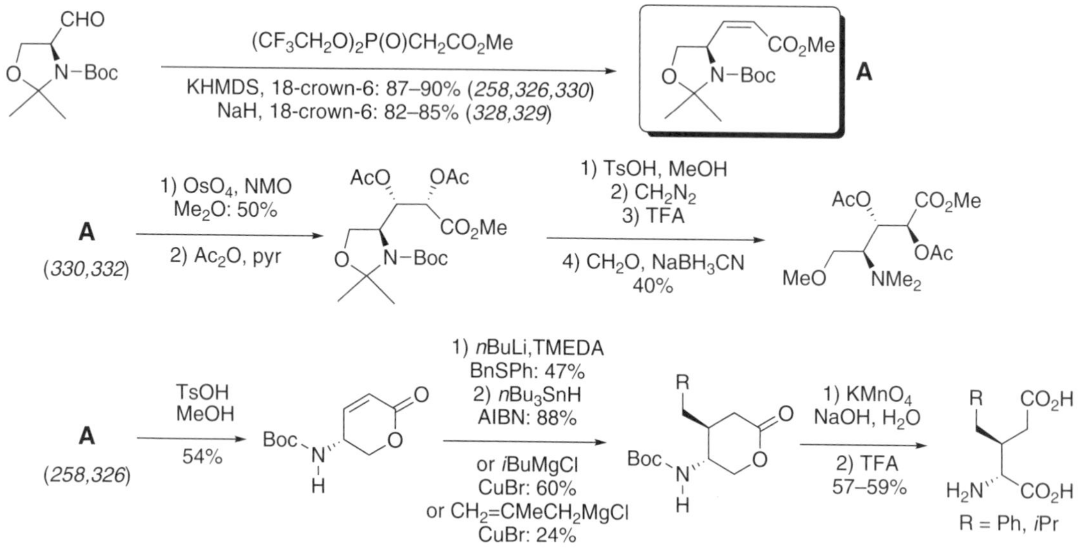

CHO
O N–Boc

(CF₃CH₂O)₂P(O)CH₂CO₂Me

KHMDS, 18-crown-6: 87–90% (*258,326,330*)
NaH, 18-crown-6: 82–85% (*328,329*)

O N–Boc   CO₂Me   **A**

A
(*330,332*)

1) OsO₄, NMO
Me₂O: 50%

2) Ac₂O, pyr

AcO OAc
O N–Boc CO₂Me

1) TsOH, MeOH
2) CH₂N₂
3) TFA

4) CH₂O, NaBH₃CN
40%

AcO CO₂Me
MeO NMe₂ OAc

A
(*258,326*)

TsOH
MeOH
54%

Boc N O O
H

1) *n*BuLi,TMEDA
BnSPh: 47%
2) *n*Bu₃SnH
AIBN: 88%

or *i*BuMgCl
CuBr: 60%
or CH₂=CMeCH₂MgCl
CuBr: 24%

R
Boc N O
H

1) KMnO₄
NaOH, H₂O

2) TFA
57–59%

R CO₂H
H₂N CO₂H
R = Ph, *i*Pr

Scheme 10.34 Synthesis and Reaction of (Z)-Unsaturated Ester Garner Aldehyde Derivative.

Z-isomer led to the γ-amino acid galantinic acid, part of the antibiotic peptide galantin I, in 2.2% overall yield from the Garner aldehyde (*329, 341*). Both allylic alcohols were used for an intramolecular cyclopropanation in the diastereoselective synthesis of conformationally restricted Glu analogs (*328, 342, 343*). The hydroxymethyl substituent on the cyclopropane ring was left as a methyl ether (*328, 342*), oxidized to an acid group (*343*), or converted to a vinyl or ethyl group (*344*). An intermolecular cyclopropanation of the Z-alkene with dibromocarbene has also been reported, but the addition was preceded by isomerization of the alkene to the E-isomer. The resulting *trans* cyclopropane was deprotected and oxidized to give *trans*-α-(2-carboxylmethylcyclopropyl)-Gly (*345*).

The benefits and drawbacks of using the Garner aldehyde for β,γ-unsaturated amino acid synthesis were both demonstrated in a 1991 report by Beaulieu et al. which described the synthesis of a series of optically active vinylglycines (*346*). The Wittig additions of stabilized and unstabilized ylides proceeded with reasonable yields (50–96%, except for methyltriphenylphosphorane, 27% yield) and good diastereoselectivity (generally >98% Z, except for the stabilized ylides). Oxazolidine opening was uneventful (75–89% yield). However, oxidation of the amino alcohols to the corresponding amino acids was problematic, resulting in variable yields (37–92% yield), racemization (82 to >95% ee), and the complete decomposition of some compounds (R = Ph, CO₂Me). The alkene bond could be reduced before oxidation to give saturated amino acids, and unsaturated nitrile adducts were transformed into unsaturated Lys analogs. To overcome the problems presented by the oxidation step, Duthaler carried out Wittig reactions on the Cys analog of the Garner aldehyde (see Scheme 10.37) (*273*).

The N-acylthiazolidines were oxidized under much milder conditions, and one example was deprotected by acid hydrolysis to give a Z-4-methylvinylglycine in 66% yield from the thiazolidine, with 94–96% ee.

A number of other Wittig reagents have been reacted with the Garner aldehyde (see Scheme 10.33, Table 10.8). (E)-Selective Wittig reactions with non-stabilized ylides were achieved when the reaction was quenched with MeOH at −78 °C, with > 10:1 (E):(Z) ratios for methyl, propyl, octyl, and undecyl substituents. Without the addition of MeOH, the reactions gave the usual (Z)-isomer, with similar selectivity (*347*). Condensation with C₆F₅CH=P(Ph)₃, followed by hydrogenation and oxidation, provided pentafluorohomophenylalanine (*348*). In 1988 Beaulieu et al. reported the successful reaction of the Cbz-protected Garner aldehyde with unstabilized alkylcyano ylides. Reduction of the cyano group, opening of the oxazolidine, oxidation of the hydroxymethyl group, and hydrogenation of the alkene generated D-α,ω-diaminoalkanoic acids in 18% overall yield and >95% ee (*264*). The ylide derived from benzyltriphenylphosphonium bromide reacted with the aldehyde to give a separable 3:1 mixture of E and Z alkenes (70% yield); the *trans* isomer was cyclopropanated with dibromocarbene in 87% yield (70% de) (*349, 350*). Oxazolidine cleavage and oxidation gave α-(2-phenylcyclopropyl)-Gly. A similar synthesis starting with propyltriphenylphosphonium bromide led to the corresponding ethyl-substituted analog, with the unreactive Z-isomer slowly converted to the E-isomer under the cyclopropanation conditions (*349*).

Wittig reaction of a β-D-galactopyranosyl- or 2-deoxy-β-D-galactopyranosyl-phosphorus ylide led to C-linked analogs of O-β-D-galactopyranosyl-L-Ser, via alkene hydrogenation, oxazolidine deprotection,

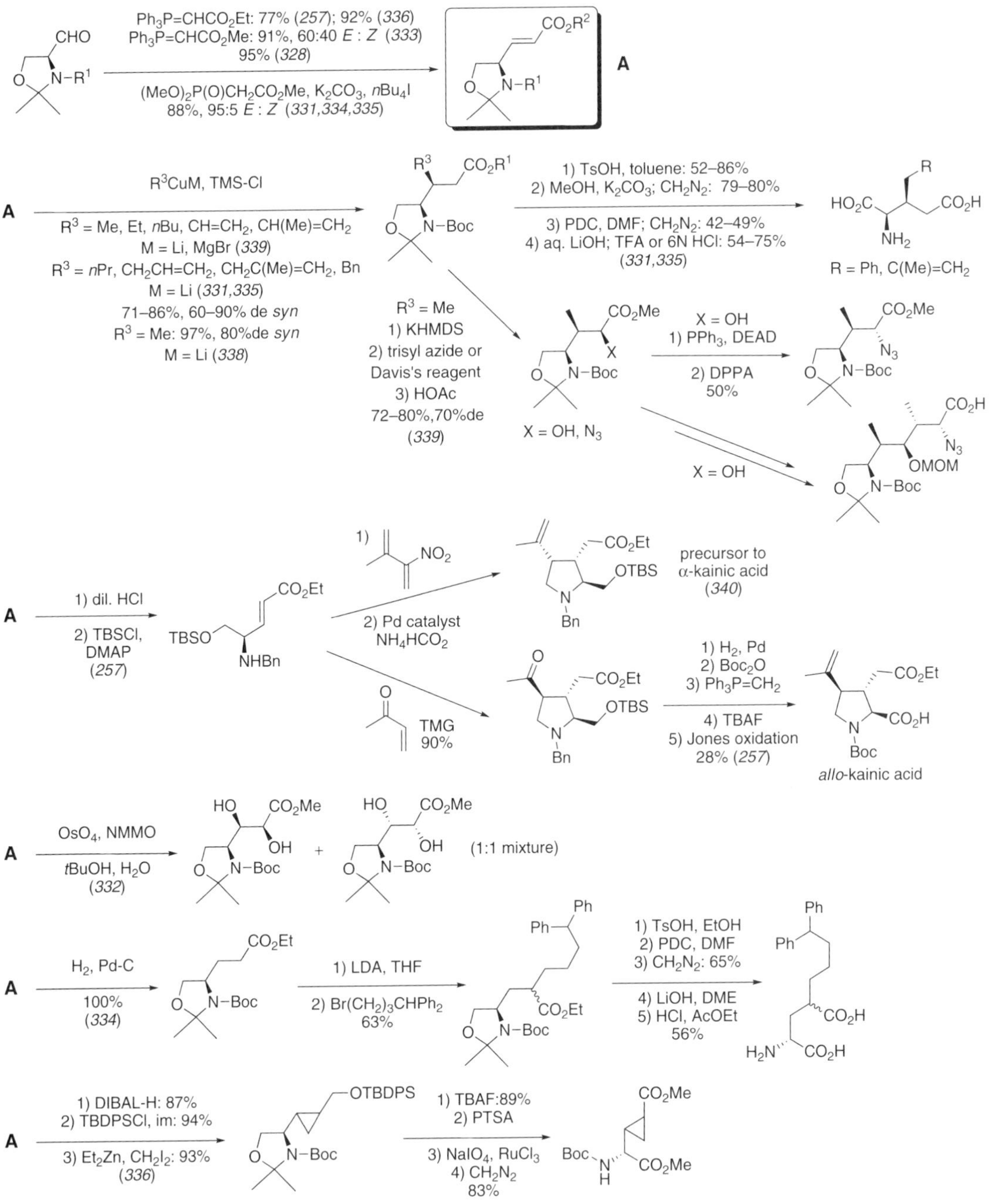

Ph₃P=CHCO₂Et: 77% (257); 92% (336)
Ph₃P=CHCO₂Me: 91%, 60:40 E : Z (333)
95% (328)

(MeO)₂P(O)CH₂CO₂Me, K₂CO₃, nBu₄I
88%, 95:5 E : Z (331,334,335)

Scheme 10.35 Synthesis and Reaction of (E)-Unsaturated Ester Ganer Aldehyde Derivative.

and hydroxymethyl oxidation (351, 352). A synthesis of (1S,3R)-1-aminocyclopentane-1,3-dicarboxylic acid employed an olefination with Ph₃P=COCH₂OTBS, followed by cyclization via a 1,5-CH insertion reaction to generate the quaternary carbon center. The side-chain and α-hydroxymethyl groups were oxidized at the same time (353). Olefination with Ph₃P=C(Br)CO₂Et, followed by Michael addition/cyclization of benzylamine,

resulted in an 85% yield mixture of four diastereomers (49:45:1:5) of aziridine-2-carboxylate with a protected 3-(1-amino-2-hydroxyethyl) substituent (354). Condensation with Ph₃P=C(Br)CO₂Et, Cu-catalyzed replacement of the bromide with a trifluoromethyl group, alkene and ester reduction, mesylation, cyclization, and hydroxymethyl oxidation led to cis- and trans-4-trifluoromethyl-Pro (355).

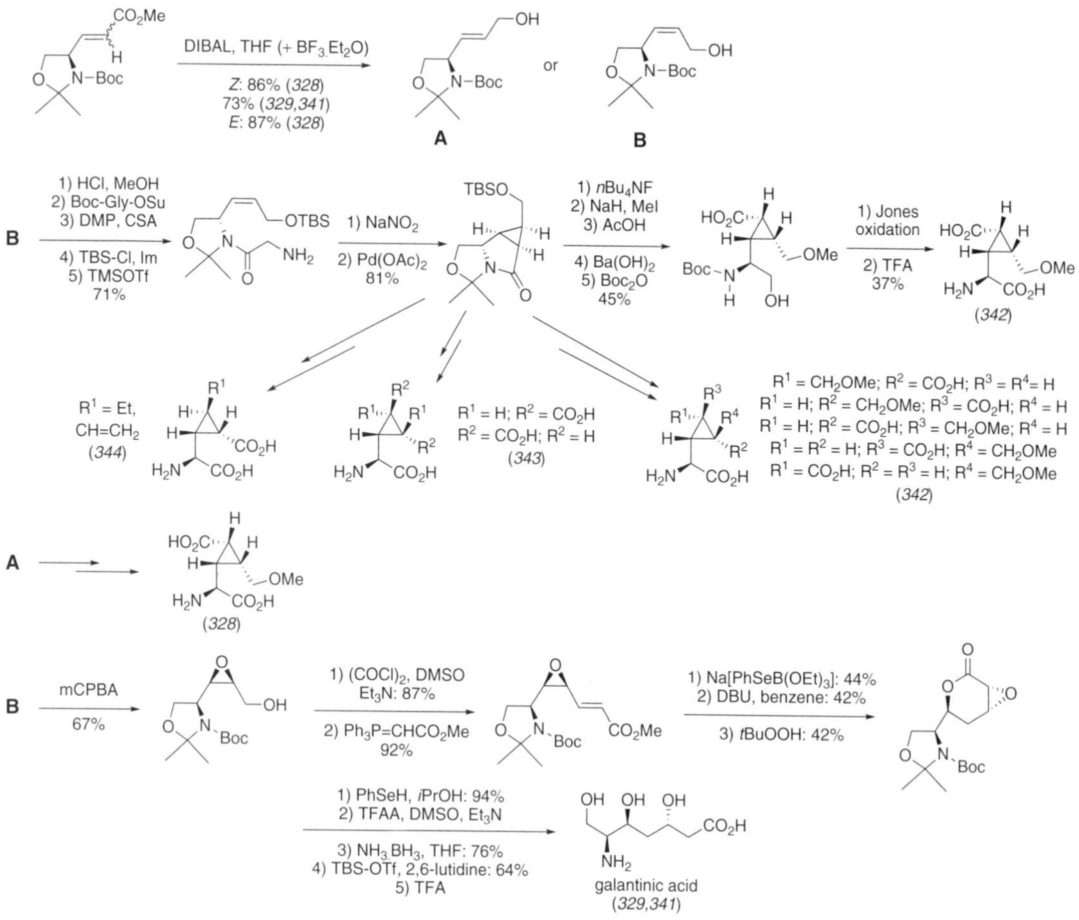

Scheme 10.36 Reduction and Reactions of α,β-Unsaturated Ester Derivatives of the Garner Aldehyde.

A Wadsworth–Horner–Emmons olefination has been used to introduce a side chain with a second amino acid moiety (*356*). Hydrogenation of the alkene proceeded with high diastereoselectivity, leading to *meso*-2,4-di-aminoglutaric acid after conventional deprotection/oxidation (*356*). The same strategy has been employed to prepare the (2*R*,4*R*) isomer of 2,4-diaminoglutaric acid (*357*). *Meso*-1,6-diaminopimelic acid has also been synthesized via the Garner aldehyde, with the second amino acid center introduced by a displacement of a bromo substituent using Schöllkopf's bislactim ether template as the nucleophile (*358*).

The aldehyde of the Garner synthon has been converted into an acetylene group by reaction with dibromoethylene triphenylphosphorane followed by base (*359, 360*) (see Scheme 10.38). Meffre et al. also transformed the aldehyde into an alkyne, but found that diazomethyl phosphonates gave better yields than the Corey–Fuchs method (*361–363*). New conditions for formation of the alkyne from the aldehyde in a one-pot reaction were described in 2002 (*364*). The preparation of the ethynylglycinol synthon from the

Garner aldehyde and its conversion into ethynylglycine and other non-natural amino acids was reviewed in 2005 (*365*). In the initial reports the alkyne substituent was used for a Stille coupling (*359*), or metallated and reacted with alkyl halides (*359*) or formaldehyde (*360*). No amino acids were prepared, but it was noted that alkynyl derivatives are readily reduced to alkenes, which have successfully been converted to β,γ-unsaturated amino acids (see above). In subsequent publications the alkyne group was again metallated and reacted with electrophiles, with fresh *n*BuLi observed to be critical for good yields (*361*). Selective oxazolidine ring cleavage was found to be difficult; thus, the Boc group was simultaneously removed during ring cleavage and the amine reprotected without isolation of the unstable fully deprotected amino alcohol. The final oxidation to generate the Boc-protected β,γ-alkynylglycines also proved to be problematic, as might be expected. A number of methods were examined, but ultimately a Jones oxidation with slow reverse addition was found to give the desired products in low but reproducible yield (26–37%). The products had 91–93% ee (similar to the Garner

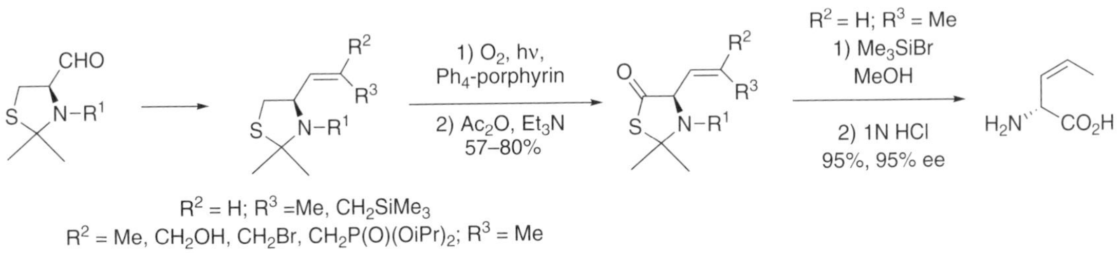

Scheme 10.37 Olefination of the Thiol Analog of the Garner Aldehyde (*273*).

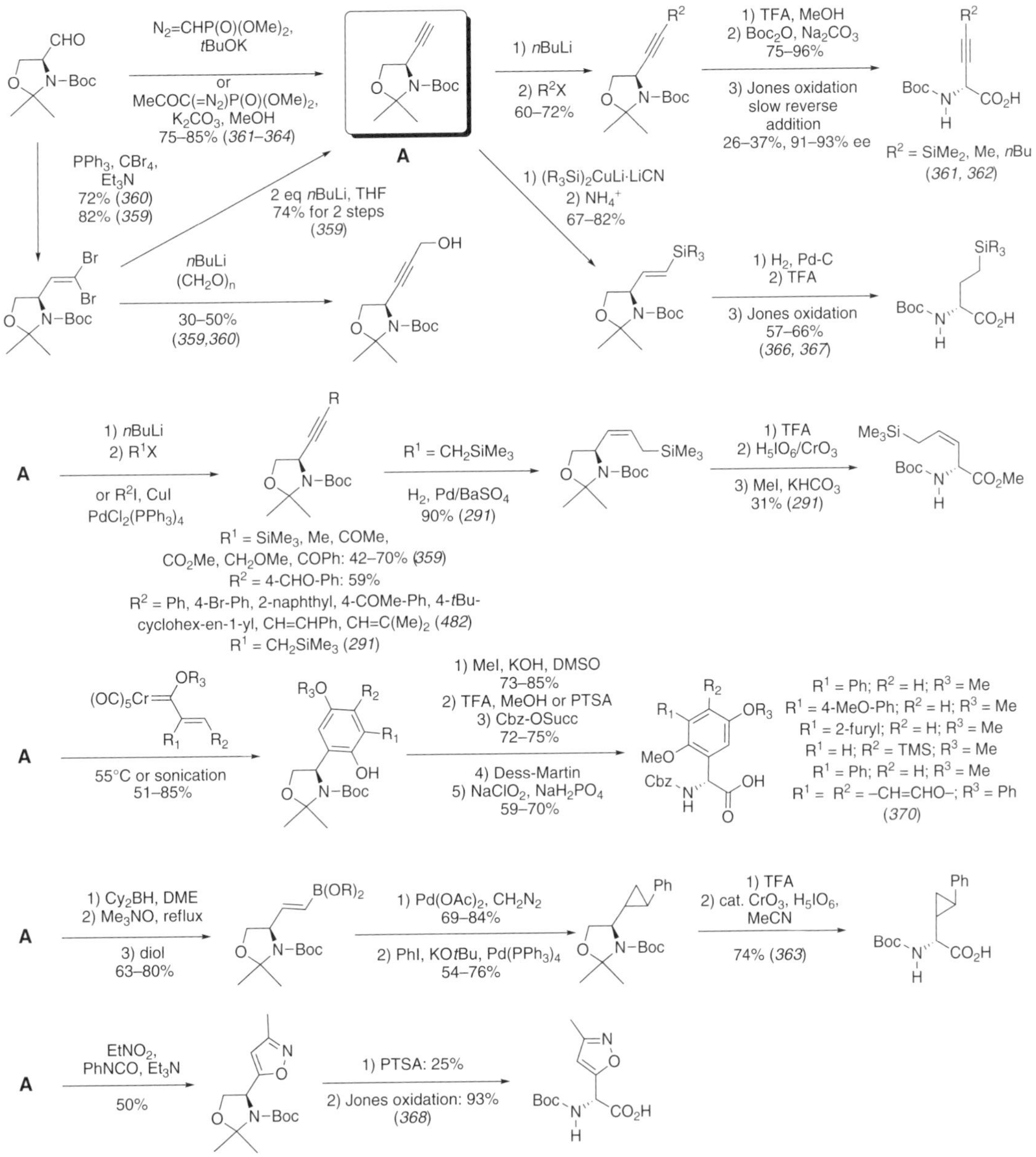

Scheme 10.38 Synthesis and Reaction of Alkynes via the Garner Aldehyde.

aldehyde starting material), with racemization during the oxidation step apparently prevented by conversion of any enolized product into an imide (*361, 362*).

The alkyne synthons were used as substrates for the addition of silyl cuprates, $(R_3Si)_2CuLi.LiCN$, with hydrogenation of the resulting alkene, oxazolidine deprotection, and hydroxymethyl oxidation providing γ-silyl-α-amino acids. Omitting the hydrogenation step resulted in the corresponding (*E*)-β,γ-unsaturated amino acids (*366, 367*). Addition of dialkylboranes to the alkyne produced alkenylboronic esters, which were then cyclopropanated with diazomethane. Suzuki couplings of the boronates with phenyl iodide led to 3,4-methanohomophenylalanine after deprotection/oxidation (*363*). (*Z*)-2-amino-5-trimethylsilylpent-3-enoic acid was obtained from the acetylene intermediate by deprotonation with *n*BuLi and alkylation with trimethylsilylmethyl bromide, followed by partial alkyne reduction (*291*). The alkyne has also been employed for cycloaddition with a nitrile oxide, leading to α-(isoxazoly-5-yl)-Gly (*368*). The α-(isoxazoly-3-yl)-Gly regioisomers were also prepared, but by a different route requiring formation of an oxime with the Garner aldehyde, conversion to a nitrile oxide, and cycloaddition with alkenes (*368, 369*). A Dötz benzannulation reaction of the alkyne with Fischer chromium carbene complexes, $(CO)_5Cr=C(OMe)C(R_1)=CHR_2$, constructed an aromatic ring, leading to 2′-hydroxy-5′-methoxy-3′,4′-disubstituted arylglycines (see Scheme 10.38) (*370*).

One-half of the complex bis-amino acid central component of asperazine (consisting of two Trp-residues asymmetrically linked through their indole 3′–7′ positions, with the 3′-linked indole also cyclized from the 2′-position to the α-nitrogen to form a Pro nucleus) was derived from the D-Ser Garner aldehyde via a carboxyacetylene intermediate (*371*).

10.6.1d Garner Aldehyde: Other Reactions

The Garner aldehyde has been used for a number of cycloaddition reactions. The aldehyde was first employed as the dienophile in a reaction with Danishefsky's diene. The resulting pyranone was oxidatively cleaved, the oxazolidine opened, and the hydroxymethyl group

oxidized, giving *threo* β-hydroxy-Glu (*251*). The same route was used in a synthesis of the 4-ethylamino sugar of calicheamicin (*372*). The Garner aldehyde was employed as a chiral auxiliary during a cycloaddition to form a β-lactam, and was eventually converted into a carboxy group. The lactam was elaborated to β-hydroxy-Asp and β-(hydroxymethyl)-Ser (*373*). As mentioned in the previous section, an oxime derivative of the Garner aldehyde was converted into a nitrone and used for cycloadditions with alkynes to give α-(isoxazoly-3-yl)-glycines (*368, 369*).

The *O*-benzyl nitrone derivative of the Garner aldehyde has also been used as a substrate for the addition of Grignard reagents, with high (>90% de) *syn* stereoselectivity. In contrast, an acyclic Ser aldehyde oxime derivative gave the *anti* isomers. *N,O*-Bond cleavage, oxazolidine deprotection, and hydroxymethyl oxidation produced β-substituted-2,3-diaminopropionic acids (*374, 375*). The aldehyde was reductively aminated with *O*-benzyl hydroxylamine and used to form a heterocycle side chain, leading to quisqualic acid (see Scheme 10.39) (*376*). The aldehyde was also reductively aminated with mono-*N*-Boc ethylene diamine enroute to 4-aza-Lys, although a more efficient route to the same product from Asn was reported in the same paper (*377, 378*). Another reductive amination, employing benzylamine/NaBH$_3$CN, led to the 2,3-diamino acid derivative D-albizzine (*379*).

Addition of HCN to the Garner aldehyde side chain formed a cyanohydrin with >99% *anti* stereoselectivity (*380*). The aldehyde side chain was also employed as the aldehyde component in three-component Biginelli cyclocondensations to construct dihydropyrimidine ring systems, leading to α-(tetrahydropyrimidin-4-yl)-glycines. The homologous aldehyde led to the corresponding β-(dihydropyrimidin-4-yl)-Ala products, while a Hantsch cyclocondensation led to β-(pyrid-2-yl)-Ala and β-(pyrid-4-yl)-Ala derivatives (*381*).

The Ser oxazolidine precursor of the Garner aldehyde has been converted into a Wittig ylide by transforming the carboxyl group into a β-ketophosphonate (*332, 382*) (see Scheme 10.40). Olefination with a number of aldehydes (*382, 383*), followed by reduction of the carbonyl (*382, 384*), gave products equivalent to the addition of allyl organometallic reagents to the Garner aldehyde, with diastereoselectivity varying from 74% de *anti* to 60% de *syn*, depending on the reducing

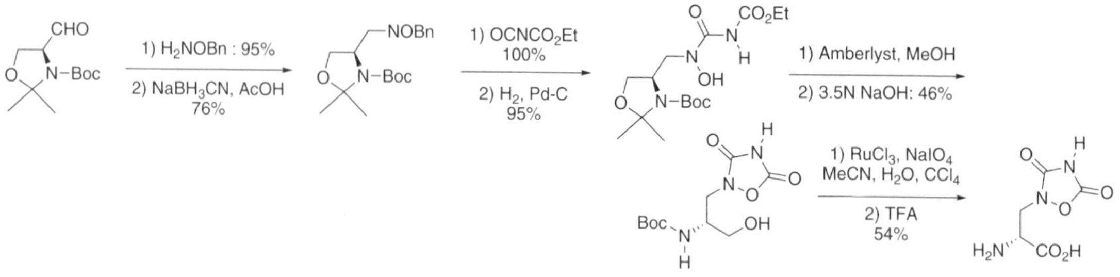

Scheme 10.39 Reductive Amination of Garner Aldehyde (*376*).

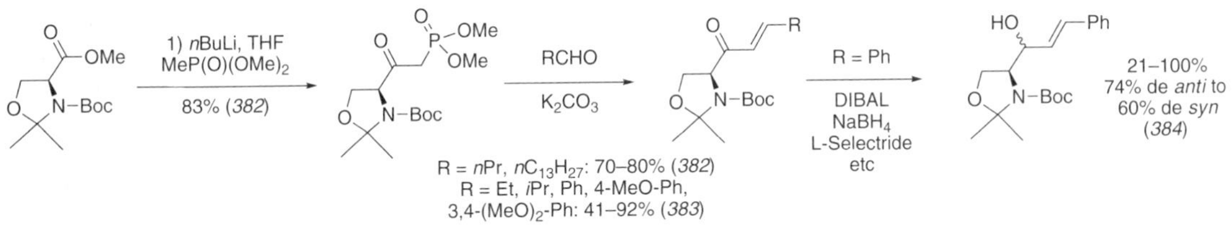

1) nBuLi, THF
MeP(O)(OMe)$_2$

83% (382)

RCHO
K$_2$CO$_3$

R = nPr, nC$_{13}$H$_{27}$: 70–80% (382)
R = Et, iPr, Ph, 4-MeO-Ph,
3,4-(MeO)$_2$-Ph: 41–92% (383)

R = Ph
DIBAL
NaBH$_4$
L-Selectride
etc

21–100%
74% de anti to
60% de syn
(384)

Scheme 10.40 Generation of Ylide from Serine Oxazolidine.

reagent (384). The products were not converted to amino acids, but this has been done for similar compounds. The Thr-derived methyl ketone equivalent of the Garner aldehyde has been employed as a homoalanine carbanion equivalent (267–269), while α-methyl serine has been converted into the α-methyl equivalent of the Garner aldehyde for use in the synthesis of α,α-disubstituted amino acids (385).

10.6.2 Rapoport's Aldehyde/Ketone

Rapoport and co-workers developed a route for the synthesis of amino acids that stemmed from earlier investigations into the reaction of organometallic reagents with the lithium carboxylates of N-protected α-amino acids, which produced α-amino ketones. No racemization was observed if appropriate N-protection was used, specifically acetyl, benzoyl, ethoxycarbonyl or phenylsulfonyl groups. It was postulated that abstraction of the nitrogen-bound proton by pretreatment with 2 equiv of alkyl lithium might be essential in preserving optical purity by discouraging α-hydrogen removal (386, 387).

In 1984 this procedure was applied to the dilithio salt of N-(phenylsulfonyl)-L-Ser (388). Reaction with a number of organometallics led to the corresponding ketones in 55–83% yield (see Scheme 10.41). The ketone group could be reduced to a methylene group via thioketal formation and reduction with Raney nickel, or by reduction with triethylsilane in TFA (76–81% yield). The ketones could also be reduced to secondary alcohols, with diastereoselectivity depending on the reducing agent (9:1 to >99:1 threo:erythro with Li- or K-Selectride; 1:2.3 to 1:6 threo:erythro with LiBH$_4$) (389).

To produce amino acids, the original Ser primary hydroxyl group was then oxidized by O$_2$/Pt at low temperature, with good discrimination observed over any secondary alcohols present. The oxidations proceeded in 55–73% yield. The N-protecting group could be removed with refluxing 48% HBr/phenol (or with Na/NH$_3$ or electrolysis for β-hydroxy compounds) to give the free D-amino acids (from L-Ser) with >99% ee (388, 389). The allyl adduct was elaborated to provide β-hydroxy-Glu, β-hydroxypipecolic acid and β-hydroxy-Lys, while the vinyl adduct provided β-hydroxy-Pro (37% overall yield of the syn diastereomer) (388, 389). An advantage of the Rapoport approach is that the Ser synthon is readily

prepared in high yields (one step, 85%). However, like the Garner aldehyde, an oxidation step is required at the end of the synthesis to regenerate the acid functionality. In addition, the N-(phenylsulfonyl) protecting group is relatively difficult to remove, although several conditions have been established (390). The aminoketone route was applied to the synthesis of 3',5'-dinitro-O-Tyr in 1998, but the product retained only 50% ee and required further optical resolution using L-amino acid oxidase (391).

Rapoport has also studied the use of the N-(9-phenylfluoren-9-yl) protecting group (PhFl) for preventing racemization of amino aldehydes prepared by reduction of amino acid carboxyl groups (392, 393). The bulky protecting group is proposed to prevent enolization by obstructing removal of the α-proton. It has been applied to the synthesis of amino aldehydes possessing side chains that increase the acidity of the α-proton, such as those derived from Ser. The PhFl group was used for reactions similar to those described in Scheme 10.41, in which an organometallic reagent was added to the carboxyl group of a Ser derivative. Further elaboration and a final oxidation of the original Ser side chain provided β-hydroxy-α-amino acids, notably including the C9 amino acid MeBmt (394). Some other applications of the PhFl group for serinal derivatives are described in Section 10.7 and Scheme 10.48 (below).

10.6.3 Other Serine Aldehyde Derivatives

The Ser carboxyl group has been reduced to an aldehyde on a number of other occasions (see Scheme 10.42), but the serinal has generally not been used to prepare amino acids. In 1973 phthalimide-protected Ser acetate was reduced to the aldehyde by hydrogenation of an acid chloride intermediate (395); organometallic addition to the aldehyde was used to prepare the amino alcohol D-erythro-sphingosine (395) and the piperidine alkaloids (−)-deoxoprosopinine and (−)-deoxoprosophylline (396). In 1981 the methyl ester of L-Ser, with the hydroxyl and amine groups protected as an oxazoline, was reduced to the aldehyde by DIBAL-H in 62% yield from Ser (397, 398). The aldehyde's instability was later noted: it was reacted without isolation and decomposed during the time required to obtain an NMR spectrum (397). Again, the aldehyde was used for organometallic additions to prepare amino alcohols,

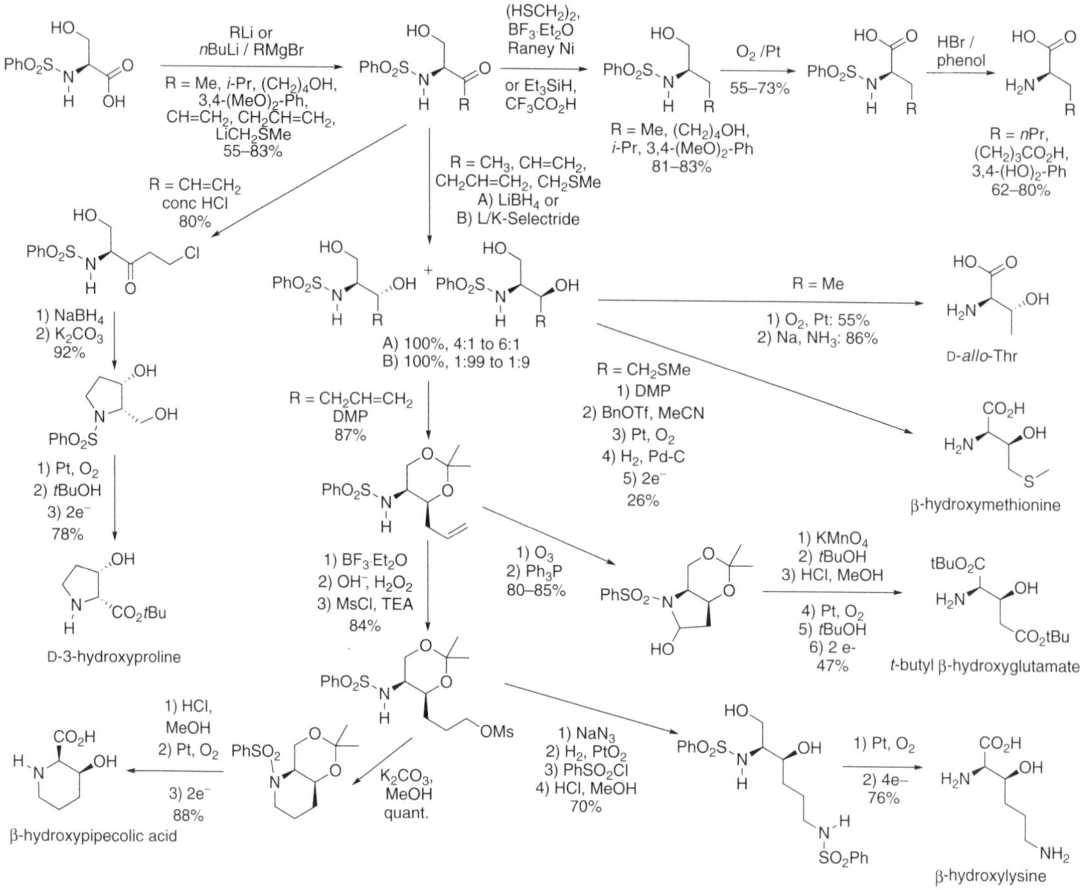

Scheme 10.41 Synthesis of Amino Acids via Addition of Organometallic to Acid Group of Serine (*388, 389*).

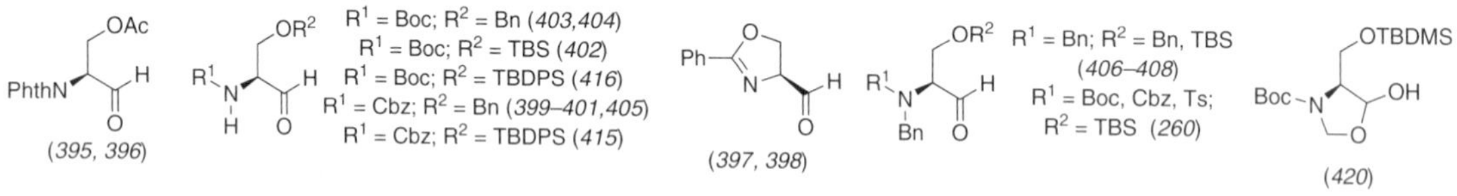

Scheme 10.42 Serinal Equivalents That Have Been Prepared by Reduction of Serine Carboxyl Group.

D-sphingosine derivatives (*397, 398*). The ester groups of Cbz-L-Ser(Bn)-OMe (*399*), Cbz-L-Ser(Bn)-OEt (*400, 401*), and Boc-L-Ser(TBS)-OMe (*402*) have also been reduced directly to the aldehyde with DIBAL-H.

N-Boc-L-Ser O-benzyl ether was converted to the corresponding aldehyde by a two-step process in 60% yield, via reduction of the acid group to an alcohol with BH₃.THF followed by PDC oxidation (*403, 404*). The N-Cbz O-benzyl D-serinal has also been prepared, but from inexpensive L-Ser instead of D-Ser, proceeding via a serinol intermediate that was oxidized to the acid (*405*). Similarly, the acid group of a series of N-monoprotected and N,N-diprotected O-TBS or

O-BOM Ser derivatives (BOM = benzyloxymethyl) were reduced to the serinol and then oxidized to the serinal. Significantly less racemization was observed with oxidation by the TEMPO procedure (82–100% ee) compared to the Swern procedure (24–100% ee). Optically pure N-Cbz, N-Boc, and N-Ts serinal were obtained by the TEMPO oxidation, while the N-Bn/N-Boc, N-Bn/N-Cbz, and N-Bn/N-Ts diprotected derivatives possessed 82–96% ee (*260*).

In Reetz's 1991 review of amino aldehydes (*406*) the preparation of an N,N-dibenzyl-protected serinal with the Ser side chain protected as a benzyl or TBDMS ether was reported, but no experimental details were given. This aldehyde has since been prepared by another group (from D-Ser) and used as a substrate for the addition of isopropyl MgCl, with the addition producing a single diastereomer. Oxidation of the original side chain of Ser gave (2S,3S)-β-hydroxy-Leu (407, 408). A similar route was employed to prepare (2R,3R)- and (2R,3S)-β-hydroxy-Nva enantiomers from N,N-dibenzyl L-Ser(TBS)-OMe (409, 410), though deprotection and oxidation of the original Ser side-chain proved problematic (409). Grignard addition of the reagent derived from 2-(2-bromoethyl)-1,3-dioxolane, followed by an intramolecular reductive amination, led to 3-hydroxypipecolic acid (411).

The aldehyde derived from Boc-L-Ser(TBS)-OMe was treated with vinylmagnesium chloride, followed by elaboration of the addition product into a number of β,γ-unsaturated amino acids by oxidation of the side-chain alcohol followed by organocuprate addition to the alkene (see Scheme 10.43) (402). The O-protecting

group can affect the stereochemistry of additions to the aldehyde. Allyltrimethylsilane added to the aldehyde of N-Cbz serinals in the presence of TiCl₄ or SnCl₄ catalyst. With O-TBS protection, up to 96% syn diastereoselectivity was obtained, but with O-BOM protection up to 94% anti diastereoselectivity was observed (412). The ester group of the benzophenone Schiff base of Ser(TBS)-OMe was reduced with iBu₅Al₂H and then reacted with alkyl Grignard reagents (R = Me, nHex, ndecyl or nC₁₅H₃₁) to give threo adducts in 60–72% yield and >90% de. The Schiff base was exchanged with an Fmoc group, the TBS protection removed, and the hydroxymethyl group oxidized with NaOCl/NaClO₂/TEMPO (413).

The Boc-Ser(TBS)-H aldehyde has also been converted into an O-benzyl nitrone. Grignard addition proceeded with high anti diastereoselectivity, in contrast to the syn products obtained from the nitrone of the Garner aldehyde. Reductive cleavage of the N–O bond,

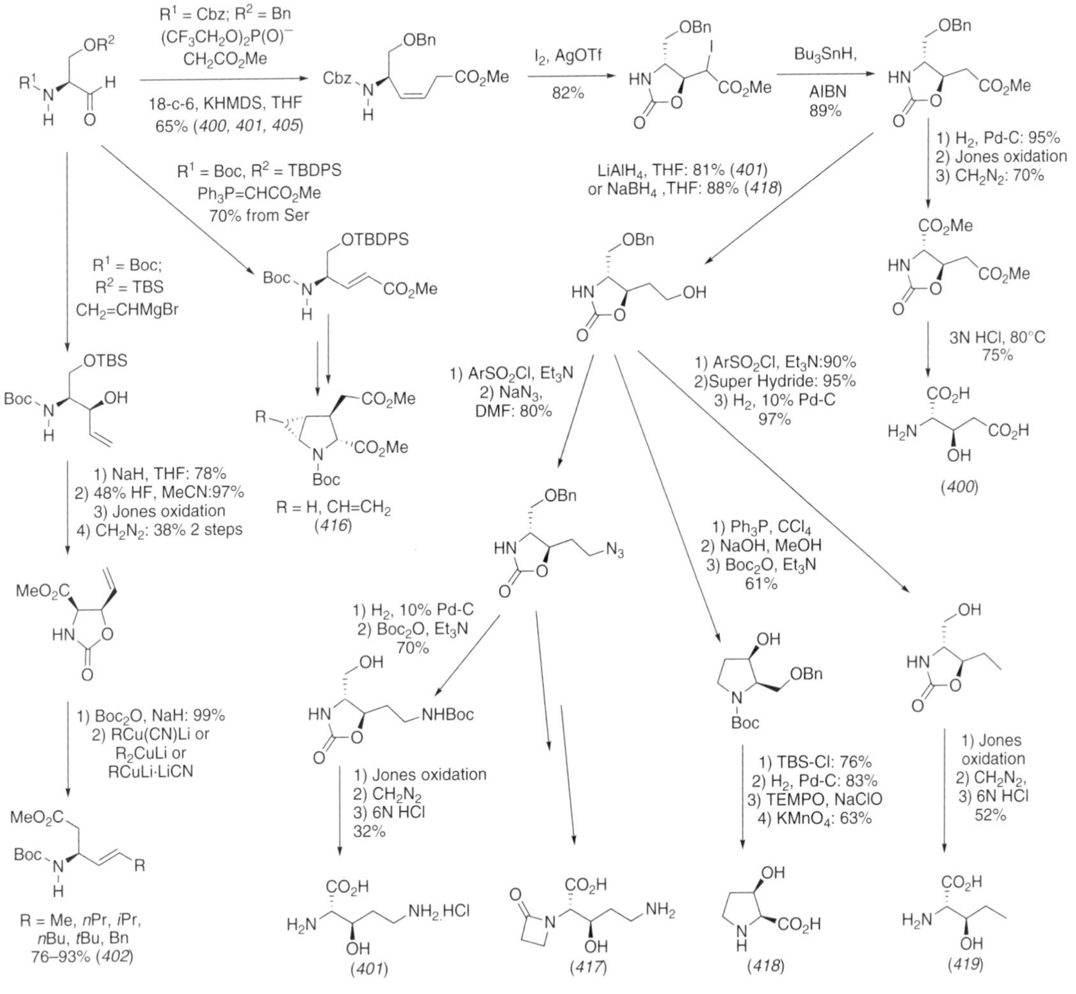

Scheme 10.43 Synthesis of Amino Acids from Acid-Reduced Serinal.

deprotection, and hydroxymethyl oxidation produced β-substituted 2,3-diaminopropionic acids (374, 375). A hetero-Diels–Alder reaction between the aldehyde of Boc-Ser(TBDPS)-H and 1-benzyloxy-2-TBSO-4-ethoxy-1,3-butadiene formed part of the synthesis of destomic acid, a polyhydroxy ε-amino acid (414, 415).

Reaction of the Boc-Ser(TBS)-H aldehyde with a Wittig reagent led to 4,5-methano congeners of α-kainic acid (see Scheme 10.43) (416). The N-Cbz benzyl ether-protected serinal has also been used for a Wittig–Horner condensation, with the cis α,β-unsaturated ester adduct converted by iodolactonization into a versatile oxazolidinone intermediate. Further derivatization led to threo-3-hydroxy-Glu(400), threo-3-hydroxy-Orn (401), proclavaminic acid (417), 3-hydroxy-Pro, and 2-hydroxymethyl-3-hydroxy-Pro (418) or the four isomers of β-hydroxy-Nva (419) (see Scheme 10.43).

In 2003, Boc-Ser(TBDMS)-OH was cyclized to an oxazolidinone by heating in paraformaldehyde with catalytic camphorsulfonic acid. Reduction of the lactone carbonyl with DIBAL-H produced a serinal derivative with the aldehyde protected as a hemiacetal, in approximately 50% overall yield from unprotected L-Ser (see Scheme 10.42). Careful analysis indicated the serinal possessed >98% ee, and was configurationally stable to a number of conditions (420).

10.6.4 Serinal Equivalents

Serinal equivalents have been prepared from sources other than Ser. In 1984 Hanessian et al. prepared a carbohydrate-based serinal equivalent, with the acid group masked as a diol (see Scheme 10.44) (421). The aldehyde was elaborated by Grignard additions, which

proceeded with reasonable stereoselectivity at the carbinol center. Diol cleavage and oxidation revealed the masked carboxyl group, giving N-protected L-threo β-hydroxy amino acids as the major diastereomer in 60% yield from the aldehyde (421).

A similar approach was used to prepare a chemically and configurationally stable N-Boc-L-serinal on a large scale from D-glucosamine hydrochloride in three steps and 70% yield (see Scheme 10.45) (422). Its stability is proposed to be due to an oligomeric/polymeric structure. It was reacted with stabilized Wittig ylides to give the corresponding alkenes in high yields, but these were not initially converted into amino acids (422). However, a different group subsequently prepared both N-Boc-L-serinal from D-glucosamine and N-Boc-D-serinal from D-mannosamine (70–72% yield), coupled the aldehydes with the Wittig reagent prepared from 4-chlorobenzyltriphenylphosphonium chloride (61–66% yield), hydrogenated the alkene (73–80% yield), and oxidized the hydroxymethyl group by Jones oxidation (64–69% yield), producing D- and L-p-chloro-homophenylalanine (423).

As mentioned earlier, vinylglycinol has been prepared from D- or L-Met. It can be employed as a serinal synthon, with ozonolysis of the alkene providing an acyclic equivalent of the Garner aldehyde (424).

10.7 Serine Aldehydes: Conversion of the Serine Side-chain Hydroxyl Group into an Aldehyde

There are very few reports of successful attempts to prepare optically active serinal derivatives (α-formyl-glycines) from Ser by oxidation of the side-chain

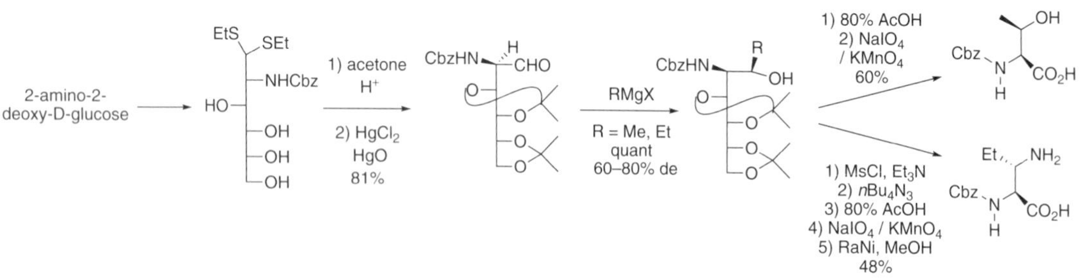

Scheme 10.44 Synthesis of Carbohydrate-Derived Serinal Equivalent (421).

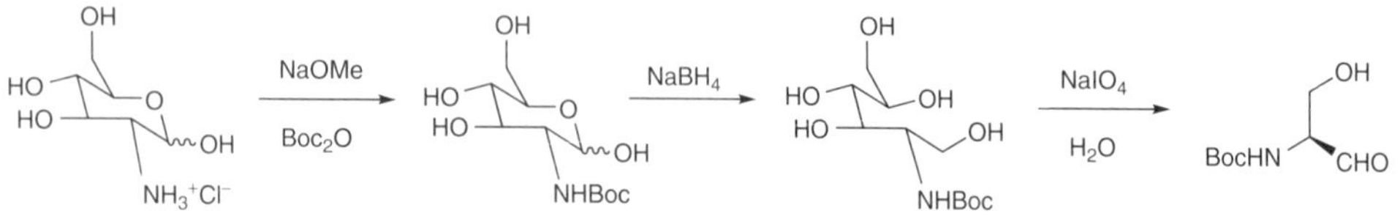

Scheme 10.45 Glucosamine-Derived Serinal Equivalent (422).

hydroxyl group. Racemic Ser aldehydes derived from other sources have been used for many years in penicillin-related syntheses, with preparation of a serinal derivative from ethylhippurate and ethyl formate described by Erlenmeyer and Stoop in 1904 (425). The same reagents were employed in a 1984 synthesis, in which the aldehyde was then reductively aminated to give N^β-substituted 2,3-diaminopropionic ester products (426). Similar serinal syntheses were published in 1954 and 1991, with the desired synthon prepared in 37–47% yield by formylation of phthalimidoacetic acid esters (427) or in 83% yield from N,N-dibenzyl Gly-OEt (428) (see Scheme 10.46). The infrared spectrum of the phthalimidoacetic acid product indicated the aldehyde existed in the enolic form (427), a tautomer that was postulated even in the original 1904 synthesis (425). A more detailed study of the product produced by formylation of N,N-dibenzyl Gly-OEt found that the resulting aldehyde existed predominantly in its enol form in equilibrium with less than 5% of the aldehyde tautomer, as indicated by integration of the ^{1}H NMR aldehyde proton resonance at δ 9.1 ppm (428). Surprisingly, serinal residues appear to exist in nature. Sulfatase enzymes, responsible for hydrolysis of sulfuric acid esters, contain a formylglycine residue (2-amino-3-oxopropanoic acid) derived from Cys. The modification is post-translational, after N-glycosylation, but the enzyme responsible has not been identified. The aldehyde may aid in cleavage of a sulfuric acid ester via nucleophilic substitution at the sulfur atom by a hydrate of the aldehyde. Humans suffering from metachromatic leukodystrophy (MLD) have a defective arylsulfatase A enzyme, apparently due to lack of oxidation of this Cys residue (429, 430).

The enol ether/aldehyde produced by formylation of a Gly enolate has been converted into a 1,3-dithiolane by reaction with ethane dithiol (431), to a dimethyl acetal with MeOH/HCl (432), or to a diethyl acetal with ethanol/triethyl formate/catalytic acid (428). Alternatively, the diethoxy acetal of protected serinal was prepared via condensation of N-formyl Gly with ethyl formate in benzene, using potassium tert-butoxide as base. The acetal was produced in 70% yield (433). An earlier synthesis by the same route in 1977 gave a 42% yield

of the diethyl acetal-protected aldehyde, which was subsequently incorporated in a synthesis of β-lactam antibiotics (434). An asymmetric synthesis of diethyl acetal-protected α-formylglycine was achieved by Ti-catalyzed alkylation of Williams' oxazinone template with triethyl orthoformate; template deprotection gave (R)-α-formylglycine diethylacetal. Attempts to couple to the amino group resulted in partial epimerization at the α-center (although formation of a Mosher amide derivative did not, perhaps due to the use of pyridine as base instead of morpholine). The diethyl acetal was deprotected by 2 N HCl/acetone, with the resulting aldehyde reacted with urea to form a 3-ureidodehydroalanine residue (435).

The preference for Ser aldehyde to exist as the enol tautomer was employed for a synthesis of O-tosyl dehydroSer, via oxidation of the Ser hydroxymethyl group and tosylation of the resulting enol tautomer (see Scheme 10.47). The tosyl group was displaced by nucleophiles to give other dehydroamino acids (436, 437). However, another group reported that attempts to oxidize the side chain of Boc- or Cbz-protected Ser esters using Swern, PDC, PCC, or Collins oxidation conditions were unsuccessful (428).

In 1989 Lubell and Rapoport used N-PhFl protection to impart configurational stability to several Ser derivatives. Introduction of the N-PhFl protecting group required temporary masking of the carboxyl and hydroxyl groups of Ser as trimethylsilyl esters/ethers, with the protected Ser produced in 75% yield (432). The PhFl-Ser-OMe side-chain was successfully oxidized under Swern conditions to generate a Ser aldehyde in 66% yield (see Scheme 10.48). Unfortunately, the aldehyde was unstable to purification on silica and to storage at 0 °C in an inert atmosphere. It also failed to give the desired products when reacted with NaBH$_4$ or MeLi, likely due to tautomerism to the unreactive enol form (as observed with the Gly enolate adducts). However, by masking the acid group as a dimethyl acetal, the acidity of the α-proton was sufficiently reduced to allow for the isolation and subsequent reaction of the aldehyde. This serinal equivalent, with >99% ee, could be reduced by LiAlH$_4$ or reacted with Wittig reagents (Ph$_3$PMeBr, Na/DMSO) without loss of chirality. Regrettably, the acetal

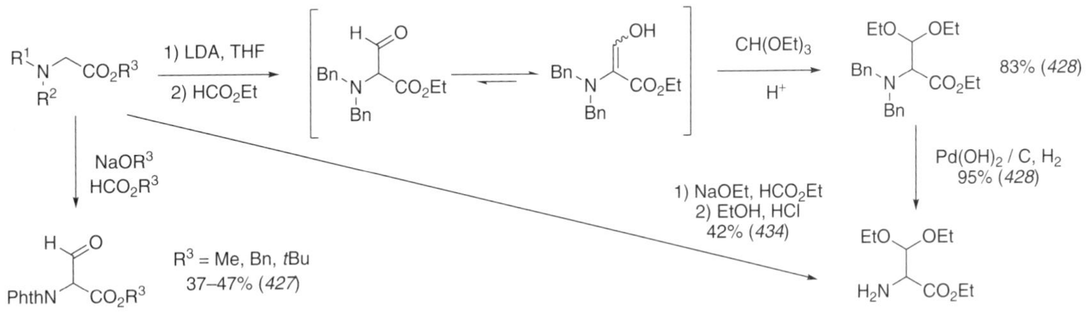

Scheme 10.46 Synthesis of Protected Serinal.

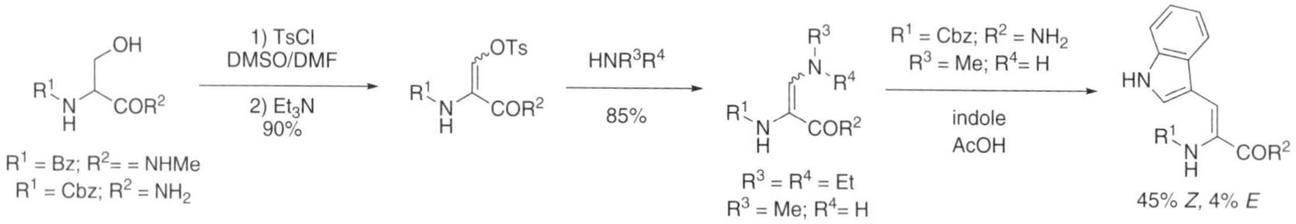

Scheme 10.47 Conversion of Serine to Dehydroserine and Displacement (436, 437).

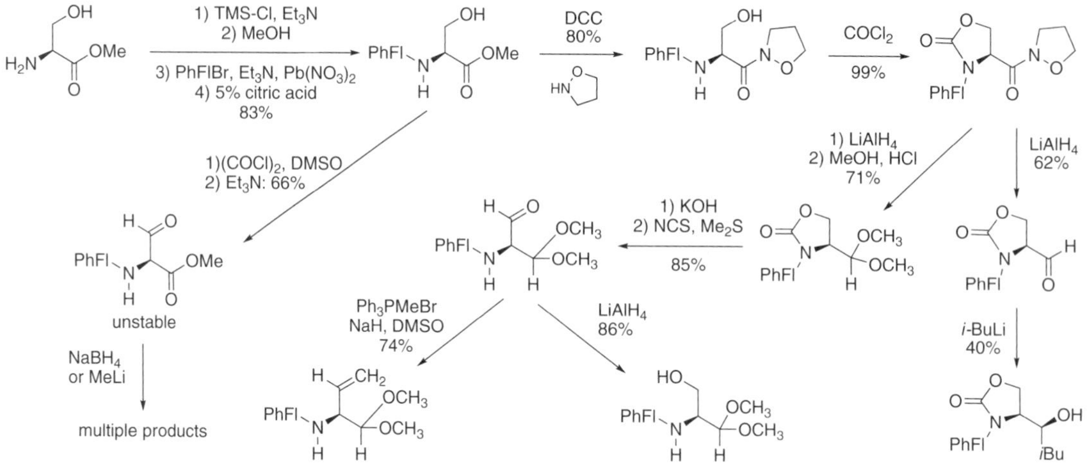

Scheme 10.48 Synthesis of Protected Serinal (432).

could not be hydrolyzed in greater than 35% yield, and no amino acids were prepared (432).

One of the intermediates in this reaction scheme, with the Ser side chain protected in a cyclic carbamate and the acid group reduced to an aldehyde, is similar to the Garner aldehyde (see Section 10.6.1). It underwent nucleophilic addition with excellent stereoselectivity, though in very poor yields (17% for iBuLi). Reaction with a Wittig ylide proceeded in 74% yield, but neither of the adducts were converted to amino acids (432). An improved reduction of the acid group to the aldehyde was reported in 2001, and addition of a Reformatsky reagent derived from bromodifluoroacetate proceeded in good yield with 71% de. However, attempts to convert the adduct to 4,4-difluoro-Glu failed at the oxazolidinone hydrolysis step. Instead, the OBO ester-protected serinal derivative was employed (see below, Scheme 10.49) (438). Another synthesis, using the Garner aldehyde, did produce 4,4-difluoro-Glu but the oxazolidine hydrolysis took 4 days and proceeded in only 30% yield, and the oxidation step gave poor 50–60% yields (316).

The most successful Ser aldehyde equivalent derived from Ser side-chain oxidation is that developed by Blaskovich and Lajoie (see Scheme 10.49) (4, 439, 440). To prevent enolization of the side-chain aldehyde, the

carboxylic acid group was protected as a base stable cyclic 4-methyl-2,6,7-trioxabicyclo[2.2.2]octane (OBO) ortho ester, reducing the acidity of the α-proton. This protecting group, formed via rearrangement of an oxetane ester intermediate, also potentially introduces steric prevention of α-proton abstraction. It has the added advantage of preventing carboxyl attack by nucleophilic reagents without altering the oxidation level of the carboxyl group. The side chain of the protected Ser derivative was successfully converted to the aldehyde under Swern conditions, giving a stable solid with 97–99% ee, as determined by both chiral shift reagent NMR measurements and by HPLC analysis after reduction of the aldehyde, deprotection, and derivatization (4, 440). The reported Swern oxidation condition are essential for obtaining high enantiopurity; a wide range of other oxidizing reagents and modified Swern reaction conditions were tested without success (440). The N-Boc- and N-Cbz-protected Ser aldehyde OBO esters have also been prepared (440), with an improved procedure for synthesizing all three synthons described in 1998 (441). Preparation of the N-Cbz aldehyde OBO ester synthon on a 10 g scale using the improved method has been reported as an Organic Syntheses procedure (442). Further improvements to the oxetane

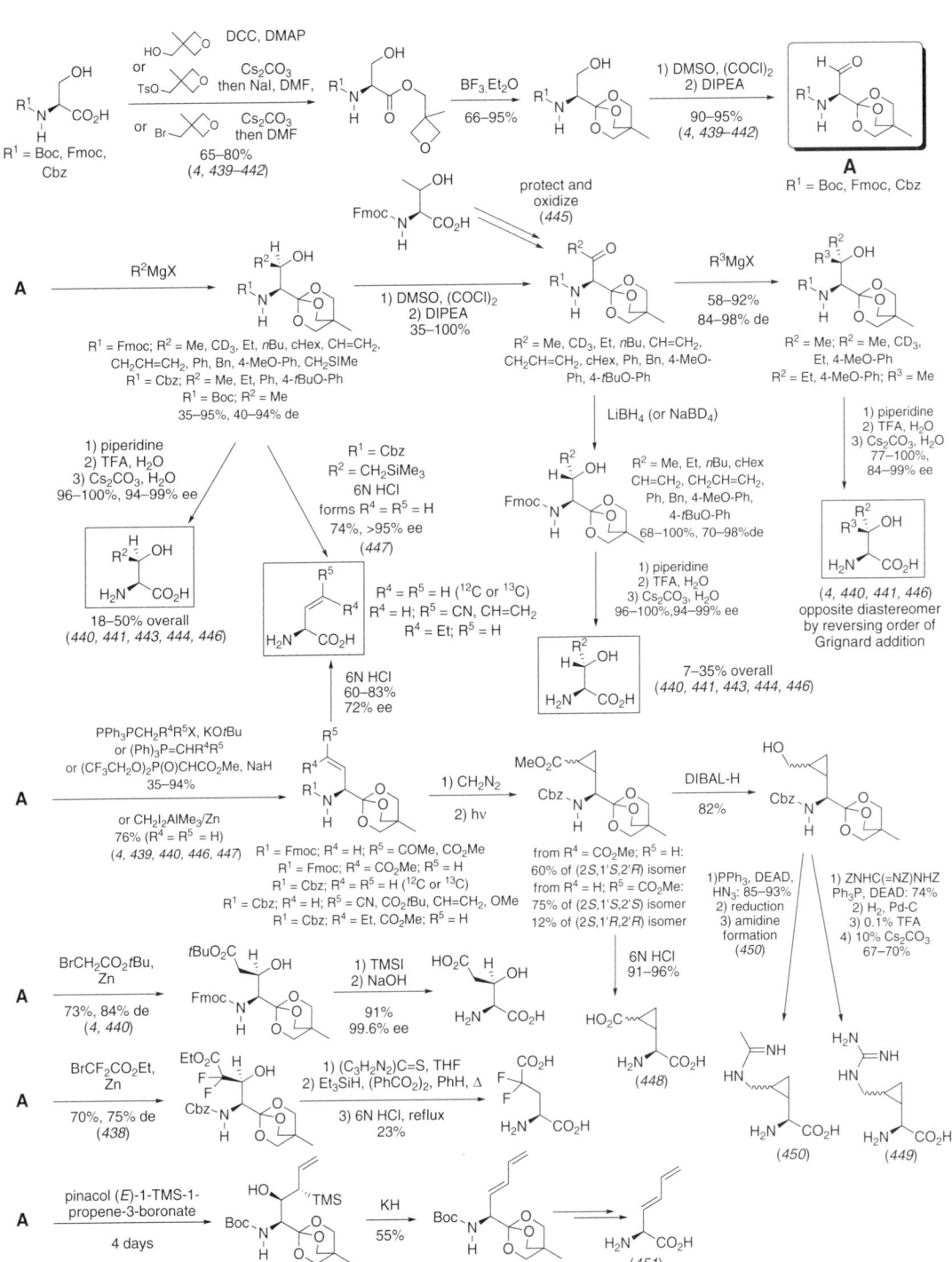

Scheme 10.49 Synthesis and Reaction of Serinal OBO Ortho Ester.

ester formation were reported in 2005, and oxidation of the β-hydroxy group with Dess–Martin periodinane was found to be somewhat easier on a large scale than the Swern oxidation procedure, but gave product with reduced enantiopurity (91–94% ee instead of 96–98% ee) (*443*). An attempt to prepare an *N*-phthaloyl analog was unsuccessful, failing to give the *ortho* ester upon attempted rearrangement of the oxetane ester (*438*). One advantage of the *N*-Fmoc derivative is that it is readily crystallized.

The aldehyde has proved to be a versatile intermediate, undergoing Grignard and Reformatsky additions to give β-hydroxy amino acids and Wittig reactions to give β,γ-unsaturated amino acids (see Scheme 10.49). The Grignard reactions proceed with reasonable stereoselectivity (40–94% de without the use of additives) to give predominantly the *threo* isomer. However, the *erythro* isomers are also readily prepared with even better diastereoselectivity (70–98% de), via reoxidation of the initial Grignard adduct to a ketone, followed by reduction with LiBH₄. Thus, by starting with ʟ- or ᴅ-Ser, all four β-hydroxy amino acid diastereomers can be selectively prepared (*4, 440, 441*). Somewhat surprisingly, the *N*-Fmoc group is stable to the Grignard addition conditions (even with excess reagent), possibly due to protective deprotonation of the urethane nitrogen. The serinal OBO method was applied to the synthesis of both (2*S*,3*R*)- and (2*S*,3*S*)-β-methoxy-Phe from the *N*-Cbz serinal OBO ester by another group. Grignard addition of PhMgBr proceeded in 61% yield with 90% de; the hydroxy group was then methylated with Me₃O⁺BF⁻₄ and proton sponge, or oxidized with Dess–Martin periodinane, reduced with LiBH₄ (74%, >95% de), and methylated (*444*). A similar procedure led to all four diastereomers of β-methoxy-Tyr, using a *tert*-butyl-protected phenol Grignard reagent (*443*). A modified route was developed to obtain (2*S*,3*R*)- or (2*S*,3*S*)-β-hydroxy-Nva. Cbz-Ser-OBO ester was converted into *N*,*N*-Bn₂ Ser-OBO ester; addition of Grignard reagents (MeMgI, EtMgBr, *i*PrMgI, PhMgBr) proceeded under chelation control with 56–76% *anti* stereoselectivity, while dialkylzinc reagents (Me₂Zn, Et₂Zn, *i*Pr₂Zn, Ph₂Zn) gave 52–100% de *syn* products in a non-chelation mode (*409*). The same paper reported the preparation of the (2*R*,3*R*)- and (2*R*,3*S*)-enantiomers from *N*,*N*-Bn₂ *O*-MEM serinal (α-carboxyl group converted to aldehyde). The advantages of keeping the carboxyl group in the desired oxidation state during the OBO ester route are apparent in this paper, as the manipulations required to oxidize the Ser side chain at the end of the synthesis from *N*,*N*-Bn₂ *O*-MEM serinal gave 35–37% yields, compared to 66–70% yields for simple deprotection of the OBO ester adducts (*409*).

The oxidation/reduction procedure was modified to prepare ᴅ- or ʟ-*allo*-Thr from ᴅ- or ʟ-Thr in 40% overall yield, via oxidation of the side-chain of suitably protected threonine to the ketone followed by reduction of the ketone; a deuterium label could be stereoselectively introduced during the reduction (*445*).

β,β-Disubstituted-β-hydroxy amino acids were prepared by adding a second Grignard reagent to the intermediate ketone produced by oxidation of an initial Grignard adduct. Complementary diastereomers could be synthesized by reversing the order of addition of the two Grignard substituents (*440, 441, 446*). A synthesis of 4,4-difluoro-Glu added the Reformatsky reagent derived from bromodifluoroacetate to the serinal derivative in 70% yield and 75% de. The β-hydroxy group was then removed by reaction with thiocarbonyl bisimidazole followed by triethylsilane in the presence of benzoyl peroxide, giving the protected difluoroester in 44% yield from the aldehyde (*438*). A similar synthesis of 4,4-difluoro-Glu from the Garner aldehyde required an oxazolidine hydrolysis that took 4 days and proceeded in only 30% yield, and a final oxidation step with low 50–60% yields, again showing the benefits of the OBO ester approach (*316*).

The Fmoc OBO ester synthon protecting groups are removed under conditions which are mild compared to most other amino acid synthesis routes, using either a two-step procedure with TMSI and aqueous NaOH, or a three-step procedure incorporating Fmoc deprotection with piperidine, mild acid hydrolysis with aqueous TFA to open the *ortho* ester, and base hydrolysis with 10% Cs₂CO₃ solution to hydrolyze the ester group. The *N*-Boc derivatives are deprotected with TFA followed by aqueous base, and the *N*-Cbz derivatives by treatment with TMSI or hydrogenolysis, followed by acid then base, or simply by refluxing with 6N HCl (*4, 441, 447, 448*). The *N*-Cbz/OBO ester protecting groups have also been removed via a sequence of OBO ring opening with 1:1:1 dioxane:HOAc:H₂O, followed by Cbz hydrogenolysis, and then finally ester cleavage by treatment of an aqueous solution in dioxane with IRA-400 (⁻OH) cation ion-exchange resin overnight (*443*).

Alkene formation from the serinal OBO esters with both stabilized and non-stabilized Wittig reagents proceeded in good yield, but racemization was evident with some reagents (*4, 440, 446, 447*). The best results were obtained with *N*-Cbz protection, although reactions with the *N*-Boc and *N*-Fmoc synthons were also successful. Both *E*- and *Z*-protected β,γ-unsaturated-Glu derivatives could be prepared with >95% ee, but unfortunately deprotection was not successful. Other efforts at preparing these compounds have shown them to be extremely labile. A number of attempts at elaborating the α,β-unsaturation (e.g., by Michael addition) were also unsuccessful, possibly due to steric crowding preventing attack. However, cyclopropanation via reaction with diazomethane, followed by photolytic decomposition of the intermediate pyrazoline adducts, produced ʟ-2-(carboxycyclopropyl)glycines with high stereoselectivity (see Scheme 10.49) (*448*). For example, the *N*-Cbz-protected *Z*-β,γ-unsaturated-Glu OBO ester gave a single (2*S*,1′*S*,2′*R*) isomer of *cis* cyclopropane product in 60% yield, which was deprotected to the free amino acid with 6 N HCl in 93% yield. The protected *E*-β,γ-unsaturated-Glu isomer produced an 87% yield of the two *trans* isomers in a 86:14

(2S,1′S,2′S):(2S,1′R,2′R) ratio (448). The acid group of the protected 3,4-methano-Glu derivatives was then reduced to give L-2-(2-hydroxymethylcycloprop-1-yl) glycines, with the alcohol used for a Mitsunobu reaction with N,N,N-tri-Cbz guanidine to form 3,4-methano-Arg (449). Alternatively, the Arg analog N-(1-iminoethyl) 3,4-methano-Orn was produced by displacement of the alcohol with azide, azide hydrogenation, and formation of the acetamidine (450).

Vinylglycine was obtained from Cbz-protected serinal OBO ester by methylenation with MePPh₃Br and KOtBu in Et₂O. A one-step deprotection with 6 N HCl gave partially racemized vinylglycine (71% ee) (440, 446, 447). Attempts at methylenation with Tebbe-type reagents were unproductive (440), although under Nozaki conditions (AlMe₃/CH₂I₂/Zn) a 76% yield of protected vinylglycine was produced. Deprotection gave vinylglycine with 86% ee (447). Vinylglycine with >95% ee could be produced by a Peterson-type olefination, via addition of trimethylsilylmethylene magnesium chloride to the aldehyde. Base-mediated elimination produced no alkene; instead a stable oxazolidinone was produced. However, under acidic conditions with 6 N HCl, deprotected vinylglycine was obtained directly (447). A Peterson-type olefination reaction was also successfully applied to the first synthesis of (S)-2-amino-(Z)-3,5-hexadienoic acid. Notably, an attempted synthesis of the same product via the Garner aldehyde was unsuccessful (451).

An added advantage of amino acid synthesis with this methodology is that isotopically labeled substituents are readily introduced at a late stage in the synthesis. Thus, NaBD₄ was used to stereoselectively prepare [3-²H]-allo-threonine (440, 445), CD₃MgI to synthesize threo- and erythro-[4,4,4-²H₂]-β-hydroxy-Val (440, 441, 446), and Ph₃P¹³CH₃Br for [4-¹³C]-vinylglycine (440, 446, 447).

The OBO ortho ester protection strategy has also been applied to Ser to prevent elimination side reactions during nucleophilic displacments of an activated side-chain hydroxyl group (105, 106). Protection of pyroglutamate as an OBO ester allowed for the synthesis and alkylation of 3,4-dehydropyroglutamate without epimerization, and with high diastereoselectivity (452, 453). The side-chain β-carboxy group of N,N-Bn₂-Asp-OBn has also been protected as an OBO ester, with the α-carboxyl group converted into an aldehyde (483).

An "improved" OBO ester protecting group with two additional methyl groups on one of the bridging methylene carbons has been described. The protecting group was formed from the oxetane 85-times faster than the OBO group (for simple acids), and was 36-times more stable towards aqueous hydrolysis. It remains to be seen whether these properties would be beneficial for amino acid synthesis (454). A closely related ABO ester, a 2,7,8-trioxabicyclo[3.2.1]octyl ring system, was employed during cyclopropanation of 3,4-dehydropyroglutamate (455). The ABO ester is purportedly easier to generate (see Scheme 9.30) and may find use in serinal

derivatives, though the presence of an additional chiral center in the protecting group may complicate NMR assignments.

10.8 Other Reactions of Serine

Ser and Thr residues within peptides can be converted into α-acetoxy-Glu residues (electrophilic Gly equivalents, see Section 3.9.1) in 95–97% yield by treatment with lead tetraacetate (430, 456–458) or (diacetoxyiodo) benzene/iodine/sunlight irradiation (459). Ser ethers have been synthesized from the side-chain hydroxy group of Boc-Ser-OMe by benzylation with photoactivatable aromatic residues (68–73% yield) (460), or by alkylation with nucleobases via a methylthiomethyl ether intermediate (64–72% yield) (461). The analogous N-Fmoc-protected nucleo amino acids were prepared by the same route in 2004 (462). Ser can be α-alkylated by derivatizing it as a cyclic oxazolidine template, using Seebach's "transfer of chirality" principle to allow asymmetric alkylation of the planar enolate (see Scheme 7.47) (9, 463).

The acid group of Boc-D-Ser(TBS) has been activated with BOP-Cl or isopropenyl chloroformate and reacted with Meldrum's acid, with heating inducing cyclization to 3-oxopyroglutaminol. Ketone and lactam reduction followed by oxidation of the original Ser hydroxymethyl group provided β-hydroxy-Pro (464, 465). The α-carboxy group of Cbz-L-Ser has been converted into an amino group via Curtius rearrangement with diphenylphosphoryl azide (DPPA), accompanied by oxazolidone ring formation via generation of a carbamate with the side-chain hydroxy group. N-Boc protection, oxazolidone ring opening, and side-chain hydroxymethyl oxidation produced (R)-Boc-Agl(Cbz)-OH (Agl = α-amino-Gly). The same route from Boc-L-Ser gave the enantiomer (466).

D-Thr has been converted into three of the unusual amino acids found in the cyclic pentapeptide protein phosphatase inhibitor motuporin (D-erythro-β-methyl-Asp, N-methyl-dehydroAbu, and the β-amino acid Adda) (467).

References

1. Smith, G.G.; Reddy, G.V. J. Org. Chem. **1989**, 54, 4529–4535.
2. Barrett, G.C. Chemistry and Biochemistry of the Amino Acids; Chapman and Hall: New York, 1985, pp 399–414.
3. Garner, P.; Park, J.M. J. Org. Chem. **1987**, 52, 2361–2364.
4. Blaskovich, M.A.; Lajoie, G.A. J. Am. Chem. Soc. **1993**, 115, 5021–5030.
5. Photaki, I. J. Am. Chem. Soc. **1963**, 85, 1123–1126.
6. Zioudrou, C.; Wilchek, M.; Patchornik, A. Biochemistry **1965**, 4, 1811–1822.
7. Fry, E.M. J. Org. Chem. **1950**, 15, 438–447.

8. Sjölin, P.; Kihlberg, J. *Tetrahedron Lett.* **2000**, *41*, 4435–4439.

9. Seebach, D.; Aebi, J.D.; Gander-Coquoz, M.; Naef, R. *Helv. Chim. Acta* **1987**, *70*, 1194–1216.

10. Pines, S.H.; Kozlowski, M.A. *J. Org. Chem.* **1972**, *37*, 292–297.

11. Wipf, P.; Miller, C.P. *J. Org. Chem.* **1993**, *58*, 1575–1578.

12. Fischer, P.M.; Sandosham, J. *Tetrahedron Lett.* **1995**, *36*, 5409–5412.

13. Fry, E.M. *J. Org. Chem.* **1949**, *14*, 887–894.

14. Seebach, D.; Aebi, J.D. *Tetrahedron Lett.* **1983**, *24*, 3311–3314.

15. Attenburrow, J.; Elliott, D.F.; Penny, G.F. *J. Chem. Soc.* **1948**, 310–318.

16. Elliott, D.F. *J. Chem. Soc.* **1949**, 589–594.

17. Elliott, D.F. *J. Chem. Soc.* **1950**, 62–68.

18. Galéotti, N.; Montagne, C.; Poncet, J.; Jouin, P. *Tetrahedron Lett.* **1992**, *33*, 2807–2810.

19. Wipf, P.; Miller, C.P. *Tetrahedron Lett.* **1992**, *33*, 6267–6270.

20. Heine, H.W. *Angew. Chem., Int. Ed. Engl.* **1962**, *1*, 528–532.

21. Heine, H.W.; Fetter, M.E.; Nicholson, E.M. *J. Am. Chem. Soc.* **1959**, *81*, 2202–2204.

22. Lown, J.W.; Itoh, T.; Ono, N. *Can. J. Chem.* **1973**, *51*, 856–869.

23. Okawa, K.; Nakajima, K.; Tanaka, T.; Neya, M. *Bull. Chem. Soc. Jpn.* **1982**, *55*, 174–176.

24. Choi, D.; Kohn, H. *Tetrahedron Lett.* **1995**, *36*, 7011–7014.

25. Hori, K.; Ishiguchi, T.; Nabeya, A. *J. Org. Chem.* **1997**, *62*, 3081–3088.

26. Laaziri, A.; Uziel, J.; Jugé, S. *Tetrahedron: Asymmetry* **1998**, *9*, 437–447.

27. Meyer, F.; Laaziri, A.; Papini, A.M.; Uziel, J.; Jugé, S. *Tetrahedron: Asymmetry* **2003**, *14*, 2229–2238.

28. Lago, M.A.; Samanen, J.; Elliot, J.D. *J. Org. Chem.* **1992**, *57*, 3493–3496.

29. Pu, Y.; Martin, F.M.; Vederas, J.C. *J. Org. Chem.* **1991**, *56*, 1280–1283.

30. Pansare, S.V.; Vederas, J.C. *J. Org. Chem.* **1989**, *54*, 2311–2316.

31. Bergel, F.; Wade, R. *J. Chem. Soc.* **1959**, 941–947.

32. Stabinsky, Y.; Fridkin, M.; Zakuth, V.; Spirer, Z. *Int. J. Pept. Prot. Chem.* **1978**, *12*, 130–138.

33. Nakajima, K.; Takai, F.; Tanaka, T.; Okawa, K. *Bull. Chem. Soc. Jpn.* **1978**, *51*, 1577–1578.

34. Theodoropoulos, D.; Schwartz, I.L.; Walter, R. *Biochemistry* **1967**, *6*, 3927–3932.

35. Monsigny, M.L.P.; Delay, D.; Vaculik, M. *Carbohydrate Res.* **1977**, *59*, 589–593.

36. Walker, M.A.; Kaplita, K.P.; Chen, T.; King, H.D. *Synlett* **1997**, 169–170.

37. Dugave, C.; Menez, A. *J. Org. Chem.* **1996**, *61*, 6067–6070.

38. Bregant, S.; Tabor, A.B. *J. Org. Chem.* **2005**, *70*, 2430–2438.

39. Panda, G.; Rao, N.V. *Synlett* **2004**, 714–716.

40. Sibi, M.P.; Rutherford, D.; Sharma, R. *J. Chem. Soc., Perkin Trans. 1* **1994**, 1675–1678.

41. Wood, J.L.; van Middlesworth, L. *J. Biol. Chem.* **1949**, *179*, 529–533.

42. Walsh, C.T.; Schonbrunn, A.; Abeles, R.H. *J. Biol. Chem.* **1971**, *246*, 6855–6866.

43. Brown, G.B.; Du Vigneaud, V. *J. Biol. Chem.* **1941**, *137*, 611–615.

44. Rachele, J.R.; Reed, L.J.; Kidwai, A.R.; Ferger, M.F.; Du Vigneaud, V. *J. Biol. Chem.* **1950**, *185*, 817–826.

45. Miller, M.J.; Mattingly, P.G.; Morrison, M.A.; Kerwin J.F., Jr. *J. Am. Chem. Soc.* **1980**, *102*, 7026–7032.

46. Hardegger, E.; Szabo, F.; Liechti, P.; Rostetter, C.; Zankowska-Jasinska, W. *Helv. Chim. Acta* **1968**, *51*, 78–85.

47. Fischer, E.; Raske, K. *Ber.* **1907**, *40*, 3717–3724.

48. Fujii, M.; Yoshida K.; Hidaka, J.; Ohtsu, T. *Bioorg. Med. Chem. Lett.* **1997**, *7*, 637–640.

49. Morin, C.; Sawaya, G. *Synthesis* **1987**, 479–480.

50. Benoiton, L. *Can. J. Chem.* **1968**, *46*, 1549–1552.

51. Kato, Y.; Fukomoto, K.; Asano, Y. *Appl. Microbiol. Biotechnol.* **1993**, *39*, 301–304.

52. Itaya, T.; Iida, T.; Shimizu, S.; Mizutani, A.; Morisue, M.; Sugimoto, Y.; Tachinaka, M. *Chem. Pharm. Bull.* **1993**, *41*, 252–261.

53. Brachwitz, H.; Ölke, M.; Bergmann, J.; Langen, P. *Bioorg. Med. Chem. Lett.* **1997**, *7*, 1739–1742.

54. Stocking, E.M.; Schwarz, J.N.; Senn, H.; Salzmann, M.; Silks, L.A. *J. Chem. Soc., Perkin Trans. 1* **1997**, 2443–2447.

55. Dugave, C.; Ménez, A. *Tetrahedron: Asymmetry* **1997**, *8*, 1453–1465.

56. Mustapa, M.F.; Harris, R.; Bulic-Subanovic, N.; Elliott, S.L.; Bregant, S.; Groussier, M.F.A.; Mould, J.; Schultz, D.; Chubb, N.A.L.; Gaffney, P.R.J.; Driscoll, P.C.; Tabor, A.B. *J. Org. Chem.* **2003**, *68*, 8185–8192.

57. Mustapa, M.F.M.; Harris, R.; Mould, J.; Chubb, N.A.L.; Schultz, D.; Driscoll, P.C.; Taboor, A.B. *Tetrahedron Lett.* **2002**, *43*, 8359–8362.

58. Maria, E.J.; Da Silva, A.D.; Fourrey, J.-L.; Machado, A.S.; Robert-Géro, M. *Tetrahedron Lett.* **1994**, *35*, 3301–3302.

59. Wilson, I.; Jackson, R.F.W. *J. Chem. Soc., Perkin Trans. 1* **2002**, 2845–2850.

60. Zhu, X.; Schmidt, R.R. *Eur. J. Org. Chem.* **2003**, 4069–4072.

61. Mayer, J.P.; Zhang, J.; Groeger, S.; Liu, C.-F.; Jarosinski, M.A. *J. Pept. Res.* **1998**, *51*, 432–436.

62. Boggs N.T., III; Gawley, R.E.; Koehler, K.A.; Hiskey, R.G. *J. Org. Chem.* **1975**, *40*, 2850–2851.

63. Erlenmeyer, E. *Ber.* **1903**, *36*, 2720–2722.

64. Zou, Y.; Fahmi, N.E.; Vialas, C.; Miller, G.M.; Hecht, S.M. *J. Am. Chem. Soc.* **2002**, *124*, 9476–9488.

65. Miyashita, K.; Murafuji, H.; Iwaki, H.; Yoshioka, E.; Imanishi, T. *Chem. Commun.* **2002**, 1922–1923.

66. Morell, J.L.; Fleckenstein, P.; Gross, E. *J. Org. Chem.* **1977**, *42*, 355–356.

67. Nakamura, Y.; Hirai, M.; Tamotsu, K.; Yonezawa, Y.; Shin, C.-G. *Bull. Chem. Soc. Jpn.* **1995**, *68*, 1369–1377.

68. Somlai, C.; Lovas, S.; Forgó, P.; Murphy, R.F.; Penke, B. *Synth. Commun.* **2001**, *31*, 3633–3640.

69. Fischer, E.; Raske, K. *Ber.* **1908**, *41*, 893–897.

70. Melchoir, J.B.; Tarver, H. *Arch. Biochem.* **1947**, *12*, 301–308.

71. Painter, E.P. *J. Am. Chem. Soc.* **1947**, *69*, 229–232.

72. Williams, L.R.; Ravve, A. *J. Am. Chem. Soc.* **1948**, *70*, 1244–1245.

73. Gieselman, M.D.; Xie, L.; van der Donk, W.A. *Org. Lett.* **2001**, *3*, 1331–1334.

74. De Marco, C.; Rinaldi, A.; Dernini, S.; Cavallini, D. *Gazz. Chim. Ital.* **1975**, *105*, 1113–1115.

75. Sadeh, T.; Davis, M.A.; Giese, R.W. *J. Pharm. Sci.* **1976**, *65*, 623–625.

76. Andreadou, I.; Menge, W.M.P.B.; Commandeur, J.N.M.; Worthington, E.A.; Vermeulen, N.P.E. *J. Med. Chem.* **1996**, 39, 2040–2046.

77. Zhou, H.; van der Donk, W.A. *Org. Lett.* **2002**, *4*, 1335–1338.

78. Knapp, S.; Myers, D.S. *J. Org. Chem.* **2001**, *66*, 3636–3638.

79. Du Vigneaud, V.; Brown, G.B. *J. Biol. Chem.* **1941**, *138*, 151–154.

80. Brown, G.B.; Du Vigneaud, V. *J. Biol. Chem.* **1941**, *140*, 767–771.

81. Kelland, J.G.; Arnold, L.D.; Palcic, M.M.; Pickard, M.A.; Vederas, J.C. *J. Biol. Chem.* **1986**, *261*, 13216–13223.

82. Du Vigneaud, V.; Brown, G.B.; Chandler, J.P. *J. Biol. Chem.* **1942**, *143*, 59–64.

83. Anslow W.P., Jr.; Simmonds, S.; Du Vigneaud, V. *J. Biol. Chem.* **1946**, *166*, 35–45.

84. Deboves, H.J.C.; Montalbetti, C.A.G.N.; Jackson, R.F.W. *J. Chem. Soc., Perkin Trans. 1* **2001**, 1876–1884.

85. Bajgrowicz, J.A.; Hallaoui, A.El.; Jacquier, R.; Pigière, Ch.; Viallefont, Ph. *Tetrahedron Lett.* **1984**, *25*, 2759–2762.

86. Bajgrowicz, J.A.; Hallaoui, A.El.; Jacquier, R.; Pigière, Ch.; Viallefont, Ph. *Tetrahedron* **1985**, *41*, 1833–1843.

87. Bernardini, A.; El Hallaoui, A.; Jacquier, R.; Pigière, Ch.; Viallefont, Ph.; Bajgrowicz, J. *Tetrahedron Lett.* **1983**, *24*, 3717–3720.

88. Rosowsky, A.; Bader, H.; Freisheim, J.H. *J. Med. Chem.* **1991**, *34*, 203–208.

89. Wojciechowska, H.; Pawlowicz, R.; Andruszkiewicz, R.; Gryzbowska, J. *Tetrahedron Lett.* **1978**, *42*, 4063–4064.

90. Arnold, L.D.; Kalantar, T.H.; Vederas, J.C. *J. Am. Chem. Soc.* **1985**, *107*, 7105–7109.

91. Mattingly, P.G.; Miller, M.J. *J. Org. Chem.* **1980**, *45*, 410–415.

92. Bose, A.K.; Sahu, D.P.; Manhas, M.S. *J. Org. Chem.* **1981**, *46*, 1229–1230.

93. Miller, M.J. *Acc. Chem. Res.* **1986**, *19*, 49–56.

94. Otsuka, M.; Kittaka, A.; Iimori, T.; Yamashita, H.; Kobayashi, S.; Ohno, M. *Chem. Pharm. Bull.* **1985**, *33*, 509–514.

95. Ibuka, T.; Nakai, K.; Habashita, H.; Hotta, Y.; Fujii, N.; Mimura, N.; Miwa, Y.; Taga, T.; Yamamoto, Y. *Angew. Chem., Int. Ed. Engl.* **1994**, *33*, 652–654.

96. Nakajima, K.; Tanaka, T.; Neya, M.; Okawa, K. *Bull. Chem. Soc. Jpn.* **1982**, *55*, 3237–3241.

97. Solomon, M.E.; Lynch, C.L.; Rich, D.H. *Synth. Commun.* **1996**, *26*, 2723–2729.

98. Turner, J.J.; Sikkema, F.D.; Filippov, D.V.; van der Marel, G.A.; van Boom, J.H. *Synlett* **2001**, 1727–1730.

99. Golding, B.T.; Howes, C. *J. Chem. Res S* **1984**, 1; *M* **1984**, 0101–0110.

100. Fabiano, E.; Golding, B.T.; Sadeghi, M.M. *Synthesis* **1987**, 190–192.

101. Boger, D.L.; Dang, Q. *J. Org. Chem.* **1992**, *57*, 1631–1633.

102. Rosenberg, S.H.; Spina, K.P.; Woods, K.W.; Polakowski, J.; Martin, D.L.; Yao, Z.; Stein, H.H.; Cohen, J.; Barlow, J.L.; Egan, D.A.; Tricarico, K.A.; Baker, W.R.; Kleinert, H.D. *J. Med. Chem.* **1993**, *36*, 449–459.

103. Cherney, R.J.; Wang, L. *J. Org. Chem.* **1996**, *61*, 2544–2546.

104. Lu, H.S.M.; Volk, M.; Kholodenko, Y.; Gooding, E.; Hochstrasser, R.M.; DeGrado, W.F. *J. Am. Chem. Soc.* **1997**, *119*, 7173–7180.

105. Luo, Y.; Blaskovich, M.A.; Lajoie, G.A. *J. Org. Chem.* **1999**, *64*, 6106–6111.

106. Turner, J.J.; Leeuwenburgh, M.A.; van der Marel, G.A.; van Boom, J.H. *Tetrahedron Lett.* **2001**, *42*, 8713–8716.

107. Lampariello, L.R.; Piras, D.; Rodriquez, M.; Taddei, M. *J. Org. Chem.* **2003**, *68*, 7893–7895.

108. Schmidt, U.; Mundinger, K.; Mangold, R.; Lieberknecht, A. *J. Chem. Soc., Chem. Commun.* **1990**, 1216–1218.

109. Pearson, C.; Rinehart, K.L.; Sugano, M. *Tetrahedron Lett.* **1999**, *40*, 411–414.

110. Horikawa, E.; Kodaka, M.; Nakahara, Y.; Okuno, H.; Nakamura, K. *Tetrahedron Lett.* **2001**, *42*, 8337–8339.

111. Nakamura, K.; Ohnishi, Y.; Horiwaka, E.; Konakahara, T.; Kodaka, M.; Okuno, H. *Tetrahedron Lett.* **2003**, *44*, 5445–5448.

112. Nakamura, K.; Isaka, T.; Toshima, H.; Kodaka, M. *Tetrahedron Lett.* **2004**, *45*, 7221–7224.

113. Siebum, A.H.G.; Woo, W.S.; Raap, J.; Lugtenburg, J. *J. Eur. J. Org. Chem.* **2004**, 2905–2913.

114. Peyman, A.; Gourvest, J.-F.; Gadek, T.R.; Knolle, J. *Angew. Chem., Int. Ed.* **2000**, *39*, 2874–2877.

115. Sibi, M.P.; Harris, B.J.; Shay, J.J.; Hajra, S. *Tetrahedron* **1998**, *54*, 7221–7228.

116. Berkowitz, D.B.; Shen, Q.; Maeng, J.-H. *Tetrahedron Lett.* **1994**, *35*, 6445–6448.

117. Porte, A.M.; van der Donk, W.A.; Burgess, K. *J. Org. Chem.* **1998**, *63*, 5262–5264.
118. Houck, D.R.; Chen, L.-C.; Keller, P.J.; Beale, J.M.; Floss, H.G. *J. Am. Chem. Soc.* **1988**, *110*, 5800–5806.
119. Lee, M.; Phillips, R.S. *Bioorg. Med. Chem. Lett.* **1992**, *2*, 1563–1564.
120. Kaiser, M.; Groll, M.; Renner, C.; Huber, R.; Moroder, L. *Angew. Chem., Int. Ed.* **2002**, *41*, 780–783.
121. Sloan, M.J.; Phillips, R.S. *Bioorg. Med. Chem. Lett.* **1992**, *2*, 1053–1056.
122. Philips, R.S.; Cohen, L.A.; Annby, U.; Wensbo, D.; Gronowitz, S. *Bioorg. Med. Chem., Lett.* **1995**, *5*, 1133–1134.
123. Walker, T.E.; Matheny, C.; Storm, C.B.; Hayden, H. *J. Org. Chem.* **1986**, *51*, 1175–1179.
124. Phillips, R.S. *Tetrahedron: Asymmetry* **2004**, *15*, 2787–2792.
125. Murakoshi, I.; Ikegami, F.; Hinuma, Y.; Hanma, Y. *Phytochemistry* **1984**, *23*, 1905–1908.
126. Antoni, A.; Omura, H.; Bergström, M.; Sundin, A.; Watanabe, Y.; Längström, B. *J. Lab. Cmpds. Radiopharm.* **1995**, *37*, 182–183.
127. Antoni, G.; Omura, H.; Sundin, A.; Takalo, R.; Valind, S.; Watanabe, Y.; Långström, B. *J. Lab. Cmpds. Radiopharm.* **1997**, *40*, 807–809.
128. Murakoshi, I.; Ikegami, F.; Yoneda, Y.; Ihara, H.; Sakata, K.; Koide, C. *Chem. Pharm. Bull.* **1986**, *34*, 1473–1478.
129. Maier, T.H.P. *Nat. Biotechnol.* **2003**, *21*, 422–427.
130. Baldwin, J.E.; Adlington, R.M.; Birch, D.J. *J. Chem. Soc., Chem. Commun.* **1985**, 256–257.
131. Baldwin, J.E.; Adlington, R.M.; Mellor, L.C. *Tetrahedron* **1994**, *50*, 5049–5066.
132. Baldwin, J.E.; Spivey, A.C.; Schofield, C.J. *Tetrahedron: Asymmetry* **1990**, *1*, 881–884.
133. Kim, B.M.; So, S.M. *Tetrahedron Lett.* **1998**, *39*, 5381–5384.
134. Kim, B.M.; So, S.M. *Tetrahedron Lett.* **1999**, *40*, 7687–7690.
135. Kuyl-Yeheskiely, E.; Lodder, M.; van der Marel, G.A.; van Boom, J.H. *Tetrahedron Lett.* **1992**, *33*, 3013–3016.
136. Wei, L.; Lubell, W.D. *Org. Lett.* **2000**, *2*, 2595–2598.
137. Cohen, S.B.; Halcomb, R.L. *J. Am. Chem. Soc.* **2002**, *124*, 2534–2543.
138. Byun, H.-S.; He, L.; Bittman, R. *Tetrahedron* **2000**, *56*, 7051–7091.
139. Meléndez, R.E.; Lubell, W.D. *Tetrahedron* **2003**, *59*, 6051–6056.
140. Ramer, S.E.; Moore, R.N.; Vederas, J.C. *Can. J. Chem.* **1986**, *64*, 706–713.
141. Lodwig, S.N.; Unkefer, C.J. *J. Lab. Cmpds. Radiopharm.* **1992**, *31*, 95–102.
142. Marzoni, G.; Kaldor, S.W.; Trippe, A.J.; Shamblin, B.M.; Fritz, J.E. *Synth. Commun.* **1995**, *25*, 2475–2482.
143. Pansare, S.V.; Huyer, G.; Arnold, L.D.; Vederas, J.C.; Jones, M.; Olson, G.L.; Coffen, D.L. *Org. Synth.* **1992**, *70*, 1–9.
144. Ratemi, E.S.; Vederas, J.C. *Tetrahedron Lett.* **1994**, *35*, 7605–7608.
145. Aggen, J.B.; Humphrey, J.M.; Gauss, C.-M.; Huang, H.-B.; Nairn, A.C.; Chamberlin, A.R. *Biorg. Med. Chem.* **1999**, *7*, 543–564.
146. Shreder, K.; Zhang, L.; Dang, T.; Yaksh, T.L.; Umeno, H.; DeHaven, R.; Daubert, J.; Goodman, M. *J. Med. Chem.* **1998**, *41*, 2631–2635.
147. Rossi, P.; Felluga, F.; Scrimin, P. *Tetrahedron Lett.* **1998**, *39*, 7159–7162.
148. Kretsinger, J.K.; Schneider, J.P. *J. Am. Chem. Soc.* **2003**, *125*, 7907–7913.
149. Warshawsky, A.M.; Patel, M.V.; Chen, T.-M. *J. Org. Chem.* **1997**, *62*, 6439–6440.
150. Elliott, J.M.; Cascieri, M.A.; Chicchi, G.; Davies, S.; Kelleher, F.J.; Kurtz, M.; Ladduwahetty, T.; Lewis, R.T.; MacLeod, A.M.; Merchant, K.J.; Sadowski, S.; Stevenson, G.I. *Biorg. Med. Chem. Lett.* **1998**, *8*, 1845–1850.
151. Zhou, H.; Schmidt, D.M.Z.; Gerlt, J.A.; van deer Donk, W.A. *ChemBioChem* **2003**, *4*, 1206–1215.
152. Epstein, W.W.; Wang, Z. *Chem. Commun.* **1997**, 863–864.
153. Shao, H.; Wang, S.H.H.; Lee, C.-W.; Ösapay, G.; Goodman, M. *J. Org. Chem.* **1995**, *60*, 2956–2957.
154. Ösapay, G.; Goodman, M. *J. Chem. Soc., Chem. Commun.* **1993**, 1599–1600.
155. Shao, H.; Lee, C.-W.; Zhu, Q.; Gantzel, P.; Goodman, M. *Angew. Chem., Int. Ed. Engl.* **1996**, *35*, 90–92.
156. Okeley, N.M.; Zhu, Y.; van der Donk, W.A. *Org. Lett.* **2000**, *2*, 3603–3606.
157. Mori, T.; Tohmiya, H.; Satouchi, Y.; Higashibayashi, S.; Hashimoto, K.; Nakata, M. *Tetrahedron Lett.* **2005**, *46*, 6423–6427.
158. Lall, M.S.; Karvellas, C.; Vederas, J.C. *Org. Lett.* **1999**, *1*, 803–806.
159. Lohse, P.A.; Felber, R. *Tetrahedron Lett.* **1998**, *39*, 2067–2070.
160. Smith, E.C.R.; McQuaid, L.A.; Paschal, J.W.; DeHoniesto, J. *J. Org. Chem.* **1990**, *55*, 4472–4474.
161. Diederichsen, U. *Angew. Chem., Int. Ed. Engl.* **1996**, *35*, 445–448.
162. Diederichsen, U. *Biorg. Med. Chem. Lett.* **1998**, *8*, 165–168.
163. Lescrinier, T.; Hendrix, C.; Kerremans, L.; Rozenski, J.; Link, A.; Samyn, B.; Van Aerschot, A.; Lescrinier, E.; Eritja, R.; Van Beeumen, J.; Herdewijn, P. *Chem. —Eur. J.* **1998**, *4*, 425–433.
164. Diederichsen, U.; Weicherding, D. *Synlett* **1999**, 917–920.
165. Wagner, T.; Davis, W.B.; Lorenz, K.B.; Michel-Beyerle, M.E.; Diederichsen, U. *Eur. J. Org. Chem.* **2003**, 3673–3679.

166. Dolman, N.P.; Troop, H.M.; More, J.C.A.; Alt, A.; Knauss, J.L.; Nistico, R.; Jack, S.; Morley, R.M.; Bortolotto, Z.A.; Roberts, P.J.; Bleakman, D.; Collingridge, G.L.; Jane, D.E. *J. Med. Chem.* **2005**, *48*, 7867–7881.

167. Arnold, L.D.; May, R.G.; Vederas, J.C. *J. Am. Chem. Soc.* **1988**, *110*, 2237–2241.

168. Pansare, S.V.; Arnold, L.D.; Vederas, J.C.; Manchand, P.S.; Mastrodonato-Delora, P.; Coffen, D.L. *Org. Synth.*, **1992**, *70*, 10–17.

169. Soucy, F.; Wernic, D.; Beaulieu, P. *J. Chem. Soc., Perkin Trans. 1* **1991**, 2885–2887.

170. Arnold, L.D.; Drover, J.C.G.; Vederas, J.C. *J. Am. Chem. Soc.* **1987**, *109*, 4649–4659.

171. Smith, N.D.; Wohlrab, A.M.; Goodman, M. *Org. Lett.* **2005**, *7*, 255–258.

172. Olma, A.; Kudaj, A. *Tetrahedron Lett.* **2005**, *46*, 6239–6241.

173. Baldwin, J.E.; Fieldhouse, R.; Russell, A.T. *Tetrahedron Lett.* **1993**, *34*, 5491–5494.

174. Baldwin, J.E.; Adlington, R.M.; Birch, D.J.; Crawford, J.A.; Sweeney, J.B. *J. Chem. Soc., Chem. Commun.* **1986**, 1339–1340.

175. Baldwin, J.E.; Adlington, R.M.; Basak, A. *J. Chem. Soc., Chem. Commun.* **1984**, 1284–1285.

176. Savithri, D. Leumann, C.; Scheffold, R. *Helv. Chim. Acta* **1996**, *79*, 288–294.

177. Adlington, R.M.; Baldwin, J.E.; Basek, A.; Kozyrod, R.P. *J. Chem. Soc., Chem. Commun.* **1983**, 944–945.

178. Adlington, R.M.; Mantell, S.J. *Tetrahedron* **1992**, *48*, 6529–6536.

179. Herpin, T.F.; Motherwell, W.B.; Weibel, J.-M. *Chem. Commun.* **1997**, 923–924.

180. Sasaki, N.A.; Hashimoto, C.; Potier, P. *Tetrahedron Lett.* **1987**, *28*, 6069–6072.

181. Pauly, R.; Sasaki, N.A.; Potier, P. *Tetrahedron Lett.* **1994**, *35*, 237–240.

182. Sasaki, N.A.; Hashimoto, C.; Pauly, R.; Potier, P. *Peptides* **1988**, 313–315.

183. Sasaki, N.A; Hashimoto, C.; Pauly, R., Potier, P. *Second Forum on Peptides,* Aubry, A.; Vitoux, B., Eds; Colloque INSERM/John Libbey Eurotext Ltd, Paris; 1989, vol 174, pp 285–289.

184. Sasaki, N.A; Hashimoto, C.; Pauly, R. *Tetrahedron Lett.* **1989**, *30*, 1943–1946.

185. Sasaki, N.A.; Pauly, R.; Fontaine, C.; Chiaroni, A.; Riche, C.; Potier, P. *Tetrahedron Lett.* **1994**, *35*, 241–244.

186. Sasaki, N.A.; Dockner, M.; Chiaroni, A.; Riche, C.; Potier, P. *J. Org. Chem.* **1997**, *62*, 765–770.

187. Sagnard, I.; Sasaki, N.A.; Chiaroni, A.; Riche, C.; Potier, P. *Tetrahedron Lett.* **1995**, *36*, 3149–3152.

188. Wang, Q.; Saski, N.A.; Potier, P. *Tetrahedron* **1998**, *54*, 15759–15780.

189. Wang, Q.; Saski, N.A.; Potier, P. *Tetrahedron Lett.* **1998**, *39*, 5755–5758.

190. Itaya, T.; Mizutani, A. *Tetrahedron Lett.* **1985**, *26*, 347–350.

191. Itaya, T.; Mizutani, A.; Iida, T. *Chem. Pharm. Bull.* **1991**, *39*, 1407–1414.

192. Bicknell, A.J.; Burton, G.; Elder, J.S. *Tetrahedron Lett.* **1988**, *29*, 3361–3364.

193. Sibi, M.P.; Renhowe, P.A. *Tetrahedron Lett.* **1990**, *31*, 7407–7410.

194. Sibi, M.P.; Christensen, J.; Li, B.; Renhowe, P.A. *J. Org. Chem.* **1992**, *57*, 4329–4330.

195. Sibi, M.; Li, B. *Tetrahedron Lett.* **1992**, *33*, 4115–4118.

196. Bertozzi, C.R.; Hoeprich P.D., Jr.; Bednarski, M.D. *J. Org. Chem.* **1992**, *57*, 6092–6094.

197. Bertozzi, C.R.; Cook, D.G.; Kobertz, W.P.; Gonzalez-Scarano, F.; Bednarski, M.D. *J. Am. Chem. Soc.* **1992**, *114*, 10639–10641.

198. Sibi, M.P.; Rutherford, D.; Renhowe, P.A.; Li, B. *J. Am. Chem. Soc.* **1999**, *121*, 7509–7516.

199. Jackson, R.F.W.; Wishart, N.; Wood, A.; James, K.; Wythes, M.J. *J. Org. Chem.* **1992**, *57*, 3397–3404.

200. Jackson, R.F.W.; James, K.; Wythes, M.J.; Wood, A. *J. Chem. Soc., Chem. Commun.* **1989**, 644–645.

201. Gair, S.; Jackson, R.F.W.; Brown, P.A. *Tetrahedron Lett.* **1997**, *38*, 3059–3062.

202. Dunn, M.; Jackson, R.F.W.; Pietruszka, J.; Turner, D. *J. Org. Chem.* **1995**, *60*, 2210–2215.

203. Dunn, M.J.; Jackson, R.F.W.; Pietruszka, J.; Wishart, N.; Ellis, D.; Wythes, M.J. *Synlett* **1993**, 499–500.

204. Rilatt, I.; Caggiano, L.; Jackson, R.F.W. *Synlett* **2005**, 2701–2719.

205. Jackson, R.F.W.; Wythes, M.J.; Wood, A. *Tetrahedron Lett.* **1989**, *30*, 5941–5944.

206. Jackson, R.F.W.; Moore, R.J.; Dexter, C.S.; Elliott, J.; Mowbray, C.E. *J. Org. Chem.* **1998**, *63*, 7875–7884.

207. Schmidt, B.; Ehlert, D.K. *Tetrahedron Lett.* **1998**, *39*, 3999–4002.

208. Dexter, C.S.; Jackson, R.F.W.; Elliott, J. *Tetrahedron* **2000**, *56*, 4539–4540.

209. Dexter, C.S.; Jackson, R.F.W. *Chem. Commun.* **1998**, 75–76.

210. Dexter, C.S.; Jackson, R.F.W.; Elliott, J. *J. Org. Chem.* **1999**, *64*, 7579–7585.

211. Dexter, C.S.; Hunter, C.; Jackson, R.F.W.; Elliott, J. *J. Org. Chem.* **2000**, *65*, 7417–7421.

212. Dow, R.L.; Bechle, B.M. *Synlett* **1994**, 293–294.

213. Smyth, M.S.; Burke T.R., Jr.; *Tetrahedron Lett.* **1994**, *35*, 551–554.

214. Swahn, B.-M.; Claesson, A.; Pelcman, B.; Besidski, Y.; Molin, H.; Sandberg, M.P.; Berge, O.-G. *Bioorg. Med. Chem. Lett.* **1996**, *6*, 1635–1640.

215. Malan, C.; Morin, C. *Synlett* **1996**, 167–168.

216. Ye, B.; Otaka, A.; Burke T.R., Jr.; *Synlett* **1996**, 459–460.

217. Kawahata, N.; Yang, M.G.; Luke, G.P.; Shakespeare, W.C.; Sundaramoorthi, R.; Wang, Y.;

Johnson, D.; Merry, T.; Violette, S.; Guan, W.; Bartlett, C.; Smith, J.; Hatada, M.; Lu, X.; Dalgarno, D.C.; Eyermann, C.J.; Bohacek, R.S.; Sawyer, T.K. *Bioorg. Med. Chem. Lett.* **2001**, *11*, 2319–2323.

218. Tabanella, S.; Valancogne, I.; Jackson, R.F.W. *Org. Biomol. Chem.* **2003**, *1*, 4254–4261.

219. Jung, M.E.; Starkey, L.S. *Tetrahedron* **1997**, *53*, 8815–8824.

220. Pearce, A.J.; Ramaya, S.; Thorn, S.N.; Bloomberg, G.B.; Walter, D.S.; Gallagher, T. *J. Org. Chem.* **1999**, *64*, 5453–5462.

221. Jackson, R.F.W.; Turner, D.; Block, M.H. *Synlett* **1996**, 862–864.

222. Jackson, R.F.W.; Wood, A.; Wythes, M.J. *Synlett* **1990**, 735–736.

223. Jackson, R.F.W.; Rettie, A.B.; Wood, A.; Wythes, M.J. *J. Chem. Soc., Perkin Trans. 1* **1994**, 1719–1726.

224. Jackson, R.F.W.; Graham, L.J.; Rettie, A.B. *Tetrahedron Lett.* **1994**, *35*, 4417–4418.

225. Jackson, R.F.W.; Turner, D.; Block, M.H. *J. Chem. Soc., Perkin Trans. 1* **1997**, 865–870.

226. Jackson, R.F.W.; Wishart, N.; Wythes, M.J. *J. Chem. Soc., Chem. Commun.* **1992**, 1587–1589.

227. Jackson, R.F.W.; Turner, D.; Block, M.H. *J. Chem. Soc., Chem. Commun.* **1995**, 2207–2208.

228. Dunn, M.J.; Gomez, S.; Jackson, R.F.W. *J. Chem. Soc., Perkin Trans.1* **1995**, 1639–1640.

229. Dunn, M.J.; Jackson, R.F.W. *Tetrahedron* **1997**, *53*, 13905–13914.

230. Dunn, M.J.; Jackson, R.F.W. *J. Chem. Soc., Chem. Commun.* **1992**, 319–320.

231. Deboves, H.J.C.; Grabowska, U.; Rizzo, A.; Jackson, R.F.W. *J. Chem. Soc., Perkin Trans. 1* **2000**, 4284–4292.

232. Rodríguez, A.; Miller, D.D.; Jackson, R.F.W. *Org. Biomol. Chem.* **2003**, *1*, 973–977.

233. IJsselstijn, M.; Kaiser, J.; van Delft, F.L.; Schoemaker, H.E.; Rutjes, F.P.J.T. *Amino Acids* **2003**, *24*, 263–266.

234. Masaki, H.; Mizozoe, T.; Esumi, T.; Iwabuchi, Y.; Hatakeyama, S. *Tetrahedron Lett.* **2000**, *41*, 4801–4804.

235. Suárez, R.M.; Sestelo, J.P.; Sarandeses, L.A. *Chem.—Eur. J.* **2003**, *9*, 4179–4187.

236. Suárez, R.M.; Sestelo, J.P.; Sarandeses, L.A. *Org. Biomol. Chem.* **2004**, *2*, 3584–3587.

237. Dunn, M.J.; Jackson, R.F.W.; Stephenson, G.R. *Synlett* **1992**, 905–906.

238. Jackson, R.F.W.; Turner, D.; Block, M.H. *Synlett* **1997**, 789–790.

239. Dorgan, B.J.; Jackson, R.F.W. *Synlett* **1996**, 859–861.

240. Jackson, R.F.W.; Wishart, N.; Wythes, M.J. *Synlett* **1993**, 219–220.

241. Fraser, J.L.; Jackson, R.F.W.; Porter, B. *Synlett* **1994**, 379–380.

242. Tamaru, Y.; Tanigawa, H.; Yamamoto, T.; Yoshida, Z.-I. *Angew. Chem., Int. Ed. Engl.* **1989**, *28*, 351–353.

243. Fraser, J.L.; Jackson, R.F.W.; Porter, B. *Synlett* **1995**, 819–820.

244. Zeng, B.; Wong, K.K.; Pompliano, D.L.; Reddy, S.; Tanner, M.E. *J. Org. Chem.* **1998**, *63*, 10081–10086.

245. Jackson, R.F.W.; Fraser, J.L.; Wishart, N.; Porter, B.; Wythes, M.J. *J. Chem. Soc., Perkin Trans. 1* **1998**, 1903–1912.

246. Acherki, H.; Alvarez-Ibarra, C.; Barrasa, A.; de Dios, A. *Tetrahedron Lett.* **1999**, *40*, 5763–5766.

247. Fu, J.-M.; Castelhano, A.L. *Biorg. Med. Chem. Lett.* **1998**, *8*, 2813–2816.

248. Kenworthy, M.N.; Kilburn, J.P.; Taylor, R.J.K. *Org. Lett.* **2004**, *6*, 19–22.

249. Harvey, J.E.; Kenworthy, M.N.; Taylor, R.J.K. *Tetrahedron Lett.* **2004**, *45*, 2467–2471.

250. Liang, X.; Andersch, J.; Bols, M. *J. Chem. Soc., Perkin Trans. 1* **2001**, 2136–2157.

251. Garner, P. *Tetrahedron Lett.* **1984**, *25*, 5855–5858.

252. Garner, G.; Park, J.M.; Takasu, M.; Yamamoto, H. *Org. Synth.* **1992**, *70*, 18–28.

253. McKillop, A.; Taylor, R.J.K.; Watson, R.J.; Lewis, N. *Synthesis* **1994**, 31–33.

254. Branquet, E.; Durand, P.; Vo-Quang, L.; Le Goffic, F. *Synth. Commun.* **1993**, *23*, 153–156.

255. Williams, L.; Zhang, Z.; Ding, X.; Joullié, M.M. *Tetrahedron Lett.* **1995**, *36*, 7031–7034.

256. Williams, L.; Zhang, Z.; Shao, F.; Carroll, P.J.; Joullié, M.M. *Tetrahedron* **1996**, *52*, 11673–11694.

257. Barco, A.; Benetti, S.; Casolari, A.; Pollini, G.P.; Spalluto, G. *Tetrahedron Lett.* **1990**, *31*, 4917.

258. Yanagida, M.; Hashimoto, K.; Ishida, M.; Shinozaki, H.; Shirahama, H. *Tetrahedron Lett.* **1989**, *30*, 3799–3802.

259. Roush, W.R.; Hunt, J.A. *J. Org. Chem.* **1995**, *60*, 798–806.

260. Jurczak, J.; Gryko, D.; Kobryzcka, E.; Gruza, H.; Prokopowicz, P. *Tetrahedron* **1998**, *54*, 6051–6064.

261. Bold, G.; Allmendinger, T.; Herold, P.; Moesch, L.; Schär, H.-P.; Duthaler, R.O. *Helv. Chim. Acta* **1992**, 865–882.

262. Campbell, A.D.; Raynham, T.M.; Taylor, R.J.K. *Synthesis* **1998**, 1707–1709.

263. Avenoza, A.; Cativiela, C.; Corzana, F.; Peregrina, J.M.; Zurbano, M.M. *Synthesis* **1997**, 1146–1150.

264. Beaulieu, P.L.; Schiller, P.W. *Tetrahedron Lett.* **1988**, *29*, 2019–2022.

265. Kumar, J.S.R.; Datta, A. *Tetrahedron Lett.* **1997**, *38*, 6779–6780.

266. Falorni, M.; Conti, S.; Giacomelli, G.; Cossu, S.; Soccolini, F. *Tetrahedron: Asymmetry* **1995**, *6*, 287–294.

267. Dondoni, A.; Massi, A.; Marra, A. *Tetrahedron Lett.* **1998**, *39*, 6601–6604.

268. Dondoni, A.; Marra, A.; Massi, A. *Chem. Commun.* **1998**, 1741–1742.

269. Dondoni, A.; Marra, A.; Massi, A. *J. Org. Chem.* **1999**, *64*, 933–944.

270. Alías, M.; Cativiela, C.; Díaz-de-Villegas, M.D.; Gálvez, J.A.; Lapeña, Y. *Tetrahedron* **1998**, *54*, 14963–14974.

271. Avenoza, A.; Busto, J.H.; Corzana, F.; Jiménez-Osés, G.; Peregrina, J.M. *Tetrahedron* **2003**, *59*, 5713–5718.

272. Tuch, A.; Sanière, M.; Le Merrer, Y.; Depezay, J.-C. *Tetrahedron: Asymmetry* **1997**, *8*, 1649–1659.

273. Duthaler, R.O. *Angew. Chem., Int. Ed. Engl.* **1991**, *30*, 705–707.

274. Dondoni, A.; Fantin, G.; Fogagnolo, M.; Medici, A. *J. Chem. Soc., Chem. Commun.* **1988**, 10–12.

275. Dondoni, A.; Fantin, G.; Fogagnolo, M.; Pedrini, P. *J. Org. Chem.* **1990**, *55*, 1439–1446.

276. Garner, P.; Park, J.M.; Malecki, E. *J. Org. Chem.* **1988**, *53*, 4395–4398.

277. Dondoni, A.; Fantin, G.; Fogagnolo, M.; Merino, P. *J. Chem. Soc., Chem. Commun.* **1990**, 854–855.

278. Nakagawa, M.; Tsuruoka, A.; Yoshida, J.; Hino, T. *J. Chem. Soc., Chem. Commun.* **1990**, 603–604.

279. Mori, K.; Matsud, H. *Liebigs Ann. Chem.* **1991**, 529–535.

280. Radunz, H.-E.; Devant, R.M.; Eiermann, V. *Liebigs Ann. Chem.* **1988**, 1103–1105.

281. Garner, P.G.; Park, J.M. *J. Org. Chem.* **1988**, *53*, 2979–2984.

282. Garner, P.; Park, J.M. *Tetrahedron Lett.* **1989**, *30*, 5065–5068.

283. Garner, P.; Park, J.M. *J. Org. Chem.* **1990**, *55*, 3772–3787.

284. Casiraghi, G.; Colombo, L.; Rassu, G.; Spanu, P. *J. Chem. Soc., Chem. Commun.* **1991**, 603–604.

285. Casiraghi, G.; Colombo, L.; Rassu, G.; Spanu, P.; Fava, G.C.; Belicchi, M.F. *Tetrahedron* **1990**, *46*, 5807–5824.

286. Casiraghi, G.; Colombo, L.; Rassu, G.; Spanu, P. *J. Org. Chem.* **1991**, *56*, 6523–6527.

287. Wagner, R.; Tilley, J.W. *J. Org. Chem.* **1990**, *55*, 6289–6291.

288. Mead, K.; MacDonald, T.L. *J. Org. Chem.* **1985**, *50*, 422–424.

289. Herold, P. *Helv. Chim. Acta* **1988**, *71*, 354–362.

290. Coleman, R.S.; Carpenter, A.J. *Tetrahedron Lett.* **1992**, *33*, 1697–1700.

291. Reginato, G.; Mordini, A.; Meffre, P.; Tenti, A.; Valacchi, M.; Cariou, K. *Tetrahedron: Asymmetry* **2006**, *17*, 922–926.

292. Koskinen, A.M.P.; Hassila, H.; Myllymäki, V.T.; Rissanen, K. *Tetrahedron Lett.* **1995**, *36*, 5619–5622.

293. East, S.P.; Joullié, M.M. *Tetrahedron Lett.* **1998**, *39*, 7211–7214.

294. Beaulieu, P.L. *Tetrahedron Lett.* **1991**, *32*, 1031–1034.

295. Husain, A.; Ganem, B. *Tetrahedron Lett.* **2002**, *43*, 8621–8623.

296. Delas, C.; Okamoto, S.; Sato, F. *Tetrahedron Lett.* **2002**, *43*, 4373–4375.

297. Casiraghi, G.; Cornia, M.; Rassu, G. *J. Org. Chem.* **1988**, *53*, 4919–4922.

298. Gruza, H.; Kiciak, K.; Krasinski, A.; Jurczak, J. *Tetrahedron: Asymmetry* **1997**, *8*, 2627–2631.

299. D'Aniello, F.; Mann, A.; Taddei, M.; Wermuth, C.-G. *Tetrahedron Lett.* **1994**, 7775–7778.

300. D'Aniello, F.; Mann, A.; Schoenfelder, A.; Taddei, M. *Tetrahedron* **1997**, *53*, 1447–1456.

301. Muller, M.; Mann, A.; Taddei, M. *Tetrahedron Lett.* **1993**, *34*, 3289–3290.

302. Kumar, J.S.R.; Datta, A. *Tetrahedron Lett.* **1997**, *38*, 473–476.

303. Okamoto, N.; Hara, O.; Makino, K.; Hamada, Y. *J. Org. Chem.* **2002**, *67*, 9210–9215.

304. Yamamoto, Y.; Asao, N. *Chem. Rev.* **1993**, *93*, 2207–2293.

305. Kiyooka, S.-I.; Nakano, M.; Shiota, F.; Fujiyama, R.; *J. Org. Chem.* **1989**, *54*, 5409–5411.

306. Hafner, A.; Duthaler, R.O.; Marti, R.; Rihs, G.; Rothe-Streit, P.; Schwarzenbach, F. *J. Am. Chem. Soc.* **1992**, *114*, 2321–2336.

307. Marshall, J.A.; Seletsky, B.M.; Coan, P.S. *J. Org. Chem.* **1994**, *59*, 5139–5140.

308. Jirgensons, A.; Marinozzi, M.; Pellicciari, R. *Tetrahedron* **2005**, *61*, 373–377.

309. Niel, G.; Roux, F.; Maisonnasse, Y.; Maugras, I.; Poncet, J.; Jouin, P. *J. Chem. Soc., Perkin Trans. 1* **1994**, 1275–1280.

310. Barrett, A.G.M.; Edmunds, J.J.; Hendrix, J.A.; Malecha, J.W.; Parkinson, C.J. *J. Chem. Soc., Chem. Commun.* **1992**, 1240–1242.

311. D'Aniello, F.; Falorni, M.; Mann, A.; Taddei, M. *Tetrahedron: Asymmetry* **1996**, *7*, 1217–1226.

312. Lombardo, M.; Gianotti, K.; Licciulli, S.; Trombini, C. *Tetrahedron* **2004**, *60*, 11725–11732.

313. Lombardo, M.; Licciulli, S.; Trombini, C. *Tetrahedron Lett.* **2003**, *44*, 9147–9149.

314. Ma, D.; Yang, J. *J. Am. Chem. Soc.* **2001**, *123*, 9706–9707.

315. Viacrio, J.L.; Rodriguez, M.; Badía, D.; Carrillo, L.; Reyes, E. *Org. Lett.* **2004**, *6*, 3171–3174.

316. Meffre, P.; Dave, R.H.; Leroy, J.; Badet, B. *Tetrahedron Lett.* **2001**, *42*, 8625–8627.

317. Jackson, M.D.; Gould, S.J.; Zabriskie, T.M. *J. Org. Chem.* **2002**, *67*, 2934–2941.

318. Dondoni, A.; Mariotti, G.; Marra, A. *J. Org. Chem.* **2002**, *67*, 4475–4486.

319. VanBrunt, M.P.; Standaert, R.F. *Org. Lett.* **2000**, *2*, 705–708.

320. Cao, B.; Park, H.; Joullié, M.M. *J. Am. Chem. Soc.* **2002**, *124*, 520–521.

321. Hanessian, S.; Bayrakdarian, M.; Luo, X. *J. Am. Chem. Soc.* **2002**, *124*, 4716–4721.

322. Moriwake, T.; Hamano, S.-I.; Saito, S.; Torii, S. *Chem. Lett.* **1987**, 2085–2088

323. Takahata, H.; Banba, Y.; Ouchi, H.; Nemoto, H.; Kato, A.; Adachi, I. *J. Org. Chem.* **2003**, *68*, 3603–3607.

324. Campbell, A.D.; Raynham, T.M.; Taylor, R.J.K. *Tetrahedron Lett.* **1999**, *40*, 5263–5266.

325. Collier, P.N.; Patel, I.; Taylor, R.J.K. *Tetrahedron Lett.* **2001**, *42*, 5953–5954.

326. Hashimoto, M.; Hashimoto, K.; Shirahama, H. *Tetrahedron* **1996**, *52*, 1931–1942.

327. Still, W.C.; Gennari, C. *Tetrahedron Lett.* **1983**, *24*, 4405–4408.

328. Shimamoto, K.; Ohfune, Y. *Tetrahedron Lett.* **1990**, *31*, 4049–4052.

329. Sakai, N.; Ohfune, Y. *Tetrahedron Lett.* **1990**, *31*, 4151–4154.

330. Koskinen, A.M.P.; Chen, J. *Tetrahedron Lett.* **1991**, *32*, 6977–6980.

331. Jako, I.; Uiber, P.; Mann, A.; Taddei, M.; Wermuth, C-G. *Tetrahedron Lett.* **1990**, *31*, 1011–1014.

332. Koskinen, A. *Pure Appl. Chem.* **1993**, *65*, 1465–1470.

333. Priepke, H.; Brückner, R.; Harms, K. *Chem. Ber.* **1990**, *123*, 555–563.

334. Wermuth, C.G.; Mann, A.; Schoenfelder, A.; Wright, R.A.; Johnson, B.G.; Burnett, J.P.; Mayne, N.G.; Schoepp, D.D. *J. Med. Chem.* **1996**, *39*, 814–816.

335. Jako, I.; Uiber, P.; Mann, A.; Wermuth, C.-G.; Boulanger, T.; Norberg, B.; Evrard, G.; Durant, F. *J. Org. Chem.* **1991**, *56*, 5729–5733.

336. Mohapatra, D.K. *J. Chem. Soc., Perkin Trans. 1* **2001**, 1851–1852.

337. Nevalainen, M.; Kauppinen, P.M.; Koskinen, A.M.P. *J. Org. Chem.* **2001**, *66*, 2061–2066.

338. Hanessian, S.; Sumi, K. *Synthesis* **1991**, 1083–1089.

339. Hanessian, S.; Wang, W.; Gai, Y. *Tetrahedron Lett.* **1996**, *37*, 7477–7480.

340. Barco, A.; Benetti, S.; Pollini, G.P.; Spalluto, G.; Zanirato, V. *J. Chem. Soc., Chem. Commun.* **1991**, 390–391.

341. Sakai, N.; Ohfune, Y. *J. Am. Chem. Soc.* **1992**, *114*, 998–1010.

342. Shimamoto, K.; Ohfune, Y. *J. Med. Chem.* **1996**, *39*, 407–423.

343. Ohfune, Y.; Shimamoto, K.; Ishida, M.; Shinozaki, H. *Bioorg. Med. Chem. Lett.* **1993**, *3*, 15–18.

344. Shimamoto, K.; Shigeri, Y.; Nakajima, T.; Yumoto, N.; Yoshikawa, S.; Ohfune, Y. *Bioorg. Med. Chem. Lett.* **1996**, *6*, 2381–2386.

345. Alcaraz, C.; Bernabé, M. *Tetrahedron: Asymmetry* **1994**, *5*, 1221–1224.

346. Beaulieu, P.L.; Duceppe, J.-S.; Johnson, C. *J. Org. Chem.* **1991**, *56*, 4196–4204.

347. Oh, J.S.; Kim, B.H.; Kim, Y.G. *Tetrahedron Lett.* **2004**, *45*, 3925–3928.

348. Babu, I.R.; Hamill, E.K.; Kumar, K. *J. Org. Chem.* **2004**, *69*, 5468–5470.

349. de Frutos, P.; Fernández, D.; Fernández-Alvarez, E.; Bernabé, M. *Tetrahedron* **1992**, *48*, 1123–1130.

350. de Frutos, P.; Fernández, D.; Fernández-Alvarez, E.; Bernabé, M. *Tetrahedron Lett.* **1991**, *32*, 541–542.

351. Dondoni, A.; Marra, A.; Massi, A. *Tetrahedron*, **1998**, *54*, 2827–2832.

352. Lieberknecht, A.; Griesser, H.; Krämer, B.; Bravo, R.D.; Colinas, P.A.; Grigera, J. *Tetrahedron* **1999**, *55*, 6475–6482.

353. Bradley, D.M.; Mapitse, R.; Thomson, N.M.; Hayes, C.J. *J. Org. Chem.* **2002**, *67*, 7613–7617.

354. Jähnisch, K.; Tittelbach, F.; Gründemann, E.; Schneider, M. *Eur. J. Org. Chem.* **2000**, 3957–3960.

355. Qiu, X.-l.; Qing, F.-l. *J. Chem. Soc., Perkin Trans. 1* **2002**, 2052–2057.

356. Avenoza, A.; Cativiela, C.; Peregrina, J.M.; Zurbano, M.M. *Tetrahedron: Asymmetry* **1996**, *7*, 1555–1558.

357. Avenoza, A.; Cativiela, C.; Peregrina, J.M.; Zurbano, M.M. *Tetrahedron: Asymmetry* **1997**, *8*, 863–871.

358. Jurgens, A.R. *Tetrahedron Lett.* **1992**, *33*, 4727–4730.

359. Reginato, G.; Mordini, A.; Degl'Innocenti, A.; Caracciolo, M. *Tetrahedron Lett.* **1995**, *36*, 8271–8274.

360. Chung, J.Y.L.; Wasicak, J.T. *Tetrahedron Lett.* **1990**, *31*, 3957–3960.

361. Meffre, P.; Gauzy, L.; Branquet, E.; Durand, P.; Le Goffic, F. *Tetrahedron* **1996**, *52*, 11215–11238.

362. Meffre, P.; Gauzy, L.; Perdigues, C.; Desanges-Levecque, F.; Branquet, E.; Durand, P.; Le Goffic, F. *Tetrahedron Lett.* **1995**, *36*, 877–880.

363. Pietruszka, J.; Witt, A.; Frey, W. *Eur. J. Org. Chem.* **2003**, 3219–3229.

364. Meffre, P.; Hermann, S. Durand, P.; Reginato, G.; Riu, A. *Tetrahedron* **2002**, *58*, 5159–5162.

365. Reginato, G.; Meffre, P.; Gaggini, F. *Amino Acids* **2005**, *29*, 81–87.

366. Reginato, G.; Mordini, A.; Valacchi, M. *Tetrahedron Lett.* **1998**, *39*, 9545–9548.

367. Reginato, G.; Mordini, A.; Valacchi, M.; Grandini, E. *J. Org. Chem.* **1999**, *64*, 9211–9216.

368. Falorni, M.; Giacomelli, G.; Spanu, E. *Tetrahedron Lett.* **1998**, *39*, 9241–9244.

369. De Luca, L.; Giacomelli, G.; Riu, A. *J. Org. Chem.* **2001**, *66*, 6823–6825.

370. Pulley, S.R.; Czakó, B.; Brown, G.D. *Tetrahedron Lett.* **2005**, *46*, 9039–9042.

371. Govek, S.P.; Overman, L.E. *J. Am. Chem. Soc.* **2001**, *123*, 9468–9469.

372. Kahne, D.; Yang, D.; Lee, M.D. *Tetrahedron Lett.* **1990**, *31*, 21–22.

373. Palomo, C.; Cabré, F.; Ontoria, J.M. *Tetrahedron Lett.* **1992**, *33*, 4819–4822.

374. Merino, P.; Lanaspa, A.; Merchan, F.L.; Tejero, T. *Tetrahedron Lett.* **1997**, *38*, 1813–1816.

375. Merino, P.; Lanaspa, A.; Merchan, F.L.; Tejero, T. *Tetrahedron: Asymmetry* **1998**, *9*, 629–646.

376. Guibourdenche, C.; Roumestant, M.L.; Viallefont, Ph. *Tetrahedron: Asymmetry* **1993**, *4*, 2041–2046.

377. Chhabra, S.R.; Mahajan, A.; Chan, W.C. *Tetrahedron Lett.* **1999**, *40*, 4905–4908.

378. Chhabra, S.R.; Mahajan, A.; Chan, W.C. *J. Org. Chem.* **2002**, *67*, 4017–4029.

379. Denis, J.-N.; Tchertchian, S.; Vallée, Y. *Synth. Commun.* **1997**, *27*, 2345–2350.

380. Marcus, J.; Vandermeulen, G.W.M.; Brussee, J.; van der Gen, A. *Tetrahedron: Asymmetry* **1999**, *10*, 1617–1622.

381. Dondoni, A.; Massi, A.; Minghini, E.; Sabbatini, S.; Bertolasi, V. *J. Org. Chem.* **2003**, *68*, 6172–6183.

382. Koskinen, A.M.P.; Krische, M.J. *Synlett* **1990**, 665–666.

383. Koskinen, A.M.P.; Koskinen, P.M. *Synlett* **1993**, 501–502.

384. Koskinen, A.M.P.; Koskinen, P.M. *Tetrahedron Lett.* **1993**, *34*, 6765–6768.

385. Avenoza, A.; Cativiela, C.; Corzana, F.; Peregrina, J.M.; Zurbano, M.M. *J. Org. Chem.* **1999**, *64*, 8220–8225.

386. Knudsen, C.G.; Rapoport, H. *J. Org. Chem.* **1983**, *48*, 2260–2266.

387. Buckley T.F., III; Rapoport, H. *J. Am. Chem. Soc.* **1981**, *103*, 6157–6163.

388. Maurer, P.J.; Takahata, H.; Rapoport, H. *J. Am. Chem. Soc.* **1984**, *106*, 1095–1098.

389. Roemmele, R.C.; Rapoport, H. *J. Org. Chem.* **1989**, *54*, 1866–1875.

390. Roemmele, R.C.; Rapoport, H. *J. Org. Chem.* **1988**, *53*, 2367–2371.

391. Sun, G.; Slavica, M.; Uretsky, N.J.; Wallace, L.J.; Shams, G.; Weinstein, D.M.; Miller, J.C.; Miller, D.D. *J. Med. Chem.* **1998**, *41*, 1034–1041.

392. Lubell, W.D.; Rapoport, H. *J. Am. Chem. Soc.* **1988**, *110*, 7447–7455.

393. Lubell, W.D.; Rapoport, H. *J. Am. Chem. Soc.* **1987**, *109*, 236–239.

394. Lubell, W.D.; Jamison, T.F.; Rapoport, H. *J. Org. Chem.* **1990**, *55*, 3511–3522.

395. Newman, H. *J. Am. Chem. Soc.* **1973**, *95*, 4098–4099.

396. Saitoh, Y.; Moriyama, Y.; Takahashi, T. *Tetrahedron Lett.* **1980**, *21*, 75–78.

397. Mori, K.; Funaki, Y. *Tetrahedron* **1985**, *41*, 2379–2386.

398. Tkaczuk, P.; Thornton, E.R. *J. Org. Chem.* **1981**, *46*, 4393–4398.

399. Angrick, M. Montash. *Chem.* **1985**, *116*, 645–649.

400. Dell'Uomo, N.; Di Giovanni, M.D.; Misiti, D.; Zappia, G.; Delle Monache, G. *Liebigs Ann. Chem.* **1994**, 641–644.

401. Di Giovanni, M.C.; Misiti, D.; Zappia, G.; Monache, G.D. *Tetrahedron*, **1993**, *34*, 11321–11328.

402. Ibuka, T.; Suzuki, K.; Habashita, H.; Otaka, A.; Tamamura, H.; Mimura, N.; Miwa, Y.; Taga, T.; Fuji, N. *J. Chem. Soc., Chem. Commun.* **1994**, 2151–2152.

403. Stanfield, C.F.; Parker, J.E.; Kanellis, P. *J. Org. Chem.* **1981**, *46*, 4797–4798.

404. Stanfield, C.F.; Parker, J.E.; Kanellis, P. *J. Org. Chem.* **1981**, *46*, 4799–4800.

405. Monache, G.D.; Di Giovanni, M.C.; Maggio, F.; Misiti, D.; Zappia, G. *Synthesis* **1995**, 1155–1158.

406. Reetz, M.T. *Angew. Chem., Int. Ed. Engl.* **1991**, *30*, 1531–1546

407. Laïb, T.; Chastanet, J.; Zhu, J. *Tetrahedron Lett.* **1997**, *38*, 1771–1772.

408. Laïb, T.; Chastanet, J.; Zhu, J. *J. Org. Chem.* **1998**, *63*, 1709–1713.

409. Andrés, J.M.; de Elena, N.; Pedrosa, R. *Tetrahedron* **2000**, *56*, 1523–1531.

410. Andrés, J.M.; Barrio, R.; Martínez, M.A.; Pedrosa, R.; Pérez-Encabo, A. *J. Org. Chem.* **1996**, *61*, 4210–4213.

411. Jourdant, A.; Zhu, J. *Tetrahedron Lett.* **2000**, *41*, 7033–7036.

412. Jurczak, J.; Prokopowicz, P. *Tetrahedron Lett.* **1998**, *39*, 9835–9838.

413. Palian, M.M.; Polt, R. *J. Org. Chem.* **2001**, *66*, 7178–7183.

414. Golebiowski, A.; Jurczak, J. *J. Chem. Soc., Chem. Commun.* **1989**, 263–264.

415. Golebiowski, A.; Kozak, J.; Jurczak, J. *J. Org. Chem.* **1991**, *56*, 7344–7347.

416. Hanessian, S.; Ninkovic, S.; Reinhold, U. *Tetrahedron Lett.* **1996**, *37*, 8971–8974.

417. Di Giovanni, M.C.; Misiti, D.; Villani, C.; Zappia, G. *Tetrahedron: Asymmetry* **1996**, *7*, 2277–2286.

418. Dell'Uomo, N.; Di Giovanni, M.C.; Misiti, D.; Zappia, G.; Delle Monache, G. *Tetrahedron: Asymmetry* **1996**, *7*, 181–188.

419. delle Monache, G.; Di Giovanni, M.C.; Misiti, D.; Zappia, G. *Tetrahedron: Asymmetry* **1997**, *8*, 231–243.

420. Yoo, D.; Seok, J.; Lee, D.-W.; Kim, Y.G. *J. Org. Chem.* **2003**, *68*, 2979–2982.

421. Hanessian, S.; Sahoo, S.P. *Can. J. Chem.* **1984**, *62*, 1400–1402.

422. Giannis, A.; Henk, T. *Tetrahedron Lett.* **1990**, *31*, 1253–1256.

423. Cessac, J.W.; Rao, P.N.; Kim, H.K. *Amino Acids* **1994**, *6*, 97–105.

424. Ohfune, Y.; Kurokawa, N. *Tetrahedron Lett.* **1984**, *25*, 1071–1074.

425. Erlenmeyer, E.; Stoop, F. *Ann.* **1904**, *337*, 236–263.

426. Bycroft, B.W.; Chhabra, S.R.; Grout, R.J.; Crowley, P.J. *J. Chem. Soc., Chem. Commun.* **1984**, 1156–1157.

427. Sheehan, J.C.; Johnson, D.A. *J. Am. Chem. Soc.* **1954**, *76*, 158–160.

428. Broxterman, H.J.G.; Liskamp, R.M.J. *Recl. Trav. Chim. Pays-Bas* **1991**, *110*, 46–52.

429. Uhlhorn-Dierks, G.; Kolter, T.; Sandhoff, K. *Angew. Chem., Int. Ed. Engl.* **1998**, *37*, 2453–2455.

430. Dierks, T.; Schmidt, B.; von Figura, K. *Proc. Natl. Acad. Sci. U.S.A.* **1997**, *94*, 11963–11968.

431. Mertes, M.P.; Ramsey, A.A. *J. Med. Chem.* **1969**, *12*, 342–343.

432. Lubell, W.; Rapoport, H. *J. Org. Chem.* **1989**, *54*, 3824–3831.

433. DiMaio, J.; Belleau, B. *J. Chem. Soc., Perkin Trans 1* **1989**, 1687–1689.

434. Doyle, T.W.; Belleau, B.; Luh, B.-Y.; Ferrari, C.F.; Cunningham, M.P. *Can. J. Chem.* **1977**, *55*, 468–483.

435. DeMong, D.E.; Williams, R.M. *J. Am. Chem. Soc.* **2003**, *125*, 8561–8565.

436. Nakazawa, T.; Suzuki, T.; Ishii, M. *Tetrahedron Lett.* **1997**, *38*, 8951–8954.

437. Groundwater, P.W.; Sharif, T.; Arany, A.; Hibbs, D.E.; Hursthouse, M.B.; Nyerges, M. *Tetrahedron Lett.* **1998**, *39*, 1433–1436.

438. Ding, Y.; Wng, J.; Abboud, K.A.; Xu, Y.; Dolbier W.R., Jr.; Richards, N.G.J. *J. Org. Chem.* **2001**, *66*, 6381–6388.

439. Blaskovich, M. A.; Lajoie, G. *Peptides: Chemistry and Biology. Proceedings of the 12th American Peptide Symposium*; Smith, J.A; Rivier, J.E., Eds; Escom Publishers: the Netherlands, 1992; pp 515–516.

440. Blaskovich, M.A. Ph.D. thesis, University of Waterloo, Waterloo, Ontario, Canada, 1993.

441. Blaskovich, M.A.; Evindar, G.; Rose, N.G.W.; Wilkinson, S.; Luo, Y.; Lajoie, G.A. *J. Org. Chem.* **1998**, *63*, 3631–3646; *63*, 4560.

442. Rose, N.G.W; Blaskovich, M.A.; Evindar, G.; Wilkinson, S.; Luo, Y.; Reid, C.; Lajoie, G.A. *Org. Synth.* **2002**, *79*, 216–227.

443. Hansen, D.B.; Wan, X.; Carroll, P.J.; Joullié, M.M. *J. Org. Chem.* **2005**, *70*, 3120–3126.

444. Hansen, D.B.; Joullié, M.M. *Tetrahedron: Asymmetry* **2005**, *16*, 3963–3969.

445. Blaskovich, M.A.; Lajoie, G.L. *Tetrahedron Lett.* **1993**, *34*, 3837–3840.

446. Blaskovich, M.A.; Lajoie, G.A. *Peptides: Chemistry and Biology. Proceedings of the 13th American Peptide Symposium*; Hodges, R.A; Smith, J.A., Eds; Escom Publishers: the Netherlands, 1994; pp 167–169.

447. Rose, N.G.W.; Blaskovich, M.A.; Wong, A.; Lajoie, G.A. *Tetrahedron* **2001**, *57*, 1497–1507.

448. Rifé, J.; Ortuño, M.; Lajoie, G.A. *J. Org. Chem.* **1999**, *64*, 8958–8961.

449. Fishlock, D.; Guillemette, J.G.; Lajoie, G.A. *J. Org. Chem.* **2002**, *67*, 2352–2354.

450. Fishlock, D.; Perdicakis, B.; Montgomery, H.J.; Guillemette, J.G.; Jervis, E.; Lajoie, G.A. *Bioorg. Med. Chem.* **2003**, *11*, 869–873.

451. Cameron, S.; Khambay, B.P.S. *Tetrahedron Lett.* **1998**, *39*, 1987–1990.

452. Herdeis, C.; Kelm, B. *Tetrahedron* **2003**, *59*, 217–229.

453. Oba, M.; Nishiyama, N.; Nishiyama, K. *Chem. Commun.* **2003**, 776–777.

454. Giner, J.-L. *Org. Lett.* **2005**, *7*, 499–501.

455. Oba, M.; Nishiyama, N.; Nishiyama, K. *Tetrahedron* **2004**, *60*, 8456–8464.

456. Apitz, G.; Jäger, M.; Jaroch, S.; Kratzel, M.; Schäffeler, L.; Steglich, W. *Tetrahedron* **1993**, *49*, 8223–8232.

457. Bogenstätter, M.; Steglich, W. *Tetrahedron* **1997**, *53*, 7267–7274.

458. Blakskjaer, P.; Gavrila, A.; Andersen, L.; Skrydstrup, T. *Tetrahedron Lett.* **2004**, *45*, 9091–9094.

459. Boto, A.; Hernández, R.; Montoya, A.; Suárez, E. *Tetrahedron Lett.* **2002**, *43*, 8269–8272.

460. Dugave, C.; Kessler, P.; Colas, C.; Hirth, C. *Tetrahedron Lett.* **1994**, *35*, 9199–9202.

461. Garner, P.; Yoo, J.U. *Tetrahedron Lett.* **1993**, *34*, 1275–1278.

462. Huang, Y.; Dey, S.; Zhang, X.; Sönnichsen, F.; Garner, P. *J. Am. Chem. Soc.* **2004**, *126*, 4626–4640.

463. Brunner, M.; Saarenketo, P.; Straub, T.; Rissanen, K.; Koskinen, A.M.P. *Eur. J. Org. Chem.* **2004**, 3879–3883.

464. Heffner, R.J.; Joullié, M.M. *Tetrahedron Lett.* **1989**, *30*, 7021–702.

465. Jiang, J.; Li, W.-R.; Przeslawski, R.M.; Joullié, M.M. *Tetrahedron Lett.* **1993**, *34*, 6705–6708.

466. Sypniewski, M.; Penke, B.; Simon, L.; Rivier, J. *J. Org. Chem.* **2000**, *65*, 6595–6600.

467. Valentekovich, R.J.; Schreiber, S.L. *J. Am. Chem. Soc.* **1995**, *117*, 9069–9070.

468. Fischer, E.; Raske, K. *Ber.* **1907**, *40*, 3717–3724.

469. Buters, J.; Rini, J.; Helquist, P. *J. Org. Chem.* **1985**, *50*, 3676–3678.

470. Nishimura, H.; Mizuguchi, A.; Mizutani, J. *Tetrahedron Lett.* **1975**, *37*, 3201–3202.

471. Roy, J.; Gordon, W.; Schwartz, I.L.; Walter, R. *J. Org. Chem.* **1970**, *35*, 510–513.

472. Erlenmeyer, E.; Stoop, F. *Ann.* **1904**, *337*, 236–263.

473. González-Roura, A.; Navarro, I.; Delgado, A.; Llebaria, A.; Casas, J. *Angew. Chem., Int. Ed.* **2004**, *43*, 862–865.

474. Nimkar, S.; Menaldino, D.; Merrill, A.H.; Liotta, D. *Tetrahedron Lett.* **1988**, 29, 3037–3040.

475. Otaka, A.; Miyoshi, K.; Burke T.R., Jr.; Roller, P.P.; Kubota, H.; Tamamura, H.; Fujii, N. *Tetrahedron Lett.* **1995**, *36*, 927–930.

476. Aoyagi, Y.; Inaba, H.; Hiraiwa, Y.; Kuroda, A.; Ohta, A. *J. Chem. Soc., Perkin Trans. 1* **1998**, 3975–3978.

477. Carrillo-Marquez, T.; Caggiano, L.; Jackson, R.F.W.; Grabowska, U.; Rae, A.; Tozer, M.J. *Org. Biomol. Chem.* **2005**, *3*, 4117–4123.

478. Atfani, M.; Wei, L.; Lubell, W.D. *Org. Lett.* **2001**, *3*, 2965–2968.

479. Smyth, M.S.; Burke, T.R., Jr. *Org. Proc. Prep. Int.* **1996**, *28*, 77–81.

480. Ye, B.; Burke T.R., Jr. *J. Org. Chem.* **1995**, *60*, 2640–2641.

481. Masaki, H.; Maeyama, J.; Kamada, K.; Esumi, T.; Iwabuchi, Y.; Hatakeyama, S. *J. Am. Chem. Soc.* **2000**, *122*, 5216–5217.

482. Crisp, G.T.; Jiang, Y.-L.; Pullman, P.J.; De Savi, C. *Tetrahedron* **1997**, *53*, 17489–17500.

483. Andrés, J.M.; Muñoz, E.M.; Pedrosa, R.; Pérez-Encabo, A. *Eur. J. Org. Chem.* **2003**, 3387–3397.

484. Dondoni, A.; Perrone, D. *Synthesis* **1997**, 527–529.

11 | Synthesis of β-, γ-, δ-, and ω-Amino Acids

11.1 Introduction

Most efforts towards amino acid synthesis have focused on α-amino acids, but in recent years there has been an intensified interest in the backbone homologated amino acids, especially β-and γ-amino acids. This chapter will provide an overview of the major methods of synthesis of these compounds, along with summaries of some of their biological properties. A monograph by M.B. Smith describing methods of non-α-amino acid synthesis (primarily acyclic C_3 to C_{10} amino acids) was published in 1995 (1) while a 1997 monograph edited by E. Juaristi reviewed enantioselective synthesis of β-amino acids, with a second edition in 2005 (2). Other reviews are discussed in the appropriate sections.

11.2 β-Amino Acids

11.2.1 Introduction

11.2.1a Nomenclature

β-Amino acids have attracted increasing attention in recent years, which can be attributed to a number of factors. β-Amino acids are found in a number of pharmacologically important molecules, including the side chain of the antitumor compound paclitaxel (Taxol, 3 and references therein), creating a strong incentive for the development of new synthetic methods. Ritalin (*threo*-methylphenidate, methyl piperidine-2-phenylacetate), the most commonly prescribed psychotropic medication for children in the United States, has a cyclic β-amino acid skeleton (4, 5). β-amino acids are also of interest as they are synthetic precursors of the medicinally important β-lactam class of compounds (6, 7). Research into

amino acid analogs for incorporation into peptides and proteins has found that the properties introduced by β-amino acids can be useful. β-Amino acids were used in structure–activity relationship (SAR) studies of peptides as early as 1966, when bioactive bradykinin analogs containing β-alanine (β-Ala) were prepared (8, 9); corticotropin analogs with β-Ala were reported in 1970 (10, 11). In recent years peptides constructed entirely of β-amino acids have been found to form unique structures (12–14), although β-peptides and their structures have been known for some time (15, 16).

A nomenclature for substituted β-amino acids has been developed, with 2- or 3-substituted β-alanines, or 2,3-, 2,2- or 3,3-disubstituted β-alanines referred to as β^2-amino acids, β^3-amino acids, $\beta^{2,3}$-amino acids, $\beta^{2,2}$-amino acids, and $\beta^{3,3}$-amino acids, respectively (see Scheme 11.1). Homologated α-amino acids, Xaa, with a methylene inserted between the α-carbon chiral center and the carboxyl group, become β^3-HXaa. An older nomenclature exists in which β-amino acids are named by attaching a β-prefix to the corresponding α-amino acid in which the amino group has been moved from the α-position to the β-position (i.e., 3-aminopropionic acid becomes, β-Ala). Thus, β-lysine (older nomenclature) is equivalent to β^3-HOrn.

11.2.1b β-Amino Acids in Natural Products

Four β-amino acids are found naturally in mammals (see Scheme 11.2) (17). The most common, β-Ala, is a catabolite of uracil (18–20) or results from metabolism of β-hydroxypropionate (21), while (R)-β-aminoisobutyric acid (2-methyl-3-aminopropionic acid, β-Aib) is a catabolite of thymine (19, 22). (S)-β-Aib is derived by transamination of a catabolite of Val (21), and β-Leu is

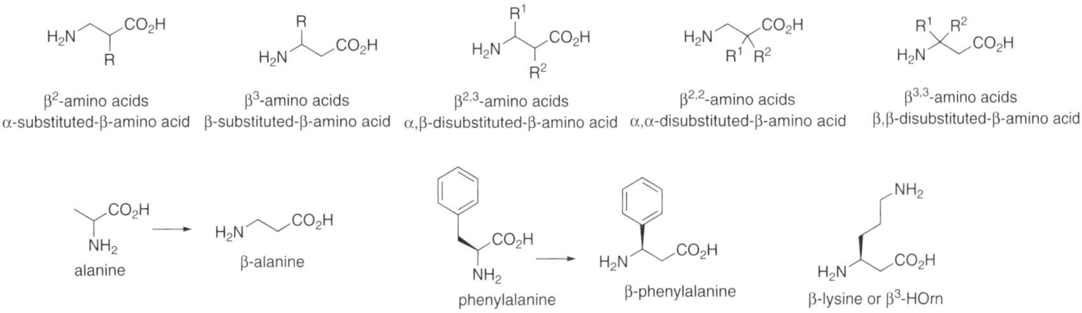

Scheme 11.1 Nomenclature of β-Amino Acids.

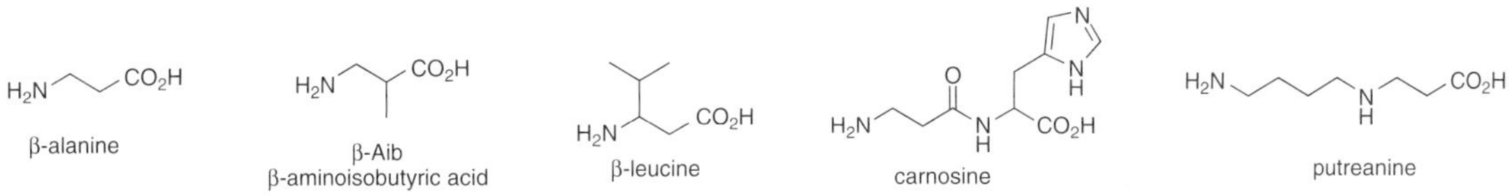

Scheme 11.2 Mammalian β-Amino Acids.

probably a precursor of Leu (17). β-Amino acid metabolism was the subject of a 1986 review (17). A number of potential neurotransmitter/neuromodulator effects of β-Ala have been observed. An orphan G-protein-coupled receptor, TGR7 (corresponding to MrgD, a sensory neuron-specific GPCR), was identified in 2004 as specifically responding to β-Ala (23). Carnosine, first isolated from extracts of muscle tissue in 1900 (24), is a β-Ala-His dipeptide (25) and represents a large endogenous store of β-Ala. Its structure was identified by synthesis in 1918 (26). Carnosine represents an important molecular marker for the presence of undesired processed animal proteins in animal feed (banned due to the potential for transfer of spongiform encephalopathies), and an anion-exchange chromatography method has been developed for its determination in feed and meat (27). β-Ala is also contained in putreanine, N-(4-aminobutyl)-β-Ala, which has been isolated from mammalian brains (with 475 mg obtained from 82 kg of bovine brain) (28). Putreanine was also detected in human brains (29) and human urine (30). β-Ala is one of the main amino acid components found in the Orgueil, Ivuna, Murray, and Murchison meteorites, and is proposed to be synthesized via an HCN polymerization reaction (31). 3,3'-Diaminoisobutanoic acid has also been identified in the Murchison meteorite (32).

β-Aib has been isolated from human urine (33) and from the urine of rats fed thymine (34). D-(−)-β- Aib was isolated from human livers (17.5 mg from 2968 g of liver) (35). The presence of elevated levels of D-β-Aib in urine is a potential diagnostic marker for cancer (17). Excretors of D-β-Aib catabolize L-β-Aib as effectively as non-excretors, suggesting different enzymes are responsible for metabolism of each enantiomer (36).

There are many other examples of naturally occurring β-amino acids (see Scheme 11.3). The natural products from lithistid sponges, which contain many β-amino acids of varying complexity, were reviewed in 1998 (37). Two β-Ala residues are contained in Barangamide A, a cyclic undecapeptide isolated from the sponge Theonella swinhoei (38), while two theonellapeptolide-related cyclic depsipeptides from an Okinawan marine sponge contain three β-Ala residues (39). A survey of the amino acids contained in 18 red algae identified traces of β-Ala in two of them (40). Flavobacterium biosynthesizes β-Ala via L-aspartic-4-carboxylase-catalyzed decarboxylation of L-Asp to give L-Ala, followed by conversion to β-Ala by β-alanine-pyruvic aminotransferase (41). β-Ala is contained in a tripeptide homolog of glutathione isolated from seedlings of Phaseolus aureus (42, 43), while a dipeptide (ophidin) isolated from snake muscles consists of β-Ala and 2'-methylhistidine (44–46). Adenopeptin, a tridecapeptide isolated from Chrysosporium sp. fungus, is an apoptosis inducer in transformed cells and therefore has potential as an anticancer agent. It contains a β-Ala residue (47). The nitro analog of β-Ala, 3-nitropropionic acid (3-NPA), has been found in extracts of several strains of fungi, and shows potent antimycobacterial activity (48).

β-Aib has been isolated in significant quantities from iris bulbs, Iris tingitana, with the major fraction bound in a dipeptide, γ-L-glutamyl-β-Aib (49). γ-L-Glutamyl-β-Ala was also isolated from iris bulbs (50). Similarly, unhydrolyzed seed extracts of the ornamental crucifer Lunaria annua yielded γ-L-glutamyl-β-Ala and γ-L-glutamyl-β-Aib (51). Crude enzyme extracts from fresh leaves of 2-week-old pea plants catalyzed the

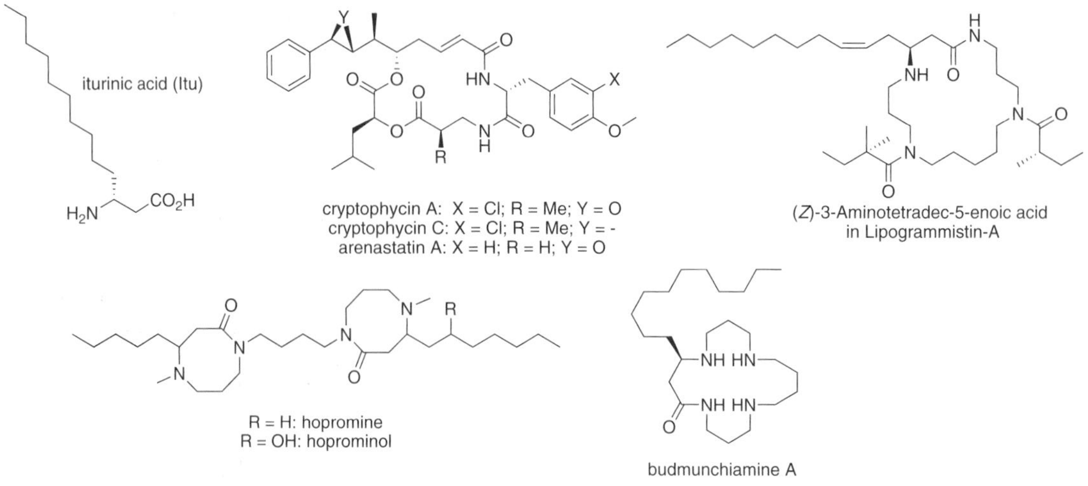

iturinic acid (Itu)

H₂N ... CO₂H

cryptophycin A: X = Cl; R = Me; Y = O
cryptophycin C: X = Cl; R = Me; Y = -
arenastatin A: X = H; R = H; Y = O

(Z)-3-Aminotetradec-5-enoic acid
in Lipogrammistin-A

R = H: hopromine
R = OH: hoprominol

budmunchiamine A

Scheme 11.3 β-Amino Acids from Natural Sources.

conversion of 4,5-dihydrouracil into *N*-carbamoyl-β-Ala and of 4,5-dihydrothymine into *N*-carbamoyl-β-Aib, respectively (*52*). 2-Methyl- and 2,2-dimethyl-3-amino-butyric acid are components of the antitumor depsipeptides cryptophycins, derived from terrestrial blue-green algae *Nostoc* sp. (*53–61*). Over 25 naturally occurring crypto-phycins have been identified (*61*). Arenastatin A, corresponding to cryptophycin-24, was also found in the Okinawa marine sponge *Dysidea arenaria* (1 mg from 6.5 kg of fresh sponge), and contains a β-Ala residue instead of the 2-methyl analogs (*60, 62*). Total syntheses of two diastereomers of cryptomycin A and four diastereomers of cryptomycin C were reported in 1996 (*63*), with syntheses of cryptophycins-1, -3, -4, -24, and -29 in 1999 (*835*), cryptophycin B in 2004 (*65*), cryptophycins-1 (*66*), -52 (*66*), and -3 (*67*) in 2005, and cryptophycin-24 in 1999 (*835*), 2000 (*68*), and 2004 (*65, 69*). A chemoenzy-matic synthesis of cryptophycin/arenastatin natural products was described in 2005, using a thioesterase to form the depsipeptide ring (*70*). Analogs have also been synthesized (*57–59, 69, 71*), including cryptophycin-1 with the depsipeptide linkage replaced by an amide (*72*). In 2002 the synthetic analog cryptophycin-52 was in phase II clinical trials for advanced non-small cell lung cancer (*71, 2187*). A review of the synthesis of the crypto-phycins appeared in 2002 (*73*).

Majusculamide C is a cyclic depsipeptide isolated from the toxic blue-green algae *Lyngbya majuscula* which contains 3-amino-2-methylpentanoic acid (*74*). Dolastatins-11 and -12 (43.6 mg and 10.6 mg, respec-tively, isolated from 1600 kg (!) of marine mollusc *Dolabella auricularia*) also contain 2-methyl-3-amino-pentanoic acid (*75, 76*). *N*-Acylated 2-methylene-β-alanines were isolated as 1–2% of the dry weight of *Fasciospongia cavernosa* sponge (*77*). A leucine 2,3-aminomutase enzyme from *Andrographis* tissue cultures interconverts the α-amino acid (2*S*)-Leu with (3*R*)-β-Leu, 3-amino-4-methylpentanoic acid (*78*).

The iturin family of antifungal cyclic peptides con-tains ituricin acid, a long-chain alkyl β-amino acid rang-ing from 13 to 17 carbons long, with the *n*-C₁₄ isomer the most common (*79–81*). Onchidin, a cytotoxic depsi-peptide with C2 symmetry found in a marine mollusc, contains 2 molecules of (2*S*,3*S*)-3-amino-2-methyl-oct-7-ynoic acid (Amo) (*82*), although a synthesis of the proposed structure gave a compound with an NMR spectra inconsistent with the natural product (*86*). 3-Amino-5-tetradecenoic acid is contained in lipogram-mistin-A, a lipophilic ichthyotoxin secreted by the skin of soapfish. The fish produces copious quantities of mucus that create a soaplike film in water (*83*): 3.0 mg of lipogrammistin-A was obtained from 0.75 g of wet mucus scraped off the skin of one 77 g *A. temmincki* fish. A total synthesis was reported in 2000 (*84, 85*). The lipopeptides lobocyclamides A–C, isolated from cyanobacteria *Lyngbya confervoides*, contain either 3-aminooctanoic acid or 3-aminodecanoic acid (*87*). An iminodicarboxylic acid, *N*-carboxymethyl 3-amino-decanoic acid, is a component of the minalemines, from the tunicate *D. rodriguesi* (*88*). The spermine alka-loids, homaline, hopromine, hoprominol, and hoproma-linol, are butane-bridged lactam rings containing combinations of 3-aminooctanoic acid, 3-aminode-canoic acid, 3-amino-5-hydroxy-decanoic acid, and β-Phe. Syntheses of (−)-(*R, R*)-hopromine and racemic homaline were reported in 2002 (*89*). 3-Aminotrade-canoic acid is found in another polyamine macrocycle, (−)-(*R*)-budmunchiamine A, with a total synthesis in 2002 (*90*).

β-Amino-β-phenylpropionic acid (β-Phe, see Scheme 11.4) is contained in a highly toxic cyclic pep-tide isolated from a mold of yellow rice (*Penicillium islandicum*) (*91*), in a cyclic tetrapeptide (roccanin) isolated from the lichen *Roccella* (*92*), in a dipeptide (γ-L-glutamyl-L-β-Phe) isolated from the Japanese azuki bean *Phaseolus angularis* (400 mg from 15 kg of pulverized bean seeds) (93, 94), and in periphylline, neoperiphylline, and dihydroperiphylline from the leaves of *Peripterygia marginata* (95, 96). D-β-Phe is a component of the antibiotics moiramide A, B, and C and andrimid, isolated from both an intracellular symbiont of *Nilaparvata lugens* (brown planthopper) and the marine bacterium *Pseudomonas fluorescens*. Moiramide B and andrimid show potent in vivo activity against antibiotic-resistant human pathogens (97). β-Phe is also contained in the astin family of cyclopentapeptides isolated from biologically active extracts of the root of *Aster tataricus*, a flowering plant used in Chinese medicinal teas (98–102), with 100 mg of astin A, 300 mg of astin B, and 500 mg of astin C obtained from 10.0 kg of *Aster tataricus* (99). Astin A was also separately isolated and named asterin, with 250 mg obtained from 5 kg of commercially available dried roots (103). The peptides have potent antitumor activity. NMR conformational analyses of astin A (104), Astin B (105), and astin C (104) have been reported, as has an X-ray crystal structure of astin B (99). All three astins contain a 3,4-dichloroproline residue, which appears to be important for antitumor activity (106). astin J is an acyclic analog of Astin C without the dichloroproline residue (107). Thioamide-containing analogs of the astins have been synthesized and conformational analyses carried out (108).

As mentioned earlier, many macrocyclic spermine alkaloids (so named as they contain a spermine, N, N′-bis(3-aminopropyl)-1,4-diaminobutane, backbone) with a β-Phe component have been identified. These include (−)-(S)-protoverbine (isolated from the leaves of *Verbascum pseudonobile* (109) with a racemic synthesis in 2000 (109) and an asymmetric synthesis in 2002 (110)), the derivatized protoverbine analogs protomethine, verbamethine, verbacine, verbamekrine, isoverbamekrine, verbasikrine, and isoverbasikrine (110, 111), and (+)-(S)-dihydroperiphylline (isolated from the leaves of the New Caledonian plant *Peripterygia marginata*, and synthesized in 2002 (112)). Closely related (S)-prelandrine (isolated from the roots of *Aphelandra squarrosa*) contains β-Tyr, and is an intermediate in the synthesis of the more complex macrobicyclic spermine alkaloids aphelandrine and orantine. Prelandrine and dihydroxyverbacine (the terminal precursor in the synthesis of aphelandrine and orantine) (113), were synthesized from the corresponding O-methyl-β-Tyr-containing analog, buchnerine (114), with a hemin oxidizing system converting dihydroxyverbacine into aphelandrine, orantine, and chaenorpine (115). (−)-(S)-Buchnerine was isolated from *Clerodendrum buchneri*, with an asymmetric synthesis in 2002 (110).

Another family of β-Phe-containing macrocyclic spermidine alkaloids includes celacinnine, celallocinnine, celafurine, and celabenzine (see Scheme 11.4) from the plants *Maytenus arbutifolia*, *Tripterygium wilfordii*, and *Pleurostylia africana* (116, 117). Cyclocelabenzine, isocyclocelabenzine, and hydroxyisocyclocelabenzine are related compounds isolated from *Maytenus mossambicensis* (118, 119). Syntheses of (±)-celabenzine (120,

121), (+)-celabenzine (established configuration as (8S)) (122), (±)-celacinnine (121), dihydrocelacinnine (established configuration as (8S)) (122), (±)-cellallocinnine (121), (+)-celacinnine (123) (±)-celafurine (121), (+)-cyclocelabenzine (established configuration as (8S,13R)) (118), (±)-isocyclocelabenzine (124), (+)-isocyclocelabenzine (established configuration as (9S, 13S)) (125), and 13-hydroxyisocyclocelabenzine (126) have been reported. Hydroxylated spermidine alkaoids have also been isolated, with 3-hydroxycelacinnine containing a 2-hydroxy-3-phenyl-3-aminopropionic acid moiety (127). As mentioned earlier, homaline, hopromine, hoprominol, and hopromalinol are butane-bridged lactam rings containing combinations of β-Phe, 3-aminooctanoic acid, 3-aminodecanoic acid, and 3-amino-5-hydroxydecanoic acid. Syntheses of (−)-(R,R)-hopromine and racemic homaline were reported in 2002 (89).

Other aryl β-amino acids include β-tyrosine (β-Tyr), a component of the bioactive marine cyclodepsipeptides (+)-jasplakinolide (128) and jaspamide (possessing antifungal, antihelminthic, insecticidal, nematocidal, and icthyotoxic acitivity) (128–131). It is also found with isoserine in the antibiotic peptides edeines A₁ and B₁, from the soil bacterium *Bacillus brevis* (132, 133). An enzyme which converts L-Tyr to β-Tyr, tyrosine α,β-amino mutase, was isolated from cells of *Bacillus brevis* (134). A 2:1 ligand:iron complex formed from (R)-β-DOPA, 3-(3,4-dihydroxyphenyl)-β-Ala, has been found to be responsible for the dark blue-violet color of the *Cortinarius violaceus* mushroom, with 4.5–7.5 mg/g dry weight content (135). β-(2-Thiazole)-β-Ala is found in the peptide antibiotics bottromycins A and B from *Streptomyces* 3668-L2 (136–138); the original structural assignments of bottromycins B₁ and B₂ were found to be incorrect during a total synthesis of the proposed structures (139). A β-heteroaryl-β-amino residue is contained in C-1027 chromophore, a natural product enediyne with potent antitumor activity. The C-1027 is produced in association with a natural delivery vehicle, an 11-kDa apoprotein, which binds C-1027 non-covalently in a cleft and kinetically prevents its decomposition. The apoprotein can deliver C-1027 through a cell to its double-stranded DNA target, acting as a drug delivery system. An engineered apoprotein provided greater stabilization of C-1027 (140). A similar natural product, kedarcidin, contains a β-(2-chloro-3-hydroxypyrid-6-yl)-β-amino acid residue, also cyclized via an aryl ether linkage (141, 142).

Several β, ω-diamino acids exist (see Scheme 11.5). D-β-Lysine (3,6-diaminohexanoic acid) is the primary component of the basic antibiotic bellenamine produced by *Streptomyces nashvillensis* (143), and is also found in the antibiotics geomycin (produced by *Streptomyces xanthophaeus*) (144), streptolin (145, 146), and the streptothricins (147). A β-(L-β-Lys)-L-β-Lys dipeptide was prepared during elucidation of the β-Lys peptide structure of the streptothricin antibiotics (147), while an isotopic labeling study with [1,2-¹³C₂]acetate was used to determine the biosynthesis of streptothricin F, including the β-Lys residue (148). The metabolism of

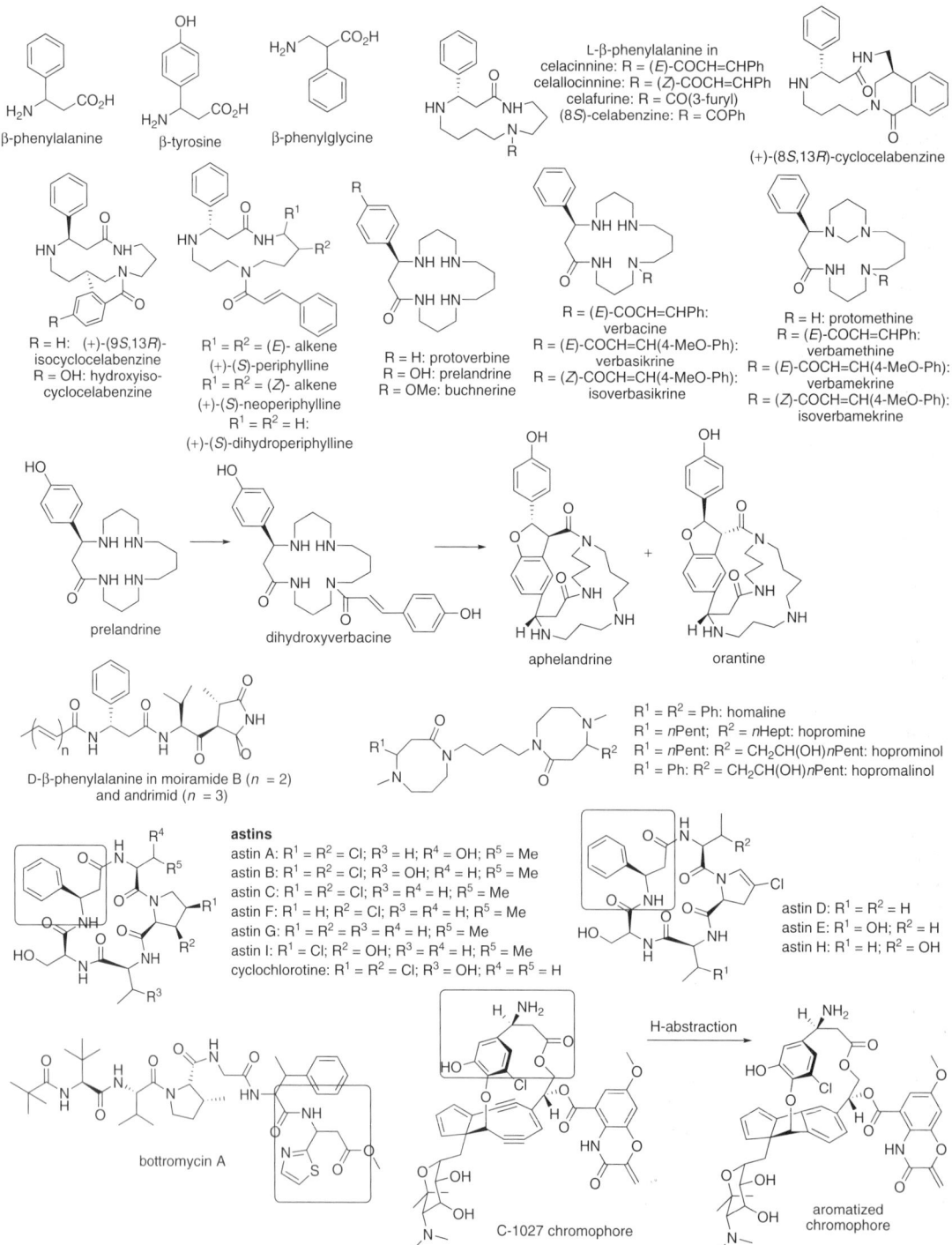

Scheme 11.4 Aryl β-Amino Acids.

L-β-Lys by *Pseudomonas* B4 has been studied (*149*). The tuberactinomycins, also called viomycins, are a family of cyclic peptide antibiotics effective against tubercularbacilli(*145,146,150–155*).Tuberactinomycins B and O contain β-Lys (*145, 146, 150–155*), while tuberactinomycins A and N have a γ-hydroxy-β-Lys residue (*152, 156–160*). The structures and stereochemistry were confirmed by total synthesis (*161*). Analogs of tuberactinomycin N have been prepared by semisynthesis in which the γ-hydroxy-β-Lys residue was

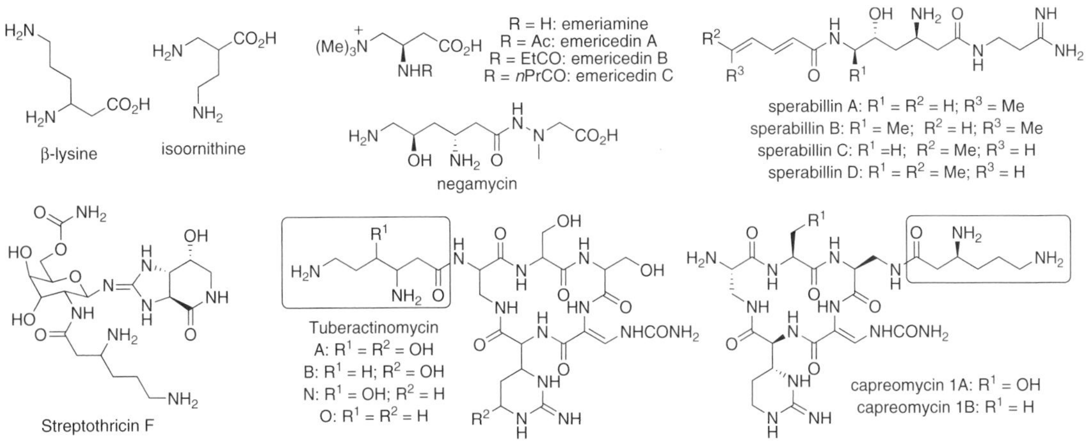

Scheme 11.5 β,ω-Diamino Acids.

removed and replaced by other α- and β-amino acids (156). Two of the closely related capreomycins (isolated from *Streptomyces capreolus*) also contain β-Lys; a total synthesis of capreomycin 1B was reported in 2003 (162).

δ-Hydroxy-β-Lys (3,6-diamino-5-hydroxyhexanoic acid) is a component of the antibiotic dipeptide negamycin (163–165) and the sperabillins A and C, novel antibiotics isolated from *Pseudomonas fluorescens* which contain the (3R,5R)-isomer (166). Sperabillins B and D contain the methylated analog (3R,5R,6R)-3,6-diamino-5-hydroxyheptanoic acid (166). Analogs of negamycin containing a variety of derivatives of the β-amino acid residue have been prepared (167). Emeriamine, (R)-3-amino-4-trimethylaminobutyric acid, is an inhibitor of fatty acid oxidation obtained by deacylation of the less potent emericedins, isolated from the culture broth of *Emericella quadrilineata* (168, 169). More active derivatives were prepared by forming *N*-palmitoyl or *N*-myristoyl derivatives of emeriamine (168). N^ε-Me-β-arginine (blastidic acid) is found in blasticidin S, an antibiotic produced by *Streptomyces griseochromogenes* that is effective against rice blast disease (170). A number of isotopically labeled Arg derivatives were employed to study the biosynthesis of the β-Arg in blasticidin S (171). An efficient asymmetric synthesis of blastidic acid was reported in 2005 (172).

α-Hydroxy-β-amino acids often possess interesting activity. Isoserine (2-hydroxy-3-aminopropionic acid) (see Scheme 11.6) is found in antibiotics and has been used to enhance the activity of other antibiotics by replacing other amino acid residues (see references cited within 173). For example, isoserine is contained in the antibiotic peptides edeines A_1 and B_1 (from the soil bacterium *Bacillus brevis*) (132, 133), in cyclocinamide (a cytotoxic hexapeptide isolated from the sponge *Psammocinia* sp.) (174), and in the keramamides (*Theonella* sponge products with antifungal, antioxidant,

and cytotoxic properties) (175, 176). α-Hydroxy-β-isobutyl-β-Ala is contained in the tripeptide aminopeptidase inhibitor amastatin (177, 178). Peptide couplings with α-hydroxy-β-amino acids (with unprotected hydroxy groups) as acyl components have been found to generate dimeric homobislactone byproducts (179). Phenylnorstatine (2-hydroxy-3-amino-4-phenylbutyric acid) has been used to construct a new polymer-supported Evans-type oxazolidinone chiral auxiliary (180).

In recent years a great deal of attention has been devoted to the antitumor compound Taxol (paclitaxel), which contains *N*-benzoyl (2S,3R)-phenylisoserine as a critical side-chain substituent on a baccatin core (3 and references therein). A brief review of Taxol's history, including an overview of six total syntheses, was published in 2001 (181). Large-scale syntheses of the phenylisoserine isomer required for the Taxol side chain have been described (182, 183). Taxotere (docetaxel) is a semisynthetic analog with a Boc group on the phenylisoserine and without an *O*-acetyl group on the baccatin. An ab initio conformational study of the phenylisoserine side chain indicates that the preferred conformation is similar to that found in the crystal structure of docetaxel, as an L-shape which minimizes steric and coulombic interactions (184). 1-Deoxypaclitaxel analogs have been prepared with phenylisoserine or 2-hydroxy-3-(2-methyl-1-propene)-3-aminopropionic acid substituents, giving compounds with similar activity to paclitaxel (185). Taxotere analogs with 3-methyl-phenylisoserine (186), a conformationally restricted phenylisoserine analog (187), 1-amino-1-(2-hydroxyacetic acid)cyclopropane (188), and 2-substituted-1-amino-1-(2-hydroxyacetic acid)cyclopropane (189) have been described. A bioconversion of taxol by *Streptomyces* sp. MA7065 hydroxylated the *para*-position of the phenylisoserine benzene ring in 10% yield (190). Abbott has developed ABT-271 as a promising

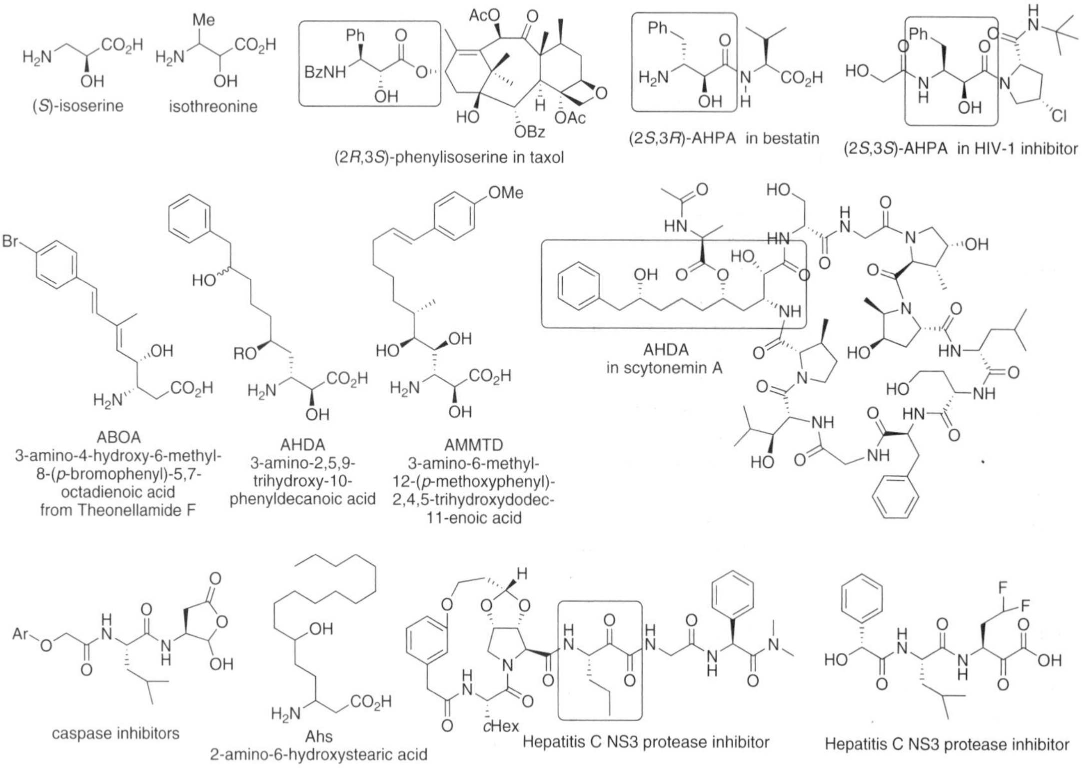

Scheme 11.6 Hydroxy-β-Amino Acids and α-Keto-β-Amino Acids.

anticancer agent, with a C9-(R)-hydroxy group on the taxane skeleton and a N-Boc norstatine (2-hydroxy-3-amino-5-methylhexanoic acid) β-amino acid moiety (191). Other taxane analogs with modified baccatin cores and a norstatine substituent were reported in 2004 (192) and 2005 (193). Taxoid analogs of calicheamicin, another anticancer antibiotic, have been prepared by attaching a phenylisoserine side chain to an enediyne core (194). Appending the phenylisoserine side chain to repin, an antitumor sesquiterpene lactone, improved its antitumor potency (195).

Other α-hydroxy-β-amino acids include Ahda (3-amino-2-hydroxydecanoic acid), found in microginin (an angiotensin-converting enzyme pentapeptide inhibitor isolated from blue-green algae) (196). The absolute configuration of C3 of Ahda was established by a modified advanced Marfey's method (197). A series of related microginins were subsequently obtained from the cyanobacterium Microcystis aeruginosa: Ahda, 10, 10-dichloro-Ahda, 10-chloro-3-aminodecanoic acid, and 10,10-dichloro-3-aminodecanoic acid (198). Bestatin is a dipeptide containing L-Val and AHPA (or AHPBA, 3-amino-2-hydroxy-4-phenylbutanoic acid) (199, 200) which is isolated from Streptomyces and inhibits aminopeptidase B (201), modifies the immune response, and reduces tumor growth (202). Bestatin and a number of analogs were prepared in 1999 via a general route to α-hydroxy-β-amino acids (203). Closely related

ubenimex is a dipeptide of L-Leu and (2S,3R)-AHPA from Streptomyces olivoreticuli fermentation broth. It enhances delayed hypersensitivity in animals, and inhibits both aminopeptidase B and leucine aminopeptidase. An aryl-[o-³H]-labeled analog was prepared to examine metabolism and pharmacokinetics (204). (2S,3S)-AHPA is the main component of a variety of HIV-1 protease inhibitors with subnanomolar IC₅₀ values and good oral availability, acting as a hydroxymethylcarbonyl isostere and transition-state mimic (205, 206).

β-Amino acids with hydroxy groups at other positions are also known. The antifungal lipopeptide nostofungicidine, isolated from field-grown terrestrial blue-green algae Nostoc commune, contains 3-amino-6-hydroxyoctadecanoic acid (stearic acid) (207). Theonellamide F, an antifungal and cytotoxic cyclic dodecapeptide from a Theonella sponge, possesses an ABOA residue, (3S, 4S,5E,7E)-3-amino-4-hydroxy-6-methyl-8-(p-bromophenyl)-5,7-octadienoic acid, in addition to β-Ala (208). The blue-green algae Scytonema produces scytonemin A, a cyclic peptide that has calcium antagonistic properties with an AHDA (3-amino-2,5,9-trihydroxy-10-phenyldecanoic acid) residue (209, 210). AMMTD, 3-amino-6-methyl-12-(p-methoxyphenyl)-2,4,5-trihydroxydodec-11-enoic acid is a component of the cyclic antifungal hexapeptides microsclerodermins A and B from Microscleroderma sponge (211, 212), while 3-amino-2,4,5-trihydroxy-8-phenyl-7-octenoic acid is found in microsclerodermins C and D (*213*), 3-amino-2,4,5-tri-hydroxy-10-(4′-ethoxyphenyl)-7,9-decadienoic acid in microsclerodermin E (*213*), 3-amino-2,4,5-trihydroxy-6-methyl-12-phenyl-7,9,11-dodecatrienoic acid in microsclerodermins F and H (*214*), and 3-amino-2,4,5-trihydroxy-6,10-dimethyl-12-phenyl-7,9,11-dodeca-trienoic acid in microsclerodermins G and I (*214*). A total synthesis of microsclerodermin E was reported in 2003 (*215*). AI-77B is an antiulcerogenic antibiotic isolated from *Bacillus pimilus* that contains 2,3-di-hydroxy-4-amino-1,6-hexanedioic acid coupled to a dihydroisocoumarin moiety (*216–224*). A total synthe-sis was reported in 2003 (*225*).

Other β-amino acids with complex side chains have been found in peptides isolated from the *Theonella swinhoei* Gray sponge, such as Adda, (2*S*,3*S*,8*S*,9*S*)-3-amino-9-methoxy-2,6,8-trimethyl-10-phenyl-4,6-decadienoic acid (see Scheme 11.7), which is con-tained in the cyclic pentapeptide motuporin (a phos-phatase-1 inhibitor)(*226–228*). Nodularin, a pentapeptide from *Nodularia spumigena* algae, also contains Adda (*229*), as do the microcystin cyclic heptapeptide toxins from the freshwater cyanobacterium *Oscillatoria agar-dhii* (*229–231*), along with *O*-demethyl-Adda (*231*). A series of analogs of microcystin containing Adda have been prepared (*232*), while the role of the 2-methyl and 3-diene groups in Adda in nodularin were examined for their effects on protein phosphatase inhibition (*233*). Another report replaced the Adda residue in nodularin and microcystin with β-Ala, 3-aryl-β-Ala, or amidated Asp (*234*). The *Anabaena* hepatotoxic bacteria was assayed for its microcystin content and composition in 2002 (*235*). A cyclic hexapeptide nostophycin from the toxic cyanobacterium *Nostoc* sp. 152, which also pro-duces microcystins, contains (2*S*,3*R*,5*R*)-3-amino-2,5-dihydroxy-8-phenyloctanoic acid (Ahoa) (*236*). HPLC methods for the separation of microcystins and nodu-larin on C$_{18}$ and amide C$_{16}$ stationary phases (*237*) or on a monolithic silica C$_{18}$ column (*238*) have been pub-lished, along with a review of the purification of micro-cystins in 2001 (*239*). A liquid chromatography–electron spray ionization–mass spectrometry (LC–ESI–MS) method for the simultaneous analysis of algal and cyanobacterial toxins extracted from phytoplankton,

including microcystins, nodularin, okadaic acid, domoic acid, saxitoxin, and anatoxin-A, was published in 2003; the method also allows for structural identification of potential new microcystins via peptide hydrolysis of fractions and chiral amino acid analysis (*240*). Other microcystin-related analytical methods include an LC–ESI–MS screening method in 2003 (*241*), an opti-mized procedure for intracellular microcystin extraction in 2005 (*242*), methods for the determination of micro-cystins in biological samples by matrix solid-phase dis-persion and LC-MS (*243*), and analysis in fish by solvent extraction and LC (*244*).

The glycopeptide antitumor antibiotic bleomycin A$_2$, isolated from *Streptomyces verticillus* in 1966 and now used clinically to treat carcinomas (*245–248*), contains pyrimidoblamic acid, β-amino-β-(4-amino-6-carboxy-5-methylpyrimidin-2-yl)propionic acid (*249*) (see Scheme 11.7). The synthesis of, and mechanistic studies with, bleomycin were reviewed in 1999 (*250*), with fur-ther studies on the chemistry of bleomycin-induced alka-li-labile DNA lesion discussed the same year (*251*) and in 2001 (*252*). A detailed study of the electronic structure of the bleomycin–Fe(III) complex and the corresponding activated ferric hydroperoxide derivative was presented in 2000 (*253*). Total syntheses of bleomycin demethyl A$_2$ (*254*), bleomycin A$_2$ (*254, 255*), deglycobleomycin A$_2$ (*256*), and decarbamoylbleomycin demethyl A$_2$ (*254*), have been reported. Catabolism of bleomycin is believed to involve hydrolysis of a 2,3-diaminopropionic acid α-amide; a total synthesis of this deamidobleomycin A$_2$ along with bleomycin demethyl A$_2$ and the correspond-ing aglycons was reported in 2002 (*257*). Analogs of deglycobleomycin A$_5$ with conformationally restricted γ-amino acid components were synthesized in 2003 (*258*), while the same year a library of 108 deglycobleo-mycin analogs was prepared by solid-phase synthesis from three bithiazole analogs, three Thr analogs, three methylvalerate analogs, and four β-hydroxy-His replace-ments (*259*). Another solid-phase synthesis prepared bleomycin A$_5$ and three monosaccharide analogs in order to investigate the role of the carbohydrate moiety in DNA/RNA cleavage (*260*). A deglycobleomycin A$_5$ analog with a modified *C*-terminal trithiazole tail pos-sessed altered sequence selectivity (*261*).

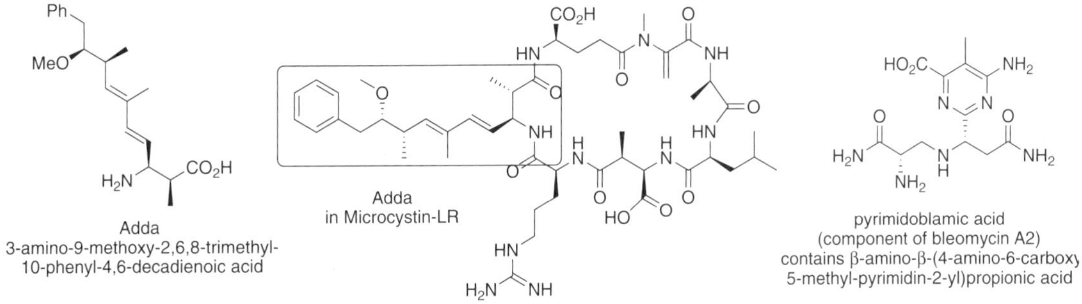

Adda
3-amino-9-methoxy-2,6,8-trimethyl-
10-phenyl-4,6-decadienoic acid

Adda
in Microcystin-LR

pyrimidoblamic acid
(component of bleomycin A2)
contains β-amino-β-(4-amino-6-carboxy
5-methyl-pyrimidin-2-yl)propionic acid

Scheme 11.7 Other Complex β-Amino Acids.

A number of α-keto-β-amino acid homologs of α-amino acids are components of metabolites from *Streptomyces* that possess prolyl endopeptidase inhibitory activity. Poststatin, from *Streptomyces viridochromogenes*, contains 3-amino-2-oxopentanoic acid (*262*) while eurystatins A and B from *Streptomyces eurythermus* contain (*S*)-3-amino-2-oxobutyric acid (*263*). Extracts of the sponges *Theonella* swinhoei and *Theonella* sp. also contain α-keto-β-amino acid homologs of α-amino acids, with the cyclotheonamides and pseudotheonamides containing an α-keto-Arg homolog (*264–267*), the orbiculamides an α-keto-Leu homolog (*268*), oriamide an α-keto-Ile homolog (3-amino-4-methyl-2-oxohexanoic acid) (*269*), and the keramamides α-keto-Leu and α-keto-Ile homologs (*175*, *270–272*). As discussed further below, α-keto-β-amino acids have found use as enzyme inhibitors in medicinal chemistry programs.

11.2.1c β-Amino Acid Peptides

Peptides formed from β-amino acids possess interesting properties. Early efforts at β-peptide synthesis are summarized in a 1977 review, including difficulties encountered in coupling reactions (*273*). The structure of β-peptides was reviewed in 2001 (*14*), with a discussion of the secondary structures of β-peptides in 2006 (*274*). A comprehensive study of the synthesis and properties of enantiomerically pure β²-, β³-, and β²,³-peptides with proteinaceous side chains was published in 1998 (*275*), while a 1997 review of the synthesis and structures of β-peptides included comparisons of β-peptides with α-peptides (*276*), and a 1993 article examined the role of β-Ala in the conformation of cyclic peptides (*277*). Short overviews of β-peptides have also been published (*278, 279*). A 2001 *Chemical Reviews* article examined the conformational properties of β-peptides and provided a summary of biologically active β-peptides (*280*, 187 references).

Difficulties encountered in synthesizing longer β-peptides include the formation of stable secondary structures during synthesis, the need for expensive building blocks that do not allow for excess reagents to be employed, and racemization of β²-amino acids. Solid-phase protocols suitable for preparing β-peptides with more than nine residues have been developed; in particular, the use of preformed β²/β³-dipeptides reduces the number of couplings required and eliminates epimerization at the β²-amino acid residue (*281*). For couplings, urethane *N*-carboxyanhydrides (UNCAs) of β-amino acids have been synthesized and reacted with amines to form amides, though peptide formation was not reported (*282*). HBTU/HOBt was employed in combination with microwave irradiation to greatly improve yields of longer sequences of β-peptides when compared to conventional heating (*283*). Microwave irradiation was also tested for the synthesis of a 100-member combinatorial library of octamers (*284*). A glass microreactor produced a β-Ala tripeptide using pentafluorophenyl-activated esters (*285*). Segments of β-peptides can be joined by native chemical ligation, in which β-peptide thioesters couple to β-peptides with an *N*-terminal β³-HCys residue. Hybrid α-/β-peptides can also be produced by coupling to α-peptides with an *N*-terminal Cys residue (*286*). Isocysteine, 2-thio-3-aminopropionic acid, has also been employed as the *N*-terminal amino acid for native chemical ligation of two peptide segments (*287*).

Homopolymers of β-Ala are quite insoluble (*288*), while poly(3-amino-3-methylbutyric acid) with molecular weight values of 7.5×10^5 forms a crystalline polymer that can be spun into filaments and has properties resembling those of natural silk (*273*). Branched β-peptides have been synthesized by using α,α-bis(aminomethyl)-β-Ala (*289*), while orthogonally protected bis(aminomethyl)malonic acid was used to prepare branched cyclic peptide conjugates (*290*). Oligomers of β-amino acids with as few as six or seven residues can form stable helical conformations in methanol (*291, 292*), compared to the 15 α-amino acid residues that are generally required for helix formation (*276*). Peptides consisting of mixtures of alkyl-mono-substituted β²-amino acids and β³-amino acids form a different type of secondary structure than do homo-β-peptides, believed to be a novel irregular helix (*293*). A β²-heptapeptide and a β²,³-hexapeptide possessed a 3_{14}-helical structure (a helix structure with interwoven 14-membered-ring hydrogen bonds), while an alternating β²/β³-hexapeptide had a novel helical structure with one central 10-membered H-bonded ring and two terminal 12-membered rings (*275*). Hexa- or hepta-β-peptides containing one or two β²,³-amino acids with two hydroxymethyl (Ser analogs) or thiomethyl (Cys analog) substituents were also found to adopt 3_{14} helical structures (*294*). The temperature dependence of helix formation of a β-hexapeptide and a β-heptapeptide was examined using nuclear megnatic resonance (NMR) and circular dicharoism(CD) spectra between 298 and 393 K. The 3_{14}-helical secondary structures did not "melt" in this temperature range, in contrast to denaturing observed with helices of α-amino acids upon heating (*295*). A β-peptide with 12/10/12/10 helical conformation was found to be very thermally stable, maintaining its structure at 80 °C (*296*).

Theoretical molecular and quantum mechanics studies of β-peptides with various substitution patterns have been used to predict secondary structures (*297, 298*). Theoretical modeling studies were carried out on six different β-dipeptide models to evaluate their tendency towards β-sheet, 14-helix, or 12-helix formation. β³-Peptides were found to have greater 14-helical propensity than β²-peptides (*299*). Predictions of the ability of β-peptides to form mixed helices with alternating hydrogen-bonding patterns found a preference for 10-membered rings involving *i* and *i+1* residues in the forward direction followed by 12-membered rings with *i* and *i+3* in the backwards direction (*300*). This pattern was identified by NMR structural analysis in hexa- and nonapeptides formed from alternating β²- and β³-amino acid residues (*301*). A 2002 theoretical study developed predictions for the propensities of different monomer substitution patterns on inducing secondary structure. Substitution at $C(\beta)$ had a higher impact on conformation than did a substituent at $C(\alpha)$ (302). Another theoretical study compared models of poly(β-ala) in various helix structures (303). Some characteristic CD spectra patterns were described in a 2000 report (304). Another 2000 report on the possible structures of a series of hexa-, hepta- and nona-β-peptides containing functionalized side chains, alternating configurations, and/or geminal backbone substituents illustrated the difficulties in ascertaining structure based on CD spectra alone (305). The dangers of relying on CD spectra to predict conformational preferences was pointed out in a 2002 study, in which two β-hexapeptides predicted to differ in secondary structure exhibited similar CD spectra (306).

A β³-heptapeptide designed to form a 3_{14}-helix was glycosylated on a β³-HSer residue, with NMR analysis indicating the helical structure was maintained, at least in methanol (307). 3-Hydroxybutanoic acid was incorporated into β-peptides as a β³-Ala analog, and used to examine the effect of deleting a backbone hydrogen bond on the stability of β-peptidic 3_{14}-helices, with significant reductions in helicity observed (308). The effect of β-substitition on intramolecular hydrogen bonding has also been examined, with alkyl substitution increasing the population of intramolecularly hydrogen-bonded conformations (309). Another theoretical conformational analysis looked at hydrazino peptides; this indicated that a variety of secondary structure types (helices, strands, and turns) were possible, with some closely resembling those formed by β-peptides, but some representing novel types (310).

The formation of 14-helical structures in β-peptides is greatly stabilized by residues with side chain branching adjacent to the β-carbon: a peptide formed from β³-residues derived by homologation of Val showed strong 14-helical conformation, while the identical peptide using Leu-derived residues had a distinctly different structure (311). β-Peptides with both acidic and basic residues could be designed to maintain short 14-helical structure (312). β³-Amino acids corresponding to homologated Lys and Glu residues were incorporated into a de novo designed monomeric helical β-peptide, with the ionized side chains stabilizing helix formation via electrostatic interactions to generate a helix in water (314). Similarly, β³-hArg and β³-hGlu were incorporated into a β³-heptapeptide to increase 3_{14}-helicity via electrostatic interactions (313). Further 3_{14}-helix stabilization is achieved by designing the helix to possess a macrodipole, with partial positive charge at the N-terminus and partial negative charge at the C-terminus (316). β³-Amino acids with thiomethyl and thioethyl side chains were incorporated into β³-penta-, hexa- and heptapeptides. Disulfide formation with a $-CH_2SSCH_2-$ bridge between residues i and i+3 produced peptides with CD spectra characteristic of helical structure, while the same linkage between i and i+2 or i and i+4 did not (317). An NMR structure of the cyclized β³-hexapeptide corresponding to hVal-hCys-hLeu-hVal-hCys-hLeu demonstrated a 3_{14}-helical structure (318).

A β-Ala-γ-aminobutyric acid sequence (equivalent to a Gly_3 tripeptide segment) can be accommodated within a regular 3_{10}- or α-helical structure in an octa- or undecapeptide (319).

β-Peptides with cyclic β-amino acids form 12-helix structures (see Section 11.3.1). It is possible to substitute some of the cyclic residues with acyclic β³-residues and retain the 12-helix conformation; using β-Lys produced β-peptides with antibacterial activity (320). β-Amino acids with sugar side chains have been described (321, 2183). Polymers prepared with β-furanosyl-β-amino acids of alternating chirality formed 10/12/10 helical structures (321). β-Peptides containing up to 12 β²,²- or β³,³-disubstituted β-amino acids do not form secondary structures similar to the other types of β-peptides. A crystal structure of a β²,²-tripeptide showed a 10-membered hydrogen-bonded turn (322). Another helical secondary structure, a (P)-2_8-helix (right-handed helix with approximately two residues per pitch and eight-membered hydrogen-bonded ring between NH of amino acid i and C=O of amino acid i−2), was assigned to a β-hexapeptide formed from (2R,3S)-3-amino-3-substituted-2-hydroxypropionic acid residues (323). A subsequent study compared two methods for the conformational interpretation of the NMR data of a β-hexapeptide (conventional analysis using NOE distance restraints and simulated annealing vs free molecular dynamic simulation), and found significant differences in the results (324).

The introduction of heteroatom (fluoro or hydroxy) substituents at the C2 position of residues within β³-peptides can have significant effects on the tertiary structure, with a single fluorine substituent preventing a β-heptapeptide from adopting a 3_{14}-helix (325). β³-Amino acids with a 2-methylene substituent, homologs of dehydroalanine, gave di-, tri-, and hexapeptides that did not form any recognizable secondary structures (326). Similarly, β-peptides constructed from achiral β²,²-3-amino-2,2-dimethylpropionic acid and/or chiral β²,²,³-3-amino-3-substituted-2,2-dimethylpropionic acid were examined by solution and solid-state NMR. No evidence for helix or sheet formation was found, indicating that geminal substitution leads to a more flexible peptide structure (327). α,β-Disubstituted β-amino acids (β²,³-amino acids) can induce formation of antiparallel β-peptide sheets (291), with two dimensional (2D) NMR spectroscopy of β-peptides formed from these disubstituted β-amino acids indicating pleated sheet and turn structures (328). Parallel sheet secondary structures have been created in β-peptides by linking two peptide segments of syn-β²,³-amino acids with a 1,2-diaminoethane moiety that allows formation of a hairpin turn (329). Homohexapeptides produced from O-tert-butyldimethylsilyl (2R,3S)-phenylisoserine residues formed highly stable β-strand-type secondary structures in chloroform solution (330). Incorporating α-substituted-β-amino acids into tetrapeptides induced a reverse-turn conformation in aprotic solvent systems (331). A tetrapeptide, Boc-β-Ala-Aib-Leu-Aib-OMe, was found to form a double-turn structure, which then self-associated into a supramolecular helix with amyloid-like fibril structure (*332*).

Cyclic β-peptides have also been synthesized, though difficulties can be encountered due to insoluble intermediates. On-resin cyclization was successful for the preparation of cyclic β-tripeptides (*333*). Cyclic β³-tetra- and pentapeptides have also been prepared by on-resin cyclization, and then characterized by X-ray crystallography and NMR and CD spectroscopy (*334*). Several linear and cyclic peptides formed from two to six β³-amino acids (Asp, and homologated Glu and Ser) were synthesized in 1998. The cyclic peptides were too insoluble for solution structural investigations, while the linear unprotected β-hexapeptides showed spectroscopic evidence of helix formation (*335*). β-Peptides containing hydrophilic side chains have been prepared to increase their water solubility. In one report homoglutamate and homolysine residues were employed, with one polar residue for every three non-polar residues required for reasonable water solubility. CD spectra of the aqueous solutions demonstrated some evidence of helical structure (*336*). Alternatively, β³-homoserine and β³-homolysine were incorporated into hexa- hepta-, and nonapeptides with two, three, and seven hydrophilic side chains, respectively. Less-pronounced secondary structures were observed in water compared with methanol (*337*).

A review containing the results of some X-ray crystallography studies of peptides containing β-Ala has been published (*338*). The crystal structure of Boc-β-Ala-NHMe shows an all-*anti*-conformation, in contrast to the folded conformation predicted by ab initio calculations (*339*). However, an X-ray crystal structure of a Boc-L-pipecolic acid-β-Ala-NHMe dipeptide maintained the β-Ala residue in a folded conformation (*340*). The solution and X-ray structures of peptides containing a β-Ala residue have been compared to a standard peptide bond (*341*). The effects of a single β-amino acid residue (homologated Val or Leu) on the conformation of tri-, tetra-, or hexapeptides formed with the α, α-disubstituted amino acids Aib or 1-aminocyclohexane-1-carboxylic acid were examined by X-ray crystallography. No obvious conformational preferences were observed (*342*). A comparison of crystal structures of Boc-β-Ala-L-Ala-NHMe and Boc-β-Ala-D-Ala-NHMe demonstrated the potential for adjacent chiral residues to switch the conformational characteristics of achiral β-Ala from right-handed to left-handed (*343*).

The research group of Dieter Seebach has prepared increasingly complex β-peptides containing residues corresponding to the 20 homologated α-amino acids. In 2000 a β³-dodecapeptide containing seven functionalized side chains (analogs of Ala, Val, Leu, Phe, Tyr, Ser, and Lys) was reported (*344*). Structural investigations using NMR and CD spectroscopy indicated a 3_{14}-helix in MeOH, but an extended conformation in water (*345*). In 2001 a β³-tetracosapeptide (24-mer) was prepared with eight different side chains (side chains of Ala, Val, Leu, Phe, Glu, Lys, Ser, and Tyr), designed as an amphipathic 3_{14}-helix. However, the CD spectrum indicated a novel, and unknown, secondary structure (*346*). The first eicosapeptide (20-mer) containing the β³-amino acid homologs of all 20 common amino acids (readily available by homologation of the corresponding α-amino acids) was prepared in 2003 (*347*). An NMR-solution structure of this peptide in methanol demonstrated a 3_{14}-helix over its whole length, forming six full turns (*348*). This is considerably longer than the helical structures normally observed with α-amino acids. The second major synthetic effort, to prepare the corresponding β²-eicosapeptide, required a 159-step synthesis in order to prepare all the β²-amino acids and successfully couple them without racemization. These efforts were summarized in a 2004 review (*349*).

The secondary structures of β-peptides have been used to design more complex structures. Cyclic β³-tetrapeptides containing 3-isobutyl or 3-(methylindole) substituents adopt flat ring shaped conformations which stack through backbone–backbone hydrogen bonding to form nanotubes. High ion transport activities were detected when tested in liposome-based proton transport assays, presumably due to the formation of transmembrane ion channels (*350*). A β-octapeptide designed with a β-hairpin turn or a β-decapeptide designed to form a 3_{14}-helix included either a β³-homohistidine and β³-homocysteine residue or two β³-homohistidine residues, respectively. Both showed evidence of Zn^{2+} complexation (*351*). Another 14-helix was designed with β³-amino acids, which included the nucleo-β-amino acids, (*S*)-γ-(guanin-9-yl)-β-homoalanine, (*S*)-γ-(cytosin-1-yl)-β-homoalanine,(*S*)-γ-(adenin-9-yl)-β-homoalanine, (*S*)-γ-(thymin-1-yl)-β-homoalanine and (*S*)-γ-(1-methyl-guanin-9-yl)-β-homoalanine (see Scheme 11.8). These base pairs were incorporated to allow for formation of an artificial helical bundle via antiparallel base pairing, and evidence for duplex formation was presented (*352*). A helical β³-pentapeptide was designed with asparagine and lysine side chains in order to complex with DNA duplexes. Initial experiments indicated that ordered interactions were present (*353*).

11.2.1d Biological Activities of β-Amino Acids and β-Peptides

A number of properties of β-peptides have been explored, as reviewed in 2001 (*14*). β-Peptides are resistant to common peptidases for at least 2 days (*276, 292*), an important consideration when designing bioactive compounds. A more comprehensive study in 2001 examined the stability of 36 linear and cyclic β- and γ-peptides (from 2- to 15-mer) to 15 different proteases, such as pepsin, chymotrypsin, elastase, β-lactamase, amidase, and 20S proteasome (a vigorous, non-specific protease). Under conditions where α-peptides were completely cleaved within 15 min, the β- and γ-peptides were stable for at least 48 h (*354*). Adding an α-hydroxy, α-fluoro, or α,α-difluoro substituent to one of the internal

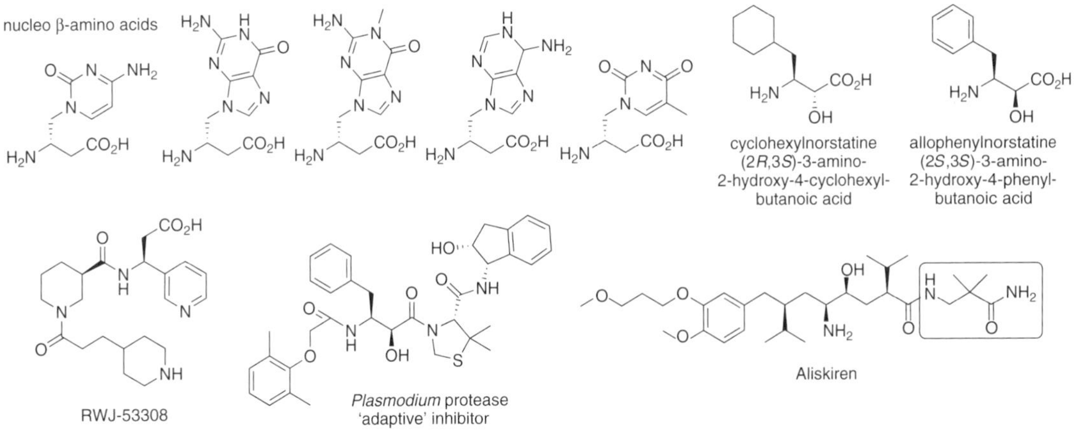

nucleo β-amino acids

cyclohexylnorstatine
(2R,3S)-3-amino-
2-hydroxy-4-cyclohexyl-
butanoic acid

allophenylnorstatine
(2S,3S)-3-amino-
2-hydroxy-4-phenyl-
butanoic acid

RWJ-53308

Plasmodium protease
'adaptive' inhibitor

Aliskiren

Scheme 11.8 β-Amino Acids in Medicinal Chemistry.

β^3-amino acid residues, which might be expected to increase the reactivity of the adjacent amide bond, did not increase proteolysis (355). A mixed tetrapeptide of α- and β-amino acids, Ac-β^3-HPhe-Trp-β^3-HLys-β^3-HThr (with potent somatostatin receptor-binding affinity), was also proteolytically resistant to pronase, proteinase K, carboxypeptidase A, chymotrypsin, and trypsin for at least 48 h (356). β-Ala was used to replace Ser as the N-terminal residue of an 18-mer analog of corticotropin, giving a hormone of increased activity. It was suggested that the increase in activity of this analog, and of other analogs where the L-Ser is replaced with D-Ser or D-Ala, was due to resistance to aminopeptidase activity (357). A direct comparison of the action of hog kidney leucine aminopeptidase on β-Ala-octadecapeptide and L-Ser-octadecapeptide confirmed the resistance of the β-Ala amide to aminopeptidases (357). Tri- and tetra-α-peptides and the corresponding β-peptides with the analogous side chains were injected into insect larvae (tobacco budworm, *Heliothis vireescens*), and the body liquids of the larvae analyzed for the presence of the peptides over time. The α-peptides disappeared quickly (<10% after 3 h), while 40–54% of the β-peptides were still detected after 24 h. In a similar manner, injection of the peptides into cultured maize cells showed <3% of α-peptides after 24 h, but 71–97% of intact β-peptides (358). Due to their proteolytic resistance, concerns have been raised about the potential persistence of β-peptides in the environment. However, sets of microorganisms obtained from soil samples were able to utilize β-tripeptides as the sole carbon and energy source, with a β-dipeptide intermediate observed. The cultures still preferred α-tripeptides as an energy source. A pure microorganism culture could not be isolated that utilized only β-peptides as the carbon source, indicating that a synergistic combination of microorganisms was required (359).

β-Peptides with the ability to translocate across a cell membrane were prepared by making analogs of the α-peptide fragment Tat 48-60, a sequence known to move spontaneously from the extracellular media into nuclei. Studies with β-peptides N-terminal acylated with fluorescein showed clear evidence of fluorescent accumulation within cells, localized to the nuclei within the cells (360). In a similar manner, β-oligoarginines prepared from β^2-homoarginine (2-aminomethyl-5-guanidyl-pentanoic acid) or β^3-homoarginine (3-amino-6-guanidyl-hexanoic acid) demonstrated cell-penetration ability for the longer oligomers (n = 8 or 10) without evidence of erythrocyte lysis (361), while fluorescently labeled fluorescein-(β^3-HArg)$_7$-NH$_2$ and fluorescein-(β^3-HLys)$_7$-NH$_2$ also entered cells without cytotoxicity (362). In a similar manner, β-oligoarginines labeled with the fluorescent dyes 7-nitrobenzo-2-oxa-1,3-diazole (NBD) or fluorescein isothiocyanate (FITC) were used to examine internalization by bacterial cells, apparently via non-endocytotic mechanisms (363). The effects of adding conformational constraints to guanidyl-substituted β-peptides on their cell-penetrating abilities has been studied (364). β-Peptides with seven or nine residues that form amphipathic 3_{14}-helical structures inhibited the transport of lipids through a mimetic system of the brush-border membrane in the small intestine of mammals (365).

Amphiphilic helical β-peptides were designed from β-Val, β-Leu, and β-Lys residues, with the peptides demonstrating length-dependent antibacterial activity that corresponded to their helical content (366). β-Peptides with potent antimicrobial activity (IC$_{50}$ 3–5 mM against *E. coli*) but low hemolytic properties were reported by Liu and DeGrado in 2001, consisting of H-(β^3-HAla-β^3-HLys-β^3-HVal)$_n$-NH$_2$ with n = 4 or 5 (367). Peptides with increased hydrophobicity possessed more undesirable activity against human erythrocytes. The same year the Seebach group employed H-(β^3-HAla-β^3-HLys-β^3-HPhe)$_n$-OH as antibacterial peptides with low hemolytic activity; curiously, the peptides stimulated growth of eukaryotic fungi, possibly acting as an energy source (368). A structure–activity study of cationic antimicrobial β-peptides was reported in 2002, examining the effect of increasing the formation/stability of the 14-helix structure by including increasing proportions of rigid *trans*-2-aminocyclohexane carboxylic acid. Circular dichroism of the 9- to13-mer peptides showed large variations in helical populations that did not correlate with changes in antibiotic activity (*369*). Other antimicrobial β-peptides are described in the Section 11.3.1 on cyclic β-amino acids. A comparison of the antibacterial and hemolytic activity of 31 different β- and γ-peptides was published in 2005 (*370*).

β³-Decapeptides with significant 14-helix character were designed to present the Trp,Trp,Ile epitope found in α-helical peptide HIV fusion inhibitors. The peptides inhibited gp41-mediated cell–cell fusion with EC_{50} = 5–27 μM activity (*371*). A cyclo-β-tetrapeptide composed of β³-residues has been synthesized as a somatostatin analog; it possessed micromolar-level activity compared to the nanomolar activity of the cyclic α-octapeptide on which it was modeled (*372, 373*). However, a mixed tetrapeptide of α- and β-amino acids, Ac-β³-HPhe-Trp-β³-HLys-β³-HThr, possessed 23 nM binding affinity at the somatostatin hsst-4 receptor, and was proteolytically resistant (*356*). Cyclic β³-tripeptides demonstrated antiproliferative activity at low μM concentrations against human cancer cell lines (*374*). Another cyclic tripeptide formed with β-Glu was synthesized as an amide analog of the cyclic triester enterobactin siderophores, and the solution structure studied by NMR spectroscopy (*375*). A cyclo-β-tetrapeptide of β-homophenylalanine has been prepared as a potential molecular scaffold (*376*). The potential of β-amino acids to act as γ-turn mimetics was explored by replacing the D-Phe-Val γ-turn region of a cyclic Arg-Gly-Asp-D-Phe-Val integrin $\alpha_v\beta_3$ antagonist with a D-β-HPhe residue; the resulting cyclic tetrapeptide retained potent (63 nM) and selective integrin affinity (*377*). Polyamides based on amino/carboxyl-substituted *N*-methylpyrrole and *N*-methylimidazole heterocycles can form sequence-specific antiparallel dimeric complexes with the minor groove of DNA. The antiparallel peptide structure was constructed within a single peptide by including α-substituted-β-amino acids as "hairpin turn" linkers (*378, 379*).

β-Amino acids have been used as replacements for α-amino acids in a number of peptides, with early efforts summarized in a 1977 review (*273*). Insertion of single β-amino acids into designed α-amino acid peptides aids the formation of β-hairpin turn structures (*380*). As mentioned earlier, bioactive bradykinin analogs containing β-Ala were prepared in 1966 (*8, 9*) and corticotropin analogs with β-Ala were reported in 1970 (*10, 11*). Pyrrolidine-2-acetic acid (β-proline, although this name has also been used for pyrrolidine-3-carboxylic acid) and β-Phe were also used in bradykinin analogs. The β-Pro bradykinin analog was equipotent to bradykinin, and resisted degradation by dipeptidylcarboxypeptidase (*381*). β-Ala and *N*-methyl β-Ala were used in tri- and tetrapeptide *C*-terminal gastrin antagonists, replacing an Asp residue (*382*). Other β-amino acids have been

incorporated in tetrapeptide gastrin antagonists (*383*). 3-Amino-4-phenylbutyric acid and 3-amino-2-benzyl-propionic acid were used to replace Phe in angiotensin II analogs, but the new compounds had greatly reduced activity (*384, 385*).

Cyclohexylnorstatine (see Scheme 11.8) was incorporated in macrocyclic peptides with potent renin inhibitory activity (*386–388*), and in an orally available human renin inhibitor (*389*). 2,2-Dimethyl-β-alanine is contained in the orally active renin inhibitor aliskiren (*390*). Allophenylnorstatine, (2*S*,3*S*)-3-amino-2-hydroxy-4-phenylbutanoic acid, has been used in HIV protease inhibitors (*391, 392*). It is also the central component of a 70 nM inhibitor of plasmepsin II, a key enzyme in the life cycle of the parasites responsible for malaria (*393*). The latter inhibitor was subsequently converted into a designed "adaptive" inhibitor of a family of *Plasmodium falciparum* proteases, targeting the most important enzyme (plasmepsin II) with high affinity (0.5 nM), but losing little affinity against four other related and essential proteases (*394*). A series of 12 different β-substituted-α,α-dimethyl-β-amino acids were prepared and incorporated into dipeptide inhibitors of chymotrypsin-like serine proteases. The 2,2-dimethyl-3-amino nonanoic acid-based inhibitor had K_i = 4.5 nM vs α-chymotrypsin (*395*).

(*S*)-3-Amino-3-(3-pyridyl)propanoic acid and piperidine-3-carboxylate are components of RWJ-53308 (see Scheme 11.8), an orally active antagonist of the platelet fibrinogen receptor glycoprotein GPIIb/IIIa. The antagonism prevents platelet aggregation and thrombus formation and is therefore potentially useful for treating thrombotic disorders such as myocardial infarction (*396*). Derivatives of homologated Orn and Lys are other potent orally active antagonists of the platelet fibrinogen receptor (*397*). β-Phe and β-Ala were also examined in orally active fibrinogen receptor antagonists (*398*), as was β-(aminomethyl)-β-alanine (*399*), 2,2-dimethyl-3-substituted-β-amino acids (*400*), and a variety of β²- and β³-amino acids (*401*). The β-substituent was determined to have little effect, but modification of the α-substituent resulted in extremely potent compounds (*401*). 3-Amino-4-pentynoic acid is used in xemilofiban, a potent orally available inhibitor of platelet aggregation which also acts via disruption of the platelet–fibrinogen interaction (*402*).

Hybrid conjugates of β-amino acids and α-amino acids have been developed as analogs of MHC (major histocompatibility complex) binding peptides (*403*), while Asp was used as a β-amino acid during the synthesis of a peptide–sugar hybrid (*404*). Replacement of the two L-Pro residues in the antibiotic gramicidin S with β-Ala produced an analog with no antibacterial activity (*405*). Dipeptides containing a 3-amino-3-biarylpropionic acid moiety were potent VLA-4 antagonists, potentially useful for treating inflammatory diseases (*406*). Antagonists of the $\alpha_v\beta_3$ integrin receptor have been prepared based on RGD mimetics with 2-aryl-β-amino acids replacing the Asp residue (*407, 408*). Ligands containing (Ala-Ala-*N*-hydroxy-β-Ala)$_n$ connected with tris(alanylaminoethyl)amine (n = 1,2) were prepared as models of ferrichrome; their iron-complexing properties were explored (*409*).

α-Keto-β-amino acids have found use as enzyme inhibitors in medicinal chemistry programs. Peptides with *C*-terminal α-keto-β-amino acids were found to be more effective then peptidyl fluoromethyl ketones as inhibitors of porcine pancreatic elastase, human neutrophil elastase, and rat and human neutrophil cathepsin G (*410*). A proline-based macrocyclic inhibitor of hepatitis C virus (HCV) NS3 protease employed a norvaline α-keto amide as the P1 active site residue (*411*), while other tripeptide or capped dipeptide inhibitors employed a *C*-terminal 2-keto-3-amino-5,5-difluoropentanoic acid residue (*412*). Dipeptide-derived cell-penetrating α-keto amide calpain inhibitors based on arylalanines were developed as a potential treatment for muscular dystrophy (*413*). The α-aldehyde derivative of Asp, 3-amino-4-oxobutanoic acid, forms a cyclic acetal that acts as an electrophile in dipeptide reversible caspase inhibitors (*414, 415*).

β-Amino acids do not need to be within a peptide exhibit biological activity. β^{2,3}-Amino acids with an α-(3-amidinobenzyl) substituent and β-(phenylvinyl) group were good inhibitors of the blood coagulation Factor Xa (*416*). *N*-Formyl-*N*-hydroxy-2-benzyl-3-aminopropanoic acid is a nanomolar inhibitor of the zinc-containing protease carboxypeptidase A (*417*), while an oxazolidinone formed from 2-benzyl-3-amino-4-hydroxybutanoic acid is a mechanism-based irreversible inhibitor of the same enzyme (*418*). A series of β-, γ-, and δ-amino acids were examined for inhibition of GABA-T (pyridoxal phosphate-dependent 4-aminobutyrate-2-oxoglutarate aminotransferase) in 1981; 4-amino-5-hexenoic acid (vinyl-GABA) was the most potent in vitro but β-difluoromethyl-β-Ala and 2,4-difluoro-3-aminobutyric acid were the most potent in vivo (*419*). The other β-amino acids evaluated included 3-amino-4-fluorobutyric acid, 3-amino-4,4,4-trifluorobutyric acid, 3-amino-4-pentenoic acid, and 3-amino-4-pentynoic acid (*419*). Thiourea derivatives of β^{2,2}-disubstituted amino acids were prepared as bioactive analogs of the HDL-elevating agent gemfibrozil (*420*).

β-Amino acids have been employed for metabolic studies. (3*S*)-[3-^{2}H$_1$]-, (3*R*)-[3-^{2}H$_1$]-, (2*R*)-[2-^{2}H$_1$]-, and (2*S*,3*RS*)-[2,3-^{2}H$_2$]-β-Ala were synthesized in order to determine the stereochemistry of the catabolism product of the RNA base [5-^{2}H]- or [6-^{2}H]-uracil (*421*). In a similar fashion, (2*S*)-3-amino-2-methylpropanoic acid and (2*RS*,3*S*)-[3-^{2}H$_1$]-3-amino-2-methylpropanoic acid helped to illustrate the stereochemistry of catabolism of the DNA base thymine (*422*), while (2*R*)-3-amino-2-fluoropropanoic acid, (2*R*,3*S*)-[3-^{2}H$_1$]-3-amino-2-fluoropropanoic acid, and (2*R*,3*R*)-[3-^{2}H$_1$]-3-amino-2-fluoropropanoic acid were used to demonstrate that catabolism of the anticancer drug 5-fluorouracil proceeded with the same absolute streochemistry as found in catabolism of thymine (*422*).

Some more esoteric macromolecules have been constructed from β-amino acids. Fourth-generation dendrimers were formed using β-alanine to link 3,5-diaminobenzoic acid branching points (*423*). Coordination polymers have been generated from zinc or cadmium salt complexes of 3-(3-pyridyl)-3-aminopropionic acid (*424*). A polyester backbone employed bis(carboxyethyl)piperazine or bis(carboxyethyl)-*N,N′*-dimethyl ethylenediamine units linked by ethylene glycol. The polymers, with tertiary amino groups in the backbone, were degradable and condensed DNA into soluble DNA/polymer particles (*425*). Oligo(3-hydroxybutanoate) was synthesized as a β-peptide surrogate (*403, 426, 427*), while β-peptoids, oligomers of *N*-substituted β-aminopropionic acid, have been reported as analogs of the α-amino acid peptoids (*428*). Poly-β-peptoids can be synthesized by a living alternating copolymerization of *N*-alkylaziridines and carbon monoxide in the presence of a Co catalyst (*429, 430*). The amide linkages in β-peptides have been replaced with thioamides, resulting in conformational changes (*431*).

β-Amino acids have been efficiently transported across a chloroform membrane by a macrocyclic dicopper(II) complex containing two 4,4′-di(aminomethyl)diphenylmethane moieties linked by dodecyloxy-substituted pyridine rings, with a clear preference for β-amino acids (β-Phe) over α-amino acids (Phe) or γ-amino acids (GABA) (*432*).

11.2.1e β-Amino Acid Synthesis

The β-amino acids can present an even more challenging synthetic target than the α-amino acids, as there are potentially at least two chiral centers in the molecule. However, the vicinal separation of the amino and carboxy moieties allows for easier introduction of these functional groups. There are four major approaches to β-amino acid synthesis: (1) amination of a propionic acid/side-chain nucleus; (2) fusion of aminomethyl and methylcarboxy components; (3) attachment of the side chain(s) to the β-amino acid backbone; and (4) elaboration of α-amino acids. Another possibility, carboxylation of the propylamine/side-chain nucleus, is comparatively rare. Many β-amino acids contain an α-hydroxy group, adding to the functional complexity. An unusual lability of α-hydroxy *O*-silyl protecting groups has been observed during saponification/mild acidification of β-amino esters. It is proposed that this is due to neighboring group participation of the deprotected carboxyl group and an NH proton (*433*).

A number of reviews relating to β-amino acid synthesis have been published, including reviews of enantioselective and stereoselective syntheses in 1996 (*434*), 1994 (*435, 436*), and 2002 (*437*), a 1995 monograph on methods of non-α-amino acid synthesis by M.B. Smith (*1*), and a 1997 monograph on enantioselective synthesis edited by E. Juaristi (with a second edition in 2005) (*2*). A microreview of the synthesis of geminally disubstituted β^{2,2}- and β^{3,3}-amino acids appeared in 2000 (*438*), while highlights of new synthetic routes towards enantiomerically pure β-amino acids were surveyed in 2003 (*439*), the same year that the 31st issue of Vol 56 of *Tetrahedron* was devoted to β-lactam chemistry.

Older syntheses of β-amino acids are addressed in Sidgwick's *The Organic Chemistry of Nitrogen* (*440*) and a 1977 review of β-amino acids included in Volume 4 of *Chemistry and Biochemistry of Amino Acids, Peptides and Proteins* (*273*). The latter review also contains a summary of earlier efforts at β-peptide synthesis. β-Peptides were discussed in a 1997 *Chemistry and Engineering News* summary (*278*) and reviewed in more detail in a 1997 *Chemical Communications* article (*276*). A chapter in *Annual Review of Biochemistry* in 1986 examined mammalian metabolism of β-amino acids, and suggested their potential use as analogs of α-amino acids in enzyme inhibitors (*17*). A comprehensive review of β-peptides (128 pages, 531 references), including discussions of natural β-amino acids, the synthesis of β-amino acids, the synthesis of β-peptides, and the properties of β-peptides, was published in 2004 (*13*). The structure and function of β-peptides was reviewed in 2001 (*14*), while syntheses of β²-amino acids, their presence in nature, and their use in β-peptides was reviewed in 2004 (249 references) (*441*).

β-amino acid analogs, such as β-amino sulfonic acids, β-amino phosphonic acids, α-aminoxy acids, and α-hydrazino acids have been prepared, but will not be discussed in this chapter. Neither will the two common α-amino acids that contain a β-amino acid backbone, Asp and 2, 3-diaminopropanoic acid (β-aminoalanine).

11.2.2 Syntheses of Achiral or Racemic β-Amino Acids (see Table 11.1)

11.2.2a Amination Reactions

One common method of β-amino acid synthesis is via conjugate addition of an amine to an α,β-unsaturated acid or ester (see Scheme 11.9). This area was reviewed in 1996 (*442*). In 1902 addition of ammonia to dimethylacrylic acid produced β-aminoisovaleric acid in 92% yield (*443*), while ammonia was reacted with ethyl methacrylate in 1952 to prepare β-aminoisobutyric acid (3-amino-2-methylpropanoic acid) (*34*). The most simple β-amino acid, β-Ala, has been synthesized by addition of ammonia or other amines to acrylonitrile, though the initial report in 1944 did not hydrolyze the cyano group (*444*). Subsequently, β-Ala was prepared in 35% overall yield via reaction of aqueous ammonia with acrylonitrile, followed by hydrolysis with concentrated HCl (*445*). Alkaline hydrolysis of the β-aminopropionitrile can also be employed (*446, 447*).

Various other β-amino acids have been synthesized by this procedure. Crotonic acid was combined with ammonia to give predominantly β-aminobutyric acid, but significant quantities of the dialkylated amine, a symmetrical imine-linked dimer of β-aminobutyric acid, were also isolated (*448*). β-Phenylalanine was prepared in 1935 via addition of ammonia to benzalmalonic acid (2-carboxy-3-phenylprop-2-enoic acid), with decarboxylation occurring upon heating (*449*). β-Methionine (*450*) and β-ethionine (*451*) were synthesized by the reaction of ammonia with γ-(methylmercapto) crotonic acid or γ-(ethylmercapto)crotonic acid, while β-homocysteine was produced via ammonia addition to γ-(benzylmercapto)crotonic acid followed by removal of the *S*-benzyl group (*452*). The β³,³-amino acid 3-amino-3-methylbutanoic acid was prepared by Michael addition of ammonia to 3-methylbut-2-enoic acid at 150 °C (*322*), while addition of ammonia to ethyl 4,4,4-trifluorocrotonate produced 2-amino-4,4,4-trifluorobutyramide (*453*). More recently, ammonia was applied to a synthesis of 3-amino-3-(5-pyrimidine)propanoic acid, with dry ammonia in *tert*-butanol adding to the α,β-unsaturated substrate in quantitative yield. The heterocyclic propenoate was obtained by a Heck coupling of 5-bromopyrimidine and *tert*-butyl acrylate (*454*). The same amination procedure was used in an earlier synthesis of β-amino-β-(4-amino-6-carboxy-5-methylpyrimidine-2-yl)propionic acid, a component of the antitumor antibiotic bleomycin II (*249*).

A variety of other amine sources have been employed as alternatives to ammonia. Improved yields of β-Ala were provided by the addition of phthalimide to acrylonitrile (100%) followed by hydrolysis (80%) (*455*). Potassium phthalimide was also added to ethyl α-fluoroacrylate enroute to α-fluoro-β-Ala (*456*). *N*-Alkyl β-alanines have been synthesized by addition of the appropriate amine hydrochloride to ethyl acrylate in the presence of triethylamine (*457*). Primary and secondary amines added to trimethylsilyl (TMS) acrylate in good yield to give *N*-alkyl or *N,N*-dialkyl β-Ala (*458*), while a number of disubstituted amines were added to ethyl β-bromopropionate, ethyl methacrylate, or acrylonitrile (*459*). Fifteen different secondary amines and anilines were added to ethyl acrylate in a solvent-free system on silica gel, with 50–99% yields (*460*). Alkylamines, including benzylamine, were also added to acrylic acid, methacrylic acid, and itaconic acid to give *N*-alkyl β-Ala, *N*-alkyl β-aminoisobutyric acid or *N*-alkyl β-carboxy-γ-aminobutyric acid in 20–88% yield. The benzyl group could be removed by hydrogenolysis (*461*). Similarly, addition of benzylamine to methyl methacrylate at 70 °C gave a 90% yield of the adduct, with hydrogenation providing methyl 3-amino-2-methylpropanoate in 90% yield (*63*). Perlmutter and

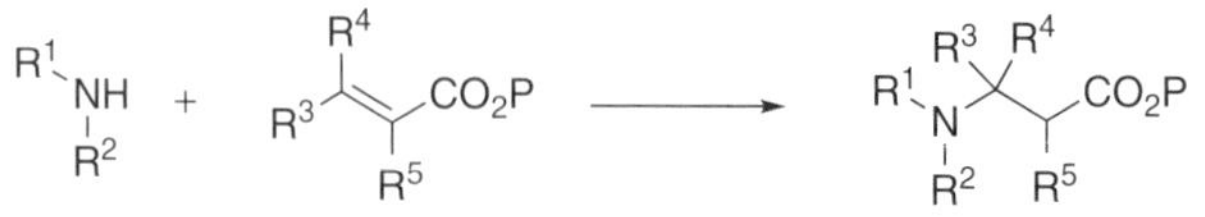

Scheme 11.9 Conjugate Addition of Amines to α,β-Unsaturated Acids.

Tabone have added benzylamine and dimethylamine to a range of α-mono-, β-mono-, and α,β-di-substituted acrylates without catalysts or high pressure in 32–97% yield, although some substrates required extended reaction times (up to 240 h) (462). Acrylonitrile was treated with *N*-acetyl 1,4-diaminobutane and sodium methoxide, with acid hydrolysis generating *N*-(4-aminobutyl) β-Ala (463).

The β-Phe moiety contained within the celacinnine family of alkaloids was generated by addition of *N*-(cyanoethyl) 1,4-diaminobutane to ethyl phenylpropionate, followed by hydrogenation (121). During model studies the same amine was also added to ethyl acrylate. Racemic β-Phe was prepared during a total synthesis of the macrocyclic spermine alkaloid protoverbine via simultaneous conjugate addition and amidation of methyl cinnamate with 2 equiv. of 1,3-diaminopropane (109). A β-amino acid analog of the herbicide phosphinothricin was prepared by addition of benzylamine to ethyl 2-[CH$_2$P(O)(Me)OH]-acrylate (466).

Catalysts have been examined in an attempt to improve addition yields; in the presence of lanthanoid (III) triflates, benzylamine added to ethyl methacrylate in high yield (95%) (464). Ytterbium triflate was combined with high pressure to induce addition of hindered amines to acrylates (465).

Other nucleophilic amines include lithium bis(trimethylsilyl)amide, which added to the α,β-unsaturated lactone contained in coumarins in 60–70% yield (467). Lithium *N*-benzyltrimethylsilylamide, LiN(SiMe$_3$)Bn, was found to add to crotonates in the desired 1,4-manner instead of the 1,2-addition normally observed with lithium amides (468). The conjugate addition adduct could be trapped with electrophiles such as methyl iodide, *n*-octyl iodide, or aldehydes to give 2-substituted-3-aminobutyric acid derivatives, with 2–78% de *anti* stereoselectivity (468, 469). The initial enolate adduct could be generated with very high Z- or E-stereoselectivity, depending on reaction conditions, as shown by trapping as the Z- or E-silyl ketene acetal. Alkylation of the E- or Z-enolates resulted in some differences in *syn:anti* product stereoselectivity, but the diastereoselectivity still varied from 6 to 80% de (469, 470).

More recently, hydrogen azide was found to add to conjugated esters and lactones in high yields when triethylamine was added as a basic catalyst (471). Arylamines were added using an aminomercuration–demercuration procedure, but this only worked with unsubstituted acrylate substrates, as substituted acrylates underwent a deaminomercuration side reaction during attempted reduction of the aminomercury adduct (472).

The imine of benzophenone has also been used as the amine nucleophile, providing β-imino esters that were readily hydrolyzed to β-amino esters in the presence of 0.1 N HCl (473). Both gaseous ammonia and hexamethyldisilazane added to 2-(trifluoromethyl)acrylic acid to give α-(trifluoromethyl)-β-Ala in excellent yield (95–100%) (474). Hydroxylamine hydrochloride was added to aryl-substituted α,β-unsaturated acids, with the β-hydroxylamine converted to an amino group by catalytic hydrogenation in 70–98% yield (475). Alternatively, hydroxylamine reduction was achieved using Zn/NaOH, as recently applied to a synthesis of 3-amino-3-(4-chlorophenyl)propanoic acid (476). In a similar manner, β-(2-chloro-3,4-dimethoxyphenyl)-β-Ala was prepared via addition of hydroxylamine to the substituted cinnamic acid in the presence of sodium ethoxide (477), while β-(2-thiazole)-β-Ala was produced from hydroxylamine and β-(2-thiazole)acrylic acid (136).

A 1949 synthesis of β-benzoylamino acids employed benzonitrile as the amine source. Reaction with α,β-unsaturated esters or β-hydroxy esters in the presence of concentrated sulfuric acid, followed by hydrolysis with KOH, gave 16 different β-amino acids, including tetrasubstituted derivatives (see Scheme 11.10) (478).

α,β-Unsaturated β-amino acids have been produced by treatment of the corresponding alkyne with ammonia (479). α,β-Unsaturated β-amino acids can also be prepared via reaction of β-ketocarboxylates with NH$_4$OAc followed by acylation, which produced mixtures of (E)- and (Z)-isomers (480–482). A 2003 report found that a compound previously described as the (E)-isomer of methyl 3-acetylamino-3-phenylacrylate was in fact methyl (Z)-3-diacetylamino-3-phenylacrylate, with the real (E)-product eluting elsewhere (483). Addition of *N*-benzylaniline to α,β,γ-allenic esters (RO$_2$C1C2=C3=C4R^2) and lactams occurs at the C3 carbon, producing enamino esters and lactams in good yield (484).

In contrast to nucleophilic aminations of α,β-unsaturated esters, nitronium tetrafluoroborate was used to nitrate α,β-unsaturated esters, giving a highly reactive α-carbonyl cation-like species in the initial β-nitro adduct. The carbocation could be trapped in a cyclopropanation reaction, and Wagner–Meerwein rearrangements of substituents were observed (485).

A different amination route to β-amino acids relies on nucleophilic aminations via displacement of β-electrophiles. 2-Hydroxy-3-aminopropionic acid (isoserine) was prepared in 70% yield in 1902 when Fischer and Leuchs combined ammonia and β-chloro-α-hydroxypropionic acid (486). Isoserine was also derived

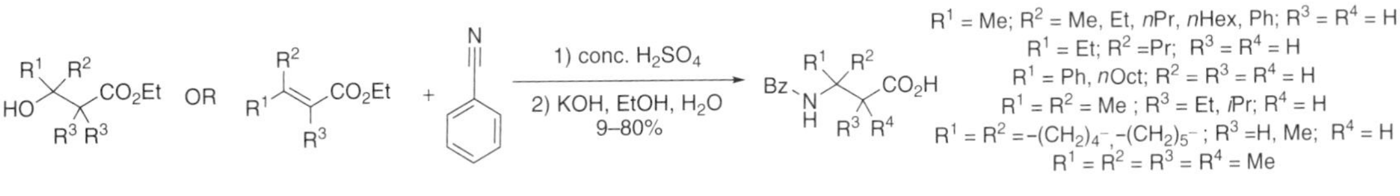

R^1 = Me; R^2 = Me, Et, *n*Pr, *n*Hex, Ph; R^3 = R^4 = H
R^1 = Et; R^2 =Pr; R^3 = R^4 = H
R^1 = Ph, *n*Oct; R^2 = R^3 = R^4 = H
R^1 = R^2 = Me ; R^3 = Et, *i*Pr; R^4 = H
R^1 = R^2 =-(CH$_2$)$_4$-,-(CH$_2$)$_5$- ; R^3 =H, Me; R^4 = H
R^1 = R^2 = R^3 = R^4 = Me

Scheme 11.10 Reaction of Benzonitrile with β-Hydroxyamino Acids (478).

from methyl bromohydroxypropionate, with the bromo-substrate prepared by reaction of methyl acrylate with aqueous bromine in the presence of silver nitrate (*487*). 2-Hydroxy-3-aminobutyric acid was reported in 1906 as a byproduct in the synthesis of 2,3-diaminobutyric acid from 2,3-dibromobutyric acid and ammonia (*488*). The first synthesis of carnosine (β-alanylhistidine) in 1918 relied on displacement of N^α-(3-iodopropionyl) His with ammonia (*26*). A Cahours-type synthesis was used to prepare 3-aminopentadecanoic acid from 3-bromopentadecanoic acid using ammonia as the aminating reagent (*79*).

Other amine nucleophiles have been employed. Methyl 2-bromomethylacylate was treated with $HNBoc_2$, $BocNHNH_2$ or BocONHBoc in the presence of K_2CO_3 to give the various *N*-substituted 2-methylene-3-aminopro-pionates in 45–99% yield (*489*). 3-Amino-4-oxopentanoic acid was derived from methyl 3-bromo-4-oxopentanoate by azide displacement followed by hydrogenation to reduce the azide group (*490*). In 2002 differentially protected tris(aminomethyl)acetate, α,α-bis(aminomethyl)-β-Ala (with *N*-Aloc, *N*-Boc, and *N*-Fmoc protection), was constructed from 2,2-bis(bromomethyl)-1,3-propanediol via displacement of the two bromides with azide, selective reduction and then protection of one azide group followed by the second, followed by a Mitsunobu displacement of one hydroxy group with phthalimide. Oxidation of the second hydroxymethyl group completed the synthesis (*491*). An earlier synthesis employed either this route, an alternative route from 2,2-bis(bromomethyl)-1,3-propanediol via Mitsunobu reaction with phthalimide followed by bis azidation, or a synthesis from 2-bromo-methyl-2-hydroxymethyl-1,3-propanediol via Mitsunobu displacements of two hydroxy groups with phthalimide and azide displacement of the bromide (*289*). In a similar manner, orthogonally protected bis(aminomethyl) malonic acid was prepared in 2001 from triprotected pentaerythritol, 2,2-bis(hydroxymethyl)propane-1,3-diol. Each amino group was introduced by Mitsunobu reaction with phthalimide, followed by phthaloyl deprotection and amine reprotection with Boc or Fmoc. The remaining hydroxymethyl groups were then sequentially oxidized, with the first protected as an allyl ester (*290*).

Phthalimide has been employed for other syntheses, such as a 1961 synthesis of α-fluoro-β-Ala which used potassium phthalimide to displace the methanesulfonate of diethyl α-hydroxymethyl-α-fluoromalonate, followed by hydrolysis/decarboxylation (*492*). 3-Aminotridec-anoic acid was prepared from substrates with a masked carboxyl group, 4-hydroxytetradec-1-ene, 4-hydroxy-pentadec-2-ene, or 2-methyl-4-hydroxypentadec-2-ene, via Mitsunobu amination with phthalimide. The alkene was then ozonolyzed in methanolic NaOH to generate the methyl ester of the final product (*493*). Several achiral *O*-substituted α,α-bis(hydroxymethyl)-β-Ala derivatives were prepared in 1999 from 2,2-bis(hydroxy-methyl)-1,3-propanediol (pentaerythritol). One hydroxy group was displaced with phthalimide via a Mitsunobu reaction and a second oxidized to form the β-amino acid. The two remaining hydroxy groups were functionalized

at various stages of the synthesis, depending on the substituent desired (H, Me, Bn or Bz) (*494*).

The amination of substrates containing both an allylic bromide and α,β-unsaturated system, $ArCH=C(CH_2Br)CO_2Me$, could be controlled between S_N2 displacement or conjugate addition/elimination S_N2' by altering the solvent, with amines in dichloromethane/7 equiv Et_3N producing exclusively the S_N2' 2-methylene-3-amino-3-arylpropionic acid product, and hexane/0.5 equiv. Et_3N favoring the S_N2 2-aminomethyl-3-arylprop-2-enoic acid adduct (*495*). β-Fluoride elimination has been combined with amine conjugate addition in the reaction of azide or amines with β-fluoro-β-perfluoroalkyl acrylates, $F(CF_2)_{n-1}C(F)$ $=CHCO_2Et$ (n = 6 or 8). The initial β-amino-α, β-unsaturated product was reduced by hydrogenation to give the perfluorinated β-amino acid products (*496*). α-Alkylidene-β-amino acids, $H_2NCH_2C(=CHR)$ CO_2Me, can be synthesized by Pd-catalyzed amination of allyl acetates, $RCH(OAc)C(=CH_2)CO_2Me$, with primary and secondary amines. Amino acids were also employed as the nucleophiles (*497*).

Another nucleophilic amination approach to β-amino acid synthesis involves amine opening of epoxides, which leads to α-hydroxy-β-amino acids from 2,3-epoxy acids if regioselective attack at C3 is obtained. A 1879 synthesis of isoserine was actually an attempt to prepare serine, via conversion of acrylic acid to α-chloro-β-hydroxypropionic acid and displacement of the α-chloro group with ammonia. However, the treatment with ammonia instead induced formation of 2,3-epoxy-propionic acid, which was then opened by ammonia to give the β-amino acid (*498, 499*). Racemic phenyliso-serine, the Taxol side chain, was prepared in 1994 via conversion of benzaldehyde to *trans*-methyl-3-phenyl-glycidate by the Darzen's procedure. The epoxide was ammonolyzed to give exclusively the *threo* phenylserine amide isomer (*500*). The racemic amino acid was resolved by crystallization using seed crystals of the desired enantiomer. The *trans*-3-phenylglycidate sub-strate has also been synthesized by a cobalt-catalyzed epoxidation of cinnamate esters and amides (*501, 502*), including by a polyaniline-supported Co(II)salen complex (*503*). Cobalt could also be employed to catalyze the subsequent epoxide opening by aniline or substituted anilines (*501–503*), with the type of aniline used controlling the *syn/anti* diastereoselectivity (*504*). In a similar fashion, both *threo* and *erythro* phenylisoserine were prepared by azide opening of the corresponding *trans* or *cis* epoxy ester, prepared by Darzen's procedure or epoxidation of (Z)-cinnamic acid, respectively (*127, 505*). The enantiomers were enzymatically resolved using lipases for hydrolysis of 2-acyloxy derivatives (*505*). The β-amino acid component of bestatin (*threo*-3-amino-2-hydroxy-4-phenylbutanoic acid) and the homologous component of amastatin (*threo*-3-amino-2-hydroxy-5-methylhexanoic acid) were synthesized by regiospecific ring opening of *cis*-2,3-epoxy acids by ammonia (*506*). The *erythro* isomer could be obtained from the *trans* epoxide. Indium trichloride catalyzed the azidolysis of 2,3-epoxy acids by NaN_3 to give the α-hydroxy-β-azido products with >99:1 regioselectivity and >99% yield (507).

β-Lactones of β-hydroxy acids have also been employed as the electrophilic; nucleophilic opening with O-benzyl hydroxylamine was immediately followed by treatment with $DIAD/PPh_3$ to generate N-benzyloxy β-lactams in a one-pot reaction. The reactions proceeded with inversion of configuration at the β-hydroxy carbon. Samarium iodide reductive cleavage of the N–O bond produced the N-unsubstituted β-lactams (508). A cyclopropyl ring has also been employed as the ring-opening substrate. A nitrene derived from ethyl azidoformate opened the cyclopropyl ring of substituted 1-alkoxy-1-siloxycyclopropanes, giving N-ethoxy-carbonyl β-amino esters in 46–71% yield (509).

Another traditional aminative route to β-amino acids involves classical acid-to-amine rearrangements, applied to one of the two acid groups of 1,4-butanedioic acid equivalents. β-Ala ethyl or methyl esters were easily prepared by a rearrangement of succinimide in aqueous NaOH and bromine, followed by a work-up in alcohol (510). Yields of 65% were obtained in this 1919 procedure. The β-Ala acid could also be directly obtained in 41–45% yield, as reported in an *Organic Syntheses* procedure (511). β-Ala was prepared in 1956 in 66% yield via a Hofmann rearrangement of the monoamide derived from 1,4-butanedioic acid anhydride (512), and in 1959 in 79% yield via a Schmidt rearrangement of the mono ester of 1,4-butanedioic acid (513). Racemic thiomalic acid was converted into N-Boc, N-Cbz or N-Fmoc isocysteine (2-mercapto-3-aminopropionic acid) via protection of the thiol and adjacent carboxyl groups with hexafluoroacetone, followed by conversion of the remaining carboxyl group to the protected amine via a Curtius-type rearrangement in the presence of tert-butanol, benzyl alcohol, or 9-fluorenyl methanol (514). A similar reaction converted homochiral malic acid to isoserine (see Scheme 11.43 in Section 11.2.3f). 2,2,3,3-Tetramethyl-β-Ala was synthesized via a modified Curtius rearrangement of monomethyl 2,2,3,3-tetramethylsuccinate (515). Another synthesis of N-Fmoc, N-Cbz and N-Boc isocysteine, also employed Curtius rearrangements (287).

Schmidt rearrangement of β-substituted-γ-keto esters resulted in preferential formation of the N-acyl β-amino ester product over the γ-carboxamide isomer (see Scheme 11.11), leading to β-Ala, 3-aminobutyric acid, 3-aminopentanoic acid, 3-amino-4-phenylbutanoic acid, and 3-amino-6-carboxyhexanoic acid. The latter product

was converted into β-Lys by rearrangement of the terminal acid group (516). The same method was applied to the preparation of β-Met (451). β-Ala was prepared in 1962 in 29–40% yield via a Beckmann or Schmidt rearrangement of cyclohexane-1,4-dione, followed by hydrolysis of the resulting bis(β-Ala) lactam (517).

An intramolecular Rh-catalyzed C–H bond insertion of a sulfamate ester derivative of 3-phenylpropanol led to a six-membered oxathiazinane through selective γ-C–H insertion of the sulfamate amino group. N-Cbz protection, ring opening with aqueous acetonitrile, and TEMPO oxidation of the hydroxymethyl group produced β-phenylalanine (518). Baylis–Hollman olefins, α-(1-hydroxybenzyl)acrylates, have been aminated via reaction with Burgess reagent to form an O-SO_2NHCO_2Me carbamate derivative. Treatment with NaH induced an allylic rearrangement to form α-(2-aminomethyl)cinnamates, while pyrolysis converted the hydroxy to a carbamate to give α-(1-aminobenzyl) acrylates (519).

Soloshonok and Kukhar have prepared β-fluoroalkyl-β-amino acids by reductive amination of β-keto esters without the need for a reducing reagent, using condensation with benzylamine followed by a [1,3]-proton shift reaction of the initial imine/enamine adduct to an N-benzylidene derivative (520–522). The 3-fluorocarbon substituent is required for the rearrangement into the aldimine; a hydrocarbon enamine is unreactive (521, 522). The nature of the benzylamine substituents affects the reaction rate (522). The same reaction can also produce α-alkyl-β-fluoroalkyl-β-amino acids, with 2-methyl-3-keto-4,4,4-trifluorobutyrate converted into 2-methyl-3-amino-4,4,4-trifluorobutyric acid. Varying the type of base employed for the isomerization altered the diastereoselectivity from 40% de *syn* to 40% de *anti* (523, 524). An earlier reaction employed an isomerization of the imine formed from ethyl acetoacetate and ammonia to produce ethyl β-aminocrotonate (525).

More conventionally, ethyl 3-keto-5,5-dimethyl-hexanoic acid was reductively aminated with ammonium acetate/sodium cyanoborohydride, forming 3-amino-5,5-dimethylhexanoic acid (526). 3-Aminopentanoic acid and 3-amino-2-methylbutanoic acid were both prepared by reduction of the corresponding β-keto esters under the same conditions (527), as was the symmetrical β-amino acid dimethyl 3-aminoglutarate (from dimethyl 3-oxoglutarate) (528). Methyl 3-ketobutanoate was reductively aminated with aniline and a dibutyltin chloride hydride complex to give 3-aminobutanoate (529).

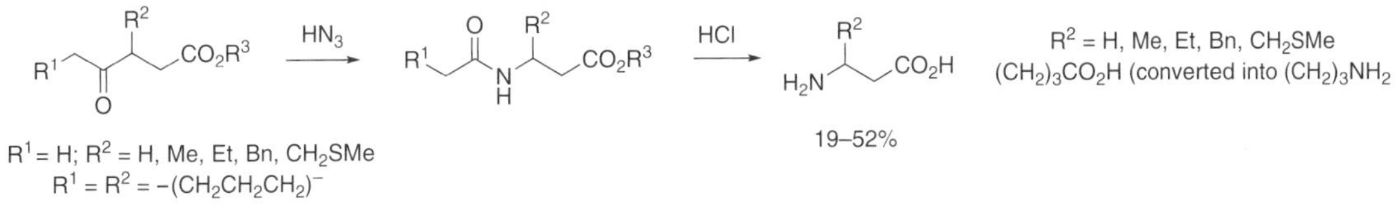

R^1 = H; R^2 = H, Me, Et, Bn, CH$_2$SMe
R^1 = R^2 = –(CH$_2$CH$_2$CH$_2$)$^-$

R^2 = H, Me, Et, Bn, CH$_2$SMe
(CH$_2$)$_3$CO$_2$H (converted into (CH$_2$)$_3$NH$_2$)

19–52%

Scheme 11.11 Rearrangement of γ-Keto-Esters (451, 516).

The asymmetric aminohydroxylation reaction, when applied to α,β-unsaturated amides with Chloramine-T as the nitrogen source, gives racemic products. However, the reaction proceeds with very little catalyst and without ligands to produce α-hydroxy-β-amino amides in high yields with short reaction times (*530*). Similarly, a range of unsaturated acids were aminohydroxylated with 4-40× less osmium catalyst and 3× less Chloramine-T than usually employed for the asymmetric version, without diol byproduct formation, and aqueous reactions conditions could be employed (*531*). Racemic α-hydroxy-β-amino amides were also produced using *tert*-butylsulfonamide, $tBuSO_2N(Cl)Na$, as the amine source, giving products with a more readily removed protecting group (*532*). Osmium-catalyzed aminohydroxylation of Baylis–Hollman olefins, α-(1-hydroxyalkyl)acrylates, produced 1,2-dihydroxy-1-(aminomethyl)alkanoic esters with good *syn* diastereoselectivity (*533*).

11.2.2b Carboxylation Reactions

Lithiated enamines were treated with diethyl carbonate or benzylchloroformate to give α,β-unsaturated β-amino esters (see Scheme 11.12). An asymmetric version of this reaction was developed by using imines with a chiral auxiliary (*534*).

Acylation of methyl ketimines with carbonyldiimidazole in the presence of $BF_3{\cdot}Et_2O$ or LDA followed by reaction with alcohols generated β-enamino esters (see Scheme 11.13) (*535*).

A Passerini-type three-component condensation of acyl cyanides (R^1COCN) with isonitriles (R^2NC) and carboxylic acids (R^3CO_2H) leads to α-acyloxy-α-cyanocarboxamides, $R^1C(CN)(OCOR^3)CONHR^2$, in 32–75% yield (24 examples). Ten of these adducts were converted into α-hydroxy-β-amino acid diamides $R^3CONHCH_2CH(OH)(R^1)CONHR^2$ by a reductive rearrangement induced by catalytic hydrogenation (30–76% yield) (*536*). Anomeric glucosyl isonitriles have been employed as a chiral isonitrile component, but induced minimal diastereoselectivity (*537*). A Ugi-like condensation of 2-phenylaziridine with *tert*-butyl isocyanide and various acids in the presence of LiOTf produced acyl derivatives of 2-phenyl-3-aminopropionic acid *tert*-butyl amide in 71–80% yield (*538*).

Coupling of carbon monoxide and *N*-alkyl aziridines proceeds in the presence of a cobalt catalyst to produce poly (*N*-alkyl-β-alanine) (*429, 430*). Alternatively, a carbonylative ring expansion of 2-arylaziridines with CO in the presence of a Rh-dendrimer catalyst produced 4-aryl-β-lactams (*539*). The carbonylative ring expansion has also been carried out with Ti or Al catalysts and $[Co(CO)_4]^-$ (*540*).

11.2.2c Fusion of Aminomethyl and Acetate Components

A number of different methods have been developed for forming the central C2–C3 bond of β-amino acids, based on either nucleophilic aminomethyl equivalents reacting with electrophilic carboxymethyl equivalents, or nucleophilic acetate equivalents adding to electrophilic aminomethyl synthons.

Cyanide has been used as an anionic aminomethyl equivalent. Displacement of 2-bromopropionic acid

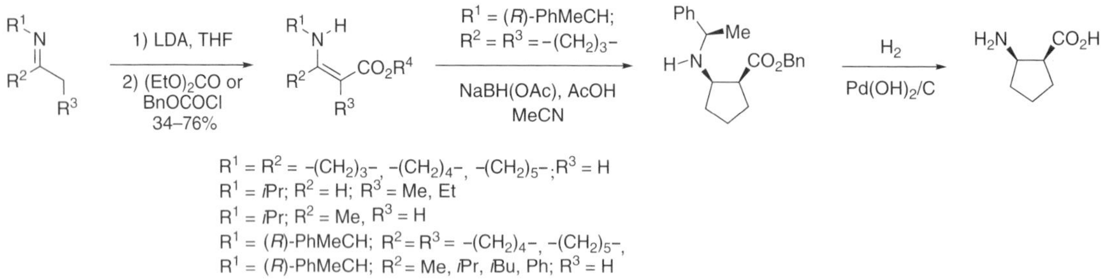
1) LDA, THF
2) (EtO)2CO or BnOCOCl
34–76%
R1 = (R)-PhMeCH; R2 = R3 = -(CH2)3-
NaBH(OAc), AcOH
MeCN
H2
Pd(OH)2/C
R1 = R2 = -(CH2)3-, -(CH2)4-, -(CH2)5-; R3 = H
R1 = iPr; R2 = H; R3 = Me, Et
R1 = iPr; R2 = Me, R3 = H
R1 = (R)-PhMeCH; R2 = R3 = -(CH2)4-, -(CH2)5-,
R1 = (R)-PhMeCH; R2 = Me, iPr, tBu, Ph; R3 = H

Scheme 11.12 Carboxylation of Lithium Enamines (*534*).

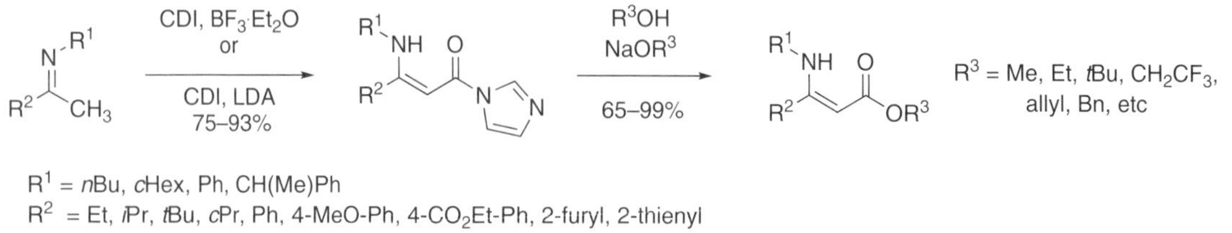
CDI, BF3·Et2O
or
CDI, LDA
75–93%
R3OH
NaOR3
65–99%
R3 = Me, Et, tBu, CH2CF3, allyl, Bn, etc
R1 = nBu, cHex, Ph, CH(Me)Ph
R2 = Et, iPr, tBu, cPr, Ph, 4-MeO-Ph, 4-CO2Et-Ph, 2-furyl, 2-thienyl

Scheme 11.13 Synthesis of β-Enamino Esters (*535*).

followed by hydrogenation of the nitrile group produced β-aminoisobutyric acid (α-methyl-β-Ala) in 30% yield (*541*). An earlier synthesis employed α-chloropropionic acid as the electrophile (*542*), while with K¹⁴CN as the nucleophile [3-¹⁴C]-β-aminoisobutyric acid was prepared (*21*). Reaction of NaCN with 2-bromobutyrate provided 3-amino-2-ethylpropanoic acid after hydrogenation of the nitrile over Raney nickel (*527*). Isoserine was prepared in two steps from ethyl glyoxylate via reaction with TMSCN to form a cyanohydrin intermediate, which was reduced to give isoserine in 85% overall yield (see Scheme 11.14) (*543*, *544*).

Nitromethane and nitroalkanes also function as aminomethyl equivalents, undergoing aldol addition to glyoxalic acid hydrate in the presence of base (NaOH) (*545*). The nitro substituent was reduced by hydrogenolysis to give isoserine or β-substituted isoserines in high yield. Lithiated N-Boc benzylamine is another nucleophilic aminomethyl equivalent, adding to the carbonyl of acrolein in 49% yield with 70% de. The carboxyl group was generated by oxidative cleavage of the alkene, resulting in the Taxol phenylisoserine side chain (*546*). Enamines were condensed with methyl bromoacetate in the presence of acetic acid and zinc or indium in moderate (23–41%) yields (see Scheme 11.15) (*547*).

Reversing the polarity of the reaction requires an electrophilic aminomethyl equivalent. One of the first was diethylaminomethylbenzamide, PhCONHCH₂NEt₂. Alkyl malonates or alkyl cyanoacetates displaced the diethylamino group to produce α-carboxy or α-cyano N-Bz β-amino acids. Hydrolysis with concentrated HBr induced decarboxylation and deprotection to the free

amino acid (*548*). Similarly, N-chloromethylphthalimide was displaced with diethyl benzylmalonate during a synthesis of 2-benzyl-3-aminopropionic acid (*384*). An analog of the γ-amino acid baclofen with the 3-(4′-chlorophenyl)-substituent moved to the 2-position was synthesized by alkylation of diethyl (4′-chlorophenyl) malonate with bromomethylphthalimide, followed by hydrolytic deprotection/decarboxylation (*549*). The dechloro analog, N-phthaloyl 3-amino-2-phenylpropanoic acid, was prepared by reaction of bromomethylphthalimide with the enolate of phenylacetic acid benzyl ester (*550*). A series of other 2-aryl-3-aminopropionic acids (Ar = 4-methoxyphenyl, 4-fluorophenyl, 3,4-dimethoxyphenyl, and 1-naphthyl) were then prepared by this method (*551*). The esters were subsequently hydrogenolyzed, converted into a ketene, and resolved by addition of (R)-pantolactone. These syntheses are more recent versions of the reaction of chloromethylphthalimide with various diethyl alkylmalonates reported in 1959 (*552*).

Other aminomethyl synthetic equivalents of $H_2NCH_2^+$ are N-alkoxycarbonyl-1-methoxyamines, which reacted with esters in the presence of LDA to give predominantly *anti* products in good yield (see Scheme 11.16) (*553*). Addition of Ti(OiPr)₄ produced an excess of the *syn* isomer (*553*). Other reports reacted ketene silyl acetals or deprotonated alkyl acetates with the same α-methoxycarbamate, RCH(OMe)NR²CO₂R³, electrophiles (*554*, *555*). Similarly, N,N-bis(trimethylsilyl) methoxymethylamine, (TMS)₂NCH₂OMe, reacted with ketene silyl acetals in the presence of trimethylsilyl trifluoromethanesulfonate to give N,N-bis(trimethylsilyl)

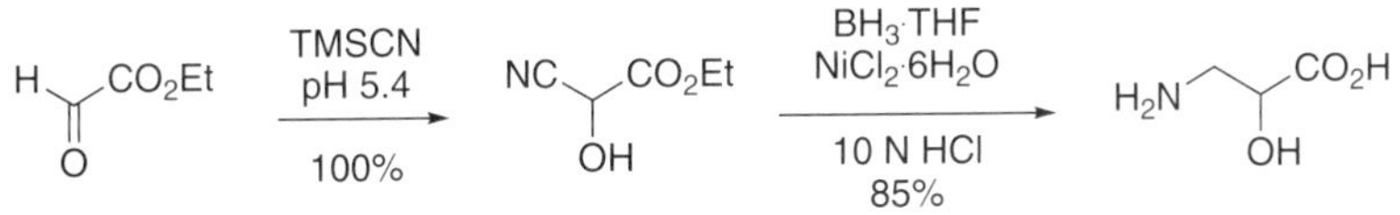

Scheme 11.14 Synthesis Isoserine via Cyanohydrin (*543*, *544*).

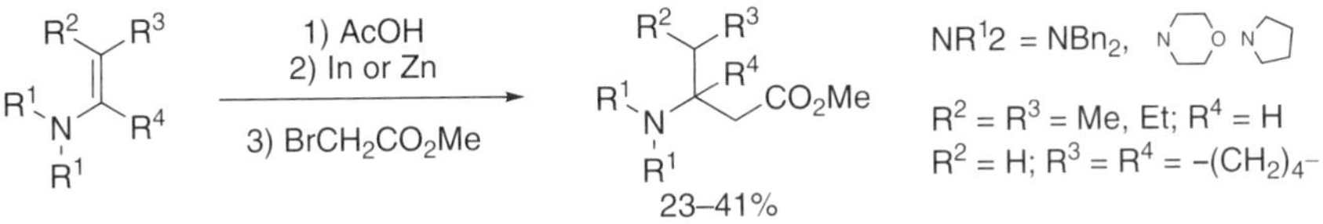

Scheme 11.15 Enamine Addition to Bromoacetate (*547*).

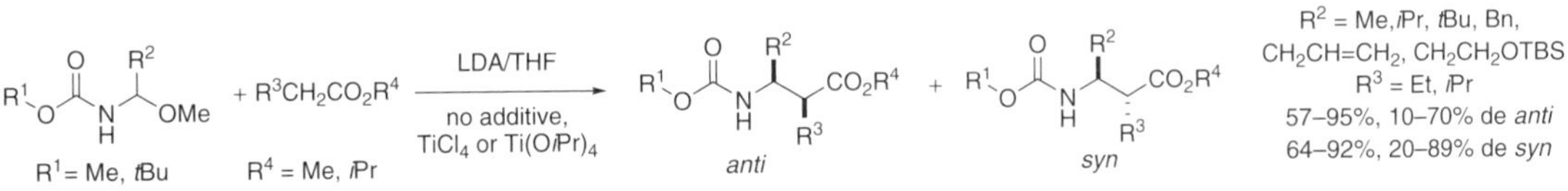

Scheme 11.16 Reaction of 1-Methoxyamines with Enolates (*553*).

methyl esters of α-substituted- or α,α-disubstituted-β-amino acids in 78–95% yield (556). This method was employed to prepare 2,2-dimethyl-β-Ala (420).

Ketene silyl acetals have also been combined with N-(alkylamino)benzotriazoles, prepared from a secondary amine, an aldehyde, and benzotriazole (see Scheme 11.17). A wide range of substituents were introduced by this procedure, although with minimal stereoselectivity (557). A subsequent report found that lithium enolates of *tert*-butyl acetates also reacted in good yield (558). In a similar fashion, Reformatsky reaction of ethyl 2-bromoalkanoates with N-(1-alkylamino)benzotriazoles produced β-amino esters (559). Displacement with a Reformatsky reagent derived from fluorinated ethyl bromoacetate provided α-fluoro-β-amino esters in 77–92% yield (560). Reformatsky reagents were also employed to open 3-benzyl-1,3-oxazinanes, producing N-(3-hydroxypropyl), N-benzyl 3-substituted-β-alanines. The 3-benzyl-1,3-oxazinanes were obtained from aldehydes and 3-(benzylamino)propanol (561). Another aminomethyl electrophile is N-Cbz vinylamine, which underwent a Pd-catalyzed reaction with sodium benzyl acetoacetate followed by carbonylation of the adduct to give 2-acetyl-3-amino-1,5-pentanedioic acid (562).

One common method of β-amino acid synthesis is to add an acetate enolate equivalent to an electrophilic imine (Mannich-type reaction), with β-lactam formation a competing reaction. Indeed, the ester enolate-imine condensation route to β-lactams has been exploited since the beginning of the century and is still widely used (563). Some of the differences in reaction pathways leading to either β-lactam or β-amino ester formation are discussed in a 1989 review, with factors such as the nature of the imine and reaction conditions (temperature, catalysis) affecting the product produced (563). Syntheses leading to β-lactams are discussed in Sections 8.5, 11.2.2f, and 11.2.3s. Addition of ketene bis(trimethylsilyl) acetals to aldimines in the presence of catalytic ZnBr$_2$ provided β-aryl-β-amino acids in excellent yields with no β-lactam formation, although diastereoselectivity was poor (see Scheme 11.18) (564). N-(Trimethylsilyl)imines were also condensed with silyl ketene acetals in the presence of ZnI$_2$ to give N-silyl β-amino esters, which could either be isolated or converted in situ into β-lactams (565, 566). Titanium tetrachloride successfully catalyzed the addition of ketene silyl acetals to Schiff bases, producing metallated β-amino ester intermediates which were hydrolyzed to the amino esters in excellent yield (567, 568). 3-Unsubstituted β-amino acids were produced by reaction of ketene silyl acetals with benzyl N-(chloromethyl)carbamates in the presence of titanium tetrachloride (569). Iminium ions have been prepared from a cyclopropanone hemiaminal under TiCl$_4$ catalysis, and reacted with ketene silyl acetals or lithium enolates to give β-amino acids with the β-carbon contained in a cyclopropyl ring (570). A TiCl$_4$/tertiary amine system provided *syn* β-phenyl-β-amino esters from the N-benzyl imine of benzaldehyde and methyl propionate or methyl phenylacetate in 78–87% yields with 46–100% de (571).

Other effective catalysts for the addition of ketene silyl acetals to various imines are FeI$_2$ or SbCl$_6$, which produced β^{2,3}-disubstituted products with good (50–88% de) *anti* diastereoselectivity (572). Trimethylsilyl trifluoromethanesulfonate catalyzed the addition of silyl ketene acetals to N-phenyl or N-(p-methoxyphenyl) imines with high *anti* stereoselectivity (70–100% de) (573). Microwave irradiation induced formation of β-amino esters in a few minutes when silyl ketene acetals and aldimines were mixed together without solvent in the presence of K$_1$- montmorillonite clay or PTSA (574). An ionic liquid, [emim]OTf, promoted the reaction of ketene silyl acetals with a number of imines in 81–100% yield without the need for an activator. A three-component reaction, combining aldehyde, 4-methoxyaniline, and silyl enolate, was also successful (575).

Other types of imines have been employed as the aminomethyl component. A polymer-supported amine was used to form imines with aldehydes. Addition of silyl ketene acetals and cleavage from the resin with

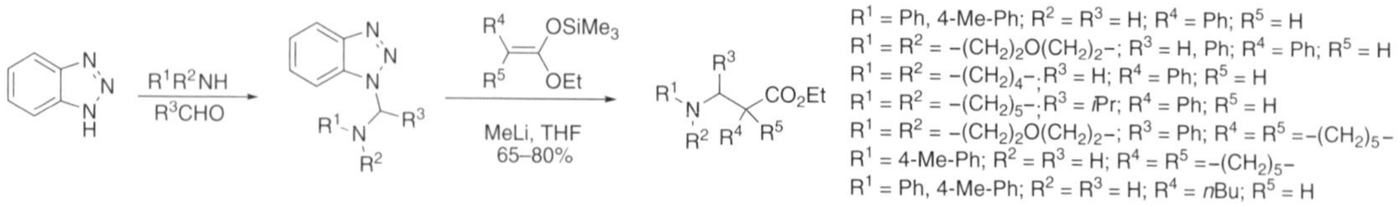

R^1 = Ph, 4-Me-Ph; R^2 = R^3 = H; R^4 = Ph; R^5 = H
R^1 = R^2 = –(CH$_2$)$_2$O(CH$_2$)$_2$–; R^3 = H, Ph; R^4 = Ph; R^5 = H
R^1 = R^2 = –(CH$_2$)$_4$–.R^3 = H; R^4 = Ph; R^5 = H
R^1 = R^2 = –(CH$_2$)$_5$–; R^3 = *i*Pr; R^4 = Ph; R^5 = H
R^1 = R^2 = –(CH$_2$)$_2$O(CH$_2$)$_2$–; R^3 = Ph; R^4 = R^5 =–(CH$_2$)$_5$–
R^1 = 4-Me-Ph; R^2 = R^3 = H; R^4 = R^5 =–(CH$_2$)$_5$–
R^1 = Ph, 4-Me-Ph; R^2 = R^3 = H; R^4 = *n*Bu; R^5 = H

Scheme 11.17 Reaction of Ketene Silyl Acetals and N-(alkylamino)benzotriazoles (557).

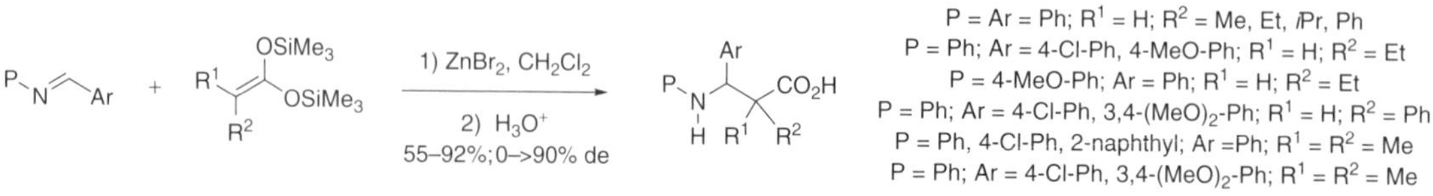

P = Ar = Ph; R^1 = H; R^2 = Me, Et, *i*Pr, Ph
P = Ph; Ar = 4-Cl-Ph, 4-MeO-Ph; R^1 = H; R^2 = Et
P = 4-MeO-Ph; Ar = Ph; R^1 = H; R^2 = Et
P = Ph; Ar = 4-Cl-Ph, 3,4-(MeO)$_2$-Ph; R^1 = H; R^2 = Ph
P = Ph, 4-Cl-Ph, 2-naphthyl; Ar =Ph; R^1 = R^2 = Me
P = Ph; Ar = 4-Cl-Ph, 3,4-(MeO)$_2$-Ph; R^1 = R^2 = Me

Scheme 11.18 Addition of Ketene Bis(trimethylsilyl) Acetal to Aldimine (564).

DDQ gave β-amino esters (576). Ketene silyl acetals have also been reacted with hexahydro-1,3,5-triazines as imine equivalents (577), with N-benzyloxyimines to give N-benzyloxy β-amino esters (578), or with the iminium ion of N-Cbz-2-(aminomethylene)-3-oxobutyrate and related substrates to give 2-carboxy-3-aminopentanedioic acids (579). The γ-carboxy group of the latter adducts could be removed by decarboxylation. LDA-deprotonated 2-arylacetates were added to the O-benzyl oxime of formaldehyde in the presence of catalytic TMS-OTf for the preparation of 2-aryl-β-amino acids (aryl = phenyl, 3-fluoro-phenyl, 3-quinolinyl, 2,3-dihydrobenzofuran-6-yl) for use in $\alpha_v\beta_3$ integrin receptor antagonists (407).

Alternatives to ketene silyl acetal nucleophilic acetate equivalents have been employed for additions to imines. A Reformatsky reagent derived from methyl bromoacetate and Zn/Cu formed exclusively β-amino ester products when an imine with an N-o-methoxyphenyl substituent was used as the electrophile. The N-o-methoxyphenyl substituent was removed by ceric ammonium nitrate in good yield (580). The Reformatsky reagent derived from tert-butyl bromoacetate was added to imines formed from aldehydes and anthracene-9-sulfonamide, providing protected β-amino esters that could be deprotected with Al/Hg followed by 1M HCl (581). The Reformatsky-imine reaction has been catalyzed by 5% NiCl(PPh₃)₃, which was also employed for a multiple-component condensation of aldehydes, amines, and methyl bromoacetate that produced β-substituted-β-amino esters in 58–97% yield on a mmol scale. A library of 64 analogs was also prepared on a μmol scale (582). Reformatsky addition of ethyl bromodifluoroacetate to N-benzyl imines, catalyzed by a Rh complex and Et₂Zn, produced the acyclic 3-amino-2,2-difluoro product instead of the β-lactam if MgSO₄ was added to the reaction (583).

A boron enolate of the tert-butyl thioester of acetic acid was added to benzylamine or β-Ala imines of propanal, isobutyraldehyde, benzaldehyde, cinnamaldehyde, or 2-formyl-6-methoxycarbonyl-pyridine in 43–80% yield (584), while a tin(II) carboxylic thioester enolate, RCH=C(StBu)OSnStBu, added to imines in the presence of stannous triflate to give 1,2-disubstituted-β-amino thioesters with high anti diastereoselectivity (585). A novel niobium enolate of propanoic acid oxazolidinone amide added to aryl aldimines in 23–80% yield with 20–80% de anti stereoselectivity (586).

Allylmagnesium bromide was employed as a masked acetate group for addition to N-tosyl imines generated in situ from the reaction of aldehydes and N-sulfinyl sulfonamides (72–85%) (587). The allyl alkene has been oxidatively cleaved to generate the acid group in other syntheses (e.g., 588). The same Grignard reagent was also added to diarylidenesulfamides (ArCH=N-SO₂-N=CHAr) in 83–92% yield (589).

β-Fluoroalkyl-β-amino-α,β-unsaturated esters were synthesized by addition of lithium ester enolates to imidoyl chlorides (imines with a chloro substituent on the carbon forming the imine), generating exclusively (Z) β-enamino esters (590, 591). The (Z)-β-enamino esters could be diastereoselectively reduced with ZnI₂/NaBH₄, producing syn α-alkyl-β-fluoroalkyl-β-amino esters with 88–96% de (591, 592). Alternatively, alkyl glycolates were reacted with the chloroimine to give an imino ether, with rearrangement in the presence of base forming α-hydroxy-β-fluoroalkyl-β-amino acids (593). 2-tert-Butylthio-3-amino-4,4,4-trifluorobutanoate was obtained by the addition of the lithium enolate of α-tert-butylthioacetate tert-butyl ester to an imidoyl chloride of trifluoroacetaldehyde, ArN=C(Cl)CF₃, followed by diastereoselective NaBH₄ reduction of the imine product (594). 2-Alkyl-Δ²-oxazolines were employed as substituted acetate equivalents, adding to nitriles or imidoyl chlorides to give masked β-enamino acids (see Scheme 11.19). The enamine was then reduced using Na/iPrOH, and the oxazoline converted into a carboxyl group by acid hydrolysis (595).

Several one-pot three-component reactions have also been developed which involve addition of an acetate equivalent to an in situ-generated imine (see earlier in this section for a few other examples). An early version of this reaction, the Rodionow reaction, employed malonic acid as the acetate equivalent, reacting with an aromatic aldehyde in the presence of ammonium acetate in refluxing alcohol to give β-aryl-β-aminoethane dicarboxylic acids. Boiling with acid induced decomposition to the β-amino acid, as well as to cinnamic acid derivatives (596). Benzaldehyde, piperonal or m-nitrobenzaldehyde were condensed with malonic acid and ammonia or methylamine in 1926 (597), while 4-bromobenzaldehyde, ammonium acetate, and malonic acid were combined in 2001 (234). Malonic acid, ammonium acetate, and 3-thienaldehyde gave 3-amino-3-(3-thienyl)propionic acid (598), while 2-furaldehyde and 2-thienylaldehyde led to 3-amino-3-(2-furyl)propionic acid and 3-amino-3-(2-thienyl)

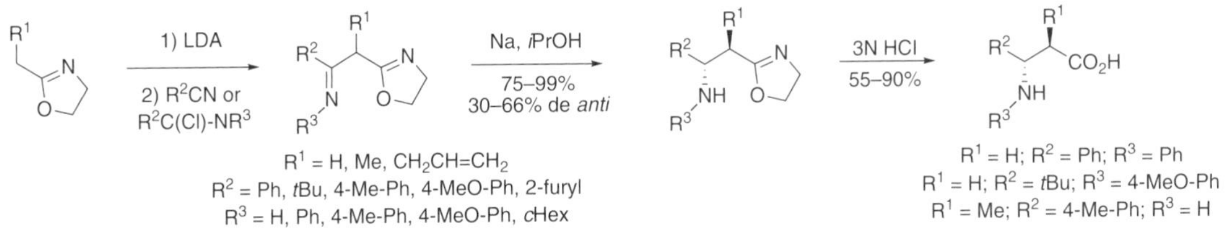

R¹ = H, Me, CH₂CH=CH₂
R² = Ph, tBu, 4-Me-Ph, 4-MeO-Ph, 2-furyl
R³ = H, Ph, 4-Me-Ph, 4-MeO-Ph, cHex

R¹ = H; R² = Ph; R³ = Ph
R¹ = H; R² = tBu; R³ = 4-MeO-Ph
R¹ = Me; R² = 4-Me-Ph; R³ = H

Scheme 11.19 β-Amino Acids from Oxazolines via Enamino Acid Equivalents (595).

propionic acid, respectively (599); these syntheses were repeated in 2002 to give substrates for enzymatic resolution (600). Dimethylamine, formaldehyde, and malonic acid were condensed in 1920 to give N,N-dimethyl 2-(dimethylaminomethyl)-β-Ala (601).

Soloshonok and co-workers (602) and other groups have regenerated interest in this reaction. Imine formation, malonate addition and decarboxylation occur in one pot, with the desired β-amino acid simply precipitating from solution. Soloshonok et al. obtained yields of 51–74% for six examples (see Scheme 11.20). Other more recent applications include syntheses of 3-(3-thienyl)-β-Ala and 3-(2-thienyl)-β-Ala (although no experimental details were reported) (603), a range of β-halothienyl-β-amino acids (604), and a number of β-aryl-β-amino acids (398). Racemic β-Phe was prepared in 75% yield by combining malonic acid, benzaldehyde, and ammonium acetate in refluxing ethanol (605). C2-substituted malonic acids can be used to give α-alkyl-β-aryl-β-amino acids (606). β-Phenyl-, β-(4-chlorophenyl)-, β-(3-nitrophenyl)-, β-(4-methoxyphenyl)-, β-tert-butyl-, and β-isopropyl-β-Ala were synthesized in 2005 (607) showing that aliphatic aldehydes can also be used.

Ytterbium triflate is an effective catalyst for the addition of silyl enolates to imines generated from an aldehyde and amine in the same vessel (608). The β-amino ester products were produced in 62–96% yield, with good syn or anti diastereoselectivity depending on if (E)- or (Z)-enolates were employed. Indium trichloride catalyzed the condensation of aldehydes, aryl amines, and silyl ketene acetal (1-methoxy-2-methyl-1-trimethylsiloxypropene) in water, giving 21–92% yields (609). A polymer-supported scandium catalyst provided yields of 73–88% for aldehydes, amines, and the ketene silyl acetal of methyl isobutyrate (610). Similarly, Sc(OTf)₃ or Cu(OTf)₂ catalyzed the three-component Mannich-type reaction of an aldehyde, o-methoxyphenylamine, and ketene silyl acetals. The reactions were carried out in water using a surfactant (SDS, sodium dodecyl sulfate) to form a micellar system (611). HBF₄ (0.1 equiv) has also catalyzed aqueous Mannich-type reactions in the presence of 1 mol % SDS (612, 613); in the absence of SDS 0.3 equiv of HBF₄ still produced the desired

products (613). Another surfactant, dodecylbenzene sulfonic acid, was itself found to act as a catalyst for the condensation of anilines, aldehydes, and ketene silyl acetals in water (614). Aqueous reactions have also been catalyzed by HBF₄ in methanol–water or isopropanol–water solutions (615). An interesting three-component TiCl₃/pyridine-catalyzed reaction produced β-amino-α-hydroxy esters in good yield and complete syn stereoselectivity by combining methyl phenylglyoxylate, aniline, and aromatic aldehydes (see Scheme 11.21) (616). A three-component condensation of aldehydes, aniline and chlorotitanium enolates of methyl 2-methoxyacetate produced 2-methoxy-3-substituted-β-amino acids in 71–94% yield with high anti stereoselectivity (617).

Another three-component condensation combined aryl aldehydes with TsNH₂ and methyl acrylate in the presence of La(OTf)₃ and either DABCO or 3-hydroxyquinuclidine base, producing α-methylene-β-aryl-β-amino acids in an aza-Baylis–Hillman reaction (618). Ti(OiPr)₄ was subsequently identified as an even better catalyst (619). A one-pot aminoalkylation of aldehydes proceeded by initial reaction of aliphatic or aromatic aldehydes with (trimethylsilyl)dialkylamines in the presence of catalytic lithium perchlorate. The presumed iminium ion intermediate was then treated with a Reformatsky organozinc reagent derived from methyl bromoacetate, giving N,N-disubstituted β-amino esters (620). The consecutive one-pot reaction was converted into a one-pot three-component reaction by combining the aldehyde and (trimethylsilyl)dialkylamine with O-silylated ketene acetals (621). Imines derived from benzylamine and aldehydes were used for a dipolar cycloaddition reaction, reacting with a Rh₂(OAc)₄-generated ylide derived from benzaldehyde and ethyl diazoacetate. The initial oxazolidine adduct was decomposed with PTSA to give α-hydroxy-β-substituted-β-amino acids with good (66–96% de) syn stereoselectivity (622).

Nitriles have also been employed as an electrophilic aminomethyl equivalent, with addition of the Reformatsky zinc enolate to the cyano group of alkylnitriles, a reaction known as the Blaise reaction, producing a range of β-amino-α,β-unsaturated esters in a one-pot reaction (see Scheme 11.22). The nitrile was

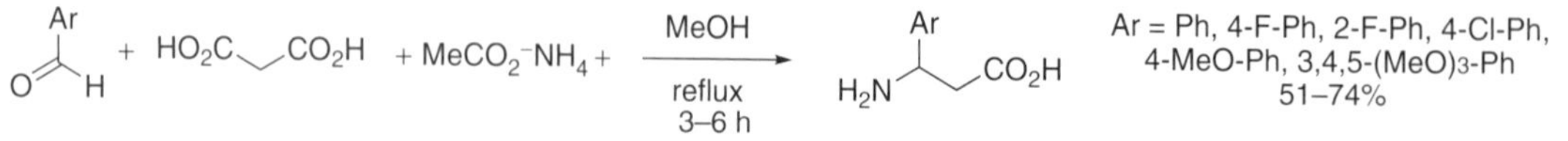

Scheme 11.20 Rodionow Reaction (602).

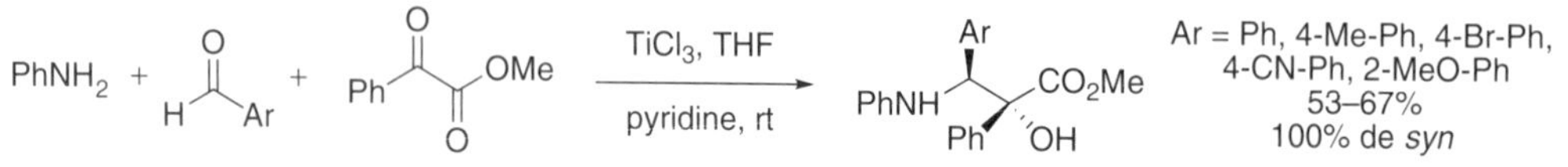

Scheme 11.21 Three Component Reaction (616).

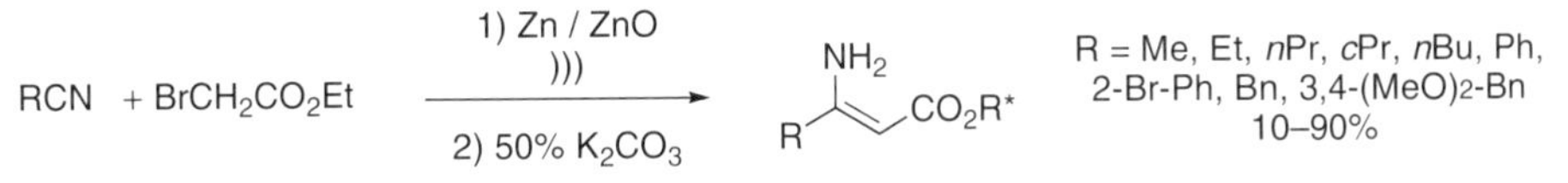

Scheme 11.22 Blaise Reaction (623).

simply combined with ethyl bromoacetetate, zinc powder, and zinc oxide in anhydrous THF and sonicated for 2 h, followed by treatment with 50% K₂CO₃ (623). Formation of a nitrile self-condensation product could be prevented by using less pure 90% zinc. A similar reaction proceeded by addition of the lithium salt of *tert*-butyl acetate to fluoroacetonitrile, followed by reduction of the initial 3-amino-4-fluoro-2-butenoate with sodium cyanoborohydride to provide 3-amino-4-fluorobutanoic acid (624). Cobalt catalysis applied to methyl acetoacetate, benzaldehydes, and acetonitrile in the presence of acetyl chloride produced N-acyl α-acetyl-β-substituted β-amino esters in 44–62% yield (625–627). Improved yields and greater *syn* selectivity were obtained by employing a polyaniline-supported cobalt catalyst (628).

Several other methods have been developed to form the C2–C3 bond of β-amino acids. A Wittig-like reaction was applied to the synthesis of a number of α,β-unsaturated amino acids, which were then hydrogenated to give the β-amino acids (629). The olefination was carried out on N-(acetyl)thioamides using methyl (triphenylphosphorylidene)acetate (see Scheme 11.23).

The carbene derived from methyl diazophenylacetate was inserted into one of the C−H bonds of triethylamine to give methyl N,N-diethyl-3-amino-2-phenylbutyrate in 60% yield (630). Isoserine, with the amino group contained within a piperidine ring, was obtained in 19% by reaction of O-tosyl-hydroxymethyl-piperidine with ethyl diazoacetate, via insertion of the alkylidene into the methylene C−O bond (631). A radical addition–cyclization of an α,β-unsaturated hydroxamate with an allylic oxime ether, H₂C=CHCON(Bn)OCH₂CH=NHOMe, led to a 5-amino-4-substituted-1,2-oxazin-3-one derivative;

hydroxamate hydrolysis and O–N hydrogenation provided 2-propyl-3-amino-4-hydroxybutanoic acid as the product from ethyl radical addition (632).

11.2.2d Elaboration of α-Amino Acids

The Arndt–Eistert homologation of α-amino acids is a common route to β³-amino acids, but it is generally applied to asymmetric syntheses and therefore is discussed in Section 11.2.3m. The Wolff rearrangement component of the Arndt–Eistert homologation converted Gly-derived diazoethylketone to α-methyl-β-Ala (633). Similarly, Wolff rearrangement of 1-phthaloylamino-3-diazo-2-alkanones (derived from amino acid chlorides and diazoalkanes), catalyzed by silver oxide or photolysis in the presence of amines, produced α- or β-substituted β-amino acid amides in good yield (see Scheme 11.24) (634).

Racemic Trp was homologated via reduction of the acid group to an alcohol, activation as the nosylate, displacement with KCN, and hydrolysis with HCl (635). A homologation of racemic 2′-bromophenylalaninal employed reaction with sodium sulfite, followed by KCN, to give the cyanohydrin adduct. Hydrolysis of the cyano group in concentrated HCl produced 2-hydroxy-3-amino-4-(2-bromophenyl)butyric acid as a mixture of four diastereomers. The product was coupled with Leu and the bromo substituent replaced with tritium to give [³H]-labeled ubenimex (204). Condensation of racemic Cbz-Asn with 4-benzyloxy-benzamide oxime gave, after hydrolysis, a mixture of β-[3-(p-hydroxyphenyl)1,2,4-oxadiazol-5-yl]-Ala (16%) and the α-[3-(p-hydroxyphenyl)1,2,4-oxadiazol-5-yl]-β-Ala isomer (32%) (636).

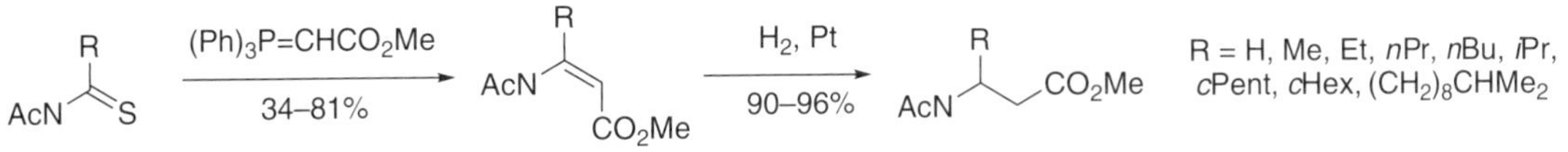

Scheme 11.23 Olefination of N-(acetyl)thioamide (629).

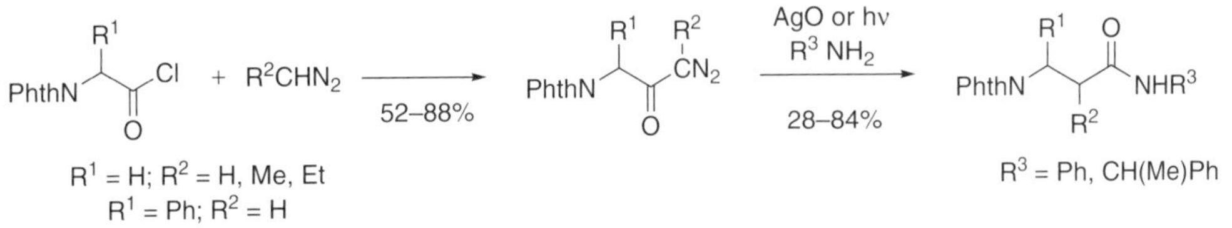

Scheme 11.24 Wolff Rearrangement of 1-Phthaloylamino-3-Diazo-2-Alkanones (634).

11.2.2e Derivatization of β-Amino Acids

An α-substituent was introduced to β-Ala by alkylation of a pyrimidinone template derivative, which was first developed to prepare racemic α-substituted-β-alanines (see Scheme 11.25) (637), but subsequently applied to asymmetric syntheses (see Scheme 11.93). Estermann and Seebach simply α-alkylated the di-deprotonated enolate of racemic N-protected 3-aminobutanoate and obtained diastereoselectivities of 75 to >99% de (anti isomer), depending on the electrophile and N-acyl group (638). This reaction was used by Cardillo et al. to prepare β-amino acid substrates for enzymatic resolution (639). A second alkylation led to the β²,²-amino acid 3-amino-2,2-dimethylpropionic acid (322). Racemic N-Bz 3-aminobutanoic acid and 3-amino-3-phenylpropionic acid methyl esters have been deprotonated with LiHMDS and alkylated with a number of electrophiles at −30 °C to −60 °C with diastereoselectivities of >98% de (640). An α-hydroxy group could then be introduced by deprotonation with NaHMDS and treatment with iodine followed by hydrolysis, with 20 to >98% de of the syn isomer (640). N-alkenyl,N-Boc β-Ala-OEt derivatives were deprotonated with LiHMDS and allylated with allyl iodide (57–73%) (641).

An allyl side chain has been introduced to the α-position of β-substituted β-alanines by free radical C-allylation reactions, which proceeded with up to >96% de anti or 90% de syn stereoselectivity, depending on reaction conditions (642). Alternatively, an Ireland–Claisen enolate rearrangement of β-Ala allyl esters also introduced an α-allyl substituent (643, 644).

Alkyl cyanoacetates have been employed as anionic β-Ala equivalents. Alkylation of ethyl cyanoacetate with acrylonitrile with NaCN as catalyst gave, after catalytic hydrogenation of both nitrile groups, 2-(3-aminopropyl)-β-Ala (146), while dimethylation of ethyl cyanoacetate or cyanoacetamide with methyl iodide followed by nitrile reduction gave 3-amino-2,2-dimethylpropanoic acid (390, 527). Methyl cyanoacetate was deprotonated and dialkylated with 1,2-dibromoethane, 1,3-dibromopropane, 1,4-dibromobutane, and 1,5-dibromopentane. Reduction of the nitrile by hydrogenation over Raney Ni provided 1-(aminomethyl)cycloalkane-1-carboxylic acid derivatives (322). Similarly, dialkylation of ethyl cyanoacetate with EtI, nPrI, Br(CH₂)₃Br or Br(CH₂)₄Br gave the adducts in 64–93% yield. The ester and cyano

groups were both reduced by LiAlH₄; then, after N-protection, the hydroxymethyl group was oxidized back to an acid (53). In the same report the nitrile group of commercially available 1-cyanocyclopropyl-1-carboxylic acid was reduced by hydrogenation over PtO₂ (53). A series of symmetrically and asymmetrically α,α-disubstituted ethyl cyanoacetates were reduced by hydrogenation over Raney nickel in 1959 (645). Reduction of ethyl cyanoacetate with lithium aluminum deuteride followed by benzoylation and oxidation of the amino alcohol gave [3,3-²H₂]-β-Ala (646), while base-catalyzed exchange in D₂O allowed [2,2-²H₂]-β-Ala and [2,2,3,3-²H₄]-β-Ala to be prepared (646). Reaction of ethyl cyanoacetate with ethanol in the presence of HCl gave β-ethoxy-β-aminoacrylate (525).

Electrophilic β-Ala equivalents have also been elaborated. Methyl 3-aminocrotonate, (methyl 3-aminobut-2-enoate) was treated with [hydroxy(tosyloxy)iodo]benzene to give a tosylate salt of methyl (E)-2-phenyliodonio-3-aminocrotonate. Displacement with nucleophiles (Br, NO₂, CN, HNEt₂, morpholine, p-TolSO₂) gave 2-substituted-3-aminocrotonates in 20–83% yield (647). The hydroxyl group of protected β-isoserine was converted into an α-bromo group by treatment with thionyl bromide. The bromine was then displaced with ¹⁸F to give α-[¹⁸F]-fluoro-β-Ala (648). Hydrogenation of ethyl (Z)-3-amino-4-methylpent-2-enoate with ³H₂ produced [2,3-³H₂]-β-Leu-OEt (649).

The alkene in N,N-Boc₂ methyl 2-methylene-3-aminopropionate was employed for radical conjugate addition reactions with alkyl iodides or for Heck reactions with aryl iodides, introducing 2-CH₂R substituents (489), while arylboronic acids coupled to benzyl α-(phthalimidomethyl) acrylate under Rh-catalysis (650). Electrophilic β-lactam templates, 1-aryl-4-substituted-3-phenylthio-3-chloroazetidinones, have been reacted with 1 or 2 equiv of an aryl nucleophile to produce 3-aryl- or 3,3-diaryl products (651). β-Enamino acids with the acid group protected within an oxazoline have been α-halogenated by treatment with N-halosuccinimides in 30–95% yield (652).

Other β-amino acid derivatives have been obtained by functionalization of the amino group. β-Ala-OEt was N-alkylated by heating with 4-phthalimido-1-bromobutane, resulting in putreanine, N-(4-aminobutyl)-3-aminopropionic acid (28). Phosphocreatine analogs were prepared by derivatization of the amino group of β-Ala with (diethylphosphono)-S-methylthioacetamidium iodide (653).

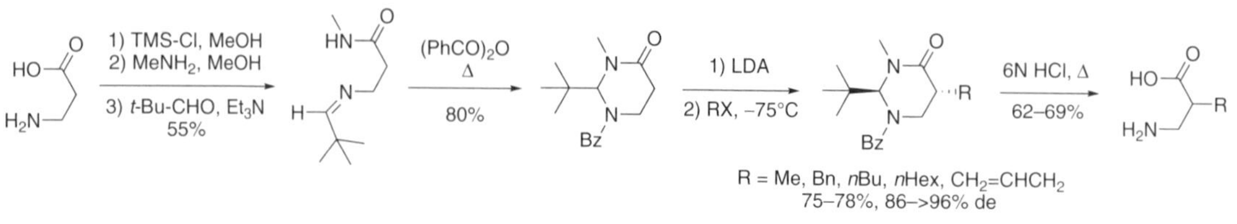

R = Me, Bn, nBu, nHex, CH₂=CHCH₂
75–78%, 86–>96% de

Scheme 11.25 α-Alkylation of β-Alanine in Tetrahydropyrimidinone Template (637).

11.2.2f Small Ring Systems

The synthesis of β-amino acids via amine opening of epoxides is discussed in Section 11.2.2a. Ring opening of aziridine-2-carboxylates by attack at the C2 center generates β-amino acids, but C3 attack produces α-amino acids (see Section 8.2). NaBr causes somewhat regioselective C2 opening (75:25 to 99:1 C2:C3) of 3-substituted-N-ethoxycarbonyl aziridine-2-carboxylates to give the 3-amino-2-bromo acids (654). 2-Substituted acrylates and cinnamates have been converted into 2- and/or 3-substituted aziridine-2-carboxylates by reaction with PhI=NSO$_2$Ar in the presence of a copper catalyst. Ring opening of the N-tosyl or N-nosyl products by hydrolysis, with hydride from NaBH$_4$/NiCl$_2$, or with the amine of α-methylbenzylamine, produced primarily the β-amino acid products (655). Reductive ring opening of N-tosyl aziridine-2-carboxylate or cis-3-phenylethyl-aziridine-2-carboxylate with polymethylhydrosiloxane in the presence of Pd-C gave β-Ala or 3-amino-5-phenylpentanoic acid (656). Racemic N-benzyl ethyl cis-3-CF$_3$-aziridine-2-carboxylate was opened with a variety of nucleophiles (Cl, OH, SPh, SBn, SEt), with the CF$_3$-group directing the ring opening to give exclusively C2 attack (657). β-Aminonitrile precursors of β-amino acids can be obtained from 2-substituted N-tosyl aziridines via ring opening with cyanotrimethylsilane in the presence of Yb(CN)$_3$ catalyst. The cyanide, acting as a carboxy equivalent, selectively attacks at the less substituted carbon of the ring (658). A general review of nucleophilic ring opening of aziridines (not focused on amino acids) appeared in 2004 (659).

Syntheses of β-amino acids via β-lactams have generally been applied to asymmetric syntheses in recent years. As mentioned earlier (Section 11.2.2c), the reaction of imines with acetate equivalents can produce either β-lactams or β-amino acids, depending on reaction conditions, with the cycloaddition of ketene and imine known as the Staudinger reaction and the acyclic addition as a Mannich-type reaction. A microreview of asymmetric syntheses of β-lactams by the Staudinger ketene–imine cycloaddition was published in 1999 (660). When ketenes are generated from acids by reaction with triphosgene and triethylamine, reaction with imines provides exclusively β-lactams (661). Several racemic α-hydroxy-β-amino acids were prepared via a Staudinger cycloaddition reaction of an imine with the ketene generated from an alkoxy acetyl chloride (662). For example, β-lactams were the exclusive product of a [2+2] cycloaddition of benzyloxyketene (generated in situ from benzyloxyacetyl chloride and triethylamine) with the imine formed from cyclohexylacetaldehyde and benzylamine, producing cis-substituted 1-benzyl-3-benzyloxy-4-cyclohexylmethylazetidinone. Lactam hydrolysis and hydrogenolysis gave racemic cyclohexylnorstatine (663). Syn-3-trifluoromethyl-isoserine methyl ester was synthesized by reaction of trifluoromethaldimine with the ketene generated from α-benzyloxyacetyl chloride, followed by lactam hydrolysis (664).

One equivalent of a transition metal (Re) benzophenone imine complex reacted with diphenylketene, phenylethylketene, or cycloheptylketene under much more mild reaction conditions than the corresponding N-silylimine, leading to the lactams of 2,2,3,3-tetraphenyl-3-aminopropionic acid, 2-ethyl-2,3,3-triphenyl-3-aminopropionic acid, or 1-(aminodiphenylmethyl)cycloheptane-1-carboxylic acid after treatment of the Re complex with triflic acid (665). Imines derived from aromatic amines and benzaldehyde reacted with acyloxy or alkyloxy acid chorides and triethylamine to form β-lactams; monocyclic aryl amines produced cis products, but polycyclic aromatics resulted in exclusively trans lactams (666). The Staudinger reaction has been carried on a solid-phase support by using Gly attached to the Wang resin, forming an imine with aldehydes, and then reacting acids in the presence of Mukaiyama's salt, 2-chloro-1-methylpyridinium iodide (667).

A range of 2,2-dimethyl-3-substituted-β-amino acids were prepared from aldehydes via formation of an N-(trimethylsilyl)imine (preformed by reaction of aldehydes with hexamethyldisilazane/nBuLi) followed by ester–imine condensation with the lithium enolate of ethyl isobutyrate. Hydrolysis of the azetidinone intermediate with 6 N HCl gave the racemic amino acids (395, 400). A β-lactam was synthesized via TiCl$_4$-catalyzed condensation of the thiopyridyl ester of 3-(3-cyanophenyl)propionic acid with the imine formed from cinnamaldehyde and p-anisidine. The lactam product was hydrolyzed and converted into Factor Xa inhibitors based on 3-amino-2-[(3-amidino phenyl-5-methyl)]-5-phenyl-hex-4-enoic acid (416). High pressure has been used to promote the cycloaddition of imines with dimethyl ketene acetals or enamines, leading to 2,2-dimethoxy-azetidines or 2-methoxy-2-aminoazetidines (often unstable) that were hydrolyzed to β-amino esters (668).

β-Lactam synthesis via reaction of olefinic substrates with chlorosulfonyl isocyanate is a useful route to β-amino acid precursors (see Scheme 11.26). A large number of β-lactams were prepared by this method in 1963, with some converted to the β-amino acids (669). The cycloaddition could also be carried out with chlorosulfonyl isocyanate, aldehydes, and ketenes (669). More products were prepared in 1970, with the N-chlorosulfonyl β-lactams deprotected by sodium sulfite (670). The method was recently applied to syntheses of 3-amino-5-phenylpentanoate (398), and cis-2-aminocyclo-heptane-, -octane-, and -dodecane-1-carboxylic acid (671).

Lactams can also be formed from the cycloaddition of lithium ynolates, LiOCCR, with N-sulfonyl imines of benzaldehydes. With more activated 2-methoxyphenyl-mines, the initial lactam enolate added to a second equivalent of imine to produce lactams of the diamino acids HO$_2$CCR[CHArNH$_2$]$_2$ (672). The β-lactam 4-acetoxyazetidine-2-one has been converted to an electrophilic β-lactam, 4-(phenylsulfonyl)azetidin-2-one. A vinyl side chain was then introduced by displacement with a Grignard reagent (673). Hydrolysis gave 3-amino-4-pentenoic acid.

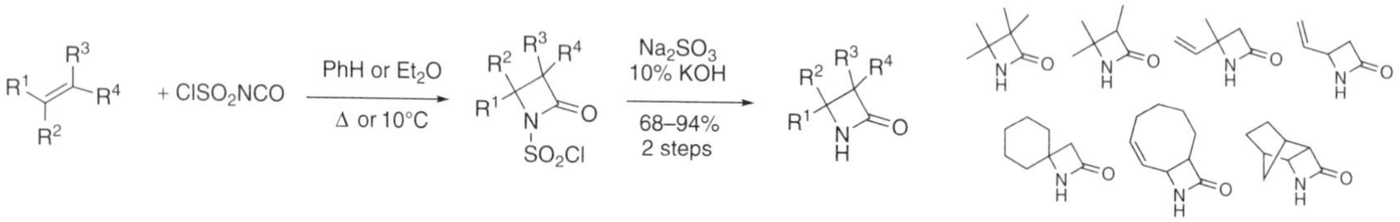

Scheme 11.26 β-Lactam Formation from Alkenes and Chlorosulfonyl Isocyanate (670).

11.2.2g Other Syntheses

Commercially available 5-substituted uracils have been converted into α-substituted-β-alanines by hydrolysis with 8 N HCl (see Scheme 11.27) (674). Other 5-substituted uracils should be accessible from the unsubstituted or 5-bromouracils by electrophilic or nucleophilic reactions.

β-Aminoacrylates have been prepared by a Rh-catalyzed decarboxylative rearrangement of diazo urethanes, prepared by an aldol reaction of ethyl trimethylsilyldiazoacetate with aldehydes (see Scheme 11.28). The structure of the rearrangement product was

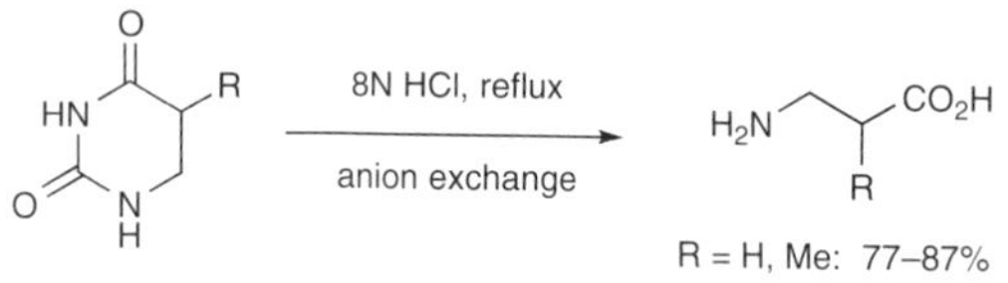

dependent on the substituent, with electron-donating aryl groups forming β-substituted products (675).

Several racemic substituted isoornithines, including isoornithine itself, were prepared via a one-pot butane-dinitrile synthesis, with the nitrile groups hydrogenated over PtO₂ to give the aminomethyl groups (see Scheme 11.29) (676). The compounds were found to be inhibitors of liver ornithine decarboxylase.

Ozonolysis of 2,3-dihydropyrroles produced β-amino esters in 60–88% yield (677). 4-Methoxypyridine has been converted into a series of substituted-2,3-dihydropyridones by addition of Grignard reagents to the N-acylpyridinium salt (see Scheme 11.30). Oxidative cleavage of the alkene with NaIO₄ produced β-substituted-β-amino acids in 83–95% yield (678).

A new Lewis acid-catalyzed allenoate-Claisen rearrangement of allylamines with 2,3-butadienoates produced 3-amino-4,5-disubstituted hepta-2,6-dienoic acids in good yields with high *syn* stereoselectivity for the 4,5-substituents (679).

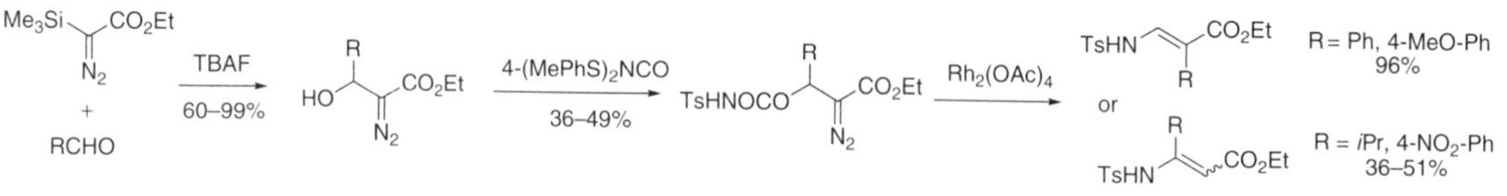

Scheme 11.27 Uracil Hydrolysis (674).

Scheme 11.28 Synthesis of β-Aminoacrylates (675).

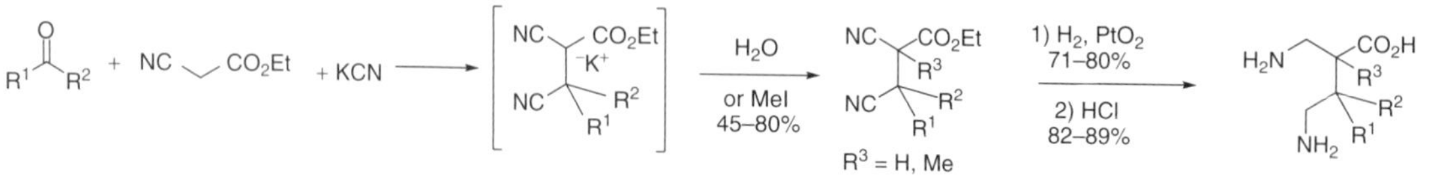

Scheme 11.29 Synthesis of Isoornithines (66).

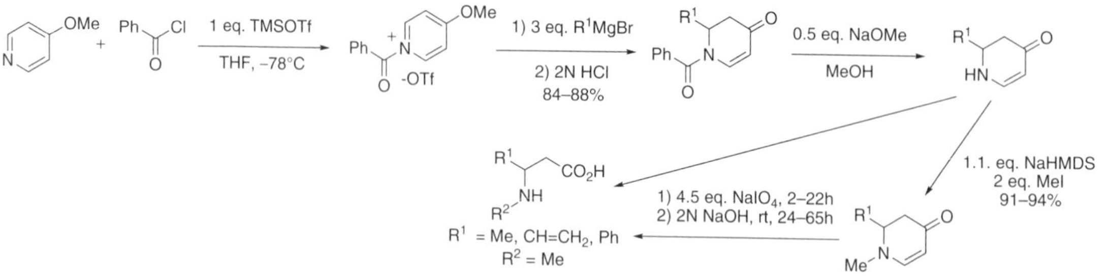

Scheme 11.30 Synthesis via Dihydropyridone (678).

Table 11.1 Syntheses of Achiral or Racemic β-Amino Acids

$$H_2N-\underset{\underset{R^4}{R^3}}{\overset{\overset{R^2}{R^1}}{C}}-CO_2H$$

Amination

R^1	R^2	R^3	R^4	Method	Yield	Reference
Me	Me	H	H	addition of ammonia to dimethylacrylic acid	92%	1902 (443)
H	H	Me	H	addition of ammonia to ethyl methacrylate, hydrolysis	20%	1952 (34)
H	H	H	H	addition of ammonia to acrylonitrile, acid hydrolysis	39% addition / 90–94% hydrolysis	1945 (445)
H	H	H	H	addition of ammonia to acrylonitrile, basic hydrolysis	90% hydrolysis	1945 (446)
H	H	H	H	hydrolysis of β-aminopropionitrile	85–90%	1947 (447)
H	Me	H	H	addition of NH_3 to crotonic acid	mixture of amino acid and imino acid dimer	1911 (448)
Ph	H	H	H	addition of NH_3 to benzalmalonic acid, decarboxylation	32%	1935 (449)
CH_2SMe	H	H	H	conjugate addition of ammonia to $MeSCH{=}CHCO_2H$	64%	1954 (450)
CH_2SEt	H	H	H	conjugate addition of ammonia to $EtSCH{=}CHCO_2H$	26%	1954 (451)
CH_2SBn / CH_2SH	H	H	H	conjugate addition of ammonia to $BnSCH{=}CHCO_2H$, removal benzyl group	78% addition / 89% debenzylation	1956 (452)
Me	Me	H	H	addition of ammonia to $Me_2C{=}CHCO_2H$	100%	1998 (322)
CF_3	H	H	H	conjugate addition of ammonia to $CF_3CH{=}CHCO_2H$	—	1964 (453)
5-pyrimidinyl	H	H	H	addition of NH_3 to 3-(5-pyrimidinyl)-2-propenoate	100%	1993 (454)
4-NH_2-5-Me-6-CO_2H-pyrimidin-2-yl	H	H	H	addition of NH_3 to 3-(5-Me-6-CO_2H-pyrimidin-2-yl)-2-propenoate	20%	1972 (249)
H	H	H	H	addition of phthalimide to acrylonitrile, hydrolysis	100% addition / 80% hydrolysis	1945 (455)
H	H	F	H	Michael addition of KNPhth to ethyl 2-F-acrylate, ester hydrolysis, Phth deprotection	78% addition and ester hydrolysis / 87% deprotection	1964 (456)

(Continued)

Table 11.1 Syntheses of Achiral or Racemic β-Amino Acids (continued)

R¹	R²	R³	R⁴	Method	Yield	Reference
Ph	H	H	H	conjugate addition of 1,3-diaminopropane to methyl cinnamate	50% addition	2000 (109)
H	H	H	H	addition of amines to ethyl acrylate	80% addition	1997 (457)
H	H	H	H	addition of R¹R²NH to TMS acrylate	37–99%	1989 (458)
HNEt₂	H	H	H	addition of piperidine, morpholine, HNEt₂, HN(nPr)₂, HN(nBu)₂, HN(nPent)₂, HN(nHex)₂ to ethyl acrylate/methacrylate or ethyl β-bromopropionate or acrylonitrile	40–88%	1945 (459)
H	H	Me	H			
H	H	H	H	addition of secondary amines and anilines to ethyl acrylate over Si gel with no solvent	50–99% addition	2004 (460)
H	H	H	H	addition of amines to acrylic acid, methacrylic acid, itaconic acid, hydrogenolysis of benzyl group	20–88% addition / 70–100% hydrogenation	1961 (461)
H	H	Me	H			
H	H	CH₂CO₂H	H			
Me	H	H	H	addition of benzylamine or dimethylamine to α,β-unsaturated esters	32–97% / 0–>90% de	1995 (462)
(CH₂)₂OH	H	CH(OH)Me	H			
Me	H	CH(OH)Me	H			
(CH₂)₂OH	H	H	H			
H	H	CH(OH)R R = Me, Et, iPr, Ph, 2-furyl	H			
H	H	H	H	addition of AcNH(CH₂)₄NH₂ to acrylonitrile, hydrolysis	79% addition / 83% hydrolysis	1978 (463)
H	H	Me	H	addition of benzylamines to methyl methacrylate, hydrogenolysis of benzyl group	90% addition / 90% hydrogenation	1996 (63)
Me	H	H	H	addition of BnNH₂ to ethyl methacrylate with catalyst	95% addition	1994 (464)
H	H	H	H	addition of iPrNH₂, tBuNH₂, Ph₂CHNH₂, iPr₂NH, iBu₂NH, iPr(Me)NH to ethyl methacrylate with catalyst at high pressure	10–100% addition	1995 (465)
Me	H	H	H			
H	H	Me	H			
Me	Me	H	H			
H	H	CH₂P(=O)(Me)OH	H	addition of BnNH₂ to H₂C=C[CH₂PO(Me)(OH)]CO₂H	—	1983 (466)
H	H	H	H	addition of N-(cyanoethyl)-1,4-diaminobutane to ethyl acrylate or ethyl phenylpropionate, hydrogenation	82–92% addition / 81% hydrogenation	1981 (121)
Ph	H	H	H			
2-HO-Ph	H	H	H	addition of LiHMDS to coumarins, desilylation with HCl/dioxane	60–70%	1994 (467)
4-MeO-2-HO-Ph	H	H	H			
4-Me-2-HO-Ph	H	H	H			
5-Me-2-HO-Ph	H	H	H			

R1	R2	R3	Conditions	Results	Year (Ref)
Me	H	H	addition of LiNBnTMS to methyl crotonate, trapping of E- or Z-adduct enolate with electrophiles	88% addition	1988 (468)
Me	Me	H		60–96% addition and trapping	1990 (470)
Me	nOct	H		6–80% de	
Me	CH(OH)Ph	H			
Me	H	Me			
Me	H	nOct			
Me	H	CH(OH)Ph			
Me	CH(OH)Me	H	addition of LiNBnTMS to methyl crotonate, trapping of E- or Z-adduct enolate with RCHO	39–73% addition and trapping	1989 (469)
Me	CH(OH)Ph	H		50–64% de $anti$	
Me	H	CH(OH)Me		or 80–86% de syn	
Me	H	CH(OH)Ph			
H	H	H	addition of hydrogen azide to α,β-unsaturated esters or lactones	70–98%	1997 (471)
Bn	H	H			
CH_2OH	H	H			
CH(OH)CH_2OH	H	H			
H	H	H	aminomercuration-demercuration of α,β-unsaturated esters with ArNH, PhNHMe Ar = Ph, 4-Cl-Ph, 4-Me-Ph, 4-MeO_2C-Ph	74–96% aminomercuration 37–47% demercuration	1981 (472)
H	H	H	addition of $Ph_2C{=}NH$ to α,β-unsaturated esters	20–100%	1989 (473)
Me	H	H			
Ph	Me	H			
Me	H	H			
H	CF_3	H	addition of NH_3 or $(Me_3Si)_2NH$ to α,β-unsaturated acid	95–10% addition	1989 (474)
CO_2Et	H	H			
H	H	H			
Ph	H	H	addition of hydroxylamine to α,β-unsaturated acids, catalytic hydrogention	80% hydroxylamine addition 70–98% reduction	1979 (475)
Bn	H	H			
2-MeO-Ph	H	H			
4-MeO-Ph	H	H			
4-NH_2-Ph	H	H			
2,4-$(MeO)_2$-Ph	H	H			
3,4-$(\text{-}OCH_2O\text{-})$Ph	H	H			
4-HO-3-MeO-Ph	H	H			
$CH_2CH(NH_2)$Ph	H	H			
4-Cl-Ph	H	H	addition of hydroxylamine to α,β-unsaturated acids, reduction using Zn/NaOH	61% addition 65% reduction	1997 (476)

(*Continued*)

Table 11.1 Syntheses of Achiral or Racemic β-Amino Acids (continued)

R^1	R^2	R^3	R^4	Method	Yield	Reference
2-Cl-3,4-(MeO)$_2$-Ph	H	H	H	addition of hydroxylamine to α,β-unsaturated acids, hydrolysis	37% addition	1957 (477)
2-Cl-3,4-(HO)$_2$-Ph	H	H	H		90% hydrolysis	
thiazol-2-yl	H	H	H	addition of hydroxylamine to β-(2-thiazole) acrylic acid	30%	1957 (136)
Me	Me	Me	Me	reaction of benzonitrile with β-hydroxy esters or α,β-unsaturated ester with H$_2$SO$_4$	9–80%	1949 (478)
Me	Me	Et	H			
Me	Me	nPr	H			
Bn	H	H	H			
nOct	H	H	H			
Et	nPr	H	H			
Me	Me	H	H			
Me	Et	H	H			
Me	nPr	H	H			
Me	nHex	H	H			
Me	Ph	H	H			
	—(CH$_2$)$_4$—	H	H			
	—(CH$_2$)$_4$—	Me	H			
	—(CH$_2$)$_5$—	H	H			
	—(CH$_2$)$_5$—	Me	H			
iBu	=	H	=	amination of β-ketocarboxylates with NH$_4$OAc	58–67%	1991 (480)
Ph	=	H	=			
Me	=	H	=	amination of β-ketocarboxylates with NH$_4$OAc	—	1999 (482)
Et	=	H	=			
nPr	=	H	=			
iPr	=	H	=			
iBu	=	H	=			
Ph	=	H	=			
Ph	=	H	=	amination of β-ketocarboxylates with NH$_4$OAc	—	2002 (918)
4-F-Ph	=	H	=			
4-Cl-Ph	=	H	=			
4-Br-Ph	=	H	=			
4-Me-Ph	=	H	=			
4-MeO-Ph	=	H	=			
2-Me-Ph	=	H	=			
2-MeO-Ph	=	H	=			
Ph	=	H	=	amination of β-ketocarboxylates with NH$_4$OAc	7–39%	2003 (483)
4-Me-Ph	=	H	=			
4-MeO-Ph	=	H	=			
4-Cl-Ph	=	H	=			

4-F-Ph	=	=	=		
2-MeO-Ph	=	=	=		
3-NO$_2$-Ph	=	=	=		
CH_2OBn	CH_2CN	=	addition of N-benzylaniline to α,β,γ-allenic esters and lactams	82–95%	1997 (484)
$CH_2C(-OCH_2CH_2O-)Me$	CH_2CN	=			
$CH_2C(-OCH_2CH_2O-)CO_2Me$	CH_2CN	=			
$CH_2C(-OCH_2CH_2O-)Me$	$CH_2CH=CH_2$	=			
H	$=CH_2$	—	addition of nitro to α,β-unsaturated esters using NO$_2$BF$_4$, trapping of carbocation (nitro group not reduced)	12–40%	1997 (485)
H	=CHMe	—			
H	=CHEt	—			
H	$-(CH_2)_2-$	—		10–22%	
H	OH	H	amination of β-chloro-α-hydroxypropionic acid	70%	1902 (486)
H	OH	H	amination of methyl 3-bromo-2-hydroxypropionate	39%	1962 (487)
Me	OH	H	byproduct of amination of 2,3-dibromobutyric acid	—	1906 (488)
H	H	H	amination of β-iodopropionyl-His	—	1918 (26)
n-C$_{13}$H$_{27}$	H	H	amination of 3-bromopentadecanoic acid with ammonia	—	1973 (79)
H	$=CH_2$	—	displacement of methyl 2-(bromomethyl) acrylate with HNBoc$_2$, BocNHNH$_2$ or BocONHBoc/K$_2$CO$_3$	45–99%	2001 (489)
COMe	H	H	azide displacement of 3-Br-4-oxopentanoate, reduction	58% displacement 90% reduction	1998 (490)
CH_2NH_2	CH_2NH_2	H	azide displacement of 2,2,bis(bromomethyl)-1,3-propanediol, consecutive azide reductions, Mitsunobu displacement of hydroxyl with phthalimide, hydroxymethyl oxidation	84% azide displacement 45% first azide reduction 82% second azide reduction 54% oxidation	2002 (491)
H	F	H	KNPhth displacement of MsOCH$_2$CF(CO$_2$Et)$_2$, hydrolysis/decarboxylation	40% displacement 41% hydrolysis	1961 (492)
C$_{10}$H$_{21}$	H	H	Mitsunobu amination of nDecCH(OH)-CH$_2$CH=CR$_2$ with phthalimide, ozonolysis	53–84% ozonolysis	1993 (493)
CH_2OR R = H, Me, Bn, Bz	CH_2OR R = H, Me, Bn, Bz	H	Mitsunobu amination of protected C(CH$_2$OH)$_4$ with HNPhth, oxidation of hydroxymethyl	55–88% Mitsunobu amination 60–83% oxidation	1999 (494)

(Continued)

Table 11.1 Syntheses of Achiral or Racemic β-Amino Acids (continued)

R^1	R^2	R^3	R^4	Method	Yield	Reference
Ph	H	=CH$_2$	—	S$_N$2 displacement of ArCH=C(CH$_2$Br)CO$_2$Me with amines in CH$_2$Cl$_2$/Et$_3$N, or conjugate addition/elimination of amines in hexane/Et$_3$N	20–95% S$_N$2′, only S$_N$2′ product, 71–99% S$_N$2, 6:1 to >99:1 S$_N$2:S$_N$2′	2005 (495)
3-Br-Ph	H	=CH$_2$	—			
4-MeO-Ph	H	=CH$_2$	—			
1-naphthyl	H	=CH$_2$	—			
2-thienyl	H	=CH$_2$	—			
2-furyl	H	=CH	—			
H	H	=CH(Ph)	—			
H	H	=CH(3-Br-Ph)	—			
H	H	=CH(4-MeO-Ph)	—			
H	H	=CH(1-naphthyl)	—			
H	H	=CH(2-thienyl)	—			
H	H	=CH(2-furyl)	—			
H	H	CH$_2$NH$_2$	CH$_2$NH$_2$	Mitsunobu displacement of 2 hydroxy groups of 2-bromomethyl-2-hydroxymethyl-1,3-propanediol, azide displacement of third; hydroxymethyl oxidation or Mitsunobu displacement of hydroxy group of 2,2-bis(bromomethyl)-1,3-propanediol, azide displacement of bromides, hydroxymethyl oxidation or azide displacement of 2,2-bis(bromomethyl)-1,3-propanediol, azide reduction, Mitsunobu displacement of hydroxyl with phthalimide, hydroxymethyl oxidation	92% Mitsunobu, 63% bis azidation, 47% oxidation, 68% bis Mitsunobu, 32% azide displacement, 39% oxidation, 49% bis azidation, 51% azide reduction, 66% Mitsubobu, 65% oxidation	2000 (289)
(CF$_2$)$_4$CF$_3$	H	H	H	addition of azide or R^{1}R^{2}NH to F(CF$_2$)$_{n-1}$C(F)=CHCO$_2$Et, hydrogenation	89–90% addition/elimination, 70–80% reduction	1999 (496)
(CF$_2$)$_6$CF$_3$	H	H	H			
H	H	CH$_2$NH$_2$	CH$_2$OH	phthalimide Mitsunobu displacement of triprotected 2,2-bis(hydroxymethyl)-1,3-propanediol, deprotect hydroxyl and repeat, deprotect hydroxyl and oxidize, deprotect hydroxyl and oxidize	98% first Mitsunobu, 96% second Mitsunobu, 56% first oxidation, 63% second oxidation	2001 (290)
H	H	CH$_2$NH$_2$	CO$_2$H			
H	H	=CHPh	—	Pd-catalyzed reaction of amines with allyl acetates RCH(OAc)-C(=CH$_2$)CO$_2$Me	57–78%	2002 (497)
H	H	=CH(4-Cl-Ph)	—			
H	H	=CH(4-MeO-Ph)	—			

H	H	=CHiBu	—	conversion of acrylic acid to α-Cl-β-OH-propionic acid, reaction with ammonia to form epoxide, open epoxide	—	1879 (498)
H	H	=CHiPr	—			1961 (499)
H	H	=CHCH=CHPh	—			
H	H	H	OH			
Ph	H	H	OH	epoxide opening with ammonia	69%	1994 (500)
Ph	H	OH	H	Co-catalyzed epoxidation of cinnamate, Co-catalyzed opening with anilines	45–66% epoxidation / 32–42% opening	1996 (501)
Ph	H	H	OH	epoxide formation by Co(II)-salen catalysis, epoxide opening with anilines	30–50% for opening, *anti* isomer	1997 (502)
Ph	H	OH	H	epoxide formation by polyaniline Co(II)-salen catalysis, epoxide opening with anilines	78–81% epoxide formation / 40–51% opening	2002 (503)
Ph	H	OH	H	Co(II) catalysis epoxide opening with anilines	33–45% for opening / *syn* or *anti* isomer	1997 (504)
H	Ph	OH	H			
Ph	H	OH	H	epoxide opening with azide	60–65% for opening / *syn* or *anti* isomer	1990 (505)
H	Ph	OH	H			
Ph	H	H	OH	epoxide opening with azide	94% for opening / *anti* isomer	2001 (127)
Bn	H	H	OH	epoxide opening with ammonia	68–72%	1980 (506)
iBu	H	H	OH			
Bn	H	OH	H			
H	Me	H	OH	epoxide opening with NaN₃/InCl₃	>99% / >99:1 regioselectivity	2001 (507)
H	nPr	H	OH			
H	Ph	H	OH			
H	Et	Me	OH			
H	Ph	Me	OH			
H	−(CH₂)₄−C2	−(CH₂)₄−C3	OH			
Me	Me	H	OH			
CH₂CH₂Ph	H	H	H	β-lactone opening with BnONH₂, one-pot cyclization to β-lactam with DIAD/PPh₃, N–O cleavage with SmI₂	45–87% one-pot lactam formation / 91–94% N–O cleavage	1999 (508)
CH₂CH₂Ph	Me	H	H			
CH₂CH₂Ph	H	Me	H			
CH₂CH₂Ph	OBn	H	H			
nHept	H	H	H			
CH₂CH(Me)OTBS	H	Me	H			
CH[−OC(Me)₂O−]CH₂	H	Me	H			
H, Me, −(CH₂)₂−, CH(Me)OSiMe₃	H, Me, Ph	H, Me, CH(Me)OSiMe₃		nitrene insertion into 1-alkoxy-1-siloxycyclopropane	46–71%	1989 (509)
H	H	H	H	Curtius rearrangement of succinimide	65%	1919 (510)

(Continued)

Table 11.1 Syntheses of Achiral or Racemic β-Amino Acids (continued)

R¹	R²	R³	R⁴	Method	Yield	Reference
H	H	H	H	Hofmann rearrangement of 1,4-butanedioic acid mono amide	66%	1956 (512)
H	H	H	H	Schmidt rearrangement of 1,4-butanedioic acid mono ester	79%	1959 (513)
H	H	SH	H	amination of thiomalic acid via Curtius rearrangement of hexafluoroacetone-protected	43–57% for amination	1996 (514)
Me	Me	Me	Me	Curtius rearrangement of monomethyl tetramethyl succinate	28% rearrangement	1971 (515)
H	H	SH	H	Curtius rearrangement of protected 2-thiosuccinic acid	77–87%	2004 (287)
H	H	H	H	Schmidt rearrangement of γ-keto esters, hydrolysis	19–50%	1953 (516)
Me	H	H	H			
Et	H	H	H			
Bn	H	H	H			
$(CH_2)_3CO_2H$	H	H	H			
$(CH_2)_3NH_2$	H	H	H			
CH_2SMe	H	H	H	Schmidt rearrangement of γ-keto esters, hydrolysis	52%	1954 (451)
H	H	H	H	Beckmann or Schmidt rearrangement of cyclohexa-1,4-dione, hydrolysis	29–40% rearrangement	1962 (517)
Ph	H	H	H	Rh-catalyzed intramolecular C—H insertion of H_2NSO_2-ester of 3-Ph-propanol, hydrolysis of resulting oxathiazinane, oxidation	91% insertion 80% hydrolysis and oxidation	2001 (518)
Ar	H	=CH₂	—	treatment of $ArCH(OH)C(=CH_2)\text{-}CO_2Me$ with $Et_3NSO_2NCO_2Me$, reaction with NaH or Δ	20–85% OH to NH	2004 (519)
H	H	=CHAr	—		23–94% allylic rearrangement	
Ar = Ph, 4-Me-Ph, 4-MeO-Ph, 3-NO₂-Ph, 4-NO₂-Ph, 4-Cl-Ph, 2-Cl-Ph		Ar = Ph, 4-Me-Ph, 4-MeO-Ph, 3-NO₂-Ph, 4-NO₂-Ph, 4-Cl-Ph, 2-Cl-Ph				
CF₃	H	H	H	reductive amination of β-keto ester using benzylamine and [1,3]-proton shift	70–81%	1994 (520)
C₂F₅	H	H	H			
C₃F₇	H	H	H			
CF₂CF₂H	H	H	H			
(CF₂)₃CF₂H	H	H	H			

774

R¹	R²	R³	Conditions	Yield	Year (Ref.)
CF_3	Me	H	reductive amination of β-keto ester using benzylamine and [1,3]-proton shift to Schiff base, hydrolysis	61–94% Schiff base / 67–84% hydrolyis%	1993 (521)
CHF_2	Me	H			
C_2F_5	Me	H			
C_3F_7	Me	H			
CF_3	H	H			
CHF_2	H	H			
C_2F_5	H	H			
C_3F_7	H	H			
CF_3	H	H	reductive amination of β-keto ester using benzylamine and [1,3]-proton shift	87–95%	1996 (522)
CF_3	Me	H	reductive amination of β-keto ester using benzylamine and [1,3]-proton shift	64–94%	1996 (523)
CF_3	H	Me		40% de syn to 40% de anti	1998 (524)
Me	=	H	imine formation, isomerization of ethyl acetoacetate	90%	1945 (525)
$CH_2C(Me)_3$	H	H	reductive amination of β-keto ester with $NaBH_3CN/NH_4OAc$	85%	1994 (526)
Et	H	H	reductive amination of β-keto ester with $NaBH_3CN/NH_4OAc$	—	1982 (527)
Me	Me	H			
CH_2CO_2H	H	H	reductive amination of β-keto ester with $NaBH_3CN/NH_4OAc$	58%	1990 (528)
Me	H	H	reductive amination of β-keto ester with aniline/Bu_2SnClH	73% reductive amination	1998 (529)
Ph	H	OH	catalytic aminohydroxylation of α,β-unsaturated amides with Chloramine-T, $K_2OsO_2(OH)_4$	95–99%	1998 (530)
4-Br-Ph	H	OH			
3-NO_2-Ph	H	OH			
H	H	OH			
Me	Me	OH			
H	H	OH			
nPr	H	OH			
H	H	OH	catalytic aminohydroxylation of α,β-unsaturated acid Na salts with Chloramine-T, $K_2OsO_2(OH)_4$.	88–98% (1.6:1 to 3.0:1 β-amino to α-amino product with β-aryl substituent)	2001 (531)
H	Me	OH			
H	CH_2CO_2H	OH			
Ph	H	OH			
4-Br-Ph	H	OH			
3-NO_2-Ph	H	OH			
C_6F_5	H	OH			
4-CO_2H-Ph	H	OH			
3-thienyl	H	OH			

(Continued)

775

Table 11.1 Syntheses of Achiral or Racemic β-Amino Acids (continued)

R^1	R^2	R^3	R^4	Method	Yield	Reference
H	H	H	OH	catalytic aminohydroxylation of α,β-unsaturated amides with $tBuSO_2N(Cl)Na$, $K_2OsO_2(OH)_4$	70–93% 1.5:1 to >99:1 regioselectivity	1999 (532)
H	H	Me	OH			
nPr	H	H	OH			
Ph	H	H	OH			
4-Me-Ph	H	H	OH			
4-Br-Ph	H	H	OH			
2,6-Me$_2$-Ph	H	H	OH			
3-NO$_2$-Ph	H	H	OH			
H	H	CH(OH)Me	OH	catalytic aminohydroxylation of α,β-unsaturated esters	65–88% amino-hydroxylation 72–84% de syn	1999 (533)

Carboxylation

R^1	R^2	R^3	R^4	Method	Yield	Reference
Me	=	H	=	carboxylation of lithium enamine to give β-enamino ester	33–76%	1995 (534)
Et	=	H	=			
H	=	Me	=			
H	=	iPr	=			
H	=	tBu	=			
H	=	Ph	=			
(CH$_2$)$_3$-C2	=	(CH$_2$)$_3$-C3	=			
(CH$_2$)$_4$-C2	=	(CH$_2$)$_4$-C3	=			
Et	=	H	=	carbonylation of enamine with CDI, then reaction with ROH to give β-enamino ester	75–93% carbonylation 65–99% ester formation	1998 (535)
iPr	=	H	=			
tBu	=	H	=			
cPr	=	H	=			
Ph	=	H	=			
4-MeO-Ph	=	H	=			
4-CO$_2$Et-Ph	=	H	=			
2-furyl	=	H	=			
2-thiophenyl	=	H	=			
H	H	Me	OH	Passerini condensation of R^{1}COCN, R^{2}NC and R^{3}CO$_2$H, reductive rearrangement	32–75% condensation 30–76% reductive rearrangement	2004 (536)
H	H	n-Hept	OH			
H	H	CH$_2$OMe	OH			
H	H	Ph	H	LiOTf catalyzed condensation of $tBuNC$ with RCO$_2$H and 2-Ph-aziridine	71–80%	2003 (538)
H	H	Ph	H	ring expansion of 2-aryl-aziridines with CO and Rh catalyst	100%	2004 (539)
H	H	4-Br-Ph	H			
H	H	4-Ph-Ph	H			

			Method	Yield	Year (Ref.)
Me	H	H	ring expansion of substituted-aziridines with $Co(CO)_4$ and Ti or Al catalyst	80–95%	2002 (540)
Me	H	Me			
$(CH_2)_4$-C2	H				

Coupling of Aminomethyl and Acetate Equivalents

			Method	Yield	Year (Ref.)
Me	H	H	displacement of 2-bromopropionic acid with CN, reduction	52% addition / 58% reduction	1960 (541)
Me	H	H	displacement of 2-chloropropionic acid with CN, reduction	20% addition / 73% reduction	1943 (542)
Me	H	H with $[3\text{-}^{14}C]$	displacement of 2-bromopropionic acid with $K^{14}CN$, reduction	73% overall	1957 (21)
Et	H	H	displacement of 2-bromobutanoic acid with CN, reduction	—	1982 (527)
OH	H	H	cyanohydrin formation from ethyl glyoxylate, reduction	85% overall	1991 (544) 1993 (543)
OH	H	H	addition of nitroalkane to glyoxalic acid	98–100% addition / 85–96% reduction	1985 (545)
$CH_2CH(OCH_2CH_2O)$	OH	H			
Ph	H	OH	addition of Boc-$NHCH_2Bn$ to acrolein, alkene oxidative cleavage	49% addition / 70% de / 77% oxidation	1993 (546)
iPr	H	H	addition of enamines to bromoacetate	23–41%	1997 (547)
$CH(Et)_2$	H	H			
$-(CH_2)_5-$	—	H			
Me	H	H	reaction of $NaC(R)(CO_2Et)_2$ with $BzHNCH_2NEt_2$, hydrolysis	38–100% alkylation / 58–80% hydrolysis	1957 (548)
Ph	H	H			
Bn	H	H			
Bn	H	H	reaction of $NaC(Bn)(CO_2Et)_2$ with $ClCH_2NPhth$, hydrolysis	76% alkylation / 70% hydrolysis	1970 (384)
Ph	H	H	reaction of $NaC(R)(CO_2Et)_2$ with $BrCH_2NPhth$, hydrolysis	87% alkylation / 70% hydrolysis	1997 (549)
4-Cl-Ph	H	H			
Ph	H	H	reaction of enolate of $PhCH_2CO_2Bn$ with $BrCH_2NPhth$	75%	1998 (550)
4-MeO-Ph	H	H	reaction of enolate of $ArCH_2CO_2Bn$ with $BrCH_2NPhth$	68–78% alkylation	2000 (551)
3,4-$(MeO)_2$-Ph					
4-F-Ph					
1-naphthyl					

(Continued)

Table 11.1 Syntheses of Achiral or Racemic β-Amino Acids (continued)

R¹	R²	R³	R⁴	Method	Yield	Reference
H	H	Me	CO₂Et	reaction of NaC(R)(CO₂Et)₂ with ClCH₂NPhth, hydrolysis	60–83%	1959 (552)
H	H	Me	H			
H	H	Et	CO₂Et			
H	H	Et	H			
H	H	Ph	CO₂Et			
H	H	Ph	H			
H	H	Me	Me	reaction of ketene silyl acetals with (TMS)₂NCH₂OMe, hydrolysis	—	2002 (420)
H	H	–(CH₂)₃–	—			
H	H	–(CH₂)₄–	—			
H	H	–(CH₂)₅–	—			
Me	H	Me	Me	reaction of α-methoxycarbamate with ketene silyl acetals	66–93%	1984 (554)
Me	H	Ph	Me			
Me	H	Me	Ph			
Me	H	nBu	H			
iPr	H	Me	Me			
iPr	H	Ph	H			
Me	H	H	H	reaction of α-methoxycarbamate with LDA enolate of alkyl acetates	62–87%	1988 (555)
iPr	H	H	H			
allyl	H	H	H			
H	H	Me	Me	reaction of ketene silyl acetals with (TMS)₂NCH₂OMe, hydrolysis	78–95% addition	1984 (556)
H	H	Me	H			
H	H	Ph	H			
H	H	–(CH₂)₄–	—			
H	H	–(CH₂)₅–	—			
H	H	Ph	H	addition of ketene silyl acetal to N-(1-alkylamino)benzotriazole	65–80%	1990 (557)
H	H	nBu	H			
Ph	H	Ph	H			
iPr	H	Ph	H			
H	H	–(CH₂)₅–	—			
Ph	H	–(CH₂)₅–	—			
H	H	H	H	addition of Li tert-butyl acetates to N-(1-alkylamino)benzotriazole	35–70%	2002 (558)
H	H	OH	H			
iPr	H	OH	H			
Bn	H	H	H			
Bn	H	OH	H			
4-Me-Ph	H	H	H			
4-Me-Ph	H	OH	H			
4-Me-Ph	H	Ph	H			

Ph	H	H	H	addition of $ZnBrCH(R)CO_2Et$ to N-(1-alkylamino)benzotriazoles	51–88%	1989 (559)
Ph	H	OH	H			
Ph	H	Ph	H			
iPr	H	H	H			
Ph	H	Me	H			
iPr	H	Me	H			
Ph	H	Me	Me			
iPr	H	Me	Me			
Ph	H	Me	Me			
Ph	H	Me	Me			
H	H	Me	Me			
Ph	H	Me	Me			
nPr	H	H	H			
H	H	F	F	Reformatsky displacement of N-(α-aminoalkyl)benzotriazoles by fluorinated ethyl bromoacetates	77–92%	1998 (560)
iPr	H	F	F			
CH_2CH_2Ph	H	F	F			
Me	H	H	H	addition of $ZnBrCH_2CO_2Et$ to 3-Bn-1,3-oxazinane	72–96%	1990 (561)
Et	H	H	H			
iPr	H	H	H			
nBu	H	H	H			
iBu	H	H	H			
CH_2CH_2Ph	H	H	H			
Ph	H	H	H			
4-Cl-Ph	H	H	H			
4-MeO-Ph	H	H	H			
CH_2CO_2Me	COMe	H		Pd-catalyzed reaction of vinylNHCbz with Na acetoacetate, then CO	92%	1989 (562)
Ph, 4-Cl-Ph, 4-MeO-Ph, or 3,4-$(MeO)_2$-Ph	Me, Et, iPr, Ph	H, Me	H	addition of ketene bis(trimethylsilyl) acetal to imine in presence of $ZnBr_2$	55–92% 0->90% de	1993 (564)
Ph	H	H		reaction of N-TMS imine with silyl ketene acetals	27–78% after in situ conversion to β-lactam	1988 (565)
Ph	Et	H				
Ph	iPr	H				
2-furyl	OPh	H				
2-furyl	Et	H				
CCPh	Me	H				
$CCSiMe_3$	Me	H				
$CCSiMe_3$	Et	H				
$CCSiMe_3$	Ph	H				
Ph	Me	Me				
2-furyl	Me	Me				
CCPh	Me	Me				
$CH=CHSiMe_3$	Me	Me				

(Continued)

Table 11.1 Syntheses of Achiral or Racemic β-Amino Acids (continued)

R¹	R²	R³	R⁴	Method	Yield	Reference
Ph	H	Me	Me	addition of ketene silyl acetal to N-TMS imine with ZnI_2 catalysis, in situ cyclization to lactam	27–82%	1985 (566)
2-furyl	H	Me	Me			
CCPh	H	Me	Me			
CH=CHTMS						
Ph	H	H	Me			
2-furyl	H	H	OPh			
Ph	H	Et	H			
CCPh	H	Me	H			
CCTMS	H	Me	H			
CCTMS	H	Ph	H			
Ph	H	Me	Me	addition of ketene bis(trimethylsilyl) acetal to imine in presence of $TiCl_4$	70–92%	1977 (567)
Ph	H	Me	H			
Ph	H	$-(CH_2)_5-$	—			
Bn	H	Me	Me			
2-furyl	H	Me	Me	addition of ketene bis(trimethylsilyl) acetal to imine in presence of $TiCl_4$	54–99%	1981 (568)
2-thienyl	H	Me	Me			
2-pyridyl	H	Me	Me			
Ph	H	OPh	Me			
Ph	H	OPh	H			
Ph	H	SPh	Me			
2-furyl	H	OPh	Me			
2-thienyl	H	OPh	Me			
2-furyl	H	SPh	Me			
2-thienyl	H	SPh	Me			
H	H	Me	Me	addition of ketene silyl acetal to $CbzNRCH_2Cl$ with $TiCl_4$	81–90%	1981 (569)
H	H	Me	H			
H	H	$-(CH_2)_5-$	—			
H	H	OPh	Me			
$-(CH_2)_2-$	—	H	H	addition of ketene silyl acetals or Li enolate to iminium ion of cyclopropanone	11–57%	1991 (570)
$-(CH_2)_2-$	—	NC	H			
Ph	H	Et	H	addition of Li enolate to N-Bn imine of PhCHO	78–87% yield	2005 (571)
Ph	H	Ph	H		46–100% de *syn*	
Ph	H	Me	H	addition of ketene silyl acetal to imine with FeI_2 or $TrtSbCl_6$ catalysis	47–99%	1990 (572)
2-thiophene	H	Me	H		48–96% de *anti*	
2-furyl	H	Me	H			
cHex	H	Me	H			

R1	R2	R3	R4	Conditions	Yield	Selectivity	Year (Ref)
Ph	H	Me	H	addition of ketene bis(trimethylsilyl) acetal to imine in presence of TFMSOTf	65–85%	70–100% de *anti*	1987 (573)
Ph	H	Ph	H				
CH=CHPh	H	Ph	H				
CC-TBDMS	H	Me	H				
CC-TBDMS	H	Ph	H				
Ph	H	Me	H	addition of ketene silyl acetal to imine with microwave irradiation	43–90%	32–56% de *anti*	1993 (574)
Ph	H	Me	Me	addition of ketene silyl acetal to imine in [emim]OTf ionic liquid	81–100% condensation	32:68 to 61:39 *syn:anti*	2005 (575)
4-Me-Ph	H	Me	Me				
4-Cl-Ph	H	Me	Me				
4-NO$_2$-Ph	H	Me	Me				
CH=CHPh	H	Me	Me				
2-thienyl	H	Me	Me				
2-furyl	H	Me	Me				
4-pyridyl	H	Me	Me				
Ph	H	OTBS	H				
Ph	H	OBn	H				
Ph	H	Me	H				
Ph	H	H	H				
Ph	H	—(CH$_2$)$_4$—					
cHex	H	Me	Me				
cHex	H	OTBS	OTBS				
Ph	H	H	H	addition of LDA-deprotonated 2-arylacetate ethyl ester to *O*-Bn oxime of formaldehyde	—		2002 (407)
3-F-Ph	H	H	H				
quinolin-3-yl	H	H	H				
2,3-dihydrobenzofuran-6-yl	H	H	H				
Ph	H	OBn	H	addition of silyl ketene acetal to resin-supported imine, cleavage	50–84%		1998 (576)
Ph	H	Me	Me				
CH$_2$CH$_2$Ph	H	Me	Me				
4-Cl-Ph	H	Me	Me				
4-Me-Ph	H	Me	Me				
2-pyridyl	H	Me	Me				
cHex	H	Me	Me				
H	H	H	H	reaction of hexahydro-1,3,5-triazines with ketene silyl acetals	20–87%		1983 (577)
H	H	Me	H				
H	H	Me	Me				
H	H	Me	Me	reaction of *N*-benzyloxyimines with ketene silyl acetals with TMSOTf catalyst	—		1983 (578)
Me	H	H	H				

(*Continued*)

Table 11.1 Syntheses of Achiral or Racemic β-Amino Acids (continued)

R^1	R^2	R^3	R^4	Method	Yield	Reference
$CH(CO_2Et)_2CH-(CO_2Et)Ph$	H	Me	H	addition of ketene silyl acetals to iminium ion of N-Cbz-2-(aminomethylene)-3-oxobutanoate, decarboxylation	90–97% yield addition 0% de	1996 (579)
$CH(CO_2Et)_2Ph$	H	H	H			
$CH(CO_2Et)COMe$	H	Me	H			
$CH(CO_2Et)COMe$	H	H	H			
CH_2COMe	H	Me	H			
CH_2COMe	H	Me	H			
Ph	H	H	H	addition of $BrZnCH_2CO_2Me$ to imine with N-o-MeO-Ph substituent	51–93%	1999 (580)
4-Me-Ph	H	H	H			
4-MeO-Ph	H	H	H			
4-Cl-Ph	H	H	H			
2-naphthyl	H	H	H			
2-furyl	H	H	H			
tBu	H	H	H			
cHex	H	H	H			
Ph	H	H	H	addition of $BrZnCH_2CO_2tBu$ to imine formed from RCHO and (9-anthracenyl)SO_2NH_2	26–66% imine formation 18–99% Reformatsky addition 63–72% deprotection	1993 (581)
4-F-Ph	H	H	H			
4-MeO-Ph	H	H	H			
4-CN-Ph	H	H	H			
2-furyl	H	H	H			
(E)-CH=CHPh	H	H	H			
4-Cl-Ph	H	H	H	one-pot multicomponent condensation of aldehydes and amines with $BrZnCH_2CO_2Me$ catalyzed by Ni(Cl)(PPh$_3$)$_3$	58–97% yield	2003 (582)
3-pyridyl	H	H	H			
CH=CHPh	H	H	H			
$CH_2OTBDMS$	H	H	H			
2-furyl	H	H	H			
tBu	H	H	H			
$CH_2CH_2OTBDMS$	H	H	H			
CH_2CH_2Ph	H	H	H			
Ph	H	F	F	addition of $BrZnCF_2CO_2Et$ to N-Bn imines with Rh catalyst and $MgSO_4$	59–65%	2005 (583)
4-MeO-Ph	H	F	F			
4-Cl-Ph	H	F	F			
Et	H	H	H	addition of boron enolate of MeCOStBu to RCH=NR′	43–80%	1981 (584)
iPr	H	H	H			
Ph	H	H	H			
CH=CHPh	H	H	H			
(6-MeO$_2$C-2)-pyridine	H	H	H			

Ph	Me	H	addition of RCH=C(S*t*Bu)OSn*t*Bu enolates to imines	60–89% addition	1986 (585)
CH-CHPh	Me	H		68–92% de	
*i*Pr	Me	H			
(CH₂)₂Ph	Me	H			
(CH₂)₂OBn	Et	H			
Ph	Me	H	addition of niobium enolate of propionic acid oxazolidinone amide enolates to aryl aldimines	23–80%	2004 (586)
4-Cl-Ph	Me	H		20–80% de *anti*	
4-MeO-Ph	Me	H			
4-NO₂-Ph	Me	H			
furyl	Me	H			
*i*Pr	H	H	addition of allylMgBr to *N*-Ts imine	72–85% addition	1990 (587)
2-naphthyl	H	H			
Ph	H	H	addition of allylMgBr to ArCH=N-SO₂-N=CHAr	72–85% addition	1986 (589)
3,4-(MeO)₂-Ph	H	H			
CF₃	=	=	addition of Li ester enolates to imidoyl chlorides	75–95%	1997 (590)
CF₂CF₃	=	=			
(CF₂)₃CF₃	=	=			
CF₃	=	=	addition of Li ester enolates to imidoyl chlorides, reduction with ZnI₂/NaBH₄	65–100% addition to imidoyl chloride	1999 (592)
CF₂Cl	=	=		36–98% reduction	
CF₂CF₃	=	=		0–96% de *syn*	
CF₃	=	=			
CF₂Cl	=	=			
CF₂CF₃	=	=			
CF₃	H	H			
CF₃	H	H			2002 (591)
CF₂Cl	H	H			
CF₂CF₃	H	H			
CF₃	Me	H			
CF₂Cl	Me	H			
CF₂CF₃	Me	H			
CF₃	Et	H			
CF₃	OH	H	reaction of alkyl glycolates with imidoyl chlorides, rearrangement	71–92% imino ether	1998 (593)
CHF₂	OH	H		81–89% rearrangement	
CF₂CF₃	OH	H			
H	S*t*Bu	CF₃	addition of enolate of *t*BuSCH₂CO₂*t*Bu to ArN=C(Cl)CF₃, reduction of imine with NaBH₄	80–92% addition	2002 (594)
CF₃	S*t*Bu	H		50% reduction (>98% de *syn*)	
				81% reduction (78% de *anti*)	

(Continued)

Table 11.1 Syntheses of Achiral or Racemic β-Amino Acids (continued)

R^1	R^2	R^3	R^4	Method	Yield	Reference
tBu	H	H	H, Me, or CH₂CH=CH₂	addition of 2-alkyloxazoline to RCN or RC(Cl)=NP, reduction of enamine, hydrolysis of oxazoline	75–98% addition and reduction 55–90% hydrolysis	1999 (595)
Ph	H	H				
4-MeO-Ph	H	H				
4-Me-Ph	H	H				
2-furyl	H	H				
Ph	H	H	H	3-component reaction of malonic acid, ArCHO, and ammonia, decomposition	50%	1929 (596)
Ph	H	H	H	3-component reaction of malonic acid, ArCHO, and ammonia or methylamine, decomposition	20–61%	1926 (597)
3-NO₂-Ph	H	H	H			
3,4-(OCH₂O)-Ph	H	H	H			
3-thiophene	H	H	H	3-component reaction of malonic acid, aldehyde, and ammonium acetate	—	1986 (598)
2-thiophene	H	H	H	3-component reaction of malonic acid, aldehyde, and ammonium acetate	28–79% yield	1979 (599)
2-furyl	H	H	H			
2-thiophene	H	H	H	3-component reaction of malonic acid, aldehyde, and ammonium acetate	—	2002 (600)
2-furyl	H	H	H			
3-thiophene	H	H	H			
3-furyl	H	H	H			
CH₂NMe₂	H	H	H	3-component reaction of malonic acid, formaldehyde, and dimethylamine	—	1920 (601)
4-F-Ph	H	H	H	Rodionow reaction of aromatic aldehyde, malonic acid, and ammonium acetate	50%	2001 (234)
Ph	H	H	H	Rodionow reaction of aromatic aldehyde, malonic acid, and ammonium acetate	51–74%	1995 (602)
4-F-Ph	H	H	H			
2-F-Ph	H	H	H			
4-Cl-Ph	H	H	H			
4-MeO-Ph	H	H	H			
3,4,5-(MeO)₃-Ph	H	H	H			
2-thienyl	H	H	H	Rodionow reaction of 2-thienyl aldehyde, malonic acid, and ammonium acetate	—	1987 (603)
3-thienyl	H	H	H			
2-thienyl	H	H	H	Rodionow reaction of 2-thienyl aldehyde, malonic acid, and ammonium acetate	38–48%	1997 (604)
2-(4-X-thienyl)	H	H	H			
2-(5-X-thienyl)	H	H	H			
2-(4,5-X₂-thienyl) X = Br, Cl	H	H	H			
Ph	H	H	H	3-component reaction of malonic acid, aryl aldehyde, and ammonium acetate	—	1995 (398)
4-OEt-Ph	H	H	H			
3-pyridyl	H	H	H			
2-OEt-5-pyridyl	H	H	H			

			Reaction	Yield	Year (Ref.)
3-MeO-Ph	H	H			
2-pyridyl	H	H			
3,4-(CH₂OCH₂)-Ph	H	H			
Ph	H	H	3-component reaction of malonic acid, PhCHO, and ammonium acetate	75%	1998 (605)
Ph or 3,4-(OCH₂O)-Ph	Me, Et, or Bn	H	3-component reaction of monosubstituted malonic acid, ArCHO, and ammonia, decomposition	6–52% condensation 13–77% decarboxylation	1929 (606)
Ph	H	H	3-component reaction of malonic acid, PhCHO, and ammonium acetate	—	2005 (607)
4-Cl-Ph	H	H			
4-MeO-Ph	H	H			
3-NO₂-Ph	H	H			
tBu	H	H			
iPr	H	H			
Ph	Me	H	3-component Yb-catalyzed reaction of aldehyde, amine, and silyl enolate	62–96% 76–86% de *syn* or *anti*	1995 (608)
Ph	OH	H			
(CH₂)₂Ph	H	OH			
nBu	H	OH			
iBu	H	OH			
Ph	Me	Me			
COPh	Me	Me			
Ph	Me	H			
COPh	H	H			
CH=CHPh	H	H			
(CH₂)₂Ph	H	H			
nBu	H	H			
nOct	H	H			
H	Me	Me	3-component InCl₃-catalyzed reaction of aldehyde, amine, and silyl enolate	21–92%	1998 (609)
Ph	Me	Me			
2-pyridyl	Me	Me			
Ph	Me	H	3-component polymer-Sc-catalyzed reaction of aldehyde, amine, and silyl enolate	73–89%	1996 (610)
2-furyl	Me	H			
CH=CHPh	Me	H			
cHex	Me	H			
Ph	Me	H	3-component Sc(OTf)₃- or Cu(OTf)₂-catalyzed reaction of aldehyde, amine, and silyl enolate in H₂O with SDS	64–88%	1999 (611)
COPh	Me	H			
CH=CHPh	Me	H			
iBu	Me	H			
(CH₂)₂Ph	Me	H			
2-furyl	Me	H			

(*Continued*)

Table 11.1 Syntheses of Achiral or Racemic β-Amino Acids (continued)

R¹	R²	R³	R⁴	Method	Yield	Reference
H	H	Me	Me	addition of silyl ketene acetal to imine of aniline and aldehyde with HBF₄ and surfactant catalysis in aqueous solution	65–98% condensation	1999 (612)
cHex	H	Me	Me			
(CH₂)₂Ph	H	Me	Me			
COPh	H	Me	Me			
CH=CHPh	H	Me	Me			
CH₂Cl	H	Me	Me			
2-furyl	H	Me	Me			
2-thienyl	H	Me	Me			
Ph	H	Me	Me			
Ph	H	H	H			
Ph	H	Me	H			
Ph	H	Ph	H			
Ph	H	Me	Me	2- or 3-component SbF₄-catalyzed reaction of aldehyde, amine, and silyl enolate in H₂O with or without SDS	53–100% 70% de *syn* to 68% de *anti*	2002 (613)
Ph	H	Bn	H			
Ph	H	Me	H			
Ph	H	Ph	H			
Ph	H	OSiPh₃	H			
Ph	H	H	H			
4-Me-Ph	H	Me	Me			
2-furyl	H	Me	Me			
PhCO	H	Me	Me			
BnOCH₂	H	Me	Me			
PhCH₂CH₂	H	Me	Me			
H	H	Me	Me	addition of silyl ketene acetal to imine of aniline and aldehyde with dodecylbenzene sulfonic acid catalyst	63–90% condensation	1999 (614)
Ph	H	Me	Me			
2-furyl	H	Me	Me			
CH=CHPh	H	Me	Me			
iBu	H	Me	Me			
Ph	H	H	Ph	addition of silyl ketene acetal to imine of aniline and aldehyde with HBF₄ catalysis in aqueous/alcohol solution	80–99% condensation	1999 (615)
Ph	H	H	Me			
CH₂CH₂Ph	H	H	Ph			
cHex	H	Me	Me			
CH₂CH₂Ph	H	Me	Me			
CH₂OBn	H	Me	Me			
Ph	H	Ph	OH	3-component Ti-catalyzed reaction of methyl phenylglyoxylate, aniline, aldehyde	53–67% 100% de *syn*	1995 (616)
4-Me-Ph	H	Ph	OH			
4-Br-Ph	H	Ph	OH			
4-CN-Ph	H	Ph	OH			

R	X	Y	Conditions	Yield	Year (Ref)
Ph	H	OMe	3-component reaction of ArCHO with 2-methoxyaniline and Ti enolate of methyl 2-methoxyacetate	71–94% / 58–98% de *anti*	2004 (617)
4-Cl-Ph	H	OMe			
4-Me-Ph	H	OMe			
2-naphthyl	H	OMe			
CH=CHPh	H	OMe			
cHexenyl	H	OMe			
Ph	H	$=\!CH_2$	3-component reaction of ArCHO with $TsNH_2$ and methyl acrylate in presence of $La(OTf)_3$ and base	50–90%	2001 (618)
3-Cl-Ph	H	$=\!CH_2$			
3-NO_2-Ph	H	$=\!CH_2$			
4-NO_2-Ph	H	$=\!CH_2$			
4-MeO-Ph	H	$=\!CH_2$			
2-naphthyl	H	$=\!CH_2$			
2-furyl	H	$=\!CH_2$			
2-pyridyl	H	$=\!CH_2$			
Ph	H	$=\!CH_2$	3-component reaction of ArCHO with $TsNH_2$ and methyl acrylate in presence of $Ti(OiPr)_4$ and base	58–94%	2002 (619)
3-Cl-Ph	H	$=\!CH_2$			
3-NO_2-Ph	H	$=\!CH_2$			
4-NO_2-Ph	H	$=\!CH_2$			
4-MeO-Ph	H	$=\!CH_2$			
2-naphthyl	H	$=\!CH_2$			
2-furyl	H	$=\!CH_2$			
2-pyridyl	H	$=\!CH_2$			
Ph	H	Et	one-pot reaction of aldehydes with $TMS\text{-}NR_2$ in presence of $LiClO_4$, then addition of $ZnBrCH_2CO_2Me$	52–73%	1997 (620)
iPr	H	H			
Ph	H	H			
3-pyridyl	H	H			
2-thienyl	H	H			
CH=CHPh	H	H			
iPr	H	H	one-pot 3-component reaction of aldehydes with $TMS\text{-}NR_2$ and ketene silyl acetal in presence of $LiClO_4$	83–87%	1994 (621)
Ph	H	Me			
Ph	H	OH	dipolar cycloaddition of ylide generated by $Rh_2(OAc)_4$ treatment of N_2CHCO_2Et/ PhCHO with BnN=CHR imine, hydrolysis of oxazolidine adduct with pTSA	61–83% yield cycloaddition and hydrolysis / 66–96% de *syn*	2005 (622)
4-NO_2-Ph	H	OH			
4-Cl-Ph	H	OH			
4-F-Ph	H	OH			
4-MeO-Ph	H	OH			
4-Me-Ph	H	OH			
2-naphthyl	H	OH			
2-furyl	H	OH			

(*Continued*)

Table 11.1 Syntheses of Achiral or Racemic β-Amino Acids (continued)

R¹	R²	R³	R⁴	Method	Yield	Reference
Me	=	H	=	Blaise reaction of bromoacetate, Zn, and alkylnitrile	10–90%	1997 (623)
Et	=	H	=			
nPr	=	H	=			
cPr	=	H	=			
nBu	=	H	=			
Ph	=	H	=			
Bn	=	H	=			
2-Br-Ph	=	H	=			
3,4-(MeO)₂-Bn	=	H	=			
CH₂F	H	COMe or CH(OH)Me	H	addition of LiCH₂CO₂tBu to FCH₂CN, NaBH₃CN reduction	79% addition, 80% reduction	1985 (624)
CH₂F	H	H	H			
4-Me-Ph	H	COMe	H	Co-catalyzed condensation of methyl acetoacetate, aldehydes, and acetonitrile, ketone reduction	44–62%, 50% de *anti*	1997 (625)
4-MeO-Ph	H	COMe	H			
4-NO₂-Ph	H	COMe	H			
4-Cl-Ph	H	COMe	H			
4-CO₂Me-Ph	H	COMe	H			
nPr	H	COMe	H			
iPr	H	COMe	H			
cHex	H	COMe	H			
4-Me-Ph	H	COMe	H	Co-catalyzed condensation of methyl acetoacetate, aldehydes, and acetonitrile	41–44%	1994 (626)
4-Cl-Ph	H	COMe	H			
4-Me-Ph	H	COMe	H	Co-catalyzed condensation of methyl acetoacetate, aldehydes, and acetonitrile	23–76%, 40% de *syn* to 80% de *anti*	2003 (627)
4-Cl-Ph	H	COMe	H			
4-F-Ph	H	COMe	H			
4-Br-Ph	H	COMe	H			
2-Br-Ph	H	COMe	H			
4-Me-Ph	H	COMe	H			
4-NO₂-Ph	H	COMe	H			
4-MeO-Ph	H	COMe	H			
Ph	H	COMe or CH(OH)Me	H	polyaniline-supported Co-catalyzed condensation of methyl acetoacetate, aldehydes, and acetonitrile, ketone reduction	51–68%, 76–82% de *anti*, 80–90% reduction	1999 (628)
4-Cl-Ph	H					
4-NO₂-Ph	H					
2-CO₂Me-Ph	H					
2-OAc-Ph	H					
3-AcO-4-MeO-Ph	H					
iBu	H					

Me	H	Et	reaction of LiCH₂CO₂R with 1-meth-oxyamines	57–95% addition	1999 (553)
iPr	H	Et		10–70% de *anti* with no additive	
tBu	H	Et		20–90% de *syn* with Ti(OiPr)₄	
Bn	H	Et			
CH₂CH=CH₂	H	Et			
CH₂CH₂OTBS	H	Et			
iPr	H	iPr			
CH₂CH₂OTBS	H	iPr			
H	Me	Et			
H	iPr	Et			
H	tBu	Et			
H	Bn	Et			
H	CH₂CH=CH₂	Et			
H	CH₂CH₂OTBS	Et			
H	iPr	iPr			
H	CH₂CH₂OTBS	iPr			
H	H	H	olefination of *N*-(acetyl)thioamides with methyl (triphenylphosphoranylidene)acetate	34–81% olefination 90–96% reduction	1981 (629)
Me	H	H			
Et	H	H			
nPr	H	H			
nBu	H	H			
iPr	H	H			
cPent	H	H			
cHex	H	H			
(CH₂)₈CHMe₂	H	H			
Me		Ph	reaction of Et₃N with methyl diazophenyl-acetate	60% (*N,N*-Et₂)	1989 (630)
H	OH	H	reaction of *O*-tosyl-1-(hydroxymethyl) piperidine with ethyl diazoacetate	19% (amino group in piperidine ring)	1979 (631)
CH₂OH	nPr	H	radical addition-cyclization of Et radical to H₂C=CHCO-N(Bn)OCH₂CH=NHOMe, oxazinone hydrolysis/hydrogenation	79% addition/cyclization 82% de *cis* 62% oxazinone hydrolysis 58% hydrogenation	2004 (632)
CH₂O-NBn	nPr	H			

From α-amino acids

H	Me	H	reaction of Phth-Gly-Cl acid chloride with diazoethane, rearrangement catalyzed by AgO in presence of amine	—	1955 (633)

(Continued)

Table 11.1 Synthesfes of Achiral or Racemic β-Amino Acids (continued)

R¹	R²	R³	R⁴	Method	Yield	Reference
H	H	H	H	reaction of Phth-amino acid chloride with diazoalkane, rearrangement catalyzed by hv or AgO in presence of amine	52–88% diazoalkanone formation; 29–84% rearrangement	1968 (634)
H	H	Me	H			
H	H	Et	H			
Ph	H	H	H			
CH₂(indol-3-yl)	H	H	H	homologation of Trp via acid reduction to alcohol, nosylation, displacement with KCN, hydrolysis	98% reduction; 81% nosylation and displacement; 65% hydrolysis	2005 (635)
3-(p-OH-Ph)-1,2,4-oxadiazol-5-yl	H	H	H	condensation of Cbz-Asn with p-BnO-Ph-C(=NOH)NH₂	32%	1977 (636)
CH₂(2-Br-Ph)	H	OH	H	reduction of 4'-Br-Phe, addition of KCN, hydrolysis, tritiation	39% reduction, KCN addition and hydrolysis	1988 (204)
CH₂(2-T-Ph)	H	OH	H			

From β-amino acids

R¹	R²	R³	R⁴	Method	Yield	Reference
H	H	Me	H	alkylation of β-Ala in pyrimidinone template	76–78% alkylation; 62–69% hydrolysis	1991 (637)
H	H	Bn	H			
H	H	nBu	H			
H	H	nHex	H			
H	H	CH₂=CHCH₂	H			
Me	H	Me	H	alkylation of dideprotonated N-Bz or N-Cbz 3-aminobutanoate	25–96% alkylation; 75–>99% de anti	1988 (638)
Me	H	Et	H			
Me	H	CH(OH)Ph	H			
Me	H	CH(OH)Me	H			
Me	H	CH₂CH=CH₂	H			
Me	H	Bn	H			
Me	H	CH(Me)CH₂NO₂	H			
Me	H	Me	H	alkylation of dideprotonated N-Bz-3-aminobutanoate	68–84% alkylation; 86–>99% de anti	1996 (639)
Me	H	Et	H			
Me	H	CH₂CH=CH₂	H			
Me	H	Bn	H			
H	H	Me	Me	consecutive deprotonations and alkylation of Boc-β-Ala-OMe	80% dialkylation	1998 (322)
Me	H	Me	H	deprotonation and alkylation of 3-aminobutanoic acid or 3-amino-3-phenylpropionic acid, deprotonation and α-hydroxylation	70–99% alkylation; >98% de; 20–99% hydroxylation; 20–>98% de	1999 (640)
Me	H	Et	H			
Me	H	CH₂CH=CH₂	H			
Me	H	Bn	H			
Ph	H	Me	H			

			Method	Result	Year (Ref.)
Ph	H	Et	alkylation of N-Boc,N-alkenyl-β-Ala-OEt	57–73% alkylation	2005 (*641*)
Ph	H	CH$_2$CH=CH$_2$			
Ph	H	Bn			
Me	OH	Me			
Me	OH	Et			
Me	OH	CH$_2$CH=CH$_2$			
Me	OH	Bn			
Ph	OH	Me			
Ph	OH	Et			
Ph	OH	CH$_2$CH=CH$_2$			
Ph	OH	Bn			
H	H	CH$_2$CH=CH$_2$	free-radical allylation of β-substituted β-Ala	72–96% yield	1996 (*642*)
Me	H	CH$_2$CH=CH$_2$		>96% de *anti* to 90% de *syn*	
iPr	H	CH$_2$CH=CH$_2$			
tBu	H	CH$_2$CH=CH$_2$			
Ph	H	CH$_2$-CH=CH$_2$			
H	H	CH(Me)CH=CH$_2$	Ireland enolate-Claisen rearrangement of β-Ala allyl esters	31–88% yield	1989 (*644*)
H	H	CH(iPr)CH=CH$_2$			1994 (*643*)
H	H	CH(Ph)CH=CH$_2$			
H	H	CH$_2$CH=CHMe			
H	H	CH(CH$_2$OH)CH=CH$_2$			
H	H	CH(CH=CH$_2$)(CH$_2$)$_2$OH			
H	H	(CH$_2$)$_3$NH$_2$	condensation of ethyl cyanoacetate with acrylonitrile, hydrogenation	33% condensation 24% reduction	1953 (*146*)
H	Me	Me	dialkylation of ethyl cyanoacetate with MeI, reduction	—	1982 (*527*)
H	Me	Me	dialkylation of cyanoacetamide with MeI, reduction	64% dialkylation 20% reduction	2005 (*390*)
OEt	=	H	addition of EtOH to ethyl cyanoacetate	71%	1945 (*525*)
H	—(CH$_2$)$_n$— n = 2,3,4,5		dialkylation of methyl cyanoacetate with Br(CH$_2$)$_n$Br, CN reduction with H$_2$/Ra Ni, hydrolysis	55–80% hydrogenation 32–89% hydrolysis	1998 (*322*)
H	Et	Et	dialkylation of ethyl cyanoacetate with Br(CH$_2$)$_n$Br, CN /ester reduction with LiAlH$_4$, hydroxymethyl oxidation	64–93% alkylation 67–87% reduction 71–87% oxidation	1999 (*53*)
H	nPr	nPr			
H	—(CH$_2$)$_n$— n = 2,4,5	—(CH$_2$)$_n$— n = 2,4,5			

(*Continued*)

Table 11.1 Syntheses of Achiral or Racemic β-Amino Acids (continued)

R^1	R^2	R^3	R^4	Method	Yield	Reference
H	H	Ph	Bn	dialkylation of ethyl cyanoacetate, cyano reduction with H_2/Raney-Ni	98% reduction (1 example)	1959 (645)
H	H	Ph	Et			
H	H	Ph	iPr			
H	H	Ph	nBu			
H	H	Ph	cHex			
H	H	Me	Me			
H	H	Et	Et			
H	H	nPr	nPr			
H	H	nBu	nBu			
^{2}H	^{2}H	^{2}H	^{2}H	$LiAlD_4$ reduction of ethyl cyanoacetate, oxidation of hydroxymethyl, D_2O exchange	42–47% reduction 70–88% oxidation 36% exchange	1988 (646)
Me	=	Br	=	reaction methyl 3-aminocrotonate with PhI(OH)OTs, displacement with Nu	73% formation 20–83% displacement	1998 (647)
Me	=	NO_2	=			
Me	=	CN	=			
Me	=	pTolSO$_2$	=			
Me	=	NEt_2	=			
Me	H	morpholine	=			
H	H	Br, ^{18}F	H	displacement of hydroxyl group of isoserine	65% bromination 75–90% radiochemical yield F displacement	1995 (648)
iPr	H	H	H	hydrogenation of 3-amino-4-methylpent-2-enoate	—	1983 (649)
iPr	^{3}H	^{3}H	H			
H	H	iBu	H	radical conjugate additions of alkyl iodide or Heck reactions of aryl iodides with N,N-Boc$_2$ 3-amino-2-methylenepropionate	70–80% radical addition 63–85% Heck coupling	2001 (489)
H	H	CH$_2$cHex	H			
H	H	(CH$_2$)$_3$CO$_2$H	H			
H	H	(CH$_2$)$_4$NH$_2$	H			
H	H	=CHPh	—			
H	H	=CH(4-MeO-Ph)	—			
H	H	CH$_2$Ph	H	Rh-catalyzed addition arylboronic acids to α-(phthalimidomethyl)–acrylate	53–77%	2004 (650)
H	H	CH$_2$(3-NO$_2$-Ph)	H			
H	H	CH$_2$(4-MeO-Ph)	H			
H	H	CH$_2$(4-Br-Ph)	H			
H	H	CH$_2$(4-Ac-Ph)	H			
H	H	CH$_2$(1-naphthyl)	H			
H	H	H	H	reaction of β-Ala-OEt with Br(CH$_2$)$_4$NPhth, hydrolysis	31%	1969 (28)

with N-(CH$_2$)$_4$NH$_2$

Ph	H	4-MeO-Ph	reaction of electrophilic β-lactam template with aryl nucleophile	36–47%	2000 (651)
Ph	4-MeO-Ph	4-MeO-Ph			
4-MeO-Ph	4-MeO-Ph	4-MeO-Ph			
Ph	1,3-(MeO)$_2$-Ph	1,3-(MeO)$_2$-Ph			
4-MeO-Ph	1,3-(MeO)$_2$-Ph	1,3-(MeO)$_2$-Ph			
4-MeO-Ph	1,4-(MeO)$_2$-Ph	1,4-(MeO)$_2$-Ph			
4-MeO-Ph (as β-lactam)	Ph	Ph			
Ph	=	Cl	halogenation of β-enamino acids with carboxy protected as oxazoline	30–95%	2004 (652)
Ph	=	Br			
4-Me-Ph	=	Cl			
4-Me-Ph	=	Br			
4-Me-Ph	=	I			
4-MeO-Ph	=	Cl			
4-MeO-Ph	=	Br			
iPr	=	Cl			
3-pyridyl	=	Br			

From Small Ring Systems

nPr	Br	H	aziridine opening with NaBr	70–100% opening	1996 (654)
nC$_{12}$H$_{25}$	Br	H			
iPr	Br	H			
cHex	Br	H			
Ph	Br	H			
H	Ph	H	aziridation of alkenes with PhI=NSO$_2$Ar, opening with H, OH or NHR	8–70% aziridation, 45–90% opening	1998 (655)
Ph	Me	H			
H	Ph	OH			
H	Ph	NHCH-(Me)Ph			
H	H	H	reductive opening of aziridines with polymethylhydro-siloxane/Pd-C	80–85%	1999 (656)
CH$_2$CH$_2$Ph	H	H			
CF$_3$	H	Cl	opening of 3-CF$_3$-aziridine-2-carboxylate with HCl, CF$_3$CO$_2$H, PhSH/CF$_3$SO$_3$H, BnSH/ CF$_3$SO$_3$H or EtSH/CF$_3$SO$_3$H	50–98%	2001 (657)
CF$_3$	H	OH			
CF$_3$	H	SPh			
CF$_3$	H	SBn			
CF$_3$	H	SEt			
nBu	H	H	2-substituted N-Ts aziridine opening with TMSCN/Yb(CN)$_3$	85–86%	1990 (658)
Ph	H	H			

(Continued)

Table 11.1 Syntheses of Achiral or Racemic β-Amino Acids (continued)

R^1	R^2	R^3	R^4	Yield	Method	Reference
Ph	H	OPh	H	78–95%	cycloaddition of ketene silyl acetal generated from acid and triphosgene with imine	2000 (661)
p-MeO-Ph	H	OPh	H			
CH=CHPh	H	OPh	H			
Ph	H	OMe	H			
p-MeO-Ph	H	OMe	H			
CH=CHPh	H	NPhth	H			
Ph	H	H	OH	60–70% cycloaddition 80–100% de	via β-lactam prepared by cycloaddition of imine with ketene	1990 (662)
Ph	H	H	OMe			
Bn	H	H	OMe			
Bn	H	H	OBn			
Bn	H	H	Ph			
CH₂cHex	H	H	OH	41–94% cycloaddition 64–89% hydrolysis/ hydrogenation	[2+2] cycloaddition of BnOCH=C=O with BnNH₂ imine of cHexCH₂CHO, lactam hydrolysis and hydrogenation	1990 (663)
CF₃	H	H	OH	50% cycloaddition 80% hydrolysis	via β-lactam prepared by cycloaddition of imine with ketene	1996 (664)
Ph	Ph	Ph	Ph	67–78% cycloaddition 64–69% decomposition	cycloaddition of Re complex of benzophenone imine with diphenylketene, decomposition of lactam Re complex with TfOH	2003 (665)
Ph	Ph	Ph	Et			
Ph	Ph	—(CH₂)₆—	—			
(as lactam)						
H	Ph	OAc	H	47–90%	Et₃N induced Staudinger reaction of ROCH₂COCl and imines formed from benzaldehyde and ArNH₂	2000 (666)
H	Ph	OPh	H			
Ph	H	OAc	H			
Ph	H	OPh	H			
2-furyl	H	OAc	H			
2-furyl	H	OPh	H			
4-MeO-Ph	H	OPh, phthaloyl, or	H	40–85%	Staudinger reaction of imines formed from aldehydes and Gly-OWang with acids in the presence of Mukaiyama's salt	2003 (667)
3,4-(MeO)₂-Ph	H	CH=CH₂	H			
2-furyl	H		H			
Ph	H	H	H	40–90%	via β-lactam prepared by cycloaddition of imine with ketene or enamine	1987 (668)
Ph	H	Me	H			
CO₂tBu	H	H	H			
CO₂tBu	H	Me	H			
CH=CHPh	H	3-[C(=NH)NH₂]-Bn	H	45% cycloaddition 54% hydrolysis 49% amidine	cycloaddition of imine, hydrolysis, amidine formation	1998 (416)
CH=CHPh	H	H	3-[C(=NH)-NH₂]-Bn			

Me	Me	H	via β-lactam prepared by cycloaddition of N-TMS-imine with Me₂C(Li)CO₂Et	15–94% cycloaddition / 41–99% hydrolysis	1998 (400)
Et	Me	H			
nPr	Me	H			
iPr	Me	H			
2-propenyl	Me	H			
nBu	Me	H			
iBu	Me	H			
nPent	Me	H			
CH₂CH₂Ph	Me	H			
2-naphthyl	Me	H			
4-Cl-Ph	Me	H			
4-BnO-Ph	Me	H			
3-Cl-Ph	Me	H			
3-BnO-Ph	Me	H			
Et	Me	H	addition of enolate of ethyl isobutyrate to N-TMS imine of RCHO	11–69% addition	1999 (395)
nPr	Me	H			
iPr	Me	H			
nBu	Me	H			
iBu	Me	H			
nPent	Me	H			
nHex	Me	H			
nHept	Me	H			
nOct	Me	H			
nDec	Me	H			
nDodec	Me	H			
CH₂CH₂Ph	Me	H			
Me	Et	H	cycloaddition of ClSO₂NCO with alkenes or aldehydes and ketenes, deprotection (β-lactam products)	65–100%	1963 (669)
Me	Me	H			
Et	Et	H			
Me	iPr	H			
Me	nPr	Me			
Me	Me	H			
Et	nBu	H			
M	neoPent	tBu			
Me	Me	H			
CH₂OMe	Me	H			
CH₂OPh e	Me	H			
Me	H	Me			
iPr	H	Me			

(Continued)

Table 11.1 Syntheses of Achiral or Racemic β-Amino Acids (continued)

R¹	R²	R³	R⁴	Method	Yield	Reference
Ph	H	Me	Me	cycloaddition of $ClSO_2NCO$ with alkenes, deprotection	68–94%	1970 (670)
Me	H	H	H			
3-Br-Ph	H	H	H			
4-Cl-Ph	H	H	H			
2,4-Cl₂-Ph	H	H	H			
2,6-Cl₂-Ph	H	H	H			
3-MeO-Ph	H	H	H			
2-NO₂-Ph	H	H	H			
3-NO₂-Ph	H	H	H			
4-NO₂-Ph	H	H	H			
3,4-(NO₂)₂-Ph	H	H	H			
3,6-(NO₂)₂-Ph	H	H	H			
2-NO₂-5-Cl-Ph	H	H	H			
Me	Me	Me	Me			
Me	Me	Me	H			
CH=CH₂	Me	H	H			
CH=CH₂	Me	H	H			
–(CH₂)₅–	—					
CH=CH(CH₂)₄-C2	H	(CH₂)₄CH=CH-C3	H			
Bn	H	H	H	cycloaddition of $ClSO_2NCO$ with alkene, hydrolysis	53% cycloaddition 100% hydrolysis	1995 (398)
Ph	H	Me	CH(Ph)NH₂	cycloaddition of LiOCCR with (2-MeO-Ph) N=CHAr followed by addition of lactam enolate to imine	52–96%	2000 (672)
1-napthyl	H	Me	CH(1-napthyl)NH₂			
2-naphthyl	H	Me	CH(2-naphthyl)NH₂			
4-CO₂Me-Ph	H	Me	CH(4-CO₂Me-Ph)NH₂			
Ph	H	nBu	CH(Ph)NH₂			
1-naphthyl	H	nBu	CH(1-naphthyl)NH₂			
2-naphthyl	H	nBu	CH(2-naphthyl)NH₂			
Ph	H	cHex	CH(Ph)NH₂			
CH=CH₂	H	H	H	via conversion 4-acetoxy- β-lactam to 4-phenylsulfonyl, vinylMgBr addition, hydrolysis	32% sulfonylation 86% Grignard addition 100% hydrolysis	1997 (673)

Other Syntheses

R1	R2	R3	Synthesis	Yield	Year (Ref)
H	H		(alkylation of uracil) uracil hydrolysis	77–87% hydrolysis	1979 (674)
H	Me		decarboxylative rearrangement of diazoure-thanes	36–96% rearrangment	1999 (675)
H	Ph	=			
H	4-MeO-Ph	=			
iPr	H	=			
4-NO2-Ph	H	=			
H	CH2CH2NH2	H	one-pot butanedinitrile synthesis, nitrile reduction	45–80% for dinitrile	1995 (676)
H	CH2CH2NH2	Me		71–80% for reduction	
H	CMe2CH2NH2	H			
H	CMe2CH2NH2	Me			
H	CHMeCH2NH2	H			
H	CHMeCH2NH2	Me			
H	H		ozonolysis of 2,3-dihydropyrroles	60–88%	1994 (677)
Ph	H				
–CH2-C2	–CH2-C3				
–CH(CO2H)-C2	–CH(CO2H)-C3				
Me	H		Grignard addition to N-acylpyridinium, oxidative alkene cleavage	84–88% addition	2004 (678)
Ph	H			83–95% oxidative cleavage	
CH=CH2 (with N-H or N-Me)	H				
CH(R1)CH(R2)CH=CH2 R1 = Me; R2 = Me, Ph, iPr R1 = Ph; R2 = Me R1 = Ph, iPr, Cl, allyl, H, R2 = Ph R1 = NHPhth; R2 = Me	H	=	Lewis acid catalyzed [3,3]sigmatropic allenoate rearrangement of allylamines with 2,3-butadienoates	75–97% yield 82–>96% de syn	2002 (679)

11.2.3 Asymmetric Syntheses of β-Amino Acids (see Table 11.2)

11.2.3a Resolutions of Racemic β-Amino Acids

There have been few enzymatic resolutions of β-amino acids compared to the number described for α-amino acids. However, a number of N-phenylacetyl β-substituted β-amino acids (680) were resolved using penicillin acylase, giving enantiopure products as determined by chiral HPLC analysis (135, 520, 602, 605, 680–682). Penicillin G acylase was immobilized and used to resolve four N-Bz 2-alkyl-3-aminobutanoic acids. Both enantiomers were isolated with 94–100% ee (639). ChiroCLEC-EC, a cross-linked penicillin G acylase, was employed to acylate racemic β-substituted-β-amino esters with methyl phenylacetate. Five amino esters were studied, with four resolved with 94 to >95% ee (R = Me, Ph, 4-MeO-Ph, 3,4-(MeO)₂-Ph) (683).

Other resolutions have used lipases. Racemic ethyl 3-aminobutyrate reacted with Candida antarctica lipase B in butyl butyrate to produce a mixture of ethyl (S)-3-aminobutyrate and butyl (S)-3-aminobutyrate, along with N-butanoyl butyl (R)-3-aminobutyrate (684). The β³-amino acid analog of Trp was also resolved via N-acylation with butyl butyrate and C. antarctica lipase B (635), as were the β³-amino acid analogs of Ala and Phg (685). Gram-scale resolutions of four β-heteroaryl β-amino esters (Ar = 2- or 3-thienyl or -furyl) were achived by acylation with ethyl butanoate and C. antarctica lipase B (600). Some examples of resolutions of β-amino acids via N-acylation by C. antarctica lipase B and penicillin acylase were included in a 2004 review of serine hydrolase acylation of chiral amines (686). C. antarctica lipase A (CAL-A) and C. antarctica lipase B (CAL-A) show low, but opposite, enantioselectivity for acylation of α-methyl-β-Ala ethyl ester with 2,2, 2-trifluoroethyl butanoate in organic solvents. However, highly pure enantiomers could be produced on a gram scale by a two-step procedure. CAL-A was allowed to convert 75% of the racemic starting material, leaving non-acylated (R)-enantiomer with 96% ee. The partially resolved N-butanoyl ethyl ester (33% ee) was then transesterified by treatment with CAL-B and butyl butanoate, leaving the (S)-N-butanoyl ethyl ester as the less reactive enantiomer, with 96% ee (687).

The lactam of 2-acetoxy-3-amino-4,4-dimethylpentanoic acid was resolved by immobilized lipase PS-30 cleavage of the acetyl group, with >48% reaction yield of >99% ee product on >100 g scales (688). Isoserine was enzymatically resolved via deacetylation of the hydroxyl group by C. cylindracea lipase (543, 544), while immobilized C. antarctica lipase was used for ester hydrolysis of N-Boc methyl β-aminoisobutyrate (3-amino-2-methylpropanoate) (63). An enzymatic desymmetrization of dimethyl 3-benzylaminoglutarate by CAL-B produced the (S)-monoamide in 76% yield. A Hofmann rearrangement then gave 3,4-diaminobutanoic acid (689). Ethyl 3-aminobutyrate was resolved by either

N-acetylation, ester aminolysis, or ester hydrolysis, which were all catalyzed by C. antarctica lipase. Optical purities of up to 94% were obtained (690). PLE (pig liver esterase) was employed to kinetically resolve N-benzoyl-3-aminobutanoate by hydrolysis of the methyl ester (638), or to desymmetrize the symmetrical β-amino acid dimethyl 3-aminoglutarate by mono-ester hydrolysis (528). Several N-unprotected β-aryl-β-amino acid ethyl esters were resolved by ester hydrolysis with Amano PS, with pH playing a significant role in enantioselectivity. At pH 8, five examples (Ar = Ph, 2Br-Ph, 4-Br-Ph, 4-F-Ph, 2-naphthyl) were hydrolyzed with 91–99% ee (691).

Enantiomerically enriched phenylisoserine, the Taxol side chain, was prepared via a lipase resolution of a β-lactam precursor, either by deacylation of an O-acetyl group, or by cleavage of the lactam ring (692). Alternatively, the same residue was obtained by microbial reduction of the carbonyl group of racemic 2-keto-3-(N-benzoylamino)-3-phenylpropionic acid. A variety of organisms were screened, with the best conditions giving 80% yields with 80% de and >98% ee (693). Baker's yeast reduction of 3-keto-4-phenyl-β-lactam produced the desired (3R,4S)-isomer of phenylisoserine β-lactam in up to 41% yield with 82% ee; Baker's yeast reduction of racemic methyl 2-keto-3-azido-3-phenylpropionate gave only the (R)-alcohol center, but no diastereoselectivity for the azido-center configuration (694).

Diastereomeric salt formation was used to resolve isoserine with brucine (695), while quinine was employed to separate the N-Cbz derivatives of 3-amino-3-(2-furyl)propionic acid and 3-amino-3-(2-thienyl) propionic acid (599). The phenylisoserine component of Taxol was resolved by crystallization using seed crystals of the desired enantiomer (500).

Few separations of covalent diastereomeric derivatives have been reported. 3-Amino-5,5-dimethyl-hexanoic acid was derivatized with (R)-(−)-α-methoxy-phenylacetic acid, with the diastereomers separated by flash chromatography (526). A chemical kinetic resolution of 2-phenyl-3-aminopropanoic acid was achieved with high diastereoselectivity (94% de) by addition of (R)-pantolactone to the ketene derived from the β-amino acid (550), with the same methodology subsequently applied to four other 2-aryl-3-aminopropionic acids (551). This methodology was initially developed for resolutions of α-amino acids (see Section 12.1.7). In a similar manner, β-substituted-β-amino acids were converted into oxazinones by treatment with isobutyl chloroformate. Kinetic resolution of the oxazinones via ring opening with allyl alcohol in the presence of a thiourea-based catalyst produced enantiomerically pure oxazinone along with β-substituted-β-amino acid allyl ester (see Scheme 12.4). Since the oxazinones are configurationally stable, dynamic resolution (with >50% yield of one enantiomer) does not occur (607).

Most attempts at chromatographic resolution of amino acids for analysis of enantiomeric purity have focused on α-amino acids. However, several reports in

1986 examined the separation of β-amino acids by HPLC on chiral stationary phases (696, 697). The use of chiral eluent HPLC (with a Cu-proline complex) was also examined (698). A glycopeptide antibiotic, A-40, 926, provided chromatographic resolution of unprotected β-amino acids when attached to silica gel microparticles, with semipreparative separations of 15–30 mg feasible (699). A ligand exchange chiral stationary phase based on N-carboxymethyl,N-undecyl-(R)-phenylglycinol attached to silica was able to resolve all 15 β-amino acids tested, using a Cu(II) eluent (700), while a quinine-derived chiral anion exchange stationary phase provided separations for nine different N-(2, 4-dinitrophenyl) β-substituted-β-amino acids (701). The enantioseparation of 18 underivatized β-amino acids was found to be, in general, better accomplished on Chirobiotic T (teicoplanin) than Chirobiotic TAG (teicoplanin aglycone) columns (702). The same report also formed diastereomeric derivatives with N-(4-nitrophenoxycarbonyl)-L-Phe methoxyethyl ester for separation on a standard C_{18} column (702). A similar study compared a Crownpak CR(+) column (chiral crown ether) with Chirobiotic T and R columns for the separation of eight underivatized β-substituted-β-amino acids, along with diastereomeric derivatization with GITC or FDAA. The best chiral column was analyte-dependent, while derivatization with FDAA was more efficient than GITC (703). Normal-phase HPLC was employed to resolve diastereomeric 10-camphorsulfonamide p-nitrobenzyl ester derivatives of β-leucine; better separation was obtained using the derivatives formed with Marfey's reagent (704).

Gas chromatography using the chiral stationary phase Chirasil-Val was applied to the separation of the enantiomers of several β-amino acids derivatized as their N-trifluoroacetyl isopropyl esters (698). Difficulties were encountered in the analysis of optical purity of α-alkyl-β-amino acids synthesized in a 2000 report.

HPLC separation of Marfey's derivatives or pseudoephedrine amide derivatives, or chiral HPLC on a Leu Pirkle column, were unsuccessful, as was ^{1}H NMR analysis of the pseudoephedrine amides or Mosher ester derivatives of acid-reduced amino alcohols. Eventually, GC analysis of N-trifluoroacetamide isopropyl ester derivatives on a Chirasil-Val column was successful (705). Derivatives of the β-amino acids β-amino-n-butyric acid and β-amino-isobutyric acid have been resolved by GC analysis using various (R)-(+)-2-alkanol esters (2-butanol, 2-pentanol, 2-hexanol, 2-heptanol, or 2-octanol) in combination with different N-protecting groups (N-trifluoroacetyl, N-pentafluoropropionyl, N-heptafluorobutyryl, or N-chlorodifluoroacetyl) (706).

(S)-2-[(R)-Fluoro(phenyl)methyl]oxirane was recently reported as a versatile general reagent for the derivatization and enantiomeric analysis of chiral amines. The diastereomeric derivatives were readily analyzed by ^{1}H, ^{19}F, or ^{13}C NMR, or separated by HPLC. Ethyl 3-aminobutyrate was among the amines tested (707). Spirocyclic Pd complexes formed from di-μ-chlorobis{[2-[1-dimethylamino-xN)ethyl]phenyl-xC}dipalladium and a range of β- or γ-amino acids provided distinct chemical shift differences in both ^{1}H and ^{13}C NMR spectra, allowing for the determination of enantiomeric purity (see Scheme 1.7) (708).

11.2.3b Amination Reactions: Conjugate Additions

One of the most common methods for the asymmetric synthesis of β-amino acids employs stereoselective Michael addition of a homochiral ammonia equivalent to an α,β-unsaturated ester (see Scheme 11.31). This area was extensively reviewed in 1996, including discussions on the effect of solvent and substituents (442, 709). The first attempts at addition employed

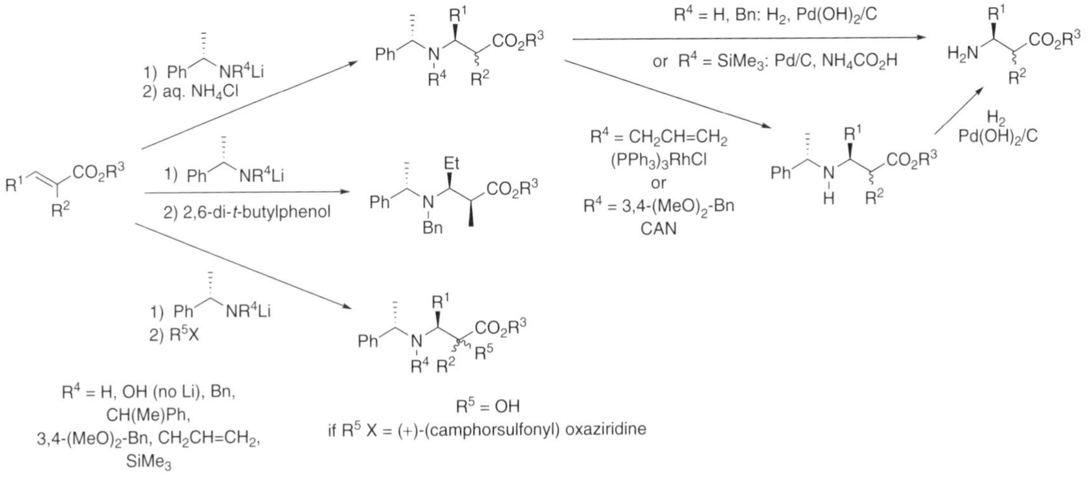

Scheme 11.31 Michael Additions of Chiral Lithium Amides and Amines to α,β-Unsaturated Esters.

amines such as (R)- and (S)-α-methylbenzylamine, which gave both low yields (9–47%) and optical purities (2–19% ee) (710, 711). The (R)- and (S)-enantiomers of 3-aminobutanoic acid were synthesized by addition of α-methylbenzylamine to methyl (638, 712) or ethyl (713) crotonate with (8–20% de), followed by chromatographic resolution of the reaction mixture to give the pure diastereomers. Homochiral α-methylbenzyl hydroxylamine has also been added to α,β-unsaturated esters, with the N-hydroxy group of the initial conjugate addition adduct cyclizing with the ester group to form 5-isoxazolidinones. Diastereoselectivities of 38–60% de were observed for β-substituted products, which were much higher than those obtained with simple α-methylbenzylamine. α-Substituted products had minimal diastereoselectivity, but could be readily separated. Hydrogenolysis of the N–O bond also cleaved the auxiliary to provide β-amino acids (714, 715).

A significant improvement was reported by Davies et al., who have made extensive use of lithium (α-methylbenzyl)benzylamide or related derivatives as nitrogen nucleophiles (716, 717). A comprehensive review of the conjugate addition of enantiomerically pure lithium amides was published in 2005, including tables of many β-amino acids produced by this route (718). Lithium (α-methylbenzyl)benzylamide is inexpensive, gives good stereoselectivity, and can be deprotected by hydrogenolysis. It was applied to a large-scale synthesis of D-β-Phe (3-amino-3-phenylpropionic acid, a component of the antibiotics moiramide B and andrimid) via addition to tert-butyl cinnamate (96% yield, >95% de on 6 g scale) followed by hydrogenation (75% yield) (97, 112, 719). The additions could also be carried out on Weinreb amide derivatives of α,β-unsaturated acids, with good yield (95–96%) and diastereoselectivity (>95% de) (720). Addition to methyl p-hydroxycinnamate O-TBS ether led to (R)-β-Tyr (131). Methyl p-benzyloxycinnamate was also employed as a substrate for the conjugate addition, leading to 3-(4-hydroxyphenyl)-β-Ala. The phenol was then converted into a triflate and used for Suzuki couplings with aryl boronic acids to form 3-biaryl-β-Ala derivatives (406). β-(5- and 6-[2,3]- dihydrobenzofuran)-β-amino acids were derived from the corresponding benzofuryl cinnamates; hydrogenolytic auxiliary removal also partially reduced the heterocyclic ring system (721). Two or three equivalents of lithiated (α-methylbenzyl) benzylamide have been added to two or three α,β-unsaturated tert-butyl esters attached to a phenyl nucleus (e.g., di-tert-butyl benzene-1,2-, -1,-3-, or -1,4-dipropenoate or tris-tert-butyl benzene-1,3,5-tripropenoate). Hydrogenolytic deprotection produced bis- or tris-β-amino acids linked at the β3-position to a common benzene ring (722).

Additions to α,β-unsaturated esters with α-alkyl substituents proceeded with complete C2 stereocontrol (syn product) in addition to the stereocontrol normally observed at the C3 position, by carrying out the addition in toluene and using the hindered acid 2,6-di-tert-butylphenol as a proton source (723–725) (note that in

the table in this reference syn/anti are mislabeled). Syn-2,3-dialkyl-β-amino acids with polar side chains were also prepared via this route (726). (2R,3S)-3-Amino-2-methylbutyrate was derived from tert-butyl (E)-2-methylbut-2-enoate in 68% yield (304). The antifungal antibiotic cispentacin, (1R,2S)-2-aminocyclopentane-1-carboxylic acid, and its cyclohexyl homolog were also prepared by this method (see Section 11.3.3b below) (716). Blastidic acid (Nε-Me-β-arginine) was prepared via addition of the lithiated (α-methylbenzyl)benzylamide to tert-butyl N-Boc,N-methyl 5-aminopent-2-enoate, followed by removal of the Boc group and guanidylation of the secondary amine (172). Addition of lithium (α-methylbenzyl)benzylamide to methyl 5-phenylpenta-2,4-dienoic acid led to 3-amino-5-phenylpentanoic acid after hydrogenation; the alkene could be reintroduced by bromination with NBS/CCl4 followed by elimination with DBU, producing 3-amino-5-phenylpent-4-enoic acid. The corresponding 2-methyl analogs were obtained by deprotonation of the conjugate addition adduct with LDA and alkylation with methyl iodide (233).

The intermediate enolates obtained by conjugate addition of lithium (α-methylbenzyl)benzylamide to α-unsubstituted substrates can be trapped with electrophiles such as alkyl halides (724, 727), acetaldehyde (728) or N-fluorobenzenesulfonimide (for fluorination) (729) to give α-substituted products. Similarly, the initial adduct of amide addition to α,β-unsaturated esters with α-alkyl substituents can also be trapped to give α,α-disubstituted products (725). The enolate trapping method led to anti products, in contrast to conjugate additions of the amide to α,β-disubstituted substrates, which resulted in the syn diastereomer (724, 727). Somewhat better diastereoselectivity was obtained by a two-step procedure in which the anion of conjugate addition was first quenched, and then regenerated by deprotonation (724, 727, 730). However, another study of the stereochemistry of the addition of lithium (α-methylbenzyl)benzylamide to α,β-unsaturated amides (instead of esters) found both tandem and stepwise conjugate addition–alkylation gave anti adducts (731).

Electrophilic hydroxylation of the α-enolate can be carried out, either by tandem reaction or a two-step procedure (732). Homochiral syn- or anti-3-phenylisoserine were prepared by the tandem amide conjugate addition–electrophilic hydroxylation approach (732–734). The diastereoselectivity of the hydroxylation was predominantly controlled by the existing chiral center, with (+)-(camphorsulfonyl)oxaziridine giving 92% de of the anti isomer of phenylisoserine as the mismatched reagent pair. The anti isomer could be epimerized to the syn diastereomer via a Mitsunobu inversion of the carbinol center. Allophenylnorstatine (3-amino-2-hydroxy-4-phenylbutanoic acid) was also synthesized by this route, although the tandem procedure led to poor diastereoselectivity (20% de syn) at the carbinol center (735, 736). Much better results (>95% de anti) were obtained by acetic acid quenching of the initial enolate generated by the amide conjugate addition, followed by in situ deprotonation with LDA and reaction with the hydroxyl-ation reagent. Alternatively, the tandem procedure gave 91% de *anti* selectivity when the matched pair of homochiral reagent/substrate was used, an exception to the lack of stereoselectivity normally induced by the reagent chirality. Conjugate addition and α-hydroxylation also led to three diastereomers (2S,3R,2S,3S, and 2R,3R) of the N-terminal component of microginin, 3-amino-2-hydroxydecanoic acid (*737, 738*). The *anti* hydroxyl-ation products were again initially obtained, but could be epimerized to the *syn* diastereomer by a Mitsunobu inversion. The conjugate addition/hydroxylation proce-dure was applied to a synthesis of *anti*-(E)-3-amino-2-hydroxy-4-hexenoate; in this case a DCC/DMAP procedure was employed to epimerize the C-α center to form the *syn* product (*739*).

Changing the metal counterion of the lithium (α-methylbenzyl)benzylamide to magnesium provided a reagent that reacted with *tert*-butyl cinnamate with similar high diastereoselectivity (>95% de) (*740*). However, trapping of the initial adduct with MeI gave a product with excellent *syn* diastereoselectivity (90% de), compared to the 30% de *anti* obtained with the lith-ium amide (*724*). Molecular modeling has been used to rationalize the high diastereoselectivites obtained with this reagent (*741*).

The major disadvantages of lithium (α-methylbenzyl) benzylamide as the nucleophilic amine are that the reductive deprotection conditions are not compatible with some side-chain functional groups (such as alk-enes), and that selective monodeprotection of one of the amine substituents is not possible. For example, attempts to prepare β-(pyridyl)-β-amino acids using (α-methyl-benzyl)benzylamide were unsuccessful due to reduction of the pyridyl ring during hydrogenolytic removal of the N-benzyl substituents. Instead, the β-(3-pyridyl)-, and (4-pyridyl)-β-amino acids were constructed by addition of lithium (R)-N-benzyl-N-α-methyl-4-methoxybenzyl-amine to the *tert*-butyl 3-(pyridyl)propenoates (84–94% de). The N-benzyl substituent was removed with 2.1 equiv of ceric ammonium nitrate (CAN), and the N-α-methyl-methoxybenzyl group with 4.0 equiv of CAN (*742*). For the β-(2-pyridyl) regioisomer, the conjugate addition proceeded with only 5% de, unless a 3-(6-sub-stituted-2-pyridyl)propenoate substrate was employed (61–83% de), presumably due to the 2-pyridyl group interfering with chelation (*742*). The same modified lith-ium amide and oxidative deprotection was used to pre-pare a series of β-haloaryl-β-amino acids, via conjugate addition to halo-substituted *tert*-butyl cinnamates, with 74–86% conjugate addition yields and 88–98% de (*743, 744*). Similarly, lithium (α-methylbenzyl)(3,4-dime-thoxybenzyl) amide was reported as a useful protected chiral ammonia equivalent. The 3,4-dimethoxybenzyl group could be selectively removed by oxidative treat-ment with CAN (*745*).

Another reagent introduced to allow for more facile N-deprotection is lithium (α-methylbenzyl)allylamide (*746, 747*). Stereoselectivity and yields were similar to the benzylamide reagent, with selective removal of the

allyl group achieved with a rhodium catalyst. This amide was added to isopropyl hepta-2,5-dienoic acid enroute to a synthesis of (3R,5R,6R)-3,6-diamino-5-hydroxy-heptanoic acid, a component of sperabillins B and D (*166*). Piperidine-2-acetic acid and pyrrolidine-2-acetic acid were obtained via addition of (S)-N-allyl-N-α-methylbenzylamide to methyl (E,E)-heptan-2,5-dienoate or *tert*-butyl (E,E)-hexa-2,4-dienoate, respectively, fol-lowed by ring-closing metathesis of the two alkene groups, and then hydrogenation (*748*). Alternatively, Michael addition of N-aryl α-methylbenzylamines to α,β-unsaturated-ω-chloroesters, followed by N-aryl deprotection, resulted in cyclization to five-, six-, and seven-membered imino acids (*749*). Another chiral lithium amide was generated from N-(trimethylsilyl)-(R)-1-phenylethylamine (*750*). Addition to ethyl *trans*-3-pyridineacrylate occurred in good yield with >96% de, and N-TMS deprotection was readily achieved. An amidocuprate reagent derived from this amine was also reported, adding with 56 to >98% de in the presence of triethyl phosphite (*751*). However, the ami-docuprate reagent derived from bis(α-methylbenzyl) amine gave much better diastereoselectivities for addi-tions to unhindered substrates (*751*). The conjugate addition adducts could be trapped with D_2O to give the α-deutero products (*751*).

Other chiral amine nucleophiles have been reported. Addition of the lithium amide of 3,5-dihydro-4H-dinaphth[2.1-c:1',2'-e]azepine (based on a binaphthyl template, see Scheme 11.32) to α,β-unsaturated esters proceeded with good diastereoselectivity (*752, 753*). The stereoselectivity has been rationalized by the use of force-field modeling (*754*). The initial adduct could be trapped by electrophiles to give α-substituted derivatives with good *threo* or *erythro* selectivity (*755*). Auxiliary removal was accomplished by hydrogenolysis.

The nitrogen contained within an optically active 3,4-diphenyloxazolidinone can also be used for Michael additions. Nearly complete stereoselectivity at the addi-tion center was observed with 2-chloro-2-cyclopropy-lidene acetates as the Michael acceptor (*756*). Enders et al. have employed a chiral hydrazone (TMS-SAMP) as a chiral ammonia equivalent, with addition to (E)-enoates proceeding in 32–67% yield and >93–97% de (see Scheme 11.33) (*757*). The TMS-SAMP also undergoes a tandem Michael addition/α-alkylation to give β-hydrazino esters in 26–68% yield, and 63 to >96% de *anti* (*758*). The hydrazine bond was cleaved with H_2/Raney-nickel to give the β-amino esters in 50–87% yield (*757, 758*).

An alternative approach to asymmetric Michael additions is to use a chiral auxiliary on the electrophile instead of the amine. Amoroso et al. added O-benzylhydroxylamine to chiral amide derivatives of crotonic acid, hex-3-enoic acid, or cinnamic acid in the presence of Lewis acids (*759, 760*). The best diastereo-selectivities (60–78% de) were obtained with an (4S,5R)-1,5-dimethyl-4-phenylimidazolidin-2-one auxiliary, but varied greatly depending on the type and quantity of Lewis acid employed. The initial adducts were later used

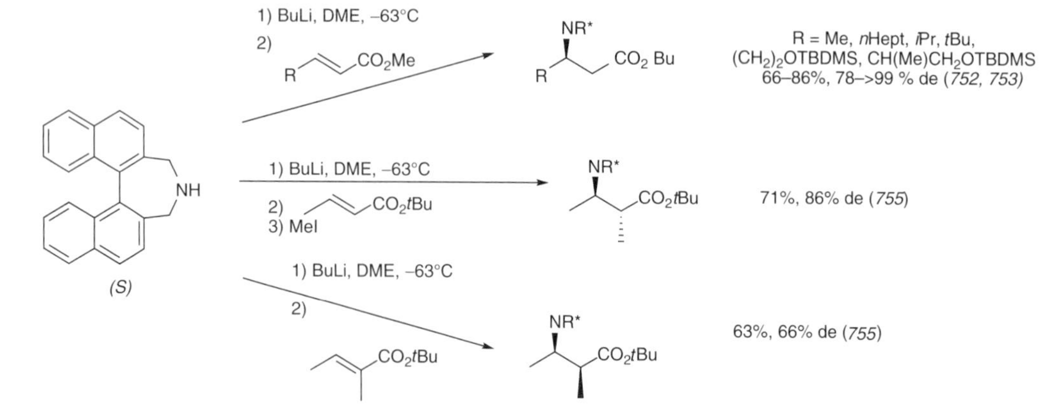

Scheme 11.32 Addition of Chiral Amine to α,β-Unsaturated Esters.

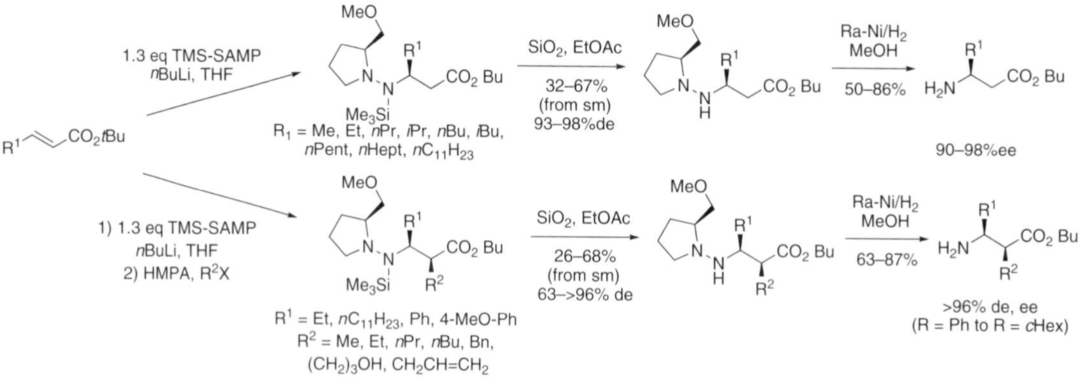

Scheme 11.33 Addition or Tandem Michael Addition/Trapping of Chiral Ammonia Equivalent (757, 758).

to form aziridine-2-carboxylates (760, 761). Phthalimide salts have also been added to acrylates derivatized with an imidazolidinone chiral amide auxiliary (762). Different salts resulted in substantially different diastereoselectivity, with Phth-MgCl giving >90% de for a β-aminobutanoic acid product. The intermediate enolate could be trapped with benzenesulfonyl bromide to give the 2-bromo-3-phthalimido imide (762).

A 5,5-dimethyloxazolidin-2-one amide auxiliary was used to induce asymmetry during conjugate additions to an N-acryloyl derivative. Several different approaches were tested (see Scheme 11.34). Addition of lithium dibenzylamide, followed by alkylation of the resulting α-anion with reactive electrophiles, produced the 2-substituted-β-amino acids with 96–97% de, though in only 8% yield with sterically hindered isopropyl idodide. The opposite diastereomers could be obtained by instead adding lithium dibenzylamide to acrylic acid derivatives with the 2-substituent already present. Some substituents gave relatively poor diastereoselectivity (e.g., 78% de for R = Ph), which could be improved to 96% de by using a chiral amine derivative of the correct (matched) chirality for the conjugate addition (763).

(S, S)-(+)-Pseudoephedrine has been employed as an amide auxiliary for addition of lithium benzylamides to 3-substituted acrylates. A range of reaction conditions were explored. Very high diastereoselectivity was obtained (>98% de in many cases) using Bn$_2$NLi or BnMeNLi in toluene at –90 °C. The diastereoselectivity could be reversed, resulting in the enantiomeric β-substituted-β-amino acid, by using O-TBS pseudoephedrine as the auxiliary (764). A synthesis of the β-Phe component of celacinnine was achieved via conjugate addition/cyclization of piperidazine to tert-butyl cinnamate derivatized with an α-(R)- or (S)-p-tolylsulfinyl chiral auxiliary substituent, producing (S)- or (R)-9-phenyl-1,6-diazabicyclo[4.3.0]nonan-7-one after removal of the p-tolylsulfinyl group. Reductive cleavage of the N–N bond produced (S)- or (R)-4-phenyl-1,5-diazacyclononan-2-one, 3-phenyl-3-amino-propionamide with a four carbon linker joining the amine and amide (123).

Chiral alcohols have been used to esterify crotonic acid; high-pressure addition of benzylamine proceeded with up to >99% de with esters of 8-(β-naphthyl)-menthol alcohol (711). A number of primary and

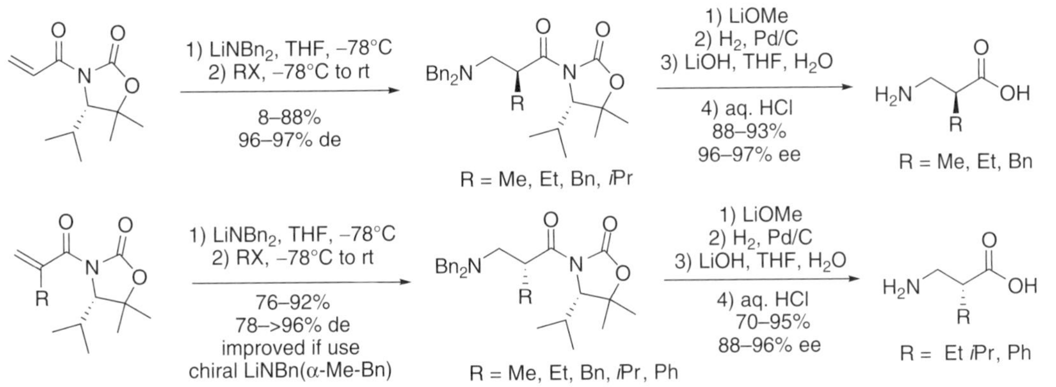

Scheme 11.34 Michael Addition to α,β-Unsaturated Acids with Oxazolidinone Chiral Auxiliary (763).

secondary amines were added to 5-alkoxy-2(5H)-fura-nones derivatized with a menthyl chiral auxiliary (765). The additions proceeded with complete stereoselectivity, but the adducts were elaborated to amino diols rather than β-amino acids. Several ester/amide auxiliaries were examined for the addition of amide cuprate reagents (a new class of nitrogen nucleophiles) to 5-phenylpenta-2,4-dienoic acid. A bornanesultam amide auxiliary induced 90% de when [Bn(TMS)N]₂CuLi was added. The initial enolate adduct could be trapped with acetaldehyde, providing a single diastereomer of all three newly generated chiral centers in 71% overall yield (766). Cinnamic acid was incorporated into a homochiral Fe complex, with lithium dimethylamide adding in 86% yield to give a β-dimethylamino complex with 97% de. Oxidative decomposition by N-bromosuccinimide provided (S)-3-dimethylamino-3-phenylpropionic acid (767).

An alternative approach for induction of asymmetry from a chiral electrophilic α,β-unsaturated acid component is the use of a removable chiral auxiliary at the C2 or C3 position. Various primary amines added to α,β-unsaturated esters possessing a chiral β-p-tolylsulfinyl group with diastereoselectivity of 49–89%. The β-auxiliary, which also helped activate the alkene for amine addition, was removed by reductive elimination with SmI₂ (768). Another approach employed a removable 1,3-oxathiane as a chiral auxiliary during the addition of benzylamine to homochiral unsaturated esters catalyzed by lanthanoid triflates (464).

Several syntheses have been based on an intramolecular addition of an amine nucleophile contained in a chiral auxiliary on the α,β-unsaturated acid component. For example, prolinamide was acylated with several α,β-unsaturated acids, and the C-terminal amide nitrogen used for an organoselenium-induced intramolecular addition. Complete diastereoselectivity and, depending on substituents, regioselectivity was observed, but yields varied from 22 to 73%. To obtain a free β-amino acid the adduct must be de-selenenylated and the seven-membered bislactam ring hydrolyzed (769, 770). A Hg-catalyzed intramolecular addition of an amino group to the β-position of 3-butenoic acid was used to

prepare 3-aminobutyric acid as a 29% de mixture of cyclic protected 6-methylperihydropyrimidin-4-one diastereomers (771). (R)-Emeriamine, 4-trimethylamino-3-aminobutanoic acid, was prepared via an intramolecular conjugate addition of the urea formed by reaction of methyl N-[(S)-1-phenylethyl)] 4-aminobut-2-enoate with tosyl isocyanate. The imidazolidinone adduct was deprotected with Li/NH₃ and hydrolyzed to give 3,4-diaminobutanoic acid, or hydrolyzed and N-trimethylated to give emeriamine (772). The N-acetoxymethyl substituent of homochiral methyl N-Boc,N-acetoxymethyl (S)-4-amino-6-methylhept-2-enoate was converted into a protected aminomethyl substituent by treatment with RCONH₂/PPTS. An intramolecular Michael addition was then induced by addition of NaH, with the resulting cyclic aminal decomposed by treatment with catalytic RuCl₃/TBHP, followed by 3 N HCl and Cs₂CO₃ in MeOH to give (3S,4S)-3,4-diamino-6-methylheptanoic acid with 80% de (773).

Other conjugate additions make use of chiral substrates instead of auxiliaries. Hydroxy/alkoxy-functionalized substrates are often employed. Benzylamine was added to homochiral 2-hydroxyalkylpropenoates with 60–64% anti de in MeOH and 40–56% syn de in THF (774). By protecting the hydroxyl group as a tert-butyldimethylsilyl ether, the additions proceeded with >90% anti de. Benzylamine was also added to homochiral unsaturated esters in the presence of lanthanoid triflate catalyst (464). Lithium amides (LiNHBn and LiN(Bn)₂) added to homochiral 4-silyloxy or 4-alkoxy alkan-2-enoates with variable diastereoselectivity (100% de syn to 40% de anti). The effects of other types of substituents were examined, as was double asymmetric induction when combined with a chiral amine nucleophile (775, 776). Double asymmetric induction was also employed during additions of chiral amines to homochiral 4-alkoxy 2,3-unsaturated acids, leading to 3-amino-4-alkoxypentanoic acid, 3-amino-4-hydroxy-4-phenylbutanoic acid, and 3-amino-4,5-dihydroxypentanoic acid. With the γ-methyl-substituted substrate, the configuration of the amine controlled the stereochemistry of the newly generated chiral center, but with γ-phenyl and γ-CH₂OtBu substrates the alkoxy γ-center chirality predominantly controlled the β-center stereochemistry (777).

A homochiral γ-lactone of 4,5-dihydroxypentenoic acid, prepared from D-mannitol via Wittig reaction of D-glyceraldehyde actonide, added benzylamine with complete stereoselectivity (778). The aminated lactone could then be deprotonated and alkylated at the α-position (778). Benzylamine was also added to acetonide derivatives of both (E)- and (Z)-4,5-dihydroxy pent-2-enoic acid methyl ester (779). Both isomers provided the same (3R)-product with complete diastereoselectivity when reacted at −50 °C, opposite to the stereochemistry of the product obtained by addition to the lactone derivative above. A series of α-glycosyl-β-amino acids were derived from sugar aldehydes via Wittig olefination with Ph₃P=CHCO₂Et, followed by Michael addition of benzylamine, using TBAF (tetra-n-butylammonium fluoride) as the basic catalyst. The N-benzyl products were obtained in 68–85% yield with 86–94% de (780).

Another potential route for asymmetric induction during addition of amines to unsaturated esters is to use a chiral catalyst. Titanium complexes of TADDOL and BINOL catalyzed the addition of O-benzylhydroxylamine to α,β-unsaturated acids amidated with a 1,3-oxazolidinone (781). However, enantioselectivity was poor (31–42% ee). Homochiral aluminum or boron complexes of N-benzylhydroxylamine added to unsaturated amides with 43–71% ee (782), while a bisoxazoline catalyst induced up to 97% ee for additions of N-benzylhydroxylamine to α,β-unsaturated amides in the presence of MgBr₂ as Lewis acid (783). By using lanthanoid triflates as the Lewis acid, the enantioselectivity was reversed (but reduced to 41–59% ee) (783). α,β-Disubstituted-β-amino acids could be obtained in a similar manner, via addition of N-benzyl hydroxylamine to a number of 2,3-disubstituted-2,3-unsaturated amide derivatives in the presence of a chiral Lewis acid bisoxazoline catalyst (93–99% de, 60–96% ee) (784). A bisoxazoline catalyst was also employed for additions of carbamates to α'-hydroxyenones, RCH=CHCOC(Me)₂OH, giving products with 88–99% ee. The β-amino acid was then generated by NaIO₄ oxidative cleavage of the acyloin moiety (785).

Other amines have been utilized. Anilines were added to 3-substituted acrylate amides in the presence of Ni(ClO₄)₂ and a bisoxazoline catalyst, with 6–87% yield and 34–90% ee (786). Aniline additions to N-alkenoylcarbamates, RCH=CHCONHCO₂tBu, with a BINAP-Pd catalyst proceeded with much better 89–97% ee for unsubstituted aniline (787). In 1999 it was found that hydrazoic acid (HN₃) added to unsaturated imides (E-RCH=CHCONHCOPh) with very high enantioselectivity (95–97% ee) and yield (96–98% ee) when catalyzed by an (S,S)-salen–Al(III)Me complex (see Scheme 11.35). One example was converted to the amino acid via hydrogenation of the azide and cleavage of the imide group by aqueous NaOH (788). Simple peptides were reasonably effective catalysts for the addition of TMSN₃ to a number of substituted acrylate pyrrolidinone amides, with 45–85% ee (789). In 2003, lithiated N-TMS benzylamine was added to a variety of α,β-unsaturated tert-butyl esters in toluene at −78 °C, giving products with 90–99% ee in the presence of a chiral 1,2-dimethoxy-1,2-diphenylethane ligand (790). Reduced enantioselectivity (75–94% ee) was observed when lithiated N-TBDMS allylamine was added to tert-butyl acrylates (791).

Modest enantioselectivity (up to 35% ee) was obtained during syntheses of 2-trifluoromethyl-β-Ala and 3-trifluoromethyl-β-Ala via Michael addition of diethylamine to the corresponding trifluoromethyl-substituted acrylates in the presence of Candida cylindracea enzyme (792). Similarly, several primary and secondary amines were added to ethyl cinnamate, p-methylcinnamate or p-methoxycinnamate in the presence of Baker's yeast and cyclodextrin, with 23–72% ee observed in the products (793).

11.2.3c Amination Reactions: Epoxide Opening

Alternative electrophilic substrates for nucleophilic amination reactions leading to β-amino acids are 2,3-epoxy acids or their equivalents, which can be readily prepared by Sharpless epoxidation of allylic alcohols followed by hydroxymethyl oxidation (see Section 8.4) (e.g., see 794, 795). However, the regioselectivity of ring opening can be a problem. Aqueous ammonia opening of unsubstituted glycidic acid produced isoserine in 33% yield, but 3-substituted epoxides tended to open at C2 to give α-amino acids (794). Oxidation of (S)-glycidol to glycidic acid, coupling to a peptide, and ring opening with NaN₃/MgSO₄ gave the desired isoserine regioisomer

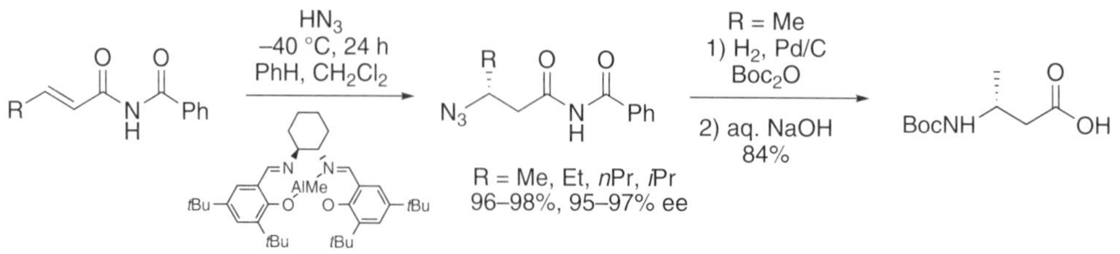

Scheme 11.35 Salen-Catalyzed Addition of Hydrazoic Acid to Unsaturated Imides (788).

in >50% yield (*175*). The regioselectivity of ring opening of 2,3-epoxy acids and 2,3-epoxy amides by secondary amines can be directed towards the C3 attack required to produce β-amino acids by the addition of Ti(O-*i*Pr)₄ (*2191*). Azide also adds with high C3 regioselectivity in the presence of this catalyst (*2191*). This method was used to synthesize the β-amino acid contained in amastatin, (2*S*,3*R*)-3-amino-2-hydroxy-5-methylhexanoic acid (*2191*). Yb(OTf)₃ also induces high C3 regioselectivity for azide opening, with a *trans*-C3-(4-methyl-phenyl)glycidic amide (prepared by an enantioselective Darzens reaction of *p*-methylbenzaldehyde with a camphor-derived sulfonium salt) opened in 79% yield (*796*). Indium trichloride catalyzed the azidolysis of racemic 2,3-epoxy acids by NaN₃ to give the α-hydroxy-β-azido products with >99:1 regioselectivity and >99% yield (*507*). Cu(NO₃)₂ not only catalyzes regioselective azide opening of 2,3-epoxy acids, but also can be employed to catalyze the subsequent azide reduction step, giving a one-pot synthesis. Phenylnorstatine, allophenyl-norstatine, and alloethylnorstatine were among the α-hydroxy-β-amino acids produced (*797*). The β-amino acid components of bestatin (2-hydroxy-3-amino-4-phenylbutyric acid) and 2-methylbestatin (2-hydroxy-2-methyl-3-amino-4-phenylbutyric acid) were both prepared via regioselective azide opening of the corresponding glycidic acid (*795*). An epoxy ester derived from L-tartaric acid was opened with azide to give D-*threo*-3-amino-2-hydroxybutanoic acid (*798*).

Other amine nucleophiles include α-methyl-benzylamine, which was employed to open a 2,3-epoxy ester leading to a complex β-amino acid with a bicyclic C2 substituent (*799, 800*). Azide or various chiral amines opened a homochiral epoxy ester with a 2-(poly-hydroxycyclopentane) substituent (*801*). A *trans*-3-furyl substituted epoxy ester was regiospecifically opened with either benzylamine or TMS-N₃, with azide producing the *syn* adduct and benzylamine the *anti* isomer (*802*). Ring openings of the closely related *trans*-3-thienyl substrate were much less regio- and stereoselective (*802*). L-Thr and L-Ser were converted into epoxides by replacement of the α-amino group with an α-bromo group, followed by intramolecular displacement with the β-hydroxy group. Epoxide opening with ammonia gave D-isothreonine and D-isoserine, respectively (*803*).

Epoxy esters can be opened with halides instead of amines, with the halide subsequently displaced by an amine equivalent. This provides access to the complementary diastereomer compared to that produced by direct epoxide opening. (2*S*,3*R*)-3-Amino-2-hydroxyde-canoic acid (Ahda) was synthesized via opening of the *trans* epoxy ester with MgBr₂ in ether, with the resulting β-bromo group displaced by azide (*804*). In a similar manner, methyl (2*S*,3*R*)-4-phenyl-2,3-epoxybutanoic acid was opened with MgBr₂ (92% yield), then displaced with NaN₃ (73% yield), and reduced (hydrogenation, 95% yield) to give (2*S*,3*R*)-3-amino-2-hydroxy-4-phenylbutanoic acid (AHPBA) (*805*). *cis*-3,4-Phenyl-2,3-epoxy-butanoic acid was treated with benzonitrile and BF₃·Et₂O to induce C3 opening, resulting in a phe-nyl-substituted oxazoline of ethyl 2-hydroxy-3-amino-4-phenylbutanoate. Hydrolysis with 6 N HCl gave the free amino acid (*806, 807*). The same approach was employed for isobutyl- and phenyl-substituted epoxides, using either benzonitrile or acetonitrile (*807*); an earlier synthesis prepared α-alkyl-α-hydroxy-β-amino acids using acetonitrile for the opening (*808*).

Several syntheses rely on ring opening of protected 2,3-epoxy alcohols rather than 2,3-epoxy acids. Sharpless epoxidation of allylic alcohols followed by opening with an amine nucleophile gave *anti*-3-amino-1,2-diols, which were then converted into both *anti*- and *syn*-α-hydroxy-β-amino acids, including phenylisoser-ine (*809*). The *anti* diastereomers were obtained by selective protection of the 2-hydroxy group followed by RuCl₃/NaIO₄ oxidation of the hydroxymethyl moiety. The *syn* isomers were produced via a Mitsunobu inversion of the 2-carbinol center, using *p*-nitrobenzoic acid as the nucleophile. Although lengthy, the latter procedure provided access to *syn*-3-amino-1,2-diols that are otherwise difficult to obtain. This methodology was applied to a synthesis of cyclohexylnorstatine, (2*R*,3*S*)-3-amino-4-cyclohexyl-2-hydroxybutanoic acid, via a Sharpless epoxidation of (*E*)-4-cyclohexyl-2-buten-1-ol followed by regioselective opening of the epoxide with either benzhydrylamine or azide in the presence of a titanium catalyst (see Scheme 11.36). Hydrogenolysis gave the aminodiol, but with undesired C2 stereochemistry at the hydroxy center. Inversion at C2 was accomplished via a Mitsunobu reaction, with the carboxyl group finally generated by oxidation of the terminal

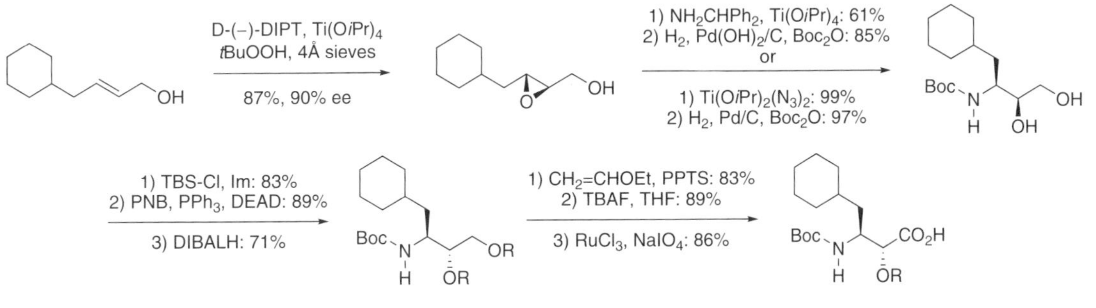

Scheme 11.36 Synthesis of Cyclohexylnorstatine via Epoxide Opening (*810*).

hydroxy group with RuCl$_3$/NaIO$_4$ (810). In a similar manner, benzhydrylamine opened the Sharpless epoxidation product of but-2-en-1-ol, with selective hydroxy protection and hydroxymethyl oxidation giving *anti* 3-amino-2-hydroxybutyrate (811). (S)-Isoserine was prepared from (S)-glycidyl p-methoxyphenyl ether (obtained by a hydrolytic kinetic resolution) via opening with azide, protection as the benzyl ether, azide reduction, p-methoxyphenyl ether removal with ceric ammonium nitrate, and oxidation with PDC. The enantiomer was also synthesized (378).

An alternative procedure converts the diol moiety of 3-amino-1,2-diols (obtained by benzhydrylamine opening of Sharpless epoxidation-derived epoxy alcohols) into a vinyl group, which is then hydroborated and oxidized (see Scheme 11.37) (812).

The *syn* phenylisoserine side chain of Taxol has been prepared by a number of research groups via opening of an epoxide with an amine nucleophile. A *cis*-substituted epoxide is required for the desired *syn* isomer product. The homochiral epoxide substrate can be derived from (Z)-cinnamyl alcohol via a Sharpless epoxidation, with the synthesis completed by oxidation of the hydroxymethyl group to an acid followed by epoxide opening with azide (813). However, Sharpless epoxidations of Z-allylic alcohols tend to proceed with much poorer enantioselectivities than do epoxidations of *E*-allylic alcohol substrates. Most routes to *syn* phenylisoserine therefore either use a different procedure to obtain the requisite *cis* α,β-epoxy ester that leads to *syn* α-hydroxy-β-amino esters, or incorporate an inversion step of the carbinol center. For example, a (salen)Mn(III) chiral complex was used for an asymmetric epoxidation of *cis*-ethyl cinnamate, followed by opening with ammonia to give the *syn* isomer (814). In earlier reports an achiral Co(II)-salen catalyst used for both epoxidation and amination steps led to either *syn* or *anti* racemic isomers, depending on the type of aniline used for epoxide opening (501, 502, 504). The cobalt catalyst has also been used on a solid support, catalyzing a one-pot epoxidation of cinnamoyl-L-Leu followed by ring opening with anilines (815). Substituted N-benzyl anilines led to the *anti* diastereomer if the phenyl ring was p-methoxy substituted, or the *syn* isomer if it was unsubstituted (504, 815). Cinnamic acid, attached to a Pro auxiliary, was epoxidized with a polyaniline-supported Co(II)salen catalyst to give the *trans* epoxide. Opening with p-bromoaniline gave (2R,3R)-phenylisoserine (816). A synthesis of the *anti* isomer of phenylisoserine proceeded via azide opening of an epoxy oxazolidine, with the epoxide chirality originating from a chiral oxazolidine

auxiliary instead of the epoxidation catalyst (817). A large-scale synthesis of both the Taxol N-benzoyl (2R,3S)-3-phenylisoserine, and the (2S,3S)-epimer, opened the chiral *trans*-epoxide ethyl ester with ammonia to give the (2S,3S)-phenylisoserine amide in 81% yield. The product was N-benzoylated and converted to the methyl ester (50%), and then epimerized at the carbinol center using SOCl$_2$/CHCl$_3$ followed by HCl/MeOH (49% yield) (818).

The *cis* glycidic ester leading to *syn*-phenylisoserine, as well as other α-hydroxy-β-amino acids, has been prepared by cyclization of *syn*-2-chloro-3-hydroxy esters induced by potassium carbonate (819). The homochiral chlorohydroxy substrate was prepared by microbial reduction of 2-chloro-3-oxoalkanoate esters. The *trans* glycidic esters were accessed by cyclization of the *anti* diastereomer, or by cyclization of either isomer in the presence of NaOEt as base. Ring opening with azide or ammonia was completely C3-specific (819). Trans-methyl cinnamate can be asymmetrically dihydroxylated and then cyclized to the *cis*-glycidic ester using TsCl/Et$_3$N followed by K$_2$CO$_3$ (820, 821), with amination by azide opening (821). The *cis* epoxy ester has also been prepared via condensation of a enolate of a bromoacetyl oxazolidinone with benzaldehyde. Azide opening of the epoxide, followed by hydrogenation, gave *syn*-phenylisoserine ester in 85% yield (822).

Several methods have been developed to invert one of the two chiral centers contained in *trans*-phenylglycidate in order to form the *syn* isomer of phenylisoserine. Asymmetric epoxidation of 1-tBu-3-phenylprop-2-enone using poly-L-Leu as catalyst gave the *trans* epoxide in 76% yield with 94% ee. Baeyer–Villiger oxidation of the ketone with *m*CPBA resulted in the same *trans* phenylglycidic ester product as produced by asymmetric epoxidation of cinnamic acid. Epoxide ring opening with HCl, followed by epoxide reformation under acidic conditions, generated the desired *cis*-substituted epoxide isomer. Epoxide opening with azide followed by hydrogenation, or opening with ammonia, gave the *syn* isomer of phenylisoserine (823). Similarly, a large-scale synthesis of N-benzoyl (2R,3S)-3-phenylisoserine opened the *trans* epoxide with chloride (70%), and then reformed a *cis* epoxide to epimerize the C2 center (61%), followed by ammonia opening (65%) and amide hydrolysis (88%) (824).

Homochiral β-phenylglycidic ester substrates have also been prepared by a lipase-mediated resolution of racemic methyl *trans*-β-phenylglycidate (825). Both enantiomers were then converted to the Taxol (2R,3S)-phenylisoserine side chain. The (2S,3R)-epoxide isomer

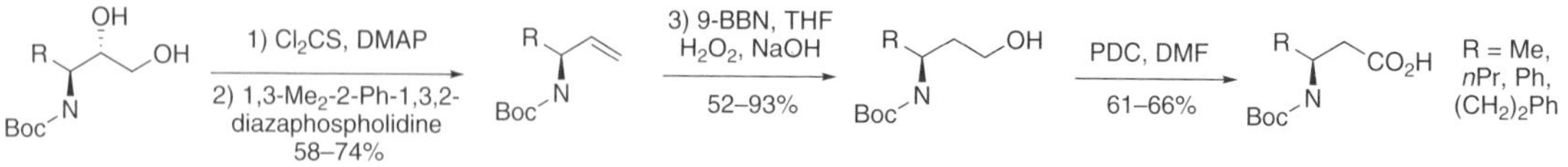

Scheme 11.37 Elaboration of 3-Amino-1,2-diol (812).

was directly opened with azide, and the hydroxy center inverted via intramolecular oxazoline formation. Alternatively, the (2R,3S)-isomer was indirectly aminated via ring opening with bromide, followed by azide displacement to invert the β-center (825). The bromide-opening/azide-displacement route was also applied to substrates derived from Sharpless epoxidations, providing both phenylisoserine and cyclohexylnorstatine (826). An unusual variation of these epoxide-opening reactions first coupled the cis- or trans-phenyl glycidic acid to protected 10-deacetylbaccatin III. Epoxide opening with azide or hydroxylamine, or with bromide followed by azide displacement, generated the desired α-hydroxy-β-amino acid (820, 827). Another enzymatic resolution was used to prepare syn-2,3-dihydroxy-3-phenylpropanoic acid methyl ester. Both enantiomers were then converted into the same phenylisoserine diastereomer required for the side chain of Taxol. The (2S,3R)-enantiomer was converted into an epoxide and opened with azide (inverting both centers), while the (2R,3S)-enantiomer was transformed into a 2-hydroxy-3-bromo derivative and displaced with azide (one center inverted twice) (828).

11.2.3d Amination Reactions: Other Nucleophilic Aminations

Another general route for nucleophilic aminations is to displace an activated β-hydroxy group on a homochiral substrate. Various procedures have been employed to prepare the homochiral β-hydroxy acid substrate required for amination, with a masked carboxyl group often present during the amination step. In one method, a Sharpless epoxidation was used to prepare the substrates for both enantiomers of (Z)-3-amino-5-tetradecenoic acid, the β-amino acid component of lipogrammistin-A, a lipophilic ichthyotoxin secreted by the skin of soapfish. (E)-Tetradec-2-en-5-yn-1-ol was epoxided, the alkyne was reduced to form the alkene, and the epoxide was opened with Red-Al to form (Z)-tetradec-5-en-1, 3-diol. The amino group was introduced by mesylation of the secondary alcohol followed by displacement with azide and azide reduction, with the acid group generated by oxidation of the terminal hydroxyl (83). 3-Amino-2-methylpropionic acid was produced by azide displacement of the tosylate of the corresponding chiral 3-hydroxy-2-methylpropionic acid methyl ester (65, 67).

Several syntheses rely on aminations of natural product-derived substrates in which the acid group is masked as an alcohol, aldehyde or alkene. Both (R)- and (S)-isoserine were synthesized from D-mannitol via direct or indirect (via epoxide) azide displacement of both terminal hydroxyl groups, followed by protecting group manipulations and oxidative cleavage of the central diol to create two molecules of the desired product (829). D-Glucose was transformed into 3-amino-4-cyclohexyl-2-hydroxybutanoate in 10% overall yield by a route which included amination via azide displacement of a mesylate, and acid group generation via a

Baeyer–Villiger oxidation (830). The complex β-amino acid Adda was prepared from a D-glucose-derived oxazoline by a route including amination via azide displacement of a triflate. The acid group was revealed by oxidation of the glucose aldehyde group (831).

Azide displacements have also been used in many other syntheses, with the β²-amino acid analog of His prepared via azide displacement of a chiral hydroxy substrate; attempted syntheses by a number of other routes were unsuccessful (832, 833). The β-amino acid component of AI-77B (2,3-dihydroxy-4-amino-1,6-hexanedioic acid) was prepared via an azide displacement of a chiral mesylate (216). A protected form of AMMTD, 3-amino-2,4,5-trihydroxy-6-methyl-12-(4′-methoxyphenyl)-11-dodecenoic acid (found in the cyclic antifungal hexapeptides microsclerodermins A and B), was prepared by a lengthy route that included amination by azide displacement of a chiral triflate. The carboxyl group was subsequently generated by oxidation of a hydroxymethyl group (212). The 3-amino-2,4,5-trihydroxy-10-(4′-ethoxyphenyl)-7,9-decadienoic acid component of microsclerodermin E employed an azide displacement of chiral tetraol substrate, followed by further elaboration of the side chain and eventual oxidation of one hydroxymethyl group (215). (2R,3R)- and (2R,3S)-3-amino-2-methylpentanoic acids were prepared from (3S,4S)- and (3R,4S)-4-methyl-5-hexen-3-ol, respectively, via azide displacement of the mesylate derivative followed by oxidative cleavage of the alkene (834). (2R)-3-Amino-2-methylpropionate was produced via azide displacement of (2S)-3-bromo-2-methylpropanol, followed by RuCl₃/NaIO₄ oxidation of the alcohol (835).

A large-scale synthesis of phenylisoserine began with an asymmetric dihydroxylation of trans-methyl cinnamate. The product was treated with trimethyl orthoacetate/p-TsOH followed by acetyl bromide to give primarily (2R,3S)-2-acetoxy-3-bromo-3-phenylpropionate. Azide displacement of the bromide and azide reduction produced phenylisoserine in 23% overall yield on a 60 g scale with 99% ee (182). The same bromination/azidation reaction sequence was applied to a dihydroxy substrate obtained in 90% yield with 99% ee via a one-pot reaction that used a bi- or tri-functional catalyst not only for the dihydroxylation reaction but also to generate the trans-cinnamate substrate via a Heck coupling of bromobenzene with methyl acrylate (836). A conformationally restricted analog of phenylisoserine, 2, 2′-methylenephenyslisoserine, was prepared via an asymmetric dihydroxylation of ethyl 1H-indene-2-carboxylate (92% ee). Bromination/azide displacement/azide reduction gave the desired cyclic β-amino acid. Attempts at asymmetric aminohydroxylation gave racemic product (187). The azido precursors of two C3 diastereomers of 2-hydroxy-3-amino-5-methylhexanoate were derived from an asymmetric dihydroxylation of ethyl 5-methylhex-2-enoate. One diastereomer was prepared by a regioselective Mitsunobu azidation of the 2,3-diol to give 2-hydroxy-3-azido-5-methyl-hexanoate, while the other isomer was obtained by regioselective Mitsunobu tosylation followed by azide displacement (837).

Mitsunobu aminations have been employed in a number of other syntheses. β-Tyr was obtained by azide displacement of homochiral ethyl 3-hydroxy-3-(4-benzyloxyphenyl)propionate under Mitsunobu conditions (130). β-Amino acids with α-benzyl substituents were prepared via Mitsunobu displacement of a chiral mono-O-acetyl 2-benzyl-1,3-diol. The monoprotected diol was obtained by enzymatic desymmetrization of a prochiral bis-O-acetyl precursor. The carboxyl group was generated after amination by oxidation of the remaining hydroxymethyl substituent (838). Adda, (2S,3S,8S,9S)-3-amino-9-methoxy-2,6,8-trimethyl-10-phenyl-4,6-decadienoic acid, was synthesized via Mitsunobu amination of a diol precursor to give 2-amino-3-methyl-4-hydroxy-butanal as a key intermediate, with the carboxyl group masked as a hydroxymethyl group. The side chain was then introduced by a Wittig coupling with the aldehyde (839). Evans' oxazolidinone chiral auxiliary was used to prepare homochiral 2-methyl-3-hydroxy-4-benzyloxybutyric acid. The amide auxiliary was then exchanged for a Weireb amide. Mitsunobu displacement of the 3-hydroxy group with HN₃, followed by hydrogenation, gave the Weinreb amide of 2-methyl-3-amino-4-hydroxybutyric acid. The Weinreb amide prevented γ-lactone formation (cyclization of the alcohol with the acid group), allowing for oxidation of the unprotected alcohol to an aldehyde. The protected β-amino-γ-keto acid was prepared as a potential synthon for the synthesis of complex β-amino acids such as Adda, via a Wittig coupling reaction (840). Two diastereomers of 2-fluoro-2-methyl-3-phenyl-3-aminopropionic acid were obtained via a Pd-catalyzed asymmetric fluorination of 2-methyl-3-keto-3-phenylpropionic acid. The ketone was then diastereoselectively reduced by phenyldimethylsilane/DMF/TBAF (>90% de syn) or triphenylsilane/TFA (>90% de anti), with the synthesis completed by Mitsunobu displacement of the alcohol with DEAD/DPPA, and then azide reduction (841).

A Mitsunobu azide displacement of a chiral polyhydroxy compound derived from L-arabinose was used in the synthesis of 2-hydroxy-3-amino-5-phenylpent-4-enoic acid (842). The β-azatyrosine residue found in the antiproliferative natural product kedarcidin, β-(2-chloro-3-hydroxy-pyrid-6-yl)-β-amino acid, was constructed via a Mitsunobu azide reaction of methyl 3-hydroxy-3-(2-chloro-3-methoxymethyloxy-pyrid-6-yl)propionate, followed by Staudinger reduction of the azide. The chiral hydroxy substrate was obtained on a large scale by an asymmetric aldol reaction (142). The diol precursor of phenylisoserine was obtained by addition of a hydroxyacetate equivalent to a chiral chromium complex of benzaldehyde, with the amine introduced by a Mitsunobu reaction with azide (843, 844). Similarly, another synthesis of the Taxol phenylisoserine component employed an aldol reaction to prepare homochiral 2,3-dihydroxyphenylpropanoic acid. Amination was achieved by Mitsunobu displacement of the 3-hydroxy group with HN₃, followed by azide reduction. The product was protected as an oxazolidine before coupling to the baccatin III skeleton during a total synthesis of Taxol (845). (2S,3R)-3-Amino-2-hydroxy-4-phenylbutanoic acid was prepared by a Mitsunobu displacement of the protected precursor with DPPA. An aldol condensation of a chiral glycolate enolate was employed to synthesize the precursor (846). The hydroxyl group formed during an aldol reaction of an acylated Evans' oxazolidinone auxiliary was displaced by azide under Mitsunobu conditions during a synthesis of β-(6-amino-4-carboxy-5-methylpyrimidin-2-yl)-β-Ala. In this case, a temporary 2-thiomethyl substituent was used during the aldol reaction to ensure transfer of chirality from the oxazolidinone auxiliary to the β-center. It was removed by Bu₃SnH reductive cleavage (847).

Intramolecular aminations have been developed. Methyl (2S,3S)-3-amino-2-methyl-7-octynoate was prepared via intramolecular displacement of a β-mesylate by the amide nitrogen of a homochiral β-hydroxy amide derivative. The homochiral β-hydroxy amide was prepared by an aldol reaction of an acylated Evans' oxazolidinone (see Scheme 11.38) (848). This is potentially a very useful route to a number of α,β-disubstituted-β-amino acids as either enantiomer, and potentially either diastereomer, can be accessed during the initial aldol reaction. In a somewhat similar amination procedure,

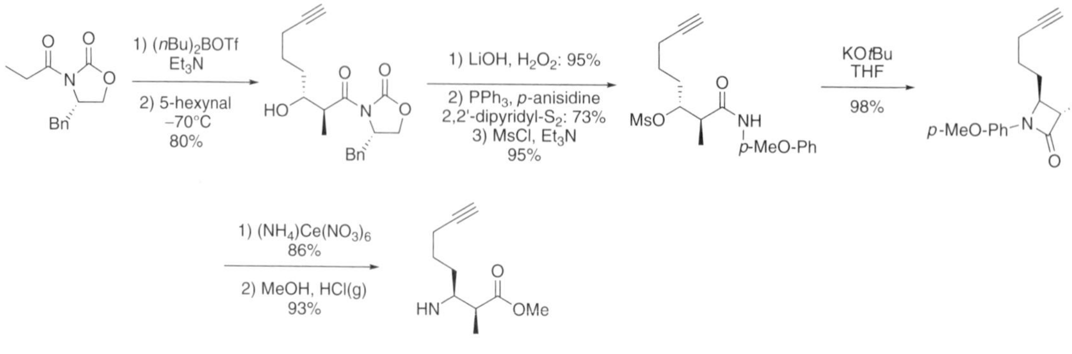

Scheme 11.38 Intramolecular Mesylate Displacement (848).

asymmetric dihydroxylation of *trans* N-benzyl or N-4-methoxyphenyl cinnamide was followed by benzylic bromination and O-acetylation to give 2-acetoxy-3-bromocarboxamides. Treatment with TBAF induced intramolecular amination to form (3R,4S)-3-hydroxy-4-phenylazetidinone, with oxidative cleavage of the N-4-methoxyphenyl group giving an activated precursor of phenylisoserine, ready for coupling to 10-deacetyl baccatin III (*849*).

An intramolecular amination was also employed during conversion of homochiral β-hydroxycarboxylic acids into the corresponding β-amino acids by a four-step procedure. The acid was coupled to O-benzyl-hydroxylamine, with an intramolecular Mitsunobu reaction generating an N-benzyloxy β-lactam, accompanied by inversion at the β-position. Acidic or basic hydrolysis provided the N-benzyloxy β-amino acid, with the N–O bond finally cleaved by hydrogenation (*850*). Chiral α-hydroxy-β, γ-unsaturated esters have been aminated by reaction of the hydroxy group with tosyl isocyanate (see Scheme 11.39). The resulting allylic carbamate cyclized upon treatment with iodine to form an oxazolidinone of 2-hydroxy-3-amino-4-iodoalkanoates with >90% de. Removal of the iodo group (by radical reduction) and the N-tosyl group (by sodium naphthalenide reduction) provided (2R,3S)- or (2S,3R)-3-amino-2-hydroxy acids, such as norstatine, cyclohexylnorstatine, and Ahda (*851*).

A large-scale (40 g) synthesis of the Taxol (2R,3S)-phenylisoserine side chain, and other aryl-substituted analogs, began with asymmetric dihydroxylation of cinnamate esters. Treatment of acetonitrile or benzonitrile solutions of the dihydroxycinnamates with concentrated sulfuric acid at −10 °C produced an oxazoline intermediate (via a benzylic carbocation that reacts with the nucleophilic nitrile nitrogen). Dilute aqueous sulfuric acid hydrolysis hydrolyzed both the oxazoline and the ester to give N-acetyl or N-benzoyl phenylisoserines with 11:1 to 20:1 *syn* stereoselectivity (*183*).

A cyclic sulfamidate was used to both aminate 2,3-dihydro-2-methylpropanoic acid, and activate the α-center for nucleophilic displacement (see Scheme 11.40). A series of β²,²-amino acids were prepared (*852*). The Sharpless asymmetric dihydroxylation has proven to be useful for the preparation of homochiral amination substrates leading to α-hydroxy-β-amino acid products. Asymmetric dihydroxylation of *cis*-methyl cinnamate, formation of a cyclic sulfate intermediate, and opening with azide gave, after azide reduction, *syn*-phenylisoserine. However, the initial dihydroxylation proceeded with only 7% ee (*853*). In contrast, *trans*-methyl cinnamate was dihydroxylated with 92% ee and then converted to *syn*-phenylisoserine via initial opening of the cyclic sulfate with bromide followed by azide displacement (*853*). Alternatively, a cyclic sulfite intermediate obtained from the asymmetric dihydroxylation product of *trans*-ethyl cinnamate was opened with azide followed by Mitsunobu inversion of the carbinol center to provide the desired stereoisomer (*854*). Enantiomerically enriched β-lactones, the cyclic esters of β-hydroxy acids, can be opened with azide in 78–95% yield or with the sodium salt of o-nitrobenzenesulfonamide in 43–83% yield, with no loss in enantiopurity (*855*).

11.2.3e Amination Reactions: Asymmetric Aminohydroxylations

Catalytic asymmetric aminohydroxylation of olefins has recently become a feasible operation. Substituted cinnamates are amenable substrates, leading directly to phenylisoserines with high regioselectivity (e.g., see Scheme 11.41). Chloramine-T trihydrate (TsNClNa·3H₂O) was initially employed as the oxidant/amine source (*856*), but removal of the resulting N-toluenesulfonyl group required 33% HBr in acetic acid with phenol at 75 °C (*856, 857*). Nevertheless, (2R,3S)-phenylisoserine was prepared in two steps in 53% yield with 82% ee (*856*). Cyclohexylnorstatine was also synthesized using this oxidant, with the aminohydroxylation of ethyl 4-cyclohexyl-2-butenoate proceeding in 60–65% yield with 89–96% ee. However, the tosyl group was not removed

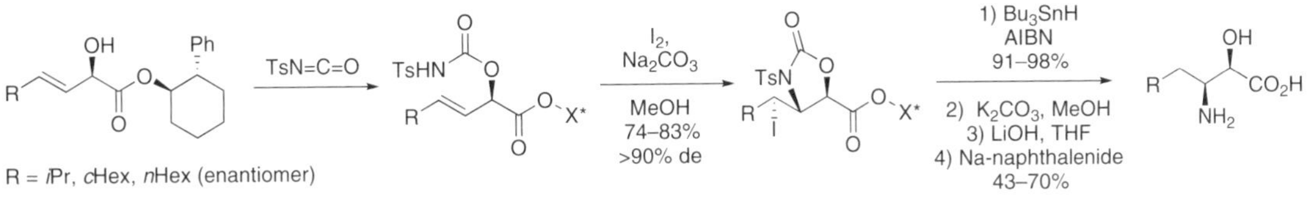

R = *i*Pr, *c*Hex, *n*Hex (enantiomer)

Scheme 11.39 Synthesis of *Syn-β-Amino-α-Hydroxy Acids* (*851*).

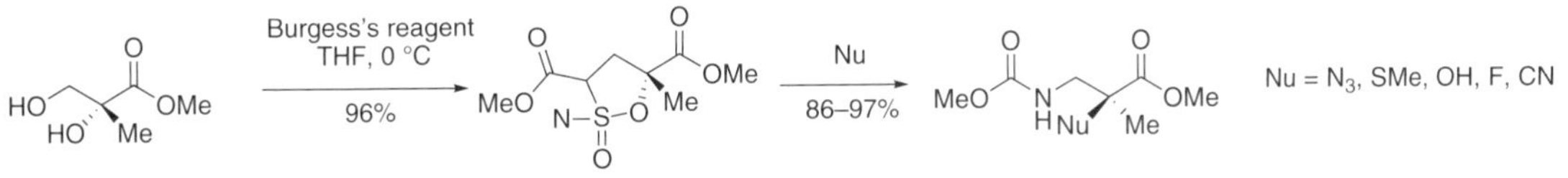

Nu = N₃, SMe, OH, F, CN

Scheme 11.40 Amination via Cyclic Sulfamidate (*852*).

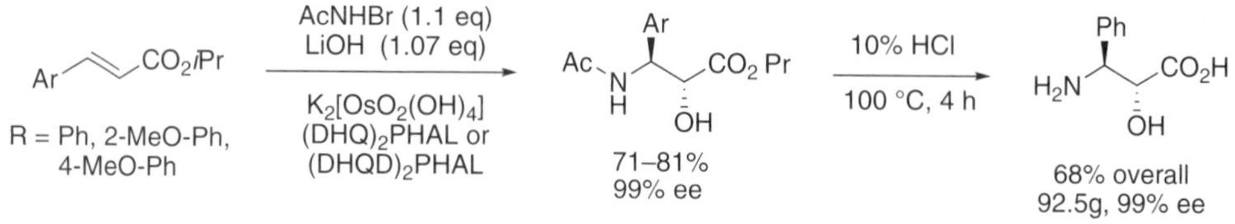

Scheme 11.41 Catalytic Asymmetric Aminohydroxylation (*872*).

(*858, 859*). Another synthesis prepared 2-hydroxy-3-aminobutanoic acid (*857*). It has subsequently been reported that the *N*-tosyl group of aminohydroxylation products can be readily removed by treatment with 2,2-dimethoxypropane in toluene with catalytic PPTS, forming an *N,O*-acetonide product (*860*). This method was used for the preparation of 2,6-dihydroxy-3-amino-hexanoic acid, an intermediate in the synthesis of the γ-amino acid vigabatrin (*861*).

Replacement of the Chloramine-T reagent with Chloramine-M gave higher enantioselectivities (*862*), while *N*-halocarbamate salts were found to give excellent enantioselectivities under convenient reaction conditions, with the added advantage of easy *N*-deprotection of the *N*-Cbz or *N*-ethoxycarbonyl products (*863*). A number of 2-hydroxy-3-aryl-β-amino acids were produced by the latter method, in addition to unsubstituted 2-hydroxy-3-aminobutyric acid (*863*). *N*-Chlorobenzyl-carbamate (CbzNHCl) was used for an aminohydroxylation of *tert*-butylcrotonate, giving (2*S*,3*R*)-2-hydroxy-3-aminobutanoate with 90% ee (*864*), and for aminohydroxylations of ethyl 3-(2-furyl)acrylate (*865*) and methyl *p*-methoxycinnamate (*866*). The same reagent, generated from benzylcarbamate/*tert*-butyl hypochlorite, was employed for aminohydroxylations of a variety of 3-(heteroaromatic)acrylates. Furan, thiophene, and indole derivatives reacted well, but pyrrole derivatives were problematic. Pyridyl groups also caused problems, but the corresponding *N*-oxide pyridines reacted smoothly, and could be reduced back to the pyridine with Raney-Ni or TiCl₃ (*867*). BocNNaCl was used for aminohydroxylation of 3,5-diisopropoxy-4-methoxycinnamate, giving the α-hydroxy-β-aryl β-amino acid in 70% yield and the regioisomeric α-amino-β-hydroxy-β-aryl ester in 11% yield. Three other byproducts were observed, a benzaldehyde (7%), benzoic acid (2%), and 2,2-diamino-3-oxo-3-arylpropionate derivative (4%), which may account for reduced chemical yields in other syntheses. The α-hydroxy group of the desired product was mesylated and displaced with cesium acetate to give the epimeric product, or with azide to give the 2-azido-3-amino product (*868*). Aminohydroxylations of nitro- or amino-substituted methyl cinnamates with BocNLiBr led to the substituted phenylisoserines with 24–55% yield and 79–86% ee (*869*).

Amino-substituted heterocycles have been identified as another nitrogen source, leading to *N*-heteroaryl β-amino acids (*N*-4,6-dimethyl-1,3,5-triazin-2-yl β-phenylserine) (*870*). *N*-Chloro-*N*-sodio-2-trimethylsilyl ethyl carbamate,

TeocN(Cl)Na, was recently used as a nitrogen source for the preparation of phenylisoserine and other arylisoserines (*871*). *N*-Bromoacetamide has also been reported as an effective nitrogen donor/oxidant (*872*). The reagent was applied to the synthesis of both enantiomers of phenylisoserine, which were obtained with >99% ee by using "pseudoenantiomeric" alkaloid ligands for complexing the potassium osmate dihydrate catalyst. The desired regioselectivity was obtained with >20:1 ratio, and complete *syn* diastereoselectivity was observed. This process gave (2*R*,3*S*)-phenylisoserine on a 100 g scale in 68% overall yield with 99% ee after hydrolysis of the *N*-acetyl group (*872*) (see Scheme 11.41). This method was employed during a total synthesis of an NK1 antagonist that used (2*R*,3*S*)-phenylisoserine as an intermediate (*873*). *N*-Bromobenzamide was also employed for a synthesis of phenylisoserine (*874*). The protected (2*R*,3*S*)-phenylisoserine produced by aminohydroxylation of isopropyl cinnamate with *N*-bromo-acetamide (81%) has been converted into both (2*R*,3*S*)- and (2*R*,3*R*)-phenylisocysteine. Protecting group exchange to the *N*-Bz benzyl ester was followed by treatment with Tf₂O to induce formation of *cis*-substituted oxazoline. Opening with thioacetate gave the (2*R*,3*S*)-product. If the *N*-Bz methyl ester was cyclized to the oxazoline via mesylation of the alcohol and treatment with DBU, the *trans*-substituted oxazoline was produced instead. Thiol opening gave the (2*R*,3*R*)-product (*875*). Osmylated macroporous resins have been developed as a safe, readily recovered/reused source of the hazardous osmium catalyst. A series of alkenes were aminohydroxylated with AcNHBr, giving phenylisoserine and *p*-methoxy-phenylisoserine (*876*).

Other catalyst ligands for the aminohydroxylation reaction have been tested. A silica gel-supported bis-cinchona alkaloid provided (2*R*,3*S*)-phenylisoserine and the *p*-methoxy-substituted analog with >99% ee, using AcNBrLi as the aminating oxidant. The catalyst could be recovered by filtration and reused (*877*). A reversal in the regioselectivity of aminohydroxylation of cinnamates was described in 1998, resulting in β-hydroxy-α-amino acids instead of β-amino-α-hydroxy acids. The reversal was induced by using cinchona ligands with an anthraquinone (AQN) core instead of a phthalazine (PHAL) core (*878*). Reversals in regioselectivity could also be obtained by altering the substrate substituents, with ethyl *O*-(*p*-methoxybenzoyl) 4-hydroxybut-2*E*-enoate giving >20:1 regioselectivity in favor of the β-amino acid, but (2-naphthyl)methyl

O-(TBDPS) 4-hydroxybut-2*E*-enoate giving a 17:1 preference for the α-amino derivative (*879*).

Aminohydroxylation of α,β-unsaturated amide substrates was found to proceed in good yield, but resulted in racemic product (*530*). Similarly, a range of unsaturated acids were aminohydroxylated with 4–40× less osmium catalyst and 3× less Chloramine-T than usually employed for the asymmetric version, without diol byproduct formation, under aqueous conditions (*531*). Acrylamide with a chiral amide auxiliary formed from 1-phenylethylamine was aminohydroxylated with 1.1 equiv Chloramine-T and 2 mol % K$_2$[OsO$_2$(OH)$_4$] to give α-hydroxy-β-amino amide with 100% regioselectivity, 84% yield, and >99% de. α- or β-substituents greatly reduced the stereo- and regioselectivity (*880*). A 1999 report highlighted the scope, limitations, and applications to synthesis of the Sharpless asymmetric aminohydroxylation reaction (*881*). Prior to the development of the catalytic asymmetric synthesis, an aminohydroxylation of a substrate with a chiral auxiliary was reported. A cinnamate ester of the taxane skeleton was oxyaminated to generate the phenylisoserine residue, but the taxane ring induced minimal stereoselectivity (*882*).

11.2.3f Amination Reactions: Rearrangements

Another traditional route to β-amino acids employs the classical acid-to-amine rearrangements. (*S*)-Isoserine was prepared in 1914 in 45% yield from L-β-malamidic acid (in which the β-carboxy group is already derivatized as an amide) via a Hofmann rearrangement (*883*), with the procedure contained in a subsequent review (*884*). More recently (1983), [bis(trifluoroacetoxy)iodo] benzene was used to induce the Hofmann rearrangement of the amide group in 82% yield (*885*). Cyclohexylnorstatine was synthesized via benzylation of a protected malate dianion, followed by selective conversion of one of the carboxyl groups to an amino group via a Hofmann rearrangement (see Scheme 11.42) (*886*).

(*S*)-Malic acid has been converted into an activated (*S*)-isoserine derivative ready for incorporation into peptides, via protection of the hydroxyl and adjacent carboxyl groups with hexafluoroacetone and conversion of the remaining carboxyl group to a urethane via a Curtius-type rearrangement (*173, 2185*) (see Scheme 11.43). This procedure was also applied to racemic thiomalic acid to prepare DL-isocysteine (*514*). A similar reaction sequence employed formaldehyde to temporarily protect the adjacent hydroxyl/carboxyl groups of malic acid, with both (*S*)-isoserine (*888*) and (*R*)-isoserine (*174*) prepared. Enzymatically deuterated malate substrates were converted to (2*S*,3*R*)-[3-^{2}H$_1$]- and (2*S*,3*S*)-[2, 3-^{2}H$_2$]-isoserine (*889, 890*). Alternatively, (*S*)-malic acid α-monomethyl or -monobenzyl esters were refluxed with DPPA to induce the Curtius rearrangement, with the transient isocyanate reacting with the α-hydroxy group to form an oxazolidin-2-one. The oxazolidinone nitrogen was protected with a Boc or Cbz group, and the oxazolidinone and ester cleaved using benzyltrimethylammonium hydroxide in THF at –40 °C, forming *N*-Boc or *N*-Cbz (*S*)-isoserine (*891*). Curtius rearrangements of chiral triacid substrates produced the β^2-amino acid analogs of Asp and Glu (*832, 892*). The side chains were amidated to produce Asn and Gln analogs (*892*).

Homochiral 2-substituted succinic acids have been prepared by alkylation of acids with *tert*-butyl bromoacetate. Evans' oxazolidinone auxiliary was employed to induce asymmetry. The *tert*-butyl ester was cleaved and a Curtius rearrangement with DPPA used to form the β-amino acid. Removal of the oxazolidinone auxiliary provided 2-phenyl-, 2-(4-hydroxyphenyl)-, 2-benzyl-, and 2-(indol-3-yl)-3-aminopropanoic acids (*893*). This route was applied to syntheses of both (*R*)- and (*S*)-β^2-homotryptophan, 2-aminomethyl-3-(indol-3-yl)propionic acid (*894*). A similar strategy was used to generate 3-substituted-β-amino acids, via Curtius rearrangement of the original acid group, rather than the introduced acetate substituent (see Scheme 11.44). β-Methyl-, -phenyl-, benzyl-, and -isopropyl-β-Ala were synthesized (*895*).

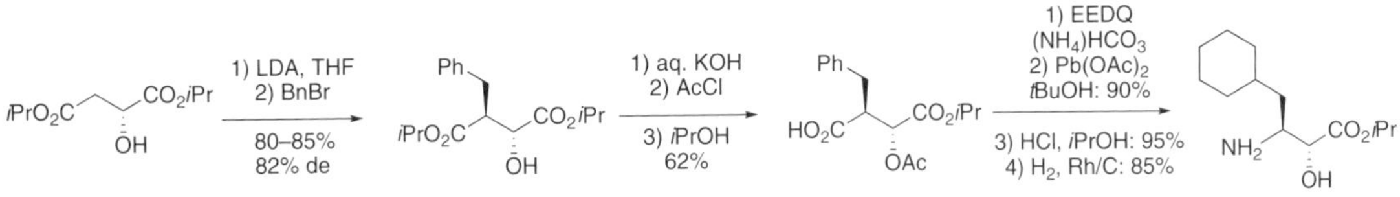

Scheme 11.42 Elaboration of Malate (*886*).

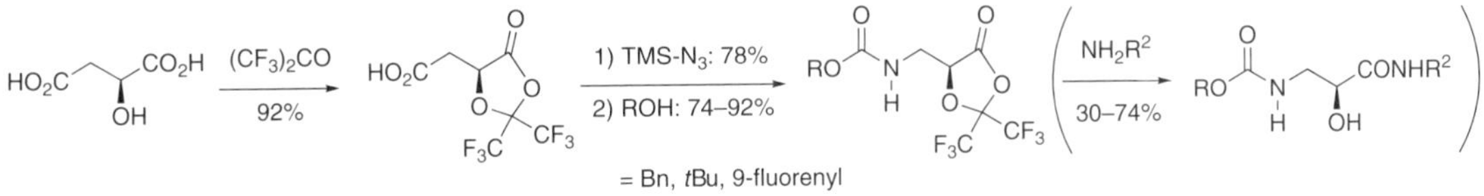

Scheme 11.43 Conversion of (*S*)-Malic Acid to (*S*)-Isoserine (*173*).

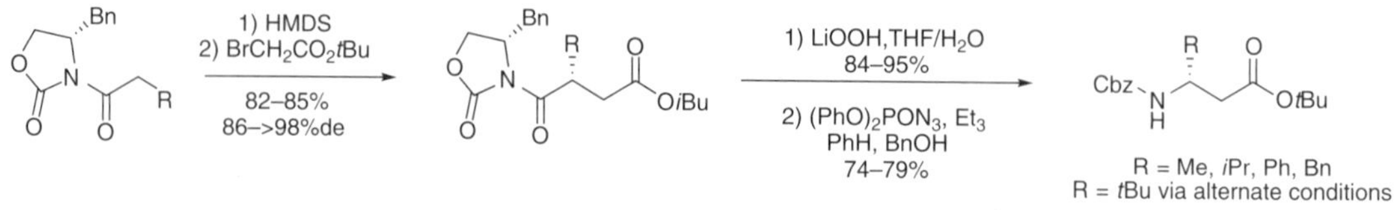

Scheme 11.44 General Route to β-Substituted-β-Amino Acids (895).

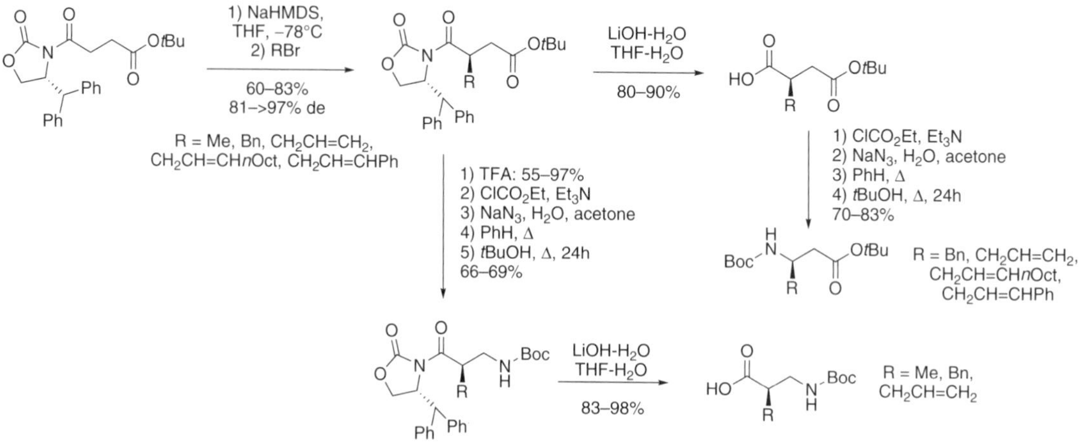

Scheme 11.45 Common Route to β²-Amino Acids and β³-Amino Acids (896).

Thus, a common chiral diacid derived by asymmetric alkylation of succinic acid can be selectively converted into either a β²-amino acid or a β³-amino acid, depending on which acid group is converted into an amine via a Curtius rearrangement (see Scheme 11.45) (896).

A similar strategy was employed with an allyl group as a masked acid group. Acids aminated with Evans' oxazolidinone auxiliary were asymmetrically allylated; the oxazolidinone was removed, the acid group amidated, the allyl group oxidatively cleaved to form the acid, and the amide converted into the amine via a Hofmann rearrangement (897). Other syntheses with alkenes (898, 899) or diphenylalkenes (898, 899) as masked carboxyl groups have been reported. The latter substrates were prepared by a chiral Pd complex-catalyzed substitution of an allylic acetate with sodium malonate. After Curtius rearrangement, the carboxyl group was revealed by oxidative cleavage using RuCl₃/NaIO₄ (898, 899). The 2,3-dihydroxy-4-amino-1, 6-hexanedioic acid component of the gastroprotective microbial agent AI-77-B was prepared by a Curtius rearrangement of protected 2-allyl-3,4-dihydroxybutanoic acid, with the terminal alcohol oxidized to an aldehyde and used for a thiazole addition to introduce the terminal carboxyl group. Alkene oxidative cleavage completed the synthesis (225). An aldehyde derived from a chiral allyllactone was protected as a dithiane, followed by Curtius rearrangement of the acid derived from the

lactone, using DPPA. Aldehyde deprotection and oxidation gave 3-amino-5-hexenoic acid (900). A synthesis of 3-aminodecanoic acid employed a Curtius rearragement of 2-allylnonanoic acid, followed by oxidative cleavage of the allyl alkene (88). Curtius rearrangements were also carried out on homochiral precursors in which the second carboxyl group was masked as an alkyne and revealed using OsO₄/NaIO₄ (901, 902).

Other aminative rearrangements include an intramolecular iodoamination of an allylic trichloroacetimidate derivative of 4-phenyl-but-3-en-1,2-diol (derived from D-glyceraldehyde acetonide). Phenylisoserine was the final product (903). 1-Amino-1-(4-methoxyphenyl)-2-carboxy-3-(3,4-methylenedioxy)-4-hydroxybutane was obtained via a diastereoselective 6-endo-trig cyclization of a homochiral substrate (904). Enantiomerically enriched allylic alcohols, produced via an asymmetric vinylzinc addition to aldehydes, underwent an Overman imidate rearrangement to give allylic amines (see Scheme 11.46). Jones oxidation provided a number of γ-unsaturated-β-amino acids (905).

An intramolecular Rh-catalyzed C–H bond insertion of a sulfamate ester derivative of (S)-3-methyl-1-pentanol led to a six-membered oxathiazinane through selective γ-C–H insertion of the sulfamate amino group. N-Cbz protection, ring opening with aqueous acetonitrile, and TEMPO oxidation of the hydroxymethyl group produced (R)-β-isoleucine (518).

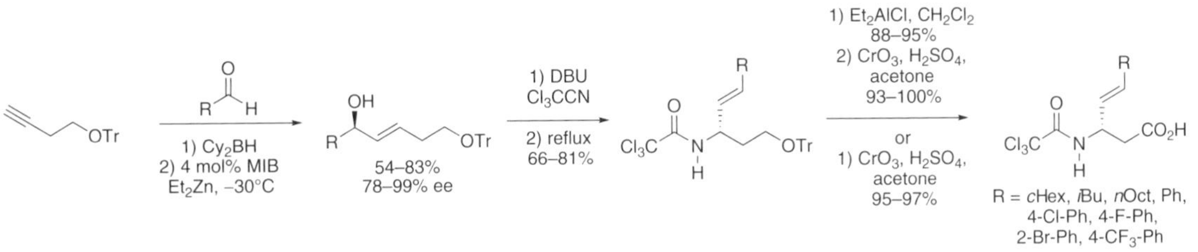

Scheme 11.46 Overman Rearrangement Leading to γ-Unsaturated-β-Amino Acids (905).

11.2.3g Amination Reactions: Reductive Aminations

β-Amino groups have been introduced via reductive amination of β-keto esters, similar to the preparation of α-amino acids from α-keto esters (see Section 5.3.9). Asymmetry can be induced by employing chiral auxiliaries on the amine or β-keto ester components, by aminating optically active substrates, or by asymmetric hydrogenation for the reduction step. One difference from α-amino acid synthesis is that the reaction of an amine with a β-keto ester forms an enamine rather than an imine. When optically active amines were used to form the intermediate enamine, poor diastereoselectivity (3–28%) was reported during reduction by catalytic hydrogenation or with sodium cyanoborohydride (906). Much better results (30–85% de) were obtained using sodium triacetoxyborohydride as the reducing reagent for β-enamino esters formed from (R)-α-methyl-benzylamine. The major diastereomer could be isolated with >99% purity in 47–73% yield (907, 908). An improvement to this procedure was reported in 2001, using benzyl ester derivatives that were deprotected during hydrogenation of the chiral auxiliary (909).

In contrast to these reports, hydrogenation over Pd(OH)$_2$/C gave higher diastereoselectivity (78% de) than sodium triacetoxyborohydride (46% de) for reduction of the enamine formed from (S)-α-methylbenzylamine and methyl 3-keto-3-(3-pyridyl)propanoate (396). Dimethyl acetonedicarboxylate was condensed with (R)-(+)-α-methylbenzylamine to give a chiral enamino ketone. The enamine was acylated with acetic anhydride or ketene, and then catalytically hydrogenated to give (2S,3'R,3S)-3-amino-2-(1-hydroxyethyl)pentanedioic acid (910). A series of enamines formed from β-aryl-β-keto esters and (S)-1-(4-methoxyphenyl)ethylamine were reduced by hydrogenation over Pd(OH)$_2$/C, with 42–80% de in the presence of BF$_3$·Et$_2$O. The chiral auxiliary was removed by treatment with Et$_3$SiH/HCO$_2$H (911). In 2004 it was reported that enamines formed from β-keto esters and phenylglycine amide could be hydrogenated over PtO$_2$ with 70–98% de; N-deprotection was accomplished by adding Pd(OH)$_2$/C after the enamine hydrogenation was complete (912).

Chiral ester auxiliaries have been employed to induce asymmetry during H$_2$/PtO$_2$ hydrogenation of 3-acetami-dobut-2-enoate, producing 3-acetamidobutanoate with up to 96% de. The enamino ester substrates were prepared from esters of acetoacetate by reaction with ammonia, followed by acetylation with acetic anhydride (913). 10-Deacetylbaccatin III was employed as a chiral ester auxiliary during the synthesis of baccatin II derivatives with various β-amino acid substituents. The C13 alcohol of 10-deacetylbaccatin III was acylated with β-keto esters via a transesterification reaction. The β-keto group of the 3-phenyl-3-oxopropionate derivative was then converted into an O-benzyl oxime. The β-oximino ester was α-hydroxylated with moderate diastereoselectivity (33% de), and the oxime reduced to an amine using ammonium formate over Pd-catalyst, producing a single anti isomer. Mitsunobu inversion of the α-hydroxyl group was required to obtain the desired syn stereochemistry of the phenylisoserine residue in paclitaxel (914, 915).

3-Amino-2-hydroxyalkanoic esters have been prepared via reductive amination of homochiral epoxy ketones, with a masked carboxylic acid group (see Scheme 11.47) (916). Intramolecular epoxide opening followed by hydroxymethyl oxidation led to syn products, while a regioselective opening using an oxygenated nucleophile provided the anti isomers.

A direct reductive amination of β-aryl-β-keto esters with ammonium acetate and H$_2$ in the presence of a chiral Rh catalyst (Ru-ClMeBIPHEP) resulted in 3-aryl-β-amino acids in 79–88% yield with 96–99% ee (917). However, β-substituted-β-amino acids are more commonly prepared by asymmetric hydrogenation of β-substituted (E)-β-(acylamino)acrylic acids (see also Section 4.3.2c), with the substrates prepared via reaction of β-ketocarboxylates with NH$_4$OAc followed by acylation, which produces mixtures of (E)- and (Z)-isomers (480, 918, 919). A 2003 report found that a compound previously described as the (E)-isomer of methyl 3-acetylamino-3-phenylacrylate was in fact methyl (Z)-3-diacetylamino-3-phenylacrylate, with the real (E)-product eluting elsewhere (483). Early attempts at reduction with a BPPM-Rh complex gave products with only 53–55% ee (920), but subsequent experiments using a BINAP-Ru(II) catalyst (see Scheme 11.48) gave the β-amino esters with 90–96% ee (480).

Rh-MeDuPHOS and Rh-BICP catalysts both provided good enantioselectivity for reductions of (E)-β-substituted-β-(acylamino)acrylates, with the DuPHOS ligand tending to give slightly better results

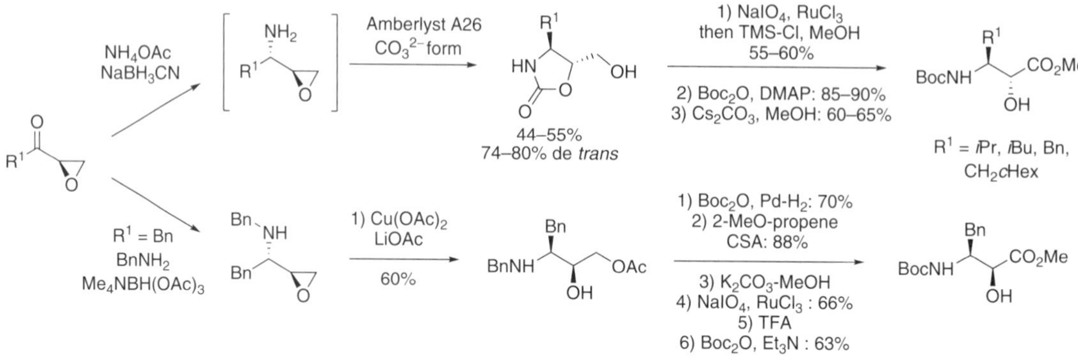

Scheme 11.47 Reductive Amination of Epoxyketones (916).

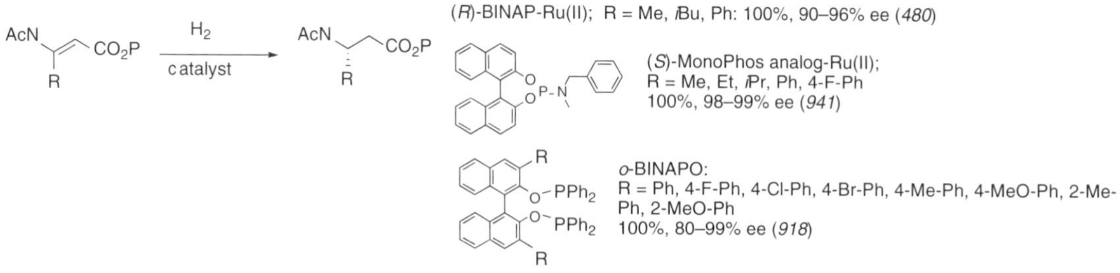

Scheme 11.48 Asymmetric Hydrogenation of β-Aminoacrylate.

(97.6–99.6% ee) than the BICP ligand (90.9–97.0% ee). In contrast, for the (Z)-isomers, the BICP (86.4–92.9% ee) was substantially better than the DuPHOS (11.2–62.4% ee) (482). A subsequent report found that polar solvents greatly improved the enantioselectivity of DuPHOS reductions of the (Z)-isomers, with up to 87.8% ee for reduction leading to 3-aminobutanoic acid (921). Temperature was found to play a crucial role in the enantioselectivity for reductions of β-acetamido-β-alkylacrylates with Et-BPE or Et-DuPHOS catalysts, (particularly for the (Z)-isomer), with the best results around 30–40 °C (922). For α-substituted-β-amido-acrylate substrates, leading to β²-amino acid products, enantioselectivities with DuPHOS and BPE only reached a maximum of 67% ee (923). Two 2002 studies compared several phosphine ligands (Me- and Et-DuPHOS, Me-BPE, Me4-BASPHOS, DIOP, HO-DIOP, DIPAMP, Et-FerroTANE) for the reduction of (Z)- and (E)- methyl β-acetylamino butenoate; Me-DuPHOS gave the best results (88%, 99% ee) for both substrates. Several other substrates were then reduced (924, 925).

A new bisphospholane ligand, MalPHOS, was compared against DuPHOS for hydrogenations of both (E)- and (Z)-isomers, with similar results for (E)-isomers (79 to >99% ee) but improved enantioselectivity for the (Z)-substrates (58–90% ee vs 4–86% ee) (926). In 2005 a ligand with two chiral phosphine centers connected by a quinoxaline backbone was prepared, QuinoxP*. Unlike other P-chirogenic ligands, this ligand is an air-stable solid. Methyl 3-acetaminobutyrate was produced with

99.2% ee (927). TangPhos, a ligand with both chiral backbone and chiral phosphine center, induced 74–99.5% ee for a variety of β-substituted amino acids from mixtures of E/Z substrates (928). A related bis(binaphthophosphepine sulfide) catalyst provided 3-methylbutyric acid with 99.2% ee from the (Z)- isomer (but only 32.7% ee from the (E)-isomer), and 96 to >99% ee for a number of (Z)-β-aryl-β-(acetylamino)acrylate substrates (929). Aryl groups have been added to TangPhos to generate DuanPhos, which reduced several β-substituted substrates with 92 to >99% ee (930).

A variety of other ligand complexes have been tested. Unprotected β-substituted enamines were reduced with a Rh-ferrocenophosphine complex to give β-amino esters with 95–97% ee (931). Et-FerroTANE provided better enantioselectivity (98 to >99% ee) than DuPHOS for reductions of β-alkyl and β-aryl (E)-isomers (483). A ferrocenyl-binaphthol phosphoramidite hybrid ligand gave 96–>99% ee for reductions of (Z)-β-aryl-β-(acylamino)acrylates, and 92–99% ee for (E)- or (Z)-β-alkyl substrates (932). A binaphthol-derived bidentate o-BINAPO ligand induced 80–99% ee for a variety of 3-aryl substrates (918), while a Ru bipyridyl complex induced 98.3–99.7% ee for reductions of (E)-β-substituted-β-acetamidoacrylates, but 68.3–82.3% ee for the (Z)-isomers (933). A biaryl diphosphine ligand with a chiral bridge between the aryl groups produced β-methyl-, β-ethyl-, β-n-propyl-, β-isopropyl, or β-tert-butyl-β-Ala with 97.7–99.8% ee (934). A bisphosphine ligand with an imidazolidin-2-one backbone, BDPMI, reduced a number of (Z)-substrates with equal or better enantioselectivity than obtained with the (E)-isomers, giving products with 75.6–97.4% ee (935). A diphosphine ligand based on a camphor backbone induced 99.3% ee for a hydrogenation leading to 3-aminobutyric acid (936). Analogs of the flexible BDDP ligand, 1,3-bis(diphenylphosphino)-1,3-diphenyl- or -dicyclohexyl-propane, provided up to 98% ee for reductions leading to 3-aminobutyrate from the (E)-substrate, and 81% ee from the (Z)-substrate, despite very poor enantioselectivities for reductions leading to α-amino acids (937, 938). An ethyl- or phenyl-bridged ligand containing two chiral phosphine substituents within benzyl-substituted five-membered rings reduced methyl-3-acetamido-2-butenoate with up to 96% ee (939). A "three-hindered quadrant" chiral ligand featuring three bulky tert-butyl groups attached to two phosphorus centers provided high enantioselectivity (98–99% ee) for a series of (E)-β-substituted-β-acetamidoacrylates. More importantly, the more common (Z)-isomers were also reduced with high stereoselectivity (92–98% ee) (940).

In recent years, monodentate ligands have been found to be effective. A Rh catalyst with a monodentate phosphoramidite ligand (an analog of binaphthol-derived MonoPhos) produced a number of 2-substituted β-amino esters with 98–99% ee (941). A comparative rate study found that the monodentate phosphoramidite catalyst could lead to higher reaction rates and/or higher enantioselectivities than state-of-the-art bidentate catalysts (942). The tBu-BisP and tBu-MiniPHOS catalysts reduced four different (E)-substrates with 95.6–99.7% ee (943). A catalyst with the MonoPhos dimethylamine replaced by a D-mannitol-derived ligand resulted in 98.7–99.9% ee for several hydrogenations (944). Other carbohydrate derivatives resulted in 93–98.4% ee (945). Several new monodentate chiral phosphonites were derived from 1,1'-spirobiindane-7,7'-diol, giving 85–98% ee for reductions leading to seven different β-aryl-β-amino acids (946). P-Chirogenic trialkylphosphonium salts induced >99% ee for reduction of (E)-β-methyl-β-acetamidoacrylate, but only 64% ee for the (Z)-isomer (947).

A parallel synthesis of 32 monodentate phosphoramidite binaphthol-derived ligands was combined with in situ screening for asymmetric hydrogenation of amino acid precursors. Different catalysts were optimal for reductions leading to 3-aminobutyrate compared to reductions leading to Phe (948). Another library screened 11 phosphite and eight phosphoramidite ligands based on a biphenol (TROPOS) unit; 3-aminobutyrate was obtained with 71% ee (949). Combining two different MonoPhos-type monodentate ligands provided much better enantioselectivity than either ligand alone. For example, reduction of ethyl β-(acetamino)-β-methacrylate with two different homogeneous ligands gave product with 54% and 80% ee, but using a mixture of the ligands gave product with 91% ee (950). A combinatorial approach towards identifying useful catalysts consisting of mixtures of chiral monodentate P-ligands found that a

mixture of BINOL-derived methyl phosphite with BINOL-derived tert-butyl phosphonite consistently gave the best results for reductions leading to 3-substituted β-amino esters, with 94–98.8% ee for 5 examples (951), or 84–97% ee for the combination of BINOL-derived tert-butyl phosphonite with an achiral (configurationally fluxional) biphenyl-derived phosphite (952).

The mechanism for asymmetric hydrogenation of β-acylamino acrylates using cationic Rh(I) complexes was compared to that of α-acylamino acrylates in 2005. The reaction sequence for both hydrogenations was determined to be the same, but for the β-acylamino acrylates the catalyst–substrate complex dominant in solution provided the major product of the reactions. In the case of α-acylamino acrylates it is the minor solution complex that leads to product, due to the minor complex possessing much greater reactivity. This extreme difference in reactivity does not appear to apply in the case of hydrogenations leading to β-amino acids (953).

Optically pure 2-substituted 1-(2, 6-dimethoxyphenyl)ethylamines were readily obtained by reductive amination of acetophenones. The aryl group acted as a masked carboxyl group, with Birch reduction and ozonolysis giving a β-substituted-α-carboxy-β-amino ester. One ester group was then removed by Krapcho decarboxylation (954). Soloshonok and Kukhar prepared β-fluoroalkyl-β-amino acids by reductive amination of β-keto esters, via condensation with benzylamine followed by a base-catalyzed [1,3]-proton shift reaction of the initial imine/enamine adduct to an N-benzylidene derivative, with the free amine readily obtained by hydrolysis (523, 955). The 3-fluorocarbon substituent is required for the rearrangement into the aldimine; a hydrocarbon enamine is unreactive (521, 522). Enantioselectivities of 29–36% were obtained by using catalytic quantities of a chiral base to induce the proton shift (523, 955). Much better enantiopurities (88–96% ee) were produced when (S)-α-phenylethylamine was employed for the initial β-keto ester condensation (956).

11.2.3h Amination Reactions: Other Aminations

2-Keto esters undergo an asymmetric amination reaction with azodicarboxylates in the presence of bis(oxazoline) copper(II) catalysts. The product racemized during purification, unless the ketone functionality was stereoselectively reduced (>90% de) with L-Selectride prior to removal of the copper catalyst. With dibenzyl azodicarboxylate, eight different α-hydroxy-β-hydrazino derivatives were produced with 90–96% ee. The hydrazino bond was cleaved in two examples to give syn-β-amino-α-hydroxy esters (957).

11.2.3i Carboxylation Reactions

Carboxylations have been accomplished by ring opening of aziridines with carboxylic acid equivalents; these reactions are discussed in Section 11.2.3r. The homologation of α-amino acids can also be considered a carboxylation reaction, but is included in Section 11.2.3m.

Homochiral N-benzoyl or N-Boc α-chloroamines (derived from amino alcohols) were deprotonated with n-butyl lithium and lithiated with lithium naphthalenide to give chiral dianionic intermediates, which were then quenched with CO_2 or $CO(OEt)_2$ to give β-amino acids or esters (see Scheme 11.49). This method provides another route for α-amino acid homologation, if the amino alcohols are derived from amino acids (958).

Treatment of lithiated enamines with diethyl carbonate or benzylchloroformate produced β-enamino esters, the same intermediates obtained during reductive aminations (see Section 11.2.3g). Asymmetry was induced during reduction of the enamine by using a chiral auxiliary on the amine, as demonstrated by the synthesis of 2-amino-1-cyclopentanecarboxylic acid (see Scheme 11.12) (534). The carboxy group of N-Boc amino acids has been transformed into an alkyne. This was converted into a stannyl-cuprate, which was carboxylated with CO_2 and then methylated to give $BocNHCH(R)C(=CHSnBu_3)$ CO_2Me. The organostannane was employed for Stille couplings with aryl iodides (959).

A chiral electrophilic carboxylation reagent has been prepared from Oppolzer's sultam auxiliary, via reaction with triphosgene to give the N-chlorocarbonyl sultam. This was used to acylate the anion of phenylacetonitrile

with up to 90% de. Reduction of the nitrile provided 2-phenyl-β-Ala (960).

Intramolecular amination of a chiral allylic sulfoxamine derivative gave a substrate that was carboxylated by deprotonation and acylation with $ClCO_2R$, with the sulfoxamine auxiliary removed with Raney Ni (see Scheme 11.50). Alternatively, the sulfoxamine was replaced with chloride and displaced with NaCN to introduce the carboxyl substituent (961).

11.2.3j Fusion of Aminomethyl and Acetate Components: Nucleophilic Aminomethyl Moiety

The central C2–C3 bond of β-amino acids can be formed from either nucleophilic aminomethyl equivalents reacting with electrophilic carboxymethyl equivalents, or from nucleophilic acetate equivalents adding to electrophilic aminomethyl synthons. There are only a few reports of asymmetric syntheses based on the first approach. In one example, nitromethane was added to glyoxylate in the presence of KF (see Scheme 11.51). Diastereoselectivity of >95% de was induced by using an (−)-8-phenylmenthyl ester auxiliary. Isoserine was produced in 50% overall yield after Raney-Ni reduction of the nitro group and ester cleavage (962). 1-Nitro-3-methylbutane was also added to this glyoxylate ester,

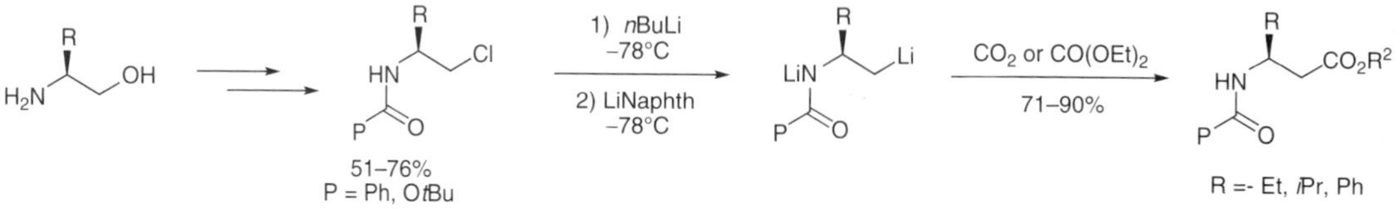

Scheme 11.49 Carboxylation of Asymmetric Lithiated Amine (958).

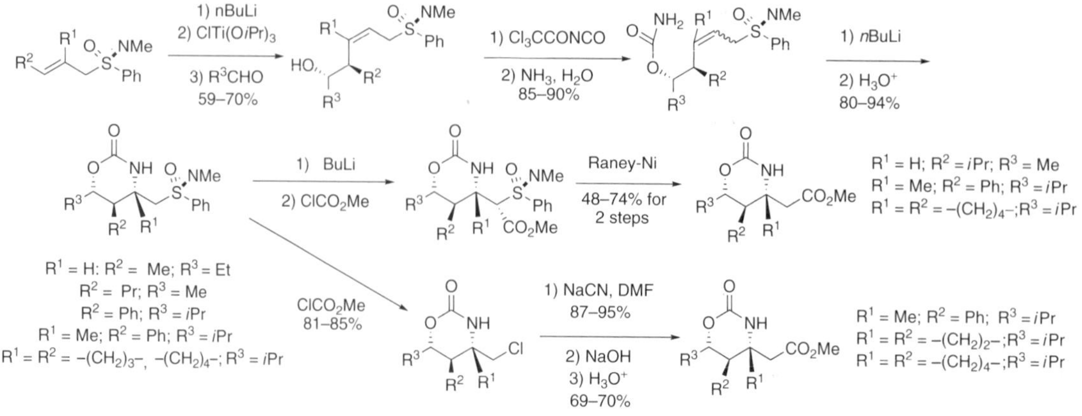

Scheme 11.50 Carboxylation of Aminated Sulfoxamine (961).

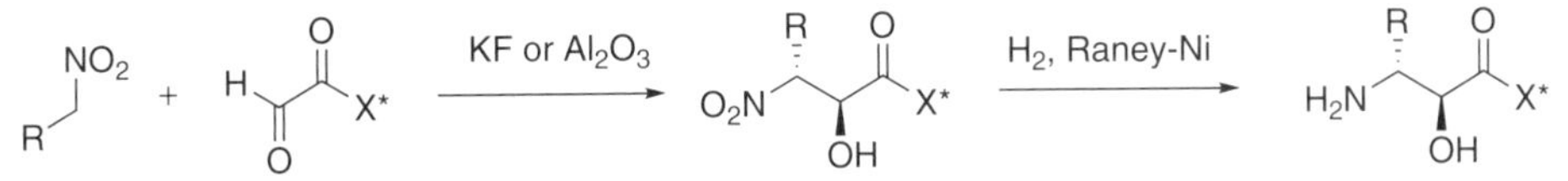

Scheme 11.51 Addition of Nitroalkanes to Glyoxylate.

giving (2S,3R)-3-amino-2-hydroxy-5-methylhexanoic acid in 53–55% overall yield (963). A subsequent report also tested a (2R)-bornane-10,2-sultam amide chiral auxiliary. Addition of 1-nitro-1-phenylmethane or 1-nitro-2-phenylethane to glyoxylate derivatized with either auxiliary led to (2S,3R)-phenylisoserine or (2S,3R)-bestatin, 2-hydroxy-3-amino-4-phenylbutanoic acid (964). A bicyclic oxazolidinone amide auxiliary produced predominantly one diastereomer from the Al₂O₅-catalyzed addition of various nitroalkanes to glyoxylic acid, though significant quantities of all four diastereomers were still produced (965).

A La(R)–BINOL complex catalyzed an asymmetric addition of 2-phenyl-1-nitroethane to ethyl glyoxylate, providing the (2S,3R)-2-hydroxy-3-nitro-4-phenylbutyrate product in 81% yield with 93% ee. Hydrogenation of the nitro group gave (−)-bestatin (966). An earlier report prepared the Taxol phenylisoserine side chain via La(R)–BINOL-catalyzed addition of phenylnitromethane to O-benzyl hydroxyacetaldehyde (80% yield, 90% ee, 100% de), followed by O-acetylation, nitro reduction and debenzylation, and hydroxymethyl oxidation (967). Nitromethane was added to several α-keto esters in the presence of a chiral bisoxazoline–copper(II) complex to produce 2-hydroxy-2-substituted-β-nitro esters in 46–99% yield with 77–93% ee (968).

A completely different strategy relied on asymmetric deprotonation of N-Boc, N-(p-methoxyphenyl) benzylamine using n-BuLi/(−)-sparteine as a chiral base (see Scheme 11.52). Alkylation with 4-bromo-2-methyl-2-butene provided access to the (S)-enantiomer of the desired β-amino acid (86% ee) after oxidative cleavage of the alkene, while quenching of the anion with methyl bromoacetate gave the (R)-enantiomer (84% ee). Other reaction conditions were reported that allowed for selective access to either enantiomer (969, 970).

11.2.3k Fusion of Aminomethyl and Acetate Components: Nucleophilic Acetate Moiety

The addition of acetate enolates to electrophilic aminomethyl equivalents is a very popular approach to synthesizing β-amino acids. Asymmetry can be introduced by employing chiral auxiliaries on either of the components, by using optically active substrates, or by employing chiral catalysts. The most common method relies on the addition of an acetate enolate to a chiral imine, with β-lactam formation often a competing reaction. As discussed in Section 11.2.2c, factors such as the nature of the imine and the reaction conditions affect whether the condensation leads to β-lactam or β-amino ester formation; these are discussed in a 1989 review (563). Asymmetric syntheses of β-amino acids via β-lactams are discussed in Section 11.2.3s.

One of the first reports of the addition of an acetate enolate to a chiral imine utilized Reformatsky reagents (BrZnCH₂CO₂R) as the acetate equivalent and imines formed from aldehydes and α-methylbenzylamine. The mixture of β-lactams and β-amino esters was hydrolyzed and hydrogenated to give β-amino acids in 10–42% yield, but with only 5–28% ee (971). A modified one-pot procedure was reported in 2003, with α-methylbenzylamine, aryl aldehydes, and the Reformatsky reagent condensed in the presence of trimethylsilylchloride and 5 M lithium perchlorate in diethyl ether. A series of 3-amino-3-arylpropionic acids were produced in 76–94% yield with 82–95% de (972).

Other acetate derivatives have been added to chiral imines derived from α-methylbenzylamine. The Taxol phenylisoserine side chain was prepared by addition of (Z)-α-methoxy trimethylsilyl ketene acetal to the imine formed from (S)-α-methylbenzylamine and benzaldehyde.

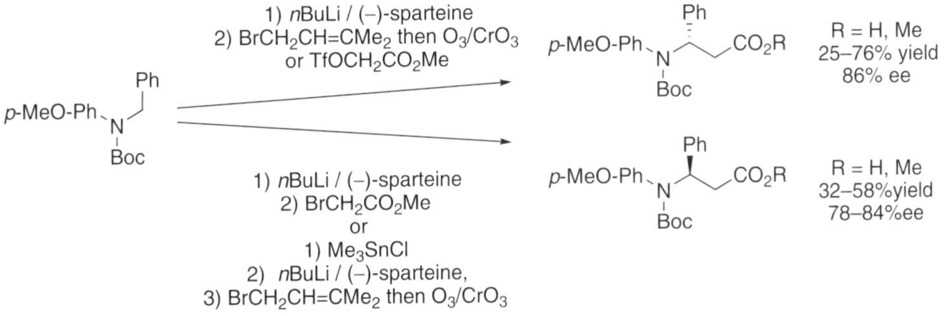

Scheme 11.52 Enantioselective Deprotonation (969).

The desired *syn* adduct was obtained with up to 66% de if the addition was carried out in the presence of the Lewis acid MgBr$_2$. The *O*-methyl group was removed with BBr$_3$, accompanied by azetidin-2-one formation, followed by auxiliary removal (*973*). A tin(II) carboxylic thioester enolate, RCH=C(S*t*Bu)OSnS*t*Bu, was added to chiral imines formed from α-methylbenzylamine to give 1,2-disubstituted-β-amino thioesters with high *anti* diastereoselectivity (*974*).

2,3,4,6-Tetra-*O*-pivaloyl-β-D-galactosylamine has been employed to form the chiral imine component with alkyl and aryl aldehydes. Silyl ketene acetals were added in the presence of zinc chloride to give the protected β-amino esters with excellent stereoselectivity (82–99% de when the ketene acetal of methyl isobutyrate was employed). The sugar auxiliary was removed from the amino group by treatment with HCl in methanol, and could be recovered if desired (*975*). Addition of prochiral bis(*O*-trimethylsilyl) ketene acetals led to exclusively *erythro* adducts with generally high (>90%) diastereoselectivity of a single isomer; removal of the chiral auxiliary revealed that the minor diastereomer obtained was the enantiomer of the major isomer. The *threo* epimers were produced exclusively by adding a lithium ester enolate instead of the silyl ketene acetal, although with these nucleophiles the net enantioselectivity was lower (*976*). Allyltributylstannane and allyltrimethylsilane have also been added to a variety of imines formed with 2,3,4,6-tetra-*O*-pivaloyl-β-D-galactosylamine (48–92% de, 16 examples), followed by hydrolysis to cleave the galactosyl auxiliary and oxidative cleavage of the alkene to generate the acid group. For example, enantiomerically pure β-Phe was produced in 41% overall yield (*977*, *978*). The use of sugars as chiral auxiliaries has been briefly reviewed (*979*).

Other chiral amines used for imine formation include an α-amino-pyrrolidine compound, a pyrrolidinopiperazinedione. Dimethylketene silyl acetal was added to the imine formed with benzaldehyde in 80% yield with 84% de. The free amino group was obtained by

hydrolysis, giving 2,2-dimethyl-β-Phe (*980*). The side-chain amino group of L-2,3-diaminopropionic amide has been used to form a chiral imine with 2-formyl-4-chloro-5-methyl-6-ethoxycarbonylpyrimidine. Addition of a boron enolate of the *tert*-butyl thioester of acetic acid produced pyrimidoblamic acid, the β-amino acid component of the bleomycin antitumor antibiotics, although with no diastereoselectivity reported (*584*). Addition of Reformatsky-type reagents to a chiral chromium tricarbonyl complex of an imine of 2-methoxybenzaldehyde produced mixtures of β-amino esters and β-lactams with >98% ee after decomplexation (*981*).

Very high diastereoselectivities were obtained by using chiral *N*-acyliminium intermediates generated from *N*-phthaloyl amino acid chlorides and aryl imines as the electrophilic aminomethyl equivalent (see Scheme 11.53). Additions of silyl ketene acetals proceeded with 56 to >99% de. The amino acid chiral auxiliary was removed by deprotection of the phthaloyl group, followed by Edman degradation (*982*). A modified version of the imine addition reaction employs an α-cyano-amino as the imine precursor, with Lewis acid treatment eliminating the cyano group. Thus, the chiral iminium ion generated in situ from 3-phenyl-2-[(*R*)-1-phenylethylamino]propanenitrile reacted with an α-methoxy trimethylsilyl ketene acetate in the presence of catalytic TMSOTf to give (2*S*,2*R*)-2-methoxy-4-phenyl-3-[(*R*)-1-phenylethylamino]butanoate; deprotection gave (2*S*,2*R*)-3-amino-2-hydroxy-4-phenylbutanoate (*983*).

An enantiopure sulfinimine has been employed as a chiral ammonia imine equivalent, with the *N*-sulfinyl auxiliary activating the C–N bond towards nucleophilic addition as well as controlling the addition stereochemistry (*984–987*) (see Scheme 11.54). The use of sulfinimines in asymmetric syntheses, including β-amino acids, was reviewed in 2004 (*988*). For example, addition of an enolate of methyl acetate to the sulfinimine of benzaldehyde led to β-Phe with >98% de, with the *N*-sulfinyl auxiliary removed by treatment with TFA/MeOH (*986*, *987*). The initial adduct can be deprotonated and

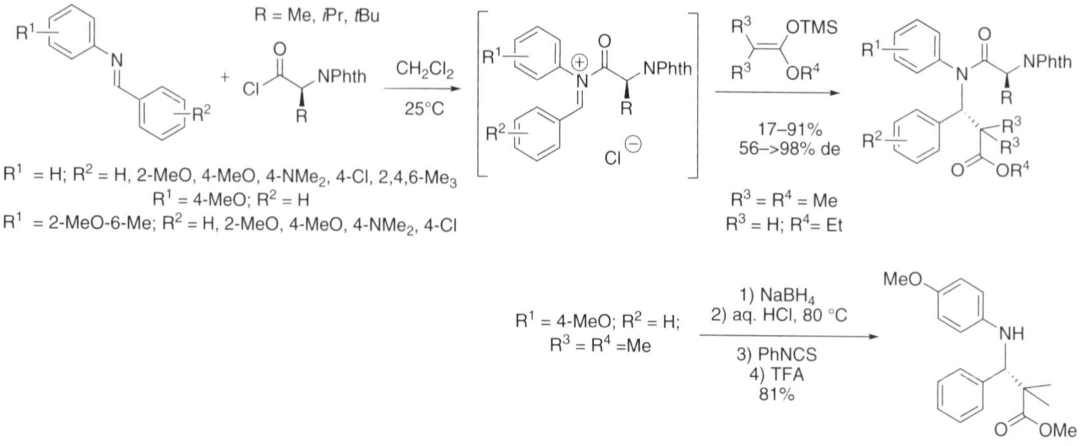

Scheme 11.53 Amino Acid as Auxiliaries for Mannich-Type Reaction (*982*).

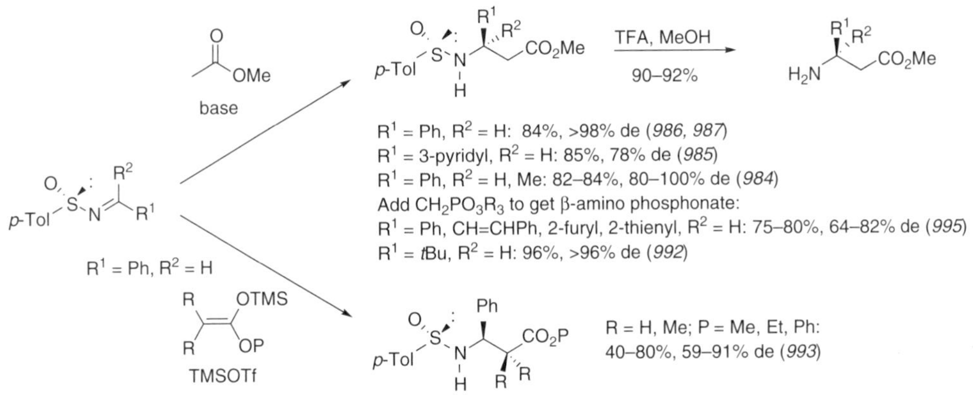

Scheme 11.54 Addition of Acetate to Enantiopure Sulfinimine.

hydroxylated, an approach that was applied to a synthesis of phenylisoserine (*984*). The anion of α-fluoroacetate was added to the sulfinimine of benzaldehyde to give a fluorinated phenylisoserine analog (*989*), while Reformatsky addition of the reagent derived from ethyl bromodifluoroacetate produced β-substituted-α, α-difluoro-β-amino acids (*990*). Addition of metal dienolates of methyl 3-butenoate produced 2-vinyl-3-phenyl-3-aminopropanoic acid in 34–90% yield, with varying diastereoselectivity of 74% de *anti* to 40% de *syn* depending on the metal used and the aryl group on the sulfinyl (*991*). The sulfinimine of pivalaldehyde reacted with the sodium enolate of methyl acetate with >96% de (*992*).

Silyl ketene acetals were added to several chiral sulfinimines in the presence of Lewis acids such as TMSOTf, giving 3-phenyl-3-aminopropanoic acid and 2,2-dimethyl-3-phenyl-3-aminopropanoic acid with up to 91% de (*993*). Vinylcuprates, generated by Michael addition of organocuprates to α,β-acetylenic esters, underwent a Yb(OTf)₃-catalyzed addition to chiral sulfinimines to form α-alkenyl-β-substituted-β-amino acids with 64 to >90% de (*994*). α-Phosphonate carbanions were also used as nucleophiles for additions to

sulfimines, giving β-aminophosphonic acids (*995*). Bis-sulfinimines have been generated from dialdehydes (isophthalaldehyde, teriphthalaldehyde, or 2,6-naphthalenedicarboxaldehyde) and a chiral N-sulfinyl auxiliary. Addition of 2 equiv of the sodium enolate of methyl acetate provided a bis(β-amino acid) linked at the β-position by a 1,3-phenyl, 1,4-phenyl, or 2,6-naphthyl moiety (*996*). A useful improvement to the sulfinimine procedure has been the development of a method to regenerate the chiral auxiliary (*986*).

More recently, *tert*-butylsulfinyl imines were reported as the electrophilic aminomethyl equivalent (see Scheme 11.55). The sulfinimines are readily formed by direct condensation of homochiral *tert*-butanesulfinamide, prepared in enantiopure form in two steps from *tert*-butyl disulfide (*997*), with aldehydes and ketones (*998*). Titanium enolates of esters added with high yields and improved diastereoselectivities compared to p-toluenesulfinyl imines. A wide range of β-substituted, α,β- and β,β-disubstituted, α,β,β- and α,α,β-trisubstituted, and α,α,β,β-tetrasubstituted β-amino acids were prepared (*999, 1000*). The *tert*-butanesulfinyl moiety is easily removed with HCl/EtOH, and can be retained as a protecting group and employed for solid-phase

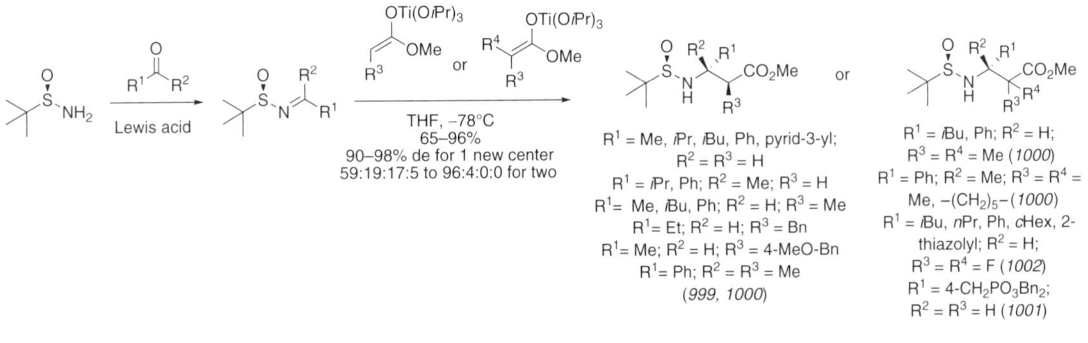

Scheme 11.55 Addition of Ester Enolates to t-Butanesulfinyl Imines.

synthesis of β-peptides (*999, 1000*). The *tert*-butylsulfinyl imine method was applied to a synthesis of the β-amino acid analog of the phosphotyrosine analog Pmp, 4′-phosphonomethyl-Phe (*1001*). Another group added the Reformatsky reagent derived from ethyl difluoroacetate to produce α,α-difluoro-β-substituted-β-amino acids (*1002*). This approach was summarized in a 2002 review of *N-tert*-butanesulfinyl imines (*1003*).

In an earlier report, homochiral sulfinimines were used as substrates for the addition of allylmagnesium bromide (82–100% de), with the allyl group oxidatively cleaved to generate the acid group (*588*). Addition of allylmagnesium bromide to chiral *N*-[α-phenyl-β-(benzyloxy)ethyl]nitrone derivatives of aryl aldehydes, ArCH=N(→O)CH(Ph)CH₂OBn, proceeded with only 20–56% de (*1005*).

Greater success was obtained with the nitrone formed from benzaldehyde and (*R*)-*N*-(α-methylbenzyl) hydroxylamine (*1006*). Zinc iodide-catalyzed addition of an achiral *E*-ketene silyl acetal derived from methyl (triethylsilyloxy)glycolate gave a protected α-hydroxy-β-hydroxylamine ester with 94% de *anti* stereoselectivity and asymmetric induction corresponding to 60% ee. Much better stereoselectivity was obtained by matching the chiral nitrone with a chiral ester auxiliary on the ketene acetal component, with the matched pair of auxiliaries giving a single diastereomer in quantitative yield. Upon treatment with TFA, the protecting groups were removed, leading to a cyclic isoxazolidinone product in 72% overall yield (addition and cyclization). The cyclic product was converted to phenylisoserine by hydrogenolysis, which cleaved the N–O bond and removed the *N*-auxiliary (*1006*). A similar procedure using *O,O*-isopropylidene D-glyceraldehyde as the aldehyde component produced the isoxazolidinone precursor of 3-amino-4,5-dihydroxypentanoic acid (*1007*). A chiral nitrone has also been prepared from the Ser-derived Garner aldehyde. Addition of *O*-methyl, *O*-TBS ketene acetal provided protected 3,4-diamino-5-hydroxypentanoic acid (*1008*).

Chiral hydrazones formed with Enders' SAMP/RAMP auxiliary were treated with an allyl cerium reagent to give protected homoallylamines in 63–99% yield and 90–99% de (*1009*). Oxidative cleavage of the alkene generated the β-amino acids (see Scheme 11.56). In a similar manner, allylmagnesium bromide was added to the chiral oxime ethers formed from ROPHy or SOPHy, (*R*)- or (*S*)-*O*-(1-phenylbutyl)hydroxylamine, and various aldehydes (80–100% yield, 86-96% de). Reductive hydroxylamine cleavage (78–98%) and oxidative cleavage of the alkene generated five different 3-substituted-β-amino acids in 33–52% yield (*1004, 1010*).

Chiral imine equivalents have been prepared by condensing aldehydes with (*R*)- or (*S*)-*N*-Bz-phenylglycinol, forming a chiral oxazolidine (see Scheme 11.57). Addition of the Reformatsky reagent derived from ethyl bromoacetate gave β-amino esters with moderate to good stereoselectivity (38–92% de), with the *N*-auxiliary readily removed by hydrogenation (*1011*) or by lead tetraacetate cleavage (*1012*). The Reformatsky reagent derived from ethyl bromodifluoroacetate added to give α,α-difluoro-β-amino acids (*919, 1013*). Ethyl tributylstannyl acetate reacted with even better diastereoselectivity (91–99% de for R = *i*Pr, *c*Hex, Ph, 4-Me-Ph, 4-Br-Ph, 1-naphthyl) under ZnCl₂/BF₃·Et₂O catalysis (*1014*). Allyl organocerium was also added (80–90% de for aryl aldehyde-derived oxazolidines), with oxidative cleavage of the alkene needed to generate the β-amino acid (*1015*).

An alternative strategy is to prepare the chiral imine or imine equivalent from a chiral aldehyde component instead of (or in addition to, as several examples above have already shown) a chiral amine component. However,

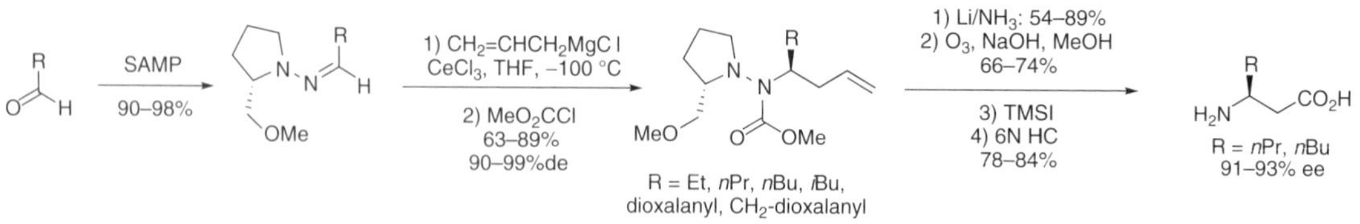

Scheme 11.56 Addition of Acetate Equivalent to Chiral Hydrazone (*1009*).

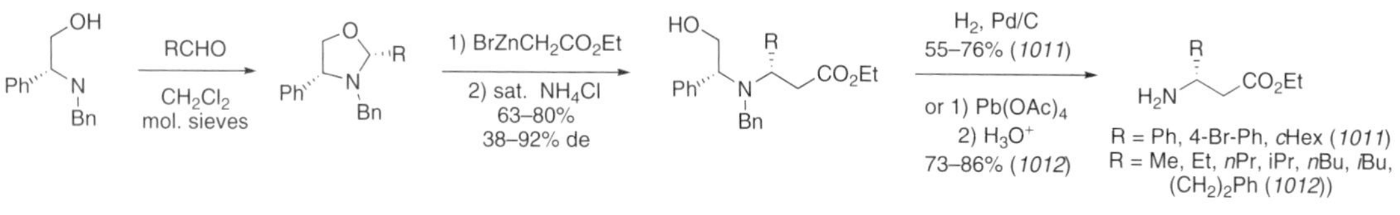

Scheme 11.57 Addition of Reformatsky Reagents to Chiral 1,3-Oxazolidines.

the chiral component remains part of the product, rather then acting as an auxiliary. An imine formed by a dioxolane ring derived from (2S,3S)-1,4-dimethoxy-2,3-butanediol induced considerable stereocontrol during additions of *tert*-butyl alkanoates, depending on which metal enolate was employed, with 100% de (3R,4S) product with Ti, 68% de (3S,4S) product with Li, and 86% de (3R,4R) product with Zn (*1016*). Chiral cyanohydrins have been obtained by the reaction of HCN and aldehydes catalyzed by (R)-oxynitrilase. *tert*-Butyl 2-bromoalkanoates were added to the nitrile in the presence of Zn dust (Blaise reaction) to give 3-amino-4-hydroxy-2-alkenoates. Unfortunately, alkene reduction by NaBH₃CN proceeded with little to moderate stereoselectivity (0–40% de) (*1017*).

Imine equivalents were prepared from D-glyceraldehyde acetonide or Cbz-prolinal by reaction with PhSO₂H and Boc-NH₂ or Cbz-NH₂, forming an α-amidoalkylphenylsulfone. Addition of lithium or zinc enolates of acetate resulted in protected 3-amino-4,5-dihydroxypentanoic acid or 3-amino-3-(pyrrolidin-2-yl)-propanoic acid, respectively (*1018*). The 2,3-dihydroxy-4-amino-1,6-hexanedioic acid component of the antiulcerogenic antibiotic AI-77B has been synthesized from D-ribose, via conversion into a isopropylidene derivative of 3,4-dihydroxy-5-acetoxy-pyrrolidin-2-one, and displacement of the acetoxy group by allyltrimethylsilane. Oxidative cleavage of the alkene completed the synthesis (*224*). A synthesis of the 3,6-diamino-5-hydroxyhexanoic acid component of the antibiotic negamycin elaborated a 3-hydroxy-4-aminobutyraldehyde derivative via imine formation and addition of an allyl group as a masked acetate group (*165*).

The chiral auxiliary can also be employed on the enolate component instead of the imine. Titanium-catalyzed

addition of a chiral silyl ketene acetal derived from N-methylephedrine-O-propionate to the benzaldehyde imine of aniline proceeded with >80% de *anti* relative stereochemistry, and absolute stereoselectivity corresponding to >94% ee (*1019*). Oppolzer's camphorsultam amide auxiliary was used to induce complete diastereoselectivity for addition of the enolate of (benzyloxy)acetyl amide to benzaldehyde N-(Boc)imine (*1020*) or N-(benzoyl)imine (*1021*) (see Scheme 11.58).

An optically pure α-sulfinyl ester enolate was added to benzaldimines containing an additional electron-withdrawing group on the nitrogen; either diastereomer could be obtained with up to 94% de, depending on the choice of N-substituent and additives (see Scheme 11.59) (*1022*).

Another chiral acetate equivalent that has been tested is the anion derived from α-diazoacetate, derivatized with an oxazolidinone amide auxiliary. This added to a number of aryl N-tosyl imines to form α-diazo-β-(tosylamino)acid amide derivatives with 52 to >90% de. The auxiliary could be removed with MeOLi and the diazo groups oxidized with dimethyldioxirane to produce α-keto-β-(tosylamino) esters. The ketone was then diastereoselectively reduced with NaBH₄ to form *anti*-α-hydroxy-β-(tosylamino) esters or by hydrogenation to give the *syn* isomers (*1023*).

α-Amido alkyl sulfones have been derived from aldehydes and employed as imine equivalents for reaction with a chiral acetate equivalent with a camphor-derived auxiliary (see Scheme 11.60). The camphor auxiliary was removed by oxidative cleavage with cerium nitrate. The route was employed to prepare glycosyl β-amino acids, but has general utility for the preparation of other β-amino acids. The C3 stereochemistry was controlled exclusively by the camphor-derived

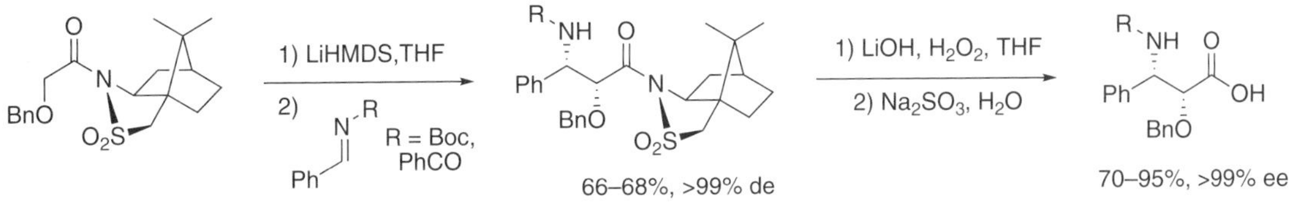

Scheme 11.58 Synthesis of Taxotere Side Chain Using Oppolzer Camphorsultam (*1020, 1021*).

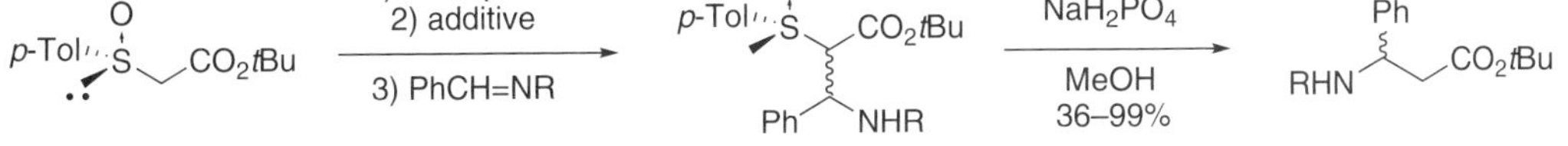

R = Ts, CO₂Me
additive = none, MgBr₂, MgI₂, CeCl₃, Et₃B, Me₄Sn
13–78%, 90% de (3S) to 94% de (3R)

Scheme 11.59 Addition of Homochiral α-Sulfinyl Ester Enolate to Benzaldimines (*1022*).

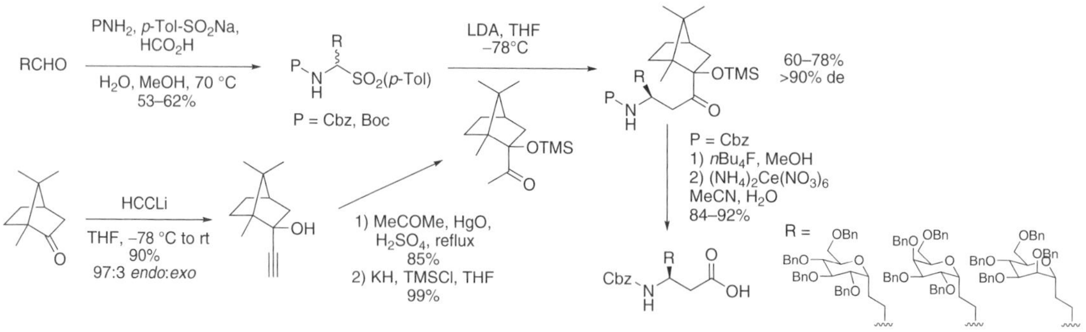

Scheme 11.60 Synthesis of Glyco β-Amino Acids (*2183*).

auxiliary; employing the methyl ketone derived from (−)-camphor produced the C3-epimeric product as the sole isomer (*2183*).

Propiolic acid esterified with chiral alcohols has been used as an acetate equivalent. HMPA-complexed DIBAL-H reduced the acetylenic moiety to an [(α-alkoxycarboyl)vinyl]aluminum species, which added to *N*-protected imines with 86 to >98% de (see Scheme 11.61). The β-phenyl β-methylene acid resulting from addition to benzaldehyde imine was coupled to the 10-desacetyl baccatin II core of taxol/taxotere and the alkene dihydroxylated, giving an α-hydroxymethyl analog of phenylisoserine (*1024*). Since the alkene is susceptible to conjugate addition, a range of other α-substituents could potentially be introduced.

A Ti homoenolate equivalent of propionaldehyde, derivatized as an acetal of homochiral 1,2-bis(cyclohexyl) ethanediol, was added to an imine formed from benzylamine and isobutyraldehyde to produce a *O*-alkyl 4-amino-5-methylhex-1-en-1-ol derivative with 88% de. Alkene oxidative ozonolysis led to 3-amino-4-methylpentanoate (*1025*).

Electrophilic aminomethyl equivalents other than imines can be employed for reactions with chiral acetate equivalents. The titanium enolate of phenylacetic acid (derivatized with an oxazolidinone chiral auxiliary) was condensed with *N*-Cbz 1-(aminomethyl)benzotriazole to give (*R*)-3-amino-2-phenylpropanoic acid after oxazolidinone removal (see Scheme 11.62) (*1026*). *N*-Cbz 1-(aminomethyl)benzotriazole was also employed as the aminomethylating reagent for alkylations of lithium enolates of 2-substituted acetates derivatized with Evans' oxazolidinone auxiliary, while titanium enolates were alkylated with *N*-Cbz aminoacetoxymethane (*893*).

N-Chloromethylbenzamide (*N*-benzoyl aminochloromethane), or the *N*-Cbz equivalent, have also been used as aminomethylation reagents, reacting with the titanium enolate of Evans' acylated oxazolidinone with 86–98% diastereoselectivity (see Scheme 11.63) (*275, 337, 1027, 1028*). This method was applied to the preparation of β-amino acids with α-methyl, α-isopropyl, α-isobutyl, or α-benzyl groups, which were converted into *N*-Fmoc protected products (*1029*). A new variant of Evans' oxazolidinone auxiliary was tested for

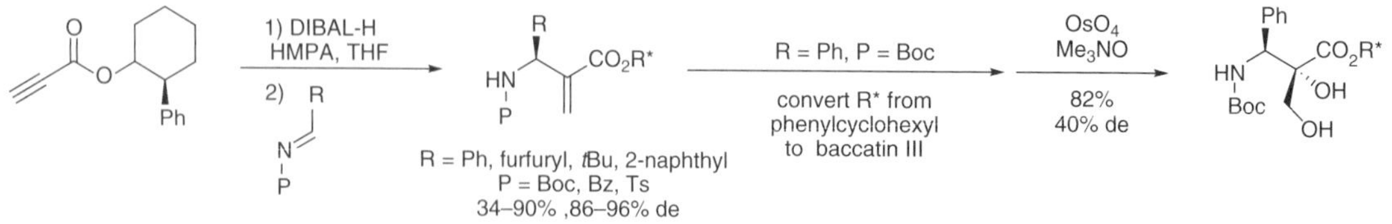

Scheme 11.61 Addition of Vinyl-aluminum to Imine (*1024*).

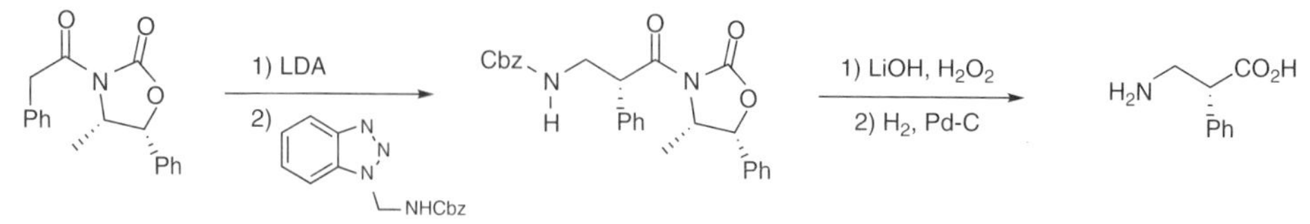

Scheme 11.62 Aminomethylation of Phenylacetic Acid (*1026*).

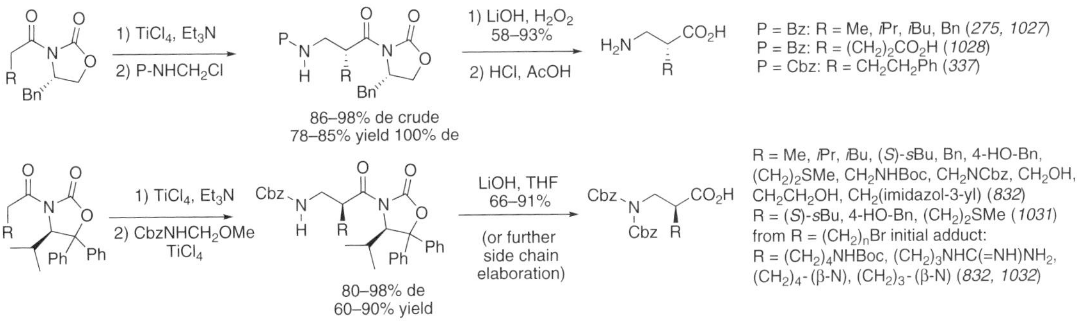

Scheme 11.63 Aminomethylation of Acylated Oxazolidinone.

aminomethylations of amide enolates using *N*-Cbz chloromethylamine, producing adducts with 84–90% de (*1030*). The new auxiliary tended to form crystalline derivatives, aiding purification. The DIOZ chiral auxiliary (4-isopropyl-5,5-diphenyloxazolidin-2-one) was subsequently employed to prepare multigram quantities of β²-amino acids corresponding to many of the proteinogenic amino acids (Ala, Val, Leu, Ile, Phe, Tyr, Met, Orn, Lys, Arg, Pro, Pip), with MeOCH₂NHCbz/TiCl₄ as the electrophile. Most of the side-chain substituents were directly derived from the initial acid that was acylated with the oxazolidinone auxiliary, but a number were obtained by further derivatization of a functionalized (e.g., bromoalkyl) side chain (*832, 1031, 1032*).

A titanium enolate of an acylated oxazolidinone was added to a nitrone to produce protected (2*S*,3*S*)-2-methyl-3-phenyl-3-aminopropionic acid with 70% de (*1033*). An acylated oxazolidinone was also alkylated with a masked aminomethyl equivalent, *tert*-butyl bromoacetate. Curtius rearrangement converted the carboxymethyl substituent into the desired aminomethyl group. This route was applied to a synthesis of β²-homoarginine (2-aminomethyl-5-guanidyl-pentanoic acid) (*361*). A chiral thiazolidinethione auxiliary was employed during additions of a propionic acid enolate to *O*-methyl oximes; a *trans*-substituted azetine intermediate was initially generated with 70 to >90% de, with *N*-acylation inducing azetine opening to provide 2-methyl-3-substituted-β-amino acids (*1034*). Anions generated from chiral 2-methyl-4-alkyloxazolines are another chiral acetate equivalent, and were condensed

with α-methoxycarbamates as the electrophile to give adducts with modest diastereoselectivity. Hydrolysis with 1 N HCl produced the β-amino esters with 44–90% ee (*555*). Lithiated homochiral 2,4-diisopropyl-2-oxazoline added to nitrones with good diastereoselectivity, with the adduct undergoing rearrangement to give a spirocyclic intermediate that was hydrolyzed to produce 5,5-dimethylisoxazolidinones (see Scheme 11.64). Hydrogenolytic cleavage of the N–O bond produced 2,2-dimethyl-3-substituted-β-amino acids (*1035*).

β-Fluoroalkyl-β-amino-α,β-unsaturated esters were synthesized by addition of lithium ester enolates to imidoyl chlorides (imines with a chloro substituent on the carbon forming the imine), generating exclusively (*Z*)-β-enamino esters (*590, 591*). By using chiral auxiliary esters, the (*Z*)-β-enamino esters could be diastereoselectively reduced with ZnI₂/NaBH₄ to produce β-fluoroallkyl-β-amino esters with 10–60% de (*591*).

Another strategy uses chiral acetate equivalents with the chirality residing in the acetate moiety, rather than a chiral auxiliary, with the asymmetric center therefore ending up in the product. A chiral ketene silyl acetal derived from (*R*)-3-hydroxybutanoate was condensed with nitrones to give various 2-(1-hydroxyethyl)-3-substituted-3-aminopropanoic acids (*1036*). Similarly, lithium enolates of other chiral hydroxy esters were added to imines to produce 2-(1-hydroxyethyl)-3-amino-3-phenyl-propionic acid (*1037*), 2-(1-hydroxyethyl)-3-amino-1,5-pentanedioic acid (*1038*) or 2-(1-hydroxyethyl)-3-amino-4-carboxypentanoic acid (*1039*). A boron enolate of the thiophenyl ester of (*R*)-3-hydroxybutanoic

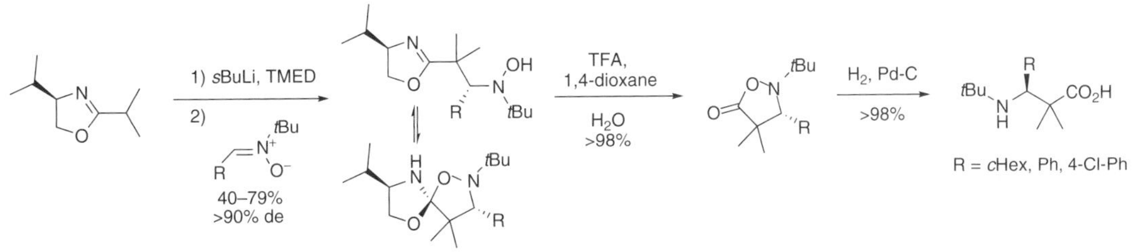

Scheme 11.64 Homochiral Oxazoline Addition to Nitrones (*1035*).

acid was also tested for additions to imines, with a reversal in stereoselectivity compared to lithium enolates (*1040*).

The third possibility for asymmetric synthesis is to condense an achiral ester and achiral imine in the presence of a homochiral catalyst. Corey et al. employed a chiral diazaborolidine during the reaction of (*S*)-*tert*-butyl thiopropionate with *N*-benzyl or *N*-allyl aldimines, giving the adducts with good diastereoselectivity (84 to >99% de) and enantioselectivity (90 to >99% ee) (*1041*). Phenylisoserine was synthesized a number of years later by the addition of an α-hydroxy thioester boron enolate to an imine of benzaldehyde, with the boron bearing chiral ligands derived from menthone. Excellent stereocontrol (>92% de *syn* or >94% de *anti*, >95% ee in both cases) was obtained by varying the thioester from S*t*Bu to SPh and, to a lesser extent, the *O*-protection from OBz to OTBS. The absolute configuration of the ligand was found to control the chirality of the carbinol center (*1042*). A chiral binaphthol-boron catalyst was employed for condensations of silyl ketene acetals with imines possessing a chiral auxiliary, generating adducts with 92–94% de for the matched pair of reagents, and 98% de *syn* to 96% de *anti* relative stereochemistry (*1043, 1044*).

A zirconium catalyst prepared with BINOL-type ligands provided 81–96% ee for the addition of various silyl enolates to 4-trifluoromethylbenzoyl hydrazone derivatives of aldehydes. β-Substituted and α,α,β-trisubstituted compounds were prepared (*1045*). A zirconium binaphthol catalyst was also used for the addition of ketene silyl acetals to 2-hydroxyaniline imines of aldehydes. *syn* 2-Hydroxy-3-amino products were preferentially obtained from an α-TBSO-ketene silyl acetal (64 to >98% de, 91–98% ee), while an α-benzyloxy ketene silyl acetal gave the *anti* adducts (6–88% de, 76–96% ee) (*1046*). A propionate ketene silyl acetal led to *anti*-2-methyl-3-substituted-3-amino products with 42–96% de and 80–96% ee (*1047*). A subsequent report generated 3-aryl-3-aminopropionic thioethyl esters using the silyl enolate of (*S*)-ethyl thioacetate (78–100% yield, 86 to >98% ee), or 2,2-dimethyl-3-aryl-3-amino-propionic methyl esters from the silyl enolate of methyl

isobutyrate (70–100%, 83–92% ee) (*1048, 1049*). Aliphatic imines generated from cyclohexanecarboxaldehyde or isovaleraldehyde worked if the 2-hydroxyaniline component was replaced with 2-amino-*m*-cresol, although yields and enantioselectivity were reduced (45–47%, 80% ee) (*1048, 1049*).

A titanium BINOL catalyst provided 80–92% ee for additions of ketene silyl acetals to *N*-benzyl oximes of aryl aldehydes (five examples) (*1050*), while a BINOL-derived chiral phosphate Brønsted acid catalyst induced 81–96% ee for reactions of various ketene silyl acetals with imines derived from 2-hydroxyaniline and aryl aldehydes. The 2-alkyl-3-aryl-β-aminopropionic acids were obtained with 72–100% *syn* stereoselectivity (*1051*). *N*-(2-Hydroxyacetyl)pyrrole has been employed as an acetate equivalent, adding to *N*-(*o*-Ts) imines in the presence of In(O*i*Pr)₃ and a dimeric BINOL ligand to give α-hydroxy-β-amino amides in 68–97% yield, 18–82% *anti* diastereoselectivity, and 89–98% ee for the major isomer (see Scheme 11.65). The pyrrole could be hydrolyzed in good yield (*1052*). An amino acid urea–Schiff base catalyst induced 86–98% ee during Mannich reactions of a silyl ketene acetal derived from isopropyl acetate with aryl aldehyde *N*-Boc imines (14 examples), producing β-aryl,β-amino esters in high yield (87–99%) (*2182*). Some catalytic asymmetric Mannich reactions leading to α-hydroxy-β-amino acids were reviewed in 2004 (*1053*), while other Mannich-type catalytic enantioselective reactions were included in a 1999 review of additions to imines (*1054*).

A one-pot, three-component imino-Reformatsky reaction employed ethyl bromoacetate and the imines generated from alkyl or aryl aldehydes and 2-methoxyaniline. When carried out with dimethylzinc, NiCl₂(PPh₃)₂ and (1*S*,2*R*)-*N*-methylephedrine as catalyst the β-substituted-β-amino esters were produced with 74–92% ee (*1055*). An aza-Baylis–Hillman reaction combined *N*-sulfonylated imines (ArCH=NHTs) with phenyl acrylate in the presence of a chiral binaphthyl phosphine Lewis base, producing α-methylene-β-aryl-β-amino acids in 53–97% yield with 52–77% ee (*1056*). β-Keto esters were added to various imines in the presence of a chiral Pd catalyst, providing the adducts

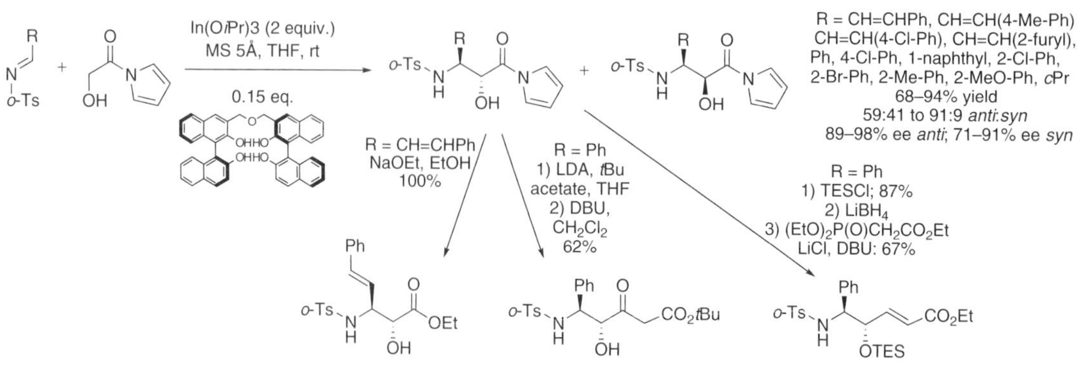

Scheme 11.65 Catalyzed Addition of *N*-(2-Hydroxyacetyl)pyrrole to Imines (*1052*).

with 86–99% ee and anywhere from 0 to 80% diastereomeric ratio for the two chiral centers generated (*1057*).

A C–H insertion reaction combined aryldiazoacetates, ArC(=N₂)CO₂Me, with *N*-Boc,*N*-benzyl, *N*-methylamine in the presence of a chiral [Rh₂(*S*-DOSP)₄] catalyst, producing *N*-benzyl β-aryl-β-amino acids with 87–96% ee in 55–66% yield (*1058*).

11.2.3l Fusion of Aminomethyl and Acetate Components: Other Methods

Several other methods of joining aminomethyl and acetate equivalents have been reported. A dipolar cycloaddition of vinyl acetate, ketene acetal, or α-chloroacrylonitrile with nitrones containing a chiral auxiliary was used to prepare β-Lys, β-Leu, β-Tyr or β-Phe, with 33–84% de obtained during the cyclization (see Scheme 11.66) (*1059*). Alternatively, the chiral auxiliary was included on amide derivatives of crotonic acid, but the induced diastereoselectivity was not as good (*1060*). The same chiral nitrone has also been used as the substrate for the addition of achiral (*E*)-ketene acetals (*1006*). An intramolecular nitrone cycloaddition of a chiral substrate (the nitrone of 3-fluoro-2-hydroxypropanal crotonic or cinnamic ester) was used to prepare a highly functionalized β-amino acid (*1061*). Nitrones formed from aldehydes and an *N*-hydroxyphenylglycinol auxiliary underwent 1,3-dipolar cycloaddition to form 1,3-isoxazolidines, with bicyclic products obtained from aldehydes linked to alkenes (see Scheme 11.66). Hydrogenation removed the auxiliary and cleaved the isoxazolidine N–O bond, with hydroxymethyl oxidation producing the β-amino acid (*1062*).

Another dipolar cycloaddition reaction employed a Rh₂(OAc)₄-generated ylide derived from benzaldehyde and ethyl diazoacetate, reacting with a chiral imine derived from (+)-α-methylbenzylamine and benzaldehyde. The initial oxazolidine adduct was decomposed with PTSA to give the Taxol phenylisoserine substituent (*622*). A hetero-Diels–Alder reaction between *N*-benzoylbenzaldimine and (*Z*)-silylketene acetals containing a chiral auxiliary was also applied to a synthesis of phenylisoserine. The dihydrooxazine cycloaddition adduct was not isolated, but hydrolyzed with dilute aqueous HCl to give *N*-benzoyl,*O*-benzyl phenylisoserine ester. Complete diastereoselectivity or enantioselectivity could be obtained by varying the auxiliary, with the best result a 75% yield with 100% ee and 86% de (*1063*). An early asymmetric synthesis of β-phenyl-α,α-dimethyl-β-aminopropionic acid was achieved by a cycloaddition reaction of a chiral Schiff base with 2 equiv of dimethylketene, giving an oxazinone intermediate that was hydrolyzed (*1064*).

A potentially useful synthetic route towards sterically congested β²,³-, β³,³- and β²,³,³-amino acids was described in 2003 (see Scheme 11.67). A 1,3-dipolar cycloaddition between various substituted oximes (which introduce the β-substitutent) and (*S*)- or (*R*)-3-alken-2-ols (which introduce the α-substituent) produced isoxazoline intermediates. The isoxazoline stereochemistry was determined by employing (*E*)- or (*Z*)-alkenes. The resulting isoxazoline was diastereoselectively reduced by hydride addition with LiAlH₄ or Grignard reagents (which introduced the second β-substitutent). LiAlH₄ reduction stereochemistry was directed by the C5 1′-hydroxyethyl group, while Grignard addition was dictated by steric constraints,

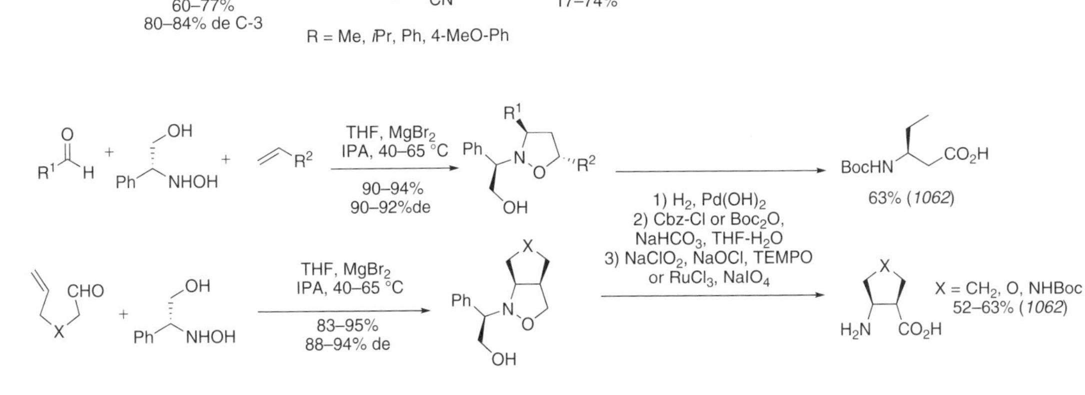

Scheme 11.66 Cycloaddition of Homochiral Nitrones.

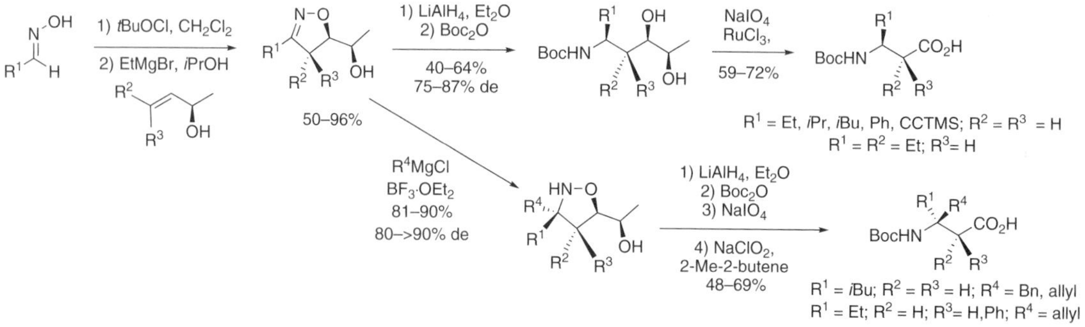

Scheme 11.67 Dipolar Cycloaddition of Oximes and Alkenols (*1065*).

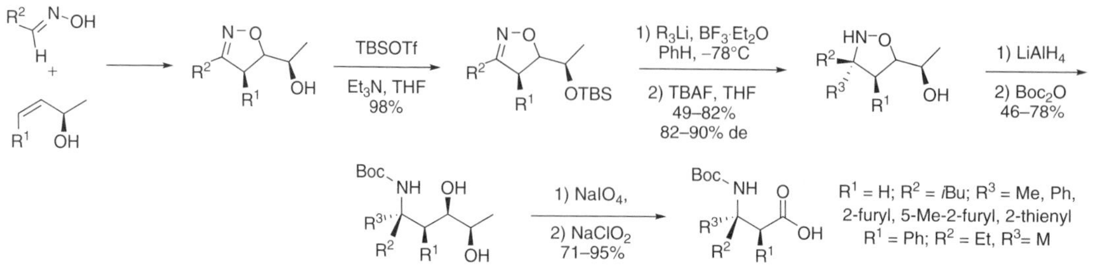

Scheme 11.68 Synthesis via Nitrone Cycloaddition (*1066*).

resulting in opposite stereoselectivity. An isoxazolidine product could not be isolated during hydride reductions; instead N−O bond cleavage directly gave the acyclic aminodiol. The Grignard additions produced a stable cyclic structure, with LiAlH$_4$ reduction or hydrogenation cleaving the N−O bond. Finally, the amino acid was generated by diol oxidative cleavage (*1065*). A subsequent report employed organolithium addition to the isoxazoline to produced 3,3-di- or 3,3,4-trisubstituted products (see Scheme 11.68). These were again converted into $\beta^{2,3}$- or $\beta^{2,3,3}$-amino acids via N−O bond cleavage and oxidative cleavage of the 1,2-diol (*1066*).

11.2.3m Elaboration of α-Amino Acids: Homologation of α-Carboxyl Group

One popular route to optically active β-substituted-β-amino acids is the Arndt–Eistert homologation (*1067*) of α-amino acids, which proceeds by initial formation of a diazo ketone from an activated amino acid and diazomethane. A Wolff rearrangement, induced by metal

catalysis or thermal or photochemical conditions in the presence of a nucleophile, generates the β-amino acid with retention of configuration (see Scheme 11.69). The rearrangement has been carried out on many amino acids, including suitably protected Gly (*421, 1068–1071*), L-Ala (*281, 288, 292, 398, 813, 1029, 1072–1082*), D-Ala (*398, 1078*), N-methyl L-Ala (*288*), L-Val (*156, 292, 1029, 1071, 1072, 1074, 1075, 1077, 1080–1086*), D-Val (*1087*), L-Nva (norvaline) (*156, 1079*), L- and D-Leu (*156, 292, 383, 1029, 1073–1075, 1081, 1084, 1085*), L-Ile (*156, 1071, 1074, 1075, 1078, 1081, 1082, 1084*), D-Ile (*1071*), L-Tle (*tert*-leucine) (*1076, 1081*), L-Pro (*381, 1071, 1078, 1088–1090*), D-Pro (*1088*), L-homopipecolic acid (*1076*), L-Ser (*335, 374, 1029, 1075, 1076*), L-Thr (*372, 1075*), L-Cys (*317, 351, 833, 1086, 1091, 1092*), L-Met (*317, 374, 715*), L- and D-Asp (*167, 169, 383, 1071, 1082, 1084, 1086, 1093*), L-Glu (*335, 1029, 1082, 1083, 1094*), L-Gln (*1095*), L-2-aminoadipic acid (*1096*), L-Lys (*281, 337, 372, 397, 1029, 1071, 1075, 1081, 1082, 1083*), L-Orn (*145, 146, 156, 337, 397, 1076, 1097*), D-Orn (*143*), L-2, 4-diaminobutyric acid (*156, 1098*), L-2,3-diaminopropionic acid

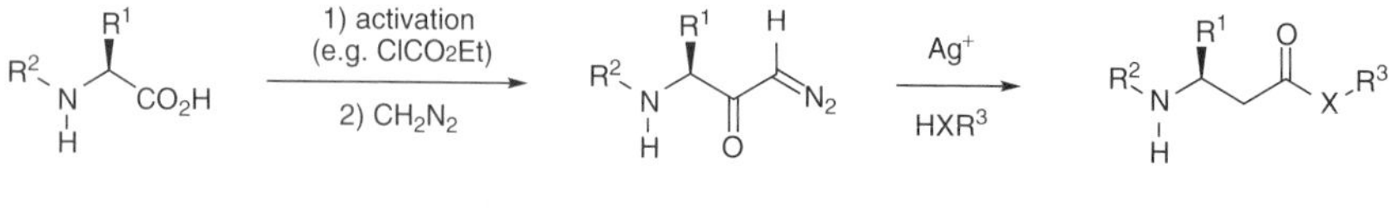

Scheme 11.69 Arndt-Eistert Homologation of α-Amino Acids.

(156, 168), L-Phe *(156, 281, 344, 372, 381, 383, 384, 1029, 1071, 1074–1076, 1080–1085, 1099–1100)*, D-Phe *(1082, 1084)*, L-Tyr *(337, 1075, 1083, 1101)*, L-DOPA *(1102)*, L-Phg (phenylglycine) *(1071, 1076, 1080)*, D-Phg *(1084, 1103)*, L-4-HO-Phg *(1104)*, L-biarylglycine (with significant racemization) *(406)*, L-homoPhe *(337)*, D-homoPhe *(416)*, L-Trp *(372, 1085, 1086)*, L-His *(351, 833, 1086)*, D-allylglycine *(1105)*, and L-2-aminohex-5-enoic acid *(1105)*.

Threo-γ-hydroxy-L-β-Lys was derived from *threo*-β-hydroxy-L-Orn *(161, 1106)* and *erythro*-γ-hydroxy-DL-β-Lys from *erythro*-β-hydroxy-DL-Orn *(161)*. Both Fmoc-Arg(Pbf) and Fmoc-Arg(Pmc) were homologated under standard conditions to the β³-hArg derivatives, or, alternatively, Fmoc-Orn(Boc) was homologated, Boc-deprotected, then guanylated *(315)*. L-2,4-Diaminobutyric acid was homologated and the δ-amino group guanylated to give L-β-Arg *(1098)*, while homologation of L-2,3-diaminopropionic acid followed by trimethylation of the γ-amino group gave emeriamine, the 3-amino analog of carnitine *(168)*. Alternatively, homologation of the L-Asp α-carboxyl group provided 3-amino-1,5-pentanedioic acid, a symmetrical molecule that is only chiral due to differential protection of the carboxyl groups. Hofmann rearrangement of either of the carboxyl moieties introduced a γ-amino group, with trimethylation producing either enantiomer of emeriamine *(169)*. Iodination of homologated L-Tyr gave 3′,5′-diiodo-β-Tyr *(1101)*. (3S)- or (3R)-[3-²H₁]-β-Ala were prepared by Arndt–Eistert homologation of N-trifluoroacetyl (3S)- or (3R)-[3-²H₁]-Gly *(421)*. The homologation procedure has been incorporated into a synthesis of 2-methyl-3-aminooct-7-ynoic acid from Lys. Cbz-Lys was converted into 6-hydroxy-Nle, homologated, and then deprotonated and α-methylated. The terminal alkyne was preared by Swern oxidation, followed by conversion of the aldehyde with the Ohira–Bestmann reagent *(86)*.

A variety of modifications to the reaction conditions have been reported. A range of N-protecting groups are tolerated. The preparation of diazoketo intermediates of a number of Phth-, Boc-, and Cbz-amino acids were described in 1970 *(1069)*, while optimized conditions for Fmoc amino acids were detailed more recently *(1029, 1078)*. N-Cbz,N-methyl amino acids also underwent the homologation, with sonication used for the diazoketone rearrangement *(1086)*. Ultrasound similarly promoted the rearrangement of Fmoc amino acids, with the enantiomeric purity of the products produced under these conditions determined by capillary electrophoresis with a chiral buffer. The homolog of Phe was produced >99% ee but the homolog of phenylglycine (Phg) had partially racemized (80.5% ee) *(1084)*. Similarly, Podlech and Seebach investigated the stereoselectivity of the rearrangement of Boc- and Cbz-protected amino acids in 1995 using modern analytical methods and found that all amino acids tested were homologated with full retention of configuration, except for carbamate-protected Phg, which was obtained with 80% ee *(1076)*. Hindered amino acids gave low yields during formation

of the diazoketone intermediate, while amino acids with heteroaromatic side chains decomposed. In contrast, yields during the rearrangement step appeared to be independent of the amino acid residue *(1076)*. Another study examining the homologation of N-Fmoc-protected amino acids again found complete retention of both chiral configuration and the N-protecting group *(1074)*, as did a 1976 homologation of Cbz-Pro *(1088)*. A 1987 examination of the rearrangement of N-Boc and N-Cbz α-amino acids determined that careful purification of the diazoketone intermediate was required in order to avoid inconsistent results for retention of optical purity during the rearrangement step. It was proposed that this was likely due to the presence of an oxazolone byproduct that would contaminate the final homochiral β-amino acid product with DL-α-amino acid *(1099)*.

Hazardous diazomethane can be replaced with trimethylsilyldiazomethane; the mixed anhydride prepared from Boc-Phe and ethyl chloroformate was treated with TMSCHN₂ to form purified diazoketone in 78% yield (compared to 76% yield using CH₂N₂) *(1107)*. Boc anhydride has been employed as an inexpensive coupling reagent with diazomethane, allowing for good yields (86–95%) of the diazomethane precursors of homologated Fmoc (Gly, Val, Asp, Lys), Boc (Pro, Ile), and Cbz (Phe, Phg) amino acids, with yields of 75–84% for the subsequent Ag-catalyzed rearrangement *(1071)*. p-Toluenesulfonyl chloride has also been used as the activating agent, with 86–94% yields of isolated diazoketone, and 80-88% yields of rearranged Fmoc-Phe, -Phg and -Val, Boc-Ala and -Val, and Cbz-Ala, -Phe, and -Phg *(1080)*. Activation with TBTU gave 86–94% yields of the diazoketones of Fmoc-Val, -Ile, -Phe, -D-Phe, -Asp, -Glu, and -Lys, Boc-Ala, -Phe, and -Val, and Cbz-Ala and -Phe *(1082)*. The rearrangement step has been improved by a microwave-assisted reaction using a slurry of silica gel with catalytic silver trifluoroacetate, with Cbz-Ala, -Val, -Ile, -Tle, -Phe, Fmoc-Leu, and Boc-Lys(Cbz) diazoketones rearranged in 15 min with 93–96% yield *(1081)*. The homologation has been carried out on amino acids attached to a Wang resin by their amino groups via a carbamate linkage, using isobutyl chloroformate for activation, followed by treatment with diazomethane and then a solution of silver benzoate *(1085)*.

Several complex multifunctional nucleophiles, including carbohydrates and nucleosides, have been examined for trapping of the rearrangement reaction. High functional group selectivity was generally observed, giving β-amino acid ester and amide derivatives *(1072)*. Sterically hindered nucleophiles (such as carbohydrates with an unprotected hydroxyl group) could be used as nucleophiles to trap the ketene intermediate if the steric hindrance was not too great, but yields were reduced *(1076)*. Oligonucleotides with a terminal 5′-amino group were also used to trap the ketene to form a β-amino acid–oligonucleotide amide-linked conjugate *(1077)*. Homologation of Boc- or Fmoc-protected amino acids with N,O-dimethylhydroxylamine as the nucleophile for the diazoketone intermediate rearrangement provided the Weinreb amide derivatives of β-amino acids, suitable for further derivatization such as formation of reduced amide dipeptide isosteres (*1075*).

Carrying out the Wolff rearrangement in the presence of an amino ester leads to dipeptide formation (*1069*). Podlech and Seebach applied the Arndt–Eistert homologation to the *C*-terminal amino acid of a peptide and trapped the intermediate ketene with the free amino group of an amino acid or dipeptide, thereby allowing for the insertion of a β-amino acid at any stage of a peptide synthesis (*1073*). Similarly, a β-Val-β-Ala-β-Leu-OMe tripeptide was constructed by sequential Arndt–Eistert homologations and concomitant peptide-bond formation, with Boc-Ala homologated and trapped with β-Leu-OMe, and then Boc-Val homologated and trapped by the β-Ala-β-Leu-OMe dipeptide. Either photochemical or silver-catalyzed transformations could be used, and their effectiveness was compared in this study (*292*). The same strategy was applied to a solid-phase β-peptide synthesis by trapping the ketene intermediate generated from Fmoc-protected amino acid diazoketones using the free amino group of peptides attached to a polymer support. α-Amino acids were then coupled by conventional methods, or additional β-amino acids were introduced by further Arndt–Eistert additions. A tetra-β-peptide was prepared in 60% crude yield (*1083*).

Several other methods have been developed for homologating the α-carboxy group of α-amino acids. One method that is suitable for large-scale conversions relies on reduction of the α-carboxyl group to a hydroxymethyl moiety (see Scheme 11.70). This was converted into an enantiopure β-amino iodide in high yield by reaction with a polymer-bound triarylphosphine–I$_2$ complex. Displacement of the iodide with tetraethylammonium cyanide and methanolysis of the

cyano group produced protected β-amino esters (eight examples) (*1108*). In a similar manner Boc-L-phenylalaninol was activated as the tosylate, displaced with NaCN, and hydrolyzed with H$_2$O$_2$/NaOH in 36% overall yield on a multigram scale (*376*). The α-carboxyl group of pyroglutamate was reduced to a hydroxymethyl moiety, activated, and displaced by cyanide. Further derivatization introduced unsaturation into the pyrrolidinone ring, with dihydroxylation of the alkene giving 4-amino-2,3-dihydroxyhexanedioic acid (*1109*). Surprisingly, His was not homologated into the corresponding β-amino acid until a 2002 report, which protected the imidazole and α-amino groups of His-OMe with 2,4,6-trimethylbenzenesulfonyl chloride, (mesitylenesulfonyl chloride, Mts-Cl), reduced the ester with NaBH$_4$, formed the mesylate of the alcohol, and then used NaCN for the displacement. Hydrolysis with HBr removed the Mts protection and hydrolyzed the nitrile (*1110*). The Arndt–Eistert homologation of His has subsequently been reported (*351, 833*).

The dimesylate of 2-aminobutane-1,4-diol was prepared via reduction of both Asp carboxyl groups (*1111, 1112*). The β-mesyl group was regioselectively displaced with cyanide, followed by functionalization of the γ-mesylate by reaction with organocuprates or azide (see Scheme 11.71). A rearrangement reaction led to β-Pro (*1111*).

A possible alternative to the Arndt–Eistert one-carbon homologation, though not yet applied to amino acid substrates reacted acid chlorides with 1-[(trimethylsilyl)methyl]benzotriazole, BtCH$_2$TMS. The resulting α-benzotriazole ketones were converted into enol triflates by reaction with triflic anhydride; treatment with NaOMe in acetonitrile, followed by hydrolysis

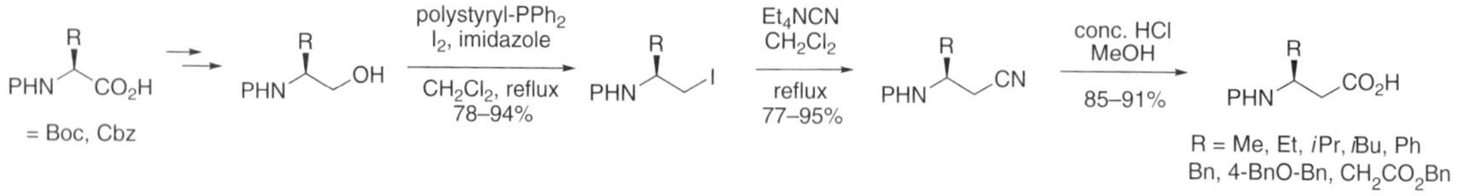

Scheme 11.70 Homologation of α-Amino Acids via Amino Alcohol (*1108*).

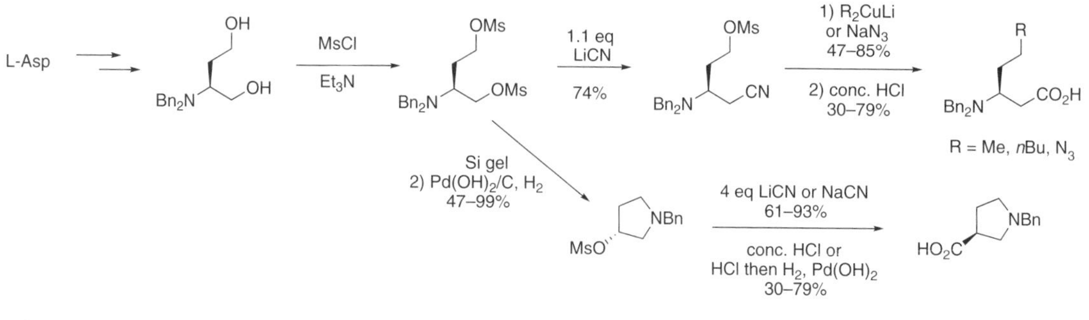

Scheme 11.71 Derivatization of Asp-derived Aminodiol (*1111, 1112*).

with HCl in methanol or ethanol, produced the homologated methyl or ethyl esters. (*1113*).

Another general method of homologation involves reduction of the α-amino acid carboxyl group to an aldehyde, with cyanide addition leading to α-hydroxy-β-amino acids. This method was applied to the synthesis of the AHPA (3-amino-2-hydroxy-4-phenylbutyric acid) β-amino acid component of bestatin in 1976. Cbz-D-Phe was reduced to phenylalaninal via a pyrazolide intermediate, and cyanide was added via a sodium bisulfite adduct. Hydrolysis gave AHPA as a mixture of diastereomers (see Scheme 11.72) (*1114, 1115*). More recently, Phe was reduced to phenylalaninal via reduction of the Weinreb amide derivative. Addition of cyanide and methanolysis gave methyl AHPA with retention of chirality at the β-center, but only 8% de. The purified diastereomer was treated with DAST to introduce a 2-fluoro group, or oxidized and treated with DAST to prepare the 2,2-difluoro derivative (*1116*). The Cbz-phenylalaninal aldehyde has also been protected as an acetal with either (2*S*,4*S*)- or (2*R*,4*R*)-pentanediol, followed by a BF$_3$·Et$_2$O-catalyzed reaction with TMSCN. The acetal auxiliary induced either 40% de *threo* or 40% de *erythro* diastereoselectivity. By starting with D- or L-Phe, all four diastereomers of AHPA could be prepared (*1117*).

Reduction of the aromatic ring of (*S*)-phenylalaninol to a cyclohexyl group followed by oxidation of the hydroxymethyl group produced cyclohexylalaninal. Addition of cyanide followed by hydrolysis gave (2*R*,3*S*)-cyclohexylnorstatine in 60% overall yield with complete diastereoselectivity (*1118*). The 3-isobutylisoserine (norstatine) component of amastatin was synthesized from D-Leu via reduction of the carboxyl group to an aldehyde, conversion to a cyanohydrin group using NaHSO$_3$ and NaCN, and then nitrile hydrolysis (*1119*), while the aldehydes derived from Boc-L-cyclohexylalanine, Boc-L-Ile, or Boc-L-Phe were hydrocyanated with NaCN-HCl at 0 °C and then hydrolyzed with 23% HCl at 80 °C to give predominantly the (2*R*,3*S*) diastereomers of cyclohexylnorstatine, norstatine, or phenylnorstatine with approximately 40% de (*389*). Another synthesis also reacted Bz-L- or D-cyclohexylalaninal with NaCN–HCl, but the intermediate cyanohydrin, as an oxazoline derivative, was treated with SOCl$_2$ in benzene to invert the C2 configuration. Hydrolysis then gave (2*S*,3*S*)- or (2*R*,3*R*)-cyclohexylnorstatine, with 50% de (*389*). Boc-alaninal, Boc-valinal, and Boc-phenylalaninal were all reacted with KCN and then hydrolyzed to give α-hydroxy acid homologs. Esterification was followed by Swern or Dess–Martin oxidation to give the corresponding α-keto

esters (*410*). β-(4-Piperidinyl)-Ala was homologated by a similar route (*1120*).

Similar strategies were applied to *N,N*-dibenzyl L-Ala, producing (2*S*,3*S*)-2-hydroxy-3-dibenzylaminobutanoate via addition of TMSCN/BF$_3$·Et$_2$O, or (2*R*,3*S*)-2-hydroxy-3-dibenzylaminobutanoate via addition of TMSCN/TiCl$_4$. DAST treatment of the (*S*,*S*) isomer resulted in the (*S*,*S*)-2-fluoro-3-aminobutanoic acid isomer (with retention of configuration, via an intermediate aziridinium ion), while oxidation followed by DAST treatment produced 2,2-difluoro-3-aminobutanoic acid (*355*). Similarly, *N,N*-dibenzyl Ala, Val, and Leu were reduced to the amino aldehydes, converted to the epimeric cyanohydrins, and then methanolyzed. Treatment with DAST, or oxidation to the ketone followed by treatment with DAST, produced the α-monofluoro- or α,α-difluoro-β-amino acids (*1121*).

Amino aldehydes derived from amino acids have been employed in Passerini three component condensations with anomeric glucosyl isonitriles (R^{2}NC) and carboxylic acids (R^{3}CO$_2$H), producing α-acetoxy-β-amino acid glycosyl amides in 31–57% yield, though with minimal diastereoselectivity at the α-acetoxy-center (*537*). Another Passerini synthesis was used to generate arylalanine-derived α-ketoamides as calpain inhibitors. Boc-phenylalanines were converted to the Weinreb amide and reduced with LiAlH$_4$ to form the amino aldehyde; reaction with an isocyanide and Boc-protected amino acids formed α-acyloxyamides, with the acyl group removed and the α-hydroxy group oxidized with Dess–Martin periodinane to form the α-ketoamide (*413*).

Modifed amino aldehyde equivalents have been used as the electrophilic component. Alkylation of Williams' oxazinone template provided the initial homochiral α-amino acid. Instead of removing the template to provide the free amino acid for further elaboration, the template lactone carbonyl was directly reduced to an aldehyde acetal (see Scheme 11.148 in Section 11.4.3g). Reaction with TMSCN led to β-amino acid derivatives, with cyano hydrolysis and removal of the template by hydrogenation providing cyclohexylnorstatine and isothreonine diastereomers (*1122*).

Other carboxyl equivalents can be used instead of cyanide. *N*-Boc 2-hydroxy-3-amino-5-methylhexanoic acid (3-isobutylisoserine, the enantiomer of the norstatine component of amastatin, and a component of the taxane-derived anticancer agent ABT-271) was prepared from L-Leu via addition of vinylmagnesium bromide to *N*-Boc leucinal derivative (66–80% de *syn*), with oxidative cleavage of the alkene generating the carboxyl group. The diastereomers could be chromatographically

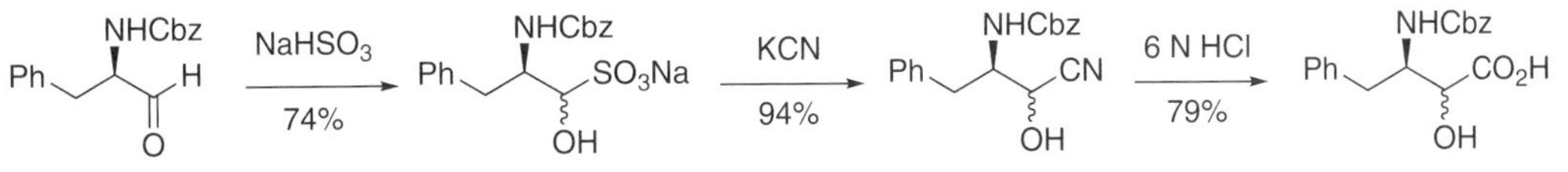

Scheme 11.72 Synthesis of β-Hydroxy-β-Amino Acids from α-Amino Aldehydes (*1114, 1115*).

separated (*191, 1123*). In a similar manner, *N*-Boc cyclo-hexylalanine methyl ester was reduced with DIBAL and the resulting aldehyde reacted with vinylmagnesium bromide to give a 6:1 mixture of carbinol diastereomers. Oxidative cleavage of the alkene gave (2*S*,3*S*)-2-hy-droxy-3-amino-4-cyclohexylbutyric acid (*388*). The car-boxyl group of (*S*)-phenylglycine was also reduced to an aldehyde. Addition of vinylmagnesium bromide gave the *syn* (*threo*) diastereomer. Oxidative cleavage of the vinyl group generated the carboxyl substituent, providing the Taxol phenylisoserine component with the desired ste-reochemistry (*1124*). Ethynylmagnesium bromide added to *N*-9-phenylfluorenyl (*R*)-phenylalaninal in 96% yield with 81% *syn* diastereoselectivity. *O*-Benzylation (97%), alkyne oxidative cleavage with KMnO$_4$ (87%), and *O*-deprotection (93%) produced AHPBA (*1125*).

2-Lithiated thiazole has been added to α-amino alde-hydes derived from amino esters, with *syn* or *anti* diaste-reoselectivity controlled by the *N*-substitution of the amino aldehyde. The thiazole substituent was converted to an aldehyde and oxidized to generate the acid moiety (*1126*). (*S*)-Phenylglycinal was converted to the Taxol phenylisoserine component via addition of 2-(trimethylsilyl)thiazole as the carboxyl equivalent (*1127*). Lithiated methoxyallene can replace lithiated thiazole for additions to *N*-diprotected α-amino alde-hydes. Diastereoselectivity was not as high (24–78% de *anti*), but unmasking the carboxyl group required only an ozonolysis step, with overall yields of 27–51% (*1126*). Both of these homologation routes were recently reviewed (*1128*).

Another general route to α-hydroxy-β-amino acids from α-amino acids begins with reaction of 2-lithiated thiazole with the α-amino esters, rather than the α-amino aldehydes (see Scheme 11.73). Stereoselective reduc-tion of the carbonyl of the resulting 2-thiazolyl-α-amino ketone produced either *syn* or *anti* α,β-amino alcohols, with the stereochemistry determined by mono- or

diprotection of the amino group. The synthesis was completed by conversion of the thiazole to an aldehyde and oxidation to the acid moiety (*1129, 1130*). Similarly, Gly was transformed into isoserine by treatment of the *N*-phthaloyl acid chloride with NaI and then CuCN, with methanolysis of the glycyl cyanide adduct providing *N*-phthaloyl 3-amino-2-oxopropanoate. The α-keto group was enantioselectively reduced with [RuCl(binap) (benzene)]Cl to give isoserine with 81% ee, enantiomer-ically pure after recrystallization and deprotection (*1131*). D-Isothreonine and L-alloisothreonine were obtained via homologation of L-Ala by addition of 3-phenylallyl Grignard or organolithium reagents, fol-lowed by stereoselective ketone reduction and protec-tion of the amino alcohol of each stereoisomer as an oxazolidinone. The alkene was then oxidatively cleaved to generate the acid group (*1132*).

None of the previous homologation methods allow for the introduction of an α-substituent, other than a hydroxyl group. Burgess et al. developed a method by which the α-substituent was added via organolithium addition to *N*-tosyl amino acids (*1133*). Homologation was then accomplished by methylenation of the result-ing ketone group via a Wittig reaction (see Scheme 11.74): *syn*- or *anti*-selective hydroboration followed by oxidation of the hydroxymethyl moiety gave α,β-disubstituted-β-amino acids.

Modifications of these methods allow for the prepa-ration of α-keto-β-amino acids. For example, the α-keto acid analog of Arg was prepared from the Weinreb amide of protected Arg, via reduction with LiAlH$_4$ to arginal, addition of the acid equivalent LiC(SMe)$_3$, hydrolysis, and oxidation of the α-hydroxy acid (*264*). Similarly, the Weinreb amide of L-Ile was displaced with vinyl-magnesium bromide, followed by NaBH$_4$ reduction of the carbonyl of the vinyl ketone. Protection of the hydroxyl group and oxidation of the alkene to an acid group gave an α-hydroxy-β-amino acid, which was

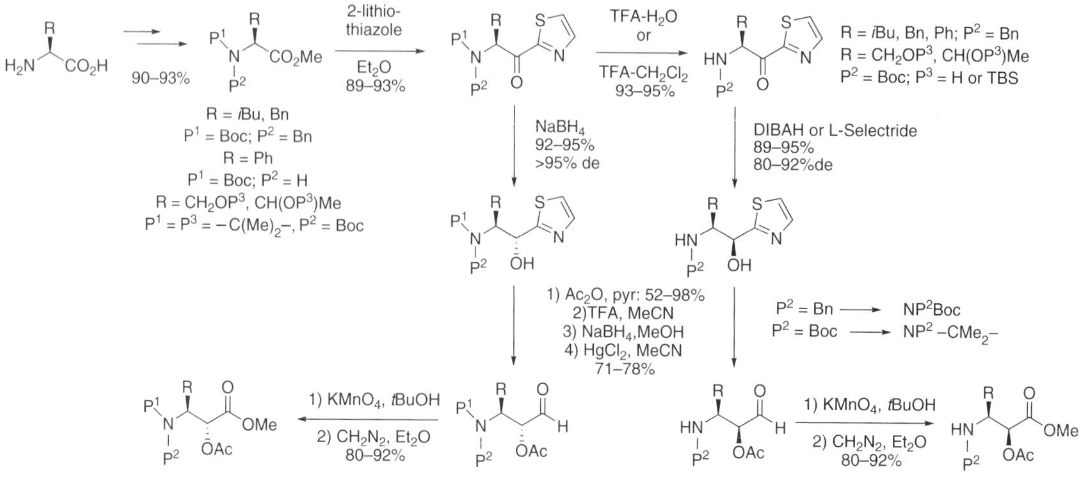

Scheme 11.73 Homologation of α-Amino Acids via α-Ketothiazole (*1129, 1130*).

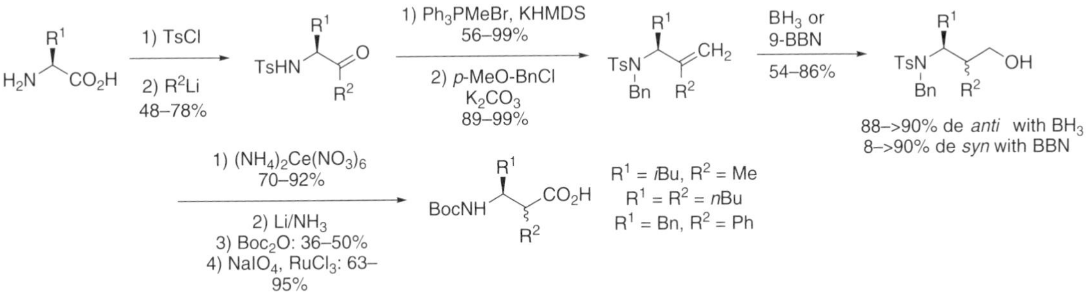

Scheme 11.74 Homologation of α-Amino Acids to α,β-Disubstituted-β-Amino Acids (*1133*).

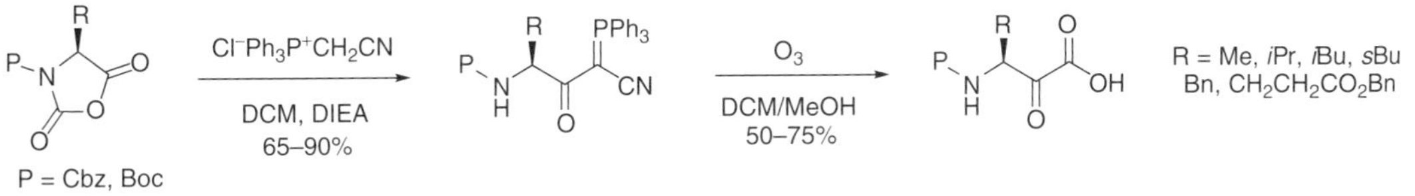

Scheme 11.75 Synthesis of β-Amino-α-Keto-Esters from UNCAs (*1134*).

oxidized to the α-keto derivative after incorporation into a peptide (*175*).

N-Protected amino acid N-carboxyanhydrides (UNCAs) were treated with cyanomethyltriphenylphosphonium chloride to provide cyanomethylene triphenylphosphoranes in 65–90% yield. Ozonolysis gave β-amino-α-keto esters in 50–75% yield (see Scheme 11.75) (*1134*). A closely related synthesis coupled Boc-L-Phe with (cyanomethylene)triphenylphosphorane, Ph₃P=CHCN, using EDCI/HOBt for activation. Ozonolysis gave the α-keto acid, which was then coupled with L-Val-OBn or L-Leu-OBn. Ketone reduction with zinc borohydride proceeded with high stereoselectivity (84–86% de) to give the β-amino-α-hydroxy amides. The peptide aminopeptidase inhibitors bestatin, phebestin, and probestin were prepared (*1135*). Other arylalanine-derived α-keto amides were prepared by this route as calpain inhibitors (*413*), while D-Abu was converted in another paper (*2184*). 2-Keto-3-amino-5,5-difluoropentanoic acid, used in HCV NS3 protease inhibitors, was constructed from the α-amino acid via the acyl cyanophosphorane, which was incorporated into di-/tripeptides before ozonolysis and methyl ester hydrolysis (*412*). The β-amino-α-keto analog of Arg, contained in the cyclotheonamides, was derived from Boc-Arg(Cbz)₂ via EDCI/DMAP coupling with Ph₃P=CHCN, followed by ozonolysis (*267*). The (cyanomethylene)triphenylphosphorane reagent has been attached to a polymer support and coupled to a number of N-Fmoc amino acids with EDC/DMAP or DIC/DMAP. Cleavage from the resin by ozonolysis followed by methanol gave the homologated α-keto esters of Leu, Phe, Orn, Asp, Asn, and Met. These could be reduced with a solid-supported trimethylammonium borohydride

(*1136*). A more mild procedure for oxidative cleavage of the cyanoketophosphorane intermediate was reported in 2001, employing dimethyldioxirane (*1137*).

In 2004 a series of N-Fmoc amino acids were coupled with (cyanomethylene)triphenylphosphorane, and ozonolyzed to give highly electrophilic diketonitriles, which were reacted in situ with α-amino esters to give α-keto amide dipeptides. The α-keto group was protected as a 1,3-dithiane, and the protected dipeptides employed for solid-phase peptide syntheses (*1138*). The pentapeptide inhibitor postatin was obtained by coupling Cbz-Abu with Ph₃P=CHCN. Cbz removal and acylation with Cbz-Val-Val-OH formed a tripeptide acyl cyanophosphorane, with ozonolysis in the presence of H-D-Leu-Val-OBn dipeptide forming the α-keto amide linkage of protected postatin (1129).

Another general homologation procedure reacted N-trityl amino acid methyl esters with excess lithium dimethyl methylphosphonate (see Scheme 11.76). The resulting β-ketophosphonates were olefinated by Horner–Wadsworth–Emmons reaction with benzaldehyde. Alkene ozonolysis generated the α-keto acids. The ketone could be reduced with NaBH₄ or Zn(BH₄)₂ with moderate *anti* stereoselectivity (31–79% de) as the N-trityl derivative, or with high *anti* selectivity (90–98% de) as the unprotected amine (*1139*). Alternatively, the Weinreb amide of N-Boc L-phenylglycine was reduced to an aldehyde and homologated by Wittig reaction with (MeO)₂POCH₂CO₂Me. The ester was reduced with DIBAL and the alcohol O-benzoylated. A Pd-catalyzed isomerization/cyclization produced (4S,5R)-2,4-diphenyl-5-vinyl-oxazoline, which was oxidized by RuCl₃/NaIO₄ to give the oxazoline of the Taxol side chain (*1140*).

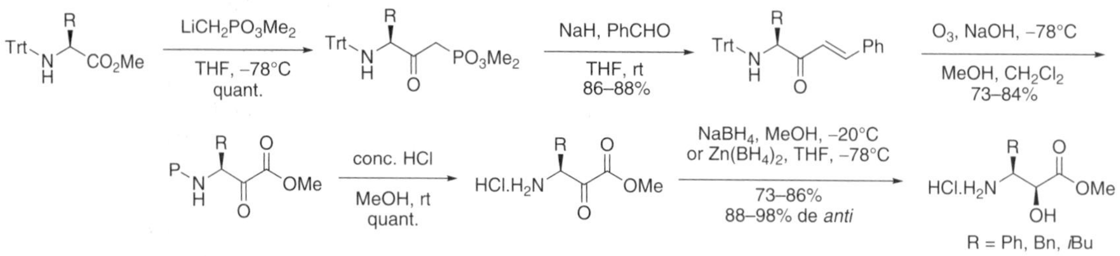

Scheme 11.76 Synthesis of β-Amino-α-Keto-Esters from Amino Esters (1139).

11.2.3n Elaboration of α-Amino Acids: Derivatization of Asp α-Carboxyl Group

Aspartic acid already contains a 3-aminopropionic acid skeleton. Modification of the α-carboxyl group can convert it into a β-substituent on the β-amino acid. The most simple modification is to remove the α-carboxyl group. Asp(OMe)-OH was decarboxylated to β-Ala by heating the hydrochloride salt with p-methoxy-acetophenone (421). This method was used to obtain (2R)-[2-²H₁]- or (2S,3RS)-[2,3-²H₂]-β-Ala from (2S,3R)-[3-²H₁]- or (2S,3S)-[2,3-²H₂]-Asp(OMe)-OH, respectively (421). (2S)-3-Amino-2-methylpropanoic acid was obtained in 24% yield by decarboxylation of (2S,3S)-3-methyl-Asp (422). Asn has also been decarboxylated within a heterocyclic template, giving a cyclic derivative of 3-aminopropenoic acid, which was then arylated using aryl iodides and Pd(0) catalysis to give (R)-β-Tyr (1141, 1142) (see Scheme 11.91 below).

Other β-substituents have been created by selective reduction of the Asp α-carboxyl group, followed by further elaboration. For example, N-Cbz (1143–1145), N-Tfa (1146) or N-Boc (1144, 1147) Asp anhydride were regioselectively reduced with sodium borohydride in 76–91% yield to give a protected γ-butyrolactone. Hydrolysis of the lactone gave β-(hydroxymethyl)-β-Ala (1144). The reduced lactone could be deprotonated and alkylated to introduce an α-substituent, with 76–82%

de syn stereoselectivity (see Scheme 11.77) (1144). If the lactone was first converted into an oxazoline or oxazolidinone, alkylation produced the anti isomers with 66–88% de (1144, 1148).

The α-carboxyl group of acyclic Cbz- or Boc-Asp(OBz)-OH was reduced via sodium borohydride reduction of the mixed anhydride formed with isobutyl chloroformate (1149), or by reduction of the N-ethoxycarbonyl Asp(OMe)-OH derivative (1150). In a similar fashion Cbz-Asp(OtBu)-OH was activated as the N-hydroxysuccinimide ester and reduced with sodium borohydride to give protected 3-amino-4-hydroxy-butanoic acid. Deprotonation and methylation gave the 2-methyl derivative with 48% de (1151). β-Benzylation of Boc-Asp(OBn)-OBn, followed by selective α-ester deprotection and reduction, provided (2S,4R)-2-benzyl-3-amino-4-hydroxybutanoic acid (418). β-(Amino-methyl)-β-Ala was prepared via α-amide formation of Boc-Asp(OMe)-OH with secondary amines, followed by borane-mediated reduction of the amide carbonyl group (399). Activation of the α-carboxyl group of Boc-Asp(Bn)-OH by treatment with CDI followed by reaction with the enolate of trimethylsilylethyl acetate gave protected 3-amino-4-ketoadipic acid; the carbonyl group was reduced with sodium borohydride (1152).

Asp has been converted into an electrophilic synthon for β-amino acid synthesis by reduction of the α-carboxyl group to a hydroxymethyl moiety followed by tosylation

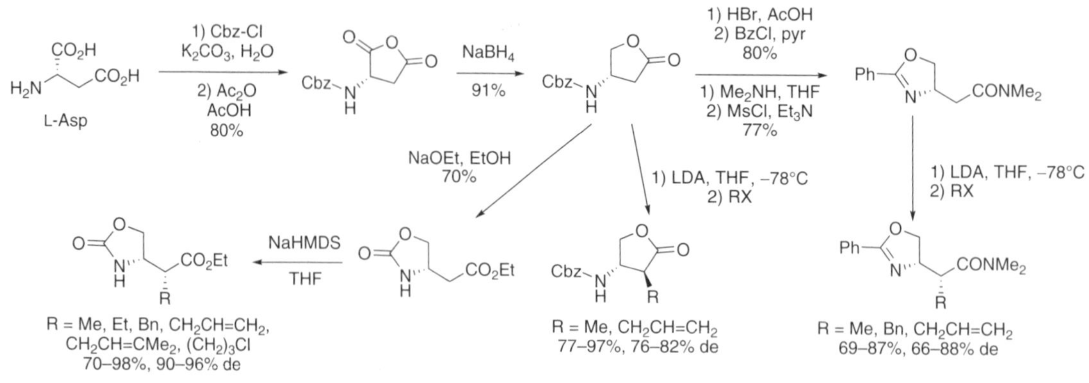

Scheme 11.77 Reduction of Asp α-Carboxyl, Alkylation (1143, 1144, 1148).

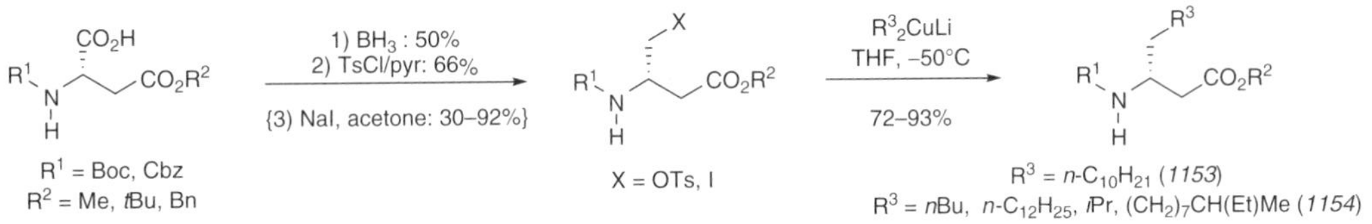

Scheme 11.78 reaction conditions:

CO_2H / CO_2R^2 with R^1, N, H

1) BH_3 : 50%
2) TsCl/pyr: 66%
{3) NaI, acetone: 30–92%}

X / CO_2R^2 with R^1, N, H   X = OTs, I

R^3_2CuLi
THF, –50°C
72–93%

R^3 / CO_2R^2 with R^1, N, H

R^1 = Boc, Cbz
R^2 = Me, tBu, Bn

R^3 = n-$C_{10}H_{21}$ (1153)
R^3 = nBu, n-$C_{12}H_{25}$, iPr, $(CH_2)_7CH(Et)Me$ (1154)

Scheme 11.78 Conversion of Asp into Electrophilic Synthon.

(see Scheme 11.78) (1153, 1154). Treatment with an organocuprate gave the Boc-protected nC$_{14}$ isomer of iturinic acid in 86% yield (1153). Other alkyl organocuprates, including the 8-methyldecylcuprate required for the natural isomer of iturinic acid, were also used for the displacement. In some cases better yields were obtained by converting the tosyl leaving group into an iodo leaving group (1154). Similarly, Ts-Asp anhydride was reduced to the lactone and opened with TMSI in EtOH to give ethyl 3-tosylamino-4-iodobutanoate. Organocuprate reactions produced 3-aminooctanoic acid and 3-aminodecanoic acid (89). The same route led to 3-aminotetradecanoic acid, (Z)-3-aminodec-7-enoic acid, and 3-amino-4-(octylthio)butanoic acid (90). Displacement of the iodide with trimethylamine produced (R)-3-aminocarnitine (from D-Asp), while reaction with trimethylphosphine produced the trimethylphosphonium iodide, and displacement with azide followed by reduction produced 3,4-diaminobutanoic acid (1155). Boc-L-Asp(OMe) was converted into $N^β$-Boc,$N^γ$-Fmoc 3,4-diaminobutanoic acid in a similar fashion (1156). The α-carboxyl group of Boc-Asp(OBn)-OH was reduced via the isobutyl chloroformate/NaBH$_4$ method, and the alcohol displaced by Mitsunobu reaction with 3-benzoyluracil or 3-benzoylthymine, or mesylated and displaced with 3-benzoylthymine or Cbz-cytosine, producing nucleo-β-amino acids (1157).

Similarly, the side-chain carboxyl group of L-Asn was dehydrated to a nitrile group and the α-carboxy group converted into an electrophilic mesylated hydroxymethyl group (see Scheme 11.79). The mesylate was displaced with organocuprates, with hydrolysis of the cyano group completing the synthesis (1158). Displacement of the O-mesyl electrophile with hydride from LiBH$_4$ led to 3-aminobutanoic acid (1159). This procedure was combined with deprotonation and alkylation of the intermediate β-aminonitriles to introduce substituents at the α-position (1160). However, in contrast to alkylations of β-amino esters (see Section 11.2.3p), diastereoselectivity was poor. For these

reactions the hydroxyl group was displaced with hydride or thioacetate.

An alternative route for converting the α-carboxyl group of Asp to an electrophile proceeded via the anhydride of N-tosyl-Asp, which was reduced, stereoselectively α-alkylated or hydroxylated, and then ring opened with TMSI to give an iodo leaving group (see Scheme 11.80; also Scheme 9.15) (1161). Displacement of the iodo group with an organocuprate gave β-amino acids (1162, 1163), including the α-hydroxy-β-amino acid components of bestatin and dolastatin (1162, 1164–1167). Another synthesis reduced the α-carboxyl group of Cbz-Asp(OtBu)-OH via the isobutyl chloroformate/NaBH$_4$ method, and then converted the resulting amino alcohol into tert-butyl N-Cbz aziridine-2-acetate by a Mitsunobu reaction with PPh$_3$/DEAD. The aziridine could be opened with amine nucleophiles to form 3,4-diaminobutyrate products, with high regioselectivity for attack of the incoming amine at the less substituted C4 position (1168).

A nucleophilic β-amino acid corresponding to a homolog of the Jackson β-(iodozinc)alanine synthon (see Section 10.5.3) was derived from Asp via the same N-Boc 3-amino-4-iodopropionic acid methyl ester intermediate used as an electrophilic synthon. The organozinc reagent was generated and then used for Pd-catalyzed couplings with aryl iodides to give homochiral 3-benzyl-β-Ala derivatives (1169). Unprotected iodophenols also coupled (1170), while aryl triflates reacted with the reagent in 44–64% yield (1171). The corresponding zinc/copper reagent coupled with allyl chloride to give 3-aminohept-6-enoic acid (1172). One problem with the iodozinc reagent is that elimination of the β-amino group is a side reaction. Surprisingly, an N-trifluoroacetyl protecting group, which would be expected to activate the amine as a better leaving group than an N-Boc group, instead resulted in reduced elimination. Aryl iodides coupled in 53–77% yield (1173). NMR studies on the organozinc reagent also demonstrated that DMA or DMF helped to

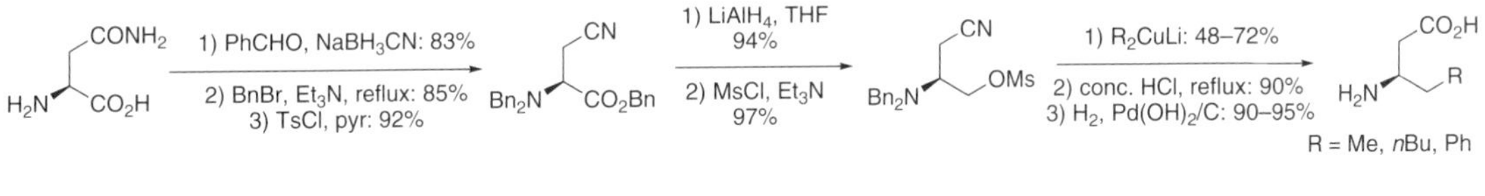

Scheme 11.79 reactions:

$CONH_2$ / CO_2H with H_2N

1) PhCHO, NaBH$_3$CN: 83%
2) BnBr, Et$_3$N, reflux: 85%
3) TsCl, pyr: 92%

CN / CO_2Bn with Bn$_2$N

1) LiAlH$_4$, THF
94%
2) MsCl, Et$_3$N
97%

CN / OMs with Bn$_2$N

1) R$_2$CuLi: 48–72%
2) conc. HCl, reflux: 90%
3) H$_2$, Pd(OH)$_2$/C: 90–95%

CO_2H / R with H_2N

R = Me, nBu, Ph

Scheme 11.79 Synthesis of β-Amino Acids from L-Asn (1158).

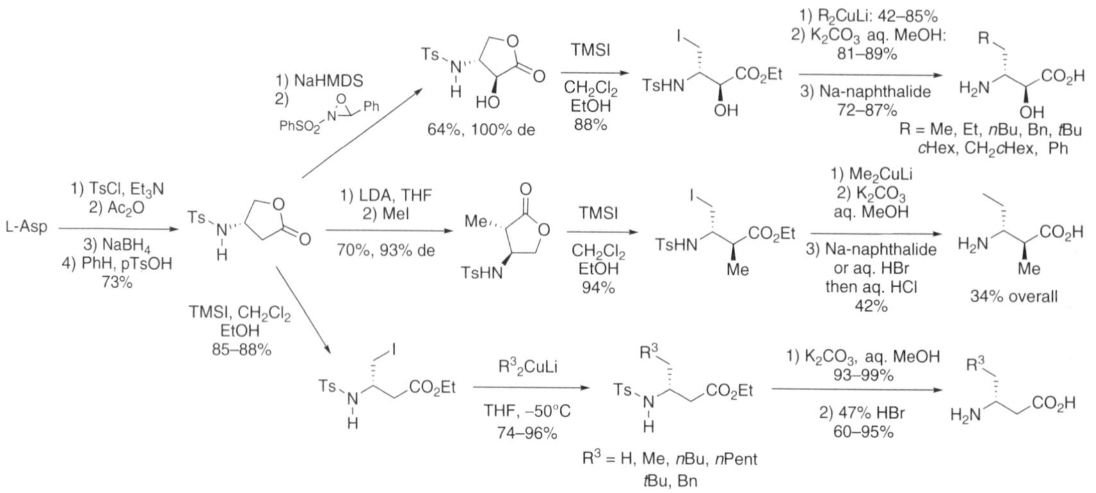

Scheme 11.80 Elaboration of Asp to β-Amino Acids (*1161–1167*).

stabilize the reagent from β-elimination decomposition, allowing for Pd-catalyzed cross-couplings with acid chlorides to produce 3-amino-5-keto-5-substituted-pentanoic acids (*1174*).

As mentioned earlier, homologation of the L-Asp α-carboxyl group provides 3-amino-1,5-pentanedioic acid, a symmetrical molecule that is chiral due to differential protection of the carboxyl groups. Hofmann rearrangement of either of the carboxyl moieties introduced a γ-amino group, with trimethylation producing either enantiomer of emeriamine (*169*). Another synthesis homologated the α-carboxyl group of Cbz-Asp(OtBu)-OH under Arndt–Eistert conditions, converted the newly formed acid group to an aldehyde, and added the Reformatsky reagent ZnBrCH$_2$CO$_2$Et. The ethyl ester was reduced to the alcohol, activated as the mesylate and displaced with azide, which was reduced to finally produce 3,7-diamino-5-hydroxyheptanoic acid (*167*). The acid group of homologated Boc-L-Asp(OBn)-OH was also reduced with borane and brominated with CBr$_4$/PPh$_3$ to give 3-amino-5-bromopentanoic acid. Displacements of the bromide with N-Cbz cytosine, 3-benzoyl-thymine, 2-amino-6-chloropurine, or N-Cbz adenine produced the four β^3-nucleoamino acids, which were then incorporated into β-peptides (*1093*).

The Asp α-carboxyl group has been homologated to a β-keto ester via conversion into a urethane N-carboxy anhydride, followed by treatment with the lithium enolate of trimethylsilylethyl acetate, giving 3-amino-4-keto-hexanedioic acid (*215*). Reduction of the α-carboxy group of Cbz-Asp(OBn)-OH with borane followed by Swern oxidation gave Asp α-aldehyde. Homologation with *tert*-butyl triphenylphosphorylidene acetate (1:1 *cis:trans* alkene product), followed by OsO$_4$-catalyzed dihydroxylation of the *cis* alkene gave the 2,3-dihydroxy-4-aminohexanedioic acid-component of the gastroprotectant AI-77-B (*220*). Another synthesis

employing Asp α-semialdehyde masked the side-chain carboxyl group of N,N-Bn$_2$-Asp by reduction to a TBDMS-protected alcohol, followed by reduction of the α-carboxyl group to an aldehyde. Addition of diethylzinc to the aldehyde gave the *syn* adduct as a single isomer in 76% yield, while EtMgBr produced the *anti* adduct in 70% yield. The side-chain carboxyl group was then reoxidized, producing both *syn* and *anti* 3-amino-4-hydroxyhexanoic acid. The side-chain carboxyl reduction/oxidation procedures could be avoided by protecting the side chain acid group as a nucleophile-resistant 2,6,7-trioxabicyclo[2.2.2]octane (OBO) ester (*1175*). The C$_{20}$ amino acid Adda was prepared by reduction of the α-carboxyl group of Boc-β-methyl-Asp(OMe)-OH to an aldehyde, followed by attachment of the remainder of the side chain via elaboration of the aldehyde to a vinylstannane and Stille coupling (see Scheme 11.81) (*1176*).

Another synthesis of Adda relied on elaboration of Ser to prepare the Boc-β-methyl-Asp aldehyde intermediate, via the Garner aldehyde (see Section 10.6.1) (*1177*). The aldehyde was iodomethylenated with CH$_2$I$_2$/CrCl$_2$ to give an (*E*)-iodo alkene (75% yield), which was then converted to a vinylstannane and coupled with the remainder of the side chain under Stille coupling conditions (58% yield). Alternatively, reduction of the α-carboxyl group of Cbz-D-Asp anhydride gave homochiral 3-amino-γ-butyrolactone, which was deprotonated and methylated with 60% de (*1178*). The predominant (*2R,3R*) isomer was epimerized to the desired (*2S,3R*) lactone by heating in benzene/Et$_3$N. Lactone saponification followed by oxidation of the hydroxymethyl side chain gave an aldehyde, with a Wittig coupling introducing the Adda side chain in 33–43% yield (see Scheme 11.82) (*1178, 1179*). A similar aldehyde equivalent was prepared by a different route, with the Adda side chain coupled in 55% by a

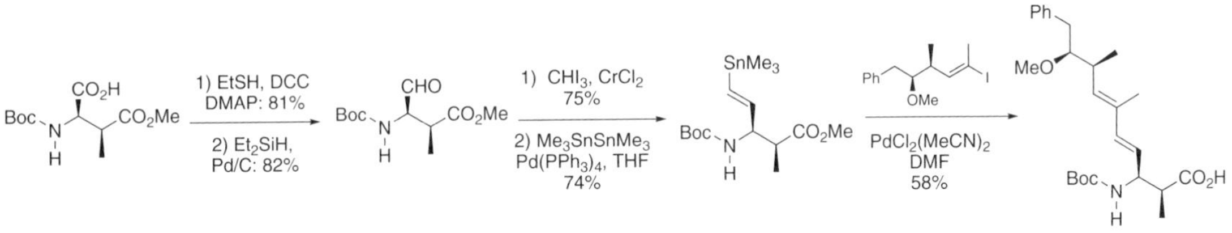

Scheme 11.81 Conversion of Asp into ADDA (*1176*).

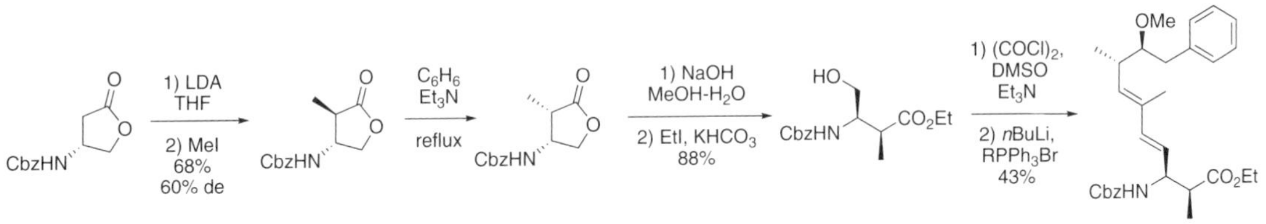

Scheme 11.82 Alkylation and Elaboration of 3-Amino-4-hydroxybutyrate (*1178, 1179*).

Wittig reaction (*839*). Asp has also been converted into Adda via cyclization to form azetidinone-3-carboxylate. The β-lactam was methylated and the carboxy substituent converted into a formyl group, which was coupled by a Wittig reaction with the remainder of the Adda side chain. Opening of the β-lactam with LiOH gave Adda, while reaction with the amino group of amino esters formed Adda–amino acid dipeptides (*1180*).

Friedel–Crafts acylations with protected Asp anhydrides have traditionally been used to prepare β-arylalanines. However, depending on the arene used and the type of Asp N-protection, substantial quantities of 3-amino-4-oxo-5-arylpentanic acids can also be isolated (*1181*). (2S,3S)-3-Amino-2-hydroxy-5-phenylpentanoic acid (allophenylnorstatine, an *erythro* α-hydroxy-β-amino acid component of HIV protease inhibitors) was synthesized from Meoc-D-Asp(OEt)-OH. The α-carboxyl group was activated as an acid chloride, and then used for a Friedel–Crafts acylation of benzene in the presence of AlCl₃. Ketone reduction, protection of the amino alcohol as an oxazolidinone, and conversion to an oxazoline gave a substrate that could be deprotonated and stereoselectively hydroxylated with 3-phenyl-2-(phenylsulfonyl)oxaziridine to introduce the 2-hydroxy group. Hydrogenation gave the target compound (*392*).

Several other simple transformations of the α-carboxyl group are possible. Asp was converted into β-(2-thiazolyl)-β-Ala via derivatization of the α-carboxyl group as a thioamide; the thioamide was condensed with bromoacetaldehyde to form the thiazole substituent (*1182*). N-Protected L-Asp α-amide (L-isoasparagine) was dehydrated with DCC in pyridine to give β-cyano-β-Ala (*1183*). A number of β-(1,2,4-oxadiazol-3-yl)-β-Ala derivatives, with substituents at the 5-position of the oxadiazole ring, were prepared by esterification of Fmoc-Asp(OtBu)-OH

with amidoximes, HON=C(R)NH₂. Treatment with a solution of sodium acetate in ethanol/water at 86 °C induced cyclization to the oxadiazole in 25–84% overall yield (*1184*).

11.2.3o Elaboration of α-Amino Acids: Other Methods

The side chains of several amino acids other than Asp have also been used to form the carboxyl group of β-amino acids. Cbz-homoserine lactone was converted into a Weinreb amide or pyrrolidinamide of homoserine and reacted with organolithium reagents to give α-amino ketones (see Scheme 11.83). The ketone was reduced to a methylene group and the side chain oxidized to form the acid (*1185*).

D-Thr was converted into the complex β-amino acid Adda via reduction of the α-carboxyl group to an aldehyde, followed by introduction of the remainder of the β-side chain by aldol addition/elimination. The carboxyl group was introduced by displacement of the original Thr hydroxyl group with cyanide (*226*). Another synthesis of Adda employed protected 2-amino-3-methyl-4-hexenoic acid, with the homochiral substituted allylglycine substrate prepared by an Ireland–Claisen rearrangement of the (R)-3-penten-2-ol ester of Boc-Gly. Reduction of the ester to a hydroxymethyl group, oxidative cleavage of the alkene, and oxidation of the hydroxymethyl group to an aldehyde gave methyl N-Boc 2-methyl-3-amino-3-formylpropanoate. Wittig coupling of the aldehyde with the Adda side-chain triphenylphosphorane produced a 2:1 E:Z-mixture of alkene product in 41% yield. The (Z) isomer could be isomerized by irradiation in the presence of catalytic iodine (*227*).

Other syntheses of β-amino acids from α-amino acids rely on moving the α-amino group to the β-position.

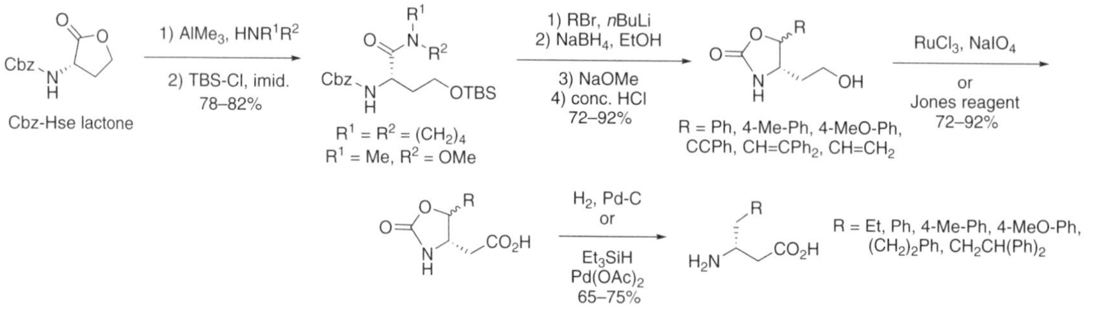

Scheme 11.83 Elaboration of Homoserine Lactone (*1185*).

Leucine was converted into β-Leu, 3-amino-4-methyl-pentanoic acid, by enzymatic treatment with leucine 2,3-aminomutase from *Clostridium sporogenes* (*1186*). (*S*)-Isoserine was prepared from L-Asn in 1976, with the α-hydroxy group derived from the α-amino group by treatment with sodium nitrite, and the β-amino group produced by a Hofmann rearrangement of the Asn side-chain amide (*1187*). *N,N*-Dibenzyl β-hydroxy-α-amino acid benzyl esters were stereospecifically converted into α-fluoro-β-amino acids upon treatment with DAST. Ser, Thr, *allo*-Thr, *threo*-3-hydroxy-Leu, and *erythro*-3-hydroxy-Leu, were all transformed in good yield (60–90%) into 2-fluoro-3-aminopropionate, 2-fluoro-3-aminobutanoate or 2-fluoro-3-amino-4-methylpen-tanoate. The reaction is proposed to proceed via fluoride opening of an aziridine intermediate (*355, 1188*). This method was applied to the preparation of (2*R*,3*S*)-[3-^{2}H$_1$]- and (2*R*,3*R*)-[2,3-^{2}H$_2$]-3-amino-2-fluoropropanoic acid from labeled *N,N*-dibenzyl Ser-OMe (*422*). (*R*)-Serine was converted into (2*S*)-3-amino-2-fluoropropionic acid by this route (*378*).

The β-amino acid component of AI-77B (2,3-dihydroxy-4-amino-1,6-hexanedioic acid) was derived from D-Glu via D-pyroglutaminol (*217, 218*). Another synthesis has been reported from homochiral 3,4,5-trihydroxynorvaline, protected as a lactone derivative (*1189*).

11.2.3p Elaboration of β-Amino Acids: Addition of Side Chain

Several routes have been developed to synthesize α-substituted-β-amino acids via α-alkylation of a β-amino acid enolate. Most of these methods rely on a cyclic template to induce high diastereoselectivity, but a number of alkylations of homochiral substrates have been described. For example, Estermann and Seebach simply α-alkylated the di-deprotonated enolate of homochiral *N*-protected 3-aminobutanoate and obtained diastereoselectivities from 75 to >99% de, depending on the electrophile and *N*-acyl group (*638, 1190*). Other homochiral β-substituted β-amino acids (obtained by Arndt–Eistert homologation of α-amino acids) have also been deprotonated and α-alkylated to give β^{2,3}-amino

acid products (*416, 1076*). A second α-alkylation to give α,α-disubstituted products (β^{2,2,3}-amino acids) is also possible (*1076*). α-Allylations of homologated Met, vinylglycine, allylglycine or 2-aminohex-5-enoic acid proceeded with complete diastereoselectivity (enroute to cyclic β-amino acids), as were α-methylations followed by α-allylations (*1105*). In another report Boc-protected methyl esters of homochiral β^3-amino acids were deprotonated with LDA and α-methylated. The reaction was not very diastereoselective, but both epimers were desired and could be resolved by column chromatography (*275*). β^{2,2,3}-Amino acids have been synthesized from *N*-Boc β^{2,3}-amino acid methyl esters via deprotonation with *n*BuLi/DIPA and methylation with MeI (*327*). Polylithiated β^3-tripeptides can be selectively alkylated on the *C*-terminal residue, with tetralithiated Boc-β-HVal-β-HAla-β-HLeu-OMe alkylated with methyl, benzyl, and allyl halides or *tert*-butyl bromoacetate in 35–80% yield (*1191*). Homologated β^3-amino acids derived from Boc-Ala, Boc-Val, and Boc-Leu were deprotonated and then alkylated with MeSCH$_2$Cl/NaI to introduce a methylthiomethyl α-substituent. Oxidation of the thioether to a sulfoxide and pyrolytic elimination at 80–90°C produced 2-methylene-3-substituted-3-aminopropionic acids, β-amino acid analogs of dehydroalanine (*326*).

Both α- and β-substituents were introduced to a β-Ala synthon prepared from Asn by dehydration of the amide to a nitrile and reduction of the Asn α-carboxyl group to a protected hydroxymethyl group. The β-amino acid α-substituent was introduced by enolate formation and alkylation adjacent to the nitrile, and the β-substituent added by nucleophilic displacement of a triflate derived from the hydroxymethyl group (as in Scheme 11.79) (*1160*). β^{2,3}-Amino acids with two hydroxymethyl (Ser analogs) or thiomethyl (Cys analog) substituents were derived from the Arndt–Eistert homologated Boc-Ser(Bn), or Boc-Cys(Bn), respectively. The L-Ser-derived (3*R*)-β^3-amino acid methyl ester was deprotonated with LDA/ZnBr$_2$ and reacted with formaldehyde to give (2*R*,3*R*)-product with >92% ee, while reaction of the enolate with methyl formate followed by NaBH$_4$ reduction gave a 1.2:1 mixture of (2*R*,3*R*):(2*S*,3*R*) isomers. In a similar manner the L-Cys derived (3*R*)-β^3-amino

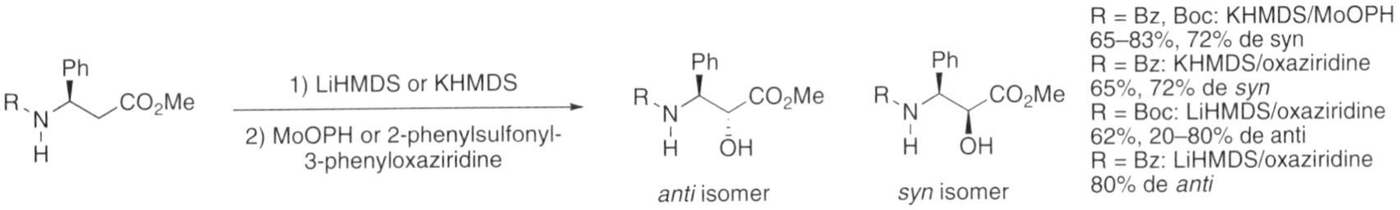

R = Bz, Boc: KHMDS/MoOPH
65–83%, 72% de syn
R = Bz: KHMDS/oxaziridine
65%, 72% de syn
R = Boc: LiHMDS/oxaziridine
62%, 20–80% de anti
R = Bz: LiHMDS/oxaziridine
80% de anti

1) LiHMDS or KHMDS

2) MoOPH or 2-phenylsulfonyl-
3-phenyloxaziridine

anti isomer   *syn* isomer

Scheme 11.84 Hydroxylation of 3-Amino-3-Phenylpropionic Acid (*1192*).

acid methyl ester was deprotonated with LDA and alkylated with BnSCH$_2$Cl to produce (2S,3R) product with 60% de (*294*).

The 3-phenylisoserine substituent of Taxol was prepared by a diastereoselective hydroxylation of the enolate of homochiral N-protected methyl 3-amino-3-phenylpropionate (see Scheme 11.84). KHMDS and MoOPH led to the correct *syn* diastereomer (72% de), while the *anti* isomer was produced with LiHMDS and 2-phenylsulfonyl-3-phenyloxaziridine (80% de) (*1192*). (2R,3S)-Phenylisoserine has also been synthesized from N-Bz methyl (3R)-3-amino-3-phenylpropionate by deprotonation with LiHMDS, iodination, and intramolecular displacement by the benzoyl carbonyl oxygen to form an oxazoline. Hydrolysis gave the acyclic product (*605*). A series of β^3-amino acids, obtained by homologation of α-amino acids (Phg, Phe, O-benzyl Ser), were N,N-diprotected with 4-methoxybenzyl chloride, and then deprotonated with KHMDS and α-hydroxylated with 2-[(4-methylphenyl)sulfonyl]-3-phenyloxaziridine. The hydroxylations proceeded with high diastereoselectivity (82–94%) and yield (89–95%) (*1193*). A fluoro analog of phenylisoserine was prepared via electrophilic fluorination of the dianion of methyl (R)-(N-benzoyl)-3-amino-3-phenylpropanoate, although the best diastereoselectivity obtained was only 62% de (*989*). Better diastereoselectivity was produced during electrophilic hydroxylation reactions (*1192*). The enolates of several β-substituted-β-amino acids were sulfenylated with >90% de *anti* diastereoselectivity using 2,4-dinitrophenyl 4-methoxybenzyl disulfide (*2188*). Similarly, treatment of the dianion of Boc-β-amino-L-butyric acid methyl ester with phenylselenyl bromide gave the (S)-α-phenylselenyl derivative, which could be exchanged with an allyl substituent with retention of stereochemistry by free-radical allylation (*331*).

Other alkylations of optically active β-amino acids have relied upon cyclic derivatives. A homochiral lactone of 3-amino-4,5-dihydroxypentanoic acid was derived from D-mannitol. Alkylation at the α-center proceeded

with 50 to >90% de (*778*). Similarly, the homochiral 3-amino-γ-butyrolactone obtained by reduction of the α-carbonyl group of Cbz-D-Asp anhydride was deprotonated and α-methylated with 60% de. The predominant (2R,3R) isomer was epimerized to the desired (2S,3R) lactone by heating in benzene/Et$_3$N (see Scheme 11.82) (*1178, 1179*). The symmetrical β-dicarboxylic acid 3-amino-1,5-pentanedioic acid was enzymatically resolved to give a monomethyl ester. The ester was reduced to an alcohol with NaBH$_4$, cyclized to a δ-lactone, and then used for an aldol addition to acetone, after which the lactone was hydrolyzed to give 2-(hydroxyisopropyl)-3-amino-5-hydroxypentanoic acid (*1194*).

Acyclic β-amino acids have been derivatized with chiral auxiliaries and alkylated. α-Methylisoserine was synthesized via methylation of the enolate of an isoserine equivalent (2-acetoxy-2-cyanopropionate) derivatized with the Oppolzer dicyclohexylsulfamoylisobornyl amide auxiliary (*1195*). The crude diastereoselectivity was only 54%, but diastereomerically pure product was isolated in 71% yield. The Oppolzer ester auxiliary was employed during alkylations of 2-cyano-2-phenylpropanoic acid, giving crude products with 60–94% de, which improved to >96% de upon recrystallization. Hydrogenation of the cyano group revealed the masked aminomethyl functionality (*1196, 1197*). Better results were obtained by using Oppolzer's sultam amide auxiliary in combination with the benzophenone Schiff base of β-Ala (see Scheme 11.85); deprotonation with LDA and benzylation, methylation, or isobutylation proceeded with complete diastereoselectivity (*1198*).

The Oppolzer auxiliary was also employed with an O-benzyl oxime of 3-oxopropionate (see Scheme 11.86). An α-substituent was introduced with high diastereoselectivity (95% de after recrystallization) by deprotonation under phase-transfer catalysis conditions. A β-substituent could then be added via a radical addition to the oxime; little diastereoselectivity (5% de) was observed in the absence of an α-substituent, but with an

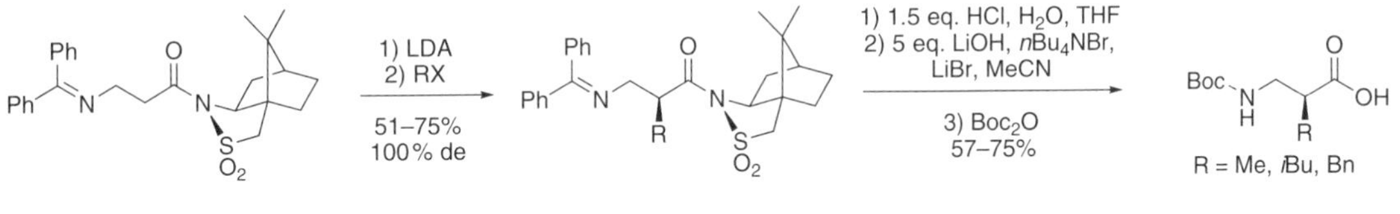

1) LDA
2) RX

51–75%
100% de

1) 1.5 eq. HCl, H$_2$O, THF
2) 5 eq. LiOH, nBu$_4$NBr,
LiBr, MeCN

3) Boc$_2$O
57–75%

R = Me, iBu, Bn

Scheme 11.85 Alkylation of β-Ala Oppolzer Sultam Amide (*1198*).

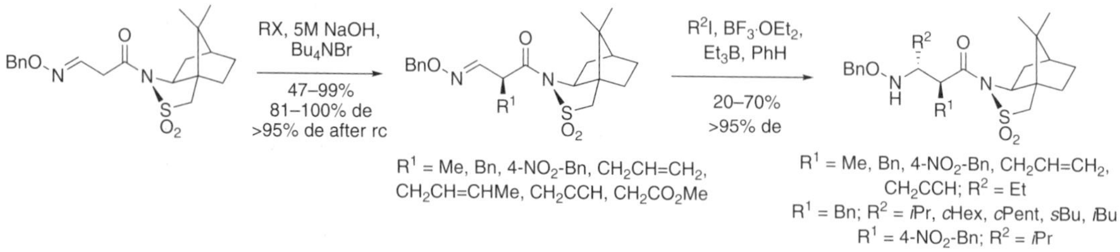

Scheme 11.86 Alkylation and Radical Addition of β-Ala Oppolzer Sultam Amide Oxime (*1199*).

α-substituent present the α,β-disubstituted-β-amino acids were produced with >95% de (*1199*).

Evans' benzyl-substituted oxazolidinone auxiliary was acylated with N-alkenyl,N-Boc β-Ala, and then deprotonated with NaHMDS and allylated with allyl iodide (68–71%, 87–89% de) (*641*). A new 4-isopropyl-5,5-diphenyl variant of Evans' oxazolidinone auxiliary (DIOZ) was tested during alkylation of N-Phth β-Ala (*1030*). The titanium enolate reacted with 1,3,5-trioxane to introduce a hydroxymethyl group, forming 2-(aminomethyl)-3-hydroxypropionic acid (the β²-analog of Ser) in 92% yield. Alkylation with ClCH₂OBn proceeded in lower yield (*832, 833, 1030*). The hydroxymethyl Ser side chain could be converted into a Cys thiomethyl substituent via Mitsunobu reaction (*832, 833*). β-Ala derivatives with the DIOZ oxazolidinone auxiliary were also employed for aldol reactions with acetaldehyde (giving a β²-Thr analog) or 3-formyl-indole, with the latter dehydroxylated to give a β²-Trp derivative (*348, 832*).

A matched pair of chiral auxiliaries was used for diastereoselective alkylation of di-deprotonated β-alanine. An N-(S)-(1-phenylethyl) substituent combined with (R,R)-bis(1-phenylethyl)amide induced 60–90% de for alkylations with methyl iodide, ethyl iodide, propyl iodide or benzyl bromide. Deprotection yields were variable (13–65%) (*1200*). Another pair of chiral auxiliaries, a menthyl ester and a 2-hydroxypinanone Schiff base, were also used for alkylation of β-alanine. Methylation proceeded in 86% yield with 84% de from the matched pair (*1201*).

A much more effective strategy for α-alkylations of β-Ala was reported by Nagula et al., with β-Ala amidated by the same pseudoephedrine chiral auxiliary successfully applied to the synthesis of α-amino acids

by Myers et al. (see Section 6.2.1b). Deprotonation and alkylation proceeded with good yield and high stereoselectivity, providing a general route to α-alkyl-β-amino acids (see Scheme 11.87). The final products were obtained by hydrolysis, with 52–74% overall yields from the β-Ala pseudoephedrine amide and 75 to >99% ee (with no purification of the intermediate diastereomers) (*705*). This route was successfully applied to the preparation of hundred gram quantities of product.

An acylated chiral oxazolidinone, with the amino group masked as a vinyl or dimethoxyphenyl substituent, provided another chiral β-Ala α-enolate equivalent (see Scheme 11.88). The oxazolidinone was deprotonated and alkylated, with the amino group revealed by oxidation to form a carboxyl group followed by a Curtius rearrangement (*1202*).

A number of cyclic β-alanine equivalents have been developed which contain chiral auxiliaries to induce asymmetry during alkylation. Juaristi and Seebach prepared a cyclic β-amino acid synthon analogous to the Boc-BMI template used for α-amino acid synthesis. Initially, a racemic 2-tert-butylperhydropyrimidin-4-one template was reported, which was alkylated with electrophiles to give α-substituted-β-amino acids with 86 to >96% de (see Scheme 11.25, above) (*637*). The analogous homochiral molecule was then prepared from (S)-Asn by condensation with pivalaldehyde, followed by removal of the Asn α-carboxyl group and N-protection (see Scheme 11.89) (*1203, 1204*). Deprotonation and alkylation with alkyl halides proceeded in 75–80% yield with >95% de. Hydrolysis with 6 N HCl in a sealed tube at 90–100 °C gave the free amino acids in 80–85% yield (*1204*). The initial *trans* adducts could be epimerized to the *cis* diastereomers by deprotonation with LDA and reprotonation, which proceeded in 85–90% yield.

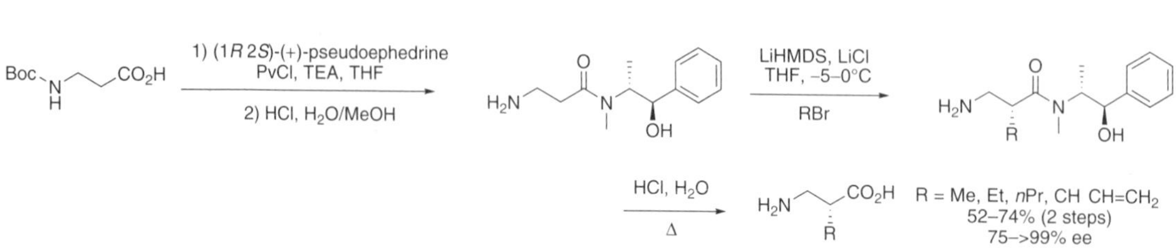

Scheme 11.87 Alkylation of β-Alanine Pseudoephedrine Amide (*705*).

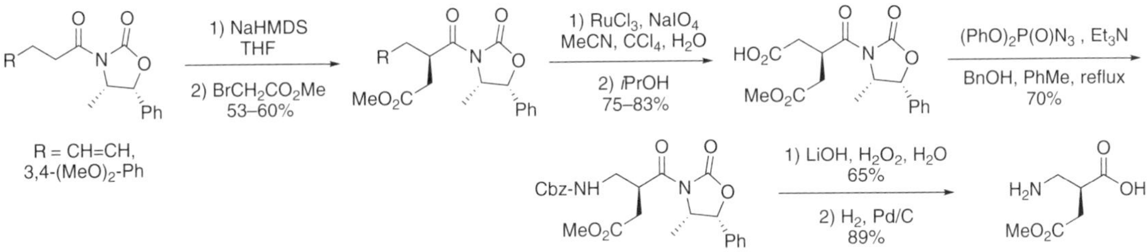

Scheme 11.88 Alkylation of β-Alanine α-Enolate Equivalent (*1202*).

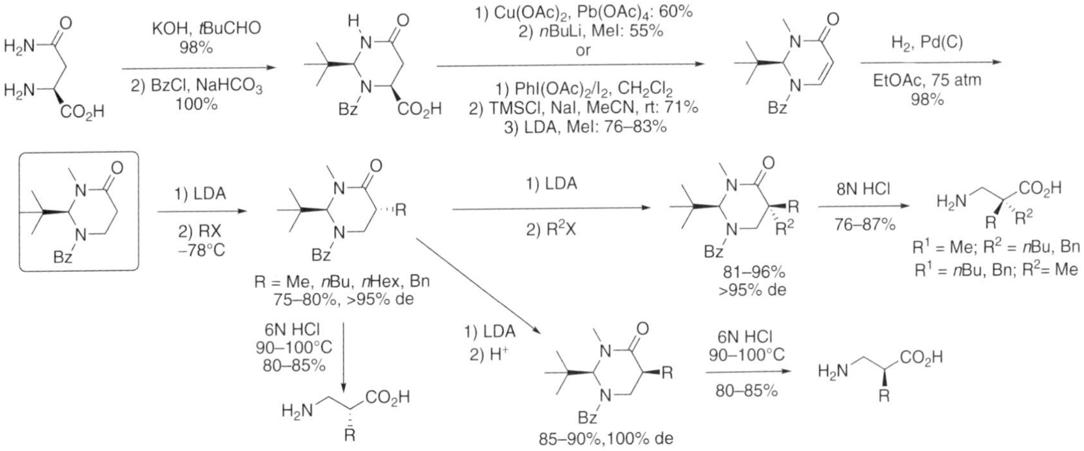

Scheme 11.89 Alkylation of Asn-Derived Tetrahydropyrimidinone Template (*1203–1207*).

Hydrolysis resulted in the enantiomeric β-amino acids to those obtained by hydrolysis of the initial alkylation adducts, with both enantiomers derived from a single template enantiomer (*1204*). α,α-Disubstituted-β-amino acids were also prepared by consecutive deprotonations and alkylations of the template (*1205*). The effect of the *N*-acyl group on conformation and stereoselectivity has been examined (*1206*). An improved synthesis of the template from L-Asn was reported in 2003, with 46–50% overall yield (*1207*).

The same 2-*tert*-butylperhydropyrimidin-4-one template, but with an additional β-methyl group, was prepared by resolving (*R*)- and (*S*)-3-aminobutanoic

acids and using them to form the templates (the *cis* template isomers predominated over the *trans* by 95:5, and were readily separated by chromatography) (*712*). Deprotonation with LDA and alkylation with MeI or BnBr gave a single *cis, trans* diastereomer in each case (see Scheme 11.90). Hydrolysis with 6 N HCl gave the α,β-disubstituted-β-amino acids (*712*). The 3-amino-butanoic acid template was also prepared without the bulky *tert*-butyl substituent by using paraformaldehyde instead of pivalaldehyde during the cyclization, but alkyl-ation diastereoselectivity was not as impressive (50% de) (*1208*). The template enolate derived from homochi-ral (*S*)-3-amino-3-phenylpropionate was hydroxylated

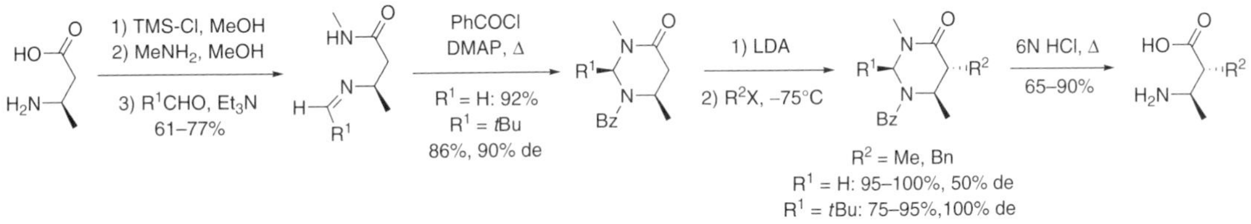

Scheme 11.90 α-Alkylation of 3-Aminobutanoic Acid in Tetrahydropyrimidinone Template (*712, 1208*).

with (+)-(camphorsulfonyl)oxaziridine to provide a single diastereomer, which was hydrolyzed to phenyliso-serine (*719*).

The 2-*tert*-butylperhydropyrimidin-4-one template has also been used for radical couplings (*1209*). An N-(2-halo-benzoyl) protecting group was employed, with the halogen ensuring a 1,5-hydrogen translocation reaction to form the radical at the β-position of the β-Ala within the template. Quenching with electophilic olefins gave the addition products in 50–59% yield with 97–98% de. An unsaturated version of this template was derived from L- or D-Asn (see Scheme 11.91). Pd(0)-catalyzed arylation with aryl iodides gave (R)-β-Tyr (a component of the marine cyclodepsipeptide (+)-jasplakinolide) (*128, 1141, 1142, 1210*); conjugate addition of organocuprates was less successful (*1142*). The C6 alkene position of a modified version (N-methyl, N'-MOM) of this heterocycle can be deprotonated and alkylated (see Schema 11.91). Reduction with sodium cyanoborohydride gave the *cis*-substituted template with high (>95%) diastereoselectivity, with concomittant reduction of the MOM group (*1211*). Demethylation was achieved by refluxing with vinyl chloroformate followed by treatment with ethanolic HCl, with acid hydrolysis liberating the free α-substituted-β-amino acids. A cyclic template containing an achiral 3-amino-4-carboxybut-2-enoic acid residue has been deproto-nated and α-alkylated (34–40% de), and then hydrogenated (>95% de), to provide 2-substituted-3-amino-4-carboxybutanoic acid products (*1212*).

β-Ala and homochiral β³-amino acids have been con-verted into a version of the 2-*tert*-butylperhydropyrimi-din-4-one template, and then treated with the Meerwein salt Me₃OBF₄ to form 1-Boc-2-*tert*-butyl-4-methoxy-5,6-dihydro-2H-pyrimidine (see Scheme 11.92). Enantiopure template could be derived from enantiopure β³-amino acid, while the unsubstituted template derived from β-Ala required resolution. Deprotonation with LDA and alkylation with alkyl halides gave the adducts in good yield with >98% de, even with the presence of a β³-substituent. A second alkylation produced β²,²-precursors, again with complete stereoselectivity. Aldol reactions and additions to imines or α,β-usaturated

esters were also successful. Deprotection was readily accomplished with 0.1 N TFA, followed by Boc protec-tion (12–91% yield). The Cbz- and Aloc-protected tem-plates could also be alkylated. A number of the examples reported used the racemic template, but all will be con-sidered as asymmetric syntheses in this summary (*1213*).

Cardillo, Tomasini, and co-workers have also reported α-alkylations of a perhydropyrimidin-4-one template with a β-substituent, leading to α,β-disubsti-tuted-β-amino acids with high *trans* stereoselectivity during the alkylation (see Scheme 11.93). The homochi-ral template was obtained by chromatographic separa-tion of diastereomers formed with an N-(S)-phenethyl substituent (*1214, 1215*). Alternatively, an Hg-catalyzed intramolecular addition of an amino group to the β-position of 3-butenoic acid was used to prepare cyclic protected 6-methylperhydropyrimidin-4-one diastereomers of 3-aminobutyric acid as a 29% de mix-ture (*771*). The template derived from 3-aminobutyric acid was employed for aldol reactions as well as alkyla-tions, with complete stereoselectivity at the template center and from 70% de *anti* to 60% *syn* selectivity at the carbinol center depending on reaction conditions (*1216*). α-Hydroxylation of the enolate proceeded with >98% de, depending on the hydroxylating reagent (*1217*). Template hydrolysis was accomplished by refluxing in 6 N HCl.

Alkylations of both acyclic and cyclic electrophilic β-Ala equivalents have also been reported. An electro-philic template derived from (S)-phenylglycinol was described in 1999. Organocuprates added with high (>95% de) stereoselectivity, resulting in β³-amino acid precursors (see Scheme 11.94) (*1218*). A β²-substituent could then be introduced by deprotonation/alkylation (*1219*). The template was decomposed via hydrogena-tion, basic hydrolysis, and a second hydrogenation.

A chiral cyanohydrin with a masked carboxyl group, produced enzymatically, was converted into α-hydroxy-β-amino acids by addition of the β-substituent (as Grignard reagent) to the cyano group, followed by reduction of the resulting imine with sodium borohydride (see Scheme 11.95). Moderate diastereoselectivity was

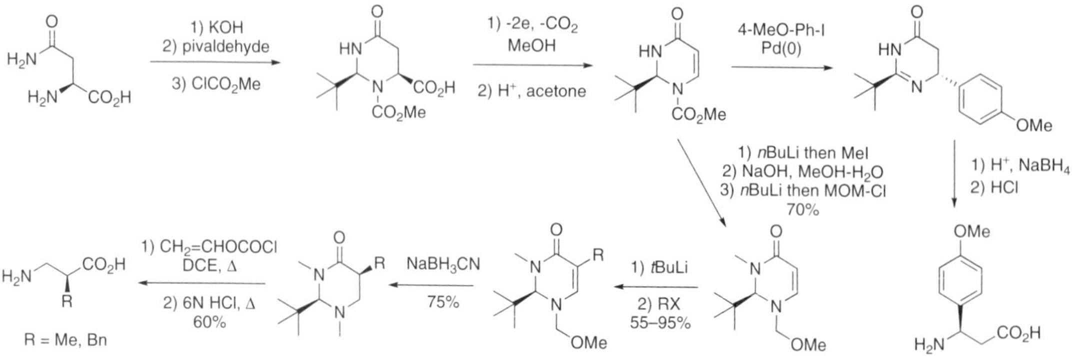

Scheme 11.91 Synthesis of β-Amino Acids via 2,3-Dihydropyrimidin-4-(1H)-one (*128, 1141, 1142, 1210, 1211*).

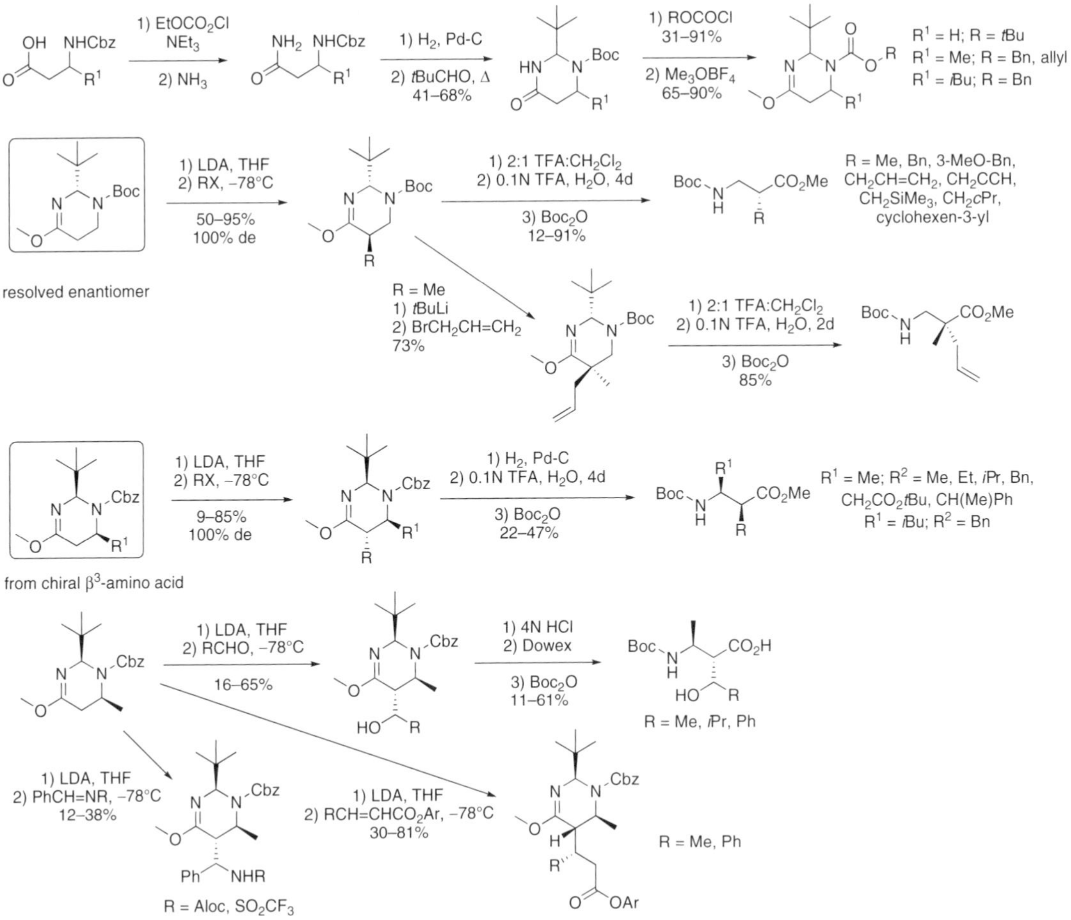

Scheme 11.92 Alkylation of β-Ala in 5,6-Dihydro-4-methoxy-2H-pyrimidine Template (*1213*).

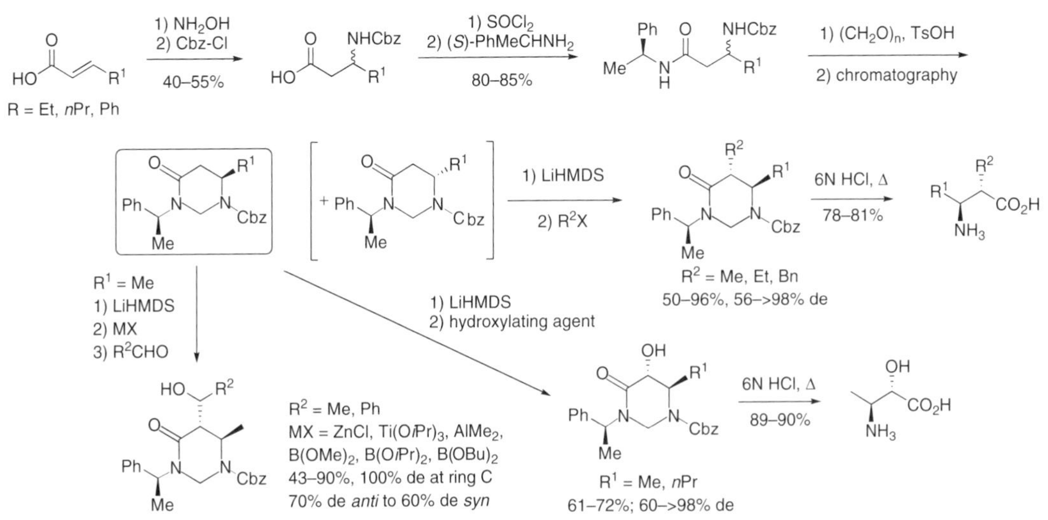

Scheme 11.93 α-Alkylation of β-Alanine in Tetrahydropyrimidinone Template (*1214–1217*).

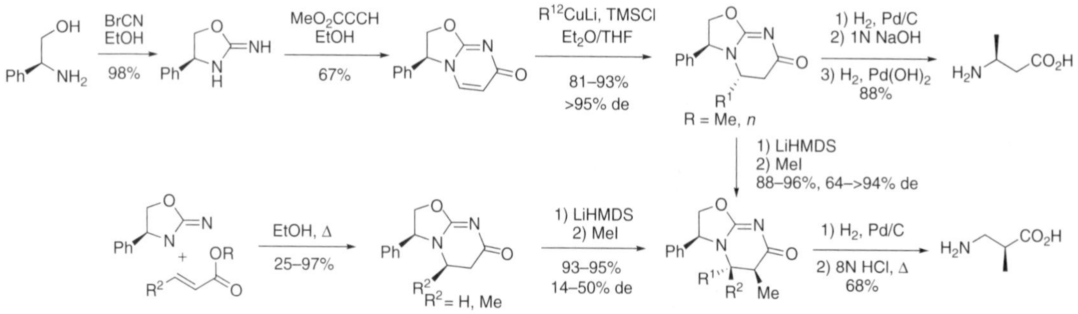

Scheme 11.94 Electrophilic β-Amino Acid Template from Phenylglycinol (*1218, 1219*).

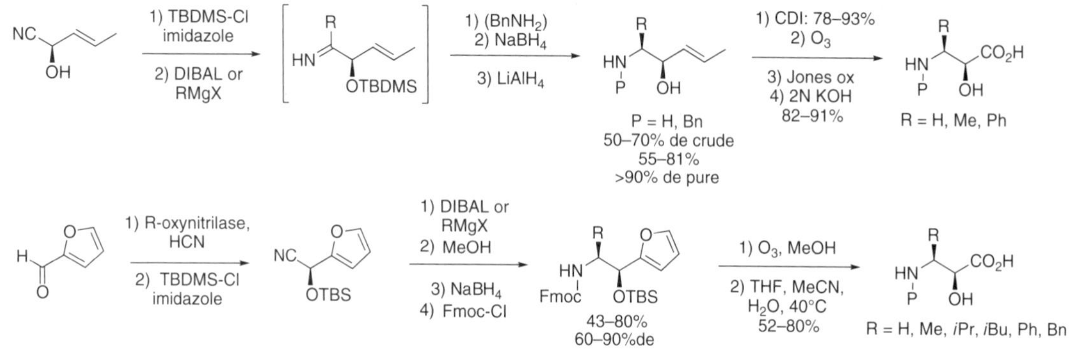

Scheme 11.95 Side Chain Addition to Chiral Cyanohydrin (*1220, 1221*).

obtained during the reduction (50–70% de), which was improved by crystallization of the crude product. The amino and hydroxyl groups were temporarily protected as an oxazolidinone during oxidative cleavage of the alkene to reveal the masked carboxylic acid (*1220*). A modification to this procedure employed a furyl group as the masked carboxyl equivalent. The chiral cyanohydrin substrate was produced using the (*R*)-oxynitrilase contained in a suspension of 30 g of defatted almond meal to convert 200 mmol of 2-furaldehyde, giving 26 g of crude product with 98.6% ee. Grignard addition/reduction proceeded with improved 60–90% de (*1221*).

Another electrophilic equivalent with a masked carboxyl group was prepared from an imine of D-glyceraldehyde (*1222*). Methylmagnesium bromide added

with complete stereoselectivity, giving 3-amino-2-hydroxybutanoic acid after protecting group manipulation and hydroxymethyl oxidation (see Scheme 11.96). An electrophilic imine prepared from 2,3-isopropylidene-D-threose and benzylamine was used for a synthesis of cyclohexylnorstatine via addition of cyclohexylmethylmagnesium bromide in the presence of cerium(III) chloride, which proceeded with complete diastereoselectivity. Addition of cyclohexylmethylcopper(I) gave the other diastereomer, also with complete stereoselectivity. Oxidative diol cleavage generated the carboxyl group at the end of the synthesis (*1223*). In a similar fashion, a benzylamine Schiff base derivative derived from L-threose was reacted with 2 equiv of phenyllithium to give the *syn* addition product with 72% de.

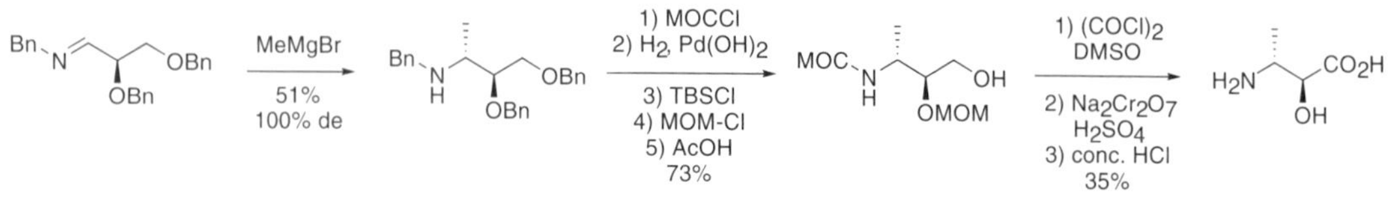

Scheme 11.96 Synthesis of α-Hydroxy-β-Amino Acids from D-Glyceraldehyde (*1222*).

Elaboration to generate the carboxy group provided (S)-β-phenyl-β-Ala (95).

Another chiral electrophilic synthon has been prepared by condensing a C2 symmetrical amine with 2-bromomethylacrylate (see Scheme 11.97). The alkene synthon underwent Michael addition of enolates, with the chiral amine substituent inducing 98% de at the α-center for the one example presented (1224).

A more versatile synthon is the nitrone of 2,3-O-isopropylidene-D-glyceraldehyde. Grignard reagents added with 56–82% de syn stereoselectivity in the presence of ZnBr₂, and 40–90% anti selectivity in the presence of Et₂AlCl. Selective oxidation of the terminal hydroxy group provided either diastereomer of (2S)-3-substituted isoserines, such as phenylisoserine (1225). Similarly, organocerium reagents added to a hydrazone prepared from 3,3-ethylenedioxypropanal and Enders' SAMP auxiliary with high (75–99% de) stereoselectivity (see Scheme 11.98). Ozonolysis of the acetal generated the carboxyl group (1226).

Both nucleophilic and radical alkylations were employed to construct 2,3-disubstituted-β-amino acids (see Scheme 11.99). A camphorsultam amide auxiliary was employed to induce asymmetry during PTC alkylations of the O-benzyl oxime of 3-oxopropionic acid, introducing the C2-substituent with >95% de. Radical addition to the oxime ether also proceeded with >95% de, allowing for selective introduction of both C2- and C3-substituents (1227).

Chiral catalysts have been employed for a number of β-amino acid alkylations. Conjugate additions of organomagnesium derivatives to enamidomalonates in the presence of a bisoxazoline chiral catalyst introduced a β-substituent with 78–93% ee (see Scheme 11.100). Decarboxylation gave the β-amino acids (1228).

Nitropropene dimethylacetals, NO₂CH=CHCH(OMe)₂, act as another masked electrophilic β-alanine equivalent. Conjugate addition of dialkylzinc reagents in the presence of chiral Cu-phosphoramidite catalysts proceeded with 88–98% ee. Nitro reduction by hydrogenation over

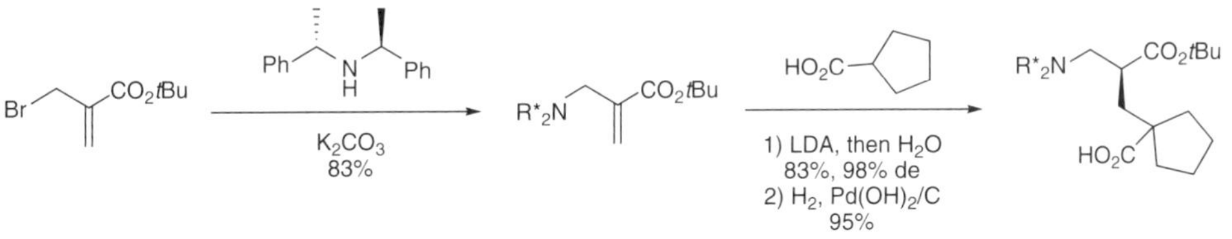

Scheme 11.97 Michael Addition to Chiral 2-Aminomethylacrylates (1224).

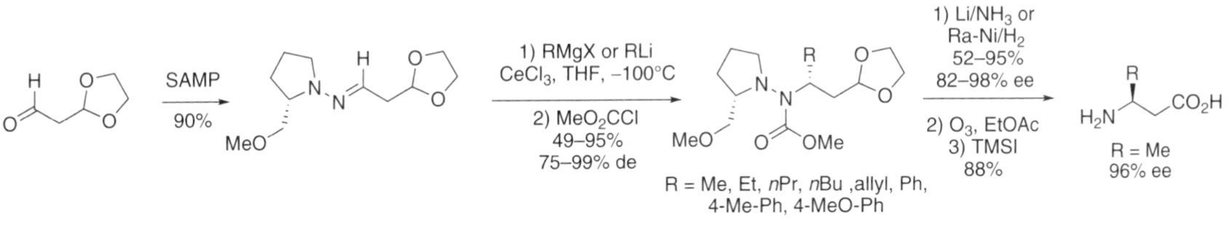

Scheme 11.98 Addition of Organocerium Reagents to Chiral Hydrazone (1226).

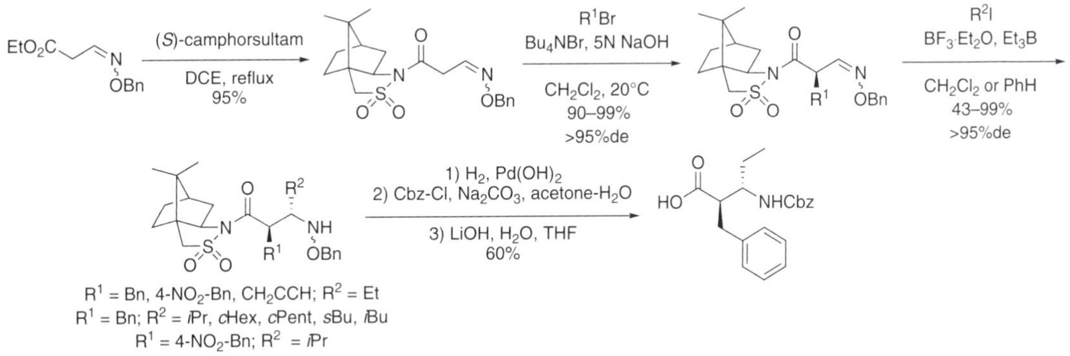

R¹ = Bn, 4-NO₂-Bn, CH₂CCH; R² = Et
R¹ = Bn; R² = iPr, cHex, cPent, sBu, iBu
R¹ = 4-NO₂-Bn; R² = iPr

Scheme 11.99 Introduction of Both C2- and C3-Substituents Using Sultam Auxiliary (1227).

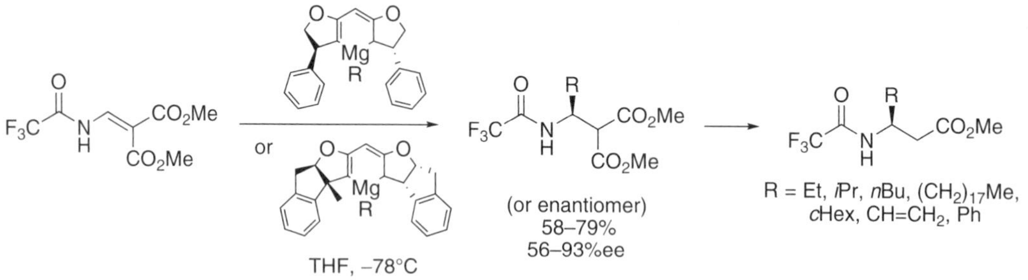

Scheme 11.100 Asymmetric Addition of Organomagnesium Reagents to Enamidomalonates (*1228*).

Raney Ni, amine protection with Boc$_2$O, acetal hydrolysis, and aldehyde oxidation produced the α-substituted-β-amino acids (*2181*). A similar report the same year added three dialkylzinc reagents to methyl nitroacrylate with 94–97% yield and 15–85% ee, avoiding the need for deprotection/oxidation (*1229*).

Racemic α-methylene-β-phthalimido esters underwent asymmetric radical additions using alkyl halides in the presence of MgI$_2$, bisoxazoline chiral catalysts, BEt$_3$ and O$_2$ (see Scheme 11.101). A series of α-substituted-β-amino esters were produced with 34–98% ee. One example was deprotected (*1230*).

11.2.3q Elaboration of β-Amino Acids: Other Methods

Modifications of β-amino acid side chains are possible. The hydroxy group of β-homoserine, 3-amino-4-hydroxybutyric acid, was displaced by nucleobases under Mitsunobu conditions to give 3-amino-4-(nucleobase)-butyric acid derivatives, analogs of peptide nucleic acids (*1231*). The alkene of 3-aminohex-5-enoic acid was iodinated with I$_2$/KI and then treated with NaN$_3$ to give a 62% yield of the lactone of 3-amino-5-hydroxy-6-azidohexanoic acid, as a mixture of isomers epimeric at the hydroxy center. The azido group was reduced to an amine, temporarily protected with 2-nitrobenzenesulfonamide, and alkylated with several alcohols under Mitsunobu conditions, or guanylated, or benzylated. Alternatively, instead of employing azide to react with the iodinated 3-aminohex-5-enoic acid, secondary amines were used, including morpholine, N-Boc piperazine, and 3-(Boc-amino)pyrrolidine (*167*).

Treatment of protected derivatives of (S,S)-2-hydroxy-3-aminobutanoic acid with DAST gave (S,S)-2-fluoro-3-aminobutanoic acid (with retention of configuration), via an intermediate aziridinium ion. Alternatively, oxidation followed by DAST treatment produced 2,2-difluoro-3-aminobutanoic acid (*355*). Dehydration of the amide group of protected L-isoasparagine (H-Asp-NH$_2$) with DCC gave L-β-cyano-β-Ala (*1232*). An enzymatic desymmetrization of dimethyl 3-benzylaminoglutarate by *Candida antarctica* lipase B produced the (S)-monoamide in 76% yield. Hofmann rearrangement then gave 3,4-diaminobutanoic acid (*689*). The S-methyl group of homologated Met has been removed by treatment with excess Na/NH$_3$, leaving a thioethyl side chain (*317*). The phenol of 3-(4-hydroxyphenyl)-β-Ala was converted into a triflate and used for Suzuki couplings with aryl boronic acids to form 3-biaryl-β-Ala derivatives (*406*). The side-chain acid group of N-Cbz 3-amino-4-methoxycarbonyl-butyric acid was reduced to an aldehyde, and then used for a Wittig reaction to give the (Z)-alkene in 3-amino-5-tetradecenoic acid, the β-amino acid contained in lipogrammistin-A, a lipophilic ichthyotoxin secreted by the skin of soapfish (*84, 85*).

Racemic 2-methylene-3-aminoalkanoates have been kinetically resolved by asymmetric hydrogenation using a chiral DIPAMP Rh catalyst, producing 2-methyl-3-aminoalkanoic acids (*1233*). In 1989 Noyori reported the preparation of enantiomerically enriched *threo*-N-benzoyl 2-(aminomethyl)-3-hydroxybutyric acid methyl ester by dynamic resolution/reduction of the ketone of racemic N-benzoyl 2-(aminomethyl)-3-ketobutyric acid methyl ester using Ru–BINAP-catalyzed hydrogenation. The ester stereomutated quickly enough, and the catalyst was selective enough, that 100% conversion to a single chiral enantiomer was possible, avoiding the 50% loss normally observed with enzymatic resolutions.

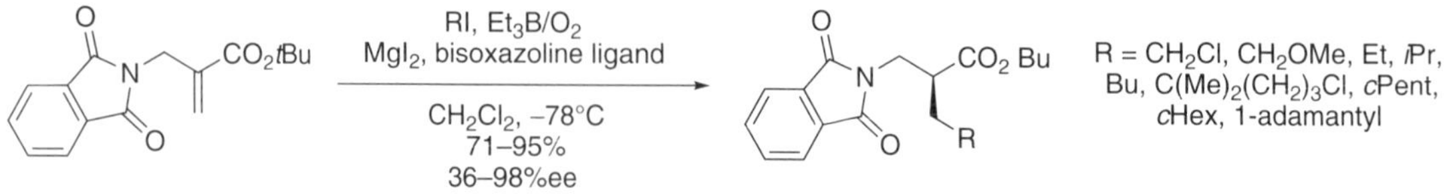

Scheme 11.101 Radical Addition to α-Methylene-β-Phthalimido Esters (*1230*).

The product was obtained with 98% ee and 88% de *syn* stereoselectivity (*1234*). A synthesis of the Taxol phenylisoserine side chain utilized a microbial reduction of the carbonyl group of racemic 2-keto-3-(*N*-benzoylamino)-3-phenylpropionic acid. A variety of organisms were screened, with the best conditions giving yields of 80% with 80% de and >98% ee (*693*). Baker's yeast reduction of 3-keto-4-phenyl-β-lactam produced the desired (3*R*,4*S*)-isomer of phenylisoserine β-lactam in up to 41% yield with 82% ee; Baker's yeast reduction of racemic methyl 2-keto-3-azido-3-phenylpropionate gave only the (*R*)-alcohol center, but no diastereoselectivity for the azido-center configuration (*694*). Bioconversion of Taxol by *Streptomyces* sp. MA7065 hydroxylated the *para*-position of the phenylisoserine benzene ring in 10% yield (*190*).

11.2.3r Small Ring Systems: Aziridines

Homochiral aziridine-2-carboxylates undergo ring opening to give β-amino acids, although regioselectivity is generally poor and favors formation of the α-amino acid product (*1235*) (see Section 8.2). However, NaBr caused somewhat regioselective C2 opening of 3-substituted-*N*-ethoxycarbonyl aziridine-2-carboxylates (75:25 to 99:1 C2:C3 attack) to give 3-amino-2-bromo acid products. The bromo atom could be removed by radical reduction (*654*). By using NaX/Amberlyst 15 in acetone at −30 °C, *N*-ethoxycarbonyl aziridine-2-carboxylates were opened to give α-halo-β-amino derivatives in good yields (*804*). This method, coupled with radical dehalogenation, was used for a synthesis of 3-aminodecanoic acid (*804*). *cis*-3-Hydroxymethylaziridine-2-carboxylate was opened with 4 N HCl or 30% HBr to give 2-halo-3-amino-4-hydroxy-butanoic acid. The bromo derivative was again radically debrominated (*1236*). Thiols such as thiophenol, cysteine, and glutathione opened aziridine-2-carboxylates with

moderate to good C2 regioselectivity when reacted in aqueous solution (*1237*). The 3-methylaziridine-2-carboxylate derived from *N*-Ts-Thr reacted with organo-cuprates predominantly at C2, although in low yield (*1238*). Reductive opening of *N*-unsubstituted-3,3-dialkyl-aziridine-2-carboxylates by H₂/Raney Ni provided 3,3-disubstituted-β-amino acids. The chiral substrates were derived from 2*H*-azirine-2-carboxylates via Grignard addition of the second 3-alkyl substituent (*1239*).

Aziridine-2-carboxylate equivalents in which the carboxyl group is cyclized to form a lactone with an alcohol contained on a C3 side chain (see Scheme 11.102) were found to exhibit considerably different reaction profiles compared to conventional aziridine-2-carboxylates (*1240*). "Soft" nucleophiles such as 1-methylindole, thiols, acetate, and halide reacted exclusively at C2 in almost all cases to give the β-amino acid products, while "hard" nucleophiles such as alcohols opened the lactone before the aziridine. The chiral aziridine was derived from D-ribose.

N-Tosyl aziridines with a masked carboxyl group, a 2-hydroxymethyl substituent, were opened with organocuprates. Oxidation of the hydroxymethyl group formed a β-amino acid, 2-ethyl-3-amino-8-hydroxy-octanoic acid (*1241*). The chiral epoxy diol 2,3-epoxy-1,5-pentanediol, obtained by a Sharpless epoxidation, has been converted into an *N*-tosyl 2,3-aziridyl-1,5-pentanediol derivative. Ring opening with an ethyl organocuprate and oxidation of one hydroxymethyl substituent gave 2-ethyl-3-amino-5-hydroxypentanoic acid (*1242*). Homochiral *N*-Boc 2-(1,2-dihydroxyethane)aziridine was opened with organocuprates in good yield, giving the *threo* diastereomer (see Scheme 11.103). Protection as an oxazolidinone and hydroxymethyl oxidation, or direct oxidation with oxygen in the presence of platinum oxide, gave the α-hydroxy-β-amino acids norstatine, cyclohexylnorstatine, AHPBA, and AHMHA (*1243*).

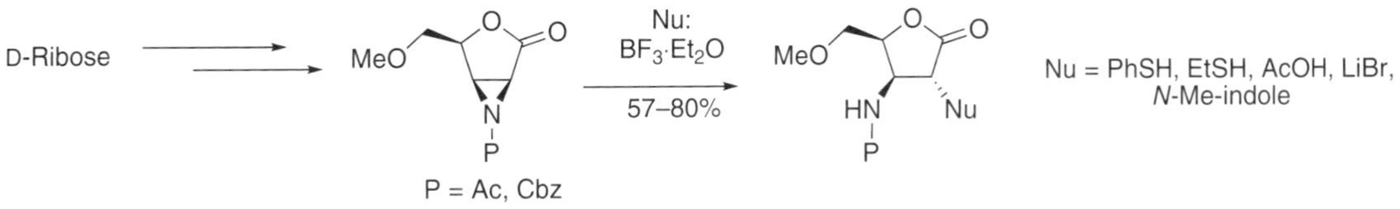

Scheme 11.102 Opening of Bicyclic Aziridine-Lactone (*1240*).

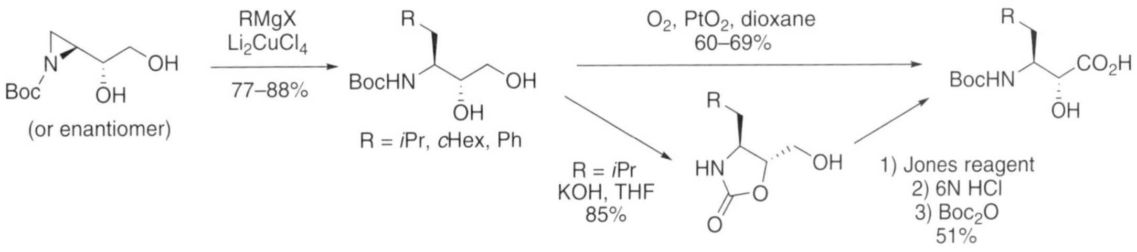

Scheme 11.103 Reaction of *N*-Boc 2-(1,2-Dihydroxyethane)-Aziridine (*1243*).

Hydride opening of 2-substituted aziridine-2-carboxy-lates proceeded with C2-regioselectivity if the acid was reduced to a hydroxymethyl group (due to coordination of the Al-based reagent by the hydroxyl), producing *syn* 1,3-amino alcohols. Oxidation regenerated the carboxyl group (*1244*). In a similar manner, Red-Al regioselec-tively opened a mono *O*-protected 2,3-bis(hydroxy-methyl)aziridine, with the unprotected alcohol oxidized to form 3-amino-4-hydroxybutyric acid (*1245*).

Another aziridine synthon has been derived from (*R*)-glyceraldehyde acetonide. Organocuprate opening of the bicyclic aziridine-oxazolidinone produced an amino alcohol oxazolidinone with a free hydroxymethyl group, which was oxidized to generate the carboxyl group (see Scheme 11.104). Coupling to Leu-O*t*Bu before oxazolidinone deprotection gave bestatin and analogs (*203*).

Another masked carboxyl group is a vinyl substitu-ent; optically active 2-vinylaziridines were reacted/opened with various aldehydes in the presence of InI and Pd(PPh$_3$)$_4$ to give *syn,syn*-2-vinyl-1,3-amino alcohols. Ozonolysis of the vinyl group gave 2-(1-hydroxy-1-arylmethyl)-3-alkyl-β-amino acids (*1246*).

An alternative approach to β-amino acid synthesis from aziridines is to open alkyl-substituted aziridines with a carboxylic acid equivalent. Homochiral *N*-tosyl 2-alkylaziridines were treated with a lithiated dithiane derivative of formyltrimethylsilane to give an acyl silane dithiane product. Dithiane removal and acyl silane hydrolysis gave *N*-tosyl 3-amino-4-phenylbutanoic acid and 3-amino-5-methylhexanoic acid (*1247*). Optically active β-aminonitrile precursors of β-amino acids were obtained from homochiral 2-substituted *N*-tosyl aziri-dines via ring opening with cyanotrimethylsilane in the presence of Yb(CN)$_3$ catalyst (*658*). The cyanide selec-tively attacked at the less-substituted carbon of the ring.

11.2.3s Small Ring Systems: β-Lactams

A number of syntheses of β-amino acids proceed via ini-tial formation of a β-lactam, followed by lactam hydro-lysis. Conversion of β-lactams to β-amino acids has often been accomplished using harsh hydrolysis condi-tions to open the lactam ring, but a recent report disclosed that ring opening of *N*-Boc azetidin-2-ones with alcohols is greatly enhanced by the presence of

NaN$_3$ or KCN. The alcoholysis occurred quickly at room temperature without racemization, even with hindered substrates (*1248*). These conditions can also be used with amino acids as the nucleophile, allowing β-lactams to be used as acylating reagents during peptide synthesis (reviewed in *1249*). For example, *N*-Boc β-lactams were opened and coupled to α-amino esters by combining the two in DMF in the presence of 1 equiv of sodium azide, allowing for dipeptide synthesis under mild conditions (*1250*). *N*-Boc 3-hydroxy-β-lactams, norstatine analogs, were effectively opened by amino esters in dichloro-methane without any catalyst, giving dipeptides in 63–94% yield (*1251*). Some syntheses of β-amino acids and their incorporation into peptides via β-lactams were reviewed in 1999 (*1252*), while the use of β-lactams as intermediates in the synthesis of α- and β-amino acids was reviewed in 2001 (*1253*).

A common route to β-lactams proceeds via reaction of an acetate enolate with a chiral imine (the ester eno-late–imine condensation route), with the cycloaddition of ketene and imine known as the Staudinger reaction. Depending upon the reaction conditions, the condensa-tion can instead directly produce β-amino esters instead of the β-lactam (see Sections 11.2.2c and 11.2.3k) (*563*). Theoretical studies of the mechanism of the Staudinger reaction have been published (*1254*), including stereo-chemical aspects of the asymmetric condensation of *N*-trialkylsilylimines with Gly-derived ketenes, with the Gly nitrogen contained within a chiral oxazolidinone ring (*1255*). A microreview of asymmetric syntheses of β-lactams by the Staudinger ketene–imine cycloaddition was published in 1999 (*660*).

Ojima and co-workers developed the β-lactam syn-thon method (reviewed in *1256, 1257*) for preparing α-hydroxy-β-amino acids via hydrolysis of 3-hydroxy-β-lactams. The β-lactams were prepared by a [2+2] cycloaddition of achiral ketenes with chiral imines, or of chiral ketenes with achiral imines. The (2*R*,3*S*)-3-phenylisoserine component of Taxol and related aryl-substituted derivatives were synthesized by this route, using an enolate with a (–)-*trans*-2-phenyl-1-cyclohexyl ester auxiliary to give the β-lactam intermediates with 96–98% ee (see Scheme 11.105) (*1258*). A range of other substituents have also been incorporated (*1259, 1260*). The β-lactams were hydrolyzed to the β-amino acids with 6 N HCl and then coupled with 7-TES-baccatin III,

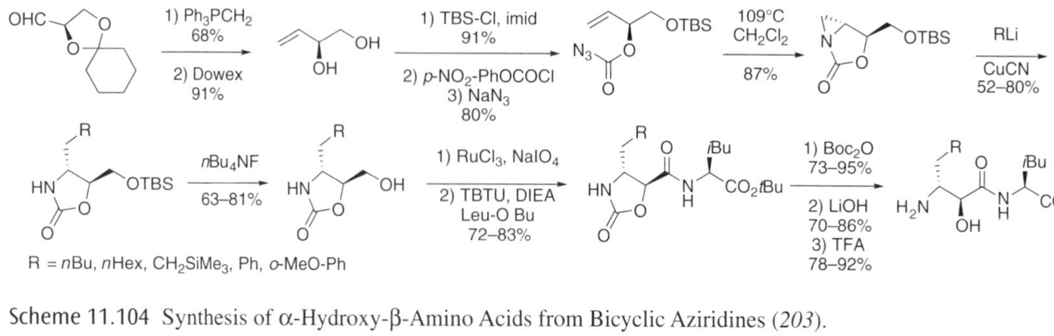

R = *n*Bu, *n*Hex, CH$_2$SiMe$_3$, Ph, *o*-MeO-Ph

Scheme 11.104 Synthesis of α-Hydroxy-β-Amino Acids from Bicyclic Aziridines (*203*).

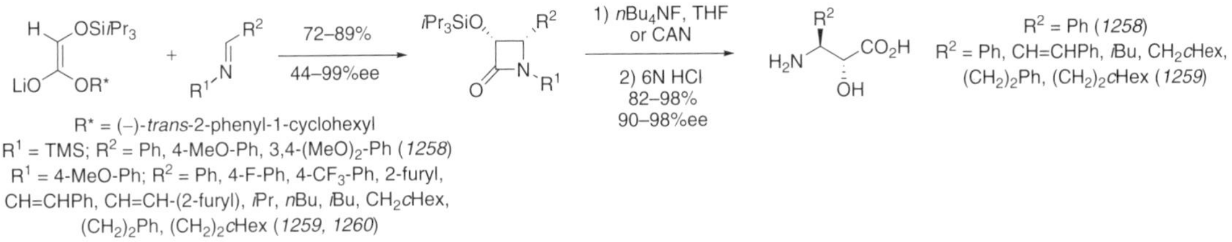
72–89%
44–99%ee
1) nBu4NF, THF or CAN
2) 6N HCl
82–98%
90–98%ee
R* = (–)-trans-2-phenyl-1-cyclohexyl
R1 = TMS; R2 = Ph, 4-MeO-Ph, 3,4-(MeO)2-Ph (1258)
R1 = 4-MeO-Ph; R2 = Ph, 4-F-Ph, 4-CF3-Ph, 2-furyl, CH=CHPh, CH=CH-(2-furyl), iPr, nBu, iBu, CH2cHex, (CH2)2Ph, (CH2)2cHex (1259, 1260)
R2 = Ph (1258)
R2 = Ph, CH=CHPh, iBu, CH2cHex, (CH2)2Ph, (CH2)2cHex (1259)

Scheme 11.105 Chiral Ester Enolate-Imine Cyclocondensation.

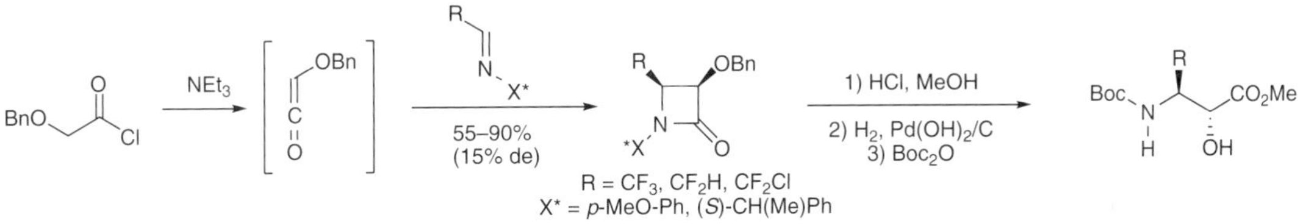
NEt3
55–90%
(15% de)
1) HCl, MeOH
2) H2, Pd(OH)2/C
3) Boc2O
R = CF3, CF2H, CF2Cl
X* = p-MeO-Ph, (S)-CH(Me)Ph

Scheme 11.106 Synthesis of β-Amino Acids via β-Lactam (*1262*).

or the lactam could be directly reacted with 7,10-bis(Troc)-baccatin II and NaH to give 10-deacetyl-Taxol derivatives (*1260*). β-Lactams have also been employed to introduce different β-substituted isoserines to analogs of 1-deoxypaclitaxel (*185*). The Ojima synthetic route was further extended by deprotonating the 3-silyloxy-4-alkyl-β-lactams with LDA and alkylating the enolate with methyl iodide or allyl bromide, stereoselectively introducing the new substituent at the carbinol center. Deprotection and lactam opening gave 2-hydroxy-2-substituted-3-alkyl-3-amino acids (*1261*).

(Fluoroalkyl)imines with a chiral auxiliary have been used for a cycloaddition reaction with the ketene generated from (benzyloxy)acetyl chloride, producing *cis*-substituted β-lactams (see Scheme 11.106). Little diastereoselectivity was observed, but the isomers were easily separated. The β-lactam was then manipulated to give *syn*-(3-fluoroalkyl)isoserines (*1262*). 2-Pyridyl thioesters of pent-4-enoic acid or hex-5-enoic acid were condensed with the α-methylbenzylamine imine of cinnamaldehyde to give the β-lactams as 3:1 or 2:1 mixtures of diastereomers. The alkenes were then cyclized by ring-closing metathesis (*1263*).

In a similar fashion, the β-lactam corresponding to the phenylisoserine component of Taxol was prepared from cycloaddition of a benzaldehyde imine formed from L-Thr as the chiral amine source, with acetoxyacetyl chloride used to generate the ketene (*1264*). The lactam was formed with 90% de, and the Thr chiral auxiliary removed by a multistep procedure. Phenylisoserines have also been prepared using α-methylbenzylamine (50–60% de) (*1265*), (*S*)-1-(4-methoxyphenyl)ethylamine (46% de) (*1266*), or β-D-galactosamine (100% de, 75% yield) (*1267*) as the chiral amine. A taxotere analog was prepared with a 3-methylphenylisoserine residue via acylation with the lactam of the β-amino acid. The lactam was prepared by Staudinger reaction of acetoxyacetyl chloride with the (*S*)-1-(4-methoxyphenyl)ethylamine imine of acetophenone (64% yield, 40% de) (*186*).

β-Lactam synthesis using other amino acids as a chiral auxiliary in the imine component has previously been reported for reaction with dimethyl ketene methyl trimethylsilyl acetal, with asymmetric induction of 14 to >98% de (*1268*). Other chiral amines were also examined with this ketene (44–78% de), but only one example was hydrolyzed and *N*-deprotected to the β-amino acid (*1269*). Protected galactopyranosylamine was used to form chiral imines for the preparation of a number of α-hydroxy-β-amino acid β-lactam precursors, but minimal diastereoselectivity was reported (*1270*). A [2+2] cycloaddition of benzyloxyketene (generated in situ from benzyloxyacetyl chloride and triethylamine) with the imine formed from cyclohexylacetaldehyde and (*S*)-α-methylphenylamine provided *cis*-substituted 1-benzyl-3-benzyloxy-4-cyclohexylmethylazetidinone, with 24% de. Lactam hydrolysis and hydrogenolysis provided cyclohexylnorstatine, 2-hydroxy-3-amino-4-cyclohexylbutyric acid (*649, 663, 1129*).

Hydrazones derived from chiral *N,N*-dialkyl hydrazines (*N*-amino-2, 5-dimethylpyrrolidine) can act as the imine component, reacting with benzyloxyacetyl chloride in the presence of triethylamine to give *N*-dialkyl *cis*-3-benzyloxy-3-substituted-β-lactams with >98% de. The chiral auxiliary was removed by oxidative N–N bond cleavage, with hydrogenolysis of the *O*-benzyl group and lactam hydrolysis with 6 N HCl producing α-hydroxy-β-substituted-β-amino acids (*1271*).

Good stereoselectivity was obtained by reacting chiral imines with the chiral center present on the aldehyde component (e.g., 2-hydroxy-2-phenylacetaldehyde) rather than the amine. The side-chain hydroxyl group of the β-lactam adduct was later removed and the aryl ring reduced to give cyclohexylnorstatine in addition to the aromatic analog (1249, 1272, 1273).

An alternative strategy employed an oxazolidinone auxiliary on the ketene component (Evans–Sjögren ketene, Gly-derived ketenes with the Gly nitrogen contained within a chiral oxazolidinone ring), which reacted with achiral or chiral imines with "virtually complete" diastereoselectivity (1274). The β-lactam adduct was deprotonated and α-hydroxylated to give an α-keto-β-lactam. Reduction of the keto group and lactam hydrolysis provided phenylisoserine. Stereochemical aspects of the asymmetric condensation of this ketene with N-trialkylsilylimines have been examined, with a 1,3-aza-diene intermediate isolated (1255). Spiro-β-lactams were synthesized by a [2+2]-cycloaddition reaction between cyclic ketenes derived from Cbz-Pro-Cl or O-protected Cbz-4-hydroxy-Pro-Cl and N-Pmp imines derived from benzaldehyde, D-glyceraldehyde acetonide or Garner's aldehyde. Lactam hydrolysis produced α-(1-amino-benzyl)-4-hydroxy-Pro, α-(1-amino-2,3-dihydroxy-propyl)-Pro, and α-(1,2-diamino-3-hydroxypropyl)-Pro (1275). The lactam precursors of a number of other analogs were reported earlier (1276). Lactam precursors of 2,3-disubstituted isoserines can be obtained by reaction of homochiral enolates of dioxolanones with achiral imines (see Scheme 11.107) (1277).

Catalytic, enantioselective β-lactam syntheses from achiral ketene and α-imino ester components have been described, with developments reviewed in 2001 (1278), 2003 (1279), and 2004 (1280). An electron-deficient imino ester, TsN=CHCO$_2$Et, reacted with a number of acid chloride-derived ketenes in the presence of benzo-ylquinine to give 2-carboxy-3-substituted-azetidin-2-ones in 36–65% yields, with 95–99% ee and >98% de cis stereoselectivity. The lactams were not opened (1281, 2189). Bifunctional catalysis, employing a chiral cin-chona alkaloid in combination with an In(III) metal cocatalyst, provided a similar series of β-lactams in 92–98% yield with 96–98% ee and 10:1 to 60:1 dr (1282). A planar-chiral ferrocenyl-like catalyst gave lactams with 81–98% ee during Staudinger reactions of symmetrical or unsymmetrical disubstituted ketenes with a range of N-tosyl imines (1283, 1284).

β-Lactams have been synthesized and derivatized in other manners. An enantioselective Cu-catalyzed coupling of alkynes with nitrones proceeded with 72–93% ee in the presence of a chiral bis(azaferrocene) ligand (1285). Asp was converted into 4-carboxyazetidin-2-one. Reduction of the carboxy substituent to an aldehyde and Corey–Fuchs homologation gave a 4-ethynyl-azetidin-2-one.

Hydrolysis provided 3-aminopent-4-ynoic acid (402). Similarly, the β-amino acid component of AI-77B (2,3-dihydroxy-4-amino-1,6-hexanedioic acid) was prepared from L-Asp-derived 4-formyl-azetidin-2-one via Wittig reaction and dihydroxylation of the alkene (219, 223). Alternatively, the carboxy substituent of 4-carboxyazeti-din-2-one was reduced to a hydroxymethyl group, and a methyl substituent stereoselectively introduced at the 3-position of the β-lactam via deprotonation and methy-lation. The 4-(hydroxymethyl) substituent was oxidized to an aldehyde and the remaining side chain of Adda added via an olefination reaction. The β-lactam could be used as an activated Adda equivalent, coupling with Gly-OMe to form a dipeptide, or it could be hydrolyzed to unprotected Adda using mild basic hydrolysis (1286).

11.2.3t Elaboration of Other Chiral Substrates

A number of optically enriched substrates for β-amino acid synthesis rely upon masked carboxylic acid groups. A carboxy group was generated from homochiral O-BPS 2-methyl-3-aminopropanol via acylation of the amino group, O-deprotection, and hydroxymethyl oxidation with RuCl$_3$/NaIO$_4$ (58). (R)-2-Methyl-3-aminobutyric acid, a component of the cryptophycin antitumor depsi-peptides, was obtained from (S)-3-hydroxy-2-methyl-propanoate. The ester was converted into an amide and reduced with borane to give an aminomethyl group, with oxidation of the hydroxymethyl group providing the acid (55). Alternatively, commercially available (S)-3-bromo-2-methyl-1-propanol was aminated by reaction with sodium azide, oxidized with Jones reagent, and hydrogenated over palladium to give the desired β-amino acid in 65% overall yield (57). 3-Amino-4-phenyl-2-methylenebutanoic acid was prepared from homochiral 2-(chloromethyl)-2-(1-amino-2-phenylethyl)oxirane via reaction with base and hydroxymethyl oxidation (1287).

D-Glucosamine was converted into an intermediate used to synthesize the β-amino acid found in the gastro-protective substance AI-77-B, 3-amino-4,5-dihydroxy-hexanedioic acid (1288). Methyl-α-D-mannopyranoside (221), methyl α-D-glucopyranoside (221), and (R)-glyceric acid (222) have also been converted into this β-amino acid. D-Glucono-1,5-lactone and D-gulono-1,4-lactone were transformed into two diastereomers of 3-amino-4,5,6-trihydroxyhexanoic acid via amination

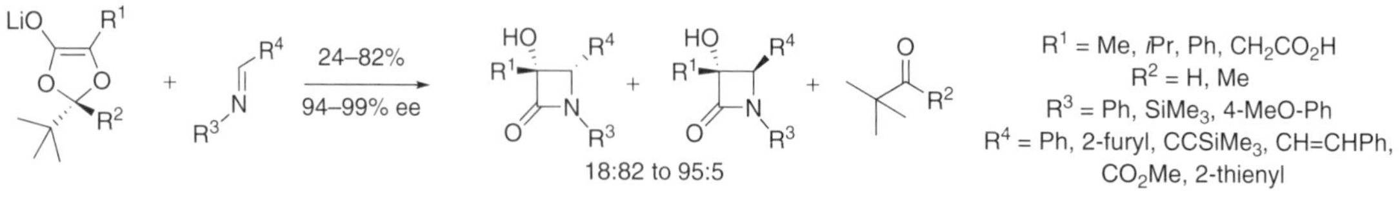

R^1 = Me, iPr, Ph, CH$_2$CO$_2$H
R^2 = H, Me
R^3 = Ph, SiMe$_3$, 4-MeO-Ph
R^4 = Ph, 2-furyl, CCSiMe$_3$, CH=CHPh, CO$_2$Me, 2-thienyl

24–82%
94–99% ee

18:82 to 95:5

Scheme 11.107 Lactam Formation from Enolates of Chiral Dioxolanones and Imines (1277).

Table 11.2 Asymmetric Syntheses of β-Amino Acids

$$H_2N,\ R^3 \diagup\hspace{-6pt}\diagdown\ R^1,\ R^2,\ CO_2H,\ R^4$$

R^1	R^2	R^3	R^4	Method	Yield, % ee	Reference
Amination: Conjugate Addition						
Me	H	H	H	addition of (R)- or (S)-α-methylbenzylamine to α,β-unsaturated esters	9–47%	1977 (710)
Ph	H	H	H		2–19% ee	
Me	H	H	H	addition of BnNH₂ to α,β-unsaturated acids derivatized as chiral esters	50–90% addition; 60–>99% de	1986 (711)
Me	H	H	H	addition of (R)- or (S)-α-methylbenzylamine to methyl crotonate	74–86%	1988 (638)
H	Me	H	H		20% de	1992 (712)
Me	H	H	H	addition of (R)- or (S)-α-methylbenzylamine to ethyl crotonate	8% de	1977 (713)
H	Me	H	H			
Me	H	H	H	addition of (R)- or (S)-α-methylbenzyl-hydroxylamine to α,β-unsaturated esters	54–91%; 38–60% de	1987 (714)
CH₂CH₂OMOM	H	H	H			
H	Me	H	H			
H	nPr	H	H			
H	iPr	H	H			
H	Ph	H	H			
H	CH₂CH₂OMOM	H	H			
H	CH₂CO₂H	H	H			
H	H	Bn	H	addition of (S)-α-methyl-benzyl-hydroxylamine to α,β-unsaturated esters	78–84% addition and cyclization to isoxazolidi-none	2003 (715)
H	H	iPr	H			
H	H	iBu	H		78–84% hydrogenation	
H	H	(CH₂)₄NH₂	H			
H	H	H	Bn			
H	H	H	iPr			
H	H	H	iBu			
H	H	H	(CH₂)₄NH₂			
–(CH₂)₃-C2	H	–(CH₂)₃-C3	H	addition of Li (α-methylbenzyl)benzylamide to α,β-unsaturated esters	65–70%; >97% de	1993 (716)
–(CH₂)₃-C2	H	–(CH₂)₄-C3	–(CH₂)₃-C3			
H	–(CH₂)₄-C2	H	H			
H	–(CH₂)₄-C2	H	–(CH₂)₄-C3			

(Continued)

Table 11.2 Asymmetric Syntheses of β-Amino Acids (continued)

R^1	R^2	R^3	R^4	Method	Yield, % ee	Reference
H	Me	H	H	addition of Li (α-methylbenzyl)benzylamide to α,β-unsaturated esters	78–82% >99% de	1991 (717)
H	4-HO-Ph	H	H	addition of Li (α-methylbenzyl)benzylamide to α,β-unsaturated ester	69% overall >95% de	1995 (719) 1998 (97)
H	Ph	H	H	addition of Li (α-methylbenzyl)benzylamide to tert-butyl cinnamate, deprotection	63% overall 92% ee	2002 (112)
Me	H	H	H	addition of Li (α-methylbenzyl)benzylamide to Weinreb amides of α,β-unsaturated acids	92–96% addition >95% de	2004 (720)
iPr	H	H	H			
Ph	H	H	H			
H	Ph	H	H			
H	[2,3]-dihydrobenzofur-5-yl	H	H	addition of Li (α-methylbenzyl)benzylamide to α,β-unsaturated ester, hydrogenolysis of N-substituents	72% amine addition >90% de	2000 (721)
H	[2,3]-dihydrobenzofur-6-yl	H	H		75% hydrogenation	
H	4-HO-Ph	H	H	addition of Li (α-methylbenzyl)benzylamide to α,β-unsaturated ester	95% addition >95% de 90% deprotection	1995 (131)
2-[CH(NH$_2$)CH$_2$CO$_2$H]-Ph	H	H	H	addition of Li (α-methylbenzyl)benzylamide to 2 or 3 tert-butyl acrylates attached at 3-position to a benzene nucleus, hydrogenolysis	78–87% conjugate addition 65–90% deprotection	2001 (722)
3-[CH(NH$_2$)CH$_2$CO$_2$H]-Ph	H	H	H			
4-[CH(NH$_2$)CH$_2$CO$_2$H]-Ph	H	H	H			
3,5-[CH(NH$_2$)CH$_2$CO$_2$H]$_2$-Ph	H	H	H			
H	Et	Me	H	addition of Li (α-methylbenzyl)benzylamide to α,β-unsaturated esters, protonation with hindered acid	65% >95% de syn 57% deprotection	1994 (723)
H	Ph	H	Me	addition of Li (α-methylbenzyl)benzylamide to α,β-unsaturated esters, protonation with hindered acid or alkylation with MeI	61–92% >94% de syn or anti 50–84% deprotection	1993 (724)
H	Ph	Me	H			
H	Me	Me	H	addition of Li (α-methylbenzyl)benzylamide to α,β-unsaturated esters, protonation with hindered acid or alkylation with RX	45–72% >95% de syn 56–69% deprotection	1994 (725)
H	Et	Me	H			
H	Ph	Me	H			
H	Ph	Bn	H			
H	Me	Me	Me			
H	Me	Me	Bn			

R¹	R²	R³	Conditions	Yield	Year (Ref)
H	Me	(CH$_2$)$_4$NH$_2$	addition of Li (α-methyl-benzyl)benzylamide to α,β-unsaturated esters, protonation with hindered acid	12–67% 86–98% de syn 14–74% deprotection 88–96% ee	2003 (726)
H	(CH$_2$)$_4$NH$_2$				
H	iBu				
H	CH$_2$cHex				
H	Me	Bn			
H	Me	4-MeO-Bn			
H	Me	3,4,5-(MeO)$_3$-Bn			
Me	H	Me	addition of Li (α-methylbenzyl)benzylamide to α,β-unsaturated ester	68% addition 99% deprotection	2000 (304)
	CH$_2$CH$_2$NHMe	H	addition of Li (α-methylbenzyl)benzylamide to α,β-unsaturated ester, guanidation of secondary amine	85% addition 74% auxiliary removal 82% guanylation	2005 (172)
	CH$_2$CH$_2$N(Me)-C(=NH)NH$_2$	H			
H	4-HO-Ph	H	addition of Li (α-methylbenzyl)benzylamide to α,β-unsaturated ester, convert phenol to triflate and use for Suzuki couplings with aryl boronic acids	66% addition 93% auxiliary removal 75% Suzuki (with Ar = Ph)	2002 (406)
H	4-Ph-Ph	H			
H	4-Ar-Ph	H			
	Ar = 2-MeO-Ph, 2-CN-Ph, 2-CF$_3$O-Ph, 2,5-(MeO)$_2$-Ph, 2-MeO-5-F-Ph, 2-MeO-6-F-Ph, 2-MeO-4-F-Ph, 2-CF$_3$O-4-F-Ph, 2-MeO-3-F-Ph, 2-F-3-MeO-Ph, 2-F-4,6-(MeO)$_2$-Ph, 3-F-5-MeO-Ph, 2,6-(MeO)$_2$-Ph, 2,6-(MeO)$_2$-5-F-Ph, 2-CF$_3$O-6-MeO-Ph, 2,6-(MeO)$_2$-3,5-F$_2$-Ph, 2-HO-6-MeO-Ph, 2,6-(OH)$_2$-Ph				
(CH$_2$)$_2$Ph	H	H	addition of Li (α-methylbenzyl)-benzylamide to methyl 5-phenyl-2,4-pentadienoic acid, deprotonation with LDA and α-methylation, auxiliary removal and alkene reduction by hydrogenation, reintroduction of 4-alkene via bromination/elimination	61% addition 57% methylation 71–77% hydrogenation 42–45% bromination/elimination	2001 (233)
CH=CHPh	Me	H			
(CH$_2$)$_2$Ph	H	H			
CH=CHPh	Me	H			

(Continued)

Table 11.2 Asymmetric Syntheses of β-Amino Acids (continued)

R¹	R²	R³	R⁴	Method	Yield, % ee	Reference
H	Me	H	Me	addition of Li (α-methylbenzyl) benzylamide to α,β-unsaturated esters, α-alkylation with RX	42–77%	1994 (727)
H	Me	H	Bn		0–90% de anti	
H	Ph	H	Me		40–75% deprotection	
H	Ph	H	Bn			
H	Ph	H	allyl			
H	(R)-CH(Me)CH₂-OTBDMS	H	H	addition of Li (α-methylbenzyl) benzylamide to α,β-unsaturated esters, trap with MeCHO	65–70%	1995 (728)
H	(R)-CH(Me)CH₂-OTBDMS	H	CH(OH)Me		>97% de	
(R)-CH(Me)CH₂-OTBDMS	H	H	H			
Me	H	F	H	addition of Li (α-methylbenzyl) benzylamide to α,β-unsaturated esters, trap with (PhSO₂)₂NF, hydrogenation	39–100% addition/trapping	2004 (729)
Ph	H	F	H		64–66% de	
					86% deprotection	
H	CH=CH₂Ph	H	CH(OH)Me	addition of Li (α-methylbenzyl) benzylamide to α,β-unsaturated esters, aldol reaction with MeCHO	81–98% conjugate addition 100% de 72–100% aldol reaction 50–82% de	1993 (730)
H	Me	H	Me	addition of Li (α-methylbenzyl) benzylamide to α,β-unsaturated amides, trap with alkyl halide	46–89% yield	1995 (731)
H	Ph	H	Me		0–>94% de	
H	Ph	H	Et			
H	Ph	H	Bn			
H	Ph	H	CH₂CH=CH₂			
H	Me	H	OH	addition of Li (α-methylbenzyl) benzylamide to α,β-unsaturated esters, trap with homochiral electrophilic hydroxylation reagent or quench, redeprotonate and hydroxylate	36–85% addition and trapping >90% de anti	1994 (732)
H	Et	H	OH			
H	Ph	H	OH			
H	Ph	H	OH	addition of Li (α-methylbenzyl) benzylamide to tert-butyl cinnamate, trap with homochiral electrophilic hydroxylation reagent, invert carbinol center	86% addition and trapping 92% de	1993 (734)
H	Ph	OH	H		70% inversion	1994 (733)

H	Bn	H	OH	addition of Li (α-methylbenzyl) benzylamide to *tert*-butyl cinnamate, trap with homochiral electrophilic hydroxylation reagent	41–63% addition and trapping; 98% de at C3; 20% de *syn* to 91% de *anti* at C2 depending on if matched/mismatched homochiral reagents	1993 (736) 1994 (735)
H	Bn	OH	H			
Bn	H	OH	H			
H	nHept	H	H	addition of Li (α-methylbenzyl) benzylamide to *tert*-butyl 2-decenoate, trap with homochiral electrophilic hydroxylation reagent	67–85% addition and trapping; 98% de at C3; 88–92% de *anti* at C2 depending on if matched/mismatched homochiral reagents *syn* via Mitsunobu inversion	1994 (738) 1995 (737)
H	nHept	H	OH			
nHept	nHept	OH	H			
nHept	H	OH	H			
H	(E)-CH=CHMe	H	OH	addition of Li (α-methylbenzyl) allylamide to *tert*-butyl hexa-2,4-dienoate, trap with homochiral electrophilic hydroxylation reagent, epimerize α-center with DCC/DMAP	63% addition and trapping; 96% de; 51% epimerization	1999 (739)
H	(E)-CH=CHMe	OH				
H	Ph	H	H	addition of MgBr (α-methylbenzyl)benzyl-amide to α,β-unsaturated esters, trap with MeI	Michael: 90%; >95% de; Michael+trap: 73%; 90% de *syn*	1994 (740)
H	Ph	H	Me			
3-pyridyl	H			addition of Li (α-methyl-4-MeO-benzyl)benzylamide to *t*Bu 3-(pyridyl)-propenoic esters, Bn removal with 2.1 eq. CAN, auxiliary removal with 4 eq. CAN	45–86% conjugate addition; 5% de (for 2-pyridyl) to 84% de; 68–86% Bn removal; 41–63% auxiliary removal	2002 (742)
4-pyridyl	H					
2-pyridyl	H					
6-Me-2-pyridyl	H					
6-Br-2-pyridyl	H					
6-TMS-2-pyridyl	H					
5-MOMO-6-Cl-2-pyridyl	H					

(*Continued*)

Table 11.2 Asymmetric Syntheses of β-Amino Acids (continued)

R¹	R²	R³	R⁴	Method	Yield, % ee	Reference
3-Cl-Ph	H	H	H	addition of Li (α-methyl-4-MeO-benzyl)benzylamide to tBu cinnamic esters. Bn removal with 2.1 eq. CAN, auxiliary removal with 4 eq. CAN	74–86% conjugate addition 88–98% de 71–86% debenzylation 48–68% auxiliary removal 94–98% ee	2001 (743)
2-Br-Ph	H	H	H			
3-Br-Ph	H	H	H			
4-Br-Ph	H	H	H			
$3,4\text{-}F_2\text{-Ph}$	H	H	H			
H	3-FPh	H	H			
H	2-I-Ph	H	H			
H	3-I-Ph	H	H			
H	4-I-Ph	H	H			
3-Cl-Ph	H	H	H	addition of Li (α-methyl-4-MeO-benzyl)benzylamide to tBu cinnamic esters, Bn removal with 2.1 eq. CAN, auxiliary removal with 4 eq. CAN	74–86% conjugate addition 88–98% de 71–86% debenzylation 48–68% auxiliary removal 94–98% ee	2000 (2179) 2001 (743)
2-Br-Ph	H	H	H			
3-Br-Ph	H	H	H			
4-Br-Ph	H	H	H			
$3,4\text{-}F_2\text{-Ph}$	H	H	H			
H	3-FPh	H	H			
H	2-I-Ph	H	H			
H	3-I-Ph	H	H			
H	4-I-Ph	H	H			
Ph	H	H	H	addition of Li (α-methyl-4-MeO-benzyl)benzylamide to tBu cinnamic esters, Bn removal with 2.1 eq. CAN, auxiliary removal with 4 eq. CAN	71–83% conjugate addition 88–92% de 77–89% debenzylation 49–64% auxiliary removal 88–>97% ee	2000 (744)
3-F-Ph	H	H	H			
3-I-Ph	H	H	H			
H	3-Cl-Ph	H	H			
H	3-Br-Ph	H	H			
Me	H	H	H	addition of Li (α-methylbenzyl) allylamide to α,β-unsaturated esters	78–97% 96% de	1995 (746)
Et	H	H	H			
iPr	H	H	H			
Ph	H	H	H			
2-furyl	H	H	H			
(E)-CH=CHMe	H	H	H			
CH=CHMe	H	H	H	addition of Li (α-methylbenzyl) allylamide to α,β-unsaturated esters	72–78% 96% de	1997 (747)
CH=CHMe	H	OH	H			
CH_2CH=CHMe	H	H	H	addition of Li (α-methylbenzyl) allylamide to α,β-unsaturated esters	69–78% conjugate addition >95% de	2002 (748)
CH=CHMe	H	H	H			
$(CH_2)_3Cl$	H	H	H	addition of Li (α-methylbenzyl) arylamide to o-Cl-α,β-unsaturated esters	60–88% addition >95% de	2000 (749)
$(CH_2)_4Cl$	H	H	H			
$(CH_2)_5Cl$	H	H	H			

H	H	CH₂CH=CHMe	H	addition of Li (α-methylbenzyl) allylamide to α,β-unsaturated esters, hydroxyamination of alkene	64% addition, 91% de, 69% deprotection	1999 (166)
H	H	CH$_2$CH(OH)CH-(I)Me	H			
H	H	CH$_2$CH(OH)CH-(NH$_2$)Me	H			
H	H	3-pyridyl	H	addition of Li (α-methylbenzyl)-TMS-amide to α,β-unsaturated esters	64%, 96% de	1993 (750)
H	H	Me	H	addition of CuLi (α-methylbenzyl)-TMS-amide or CuLi-bis(α-methylbenzyl) amide to α,β-unsaturated esters, trapping with D$_2$O	49–62%, 56–>98% de	1995 (751)
H	H	iPr	H			
H	H	Ph	H			
H	H	4-BnO-Ph	H			
H	H	Me	D			
H	H	iPr	D			
H	H	Ph	D			
H	H	4-BnO-Ph	D			
H	H	Me	H	addition of Li binaphthylamine to α,β-unsaturated esters	81–88%, 97% de	1986 (752)
H	H	Me	H	addition of Li binaphthylamine to α,β-unsaturated esters	66–86%, 78–>99% de	1992 (753)
H	H	nHept	H			
H	H	iBu	H			
H	H	iPr	H			
H	H	(CH$_2$)$_2$-OTBDMS	H			
H	H	CH(Me)CH$_2$-OTBDMS	H			
H	H	Me	Me	addition of Li binaphthylamine to α,β-unsaturated esters	63–71%, 66–86% de	1994 (755)
H	H	Me	H			
−CH$_2$CH(CH$_2$R)CH$_2$−C3, R = H, Me, OMe, OBn, SMe, N$_3$, NH$_2$	H			addition of K 4,5-diphenyloxazolidinone to α,β-unsaturated esters	48–88% addition, 74–95% auxiliary removal	1993 (756)
Me	H			addition of chiral hydrazine TMS-SAMP to α,β-unsaturated acids	32–67% yield addition, >93–>99% de, 50–86% deprotection, >90–98% ee	1995 (757)
Et	H					
nPr	H					
iPr	H					
nBu	H					
iBu	H					
nPent	H					
nHept	H					
nC$_{11}$H$_{23}$	H					

(Continued)

Table 11.2 Asymmetric Syntheses of β-Amino Acids (continued)

R¹	R²	R³	R⁴	Method	Yield, % ee	Reference
$nC_{11}H_{23}$	H	nPr	H	addition of chiral hydrazine TMS-SAMP to α,β-unsaturated acids, trapping with electrophile	26–68% yield addition and trapping	1994 (758)
Ph	H	nPr	H		63–>96% de	
4-MeO-Ph	H	Me	H		63–87% deprotection	
Ph	H	Et	H		>96% de,ee	
Ph	H	nBu	H			
Ph	H	allyl	H			
Ph	H	Bn	H			
Et	H	Me	H			
Et	H	Et	H			
Et	H	nPr	H			
Et	H	Bn	H			
Et	H	(CH₂)₃OH	H			
Me	H	H	H	addition of BnONH₂ to α,β-unsaturated acids derivatized as chiral amide	38–92% addition	1993 (759)
nPr	H	H	H		10–78% de	
H	Me	H	H			
H	nPr	H	H			
Ph	H	H	H	addition of BnONH₂ to α,β-unsaturated acids derivatized as chiral amide	50–98% addition	1999 (760)
H	Ph	H	H		20–60% de	
H	Me	H	H	addition of MgCl-phthalimide to α,β-unsaturated acids derivatized with chiral imidazolidinone, trapping with BnSO₂Br	90% addition	1994 (762)
H	H	Br	H		90% de	
H	Me	Br	H		95–98% addition + trapping	
H	nPr	Br	H		20–60% de	
H	Ph	Br	H			
H	H	Me	H	conjugate addition of LiNBn₂ to acrylate amidated with chiral oxazolidinone, trap anion with electrophile OR conjugate addition to 2-substituted acrylate	8–88% addition and trapping	2004 (763)
H	H	Et	H		96–97% de	
H	H	Bn	H		76–92% addition to 2-substituted	
H	H	H	Et		78–>96% de	
H	H	H	iPr		70–95% deprotection	
H	H	H	Ph		88–97% ee	
H	Me	H	H	addition of LiNBn₂ to RCH=CHCO₂H derivatized with pseudoephedrine, auxiliary removal	15–85% addition	2005 (764)
H	Et	H	H		46–>98% de	
H	iPr	H	H		57–98% deprotection	
H	iBu	H	H			
H	Ph	H	H			

R1	R2	R3	Conditions	Results	Year (Ref)
H	Ph	H	conjugate addition of piperadizine to chiral vinyl sulfoxide derivative of tert-butyl cinnamate, N–N bond cleavage	73% yield addition/ auxiliary removal; 94% N–N cleavage	2001 (123)
Ph	H				
CH=CHPh	H	H	addition of [Bn(TMS)N]$_2$CuLi to PhCH=CHCH=CHCONR* with borname sultam auxiliary, trapping with MeCHO	78% addition, 90% de; 71% addition/ trapping, >99% de	1992 (766)
CH=CHPh	CH(OH)Me	H			
H	Ph	H	addition of LiNMe$_2$ to chiral Fe complex of cinnamic acid, complex decomposition	86% addition, 97% de, 82% decomposition	1990 (767)
Ph	H	H	addition of primary amines to α,β-unsaturated esters with chiral β-p-tolylsulfinyl substituent	37–89% addition and auxiliary removal, 17–89% ee	1997 (768)
H	Ph	H			
CH(OBn)Me	H	H	addition of BnNH$_2$ to homochiral α,β-unsaturated acids	37–78% addition, 5–94% de	1994 (464)
CH(–OR*S–)	H	H			
CH$_2$OH	H	H			
H	Ph	H	Se-catalyzed intramolecular addition of prolinamide auxiliary to α,β-unsaturated acid	22–73% addition yield, 100% de; 87% deselenation	1997 (769) 1998 (770)
Me	Me	H			
Me	Me	H	Hg-catalyzed intramolecular addition of amine to 3-butenoic acid	70% yield, 26% de	1992 (771)
H	H	H			
CH$_2$NH$_2$	H	H	reaction of N-CH(Ph)Me 4-aminobut-2-enoic acid with tosyl isocyanate, hydrolysis, N-trimethylation	90% cyclization, 0% de; quant. hydrolysis and methylation	1990 (772)
CH$_2$NMe$_2$	H	H			
H	CH(iBu)NH$_2$	H	intramolecular conjugate addition of N-aminomethyl group of 4-amino-6-methyl-hept-2-enoic acid, aminal deprotection	78% conjugate addition, 60% aminal deprotection	2005 (773)
H	CH(OH)Me	H	addition of BnNH$_2$ to homochiral 2-alkoxyalkyl-α,β-unsaturated acids	52–98% addition, 40% de syn to >90% de anti	1988 (774)
H	CH(OH)Ph	H			

(Continued)

Table 11.2 Asymmetric Syntheses of β-Amino Acids (continued)

R¹	R²	R³	R⁴	Yield, % ee	Method	Reference
H	CH(OR)Me	H	H	74–99% 100% de *syn* to 100% de *anti*	addition of LiNR¹R² to homochiral α,β-unsaturated acids	1997 (776)
H	CH(OR)Ph	H	H			
H	CH(Me)Ph	H	H			
H	CH(Me)CH₂OH	H	H			
CH(OMOM)Me	H	H	H	27–95% 36–98% de	conjugate addition of chiral amine to homochiral γ-alkoxy-α,β-unsaturated ester	1998 (777)
CH(OMOM)CH₂OtBu	H	H	H			
H	CH(OMOM)Me	H	H	43% yield amination 100% de 51–89% alkylation 50–>90% de C2	addition of BnNH to homochiral unsaturated lactone, deprotonation, alkylation	1995 (778)
H	CH(OMOM)Ph	H	H			
H	CH(OH)CH₂OH	H	H			
H	CH(OH)CH₂OH	Et	H			
H	CH(OH)CH₂OH	C(OH)Me₂	H			
H	CH(OH)CH₂OH	CH(OH)Ph	H			
H	CH(OH)CH₂OH	C(OH)(Me)Ph	H			
CH(OH)CH₂OH	H	H	H	85% yield amination 100% de	addition of BnNH to homochiral unsaturated ester diol	1983 (779)
H	(sugar structures; R' = Me, Bn)	H	H	68–85% 86–94% de	conjugate addition of benzyl-amine to chiral β-(sugar) unsaturated ester	2002 (780)
Me	H	H	H	54–94% yield 31–42% ee	addition of O-benzylhydroxylamine to α,β-unsaturated acids using chiral catalyst	1996 (781)
nPr	H	H	H			
Ph	H	H	H			
Bn	H	H	H	40–77% yield 43–71% ee	addition of N-benzylhydroxylamine to α,β-unsaturated amides using chiral catalyst	1998 (782)
H	Me	H	H			
H	Bn	H	H			
H	Me	H	H	21–80% yield 83–96% ee one enantiomer with MgBr₂ 41–59% ee other enantiomer with Yb(OTf)₂	addition of N-benzylhydroxylamine to α,β-unsaturated amides using chiral catalyst	1998 (783)
Me	H	H	H			
H	Et	H	H			
H	CH₂cHex	H	H			
H	Bn	H	H			
H	iPr	H	H			
H	Ph	H	H			

			Results	Conditions	Year (Ref.)
H	Me	Me	28–95% yield 93–99% de 60–96% ee	addition of N-benzylhydroxylamine to α,β-unsaturated amides using MgBr₂ chiral bisoxazoline catalyst	2003 (784)
H	Me	Et			
H	Me	Br			
H	Me	Ph			
H	Et	Me			
H	nPr	Me			
H	iPr	Me			
H	iBu	Me			
H	nHept	Me			
H	Et	Et			
H	Ph	Me			
H	Ph	Ph			
H	Et	H	51–92% addition 88–99% ee >90% oxidative cleavage	addition of carbamates to α,β-unsaturated-α′-hydroxy enones. RCH=CHCOC(Me)₂OH, using MgBr₂ chiral bisoxazoline catalyst, NaIO₄ oxidative cleavage of acyloin	2004 (785)
H	iPr	H			
H	iBu	H			
H	tBu	H			
H	nHex	H			
H	cHex	H			
H	(CH₂)₂Ph	H			
H	Me	H	6–93% addition 34–96% ee	conjugate addition anilines to 3-substituted acylates amides in presence of chiral catalyst	2001 (786)
H	Pr	H			
H	Me	H	94–>99% 89–97% ee for unsubstituted aniline	conjugate addition anilines to 3-substituted acylates amides in presence of chiral BINAP–Pd catalyst	2004 (787)
H	Et	H			
H	nPr	H			
H	H	Me	96–98% addition 95–97% ee 84% hydrogenation and hydrolysis	addition of HN₃ to RCH=CHCONHPh catalyzed by chiral salen-AlMe complex, azide hydrogenation and imide hydrolysis	1999 (788)
H	H	Et			
H	H	nPr			
H	H	iPr			
H	Me	H	79–97% 45–85% ee	addition of TMSN₃ to RCH=CHCO(pyrrolidin-2-one) catalyzed by peptides	2000 (789)
H	cHex	H			
H	iPr	H			
H	Et	H			
H	piperidin-4-yl	H			
H	Me	H	70–99% addition 92–97% ee	addition of LiN(TMS)Bn to RCH=CHCO₂tBu in presence of chiral MeOCH(Ph)CH(Ph)OMe ligand	2003 (790)
H	iPr	H			
H	CH=CHMe	H			
H	1-naphthyl	H			
H	2-naphthyl	H			
H	(CH₂)₃-C2	H			
H	(CH₂)₃-C3				

(Continued)

Table 11.2 Asymmetric Syntheses of β-Amino Acids (continued)

R¹	R²	R³	R⁴	Method	Yield, % ee	Reference
Me	H	H	H	addition of LiN(TBMS)allyl to RCH=CHCO₂tBu in presence of chiral MeOCH(Ph)CH(Ph)OMe ligand	11–90% addition 75–94% ee	2004 (791)
iPr	H	H	H			
CH=CHMe	H	H	H			
1-naphthyl	H	H	H			
(CH₂)₃-C2	H	H	(CH₂)₃-C3			
CF₂	H	H	H	addition of HNEt₂ to α,β-unsaturated ester using enzymatic catalyst	47–82% 35–37% ee	1987 (792)
H	H	CF₃	H			
Ph	H	H	H	addition of HNEt₂ to α,β-unsaturated ester using enzymatic catalyst and cyclodextrin	41–72% 23–72% ee	1991 (793)
4-Me-Ph	H	H	H			
4-MeO-Ph	H	H	H			

Amination: Epoxide Opening

R¹	R²	R³	R⁴	Method	Yield, % ee	Reference
H	H	H	OH	Sharpless epoxidation allyl-OH, oxidation, opening with ammonia	45% epoxidation 65–70% oxidation 33% opening	1990 (794)
H	H	OH	H			
H	Bn	OH	H	Sharpless epoxidation allyl-OH, oxidation, opening with azide	83–90% epoxidation 91% ee 39–57% oxidation and opening	1983 (795)
H	Bn	OH	Me			
iPr	H	OH	H	trans- and cis-epoxide opening with dialkylamines, azide in presence of Ti(OiPr)₄	71–98%	1985 (2191)
nHept	H	OH	H			
cHex	H	OH	H			
CH₂OH	H	OH	H			
H	nHept	OH	H			
H	cHex	OH	H			
H	CH₂OH	OH	H			
H	4-Me-Ph	H	OH	trans-epoxide opening with azide in presence of Yb(OTf)₃	79%	2002 (796)
Me	H	OH	H	epoxide opening with azide in presence of Cu(NO₃)₂, azide reduction with same catalyst	79–91% azide opening and azide reduction	2003 (797)
nPr	H	OH	H			
Ph	H	OH	H			
Et	H	OH	Me			
Me	Me	OH	Me			
Ph	H	OH	Me			
H	Me	OH	H	epoxide opening with azide	56–86% for opening	1988 (798)
H	Me	H	OH			

R1	R2	R3	R4	Synthesis	Yield	Year (Ref.)
H	H	H	OH	oxidation of (S)-glycidol, coupling with peptide, opening with azide	36% oxidation and coupling; 48% azide opening and reduction	1999 (175)
H	H	OH	3-(1-CO$_2$H-4-OH-bicyclo[3.1.0]hexane)	epoxide opening with benzyl-amine	88%	1995 (799)
H	H	OH	3-(1-CO$_2$H-4-OH-bicyclo[3.1.0]hexane)	epoxide opening with α-methylbenzylamine	80–83%	1996 (800)
H	H	2,3-(HO)$_2$-3-CO$_2$H-cPent	HO	epoxide opening with azide, chiral amine	24–86%	1995 (801)
H	H	OH	2,3-(HO)$_2$-3-CO$_2$H-cPent			
2-furyl	2-furyl	OH	OH	epoxide opening with azide or benzylamine	50% for opening; 100% *syn* or *anti*	1995 (802)
H	H	OH	H	epoxide formation from L-Ser, L-Thr, opening with ammonia	31–58% from amino acid	1979 (803)
H	Me	OH	H			
H	nHept	OH	H	epoxide opening with bromide, displacement with azide, reduction	98% opening; 75% displacement; 62% reduction	1997 (804)
H	Bn	OH	H	epoxide opening with bromide, displacement with azide, reduction	92% opening; 73% displacement; 95% reduction	2003 (805)
H	Bn	OH	H	epoxide opening with PhCN/BF$_3$·Et$_2$O, hydrolysis of oxazoline	78% opening; 83% oxazoline hydrolysis	2005 (806)
iBu	H	H	OH	epoxide opening with PhCN or MeCN/BF$_3$·Et$_2$O, hydrolysis of oxazoline	95–98% opening; 95–98% oxazoline hydrolysis	2005 (807)
Ph	H	H	OH			
Bn	H	H	OH			
H	Me	Me	OH	epoxide opening with MeCN/BF$_3$·Et$_2$O, hydrolysis of oxazoline	60–92% opening	2002 (808)
H	Me	nPr	OH			
H	Me	iPr	OH			
Me	H	iPr	OH			
H	Me	Me	OH			
H	Me	H	OH			
Me	Me	H	OH	conversion of 3-amino-1,2-diol obtained by Sharpless epoxidation, amine opening	19–34% from aminodiol	1996 (809)
Ph	Ph	H	OH			
Me	Me	OH	H			
Ph	Ph	OH	H			

(Continued)

Table 11.2 Asymmetric Syntheses of β-Amino Acids (continued)

R^1	R^2	R^3	R^4	Method	Yield, % ee	Reference
CH$_2$cHex	H	H	OH	epoxide opening with azide/ amine, oxidation of hydroxymethyl group	61–99% epoxide opening 86% oxidation 94% ee	1996 (810)
CH$_2$cHex	H	OH	H			
Me	H	OH	H	epoxide opening with benzhydrylamine, oxidation of hydroxymethyl group	68–74% epoxide opening 67–70% oxidation	1995 (811)
Me	H	H	H	epoxide opening with benzhydrylamine, conversion of 1,2-diol to methylene-carboxy	15–38% for diol conversion	1994 (812)
*n*Pr	H	H	H			
Ph	H	H	H			
(CH$_2$)$_2$Ph	H	H	H			
H	Me	H	H			
H	*n*Pr	H	H			
H	Ph	H	H			
H	(CH$_2$)$_2$Ph	H	H			
H	H	OH	H	epoxide opening with azide, oxidation of hydroxymethyl group	98% azide opening 46% azide reduction 38% oxidation	2000 (378)
H	H	H	OH			
Ph	H	H	OH	epoxide opening with azide	90% for opening	1986 (813)
Ph	H	H	OH	epoxide formation by Mn(II)-salen catalysis, *cis* epoxide opening with ammonia	65% for opening	1992 (814)
Ph	H	OH	H	epoxide formation by Co(II)-salen polymer catalysis, epoxide opening with anilines, chiral amide auxiliary	28–49%	1997 (815) 1997 (504)
H	Ph	OH	H			
H	Ph	H	OH	epoxide formation by Co(II)-salen polymer catalysis, epoxide opening with anilines, chiral amide auxiliary	68–71% epoxidation 72–76% opening	2002 (816)
H	Ph	H	OH	azide opening of epoxy-oxazolidine, aldehyde oxidation	71% opening 54% oxidation	1997 (817)
Ph	H	OH	H	*trans* epoxide ester opening with ammonia, hydrolysis of resulting amide and N-benzoylation, epimerization of carbinol	81% opening 50% amide to ester and N-benzoylation 49% carbinol epimerization	2003 (818)
Ph	H	H	OH			

Ph	H	OH	cis or trans epoxide formation from 2-chloro-3-amino acids, opening with ammonia or azide	50–85% for opening	1995 (819)
Ph	OH	H		0% de to 100% de syn or anti depending on amine	
Ph	H	OH	epoxide formation via dihydroxylation, cyclization; opening with azide	65% dihydroxylation 78% epoxide formation 78% azide opening	1998 (820)
Ph	H	OH	epoxide opening with azide	95% for opening	1990 (821)
Ph	H	OH	epoxide opening with azide	85% for opening, azide reduction	1992 (822)
H	Ph	OH	epoxidation of PhCH=CHCOtBu, mCPBA rearrangement, HCl opening, epoxide formation, azide or ammonia opening	76% epoxidation 94% ee 94% rearrangement 66% HCl opening 80% cis-epoxide formation 49% opening	1997 (823)
Ph	H	OH	trans epoxide ester opening with HCl, reformation of cis epoxide, opening with ammonia, amide hydrolysis	70% chloride opening 61% cis epoxide 65% ammonia opening 88% amide hydrolysis	2001 (824)
Ph	H	OH	epoxide opening with azide, or with bromide followed by azide displacement	72–95% for amination	1993 (825)
Et	H	OH	asymmetric epoxidation, opening with bromide then azide displacement	98% bromide opening 70–99% azide displacement 80–88% azide reduction	1996 (826)
nPr	H	OH			
Ph	H	OH			
CH2cHex	H	OH			
Ph	H	OH	epoxide opening with azide or hydroxyamine, or Br then azide; epoxide attached to 10-deacetylbaccatin III	20–78% opening	1999 (827)
Ph	H	OH	enzymatic resolution of 2,3-diol, conversion of one enantiomer via epoxide, azide opening; other enantiomer via bromination, azide displacement	83% epoxide formation 94% azide opening 89% bromination 85% azide displacement 88% azide reduction	1998 (828)

(Continued)

Table 11.2 Asymmetric Syntheses of β-Amino Acids (continued)

R¹	R²	R³	R⁴	Method	Yield, % ee	Reference
Amination: Other Nucleophilic Reactions						
(Z)-CH₂CH=CH-nOct	H	H	H	Sharpless epoxidation of (E)-tetrade-2-en-5-yn-1-ol, alkyne reduction, epoxide opening with Red-Al, mesylation of secondary ROH, displacement with azide, azide reduction, hydroxymethyl oxidation	76% epoxidation 90% reduction 99% epoxide opening 74% mesylation/azide displacement 84% azide reduction 90% oxidation	1998 (83)
H	H	H	Me	azide displacement of chiral tosylate, reduction	80% displacement 90% hydrogenation	2005 (67)
H	H	H	Me	azide displacement of chiral tosylate, reduction	99% tosylate formation 100% displacement, hydrogenation and N-Boc protection	2004 (65)
2-Cl-3-MOMO-pyrid-6-yl	H	H	H	Mitsunobu azide displacement of chiral 3-Ar-3-OH-propionic acid, azide reduction, azide reduction with PPh₃/H₂O	77% azidation and reduction	2002 (142)
H	CH(OH)-CH(OH)CO₂H	H	H	azide displacement of chiral substrate	66% for azide displacement	1997 (216)
H	CH(OH)CH(OH)CH(Me)-(CH₂)₄CH=CH=CH(4-MeO-Ph)	OH	H	azide displacement of chiral substrate, generate carboxy from hydroxymethyl	60% for triflate formation and azide displacement, 75% hydroxymethyl oxidation	1999 (212)
H	CH(OH)CH(OH) CH₂CH=CHCH=CH (4-EtO-Ph)	OH	H	azide displacement of chiral substrate, elaborate side chain, generate carboxy from hydroxymethyl	97% mesylate 83% azide displacement	2003 (215)
H	H	H	OH	amination of D-mannitol, oxidative diol cleavage	34–50% azide displacement	1987 (829)
H	H	OH	H		75% oxidative diol cleavage 92% azide reduction	
CH₂cHex	H	H	OH	amination of D-glucose via azide displacement of mesylate, carboxyl generation by Baeyer–Villiger oxidation	10% overall	1991 (830)

CH=CH-C(Me)=CH-CH(OMe)Bn	Me	H	azide displacement of D-glucose-derived triflate, aldehyde oxidation	77% amination 91% oxidation	1996 (831)
H	CH₂(imidazol-4-yl)	H	Mitsunobu azide displacement of chiral 3-Ar-3-OH-propionic acid, azide reduction, azide reduction with PPh₃/H₂O	24% displacement, azide reduction and deprotection	2003 (832) 2004 (833)
H Et	Et H	Me Me	azide displacement of chiral mesylate, oxidative cleavage of alkene	97% azide displacement 68% oxidation	1993 (834)
H	H	Me	azide displacement of chiral 3-Br-2-Me-propanol, oxidation	82% azide displacement 60% azide reduction 98% oxidation	1999 (835)
Ph	H	OH	dihydroxylation of cinnamate substrate obtained via Heck coupling of bromobenzene with methyl acrylate in one-pot, bromination, azide displacement	90% for Heck and dihydroxylation 99% ee 89% bromination 85% azide displacement 88% azide reduction	2003 (836)
iBu H	H iBu	OH OH	asymmetric dihydroxylation of iBuCH=CHCO₂Et, regioselective Mitsunobu azidation, or Mitsunobu inversion with TsOH followed by azide displacement	96% dihydroxylation 99% ee 82% Mitsunobu azidation 70% Mitsunobu inversion 90% azide displacement	2002 (837)
Ph	H	OH	dihydroxylation of cinnamate, bromination, azide displacement	74% for amination 23% overall 99% ee	1994 (182)
Ph 4-Me-Ph 4-F-Ph	H H H	OH OH OH	asymmetric dihydroxylation of alkyl cinnamate, amination via oxazoline formation with H₂SO₄ in MeCN or PhCN, aqueous hydrolysis	52–86% 84–90% de syn	2003 (183)
H H H	3-(CH₂-PO₃Et₂)-Bn 4-(CH₂-PO₃Et₂)-Bn 4-(CF₂-PO₃Et₂)-Bn	H H H	Mitsunobu amination of mono-O-Ac 2-Bn-1,3-propanediol, oxidation	63–65% desymmetrization 95–98% ee 68–98% Mitsunobu amination 84–91% oxidation	1998 (838)

(Continued)

Table 11.2 Asymmetric Syntheses of β-Amino Acids (continued)

R¹	R²	R³	R⁴	Method	Yield, % ee	Reference
CH=CHC(Me)=CHCH(Me)-CH(OMe)Bn	H	Me	H	Mitsunobu amination of chiral 1,2,4-trihydroxy-4-methylbutane oxidation and Wittig reaction of hydroxymethyl group	72% amination 90% oxidation 55% Wittig	1998 (839)
CH₂OH	H	Me	H	aldol reaction of acylated Evans' oxazolidinone, conversion to Weinreb amide, Mitsunobu azide displacement of hydroxyl, oxidation	98% aldol 89% amide exchange 92% Mitsunobu 87% oxidation	1999 (840)
CHO	H	Me	H			
Ph	H	Me	F	Mitsunobu amination with DEAD/DPPA of chiral substrate obtained by asymmetric fluorination of PhCOCH(Me)CO₂*t*Bu and diastereoselective ketone reduction; azide reduction	92–96% fluorination 91% ee 75–83% ketone reduction >90% de *syn* or *anti* 73–79% Mitsunobu amination 57–80% azide reduction	2002 (841)
H	Ph	Me	F			
4-HO-Ph	H	H	H	Mitsunobu displacement of homochiral hydroxy with azide, reduction	80% displacement	1993 (130)
CH=CHPh	H	H	OH	Mitsunobu azidation of chiral-protected polyhydroxy compound derived from L-arabinose	—	1997 (842)
Ph	H	H	OH	formation of diol by addition of hydroxyacetate to Cr complex of benzaldehyde, azide displacement	63% for amination	1992 (843) 1993 (844)
Ph	H	H	OH	Mitsunobu displacement of homochiral hydroxy with azide, reduction	82% displacement 90% reduction	1999 (845)
H	Bn	OH	H	Mitsunobu azide displacement of chiral alcohol	—	1989 (846)
6-amino-4-carboxy-5-methylpyrimidin-2-yl	H	H	H	aldol reaction of acylated Evans' oxazolidinone, Mitsunobu azide displacement of hydroxyl	52% Mitsunobu 82% azide reduction	1994 (847)

R^1	R^2	R^3	R^4	Reaction	Yield	Year (Ref)
$(CH_2)_3CCH$	Me	H	H	intramolecular displacement of homochiral mesylate by amide nitrogen	95–98% for displacement	1997 (848)
H	H					
Ph	$(CH_2)_3CCH$	OH		asymmetric dihydroxylation of cinnamide, bromination, intramolecular amination, oxidative cleavage of N-4-MeO-Ph	88% dihydroxylation; 94% ee; 95% bromination; 94% amination; 80% deprotection	1998 (849)
H	Me			intramolecular amination of β-hydroxy acid O-Bn hydroxylamine amide, hydrolysis, N–O cleavage	80–89% overall	1998 (850)
H	iPr					
H	iBu					
H	Bn					
Me	H					
iBu		OH	H	reaction of chiral α-OH-β-γ-unsaturated ester with TsNCO, iodocyclization, radical I removal, deprotection	74–83% carbamate formation and cyclization; 91–98% I removal; 43–70% deprotection	1997 (851)
CH_2cHex		OH	H			
H	nHept	H	OH			
H	N_3	Me		amination via formation of cyclic sulfamidate from chiral diol using Burgess reagent, opening with nucleophiles	96% sulfamidate formation; 86–97% opening	2004 (852)
H	SMe	Me				
H	OH	Me				
H	F	Me				
H	CN	Me				
Ph	H	OH		dihydroxylation of cinnamate, cyclic sulfate opening with azide, or with bromide followed by azide displacement	69–81% for amination	1994 (853)
Ph	H	OH		dihydroxylation of cinnamate, cyclic sulfite opening with azide followed by Mitsunobu inversion of carbinol	70–76% for amination; 65% for inversion	1995 (854)
nPr	H			opening of enantiomerically enriched β-lactones with azide or $2\text{-}NO_2\text{-}PhSO_2NHNa$	78–95% opening with azide; 43–83% opening with sulfonamide	2002 (855)
nBu	H					
iBu	H					
$(CH_2)_8 iPr$	H					
cHex	H					
$4\text{-}NO_2\text{-}Ph$	H					
CH_2CH_2Ph	H					
CH_2OBn	H					

(Continued)

Table 11.2 Asymmetric Syntheses of β-Amino Acids (continued)

Amination: Asymmetric Aminohydroxylation

R¹	R²	R³	R⁴	Method	Yield, % ee	Reference
Ph	H	H	OH	catalytic asymmetric aminohydroxylation of methyl cinnamate using Chloramine-T, hydrolysis	69% aminohydroxylation 82% ee 77% hydrolysis	1996 (856)
Ph	H	H	OH	catalytic asymmetric aminohydroxylation of methyl cinnamate/ ethyl crotonate using Chloramine-T	52–64% aminohydroxylation 60–82% ee	1996 (857)
Me	H	H	OH			
CH₂cHex	H	H	OH	catalytic asymmetric aminohydroxylation of α,β-unsaturated ester using Chloramine-T	60% aminohydroxylation 96% ee	1997 (858)
CH₂cHex	H	H	OH	catalytic asymmetric aminohydroxylation of α,β-unsaturated ester using Chloramine-T	65% aminohydroxylation 89% ee	1999 (859)
(CH₂)₃OH	H	H	OH	asymmetric aminohydroxylation of 6-HO-2-hexenoate using Chloramine-T	70% aminohydroxylation 85% ee 64% ester to alkene 79% oxidation	1998 (861)
Ph	H	H	OH	catalytic asymmetric aminohydroxylation of methyl cinnamate/ethyl crotonate using Chloramine-M	65% aminohydroxylation 95% ee	1996 (862)
H	H	H	OH	catalytic asymmetric aminohydroxylation of α,β-unsaturated esters using N-halocarbamates	61–89% aminohydroxylation 84–99% ee	1996 (863)
Ph	H	H	OH			
2,6-Me₂-Ph	H	H	OH			
4-MeO-Ph	H	H	OH			
4-NO₂-Ph	H	H	OH			
1-naphthyl	H	H	OH			
H	Me	OH	H	catalytic asymmetric aminohydroxylation of crotonate using CbzNCl	90% ee	1998 (864)
2-furyl	H	H	OH	catalytic asymmetric aminohydroxylation of crotonate using CbzNCl	41% >86% ee	1999 (865)

R1	R2	R3	R4	Reaction	Yield, ee	Year (Ref.)
H	4-MeO-Ph	OH	H	catalytic asymmetric aminohydroxylation of methyl p-MeO-cinnamate using CbzNCl	71% 99% ee	2003 (866)
Ar		H	OH	catalytic asymmetric aminohydroxylation of aryl acrylates using CbzNH(Cl)	45–75% 89–99% ee	1999 (867)
H	Ar = 2-furyl, 5-Me-2-furyl, 3-furyl, 2-thiophene, 5-Me-2-thiophene, 3-thiophene, 3-indolyl, 1-pyridyl, 2-pyridyl, 3-pyridyl	OH	H			
H	3,5-iPrO$_2$-4-MeO-Ph	H	H	catalytic asymmetric aminohydroxylation of aryl acrylates using BocNH(Cl), mesylation of alcohol and displacement with acetate or azide	70% aminohydroxylation; 98% mesylation; 54% acetate displacement; 91% azide displacement	2005 (868)
H	3,5-iPrO$_2$-4-MeO-Ph	OH	H			
H	3,5-iPrO$_2$-4-MeO-Ph	N$_3$	H			
Ph		H	OH	catalytic asymmetric aminohydroxylation of methyl cinnamates using BocNH(Br)	24–79% 79–86% ee	2002 (869)
2-NO$_2$-Ph		H	OH			
3-NO$_2$-Ph		H	OH			
4-NO$_2$-Ph		H	OH			
4-[NHCO(CH$_2$)$_3$CO$_2$Bn]-Bn		H	OH			
Ph		H	OH	catalytic asymmetric amino-hydroxylation of isopropyl cinnamate with 2-aminotriazine amine source	97–99% ee	1999 (870)
Ph	H	OH	H	catalytic asymmetric amino-hydroxylation of cinnamates using TeocN(Cl)Na	70–86% 98–99% ee	1998 (871)
3-NO$_2$-Ph	H	OH	H			
1-naphthyl	H	OH	H			
H	Ph	H	OH			
H	3-NO$_2$-Ph	H	OH			
H	1-naphthyl	H	OH			
Ph	H	OH	H	catalytic asymmetric amino-hydroxylation of cinnamates using bromoacetamide	71–81% 99% ee	1997 (872)
2-MeO-Ph	H	OH	H			
4-MeO-Ph	H	OH	H			
H	Ph	H	OH			
H	2-MeO-Ph	H	OH			
H	4-MeO-Ph	H	OH			

(Continued)

Table 11.2 Asymmetric Syntheses of β-Amino Acids (continued)

R¹	R²	R³	R⁴	Method	Yield, % ee	Reference
Ph	H	H	OH	catalytic asymmetric amino-hydroxylation of isopropyl cinnamate using bromoacetamide	76%	2005 (873)
Ph	H	H	OH	catalytic asymmetric amino-hydroxylation of cinnamates using bromobenzamide	46% 97% ee	1999 (874)
Ph	H	H	OH	catalytic asymmetric amino-hydroxylation of isopropyl cinnamate using bromoacetamide, formation of *cis*- or *trans*-substituted oxazoline derivative, opening with AcSH	81% aminohydroxylation 74% *trans* oxazoline 80% *cis* oxazoline 78–89% thiol opening	2002 (875)
Ph	H	H	SH			
Ph	H	SH	H			
Ph	H	H	OH	catalytic asymmetric amino-hydroxylation of isopropyl cinnamates using bromoacetamide and osmylated macroporous resin	80–93% amino-hydroxylation >99% ee	2003 (876)
4-MeO-Ph	H	H	OH			
Ph	H	H	OH	catalytic asymmetric amino-hydroxylation of cinnamates using polymer-suppoted ligand and AcNBrLi	30–71% 88–>99% ee	1998 (877)
4-MeO-Ph	H	H	OH			
4-NO₂-OH	H	H	OH			
H	H	OH	H	catalytic asymmetric amino-hydroxylation of α,β-unsaturated ester using *N*-Br-acetamide	65–79% >95% ee	1999 (879)
H	Ph	OH	H			
H	CH₂OBz	OH	H			
Ph	H	H	OH	catalytic aminohydroxylation of acrylate 1-phenylethyl amides	77–96% 0–100% regioselectivity 0–100% de	2005 (880)
H	H	OH	H			
H	H	Me	OH			
Me	H	H	OH			
Ph	H	H	OH			
Ph	H	H	OH	catalytic asymmetric aminohydroxylation of cinnamate ester of taxane	40–85%	1989 (882)
H	Ph	H	OH			

Amination: Rearrangement

R¹	R²	R³	R⁴	Method	Yield, % ee	Reference
H	H	OH	H	Hofmann rearrangement of L-β-malamidic acid	45% overall	1914 (883) 1946 (884)

H	H	OH	H	Hofmann rearrangement of L-β-malamidic acid	82% overall	1983 (885)
H	Bn	H	OH	alkylation of malic ester, selective Hofmann degradation	52% alkylation	1992 (886)
CH$_2$cHex	H	H	OH		90% Hofmann	
H	H	OH	H	amination of malic acid via Curtius rearrangement	58–72% for amination	1995 (173); 1995 (2185)
H	H	OH	H	amination of malic acid via Curtius rearrangement	46% overall	1988 (888)
H	H	OH	H	amination of malic acid via Curtius rearrangement	29% overall	1998 (174)
D	H	OH	H	amination of deuterated malic acid via Curtius rearrangement	—	1991 (890)
H	D	OH	D	amination of malic acid via Curtius rearrangement		1994 (889)
H	H	OH	H	amination of malic acid via Curtius rearrangement	31–32% overall	2002 (891)
H	CH$_2$CO$_2$H	H	H	Curtius rearrangement of chiral triacid, amidation of side chain	75–76% Curtius rearrangement	2003 (832)
H	(CH$_2$)$_2$CO$_2$H	H	H		72–83% amidation	2004 (892)
H	CH$_2$CONH$_2$	H	H			
H	(CH$_2$)$_2$CONH$_2$	H	H			
H	H	Ph	H	Curtius rearrangement of 2-substituted succinic acids prepared via alkylation of acids derivatized with Evans' oxazolidinone with BrCH$_2$CO$_2$tBu	49–69% alkylation	1998 (893)
H	H	4-HO-Ph	H		40–58% rearrangement	
H	H	Bn	H		65–84% deprotection	
H	H	indol-3-yl	H			
H	H	CH$_2$(indol-3-yl)	H	Curtius rearrangement of 2-substituted succinic acids prepared via alkylation of acids derivatized with Evans' oxazolidinone with BrCH$_2$CO$_2$tBu	76% alkylation	2002 (894)
H	H	H	CH$_2$(indol-3-yl)		>95% de	
					100% ester hydrogenation	
					74% Curtius	
H	H	Me	H	Curtius rearrangement of 2-substituted succinic acids prepared via alkylation of acids derivatized with Evans' oxazolidinone with BrCH$_2$CO$_2$tBu	82–89% alkylation	1999 (895)
H	H	iPr	H		86–>98% de	
H	H	tBu	H		84–95% hydrolysis	
H	H	Ph	H		74–79% rearrangement	
H	H	Bn	H			

(Continued)

Table 11.2 Asymmetric Syntheses of β-Amino Acids (continued)

R¹	R²	R³	R⁴	Method	Yield, % ee	Reference
H	Bn	H	H	asymmetric alkylation of succinic acid mono tBu ester with oxazolidinone chiral auxiliary, auxiliary removal and Curtius rearrangement, or tBu ester removal, Curtius rearrangement, and auxiliary removal	60–83% allylation	2000 (896)
H	CH₂CH=CH₂	H	H		81–>97% de	
H	CH₂CH=CHnOct	H	H		80–90% auxiliary removal	
H	CH₂CH=CHPh	H	H		70–83% Curtius	
H	H	H	Me		or	
H	H	H	Bn		55–97% tBu ester removal	
H	H	H	CH₂CH=CH₂		66–69% Curtius	
					83–98% auxiliary removal	
H	Me	H	H	asymmetric allylation of acids with oxazolidinone chiral auxiliary, auxiliary removal, amidation, allyl oxidative cleavage, Hofmann rearrangement	60–73% allylation	2002 (897)
H	iPr	H	H		84–95% oxazolidinone hydrolysis	
H	nOct	H	H		91–96% amidation	
H	C₁₆H₃₃	H	H		49–63% alkene oxidative cleavage	
H	Ph	H	H		61–70% Hofmann rearrangement	
H	2,4,6-Me₃-Ph	H	H	Curtius rearrangement of chiral acid with 2nd carboxyl masked as diphenylalkene	49–52% Curtius	1996 (898)
H	H	H	Me		60–65% alkene oxidative cleavage	1997 (899)
H	H	H	Ph			
H	H	H	2,4,6-Me₃-Ph			
H	H	H	1-naphthyl			
H	CH₂CH=CH₂	H	H	Curtius rearrangement of chiral acid with 2nd carboxyl masked as dithiane-protected aldehyde, aldehyde oxidation	72% rearrangement	1982 (900)
					92% deprotection/ oxidation	
H	Me	Et	H	Curtius rearrangement of chiral acid with 2nd carboxyl masked as alkyne	62–83% Curtius	1993 (901)
H	Me	iPr	H		64–97% alkyne oxidative cleavage	1993 (902)
H	Me	CH(OBn)-Me	H			
H	CH(OH)CH(OH)CO₂H	H	H	Curtius rearrangement of 2-allyl-3,4-dihydroxybutanoic acid with 2nd carboxyl masked as allyl group	92% Curtius	2003 (225)
					80% oxidative cleavage	
nHept	H	H	H	Curtius rearrangement of chiral acid with 2nd carboxyl masked as alkene, oxidative alkene cleavage	83% rearrangement	2001 (88)
					93–99% oxidative cleavage	

Ph	H	H	OH	intramolecular iodoamidation of trichloroacetimidate derivative of 4-Ph-but-3-ene-1,2-diol	69% iodoamidation	1999 (903)
H	4-MeO-Ph	H	CH[3,4-(OCH$_2$O)-Ph]CH$_2$OH	6-*endo-trig* cyclization of homochiral substrate	97% cyclization/amination	1999 (904)
CH=CHPh	H	H	H	Overman rearrangement of allylic alcohol, alcohol oxidation	66–72% rearrangement, 93–100% oxidation	2003 (905)
H	CH=CHPh	H	H			
H	CH=CH(4-CF$_3$-Ph)	H	H			
H	CH=CH(2-Br-Ph)	H	H			
H	CH=CH(4-Cl-Ph)	H	H			
H	CH=CH(4-F-Ph)	H	H			
H	CH=CH(*c*Hex)	H	H			
H	CH=CH(*t*Bu)	H	H			
H	CH=CH(*n*Oct)	H	H			
Me	Et	H		Rh catalyzed intramolecular C–H insertion of H$_2$NSO$_2$-ester of (S)-3-Me-1-pentanol, hydrolysis of resulting oxathiazinane, oxidation	91% insertion, 81% hydrolysis and oxidation	2001 (518)
Amination: Reductive Amination						
Me	H			reductive amination of β-keto ester using chiral amine and hydrogenation or NaBH$_3$CN	23–44%, 3–28% ee	1979 (906)
Ph	H					
Me	H			reductive amination of β-keto ester using chiral amine and NaBH(OAc)$_3$	30–85% de crude; 47–73% yield with >99% de; 71–86% auxiliary removal	1994 (908); 1996 (907)
*n*Pr	H					
*i*Pr	H					
*t*Bu	H					
Ph	H					
Me	(CH$_2$)$_2$OH					
Me	(CH$_2$)$_2$-β-N					
Me	Me					
-(CH$_2$)$_3$-C2	-(CH$_2$)$_3$-C3					
(CH$_2$)$_4$-C2	-(CH$_2$)$_4$-C3					
Me	H			reductive amination of β-keto benzyl ester using chiral amine and NaBH(OAc)$_3$	6–89% yield purified major isomer; 83–93% auxiliary removal	2001 (909)
*i*Pr	H					
Ph	H					
-(CH$_2$)$_3$-C2	-(CH$_2$)$_3$-C3					
-(CH$_2$)$_4$-C2	-(CH$_2$)$_4$-C3					

(Continued)

Table 11.2 Asymmetric Syntheses of β-Amino Acids (continued)

R¹	R²	R³	R⁴	Method	Yield, % ee	Reference
H	3-pyridyl	H	H	reductive amination of β-keto ester using chiral amine and hydrogenation	79% hydrogenation 78% de 94% deprotection	1999 (396)
H	CH₂CO₂H	H	CH(OH)Me	enamine formation from acetonedicarboxylate and α-methylbenzylamine, acetylation, catalytic hydrogenation	80–95% enamine and acetylation 63% hydrogenation	1986 (910)
H	Ph	H	H	enamine formation from β-keto esters and Phg-NH₂, catalytic hydrogenation over Pd(OH)₂/C + BF₃·Et₂O, auxiliary removal using Et₃SiH/HCO₂H	75–95% enamine formation 42–80% de enamine reduction 75–95% auxiliary removal	2002 (911)
H	4-F-Ph	H	H			
H	4-HO-Ph	H	H			
H	4-MeO-Ph	H	H			
H	4-CO₂Me-Ph	H	H			
H	2-pyridyl	H	H			
H	3-pyridyl	H	H			
H	4-pyridyl	H	H			
H	Me	H	H	enamine formation from β-keto esters and Phg-NH₂, catalytic hydrogenation over PtO₂, auxiliary removal using Pd(OH)₂/C	24–99% hydrogenation 70–99% de	2004 (912)
H	iPr	H	H			
H	Bn	H	H			
H	4-MeO-Ph	H	H			
H	4-CF₃-Ph	H	H			
Me	H	H	H	hydrogenation of 3-aminoacrylate chiral ester auxiliaries	96% de	1990 (913)
Ph	H	OH	H	oxime formation of β-keto esters of 10-deacetylbaccatin III, α-hydroxylation, oxime reduction, α-OH inversion	74% oxime and α-hydroxylation 33% de 78% oxime reduction 100% de 72% inversion	2000 (914) 2000 (915)
Ph	H	H	OH			
iPr	H	H	OH	reductive amination of β-keto epoxide using benzylamine or ammonium acetate, epoxide opening with O nucleophile, hydroxymethyl oxidation	44–55% reductive amination and epoide opening 55–60% oxidation	1996 (916)
iBu	H	H	OH			
Bn	H	H	OH			
CH₂cHex	H	H	OH			
Bn	H	OH	H			

			Reaction	Results	Year (Ref.)
Ph	H		reductive amination of β-aryl-β-keto esters with ammonium acetate and H$_2$ using Rh-ClMeBIPHEP catalyst	79–88% 96–99% ee	2005 (917)
3-Cl-Ph	H				
3-MeO-Ph	H				
4-F-Ph	H				
4-MeO-Phh	H				
4-Cl-Ph	H				
H	Ph	H	hydrogenation of 3-aminoacrylates with o-BINAPO-Ru(II) catalyst	100% 80–99% ee	2002 (918)
H	4-F-Ph	H			
H	4-Cl-Ph	H			
H	4-Br-Ph	H			
H	4-Me-Ph	H			
H	4-MeO-Ph	H			
H	2-Me-Ph	H			
H	2-MeO-Ph	H			
Me	H	H	hydrogenation of 3-aminoacrylates with BPPM-Rh(II) catalyst	91–100% 53–55% ee	1978 (920)
Ph	H	H			
Me	H	H	hydrogenation of (Z)-3-substituted-3-aminoacrylates with DuPHOS in polar solvent	87.8% ee	2001 (921)
Me	H	H	hydrogenation of 2-substituted-3-aminoacrylates with DuPHOS or BPE at different temperatures	99% ee for (E) 82–90% ee for (Z)	2003 (922)
H	Me	H	hydrogenation of 2-substituted-3-aminoacrylates with DuPHOS or BPE	10–67% ee	2005 (923)
H	iPr	H			
H	iBu	H			
H	Ph	H			
Me	H	H	hydrogenation of 3-aminoacrylates with Et-DuPHOS ligand	97–>99% ee from E 68–87% ee from Z	2002 (924) 2002 (925)
Et	H	H			
iPr		H			
Me	H	H	hydrogenation of 3-acetamido-acrylates with MalPHOS catalyst	79–>99.5% ee from (E) 63–90% ee from (Z)	2003 (926)
Et	H	H			
iPr	H	H			
Ph	H	H			
H	Me	H			
H	Et	H			
H	iPr	H			
H	Ph	H			

(Continued)

Table 11.2 Asymmetric Syntheses of β-Amino Acids (continued)

R¹	R²	R³	R⁴	Method	Yield, % ee	Reference
Me	H	H	H	catalytic asymmetric hydrogenation with QuinoxP* catalyst	99.2% ee	2005 (927)
Me	H	H	H	hydrogenation of 3-aminoacrylates with Rh-TangPhos catalyst	74.3–99.6% ee	2002 (928)
Et	H	H	H			
nPr	H	H	H			
iBu	H	H	H			
Ph	H	H	H			
4-F-Ph	H	H	H			
4-Cl-Ph	H	H	H			
4-Br-Ph	H	H	H			
4-Me-Ph	H	H	H			
4-MeO-Ph	H	H	H			
4-BnO-Ph	H	H	H			
2-Me-Ph	H	H	H			
2-MeO-Ph	H	H	H			
Me	H	H	H	hydrogenation of 3-aminoacrylates with Rh catalyst with bis(binaphtho-phosphepine sulfide) ligand	96–>99% ee from E substrate	2003 (929)
Ph	H	H	H			
4-F-Ph	H	H	H			
4-Cl-Ph	H	H	H			
4-Br-Ph	H	H	H			
4-Me-Ph	H	H	H			
4-MeO-Ph	H	H	H			
4-BnO-Ph	H	H	H			
2-Me-Ph	H	H	H			
2-MeO-Ph	H	H	H			
3-pyridyl	H	H	H			
Et	H	H	H	hydrogenation of 3-aminoacrylates with Rh-DuanPhos catalyst	92–>99% ee	2005 (930)
4-MeO-Ph	H	H	H			
4-Cl-Ph	H	H	H			
Ph	H	H	H	hydrogenation of 3-aminoacrylates with Rh-ferrocenophosphines	85–98% 96–97% ee	2004 (931)
4-MeO-Ph	H	H	H			
4-F-Ph	H	H	H			
3-pyridyl	H	H	H			
Bn	H	H	H			

Me	H	H	hydrogenation of 3-aminoacrylates with ferrocenyl-binaphthol phosphoamidite catalyst	92–99% ee	2005 (932)
H	Me	H			
Et	H	H			
H	Et	H			
iPr	H	H			
H	iPr	H			
Ph	H	H			
4-Me-Ph	H	H			
4-MeO-Ph	H	H			
4-Cl-Ph	H	H			
4-F-Ph	H	H			
3-MeO-Ph	H	H			
Me	H	H	hydrogenation of 3-acetamido-acrylates with Ru-bipyridyl catalyst	>98% 97.3–99.7% ee from (E) 68.3–82.3% ee from (Z)	2003 (933)
Et	H	H			
iPr	H	H			
nPr	H	H			
tBu	H	H			
H	Me	H			
H	Et	H			
H	iPr	H			
H	nPr	H			
H	tBu	H			
Me	H	H	hydrogenation of 3-acetamido-acrylates with Ru-biaryl catalyst	>98% 96.0–99.8% ee from (E)	2004 (934)
Et	H	H			
iPr	H	H			
nPr	H	H			
tBu	H	H			
Me	H	H	hydrogenation of 3-aminoacrylates with Rh-BDPMI	75.6–97.4% ee	2002 (935)
Et	H	H			
iPr	H	H			
iBu	H	H			
Ph	H	H			
Me	H	H	hydrogenation of 3-aminoacrylates with camphor-based bisphosphine	99.3% ee	2005 (936)
Me	H	H	hydrogenation of 3-aminoacrylates with 1,3-biphenyl-1,3-bis(diphenylphosphine)-propane ligand	98% ee	2003 (938) 2005 (937)

(Continued)

Table 11.2 Asymmetric Syntheses of β-Amino Acids (continued)

R^1	R^2	R^3	R^4	Method	Yield, % ee	Reference
Me	H	H	H	hydrogenation of 3-aminoacrylates with P-chirogenic bisphospholane ligands	83–96% ee	2004 (939)
Me	H	H	H	hydrogenation of 3-acetamido-acrylates with three-hindered quadrant-Ru(II) catalyst	100%	2004 (940)
iPr	H	H	H		98–99% ee from (E)	
iBu	H	H	H		92–98% ee from (Z)	
tBu	H	H	H			
Ph	H	H	H			
-(CH₂)₃-C2	H	-(CH₂)₃-C3				
H	Me	H	H	hydrogenation of 3-aminoacrylates with BINAP-Ru(II) catalyst	100%	1991 (480)
H	iBu	H	H		90–96% ee	
H	Ph	H	H			
H	Me	H	H	hydrogenation of 3-aminoacrylates with MonoPhos-Ru(II) catalyst analog	100%	2002 (941)
H	Et	H	H		98–99% ee	
H	iPr	H	H			
H	Ph	H	H			
H	4-F-Ph	H	H			
Me	H	H	H	hydrogenation of 3-aminoacrylates with Rh-tBu-miniPHOS or tBu-BisP	95.6–99.7% ee	2001 (943)
Et	H	H	H			
nPr	H	H	H			
H	Me	H	H	hydrogenation of 3-aminoacrylates with monodentate phosphite binaphthol-mannitol-Ru(II) catalyst analog	93–100%	2004 (944)
H	Et	H	H		98.7–99.9% ee	
H	iPr	H	H			
H	Me	H	H	hydrogenation of 3-aminoacrylates with monodentate phosphite binaphthol-carbohydrate Ru(II) catalyst analog	40–92%	2004 (945)
H	Et	H	H		93.0–97.5% ee	
H	Ph	H	H			
H	Ph	H	H	hydrogenation of 3-aminoacrylates with chiral Rh catalyst using monodentate spirophosphonite ligand	100%	2004 (946)
H	4-Me-Ph	H	H		98–99% ee	
H	2-MeO-Ph	H	H			
H	4-MeO-Ph	H	H			
H	4-Cl-Ph	H	H			
H	3-Br-Ph	H	H			
H	4-Br-Ph	H	H			
Me	H	H	H	hydrogenation of 3-aminoacrylates with P-chirogenic trialkylphosphonium salts	>99% ee for (E)-substrate 64% ee for (Z)-substrate	2003 (947)

Me	R	R	hydrogenation of 3-aminoacrylates with chiral Rh catalyst using monodentate phosphoramidite ligand	93% ee	2004 (950)
Me	H	H	hydrogenation of 3-aminoacrylates with chiral Rh catalyst using combinations of monodentate P-ligands	100% 94.4–98.8% ee	2004 (951)
Et	H	H			
nPr	H	H			
Ph	H	H			
Me	H	H	hydrogenation of 3-aminoacrylates with chiral Rh catalyst using combination of monodentate P-ligand with achiral ligand	69–94% 84–97% ee	2005 (952)
Et	H	H			
nPr	H	H			
Ph	H	H			
Me	H	H	hydrogenation of (E)- or (Z)-3-substituted 3-aminoacrylates with DuPHOS or BICP-Ru(II) catalyst	93–100% (E) substrate: DuPHOS: 97.6–99.6% ee BICP: 90.9–97.0% ee (Z) substrate: DuPHOS: 11.2–62.4% ee BICP: 86.4–92.9% ee	1999 (482)
Et	H	H			
nPr	H	H			
iPr	H	H			
iBu	H	H			
Ph	H	H			
Me	H	H	hydrogenation of 3-aminoacrylates with FerroTANE	98–>99% ee	2003 (483)
iPr	H	H			
Ph	H	H			
4-Me-Ph	H	H			
4-MeO-Ph	H	H			
4-Cl-Ph	H	H			
4-F-Ph	H	H			
2-MeO-Ph	H	H			
3-NO$_2$-Ph	H	H			
Me	CO$_2$Me	H	conversion of aryl group to malonate	77% aryl reduction, ozonolysis 86% decarboxylation	1991 (954)
Me	H	H			
CF$_3$	H	H	reductive amination of β-keto ester using benzylamine and [1,3]-proton shift with chiral base, hydrolysis	67–89% 15–36% ee 88–93% hydrolysis	1994 (955) 1996 (523)
C$_2$F$_5$	H	H			
C$_3$F$_7$	H	H			
CF$_2$CF$_2$H	H	H			
(CF$_2$)$_3$CF$_2$H	H	H			
CF$_3$	H	H	reductive amination of β-keto ester using α-phenylethyl-amine and base-catalyzed [1,3]-proton shift	57–88% 88–96% ee	1997 (956)
C$_3$F$_7$	H	H			

(Continued)

Table 11.2 Asymmetric Syntheses of β-Amino Acids (continued)

R¹	R²	R³	R⁴	Method	Yield, % ee	Reference
Amination: Other						
Me	H	H	OH	asymmetric amination via reaction with dibenzyl azodicarboxylate in presence of chiral Cu bis(oxazoline) catalyst, hydrazino reduction	39–78% amination / 90–96% ee / 33–89% reduction	2002 (957)
nPent	H	H	OH			
iPr	H	H	OH			
iBu	H	H	OH			
CH₂CH=CH₂	H	H	OH			
CH₂cHex	H	H	OH			
Bn	H	H	OH			
Carboxylation						
H	Et	H	H	conversion of chiral amino alcohol into N-acyl α-chloroamine, di-deprotonation, quenching with CO₂ or CO(OEt)₂	51–76% conversion / 64–90% quenching	1996 (958)
H	iPr	H	H			
H	Ph	H	H			
Me	H	H	H	carboxylation of lithium enamine to give β-enamino ester, reduction (only 1 reduced)	33–76%	1995 (534)
Et	H	H	H			
H	H	Me	H			
H	H	iPr	H			
H	H	tBu	H			
H	H	Ph	H			
-(CH₂)₃-C2	H	-(CH₂)₃-C3	H			
-(CH₂)₄-C2	H	-(CH₂)₄-C3	H			
Bn	H	=CHSnBu₃	—	conversion of amino acid into amino alkyne, formation of organostannane, Stille coupling of stannane, alkene hydrogenation	45–55% carboxylation / 52–74% Stille coupling	2002 (959)
iBu	H	=CHSnBu₃	—			
H	H	=CHSnBu₃	—			
H	H	=CHPh	—			
H	H	=CH(4-MeO-Ph)	—			
Bn	H	=CHPh	—			
Bn	H	=CH(4-MeO-Ph)	—			
iBu	H	=CHPh	—			
iBu	H	=CH(4-MeO-Ph)	—			
Bn	H	Bn	H			
Bn	H	4-MeO-Bn	H			
H	H	Ph	H	carboxylation of anion of PhCH₂CN with N-(ClCO)-Oppolzer's sultam, reduction CN/amide, oxidation	90% acylation / 90% de / 20% reduction/oxidation	1998 (960)
H	H	H	Ph			

CH(iPr)CH(OH)Me	H	H	intramolecular amine addition to chiral allylic sulfoxamine, deprotonation and carboxylation or conversion to chloride and NaCN displacement	80–94% intramolecular amination; 48–74% carboxylation and auxiliary removal	2003 (961)
CH(Ph)CH(OH)iPr	H	Me			

Fusion of Aminomethyl and Acetate: Nucleophilic Aminomethyl

H	OH	H	addition of nitromethane to glyoxylate with chiral auxiliary, reduction	90% addition; >95% de; 50% overall	1988 (962)
H	OH	iBu	addition of nitroalkane to glyoxylate with chiral auxiliary	90% addition; 80% de C2; 74% de C3; 53–55% overall	1990 (963)
H	OH	Ph	addition of nitroalkane to glyoxylate with chiral auxiliary, nitro reduction	50–98% yield addition; 18–84% ds major diastereomer	2004 (964)
H	OH	Bn			
H	OH	H	addition of nitroalkane to glyoxylate with chiral bicyclic oxazolidinone auxiliary, nitro reduction	72–93% yield addition; 10–30% ds major diastereomer	2004 (965)
H	OH	cPent			
H	OH	CH(OEt)$_2$			
H	OH	CO$_2$Et			
H	OH	Ph			
H	OH	Bn			
H	OH	Bn	addition of nitroalkane to glyoxylate with chiral La catalyst, nitro reduction	81% yield addition; 93% ee; 60% hydrogenation	2005 (966)
Ph	H	OH	addition of phenylnitromethane to O-benzyl hydroxyacetaldehyde with chiral La catalyst, nitro reduction, hydroxymethyl oxidation	80% yield addition; 98% ee; 100% de; 80% hydrogenation; 82% oxidation; 96% ee	2004 (967)
H	Me		addition of nitromethane to α-keto esters using chiral bisoxazoline-Cu(II) catalyst	46–99% addition; 77–93% ee	2001 (968)
H	Et				
H	CH$_2$CH$_2$Ph				
H	Ph				
H	4-NO$_2$-Ph				
(as nitro derivative)					
Ph	H		asymmetric deprotonation of benzylamine, quenching with methylenecarboxy equivalent	25–76% yield; 87% ee (S) to 84% ee (R)	1997 (969)
H	Ph				

(Continued)

Table 11.2 Asymmetric Syntheses of β-Amino Acids (continued)

R¹	R²	R³	R⁴	Method	Yield, % ee	Reference
H	Ph	H	H	alkylation of *N*-protected benzylamine with 4-Br-2-Me-2-butene using *n*-BuLi/(−)-sparteine, ozonolysis	76% alkylation and ozonolysis 86% ee	1999 (970)

Fusion of Aminomethyl and Acetate: Nucleophilic Acetate

R¹	R²	R³	R⁴	Method	Yield, % ee	Reference
Me	H	H	H	addition of ZnBrCH₂CO₂R to chiral imine	10–40%	1978 (971)
Ph	H	H	H		3–28% ee	
Ph	H	H	H	one-pot addition of ZnBrCH₂CO₂R to imines formed from α-methylbenzylamine and aryl aldehydes	76–94%	2003 (972)
4-Cl-Ph	H	H	H		82–95% de	
2-Cl-Ph	H	H	H			
2,4-Cl₂-Ph	H	H	H			
2-NO₂-Ph	H	H	H			
3-pyridyl	H	H	H			
4-MeO-Ph	H	H	H			
Ph	H	H	OH	addition of α-MeO-TMS ketene acetal to chiral imine of benzaldehyde	93% addition 66% de *syn* (25–30% overall, including deprotection)	1998 (973)
H	CO₂Et	Me	H	addition of RCH=C(S*t*Bu)OSn*t*Bu enolates to chiral imines	78–85%	1987 (974)
H	CO₂Et	Et	H		50–90% de	
H	CO₂Et	*i*Pr	H		70–84% ee	
*n*Pr	H	Me	Me	addition of silyl ketene acetal to chiral imine formed from galactosylamine	82–92% yield	1989 (975)
Ph	H	Me	Me		82–99% de	
2-Cl-Ph	H	Me	Me		95% hydrolysis	
3-Cl-Ph	H	Me	Me			
4-Cl-Ph	H	Me	Me			
4-F-Ph	H	Me	Me			
2-naphthyl	H	Me	Me			
H	Ph	H	Me	addition of silyl ketene acetal or lithium ester enolate to chiral imine formed from galactosylamine	45–97% addition yield	1997 (976)
H	3-Cl-Ph	H	Me		100% de *erythro* or *threo*	
H	4-F-Ph	H	Me		50–>90% ee	
H	4-Me-Ph	H	Me		88% hydrolysis	
H	2-naphthyl	H	Me			
H	4-Cl-Ph	H	Et			
H	*n*Pent	H	Ph			
H	Ph	H	Ph			
H	4-Cl-Ph	H	Ph			
H	Ph	Ph	H			

nPr	H	H	addition of allyl trimethylsilane or allylstannane to chiral imine formed from galactosylamine, alkene oxidative cleavage	26–82% addition yield	1990 (977)
nNon	H	H		48–92% de	1991 (978)
Ph	H	H		27–87% alkene oxidation	
2-Cl-Ph	H	H			
3-Cl-Ph	H	H			
4-Cl-Ph	H	H			
2-NO₂-Ph	H	H			
4-NO₂-Ph	H	H			
4-Me-Ph	H	H			
2-MeO-Ph	H	H			
4-MeO-Ph	H	H			
4-F-Ph	H	H			
4-CN-Ph	H	H			
3-pyridyl	H	H			
2-naphthyl	H	H			
CH=CHPh	H	H			
Ph	Me	Me	addition of dimethylketene silyl acetal to imines prepared from chiral aminopyrrolidine	80% yield addition 84% de	1997 (980)
2-(MeO)-Ph	Me, H	Me, H	addition of acetate to chiral Cr(CO)₃ imine complex	35–62%	1996 (981)
4-Cl-5-Me-6-EtO₂C-2-pyrimidine	H	H	addition of acetate thioester boron enolate to chiral imine prepared from 2-CHO-pyrimidine derivative and side chain of L-Dap	40% addition	1981 (584)
Ph	H	H	addition of silyl ketene acetal to acyliminium ion derived from arylamine and amino acid	17–91% condensation 56–>98% de 81% deprotection (1 example)	1999 (982)
2-MeO-Ph	H	H			
4-MeO-Ph	H	H			
Ph	Me	Me			
2-MeO-Ph	Me	Me			
4-MeO-Ph	Me	Me			
4-NMe₂-Ph	Me	Me			
4-Cl-Ph	Me	Me			
2,4,6-Me₂-Ph	Me	Me			
Bn	OH	H	addition of silyl ketene acetal to chiral imine generated in situ from α-cyanoamine	97% addition 67% de	1999 (983)
Me	Ph	H	addition of acetate to homochiral sulfinimine, enolate generation and hydroxylation	82–84% addition 80–100% de 58% hydroxylation 72% de	1992 (984)
Me	Ph	H			
H	H	OH			

(Continued)

Table 11.2 Asymmetric Syntheses of β-Amino Acids (continued)

R^1	R^2	R^3	R^4	Method	Yield, % ee	Reference
H	3-pyridyl	H	H	addition of acetate to homochiral sulfinimine	85% addition / 78% de / 90% deprotection	1996 (985)
Ph	H	H	H	addition of acetate to homochiral sulfinimine	73–84% addition / 98% de / 92% deprotection	1994 (987) / 1995 (986)
Ph	Ph	H	F	addition of fluoroacetate to homochiral sulfinimine	37% addition / 16% de syn	1994 (989)
Ph	Ph	F	H			
nPent	H	F	F	addition of Reformatsky reagent derived from BrCF$_2$CO$_2$Et to chiral sulfinimines	59–85% addition / 72–>98% de / 56–96% auxiliary removal	2003 (990)
iPr	H	F	F			
tBu	H	F	F			
Ph	H	F	F			
4-MeO-Ph	H	F	F			
4-F-Ph	H	F	F			
4-Cl-Ph	H	F	F			
4-CF$_3$-Ph	H	F	F			
2-furyl	H	F	F			
Ph	H	CH=CH$_2$	H	addition of dienolate of Me 3-butenoate to chiral benzaldehyde N-sulfinimine	34–90% addition / 74% de anti to 40% de syn	1998 (991)
Ph	H	H	CH=CH$_2$			
tBu	H	H	H	addition of acetate to chiral benzaldehyde N-sulfinimine	96% addition / >96% de	2003 (992)
Ph	H	H	H	addition of silyl ketene acetals to chiral imines in presence of Lewis acids such as TMSOTf	40–80% addition / 59–91% de	1999 (993)
Ph	H	Me	Me			
Ph	H	=CR^{1}R^2	—	addition of organocuprates generated from R$_2$CuLi and RCCCO$_2$Et to homochiral sulfinyl imines	51–64% / 64–>90% de	1999 (994)
4-Cl-Ph	H	R^1 = H; R^2 = Ph	—			
4-F-Ph	H	R^1 = Ph; R^2 = Me, Ph	—			
2-furyl	H		—			
2-thienyl	H		—			
Ph	H	H	H	addition of α-phosphonate carbanion to homochiral sulfinimine	75–80% addition / 64–82% de	1996 (995)
2-furyl	H	H	H			
2-thienyl	H	H	H			
CH=CHPh (with PO$_3$R$_2$ instead of CO$_2$H)	H	H	H			

Substrate R¹	R²	R³	Method	Yields	Year (ref)
3-R-Ph	H	H	addition of acetate to homochiral bis(sulfinimine) formed from dialdehyde	40–46% sulfinimine formation; 24–46% addition; 20–40% de; 82–93% deprotection	1998 (*996*)
4-R-Ph	H	H			
6-R-2-naphthyl	H	H			
R = $CH(NH_2)$-CH_2CO_2H					
Me	H	H	addition of Ti ester enolates to homochiral *tert*-butanesulfinyl imines	65–96% addition; 90–98% de	1999 (*999*); 2002 (*1000*)
Me	Me	H			
iPr	H	H			
iBu	H	H			
Ph	H	H			
Ph	Me	H			
pyrid-3-yl	H	H			
Me	Me	Me			
iBu	Me	Me			
Ph	Me	Me			
Et	Bn	H			
Me	Me	H			
Me	4-MeO-Bn	H			
Ph	Me	Me			
iBu	Me	Me			
Ph	Me	Me			
Ph	Me	Me			
Ph	Me	Me			
-$(CH_2)_5$-		—			
4-$(CH_2PO_3Bn_2)$-Ph	H	H	addition of Ti ester enolates to homochiral *tert*-butanesulfinyl imines	79% imine formation; 83% addition; >90% de	1999 (*999*); 2002 (*1000*)
iBu	F	F	addition of ethyl difluoroacetate Reformatsky reagent to homochiral *tert*-butanesulfinyl imines	51–82% addition; 60–90% de	2002 (*1002*)
nPr	F	F			
Ph	F	F			
cHex	F	F			
2-thiazolyl	F	F			
Ph	Me	H	addition of allylMgBr to homochiral sulfinimine, oxidative cleavage of alkene	92–98% addition; 82–100% de; 56–60% oxidation and deprotection	1991 (*588*)
Ph	nBu	H			
Ph	H	H	addition of allylMgBr to oxime formed from RCHO and H_2NOCH(nBu)Ph, N–O reductive cleavage, alkene oxidation	54–100% oxime formation; 78–100% addition; 86–>96% de; 47–75% reductive cleavage; 33–52% oxidation	1998 (*1004*)
4-MeO-Ph	H	H			
cHex	H	H			
iPr	H	H			
$CHEt_2$	H	H			

(*Continued*)

Table 11.2 Asymmetric Syntheses of β-Amino Acids (continued)

R¹	R²	R³	R⁴	Method	Yield, % ee	Reference
H	Ph	H	H	addition of allylMgBr to homochiral nitrone derivative of aldehyde	50–87% addition	1990 (1005)
H	nPent	H	H		20–56% de	
Ph	H	OH	H	addition of silyl ketene acetal with chiral auxiliary to nitrone of benzaldehyde with chiral auxiliary, hydrogenolysis	98% addition >95% de anti >95% ee 60% deprotection, hydrogenolysis	1997 (1006)
H	CH(OH)CH$_2$OH	H	H	addition of α-Cl-acrylonitrile to nitrone of isopropylidene-D-glyceraldehyde	72% nitrone formation 79% cycloaddition	1990 (1007)
H	CH(NH)CH$_2$OH	H	H	addition of silyl ketene acetal to nitrone of Garner aldehyde	92% addition 86% de	1998 (1008)
Me	H	H	H	addition of acetate equivalent (allyl cerium) to chiral SAMP hydrazone	63–99% addition	1994 (1009)
nPr	H	H	H		90–99% de	
nBu	H	H	H		54–89% SAMP removal	
iBu	H	H	H		66–74% oxidative cleavage	
iPr	H	H	H	addition of acetate equivalent (allylMgBr) to chiral oxime formed from ROPHy or SOPHy, hydroxylamine cleavage, alkene oxidative cleavage	78–100% addition	1999 (1010)
CH(Et)$_2$	H	H	H		86–96% de	
cHex	H	H	H		78–98% N-O cleavage	
Ph	H	H	H		33–52% oxidative cleavage	
4-MeO-Ph	H	H	H			
H	Me	H	H	addition of Reformatsky reagent to chiral oxazolidinone prepared from RCHO and N-Bz-phenylglycinol, deprotection by hydrogenation	63–80%	1992 (1011)
Me	H	H	H		38–92% de	
Et	H	H	H			
nPr	H	H	H			
iPr	H	H	H			
nBu	H	H	H			
iBu	H	H	H			
(CH$_2$)$_2$Ph	H	H	H			
Ph	H	H	H	addition of Reformatsky reagent to chiral oxazolidinone prepared from RCHO and N-Bz-phenylglycinol, deprotection by Pb(OAc)$_4$	73–87% deprotection	1993 (1012)
4-Br-Ph	H	H	H			
cHex	H	H	H			

			Method	Results	Year (Ref)
*n*Pent	F	F	addition of Reformatsky reagent from BrCF$_2$CO$_2$Et to chiral oxazolidinone, deprotection using Tf$_2$O/Pyr, then KO*t*Bu, then 6N HCl	32–69% addition; 85–>99% ee; 60–70% deprotection	1999 (*919*); 2004 (*1013*)
(*E*)-CH=CH(Me)	F	F			
Ph	F	F			
2-furyl	F	F			
*i*Pr	H	H	addition of Bu$_3$SnCH$_2$CO$_2$Et to chiral oxazolidinone prepared from RCHO and *N*-Bz-phenylglycinol	33–71%; 91–99% de; 73–87% deprotection	1993 (*1014*)
Ph	H	H			
4-Br-Ph	H	H			
4-Me-Ph	H	H			
1-naphthyl	H	H			
*c*Hex	H	H			
Ph	H	H	addition of allylCeCl$_2$ to chiral oxazolidinone prepared from RCHO and *N*-Bz-phenylglycinol	52–98%; 80–90% de	1990 (*1015*)
4-Br-Ph	H	H			
4-Me-Ph	H	H			
1-naphthyl	H	H			
3-pyridyl	H	H			
C(Me)[–CH(CH$_2$OMe)–CH(CH$_2$OMe)–]	Me	H	addition of acetate enolate to chiral imine prepared from chiral aldehyde	72–87%; 68–100% de *syn* or *anti* depending on enolate metal	1993 (*1016*)
	Et	H			
	H	Me			
	Et	Et			
CH(*n*Pr)OH	H	H	Blaise reaction of 2-bromo-alkanoates to chiral cyanohydrins, alkene reduction	54–80% addition; 72–95% ee; 71–88% reduction; 0–40% de	1998 (*1017*)
CH(Ph)OH	H	H			
CH(*n*Oct)OH	H	H			
2,3-dehydro with CH(R)OH, R = *n*Pr; Ph, *n*Oct	2,3-dehydro	2,3-dehydro with H, Me, Et			
H	CH(OH)CH$_2$OH	H	preparation of α-amidoalkylphenylsulfone from glyceraldehyde or prolinal, reaction with acetate enolate	65–90% amidosulfone formation; 90% de; 91–95% enolate addition; 80% de *anti*	2004 (*1018*)
pyrrolidin-2-yl	H	H			
CH$_2$CH(OH)-CH$_2$OH	H	H	addition of allyl group to imine formed from 3-hydroxy-4-aminobutyraldehyde and benzylamine, alkene oxidative cleavage	96% addition; 80% de; 71% oxidative cleavage	2002 (*165*)

(*Continued*)

Table 11.2 Asymmetric Syntheses of β-Amino Acids (continued)

R¹	R²	R³	R⁴	Method	Yield, % ee	Reference
H	CH(OH)CH(OH)CO₂H	H	H	conversion of D-ribose into 3,4-dihydroxy-5-acetoxypyrrolidin-2-one, acetoxy displacement with allylTMS/BF₃·Et₂O, alkene oxidative cleavage	100% allyl group introduction; 85% oxidative cleavage	1999 (224)
H	Ph	Me	H	addition of propionate silyl ketene acetal with chiral auxiliary to imine with Ti catalysis	75%; >80% de; >94% ee	1987 (1019)
Ph	H	H	OBn	addition of acylated Oppolzer camphorsultam to imine	66% yield; >99 % de	1994 (1020)
Ph	H	H	4-MeO-Bn	addition of acylated Oppolzer camphorsultam to imine	68% yield; >99 % de	1994 (1021)
Ph	H	H	H	addition of homochiral α-sulfinyl ester enolate to benzaldimine	13–78% yield	1994 (1022)
H	Ph	H	H		90% de (3S) to 94% de (3R); 36–99% auxiliary removal	
H	Ar₁	=N₂	—	addition of enolate of chiral acetate equivalent (oxazolidinone amide of α-diazoacetate) to N-tosyl imine, oxazolidinone removal, diazo oxidation, ketone reduction with NaBH₄, or H₂/Pd	73–94% addition	2004 (1023)
H	Ar₂	=O	—		52–>90% de	
H	Ar₂	OH	H		56–83% oxazolidinone hydrolysis	
H	Ar₂ (Ar₁ = Ph, 4-Ph-Ph, 4-Cl-Ph, 4-F-Ph, 4-MeO-Ph, 3-CN-Ph, 3-Br-Ph, CH=CHPh, 5-Br-2-thienyl, 2-furyl, nPr Ar₂ = Ph, 4-Ph-Ph, 4-Cl-Ph)	H	OH		83–>99% ee; 72–92% diazo oxidation and ketone reduction	
CH₂CH₂(α-Bn₄-glucose)	H	H	H	addition of enolate of chiral acetate equivalent derived from camphor to α-amido alkyl sulfone derived from glycosyl aldehyde, Cbz-NH₂ and pTol-SO₂Na, oxidative cleavage of auxiliary	53–62% formation of α-amido sulfone	2002 (2183)
CH₂CH₂(α-Bn₄-galactose)	H	H	H		60–78% acetate addition	
CH₂CH₂(α-Bn₄-mannose)	H	H	H		84–92% auxiliary cleavage	
H	CH₂CH₂(α-Bn₄-glucose)	H	H			
Ph	H	=CH₂	—	addition of hydroaluminated chiral ester of propiolic acid to benzaldehyde imine, dihydroxylation of alkene	34–90% addition	1996 (1024)
tBu	H	=CH₂	—		86–>98% de	
2-naphthyl	H	=CH₂	—		82% dihydroxylation	
fufuryl	H	=CH₂	—		40% de	
Ph	H	CH₂OH	OH			

iPr	H	H	addition of chiral acetal Ti homoenolate equivalent of propionaldehyde to benzylamine imine of butyraldehyde, enol ether oxidative cleavage	85% addition / 88% de / 65% oxidative cleavage	1999 (1025)
H	H	Ph	aminomethylation of phenyl-acetic acid with Evans' oxazolidinone auxiliary	68%	1995 (1026)
H	H	Ph	aminomethylation of acylated Evans' oxazolidinone auxiliary	0–65% / 65–84% deprotection	1998 (893)
H	H	4-HO-Ph			
H	H	Bn			
H	H	indol-3-yl			
H	H	Me	aminomethylation of acylated Evans' oxazolidinone auxiliary	86–98% de crude / 78–85% yield / 100% de	1997 (1027)
H	H	iPr			
H	H	iBu			
H	H	(CH$_2$)$_2$-CO$_2$Me	aminomethylation of acylated Evans' oxazolidinone auxiliary	87% / 92% de	1990 (1028)
Me	H	H	aminomethylation of acylated Evans' oxazolidinone auxiliary	55–85% aminomethylation / 86–98% de / 58–93% auxiliary removal	1998 (275)
iPr	H	H			
iBu	H	H			
Bn	H	H			
H	H	Me			
H	H	iPr			
H	H	iBu			
H	H	Bn			
CH$_2$CH$_2$Ph	H	H	aminomethylation of acylated Evans' oxazolidinone auxiliary	44% alkylation / 44% deprotection	1998 (337)
Me	H	H	aminomethylation of acylated Evans' oxazolidinone auxiliary, deprotection, conversion to Fmoc-amino acid	65–78% Fmoc protection	1998 (1029)
iPr	H	H			
iBu	H	H			
Bn	H	H			
H	H	Me	aminomethylation of acylated new Evans' oxazolidinone auxiliary (DIOZ)	30–78% alkylation / 84–90% de	1998 (1030)
H	H	iPr			
H	H	iBu			
H	H	Bn			
H	H	(CH$_2$)$_2$CO$_2$Me			
H	H	CH$_2$NPhth			
H	H	4-BnO-Bn			

(Continued)

Table 11.2 Asymmetric Syntheses of β-Amino Acids (continued)

R¹	R²	R³	R⁴	Method	Yield, % ee	Reference
H	H	(S)-s-Bu	H	aminomethylation of acylated new Evans' oxazolidinone auxiliary (DIOZ)	33–66% aminomethylation	2003 (832)
H	H	4-HO-Bn	H		91–96% de	2003 (1031)
H	H	(CH₂)₂SMe	H			
H	H	Me	H	aminomethylation of acylated new Evans' oxazolidinone auxiliary (DIOZ), functionalization of bromoalkyl side chain	60–90%	2003 (832)
H	H	iPr	H		80–98% de	
H	H	iBu	H			
H	H	Bn	H			
H	H	CH₂NHBoc	H			
H	H	CH₂NCbz	H			
H	H	(CH₂)₃NHC(=NH)-NH₂	H			
H	H	(CH₂)$_n$Br n = 2,3,4	H	aminomethylation of acylated new Evans' oxazolidinone auxiliary (DIOZ), functionalization of bromoalkyl side chain via thiomethyl or azide displacement	77–80% aminomethylation	2003 (832)
H	H	(CH₂)₂SMe	H		78–88% de	2005 (1032)
H	H	(CH₂)₃NHC(=NH)-NH₂	H		53% displacement with NaSMe	
H	H	(CH₂)₄NH₂	H		95% displacement with NaN₃	
					69% azide reduction	
H	Ph	Me	H	addition of Ti enolate of acylated oxazolidinone auxiliary to nitrone, auxiliary removal and N–O cleavage	65% addition	1999 (1033)
					70% de	
H	H	H	(CH₂)₃NHC(=NH)NH₂	alkylation of acylated oxazolidinone with BrCH₂CO₂tBu, Curtius rearrangement	82% alkylation	2004 (361)
					92% Curtius rearrangement	
H	iPr	H	H	addition of chiral lithiated 2-methyl-4-alkyloxazoline to α-methoxycarbamate, hydrolysis	40–63% condensation	1988 (555)
H	Me	H	H		44–90% ee of hydrolyzed product	
H	CH₂CH=CH₂	H	H			
H	CH₂CH₂OH	H	H			
H	Ph	H	Me	addition of enolate of propionic acid amidated with thiazolidinethione auxiliary to O-methyl oximes, N-acylation of azetine product	31–78% addition	2003 (1034)
H	1-naphthyl	H	Me		71–>90% de	
H	2-thienyl	H	Me		38–78% azetine opening	
H	cHex	H	Me			
H	CH₂CH₂Ph	H	Me			
Ph	H	Me	Me	addition of homochiral lithiated 2,4-diisopropyl-oxazoline to nitrone, rearrangement, hydrogenation	40–79% addition	2003 (1035)
4-Cl-Ph	H	Me	Me		96% de	
cHex	H	Me	Me		>98% hydrogenation	

R¹	R²	R³	R⁴	Reaction	Results	Year (Ref.)
H	H		$CH(OH)Me$	addition of chiral silyl ketene acetal to nitrone	71–89%; 100% de	1999 (1036)
Ph	H		$CH(OH)Me$			
$(CH_2)_3$-(α-N)			$CH(OH)Me$			
$(CH_2)_4$-(α-N)			$CH(OH)Me$			
2-[$(CH_2)_2$-(α-N)]-Ph			$CH(OH)Me$			
H		Ph	$CH(OH)Me$	addition of chiral enolate to imine	43% addition; 90% de	1984 (1037)
H	$CH{=}CHSPh$	$CH(OH)Me$	H	addition of chiral enolate to imine, oxidation, carboxy inversion	96% additon; 33–78% de	1987 (1038)
H	CH_2CO_2H	$CH(OH)Me$	H			
H	CH_2CO_2H		$CH(OH)Me$			
H	$CH(Me)CO_2H$		H	addition of chiral enolate to imine, carboxy inversion	87–95% additon; 33–60% de	1987 (1039)
H	$CH(Me)CO_2H$		$CH(OH)Me$			
H	CH_2CH_2OBn		$CH(OH)Me$	addition of chiral boron enolate to imine	36–73% addition; 72–80% de	1986 (1040)
H	$CH{=}CHPh$		$CH(OH)Me$			
H	CCTMS		$CH(OH)Me$			
H	CCTMS		$CH(OH)Et$			
CF_3			H	addition of Li ester enolates to imidoyl chlorides, reduction with $ZnI_2/NaBH_4$	65–100% addition to imidoyl chloride; 75–93% reduction; 10–60% de	2002 (591)
CF_2Cl			H			
H	Ph		Me	addition of propionate enolate to imine in presence of homochiral diazaborolidine	67–77%; 84–>98% de; 90–>99% ee	1991 (1041)
H	1-naphthyl		Me			
H	2-naphthyl		Me			
H	$CH{=}CHPh$		Me			
H	Ph	OH	H	addition of hydroxyacetate enolate to imine in presence of homochiral boron complex	60–71%; >92% de *syn* to 94% de *anti*; >95% ee	1997 (1042)
H			OH			
nPr			H	addition of silyl ketene acetal to chiral imine using chiral catalyst	50–99%	1993 (1043)
Ph			H			
H	CCTMS		Et			
Ph			OH		92–98% de matched pair; 96% de *anti* to 98% de *syn*	1994 (1044)
cHex			OH			
iBu			OH			
Ph			H			
cHex			H			
iBu			H			

(Continued)

Table 11.2 Asymmetric Syntheses of β-Amino Acids (continued)

R¹	R²	R³	R⁴	Method	Yield, % ee	Reference
H	CH₂CH₂Ph	Me	Me	addition of silyl enolates to hydrazone in presence of homochiral Zr catalyst	39–66% addition	1998 (1045)
H	CH₂CH₂Ph	H	H		81–96% ee	
H	nHex	Me	Me			
H	nHex	H	H			
H	Ph	Me	Me			
H	CH₂Cl	Me	Me			
H	nC₁₂H₂₅	Me	Me			
H	Ph	OH	H	addition of silyl ketene acetals to imines in presence of homochiral Zr catalyst	41–100% addition	1998 (1046)
H	1-naphthyl	OH	H		>98% de syn to 88% de anti	
H	2-furyl	OH	H			
H	4-Cl-Ph	OH	H		76–96% ee	
H	Ph	H	OH			
H	1-naphthyl	H	OH			
H	2-furyl	H	OH			
H	4-Cl-Ph	H	OH			
H	cHex	H	OH			
H	Ph	H	Me	addition of propionate silyl ketene acetal to imines in presence of homochiral Zr catalyst, cleavage of N-aryl group with CAN	54–96% addition	2002 (1047)
H	4-Cl-Ph	H	Me		42–96% de anti	
H	2-Me-Ph	H	Me		80–96% ee	
H	1-naphthyl	H	Me		44% N-aryl cleavage	
H	2-furyl	H	Me		(2 examples)	
H	iBu	H	Me			
H	nPent	H	Me			
H	(CH₂)₂OTBS	H	Me			
H	cHex	H	Me			
H	(CH₂)₃CCH	H	Me			
H	(CH₂)₂OBn	H	Et			
H	Ph	H	H	addition of silyl ketene acetals to imines in presence of homochiral Zr catalyst	45–100% addition	2000 (1049)
H	1-naphthyl	H	H		80–>98% ee	2004 (1048)
H	2-furyl	H	H			
H	4-Cl-Ph	H	H			
H	cHex	H	H			
H	iBu	H	H			
H	Ph	Me	Me			
H	1-naphthyl	Me	Me			
H	4-Cl-Ph	Me	Me			

Ph	H	H	addition of silyl ketene acetals to N-Bn oximes in presence of BINOL Ti catalyst	74–99% addition 80–92% ee	2002 (*1050*)
2-naphthyl	H	H			
4-Me-Ph	H	H			
3-pyridyl	H	H			
3,4-(OCH$_2$O)-Ph	H	H			
Ph	H	H	addition of silyl ketene acetal of isopropyl acetate to N-Boc aryl imines in presence of chiral Tle-urea-Schiff base catalyst	84–99% yield 86–98% ee	2002 (*2182*)
2-Me-Ph	H	H			
3-Me-Ph	H	H			
4-Me-Ph	H	H			
4-MeO-Ph	H	H			
4-F-Ph	H	H			
3-Br-Ph	H	H			
4-Br-Ph	H	H			
1-naphthyl	H	H			
2-naphthyl	H	H			
2-furyl	H	H			
2-thienyl	H	H			
3-quinolinyl	H	H			
3-pyridyl	H	H			
Ph	Me	Me	addition of ketene silyl enolates to imine from o-HO-aniline/RCHO in presence of chiral BINOL-derived phosphate Bronsted acid	65–100% yield 72–100% de syn 81–96% ee	204 (*1051*)
4-Me-Ph	Me	Me			
4-F-Ph	Me	Me			
4-Cl-Ph	Me	Me			
Ph	H	H			
4-MeO-Ph	H	H			
4-F-Ph	H	H			
4-Cl-Ph	H	H			
4-Me-Ph	H	H			
2-thienyl	H	H			
CH=CHPh	H	H			
Ph	H	Bn			
4-MeO-Ph	H	Bn			
CH=CHPh	H	Bn			
Ph	H	OSiPh$_3$			
CH=CHPh	H	OH	Mannich-type condensation of RCH=NHTs with N-(hydroxyacetyl)pyrrole in presence of chiral catalyst, pyrrole hydrolysis	68–94% yield 59:41 to 91:9 anti:syn 89–98% ee anti; 71–91% ee syn 100% hydrolysis	2005 (*1052*)
CH=CH(4-Me-Ph)	H	OH			
CH=CH(4-Cl-Ph)	H	OH			
CH=CH(2-furyl)	H	OH			
Ph	H	OH			

(*Continued*)

Table 11.2 Asymmetric Syntheses of β-Amino Acids (continued)

R¹	R²	R³	R⁴	Method	Yield, % ee	Reference
4-Cl-Ph	H	H	OH			
1-naphthyl	H	H	OH			
2-Cl-Ph	H	H	OH			
2-Br-Ph	H	H	OH			
2-Me-Ph	H	H	OH			
2-MeO-Ph	H	H	OH			
cPr	H	H	OH			
Ph	H	H	H	Reformatsky addition of BrCH$_2$O$_2$Et to imine formed from 2-MeO-PhNH$_2$ and RCHO in presence of Me$_2$Zn, [NiCl$_2$(PPh$_3$)$_2$], and (1S,2R)-N-Me-ephedrine	30–90% yield	2005 (1055)
4-Cl-Ph	H	H	H		74–90% ee	
4-CF$_3$-Ph	H	H	H			
C$_6$F$_5$	H	H	H			
4-MeO-Ph	H	H	H			
4-tBu-Ph	H	H	H			
2-naphthyl	H	H	H			
2-thiophene	H	H	H			
CH=CHPh	H	H	H			
iPr	H	H	H			
cHex	H	H	H			
Ph	H	=CH$_2$	—	aza-Baylis–Hillman condensation of ArCH=NHTs with H$_2$C=CHCO$_2$Ph in presence of chiral phosphine catalyst	60–97% yield	2005 (1056)
4-Cl-Ph	H	=CH$_2$	—		52–77% ee	
3-Cl-Ph	H	=CH$_2$	—			
4-Et-Ph	H	=CH$_2$	—			
4-F-Ph	H	=CH$_2$	—			
4-Br-Ph	H	=CH$_2$	—			
3-F-Ph	H	=CH$_2$	—			
4-NO$_2$-Ph	H	=CH$_2$	—			
3-NO$_2$-Ph	H	=CH$_2$	—			
Ph	H	COMe	Me	addition of β-keto ester to imine in presence of chiral Pd catalyst	61–99% yield	2005 (1057)
4-Me-Ph	H	COMe	Me		0–>90% de	
2-furyl	H	COMe	Me		86–99% ee	
4-Cl-Ph	H	COMe	Me			
CH=CHPh	H	COMe	Me			
Ph	H	CO(CH$_2$)$_3$-C2	—			
4-Me-Ph	H	CO(CH$_2$)$_3$-C2	—			
2-furyl	H	CO(CH$_2$)$_3$-C2	—			
Ph	H	CO(CH$_2$)$_4$-C2	—			

Ph	H	H	C–H insertion reaction of ArC(=N$_2$)CO$_2$Me with Bn(Boc)NMe catalyzed by [Rh$_2$(S-DOSP)$_4$]	55–67% / 87–96% ee	2002 (*1058*)
4-CF$_3$-Ph	H	H			
4-MeO-Ph	H	H			
4-Me-Ph	H	H			
2-naphthyl	H	H			
4-Cl-Ph	H	H			
3-thienyl	H	H			
CH=CHPh	H	H			

Fusion of Aminomethyl and Acetate: Other Methods

(CH$_2$)$_3$NH$_2$	H	H	cycloaddition of nitrone with chiral auxiliary and vinyl acetate or α-chloroacrylonitrile	30–81% cycloaddition / 33–84% de	1991 (*1059*)
iPr	H	H			
Ph	H	H			
4-MeO-Ph	H	H			
H	(CH$_2$)$_3$NH$_2$	H			
H	iPr	H			
H	Ph	H			
H	4-MeO-Ph	H			
(CH$_2$)$_3$-β-N	CH(OH)Me	H	cycloaddition of nitrone and crotonic acid with chiral amide auxiliary	86–99% cycloaddition / 2–8% de	1993 (*1060*)
(CH$_2$)$_4$-β-N	CH(OH)Me	H			
H	CH(OH)CH$_2$F	CH(OH)Me	intramolecular cycloaddition of nitrone of 3-fluoro-2-hydroxypropanal crotonic and cinnamic esters	63–69%	1997 (*1061*)
H	CH(OH)CH$_2$F	CH(OH)Ph			
Et	H	H	1,3-dipolar cycloaddition of nitrone formed from phenylglycinol and propanal with allyl alcohol, hydrogenation to cleave auxiliary and isoxazolidine N–O bond, hydroxymethyl oxidation	94% cycloaddition / 90% de / 63% hydrogenation and oxidation	2003 (*1062*)
Ph	H	OH	dipolar cycloaddition of ylide generated by Rh$_2$(OAc)$_4$ treatment of N$_2$CHCO$_2$Et/PhCHO with PhCH(Me)N=CHR imine, hydrolysis of oxazolidine adduct with pTSA	61–83% yield cycloaddition and hydrolysis / 66–96% de *syn*	2005 (*622*)

(Continued)

Table 11.2 Asymmetric Syntheses of β-Amino Acids (continued)

R¹	R²	R³	R⁴	Method	Yield, % ee	Reference
Ph	H	H	OH	hetero-Diels–Alder of N-benzoylbenzaldimine with Z ketene acetal with chiral auxiliary	47–75% yield	1993 (1063)
H	Ph	OH	H		0–100% ee	
					69–100% de	
Ph	H	Me	Me	cycloaddition of chiral Schiff base with 2 equiv of dimethylketene	56–68%	1977 (1064)
H	Ph				47–53% ee	
Et	H	H	H	1,3-dipolar cycloaddition of R¹CH=NOH with (R)-R²CH=CHCH(OH)Me, isoxazolidine reduction with LiAlH₄ or R³MgX addition then LiAlH₄ reduction, diol oxidative cleavage	50–96% cycloaddition	2003 (1065)
Et	H	H	Et		40–64% reduction	
iPr	H	H	H		75–87% de	
iBu	H	H	H		81–95% Grignard addition	
Ph	H	H	H		80–>90% de	
CCTMS	H	H	H		48–72% oxidative cleavage	
allyl	iBu	H	H			
Bn	iBu	H	H			
allyl	Et	H	H			
allyl	Et	Ph	H			
Me	iBu	H	H	1,3-dipolar cycloaddition of R¹CH=NOH with (R)-(Z)-R²CH=CHCH(OH)Me, TBS protection, R³Li/BF₃·Et₂O addition then LiAlH₄ reduction, diol oxidative cleavage	98% TBS protection	2004 (1066)
Ph	iBu	H	H		49–82% organolithium addition	
2-furyl	iBu	H	H		82–90% de	
5-Me-2-furyl	iBu	H	H		46–78% N–O cleavage	
2-thienyl	iBu	H	H		70–95% oxidative cleavage	
Me	Et	Ph	H			
Ph	Et	Ph	H			

Elaboration of α-Amino Acids: Homologation

R¹	R²	R³	R⁴	Method	Yield, % ee	Reference
H	H	H	H	Arndt–Eistert homologation of Phth-Gly	26%	1951 (16)
4-HO-Bn	H	H	H	Arndt–Eistert homologation of Phth-L-Tyr, iodination	93% acid chloride	1951 (1101)
3,5-I₂-4-HO-Bn	H	H	H		78% azide	
					67% rearrangement and hydrolysis	
					63% iodination	
Me	H	H	H	Arndt–Eistert homologation of Ala	45% overall	1952 (1070)
CH₂SBn	H	H	H	Arndt–Eistert homologation of S-Bn-L-Cys	26%	1952 (1091)
						1956 (1092)
(CH₂)₃NH₂	H	H	H	Arndt–Eistert homologation of Phth-L-Orn(Phth)	45% rearrangement	1952 (145)
					46% hydrolysiss	1953 (146)

H	iPr	H	Arndt–Eistert homologation of Phth-D-Val	89% acid chloride 89% diazoketone 72% Ag-catalyzed rearrangement 93% deprotection	1954 (*1087*)
$CH_2CH_2CO_2H$	H	H	Arndt–Eistert homologation of Ts-pyroGlu	88% diazoketone 67% rearrangement and hydrolysis	1963 (*1094*)
H	H	H	Arndt–Eistert homologation of Phth-Gly	30% diazoketone 65% rearrangement	1970 (*1069*)
diazo precursors of: Me iPr iBu sBu Bn CH(MeOH) $(CH_2)_3$-(β-N) CH_2(indol-3-yl)	H		prep diazo intermediates of other Phth-, Cbz- or Boc-amino acids	30–70% diazoketones	
Bn	H	H	Arndt–Eistert homologation of Cbz-L-Phe	57% overall	1970 (*384*)
H	Ph	H	Arndt–Eistert homologation of D-Phg	54%	1973 (*1103*)
CH(OH)$CH_2CH_2NH_2$	H	H	Arndt–Eistert homologation of *threo*- or *erythro*-β-hydroxy-DL-Orn	10–38%	1974 (*161*)
$(CH_2)_3$-(β-N)	H	H	Arndt–Eistert homologation of Cbz-L-Pro	91% diazoketone 95% rearrangement	1975 (*1089*)
$(CH_2)_3$NHP P = Cbz, Boc	H	H	Arndt–Eistert homologation of Boc-L-Orn(Boc) or Cbz-L-Orn(Cbz)	—	1975 (*1097*)
Bn	H	H	Arndt–Eistert homologation of Boc-L-Phe or -Pro	33–40% overall	1975 (*381*)
$(CH_2)_3$-(β-N)	H	H	Arndt–Eistert homologation of Cbz-L-Pro, Cbz-D-Pro	53–65% diazoketone 98% rearrangement	1976 (*1088*)
$(CH_2)_3$-(β-N)	$(CH_2)_3$-(β-N)	H			
H	H	$(CH_2)_3$-(β-N)			
Bn	H	H	Arndt–Eistert homologation of Cbz-L-Phe	92% rearrangement	1977 (*1100*)

(*Continued*)

Table 11.2 Asymmetric Syntheses of β-Amino Acids (continued)

R¹	R²	R³	R⁴	Method	Yield, % ee	Reference
*i*Pr	H	H	H	Arndt–Eistert homologation of Boc-L-Orn(Boc), Boc-L-Lys(Cbz), Cbz-L-Dap(Cbz), Cbz-L-Val, Cbz-L-Leu, Cbz-L-Ile, Cbz-L-Nva, Cbz-L-Phe	70–91%	1977 (*156*)
*i*Bu	H	H	H			
*s*Bu	H	H	H			
*n*Pr	H	H	H			
Bn	H	H	H			
CH_2NH_2	H	H	H			
$CH_2CH_2NH_2$	H	H	H			
$(CH_2)_3NH_2$	H	H	H			
$CH_2CH_2NH_2$	H	H	H	Arndt–Eistert homologation of L-2,4-diaminobutyric acid, guanidation	75% homologation 86% guanidation	1978 (*1098*)
$CH_2CH_2NHC(=NH)NH_2$	H	H	H			
$CH(OH)CH_2-CH_2NH_2$	H	H	H	Arndt–Eistert homologation of *threo*-β-hydroxy-L-Orn	30%	1980 (*1106*)
D	D	H	H	Arndt–Eistert homologation of labeled Tfa-Gly	26%	1985 (*421*)
H	D	H	H			
H	H	H	H			
Bn	H	H	H	Arndt–Eistert homologation of Cbz- or Boc-L-Phe	75% diazoketone 72–75% rearrangement	1987 (*1099*)
CH_2NH_2	H	H	H	Arndt–Eistert homologation of Cbz-L-Dap(Boc), methylation	73% homologation 11% trimethylation	1987 (*168*)
$CH_2N(Me)_3$	H	H	H			
4-HO-Ph	H	H	H	Arndt–Eistert homologation of L-4-HO-Phg	49%	1988 (*1104*)
*i*Bu	H	H	H	Arndt–Eistert homologation of Cbz-L-Phe, -L- and -D-Leu, -L- and -D-Asp(O*t*Bu)	65–75%	1989 (*383*)
$(CH_2)_2CO_2tBu$	H	H	H			
Bn	H	H	H			
H	*i*Bu	H	H			
H	$(CH_2)_2CO_2tBu$	H	H			
Me	H	H	H	Arndt–Eistert homologation of Ala, Nva with N in pyrrole ring	59–75% overall	1991 (*1079*)
*n*Pr	H	H	H			
H	$(CH_2)_3NH_2$	H	H	Arndt–Eistert homologation of D-Orn	—	1992 (*143*)
H	$CH_2N(Me)_3$	H	H	Arndt–Eistert homologation of L-Asp, Hofmann rearrangement of β-carboxy group, trimethylation	22–25% overall	1992 (*169*)
$CH_2N(Me)_3$	H	H	H			

R¹	R²	R³	Reaction	Yield	Year (Ref)
Me	H	H	Arndt–Eistert homologation of Ala or Leu (Cbz-amino acid or C-terminal of dipeptide, tripeptide) combined with trapping by amino acid or dipeptide	41–95% overall	1995 (1073)
iBu	H	H			
Me	H	H	Arndt–Eistert homologation of Cbz- and Boc-L-Ala, Phe, Tle, Ser(OtBu), Phg, Orn(Boc), homopipecolic acid	52–89% diazoketone; 73–95% ketene generation and trapping with MeOH; 23–76% trapping with carbohydrates	1995 (1076)
tBu	H	H			
Ph	H	H			
Bn	H	H			
CH₂OtBu	H	H			
(CH₂)₃NHBoc	H	H			
(CH₂)₅-β-N	H	H			
Me	H	H	Arndt–Eistert homologation of Boc-L-Ala, Boc-D-Ala	87% diazoketone and rearrangement	1995 (398)
H	Me	H			
Me	H	H	Arndt–Eistert homologation of Ala and AVal using Ag catalysis, with multifunctional nucleophiles	30–74%	1996 (1072)
iPr	H	H			
(CH₂)₃CO₂Et	H	H	Arndt–Eistert homologation of L-2-aminoadipic acid	58%	1996 (1096)
Me	H	H	Arndt–Eistert homologation of Boc-L-Ala, -L-Val and -L-Leu, trapping with MeOH, H₂O or amino acid	62–85%	1996 (292)
iPr	H	H			
iBu	H	H			
Me	H	H	Arndt–Eistert homologation of Fmoc-L-Ala, Val, Leu, Ile, Phe	93–97% diazoketone; 50–90% ketene generation and trapping with H₂O	1997 (1074)
iPr	H	H			
iBu	H	H			
sBu	H	H			
Bn	H	H			
Me	H	H	Arndt–Eistert homologation of Boc- or Cbz-L-Ala or -L-Val, trapping with MeOH, or 5-amino oligonucleotide	70–93%	1997 (1077)
iPr	H	H			
iPr	H	H	Arndt–Eistert homologation of Fmoc-amino acid or combined with trapping by amino acid on polymer support	60–95% crude yield of di- to tetra-peptides	1997 (1083)
Bn	H	H			
4-HO-Bn	H	H			
(CH₂)₂CO₂H	H	H			
(CH₂)₄NH₂	H	H			

(Continued)

Table 11.2 Asymmetric Syntheses of β-Amino Acids (continued)

R^1	R^2	R^3	R^4	Method	Yield, % ee	Reference
Me	H	H	H	Arndt–Eistert homologation of Boc-L-Ala and Boc-N-Me-L-Ala	72–73%	1997 (288)
$(CH_2)_3NH_2$	H	H	H	Arndt–Eistert homologation of Boc-L-Orn(Cbz), Boc-L-Lys(Cbz)	70–80%	1997 (397)
$(CH_2)_4NH_2$	H	H	H			
Me	H	H	H	Arndt–Eistert homologation of Fmoc-L-Ala, Fmoc-L-Val, Fmoc-L-Leu, Fmoc-L-Phe, Fmoc-L-Lys(Boc), Fmoc-L-Glu(OtBu), and Fmoc-L-Ser(tBu)	68–93% diazoketone	1998 (1029)
iPr	H	H	H		38–75% rearrangement	
iBu	H	H	H			
Bn	H	H	H			
$(CH_2)_3NH_2$	H	H	H			
$(CH_2)_2CO_2H$	H	H	H			
CH_2OH	H	H	H			
Me	H	H	H	Arndt–Eistert homologation of Boc-Ala, Boc-Val, Boc-Phe, Boc-Ile, Boc-Leu, Boc-Thr(Bz), Boc-Tyr(Bz), Boc-Lys(2-Cl-Z), Boc-Ser(OBz), Fmoc-Phe, or Fmoc -Ser(OtBu) to give Weinreb amide derivatives	86–98% diazoketone	1998 (1075)
iPr	H	H	H		84–97% Weinreb amide	
iBu	H	H	H			
sBu	H	H	H			
Bn	H	H	H			
4-HO-Bn	H	H	H			
CH_2OH	H	H	H			
CH(OH)Me	H	H	H			
$(CH_2)_4NH_2$	H	H	H			
Me	H	H	H	Arndt–Eistert homologation of Boc-L-Ala, Fmoc-D-Ala, Fmoc-L-Ile, and Fmoc-L-Pro	82–100% diazoketone	1998 (1078)
H	Me	H	H		83–100% rearrangement	
sBu	H	H	H			
$(CH_2)_3$-(β-N)	H	H	H			
H	Ph	H	H	Arndt–Eistert homologation of Fmoc-D-Phg, Fmoc-L-Phe, Fmoc-D-Phe, Fmoc-L-Asp, Fmoc-L-Val, Fmoc-L-Ile, and Fmoc-L-Leu	65–82% overall	1998 (1084)
Bn	H	H	H		80–>99% ee	
H	Bn	H	H			
sBu	H	H	H			
iBu	H	H	H			
iPr	H	H	H			
CH_2CO_2H	H	H	H			
$(CH_2)_2CONH_2$	H	H	H	Arndt–Eistert homologation of Boc-L-Gln(Trt)-OH	75%	1998 (1095)
CH_2OH	H	H	H	Arndt–Eistert homologation of Boc-L-Ser(Bn)-OH or Boc-L-Glu(OBn)-OH	48–58%	1998 (335)
$CH_2CH_2CO_2H$	H	H	H			

(CH₂)₂Ph	H	Arndt–Eistert homologation of Boc-L-homoPhe, Boc-L-Lys(2-Cl-Cbz), Fmoc-L-Orn(Boc), Fmoc-L-Tyr(tBu)	60–89% diazoketone 63–75% rearrangement	1998 (337)
(CH₂)₂NH₂	H			
(CH₂)₂NH₂	H			
4-HO-Bn	H			
H	(CH₂)₂Ph	Arndt–Eistert homologation of Boc-D-homoPhe	80%	1998 (416)
(CH₂)₃-(β-N)	H	Arndt–Eistert homologation of Boc-L-Pro	56–77% diazoketone 61–76% rearrangement	1999 (1090)
(CH₂)₂SMe	H	Arndt–Eistert homologation of Boc-L-Met, or Boc-L-Cys(Bn)	61–73% rearrangement	1999 (317)
CH₂SH	H			
Bn	H	Arndt–Eistert homologation of Boc-L-Xaa	50–69% diazoketone 88–94% rearrangement	1999 (372)
(CH₂)₂NH₂	H			
CH₂(indol-3-yl)	H			
CH(OH)Me	H			
Bn	H	Arndt–Eistert homologation of Boc-L-Phe	70% from diazo intermediate	2000 (344)
CH₂OBn	H	Arndt–Eistert homologation of Boc-Met or Boc-Ser(OBn)	71–79% diazoketone 78–95% rearrangement	2001 (374)
CH₂CH₂SMe	H			
H	H	Arndt–Eistert homologation of Fmoc-Gly, Fmoc-Val, Fmoc-Asp(OBu), Fmoc-Lys(Boc), Cbz-Phe, Cbz-Phg, Boc-L-Pro, Boc-L-Ile, or Boc-D-Ile using Boc2O for activation and coupling with diazomethane	86–95% diazomethane intermediate 75–84% rearrangement	2002 (1071)
iPr	H			
CH₂CO₂H	H			
(CH₂)₄NH₂	H			
(CH₂)₃-(β-N)	H			
sBu	H			
Bn	H			
Ph	H			
H	sBu			
H	H	Arndt–Eistert homologation of Fmoc-Val, Fmoc-Phe, Fmoc-Phg, Cbz-Ala, Cbz-Phe, Cbz-Phg, Boc-Ala or Boc-Val using p-toluenesulfonyl chloride for activation and coupling with diazomethane	86–94% diazomethane intermediate 80–88% rearrangement	2002 (1080)
iPr	H			
CH₂CO₂H	H			
(CH₂)₄NH₂	H			
(CH₂)₃-(β-N)	H			
sBu	H			
Bn	H			
Ph	H			
H	sBu			
4-Ph-Ph	H	Arndt–Eistert homologation of L-4-Ph-Phg	45% 50% ee	2002 (406)

(Continued)

Table 11.2 Asymmetric Syntheses of β-Amino Acids (continued)

R^1	R^2	R^3	R^4	Method	Yield, % ee	Reference
Me	H	H	H	Arndt–Eistert homologation of Fmoc-Val, Fmoc-Ile, Fmoc-Phe, Fmoc-D-Phe, Fmoc-Asp(OtBu), Fmoc-Glu(OtBu), Fmoc-Lys(Boc), Cbz-Ala, Cbz-Phe, Boc-Ala, Boc -Phe or Boc-Val using TBTU for activation and coupling with diazomethane	86–94% diazomethane intermediate 87–93% rearrangement	2003 (1082)
iPr	H	H	H			
sBu	H	H	H			
CH$_2$CO$_2$H	H	H	H			
CH$_2$CH$_2$CO$_2$H	H	H	H			
(CH$_2$)$_4$NH$_2$	H	H	H			
Bn	H	H	H			
H	Bn	H	H			
CH$_2$CO$_2$H	H	H	H	Arndt–Eistert homologation of Boc-L-4-Asp(OBn)-OH	95% diazoketone 88% rearrangement	2003 (1093)
H	CH$_2$CO$_2$H	H	H	Arndt–Eistert homologation of Cbz-L-Asp(OtBu)-OH	79% homologation	2003 (167)
Me	H	H	H	Arndt–Eistert homologation of Boc- or Fmoc-L-Ala, -Phe and -Lys	65–84% diazoketone 80–96% rearrangement	2003 (281)
Bn	H	H	H			
(CH$_2$)$_4$NH$_2$	H	H	H			
CH$_2$SH	H	H	H	Arndt–Eistert homologation of Boc-His(Tos)-OH or Fmoc-Cys(Pmb)-OH	56–86% diazoketone 85–91% rearrangement	2003 (351) 2004 (833)
CH$_2$(imidazol-3-yl)	H	H	H			
iPr	H	H	H	Arndt–Eistert homologation of resin-linked Phe, Trp, Val, Leu	56–92%	2004 (1085)
iBu	H	H	H			
Bn	H	H	H			
CH$_2$(indol-3-yl)	H	H	H			
(CH$_2$)$_2$SMe	H	H	H	Arndt–Eistert homologation of Cbz-L-Met, Cbz-D-allylGly or Boc-L-2-aminohex-5-enoic acid	93–96%	2004 (1105)
H	CH$_2$CH=CH$_2$	H	H			
(CH$_2$)$_2$CH=CH$_2$	H	H	H			
(CH$_2$)$_3$C(=NH)NH$_2$	H	H	H	Arndt–Eistert homologation or Fmoc-Arg(Pmc) or Fmoc-Arg(Pbf) or homologation of Fmoc-Orn(Boc), then guanylation	55–63% diazoketone 83–91% rearrangement 70% guanylation of hOrn	2004 (315)

(CH₂)₂OH	H	H	conversion of Lys to 6-hydroxy-Nle, Arndt–Eistert homologation, α-methylation, side-chain oxidation and conversion to alkyne	27% 6-hydroxy-Nle 63% homologation 75% methylation 82% oxidation 93% alkyne	2004 (86)
(CH₂)₄OH	H	Me			
(CH₂)₃CHO	H	Me			
(CH₂)₃CCH	H	Me			
Me	H	H	Arndt–Eistert homologation of Cbz-Ala, Cbz-Val, Cbz-Ile, Cbz-Tle, Cbz-Phe, Fmoc-Leu, or Boc-Lys(Cbz) diazoketones using microwave reaction with Si gel slurry and catalytic Ag trifluoroacetate	92–97% rearrangement	2005 (1081)
iPr	H	H			
iBu	H	H			
sBu	H	H			
tBu	H	H			
Bn	H	H			
(CH₂)₄NH₂	H	H			
iPr	H	H	Arndt–Eistert homologation of N-methyl Cbz-Val, -Asp(NBn₂), -Cys(Bn), -Trp(CHO), -His(DNP) with sonication for rearrangement	50–65% diazomethane 68–87% rearrangement	2005 (1086)
(CH₂)₂CONBn₂	H	H			
CH₂SH	H	H			
CH₂(indol-3-yl)	H	H			
CH₂(imidazol-4-yl)	H	H			
3,4-(BnO)₂-Bn	H	H	Arndt–Eistert homologation of Boc-DOPA(Bn)₂	90% diazoketone 61% rearrangement	2005 (1102)
3,4-(OH)₂-Bn	H	H			
Me	H	H	homologation of amino acid via reduction carboxyl to hydroxymethyl, displacement with iodide, cyanide, hydrolysis	87–96% reduction 78–94% iodination 77–95% cyanidation 85–91% hydrolysis	1995 (1108)
Et	H	H			
iPr	H	H			
iBu	H	H			
Ph	H	H			
Bn	H	H			
4-BnO-Bn	H	H			
CH₂OBn	H	H			
H	CH(OH)CH(OH)-CO₂H	H	homologation of pyroglutamate via reduction carboxyl to hydroxymethyl, displacement with cyanide, hydrolysis	43% yield displacement 78% hydrolysis	1989 (1109)
H	Bn	H	homologation of Phe via reduction carboxyl to hydroxymethyl, displacement with cyanide, hydrolysis	62% cyanidation 59% hydrolysis	1999 (376)
H	CH₂(imidazol-4-yl)	H	homologation of His-OMe via reduction carboxyl to hydroxymethyl, displacement mesylate derivative with cyanide, hydrolysis	58% ester reduction 81% mesylate 63% cyanidation 23% hydrolysis and Boc protection	2002 (1110)

(Continued)

Table 11.2 Asymmetric Syntheses of β-Amino Acids (continued)

R^1	R^2	R^3	R^4	Method	Yield, % ee	Reference
nPr	H	H	H	reduction and mesylation of both Asp carboxyl groups, displacement with CN, then nucleophile, hydrolysis	74% CN displacement	1995 (1111)
nHex	H	H	H		47–85% for Nu	1998 (1112)
$(CH_2)_2N_3$	H	H	H		30–79% hydrolysis	
H	Bn	OH	H	homologation of Phe via reduction carboxyl to aldehyde, cyanide addition, hydrolysis	61% reduction	1976 (1114)
H	Bn	H	OH		94% cyanidation 79% hydrolysis	1977 (1115)
Bn	H	OH	H	homologation of Phe via reduction carboxyl to aldehyde, cyanide addition, fluorination of hydroxyl	87% reduction	1996 (1116)
Bn	H	H	OH		80% cyanidation	
Bn	H	F	H		8% de	
Bn	H	H	F		74% hydrolysis	
Bn	H	F	F		31–55% fluorination	
Bn	H	OH	H	homologation of Phe via reduction carboxyl to aldehyde, protection as chiral acetal, TMSCN addition, hydrolysis	95% acetal formation	1989 (1117)
Bn	H	H	OH		75% TMSCN addition	
H	Bn	OH	H		40% de *threo* or 40% de *erythro*	
H	Bn	H	OH			
Me	H	OH	H	homologation of Bn₂-Ala via reduction to aldehyde, stereoselective addition of TMSCN/BF$_3$·Et$_2$O or TMSCN/TiCl$_4$, DAST treatment of 2-hydroxy-3-aminobutanoic acid, or oxidation, then DAST	23–38% CN, addition	2004 (355)
Me	H	H	OH		76–88% hydrolysis	
Me	H	F	H			
Me	H	F	F			
CH$_2$cHex	H	H	OH	reduction of Ph ring of phenylalaninol, oxidation to aldehyde, CN addition, hydrolysis	60% overall	1989 (1118)
H	iBu	H	OH	reduction of D-Leu to aminoaldehyde, cyanohydrin formation, hydrolysis	18% overall	1979 (1119)
H	iBu	OH	H			
CH$_2$(4-piperidinyl)	H	OH	H	reduction of β-(4-piperidinyl)Ala to aldehyde, addition of NaCN, methanolysis, oxidation	100% reduction 100% CN addition 89% methanolysis 100% oxidation	1999 (1120)
CH$_3$(4-piperidinyl)	H	=O	=O			

Me	H	–OH	H	addition of CN to Boc-amino acid-derived aldehyde, hydrolysis, oxidation	56–63% CN addition and hydrolysis	1990 (410)
Me	H	=O	—			
iPr	H	–OH	H			
iPr	H	=O	—			
Bn	H	–OH	H			
Bn	H	=O	—			
Me	H	OH	H	addition of CN to N,N-Bn₂ amino acid-derived aldehyde, methanolysis, treatment with DAST or oxidation, then DAST	—	2003 (1121)
Me	H	H	OH			
Me	H	F	H			
Me	H	F	F			
Me	H	=O	—			
iPr	H	OH	H			
iPr	H	H	OH			
iPr	H	F	H			
iPr	H	F	F			
iPr	H	=O	—			
iBu	H	OH	H			
iBu	H	H	OH			
iBu	H	F	H			
iBu	H	F	F			
iBu	H	=O	—			
CH2cHex	H	OH	OH	CN addition to aldehyde from Boc-Cha, Boc-Leu, or Boc-Phe, hydrolysis, or addition to aldehyde from Bz-Cha, oxazoline inversion, hydrolysis	50–55% from amino alcohol; 40–50% de	1990 (389)
CH2cHex	CH2cHex	H	OH			
H	Bn	H	OH			
Bn	iBu	H	OH			
iBu	H	H	OH	oxidation Boc-leucinol, addition vinylMgCl, protection, oxidative alkene cleavage	77% oxidation and addition; 66% de; 91% protection; 85% oxidation	2001 (191)
H	H	OAc	H	Passerini condensation of amino acid-derived aldehydes with acids and glycosyl isonitriles	31–57% condensation; 4–20% de	1999 (537)
Bn	H	OAc	H			
Bn	H	OCOR	H	Passerini condensation of amino acid-derived aldehydes with acids and isonitriles acyl removal, oxidation	60–95% amino aldehyde; 40–55% Passerini; 25–85% oxidation	2005 (413)
4-Cl-Bn	H	OCOR	H			
4-Br-Bn	H	OCOR	H			
Bn	H	=O	—			
4-Cl-Bn	H	=O	—			
4-Br-Bn	H	=O	—			

(Continued)

Table 11.2 Asymmetric Syntheses of β-Amino Acids (continued)

R¹	R²	R³	R⁴	Method	Yield, % ee	Reference
Me	H	H	OH	TMSCN addition to amino aldehyde equivalent derived from Williams' alkylated oxazinone template	93–95% cyanation 10–>90% de 88–97% hydrolysis 95–97% hydrogenation	2001 (1122)
H	Me	OH	H			
CH₂cHex	H	H	OH			
H	CH₂cHex	OH	H			
CH₂cHex	H	H	OH	addition of vinylMgBr to cyclohexylalaninal, oxidative cleavage of alkene	55–65% reduction of Cha, RMgBr addition 70% de 95% oxidation	1991 (388)
H	Bn	OH	H	addition of ethynylMgBr to N-PhFl-phenylalaninal, O-benzylation, oxidative cleavage of alkene with KMnO₄	96% addition 81% de 97% benzylation 87% oxidative cleavage	2003 (805)
iBu	H	H	OH	reduction of L-Leu to aminoaldehyde, vinylMgBr addition, alkene oxidative cleavage	59% addition 80% de 63% oxidation	1997 (1123)
Ph	H	H	OH	addition of vinylL MgBr to phenylglycinal, oxidative cleavage of alkene	62% addition yield 68–82% oxidative cleavage	1991 (1124)
Me	H	OH	H	homologation of amino acids via reduction to amino aldehyde, lithiated methoxyallene addition, ozonolysis	7–18% reduction 27–51% addition, ozonolysis 48–76% de	1994 (1126)
iBu	H	OH	H			
Bn	H	OH	H			
CH₂OH	H	OH	H			
Ph	H	H	OH	homologation of Phg using 2-(trimethylsilyl)thiazole addition to phenylglycinal	70–75% for addition	1995 (1127)
iBu	H	OH	H	homologation of amino acids via thiazole addition, stereoselective ketone reduction, conversion of thiazole to acid	89–93% thiazole addition 85–95% reduction, >95% de syn to 95% de anti 71–78% aldehyde 80–92% oxidation	1991 (1130) 1993 (1129)
iBu	H	H	OH			
Ph	H	OH	H			
Ph	H	H	OH			
Bn	H	OH	H			
Bn	H	H	OH			
CH₂OH	H	OH	H			
CH₂OH	H	H	OH			
CH(OH)Me	H	OH	H			
CH(OH)Me	H	H	OH			

H	H	OH	H	homologation of glycine to 3-amino-2-oxopropanoate, asymmetric ketone reduction	30% homologation; 85% reduction, 81% ee; 64% recrystallize, 100% ee; 78% deprotection	1993 (1131)
Me	H	H	OH	homologation of Ala via addition of allyl reagent, ketone reduction, alkene oxidative cleavage	42–57% oxidation	1987 (1132)
Me	OH	OH	H			
iBu	H	Me	H	homologation of amino acids via organolithium addition, ketone methylenation, alkene hydroboration, hydroxy-methyl oxidation	48–78% RLi addition; 56–99% methylenation; 54–85% hydroboration; >90% de *syn* to >90% de *anti*; 63–95% oxidation	1993 (1133)
iBu	H	H	Me			
nBu	H	nBu	H			
nBu	H	H	nBu			
Bn	H	Ph	H			
$(CH_2)_3NH\text{–}C(=NH)NH_2$	H	OH	H	homologation of Arg via reduction to arginal, addition of acid equivalent, oxidation of α-hydroxy acid	48% reduction, addition; 89% hydrolysis	1992 (264)
$(CH_2)_3NH\text{–}C(=NH)NH_2$	H	=O	—			
CH(Me)Et	H	H	OH	homologation of Ile via addition of vinylMgBr to Weinreb amide, ketone reduction, oxidation of α-hydroxy acid	76% ketone reduction; 66% alkene oxidation	1999 (175)
CH(Me)Et	H	=O	—			
Me	H	=O	—	homologation of amino acid UNCAs via reaction with $ClPh_3PCH_2CN$, then ozonolysis	65–90% UNCA reaction; 50–75% ozonolysis	1998 (1134)
iPr	H	=O	—			
iBu	H	=O	—			
sBu	H	=O	—			
Bn	H	=O	—			
$CH_2CH_2CO_2Bn$	Bn	=O	—	homologation of amino acid via activation with EDCI/HOBt, reaction with $ClPh_3PCH_2CN$, then ozonolysis, then ketone reduction	88% coupling; 58–62% ozonolysis and amide formation; 82–85% ketone reduction; 84–86% de	1999 (1135)
$CH_2CH_2CO_2Bn$	Bn	OH	H			
Bn	H	=O	—	homologation of amino acid via activation with EDCI/HOBt, reaction with $ClPh_3PCH_2CN$, then ozonolysis	48–83% coupling; 40–85% ozonolysis and amide formation	2005 (413)
4-Cl-Bn	H	=O	—			
4-Br-Bn	H	=O	—			

(Continued)

Table 11.2 Asymmetric Syntheses of β-Amino Acids (continued)

R¹	R²	R³	R⁴	Method	Yield, % ee	Reference
H	Et	=O	—	homologation of amino acid via activation with EDCI/HOBt, reaction with ClPh₃PCH₂CN, then ozonolysis	88% coupling 32–50% ozonolysis and amide formation	1997 (2184)
(CH₂)₃NHC(=NH)H₂	H	=O	—	homologation of Boc-Arg(Cbz)₂ via activation with EDCI/DMAP, reaction with ClPh₃PCH₂CN, then ozonolysis	86% coupling 75% ozonolysis and amide formation	2002 (267)
iBu	H	=O	—	homologation of Fmoc amino acid via activation with EDCI/DMAP, reaction with polymer-BrPh₂PCH₂CN, then oxidative cleavage from resin with ozone; ketone reduction with polymer-supported NMe₃BH₄	50–65% α-keto ester 76–>95% reduction 0% de	2003 (1136)
Bn	H	=O	—			
(CH₂)₃NH₂	H	=O	—			
CH₂CO₂H	H	=O	—			
CH₂CONH₂	H	=O	—			
(CH₂)₂SMe	H	=O	—			
Bn	H	OH	H			
Bn	H	H	OH			
CH₂CONH₂	H	OH	H			
CH₂CONH₂	H	H	OH			
H	Bn	=O	—	homologation of amino acid via activation with EDCI/HOBt, reaction with ClPh₃PCH₂CN, then oxidative cleavage with dimethyldioxirane	62% oxidative cleavage	2001 (1137)
Me	H	=O	—	homologation of amino acid via activation with EDCI/HOBt, reaction with ClPh₃PCH₂CN, then ozonolysis	79–83% coupling 67–85% ozonolysis and amide formation	2004 (1138)
iBu	H	=O	—			
Bn	H	=O	—			
CH₂CF₂H	H	=O	H	homologation of difluoroAbu via activation, reaction with ClPh₃PCH₂CN, then ozonolysis and Me ester hydrolysis	64% ozonolysis 69% ester hydrolysis	2002 (412)

Ph	H	$=O$	—	homologation of *N*-Trt amino esters via addition of $LiCH_2PO_3Me_2$, olefination with PhCHO, ozonolysis, ketone reduction	100% phosphonate addition	2003 (*1139*)
Bn	H	$=O$	—		86–88% olefination	
iBu	H	$=O$	—		73–84% ozonolysis	
Ph	H	OH	H		73–86% reduction	
Bn	H	OH	H		90–98% de *anti*	
iBu	H	OH	H			
Ph	H	H	OH	homologation of Phg Weinreb amide via aldehyde formation, Wittig reaction, ester reduction, Pd-catalyzed oxazoline formation, oxidation	80% reduction and Wittig; 89% ester reduction; 78% cyclization; 78% oxidation	1998 (*1140*)

Elaboration of α-Amino Acids: Asp

H	H	H	decarboxylation of labeled Asp(OMe)-OH	62%	1985 (*421*)
H	H	D			
D	D	H			
H	Me	H	decarboxylation of 3-Me- Asp	24%	1985 (*422*)
4-MeO-Ph	H	H	arylation of α,β-unsaturated β-Ala derived from Asn within chiral dihydropyrimidinone template	55–78% arylation	1991 (*1210*); 1991 (*128*); 1992 (*1142*); 1996 (*1141*)
CH_2OH	Me	H	reduction of Cbz-Asp anhydride, deprotonation and alkylation	91% reduction; 69–97% alkylation; 76–82% de *syn*, 66–88% de *anti*	1983 (*1143*)
CH_2OH	$CH_2CH{=}CH_2$	H			
CH_2OH	H	Me			
CH_2OH	H	Bn			
CH_2OH	H	$CH_2\text{-}CH{=}CH_2$			
CH_2OH	H		reduction of Cbz- or Boc-Asp anhydride	76% reduction	1997 (*1144*)
CH_2OH	H		reduction of Cbz-Asp anhydride	—	2003 (*1145*)
CH_2OH	H		reduction of Cbz- or Boc-Asp anhydride	92% reduction	1990 (*1146*)
CH_2OH	H		reduction of Boc-Asp(OBn) anhydride	95% reduction	2002 (*1147*)
CH_2OH	Me	H	alkylation of Asp-derived oxazolidinone	70–92% alkylation; 90–96% de *anti*	1996 (*1148*)
CH_2OH	Et	H			
CH_2OH	Bn	H			
CH_2OH	$CH_2\text{-}CH{=}CH_2$	H			
CH_2OH	$CH_2\text{-}CH{=}C(Me)_2$	H			
CH_2OH	$(CH_2)_3Cl$	H			

(Continued)

Table 11.2 Asymmetric Syntheses of β-Amino Acids (continued)

R^1	R^2	R^3	R^4	Method	Yield, % ee	Reference
H	CH_2OH	H	H	reduction of Cbz- or Boc-Asp(OBzl) α-carboxyl	61–84%	1991 (1149)
CH_2OH	H	H	H			
CH_2OH	H	H	H	reduction of EtO₂C-Asp(OMe) α-carboxyl	95%	1992 (1150)
H	CH_2OH	H	H	reduction of Asp α-carboxyl, deprotonation, and methylation	93% reduction	2000 (1151)
H	CH_2OH	H	Me		97% methylation	
H	CH_2OH	Me	H		48% de	
CH_2NMe_2	H	H	H	amide formation of Boc-Asp(OMe) α-carboxyl, carbonyl reduction with borane	—	1997 (399)
$CH_2N[\text{-}(CH_2)_4\text{-}]$	H	H	H			
CH_2OH	H	Bn	H	β-benzylation of Boc-Asp(OBn)-OBn, selective α-ester deprotection, reduction	46% benzylation; 90% ester hydrolysis; 33% ester reduction	1998 (418)
H	$COCH_2CO_2H$	H	H	activation of Boc-Asp(OBn)-OH α-carboxyl with CDI, addition of LiCH₂CO₂TMSE, ketone reduction	92% activation and addition	1999 (1152)
H	$CH(OH)CH_2\text{-}CO_2H$	H	H		82% reduction	
H	CH_2CO_2H	H	H	Arndt–Eister homologation of α-carboxyl group of Cbz-Asp(OtBu)-OH, conversion to aldehyde, Reformatski addition, ester reduction to alcohol, activation and azide displacement, azide reduction	79% homologation	2003 (167)
H	CH_2CH_2OH	H	H		68% reduction	
H	CH_2CHO	H	H		93% oxidation	
H	$CH_2CH(OH)CH_2CO_2H$	H	H		57% Reformatski	
H	$CH_2CH(OH)CH_2CH_2OH$	H	H		69% azide displacement	
H	$CH_2CH(OH)CH_2CH_2N_3$	H	H		92% azide reduction	
H	$CH_2CH(OH)CH_2CH_2NH_2$	H	H			
H	CH_2CO_2H	H	H	Arndt–Eistert homologation of Boc-L-Asp(OBn)-OH, borane reduction of acid, bromination, bromide displacement with nucleobases	84% homologation	2003 (1093)
H	CH_2CH_2OH	H	H		73% reduction	
H	CH_2CH_2Br	H	H		79% bromination	
H	$CH_2CH_2(cytosin\text{-}1\text{-}yl)$	H	H		44–92% nucleobase displacement	
H	$CH_2CH_2(adenin\text{-}9\text{-}yl)$	H	H			
H	$CH_2CH_2(thymin\text{-}1\text{-}yl)$	H	H			
H	$CH_2CH_2(2\text{-}amino\text{-}6\text{-}chloropurin\text{-}9\text{-}yl)$	H	H			
H	$CH_2CH_2(guanin\text{-}9yl)$	H	H			
H	$nC_{10}H_{21}$	H	H	organocuprate addition to Asp derivative	86% for addition	1995 (1153)

H	H	nBu	organocuprate addition to Asp derivative	72–93% for addition	1992 (1154)
H	H	nC$_{12}$H$_{25}$			
H	H	iPr			
H	H	(CH$_2$)$_7$CH(Et)-Me			
H	H	nPent	reduction Ts-Asp anhydride to lactone, opening with TMSI/EtOH to give 3-TsNH-4-I-butanoate, organocuprate with BuI or HexI	63% anhydride reduction 88% iodination 66–79% organocuprate	2002 (89)
H	H	nHept			
H	H	(Z)-(CH$_2$)$_3$CH=CHEt	reaction of 3-TsNH-4-I-butanoate, with organocuprate or octane-1-thiol, thiol oxidation	75–89% organocuprate or thiol	2002 (90)
H	H	n-Undecane			
H	H	CH$_2$SnOct			
H	H	CH$_2$SOnOct			
H	H	CH$_2$SO$_2$nOct			
H	H	CH$_2$I	reduction Ts-D-Asp anhydride to lactone, opening with TMSI/EtOH to give 3-TsNH-4-I-butanoate, iodide displacement with NMe$_3$, Me$_3$P or NaN$_3$, azide reduction, Ts deprotection	80% lactone reduction 70% lactone iodination 78–98% iodide displacement 78–95% Ts removal	2003 (1155)
H	H	CH$_2$NMe$_3$			
H	H	CH$_2$PMe$_3$			
H	H	CH$_2$N$_3$			
H	H	CH$_2$NH$_2$			
H	H	CH$_2$NH$_2$	reduction of Boc-Asp(OMe)-OH α-carboxyl, activation and displacement with azide, azide reduction	48% azide reduction and N-Fmoc protection	2004 (1156)
H	H	CH$_2$(1-uracil)	reduction of Boc-Asp(OBn)-OH α-carboxyl with iBuOCOCl/NaBH$_4$, Mitsunobu displacement with 3-Bz-uracil or 3-Bz-thymine, or mesylation and displacement with 3-Bz-thymine or Cbz-cytosine	95% reduction 65–70% Mitsunobu 79% mesylation 23–66% displacement 82–86% deprotecton	2002 (1157)
H	H	CH$_2$(1-thymine)			
H	H	CH$_2$(1-cytosine)			
H	H	Et	conversion of Asn via β-amide dehydration to nitrile, reduction of α-carboxy to CH$_2$OMs, organocuprate displacement	91% reduction/ mesylation 48–72% organocuprate displacement >99% ee	1990 (1158)
H	H	nPent			
H	H	Bn			
H	H	Me	reduction of Asn α-carboxyl, amide hydrolysis	35% from mesylate	1991 (1159)

(Continued)

Table 11.2 Asymmetric Syntheses of β-Amino Acids (continued)

R¹	R²	R³	R⁴	Method	Yield, % ee	Reference
Me	H	Me	H	conversion of Asn via β-amide dehydration to nitrile, reduction of α-carboxy to CH₂OR, α-deprotonation and alkylation, nucleophilic displacement of hydroxyl	85–92% deprotonation/ alkylation 0% de 37–70% displacement	1995 (1160)
Me	H	Bn	H			
H	CH₂OH	H	H	reduction of Asp anhydride α-carboxyl, α-alkylation	91% reduction	1986 (1161)
H	CH₂OH	Me	H		77% allkylation (84% de)	
H	Me	H	H	organocuprate addition to Asp derivative, or α-hydroxylation followed by organocuprate addition	74–96% for addition 64–68% for hydroxylation, then 71–90% for addition (syn)	1993 (1163) 1994 (1165) 1996 (1162)
H	Et	H	H			
H	nPent	H	H			
H	CH₂tBu	H	H			
H	(CH₂)₂Ph	H	H			
H	Et	OH	H			
H	nPr	OH	H			
H	nPent	OH	H			
H	CH₂tBu	OH	H			
H	Bn	OH	H			
H	(CH₂)₂Ph	OH	H			
H	CH₂cHex	OH	H			
H	(CH₂)₂cHex	OH	H			
H	Et	Me	H	β-alkylation, organocuprate addition to Asp derivative	—	1996 (1164)
H	Ph	OH	H			
H	Et	Me	H	reduction of Asp anhydride α-carboxyl, α-alkylation, lactone opening to form electrophile, organocuprate displacement	75% reduction 70% allkylation (93% de) 79% methylation	1994 (1166)
H	Et	OH	H	reduction of Asp anhydride α-carboxyl, α-hydroxylation, lactone opening to form electrophile, organocuprate displacement	74% reduction 64% hydroxylation (93% de) 42–85% organocuprate	1993 (1167)
H	nPr	OH	H			
H	nPent	OH	H			
H	(CH₂)₂Ph	OH	H			
H	CH₂tBu	OH	H			
H	CH₂cHex	OH	H			
H	CH₂NHR R = Bn, CH₂CH=CH₂, BnO, CbzNH, (CH₂)₄CH-(NHBoc)-CO₂Me	H	H	reduction of α-carboxy group, of Cbz-Asp(OtBu)-OH, Mitsunobu cyclization of alcohol, aziridine opening with alkylamines	60–90% aziridine 60–83% aziridine opening	2004 (1168)

H	H	Bn	conversion of Asp to 3-amino-4-iodobutyric acid, form organoZn reagent, Pd-catalyzed coupling with ArI	20–89% coupling	1998 (*1169*) 1999 (*1172*)
H	H	CH₂(1-naphthyl)			
H	H	CH₂(4-Me-Ph)			
H	H	CH₂(2-MeO-Ph)			
H	H	CH₂(4-MeO-Ph)			
H	H	CH₂(2-NH₂-Ph)			
H	H	CH₂(4-Br-Ph)			
H	H	CH₂(2- F-Ph)			
H	H	CH₂(4-F-Ph)			
H	H	CH₂(2-NO₂-Ph)			
H	H	CH₂(3-NO₂-Ph)			
H	H	CH₂(4-NO₂-Ph)			
H	H	CH₂CH=CH₂			
H	H	CH₂(2-HO-Ph)	conversion of Asp to Boc-3-amino-4-iodobutyric acid Me ester, form organoZn reagent, Pd-catalyzed coupling with unprotected iodoophenol	42–85% coupling	2004 (*1170*)
H	H	CH₂(3-HO-Ph)			
H	H	CH₂(4-HO-Ph)			
H	H	Bn	conversion of Asp to 3-amino-4-iodobutyric acid, form organoZn reagent, Pd-catalyzed coupling with ArOTf	44–64% coupling	2000 (*1171*)
H	H	CH₂(1-naphthyl)			
H	H	CH₂(4-NO₂-Ph)			
H	H	CH₂COPh	conversion of Asp to Boc-3-amino-4-iodobutyric acid Me ester, form organoZn reagent, Pd-catalyzed coupling with RCOCl	20–59% coupling	2000 (*1174*)
H	H	CH₂COCH=CH₂			
H	H	CH₂COCH₂OAc			
H	H	CH₂COnPent			
H	H	CH₂CO(2-furyl)			
H	H	Bn	conversion of Asp to TFA-3-amino-4-iodobutyric acid Me ester, form organoZn reagent, Pd-catalyzed coupling with ArI	53–77% coupling	2003 (*1173*)
H	H	CH₂(1-naphthyl)			
H	H	CH₂(4-Me-Ph)			
H	H	CH₂(4-MeO-Ph)			
H	H	CH₂(4-Br-Ph)			
H	H	CH₂(4-F-Ph)			
H	H	CH₂(4-NO₂-Ph)			
H	H	CH₂(4-CN-Ph)			
H	H	CH=CHCO₂H	conversion Cbz-Asp(OBn)-OH α-carboxy to aldehyde, Wittig homologation, dihydroxylation	59% aldehyde 70% Wittig (1:1 *E:Z*) 50% dihydroxylation	1992 (*220*)
H	H	CH(OH)CH(OH)CO₂H			

(*Continued*)

Table 11.2 Asymmetric Syntheses of β-Amino Acids (continued)

R^1	R^2	R^3	R^4	Method	Yield, % ee	Reference
H	syn and anti CH(OH)Et	H	H	reduction of N,N-Bn$_2$-Asp γ-carboxy group to alcohol and protection as TBDMS ether, reduction of α-carboxyl group to aldehyde, addition of Et$_2$Zn or EtMgBr, reoxidation of side-chain carboxyl OR same procedure with side-chain protected as OBO ester or simple reduction of Cbz-Asp(OMe)-OH α-carboxyl	76% Et$_2$Zn addition; 100% de syn; 70% EtMgBr addition; 100% de anti; 71–76% reoxidation OR; 52–64% OBO cleavage; 48% α-carboxyl reduction	2003 (1175)
H	CH$_2$OH	H	H			
CH=CHC(Me)=CHCH(Me)- CH-(OMe)Bn	H	Me	H	organostannane formation from Asp α-aldehyde derivative, Stille coupling	66% for aldehyde; 56% for stannane; 58% for Stille	1996 (176)
CH=CHC(Me)=CHCH(Me)- CH-(OMe)Bn	H	Me	H	conversion of serine into β-methyl-Asp-aldehyde, iodomethylenation. Stille coupling	75% iodo-methylenation; 58% Stille	1997 (1177)
CH$_2$OH	H	Me	H	methylation of enolate of 3-amino-4-hydroxybutyrate lactone, oxidation and Wittig reaction of hydroxymethyl group	68% alkylation; 60% de; 43% oxidation and Wittig	1989 (1178)
CH=CHC(Me)=CHCH(Me)- CH-(OMe)Bn	H	Me	H			
CH=CHC(Me)=CHCH(Me)- CH-(OMe)Bn	H	Me	H	Mitsunobu amination of chiral 1,2,4-trihydroxy-4-methyl-butane oxidation and Wittig reaction of hydroxymethyl group	72% amination; 90% oxidation; 55% Wittig	1990 (839)
CH=CHC(Me)=CHCH(Me)- CH-(OMe)Bn	H	Me	H	organostannane formation from Asp α-aldehyde derivative, Wittig coupling	7% for aldehyde; 30% for Wittig	1992 (1179)
CH$_2$OH	H	Me	H	conversion of Asp to β-lactam, methylation, reduction of carboxyl, Wittig reaction, lactam opening	50–60% β-lactam; 50–55% methylation; 90% aldehyde; 10–45% Wittig; 76–95% lactam opening	1998 (1180)
CH=CHC(Me)=CHCH(Me)- CH-(OMe)Bn	H	Me	H			

H	COPh	H	Friedel–Crafts acylation using Asp anhydride	38–45%	1997 (1181)
H	CO(4-Me-Ph)	H			
H	CO(2,3-Me_2-Ph)	H	Friedel–Crafts arylation of Meoc-Asp(OEt)-Cl, reduction, protection, enolate formation and hydroxylation, deprotection	57% arylation 87% reduction 62% hydroxylation 100% de 86% deprotection	1999 (392)
COPh	H	H			
CH(OH)Ph	H	H			
Bn	OH	H			
H	2-thiazolyl	H	conversion of Asp α-carboxyl to thioamide, cyclization to thiazole	57% thioamide 77% thiazole	1974 (1182)
CN	H	H	dehydration of isoasparagine	50–60%	1986 (1183)
CH_2(5-R-1,2,4-oxadiazol-3-yl), R = Me, iPr, nBu, tBu, 4-Me-Ph, CH_2CONH_2, $CH_2CH_2NH_2$	H	H	esterification of Fmoc-Asp(OtBu)-OH with HON=C(R)NH_2, cyclization in NaOAc/EtOH/H_2O	25–84% overall	2003 (1184)
$COCH_2CO_2H$	H	GH	formation of Asp UNCA. opening with $LiCH_2CO_2$TMSE	90% UNCA formation 80% opening	2003 (215)
Elaboration of α-Amino Acids: Other					
nPr	H	H	elaboration of homoserine via RLi addition to Weinreb amide, side-chain oxidation, ketone reduction	72–92% RLi addition 72–92% oxidation 65–78% reduction	1996 (1185)
Bn	H	H			
4-Me-Bn	H	H			
4-MeO-Bn	H	H			
$(CH_2)_3$Ph	H	H			
$(CH_2)_2CH(Ph)_2$	H	H			
CH=CHC(Me)=CHCH(Me)-CH(OMe)Bn	Me	H	elaboration of D-Thr	38% overall	1995 (226)
CH_2OH	Me	H	ester reduction of 2-amino-3-Me-4-hexenoic acid, alkene oxidative cleavage, hydroxymethyl oxidation, Wittig reaction	86% ester reduction 90% alkene oxidation 84% hydroxymethyl oxidation 41% Wittig	1999 (227)
CHO	Me	H			
CH=CHC(Me)=CHCH(Me)-CH(OMe)Bn	Me	H			
iPr	H	H	conversion of Leu by Leu 2,3-aminomutase	—	1976 (1186)
H	H	OH	conversion of Asn via reaction with $NaNO_3$ then Hofmann rearrangement	23% overall	1976 (1187)

(Continued)

Table 11.2 Asymmetric Syntheses of β-Amino Acids (continued)

R¹	R²	R³	R⁴	Method	Yield, % ee	Reference
H	H	H	F	treatment of N,N-Bn₂-β-hydroxy-α-amino acids with DAST	60–90%	1982 (1188)
H	Me	H	F			
H	iPr	H	F			
Me	H	H	F			
iPr	H	H	F			
H	H	H	F	treatment of labeled N,N-Bn₂-Ser-OMe with DAST	78–86%	1985 (422)
D	D	H	F			
H	D	D	F			
H	H	F	H	treatment of N,N-Bn₂-D-Ser-OMe with DAST	—%	2000 (378)
H	H	H	F	treatment of N,N-Bn₂-β-hydroxy-α-amino acids with DAST	60–90%	2004 (355)
Me	H	H	F			
H	CH(OH)CH(OH)CO₂H	H	H	elaboration of D-Glu	17% from pyroglutaminol	1989 (217) 1991 (218)
H	CH(OH)CH(OH)CO₂H	H	H	elaboration of 3,4,5-trihydroxy-Nva	—	1990 (1189)

Elaboration of β-Amino Acids: Addition of Side Chain

R¹	R²	R³	R⁴	Method	Yield, % ee	Reference
H	Me	H	Me	alkylation of homochiral di-deprotonated N-Bz or N-Cbz-3-aminobutanoate	38–96% alkylation	1987 (1190)
H	Me	H	Et		75–>99% de	1988 (638)
H	Me	H	CH(OH)Ph			
H	Me	H	CH₂-CH=CH₂			
H	Me	H	Bn			
H	Me	H	N(Boc)NH-Boc			
iBu	H	Me	H	polydeprotonation and alkylation of β³-HLeu residue in β-tripeptide	35–75%	1998 (1191)
iBu	H	Bn	H		60–>94% de	
iBu	H	CH₂CH=CH₂	H			
iBu	H	CH₂CO₂tBu	H			
Bn	H	Me	H	alkylation or sequential dialkylation of homochiral dideprotonated N-Cbz-3-amino-3-phenylpropanoate	23–95% alkylation	1995 (1076)
Bn	H	Bn	H		66–>98% de	
Bn	H	N(Boc)NH-Boc	H			
Bn	H	Et	Me			
Bn	H	CH₂CH=CH₂	Me			
(CH₂)₂SMe	H	CH₂CH=CH₂	H	alkylation or sequential dialkylation of homochiral di-deprotonated N-Cbz homologated Met, vinylglycine, allylglycine or 2-aminohex-6-enoic acid	42–86% alkylation	2004 (1105)
CH=CH₂	H	CH₂CH=CH₂	H			
H	CH₂CH=CH₂	H	CH₂CH=CH₂			
(CH₂)₂CH=CH₂	H	CH₂CH=CH₂	H			
(CH₂)₂SMe	H	Me	H			
(CH₂)₂SMe	H	CH₂CH=CH₂	Me			

$CH=CH_2$	$CH_2CH=CH_2$	Me			
H	H	Me			
H	H	Et			
H	Me	$CH_2CH=CH_2$			
H	Et	$CH_2CH=CH_2$			
H	H	$CH_2CH=CH_2$			
H	$CH_2CH=CH_2$	Me			
H	$(CH_2)_2Ph$	3-CN-Bn	alkylation of homochiral di-deprotonated N-Ac-3-amino-5-Ph-pentanoate, convert CN to amidine	50% alkylation, >95% de, 60% amidine formation	1998 (416)
H	$(CH_2)_2Ph$	3-[C(=NH)-NH$_2$]-Bn			
Me	H	H	deprotonation and methylation of homochiral Boc-β-substituted-β-amino esters	75–90% methylation, 23–50% de	1998 (275)
iPr	H	H			
iBu	H	H			
Bn	H	H			
Me	H	Me			
iPr	H	Me			
iBu	H	Me			
Bn	H	Me			
CH_2OH	H	H	deprotonation and alkylation of Ser- or Cys-homologated β³-amino acids	24–46% alkylation, 10–>92% de	2000 (294)
CH_2OH	H	CH_2OH			
CH_2SH	H	H			
Me	Me	Me	deprotonation and alkylation of Boc-β²,³-amino acid methyl esters		2002 (327)
iPr	Me	Me			
iBu	Me	Me			
Me	CH_2SMe	H	polydeprotonation and alkylation of Boc-β³-HLeu, of Boc-β³-HLeu, or of Boc-β³-HLeu with ClCH$_2$SMe, oxidation, pyrolytic elimination	60–89% alkylation, 74–89% oxidation and elimination	2004 (326)
iPr	CH_2SMe	H			
iBu	CH_2SMe	H			
Me	$=CH_2$	—			
iPr	$=CH_2$	—			
iBu	$=CH_2$	—			
Me	Me	H	conversion of Asn via β-amide dehydration to nitrile, reduction of α-carboxy to CH$_2$OR, α-deprotonation and alkylation, nucleophilic displacement of hydroxyl	85–92% deprotonation/ alkylation, 0% de, 37–70% displacement	1995 (1160)
Me	Bn	H			
Ph	H	OH	hydroxylation of enolate of 3-amino-3-phenylpropionic acid	65–83%, 72% de syn to 80% de anti	1996 (1192)
Ph	OH	H			

(Continued)

Table 11.2 Asymmetric Syntheses of β-Amino Acids (continued)

R^1	R^2	R^3	R^4	Method	Yield, % ee	Reference
Ph	H	H	F	fluorination of homochiral di-deprotonated N-Bz-3-amino-3-phenylbutanoate	41–94% fluorination 4–62% de	1994 (989)
Ph	H	F	H			
Ph	H	H	OH	hydroxylation of Bz-3-Ph-β-Ala via iodination, oxazoline formation	95% iodination/ oxazoline 85% hydrolysis	1998 (605)
H	H	OH	H	N,N-diprotection of amino acid with 4-MeO-BnCl, deprotonation with KHMDS, hydroxylation with 2-[(4-Me-Ph)sulfonyl]-3-Ph-oxaziridine	89–95% yield 82–94% de	2002 (1193)
Ph	H	OH	H			
Bn	H	OH	H			
CH$_2$OBn	H	OH	H			
iBu	H	SH	H	sulfenylation of enolate of 3-amino-3-alkylpropionic acid	20–69% >90% de anti	1997 (2188)
Bn	H	SH	H			
(CH$_2$)$_2$OH	H	SH	H			
(CH$_2$)$_2$SO$_2$NH$_2$	H	SH	H			
(CH$_2$)$_4$NH$_2$	H	SH	H			
(CH$_2$)$_2$CO$_2$H	H	SH	H			
Me	H	SePh	H	phenylselenation of enolate of 3-aminobutyric acid, free radical allylation	58–78% >96% de anti	1997 (331)
Me	H	CH$_2$-CH=CH$_2$	H			
CH=CHC(Me)=CHCH(Me)-CH(OMe)Bn	H	Me	H	methylation of enolate of 3-amino-4-hydroxybutyrate lactone, oxidation and Wittig reaction of hydroxymethyl group	68% alkylation 60% de 43% oxidation and Wittig	1989 (1178)
H	CH(OH)CH$_2$OH	H	H	addition of BnNH to homochiral unsaturated lactone, deprotonation, alkylation	43% amination 100% de 51–89% alkylation 50–>90% de C2	1995 (778)
H	CH(OH)CH$_2$OH	Et	H			
H	CH(OH)CH$_2$OH	C(OH)Me$_2$	H			
H	CH(OH)CH$_2$OH	CH(OH)Ph	H			
H	CH(OH)CH$_2$OH	C(OH)(Me)-Ph	H			
H	(CH$_2$)$_2$OH	C(Me)$_2$OH	H	reduction of carboxymethyl group of 3-amino-4-(carboxymethyl)-butyric acid, lactone formation, aldol reaction, hydrolysis	65% reduction and lactone formation 77% aldol 88% hydrolysis	1983 (1194)
H	H	Me	OH	alkylation of isoserine equivalent with Oppolzer chiral amide auxiliary	92% yield crude, 54% de 71% pure, >99% de	1996 (1195)

R1	R2	R3	R4	Method	Results	Year (Ref)
H	H	Bn	Me	alkylation of 3-phenyl-2-cyano-propanoate, reduction of cyano group	94–97% alkylation	1992 (1197)
H	H	Bn	CH₂-CH=CH₂		60–94% de crude	1993 (1196)
H	H	Bn	CH₂-C(Me)=CH₂		>96% de recrystallized	
H	H	Bn	CH₂-CH=C(Me)₂		89–93% reduction	
H	H	Bn	CH₂-CH=CHPh			
H	H	Bn	CH₂CCH			
H	H	H	Me	alkylation of benzophenone Schiff base of β-Ala with Oppolzer sultam amide chiral auxiliary	51–772% alkylation	2000 (1198)
H	H	H	iBu		100% de	
H	H	H	Bn		57–75% deprotection	
H	H	H	Me	alkylation of BnON=CHCH₂CO₂H with Oppolzer sultam amide chiral auxiliary under PTC conditions, radical addition to oxime	47–99% alkylation	2003 (1199)
H	H	H	Bn		80–100% de	
H	H	H	4-NO₂-Bn		>95% de after rc	
H	H	H	CH₂-CH=CH₂		20–70% radical addition	
H	H	H	CH₂CH=CHMe		>95% de if α-substituent present	
H	H	H	CH₂CCH			
H	H	H	CH₂CO₂Me			
H	H	Et	Me			
H	H	Et	Bn			
H	H	Et	4-NO₂-Bn			
H	H	Et	CH₂-CH=CH₂			
H	H	Et	CH₂CCH			
H	H	iPr	Bn			
H	H	cHex	Bn			
H	H	cPent	Bn			
H	H	sBu	Bn			
H	H	iBu	Bn			
H	H	iPr	4-NO₂-Bn			
H	H	H	CH₂CH=CH₂	alkylation of N-Boc,N-alkenyl-β-Ala with chiral oxazolidinone auxiliary	68–71% alkylation; 87–89% de	2005 (641)
H	CH₂OH	H	H	alkylation of Phth-β-Ala with DIOZ chiral oxazolidinone auxiliary	92% alkylation	1998 (1030)
H	CH(OH)Me	H	H	aldol reaction of Phth-β-Ala with DIOZ chiral oxazolidinone auxiliary, dehydroxylation	67–68% aldol	2003 (832)
H	CH(OH)(indol-3-yl)	H	H		75% dehydroxylation	2005 (348)
H	CH₂(indol-3-yl)	H	H			

(Continued)

Table 11.2 Asymmetric Syntheses of β-Amino Acids (continued)

R^1	R^2	R^3	R^4	Method	Yield, % ee	Reference
H	H	CH_2OH	H	reaction of Phth-β-Ala with	89% aldol	2003 (832)
H	H	CH_2SH	H	DIOZ chiral oxazolidinone auxiliary with trioxane, conversion of hydroxy to thiol via Mitsunobu	71% Mitsunobu	2004 (833)
H	H	Me	H	alkylation of (S,R,R)-PhCH(Me) NHCH$_2$CH$_2$CON[CH(Me) Ph]$_2$	29–86% alkylation 60–90% de 13–63% deprotection	2002 (1200)
H	H	Et	H			
H	H	nPr	H			
H	H	Bn	H			
H	H	Me	H	alkylation of β-Ala with 2-hydroxypinanone Schiff base and menthol ester chiral auxiliaries	86% alkylation 84% de 79% deprotection	1994 (1201)
H	H	H	Me	alkylation of β-Ala pseudoephedrine amide, hydrolysis	52–74% alkylation and hydrolysis 75->99% ee	2000 (705)
H	H	H	Et			
H	H	H	nPr			
H	H	H	CH$_2$CH=CH$_2$			
H	H	CH_2CO_2H	H	alkylation of β-Ala with masked amino group and chiral oxazolidinone auxiliary	53–60% alkylation >95% de 52–58% revealing of amino group	1996 (1202)
H	H	H	Me	alkylation of 3-aminopropanoic acid in 2-tBu-3,6-dimethyl-perhydropyrimidin-4-one template	—	1992 (1203)
H	H	Me	H	alkylation of β-Ala in 2-tBu-3-methylperhydropyrimidin-4-one template	75–80% alkylation >95% de 85–90% deprotonation to other diastereomer 80–85% hydrolysis	1996 (1204)
H	H	nBu	H			
H	H	nHex	H			
H	H	Bn	H			
H	H	H	Me			
H	H	H	nBu			
H	H	H	nHex			
H	H	H	Bn			
H	H	Me	Bn	consecutive alkylations of β-Ala in 2-tBu-3-methyl-perhydro-pyrimidin-4-one template	81–96% second alkylation >95% de 76–87% hydrolysis	1998 (1205)
H	H	Me	nBu			
H	H	nBu	Me			
H	H	Bn	Me			

R¹	R²	R³	Reaction	Results	Year (Ref.)
H	Me	H	alkylation of 3-aminobutanoic acid in 2-*t*-Bu-3,6-dimethylperhydropyrimidin-4-one template	75–95% alkylation 100% de 50–90% hydrolysis	1992 (*712*)
H	Me	Bn			
Me	Me	Me	alkylation of 3-aminobutanoic acid in 3,6-dimethylperhydropyrimidin-4-one template	95–100% alkylation 50% de 70–84% hydrolysis	1993 (*1208*)
Me	Me	Bn			
H	Ph	OH	hydroxylation of 3-amino-3-phenylpropanoic acid in 3,6-dimethylperhydropyrimidin-4-one template	71% alkylation 100% de 41% hydrolysis	1995 (*719*)
(CH₂)₂CO₂H	H	H	radical reaction of β-Ala in 2-*t*-Bu-3-methylperhydropyrimidin-4-one template	50–57% yield addition 97–98% de 45–80% hydrolysis	1996 (*1209*)
(CH₂)₂CN	H	H			
(CH₂)₂SO₂Ph	H	H			
4-MeO-Ph	H	H	arylation of α,β-unsaturated β-Ala within chiral dihydropyrimidinone template	55–78% arylation	1991 (*1210*) 1991 (*128*) 1992 (*1142*) 1996 (*1141*)
Me	H	H	alkylation of α,β-unsaturated β-Ala within chiral template, hydrogenation	55–95% alkylation 75% reduction 60% deprotection	1993 (*1211*)
Bn	H	H			
H	CH₂CO₂H	H	alkylation of H₂NC(CH₂CO₂Me)=CHCO₂H within cyclic template, hydrogenation	46–72% alkylation 34–40% de 37–53% hydrogenation >95% de	2005 (*1212*)
H	CH₂CO₂H	Me			
H	CH₂CO₂H	Bn			
Me	H	H	α-alkylation or dialkylation of β-Ala or β-substituted β-Ala in 1-Boc-2-*t*-Bu-5,6-dihydro-4-methoxy-2*H*-pyrimidine template, deprotection and N-Boc protection	47–97% alkylation of β-Ala 100% de 73% second alkylation 9–78% alkylation of β³-Ala 5–65% aldol 12–38% additon to imine 30–81% additon to α,β-unsaturated ester 12–91% deprotection	1999 (*1213*)
Bn	H	H			
CH₂CH=CH₂	H	H			
CH₂CCH	H	H			
CH₃SiMe₃	H	H			
CH₂cPr	H	H			
3-MeO-Bn	H	H			
cyclohexen-3-yl	H	H			
CH₂CH=CH₂	Me	H			
Me	Me	H			
Me	Et	H			
Me	iPr	H			
Me	Bn	H			
Me	CH₂CO₂Bu	H			
Me	CH(Me)Ph	H			

(Continued)

Table 11.2 Asymmetric Syntheses of β-Amino Acids (continued)

R^1	R^2	R^3	R^4	Method	Yield, % ee	Reference
iBu	H	Bn	H	α-alkylation of β-substituted β-Ala in pyrimidinone template	50–96% alkylation	1992 (1215)
Me	H	CH(OH)Me	H		56–>98% de	1994 (1214)
Me	H	CH(OH)nBu	H		78–81% hydrolysis	
Me	H	CH(OH)Ph	H			
not deprotected:						
Me	H	CH(Ph)NHAloc	H			
Me	H	CH(Me)CH$_2$CO$_2$R	H			
Me	H	CH(Ph)CH$_2$CO$_2$R	H			
Me	H	Et, nPr, Ph	H	aldol reaction of 3-amino-butanoic acid in pyrimidinone template	43–90% yield	1993 (1216)
Et	H	Et, nPr, Ph	H		100% de C2	
Bn	H	Et, nPr, Ph	H		70% anti to 60% syn	
H	Me	H	Et, nPr, Ph		78–81% hydrolysis	
H	Et	H	Et, nPr, Ph			
H	Bn	H	Et, nPr, Ph			
Me	H	CH(OH)Ph	H	α-hydroxylation of β-substituted β-Ala in pyrimidinone template	61–72% hydroxylation	1995 (1217)
Me	H	CH(OH)Me	H		60–>98% de	
H	Me	H	CH(OH)Ph		88–90% hydrolysis	
H	Me	H	CH(OH)Me			
H	Me	H	OH			
H	nPr	H	OH			
H	Me	OH	H			
H	nPr	OH	H			
Me	H	H	H	addition of organocuprate to electrophilic β-Ala equivalent in chiral template, template decomposition	81–86% addition	1999 (1218)
nBu	H	H	H		>95% de	
Ph	H	H	H		88% template removal	
H	H	Me	H	deprotonation and α-alkylation of β-Ala equivalent in chiral template, template decomposition	88–96% methylation	1999 (1219)
Me	H	Me	H		14–>94% de	
H	Me	Me	H		68% deprotection	
nBu	H	Me	H			
H	H	OH	H	addition of Grignard reagent to chiral cyanohydrin with masked carboxyl group (alkene), imine reduction, alkene oxidative cleavage	55–81% addition and reduction	1996 (1220)
Me	H	OH	H		50–70% de of crude	
Ph	H	OH	H		82–91% oxidative cleavage	

			Method	Results	Year (Ref)
H	OH	H	addition of Grignard reagent to chiral cyanohydrin with masked carboxyl group, imine reduction, furan oxidative cleavage	43–80% addition and imine reduction 60–90% de 52–80% oxidation	2003 (1221)
Me	OH	H			
iPr	OH	H			
iBu	OH	H			
Ph	OH	H			
Bn	OH	H			
Me	OH	H	addition of Grignard reagent to imine of glyceraldehyde with masked carboxyl group, imine reduction, alkene oxidative cleavage	51% addition 100% de 41% oxidation	1996 (1222)
cHexCH₂	H	OH	addition of cHexCH₂ Grignard or organocuprate reagent to electrophilic β-Ala equivalent derived from D-threose	52–75% 100% de	1990 (1223)
H	cHexCH₂	OH			
H	Ph	H	addition of PhLi to electrophilic β-Ala equivalent derived from L-threose	68% addition 72% de 18% formation of carboxy	1989 (95)
H	CH₂Cl[-(CH₂)₅]-CO₂H	H	Michael addition of enolate to chiral electrophilic 2-Me-3-aminopropionate equivalent	83% 98% de	1993 (1224)
Me	OH	H	addition of Grignard reagent to nitrone of glyceraldehyde with masked carboxyl group in presence of ZnBr₂ or Et₂AlCl, deprotection, hydroxy oxidation	72–86% addition 56–82% de syn or 40–90% de anti 50–60% oxidation	1998 (1225)
Et	OH	H			
Ph	OH	H			
Me	OH	H			
Et	OH	H			
Ph	OH	H			
Me	OH	H	addition of organocerium reagent to SAMP hydrazone of 3,3-ethylenedioxypropanal	49–95% addition 75–99% de 52–95% hydrazone cleavage 82–98% ee 88% oxidation (only R = Me)	1993 (1226)
Et	OH	H			
nPr	OH	H			
nBu	OH	H			
allyl	OH	H			
Ph	OH	H			
4-Me-Ph	OH	H			
4-MeO-Ph	OH	H			

(Continued)

Table 11.2 Asymmetric Syntheses of β-Amino Acids (continued)

R¹	R²	R³	R⁴	Method	Yield, % ee	Reference
Et	H	H	Bn	PTC alkylation of O-Bn oxime of 3-oxopropionic acid with camphor sultam auxiliary, radical addition to oxime, deprotection	90–99% alkylation >95% de 20–99% radical addition >95% de 60% deprotection (1 example)	1999 (1227)
Et	H	H	4-NO₂-Bn			
Et	H	H	CH₂CCH			
iPr	H	H	Bn			
iBu	H	H	Bn			
sBu	H	H	Bn			
cPent	H	H	Bn			
cHex	H	H	Bn			
iPr	H	H	4-NO₂-Bn			
H	Et	H	H	conjugate addition of organo-magnesium reagents complexed with chiral bis-oxazoline ligand to CF₃CONHCH=C(CO₂Me)₂, decarboxylation	58–79% addition 56–93% ee	2001 (1228)
H	iPr	H	H			
H	nBu	H	H			
H	cHex	H	H			
H	C₁₈H₃₇	H	H			
H	CH=CH₂	H	H			
H	Ph	H	H			
H	Et	H	H			
H	iPr	H	H			
H	nBu	H	H			
H	cHex	H	H			
H	C₁₈H₃₇	H	H			
H	CH=CH₂	H	H			
H	Ph	H	H			
H	H	H	Me	conjugate addition of dialkylzinc to NO₂CH=CHCO₂Me, nitro reduction	94–97% addition 18–85% ee 60–65% overall including nitro hydrogenation	2003 (1229)
H	H	H	Et			
H	H	H	iBu			
H	H	(CH₂)₂Cl	H	radical addition of RI to PhthNCH₂C(=CH₂)CO₂tBu catalyzed by chiral oxazoline catalyst	71–95% 36–95% ee	2004 (1230)
H	H	(CH₂)₂OMe	H			
H	H	nPr	H			
H	H	iBu	H			
H	H	CH₂tBu	H			
H	H	CH₂cPent	H			
H	H	CH₂cHex	H			
H	H	CH₂(1-adamantyl)	H			
H	H	CH₂C(Me)(CH₂)₃Cl	H			

R^1	R^2	R^3	Method	Yield	Year (ref)
H	H	Me	conjugate addition of dialkylzinc to $NO_2CH=CHCH(OMe)_2$ in presence of chiral catalyst, nitro reduction and amine protection, acetal hydrolysis and oxidation	74–86% addition	2003 (2181)
H	H	Et		88–98% ee	
H	H	nBu		64% reduction and Boc protection (1 example)	
H	H	$(CH_2)_6CO_2H$		82% acetal hydrolysis and oxidation (1 example)	

Elaboration of β-Amino Acids: Other

R^1	R^2	R^3	Method	Yield	Year (ref)
CH_2(adenin-9-yl)	H	H	Mitsunobu reaction of β-homoserine	27–40% Mitsunobu	1998 (1231)
Me	F	H	DAST treatment of 2-hydroxy-3-aminobutanoic acid, or oxidation, then DAST		2004 (355)
Me	F	F			
CN	H	H	dehydration of isoAsn	50–60%	1971 (1232)
H	H	Me	asymmetric hydrogenation/ kinetic resolution of 2-methylene-3-amino esters	56–61%	1987 (1233)
H	H	Et		95–98% ee	
CH_2CH_2SH	H	H	demethylation of homologated Met $(CH_2)_2SMe$ side chain	—	1999 (317)
4-Ar-Ph Ar = Ph, 2-MeO-Ph, 2-CN-Ph, 2-CF$_3$O-Ph, 2,5-(MeO)$_2$-Ph, 2-MeO-6-F-Ph, 2-MeO-4-F-Ph, 2-CF$_3$-O-4-F-Ph, 2-MeO-3-F-Ph, 2-F-3-MeO-Ph, 2-F-4,6-(MeO)$_2$-Ph, 3-F-5-MeO-Ph, 2,6-(MeO)$_2$-Ph, 2,6-(MeO)$_2$-5-F-Ph, 2-CF$_3$-O-6-MeO-Ph, 2,6-(MeO)$_2$-3,5-F$_2$-Ph, 2-HO-6-MeO-Ph, 2,6-(OH)$_2$-Ph	H	H	convert phenol of 3-(4-hydroxyphenyl)-β-Ala to triflate and use for Suzuki couplings with aryl boronic acids	75% Suzuki (with Ar = Ph)	2002 (406)

(Continued)

Table 11.2 Asymmetric Syntheses of β-Amino Acids (continued)

R^1	R^2	R^3	R^4	Method	Yield, % ee	Reference
H	H	CH(OH)Me	H	asymmetric hydrogenation of BzNHCH$_2$CH-(COMe) CO$_2$Me with dynamic catalytic resolution	92–98% ee 98% de *threo*	1989 (1234)
Ph	H	H	OH	Bakers yeast reduction of 2-keto-3-azido-3-phenylpri-onate or 3-keto-4-phenyl-β-lactam	67% ketone reduction 41% lactam reduction 82% ee	1999 (694)
4-HO-Ph	H	H	OH	bioconversion of taxol by *Streptomyces* sp. MA7065	10%	2001 (190)
H H H	CH$_2$CH(OH)CH$_2$N$_3$ CH$_2$CH(OH)CH$_2$NH$_2$ CH$_2$CH(OH)CH$_2$NHR R = Me, Et, nPr, (CH$_2$)$_3$iPr, CONH$_2$, C(=NH)NH$_2$, Bn, 3,4-(OH)$_2$-Bn CH$_2$CH(OH)CH$_2$NR$_2$ R = –CH$_2$CH$_2$OCH$_2$CH$_2$–, –CH$_2$CH$_2$NHCH$_2$CH$_2$–, –CH$_2$CH(NH$_2$)CH$_2$CH$_2$–	H H H	H H H	hydroxyazidation-hydroxylation or hydroxyamination-hydroxylation of alkene in 3-aminohex-5-enoic acid, azide reduction, temporary protection and *N*-alkylation	62% hydroxyazidation 38–65% hydroxyamina-tion	2003 (167)
CH$_2$CONH$_2$ CH$_2$NH$_2$	H H	H H	H H	enzymatic symmetrization/monoamidation of dimethyl 3-benzylaminoglutarate, Hoffman rearrangement	76% monoamidation	2004 (689)
Ph	H	H	OH	microbial reduction of 2-keto-3-amino-3-phenylpro-pionic acid	31–80% 26–99% ee 60–80% de	1993 (693)
CH$_2$CHO (Z)-CH$_2$CH=CHnOct	H H	H H	H H	reduction of acid side chain to aldehyde, Wittig coupling	52% aldehyde and Wittig	2000 (85) 2002 (84)

Small Ring Systems: Aziridines

R^1	R^2	R^3	R^4	Method	Yield, % ee	Reference
H H H	H H H	H H H	Cl SBn indole	aziridine-2-carboxylate opening with nucleophiles	20–40% 97–90% ee	1995 (1235)
H H H H H H H	nHept nHept nPr nC$_{12}$H$_{25}$ iPr cHex Ph	H H H H H H H	H Br Br Br Br Br Br	aziridine-2-carboxylate opening with NaBr, radical reduction (only nHept chiral)	70–100% opening 80% reduction	1996 (654)

nHept	H	H	aziridine-2-carboxylate opening with NaBr, radical reduction	70% opening 80% reduction	1997 (804)
CH$_2$OH	Cl	H	HCl or HBr opening of 3-hydroxymethylaziridine-2-carboxylate, radical debromination	81–89% opening 62% debromination	2000 (236)
CH$_2$OH	Br	H			
CH$_2$OH	H	H			
H or Me	SPh	H	aziridine-2-carboxylate opening with thiols	63–86%	1987 (1237)
H or Me	SCH$_2$CH(NH$_2$)CO$_2$H	H			
H or Me	S-glutathione	H			
Me	H	Me, nBu	aziridine-2-carboxylate opening with organocuprate	30–34%	1995 (1238)
(CH$_2$)$_4$Ph	H	Me	aziridine-2-carboxylate opening with H$_2$/Raney Ni	71–86% opening	2002 (1239)
Me	H	H			
iPr	H	H			
nBu	H	H			
CH(OH)CH$_2$OMe	H	SEt	bicyclic aziridine-lactone opening with soft nucleophiles	57–80%	1995 (1240)
CH(OH)CH$_2$OMe	H	SPh			
CH(OH)CH$_2$OMe	H	OAc			
CH(OH)CH$_2$OMe	H	Br			
CH(OH)CH$_2$OMe	H	3-(N-Me)-indolyl			
(CH$_2$)$_5$OH	H	Et	aziridine opening with organocuprate, hydroxymethyl oxidation	40% overall	1993 (1241)
CH$_2$CH$_2$OH	H	Et	aziridine formation from chiral epoxydiol, opening with LiEt$_2$Cu, hydroxymethyl oxidation	84% aziridine formation 80% opening 82% oxidation	1988 (1242)
iBu	H	OH	opening of N-Boc 2-(1,2-dihydroxyethyl)aziridine with RMgX/Cu, tosylation and displacement with NaCN, hydrolysis	77–88% opening 77% displacement 93% opening	1995 (1243)
Bn	H	OH			
CH$_2$cHex	H	OH			
Et	H	Me	aziridine 2-carboxylate opening with LiAlH$_2$, hydroxymethyl oxidation	90–95% opening 81–90% oxidation	1997 (1244)
Ph	H	Me			
CH$_2$OH	H	H	aziridine opening with Red-Al, hydroxymethyl oxidation	92% opening 90% oxidation	1987 (1245)

(Continued)

Table 11.2 Asymmetric Syntheses of β-Amino Acids (continued)

R¹	R²	R³	R⁴	Method	Yield, % ee	Reference
H	nBu	OH	H	bicyclic aziridine-oxazolidinone opening with organocuprates, hydroxymethyl oxidation	52–80% opening / 72–83% oxidation	1999 (*203*)
H	nHe	OH	H			
H	CH₂SiMe₃	OH	H			
H	Ph	OH	H			
H	4-MeO-Ph	OH	H			
iPr	H	CH(OH)Ph	H	opening of 2-vinylaziridine with RCHO/InI/Pd(PPh₃)₄, vinyl ozonolysis	76–88% opening/ aldol / 58–89% ozonolysis/ oxidation	2002 (*1246*)
iPr	H	CH(OH)(4-MeO-Ph)	H			
iPr	H	CH(OH)(4-Cl-Ph)	H			
Bn	H	H	H	aziridine opening with dithiane of HCOSiMe₃, dithiane removal, hydrolysis	67–90% aziridine opening / 50–53% dithiane removal and hydrolysis	1993 (*1247*)
iBu	H	H	H			
Bn	H	H	H	2-substituted N-Ts aziridine opening with TMSCN/Yb(CN)₃	87–93%	1990 (*658*)
CH₂CH₂SMe	H	H	H			

Small Ring Systems: β-Lactams

R¹	R²	R³	R⁴	Method	Yield, % ee	Reference
Ph	H	H	OH	cycloaddition of enolate with chiral auxiliary with imine	80–85% cycloaddition / 96–98% ee / 97% hydrolysis	1991 (*1258*)
4-MeO-Ph	H	H	OH			
3,4-(MeO)₂-Ph	H	H	OH			
Ph	H	H	OH	cycloaddition of enolate with chiral auxiliary with imine	72–89% cycloaddition / 90–98% ee / 82–91% hydrolysis	1992 (*1260*) / 1992 (*1259*)
4-MeO-Ph	H	H	OH			
3,4-(MeO)₂-Ph	H	H	OH			
4-F-Ph	H	H	OH			
4-CF₃-Ph	H	H	OH			
2-furyl	H	H	OH			
CH=CHPh	H	H	OH			
CH=CH(2-furyl)	H	H	OH			
iBu	H	H	OH			
CH₂cHex	H	H	OH			
(CH₂)₂Ph	H	H	OH			
(CH₂)₂cHex	H	H	OH			
iBu	H	Me	OH	deprotonation and alkylation of 3-silyloxy-4-alkyl β-lactam, hydrolysis	56–84% alkylation / 16–70% N-deprotection / 79–82% hydrolysis	1998 (*1261*)
Ph	H	Me	OH			
CH₂C(Me)=CH₂	H	Me	OH			
iBu	H	CH₂CH=CH₂	OH			
Ph	H	CH₂CH=CH₂	OH			
CH₂C(Me)=CH₂	H	CH₂CH=CH	OH			

CF$_3$	H	OH	cycloaddition of imine with chiral auxiliary with ketene	55–90% cycloaddition 15% de	1991 (1262)
CH=CHPh	CH$_2$CH=CH	H	cycloaddition of imine derived from chiral α-methylbenzylamine pyridyl thioester of alkenoic acids in presence of Lewis acid SnCl$_4$	44–63% cycloaddition	2003 (1263)
CH=CHPh	CH$_2$CH$_2$CH=CH$_2$	H			
H	CH=CHPh	H			
H	CH=CHPh	H			
Ph	H	OH	cycloaddition of imine with chiral auxiliary (Thr) with ketene	65% cycloaddition 90% de	1992 (1264)
Ph	H	OH	cycloaddition of imine with chiral auxiliary (α-methylbenzylamine) with ketene	47–74% cycloaddition 50–60% de 84–89% hydrolysis 48–72% auxiliary removal	1993 (1265)
4-F-Ph	H	OH			
4-NMe$_2$	H	OH			
Ph	H	H	cycloaddition of imine with chiral auxiliary (α-methyl(4-MeO-benzyl)amine) with ketene	46% de	1998 (1266)
Ph	Me	OH	cycloaddition of acetophenone imine with chiral auxiliary (α-methyl(4-MeO-benzyl)amine) with ketene	64% cycloaddition 40% de	2002 (186)
H	Ph	OH	cycloaddition of imine with chiral auxiliary (D-galactosamine) with ketene	75% cycloaddition 100% de 84% hydrolysis	1991 (1267)
Et	Me	Me	cycloaddition of imine with chiral auxiliary (α-methylbenzylamine) with ketene	26–73% cycloaddition 44–78% de only R = tBu hydrolyzed	1980 (1269)
nPr	Me	Me			
iPr	Me	Me			
nBu	Me	Me			
iBu	Me	Me			
Ph	H	OH	cycloaddition of imine with chiral auxiliary (galactopyranosylamine) with ketene	32–93% cycloaddition 0–33% de	1992 (1270)
4-Cl-Ph	H	OH			
4-NO$_2$-Ph	H	OH			
4-MeO-Ph	H	OH			
4-Me-Ph	H	OH			
CH=CHPh	H	OH			
CH$_2$cHex	H	OH	[2+2] cycloaddition of BnOCH=C=O with (S)-PhCH(Me)NH$_2$ imine of cHexCH$_2$CHO, lactam hydrolysis and hydrogenation	84% cycloaddition 68% de cis lactam 34% de 48% hydrolysis/ hydrogenation	1990 (663)

(Continued)

Table 11.2 Asymmetric Syntheses of β-Amino Acids (continued)

R¹	R²	R³	R⁴	Method	Yield, % ee	Reference
iPr	H	H	OH	cycloaddition of chiral hydrazone with benzyloxyketene, N–N bond cleavage, β-lactam hydrolysis	70–96% cycloaddition 90–>98% de cis, >98% de 84–95% N–N cleavage 72–85% hydrogenation 100% hydrolysis	2002 (1271)
iBu	H	H	OH			
Ph	H	H	OH			
(CH₂)₂Ph	H	H	OH			
Bn	H	H	OH	cycloaddition of chiral Schiff base with benzyloxyketene, β-lactam hydrolysis	84–90% cycloaddition 72–>95% de, 24–88% de syn	1990 (1273)
CH₂cHex	H	H	OH			1992 (1272)
CH(OH)Ph	H	H	H			
	Bn	OH	OH			
Ph	H	H	OH	cycloaddition of ketene with chiral auxiliary with imine, α-hydroxylation, reduction, hydrolysis	83% cycloaddition 100% de 70% hydrolysis	1993 (1274)
H	Ph	–NHCH₂CH(OH)–CH₂–	—	Staudinger cycloaddition of ketene derived from Pro or 4-hydroxy-Pro with imines derived from benzaldehyde, D-glyceraldehyde or the Garner aldehyde, hydrolysis of resulting spiro-β-lactam	40–70% cycloaddition 62–90% hydrolysis	2004 (1275)
H	CH(NH₂)CH₂OH	–NH(CH₂)₃–	—			
CH₂OH	H	–NH(CH₂)₃–	—			
CH(OH)CH₂OH	H	–NH(CH₂)₃–				
H	Ph	–NH(CH₂)₃–	—	Staudinger cycloaddition of ketene derived from Pro or 4-hydroxy-Pro with imines derived from various aldehydes to give spiro-β-lactam	53–71% cycloaddition	2004 (1276)
H	4-F-Ph	–NH(CH₂)₃–	—			
H	4-MeO-Ph	–NH(CH₂)₃–	—			
H	Me	–NH(CH₂)₃–	—			
H	cHex	–NH(CH₂)₃–	—			
Ph	H	Me	OH	β-lactam formation from R¹N=CHR² imine and enolate of chiral dioxolanone	24–82% 94–99% ee 90% de cis to 64% de trans	1999 (1277)
Ph	H	Ph	OH			
Ph	H	iPr	OH			
Ph	H	CH₂CO₂H	OH			
2-furyl	H	Me	OH			
CCSiMe₃	H	Me	OH			
CCSiMe₃	H	iPr	OH			
CH=CHPh	H	Me	OH			
CO₂Me	H	Me	OH			
2-thienyl	H	Me	OH			
CO₂Me	H	iPr	OH			

H	Ph	Me	OH		
H	Ph	Ph	OH		
H	Ph	iPr	OH		
H	Ph	CH$_2$CO$_2$H	OH		
H	2-furyl	Me	OH		
H	CCSiMe$_3$	Me	OH		
H	CCSiMe$_3$	iPr	OH		
H	CH=CHPh	Me	OH		
H	CO$_2$Me	Me	OH		
H	2-thienyl	Me	OH		
H	CO$_2$Me	iPr	OH		
H	CO$_2$Et	Ph	Ph	β-lactam formation from TsN=CHCO$_2$Et and RCOCl-derived ketenes in presence of chiral benzoylquinine catalyst	2000 (2189) 2002 (1281) 36–65% lactam formation >98% de cis 95–99% ee
H	CO$_2$Et	H	Ph		
H	CO$_2$Et	H	Et		
H	CO$_2$Et	H	OPh		
H	CO$_2$Et	H	OBn		
H	CO$_2$Et	H	OAc		
H	CO$_2$Et	H	CH=CH$_2$		
H	CO$_2$Et	H	N$_3$		
H	CO$_2$Et	H	Br		
H	CO$_2$Et	H	CH$_2$OPh		
H	CO$_2$Et	H	Bn		
(as lactam)					
H	CO$_2$Et	H	Ph	β-lactam formation from TsN=CHCO$_2$Et and RCOCl-derived ketenes in presence of chiral benzoylquinine catalyst and In(OTf)$_3$	2005 (1282) 92–98% lactam formation 82–96% de cis 96–98% ee
H	CO$_2$Et	H	Bn		
H	CO$_2$Et	H	OPh		
H	CO$_2$Et	H	OBn		
H	CO$_2$Et	H	OAc		
H	CO$_2$Et	H	CH=CH$_2$		
H	CO$_2$Et	H	Br		
H	CO$_2$Et	H	CH$_2$OPh		
(as lactam)					

(Continued)

Table 11.2 Asymmetric Syntheses of β-Amino Acids (continued)

R^1	R^2	R^3	R^4	Method	Yield, % ee	Reference
Ph	H	$-(CH_2)_6-$	—	β-lactam formation from $TsN=C(R^1)CO_2Et$ imine and $R^2R^3C=C=O$ ketenes in presence of chiral ferrocenyl-like catalyst	76–98% lactam formation	2002 (1283)
2-furyl	H	$-(CH_2)_6-$	—		77–88% de cis	
2-furyl	H	Et	—		81–98% ee	
CH=CHPh	H	$-(CH_2)_6-$	—			
CH=CHPh	H	Et	—			
cPr	H	$-(CH_2)_6-$	—			
cHex	H	$-(CH_2)_6-$	—			
Ph	H	iBu	Ph			
2-furyl	H	Et	Ph			
2-furyl	H	iBu	Ph			
cPr	H	Et	Ph			
cPr	H	iBu	Ph			
CH=CHPh (as lactam)	H	iBu	Ph			
Ph	H	H	Ph	Cu-catalyzed reaction of alkynes with nitrones in presence of bis(azaferrocene) chiral ligand	42–65% yield	2002 (1285)
Ph	H	H	4-CF$_3$-Ph		72–93% ee	
Ph	H	H	4-MeO-Ph		42–>90% de cis	
Ph	H	H	Bn			
4-MeO-Ph	H	H	Ph			
4-MeO-Ph	H	H	1-cyclohexenyl			
4-CF$_3$-Ph	H	H	Ph			
3,5-Me$_2$-Ph	H	H	Ph			
COPh (as lactam)	H	H	Ph			
H	CCH	H	H	conversion Asp to 4-carboxyazetidin-2-one, conversion carboxy to ethyne, hydrolysis	59% carboxy to alkyne 74% hydrolysis	1998 (402)
H	CH(OH)CH(OH)-CO$_2$H	H	H	conversion 4-formyl-azetidin-2-one	45–76% Wittig 75% dihydroxylation	1991 (223) 1999 (219)
CH=CHC(Me)=CH-CH(Me)CH-(OMe)Bn	H	Me	H	reduction of carboxyl of 4-carboxy-β-lactam, deprotonation and methylation, oxidation to aldehyde, olefination, lactam hydrolysis	62% reduction 66% methylation 90% oxidation 40–45% coupling	1998 (1286)

Elaboration of Other Chiral Substrates

			reaction	yield	year (ref.)
H	H	Me	oxidation of homochiral 3-aminopropanol	83%	1997 (58)
H	H	Me	amidation of ester of (S)-3-HO-2-Me-propanoate, amide reduction, hydroxymethyl oxidation	66% amidation, 77% reduction, 74% oxidation	1995 (55)
H	H	Me	amidation of ester of (S)-3-Br-2-Me-1-propanol, azide displacement, hydroxymethyl oxidation, azide hydrogenation	65% overall	1996 (57)
Bn	H	=CH$_2$	reaction of 2-(chloromethyl)-2-(1-amino-2-Ph-ethyl)oxirane	70–75% oxirane reaction, 70% oxidation	1999 (1287)
H	CH(OH)CH(OH)CO$_2$H	H	conversion of D-glucosamine	—	1996 (1288)
CH(OH)CH(OH)CH$_2$OH	H	H	amination of D-glucono-1,5-lactone or D-gulono-1,4-lactone via aziridine, reductive opening with hydrazine	26–40% aziridine formation, 74–100% aziridine opening and deprotection	1998 (1289)
H	CH(OH)CH(OH)CH$_2$OH	H	conversion of synthon derived from D-isoascorbic acid	synthon in 38%	1996 (1290)
H	nHept	OH	conversion of synthon derived from D-isoascorbic acid	34% from synthon	1996 (1291)
H	CH(OH)CH(OH)CO$_2$H	H	conversion of methyl α-D-glucopyranoside or methyl α-D-mannopyranoside	—	1989 (221)
H	CH(OH)CH(OH)CO$_2$H	H	conversion of (R)-glyceric acid	—	1988 (222)
CH=CHC(Me)=CHCH(Me)-CH(OMe)Bn	Me	H	functionalization of β-silyl ester, Pd(0)-catalyzed cross-coupling	27 % overall, 11 steps	1997 (1292)
CH=CHC(Me)=CHCH(Me)-CH(OMe)Bn	Me	H	Pd(0)-catalyzed cross-coupling of alkyne-Zn reagent with 2-Me-3-NHBoc-5-iodopent-4-enol, oxidation	84% coupling, 86% oxidation	2002 (228)
H	CH$_2$CH(OH)-CH$_2$NH$_2$	H	Pd(II)-catalyzed alkylation of optically active ene carbamate, elaboration	13–20% overall	1993 (164)
H	nHept	OH	addition of nHexCeCl$_2$ to 4-formyl-5-vinyl-2-oxazolidinone, dehydroxylation of adduct, oxidative cleavage of vinyl group	78% addition, 54% dehydroxylation, 69% oxidative cleavage	2003 (1293)

using a aziridine intermediate that was reductively opened with refluxing hydrazine (*1289*). A Wittig reaction introduced a side chain to homochiral-protected 2-amino-3,4-dihydroxybutanal, itself derived by oxidation of a synthon prepared from D-isoascorbic acid (*1290*). The terminal hydroxymethyl group was then oxidized to the acid moiety, and the alkene hydrogenated, giving Ahda (3-amino-2-hydroxydecanoic acid), the *N*-terminal amino acid of microginin (*1291*).

The complex β-amino acid Adda, 3-amino-9-methoxy-2,6,8-trimethyl-10-phenyldeca-4,6-diencoic acid, has generally been prepared via coupling of the side chain to a β-amino acid equivalent core, with the core derived via an amination reaction or derivatization of an amino acid. A more convergent synthesis prepared the β-amino acid core and first alkene of the side chain from a chiral β-silyl ester, with the remainder of the side chain subsequently added via a Pd(0)-catalyzed cross-coupling reaction (*1292*). The amino acid was obtained in 11 steps and 27% overall yield. A modified version of this route was employed during a total synthesis of the Adda-containing cyclic pentapeptide motuporin, with the coupling carried on on an acid-reduced *N*-Boc amino alcohol, or an *N*-Val-amino alcohol (*228*).

(+)-Negamycin and (−)-5-*epi*-negamycin, containing (3*R*,5*R*)- or (3*R*,5*S*)-3,6-diamino-5-hydroxyhexanoic acid, respectively, were synthesized in 13–20% overall yield via a Pd(II)-assisted alkylation of an optically active ene carbamate, followed by further elaboration (*164*). (2*S*,3*R*)-3-Amino-2-hydroxydecanoic acid was obtained from a 4-formyl-5-vinyl-2-oxazolidinone chiral substrate, via addition of a hexylCeCl$_2$ reagent to the aldehyde, dehydroxylation of the adduct, and oxidative cleavage of the vinyl substituent (*1293*).

11.3 Cyclic β-Amino Acids

11.3.1 Introduction

Cyclic β-amino acids provide considerable conformational constraints on both the β-amino acid backbone and the orientation of side chains. Several cyclic β-amino acids have been isolated from natural sources (see Scheme 11.108). The (1*R*,2*S*) isomer of 2-aminocyclopentane-1-carboxylic acid (cispentacin) is an antifungal antibiotic, and has been isolated from both *Streptomyces setoni* and *Bacillus cereus* (*716, 1294–1296*). It is also found in the antibiotic amipurimycin, isolated from *Streptomyces novoguineensis* (*1297*). Oxetin, (2*R*,3*S*)-2-carboxy-3-amino-oxetane, is an antibiotic isolated from *Streptomyces* which exhibits herbicidal activity and inhibits glutamine synthetase from spinach leaves (*1298, 1299*). The other three stereoisomers were synthesized and compared for biological activities; they

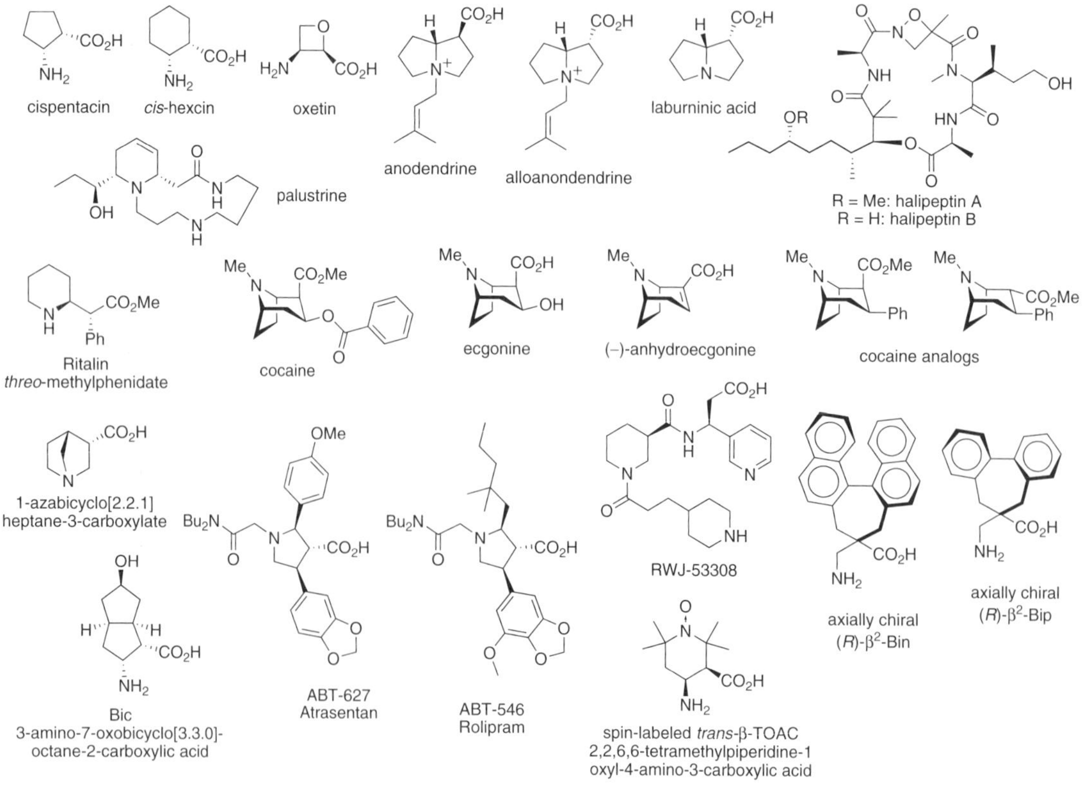

Scheme 11.108 Structures of Cyclic β-Amino Acids.

were found to be inactive against *Bacillus subtilis* (*1300*). Two bicyclic quaternary β-amino acids, anodendrine and alloanodendrine, were isolated from the leaves and stem of the plant *Anodendron affine*. Laburninic acid is a possible *N*-unsubstituted precursor (*1301*). An unusual cyclic β-amino acid, 1,2-oxazetidine-4-methyl-4-carboxylic acid, was identified in the cyclic depsipeptides halipeptins A and B, potent anti-inflammatory metabolites (60% inhibition of edema in mice at 300 μg/kg) isolated from the marine sponge *Haliclona* sp. (*1252*). The palustrine class of alkaloids contains a 6-substituted-3,4-dehydro-piperidine-2-acetic acid moiety (*1302*). Cocaine is a bicyclic β-amino acid derivative with an ethanobridge across the 2,6-positions of an *N*-methylpiperidine-3-carboxylate. Ecgonine is the non-benzoyl free acid analog of cocaine. The NMR spectra of various isomers have been reported (*1303*), while an analog with the phenyl ring directly attached to the tropane skeleton has been synthesized (*1304*).

A number of medically important drugs possess a cyclic β-amino acid skeleton. Ritalin (*threo*-methylphenidate, methyl piperidine-2-phenylacetate), the most commonly prescribed pyschotropic medication for children in the United States, is sold as a racemic mixture of *threo* diastereomers (*4, 5, 1305*). Detailed solution and solid-state conformational and structural analysis of the *N*-methyl derivatives of both *threo* and *erythro* isomers of methylphenidate, and of *threo* *p*-methyl-methylphenidate have been carried out in order to determine bioactive conformations (*1306*). The Abbott antidepressant Rolipram (ABT-546), an endothelin A antagonist, possesses a pyrrolidine-3-carboxylate core with C2-alkyl and C4-aryl substituents (*1307*). The analog ABT-627 (Atrasentan), in clinical trials for prostate cancer, contains a 2,4-diaryl-pyrrolidine-3-carboxylic acid core (*904, 1308*). Thiourea derivatives of cyclic β²,²-disubstituted amino acids were prepared as bioactive analogs of the HDL-elevating agent gemfibrozil (*420*). Analogs of Taxol have been prepared with the phenylisoserine moiety replaced with 1-amino-1-(2-hydroxyacetic acid)cyclopropane β-amino acid (*188*) and 2-substituted-1-amino-1-(2-hydroxyacetic acid)cyclopropane β-amino acids (*189*). A review of alicyclic β-amino acids in medicinal chemistry was published in 2005 (*1309*).

The steric constraints of cyclic β-amino acids produce unique properties. β-Peptides formed from (1R,2R)-*trans*-2-aminocyclopentane-1-carboxylic acid (ACPC) oligomers were synthesized and characterized. The crystal structures of a hexamer and octamer were obtained, showing a unique "12-helix" defined by a series of interwoven 12-membered ring hydrogen bonds, with each hydrogen bond linking a carbonyl atom to an amide proton three residues towards the *C*-terminus. Far-UV circular dichroism (CD) studies of various oligomers (*n* = 1,2,3,4,6,8) showed a length-dependent conformational preference similar to that observed with α-helices formed by α-amino acids (*1310, 1311*). However, the danger of relying on CD spectra to predict conformational preferences was pointed out in a

2002 study, in which two β-hexapeptides predicted to differ in secondary structure exhibited similar CD spectra (*306*). NMR spectroscopy was employed to examine the solution conformations of a hexamer of *trans*-2-aminocyclopentane-1-carboxylic acid or an octamer of *trans*-2-aminocyclohexane-1-carboxylic acid, with both showing a 12-helix structure, while a tetramer of *trans*-2-aminocyclohexane-1-carboxylic acid formed a 14-helix structure (*1312*). (1R,2S)-*cis*-2-Aminocyclopentane-1-carboxylic acid oligomers (trimer, pentamer, and heptamer) were also studied by solution NMR, with the pentamer and hexamer found to adopt a six-strand sheet-like structure, distinctly different from the helices observed with (1R,2R)-*trans*-isomers (*1313*).

Hexamers formed from alternating (3R,4S)-*trans*-4-aminopyrrolidine-3-carboxylic acid residues and the same residues sulfonylated on the imino nitrogen also formed 12-helical structures, as determined by NMR analysis (*1314*). Theoretical calculations of the conformations of the monomer, dimer, and hexamer of pyrrolidine-3-carboxylic acid were reported in 2003, with the hexamer showing preferences for left-handed helices with *cis*-amide bonds, or right-handed helices with *trans*-peptide bonds (*1315*). Polymers of 2,2-disubstituted-pyrrolidine-4-carboxylic acid were prepared in an attempt to generate unusual secondary structures, such as the polyproline helices that lack intramolecular hydrogen bonds. Preliminary results indicate that the β-foldamers possess distinct conformational preferences (*1316*). Hexamers of 3-substituted *trans*-2-aminocyclopentanecarboxylic acids again formed 12-helix structures, as determined by both CD and NMR spectra. The additional substituents dispersed the NMR signals, allowing for assignment of most NOE signals (*1317*). The tolerance of 12-helix structures formed from β-peptide heptamers of ACPC towards replacement with acyclic β²-amino acid residues was examined in 2003. One or two substitutions still produced pronounced 12-helix formation (*1318*).

In contrast to the 12-helical structures formed by cyclopentane-based β-amino acid oligomers, tetramer and hexamers of (1R,2R)-*trans*-2-aminocyclohexane-1-carboxylic acid were found to possess a 14-helical structure, with 14-membered-ring hydrogen bonds between a carbonyl oxygen and the amide portion of the second residue towards the *N*-terminus. Both crystal structures and CD spectra were obtained (*1311, 1319–1321*). β-Peptides with both acidic and basic residues could be designed to maintain short 14-helical structure by incorporating a number of *trans*-2-aminocyclohexane-1-carboxylic acid residues (*312*). Both *cis*- and *trans*-ACHC were incorporated into cyclic penta or hexapeptides of α-amino acids, and their structures examined by NMR (*1322*). Several peptides have been constructed from alternating α- and β-amino acids, using Ala as the α-amino acid. With *trans*-2-aminocyclohexane-1-carboxylic acid as the β-amino acid component, no secondary structure was observed, but with *trans*-2-aminocyclopentane-1-carboxylic acid a strong propensity for helical secondary structure was observed, with an apparent rapid conversion between two helical conformations (*1323*). Further optimization of the α-amino acid components produced an α/β-octamer peptide forming an 11-helix, for which a crystal structure was obtained (*1324*).

Oligomers (up to octadecamer) of β²-homoproline (piperidine-3-carboxylic acid, nipecotic acid) and β³-homoproline (pyrrolidine-2-acetic acid) are unable to form hydrogen bonds, but show strong and unique CD spectra (*1090*). Formation of a hairpin turn in a β-peptide was promoted by a dinipecotic acid (consecutive piperidine-3-carboxylic acid residues) sequence (*1325, 1326*), and an Asn-Pro reverse-turn sequence within RNase A was successfully replaced with (*R*)-nipecotic acid-(*S*) nipecotic acid. The modified RNase A retained catalytic activity and enhanced the enzyme's conformational stability, whereas an (*R*)-nipecotic acid-(*R*)-nipecotic acid sequence gave inactive enzyme (*1327*). *endo*-(2*S*,3*R*)-2-Amino-3-carboxy-norborn-5-ene was also incorporated into peptides as a turn inducer (*1328*). Conformational analysis by NMR spectroscopy demonstrated that an antiparallel β-sheet was formed by the peptide segments attached to either side of the constrained β-amino acid (*1329*). Cyclic β-amino acid analogs of the spin-labeled α-amino acid TOAC, *cis*- and *trans*-4-amino-1-oxyl-2,2,6,6-tetramethylpiperidine-3-carboxylic (β-TOAC) and *trans*-3-amino-1-oxyl-2,2,5,5-tetramethylpyrrolidine-4-carboxylic acid, have been prepared to study the conformation of peptides (*1330–1332*).

Dipeptides of *cis*-2-aminocyclobutane-1-carboxylic acid have a highly rigid structure (*1333*). However, the 2-amino-1-cyclobutanecarboxylic acid skeleton has been found to undergo a facile retro-Mannich-type ring opening in solution under mild conditions, such as those used for hydrogenolysis, leading to byproducts such as 5-aminopentanoic acid and *N*-(carboxybutyl) 5-aminopentanoic acid (*1334*). With *cis*-β-aminocyclopropanecarboxylic acid as the β-amino acid component, peptides as short as pentapeptides were found to form helical structures (*1335*). A polypeptide has been constructed from 1-(aminomethyl)cyclopropane-1-carboxylic acid residues, with X-ray crystal structures showing unusual H-bonded eight-membered rings (*1336*).

Short β-peptides composed of (3*R*,4*S*)-*trans*-4-aminopyrrolidine-3-carboxylic acid (APC) and (1*R*,2*R*)-*trans*-2-aminocyclopentanecarboxylic acid (ACPC) were designed to form an amphiphilic 12-helix structure (*1337*). A longer 17-mer peptide showed potent antimicrobial activity against four different bacteria (including antibiotic-resistant varieties) with better activity and less hemolytic activity against red blood cells than the α-amino acid-based peptide antibiotic magainin (*1338*). Analogs were prepared to examine the factors important for activity, using APC, ACPC, and (3*S*,4*R*)-*trans*-3-aminopyrrolidine-4-carboxylic acid (AP); a 12-helical structure with 40% cationic face appeared ideal. The β-peptides were resistant to proteolysis by trypsin or pronase (*1339*). It was possible to substitute some cyclic residues with acyclic β³-residues and retain the 12-helix conformation; using β-Lys produced β-peptides with antibacterial activity (*320*). Another structure–activity study of cationic antimicrobial β-peptides was reported

in 2002, examining the effect of increasing the formation/stability of 14-helix structure on antimicrobial activity by including increasing proportions of rigid *trans*-2-aminocyclohexanecarboxylic acid. Circular dichroism of the 9- to 13-mer peptides showed large variations in helical populations that did not correlate with changes in antibiotic activity (*369*). Potential antimicrobial peptides composed of alternating *trans*-2-aminocyclohexane-1-carboxylic acid (or piperidine-3-carboxylic acid) and α-amino acid residues were designed to form amphiphilic helices based on putative 11-helix or 14-helix formation. Again, antimicrobial activity did not correlate with apparent helix formation (*1340*).

Cyclic β-amino acids have been incorporated into peptides for SAR studies. The four stereoisomers of 2-aminocyclopentane-1-carboxylic acid (2-Ac⁵c) were used in Asp-2-Ac⁵c dipeptides to investigate the validity of a molecular model proposed to account for the conformation required for a sweet-tasting molecule. A planar "L"shape, as produced by the *trans*-(1*R*,2*R*)- and *cis*-(1*S*,2*R*)-cyclic amino acid isomers, gave sweet derivatives as predicted (*1341*). The crystal structures of the cyclic β-amino acids were reported (*1341*). 2-Aminocyclopentanecarboxylic acids have also been used as proline replacements in HIV-protease inhibitors (*1342*), while an enkephalinase inhibitor was synthesized with *cis*-hexacin, (1*S*,2*R*)-2-aminocyclohexane-1-carboxylic acid (*1343*). Pyrrolidine-2-acetic acid (β-proline, although this name has also been applied to pyrrolidine-3-carboxylic acid) was incorporated into bradykinin analogs (*381, 1089*). The β-Pro bradykinin analog was equipotent to bradykinin, and resisted degradation by dipeptidylcarboxypeptidase (*381*). Similarly, analogs of the μ-opioid receptor agonist endomorphin-1 (Tyr-Pro-Trp-Phe-NH₂), with the Pro replaced by β-Pro (pyrrolidine-3-carboxylic acid), showed good resistance to enzymatic hydrolysis of the Pro–Trp bond (*1344*). Piperidine-3-carboxylate is contained within RWJ-53308, an orally active antagonist of the platelet fibrinogen receptor glycoprotein GPIIb/IIIa. The antagonist prevents platelet aggregation and thrombus formation and has potential for treating thrombotic disorders such as myocardial infarction (*396*). Other constrained β-amino acids have been examined for their antagonistic activity towards GPIIb/IIIa (*457*). A bicyclic β-amino acid, Bic (3-amino-7-hydroxybicyclo[3.3.0]octane-2-carboxylic acid) was incorporated into an analog of GnRH (gonadotropin-releasing hormone) as a rigid β-reverse-turn inducer, producing a derivative with equipotent activity (*1345*).

Cyclic β-amino acids by themselves have been evaluated for biological activity. A number of β-proline analogs were tested for agonist activity at the strychnine-sensitive glycine receptor. 3-Carboxy-3,4-dehydropyrrolidine bound as strongly as glycine (*1346*). 2,4-Disubstituted-3-carboxypyrrolidines were examined as endothelin receptor antagonists, with the substituents affecting selectivity over the ETₐ and ET_B types of receptors (*1347–1349*).

Other cyclic β-amino acids have been prepared with unusual structures. 3-Carboxy-4-hydroxypiperidine was used in a combinatorial synthesis to prepare amino acid-based carbohydrate mimics (*1350*). Carbohydrate mimetics called carbopeptoids have been constructed from carbohydrate-based cyclic β-amino acids. For example, Suhara, Hildreth, and Ichikawa used a β-amino acid monomer derived from D-glucosamine by carboxylation at the C1 position (*1351*). A tetrameric "carbopeptoid" was assembled in 59% yield (*350*). In these oligosaccharide mimetics, the amide bond replaces the glycosidic linkage. Two unusual axially chiral α,α-disubstituted-β-amino acids based on a binaphthyl or biphenyl chiral axis have been synthesized (see Scheme 11.108); the binaphthyl derivative could be isolated in enantiomerically pure form, but the biphenyl derivative interconverted between enantiomers at room temperature (*1352, 1353*). A variety of bicyclic β-amino acids have been synthesized; some of these are summarized in Scheme 11.109.

The chemistry of 2-aminocycloalkane-1-carboxylic acids was reviewed in a 2001 *Chemical Reviews* article (312 references) (*1354*), with a review of β-amino-carboxylic acids containing a cyclopropyl ring appearing in 2003 (*1355*), and syntheses of 2-aminocyclobutane-1-carboxylic acids described in a 1993 *Russian Chemical Reviews* paper (*1356*). A comprehesive review of β-peptides (128 pages, 531 references), including discussions of natural β-amino acids, the synthesis of β-amino acids, the synthesis of β-peptides, and properties of β-peptides, was published in 2004 (*13*). 5-Azaprolines and 6-azapipecolic acids, which can be considered as either α- or β-imino acids, are not discussed in this chapter.

11.3.2 Syntheses of Achiral or Racemic Cyclic β-Amino Acids (see Table 11.3)

11.3.2a Amination Reactions

Several syntheses of cyclic β-amino acids proceed via conjugate addition of an amine nucleophile to cyclic unsaturated substrates. Addition of ammonia to cyclopentene-1-carboxylic acid under pressure

and at elevated temperatures gave a mixture of *cis*- and *trans*-2-aminocyclopentane-1-carboxylic acid (*1357*), while *trans*-2-aminocyclohexane-1-carboxylic acid was obtained by addition of ammonia to cyclohexene-1-carboxylic acid (*1358*). The imine of benzophenone has also been used as an amine nucleophile, adding to methyl cyclopropylideneacetate or α-halogenated derivatives to provide β-imino esters that were readily hydrolyzed to β-amino esters by 0.1 N HCl (*473*). An intramolecular Michael addition was used to prepare *cis*- or *trans*-6-methylpiperidine-2-acetic acid from an acyclic 7-amino-oct-2-enoate substrate, with the relative stereochemistry controlled by the alkene configuration. The substrate was prepared by reductive amination of 7-oxooct-2-enoic acid ester (*1359*). Protected 1-amino-1-(2-hydroxyacetic acid)-2-substituted-cyclopropanes were obtained by reaction of benzamide with 2-chloro-2-(2'-R-cyclopropylidene)acetate (R = H, Me, iPr) (*189*).

Most aminative syntheses of cyclic β-amino acids employ rearrangements of differentially protected cycloalkane-1,2-dicarboxylic acids. *cis*-2-Aminocyclohexane-1-carboxylic acid was prepared via a Hofmann rearrangement of the monoamide of *cis*-cyclohexane-1,2-dicarboxylic acid (*1358*), while an electrochemically induced Hofmann rearrangement of *trans*-cyclohexane-1,2-dicarboxylic acid monoamide gave *trans*-2-aminocyclohexane-1-carboxylic acid in 74% yield (*1360*). A Hofmann rearrangement was also applied to a synthesis of *cis*-2-amino-cyclopentane-1-carboxylic acid, starting from the anhydride of *cis*-cyclopentane-1,2-carboxylic acid (*1361*). *endo*-Norborn-5-ene-2,3-dicarboxylic anhydride, produced by a Diels–Alder reaction of cyclopentadiene and maleic anhydride, was opened with ammonium hydroxide, and the resulting monoamide converted into an amine via a Hofmann rearrangement, providing 2-*endo*-amino-5-norbornene-3-*endo*-carboxylic acid (*1362*). The bicyclic anhydride was also opened with methanol and the resulting free monoacid was converted into an amine via Curtius rearrangement, also giving racemic *endo*-2-amino-3-carboxy-norborn-5-ene. Oxidative cleavage of the alkene produced all-*cis* 1-aminocyclopentane-2,3,5-tricarboxylic acid (*1363*).

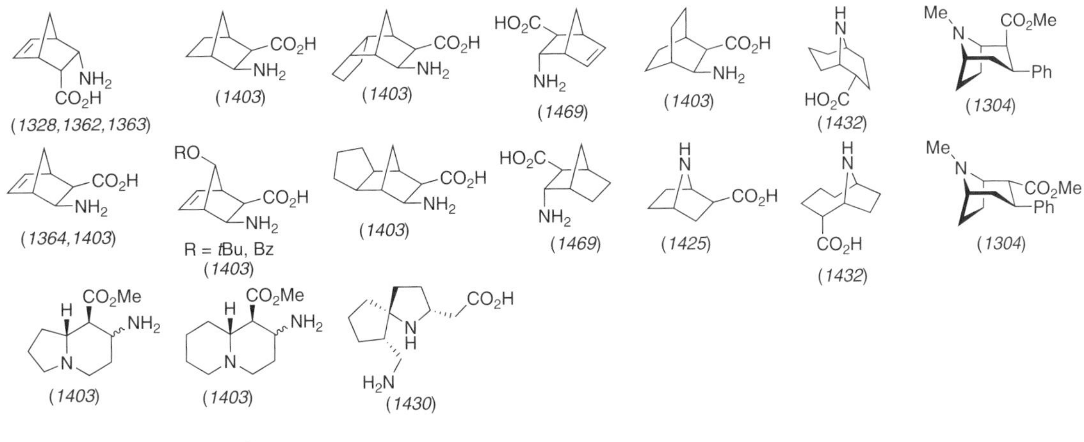

Scheme 11.109 Bicyclic β-Amino Acids.

Both *trans*-1-amino-2-carboxycyclopropane and *trans*-1-amino-2-carboxycyclobutane were prepared via Curtius rearrangements of the corresponding monoester *trans*-diacids (*1364*). The *cis*- or *trans*-1-amino-2-carboxycyclopropane could be prepared from the corresponding dicarboxylic acids via a Curtius rearrangement, as long as acid hydrolysis of the isocyanate intermediate was avoided (*1365*). Cyclobutane-1,2-dicarboxylic acid, cyclohexane-1,2-dicarboxylic acid, or pyridine-2,3-dicarboxylic acid were attached to a polystyrene resin via an amide linkage to one of the carboxy groups. Curtius rearrangement (DPPA, Et$_3$N, and then heating) gave the amino amides in 70–85% yield (*1366*). A Curtius rearrangement of selectively protected 2,3-pyrazinedicarboxylic acid gave 3-amino-pyrazinecarboxylic acid in 80% overall yield from the diacid (*1367*). Both regioisomers of a cyclic β-amino acid based on a 1,3-diphenylpyrazole nucleus were also prepared via a Curtius rearrangement (*1368*). Monoesters of bicyclo[2.2.1]hept-5-ene-2-*exo*,3-*exo*-dicarboxylate, bicyclo[2.2.1]hept-5-ene-2-*endo*,3-*endo*-dicarboxylate, 3-methylbicyclo[2.2.1]hept-5-ene-2-*endo*,3-*endo*-dicarboxylate and 7-oxabicyclo[2.2.1]hept-5-ene-2-*exo*, 3-*exo*-dicarboxylate, were obtained via cycloadditions of furan or cyclopentadiene. Curtius rearrangement gave the corresponding N-Boc bicyclic β-amino acids with 3-amino-2-methoxycarbonyl substituents (*1369*).

Other amination reactions include conversion of the hydroxy group of 2-*cis*-hydroxy-3-*trans*-4-*trans*-diphenylcyclobutane-1-carboxylic acid into an amine to form 2-*cis*-amino-3-*trans*-4-*trans*-diphenylcyclobutane-1-carboxylic acid (*1370*). A series of 2-substituted-3-carboxy-4-(1,3-benzodioxol-5-yl)pyrrolidines were derived from a nitro-ketone substrate, RCOCH(CO$_2$Et)CH(1,3-benzodioxol-5-yl)CH$_2$NO$_2$, by hydrogenation of the nitro group and intramolecular reductive amination (*1347–1349*). This method was employed to construct the pyrrolidine-3-carboxylate core of the Abbott antidepressant Rolipram (ABT-546), an endothelin A antagonist (*1307*).

11.3.2b Carboxylation Reactions

N-Tosyl or N-methoxycarbonyl-2-piperidone were carboxylated by deprotonation and reaction of the enolate with ClCO$_2$Bn (*1371*), producing 3-carboxy-2-piperidone. Unsaturation was then introduced within the ring and used for Diels–Alder reactions to generate a number of polycyclic β-amino acids. A kinetic deprotonation of racemic O-protected N-benzylpiperidine-2-methanol was achieved with TMEDA and sBuLi. Carboxylation using methyl chloroformate provided one diastereomer of piperidine-2-(hydroxyacetic acid) with 70–84% de (*1372*).

1-(Aminomethyl)cyclopropane-1-carboxylic acid was synthesized by carboxylation of 1-cyano-1-trimethyl-silylcyclopropane using CO$_2$ and CsF, followed by reduction of the cyano group to generate the amino-methyl substituent. The cyanosilyl substrate was prepared by bis-alkylation of trimethylsilylacetonitrile with

1,2-dibromoethane (*1373*). Azetidine-3-carboxylic acid was prepared by displacement of the mesylate of 1-diphenylmethane-3-hydroxyazetidine with NaCN followed by hydrolysis and hydrogenation. The substrate was prepared from epichlorohydrin and benzhydryl-amine (*1374*).

A carbonylative ring expansion of N-benzyl cyclohexane aziridine has been carried out with Ti or Al catalysts and [Co(CO)$_4$]$^-$, producing the β-lactam of 1-aminocyclohexane-2-carboxylic acid (*540*).

11.3.2c Fusion of Aminomethyl and Acetate Components

The acyclic β-amino acid synthetic strategy of addition of an acetate nucleophile to an electrophilic amino-methyl equivalent has also been applied to syntheses of cyclic β-amino acids. An iminium ion prepared from a cyclopropanone hemiaminal under TiCl$_4$ catalysis was reacted with ketene silyl acetals or lithium enolates to give β-amino acids with the β-carbon contained in a cyclopropyl ring (*570*). N-Tosyl iminium ions were generated from N-tosyl 2-methoxypyrrolidine and N-tosyl 2-methoxypiperidine upon treatment with TMSOTf. Addition of the silyl ketene acetal of ethyl acetate provided N-tosyl pyrrolidine-2-acetic acid or N-tosyl piperidine-2-acetic acid (*1375*). Pyrrolidine-2-dimethylacetic acid was obtained by reaction of a trispyrrolidine hexahydro-1,3,4-triazine with the ketene silyl acetal of 2-methylpropionic acid (*577*). The silyl ketene acetal of methyl acetate added to an N-Cbz 2,3-dehydro-5-phenylseleno-6-isopropylpiperidin-4-one in the presence of BF$_3$·Et$_2$O with complete *cis*-stereoselectivity. The phenylseleno group could be removed under radical conditions to give 6-isopropylpiperidin-4-one-2-acetic acid, or the ketone reduced and phenylseleno/hydroxy groups eliminated to form 4,5-dehydro-6-isopropylpiperidine-2-acetic acid (*1376*).

N, N-Bis(trimethylsilyl)methoxymethylamine, (TMS)$_2$ NCH$_2$OMe, is another synthetic equivalent of H$_2$NCH$_2^+$, reacting with ketene silyl acetals in the presence of trimethylsilyl trifluoromethanesulfonate to give N,N-bis (trimethylsilyl) methyl esters of 1-aminomethylcyclo-alkane-1-carboxylic acid (ring size n = 4,5,6) in 78–95% yield (*557*). A similar reaction with N-(1-aminoalkyl) benzotriazoles gave 1-aminomethylcyclohexane-1-carboxylic acid and 1-(1-aminobenzyl)cyclohexane-1-carboxylic acid (*557*).

Addition of a nucleophilic aminomethyl moiety to an electrophilic acetate equivalent is also possible. The benzophenone Schiff base of 3-(aminomethyl)pyridine was reacted with diethylfumarate under PTC conditions. The initial Michael adduct cyclized to a pyrrolidinone after acid catalysis (see Scheme 11.110) (*1377*). Reaction of pyrrolidin-2-thione with BrC(Ph)CO$_2$Me formed a thioimine, 1,2-dehydropyrrolidine-2-SCH(Ph)CO$_2$Me. Treatment with Ph$_3$P induced an Eschenmoser sulfide contraction to pyrrolidine-2-[=C(Ph)CO$_2$Me], with hydrogenation of the alkene giving the methylphenidate analog pyrrolidine-2-phenylacetic acid (*1378*).

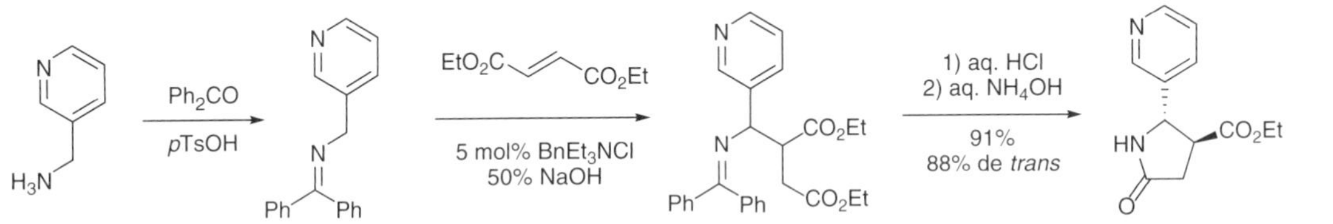

Scheme 11.110 Schiff Base Michael Addition (*1377*).

Various cycloaddition reactions have been employed to form the C2–C3 bond, generating three-, four-, five-, or six-membered ring systems. *N*-Boc pyrrole was monocyclopropanated with methyl diazoacetate in 45% yield. The remaining alkene bond in the pyrrole ring was then oxidatively cleaved, leaving 1-amino-2,3-dicarboxycyclopropane (*1379*). 1-Amino-2,3-dicarboxycyclopropanes are unstable unless an electron-withdrawing group protects the amino functionality, making peptide synthesis difficult as the unprotected amine that is required undergoes rapid ring opening (*1380*). However, applying the synthetic route to pyrroles that were already *N*-acylated with an amino acid allowed the *cis*-β-aminocyclopropanecarboxylic acid residue to be incorporated into peptides. No diastereoselectivity was observed in these cyclopropanations, but the diastereomers were easily resolved (*1381*).

Most 2-aminocyclobutane-1-carboxylic acids have been prepared by [2+2] cycloadditions of enamines with electrophilic olefins such as acrylonitrile or acrylic esters (*1356*). For example, *N*,*N*-dimethylisobutenylamine was heated with methyl acrylate to give a 75% yield of methyl 2-dimethylamino-3,3-dimethylcyclobutane-1-carboxylate (*1382*). Cycloaddition of *N*,*N*,*N'*,*N'*,2-pentamethyl-1-propene-1,1-diamine was accompanied by elimination of one dimethylamino group to give 2-dimethylamino-3,3-dimethylcyclobutene-1-carboxylate in low yield (*1383*). A large number of *N*,*N*-dialkyl enamines were reacted with methyl acrylate or diethyl fumarate in 1964 to give 2-dialkylamino-3,3-dialkylcyclobutane-1-carboxylates or 2-dialkylamino-4,4-dialkylcyclobutane-1,3-dicarboxylates (*1384*). A cycloaddition was also used to prepare the antibiotic oxetin, an oxetane-based cyclic amino acid. Photochemical Paterno–Büchi reaction of *n*-butyl glyoxylate and *N*-Bn,*N*-Boc vinylamine gave the protected amino acid in 28–35% yield, with >80% de (*1298*).

A constrained analog of the γ-amino acid gabapentin was prepared by a [3+2] cycloaddition reaction between an alkene and a transient azomethine ylide derived from an amino acid. *N*-Benzyl Gly was condensed with para-formaldehyde and a cyclohexylidene derivative, followed by decarboxylation and deprotection (see Scheme 11.111). The cyclic β-imino acid was resolved via amidation with (*R*)-(+)-1-(2-naphthyl)ethylamine, followed by flash chromatography and hydrolysis (*1385*).

Condensation of *N*-methyl Gly, paraformaldehyde, and either ethyl 2-propynoate, ethyl 3-phenylpropynoate, methyl 2-butynoate, or ethyl 2-hexynoate produced 3,4-dehydro-pyrrolidine-3-carboxylic acid with 1-methyl-, 1-methyl-4-phenyl-, 1,4-dimethyl-, or 1-methyl-4-pentyl-substituents. The *N*-demethylated products were also obtained (*1346*). Similarly, *N*-methyl Ala, paraformaldehyde, and ethyl propiolate were combined to give a mixture of 1,2-dimethyl-3,4-dehydro-pyrrolidine-3-carboxylic acid and 1,5-dimethyl-3,4-dehydro-pyrrolidine-3-carboxylic acid. The isomers were separated, and then demethylated by treatment with 1-chloroethyl chloroformate to give 2-methyl- or 5-methyl-3,4-dehydropyrrolidine-3-carboxylic acid (*1346*). Schiff bases of α-aminophosphonates underwent a cycloaddition reaction with ethyl acrylates in the presence of base to give, with aqueous quenching, 4-ethoxycarbonyl pyrrolidine-2-phosphonates (see Scheme 11.112). Depending on reaction conditions and the steric bulk of substituents, an alternative reaction led to 1,2-dehydropyrrolidine-4-carboxylates via elimination of the phosphonate. These products could be reduced with sodium borohydride to give pyrrolidine-3-carboxylates (*1386*). Reaction of vinyldiazoacetates with imines in the presence of Rh catalysts produced 2,3-dehydro-Pro derivatives. However, in the presence of a Cu catalyst, the regioisomeric 2,5-dihydropyrrole-3-carboxylic acids were generated (*1387*).

A carbenoid generated from a vinyl-substituted diazoacetate reacted with pyrroles to give bicyclic β-amino acids as the predominant product (see Scheme 11.113) (*1388*).

Both *exo*- and *endo*-2-(methoxycarbonyl)-7-azabicyclo[2.2.1]heptane were synthesized via cycloaddition

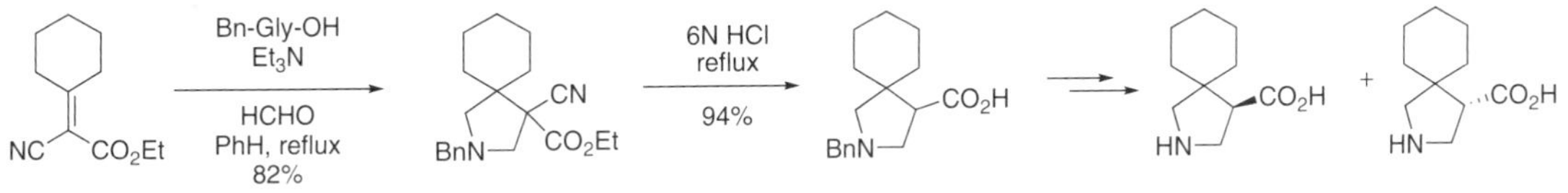

Scheme 11.111 Synthesis of Gabapentin Analog (*1385*).

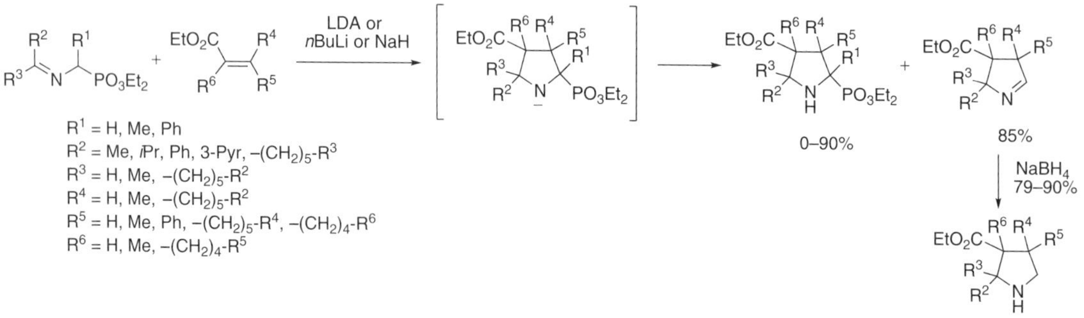

R^1 = H, Me, Ph
R^2 = Me, iPr, Ph, 3-Pyr, –(CH$_2$)$_5$-R^3
R^3 = H, Me, –(CH$_2$)$_5$-R^2
R^4 = H, Me, –(CH$_2$)$_5$-R^2
R^5 = H, Me, Ph, –(CH$_2$)$_5$-R^4, –(CH$_2$)$_4$-R^6
R^6 = H, Me, –(CH$_2$)$_4$-R^5

0–90%

85%

NaBH$_4$
79–90%

Scheme 11.112 Formation of Pyrrolidine-3-Carboxylates (1386).

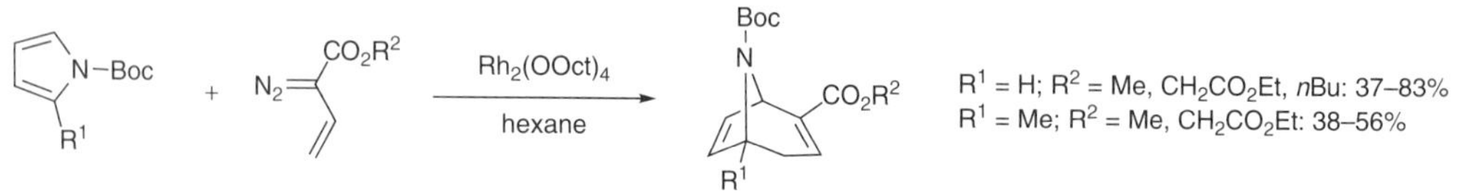

R^1 = H; R^2 = Me, CH$_2$CO$_2$Et, nBu: 37–83%
R^1 = Me; R^2 = Me, CH$_2$CO$_2$Et: 38–56%

Scheme 11.113 Reaction of Carbene with Pyrrole (1388).

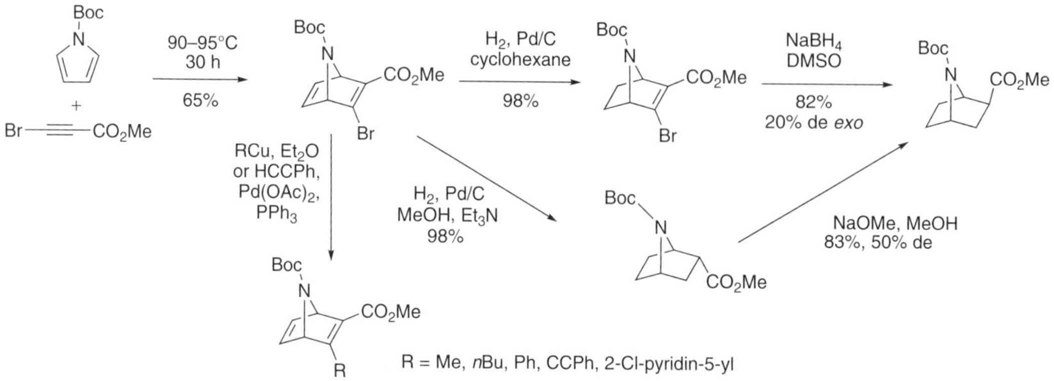

R = Me, nBu, Ph, CCPh, 2-Cl-pyridin-5-yl

Scheme 11.114 Cycloaddition of Pyrrole (1389, 1390).

of N-Boc pyrrole with methyl 3-bromopropiolate (see Scheme 11.114) (1389). The bromo substituent was used to introduce alkyl and aryl groups via organocopper coupling reactions (1390).

Trimethylsilylketene was reacted with acyl isocyanates to give a labile [4+2] cycloaddition product, which was then treated with dimethyl acetylene dicarboxylate or methyl propiolate to form 5-carboxy-2-pyridones (1391).

11.3.2d Elaboration of α-Amino Acids

Racemic 1,2,3,4-tetrahydroquinoline-2-carboxylic acid was homologated under Arndt–Eistert conditions to give 1,2,3,4-tetrahydroquinoline-2-acetic acid. The methyl ester of the product was resolved using Novozym 435 (1392). A racemic α-(3-methoxyphenoxymethyl)-Asp

derivative, prepared via a hydantoin intermediate, was cyclized by a Friedel–Crafts acylation of the side-chain aromatic ring with the α-carboxyl group to give a 1-aminomethyl-2-oxocyclohexane-1-carboxylic acid with a fused aromatic ring at the 3,4-positions (1393).

11.3.2e Derivatization of β- or γ-Amino Acids

Nucleophilic β-alanine equivalents have been dialkylated with bis-electrophiles to form ring systems. Methyl cyanoacetate was deprotonated and dialkylated with 1,2-dibromoethane, 1,3-dibromopropane, 1,4-dibromobutane, or 1,5-dibromopentane. Reduction of the nitrile by hydrogenation over Raney Ni provided 1-(aminomethyl)cycloalkane-1-carboxylic acid derivatives (322). Similarly, dialkylation of ethyl cyanoacetate with Br(CH₂)₃Br or Br(CH₂)₄Br gave the adducts in 78–82% yield. The ester and cyano groups were both reduced by LiAlH₄, and then after *N*-protection the hydroxymethyl group was oxidized back to an acid (*53*). In the same report the nitrile group of commercially available 1-cyanocyclopropyl-1-carboxylic acid was reduced by hydrogenation over PtO₂ (*53*). An unusual α,α-disubstituted-β-amino acid with potential axial chirality was prepared by bis-alkylation of ethyl cyanoacetate with 2,2′-bis-(bromomethyl)-1,1′-diphenyl, followed by regioselective reduction of the cyano group with sodium borohydride. The biphenyl derivative interconverted between enantiomers at room temperature and so was racemic, in contrast to the binaphthyl-based analog which could be isolated in enantiomerically pure form (*1352, 1353*). Chiral analysis of the binaphthyl derivative is possible on a β-cyclodextrin HPLC column as the free amino acid, the ethyl ester, and the *N*-Boc derivative. The amino acid could also be derivatized with Marfey's reagent, FDAA, and separated on a normal C₁₈ column (*1394*).

Ethyl (*E*)-3-nitroacrylate undergoes Diels–Alder reaction with furan to give a 90% yield of a 2:1 mixture of *endo*- and *exo*-5-nitro-6-carboxy-7-oxabicyclo[2.2.1]hept-2-ene; nitro reduction gave the bicyclic β-amino acids, while treatment with KHMDS eliminated the oxygen bridge to provide 2-amino-3-carboxycyclohexa-3,5-dien-1-ol. The diene could then be hydrogenated, or epoxidized and ring-opened to give further functionalized cyclohexane derivatives (*1395*). *N*-alkenyl,*N*-Boc β-Ala-OEt derivatives were deprotonated with LiHMDS and allylated with allyl iodide (57–73%). Ring-closing metathesis and hydrogenation resulted in seven-, eight- or nine-membered 1-azacycloheptane-, -cyclooctane-, or -cyclononane-3-carboxylic acids (*641*).

N-Boc piperidine-3-carboxylate ethyl ester was deprotonated with LDA and alkylated with methyl iodide in 90% yield (*420*). Deprotonation of ethyl *N*-Boc 4-keto-piperidine-3-carboxylate and alkylation with ROCH₂Cl gave an enol ether product. Reduction of the enol by catalytic hydrogenation or with other reducing agents produced mixtures of 4-*cis*-hydroxy-5-*trans*-hydroxymethyl- and 4-*cis*-hydroxy-5-*cis*-hydroxymethyl-piperidine-3-carboxylate (*1396*). The *N*-ethoxycarbonyl lactam of γ-aminobutyric acid was deprotonated and acylated with Cbz-Cl and then deprotected and reduced to give pyrrolidine-2,3-dehydro-3-carboxylate (*1397*). A domino 1,4-addition/carbocyclization reaction of *N*-allyl, *N*-benzyl α-methylene-β-amino esters, initiated by reaction with organozinc reagents, produced 2,3,4,5-tetrasubstituted-3-carboxypyrrolidines (*1398*).

11.3.2f Small Ring Systems

The β-aminonitrile precursor of *trans*-1-aminocyclohexane-2-carboxylic acid was obtained from *N*-tosyl cyclohexylaziridine via ring opening with cyanotrimethylsilane in the presence of Yb(CN)₃ catalyst (*658*). 2-Substituted aziridines underwent Michael addition to methyl propiolate to give *N*-(2-acryloyl)-aziridines. Aziridine ring opening by benzeneselenolate generated MeO₂CCH=CHNHCH(R)CH₂SePh vinylogous carbamates, substrates for a radical carbonylation/reductive cyclization that produced 5-substituted-pyrrolidin-3-one-2-acetic acids in 65–85% yield (*1399*). Azirine-2-carboxylate was employed as a dienophile in a Diels–Alder reaction with cyclohexa-1,4-diene; halide opening of the resulting tricyclic aziridine product produced 1-aza-3-halobicyclo[3.2.1]oct-4-ene-3-carboxylate (*1400*).

β-Lactam synthesis via reaction of olefinic substrates with chlorosulfonyl isocyanate is a useful route to β-amino acid precursors (see Scheme 11.26 earlier). The method was recently applied to syntheses of *cis*-2-aminocycloheptane-1-carboxylic acid, *cis*-2-aminocyclooctane-1-carboxylic acid, and *cis*-2-aminocyclododecane-1-carboxylic acid (*671*). Reaction of the alkene group of methylene-substituted cyclobutane, cyclopentane or cyclohexane with chlorosulfonyl isocyanate (CSI) formed bicyclic β-lactam derivatives. Ring opening with amino esters in the presence of KCN produced dipeptides with a 1-aminocycloalkyl-1-acetic acid residue (*1401*). This method was applied to the synthesis of racemic cispentacin, 2-aminocyclopentane-1-carboxylic acid, via cycloaddition of cyclopentene and chlorosulfonyl isocyanate to give the bicyclic β-lactam derivative. Removal of the *N*-chlorosulfonyl group and lactam hydrolysis with concentrated HCl gave cispentacin in 10% overall yield (*1296*). It was resolved by salt formation with (+)-dehydroabietylamine. Similarly, cycloaddition of norbornadiene and chlorosulfonyl isocyanate gave a tricyclic adduct (see Scheme 11.115). The chlorosulfonyl group was removed with sodium sulfite and the azetidinone ring hydrolyzed to provide 3-*exo*-aminobicyclo[2.2.1]hept-5-ene-2-*exo*-carboxylic acid in 43% yield (*1402*). A series of other bi- and tricyclic alkenes (norbornene, norbornadiene, bornylene, 7-benzyloxy-norbornadiene, 7-*tert*-butoxy-norbornadiene, bicyclo[2.2.2]octene, and both *endo*- and *exo*-tricyclic dicyclopentadiene) were also reacted with chlorosulfonyl isocyanate and converted into the β-amino acids (*1403*). The Staudinger ketene–imine cycloaddition

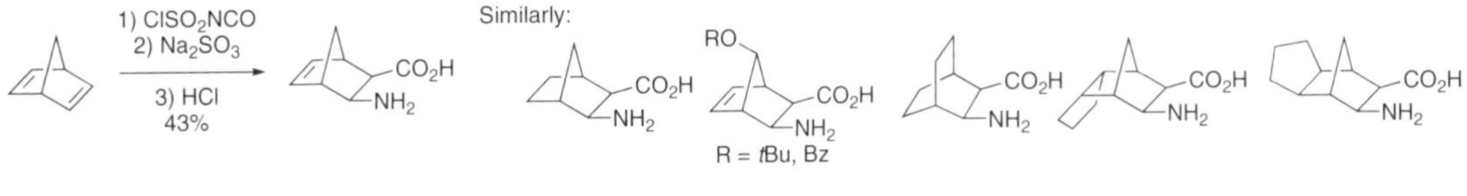

Scheme 11.115 Synthesis of 3-*Exo*-aminobicyclo[2.2.1]hept-5-ene-2-*exo*-carboxylic Acid and Others (*1364, 1402, 1403*).

lactam synthesis was employed to prepare tetrahydro-furan-derived spiro-β-lactams, using the acid chlorides of tetrahydrofuran-2- or -3-carboxylic acids for reaction with a variety of arylaldehyde-derived N-aryl imines (*1404*).

11.3.2g Syntheses from Heterocycles

Hydrogenation of pyridine-3,5-dicarboxylate dimethyl ester produced a 1:1 mixture of *cis*- and *trans*-piperidine-3,5-dicarboxylate. Addition of LDA to a THF solution of this product in the presence of MeI resulted in a mixture of *trans* and *cis* isomers of the dimethylated 3,5-dimethyl-3,5-piperidinedicarboxylic acid, in a 4:1 ratio (*1405*). Pyridine-3-carboxylic acid (nicotinic acid), ester or amide was hydrogenated with a RuO_2 catalyst to give piperidine-3-carboxylic acid derivatives (*1406*). Lower pressures (3 atm instead of 70 atm) could be used with Rh on carbon as catalyst (*1407*). Reduction with a Pt catalyst was reported in 1917 (*1408*).

Birch reduction of pyrrole-3,4-dicarboxylates in the presence of electrophiles resulted in a double reductive alkylation to give *cis*-3,4-disubstituted-pyrrolidine-3,4-dicarboxylates (*1409*), while ozonolysis of substituted 2,3-dihydropyrroles resulted in 2-aminocyclopropane-1-carboxylic acid derivatives (*677*).

A number of complex heterocycles have been synthesized that contain amino and carboxy groups in a 1,2-relationship (see Scheme 11.116). Examples include 4-aryl-6-methyl-4,7-dihydro-1H-pyrazolo-[3,4-b]pyridine-5-carboxylates (*1410*), 1-amino-3,4-dihydro-2-naphthalenecarboxylic acids (*1411, 1412*), 1-amino-2-napthalenecarboxylic acids (*1411, 1412*), 3-amino-2-ethoxycarbonylimidazo[1,2-a]pyridine (*1413*), carboxy-pyridones (*1414*), carboxy-indolizidines (*1415*), and carboxy-quinolizidines (*1415*). 3-Aminopyrrole-2,4-dicarboxylates were prepared via a cyclization reaction (*1416*), while condensation of thioaroylketene S,N-acetals with β-keto esters produced 3-amino-5-aryl-thiophene-2-carboxylates (*1417*). Complex 2-amino-5-aryl-pyrrole-3,4-dicarboxylates were synthesized by a three-component condensation of tosylimines with dimethyl acetylene dicarboxylate and alkyl isocyanides (*1418*). Deprotonated N-Boc-protected 3-amino-4-carboxymethylthiophene was regioselectively alkylated next to the amino group under kinetic conditions, or next to the carboxy group under thermodynamic conditions (*1419*).

11.3.2h Other Syntheses

Many other disparate routes to cyclic β-amino acids have been reported. Some of these involve cycloadditions. A thermal hetero[3+2] cycloaddition of O-benzyl oximes with dipolar trimethylenemethane derivatives derived from alkylidenecyclopropane acetals led to predominantly *trans*-5-substituted-pyrrolidine-3-carboxylates after hydrolysis (see Scheme 11.117) (*1420*).

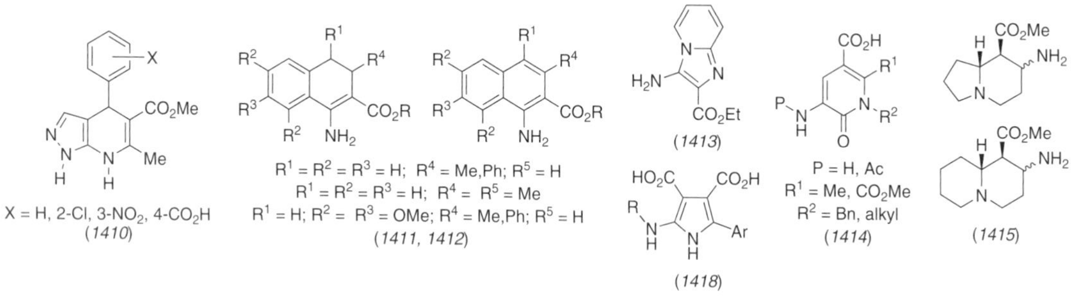

X = H, 2-Cl, 3-NO₂, 4-CO₂H
(*1410*)

R¹ = R² = R³ = H; R⁴ = Me,Ph; R⁵ = H
R¹ = R² = R³ = H; R⁴ = R⁵ = Me
R¹ = H; R² = R³ = OMe; R⁴ = Me,Ph; R⁵ = H
(*1411, 1412*)

(*1413*)

(*1418*)

P = H, Ac
R¹ = Me, CO₂Me
R² = Bn, alkyl
(*1414*)

(*1415*)

Scheme 11.116 Some Heterocyclic β-Amino Acids.

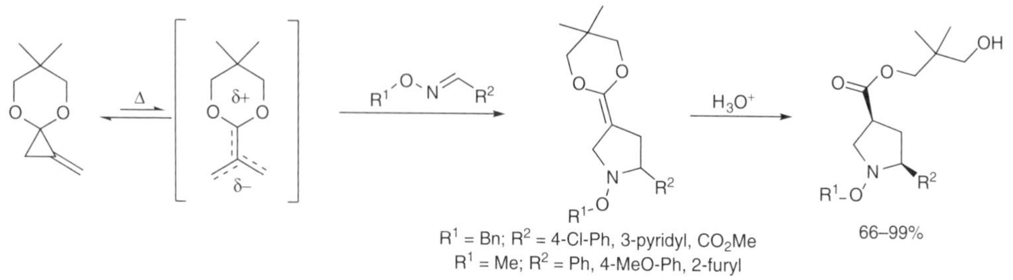

R¹ = Bn; R² = 4-Cl-Ph, 3-pyridyl, CO₂Me
R¹ = Me; R² = Ph, 4-MeO-Ph, 2-furyl

66–99%

Scheme 11.117 [3+2] Cycloaddition of Dipolar Trimethylenemethane with Oximes (*1420*).

1-Aminomethylcyclopropane-1-carboxylic acids were obtained by a 1,3-dipolar cycloaddition reaction of nitrone derivatives with bicyclopropylidene, initially generating a five-membered bispirocyclopropanated isoxazolidine that rearranged into a spirocyclopropanated β-lactam upon treatment with TFA (1421).

Ritalin and a number of related analogs were prepared from aryl glyoxylates via reaction with cyclic secondary imines to form an α-ketoamide, followed by condensation with tosylhydrazine to give a tosylhydrazone (see Scheme 11.118). Refluxing in toluene with KOtBu induced an intramolecular reaction to form a bicyclic lactam, with hydrolysis generating the desired product (4).

β-Amino acids based on the tropane alkaloid nucleus have been synthesized via a 2-carbomethoxy-3-tropinone intermediate (see Scheme 11.119), obtained by Willstätter condensation of acetonedicarboxylic acid monomethyl ester with methylamine and succindialdehyde, or by condensation of 3-tropinone with dimethylcarbonate (1303). The racemic product could be resolved. The ketone group was then reduced, eliminated, or hydrogenated to give various derivatives (1422). Treatment of the products with vinyl chloroformate induced N-demethylation (1422). A more recent synthesis prepared the tropane skeleton via a 1,3-dipolar cycloaddition of 1-methyl-3-oxidopyridinium and methyl acrylate, followed by alkene hydrogenation and ketone removal (1423, 1424). A ring contraction of dealkylated tropinone via a Favorskii rearrangement yielded 7-azabicyclo[2.2.1]hexane-2-exo-carboxylic acid (1425).

Other syntheses of cyclic β-amino acids involve cyclization steps. A stereoselective synthesis of trans-3-hydroxypyrrolidine-2-acetic acids (and their bicyclic lactones) was achieved via SmI₂-mediated 5-exo-trig cyclization of N-Cbz,N-(3-propanal)-3-aminoacrylates (1426). Cispentacin, cis-1-aminocyclopentane-2-carboxylic acid, was prepared via a sulfanyl radical addition–cyclization reaction of the O-methyl oxime of hex-5-enal, which produced S-phenyl cis-1-methoxyamino-2-thiomethylcyclopentane in 49% yield with 53% de. Conversion of the thiomethyl substituent to an acid and N–O cleavage proceeded in 31% yield (1427, 1428). The trans isomer could also be generated, as could the pyrrolidine analogs, trans-4-amino-3-pyrrolidinecarboxylic acid and 5-phenyl-4-amino-3-pyrrolidinecarboxylic acid (1428). An SmI₂-mediated radical cyclization of the imine adducts formed from secondary amines/benzotriazole and HC(=O)CH₂CH₂N(Cbz)CH=CHCO₂Et led to 3-aminopyrrolidine-2-acetic acid (1429). Cyclization of a hydrazone of 6-keto-undeca-2,10-dienoic acid via intramolecular conjugate addition was followed by an azomethine imine [3+2] cycloaddition to give a tricyclic pyrazoline. Hydrogenation generated pyrrolidine-2-acetic acid with a 5,5-spiro [2-(aminomethyl)cyclopentane] substituent (1430). 5-Hydroxy-2,3,4,5-tetrahydro-4H-indole-3a-carboxylic acid was obtained via a cyclization of N-(2-furyl) 2-methylene-4-aminobutyric acid (1431).

A Pd-catalyzed aminocarbonylation reaction of 5-acetamidocyclooctene with an intramolecular amine source and CO for the carboxyl source provided a mixture of 9-azabicyclo[4.2.1]nonane-2-carboxylate and 9-azabicyclo[3.3.1]nonane-2-carboxylate in ratios from 72:28 to 3:97, depending on reaction conditions (1432). 2-Azabicyclo[2.1.1]hexane-5-carboxylic acid, the β-isomer of 2,4-methano-Pro, has been obtained via a fairly lengthy synthesis that began with construction of the 2-azabicyclo[2.1.1]hexane ring system via base-induced cyclization of 1-phenylseleno-2-bromo-3,4-bis(aminomethyl)cyclobutane (1433).

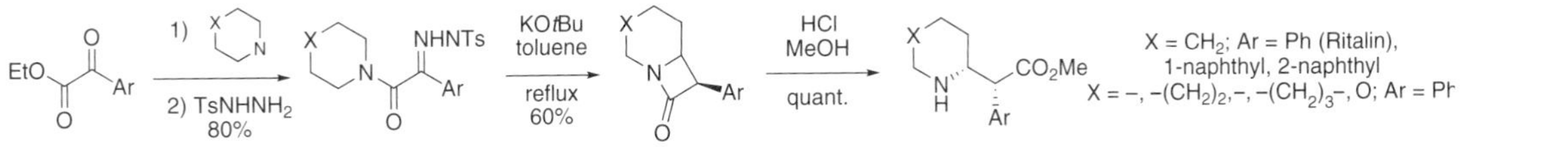

Scheme 11.118 Synthesis of Ritalin and Analogs (4).

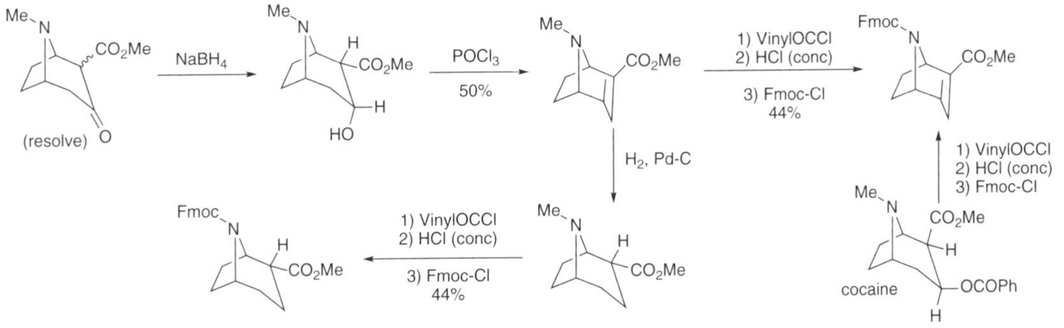

Scheme 11.119 Synthesis of Tropane Amino Acids (1422).

Table 11.3 Syntheses of Racemic or Achiral Cyclic β-Amino Acids

Amination

Ring Size (n)	Substituents	Method	Yield	Reference
5	1-NH$_2$-2-CO$_2$H	addition of ammonia to cyclopent-1-ene carboxylic acid	23% / 70% de	1960 (1357)
6	1-NH$_2$-2-CO$_2$H	addition of ammonia to cyclohexene-1-carboxylic acid	54%	1959 (1358)
3	1-NH$_2$-1-CH$_2$CO$_2$H / 1-NH$_2$-1-CH(Cl)CO$_2$H / 1-NH$_2$-1-CH(Br)CO$_2$H	addition of Ph$_2$C=NH to α,β-unsaturated esters	20–100%	1989 (473)
6	1-aza-2-CH$_2$CO$_2$H-6-methyl	intramolecular addition of 7-amino group (prepared by reductive amination of ketone) to α,β-unsaturated ester	73–86% for reductive amination and addition	1996 (1359)
3	1-NH$_2$-1-CH(OH)CO$_2$H-2-R R = H, Me, iPr	reaction of benzamide with 2-R-cPr=C(Cl)CO$_2$Me	16%	2005 (189)
6	1-NH$_2$-2c-CO$_2$H	Hofmann rearrangement of monoamide of cis-cyclohexane-1,2-dicarboxylic acid	70%	1959 (1358)
6	1-NH$_2$-2-CO$_2$H	electrochemical Hofmann rearrangement of 1,2-cyclohexanedicarboxylate monoamide	74%	1997 (1360)
5	1-NH$_2$-2-CO$_2$H	Hofmann rearrangement of anhydride of 1,2-cyclopentanedicarboxylate	—	1972 (1361)
6	1-NH$_2$-2-CO$_2$H-3,6-(—CH$_2$—)-4,5-dehydro	cycloaddition of maleic anhydride and cyclopentadiene, anhydride opening with NH$_4$OH, Hofmann rearrangemen	83% Diels–Alder / 69% Hofmann	1981 (1362)
6 / 5	1-NH$_2$-2-CO$_2$H-3,6-(—CH$_2$—)-4,5-dehydro / 1-NH$_2$-2c,3c,5c-(CO$_2$H)$_2$	Curtius rearrangement of endo-norborn-5-ene-2,3-dicarboxylate, alkene oxidative cleavage	75% Curtius / 68% alkene oxidation	1994 (1363)
3 / 4	trans-1-NH$_2$-2-CO$_2$H / trans-1-NH$_2$-2-CO$_2$H	Curtius rearrangement of monoester of trans-1,2-dicarboxycycloalkanes	58–66% Curtius	1971 (1364)
3	1-NH$_2$-2-CO$_2$H	Curtius rearrangement of cis- or trans 1,2-dicarboxycyclopropane	36–70%	1975 (1365)
4 / 6 / 6 (pyridine)	1-NH$_2$-2-CO$_2$H / 1-NH$_2$-2-CO$_2$H / 3-NH$_2$-2-CO$_2$H	Curtius rearrangement of resin-linked dicarboxylic acids	70–85%	1998 (1366)
6	2-NH$_2$-3-CO$_2$H on pyrazine nucleus	Curtius rearrangement of monoprotected 2,3-pyrazinedicarboxylic acid	80% overall	1996 (1367)
5	3-NH$_2$-4-CO$_2$H on 1,3-diphenylpyrazole nucleus	Curtius rearrangement of 2,3-pyrazoledicarboxylic acid	46–54% overall	1995 (1368)
6	1-NH$_2$-2c-CO$_2$H-3c,6c-(—CH$_2$—)-4,5-dehydro / 1-NH$_2$-2c-CO$_2$H-3t,6t-(—CH$_2$—)-4,5-dehydro / 1-NH$_2$-1-Me-2c-CO$_2$H-3t,6t-(—CH$_2$—)-4,5-dehydro / 1-NH$_2$-2c-CO$_2$H-3t,6t-(—O—)-4,5-dehydro	Curtius rearrangement of bicyclic monoester diacids	75–85% rearrangement	1993 (1369)
4	1-NH$_2$-2c-CO$_2$H-3t-Ph-4t-Ph	amination of 1-OH-2c-CO$_2$H-3t-Ph-4t-Ph-cyclobutane	80%	1927 (1370)

Ring	Substituents	Yield	Method	Year (Ref)
5	1-aza-2-R-3-CO_2H-4-(3,4-OCH_2O-Ph), R = H, Me, Et, CF_3, tBu, nPr, nBu, cPent, cHex, cHept, iPr, iPent, sPent, 4-Me-cHex, CH_2cHex, CH_2(4-Me-cHex), 4-MeO-cHex, CH_2OH, CH_2OEt, CH_2OnPr, CH_2OnBu, $CH_2OCH_2CH_2OMe$, CH_2OiBu, $CH_2OC(Me)_2Et$, CH_2OPh, CH_2OBn, CH_2CH_2OMe, CH_2CH_2OEt, OMe	—	cyclization of nitro ketone via nitro reduction and reductive amination	1999 (1347)
5	1-aza-2-(4-MeO-Ph)-3-carboxy-4-(3,4-OCH_2O-Ph)	60% nitro reduction; 75% reductive cyclization	cyclization of nitro ketone via nitro reduction and reductive amination	1996 (1348)
5	1-aza-2-Ar-3-CO_2H-4-(3,4-OCH_2O-Ph), Ar = 4-MeO-Ph, 4-$MeOCH_2O$-Ph, 4-nPrO-Ph, 3-F-4-MeO-Ph, 3-F-4-EtO-Ph, 3,4-F_2-Ph, 3,4-$(MeO)_2$-Ph	—	cyclization of nitro ketone via nitro reduction and reductive amination	1997 (1349)
5	1-aza-2-[$CH_2C(Me)_2n$Pr]-3-CO_2H-4-(3,4-OCH_2O-5-MeO-Ph)	91%	cyclization of nitro ketone via nitro reduction and reductive amination	2002 (1307)

Carboxylation

Ring	Substituents	Yield	Method	Year (Ref)
6	1-aza-2-oxo-3-CO_2H-4,5-dehydro; 1-aza-2-oxo-3-CO_2H-4,5-(-R-); (-R-) = (-$CH(OMe)CH_2COCH_2$-), (-CH=$CHCOCH_2$-), (-$CH(OR^1)$CH=$CHCH_2$-) with R^1 = TMS, Me, Ac; (-CH_2C(Me)=C(Me)CH_2-), (-CH(Me)CH=CHCH(Me)-), (-CH_2CH=C(Me)CH_2-)	31–89% Diels–Alder	carboxylation of enolate of 2-piperidone with $ClCO_2Bn$, alkene formation, Diels–Alder	1997 (1371)
6	1-aza-2-CH(OH)CO_2H	49–80%; 70–84% de	kinetic deprotonation of protected piperidine-2-methanol with TMEDA/sBuLi; carboxylation with $ClCO_2Me$	1999 (1372)
3	1-CH_2NH_2-1-CO_2H	100% carboxylation; 67% reduction	carboxylation of 1-cyano-1-TMS-cyclopropane, reduction cyano	1991 (1373)
4	1-aza-3-CO_2H	—	CN displacement of mesylate of 3-HO-azetidine	1990 (1374)
6	1-NH_2-2-CO_2H	80%	ring expansion of substituted-aziridines with $Co(CO)_4$ and Ti or Al catalyst	2002 (540)

Fusion of Aminomethyl and Methylcarboxy

Ring	Substituents	Yield	Method	Year (Ref)
3	1-NH_2-1-CH_2CO_2H; 1-NH_2-1-$CH(CN)CO_2H$	11–57%	addition of ketene silyl acetals or Li enolate to iminium ion of cyclopropanone	1991 (570)
5	1-aza-2-CH_2CO_2H	71–82%	reaction of N-Ts 2-MeO-pyrrolidine or piperidine with TMSOTf, H_2C=C(OTBS)OEt	1992 (1375)
6	1-aza-2-CH_2CO_2H			
5	1-aza-2-$C(Me)_2CO_2H$	20%	reaction of pyrrole hexahydro-1,3,5-triazine with ketene silyl acetal	1983 (577)
6	1-aza-2-CH_2CO_2H-4-keto-5-SePh-6-iPr; 1-aza-2-CH_2CO_2H-4-keto-6-iPr; 1-aza-2-CH_2CO_2H-4,5-dehydro-6-iPr	84–85% addition; 93% radical PhSe removal; 86–88% ketone reduction; 100% elimination	addition of ketene silyl acetal in presence of $BF_3 \cdot Et_2O$ to 2,3-dehydro-5-PhSe-6-iPr-piperidin-4-one, radical deselenation or ketone reduction and elimination	1999 (1376)

(Continued)

Table 11.3 Syntheses of Racemic or Achiral Cyclic β-Amino Acids (continued)

Ring Size (n)	Substituents	Method	Yield	Reference
5 6	1-CH$_2$NH$_2$-1-CO$_2$H 1-CH$_2$NH$_2$-1-CO$_2$H	reaction of ketene silyl acetals with (TMS)$_2$NCH$_2$OMe, hydrolysis	78–95% addition	1984 (556)
4 5 6	1-CH$_2$NH$_2$-1-CO$_2$H 1-CH$_2$NH$_2$-1-CO$_2$H 1-CH$_2$NH$_2$-1-CO$_2$H	reaction of ketene silyl acetals with (TMS)$_2$NCH$_2$OMe, hydrolysis	—	2002 (420)
5	1-aza-2-(3-pyridyl)-3-CO$_2$H-5-keto	addition of Schiff base to diethyl fumarate, deprotection and cyclization	91% overall, 88% de	1997 (1377)
5	1-aza-2=C(Ph)CO$_2$H 1-aza-2-CH(Ph)CO$_2$H	reaction of pyrrolidin-2-thione with BrCH(Ph)CO$_2$Me/DBU, treatment with Ph$_3$P, hydrogenation	91% thiobromination 86% thio contraction 95% hydrogenation	2004 (1378)
3 6	1-NH$_2$-2,3-(CO$_2$H)$_2$ 1-aza-3-CO$_2$H-2,4-(–)-5,6-dehydro	cyclopropanation of Boc- or N-acyl pyrrole, oxidative cleavage of alkene	24–45% cyclopropanation (0% de) 72–84% oxidative cleavage	1997 (1379) 1997 (1381)
4	1-NH$_2$-2-CO$_2$H-4,4-Me$_2$	cycloaddition of NMe$_2$-CH=C(Me)$_2$ with methyl acrylate	75%	1961 (1382)
4	1-NH$_2$-2-CO$_2$H-1,2-dehydro-4,4-Me$_2$	cycloaddition of (NMe$_2$)$_2$C=C(Me)$_2$ with methyl acrylate	45%	1964 (1383)
4	1-NH$_2$-2-CO$_2$H-4,4-Me$_2$ 1-NH$_2$-2-CO$_2$H-4,4-[–(CH$_2$)$_5$–] 1-NH$_2$-2,3-CO$_2$H-4,4-Me$_2$ 1-NH$_2$-2,3-CO$_2$H-4,4-[–(CH$_2$)$_5$–]	cycloaddition of N(alkyl)$_2$-CH=C(R)$_2$ with methyl acrylate or diethyl fumarate	18–72%	1964 (1384)
4	1-oxo-2-CO$_2$H-3-NH$_2$	Paterno–Büchi cycloaddition of nBu glyoxylate and BnN(Boc)CH=CH$_2$	28–35% for cyclization	1997 (1298)
5	1-aza-3-CO$_2$H-4,4-[–(CH$_2$)$_5$–]	reaction of Gly, paraformaldehyde, and cyclohexylidene derivative, hydrolysis	82% condensation 94% hydrolysis	1999 (1385)
5	1-aza-1-Me-3-CO$_2$H-3,4-dehydro-4-R 1-aza-3-CO$_2$H-3,4-dehydro-4-R R = H, Me, Ph, nPent 1-aza-3-CO$_2$H-1,2-Me$_2$-3,4-dehydro 1-aza-3-CO$_2$H-1,5-Me$_2$-3,4-dehydro 1-aza-3-CO$_2$H-2-Me$_2$-3,4-dehydro 1-aza-3-CO$_2$H-5-Me$_2$-3,4-dehydro	reaction of N-methyl-Gly or Ala with paraformaldehyde and 2-alkynoates, demethylation	10–41% condensation 10–59% demethylation	1992 (1346)
5	1-aza-2-R^2-2-R^3-3-CO$_2$H-3-R^6-4-R^4-4-R^4-5-R^1 1-aza-2-R^2-2-R^3-3-CO$_2$H-3-R^6-4-R^4-4-R^4-5-R^1-5-PO$_3$H$_2$ R^1 = H, Me, Ph R^2 = Me, iPr, Ph, 3-Pyr, -(CH$_2$)$_5$-R^3 R^3 = H, Me, -(CH$_2$)$_5$-R^2 R^4 = H, Me, -(CH$_2$)$_5$-R^5 R^5 = H, Me, Ph, -(CH$_2$)$_5$-R^4, -(CH$_2$)$_4$-R^6 R^6 = H, Me, -(CH$_2$)$_5$-R^5	cycloaddition or cycloaddition/elimination of anion of α-aminophosphonate Schiff bases with ethyl acrylates, reduction	5–90% cycloaddition 0–85% cycloaddition/ elimination	1988 (1386)
6	1-aza-2,6-(–CH=CH–)-3-CO$_2$H-3,4-dehydro 1-aza-2,6-(–CH=CH–)-3-CO$_2$H-3,4-dehydro-6-Me	Rh(II)-catalyzed decomposition of vinyldiazomethanes in presence of N-Boc-pyrroles	38–62%	1995 (1388)

5	1-aza-2-Ar-3-CO₂H-3,4-dehydro-4-*trans*-Ph Ar = Ph, 4-NO₂-Ph, 4-Cl-Ph, 4-Me-Ph, 4-MeO-Ph	Cu(II)-catalyzed reaction of methyl styryldiazoacetate with benzylamines	40–74% 66–96% de	2003 (1387)
5	1-aza-3-CO₂H-2,5-(−CH₂CH₂−)	cycloaddition of N-Boc pyrrole with methyl 3-bromopropiolate, reduction	60% cycloaddition 80–98% reduction	1997 (1389)
5	1-aza-3-CO₂H-3,4-dehydro-4-R-2,5-(−CH=CH₂) R = Br, Me, nBu, Ph, 2-Cl-pyrid-5-yl, CCPh	cycloaddition of N-Boc pyrrole with methyl 3-bromopropiolate, organocuprate reaction	45–60% cycloaddition 40–78% RCu addition	1998 (1390)
6	1-aza-2-R¹-3-CO₂H-4-R²-2,3,4,5-dehydro-6-keto R¹ = Ph, 4-Me-Ph, 4-NO₂-Ph, 4-MeO-Ph, 2-furyl, 2-thienyl R² = H, CO₂Me	cycloaddition of TMSketene with acyl isocyanates and acetylene carboxylates	22–99%	1996 (1391)

Elaboration of α-Amino Acids

6	1-aza-2-CH₂CO₂H-5,6-benzo	homologation of 5,6-benzopipecolic acid	48%	1998 (1392)
6	1-CH₂NH₂-1-CO₂H-2-oxo-3,4-(4-MeO-benzo)-5-oxa	cyclization of α-substituted Asp by intramolecular Friedel–Crafts acylation using α-carboxyl group	—	1940 (1393)

Elaboration of β-Amino Acids

3	1-CH₂NH₂-1-CO₂H	dialkylation of methyl cyanoacetate with Br(CH₂)ₙBr, CN reduction with H₂/Ra Ni, hydrolysis	55–80% hydrogenation 32–89% hydrolysis	1998 (322)
4	1-CH₂NH₂-1-CO₂H			
5	1-CH₂NH₂-1-CO₂H			
6	1-CH₂NH₂-1-CO₂H			
3	1-CH₂NH₂-1-CO₂H	dialkylation of ethyl cyanoacetate with Br(CH₂)ₙBr, CN /ester reduction with LiAlH₄, hydroxymethyl oxidation	64–93% alkylation 67–87% reduction 71–87% oxidation	1999 (53)
5	1-CH₂NH₂-1-CO₂H			
6	1-CH₂NH₂-1-CO₂H			
7	1-CH₂NH₂-1-CO₂H-3,4-benzo-5,6-benzo	dialkylation of ethyl cyanoacetate with 2,2'-(CH₂Br)₂-1,1'-biphenyl, regioselective CN reduction	82% alkylation 76% reduction	1998 (1352) 2000 (1353)
6	1-amino-2-CO₂H-3,6-(−O−)-4,5-dehydro 1-amino-2-CO₂H-6-HO-2,3,4,5-dehydro 1-amino-2-CO₂H-6-HO-2,3-dehydro-4,5-(−O−) 1-amino-2-CO₂H-5,6-(HO)₂	Diels-Alder cycloaddition of furan with 3-nitroacrylate, basic elimination of oxa bridge, alkene hydrogenation or epoxidation, reductive epoxide opening	90% Diels–Alder 2:1 endo:exo 75–77% nitro reduction 69–71% oxa elimination 98% alkene hydrogenation 77–82% epoxidation 84–98% reductive epoxide opening	2004 (1395)
7	1-aza-3-CO₂H-5,6-dehydro	alkylation of N-Boc,N-alkenyl-β-Ala-OEt, ring closing metathesis, hydrogenation	57–73% alkylation 87–89% de 24–98% metathesis 88–95% reduction	2005 (641)
7	1-aza-3-CO₂H			
8	1-aza-3-CO₂H-5,6-dehydro			
8	1-aza-3-CO₂H			
9	1-aza-3-CO₂H-5,6-dehydro			
9	1-aza-3-CO₂H			

(Continued)

Table 11.3 Syntheses of Racemic or Achiral Cyclic β-Amino Acids (continued)

Ring Size (n)	Substituents	Method	Yield	Reference
6	1-aza-3-Me-3-CO$_2$H	deprotonation and alkylation of Boc-piperidine-3-carboxylate ethyl ester	90%	2002 (420)
6	1-aza-3-CO$_2$H-4-OH-5-CH$_2$OH	deprotonation and alkylation of 4-ketopiperidine-3-carboxylate, reduction of enol ether	57–95% alkylation 66–96% reduction 0–50% de	1999 (1396)
6	1-aza-3-CO$_2$H-4-hydroxy	reduction of ketone of 3-carboxyethyl-4-piperidone	23–47% reduction	1997 (1350)
5	1-aza-2,3-dehydro-3-CO$_2$H	deprotonation and acylation of GABA lactam with Cbz-Cl, deprotection, lactam reduction with NaBH$_4$	72% carboxylation and reduction	2002 (1397)
5	1-aza-2-Me-3-CO$_2$H-3-nPent-4-Me 1-aza-3-CO$_2$H-3-nPent-4-Me 1-aza-3-CO$_2$H-3-nPent-4-Me-5-Me 1-aza-3-CO$_2$H-3-nPent-4-Me-5-iPr 1-aza-3-CO$_2$H-3-nPent-4-Me-5-Ph	domino reaction of H$_2$C=CHCH=CHCH(R^1)N(Bn)CH(R^2)C(=CH$_2$)CO$_2$Me initiated by addition of alkylzinc	49–64% mainly cis	2003 (1398)

Small Ring Systems

Ring Size (n)	Substituents	Method	Yield	Reference
6	1-NH$_2$-2-CO$_2$H	cHex N-Ts aziridine opening with TMSCN/Yb(CN)$_3$	90%	1990 (658)
5	1-aza-2-CH$_2$CO$_2$H-3-(=O)-5-R R = Bn, CH$_2$OPh, nHex, tBu, Me$_2$	Michael addition of 2-substituted aziridines to methyl propiolate, aziridine ring opening with PhSeNa, radical carbonylation/cyclization or PhSeCH$_2$CH(R) NHCH=CHCO$_2$Me substrates	100% Michael addition 92–98% aziridine opening 65–85% cyclization	2003 (1399)
6	1-aza-3-CO$_2$H-3-X-4,6-(CH=CH) X = Cl, Br, I	Diels–Alder cycloaddition of cyclohexadiene with azirine-2-carboxylate, aziridine opening with halide	100% cycloaddition 10–31% de 45–50% halide	2003 (1400)
4	1-NH$_2$-1-CH$_2$CO$_2$H	cycloaddition reaction of methylene cycloalkanes with chlorosulfonyl isocyanate, ring opening of β-lactam	65–83% bicyclic β-lactam formation 86–98% opening	1998 (1401)
5	1-NH$_2$-1-CH$_2$CO$_2$H			
6	1-NH$_2$-1-CH$_2$CO$_2$H			
7	cis-1-NH$_2$-2-CO$_2$H	cycloaddition reaction of cycloalkenes with chlorosulfonyl isocyanate, ring opening of β-lactam	60–63% bicyclic β-lactam formation 72–74% opening	2003 (671)
8	cis-1-NH$_2$-2-CO$_2$H			
12	cis-1-NH$_2$-2-CO$_2$H			
12	trans-1-NH$_2$-2-CO$_2$H			
5	1-NH$_2$-2-CO$_2$H	reaction of cyclopentene with chlorosulfonyl isocyanate, deprotection and lactam hydrolysis	10% overall	1989 (1296)
5	1-NH$_2$-2-CO$_2$H-3,5-(–CH=CH–)	cycloaddition of ClSO$_2$NCO and norbornadiene, reduction, hydrolysis	43%	1984 (1402)
5	1-NH$_2$-2-CO$_2$H-3,5-(–CH=CH–) 1-NH$_2$-2-CO$_2$H-3,5-(–CH=CH–)-4-OtBu 1-NH$_2$-2-CO$_2$H-3,5-(–CH=CH–)-4-OBz 1-NH$_2$-2-CO$_2$H-3,5-(–CH$_2$CH$_2$–) 1-NH$_2$-2-CO$_2$H-3,6-(–CH$_2$–)-4,5-	cycloaddition of ClSO$_2$NCO and bicyclic alkenes, reduction, hydrolysis	11–94%	1968 (1403)

5	1-CH(R)NH$_2$-1-CO$_2$H-2-oxa 1-CH(R)NH$_2$-1-CO$_2$H-3-oxa R = Ph, 4-MeO-Ph, 4-NO$_2$-Ph, 2-furyl, 3-pyridyl, CH=CHPh	Staudinger cycloaddition of ketenes derived from acid chloride of tetrahydrofuran-2- or -3-carboxylic acid with N-aryl imines derived from substituted benzaldehydes	49–77% cycloaddition	2002 (1404)

Syntheses from Heterocycles

6	1-aza-3,5-(CO$_2$H)$_2$ 1-aza-3,5-dimethyl-3,5-(CO$_2$H)$_2$	hydrogenation of pyridine-3,5-dicarboxylate, deprotonation, and dimethylation	100% hydrogenation 0% de 70% dimethylation 60% de *trans*	1994 (1405)
6	1-aza-3-CO$_2$H	hydrogenation of pyridine-3-carboxylate with H$_2$/RuO$_2$ at 70–100 atm	86–90% hydrogenation	1961 (1406)
6	1-aza-3-CO$_2$H	hydrogenation of pyridine-3-carboxylic acid, ester or amide with H$_2$/Ru-C at 3 atm	42–87% hydrogenation	1962 (1407)
6	1-aza-3-CO$_2$H	hydrogenation of pyridine-3-carboxylic acid with Pt catalyst	92% hydrogenation	1917 (1408)
5	1-aza-3,4-(CO$_2$H)$_2$-3,4-R$_2$ R = Me, Et, *t*Bu, CH$_2$CH=CH$_2$	double Birch reduction/alkylation of pyrrole-3,4-dicarboxylate	70–82% reductive alkylation	1999 (1409)
3	1-NH$_2$-2-CO$_2$H 1-NH$_2$-2,3-(CO$_2$H)$_2$	ozonolysis of 2,3-dihydropyrroles	60–88%	1994 (677)
5	3-NH$_2$-2,4-(CO$_2$H)$_2$-pyrrole	cyclization reaction	17–90%	1998 (1416)
5	2-CO$_2$H-3-NH$_2$-5-Ar-thiophene Ar = Ph, 3-MeO-Ph, 3-Cl-Ph	condensation of thioaroylketene S,N-acetal with β-keto ester	82–94%	1998 (1417)
5	pyrrole 3,4-(CO$_2$H)$_2$ with 2-NHR and 5-Ar	condensation of tosylimines with alkyl isocyanide and dimethyl acetylenedicarboxylate	72–94%	2001 (1418)
5	2-R-3-NH$_2$-4-CO$_2$H-thiophene 3-amino-4-carboxy-5-R thiophene R = Me, TMS, CHO, CO$_2$Me	deprotonation and alkylation of N-Boc-3-amino-4-carboxymethylthiophene	43–74%	1997 (1419)

Other Syntheses

5	1-aza-3-CO$_2$H-5-R R = Ph, 4-Cl-Ph, 4-MeO-Ph, 3-pyridyl, 2-furyl, CO$_2$Me	dipolar cycloaddition of trimethylenemethane with oxime, hydrolysis	66–99%	1998 (1420)
3	1-CH(R)NH$_2$-1-CO$_2$H R = Ph, 2-pyridyl, CN	dipolar cycloaddition of nitrone with bicyclopropylidene, rearrangement of bisspirocyclopropanated isoxazolidine into 3-spirocyclopropanated β-lactam	71–100% isoxazolidine 75–96% β-lactam	2004 (1421)
6	1-aza-2-CH(Ph)CO$_2$H	amidation of arylglyoxylate with secondary amine, formation of tosylhydrazone, bicyclic β-lactam formation, hydrolysis	80% amidation and hydrazone formation 60% lactam formation 100% hydrolysis	1998 (4)
6	1-aza-2-CH(1-naphthyl)CO$_2$H			
6	1-aza-2-CH(2-naphthyl)CO$_2$H			
6	1-aza-2-CH(Ph)CO$_2$H-4-oxa			
7	1-aza-2-CH(Ph)CO$_2$H			
8	1-aza-2-CH(Ph)CO$_2$H			

(*Continued*)

Table 11.3 Syntheses of Racemic or Achiral Cyclic β-Amino Acids (continued)

Ring Size (n)	Substituents	Method	Yield	Reference
6	1-aza-3-CO$_2$H-2,6-(−CH$_2$CH$_2$−) 1-aza-3-CO$_2$H-2,6-(−CH$_2$CH$_2$−)-2,3-dehydro	carboxylation of 3-tropinone, elaboration	—	1997 (1422)
5	1-aza-3-CO$_2$H-2,5-[−(CH$_2$)$_3$−]	1,3-dipolar cycloaddition of 1-Me-3-oxidopyridinium and Me acrylate, alkene reduction, ketone removal	96% reduction 74% ketone removal	1998 (1423)
5	1-aza-3-CO$_2$H-2,5-[−(CH$_2$)$_3$−] 1-aza-3-CO$_2$H-2,5-[−CH$_2$CH$_2$CO−] 1-aza-3-CO$_2$H-2,5-[−CH$_2$CH$_2$CH(OH)−] 1-aza-3-CO$_2$H-2,5-[−CH$_2$CH$_2$CH(OAc)−] 1-aza-3-CO$_2$H-2,5-[−CH$_2$CH$_2$CH(SEt)−]	1,3-dipolar cycloaddition of 1-Me-3-oxidopyridinium and Me acrylate, alkene reduction, ketone removal	47% cycloaddition 95% reduction 24% ketone removal	1998 (1424)
5	1-aza-3-CO$_2$H-2,4-(−CH$_2$CH$_2$−)	Favorskii rearrangement of tropinone	—	1998 (1425)
5	1-aza-2-CH$_2$CO$_2$H-3-OH 1-aza-2-CH(nPr)CO$_2$H-3-OH	SmI$_2$ mediated cyclization of HCOCH$_2$CH$_2$N(Cbz)-CH=C(R)CO$_2$Et	48–53%	1998 (1426)
5	cis-1-NH$_2$-2-CO$_2$H trans-1-NH$_2$-2-CO$_2$H trans-1-NH$_2$-2-CO$_2$H-4-aza trans-1-NH$_2$-2-CO$_2$H-4-aza-5-Ph	sulfanyl radical addition-cyclization of H$_2$C=CH(CH$_2$)$_3$CH=NOMe, conversion PhSCH$_2$ to CO$_2$H, N–O cleavage	20–76% cyclization 31% conversion and cleavage for cispentacin	1998 (1427) 2002 (1428)
5	1-aza-2-CH$_2$CO$_2$H-3-NR^{1}R^2 R^{1}R^2 = −(CH$_2$)$_2$O(CH$_2$)$_2$−, −(CH$_2$)$_2$N(Ph)(CH$_2$)$_2$−, −(CH$_2$)$_5$−, Me/Bn, 3,4-(MeO)$_2$-PhEt/allyl	SmI$_2$ radical cyclization of HCOH CH$_2$CH$_2$N(Ts) CH=CHCO$_2$Et in presence of amine and benzotriazole	52–84% 14–20% de	2002 (1429)
5	1-aza-2-CH$_2$CO$_2$H-5,5-spiro-[2-(aminomethyl)cyclopentane]	cyclization of hydrazone of H$_2$C=CH(CH$_2$)$_3$CO(CH$_2$)$_2$CH=CHCO$_2$Me, hydrazine hydrogenolysis	75% cyclization 87% hydrogenation	1999 (1430)
4	1-aza-3-CO$_2$H-2,3-[=CHCH(OBn)CH(OH)CH$_2$−] 1-aza-3-CO$_2$H-2,3-[=CHCH$_2$CH(OH)CH$_2$−]	cyclization of N-(2-furyl) 2-methylene-4-aminobutyric acid	62% cyclization	2001 (1431)
6 7	1-aza-3-CO$_2$H-2,6-[−(CH$_2$)$_3$−] 1-aza-3-CO$_2$H-2,7-[−(CH$_2$)$_2$−]	Pd-catalyzed aminocarboxylation of 5-acetamidocyclooctene	47–66%	1998 (1432)
5	1-aza-3-CO$_2$H-2,4-(−CH$_2$−)	cyclization of 1-PhSe-2-Br-3,4-(CH$_2$NH$_2$)$_2$ to bicyclic system, conversion of CH$_2$NH$_2$ to CO$_2$H	71% cyclization 77% deselenation 18% conversion of CH$_2$NH$_2$ to CO$_2$H	2001 (1433)

11.3.3 Asymmetric Syntheses of Cyclic β-Amino Acids (see Table 11.4)

11.3.3a Resolutions of Racemic Cyclic β-Amino Acids

Resolutions of cyclic β-amino acids via diastereomeric salt formation include the combinations of N-Boc cis-2-aminocyclopentane-1-carboxylic acid with (−)-ephedrine (1341, 1342), cis-2-benzamidocyclohexane-1-carboxylic acid with cinchonidine (1434), cispentacin (2-aminocyclopentane-1-carboxylic acid) with (+)-dehydroabietylamine (1296), and cis-2-aminocyclohexane-1-carbohydrazide with dibenzoyl (+)-tartaric acid (1434). The resolved (1R,2S)- and (1S,2R)-cyclohexane enantiomers from the latter resolution were subsequently epimerized at the carboxyl center with hot concentrated HCl to give the (1S,2S)- and (1R,2R)-trans-enantiomers, respectively (1434). N-Benzoyl trans-2-aminocyclohexane-1-carboxylic acid was resolved by seed-induced enantioselective crystallization, with the small amount of seed crystal first obtained by quinine resolution (1319). The same procedure was used for the cis isomer (1435), and for 3-endo-benzamido-5-norbornene-2-endo-carboxylic acid (1362). The Abbott antidepressant Rolipram (ABT-546), a pyrrolidine-3-carboxylate core with C2-alkyl and C4-aryl substituents, was resolved with D-tartaric acid (1307), while diastereomeric salt separation of threo-methylphenidate (methyl piperidine-2-phenylacetate: Ritalin, used to treat attention deficit hyperactivity disorder) employed O,O′-dibenzoyl-D-(+)-tartaric acid in the presence of 4-methylmorpholine (1436). Tartaric acid was also used to resolve ethyl nipecotate (β²-homoproline, piperidine-3-carboxylic acid) (1090, 1326,), while the unprotected amino acid was separated with camphorsulfonic acid (1326). The purified enantiomers had low optical rotation, thus, enantiopurity was assessed by formation of the N-2,4-dinitrophenyl derivative and chromatography on a Chiracel-OD column (1090).

The HPLC enantioseparation of 12 underivatized cyclic β-amino acids was compared using Chirobiotic T (teicoplanin) and Chirobiotic TAG (teicoplanin aglycone) columns, with the TAG column tending to give better resolution (1437). In another study, Chirobiotic T gave resolutions of 10 different cyclic and bicyclic β-amino acids, using water–methanol mobile phase (1438). Enantiomers of derivatized nipecotic acid, piperidine-3-carboxylic acid, have been separated by a cyclodextrin-based column during GC analysis (1439). An HPLC column with a chiral crown ether stationary phase, Crownpak CR(+), provided resolutions for some of the isomers of bicyclo[2.2.1]heptane and heptene-based β-amino acids (1440). Diastereomeric derivatives formed with FDAA (Marfey's reagent) or GITC (tetra-O-acetyl-β-D-glucopyranosyl isothiocyanate) gave complementary results for the compounds that were not resolved by the chiral column (1440). The novel atropic α,α-disubstituted-β-amino acid Bin was resolved on a β-cyclodextrin HPLC column as the free amino acid, the

ethyl ester, and the N-Boc derivative. The amino acid could also be derivatized with Marfey's reagent, FDAA, and separated on a normal C_{18} column (1394). The β-amino acid spin-labeled analogs of TOAC, cis-4-amino-1-oxyl-2,2,6,6-tetramethylpiperidine-3-carboxylic and trans-3-amino-1-oxyl-2,2,5,5-tetramethylpyrrolidine-4-carboxylic acid, have been enantioseparated by HPLC using a Chiralcel OD-RH column (1332).

A variety of approaches to enzymatic resolution have been reported. Ten different cyclic β-amino esters were resolved via acylation with 2,2,2-trifluoroethyl acetate catalyzed by lipase SP 526 from Candida antarctica or lipase PS from Pseudomonas cepacia. Other acyl donors such as 2,2,2-trifluoroethyl chloroacetate or 2,2,2-trifluoroethyl hexanoate were also examined (1441). Gram-scale resolutions of methyl cis-2-aminocycloheptane-1-carboxylic acid, cis-2-aminocyclooctane-1-carboxylic acid, and cis-2-aminocyclododecane-1-carboxylic acid were accomplished by N-acylation with 2,2,2-trifluoroethyl butanoate as the acyl donor and Candida antarctica lipase A in diisopropyl ether, or by acylation of the hydroxymethyl group of the N-hydroxymethyl-β-lactam derivatives, using Pseudomonas cepacia lipase in dry acetone (671). The N-hydroxymethyl β-lactam precursors of 2-aminocyclopentane-1-carboxylic acid and 3-aminobicyclo[2.2.1]heptane-1-carboxylic acid were also resolved via a lipase-mediated acylation of the hydroxymethyl substituent. Acid hydrolysis cleaved the β-lactam and N-acylhydroxymethyl substituents to give the cyclic β-amino acids (1442). The β-lactam derivative of 1-aminocyclopent-4-ene-1-carboxylic acid (6-azabicyclo[3.2.0]hept-3-en-7-one) was resolved by lactam hydrolysis using whole cells of Rhodococcus equi. The non-hydrolyzed lactam was hydrogenated and then hydolyzed to give (−)-cispentacin, (1R,2S)-cis-1-amino-cyclopentane-2-carboxylic acid (1443). Threo-methylphenidate was resolved by ester hydrolysis with α-chymotrypsin or subtilisin Carlsberg (5). Methyl and ethyl esters of N-Boc homoproline (pyrrolidine-2-acetic acid), homopipecolic acid (piperidine-2-acetic acid), and 3-carboxymethylmorpholine were resolved by hydrolysis with Burkholderia cepacia lipase to give the acids and residual esters with 98 to >99% ee (1444).

Reduction of the ketone group of racemic ethyl N-Boc 4-oxopiperidine-3-carboxylate by Baker's yeast produced a single (cis) diastereomer of 4-hydroxypiperidine-3-carboxylate in 78% yield with 93% ee, as determined by both chiral shift reagent ^{1}H NMR analysis and by conversion to a known chiral derivative (1445, 1446). However, another publication was unable to duplicate the high enantiomeric excess, obtaining product with the same optical rotation but only 24% ee, as determined by chiral HPLC analysis of a derivative (1350). Enantioselective hydrogenation of 2-acetylaminocycloalk-1-ene-1-carboxylic acid ethyl esters using a Ru catalyst with a (S)-C3-TunaPhos ligand produced the corresponding cis-substituted cycloalkyl β-amino esters, with enantioselectivity decreasing with increasing ring size (99% ee for cyclopentyl to 44% ee for cyclooctyl). The cis cyclopentyl product was epimerized in NaOMe/MeOH to form the *trans* isomer, although the enantiomeric purity was not reported (*1447*). *N*-Protected pyrrolidine enamino acid methyl esters were reduced by asymmetric hydrogenation with a variety of catalysts. Me-DuPHOS produced pyrrolidine-2-acetic acid with >99% ee but only 37% conversion, while Me-BDPMI gave 100% conversion with 96% ee (*1448*).

11.3.3b Amination Reactions

Davies et al. have made extensive use of lithium (α-methylbenzyl)benzylamide as a chiral amine nucleophile for Michael additions to acyclic conjugated acids (see Scheme 11.31 earlier). The same procedure was employed with cycloalkene-1-carboxylate substrates. The antifungal antibiotic cispentacin, (1*R*,2*S*)-2-aminocyclopentane-1-carboxylic acid, and the cyclohexyl homolog hexacin, were synthesized by this method. In both cases the desired *cis* isomer was produced with >98% de (*602, 716, 746, 1294*). Treatment with KO*t*Bu induced epimerization to the more stable *trans* epimer (see Scheme 11.120) (*716, 1294*). The same strategy was applied to the syntheses of both *cis*- and *trans*-4-aminopyrrolidine-3-carboxylic acid, 4-amino-tetrahydrofuran-3-carboxylic acid, and 4-amino-tetrahydrothiophene-3-carboxylate (*1449*). For the preparation of (1*R*,2*S*,3*R*)-3-methylcispentacin and (1*S*,2*S*,3*R*)-3-methyltranspentacin, racemic *tert*-butyl 3-methylcyclopent-1-ene-1-carboxylate was reacted with lithium (*S*)-*N*-benzyl-*N*-α-methylbenzylamide, resulting in three diastereomers in 95.5:1.7:2.8 ratio, with a 39% isolated yield of the (1*R*,2*S*,3*R*)-isomer (>99% de). Treatment with KO*t*Bu induced epimerization to the (1*S*,2*S*,3*R*)-isomer in quantitative yield, and also with >99% de. Deprotection of both isomers by hydrogenation proceeded in 64–69% yield (*1450*). In contrast, a mixture of (1*R*,2*R*,5*R*)- and (1*R*,2*R*,5*S*)-2-amino-5-carboxymethyl-cyclopentane-1-carboxylic acids (5-carboxymethyl-transpentacin) was obtained by a tandem addition/cyclization to dialkyl (*E,E*)-octa-2,6-dienedioic acid, with the (1*R*,2*R*,5*R*)-isomer isolated in 77–83% yield and the (1*R*,2*R*,5*R*)-isomer in 6–7% yield, depending on ester protecting group. Since the *cis* isomers were not produced by this route, the initial adducts were oxidatively eliminated to give 5-carboxymethyl-cyclopent-1-ene-1-carboxylic acid, with lithium

(α-methylbenzyl)benzylamide addition now producing the 5-carboxymethylcispentacin diastereomers (*1451*).

Piperidine-2-acetic acid and pyrrolidine-2-acetic acid were obtained via addition of (*S*)-*N*-allyl-*N*-α-methylbenzylamide to methyl (*E,E*)-heptan-2,5-dienoate or *tert*-butyl (*E,E*)-hexa-2,4-dienoate, respectively, followed by ring-closing metathesis of the two alkene groups, and then hydrogenation (*748*). Alternatively, Michael addition of *N*-aryl α-methylbenzylamines to α,β-unsaturated-ω-chloroesters, followed by *N*-aryl deprotection, resulted in cyclization to five-, six- and seven-membered imino acids (*749*). Michael addition of (*R*)-α-methylbenzylamine to *N*-Boc ethyl 3-carboxy-2,5-dihydropyrrole led to *trans*-3-aminopyrrolidine-4-carboxylic acid (*1337*). A more efficient synthesis employed the same *N*-Boc ethyl 3-carboxy-4-oxopyrrolidine starting material, but utilized reductive amination with (*R*)-(+)-α-methylbenzylamine/NaBH₃CN; diastereomerically pure product was obtained in 38% yield after one recrystallization (*1452*). A comprehensive review of the conjugate addition of enantiomerically pure lithium amides was published in 2005, including tables of many β-amino acids produced by this route (*718*).

Addition of Enders' TMS-SAMP chiral amine equivalent to ω-halide-substituted α,β-unsaturated esters resulted in either 1-amino-2-carboxycycloalkanes (via intramolecular alkylation of the initial Michael adduct enolate) or 1-aza-2-(acetic ester)cycloalkanes (via quenching of the initial adduct followed by an intramolecular *N*-alkylation) (see Scheme 11.121). The 1-amino-2-carboxyalkane series was generated with *trans* stereochemistry, and both procedures gave products with >97% ee (*1453, 1454*). A similar tandem Michael addition/cyclization quench was employed for an asymmetric synthesis of (1*S*,2*R*)-1-amino-2,3-dihydro-1*H*-indene-2-carboxylic acid, via addition of the lithium (α-methylbenzyl)benzylamide reagent to *tert*-butyl *o*-bromomethylcinnamate (*1455*).

Chiral auxiliaries have also been employed on the electrophilic propionate equivalent. Lithium amides were added to naphthalene substituted in either the 1- or 2-position with a carboxyl group protected as a chiral oxazoline, forming rigid bicyclic β-amino acids with >97% de. Quenching the amide addition reaction with an electrophile introduced an additional substituent adjacent to the carboxyl group (see Scheme 11.122) (*1456*).

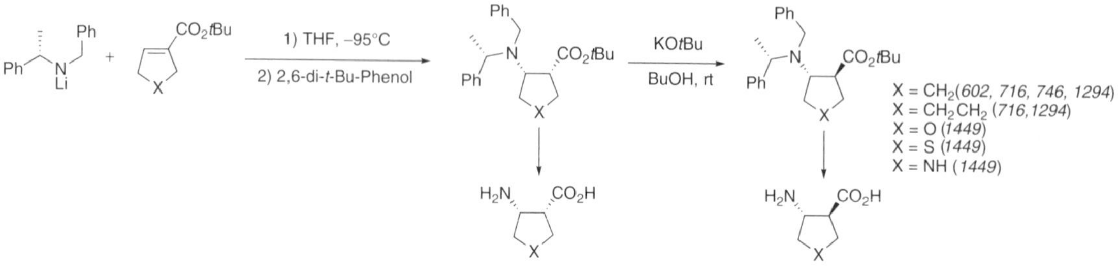

Scheme 11.120 Conjugate Additions of Lithium Amides and Epimerization.

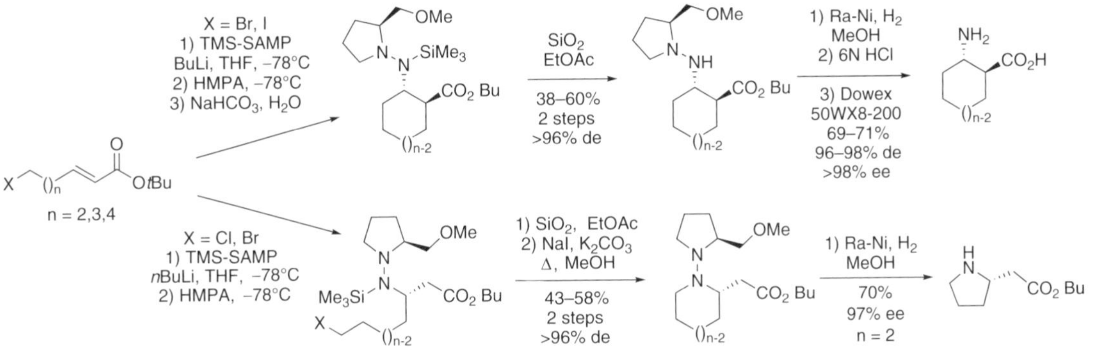

Scheme 11.121 Addition of TMS-SAMP to ω-Halo-α,β-Unsaturated Esters (*1453, 1454*).

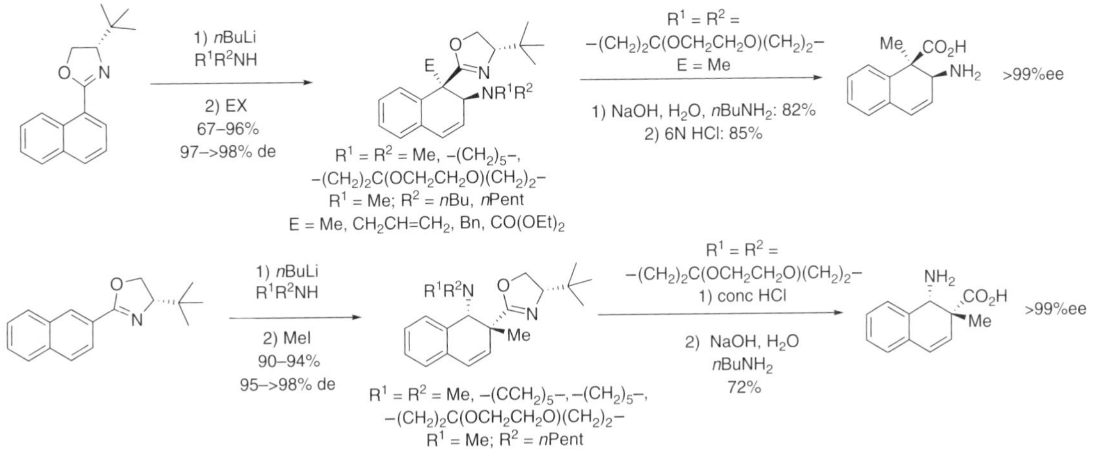

Scheme 11.122 Conjugate Additions of Lithium Amides to Chiral Naphthyloxazolines (*1456*).

(1S,2S,3S)-1-Amino-2-carboxy-3-(carboxymethyl) cyclohexane was prepared via a tandem conjugate addition/cyclization reaction of 2,7-nonadienedioic acid with LiZn[Bn(TMS)N]₃, with a bis(menthyl ester) auxiliary employed to induce enantioselectivity (*1457*). Nitrogen nucleophiles such as hydroxylamines and hydrazines were added to a homochiral α, β-unsaturated-α-carboxy-γ-substituted-pyrrolidinone with moderate to good diastereoselectivity (*1458*). 2,5-Dihydropyrrole-2-acetic acid derivatives were prepared via an intramolecular Michael addition of a homochiral tricyclic substrate followed by flash thermolysis (*1459*).

Amination has also been achieved by nucleophilic displacements. A bicyclic β-amino acid, Bic (3-amino-7-hydroxybicyclo[3.3.0]octane-2-carboxylic acid) was prepared via azide displacement of a nosylate (*1345*). A synthesis of cispentacin, (1R,2S)-2-aminocyclopentane-1-carboxylic acid, employed Mitsunobu displacement of a chiral cyclopentanol substrate with phthalimide. The cyclopentanol substrate, which was resolved by a lipase-catalyzed kinetic resolution, contained a protected

2-(hydroxymethyl) substituent that was later oxidized to generate the carboxyl group (*1460*). The *trans* isomer, (1R,2R)-2-aminocyclopentane-1-carboxylic acid, was prepared from (1R,2S)-cis-2-hydroxycyclopentane-1-carboxylate, itself obtained via Baker's yeast reduction of racemic ethyl 2-oxocyclopentanecarboxylate. The hydroxyl group was aminated by Mitsunobu reaction with HN₃, and then reduced (*1310*). A somewhat-related synthesis employed (R)-O-benzyl 3-(hydroxymethyl) cyclopentene as the starting material. *cis*-Epoxidation, followed by epoxide opening with azide, mesylation, azide reduction, and aziridine formation produced the *trans*-aziridine   2-hydroxymethyl-6-azabicyclo[3.1.0] hexane. Aziridine opening with alcohols or cyanide, followed by oxidation of the hydroxymethyl group, produced 3-substituted-2-aminocyclopentanecarboxylic acids with methoxy, phenoxy, or aminomethyl substituents (*1317*).

Both enantiomers of β-proline have been prepared from a common homochiral hydroxymethyl lactam intermediate, itself derived from 3-[(benzyloxymethyl)

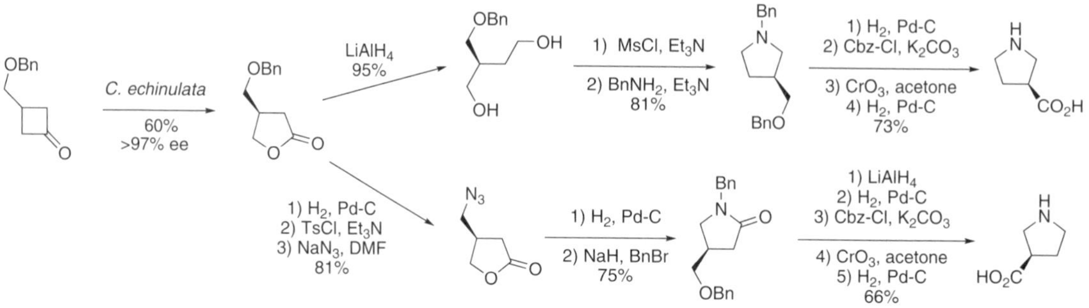

Scheme 11.123 Synthesis of β-Proline (1461).

methyl]cyclobutanone by an enzymatic Baeyer–Villiger oxidation (see Scheme 11.123). The proline ring was formed by azide displacement of a tosylate, followed by azide reduction, or by displacement of a dimesylate with benzylamine (1461).

A conformationally restricted analog of phenylisoserine, 2,2'-methylenephenylisoserine, was synthesized via asymmetric dihydroxylation of ethyl 1H-indene-2-carboxylate (92% ee). Bromination/azide displacement/azide reduction gave the desired cyclic β-amino acid (187). Azide opening of a chiral epoxide (itself prepared via a Sharpless epoxidation reaction), followed by treatment of the azide with triphenylphosphine, generated an aziridine which was opened with samarium iodide to form 3-amino-1,2,3,4-tetrahydro-2-naphthoic acid (1462). 1,2-Epoxy-cyclohexane-1-carboxylic acid was regioselectively opened with azide in the presence of catalytic Cu(NO₃)₂, which not only catalyzed the epoxide opening but also catalyzed the subsequent azide reduction step, giving a one-pot synthesis of 2-hydroxy-3-amino-cyclohexane-1-carboxylic acid (797).

Rearrangements of acids or esters to amines have been used in a number of synthetic routes. Dimethyl cis-cyclohex-4-ene-1,2-dicarboxylate was enzymatically resolved via hydrolysis of one of the ester groups with pig liver esterase. Curtius rearrangement of the free acid group produced cis-1-aminocyclopent-4-ene-2-carboxylic acid. The enantiomeric cis isomer was obtained by

tert-butylation of the free acid group, hydrolysis of the remaining methyl ester, and rearrangement of the free acid group. Refluxing either cis enantiomer with sodium methoxide epimerized the carboxyl center to provide the more stable trans isomer (1463). The Curtius rearrangement has also been applied to a synthesis of endo-(2S,3R)-2-amino-3-carboxy-norborn-5-ene, trapping the intermediate isocyanate with alcohols or the amino group of amino acids (1464). The chiral monoprotected diacid substrate was obtained by a desymmetrization of meso cis-5-norbornene-endo-2,3-dicarboxylic anhydride using methyl (S)-prolinate as a chiral reagent (1328, 1464). The same procedure was used to prepare (1S,2R)-2-aminocyclopropane-1-carboxylic acid (see Scheme 11.124) (1465). This amino acid is normally difficult to incorporate into peptides due to facile cyclopropyl ring opening, but by using the amino group of a peptide to trap the isocyanate intermediate a pseudopeptide with a urea peptide isostere could be prepared. Another synthesis of (1S,2R)-trans-2-aminocyclopropane-1-carboxylate also employed a Curtius rearrangement, with the chiral cyclopropane-1,2-dicarboxylate prepared via an asymmetric cyclopropanation reaction (1466).

In a similar manner, both (+)- and (−)-cis-2-amino-cyclobutane-1-carboxylic acids were prepared from a common chiral monomethyl cis-cyclobutane-1,2-dicar-boxylic acid via Curtius rearrangements (1333). Another monoprotected diacid prepared by enzymatic hydrolysis

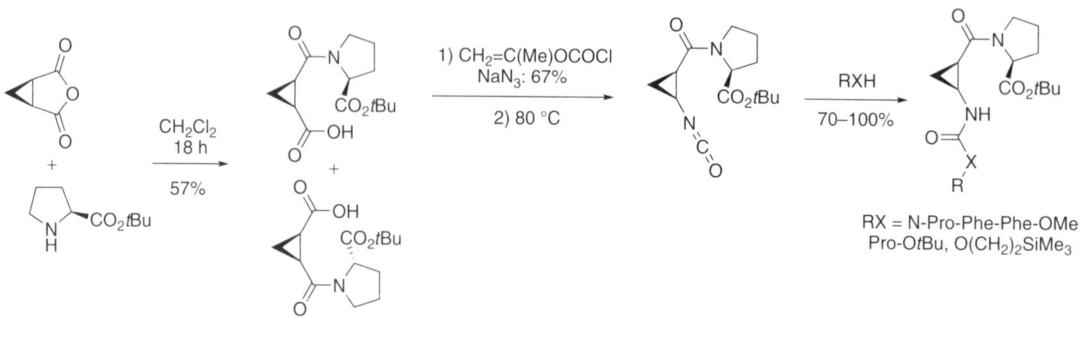

Scheme 11.124 Curtius Rearrangement of De-Symmetrized Anhydrides (1465).

of *trans*-(3R,4R)-bis(methoxycarbonyl)cyclopentanone led to *trans*-2-aminocyclopentane-1-carboxylic acid after rearrangement, with the ketone removed by thioketal reduction (*1342*). Enzymatic hydrolysis of a diester also led to enantiomerically enriched mono-methyl *cis*-cyclobutane-1,2-dicarboxylic acid. Curtius rearrangement gave (1R,2S)-2-aminocyclobutane-1-car-boxylic acid with 91% ee (*1467*). One diastereomer of the bicyclic β-amino acid 2-amino-3-carboxy-7-oxa-bicyclo[2.2.1]hept-5-ene was prepared by Curtius rear-rangement of monoacid/monoester substrate, with a second diastereomer obtained by epimerization of the α-carboxy center. The product β-amino acids were used as catalysts for Ugi 4CC condensations (*1468*). An asymmetric Diels–Alder reaction of a chiral fumarate ester with cyclopentadiene gave a monoester of 5-nor-bornene-*endo*-2-*exo*-3-dicarboxylic acid, with Curtius rearrangement used to prepare the β-amino ester. The alkene was reduced by hydrogenation (*1469*).

Reductive amination of β-keto esters produces β-amino acids in a manner similar to the preparation of α-amino acids from α-keto esters (see Section 5.3.9), generally using a chiral amine component. Several 1-amino-2-carboxycycloalkanes and 2-methyl-3-carboxy-tetrahydropyrroles were obtained with 67–85% de by sodium triacetoxyborohydride reduction of β-enamino esters containing a chiral auxiliary. The major diastere-omer could be isolated with >99% purity in 66–73% yield. The β-enamino ester substrates were formed by condensation of the corresponding β-keto esters with (R)-α-methylbenzylamine, or by acylation of lithium imines with chloroformates or carbonates (*907, 908*). An improvement to this procedure was reported in 2001, using benzyl ester derivatives that were deprotected during hydrogenation of the chiral auxiliary (*909*). Similar conditions were applied to a large-scale synthe-sis of ethyl *cis*-2-amino-1-cyclohexanecarboxylate from 2-oxo-cyclohexanecarboxylate, with sodium borohy-dride used for the reduction. The major diastereomer was isolated in 67% overall yield with >99% ee on a 71 g scale, with purification by crystallization (*1470*). Likewise, N-Boc 4-ketopiperidine-3-carboxylic acid was reductively aminated with α-methylbenzylamine, giving mainly *cis* diastereomers. Treament with Na/EtOH induced epimerization to predominantly *trans* isomers, with recrystallization and auxiliary cleavage providing *trans*-4-aminopiperidine-3-carboxylic acid with >99% de. The same strategy was applied to ethyl cyclohexanone-2-carboxylate, resulting in *trans*-ACHC (*1471*).

Either enantiomer of N-Fmoc *trans*-2-aminocyclo-pentane-1-carboxylic acid was prepared via formation of an imine of ethyl 2-carboxycyclopentanone with (R)- or (S)-α-methylbenzylamine, followed by reduction with NaBH₃CN, ester hydrolysis, reductive auxiliary removal, and Fmoc protection (*1472*). In a similar manner, the spin-labeled analog of *trans*-ACHC, *trans*-4-amino-1-oxyl-2,2,6,6-tetramethylpiperidine-3-carboxylic acid, was synthesized via reductive amina-tion with α-methylbenzylamine, with only the *cis*-(3S,4S)

and *trans*-(3R,4S) isomers formed as a 1:1 mixture (*1330, 1331*). In 2004 the enamine formed from 2-meth-oxycarbonylcyclohexanone and phenylglycine amide was hydrogenated over PtO₂ with 98% de. N-deprotection was accomplished by adding Pd(OH)₂/C after the enam-ine hydrogenation was complete (*912*). A synthesis of *trans*-3-aminopyrrolidine-4-carboxylic acid reductively aminated N-Boc ethyl 3-carboxy-4-oxopyrrolidine with (R)-(+)-α-methylbenzylamine/NaBH₃CN; diastereomer-ically pure product was obtained in 38% yield after one recrystallization (*1452*). Esters of pyrrolidine-2-acetic acid were also obtained by reduction of enamino esters (*1473, 1474*).

Chiral substrates have been employed for the reduc-tive amination. Optically active 2-benzylcyclopen-tanone-2-carboxylic acid, prepared via an asymmetric alkylation, was reductively aminated with benzylamine to produce *trans*-2-aminocyclopentane-1-phenyl-1-car-boxylic acid with 68% de (*1475*).

Amination via nitration was employed for the syn-thesis of a ferrocene-based planar analog of *cis*-pentacin from a ferrocene-carboxylate analog in which the car-boxy group was masked as a chiral (S)-4-(1-methylethyl) oxazoline. Nitration with N₂O₄ gave the (pS)-2-nitrofer-rocene derivative. By first blocking the 2-position with BuLi/TMS-Cl, nitration was directed to the 6-position, providing the (pR)-2-nitroferrocene derivative after silyl deprotection. The oxazoline of both diastereomers was hydrolyzed and the nitro group reduced to give both enantiomers of 2-aminoferrocenecarboxylic acid (*1476*).

11.3.3c Carboxylation Reactions

(S)- and (R)-β-proline, pyrrolidine-3-carboxylic acid, were obtained from N-Cbz (R)- or (S)-3-hydroxy-pyrrolidine via tosylation, displacement with KCN, and hydrolysis (*1344*). A synthesis of 2-aminocyclopentane-1-carboxylic acid proceeded via carboxylation of lithiated enamines with diethyl carbonate or benzylchloroformate, employing a chiral auxiliary on the amine to induce asymmetry (see Scheme 11.12 above). The resulting α,β-unsaturated β-amino ester was reduced and hydrog-enolyzed to give the β-amino acid (*534*).

Kinetic deprotonation of racemic O-protected N-benzylpiperidine-2-methanol by (−)-sparteine and sBuLi, followed by carboxylation with methyl chloro-formate, provided one diastereomer of piperidine-2-(hydroxyacetic acid) with >96% de and 86–94% ee (*1372*). Carboxyl groups were generated from ω-azabicyclo[3.n.1]alkan-3-ones via an asymmetric deprotonation of the ketone using a chiral base, trapping as a silyl enol ether, and ozonolysis to cleave the alkene and form the carboxyl moiety (*1477*).

An intramolecular amination of chiral allylic sulfox-amine derivatives gave a substrate that was carboxylated by deprotonation and acylation with ClCO₂R, with the sulfoxamine auxiliary removed with Raney Ni (see Scheme 11.50 above). Alternatively, the sulfoxamine was replaced with chloride and displaced with NaCN to introduce the carboxyl substituent (*961*).

11.3.3d Fusion of Aminomethyl and Acetate Components

There are relatively few examples of addition of an acetate equivalent to an electrophilic aminomethyl group for preparing cyclic β-amino acids. A titanium enolate of an acylated oxazolidinone was added to a nitrone derived from 1-pyrroline N-oxide to produce protected (2S,2′S)-2-(2′-propionic acid)-pyrrolidine with 96% de (*1033*). Optically active *threo*-methylphenidate (Ritalin, methyl 2-phenyl-2-(2′-piperidyl)acetate) was obtained by reaction of the titanium enolate of phenylacetic acid, amidated with Evans' oxazolidinone chiral auxiliary, with N-methoxycarbonyl 2-methoxy-piperidine (obtained by electrochemical oxidation of piperidine) (see Scheme 11.125). The C–C bond-forming reaction proceeded with 90–96% *threo* diastereoselectivity, producing the *threo* isomer with 82–99.6% ee, depending on the oxazolidinone auxiliary. Aryl-substituted phenylacetic acids were also reacted, as were N-methoxycarbonyl 2-methoxy-pyrrolidine and N-methoxycarbonyl 2-methoxy-hexamethyleneimine (*1478*). Cyclic β-keto esters were added to various imines in the presence of a chiral Pd catalyst, providing the 1-carboxy-1-(aminomethyl)cycloalkan-2-one adducts with anywhere from 0 to 80% diastereomeric ratio at the two chiral centers generated, and with 86–99% ee (*1057*).

In contrast, several different cycloaddition or cyclization reactions have been employed to form the C2–C3 bond of β-amino acids. A dipolar cycloaddition of cyclic nitrones with crotonic acid derivatives containing an amide chiral auxiliary was used to introduce a substituted acetate side chain to pyrrolidine and piperidine rings, although diastereoselectivity was poor (see Scheme 11.66 earlier for acyclic version) (*1060*). Nitrones formed from aldehydes and an N-hydroxy phenylglycinol auxiliary underwent 1,3-dipolar cycloaddition with alkenes to form 1,3-isoxazolidines, with bicyclic products obtained from aldehydes linked to alkenes (see Scheme 11.66). Hydrogenation removed the auxiliary and cleaved the isoxazolidine N–O bond, with hydroxymethyl oxidation producing 2-aminocyclopentane-1-carboxylic acid, 3-amino-4-carboxy-tetrahydrofuran, and 3-amino-4-carboxy-pyrrolidine (*1062*). 4-Aryl-pyrrolidine-3-carboxylates were prepared by a [3+2] cycloaddition of an azomethine ylide with chiral amide derivatives of substituted cinnamates. Using a nitrone for the cycloaddition gave 5-aryl-isoxazolidine-4-carboxylates (*1479*).

An asymmetric 1,3-dipolar cycloaddition using a chiral auxiliary on 5-hydroxy-pent-2Z-enoic acid gave 4-(2-hydroxyethyl)pyrrolidine-3-carboxylate, which was cyclized to produce 1-azabicyclo[2.2.1]heptane-3-carboxylate (*1480*).

Bicyclic β-amino acids (2-carboxymethyl-7-azabicyclo[2.2.1]heptane, 6-carboxymethyl-8-azabicyclo[3.2.1]octane, and 7-carboxymethyl-9-azabicyclo[4.2.1]nonane) were produced via [3+2]cycloaddition reactions between acrylate and N-alkyl 2,5-bis(trimethylsilyl) pyrrolidine, 2,5-bis(trimethylsilyl)piperidine, or 2,5-bis(trimethylsilyl)azepine. A sultam amide chiral auxiliary on the acrylate component induced asymmetry (*1481*). Diels–Alder reaction between methyl acrylate and 1-aminobutadiene, with the amine contained within a 4-phenyloxazolidin-2-one or 4-phenyloxazolidin-2-thione chiral auxiliary, produced 2-aminocyclohex-3-ene-1-carboxylic acid with 96% ee. Dimethyl fumarate and dimethyl butynedioic acid reacted with reduced enantioselectivity (*1482*).

β-Lactams containing a 2-acetic acid substituent (4-oxo-2-azetidineacetic acids) were prepared by a tributylstannane-mediated radical cyclization of N-vinyl α-bromo amides with a phenyl substituent on the vinyl group. Asymmetry was induced by using a chiral auxiliary on the amide nitrogen, with the acid group generated by ruthenium tetraoxide oxidation of the phenyl group (*1483*). A coupling reaction of 1-((1S)-phenethyl) pyrrolidine-2-thione with α-bromoesters using triethylamine and triphenylphosphine gave enamino esters in 60–70% yield as predominantly E-isomers. Hydrogenation of the alkene produced pyrrolidine-2-acetic acids with an alkyl substituent on the acetic acid side chain. The N-phenethyl auxiliary induced moderate to good stereoselectivity during the reduction (*1484*).

A regio-, diastereo-, and enantioselective C–H insertion reaction of methyl aryldiazoacetates into N-Boc-protected pyrrolidine, piperidine or 3,4-dehydropiperidine, catalyzed by a chiral Rh₂ complex, produced pyrrole- or piperidine-2-(2-arylacetic acids). An excess of diazoacetate created a symmetrical pyrrole-2,5-bis(2-arylacetic acid). Among the imino acids synthesized was *threo*-methylphenidate, from N-Boc piperidine and phenyldiazoacetate (*1485*). The subsequent paper in the same journal reported a similar synthesis of D-*threo*-methylphenidate using different Rh catalysts (*1486*). N-Boc pyrrolidine and phenyldiazoacetate produced pyrrolidine-2-(1-phenylacetic acid) (*1487*).

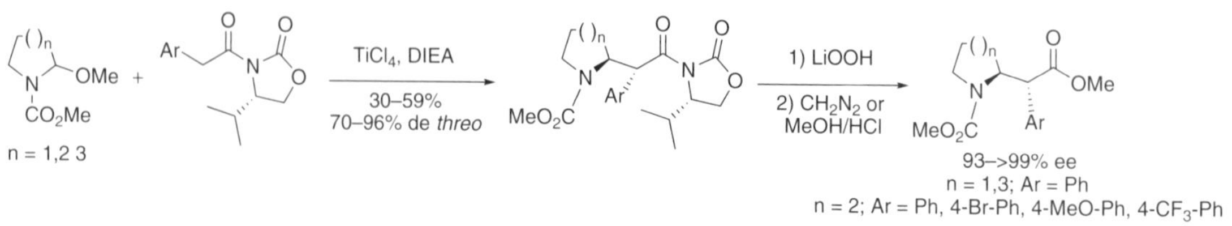

Scheme 11.125 Reaction of Enolates with α-Methoxy-Imines (*1478*).

11.3.3e Elaboration of α-Amino Acids

The acid groups of azetidine-, pyrrolidine-, and piperidine-2-carboxylates were homologated to give the corresponding 2-acetic acids under Arndt–Eistert reaction conditions (457). Both L-Pro (381, 1071, 1078, 1088–1090) and D-Pro (1088) were among those converted in this, and other, reports. Boc anhydride has been employed as an inexpensive coupling reagent for production of the diazomethane intermediate (86%) leading to homologated Boc-Pro (80% for the Ag-catalyzed rearrangement) (1071). The 6-(1-hydroxypropyl)-piperidine-2-acetic acid moiety of the palustrine alkaloids was prepared via an Arndt–Eistart homologation of a protected pipecolic acid (1302).

Ring-closing metatheses of α-amino-acid derived alkenes were employed to prepare five-, six- and seven-membered cyclic β-amino esters (see Scheme 11.126). For one synthesis, Met was homologated and the side chain eliminated to give a homologated vinylglycine. This was α-allylated with complete diastereoselectivity. The same α-allyl-β-vinyl-amino acid was obtained in better yield by α-allylation of homologated Met, followed by the elimination reaction. Ru-catalyzed metathesis produced 2-aminocyclopent-3-ene-1-carboxylic acid, with hydrogenation giving the saturated analog. The homologated Met substrate could also be α-methylated before the α-allylation reaction, leading to 1-methyl-2-aminocyclopentane-1-carboxylic acid. In a similar fashion, allylglycine or 2-aminohex-5-enoic acid were homologated, α-allylated, cyclized, and hydrogenated to give 2-aminocyclohexane-1-carboxylic acid and 2-aminocycloheptane-1-carboxylic acid (1105).

The α-carboxyl group of Cbz-Asp was reduced to a hydroxymethyl group and protected within an oxazolidinone along with the amino group. Deprotonation and alkylation adjacent to the side-chain carboxyl group with 1-iodo-3-chloropropane, followed by cyclization with the amino group, gave 2-hydroxymethyl-3-carboxy-piperidine (1148). Asp has also been converted into 2-amino-1,4-butanediol, which was mesylated and rearranged to give 3-mesyloxy-pyrrolidine (see Scheme 11.71 earlier). Mesylate displacement with cyanide followed by hydrolysis gave β-proline, pyrrolidine-3-carboxylic acid (1111, 1112). Trans-D- or -L-4-hydroxy-proline were converted to enantiomers of the same product, (R)- and (S)-pyrrolidine-3-carboxylic acid, via a similar intermediate. The prolines were decarboxylated, tosylated on the hydroxyl group, displaced with cyanide, and the nitrile hydrolyzed (457). A synthesis of aziridine-2-acetic acid began with β-alkylation of Boc-Asp(OBn)-OBn with benzyl bromide, methyl iodide or methyl bromoacetate. The α-carboxy group was reduced to a hydroxymethyl substituent, and then cyclized with the α-amino group via Mitsunobu reaction to form the aziridine group. The epimeric product was obtained by β-alkylation of an aspartaminol lactam (1488). Another synthesis reduced the α-carboxyl group of Cbz-Asp(OtBu)-OH via the isobutyl chloroformate/NaBH₄ method, and then converted the resulting amino alcohol into tert-butyl N-Cbz aziridine-2-acetate by a Mitsunobu reaction with PPh₃/DEAD (1168).

L-Pyroglutamate was converted into 3-trans-hydroxy-L-prolinol (which could be epimerized to 3-cis-hydroxy-L-prolinol). The primary hydroxyl group was converted to a nitrile via in situ chlorination/KCN

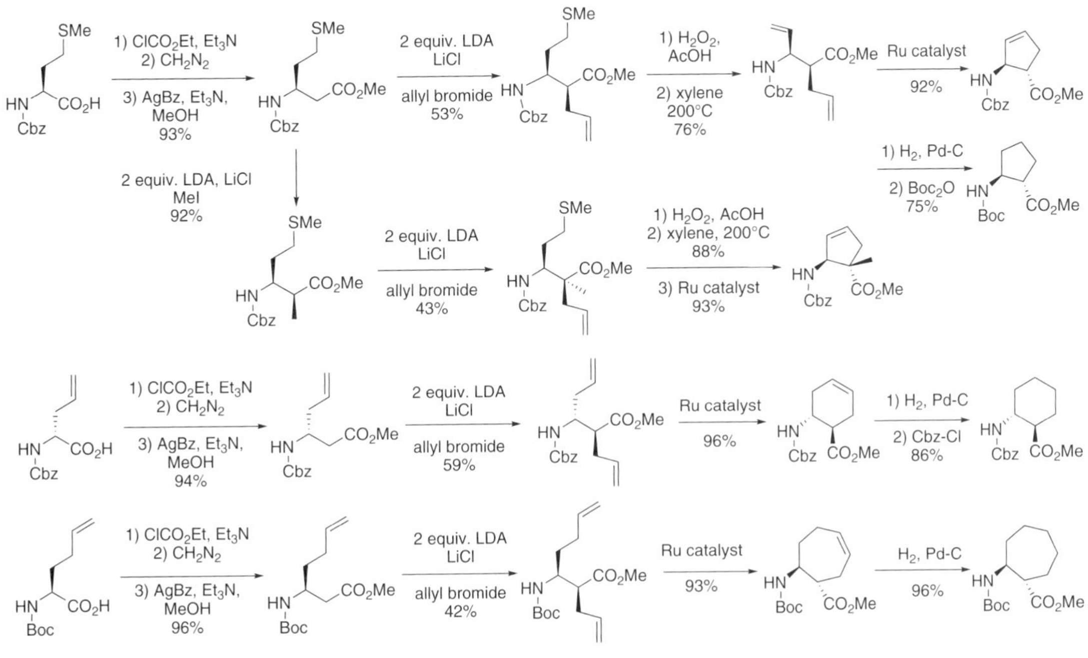

Scheme 11.126 Homologation, Alkylation and Ring-Closing Metathesis of α-Amino Acids (1105).

displacement, but this reaction was only successful for the *trans* diastereomer. Hydrolysis provided 3-hydroxy-pyrrolidine-2-acetic acid. Mitsunobu reaction inverted the C3 carbinol via formation of the lactone, producing the *cis*-hydroxy isomer (*1489*). Another synthesis deprotonated pyroglutaminol and carboxylated it with Cbz-Cl, with lactam carbonyl reduction providing pyrrolidine-2,3-dehydro-5-hydroxymethyl-3-carboxylate (*1397*). Mitsunobu reaction of a Glu-derived 2,5-dihydroxypentanoic acid with *N*-hydroxy-phthalimide gave the α-aminoxy acid analog of 5-hydroxy-Nva. Deprotection of the phthalimide group, reprotection of the amine with a urethane group, and cyclization with the 5-hydroxy group under Mitsunobu conditions provided 2-oxa-piperidine-3-carboxylic acid (*1490*). L-Serine was elaborated into 4-hydroxy-5-amino-piperidine-2-acetic acid via functionalization of both side-chain hydroxy and α-carboxy groups (*1491*).

Pipecolic acid has been used for an asymmetric synthesis of the four diastereomers of methyl piperidine-2-phenylacetate (Ritalin, methylphenidate), as well as 4′-aryl substituted derivatives. The Weinreb amide of Boc-D-Pip-OH or Boc-L-Pip-OH was treated with PhLi or 4-substituted PhLi to give aryl ketone intermediates. The ketone was methylenated by a Wittig reaction and stereoselectively hydroborated to generate a hydroxymethyl substituent. With (+)-IPC-BH$_2$ 100% de *threo* selectivity was achieved; up to 50% de of the *erythro* isomer was provided by (−)-IPC-BH$_2$, although the yields were only 40–55%. Improved overall yields of the diastereomers were obtained with BH$_3$·THF and BH$_3$·Me$_2$S, respectively, with a 64% yield of the *threo* isomer and 30% yield of the *erythro* product after separation. Oxidation of the hydroxymethyl group and esterification completed the synthesis (*1305*).

A 6-hydroxyethyl-substituted piperidine-2-carboxylate was converted into a substituted piperidine-2-acetic acid via reduction of the carboxyl group and oxidation of the hydroxyethyl group (*1492*). As discussed earlier, the ketone group of racemic ethyl *N*-Boc 4-oxopiperidine-3-carboxylate was reduced with Baker's yeast to give a single diastereomer of 4-hydroxypiperidine-3-carboxylate, (3*R*,4*S*)-*cis*-4-hydroxypiperidine-3-carboxylate, in 78% yield with 93% ee, as determined by both chiral shift reagent ¹H NMR analysis and conversion to a known chiral derivative (*1445, 1446*). However, another group was unable to duplicate the high enantiomeric excess, obtaining product with the same optical rotation but only 24% ee, as determined by chiral HPLC analysis of a derivative (*1350*).

Lithiated *N*-acyl 6,7-dimethoxytetrahydroisoquinoline-3-carboxylic acid underwent an unusual 1,2-migration rearrangement in the presence of methyl iodide to give enantiomerically pure 5,6-dimethoxy-2-methyl-*cis*-1-aminoindane-2-carboxylic acid (*1493*).

11.3.3f Elaboration of β-Amino Acids

Cyclic β-amino acids have been derived from acyclic β-amino acids via a number of cyclization reactions.

The DIOZ chiral auxiliary (4-isopropyl-5,5-diphenyloxazolidin-2-one) was employed during α-aminomethylations of 4-bromobutanoic acid and 5-bromopentanoic acid, giving the β²-bromoalkyl-β-amino acids (see Scheme 11.63 earlier). Heating induced cyclization to form pyrrolidine-3-carboxylic acid or piperidine-3-carboxylic acid (*832, 1032*).

The side chains of the β-lactams of 2-allyl-3-amino-5-phenylpent-4-enoic acid or 2-homoallyl-3-amino-5-phenylpent-4-enoic acid were cyclized by ring-closing metathesis to produce *trans*-1-amino-2-cyclopent-4-enoic acid or 1-amino-2-cyclohex-5-enoic acid, with alkene reduction producing the saturated cyclic β-amino acids (*1263*). *N*-Allyl 2-aminohept-5-enoate or 2-aminohex-4-enoate, obtained via addition of (*S*)-*N*-allyl-*N*-α-methylbenzylamide to methyl (*E,E*)-heptan-2,5-dienoate or *tert*-butyl (*E,E*)-hexa-2,4-dienoate, respectively, were converted into piperidine-2-acetic acid or pyrrolidine-2-acetic acid via ring-closing metathesis of the two alkene groups, and then hydrogenation (*748*). Alternatively, 3-amino-ω-chloroesters, obtained by Michael addition of *N*-aryl α-methylbenzylamines to α,β-unsaturated-ω-chloroesters followed by *N*-aryl deprotection, were cyclized to five-, six- and seven-membered imino acids (*749*). Evans' benzyl-substituted oxazolidinone auxiliary was acylated with *N*-alkenyl,*N*-Boc β-Ala, and then deprotonated with NaHMDS and allylated with allyl iodide (68–71%, 87–89% de). Ring-closing metathesis and hydrogenation resulted in seven- or eight-membered 1-azacycloheptane- or -cyclooctane-3-carboxylic acids (*641*).

3-Substituted-β-amino acids have been added to methyl acrylate to give *N*-carboxyethyl-substituted β-amino acids. Deprotonation with Na in EtOH induced an intramolecular Dieckmann condensation to give 3,4-dehydro-4-hydroxy-6-substituted-piperidine-3-carboxylates as the major regioisomer. Hydrogenation over Raney Ni produced the 3,4-*cis*, 3,6-*trans*-diastereomers (*1494*). 1-Amino-1-(4-methoxyphenyl)-2-carboxy-3-(3,4-methylenedioxy)-4-hydroxybutane was cyclized via bromination/displacement of the hydroxy group, generating a 2,4-diaryl-pyrrolidine-3-carboxylic acid (*904*). Prochiral dimethyl β-aminoglutarate was converted into a homochiral monoester by enzymatic ester hydrolysis, and then cyclized to give methyl azetidinone-4-acetic acid (*1495*). β-Enamino esters with a chiral auxiliary on the nitrogen underwent an aza-annulation reaction with acryloyl chloride to form two diastereomers of a bicyclic piperidone system (see Scheme 11.127). The amide carbonyl was reduced and the acetal cleaved upon treatment with BH$_3$·Me$_2$S. Auxiliary removal by hydrogenation completed the synthesis of 2,3-*cis*- or 2,3-*trans*-2-methylpiperidine-3-carboxylate (*1496*).

An unusual α,α-disubstituted-β-amino acid possessing axial chirality was prepared by bis-alkylation of the β-Ala equivalent ethyl cyanoacetate with (*R*)-2,2′-*bis*-(bromomethyl)-1,1′-binaphthyl, followed by regioselective reduction of the cyano group by sodium borohydride. The binaphthyl derivative could be isolated in enantiomerically pure form, in contrast to the biphenyl-based analog, which interconverted between enantiomers at room temperature (*1352*). Chiral analysis of the binaphthyl derivative was possible on a β-cyclodextrin HPLC column as the free amino acid, the ethyl ester, and the *N*-Boc derivative. The amino acid could also be derivatized with Marfey's reagent, FDAA, and separated on a normal C_{18} column (*1394*).

An unexpected side reaction occurred during attempted *N*-methylation of Fmoc-β³-hPhe using microwave heating of the amino acid with paraformaldehyde and *p*TSA in acetonitrile. Instead of the expected oxazinanone, *N*-Fmoc 1,2,3,4-tetrahydroisoquinoline-3-acetic acid was produced in 88% yield (*1497*).

Other derivatizations of β-amino acids start with a cyclic β-amino acid core. Non-chiral 2-acetylaminocycloalk-1-ene-1-carboxylic acid ethyl esters could be enantioselectively reduced to the corresponding *cis*-substituted cycloalkyl β-amino esters, as discussed in Section 11.3.3a (*1447*). The alkene of both *cis*- and *trans*-2-amino-4-cyclohexenecarboxylic acid was iodolactonized with KI/I_2 to form 4-iodo-5-hydroxy lactone products, with the iodo group reductively removed by tributyltin hydride. The 4-hydroxy products were obtained by a similar sequence, starting with *N*-iodosuccinimide treatment to induce formation of an oxazinine with the amine acyl carbonyl (*1498*). Pyrrolidine-2-acetic esters (homoprolines), prepared by reduction of pyrrolidine enamino esters, were deprotonated and alkylated α to the carboxy group. A mixture of four diastereomers was obtained with sodium methoxide in refluxing methanol as base, due to a reversible Michael reaction epimerizing the stereocenters. Alternatively, by employing LDA at −70 °C, single diastereomers were produced (*1474*). Similarly, homochiral piperidine-2-acetic acid derivatives were alkylated α to the carboxyl group, with a number of electrophiles in good yield (45–92%) and high stereoselectivity (90–100% de) (*1499*). *N*-Boc (*S*)-homoproline was esterified with allyl alcohol and then α-allylated via a Claisen enolate rearrangement to give a mixture of diastereomers of α-allylhomoproline, pyrrolidine-2-allylacetic acid. Direct allylation of the enolate of *N*-Boc homoproline methyl ester with allyl bromide gave the same product with a similar diastereoselectivity (*1500*).

β-Amino acids based on the tropane alkaloid nucleus have been synthesized from cocaine (*1422*) (see Scheme 11.119 earlier). A racemic synthon was also prepared, resolved using tartaric acid, and then elaborated into the same products. The chiral tropane-carboxylic acid nucleus is also contained in the dehydroamino acid anhydroecgonine; a cocaine analog was prepared via Michael addition of phenylmagnesium bromide (*1304*).

11.3.3g Small Ring Systems

β-Lactam synthesis via reaction of olefinic substrates with chlorosulfonyl isocyanate is a useful route to β-amino acid precursors (see Scheme 11.26 earlier). The alkene of (+)-3-carene, 3,7,7-trimethylbicyclo[4.1.0]hept-3-ene, was converted to the β-lactam (62%) and hydrolyzed (94%) to give (1*R*,3*R*,4*S*,6*S*)-4-amino-4,7,7-trimethylbicyclo[4.1.0]heptane-3-carboxylic acid (*1501*). Spiro-β-lactams were synthesized by a [2+2]-cycloaddition reaction between cyclic ketenes derived from Cbz-Pro-Cl or *O*-protected Cbz-4-hydroxy-Pro-Cl and *N*-Pmp imines derived from benzaldehyde, D-glyceraldehyde acetonide or Garner's aldehyde. Lactam hydrolysis produced α-(1-aminobenzyl)-4-hydroxy-Pro, α-(1-amino-2,3-dihydroxypropyl)-Pro, and α-(1,2-diamino-3-hydroxypropyl)-Pro (*1275*). The lactam precursors of a number of other analogs were reported earlier (*1276*).

11.3.3h Elaboration of Other Chiral Substrates

D-Glucosamine has been converted into cyclic sugar-based β-amino acids. Polymers of these amino acids form oligosaccharide analogs, carbopeptoids. Three different stereoisomers of oxetin, (2*R*,3*S*)-2-carboxy-3-amino-oxetane, were all derived from 2*O*,3*O*-isopropylidene-4*O*-benzyl-D-glucose, which was used as a masked chiral carboxyl group during the syntheses (*1300*). Other sugar β-amino acids were obtained from carbohydrate lactones by addition of lithium *tert*-butyl acetate to the lactone, followed by a Ritter amination using a nitrile under Lewis acid-catalyzed conditions (*1502*).

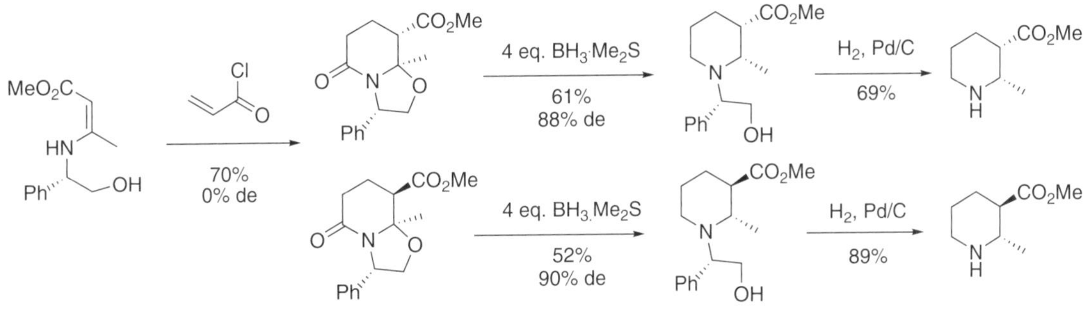

Scheme 11.127 Aza-annulation of Enamino Esters (*1496*).

The imines formed from benzylamine and homochiral 4,5-dimethyl-7-oxohept-2-enoic acids were treated with TMSCN to form 3,4-dimethyl-6-cyano-piperidine-2-acetic acid products. Reaction of the cyano piperidine with silver triflate formed an electrophilic iminium salt, with organozinc reagents adding to provide 6-alkyl-substituted piperidine-2-acetic acids (*1503*).

(*S*)-3-Ethoxycarbonyl-3-benzylpiperidine was derived from 2-ethoxycarbonyl-2-benzyl-5-aminopentanoic acid via lactam formation and amide reduction. The homochiral substrate was synthesized via Michael addition of diethyl benzylmalonate to acrylonitrile, followed by cyano reduction and enantioselective ester hydrolysis by porcine liver esterase (PLE) (*1504*).

11.3.3i Other Syntheses

Diels–Alder cycloaddition of *E*-2-cyanocinnamate and butadiene generated 1-cyano-2-phenyl-cyclohex-4-enyl-1-carboxylate, with a chiral ester auxiliary introducing diastereoselectivity. Reduction of the cyano/alkene groups with Raney Ni provided 1-(aminomethyl)-2-phenylcyclohexane-1-carboxylic acid (*1505*). An asymmetric Diels–Alder reaction was also employed to prepare functionalized *cis*- and *trans*-2-aminocyclohexanecarboxylic acid derivatives. The *cis* isomers were prepared with the carboxyl group masked as a hydroxymethyl substituent on the diene, component while the *trans* products employed a 3-furyl substituent on the dienophile (see Scheme 11.128). In both cases the amino group was derived from a nitro substituent on the dienophile, with asymmetry induced by a pyrrolidine substituent on the diene (*1506*). Syntheses of (−)-cispentacin, (1*R*,2*S*)-2-aminocyclopentane-1-carboxylic acid, and *cis*-(3*R*,4*R*)-4-aminopyrrolidine-3-carboxylic acid employed an intramolecular [2+3] cycloaddition between a chiral ketene dithioacetal dioxide and a nitrone, followed by deprotection (*1507*).

A synthesis of the 2-(4-methoxyphenyl)-4-(1,3-benzodioxo-5-yl)-pyrrolidine-3-carboxylic acid core of ABT-627 (Atrasentan) proceeded via a chiral oxazine, which was hydrogenated to induce a ring contraction to form the pyrrolidine ring (*1308*). A tricyclic β-amino acid, 3-carboxy-4-hydroxy-1,2,3,4-tetrahydroquinoline with a cyclopentyl ring fused at the 3,4-positions, was the unexpected product of the reaction between 2-carboethoxycyclopentanone and benzyl azide in the presence of TiCl₄ (*1508*).

Pyrroles that have been *N*-acylated with amino acids can be mono-cyclopropanated with methyl diazoacetate. Oxidative cleavage of the remaining alkene bond in the pyrrole ring leaves a 1-amino-2,3-dicarboxycyclopropane structure. This method allows for the *cis*-β-aminocyclopropanecarboxylic acid moiety to be incorporated into peptides, normally a difficult procedure due to rapid ring opening. No diastereoselectivity was induced by the amino acid *N*-acyl substituent in these cyclopropanations, but the diastereomers were easily resolved (*1379, 1381*).

11.4 γ-Amino Acids

11.4.1 Introduction

There are a number of biologically important γ-amino acids (see Scheme 11.129). γ-Aminobutyric acid (GABA) is one of the two major neurotransmitters that regulate neuronal activity in the brain (L-Glu is the other). GABA is an inhibitory neurotransmitter, while L-Glu is an excitatory neurotransmitter (*1509*). GABA is found in mammalian brain tissue, but is also present in yeast and almost all plants (*440*). Many amino acids undoubedly exist in natural sources in quantities too small to be detected; to demonstrate this possibility the nitrogenous fraction resulting from large-scale industrial processing of sugar beet was studied. γ-L-Glutamyl-GABA was among those isolated (*1510*). A survey of the amino acids contained in 18 red algae identified traces of GABA in *Scinaia furcellata* (*1511*). The metabolism of L-β-lysine to GABA by *Pseudomonas* B4 has been studied (*149*). Biological levels of GABA, Glu, and Ala can be monitored by derivatization with naphthalene-2,3-dicarboxaldehyde (NDA) followed by separation by mixed micellar electrokinetic chromatography, with fluorescence detection. Levels of GABA in plant tissues were then determined (*1512*). A sensitive method was employed to measure GABA concentrations in

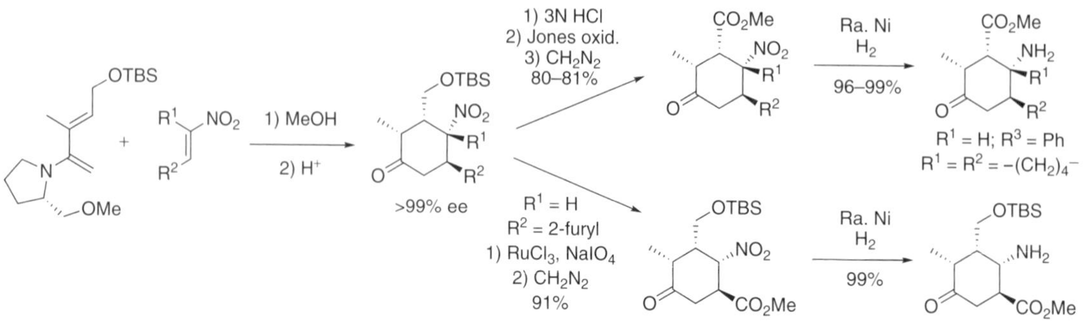

Scheme 11.128 Synthesis of 2-Aminocyclohexane-1-Carboxylic Acids (*1506*).

Table 11.4 Asymmetric Syntheses of Cyclic β-Amino Acids

Ring Size (n)	Substituents	Method	Yield, % ee	Reference
Amination				
5	1-NH$_2$-2c-CO$_2$H	addition of Li (α-methylbenzyl)benzylamide to α,β-unsaturated esters, KOtBu epimerization to $trans$ isomer	65–70%	1993 (716)
5	1-NH$_2$-2t-CO$_2$H		>97% de	1994 (1294)
6	1-NH$_2$-2c-CO$_2$H			
6	1-NH$_2$-2t-CO$_2$H			
5	1-NH$_2$-2c-CO$_2$H-4-oxa	addition of Li (α-methylbenzyl)benzylamide to α,β-unsaturated esters, KOtBu epimerization to $trans$ isomer	initial adduct	2004 (1449)
	1-NH$_2$-2t-CO$_2$H-4-oxa		>98:<2 to 69:31 cis: $trans$	
	1-NH$_2$-2c-CO$_2$H-4-thia		cis isolated in 46–79% yield with	
	1-NH$_2$-2t-CO$_2$H-4-thia		>98% de	
	1-NH$_2$-2c-CO$_2$H-4-aza		quant. epimerization	
	1-NH$_2$-2t-CO$_2$H-4-aza		>98% de	
			66–71% deprotection	
6	1-NH$_2$-2c-CO$_2$H	addition of Li (α-methylbenzyl)benzylamide to cyclohex1-ene-1-carboxylate	47% addition	1998 (1343)
			>96% de	
			86% hydrogenation	
			96% ee	
5	1-NH$_2$-2c-CO$_2$H-5-t-Me	addition of Li (α-methylbenzyl)benzylamide to racemic 3-Me-cyclopent-1-ene-1-carboxylate, KOtBu epimerization	95.5:1.7:2.8 initial diastereomers	2003 (1450)
	1-NH$_2$-2t-CO$_2$H-5-t-Me		isolate main isomer with >99% de	
			in 39% yield	
			100% epimerization	
			>99% de	
			64–69% deprotection	
6	1-aza-2-CH$_2$CO$_2$H-4,5-dehydro	addition of Li (α-methyl- benzyl)allylamide to hept-2,5-dienoic acid or hexa-2,4-dienoic acid, ring-closing metathesis, reduction	69–78% conjugate addition	2002 (748)
6	1-aza-2-CH$_2$CO$_2$H		>95% de	
5	1-aza-2-CH$_2$CO$_2$H-3,4-dehydro		49–77% metathesis	
5	1-aza-2-CH$_2$CO$_2$H		86–93% reduction	
5	1-aza-2-CH$_2$CO$_2$H	addition of Li (α-methyl- benzyl)arylamide to ω-Cl-α,β-unsaturated esters, N-aryl deprotection, cyclization, auxiliary removal	60–88% addition	2000 (749)
6	1-aza-2-CH$_2$CO$_2$H		>95% de	
7	1-aza-2-CH$_2$CO$_2$H		66–70% deprotection/cyclization	
			41% auxiliary removal	
5	1-NH$_2$-2c-CO$_2$H-3-t-CH$_2$CO$_2$H	tandem addition/cyclization via addition of Li (α-methylbenzyl)benzylamide to dialkyl (E,E)-octa-2,6-diendioic acid, deprotection; cis isomer via oxidative elimination of amine group, addition of Li (α-methylbenzyl)benzylamide	77–83% major isomer from addition/cyclization	2004 (1451)
	1-NH$_2$-2c-CO$_2$H-3-c-CH$_2$CO$_2$H		6–7% minor isomer	
	1-NH$_2$-2t-CO$_2$H-3-t-CH$_2$CO$_2$H		74–83% deprotection	
	1-NH$_2$-2t-CO$_2$H-3-c-CH$_2$CO$_2$H		65–85% oxidative elimination	
			62–68% addition	
			50–100% de	

(Continued)

Table 11.4 Asymmetric Syntheses of Cyclic β-Amino Acids (continued)

Ring Size (n)	Substituents	Method	Yield, % ee	Reference
5 5 6 6 7	1-NH$_2$-2-CO$_2$H 1-aza-2-CH$_2$CO$_2$H 1-NH$_2$-2-CO$_2$H 1-aza-2-CH$_2$CO$_2$H$_1$-NH$_2$-2-CO$_2$H 1-aza-2-CH$_2$CO$_2$H	addition of TMS-SAMP to ω-halo-α,β-unsaturated esters, intramolecular alkylation	38–60% addition and cyclization 69–71% auxiliary removal	1995 (*454*) 1997 (*453*)
5	1-NH$_2$-2-CO$_2$H-4,5-benzo	Michael addition of Li (α-methylbenzyl) benzylamide to *o*-bromomethylcinnamate, intramolecular quenching	80% addition/cyclization 86% de	1999 (*455*)
6	1-NH$_2$-2-CO$_2$H-2-Me-3,4-benzo-4,5-dehydro 1-NH$_2$-2-CO$_2$H-2-Me-3,4-dehydro-4,5-benzo	Michael addition of Li (α-methylbenzyl) benzylamide to naphthyl derivative	67–96% addition 97–98% de	1995 (*456*)
5	1-aza-3-NH$_2$-4-CO$_2$H	Michael addition of α-methylbenzylamine to *N*-Boc ethyl 3-carboxy-2,5-dihydropyrrole, hydrogenation	13% addition and chromatography 90% hydrogenation	2000 (*337*)
6	1-NH$_2$-2*t*-CO$_2$H-3*c*-CH$_2$CO$_2$H	tandem conjugate addition/cyclization of LiZn[Bn(TMS)N]$_3$ to RO$_2$CCH=CHCH$_2$CH$_2$CH=CHCO$_2$R, R = menthyl ester	40–87% 54–90% de	1992 (*457*)
5	1-aza-2-CO$_2$H-3-NH$_2$-4-CH$_2$OH-5-keto	addition of amine nucleophiles to homochiral α,β-unsaturated-α-carboxyl pyrrolidinone	90–99% 14–84% de	1997 (*1458*)
5	1-aza-2-CH$_2$CO$_2$H-3,4-dehydro	intramolecular Michael addition of chiral substrate, flash thermolysis	72–85% Michael addition 71–88% flash thermolysis	1996 (*1459*)
5	1-NH$_2$-2*t*-CO$_2$H	Mitsunobu phthalimide displacement of chiral alcohol, hydroxymethyl oxidation	70% displacement 92% oxidation	1996 (*1460*)
5	1-NH$_2$-2-CO$_2$H	Mitsunobu azide displacement of *cis*-2-hydroxycyclopentane-2-carboxylate	18% azide displacement and azide reduction	1999 (*1310*)
5	1-NH$_2$-2-CO$_2$H-5-OMe 1-NH$_2$-2-CO$_2$H-5-OPh 1-NH$_2$-2-CO$_2$H-5-CH$_2$NHBoc	azide opening of *cis*-epoxide derived from 3-hydroxymethylcyclopentene, azide reduction and cyclization to aziridine, aziridine opening with ROH or CN, hydroxymethyl oxidation	85% epoxidation 85% azide opening 84% aziridine formation 71–90% aziridine opening 73–89% oxidation	2002 (*1317*)
5	1-NH$_2$-2-CO$_2$H-3,4-[–CH$_2$CH(OH)CH$_2$–]	azide displacement of nosylate	95% displacement 82% azide reduction	2002 (*1345*)
5	1-aza-3-CO$_2$H	azide displacement of tosylate or benzylamine displacement of dimesylate of homochiral hydroxymethyllactone	90% azide displacement 85% BnNH$_2$ reaction	1997 (*1461*)
5	1-NH$_2$-2-CO$_2$H-2-OH-4,5-benzo	asymmetric dihydroxylation of ethyl 1*H*-indene-2-carboxylate, bromination and azide displacement	65% dihydroxylation 92% ee 16% bromination, azide displacement, reduction	1998 (*187*)

	Compound	Method	Results	Year	Ref.
6	1-NH₂-2-CO₂H-4,5-benzo	azide opening of chiral epoxide, aziridine formation, opening with SmI₂	67% azide opening and aziridine formation; 66% aziridine opening; 24% de cis	1999	(1462)
6	1-NH₂-2c-CO₂H-2t-OH	epoxide opening with azide in presence of Cu(NO₃)₂, azide reduction with same catalyst	80% azide opening and azide reduction	2003	(797)
6	1-NH₂-2c-CO₂H-4,5-dehydro; 1-NH₂-2t-CO₂H-4,5-dehydro	Curtius rearrangement of monoprotected diacid, acid epimerization	89–91% rearrangement; 72–76% epimerization	1984	(1463)
6	1-NH₂-2-CO₂H-3,6-(-(CH₂-)-4,5-dehydro	Curtius rearrangement of desymmetrized endo-norborn-5-ene-2,3-dicarboxylate	57–60% azide formation; 55–100% rearrangement	1997; 1998	(1464); (1328)
3	1-NH₂-2-CO₂H	Curtius rearrangement of monoprotected diacid	47–67% rearrangement	1997	(1465)
3	1-NH₂-2-CO₂H	Curtius rearrangement of monoprotected diacid	94% rearrangement	2003	(1466)
4	cis-1-NH₂-2-CO₂H	Curtius rearrangement of monoester of cis-1,2-dicarboxycyclobutane	82–92% Curtius	2005	(1333)
4	1-NH₂-2-CO₂H	Curtius rearrangement of monoprotected diacid	63% rearrangement	1998	(1467)
6	1-NH₂-2t-CO₂H-3c,6c-(-O-)-4,5-dehydro; 1-NH₂-2c-CO₂H-3c,6c-(-O-)-4,5-dehydro	Curtius rearrangement of monoprotected diacid	84% rearrangement; 90% epimerization	2005	(1468)
6	1-NH₂-2t-CO₂H-3t,6t-(-CH₂-)-4,5-dehydro; 1-NH₂-2t-carboxy-3t,6t-(-CH₂-)	Curtius rearrangement of homochiral o-norborn-5-ene-2,3-dicarboxylate, hydrogenation	—	1987	(1469)
5	1-NH₂-2-CO₂H	Curtius rearrangement of monoprotected diacid	73% rearrangement	1997	(1342)
5	1-NH₂-2-CO₂H	reductive amination of β-keto ester via β-enamino ester using chiral amine and NaBH(OAc)₃	67–81% de crude	1994	(908)
6	1-NH₂-2-CO₂H		66–73% yield with >99% de; >99% de; 74% auxiliary removal	1996	(907)
5	1-aza-2-methyl-3-CO₂H	reductive amination of β-keto ester via β-enamino benzyl ester using chiral amine and NaBH(OAc)₃	6–83% yield purified major isomer; 83–89% auxiliary removal	2001	(909)
5	1-NH₂-2-CO₂H				
6	1-aza-2-methyl-3-CO₂H	reductive amination of β-keto ester using chiral amine and NaBH₄	67% overall; >99% ee	1997	(1470)
6	cis-1-NH₂-2-CO₂H; trans-1-NH₂-2-CO₂H; trans-1-NH₂-2-CO₂H-4-aza	reductive amination of β-keto ester using chiral amine and NaBH₄, epimerization to trans with Na/EtOH	71–87% imine formation; 16–25% reduction and epimerization; 77–90% auxiliary removal	2003	(1471)
5	1-NH₂-2-CO₂H	reductive amination of β-keto ester using chiral amine and NaBH₃CN	40% reductive amination; 85% ester hydrolysis and auxiliary removal	2001	(1472)
6	cis and trans-1-NH₂-2-CO₂H-3,3,5,5-Me₄-4-aza-4-O	reductive amination of β-keto ester using chiral amine and NaBH₃CN	59% reductive amination; 2 of 4 isomers	2003; 2005	(1331); (1330)
5	1-aza-3-NH₂-4-CO₂H	reductive amination of β-keto ester using chiral amine and NaBH₃CN	38% reductive amination; 72% auxiliary removal and ester hydrolysis	2001	(1452)

(Continued)

Table 11.4 Asymmetric Syntheses of Cyclic β-Amino Acids (continued)

Ring Size (m)	Substituents	Method	Yield, % ee	Reference
5	1-NH$_2$-2-Ph-2-CO$_2$H	reductive amination of optically active β-keto ester with benzylamine and NaBH$_3$CN	98% yield 68% de	2003 (1475)
5	1-aza-2-CH$_2$CO$_2$Et 1-aza-2-CH(Me)CO$_2$Et 1-aza-2-CH(Et)CO$_2$Et 1-aza-2-CH(nPr)CO$_2$Et 1-aza-2-CH(nBu)CO$_2$Et	reduction of β-enamino ester with chiral amine auxiliary	80–98% 68–>95% de	1997 (1473) 1997 (1474)
6	1-NH$_2$-2-CO$_2$H	enamine formation from β-keto ester and Phg-NH$_2$, catalytic hydrogenation over PtO$_2$, auxiliary removal using Pd(OH)$_2$/C	97% hydrogenation 98% de	2004 (912)
5	1-NH$_2$-2-CO$_2$H-ferrocene	nitration of ferrocene-1-carboxylate with chiral oxazoline carboxyl protection	74–86% nitration 34–64% oxazoline hydrolysis 99% nitro reduction	1998 (1476)

Carboxylation

Ring Size (m)	Substituents	Method	Yield, % ee	Reference
5	1-aza-3-CO$_2$H	tosylation of N-Cbz (R)- or (S)-3-hydroxypyrrolidine, displacement with KCN, hydrolysis	70–80% tosylation 70–72% CN displacement 83–85% hydrolysis	2003 (1344)
5	1-NH$_2$-2-CO$_2$H	carboxylation of lithium enamine with chiral to give β-enamino ester, reduction	33–76%	1995 (534)
6	1-aza-2-CH(OH)CO$_2$H	kinetic deprotonation of protected piperidine-2-methanol with (−)-sparteine/sBuLi, carboxylation with ClCO$_2$Me	38–41% 92–>96% de 86–94% ee	1999 (1372)
5	1-aza-2-CH$_2$CO$_2$H-5-CH$_2$OH	enantioselective deprotonation of azabicyclo[3.n.1]alkan-3-one, trap as silyl enol ether, ozonolysis	53–56% deprotonation and ozonolysis	1997 (1477)
6	1-aza-2-CH$_2$CO$_2$H-6-CH$_2$OH			
7	1-aza-2-CH$_2$CO$_2$H-7-CH$_2$OH			
6	1-aza-2-CH$_2$CO$_2$H-6-methyl			
5	1-NH$_2$-1-CH$_2$CO$_2$H-2-CH(OH)iPr	intramolecular amine addition to chiral allylic sulfoxamine, deprotonation and carboxylation or conversion to chloride and NaCN displacement	80–94% intramolecular amination 48–74% carboxylation and auxiliary removal	2003 (961)
6	1-NH$_2$-1-CH$_2$CO$_2$H-2-CH(OH)iPr			

Fusion of Aminomethyl and Acetate Components

Ring Size (m)	Substituents	Method	Yield, % ee	Reference
5	1-aza-2-CH(Me)CO$_2$H	addition of Ti enolate of acylated oxazolidinone auxiliary to nitrone, auxiliary removal and N–O cleavage	84% addition 96% de	1999 (1033)
5	1-aza-2-CH(Ph)CO$_2$H	addition of Ti enolate of arylacetic acids with oxazolidinone auxiliary to 2-methoxy-pyrrolidine, -piperidine, or -hexamethyleneimine	30–59% addition 70–96% de $threo$ 93–>99% ee	1999 (1478)
6	1-aza-2-CH(Ph)CO$_2$H			
6	1-aza-2-CH(4-MeO-Ph)CO$_2$H			
6	1-aza-2-CH(4-Br-Ph)CO$_2$H			
6	1-aza-2-CH(4-CF$_3$-Ph)CO$_2$H			

5 6	1-CH(Ar)NH₂-1-CO₂H-2-=O 1-CH(Ar)NH₂-1-CO₂H-2-=O Ar = Ph, 4-Me-Ph, 2-furyl	addition of β-keto ester to imine in presence of chiral Pd catalyst	61–99% yield 0–>90% de 86–99% ee	2005 (1057)
5 6	1-aza-2-CH(CHMeOH)CO₂H 1-aza-2-CH(CHMeOH)CO₂H	cycloaddition of nitrone and crotonic acid with chiral amide auxiliary	86–99% cycloaddition 2–8% de	1993 (1060)
5	1-NH₂-2-CO₂H 1-NH₂-2-CO₂H-4-oxa 1-NH₂-2-CO₂H-4-aza	intramolecular 1,3-dipolar cycloaddition of nitrone formed from phenylglycinol and alkenes linked to aldehydes, hydrogenation to cleave auxiliary and isoxazolidine N–O bond, hydroxymethyl oxidation	89–95 cycloaddition 88–92% de 52–63% hydrogenation and oxidation	2003 (1062)
5	1-aza-3-CO₂H-4-(3-CN-Ph) 1-aza-3-CO₂H-4-[3-[C(=NH)NH₂]-Ph] 1-aza-3-CO₂H-4-(3-CN-Ph)-5-oxa 1-aza-3-CO₂H-4-[3-[C(=NH)NH₂]-Ph]-5-oxa	cycloaddition of azomethine ylide or nitrone with cinnamic acid with chiral amide auxiliary	75–78% cycloaddition	1999 (1479)
5	1-aza-3-CO₂H-4-(CH₂)₂OH 1-aza-3-CO₂H-4-(CH₂)₂-(1-N)	cycloaddition of N-MeOCH₂,N-TMSCH₂ benzylamine with 5-hydroxypent-2Z-enoic acid with chiral amide auxiliary, cyclization of hydroxyethyl substituent	73% cycloaddition 85% cyclization	2001 (1480)
5	1-aza-3-CO₂H-2,5-[-(CH₂)₂-] 1-aza-3-CO₂H-2,5-[-(CH₂)₃-] 1-aza-3-CO₂H-2,5-[-(CH₂)₄-]	[3+2]cycloaddition between acrylate with chiral sultam amide auxiliary, and N-alkyl 2,5-bis(TMS)-pyrrolidine, 2,6-bis(TMS)-piperidine, or 2,7-bis(TMS)-azepine	58–68% cycloaddition 80:20 to 98:2 exo:endo	2002 (1481)
6	1-NH₂-2-CO₂H-5,6-didehydro 1-NH₂-2-CO₂H-2,3,5,6-tetradehydro 1-NH₂-2,3-(CO₂H)₂-5,6-didehydro 1-NH₂-2-CO₂H-3-PO₃H₂-5,6-didehydro	Diels-Alder cycloaddition of 1-aminobutadiene (with amino group contained within 4-Ph-oxazolidin-2-one or 4-Ph-oxazolidin-2-thione) and methyl acrylate or α-phosphonoacrylate or dimethyl fumarate	83–100% conversion 51–>99% de	2003 (1482)
4	1-aza-2-CH₂CO₂H-3-Et-4-keto 1-aza-2-CH₂CO₂H-3-CH(OH)Me-4-keto	radical cyclization of N-styryl α-bromoamide, oxidation of β-phenyl group	—	1996 (1483)
5	1-aza-2-CH(Me)CO₂H 1-aza-2-CH(Et)CO₂H 1-aza-2-CH(nPr)CO₂H 1-aza-2-CH(nBu)CO₂H 1-aza-2-CH(Ph)CO₂H	enamino ester formation from pyrrolidine-2-thione and α-bromoesters, hydrogenation	60–70% enamino ester 8–94% de hydrogenation	1999 (1484)
5 5 6 6	1-aza-2-CH(Ar)CO₂H 1-aza-2,5-[CH(Ar)CO₂H]₂ 1-aza-2-CH(Ar)CO₂H 1-aza-2-CH(Ar)CO₂H-3,4-dehydro Ar = Ph, 4-Cl-Ph, 4-Me-Ph, 4-MeO-Ph, 2-naphthyl	chiral Rh-complex-catalyzed C–H insertion of methyl aryldiazoacetates into Boc-pyrrolidine, Boc-piperidine, or Boc-3,4-dehydropiperidine	40–86% 55–94% ee 42–94% de	1999 (1485)

(Continued)

Table 11.4 Asymmetric Syntheses of Cyclic β-Amino Acids (continued)

Ring Size (n)	Substituents	Method	Yield, % ee	Reference
6	1-aza-2-CH(Ph)CO$_2$H	chiral Rh-complex-catalyzed C–H insertion of methyl phenyldiazoacetate into Boc-piperidine	69% ee 94% de *threo*	1999 (*1486*)
5	1-aza-2-CH(Ph)CO$_2$H	chiral Rh-complex-catalyzed C–H insertion of methyl phenyldiazoacetate into Boc-pyrrolidine	70% yield 88% ee 90% de *threo*	2003 (*1487*)

Elaboration of α-Amino Acids

Ring Size (n)	Substituents	Method	Yield, % ee	Reference
5	1-aza-2-CH$_2$CO$_2$H	Arndt–Eistert homologation of Cbz-L-Pro	91% diazoketone 95% rearrangement	1975 (*1089*)
5	1-aza-2-CH$_2$CO$_2$H	Arndt–Eistert homologation of Boc-L-Pro	33% overall	1975 (*381*)
5	1-aza-2-CH$_2$CO$_2$H	Arndt–Eistert homologation of Cbz-L-Pro, Cbz-D-Pro	53–65% diazoketone 98% rearrangement	1976 (*1088*)
4 5 6	1-aza-2-CH$_2$CO$_2$H 1-aza-2-CH$_2$CO$_2$H 1-aza-2-CH$_2$CO$_2$H	Arndt–Eistert homologation of 2-carboxy-azetidine, -pyrrolidine, and -piperidine	—	1997 (*457*)
5	1-aza-2-CH$_2$CO$_2$H	Arndt–Eistert homologation of Fmoc-L-Pro	82–100% diazoketone 83–100% rearrangement	1998 (*1078*)
5	1-aza-2-CH$_2$CO$_2$H	Arndt–Eistert homologation of Boc-L-Pro	56–77% diazoketone 61–76% rearrangement	1999 (*1090*)
5	1-aza-2-CH$_2$CO$_2$H	Arndt–Eistert homologation of Boc-L-Pro using Boc$_2$O for activation and coupling with diazomethane	86% diazoketone 80% rearrangement	2002 (*1071*)
5 5 5 6 6 6 6 6 7 7	1-NH$_2$-2-CO$_2$H-4,5-dehydro 1-NH$_2$-2-Me-2-CO$_2$H-4,5-dehydro 1-NH$_2$-2-CO$_2$H 1-NH$_2$-2-CO$_2$H-4,5-dehydro 1-NH$_2$-2-CO$_2$H-4,5-dehydro 1-NH$_2$-2-Me-2-CO$_2$H-4,5-dehydro 1-NH$_2$-2-Et-2-CO$_2$H-4,5-dehydro 1-NH$_2$-2-CO$_2$H 1-NH$_2$-2-CO$_2$H-4,5-dehydro 1-NH$_2$-2-CO$_2$H	Arndt–Eistert homologation of Cbz-L-Met, Cbz-D-allylGly or Boc-L-2-aminohex-5-enoic acid, alkylation or sequential dialkylation of homochiral di-deprotonated *N*-Cbz homologated Met, vinylglycine, allylglycine or 2-aminohex-6-enoic acid; Ru-catalyzed metathesis, hydrogenation	93–96% homologation 42–86% alkylation 91–96% metathesis 75–96% hydrogenation	2004 (*1105*)
6	1-aza-2-CH$_2$CO$_2$H-6-CH(OH)Et	Arndt–Eistert homologation of protected 6-substituted pipecolic acid derivative	61% homologation	2004 (*1302*)
6	1-aza-2-CH$_2$OH-3-CO$_2$H	reduction of Asp carboxyl, alkylation of oxazolidinone derivative with I(CH$_2$)$_3$Cl, cyclization	80% alkylation 94% de 73% cyclization	1996 (*1148*)
5	1-aza-3-CO$_2$H	reduction of Asp carboxyl, mesylation, rearrangement, CN displacement, hydrolysis	60% for mesylation/rearrangement 61% displacement	1995 (*1111*)
5	1-aza-3-CO$_2$H	reduction of Asp carboxyl, mesylation, rearrangement, CN displacement, hydrolysis	99% for mesylation/rearrangement 93% displacement	1998 (*1112*)

n	Amino acid	Method	Yield	Year (Ref.)
5	1-aza-3-CO₂H	decarboxylation of 4-hydroxy-Pro, tosylation, CN displacement, hydrolysis	70% decarboxylation 72% CN displacement 65% hydrolysis	1995 (457)
3	1-aza-2-CH₂CO₂H	reduction of α-carboxy group, of Cbz-Asp(OtBu)-OH, Mitsunobu cyclization of alcohol	60–90% overall	2004 (1168)
3	1-aza-2-CH(R)CO₂H R = Me, Bn, CH₂CO₂H	β-alkylation of Asp, reduction of α-carboxy group, Mitsunobu cyclization of alcohol	38–73% alkylation 69–75% reduction 76–83% cyclization	2001 (1488)
5	1-aza-2-CH₂CO₂H-3-OH	conversion of pyroglutamate to 3-HO-prolinol, chlorination/CN displacement, hydrolysis	76% nitrile displacement and hydrolysis	1988 (1489)
5	1-aza-2,3-dehydro-3-CO₂H-5-CH₂OH	deprotonation and acylation of pyroglutaminol with Cbz-Cl, deprotection, lactam reduction with NaBH₄	89% carboxylation 84% reduction	2002 (1397)
6	1-aza-2-oxa-3-CO₂H	conversion of α-amino of Glu to α-hydroxy, reduction of γ-carboxy, Mitsunobu reaction with PhthNOH, deprotection, intramolecular	93% Mitsunobu 93% deprotection/reprotection 63% cyclization	2000 (1490)
6	1-aza-2-CH(Ph)CO₂H 1-aza-2-CH(4-Br-Ph)CO₂H 1-aza-2-CH(4-MeO-Ph)CO₂H	arylLi addition to Weinreb amide of Boc-pipecolic acid, Wittig methylenation of ketone, stereoselective hydroboration, hydroxymethyl oxidation	28–56% addition 9–98% methylenation 30–68% hydroboration *threo* or *erythro* 58–73% oxidation	1998 (1305)
6	*cis*-1-aza-3-CO₂H-4-OH	Baker's yeast reduction of ketone	78% 100% de 93% ee	1993 (1445) 1998 (1446)
6	*cis*-1-aza-3-CO₂H-4-OH	Baker's yeast reduction of ketone	100% de 24% de	1997 (1350)
6	1-aza-2-CH₂CO₂H-4-OH-5-NH₂	elaboration of L-Ser	—	2005 (1491)
6	1-aza-2-CH₂CO₂H-6-Me 2-CH₂CO₂H-5-OH-6-Me	reduction of carboxyl of 2-carboxy-3-acetoxy-6-hydroxyethylpiperidine, oxidation of hydroxyethyl	56% carboxyl reduction 66% hydroxyethyl oxidation	1997 (1492)
5	*cis*- 1-amino-a2-CO₂H-2-Me-4,5-(3,4-MeO2-Ph)	treatment of 6,7-MeO₂-tetrahydroisoquinoline-2-carboxylate with tBuLi/MeI	35%	1993 (1493)

Elaboration of β-Amino Acids

n	Amino acid	Method	Yield	Year (Ref.)
5	1-aza-3-CO₂H	aminomethylation of 4-bromobutanoic acid or 5-bromopentanoic acid with DIOZ oxazolidinone auxiliary, cyclization, deprotection	79–80% aminomethylation 78–86% de	2003 (832)
6	1-aza-3-CO₂H		59–80% cyclization 85% auxiliary removal	2005 (1032)

(Continued)

Table 11.4 Asymmetric Syntheses of Cyclic β-Amino Acids (continued)

Ring Size (n)	Substituents	Method	Yield, % ee	Reference
5	*trans*-1-amino-2-carboxy-4,5-dehydro	ring-closing metathesis of 2-allyl or 2-homoallyl-3-amino-5-phenylpent-4-enoic acid, hydrogenation	46–56% metathesis; 68–79% hydrogenation	2003 (*1263*)
6	*trans*-1-amino-2-carboxy-5,6-dehydro			
5	*trans*-1-amino-2-carboxy			
6	*trans*-1-amino-2-carboxy			
6	1-aza-2-CH_2CO_2H-4,5-dehydro	addition of Li (α-methyl benzyl)allylamide to hept-2,5-dienoic acid or hexa-2,4-dienoic acid, ring-closing metathesis, reduction	69–78% conjugate addition; >95% de; 49–77% metathesis; 86–93% reduction	2002 (*748*)
6	1-aza-2-CH_2CO_2H			
5	1-aza-2-CH_2CO_2H-3,4-dehydro			
5	1-aza-2-CH_2CO_2H			
5	1-aza-2-CH_2CO_2H	addition of Li (α-methyl benzyl)arylamide to ω-Cl-α,β-unsaturated esters, N-aryl deprotection, cyclization, auxiliary removal	60–88% addition; >95% de; 66–70% deprotection / cyclization; 41% auxiliary removal	2000 (*749*)
6	1-aza-2-CH_2CO_2H			
7	1-aza-2-CH_2CO_2H			
7	1-aza-3-CO_2H-5,6-dehydro	alkylation of N-Boc,N-alkenyl-β-Ala with chiral oxazolidinone auxiliary, ring-closing metathesis, hydrogenation	68–71% alkylation; 87–89% de; 96–97% metathesis; 87–93% auxiliary removal; 100% reduction	2005 (*641*)
7	1-aza-3-CO_2H			
8	1-aza-3-CO_2H-5,6-dehydro			
8	1-aza-3-CO_2H			
6	1-aza-3-CO_2Me-3,4-dehydro-4-HO-6-R; 1-aza-3-CO_2Me-4-HO-6-R; R = nPr, $nC_{15}H_{31}$, Bn, $(CH_2)_6$OBn	N-alkylation of 3-substituted-β-amino acids via addition to methyl acrylate, intramolecular Dieckmann condensation, hydrogenation of alkene	63–71% N-alkylation; 44–59% major isomer from cyclization and hydrogenation	2000 (*1494*)
4	1-aza-2-CH_2CO_2H-4-oxo	enzymatic hydrolysis of dimethyl β-aminoglutarate, cyclization	94% hydrolysis; 84% cyclization	1981 (*1495*)
6	2,3-*cis*-1-aza-2-Me-3-CO_2H; 2,3-*trans*-1-aza-2-Me-3-CO_2H	aza-annulation of enamino esters with chiral amino auxiliary with acryloyl chloride, BH_3 reduction of resulting piperidone, hydrogenation	70% condensation; 0% de; 52–61% borane reduction; 88–90% de; 69–89% hydrogenation	2002 (*1496*)
6	2-CH_2CO_2H-4,5-fused phenyl	treatment of Fmoc β³-hPhe with paraformaldehyde, PTSA, microwave heating	88%	2006 (*1497*)
5	*cis*-1-amino-2-CO_2H	asymmetric hydrogenation of 1,2-dehydro analog using chiral Ru catalyst, epimerization of *cis* product	100%; 44–99% ee; 80% epimerization	2003 (*1447*)
5	*trans*-1-amino-2-CO_2H			
5	*cis*-4-aza-1-amino-2-CO_2H			
6	*cis*-1-amino-2-CO_2H			
7	*cis*-1-amino-2-CO_2H			
8	*cis*-1-amino-2-CO_2H			

6	cis-1-amino-2-CO₂H-cis-4-hydroxy-trans-5-iodo cis-1-amino-2-CO₂H-cis-4-hydroxy cis-1-amino-2-CO₂H-trans-4-iodo-cis-5-hydroxy cis-1-amino-2-CO₂H-cis-5-hydroxy trans-1-amino-2-CO₂H-cis-4-hydroxy-trans-5-iodo trans-1-amino-2-CO₂H-cis-4-hydroxy trans-1-amino-2-CO₂H-trans-4-iodo-cis-5-hydroxy trans-1-amino-2-CO₂H-cis-5-hydroxy	treatment of *cis*- or *trans*-2-amino-4-cyclohexenecarboxylic acid with NIS or KI/I₂, dehydrohalogenation with Bu₃SnH	80–83% iodo-oxazine formation 78–84% iodolactonization 60–75% dehalogenation	2005 (*1498*)
5	1-aza-2-CH₂CO₂Et	asymmetric hydrogenation of 1,2-dehydro analog using chiral Ru catalyst	100% yield 96% ee	2004 (*1448*)
5	1-aza-2-CH(Me)CO₂Et 1-aza-2-CH(Et)CO₂Et 1-aza-2-CH(*n*Pr)CO₂Et 1-aza-2-CH(*n*Bu)CO₂Et 1-aza-2-CH(*n*Hex)CO₂Et 1-aza-2-CH(Bn)CO₂Et 1-aza-2-CH(CH₂CH=CH₂)CO₂Et 1-aza-2-CH(CH₂CH=CHMe)-CO₂Et	deprotonation, alkylation	9–84% 100% de	1997 (*1474*)
6	1-aza-2-CH(Me)CO₂Et 1-aza-2-CH(Et)CO₂Et 1-aza-2-CH(*n*Pr)CO₂Et 1-aza-2-CH(*n*Bu)CO₂Et 1-aza-2-CH(*n*Hex)CO₂Et 1-aza-2-CH(Bn)CO₂Et 1-aza-2-CH(CH₂CH=CH₂)CO₂Et 1-aza-2-CH(CH₂CH=CHMe)-CO₂Et	deprotonation, alkylation	45–92% 90–100% de	1999 (*1499*)
5	1-aza-2-CH(CH₂CH=CH₂)CO₂H	allylation of homoproline via enolate formation and allylation or Claisen rearrangement of allyl ester	78% Claisen rearrangement 84% direct allylation	1991 (*1500*)
7	1-CH₂NH₂-1-CO₂H-3,4-naphtho-5,6-naphtho	dialkylation of ethyl cyanoacetate with 2,2′-(CH₂Br)₂-1,1′-binaphthyl, regioselective CN reduction	79% alkylation 59% reduction	1998 (*1352*)
5	1-aza-2-(4-MeO-Ph)-3-CO₂H-4-[3,4-(OCH₂O)-Ph]	cyclization of 1-amino-2-carboxy-4-hydroxy-butane derivative	55% cyclization	1999 (*904*)
6	1-aza-3-CO₂H-2,6-(–CH₂CH₂–) 1-aza-3-CO₂H-2,6-(–CH₂CH₂–)-2,3-dehydro	elaboration of cocaine or via resolved synthon	—	1997 (*1422*)

(Continued)

Table 11.4 Asymmetric Syntheses of Cyclic β-Amino Acids (continued)

Ring Size (n)	Substituents	Method	Yield, % ee	Reference
6	1-aza-3-CO$_2$H-4-Ar-2,6-(−CH$_2$CH$_2$−) Ar = Ph, 4-F-Ph, 4-MeO-Ph, 3-MeO-Ph	PhMgBr addition to (−)-anhydroecgonine	16–92%	1973 (*1304*)

Elaboration of Chiral Substrates

4	1-NH$_2$-2-CO$_2$H-3-oxa	elaboration of D-glucose	—	1986 (*1300*)
6	1-NH$_2$-1-Me-2-CO$_2$H-4,5-[−C(Me)$_2$−]	cycloaddition of alkene of (+)-3-carene with chlorosulfonyl isocyanate, lactam hydrolysis	62% cycloaddition 94% lactam hydrolysis	2003 (*1501*)
6	1-oxa-2-NH$_2$-2-CH$_2$CO$_2$H-3,4,5-(OH)$_3$-6-CH$_2$OH	addition of Li acetate to lactone, Ritter amination using nitriles	84% acetate addition 60–70% amination	2001 (*1502*)
6	1-aza-2-CH$_2$CO$_2$H-3,4-Me$_2$-6-CN 1-aza-2-CH$_2$CO$_2$H-3,4-Me$_2$-6-CH$_2$CH=CH$_2$ 1-aza-2-CH$_2$CO$_2$H-3,4-Me$_2$-6-CH$_2$CCH	TMSCN addition/cyclization of BnNH$_2$ imine of 3,4-dimethyl-7-oxo-hept-2-enoate, iminium ion formation, addition of RZnBr	75–81% cyclization 69–78% RZnBr addition	1999 (*1503*)
6	1-aza-3-CO$_2$H-3-Bn	addition of diethyl benzylmalonate to acrylonitrile, reduction, enzymatic hydrolysis, lactam formation, reduction	62% addition and reduction 86% hydrolysis, 99% ee 70% lactam formation 78% reduction	1998 (*1504*)

Small Ring Systems

5	1-aza-2-CO$_2$H-2-CH(R)NH$_2$, R = CH$_2$OH, CH(OH)CH$_2$OH, CH(NH$_2$)CH$_2$OH 1-aza-2-CO$_2$H-2-CH(Phh)NH$_2$-4-OH	Staudinger cycloaddition of ketene derived from Pro or 4-hydroxy-Pro with imines derived from benzaldehyde, D-glyceraldehyde or the Garner aldehyde, hydrolysis of resulting spiro-β-lactam	40–70% cycloaddition 62–90% hydrolysis	2004 (*1275*)

Other Syntheses

6	1-(CH$_2$NH$_2$)-2-CO$_2$H-3-Ph	Diels–Alder cycloaddition of 2-cyanocinnamate and butadiene	81% cycloaddition 100% de 92% reduction	1995 (*1505*)
6	1-NH$_2$-2-CO$_2$H-4-keto-5-Me-6-CH$_2$OH 1-NH$_2$-2-CO$_2$H-3-Me-4-keto-6-Ph 1-NH$_2$-2-CO$_2$H-3-Me-4-keto-6,1-[−(CH$_2$)$_4$−]	Diels–Alder cycloaddition of nitro alkene with diene containing chiral auxiliary, carboxyl masked as hydroxymethyl of furyl	80–91% carboxyl generation 96–99% nitro reduction	1998 (*1506*)
5	1-aza-2-(4-MeO-Ph)-3-CO$_2$H-4-(1,3-benzodioxol-5-yl)	contraction of substituted oxazine	66%	2003 (*1308*)
5	1-NH$_2$-2c-CO$_2$H 1-NH$_2$-2c-CO$_2$H-4-aza	intramolecular dipolar cycloaddition between chiral ketene dithioacetal dioxide and a nitrone	70–73% cycloaddition 65% deprotection	2003 (*1507*)
6	1-aza-3-CO$_2$H-4-OH-3,4-[−(CH$_2$)$_3$−]-5,6-benzo	TiCl$_4$-catalyzed reaction of 2-ethoxycarbonyl-cyclopentanone with benzyl azide	88%	2000 (*1508*)
3	1-NH$_2$-2,3-(CO$_2$H)$_2$	cyclopropanation of Boc- or N-acyl pyrrole, oxidative cleavage of alkene	24–45% cyclopropanation (0% de) 72–84% oxidative cleavage	1997 (*1379*) 1997 (*1381*)

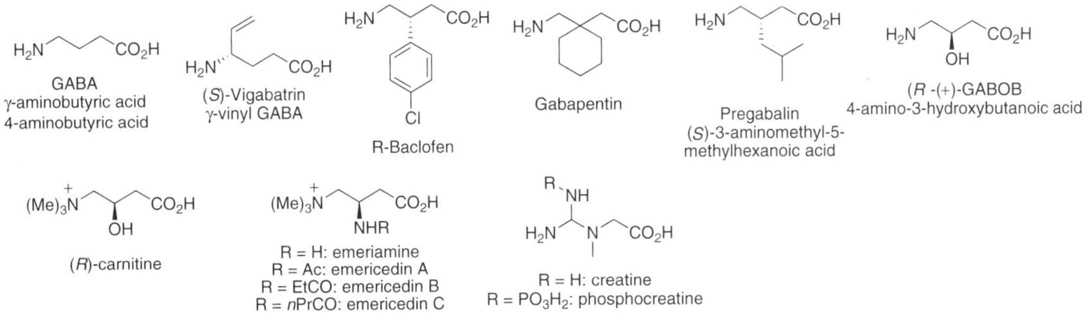

GABA
γ-aminobutyric acid
4-aminobutyric acid

(S)-Vigabatrin
γ-vinyl GABA

R-Baclofen

Gabapentin

Pregabalin
(S)-3-aminomethyl-5-
methylhexanoic acid

(R-(+)-GABOB
4-amino-3-hydroxybutanoic acid

(R)-carnitine

R = H: emeriamine
R = Ac: emericedin A
R = EtCO: emericedin B
R = nPrCO: emericedin C

R = H: creatine
R = PO₃H₂: phosphocreatine

Scheme 11.129   Structures of γ-Amino Acids.

individual nerve cells of the cat central nervous system; concentrations of 0.2–6.6 mM were obtained (1513).

An imbalance of GABA and L-Glu in the brain can lead to convulsions, and GABA deficiency is associated with Parkinson's and Huntington's diseases (1514, 1515). Some effects of the clinical and recreational drug γ-hydroxybutyric acid (GHB) are believed to result through indirect action on the GABA_B receptor via formation of GABA (1516). The brain concentration of GABA can be increased by inhibiting GABA transaminase (GABA-T, γ-aminobutyric acid-α-ketoglutaric acid aminotransferase), the enzyme that degrades GABA. 4-Amino-5-hexenoic acid (γ-vinyl-GABA, or vigabatrin, marketed as the anticonvulsant Sabril) is a highly selective mechanism-based irreversible inhibitor of pyridoxal phosphate-dependent GABA-T (419, 1515, 1517–1521). Only the (S)-isomer of γ-vinyl-GABA is active, although Sabril is racemic. Two divergent mechanisms of inactivation are proposed (1518, 1522). A constrained analog, cis-3-aminocyclohex-4-ene-1-carboxylic acid, was successfully designed to inhibit via one of these pathways, via an enamine (1522). The alkyne analog of vigabatrin, 4-aminohex-5-ynoic acid acid (γ-ethynyl-GABA), is also an effective inhibitor, and its mechanism of inactivation has been investigated (419, 1520, 1523). The (R)-4-aminohex-5-ynoic acid enantiomer is also an irreversible inhibitor of bacterial glutamic acid decarboxylase, with mechanistic studies employing [4-²H]- or [2-³H]-labeled derivatives (1524).

The discovery of γ-vinyl-GABA stemmed from investigations of a series of β-, γ-, and δ-amino acids, which were examined for inhibition of GABA-T in 1981. Vinyl-GABA was the most potent in vitro but β-difluoromethyl-β-Ala and 2,4-difluoro-3-aminobutyric acid were the most potent in vivo (419). The other γ-amino acids evaluated included 4-amino-5-hexynoic acid, 4-amino-5-fluoropentanoic acid, 4-amino-5,5-difluoropentanoic acid, and 4-amino-5,5,5-trifluoropentanoic acid (419). Other irreversible inactivators of GABA-T include 4-amino-5-halopentanoic acids (419, 1525). However, 4-amino-3-halobutanoic acids and 4-amino-5-hydroxypentanoic acid were substrates instead of inhibitors (1526). [4-²H]- and [U-¹⁴C]-labeled

4-amino-5-halopentanoic acids were used to study the mechanism of inactivation (1527). The γ-fluoro compound was 12-times more potent than the γ-chloro derivative, while the γ-bromo analog was only weakly active (1525). 4-Amino-2-(hydroxymethyl)-2-butenoic acid, and the corresponding chloro and fluoro analogs, are also potent competitive reversible inhibitors of GABA-T (1528), while (Z)-4-amino-2-fluorobut-2-enoic acid was determined to be a mechanism-based inactivator of pig brain GABA-T (1529).

3-Alkyl-4-aminobutyric acids and related alkyl-substituted 4-aminobutyric acid derivatives were tested as alternative substrates/inhibitors of GABA-T (1530), as were 4-amino-3-arylbutyric acids (1531). The 3-alkyl-4-aminobutyric acids were identified as anticonvulsant agents, but they acted via activation of L-glutamic acid decarboxylase (1532). (S)-(+)-3-Isobutyl GABA was found to be a potent anticonvulsant lipophilic GABA analog that presumably can cross the blood–brain barrier (1514). Other GABA analogs include R-baclofen, (3R)-4-amino-3-(4-chlorophenyl)butanoic acid, a high-affinity agonist of the GABA_B-receptor which is in clinical use for the treatment of spasticity caused by disease of the spinal cord (1519, 1533). Baclofen is also sold as the racemate. R-Baclofen was employed to help identify a new GABA receptor (1534). Gabapentin (Neurontin), 1-(aminomethyl)cyclohexane-1-acetic acid, is a constrained cyclohexyl-based γ-amino acid anticonvulsant. It was designed as a lipophilic analog of GABA, but does not bind to any of the GABA receptors (1535). The anticonvulsant effect of gabapentin is proposed to be related to its high-affinity binding to a calcium channel site (1519). (S)-3-Aminomethyl-5-methylhexanoic acid, pregabalin, is another potent anticonvulsant (1536). A commentary in 1979 reviewed inhibitors of GABA metabolism such as GABA uptake inhibitors, inactivators of 4-aminobutyrate:2-oxoglutarate aminotransferase, and glutamic acid decarboxylase (1537).

GABOB (3-hydroxy-4-aminobutanoic acid) is a neuromodulator in the mammalian nervous system, with the (R)-enantiomer having greater biological activity than the (S)-isomer (references within 1538). An analytical method to determine GABOB concentrations in finished pharmaceutical products relied on HPLC separation of the underivatized amino acid on a C_{18} column, using an aqueous sodium heptasulfonate mobile phase and UV detection at 210 nm (*1539*). The amino acid is also contained in microsclerodermins A–I, anti-fungal cyclic peptides isolated from the lithistid sponge *Microscleroderma* sp. (*211, 213, 214*). A total synthesis of microsclerodermin E was reported in 2003 (*215*).

The *N*-trimethylated derivative of GABOB, carnitine, plays a role in the transportation of fatty acids through mitochondrial membranes (see references within *1538, 1540*). It was first isolated from a meat extract in 1905 and its structure reported in 1927. The absolute configuration of (−)-*l*-carnitine was established as (*R*) in 1962 via α-decarboxylation of *threo*-β-hydroxy-L-Glu and *N*-methylation (*1541*). (−)-Carnitine can be isolated by alcoholic extraction of beef extract, which contains 2–3% of the amino acid (*1542*). Large-scale bioreactor production of L-(−)-carnitine from *E. coli* is possible (*1543*). Carnitine is biosynthesized from Lys, with N^{ε}-trimethyl-Lys a key intermediate. Rat liver slices fed [methyl-^{3}H]N^{ε}-trimethyl-Lys produced labeled carnitine, but tissue slices of heart, skeletal muscle, kidney or testes did not (*1544*). Feeding experiments with [1-^{14}C]N^{ε}-trimethyl-Lys demonstrated that Gly was extruded as the 2-carbon fragment during the biosynthesis, as labeled hippuric acid was produced (*1545*). Three analogs of carnitine, 1-(trimethylammonio)-cyclopropanepropionic acid, 1-(2-trimethylammonioethyl)-cyclopropanecarboxylic acid, and 4-(trimethylammonio)-3-cyclopropylbutyric acid, were found to inhibit γ-butyrobetaine hydroxylase, which catalyzes the final step in carnitine biosynthesis from Lys (*1546*). Cyclic analogs of carnitine have also been prepared as inhibitors of carnitine acetyltransferase (*1547*).

Carnitine and acetyl carnitine are involved in transporting metabolic energy between different organs; plasma levels of both have been found to vary with body muscular mass (*1548*). Some diseases result in an accumulation of acyl carnitines, giving importance to analysis of carnitine and acyl carnitines in biological fluids. Carnitine-deficient patients are given L-carnitine, known as levocarnitine, or Carnitor. A number of analyses of carnitine have been developed, including a capillary electrophoresis (CE) analysis of carnitine and acylcarnitines in urine samples following derivatization of the carboxyl group with 4′-bromophenacyl trifluoromethanesulfonate (*1549*). Conductivity detection was employed for the analysis of carnitine in food supplement formulations, with ion-pair chromatography separation using octanesulfonate eluent (*1550*). Capillary zone electrophoresis and capillary isotachophoresis separations were also employed for carnitine analysis in food supplements, with UV or conductivity detection, respectively (*1551*). A GC-based resolution of carnitine enantiomers utilized online conversion of carnitine into β-hydroxy-γ-butyrolactone (based on thermal intramolecular nucleophilic displacement of the trimethylammonium group by the carboxylate oxygen), followed by resolution on a β-cyclodextrin-based column (*1552*). An enantiomeric separation of carnitine and

O-acylcarnitine enantiomers is possible by HPLC using a teicoplanin chiral stationary phase (*1553*), while analysis of the enantiomeric purity of *O*-acetyl carnitine was accomplished on a normal reversed-phase column via formation of diastereomeric derivatives by amidation with L-Ala-β-naphthylamide (*1554*). A CE resolution employed online derivatization with Fmoc-Cl, followed by separation using a modified cyclodextrin chiral selector (*1555*).

Emeriamine, (*R*)-3-amino-4-trimethylaminobutyric acid, is an analog of carnitine with an amino group replacing the hydroxyl. It is an inhibitor of fatty acid oxidation (*168, 169*). Emeriamine is derived by deacylation of the emericedins, less potent inhibitors of fatty acid oxidation isolated from the culture broth of *Emericella quadrilineata* (*168*). More potent analogs could be prepared by forming *N*-palmitoyl or *N*-myristoyl derivatives of emeriamine (*168*). Creatine and phospho-creatine contain guanyl functions with an amino group γ to the carboxyl group. The most simple γ-amino acid, 4-aminobutanoic acid, was identified in 2003 in a novel cytotoxic cyclodepsipeptide isolated from a New Zealand fungi (*1556*). 4,4′-Diaminoisopentanoic acid was identified in the Murchison meteorite (*32*).

A number of biologically active compounds contain β-hydroxy-γ-amino acids (see Scheme 11.130). Two molecules of statine, (3*S*,4*S*)-(−)-4-amino-3-hydroxy-6-methylheptanoic acid (AHMHA), are found in the peptide antibiotic pepstatin, which was isolated from various species of actinomycetes (*1557, 1558*). Ahpatinin C, obtained from a soil isolate, *Streptomyces* sp. WK-142, was found to be identical to pepstatin A (*1559*). The peptides were identified during screening tests for enzyme inhibitors (*1558*), and found to inhibit renin (*1560*), pepsin (preventing stomach ulceration) (*1558, 1560*), cathepsin D (*1560*), and other acid proteases. Both pepstatin and a related analog, leupeptin, delay degeneration of muscle tissue caused by muscular dystrophy, presumably via inhibition of the proteases involved in catabolism (*1561*). Statine was also identified as a component of miraziridine A, a cysteine protease (cathepsin B) inhibitor isolated from the marine sponge *Theonella mirabilis* (*1562*).

The four isomers of statine have been prepared (*1563–1566*) and incorporated into pepstatin analogs. The (3*S*,4*S*)-statine isomer possessed the greatest activity (*1566*) and was identified as the diastereomer present in pepstatin (*1564, 1565*). The absolute configuration of natural (−)-statine was also established by X-ray crystallography. The conformational preferences of statine have been discussed, based on crystal structures of statine, statine-containing peptides, and statine-containing inhibitors within enzymes (*1567*). The crystal structures of peptides containing statine were reviewed in 1996. Both folded and extended structures are observed, suggesting statine may be able to alter its conformation to accommodate the environment (*1568*). Evidence for and against the statine residues within pepstatin acting as transition-state analogs was discussed in a 1985 article (*1569*).

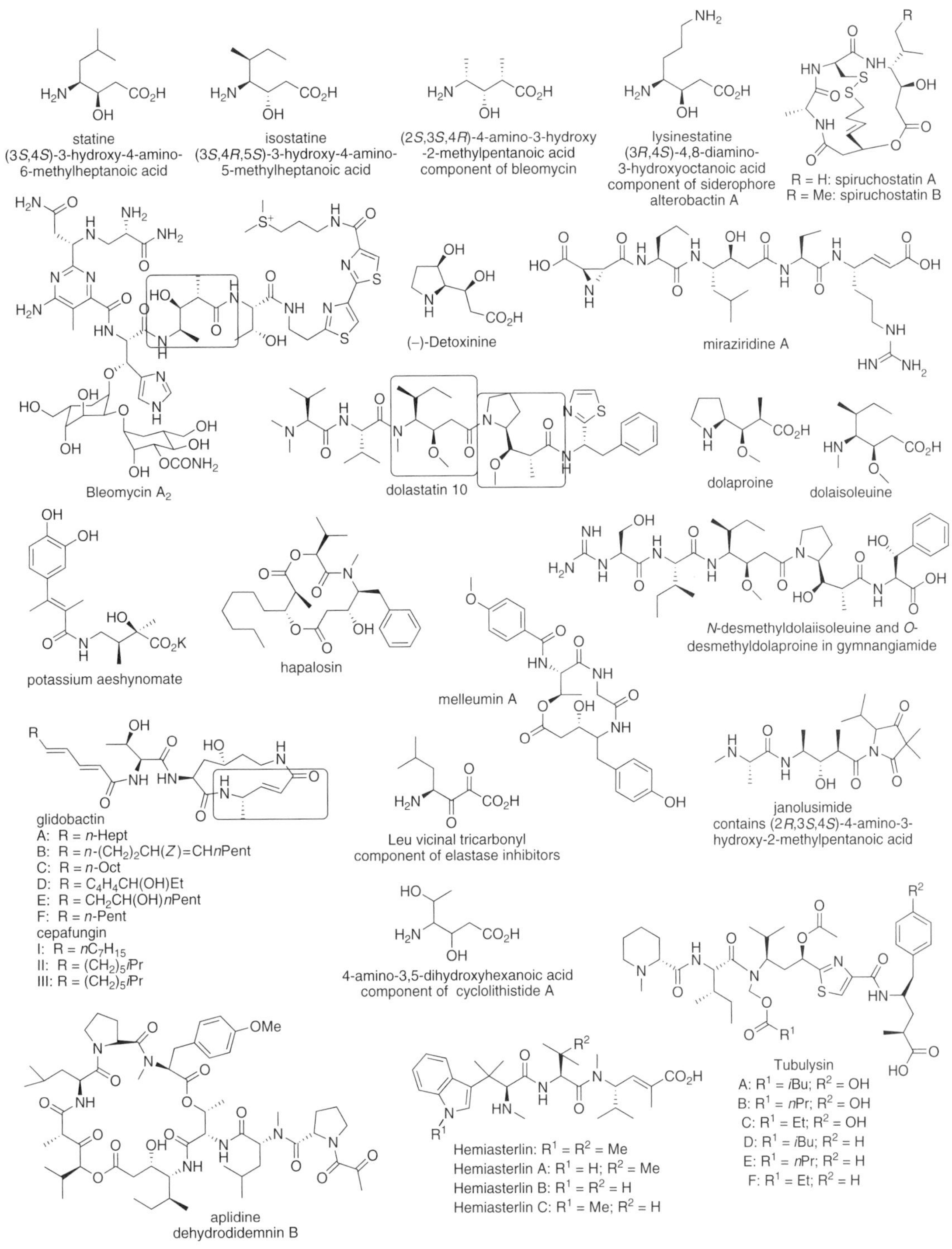

statine
(3S,4S)-3-hydroxy-4-amino-
6-methylheptanoic acid

isostatine
(3S,4R,5S)-3-hydroxy-4-amino-
5-methylheptanoic acid

(2S,3S,4R)-4-amino-3-hydroxy
-2-methylpentanoic acid
component of bleomycin

lysinestatine
(3R,4S)-4,8-diamino-
3-hydroxyoctanoic acid
component of siderophore
alterobactin A

R = H: spiruchostatin A
R = Me: spiruchostatin B

Bleomycin A₂

(−)-Detoxinine

miraziridine A

dolastatin 10

dolaproine

dolaisoleuine

potassium aeshynomate

hapalosin

melleumin A

N-desmethyldolaiisoleuine and O-
desmethyldolaproine in gymnangiamide

glidobactin
A: R = n-Hept
B: R = n-(CH₂)₂CH(Z)=CHnPent
C: R = n-Oct
D: R = C₄H₄CH(OH)Et
E: R = CH₂CH(OH)nPent
F: R = n-Pent
cepafungin
I: R = nC₇H₁₅
II: R = (CH₂)₅iPr
III: R = (CH₂)₅iPr

Leu vicinal tricarbonyl
component of elastase inhibitors

4-amino-3,5-dihydroxyhexanoic acid
component of cyclolithistide A

janolusimide
contains (2R,3S,4S)-4-amino-3-
hydroxy-2-methylpentanoic acid

Tubulysin
A: R¹ = iBu; R² = OH
B: R¹ = nPr; R² = OH
C: R¹ = Et; R² = OH
D: R¹ = iBu; R² = H
E: R¹ = nPr; R² = H
F: R¹ = Et; R² = H

aplidine
dehydrodidemnin B

Hemiasterlin: R¹ = R² = Me
Hemiasterlin A: R¹ = H; R² = Me
Hemiasterlin B: R¹ = R² = H
Hemiasterlin C: R¹ = Me; R² = H

Scheme 11.130 Structures of More Complex γ-Amino Acids.

A variety of statine analogs have been synthesized. Removal of the statine hydroxy groups from pepstatin resulted in a 2000-fold loss in activity for inhibition of porcine pepsin, indicating an important role for the 3-hydroxy group (*1570*). The ketone analog of pepstatin,

in which the statin β-hydroxy group is oxidized to a ketone, is 50-fold less active at inhibition of renin. However, the difluoroketone analog, 2,2-difluoro-3-keto-4-amino-6-methylheptanoic acid, is an extremely potent inhibitor with K_i = 0.06 nM, 18-fold more active than pepstatin (*1571, 1572*). Other peptidic pepstatin analogs have been prepared containing statine (*1573, 1574*), GABOB (*1573, 1574*), 2,2-difluorostatine and related derivatives (*1575*) or (3S,4S)-3-hydroxy-4-amino-5-phenylpentanoic acid (*1576*). Potent, orally available renin inhibitors have been prepared containing statine and various other (3S,4S)-3-hydroxy-4-amino-4-substituted butanoic acids (*1577*). Statine has been incorporated into an affinity-based probe of aspartic proteases (*1578*), and used as the basis of several large (13,000–18,900-member) combinatorial libraries employed to find selective inhibitors of the aspartyl proteases cathepsin D and malarial protease plasmepsin II (*1579, 1580*). Statine analogs, 2-heterosubstituted-4-amino-3-hydroxy-5-phenylpentanoic acids, have been used in inhibitors of HIV-1 protease (*1581*).

Isostatine, (3S,4R,5S)-4-amino-3-hydroxy-5-methylheptanoic acid, is found in the didemnins, cyclodepsipeptides with antitumor, antiviral, and immunosuppressive activity that were isolated from marine tunicates (*1582–1590*). Closely related tamandarin A also contains this residue, while tamandarin B has (3S,4R)-4-amino-3-hydroxy-5-methylhexanoic acid. Total synthesis of both tamandarins and related analogs were reported in 1999 (*1591*) and 2001 (*1592*). Didemnin B analogs with *N*-Me Leu or *N*-Me Phe replacing an *N,O*-Me₂-Tyr residue retained biological activity (*1593*). A constrained didemnin analog was prepared using 3-amino-2-hydroxycyclohexane-1-carboxylic acid to replace the isostatine unit, but potency was reduced (*1594, 1595*). Dehydrodidemnin B (aplidine), in Phase II clinical trials in 2003, exists in DMSO as a mixture of four interconverting isomers that arise due to isomerization around the *N*-methyl-Leu⁷-Pro⁸ and Pro⁸-pyruvate amide bonds, as measured by NMR spectroscopy (*1596, 1597*). A didemnin derivative with a dimethylaminocoumarin fluorophore linked via a Gly tether was constructed in order to study cellular localization (*1598*). Similarly, (3S,4R)-4-amino-3-hydroxy-5-methylhexanoic acid is found in spiruchostatin A, and 4-amino-3-hydroxy-5-methylheptanoic acid in spiruchostatin B. The spiruchostatins, isolated from a *Pseudomonas* extract, are potent histone deacetylase inhibitors with potential as anticancer agents. A total synthesis of spiruchostatin A was reported in 2004 (*1599*).

Bleomycin A₂, the major constituent of the clinical antitumor drug Blenoxane, contains (2S,3S,4R)-4-amino-3-hydroxy-2-methylpentanoic acid (references within *247, 1600, 1601*). The stereochemistry was identified by X-ray studies and chemical synthesis (*1602*). Other bleomycins in the family of glycopeptide-derived antibiotics isolated from *Streptomyces verticillus* contain this γ-amino acid, as do the closely related cleomycins, produced by a mutant bacterial strain (*1603*). The antitumor activity of bleomycin is apparently due to its ability to mediate DNA degradation; the possible chemistry of this reaction was reviewed in 1986 (*1604*). Some reverse transcriptase inhibition has also been reported (*1605*). Total syntheses of both the bleomycin aglycon (*1606*) and bleomycin itself (*1607*), as well as bleomycin

demethyl A₂ (*254*), bleomycin A₂ (*254, 255*), deglycobleomycin A₂ (*256*), and decarbamoylbleomycin demethyl A₂ (*254*) have been reported, as have partial syntheses incorporating the γ-amino acid (*1601, 1608*). Analogs of deglycobleomycin A₅ with various conformationally restricted γ-amino acid components, joining the C2–C4 backbone carbons with a butyl or 2-butene linker (diastereomers of 2-hydroxy-3-aminocycloheptyl-1-carboxylic acid), were synthesized in 2003 (*258, 1609*). The same year a library of 108 deglycobleomycin analogs was prepared by solid-phase synthesis from three bithiazole analogs, three Thr analogs, three γ-aminomethylvalerate analogs, and four β-hydroxy-His replacements (*259*). A deglycobleomycin A₅ analog with a modified *C*-terminal trithiazole tail possessed altered sequence selectivity (*261*). The synthesis and mechanistic studies of bleomycin were reviewed in 1999 (*250*).

The tripeptide marine toxin janolusimide (129 mg from 300 specimens of the nudibranch mollusk *Janolus cristatus*) (*1610*) also contains 4-amino-3-hydroxy-2-methylpentanoic acid, but synthesis of all four possible diastereomers revealed this natural product contained the (2R,3S,4S)-isomer (*1611*). Butirosin is an aminoglycoside antibiotic produced by mucoid strains of *Bacillus circulans*, with a (S)-(−)-4-amino-2-hydroxybutyric acid component (*1612, 1613*). α-Hydroxy-β-keto-γ-aminobutyric acid has been isolated from normal human urine (*1614*). Lysinestatine, (3R,4S)-4,8-diamino-3-hydroxyoctanoic acid, is found in the marine siderophore alterobactin A, which has possibly the highest affinity for the ferric ion of any known siderophore (*1615*). Ahpatinins A, B, E, and G, isolated from a soil isolate, *Streptomyces* sp. WK-142, contain AHPPA, 4-amino-3-hydroxy-5-phenylpentanoic acid (*1559*). Ahpatinin E, an analog of pepstatin A with AHPPA replacing statine, demonstrated higher renin-binding activity than pepstatin A. Ahpatinin C was found to be identical to pepstatin A, containing statine (*1559*). Hapalosin is a cyclic depsipeptide which reverses multidrug resistance in tumor cell chemotherapy. It contains (3R,4S)-4-methylamino-3-hydroxy-5-phenylpentanoic acid (*1616, 1617*). Total syntheses of hapalosin (*1617–1619*) and analogs containing *N*-desmethyl and deoxy-γ-amino acid analogs (*1617*) have been reported. Melleumin A, a peptide lactone from the myxomycete *Physarum melleum*, contains a 3-hydroxy-4-amino-5-(4-hydroxyphenyl)pentanoic acid residue (*1620*). 4-Amino-3-hydroxy-5-(indol-3-yl)-pentanoic acid is a component of the antialgal peptide kasumigamide, from the cyanobacterium *Microcystis aeruginosa* (*1621*).

Potassium aeshynomate was isolated from the nyctinastic plant (plants that close their leaves in the evening and open them in the morning) *Aeshynomene indica* (8.7 mg from 7.7 kg of fresh plant) and found to be effective at inducing leaf opening. The compound contains 2,3-dihydroxy-2-methyl-4-aminobutyric acid (*1622*). Calyculins A–H are cytotoxic metabolites (selective phosphatase inhibitors) from the marine sponge *Discodermia calyx* which also contain a polyhydroxy γ-amino acid, 2,3-dihydroxy-4-dimethylamino-5-methoxypentanoic acid

(129, 1623–1628). The absolute configuration of the calyculins was established by a stereoselective synthesis *(1627)*. A closely related analog was found in AI-77-B, an antiulcer substance isolated from the culture broth of *Bacillus pumilus (1288)*. The antifungal cyclodepsipeptide cyclolithistide A, isolated from the sponge *Theonella swinhoei*, contains 4-amino-3,5-dihydroxyhexanoic acid *(1629)*.

The pentapeptide dolastatin 10, isolated from the Indian Ocean sea hare *Dolabella auricularia*, has potent antineoplastic properties (references within *1630–1632*), with only 28.7 mg obtained from 1600 kg of *Dolabella auricularia (75)*! It contains two β-methoxy-γ-amino acids, dolaisoleuine (also called dolaisoleucine) and dolaproine (see Scheme 11.130). Dolastatins-11 and -12 (43.6 mg and 10.6 mg, respectively, from 1600 kg of mollusc) contain 2,2,-dimethyl-3-keto-4-aminopentanoic acid *(75, 76)*. An analog of dolastatin 10 was prepared and was cyclized via an ester linkage between the modified *C*- and *N*-termini *(1633)*. Both *N*-desmethyldolaisoleuine and *O*-desmethyldolaproine are components of gymnangiamide, a cytotoxic peptide from the marine hydroid *Gymnangium regae (1634)*. The tubulysins are extremely potent microtubule-perturbing compounds with 10-fold higher cytostatic activity than the dolastatins. They contain a 2-methyl-4-amino-5-arylpentanoic acid residue, with the aryl group either a phenyl (tubuphenylalanine) or a *p*-hydroxyphenyl group (tubutyrosine) *(1635)*.

Other γ-amino acids isolated from bioactive natural products include a vinylogous derivative of Tyr, 2,3-dehydro-4-amino-4-(4'hydroxyphenyl)pentanoic acid, found in the cyclotheonamides, cyclic pentapeptides from *Theonella swinhoei* sponge which inhibit thrombin, trypsin, and plasmin *(129, 264, 265, 267)*. Pseudotheonamides C and D also contain this residue *(266)*, while the vinylogous homolog of Arg was identified as a component of miraziridine A, a cysteine protease (cathepsin B) inhibitor isolated from the marine sponge *Theonella mirabilis (1562)*. (4*S*)-4-Amino-2(*E*)-pentenoic acid is a component of the antitumor antibiotics glidobactins A–F and the closely related cepafungins I–III from *Polyangium brachysporum (1636, 1637)*. The hemiasterlins, milnamides, and criamides are cytotoxic tripeptides isolated from marine sponges that contain a 4-amino-2,5-dimethylhex-2-enoic acid *(1638)*.

γ-Amino acids have been synthesized for use in drug discovery. An Arg homolog, 4-amino-3-oxo-7-guanidinoheptanoic acid, was prepared for incorporation in hirudin-like thrombin inhibitors *(1639)*, while a Leu homolog with a vicinal tricarbonyl moiety is contained in two cyclic depsipeptide elastase inhibitors, YM-47141 and YM-47142 *(1640)*.

The structural properties of γ-amino acids, like β-amino acids, have been investigated for their potential to form unique structures compared to α-amino acids. As with synthetic β-amino acids, a nomenclature for substitution patterns has been developed, with, for example, a γ^{2,3,4}-amino acid representing a γ-amino acid with side chains in the 2-,3-, and 4-positions. GABA has been used in a dipeptide with β-alanine as a homolog of

(Gly)$_3$ and inserted into octa- and undecapeptides. The peptides retained their helical motifs with minimal perturbations *(319)*. A conformational analysis of amide derivatives of 4-amino-2-methylpentanoic acid indicated two predominant conformers *(1641)*. Peptides consisting of homochiral γ-substituted γ-amino acids have also been examined. A γ-hexapeptide, the homolog of H-(Val-Ala-Leu)$_2$-OH, formed a right-handed helical structure with surprising stability, in contrast to the left-handed helix of a corresponding β-peptide and the right-handed helix of α-peptides *(1642)*. A homologated analog of Ala-Val-Ala-Val was sufficient to form a right-handed helix. α-Substitution in the γ-peptides further stabilized the helical structure, as long as they were of the correct configuration at the α-center *(1643)*. One-dimensional (1D)- and two- dimensional (2D)-NMR studies of oligopeptides formed from α,γ-disubstituted-γ-amino acids provided evidence that they formed γ-turn and right-handed helical structures *(1644)*. γ-Tetra- and γ-hexapeptides formed from γ^{2,3,4}-amino acids were shown to form left-handed 2.6$_{14}$ helices by 2D-NMR and X-ray structural analysis *(1645, 1646)*, while the structures of γ^2- or γ^3-hexapeptides were unable to be determined *(1646)*. Introduction of hydroxy groups at the C2 or C3 position in γ^4-tri- or -hexapeptides resulted in changes in conformation, as measured by CD spectra *(1647)*. Crystal structures of gabapentin oligomers have shown C$_9$ helix and ribbon structures *(1648)*. Theoretical quantum mechanics studies of γ-peptides have examined the influences of substituents on secondary helical structures *(298)*.

Schreiber and co-workers. have developed polypeptides of γ-substituted-α,β-unsaturated-γ-amino acids (vinylogous polypeptides) as an alternative peptide backbone, and identified possible parallel and antiparallel sheet secondary structures, in addition to a novel helical sheet structure *(1649)*. The vinylogous analogs of Pro-Gly or Pro-Ala induced formation of β-hairpin structures when incorporated into peptides *(1650)*. Vinylogous amino acids have been used as dipeptide isosteres, and as Michael acceptors which inactivate human rhinovirus 3C protease *(1651–1656)*. A theoretical study in 2003 examined all possible periodic structures with hydrogen-bonding for oligomers of γ-amino acids or the corresponding vinylogues, looking at pseudocycles with ring sizes between C$_7$ and C$_{24}$. Structures with 9- and 14-membered rings were found to be most stable, with (*E*)-alkene bonds in the backbone increasing the size of the preferred pseudocycle *(1657)*.

GABA was used to replace Gly in a Piv-Pro-Gly-NHME sequence known to form a classical Type II β-turn; a crystal structure showed a novel 10-membered turn structure involving a C–H⋯O interaction between one of the α-methylene hydrogen atoms and the carbonyl of the Piv group *(1658)*. A crystal structure of an *N*-acyl γ-dipeptide demonstrated a type βII'-turn conformation *(1659)*. The helical and other secondary structures of γ-peptides were discussed in a 2006 article *(274)*.

Like β-peptides, γ-peptides have potential therapeutic use if they are more stable to proteolysis than peptides made from α-amino acids. A study in 2001 examined the stability of five linear γ-peptides (3- or 6-mer) to 15 different proteases, such as pepsin, chymotrypsin, elastase, β-lactamase, amidase, and 20S proteasome (a vigorous, non-specific protease). Under conditions where α-peptides were completely cleaved within 15 min, the γ-peptides were stable for at least 48 h (*354*). A comparison of the antibacterial and hemolytic activity of 31 different β- and γ-peptides was published in 2005 (*370*). GABA was used to replace the *N*-terminal amino acid in corticotropin analogs in 1970 (*10*). A tetrapeptide of (*S*)-4-amino-5-hydroxypentanoic acid was used to replace a heptapeptide sequence in the 31-residue peptide hormone β-endorphin. The substitution was made in a proposed linker region joining the *N*-terminal Met-enkephalin (residues 1–5) and a *C*-terminal helix (residues 13–31). The endorphin analog maintained similar analgesic potency upon intracerebroventricular (icv) injection into mice (*1660*).

Dendrimers have been formed from aminotri (carboxyethyl)methane (*1661*), including a third generation dendrimer attached to a fluorescent dansyl core (*1662*). A fluorescent receptor for zwitterionic amino acids was based on triaza-18-crown-6 ether, with two pendant guanidinium groups and an anthryl fluorophore. Binding of Lys and GABA induced a very strong fluorescence, while Gly gave a small increase, and other amino acids (Ala, Phe, Val, Ser, Glu, Arg) showed no response (*1663*).

A monograph describing methods of non-α-amino acid synthesis (including γ-amino acids) was published in 1995 by M.B. Smith (*1*). β-Fluoro-substituted α-, β-, and γ-amino acids were reviewed in 1990 (*1664*), and the synthesis of β-oxygenated-γ-amino acids from α-amino acids in 1992 (*1665*). A comprehensive 2004 review (128 pages, 531 references) focused on the synthesis and properties of β-peptides and also included extensive discussions on the synthesis, structures, and biological properties of γ-peptides (*13*). The stereoselective synthesis of γ-amino acids was thoroughly reviewed in a 2007 *Tetrahedron: Asymmetry* report (417 references) (*1666*).

11.4.2 Syntheses of Achiral or Racemic γ-Amino Acids (see Table 11.5)

11.4.2a. Amination Reactions

Several nucleophilic aminations have employed phthalimide as the aminating reagent. GABA was prepared in 1958 by ring opening γ-butyrolactone with potassium phthalimide; hydrolysis with HCl gave the product in 50% overall yield (*1667*). Potassium phthalimide was also used to open 2-methylbutyrolactone, giving a 70% yield of *N*-phthaloyl 2-methyl-4-aminobutyric acid (*1668*). An early synthesis of GABOB reacted phthalimide with γ-chloro-β-hydroxybutyric acid, itself prepared from 3-hydroxy-4-chlorobutyronitrile (*1669*). Similarly, β-methyl-GABA was derived from 3-methyl-4-chlorobutyronitrile via

displacement with phthalimide and hydrolysis (*1670*). Potassium phthalimide and ethyl 2-fluoro-4-bromocrotonate provided 2-fluoro-4-aminocrotonic acid (*1671*). More recently, phthalimide was used in a Mitsunobu displacement of the 4-hydroxy group of (7-*O*-tetrahydropyranyl)-1-fluoro-4,7-dihydroxy-hepta-1,2-diene. Deprotection and oxidation of the terminal hydroxy group gave 4-(3′-fluoroallenyl)-GABA (*1672*).

Ammonia has also been used for nucleophilic aminations. Several alkyl α,β-unsaturated acids were converted into substituted 4-aminobut-2-enoic acids via bromination at the allylic position using NBS followed by amination in liquid ammonia (*1673, 1674*). Addition of ammonia to the double bond of triethyl aconitate (3-carboxypent-2-enedioic acid) gave a tetraamide of the aminotriacid diketopiperazine, with saponification yielding aminotricarballylic acid (2-amino-3-carboxypentanedioic acid) in 12% yield (*1675*). A GABA analog with tetronic acid [furan-2,4(3*H*,5*H*)-dione] as an isostere of the carboxyl group was synthesized by displacement of a bromo precursor with an ammonia equivalent, potassium bis(Boc)amide (*1676*). (η³-Allyl)dicarbonylnitrosyliron complexes containing a methoxycarbonyl substituent were reacted with various amines to produce γ-amino-α,β-unsaturated esters in 63–95% yields (*1677*). Chiral ester auxiliaries could be used to induce asymmetry.

Reductive aminations provide another route to γ-amino acids. 4-Aminopentanoic acid was prepared in 1911 via reductive amination of the corresponding ketone using phenylhydrazine and aluminum amalgam (*1678*), while in 1927 4-amino-3-hydroxypentanoic acid was derived from 3-hydroxy-4-ketopentanoic acid (β-hydroxylevulinic acid) using hydroxylamine and sodium amalgam (*1679*). Several γ-aryl-GABA derivatives were prepared from the corresponding 4-keto acids by reductive amination via hydrogenation of a phenylhydrazone intermediate (*1680*). *N*-Substituted 4-aminopentanoic acid could be prepared by reductive amination of the ketone with norepinephrine (*1681*).

Several types of rearrangements have been employed to introduce the amino group. A Curtius rearrangement of 2,4-dimethylglutaric monomethyl ester provided 4-amino-2-methylpentanoic acid (*1641*), while an aza-Cope rearrangement was employed for amination during a synthesis of γ-allenic GABA (4-amino-hepta-5,6-dienoic acid) and several alkyl-substituted derivatives. The product is a potent inhibitor of mamalian GABA transaminase (*1682*). The diamino acid 4,6-diaminohexanoic acid was prepared in 1953 via a Schmidt rearrangement of 5-(β-carbomethoxyethyl)-2-pyrrolidone, cyclized 4-aminoheptanedioic acid (*146*). Diethyl arylmalonates were combined with acrylonitrile to give 2-aryl-4-cyanobutanoic acids after hydrolysis/decarboxylation. These versatile intermediates could be converted into 4-aryl-GABA products via partial hydrolysis of the cyano group to an amide followed by Hofmann rearrangement, while complete hydrolysis of the cyano group and selective Hofmann rearrangement of the carboxyl adjacent to the aryl group gave 2-aryl-GABA regioisomers (*1683*). 4-Aminohex-2-enoic acid was synthesized by palladium-catalyzed rearrangement of an allylic sulfoximine, a procedure that is potentially a general route to γ-amino-α,β-unsaturated acids. The rearrangement substrates were prepared by a Knoevengel-type condensation of α-sulfonimidoyl methyl acetate with aldehydes (*1684*). γ-Aminophosphonates have also been synthesized via introduction of the amine group using a rearrangement (*1685*).

Pd(0)-catalyzed azidation of allylic cyanohydrin carbonates, RCH=CHCH(CN)OCO₂Et, produced RCH(N₃) CH=CH(CN) γ-azido-α,β-unsaturated nitrile precursors of γ-amino acids in 85–90% yield. The cyanocarbonate substrates were obtained from α,β-unsaturated aldehydes by reaction with NCCO₂Et/DABCO (*1686*). An intramolecular Rh-catalyzed C–H bond insertion of sulfamate ester derivatives of 2-hydroxy esters led to six-membered oxathiazinane products through selective γ-C–H insertion of the sulfamate amino group. Oxathiazinane ring opening with KOAc in DMSO produced 2-acetoxy-4-aminoesters with high *syn* stereoselectivity (*518*).

11.4.2b Carboxylation Reactions

A synthesis of GABOB introduced the carboxyl group via NaCN displacement of a protected derivative of 1-chloro-2-hydroxy-3-aminopropane, derived via condensation of benzaldehyde, ammonia, and epichlorohydrin (*1687*). Reaction with methyl iodide gave carnitine (*1687*). GABA analogs with the carboxyl group replaced with an isostere, squaric acid, were synthesized by reaction of monoprotected 1,2-diaminoethane or 1,3-diaminopropane with diethyl squarate. However, little biological activity was observed (*1688*). A β-keto ester desmethyl precursor of carnitine was prepared by carboxylation of the enolate of dimethylamino acetone with ethyl carbonate (*1689*).

Rearrangement of substituted 1-amino-2,3-dehydro-butan-4-ols in the presence of ethyl or methyl orthoacetate led to substituted 3-vinyl-4-aminobutyric acid derivatives. The alkene could be ozonolyzed and the aldehyde used for Wittig couplings to give other substituted 3-vinyl-4-aminobutyric acids (*1690*).

11.4.2c Fusion of Amino-Containing and Carboxy-Containing Components

Several different approaches have been developed to join amino- and carboxy-containing components of γ-amino acids, via formation of either the C2−C3 or C3−C4 bonds. For C3−C4 bond formation, nucleophilic aminomethyl groups can be reacted with electrophilic propionate equivalents. 4-Amino-2-methylbutanoic acid and 4-amino-3-methylbutanoic acid were obtained by conjugate addition of nitromethane to methyl methacrylate or ethyl crotonate, followed by Raney nickel hydrogenation (*527*), while addition to ethyl 3,3-dimethylacrylate with DBU as base led to 4-amino-3,3-dimethylbutanoic

acid (*1530*), and addition to methyl (*E*)-3-cyclopropylacrylate followed by nitro reduction and ester hydrolysis and resulted in 4-amino-3-cyclopropyl-butanoic acid. The amino group was then trimethylated (*1546*). Nitromethane was also added to a number of other 2-alkenoic esters in the presence of TMG or DBU in 40–74% yield, giving 3-alkyl-4-aminobutanoic acids after hydrogenation and hydrolysis (R = Me, Et, *n*Pr, *i*Pr, *n*Bu, *s*Bu, *i*Bu, *t*Bu) (*1691*). Nitromethane has also been added to acrylates attached via an ester linkage to a dendritic polyglycerol dendritic support. Nitro reduction was accompanied by cyclization with concomitant cleavage from the support to form γ-lactams (*1692*). A γ-aminotricarboxylic acid suitable for constructing dendritic polymers, tri-*tert*-butyl aminotri(carboxyethyl) methane, was prepared by the Michael addition of nitromethane to three equivalents of *tert*-butyl acrylate in dimethoxyethane, followed by reduction of the nitro group to give the aminotriester (*1661, 1693*). Both 3-phenyl-GABA and 3-(4-chlorophenyl)-GABA were synthesized by a chemoenzymatic route which began with addition of nitromethane to the methyl cinnamates. An enzymatic resolution was carried out before nitro reduction (*1694*). Selective catalytic hydrogenations of γ-nitro esters have been optimized for process synthesis (*1695*).

A variety of nitrones, RCH=N⁺(Bn)O⁻, were added to ethyl acrylate or other substituted acrylates in 2003 via a reductive conjugate addition in the presence of 2 equiv of SmI₂. Hydrogenation of the resulting *N*-hydroxy, *N*-benzyl γ-amino esters over Raney Ni in the presence of Boc₂O produced the Boc-protected γ-amino esters in good yield (*1696*). 4-Aminohex-5-ynoic acid was obtained by Michael addition of a substituted aminomethyl equivalent, the enolate of the benzaldehyde imine of 1-amino-3-trimethylsilylprop-2-yne, to methyl acrylate (*1520*). A precursor of GABOB was prepared by addition of cyanide (a masked aminomethyl equivalent, from TMSCN) to ethyl formyl acetate, generated in situ from ethyl 3,3-diethoxypropionate. The cyano group of the resulting ethyl 3-cyano-3-hydroxypropionate was reduced with borane to give GABOB, after resolution of the cyanohydrin (*543, 1697*). A similar reaction sequence led to racemic *threo*-statine in 32% overall yield with 68% de. An oxazoline-protected formyl acetate equivalent, prepared in situ from 2,4,4-trimethyl-2-oxazoline, was reacted with TMSCN to give a cyanohydrin intermediate. Addition of isobutylmagnesium chloride (not isopropyl, as stated in the text) to the cyanide group followed by borane reduction and hydrolysis gave statine (see Scheme 11.131) (*543*).

Alternatively, γ-amino acids can be synthesized by reversing the reactant polarity of C3−C4 bond formation, using nucleophilic propionate and electrophilic aminomethyl equivalents. An organozinc reagent prepared from methyl 3-bromopropionate was added to an imine intermediate generated in situ from benzaldehyde and (trimethylsilyl)dimethylamine, giving *N, N*-dimethyl-4-amino-4-phenylbutanoate in 59% yield (*620*). Racemic statine was prepared by addition of γ-oxygenated allyltin to an "activated imine," which proceeded with high *syn*

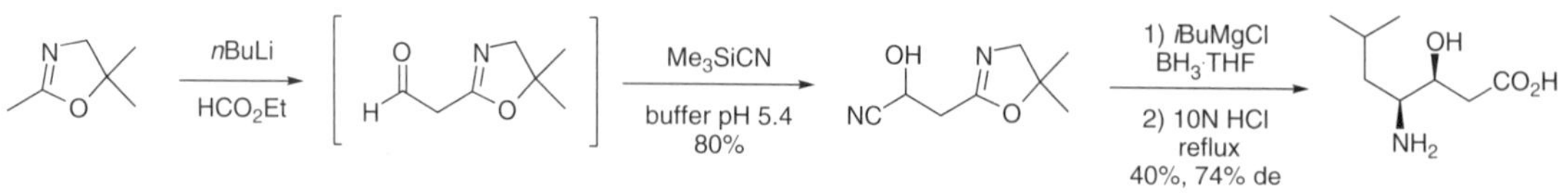

Scheme 11.131 Synthesis of Statine via Cyanohydrin (543).

selectivity. Hydroboration and oxidation of the alkene generated the carboxyl group, although in low yield (1698). Addition of the lithium anions of ethyl or tert-butyl propiolate to nitrones gave 4-substituted-4-hydroxylamine acetylenic esters. Reduction of the hydroxylamine and alkyne functionalities resulted in γ-substituted γ-amino-α,β-unsaturated butyric esters as mixtures of E- and Z-isomers. The alkene bond could then be dihydroxylated (1699). Electrophilic 2,2-dimethylaziridine was opened with ethyl cyanoacetate (or other substituted cyanoacetates) in the presence of sodium ethoxide, giving 2-cyano-3,3-dimethyl-GABA (1700).

Similarly, C2−C3 bonds have been generated from nucleophilic acetate and electrophilic aminoethyl equivalents. An analog of Baclofen with the 3-(4'-chlorophenyl)-substituent moved to the 2-position was synthesized by alkylation of diethyl (4'-chlorophenyl) malonate with bromoacetonitrile, followed by hydrogenation of the nitrile and decarboxylation by hydrolysis. The dechloro analog was also synthesized, as were γ-aminophosphonic acid analogs (549). 3-Cyano-2-phenylpropionic acid was obtained by treatment of the LiHMDS enolate of methyl phenylacetate with bromoacetonitrile. Reduction with LiEt₃BH gave the 4-amino-2-phenylbutyric acid product, which was subsequently converted into an optically enriched product via enantioselective protonation (1701). Other methyl aryl acetates were also deprotonated with LiHMDS and alkylated with bromoacetonitrile; reduction of the nitrile was accomplished after resolution of the cyanopropionate (1702). Other C2−C3 bond-forming reactions rely on amino acid-derived aminoethyl components, as discussed in Section 11.4.2d.

A cyclopropanation of ketene acetals with diazoacetonitrile produced cyclopropylnitriles; a two-step hydrogenation first removed the ketal protection and cleaved

the cyclopropyl ring, and then reduced the cyano group to give γ-amino acids (see Scheme 11.132). By carrying out the cyclopropyl cleavage with deuterium, 3-deutero products were obtained (1703).

11.4.2d Elaboration of α-Amino Acids and β-Amino Acids

Glutamic acid already contains a 4-aminobutyric acid skeleton, and so is useful for preparing γ-amino acids, just as aspartic acid is convenient for β-amino acid synthesis. [¹³N]-GABA suitable for PET imaging was obtained by incorporation of [¹³N]-ammonia into α-oxoglutaric acid using glutamate dehydrogenase, giving [¹³N]-Glu. This was regioselectively α-decarboxylated by glutamate decarboxylase (1704). The same procedure was used many years earlier in a synthesis of the [¹⁵N]-GABA analog (1705). Racemic [4-²H]-4-amino-5-bromopentanoic acid was prepared from DL-[2-²H]-Glu via cyclization to pyroGlu-OEt, reduction of the α-carboxyl group to 5-(hydroxymethyl)pyrrolidinone, bromination with PPh₃/CBr₄, and lactam hydrolysis (1527).

Derivatization of other α-amino acids requires bis-homologation of the α-carboxy group. Treatment of the acid chloride of Cbz-Gly with ethyl benzoyloxyacetate under basic conditions gave, after hydrogenation and hydrolysis, 2-hydroxy-3-keto-4-aminobutyric acid (1614). Boc-Sar (N-methyl-Gly) was activated as the N-hydroxysuccinimide ester and reacted with methyl or ethyl cyanoacetate in the presence of KOtBu, producing the enol tautomer of the adduct, 2-cyano-3-hydroxy-4-aminobute-2-enoate, as evidenced by NMR and X-ray analysis (1706). Glycinal, attached to the C-terminus of a resin-linked peptide, was employed for a Horner–Emmons reaction with (EtO)₂P(O)CH₂CO₂allyl in the presence of LiHMDS to give 4-aminobut-2-enoic

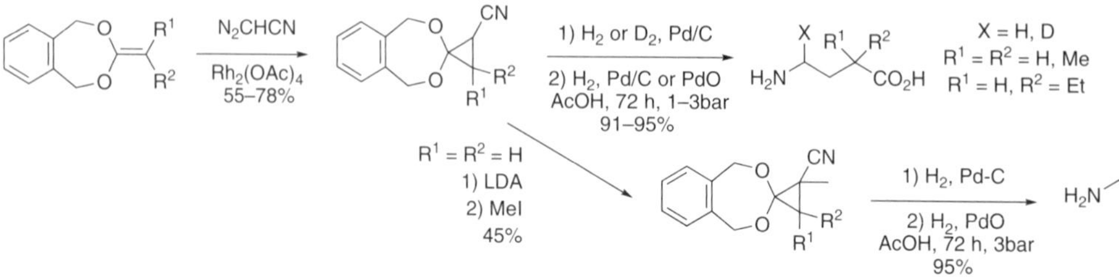

Scheme 11.132 Synthesis of γ-Amino Acids from Cyanocyclopropanes (1703).

acid (*1707*). The same homologation was carried out on glycinal derivatives in solution (*1708*). 5-Amino-3-carboxypentanoic acid was synthesized by a Wittig reaction of *N*-benzoyl glycinal with a phosphorane derivative of monomethyl succinate, with hydrogenation and hydrolysis giving the amino diacid (*1709*). An early attempt at homologation of L-Leu-OEt employed an intramolecular NaH-induced Dieckmann condensation of the *N*-(ethoxycarbonylacetyl) derivative. This provided a pyrrolin-2-one intermediate that was decarboxylated to give the lactam of racemic 3-keto-4-amino-6-methylheptanoic acid. Reduction of the ketone (H₂-Pt) followed by lactam hydrolysis with 48% HBr produced *threo*-3-hydroxy-4-amino-6-methylheptanoic acid (*1710*). The same procedure was applied to β-(3-pyridyl)-Ala, but with *N*-(2-ethoxycarbonylpropionyl) as the nucleophile to introduce a 2-methyl substituent in the product, 2-methyl-3-hydroxy-4-amino-5-(3-pyridyl)-pentanoic acid (*1710*). Another homologation procedure employed the radical generated by decarboxylative reduction of selenoesters of *N*-protected amino acids to add to methyl acrylate or [2-(methoxycarbonyl)propenyl]tributylstannane, providing γ-substituted γ-amino esters (*1711*).

Other modifications of α-amino acids include using the amino group of Gly to displace the SMe group of 2-(diethylphosphono)-*S*-methylthioacetamidinium iodide to give an analog of phosphocreatine with the external amidine nitrogen γ to the carboxyl group (*653*).

β-Amino acids can also be homologated, with Cbz-β-Ala converted into 2-keto-GABA via coupling with cyanomethylenetriphenylphosphorane in the presence of EDC/DMAP, followed by ozonolysis (*1712*), or by reaction of the ethyl ester with dimesylsodium followed by a Pummerer-type rearrangement of the β-keto-sulfoxide (*1713*).

11.4.2e Elaboration of γ-Amino Acids

A number of alkylations of γ-aminobutyric acids have been reported. The *N*-Boc-protected lactam of GABA was deprotonated and α-dialkylated with iodomethane; lactam hydrolysis with NaOH provided 2,2-dimethyl-GABA (*1714*). Selenation of the α-anion of Cbz-GABA-O*t*Bu with phenylselenylbromide gave 4-amino-2-(phenylseleno)butanoic acid, which was deprotonated again and reacted with formaldehyde to introduce an α-(hydroxymethyl)group. Oxidative elimination of the seleno group provided 4-amino-2-(hydroxymethyl)-2-butenoic acid, which could be converted into 2-(fluoromethyl)-, -(chloromethyl)-, or -(bromomethyl)-derivatives (*1528*). *N*-Boc 4-amino-3-methylbutanoic acid *tert*-butyl ester was also α-selenated with phenylselenylbromide. Oxidative elimination of the seleno substituent provided (*E*)-4-amino-3-methyl-2-butenoic acid (*1530*). The same reaction sequence was applied to the synthesis of (*E*)-4-amino-3-phenyl-2-butenoic acid and (*E*)-4-amino-3-(4-chlorophenyl)-2-butenoic acid (*1531*). Allylation of the lactam 5-ethoxy-2-pyrrolidone

with allyltrimethylsilane in the presence of titanium tetrachloride produced 5-allyl-2-pyrrolidone. Hydrolysis gave 4-amino-5-heptenoic acid, which could be ozonolyzed to 4-amino-6-oxohexanoic acid (*1523*).

Other derivatizations of γ-aminobutyric acids include a synthesis of 4-amino-3-chlorobutanoic acid in 45% yield by photochlorination of GABA (*1526*). Racemic GABOB was converted to 3-fluoro-4-aminobutyric acid by treatment with sulfur tetrafluoride in liquid HF (*1715*). The alkene of 3-vinyl-4-aminobutyric acid was ozonolyzed and the aldehyde used for Wittig couplings to give vinyl-substituted 3-vinyl-4-aminobutyric acid derivatives (*1690*). Racemic [5,6-³H₂]-4-aminohex-5-enoic acid was obtained by Lindlar reduction of racemic 4-aminohex-5-ynoic acid (*1716*).

γ-Guanidino-β-hydroxybutyric acid was prepared by formation of a guanyl group using the amino terminus of GABOB (*1717*), while carnitine was derived from GABOB via trimethylation of the amino group (*1687*).

11.4.2f Other Syntheses

Ozonolysis of 2-substituted 1,2,3,4-tetrahydropyridines provided 4-substituted γ-amino methyl esters in 40–71% yield, with the γ-amino aldehyde as the major byproduct (*677*). Lactams of GABA and 3-ethyl-4-aminobutyric acid were produced from ethane-1,2-dinitrile and butane-1,2-dinitrile, respectively. One nitrile group was selectively hydrolyzed using a microbial cell catalyst with aliphatic nitrilase activity, producing β-cyanocarboxylic acids. The second cyano group was then hydrogenated, accompanied by cyclization to form the lactam (*1718*).

11.4.3 Asymmetric Syntheses of γ-Amino Acids (see Table 11.6)

11.4.3a Resolutions of Racemic γ-Amino Acids

There have been comparatively few resolutions reported for γ-amino acids. Quinine was used to resolve *N*-phthaloyl 2-methyl-4-aminobutyric acid (*1668*) and *N*-benzoyl 4-aminopentanoic acid (*1678*) via diastereomeric salt formation, while the enantiomers of 2-aminohex-5-ynoic acid were separated using (+)- and (−)-binaphthylphosphoric acid (*1524*). Immobilized benzylpenicillin acylase enzymatically resolved *N*-phenacetyl 3-hydroxy-4-aminobutyronitrile, with *N*-methylation and nitrile hydrolysis providing (*S*)-carnitine (*1719*). Other resolutions of carnitine were summarized in 1987 (*1540*). γ-Substituted-γ-aminobutyric acids (with ethynyl, allenyl, and vinyl substituents) have also been resolved by immobilized penicillin acylase-catalyzed hydrolysis of their *N*-phenylacetyl derivatives. The hydrolyzed (*R*)-enantiomer was produced with 78–96% ee, with the remaining (*S*) enantiomer possessing 83-99% ee (*1720*). Both 3-phenyl-GABA and 3-(4-chlorophenyl)-GABA were synthesized by a

Table 11.5 Syntheses of Achiral or Racemic γ-Amino Acids

H_2N-R-CO_2H	Method	Yield	Reference
Amination			
$-CH_2CH_2CH_2-$	opening of γ-butyrolactone with KNPhth, hydrolysis	73% opening 68% hydrolysis	1958 (*1667*)
$-CH_2CH_2CH(Me)-$	KNPhth opening of 2-Me-butyrolactone	70%	1959 (*1668*)
$-CH_2CH(OH)CH_2-$	displacement of 4-Cl-3-HO-butyric acid with KNPhth, hydrolysis	—	1935 (*1669*)
$-CH_2CH(Me)CH_2-$	displacement of 4-Cl-3-Me-butyronitrile with KNPhth, hydrolysis	40%	1945 (*1670*)
$-CHFCH=CH-$	KNPhth displacement of 2-F-4-Br-crotonate, hydrolysis	75% amination 55% hydrolysis	1967 (*1671*)
$-CH(CH=C=CHF)CH_2CH_2-$	amination of HOCH(CH=C=CHF) CH_2CH_2OTHP via Mitsunobu of with HNPhth, deprotection and oxidation	40–45% amination and O-deprotection 70% oxidation	1987 (*1672*)
$-CH(Me)CH=CH-$ $-C(Me)_2CH=CH-$	bromination of α,β-unsaturated acid, amination with ammonia	52–96% bromination and amination	1979 (*1673*)
$-CH_2CH=CH-$ $-CH_2CH=C(Me)-$ $-CH_2CH=C(Br)-$ $-CH_2CH=C(Cl)-$ $-CH(Me)CH=CH-$ $-CH(Br)CH=CH-$	bromination of α,β-unsaturated acid, amination with ammonia	7–29% bromination and amination	1978 (*1674*)
$-CH(CO_2H)CH(CO_2H)CH_2-$	amination of triethyl aconitate, saponification	12% overall	1935 (*1675*)
$-CH_2CH=CH-$ $-CH(Me)CH=CH-$ $-CH(nHex)CH=CH-$ $-C(Me)_2CH=CH-$	amination of $(\eta^3$-allyl)Fe(CO)$_2$(NO) complex	63–95%	1998 (*1677*)
$-CH(Me)CH_2CH_2-$	reductive amination of 4-ketovaleric acid via phenylhydrazone	—	1911 (*1678*)
$-CH(Me)CH(OH)CH_2-$	reductive amination of β-hydroxylevulinic acid via oxime	—	1927 (*1679*)
$-CH(Ph)CH_2CH_2-$ $-CH(4-MeO-Ph)CH_2CH_2-$ $-CH(2-naphthyl)CH_2CH_2-$	reductive amination of 4-keto-4-aryl-butyric acid via hydrazone	50–96% reduction	1977 (*1680*)
$-CH(Me)CH_2CH_2-$	reductive amination of 4-keto-pentanoic acid with norepinephrine	51%	1983 (*1681*)
$-CH(Me)CH_2CH(Me)-$	Curtius rearrangement of monomethyl ester of 2,4-dimethylglutaric acid	—	1999 (*1641*)
$-CH(CH_2CH_2NH_2)CH_2CH_2-$	Schmidt rearrangement of 5-(2-carbomethoxyethyl)-2-pyrrolidone	29%	1953 (*146*)
$-CH_2CH_2CH(4-iPrO-Ph)-$ $-CH(4-iPrO-Ph)CH_2CH_2-$	reaction of CH_2=CCN with NaC(R)(CO$_2$Et)$_2$, hydrolysis and Hofmann rearrangement	70–97% Hofmann rearrangement	1996 (*1683*)
$-CH(R)CH_2CH_2-$ R = CH=C=CH$_2$, C(Me)=C=CH$_2$, CH=C=CHMe	aza-Cope rearrangement	35–80% rearrangement	1984 (*1682*)
$-CH(Et)CH=CH-$	Pd-catalyzed rearrangement of allylic sulfoximine	87% allylic sulfoximine 57% rearrangement	1996 (*1684*)
$-CH(Me)CH=CH-$ $-CH(nPr)CH=CH-$ $-CH(CH_2OBn)CH=CH-$ as γ-azidocyano precursor	Pd(0)-catalyzed azidation of allylic cyanohydrin carbonate, RCH=CHCH(CN)OCO$_2$Et	85–90%	2001 (*1686*)
$-CH(Ph)CH_2CH(OH)-$ $-CH(nPr)CH_2CH(OH)-$ $-C(Me)_2CH_2CH(OH)-$	Rh-catalyzed intramolecular C–H insertion of H$_2$NSO$_2$ ester of 2-HO esters, hydrolysis of resulting oxathiazinane	78–91% rearrangement 86% hydrolysis	2001 (*518*)
Carboxylation			
$-CH_2CH(OH)CH_2-$ with NH$_2$, NMe$_3$	CN displacement of 1-Cl-2-OH-3-NH$_2$-propane prepared from PhCHO, NH$_3$, ClCH$_2$CH(−O−)CH$_2$, methylation	69% Cl precursor 78% methylation	1953 (*1687*)

Table 11.5 Syntheses of Achiral or Racemic γ-Amino Acids (continued)

H_2N-R-CO_2H	Method	Yield	Reference
$-CH_2CH_2N-$ (with squaric acid replacing carboxyl)	reaction of monoprotected diamine with diethyl squarate	82–87%	1995 (1688)
$-CH_2COCH_2-$ with NMe_2	reaction of $NaCH_2COCH_2NMe_2$ with $(EtO)_2CO$	—	1991 (1689)
$CH(CT=CHT)CH_2 CH_2-$	Lindlar tritiation of 4-aminohex-5-ynoic acid	—	2003 (1716)
$-CH_2CH(CH=CH_2)CH_2-$ $-CH_2CH(CMe=CHMe)CH_2-$ $-CH_2CH(CH=CHMe)CH_2-$ $-CH_2CH(CH=CHEt)CH_2-$	rearrangement of 1-amino-2,3-dehydro-butanols	27–80% rearrangement	1998 (1690)
$-CH(iBu)CH(NH_2)CH_2-$	intramolecular conjugate addition of N-aminomethyl group of 4-amino-6-methylhept-2-enoic acid, aminal deprotection	78% conjugate addition 60% aminal deprotection	2005 (773)

Fusion of Amino and Carboxyl Components

H_2N-R-CO_2H	Method	Yield	Reference
$-CH_2CH_2CH(Me)-$ $-CH_2CH(Me)CH_2-$	Michael addition of nitromethane to 2-alkenoic esters, hydrogenation and hydrolysis	—	1982 (527)
$-CH_2CH(Me)CH_2-$ $-CH_2CH(Et)CH_2-$ $-CH_2CH(nPr)CH_2-$ $-CH_2CH(iPr)CH_2-$ $-CH_2CH(nBu)CH_2-$ $-CH_2CH(iBu)CH_2-$ $-CH_2CH(sBu)CH_2-$ $-CH_2CH(tBu)CH_2-$	Michael addition of nitromethane to 2-alkenoic esters, hydrogenation and hydrolysis	40–74% addition 78–86% hydrogenation and deprotection	1989 (1691)
$-CH_2C(Me)_2CH_2-$	Michael addition of nitromethane to ethyl 3,3-dimethylacrylate, hydrogenation	48% addition 80% reduction	1990 (1530)
$-CH_2CH(cPr)CH_2-$	addition of nitromethane to 3-cPr-acrylate, nitro reduction/hydrolysis	76% Michael addition 55% nitro reduction 64% hydrolysis 53% trimethylation	1990 (1546)
$-C(CH_2CH_2CO_2H)_3$	addition of nitromethane to 3 eq. of tert-butyl acrylate, reduction	75–81% Michael addition 89–93% reduction	1996 (1693)
$-CH_2CH(Ph)CH_2-$ $-CH_2CH(4-Cl-Ph)CH_2-$	addition of nitromethane to methyl cinnamates	84–89% addition	2005 (1694)
$-C(CH_2CH_2CO_2H)_2CH_2CH_2-$	addition of nitromethane to 3 equiv of tert-butyl acrylate, nitro reduction	72% addition 88% nitro reduction	1991 (1661)
$-CH_2CH(Ph)CH_2-$ $-CH_2CH(4-Br-Ph)CH_2-$ $-CH_2CH(4-F-Ph)CH_2-$ $-CH_2CH(2-thienyl)CH_2-$ $-CH_2C[-(CH_2)_5-]CH_2-$ $-CH_2C[-(CH_2)_2CH(tBu)(CH_2)_2-]CH_2-$	addition of nitromethane to acrylates attached to dendritic polyglycerol support, nitro reduction with cyclization/cleavage from support	24–72% addition 35–70% reduction/cyclization	2002 (1692)
$-CH(Et)CH_2CH_2-$ $-CH(iPr)CH_2CH_2-$ $-CH(iBu)CH_2CH_2-$ $-CH[(CH_2)_6OH]CH_2CH_2-$ $-C[-(CH_2)_5-]CH_2CH_2-$ $-CH(Et)CH_2CH(Me)-$ $-CH(iPr)CH_2CH(Me)-$ $-CH(iPr)CH(Me)CH_2-$ $-CH(iPr)CH=CH-$	reductive conjugate addition of nitrones R CH=N^+(Bn)O^- to acrylates in presence of SmI_2 deprotection via hydrogenation over Raney Ni	55–81% 62–82% hydrogenation	2003 (1696)
$-CH(CCH)CH_2CH_2-$	Michael addition of enolate of $PhCH=NCH_2CCSiMe_3$ to methyl acrylate	53% addition	1975 (1520)

(Continued)

Table 11.5 Syntheses of Achiral or Racemic γ-Amino Acids (continued)

H₂N-R-CO₂H	Method	Yield	Reference
$-CH(iBu)CH(OH)CH_2-$	cyanohydrin formation from ethyl formyl acetate equivalent, Grignard addition	80% cyanohydrin formation 40% addition 64% de *threo*	1993 (*543*)
$-CH_2CH(OH)CH_2-$	CN addition to ethyl formyl acetate, reduction	100% addition 93–100% reduction	1990 (*1697*) 1993 (*543*)
$-CH(Ph)(CH_2)_2-$	one-pot reaction of ZnBr(CH₂)₂CO₂Et with imine of PhCHO and Me₂NSiMe₃	59%	1997 (*620*)
$-CH(iBu)CH(OH)CH_2-$	γ-oxy allyltin addition to activated imine, alkene hydroboration/oxidation	84% addition 100% de *syn* 46% hydroboration 27% deprotection	1989 (*1698*)
$-CH(Me)CCH-$ $-CH(Et)CCH-$ $-CH(iPr)CCH-$ $-CH(tBu)CCH-$ $-CH(Ph)CCH-$ (all with N-OH) $-CH(iPr)CH=CH-$	addition of LiCCCO₂R to nitrone, reduction	10–94% addition 40–75% reduction	1997 (*1699*)
$-CH_2C(Me)_2C(Ph)(CN)-$	opening of 2,2-dimethyl aziridine with ethyl cyanoacetates	60%	1987 (*1700*)
$-CH_2CH_2CH(Ph)-$	reaction of BrCH₂CN with LiCH(Ph)CO₂Me, CN reduction	77% CN formation 25% reduction	2002 (*1701*)
$-CH_2CH_2CH(Ph)-$ $-CH_2CH_2CH(4-iBu-Ph)-$ $-CH_2CH_2CH(6-MeO-2-naphthyl)-$	reaction of BrCH₂CN with LiCH(Ar)CO₂Me, CN reduction	73–76% alkylation	2004 (*1702*)
$-CH_2CH_2CH(Ph)-$ $-CH_2CH_2CH(4-Cl-Ph)-$	reaction of BrCH₂CN with NaC(R)(CO₂Et)₂, hydrogenation, hydrolysis	82% alkylation 84% reduction 81% hydrolysis	1997 (*549*)
$-(CH_2)_3-$ $-CH_2CH_2C(Me)_2-$ $-CH_2CH_2CH(Et)-$ $-CH_2CHDCH(Et)-$ $-CH_2CH(Me)CH_2-$	cyclopropanation of ketene acetal with diazoacetonitrile, (deprotonation and alkylation), hydrogenolytic cleavage of cyclopropane ring then cyano reduction	55–78% cyclopropanation 45% alkylation 91–95% hydrogenation	2001 (*1703*)

Elaboration of α- and β-Amino Acids

$-(CH_2)_3-$ with [¹³N]	enzymatic amination of α-oxoglutaric acid with glutamate dehydrogenase, decarboxylation with glutamate decarboxylase	—	1985 (*1704*)
$-(CH_2)_3-$ with [¹⁵N]	enzymatic amination of α-oxoglutaric acid with glutamate dehydrogenase, decarboxylation with glutamate decarboxylase	80% amination 68% decarboxylation	1975 (*1705*)
$-CD(CH_2Br)(CH_2)_2-$	cyclization of [2-²H]-Glu to pyroGlu, carboxyl reduction, bromination, hydrolysis	quant. cyclization and reduction 50% bromination	1981 (*1527*)
$-CH_2COCH(OH)-$	addition of anion of BzOCH₂CO₂Et to Cbz-Gly-Cl, deprotection	12% overall	1978 (*1614*)
$-CH_2C(OH)=C(CN)-$	activation of Boc-Sar as NHS ester, reaction with Me or Et cyanoacetate and KOtBu	62–75%	2001 (*1706*)
$-CH_2CH=CH-$	Horner–Emmons reaction of resin-linked glycinal with (EtO)₂P(O)CH₂CO₂allyl/ LiHMDS	—	2004 (*1707*)
$-CH_2CH=CH-$	Horner–Emmons reaction of protected glycinal with (EtO)₂P(O)CH₂CO₂Etl/ LiHMDS	57–82%	2004 (*1707*)

Table 11.5 Syntheses of Achiral or Racemic γ-Amino Acids (continued)

H₂N-R-CO₂H	Method	Yield	Reference
$-(CH_2)_2CH(CH_2CO_2H)-$	Wittig reaction of Bz-glycinal and succinate phosphorane, reduction, hydrolysis	45% Wittig 71% reduction 61% hydrolysis	1993 (*1709*)
$-CH(iBu)CH(OH)CH_2-$ $-CH[CH_2(3-pyridyl)])CH(OH)$ $CH(Me)_2-$	homologation of Leu or β-(3-pyridyl)Ala via intramolecular Dieckmann condensation of N-COCHRCO₂Et derivative, decarboxylation, ketone reduction, lactam hydrolysis	41–85% cyclization 76–100% decarboxyla-tion 82–94% reduction 62–74% hydrolysis	1976 (*1710*)
$-CH(Bn)CH_2CH_2-$ $-CH(Bn)CH_2CH(=CH_2)-$	radical decarboxylation of amino acid, addition to acrylates	81–91% selenoester 53–58% decarboxyla-tion/addition	1997 (*1711*)
$-CH(CH_2PO_3H_2)NHCH_2-$ $-CH(CH_2PO_3H_2)N(Me)CH_2-$	reaction of Gly, Sar with MeSC(=NH) CH₂PO₃Et₂	44–99%	1997 (*653*)
$-CH_2CH_2CO-$	homologation of β-Ala	—	1999 (*1712*)
$-CH_2CH_2CO-$	homologation of β-Ala using NaCH₂SOMe then Pummerer rearrangement	32%	1995 (*1713*)

Elaboration of γ-Amino Acids

$-CH_2CH_2C(Me)_2-$	dialkylation of GABA lactam, hydrolysis	44% dialkylation 85% hydrolysis	1993 (*1714*)
$-CH(Me)CH_2CH_2-$	hydrolysis of 5-Me-2-pyrrolidinone	—	1982 (*527*)
$-CH_2CH_2CH(SePh)-$ $-CH_2CH=C(CH_2OH)-$ $-CH_2CH=C(CH_2Cl)-$ $-CH_2CH=C(CH_2Br)-$ $-CH_2CH=C(CH_2F)-$	deprotonation and phenylselenation of Cbz-GABA-OtBu, deprotonation and hydroxymethylation, oxidative elimination, halogenation	79% selenation 49% hydroxymethyla-tion 55% elimination 55–70% halogenation	1986 (*1528*)
$-CH_2CH(Me)CH(SePh)-$ $-CH_2C(Me)=CH-$	α-deprotonation and selenation of 4-amino-3-methylbutanoic acid, oxidative elimination	85% selenation 40% oxidative elimination	1990 (*1530*)
$-CH_2CH(Ph)CH(SePh)-$ $-CH_2CH(Ph)CH(SePh)-$ $-CH_2C(4-Cl-Ph)=CH-$ $-CH_2C(4-Cl-Ph)=CH-$	α-deprotonation and selenation of 4-amino-3-arylbutanoic acid, oxidative elimination	70% selenation 64% oxidative elimination	1987 (*1531*)
$-CH(CH_2CH=CH_2)CH_2CH_2-$ $-CH(CH_2CHO)CH_2CH_2-$	allylation of 5-EtO-2-pyrrolidone, hydrolysis, ozonolysis	72% allylation 6% hydrolysis 35% ozonolysis	1991 (*1523*)
$-CH_2CH(Cl)CH_2-$	photochlorination of GABA	45%	1981 (*1526*)
$-CH_2CHFCH_2-$	fluorodehydroxylation of GABOB	50%	1979 (*1715*)
$-CH_2CH(OH)CH_2-$ with N-C(=NH)NH₂	guanylation of GABA	—	1935 (*1717*)
$-CH_2CH(OH)CH_2-$ with NMe₃	methylation of GABA	78% methylation	1953 (*1687*)
$-CH_2CH(CH=CMe)CH_2-$ $-CH_2CH(CMe=CHEt)CH_2-$ $-CH_2CH(CH=CHnPr)CH_2-$ $-CH_2CH(CH=CHiPr)CH_2-$ $CH_2CH(CH=CMe_2)CH_2-$ $CH_2CH(iPr)CH_2-$	ozonolysis of 3-vinyl-4-aminobutyric acid, Wittig reaction, hydrogenation	88% ozonolysis 29–51% Wittig 20% hydrogenation	1998 (*1690*)

Other Syntheses

$-CH(Me)CH_2CH_2-$ $-CH(Et)CH_2CH_2-$ $-CH(Ph)CH_2CH_2-$	ozonolysis of 2-substituted-1,2,3,4-tetrahydropyridines	40–71%	1994 (*677*)
$-CH_2CH_2CH_2-$ $-CH_2CH(Et)CH_2-$	aliphatic nitrilase hydrolysis of one cyano group of ethane-1,2-dinitriles, hydrogenation	99–100% hydrolysis 80–91% reduction	198 (*1718*)

chemoenzymatic route that began with addition of nitromethane to the methyl cinnamates. An enzymatic resolution was carried on the γ-nitro esters using α-chymotrypsin, followed by nitro reduction with Raney nickel (*1694*).

Various 4-substituted GABA analogs, derivatized as their *N*-pentafluoropropionyl ethyl esters, were successfully separated by gas chromatography with Chirasil-Val as a chiral stationary phase. Attempts at resolution by ligand-exchange reversed-phase HPLC were unsuccessful (*698*). A GC-based analysis of carnitine enantiomers employed online conversion of carnitine into β-hydroxy-γ-butyrolactone (based on thermal intramolecular nucleophilic displacement of the trimethylammonium group by the carboxylate oxygen), followed by resolution on a β-cyclodextrin-based column (*1552*). Carnitine and *O*-acyl carnitine enantiomers were resolved by HPLC using a teicoplanin chiral stationary phase (*1553*). Analysis of the enantiomeric purity of *O*-acetyl carnitine has also been accomplished on a normal reversed-phase column via formation of diastereomeric derivatives by amidation with L-Ala-β-naphthylamide (*1554*). Spirocyclic Pd complexes formed from di-μ-chlorobis{[2-(1-dimethylamino-*xN*) ethyl]phenyl-*xC*}dipalladium and a range of γ-amino acids provided distinct chemical shift differences in both ^{1}H and ^{13}C NMR spectra, allowing for the determination of enantiomeric purity (see Scheme 1.7) (*708*).

Several chemical dynamic resolutions have been reported. Racemic *N*-phthaloyl 4-amino-2-phenylbutyric acid was treated with oxalyl chloride and triethylamine to form a ketene intermediate, with addition of the chiral alcohol (*R*)-pantolactone generating predominantly the *N*-phthaloyl (*R*)-4-amino-2-phenylbutyric acid ester as an 85:15 diastereomeric mixture (*1701*). 2-Aryl-3-cyanopropionic acids, obtained via alkylation of methyl arylacetates with bromoacetonitrile, were activated as the acid chloride and then esterified with (*R*)- or (*S*)-*N*-phenylpantolactam. Under the reaction conditions, the derivative epimerized to give predominantly one diastereomer, with 74–86% de. After column chromatography to purify one diastereomer (>96% de), the ester was cleaved and the nitrile hydrogenated to give 2-aryl-γ-amino acids (Ar = Ph, 4-*i*Bu-Ph, 6-MeO-2-naphthyl) with >99% ee (*1702*).

11.4.3b Amination Reactions

Several nucleophilic amination reactions leading to γ-amino acids employ optically enriched electrophilic substrates. Epoxide substrates provide β-hydroxy-γ-amino acids. (*R*)-GABOB was synthesized during an early application of the Sharpless epoxidation procedure to 4-hydroxy-1-butene. After epoxidation (25% yield, only 55% ee) the hydroxyl group was oxidized to an acid, and the epoxide opened with ammonium hydroxide in THF (*795, 1721*). The GABOB derivative (*R*)-(−)-carnitine has also been prepared by epoxide opening (using trimethylamine), but in this case the

chiral epoxide was prepared by enzymatic resolution of alkyl 3,4-epoxybutyrates (*1722*). A synthesis of (*S*)-vigabatrin began with homochiral 2,3-epoxy-5-phenyl-1-pentanol, which contains the carboxy group masked as a phenyl group. The amino group was introduced by epoxide opening with benzylamine, the carboxyl group revealed by oxidative cleavage of the phenyl ring, and the 1,2-dihydroxyethane side chain converted to the vinyl group (*1517*).

Nucleophilic aminations of other γ-electrophiles have been reported. Chemoselective opening of the γ-lactone of homochiral 3,4-dihydroxybutanoic acid by trimethylsilyliodide in ethanol gave ethyl γ-iodo-β-hydroxybutanoate. Iodide displacement with azide, saponification, and hydrogenolysis of the azide provided (*R*)-GABOB (*2192*). (*S*)-GABOB was derived from ethyl (*S*)-4-chloro-3-hydroxybutyrate via displacement with NaN$_3$ and hydrogenation over Pd/C (*1724*). Methyl (*R*)-3,4-dihydroxybutanoate was selectively tosylated on the primary hydroxy group, displaced with azide, and eventually reduced to give GABOB (*215*). Both enantiomers of GABOB were synthesized via displacement of ethyl 4-bromo-2-(*E*)-butenoate with (*S*)-phenylethylamine. The secondary amine was Cbz-protected and the 3-hydroxy-group introduced by iodocyclization with the urethane carbonyl oxygen, producing a 1:1 diastereomeric mixture of oxazolidin-2-ones. Chromatographic resolution, radical deiodination, reductive removal of the phenethyl auxiliary, and hydrolysis of the oxazolidinone completed the syntheses (*1725*). (3*R*)-4-Amino-3-(4-chlorophenyl)butanoic acid, (*R*-baclofen), was obtained by azide displacement of a chiral iodo precursor (*1726*), while azide displacement of homochiral tosylated benzyl 3-(hydroxymethyl)-5-methylhexanoic acid gave (*S*)-(+)-3-isobutyl-GABA. The chiral substrate for the latter synthesis was prepared by asymmetric alkylation of 5-methylpentanoic acid (derivatized with Evans' oxazolidinone auxiliary) with benzyl bromoacetate, followed by auxiliary removal and reduction of the free acid to a hydroxymethyl group (*1514*). Another nucleophilic reaction aminated homochiral ethyl 3-methyl-4-bromobutyrate with *N*-Boc,*O*-benzyl hydroxyamine (*1727*).

(*R*)-(−)-Carnitine was prepared by reaction of homochiral 1-chloro-2-hydroxy-4-pentene with aqueous trimethylamine, followed by oxidative cleavage of the alkene to generate the acid group. The substrate was readily prepared from commercially available (*R*)-epichlorohydrin by opening with vinylmagnesium bromide (*1728*). Another synthesis of carnitine relied on amination of 3-hydroxy-4-chlorobutanoic acid using trimethylamine, with the chiral substrate prepared by reduction of 4-chloro-3-ketobutyrate with Baker's yeast (*1729*). Alternatively, asymmetric hydrogenation of ethyl 3-keto-4-chlorobutyrate with (*S*)-BINAP-Ru as catalyst gave the precursor in 97% yield with 97% ee (*1730*). Carnitine has also been prepared by enzymatic resolution of diethyl 3-hydroxyglutarate, with the trimethylamino group added by displacement of a bromo intermediate obtained by a Hunsdiecker rearrangement (*1731*).

E-Vinylogous (*R*)-amino acids were produced via azide displacement of homochiral acylated vinylcyano-hydrins, RCH=CHCH(OCO₂Et)CN, or the corresponding esters, RCH=CHCH(OCO₂Et)CO₂Et. The homochiral cyanohydrin substrates were obtained from unsaturated aldehydes, RCH=CHCHO, via enzymatic reaction with HCN and oxynitrilase (>95% ee), followed by acylation with ethyl chloroformate. Reaction of the vinyl cyanohydrins with TMSN₃/Pd(PPh₃)₄ produced the unsaturated azido nitriles, RCH(N₃)CH=CHCN, with 81–85% ee, while the esters reacted to give products with 95% ee. Azide reduction completed the synthesis (*1732*).

Homochiral allylic alcohols with a protected diol as a masked carboxyl group have been derived from (*R*)-isopropylidene glyceraldehyde, ethyl (*S*)-lactate or D-mannitol. The allylic alcohol undergoes a Mitsunobu reaction with phthalimide by two possible pathways; direct S_N2-inversion to give a β, γ-unsaturated-α-amino acid equivalent, or S_N2′ allylic reaction to give γ-amino-α,β-unsaturated acids, with the product distribution depending on the substrate structure. Oxidative cleavage of the diol by Pb(OAc)₂ generated the acid group, with the phthalimide converted to an amino group by hydrazinolysis (*1733*). γ-Amino-α,β-unsaturated acids can also be obtained by a Mitsunobu reaction of homochiral γ-hydroxy-α,β-unsaturated esters using DEAD and phthalimide (*1733*), or by amine treatment of (η³-allyl)dicarbonylnitrosyliron complexes containing a chiral ester substituent (28–82% yields with 79 to >98% de) (*1677*).

A number of rearrangement reactions have been reported. (*R*)-GABOB was synthesized by a Curtius rearrangement of enzymatically resolved monoethyl 3-hydroxyglutarate (*1731*), or by a Hofmann rearrangement of a homochiral monoamide monomethyl ester of 3-hydroxyglutaric acid, prepared by an enzymatic ammonolysis of dimethyl 3-hydroxyglutarate (*1734*). The thiol analog of GABOB was prepared in a similar manner, with dimethyl 3-*p*-methoxybenzylthio-glutarate obtained from dimethyl 3-hydroxyglutarate by mesylation and displacement with the sodium salt of the thiophenol. Enzymatic monodemethylation provided optically active half-ester with up to 71% ee. Curtius rearrangement and *S*-deprotection gave (*R*)-3-mercapto-4-aminobutyric acid (*1724*). Likewise, homochiral monomethyl (*R*)-3-methylglutarate was obtained by enzymatic hydrolysis of the diester. Curtius rearrangement with DPPA produced (*S*)-4-amino-3-methyl-butanoic acid. Alternatively, treatment of the same ester intermediate with ammonia produced a monoamide, which was used for a Hofmann rearrangement to produce the enantiomeric (*R*)-4-amino-3-methylbutanoic acid (*1735*). The same procedure was used to prepare (*R*)- and (*S*)-4-amino-3-ethylbutanoic acid (*1530*), and both enantiomers of baclofen (*1736*), with the substrate for the latter products obtained from 4-chlorocinnamic acid via conversion into dimethyl 3-(4-chlorophenyl) glutarate followed by enzymatic hydrolysis of one of the methyl esters with chymotrypsin. Beckmann ring

expansion of a homochiral 2,2-dichloro-3-hydroxy-4-benzylcyclobutanone derivative (prepared via an asymmetric [2+2]cycloaddition reaction) induced with *O*-(mesitylenesulfonyl)hydroxyamine and Al₂O₃, followed by reduction with Zn–Cu in methanol, provided the lactam of (3*S*,4*S*)-3-hydroxy-4-amino-5-phenylpentanoic acid. Hydrolysis gave (−)-AHPPA (*1737*). The same route was used to prepare (−)-statine (*1738*).

Asymmetric reductive aminations are possible. Isopropyl 2,5-dimethoxyphenyl ketone was converted into statine via a diastereoselective reductive amination of the ketone using (*S*)-1-phenylethylamine and catalytic hydrogenation (94% de). The auxiliary was removed from the amine, and the aromatic ring converted to a cyclohexadiene ring, by a Birch reduction. Ozonolysis produced the β-keto-γ-amino ester precursor of statine, ethyl 3-keto-4-amino-6-methylheptanoate, with stereoselective reduction of the ketone leading to either diastereomer (*1739*).

The anticonvulsant (*S*)-3-aminomethyl-5-methyl-hexanoic acid (pregabalin) was prepared by hydrogenation of the cyano group of homochiral 3-cyano-5-methylhexanoic acid, itself prepared via an asymmetric hydrogenation of 3-cyano-5-methylhex-3-enoic acid (*1536*).

11.4.3c Carboxylation Reactions

α-Phenyl-GABA was prepared via carboxylation of *N*-aryl,*N*-Boc cinnamylamine, with enantioselectivity induced by using a *n*BuLi/(−)-sparteine complex for deprotonation. With carbon dioxide as the electrophile, the (*S*)-isomer was produced with 92% ee, while methyl chloroformate led to the (*R*)-enantiomer with 84% ee. Hydrogenation of the initial 3,4-dehydro adduct gave the desired product (*970*).

Nucleophilic carboxylations have also been developed. A vinyl group was used as a masked carboxyl group in a synthesis of the *N*-methyl 3-hydroxy-4-amino-5-phenylpentanoic acid component of hapalosin. (Vinyl)₂CuMgBr opened homochiral *N*-methyl,*N*-Boc 1,2-epoxy-3-amino-4-phenylbutane, followed by oxidative cleavage of the alkene (*1618*). Cyanide was also employed as the nucleophilic carboxyl equivalent for this synthesis (*1740*). A more general synthetic route opened homochiral *N*-Boc 2-(1,2-dihydroxyethane) aziridine with organocuprates in good yield, giving the *threo* diastereomers (see Scheme 11.133). Activation of the primary hydroxyl as a tosylate, displacement with NaCN, and hydrolysis gave the γ-amino acids (*1243*).

11.4.3d Amination Combined with Carboxylation

Several syntheses have employed both amination and carboxylation reaction steps. Trimethylamine opening of racemic epichlorohydrin gave 1-chloro-2-hydroxy-3-trimethylaminopropane. Resolution with (+)-L-tartrate, displacement of the chloride with cyanide, and hydrolysis produced (*R*)-(−)-L-carnitine (*1540*). Alternatively,

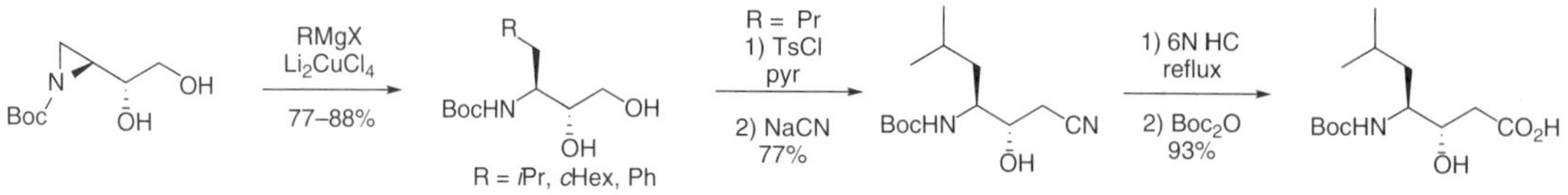

Scheme 11.133 Reaction of N-Boc 2-(1,2-Dihydroxyethane)-Aziridine (1243).

(R)-(−)-carnitine was derived from glycerol via an asymmetric mesylation of one primary hydroxyl group and displacement with trimethylamine, followed by conversion of the second primary hydroxy group to a bromide, displacement with cyanide, and hydrolysis (1741).

GABOB has also been prepared from (R)-epichlorohydrin, with a masked carboxyl group introduced by epoxide opening with phenyllithium, and the amino group by azide displacement of the chloro group (1742). Homochiral epoxides, derived from allylic alcohols via Sharpless epoxidation, were converted into anti 3-amino-1,2-diols by titanium-promoted oxirane opening with benzhydrylamine (see Scheme 11.134). The carboxyl group was then introduced by a stereodivergent introduction of cyanide, giving either syn or anti β-hydroxy-γ-amino acids (1743). Simple hydrolysis of the cyano group gave poor yields, and so the hydroxyl and amino groups were protected in an oxazolidine, and the nitrile converted to a carboxyl using non-hydrolytic conditions.

A similar synthetic route was used to prepare statine, (3S,4S)-3-hydroxy-4-amino-6-methylheptanoic acid, and its 3-epimer, with azide employed for the epoxide opening (1744). The azide was not reduced until after hydrolysis of the nitrile, allowing for reasonable hydrolysis yield (63%) compared to the poor yields observed with the previous sequence. All four diastereomers of statine amide were also prepared by a slightly different route, in which 3-hydroxy-5-methyl-1-hexene was used as the initial asymmetric epoxidation substrate. The azide group was then introduced by displacement of the

central hydroxyl group and the epoxide opened with KCN. Problems during hydrolysis of the azido cyanides were again reported (1745, 1746).

Asymmetric dihydroxylations also give substrates suitable for amination and carboxylation. Both statine and its C4 epimer were derived from an asymmetric dihydroxylation of ethyl 5-methylhex-2-enoate, with one diastereomer prepared by a regioselective Mitsunobu azidation of the 2,3-diol to give 2-hydroxy-3-azido-5-methylhexanoate and the other isomer by regioselective Mitsunobu tosylation followed by azide displacement. The carboxyl group was homologated by Arndt–Eistert reaction to form the statine products (837). Allyl bromide was asymmetrically dihydroxylated to give (S)-3-bromopropane-1,2-diol with 72% ee. The primary hydroxy group was converted to a chloride, and the bromide displaced with cyanide to introduce the carboxyl group. The chloride was then displaced with ammonia or trimethylamine to give, after hydrolysis, (R)-GABOB or (R)-carnitine with 90–95% ee (1747).

(R)-(−)-Baclofen was derived from racemic 2-(4-chlorophenyl)-4-pentenenitrile, via enzymatic (Rhodococcus) nitrile hydrolysis to give the resolved amide with >99.5% ee. Amide reduction with LiAlH$_4$ generated the amino group (77% yield), while the carboxyl group was derived by OsO$_4$/Jones oxidation of the alkene (53% yield). However, the final product possessed only 91% ee (1748).

A number of other syntheses introduce carboxyl and amino groups to chiral substrates; these are described in Section 11.4.3q below.

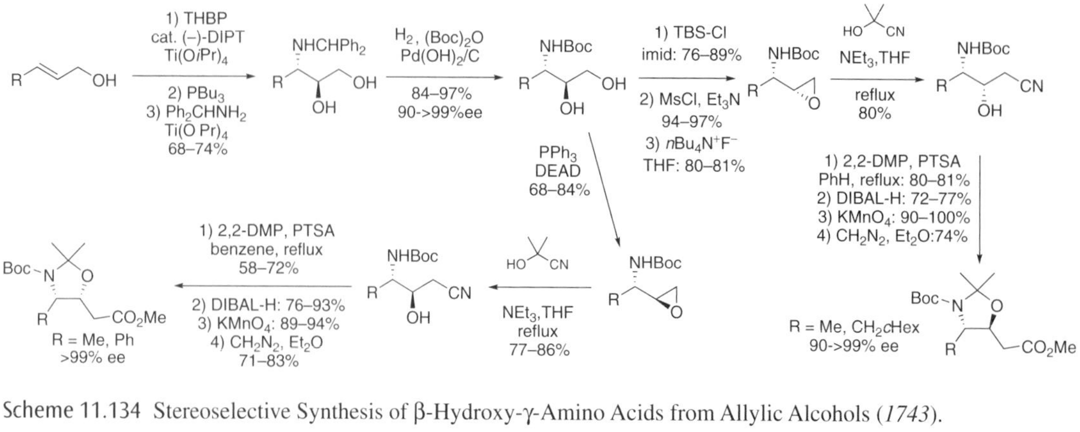

Scheme 11.134 Stereoselective Synthesis of β-Hydroxy-γ-Amino Acids from Allylic Alcohols (1743).

11.4.3e Fusion of Amino- and Carboxy-Containing Components: C2–C3 Formation

One method of forming the C2–C3 bond of γ-amino acids is to react a nucleophilic acetate equivalent with an electrophilic aminoethyl group. This approach is often used for elaborations of α-amino acids, with these derivatizations discussed further in Sections 11.4.3g–11.4.3l below. However, syntheses not relying on amino acids are also possible. A general route to 2-substituted 4-aminobutanoic acids proceeded via stereoselective introduction of a cyanomethyl group (aminoethyl equivalent) to the α-position of deprotonated alkyl and aryl acids derivatized with a chiral oxazolidinone auxiliary (see Scheme 11.135). Bromoacetonitrile was used as the electrophile. Only moderate stereoselectivity was observed, but the pure diastereomers were generally easily isolated. Oxazolidinone auxiliary removal and cyano group reduction gave the γ-amino acids (*1749*).

Nitroalkenes provide another electrophilic amino-ethyl group. Several alkanoic acids, derivatized with a chiral oxazolidinone auxiliary, were deprotonated and added to nitroethene, nitrostyrene, and other nitroolefins (see Scheme 11.136). Products with α-substituents (from the acid moiety) were produced with >90% de, but products with β-substituents (from the nitroalkene) had reduced stereoselectivity (72–92% de). α,β-Disubstituted 2-methyl-3-phenyl-4-nitrobutanoic acid was obtained from propionic acid and nitrostyrene (>98% de), while propionic acid and (2-nitropropenyl)benzene led to trisubstituted 2-methyl-3-phenyl-4-nitropentanoic acid (60% de at the γ-position). Catalytic

hydrogenation of the γ-nitro products led to γ-lactams via nitro reduction and oxazolidinone cleavage/cyclization, with the lactams opened using LiOH (*1645, 1750*). 2-Methyl, 2-isopropyl, and 2-isobutyl-3-methyl-4-aminopentanoic acid were also prepared by this route (*1646*).

In a similar manner, an enolate of a homochiral 1,3-dioxolan-4-one derived from (S)-lactic acid added to nitropropene with high diastereoselectivity to give, after nitro reduction and auxiliary hydrolysis, 2-hydroxy-2,3-dimethyl-4-aminobutyric acid (*1751*). This strategy was applied to more complex derivatives, using chiral cyclic templates derived from 3-substituted-3-hydroxy-propanoic acids to add to 2-substituted nitroalkenes (*1752*). Alternatively, a chiral Rh catalyst induced 58–98% ee for Michael additions of dialkyl malonates or β-keto esters to nitrostyrenes, producing 2-acyl-3-aryl-4-nitrobutanoate precursors of γ-amino acids. Minimal diastereoselectivity was observed for reactions generating two chiral centers (*1753*). A Fischer carbene anion with a chiral auxiliary, MeC[=Cr(CO)₅]-N*, added to *trans-p*-chloro-nitrostyrene to give the γ-nitrocarbene complex adduct in 90% yield and 76% de. The complex was converted into (R)-baclofen via oxidation of the carbene with CAN and nitro reduction with H₂/Raney Ni (*1754*).

Homochiral *N*-tosyl 2-alkylaziridines were opened with a lithiated dithiane derivative of glyoxalic acid to give the dithiane derivative of 2-keto-4-aminoalkanoic acids (*1247*).

Nucleophilic aminoethyl equivalents and electrophilic acetate equivalents can also be combined. Enders and Reinhold applied the SAMP hydrazone auxiliary

Scheme 11.135 Enantioselective Synthesis of 2-Substituted 4-Aminobutanoic Acid (*1749*).

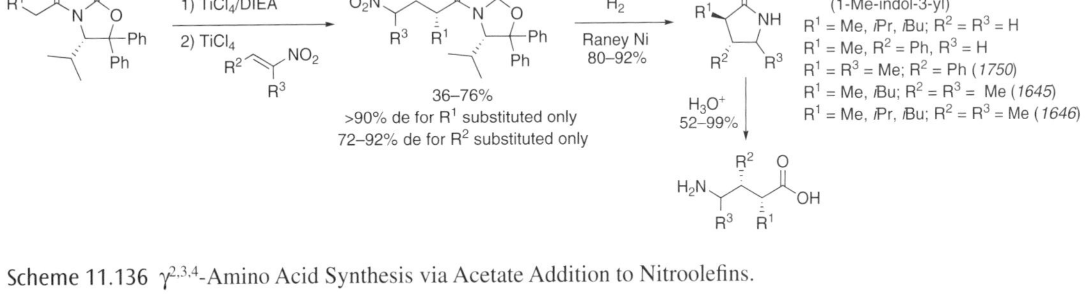

Scheme 11.136 γ²,³,⁴-Amino Acid Synthesis via Acetate Addition to Nitroolefins.

analog SADP to a synthesis of (*R,R*)-statine, using allyl bromide as a masked electrophilic acetate group (see Scheme 11.137). Allylation of the SADP hydrazone of *O*-Bn hydroxyacetaldehyde was followed by stereo-selective Grignard addition to the hydrazone bond to introduce the isobutyl substituent. Auxiliary removal, alkene ozonolysis, and deprotection completed the synthesis (*1755, 1756*). The precursors of cyclohexylstatine and AHPPA were also prepared.

Another synthesis with an allyl group as a masked acetate also employed a nitrile as a masked aminomethyl group. Alkylation of methyl cyanoacetate with allylic acetates in the presence of a chiral Pd catalyst produced 2-cyano-3-substituted-5,5-diphenylpent-4-enoate with 64–96% ee (see Scheme 11.138). Decarboxylation, nitrile reduction, and alkene oxidative cleavage gave 2-substituted-4-amino acids (*1757*).

Other methods of forming the C2–C3 bond have been described. The γ-lactam of homochiral 3-alkyl-4-aminobutyric acids was generated via an intramolecular Rh-catalyzed C–H insertion reaction of α-diazo derivatives of monoamides of succinic acid. A chiral

ester auxiliary induced 30–98% de for the cyclization step (see Scheme 11.139). Decarboxylation proceeded in good yield (*1758*).

The same reaction was used to prepare (*R*)-(−)-Baclofen, 3-(4-chlorophenyl)-GABA, and other 3-substituted-GABA analogs, but for this synthesis a chiral dirhodium(II) catalyst promoted the C–H insertion reaction of an α-methoxycarbonyl-α-diazoacetanilide (see Scheme 11.140). The initial product, an *N*-aryl 3-carboxymethyl-4-arylpyrrolidin-2-one (33–82% ee) was decarboxylated and *N*-deprotected to give a γ-lactam, which was opened with 6 N HCl. Recrystallization of the lactam of (*R*)-(−)-baclofen before hydrolysis gave optically pure product (*1759*).

The C2–C3 bond of protected 2,4-dichloro-2-substituted-3-hydroxy-4-aminobutanoic acid was created via a Ru-initiated intramolecular radical addition of a 2,2-dichloropropionyl moiety, linked by a chiral apocamphane tether to a 2-oxazolone group (see Scheme 11.141). Methanolysis replaced the 4-chloro group with a methoxy substituent, and the 2-chloro group was reductively removed using (TMS)$_3$SiH, resulting in a

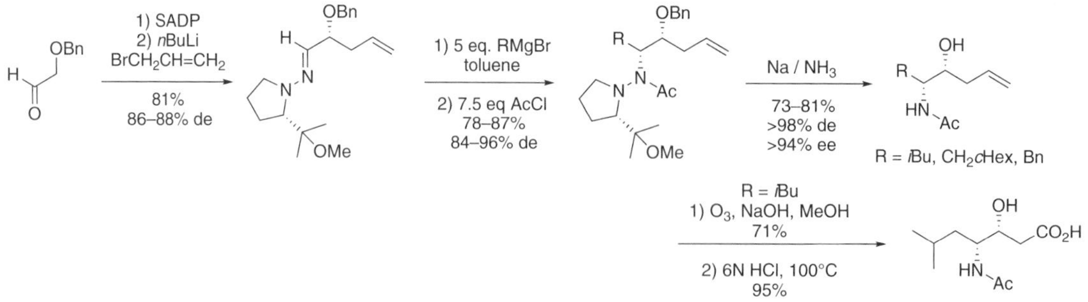

Scheme 11.137 Synthesis of γ-Amino Acids using SADP Hydrazones (*1755, 1756*).

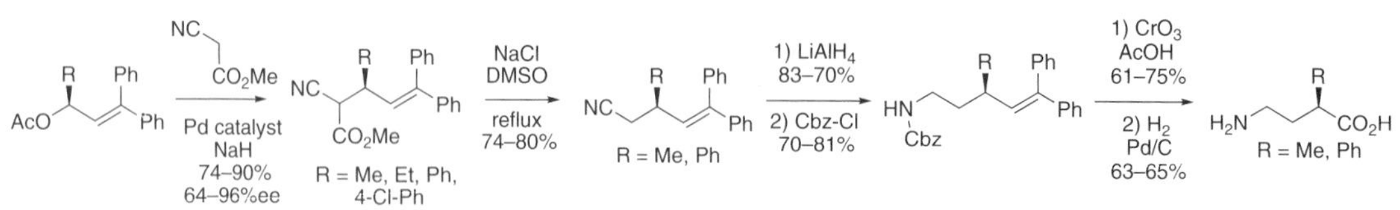

Scheme 11.138 Synthesis of 2-Substituted-4-Amino Acids (*1757*).

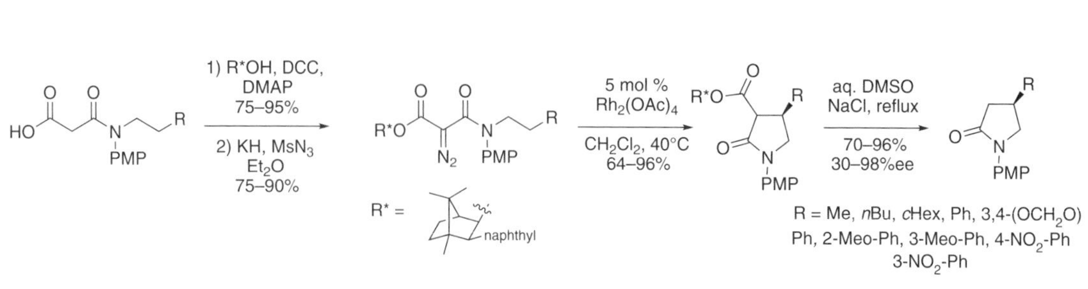

Scheme 11.139 Rh-Catalyzed C–H Insertion in Chiral Ester Diazoanilides (*1758*).

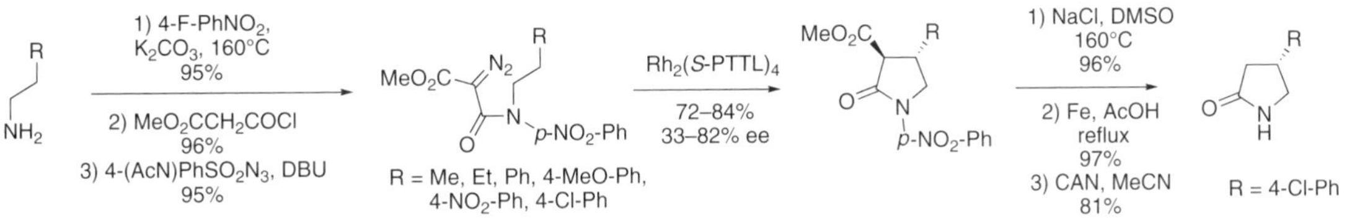

Scheme 11.140 Synthesis of Baclofen and Analogs (*1759*).

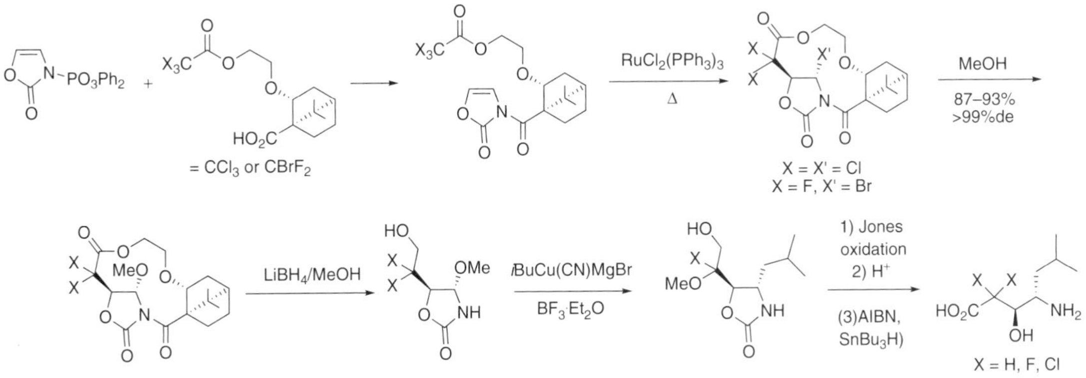

Scheme 11.141 Synthesis of Statine and Analogs (*1761*).

single diastereomer. Further manipulation led to three different isomers of 2-methyl-3-hydroxy-4-amino-butanoic acid (*1760*). The same methodology was used to synthesize statine (4-amino-3-hydroxy-6-methylheptanoic acid) and its 2,2-dichloro and 2,2-difluoro analogs (*1761*).

11.4.3f Fusion of Amino- and Carboxy-Containing Components: C3–C4 Formation

Nucleophilic aminomethyl equivalents have been added to electrophilic propionic acids. Enzymatic kinetic resolution of a racemic cyanohydrin adduct led to both enantiomers of GABOB and carnitine, and is potentially applicable to other γ-amino-β-hydroxy acids such as

statine (*543*, *1697*). The cyanohydrin precursor was formed from an ethyl formyl acetate equivalent and TMSCN (see Scheme 11.142). Acetylation followed by resolution using lipase (CCL) gave a 32% yield of the (*R*)-isomer after 40% conversion. Treatment of the remaining starting material with a different lipase (PPL) gave a 48% yield of the (*S*)-enantiomer after 60% conversion. Cyano reduction resulted in GABOB, with carnitine provided by *N*-methylation. Statine could be synthesized by addition of isobutylmagnesium chloride to the cyano group, but the ester functionality required protection as an oxazoline group (see Scheme 11.131 earlier). However, for this synthesis, the statine cyanohydrin precursor was not resolved (*543*).

Other combinations of nucleophilic aminomethyl and electrophilic propionate equivalents have been

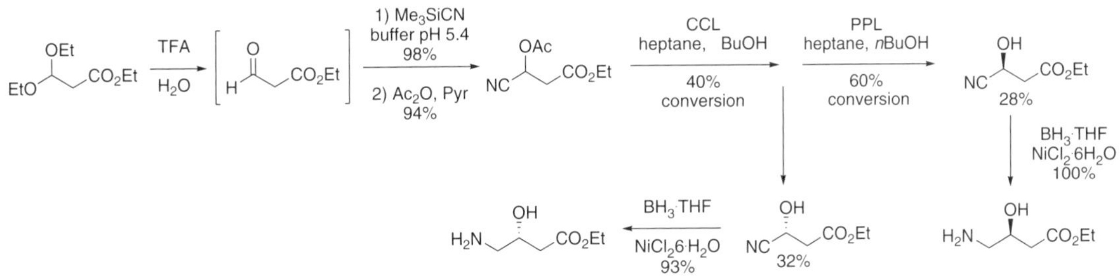

Scheme 11.142 Synthesis of GABOB via Cyanohydrin (*543*).

developed. 4-Phenyl-4-aminobutyric acid was prepared by an asymmetric conjugate addition of the anion of N-aryl,N-Boc-benzylamine to acrolein, with enantioselectivity induced by an asymmetric deprotonation using nBuLi/(−)-sparteine. Oxidation of the aldehyde with Jones reagent gave the amino acid with 94% ee. The opposite enantiomer (92% ee) could be accessed by a lithiation–stannylation–transmetalation protocol (969). An enantioselective Michael addition of nitromethane to substituted acrylates in the presence of a chiral Ni catalyst proceeded with 77–97% ee, producing 12 different β-substituted-γ-nitro products in high yield (1762). Nitroalkanes have also been added to alkylidenemalonates in the presence of a chiral quaternary ammonium phase transfer catalyst, leading to 3,4-disubstituted-γ-nitro acids with high enantioselectivity (88–99% ee) and good diastereoselectivity (42–90% de anti). Nitro reduction (H₂/Raney Ni) and decarboxylation (6 N HCl) produced the γ-amino acids (1763). Nitromethane was added to a chiral α,β-unsaturated ester derived from (S)-verbenone, and then hydrogenated to give a carboxycyclobutyl-substituted GABA analog, 3-amino-2-(2,2-dimethyl-3-carboxycyclobut-1-yl)propanoic acid (1764). Selective catalytic hydrogenations of γ-nitro esters have been optimized for process synthesis (1695).

Chiral nitrones derived from D-glyceraldehyde, R*CH=N⁺(Bn)O⁻, were added to ethyl acrylate in 2003 via a reductive conjugate addition in the presence of 2 equiv of SmI₂. Hydrogenation of the resulting N-hydroxy, N-benzyl γ-amino esters over Raney Ni in the presence of Boc₂O produced the Boc-protected γ-amino esters in good yield (1696, 1765). Alternatively, nitrones bearing a chiral auxiliary on the nitrogen center were prepared and added to ethyl acrylate (see Scheme 11.143). A 1-(triisopropylphenyl)ethyl auxiliary induced excellent (>90%) diastereoselectivity (1696). A carbohydrate auxiliary on the nitrone nitrogen also gave >90% de for six different nitrone derivatives added to butyl acrylate (1766).

The polarity of this C3–C4 bond-forming reaction has been reversed. Stereoselective addition of allylmagnesium bromide (a masked nucleophilic propionate equivalent) to chiral N-benzylidene-p-toluenesulfinamides resulted in 3-amino-3-phenyl-3-substituted-propenes after auxiliary removal. Hydroboration of the alkene followed by oxidation furnished the carboxyl

group (588). Another synthesis added lithiated (R)-p-tolyl-γ-butenyl sulfoxide (a masked 3-hydroxypropanoic acid group) to an imine of trifluoroacetaldehyde (46% de). Pummerer rearrangement converted the sulfoxide auxiliary into a hydroxy substituent, with oxidative cleavage of the alkene providing (3S,4R)-3-hydroxy-4-trifluoromethyl-4-aminobutyric acid (1767). A similar route was employed for syntheses of (+)- and (−)-statine, 4-amino-3-hydroxy-6-methylheptanoic acid (1768).

A Ti homoenolate equivalent of propionaldehyde, derivatized as an acetal of homochiral 1,2-bis(cyclohexyl) ethanediol, was added to an imine formed from benzylamine and isobutyraldehyde to produce an O-alkyl 4-amino-5-methylhex-1-en-1-ol derivative with 88% de. Conversion of the enol ether to an aldehyde and oxidation provided 4-amino-5-methylhexanoate. A variety of other imines were also reacted (1025, 1769). Alternatively, an achiral (γ-alkoxyallyl)titanium complex was added to chiral imines formed from aldehydes and α-methylbenzylamine to give syn 1-vinyl-2-amino alcohols with 60–66% diastereoselectivity for the major isomer. The alkene could be hydroxylated and oxidized to form the acid group of β-hydroxy-γ-amino acids, including statine (1770). A chiral Pd complex catalyzed the reaction of aryl imines with ethyl 2-(tributylstannylmethyl)-acrylate, producing 2-methylene-4-aryl-4-aminobutanoic acids with 60–89% ee (1771).

Nitrones have been used as the electrophilic component. Lithiated ethyl or tert-butyl propiolate added to a chiral nitrone prepared from the Ser-derived Garner aldehyde to give 4-substituted-4-hydroxyamino acetylenic esters with complete syn stereoselectivity. A reversal of stereoselectivity was obtained by reacting the analogous Cys-derived aldehyde in the presence of MgBr₂, giving the anti product with 74% de. Reduction of the hydroxylamine and alkyne functionalities with Zn in acidic solution gave protected 4,5-diamino-6-hydroxyhex-2-enoic acid with high Z-selectivity (1699). The alkene bond could be dihydroxylated.

11.4.3g Elaboration of α-Amino Acids: Aldol Reactions of Amino Aldehydes

A number of stereoselective syntheses of statine analogs, 3-hydroxy-4-substituted-4-amino acids, have

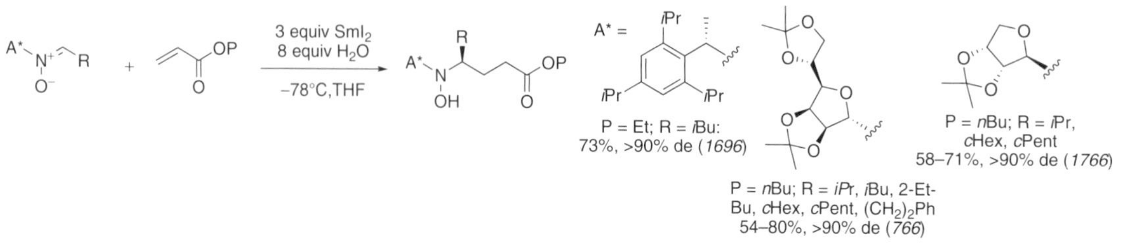

Scheme 11.143 Reductive Conjugate Addition of Nitrones to Acrylates.

relied on aldol reactions of an acetate enolate with chiral aldehydes derived from amino acids. These reactions generally rely on substrate control (with the amino acid chirality controlling the stereoselectivity), but it is also possible to use chiral acetate enolate equivalents to induce stereoselectivity (reagent control). A good summary of stereochemical control has been published (1772). The synthesis of β-oxygenated-γ-amino acids from α-amino acids was reviewed in 1992 (1665).

An early synthesis of all four isomers of statine, (3S,4S)-4-amino-3-hydroxy-6-methylheptanoic acid, added the zinc enolate of tert-butyl acetate to Leu-derived N-Phth L- or D-leucinal with 8–10% de. The diastereomers were resolved by preparative LC before deprotection (1566, 1773). Similarly, N-Boc leucinal was treated with the lithium enolate of ethyl acetate to give a mixture of diastereomers of protected statine (73–80% yield, 4–20% de) (see Scheme 11.144) (1774–1776). The same condensation with Boc-L-phenylalaninal (55% yield, de not disclosed) produced the aromatic analog of statine, (3S,4S)-3-hydroxy-4-amino-5-phenylpentanoic acid (AHPPA) (1576, 1777). Eight different steroisomers of isostatine (4-amino-3-hydroxy-5-methylheptanoic acid) resulted from reaction of D- or L-isoleucinal or -alloisoleucinal with ethyl lithioacetate, followed by separation of the two diastereomers produced by each addition. The products were used to determine the absolute configuration of the natural (3S,4R,5S)-isomer found in the didemnins (1588). Steroselective synthesis of the natural (3S,4R,5S)-isostatine isomer was achieved via condensation of N-Boc (R)-alloisoleucinal with ethyl lithioacetate, giving a mixture of diastereomers epimeric at the carbinol center. Oxidation to the ketone and reduction with sodium borohydride gave the desired carbinol center stereochemistry with >82% de (1585). Dolaisoleuine, the β-methoxy-γ-amino acid component of dolastatin 10, was derived from (S)-Ile via N-methylation, reduction to isoleucinal, addition of the lithium enolate of tert-Bu acetate, and O-methylation (1632). Addition of lithiated ethyl acetate to the aldehyde derived from orthogonally protected 2,3-diaminopropionic acid, Cbz-Dap (Boc)-OH, led to (3S,4S)-4,5-diamino-3-hydroxypentanoic acid with 38% de. The (3R,4S)-diastereomer was obtained preferentially via a different route (1778).

A mixture of diastereomers of 2,2-difluorostatine was obtained by the addition of the Reformatsky reagent derived from BrF$_2$CCO$_2$Et to Boc-L-leucinal, with diastereoselectivity ranging from 70:30 at room temperature to 100:0 at reflux. The secondary alcohol of

either isomer could be oxidized into a ketone by Swern oxidation after incorporation of the amino acid into a peptide (1572, 1575). The amino aldehydes derived from L-Phe and L-cyclohexylalanine have also been elaborated with this reagent (481, 1575). Other 2-substituted statine analogs were prepared by addition of the enolate of dimethyl 3-methylglutaconate [MeO$_2$CCH=C(Me)CH$_2$CO$_2$Me] to amino acid-derived N, N-dibenzyl amino aldehydes. With amino aldehydes derived from Ala, Phe, and Leu, only one of the possible four diastereomers predominated (1779).

In order to overcome the poor diastereoselectivity of achiral acetic acid enolate additions, an O-methyl-O-trimethylsilyl ketene acetal was employed as the acetate equivalent. Addition to N-carbamate-protected aldehydes derived from Leu or cyclohexylalanine in the presence of TiCl$_4$ proceeded with 88–90% de syn stereoselectivity (89–95%). Hydrolysis gave (3S,4S)-statine or cyclohexylstatine in 88–93% yield (1780). Reaction of N-Boc leucinal with the trimethylsilyl ketene acetal of ethyl acetate using SnCl$_4$ as Lewis acid catalyst gave the syn adduct with 82% de, while reaction of N,N-dibenzyl leucinal with the same ketene silyl acetal in the presence of EtAlCl$_2$ gave the anti adduct with 88% de. The silyl ketene acetal of benzyl acetate improved the stereoselectivity to >99% de (1781). A similar stereoselective reaction of a silyl ketene acetal with N,N-dibenzyl alaninal in the presence of lithium perchlorate catalyst produced 3-hydroxy-4-aminopentanoic acid in 58% yield with >98% de for the anti product (1782).

An allyl group has been used as a masked acetate equivalent. The amino aldehydes derived from Boc-Nle or Boc-Lys(Cbz) were treated with allyltrimethylsilane in the presence of tin tetrachloride to give allyl alcohols with good threo selectivity; oxidation of the alkene with catalytic ruthenium trichloride and sodium periodate provided statine analogs (1783). Similarly, N-Boc leucinal was converted to statine by a syn selective addition of allylmagnesium bromide (53% yield, 76% de), followed by oxidative cleavage of the alkene (1123). The amino alcohol obtained via addition of allylmagnesium bromide to Cbz-L-leucinal was epimerized at the carbinol center via oxazolidinone formation. Oxidative cleavage of the alkene produced (3S,4S)-statine (1784). Addition of allylmagnesium chloride to a homochiral α-aminotrifluoropropanal equivalent led to (3R,4S)-3-hydroxy-4-trifluoromethyl-4-aminobutyric acid (1767).

Another amino acid homologation added a vinyl anion to amino acid-derived aldehydes. Conversion of the resulting amino alcohol to an oxazolidinone was

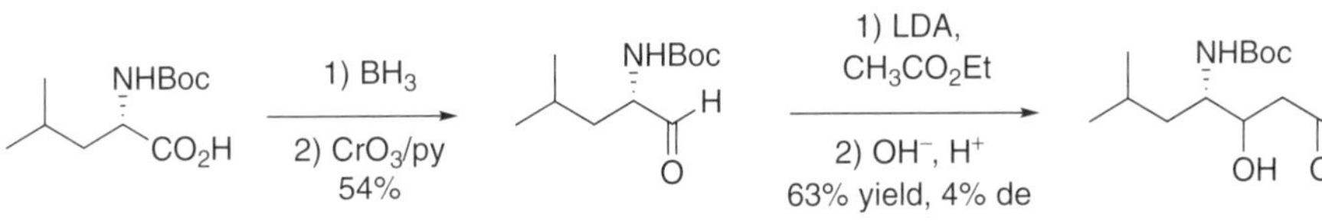

Scheme 11.144 Synthesis of β-Hydroxy-γ-amino Acids from α-Amino Aldehydes (1774).

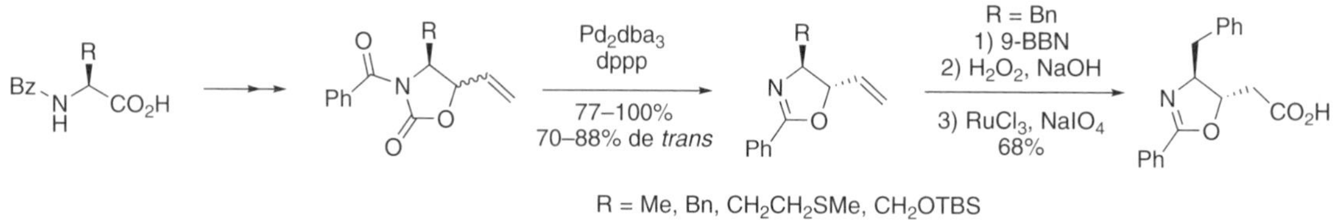

Scheme 11.145 Synthesis of β-Hydroxy-γ-Amino Acids from Amino Acids (*1785*).

followed by Pd-catalyzed isomerization to a 5-vinylox-azoline (see Scheme 11.145). Hydroboration and oxida-tion of the vinyl group generated the acid substituent, giving the protected precursor of β-hydroxy-γ-amino acids such as AHPPA, (3S,4S)-4-amino-3-hydroxy-5-phenylpentanoic acid (*1785*).

Several syntheses have employed chiral enolates in combination with chiral aldehydes in order to improve the induction of stereoselectivity. The magnesium eno-late of a chiral acetate ester, 2-acetoxy-1,1,2-triphenyl-ethanol, was added to Boc-L-leucinal. The (S)-ester produced the (3S,4S)-precursor of statine with 86% de, while the (R)-ester resulted in product epimeric at C3 with 50% de (*1786*). The same set of chiral enolates were added to Cbz-glycinal, producing protected (S)- or (R)-GABOB with 82% de. Deprotection and recrystal-lization gave either enantiomer of GABOB with 98% ee (*1787*). In the same manner, lithium acetate enolates with chiral ester auxiliaries were added to L-phenyla-laninal or L-cyclohexylalaninal, with 80–82% de (*1788*). Dolaisoleuine, (3R,4S,5S)-4-(methylamino)-3-methoxy-5-methyl heptanoic acid, was synthesized from Cbz-protected N-methyl isoleucinal via a stereoselective aldol condensation with a homochiral α-(methylsulfanyl) acetyloxazolidinone. The synthesis was completed by reductive desulfurization, O-methylation, and oxazolidi-none cleavage (*1630*). An earlier synthesis of statine from N-Boc-L-Leu used the same procedure (*1789*).

Another chiral acetate enolate equivalent was pre-pared from diethylaluminum enolates of an iron acetyl complex. Addition of the (R)-enolate to the amino alde-hyde derived from N,N-dibenzyl (S)-Val produced pre-dominantly the (S,R,R)- and (S,S,R)-isomers in a 1:4 ratio; while the (S)-enolate added to give the (S,R,S)- and (S,S,S)-isomers in a 23:1 ratio. Similarly, addition of the (S)-enantiomer to (S)-leucinal gave mainly the (S,R,S)-diastereomer (92% de), which, after removal of the iron auxiliary with Br$_2$ in EtOH, provided (3R,4S)-statine (*1790*). A camphor-derived chiral lithium acetate enolate reagent was added to a variety of Boc- or Cbz-protected N-methyl amino aldehydes with complete diastereoselectivity, producing one isomer in 55–75% yield. The auxiliary was removed by desilylation and oxidative treatment with CAN (*1619*). The enolate of acetylated pseudoephedrine was added to the Ser-derived Garner aldehyde, with the matched pair of enantiomers giving the aldol adduct with 96% de and the mismatched pair with 12% de. Hydrolysis gave 3,5-dihydroxy-4-aminopentanoic acid (*1791*).

C2–substituents can be introduced. A synthesis of 2,3-dihydroxy-4-dimethylamino-5-methoxypentanoic acid (the component of calyculin A) added a chiral eno-late derivative of glycolic acid (using an oxazolidinone auxiliary) to an N-Cbz N,O-dimethyl D-serinal interme-diate, providing the desired *anti* aldol adduct in 60% isolated yield (*1624*, *1625*). The boron (Z)-enolate derived from propionic acid derivatized with Evans' oxazolidinone was added to N-Boc-D-alaninal to give the *syn* aldol adduct in 73% yield. Hydrolysis of the chiral auxiliary resulted in the (2S,3S,4R)-4-amino-3-hydroxy-2-methylpentanoic acid component of bleomy-cin A$_2$ (*247*, *1792*). The same procedure was applied to N-Boc-D-leucinal and N-Boc-D-phenylalaninal (*259*). Constrained analogs of this bleomycin γ-amino acid component, joining the C2–C4 backbone carbons with a butyl or 2-butene linker, were synthesized from allyl-glycinal, via addition of pent-4-enoic acid derivatized with Evans' oxazolidinone (see Scheme 11.146). The two allyl substituents of the resulting 2,4-di(allyl)-3-hydroxy-4-aminobutanoic acid derivative were then joined by a metathesis reaction, with the resulting alkene hydrogenated if desired. The diallyl intermediate was also deprotected (*258*, *1609*).

Gennari et al. reported the use of chiral boron eno-lates of *tert*-butyl thioacetate to give highly stereoselec-tive *syn* or *anti* products (91 to >98% de) upon addition to N,N-dibenzyl amino aldehydes. The diastereoselec-tivity depended entirely on the configuration of the chiral boron ligands (see Scheme 11.147). The amino aldehydes derived from Ala, Val, Leu, Ile and Phe were employed as substrates, with the adduct from Leu con-verted to statine (*1772*, *1793*, *1794*). The boron auxiliary also controlled the stereochemistry at the C2 position if an enolate of propionic thioester was added (*1794*). An earlier synthesis employed achiral boron enolates of thioesters of propionic acid (*1795*). A good summary of the stereoselectivity of acetate enolate addition to amino aldehydes was contained in one of these papers (*1772*).

The four diastereomers of 4-amino-3-hydroxy-2-methylpentanoic acid were prepared from Boc-L-alaninal via diastereoselective addition of four organoborane reagents prepared from (+)- or (−)-methoxydiisopi-nocamphenylborane and (Z)- or (E)-2-butene. Three of the diastereomers of 2-amino-3-hydroxy-4-methylhex-5-ene were obtained with very high stereoselectivity (>95% de). After protecting group manipulation, the alkene was oxidatively cleaved to generate the carboxy

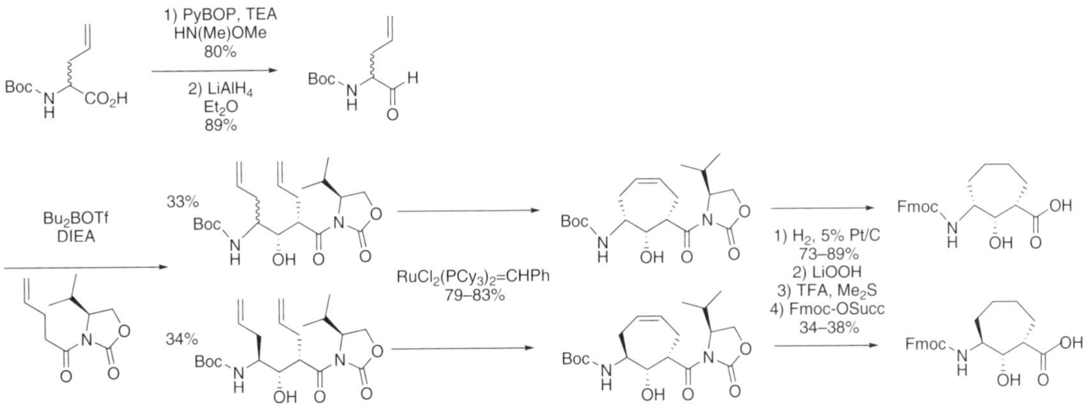

Scheme 11.146 Reaction of N-Boc Allylglycinal With Enolate (258).

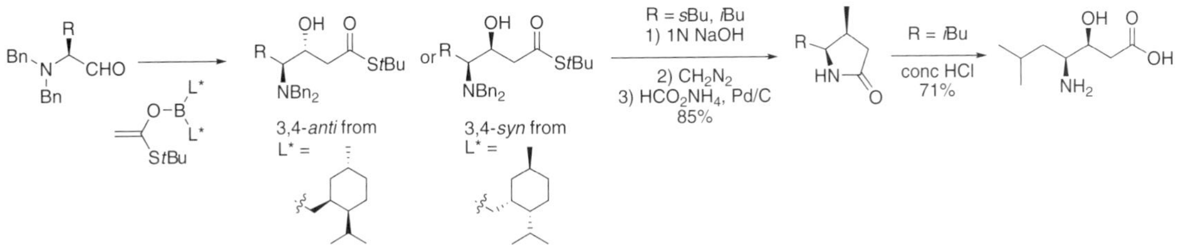

Scheme 11.147 Addition of Chiral Boron Enolates to α-Amino Aldehydes (1772, 1793, 1794).

group (1611). A synthesis of 2,3-dihydroxy-4-dimethyl-amino-5-methoxypentanoic acid (the component of calyculin A) also employed addition of a chiral allyl boron enolate for homologation of the Garner aldehyde (1796).

Modifed amino aldehyde equivalents have been used as the electrophilic component. A new approach to statine-like γ-amino acids relied on alkylation of Williams' oxazinone template to prepare the initial homochiral α-amino acid. Instead of removing the template to provide the free amino acid for further elaboration, the template lactone carbonyl was directly reduced to an aldehyde acetal (see Scheme 11.148). Homologation of the amino acid was carried out by addition of allyltrimethylsilane or a silyl ketene acetal of methyl acetate, which proceeded with variable stereoselectivity (1797, 1798). The allyl adduct also allowed for elaboration to a γ-hydroxy-δ-amino acid (1798). Reaction of the template acetal with TMSCN led to β-amino acid derivatives (1122).

11.4.3h Elaboration of α-Amino Acids: Olefination Reactions of Amino Aldehydes

Another general approach to γ-amino acids from α-amino acid-derived amino aldehydes relies on olefination reactions to give γ-amino-α,β-unsaturated esters. Deoxystatin was prepared from N-Phth leucinal by Horner–Emmons homologation with ethyl diethylphosphonoacetate. Catalytic hydrogenation and acid hydrolysis completed the synthesis (1570). Seebach and co-workers also relied on this approach to homologate N-Boc amino acids, after attempted double Arndt–Eistert homologation gave poor results. The N-Boc α-amino acids were converted into Weinreb amides, reduced to the aldehyde with LiAlH₄, and condensed with (PhO)₂P(O)CH₂CO₂Me to give cis/trans mixtures of γ-amino-alkenes (see Scheme 11.149). Hydrogenation and saponification gave the desired homochiral γ-substituted γ-amino acids in 55–72% overall yield from the α-amino acid (1642). The group of Schreiber et al. refined this procedure in order to prepare monomers of vinylogous polypeptides. The N-Boc amino acids (Ala, Val, Phe) were again converted to the Weinreb amides (78–99%) and reduced with LiAlH₄ to give the amino aldehydes. Homologation with Ph₃P=CHCO₂Me or Ph₃P=C(Me)CO₂Et gave the (E)-γ-substituted-γ-amino esters or (E)-substituted-β-methyl-γ-amino esters in 69–92% from the Weinreb amide (1649).

This sequence was applied to the synthesis of the vinylogous amino acid derivatives of Val (1635), Met (1651, 1652), Glu (1652), Ser (1652), Gln (1651–1653), Pro (1799), lactam analogs of Gln (1655), 2,3-diaminopropionic acid (1652), and the Tyr analog found in the

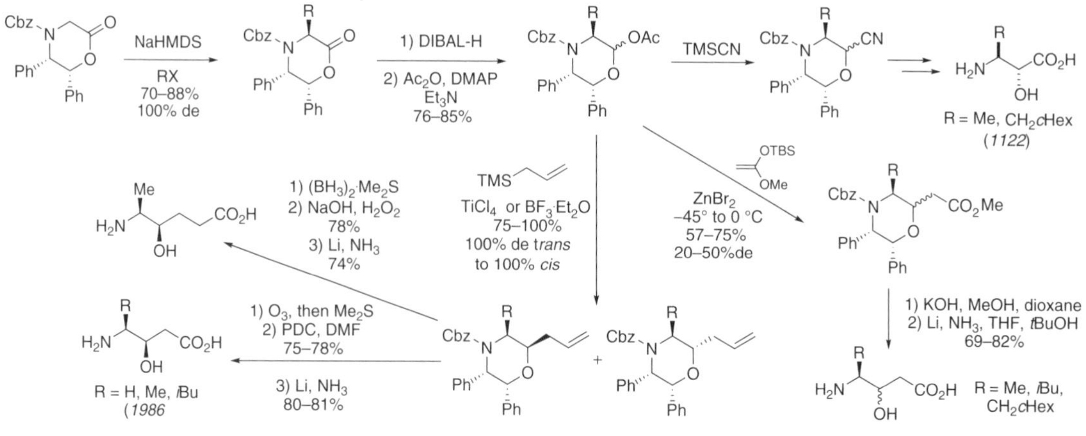

Scheme 11.148 Elaboration of Williams' Oxazinone to γ-Amino-β-Hydroxy Acids (*1797, 1798*).

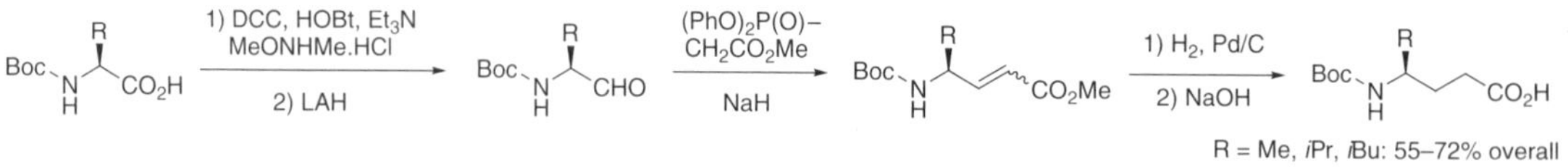

Scheme 11.149 Synthesis of γ-Amino Acids from α-Amino Acids (*1642*).

cyclotheonamides (*264, 267*). (*E*)-4-Amino-5-hydroxy-pent-2-enoic acid was prepared by Wittig reaction of the Garner aldehyde (*1623, 1800*), or via reduction of *O*-TBDPS Boc-Ser-OMe to an aldehyde and Wittig reaction with Ph$_3$P=CHCO$_2$Et (*1801*). 4-Amino-2,5-dimethylhex-2-enoic acid, a component of the marine cytotoxic tripeptides hemiasterlins, milnamides, and criamides, was obtained via a Wittig reaction of valinal with Ph$_3$P=C(Me)CO$_2$Et (*1638*), as was a 2-methyl vinylogous Gln derivative (*1653*). A vinylogous Gln-Gly γ-amino ester, employed as an electrophilic Michael acceptor in inhibitors of the human rhinovirus 3C protease, was prepared by reduction of Cbz-Gln(Trt) to a protected alcohol, *N*-deprotection, and then acylation with a dipeptide analog, followed by alcohol oxidation and coupling with Ph$_3$P=CHCO$_2$Et (*1656*). An interesting variation of the olefination homologation employed racemic *N,N*-diprotected amino aldehydes and homochiral Horner–Wadsworth–Emmons reagents to induce stereoselectivity at the 4-position. The amino aldehyde was coupled under dynamic kinetic resolution conditions, allowing for up to 90% de for bulky substituents (*1802*).

The alkenes obtained from olefination of α-amino aldehydes can be derivatized in a number of ways. Alkene hydrogenation produces γ-substituted γ-amino acids. Hydrogenation conditions have been reported which allow for chemoselective alkene reduction in the presence of *N*-Cbz or benzyl ester/ether protecting groups (*1803*). The vinylogous amino acid derived fom Cbz-valinal was reduced to 4-amino-5-methylhexanoic

acid derivative, and then diastereoselectively hydroxylated at the α-position to give 2-hydroxy-4-amino-5-methylhexanoic acid (*1635*). In the same paper, Boc-phenylalaninol was oxidized and coupled to Ph$_3$P=C(Me)CO$_2$Et. Again the alkene was reduced, giving the 2-methyl-4-amino-5-phenylpentanoic acid component of the tubulysin cytostatic peptides (*1635*). Both L- and D-Met were converted into enantiomers of vigabatrin, 4-amino-5-hexenoic acid. A one-pot homologation of the *N*-protected amino ester via DIBAL reduction in the presence of trialkylphosphonoacetate/*t*BuLi gave methyl 6-methylthio-4-aminohex-2-enoate. Mg-methanol reduced the 2,3-alkene bond, with the thiol ether then oxidized to a sulfoxide and thermally eliminated to generate the desired 5,6-alkene (*1804, 1805*).

Other alkene functionalizations include an OsO$_4$-induced dihydroxylation of the (*Z*)-alkene obtained by homologation of *N*-Boc,*O*-methyl serinal, producing a protected derivative of the 2,3-dihydroxy-4-dimethylamino-5-methoxypentanoic acid component of calyculin A (*1806*). The same product was obtained from the Ser-derived Garner aldehyde via homologation by a Wittig reaction (*1623, 1800*), followed by *cis*-dihydroxylation of the alkene bond. *O*- and *N*-methylation completed the synthesis (*1623*). The (*E*)-alkene obtained from *N*-Boc phenylglycinal and Ph$_3$P=CHCO$_2$Et was epoxidized with *m*CPBA to give predominantly the (2*S*,3*R*,4*S*)-*syn* isomer with 80% de. The epoxide was then opened with heteroatom nucleophiles (amines and thiols) to provide 2-heterosubstituted-3-hydroxy-4-amino-5-phenylpentanoic acids, which were incorporated into inhibitors of HIV-1 protease (1581). An epoxidation was also carried out on the (Z)-alkene derived from Cbz-Leu and (CF₃CH₂O)₂P(O)CH₂CO₂Me. The carboxyl group was temporarily reduced to give an allylic alcohol, which was epoxidized by mCPBA with almost exclusive syn selectivity. Epoxide opening with Red-Al and reoxidation of the hydroxymethyl group gave statine (see Scheme 11.150) (887).

Reetz and Röhrig employed Michael addition of organocuprates to (E)-α,β-unsaturated-γ-amino esters to give 3,4-disubstituted γ-amino acids with high stereoselectivity (84 to >90% de) (see Scheme 11.151) (1807).

(3S, 4S)-Statine and AHPPA were derived from N-Cbz leucinal or phenylalaninal via olefination with NaH/(MeO)₂POCH₂CO₂Me, followed by a stereoselective iodocyclocarbamation induced by treatment with iodine. Radical deiodination and alkaline hydrolysis gave the desired products (1808). The ester group of the Wittig adducts of Boc-leucinal or Boc-phenylalaninal was reduced and chlorinated to give a (chloromethyl) allylamine, which was N-protected as a silyl carbamate (see Scheme 11.152). Treatment with AgF formed a cyclic carbamate of a vicinal syn amino alcohol. Hydroboration and oxidation of the terminal alkene

provided statine and AHPPA (4-amino-3-hydroxy-5-phenylpentanoic acid) in 27% and 19% overall yields from Boc-leucinal and Boc-phenylalaninal, repectively (1809, 1810).

Reetz et al. employed a Wittig reaction of amino acid-derived amino aldehydes to prepare, after ester reduction and O-alkylation, 4-amino-allyloxyacetates (see Scheme 11.153). Addition of LDA and tetramethylethylenediamine initiated a [2,3]sigmatropic Wittig rearrangement, producing α-hydroxy-β-vinyl-γ-substituted γ-amino acids with good stereoselectivity and no racemization (1811).

11.4.3i Elaboration of α-Amino Acids: Other Reactions of Amino Aldehydes

Optically active α-amino aldehydes have been converted into γ-amino acids by several other methods. The N-Boc α-amino aldehydes derived from Boc-L-Val and Boc-L-Phe were transformed into 4-(aminoalkyl)-substituted 3,3-dichloro-β-lactones by a diastereoselective [2+2] cycloaddition with dichloroketene, giving the (3S,4S)-isomer exclusively (see Scheme 11.154). The lactones act as activated equivalents of γ-amino-β-hydroxyamino

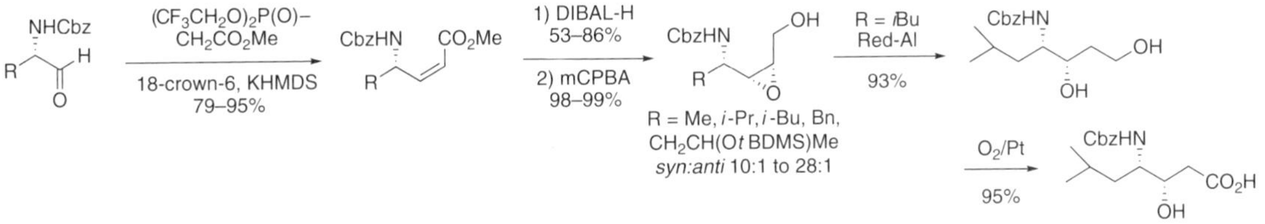

Scheme 11.150 Synthesis of β-Hydroxy-γ-amino Acids from α-Amino Aldehydes (887).

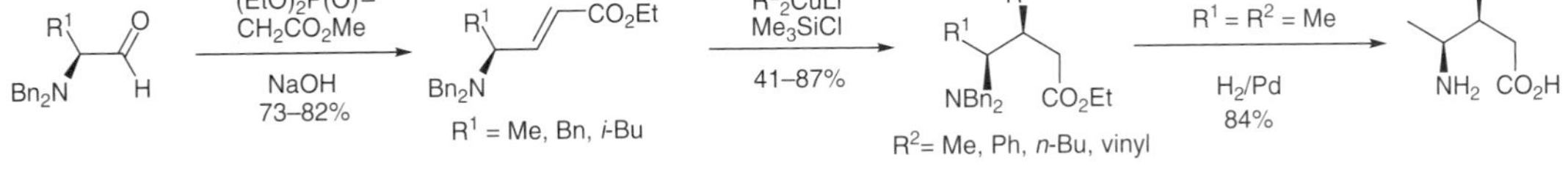

Scheme 11.151 Synthesis of β-Substituted-γ-amino Acids from α-Amino Aldehydes (1807).

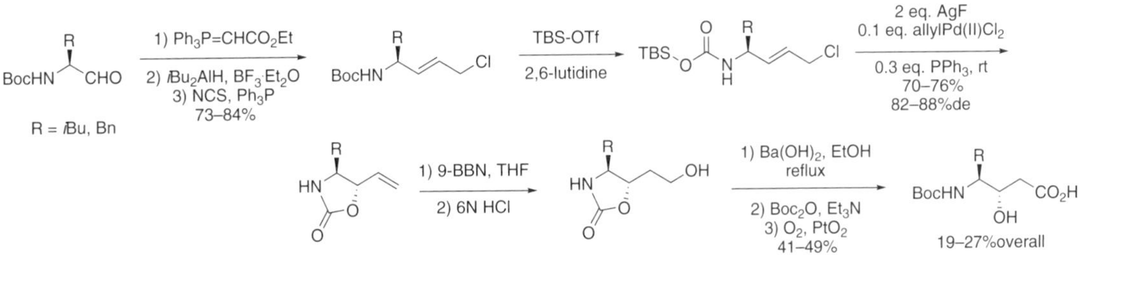

Scheme 11.152 Cyclic Carbamate Formation of (Chloromethyl)homoallylamines (1809, 1810).

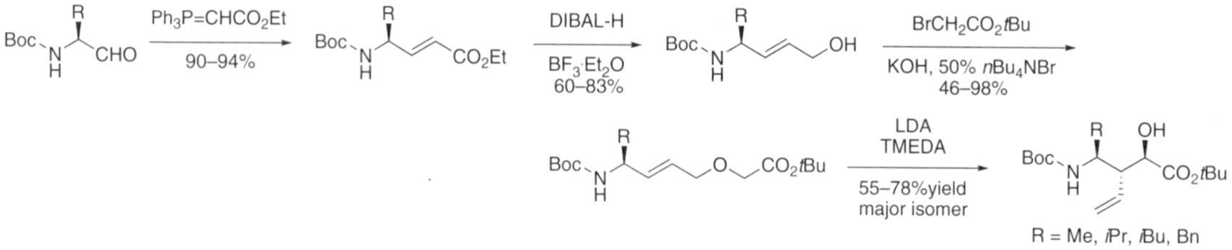

Scheme 11.153 Synthesis of α-Hydroxy-γ-Amino Acids (*1811*).

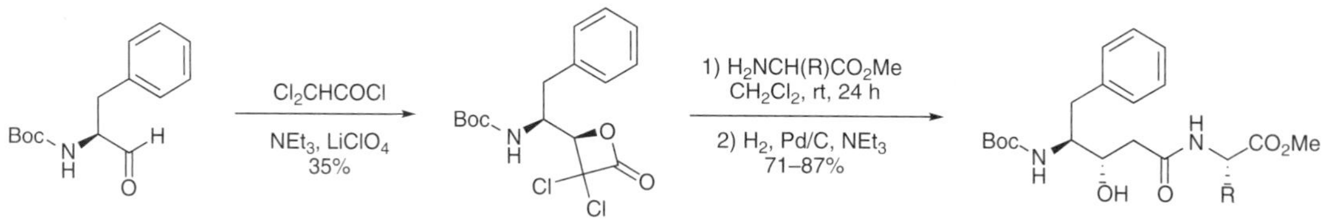

Scheme 11.154 Stereoselective Synthesis of β-Hydroxy-γ-Amino Acids via Cycloaddition of α-Amino Aldehyde (*1249, 1812, 1813*).

acids that can be coupled with amino esters to give dipeptides (opening with alcohols gives esters). The chloro substituents were removed by extended hydrogenolysis. Only AHPPA, (3*S*),4*S*)-4-amino-3-hydroxy-5-phenylpentanoic acid, was prepared as the amino acid, but the procedure should be generally applicable (*1249, 1812, 1813*).

A hetero-Diels–Alder reaction of homochiral amino aldehydes with 1,3-dimethoxy-1-(silyloxy)butadiene under Lewis acid catalysis produced lactones of 6-amino-3-methoxy-5-hydroxyhex-2-eneoate with either *threo* or *erythro* selectivity, depending on catalyst (see Scheme 11.155) (*1814*). Ozonolysis provided statine analogs.

A synthesis of γ-substituted α,β-acetylenic-γ-amino acids from amino acid-derived amino aldehydes employed the Corey–Fuchs reaction to generate an alkyne linkage (see Scheme 11.156) (*1815, 1816*).

Alternatively, the aldehydes were converted into unsubstituted alkynes by reaction with a diazophosphonate reagent, with the alkyne then deprotonated with *n*BuLi and carboxylated with CO_2 or $ClCO_2Me$. The alkynes were employed for a Ru-catalyzed Alder ene reaction with monosubstituted alkenes, producing (*Z*)-γ-amino-α,β-unsaturated-α-substituted esters (*1817*).

11.4.3j Elaboration of α-Amino Acids: Amino Alcohols

Several syntheses of γ-amino acids rely upon further reduction of the acid group of α-amino acids to give amino alcohol intermediates. Leu was homologated by reduction to *N,O*-ditosyl leucinol, followed by displacement with KO*t*Bu/diethyl malonate. Treatment with

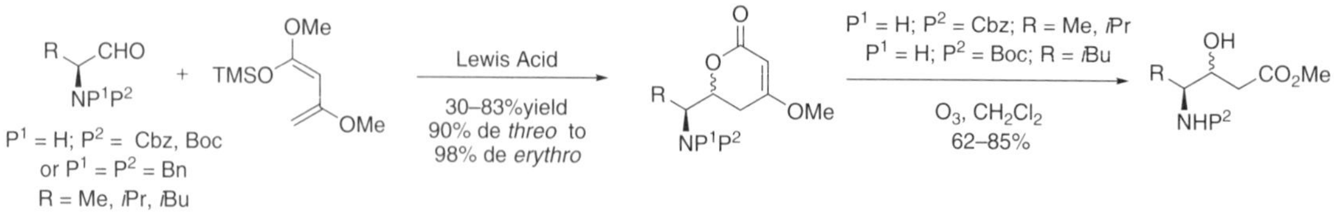

Scheme 11.155 Synthesis of Statine Analogs via Hetero-Diels–Alder Reactions (*1814*).

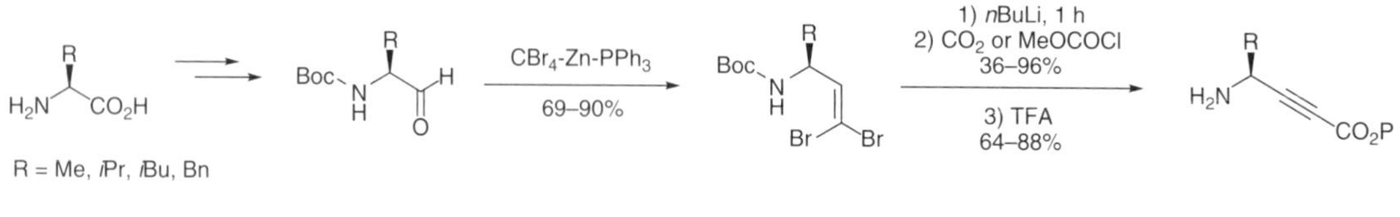

Scheme 11.156 Synthesis of Acetylenic γ-Amino Acids (*1815, 1816*).

47% HBr induced hydrolysis, decarboxylation, and N-Ts deprotection to give the γ-substituted γ-amino acid (*1723*). The same procedure was applied to the reduced amino alcohols derived from Val, Phe, and Pro (*1818*). 2-Vinyloxirane was elaborated into enantiomerically pure vigabatrin (4-amino-5-hexenoic acid) in four steps and 59% overall yield via a vinylglycinol intermediate. The butadiene monoepoxide was opened with phthalimide under dynamic kinetic conditions in the presence of a palladium catalyst with chiral ligands. The protected N-Phth vinylglycinol product was obtained with extremely high regioselectivity (75:1), enantioselectivity (98% ee), and yield (99%). Homologation was achieved by displacement of a triflate derivative with the sodium salt of dimethyl malonate, with acid hydrolysis removing the phthalimide and ester protection and inducing decarboxylation (*1515, 1819*).

Amino alcohols derived from amino acids were converted into homochiral N-benzoyl or N-Boc α-chloroamines, and then di-deprotonated with n-butyl lithium followed by lithium naphthalenide to give chiral dianionic intermediates (see Scheme 11.49 earlier). One example was quenched with BrCH₂CO₂R to give N-Bz 4-aminohexanoic acid (*958*).

A chemoenzymatic synthesis of homochiral 4-sub-stituted-4-amino-2-hydroxy acids was reported in 1998. Amino acids were converted into α-keto-γ-amino acids by a multistep reaction sequence proceeding via amino alcohol and β-amino acid intermediates (see Scheme 11.157). The ketone was then enzymatically reduced using mutant dehydrogenases (*1820*).

11.4.3k Elaboration of α-Amino Acids: Reactions of Activated Carboxyl Group

Another approach for the homologation of α-amino acids to γ-amino acids reacts activated amino acids with acetate nucleophiles to give β-keto ester intermediates. Stereoselective reduction of the β-keto ester results in the same β-hydroxy-γ-amino acids provided by enolate addition to amino aldehydes (see Section 11.4.3g), but often with improved stereocontrol. Asymmetric ketone reductions are also possible. Chiral catalysts were employed for asymmetric hydrogenation of a carnitine precursor (*1689*), while BINAP-Ru complexes stereo-selectively provided *threo* or *erythro* statine derivatives via asymmetric reduction of 3-keto-4-amino esters (*1821*).

The most common method for synthesis of the β-keto ester intermediates is reaction of an alkyl lithioacetate with the activated derivative obtained from an N-protected amino acid and carbonyldiimidazole (CDI). For example, Boc- or Cbz-D-*allo*-Ile were converted to isostatine by activation of the acid group with CDI followed by reaction with the lithium enolate of methyl acetate, giving a γ-amino-β-keto ester derivative (see Scheme 11.158). The ketone was stereoselectively reduced with potassium borohydride in methanol to give the desired stereoisomer (*1822*). In a similar manner, ethyl lithioacetate was added to CDI-activated Boc-D-*allo*-Ile (*1585, 1823*), while (3S,4R)-statine or (3S,4R,5S)-isostatine were prepared from Boc-D-Leu or Boc-D-allo Ile (*1823*). For the latter syntheses, the CDI-activated acid was reacted with the lithium enolate of *tert*-butyl acetate. A range of reagents, both achiral and chiral, were examined for stereoselective reduction of the β-keto group to generate the statine carbinol center. The desired *erythro* product, with (3S,4R) stereochemistry, was obtained as the major isomer with LiBH₄, NaBH₄, KBH₄, or Zn(BH₄)₂, with up to 10:1 stereoselectivity. The diastereomer derived from Boc-L-Ile was also prepared (*1823*). In another synthesis of (3R,4R)-statine, Boc-D-Leu was again activated with CDI, displaced with lithiated ethyl or *tert*-butyl acetate,

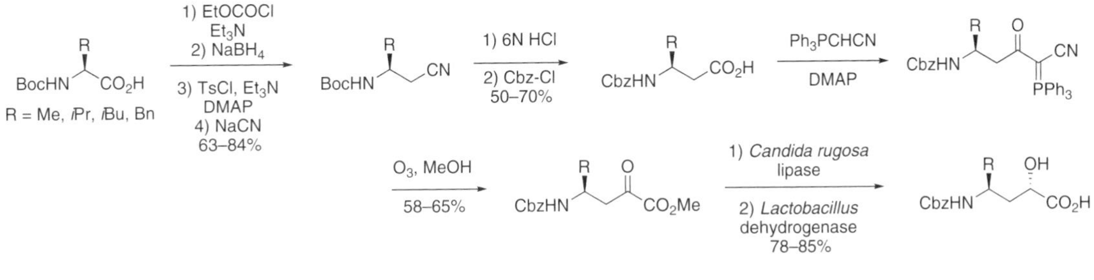

Scheme 11.157 Chemoenzymatic Synthesis of 4-Amino-2-Hydroxy Acids from Amino Acids (*1820*).

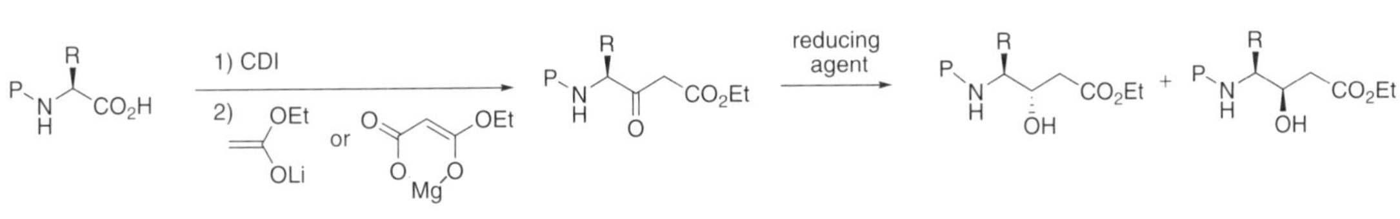

Scheme 11.158 Synthesis of Statines from Amino Acids via Acetate or Malonate.

and then reduced with a chiral reducing agent derived from lithium borohydride and *N,N'*-dibenzoyl-D-Cys with 72% de (*1824*).

A closely related synthesis prepared (3*S*,4*R*,5*S*)-isostatine via activation of Cbz-D-*allo* Ile as a pentafluorophenol ester, reaction with lithiated methyl acetate, and reduction with KBH₄ (11:1 stereoselectivity) (*1591, 1592*). Cbz-D-Val (*1591, 1592*) and Boc-D-Val (*1599*) were also converted by this procedure into (3*S*,4*R*)-isostatine, (3*S*,4*R*)-4-amino-3-hydroxy-5-methylhexanoic acid. A slight variation of the syntheses leading to isostatine reacted the lithium enolate of benzyl acetate with CDI-activated homochiral β-methylallylglycine (prepared by an asymmetric Claisen rearrangement of Gly allyl ester). NaBH₄ reduction of the ketone and hydrogenation of the ester/alkene gave the desired product (*1825*).

The 3-hydroxy-4-methylamino-5-phenylpentanoic acid constituent of hapalosin was prepared by reaction of lithiated ethyl acetate with CDI-activated Boc-L-Phe. Ketone reduction with NaBH₄ gave the *anti* isomer with 80% de. *N*-Methylation completed the synthesis. Basic conditions induced dehydration to 4-methylamino-5-phenylpent-2-enoic acid, which was hydrogenated to form the saturated analog (*1617*). Reaction of lithiated ethyl acetate with CDI-activated orthogonally protected 2,3-diaminopropionic acid, Cbz-Dap(Boc)-OH, followed by NaBH₄ reduction, provided (3*R*,4*S*)-4,5-diamino-3-hydroxypentanoic acid with 32% de. The (3*S*,4*S*)-diastereomer was obtained preferentially via a different route (*1778*).

Non-activated esters can be employed. The β-keto ester derived from the lithium enolate of ethyl acetate and L-Ser (protected as an *N*-Boc oxazolidine methyl ester) was used for Pd-catalyzed Stille or Sonogashira cross-coupling reactions with organostannanes, resulting in 3-aryl or 3-alkynyl-4-amino-5-hydroxypent-2-enoic acid derivatives (*1826*). Hofman and Tao found that *N,N*-dibenzyl α-amino acids could be prepared by displacement of scalemic α-hydroxy methyl esters with dibenzylamine, allowing for a greater diversity of side-chain substituents than provided by the common amino acids. Treatment of the amino esters with the lithium enolate of *tert*-butyl acetate gave β-keto esters in one step, but the reaction was slow, and, in the case of sterically hindered β-branched substrates, unsuccessful. Racemization also occurred. To overcome these limitations, the dibenzylamino methyl esters were hydrolyzed using LiI/NaCN, with the acid groups then activated with CDI before treatment with lithiated *tert*-butyl acetate. The β-keto esters were almost all obtained with >97% ee, and reduction to the *syn* diastereomer with sodium borohydride proceeded with 88-98% de (*1827*).

Malonates have been employed as the acetate nucleophile. Lysinestatine, (3*R*,4*S*)-4,8-diamino-3-hydroxyoctanoic acid (a component of the marine siderophore alterobactin A), was prepared from Cbz-Lys(Boc) by activation with CDI and displacement with the magnesium enolate of ethyl hydrogen malonate, which was accompanied by decarboxylation (see Scheme 11.158). The β-keto group was reduced with NaBH₄ to give the desired diastereomer with 66% de (*1615*). Boc-Leu, Boc-Lys(Cbz), Boc-Phe, Boc-Cys(Me), and Boc-His(Ts) were all converted to the β-keto esters by the same method in a 1988 report. A variety of ketone reduction conditions were then examined, with high diastereoselectivity (but low yields) obtained with K-*sec*-Bu₃BH (*1828*). Reetz et al. reported in 1989 that the β-keto esters derived from *N,N*-dibenzyl α-amino acids by activation with CDI and reaction with the magnesium enolate of malonic acid monomethyl ester could be reduced with NaBH₄ in methanol with 80–88% de *syn* selectivity (*1829*). (3*S*,4*S*)-4-Amino-3-hydroxy-6-methylheptanoic acid was obtained from CDI-activated Boc-L-Leu and the magnesium salt of ethyl malonate, followed by ketone reduction using H₂/Raney Ni or KBH₄/LiCl (*1776*). The CDI activation/malonate displacement homologation has also been applied to Cbz-Gly (*1830*), Boc-Gly (*1831*), Boc-Ala (*1831*), Boc-Abu (*1831*), Boc-Val (*1831*), Boc-Leu (*1831*), Boc-Ile (*1831*), Boc-Met (*1831*), Boc-Cys (*1831*), Boc-Cys(Me) (*1831*), Boc-Ser (*1831*), Boc-Thr (*1831*), Boc-vinylglycine (*1831*), Boc-Trp(CHO) (*1621*), Boc-3-hydroxy-4-azido-Abu (*1832*), and L-cyclohexylalanine (*1833*). The ketone of the latter adduct was reduced with sodium cyanoborohydride to give a nearly 1:1 mixture of 4-amino-5-cyclohexyl-3-hydroxypentanoic acid epimers (*1833*).

Boc-L-Phe was activated with CDI and then reacted with lithiated ethyl acetate or the magnesium salt of monoethyl malonate to give the β-keto ester; hydrogenation of the ketone with a chiral Ru catalyst produced the *threo* (3*S*,4*S*)-diastereomer with >98% de. The aromatic ring was then hydrogenated to a cyclohexyl ring (*1577*). Statine analogs were also derived from amino acids via activation with CDI and reaction with (PhMe₂Si)₂CuCNLi₂, generating an acylsilane intermediate (see Scheme 11.159). The two-carbon homologation was achieved by aldol addition of *O*-ethyl-*O-tert*-butyldimethylsilyl ketene acetal, which proceeded with high *syn* selectivity. Desilylation with TBAF gave the statine products (*1834*).

Amino acids have been converted into acid chlorides for activation. Two statine diastereomers were synthesized by reaction of Phth-L-Leu acid chloride with the magnesium derivative of diethyl malonate to give the β-keto derivative. Ketone reduction by NaBH₄ preceded the decarboxylation step. The diastereomers were separated by ion-exchange chromatography (*1563*). A closely related synthesis prepared 4-amino-3-hydroxy-2-methylpentanoic acid, the component of the bleomycin antibiotics, via reaction of the magnesium enolate of monomethyl methylmalonate with the acid chloride of Phth-D-Ala, producing directly the decarboxylated β-keto ester. Reduction of the ketone (NaBH₄) gave a mixture of four diastereomers, from which the (2*S*,3*S*,4*R*)-isomer was isolated in 12% yield on a sulfonic acid resin (*1602*). A synthesis of (3*S*,4*R*,5*S*)-isostatine employed the acid chloride of Fmoc-(2*R*,3*S*)-D-*allo*-Ile and the lithium enolate of monomethyl monotrimethylsilyl malonate to give the β-keto ester (*1590, 1835*). Ketone reduction with sodium cyanoborohydride

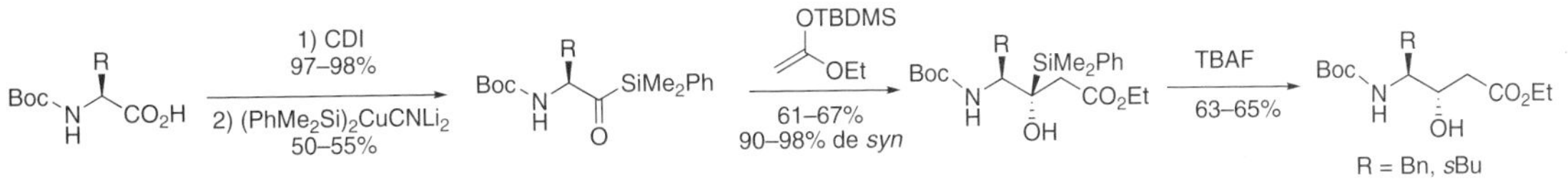

Scheme 11.159 Synthesis of Statines from Amino Acids via Acylsilane (*1834*).

gave a 77:23 mixture of diastereomers that were separated by crystallization (*1590*). Alternatively, the same γ-amino-β-keto ester intermediate was reduced with NaBH₄ to give the same (3S,4R,5S)-isomer of isostatine, but with 80% de (*1835*). (3R,4S)-Statine and (3S,4R,5S)-isostatine were both prepared in 1990 by reaction of Fmoc-L-Leu or Fmoc-D-allo Ile with thionyl chloride to form the activated acid chlorides. Condensation with excess lithiated *tert*-butyl acetate gave the β-keto esters, which were reduced with KBH₄ in ethanol (90% de, and 50% de, respectively). Crystallization gave diastereomerically pure product (*1836*).

Amino acid anhydrides have also been reacted. The enolate of propionic acid, derivatized with Evans' oxazolidinone auxiliary, was acylated with Boc-D-Ala anhydride. Ketone reduction by Zn(BH₄)₂ gave the desired *syn* stereochemistry of the (2S,3S,4R)-4-amino-3-hydroxy-2-methylpentanoic acid component found in bleomycin A₂ (*1601*). An Arg homolog, 4-amino-3-oxo-7-guanidinoheptanoic acid, was prepared for incorporation in hirudin-like thrombin inhibitors via reaction of a mixed anhydride of Boc-Arg(Ts) with lithioacetate (*1639*). More recently, a number of Boc- and Cbz-protected amino acid N-carboxyanhydride derivatives (UNCAs) were treated with the lithium enolate of ethyl acetate in

THF, giving γ-amino-β-keto ester products in 50–70% yield (see Scheme 11.160). Reduction of the ketone with NaBH₄ provided γ-amino-β-hydroxy ester derivatives with moderate stereoselectivity (*1837*). By employing substituted lithium enolates for the initial UNCA addition, α-alkyl-β-keto-γ-amino esters were produced in 30–81% yield (*1838*).

N-Boc and N-Cbz amino acids have also been activated as 2-acyl derivatives of 3,5-dioxo-1,2,4-oxadiazolidines (MODD). Lithium enolates of ethyl acetate or the *tert*-butyl thioester of propionate displaced the MODD to give γ-amino-β-keto esters, with much improved yields compared to displacements of Weinreb amide derivatives. The ketones were diastereoselectively reduced with NaBH₄ to give statine analogs (*1839*).

11.4.3l Elaboration of α-Amino Acids: Other Reactions of Carboxyl Group

A Wittig-based homologation has been applied to amino acids, instead of amino aldehydes as described in Section 11.4.3h. The lactone carbonyl of N-Cbz amino acid-derived oxazolidinones, formed from the N-Cbz amino acid and paraformaldehyde (see Scheme 11.161), were

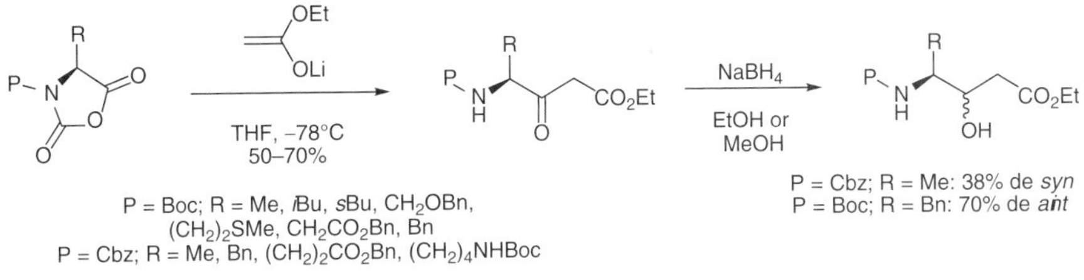

Scheme 11.160. Synthesis of γ-Amino-β-Keto Esters from UNCAs (*1837*).

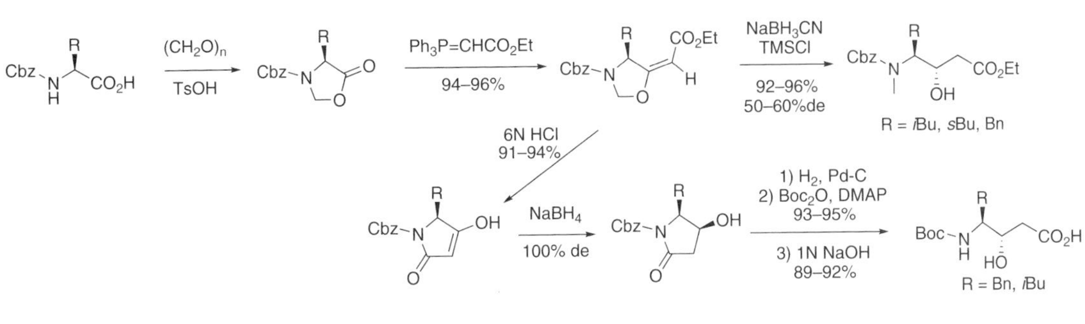

Scheme 11.161 Elaboration of Amino Acid Oxazolidinones (*1650, 1840*).

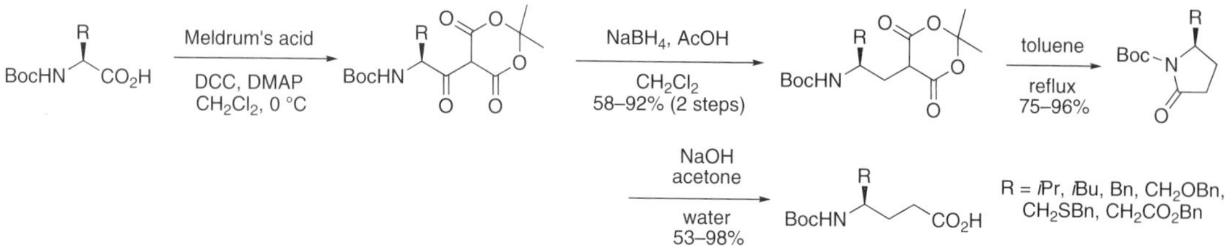

Scheme 11.162 Homologation of α-Amino Acids Using Meldrum's Acid (*1841*).

homologated with Ph₃P=CHCO₂Et. Treatment of the resulting *O*-alkyl enol with sodium cyanoborohydride/ TMSCl produced *N*-methyl γ-amino-β-hydroxy acids in high yield (*1650*). Alternatively, hydrolysis induced oxazolidinone cleavage and lactam formation. Alkene reduction with NaBH₄ proceeded with complete stereoselectivity, with lactam hydrolysis giving (3*S*,4*S*)-statine and (3*S*,4*S*)-AHPPA (*1840*).

Another general route to γ-substituted-γ-amino acids from α-amino acids relied on homologation using an initial coupling of Meldrum's acid (see Scheme 11.162). Reduction with sodium triacetoxyborohydride selectively removed the α-carbonyl group in 60–90% yield. Decarboxylative ring closure gave a 5-substituted pyrrolidinone, which was opened using basic hydrolysis (*1841*).

A variety of statine analogs were prepared by condensation of *N*-protected amino acids with Meldrum's acid, leading to a tetramic acid intermediate after heating (see Scheme 11.163, top). Reduction and hydrolysis of the pyrrolidinone provided the desired *N*-protected statines (*1842, 1843*). Tetramic acid intermediates have also been synthesized via *N*-acylation of chiral amino esters with ethyl malonyl chloride (see Scheme 11.163, bottom). Intramolecular Dieckmann condensation in the presence of NaOEt formed the tetramic acids, which

were decarboethoxylated to a keto lactam, stereospecifically hydrogenated, and hydrolyzed (*1844*).

A general route to γ-substituted α,β-acetylenic γ-amino acids began with carbamate-protected amino acids, which were converted to stabilized phosphorus ylides, e.g., CbzNHCH(R)C(=O)C(=PPh₃)CO₂Et, by EDCI-mediated coupling with Ph₃P=CHCO₂Et (45–51% for seven different amino acids) (see Scheme 11.164). Flash vacuum pyrolysis at 600 °C generated the acetylenic amino esters in 29–58% via extrusion of Ph₃PO (*1845, 1846*). Alkyne hydrogenation provided γ-substituted-γ-amino acids; hydrobromination was also possible.

A vicinal tricarbonyl derivative of Leu was prepared by activation with CDI, reaction with a phosphorane, and ozonolysis of the resulting phosphoranylidene ylide (*1640*).

Reduction of the α-carboxyl group of Cbz-Asp(O*t*Bu)-OH via the isobutyl chloroformate/NaBH₄ method produced an amino alcohol that was converted into *tert*-butyl *N*-Cbz aziridine-2-acetate by a Mitsunobu reaction with PPh₃/DEAD. The aziridine could be opened with amine nucleophiles to form 3,4-diaminobutyrate products, with high regioselectivity for attack of the incoming amine at the less substituted C4 position (*1168*).

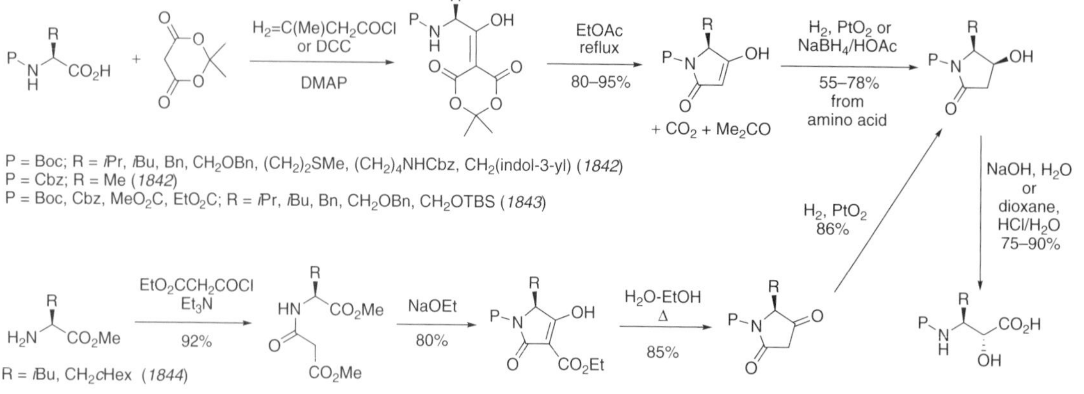

P = Boc; R = *i*Pr, *i*Bu, Bn, CH₂OBn, (CH₂)₂SMe, (CH₂)₄NHCbz, CH₂(indol-3-yl) (*1842*)
P = Cbz; R = Me (*1842*)
P = Boc, Cbz, MeO₂C, EtO₂C; R = *i*Pr, *i*Bu, Bn, CH₂OBn, CH₂OTBS (*1843*)

R = *i*Bu, CH₂cHex (*1844*)

Scheme 11.163 Synthesis of Statines from Amino Acids via Tetramic Acid.

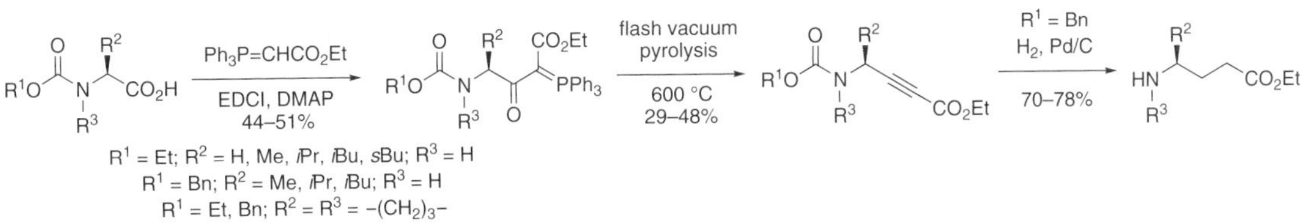

Scheme 11.164 Synthesis of Acetylenic γ-Amino Acids (*1845*, *1846*).

11.4.3m Elaboration of α-Amino Acids: Derivatization of Glu α-Carboxyl Group

Glutamic acid already contains a 4-aminobutyric acid skeleton, and so is a useful synthon for preparing γ-amino acids, just as aspartic acid is convenient for β-amino acid synthesis. Regioselective reduction of the α-carboxy group of L-Glu to an hydroxymethyl group provided (S)-4-amino-5-hydroxypentanoic acid (*1847–1849*). The same product was obtained by reduction of the ester group of Cbz-L-Gln-OMe, followed by hydrogenation and hydrolysis (*1660*). Conversion of the hydroxy group of the reduction product to an iodo leaving group was followed by displacement with alkyl organocuprates to give γ-substituted-γ-amino acids (see Scheme 11.165) (*1154*), while displacement with azide led to 4,5-diaminopentanoic acid (*1850*).

A similar N-Boc 4-amino-5-iodopropionic acid methyl ester intermediate was converted into a nucleophilic iodozinc derivative corresponding to an α-carboxy-bis(homologated) version of the Jackson β-(iodozinc)alanine synthon (see Sections 10.5.3 and 11.2.3n). Pd-catalyzed coupling of the zinc reagent with aryl iodides provided homochiral 4-benzyl-γ-aminobutyric acid derivatives (*1169*, *1172*). Unprotected iodophenols also coupled (*1170*), as did aryl triflates, although with reduced yields (29–49%) (*1171*). The corresponding zinc/copper reagent was reacted with allyl chloride to give 4-aminooct-7-enoic acid (*1172*). NMR studies on the organozinc reagent demonstrated that DMA or DMF stabilized the reagent from β-elimination decomposition, allowing for Pd-catalyzed cross-couplings with acid chlorides to produce 4-amino-6-keto-6-substituted-hexanoic acids (*1174*).

The α-carboxyl group of N-unprotected ethyl pyroglutamate has been reduced to a hydroxymethyl group to give pyroglutaminol, with the hydroxyl replaced with chloride, bromide, fluoride or cyanide before opening of the lactam by hydrolysis (*1851*). (S)-[4-²H]-4-amino-5-chloropentanoic acid and (S)-[U-¹⁴C]-4-amino-5-chloropentanoic acid were prepared from L-[2-²H]-Glu or L-[U-¹⁴C]-Glu via cyclization to pyroGlu-OEt, reduction to pyroglutaminol, chlorination with PPh₃/CCl₄, and hydrolysis (*1527*). (R)- or (S)-pyroglutaminol were O-mesylated and treated with sodium iodide/zinc dust to give 5-methyl-2-pyrrolidone. Lactam hydrolysis provided (S)- or (R)-4-aminopentanoic acid (*1530*). Protected pyroglutaminol was carboxylated at the 3-position, alkylated with *tert*-butyl bromoacetate, and then decarboxylated to give the lactam of 3-carboxy-6-amino-6-hydroxyhexanoic acid (*1852*). Alternatively, the side-chain carboxyl group of N,N-Bn₂-Glu was masked by reduction to a TBDMS-protected alcohol, and the α-carboxyl group reduced to an aldehyde. Addition of diethylzinc to the aldehyde gave the *syn* adduct with 74% de, while EtMgBr produced the *anti* adduct in 73% yield with 84% de. The side-chain carboxyl group was then reoxidized, producing either *syn* or *anti* 4-amino-5-hydroxyhexanoic acid (*1175*).

S-Glu has been converted into 3,4-dehydropyroglutamate, and the α-carboxyl group reduced to a hydroxymethyl group to provide 3,4-dehydropyroglutaminol. Michael addition of a *p*-chlorophenyl organocuprate reagent to the alkene was followed by removal of the hydroxymethyl substituent via oxidation to an acid and Barton decarboxylation. Hydrolysis of the lactam bond gave (R)-baclofen, (3R)-4-amino-3-(4-chlorophenyl)butanoic acid (*1533*). Other organocuprates could be employed for the addition step (*1533*). 2,5-Dihydroxy-3-ethyl-4-aminopentanoic acid was prepared by addition of an ethyl organocuprate to 3,4-dehydropyroglutaminol, followed by deprotonation and hydroxylation of the lactam before hydrolysis (*1853*). The 2,3-dihydroxy-4-dimethylamino-5-methoxypentanoic acid component of calyculin was obtained from pyroglutaminol via O-methylation of the hydroxymethyl group, formation

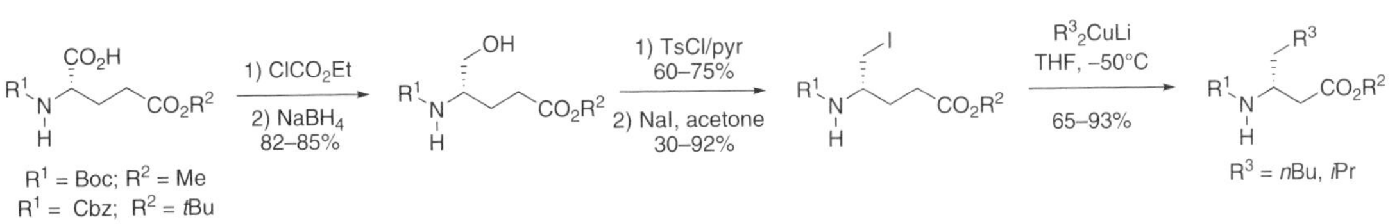

Scheme 11.165 Conversion of Glu into Electrophilic Synthon (*1154*).

and dihydroxylation of a 3,4-dehydropyroglutaminol intermediate, and lactam hydrolysis (*1627*). Simple dihydroxylation of 3,4-dehydropyroglutaminol provides 2,3,5-trihydroxy-4-aminopentanoic acid (*1854*).

The hydroxy group of L-glutaminol was oxidized to an aldehyde and methylenated with $Ph_3P^+MeBr^-$/ NaHMDS to give (*S*)-4-amino-5-hexenoic acid, (*S*)-vigabatrin (*1849*). (*S*)-Vigabatrin was also obtained from *N*-Boc methyl L-pyroglutamate, via reduction and methylenation of the α-carboxy group. Hydrolysis with 5 N HCl gave the desired product (*1855*). Alternatively, pyroglutamate was *N*-protected with a butenyl group, and the α-ester group reduced to an aldehyde. Wittig methylenation and pyrrolidinone hydrolysis provided (*S*)-4-aminohex-5-enoic acid in six steps and an overall yield of 33% from L-Glu (*1856*). Similarly, the α-carboxy group of *N*-aryl (*S*)-ethyl pyroglutamate was first reduced to an aldehyde, and then converted to an acetylene group using diethylmethyldiazophosphonate, leading to (*S*)-4-amino-5-hexynoic acid (*1857*).

Another synthesis beginning with Glu dehydrated Cbz-L-Glu(OMe)-NH₂ (L-isoglutamine methyl ester) to give Cbz-L-γ-cyano-γ-aminobutyric acid after ester hydrolysis (*1183*). (*R*)-GABOB was synthesized in 1962 via decarboxylation of *threo*-β-hydroxy-L-Glu with *E. coli* glutamic acid decarboxylase. It was then *N*-trimethylated with methyl iodide to give (−)-*l*-carnitine (*1541*).

11.4.3n Elaboration of α-Amino Acids: Other Syntheses

(*S*)-(−)-4-Amino-2-hydroxybutyric acid was prepared from L-2,4-diaminobutyric acid dihydrochloride via regioselective conversion of the α-amino group to a hydroxy group upon treatment with sodium nitrite (*1613*). In a similar fashion, L-asparagine was converted into L-4-amino-2-hydroxybutyric acid by treatment with sodium nitrite, followed by dehydration of the amide with acetic anhydride in pyridine, and hydrogenation of the cyano group (*1858*). Both enantiomers of emeriamine, 3-amino-4-trimethylammoniumbutanoic acid, were prepared from Cbz-L-Asp(O*t*Bu) via one-carbon homologation of the α-carboxyl group to give an asymmetrically protected 3-aminopentanedioic acid intermediate. The γ-trimethylammonium group was introduced by regioselective deprotection of either of the carboxyl groups, followed by Curtius rearrangement and trimethylation (*169*).

N-Acetyl (2*S*,4*R*)-4-hydroxyproline was converted into (*R*)-GABOB by an electrochemical decarboxylation, with the resulting 2-methoxy-4-hydroxypyrrolidine oxidized with peracetic acid or *m*CPBA to give the lactam of GABOB. Hydrolysis in 4 N HCl gave the amino acid in 65% overall yield (*1859*). *N*-Bz *trans*-4-hydroxy-L-Pro-OMe was transformed into (*S*)-4-amino-3-(4-chlorophenyl) butyric acid ((*S*)-baclofen), regioisomeric (*R*)-4-amino-2-(4-chlorophenyl)butyric acid, and their enantiomers. The aromatic group was introduced by addition of (4-Cl-Ph) MgBr to 4-keto-Pro, followed by dehydroxylation,

decarboxylation, nonregioselective oxidation of the 3-(4-chlorophenyl)pyrrolidine to a mixture of 3- and 4-(4-chlorophenyl)pyrrolidin-2-one, and hydrolysis of the lactam (*1860*).

Treatment of histidine benzyl ester with excess benzoyl chloride opens the imidazole ring to form tribenzoyl 2,4,5-triaminopent-4-enoic acid, with hydrolysis generating the free triamino acid (*1675*).

Methyl 2-azido-3-(dimethylphenylsilyl)hex-4-enoate was treated with the formaldehyde equivalent 1,3,5-trioxane-trioxane in the presence of Lewis acids to give, after an allylic azide isomerization, methyl 4-azido-5-methyl-6-hydroxyhex-2-enoate. Azide reduction by $SnCl_2$ gave the amino ester (*228*).

11.4.3o Elaboration of β-Amino Acids

A general route to γ-substituted-γ-amino acids proceeds via Arndt–Eistert homologation of β-amino acids, themselves obtained by Arndt–Eistert homologation of α-amino acids. Low yields were encountered during attempted bis(homologation) of Ala, Val, and Leu (*1642*), but successful bis-homologation was reported for Cbz-L-Pro and Cbz-L-Phe (*1100*). Arndt–Eistert homologation of L-2,3-diaminopropionic acid, followed by trimethylation of the γ-amino group, gave emeriamine (*168*).

Homochiral β-amino acids have also been elaborated into γ-amino acid derivatives via a stereospecific ring expansion of β-lactams to γ-lactams. A diazoketone intermediate was obtained in 44–82% yield by treatment of β-lactams with trimethylsilyldiazomethane, with photolytic rearrangement to the γ-lactam proceeding in 62–82% yield (*1861*). Cbz-β-Ala-OMe was homologated to give protected 2-keto-4-aminobutyric acid via addition of dimsyl sodium, α-bromination, and acid methanolysis. The ketone was stereoselectively reduced with Baker's yeast to provide (*S*)-4-amino-2-hydroxybutanoic acid, a component of the butirosin antibiotics, but with only 49% ee (*1862*).

An enzymatic desymmetrization of dimethyl 3-benzylaminoglutarate by *Candida antarctica* lipase B (CAL-B) produced the (*S*)-monoamide in 76% yield. A Hofmann rearrangement then gave 3,4-diaminobutanoic acid (*689*).

11.4.3p Elaboration of γ-Amino Acids

α-Substituted-γ-amino acids can be prepared by deprotonation and alkylation of γ-amino acid derivatives. Hanessian and Schaum reported that homochiral *N*-protected 4-substituted γ-amino esters or amides, prepared by homologation of α-amino esters, could be dideprotonated and alkylated at the α-position with high stereoselectivity. The 1,3-asymmetric induction was generated under a variety of conditions, and in all cases the *anti* isomer was obtained with >90% de (*1863*). This strategy was applied to α-methylation or α-benzylation of the bis-homologated γ-amino acids derived from

Boc-Val or Boc-Ala, respectively (1659). The pyrrolidinones formed by cyclizing γ-substituted-γ-amino acids (again derived from α-amino acids) could be deprotonated and stereoselectively alkylated with cinnamyl bromide, giving α-cinnamyl-γ-substituted-γ-amino acids (1644). Homochiral 4-substituted GABA amides have been α-iodinated and then used for a free-radical C-allylation reaction to provide α-allyl derivatives with *anti* stereoselectivity (50–78% de) (642).

The acid groups of γ-aminobutyric acid or N-methyl GABA were converted into a homochiral oxazoline by reaction with an amino alcohol. Deprotonation and alkylation, followed by hydrolysis, gave α-substituted-GABA with excellent diastereoselectivity (>90%), depending on the amino alcohol employed (177). The deprotonated oxazoline derivatives have also been sulfenylated to give γ-amino-α-thiohydroxycarboxylic acids (1864). Racemic N-phthaloyl 4-amino-2-phenylbutyric acid was treated with oxalyl chloride and triethylamine to form a ketene intermediate, with addition of the chiral alcohol (R)-pantolactone to the ketene generating predominantly the N-phthaloyl (R)-4-amino-2-phenylbutyric acid ester as an 85:15 diastereomeric mixture (1701). (S)-4-Amino-5-fluoropentanoic acid was converted into (S,E)-4-amino-5-fluoropent-2-enoic acid via deprotonation, α-selenation, and oxidative elimination (1865).

N-Boc-2-(TBS)-pyrrole (TBSOP) has been developed as an anionic GABA equivalent (see Scheme 11.166). Reaction with chiral aldehydes in the presence of Lewis acids produced polyhydroxy γ-amino acids, hybrids of amino acids and carbohydrates, as diastereomerically pure compounds. The stereochemistry at both the newly generated carbinol center and the amino-substituted center was controlled by the configuration of the aldehyde. The lactam center could be epimerized by treatment with triethylamine/DMAP. The γ-amino acid was revealed by catalytic hydrogenation of the unsaturated lactam, followed by ring opening with refluxing 6 N HCl (1866). TBSOP was also added to tridecanal in the presence of Ti(OiPr)₄ and (R)-Binol catalyst to give the aldol adduct in 83% yield, with >98% de *threo*

diastereoselectivity and up to 68% ee. Hydrogenation generated the lactam of 4-amino-5-hydroxyheptadecanoic acid, a non-natural aza-analog of the bioactive acetogenin (+)-muricatacin (1867).

Unsaturated γ-aminobutyric acids can be derivatized. Conjugate addition of organocuprates to 4-aminobut-2-enoic acid, with the amino group contained in a chiral oxazolidinone, proceeded with 50–78% de. The diastereomers were chromatographically separated before removal of the auxiliary (1868). (−)-(R)-Baclofen and other 3-aryl-4-amino acids were obtained from N-protected ethyl (E)-4-aminobut-2-enoate via an asymmetric conjugate addition of arylboronic acids in the presence of a chiral Rh(I)-BINAP catalyst, with 86–92% ee (1708). (R)-Emeriamine, 4-trimethylamino-3-aminobutanoic acid, was prepared via an intramolecular conjugate addition of the urea formed by reaction of methyl N-[(S)-1-phenylethyl)]4-aminobut-2-enoate with tosyl isocyanate. The imidazolidinone adduct was deprotected with Li/NH₃ and hydrolyzed to give 3,4-diaminobutanoic acid, or hydrolyzed and N-trimethylated to give emeriamine (772). Homochiral N-protected 4-substituted-4-aminobut-2-enoic acid esters, derived from amino acids, were dihydroxylated under Sharpless asymmetric dihydroxylation conditions. With N-Boc protection, the "matched" reagent gave the (2R,3S,4S)-diastereomers with 94–97% de, but the "mismatched' reagent gave only 26–82% de. However, by using N,N-dibenzyl protection the (2S,3R,4S)-isomer was obtained as the "matched" product, with 72–90% de (1869).

Other γ-aminobutyric acids have been modified by derivatizing their side chains. The amide group of protected L-isoglutamine (Glu α-amide) was dehydrated with DMF/SOCl₂ to give L-γ-cyano-γ-aminobutyric acid (1232). Nucleo γ-amino acids, with the nucleobase attached to the γ-carbon by a methyl linker, were synthesized from (S)-N-Boc-5-(hydroxymethyl)-2-pyrrolidinone. O-Tosyl derivatives were displaced with cytosine or adenine, followed by lactam hydrolysis (1870).

Several syntheses convert achiral or racemic γ-aminobutyric acids into enantiomerically enriched

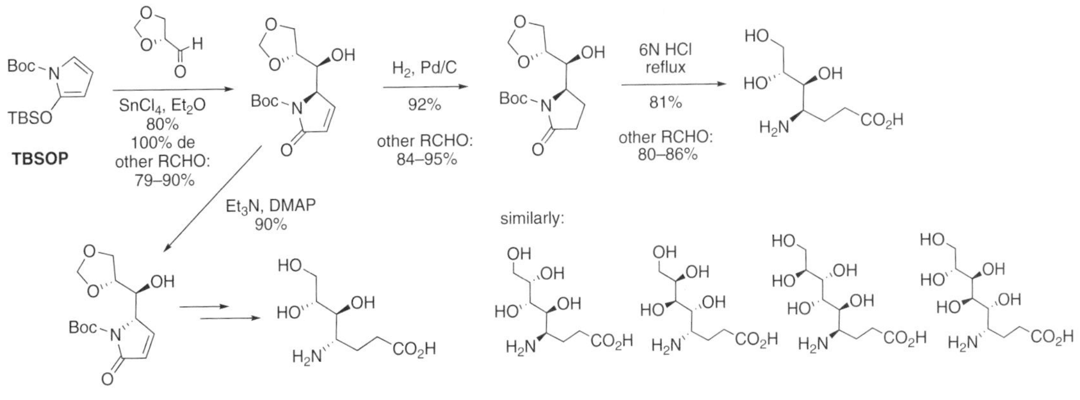

Scheme 11.166 Addition of TBSOP to Aldehydes (1866).

compounds by some form of asymmetric transformation. The statine analog AHPPA, 4-amino-3-hydroxy-5-phenylpentanoic acid, was prepared from achiral 3-keto-4-amino-5-phenylpent-4-enoic acid via a one-pot sequential asymmetric hydrogenation. In a stepwise synthesis, a Rh(I) catalyst (DuPHOS or BINAP ligand) reduced the alkene, producing the (4R)-isomer with >99% ee. Next, a Ru-BINAP catalyst reduced the β-keto ester to provide the (3R,4R)-diastereomer with >95% de. Both catalysts could be combined in a one-pot reaction to provide only one product with >95% ee. The best results were obtained with BINAP ligands for both the Rh and Ru catalysts (1871).

The γ-lactam of racemic N-benzyl 2-hydroxy-3,3-dimethyl-4-aminobutanoic acid has been oxidized to the α-keto lactam. Asymmetric hydrogenation of the ketone regenerated 2-hydroxy-3,3-dimethyl-4-amino-butanoic acid, but as a homochiral product (1872). A β-keto ester precursor of (R)-(−)-carnitine was prepared by carboxylation of the enolate of dimethylamino acetone with ethyl carbonate. The ketone group was enantioselectively reduced by catalytic asymmetric hydrogenation in the presence of a chiral rhodium complex, giving norcarnitine hydrochloride with up to 85% ee. Quaternization of the dimethylamino group with methyl iodide gave carnitine (1689). Enzymatic reductions of alkyl 3-oxo-4-azidobutyrates using Baker's yeast gave the azido precursors of GABOB and carnitine (1873). Achiral N-Cbz 4-amino-2-oxobutanoic acid was enantioselectively reduced by Bacillus stearothermophilus—or Staphylococcus epidermis—lactate dehydrogenase to provide (S)- or (R)-2-hydroxy-4-aminobutyric acid, respectively (1713).

GABA lactam can be condensed with aldehydes under basic conditions to introduce an α-alkylidene substituent. Asymmetric hydrogenation with a chiral 2,4-bis(diphenylphosphine)pentane-Ir catalyst produced the lactam of 2-(p-fluorophenyl)-4-aminobutanoic acid or 2-(isobutyl)-4-aminobutanoic acid with 82–89% ee (1874).

The N-acetoxymethyl substituent of homochiral methyl N-Boc,N-acetoxymethyl (S)-4-amino-6-methylhept-2-enoate was converted into a protected aminomethyl substituent by treatment with RCONH₂/PPTS. An intramolecular Michael addition was then induced by addition of NaH, with the resulting cyclic aminal decomposed by treatment with catalytic RuCl₃/TBHP, followed by 3 N HCl and Cs₂CO₃ in MeOH, to give (3S,4S)-3,4-diamino-6-methylheptanoic acid with 80% de (773).

11.4.3q Elaboration of Chiral Substrates

(R)-GABOB was prepared from monobenzyl L-malic acid via treatment with DPPA, with a Curtius rearrangement forming an oxazolidin-2-one intermediate (see Scheme 11.167). After hydrogenolytic removal of the benzyl group, the free carboxyl group was homologated by an Arndt–Eistert reaction, with acid hydrolysis giving the free amino acid in 30% overall yield (1538). An alternative synthesis from the same L-malic acid substrate reduced both carboxy groups to hydroxymethyl groups, with the 1,2-diol converted to an epoxide and opened with azide. Azide reduction and oxidation of the remaining hydroxy group formed (R)-GABOB (1875). A third synthesis converted the α-hydroxy acid group to an amide and the other carboxylic acid to an ester. Lithium aluminum hydride simultaneously reduced both the ester (to a hydroxymethyl group) and the amide (to an aminomethyl substituent). After N-protection, the hydroxymethyl group was reoxidized to an acid by zinc permanganate, giving GABOB in 25% overall yield (1876). Yet another route selectively reduced the 1-carboxyl group of dimethyl malate to give (3R)-3,4-dihydroxybutanoate. The amino group was introduced by regioselective tosylation of the primary alcohol and displacement with azide, followed by hydrogenation (1152). In a similar fashion, carnitine was prepared from malic acid via chemoselective reduction of the 1-carboxyl group to a hydroxymethyl moiety, tosylation, and displacement with trimethylamine (1877).

(R)-GABOB and (S)-GABOB have both been obtained from L- and D-arabinose, respectively, as have (R)-carnitine and (S)-carnitine. Oxidation of arabinose gave 2,3,4,5-tetrahydroxypentanoic acid, with bromination by HBr in acetic acid yielding 2,4-dibromo-3-hydroxypentanoic acid. Selective reduction of the 2-bromo group by hydrogenation over Pd-C and displacement of the terminal bromo group with sodium azide gave the GABOB precursors, while displacement with trimethylamine gave the carnitine enantiomers. A similar strategy was employed to prepare (R)-GABOB and (R)-carnitine from L-ascorbic acid (1878). (R)-GABOB has also been prepared from ascorbic acid via an (R)-glycerol acetonide intermediate. The free hydroxyl was tosylated and displaced with KCN to introduce the carboxyl group, with azide displacement of the tosylated or mesylated second terminal hydroxy group (after protecting group manipulation) introducing the amino group (1879). Another source of (R)-GABOB and (R)-carnitine is D-mannitol, via an (S)-glycerol

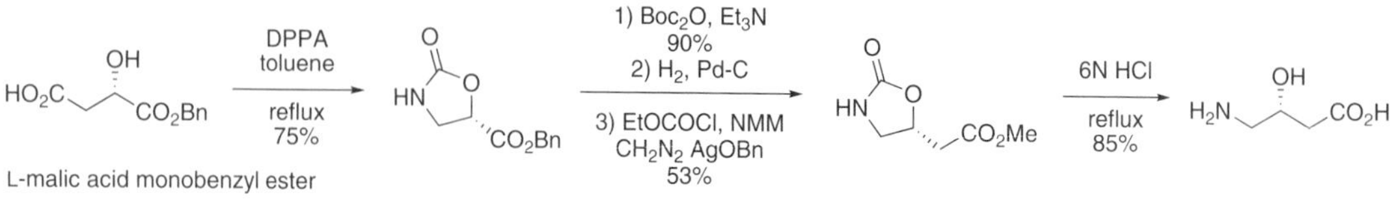

Scheme 11.167 Synthesis of (R)-GABOB from L-Malic Acid (1538).

acetonide intermediate. The amino group was introduced via Mitsunobu displacement of the free hydroxyl with phthalimide, with carboxylation achieved by KCN opening of a cyclic sulfite derivative (*1880*).

(*R*)-Carnitine was derived from D-galactono-1,4-lactone by two routes. In one procedure the amino group was introduced by reductive amination of an aldehyde with methylamine, while the other method formed an amide with dimethylamine and then reduced the amide carbonyl. Methyl iodide or dimethyl sulfate were employed to generate the desired quaternary amino group (*1881*). Both GABOB and carnitine have been prepared from (*S*)-3-hydroxybutyrolactone, which is readily available in large quantities but usually leads to the wrong enantiomer of GABOB. However, by using the carboxy group as a masked amino group and a cyano group as a masked carboxy group, the correct isomer could be prepared (see Scheme 11.168) (*1882*).

Malic acid has also been employed to prepare statin analogs. A cyclic imide was formed upon refluxing with benzylamine, and one carbonyl group reduced to form an acetoxy leaving group. This was stereoselectively displaced with trimethylallylsilane or tributylallyltin to give either the *trans*- or *cis*-substituted lactam of γ-allyl-γ-aminobutyric acid, with the stereoselectivity depending on reaction additives. Hydrogenation, deprotection, and hydrolysis of the *cis* derivative gave the *n*-propyl analog of statine (*1883*). 4-*Epi*-statine was synthesized by using methallyltrimethylsilane in the presence of boron trifluoride etherate for the alkylation reaction, giving the *trans* cyclic diastereomer with 80% de (*1884*). The *cis* isomer was obtained from the same acetoxy-substituted electrophile via a radical cyclization reaction (*1884*). A slightly different reaction sequence was used in another synthesis of (−)-statine. The malimide prepared from malic acid was alkylated with methallyl magnesium chloride and then deoxygenated by hydrogenation (*1885*).

A synthesis of (3*S*,4*R*)-4-methylamino-3-hydroxy-5-phenylpentanoic acid, the enantiomer of the component of hapalosin, also began with malic acid. Aminative cyclization, acetylation, and methylation provided 1-methyl-3-acetoxy-2,5-pyrrolidinedione, which was converted to the 3-benzyloxy derivative. The Grignard reagent BnMgCl reacted regioselectively with the lactam carbonyl adjacent to the hydroxy substituent; deoxygenation of the adduct provided 1-methyl-4-benzyloxy-5-benzyl-2-pyrrolidinone with *trans* stereochemistry. Hydrogenation and hydrolysis gave the γ-amino acid (*1886*).

An early synthesis of all four isomers of statine (4-amino-3-hydroxy-6-methylheptanoic acid) proceeded via Grignard addition of the isobutyl side chain to protected 3-deoxy-α-D-*erythro*-pentodialdo-1,4-furanose or 3-deoxy-β-L-*threo*-pentodialdo-1,4-furanose, with one diastereomer produced in the first reaction, and a mixture of two in the second. The carbinol center of the first adduct could be inverted by tosylation and displacement with sodium benzoate to provide the fourth isomer. To complete the synthesis, the hydroxy group was activated and displaced with azide, while the diol was oxidatively cleaved to form the acid group (*1564, 1565*). Another synthesis leading to all four statine isomers relied on condensation of benzyl isocyanate with methyl (*S*)-α-hydroxy-β-phenylpropionate, followed by DIBAL reduction to give a 3,5-dibenzyl-4-ethoxyoxazolidin-2-one intermediate. An isobutenylation reaction with β-methallyltriphenylstannane introduced the side chain, with the carboxyl group unmasked at the end of the synthesis by oxidation of the phenyl substituent (*1887*).

Another synthesis of statine (and analogs) proceeded from protected D-glucofuranose (see Scheme 11.169). An epoxide derivative was opened with Grignard reagents in the presence of CuI, with the amino group introduced by mesylation and displacement with azide. The acid group was generated by oxidative diol cleavage (*1888*). A similar approach elaborated (*R*)-2,3-*O*-isopropylidene glyceraldehyde into statine in nine steps and 13% overall yield, via addition of diallylzinc to the aldehyde, conversion of the protected diol to an epoxide, epoxide opening with *i*PrMgBr, hydroxyl displacement with phthalimide, and oxidative cleavage of the alkene (*1889*).

(3*S*,4*S*)-Statine and (3*S*,4*S*)-4-amino-3-hydroxy-5-phenylpentanoic acid (AHPPA) were derived from D-glucosamine through a common intermediate that could be applicable to the preparation of other γ-substituted-β-hydroxy-γ-amino acids (*1890, 1891*). The desired synthon, essentially γ-formyl-β-hydroxy-γ-aminobutyric acid with a masked carboxy group, was prepared in nine steps and 27% overall yield via C6 carbon degradation and C4-hydroxy group elimination (see Scheme 11.170). The aldehyde group was reacted with a Wittig or Grignard reagent to introduce the γ-substituent and the vinyl group was hydroborated and then oxidized to reveal the acid. D-Glucosamine has also been elaborated to give the γ-amino acid component of calyculin, 2,3-dihydroxy-4-dimethylamino-5-methoxy-pentanoic acid, as well as one of the dihydroxyamino

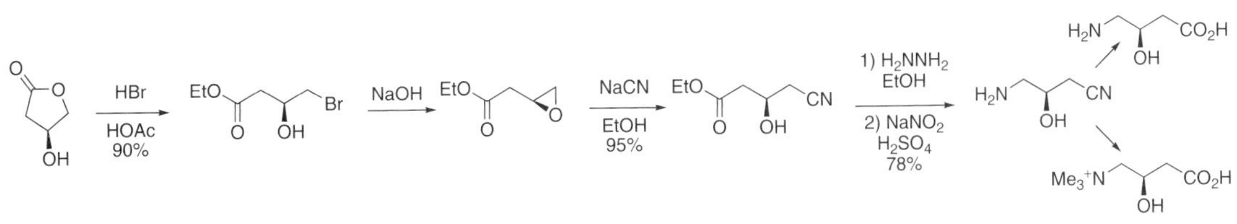

Scheme 11.168 Synthesis of GABOB and Carnitine from (*S*)-3-Hydroxybutyrolactone (*1882*).

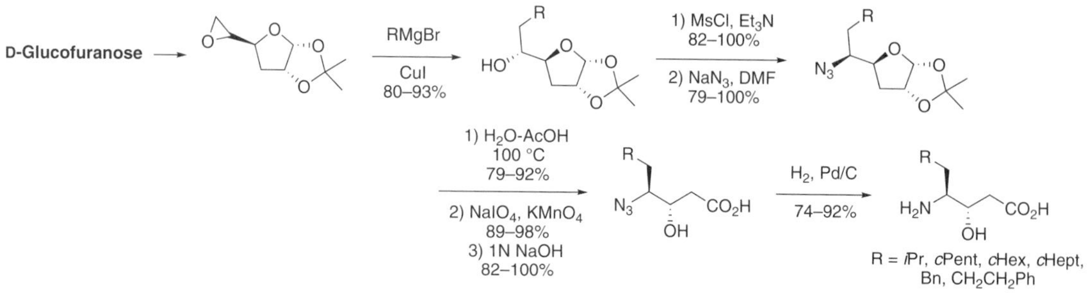

Scheme 11.169 Elaboration of D-Glucofuranose (*1888*).

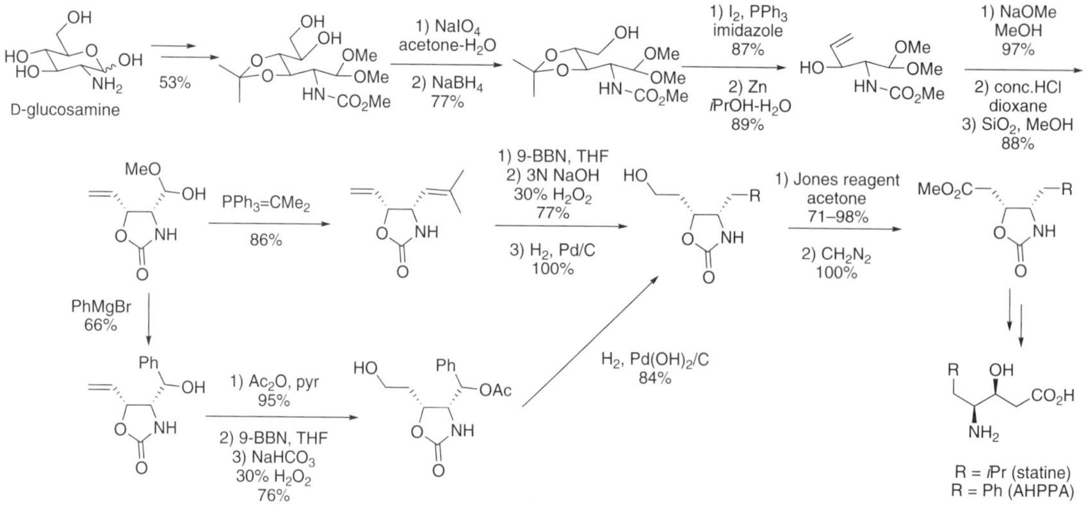

Scheme 11.170 Synthesis of *threo*-β-Hydroxy-γ-Amino Acids from D-Glucosamine (*1890, 1891*).

acid components of AI-77-B, the lactone of 2-acetoxy-3-hydroxy-4-amino-hexanedioic acid (*1288*). The same amino acid was also prepared via amination of a gulonolactone-derived intermediate (*1628*).

1,2-*O*-Isopropylidene-α-D-xylofuranoside and 1,2,4,5-di-*O*-isopropylidene-α-D-allofuranoside were converted into two isomers of 2,3-dihydroxy-4-aminobutanoic acid by tosylation and azide displacement of the hydroxymethyl group, followed by oxidative cleavage of the diol group to generate the acid (*1892*). (2*S*,3*S*,4*R*)-4-Amino-3-hydroxy-2-methylvalerate, a component of the antitumor antibiotic bleomycin, was derived from L-rhamnose, with the amino group introduced by azide displacement of a mesyl derivative (*1600*).

11.4.3r Other Syntheses

A potentially versatile route to α,γ-disubstituted-γ-amino acids was applied to the synthesis of tubuphenylalanine and the γ-amino acid precursor of tubuvaline. The γ-center chirality was introduced by a stereoselective Mn-mediated addition to a hydrazone, with the

α-substituent center present on either one of the reacting components. The acid group was masked as an alcohol until the end of the synthesis (see Scheme 11.171) (*1893*).

An asymmetric functionalization of 2-oxazolone generated a protected 2-oxazolidinone synthon corresponding to an activated GABOB derivative (see Scheme 11.172). Reaction with nucleophilic reagents introduced a 4-substituent, with the synthesis completed by oxidative cleavage of the alkene group to form the acid, followed by oxazolidine ring opening. Statine and analogs such as cyclohexylstatine were prepared using this method (*1894, 1895*).

A stereoselective iodolactonization reaction of *N*-[(*R*)-1-phenylethyl]-3-hydroxypent-4-enethioacetamide, via a thioimidate intermediate, provided (4*S*,5*R*)-4-hydroxy-5-(iodomethyl)pyrrolidine-2-one. The iodo group was displaced by Grignard reagents in the presence of Cu(I) salts, with isopropyl or cyclohexyl reagents resulting in the lactams of statine or 4-amino-5-cyclohexyl-3-hydroxypentanoic acid (ACHPA), respectively. The *N*-substituted chiral auxiliary was removed with

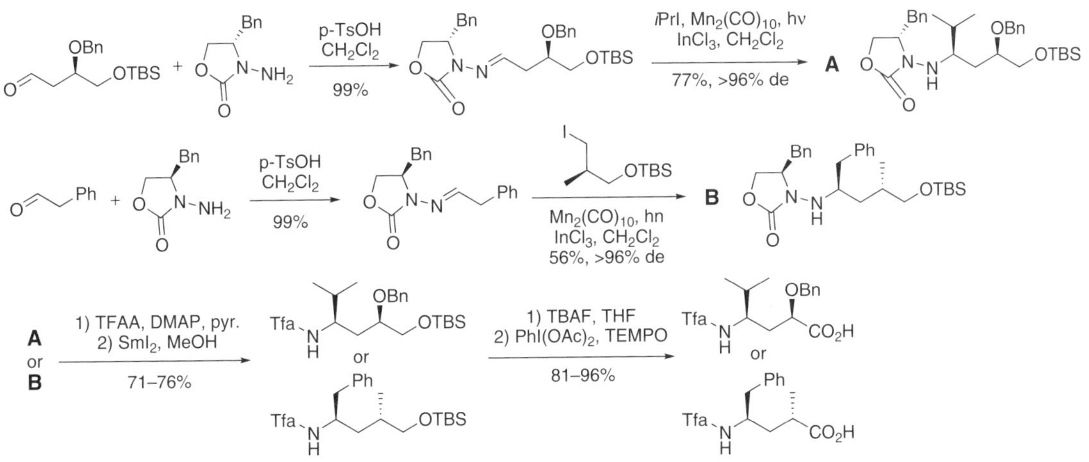

Scheme 11.171 Synthesis of α,γ-Disubstituted-γ-Amino Acid via Hydazone (*1893*).

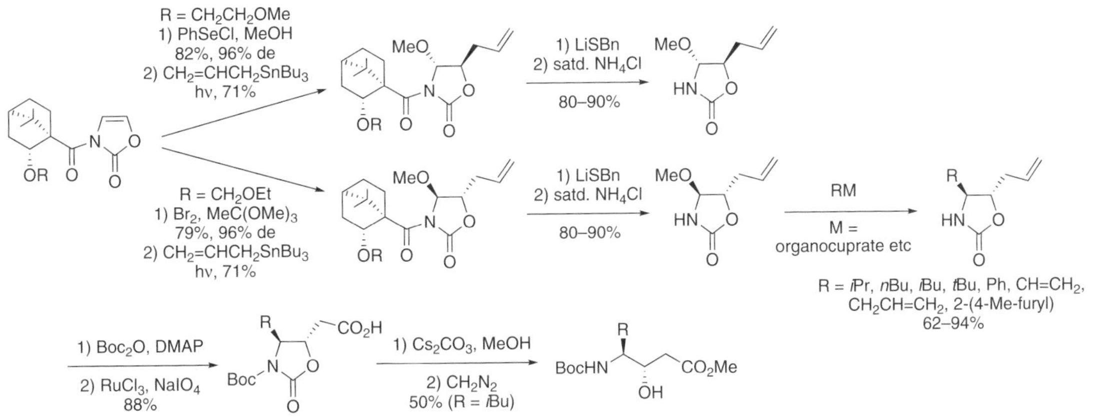

Scheme 11.172 Synthesis From 2-Oxazolone (*1894, 1895*).

Na/NH$_3$, and the lactam ring opened with sodium methoxide (*1896*).

A synthesis of (R)-GABOB from 3-hydroxycyclobutanone converted the ketone to an imine of (S)-α-methylbenzylamine, with oxidation by mCPBA generating a mixture of oxaziridine diastereomers. Photolysis-induced ring expansion gave a mixture of γ-lactam diastereomers, with the desired precursor of (−)-GABOB obtained in 40% yield. Deprotection and hydrolysis produced GABOB in 36% yield (*1897*). (R)-GABOB was also prepared from 3-hydroxypyridine via reduction with sodium borohydride to give 1,2,3,4-tetrahydro-3-hydroxypyridine, resolution using lipase PS, and oxidative cleavage of the alkene bond with ruthenium tetraoxide (*1898*). Another synthesis of (R)-GABOB and (R)-carnitine employed β-pinene as a masked chiral acetate group. An ene-reaction with methyl glyoxylate generated the chiral carbinol center, with the glyoxylate ester converted into a chloromethyl group. Oxidative cleavage of the pinene auxiliary revealed the carboxyl group, with halide displacement using ammonia or trimethylamine leading to GABOB or carnitine, respectively (*1899*).

γ-Vinyl-GABA, the anticonvulsant drug (S)-vigabatrin, was prepared from ethyl 6-hydroxyhex-2E-enoate via an asymmetric aminohydroxylation reaction. The α-hydroxyester moiety of the initial α-hydroxy-β-amino ester adduct was converted into a vinyl group, and the terminal hydroxyl group oxidized to generate the desired γ-amino acid (*861*). (3R,4S)-4-Methylamino-3-hydroxy-5-phenylpentanoic acid, the γ-amino acid component of hapalosin, was constructed via an aldol reaction of 3-phenylpropanoic acid with acrolein, using the Evans' oxazolidinone auxiliary to control diastereoselectivity. The carboxyl group was converted into an amine by a Curtius rearrangement, while the alkene was hydroborated and oxidized to form the acid (*1616*).

Table 11.6 Asymmetric Syntheses of γ-Amino Acids

H₂N-R-CO₂H	Method	Yield, % ee	Reference
Amination			
−CH₂CH(OH)CH₂−	asymmetric epoxidation of 4-hydroxybutene, oxidation, epoxide opening with NH₄OH	25% epoxidation 55% ee 66% oxidation and amination	1983 (795) 1984 (1721)
−CH₂CH(OH)CH₂− (with NMe₃)	enzymatic hydrolysis of alkyl 3,4-epoxybutyrate, epoxide opening with NMe₃	35% resolution 95% ee 75–80% amination	1988 (1722)
−CH(CH=CH₂)CH₂CH₂−	opening of homochiral 2,3-epoxy-5-Ph-1-pentanol with BnNH₂, oxidation of Ph, conversion of diol to vinyl	70% epoxide opening 63–70% Ph oxidation 40% side chain conversion	1997 (1517)
−CH₂CH(OH)CH₂−	opening of lactone of 3,4-dihydroxybutanoic acid with TMSI, azide displacement, reduction	80% opening 100% azide displacement 75% reduction	1990 (2192)
−CH₂CH(OH)CH₂−	azide displacement of ethyl 4-Cl-3-OH-butyrate, hydrogenation	100% displacement	1999 (1724)
−CH₂CH(OH)CH₂−	tosylation and azide displacement of methyl 3,4-(OH)₂-butyrate, reduction	60% tosylation 75% displacement	2003 (215)
−CH₂CH(OH)CH₂−	displacement of BrCH₂CH=CHCO₂Et with PhCH(Me)NH₂, N-Cbz protection, iodocyclization, radical deiodination, deprotection	80% displacement 90% iodocyclization 0% de 85% deiodination 64% deprotection	1987 (1725)
−CH₂CH(4-Cl-Ph)CH₂−	azide displacement of homochiral iodo precursor	95% displacement 80% reduction/deprotection	1997 (1726)
−CH₂CH(iBu)CH₂−	azide displacement of tosylated homochiral precursor, reduction	—	1994 (1514)
−CH₂CH(Me)CH₂−	BocNHOBn displacement of homochiral BrCH₂CH(Me)CH₂CO₂Et	—	1992 (1727)
−CH₂CH(OH)CH₂− with NMe₃	displacement of 1-chloro-2-hydroxy-4-pentene with NMe₃, oxidative cleavage of alkene	99% displacement 81% oxidation	1997 (1728)
−CH₂CH(OH)CH₂− with NMe₃	displacement of 1-chloro-2-hydroxy-4-butyric acid with NMe₃, substrate by Baker's yeast reduction of β-keto ester	45% overall yield	1983 (1729)
−CH₂CH(OH)CH₂− with NH₂ or NMe₃	asymmetric reduction of ethyl 3-oxo-4-chlorobutyrate by asymmetric hydrogenation	97% yield 97% ee	1988 (1730)
−CH₂CH(OH)CH₂− with NH₂ or NMe₃	enzymatic resolution of diethyl 3-hydroxyglutarate, amination via Curtius or trimethylamine displacement of bromo derivative	36% overall GABOB 50% amination carnitine	1984 (1731)
−CH(Me)CH=CH− CH(nPr)CH=CH−	azide addition to RCH=CHC(OCO₂Et)CO₂Et, prepared from aldehyde using oxynitrilase, azide reduction	77–91% cyanohydrin >95% ee 87–95% acylation 85–88% nitrile hydrolysis 84–85% azide addition 95% ee 80–86% azide reduction	2005 (1732)
−CH₂CH=CH− −CH(Me)CH=CH− −CH(Me)CMe=CH−	Mitsunobu reaction of homochiral allylic alcohols with masked carboxyl group via Sₙ2′ with HNPhth, Pb(OAc)₄ cleavage, hydrazinolysis	25–62% Mitsunobu 50–80% de	1995 (1733)

Table 11.6 Asymmetric Syntheses of γ-Amino Acids (continued)

H$_2$N-R-CO$_2$H	Method	Yield, % ee	Reference
−CH(Me)CH=CH− −CH(Me)CMe=CH− −CH(Me)CH=CMe−	Mitsunobu reaction of homochiral γ-OH-α,β-unsaturated ester with HNPhth, hydrazinolysis	10–30% overall >99% ee	1995 (*1733*)
−CH(Me)CH=CH−	amination of (η³-allyl)Fe(CO)$_2$(NO) complex with chiral ester auxiliary	28–82% 79–>98% de	1998 (*1677*)
−CH$_2$CH(OH)CH$_2$−	enzymatic ammonolysis of 3-hydroxyglutarate, Hofmann rearrangement of monoamide	49% rearrangement 34% overall from dimethyl 3-hydroxyglutarate	1996 (*1734*)
−CH$_2$CH[S(4-MeO-Bn)]CH$_2$− −CH$_2$CH(SH)CH$_2$−	thiol displacement of mesylate of glutarate, enzymatic monodem-ethylation, Curtius rearrangement, S-deprotection	98% displacement 81% monodemethylation 71% ee 73–91% rearrangement	1999 (*1724*)
−CH$_2$CH(Me)CH$_2$−	enzymatic resolution of dimethyl 3-methylglutarate, amination via Curtius of acid or conversion of ester to amide and Hofmann rearrangement	71% ammonolysis 84% Hofmann 94% Curtius	1990 (*1735*)
−CH$_2$CH(Et)CH$_2$−	enzymatic resolution of dimethyl 3-ethylglutarate, amination via Curtius of acid or conversion of ester to amide and Hofmann rearrangement	46–57% from 3-ethylglutaric acid	1990 (*1530*)
−CH(CH$_2$OH)CH$_2$CH(CH$_2$CO$_2$H)− as lactam	carboxylation, then carboxymethyla-tion of protected pyroglutaminol, decarboxylation	82–90% overall	2002 (*1852*)
−CH$_2$CH(4-Cl-Ph)CH$_2$−	enzymatic resolution of dimethyl 3-(4-chlorophenyl)glutarate, amination via Curtius of acid or conversion of ester to amide and Hofmann rearrangement	85% enzymatic resolution 60% ammonolysis and Hofmann 40% acyl azide and Curtius	1991 (*1736*)
−CH(Bn)CH(OH)CH$_2$−	Beckmann ring expansion/amination of homochiral cyclobutanone,	82% ring expansion 60% hydrolysis	1998 (*1737*)
−CH(iBu)CH(OH)CH$_2$−	Beckmann ring expansion/amination of homochiral cyclobutanone,	40% ring expansion 60% hydrolysis	1996 (*1738*)
−CH(iBu)COCH$_2$− −CH(iBu)CH(OH)CH$_2$−	reductive amination of iPrCO[2,5-(MeO)$_2$-Ph] with PhCH(Me)NH$_2$, Birch reduction, ozonolysis, ketone reduction	71% reductive amination 94% de 97% Birch reduction 84% ozonolysis	1990 (*1739*)
−CH$_2$CH(iBu)CH$_2$−	H$_2$/Ni hydrogenation of homochiral 3-CN-5-Me-hexanoic acid	61% reduction	2003 (*1536*)
Carboxylation			
−CH=CHCH(Ph)− −CH$_2$CH$_2$CH(Ph)−	alkylation of N-protected cinnamylamine with CO$_2$ or ClCO$_2$Me using n-BuLi/(−)-sparteine, hydrogenation	81–92% carboxylation and hydrogenation 84–92%ee	1999 (*970*)
−CH(Bn)CH(OH)CH$_2$−	vinyl$_2$CuMgBr opening of homochiral epoxide, alkene oxidative cleavage	88% epoxide opening 67–80% oxidative cleavage	1999 (*1618*)
−CH(Bn)CH(OH)CH$_2$−	CN opening of homochiral epoxide, CN hydrolysis	52–54% epoxide opening 88–100% CN hydrolysis	1999 (*1740*)
−CH(iBu)CH(OH)CH$_2$−	opening of N-Boc 2-(1,2-dihydroxy-ethyl)aziridine with RMgX/Cu, tosylation and displacement with NaCN, hydrolysis	77–88% opening 77% displacement 93% opening	1995 (*1243*)
Amination and Carboxylation			
−CH$_2$CH(OH)CH$_2$− (with NMe$_3$)	amination of epichlorohydrin with NMe$_3$, resolution with tartrate, displacement with CN, hydrolysis	27% epichlorohydrin amination and resolution 88% displacement with CN 82% hydrolysis	1987 (*1540*)

Table 11.6 Asymmetric Syntheses of γ-Amino Acids (continued)

H$_2$N-R-CO$_2$H	Method	Yield, % ee	Reference
−CH$_2$CH(OH)CH$_2$− (with NMe$_3$)	asymmetric mesylation of glycerol, displacement with trimethylamine, bromination of second hydroxyl, displacement with CN, hydrolysis	90–95% mesylation 97–99% amination 82% bromination 99% cyanide 80% hydrolysis	2000 (1741)
−CH$_2$CH(OH)CH$_2$−	carboxylation of epichlorohydrin by opening with PhLi, amination by displacement with N$_3$, oxidation of Ph	93% Ph opening 93% azidation 81% Ph oxidation	1987 (1742)
−CH(Me)CH(OH)CH$_2$− −CH(Ph)CH(OH)CH$_2$− −CH(CH$_2$cHex)CH(OH)CH$_2$−	epoxidation of allylic alcohol, amination, epoxide formation, opening with CN, deprotection	90–99% ee syn or anti	1996 (1743)
−CH(iBu)CH(OH)CH$_2$−	epoxidation of allylic alcohol, azidation, epoxide formation, opening with CN, hydrolysis	86% ee epoxidation 82% CN epoxide opening 63% hydrolysis syn or anti	1993 (1744)
−CH(iBu)CH(OH)CH$_2$−	epoxidation of allylic alcohol, azidation, epoxide formation, opening with CN, hydrolysis	96–98% epoxidation 84–95% azidation 82–88% CN epoxide opening 90–95% hydrolysis 92–99% azide reduction syn or anti	1992 (1745)
−CH(iBu)CH(OH)CH$_2$−	asymmetric epoxidation of H$_2$C=CCH(OH)iBu, azide displacement, epoxide opening with KCN, hydrolysis and azide reduction	—	1991 (1746)
−CH$_2$CH(OH)CH$_2$− with NH$_3$ or NMe$_3$	asymmetric dihydroxylation of allyl bromide, conversion 1° OH to Cl, Br displacement with CN, Cl displacement with NH$_3$ or NMe$_3$	61–74% dihydroxylation 72% ee 88–93% bromination 70–82% CN displ. 47–49% amine displ. and hydrolysis 90–95% ee	1993 (1747)
−CH$_2$CH(4-Cl-Ph)CH$_2$−	enzymatic hydrolysis of racemic 2-(4-Cl-Ph)-4-butenenitrile, amide reduction wiith LiAlH$_4$, alkene oxidative cleavage	50% nitrile hydrolysis/ resolution >99.5% ee 77% reduction 53% oxidative cleavage 91% ee	2002 (1748)
−CH(iBu)CH(OH)CH$_2$−	asymmetric dihydroxylation of iBuCH=CHCO$_2$Et, regioselective Mitsunobu azidation, or Mitsunobu inversion with TsOH followed by azide displacement; homologation of carboxy group by Arndt–Eistert reaction, azide reduction	96% dihydroxylation 99% ee 82% Mitsunobu azidation 70% Mitsunobu inversion 90% azide displacement 53–55% homologation 82–86% azide reduction	2002 (837)

Fusion of Amino and Carboxyl Components: C2–C3 Formation

−(CH$_2$)$_2$CH(Me)− −(CH$_2$)$_2$CH(Et)− −(CH$_2$)$_2$CH(Ph)− −(CH$_2$)$_2$CH(CH$_2$CO$_2$H)− −(CH$_2$)$_2$CH[3,4-(MeO)$_2$-Bn]−	cyanomethylation of homochiral acyl ated oxazolidinones, auxiliary removal, reduction	59–77% de crude 59–81% yield >92% de 83–97% auxiliary removal 43–86% reduction	1996 (1749)

Table 11.6 Asymmetric Syntheses of γ-Amino Acids (continued)

H₂N-R-CO₂H	Method	Yield, % ee	Reference
−CH₂CH(Me)CH₂− −CH₂CH(iPr)CH₂− −CH₂CH(iBu)CH₂− −CH₂CH(Ph)CH₂− −CH₂CH(1-Me-indol-3-yl)CH₂− −CH₂CH₂CH(Me)− −CH₂CH₂CH(iPr)− −CH₂CH₂CH(iBu)− −CH₂CH(Ph)CH(Me)− −CH(Me)CH(Ph)CH(Me)−	Michael addition of acids derivatized with oxazolidinone auxiliary to nitroalkenes, nitro reduction	36–76% addition >90% de for α-substituted 72–92% de for β-substituted 80–92% nitro reduction/ lactam formation 52–99% lactam hydrolysis	1999 (1750) 2001 (1645)
−CH₂CH(Me)C(OH)(Me)−	addition of chiral 1,3-dioxolan-4-one to nitropropene, reduction, hydrolysis	58% yield addition 93% de 95% reduction 39% hydrolysis	1985 (1751)
−CH₂CH(Bu)CH[CH(Me)]− −CH₂CH(Ph)CH[CH(Me)OH]− −CH₂CH(Bu)CH[CH(CF₃)OH]− −CH₂CH(Ph)CH[CH(CF₃)OH]− −CH₂CH(CH₂OH)CH[CH(Ph)OH]−	Michael addition of cyclic dioxanone derived from RCH(OH)CH₂CO₂H to RCH=CHNO₂	86–97% addition 0–70% de	2003 (1752)
−CH(Me)CH(Me)CH(Me)− −CH(Me)CH(Me)CH(iPr)− −CH(Me)CH(Me)CH(iBu)−	Michael addition of acids derivatized with oxazolidinone auxiliary to nitroalkenes, nitro reduction	47–54% addition 84–>94% dr 91–94% nitro reduction/ lactam formation 67–86% lactam hydrolysis	2002 (1646)
−CH₂CH(Ar)CH(CO₂R)− Ar = Ph, 4-Me-Ph, 4-Cl-Ph, 4-F-Ph, 3,4-OCH₂O-Ph, 2-thienyl, 2-furyl −CH₂CH(Ph)C(Me)(CO₂R)− −CH₂CH(Ph)CH(COR)− R = Me, Et, iPr, Ph	Michael addition of malonates or β-keto esters to nitrostyrenes catalyzed by chiral Ru complex	93–99% 50–98% ee	2004 (1753)
−CH₂CH(4-Cl-Ph)CH₂−	Michael addition of chiral Fischer carbene to nitrostyrene, carbene oxidation and nitro reduction	90% addition 76% de 90% oxidation 95% reduction	2000 (1754)
−CH(Bn)CH₂C(-SCH₂CH₂CH₂S-)− −CH(iBu)CH₂C(-SCH₂CH₂CH₂S-)−	aziridine opening with dithiane of glyoxylic acid	89–92% aziridine opening	1993 (1247)
−CH(iBu)CH(OH)CH₂− −CH(CH₂cHex)CH(OH)CH₂− −CH(Bn)CH(OH)CH₂−	allylation of SADP hydrazone of BnOCH₂CHO, Grignard addition to hydrazone, auxiliary removal, ozonolysis	81% allylation 78–87% Grignard addition >84% de 73–81% deprotection >96% de, >94% ee 71% ozonolysis	1995 (1755) 1996 (1756)
−CH₂CH₂CH(Me)− −CH₂CH₂CH(Et)− −CH₂CH₂CH(Ph)− −CH₂CH₂CH(4-Cl-Ph)−	Pd-catalyzed allylation of methyl cyanoacetate, decarboxylation, nitrile reduction, akene oxidation	74–90% allylation 64–96% ee 74–80% decarboxylation 70–83% nitrile reduction 61–75% alkene oxidation	1998 (1757)
−CH₂CH(Me)CH₂− −CH₂CH(nBu)CH₂− −CH₂CH(cHex)CH₂− −CH₂CH(Ph)CH₂− −CH₂CH(2-MeO-Ph)CH₂− −CH₂CH(3-MeO-Ph)CH₂− −CH₂CH(3-NO₂-Ph)CH₂− −CH₂CH(4-NO₂-Ph)CH₂− −CH₂CH[3,4-(OCH₂O)-Ph]CH₂−	Rh-catalyzed C–H insertion of chiral ester diazoanilides, decarboxylation	64–96% insertion 70–98% decarboxylation 30–98% ee	1998 (1758)

(Continued)

Table 11.6 Asymmetric Syntheses of γ-Amino Acids (continued)

H₂N-R-CO₂H	Method	Yield, % ee	Reference
−CH₂CH(Me)CH₂− −CH₂CH(Et)CH₂− −CH₂CH(Ph)CH₂− −CH₂CH(4-Cl-Ph)CH₂− −CH₂CH(4-MeO-Ph)CH₂− −CH₂CH(4-NO₂-Ph)CH₂−	Rh-catalyzed insertion reaction of α-MeO₂C-α-diazoacetanilides, decarboxylation, deprotection, lactam hydrolysis	72–84% insertion 33–82% ee 75% decarboxylation/ deprotection 74% hydrolysis	1998 (1759)
−CH(Me)CH(OH)CH(Me)−	Rh-initiated intramolecular radical addition of dichloroacyl group to oxazolone, further derivatization	75–83% cycloaddition >99% dechlorination >99% de	1998 (1760)
−CH(iBu)CH(OH)CH₂− −CH(iBu)CH(OH)CF₂− −CH(iBu)CH(OH)CCl₂−	Rh-initiated intramolecular radical addition of dihaloacyl grop to oxazolone, further derivatization	87–100% cycloaddition >99% de	1993 (1761)

Fusion of Amino and Carboxyl Components: C3-C4 Formation

H₂N-R-CO₂H	Method	Yield, % ee	Reference
−CH₂CH(OH)CH₂−	cyanohydrin formation from ethyl formyl acetate equivalent, enzymatic resolution, reduction	98% cyanohydrin formation 32% one enantiomer 28% second enantiomer 93–100% reduction	1990 (1697) 1993 (543)
−CH(Ph)CH₂CH₂−	enantioselective deprotonation of ArN(Boc)Bn, addition to acrolein, oxidation	61–72% Michael addition 92–94% ee 75–77% oxidation	1997 (969)
−CH₂CH(Me)CH₂− −CH₂CH(Et)CH₂− −CH₂CH(nPr)CH₂− −CH₂CH(iPr)CH₂− −CH₂CH(cHex)CH₂− −CH₂CH(tBu)CH₂− −CH₂CH(CH=CHMe)CH₂− −CH₂CH(CO₂Me)CH₂− −CH₂CH(Ph)CH₂− −CH₂CH(3,4-OCH₂O-Ph)CH₂− −CH₂CH(2-furyl)CH₂− −CH₂CH(2-thienyl)CH₂−	asymmetric addition of nitromethane to RCH=CHCO(3,5-dimethyl-1-pyrazolyl) in presence of chiral Ni catalyst	39–97% yield 77–97% ee	2002 (1762)
−CH(Et)CH(Ph)CH₂− −CH(Et)CH(4-F-Ph)CH₂− −CH(Et)CH(4-MeO-Ph)CH₂− −CH(Et)CH(2-naphthyl)CH₂− −CH(Et)CH(cHex)CH₂− −CH(Et)CH(iBu)CH₂− −CH(Me)CH(Ph)CH₂− −CH(iPr)CH(Ph)CH₂− −CH[(CH₂)₂OBn]CH(Ph)CH₂− −CH₂CH(Ph)CH₂−	asymmetric addition of nitroalkanes to RCH=C(CO₂R)₂ in presence of chiral-phase transfer catalyst, nitro reduction and decarboxylation	97–99% yield addition 88–99% ee 42–90% de anti 75% nitro reduction and decarboxylation (1 example)	2004 (1763)
−CH₂CH(2,2-Me₂-3-CO₂H-cyclobut-1-yl) CH₂−	nitromethane addition to chiral α,β-unsaturated ester derived from (S)-verbenone, nitro reduction	75–80% nitromethane addition 68–70% nitro reduction	2002 (1764)
−CH(Ph)CH₂CH₂− −CH(iPr)CH₂CH₂− −CH(iBu)CH₂CH₂− −CH[(CH(OH)CH₂OH]CH₂CH₂−	reductive conjugate addition of nitrones RCH=N⁺(R*)O⁻ with chiral auxiliary or chiral substituent to acrylates in presence of SmI₂, deprotection via hydrogenation over Raney Ni	36–96% 20–>90% de 62–82% hydrogenation	2003 (1696)
−CH[(CH(OH)CH₂OH]CH₂CH₂−	reductive conjugate addition of glyceraldehyde-derived N-aryl nitrones RCH=N⁺(Ar*)O⁻ to acrylates in presence of SmI₂, deprotection via hydrogenation over Raney Ni	70–77% 80% de 82% hydrogenation	2003 (1765)

Table 11.6 Asymmetric Syntheses of γ-Amino Acids (continued)

H$_2$N-R-CO$_2$H	Method	Yield, % ee	Reference
−CH(CH$_2$CHEt$_2$)CH$_2$CH$_2$− −CH(iPr)CH$_2$CH$_2$− −CH(iBu)CH$_2$CH$_2$− −CH(cHex)CH$_2$CH$_2$− −CH(iPent)CH$_2$CH$_2$− −CH(CH$_2$CH$_2$Ph)CH$_2$CH$_2$−	reductive conjugate addition of nitrones RCH=N$^+$(R*)O$^-$ with carbohydrate chiral auxiliary to n-butyl acrylate in presence of SmI$_2$	54–80% >90% de	2004 (1766)
−C(Me)(Ph)CH$_2$CH$_2$− −C(nBu)(Ph)CH$_2$CH$_2$−	addition of allylMgBr to chiral sulfinamide, hydroboration and oxidation of alkene	84–98% addition 55–60% hydroboration 65–80% oxidation	1991 (588)
−CH(CF$_3$)CH(OH)CH$_2$−	addition of chiral γ-butenyl sulfoxide to imine of trifluoroacetaldehyde, Pummerer rearrangement, alkene oxidation	100% addition 46% de 94% Pummerer 89% oxidative cleavage	1998 (1767)
−CH(iBu)CH(OH)CH$_2$−	addition of chiral γ-butenyl sulfoxide to imine of isopentanal, Pummerer rearrangement, alkene oxidation	97% addition 63:16:16:6 diastereomers 76% Pummerer 48% oxidative cleavage	2001 (1768)
−CH(iPr)CH$_2$CH$_2$− enol ether precursors of: −CH(Me)CH$_2$CH$_2$− −CH(nPr)CH$_2$CH$_2$− −CH(Ph)CH$_2$CH$_2$−	addition of chiral acetal Ti homoeno-late equivalent of propionaldehyde to benzylamine imine of butyralde-hyde, enol ether deprotection and oxidation	71–85% addition 83–88% de 87% oxidation	1999 (1025) 2001 (1769)
−CH(Me)CH(OH)CH$_2$− −CH(Et)CH(OH)CH$_2$− −CH(nHept)CH(OH)CH$_2$− −CH(iBu)CH(OH)CH$_2$−	addition of achiral (γ-alkoxyallyl)Ti complex to chiral imine, alkene hydroxylation and oxidation	60–76% addition 60–66% de major syn isomer 69% hydroxylation (one example)	2000 (1770)
−CH(Ph)CH$_2$C(=CH$_2$)− −CH(2-naphthyl)CH$_2$C(=CH$_2$)− −CH(4-MeO-Ph)CH$_2$C(=CH$_2$)− −CH(2-thienyl)CH$_2$C(=CH$_2$)− −CH(2-MeO-Ph)CH$_2$C(=CH$_2$)− −CH(cHex)CH$_2$C(=CH$_2$)− −CH(3,4-OCH$_2$O-Ph)CH$_2$C(=CH$_2$)−	addition of Bu$_3$SnCH$_2$C(=CH$_2$)CO$_2$Et to ArCH=NR imine	59–81% addition 60–89% ee	2004 (1771)
−CH(R)CCH− (with N-OH) −CH(R)CH=CH− −CH(R)CH(OH)CH(OH)− R = CH(−CH$_2$OCMe$_2$NBoc−)	addition of LiCCCO$_2$R to nitrone, reduction, dihydroxylation	78% addition 100% de 86% reduction (Z only) 55% dihydroxylation	1997 (1699)

Elaboration of α-Amino Acids: Aldol of Amino Aldehydes

H$_2$N-R-CO$_2$H	Method	Yield, % ee	Reference
−CH(iBu)CH(OH)CH$_2$−	aldol reaction of Phth-leucinal with ZnBrCH$_2$CO$_2$tBu, resolution, deprotection	57% addition 8% de 94% deprotection	1978 (1566)
−CH(iBu)CH(OH)CH$_2$−	aldol reaction of Phth-leucinal with ZnBrCH$_2$CO$_2$tBu, deprotection	38% addition and deprotection 10% de	1978 (1773)
−CH(iBu)CH(OH)CH$_2$−	aldol reaction of Boc-leucinal with LiCH$_2$CO$_2$Et	73% aldol 4% de	1982 (1774)
−CH(iBu)CH(OH)CH$_2$−	aldol reaction of Boc-leucinal with LiCH$_2$CO$_2$Et	80% aldol 20% de	1978 (1775)
−CH(iBu)CH(OH)CH$_2$−	activation of Boc-Leu with CDI, reduction with LiAlH$_4$, addition of lithiated Et acetate, hydrolysis	82% reduction 80% LiCH$_2$CO$_2$Et addition 94% hydrolysis	1975 (1776)

(Continued)

Table 11.6 Asymmetric Syntheses of γ-Amino Acids (continued)

H$_2$N-R-CO$_2$H	Method	Yield, % ee	Reference
—CH$_2$COCH$_2$— —CH(Me)COCH$_2$— —CH(Et)COCH$_2$— —CH(iPr)COCH$_2$— —CH(iBu)COCH$_2$— —CH(sBu)COCH$_2$— —CH(CH$_2$CH$_2$SMe)COCH$_2$— —CH(CH$_2$SMe)COCH$_2$— —CH(CH$_2$SH)COCH$_2$— —CH(CH$_2$OH)COCH$_2$— —CH[CH(OH)Me]COCH$_2$— —CH(CH=CH$_2$)COCH$_2$—	activation of Boc-Xaa with CDI, addition of KO$_2$CCH$_2$CO$_2$Me, hydrolysis	60–76% CDI activation and malonate addition	2004 (1831)
—CH(Bn)CH(OH)CH$_2$—	aldol reaction of Boc-phenylalaninal with LiCH$_2$CO$_2$Et, resolution, deprotection	55% addition de not disclosed	1980 (1777)
—CH(Bn)CH(OH)CH$_2$—	aldol reaction of Boc-phenylalaninal with LiCH$_2$CO$_2$Et	97% Phe reduction 55–60% aldol	1980 (1576)
—CH(sBu)CH(OH)CH$_2$—	addition of Li acetate to D- or L-leucinal or alloleucinal	41–99% addition	1992 (1588)
—CH(sBu)COCH$_2$— —CH(sBu)CH(OH)CH$_2$—	reaction of Boc-alloisoleucinal with LiCH$_2$CO$_2$Et, carbinol oxidation, or reaction of CDI derivative of Boc-aIle with LiCH$_2$CO$_2$Et, reduction	51% addition 80% oxidation or 78% addition of CDI derivative 65–73% reduction 82% de	1989 (1585)
—CH(CH$_2$NH$_2$)CH(OH)CH$_2$—	reaction of Cbz-Dap(Boc)-H aldehyde with LiCH$_2$CO$_2$Et	48% addition 38% de	2003 (1778)
—CH(sBu)CH(OMe)CH$_2$— with N-Me	N-methylation of Ile, reduction to aldehyde, reaction with LiCH$_2$CO$_2$Et, O-methylation	95% methylation 78% reduction 56% aldol 18% de 67% methylation	1989 (1632)
—CH(iBu)CH(OH)CF$_2$— —CH(iBu)COCH$_2$—	reaction of Boc-leucinal with ZnBrCF$_2$CO$_2$Et, oxidation	80% addition 40–100% de	1985 (1572)
—CH(iBu)CH(OH)CF$_2$— —CH(Bn)CH(OH)CF$_2$— —CH(CH$_2$$c$Hex)CH(OH)CF$_2$— —CH($i$Bu)COCF$_2$— —CH(Bn)COCF$_2$— —CH(CH$_2$$c$Hex)COCF$_2$—	reaction of Boc-leucinal with ZnBrCF$_2$CO$_2$Et, oxidation	60–97% addition 40–100% de	1986 (1575)
—CH(CH$_2$$c$Hex)CH(OH)CF$_2$—	reaction of cyclohexylalaninal with Reformatsky reagent derived from ethyl bromodifluoroacetate	70% addition 33% de	1990 (481)
—CH(Me)CH(OH)CH[C(Me)=CHCO$_2$Me]— —CH(iBu)CH(OH)CH[C(Me)=CHCO$_2$Me]— —CH(Bn)CH(OH)CH[C(Me)=CHCO$_2$Me]—	addition of Li enolate of dimethyl 3-methylglutaconate to amino acid-derived amino aldehyde	65–77% addition 36–70% de major diastereomer	1997 (1779)
—CH(iBu)CH(OH)CH$_2$— —CH(CH$_2$$c$Hex)CH(OH)CH$_2$—	aldol reaction of amino acid-derived amino aldehyde with O-Me-OTMS ketene acetal, deprotection	89–95% addition 88–90% de syn 88–93% deprotection	1990 (1780)
—CH(iBu)CH(OH)CH$_2$—	reaction of Boc-leucinal or Bn$_2$-leucinal with ketene silyl acetal	82–94% 88–>98% de $anti$ with EtAlCl$_2$ catalyst 60% 82% de syn with SnCl$_4$ catalyst	1990 (1781)

Table 11.6 Asymmetric Syntheses of γ-Amino Acids (continued)

H₂N-R-CO₂H	Method	Yield, % ee	Reference
$-CH(Me)CH(OH)CH_2-$	aldol reaction of N,N-Bn₂ alaninal with silyl ketene acetal in presence of LiClO₄	58% aldol >98% de *anti*	1993 (*1782*)
$-CH(nBu)CH(OH)CH_2-$ $-CH[(CH_2)_4NH_2]CH(OH)CH_2-$	addition of allylM to Boc-aminoaldehyde, alkene oxidation	80–85% allyl addition 82–84% de	1990 (*1783*)
$-CH(iBu)CH(OH)CH_2-$	reaction of Boc-leucinal with allylMgBr, oxidative cleavage	53% addition 76% de 58% oxidative cleavage	1997 (*1123*)
$-CH(iBu)CH(OH)CH_2-$	inversion of hydroxy center of amino alcohol corresponding to allyl addition to Cbz-leucinal, oxidative cleavage of alkene	67% inversion 74% oxidative cleavage	1987 (*1784*)
$-CH(CF_3)CH(OH)CH_2-$	addition of allylMgBr to chiral α-aminotrifluoropropanal equivalent, alkene oxidation	85% addition 64% de	1998 (*1767*)
$-CH(Bn)CH(OH)CH_2-$ oxazoline precursors: $-CH(Me)CH(OH)CH_2-$ $-CH(CH_2CH_2SMe)CH(OH)CH_2-$ $-CH(CH_2OH)CH(OH)CH_2-$	addition of MCH=CH₂ to amino aldehyde, oxazolidinone formation, Pd-catalyzed rearrangement to oxazoline, alkene hydroboration/oxidation	77–100% rearrangement 70–88% de 68% alkene oxidation	1998 (*1785*)
$-CH(iBu)CH(OH)CH_2-$	aldol reaction of chiral Mg enolate of MeCO₂R with Boc-L-leucinal, hydrolysis	48–100% aldol 87% de *syn* to 50% de *anti* 81–90% hydrolysis	1989 (*1786*)
$-CH_2CH(OH)CH_2-$	aldol reaction of chiral Mg enolate of MeCO₂R with Cbz-glycinal, hydrolysis	61% aldol 82% de 47% deprotection 98% ee	1989 (*1787*)
$-CH(Bn)CH(OH)CH_2-$ $-CH(CH_2cHex)CH(OH)CH_2-$	addition of Li acetate with chiral auxiliary to amino acid-derived aldehyde	49–61% addition 80–82% de	1988 (*1788*)
$-CH(sBu)CH(OMe)CH_2-$	aldol reaction of α-(methylsulfanyl) acetyloxazolidinone with Cbz-N-Me-isoleucinal, desulfurization, O-methylation, hydrolysis	70–75% aldol 76–79% reduction 70–79% methylation 97–99% hydrolysis	1996 (*1630*)
$-CH(iBu)CH(OH)CH_2-$	aldol reaction of α-(methylsulfanyl) acetyloxazolidinone with Boc-leucinal, desulfurization, hydrolysis	24% overall	1985 (*1789*)
$-CH(iPr)CH(OMe)CH_2-$ $-CH(iBu)CH(OMe)CH_2-$	addition of chiral Fe complex acetate enolate to N,N-Bn₂-valinal or leucinal, auxiliary removal	44–71% addition 60–92% de 65% auxiliary removal	1993 (*1790*)
$-CH(CH_2OMe)CH(OH)CH(OH)-$ (with N-Me₂)	alkylation of Sar with chiral auxiliary, reduction to give N,O-Me₂-D-serinal, aldol addition of chiral enolate	80% Sar alkylation 96% de 60% reduction and aldol	1992 (*1624*) 1992 (*1625*)
$-CH(iPr)CH(OH)CH_2-$ $-CH(iBu)CH(OH)CH_2-$ $-CH(sBu)CH(OH)CH_2-$ $-CH(Bn)CH(OH)CH_2-$ (with N-methyl) $-CH[(CH_2)_3-(\gamma-N)]CH(OH)CH_2-$	aldol addition of champro-derived chiral Li acetate enolate to N-methyl amino aldehydes	55–78% yield 100% de 63–86% auxiliary deprotection	2004 (*1619*)
$-CH(CH_2OH)CH(OH)CH_2-$	aldol reaction of acetate enolate with pseudoephedrine chiral auxiliary with Garner aldehyde	79% aldol 96% de for matched pair 61% aldol 12% de for mismatched pair	2004 (*1791*)
$-CH(Me)CH(OH)CH(Me)-$	aldol reaction of D-alaninal with propionic acid derivatized with Evans' oxazolidinone auxiliary	73% aldol 100% *syn* 88% deprotection	1992 (*1792*)

(*Continued*)

Table 11.6 Asymmetric Syntheses of γ-Amino Acids (continued)

H$_2$N-R-CO$_2$H	Method	Yield, % ee	Reference
−CH(Me)CH(OH)CH(Me)− −CH(iBu)CH(OH)CH(Me)− −CH(Bn)CH(OH)CH(Me)−	aldol reaction of D-alaninal, leucinal or phenylalaninal with propionic acid derivatized with Evans' oxazolidinone auxiliary	27–71% aldol 100% syn 84–88% deprotection	2003 (259)
−CH(Me)CH(OH)CH(Me)−	aldol reaction of chiral boron enolate of EtCO$_2$H with Boc-D-alaninal, hydrolysis	73% aldol >98% de syn 88% hydrolysis	1994 (247)
−CH(allyl)CH(OH)CH(allyl)− −CH(CH$_2$CH=CHCH$_2$-C2)CH(OH)CH(-)− −CH[(CH$_2$)$_4$-C2]CH(OH)CH(-)−	reduction of Boc-allylglycine to Boc-allylglycinal, addition of enolate of 4-pentenoic acid with Evans' oxazolidinone auxiliary, Ru-metathesis, hydrogenation, deprotection	71% reduction 67% aldol reaction, 0% de 79–83% metathesis 73–89% hydrogenation 34–38% deprotection	2001 (1609) 2003 (258)
−CH(Me)CH(OH)CH$_2$− −CH(iPr,CH(OH)CH$_2$− −CH(iBu)CH(OH)CH$_2$− −CH(sBu)CH(OH)CH$_2$− −CH(Bn)CH(OH)CH$_2$−	aldol reaction of chiral boron enolate of MeCOStBu with amino acid-derived Bn$_2$NCH(R)CHO, hydrolysis, hydrogenation	71–80% aldol 91–>98% de syn or anti 85% hydrolysis and hydrogenation	1995 (1793) 1997 (1772)
−CH(iBu)CH(OH)CH$_2$− −CH(Bn)CH(OH)CH$_2$− −CH(iBu)CH(OH)CH(Me)− −CH(Bn)CH(OH)CH(Me)−	reaction of Bn$_2$-amino aldehydes with thioester-derived boron enolates	50–85% 91% de syn to 88% de anti	1990 (1794)
−CH(Me)CH(OH)CH(Me)− four diastereomers	addition of chiral organoborane of (E)-or (Z)-2-butene to Boc-L-alaninal, alkene oxidative cleavage	31–45% addition and protection 26–>95% de 62–73% oxidation	1999 (1611)
−CH(Me)CH(OH)CH(Me)−	aldol reaction of boron enolate of EtCOSAr with Boc-D-alaninal, hydrolysis	60–77% aldol 75–94% de syn	1982 (1795)
−CH(CH$_2$OMe)CH(OH)CH(OH)− (with N-Me$_2$)	aldol additon to Ser-derived Garner aldehyde, methylation, alkene oxidative cleavage	—	1992 (1796)
−CH$_2$CH(OH)CH$_2$− −CH(Me)CH(OH)CH$_2$− −CH(CH$_2$cHex)CH(OH)CH$_2$− −CH(iBu)CH(OH)CH$_2$−	alkylation of Williams' oxazinone, lactone reduction, addition of ketene silyl acetal or allylTMS, deprotection	70–88% alkylation 76–85% reduction 57–75% acetate addition 20–50% de 69–82% deprotection	1994 (1797) 1998 (1798)

Elaboration of α-Amino Acids: Olefination of Amino Aldehydes

−CH(iBu)CH=CH− −CH(iBu)CH$_2$CH$_2$−	Horner–Emmons coupling of (EtO)$_2$PO=CHCO$_2$Et to Phth-Leu-H, hydrogenation and hydrolysis	90% aldehyde 65% olefin 100% hydrogenation 52% hydrolysis	1977 (1570)
−CH(Me)CH=CH− −CH(iPr)CH=CH− −CH(iBu)CH=CH− −CH(Me)CH$_2$CH$_2$− −CH(iP)CH$_2$CH$_2$− −CH(iBu)CH$_2$CH$_2$−	Wittig coupling of (PhO)$_2$P(O)CHCO$_2$Me to amino acid-derived aldehyde, hydrogenation, hydrolysis	55–72% overall	1998 (1642)
−CH(Me)CH=CH− −CH(iPr)CH=CH− −CH(Bn)CH=CH− −CH(Me)CH=C(Me)− −CH(iPr)CH=C(Me)− −CH(Bn)CH=C(Me)−	Wittig coupling of Ph$_3$P=CHCO$_2$Me or Ph$_3$P=C(Me)CO$_2$Et to amino acid-derived aldehyde	78–99% Weinreb amide 69–92% reduction and Wittig	1992 (1649)
−CH(R)CH=CH− R = (CH$_2$)$_2$SMe, (CH$_2$)$_2$CONH$_2$,	Wittig coupling of Ph$_3$P=CHCO$_2$Me to amino acid-derived aldehyde	75% Weinreb amide 88% reduction 70–83% Wittig	1998 (1651)

Table 11.6 Asymmetric Syntheses of γ-Amino Acids (continued)

H_2N-R-CO_2H	Method	Yield, % ee	Reference
$-CH(R)CH=CH-$ R = $(CH_2)_2SMe$, $(CH_2)_2CONH_2$, $(CH_2)_2CONHMe$, $(CH_2)_2CONMe_2$, $(CH_2)_2CO_2H$, $(CH_2)_2COMe$, $(CH_2)_2CH(OH)Me$, $(CH_2)_2SOMe$, $(CH_2)_2SO_2Me$, CH_2OH, CH_2OCONH_2, CH_2NHAc, CH_2NHONH_2	Wittig coupling of $Ph_3P=CHCO_2Me$ to amino acid-derived aldehyde	41–63% Wittig	1998 (1652)
$-CH(R)CH=CH-$ $-CH(R)CH=CMe-$ R = $(CH_2)_2CONH_2$	Wittig or Horner–Emmons coupling of amino acid-derived aldehyde	35–88% Wittig	1998 (1653)
$-CH(iPr)CH=CH-$ $-CH(iPr)CH_2CH_2-$ $-CH(iPr)CH_2CH(OH)-$	Wittig or Horner–Emmons coupling of amino acid-derived aldehyde, alkene reduction, α-hydroxylation	64% oxidation and Wittig 80% reduction 66% hydroxylation 100% de	2004 (1635)
$-CH[CH_2)_3-(\gamma-N)]CH=CH-$ $-CH[CH_2)_3-(\gamma-N)]CH=C(Me)-$	Wittig or Horner–Emmons coupling of Boc-prolinal	—	2003 (1799)
$-CH(Bn)CH=C(Me)-$ $-CH(Bn)CH_2CH(Me)-$	Wittig or Horner–Emmons coupling of amino acid-derived aldehyde, alkene reduction, α-hydroxylation	69% oxidation and Wittig 72% reduction 50% de	2004 (1635)
$-CH[CH_2(4-HO-Ph)]CH=CH-$	Wittig coupling of $Ph_3P=CHCO_2Et$ to amino acid-derived aldehyde	89% Weinreb reduction and Wittig	1992 (264)
$-CH[CH_2(4-HO-Ph)]CH=CH-$	Wittig coupling of $Ph_3P=CHCO_2Et$ to amino acid-derived aldehyde	90% Wittig reaction	2002 (267)
$-CH(R)CH=CH-$ R = $CH_2CH(-CONHCH_2CH_2-)$, $CH_2N(-CONHCH_2CH_2-)$, CH_2CH $(-CONHCH_2CH_2CH_2-)$	Wittig or Horner–Emmons coupling of amino aldehyde	35–88% Wittig	1999 (1655)
$-CH(CH_2OH)CH(OH)CH(OH)-$ $-CH(CH_2OMe)CH(OH)CH(OH)-$ (with N-Me_2)	Wittig reaction of Ser-derived Garner aldehyde, dihydroxylation	90% Wittig 50% dihydroxylation 84% O-methylation 48% N-dimethylation	1991 (1623)
$-CH(CH_2OH)CH=CH-$	Wittig reaction of Ser-derived Garner aldehyde	—	1990 (1800)
$-CH(CH_2OH)CH=CH-$	reduction of Ser carboxyl to aldehyde, Wittig reaction	74% reduction and Wittig	1999 (1801)
$-CH(iPr)CH=C(Me)-$	oxidation of Bts-valinol, Wittig reaction with $Ph_3P=C(Me)CO_2Et$	87% oxidation 87% Wittig	2001 (1638)
$-CH[(CH_2)_2CONH_2]CH=CH-$ $-CH[CH_2CH(-CH_2CH_2NHCO-)]$ $CH=CH-$	reduction of Cbz-Gln(Trt) to alcohol, N-deprotection and acylation, oxidation, Wittig	71% reduction 63–77% oxidation and Wittig	2002 (1656)
$-CH(Me)CH=CH-$ $-CH(Bn)CH=CH-$ $-CH(Me)CH=C(Me)-$ $-CH(Bn)CH=C(Me)-$	dynamic kinetic resolution of Horner–Wadsworth–Emmons reaction of homochiral phospho- nate with racemic amino aldehyde	4–90% de 68–90% de Z	1995 (1802)
$-CH(Me)CH_2CH_2-$ $-CH(CH_2OBn)CH_2CH_2-$ $-CH(CH_2CH_2CO_2Bn)CH_2CH_2-$ $-CH(iBu)CH_2CH_2-$ $-CH(Bn)CH_2CH_2-$ $-CH(CH_2CO_2tBu)CH_2CH_2-$ $-CH[(CH_2)_3-\gamma-N]CH_2CH_2-$	hydrogenation of amino acid-derived α,β-unsaturated-γ-amino esters	60–96%	1999 (1803)
$-CH(CH_2SMe)CH_2CH_2-$ $-CH(CH=CH_2)CH_2CH_2-$	homologation of Met via reduction to amino aldehyde, Wittig, alkene reduction, oxidation, and thermal elimination	62–78% homologation 92–95% reduction 56% oxidation/elimination	1993 (1805) 1994 (1804)

(Continued)

Table 11.6 Asymmetric Syntheses of γ-Amino Acids (continued)

H₂N-R-CO₂H	Method	Yield, % ee	Reference
$-CH(Bn)CH=CH-$ $-CH(Bn)CH(-O-)CH-$ $-CH(Bn)CH(OH)CH(R)-$ R = NHBn, SBn, NHPh, NHCH₂CH₂Ph, NHnBu, NHcHex, NHCH₂(1-naphthyl), NH(4-Ph-Ph), NHCH₂CH₂(indol-3-yl), NHCH₂CH₂(2-pyridyl), NH(4-MeO-Bn), NH(4-Cl-Bn), NH(4-Br-Bn)	coupling of Ph₃P=CHCO₂Et Boc-Phe-H, epoxidation, opening with nucleophiles	82% Wittig 62–65% epoxidation 80% de 50–90% opening	1994 (*1581*)
$-CH(iBu)CH=CH-$ $-CH(iBu)CH(-O-)CH-$ $-CH(iBu)CH(OH)CH₂-$	coupling of (CF₃CH₂O)₂P(O) CH₂CO₂Me with Leu-derived aldehyde, carboxyl reduction, epoxidation, hydride opening, oxidation	86% coupling 75% reduction 98% epoxidation 93% hydride opening 95% oxidation	1987 (*887*)
$-CH(Me)CH=CH-$ $-CH(iBu)CH=CH-$ $-CH(Bn)CH=CH-$ $-CH(Me)CH(R)CH₂-$ $-CH(iBu)CH(R)CH₂-$ $-CH(Bn)CH(R)CH₂-$ R = Me, CH=CH₂, nBu, Ph	Wittig coupling of (EtO)₂P(O) CH₂CO₂Et to amino acid-derived aldehyde, organocuprate addition	73–82% Wittig 41–84% Michael addition 84–>90% de	1989 (*1807*)
$-CH(iBu)CH(OH)CH₂-$ $-CH(Bn)CH(OH)CH₂-$	homologation of Cbz-Leu or Phe via reduction to amino aldehyde, Horner–Emmons olefination, I₂-induced iodocyclocarbamation, radical deiodination, hydrolysis	75% olefination 85% iodocyclocarbamation 85% deiodination (45–47% overall)	1990 (*1808*)
$-CH(iBu)CH(OH)CH₂-$ $-CH(Bn)CH(OH)CH₂-$	Wittig coupling of Ph₃P=CHCO₂Et to amino acid-derived aldehyde, reduction, chlorination, silyl carbamate cyclization, hydroboration, oxidation	19–27% overall	1987 (*1810*) 1990 (*1809*)
$-CH(Me)CH(CH=CH₂)CH(OH)-$ $-CH(iPr)CH(CH=CH₂)CH(OH)-$ $-CH(iBu)CH(CH=CH₂)CH(OH)-$ $-CH(Bn)CH(CH=CH₂)CH(OH)-$	Wittig reaction of amino acid-derived amino aldehyde, reduction and O-alkylation, Wittig rearrangement	90–94% Wittig 60–83% reduction 46–96% alkylation 55–78% rearrangement (major isomer yield)	1995 (*1811*)

Elaboration of α-Amino Acids: Other Reactions of Amino Aldehydes

$-CH(Bn)CH(OH)CH₂-$	cycloaddition of dichloroketene with α-amino aldehyde, opening with amino acid, reduction	35% cycloaddition 100% de 70–87% coupling and reduction	1996 (*1812*) 1996 (*1249*)
$-CH(Bn)CH(OH)CH₂-$ $-CH(iBu)CH(OH)CH₂-$	[2+2] cycloaddition of amino acid-derived aldehyde with dichloroketene, hydrogenolysis	35–44% cycloaddition 81% hydrogenolysis	1995 (*1813*)
$-CH(Me)CH(OH)CH₂-$ $-CH(iPr)CH(OH)CH₂-$ $-CH(iBu)CH(OH)CH₂-$	hetero-Diels–Alder of RCH(NHP) CHO and Brassard's diene, ozonolysis	30–83% Diels–Alder 92% de *threo* to 98% de *erythro* 62–85% ozonolysis	1989 (*1814*)
$-CH(Me)-CC-$ $-CH(iPr)-CC-$ $-CH(iBu)-CC-$ $-CH(Bn)-CC-$	Corey–Fuchs reaction of amino acid-derived amino aldehyde	69–90% olefination 36–96% alkyne formation 64–88% deprotection	1996 (*1815*) 1998 (*1816*)

Table 11.6 Asymmetric Syntheses of γ-Amino Acids (continued)

H₂N-R-CO₂H	Method	Yield, % ee	Reference
–CH₂-CC– –CH(Me)-CC– –CH(iPr)-CC– –CH(tBu)-CC– –CH(Bn)-CC– –CH₂CH=C(R)– –CH(Me)CH=C(R)– –CH(iPr)CH=C(R)– –CH(tBu)CH=C(R)– –CH(Bn)CH=C(R)– R = CH₂CH=CH(CH₂)₇CO₂Me	conversion of amino aldehyde CHO in alkyne CCH with diazophosphonate reagent, deprotonation and carboxylation of alkyne, Ru-catalyzed ene reaction with monosubstituted alkenes	55–71% alkyne formation 66–91% carboxylation 36–81% ene reaction 2:1 to 10:1 Z:E	1999 (*1817*)

Elaboration of α-Amino Acids: Amino Alcohols

–CH(iBu)(CH₂)₂–	reduction of Leu to N,O-ditosyl leucinol, displacement with K diethyl malonate, hydrolysis	72% tosylation 66% malonate addition 90% hydrolysis	1989 (*1723*)
–CH(iPr)CH₂CH₂– –CH(Bn)CH₂CH₂– –CH[(CH₂)₃-(γ-N)]CH₂CH₂–	reduction of amino acid to alcohol, ditosylation, displacement with KCH(CO₂Et)₂, decarboxylation	75–94% tosylation 25–71% malonate addition 27–49% decarboxylation	1977 (*1818*)
–CH(CH=CH₂)CH₂CH₂–	opening of butadiene monoepoxide with phthalimide and chiral catalyst, form triflate, displace with Na malonate, hydrolysis with 6N HCl	99% opening 75:1 regioselectivity 98% ee 97% triflate 64% malonate displacement 96% hydrolysis	1996 (*1515*) 2000 (*1819*)
–CH(Et)CH₂CH₂–	conversion of chiral amino alcohol into N-acyl α-chloroamine, dideprotonation, quenching with BrCH₂CO₂Et	76% conversion 25% quenching	1996 (*958*)
–CH(Me)CH₂CO– –CH(iPr)CH₂CO– –CH(iBu)CH₂CO– –CH(Ph)CH₂CO– –CH(Me)CH₂CH(OH)– –CH(iPr)CH₂CH(OH)– –CH(iBu)CH₂CH(OH)– –CH(Ph)CH₂CH(OH)–	conversion of amino acid to β-keto-γ-amino acid, enzymatic ketone reduction	18–38% γ-amino-β-keto formation 78–90% ketone reduction	1998 (*1820*)

Elaboration of α-Amino Acids: Reactions of Activated Carboxy Group

–CH(Bn)CH(OH)CH₂– –CH(iBu)CH(OH)CH₂– –CH(CH₂cHex)CH(OH)CH₂–	asymmetric hydrogenation of 3-keto-4-amino esters	96–99% hydrogenation 97–100% ee 82–98% de	1988 (*1821*)
–CH(sBu)COCH₂– –CH(sBu)CH(OH)CH₂–	reaction of CDI derivative of Boc-aIle with LiCH₂CO₂Et, reduction	78% addition of CDI derivative 65–73% reduction 82% de	1989 (*1585*)
–CH(sBu)COCH₂– –CH(sBu)CH(OH)CH₂–	activation of Cbz- or Boc-D-aIle with CDI, reaction with Li enolate of methyl acetate, reduction of ketone	57% acylation 83% reduction 100% de	1994 (*1822*)
–CH(sBu)COCH₂– –CH(sBu)CH(OH)CH₂– –CH(iBu)COCH₂– –CH(iBu)CH(OH)CH₂–	reaction of CDI derivative of Boc-aIle, Boc-Ile or Boc-Leu with LiCH₂CO₂Et, ketone reduction	76–86% addition 80–89% reduction 58–82% de	1988 (*1823*)
–CH(sBu)COCH₂– –CH(sBu)CH(OH)CH₂– –CH(iPr)COCH₂– –CH(iPr)CH(OH)CH₂–	reaction of Pfp ester of Cbz-D-aIle or Cbz-Val LiCH₂CO₂Et, ketone reduction with KBH₄	80–83% addition 77–99% reduction 84% de	1999 (*1591*) 2001 (*1592*)

(Continued)

Table 11.6 Asymmetric Syntheses of γ-Amino Acids (continued)

H₂N-R-CO₂H	Method	Yield, % ee	Reference
−CH(iPr)COCH₂− −CH(iPr)CH(OH)CH₂−	reaction of Pfp ester of Boc-D-Val LiCH₂CO₂Et, ketone reduction with KBH₄	66% addition 70% reduction 84% de	2004 (1599)
−CH(iBu)COCH₂− −CH(iBu)CH(OH)CH₂−	activation of Boc-D-Leu with CDI, reaction with Li enolate of ethyl or tert-butyl acetate, reduction of ketone with chiral LiBH₄	82–88% CDI activation/ displacement 81% reduction 72% de	1987 (1824)
−CH(CHMeCH=CH₂)COCH₂− −CH(CHMeCH=CH₂)CH(OH)CH₂− −CH(sBu)CH(OH)CH₂−	activation of TFA-β-Me-allylGly with CDI, reaction with Li enolate of benzyl acetate, reduction of ketone, hydrogenation of alkene	78% acylation and acetate addition 84% ketone reduction 84% hydrogenation	1999 (1825)
−CH(Ph)CH(OH)CH₂− −CH(Ph)CH=CH− −CH(Ph)CH₂CH₂− with N-Me	reaction of CDI derivative of Boc-Phe with LiCH₂CO₂Et, ketone reduction, N-methylation, elimination, hydrogenation	92% activation/displacement 74% reduction (80% de anti crude) 62% N-methylation	1999 (1617)
−CH(CH₂NH₂)COCH₂− −CH(CH₂NH₂)CH(OH)CH₂−	reaction of CDI derivative of Cbz-Dap(Boc)-OH aldehyde with LiCH₂CO₂Et, ketone reduction	75% activation/displacement 44% reduction 32% de	2003 (1778)
−CH₂COCH(CH₂OH)− −CH=C(3,4-OCH₂O-Ph)CH(CH₂OH)− −CH=C(CCTMS)CH(CH₂OH)−	addition of Li enolate of EtOAc to protected Ser, Pd-catalyzed reaction with organostannanes	76% β-keto formation 44–64% coupling	1999 (1826)
−CH(Me)COCH₂− −CH(iPr)COCH₂− −CH(iBu)COCH₂− −CH(sBu)COCH₂− −CH(cHex)COCH₂− −CH(CH₂cHex)COCH₂− −CH(Bn)COCH₂− −CH(CH₂Bn)COCH₂− −CH(Me)CH(OH)CH₂− −CH(iPr)CH(OH)CH₂− −CH(iBu)CH(OH)CH₂− −CH(sBu)CH(OH)CH₂− −CH(cHex)CH(OH)CH₂− −CH(CH₂cHex)CH(OH)CH₂− −CH(Bn)CH(OH)CH₂− −CH(CH₂Bn)CH(OH)CH₂− −CH(R)CHCH(OH)CH₂−	activation of Bn₂-amino acids with CDI, reaction with Li enolate of tBu acetate, reduction of ketone with NaBH₄	73–80% activation and enolate addition 70–>97% ee 87–92% reduction 88–98% de syn	1997 (1827)
−CH[(CH₂)₄NH₂]COCH₂− −CH[(CH₂)₄NH₂]CH(OH)CH₂−	activation of Cbz-Lys(Boc) with CDI, reaction with Mg enolate of malonic acid monoethyl ester, reduction of ketone with NaBH₄	83% activation and enolate addition 94% reduction 66% de syn	1998 (1615)
−CH(iBu)COCH₂− −CH(Bn)COCH₂− −CH[(CH₂)₄NH₂]COCH₂− −CH(CH₂SMe)COCH₂− −CH[CH₂(imidazol-4-yl)]COCH₂− −CH(iBu)CH(OH)CH₂− −CH(Bn)CH(OH)CH₂− −CH[(CH₂)₄NH₂]CH(OH)CH₂− −CH(CH₂SMe)CH(OH)CH₂− −CH[CH₂(imidazol-4-yl)]CH(OH)CH₂−	activation of Boc-Phe, -Leu, -Cys(Me), -His(Ts) or -Lys(Cbz) with CDI, reaction with Mg enolate of malonic acid monoethyl ester, ketone reduction	70–88% activation and enolate addition 24–97% ketone reduction 92% de one iosmer to 66% de the other	1988 (1828)
−CH(iBu)COCH₂− −CH(CH₂cHex)COCH₂− −CH(Bn)COCH₂− −CH(iBu)CH(OH)CH₂− −CH(CH₂cHex)CH(OH)CH₂− −CH(Bn)CH(OH)CH₂−	activation of Bn₂-amino acids with CDI, reaction with Mg enolate of malonic acid monomethyl ester, reduction of ketone with NaBH₄	71–74% activation and enolate addition 80–88% de syn reduction	1989 (1829)

Table 11.6 Asymmetric Syntheses of γ-Amino Acids (continued)

H₂N-R-CO₂H	Method	Yield, % ee	Reference
$-CH_2COCH_2-$	activation of Cbz-Gly with CDI, reaction with Mg enolate of malonic acid monoethyl ester	71% activation and enolate addition	1985 (*1830*)
$-CH[CH(OH)H_2N_3]COCH_2-$	activation of 3-HO-4-N₃-Abu with CDI, reaction with Mg enolate of malonic acid monoethyl ester	88% activation and enolate addition	1985 (*1832*)
$-CH(CH_2cHex)COCH_2-$ $-CH(CH_2cHex)CH(OH)CH_2-$	activation of Boc-Cha with CDI, reaction with Mg enolate of malonic acid monoethyl ester, ketone reduction	62% activation and enolate addition 83% ketone reduction 4% de	1988 (*1833*)
$-CH(CH_2indol-3-yl)COCH_2-$ $-CH(CH_2indol-3-yl)CH(OH)CH_2-$	activation of Boc-Trp(CHO) with CDI, reaction with Mg enolate of malonic acid monoethyl ester, ketone reduction	—	2000 (*1621*)
$-CH(iBu)COCH_2-$ $-CH(iBu)CH(OH)CH_2-$	activation of Boc-Leu with CDI, addition of Mg salt of ethyl malonate, ketone reduction with H₂/Raney Ni or NaBH₄/LiCl	37% malonate addition/ decarboxylation 90–98% ketone reduction	1975 (*1776*)
$-CH(Bn)COCH_2-$ $-CH(Bn)CH(OH)CH_2-$ $-CH(CH_2cHex)CH(OH)CH_2-$	activation of Boc-L-Phe with CDI, reaction with Mg enolate of monoethyl malonate or LiCH₂CO₂Et, asymmetric ketone hydrogenation with Rh catalyst, Ph hydrogenation, reduction of ketone	80–90% β-keto ester 97% ketone reduction >98% de 92% aryl reduction	1990 (*1577*)
$-CH(Bn)CH(OH)CH_2-$ $-CH(sBu)CH(OH)CH_2-$	activation of Boc-amino acids with CDI, reaction with (PhMe₂Si)₂CuCNLi₂, addition of acetate enolate to acylsilane, desilylation	97–98% CDI 50–55% acylsilane 61–67% acetate addition 90–98% de *syn* 63–65% desilylation	1999 (*1834*)
$-CH(iBu)CH(OH)CH_2-$	reaction of acid chloride of Phth-L-Leu with MgCH₂(CO₂Et)₂, ketone reduction, decarboxylation	82% malonate addition 0% de reduction	1973 (*1563*)
$-CH(Me)CH(OH)CH(Me)-$	reaction of acid chloride of Phth-D-Ala with MgCH₂(CO₂Et)CO₂H/ decarboxylation/ ketone reduction	43% malonate addition/ decarboxylation 95% ketone reduction	1974 (*1602*)
$-CH(sBu)COCH_2-$ $-CH(sBu)CH(OH)CH_2-$	reaction of Fmoc-alloIle chloride with LiCH(CO₂SiMe₃)CO₂Me, ketone reduction	100% addition 90% reduction 54% de	1988 (*1590*)
$-CH(sBu)COCH_2-$ $-CH(sBu)CH(OH)CH_2-$	activation of Fmoc-D-alle with SOCl₂, reaction with Li enolate of MeO₂CCH₂CO₂SiMe₃, reduction of ketone	96% chlorination 95% acylation 93% reduction 80% de	1989 (*1835*)
$-CH(iBu)COCH_2-$ $-CH(sBu)COCH_2-$ $-CH(iBu)CH(OH)CH_2-$ $-CH(sBu)CH(OH)CH_2-$	activation of Fmoc- L-Leu or D-alle with SOCl₂, reaction with Li enolate of *t*Bu acetate, reduction of ketone	75–76% acylation and reduction 50–90% de	1990 (*1836*)
$-CH(Me)CH(OH)CH(Me)-$	acylation reaction of propionic acid derivatized with Evans' oxazolidinone auxiliary with D-Ala, ketone reduction	78% acylation 95% ketone reduction 60–72% deprotection	1983 (*1601*)
$-CH[(CH_2)_3NHC(=NH)NH_2]COCH_2-$	homologation of Arg via addition of Li enolates to mixed anhydride	43%	1992 (*1639*)

(*Continued*)

Table 11.6 Asymmetric Syntheses of γ-Amino Acids (continued)

H₂N-R-CO₂H	Method	Yield, % ee	Reference
$-$CH(Me)COCH$_2-$	reaction of amino acid UNCAs with Li	50–70% homologation	1996 (*1837*)
$-$CH(*i*Bu)COCH$_2-$	enolate of ethyl acetate, reduction	38–70% de reduction	
$-$CH(*s*Bu)COCH$_2-$	of ketone		
$-$CH[(CH$_2$)$_2$SMe]COCH$_2-$			
$-$CH(CH$_2$OBn)COCH$_2-$			
$-$CH(CH$_2$CO$_2$Bn)COCH$_2-$			
$-$CH[(CH$_2$)$_4$NHBoc]COCH$_2-$			
$-$CH[(CH$_2$)$_2$CO$_2$Bn]COCH$_2-$			
$-$CH(Bn)COCH$_2-$			
$-$CH(Me)CH(OH)CH$_2-$			
$-$CH(*i*Bu)CH(OH)CH$_2-$			
$-$CH(*s*Bu)CH(OH)CH$_2-$			
$-$CH[(CH$_2$)$_2$SMe]CH(OH)CH$_2-$			
$-$CH(CH$_2$OBn)CH(OH)CH$_2-$			
$-$CH(CH$_2$CO$_2$Bn)CH(OH)CH$_2-$			
$-$CH[(CH$_2$)$_4$NHBoc]CH(OH)CH$_2-$			
$-$CH[(CH$_2$)$_2$CO$_2$Bn]CH(OH)CH$_2-$			
$-$CH(Bn)CH(OH)CH$_2-$			
$-$CH(Me)COCH(*i*Pr)$-$	homologation of α–amino acids via	30–81%	1998 (*1838*)
$-$CH(Bn)COCH(*i*Pr)$-$	addition of Li enolates to UNCA		
$-$CH(*i*Pr)COCH(*i*Pr)$-$			
$-$CH(*i*Bu)COCH(*i*Pr)$-$			
$-$CH(CH$_2$OBn)COCH(*i*Pr)$-$			
$-$CH(CH$_2$CO$_2$Bn)COCH(*i*Pr)$-$			
$-$CH(CH$_2$CH$_2$CO$_2$Bn)COCH(*i*Pr)$-$			
$-$CH(Me)COCH(*i*Bu)$-$			
$-$CH(Bn)COCH(*i*Bu)$-$			
$-$CH(*i*Pr)COCH(*i*Bu)$-$			
$-$CH(*i*Bu)COCH(*i*Bu)$-$			
$-$CH(*s*Bu)COCH(*i*Bu)$-$			
$-$CH(CH$_2$CH$_2$SMe)COCH(*i*Bu)$-$			
$-$CH(CH$_2$CO$_2$Bn)COCH(*i*Bu)$-$			
$-$CH(CH$_2$CH$_2$CO$_2$Bn)COCH(*i*Bu)$-$			
$-$CH(Me)COCH(Bn)$-$			
$-$CH(Bn)COCH(Bn)$-$			
$-$CH(*i*Pr)COCH(Bn)$-$			
$-$CH(*i*Bu)COCH(Bn)$-$			
$-$CH(CH$_2$OBn)COCH(Bn)$-$			
$-$CH(CH$_2$CH$_2$CO$_2$Bn)COCH(Bn)$-$			
$-$CH(Me)COC(Me)$_2-$			
$-$CH(Bn)COC(Me)$_2-$			
$-$CH(*i*Pr)COC(Me)$_2-$			
$-$CH(*i*Bu)COC(Me)$_2-$			
$-$CH(CH$_2$OBn)COC(Me)$_2-$			
$-$CH(CH$_2$CO$_2$Bn)COC(Me)$_2-$			
$-$CH(CH$_2$CH$_2$CO$_2$Bn)COC(Me)$_2-$			
$-$CH(Bn)COCH$_2-$	activation of Boc- or Cbz-Phe, -Leu,	75–95% enolate displace-	1988 (*1839*)
$-$CH(*s*Bu)COCH$_2$-CH(*i*Pr)COCH$_2-$	-Val, -Pro as MODD amide,	ment	
$-$CH(*i*Pr)COCH(Me)$-$	displacement with Li enolate of	>96% de reduction	
$-$CH[(CH$_2$)$_3$-(γ-N)]COCH(Me)$-$	ethyl acetate or propionate, ketone		
$-$CH(*s*Bu)CH(OH)CH$_2-$	reduction		

Elaboration of α-Amino Acids: Other Reactions of Carboxy Group

H₂N-R-CO₂H	Method	Yield, % ee	Reference
$-$CH(*i*Bu)CH(OH)CH$_2-$	Wittig reaction of amino acid-derived	94–97% Wittig	1999 (*1650*)
$-$CH(*s*Bu)CH(OH)CH$_2-$	oxazolidinone, reduction	92–96% reduction	
$-$CH(Bn)CH(OH)CH$_2-$		50–60% de	
$-$CH(*i*Bu)CH(OH)CH$_2-$	Wittig reaction of amino acid	91–94% hydrolysis	1999 (*1840*)
$-$CH(Bn)CH(OH)CH$_2-$	oxazolidinone, hydrolysis,	89–92% saponification	
	reduction, saponification		

Table 11.6 Asymmetric Syntheses of γ-Amino Acids (continued)

H$_2$N-R-CO$_2$H	Method	Yield, % ee	Reference
−CH(CH$_2$OH)(CH$_2$)$_2$−	coupling of amino acid with Meldrum's acid, ketone reduction, decarboxylative cyclization, hydrolysis	58–92% coupling and reduction 75–96% cyclization 53–98% hydrolysis	1997 (*1841*)
−CH(Me)CH(OH)CH$_2$− −CH(*i*Pr)CH(OH)CH$_2$− −CH(*i*Bu)CH(OH)CH$_2$− −CH(Bn)CH(OH)CH$_2$− −CH(CH$_2$OBn)CH(OH)CH$_2$− −CH(CH$_2$CH$_2$SMe)CH(OH)CH$_2$− −CH[(CH$_2$)$_4$NHCbz]CH(OH)CH$_2$− −CH[CH$_2$(indol-3-yl)]CH(OH)CH$_2$−	reaction of amino acid with Meldrum's acid, cyclization, reduction, hydrolysis	55–78% addition, cyclization and reduction 75–92% hydrolysis	1987 (*1842*)
−CH(*i*Pr)CH(OH)CH$_2$− −CH(*i*Bu)CH(OH)CH$_2$− −CH(Bn)CH(OH)CH$_2$− −CH(CH$_2$OBn)CH(OH)CH$_2$− −CH(CH$_2$OTBS)CH(OH)CH$_2$−	reaction of amino acid with Meldrum's acid, cyclization, reduction, hydrolysis	81–96% addition and cyclization 55–80% reduction	1996 (*1843*)
−CH(*i*Bu)CH(OH)CH$_2$− −CH(CH$_2$*c*Hex)CH(OH)CH$_2$−	acylation of amino ester with ethyl malonyl chloride, cyclization to tetramic acid, decarboxylation, hydrogenation, hydrolysis	92% acylation 80% cyclization 85% decarboxylation 96% hydrogenation 60–80% hydrolysis	1989 (*1844*)
−CH$_2$CC− −CH(Me)CC− −CH(*i*Pr)CC− −CH(*i*Bu)CC− −CH(*s*Bu)CC− −CH[(CH$_2$)$_3$-(γ-N)]CC− −CH(Me)CH$_2$CH$_2$− −CH(*i*Pr)CH$_2$CH$_2$− −CH(*i*Bu)CH$_2$CH$_2$− −CH[(CH$_2$)$_3$-(γ-N)]CH$_2$CH$_2$− −CH(Me)CH=C(Br)−	EDC coupling of Cbz-amino acid with Ph$_3$P=CHCO$_2$Et, flash vacuum pyrolysis, akyne hydrogenation or hydrobromination	45–51% coupling 29–58% pyrolysis 70–78% hydrogenaiton 80% hydrobromination	1996 (*1845*) 2002 (*1846*)
−CH(*i*Bu)COCO−	activation of Leu with CDI, reaction with Ph$_3$PCH$_2$CONHR, ozonolysis	96% coupling 89–92% ozonolysis	1999 (*1640*)
−CH$_2$CH(NH$_2$)CH$_2$−	reduction of α-carboxy group of Cbz-Asp(O*t*Bu)-OH, Mitsunobu cyclization of alcohol, aziridine opening with alkylamines	60–90% aziridine 60–83% aziridine opening	2004 (*1168*)

Elaboration of α-Amino Acids: Glu

H$_2$N-R-CO$_2$H	Method	Yield, % ee	Reference
−CH(CH$_2$OH)CH$_2$CH$_2$−	reduction of Trt-Glu(OMe)-OBt or Trt-Glu-OMe α-carboxyl, deprotection	72–77% overall	1987 (*1847*)
−CH(CH$_2$OH)CH$_2$CH$_2$−	reduction of Boc-Glu(OMe)-OSu α-carboxyl, deprotection	83% overall	1991 (*1848*)
−CH(CH$_2$OH)CH$_2$CH$_2$− −CH(CHO)CH$_2$CH$_2$− −CH(CH=CH$_2$)CH$_2$CH$_2$−	reduction of MeO$_2$C-Glu(OEt)-OH α-carboxyl, oxidation, methylenation	72% reduction 64% oxidation and methylenation 89% deprotection	1993 (*1849*)
−CH(CH$_2$OH)CH$_2$CH$_2$−	reduction of Cbz-Gln-OBn α-carboxyl, hydrogenation and hydrolysis	81% reduction 65% deprotection	1986 (*1660*)
−CH(*i*Pr)CH$_2$CH$_2$− −CH(*n*Bu)CH$_2$CH$_2$−	reduction of Glu α-carboxyl, conversion to electrophile, displacement with organocuprate	82–85% reduction 75–90% tosylation 30–92% iodination 65–93% displacement	1992 (*1154*)
CH(CH$_2$=CH$_2$)CH$_2$CH$_2$−	oxidation and methylenation of pyroglutaminol	64% oxidation and methylenation 89% deprotection	1993 (*1849*)

(Continued)

Table 11.6 Asymmetric Syntheses of γ-Amino Acids (continued)

H₂N-R-CO₂H	Method	Yield, % ee	Reference
$-CH(CH_2NH_2)CH_2CH_2-$	reduction of Glu α-carboxyl, azide displacement, azide reduction	78% reduction 86% displacement 62–91% reduction	1996 (*1850*)
$-CH(CH_2Ar)CH_2CH_2-$ Ar = Ph, 4-Me-Ph, 2- or 4-MeO-Ph, 2-NH₂-Ph, 2-F-Ph, 4-NO₂-Ph, CH₂CH=CH₂	conversion of Glu to 4-amino-5-iodobutyric acid, form organoZn reagent, Pd-catalyzed coupling with ArI	34–80% coupling	1998 (*1169*) 1999 (*1172*)
$-CH(CH_2COR)CH_2CH_2-$ R = Ph, 2-furyl, CH₂OAc, CH=CH₂	conversion of Glu to 4-amino-5-iodobutyric acid, form organoZn reagent, Pd-catalyzed coupling with RCOCl	45–52% coupling	2000 (*1174*)
$-CH(CH_2Ar)CH_2CH_2-$ Ar = Ph, 4-NO₂-Ph, 1-naphthyl	conversion of Glu to 4-amino-5-iodobutyric acid, form organoZn reagent, Pd-catalyzed coupling with ArOTf	29–49% coupling	2000 (*1171*)
$-CH[CH_2(4-HO-Ph)]CH_2CH_2-$	iodobutyric acid, form organoZn reagent, Pd-catalyzed coupling with unprotected iodophenol	64% coupling	2004 (*1170*)
$-CH(CH_2OH)CH_2CH_2-$ $-CH(CH_2F)CH_2CH_2-$ $-CH(CH_2Cl)CH_2CH_2-$ $-CH(CH_2Br)CH_2CH_2-$ $-CH(CH_2CN)CH_2CH_2-$	reduction of pyroGlu-OEt ester to hydroxymethyl, hydroxyl displacement with halide, lactam hydrolysis	88% reduction 74–86% hydroxyl conversion	1980 (*1851*)
$-CH(CH_2Cl)CH_2CH_2-$ uniformly labelled with ¹⁴C $-CD(CH_2Cl)CH_2CH_2-$	cyclization of [2-²H]- or [U-¹⁴C]-Glu to pyroGlu, carboxyl reduction, chlorination, hydrolysis	quant. cyclization and reduction 39% chlorination and hydrolysis	1981 (*1527*)
$-CH(Me)CH_2CH_2-$	reduction of pyroGlu carboxyl group, mesylation, reduction, lactam hydrolysis	67–75% mesylation 60–61% reduction 58–68% hydrolysis	1990 (*1530*)
$-CH[CH(OH)Et]CH_2CH_2-$ *syn* and *anti*	reduction of N,N-Bn₂-Glu γ-carboxy group to alcohol and protection as TBDMS ether, reduction of α-carboxyl group to aldehyde, addition of Et₂Zn or EtMgBr, reoxidation of side-chain carboxyl	67% Et₂Zn addition 74% de *syn* 73% EtMgBr addition 84% de *anti* 62–68% reoxidation	2003 (*1175*)
$-CH_2CH(4-Cl-Ph)CH_2-$	conversion of Glu to dehydropyroGlu, carboxyl reduction, organocuprate addition, hydroxymethyl oxidation, decarboxylation, hydrolysis	60–70% organocuprate addition 80% oxidation 60–65% decarboxylation 64% hydrolysis	1992 (*1533*)
$-CH(CH_2OH)CH(Et)CH(OH)-$	conversion of Glu to dehydropyroGlu, carboxyl reduction, organocuprate addition, hydroxylation, hydrolysis	—	1991 (*1853*)
$-CH(CH_2OMe)CH(OH)CH(OH)-$	O-methylation of pyroglutaminol, formation of 3,4-dehydro derivative, dihydroxylation, lactam hydrolysis	75% methylation 49% 3,4-dehydro 67% dihydroxylation 78% lactam hydrolysis	1991 (*1627*)
$-CH(CH_2OH)CH(OH)CH(OH)-$	dihydroxylation of 3,4-dehydropyro-Glu derivative	78% dihydroxylation	1989 (*1854*)
$-CH(CH_2=CH_2)CH_2CH_2-$	reduction of pyroGlu carboxyl group, methylenation, hydrolysis	46% reduction and methylenation quant. hydrolysis	1994 (*1855*)
$-CH(CH=CH_2)CH_2CH_2-$	conversion of Glu to pyroGlu, N-protection, ester reduction to aldehyde, methylenation, N-deprotection, pyrrolidinone hydrolysis	33% overall	1992 (*1856*)

Table 11.6 Asymmetric Syntheses of γ-Amino Acids (continued)

H₂N-R-CO₂H	Method	Yield, % ee	Reference
$-CH(CCH)(CH_2)_2-$	reduction of pyroGlu α-carboxyl, conversion to alkyne	73% reduction 64% alkyne formation 36% deprotection	1995 (*1857*)
$-CH(CN)CH_2CH_2-$ $-CH_2CH(OH)CH_2-$ with NH₂ or NMe₃	dehydration of Cbz-Glu(OMe)-NH₂ enzymatic decarboxylation of Glu, N-methylation	69–93% dehydration 40% decarboxylation 40% methylation	1971 (*1183*) 1962 (*1541*)

Elaboration of α-Amino Acids: Other Amino Acids

$-CH_2CH_2CH(OH)-$	partial deamination of L-2,4-diamin-obutyric acid with NaNO₂	—	1971 (*1613*)
$-CH_2CH_2CH(OH)-$	conversion of Asn α-NH₂ to OH, dehydration of amide to nitrile, reduction	30% overall	1978 (*1858*)
$-CH_2CH(NH_2)CH_2-$ $-CH_2CH(NMe_3)CH_2-$	homologation of Cbz-Asp(O*t*Bu) α-carboxyl, selective deprotection, Curtius rearrangement, methylation	22% overall	1992 (*169*)
$-CH_2CH(OH)CH_2-$	electrochemical decarboxylation of 4-hydroxyproline, hydrolysis	65% overall	1986 (*1859*)
$-CH_2CH(4-Cl-Ph)CH_2-$ $-CH_2CH_2CH(4-Cl-Ph)-$	conversion of 4-HO-Pro	13–20%	1995 (*1860*)
$-C(=CHNH_2)CH_2CH(NH_2)-$	reaction of His with excess benzoyl chloride, hydrolysis	60% hydrolysis	1935 (*1675*)
$-CH[(CH(Me)CH_2OH)CH=CH-$	Lewis-catalyzed reaction of methyl 2-azido-3-SiMe₂Ph-hex-4-enoate with trioxane, rearrangement, azide reduction	85% reaction/rearrangement 84% reduction	2002 (*228*)

Elaboration of β-Amino Acids

$-CH(Bn)CH_2CH_2-$ $CH[CH_2CH_2CH_2-(γ-N)]CH_2CH_2-$	homologation of α-amino acids via consecutive Arndt–Eistert reactions	56–73% second homologation	1977 (*1100*)
$-CH_2CH(NH_2)CH_2-$ with NMe₃	Arndt–Eistert homologation of Cbz-L-Dap(Boc), methylation	73% homologation 11% trimethylation	1987 (*168*)
$-CH(Ph)CH(iPr)CH_2-$ $-CH(Ph)CH(Et)CH_2-$ $-CH(4-MeO-Ph)CH(OBn)CH_2-$ $-CH(CH_2CH_2Ph)CH(iPr)CH_2-$	homologation of β–amino acids via TMSCH₂N₂ with β-lactam, photolysis	44–81% diazoketone 62–82% γ-lactam	1998 (*1861*)
$-CH_2CH_2CH(OH)-$	homologation of Cbz-β-Ala-OMe, ketone reduction	56% homologation 47% reduction 49% ee	1982 (*1862*)
$-CH_2CH(NH_2)CH_2-$	enzymatic symmetrization/monoami-dation of dimethyl 3-benzylamino-glutarate, Hofmann rearrangement	76% monoamidation	2004 (*689*)

Elaboration of γ-Amino Acids

$-CH_2CH_2CH(Me)-$ $-CH_2CH_2CH(Et)-$ $-CH_2CH_2CH(iPr)-$	formation of homochiral oxazoline of GABA, alkylation	52–58% alkylation >90% de 50–95% hydrolysis	1996 (*177*)
$-CH(Me)CH_2CH(CH_2CH=CH_2)-$ $-CH(Me)CH_2CH(CH=CHPh)$ $-CH(Bn)CH_2CH(Me)-$ $-CH(Bn)CH_2CH(CH_2CH=CH_2)-$ $-CH(Bn)CH_2CH(CH=CHPh)-$ $-CH(Bn)CH_2CH(Bn)-$ $-CH(iPr)CH_2CH(CH_2CH=CH_2)-$ $-CH(iBu)CH_2CH(CH_2CH=CH_2)-$	di-deprotonation of homochiral N-protected 4-substituted-GABA esters/amides, alkylation with R₂X	45–90% alkylation >90% de	1997 (*1863*)
$-CH_2CH_2CH(SMe)-$ $-CH_2CH_2CH(SPh)-$	formation of homochiral oxazoline of GABA, sulfenylation	32–69% sulfenylation	1997 (*1864*)
$-CH(CH_2F)C=CH-$	α-deprotonation and selenation of 4-amino-5-F-pentanoic acid, oxidative elimination	64% selenation 76% elimination	1986 (*1865*)

Table 11.6 Asymmetric Syntheses of γ-Amino Acids (continued)

H$_2$N-R-CO$_2$H	Method	Yield, % ee	Reference
−CH(Me)CH$_2$CH(Bn)− −CH(iPr)CH$_2$CH(Me)−	di-deprotonation of homochiral Boc-4-substituted-GABA Me esters, alkylation with R^{2}X	49–87% alkylation 50–>90% de	2001 (1659)
−CH(Me)CH$_2$CH(CH$_2$CH=CHPh)− −CH(iPr)CH$_2$CH(CH$_2$CH=CHPh)−	conversion of amino acid to γ-substituted γ-amino acid, cyclization to pyrrolidinone, deprotonation and cinnamylation, hydrolysis	77–85% alkylation 84–87% hydrolysis	1999 (1644)
−CH$_2$CH$_2$CH(Ph)−	formation of ketene of Phth-γ-amino acid, addition of (R)-pantolactone	75% ketene formation and ester formation 70% de	2002 (1701)
−CH[CH(OH)CH(OH)R]CH$_2$CH$_2$− R = CH$_2$OH, CH(OH)CH$_2$OH, CH(OH)CH(OH)CH$_2$OH	addition of GABA equivalent, TBSOP, to chiral aldehydes, reduction, hydrolysis	79–90% addition 100% de 84–95% reduction 80–88% hydrolysis	1993 (1866)
−CH[CH(OH)nC$_{12}$H$_{25}$]CH$_2$CH$_2$−	addition of GABA equivalent, TBSOP, to aldehydes with chiral catalyst, reduction	52% addition >98% de, 68% ee 83% reduction	1998 (1867)
−CH(R)CH$_2$CH(CH$_2$CH=CH$_2$)−	radical allylation of homochiral 4-substituted GABA amide	72–97% 50–78% de anti	1996 (642)
−CH$_2$CH(Me)CH$_2$− −CH$_2$CH(Et)CH$_2$− −CH$_2$CH(nPr)CH$_2$− −CH$_2$CH(nBu)CH$_2$−	organocuprate addition to 4-amino-2-butenoate with amino group in chiral oxazolidinone	75–80% addition 50–75% de	1993 (1868)
−CH$_2$CH(Ph)CH$_2$− −CH$_2$CH(4-F-Ph)CH$_2$− −CH$_2$CH(4-Cl-Ph)CH$_2$−	asymmetric conjugate addition of arylboronic acids to N-protected ethyl 4-aminobut-2-enoate in presence of chiral Rh(I)-BINAP catalyst	46–74% addition 86–92% ee	2003 (1708)
−CH(Me)CH(OH)CH(OH)− −CH(Bn)CH(OH)CH(OH)− −CH(iPr)CH(OH)CH(OH)− −CH(iBu)CH(OH)CH(OH)−	asymmetric dihydroxylation of γ-amino α,β-unsaturated esters	7–88% 26–94%de	1996 (1869)
−CH$_2$CH(NH$_2$)CH$_2$− with NH$_2$ or NMe$_3$	intramolecular conjugate addition of urea formed from N-CH(Ph)Me 4-aminobut-2-enoic acid and tosyl isocyanate, hydrolysis, N-trimethylation	90% cyclization 0% de quant. hydrolysis and methylation	1990 (772)
−CH(CN)]CH$_2$CH$_2$−	dehydration of isoGln	70%	1971 (1232)
−CH[CH$_2$(cytosin-1-yl)]CH$_2$CH$_2$− −CH[CH$_2$(adenin-1-yl)]CH$_2$CH$_2$−	tosylation and displacement of lactam of 4-amino-5-hydroxypentanoic acid, lactam hydrolysis	52–73% displacement 70–81% hydrolysis	1991 (1870)
−CH(Bn)COCH$_2$− −CH(4-Cl-Bn)COCH$_2$− −CH(Bn)CH(OH)CH$_2$− −CH(4-Cl-Bn)CH(OH)CH$_2$−	consecutive or one-pot asymmetric hydrogenation of 3-keto-4-amino-5-Ph-pent-4-enoate with Rh(I) and Ru(II) catalysts	>99% ee alkene reduction >95% de ketone reduction	1998 (1871)
−CH$_2$C(Me)$_2$CH(OH)−	oxidation then asymmetric reduction of racemic γ-lactam	94% oxidation 100% reduction 76% ee 62% after crystallization >99% ee	1999 (1872)
−CH$_2$COCH$_2$− −CH$_2$CH(OH)CH$_2$− with NMe$_2$, NMe$_3$	reaction of NaCH$_2$COCH$_2$NMe$_2$ with (EtO)$_2$CO, asymmetric hydrogenation	100% reduction 85% ee	1991 (1689)
−CH$_2$CH$_2$CH(OH)−	enzymatic reduction of N-Cbz 2-keto-4-aminobutanoic acid	91–95% reduction >99% ee	1995 (1713)
−CH$_2$CH(OH)CH$_2$−	asymmetric reduction of alkyl 3-oxo-4-azidobutyrate with Baker's yeast	70–80%	1985 (1873)

Table 11.6 Asymmetric Syntheses of γ-Amino Acids (continued)

H₂N-R-CO₂H	Method	Yield, % ee	Reference
−CH₂CH₂CH(4-F-Bn)− −CH₂CH₂CH(iBu)− (lactam)	condensation of lactam with RCHO, alkene reduction by asymmetric hydrogenation with chiral Ir catalyst	98–100% reduction 82–89% ee	2002 (*1874*)

Elaboration of Chiral Substrates

H₂N-R-CO₂H	Method	Yield, % ee	Reference
−CH₂CH(OH)CH₂−	derivatization of malic acid via amination and homologation	30% overall	1995 (*1538*)
−CH₂CH(OH)CH₂−	derivatization of malic acid via reduction, epoxide formation, azide displacement, oxidation	63% epoxide formation 87% azide opening 72% azide reduction 60% oxidation	1995 (*1875*)
−CH₂CH(OH)CH₂−	derivatization of malic acid via amidation, reduction, oxidation	25% overall	1985 (*1876*)
−CH₂CH(OH)CH₂−	derivatization of dimethyl malate via chemoselective reduction, tosylation, amination	64% tosylation 83% azide displacement 87% hydrogenation	1999 (*1152*)
−CH₂CH(OH)CH₂− (with NMe₃)	derivatization of malic acid via chemoselective reduction, tosylation, amination	60–70% reduction 65–70% tosylation 90% amination 80–95% deprotection	1990 (*1877*)
−CH₂CH(OH)CH₂− with NH₂ or NMe₃	derivatization of D- or L-arabinose or L-ascorbic acid, via oxidation, dibromination, Br reduction by hydrogenation, Br displacement with azide or NMe₃, azide reduction	50–79% oxidation 65–76% bromination 70–89% Br reduction 89% azide displacement 58% reduction 65% displacement with NMe₃	1983 (*1878*)
−CH₂CH(OH)CH₂−	derivatization of ascorbic acid via (R)-glycerol acetonide, CN displacement of tosylate, azide displacement of mesylate	65% CN displacement 63–77% azide displacement 97% azide reduction 96% CN hydrolysis	1980 (*1879*)
−CH₂CH(OH)CH₂−	derivatization of D-mannitol via (S)-glycerol acetonide, HNPhth Mitsunobu amination, CN opening of cyclic sulfite	68% amination 70% CN sulfite opening 84% CN hydrolysis	1996 (*1880*)
−CH₂CH(OH)CH₂− (with NMe₃)	derivatization of D-galactono-1,4-lactone via amidation with NMe₂, carbonyl reduction or reductive amination of aldehyde with MeNH₂	28–47% overall	1992 (*1881*)
−CH₂CH(OH)CH₂−	opening of 3-hydroxybutyrolactone with HBr, epoxide formation, opening with NaCN, Curtius rearrangement	85% to cyano derivative 78% rearrangement	1999 (*1882*)
−CH(nPr)CH(OH)CH₂−	derivatization of malic acid via imide formation, reduction, allyl addition, deprotection, hydrolysis	70% imide formation 85–99% allyl addition 50% de *trans* to 60% de *cis*	1990 (*1883*)
−CH(iBu)CH(OH)CH₂−	derivatization of malic acid via imide formation, reduction, methallyl addition or radical reaction, deprotection, hydrolysis	83% imide formation 77% reduction 95% allyl addition 80% de *trans* or 75% radical cyclization 100% de *cis*	1991 (*1884*)
−CH(iBu)CH(OH)CH₂−	derivatization of malic acid via imide formation, methallylMgBr addition, reduction, deprotection, hydrolysis	46% allyl addition 73% hydrogenation 50% de *cis* 40% deprotection	1990 (*1885*)

(Continued)

Table 11.6 Asymmetric Syntheses of γ-Amino Acids (continued)

H₂N-R-CO₂H	Method	Yield, % ee	Reference
$-CH(Bn)CH(OH)CH_2-$ with *N*-Me	derivatization of malic acid via cyclization to pyrrolidinedione, addition of BnMgCl, deoxygenation, hydrogenation/hydrolysis	91% addition 85% deoxygenation 94% hydrogenation	1998 (*1886*)
$-CH(iBu)CH(OH)CH_2-$	derivatization of 3-deoxy-1,2-*O*-isopropylidene-α-D-*erythro*-pentodialdo-1,4-furanose	67% Grignard addition 52% carbinol inversion 85% azide displacement 96% diol cleavage 63–76% oxidation 93% azide reduction	1973 (*1564*) 1975 (*1565*)
$-CH(iBu)CH(OH)CH_2-$	condensation of benzyl isocyanate with α-OH-β-Ph-propionate	92% condensation 95% reduction 84% isobutenylation 85% Ph oxidation	1988 (*1887*)
$-CH(iPr)CH(OH)CH_2-$ $-CH(cPent)CH(OH)CH_2-$ $-CH(cHex)CH(OH)CH_2-$ $-CH(cHept)CH(OH)CH_2-$ $-CH(Bn)CH(OH)CH_2-$ $-CH(CH_2CH_2Ph)CH(OH)CH_2-$	elaboration of D-glucofuranose	80–93% organocuprate addition 79–92% azide displacement 89–98% oxidative cleavage	1989 (*1888*)
$-CH(iBu)CH(OH)CH_2-$	conversion of (*R*)-2,3-isopropylidene glyceraldehyde via addition of diallylzinc, formation of epoxide, opening with *i*PrMgBr, amination with NaNPhth, alkene oxidative cleavage	13% overall	1988 (*1889*)
$-CH(iBu)CH(OH)CH_2-$ $-CH(Bn)CH(OH)CH_2-$	formation of synthon from D-glucosamine, addition of side chain, oxidation to form carboxyl	27% yield of synthon 66–86% Wittig or Grignard 54–75% carboxyl formation	1992 (*1891*) 1996 (*1890*)
$-CH(CH_2CO_2H)CH(OH)CH(OAc)-$ $-CH(CH_2OMe)CH(OH)CH(OH)-$	elaboration of D-glucosamine	—	1996 (*1288*)
$-CH_2CH(OR)CH(OR)-$ R = H, Me, Bn	tosylation, azide displacement, oxidative diol cleavage of D-xylofuranoside, D-allofuranoside	—	1992 (*1892*)
$-CH(Me)CH(OH)CH(Me)-$	elaboration of L-rhamnose		1981 (*1600*)
$CH(CH_2OMe)CH(OHCH(OH)-$	amination of chiral 2,3,4,5-(OH)₂-pentanoic acid lactone derivative	99% amination	1992 (*1628*)

Other Syntheses

H₂N-R-CO₂H	Method	Yield, % ee	Reference
$-CH(iPr)CH(Me)CH_2-$ $-CH(iPr)CH(Me)CH_2-$	Mn-mediated addition to chiral hydrazone derivative of 3,4-dihydroxybutyraldehyde or phenylacetaldehyde, hydrazone reduction, hydroxymethyl oxidation	56–77% addition >96% de 71–76% hydrazone cleavage 81–96% oxidation	2004 (*1893*)
$-CH(iBu)CH(OH)CH_2-$ $-CH(cHex)CH(OH)CH_2-$	organocuprate displacement of oxazolidine of 2-OH-1-NH₂-1-OMe-pent-4-ene, oxazolidinone hydrolysis, alkene oxidative cleavage	60–94% organocuprate displacement 53–57% oxazolidinone hydrolysis 54–82% alkene oxidative cleavage	1991 (*1895*) 1993 (*1894*)
$-CH(CH_2I)CH(OH)CH_2-$ $-CH(iBu)CH(OH)CH_2-$ $-CH(CH_2cHex)CH(OH)CH_2-$	iodolactonization of H₂C=CHCH(OH)-CH₂C(SMe)=NCH(Me)Ph, organocuprate displacement of I, auxiliary removal, lactam opening	66% iodolactonization 83–91% displacement 41–76% auxiliary removal 50–88% lactam opening	1990 (*1896*)
$-CH_2CH(OH)CH_2-$	conversion of 3-hydroxycyclobutanone to chiral imine, oxaziridine formation, photolysis, deprotection	70–79% oxaziridine 40% photolysis 36% deprotection	1991 (*1897*)
$-CH_2CH(OH)CH_2-$	reduction of 3-hydroxypyridine, resolution, oxidative cleavage	71% reduction 55% oxidative cleavage	1997 (*1898*)

Table 11.6 Asymmetric Syntheses of γ-Amino Acids (continued)

H₂N-R-CO₂H	Method	Yield, % ee	Reference
–CH₂CH(OH)CH₂– with NH₂ or NMe₃	ene reaction of β-pinene with ethyl glyoxylate, ester reduction and chlorination, pinene oxidative cleavage, Cl displacement with amine	60% ene reaction 84% reduction 53% chlorination 48% oxidative cleavage 56–82% amination	1985 (*1899*)
–CH(CH=CH₂)CH₂CH₂–	asymmetric aminohydroxylation of 6-HO-2-hexenoate, ester reduction, hydroxy/amine elimination, hydroxy oxidation	70% aminohydroxylation 85% ee 64% ester to alkene 79% oxidation	1998 (*861*)
–CH(Bn)CH(OH)CH₂– with *N*-Me	aldol reaction of 3-Ph-pentanoic acid with Eavans' oxazolidinone auxiliary with acrolein, Curtius rearrangement, hydroboration, and oxidation	79% aldol 80% Curtius 75% hydroboration 80% oxidation	1999 (*1616*)

11.5 Cyclic γ-Amino Acids

11.5.1 Introduction

There are a number of naturally occurring cyclic γ-amino acids. (–)-Detoxinine, 3-hydroxypyrrolidine-2-(3-hydroxy-3-propionic acid) (see Scheme 11.173), is a constituent of the depsipeptide detoxin D₁ isolated from *Streptomyces caespitosus*. The peptide is a potent antagonist of the cytotoxicity of the antibiotic blasticidin S, a fungicide used to treat rice blast disease (*1900–1903*). The absolute configuration of detoxinine was revised in 1980 (*1904*). Several syntheses of detoxinine have been described (*1900–1903, 1905–1907*). A number of different detoxin D group depsipeptides contain this

O-acyl 3-hydroxypyrrolidine-2-(3-hydroxypropionic acid) moiety (*1908*).

The cyclic antifungal hexapeptides microsclerodermins A–I from *Theonella* and *Microscleroderma* sponges contain a γ-amino acid with the backbone contained in a pyrrolidinone ring system (*211, 213, 214*). A total synthesis of microsclerodermin E was reported in 2003 (*215*). Dolastatin-10 is a pentapeptide isolated from the Indian Ocean sea hare *Dolabella auricularia* which has potent antineoplastic properties (references within *1630–1632*). Only 28.7 mg was obtained from 1600 kg of *Dolabella auricularia* (*75*). The peptide contains two β-methoxy-γ-amino acids, acyclic dolaisoleuine and cyclic dolaproine. A cyclic analog of dolastatin-10 was prepared with an ester linkage between the modified

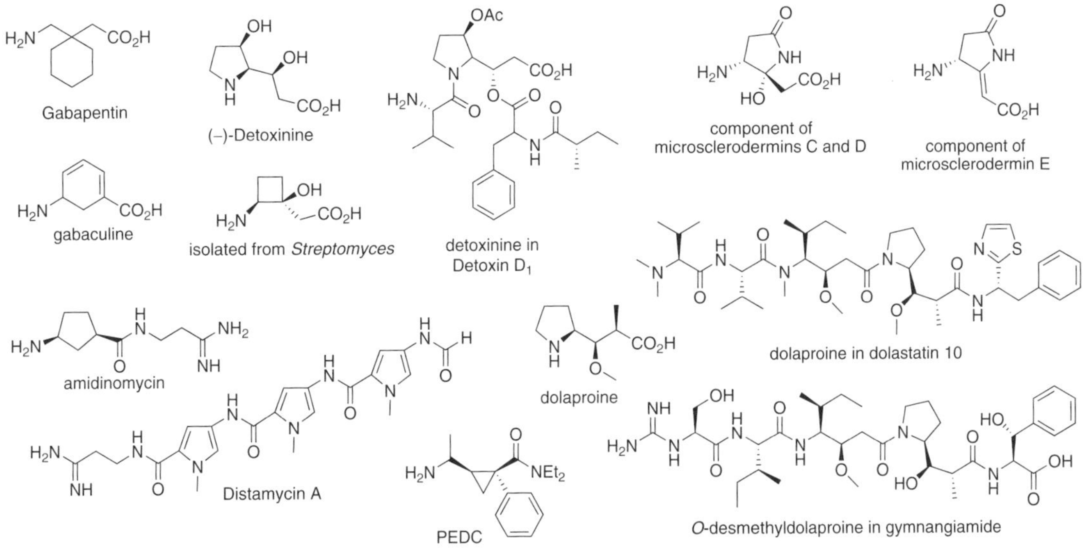

Scheme 11.173 Cyclic γ-Amino Acids.

C- and *N*-termini (*1633*). Both *N*-desmethyldolaisoleuine and *O*-desmethyldolaproine are components of gymnangiamide, a cytotoxic peptide from the marine hydroid *Gymnangium regae* (*1634*).

An antibiotic dipeptide containing (1*S*,2*S*)-1-hydroxy-2-aminocyclobutane-1-acetic acid was extracted from a *Streptomyces* species and shown to inhibit the growth of several Gram-positive organisms, possibly via interference with Cys/Met metabolism (*1909, 1910*). Amidinomycin, an antiviral antibiotic metabolite from another *Streptomyces* species, consists of *N-cis*-(2-amidinoethyl)-3-aminocyclopentanecarboxamide (*1911*). Gabaculine (5-amino-1,3-cyclohexadienyl-1-carboxylic acid) was isolated from *Streptomyces toyocaensis* (*1912*), and is a potent irreversible inhibitor of GABA-α-ketoglutaric acid transaminase (*1912–1915*). An analog, (*S*)-4-amino-4,5-dihydro-2-thiophenecarboxylic acid, is also an irreversible inhibitor. Mechanistic studies show the inactivation results from reaction with the pyridoxal 5'-phosphate (PLP) cofactor, and results in aromatization of the thiophene nucleus (*1916*).

Distamycin A, isolated from the fermentation broth of *Streptomyces distallicus*, possesses antiviral and oncolytic properties (*1917*). Distamycin is a sequence-specific DNA-binding peptide made up of *N*-methyl 2-carboxy-4-aminopyrrole monomers (*1918*). Oligomers of this monomer were prepared as possible antitumor antibiotics (*1919*) as well as for a total synthesis of distamycin A (*1917*). Attachment of an EDTA molecule to one end converted the sequence-specific DNA binding antibiotic into a sequence-specific DNA-cleaving molecule (*1920–1922*); the alkylating subunit of the DNA cross-linking agent isochrysohermidin has also been attached (*1923*). 4-Amino-1-methylimidazole-2-carboxylic acid (*1924, 1925*), and 3-amino-1-methylpyrazole-5-carboxylic acid (*1926, 1927*) were synthesized as analogs of the distamycin monomer. Another analog employed a 2-carboxy-4-amino-thiophene ring to replace the aromatic heterocycle, with a uracil nucleobase at the 3-position. However, a tetramer of this analog showed no binding to complementary DNA or RNA (*1928, 1929*).

A number of synthetic cyclic γ-amino acids possess biological activity. Milnacipran, racemic (*Z*)-1-phenyl-2-aminomethyl-*N,N*-diethylcyclopropanecarboxamide, is a clinically used antidepressant which acts by inhibition of serotonin reuptake. It is also a non-competitive NMDA receptor antagonist. The 1-phenyl-2-[(*S*)-1-aminoethyl]-*N,N*-diethylcyclopropanecarboxamide enantiomer (PEDC) was identified as the potent isomer (*1930*). A series of 3-substituted derivatives were prepared, and also had very potent affinities for the NMDA receptor (*1931*). A proline-derived γ-imino acid, 4-hydroxypyrrolidine-2-(3-hydroxy-3-propionic acid), has been incorporated into analogs of the cyclic depsipeptide antitumor agent hapalosin. Two of the analogs possessed greater activity (*1932*). Analogs of deglycobleomycin A5 (a glycopeptide antitumor antibiotic) with various conformationally restricted γ-amino acid

components, joining the C2–C4 backbone carbons with a butyl or 2-butene linker (diastereomers of 2-hydroxy-3-aminocycloheptyl-1-carboxylic acid), were synthesized in 2003 (*258, 1609*).

Cyclic analogs of carnitine were prepared as inhibitors of carnitine acetyltransferase (*1547*), while piperidine-4-carboxylic acid was found to be an effective component of ligands for the SH3 domain of the protein tyrosine kinase Src, identified via combinatorial libraries (*1933*). 3-Amino-2-hydroxycyclohexane-1-carboxylic acid was used to replace the isostatine unit contained in the didemnins (cyclodepsipeptides isolated from marine tunicates with antitumor, antiviral, and immunosuppressive activity), but potency was reduced (*1594, 1595*). A number of potent carbocyclic inhibitors of influenza neuraminidase based on the structure of *N*-acetylneuraminic acid but containing an additional amino group in the γ-position have been synthesized (*1934–1936*). They have potential for protection against influenza infection.

The best known cyclic γ-amino acid is gabapentin (Neurontin, 1-(aminomethyl)cyclohexane-1-acetic acid), a constrained γ-amino acid anticonvulsant (see Scheme 11.173). It was designed as a lipophilic analog of GABA, but does not bind to any of the GABA receptors (*1535*). The anticonvulsant effect of gabapentin is proposed to be related to its high-affinity binding to a calcium channel site (*1519*). Alkylated gabapentin analogs have been prepared, with two compounds showing higher affinity for the calcium channel binding site (*1519*). A constrained Gabapentin analog, with a methylene bridge between the amino group and acetic acid substituent methylene group, was used to investigate the possible binding conformation of gabapentin (*1385*).

Several cyclic γ-amino acids are mechanism-based inactivators of γ-aminobutyric acid aminotransferase (GABA-AT). Inhibition of this enzyme is a therapy for the treatment of epilepsy, as it raises the GABA levels in the brain. *cis*-3-Aminocyclohex-4-ene-1-carboxylic acid, a constrained analog of 4-amino-5-hexenoic acid (γ-vinyl-GABA, or vigabatrin, marketed as the anticonvulsant Sabril), was successfully designed to inhibit via only one of two possible divergent mechanisms of inactivation (*1522*). Similarly, (1*R*,3*S*,4*S*)-3-amino-4-fluorocyclopentane-1-carboxylic acid is a time-dependent, irreversible inactivator of GABA-T. An X-ray crystal structure demonstrated reaction via an enamine mechanism (*1937*).

Both *cis*- and *trans*-2-(aminomethyl)-cyclopropane-1-carboxylic acid are conformationally restricted analogs of GABA. The enantiomers of both diastereomers were resolved via esterification with (*R*)-pantolactone, and then conformational analysis was carried out using ^{1}H NMR in addition to semiempirical and ab initio calculations (*1938*). Also, *cis*- and *trans*-4-aminocyclopent-2-ene-1-carboxylic acid, *cis*- and *trans*-3-aminocyclopentane-1-carboxylic acid, 4-aminocyclopent-1-ene-1-carboxylic acid, and 3-aminocyclopent-1-ene-1-carboxylic acid were evaluated as constricted GABA analogs (*1939, 1940*), as were (*R*)- and (*S*)-pyrrolidine-3-acetic acid (homo-β-proline) (*1941*). The (*R*)-enantiomer of the latter compound was an inhibitor of GABA$_A$ receptor binding, while the (*S*)-enantiomer bound to the GABA$_B$ receptor (*1941*).

Crystal structures of gabapentin oligomers have shown C$_9$ helix and ribbon structures (*1648*). Gabapentin has been incorporated into dipeptides with an *N*-terminal α-amino acid, with crystal structures of four dipeptides showing a 10-atom β-turn mimetic conformation (*1942*). A parallel secondary sheet structure is formed by a hexapeptide containing four *trans*-3-amino-cyclopentane-1-carboxylic acid residues (*1943*). 5-Amino-2-methoxybenzoic acid was incorporated into several different β-strand peptide mimetics, with a hydrogen bond between the oxygen of the methoxy group and NH proton of the amidated carboxy group producing the desired geometry (*1944*). A cyclic hexa-peptide composed of three (1*R*,3*S*)-3-aminocyclohexane-carboxylic acid residues formed a flat planar cyclic structure that can self-assemble into a stacked dimeric structure, and potentially can stack to form peptide nanotubes with a hydrophobic cavity (*1945*). It was found that 32-membered cyclic octapeptides formed from alternating α-amino acids and *cis*-3-aminocyclohexane-1-carboxylic acids self-assembled into dimeric peptide tubelets with a 7 Å partially hydrophobic pore, as shown by crystal structures (*1946*). 5-(Alkanoylamino)-3-aminobenzoic acid was employed to prepare cyclic peptide artificial ion channels (*1947*).

Syntheses of 3-aminocyclobutane-1-carboxylic acids were reviewed in 1993 (*1356*).

11.5.2 Syntheses of Achiral or Racemic Cyclic γ-Amino Acids (see Table 11.7)

11.5.2a Amination Reactions

Several syntheses employing nucleophilic aminations have been described. A series of conformationally constricted GABA analogs, consisting of *cis*- or *trans*-2-aminomethylcycloalkane-1-carboxylic acids with three-, four-, or six-membered rings, were prepared from *cis*- or *trans*-cycloalkane-1,2-dicarboxylates. One carboxyl group was reduced to a hydroxymethyl substituent, and then converted into a lactone, tosylated, or brominated. The amino group was introduced via phthalimide-induced lactone opening or electrophile displacement (*1832*). Both (2-aminocyclohexylidene)acetic acid and (2-aminocyclopentylidene)acetic acid were prepared by NBS-induced bromination at the allylic position of cyclohexylideneacetic acid or cyclopentylideneacetic acid. The amine group was then introduced by amination in liquid ammonia (*1673*). In a similar manner, cyclopentene-1-carbonitrile was brominated, displaced with ammonia, and then hydrolyzed to give racemic 3-aminocyclopent-1-ene-1-carboxylic acid (*1939*). Amination of 3-*cis*-hydroxy-2-*cis*,4-*trans*-diphenylcy-clobutane-1-carboxylic acid led to 3-*cis*-amino-2-*cis*, 4-*trans*-diphenylcyclobutane-1-carboxylic acid (*1370*),

while racemic *cis*-3-aminocyclohex-4-ene-1-carboxylic acid was obtained from cyclohex-3-ene-1-carboxylic acid via a 3-phthalimidocyclohex-4-ene-1-carboxylic acid intermediate (*1522*).

Rearrangement amination reactions have also been employed. The rigid bicyclic GABA analog 3-aminobicyclo[1.1.1]pentane-1-carboxylic acid was produced via a Curtius rearrangement of the diacid monomethyl ester (*1948*). A modified Curtius rearrangement of 1-(1-but-3-enyl)cyclopropanecarboxylic acid, followed by oxidative cleavage of the alkene, provided 1-aminocyclopropane-1-propionic acid. The amino group was then trimethylated (*1546*). The isomeric 1-(2-aminoethyl)cyclopropane-1-carboxylic acid was produced by a Curtius rearrangement of the corresponding mono ester diacid (*1546*). 1,1-Bis(carboxymethyl)cyclo-pentane has been attached to a polystyrene resin via an amide linkage to one of the carboxy groups. Curtius rearrangement (DPPA, Et$_3$N, then heating) and cleavage from the resin gave 1-(aminomethyl)-1-(carboxylmethyl) cyclopentane amide in 78% yield (*1366*).

Other amination methods include nitration of *N*-methyl pyrrole-2-carboxylate followed by hydrogenation, resulting in the *N*-methyl 2-carboxy-4-aminopyrrole residue contained in distamycin A (*1918*). Oligopyrrole–peptide conjugates were prepared (*1918*, *1949*). Racemic gabaculine (5-amino-1,3-cyclohexadie-nylcarboxylic acid) (*1912*) and its [2-^{3}H]- and [4,5-^{2}H$_2$]-labeled analogs (*1915*) were prepared by amination of the corresponding methyl 1,4-cyclohexadienylcarboxy-lates. One alkene was selectively aminated by treatment with iodoisocyanate followed by dibutyltin dilaurate and then DABCO. 3-Amino-cyclobutane-1-carboxylic acid was obtained by reductive amination of cyclobutanone-3-carboxylate, via oxime formation with hydroxylamine and hydrogenation over PtO$_2$ (*1950*). Similarly, *cis*- and *trans*-1-aminocycloalkane-2-acetic acids with cyclopen-tyl or cyclohexyl rings were prepared by hydrogenation of the oximes of 2-ketocycloalkyl-1-acetic esters, with the *cis* and *trans* isomers separated after lactam formation (*1951*).

11.5.2b Carboxylation Reactions

A 1928 synthesis of piperidine-3-acetic acid proceeded via conversion of 3-(hydroxymethyl)piperidine to 3-(bromomethyl)piperidine, followed by bromide displacement with NaCN and then hydrolysis (*1952*). *N*-Boc *cis*-4-aminocyclopent-2-en-1-ol was activated as an ethyl carbonate and the hydroxy group replaced with a nitromethane group via Pd-catalyzed allylic reaction. The nitro group was transformed into a carboxylate by treatment with NaNO$_2$ and AcOH in DMF, producing 3-aminocyclopent-4-ene-1-carboxylic acid (*1953*). *N*-Benzyl,*N*-but-3-enyl aminomethyltributylstannane was cyclized by treatment with *n*BuLi, producing an organolithium pyrrolidine intermediate which was quenched with ethyl chloroformate to give protected pyrrolidine-3-acetic acid (*1954*).

11.5.2c Amination Combined
with Carboxylation

A series of conformationally constricted GABA analogs were prepared from *cis*- or *trans*-cycloalkane-1,2-dicarboxylates. One carboxyl group was reduced to a hydroxymethyl substituent, tosylated or brominated, and then displaced with cyanide. The remaining carboxyl group was converted into an amine via Curtius rearrangement, with hydrolysis of the nitrile giving *cis*- and *trans*-1-aminocycloalkane-2-acetic acids with cyclopropyl or cyclobutyl rings (*1951*). The corresponding five- and six-membered ring systems were prepared by an alternative method (*1951*).

N-Methylpyrrole and N-methylimidazole were carboxylated and nitrated to give, after reduction of the nitro group, the heteroaromatic γ-amino acids 4-amino-1-methylpyrrole-2-carboxylate and 4-amino-1-methyl-imidazole-2-carboxylate (*1955*). The method is suitable for large-scale syntheses.

A Diels–Alder reaction of tosyl nitrile (TsCN) with cyclopentadiene gave a bicyclic adduct, which was not isolated but hydrolyzed to give *cis*-3-aminocyclopent-4-ene-1-carboxylic acid. Regioisomerization with 2 N NaOH formed 3-aminocyclopent-5-ene-1-carboxylic acid (*1939, 1940*), while N-protection with phthalic anhydride and thermal isomerization gave a mixture of *cis*- and *trans*-3-phthalimidocyclopent-4-ene-1-carboxylic acids. Both *cis* and *trans* isomers could be hydrogenated to the 3-aminocyclopentane-1-carboxylic acid derivatives (*1940*).

11.5.2d Fusion of Amino-Containing
and Carboxy-Containing Components

Methyl-substituted analogs of gabapentin (1-amino-methyl-cyclohexane-1-acetic acid) were synthesized by Horner–Wadsworth–Emmons reaction of 4-methyl cyclohexanone or 3,5-dimethylcyclohexanone with $(EtO)_2P(O)CH_2CO_2Et$, followed by Michael addition of nitromethane to the alkene (see Scheme 11.174). The opposite diastereomers were created by an initial aldol reaction/elimination of ethyl cyanoacetate, followed by Michael addition of cyanide, and selective hydrolysis/reduction of the two nitrile groups (*1535*). The same

sequence of reactions was employed to prepare other gabapentin analogs (*1519*). Nitromethane has also been added to cyclic acrylates attached via an ester linkage to a dendritic polyglycerol dendritic support. Nitro reduction was accompanied by lactamization and concomitant cleavage from the support to form γ-lactams of 1-aminomethyl-cyclohexane-1-acetic acid and 4-*tert*-butyl-1-aminomethyl-cyclohexane-1-acetic acid (*1692*).

Several other syntheses of racemic gabapentin were reported in 1991, based upon the addition of a malonate equivalent to cyclohexanone to form a (cyclohexylidene) malonate, to which was added cyanide (see Scheme 11.175). Malonate hydrolysis/decarboxylation and nitrile reduction produced the desired product as the HCl salt; a route to the free amino acid had poor yield in the final step (*1956*). Azetidin-3-ylacetic acid was formed from N-CHPh$_2$ 3-hydroxyazetidine (derived from epichlorohydrin and benzhydrylamine) via conversion of the hydroxy group to a bromide leaving group and displacement with malonate anion. It was necessary to replace the N-benzhydryl group with a Boc group before saponification and decarboxylation, with TFA deprotection giving the free imino acid (*1957*).

(1S,2S)-1-Hydroxy-2-aminocyclobutane-1-acetic acid was obtained by Reformatsky addition of *tert*-butyl bromoacetate to 2-(dibenzylamino)cyclobutanone, producing predominantly the *cis* alcohol (80% de). The N-deprotected product was resolved by coupling with Boc-L-Val, followed by chromatographic separation (*1909*). Addition of lithiated methyl propargylate to a cyclopropiminium ion derived from 1-dibenzylamino-1-cyclopropanol produced racemic 1-aminocyclopropyl-1-(2-carboxyacetylene) in 35% yield (*570*).

Cis- and *trans*-2-(aminomethyl)cyclopropane-1-carboxylic acids were obtained via Rh(OAc)$_4$-catalyzed cyclopropanation of N,N-bis(trimethylsilyl) allylamine with methyl diazoacetate (*1958*). The *cis* isomers were readily separated via lactam formation upon treatment with NaH in MeOH. The cyclopropanation was also carried out on N,N-bis(trimethylsilyl) 3-amino-2-methyl-1-propene to give *cis*- and *trans*-2-(aminomethyl)-2-methylcyclopropane-1-carboxylic acid (*1958*). The enantiomers of both diastereomers have been resolved via esterification with (R)-pantolactone (*1938*). N-Benzylmaleimide was cyclopropanated with ethyl

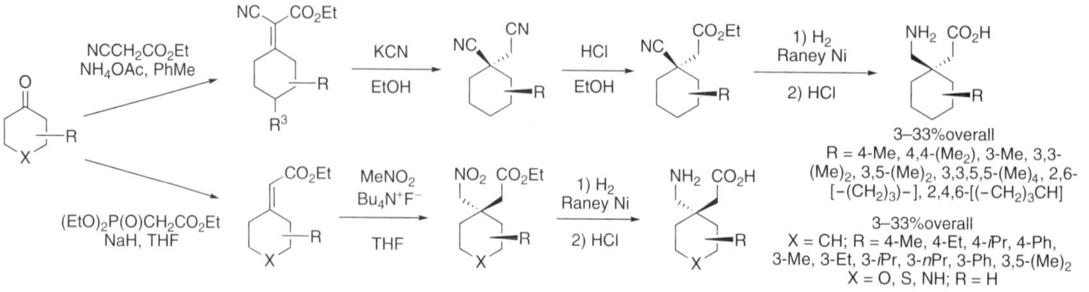

Scheme 11.174 Synthesis of Gabapentin Analogs (*1519, 1535*).

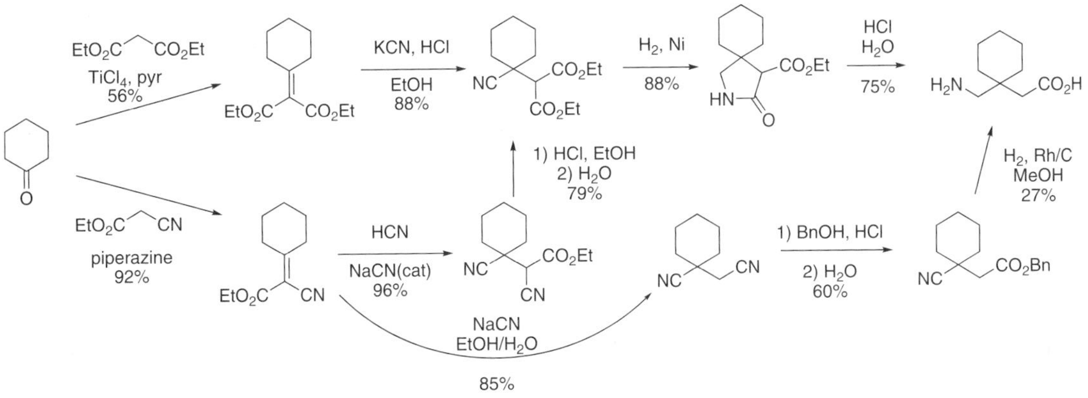

Scheme 11.175 Synthesis of Gabapentin (*1956*).

diazoacetate to give a 3-azabicyclo[3.1.0]hexane ring system containing succinimide. Reduction of the carbonyl and ester groups, followed by reoxidation of the hydroxyethyl group, produced 6-carboxy-3-azabicyclo[3.1.0]hexane (*1959*).

11.5.2e Elaboration of α-Amino Acids

1-Aminocyclopropanecarboxylic acid was homologated to (*E*)-3-(1-aminocyclopropyl)-2-propenoic acid via conversion to an amino aldehyde followed by Wittig reaction (*1865*). Activation of Cbz-Pro with EDC and coupling with Ph₃P=CHCO₂Et produced a stabilized phosphorus ylide (e.g., Cbz-Pro-C(=PPh₃)CO₂Et) in 44% yield. Flash-vacuum pyrolysis at 600 °C led to extrusion of Ph₃PO to produce an acetylenic γ-amino acid, Cbz-Pro-CCO₂Et, in 48–49% yield, on a 100–500 mg scale. The alkyne could be hydrogenated, producing (*S*)-3-(pyrrolidin-2-yl)propionic acid (*1846*). The carboxyl group of racemic *cis*-3-hydroxyproline has been used to acylate Meldrum's acid, leading to a synthesis of 3-hydroxypyrrolidine-2-(3-hydroxy-3-propionic acid) (*rac*-detoxinin), its C3-epimer, and a Z-2,3-unsaturated analog (*1960*). An intramolecular Dieckmann cyclization of *N*-(3-carbomethoxypropyl),*N*-Boc Gly-OMe provided 3-keto-4-carboxypiperidine in 47% yield (*1446*).

11.5.2f Elaboration of γ-Amino Acids

Several acyclic γ-amino acids have been cyclized. A cyclic analog of carnitine, 2,6-bis(carboxymethyl)-4,4-dimethylmorpholine, was prepared by condensing sodium norcarnitine with methyl (*E*)-4-bromo-2-butenoate (*1547*). Racemic 5,5-disubstituted-morpholine-2-acetic acid derivatives were synthesized by amination of ethyl 4-bromocrotonate with substituted aminoethanols. An intramolecular Michael addition/cyclization of the hydroxy group induced by treatment with DBU generated the morpholine derivative (*1961*). A γ-amino, γ-imino acid, *trans*-3-aminopyrrolidine-2-acetic

acid, was obtained via intramolecular Michael addition of the terminal amino group of (*E*)-4-phthalimido-6-amino-hex-2-enoate. An α-allylated derivative was also prepared (*1962*). Intramolecular thermal ene cyclization of *N*-CH₂CH=CMe₂,*N*-Tfa 4-aminobut-2-enoate provided a mixture of *cis*- and *trans*-4-isopropylidene-pyrrolidine-3-acetic acid (*1963*). Ru-catalyzed metathesis of *N*,*S*-diallyl 3-thio-4-aminobutanoic acid required the thiol to be in the sulfoxide or sulfone oxidation state, but proceeded in good yield to give eight-membered 3,4,5,8-tetrahydro-2*H*-1,4-thiazocine-2-acetic acid ring systems (*1964*).

Several derivatizations of cyclic γ-amino acids have been reported. *cis*-3-Aminocyclohexane-1-carboxylic acid was produced by hydrogenation of the sodium salt of *m*-aminobenzoic acid in the presence of alkaline Raney nickel (*1946*). Pyridine-4-carboxylic acids or esters were hydrogenated with a RuO₂ catalyst to give piperidine-4-carboxylic acid derivatives (*1406*). Lower pressures (3 atm instead of 70 atm) could be used with Rh on carbon as catalyst (*1407*). Racemic *cis*-4-hydroxy-3-carboxypiperidine was prepared by reduction of 4-oxo-3-carboxypiperidine; attempts to duplicate an asymmetric reduction were unsuccessful. The product was used as a carbohydrate mimic (*1350*). The alkene of *cis*-1-aminocyclopent-2-ene-4-carboxylic acid lactam was dihydroxylated, with lactam hydrolysis, providing 1-*cis*-amino-2-*trans*,3-*trans*-dihydroxycyclopentane-4-carboxylic acid. The diastereomers epimeric at the amino or hydroxyl centers were also prepared (*1965*). The amide bond of 2-azabicyclo[2.2.1]hept-5-en-3-one was hydrolyzed to produce *cis*-4-aminocyclopent-2-ene-1-carboxylic acid (*1966*).

11.5.2g Other Syntheses

Racemic *cis*-2,3-methano-GABA derivatives were prepared from *N*-allyl amides of α-bromo acids. Treatment of the bromo acids with a Cu catalyst induced a radical cyclization to give 4-(bromomethyl)-2-pyrrolidones. The cyclopropyl ring system was then formed by dehydrobromination using DBU at 100–120 °C, with hydrolysis of the lactam giving the γ-amino acid (*1967*). Another radical cyclization employed α-aminoalkyl radicals generated from *N*-(α-benzotriazolyl)homoallylamines by treatment with SmI$_2$, producing 2,4-disubstituted-pyrrolidine-2-acetic acid derivatives (*1968*). Similar cyclization conditions with different substrates

led to 3-aminopyrrolidine-4-acetic acid and 3-aminopyrrolidine-2-acetic acid products (*1429*).

Reaction of *N*-allyl, *N*-propargyl amides with a molybdenum carbene complex generated 3,4-methano-pyrrolidine-3-acetic acid (*1969*). Ozonolysis of 2,3-methano-5,6-dehydropiperidine gave a 62% yield of 1-amino-2-(carboxymethyl)cyclopropane methyl ester (*677*).

Table 11.7 Syntheses of Racemic or Achiral Cyclic γ-Amino Acids

Ring Size (n)	Substituents	Method	Yield	Reference
Amination				
3	*cis*-1-CH$_2$NH$_2$-2-CO$_2$H	amination via KNPhth displacement	30–66% displacement	1985 (*1832*)
3	*trans*-1-CH$_2$NH$_2$-2-CO$_2$H			
4	*cis*-1-CH$_2$NH$_2$-2-CO$_2$H			
4	*trans*-1-CH$_2$NH$_2$-2-CO$_2$H			
6	*cis*-1-CH$_2$NH$_2$-2-CO$_2$H			
6	*trans*-1-CH$_2$NH$_2$-2-CO$_2$H			
5	1-amino-2-(=CHCO$_2$H)	bromination of α,β-unsaturated acid, amination with ammonia	19–22% bromination and amination	1979 (*1673*)
6	1-amino-2-(=CHCO$_2$H)			
5	1-NH$_2$-2,3-dehydro-3-CO$_2$H	bromination of cyclopentene-1-carbonitrile, amination, hydrolysis	18% bromination, amination and hydrolysis	1980 (*1939*)
6	1-NH$_2$-5,6-dehydro-3-CO$_2$H	reaction with phthalimide, deprotection with hydrazine	48% deprotection	2002 (*1522*)
4	1-NH$_2$-3c-CO$_2$H-2c-Ph-4t-Ph	amination of 1-OH-3c-CO$_2$H-2c-Ph-4t-Ph-cyclobutane	80%	1927 (*1370*)
6	1- NH$_2$-3-CO$_2$H-3,4,5,6-tetradehydro	amination of methyl 1,4-cyclohexadienylcarboxylate	—	1976 (*1912*)
6	1- NH$_2$-3-CO$_2$H-3,4,5,6-tetradehydro, with [2-^{3}H] or [4,5-^{2}H$_2$]	amination of labeled methyl 1,4-cyclohexadienyl-carboxylate	—	1977 (*1915*)
5	1-aza-1-Me-2-CO$_2$H-4-NH$_2$-2,3,4,5-tetradehydro	nitration of pyrrole-2-carboxylate, hydrogenation	23% overall	1999 (*1918*)
4	1- NH$_2$-3-CO$_2$H-1,3-(-CH$_2$-)	Curtius rearrangement of diacid monomethyl ester	48–83%	2004 (*1948*)
4	1- NH$_2$-3- CO$_2$H	reductive amination of cyclobutanone-3-carboxylate	84% oxime formation 37% hydrogenation	1957 (*1950*)
5	*cis*-1-NH$_2$-2-CH$_2$CO$_2$H	oxime reduction	—	1982 (*1951*)
5	*trans*-1-NH$_2$-2-CH$_2$CO$_2$H			
6	*cis*-1-NH$_2$-2-CH$_2$CO$_2$H			
6	*trans*-1-NH$_2$-2-CH$_2$CO$_2$H			
3	1- CH$_2$CH$_2$NH$_2$-1-CO$_2$H	amination via Curtius rearrangement	52% Curtius	1990 (*1546*)
3	1- NH$_2$-1-CH$_2$CH$_2$CO$_2$H	amination via Curtius rearrangement, acid from oxidative cleavage of alkene	69% Curtius 58% oxidative cleavage	1990 (*1546*)
5	1- CH$_2$NH$_2$-1-CH$_2$CO$_2$H	Curtius rearrangement of resin-linked dicarboxylic acid	78%	1998 (*1366*)
Carboxylation				
6	1-aza-3-CH$_2$CO$_2$H	bromination of 3-(CH$_2$OH) piperidine, displacement with CN, hydrolysis	33–38% bromination and displacement 30% hydrolysis/ esterification	1928 (*1952*)

Table 11.7 Syntheses of Racemic or Achiral Cyclic γ-Amino Acids (continued)

Ring Size (n)	Substituents	Method	Yield	Reference
5	1-amino-3-CO$_2$H-4,5-dehydro	Pd-catalyzed replacement of O-CO$_2$Et 3-aminocyclopent-4-enol with MeNO$_2$, oxidation with NaNO$_2$/AcOH/DMF	70% Pd-catalyzed replacement 58% nitro oxidation	2003 (*1953*)
5	1-aza-3-CH$_2$CO$_2$H	anionic cyclization of BnN(CH$_2$SnBu$_3$) CH$_2$CH$_2$CH=CH$_2$, quenching with ClCO$_2$Et	68% cyclization and quench	1996 (*1954*)

Amination and Carboxylation

Ring Size (n)	Substituents	Method	Yield	Reference
3	*cis*-1-NH$_2$-2-CH$_2$CO$_2$H	reduction of one acid group of cycloalkane-1,2-dicarboxy-late, activation and displacement with KCN, Curtius rearrangement of second carboxyl	29–78% hydrazide formation 22–78% rearrange-ment	1982 (*1951*)
3	*trans*-1-NH$_2$-2-CH$_2$CO$_2$H			
4	*cis*-1-NH$_2$-2-CH$_2$CO$_2$H			
4	*trans*-1-NH$_2$-2-CH$_2$CO$_2$H			
5 pyrrole or imidazole nucleus	1-Me-2-CO$_2$H-4-amino	carboxylation and nitration of N-Me pyrrole or imidazole, reduction	18–25% carboxylation and nitration 78–82% reduction	1996 (*1955*)
5	1-*cis*-NH$_2$-2,3-dehydro-4-CO$_2$H 1-NH$_2$-3,4-dehydro-4-CO$_2$H 1-*trans*-NH$_2$-2,3-dehydro-4-CO$_2$H 1-*cis*-NH$_2$-3-CO$_2$H 1-*trans*-NH$_2$-3-CO$_2$H	cycloaddition of TsCN and cyclopentadiene, hydrolysis, basic isomerization, isomerization with phthalic anhydride, hydrogenation	84% cycloaddition and hydrolysis 91% alkene isomerization 37% isomerization 78–98% hydrogena-tion	1980 (*1939*) 1986 (*1940*)

Fusion of Amino- and Carboxy-Containing Components

Ring Size (n)	Substituents	Method	Yield	Reference
6	1-CH$_2$NH$_2$-1-CH$_2$CO$_2$H-4-Me 1-CH$_2$NH$_2$-1-CH$_2$CO$_2$H-3,5-(Me)$_2$	Wittig reaction or aldol reaction of substituted cyclohexanone, Michael addition of nitromethane or KCN, reduction	10–15% overall	1997 (*1535*)
6	1-CH$_2$NH$_2$-1-CH$_2$CO$_2$H-4-oxa 1-CH$_2$NH$_2$-1-CH$_2$CO$_2$H-4-thia 1-CH$_2$NH$_2$-1-CH$_2$CO$_2$H-4-aza 1-CH$_2$NH$_2$-1-CH$_2$CO$_2$H-3-Me 1-CH$_2$NH$_2$-1-CH$_2$CO$_2$H-3-Et 1-CH$_2$NH$_2$-1-CH$_2$CO$_2$H-3-*i*Pr 1-CH$_2$NH$_2$-1-CH$_2$CO$_2$H-3-*n*Pr 1-CH$_2$NH$_2$-1-CH$_2$CO$_2$H-3-Ph 1-CH$_2$NH$_2$-1-CH$_2$CO$_2$H-3,3-Me$_2$ 1-CH$_2$NH$_2$-1-CH$_2$CO$_2$H-3,5-Me$_2$ 1-CH$_2$NH$_2$-1-CH$_2$CO$_2$H-3,3,5,5-Me$_4$ 1-CH$_2$NH$_2$-1-CH$_2$CO$_2$H-4-Me 1-CH$_2$NH$_2$-1-CH$_2$CO$_2$H-4-Et 1-CH$_2$NH$_2$-1-CH$_2$CO$_2$H-4-*i*Pr 1-CH$_2$NH$_2$-1-CH$_2$CO$_2$H-4-Ph 1-CH$_2$NH$_2$-1-CH$_2$CO$_2$H-4,4-Me$_2$ 1-CH$_2$NH$_2$-1-CH$_2$CO$_2$H-2-MeO 1-CH$_2$NH$_2$-1-CH$_2$CO$_2$H-2-Me 1-CH$_2$NH$_2$-1-CH$_2$CO$_2$H-2-*c*Hex 1-CH$_2$NH$_2$-1-CH$_2$CO$_2$H-2,6-[−(CH$_2$)$_3$−] 1-CH$_2$NH$_2$-1-CH$_2$CO$_2$H-2,4,6- [(−CH$_2$)$_3$CH]	Wittig reaction or aldol reaction of substituted cyclohexanone, Michael addition of nitromethane or KCN, reduction	3–33% overall	1998 (*1519*)

Table 11.7 Syntheses of Racemic or Achiral Cyclic γ-Amino Acids (continued)

Ring Size (n)	Substituents	Method	Yield	Reference
6	1-CH$_2$NH$_2$-1-CH$_2$CO$_2$H 1-CH$_2$NH$_2$-1-CH$_2$CO$_2$H-4-tBu	addition of nitromethane to acrylates attached to dendritic polyglycerol support, nitro reduction with cyclization/ cleavage from support	48–55% addition 58–70% reduction/ cyclization	2004 (*1692*)
6	1-CH$_2$NH$_2$-1-CH$_2$CO$_2$H	addition of malonate equivalent to cyclohexanone, addition of CN, hydrolysis and CN reduction	56–92% malonate addition 88–96% CN addition 60–75% hydrolysis 27–88% reduction	1991 (*1956*)
4	1-NH$_2$-2c-OH-2t-CH$_2$CO$_2$H	Reformatsky addition of t-butyl bromoacetate to 2-(NBn$_2$)- cyclobutanone, N-deprotection, resolution after coupling with Boc-L-Val	70% addition 80% de 68% N-deprotection 90% acylation	1983 (*1909*)
4	1-aza-3-CH$_2$CO$_2$H	displacement of 3-Br-azetidine with malonate, decarboxylation	51% displacement 20% decarboxylation	1996 (*1957*)
3	1-NH$_2$-1-CCCO$_2$Me	addition of LiCCCO$_2$Me to cyclopropiminium ion	35%	1991 (*570*)
3	1-CO$_2$H-2-CH$_2$NH$_2$ 1-CO$_2$H-2-Me-2-CH$_2$NH$_2$	Rh(II)-catalyzed cyclopropena- tion of N,N-(TMS)$_2$- allylamine with methyl diazoacetate	67–88% cyclopro- panation 20–32% de *trans*	1991 (*1958*)
6	1-aza-4-CO$_2$H-3,5-(–) (3-azabicyclo[3.1.0]hexane system)	cyclopropanation of N-Bn maleimide with diazoacetate, carbonyl reduction, reoxidation	35% cyclopropanation 96% reduction 86% oxidation	1996 (*1959*)

From α-Amino Acids

3	1-NH$_2$-1-CH=CHCO$_2$H	conversion of 1-aminocyclopro- pane-1-carboxylic acid to aldehyde, Wittig reaction	45% overall	1986 (*1865*)
5	1-aza-2-CH(OH)CH$_2$CO$_2$H 1-aza-2-CH=CHCO$_2$H	acylation of Meldrum's acid with lactone of 3-HO-Pro, alkene reduction, lactone hydrolysis	44% Meldrum's acylation 86–90% reduction 63–85% hydrolysis	1983 (*1960*)
6	1-aza-3-keto-4-carboxy	Dieckmann cyclization of N-(3-carbomethoxypropyl), N-Boc-Gly-OMe	47%	1998 (*1446*)
5	1-aza-2-CCCO$_2$H 1-aza-2-CH$_2$CH$_2$CO$_2$H	EDC coupling of Cbz-amino acid with Ph$_3$P=CHCO$_2$Et, flash vacuum pyrolysis, akyne hydrogenation	44–49% coupling 48–49% pyrolysis 78% hydrogenation	2002 (*1846*)

From γ-Amino Acids

6	1-aza-1,1-Me$_2$-3,5-(CH$_2$CO$_2$H)$_2$-4-oxa	condensation of Na norcarnitine with BrCH$_2$CH=CHCO$_2$Me	83% condensation	1987 (*1547*)
6	1-aza-3-(CH$_2$CO$_2$H)-4-oxa-6,6-R^{1}R^2 R^1 = H; R^2 = H, Me, Et R^1 = Me, Et; R^2 = CH$_2$OH R^1 = R^2 = –(CH$_2$)$_2$–, –(CH$_2$)$_4$–	N-alkylation of substituted aminoethanol with ethyl 4-bromocrotonate, intramo- lecular Michael addition	5–67% cyclization, hydrolysis	1996 (*1961*)
5	1-aza-2-CH$_2$CO$_2$H-3-NH$_2$ 1-aza-2-CH(CH$_2$CH=CH$_2$)CO$_2$H-3-NH$_2$	intramolecular Michael addition, deprotonation and allylation	69–87% cyclization 64–87% allylation	1998 (*1962*)
5	1-aza-3-CH$_2$CO$_2$H-4-C(Me)=CH$_2$	ene cyclization of N-CH$_2$CH=C(Me)$_2$ 4-aminobutenoate	81% cyclization	1980 (*1963*)

Table 11.7 Syntheses of Racemic or Achiral Cyclic γ-Amino Acids (continued)

Ring Size (n)	Substituents	Method	Yield	Reference
8	1-aza-3-CH$_2$CO$_2$H-4-SO-6,7-dehydro 1-aza-3-CH$_2$CO$_2$H-4-SO$_2$-6,7-dehydro	Ru-catalyzed metathesis of *N,S*-diallyl 3-thio-4-amino-butanoic acid	95–100%	2004 (*1964*)
6	1-NH$_2$-3-CO$_2$H	hydrogenation of *m*-aminoben-zoic acid with H$_2$/Raney Ni at 90–100 atm and 150 °C	80% hydrogenation	2005 (*1946*)
6	1-aza-4-CO$_2$H	hydrogenation of pyridine-4-carboxylic acid with H$_2$/RuO$_2$ at 70–100 atm	86–91% hydrogena-tion	1961 (*1406*)
6	1-aza-4-CO$_2$H	hydrogenation of pyridine-4-carboxylic acid with H$_2$/Ru-C at 3 atm	87% hydrogenation	1962 (*1407*)
6	1-aza-3-CO$_2$H-4-OH	reduction of 3-carboxy-4-oxopiperidine	47%	1997 (*1350*)
5	1-NH$_2$-2,3-dehydro-4*c*-CO$_2$H 1-NH$_2$-2,3-dehydro-4*t*-CO$_2$H 1-NH$_2$-2*t*-OH-3*t*-OH-4*c*-CO$_2$H 1-NH$_2$-2*t*-OH-3*t*-OH-4*t*-CO$_2$H 1-NH$_2$-2*c*-OH-3*c*-OH-4*t*-CO$_2$H	dihydroxylation, epimerization and hydrolysis of lactam of *cis* 1-aminocyclopent-2-ene-4-carboxylic acid	85–91% dihydroxyla-tion	1981 (*1965*)
5	1-NH$_2$-2,3-dehydro-4-CO$_2$H	hydrolysis of lactam of 2-azabicyclo[2.2.1]hept-5-en-3-one	90%	1976 (*1966*)

Other Syntheses

3	1-CO$_2$H-1-Me-2-CH$_2$NH$_2$ 1-CO$_2$H-1-*i*Pr-2-CH$_2$NH$_2$ 1-CO$_2$H-1-Ph-2-CH$_2$NH$_2$ 1-CO$_2$H-1-Me-2-CH$_2$NH$_2$-2-Me 1-CO$_2$H-1-*i*Pr-2-CH$_2$NH$_2$-2-Me	radical cyclization of H$_2$C=C(R^2) CH$_2$NPCOCHBrR^1 dehydrobromination lactam hydrolysis	91–96% cyclization 84–88% hydrolysis	1999 (*1967*)
5	1-aza-3-CH$_2$CO$_2$H-5-*n*Pr 1-aza-3-CH$_2$CO$_2$H-5-Ph 1-aza-3-CH$_2$CO$_2$H-5-(3-pyridyl) 1-aza-3-CH$_2$CO$_2$H-4-Me 1-aza-2-*n*Pr-3-CH$_2$CO$_2$H-4-Me	SmI$_2$ radical cyclization of R^{1}CH(Bt)N(Bn)CH(R^2) CH(R^3)CH=CHCO$_2$Et	66–79% cyclization 10–80% ds main isomer	2002 (*1968*)
5	1-amino-2-CH$_2$CO$_2$H-4-aza 1-amino-2-CH$_2$CO$_2$H-3-aza	SmI$_2$ radical cyclization of HCOCHCH$_2$N(Ts) CH$_2$CH=CH-CO$_2$Et or HCOCHCH$_2$CH$_2$N(Ts)-CH=CHCO$_2$Et in the presence of amine and benzotriazole	47–84% 10–20% de	2002 (*1429*)
5	1-aza-3-CH$_2$CO$_2$H-3,4-(−CH$_2$−)	reaction of *N*-allyl,*N*-propargylamide with Mo–carbene complex	91%	1996 (*1969*)
3	1-amino-2-CH$_2$CO$_2$H	ozonolysis of 2,3-methano-5,6-dehydropiperidine	62%	1994 (*677*)

11.5.3 Asymmetric Syntheses of Cyclic γ-Amino Acids (see Table 11.8)

11.5.3a Resolutions of Racemic Cyclic γ-Amino Acids

Enzymatic hydrolysis of the lactam of racemic *cis*-1-aminocyclopent-2-ene-4-carboxylic acid by various lactamases gave either enantiomer of the cyclic γ-amino acid (*1970*). Alternatively, PLE or the lipase from *Candida cylindracea* was used for ester hydrolysis of the racemic *N*-acetyl methyl ester (*1971*). *N*-Phthaloyl derivatives of *trans*-4-aminocyclopent-2-ene-1-carboxylic acid, 4-aminocyclopent-1-ene-1-carboxylic acid, and *trans*-3-aminocyclopentane-1-carboxylic acid were resolved via crystallization of their D-pantolactone or 2,3-isopropylidine-D-ribonolactone esters (*1940*), while both *cis*- and *trans*-2-(aminomethyl)-2-methylcyclopro-pane-1-carboxylic acid were resolved via esterification with (*R*)-pantolactone (*1938*). The diastereomers formed by coupling (1*S*,2*S*)-1-hydroxy-2-aminocyclobutane-1-acetic acid with Boc-L-Val were chromatographically separated (*1909*). *N*-Boc *cis*-3-aminocyclohexane-1-carboxylic acids was resolved to >95% ee by crystallization with (+)-1-phenylethylamine (*1946*).

11.5.3b Amination Reactions

A number of nucleophilic aminations of chiral electrophilic substrates have been described. Azide opening of homochiral 3,4-*cis*-epoxy-cyclopentane-1-carboxylate, attached to a solid support resin, provided *cis*-3-amino-*trans*-4-hydroxycyclopentane-1-carboxylic acid after azide reduction. The carbinol epimer could be produced by Mitsunobu inversion (*1972*). Benzylamine opening of an epoxide derivative of (–)-quinic acid gave (1*R*,3*R*, 4*S*,5*R*)-5-amino-1,3,4-trihydroxycyclohexane-1-carboxylic acid (aminoquinic acid) (*1973*). A conformationally restricted analog of statine, (1*S*,2*S*,3*S*)-2-hydroxy-3-aminocyclohexane-1-carboxylic acid, was prepared by azide displacement of a homochiral triflate precursor, itself obtained via an asymmetric Diels–Alder reaction (*1974*). Replacement of the 3-hydroxy group of (–)-shikimic acid with bromide, followed by displacement with azide and azide reduction, produced (–)-(3*R*)-amino-(4*R*,5*R*)-dihydroxy-1-cyclohexene-1-carboxylic acid (*1975, 1976*). 3-Substituted-2-phenyl-2,3-methano-γ-amino acids, analogs of the NMDA antagonist milnacipran, were derived from optically active *cis* 1-phenyl-2-hydroxymethylcyclopropane-1-carboxylic acid, via oxidation of the hydroxymethyl group to an aldehyde, addition of Grignard reagents to introduce the γ-substituent, and azide displacement of the resulting alcohol to introduce the γ-amino group (*1931*). Dimethyl *cis*-cyclopropane-, -cyclobutane-, and -cyclopentane-1,2-dicarboxylates were desymmetrized by enzymatic ester hydrolysis, with the free acid group reduced with BH$_3$·SMe$_2$ in THF. The resulting alcohol was brominated and displaced with azide, or converted to the lactone and then opened with potassium phthalimide to give the 2-aminomethyl-cycloalkane-1-carboxylates (*1977*).

Reductive amination of homochiral 2-methoxy-3-methoxycarbonyl-cyclohexanone produced a cyclohexane-based analog of statine/isostatine. When sodium triacetoxyborohydride and benzylamine were used for the reduction, the yield was excellent (99%) but no diastereoselectivity was observed (*1582*). Hydrogenation of an oxime derivative proceeded with 50% de, but the yield decreased to 55% (*1978*). Reductive amination of (1*S*,2*S*)-2-methoxy-3-ketocyclohexane gave a mixture of (1*S*,2*S*,3*R*)- and (1*S*,2*S*,3*S*)-isomers of 2-methoxy-3-aminocyclohexane-1-carboxylic acid (*1594*). A more efficient synthesis was reported in 2001, using azide displacement of a triflate derivative of methyl 2-benzyloxy-3-hydroxy-cyclohexane-1-carboxylate to give the (*S*,*S*,*S*)-isomer (*1595*). 1-Phenyl-2-[(*S*)-1-aminoethyl]-*N*,*N*-diethylcyclopropanecarboxamide, PEDC, was prepared from a chiral cyclopropyl aldehyde via reductive amination using *N*-benzyl hydroxylamine to form the

nitrone, followed by addition of PhMgBr and hydrogenation of the *N*-benzyl group (*1930*).

Curtius rearrangement of derivatives of *d*-camphoric acid led to isoaminodihydrocampholytic acid in 1914 (*1979*) and isoaminocamphonanic acid in 1917 (*1980*), with similar reactions reported in earlier years on related substrates (e.g., *1981*). Either enantiomer of *cis*-3-amino-cyclopentane-1-carboxylic acid was prepared by Curtius rearrangement of the 1,3-diacid monoesters. The homochiral substrates were obtained by an enzymatic desymmetrization step (*1911, 1982*).

11.5.3c Amination Combined with Carboxylation

An asymmetric azide displacement of one of the hydroxy groups of *meso*-2-cyclopentene-1,4-diol was achieved by using a chiral Pd catalyst, giving protected *cis*-3-azido-2-cyclopentenol with 80–90% ee. A Pd-catalyzed allylic alkylation introduced the carboxyl group following azide reduction, using phenylsulfonylnitromethane as a methoxycarbonyl surrogate. Homochiral methyl *cis*-3-aminocyclopent-4-ene-1-carboxylic acid was obtained in four steps and 31% overall yield (*1983*).

Conjugate addition of (*R*)-(+)-1-phenylethylamine to itaconic acid (2-methylenesuccinic acid), followed by lactam formation, produced two diastereomers of *N*-(1-phenylethyl) pyrrolidin-2-one-4-carboxylate. The lactam carbonyl and methyl ester were reduced with LiAlH$_4$ to give 3-(hydroxymethyl)pyrrolidine. Chlorination of the hydroxymethyl group, displacement with NaCN, hydrolysis, and hydrogenation gave (*R*)- or (*S*)-pyrrolidine-3-acetic acid, homo-β-proline (*1941*).

11.5.3d Fusion of Amino-Containing and Carboxy-Containing Components

A Rh(II)-catalyzed asymmetric cyclopropenation reaction of *N*,*N*-diprotected propargylamines with ethyl diazoacetate yielded 1-aminomethyl-cyclopropene-3-carboxylic acids in up to 85% yield with >97% ee. Hydrogenation provided the corresponding *cis*-2-amino-ethyl-cyclopropane-1-carboxylic acids (*1984*). Both enantiomers of *cis*-2-aminomethylcyclobutane-1-carboxylic acid were prepared via cyclization of a malonate derivatized with a chiral amide auxiliary. Monomethyl malonate *N*-(1-phenylethyl),*N*-(ethyl 2-butenoate) amide underwent intramolecular Michael addition to form, after decarboxylation, a 3-carboxypyrrolidin-2-one-4-acetic acid skeleton. The carboxymethyl substituent was converted into an iodoethane moiety, and then cyclized to form the cyclobutane ring, with *N*-auxiliary removal and lactam hydrolysis completing the synthesis (*1985*).

11.5.3e Elaboration of α-Amino Acids

Most syntheses of cyclic γ-amino acids from α-amino acids rely upon derivatization of proline. Bis-homologation of Cbz-L-Pro via consecutive Arndt–Eistert reactions produced (S)-pyrrolidine-2-propionic acid (*1100*). The same product was obtained by displacement of the N,O-ditosylate of prolinol with diethyl potassium malonate, followed by decarboxylation (*1818*). Wittig reaction of homochiral *trans*-3-benzylaminoprolinal followed by hydrogenation led to 3-amino-pyrrolidine-2-propionic acid (*1986*). Vinylogous analogs of Pro were obtained by homologation of Boc-prolinal with $Ph_3P=CHCO_2Me$ or $Ph_3P=C(Me)CO_2Et$ (*1799*). An interesting variation of the olefination homologation of chiral amino aldehydes coupled racemic N-protected 2-formyl-piperidine with homochiral Horner–Wadsworth–Emmons reagents to induce stereoselectivity at the 4-position in the product γ-amino acid. The condensation was carried out under dynamic kinetic resolution conditions, allowing for up to 94% de in the piperidine-2-propenoic acid products (*1802*). Alternatively, D- or L-pipecolic acid were converted into homochiral 2-formylpiperidines and coupled with $Ph_3P=CHCO_2Et$. Asymmetric dihydroxylation of the alkene provided piperidine-2-(2,3-dihydroxy-3-propionic acid) (*1987*).

Addition of lithiated ethyl acetate to Boc-L-prolinal gave predominantly the (3R,4S)-statine analog, pyrrolidine-2-[3-(1-hydroxypropionic acid)], with 60% de (*1988*). Aldol reaction of (S)-prolinal with the enolate of acetate contained in a chiral iron complex produced the same product, with either (S,R)- or (S,S)-stereochemistry depending on the metal counterion used for the enolate (*1989*). 4-Hydroxyprolinal was reacted with an allylborane reagent. The alkene was converted to an acid, giving 2-(1-hydroxypropionic)-3-hydroxypyrrolidine. This was incorporated into analogs of hapalosin, a cyclic depsipeptide that reverses tumor multidrug resistance (*1932*). Dolaproine, the γ-imino acid component of dolastatin-10, was prepared from two chiral components. Boc-Pro was reduced to prolinal, and a chiral enolate derived from the (2S)-1,1,2-triphenylethanol ester of propionic acid was added to give predominantly one diastereomer in 47% isolated yield. This was the wrong stereoisomer at the C2 center, but epimerization (after O-methylation of the alcohol) gave dolaproine (*1632*). A precursor of (2R,3R,4S)-dolaproine was prepared via addition of a chiral (Z)-crotylorganoboron compound to Boc-prolinal. Excellent diastereoselectivity was observed for the matched reagent/substrate pair. Oxidative cleavage of the alkene group and O-methylation was still required to complete the synthesis of dolaproine (*1990*).

(−)-Detoxinine has been prepared by aldol condensation of protected racemic *cis*-3-hydroxyprolinal with a chiral enolate derived from S-mandelic acid (*1902*). It was also synthesized from 4-*cis*-iodoproline, via reduction to prolinal, Wittig reaction, and epoxidation of the alkene. The 3-hydroxy group on the pyrrolidine ring was obtained by elimination of the 4-iodo group and intramolecular bromolactonization (*1905*). Another route to detoxinine began with allylic oxidation of allylglycine to β-hydroxyallylglycine, followed by reduction to allylglycinal, and *threo*-selective aldol condensation with lithium *tert*-butyl acetate. The vinyl group was converted to an hydroxyethyl group, activated, and cyclized

with the amino group to form the pyrrolidine ring (*1900*). In another synthesis, Boc-D-Ser(TBS)-OH was treated with isopropenyl chloroformate and Meldrum's acid, heated in ethyl acetate, and then reduced with sodium borohydride to give 3-*cis*-hydroxy-L-pyroglutaminol. Further functionalization gave 3-*cis*-hydroxyprolinal, with an asymmetric aldol condensation with the enolate of 2-acetoxy-1,1,2-triphenylethanol leading to detoxinine (*1903*).

Asp has been converted into an N-protected aziridine, 5,6-iminohexanoate. Deprotonation with LDA induced an intramolecular aziridine opening, giving protected 3-aminocyclopentanecarboxylate in 65% yield as a 2:1 mixture of diastereomers (*1991*). Another synthesis from Asp formed a 2-amino-1,4-butanediol intermediate, which was mesylated and rearranged to give a mesylate of 3-hydroxypyrrolidine. Mesylate displacement with diethyl malonate followed by hydrolysis/decarboxylation generated homo-β-proline, pyrrolidine-3-acetic acid (*1112*). The complex pyrrolidinone γ-amino acid found in microsclerodermin E was synthesized from D-Asp in 2003. The α-carboxy group was homologated to a β-keto ester, and then treated with ammonia to induce cyclization/elimination to form the desired 4-amino-5-carboxymethylene-pyrrolidin-2-one system (*215*). An intramolecular Dieckmann cyclization of (2S)-2-aminoadipic acid was achieved via activation of the α-carboxyl group with CDI followed by treatment with KHMDS, providing 1-aminocyclopentan-2-one-3-carboxylic acid. Ketone reduction by $NaBH_4$, hydroxyl mesylation, and elimination gave 1-aminocyclopent-2-ene-3-carboxylic acid. Hydrogenation of the alkene gave a 1:1 to a 1:10 mixture of *cis*:*trans* 1-aminocyclopentane-3-carboxylic acid (*64*).

Constrained analogs of the bleomycin γ-amino acid component, joining the C2–C4 backbone carbons with a butyl or 2-butene linker, were synthesized from allylglycinal, via addition of pent-4-enoic acid derivatized with Evans' oxazolidinone (see Scheme 11.146). The two allyl substituents of the resulting 2,4-di(allyl)-3-hydroxy-4-aminobutanoic acid derivative were joined by a Rh-catalyzed metathesis, with the resulting alkene hydrogenated if desired (*258, 1609*). Homochiral azetidin-3-ones have been prepared from amino acids via Rh-catalyzed cyclization of diazoketone derivatives. The lithium enolate of *tert*-butyl actetate added to the carbonyl group to give β-hydroxy-γ-imino acids with the nitrogen contained in an azetidine ring (*1992*). Pyrrolidine γ-amino esters were produced by addition of radicals generated by decarboxylative reduction of N-protected proline or 4-hydroxyproline derivatives to methyl acrylate or [2-(methoxycarbonyl)propenyl]tributylstannane (*1711*).

11.5.3f Elaboration of γ-Amino Acids

cis-2,3-Methano-GABA was prepared by an intramolecular displacement of a homochiral mesylate derivative (*1993*). A radical cyclization of optically active N-acrylyl 4-amino-5-phenyl-pent-2,3-enoate gave a lactam intermediate that was then converted to 2-benzyl-3-carboxymethyl-4,5-methanopyrrolidine in 42% yield (*1994*). Stereoselective reduction of the ketone group of racemic ethyl N-benzyl-3-oxopiperidine-4-carboxylate with Baker's yeast gave ethyl *cis*-3-hydroxypiperidine-4-carboxylate in 65% yield with 95% ee and 73% de (*1445*), while reduction of methyl N-Boc 3-oxopiperidine-4-carboxylate provided exclusively *cis*-substituted (3R,4R)-3-hydroxypiperidine-4-carboxylate in 89% yield with 78% ee, and reduction of ethyl N-methoxycarbonyl 3-oxopiperidine-4-carboxylate led to (3R,4R) *cis*-3-hydroxypiperidine-4-carboxylate in 81% yield (*1446*).

11.5.3g Elaboration of Chiral Substrates

D-Mannitol (*1906*) and D-glucose (*1901*) were both elaborated to provide (+)-detoxinine via conversion to a 3-*cis*-hydroxyprolinal intermediate, to which was added the enolate of *tert*-butyl acetate (*1901*, *1906*). Two diastereomers of detoxinine were prepared from D-glucose in 1980 (*1904*). A number of potent carbocyclic inhibitors of influenza neuraminidase are based on the structure of N-acetylneuraminic acid but contain an additional amino group in the γ-position. They have been synthesized from chiral substrates by several methods (*1934, 1935, 1995*). Side-chain analogs in which an ether linkage is replaced by a sulfur or carbon atom were derived from (−)-quinic acid (*1935*), while other analogs were prepared from shikimic acid (*1975, 1976, 1996*). 3,4,5-Triaminocyclohex-1-ene-1-carboxylic acid (*1997*) and 5-amino-1,3,4-trihydroxycyclohexane-1-carboxylic acid (aminoquinic acid) (*1973*) have also been derived from (−)-quinic acid. Some of the most potent N-acetylneuraminic analogs are 3-alkoxy-4-acetylamino-5-amino-cyclohex-1-ene-1-carboxylic acids, also derived from shikimic acid (*1936*). Sugar γ-amino acids were obtained from carbohydrate lactones by addition of lithium *tert*-butyl acetate to the lactone, followed by introduction of an aminomethyl group to the anomeric center using TMSCN/TMSOTf (*2186*).

Another synthesis of detoxinine stemmed from (S)-malic acid, which was converted into a 2-(1,3-dihydroxypropyl)-3-hydroxypyrrolidine derivative before oxidation of the terminal hydroxy group (*1907*).

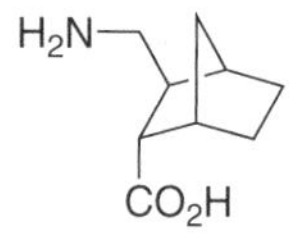

Scheme 11.177 Norbornane-Based γ-Amino Acid (*1999*).

Both enantiomers of a cyclobutane-linked γ-amino acid, 2,2-dimethyl-3-aminocyclobutane-1-carboxylate, were constructed from (+)-α-pinene by oxidative cleavage of the alkene bond followed by a either a Curtius rearrangement of the carboxyl group or oxidation of the ketone (see Scheme 11.176) (*1998*).

Norbornene dicarboxylic acid was converted into a norbornane-based γ-amino acid, 2-carboxy-3-aminomethyl-bicyclo[2.2.1]heptane (see Scheme 11.177), via selective conversion of one carboxy group to a cyano group, followed by selective cyano reduction (*1999*).

Several chiral aziridine-2-propenoic acid derivatives (both E and Z) were prepared by a variety of methods, starting with aziridine-2-carboxylate, aziridine-2-hydroxymethyl, or aziridine-2-vinyl derivatives (*2000*).

11.5.3h Other Syntheses

(R)- and (S)- pyrrolidine-3-acetic acids were obtained by Mn(III)-mediated oxidative cyclization of N-(2-butenyl), N-[(S)-1-phenylethyl] acetoacetamide, providing N-(1-phenylethyl) 3-methoxycarbonyl-4-vinylpyrrolidine-2-one as the initial adduct. Decarboxylation of the 3-methoxycarbonyl substituent, hydroboration and oxidation of the vinyl group, lactam carbonyl reduction, and hydrogenation of the N-(1-phenylethyl) group gave the desired product (*2001, 2002*).

Both (+)- and (−)-*cis*-3-aminocyclopentane-1-carboxylic acid have been prepared from racemic 2-azabicyclo[2.2.1]hept-5-en-3-one via enantiospecific hydrolysis of the lactam followed by alkene hydrogenation; one lactam alkene isomer was inverted into the other reduced enantiomer via treatment with bromine and rearrangement of a bromonium ion intermediate, followed by reduction of the dibromoproduct (*2003*).

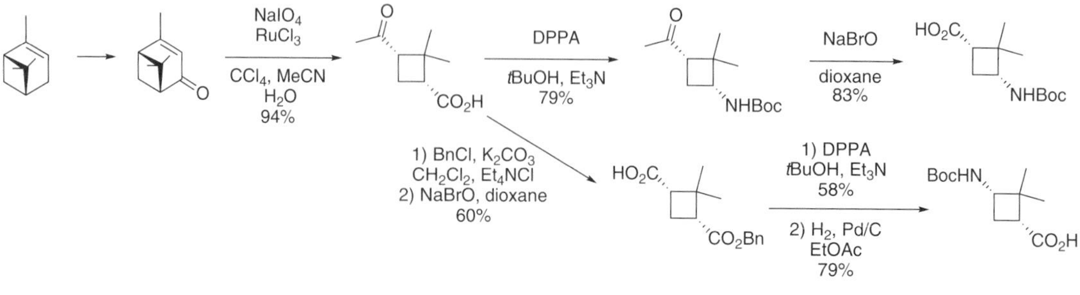

Scheme 11.176 Elaboration of (+)-α-Pinene (*1998*).

Table 11.8 Asymmetric Syntheses of Cyclic γ-Amino Acids

Ring Size (n)	Substituents	Method	Yield, % ee	Reference
Amination				
5	1-NH$_2$-3c-CO$_2$H-5t-OH 1-NH$_2$-3c-CO$_2$H-5c-OH	azide opening of 3,4-epoxy-cyclo-pentane-1-carboxylate attached to resin, inversion of carbinol by Mitsunobu	95% opening 92% ee 90% inversion	1998 (*1972*)
6	1-NH$_2$-3-CO$_2$H-3,5,6-(OH)$_3$	amination of (−)-quinic acid derived epoxide	40% amination	1998 (*1973*)
6	1-NH$_2$-2-OH-3-CO$_2$H	azide displacement of homochiral triflate	62%	1997 (*1974*)
6	1- NH$_2$-2,3-dehydro-3-CO$_2$H-5,6-(OH)$_2$	replacement of 3-hydroxy group of (−)-shikimic acid	62% bromination 61% azide displace-ment	1996 (*1975*) 1996 (*1976*)
3	1c-CO$_2$H-1-Ph-2-C(CCH)RNH$_2$ 1c-CO$_2$H-1-Ph-2-CH(Et)NH$_2$ 1c-CO$_2$H-1-Ph-2-CH(CN)NH$_2$ 1c-CO$_2$H-1-Ph-2-CH(CO$_2$H)NH$_2$ 1c-CO$_2$H-1-Ph-2-CH(CH$_2$OH)NH$_2$ 1c-CO$_2$H-1-Ph-2-CH(CHO)NH$_2$ 1c-CO$_2$H-1-Ph-2-CH(CH=CH$_2$)NH$_2$	addition of Grignard reagent to 1-Ph-2-CHO-cyclopropane-1-carboxylic amide, azide displacement	—	1998 (*1931*)
3 4 5	1-CO$_2$H-2c-CH$_2$NH$_2$ 1-CO$_2$H-2c-CH$_2$NH$_2$ 1-CO$_2$H-2c-CH$_2$NH$_2$	desymmetrization of dimethyl cycloalkane-1,2-dicarboxylate by enzymatic monoester hydrolysis, acid reduction with BH$_3$·SMe$_2$, bromination and azide displacement or lactone formation and phthalimide opening	57–88% acid reduction and lactone formation 85% bromination 84% azide displace-ment 95% azide reduction 24–70% lactone opening wiith KNPhth	2002 (*1977*)
6	1- NH$_2$-2-OMe-3-CO$_2$H	reductive amination of chiral 2-methoxy-3-methoxycarbonyl-cyclohexanone	55–99% 0–50% de	1994 (*1582*) 1994 (*1978*)
6	1-NH$_2$-2-OMe-3-CO$_2$H	reductive amination of homochiral ketone	60–98%	1996 (*1594*)
6	1-NH$_2$-2-OH-3-CO$_2$H	azide displacement of homochiral triflate	56% azide displace-ment 81% azide reduction	2001 (*1595*)
3	1-CH(Me)NH$_2$-2-Ph-2-CO$_2$H	reductive amination of chiral aldehyde with N-benzyl hydroxylamine, Grignard addition and hydrogenation	81% Grignard addition 96% de 98% hydrogenation	2000 (*1930*)
5	1-NH$_2$-1-Me-2,2-Me$_2$-3-CO$_2$H	Curtius rearrangement of derivative of *d*-camphoric acid	—	1914 (*1979*)
5	1-NH$_2$-2,2-Me$_2$-3-Me-3-CO$_2$H	Curtius rearrangement of derivative of *d*-camphoric acid	70% rearrangement	1917 (*1980*)
5	1-CO$_2$H-3c-NH$_2$	Curtius rearrangement of homochiral monoester	59–75% rearrange-ment	1992 (*1982*)
5	1-CO$_2$H-3c-NH$_2$	Curtius rearrangement of homochiral monoester	74% rearrangement	1994 (*1911*)
Amination and Carboxylation				
5	1-NH$_2$-3- CO$_2$H-4,5-dehydro	asymmetric azide displacement of hydroxyl, carboxyl addition	31% overall	1995 (*1983*)
5	1-aza-3-CH$_2$CO$_2$H	reaction of itaconic acid with α-Me-BnNH$_2$, lactam/ester reduction, chlorination, displacement with NaCN, hydrolysis and hydrogenation	80–85% carboxyl/lactam reduction 92–98% chlorination 60–67% CN displacement 42–62% hydrolysis	1990 (*1941*)

(*Continued*)

Table 11.8 Asymmetric Syntheses of Cyclic γ-Amino Acids (continued)

Ring Size (n)	Substituents	Method	Yield, % ee	Reference
Fusion of Amino- and Carboxy-Containing Components				
3	1-CO$_2$H-1,2-dehydro-3-CH$_2$NH$_2$ 1-CO$_2$H-2-CH$_2$NH$_2$	Rh(II)-catalyzed asymmetric cyclopropenation of propargylamines with ethyl diazoacetate, hydrogenation	79–85% cyclopropenation >95% ee 68% hydrogenation	1998 (*1984*)
4	1-CO$_2$H-2c-CH$_2$NH$_2$	intramolecular Michael addition, reduction carboxymethyl to iodoethyl, cyclization	80% Michael addition 86–89% reduction 88–91% iodination 83–85% cyclization	1999 (*1985*)
Elaboration of α-Amino Acids				
5	1-aza-2-CH$_2$CH$_2$CO$_2$H	homologation of Pro via consecutive Arndt–Eistert reactions	73% second homologation	1977 (*1100*)
5	1-aza-2-CH$_2$CH$_2$CO$_2$H	reduction of Pro to alcohol, ditosylation, displacement with KCH(CO$_2$Et)$_2$, decarboxylation	75% tosylation 43% malonate addition 27% decarboxylation	1977 (*1818*)
5	1-aza-2-CH(OH)CH$_2$CO$_2$H	addition of lithiated ethyl acetate to Boc-prolinal	70% addition 60% de	1986 (*1988*)
5	1-aza-2-CH(OH)CH$_2$CO$_2$H	homologation of Pro via addition of chiral acetate enolate to prolinal	79-81% addition 94% de	1992 (*1989*)
5	1-aza-2-CH(OH)CH$_2$CO$_2$H-3-OH	addition of allyl boronate to 4-hydroxyPro, ozonolysis/ oxidation	37% addition 60% alkene cleavage	1997 (*1932*)
5	1-aza-2-CH(OMe)CH(Me)CO$_2$H	reduction Boc-Pro, aldol reaction with chiral enolate, epimerization, *O*-methylation	75% reduction 47% aldol 57% epimerization	1989 (*1632*)
5	1-aza-2-CH(OMe)CH(Me)CO$_2$H	addition of chiral crotylorganoborane to Boc-prolinal	50% yield 98% de *anti*	1994 (*1990*)
5	1-aza-2-(CH=CHCO$_2$H)-3-NH$_2$ 1-aza-2-CH$_2$CH$_2$CO$_2$H-3-NH$_2$	Wittig reaction of 3-benzylamino-prolinal, hydrogenation	79% Wittig 97% hydrogenation	1998 (*1986*)
5	1-aza-2-CH=CHCO$_2$H 1-aza-2-CH=C(Me)CO$_2$H	Wittig or Horner–Emmons coupling of Boc-prolinal	—	2003 (*1799*)
5	1-aza-2-CH(OH)CH$_2$CO$_2$H-3-OH	aldol reaction of enolate derived from *S*-mandelic acid and *cis*-3-OH-prolinal	40% aldol	1986 (*1902*)
5	1-aza-2-CH(OH)CH$_2$CO$_2$H-3-HO	conversion of L-4-I-proline to prolinal, Wittig, epoxidation, HI elimination, bromolactonization, dehalogenation	79% aldehyde and Wittig 94% epoxidation 92% elimination 74% epoxide reduction 95% bromolactone 97% dehalogenation	1990 (*1905*)
5	1-aza-2-CH(OH)CH$_2$CO$_2$H	hydroxylation of vinylglycine, reduction to aldehyde, addition of LiCH$_2$CO$_2$tBu, alkene hydroxylation, pyrrolidine formation	96% yield aldol 74% de *threo*	1984 (*1900*)
5	1-aza-2-CH(OH)CH$_2$CO$_2$H-3-HO	conversion of D-Ser to 3-OH-prolinal, aldol addition of enolate of chiral ester of acetate	25% formation of prolinal 48% aldol reaction 50% de	1988 (*1903*)
6	1-aza-2-CH=CHCO$_2$H 1-aza-2-CH=C(Me)CO$_2$H	dynamic kinetic resolution of Horner–Wadsworth-Emmons reaction of homochiral phosphonate with racemic amino aldehyde	6–94% de 99% de *E* to 74% de *Z*	1995 (*1802*)

Table 11.8 Asymmetric Syntheses of Cyclic γ-Amino Acids (continued)

Ring Size (n)	Substituents	Method	Yield, % ee	Reference
6	1-aza-2-CH=CHCO$_2$H 1-aza-2-CH(OH)CH(OH)CO$_2$H	reduction of pipecolic acid carboxyl to aldehyde, Wittig homologation, asymmetric dihydroxylation	—	1994 (*1987*)
7	1-NH$_2$-2-OH-3-CO$_2$H-5,6-dehydro 1-NH$_2$-2-OH-3-CO$_2$H	reduction of Boc-allylglycine to Boc-allylglycinal, addition of enolate of 4-pentenoic acid with Evans' oxazolidinone auxiliary, Ru-metathesis, hydrogenation, deprotection	71% reduction 67% aldol reaction, 0% de 79–83% metathesis 73–89% hydrogenation 34–38% deprotection	2001 (*1609*) 2003 (*258*)
5	1-NH$_2$-3-CO$_2$H	conversion of Asp to 5,6-imino-hexanoate, intramolecular aziridine opening	22% aziridine 65% cyclization 33% de	1993 (*1991*)
5	1-aza-3-CH$_2$CO$_2$H	reduction of Asp carboxyl, mesylation, rearrangement, displacement with malonate, hydrolysis/decarboxylation	99% for mesylation/rearrangement 58% displacement 49% deprotection/decarboxylation	1998 (*1112*)
5	1-NH$_2$-2-=O-3-CO$_2$H 1-NH$_2$-2-OH-3-CO$_2$H 1-NH$_2$-2,3-dehydro-3-CO$_2$H 1-NH$_2$-3c-CO$_2$H 1-NH$_2$-3t-CO$_2$H	Dieckmann cyclization of 2-amino-adipic acid, ketone reduction, hydroxy elimination, alkene reduction	93% cyclization 85% ketone reduction 93% elimination 72–87% reduction 1:1 to 1:10 *cis:trans*	1993 (*64*)
4	1-NH$_2$-2-=CHCO$_2$H-3-aza-4-=O	conversion of Asp α-carboxyl group to α-keto ester, treatment with ammonia	80% keto ester 59% amination/cyclization/elimination	2003 (*215*)
4	1-aza-2-Me-3-OH-3-CH$_2$CO$_2$tBu 1-aza-2-Bn-3-OH-3-CH$_2$CO$_2$tBu	cyclization of diazoketone derivative of amino acids, addition of *t*-butyl acetate enolate	59–63% azetidine formation 50–64% addition	1995 (*1992*)
5	1-aza-2-CH$_2$C(=CH$_2$)CO$_2$H 1-aza-2-CH$_2$C(=CH$_2$)CO$_2$H-4-OH	radical decarboxylation of Pro or 4-HO-Pro, addition to acrylates	60–79% selenoester 62–69% decarboxylation/ addition	1997 (*1711*)

Elaboration of γ-Amino Acids

3	1-CO$_2$H-2-CH$_2$NH$_2$	cyclization of homochiral mesylate	90% cyclization	1997 (*1993*)
5	1-aza-2-Bn-3-CH$_2$CO$_2$Me-4,5-(−CH$_2$−)	radical cyclization of *N*-acrylyl-4-aminopentenoate	42%	1996 (*1994*)
6	1-aza-3-OH-4-CO$_2$H	reduction of 3-ketopiperidine-2-carboxylate with Baker's yeast	65% yield 95% ee 73% de	1993 (*1445*)
6	1-aza-3-OH-4-CO$_2$H	Bakers' yeast reduction of 3-oxopiperidine-4-carboxylate	81–89% 78% ee	1998 (*1446*)

Elaboration of Chiral Substrates

5	1-aza-2-CH(OH)CH$_2$CO$_2$H-3-OH	elaboration of D-mannitol	21% overall	1996 (*1906*)
5	1-aza-2-CH(OH)CH$_2$CO$_2$H-3-OH	elaboration of D-glucose	—	1993 (*1901*)
5	1-aza-2-CH(OH)CH$_2$CO$_2$H-3-OH	elaboration of D-glucose	—	1980 (*1904*)
6	1,2-diamino-4-carboxy-4,5-dehydro-6-OR R = *n*Pr, CH$_2$OMe, (CH$_2$)$_2$CF$_3$, CH(Et)$_2$, CH$_2$CH=CH$_2$, *c*Pent, *c*Hex, Ph	derivatization of chiral substrates	—	1997 (*1934*)
6 6	1,2-diamino-4-CO$_2$H-4,5-dehydro-6-S*n*Pr 1,2-diamino-4-CO$_2$H-4,5-dehydro-6-S*n*Bu	derivatization of (−)-quinic acid	—	1997 (*1935*)

(Continued)

Table 11.8 Asymmetric Syntheses of Cyclic γ-Amino Acids (continued)

Ring Size (n)	Substituents	Method	Yield, % ee	Reference
6	1,2-diamino-4-CO$_2$H-4,5-dehydro-5-R-6-OCH(Et)$_2$ R = Cl, SMe, Me	derivatization of shikimic acid	—	1997 (*1996*)
6	1-CO$_2$H-1,2-dehydro-3,4,5-(NH$_2$)$_3$	derivatization of (–)-quinic acid	—	1998 (*1997*)
4	1-NH$_2$-2,2-Me$_2$-3-CO$_2$H	derivatization of (+)-α-pinene	32–62%	1997 (*1998*)
6	1-NH$_2$-3-CO$_2$H-3,4-dehydro-5-RO-6-AcNH	derivatization of shikimic acid	—	1998 (*1936*)
6	1-oxa-2-CH$_2$NH$_2$-2-CH$_2$CO$_2$H-3,4,5-(OH)$_3$-6-CH$_2$OH	addition of Li acetate to glycosyl lactone, addition of TMSCN/ TMSOTf, nitrile reduction	85–95% acetate addition 50–55% CN addition 80–90% reduction	2001 (*2186*)
5	1-aza-2-[CH(OH)CH$_2$CO$_2$H]-3-OH	conversion of (*S*)-malic acid into detoxinine	—	2003 (*1907*)
6	1-CH$_2$NH$_2$-2-CO$_2$H-3,6-(–CH$_2$–)	conversion of carboxyl group of norbornene dicarboxylic acid to aminomethyl	44% cyano formation 59% cyano reduction	1999 (*1999*)
3	1-aza-2-CH=CHCO$_2$H-3-Me 1-aza-2-CH=CMeCO$_2$H-3-Me 1-aza-2-CH=CMeCO$_2$H-3-Ph 1-aza-2-CH=CMeCO$_2$H-3-Bn	derivatization of chiral aziridines	—	1997 (*2000*)

Other syntheses

5	1-aza-3-CH$_2$CO$_2$H	oxidative cyclization of MeO$_2$CCH$_2$CON (CH$_2$CH=CHMe)CH(Me)Ph, decarboxylation, alkene hydroboration, oxidation, lactam reduction, deprotection	53% cyclization 40% de 79% decarboxylation 69% hydroboration 73% oxidation 55% lactam reduction 54% hydrogenation	1996 (*2001*) 1995 (*2002*)
5	1-CO$_2$H-3*c*-NH$_2$	enzymatic resolution of 2-azabicyclo[2.2.1]hept-5-en-3-one, hydrogenation; inversion via bromination and reduction	94% hydrogenation 79% bromination 87% reduction	1991 (*2003*)

11.6 δ-Amino Acids

11.6.1 Introduction

δ-Amino acids, or 5-aminopentanoic acids, correspond to the length of a dipeptide (see Scheme 11.178). Many δ-amino acids have been synthesized as non-hydrolyz-able dipeptide surrogates, with various functional groups acting as isosteres for the missing peptide bond. Conformationally restricted cyclic and bicyclic peptido-mimetics also often have the amino and carboxy groups in a 1,4-relationship, such as indolizidinone amino acid (2-oxo-3-amino-1-azabicyclo[4.3.0]nonane-9-carboxy-late) (*2004*) and an oxa-substituted analog (*2005*). An excellent review of these constrained bicyclic δ-amino acids was published in 1997 (*2006*). A number of carbo-hydrate-derived sugar amino acids have also been designed as peptidomimetics (e.g., *404, 2007*). The fol-lowing section will restrict itself somewhat arbitrarily to those δ-amino acids which have not been specifically designed as peptide isosteres, and which therefore

generally lack functional groups attempting to mimic the central amide bond of a dipeptide.

The difficulty of distinguishing δ-amino acid pepti-domimetics from other δ-amino acids is illustrated by the use of even the most simple δ-amino acid, δ-aminovaleric acid (5-aminopentanoic acid), as a dipeptide mimetic. For example, it replaced Gly within the turn segment of a designed β-hairpin peptide, main-taining the registry of the antiparallel β-strand segments and connecting them by a three-residue loop instead of the two-residue hairpin turn (*2008*). It also replaced the central Gly-Gly sequence of a helical octapeptide, Boc-Leu-Aib-Val-Gly-Gly-Leu-Aib-Val-OMe. CD spectros-copy of the modified peptide in trifluoroethanol showed no evidence for helix formation, although NMR confor-mational analysis in DMSO indicated a folded helical structure (*2009*). Tripeptides containing a δ-aminovaleric acid residue have been found to self-assemble to form supramolecular β-sheets, which aggregate to generate fibrils resembling the amyloid fibrils believed to cause neurodegenerative disease (*2010*). The achiral δ-amino

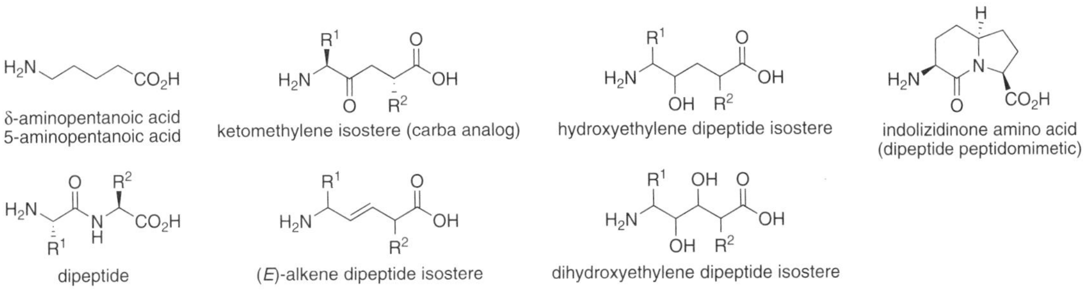

Scheme 11.178 δ-Amino Acids as Dipeptide Mimetics.

acid *trans*-5-amino-3,4-dimethylpent-3-enoate promotes β-hairpin formation when incorporated into peptides (*2011, 2012*). A systematic conformational analysis of possible helix types formed by oligomers of δ-amino acids, using ab initio MO theory, predicted a variety of novel helical structures with hydrogen-bonded ring systems of various sizes (*2013*).

5-Aminopentanoic acid was recently identified in a novel cytotoxic cyclodepsipeptide isolated from a New Zealand fungi (*1556*). However, the δ-amino acid of greatest natural importance is δ-aminolevulinic acid (ALA, 4-oxo-5-aminopentanoic acid) (see Scheme 11.179), which is the first common intermediate in the tetrapyrrole biosynthetic pathway leading to the hemes, chlorophylls, and vitamin B$_{12}$ (*2014*). δ-Aminolevulinic acid is biosynthesized by condensation of succinate and glycine to give α-amino-β-ketoadipic acid, followed by decarboxylation (*2015*). Vitamin B$_{12}$ biosynthesis was studied by employing [1,1,4-^{18}O$_3$, 4-^{13}C,]-5-aminolevulinic acid (*2016*), while [1,4-^{13}C$_2$]-, [1-^{13}C]-, [5-^{13}C]-, [1,4-^{13}C$_2$,1,1,4-^{18}O$_3$], or [1,4-^{13}C$_2$,1,1,4-^{18}O$_3$]-5-aminolevulinic acid were used to investigate the biosynthesis of porphyrins and related macrocycles (*2017, 2018*), including bacteriochlorophyll in *Rhodopseudomonas sphaeroides* (*2019*). The conversion of [1-^{13}C]-, [2-^{13}C]-, [3-^{13}C]-, [4-^{13}C]-, and [5-^{13}C]-5-aminolevulinic acid into porphobilinogen was examined in cell-free enzymatic

systems (*2020, 2021*). [5-^{13}C]-5-Aminolevulinic acid was also employed to demonstrate that a strong red fluorescence component of *Saccharopolyspora erythraea* was uroporphyrin (*2022*). 3-Oxa- and 3-thia-analogs of 5-aminolevulinic acid are potent mechanism-based inhibitors of porphobilinogen synthase, probably inactivating the active-site lysine residue (*2023*). Peptides containing ALA have been prepared as potential novel prodrugs for photodynamic therapy, releasing ALA within cells, where it is converted into the fluorescent photosensitizer protoporphyrin IX. Visible light photodynamic therapy is then able to destroy the cell (*2024, 2025*). A capillary electrophoresis (CE) method for the analysis of 5-aminolevulinic acid, porphobilinogen, levulinic acid, and Gly in culture broth has been reported (*2026*), as has a CE method for determination of ALA and its degradation products (*2027*).

A series of β-,γ-, and δ-amino acids were examined for inhibition of GABA-T (pyridoxal phosphate-dependent 4-aminobutyrate-2-oxoglutarate aminotransferase) in 1981, including 5-amino-6-fluorohexanoic acid and 5-amino-6,6-difluorohexanoic acid. They were not the most potent of those tested either in vitro or in vivo (*419*). The orally active renin inhibitor aliskiren contains a 2,7-diisopropyl-4-hydroxy-5-amino-8-(3-methoxypropyloxy-4-methoxyphenyl)octanoic acid residue (*390*). 5-Aminopentanoic acid has been *N*-alkylated

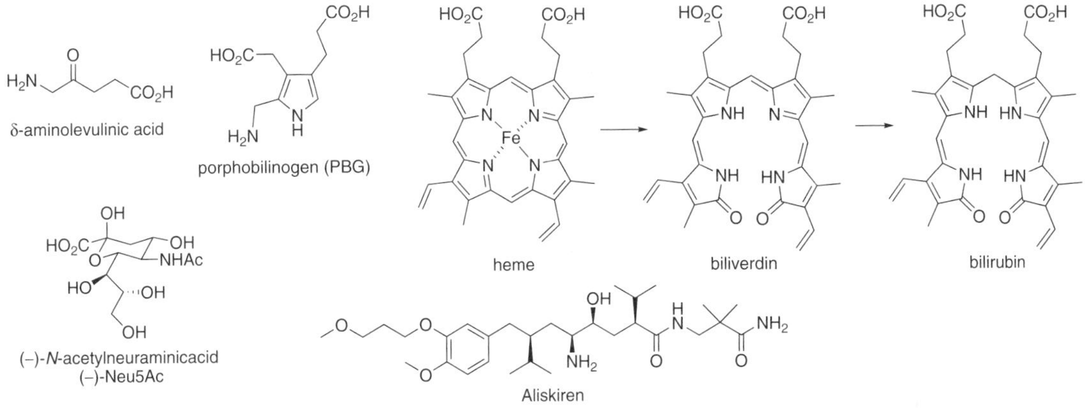

Scheme 11.179 Structures of δ-Amino Acids.

with ethane-linked thymine, giving a peptide nucleic acid (PNA) monomer (*2028*).

A monograph which describes methods of non-α-amino acid synthesis, including δ-amino acids, was published in 1995 by M.B. Smith (*1*). The synthesis of γ-oxygenated-δ-amino acids from α-amino acids was reviewed in 1992 (*1665*).

11.6.2 Syntheses of Achiral or Racemic δ-Amino Acids (see Table 11.9)

11.6.2a Amination Reactions

The most simple δ-amino acid, 5-aminopentanoic acid, was synthesized in 1995 via a Beckmann rearrangement of an in situ-generated oxime of cyclopentanone. Cyclopentanone and hydroxylamine-*O*-sulfonic acid were reacted over silica in a microwave oven, forming valerolactam in 60% yield. Hydrolysis of the lactam gave the ring-opened δ-aminovalerate salt (see Scheme 11.180) (*2029*). This is an updated version of one of the oldest syntheses of 5-aminopentanoic acid, with the oxime of cyclopentanone converted to the lactam in 1900 (*2030*). The lactam was obtained in 60% yield, with hydrolysis to the amino acid in 80% yield (*2031*). 2-Methylpentanone was also rearranged in the 1900 report, producing 2-methyl-5-aminopentanoic acid (*2030*). A 1941 synthesis employed hydroxylamine sulfate to form the cyclopentanone oxime and concentrated sulfuric acid for the rearrangement. 5-Aminopentanoic acid was isolated as the *N*-benzoyl derivative in 66% overall yield (*2032*). Gaudry and Berlinguet also reported a synthesis from cyclopentanone in 1950, with 60% yield for the rearrangement (*2033*).

5-Aminopentanoic acid has also been prepared via a Beckmann or Schmidt rearrangement of cyclodecan-1,6-dione, followed by hydrolysis of the resulting bis (5-aminopentanoic acid) lactam (*517*). A Hofmann rearrangement of the monomethyl ester monoamide of adipic acid produced *N*-carboxymethyl 5-aminopentanoic acid in 85% yield, with hydrolysis to the free amino acid in 95% yield (*512*). Schmidt rearrangement of adipic acid monoester also produced 5-aminopentanoic acid, with 74% yield (*513*).

δ-Aminolevulinic acid (ALA, 4-oxo-5-aminopentanoic acid) was synthesized in 75% yield via regioselective bromination of levulinic acid methyl ester, followed by displacement with azide, azide reduction, and ester hydrolysis (*2034*). Gabriel condensation of

potassium [^{15}N]-phthalimide with methyl 5-chloro-4-oxopentanoate gave [^{15}N]-labeled δ-aminolevulinic acid (*2015, 2035*). Alternatively, potassium [^{15}N]-phthalimide was reacted with tetrahydrofurfuryl bromide. Oxidative cleavage of the tetrahydrofuran moiety revealed the masked carboxy group and introduced the ketone in 95% yield, with hydrolysis of the phthalimido group (6 N HCl) proceeding in 93% yield (*2036*). A 2-fluoro analog of 5-aminolevulinic acid was prepared by an iodolactonization of 2-fluoropent-4-enoic acid, followed by azide displacement of the iodo group (*2023*).

Diethyl β-ketoadipic acid was α-aminated by treatment with ethyl nitrate to give the α-oximino derivative, with reduction by $SnCl_2$ or catalytic hydrogenation providing diethyl 2-amino-3-ketoadipic acid. Refluxing with 6 N HCl effected ester hydrolysis and decarboxylation to δ-aminolevulinic acid (*2015, 2037*). The same procedure was used to prepare [^{15}N]- and [5-^{14}C]-δ-aminolevulinic acid (*2015*). A synthesis of various backbone [^{13}C]-labeled ALA derivatives, including [1,2,3, 4,5-[13]C$_5$]-ALA, [2-^{13}C]-ALA, [3-^{13}C]-ALA, and [^{15}N]-ALA, was achieved by amination of labeled monoethyl 3-oxoadipic acid with sodium nitrite in acetic acid, followed by reductive acetylation, and then hydrolysis/decarboxylation (*2038*). In 1999 δ-aminolevulinic acid was synthesized from dimethyl 3-oxohexanedioate, which was obtained from monomethyl succinate via activation with CDI and homologation with potassium monomethyl malonate. Amination via oxime formation/reduction followed by hydrolysis/decarboxylation gave the product in 44–48% overall yield (*2039*). 4,4-Dimethyl-5-aminopentanoic acid was also prepared by reductive amination of methyl 4,4-dimethyl-5-ketopentanoic acid via oxime formation and hydrogenation (*2040*), while an *N*-substituted 5-amino-hexanoic acid was prepared by a reductive amination of the ketone with norepinephrine (*1681*).

11.6.2b Carboxylation or Amination Combined with Carboxylation

Attempts to homologate Cbz-GABA to 2-keto-5-aminopentanoic acid via coupling with cyanomethylenetriphenylphosphorane in the presence of EDC/DMAP, followed by ozonolysis, gave only a 35% yield of the desired phosphorane intermediate (*1712*). However, by using a derivative with the amino group masked as a nitro group, the homologation proceeded in good yield (*1712*).

Both amino and carboxyl groups were introduced during a synthesis of *trans*-5-amino-3,4-dimethylpent-3-enoate

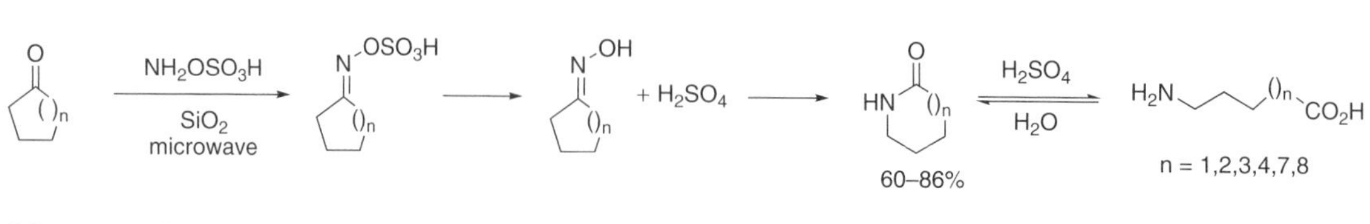

Scheme 11.180 Beckmann Rearrangement of Cycloalkanones (*2029*).

by sequential displacements of *trans*-1,4-dibromo-2,3-dimethylbut-2-ene with nucleophilic amino (azide) and carboxyl (2-lithio-1,3-dithiane) equivalents (*2011, 2012*).

11.6.2c Fusion of Amino-Containing and Carboxy-Containing Components: C2−C3 Formation

Formation of the C2–C3 bond can be achieved by coupling a nucleophilic acetate equivalent with an electrophilic aminopropyl equivalent. One of the first syntheses of δ-amino acids was reported in 1890, when 3-bromopropylphthalimide was condensed with various α-substituted dimethyl sodium malonates. Hydrolysis/decarboxylation gave 2-methyl-, 2-ethyl-, 2-benzyl-, or 2-propyl-5-aminopentanoic acid (*2041, 2042*). Much more recently, an analog of the γ-amino acid baclofen was synthesized by alkylation of diethyl(4′-chlorophenyl) malonate with acrylonitrile, followed by hydrogenation of the nitrile and decarboxylation by hydrolysis (*549*). The dechloro analog was also prepared. 5-Amino-3-oxo-pentanoic acid was prepared via activation of Cbz-β-Ala with CDI and displacement with the magnesium enolate of ethyl hydrogen malonate (*1830*). Reduction of *N*-Boc 3-methyl-3-aminobutanoic acid to an aldehyde and homologation via a Horner–Emmons reaction gave (*E*)-5-amino-5-methylhex-2-enoic acid (*2043*).

The polarity of the coupling reaction has been reversed. Ethyl [1-^{13}C]-bromoacetate, ethyl phthalimidoacetoacetate, and sodium hydride were combined to give ethyl [1-^{13}C]-3-ethoxycarbonyl-5-phthalimidolevulinate. Acidic hydrolysis/decarboxylation and *N*-deprotection produced [1-^{13}C]-5-aminolevulinic acid (*2017, 2021*). Ethyl [2-^{13}C]-bromoacetate led to [2-^{13}C]-5-aminolevulinic acid in the same manner (*2021*). *N*-Phth [1-^{13}C]-Gly or [2-^{13}C]-Gly were converted into acid chlorides and treated with Meldrum's acid to give the labeled ethyl 4-phthalimido-3-ketobutyrate aminopropyl equivalents. Deprotonation with NaH and alkylation with ethyl bromoacetate generated ethyl 3-ethoxycarbonyl-5-phthalimidolevulinates, which were converted into [4-^{13}C]-5-aminolevulinic acid or [5-^{13}C]-5-aminolevulinic acid (*2021*). [3-^{13}C]-5-Aminolevulinic acid was produced by the above route from unlabeled glycine and [2-^{13}C]-Meldrum's acid (*1884*), while unlabeled glycine, Meldrum's acid or [5-^{13}C]-Meldrum's acid, and [1,2-^{13}C$_2$]- or [2-^{13}C]-ethyl bromoacetate led to [1,2-^{13}C$_2$]- or [2,3-^{13}C$_2$]-δ-aminolevulinic acids (*2044*). A 3-oxa-analog of 5-aminolevulinic acid corresponding to a glycine ester of glycolic acid was prepared by *O*-alkylation of Boc-Gly with *tert*-butyl chloroacetate (*2023*).

11.6.2d Fusion of Amino-Containing and Carboxy-Containing Components: C3–C4 Formation

Electrophilic aminoethyl equivalents react with nucleophilic propionic acid equivalents to form the C3–C4 bond. *N*-Phth Gly was used as the source of the aminoethyl

component of δ-aminolevulinic acid, via conversion to an acid chloride and Pd-catalyzed coupling with a zinc homoenolate of ethyl propionate (*2014*). Various [1-^{13}C], [2-^{13}C], and [^{15}N]-labeled Gly substrates were employed to give the corresponding [4-^{13}C]-, [5-^{13}C], and [^{15}N]-labeled δ-aminolevulinic acid products. An earlier synthesis coupled *N*-Phthl Gly acid chloride with the sodio derivative of benzyl ethane-1,1,2-tricarboxylate, followed by hydrogenolytic debenzylation and thermally induced decarboxylation (*2035*). 5-Amino-3-carboxy-pentanoic acid was synthesized by a Wittig reaction of *N*-benzoyl glycinal with a phosphorane derivative of monomethyl succinate, with hydrogenation and hydrolysis giving the amino diacid (*1709*).

A 3-thia-analog of 5-aminolevulinic acid corresponding to a glycine thioester of thioacetic acid was prepared by activation of Boc-Gly with ethyl chloroformate and reaction with mercaptoacetic acid (*2023*). The amino group of β-alanine displaced the SMe group of 2-(diethylphosphono)-*S*-methylthioacetamidinium iodide to give an analog of phosphocreatine with the external amidine nitrogen δ to the carboxyl group (*653*).

A reversal of the normal reaction polarity was achieved via copper-catalyzed coupling of an iodozinc reagent derived from *N*-Boc 2-amino-1-iodoethane with ethyl 2-(bromomethyl)acrylate, producing 2-methylene-5-aminopentanoic acid in 78% yield (*2045*).

11.6.2e Fusion of Amino-Containing and Carboxy-Containing Components: C4–C5 Formation

Combining an aminomethyl equivalent with a butanoic acid equivalent generates the C4–C5 bond. Succinic acid derivatives are commonly employed as electrophilic butanoic acid equivalents. δ-Aminolevulinic acid was prepared by acylation of di-*tert*-butyl acetamidomalonate with ClCOCH$_2$CH$_2$CO$_2$Me (monomethyl succinyl chloride), followed by hydrolysis/decarboxylation of both malonate carboxyl groups (*2046*). In 1999 ethyl succinyl chloride was treated with imidazole and the imidazole displaced by the anion of nitromethane, producing ethyl 5-nitro-4-oxopentanoate in 60% yield. Hydrogenation and hydrolysis gave the δ-aminolevulinic acid product in 94% yield (*2039*). Ring opening of dilabeled succinic anhydride with Bz-Gly-OEt in the presence of LDA provided [1,4-^{13}C$_2$]-5-aminolevulinic acid (*2017*). All three oxygen atoms were then exchanged for ^{18}O using acid catalysis with H$_2$^{18}O at 120 °C to give [1,4-^{13}C$_2$,1,1,4-^{18}O$_3$]-5-aminolevulinic acid (*2017*).

A simple, high-yielding synthesis of δ-aminolevulinic acid reacted monomethyl succinyl chloride with CuCN to give methyl succinyl cyanide (methyl 4-keto-4-cyanobutyric acid) in 70–80% yield. Selective reduction of the acyl cyanide to an *N*-acetyl aminomethylketone proceeded in 83% yield using zinc in acetic acid/acetic anhydride. Ester and acetyl hydrolysis with 2 N HCl produced δ-aminolevulinic acid in >55% overall yield (*2047*). A similar strategy, with Cu^{13}CN, gave [5-^{13}C]-5-aminolevulinic acid (*2021*). By employing 6 N DCl for the hydrolysis step, [3,3,5,5-^{2}H$_4$-5-^{13}C]-5-aminolevulinic acid was obtained (*2021*). Another synthesis of δ-aminolevulinic acid added cyanide to the aldehyde of ethyl 4-oxobutyrate. Acetylation of the cyanohydrin hydroxyl group, hydrogenation of the cyano group, deacetylation, and lactam formation resulted in 5-hydroxy-2-piperidone. The hydroxyl group was oxidized and the lactam hydrolyzed to give the desired product in 60% overall yield from sodium cyanide (*2018*). By employing K^{13}CN, [5-^{13}C]-δ-aminolevulinic acid was obtained (*2018*).

Again, the polarity of bond formation has been reversed. An organozinc reagent prepared from methyl 4-bromobutanoate was added to an imine intermediate generated in situ from benzaldehyde and N-trimethylsilyl morpholine, giving N,N-disubstituted 5-amino-5-phenylpentanoate in 57% yield (*620*). Similarly, the organozinc reagent derived from 4-bromo-2-butenoate was added to iminium ions generated from aldehydes, secondary amines, and benzotriazole to give 5-substituted-5-aminopent-2-enoates in 39–73% yield (*2048*). Another route added vinylic ketene bis(trimethysilyl) acetals to aromatic aldimines in the presence of ZnBr$_2$ to give δ-phenyl-δ–amino acids in good yield (see Scheme 11.181) (*564*).

11.6.2f Elaboration of δ-Amino Acids

N-Benzoyl 5-aminopentanoic acid was α-brominated in 1935 (*2031*) and 1941 (*2032*). Displacement of the bromide with potassium ethyl xanthogenate (KSSCOEt), followed by reduction, gave N-benzoyl 2-thio-5-aminopentanoic acid in 75% yield (*1675*). N-(3-Nitrobenzoyl) 2-bromo-5-aminopentanoic acid, prepared by bromination of N-(3-nitrobenzoyl) 5-aminopentanoic acid, was treated with alkali to give the 2-hydroxy derivative (*2049*). N-Phth-5-aminopentanoic acid has also been both α-brominated and chlorinated (*2033*).

[1-^{13}C,1,1,4-^{18}O$_3$]-5-aminolevulinic acid was obtained by autoclaving [1-^{13}C]-5-aminolevulinic acid in [^{18}O]water with a trace of HCl (*2019*); a similar procedure was employed to prepare [1-^{13}C,1,1,4-^{18}O$_3$]- and [1,4-^{13}C$_2$,1,1,4-^{18}O$_3$]-5-aminolevulinic acid by acid catalysis with H$_2$^{18}O at 120 °C (*2017*).

11.6.2g Other Syntheses

Furfurylamine (2-aminomethylfuran) was converted into δ-aminolevulinic acid in a 1963 *Biochemical*

Preparations procedure, with the N-phthaloyl furfurylamine derivative transformed to a 2,5-dimethoxytetrahydrofuran ring, which was then oxidatively cleaved with chromium trioxide (*2050*). An earlier version of this reaction employed N-benzoyl furfurylamine for the synthesis; 5-amino-4-keto-pent-2-enoic acid was also prepared (*2051*). A 1991 paper reported conversion of N-phthaloyl tetrahydrofurfurylamine into 5-phthalimidyl-levulinic acid in 59% via oxidation with NaIO$_4$/RuCl$_3$, with hydrolysis of the phthalimido group in 64% yield using 6 N HCl (*2052*).

Lactams of 5-aminopentanoic acid and 4-methyl-5-aminopentanoic acid have been produced from propane-1,3-dinitrile, and butane-1,3-dinitrile, respectively. One nitrile group was hydrolyzed using a microbial cell catalyst with aliphatic nitrilase activity, producing γ-cyanocarboxylic acids. The second cyano group was then hydrogenated, with accompanying cyclization, to form the lactam (*1718*).

11.6.3 Asymmetric Syntheses of δ-Amino Acids (see Table 11.10)

11.6.3a Amination Reactions

Nucleophilic aminations of homochiral electrophilic substrates have been employed. Methyl (*R*)-5-bromo-4-methylpentanoate was displaced with potassium phthalimide and hydrolyzed to (*R*)-4-methyl-5-phthalimidopentanoic acid, and then α-brominated to give a mixture of (2*RS*,4*R*)-2-bromo-4-methyl-5-phthalimidopentanoic acid diastereomers. These could be separated as the brucine salt. The amino acid hydrochloride was obtained by refluxing in 6 N HCl (*2053*). 5-Amino-6-hydroxyhexanoic acid was prepared by azide displacement of 5-mesyloxy-6-hydroxyhexanoic acid. The diol substrate was obtained by an asymmetric dihydroxylation reaction of the 4-methoxyphenylmethyl ester of 5-hexenoic acid (*2054*). Azide displacement of a chiral tosylated substrate led to 5-azido-6-methyl-7,8-dihydroxy-8-phenyloct-2-enoic acid, which was used in an analog of cryptophycin 1 (*72*). Evans' oxazolidinone auxiliary has been acylated with 5-bromopentanoyl chloride, and the bromo group displaced with azide. Deprotonation and alkylation with benzyl bromide, followed by hydrolysis and hydrogenation, gave 2-benzyl-5-aminopentanoic acid with 100% ee (*2055*).

The lactone of (4*S*)-4-hydroxy-5-aminopentanoic acid was synthesized by a Mitsunobu azidation of the homochiral diol lactone substrate with hydrazoic acid. The azide was reduced by triphenylphosphine (*2056*).

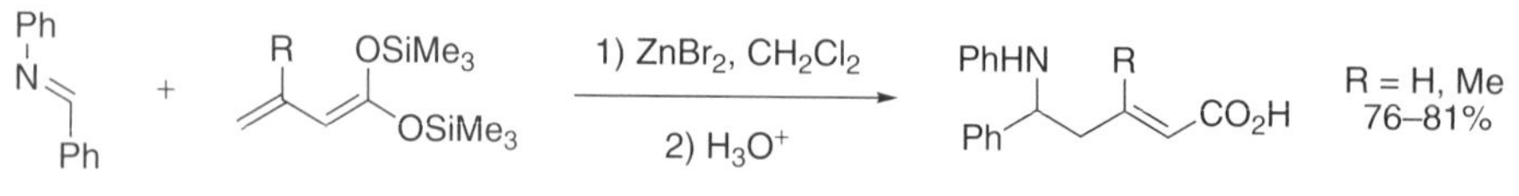

Scheme 11.181 Addition of Vinylic Ketene bis(TMS) Acetals to Aldimines (*564*).

Table 11.9 Syntheses of Racemic δ-Amino Acids

H$_2$N-R-CO$_2$H	Method	Yield	Reference
Amination			
$-$(CH$_2$)$_4-$	Beckmann rearrangement of cyclopentanone, lactam hydolysis	60%	1995 (2029)
$-$(CH$_2$)$_4-$ $-$(CH$_2$)$_3$CH(Me)$-$	Beckmann rearrangement of cyclopentanone or 2-Me-cyclopentanone	—	1900 (2030)
$-$(CH$_2$)$_4-$	Beckmann rearrangement of oxime of cyclopentanone	60% rearrangement 80% hydrolysis	1935 (2031)
$-$(CH$_2$)$_4-$	Beckmann rearrangement of oxime of cyclopentanone	66% rearrangement and hydrolysis	1941 (2032)
$-$(CH$_2$)$_4-$	Beckmann rearrangement of oxime of cyclopentanone	60% rearrangement	1950 (2033)
$-$(CH$_2$)$_4-$	Beckmann or Schmidt rearrangement of cyclodeca-1,6-dione, hydrolysis	61% rearrangement	1962 (517)
$-$(CH$_2$)$_4-$	Schmidt rearrangement of diacid monoester	74%	1959 (513)
$-$(CH$_2$)$_4-$	Hofmann rearrangement of adipic acid monomethyl ester monoamide, hydrolysis	85% rearrangement 95% hydrolysis	1956 (512)
$-$CH$_2$COCH$_2$CH$_2-$	bromination of methyl levulinate, azide displacement, reduction/hydrolysis	64–86% bromination 75% displacement, reduction, and hydrolysis	1994 (2034)
$-$CH$_2$COCH$_2$CH(F)$-$	iodolactonization of 2-F-pent-4-enoic acid, azide displacement	30%	1998 (2023)
$-$CH$_2$COCH$_2$CH$_2-$ with ^{15}N	reaction of K[^{15}N]Phth with methyl 5-chloro-4-oxopentanoate, hydrolysis	80% displacement quant. hydrolysis	1954 (2035) 1955 (2015)
$-$CH$_2$COCH$_2$CH$_2-$ with ^{15}N	reaction of K[^{15}N]Phth with tetrahydrofurfuryl bromide, oxidative cleavage, hydrolysis	89% displacement 95% oxidative cleavage 93% hydrolysis	1997 (2036)
$-$CH$_2$COCH$_2$CH$_2-$	α-amination of diethyl β-ketoadipic acid via α-oximino formation with ethyl nitrite, catalytic hydrogenation, decarboxylation	100% α-oximino 65% reduction 80% hydrolysis/ decarboxylation	1956 (2037)
$-$CH$_2$COCH$_2$CH$_2-$ $-$CH$_2$COCH$_2$CH$_2-$ with ^{15}N $-$^{13}CH$_2$^{13}CO^{13}CH$_2$^{13}CH$_2-$ with ^{13}CO$_2$ $-$CH$_2$COCH$_2$^{13}CH$_2-$ $-$CH$_2$CO^{13}CH$_2$CH$_2-$	amination of labeled 3-oxoadipic acid monoethyl ester with NaNO$_2$, oxime reduction, hydrolysis/decarboxylation	82% oxime 84% reduction 73% decarboxylation	2003 (2038)
$-$CH$_2$COCH$_2$CH$_2-$ $-$^{15}CH$_2$COCH$_2$CH$_2-$ $-$^{15}CH$_2$COCH$_2$CH$_2-$ with ^{15}N	reaction of β-ketoadipic acid with EtONO, reduction with SnCl$_2$	—	1955 (2015)
$-$CH$_2$COCH$_2$CH$_2-$	displacement of imidazole-activated monomethyl succinate with monomethylmalonate, electrophilic amination, decarboxylation	44–48% overall	1999 (2039)
$-$CH(Me)CH$_2$CH$_2$CH$_2-$	reductive amination of 5-keto-hexanoic acid with norepinephrine	50%	1983 (1681)
$-$CH$_2$C(Me)$_2$CH$_2$CH$_2-$	reductive amination of HCOC(Me)$_2$CH$_2$CH$_2$CO$_2$Me	42% oxime 60% reduction	1964 (2040)
Carboxylation or Amination and Carboxylation			
$-$CH$_2$C(Me)=C(Me)CH$_2-$	2-Li-1,3-dithiane then azide displacement of BrCH$_2$C(Me)=C(Me)CH$_2$Br, reduction, oxidation	2% overall	1995 (2011) 1999 (2012)
$-$(CH$_2$)$_3$CO$-$	homologation of GABA analog	51%	1999 (1712)

(Continued)

Table 11.9 Syntheses of Racemic δ-Amino Acids (continued)

H$_2$N-R-CO$_2$H	Method	Yield	Reference
Fusion of Amino- and Carboxy-Components: C2–C3 Formation			
−(CH$_2$)$_3$CH(Me)− −(CH$_2$)$_3$CH(Et)− −(CH$_2$)$_3$CH(nPr)− −(CH$_2$)$_3$CH(Bn)−	alkylation of dimethyl Na α-R-malonate with Br(CH$_2$)$_3$NPhth, hydrolysis	—	1890 (*2041*) 1891 (*2042*)
−(CH$_2$)$_3$CH(Ph)− −(CH$_2$)$_3$CH(4-Cl-Ph)−	reaction of NaC(R)(CO$_2$Et)$_2$ with acrylonitrile, hydrogenation, hydrolysis	82% alkylation 65–70% reduction 78–80% hydrolysis	1997 (*549*)
−CH$_2$CH$_2$COCH$_2$−	activation of Cbz-β-Ala with CDI, reaction with Mg enolate of malonic acid monoethyl ester	91% activation and enolate addition	1985 (*1830*)
−C(Me)$_2$CH$_2$CH=CH−	olefination of aldehyde from BocNHC(Me)$_2$CH$_2$CO$_2$H	50% aldehyde 79% olefination	1998 (*2043*)
−CH$_2$COCH$_2$^{13}CH$_2$− −CH$_2$COCH$_2$CH$_2$− with [1-^{13}C]	reaction of [1-^{13}C]- or [2-^{13}C]-bromoacetate with ethyl phthalimidoacetoacetate, hydrolysis/decarboxylation	74% alkylation 71% hydrolysis	1989 (*2021*)
−CH$_2$CO^{13}CH$_2$CH$_2$− −CH$_2$^{13}COCH$_2$CH$_2$− −^{13}CH$_2$COCH$_2$CH$_2$−	reaction of Gly, [1-^{13}C]- or [2-^{13}C]-Gly acid chloride with Meldrum's acid or [2-^{13}C]-Meldrum's acid, alkylation with bromoacetate, hydrolysis/decarboxylation	65–80% acylation of Meldrum's acid 27–28% alkylation, hydrolysis/decarboxylation	1989 (*2021*)
−CH$_2$COCH$_2$CH$_2$− −CH$_2$^{13}COCH$_2$CH$_2$− with 1,1,4-^{18}O$_3$-1-^{13}C	reaction of PhthNCH$_2$COCH$_2$CO$_2$Et with BrCH$_2$^{13}CO$_2$Me, decarboxylation, oxygen exchange	89% alkylation 69% deprotection/ decarboxylation 51% alkylation with succinic anhydride 98% deprotection >90% O exchange	1993 (*2017*)
−CH$_2$CO^{13}CH$_2$^{13}CH$_2$− −CH$_2$COCH$_2$^{13}CH$_2$− with ^{13}CO$_2$H	coupling of Phth-Gly-Cl with Meldrum's acid (unlabelled or 5-^{13}C), deprotonation and akylation with [1,2-^{13}C$_2$] or [2-^{13}C]ethyl bromoacetate, hydrolysis	83–88% coupling 69–70% alkylation 76% hydrolysis	1997 (*2044*)
−CH$_2$C(O)OCH$_2$−	coupling of Gly with chloroacetic acid	58–62% coupling	1998 (*2023*)
Fusion of Amino- and Carboxy-Components: C3–C4 Formation			
−CH$_2$COCH$_2$CH$_2$− −^{13}CH$_2$COCH$_2$CH$_2$− −CH$_2$^{13}COCH$_2$CH$_2$− with or without ^{15}N	coupling of Phth-Gly-Cl with Zn(CH$_2$CH$_2$CO$_2$Et)$_2$, hydrolysis	100% coupling 95% hydrolysis	1997 (*2014*)
−CH$_2$COCH$_2$CH$_2$−	coupling of Phth-Gly-Cl with NaC(CH$_2$CO$_2$Bn)(CO$_2$Bn)$_2$, hydrogenation, decarboxylation, hydrolysis	34% coupling, hydrogenation and decarboxylation quant. hydrolysis	1954 (*2035*)
−(CH$_2$)$_2$CH(CO$_2$H)CH$_2$−	Wittig reaction of Bz-glycinal and succinate phosphorane, reduction, hydrolysis	45% Wittig 71% reduction 61% hydrolysis	1993 (*1709*)
−CH$_2$CH$_2$CH$_2$C(=CH$_2$)−	Cu-catalyzed coupling of BocNHCH$_2$CH$_2$ZnI with BrCH$_2$C(=CH$_2$)CO$_2$Et	78% coupling	2001 (*2045*)
−CH$_2$COXCH$_2$− X = O, S	coupling of Gly with chloroacetic acid or mercaptoacetic acid	58–62% coupling	1998 (*2023*)
−CH(CH$_2$PO$_3$H$_2$)NHCH$_2$CH$_2$−	reaction of β-Ala with MeSC(=NH)CH$_2$PO$_3$Et$_2$	76%	1997 (*653*)
Fusion of Amino- and Carboxy-Components: C4–C5 Formation			
−CH$_2$COCH$_2$CH$_2$−	acylation of di-tBu acetamidomalonate with ClCOCH$_2$CH$_2$CO$_2$Me, hydrolysis/decarboxylation	10% overall	1958 (*2046*)

Table 11.9 Syntheses of Racemic δ-Amino Acids (continued)

H$_2$N-R-CO$_2$H	Method	Yield	Reference
−CH$_2$COCH$_2$CH$_2$−	displacement of imidazole-activated monoethyl succinate with nitromethane, nitro reduction	56% overall	1999 (*2039*)
−CH$_2$COCH$_2$CH$_2$− −CH$_2$^{13}COCH$_2$CH$_2$− with 1,1,4-^{18}O$_3$-1-^{13}C	reaction of Bz-Gly-OEt with dilabeled succinic anhydride, oxygen exchange	51% alkylation with succinic anhydride 98% deprotection >90% O exchange	1993 (*2017*)
−CH$_2$COCH$_2$CH$_2$−	CN acylation of ClCOCCH$_2$CH$_2$CO$_2$H, selective CN reduction, hydrolysis	>55% overall	1984 (*2047*)
−^{13}CH$_2$COCH$_2$CH$_2$− −^{13}CD$_2$COCD$_2$CH$_2$−	^{13}CN acylation of ClCOCCH$_2$CH$_2$CO$_2$H, selective CN reduction, hydrolysis or deuterolysis	90% CN addition and reduction	1989 (*2021*)
−CH$_2$COCH$_2$CH$_2$− −^{13}CH$_2$COCH$_2$CH$_2$−	CN addition to OHCCH$_2$CH$_2$CO$_2$H, acetylation, hydrogenation, deacetylation, cyclization, oxidation, hydrolysis	57–60% overall	1973 (*2018*)
−CH(Ph)(CH$_2$)$_3$−	one-pot reaction of ZnBr(CH$_2$)$_3$CO$_2$Et with imine of PhCHO and Me$_2$NSiMe$_3$	57%	1997 (*620*)
−CH$_2$CH$_2$CH=CH− −CH(*i*Pr)CH$_2$CH=CH− −CH(*n*Pr)CH$_2$CH=CH− −CH(3-pyridyl)CH$_2$CH=CH−	reaction of ZnBrCH$_2$CH=CHCO$_2$R with iminium ion formed from RCHO, R^{1}NHR^2, and benzotriazole	39–73%	2003 (*2048*)
−CH(Ph)CH$_2$CH=CH− −CH(Ph)CH$_2$C(Me)=CH−	addition of vinylic ketene bis(TMS) acetal to PhCH=NPh	76–81%	1993 (*564*)

Elaboration of δ-Amino Acids

−(CH$_2$)$_3$CH(Br)−	bromination of *N*-Bz-5-aminopentanoic acid with P, Br$_2$	—	1935 (*2031*) 1941 (*2032*)
−(CH$_2$)$_3$CH(SH)−	reaction of *N*-Bz-2-bromo-5-aminopentanoate with KSSCOEt, reduction	75%	1935 (*1675*)
−(CH$_2$)$_3$CH(OH)−	reaction of *N*-3-NO$_2$Bz-2-bromo-5-aminopentanoate with OH	—	1909 (*2049*)
−(CH$_2$)$_3$CH(Cl)− −(CH$_2$)$_3$CH(Br)−	α-halogenation of *N*-Phth-5-aminopentanoic acid with P, X$_2$	97–98%	1950 (*2033*)
−CH$_2$COCH$_2$CH$_2$− with 1,1,4-^{18}O$_3$-1-^{13}C	oxygen exchange with H$_2$^{18}O	—	1985 (*2019*)
−CH$_2$COCH$_2$CH$_2$− −CH$_2$^{13}COCH$_2$CH$_2$− with 1,1,4-^{18}O$_3$-1-^{13}C	oxygen exchange with H$_2$^{18}O	>90% O exchange	1993 (*2017*)

Other syntheses

−CH$_2$COCH$_2$CH$_2$−	conversion of furfurylamine	90% *N*-Phth formation 61–69% oxidative cleavage 91% hydrolysis	1963 (*2050*)
−CH$_2$COCH=CH− −CH$_2$COCH$_2$CH$_2$−	conversion of furfurylamine	30% electrolytic methoxylation 95% reduction 85% oxidative cleavage	1958 (*2051*)
−CH$_2$COCH$_2$CH$_2$−	conversion of tetrahydrofurfurylamine	95% *N*-Phth formation 59% oxidative cleavage 64% hydrolysis	1991 (*2052*)
−CH$_2$CH$_2$CH$_2$CH$_2$− −CH$_2$CH(Me)CH$_2$CH$_2$−	aliphatic nitrilase hydrolysis of one cyano group of propane-1,3-dinitriles, hydrogenation	92–99% hydrolysis 94–96% reduction	1998 (*1718*)

Phthalimide and toluenesulfonamide were employed for allylic reaction with methyl 2-methyl-2-carboxymethyl-3-acetoxy-5-phenyl-4-pentenoate or methyl 2-methyl-2-carboxymethyl-3-acetoxy-4-hexenoate, respectively, producing 2-methyl-2-carboxymethyl-5-aminopent-4-ene products (*2057*).

11.6.3b Carboxylation Reactions

Cyanide was used as a carboxyl equivalent to open homochiral 1,2-epoxy-3-hydroxy-4-aminoalkanes. Potassium cyanide gave a mixture of regioisomers, but diethylaluminum cyanide cleanly gave the desired 1-cyano-4-amino-2,3-diols in 60–62% yield with >98% regioselectivity (*2058*).

11.6.3c Fusion of Amino-Containing and Carboxy-Containing Components: C2–C3 Formation

β-Amino acids can be used as aminopropyl equivalents to couple with acetate equivalents. A number of non-racemic β-substituted β-amino acids (obtained by Arndt–Eistert homologation of α-amino acids) were reduced to the γ-amino alcohols via NaBH₄ reduction of the mixed anhydride. Oxidation by manganese dioxide in acetonitrile produced the corresponding amino aldehydes, which were trapped in situ with Ph₃P=CHCO₂Me to give unsaturated δ-amino acids in 64–91% yield (*2059*). The carboxyl group of the β-amino acid derived from L-Leu has also been reduced to give an aldehyde substrate suitable for coupling with a Horner–Wadsworth–Emmons reagent, triethyl 2-benzyl-2-phosphonoacetate. 2-Benzyl-5-amino-7-methyloct-2-enoic acid was initially produced, with the alkene then hydrogenated (*2060*).

4-Hydroxy-5-amino-5-phenylpent-2-enoic acid, a vinylog of the Taxol side chain, was prepared from an oxazoline derivative of 2-hydroxy-3-amino-3-phenyl-propanoic acid, via reduction of the carboxyl group to a formyl group and homologation via Wittig reaction. The β-amino acid oxazoline substrate was derived from homochiral methyl 3-phenylglycidate (*842*). A general route for the synthesis of γ-substituted α,β-acetylenic γ-amino acids (see Scheme 11.164 above) has also been used to prepare 5-aminopen-2-ynoic acid by starting with β-alanine instead of α-amino acids. EDCI-mediated coupling with Ph₃P=CHCO₂Et gave an ylide that was converted to the acetylenic amino ester by flash vacuum pyrolysis (*1845*).

The polarity of the bond formation has been reversed. Boc-Ala-OMe was converted into a β-ketosulfone by reaction with dilithio phenyl methyl sulfone, alkylated with ethyl bromoacetate, and then reduced and desulfonylated to give 5-aminohexanoic acid (*2061*).

11.6.3d Fusion of Amino-Containing and Carboxy-Containing Components: C3–C4 Formation

A number of syntheses of δ-amino acids rely upon homologation of α-amino acids, forming the C3–C4 bond in the final product. The aldehyde derived from protected Phe was homologated in one step by Wittig reaction with PPh₃P=CH-CC-TMS, leading to 5-amino-6-phenyl-hex-3-enoic acid after alkyne-to-acetate conversion (*2062*). The alkene was epoxidized and opened with fluoride ion to provide 4-hydroxy-5-amino-6-phenyl-hex-2-enoic acid. The hydroxy group could be displaced with azide or fluoride. A new approach to γ-hydroxy-δ-substituted-δ-amino acids alkylated Williams' oxazinone template to generate unusual α-amino acids. Homologation of the amino acid was then carried out by reduction of the lactone carbonyl, followed by addition of allyltrimethylsilane to the template (see Scheme 11.148). The homologation proceeded with variable stereoselectivity. Hydroboration and oxidation of the allyl substituent formed the carboxyl group (*1798*).

The 2,7-diisopropyl-4-hydroxy-5-amino-8-(3-meth-oxypropyloxy)-4-methoxyphenyl)octanoic acid residue found in the orally active renin inhibitor aliskiren was derived from 2-amino-4-isopropyl-5-(3-methoxypropy-loxy-4-methoxyphenyl)pentanoic acid via reduction of the acid group to an aldehyde followed by Grignard addition of the isopropyl-substituted propionate unit (*390*). A samarium-mediated ketyl–alkene coupling reaction between amino acid-derived aldehydes and alkyl acrylates produced lactones of γ-hydroxy-δ-amino acids, with a chiral ester auxiliary helping to improve diastereoselectivity (*2063*).

Amino acid S-aryl thioesters produce an acyl radical species without decarbonylation upon reduction with SmI₂; interception of the radical with acrylates or acrylamides produced δ-amino acid species (see Scheme 11.182) (*2064*). A copper-catalyzed coupling of the Ala-derived iodozinc reagent N-Boc 2-amino-1-iodozincpropane with ethyl 2-(bromomethyl)acrylate produced 2-methyl-ene-5-aminohexanoic acid in 69% yield (*2045*).

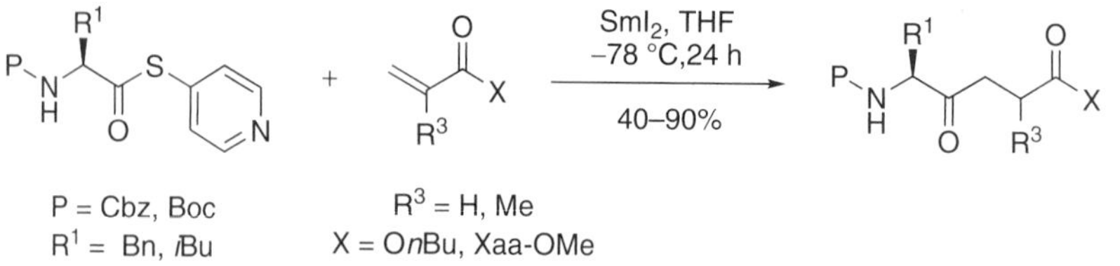

P = Cbz, Boc   R³ = H, Me
R¹ = Bn, *i*Bu   X = O*n*Bu, Xaa-OMe

Scheme 11.182 Radical Reaction of Amino Acid-Derived Acyl Radical with Acrylates (*2064*).

11.6.3e Elaboration of α-Amino Acids

A number of conversions of α-amino acids into δ-amino acids, other than the homologations described above, have been developed. Nucleic δ-amino acids, with a backbone ether linkage and side-chain nucleobase, have been prepared from Ser. The α-carboxyl group was reduced to an hydroxymethyl moiety, which was O-alkylated with tBu bromoacetate or tBu 2-bromopropionate. The side-chain hydroxyl was then displaced by a Mitsunobu reaction with the imino group of thymine or uracil, with the latter further elaborated to give a cytosine derivative (2065). The same reaction sequence was carried out on homoserine (2065). Other nucleic amino acids have been synthesized from Ser via β-bromo Ala, which was converted into 2-aminoadipic acid (this residue, like Asp for β-amino acids and Glu for γ-amino acids, contains the δ-amino acid skeleton). Homologation of the α-carboxyl group gave 3-amino-heptanedioic acid. The homologated carboxymethyl group was then reduced to a hydroxyethyl group, brominated, and displaced with thymine to give a peptide–nucleic acid monomer based on a flexible 5-substituted δ-amino acid (1096).

The α-carboxy group of Glu has also been reduced to a hydroxymethyl group, with hydroxy group displacement and azide reduction giving 4,5-diaminopentanoic acid (1850). 5-Amino-4-hydroxypentanoic acid, the homolog of GABOB, was prepared from L- or D-Glu. Reaction with NaNO₂ converted the α-amino group into a hydroxyl group. The α-carboxyl group was then reduced to a hydroxymethyl group, mesylated, and displaced with azide. Azide reduction provided the desired (4R)- or (4S)-4-hydroxy-5-aminopentanoic acid (2066). Similarly, O-protected Glu was deaminated and converted to a hydroxymethyl-substituted lactone, with tosylation/azide displacement leading to 4-methoxy-5-amino-pentanoic acid (2067).

Diazotization of L-Orn provided (S)-5-amino-2-hydroxypentanoic acid (1612), while treatment of His-OBn with excess benzoyl chloride opened the imidazole ring to give tribenzoyl 2,4,5-triaminopent-4-enoic acid. Hydrolysis gave the free triamino acid (1675).

11.6.3f Elaboration of β-, γ-, and δ-Amino Acids

Several homologations of β-amino acids have been reported. The Weinreb amides of three β³-amino acids were reduced with DIBAL-H to give the amino aldehydes, which were then homologated by Wadsworth–Emmons reaction with (EtO)₂POCH₂CO₂tBu, Hydrogenation provided 5-substituted-5-amino acids (R = Me, iPr, Ph) (720). N-Tosyl α-hydroxy-β-amino acid pyrrole esters, synthesized by an asymmetric addition of N-(2-hydroxyacetyl)pyrrole to N-(o-Ts) imines in the presence of In(OiPr)₃ and a dimeric binol ligand, were homologated by displacement of the pyrrole with lithium tert-butyl acetate, or by reduction to an aldehyde

with LiBH₄ followed by Horner–Wadsworth homologation (see Scheme 11.65 above) (1052). Protected homochiral 3-amino-4,4-dimethylpentanoic acid methyl ester was homologated to 3-oxo-5-amino-6,6-dimethylheptanoic acid via addition of the sodium enolate of methyl acetate. The analogous 5-amino-6,6-dimethylheptanoic acid was obtained from the same substrate via ester reduction to the aldehyde followed by Wittig reaction and alkene hydrogenation (992).

γ-Amino acids have also been converted into δ-amino acids. Homochiral N-phthaloyl 2-methyl-4-aminobutyric acid was transformed into 3-methyl-5-aminopentanoic acid via an Arndt–Eistert homologation (1668). Another homologation employed a nitro analog of GABA, with the substrate prepared by addition of nitromethane to tert-butyl acrylate. Coupling with cyanomethylenetriphenylphosphorane in the presence of EDC/DMAP, followed by ozonolysis, gave 2-keto-5-nitropentanoic acid. The ketone was enantioselectively reduced using a lactate dehydrogenase from Bacillus stearothermophilus or Staphylococcus epidermis to give (2S)- or (2R)-2-hydroxy-5-nitropentanoic acids, respectively. Nitro reduction by hydrogenation was accompanied by cyclization to a a δ-lactam (1712). A similar chemoenzymatic route provided the lactams of (2S,3R)- and (2S,3S)-2-hydroxy-4-methyl-5-aminopentanoic acid (2068).

Functionalizations of δ-amino acids are possible. 5-Azidopentanoic acid was derivatized with Evans' oxazolidinone chiral auxiliary. Deprotonation and benzylation gave the α-benzyl derivative with >90% de. The auxiliary was removed by hydroperoxide hydrolysis (with the enantiomeric purity of the azido acid determined by NMR analysis of the (S)-methyl mandelate ester) and the azide group reduced by hydrogenation (2055, 2069). Valerolactam was condensed with aldehydes under basic conditions to introduce an α-alkylidene substituent. Asymmetric hydrogenation of the 3-alkylidene-2-piperidone substrates with a chiral 2,4-bis (diphenylphosphine)pentane-Ir catalyst produced the lactam of 2-substituted-5-aminopentanoic acids with 81–95% ee (seven examples) (1874). The alkene of 2-methylene-4-hydroxy-5-aminoalkanoic acids, protected as the γ-butyrolactone, can be used for Michael additions of C-, S-, N-, and O-nucleophiles, resulting in a range of highly functionalized δ-amino acids (2070). Nucleo δ-amino acids, with the nucleobase attached to the δ-carbon by a methyl linker, have been synthesized from the lactam of 5-amino-6-hydroxyhexanoic acid, (S)-N-Boc-6-(hydroxymethyl)-2-piperidone. O-Tosyl derivatives were displaced with cytosine or adenine, followed by lactam hydrolysis (1870).

11.6.3g Elaboration of Chiral Substrates

Homostatine analogs were prepared from protected D-glucofuranose via epoxide formation, opening with organocuprates, azide displacement, hydroxyl removal, and oxidation (see Scheme 11.183) (1888).

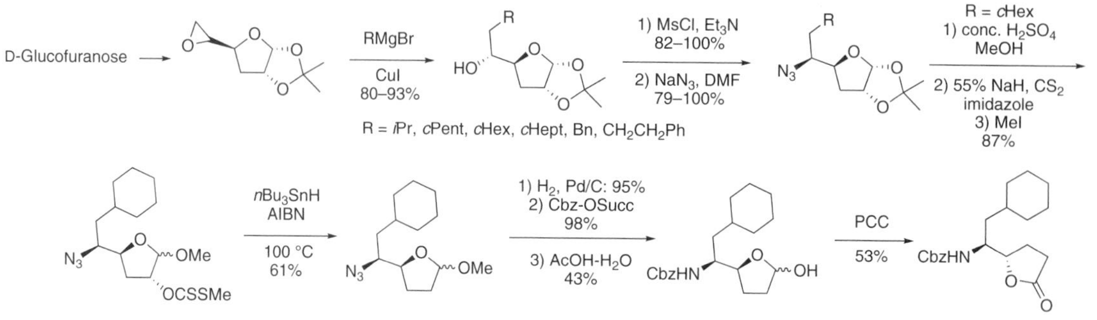

Scheme 11.183 Elaboration of D-Glucofuranose (*1888*).

The isopropylidene derivative of D-ribono-1,4-lactone was converted into a lactone of 2,3,4-trihydroxy-5-aminopentanoic acid by azide displacement of a tosylate, followed by hydrogenation (*2071*). Similarly, acyclic O-methyl 2,3,4-trihydroxy δ-amino acids were prepared from L-arabinose and D-xylose via azide displacement of tosylated derivatives (*2067*), while 3,4-dialkoxy-5-aminopentanoic acid was derived from methyl-2-deoxy-D-ribofuranoside (*1892*). N-Methyl 2,4-dihydroxy-5-aminopentanoic acid was produced from D-galactono-1,4-lactone during a synthesis of the γ-amino acid (*R*)-carnitine via reductive amination of an aldehyde with methylamine (*1881*).

Homochiral 2-(carboxyalkenyl)aziridines, derived from epoxy alcohols, were reductively opened by formic acid in the presence of a palladium catalyst to give α,β-unsaturated- or β,γ-unsaturated-δ-amino esters, depending on the reaction conditions (see Scheme 11.184) (*2000, 2072*). In all cases regioselective opening generated the δ-amino acid. Alternatively, the (*E*)-β,γ-enoate could also be selectively produced by addition of an organocuprate to the α,β-unsaturated ester aziridine, with the organocuprate substituent adding adjacent to the ester group (*2073, 2074*).

11.7 Cyclic δ-Amino Acids (see Table 11.11)

Many cyclic δ-amino acids are achiral compounds based upon an aromatic ring system. Porphobilinogen (PBG), 2-(aminomethyl)-3-(carboxymethyl)-4-(carboxyethyl)-pyrrole (see Scheme 11.179 above), is the precursor of all natural tetrapyrrole systems, including chlorophyll, vitamin B$_{12}$, and the hemes. The free amino group is in a δ-position to the carboxyl group. Interest in PBG has increased in recent years due to the potential uses of porphyrins in photodynamic therapies for the treatment of cancers. PBG is derived from 5-amino-levulinic acid; [1-^{13}C]-, [3-^{13}C]-, and [5-^{13}C]-5-amino-levulinic acid were used in cell-free enzymatic systems to examine the conversion into porphobilinogen (*2020*). A biomimetic synthesis has been reported (*2075*). Traditional methods of synthesis include preparation from unsubstituted pyrroles (*2076, 2077*), or from tri- or tetra-substituted pyrroles (*2078–2086*). PBG and a number of analogs were prepared from 1*H*-pyrrolo [2, 3-*c*]pyridine, while a methyl-substituted analog was derived from 2-(carboxy)-3-(carboxymethyl)-4-(carboxyethyl)-5-methylpyrrole (*2087*). PBG labeled with [^{13}C] on the aminomethyl substituent was prepared

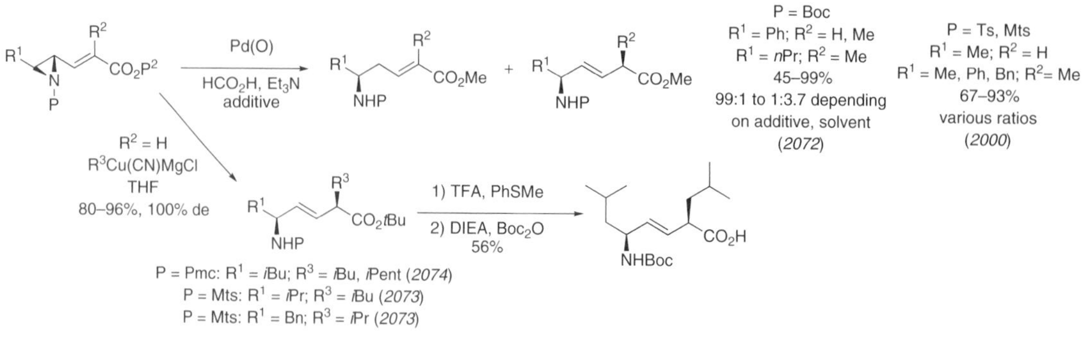

Scheme 11.184 Aziridine Opening.

Table 11.10 Asymmetric Syntheses of δ-Amino Acids

H$_2$N-R-CO$_2$H	Method	Yield, % ee	Reference
Amination			
−CH$_2$CH(Me)CH$_2$CH$_2$−	displacement of homochiral	98% displacement	1962 (*2053*)
−CH$_2$CH(Me)CH$_2$CH(Br)−	5-Br-4-Me-pentanoic acid with	72% bromination	
	KNPhth, α-bromination	45% hydrolysis	
−CH(CH$_2$OH)(CH$_2$)$_3$−	azide displacement of mesylate of	82% displacement	1998 (*2054*)
	5,6-dihydroxyhexanoate	84% azide reduction	
−(CH$_2$)$_3$CH(Bn)−	azide displacement of	97% azidation	1993 (*2055*)
	ω-bromoalkanoic acid derivatized	83% benzylation	
	with Evans' oxazolidinone,	49% reduction/	
	benzylation	hydrolysis	
−(CH$_2$)$_2$CH(OH)CH$_2$−	Mitsunobu azidation of homochiral	40%	1987 (*2056*)
	diol lactone, reduction		
−CH(Ph)CH=CHC(Me)(CO$_2$Me)−	allylic amination of chiral substrate	81–93%	2001 (*2057*)
−CH(*n*Pr)CH=CHC(Me)(CO$_2$Me)−	with phthalimide or MePhSO$_2$NH$_2$		
−CH[CH(Me)CH(OH)CH(OH)Ph]CH$_2$CH=CH−	azide displacement of tosylate	69%	2000 (72)
Carboxylation			
−CH(*i*Pr)CH(OH)CH(OH)CH(*i*Pr)−	Et$_2$AlCN opening of 1,2-epoxy-3-	60–62% opening	1999 (*2058*)
−CH(Bn)CH(OH)CH(OH)CH(Bn)−	hydroxy-4-aminoalkanes		
Fusion of Amino- and Carboxyl-Containing Components: C2–C3 Formation			
−CH(Me)CH$_2$CH=CH−	homologation of α-amino acids via	70–85% reduction	1999 (*2059*)
−CH(*i*Pr)CH$_2$CH=CH−	Arndt–Eistert conversion to	64–91% oxidation/	
−CH(*n*Pr)CH$_2$CH=CH−	β-amino acid, reduction of mixed	Wittig	
−CH(*i*Bu)CH$_2$CH=CH−	anhydride, oxidation/Wittig		
−CH(CH$_2$CH$_2$SMe)CH$_2$CH=CH−			
−CH(Bn)CH$_2$CH=CH−			
−CH[(CH$_2$)$_3$-(δ-N)]CH$_2$CH=CH−			
−CH(*i*Bu)CH$_2$CH=CH(Bn)−	homologation of Leu via reduction to	83% coupling	1990 (*2060*)
−CH(*i*Bu)CH$_2$CH$_2$CH(Bn)−	aldehyde, coupling with		
	EtO$_2$CCH(Bn)PO$_3$Et$_2$, reduction		
−CH(Ph)CH(OH)CH=CH−	3-phenyl glycidate epoxide opening,	87% overall	1997 (*842*)
	carboxy homologation via		
	reduction, Wittig reaction		
−(CH$_2$)$_2$-CC−	coupling of Ph$_3$P=CHCO$_2$Et with	52% coupling	1996 (*1845*)
	β-Ala, flash vacuum pyrolysis	49% pyrolysis	
−CH(Me)CH$_2$CH$_2$CH$_2$−	reaction of Boc-Ala-OMe with	66% sulfone addition	1992 (*2061*)
	PhSO$_2$CHLi$_2$, alkylation with	86% alkylation	
	BrCH$_2$CO$_2$Et, reduction/	80% reduction/	
	elimination	elimination	
Fusion of Amino- and Carboxyl-Containing Components: C3–C4 Formation			
−CH(Bn)CH=CHCH$_2$−	homologation of Phe, alkene	70% epoxidation	1992 (*2062*)
−CH(Bn)CH(OH)CH=CH−	epoxidation, epoxide opening	94% opening	
−CH(Bn)CH(F)CH=CH−	with F, displacement of OH	89% azidation	
−CH(Bn)CH(N$_3$)CH=CH−		40% fluorination	
−CH(Me)CH(OH)CH$_2$CH$_2$−	alkylation of Williams' oxazinone,	88% alkylation	1998 (*1798*)
	lactone reduction, addition of	76% reduction	
	allylTMS, hydroboration and	100% allylTMS	
	oxidation	addition	
		52–56% de	
		38% oxidation and	
		deprotection	
−CH[CH$_2$CH(*i*Pr)CH$_2$(3-MeOPrO-4-MeO-Ph)]	elaboration of complex α-amino acid	23% Grignard addition	2005 (*390*)
CH(OH)CH$_2$CH(*i*Pr)−	via acid reduction to aldehyde,	38% oxidation	
	Grignard addition of propionate		
	equivalent		

(Continued)

Table 11.10 Asymmetric Syntheses of δ-Amino Acids (continued)

H₂N-R-CO₂H	Method	Yield, % ee	Reference
$-CH(Me)CH(OH)CH_2CH_2-$ $-CH(iPr)CH(OH)CH_2CH_2-$ $-CH(iBu)CH(OH)CH_2CH_2-$ $-CH(Bn)CH(OH)CH_2CH_2-$	Sm-mediated ketyl–alkene coupling reaction of amino aldehydes with alkyl acrylates	32–83% 68–80% de	2003 (*2063*)
$-CH(Bn)C(=O)CH_2CH_2-$ $-CH(iBu)C(=O)CH_2CH_2-$ $-CH(Bn)C(=O)CH_2CH(Me)-$	radical reaction of amino acid pyridyl thioester generated from SmI₂ with acylates	40–90%	2003 (*2064*)
$-CH(Me)CH_2CH_2C(=CH_2)-$	Cu-catalyzed coupling of BocNHCH(Me)CH₂ZnI with BrCH₂C(=CH₂)CO₂Et	69% coupling	2001 (*2045*)

Elaboration of α-Amino Acids

H₂N-R-CO₂H	Method	Yield, % ee	Reference
$-CH(CH_2R)CH_2OCH_2-$ $-CH(CH_2CH_2R)CH_2OCH_2-$ $-CH(CH_2R)CH_2OCH(Me)-$ $-CH(CH_2R)CH_2OCH(Me)-$ R = thymin-1-yl, uracil, cytosin-1-yl	reduction of Ser or Hse α-carboxyl, alkylation with bromoacetate or propionate, Mitsunobu displacement of side chain with thymine, uracil	85–90% alkylation 33–75% Mitsunobu	1997 (*2065*)
$-CH[CH_2(thymine)](CH_2)_3-$	conversion of Ser to 2-aminoadipic acid, α-carboxy homologation, reduction, bromination, displacement with thymine	58% homologation 79% reduction 73% bromination 42% thymine displacement	1996 (*1096*)
$-CH(NH_2)(CH_2)_3-$	reduction of Glu α-carboxyl, azide displacement, azide reduction	78% reduction 86% displacement 62–91% reduction	1996 (*1850*)
$-CH_2CH(OH)CH_2CH_2-$	conversion of Glu amino group to hydroxy, reduction of α-carboxyl group, azide displacement and reduction	55% hydroxyl 72% carboxyl reduction 82% azide displacement 77% azide reduction	1986 (*2066*)
$-CH_2CH(OMe)(CH_2)_2-$	elaboration of Glu	74% azide displacement, reduction	1996 (*2067*)
$-(CH_2)_3CH(OH)-$	diazotization of L-Orn	—	1973 (*1612*)
$-CH=C(NH_2)CH_2CH(NH_2)-$	reaction of His with excess benzoyl chloride, hydrolysis	60% hydrolysis	1935 (*1675*)

Elaboration of γ- and δ-Amino Acids

H₂N-R-CO₂H	Method	Yield, % ee	Reference
$-CH(R)CH_2CH=CH-$ $-CH(R)(CH_2)_3-$ R = Me, iPr, Ph	homologation of 3-substituted-β-amino acids via Weinreb amide reduction to aldehyde, condensation with (EtO)₂POCH₂CO₂tBu/ nBuLi, hydrogenation	84–87% reduction/ homologation 94–98% hydrogenation	2004 (*720*)
$-CH_2CH_2CH(Me)CH_2-$	homologation of N-Phth 2-methyl-4-aminobutyric acid	17% homologation 45% hydrolysis	1959 (*1668*)
$-(CH_2)_3CH(OH)-$	homologation of 4-nitro-pentanoic acid, enzymatic reduction of ketone, hydrogenation of nitro	51% homologation 93% nitro reduction	1999 (*1712*)
$-CH_2CH(Me)CH_2CH(OH)-$	homologation of 4-nitro-3-methyl-pentanoic acid, enzymatic reduction of ketone, hydrogenation of nitro	60% homologation 60–90% ketone reduction 0–50% de	2000 (*2068*)
$-(CH_2)_3CH(Bn)-$	derivatization of 5-azidopentanoic acid with Evans' oxazolidinone, alkylation, hydrolysis, azide reductrion	50% oxazolidinone coupling/alkylation >90% de 86% hydrolysis	1993 (*2069*)
$-(CH_2)_3CH(Bn)-$	azide displacement of ω-bromoalkanoic acid derivatized with Evans' oxazolidinone, benzylation	97% azidation 83% benzylation 49% reduction/ hydrolysis	1993 (*2055*)

Table 11.10 Asymmetric Syntheses of δ-Amino Acids (continued)

H₂N-R-CO₂H	Method	Yield, % ee	Reference
$-$CH[CH₂(cytosin-1-yl)]CH₂CH₂CH₂$-$ $-$CH[CH₂(adenin-1-yl)]CH₂CH₂CH₂$-$	tosylation and displacement of lactam of 4-amino-5-hydroxypentanoic acid, lactam hydrolysis	31–88% displacement 64–86% hydrolysis	1991 (*1870*)
$-$CH(*i*Pr)CH(OH)CH₂CH(*n*Pent)$-$ $-$CH(*i*Pr)CH(OH)CH₂CH(CH₂Ph)$-$ $-$CH(*i*Pr)CH(OH)CH₂CH(CH₂CH=CH₂)$-$ $-$CH(*i*Pr)CH(OH)CH₂CH[(CH₂)₂CH=CH₂)]$-$ $-$CH(*i*Pr)CH(OH)CH₂CH[CH₂CH(CO₂Me)₂]$-$ $-$CH(*i*Pr)CH(OH)CH₂CH(CH₂CN)$-$ $-$CH(Bn)CH(OH)CH₂CH(CH₂CN)$-$ $-$CH(*i*Pr)CH(OH)CH₂CH(CH₂SBn)$-$ $-$CH(Bn)CH(OH)CH₂CH(CH₂SBn)$-$ $-$CH(Bn)CH(OH)CH₂CH[CH₂N(OH)Bn]$-$ $-$CH(*i*B)CH(OH)CH₂CH(CH₂OMe)$-$ $-$CH(Bn)CH(OH)CH₂CH(CH₂OMe)$-$ $-$CH(*i*Pr)CH(OH)CH₂CH(CH₂*t*Bu)$-$ $-$CH(*i*Pr)CH(OH)CH₂CH(Me)$-$ $-$CH(*i*Bu)CH(OH)CH₂CH(Me)$-$ $-$CH(*i*Pr)CH(OH)CH=C(Me)$-$	Michael addition of nucleophiles or radical addition to alkene in CbzNH-CH(R)CH(OH)-CH₂C(=CH₂)CO₂H butyrolactone derivative, or alkene hydrogenation or isomerization	41–99% addition 20–>90% de	2002 (*2070*)
$-$CH₂CH₂CH₂CH(4-F-Bn)$-$ $-$CH₂CH₂CH₂CH(4-CF₃-Bn)$-$ $-$CH₂CH₂CH₂CH(4-MeO-Bn)$-$ $-$CH₂CH₂CH₂CH[CH₂(2-furyl)]$-$ $-$CH₂CH₂CH₂CH[CH₂(3-furyl)]$-$ $-$CH₂CH₂CH₂CH(*n*Bu)$-$ $-$CH₂CH₂CH₂CH(*i*Bu)$-$ (lactam)	condensation of lactam with RCHO, alkene reduction by asymmetric hydrogenation with chiral Ir catalyst	92–100% reduction 81–95% ee	2002 (*1874*)
$-$CH(*t*Bu)CH₂COCH₂$-$ $-$CH(*t*Bu)CH₂CH=CH$-$ $-$CH(*t*Bu)CH₂CH₂ CH₂$-$	homologation of homochiral pTol-SO-NHCH(*t*Bu)CH₂CO₂Me via addition of Me acetate or ester reduction to aldehyde and Wittig reaction, reduction	85% acetate addition 85% ester reduction, Wittig reaction and alkene hydrogenation	2003 (*992*)
$-$CH(Ph)CH(OH)CH=CH$-$ $-$CH(Ph)CH(OH)COCH₂$-$	homologation of pyrrole amide of oTol-SO-NHCH(Ph)CH(OH) CO₂H via displacement with Li *t*Bu acetate or reduction with LiBH₄, then olefination with (EtO)₂P(O)CH₂CO₂Et	62% acetate displace- ment 67% reduction and olefi- nation	2005 (*1052*)

Elaboration of Chiral Substrates

H₂N-R-CO₂H	Method	Yield, % ee	Reference
$-$CH(*i*Pr)CH(OH)CH₂ CH₂$-$ $-$CH(*c*Pent)CH(OH)CH₂ CH₂$-$ $-$CH(*c*Hex)CH(OH)CH₂ CH₂$-$ $-$CH(*c*Hept)CH(OH)CH₂ CH₂$-$ $-$CH(Bn)CH(OH)CH₂ CH₂$-$ $-$CH(CH₂CH₂Ph)CH(OH)CH₂ CH₂$-$	elaboration of D-glucofuranose	80–93% organocuprate addition 79–92% azide displacement 87% oxidation (R = *c*Hex)	1989 (*1888*)
$-$CH₂CH(OH)CH(OH)CH(OH)$-$	azide displacement of tosylate of D-ribono-1,4-lactone isopropy-lidene, hydrogenation	—	1998 (*2071*)
$-$CH₂CH(OMe)CH(OMe)CH(OMe)$-$	tosylation, azide displacement of L-arabinose, D-xylose	24–35% overall	1996 (*2067*)
$-$CH₂CH(OH)CH(OH)CH₂$-$ $-$CH₂CH(OH)CH(OMe)CH₂$-$ $-$CH₂CH(OH)CH(OBn)CH₂$-$	tosylation, azide displacement of D-ribofuranoside	70% azidation	1992 (*1892*)
$-$CH₂CH(OH)CH₂CH(OHO)$-$ (with NMe)	derivatization of D-galactono-1,4-lactone via reductive amination of aldehyde with MeNH₂	63% overall	1992 (*1881*)

Table 11.10 Asymmetric Syntheses of δ-Amino Acids (continued)

H₂N-R-CO₂H	Method	Yield, % ee	Reference
−CH(Ph)CH₂CH=CH− −CH(Ph)CH₂CH=C(Me)− −CH(nPr)CH₂CH=C(Me)− −CH(Ph)CH=CHCH₂− −CH(Ph)CH=CHCH(Me)− −CH(nPr)CH=CHCH(Me)−	aziridine opening	45–99%	1995 (2072)
−CH(Me)CH₂CH=C(Me)− −CH(Ph)CH₂CH=C(Me)− −CH(Bn)CH₂CH=C(Me)− −CH(Me)CH=CHCH(Me)− −CH(Ph)CH=CHCH(Me)− −CH(Bn)CH=CHCH(Me)−	Pd-catalyzed reductive ring opening of α,β-unsaturated ester aziridine	67–93% yield	1997 (2000)
−CH(iPr)CH=CHCH(iBu)−, −CH(Bn)CH=CHCH(iPr)−,	Pd-catalyzed isomerization of α,β-unsaturated ester aziridine, opening with organocuprate	80–95% opening	1997 (2073) 1997 (2074)

from 3-carboxypropyl-4-carboxymethylpyrrole-2-carboxylate (2018, 2088), while PBG with a [2,3-²H₂]-label on the propionic substituent was prepared by a synthesis from 4-methyl-5-nitropyrimidone (2089): both were used to investigate the biosynthesis of protoporphyrin-IX.

Condensation of an α-acetoxynitro derivative with isocyanoacetonitrile provided 2-cyano-3-(2′-hydroxyethyl)-4-(2-methoxycarbonylethyl)pyrrole, which was readily converted into PBG via oxidation of the hydroxyethyl group and hydrolysis of the nitrile (2090). A pyrrole precursor of PBG was constructed by the base-induced cyclization of tosylmethyl isocyanide (CNCHTs) with diethyl gluconate (EtO₂C-CH=CH-CH₂CO₂Et), giving diethyl 3-carboxy-4-acetic acid-pyrrole in 72% yield (see Scheme 11.185). A 2-formyl group was then introduced using the Vilsmeier reagent, with reductive amination introducing the amino group. The final step involved transformation of the carboethoxy group to a propionic acid side chain, with reduction to an alcohol giving a substrate suitable for an elimination/addition reaction with malonate. The lactam precursor of PBG was obtained in eight steps and 17–22% overall yield (2091).

Bilirubin is another natural product containing cyclic δ-amino acids—in this case, cross-linked 3-carboxyethyl-4-methylpyrrole derivatives derived from heme. A 10-oxo-bilirubin analog has been synthesized (2092).

One of the first syntheses of a cyclic amino δ-amino acid was the preparation of 3-(aminomethyl)benzoic acid in 1901, starting from cyanobenzyl chloride (2093). 3-(Aminomethyl)benzoic acid is a phenylogous dipeptide mimetic. Both 3-(aminomethyl)benzoic acid and the 2-(aminomethyl)benzoic acid isomers were incoporated into peptidic growth hormone releasers (2094). Replacement of the 3-(aminomethyl)benzoic acid residue with (E)-5-methyl-5-aminohex-2-enoic acid gave orally available compounds with reduced weight (2043). 3-(Aminomethyl)benzoic acid was also used as a spacer between Cys and Met in inhibitors of P21ras farnesyltransferase (2095), and to constrain the tripeptide Arg-Gly-Asp in a high-affinity ligand of platelet glycoprotein IIb/IIIa (2096).

Other cyclic δ-amino acids have chiral centers. Neuraminic acid (sialic acid) and its N-acetyl derivative (Neu5Ac), are very important cyclic δ-amino acid amino sugars found in mammalian tissues (see

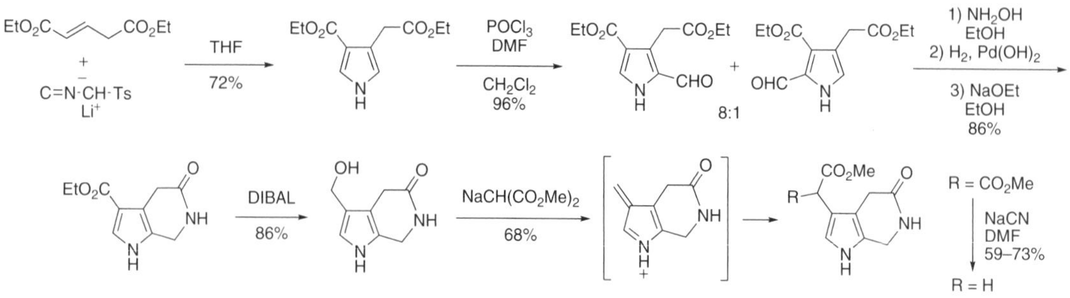

Scheme 11.185 Synthesis of Porphobilinogen (2091).

Scheme 11.179 above). A number of potent carbocyclic inhibitors of influenza neuraminidase based on the structure of N-acetylneuraminic acid but containing an additional amino group in the γ-position have been synthesized (*1934, 1935, 1995, 1996*). C-4-Disubstituted Neu5Ac analogs have been prepared via opening of a 4-epoxide derivative (*2097*), while 3,4,5-triamino-cyclohex-1-ene-1-carboxylic acid was derived from (−)-quinic acid (*1997*). These and other neuramic acid analogs (e.g., *2098*) and other related sugar amino acids have been designed as peptidomimetics (*404, 2007*).

Another constrained δ-amino acid is designed to mimic pyranosyl-RNA, and is based on 1-amino-2-nucleobase-cyclohexane-5-acetic acid monomers. An NMR study of the duplex formed by an AATAT nucleo-δ-peptide identified an antiparallel arrangement of two complementary strands (*2190*). A 1-amino-3-carboxy-methylcyclohexane core was employed in a δ-peptide analog of RNA (nucleo-δ-peptide, NDP), with the nucleobase substituted at the 6-position (*2099*). NMR studies of the solution structure have been carried out (*2099*).

Racemic 3-(aminomethyl)-5-(4-methoxybenzyl) cyclohexane-1-carboxylic acid was synthesized from trimesic acid (3,5-dicarboxybenzoic acid), which was selectively diprotected and hydrogenated to give a tricarboxycyclohexane derivative. One carboxyl was converted into the benzyl substituent, while a second was reduced to a hydroxymethyl group and reacted with phthalimide under Mitsunobu conditions to introduce the amino group (*2100, 2101*). Racemic dihydroxy-bicyclo[2.2.2]octane, substituted on the bridgehead carbons with the amino and carboxy groups, was prepared by a tandem cyclization of a 3-carboxy-3-(3-nitropropyl) pentanedianal intermediate (*2102*). A bicyclo[3.2.1]

octene derivative was also prepared (*2103*). Oxazolines were used as a masked carboxyl group in 1-(4′, 4′-dimethyloxazolin-2′-yl)naphthalene, derived from 1-naphthoic acid (see Scheme 11.186). A diastereoselective 1,6-addition of lithium dialkylamides, followed by enolate trapping with alkyl halides, led to 2-alkyl-2-carboxy-4-amino-1,4-dihydronaphthalenes after oxazoline hydrolysis and N-deprotection. Monoalkyl amides tended to give the 1,4-addition products. The *trans* isomer was formed preferentially (*2104, 2105*).

1-Aminocyclopentane-3-acetic acid and 1-amino-cyclopentane-3-(2-propionic acid) were both synthesized from L-Glu via conversion into a 2-amino-5-hexenyl derivative followed by radical cyclization (*2106*). A chiral N-acrylyl 5-amino-6-phenyl-hex-2, 3-enoate was also cyclized under radical conditions to give a lactam intermediate that was then converted to 2-benzyl-4-carboxymethyl-5,6-methanopiperidine in 38% yield (*1994*). A constrained δ-amino acid with an ether linkage in the backbone, *trans*-1-carboxymethyloxy-2-aminocyclohexane, was obtained by azide opening of optically active cyclohexene epoxide followed by alkylation of the alcohol with *tert*-butyl bromoacetate. It was incorporated into an analog of gramicidin A as a synthetic ion channel (*2107*). A 2-carboxy-8-amino-quinoline derivative was prepared as a δ-amino acid analog for generating helical structures; indeed, a homooctamer was shown by crystallography to form a very stable helix (*2108*).

A cyclobutane-linked δ-amino acid, 2,2-dimethyl-3-aminomethylcyclobutane-1-carboxylic acid, was constructed from (+)-α-pinene by oxidative cleavage of the alkene bond followed by a Curtius rearrangement of the carboxyl group and oxidation of the ketone (see Scheme 11.187) (*1998*). An earlier preparation of the

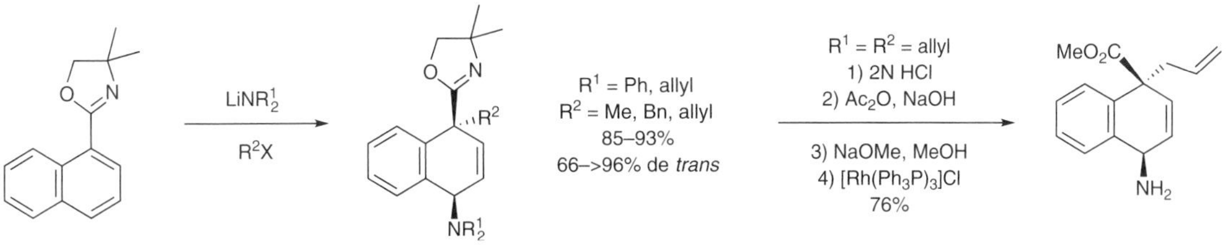

Scheme 11.186 1,6-Addition of Lithium Dialkylamides to Naphthyloxazolines (*2104, 2105*).

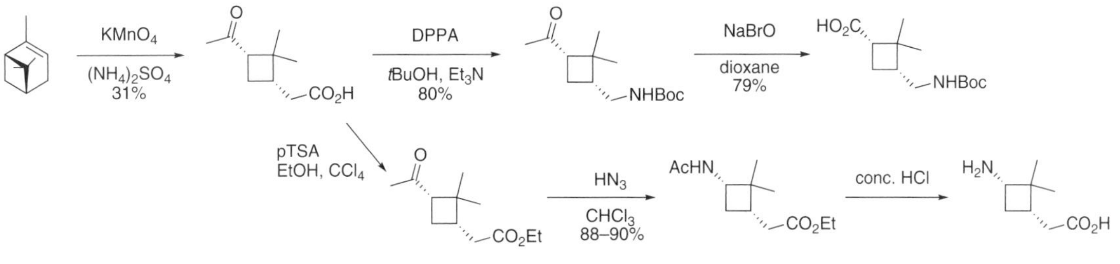

Scheme 11.187 Elaboration of (+)-α-Pinene (*1998, 2109*).

1-2,2-dimethyl-3-aminocyclobutane-1-acetic acid isomer employed a Schmidt rearrangement for the amination step (*2109*). Several cyclic δ-amino acids are included in a 1993 review of amino acids of the cyclobutane series (*1356*). A Beckmann rearrangement of camphor, via microwave solid-phase reaction of camphor with hydroxylamine hydrochloride and P_2O_5/SiO_2, produced the lactam of 1,2,2-trimethyl-3-aminomethylcyclo-pentane-1-carboxylic acid with 75% isolated yield (*2110*).

Table 11.11 Syntheses of Cyclic δ-Amino Acids

Ring Size (Aromatic Template)	Substituents	Method	Yield, % ee	Reference
5 pyrrole nucleus	1-aza-2-CH_2NH_2-3-CH_2CO_2H-4-$(CH_2)_2CO_2H$	Mukaiyama aldol reaction with acylnitrile, nitrile reduction/cyclization	26% overall	1998 (*2075*)
5 pyrrole nucleus	1-aza-2-CH_2NH_2-3-CH_2CO_2H-4-$(CH_2)_2CO_2H$-5-R R = H, CO_2H	conversion of unsubstituted pyrrole	—	1983 (*2076*)
5 pyrrole nucleus	1-aza-2-CH_2NH_2-3-CH_2CO_2H-4-$(CH_2)_2CO_2H$-5-R R = H, CO_2H	conversion of unsubstituted pyrrole	11% overall	1984 (*2077*)
5 pyrrole nucleus	1-aza-2-CH_2NH_2-3-CH_2CO_2H-4-$(CH_2)_2CO_2H$	derivatization of substituted pyrroles	—	1956 (*2078*)
5 pyrrole nucleus	1-aza-2-CH_2NH_2-3-CH_2CO_2H-4-$(CH_2)_2CO_2H$	derivatization of substituted pyrrole prepared from nitropyridine and diethyl oxalate	—	1969 (*2079*)
5 pyrrole nucleus	1-aza-2-CH_2NH_2-3-CH_2CO_2H-4-$(CH_2)_2CO_2H$	derivatization of substituted pyrrole	—	1977 (*2080*)
5 pyrrole nucleus	1-aza-2-CH_2NH_2-3-CH_2CO_2H-4-$(CH_2)_2CO_2H$	pyrrole synthesis and reduction	41% pyrrole synthesis 54% derivatization 51% reduction	1995 (*2081*)
5 pyrrole nucleus	1-aza-2-CH_2NH_2-3-CH_2CO_2H-4-$(CH_2)_2CO_2H$	pyrrole synthesis further derivatization	50% pyrrole formation	1973 (*2082*)
5 pyrrole nucleus	1-aza-2-CH_2NH_2-3-CH_2CO_2H-4-$(CH_2)_2CO_2H$	pyrrole synthesis and reduction	—	1976 (*2083*)
5 pyrrole nucleus	1-aza-2-CH_2NH_2-3-CH_2CO_2H-4-$(CH_2)_2CO_2H$-5-CO_2H	conversion of tetrasubstituted pyrrole	—	1957 (*2084*)
5 pyrrole nucleus	1-aza-2-CH_2NH_2-3-CH_2CO_2H-4-$(CH_2)_2CO_2H$-5-R R = H, CO_2H	conversion of tetrasubstituted pyrrole	—	1961 (*2085*)
5 pyrrole nucleus	1-aza-2-CH_2NH_2-3-CH_2CO_2H-4-$(CH_2)_2CO_2H$ 1-aza-2-CH_2NH_2-3-$(CH_2)_2CO_2H$-4-CH_2CO_2H	conversion of disubstituted pyrrole	—	1978 (*2086*)
5 pyrrole nucleus	1-aza-2-CH_2NH_2-3-CH_2CO_2H-4-R^1-5-R^2 R^1 = $(CH_2)_2CO_2H$, $CH=CHCO_2H$, CH_2CHFCO_2H; R^2 = H R^1 = $(CH_2)_2CO_2H$; R^2 = Me	derivatization of substituted pyrroles	—	1996 (*2087*)
5 pyrrole nucleus	1-aza-2-$^{13}CH_2NH_2$-3-CH_2CO_2H-4-$(CH_2)_2CO_2H$	formylation, reduction, chlorination, azide displacement, reduction of 3-carboxypropyl-4-carboxymethyl-pyrrole-2-carboxylate	—	1972 (*2088*)

Table 11.11 Syntheses of Cyclic δ-Amino Acids (continued)

Ring Size (Aromatic Template)	Substituents	Method	Yield, % ee	Reference
5 pyrrole nucleus	1-aza-2-^{13}CH$_2$NH$_2$-3-CH$_2$CO$_2$H-4-(CH$_2$)$_2$CO$_2$H	methylation of 3-CH$_2$CH$_2$CO$_2$Et, 4-CH$_2$CO$_2$Et-pyrrole-2-CO$_2$Et, chlorination, azide displacement, reduction	66% methylation 43% amination	1973 (2018)
5 pyrrole nucleus	1-aza-2-CH$_2$NH$_2$-3-CH$_2$CO$_2$H-4-(CHD)$_2$CO$_2$H	derivatization of 4-Me-5-NO$_2$-pyrimidin-2-one	—	1975 (2089)
5 pyrrole nucleus	1-aza-2-CH$_2$NH$_2$-3-CH$_2$CO$_2$H-4-(CH$_2$)$_2$CO$_2$H	pyrrole synthesis	17–22% overall	1997 (2091)
5 pyrrole nucleus	1-aza-2-CH$_2$NH$_2$-3-CH$_2$CO$_2$H-4-(CH$_2$)$_2$CO$_2$H	pyrrole synthesis	12% overall	1996 (2090)
6 Ph nucleus	1-CO$_2$H-3-CH$_2$NH$_2$	from 3-cyanobenzylchloride	—	1901 (2093)
6 optically active	1-CO$_2$H-1,2-dehydro-3,4,5-(NH$_2$)$_3$	from (-)-quinic acid	—	1998 (1997)
6 racemic	1-CO$_2$H-3-CH$_2$NH$_2$-5-(4-MeO-Bn)	derivatization of trimesic acid	4–24% overall	1997 (2100) 1998 (2101)
6 racemic	1-amino-2,6-(OH)$_2$-1,4-(−CH$_2$CH$_2$−)-4-CO$_2$H	tandem cyclization of OHCCH$_2$C(CO$_2t$Bu)(CH$_2$CH$_2$CH$_2$NO$_2$)CH$_2$CHO	48% cyclization	1999 (2102)
6 racemic	1-amino-2,6-(OH)$_2$-1,3-(−CH$_2$CH$_2$−)-4,5-dehydro-4-CO$_2$H	tandem Michael addition/aldol reaction	64% cyclization	1999 (2103)
6 racemic	1-NH$_2$-2,3-(fused Ph)-4-CO$_2$H-4-R-5,6-dehydro R = Me, Bn, CH$_2$CH=CH$_2$	1,6-addition of dialkylamides to naphthyloxazolines	85–93% addition 66–>96% de anti 97–99% hydrolysis 79–86% N-deprotection	1996 (2104) 1998 (2105)
5 optically active	1-NH$_2$-3-CH$_2$CO$_2$H 1-NH$_2$-3-CH(Me)CO$_2$H	conversion of L-Glu	27%	1994 (2106)
6 optically active	1-NH$_2$-2-OCH$_2$CO$_2$H	azide opening of optically active cyclohexene epoxide, alkylation with tert-butyl bromoacetate	89% opening and alkylation 95% azide reduction	2001 (2107)
4 optically active	1-CO$_2$H-2,2-Me$_2$-3-CH$_2$NH$_2$	derivatization of (+)-α-pinene	20%	1997 (1998)
4 optically active	1-NH$_2$-2,2-Me$_2$-3-CH$_2$CO$_2$H	derivatization of (+/−)-α-pinene	88–90% Schmidt rearrangement	1958 (2109)
5 optically active	1-CO$_2$H-1,2,2-Me$_3$-3-CH$_2$NH$_2$	Beckmann rearrangement of camphor using microwave solid phase with hydroxylamine hydrochloride and P$_2$O$_5$/SiO$_2$	75% rearrangement to lactam	2003 (2110)
6 optically active	1-aza-2-Bn-4-CH$_2$CO$_2$Me-5,6-(−CH$_2$−)	radical cyclization of N-acrylyl-5-aminohexenoate	38%	1996 (1994)

11.8 ω-Amino Acids

11.8.1 Introduction

Several naturally occurring amino acids have the amino and carboxyl group separated by more than four carbons (see Scheme 11.188). Galantinic acid, 6-amino-3,5,

7-trihydroxyheptanoic acid (2111–2115), and galantinamic acid, 6,10-diamino-2,3,5-trihydroxydecanoic acid (2114, 2116) are components of the peptide antibiotic galantin I, isolated from Bacillus pulvifaciens (2117, 2118). The original proposed structure for galantinic acid, 3-amino-4-hydroxypyranose-6-acetic acid (2119), was later found to be an artifact resulting from

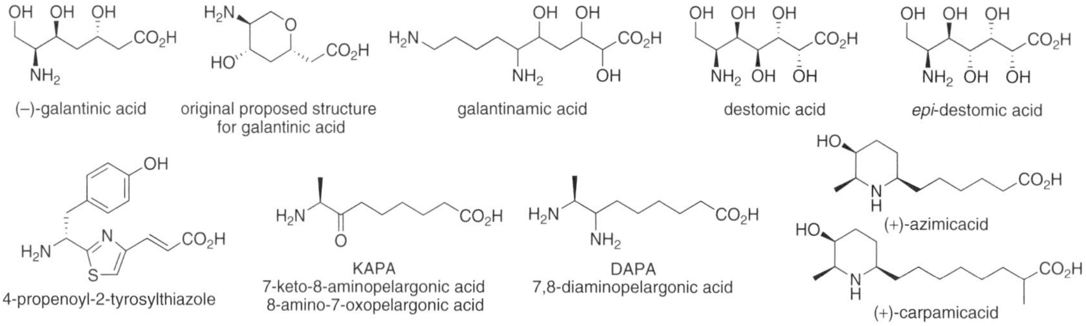

Scheme 11.188 Structures of ω-Amino Acids.

chemical degradation studies of galantin I. The absolute configuration of galantinamic acid was established to be (2R,3S,5S,6R) via stereoselective synthesis of eight of the possible diastereoisomers (2120). Other polyhydroxylated ω-amino acids include destomic acid and epi-destomic acid, isomers of 6-amino-2,3,4,5,7-penta-hydroxyheptanoic acid, which are components of the antibiotic natural products destomycin A (2121), destomycin B (2121), and hygromycin (2122). The absolute configuration of destomic acid was determined to be 6-amino-6-deoxy-L-glycero-D-galacto-heptanoic acid, while epi-destomic acid was the C4 epimer, 6-amino-6-deoxy-L-glycero-D-gluco-heptanoic acid (2121).

An unusual amino acid with a 4-propenoyl-2-tyrosylthiazole nucleus is contained within oriamide, a cyclic peptide isolated from the marine sponge Theonella sp. The thiazole forms part of the amino acid backbone in a similar fashion to the thiazole dipeptide isosteres (269). An analogous (O-methylseryl)thiazole amino acid was identified in keramamide F, another cyclic peptide isolated from the same type of sponge (271). Azimic acid, from Azima tetracantha, is 2-methyl-3-hydroxy-6-(5-carboxypentane)piperidine (2123, 2124), while carpimic acid is the homologated 2-methyl-3-hydroxy-6-(5-carboxyheptane)piperidine, with a total synthesis reported in 2004 (2125). The vitamin biotin, an essential cofactor for carboxylase-catalyzed reactions, is synthesized from pimeloyl CoA via KAPA, 7-keto-8-amino-pelargonic acid, and DAPA, 7,8-diaminopelargonic acid (1831, 2126). A number of analogs have been synthesized as potential herbicides that act via inhibition of biotin synthesis (1831).

11-Aminoundecanoic acid and a triethylene glycol analog were incorporated into analogs of human atrial natriuretic factor as a replacement for an octapeptide sequence, but caused loss of activity (2127). Several isomers of 2,8-dibenzyl-3,7-dihydroxy-8-aminooct-4-enoic acid were prepared in 2002 as high-affinity mu-opioid receptor ligands, acting as a Tyr-Pro-Phe tripeptide isostere (2128). A conformational analysis of all isomers of 2,7-dimethyl-3,6-dihydroxy-7-aminohep-4-Z-enoic acid was reported the same year. The 16 stereoisomers of the corresponding 2,7-diisobutyl analog were substituted

into a renin inhibitor, replacing two (L-V) or three (L-V-I) residues of the residues flanking the scissile bond. Activities within both series varied from 10 to >1000 µM IC$_{50}$ (2129). 3-(4-Piperidinyl)propionic acid, 4-(4-piperidinyl)butyric acid and 5-(4-piperidinyl)pentanoic acid replaced Arg in orally active fibrinogen receptor antagonists (2130), while piperidine-4-propionic acid and several unsaturated analogs were tested as GABA analogs (2131).

Substituted 6-aminocaproic acids have been used in β-turn mimetic cyclic peptides (2132), while five monobenzylated 6-aminocaproic acid analogs with the benzyl group on each of the five methylene groups were tested as constraints for cyclic dipeptides (2133). The biphenyl-containing amino acids 2'-(aminomethyl) biphenyl-2-carboxylic acid and 2'-(aminomethyl)biphenyl-2-acetic acid act as rigid turn spacers to allow for antiparallel β-sheet formation (2134). 3'-(2-Aminoethyl)-2-biphenylpropionic acid (see Scheme 11.189) induced a β-hairpin structure when incorporated into peptides (2135). A (Z)-11-aminoundec-6-en-4,8-diynoic acid derivative has been used as a turn motif in forming a β-sheet nucleus (2136).

A cyclic tetrapeptide has been formed from alternating units of Phe and an ε-amino acid, 2-(4-amino-methyl-1,2,3-triazol-1-yl)-4-methylpentanoic acid. The macrocycles self-assembled into stacks that formed an open-ended hollow tubular structure, with a crystal structure obtained of the resulting solvent-filled nanotube (2137). A series of N-acyl ω-amino acids (ranging from 6-aminohexanoic acid to 13-amino-tridecanoic acid) were found to be potent gelators in DMF at

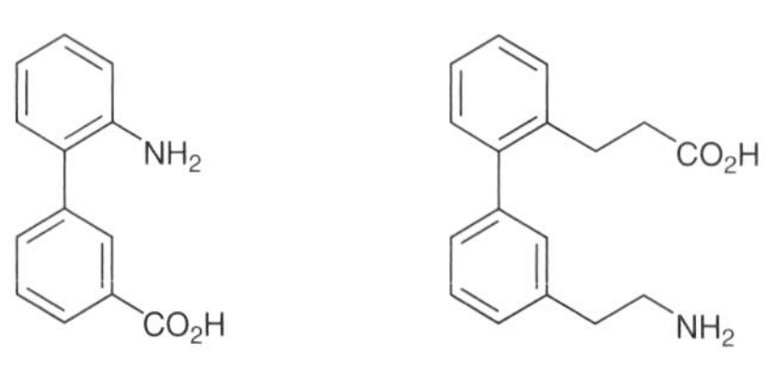

Scheme 11.189 Biphenyl-Based Amino Acids (2151).

concentrations as low as 0.4 mg/mL (0.04 wt%), producing a thermoreversible gel that was stable and transparent for more than a year in a stoppered tube (*2138*). 11-Aminoundecanoic acid has been used to form a number of derivatives that gelated both water and organic solvents at concentrations of 1–20 g/L (*2139, 2140*). When also incorporating a chiral α-amino acid unit, some racemic salts could gelate much greater quantities (up to 16-times more) of solvent than the pure enantiomers (*2140*).

The polymer nylon 6 (Perlon) is a linear polymer of 6-aminohexanoic acid, formed by ring opening of caprolactam. Macrocyclic oligomers composed of alternating 6-aminohexanoic acid and 2,3,4,5-tetrahydroxy-6-aminohexanoic acid units, with 140-, 28-, and 42-membered lactam rings, were prepared in 2004 (*2141*). Linear oligomers (up to octamer) of a fully hydroxylated analog of nylon 6 have been assembled from a protected 2,3,4,5-tetrahydroxy-6-aminohexanoic acid derived from D-ribose (*2142*).

A monograph that describes methods of non-α-amino acid synthesis (focusing on acyclic amino acids of C_3–C_{10}) was published in 1995 by M.B. Smith (*1*).

11.8.2 Syntheses of Achiral or Racemic ω-Amino Acids (see Table 11.12)

11.8.2a Amination Reactions

The simple unsubstituted ω-amino acids 6-aminohexanoic acid, 7-aminoheptanoic acid, 8-aminooctanoic acid, 11-aminoundecanoic acid, and 12-aminododecanoic acid have been synthesized via a Beckmann rearrangement of an in situ-generated oxime of the corresponding cycloalkanone (see Scheme 11.180 above). The lactams of the ω-amino acids were formed upon reaction of the cycloalkanone with hydroxylamine-O-sulfonic acid over silica in a microwave oven. Lactam hydrolysis gave the ω-aminoalkanoate salts (*2029*). Similarly, microwave solid-phase reaction of cyclohexanone with hydroxylamine hydrochloride and P_2O_5/SiO_2 produced the lactam of 6-aminohexanoic acid with 88% isolated yield (*2110*). A catalytic Beckmann rearrangement of cyclohexanone oxime in ionic liquids (bmiPF$_6$) proceeded in the presence of 10% P_2O_5 at only 75 °C with >95% yields (*2143*). An environmentally friendly production of ε-caprolactam employed supercritical water in a microreaction system, producing the caprolactam from cyclohexanone oxime in nearly 100% yield within 1 s (*2144*). A mixture of air and ammonia reacted with cyclohexanone in the presence of a transition metal aluminophosphate molecular sieve catalyst to produce up to 45% of the ε-caprolactam (*2145*). An asymmetric version of this reaction, employing 4-substituted cyclohexanone substrates, has been described (*2146*).

These are updated procedures for a much older reaction. N-Nonoyl 9-aminononanoic acid was prepared in 1894 via a Beckmann rearrangement of the oxime of 10-ketooctadecanoic acid (*2147*), while 8-aminooctanoic

acid was reported the same year by hydrolysis of the rearrangement products of 9-ketoheptadecanoic acid (*2148*). 8-Aminooctanoic acid was again prepared from cyclooctanone in 1996, via treatment with hydroxyamine-O-sulfonic acid. The resulting 2-azacyclononanone (65% yield) was hydrolytically opened with 5 N NaOH, with the N-Boc-protected derivative isolated in 63% yield (*2149*). The oximes of cyclohexanone, 2-methylcyclohexanone, menthone, tetrahydrocarvone, and heptanone were rearranged in 1900 (*2030*) and the bicyclic ketones bicyclo[3.3.1]nonan-3-one and nopinone in 1963 (*2150*). The latter substrates resulted in the bicyclic lactams of 3-aminomethyl-cyclohexane-1-acetic acid and 2,2-dimethyl-3-aminocyclobutane-1-propionic acid.

In 2000, a series of substituted cyclohexanones were treated with benzyl azide or n-hexyl azide in the presence of $TiCl_4$, giving the Schmidt rearrangement N-substituted lactams in variable yield depending on substituents, with a major side product being aminomethyl-substituted cyclohexanones resulting from a Mannich reaction. Cycloheptanone and cyclooctanone gave exclusively the aminomethyl-substituted byproduct ketone, while cyclopentanone gave no product. 2-Adamantanone, 2-decalone, 5α-cholestanone, and norbornanone gave the bicyclic lactams in 40–100% yield, although regioselectivity was poor (*1508*).

5-Aminopentanoic acid, 8-aminooctanoic acid, and 9-aminononanoic acid have also been synthesized by Schmidt rearrangements of the corresponding diacid monoesters (*513*). Similarly, Hofmann rearrangement of the monomethyl ester monoamides of alkane dicarboxylic acids produced N-carboxymethyl 6-aminohexanoic acid, 7-aminoheptanoic acid, 8-aminooctanoic acid, 9-aminononanoic acid, and 13-aminotridecanoic acid, with hydrolysis providing the free amino acids (*512*). The amino groups of two biphenyl-based amino acids, 3′-(2-aminoethyl)-2-biphenylpropionic acid and 2-amino-3′-biphenylcarboxylic acid (see Scheme 11.189) were introduced by a Curtius rearrangement (*2151*). The first of these derivatives induced a β-hairpin structure when incorporated into peptides (*2135*).

Other aminative syntheses include the preparation of several steroidal amino acids, with the amino and carboxyl groups bridged by a steroid nucleus, by a Mitsunobu azide displacement of the 3-hydroxyl group of 7,12-dideoxycholic, 7-deoxycholic, and cholic acid methyl esters (*2152, 2153*). A reductive amination of 5-benzoylvaleric acid, via hydrogenation of a phenylhydrazone intermediate, produced 6-amino-6-phenylhexanoic acid (*1680*). N-Substituted 6-aminoheptanoic acid and 7-aminooctanoic acid were prepared by reductive amination of the corresponding ketone with racemic norepinephrine (*1681*).

11.8.2b Carboxylation Reactions

6-Aminohexanoic acid was prepared in 1907 from N-benzoyl piperidine via conversion to N-benzoyl 5-aminopentyl chloride, displacement with cyanide, and hydrolysis (*2154*). The synthesis was repeated with modifications in 1924 (*2155*). A similar procedure was used to prepare 4-(2-aminophenyl)butanoic acid (*2154*).

11.8.2c Amination and Carboxylation

7-Amino-2-pyrenecarboxylic acid has been prepared by partial reduction of pyrene, Friedel–Crafts acylation at the 2-position and conversion to a carboxyl group, oxidation back to a pyrene system, and nitration/reduction to introduce the amino group (*2156*).

11.8.2d Fusion of Amino-Containing and Carboxy-Containing Components

7-Aminoheptanoic acid and 4-(2-aminophenyl)pentanoic acid were prepared in 1907 via alkylation of diethyl malonate with the corresponding *N*-benzoyl alkyl iodide, followed by decarboxylation (*2154*), while alkylation with *N*-cyano piperidine-2-(3-bromopropane) led to piperidine-2-pentanoic acid (*2157*). Activation of Cbz-γ-aminobutyric acid or *N*-Cbz-5-aminopentanoic acid with CDI and displacement with the magnesium enolate of ethyl hydrogen malonate formed 6-amino-3-oxohexanoic acid and 7-amino-3-oxoheptanoic acid (*1830*).

3-(4-Piperidinyl)propionic acid was prepared by homologation of *N*-Boc-piperidine-4-carboxylic acid via reduction to the aldehyde and Wittig reaction with methyl triphenylphosphorylanylideneacetate, followed by hydrogenation. Repeating the reaction sequence generated the 5-(4-piperidinyl)pentanoic acid bishomolog, while 4-(4-piperidinyl)butyric acid was obtained by addition of diethyl malonate to 4-vinylpyridine, followed by decarboxylation and pyridine reduction (*2130*). An analog of 11-aminoundecanoic acid with a triethylene glycol backbone was constructed by reacting 2-(chloroethoxy)ethoxyethanol with ethyl diazoacetate to introduce an *O*-carboxymethyl group. The chloro group was then displaced with azide, with hydrogenolysis giving the amino acid (*2127*).

A photoactivable amino acid containing an azabenzene moiety was synthesized by coupling the nitroso-derivative

of *tert*-butyl 4-nitrobenzoate with Fmoc-protected 4-(aminomethyl)aniline (*2158*). Stille coupling of 2-trimethylstannyl-5-nitropyridine and methyl 2-triflylpyridine-5-carboxylate provided the metal-chelating bipyridyl-based amino acid 5′-amino-2,2′-bipyridine-5-carboxylic acid (*2159*). (Z)-11-Aminoundec-6-en-4,8-diynoic acid was constructed by Pd(0) Sonogashira coupling of *N*-acyl (Z)-1-chloro-7-aminohept-1-en-3-yne with amidated pent-4-ynoic acid (*2136*).

An ε-amino acid, 2-(4-aminomethyl-1, 2, 3-triazol-1-yl)-4-methylpentanoic acid, was obtained via a 1,3-dipolar cycloaddition reaction of a D-Leu-derived α-azido acid with *N*-Fmoc propargylamine (*2137*).

11.8.2e Elaboration of ω-Amino Acids

α-Chlorination of *N*-Bz 7-aminoheptanoic acid with sulfuryl chloride and chlorine, followed by hydrolysis with NaOH, formed the 2-hydroxy derivative (*1612*). *N*-Bz 2-bromo-6-aminohexanoic acid was also treated with alkali to give the 2-hydroxy derivative (*2049*). *N*-Boc-protected cyclic lactams formed from 6-aminohexanoic acid, 7-aminoheptanoic acid, and 8-aminooctanoic acid have been α-deprotonated and alkylated with a number of electrophiles (see Scheme 11.190). Hydrogen peroxide oxidation of the adducts obtained with PhSeCl generated α,β-unsaturated lactams, which were then employed for conjugate addition of lower-order organocuprates to give β-substituted derivatives. The lactams were converted to acyclic ω-amino acids by ring opening with nucleophiles, including hydroxide (*2160*).

Reduction of pyridine-4-propenoic acid with sodium borohydride gave 1,2,5,6-tetrahydropyridine-4-propenoic acid, with hydrogenation providing the saturated piperidine-4-propionic acid (*2131*). Similarly, pyridine-4-propanoic acid was reduced to 1,2,5,6-tetrahydropyridine-4-propanoic acid (*2131*). Pyridine-4-propenoic acid was also prepared, via another route (*2131*).

11.8.2f Other Syntheses

An efficient synthesis of 9-aminononanoic acid suitable for industrial synthesis was developed from

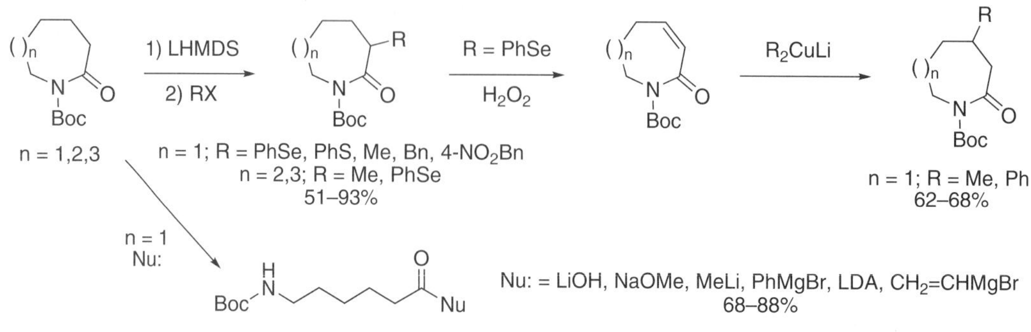

Scheme 11.190 Elaboration of ω-Lactams (*2160*).

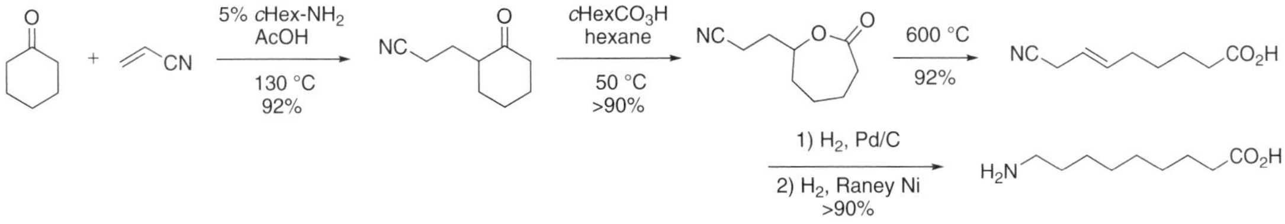

Scheme 11.191 Synthesis of 9-Aminononanoic Acid (*2161*).

cyclohexanone and acrylonitrile (see Scheme 11.191) (*2161*). Catalytic cyclohexylamine/acetic acid generated an imine intermediate of cyclohexanone, which isomerized to the enamine and underwent Michael addition to acrylonitrile, producing 2-(2-cyanoethyl)cyclohexanone in 92% yield. Baeyer–Villiger rearrangement gave the lactone, with pyrolytic ring opening and catalytic hydrogenation completing the synthesis (*2161*).

6-Aminohexanoic acid has been produced from butane-1,4-dinitrile. One nitrile group was hydrolyzed using a microbial cell catalyst with aliphatic nitrilase activity, producing δ-cyanocarboxylic acids. The second cyano group was then hydrogenated (*1718*). A series of ω-amino acids with the amino nitrogen contained in an imidazole ring were prepared by elaboration of 4-substituted imidazoles (*2162*).

Table 11.12 Syntheses of Racemic or Achiral ω-Amino Acids

Carbon Chain Length (n)	H₂N-R-CO₂H	Method	Yield	Reference
Amination				
4	$-(CH_2)_4-$	Beckmann rearrangement of	60–86% rearrange-	1995 (*2029*)
5	$-(CH_2)_5-$	cycloalkanones, hydrolysis	ment	
6	$-(CH_2)_6-$			
7	$-(CH_2)_7-$			
10	$-(CH_2)_{10}-$			
11				
5	$-(CH_2)_5-$	Beckmann rearrangement of cyclohexanone using microwave solid phase with hydroxylamine hydrochloride and P_2O_5/SiO_2	88% rearrangement to lactam	2003 (*2110*)
5	$-(CH_2)_5-$	Beckmann rearrangement of cyclohexanone oxime using P_2O_5 in ionic liquid	95% rearrangement to lactam	2001 (*2143*)
5	$-(CH_2)_5-$	reaction of cyclohexanone with air and ammonia in presence of transition metal molecular sieve catalyst	455	2001 (*2145*)
8	$-(CH_2)_8-$	Beckmann rearrangement of 10-ketooctadecanoic acid	20%	1894 (*2147*)
7	$-(CH_2)_7-$	Beckmann rearrangement of 9-ketoheptadecanoic acid	—	1894 (*2148*)
7	$-(CH_2)_7-$	Beckmann rearrangement of cycloalkanones, hydrolysis	41% overall	1996 (*2149*)
5 (as lactam)	$-(CH_2)_5-$ $-(CH_2)_2CH(tBu)(CH_2)_2-$ $-(CH_2)_2CH(Ph)(CH_2)_2-$ $-CH(Me)(CH_2)_4-$ $-(CH_2)_4CH(Me)-$ $-CH_2CH(Me)(CH_2)_3-$ $-(CH_2)_3CH(Me)CH_2-$	TiCl₄-catalyzed Schmidt rearrangement of cycloalkanones with benzyl azide or n-hexylazide	15–100%	2000 (*1508*)

(Continued)

Table 11.12 Syntheses of Racemic or Achiral ω-Amino Acids (continued)

Carbon Chain Length (n)	H₂N-R-CO₂H	Method	Yield	Reference
	$-CH_2CH[(CH_2)_3-]CH(-)(CH_2)_2-$			
	$-(CH_2)_2CH[(CH_2)_3-]CH(-)CH_2-$			
	$-CH_2CH(CH_2-)(CH_2)_2CH(-)-$			
	$-CH(CH_2-)(CH_2)_2CH(-)CH_2-$			
	3-amino-bicyclo[2.2.1]nonane-7-carboxylic acid			
5	$-(CH_2)_5-$	Beckmann rearrangement of	—	1900 (2030)
5	$-(CH_2)_4CH(Me)-$	cyclohexanone, 2-Me-		
5	$-CH_2CH(iPr)(CH_2)_2CH(Me)-$	cyclohexanone, methone,		
5	$-CH_2CH(Me)(CH_2)_2CH(iPr)-$	tetrahydrocarvone,		
6	$-(CH_2)_6-$	cycloheptanone		
5	$-CH_2(1,3\text{-cyclohexane})CH_2-$	Beckmann rearrangement of	71% oxime	1963 (2150)
	$-(2,2\text{-Me}_2\text{-}1,3\text{-cyclobutane})CH_2CH_2-$	bicyclic ketones	formation	
			43–56% rearrangement	
4	$-(CH_2)_4-$	Schmidt rearrangement of	74–83%	1959 (513)
7	$-(CH_2)_7-$	diacid monoester		
8	$-(CH_2)_8-$			
5	$-(CH_2)_5-$	Hofmann rearrangement of	—	1956 (512)
6	$-(CH_2)_6-$	adipic acid monomethyl		
7	$-(CH_2)_7-$	ester mono amide,		
8	$-(CH_2)_8-$	hydrolysis		
12	$-(CH_2)_{12}-$			
5	$-(1,2\text{-Ph})\text{-}(1,3\text{-Ph})-$	Suzuki coupling of aryl	69–92% coupling	1996 (2151)
9	$-(CH_2)_2\text{-}(1,3\text{-Ph})\text{-}(1,2\text{-Ph})\text{-}(CH_2)_2-$	derivatives, Curtius	46–70% Curtius	
		rearrangement		
12	3-amino-cholan-24-oic acid, and 12-acetoxy and 7,12-diacetoxy derivatives	Mitsunobu azidation of cholic acid derivatives	80% Mitsunobu 80% azide reduction	1994 (2153) 1997 (2152)
5	$-CH(Ph)(CH_2)_4-$	reductive amination of 6-keto-6-Ph-hexanoic acid via hydrazone	100% reduction	1977 (1680)
5	$-CH(Me)(CH_2)_4-$	reductive amination of	65–100%	1983 (1681)
6	$-CH(Me)(CH_2)_5-$	keto-alkanoic acid with norepinephrine		

Carboxylation

5	$-(CH_2)_5-$	displacement of N-Bz	—	1907 (2154)
5	$-(1,2\text{-Ph})\text{-}(CH_2)_3-$	aminoalkyliodide with CN, hydrolysis		
5	$-(CH_2)_5-$	conversion of Bz-piperidine to BzN(CH₂)₅I, displacement with CN, hydrolysis	—	1924 (2155)

Amination and Carboxylation

8	-2,7-pyrene-	Friedels–Crafts acylation of pyrene derivative, conversion to carboxyl, nitration	16% overall	1996 (2156)

Fusion of Amino- and Carboxy-Containing Components

5	$-(CH_2)_6-$	alkylation of diethyl malonate	—	1907 (2154)
6	$-(1,2\text{-Ph})\text{-}(CH_2)_4-$	with N-Bz aminoalkyliodide, hydrolysis		
5	$-CH[(CH_2)_4-(\omega\text{-N})]-(CH_2)_4-$	alkylation of diethyl malonate with 1-cyanopiperidine-2-(3-bromopropane), hydrolysis	67% alkylation 50% hydrolysis	1959 (2157)

Table 11.12 Syntheses of Racemic or Achiral ω-Amino Acids (continued)

Carbon Chain Length (n)	H$_2$N-R-CO$_2$H	Method	Yield	Reference
5 6	−(CH$_2$)$_3$COCH$_2$− −(CH$_2$)$_4$COCH$_2$−	activation of Cbz-GABA or Cbz-5-aminopentanoic acid with CDI, reaction with Mg enolate of malonic acid monoethyl ester	74–90% activation and enolate addition	1985 (1830)
5 7	−(CH$_2$)$_2$CH[(CH$_2$)$_2$-(ω-N)](CH$_2$)$_2$− −(CH$_2$)$_2$CH[(CH$_2$)$_2$-(ω-N)](CH$_2$)$_4$−	Wittig reaction of N-Boc piperidine-4-carboxalde-hyde with PPh$_3$P=CHCO$_2$Me; hydrogenation, reduction and repeat homologation	81% first Wittig 100% hydrogenation	1998 (2130)
6	−(CH$_2$)$_2$CH[(CH$_2$)$_2$-(ω-N)](CH$_2$)$_3$−	addition of diethyl malonate to 4-vinylpyridine, hydroge-nation	65% addition 98% hydrogenation	1998 (2130)
10	−(CH$_2$CH$_2$O)$_3$CH$_2$−	reaction of Cl(CH$_2$CH$_2$O)$_3$H with ethyl diazoacetate, then azide, hydrolysis, reduction	85% carboxymethy-lation 96% azide displacement 90% hydrolysis 100% reduction	1997 (2127)
10	−CH$_2$CH$_2$-CC-CH=CH-CC-CH$_2$CH$_2$−	Pd(0) Sonogashira coupling of N-acyl (Z)-1-chloro-7-aminohept-1-en-3-yne with amidated pent-4-ynoic acid	55–65% coupling	2002 (2136)
11	−(CH$_2$)$_2$−(1,4-Ph)-N=N-(1,4-Ph)−	coupling of 4-nitrobenzoic acid and 4-(aminomethyl) aniline	93% coupling	1994 (2158)
8	-(pyrid-5-yl)-(pyridy-2-yl)-	Stille coupling of 5-NO$_2$-2-SnMe$_3$-pyridine and 5-TfO-2-CO$_2$Me-pyridine, reduce nitro	73% coupling 89% nitro reduction	1996 (2159)
5	−CH$_2$(4-1,2,3-triazol-1-yl)-CH(iBu)−	1,3-dipolar cycloaddition of azido-Leu and N-Fmoc propargylamine	97%	2003 (2137)

Elaboration of ω-Amino Acids

6 6	−(CH$_2$)$_5$CH(Cl)− −(CH$_2$)$_5$CH(OH)−	chlorination of N-Bz-7-aminoheptanoate, displacement with OH	—	1973 (1612)
5	−(CH$_2$)$_4$CH(OH)−	reaction of N-Bz-2-bromo-6-aminohexanoate with OH	—	1909 (2049)
5 5 5 5 5 6 6 7 7 5 5	−(CH$_2$)$_4$CH(SePh)− −(CH$_2$)$_4$CH(Me)− −(CH$_2$)$_4$CH(SPh)− −(CH$_2$)$_4$CH(Bn)− −(CH$_2$)$_4$CH(4-NO$_2$-Bn)− −(CH$_2$)$_5$CH(SePh)− −(CH$_2$)$_5$CH(Me)− −(CH$_2$)$_6$CH(SePh)− −(CH$_2$)$_6$CH(Me)− −(CH$_2$)$_3$CH(Me)CH$_2$− −(CH$_2$)$_3$CH(Ph)CH$_2$−	alkylation of N-Boc lactam, elimination and conjugate addition, nucleophilic ring opening	51–93% alkylation 62–84% conjugate addition 68–88% ring opening	1990 (2160)
5 5 5 5	−CH$_2$CH$_2$CH[CH$_2$CH$_2$-(ω-N)]CH$_2$CH$_2$− −CH$_2$CH=C[CH$_2$CH$_2$-(ω-N)]CH$_2$CH$_2$− −CH$_2$CH$_2$CH[CH$_2$CH$_2$-(ω-N)]CH=CH− −CH$_2$CH=C[CH$_2$CH$_2$-(ω-N)]CH=CH−	reduction of pyridine-4-propanoic acid or pyridine-4-propenoic acid	49–68% pyridine reduction 66–69% hydrogena-tion	1995 (2131)

(Continued)

Table 11.12 Syntheses of Racemic or Achiral ω-Amino Acids (continued)

Carbon Chain Length (n)	H₂N-R-CO₂H	Method	Yield	Reference
Other Syntheses				
8	$-(CH_2)_8-$	Michael addition of cyclo-hexanone to acrylonitrile, Baeyer–Villiger oxidation, ring opening, reduction	>71% overall	1997 (*2161*)
5	$-(CH_2)_5-$	aliphatic nitrilase hydrolysis of one cyano group of butane-1,4-dinitrile, hydrogenation	<50% hydrolysis 95% reduction	1998 (*1718*)
3,4,5,6	(3-imidazole)(CH₂)ₙCH₂– (3-imidazole)(CH₂)ₙCH(Ph)– n = 0,1,2,3 (N in imidazole)	elaboration of 4-(substituted) imidazoles	7–59%	1997 (*2162*)

11.8.3 Asymmetric Syntheses of ω-Amino Acids (see Table 11.13)

11.8.3a. Amination Reactions

Evans' oxazolidinone auxiliary has been acylated with ω-bromoalkanoyl chlorides, followed by displacement of the terminal bromo group with azide. Deprotonation and α-alkylation with benzyl bromide, followed by hydrolysis and hydrogenation, gave 2-benzyl-8-aminooctanoic acid, 2-benzyl-10aminodecanoic acid or 2-benzyl-11-aminoundecanoic acids with 100% ee (*2055*). A synthesis of galantinic acid, 6-amino-3,5,7-trihydroxyheptanoic acid, employed an azide displacement of a optically enriched 2-bromo-1,3,5-trihydroxy-6-phenylsulfinyl-hexane. The sulfinyl group was converted into an epoxide and opened with cyanide to introduce the carboxyl group (*2163*).

A stereospecific oxaziridine-mediated nitrogen ring expansion was employed to prepare all four stereoisomers of 3,5-dimethyl-6-aminohexanoic acid from cis- or trans-3,5-dimethylcyclohexanone via imine formation with (S)- or (R)-α-methylbenzylamine, oxidation with m-CPBA, photolysis, auxiliary removal, and lactam hydrolysis (*2132*). The same reaction was applied to syntheses of 3-, 4-, and 5-benzyl-6-amino-hexanoic acids (*2133*). The Schmidt rearrangement of cyclic ketones has been converted into an asymmetric reaction by reacting 4-substituted cyclohexanones with chiral 1,3-hydroxyalkyl azides, producing N-alkyl caprolactams with up to 92% de (*2146*). Alternatively, Schmidt rearrangement of chiral 3,5-dimethylcyclo-hexanone with benzyl azide or n-hexyl azide in the presence of TiCl₄ gave the N-substituted lactams of 6-amino-3,5-dimethylhexanoic acid in 52–62% yields (*1508*).

Two isomers of ω-amino acids with a chiral [2.2] paracyclophane backbone have been reported, (+)-4-amino-13-carboxy-[2.2]paracyclophane and (+)-4-amino-12-carboxy-[2.2]paracyclophane. They were prepared by nitration of (+)-4-carbomethoxy-[2.2] paracyclophane (*2164*). Reductive amination of a chiral bicyclic ketone with a 7-hept-5-ynoic acid substituent produced 7-[(1R,2R,3R,5S)-2-amino-6,6-dimethylbicy-clo-[3.1.1]hept-3-yl]hept-5-ynoic acid (*2165*).

11.8.3b Fusion of Amino-Containing and Carboxy-Containing Components

The originally proposed (incorrect) structure of (+)-galan-tinic acid, 3-amino-4-hydroxypyranose-6-acetic acid, was synthesized via a Wittig reaction of a hemiacetal of protected 4-amino-3,5-dihydroxy-pentanal with methyl (tri-phenylphosphorylidene)acetate. Intramolecular Michael addition induced by potassium carbonate provided the cyclic ether structure (*2166*). The revised acyclic (−)-gal-antinic acid isomer was obtained from D-ribonolactone via a 3-cis-hydroxyprolinol intermediate; oxidation to the pyroglutaminol and ring opening with lithio tert-butylac-etate produced the desired skeleton (*2112*).

Several isomers of 2,8-dibenzyl-3,7-dihydroxy-8-aminooct-4-enoic acid were prepared in 2002 as high-affinity mu-opioid receptor ligands, acting as a Tyr-Pro-Phe tripeptide isostere. The analog was constructed using a cross-metathesis to form the alkene linkage (*2128*).

A synthesis of azimic acid, 2-methyl-3-hydroxy-6-(5-carboxypentane)piperidine, began with an asymmetric dipolar cycloaddition reaction of a homochiral allyl alcohol derived from methyl (−)-lactate with the Z-nitrone derived from N-benzylhydroxylamine and methyl 7-oxoheptanoate. Hydrogenolysis was used to cleave the N–O bond and give protected 6-amino-8,9,

10-trihydroxyundecanoic acid. The 10-hydroxy group was activated and cyclized to form the piperidine ring (*2123, 2124*).

11.8.3c Elaboration of α-Amino Acids

Amino aldehydes derived from amino acids have been elaborated into several ω-amino acids. Reetz et al. have used Wittig reactions of amino acid-derived amino aldehydes to prepare, after ester reduction and *O*-alkylation, ether-linked 4-amino-allyloxyacetates (see Scheme 11.153 above) (*1811*). A Wittig reaction of Boc-phenylalaninal was also used to prepare 6-amino-7-phenylheptanoic acid (*2133*). Hetero-Diels–Alder reaction of homochiral amino aldehydes with 1,3-dimethoxy-1-(silyloxy)butadiene under Lewis acid catalysis gave lactones of 6-amino-3-methoxy-5-hydroxyhex-2-eneoate (see Scheme 11.155 above). These were ozonolyzed to γ-amino acids, but presumably could also be derivatized to various 6-aminohexanoic acids (*1814*).

(*R*)- and (*S*)-carboxymethyl-substituted (*E*)-crotylsilanes were added to α-aminoaldehydes in the presence of boron trifluoride etherate to give good yields of 7-aminoheptanoic acid or 7-aminooctanoic acid derivatives, with varying stereoselectivity (*2167*). The aldehyde derived from *N*-Boc,*O*-benzyl Tyr was treated with vinylzinc chloride, followed by acetylation, to give an allylic acetate. Pd-catalyzed coupling with methyl (3-phenylsulfone)acetate and desulfonylation gave 6-amino-7-(4-benzyloxyphenyl)hept-4-enoic acid (*2168*). ε-Amino acids designed to mimic a polyketide synthesis were prepared by an aldol reaction of a (*Z*)-trichlorotitanium enolate of 3-keto-2-methylpentanoic acid (amidated with Evans' oxazolidinone auxiliary) with Fmoc-glycinal, producing 2,4-dimethyl-3-keto-5-hydroxy-6-aminohexanoic acid. Stereoselective ketone reduction with $Zn(BH_4)_2$ or $Me_4N(OAc)_3BH$ led to the two diastereomers of 2,4-dimethyl-3,5-dihydroxy-6-aminohexanoic acid (*2169*).

A total synthesis of carpimic acid, 2-methyl-3-hydroxy-6-(5-carboxyheptane)piperidine, began with addition of vinylMgBr to Cbz-alaninal. The vinyl group was employed for a Ru-catalyzed cross-metathesis reaction with methyl 9-ketoundec-10-enoate, with hydrogenation inducing *N*-deprotection, intramolecular reductive amination, and alkene reduction to produce the desired product (*2125*). Boc-alaninal was also used for a synthesis of azimic acid, 2-methyl-3-hydroxy-6-(5-carboxypentane)piperidine, which began with addition of $H_2C=CH(CH_2)_2MgBr$ to the aldehyde. Further side-chain elaboration provided 7-keto-10-hydroxy-11-amino-dodecanoic acid, which was cyclized by intramolecular reductive amination to form the desired piperidine ring product (*2170*). L-Ala was also elaborated into an unusual amino acid with the amino and carboxy termini linked by a series of three tetrahydrofuranyl rings (*2171*).

The Ser-derived Garner aldehyde was used as the source of the amino terminus of 6-amino-3,5,

7-trihydroxyheptanoic acid, galantinic acid. Two successive Wittig reactions and epoxidations were used to introduce the remainder of the molecule (*2113, 2114*). A more recent synthesis relied on an addition of allylMgBr to *O*-TBS Boc-serinal. The alkene was oxidatively cleaved to give an aldehyde, with a Reformatsky addition of ethyl zincbromoacetate introducing the *C*-terminal carboxymethyl group. The addition proceeded with little diastereoselectivity, so the carbinol center was oxidized to a ketone then reduced back to an alcohol (84% de). Galantinic acid was obtained in 16% overall yield from L-Ser, compared to 2% for the previous route (*2172*). An efficient synthesis of galantinic acid began with a diastereoselective aldol reaction of *N*-Cbz,*O*-TBS L-serinal with the ketene silyl acetal of phenyl acetate. The ester group was reduced to an aldehyde and a second aldol reaction with an acetate equivalent generated the polyhydroxylated carbon skeleton with the desired stereochemistry (*2115*). *N*-Cbz, *O*-TBDPS L-serinal was employed to prepare both destomic acid (6-amino-2,3,4,5,7-pentahydroxyheptanoic acid) and anhydrogalantinic acid via a chelation-controlled [4+2] cycloaddition with 1-benzyloxy-2-TBSO-4-ethoxy-1,3-butadiene, or 1-ethoxy-3-TMSO-1,3-butadiene respectively, followed by alkene dihydroxylation or intramolecular hydroxy Michael addition (*2173, 2174*).

The absolute configuration of galantinamic acid, 6,10-diamino-2,3,5-trihydroxydecanoic acid, was established to be (2*R*,3*S*,5*S*,6*R*) via stereoselective synthesis of eight isomers from L-Lys, with the natural isomer produced from D-Lys (*2120*). The Lys carboxyl group was reduced to an aldehyde, olefinated to give a hydroxymethylvinyl substituent, epoxidized, further homologated by a Wittig reaction, and then dihydroxylated. The cyclic ether version of galantinic acid, with an ether linkage between C3 and C7 instead of the free hydroxyl groups, was prepared by asymmetric epoxidation of vinylglycinol, followed by organocuprate opening of the epoxide to form 5,6-dihydroxy-6-amino-hept-2-enoic acid. Deprotection led to spontaneous cyclization (*2175*).

Both enantiomers of KAPA (7-keto-8-aminopelargonic acid) have been prepared by a Horner–Wadsworth–Emmons reaction of an Ala-derived phosphonate with benzyl 4-formylbutanoate (see Scheme 11.192) (*2126*). A series of KAPA analogs were derived from 12 different protected amino acids via activation of the carboxyl group with CDI and displacement with monomethylmalonate, followed by decarboxylation to give PNHCH(R)COCH$_2$CO$_2$Me. Alkylation with $I(CH_2)_4CO_2Et$ and hydrolysis/decarboxylation produced the 7-keto-8-amino-8-substituted octanoic acids. The ketone could be reductively aminated to give the corresponding 7,8-diamino-8-substituted octanoic acids (*1831*).

Lys contains an ε-amino acid skeleton. The α-amino group of Lys was deaminated with retention of configuration to give (*S*)-6-amino-2-hydroxypentanoic acid, with two reports 25 years apart (*1612, 2176*).

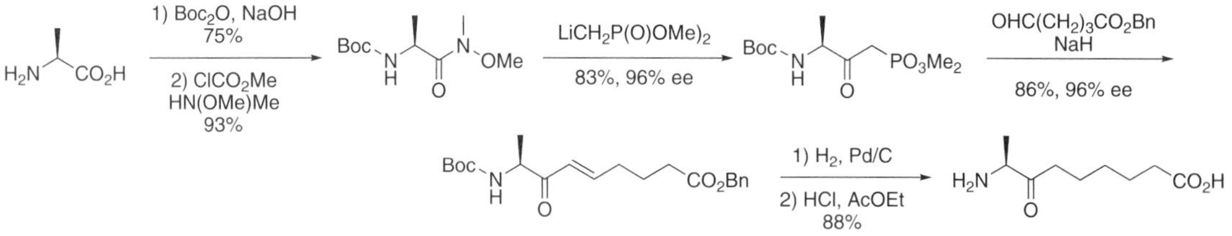

Scheme 11.192 Synthesis of 8-Amino-7-oxopelargonic Acid (*2126*).

11.8.3d Elaboration of ω-Amino Acids

The acid groups of *N*-methyl 6-aminohexanoic acid and 12-aminododecanoic acid were converted into homochiral oxazolines by reaction with an amino alcohol. Deprotonation and α-alkylation, followed by hydrolysis of the oxazoline, produced α-substituted-ω-aminoalkanoic acids with excellent diastereoselectivity (>90%), depending on the amino alcohol employed (*177*). Similarly, the oxazoline derivative of *N*-methyl 6-aminohexanoic acid was deprotonated and sulfenylated to give 2-thiomethyl-6-aminohexanoic acid or 2-thiophenyl-6-aminohexanoic acid (*1864*). The azido analog of 6-aminohexanoic acid, derivatized with Evans' oxazolidinone auxiliary, was deprotonated and α-benzylated with benzyl bromide to give, after auxiliary hydrolysis and azide reduction, 2-benzyl-6-aminohexanoic acid (*2133*).

Condensation of the lactam of 6-aminohexanoic acid with *p*-fluorobenzaldehyde under basic conditions introduced an α-alkylidene substituent. Asymmetric hydrogenation with a chiral 2,4-bis(diphenylphosphine) pentane-Ir catalyst produced the lactam of 2-(*p*-fluorobenzyl)-6-aminohexanoic acid with 67% ee (*1874*).

11.8.3e Elaboration of Chiral Substrates

Destomic acid (6-amino-2,3,4,5,7-pentahydroxyheptanoic acid) was synthesized from α-D-galactohexodialdo-1,5-pyranose diacetonide (see Scheme 11.193). The amino and hydroxymethyl moieties were introduced via thiazole addition to a nitrone intermediate (*2177, 2178*). Protected 6-amino-2,3,4,5-tetrahydroxyhexanoic acid was derived from D-ribose via addition and hydrolysis of NaCN, protection of the secondary hydroxy groups, and tosylation/azide displacement of the primary hydroxy group. Polymerization provided linear oligomers (up to octamer) as fully hydroxylated analogs of nylon 6 (*2142*). D-Galactono-1,4-lactone was converted into the lactam of 6-amino-2,3,4,5-tetrahydroxyhexanoic acid, via mesylation and azide displacement of the terminal hydroxyl group (*2179*). As mentioned earlier, a synthesis of the correct acyclic (−)-galantinic acid structure was achieved from D-ribonolactone (*2112*).

Axially asymmetric biaryl-based ω-amino acids (*R*)-2′-aminomethyl-6,6′-dimethyl-1,1′-biphenyl-2-carboxylic acid and (*S*)-2′-aminomethyl-6,6′-dimethyl-1,1′-biphenyl-2-acetic acid have been prepared from the chiral biaryl substrates (*2180*).

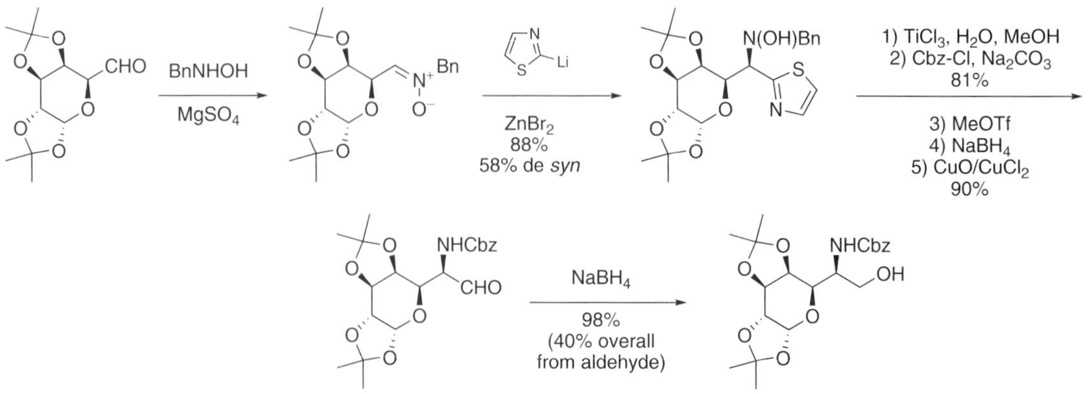

Scheme 11.193 Synthesis of Destomic Acid (*2177, 2178*).

Table 11.13 Asymmetric Syntheses of ω-Amino Acids

Carbon Chain Length (n)	H$_2$N-R-CO$_2$H	Method	Yield, % ee	Reference
Amination				
7	–(CH$_2$)$_6$CH(Bn)–	azide displacement of ω-bromoalkanoic acid derivatized with Evans' oxazolidinone, benzylation	97% azidation 83% benzylation 49% reduction/hydrolysis	1993 (2055)
9	–(CH$_2$)$_8$CH(Bn)–			
10	–(CH$_2$)$_9$CH(Bn)–			
5	–CH(CH$_2$OH)CH(OH)CH$_2$CH(OH)CH$_2$–	azide displacement of homochiral bromide substrate, CN opening of epoxide for carboxyl group	75% azide displacement 80% CN opening of epoxide 70% CN hydrolysis 80% azide reduction	2003 (2163)
5	–CH$_2$CH(Me)CH$_2$CH(Me)CH$_2$–	stereospecific ring expansion of 3,5-dimethylcyclohexanone via imine formation with α-Me-benzylamine, oxidation, photolysis, auxiliary removal, hydrolysis	70–83% imine and oxidation 53–60% photolysis 81–89% auxiliary removal	1995 (2132)
5	–CH$_2$CH$_2$CH$_2$CH(Bn)CH$_2$– –CH$_2$CH$_2$CH(Bn)CH$_2$CH$_2$– –CH$_2$CH(Bn)CH$_2$CH$_2$CH$_2$–	stereospecific ring expansion of 3- or 4-benzyl-cyclo-hexanones via imine formation with α-Me-benzylamine, oxidation, photolysis, auxiliary removal, hydrolysis	51–77% oxaziridine formation 79% ring expansion 76–85% deprotection 100% hydrolysis	2001 (2133)
5 (as lactam)	–CH$_2$CH(Me)CH$_2$CH(Me)CH$_2$–	TiCl$_4$-catalyzed Schmidt rearrangement of chiral 3,5-dimethyl-ylcyclohexanone with benzyl azide or n-hexylazide	15–100%	2000 (1508)
5 6	4-amino-13-carboxy-[2.2]paracyclophane 4-amino-12-carboxy-[2.2]paracyclophane	nitration of 4-carbomethoxy-[2.2]paracyclophane, reduction	23% nitration 83–86% nitro reduction	1997 (2164)
8	–CH[–CH$_2$–)C(Me)$_2$CH(–)CH$_2$–]CH(–)(CH$_2$CC(CH$_3$)$_3$–	reductive amination of chiral bicyclic ketone	75%	2003 (2165)
6	–CH$_2$CH$_2$CH(R)CH$_9$CH$_9$– R = Me, Ph, tBu	Schmidt rearrangement of 4-substituted cyclohexanones with chiral 1,2- or 1,3-azido alcohols	98–100% 86–92% de	2003 (2146)
Fusion of Amino- and Carboxy-Containing Components				
5	–CH(CH$_2$O–)CH(OH)CH$_2$CH(–)CH$_2$–	Wittig reaction of protected NH$_2$CH(CH$_2$OH)CH(OH)CH$_2$CHO, cyclization	49% Wittig 25% cyclization	1991 (2166)
5	–CH(CH$_2$OH)CH(OH)CH$_2$CH(OH)CH$_2$–	ring opening of 3-HO-pyroglutaminol with Li tBu acetate, ketone reduction	49% Wittig 25% cyclization	1991 (2112)
7	–CH(4-HO-BnCH(OH)CH$_2$CH=CHCH(OH)CH(Bn)–	cross-coupling olefin metathesis from two chiral alkenes	24–38%	2002 (2128)
5	–CH[CH$_2$CH(OH)CH(OH)CH(OH)Me](CH$_2$)$_4$	dipolar cycloaddition of nitrone of 7-oxoheptanoate and methyl lactate-derived allylic alcohol, opening and cyclization	83% cycloaddition	1996 (2123)
5	–CH[(CH$_2$CH(OH)CH(OH)CH(Me)–(ω-N)](CH$_2$)$_4$		62–93% ring opening and cyclization	1998 (2124)
5	–CH[(CH$_2$)$_2$CH(OH)CH(Me)–(ω-N)](CH$_2$)$_5$			

(*Continued*)

Table 11.13 Asymmetric Syntheses of α-Amino Acids (continued)

Carbon Chain Length (n)	$H_2N-R-CO_2H$	Method	Yield, % ee	Reference
Elaboration of α-Amino Acids				
5	$-CH_2CH(OH)CH(Me)C(Me)=CHCH_2-$	addition of (E)-crotylsilanes to α-amino aldehydes	46–94% yield	1997 (2167)
5	$-CH_2CH(OH)CH(Me)C(iBu)=CHCH_2-$		0–94% de	
5	$-CH_2CH(OH)CH(Me)C(Bn)=CHCH_2-$			
5	$-CH_2CH(OH)CH(Me)C(CH_2OTBS)=CHCH_2-$			
5	$-CH(Me)CH(OH)CH(Me)C(Bn)=CHCH_2-$			
5	$-CH(4-BnO-Bn)CH=CHCH_2CH_2-$	addition of vinylZnCl to tyrosinal, Pd-catalyzed reaction with $PhO_2SCH_2CO_2Me$, desulfonylation	>60% vinyl addition	1990 (2168)
			80% Pd-catalyzed coupling	
			61% desulfonylation	
6	$-CH(Me)CH=CHCH_2OCH_2-$	Wittig reaction of amino acid-derived amino aldehyde, reduction and O-alkylation	90–94% Wittig	1995 (1811)
6	$-CH(iPr)CH=CHCH_2OCH_2-$		60–83% reduction	
6	$-CH(iBu)CH=CHCH_2OCH_2-$		46–96% alkylation	
6	$-CH(Bn)CH=CHCH_2OCH_2-$			
5	$-CH(Bn)CH=CHCH_2CH_2-$	Wittig reaction of Phe-derived amino aldehyde, reduction	69% Wittig	2001 (2133)
5	$-CH(Bn)CH_2CH_2CH_2-$		100% reduction	
5	$-CH(Me)CH(OH)CH_2C(OMe)=CH-$	hetero-Diels-Alder of RCH(NHP)CHO and Brassard's diene	30–83%	1989 (1814)
5	$-CH(iPr)CH(OH)CH_2C(OMe)=CH-CH(iBu)CH(OH)CH_2C(OMe)=CH-$			
8	$-CH[-CH_2CH_2CH(OH)CH(Me)-(\omega-N)](CH_2)_7-$	addition of vinyl-MgCl to Cbz-alaninal, Ru metathesis with $H_2C=CHCO(CH_2)_7CO_2Me$, intramolecular reductive amination/alkene reduction	64% addition	2004 (2125)
			78% metathesis	
			100% reduction	
5	$-CH_2CH(OH)CH(Me)COCH(Me)-$	aldol reaction of (Z) $TiCl_3$ enolate of $EtCOCH(Me)CO$(Evans' oxazolidinone) with Fmoc-glycinal; stereoselective ketone reduction	81% aldol	2004 (2169)
5	$-CH_2CH(OH)CH(Me)CH(OH)CH(Me)-$		76% de	
			42–61% ketone reduction	
5	$-CH(CH_2OH)CH(OH)CH(OH)CH_2CH(OH)CH_2-$	elaboration of Ser-derived Garner aldehyde	2% overall	1990 (2113)
				1992 (2114)
5	$-CH(CH_2OH)CH(OH)CH_2C(=O)CH_2-$	elaboration of serinal	19% overall	1999 (2172)
5	$-CH(CH_2OH)CH(OH)CH_2CH(OH)CH_2-$			
5	$-CH(CH_2OH)CH(OH)CH(OH)CH(OH)CH(OH)-$	Diels–Alder of serinal aldehyde with 1-BnO-2-TBS-O-4-EtO-butadiene, dihydroxylation, oxidation or intramolecular Michael addition	—	1989 (2174)
5	$-CH(CH_2O-C3)CH(OH)CH_2CH(OCH_2-C6)CH_2-$			1991 (2173)
5	$-CH[(CH_2)_4NH_2]CH(OH)CH_2CH(OH)CH(OH)-$	elaboration of Lys	—	1988 (2120)
5	$-CH(CH_2OH)CH(OH)CH_2CH=CH-$	epoxidation of vinylglycinol, organocuprate opening, oxidation, deprotection/cyclization	60% epoxidation	1984 (2175)
5	$-CH(CH_2O-C3)CH(OH)-CH_2CH(OCH_2-C6)CH_2-$		51% opening	
			75% oxidation	

n	Structure	Method	Yield	Year (Ref)
7	—CH(Me)CO(CH₂)₅—	Horner–Wadsworth–Emmons coupling of Ala-derived phosphonate	86% coupling 96% ee	1996 (2126)
7	—CH₂CO(CH₂)₅— CH(Me)CO(CH₂)₅— CH(Et)CO(CH₂)₅— CH(iPr)CO(CH₂)₅— CH(Bu)CO(CH₂)₅— CH(sBu)CO(CH₂)₅— CH(CH₂CH₂SMe)CO(CH₂)₅— CH(CH₂SMe)CO(CH₃)₅— CH(CH₂SH)CO(CH₂)₅— CH(CH₂OH)CO(CH₂)₅— CH(CH(OH)Me[CO(CH₃)₅— CH(CH=CH₂)CO CO(CH₂)₅—	activation of Boc-Xaa with CDI, addition of KO₂CCH₂CO₂Me, hydrolysis, alkylation with I(CH₂)₄CO₂Et, hydrolysis/ decarboxylation, reductive amination	60–76% CDI activation and malonate addition 55% alkylation 84–96% hydrolysis/ decarboxylation 74% reductive amination	2004 (1831)
6	—CH(CH₂NH₂)(CH₂)₅— CH[CH(Me)NH₂](CH₂)₅— CH[CH(Et)NH₂](CH₂)₅— CH[CH(iPr)NH₂](CH₂)₅— CH[CH(Bu)NH₂](CH₂)₅— CH[CH(sBu)NH₂](CH₂)₅— CH[CH(CH₂CH₂SMe)NH₂](CH₂)₅— CH[CH(CH₂SMe)NH₂](CH₂)₅— CH[CH(CH₂OH)NH₂](CH₂)₅— CH[CH(CH=CH₂)NH₂](CH₂)₅— . CH[CH[CH(OH)Me]NH₂](CH₂)₅—	elaboration of Boc-Ala via alaninal, addition of Grignard reagent, elaboration to 7-keto-10-hydroxy-11-amino-dodecanoic acid, intramolecular reductive amination	38% alaninal formation and Grignard addition 42% elaboration 55% reductive amination	1999 (2170)
6	—CH[CH₂)₃CH(OH)CH(OH)CH(Me)-(ω-N)](CH₂)₅—			
11	—CH(Me)-[(2,5-tetrahydrofuran)₃]-CH(Me)—	elaboration of Ala	3% overall	1996 (2171)
5	—(CH₂)₃CH(OH)—	nitrous deamination of Lys	—	1999 (2176)
5	—(CH₂)₄CH(OH)—	diazotization of L-Lys	—	1973 (1612)
5	—CH(CH₂OH)CH(OH)CH(OH)CH₂CH(OH)CH₂—	aldol reaction of Ser aldehyde with acetate silyl enol, acid reduction to aldehyde, second aldol	74% aldol with serinal 83% reduction to aldehyde 76% aldol	2000 (2115)

Elaboration of ω-Amino Acids

n	Structure	Method	Yield	Year (Ref)
5	—(CH₂)₄CH(Me)—	formation of homochiral oxazoline, alkylation	54–68% alkylation >90% de	1996 (177)
12	—(CH₂)₁₁CH(nBu)—		66–93% hydrolysis	

(*Continued*)

Table 11.13 Asymmetric Syntheses of ω-Amino Acids (continued)

Carbon Chain Length (n)	H$_2$N-R-CO$_2$H	Method	Yield, % ee	Reference
5	-(CH$_2$)$_4$CH(SMe)-	formation of homochiral oxazoline, sulfenylation	44–73% sulfenylation	1997 (1864)
5	-(CH$_2$)$_4$CH(SPh)-			
5	-CH$_2$CH$_2$CH$_2$CH$_2$CH(Bn)-	benzylation of 6-azidohexanoic acid derivatized with Evans' oxazolidinone, hydrolysis, azide reduction	73% benzylation 90% auxiliary removal 80% azide reduction	2001 (2133)
5	-CH$_2$CH$_2$CH$_2$CH$_2$CH(4-F-Bn)- (lactam)	condensation of lactam with RCHO, alkene reduction by asymmetric hydrogenation with chiral Ir catalyst	97% reduction 67% ee	2002 (1874)

Elaboration of Chiral Substrates

Carbon Chain Length (n)	H$_2$N-R-CO$_2$H	Method	Yield, % ee	Reference
5	-CH(CH$_2$OH)CH(OH)CH(OH)CH(OH)CH(OH)-	elaboration of sugar derivative via thiazole additon to nitrone, conversion of thiazole to aldehyde, aldehyde reduction	68–90% thiazole addition 58% de *syn* to 82% de *anti* 86–90% thiazole to aldehyde 95% aldehyde reduction	1993 (2178) 1995 (2177)
4	-CH$_2$CH(OH)CH(OH)CH(OH)CH(OH)-	elaboration of D-ribose via NaCN addition, hydrolysis, protection of secondary hydroxyl, displacement of tosylate of primary hydroxyl with azide	30% NaCN addition/ hydrolysis 73% tosylation 81% azide displacement	2003 (2142)
5	CH(OH)CH(OH)CH(OH)CH(OH)CH$_2$-	elaboration of sugar derivative, amination via azide displacement of mesylate	86% azide displacement	1999 (2179)
5	-CH$_2$(3-Me-1,2-Ph)-(3-Me-1,2-Ph)-	elaboration of chiral biphenyl derivatives	—	1998 (2180)
6	-CH$_2$(3-Me-1,2-Ph)-(3-Me-1,2-Ph)-CH$_2$-			

References

1. Smith, M.B. *Methods of Non-α-Amino Acid Synthesis*; Marcel Dekker: New York, 1995.
2. *Enantioselective Synthesis of β-Amino Acids*; Juaristi, E., Ed.; VCH Publishers (John Wiley and Sons): New York, 1997; 2nd edn, Juaristi, E.; Soloshonok, V.A., Eds; Wiley-VCH: New York, 2005.
3. Nicolaou, K.C.; Dai, W.-M.; Guy, R.K. *Angew. Chem., Int. Ed. Engl.* **1994**, *33*, 15.
4. Axten, J.M.; Krim, L.; Kung, H.F.; Winkler, J.D. *J. Org. Chem.* **1998**, *63*, 9628–9629.
5. Prashad, M.; Har, D.; Repic, O.; Blacklock, T.J.; Giannousis, P. *Tetrahedron: Asymmetry* **1998**, *9*, 2133–2136.
6. Hart, D.J.; Ha, D.-C. *Chem. Rev.* **1989**, *89*, 1447.
7. *The Organic Chemistry of β-Lactams*; George, G.I., Ed.; VCH Publishers: New York, 1992.
8. Okada, Y.; Tsuda, Y.; Yagyu, M. *Chem. Pharm. Bull.* **1980**, *28*, 310–313.
9. Suzuki, K.; Abiko, T. *Chem. Pharm. Bull.* **1966**, *14*, 1017–1023.
10. Fujino, M.; Hatanaka, C.; Nishimura, O. *Chem. Pharm. Bull.* **1970**, *18*, 1288–1291.
11. Fujino, M.; Hatanaka, C.; Nishimura, O. *Chem. Pharm. Bull.* **1970**, *18*, 771–778.
12. Iveerson, B.L. *Nature* **1997**, *385*, 113–115.
13. Seebach, D.; Beck, A.K.; Bierbaum, D.J. *Chem. Biodiv.* **2004**, *1*, 1111–1239.
14. Cheng, R.P.; Gellman, S.H.; DeGrado, W.F. *Chem. Rev.* **2001**, *101*, 3219–3232.
15. Drey, C.N.C.; Mtetwa, E. *J. Chem. Soc., Perkin Trans. 1* **1982**, 1587–1592.
16. Chen, F.; Lepore, G.; Goodman, M. *Macromolecules* **1974**, *7*, 779–783.
17. Griffith, O.W. *Ann. Rev. Biochem.* **1986**, *55*, 855–878.
18. Fritzson, P.; Pihl, A. *J. Biol. Chem.* **1957**, *226*, 229–235.
19. Canellakis, E.S. *J. Biol. Chem.* **1956**, *221*, 315–322.
20. Wallach, D.P.; Grisolia, S. *J. Biol. Chem.* **1958**, *231*, 357–365.
21. Kupiecki, F.P.; Coon, M.J. *J. Biol. Chem.* **1957**, *229*, 743–754.
22. Fink, K.; Cline, R.E.; Henderson, R.B.; Fink, R.M. *J. Biol. Chem.* **1956**, *221*, 425–433.
23. Shinohara, T.; Harada, M.; Ogi, K.; Maruyama, M.; Fujii, R.; Tanake, H.; Fukusumi, S.; Komatsu, H.; Hosoya, M.; Noguchi, Y.; Watanabe, T.; Moriya, T.; Itoh, Y.; Hinuma, S. *J. Biol. Chem.* **2004**, *279*, 23559–23564.
24. Gulewitsch, W.; Amiradzibi, S. *Ber.* **1900**, *33*, 1902–1903.
25. Barger, G.; Tutin, F. *Biochem J.* **1918**, *12*, 402–407.
26. Baumann, L.; Ingvaldsen, T. *J. Biol. Chem.* **1918**, *35*, 263–276.
27. Nardiello, D.; Cataldi, T.R.I. *J. Chromatogr., A* **2004**, *1035*, 285–289.
28. Kakimoto, Y.; Nakajima, T.; Kumon, A.; Matsuoka, Y.; Imaoka, N.; Sano, I.; Kanazawa, A.; Shiba, T.; Kaneko, T. *J. Biol. Chem.* **1969**, *244*, 6003–6007.
29. Nakajima, T.; Kakimoto, Y.; Sano, I. *J. Neurochem.* **1970**, *17*, 1427–1428.
30. Nakajima, T.; Matsuoka, Y.; Akazawa, S. *Biochim. Biophys. Acta* **1970**, *222*, 405–408.
31. Ehrenfreund, P.; Glavin, D.P.; Botta, O.; Cooper, G.; Bada, J.L. *Proc. Natl. Acad. Sci.U.S.A* **2001**, *98*, 2138–2141.
32. Meierhenrich, U.J.; Muñoz Caro, G.M.; Bredehöft, J.H.; Jessberger, E.K.; Thiemann, W.H.-P. *Proc. Notl. Aead. Sci. U.S.A* **2004**, *101*, 9182–9186.
33. Crumpler, H.R.; Dent, C.E.; Harris, H.; Westall, R.G. *Nature* **1951**, *167*, 307–308.
34. Fink, K.; Henderson, R.B.; Fink, R.M. *J. Biol. Chem.* **1952**, *197*, 441–452.
35. Kakimoto, Y.; Kanazawa, A.; Sano, I. *Biochim. Biophys. Acta* **1965**, *97*, 376–377.
36. Kakimoto, Y.; Kanazawa, A.; Taniguchi, K.; Sano, I. *Biochim. Biophys. Acta* **1968**, *156*, 374–380.
37. Bewley, C.A.; Faulkner, D.J. *Angew. Chem., Int. Ed. Engl.* **1998**, *37*, 2162–2178.
38. Roy, M.C.; Ohtani, I.I.; Tanaka, J.; Higa, T.; Satari, R. *Tetrahedron Lett.* **1999**, *40*, 5373–5376.
39. Tsuda, M.; Shimbo, K.; Kubota, T.; Mikami, Y.; Kobayashi, J. *Tetrahedron* **1999**, *55*, 10305–10314.
40. Impellizzeri, G.; Mangiafico, S.; Oriente, G.; Piattelli, M.; Sciuto, S.; Fattorusso, E.; Magno, S.; Santacroce, C.; Sica, D. *Phytochemistry* **1975**, *14*, 1549–1557.
41. Durham, N.N.; Jacobs, C.D.; Ferguson, D. *J. Bacteriol.* **1964**, *88*, 1525–1526.
42. Carnegie, P.R. *Biochem. J.* **1963**, *89*, 459–471.
43. Carnegie, P.R. *Biochem. J.* **1963**, *89*, 471–478.
44. Ono, T.; Hirohata, R. *Z. Physiol. Chem.* **1956**, *304*, 77–81.
45. Kendo, K. *J. Biochem.* **1944**, *36*, 265–276.
46. Imamura, H. *J. Biochem.* **1939**, *30*, 479–490.
47. Hayakawa, Y.; Adachi, H.; Kim, J.W.; Shin-ya, K.; Seto, H. *Tetrahedron* **1998**, *54*, 15871–15878.
48. Chomcheon, P.; Wiyakrutta, S.; Sriubolmas, N.; Ngamrojanavanich, N.; Isarangkul, D.; Kittakoop, P. *J. Nat. Prod.* **2005**, *68*, 1103–1105.
49. Morris, C.J.; Thompson, J.F.; Asen, S.; Irreverre, F. *J. Biol. Chem.* **1961**, *236*, 1181–1182.
50. Morris, C.J.; Thompson, J.F.; Asen, S.; Irreverre, F. *J. Biol. Chem.* **1962**, *237*, 2180–2181.
51. Larsen, P.O. *Acta Chem. Scand.* **1962**, *16*, 1511–1518.
52. Mazus, B.; Buchowicz, J. *Acta Biochimi. Pol.* **1966**, *13*, 267–273.
53. Varie, D.L.; Shih, C.; Hay, D.A.; Andis, S.L.; Corbett, T.H.; Gossett, L.S.; Janisse, S.K.; Martinelli, M.J.; Moher, E.D.; Schultz, R.M.; Toth, J.E. *Biorg. Med. Chem. Lett.* **1999**, *9*, 369–374.
54. Norman, B.H.; Hemscheidt, T.; Schultz, R.M.; Andis, S.L. *J. Org. Chem.* **1998**, *63*, 5288–5294.
55. Barrow, R.A.; Hemscheidt, T.; Liang, J.; Paik, S.; Moore, R.E.; Tius, M.A. *J. Am. Chem. Soc.* **1995**, *117*, 2479–2490.

56. Trimurtulu, G.; Ohtani, I.; Patterson, G.M.L.; Moore, R.E.; Corbett, T.H.; Valeriote, F.A.; Demchik, L. *J. Am. Chem. Soc.* **1994**, *116*, 4729–4737.

57. Rej, R.; Nguyen, D.; Go, B.; Fortin, S.; Lavallée, J.-F. *J. Org. Chem.* **1996**, *61*, 6289–6295.

58. Gardinier, K.M.; Leahy, J.W. *J. Org. Chem.* **1997**, *62*, 7098–7099.

59. de Muys, J.-M.; Rej, R.; Nguyen, D.; Go, B.; Fortin, S.; Lavallée, J.-F. *Biorg. Med. Chem. Lett.* **1996**, *6*, 1111–1116.

60. Ali, S.M.; Georg, G.I. *Tetrahedron Lett.* **1997**, *38*, 1703–1706.

61. Golakoti, T.; Ogino, J.; Heltzel, C.E.; Husebo, T.L.; Jensen, C.M.; Larsen, L.K.; Patterson, G.M.L.; Moore, R.E.; Mooberry, S.L.; Corbett, T.H.; Valeriote, F.A. *J. Am. Chem. Soc.* **1995**, *117*, 12030–12049.

62. Kobayashi, M.; Aoki, S.; Ohyabu, N.; Kurosu, M.; Wang, W.; Kitagawa, I. *Tetrahedron Lett.* **1994**, *35*, 7969–7972.

63. Salamonczyk, G.M.; Han, K.; Guo, Z.-W.; Sih, C.J. *J. Org. Chem.* **1996**, *61*, 6893–6900.

64. Bergmeier, S.C.; Cobás, A.A.; Rapoport, H. *J. Org. Chem.* **1993**, *58*, 2369–2376.

65. Ghosh, A.K.; Bischoff, A. *Eur. J. Org. Chem.* **2004**, 2131–2141.

66. Mast, C.A.; Eißler, S.; Stoncius, A.; Stammler, H.-G.; Neumann, B.; Sewald, N. *Chem.—Eur. J.* **2005**, *11*, 4667–4677.

67. Danner, P.; Bauer, M.; Phukan, P.; Maier, M.E. *Eur. J. Org. Chem.* **2005**, 317–325.

68. Eggen, M.; Mossman, C.J.; Buck, S.B.; Nair, S.K.; Bhat, L.; Ali, S.M.; Reiff, E.A.; Boge, T.C.; Georg, G.I. *J. Org. Chem.* **2000**, *65*, 7792–7799.

69. Vidya, R.; Eggen, M.; Nair, S.K.; Georg, G.I.; Himes, R.H. *J. Org. Chem.* **2003**, *68*, 9687–9693.

70. Beck, Z.Q.; Aldrich, C.C.; Magarvey, N.A.; Georg, G.I.; Sherman, D.H. *Biochemistry* **2005**, *44*, 13457–13466.

71. Ghosh, A.K.; Swanson, L. *J. Org. Chem.* **2003**, *68*, 9823–9826.

72. Barrow, R.A.; Moore, R.E.; Li, L.-H.; Tius, M.A. *Tetrahedron* **2000**, *56*, 3339–3351.

73. Tius, M.A. *Tetrahedron* **2002**, *58*, 4343–4367.

74. Carter, D.C.; Moore, R.E.; Mynderse, J.S.; Niemczura, W.P.; Todd, J.S. *J. Org. Chem.* **1984**, *49*, 236–241.

75. Pettit, G.R.; Kamano, Y.; Herald, C.L.; Fujii, Y.; Kizu, H.; Boyd, M.R.; Boettner, F.E.; Doubek, D.L.; Schmidt, J.M.; Chapuis, J.-C.; Michel, C. *Tetrahedron* **1993**, *49*, 9151–9170.

76. Pettit, G.R.; Kamano, Y.; Kizu, H.; Dufresne, C.; Herald, C.L.; Bontems, R.J.; Schmidt, J.M.; Boettner, F.E.; Nieman, R.A. *Heterocycles* **1989**, *28*, 553–558.

77. Kashman, Y.; Fishelson, L.; Ne'eman, I. *Tetrahedron* **1973**, *29*, 3655–3657.

78. Freer, I.; Pedrocchi-Fantoni, G.; Picken, D.J.; Overton, K.H. *J. Chem. Soc., Chem. Commun.* **1981**, 80–82.

79. Peypoux, F.; Guinand, M.; Michel, G.; Delcambe, L.; Das, B.C.; Varenne, P.; Lederer, E. *Tetrahedron* **1973**, *29*, 3455–3459.

80. Isogai, A.; Takayama, S.; Murakoshi, S.; Suzuki, A. *Tetrahedron Lett.* **1982**, *23*, 3065–3068.

81. Nagai, U. *Tetrahedron Lett.* **1979**, *25*, 2359–2360.

82. Rodríguez, J.; Fernández, R.; Quiñoá, E.; Riguera, R.; Debitus, C.; Bouchet, P. *Tetrahedron Lett.* **1994**, *35*, 9239–9242.

83. Onuki, H.; Ito, K.; Kobayashi, Y.; Matsumori, N.; Tachibana, K.; Fusetani, N. *J. Org. Chem.* **1998**, *63*, 3925–3932.

84. Kan, T.; Fujiwara, A.; Kobayashi, H.; Fukuyama, T. *Tetrahedron* **2002**, *58*, 6267–6276.

85. Fujiwara, A.; Kan, T.; Fukuyama, T. *Synlett* **2000**, 1667–1669.

86. Peng, Y.; Pang, H.W.; Ye, T. *Org. Lett.* **2004**, *6*, 3781–3784.

87. MacMillan, J.B.; Ernst-Russell, M.A.; de Ropp, J.S.; Molinski, T.F. *J. Org. Chem.* **2002**, *67*, 8210–8215.

88. Expósito, A.; Fernández-Suárez, M.; Iglesias, T.; Muñoz, L.; Riguera, R. *J. Org. Chem.* **2001**, *66*, 4206–4213.

89. Ensch, C.; Hesse, M. *Helv. Chim. Acta* **2002**, *85*, 1659–1673.

90. Detterbeck, R.; Guggisberg, A.; Popaj, K.; Hesse, M. *Helv. Chim. Acta* **2002**, *85*, 1742–1758.

91. Sato, M.; Tatsuno, T. *Chem. Pharm. Bull.* **1968**, *16*, 2182–2190.

92. Bohman-Lindgren, G.; Ragnarsson, U. *Tetrahedron* **1972**, *28*, 4631–4634.

93. Koyama, M.; Obata, Y. *Agric. Biol. Chem.* **1966**, *30*, 472–477.

94. Koyama, M.; Obata, Y. *Agric. Biol. Chem.* **1967**, *31*, 738–742.

95. Kaseda, T.; Kikuchi, T.; Kibayashi, C. *Tetrahedron Lett.* **1989**, *30*, 4539–4542.

96. Sergeyev, S.; Hesse, M. *Helv. Chim. Acta* **2003**, *86*, 465–473.

97. Davies, S.G.; Dixon, D.J. *J. Chem. Soc., Perkin Trans. 1* **1998**, 2635–2643.

98. Schumacher, K.K.; Hauze, D.B.; Jiang, J.; Szewczyk, J.; Reddy, R.E.; Davis, F.A.; Joullié, M.M. *Tetrahedron Lett.* **1999**, *40*, 455–458.

99. Morita, H.; Nagashima, S.; Takeya, K.; Itokawa, H. *Tetrahedron* **1995**, *51*, 1121–1132.

100. Morita, H.; Nagashima, S.; Takeya, K.; Itokawa, H. *Chem. Pharm. Bull.* **1993**, *41*, 992–993.

101. Morita, H.; Nagashima, S.; Shirota, O.; Takeya, K.; Itokawa, H. *Chem. Lett.* **1993**, 1877–1880.

102. Morita, H.; Nagashima, S.; Takeya, K.; Itokawa, H. *Chem. Lett.* **1994**, 2009–2010.

103. Kosemura, S.; Ogawa, T.; Totsuka, K. *Tetrahedron Lett.* **1993**, *34*, 1291–1294.

104. Morita, H.; Nagashima, S.; Takeya, K.; Itokawa, H. *Chem. Pharm. Bull.* **1995**, *43*, 1395–1397.

105. Morita, H.; Nagashima, S.; Takeya, K.; Itokawa, H. *Tetrahedron* **1994**, *50*, 11613–11622.

106. Morita, H.; Nagashima, S.; Uchimi, Y.; Kuroki, O.; Takeya, K.; Itokawa, H. *Chem. Pharm. Bull.* **1996**, *44*, 1026–1032.

107. Morita, H.; Nagashima, S.; Takeya, K.; Itokawa, H. *Chem. Pharm. Bull.* **1995**, *43*, 271–273.

108. Morita, H.; Nagashima, S.; Takeya, K.; Itokawa, H. *J. Chem. Soc., Perkin Trans. 1* **1995**, 2327–2331.

109. Guggisberg, A.; Drandarov, K.; Hesse, M. *Helv. Chim. Acta* **2000**, *83*, 3035–3042.

110. Drandarov, K.; Guggisberg, A.; Hesse, M. *Helv. Chim. Acta* **2002**, *85*, 979–989.

111. Youhnovski, N.; Drandarov, K.; Guggisberg, A.; Hesse, M. *Helv. Chim. Acta* **1999**, *82*, 1185–1194.

112. Sergeyev, S.A.; Hesse, M. *Helv. Chim. Acta* **2002**, *85*, 161–167.

113. Nezbedová, L.; Hesse, M.; Drandarov, K.; Werner, C. *Tetrahedron Lett.* **2000**, *41*, 7859–7862.

114. Nezbedova, L.; Drandarov, K.; Werner, C.; Hesse, M. *Helv. Chim. Acta* **2000**, *83*, 2953–2974.

115. Dimitrov, V.; Geneste, H.; Guggisberg, A.; Hesse, M. *Helv. Chim. Acta* **2001**, *84*, 2108–2118.

116. Kupchan, S.M.; Hintz, H.P.J.; Smith, R.M.; Karim, A.; Cass, M.W.; Court, W.A.; Yatagai, M. *J. Org. Chem.* **1977**, *42*, 3660–3664.

117. Kupchan, S.M.; Hintz, H.P.J.; Smith, R.M.; Karim, A.; Cass, M.W.; Court, W.A.; Yatagai, M. *J. Chem. Soc., Chem. Commun.* **1974**, 329–330.

118. Schultz, K.; Hesse. M. *Helv. Chim Acta* **1996**, *79*, 1295–1304.

119. Wagner, H.; Burghart, J.; Hull, W.E. *Tetrahedron Lett.* **1978**, 3893–3896.

120. Wasserman, H.H.; Robinson, R.P.; Matsuyama, H. *Tetrahedron Lett.* **1980**, *21*, 3493–3496.

121. Yamamoto, H.; Maruoka, K. *J. Am. Chem. Soc.* **1981**, *103*, 6133–6136.

122. Iida, H.; Fukuhara, K.; Machiba, M.; Kikuchi, T. *Tetrahedron Lett.* **1986**, *27*, 207–210.

123. Matsuyama, H.; Itoh, N.; Matsumoto, A.; Ohira, N.; Hara, K.; Yoshida, M.; Iyoda, M. *J. Chem. Soc., Perkin Trans. 1* **2001**, 2924–2930.

124. Iida, H.; Fukuhara, K.; Murayama, Y.; Machiba, M.; Kikuchi, T. *J. Org. Chem.* **1986**, *51*, 4701–4703.

125. Schultz, K.; Hesse, M. *Tetrahedron* **1996**, *52*, 14189–14198.

126. Li, Y.; Linden, A.; Hesse, M. *Helv. Chim. Acta* **2003**, *86*, 579–591.

127. Khanjin, N.A.; Hesse, M. *Helv. Chim. Acta* **2001**, *84*, 1253–1267.

128. Chu, K.S.; Negrete, G.R.; Konopelski, J.P. *J. Org. Chem.* **1991**, *56*, 5196–5202.

129. Fusetani, N; Matsunaga, S. *Chem. Rev.* **1993**, *93*, 1793–1806.

130. Rama Rao, A.V.; Gurjar, M.K.; Nallaganchu, B.R.; Bhandari, A. *Tetrahedron Lett.* **1993**, *34*, 7085–7088.

131. Ashworth, P.; Broadbelt, B.; Jankowski, P.; Kocienski, P.; Pimm, A.; Bell, R. *Synthesis* **1995**, 199–206.

132. Hettinger, T.P.; Craig, L.C. *Biochemistry* **1970**, *9*, 1224–1232.

133. Hettinger, T.P.; Kurylo-Borowska, Z.; Craig, L.C. *Biochemistry* **1968**, *7*, 4151–4160.

134. Kurylo-Borowska, Z.; Abramsky, T. *Biochim. Biophys. Acta* **1972**, *264*, 1–10.

135. von Nussbaum, F.; Spiteller, P.; Rüth, M.; Steglich, W.; Wanner, G.; Gamblin, B.; Stievano, L.; Wagner, F.E. *Angew. Chem., Int. Ed. Engl.* **1998**, *37*, 3292–3295.

136. Waisvisz, J.M.; van der Hoeven, M.G.; te Nijenhais, B. *J. Am. Chem. Soc.* **1957**, *79*, 4524–4527.

137. Nakamura, S.; Chikaike, T.; Yonehara, H.; Umezawa, H. *J. Antibiotics* **1965**, *18*, 60–61.

138. Nakamura, S.; Chikaike, T.; Yonehara, H.; Umezawa, H. *Chem. Pharm. Bull.* **1965**, *13*, 599–602.

139. Yamada, T.; Takashima, K.; Miyazawa, T.; Kuwata, S.; Watanabe, H. *Bull. Chem. Soc. Jpn.* **1978**, *51*, 878–883.

140. Usuki, T.; Inoue, M.; Hirama, M.; Tanaka, T. *J. Am. Chem. Soc.* **2004**, *126*, 3022–3023.

141. Myers, A.G.; Hurd, A.R.; Hogan, P.C. *J. Am. Chem. Soc.* **2002**, *124*, 4583–4585.

142. Myers, A.G.; Hogan, P.C.; Hurd, A.R.; Goldberg, S.D. *Angew. Chem., Int. Ed.* **2002**, *41*, 1062–1067.

143. Ikeda, Y.; Ikeda, D.; Kondo, S. *J. Antibiotics* **1992**, *45*, 1677–1680.

144. Brockmann, H.; Musso, H. *Ber.* **1955**, *88*, 648–661.

145. van Tamelen, E.E.; Smissman, E.E. *J. Am. Chem. Soc.* **1952**, *74*, 3713–3714.

146. van Tamelen, E.E.; Smissman, E.E. *J. Am. Chem. Soc.* **1953**, *75*, 2031–2035.

147. Taniyama, H.; Sawada, Y.; Miyazeki, K.; Miyoshi, F. *Chem. Pharm. Bull.* **1972**, *20*, 601–604.

148. Gould, S.J.; Martinkus, K.J.; Tann, C.-H. *J. Am. Chem. Soc.* **1981**, *103*, 2871–2872.

149. Ohsugi, M.; Kahn, J.; Hensley, C.; Chew, S.; Barker, H.A. *J. Biol. Chem.* **1981**, *256*, 7642–7651.

150. Teshima, T.; Nomoto, S.; Wakamiya, T.; Shiba, T. *Bull. Chem. Soc. Jpn.* **1977**, 50, 3372–3380.

151. Yoshioka, H.; Aoki, T.; Goko, H.; Nakatsu, K.; Noda, T.; Sakakibara, H.; Take, T.; Nagata, A.; Abe, J.; Wakamiya, T.; Shiba, T.; Kaneko, T. *Tetrahedron Lett.* **1971**, *23*, 2043–2046.

152. Wakamiya, T.; Shiba, T. *J. Antibiotics* **1975**, *28*, 292–297.

153. Izumi, R.; Noda, T.; Ando, T.; Take, T.; Nagata, A. *J. Antibiotics* **1972**, *25*, 201–207.

154. Noda, T.; Take, T.; Nagata, A.; Wakamiya, T.; Shiba, T. *J. Antibiotics* **1972**, *25*, 427–428.

155. Haskell, T.H.; Fusari, S.A.; Frohardt, R.P.; Bartz, Q.R. *J. Am. Chem. Soc.* **1952**, *74*, 599–602.

156. Wakamiya, T.; Teshima, T.; Sakakibara, H.; Fukukawa, K.; Shiba, T. *Bull. Chem. Soc. Jpn.* **1977**, *50*, 1984–1989.

157. Nagata, A.; Ando, T.; Izumi, R.; Sakakibara, H.; Take, T.; Hayano, K.; Abe, J.-N. *J. Antibiotics* **1968**, *21*, 681–687.

158. Ando, T.; Matsuura, K.; Izumi, R.; Noda, T.; Take, T.; Nagata, A.; Abe, J.-N. *J. Antibiotics* **1971**, *24*, 680–686.

159. Wakamiya, T.; Shiba, T. *J. Antibiotics* **1974**, *27*, 900–902.

160. Wakamiya, T.; Shiba, T.; Kaneko, T. *Bull. Chem. Soc. Jpn.* **1972**, *45*, 3668–3672.

161. Wakamiya, T.; Teshima, T.; Kubota, I.; Shiba, T.; Kaneko, T. *Bull. Chem. Soc. Jpn.* **1974**, *47*, 2292–2296.

162. DeMong, D.E.; Williams, R.M. *J. Am. Chem. Soc.* **2003**, *125*, 8561–8565.

163. Kondo, S.; Shibahara, S.; Takahashi, S.; Maeda, K.; Umezawa, H.; Ohno, M. *J. Am. Chem. Soc.* **1971**, *93*, 6305–6306.

164. Masters, J.J.; Hegedus, L.S. *J. Org. Chem.* **1993**, *58*, 4547–4554.

165. Jain, R.P.; Williams, R.M. *J. Org. Chem.* **2002**, *67*, 6361–6365.

166. Davies, S.GZ.; Ichihara, O. *Tetrahedron Lett.* **1999**, *40*, 9313–9316.

167. Raju, B.; Mortell, K.; Anandan, S.; O'Dowd, H.; Gao, H.; Gomez, M.; Hackbarth, C.; Wu, C.; Wang, W.; Yuan, Z.; White, R.; Trias, J.; Patel, D.V. *Bioorg. Med. Chem. Lett.* **2003**, *13*, 2413–2418.

168. Shinagawa, S.; Kanamaru, T.; Harada, S.; Asai, M.; Okazaki, H. *J. Med. Chem.* **1987**, *30*, 1458–1463.

169. Misiti, D.; Santaniello, M.; Zappia, G. *Bioorg. Med. Chem. Lett.* **1992**, *2*, 1029–1032.

170. Otake, N.; Takeuchi, S.; Endo, T.; Yonehara, H. *Tetrahedron Lett.* **1965**, *19*, 1411–1419.

171. Prabhakaran, P.C.; Woo, N.-T.; Yorgey, P.S.; Gould, S.J. *J. Am. Chem. Soc.* **1988**, *110*, 5785–5791.

172. Bischoff, R.; McDonald, N.; Sutherland, A. *Tetrahedron Lett.* **2005**, *46*, 7147–7149.

173. Burger, K.; Windeisen, E.; Pires, R. *J. Org. Chem.* **1995**, *60*, 7641–7645.

174. Grieco, P.A.; Reilly, M. *Tetrahedron Lett.* **1998**, *39*, 8925–8928.

175. Sowinski, J.A.; Toogood, P.L. *Chem. Commun.* **1999**, 981–982.

176. Uemoto, H.; Yahiro, Y.; Shigemori, H.; Tsuda, M.; Takao, T.; Shimonishi, Y.; Kobayashi, J. *Tetrahedron* **1998**, *54*, 6719–6724.

177. Rottmann, A.; Liebscher, J. *Tetrahedron Lett.* **1996**, *37*, 359–362.

178. Aoyagi, T.; Tobe, H.; Kojima, F.; Hamad, M.; Takeuchi, T.; Umezawa, H. *J. Antibiotics* **1978**, *31*, 636–638.

179. Hayashi, Y.; Kinoshita, Y.; Hidaka, K.; Kiso, A.; Uchibori, H.; Kimura, T.; Kiso, Y. *J. Org. Chem.* **2001**, *66*, 5537–5544.

180. Kotake, T.; Rajesh, S.; Hayashi, Y.; Mukai, Y.; Udea, M.; Kimura, T.; Kiso, Y. *Tetrahedron Lett.* **2004**, *45*, 3651–3654.

181. Kingston, D.G.I. *Chem. Commun.* **2001**, 867–880.

182. Wang, Z.-M.; Kolb, H.C.; Sharpless, K.B. *J. Org. Chem.* **1994**, *59*, 5104–5105.

183. Vronkov, M.V.; Gontcharov, A.V.; Wang, Z.-M. *Tetrahedron Lett.* **2003**, *44*, 407–409.

184. Milanesio, M.; Ugliengo, P.; Viterbo, D. *J. Med. Chem.* **1999**, *42*, 291–299.

185. Kingston, D.G.I.; Chordia, M.D.; Jagtap, P.G.; Liang, J.; Shen, Y.-C.; Long, B.H.; Fairchild, C.R.; Johnston, K.A. *J. Org. Chem.* **1999**, *64*, 1814–1822.

186. Lucatelli, C.; Viton, F.; Gimbert, Y.; Greene, A.E. *J. Org. Chem.* **2002**, *67*, 9468–9470.

187. Barboni, L.; Lambertucci, C.; Ballini, R.; Appendino, G.; Bombardelli, E. *Tetrahedron Lett.* **1998**, *39*, 7177–7180.

188. Liu, C.; Tamm, M.; Nötzel, M.W.; de Meijere, A.; Schilling, J.K.; Kingston, D.G.I. *Tetrahedron Lett.* **2003**, *44*, 2049–2052.

189. Liu, C.; Tamm, M.; Nötzel, M.W.; Rauch, K.; de Meijere, A.; Schilling, J.K.; Lakdawala, A.; Snyder, J.P.; Bane, S.L.; Shanker, N.; Ravindra, R.; Kingston, D.G.I. *Eur. J. Org. Chem.* **2005**, 3962–3972.

190. Chen, T.S.; Li, X.; Bollag, D.; Liu, Y.-c.; Chang, C.-j. *Tetrahedron Lett.* **2001**, *42*, 3787–3789.

191. DeMattei, J.A.; Leanna, M.R.; Li, W.; Nichols, P.J.; Rasmussen, M.W.; Morton, H.E. *J. Org. Chem.* **2001**, *66*, 3330–3337.

192. Baldelli, E.; Battaglia, A.; Bombardelli, E.; Carenzi, G.; Fontana, G.; Gelmi, M.L.; Guerrini, A.; Pocar, D. *J. Org. Chem.* **2004**, *69*, 6610–6616.

193. Barboni, L.; Ballini, R.; Giarlo, G.; Appendino, G.; Fontana, G.; Bombardelli, E. *Bioorg. Med. Chem. Lett.* **2005**, *15*, 5182–5186.

194. Py, S.; Harwig, C.W.; Banerjee, S.; Brown, D.L.; Fallis, A.G. *Tetrahedron Lett.* **1998**, *39*, 6139–6142.

195. Bruno, M.; Rosselli, S.; Maggio, A.; Raccuglia, R.A.; Bastow, K.F.; Lee, K.-H. *J. Nat. Prod.* **2005**, *68*, 1042–1046.

196. Okina, T; Matsuda, H.; Murakami, M.; Yamaguchi, K. *Tetrahedron Lett.* **1993**, *34*, 501–504.

197. Harada, K.-i.; Shimizu, Y.; Fujii, K. *Tetrahedron Lett.* **1998**, *39*, 6245–6248.

198. Ishida, K.; Matsuda, H.; Murakami, M. *Tetrahedron* **1998**, *54*, 13475–13484.

199. Nakamura, H.; Suda, H.; Takita, T.; Aoyagi, T.; Umezawa, H. *J. Antibiotics* **1976**, *29*, 102–103.

200. Suda, H.; Takita, T.; Aoyagi, T.; Umezawa, H. *J. Antibiotics* **1976**, *29*, 100–101.

201. Umezawa, H.; Aoyagi, T.; Suda, H.; Hamada, M.; Takeuchi, T.; *J. Antibiotics* **1976**, *29*, 97–99.

202. Ishizuka, M.; Masuda, T.; Kanbayashi, N.; Fukasawa, S.; Takeuchi, T.; Aoyagi, T.; Umezawa, H. *J. Antibiotics* **1980**, *33*, 642–643.

203. Bergmeier, S.C.; Stanchina, D.M. *J. Org. Chem.* **1999**, *64*, 2852–2859.

204. Koyama, M.; Saino, T. *J. Lab. Compds. Radiopharm.* **1988**, *25*, 1299–1306.

205. Takashiro, E.; Watanabe, T.; Nitta, T.; Kasuya, A.; Miyamoto, S.; Ozawa, Y.; Yagi, R.; Nishigaki, T.; Shibayama, T.; Nakagawa, A.; Iwamoto, A.; Yabe, Y. *Biorg. Med. Chem.* **1998**, *6*, 595–604.

206. Kiso, Y.; Matsumoto, H.; Mizumoto, S.; Kimura, T.; Fujiwara, Y.; Akaji, K. *Biopolymers (Pept. Sci.)* **1999**, *51*, 59–68.

207. Kajiyama, S.-I.; Kanzaki, H.; Kawazu, K.; Kobayashi, A. *Tetrahedron Lett.* **1998**, *39*, 3737–3740.

208. Matsunaga, S.; Fusetani, N.; Hashimoto, K.; Wälchli, M. *J. Am. Chem. Soc.* **1989**, *111*, 2582–2588.

209. Helms, G.L.; Moore, R.E.; Niemczura, W.P.; Patterson, G.M.L.; Tomer, K.B.; Gross, M.L. *J. Org. Chem.* **1988**, *53*, 1298–1307.

210. See Reference *209*.

211. Bewley, C.A.; Debitus, C.; Faulkner, D.J. *J. Am. Chem. Soc.* **1994**, *116*, 7631–7636.

212. Sasaki, S.; Hamada, Y.; Shioiri, T. *Tetrahedron Lett.* **1999**, *40*, 3187–3190.

213. Schmidt, E.W.; Faulkner, D.J. *Tetrahedron* **1998**, *54*, 3043–3056.

214. Qureshi, A.; Colin, P.L.; Faulkner, D.J. *Tetrahedron* **2000**, *56*, 3679–3685.

215. Zhu, J.; Ma, D. *Angew. Chem., Int. Ed.* **2003**, *42*, 5348–5351.

216. Mukai, C.; Miyakawa, M.; Hanaoka, M. *J. Chem. Soc., Perkin Trans. 1* **1997**, 913–917.

217. Hamada, Y.; Kawai, A.; Kohno, Y.; Hara, O.; Shiori, T. *J. Am. Chem. Soc.* **1989**, *111*, 1524–1525.

218. Hammda, Y.; Hara, O.; Kawai, A.; Kohno, Y.; Shioiri, T. *Tetrahedron* **1991**, *47*, 8635–8652.

219. Broady, S.D.; Rexhausen, J.E.; Thomas, E.J. *J. Chem. Soc., Perkin Trans. 1* **1999**, 1083–1094.

220. Ward, R.A.; Procter, G. *Tetrahedron Lett.* **1992**, *33*, 3359–3362.

221. Gesson, J.P.; Jacquesy, J.C.; Mondon, M. *Tetrahedron Lett.* **1989**, *30*, 6503–6506.

222. Kawai, A.; Hara, O.; Hamada, Y.; Shioiri, T. *Tetrahedron Lett.* **1988**, *29*, 6331–6334.

223. Broady, S.D.; Rexhausen, J.E.; Thomas, E.J. *J. Chem. Soc., Chem. Commun.* **1991**, 708–710.

224. Kotsuki, H.; Araki, T.; Miyazaki, A.; Iwasaki, M.; Datta, P.K. *Org. Lett.* **1999**, *1*, 499–502.

225. Ghosh, A.K.; Bischoff, A.; Cappiello, J. *Eur. J. Org. Chem.* **2003**, 821–832.

226. Valentekovich, R.J.; Schreiber, S.L. *J. Am. Chem. Soc.* **1995**, *117*, 9069–9070.

227. Samy, R.; Kim, H.Y.; Brady, M.; Toogood, P.L. *J. Org. Chem.* **1999**, *64*, 2711–2728.

228. Hu, T.; Panek, J.S. *J. Am. Chem. Soc.* **2002**, *124*, 11368–11378.

229. Rinehart, K.L.; Harada, K.-I.; Namikoshi, M.; Chen, C.; Harvis, C.A.; Munro, M.H.G.; Blunt, J.W.; Mulligan, P.E.; Beasley, V.R.; Dahlem, A.M.; Carmichael, W.W. *J. Am. Chem. Soc.* **1988**, *110*, 8557–8558.

230. Sano, T.; Kaya, K. *Tetrahedron* **1998**, *54*, 463–470.

231. Namikoshi, M.; Rinehart, K.L.; Sakai, R.; Stotts, R.R.; Dahlem, A.M.; Beasley, V.R.; Carmichael, W.W.; Evans, W.R. *J. Org. Chem.* **1992**, *57*, 866–872.

232. Aggen, J.B.; Humphrey, J.M.; Gauss, C.-M.; Huang, H.-B.; Nairn, A.C.; Chamberlin, A.R. *Biorg. Med. Chem.* **1999**, *7*, 543–564.

233. O'Donnell, M.E.; Sanvoisin, J.; Gabi, D. *J. Chem. Soc., Perkin Trans. 1* **2001**, 1696–1708.

234. Webster, K.L.; Maude, A.B.; O'Donnell, M.E.; Mehrotra, A.P.; Gani, D. *J. Chem. Soc., Perkin Trans. 1* **2001**, 1673–1695.

235. Fujii, K.; Sivonen, K.; Nakano, T.; Harada, K.-I. *Tetrahedron* **2002**, *58*, 6863–6871.

236. Fujii, K.; Sivonen, K.; Kashiwagi, T.; Hirayama, K.; Harada, K.-I. *J. Org. Chem.* **1999**, *64*, 5777–5782.

237. Spoof, L.; Karlsson, K.; Meriluoto, J. *J. Chromatogr., A* **2001**, *909*, 225–236.

238. Spoof, L.; Meriluoto, J. *J. Chromatogr., A* **2002**, *947*, 237–245.

239. Lawton, L.A.; Edwards, C. *J. Chromatogr., A* **2001**, *912*, 191–209.

240. Dahlmann, J.; Budakowski, W.R.; Luckas, B. *J. Chromatogr., A* **2003**, *994*, 45–57.

241. Spoof, L.; Vesterkvist, P.; Lindholm, T.; Meriluoto, J. *J. Chromatogr., A* **2003**, *1020*, 105–119.

242. Barco, M.; Lawton, L.A.; Rivera, J.; Caixach, J. *J. Chromatogr., A* **2005**, *1074*, 23–30.

243. Ruiz, M.J.; Cameán, A.M.; Moreno, I.M.; Picó, Y. *J. Chromatogr., A* **2005**, *1073*, 257–262.

244. Moreno, I.M.; Molina, R.; Jos, A.; Picó, Y.; Cameán, A.M. *J. Chromatogr., A* **2005**, *1080*, 199–203.

245. Kittaka, A.; Sugano, Y.; Otsuka, M.; Ohno, M. *Tetrahedron* **1988**, *44*, 2811–2820.

246. Boger, D.L.; Ramsey, T.M.; Cai, H. *Bioorg. Med. Chem.* **1997**, *5*, 195–207.

247. Boger, D.L.; Colletti, S.L.; Honda, T.; Menezes R.F. *J. Am. Chem. Soc.* **1994**, *116*, 5607–5618.

248. Takita, T.; Yoshioka, T.; Muraoka, Y.; Maeda, K.; Umezawa, H. *J. Antibiotics* **1971**, *24*, 795–796.

249. Yoshioka, T.; Muraoka, Y.; Takita, T.; Maeda, K.; Umezawa, H. *J. Antibiotics* **1972**, *25*, 625–626.

250. Boger, D.L.; Cai, H. *Angew. Chem., Int. Ed. Engl.* **1999**, *38*, 448–476.

251. Aso, M.; Kondo, M.; Suemune, H.; Hecht, S.M. *J. Am. Chem. Soc.* **1999**, *121*, 9023–9033.

252. Keck, M.V.; Manderville, R.A.; Hecht, S.M. *J. Am. Chem. Soc.* **2001**, *123*, 8690–8700.

253. Neese, F.; Zaleski, J.M.; Zaleski, K.L.; Solomon, E.I. *J. Am. Chem. Soc.* **2000**, *122*, 11703–11724.

254. Katano, K.; An, H.; Aoyagi, Y.; Overhand, M.; Sucheck, S.J.; Stevens, Jr. W.C.; Hess, C.D.; Zhou, X.; Hecht, S.M. *J. Am. Chem. Soc.* **1998**, *120*, 11285–11296.

255. Takita, T.; Umezawa, Y.; Saito, S.-i.; Morishima, H.; Naganawa, H.; Umezawa, H.; Tsuchiya, T.; Miyake, T.; Kageyama, S.; Umezawa, S.; Muraoka, Y.; Suzuki, M.; Otsuka, M.; Narita, M.; Kobayashi, S.; Ohno, M. *Tetrahedron Lett.* **1982**, *23*, 1521–524.

256. Saito, S.-I.; Umezawa, Y.; Morishima, H.; Takita, T.; Umezawa, H.; Narita, M.; Otsuka, M.; Kobayashi, S.; Ohno, M. *Tetrahedron Lett.* **1982**, *23*, 529–532.

257. Zou, Y.; Fahmi, N.E.; Vialas, C.; Miller, G.M.; Hecht, S.M. *J. Am. Chem. Soc.* **2002**, *124*, 9476–9488.

258. Rishel, M.J.; Thomas, C.J.; Tao, Z.-F.; Vialas, C.; Leitheiser, C.J.; Hecht, S.M. *J. Am. Chem. Soc.* **2003**, *125*, 10194–10205.

259. Leitheiser, C.J.; Smith, K.L.; Rishel, M.J.; Hashimoto, S.; Konishi, K.; Thomas, C.J.; Li, C.; McCormick, M.M.; Hecht, S.M. *J. Am. Chem. Soc.* **2003**, *125*, 8218–8227.

260. Thomas, C.J.; Chizhov, A.O.; Leitheiser, C.J.; Rishel, M.J.; Konishi, K.; Tao, Z.-F.; Hecht, S.M. *J. Am. Chem. Soc.* **2002**, *124*, 12926–12927.

261. Thomas, C.J.; McCormick, M.M.; Vialas, C.; Tao, Z.-F.; Leitheiser, C.J.; Rishel, M.J.; Wu, X.; Hecht, S.M. *J. Am. Chem. Soc.* **2002**, *124*, 3875–3884.

262. Nagai, M.; Ogawa, K.; Muraoka, Y.; Naganawa, H.; Aoyagi, T.; Takeuchi, T. *J. Antibiotics* **1991**, *44*, 956–961.

263. Toda, S.; Kotake, C.; Tsuno, T.; Narita, Y.; Yamasaki, T.; Konishi, M. *J. Antibiotics* **1992**, *45*, 1580–1586.

264. Hagihara, M.; Schreiber, S.L. *J. Am. Chem. Soc.* **1992**, *114*, 6570–6571.

265. Fusetani, N.; Matsunaga, S.; Matsumoto, H.; Takebayashi, Y. *J. Am. Chem. Soc.* **1990**, *112*, 7053–7054.

266. Nakao, Y.; Masuda, A.; Matsunaga, S.; Fusetani, N. *J. Am. Chem. Soc.* **1999**, *121*, 2425–2431.

267. Wasserman, H.H.; Zhang, R. *Tetrahedron* **2002**, *58*, 6277–6283.

268. Fusetani, N.; Sugawara, T.; Matsunaga, S.; Hirota, H. *J. Am. Chem. Soc.* **1991**, *113*, 7811–7812.

269. Chill, L.; Kashman, Y.; Schleyer, M. *Tetrahedron* **1997**, *53*, 16147–16152.

270. Kobayashi, J.; Itagaki, F.; Shigemori, H.; Ishibashi, M.; Takahashi, K.; Ogura, M.; Nagasawa, S.; Nakamura, T.; Hirota, H.; Ohta, T.; Nozoe, S. *J. Am. Chem. Soc.* **1991**, *113*, 7812–7813.

271. Itagaki, F.; Shigemori, H.; Ishibashi, M.; Nakamura, T.; Sasaki, T.; Kobayashi J. *J. Org. Chem.* **1992**, *57*, 5540–5542.

272. Tsuda, M.; Ishiyama, H.; Masuko, K.; Takeo, T.; Shimonishi, Y.; Kobayashi, J. *Tetrahedron* **1999**, *55*, 12543–12548.

273. Drey, C.N.C. In *Chemistry and Biochemistry of Amino Acids, Peptides and Proteins. A Survey of Recent Developments, Volume 4*; Weinstein, B. Ed.; Marcel Dekker: New York, **1977**, pp 241–299.

274. Seebach, D. Hook, D.F.; Glättli, A. *Biopolymers (Pept. Sci.)* **2006**, *84*, 23–37.

275. Seebach, D.; Abele, S.; Gademann, K.; Guichard, G.; Hintermann, T.; Jaun, B.; Matthews, J.L.; Schreiber, J.V.; Oberer, L.; Hommel, U.; Widmer, H. *Helv. Chim. Acta* **1998**, *81*, 932–982.

276. Seebach, D.; Matthews, J.L. *Chem. Commun.* **1997**, 2015–2022.

277. Di Blasio, B.; Pavone, V.; Lombardi, A.; Pedone, C.; Benedetti, E. *Biopolymers* **1993**, *33*, 1037–1049.

278. Borman, *Chem. Eng. News* **1997**, *June 16*, 32–35.

279. Koert, U. *Angew. Chem., Int. Ed. Engl.* **1997**, *36*, 1836–1837.

280. Cheng, R.P.; Gellmen, S.H.; deGrado, W.F. *Chem. Rev.* **2001**, *101*, 3219–3232.

281. Arvidsson, P.I.; Frackenpohl, J.; Seebach, D. *Helv. Chim. Acta* **2003**, *86*, 1522–1553.

282. McKiernan, M.; Huck, J.; Fehrentz, J.-A.; Roumestant, M.-L.; Viallefont, P.; Martinez, J. *J. Org. Chem.* **2001**, *66*, 6541–6544.

283. Murray, J.K.; Gellman, S.H. *Org. Lett.* **2005**, *7*, 1517–1520.

284. Murray, J.K.; Farooqi, B.; Sadowsky, J.D.; Scalf, M.; Freund, W.A.; Smith, L.M.; Chen, J.; Gellman, S.H. *J. Am. Chem. Soc.* **2005**, *127*, 13271–13280.

285. Watts, P.; Wiles, C.; Haswell, S.J.; Pombo-Villar, E. *Tetrahedron* **2002**, *58*, 5427–5439.

286. Kimmerlin, T.; Seebach, D.; Hilvert, D. *Helv. Chim. Acta* **2002**, *85*, 1812–1826.

287. Dose, C.; Seitz, O. *Org. Biomol. Chem.* **2004**, *2*, 59–65.

288. Matthews, J.L.; Overhand, M.; Kühnle, F.N.M.; Ciceri, P.E.; Seebach, D. *Liebigs Ann./Recueil* **1997**, 1371–1379.

289. Heinonen, P.; Rosenberg, J.; Lönnberg, H. *Eur. J. Org. Chem.* **2000**, 3647–3652.

290. Virta, P.; Rosenberg, J. Karskela, T.; Heinonen, P.; Lönnberg, H. *Eur. J. Org. Chem.* **2001**, 3467–3473.

291. Krauthäuser, S.; Christianson, L.A.; Powell, D.R.; Gellman, S.H. *J. Am. Chem. Soc.* **1997**, *119*, 11719–11720.

292. Seebach, D.; Overhand, M.; Kühnle, F.N.M.; Martinoni, B.; Oberer, L.; Hommel, U.; Widmer, H. *Helv. Chim. Acta* **1996**, *79*, 913–941.

293. Seebach, D.; Gademann, K.; Schreiber, J.V.; Matthews, J.L.; Hintermann, T.; Jaun, B.; Oberer, L.; Hommel, U.; Widmer, H. *Helv. Chim. Acta* **1997**, *80*, 2033–2038.

294. Seebach, D.; Jacobi, A.; Rueping, M.; Gademann, K.; Ernst, M.; Jaun, B. *Helv. Chim. Acta* **2000**, *83*, 2115–2140.

295. Gademann, K.; Jaun, B.; Seebach, D.; Perozzo, R.; Scapozza, L.; Folkers, G. *Helv. Chim. Acta* **1999**, *82*, 1–11.

296. Hamm, P.; Woutersen, S.; Rueping, M. *Helv. Chim. Acta* **2002**, *85*, 3883–3894.

297. Wu, Y-D.; Wang, D.-P. *J. Am. Chem. Soc.* **1999**, *121*, 9352–9362.

298. Baldauf, C.; Günther, R.; Hofmann, H.-J. *Biopolymers (Pept. Sci.)* **2005**, *80*, 675–687.

299. Wu, Y.-D.; Wang, D.-P. *J. Am. Chem. Soc.* **1998**, *120*, 13485–13493.

300. Baldauf, C.; Günther, R.; Hogmann, H.-J. *Angew. Chem., Int. Ed.* **2004**, *43*, 1594–1597.

301. Rueping, M.; Schreiber, J.V.; Lelais, G.; Jaun, B.; Seebach, D. *Helv. Chim. Acta* **2002**, *85*, 2577–2593.

302. Günther, R.; Hofmann, H.-J. *Helv. Chim. Acta* **2002**, *85*, 2149–2168.

303. Wu, Y.-D.; Lin, J.-Q.; Zhao, Y.-L. *Helv. Chim. Acta* **2002**, *85*, 3144–3160.

304. Seebach, D.; Schreiber, J.V.; Abele, S.; Daura, X.; van Gunsteren, W.F. *Helv. Chim. Acta* **2000**, *83*, 34–57.

305. Seebach, D.; Sifferlen, T.; Mathieu, P.A.; Häne, A.M.; Krell, C.M.; Bierbaum, D.J.; Abele, S. *Helv. Chim. Acta* **2000**, *83*, 2849–2864.

306. Glättli, A.; Daura, X.; Seebach, D.; van Gunsteren, W.F. *J. Am. Chem. Soc.* **2002**, *124*, 12972–12978.

307. Norgren, A.S.; Arvidsson, P.I. *Org. Biomol. Chem.* **2005**, *3*, 1359–1361.

308. Seebach, D.; Mahajan, Y.R.; Senthilkumar, R.; Rueping, M.; Jaun, B. *Chem. Commun.* **2002**, 1598–1599.

309. Gung, B.W.; MacKay, J.A.; Zou, D. *J. Org. Chem.* **1999**, *64*, 700–706.

310. Günther, R.; Hofmann, H.-J. *J. Am. Chem. Soc.* **2001**, *123*, 247–255.

311. Raguse, T.L.; Li, J.R.; Gellman, S.H. *Helv. Chim. Acta* **2002**, *85*, 4154–4164.

312. Raguse, T.L.; Lai, J.R.; Gellman, S.H. *J. Am. Chem. Soc.* **2003**, *125*, 5592–5593.

313. Arvidsson, P.I.; Rueping, M.; Seebach, D. *Chem. Commun.* **2001**, 649–650.

314. Cheng, R.P.; deGrado, W.F. *J. Am. Chem. Soc.* **2001**, *123*, 5162–5163.

315. Rueping, M.; Mahajan, Y.R.; Jaun, B.; Seebach, D. *Chem.—Eur. J.* **2004**, *10*, 1607–1615.

316. Hart, S.A.; Bahadoor, A.B.F.; Matthews, E.E.; Qiu, X.J.; Schepartz, A. *J. Am. Chem. Soc.* **2003**, *125*, 4022–4023.

317. Jacobi, A.; Seebach, D. *Helv. Chim. Acta* **1999**, *82*, 1150–1172.

318. Rueping, M.; Jaun, B.; Seebach, D. *Chem. Commun.* **2000**, 2267–2268.

319. Karle, I.L.; Pramanik, A.; Banerjee, A.; Bhattacharjya, S.; Balaram, P. *J. Am. Chem. Soc.* **1997**, *119*, 9087–9095.

320. LePlae, P.R.; Fisk, J.D.; Porter, E.A.; Weisblum, B.; Gellman, S.H. *J. Am. Chem. Soc.* **2002**, *124*, 6820–6821.

321. Sharma, G.V.M.; Reddy, K.R.; Krishna, P.R.; Sankar, A.R.; Narsimulu, K.; Kumar, S.K.; Jayaprakash, P.; Jagannadh, B.; Kunwar, A.C. *J. Am. Chem. Soc.* **2003**, *125*, 13670–13671.

322. Seebach, D.; Abele, S.; Sifferlen, T.; Hänggi, M.; Gruner, S.; Seiler, P. *Helv. Chim. Acta* **1998**, *81*, 2218–2243.

323. Gademann, K.; Häne, A.; Rueping, M.; Jaun, B.; Seebach, D. *Angew. Chem., Int. Ed.* **2003**, *42*, 1534–1537.

324. Glättli, A.; van Gunsteren, W.F. *Angew. Chem., Int. Ed.* **2004, *43*, 6312–6316.

325. Mathad, R.I.; Gessier, F.; Seebach, D.; Jaun, B. *Helv. Chim. Acta* **2005**, *88*, 266–280.

326. Bierbaum, D.J.; Seebach, D. *Aust. J.Chem.* **2004**, *57*, 859–863.

327. Seebach, D.; Sifferlen, T.; Bierbaum, D.J.; Rueping, M.; Jaun, B.; Schweizer, B.; Schaefer, J.; Mehta, A.K.; O'Connor, R.D.; Meier, B.H.; Ernst, M.; Glättli, A. *Helv. Chim. Acta* **2002**, *85*, 2877–2917.

328. Seebach, D.; Abele, S.; Gademann, K.; Jaun, B. *Angew. Chem., Int. Ed. Engl.* **1999**, *38*, 1595–1597.

329. Langenhan, J.M.; Guzei, I.A.; Gellman, S.H. *Angew. Chem., Int. Ed.* **2003**, *42*, 2402–2405.

330. Motorina, I.A.; Huel, C.; Quiniou, E.; Mispelter, J.; Adjadj, E.; Grierson, D.S. *J. Am. Chem. Soc.* **2001**, *123*, 8–17.

331. Hanessian, S.; Yang, H. *Tetrahedron Lett.* **1997**, *38*, 3155–3158.

332. Banerjee, A.; Maji, S.K.; Drew, M.G.B.; Haldar, D.; Banerjee, A. *Tetrahedron Lett.* **2003**, *44*, 699–702.

333. Büttner, F.; Erdélyi, M.; Arvidsson, P.I. *Helv. Chim. Acta* **2004**, *87*, 2735–2741.

334. Büttner, F.; Norgren, A.S.; Zhang, S.; Prabpai, S.; Kongsaeree, P.; Arvidsson, P.I. *Chem.—Eur. J.* **2005**, *11*, 6145–6158.

335. Matthews, J.L.; Gademann, K.; Jaun, B.; Seebach, D. *J. Chem. Soc., Perkin Trans. 1* **1998**, 3331–3340.

336. Gung, B.W.; Zou, D.; Stalcup, A.M.; Cottrell, C.E. *J. Org. Chem.* **1999**, *64*, 2176–2177.

337. Abele, S.; Guichard, G.; Seebach, D. *Helv. Chim. Acta* **1998**, *81*, 2141–2156.

338. Benedetti, E. *Biopolymers* **1996**, *40*, 3–44.

339. Thakur, A.K.; Kishore, R. *Tetrahedron Lett.* **1999**, *40*, 5091–5094.

340. Thakur, A.K.; Kishore, R. *Tetrahedron Lett.* **1998**, *39*, 9553–9556.

341. Aubry, A.; Marraud, M. *Biopolymers* **1989**, *28*, 109–122.

342. Romanelli, A.; Garella, I.; Menchise, V.; Iacovino, R.; Saviano, M.; Montesarchio, D.; Didierjean, C.; Di Lello, P.; Rossi, F.; Benedetti, E. *J. Pept. Sci.* **2001**, *7*, 15–26.

343. Ashish; Banumathi, S.; Velmurugan, D.; Anushree; Kishore, R. *Tetrahedron* **1999**, *55*, 13791–13804.

344. Schreiber, J.V.; Seebach, D. *Helv. Chim. Acta* **2000**, *83*, 3139–3152.

345. Etezady-Esfarjani, T.; Hilty, C.; Wüthrich, K.; Rueping, M.; Schreiber, J.; Seebach, D. *Helv. Chim. Acta* **2002**, *85*, 1197–1209.

346. Seebach, D.; Schreiber, J.V.; Arvidsson, P.I.; Frackenpohl, J. *Helv. Chim. Acta* **2001**, *84*, 271–279.

347. Kimmerlin, T.; Seebach, D. *Helv. Chim. Acta* **2003**, *86*, 2098–2103.

348. Seebach, D.; Mathad, R.I.; Kimmerlin, T.; Mahajan, Y.R.; Bindschädler, P.; Rueping, M.; Jaun, B.; Hilty, C.; Etezady-Esfarjani, T. *Helv. Chim. Acta* **2005**, *88*, 1969–1982.

349. Seebach, D.; Kimmerlin, T.; Sebesta, R.; Campo, M.A.; Beck, A.K. *Tetrahedron* **2004**, *60*, 7455–7506.

350. Clark, T.D.; Buehler, L.K.; Ghadiri, M.R. *J. Am. Chem. Soc.* **1998**, *120*, 651–656.

351. Rossi, F.; Lelais, G.; Seebach, D. *Helv. Chim. Acta* **2003**, *86*, 2653–2661.

352. Brückner, A.M.; Chakraborty, P.; Gellman, S.H.; Diederichsen, U. *Angew. Chem., Int. Ed.* **2003**, *42*, 4395–4399.

353. Kimmerlin, T.; Namoto, K.; Seebach, D. *Helv. Chim. Acta* **2003**, *86*, 2104–2109.

354. Frackenpohl, J.; Arvidsson, P.I.; Schreiber, J.V.; Seebach, D. *ChemBioChem* **2001**, *2*, 445–455.

355. Hook, D.F.; Gessier, F.; Noti, C.; Kast, P.; Seebach, D. *ChemBioChem* **2004**, *5*, 691–706.

356. Seebach, D.; Rueping, M.; Arvidsson, P.I.; Kimmerlin, T.; Mieuch, P.; Noti, C.; Langenegger, D.; Hoyer, D. *Helv. Chim. Acta* **2001**, *84*, 3503–3510.

357. Inouye, K.; Tanaka, A.; Otsuka, H. *Bull. Chem. Soc. Jpn.* **1970**, *43*, 1163–1172.

358. Lind, R.; Greenhow, D.; Perry, S.; Kimmerlin, T.; Seebach, D. *Chem. Biodiv.* **2004**, *1*, 1391–1400.

359. Schreiber, J.V.; Frackenpohl, J.; Moser, F.; Fleischmann, T.; Kohler, H.-P.E.; Seebach, D. *ChemBioChem* **2002**, 424–432.

360. Umezawa, N.; Gelman, M.A.; Haigis, M.C.; Raines, R.T.; Gellman, S.H. *J. Am. Chem. Soc.* **2002**, *124*, 368–369.

361. Seebach, D.; Namoto, K.; Mahajan, Y.R.; Bindschädler, P.; Sustman, R.; Kirsch, M.; Ryder, N.S.; Weiss, M.; Sauer, M.; Roth, C.; Wener, S.; Beer, H.-D.; Munding, C.; Walde, P.; Voser, M. *Chem. Biodiv.* **2004**, *1*, 65–97.

362. Rueping, M.; Mahajan, Y.; Sauer, M.; Seebach, D. *ChemBioChem* **2002**, *2*, 257–259.

363. Geueke, B.; Namoto, K.; Agarkova, I.; Perriard, J.-C.; Kohler, H.-P.E.; Seebach, D. *ChemBioChem* **2005**, *6*, 982–985.

364. Potocky, T.B.; Menon, A.K.; Gellman, S.H. *J. Am. Chem. Soc.* **2005**, *127*, 3686–3687.

365. Werder, M.; Hauser, H.; Abele, S.; Seebach, D. *Helv. Chim. Acta* **1999**, *82*, 1774–1783.

366. Hamuro, Y.; Schneider, J.P.; DeGrado, W.F. *J. Am. Chem. Soc.* **1999**, *121*, 12200–12201.

367. Liu, D.; DeGrado, W.F. *J. Am. Chem. Soc.* **2001**, *123*, 7553–7559.

368. Arvidsson, P.I.; Frackenpohl, J.; Ryder, N.S.; Liechty, B.; Petersen, F.; Zimmermann, H.; Camenisch, G.P.; Woessner, R.; Seebach, D. *ChemBioChem* **2001**, 771–773.

369. Raguse, T.L.; Porter, E.A.; Weisblum, B.; Gellman, S.H. *J. Am. Chem. Soc.* **2002**, *124*, 12774–12785.

370. Arvidsson, P.I.; Ryder, N.S.; Weiss, H.M.; Hook, D.F.; Escalante, J.; Seebach, D. *Chem. Biodiv.* **2005**, *2*, 401–420.

371. Stephens, O.M.; Kim, S.; Welch, B.D.; Hodsdon, M.E.; Kay, M.S.; Schepartz, A. *J. Am. Chem. Soc.* **2005**, *127*, 13126–13127.

372. Gademann, K.; Ernst, M.; Hoyer, D.; Seebach, D. *Angew. Chem., Int. Ed. Engl.* **1999**, *38*, 1223–1226.

373. Gademann, K.; Ernst, M.; Seebach, D.; Hoyer, D. *Helv. Chim. Acta* **2000**, *83*, 16–33.

374. Gademann, K.; Seebach, D. *Helv. Chim. Acta* **2001**, *84*, 2924–2937.

375. Gademann, K.; Seebach, D. *Helv. Chim. Acta* **1999**, *82*, 957–962.

376. Sutton, P.W.; Breadley, A.; Elsegood, M.R.J.; Farràs, J.; Jackson, R.F.W.; Romea, P.; Urpé, F.; Vilarrasa, J. *Tetrahedron Lett.* **1999**, *40*, 2629–2632.

377. Schumann, F.; Müller, A.; Koksch, M.; Müller, G.; Sewald, N. *J. Am. Chem. Soc.* **2000**, *122*, 12009–12010.

378. Floreancig, P.E.; Swalley, S.E.; Trauger, J.W.; Dervan, P.B. *J. Am. Chem. Soc.* **2000**, *122*, 6342–6350.

379. Woods, C.R.; Ishii, T.; Wu, B.; Bair, K.W.; Boger, D.L. *J. Am. Chem. Soc.* **2002**, *124*, 2148–2152.

380. Gopi, H.N.; Roy, R.S.; Ragbothama, S.R; Karle, I.I.; Balaram, P. *Helv. Chim. Acta* **2002**, *85*, 3313–3330.

381. Ondetti, M.A.; Engel, S.L. *J. Med. Chem.* **1975**, *18*, 761–763.

382. Yasui, A.; Douglas, A.J.; Walker, B.; Magee, D.F.; Murphy, R.F. *Int. J. Peptide Prot. Res.* **1990**, *35*, 301–305.

383. Rodriguez, M.; Fulcrand, P.; Laur, J.; Aumelas, A.; Bali, J.P.; Martinez, J. *J. Med. Chem.* **1989**, *32*, 522–528.

384. Chaturvedi, N.C.; Park, W.K.; Smeby, R.R.; Bumpus, F.M. *J. Med. Chem.* **1970**, *13*, 177–181.

385. Khosla, M.C.; Leese, R.A.; Maloy, W.L.; Ferreira, A.T.; Smeby, R.R.; Bumpus, F.M. *J. Med. Chem.* **1972**, *15*, 792–795.

386. Dhanoa, D.S.; Parsons, W.H.; Greenlee, W.J.; Patchett, A.A. *Tetrahedron Lett.* **1992**, *33*, 1725–1728.

387. Yang, L.; Weber, A.E.; Greenlee, W.J.; Patchett, A.A. *Tetrahedron Lett.* **1993**, *34*, 7035–7038.

388. Weber, A.E.; Halgren, T.A.; Doyle, J.J.; Lynch, R.J.; Siegl, P.K.S.; Parsons, W.H.; Greenlee, W.J.; Patchett, A.A. *J. Med. Chem.* **1991**, *34*, 2692–2701.

389. Iizuka, K.; Kamijo, T.; Harada, H.; Akahane, K.; Kubota, T.; Umeyama, H.; Ishida, T.; Kiso, Y. *J. Med. Chem.* **1990**, *33*, 2707–2714.

390. Dong, H.; Zhang, Z.-L.; Huang, J.-H.; Ma, R.; Chen, S.-H.; Li, G. *Tetrahedron Lett.* **2005**, *46*, 6377–6340.

391. Mimoto, T.; Kato, R.; Takaku, H.; Nojima, S.; Terashima, K.; Misawa, S.; Fukazawa, T.; Ueno, T.; Sato, H.; Shintani, M.; Kiso, Y.; Hyashi, H. *J. Med. Chem.* **1999**, *42*, 1789–1802.

392. Seki, M.; Matsumoto, K. *Synthesis* **1999**, 924–926.

393. Nezami, A.; Luque, I.; Kimura, T.; Kiso, A.; Freire, E. *Biochemistry* **2002**, *41*, 2273–2280.

394. Nezami, A.; Kimura, T.; Hidaka, K.; Kiso, A.; Liu, J.; Kiso, Y.; Goldberg, D.E.; Freire, E. *Biochemistry* **2003**, *42*, 8459–8464.

395. Iijima, K.; Katada, J.; Yasuda, E.; Uno, I.; Hayashi, Y. *J. Med. Chem.* **1999**, *42*, 312–323.

396. Zhong, H.M.; Cohen, J.H.; Abdel-Magid, A.F.; Kenney, B.D.; Maryanoff, C.A.; Shah, R.D.; Villani, F.J., Jr.; Zhang, F.; Zhang, X. *Tetrahedron Lett.* **1999**, *40*, 7721–7725.

397. Kottirsch, G.; Zerwes, H.-G.; Cook, N.S.; Tapparelli, C. *Bioorg. Med. Chem. Lett.* **1997**, *7*, 727–732.

398. Zablocki, J.A.; Tjoeng, F.S.; Bovy, P.R.; Miyano, M.; Garland, R.B.; Williams, K.; Schretzman, L.; Zupec, M.E.; Rico, J.G.; Lindmark, R.J.; Toth, M.V.; McMackins, D.E.; Adams, S.P.; Panzer-Knodle, S.G.; Nicholson, N.S.; Taite, B.B.; Salyers, A.K.; King, L.W.; Campion, J.G.; Feigen, L.P. *Biorg. Med. Chem.* **1995**, *3*, 539–551.

399. Xue, C.-B.; Wityak, J.; Sielecki, T.M.; Pinto, D.J.; Batt, D.G.; Cain, G.A.; Sworin, M.; Rockwell, A.L.; Roderick, J.J.; Wang, S.; Orwat, M.J.; Frietze, W.E.; Bostrom, L.L.; Liu, J.; Higley, A.; Rankin, F.W.; Tobin, A.E.; Emmett, G.; Lalka, G.K.; Sze, J.Y.; Di Meo, S.V.; Mousa, S.A.; Thoolen, M.J.; Racanelli, A.L.; Hausner, E.A.; Reilly, T.M.; DeGrado, W.F.; Wexler, R.R.; Olson, R.E. *J. Med. Chem.* **1997**, *40*, 2064–2084.

400. Hayashi, Y.; Katada, J.; Harada, T.; Tachiki, A.; Iijima, K.; Takiguchi, Y.; Muramatsu, M.; Miyazaki, H.; Asari, T.; Okazaki, T.; Sato, Y.; Yasuda, E.; Yano, M.; Uno, I.; Ojima, I. *J. Med. Chem.* **1998**, *41*, 2345–2360.

401. Su, T.; Naughton, M.A.H.; Smyth, M.S.; Rose, J.W.; Arfsten, A.E.; McCowan, J.R.; Jakubowski, J.A.; Wyss, V.L.; Ruterbories, K.J.; Sall, D.J.;

Scarborough, R.M. *J. Med. Chem.* **1997**, *40*, 4308–4318.

402. Boys, M.L. *Tetrahedron Lett.* **1998**, *39*, 3449–3450.

403. Seebach, D.; Poenaru, S.; Folkers, G.; Rognan, D. *Helv. Chim. Acta* **1998**, *81*, 1181–1200.

404. Suhara, Y.; Izumi, M.; Ichikawa, M.; Penno, M.B.; Ichikawa, Y. *Tetrahedron Lett.* **1997**, *38*, 7167–7170.

405. Matsuura, S.; Waki, M.; Makisumi, S.; Izumiya, N. *Bull. Chem. Soc. Jpn.* **1970**, *43*, 1197–1202.

406. Kopka, I.E.; Lin, L.S.; Mumford, R.A.; Lanza, T., Jr.; Magriotis, P.A.; Young, D.; DeLaszlo, S.E.; MacCoss, M.; Mills, S.G.; Van Riper, G.; McCauley, E.; Lyons, K.; Vincent, S.; Egger, L.A.; Kidambi, U.; Stearns, R.; Colletti, A.; Teffera, Y.; Tong, S.; Owens, K.; Levorse, D.; Schmidt, J.A.; Hagmann, W.K. *Bioorg. Med. Chem. Lett.* **2002**, *12*, 2415–2418.

407. Brashear, K.M.; Hunt, C.A.; Kucer, B.T.; Duggan, M.E.; Hartman, G.D.; Rodan, G.A.; Rodan, S.B.; Leu, C.-T.; Prueksaritanont, T.; Fernandez-Metzler, C.; Barrish, A.; Homnick, C.F.; Hutchinson, J.H.; Coleman, P.J. *Bioorg. Med. Chem. Lett.* **2002**, *12*, 3483–3486.

408. Coleman, P.J.; Askew, B.C.; Hutchinson, J.H.; Whitman, D.B.; Perkins, J.J.; Hartman, G.D.; Rodan, G.A.; Leu, C.-T.; Prueksaritanont, T.; Fernandez-Metzler, C.; Merkle, K.M.; Lynch, R.; Lynch, J.J.; Rodan, S.B.; Dugan, M.E. *Bioorg. Med. Chem. Lett.* **2002**, *12*, 2463–2465.

409. Hara, Y.; Akiyama, M. *J. Am. Chem. Soc.* **2001**, *123*, 7247–7256.

410. Peet, N.P.; Burkhart, J.P.; Angelastro, M.R.; Giroux, E.L.; Mehdi, S.; Bey, P.; Kolb, M.; Neises, B.; Schirlin, D. *J. Med. Chem.* **1990**, *33*, 394–407.

411. Chen, K.X.; Njoroge, F.G.; Vibulbhan, B.; Prongay, A.; Pichardo, J.; Madison, V.; Buevich, A.; Chan, T.-M. *Angew. Chem., Int. Ed.* **2005**, *44*, 7024–7028.

412. Nizi, E.; Koch, U.; Ponzi, S.; Matassa, V.G.; Gardelli, C. *Bioorg. Med. Chem. Lett.* **2002**, *12*, 3325–3328.

413. Lescop, C.; Herzner, H.; Siendt, H.; Bolliger, R.; Henneböhle, M.; Weyermann, P.; Briguet, A.; Courdier-Fruh, I.; Erb, M.; Foster, M.; Meier, T.; Magyar, J.P.; von Sprecher, A. *Bioorg. Med. Chem. Lett.* **2005**, *15*, 5176–5181.

414. Linton, S.D.; Karanewsky, D.S.; Ternansky, R.J.; Wu, J.C.; Pham, B.; Kodandapani, L.; Smidt, R.; Diaz, J.-L.; Fritz, L.C.; Tomaselli, K.J. *Bioorg. Med. Chem. Lett.* **2002**, *12*, 2969–2971.

415. Linton, S.D.; Karanewsky, D.S.; Ternansky, R.J.; Chen, N.; Guo, X.; Jahangiri, K.G.; Kalish, V.J.; Meduna, S.P.; Robinson, E.D.; Ullman, B.R.; Wu, J.C.; Pham, B.; Kodandapani, L.; Smidt, R.; Diaz, J.-L.; Fritz, L.C.; von Krosigk, U.; Roggo, S.; Schmitz, A.; Tomaselli, K.J. *Bioorg. Med. Chem. Lett.* **2002**, *12*, 2973–2975.

416. Klein, S.I.; Czekaj, M.; Gardner, C.J.; Guertin, K.R.; Cheny, D.L.; Spada, A.P.; Bolton, S.A.; Brown, K.; Colussi, D.; Heran, C.L.; Morgan, S.R.; Leadley, R.J.; Dunwiddie, C.T.; Perrone, M.H.; Chu, V. *J. Med. Chem.* **1998**, *41*, 437–450.

417. Kim, D.H.; Jin, Y. *Biorg. Med. Chem. Lett.* **1999**, *9*, 691–696.

418. Kim, D.H.; Chung, S.J.; Kim, E.-J.; Tian, G.R. *Biorg. Med. Chem. Lett.* **1998**, *8*, 859–864.

419. Bey, P.; Jung, M.J.; Gerhart, F.; Schirlin, D.; Van Dorsselaer, V.; Casara, P. *J. Neurochem.* **1981**, *37*, 1341–1344.

420. Coppola, G.M.; Damon, R.E.; Eskesen, J.B.; France, D.S.; Paterniti, J.R. Jr., *Bioorg. Med. Chem. Lett.* **2002**, *12*, 2439–2442.

421. Gani, D.; Young, D.W. *J. Chem. Soc., Perkin Trans. I* **1985**, 1355–1362.

422. Gani, D.; Hitchcock, P.B.; Young, D.W. *J. Chem. Soc., Perkin Trans. I* **1985**, 1363–1372.

423. Mong, T.K.-K.; Niu, A.; Chow, H.-F.; Wu, C.; Li, L.; Chen, R. *Chem.—Eur. J.* **2001**, *7*, 686–699.

424. Qu, Z.-R.; Zhao, H.; Wang, Y.-P.; Wang, X.-S.; Ye, Q.; Li, Y.-H.; Xiong, R.G.; Abrahams, B.F.; Liu, Z.-G.; Xue, Z.-L.; You, X.-Z. *Chem.—Eur. J.* **2004**, *10*, 53–60.

425. Lynn, D.M.; Langer, R. *J. Am. Chem. Soc.* **2000**, *122*, 10761–10768.

426. Bachmann, B.M.; Seebach, D. *Helv. Chim. Acta* **1998**, *81*, 2430–2461.

427. Fritz, M.G.; Seebach, D. *Helv. Chim. Acta* **1998**, *81*, 2414–2429.

428. Hamper, B.C.; Kolodziej, S.A.; Scates, A.M.; Smith, R.G.; Cortez, E. *J. Org. Chem.* **1998**, *63*, 708–718.

429. Jia, L.; Sun, H.; Shay, J.T.; Allgeier, A.M.; Hanton, S.D. *J. Am. Chem. Soc.* **2002**, *124*, 7282–7283.

430. Darensbourg, D.J.; Phelps, A.L.; Le Gall, N.; Jia, L. *J. Am. Chem. Soc.* **2004**, *126*, 13808–13815.

431. Siefferlen, T.; Rueping, M.; Gademann, K.; Jaun, B.; Seebach, D. *Helv. Chim. Acta* **1999**, *82*, 2067–2093.

432. Pichler, U.; Scrimin, P.; Tecilla, P.; Tonellato, U.; Veronese, A.; Verzini, M. *Tetrahedron Lett.* **2004**, *45*, 1643–1646.

433. Greco, M.N.; Zhong, H.M.; Maryanoff, B.E. *Tetrahedron Lett.* **1998**, *39*, 4959–4962.

434. Cardillo, G.; Tomasini, C. *Chem. Soc. Rev.* **1996**, 117–128.

435. Juaristi, E.; Quintana, D.; Escalante, J. *Aldrichimica Acta* **1994**, *27*, 3–10.

436. Cole, D.C. *Tetrahedron* **1994**, *50*, 9517–9582.

437. Liu, M.; Sibi, M.P. *Tetrahedron* **2002**, *58*, 7991–8035.

438. Abele, S.; Seebach, D. *Eur. J. Org. Chem.* **2000**, 1–15.

439. Sewald, N. *Angew. Chem., Int. Ed.* **2003, *42*,** 5794–5795.

440. Sidgwick, N.V.; Millar, I.T.; Springall, H.D. *The Organic Chemistry of Nitrogen*, 3rd edn; Oxford University Press: Toronto, Canada, 1966, pp 195–222.

441. Lelais, G.; Seebach, D. *Biopolymers (Pept. Sci.)* **2004**, *76*, 206–243.

442. Romnova, N.N.; Gravis, A.G.; Bundel, Y.G. *Russ. Chem. Rev.* **1996**, *65*, 1083–1092.

443. Slimmer, M.D. *Ber.* **1902**, *35*, 400–410.

444. Whitmore, F.C.; Mosher, H.S.; Adams, R.R.; Taylor, R.B.; Chapin, E.C.; Weisel, C.; Yanko, W. *J. Am. Chem. Soc.* **1944**, *66*, 725–731.

445. Buc, S.R.; Ford, J.H.; Wise, E.C. *J. Am. Chem. Soc.* **1945**, *67*, 92–94.

446. Ford, J.H. *J. Am. Chem. Soc.* **1945**, *67*, 876–877.

447. Ford, J.H.; Adkins, H.; Caffrey, J.M. *Org. Synth.* **1947**, *27*, 1–3.

448. Stadnikoff, G. *Ber.* **1911**, *44*, 44–52.

449. Scudi, J.V. *J. Am. Chem. Soc.* **1935**, *57*, 1279.

450. Birkofer, L.; Hartwig, I. *Ber.* **1954**, *87*, 1189–1198.

451. Birkofer, L.; Storch, I. *Ber.* **1954**, *87*, 571–577.

452. Birkofer, L.; Birkofer, A. *Ber.* **1956**, *89*, 1226–1229.

453. Loncrini, D.F.; Walborsky, H.M. *J. Med. Chem.* **1964**, *7*, 369–370.

454. Bovy, P.R.; Rico, J.G. *Tetrahedron Lett.* **1993**, *34*, 8015–8018.

455. Galat, A. *J. Am. Chem. Soc.* **1945**, *67*, 1414–1415.

456. Tolman, V.; Veres, K. *Collect. Czech. Chem. Commun.* **1964**, *29*, 234–238.

457. Klein, S.I.; Cekaj, M.; Molino, B.F.; Chu, V. *Bioorg. Med. Chem. Lett.* **1997**, *7*, 1773–1778.

458. Kwiatkowski, S.; Jeganathan, A.; Tobin, T.; Watt, D.S. *Synthesis* **1989**, 946–949.

459. Weisel, C.A.; Taylor, R.B.; Mosher, H.S.; Whitmore, F.C. *J. Am. Chem. Soc.* **1945**, *67*, 1071–1072.

460. Basu, B.; Das, P.; Hossain, I. *Synlett* **2004**, 2630–2632.

461. Zilkha, A.; Rachman, E.S.; Rivlin, J. *J. Org. Chem.* **1961**, *26*, 376–380.

462. Perlmutter, P.; Tabone, M. *J. Org. Chem.* **1995**, *60*, 6515–6522.

463. Andruszkiewicz, R.; Gryzbowska, J.; Wojciechowska, H. *Polish J. Chem.* **1978**, *52*, 2251–2254.

464. Matsubara, S.; Yoshioka, M.; Utimoto, K. *Chem. Lett.* **1994**, 827–830.

465. Jenner, G. *Tetrahedron Lett.* **1995**, *36*, 233–236.

466. Maier, L.; Rist, G.; Lea, P.J. *Phosphorus and Sulfur* **1983**, *18*, 349–352.

467. Rico, J.G. *Tetrahedron Lett.* **1994**, *35*, 6599–6602.

468. Asao, N.; Uyehara, T.; Yamamoto, Y. *Tetrahedron* **1988**, *44*, 4173–4180.

469. Uyehara, T.; Asao, N.; Yamamoto, Y. *J. Chem. Soc., Chem. Commun.* **1989**, 753–754.

470. Asao, N.; Uyehara, T.; Yamamoto, Y. *Tetrahedron* **1990**, *46*, 4563–4572.

471. Lakshmipathi, P.; Rao, A.V.R. *Tetrahedron Lett.* **1997**, *38*, 2551–2552.

472. Barluenga, J.; Villamaña, J.; Yus, M. *Synthesis* **1981**, 375–376.

473. Wessjohann, L.; McGaffin, G.; de Meijere, A. *Synthesis* **1989**, 359–363.

474. Ojima, I.; Kato, K.; Nakahashi, K.; Fuchikami, T.; Fujita, M. *J. Org. Chem.* **1989**, *54*, 4511–4522.

475. Basheeruddin, K.; Siddiqui, A.A.; Khan, N.H.; Saleha, S. *Synth. Commun.* **1979**, *9*, 705–712.

476. Abbenante, G.; Hughes, R.; Prager, R.H. *Aust. J. Chem.* **1997**, *50*, 523–527.

477. Kaiser, C.; Burger, A. *J. Am. Chem. Soc.* **1957**, *79*, 4365–4370.

478. Hartzel, L.W.; Ritter, J.J. *J. Am. Chem. Soc.* **1949**, *71*, 4130–4131.

479. Aberhart, D.J.; Lin, H.-J. *J. Org. Chem.* **1981**, *46*, 3749–3751.

480. Lubell, W.D.; Kitamura, M.; Noyori, R. *Tetrahedron: Asymmetry* **1991**, *2*, 543–554.

481. Sham, H.L.; Rempel, C.A.; Stein, H.; Cohen, J. *J. Chem. Soc., Chem. Commun.* **1990**, 904–905.

482. Zhu, G.; Chen, Z.; Zhang, X. *J. Org. Chem.* **1999**, *64*, 6907–6910.

483. You, J.; Drexler, H.-J.; Zhang, S.; Fischer, C.; Heller, D. *Angew. Chem., Int. Ed.* **2003**, *42*, 913–916.

484. Ibrahim-Ouali, M.; Sinibaldi, M.-E.; Trooin, Y.; Gardette, D.; Gramain, J.-C. *Synth. Commun.* **1997**, *27*, 1827–1848.

485. Hewlins, S.A.; Murphy, J.A.; Lin, J.; Hibbs, D.E.; Hursthouse, M.B. *J. Chem. Soc., Perkin Trans. 1* **1997**, 1559–1570.

486. Fischer, E.; Leuchs, H. *Ber.* **1902**, *35*, 3787–3805.

487. Leibman, K.C.; Fellner, S.K. *J. Org. Chem.* **1962**, *27*, 438–440.

488. Neuberg, C. *Biochem. Z.* **1906**, *1*, 282–298.

489. Huck, J.; Receveur, J.-M.; Roumestant, M.-L.; Martinez, J. *Synlett* **2001**, 1467–1469.

490. Collins, J.L; Blanchard, S.G.; Boswell, E.G.; Charifson, P.S.; Cobb, J.E.; Henke, B.R.; Hull-Ryde, E.A.; Kazmierski, W.M.; Lake, D.H.; Leesnitzer, L.M.; Lehmannm, J.; Lenhard, J.M.; Orband-Miller, L.A.; Gray-Nunez, Y.; Parks, D.J.; Plunket, K.D.; Tong, W.-Q. *J. Med. Chem.* **1998**, *41*, 5037–5054.

491. Katajisto, J.; Karskela, T.; Heinonen, P.; Lönnberg, H. *J. Org. Chem.* **2002**, *67*, 7995–8001.

492. Bergmann, E.D.; Cohen, S. *J. Chem. Soc.* **1961**, 4669–4675.

493. Marshall, J.A.; Garofalo, A.W. *J. Org. Chem.* **1993**, *58*, 3675–3680.

494. Heinonen, P.; Virta, P.; Lönnberg, H. *Tetrahedron* **1999**, *55*, 7613–7624.

495. Chen, H.-Y.; Patkar, L.N.; Ueng, S.-H.; Lin, C.-C.; Lee, A.S.-Y. *Synlett* **2005**, 2035–2038.

496. Özer, M.S.; Gérardin-Charbonnier, C.; Thiébaut, S.; Rodehüser, L.; Selve, C. *Amino Acids* **1999**, *16*, 381–389.

497. Rajesh, S.; Banergi, B.; Iqbal, J. *J. Org. Chem.* **2002**, *67*, 7852–7857.

498. Melikoff, P. *Ber.* **1879**, *12*, 2227–2228.

499. Greenstein, J.P.; Winitz, N. *Chemistry of the Amino Acids, Volume 3*; Wiley: New York, 1961, p 2215.

500. Srivastava, R.P.; Zjawiony, J.K.; Peterson, J.R.; McChesney, J.D. *Tetrahedron: Asymmetry* **1994**, *5*, 1683–1688.

501. Bhatia, B.; Jain, S.; De, A.; Bagchi, I.; Iqbal, J. *Tetrahedron Lett.* **1996**, *37*, 7311–7314.

502. Das, B.C.; Iqbal, J. *Tetrahedron Lett.* **1997**, *38*, 2903–2906.

503. Prabhakaran, E.N.; Rao, I.N.; Boruah, A.; Iqbal, J. *J. Org. Chem.* **2002**, *67*, 8247–8250.

504. De, A.; Ghosh, S.; Iqbal, J. *Tetrahedron Lett.* **1997**, *38*, 8379–8382.

505. Hönig, H.; Seufer-Wasserthal, P.; Weber, H. *Tetrahedron* **1990**, *46*, 3841–3850.

506. Kato, K.; Saino, T.; Nishizawa, R.; Takita, T.; Umezawa, H. *J. Chem. Soc., Perkin Trans. 1* **1980**, 1618–1621.

507. Fringuelli, F.; Pizzo, F; Vaccaro, L. *J. Org. Chem.* **2001**, *66*, 3554–3558.

508. Yang, H.W.; Romo, D. *J. Org. Chem.* **1999**, *64*, 7657–7660.

509. Mitani, M.; Tachizawa, O.; Takeuchi, H.; Koyama, K. *J. Org. Chem.* **1989**, *54*, 5397–5399.

510. Hale, W.J.; Honan, E.M. *J. Am. Chem. Soc.* **1919**, *41*, 770–776.

511. Clarke, H.T.; Behr, L.D.; Carothers, W.H.; McEwen, W.L. *Org. Synth.* **1948**, *Coll. Vol. 2*, 19–21.

512. Treibs, W.; Hauptmann, S. *Ber.* **1956**, *89*, 117–120.

513. Takagi, S.; Hayashi, K. *Chem. Pharm. Bull.* **1959**, *7*, 99–102.

514. Pires, R.; Burger, K. *Tetrahedron Lett.* **1996**, *37*, 8159–8160.

515. Shadbolt, R.S.; Stephens, F.F. *J. Chem. Soc. C* **1971**, 1665–1666.

516. Birkofer, L.; Storch, I. *Ber.* **1953**, *86*, 749–755.

517. Rothe, M. *Chem. Ber.* **1962**, *95*, 783–794.

518. Espino, C.G.; Wehn, P.M.; Chow, J.; Du Bois, J. *J. Am. Chem. Soc.* **2001**, *123*, 6935–6936.

519. Mamaghani, M.; Badrian, A. *Tetrahedron Lett.* **2004**, *45*, 1547–1550.

520. Soloshonok, V.A.; Kirilenko, A.G.; Fokina, N.A.; Kukhar, V.P.; Galushko, S.V.; Svedas, V.K.; Resnati, G. *Tetrahedron: Asymmetry* **1994**, *5*, 1225–1228.

521. Soloshonok, V.A.; Kirilenko, A.G.; Kukhar, V.P. *Tetrahedron Lett.* **1993**, *35*, 3621–3624.

522. Soloshonok, V.A.; Ono, T. *Tetrahedron* **1996**, *52*, 14701–14712.

523. Soloshonok, V.A.; Kukhar, V.P. *Tetrahedron* **1996**, *52*, 6953–6964.

524. Soloshonok, V.A.; Soloshonok, I.V.; Kukhar, V.P.; Svedas, V.K. *J. Org. Chem.* **1998**, *63*, 1878–1884.

525. Glickman, S.A.; Cope, A.C. *J. Am. Chem. Soc.* **1945**, *67*, 1017–1020.

526. Saeed, A.; McMillin, J.B.; Wolkowicz, P.E.; Brouillette, W.J. *J. Med. Chem.* **1994**, *37*, 3247–3251.

527. Cronin, J.R.; Yuen, G.U.; Pizzarello, S. *Anal. Biochem.* **1982**, *124*, 139–149.

528. Crossley, M.J.; Fisher, M.L.; Potter, J.J.; Kuchel, P.W.; York, M.J. *J. Chem. Soc., Perkin Trans 1* **1990**, 2363–2369.

529. Shibata, I.; Suwa, T.; Sugiyama, E.; Baba, A. *Synlett* **1998**, 1081–1082.

530. Rubin, A.E.; Sharpless, K.B. *J. Am. Chem. Soc.* **1998**, *120*, 2637–2640.

531. Fokin, V.V.; Sharpless, K.B. *Angew. Chem., Int. Ed.* **2001**, *40*, 3455–3457.

532. Gontcharov, A.V.; Liu, H.; Sharpless, K.B. *Org. Lett.* **1999**, *1*, 783–786.

533. Pringle, W.; Sharpless, K.B. *Tetrahedron Lett.* **1999**, *40*, 5151–5154.

534. Bartoli, G.; Cimarelli, C.; Dalpozzo, R.; Palmieri, G. *Tetrahedron* **1995**, *51*, 8613–8622.

535. Fustero, S.; de la Torre, M.G.; Jofré, V.; Carlón, R.P.; Navarro, A.; Fuentes, A.S.; Carrió, J.S. *J. Org. Chem.* **1998**, *63*, 8825–8836.

536. Oaksmith, J.M.; Peters, U.; Ganem, B. *J. Am. Chem. Soc.* **2004**, *126*, 13606–13607.

537. Ziegler, T.; Kaisers, H.-J.; Schlömer, R.; Koch, C. *Tetrahedron* **1999**, *55*, 8397–8408.

538. Kern, O.T.; Motherwell, W.B. *Chem. Commun.* **2003**, 2988–2989.

539. Lu, S.-M.; Alper, H. *J. Org. Chem.* **2004**, *69*, 3558–3561.

540. Mahadevan, V.; Getzler, Y.D.Y.L.; Coates, G. *Angew. Chem., Int. Ed.* **2002**, *41*, 2781–2784.

541. Kupiecki, F.P.; Coon, M.J.; Khettry, A.K. *Biochem. Prep.* **1960**, *7*, 20–22.

542. Pollack, M.A. *J. Am. Chem. Soc.* **1943**, *65*, 1335–1339.

543. Lu, Y.; Miet, C.; Kunesch, N.; Poisson, J.E. *Tetrahedron: Asymmetry* **1993**, 893–902.

544. Lu, Y.; Miet, C.; Kunesch, N.; Poisson, J.E. *Tetrahedron: Asymmetry* **1991**, *2*, 871–872.

545. Williams, T.M.; Crumbie, R.; Mosher, H.S. *J. Org. Chem.* **1985**, *50*, 91–97.

546. Kanazawa, A.M.; Correa, A.; Denis, J.-N.; Luche, M.-J.; Greene, A.E. *J. Org. Chem.* **1993**, *58*, 255–257.

547. Tussa, L.; Lebreton, C.; Mosset, P. *Chem.—Eur. J.* **1997**, *3*, 1064–1070.

548. Hellman, H.; Haas, G. *Ber.* **1957**, *90*, 1357–1363.

549. Prager, R.H.; Schafer, K. *Aust. J. Chem.* **1997**, *50*, 813–823.

550. Calmes, M.; Escale, F. *Tetrahedron: Asymmetry* **1998**, *9*, 2845–2850.

551. Calmès, M.; Escale, F.; Glot, C.; Rolland, M.; Martinez, J. *Eur. J. Org. Chem.* **2000**, 2459–2466.

552. Böhme, H.; Broese, R.; Eiden, F. *Ber.* **1959**, *92*, 1258–1262.

553. Kise, N.; Ueda, N. *J. Org. Chem.* **1999**, *64*, 7511–7514.

554. Shono, T.; Tsubata, K.; Okinaga, N. *J. Org. Chem.* **1984**, *49*, 1056–1059.

555. Shono, T.; Kise, N.; Sanda, F.; Ohi, S.; Tsubata, K. *Tetrahedron Lett.* **1988**, *29*, 231–234.

556. Okano, K.; Morimoto, T.; Sekiya, M. *J. Chem. Soc., Chem. Commun.* **1984**, 883–884.

557. Katritzky, A.R.; Shobana, N.; Harris, P.A. *Tetrahedron Lett.* **1990**, *31*, 3999–4002.

558. Katritzky, A.R.; Kirichenko, K.; Elsayed, A.M.; Ji, Y.; Fang, Y. *J. Org. Chem.* **2002**, *67*, 4957–4959.

559. Katritzky, A.R.; Yannakopoulou, K. *Synthesis* **1989**, 747–751.

560. Katritzky, A.R.; Nichols, D.A.; Qi, M. *Tetrahedron Lett.* **1998**, *39*, 7063–7066.

561. Alberola, A.; Alvarez, M.A.; Andrés, C.; González, A.; Pedrosa, R. *Synthesis* **1990**, 1057–1058.

562. Wieber, G.M.; Hegedus, L.S.; Åkermark, B.; Michalson, E.T. *J. Org. Chem.* **1989**, *54*, 4649–4653.

563. Hart, D.J.; Ha, D.-C. *Chem. Rev.* **1989**, *89*, 1447–1465.

564. Mladenova, M.; Bellassoued, M. *Synth. Commun.* **1993**, *23*, 725–736.

565. Colvin, E.W.; McGarry, D.; Nugent, M.J. *Tetrahedron* **1988**, *44*, 4157–4172.

566. Colvin, E.W.; McGarry, D.G. *J. Chem. Soc., Chem. Commun.* **1985**, 539–540.

567. Ojima, I.; Inaba, S.-i.; Yoshida, K. *Tetrahedron Lett.* **1977**, *41*, 3643–3646.

568. Ojima, I.; Inabe, S.-i.; Nagai, M. *Synthesis* **1981**, 545–547.

569. Ikeda, K.; Terao, Y.; Sekiya, M. *Chem. Pharm. Bull.* **1981**, *29*, 1747–1749.

570. Mertin, A.; Thiemann, T.; Hanss, de Meijere, A. *Synlett* **1991**, 87–89.

571. Periasamy, M.; Suresh, S.; Ganesan, S.S. *Tetrahedron Lett.* **2005**, *46*, 5521–5524.

572. Mukaiyama, T.; Akamatsu, H.; Han, J.S. *Chem. Lett.* **1990**, 889–892.

573. Guanti, G.; Narisano, E.; Banfi, L. *Tetrahedron Lett.* **1987**, *28*, 4331–4334.

574. Texier-Boullet, F.; Latouche, R.; Hamelin, J. *Tetrahedron Lett.* **1993**, *34*, 2123–2126.

575. Akiyama, T.; Suzuki, A; Fuchibe, K. *Synlett* **2005**, 1024–1026.

576. Kobayashi, S.; Aoki, Y. *Tetrahedron Lett.* **1998**, *39*, 7345–7348.

577. Ikeda, K.; Achiwa, K.; Sekiya, M. *Tetrahedron Lett.* **1983**, *24*, 913–916.

578. Ikeda, K.; Achiwa, K.; Sekiya, M. *Tetrahedron Lett.* **1983**, *24*, 4707–4710.

579. Saito, S.; Uedo, E.; Kato, Y.; Murakami, Y.; Ishikawa, T. *Synlett* **1996**, 1103–1105.

580. Adrian, Jr., J.C.; Barkin, J.L.; Hassib, L. *Tetrahedron Lett.* **1999**, *40*, 2457–2460.

581. Robinson, A.J.; Wyatt, P.B. *Tetrahedron* **1993**, *49*, 11329–11340.

582. Adrian, Jr., J.C.; Snapper, M.L. *J. Org. Chem.* **2003**, *68*, 2143–2150.

583. Sato, K.; Tarui, A.; Matsuda, S.; Omote, M.; Ando, A.; Kumadaki, I. *Tetrahedron Lett.* **2005**, *46*, 7679–7681.

584. Otsuka, M.; Yoshida, M.; Kobayashi, S.; Ohno, M.; Umezawa, Y.; Morishima, H. *Tetrahedron Lett.* **1981**, *22*, 2109–2112.

585. Yamasaki, N.; Murakami, M.; Mukaiyama, T. *Chem. Lett.* **1986**, 1013–1016.

586. Andrade, C.K.Z.; Kalil, P.P.; Rocha, R.O.; Alves, L.M.; Panisset, C.M.A. *Lett. Org. Chem.* **2004**, *1*, 109–111.

587. Sisko, J.; Weinreb, S.M. *J. Org. Chem.* **1990**, *55*, 393–395.

588. Hua, D.H.; Miao, S.W.; Chen, J.S.; Iguchi, S. *J. Org. Chem.* **1991**, *56*, 4–6.

589. Davis, F.A.; Giangiordano, M.A.; Starner, W.E. *Tetrahedron Lett.* **1986**, *27*, 3957–3960.

590. Fustero, S.; Pina, B.; Simón-Fuentes, A. *Tetrahedron Lett.* **1997**, *38*, 6771–6774.

591. Fustero, S.; Pina, B.; Salavert, E.; Navarro, A.; Ramírez de Arellano, M.C.; Fuentes, A.S. *J. Org. Chem.* **2002**, *67*, 4667–4679.

592. Fustero, S.; Pina, B.; de la Torre, M.; Navarro, A.; de Arellano, C.R.; Simón, A. *Org. Lett.* **1999**, *1*, 977–980.

593. Uneyama, K.; Hao, J.; Amii, H. *Tetrahedron Lett.* **1998**, *39*, 4079–4082.

594. Ohkura, H.; Handa, M.; Katagiri, T.; Uneyama, K. *J. Org. Chem.* **2002**, *67*, 2692–2695.

595. Fustero, S.; Díaz, M.D.; Navarro, A.; Salavert, E.; Aguilar, E. *Tetrahedron Lett.* **1999**, *40*, 1005–1008.

596. Rodionow, W.M. *J. Am. Chem. Soc.* **1929**, *51*, 847–852.

597. Rodionow, W.M.; Malewinskaja, E.T. *Ber.* **1926**, *59*, 2952–2958.

598. Dallemagne, P.; Rault, S.; de Sévricourt, M.C.; Hassan, K.M.; Robba, M. *Tetrahedron Lett.* **1986**, *27*, 2607–2610.

599. Kuwata, S.; Yamada, T.; Shinogi, T.; Yamagami, N.; Kitabashi, F.; Miyazawa, T.; Watanabe, H. *Bull. Chem. Soc. Jpn.* **1979**, *52*, 3326–3328.

600. Solymár, M.; Fülöp, F.; Kanerva, L.T. *Tetrahedron Asymmetry* **2002**, *135*, 2383–2388.

601. Mannich, C.; Kather, B. *Ber.* **1920**, *53*, 1368–1371.

602. Soloshonok, V.A.; Fokina, N.A.; Rybakova, A.V.; Shishkina, I.P.; Galushko, S.V.; Sorochinsky, A.E.; Kukhar, V.P.; Savchenko, M.V.; Svedas, V.K. *Tetrahedron: Asymmetry* **1995**, *6*, 1601–1610.

603. Dallemagne, P.; Rault, S.; Gordaliza, M.; Robba, M. *Heterocycles* **1987**, *26*, 3233–3237.

604. Renault, O.; Dallemagne, P.; Rault, S. *Org. Proc. Prep. Int.* **1997**, *29*, 488–494.

605. Cardillo, G.; Gentilucci, L.; Tolomelli, A.; Tomasini, C. *J. Org. Chem.* **1998**, *63*, 2351–2353.

606. Rodionow, W.M. *J. Am. Chem. Soc.* **1929**, *51*, 841–847.

607. Berkessel, A.; Cleemann, F.; Mukherjee, S. *Angew. Chem., Int. Ed.* **2005**, *44*, 7466–7469.

608. Kobayashi, S.; Mitsuharu, A.; Yasuda, M. *Tetrahedron Lett.* **1995**, *36*, 5773–5776.

609. Loh, T.-P.; Wei, L.-L. *Tetrahedron Lett.* **1998**, *39*, 323–326.

610. Kobayashi, S.; Nagayama, S.; Busujima, T. *Tetrahedron Lett.* **1996**, *37*, 9221–9224.

611. Kobayashi, S.; Busujima, T.; Nagayama, S. *Synlett* **1999**, *5*, 545–546.

612. Akiyama, T.; Takaya, J.; Kagoshima, H. *Synlett* **1999**, 1426–1428.

613. Akiyama, T.; Itoh, J.; Fuchibe, K. *Synlett* **2002**, 1269–1272.

614. Manbe, K.; Mori, Y.; Kobayashi, S. *Synlett* **1999**, 1401–1402.

615. Akiyama, T.; Takaya, J.; Kagoshima, H. *Synlett* **1999**, 1045–1048.

616. Clerici, A.; Clerici, L.; Porta, O. *Tetrahedron Lett.* **1995**, *36*, 5955–5958.

617. Joffe, A.L.; Thomas, T.M.; Adrian, Jr., J.C. *Tetrahedron Lett.* **2004**, *45*, 5007–5009.

618. Balan, D.; Adolfsson, H. *J. Org. Chem.* **2001**, *66*, 6498–6501.

619. Balan, D.; Adolfsson, H. *J. Org. Chem.* **2002**, *67*, 2329–2334.

620. Saidi, M.R.; Khalaji, H.R.; Ipaktschi, J. *J. Chem. Soc., Perkin Trans. 1* **1997**, 1983–1986.

621. Saidi, M.R.; Heydari, A.; Ipaktschi, J. *Chem. Ber.* **1994**, *127*, 1761–1764.

622. Torssell, S.; Kienle, M.; Somfai, P. *Angew. Chem., Int. Ed.* **2005**, *44*, 3096–3099.

623. Lee, A.S.-Y.; Cheng, R.-Y.; Pan, O.-G. *Tetrahedron Lett.* **1997**, *38*, 443–446.

624. Mathew, J.; Invergo, B.J.; Silverman, R.B. *Synth. Commun.* **1985**, *15*, 377–383.

625. Mukhopadhyay, M.; Bhatia, B.; Iqbal, J. *Tetrahedron Lett.* **1997**, *38*, 1083–1086.

626. Bhatia, B.; Reddy, M.M.; Iqbal, J. *J. Chem. Soc., Chem. Commun.* **1994**, 713–714.

627. Rao, I.N.; Prabhakaran, E.N.; Das, S.K.; Iqbal, J. *J. Org. Chem.* **2003**, *68*, 4079–4082.

628. Prabhakaran, E.N.; Iqbal, J. *J. Org. Chem.* **1999**, *64*, 3339–3341.

629. Slopianka, M.; Gossauer, A. *Liebigs Ann. Chem.* **1981**, 2258–2265.

630. Tomioka, H.; Suuki, K. *Tetrahedron Lett.* **1989**, *30*, 6353–6356.

631. Akasaka, Y.; Morimoto, T.; Sekiya, M. *Chem. Pharm. Bull.* **1979**, *27*, 803–805.

632. Miyata, O.; Namba, M.; Ueda, M.; Naito, T. *Org. Biomol. Chem.* **2004**, *2*, 1274–1276.

633. Balenovic, K.; Jambresic, I. *Chem. Ind.* **1955**, 1673.

634. Jugelt, W.; Falck, P. *J. Prakt. Chem.* **1968**, *38*, 88–100.

635. Li, X.-G.; Kanerva, L.T. *Tetrahedron: Asymmetry* **2005**, *16*, 1709–1714.

636. Moussebois, C.; Heremans, J.F.; Merenyi, R.; Rennerts, W. *Helv. Chim. Acta* **1977**, *60*, 237–242.

637. Juaristi, E.; Quintana, D.; Lamatsch, B.; Seebach, D. *J. Org. Chem.* **1991**, *56*, 2553–2557.

638. Estermann, H.; Seebach, D. *Helv. Chim. Acta* **1988**, *71*, 1824–1839.

639. Cardillo, G.; Tolomelli, A.; Tomasini, C. *J. Org. Chem.* **1996**, *61*, 8651–8654.

640. Nocioni, A.M.; Papa, C.; Tomasini, C. *Tetrahedron Lett.* **1999**, *40*, 8453–8456.

641. Yamanaka, T.; Ohkubo, M.; Kato, M.; Kawamura, Y.; Nishi, A.; Hosokawa, T. *Synlett* **2005**, 631–634.

642. Hanessian, S.; Yang, H.; Schaum, R. *J. Am. Chem. Soc.* **1996**, *118*, 2507–2508.

643. Dell, C.P.; Khan, K.M.; Knight, D.W. *J. Chem. Soc., Perkin Trans. 1* **1994**, 341–347.

644. Dell, C.P.; Khan, K.M.; Knight, D.W. *J. Chem. Soc., Chem. Commun.* **1989**, 1812–1814.

645. Testa, E.; Fontanella, L. *Liebigs Ann. Chem.* **1959**, *625*, 95–98.

646. Hanai, K.; Kuwae, A. *J. Lab. Cmpds. Radiopharm.* **1988**, *25*, 217–224.

647. Papoutsis, I.; Spyroudis, S.; Varvoglis, A. *Tetrahedron*, **1998**, *54*, 1005–1012.

648. Sergis, A.N.; Brady, F.; Luthra, S.K.; Prenant, C.; Waters, S.L.; Steel, C.J.; Osman, S.; Price, P.M. *J. Lab. Cmpds. Radiopharm.* **1995**, *37*, 586–588.

649. Aberhart, D.J.; Lin, H.-J. *J. Lab. Cmpds. Radiopharm.* **1983**, *20*, 611–617.

650. Wadsworth, K.J.; Wood, F.K.; Chapman, C.J.; Frost, C.G. *Synlett* **2004**, 2022–2024.

651. Madan, S.; Arora, R.; Venugopalan, P.; Bari, S.S. *Tetrahedron Lett.* **2000**, *41*, 5577–5581.

652. Fustero, S.; Salavert, E.; Sanz-Cervera, J.F.; Román, R.; Fernández-Gutiérrez, B.; Asensio, A. *Lett. Org. Chem.* **2004**, *1*, 163–167.

653. Bergnes, G.; Kaddurah–Daouk, R. *Bioorg. Med. Chem. Lett.* **1997**, *7*, 1021–1026.

654. Righi, G.; D'Achille, R.; Bonini, C. *Tetrahedron Lett.* **1996**, *37*, 6893–6896.

655. Dauban, P.; Dodd, R.H. *Tetrahedron Lett.* **1998**, *39*, 5739–5742.

656. Chandrasekhar, S.; Ahmed, M. *Tetrahedron Lett.* **1999**, *40*, 9325–9327.

657. Crousse, B.; Narizuka, S.; Bonnet–Delpon, D.; Bégué, J. -P. *Synlett* **2001**, 679–681.

658. Matsubara, S.; Kodama, T.; Utimoto, K. *Tetrahedron Lett.* **1990**, *31*, 6379–6380.

659. Hu, X.E. *Tetrahedron* **2004**, *60*, 2701–2743.

660. Palomo, C.; Aizpurua, J.M.; Ganboa, I.; Oiarbide, M. *Eur. J. Org. Chem.* **1999**, 3223–3235.

661. Krishnaswamy, D.; Bhawal, B.M.; Deshmukh, A.R.A.S. *Tetrahedron Lett.* **2000**, *41*, 417–419.

662. Palomo, C.; Arrieta, A; Cossio, F.P.; Aizpurua, J.M.; Mielgo, A.; Aurrekoetxea, N. *Tetrahedron Lett.* **1990**, *31*, 6429–6432.

663. Ito, Y.; Kamijo, T.; Harada, H.; Terashima, S. *Heterocycles* **1990**, *30*, 299–302.

664. Abouabdellah, A.; Bégué, J.-P.; Bonnet-Delpon, D. *Synlett*, **1996**, 399–400.

665. Hevia, E.; Pérez, J.; Riera, V.; Miguel, D.; Campomanes, P.; Menéndez, M.I.; Sordo, T.L.; García-Granda, S. *J. Am. Chem. Soc.* **2003**, *125*, 3706–3707.

666. Banik, B.K.; Becker, F.F. *Tetrahedron Lett.* **2000**, *41*, 6551–6554.

667. Delpiccolo, C.M.L.; Fraga, M.A.; Mata, E.G. *J. Comb. Chem.* **2003**, *5*, 208–210.

668. Aben, R.W.M.; Smit, R.; Scheeren, J.W. *J. Org. Chem.* **1987**, *52*, 365–370.

669. Graf, R. *Liebigs Ann. Chem.* **1963**, *661*, 111–157.

670. Durst, T.; O'Sullivan, M.J. *J. Org. Chem.* **1970**, *35*, 2043–2044.

671. Gyarmati, Z.C.; Liljeblad, A.; Rintola, M.; Bernáth, G.; Kanerva, L.T. *Tetrahedron Asymmetry* **2003**, *14*, 3805–3814.

672. Shindo, M.; Oya, S.; Murakami, R.; Sato, Y.; Shishido, K. *Tetrahedron Lett.* **2000**, *41*, 5934–5946.

673. Cheung, K.-M.; Shoolingin-Jordan, P.M. *Tetrahedron* **1997**, *53*, 15807–15812.

674. Dietrich, R.F.; Sakurai, T.; Kenyon, G.L. *J. Org. Chem.* **1979**, *44*, 1894–1896.

675. Kanemasa, S.; Araki, T.; Kanai, T.; Wada, E. *Tetrahedron Lett.* **1999**, *40*, 5059–5062.

676. Aizencang, G.; Frydman, R.B.; Giorgieri, S.; Sambrotta, L.; Guerra, L. Frydman, B. *J. Med. Chem.* **1995**, *38*, 4337–4341.

677. Bubert, C.; Voigt, J.; Biasetton, S.; Reiser, O. *Synlett* **1994**, 675–677.

678. Ege, M.; Wanner, K.T. *Org. Lett.* **2004**, *6*, 3553–3556.

679. Lambert, T.H.; MacMillan, D.W.C. *J. Am. Chem. Soc.* **2002**, *124*, 13646–13647.

680. Soloshonok, V.A.; Svedas, V.K.; Kukhar, V.P.; Kirilenko, A.G.; Rybakova, A.V.; Solodenko, V.A.; Fokina, N.A.; Kogut, O.V.; Galaev, I.Y.; Kozlova, E.V.; Shishkina, I.P.; Galushko, S.V. *Synlett* **1993**, 339–341.

681. See Reference *680*.

682. Kukhar, V.P.; Svedas, V.K.; Kozlova, E.V. *Tetrahedron: Asymmetry* **1994**, *5*, 1119–1126.

683. Roche, D.; Prasad, K.; Repic, O. *Tetrahedron Lett.* **1999**, *40*, 3665–3668.

684. Gedey, S.; Liljeblad, A.; Fülöp, F.; Kanerva, L.T. *Tetrahedron: Asymmetry* **1999**, *10*, 2573–2581.

685. Flores-Sánchez, P.; Escalante, J.; Castillo, E. *Tetrahedron: Asymmetry* **2005**, *16*, 629–634.

686. van Rantwijk, F.; Sheldon, R.A. *Tetrahedron* **2004**, *60*, 501–519.

687. Solymár, M.; Liljeblad, A.; Lázár, L.; Fülöp, F.; Kanerva, L.T. *Tetrahedron Asymmetry* **2002**, *13*, 1923–1928.

688. Patel, R.N.; Howell, J.; Chidambaram, R.; Benoit, S.; Kant, J. *Tetrahedron Asymmetry* **2003**, *14*, 3673–3677.

689. López-Garciá, M.; Alfonso, I.; Gotor, V. *Lett. Org. Chem.* **2004**, *1*, 254–256.

690. Sánchez, V.M.; Rebolledo, F.; Gotor, V. *Tetrahedron: Asymmetry* **1997**, *8*, 37–40.

691. Faulconbridge, S.J.; Holt, K.E.; Sevillano, L.G.; Lock, C.J.; Tiffin, P.D.; Tremayne, N.; Winter, S. *Tetrahedron Lett.* **2000**, *41*, 2679–2681.

692. Brieva, R.; Crich, J.Z.; Sih, C.J. *J. Org. Chem.* **1993**, *58*, 1068–1075.

693. Patel, R.N.; Banerjee, A.; Howell, J.M.; McNamee, C.G.; Brozozowski, D.; Mirfakhrae, D.; Nanduri, V.; Thottathil, J.K.; Szarka, L.J. *Tetrahedron: Asymmetry* **1993**, *4*, 2069–2084.

694. Kayser, M.M.; Mihovilovic, M.D.; Kearns, J.; Feicht, A.; Stewart, J.D. *J. Org. Chem.* **1999**, *64*, 6603–6608.

695. Fischer, E.; Jacobs, W.A. *Ber.* **1907**, *40*, 1057–1070.

696. Griffith, O.W.; Campbell, E.B.; Pirkle, W.H.; Tsipouras, A.; Hyun, M.H. *J. Chromatogr.* **1986**, *362*, 345–352.

697. Pirkle, W.H.; Pochapsky, T.C.; Mahler, G.S.; Corey, D.E.; Reno, D.S.; Alessi, D.M. *J. Org. Chem.* **1986**, *51*, 4991–5000.

698. Wagner, J.; Wolf, E.; Heintzelmann, B.; Gaget, C. *J. Chromatogr.* **1987**, *392*, 211–224.

699. D'Acquarica, I.; Gasparrini, F.; Misiti, D.; Zappia, G.; Cimarelli, C.; Palmieri, G.; Carotti, A.; Cellamare, S.; Villani, C. *Tetrahedron: Asymmetry* **2000**, *11*, 2375–2385.

700. Hyun, M.H.; Han, S.C.; Whangbo, S.H. *J. Chromatogr., A* **2003**, *992*, 47–56.

701. Péter, A. *J. Chromatogr., A* **2002**, *955*, 141–150.

702. Péter, A.; Árki, A.; Vékes, E.; Tourwé, D.; Lázár, L.; Fülöp, F.; Armstrong, D.W. *J. Chromatogr., A* **2004**, *1031*, 171–178.

703. Péter, A.; Lázár, L.; Fülöp, F.; Armstrong, D.W. *J. Chromatogr., A* **2001**, *926*, 229–238.

704. Aberhart, D.J.; Cotting, J.-A.; Lin, H.-J. *Analyt. Biochem.* **1985**, *151*, 88–91.

705. Nagula, G.; Huber, V.J.; Lum, C.; Goodman, B.A. *Org. Lett.* **2000**, *2*, 3527–3529.

706. Pollock, G.E. *Anal. Chem.* **1972**, *44*, 2368–2372.

707. Rodríguez-Escrich, S.; Popa, D.; Jimeno, C.; Vidal-Ferran, A.; Pericàs, M.A. *Org. Lett.* **2005**, *7*, 3829–3832.

708. Böhm, A.; Seebach, D. *Helv. Chim. Acta* **2000**, *83*, 3262–3278.

709. Sewald, N. *Amino Acids* **1996**, *11*, 397–408.

710. Furukawa, M.; Okawara, T.; Terawaki, Y. *Chem. Pharm. Bull.* **1977**, *25*, 1319–1325.

711. d'Angelo, J.; Maddaluno, J. *J. Am. Chem. Soc.* **1986**, *108*, 8112–8114.

712. Juaristi, E.; Escalante, J.; Lamatsch, B.; Seebach, D. *J. Org. Chem.* **1992**, *57*, 2396–2398.

713. Kinas, R.; Pankiewicz, K.; Stec, W.J.; Farmer, P.B.; Foster, A.B.; Jarman, M. *J. Org. Chem.* **1977**, *42*, 1650–1652.

714. Baldwin, S.W.; Aubé, J. *Tetrahedron Lett.* **1987**, *28*, 179–182.

715. Lee, H.-S.; Park, J.-S.; Kim, B.M.; Gellman, S.H. *J. Org. Chem.* **2003**, *68*, 1575–1578.

716. Davies, S.G.; Ichihara, O.; Walters, I.A.S. *Synlett* **1993**, 461–462.

717. Davies, S.G.; Ichihara, O. *Tetrahedron: Asymmetry* **1991**, *2*, 183–186.

718. Davies, S.G.; Smith, A.D.; Price, P.D. *Tetrahedron: Asymmetry* **2005**, *16*, 2833–2891.

719. Escalante, J.; Juaristi, E. *Tetrahedron Lett.* **1995**, *36*, 4397–4400.

720. Burke, A.J.; Davies, S.G.; Garner, A.C.; McCarthy, T.D.; Roberts, P.M.; Smith, A.D.; Rodriguez-Solla, H.; Vickers, R.J. *Org. Biomol. Chem.* **2004**, *2*, 1387–1394.

721. Coleman, P.J.; Hutchinson, J.H.; Hunt, C.A.; Lu, P.; Delaporte, E.; Rushmore, T. *Tetrahedron Lett.* **2000**, *41*, 5803–5806.

722. Bull, S.D.; Davies, S.G.; Smith, A.D. *J. Chem. Soc., Perkin Trans. 1* **2001**, 2931–2938.

723. Davies, S.G.; Ichihara, O.; Walters, I.A.S. *Synlett*, **1994**, 117–118.

724. Davies, S.G.; Garrido, N.M.; Ichihara, O.; Walters, I.A.S. *J. Chem. Soc., Chem. Commun.* **1993**, 1153–1155.

725. Davies, S.G.; Ichihara, O.; Walters, I.A.S. *J. Chem. Soc., Perkin Trans. 1* **1994**, 1141–1147.

726. Langenhan, J.M.; Gellman, S.H. *J. Org. Chem.* **2003**, *68*, 6440–6443.

727. Davies, S.G.; Walters, I.A.S. *J. Chem. Soc., Perkin Trans. 1* **1994**, 1129–1139.

728. Tsukada, N.; Shimada, T.; Gyoung, Y.S.; Asao, N.; Yamamoto, Y. *J. Org. Chem.* **1995**, *60*, 143–148.

729. Andrews, P.C.; Bhaskar, V.; Bromfield, K.M.; Dodd, A.M.; Duggan, P.J.; Duggan, S.A.M.; McCarthy, T.D. *Synlett* **2004**, 791–794.

730. Asao, N.; Tsukada, N.; Yamamoto, Y. *J. Chem. Soc., Chem. Commun.* **1993**, 1660–1662.

731. Davies, S.G.; Edwards, A.J.; Walters, I.A.S. *Rec. Trav. Chim. Pay-Bas* **1995**, *114*, 175–183.

732. Bunnage, M.E.; Chernega, A.N.; Davies, S.G.; Goodwin, C.J. *J. Chem. Soc., Perkin Trans. 1* **1994**, 2373–2384.

733. Bunnage, M.E.; Davies, S.G.; Goodwin, C.J. *J. Chem. Soc., . Perkin Trans. 1* **1994**, 2385–2391.

734. Bunnage, M.E.; Davies, S.G.; Goodwin, C.J. *J. Chem. Soc. Perkin Trans. 1* **1993**, 1375–1376.

735. Bunnage, M.E.; Davies, S.G.; Goodwin, C.J.; Ichihara, O. *Tetrahedron* **1994**, *50*, 3975–3986.

736. Bunnage, M.E.; Davies, S.G.; Goodwin, C.J. *Synlett* **1993**, 731–732.

737. Bunnage, M.E.; Burke, A.J.; Davies, S.G.; Goodwin, C.J. *Tetrahedron: Asymmetry* **1995**, *6*, 165–176.

738. Bunnage, M.E.; Burke, A.J.; Davies, S.G.; Goodwin, C.J. *Tetrahedron: Asymmetry* **1994**, *5*, 203–206.

739. Brackenridge, I.; Davies, S.G.; Fenwick, D.R.; Ichihara, O.; Polywka, M.E.C. *Tetrahedron* **1999**, *55*, 533–540.

740. Bunnage, M.E.; Davies, S.G.; Goodwin, C.J.; Walters, I.A.S. *Tetrahedron: Asymmetry* **1994**, *5*, 35–36.

741. Costello J.F.; Davies, S.G.; Ichihara, O. *Tetrahedron: Asymmetry* **1994**, *5*, 1999–2008.

742. Bull, S.D.; Davies, S.G.; Fox, D.J.; Gianotti, M.; Kelly, P.M.; Pierres, C.; Savory, E.D.; Smith, A.D. *J. Chem. Soc., Perkin Trans. 1* **2002**, 1858–1868.

743. Bull, S.D.; Davies, S.G.; Delgado-Ballester, S.; Kelly, P.M.; Kotchie, L.J.; Gianotti, M.; Laderas, M.; Smith, A.D. *J. Chem. Soc., Perkin Trans. 1* **2001**, 3112–3121.

744. Bull, S.D.; Davies, S.G.; Delgado-Ballester, S.; Fenton, G.; Kelly, P.M.; Smith, A.D. *Synlett* **2000**, 1257–1260.

745. Davies, S.G.; Ichihara, O. *Tetrahedron Lett.* **1998**, *39*, 6045–6048.

746. Davies, S.G.; Fenwick, D.R. *J. Chem. Soc., Chem. Commun.* **1995**, 1109–1110.

747. Davies, S.G.; Fenwick, D.R.; Ichihara, O. *Tetrahedron: Asymmetry* **1997**, *8*, 3387–3391.

748. Davies, S.G.; Iwamoto, K.; Smethurst, C.A.P.; Smith, A.D.; Rodriguez-Sola, H. *Synlett* **2002**, 1146–1148.

749. O'Brien, P.; Porter, D.W.; Smith, N.M. *Synlett* **2000**, 1336–1338.

750. Rico, J.G.; Lindmark, R.J.; Rogers, T.E.; Bovy, P.R. *J. Org. Chem.* **1993**, *58*, 7948–7951.

751. Sewald, N.; Hiller, K.D.; Helmreich, B. *Liebigs Ann.* **1995**, 925–928.

752. Hawkins, J.M.; Fu, G.C. *J. Org. Chem.* **1986**, *51*, 2820–2822.

753. Hawkins, J.M.; Lewis, T.A. *J. Org. Chem.* **1992**, *57*, 2114–2121.

754. Rudolf, K.; Hawkins, J.M.; Loncharich, R.J.; Houk, K.N. *J. Org. Chem.* **1988**, *53*, 3879–3882.

755. Hawkins, J.M.; Lewis, T.A. *J. Org. Chem.* **1994**, *59*, 649–652.

756. Es-Sayed, M.; Gratkowski, C.; Krass, N.; Meyers, A.I.; de Meijere, A. *Tetrahedron Lett.* **1993**, *34*, 289–292.

757. Enders, D.; Wahl, H.; Bettray, W. *Angew. Chem., Int. Ed. Engl.* **1995**, *34*, 455–457.

758. Enders, D.; Bettray, W.; Raabe, G.; Runsink, J. *Synthesis* **1994**, 1322–1326.

759. Amoroso, R.; Cardillo, G.; Sabatino, P.; Tomasini, C.; Trerè, A. *J. Org. Chem.* **1993**, *58*, 5615–5619.

760. Cardillo, G.; Gentilucci, L.; Tolomelli, A. *Tetrahedron Lett.* **1999**, *40*, 8261–8264.

761. Cardillo, G.; Casolari, S.; Gentilucci, L.; Tomasini, C. *Angew. Chem., Int. Ed. Engl.* **1996**, *35*, 1848–1849.

762. Cardillo, G.; De Simone, A.; Gentilucci, L.; Sabatino, P.; Tomasini, C. *Tetrahedron Lett.* **1994**, *35*, 5051–5054.

763. Beddow, J.E.; Davies, S.G.; Smith, A.D.; Russell, A.J. *Chem. Commun.* **2004**, 2778–2779.

764. Etxebarria, J.; Vicario, J.L.; Badia, D.; Carrillo, L.; Ruiz, N. *J. Org. Chem.* **2005**, *70*, 8790–8800.

765. de Lange, B.; van Bolhuis, F.; Feringa, B.L. *Tetrahedron* **1989**, *45*, 6799–6818.

766. Yamamoto, Y.; Asao, N.; Uyehara, T. *J. Am. Chem. Soc.* **1992**, *114*, 5427–5429.

767. Davies, S.G.; Dupont, J.; Easton, R.J.C. *Tetrahedron: Asymmetry* **1990**, *1*, 279–280.

768. Matsuyama, H.; Itoh, N.; Yoshida, M.; Kamigata, N.; Sasaki, S.; Iyoda, M. *Chem. Lett.* **1997**, 375–376.

769. Chung, S.-K.; Jeong, T.-H.; Kang, D.-H. *Tetrahedron: Asymmetry* **1997**, *8*, 5–9.

770. Chung, S.-K.; Jeong, T.-H.; Kang, D.-H. *J. Chem. Soc., Perkin Trans. 1* **1998**, 969–976.

771. Amoroso, R.; Cardillo, G.; Tomasini, C. *Heterocycles* **1992**, *34*, 349–355.

772. Cardillo, G.; Orena, M.; Penna, M.; Sandri, S.; Tomasini, C. *Synlett* **1990**, 543–544.

773. Yioo, D.; Kwon, S.; Kim, Y.G. *Tetrahedron: Asymmetry* **2005**, *16*, 3762–3766.

774. Perlmutter, P.; Tabone, M. *Tetrahedron Lett.* **1988**, *29*, 949–952.

775. Asao, N.; Shimada, T.; Sudo, T.; Tsukada, N.; Yazawa, K.; Gyoung, Y.S.; Uyehara, T.; Yamamoto, Y. *J. Org. Chem.* **1997**, *62*, 6274–6282.

776. See Reference *775*.

777. Körner, M.; Findeisen, M.; Sewald, N. *Tetrahedron Lett.* **1998**, *39*, 3463–3464.

778. Collis, M.P.; Hockless, D.C.R.; Perlmutter, P. *Tetrahedron Lett.* **1995**, *36*, 7133–7136.

779. Matsunaga, H.; Sakamaki, T.; Nagaoka, H.; Yamada, Y. *Tetrahedron Lett.* **1983**, *24*, 3009–3012.

780. Sharma, G.V.M.; Reddy, V.G.; Chander, A.S.; Reddy, K.R. *Tetrahedron: Asymmetry* **2002**, *13*, 21–24.

781. Falborg, L.; Jørgensen, K.A. *J. Chem. Soc., Perkin Trans. 1* **1996**, 2823–2826.

782. Ishikawa, T.; Nagai, K.; Kudoh, T.; Saito, S. *Synlett* **1998**, 1291–1293.

783. Sibi, M.P.; Shay, J.J.; Liu, M.; Jasperse, C.P. *J. Am. Chem. Soc.* **1998**, *120*, 6615–6616.

784. Sibi, M.P.; Prabagaran, N.; Ghorpade, S.G.; Jasperse, C.P. *J. Am. Chem. Soc.* **2003**, *125*, 11796–11797.

785. Palomo, C.; Oiarbide, M.; Halder, R.; Kelso, M.; Gómez-Bengoa, E.; García, J.M. *J. Am. Chem. Soc.* **2004**, *126*, 9188–9189.

786. Zhuang, W.; Hazell, R.G.; Jørgenen, K.A. *Chem. Commun.* **2001**, 1240–1241.

787. Li, K.; Cheng, X.; Hii, K.K. *J. Eur. J. Org. Chem.* **2004**, 959–964.

788. Myers, J.K.; Jacobsen, E.N. *J. Am. Chem. Soc.* **1999**, *121*, 8959–8960.

789. Horstmann, T.E.; Guerin, D.J.; Miller, S.J. *Angew. Chem., Int. Ed.* **2000**, *39*, 3635–3638.

790. Doi, H.; Sakai, T.; Iguchi, M.; Yamada, K.-i.; Tomioka, K. *J. Am. Chem. Soc.* **2003**, *125*, 2886–2887.

791. Doi, H.; Sakai, T.; Yamada, K.-i.; Tomioka, K. *Chem. Commun.* **2004**, 1850–1851.

792. Kitazume, T.; Murata, K. *J. Fluorine Chem.* **1987**, *36*, 339–349.

793. Rao, K.R.; Nageswar, Y.V.D.; Kumar, H.M.S. *Tetrahedron Lett.* **1991**, *32*, 6611–6612.

794. Pons, D.; Savignac, M.; Genet, J.-P. *Tetrahedron Lett.* **1990**, *31*, 5023–5026.

795. Sharpless, K.B.; Behrens, C.H.; Katsuki, T.; Lee, A.W.M.; Martin, V.S.; Takatani, M.; Viti, S.M.; Walker, F.J.; Woodard, S.S. *Pure Appl. Chem.* **1983**, *55*, 589–604.

796. Aggarwal, V.K.; Hynd, G.; Picoul, W.; Vasse, J.-L. *J. Am. Chem. Soc.* **2002**, *124*, 9964–9965.

797. Fringuelli, F.; Pizzo, F.; Rucci, M.; Vaccaro, L. *J. Org. Chem.* **2003**, *68*, 7041–7045.

798. Umemura, E.; Tsuchiya, T.; Umezawa, S. *J. Antibiotics* **1988**, *41*, 530–537.

799. Dîaz, M.; Ortuño, R.M. *Tetrahedron: Asymmetry* **1995**, *6*, 1845–1848.

800. Díaz, M.; Ortuño, R.M. *Tetrahedron: Asymmetry* **1996**, *7*, 3465–3478.

801. Díaz, M.; Branchadell, V.; Oliva, A.; Ortuño, R.M. *Tetrahedron* **1995**, *51*, 11841–11854.

802. Alcaide, B.; Biurrun, C.; Martínez, A.; Plumet, J. *Tetrahedron Lett.* **1995**, *36*, 5417–5420.

803. Shimohigashi, Y.; Waki, M.; Izumiya, N. *Bull. Chem. Soc. Jpn.* **1979**, *52*, 949–950.

804. Righi, G.; Chionne, A.; D'Achille, R.; Bonini, C. *Tetrahedron: Asymmetry* **1997**, *8*, 903–907.

805. Righi, G.; D'Achille, C.; Pescatore, G.; Bonini, C. *Tetrahedron Lett.* **2003**, *44*, 6999–7002.

806. Feske, B.D.; Steart, J.D. *Tetrahedron: Asymmetry* **2005**, *16*, 3124–3127.

807. Rodrigues, J.A.R.; Milagre, H.M.S.; Milagre, C.D.F.; Moran, P.J.S. *Tetrahedron: Asymmetry* **2005**, *16*, 3099–3106.

808. García Ruano, J.L.; García Paredes, C. *Tetrahedron Lett.* **2000**, *41*, 5357–5361.

809. Pastó, M.; Moyano, A.; Pericàs, M.A.; Riera, A. *Tetrahedron: Asymmetry* **1996**, *7*, 243–262.

810. Pastó, M.; Castejøn, P.; Moyano, A.; Pericàs, M.A.; Riera, A. *J. Org. Chem.* **1996**, *61*, 6033–6037.

811. Pastó, M.; Moyano, A.; Pericàs, M.A.; Riera, A. *Tetrahedron: Asymmetry* **1995**, *6*, 2329–2342.

812. Alcón, M.; Canas, M.; Poch, M.; Moyano, A.; Pericàs, M.A.; Riera, A. *Tetrahedron Lett.* **1994**, *35*, 1589–1592.

813. Denis, J.-N.; Greene, A.E.; Serra, A.A.; Luche, M.-J. *J. Org. Chem.* **1986**, *51*, 46–50.

814. Deng, L.; Jacobsen, E.N. *J. Org. Chem.* **1992**, *57*, 4320–4323.

815. De, A.; Basak, P.; Iqbal, J. *Tetrahedron Lett.* **1997**, *38*, 8383–8386.

816. Saha, B.; Nandy, J.P.; Shukla, S.; Siddiqui, I.; Iqbal, J. *J. Org. Chem.* **2002**, *67*, 7858–7860.

817. Agami, C.; Couty, F.; Hamon, L.; Venier, O. *J. Org. Chem.* **1997**, *62*, 2106–2112.

818. Zhou, Z.; Mei, X. *Synth. Commun.* **2003**, *33*, 723–728.

819. Cabon, O.; Buisson, D.; Larcheveque, M.; Azerad, R. *Tetrahedron: Asymmetry* **1995**, *6*, 2211–2218.

820. Yamaguchi, T.; Harada, N.; Ozaki, K.; Hashiyama, T. *Tetrahedron Lett.* **1998**, *39*, 5575–5578.

821. Denis, J.-N.; Correa, A.; Greene, A.E. *J. Org. Chem.* **1990**, *55*, 1957–1959.

822. Commerçon, A.; Bézard, D.; Bernard, F.; Bourzat, J.D. *Tetrahedron Lett.* **1992**, *33*, 5185–5188.

823. Adger, B.M.; Barkley, J.V.; Bergeron, S.; Cappi, M.W.; Flowerdew, B.E.; Jackson, M.P.; McCague, R.; Nugent, T.C.; Roberts, S.M. *J. Chem. Soc., Perkin Trans. 1* **1997**, 3501–3507.

824. Zhou, Z.; Mei, X.; Chang, J.; Feng, D. *Synth. Commun.* **2001**, *31*, 3609–3615.

825. Gou, D.-M.; Liu, Y.-C.; Chen, C.-S. *J. Org. Chem.* **1993**, *58*, 1287–1289.

826. Righi, G.; Rumboldt, G.; Bonini, C. *J. Org. Chem.* **1996**, *61*, 3557–3560.

827. Yamagichi, T.; Harada, N.; Ozaki, K.; Hayashi, M.; Arakawa, H.; Hashiyama, T. *Tetrahedron* **1999**, *55*, 1005–1016.

828. Lee, D.; Kim, M.-J. *Tetrahedron Lett.* **1998**, *39*, 2163–2166.

829. Dureault, A.; Tranchepain, I.; Depezay, J.C. *Synthesis* **1987**, 491–493.

830. Inokuchi, T.; Tanigawa, S.; Kanazaki, M.; Torii, S. *Synlett*, **1991**, 707–708.

831. Sin, N.; Kallmerten, J. *Tetrahedron Lett.* **1996**, 37, 5645–5648.

832. Seebach, D.; Schaeffer, L.; Gessier, F.; Bindschädler, P.; Jäger, C.; Josien, D.; Kopp, S.; Lelais, G.; Mahajan, Y.R.; Miuech, P.; Sebesta, R.; Schweizer, B.W. *Helv. Chim. Acta* **2003**, *86*, 1852–1861.

833. Lelais, G.; Mieuch, P.; Josien-Lefebvre, D.; Rossi, F.; Seebach, D. *Helv. Chim. Acta* **2004**, *87*, 3131–3159.

834. Bates, R.B.; Gangwar, S. *Tetrahedron: Asymmetry* **1993**, *4*, 69–72.

835. White, J.D.; Hong, J.; Robarge, L.A. *J. Org. Chem.* **1999**, *64*, 6206–6216.

836. Choudary, B.M.; Chowdari, N.S.; Madhi, S.; Kantam, M.L. *J. Org. Chem.* **2003**, *68*, 1736–1746.

837. Ko, S.Y. *J. Org. Chem.* **2002**, *67*, 2689–2691.

838. Yokomatsu, T.; Minowa, T.Y.; Murano, T.; Shibuya, S. *Tetrahedron* **1998**, *54*, 9341–9356.

839. Chakraborty, T.K.; Joshi, S.P. *Tetrahedron Lett.* **1990**, *31*, 2043–2046.

840. Pearson, C.; Rinehart, K.L.; Sugano, M. *Tetrahedron Lett.* **1999**, *40*, 411–414.

841. Hamashima, Y.; Yagi, K.; Takano, H.; Tamás, L.; Sodeoka, M. *J. Am. Chem. Soc.* **2002**, *124*, 14530–14531.

842. Yadav, J.S.; Chandrasekhar, S.; Sasmal, P.K. *Tetrahedron Lett.* **1997**, *38*, 8765–8768.

843. Mukai, C.; Kim, I.J.; Hanaoka, M. *Tetrahedron: Asymmetry* **1992**, *3*, 1007–1010.

844. Mukai, C.; Kim, I.J.; Furu, E.; Hanaoka, M. *Tetrahedron* **1993**, *49*, 8323–8336.

845. Mukaiyama, T.; Shiina, I.; Iwadare, H.; Saitoh, M.; Nishimura, T.; Ohkawa, N.; Sakoh, H.;

Nishimura, K.; Tani, Y.; Hasegawa, M.; Yamada, K.; Saitoh, K. *Chem.— Eur. J.* **1999**, *5*, 121–161.

846. Pearson, W.H.; Hines, J.V. *J. Org. Chem.* **1989**, *54*, 4235–4237.

847. Boger, D.L.; Honda, T.; Dang, Q. *J. Am. Chem. Soc.* **1994**, *116*, 5619–5630.

848. Fernández-Suárez, M.; Muñoz, L.; Fernández, R.; Riguera, R. *Tetrahedron: Asymmetry* **1997**, *8*, 1847–1854.

849. Song, C.E.; Lee, S.W.; Roh, E.J.; Lee, S.-g.; Lee, W.-K. *Tetrahedron: Asymmetry* **1998**, *9*, 983–992.

850. Jin, Y.; Kim, D.H. *Synlett* **1998**, 1189–1190.

851. Sugimura, H.; Miura, M.; Yamada, N. *Tetrahedron: Asymmetry* **1997**, *8*, 4089–4099.

852. Avenoza, A.; Busto, J.H.; Corzana, F.; Jiménez-Osés, G.; Peregrina, J.M. *Chem. Commun.* **2004**, 980–981.

853. Koskinen, A.M.P.; Karvinen, E.K.; Siirilä, J.P. *J. Chem. Soc., Chem. Commun.* **1994**, 21–22.

854. Lohray, B.B.; Bhushan, V. *Ind. J. Chem.* **1995**, *34B*, 471–473.

855. Nelson, S.G.; Spencer, K.L.; Cheung, W.S.; Mamie, S.J. *Tetrahedron* **2002**, *58*, 7081–7091.

856. Li, G.; Sharpless, K.B. *Acta Chem. Scand.* **1996**, *50*, 649–651.

857. Sharpless, K.B. *Angew. Chem., Int. Ed. Engl.* **1996**, *35*, 451–454.

858. Upadhya, T.T.; Sudalai, A. *Tetrahedron: Asymmetry* **1997**, *8*, 3685–3689.

859. Chandrasekhar, S.; Mohapatra, S.; Yadav, J.S. *Tetrahedron* **1999**, *55*, 4763–4768.

860. Chandrasekhar, S.; Mohapatra, S. *Tetrahedron Lett.* **1998**, *39*, 695–698.

861. Chandrasekhar, S.; Mohapatra, S. *Tetrahedron Lett.* **1998**, *39*, 6415–6418.

862. Rudolph, J.; Sennhenn, P.C.; Vlaar, C.P.; Sharpless, K.B. *Angew. Chem., Int. Ed. Engl.* **1996**, 35, 2810–2813.

863. Li, G.; Angert, H.H.; Sharpless, K.B. *Angew. Chem., Int. Ed. Engl.* **1996**, 35, 2813–2817.

864. Han, H.; Yoon, J.; Janda, K.D. *J. Org. Chem.* **1998**, *63*, 2045–2048.

865. Bushey, M.L.; Haukaas, M.H.; O'Doherty, G.A. *J. Org. Chem.* **1999**, *64*, 2984–2985.

866. Jiang, W.; Wanner, J.; Lee, R.J.; Bounaud, P.-Y.; Boger, D.L. *J. Am. Chem. Soc.* **2003**, *125*, 1877–1887.

867. Raatz, D.; Innertsberger, C.; Reiser, O. *Synlett* **1999**, 1907–1910.

868. Liu, Z.; Ma, N.; Jia, Y.; Bois-Choussy, M.; Malabarba, A.; Zhu, J. *J. Org. Chem.* **2005**, *70*, 2847–2850.

869. Montiel-Smith, S.; Cervantes-Mejía, V.; Dubois, J.; Guénard, D.; Guéritte, F.; Sandoval-Ramírez, J. *Eur. J. Org. Chem.* **2002**, 2260–2264.

870. Goossen, L.J.; Liu, H.; Dress, K.R.; Sharpless, K.B. *Angew. Chem., Int. Ed. Engl.* **1999**, *38*, 1080–1083.

871. Reddy, K.L.; Dress, K.R.; Sharpless, K.B. *Tetrahedron Lett.* **1998**, *39*, 3667–3670.

872. Bruncko, M.; Schlingloff, G.; Sharpless, K.B. *Angew. Chem., Int. Ed. Engl.* **1997**, *36*, 1483–1486.

873. Kandula, S.R.V.; Kumar, P. *Tetrahedron: Asymmetry* **2005**, *16*, 3579–3583.

874. Song, C.E.; Oh, C.R.; Roh, E.J.; Lee, S.-G.; Choi, J.H. *Tetrahedron: Asymmetry* **1999**, *10*, 671–674.

875. Lee, S.-H.; Qi, X.; Yoon, J.; Nakamura, K.; Lee, Y.-S. *Tetrahedron* **2002**, *58*, 2777–2787.

876. Jo, C.H.; Han, S.-H.; Yang, J.W.; Roh, E.J.; Shin, U.-S.; Song, C.E. *Chem. Commun.* **2003**, 1312–1313.

877. Song, C.E.; Oh, C.R.; Lee, S.W.; Lee, S.-G.; Canali, L.; Sherrington, D.C. *Chem. Commun.* **1998**, 2435–2436.

878. Tao, B.; Schlingloff, G.; Sharpless, K.B. *Tetrahedron Lett.* **1998**, *39*, 2507–2510.

879. Han, H.; Cho, C.-W.; Janda, K.D. *Chem. —Eur. J.* **1999**, *5*, 1565–1569.

880. Streuff, J.; Osterath, B.; Nieger, M.; Muñiz, K. *Tetrahedron: Asymmetry* **2005**, *16*, 33492–3496.

881. O'Brien, P. *Angew. Chem., Int. Ed. Engl.* **1999**, *38*, 326–329.

882. Mangatal, L.; Adeline, M.-T.; Guénard, D.; Guéritte-Voegelein, F.; Potier, P. *Tetrahedron* **1989**, *45*, 4177–4190.

883. Freudenberg, K. *Ber.* **1914**, *47*, 2027–2037.

884. Wallis, E.S.; Lane, J.F. *Org. React.* **1946**, *3*, 267–306.

885. Andruszkiewicz, R.; Czerwinski, A.; Grzybowska, J. *Synthesis* **1983**, 31.

886. Dugger, R.W.; Ralbovsky, J.L.; Bryant, D.; Commander, J.; Massett, S.S.; Sage, N.A.; Selvidio, J.R. *Tetrahedron Lett.* **1992**, *33*, 6763–6766.

887. Kogen, H.; Nishi, T. *J. Chem. Soc., Chem. Commun.* **1987**, 311–312.

888. Milewska, M.J.; Polonski, T. *Synthesis* **1988**, 475.

889. Axelsson, B.S.; O'Toole, K.J.; Spencer, P.A.; Young, D.W. *J. Chem. Soc., Perkin Trans. 1* **1994**, 807–815.

890. Axelsson, B.S.; O'Toole, K.J.; Spencer, P.A.; Young, D.W. *J. Chem. Soc., Chem. Commun.* **1991**, 1085–1086.

891. Andruszkiewicz, R.; Wyszogrodzka, M. *Synlett* **2002**, 2101–2103.

892. Lelais, G.; Campo, M.A.; Kopp, S.; Seebach, D. *Helv. Chim. Acta* **2004**, *87*, 1545–1560.

893. Arvanitis, E.; Ernst, H.; Ludwig, A.A.; Robinson, A.J.; Wyatt, P.B. *J. Chem. Soc., Perkin Trans. 1* **1998**, 521–528; corrigendum **1998**, 1459.

894. Micuch, P.; Seebach, D. *Helv. Chim. Acta* **2002**, *85*, 1567–1577.

895. Evans, D.A.; Wu, L.D.; Wiener, J.J.M.; Johnson, J.S.; Ripin, D.H.B.; Tedrow, J.S. *J. Org. Chem.* **1999** *64*, 6411–6417.

896. Sibi, M.; Deshpande, P.K. *J. Chem. Soc., Perkin Trans. 1* **2000**, 1461–1466.

897. Chakraborty, T.K.; Ghosh, A. *Synlett* **2002**, 2039–2040.

898. Bower, J.F.; Williams, J.M.J. *Synlett* **1996**, 685–686.

899. Bower, J.F.; Jumnah, R.; Williams, A.C.; Williams, J.M.J. *J. Chem. Soc., Perkin Trans 1* **1997**, 1411–1420.

900. Takano, S.; Kasahara, C.; Ogasawara, K. *Chem. Lett.* **1982**, 631–634.

901. Jacobi, P.A.; Zheng, W. *Tetrahedron Lett.* **1993**, *34*, 2585–2588.

902. Jacobi, P.A.; Zheng, W. *Tetrahedron Lett.* **1993**, *34*, 2581–2584.

903. Kang, S.H.; Kim, C.M.; Youn, J.-H. *Tetrahedron Lett.* **1999**, *40*, 3581–3582.

904. Wittenberger, S.J.; McLaughlin, M.A. *Tetrahedron Lett.* **1999**, *40*, 7175–7178.

905. Lurain, A.E.; Walsh, P.J. *J. Am. Chem. Soc.* **2003**, *125*, 10677–10683.

906. Furukawa, M.; Okawara, T.; Noguchi, Y.; Terawaki, Y. *Chem. Pharm. Bull.* **1979**, *27*, 2223–2226.

907. Cimarelli, C.; Palmieri, G. *J. Org. Chem.* **1996**, *61*, 5557–5563.

908. Cimarelli, C.; Palmieri, G.; Bartoli, G. *Tetrahedron: Asymmetry* **1994**, *5*, 1455–1458.

909. Cimarelli, C.; Palmieri, G.; Volpini, E. *Synth. Commun.* **2001**, *31*, 2943–2953.

910. Melillo, D.G.; Cvetovich, R.J.; Ryan, K.M.; Sletzinger, M. *J. Org. Chem.* **1986**, *51*, 1498–1504.

911. Cohen, J.H.; Abdel-Magid, A.F.; Almond, H.R., Jr.; Maryanoff, C.A. *Tetrahedron Lett.* **2002**, *43*, 1977–1981.

912. Ikemoto, N.; Tellers, D.M.; Dreher, S.D.; Liu, J.; Huang, A.; Rivera, N.R.; Njolito, E.; Hsiao, Y.; McWilliams, J.C.; Williams, J.M.; Armstrong, J.D., III; Sun, Y.; Mathre, D.J.; Grabowski, E.J.J.; Tillyer, R.D. *J. Am. Chem. Soc.* **2004**, *126*, 3048–3049.

913. Potin, D.; Dumas, F.; d'Angelo, J. *J. Am. Chem. Soc.* **1990**, *112*, 3483–3486.

914. Mandai, T.; Kuroda, A.; Okumoto, H.; Nakanishi, K.; Mikuni, K.; Hara, K.; Hara, K. *Tetrahedron Lett.* **2000**, *41*, 239–242.

915. Mandai, T.; Kuroda, A.; Okumoto, H.; Nakanishi, K.; Mikuni, K.; Hara, K.; Hara, K. *Tetrahedron Lett.* **2000**, *41*, 243–246.

916. Pégorier, L.; Haddad, M.; Larchevêque, M. *Synlett* **1996**, 585–586.

917. Bunlaksananusorn, T.; Rampf, F. *Synlett* **2005**, 2682–2684.

918. Zhou, Y.-G.; Tang, W.; Wang, W.-B.; Li, W.; Zhang, X. *J. Am. Chem. Soc.* **2002**, *124*, 4952–4953.

919. Marcotte, S.; Pannecoucke, X.; Feasson, C.; Quirion, J.-C. *J. Org. Chem.* **1999**, *64*, 8461–8484.

920. Achiwa, K.; Soga, T. *Tetrahedron Lett.* **1978**, *13*, 1119–1120.

921. Heller, D.; Holz, J.; Drexler, H.-J.; Lang, J.; Drauz, K.; Krimmer, H.-P.; Börner, A. *J. Org. Chem.* **2001**, *66*, 6816–6817.

922. Jerphagnon, T.; Renaud, J.-L.; Demonchaux, P.; Ferreira, A.; Bruneau, C. *Tetrahedron: Asymmetry* **2003**, *14*, 1973–1977.

923. Elaridi, J.; Thaqi, A.; Prosser, A.; Jackson, W.R.; Robinson, A.J. *Tetrahedron: Asymmetry* **2005**, *16*, 1309–1319.

924. Heller, D.; Holz, J.; Komarov, I.; Drexler, H.-J.; You, J.; Drauz, K.; Börner, A. *Tetrahedron: Asymmetry* **2002**, *13*, 2735–2741.

925. Heller, D.; Drexler, H.-J.; You, J.; Baumann, W.; Drauz, K.; Krimmer, H.-P.; Börner, A. *Chem.—Eur. J.* **2002**, *8*, 5196–5203.

926. Holz, J.; Monsees, A.; Jiao, H.; You, J.; Komarov, I.V.; Fischer, C.; Frauz, K.; Börne, A. *J. Org. Chem.* **2003**, *68*, 1701–1707.

927. Imamoto, T.; Sugita, K.; Yoshida, K. *J. Am, . Chem. Soc.* **2005**, *127*, 11934–11935.

928. Tang, W.; Zhang, X. *Org. Lett.* **2002**, *4*, 4159–4161.

929. Tang, W.; Wang, W.; Chi, Y.; Zhang, X. *Angew. Chem., Int. Ed.* **2003**, *42*, 3509–3511.

930. Liu, D.; Zhang, X. *Eur. J. Org. Chem.* **2005**, 646–649.

931. Hsiao, Y.; Rivera, N.R.; Rosner, T.; Krska, S.W.; Njolito, E.; Wang, F.; Sun, Y.; Armstrong, J.D., III; Grabowski, E.J.J.; Tillyer, R.D.; Spindler, F.; Malan, C. *J. Am. Chem. Soc.* **2004**, *126*, 9918–9919.

932. Hu, X.-P.; Zheng, Z. *Org. Lett.* **2005**, *7*, 419–422.

933. Wu, J.; Chen, X.; Guo, R.; Yeung, C.-h.; Chan, A.S.C. *J. Org. Chem.* **2003**, *68*, 2490–2493.

934. Qiu, L.; Wu, J.; Chan, S.; Au-Yeung, T.T.-L.; Ji, J.-X.; Guo, R.; Pai, C.-C.; Zhou, Z.; Li, X.; Fan, Q.-H.; Chan, A.S.C. *Proc. Natl. Acad. Sci. U.S.A.* **2004**, *101*, 5815–5820.

935. Lee, S.-g.; Zhang, Y.J. *Org. Lett.* **2002**, *4*, 2429–2431.

936. Kadyrov, R.; Ilaldinov, I.Z.; Almena, J.; Monsees, A.; Riermeier, T.H. *Tetrahedron Lett.* **2005**, *46*, 7397–7400.

937. Dubrovina, N.V.; Tararov, V.I.; Monsees, A.; Spannenberg, A.; Kostas, I.D.; Börner, A. *Tetrahedron: Asymmetry* **2005**, *16*, 3640–3649.

938. Dubrovina, N.V.; Tararov, V.I.; Monsees, A.; Kadyrov, R.; Fischer, C.; Börner, A. *Tetrahedron: Asymmetry* **2003**, *14*, 2739–2745.

939. Hoge, G.; Samas, B. *Tetrahedron: Asymmetry* **2004**, *15*, 2155–1257.

940. Wu, H.-P.; Hoge, G. *Org. Lett.* **2004**, *6*, 3645–3647.

941. Peña, D.; Minnaard, A.J.; de Vries, J.G.; Feringa, B.L. *J. Am. Chem. Soc.* **2002**, *124*, 14552–14553.

942. Peña, D.; Minnaard, A.J.; de Vries, A.H.M.; de Vries, J.G.; Feringa, B.L. *Org. Lett.* **2003**, *5*, 475–478.

943. Yasutake, M.; Gridnev, I.D.; Higashi, N.; Imamoto, T. *Org. Lett.* **2001**, *3*, 1701–1704.

944. Huang, H.; Zheng, Z.; Luo, H.; Bai, C.; Hu, X.; Chen, H. *J. Org. Chem.* **2004**, *69*, 2355–2361.

945. Huang, H.; Liu, X.; Chen, S.; Chen, H.; Zheng, Z. *Tetrahedron: Asymmetry* **2004**, *15*, 2011–2019.

946. Fu, Y.; Hou, G.-H.; Xie, J.-H.; Xing, L.; Wang, L.-X.; Zhou, Q.-L. *J. Org. Chem.* **2004**, *69*, 8157–8160.

947. Danjo, H.; Sasaki, W.; Miyazaki, T.; Imamoto, T. *Tetrahedron Lett.* **2003**, *44*, 3467–3469.

948. Lefort, L.; Boogers, J.A.F.; de Vries, A.H.M.; de Vries, J.G. *Org. Lett.* **2004**, *6*, 1733–1735.

949. Monti, C.; Gennari, C.; Piarulli, U.; de Vries, J.G.; de Vries, A.H.M.; Lefort, L. *Chem.—Eur. J.* **2005**, *11*, 6701–6717.

950. Peña, D.; Minnaard, A.J.; Boogers, J.A.F.; de Vries, A.H.M.; de Vries, J.G.; Feringa, B.L. *Org. Biomol. Chem.* **2003**, *1*, 1087–1089.

951. Reetz, M.T.; Li, X. *Tetrahedron* **2004**, *60*, 9709–9714.

952. Reetz, M.T.; Li, X. *Angew. Chem., Int. Ed.* **2005**, *44*, 2959–2962.

953. Drexler, H.-J.; Baumann, W.; Schmidt, T.; Zhang, S.; Sun, A.; Spannenberg, A.; Fischer, C.; Buschmann, H.; Heller, D. *Angew. Chem., Int. Ed.* **2005**, *44*, 1184–1188.

954. Bringmann, G.; Geuder, T. *Synthesis* **1991**, 829–831.

955. Soloshonok, V.A.; Kirilenko, A.G.; Galushko, S.V.; Kukhar, V.P. *Tetrahedron Lett.* **1994**, *35*, 5063–5064.

956. Soloshonok, V.A.; Ono, T.; Soloshonok, I.V. *J. Org. Chem.* **1997**, *62*, 7538–7539.

957. Juhl, K.; Jørgensen, K.A. *J. Am. Chem. Soc.* **2002**, *124*, 2420–2421.

958. Foubelo, F.; Yus, M. *Tetrahedron: Asymmetry* **1996**, *7*, 2911–2922.

959. Reginato, G.; Mordini, A.; Valacchi, M.; Piccardi, R. *Tetrahedron: Asymmetry* **2002**, *13*, 595–600.

960. Ponsinet, R.; Chassaing, G.; Lavielle, S. *Tetrahedron: Asymmetry* **1998**, *9*, 865–871.

961. Gais, H.-J.; Loo, R.; Roder, D.; Das, P.; Raabe, G. *Eur. J. Org. Chem.* **2003**, 1500–1526.

962. Solladie-Cavallo, A.; Khiar, N. *Tetrahedron Lett.* **1988**, *28*, 2189–2192.

963. Solladie-Cavallo, A.; Khiar, N. *J. Org. Chem.* **1990**, 55, 4750–4754.

964. Kudyba, I.; Raczko, J.; Jurczak, J. *J. Org. Chem.* **2004**, *69*, 2844–2859.

965. Kudyba, I.; Raczko, J.; Jurczak, J. *Helv. Chim. Acta* **2004**, *87*, 1724–1736.

966. Gogoi, N.; Boruwa, J.; Barua, N.C. *Tetrahedron Lett.* **2005**, *46*, 7581–7582.

967. Borah, J.C.; Gogoi, S.; Boruwa, J.; Kalita, B.; Barua, N.C. *Tetrahedron Lett.* **2004**, *45*, 3689–3691.

968. Christensen, C.; Juhl, K.; Jørgensen, K.A. *Chem. Commun.* **2001**, 2222–2223.

969. Park, Y.S.; Beak, P. *J. Org. Chem.* **1997**, *62*, 1574–1575.

970. Kim, B.J.; Park, Y.S.; Beak, P. *J. Org. Chem.* **1999**, *64*, 1705–1708.

971. Furukawa, M.; Okawara, T.; Noguchi, Y.; Terawaki, Y. *Chem. Pharm. Bull.* **1978**, *26*, 260–263.

972. Saidi, M.R.; Azizi, N. *Tetrahedron Asymmetry* **2003**, *14*, 2523–2527.

973. Ha, H.-J.; Park, G.-S.; Ahn, Y.-G.; Lee, G.S. *Biorg. Med. Chem. Lett.* **1998**, *8*, 1619–1622.

974. Yamada, T.; Suzuki, H.; Mukaiyama, T. *Chem. Lett.* **1987**, 293–296.

975. Kunz, H.; Schanzenbach, D. *Angew. Chem., Int. Ed. Engl.* **1989**, *28*, 1068–1069.

976. Kunz, H.; Burgard, A.; Schanzenbach, D. *Ang. Chem., Int. Ed. Engl.* **1997**, *36*, 386–387.

977. Laschat, S.; Kunz, H. *Synlett*, **1990**, 51–52.

978. Laschat, S.; Kunz H. *J. Org. Chem.* **1991**, *56*, 5883–5889.

979. Reissig, H.-U. *Angew. Chem., Int. Ed. Engl.* **1992**, *31*, 288–290.

980. Guenoun, F.; Zair, T.; Lamaty, F.; Pierrot, M.; Lazaro, R.; Viallefont, P. *Tetrahedron Lett.* **1997**, *38*, 21563–1566.

981. Baldoli, C.; Buttero, P.D.; Licandro, E.; Papagni, A. *Tetrahedron* **1996**, *52*, 4849–4856.

982. Müller, R.; Goesmann, H.; Waldmann, H. *Angew. Chem., Int. Ed. Engl.* **1999**, *38*, 184–187.

983. Ha, H.-J.; Ahn, Y.-G.; Lee, G.S. *Tetrahedron: Asymmetry* **1999**, *10*, 2327–2336.

984. Davis, F.A.; Reddy, R.T.; Reddy, R.E. *J. Org. Chem.* **1992**, *57*, 6387–6389.

985. Davis, F.A.; Szewczyk, J.M.; Reddy, R.E. *J. Org. Chem.* **1996**, *61*, 2222–2225.

986. Davis, F.A.; Reddy, R.E.; Szewczyk, J.M. *J. Org. Chem.* **1995**, *60*, 7037–7039.

987. Jiang, J.; Schumacher, K.K.; Joullié, M.M.; Davis, F.A.; Reddy, R.E. *Tetrahedron Lett.* **1994**, *35*, 2121–2124.

988. Zhou, P.; Chen, B.-C.; Davis, F.A. *Tetrahedron* **2004**, *60*, 8003–8030.

989. Davis, F.A.; Reddy, R.E. *Tetrahedron: Asymmetry* **1994**, *5*, 955–960.

990. Sorochinsky, A.; Voloshin, N.; Markovsky, A.; Belik, M.; Yasuda, N.; Uekusa, H.; Ono, T.; Berbasov, D.O.; Soloshonok, V.A. *J. Org. Chem.* **2003**, *68*, 7448–7454.

991. García Ruano, J.L.G.; Fernández, I.; del Prado Catalina, M.; Hermoso, J.A.; Sanz-Aparicio, J.; Martínez-Ripoll, M. *J. Org. Chem.* **1998**, *63*, 7157–7161.

992. Davis, F.A.; Yang, B.; Deng, J. *J. Org. Chem.* **2003**, *68*, 5147–5152.

993. Kawecki, R. *J. Org. Chem.* **1999**, *64*, 8724–8727.

994. Li, G.; Wei, H.-X.; Hook, J.D. *Tetrahedron Lett.* **1999**, *40*, 4611–4614.

995. Mikolajczyk, M.; Lyzwa, P.; Drabowicz, J.; Wieczorek, M.W.; Blaszczyk, J. *Chem. Commun.* **1996**, 1503–1504.

996. Adamczyk, M.; Reddy, R.E. *Tetrahedron: Asymmetry* **1998**, *9*, 3919–3921.

997. Cogan, D.A.; Liu, G.; Kim, K.; Backes, B.J.; Ellman, J.A. *J. Am. Chem. Soc.* **1998**, *120*, 8011–8019.

998. Liu, G.; Cogan, D.A.; Owens, T.D.; Tang, T.P.; Ellman, J.A. *J. Org. Chem.* **1999**, *64*, 1278–1284.

999. Tang, T.P.; Ellman, J.A. *J. Org. Chem.* **1999**, *64*, 12–13.

1000. Tang, T.P.; Ellman, J.A. *J. Org. Chem.* **2002**, *67*, 7819–7832.

1001. Lee, K.; Zhang, M.; Yang, D.; Burke, T.R., Jr., *Bioorg. Med. Chem. Lett.* **2002**, *12*, 3399–3401.

1002. Staas, D.D.; Savage, K.L.; Homnick, C.F.; Tsou, N.N.; Ball, R.G. *J. Org. Chem.* **2002**, *67*, 8276–8279.

1003. Ellman, J.A.; Owens, T.D.; Tang, T.P. *Acc. Chem. Res.* **2002**, *35*, 984–995.

1004. Moody, C.J.; Hunt, J.C.A. *Synlett* **1998**, 733–734.

1005. Chang, Z.-Y.; Coates, R.M. *J. Org. Chem.* **1990**, *55*, 3475–3483.

1006. Jost, S.; Gimbert, Y.; Greene, A.E.; Fotiadu, F. *J. Org. Chem.* **1997**, *62*, 6672–6677.

1007. Freer, A.; Overton, K.; Tomanek, R. *Tetrahedron Lett.* **1990**, *31*, 1471–1474.

1008. Merino, P.; Franco, S.; Merchan, F.L.; Tejero, T. *Tetrahedron Lett.* **1998**, *39*, 6411–6414.

1009. Enders, D.; Schankat, J.; Klatt, M. *Synlett* **1994**, 795–797.

1010. Hunt, J.C.A.; Lloyd, C.; Moody, C.J.; Slawin, A.M.Z.; Takle, A.K. *J. Chem. Soc., Perkin Trans. 1* **1999**, 3443–3454.

1011. Andrés, C.; González, A.; Pedrosa, R.; Pérez-Encabo, A. *Tetrahedron Lett.* **1992**, *33*, 2895–2898.

1012. Mokhallalati, M.K.; Pridgen, L.N. *Synth. Commun.* **1993**, *23*, 2055–2064.

1013. Gouge, V.; Jubault, P.; Quirion, J.-C. *Tetrahedron Lett.* **2004**, *45*, 773–776.

1014. Mokhallalati, M.K.; Wu, M.-J.; Pridgen, L.N. *Tetrahedron Lett.* **1993**, *34*, 47–50.

1015. Wu, M-J.; Pridgen, L.N. *Synlett* **1990**, 636.

1016. Fujisawa, T.; Ichikawa, M.; Ukaji, Y.; Shimizu, M. *Tetrahedron Lett.* **1993**, *34*, 1307–1310.

1017. Syed, J.; Förster, S.; Effenberger, F. *Tetrahedron: Asymmetry* **1998**, *9*, 805–815.

1018. Giri, N.; Petrini, M.; Profeta, R. *J. Org. Chem.* **2004**, *69*, 7303–7308.

1019. Gennari, C.; Venturini, I.; Gislon, G.; Schimperna, G. *Tetrahedron Lett.* **1987**, *28*, 227–230.

1020. Kanazawa, A.M.; Denis, J.-N.; Greene, A.E. *J. Org. Chem.* **1994**, *59*, 1238–1240.

1021. Kanazawa, A.M.; Denis, J.-N.; Greene, A.E.; *J. Chem. Soc., Chem. Commun.* **1994**, 2591–2592.

1022. Shimizu, M.; Kooriyama, Y.; Fujisawa, T. *Chem. Lett.* **1994**, 2419–2422.

1023. Zhao, Y.; Ma, Z.; Zhang, X.; Zou, Y.; Jin, X.; Wang, J. *Angew. Chem., Int. Ed.* **2004**, *43*, 5977–5980.

1024. Génisson, Y.; Massardier, C.; Gautier-Luneau, I.; Greene, A.E. *J. Chem. Soc., Perkin Trans. 1* **1996**, 2869–2872.

1025. Teng, X.; Takayama, Y.; Okamoto, S.; Sato, F. *J. Am. Chem. Soc.* **1999**, *121*, 11916–11917.

1026. D'Souza, A.A.; Motevalli, M.; Robinson, A.J.; Wyatt, P.B. *J. Chem. Soc., Perkin Trans. 1* **1995**, 1–2.

1027. Hintermann, T.; Seebach, D. *Synlett* **1997**, 437–438.

1028. Evans, D.A.; Urpí, F.; Somers, T.C.; Clark, J.S.; Bilodeau, M.T. *J. Am. Chem. Soc.* **1990**, *112*, 8215–8216.

1029. Guichard, G.; Abele, S.; Seebach, D. *Helv. Chim. Acta* **1998**, *81*, 187–206.

1030. Hintermann, T.; Seebach, D. *Helv. Chim. Acta* **1998**, *81*, 2093–2126.

1031. Sebesta, R.; Seebach, D. *Helv. Chim. Acta* **2003**, *86*, 4061–4072.

1032. Gessier, F.; Schaeffer, L.; Kimmerlin, T.; Flögel, O.; Seebach, D. *Helv. Chim. Acta* **2005**, *88*, 2235–2249.

1033. Kawakami, T.; Ohtake, H.; Arakawa, H.; Okachi, T.; Imada, Y.; Murahashi, S.-I. *Org. Lett.* **1999**, *1*, 107–110.

1034. Ambhaikar, N.B.; Snyder, J.P.; Liotta, D.C. *J. Am. Chem. Soc.* **2003**, *125*, 3690–3691.

1035. Luisi, R.; Capriati, V.; Florio, S.; Vista, T. *J. Org. Chem.* **2003**, *68*, 9861–9864.

1036. Ohtake, H.; Imada, Y.; Murahashi, S.-I. *J. Org. Chem.* **1999**, *64*, 3790–3791.

1037. Georg, G.I. *Tetrahedron Lett.* **1984**, *25*, 3779–3782.

1038. Hatanaka, M.; Nitta, H. *Tetrahedron Lett.* **1987**, *28*, 69–72.

1039. Hatanaka, M. *Tetrahedron Lett.* **1987**, *28*, 83–86.

1040. Iimori, T.; Ishida, Y.; Shibasaki, M. *Tetrahedron Lett.* **1986**, *27*, 2153–2156.

1041. Corey, E.J.; Decicco, C.P.; Newbold, R.C. *Tetrahedron Lett.* **1991**, *32*, 5287–5290.

1042. Gennari, C.; Carcano, M.; Donghi, M.; Mongelli, N.; Vanotti, E.; Vulpetti, A. *J. Org. Chem.* **1997**, *62*, 4726–4755.

1043. Hattori, K.; Miyata, M.; Yamamoto, H. *J. Am. Chem. Soc.* **1993**, *115*, 1151–1152.

1044. Hattori, K.; Yamamoto, H. *Tetrahedron* **1994**, *50*, 2785–2792.

1045. Kobayashi, S.; Hasegawa, Y.; Ishitani, H. *Chem. Lett.* **1998**, 1131–1132.

1046. Kobayashi, S.; Ishitani, H.; Ueno, M. *J. Am. Chem. Soc.* **1998**, *120*, 431–432.

1047. Kobayashi, S.; Kobayashi, J.; Ishiani, H.; Ueno, M. *Chem. —Eur. J.* **2002**, *8*, 4185–4190.

1048. Kobayashi, S.; Ueno, M.; Saito, S.; Mizuki, Y.; Ishitani, H. Yamashita, Y. *Proc. Natl. Acad. Sci U.S.A.* **2004**, *101*, 5476–5481.

1049. Ishitani, H.; Ueno, M.; Kobayashi, S. *J. Am. Chem. Soc.* **2000**, *122*, 8180–8186.

1050. Murahashi, S.-i.; Imada, Y.; Kawakami, T.; Harada, K.; Yonemushi, Y.; Tomita, N. *J. Am. Chem. Soc.* **2002**, *124*, 2888–2889.

1051. Akiyama, T.; Itoh, J.; Yokota, K.; Fuchibe, K. *Angew. Chem., Int. Ed.* **2004, *43*,** 1566–1568.

1052. Harada, S.; Handa, S.; Matsunaga, S.; Shibasaki, M. *Angew. Chem., Int. Ed.* **2005, *44*,** 4365–4368.

1053. Córdova, A. *Acc. Chem. Res.* **2004**, *37*, 102–112.

1054. Kobayashi, S.; Ishitani, H. *Chem. Rev.* **1999**, *99*, 1069–1094.

1055. Cozzi, P.G.; Rivalta, E. *Angew. Chem., Int. Ed.* **2005, *44*,** 3600–3603.

1056. Shi, M.; Chen, L.-H.; Li, C.-Q. *J. Am. Chem. Soc.* **2005**, *127*, 3790–3800.

1057. Hamashima, Y.; Sasamoto, N.; Hotta, D.; Somei, H.; Umebayashi, N.; Sodeoka, M. *Angew. Chem., Int. Ed.* **2005**, *44*, 1525–1529.

1058. Davies, H.M.L.; Venkataramani, C. *Angew. Chem., Int. Ed.* **2002**, *41*, 2197–2199.

1059. Keirs, D.; Moffat, D.; Overton, K.; Tomanek, R. *J. Chem., Soc., Perkin Trans. 1* **1991**, 1041–1051.

1060. Murahashi, S.-I.; Imada, Y.; Kohno, M.; Kawakami, T. *Synlett* **1993**, 395–396.

1061. Arnone, A.; Blasco, F.; Resnati, G. *Tetrahedron* **1997**, *53*, 17513–17518.

1062. Hanselmann, R.; Zhou, J.; Ma, P.; Confalone, P.N. *J. Org. Chem.* **2003**, *68*, 8739–8741.

1063. Swindell, C.S.; Tao, M. *J. Org. Chem.* **1993**, *58*, 5889–5891.

1064. Furukawa, M.; Okawara, T.; Noguchi, H.; Terawaki, Y. *Heterocycles* **1977**, *6*, 1323–1328.

1065. Minter, A.R.; Fuller, A.A.; Mapp, A.K. *J. Am. Chem. Soc.* **2003**, *125*, 6846–6847.

1066. Fuller, A.A.; Chen, B.; Minter, A.R.; Mapp, A.K. *Synlett* **2004**, 1409–1413.

1067. Arndt, F.; Eistert, B. *Ber.* **1936**, *69*, 1805.

1068. Balenovic, K.; Bregant, N.; Cerar, D.; Tkalcic, M. *J. Org. Chem.* **1951**, *16*, 1308–1310.

1069. Penke, B.; Czombos, J.; Baláspiri, L.; Petres, J.; Kovács, K. *Helv. Chim. Acta* **1970**, *53*, 1057–1061.

1070. Balenovíc, K.; Cerar, D.; Fuks, Z. *J. Chem. Soc.* **1952**, 3316–3317.

1071. Vasanthakumar, G.-R.; Patil, B.S.; Babu, V.V.S. *J. Chem. Soc., Perkin Trans. 1* **2002**, 2087–2089.

1072. Guibourdenche, C.; Podlech, J.; Seebach, D. *Liebigs Ann. Chem.* **1996**, 1121–1129.

1073. Podlech, J.; Seebach, D. *Angew. Chem., Int. Ed. Engl.* **1995**, 34, 471–472.

1074. Leggio, A.; Liguori, A.; Procopio, A.; Sindona, G. *J. Chem. Soc., Perkin Trans. 1* **1997**, 1969–1971.

1075. Limal, D.; Quesnel, A.; Briand, J.-P. *Tetrahedron Lett.* **1998**, *39*, 4239–4242.

1076. Podlech, J.; Seebach, D. *Liebigs Ann.* **1995**, 1217–1228.

1077. Guibourdenche, C.; Seebach, D.; Natt, F. *Helv. Chim. Acta.* **1997**, *80*, 1–13.

1078. Ellmerer-Müller, E.P.; Brössner, D.; Maslouh, N.; Takó, A. *Helv. Chim. Acta* **1998**, *81*, 59–65.

1079. Jefford, C.W.; Tang, Q.; Zaslona, A. *J. Am. Chem. Soc.* **1991**, *113*, 3513–3518.

1080. Vasanthakumar, G.R.; Babu, V.V.S. *Synth. Commun.* **2002**, *32*, 651–657.

1081. Koch, K.; Podlech, J. *Synth. Commun.* **2005**, *35*, 2789–2794.

1082. Patil, B.S.; Vasanthakumar, G.-R.; Babu, V.V.S. *Synth. Commun.* **2003**, *33*, 3089–3096.

1083. Marti, R.E.; Bleicher, K.H.; Bair, K.W. *Tetrahedron Lett.* **1997**, *38*, 6145–6148.

1084. Müller, A.; Vogt, C.; Sewald, N. *Synthesis* **1998**, 837–841.

1085. Cantel, S.; Martinez, J.; Fehrentz, J.-A. *Synlett* **2004**, 2791–2793.

1086. Hughes, A.B.; Sleebs, B.E. *Aust. J.Chem.* **2005**, *58*, 778–784.

1087. Balenovic, K.; Dvornik, D. *J. Chem. Soc.* **1954**, 2976.

1088. Cassal, J.-M.; Fürst, A.; Meier, W. *Helv. Chim. Acta* **1976**, *59*, 1917–1924.

1089. Baláspiri, L.; Penke, B.; Papp, G.; Dombi, G.; Kovács, K. *Helv. Chim. Acta* **1975**, *58*, 969–973.

1090. Abele, S.; Vögtli, K.; Seebach, D. *Helv. Chim. Acta* **1999**, *82*, 1539–1558.

1091. Balenovic, K.; Fles, D. *J. Org. Chem.* **1952**, *17*, 347–349.

1092. Balenovic, K; Jambresic, I.; Gaspert, B.; Cerar, D. *Rec. Trav. Chim. Pay-Bas* **1956**, *75*, 1252–1258.

1093. Brückner, A.M.; Garcia, M.; Marsh, A.; Gellman, S.H.; Diederichsen, U. *Eur. J. Org. Chem.* **2003**, 3555–3561.

1094. Rudinger, J.; Farkasová, H. *Collect. Czech. Chem. Commun.* **1963**, *28*, 2941–2952.

1095. Martin, L.; Cornille, F.; Coric, P.; Roques, B.P.; Fournié-Zaluski, M.-C. *J. Med. Chem.* **1998**, *41*, 3450–3460.

1096. Savithri, D. Leumann, C.; Scheffold, R. *Helv. Chim. Acta* **1996**, *79*, 288–294.

1097. Wakamiya, T.; Uratani, H.; Teshima, T.; Shiba, T. *Bull. Chem. Soc. Jpn.* **1975**, *48*, 2401–2402.

1098. Nomoto, S.; Shiba, T. *Chem. Lett.* **1978**, 589–590.

1099. Plucinska, K.; Liberek, B. *Tetrahedron* **1987**, *43*, 3509–3517.

1100. Buchschacher, P.; Cassal, J.-M.; Fürst, A.; Meier, W. *Helv. Chim. Acta* **1977**, *60*, 2727–2755.

1101. Balenovic, K.; Thaller, V.; Filipovic, L. *Helv. Chim. Acta* **1951**, *34*, 744–747.

1102. Gaucher, A.; Dutot, L.; Barbeau, O.; Hamchaoui, W.; Wakselman, M.; Mazaleyrat, J.-P. *Tetrahedron: Asymmetry* **2005**, *16*, 857–864.

1103. Païs, M.; Sarfati, R.; Jarreau, F.-X. *Bull. Soc. Chim. Fr.* **1973**, 331–334.

1104. Grieco, P.A.; Hon, Y.-S.; Perez-Medrano, A. *J. Am. Chem. Soc.* **1988**, *110*, 1630–1631.

1105. Gardiner, J.; Anderson, K.H.; Downard, A.; Abell, A.D. *J. Org. Chem.* **2004**, *69*, 3375–3382.

1106. Teshima, T.; Ando, T.; Shiba, T. *Bull. Chem. Soc. Jpn.* **1980**, *53*, 1191–1192.

1107. Cesar, J.; Dolenc, M.S. *Tetrahedron Lett.* **2001**, *42*, 7099–7102.

1108. Caputo, R.; Cassano, E.; Longobardo, L.; Palumbo, G. *Tetrahedron* **1995**, *51*, 12337–12350.

1109. Ikota, N.; Hanaki, A. *Chem. Pharm. Bull.* **1989**, *37*, 1087–1089.

1110. Kumar, A.; Ghilagaber, S.; Knight, J.; Wyatt, P.B. *Tetrahedron Lett.* **2002**, *43*, 6991–6994.

1111. Gmeiner, P.; Orecher, F.; Thomas, C.; Weber, K. *Tetrahedron Lett.* **1995**, *36*, 381–382.

1112. Thomas, C.; Orecher, F.; Gmeiner, P. *Synthesis* **1998**, 1491–1496.

1113. Katritzky, A.R.; Zhang, S.; Hussein, A.H.M.; Fang, Y.; Steel, P.J. *J. Org. Chem.* **2001**, *66*, 5606–5612.

1114. Suda, H.; Takita, T.; Aoyagi, T.; Umezawa, H. *J. Antibiotics* **1976**, *29*, 600–601.

1115. Nishizawa, R.; Saino, T.; Takita, T.; Suda, H.; Aoyagi, T.; Umezawa, H. *J. Med. Chem.* **1977**, *20*, 510–515.

1116. Ohba, T.; Ikeda, E.; Takei, H. *Bioorg. Med. Chem. Lett.* **1996**, *6*, 1875–1880.

1117. Herranz, R.; Castro-Pichel, J.; Vinuesa, S.; García-López, M.T. *J. Chem. Soc., Chem. Commun.* **1989**, 938–939.

1118. Iizuka, K.; Kamijo, T.; Harada, H.; Akahane, K.; Kubota, T.; Umeyama, H.; Kiso, Y. *J. Chem. Soc., Chem. Commun.* **1989**, 1678–1680.

1119. Tobe, H.; Morishima, H.; Naganawa, H.; Takita, T.; Aoyagi, T.; Umezawa, H. *Agric. Biol. Chem.* **1979**, *43*, 591–596.

1120. Adang, A.E.P.; Peters, C.A.M.; Gerritsma, S.; de Zwart, E.; Veeneman, G. *Biorg. Med. Chem. Lett.* **1999**, *9*, 1227–1232.

1121. Gessier, F.; Noti, C.; Rueping, M.; Seebach, D. *Helv. Chim. Acta* **2003**, *86*, 1862–1870.

1122. Aoyagi, Y.; Jain, R.P.; Williams, R.M. *J. Am. Chem. Soc.* **2001**, *123*, 3472–3477.

1123. Veeresha, G.; Datta, A. *Tetrahedron Lett.* **1997**, *38*, 5223–5224.

1124. Denis, J.-N.; Correa, A.; Greene, A.E. *J. Org. Chem.* **1991**, *56*, 6939–6942.

1125. Lee, B.W.; Lee, J.H.; Jang, K.C.; Kang, J.E.; Kim, J.H.; Park, K.-M.; Park, K.H. *Tetrahedron Lett.* **2003**, *44*, 5905–5907.

1126. Hormuth, S.; ReiBig, H.-U.; Dorsch, D. *Liebigs Ann. Chem.* **1994**, 121–127.

1127. Dondoni, A.; Perrone, D.; Semola, T. *Synthesis* **1995**, 181–186.

1128. Dondoni, A.; Perrone, D. *Aldrichim. Acta* **1997**, *30*, 35–46.

1129. Dondoni, A.; Perrone, D. *Synthesis* **1993**, 1162–1176.

1130. Dondoni, A.; Perrone, D.; Merino, P. *J. Chem. Soc., Chem. Commun.* **1991**, 1313–1314.

1131. Nozaki, K.; Sato, N.; Takaya, H. *Tetrahedron: Asymmetry* **1993**, *4*, 2179–2182.

1132. Wolf, J.-P.; Pfander, H. *Helv. Chim. Acta* **1987**, *70*, 116–120.

1133. Burgess, K.; Liu, L.T.; Pal, B. *J. Org. Chem.* **1993**, *58*, 4758–4763.

1134. Paris, M.; Pothion, C.; Michalak, C.; Martinez, J.; Fehrentz, J.-A. *Tetrahedron Lett.* **1998**, *39*, 6889–6890.

1135. Wasserman, H.H.; Xia, M.; Petersen, A.K.; Jorgensen, M.R.; Curtis, E.A. *Tetrahedron Lett.* **1999**, *40*, 6163–6166.

1136. Weik, S.; Rademann, J. *Angew. Chem., Int. Ed.* **2003**, *42*, 2491–2494.

1137. Wong, M.-K.; Yu, C.-W.; Yuen, W.-H.; Yang, D. *J. Org. Chem.* **2001**, *66*, 3606–3609.

1138. Papanikos, A.; Meldal, M. *J. Comb. Chem.* **2004**, *6*, 181–195.

1139. Lee, J.-M.; Lim, H.-S.; Seo, K.-C.; Chung, S.-K. *Tetrahedron Asymmetry* **2003**, *14*, 3639–3641.

1140. Lee, K.-Y.; Kim, Y.-H.; Park, M.-S.; Ham, W.-H. *Tetrahedron Lett.* **1998**, *39*, 8129–8132.

1141. Lakner, F.J.; Chu, K.S.; Negrete, G.R.; Konopelski, J.P.; Madan, P.B.; Yiannikouros, G.P.; Coffen, D.L. *Org. Syntheses* **1996**, *73*, 201–213.

1142. Chu, K.S.; Negrete, G.R.; Koopelski, J.P.; Lakner, F.J.; Woo, N.-T.; Olmstead, M.M. *J. Am. Chem. Soc.* **1992**, *114*, 1800–1812.

1143. McGarvey, G.J.; Hiner, R.N.; Matsubara, Y.; Oh, T. *Tetrahedron Lett.* **1983**, *24*, 2733–2736.

1144. Nitta, H.; Ueda, I.; Hatanaka, M. *J. Chem. Soc., Perkin Trans. 1* **1997**, 1793–1798.

1145. Luppi, G.; Villa, M.; Tomasini, C. *Org. Biomol. Chem.* **2003**, *1*, 247–250.

1146. Gong, B.; Lynn, D.G. *J. Org. Chem.* **1990**, *55*, 4763–4765.

1147. Sibrian-Vazquez, M.; Spivak, D.A. *Synlett* **2002**, 1105–1106.

1148. Ha, D.-C.; Kil, K.-E.; Choi, K.-S.; Park, H.-S. *Tetrahedron Lett.* **1996**, *37*, 5723–5726.

1149. Rodriguez, M.; Llinares, M.; Doulut, S.; Heitz, A.; Martinez, J. *Tetrahedron Lett.* **1991**, *32*, 923–926.

1150. Cooper, J.; Knight, D.W.; Gallagher, P.T. *J. Chem. Soc., Perkin Trans 1* **1992**, 553–559.

1151. McAlpine, I.J.; Armstrong, R.W. *Tetrahedron Lett.* **2000**, *41*, 1849–1853.

1152. Sasaki, S.; Hamada, Y.; Shioiri, T. *Synlett* **1999**, 453–455.

1153. Bland, J.M. *Synth. Commun.* **1995**, *25*, 467–477.

1154. El Marini, A.; Roumestant, M.L.; Viallefont, P.; Razafindramboa, D.; Bonato, M.; Follet, M. *Synthesis* **1992**, 1104–1108.

1155. Calvisi, G.; Dell-Uomo, N.; De Angelis, F.; Dejas, R.; Giannessi, F.; Tinti, M.O. *Eur. J. Org. Chem.* **2003**, 4501–4505.

1156. Morin, C.; Thimon, C. *Eur. J. Org. Chem.* **2004**, 3828–3832.

1157. Brückner, A.M.; Schmidt, H.W.; Diederichsen, U. *Helv. Chim. Acta* **2002**, *85*, 3855–3866.

1158. Gmeiner, P. *Tetrahedron Lett.* **1990**, *31*, 5717–5720.

1159. Gmeiner, P. *Liebigs Ann. Chem.* **1991**, 501–502.

1160. Gmeiner, P.; Hummel, E.; Haubmann, C. *Liebigs Ann.* **1995**, 1987–1992.

1161. McGarvey, G.J.; Williams, J.M.; Hiner, R.N.; Matsubara, Y.; Oh, T. *J. Am. Chem. Soc.* **1986**, *108*, 4943–4952.

1162. Jefford, C.W.; McNulty, J.; Lu, Z.-H.; Wang, J.B. *Helv. Chim. Acta* **1996**, *79*, 1203–1216.

1163. Jefford, C.W.; Wang, J. *Tetrahedron Lett.* **1993**, *34*, 1111–1114.

1164. Jefford, C.W. *Pure Appl. Chem.* **1996**, *68*, 799–804.

1165. Jefford, C.W.; Lu, Z.-H.; Wang, J. B. *Pure Appl. Chem.* **1994**, *66*, 2075–2078.

1166. Jefford, C.W.; McNulty, J. *Helv. Chim. Acta* **1994**, *77*, 2142–2146.

1167. Jefford, C.W.; Wang, J.B.; Lu, Z.-H. *Tetrahedron Lett.* **1993**, *34*, 7557–7560.

1168. Thierry, J.; Servajean, V. *Tetrahedron Lett.* **2004**, *45*, 821–823.

1169. Dexter, C.S.; Jackson, R.F.W. *Chem. Commun.* **1998**, 75–76.

1170. Jackson, R.F.W.; Rilatt, I.; Murray, P.J. *Org. Biomol. Chem.* **2004**, *2*, 110–113.

1171. Dexter, C.S.; Jackson, R.F.W.; Elliott, J. *Tetrahedron* **2000**, *56*, 4539–4540.

1172. Dexter, C.S.; Jackson, R.F.W.; Elliott, J. *J. Org. Chem.* **1999**, *64*, 7579–7585.

1173. Jackson, R.F.W.; Rilatt, I.; Murray, P.J. *Chem. Commun.* **2003**, 1242–1243.

1174. Dexter, C.S.; Hunter, C.; Jackson, R.F.W.; Elliott, J. *J. Org. Chem.* **2000**, *65*, 7417–7421.

1175. Andrés, J.M.; Muñoz, E.M.; Pedrosa, R.; Pérez-Encabo, A. *Eur. J. Org. Chem.* **2003**, 3387–3397.

1176. D'Aniello, F.; Mann, A.; Taddei, M. *J. Org. Chem.* **1996**, *61*, 4870–4871.

1177. D'Aniello, F.; Mann, A.; Schoenfelder, A.; Taddei, M. *Tetrahedron* **1997**, *53*, 1447–1456.

1178. Namikoshi, M.; Rinehart, K..; Dahlem, A.M.; Beasley, V.R.; Carmichael, W.W. *Tetrahedron Lett.* **1989**, *30*, 4349–4352.

1179. Beatty, M.F.; Jennings-White, C.; Avery, M.A. *J. Chem. Soc., Perkin Trans. 1* **1992**, 1637–1641.

1180. Cundy, D.J.; Donohue, A.C.; McCarthy, T.D. *Tetrahedron Lett.* **1998**, *39*, 5125–5128.

1181. Griesbeck, A.G.; Heckroth, H. *Synlett* **1997**, 1243–1244.

1182. Seto, Y.; Torii, K.; Bori, K.; Inabata, K.; Kuwata, S.; Watanabe, H. *Bull. Chem. Soc. Jpn.* **1974**, *47*, 151–155.

1183. Ressler, C.; Nagarajan, G.R.; Kirisawa, M.; Kashelikar, D.V. *J. Org. Chem.* **1971**, *36*, 3960–3965.

1184. Hamzé, A.; Hernandez, J.-F.; Fulcrand, P.; Martinez, J. *J. Org. Chem.* **2003**, *68*, 7316–7321.

1185. Seki, M.; Matsumoto, K. *Tetrahedron Lett.* **1996**, *37*, 3165–3168.

1186. Poston, J.M. *J. Biol. Chem.* **1976**, *251*, 1859–1863.

1187. Miyazawa, T.; Akita, E.; Ito, T. *Agric. Biol. Chem.* **1976**, *40*, 1651–1652.

1188. Somekh, L.; Shanzer, A. *J. Am. Chem. Soc.* **1982**, *104*, 5836–5837.

1189. Hamada, Y.; Kawai, A.; Matsui, T.; Hara, O.; Shiori, T. *Tetrahedron* **1990**, *46*, 4823–4846.

1190. Seebach, D.; Estermann, H. *Tetrahedron Lett.* **1987**, *28*, 3103–3106.

1191. Hintermann, T.; Mathes, C.; Seebach, D. *Eur. J. Org. Chem.* **1998**, 2379–2387.

1192. Hanessian, S.; Sancéau, J.-Y. *Can. J. Chem.* **1996**, 74, 621–624.

1193. Caputo, R.; Cecere, G.; Guaragna, A.; Palumbo, G.; Pedatella, S. *Eur. J. Org. Chem.* **2002**, 3050–3054.

1194. Iimori, T.; Takahashi, Y.; Izawa, T.; Kobayashi, S.; Ohno, M. *J. Am. Chem. Soc.* **1983**, *105*, 1659–1660.

1195. Cativiela, C.; Diaz-de-Villegas, M.D.; Gálvez, J.A. *Tetrahedron* **1996**, *52*, 687–694.

1196. Cativiela, C.; Díaz-de-Villegas, M.D.; Gálvez, J.A. *Tetrahedron: Asymmetry* **1993**, *4*, 229–238.

1197. Cativiela, C.; Diaz-de-Villegas, M.D.; Galvez, J.A. *Tetrahedron: Asymmetry* **1992**, *3*, 1141–1144.

1198. Ponsinet, R.; Chassaing, G.; Vaissermann, J.; Lavielle, S. *Eur. J. Org. Chem.* **2000**, 83–90; correction **2000**, 3807.

1199. Miyabe, H.; Fujii, K.; Naito, T. *Org. Biomol. Chem.* **2003**, *1*, 381–390.

1200. Gutiérrez-García, V.M.; Reyes-Rangel, G.; Muñoz-Muñiz, O.; Juaristi, E. *Helv. Chim. Acta* **2002**, *85*, 4189–4199.

1201. Akssira, M.; Boumzebra, M.; Kasmi, H.; Roumestant, M.L.; Viallefont, Ph. *Amino Acids* **1994**, *7*, 79–81.

1202. Arvanitis, E.; Motevalli, M.; Wyatt, P.B. *Tetrahedron Lett.* **1996**, *37*, 4277–4280.

1203. Juaristi, E.; Quintana, D. *Tetrahedron: Asymmetry* **1992**, *3*, 723–726.

1204. Juaristi, E.; Quintana, D.; Balderas, M.; García-Pérez, E. *Tetrahedron: Asymmetry* **1996**, *7*, 2233–2246.

1205. Juaristi, E.; Bladeras, M.; Ramírez-Quirós, Y. *Tetrahedron: Asymmetry* **1998**, *9*, 3881–3888.

1206. Seebach, D.; Lamatsch, B.; Amstutz, R.; Beck, A.K.; Dobler, M.; Egli, M.; Fitzi, R.; Gautschi, M.; Herradón, B.; Hidber, P.C.; Irwin, J.J.; Locher, R.; Maestro, M.; Maetzke, T.; Mouriño, A.; Pfammatter, E.; Plattner, D.A.; Schickli, C.;

Schweizer, W.B.; Seiler, P.; Stucky, G.; Petter, W.; Escalante, J.; Juaristi, E.; Quintana, D.; Miravitlles, C.; Molins, E. *Helv. Chim. Acta* **1992**, *75*, 913–934.

1207. Iglesias-Arteaga, M.A.; Castellanos, E.; Juaristi, E. *Tetrahedron: Asymmetry* **2003**, *14*, 577–580.

1208. Juaristi, E.; Escalante, J.; Lamatsch, B.; Seebach, D. *J. Org. Chem.* **1993**, *58*, 2282–2285.

1209. Beaulieu, F.; Arora, J.; Veith, U.; Taylor, N.J.; Chapell, B.J.; Snieckus, V. *J. Am. Chem. Soc.* **1996**, *118*, 8727–8728.

1210. Konopelski, J.P.; Chu, K.S.; Negrete, G.R. *J. Org. Chem.* **1991**, *56*, 1355–1357.

1211. Chu, K.S.; Konopelski, J.P. *Tetrahedron* **1993**, *49*, 9183–9190.

1212. Alladoum, J.; Dechoux, L. *Tetrahedron Lett.* **2005**, *46*, 8203–8205.

1213. Seebach, D.; Boog, A.; Schweizer, W.B. *Eur. J. Org. Chem.* **1999**, 335–360.

1214. Braschi, I.; Cardillo, G.; Tomasini, C.; Venezia, R. *J. Org. Chem.* **1994**, *59*, 7292–7298.

1215. Amoroso, R.; Cardillo, G.; Tomasini, C. *Tetrahedron Lett.* **1992**, *33*, 2725–2728.

1216. Amoroso, R.; Cardillo, G.; Mobbili, G.; Tomasini, C. *Tetrahedron: Asymmetry* **1993**, *4*, 2241–2254.

1217. Cardillo, G.; Tolomelli, A.; Tomasini, C. *Tetrahedron* **1995**, *51*, 11831–11840.

1218. Agami, C.; Cheramy, S.; Dechoux, L.; Kadouri-Puchot, C. *Synlett* **1999**, *5*, 727–728.

1219. Agami, C.; Cheramy, S.; Dechoux, L. *Synlett* **1999**, 1838–1840.

1220. Warmerdam, E.G.J.C.; van Rijn, R.D.; Brussee, J.; Kruse, C.G.; van der gen, A. *Tetrahedron: Asymmetry* **1996**, *7*, 1723–1732.

1221. Tromp, R.A.; van der Hoeven, M.; Amore, A.; Brussee, J.; Overhand, M.; van der Marel, G.A.; van der Gen, A. *Tetrahedron: Asymmetry* **2003**, *14*, 1645–1652.

1222. Cativiela, C.; Diaz-de-Villegas, M.D.; Gálvez, J.A. *Tetrahedron: Asymmetry* **1996**, *7*, 529–536.

1223. Matsumoto, T.; Kobayashi, Y.; Takemoto, Y.; Ito, Y.; Kamijo, T.; Harada, H.; Terashima, S. *Tetrahedron Lett.* **1990**, *31*, 4175–4176.

1224. Barnish, I.T.; Corless, M.; Dunn, P.J.; Ellis, D.; Finn, P.W.; Hardstone, J..D.; James, K. *Tetrahedron Lett.* **1993**, *34*, 1323–1326.

1225. Merino, P.; Castillo, E.; Franco, S.; Merchán, F.L.; Tejero, T. *Tetrahedron* **1998**, *54*, 12301–12322.

1226. Enders, D.; Klatt, M.; Funk, R. *Synlett* **1993**, 226–228.

1227. Miyabe, H.; Fujii, K.; Naito, T. *Org. Lett.* **1999**, *1*, 569–572.

1228. Sibi, M.P.; Asano, Y. *J. Am. Chem. Soc.* **2001**, *123*, 9708–9709.

1229. Eilitz, U.; Leβmann, F.; Seidelmann, O.; Wendisch, V. *Tetrahedron Asymmetry* **2003**, *14*, 189–191.

1230. Sibi, M.P.; Patil, K. *Angew. Chem., Int. Ed.* **2004**, *43*, 1235–1238.

1231. Diederichsen, U.; Schmitt, H.W. *Angew. Chem., Int. Ed. Engl.* **1998**, *37*, 302–305.

1232. Ressler, C.; Nagarajan, G.R.; Kirisawa, M.; Kashelikar, D.V. *J. Org. Chem.* **1971**, *36*, 3960–3966.

1233. Brown, J.M.; James, A.P.; Prior, L.M. *Tetrahedron Lett.* **1987**, *28*, 2179–2182.

1234. Noyori, R.; Ikeda, T.; Ohkuma, T.; Widham, M.; Kitamura, M.; Takaya, H.; Akutagawa, S.; Sayo, N.; Saito, T.; Taketomi, T.; Kumobayashi, H. *J. Am. Chem. Soc.* **1989**, *111*, 9134–9135.

1235. Bucciarelli, M.; Forni, A.; Moretti, I.; Prati, F.; Torre, G. *Tetrahedron: Asymmetry* **1995**, *6*, 2073–2080.

1236. Hanessian, S.; Cantin, L.-D. *Tetrahedron Lett.* **2000**, *41*, 787–790.

1237. Hata, Y.; Watanabe, M. *Tetrahedron* **1987**, *43*, 3881–3888.

1238. Church, N.J.; Young, D.W. *Tetrahedron Lett.* **1995**, *36*, 151–154.

1239. Davis, F.A.; Deng, J.; Zhang, Y.; Haltiwanger, R.C. *Tetrahedron* **2002**, *58*, 7135–7143.

1240. Dauben, P.; Dubois, L.; Tran Huu Dau, M.E.; Dodd, R.H. *J. Org. Chem.* **1995**, *60*, 2035–2043.

1241. Tanner, D. *Pure Appl. Chem.* **1993**, *65*, 1319–1328.

1242. Tanner, D.; Somfai, P. *Tetrahedron* **1988**, *44*, 619–624.

1243. Kang, S.H.; Ryu, D.H. *Biorg. Med. Chem. Lett.* **1995**, *5*, 2959–2962.

1244. Davis, F.A.; Reddy, G.V.; Liang, C.-H. *Tetrahedron Lett.* **1997**, *38*, 5139–5142.

1245. Tanner, D.; Somfai, P. *Tetrahedron Lett.* **1987**, *28*, 1211–1214.

1246. Anzai, M.; Yanada, R.; Fuji, N.; Ohno, H.; Ibuka, T. Takemoto, Y. *Tetrahedron* **2002**, *58*, 5231–5239.

1247. Osborn, H.M.I.; Sweeney, J.B.; Howson, B. *Synlett* **1993**, 675–676.

1248. Palomo, C.; Aizpurua, J.M.; Cuevas, C.; Mielgo, A.; Galarza, R. *Tetrahedron Lett.* **1995**, 36, 9027–9030.

1249. Palomo, C.; Aizpurua, J.M.; Ganboa, I. *Russ. Chem. Bull.* **1996**, *45*, 2463–2483.

1250. Palomo, C.; Aizpurua, J.M.; Cuevas, C. *J. Chem. Soc., Chem. Commun.* **1994**, 1957–1958.

1251. Ojima, I.; Sun, C.M.; Park, Y.H. *J. Org. Chem.* **1994**, *59*, 1249–1250.

1252. Paloma, C.; Aizpurua, J.M.; Ganboa, I.; Oiarbide, M. *Amino Acids* **1999**, *16*, 321–343.

1253. Palomo, C.; Aizpurua, J.M.; Ganboa, I.; Oiarbide, M. *Synlett* **2001**, 1813–1826.

1254. Venturini, A.; González, J. *J. Org. Chem.* **2002**, *67*, 9089–9092.

1255. Bongini, A.; Panunzio, M.; Piersanti, G.; Bandini, E.; Martelli, G.; Spunta, G.; Venturini, A. *Eur. J.Org. Chem.* **2000**, 2379–2390.

1256. Ojima, I.; Delaloge, F. *Chem. Soc. Rev.* **1997**, *26*, 377–386.

1257. Ojima, I. *Acc. Chem. Res.* **1995**, *28*, 383–389.

1258. Ojima, I.; Habus, I.; Zhao, M. *J. Org. Chem.* **1991**, *56*, 1681–1683.

1259. Ojima, I.; Park, Y.H.; Sun, C.M.; Brigaud, T.; Zhao, M. *Tetrahedron Lett.* **1992**, *33*, 5737–5740.

1260. Ojima, I.; Habus, I.; Zhao, M.; Zucco, M.; Park, Y.H.; Sun, C.M.; Brigaud, T. *Tetrahedron* **1992**, *48*, 6985–7012.

1261. Ojima, I.; Wang, T.; Delaloge, F. *Tetrahedron Lett.* **1998**, *39*, 3663–3666.

1262. Abouabdellah, A.; Bégué, J.-P.; Bonnet-Delpon, D.; Nga, T.T.T. *J. Org. Chem.* **1997**, *62*, 8826–8833.

1263. Perlmutter, P.; Rose, M.; Vounatsos, F. *Eur. J. Org. Chem.* **2003**, 756–760.

1264. Farina, V.; Hauck, S.I.; Walker, D.G. *Synlett* **1992**, 761–763.

1265. Bourzat, J.D.; Commerçon, A. *Tetrahedron Lett.* **1993**, *34*, 6049–6052.

1266. Brown, S.; Jordan, A.M.; Lawrence, N.J.; Pritchard, R.G.; McGown, A.T. *Tetrahedron Lett.* **1998**, *39*, 3559–3562.

1267. Georg, G.I.; Mashava, P.M.; Akgün, E.; Milstead, M.W. *Tetrahedron Lett.* **1991**, *32*, 3151–3154.

1268. Ojima, I.; Inabe, S.-i. *Tetrahedron Lett.* **1980**, *21*, 2081–2084.

1269. Ojima, I.; Inabe, S.-i. *Tetrahedron Lett.* **1980**, *21*, 2077–2080.

1270. Georg, G.I.; Akgün, E.; Mashava, P.M.; Milstead, M.; Ping, H.; Wu, Z.-J.; Vander Velde, D.; Takusawaga, F. *Tetrahedron Lett.* **1992**, *33*, 2111–2114.

1271. Fernández, R.; Ferrete, A.; Lassaletta, J.M.; Llera, J.M.; Martín-Zamora, E. *Angew. Chem., Int. Ed.* **2002,** *41*, 831–833.

1272. Kobayashi, Y.; Takemoto, Y.; Kamijo, T.; Harada, H.; Ito, Y.; Terashima, S. *Tetrahedron* **1992**, *48*, 1853–1868.

1273. Kobayashi, Y.; Takemoto, Y.; Ito, Y.; Terashima, S. *Tetrahedron Lett.* **1990**, *31*, 3031–3034.

1274. Palomo, C.; Aizpurua, J.M.; Miranda, J.I.; Mielgo, A.; Odriozola, J.M. *Tetrahedron Lett.* **1993**, *34*, 6325–6328.

1275. Macías, A.; Alonso, E.; del Pozo, C.; Venturini, A.; González, J. *J. Org. Chem.* **2004**, *69*, 7004–7012.

1276. Khasanov, A.B.; Ramirez-Weinhouse, M.M.; Webb, T.; Thiruvazhi, M. *J. Org. Chem.* **2004**, *69*, 5766–5769.

1277. Barbaro, G.; Battaglia, A.; Guerrini, A.; Bertucci, C. *J. Org. Chem.* **1999**, *64*, 4643–4651.

1278. Magriotis, P.A. *Angew. Chem., Int. Ed.* **2001,** *40*, 4377–4379.

1279. Taggi, A.E.; Hafez, A.M.; Lectka, T. *Acc. Chem. Res.* **2003**, *36*, 10–19.

1280. France, S.; Weatherwax, A.; Taggi, A.E.; Lectka, T. *Acc. Chem. Res.* **2004**, *37*, 592–600.

1281. Taggi, A.E.; Hafez, A.M.; Wack, H.; Young, B.; Ferraris, D.; Lectka, T. *J. Am. Chem. Soc.* **2002**, *124*, 6626–6635.

1282. France, S.; Shah, M.H.; Waetherwax, A.; Wack, H.; Roth, J.P.; Lectka, T. *J. Am. Chem. Soc.* **2005**, *127*, 1206–1215.

1283. Hodous, B.L.; Fu, G.C. *J. Am. Chem. Soc.* **2002**, *124*, 1578–1579.

1284. Fu, G.C. *Acc. Chem. Res.* **2004**, *37*, 542–547.

1285. Lo, M.M.-C.; Fu, G.C. *J. Am. Chem. Soc.* **2002**, *124*, 4572–4573.

1286. Cundy, D.J.; Donohue, A.C.; McCarthy, T.D. *J. Chem., Soc., Perkin Trans. 1* **1998**, 559–567.

1287. Barluenga, J.; Baragaña, B.; Concellón, J.M. *J. Org. Chem.* **1999**, *64*, 2843–2846.

1288. Shinozaki, K.; Mizuno, K.; Wakamatsu, H.; Masaki, Y. *Chem. Pharm. Bull.* **1996**, *44*, 1823–1830.

1289. Jørgensen, C.; Pedersen, C.; Søtofte, I. *Synthesis* **1998**, 325–328.

1290. Tuch, A.; Sanière, M.; Le Merrer, Y.; Depezay, J.-C. *Tetrahedron: Asymmetry* **1996**, *7*, 897–906.

1291. Tuch, A.; Sanière, M.; Le Merrer, Y.; Depezay, J.-C. *Tetrahedron: Asymmetry* **1996**, *7*, 2901-2909.

1292. Panek, J.S.; Hu, T. *J. Org. Chem.* **1997**, *62*, 4914–4915.

1293. Wee, A.G.H.; McLeod, D.D. *J. Org. Chem.* **2003**, *68*, 6268–6273, 9532.

1294. Davies, S.G.; Ichihara, O.; Lenoir, I.; Walters, I.A.S *J. Chem. Soc., Perkin Trans. 1* **1994**, 1411–1415.

1295. Oki, T.; Hirano, M.; Tomatsu, K.; Numata, K.-i.; Kamei, H. *J. Antibiotics* **1989**, *42*, 1756–1762.

1296. Konishi, M.; Nishio, M.; Saitoh, K.; Miyaki, T.; Oki, T.; Kawaguchi, H. *J. Antibiotics* **1989**, *42*, 1749–1755.

1297. Goto, T.; Toya, Y.; Ohgi, T.; Kondo, T. *Tetrahedron Lett.* **1982**, *23*, 1271–1274.

1298. Bach, T.; Schröder, J. *Leibigs Ann. / Recueil* **1997**, 2265–2267.

1299. Omura, S.; Murata, M.; Imamura, N.; Iwai, Y.; Tanaka, H. *J. Antibiotics* **1984**, *37*, 1324–1332.

1300. Kawahata, Y.; Takatsuto, S.; Ikekawa, N.; Murata, M.; Omura, S. *Chem. Pharm. Bull.* **1986**, *34*, 3102–3110.

1301. Sasaki, K.; Hirata, Y. *Tetrahedron* **1970**, *26*, 2119–2126.

1302. Touré, B.B.; Hall, D.G. *Angew. Chem., Int. Ed.* **2004,** *43*, 2001–2004.

1303. Carroll, F.I.; Coleman, M.L.; Lewin, A.H. *J. Org. Chem.* **1982**, *47*, 13–19.

1304. Clarke, R.L.; Daum, S.J.; Gambino, A.J.; Acetyo, M.D.; Pearl, J.; Levitt, M.; Cumiskey, W.R.; Bogado, E.F. *J. Med. Chem.* **1973**, *16*, 1260–1267.

1305. Thai, D.L.; Sapko, M.T.; Reiter, C.T.; Bierer, D.E.; Perel, J.M. *J. Med. Chem.* **1998**, *41*, 591–601.

1306. Glaser, R.; Adin, I.; Shiftan, D.; Shi, Q.; Deutsch, H.M.; George, C.; Wu, K.-M.; Froimowitz, M. *J. Org. Chem.* **1998**, *63*, 1785–1794.

1307. Barnes, D.M.; Ji, J.; Fickes, M.G.; Fitzgerald, M.A.; King, S.A.; Morton, H.E.; Plagge, F.A.; Preskill, M.; Wagaw, S.H.; Wittenberger, S.J.; Zhang, J. *J. Am. Chem. Soc.* **2002**, *124*, 13097–13105.

1308. Buchholz, M.; Reißig, H.-U. *Eur. J. Org. Chem.* **2003**, 3524–3533.

1309. Kuhl, A.; Hahn, M.G.; Dumic, M.; Mittendorf, J. *Amino Acids* **2005**, *29*, 89–100.

1310. Appella, D.H.; Christianson, L.A.; Klein, D.A.; Richards, M.R.; Powell, D.R.; Gellman, S.H. *J. Am. Chem. Soc.* **1999**, *121*, 7574–7581.

1311. Appella, D.H.; Christianson, L.A.; Klein, D.A.; Powell, D.R.; Huang, X.; Barchi J.J., Jr.; Gellman, S.H. *Nature* **1997**, *387*, 381–384.

1312. Barchi, J.J., Jr.; Huang, X.; Appella, D.H.; Christianson, L.A.; Durell, S.R.; Gellman, S.H. *J. Am. Chem., Soc.* **2000**, *122*, 2711–2718.

1313. Martinek, T.A.; Tóth, G.K.; Vass, E.; Hollósi, M.; Fülöp, F. *Angew. Chem., Int. Ed.* **2002**, *41*, 1718–1721.

1314. Lee, H.-S.; Syud, F.A.; Wang, X.; Gellman, S.H. *J. Am. Chem. Soc.* **2001**, *123*, 7721–7722.

1315. Sandvoss, L.M.; Carlson, H.A. *J. Am. Chem. Soc.* **2003**, *125*, 15855–15862.

1316. Huck, B.R.; Fisk, J.D.; Guzei, I.A.; Carlson, H.A.; Gellman, S.H. *J. Am. Chem. Soc.* **2003**, *125*, 9035–9037.

1317. Woll, M.G.; Fisk, J.D.; LePlae, P.R.; Gellman, S.H. *J. Am. Chem. Soc.* **2002**, *124*, 12447–12452.

1318. Park, J.-S.; Lee, H.-S.; Lai, J.R.; Kim, B.M.; Gellman, S.H. *J. Am. Chem. Soc.* **2003**, *125*, 8539–8545.

1319. Appella, D.H.; Christianson, L.A.; Karle, I.L.; Powell, D.R.; Gellman, S.H. *J. Am. Chem. Soc.* **1999**, *121*, 6206–6212.

1320. Appella, D.H.; Christianson, L.A.; Karle, I.L.; Powell, D.R.; Gellman, S.H. *J. Am. Chem. Soc.* **1996**, *118*, 13071–13072.

1321. Appella, D.H.; Barchi, J.J., Jr.; Durell, S.R.; Gellman, S.H. *J. Am. Chem. Soc.* **1999**, *121*, 2309–2310.

1322. Strijowski, U.; Sewald, N. *Org. Biomol. Chem.* **2004**, *2*, 1105–1109.

1323. Hayen, A.; Schmitt, M.A.; Ngassa, F.N.; Thomasson, K.A.; Gellman, S.H. *Angew. Chem., Int. Ed.* **2004**, *43*, 505–510.

1324. Schmitt, M.A.; Choi, S.H.; Guzei, I.A.; Gellman, S.H. *J. Am. Chem. Soc.* **2005**, *127*, 13130–13131.

1325. Chung, Y.J.; Christianson, L.A.; Stanger, H.E.; Powell, D.R.; Gellman, S.H. *J. Am. Chem. Soc.* **1998**, *120*, 10555–10556.

1326. Chung, Y.J.; Huck, B.R.; Christianson, L.A.; Stanger, H.E.; Krauthäuser, S.; Powell, D.R.; Gellman, S.H. *J. Am. Chem. Soc.* **2000**, *122*, 3995–4004.

1327. Arnold, U.; Hinderaker, M.P.; Nilsson, B.L.; Huck, B.R.; Gellman, S.H.; Raines, R.T. *J. Am. Chem. Soc.* **2002**, *124*, 8522–8523.

1328. Hibbs, D.E.; Hursthouse, M.B.; Jones, I.G.; Jones, W.; Malik, K.M.A.; North, M. *J. Org. Chem.* **1998**, *63*, 1496–1504.

1329. Jones, I.G.; Jones, W.; North, M. *J. Org. Chem.* **1998**, *63*, 1505–1513.

1330. Wright, K.; de Castries, A.; Sarciaux, M.; Formaggio, F.; Toniolo, C.; Toffoletti, A.; Wakselman, M.; Mazaleyrat, J.-P. *Tetrahedron Lett.* **2005**, *46*, 5573–5576.

1331. Wright, K.; Crisma, M.; Toniolo, C.; Török, R.; Péter, A.; Wakselman, M.; Mazaleyray, J.-P. *Tetrahedron Lett.* **2003**, *44*, 3381–3384.

1332. Péter, A.; Török, R.; Wright, K.; Wakselman, M.; Mazaleyrat, J.P. *J. Chromatogr., A* **2003**, *1021*, 1–10.

1333. Izquierdo, S.; Rúa, F.; Sbai, A.; Parella, T.; Álvarez-Larena, A.; Branchadell, V.; Ortuño, R.M. *J. Org. Chem.* **2005**, *70*, 7963–7971.

1334. Aitken, D.J.; Gauzy, C.; Pereira, E. *Tetrahedron Lett.* **2004**, *45*, 2359–2361.

1335. De Pol, S.; Zorn, C.; Klein, C.D.; Zerbe, O.; Reiser, O. *Angew. Chem., Int. Ed.* **2004**, *43*, 511–514.

1336. Abele, S.; Seifer, P.; Seebach, D. *Helv. Chim. Acta* **1999**, *82*, 1559–1571.

1337. Wang, X.; Espinosa, J.F.; Gellman, S.H. *J. Am. Chem. Soc.* **2000**, *122*, 4821–4822.

1338. Porter, E.A.; Wang, X.; Lee, H.-S.; Weisblum, B.; Gellman, S.H. *Nature* **2000**, *404*, 565.

1339. Porter, E.A.; Weisblum, B.; Gellman, S.H. *J. Am. Chem. Soc.* **2002**, *124*, 7324–7330.

1340. Schmitt, M.A.; Weisblum, B.; Gellman, S.H. *J. Am. Chem. Soc.* **2004**, *126*, 6848–6849.

1341. Yamazaki, T.; Zhu, Y.-F.; Probstl, A.; Chadha, R.K.; Goodman, M. *J. Org. Chem.* **1991**, *56*, 6644–6655.

1342. Nöteberg, D.; Brånalt, J.; Kvarnström, I.; Classon, B.; Samuelsson, B.; Nillroth, U.; Danielson, U.H.; Karlén, A.; Hallberg, A. *Tetrahedron* **1997**, *53*, 7975–7984.

1343. Davies, S.G.; Dixon, D.J. *J. Chem. Soc., Perkin Trans. 1* **1998**, 2629–2634.

1344. Cardillo, G.; Gentilucci, L.; Tolomelli, A.; Calienni, M.; Qasem, A.R.; Spampinato, S. *Org. Biomol. Chem.* **2003**, *1*, 1498–1502.

1345. Langer, O.; Kählig, H.; Zierler-Gould, K.; Bats, J.W.; Mulzer, J. *J. Org. Chem.* **2002**, *67*, 6878–6883.

1346. Johnson, G.; Drummond, J.T.; Boxer, P.A.; Bruns, R.F. *J. Med. Chem.* **1992**, *35*, 233–241.

1347. Boyd, S.A.; Mantei, R.A.; Tasker, A.S.; Liu, G.; Sorensen, B.K.; Henry, K.J., Jr.; von Geldern, T.W.; Winn, M.; Wu-Wong, J.R.; Chiou, W.J.; Dixon, D.B.; Hutchins, C.W.; Marsh, K.C.; Nguyen, B.; Opgenorth, T.J. *Biorg. Med. Chem.* **1999** *7*, 991–1002.

1348. Winn, M.; von Geldern, T.W.; Opgenorth, T.J.; Jae, H.-S.; Tasker, A.S.; Boyd, S.A.; Kester, J.A.;

Mantei, R.A.; Bal, R.; Sorensen, B.K.; Wu-Wong, J.R.; Chiou, W.J.; Dixon, D.B.; Novosad, E.I.; Hernandez, L.; Marsh, K.C. *J. Med. Chem.* **1996**, 39, 1039–1048.

1349. Jae, H.-S.; Winn, M.; Dixon, D.B.; Marsh, K.C.; Nguyen, B.; Opgenorth, T.J.; von Geldern, T.W. *J. Med. Chem.* **1997**, 40, 3217–3227.

1350. Byrgesen, E.; Nielsen, J.; Willert, M.; Bols, M. *Tetrahedron Lett.* **1997**, 38, 5697–5700.

1351. Suhara, Y.; Hildreth, J.E.K.; Ichikawa, Y. *Tetrahedron Lett.* **1996**, 37, 1575–1578.

1352. Gaucher, A.; Bintein, F.; Wakselman, M.; Mazaleyrat, J.-P. *Tetrahedron Lett.* **1998**, 39, 575–578.

1353. Gaucher, A.; Wakselman, M.; Mazaleyrat, J.-P.; Crisma, M.; Formaggio, F.; Toniolo, C. *Tetrahedron* **2000**, 56, 1715–1723.

1354. Fülöp, F. *Chem. Rev.* **2001**, 101, 2181–2204.

1355. Gand, F.; Reiser, O. *Chem. Rev.* **2003**, 103, 1603–1623.

1356. Avotins, F. *Russ. Chem. Rev.* **1993**, 62, 897–906.

1357. Conners, T.A.; Ross, W.C.J. *J. Chem. Soc.* **1960**, 2119–2132.

1358. Plieninger, H.; Schneider, K. *Ber.* **1959**, 92, 1594–1599.

1359. Banwell, M.G.; Bui, C.T.; Pham, H.T.T.; Simson, G.W. *J. Chem. Soc., Perkin Trans. 1* **1996**, 967–969.

1360. Matsumura, Y.; Maki, T.; Satoh, Y. *Tetrahedron Lett.* **1997**, 38, 8879–8882.

1361. Nativ, E.; Rona, P. *Israel. J. Chem.* **1972**, 10, 55–58.

1362. Saigo, K.; Okuda, Y.; Wakabayashi, S.; Hoshiko, T.; Nohira, H. *Chem. Lett.* **1981**, 857–860.

1363. Pátek, M.; Drake, B.; Lebl, M. *Tetrahedron Lett.* **1994**, 35, 9169–9172.

1364. Shroff, C.C.; Stewart, W.S.; Uhm, S.J.; Wheeler, J.W. *J. Org. Chem.* **1971**, 36, 3356–3361.

1365. Cannon, J.G.; Garst, J.E. *J. Org. Chem.* **1975**, 40, 182–184.

1366. Richter, L.S.; Andersen, S. *Tetrahedron Lett.* **1998**, 39, 8747–8750.

1367. Chen, J.J.; Hinkley, J.M.; Wise, D.S.; Townsend, L.B. *Synth. Commun.* **1996**, 26, 617–622.

1368. El Mahdi, O.; Lavergne, J.-P.; Viallefont, Ph.; Akssira, M. *Bull. Soc. Chim. Belg.* **1995**, 104, 31–37.

1369. Canonne, P.; Akssira, M.; Dahdouh, A.; Kasmi, H.; Boumzebra, M. *Tetrahedron* **1993**, 49, 1985–1993.

1370. Stoermer, R.; Schenck, F. *Ber.* **1927**, 60, 2566–2591.

1371. Casamitjana, N.; Jorge, A.; Pérez, C.G.; Bosch, J.; Espinosa, E.; Molins, E. *Tetrahedron Lett.* **1997**, 38, 2295–2298.

1372. Weber, B.; Schwerdtfeger, J.; Fröhlich, R.; Göhrt, A.; Hoppe, D. *Synthesis* **1999**, 1915–1924.

1373. Ohno, M.; Tanaka, H.; Komatsu, M.; Ohshiro, Y. *Synlett* **1991**, 919–920.

1374. Archibald, T.G.; Baum, K.; Garver, L.C. *Synth. Commun.* **1990**, 20, 407–411.

1375. Åhman, J.; Somfai, P. *Tetrahedron* **1992**, 48, 9537–9544.

1376. Kuethe, J.T.; Comins, D.L. *Org. Lett.* **1999**, 1, 1031–1033.

1377. Yee, N.K. *Tetrahedron Lett.* **1997**, 38, 5091–5094.

1378. Russowsky, D.; da Silveira Neto, B.A. *Tetrahedron Lett.* **2004**, 45, 1437–1440.

1379. Bubert, C.; Cabrele, C.; Reiser, O. *Synlett* **1997**, 827–829.

1380. Zorn, C.; Gnad, F.; Salmen, S.; Herpin, T.; Reiser, O. *Tetrahedron Lett.* **2001**, 42, 7049–7053.

1381. Voigt, J.; Noltemeyer, M.; Reiser, O. *Synlett* **1997**, 202–204.

1382. Brannock, K.C.; Bell, A.; Burpitt, R.D.; Kelly, C.A. *J. Org. Chem.* **1961**, 26, 625–626.

1383. Brannock, K.C.; Burpitt, R.D.; Thweatt, J.G. *J. Org. Chem.* **1964**, 29, 940–941.

1384. Brannock, K.C.; Bell, A.; Burpitt, R.D.; Kelly, C.A. *J. Org. Chem.* **1964**, 29, 801–812.

1385. Bryans, J.S.; Horwell, D.C.; Ratcliffe, G.S.; Receveur, J.-M.; Rubin, J.R. *Biorg. Med. Chem.* **1999**, 7, 715–721.

1386. Dehnel, A.; Kanabus-Kaminska, J.M. *Can. J. Chem.* **1988**, 66, 310–318.

1387. Doyle, M.P.; Yan, M.; Hu, W.; Gronenberg, L.S. *J. Am. Chem. Soc.* **2003**, 125, 4692–4693.

1388. Davies, H.M.W.; Matasi, J.J.; Thornley, C. *Tetrahedron Lett.* **1995**, 7205–7208.

1389. Singh, S.; Basmadjian, G.P. *Tetrahedron Lett.* **1997**, 38, 6829–6830.

1390. Zhang, C.; Trudell, M.L. *Tetrahedron* **1998**, 54, 8349–8354.

1391. Takaoka, K.; Aoyama, T.; Shioiri, T. *Tetrahedron Lett.* **1996**, 37, 4973–4976.

1392. Katayama, S.; Ae, N.; Nagata, R. *Tetrahedron: Asymmetry* **1998**, 9, 4295–4299.

1393. Pfeiffer, P.; Heinrich, E. *J. Prak. Chem.* **1940**, 156, 241–259.

1394. Török, G.; Péter, A.; Gaucher, A.; Wakselman, M.; Mazaleyrat, J.-P.; Armstrong, D.W. *J. Chromatogr., A* **1999**, 846, 83–91.

1395. Masesane, I.B.; Steel, P.G. *Tetrahedron Lett.* **2004**, 45, 5007–5009.

1396. Bach, P.; Lohse, A.; Bos, M. *Tetrahedron Lett.* **1999**, 40, 367–370.

1397. Humphrey, J.M.; Liao, Y.; Ali, A.; Rein, T.; Wong, Y.-L.; Chen, H.-J.; Courtney, A.K.; Martin, S.F. *J. Am. Chem. Soc.* **2002**, 124, 8584–8592.

1398. Denes, F.; Chemla, F.; Normant, J.F. *Angew. Chem., Int. Ed.* **2003**, 42, 4043–4046.

1399. Berlin, S.; Ericsson, C.; Engman, L. *J. Org. Chem.* **2003**, 68, 8386–8396.

1400. Timén, A.S.; Somfai, P. *J. Org. Chem.* **2003**, 68, 9958–9963.

1401. Palomo, C.; Oiarbide, M.; Bindi, S. *J. Org. Chem.* **1998**, 63, 2469–2474.

1402. Stájer, G.; Mód, L.; Szabó, A.E.; Fülöp, F.; Bernáth, G. *Tetrahedron* **1984**, *40*, 2385–2393.

1403. Moriconi, E.J.; Crawford, W.C. *J. Org. Chem.* **1968**, *33*, 370–378.

1404. Alonso, E.; del Pozo, C.; González, J. *J. Chem. Soc., Perkin Trans. 1* **2002**, 571–576.

1405. Curran, T.P.; Smith, M.B.; Pollastri, M.P. *Tetrahedron Lett.* **1994**, *35*, 4515–4518.

1406. Freifelder, M.; Stone, G.R. *J. Org. Chem.* **1961**, *26*, 3805–3807.

1407. Freifelder, M.; Robinson, R.M.; Stone, G.R. *J. Org. Chem.* **1962**, *27*, 284–286.

1408. Hess, K.; Leibbrandt, F. *Ber.* **1917**, *50*, 385–389.

1409. Donohoe, T.J.; Harji, R.R.; Cousins, R.P.C. *Chem. Commun.* **1999**, 141–142.

1410. Verdecia, Y.; Suárez, M.; Morales, A.; Rodríguez, E.; Ochoa, E.; González, L.; Martín, N.; Quinteiro, M.; Seoane, C.; Soto, J.L. *J. Chem. Soc., Perkin Trans. 1* **1996**, 947–951.

1411. Kobayashi, K.; Takada, K.; Tanaka, H.; Uneda, T.; Kitamura, T.; Morikawa, O.; Konishi, H. *Chem. Lett.* **1996**, 25–26.

1412. Kobayashi, K.; Uneda, T.; Takada, K.; Tanaka, H.; Kitamura, T.; Morikawa, O.; Konishi, H. *J. Org. Chem.* **1997**, *62*, 664–668.

1413. Moutou, J.-L.; Schmitt, M.; Collot, V.; Bourguignon, J.-J. *Tetrahedron Lett.* **1996**, *37*, 1787–1790.

1414. Beholz, L.G.; Benovsky, P.; Ward, D.L.; Barta, N.S.; Stille, J.R. *J. Org. Chem.* **1997**, *62*, 1033–1042.

1415. Cordero, F.M.; Machetti, F.; De Sarlo, F.; Brandi, A. *Gazz. Chim. Ital.* **1997**, *127*, 25–29.

1416. Selic, L.; Stanovnik, B. *Helv. Chim. Acta* **1998**, *81*, 1634–1639.

1417. Kim, B.S.; Choi, K.S.; Kim, K. *J. Org. Chem.* **1998**, *63*, 6086–6087.

1418. Nair, V.; Vinod, A.U.; Rajesh, C. *J. Org. Chem.* **2001**, *66*, 4427–4429.

1419. Carroll, W.A.; Zhang, X. *Tetrahedron Lett.* **1997**, *38*, 2637–2640.

1420. Yamago, S.; Nakamura, M.; Wang, X.Q.; Yanagawa, M.; Tokumitsu, S.; Nakamura, E. *J. Org. Chem.* **1998**, *63*, 1694–1703.

1421. Zanobini, A.; Gensini, M.; Magull, J.; Vidovic, D.; Kozhushkov, A.I.; Brandi, A.; de Meijere, A. *Eur. J. Org. Chem.* **2004**, 4158–4166.

1422. Thompson, P.E.; Hearn, M.T. *Tetrahedron Lett.* **1997**, *38*, 2907–2910.

1423. Sawa, M.; Imaeda, Y.; Hiratake, J.; Fujii, R.; Umeshita, R.; Watanabe, M.; Kondo, H.; Oda, J. *Biorg. Med. Chem. Lett.* **1998**, *8*, 647–652.

1424. Pei, X.-F.; Gupta, T.H.; Badio, B.; Padgett, W.L.; Daly, J.W. *J. Med. Chem.* **1998**, *41*, 2047–2055.

1425. Thompson, P.E.; Steer, D.L.; Aguilar, M.-I.; Hearn, M.T.W. *Biorg. Med. Chem. Lett.* **1998**, *8*, 2699–2704.

1426. Macdonald, S.J.F.; Mills, K.; Spooner, J.E.; Upton, R.J.; Dowle, M.D. *J. Chem. Soc. Perkin Trans. 1* **1998**, 3931–3936.

1427. Miyata, O.; Muroya, K.; Koide, J.; Naito, T. *Synlett* **1998**, 271–272.

1428. Miyata, O.; Muroya, K.; Kobayashi, T.; Yamanaka, R.; Kajisa, S.; Koide, J.; Naito, T. *Tetrahedron* **2002**, *58*, 4459–4479.

1429. Suero, R.; Gorgojo, J.M.; Aurrecoechea, J.M. *Tetrahedron* **2002**, *58*, 6211–6221.

1430. Dolle, R.E.; Barden, M.C.; Brennan, P.E.; Ahmed, G.; Tran, V.; Ho, D.M. *Tetrahedron Lett.* **1999**, *40*, 2907–2908.

1431. Padwa, A.; Brodney, M.A.; Lynch, S.M. *J. Org. Chem.* **2001**, *66*, 1716–1724.

1432. Oh, C.-Y.; Kim, K.-S.; Ham, W. *Tetrahedron Lett.* **1998**, *39*, 2133–2136.

1433. Lescop, C.; Mévellec, L.; Huet, F. *J. Org. Chem.* **2001**, *66*, 4187–4193.

1434. Armarego, W.L.F.; Kobayashi, T. *J. Chem. Soc. C* **1970**, 1597–1600.

1435. Nohira, H.; Watanabe, K.; Kurokawa, M. *Chem. Lett.* **1976**, 299–300.

1436. Prashad, M.; Har, D.; Repic, O.; Blacklock, T.J.; Giannousis, P. *Tetrahedron: Asymmetry* **1999**, *10*, 3111–3116.

1437. Péter, A.; Árki, A.; Vékes, E.; Tourwé, D.; Forró, E.; Fülöp, F.; Armstrong, D.W. *J. Chromatogr., A* **2004**, *1031*, 159–170.

1438. Péter, A.; Török, G.; Armstrong, D.W. *J. Chromatogr., A* **1998**, *793*, 283–296.

1439. McGachy, N.T.; Grinberg, N.; Variankaval, N. *J. Chromatogr., A* **2005**, *1064*, 193–204.

1440. Török, G.; Péter, A.; Csomós, P.; Kanerva, L.T.; Fülöp, F. *J. Chromatogr., A* **1998**, *797*, 177–186.

1441. Kanerva, L.T.; Csomós, P.; Sundholm, O.; Bernáth, G.; Fülöp, F. *Tetrahedron: Asymmetry* **1996**, *7*, 1705–1716.

1442. Csomós, P.; Kanerva, L.; Bernáth, G.; Fülöp, F. *Tetrahedron: Asymmetry* **1996**, *7*, 1789–1796.

1443. Evans, C.; McCague, R.; Roberts, S.M.; Sutherland, A.G.; Wisdom, R. *J. Chem. Soc., Perkin Trans. 1* **1991**, 2276–2277.

1444. Pousset, C.; Callens, R.; Haddad, M.; Larchevêque, M. *Tetrahedron Asymmetry* **2004**, *15*, 3407–3412.

1445. Knight, D.W.; Lewis, N.; Share, A.C. *Tetrahedron: Asymmetry* **1993**, *4*, 625–628.

1446. Knight, D.W.; Lewis, N.; Share, A.C.; Haigh, D. *J. Chem. Soc., Perkin Trans. 1* **1998**, 3673–3683.

1447. Tang, W.; Wu, S.; Zhang, X. *J. Am. Chem. Soc.* **2003**, *125*, 9570–9571.

1448. Zhang, Y.J.; Park, J.H.; Lee, S.-g. *Tetrahedron: Asymmetry* **2004**, *15*, 2209–2212.

1449. Bunnage, M.E.; Davies, S.G.; Roberts, P.M.; Smith, A.D.; Withey, J.M. *Org. Biomol. Chem.* **2004**, *2*, 2763–2776.

1450. Bunnage, M.E.; Chippindale, A.M.; Davies, S.G.; Parkin, R.M.; Smith, A.D.; Withey, J.M. *Org. Biomol. Chem.* **2003**, *1*, 3698–3707.

1451. Urones, J.G.; Garrido, N.M.; Díez, D.; El Hammoumi, M.M.; Dominguez, S.H.; Casaseca, J.A.; Davies, S.G.; Smith, A.D. *Org. Biomol. Chem.* **2004**, *2*, 364–372.

1452. Lee, H.-S.; LePlae, P.R.; Porter, E.A.; Gellman, S.H. *J. Org. Chem.* **2001**, *66*, 3597–3599.

1453. Enders, D.; Wiedemann, J. *Liebigs Ann. / Recueil* **1997**, 699–706.

1454. Enders, D.; Wiedemann, J.; Bettray, W. *Synlett*, **1995**, 369–371.

1455. Price, D.A. *Synlett* **1999**, 1919–1920.

1456. Shimano, M.; Meyers, A.I. *J.Org. Chem.* **1995**, *60*, 7445–7455.

1457. Shida, N.; Uyehara, T.; Yamamoto, Y. *J. Org. Chem.* **1992**, *57*, 5049–5051.

1458. Chan, P.W.H.; Cottrell, I.F.; Moloney, M.G. *Tetrahedron Lett.* **1997**, *38*, 5891–5894.

1459. Cinquin, C.; Bortolussi, M.; Bloch, R. *Tetrahedron* **1996**, *52*, 6943–6952.

1460. Theil, F.; Ballschuh, S. *Tetrahedron: Asymmetry* **1996**, *7*, 3565–3572.

1461. Mazzini, C.; Lebreton, J.; Alphand, V.; Furstoss, R. *J. Org. Chem.* **1997**, *62*, 5215–5218.

1462. Kawahata, N.H.; Goodman, M. *Tetrahedron Lett.* **1999**, *40*, 2271–2274.

1463. Kobayashi, S.; Kamiyama, K.; Iimori, T.; Ohno, M. *Tetrahedron Lett.* **1984**, *25*, 2557–2560.

1464. Jones, I.G.; Jones, W.; North, M. *Synlett* **1997**, 63–65.

1465. Hibbs, D.E.; Hursthouse, M.B.; Jones, I.G.; Jones, W.; Malik, K.M.A.; North, M. *Tetrahedron* **1997**, *53*, 17417–17424.

1466. Miller, J.A.; Hennessy, E.J.; Marshall, W.J.; Scialdone, M.A.; Nguyen, S.T. *J. Org. Chem.* **2003**, *68*, 7884–7886.

1467. Martín-Vilà, M.; Minguillón, C.; Ortuño, R.M. *Tetrahedron: Asymmetry* **1998**, *9*, 4291–4294.

1468. Basso, A.; Banfi, L.; Riva, R.; Guanti, G. *J. Org. Chem.* **2005**, *70*, 575–579.

1469. Furuta, K.; Hayashi, S.; Miwa, Y.; Yamamoto, H. *Tetrahedron Lett.* **1987**, *28*, 5841–5844.

1470. Xu, D.; Prasad, K.; Repic, O.; Blacklock, T.J. *Tetrahedron: Asymmetry* **1997**, *8*, 1445–1451; **1998**, *9*, 1635.

1471. Schinner, M.; Murray, J.K.; Langenhan, J.M.; Gellman, S.H. *Eur. J. Org. Chem.*, **2003**, 721–726.

1472. LePlae, P.R.; Umezawa, N.; Lee, H.-S.; Gellman, S.H. *J. Org. Chem.* **2001**, *66*, 5629–5632.

1473. Blot, J.; Bardou, A.; Bellec, C.; Fargeau-Bellassoued, M.C.; Célérier, J.P.; Lhommet, G.; Gardette, D.; Gramain, J.-C. *Tetrahedron Lett.* **1997**, *38*, 8511–8514.

1474. Bardou, A.; Célérier, J.P.; Lhommet, G. *Tetrahedron Lett.* **1997**, *38*, 8507–8510.

1475. Ooi, T.; Miki, T.; Taniguchi, M.; Shiraishi, M.; Takeuchi, M.; Maruoka, K. *Angew. Chem., Int. Ed.* **2003**, *42*, 3796–3798.

1476. Salter, R.; Pickett, T.E.; Richards, C.J. *Tetrahedron: Asymmetry* **1998**, *9*, 4239–4247.

1477. Momose, T.; Toshima, M.; Toyooka, N.; Hirai, Y.; Eugster, C.H. *J. Chem. Soc., Perkin Trans. 1* **1997**, 1307–1313.

1478. Matsumura, Y.; Kanda, Y.; Shirai, K.; Onomura, O.; Maki, T. *Org. Lett.* **1999**, *1*, 175–178.

1479. Fevig, J.M.; Buriak, Jr., J.; Stouten, P.F.W.; Knabb, R.M.; Lam, G.N.; Wong, P.C.; Wexler, R.R. *Biorg. Med. Chem. Lett.* **1999**, *9*, 1195–1200.

1480. Carey, J.S. *J. Org. Chem.* **2001**, *66*, 2526–2529.

1481. Pandey, G.; Laha, J.K.; Lakshmaiah, G. *Tetrahedron* **2002**, *58*, 3525–3534.

1482. Robiette, R.; Cheboub-Benchaba, K.; Peeters, D.; Marchand-Brynaert, J. *J. Org. Chem.* **2003**, *68*, 9809–9812.

1483. Ishibashi, H.; Kodama, K.; Kameoka, C.; Kawanami, H.; Ikeda, M. *Tetrahedron* **1996**, *52*, 13867–13880.

1484. David, O.; Blot, J.; Bellec, C.; Fargeau-Bellassoued, M.-C.; Haviari, G.; Célérier, J.-P.; Lhommet, G.; Gramain, J.-C.; Gardette, D. *J. Org. Chem.* **1999**, *64*, 3122–3131.

1485. Davies, H.M.L.; Hansen, T.; Hopper, D.W.; Panaro, S.A. *J. Am. Chem. Soc.* **1999**, *121*, 6509–6510.

1486. Axten, J.M.; Ivy, R.; Krim, L.; Winkler, J.D. *J. Am. Chem. Soc.* **1999**, *121*, 6511–6512.

1487. Davies, H.M.L.; Walji, A.M. *Org. Lett.* **2003**, *5*, 479–482.

1488. Park, J.-i.; Tian, G.R.; Kim, D.H. *J. Org. Chem.* **2001**, *66*, 3696–3703.

1489. Ikota, N.; Hanaki, A. *Heterocycles* **1988**, *27*, 2535–2537.

1490. Shin, I.; Lee, M.-r.; Lee, J.; Jung, M.; Lee, W.; Yoon, J. *J. Org. Chem.* **2000**, *65*, 7667–7675.

1491. Haddad, M.; Larchevêque, M.; Tong, H.M. *Tetrahedron Lett.* **2005**, *46*, 6015–6017.

1492. Momose, T.; Toyooka, N.; Jin, M. *J. Chem. Soc., Perkin Trans. 1* **1997**, 2005–2013.

1493. Gees, T.; Schweizer, W.B.; Seebach, D. *Helv. Chim. Acta* **1993**, *76*, 2640–2653.

1494. Ma, D.; Sun, H. *J. Org. Chem.* **2000**, *65*, 6009–6016.

1495. Ohno, M.; Kobayashi, S.; Iimori, T.; Wang, Y.-F.; Izawa, T. *J. Am. Chem. Soc.* **1981**, *103*, 2405–2406.

1496. Agami, C.; Dechoux, L.; Ménard, C.; Hebbe, S. *J. Org. Chem.* **2002**, *67*, 7573–7576.

1497. Govender, T.; Arvidsson, P.I. *Tetrahedron Lett.* **2006**, *47*, 1691–1694.

1498. Fülöp, F.; Palko, M.; Forró, E.; Dervarics, M.; Martinek, T.A.; Sillanpää, R. *Eur. J. Org. Chem.* **2005**, 3214–3220.

1499. Ledoux, S.; Célérier, J.-P.; Lhommet, G. *Tetrahedron Lett.* **1999**, *40*, 9019–9020.

1500. Knight, D.W.; Share, A.C.; Gallagher, P.T. *J. Chem. Soc., Perkin Trans. 1* **1991**, 1615–1616.

1501. Gyónfalvi, S.; Szakonyi, Z.; Fülöp, F. *Tetrahedron Asymmetry* **2003**, *14*, 3965–3972.

1502. Schweizer, F.; Lohse, A.; Otter, A.; Hindsgaul, O. *Synlett* **2001**, 1434–1436.

1503. Schneider, C.; Börner, C. *Synlett* **1998**, 652–654.

1504. Maligres, P.E.; Chartrain, M.M.; Upadhyay, V.; Cohen, D.; Reamer, R.A.; Askin, D.; Volante, R.P.; Reider, P.J. *J. Org. Chem.* **1998**, *63*, 9548–9551.

1505. Avenoza, A.; Cativiela, C.; París, M.; Peregrina, J.M. *Tetrahedron: Asymmetry* **1995**, *6*, 1409–1418.

1506. Barluenga, J.; Aznar, F.; Ribas, C.; Valdés, C. *J. Org. Chem.* **1998**, *63*, 10052–10056.

1507. Aggarwal, V.K.; Roseblade, S.; Alexander, R. *Org. Biomol. Chem.* **2003**, *1*, 684–691.

1508. Desai, P.; Schildknegt, K.; Agrios, K.A.; Mossman, C.; Milligan, G.L.; Aubé, J. *J. Am. Chem. Soc.* **2000**, *122*, 7226–7232.

1509. *Glutamine, Glutamate, and GABA in the Central Nervous System*; Hertz, L; Kvamme, E.; McGeer, E.G.; Schousboe, A. Eds.; Alan R Liss.: New York, 1983.

1510. Fowden, L. *Phytochemistry* **1972**, *11*, 2271–2276.

1511. Impellizzeri, G.; Mangiafico, S.; Oriente, G.; Piattelli, M.; Sciuto, S.; Fattorusso, E.; Magno, S.; Santacroce, C.; Sica, D. *Phytochemistry* **1975**, *14*, 1549–1557.

1512. Zhang, L.-Y.; Sun, M.-X. *J. Chromatogr., A* **2005**, *1095*, 185–188.

1513. Otsuka, M.; Obata, K.; Miyata, Y.; Tanaka, Y. *J. Neurochem.* **1971**, *18*, 287–295.

1514. Yuen, P.-O.; Kanter, G.D.; Taylor, C.P.; Vartanian, M.G. *Bioorg. Med. Chem. Lett.* **1994**, *4*, 823–826.

1515. Trost, B.M.; Lemoine, R.C. *Tetrahedron Lett.* **1996**, *37*, 9161–9164.

1516. Wong, C.G.T.; Gibson, K.M.; Snead, O.C., III, *Trends Pharmacol. Sci.* **2004**, *25*, 29–34.

1517. Alcón, M.; Poch, M.; Moyano, A.; Pericàs, M.A.; Riera, A. *Tetrahedron: Asymmetry* **1997**, *8*, 2967–2974.

1518. Nanavati, S.M.; Silverman, R.B. *J. Am. Chem. Soc.* **1991**, *113*, 9341–9349.

1519. Bryans, J.S.; Davies, N.; Gee, N.S.; Dissanayake, V.U.K.; Ratcliffe, G.S.; Horwell, D.C.; Kneen, C.O.; Morrell, A.I.; Oles, R.J.; O'Toole, J.C.; Perkins, G.M.; Singh, L.; Suman-Chauhan, N.; O'Neill, J.A. *J. Med. Chem.* **1998**, *41*, 1838–1845.

1520. Metcalf, B.W.; Casara, P. *Tetrahedron Lett.* **1975**, *38*, 3337–3340.

1521. Lippert, B.; Metcalf, B.W.; Jung, M.J.; Casara, P. *Eur. J. Biochem.* **1977**, *74*, 441–445.

1522. Choi, S.; Storici, P.; Schirmer, T.; Silverman, R.B. *J. Am. Chem. Soc.* **2002**, *124*, 1620–1624.

1523. Burke, J.R.; Silverman, R.B. *J. Am. Chem. Soc.* **1991**, *113*, 9329–9340.

1524. Jung, M.J.; Metcalf, B.W.; Lippert, B.; Casara, P. *Biochemistry* **1978**, *17*, 2628–2631.

1525. Silverman, R.B.; Levy, M.A. *Biochem. Biophys. Res. Commun.* **1980**, *95*, 250–255.

1526. Silverman, R.B.; Levy, M.A. *J. Biol. Chem.* **1981**, *256*, 11565–11568.

1527. Silverman, R.B.; Levy, M.A. *Biochemistry* **1981**, *20*, 1197–1203.

1528. Silverman, R.B.; Durkee, S.C.; Invergo, B.J. *J. Med. Chem.* **1986**, *29*, 764–770.

1529. Silverman, R.B.; George, C. *Biochemistry* **1988**, *27*, 3285–3289.

1530. Andruszkiewicz, R.; Silverman, R.B. *J. Biol. Chem.* **1990**, *265*, 22288–22291.

1531. Silverman, R.B.; Invergo, B.J.; Levy, M.A.; Andrew, C.R. *J. Biol. Chem.* **1987**, *262*, 3192–3195.

1532. Silverman, R.B.; Andruszkiewicz, R.; Nanavati, S.M.; Taylor, C.P.; Vartanian, M.G. *J. Med. Chem.* **1991**, *34*, 2295–2298.

1533. Herdeis, C.; Hubmann, H.P. *Tetrahedron: Asymmetry* **1992**, *3*, 1213–1221.

1534. Bowery, N.G.; Hill, D.R.; Hudson, A.L.; Doble, A.; Middlemiss, D.N.; Shaw, J.; Turnbull, M. *Nature* **1980**, *283*, 92–94.

1535. Bryans, J.S.; Davies, N.; Gee, N.S.; Horwell, D.C.; Kneen, C.O.; Morrell, A.I.; O'Neill, J.A.; Ratcliffe, G.S. *Bioorg. Med. Chem. Lett.* **1997**, *7*, 2481–2484.

1536. Burk, M.J.; de Koning, P.D.; Grote, T.M.; Hoekstra, M.S.; Hoge, G.; Jennings, R.A.; Kissel, W.S.; Le, T.V.; Lennon, I.C.; Mulhern, T.A.; Ramsden, J.A.; Wade, R.A. *J. Org. Chem.* **2003**, *68*, 5731–5734.

1537. Metcalf, B.W. *Biochem. Pharmacol.* **1979**, *28*, 1705–1712.

1538. Misiti, D.; Zappia, G.; Monache, G. *Synth. Commun.* **1995**, *25*, 2285–2294.

1539. Candela, M.; Ruiz, A.; Feo, F.J. *J. Chromatogr., A* **2000**, *890*, 273–280.

1540. Voeffray, R.; Perlberger, J.-C.; Tenud, L.; Gosteli, J. *Helv. Chim. Acta* **1987**, *70*, 2058–2064.

1541. Kaneko, T.; Yoshida, R. *Bull. Chem. Soc. Jpn.* **1962**, *35*, 1153–1155.

1542. Friedman, S.; McFarlane, J.E.; Bhattacharyya, P.K.; Fraenkel, G.; Berger, C.R.A. *Biochem. Prep.* **1960**, *7*, 26–30.

1543. Sevilla, A.; Vera, J.; Dîaz, Z.; Cánovas, M.; Torres, N.V.; Iborra, J.L. *Biotechnol. Prog.* **2005**, *21*, 329–337.

1544. Haigler, H.T.; Broquist, H.P. *Biochem. Biophys. Res. Commun.* **1974**, *56*, 676–681.

1545. Hochalter, J.B.; Henderson, L.M. *Biochem. Biophys. Res. Commun.* **1976**, *70*, 364–366.

1546. Petter, R.C.; Banerjee, S.; Englard, S. *J. Org. Chem.* **1990**, *55*, 3088–3097.

1547. Colucci, W.J.; Gandour, R.D.; Fronczek, F.R.; Brady, P.S.; Brady, L.J. *J. Am. Chem. Soc.* **1987**, *109*, 7915–7916.

1548. Gatti, R.; De Palo, C.B.; Spinella, P.; De Palo, E.F. *Amino Acids* **1998**, *14*, 361–369.

1549. Vernez, L.; Thormann, W.; Krähenbühl, S. *J. Chromatogr., A* **2000**, *895*, 309–316.

1550. Kakou, A.; Megoulas, N.C.; Koupparis, M.A. *J. Chromatogr., A* **2005**, *1069*, 209–215.

1551. Prokorátová, V.; Kvasnicka, F.; Sevčík, R.; Voldrich, M. *J. Chromatogr., A* **2005**, *1081*, 60–64.

1552. Di Tullio, A.; D'Acquarica, I.; Gasparrini, F.; Desiderio, P.; Giannessi, F.; Muck, S.; Piccirilli, F.; Tinti, M.O.; Villani, C. *Chem. Commun.* **2002**, 474–475.

1553. D'Acquarica, I.; Gasparrini, F.; Misiti, D.; Villani, C.; Carotti, A.; Cellamare, S.; Muck, S. *J. Chromatogr., A* **1999**, *857*, 145–155.

1554. Kagawa, M.; Machida, Y.; Nishi, H. *J. Chromatogr., A* **1999**, *857*, 127–135.

1555. Mardones, C.; Ríos, A.; Valcárcel, M.; Cicciarelli, R. *J. Chromatogr., A* **1999**, *849*, 609–616.

1556. Feng, Y.; Blunt, J.W.; Cole, A.L.J.; Cannon, J.F.; Robinson, W.T.; Munro, M.H.G. *J. Org. Chem.* **2003**, *68*, 2002–2005.

1557. Morishima, H.; Takita, T.; Aoyagi, T.; Takeuchi, T.; Umezawa, H. *J. Antibiotics* **1970**, *23*, 263–264.

1558. Umezawa, H.; Aoyagi, T.; Morishima, H.; Matsuzaki, M.; Hamada, H.; Takeuchi, T.; *J. Antibiotics* **1970**, *23*, 259–262.

1559. Omura, S.; Imamura, N.; Kawakita, K.; Mori, Y.; Yamazaki, Y.; Masuma, R.; Takahashi, Y.; Tanaka, H.; Huang, L.-Y.; Woodruff, H.B. *J. Antibiotics* **1986**, *39*, 1079–1085.

1560. Marciniszyn, J., Jr.; Hartsuck, J.A.; Tang, J. *J. Biol. Chem.* **1976**, *251*, 7088–7094.

1561. Stracher, A.; McGowan, E.B.; Shafiq, S.A. *Science* **1978**, *200*, 50–51.

1562. Nakao, Y.; Fujita, M.; Warabi, K.; Matsunaga, S.; Fusetani, N. *J. Am. Chem. Soc.* **2000**, *122*, 10462–10463.

1563. Morishima, H.; Takita, T.; Umezawa, H. *J. Antibiotics* **1973**, *26*, 115–116.

1564. Kinoshita, M.; Aburaki, S.; Hagiwara, A.; Imai, J. *J. Antibiotics* **1973**, *26*, 249–251.

1565. Kinoshita, M.; Hagiwara, A.; Aburaki, S. *Bull. Chem. Soc. Jpn.* **1975**, *48*, 570–575.

1566. Liu, W.-S.; Smith, S.C.; Glover, G.I. *J. Med. Chem.* **1979**, *22*, 577–579.

1567. Precigoux, G. *Biopolymers* **1991**, *31*, 683–689.

1568. Marraud, M.; Aubry, A. *Biopolymers* **1996**, *40*, 45–83.

1569. Rich, D.H. *J. Med. Chem.* **1985**, *28*, 263–273.

1570. Rich, D.H.; Sun, E.; Singh, J. *Biochem. Biophys. Res. Commun.* **1977**, *74*, 762–767.

1571. Gelb, M.H.; Svaren, J.P.; Abeles, R.H. *Biochemistry* **1985**, *24*, 1813–1917.

1572. Thaisrivongs, S.; Pals, D.T.; Kati, W.M.; Turner, S.R.; Thomasco, L.M. *J. Med. Chem.* **1985**, *28*, 1553–1555.

1573. Okada, K.; Kurosawa, Y.; Nagai, S. *Chem. Pharm. Bull.* **1979**, *27*, 2163–2170.

1574. Matsushita, Y.; Tone, H.; Hori, S.; Yagi, Y.; Takamatsu, A.; Morishima, H.; Aoyagi, T.; Takeuchi, T.; Umezawa, H. *J. Antibiotics* **1975**, *28*, 1016–1018.

1575. Thaisrivongs, S.; Pals, D.T.; Kati, W.M.; Turner, S.R.; Thomasco, L.M.; Watt, W. *J. Med. Chem.* **1986**, *29*, 2080–2087.

1576. Rich, D.H.; Sun, E.T.O.; Ulm, E. *J. Med. Chem.* **1980**, *23*, 27–33.

1577. Nishi, T.; Saito, F.; Nagahori, H.; Kataoka, M.; Morisawa, Y.; Yabe, Y.; Sakurai, M.; Higashida, S.; Shoji, M.; Matsushita, Y.; Iijima, Y.; Ohizumi, K.; Koike, H. *Chem. Pharm. Bull.* **1990**, *38*, 103–109.

1578. Chattopadhaya, S.; Chan, E.W.S.; Yao, S.Q. *Tetrahedron Lett.* **2005**, *46*, 4053–4056.

1579. Carroll, C.D.; Johnson, T.O.; Tao, S.; Lauri, G.; Orlowski, M.; Gluzman, I.Y.; Goldberg, D.E.; Dolle, R.E. *Biorg. Med. Chem. Lett.* **1998**, *8*, 3203–3206.

1580. Carroll, C.D.; Patel, H.; Johnson, T.O.; Guo, T.; Orlowski, M.; He, Z.-M.; Cavallaro, C.L.; Guo, J.; Oksman, A.; Gluzman, I.Y.; Connelly, J.; Chelsky, D.; Goldberg, D.E.; Dolle, R.E. *Biorg. Med. Chem. Lett.* **1998**, *8*, 2315–2320.

1581. Scholz, D.; Billich, A.; Charpiot, B.; Ettmayer, P.; Lehr, P.; Rosenwirth B.; Schreiner, E.; Gstach, H. *J. Med. Chem.* **1994**, *37*, 3079–3089.

1582. Mayer, S.C.; Pfizenmayer, A.J.; Cordova, R.; Li, W.-R.; Joullié, M.M. *Tetrahedron: Asymmetry* **1994**, *5*, 519–522.

1583. Rinehart, K.L.; Kishore, V.; Nagarajan, S.; Lake, R.J.; Gloer, J.B.; Bozich, F.A.; Li, K.-M.; Maleczka, R.E., Jr.; Todsen, W.L.; Munro, M.H.G.; Sullins, D.W.; Sakai, R. *J. Am. Chem. Soc.* **1987**, *109*, 6846–6848.

1584. Rinehart, K.L., Jr.; Gloer, J.B.; Cook, J.C., Jr., *J. Am. Chem. Soc.* **1981**, *103*, 1857–1859.

1585. Hamada, Y.; Kondo, Y.; Shibata, M.; Shiori, T. *J. Am. Chem. Soc.* **1989**, *111*, 669–673.

1586. Banaigs, B.; Jeanty, G.; Francisco, C.; Jouin, P.; Poncet, J.; Heitz, A.; Cave, A.; Prome, J.C.; Wahl, M.; Lafargue, F. *Tetrahedron* **1989**, *45*, 181–190.

1587. Mayer, S.C.; Ramanjulu, J.; Vera, M.D.; Pfizenmayer, A.J.; Joullié, M.M. *J. Org. Chem.* **1994**, *59*, 5192–5205.

1588. Rinehart, K.L.; Sakai, R.; Kishore, V.; Sullins, D.W.; Li, K.-m. *J. Org. Chem.* **1992**, *57*, 3007–3013.

1589. Rinehart, K.L., Jr.; Gloer, J.B.; Hughes, R.G., Jr.; Renis, H.E.; McGovren, J.P.; Swynenberg, E.B.; Stringfellow, D.A.; Kuentzel, S.L.; Li, L.H. *Science* **1981**, *212*, 933–935.

1590. Schmidt, U.; Kroner, M.; Griesser, H. *Tetrahedron Lett.* **1988**, *29*, 3057–3060.

1591. Liang, B.; Portonovo, P.; Vera, M.D.; Xiao, D.; Joullié, M.M. *Org. Lett.* **1999**, *1*, 1319–1322.

1592. Liang, B.; Richard, D.J.; Portonovo, P.S.; Joullié, M.M. *J. Am. Chem. Soc.* **2001**, *123*, 4469–4474.

1593. Pfizenmayer, A.J.; Ramanjulu, J.M.; Vera, M.D.; Ding, X.; Xiao, D.; Chen, W.-C.; Joullié, M.M. *Tetrahedron* **1999**, *55*, 313–334.

1594. Mayer, S.C.; Pfizenmayer, A.J.; Joullié, M.M. *J. Org. Chem.* **1996**, *61*, 1655–1664.

1595. Xiao, D.; Vera, M.D.; Liang, B.; Joullié, M.M. *J. Org. Chem.* **2001**, *66*, 2734–2742.

1596. Cárdenas, F.; Caba, J.M.; Eliz, M.; Lloyd-Williams, P.; Giralt, E. *J. Org. Chem.* **2003**, *68*, 9554–9562.

1597. Cárdenas, F.; Thormann, M.; Feliz, M.; Caba, J.-M.; Lloyd-Williams, P.; Giralt, E. *J. Org. Chem.* **2001**, *66*, 4580–4584.

1598. Protonovo, P.; Ding, X.; Leonard, M.S.; Joullié, M.M. *Tetrahedron* **2000**, *56*, 3687–3690.

1599. Yurek-George, A.; Habens, F.; Brimmell, M.; Packham, G.; Ganesan, A. *J. Am. Chem. Soc.* **2004**, *126*, 1030–1031.

1600. Ohgi, T.; Hecht, S.M. *J. Org. Chem.* **1981**, *46*, 1232–1234.

1601. DiPardo, R.M.; Bock, M.G. *Tetrahedron Lett.* **1983**, *24*, 4805–4808.

1602. Yoshioka, T.; Hara, T.; Takita, T.; Umezawa, H. *J. Antibiotics* **1974**, *27*, 356–357.

1603. Umezawa, H.; Muraoka, Y.; Fujii, A.; Naganawa, H.; Takita, T. *J. Antibiotics* **1980**, *33*, 1079–1082.

1604. Hecht, S.M. *Acc. Chem. Res.* **1986**, *19*, 383–391.

1605. Inouye, Y.; Take, Y.; Nakamura, S.; Nakashima, H.; Yamamoto, N.; Kawaguchi, H. *J. Antibiotics* **1987**, *40*, 100–104.

1606. Aoyagi, Y.; Suguna, H.; Murugesan, N.; Ehrenfeld, G.M.; Chang, L.-H.; Ohgi, T.; Shekhani, M.S.; Kirkup, M.P.; Hecht, S.M. *J. Am. Chem. Soc.* **1982**, *104*, 5237–5239.

1607. Aoyagi, Y.; Katano, K.; Suguna, H.; Primeau, J.; Chang, L.-H.; Hecht, S.M. *J. Am. Chem. Soc.* **1982**, *104*, 5537–5538.

1608. Levin, M.S.; Subrahamanian, K.; Katz, H.; Smith, M.B.; Burlett, D.J.; Hecht, S.M. *J. Am. Chem. Soc.* **1980**, *102*, 1452–1453.

1609. Rishel, M.J.; Hecht, S.M. *Org. Lett.* **2001**, *3*, 2867–2869.

1610. Sodano, G.; Spinella, A. *Tetrahedron Lett.* **1986**, *27*, 25605–2508.

1611. Giordano, A.; Spinella, A.; Sodano, G. *Tetrahedron: Asymmetry* **1999**, *10*, 1851–1854.

1612. Haskell, T.H.; Rodebaugh, R.; Plessas, N.; Watson, D.; Westland, R.D. *Carbohydrate Res.* **1973**, *28*, 263–280.

1613. Woo, P.W.K.; Dion, H.W.; Bartz, Q.R. *Tetrahedron Lett.* **1971**, *28*, 2617–2620.

1614. Kinuta, M. *Biochim. Biophys. Acta* **1978**, *542*, 56–62.

1615. Deng, J.; Hamada, Y.; Shiori, T. *Synthesis* **1998**, 627–638.

1616. Pais, G.C.G.; Maier, M.E. *J. Org. Chem.* **1999**, *64*, 4551–4554.

1617. Wagner, B.; Gonzalez, G.I.; Tran Hun Dau, M.E.; Zhu, J. *Biorg. Med. Chem.* **1999**, *7*, 737–747.

1618. Haddad, M.; Botuha, C.; Larchevêque, M. *Synlett* **1999**, 1118–1120.

1619. Palomo, C.; Oiarbide, M.; García, J.M.; González, A.; Pazos, R.; Odriozola, J.M.; Bañuelos, P.; Tello, M.; Linden, A. *J. Org. Chem.* **2004**, *69*, 4126–4134.

1620. Nakatani, S.; Kamata, K.; Sato, M.; Onuki, H.; Hirota, H.; Matsumoto, J.; Ishibashi, M. *Tetrahedron Lett.* **2005**, *46*, 267–271.

1621. Ishida, K.; Murakami, M. *J. Org. Chem.* **2000**, *65*, 5898–5900.

1622. Ueda, M.; Hiraoka, T.; Niwa, M.; Yamamura, S. *Tetrahedron Lett.* **1999**, *40*, 6777–6780.

1623. Koskinen, A.M.P.; Chen, J. *Tetrahedron Lett.* **1991**, *32*, 6977–6980.

1624. Evans, D.A.; Gage, J.R.; Leighton, J.L. *J. Am. Chem. Soc.* **1992**, *114*, 9434–9453.

1625. Evans, D.A.; Gage, J.R.; Leighton, J.L.; Kim, A.S. *J. Org. Chem.* **1992**, *57*, 1961–1963.

1626. Kato, Y.; Fusetani, N.; Matsunaga, S.; Hashimoto, K.; Koseki, K. *J. Org. Chem.* **1988**, *53*, 3930–3932.

1627. Hamada, Y.; Tanada, Y.; Yokokawa, F.; Shioiri, T. *Tetrahedron Lett.* **1991**, *32*, 5983–5986.

1628. Vaccaro, H.A.; Levy, D.E.; Sawabe, A.; Jaetsch, T.; Masamune, S. *Tetrahedron Lett.* **1992**, *33*, 1937–1940.

1629. Clark, D.P.; Carroll, J.; Naylor, S.; Crews, P. *J. Org. Chem.* **1998**, *63*, 8757–8764.

1630. Pettit, G.R.; Burkett, D.D.; Williams, M.D. *J. Chem. Soc., Perkin Trans. 1* **1996**, 863–858.

1631. Pettit, G.R.; Kamano, Y.; Herald, C.L.; Tuinman, A.A.; Boettner, F.E.; Kizu, H.; Schmidt, J.M.; Baczynskyj, L.; Tomer, K.B.; Bontems, R.J. *J. Am. Chem. Soc.* **1987**, *109*, 6883–6885.

1632. Pettit, G.R.; Singh, S.B.; Hogan, F.; Lloyd-Williams, P.; Herald, D.L.; Burkett, D.D.; Clewlow, P.J. *J. Am. Chem. Soc.* **1989**, *111*, 5463–5465.

1633. Poncet, J.; Hortala, L.; Busquet, M.; Guéritte-Voegelein, F.; Thoret, S.; Pierré, A.; Atassi, G.; Jouin, P. *Biorg. Med. Chem. Lett.* **1998**, *8*, 2849–2854.

1634. Milanowski, D.J.; Gustafson, K.R.; Rashid, M.A.; Pannell, L.K.; McMahon, J.B.;Boyd, M.R. *J. Org. Chem.* **2004**, *69*, 3036–3042.

1635. Wipf, P.; Takada, T.; Rishel, M.J. *Org. Lett.* **2004**, *6*, 4057–4060.

1636. Oka, M.; Yaginuma, K.; Numata, K.; Konishi, M.; Oki, T.; Kawaguchi, H. *J. Antibiotics* **1988**, *41*, 1338–1350.

1637. Schmidt, U.; Kleefeldt, A.; Mangold, R. *J. Chem. Soc., Chem. Commun.* **1992**, 1687–1689.

1638. Vedejs, E.; Kongkittingam, C. *J. Org. Chem.* **2001**, *66*, 7355–7364.

1639. DiMaio, J.; Gibbs, B.; Lefebvre, J.; Konishi, Y.; Munn, D.; Yue, S.Y.; Hornberger, W. *J. Med. Chem.* **1992**, *35*, 3331–3341.

1640. Wasserman, H.H.; Chen, J.-H.; Xia, M. *J. Am. Chem. Soc.* **1999**, *121*, 1401–1402.

1641. Hoffman, R.W.; Lazaro, M.A.; Caturla, F.; Framery, E.; Valancogne, I.; Montalbetti, G.N. *Tetrahedron Lett.* **1999**, *40*, 5983–5986.

1642. Hintermann, T.; Gademann, K.; Jaun, B.; Seebach, D. *Helv. Chim. Acta* **1998**, *81*, 983–1002.

1643. Hanessian, S.; Luo, X.; Schaum, R.; Michnick, S. *J. Am. Chem. Soc.* **1998**, *120*, 8569–8570.

1644. Hanessian, S.; Luo, X.; Schaum, R. *Tetrahedron Lett.* **1999**, *40*, 4925–4929.

1645. Seebach, D.; Brenner, M.; Rueping, M.; Schweizer, B.; Jaun, B. *Chem. Commun.* **2001**, 207–208.

1646. Seebach, D.; Brenner, M.; Rueping, M.; Jaun, B. *Chem.—Eur. J.* **2002**, *8*, 573–584.

1647. Brenner, M.; Seebach, D. *Helv. Chim. Acta* **2001**, *84*, 1181–1189.

1648. Vasudev, P.G.; Shamala, N.; Ananda, K.; Balaram, P. *Angew. Chem., Int. Ed.* **2005, *44*,** 4972–4975.

1649. Hagihara, M.; Anthony, N.J.; Stout, T.J.; Clardy, J.; Schreiber, S.L. *J. Am. Chem. Soc.* **1992**, 114, 6568–6570.

1650. Reddy, G.V.; Rao, G.V.; Iyengar, D.S. *Chem. Commun.* **1999**, 317–318.

1651. Kong, J.-s.; Venkatraman, S.; Furness, K.; Nimkar, S.; Shepherd, T.A.; Wang, Q.M.; Aubé, J.; Hanzlik, R.P. *J. Med. Chem.* **1998**, *41*, 2579–2587.

1652. Dragovich, P.S.; Webber, S.E.; Babine, R.E.; Fuhrman, S.A.; Patick, A.K.; Matthews, D.A.; Reich, S.H.; Marakovits, J.T.; Prins, T.J.; Zhou, R.; Tikhe, J.; Littlefield, E.S.; Bleckman, T.M.; Wallace, M.B.; Little, T.A.; Ford, C.E.; Meador, J.W., III; Ferre, R.A.; Brown, E.L.; Binford, S.L.; DeLisle, D.M.; Worland, S.T. *J. Med. Chem.* **1998**, *41*, 2819–2834.

1653. Dragovich, P.S.; Webber, S.E.; Babine, R.E.; Fuhrman, S.A.; Patick, A.K.; Matthews, D.A.; Lee, C.A.; Reich, S.H.; Prins, T.J.; Marakovits, J.T.; Littlefield, E.S.; Zhou, R.; Tikhe, J.; Ford, C.E.; Wallace, M.B.; Meador, J.W. III; Ferre, R.A.; Brown, E.L.; Binford, S.L.; DeLisle, D.M.; Worland, S.T. *J. Med. Chem.* **1998**, *41*, 2806–2818.

1654. Dragovich, P.S.; Prins, T.J.; Zhou, R.; Fuhrman, S.A.; Patick, A.K.; Matthews, D.A.; Ford, C.E.; Meador, J.W. III; Ferre, R.A.; Worland, S.T. *J. Med. Chem.* **1999**, *42*, 1203–1212.

1655. Dragovich, P.S.; Prins, T.J.; Zhou, R.; Webber, S.E.; Marakovits, J.T.; Fuhrman, S.A.; Patick, A.K.; Matthews, D.A.; Lee, C.A.; Ford, C.E.; Burke, B.J.; Rejto, P.A.; Hendrickson, T.F.; Tuntland, T.; Brown, E.L.; Meador J.W., III; Ferre, R.A.; Harr, J.E.V.; Kosa, M.B.; Worland, S.T. *J. Med. Chem.* **1999**, *42*, 1213–1224.

1656. Dragovich, P.S.; Prins, T.J.; Zhou, R.; Brown, E.L.; Maldonado, F.C.; Fuhrman, S.A.; Zalman, L.S.; Tuntland, T.; Lee, C.A.; Patick, A.K.; Matthews, D.A.; Hendrickson, T.F.; Kosa, M.B.; Liu, B.; Batugo, M.R.; Gleeson, J.-P.R.; Sakata, S.K.; Chen, L.; Guzman, M.C.; Meador, J.W., III; Ferre, R.A.; Worland, S.T. *J. Med. Chem.* **2002**, *45*, 1607–1623.

1657. Baldauf, C.; Günther, R.; Hofmann, H-J. *Helv. Chim. Acta* **2003**, *86*, 2573–2588.

1658. Cheung, E.Y.; McCabe, E.E.; Harris, K.D.M.; Johnston, R.L.; Tedesco, E.; Raja, K.M.P.; Balaram, P. *Angew. Chem., Int. Ed.* **2002**, *41*, 494–496.

1659. Brenner, M; Seebach, D. *Helv. Chim. Acta* **2001**, *84*, 2155–2166.

1660. Rajashekhar, B.; Kaiser, E.T. *J. Biol. Chem.* **1986**, *261*, 13617–13623.

1661. Newkome, G.R.; Behera, R.K.; Moorefield, C.N.; Baker, G.R. *J. Org. Chem.* **1991**, *56*, 7162–7167.

1662. Cardona, C.M.; Alvarez, J.; Kaifer, A.E.; McCarley, T.D.; Pandey, S.; Baker, G.A.; Bonzagni, N.J.; Bright, F.V. *J. Am. Chem. Soc.* **2000**, *122*, 6139–6144.

1663. Sasaki, S.-i.; Hashizume, A.; Citterio, D.; Fujii, E.; Suzuki, K. *Tetrahedron Lett.* **2002**, *43*, 7243–7245.

1664. Kukhar, V.P.; Yagupolskii, Y.L.; Soloshonok, V.A. *Russ. Chem. Rev.* **1990**, *59*, 89–102.

1665. Yokomatsu, T.; Yuasa, Y.; Shibuya, S. *Heterocycles* **1992**, *33*, 1051–1078.

1666. Ordóñez, M.; Cativiela, C. *Tetrahedron: Asymmetry* **2007**, *18*, 3–99.

1667. Talbot, G.; Gaudry, R.; Berlinguet, L. *Can. J. Chem.* **1958**, *36*, 593–596.

1668. Adams, R.; Fles, D. *J. Am. Chem. Soc.* **1959**, *81*, 4946–4951.

1669. Tomita, M.; Nakashima, M. *Z. Physiol. Chem.* **1935**, *231*, 199–201.

1670. Cloke, J.B.; Stehr, E.; Steadman, T.R.; Westcott, L.C. *J. Am. Chem. Soc.* **1945**, *67*, 1587–1591.

1671. Tolman, V.; Veres, K. *Collect. Czech. Chem. Commun.* **1967**, *32*, 4460–4469.

1672. Castelhano, A.L.; Krantz, A. *J. Am. Chem. Soc.* **1987**, *109*, 3491–3493.

1673. Allan, R.D. *Aust. J. Chem.* **1979**, *32*, 2507–2516.

1674. Allan, R.D.; Twitchin, B. *Aust. J. Chem.* **1978**, *31*, 2283–2289.

1675. Greenstein, J.P. *J. Biol. Chem.* **1935**, *109*, 529–540.

1676. Allan, R.D.; Johnston, G.A.R.; Kazlauskas, R.; Tran, H.W. *J. Chem. Soc., Perkin Trans. 1* **1983**, 2983–2985.

1677. Nakanishi, S.; Okamoto, K.; Yamaguchi, H.; Takata, T. *Synthesis* **1998**, 1735–1741.

1678. Fischer, E.; Groh, R. *Ann.* **1911**, *383*, 363–372.

1679. Osterberg, A.E. *J. Am. Chem. Soc.* **1927**, *49*, 538–540.

1680. Khan, N.H.; Siddiqui, A.A.; Kidwai, A.R. *Ind. J. Chem.* **1977**, *15B*, 573–574.

1681. Jacobsen, K.A.; Marr-Leisy, D.; Rosenkranz, R.P.; Verlander, M.S.; Melmon, K.L.; Goodman, M. *J. Med. Chem.* **1983**, *26*, 492–499.

1682. Castelhano, A.L.; Krantz, A. *J. Am. Chem. Soc.* **1984**, *106*, 1877–1879.

1683. Avetisyan, S.A.; Kocharov, S.L.; Azaryan, L.V. *Russ. J. Org. Chem.* **1996**, *32*, 1667–1671.

1684. David, D.M.; O'Meara, G.W.; Pyne, S.G. *Tetrahedron Lett.* **1996**, *37*, 5417–5420.

1685. Öhler, E.; Kotzinger, S. *Liebigs Ann. Chem.* **1993**, 269–280.

1686. Deardorff, D.R.; Taniguchi, C.M.; Tafti, S.A.; Kim, H.Y.; Choi, S.Y.; Downey, K.J.; Nguyen, T.V. *J. Org. Chem.* **2001**, *66*, 7191–7194.

1687. Carter, H.E.; Bhattacharyya, P.K. *J. Am. Chem. Soc.* **1953**, *75*, 2503–2504.

1688. Chan, P.C.M.; Roon, R.J.; Koerner, J.F.; Taylor, N.J.; Honek, J.F. *J. Med. Chem.* **1995**, *38*, 4433–4438.

1689. Takeda, H.; Hosokawa, S.; Aburatani, M.; Achiwa, K. *Synlett* **1991**, 193–194.

1690. Serfass, L.; Casara, P.J. *Biorg. Med. Chem. Lett.* **1998**, *8*, 2599–2602.

1691. Andruszkiewicz, R.; Silverman, R.B. *Synthesis* **1989**, 953–955.

1692. Roller, S.; Siegers, C.; Haag, R. *Tetrahedron* **2004**, *60*, 8711–8720.

1693. Newkome, G.R.; Weis, C.D. *Org. Proc. Prep. Int.* **1996**, *28*, 495–498.

1694. Felluga, F.; Gombac, V.; Pitacco, G.; Valentin, E. *Tetrahedron: Asymmetry* **2005**, *16*, 1341–1345.

1695. Hoogenraad, M.; van der Linden, J.B.; Smith, A.A.; Hughes, B.; Derrick, A.M.; Harris, L.J.; Higginson, P.D.; Pettman, A.J. *Org. Proc. Res. Devel.* **2004**, *8*, 469–476.

1696. Masson, G.; Cividino, P.; Py, S.; Vallée, Y. *Angew. Chem., Int. Ed.* **2003**, *42*, 2265–2268.

1697. Lu, Y.; Miet, C.; Kunesch, N.; Poisson, J. *Tetrahedron: Asymmetry* **1990**, *1*, 707–710.

1698. Yamamoto, Y.; Schmid, M. *J. Chem. Soc., Chem. Commun.* **1989**, 1310–1312.

1699. Denis, J.-N.; Tchertchian, S.; Tomassini, A.; Vallée, Y. *Tetrahedron Lett.* **1997**, *38*, 5503–5506.

1700. Buchholz, B.; Stamm, H. *Chem. Ber.* **1987**, *120*, 1239–1244.

1701. Calmès, M.; Escale, F.; Martinez, J. *Tetrahedron: Asymmetry* **2002**, *13*, 293–296.

1702. Camps, P.; Muñoz-Terrero, D.; Sánchez, L. *Tetrahedron: Asymmetry* **2004**, *15*, 311–321.

1703. Royer, F.; Felpin, F.-X.; Doris, E. *J. Org. Chem.* **2001**, *66*, 6487–6489.

1704. Lambrecht, R.H.B., Slegers, G.; Mannens, G.; Claeys, A. *J. Lab. Cmpds. Radiopharm.* **1985**, *23*, 1114–1115.

1705. Greenaway, W.; Whatley, F.R.; *J. Lab. Cmpds.* **1975**, *11*, 395–400.

1706. Detsi, A.; Gavrielatos, E.; Adam, M.-A.; Igglessi-Markopoulou, O.; Markopoulos, J.; Theologitis, M.; Reis, H.; Papadopoulos, M. *Eur. J. Org. Chem.* **2001**, 4337–4342.

1707. Bang, J.K.; Hasegawa, K.; Kawakami, T.; Aimoto, S.; Akaji, K. *Tetrahedron Lett.* **2004**, *45*, 99–102.

1708. Meyer, O.; Becht, J.-M.; Helmchen, G. *Synlett* **2003**, 1539–1541.

1709. Malik, S.; Wyatt, P.B. *Synth. Commun.* **1993**, *23*, 1047–1051.

1710. Katsuki, T.; Yamaguchi, M. *Bull. Chem. Soc. Jpn.* **1976**, *49*, 3287–3290.

1711. Stojanovic, A.; Renaud, P. *Synlett* **1997**, 181–182.

1712. Gibbs, G.; Hateley, M.J.; McLaren, L.; Welham, M.; Willis, C.L. *Tetrahedron Lett.* **1999**, *40*, 1069–1072.

1713. Bentley, J.M.; Wadsworth, H.J.; Willis, C.L. *J. Chem. Soc., Chem. Commun.* **1995**, 231–232.

1714. Scheinmann, F.; Stachulski, A.V. *J. Chem. Res. S* **1993**, 414–415.

1715. Kollonitsch, J.; Marburg, S.; Perkins, L.M. *J. Org. Chem.* **1979**, *44*, 771–777.

1716. Ahern, D.G.; Laseter, A.G.; Filer, C.N. *Synth. Commun.* **2003**, *33*, 3327–3330.

1717. Fukagawa, T. *Z. Physiol. Chem.* **1935**, *231*, 202–204.

1718. Gavagan, J.E.; Fager, S.K.; Fallon, R.D.; Folsom, P.W.; Herkes, F.E.; Eisenberg, A.; Hann, E.C.; DiCosimo, R. *J. Org. Chem.* **1998**, *63*, 4792–4801.

1719. Fuganti, C.; Grasselli, P.; Seneci, P.F.; Servi, S.; Casati, P. *Tetrahedron Lett.* **1986**, *27*, 2061–2062.

1720. Margolin, A.L. *Tetrahedron Lett.* **1993**, *34*, 1239–1242.

1721. Rossiter, B.E.; Sharpless, K.B. *J. Org. Chem.* **1984**, *49*, 3707–3711.

1722. Bianchi, D.; Cabri, W.; Cesti, P.; Francalanci, F.; Ricci, M. *J. Org. Chem.* **1988**, *53*, 104–107.

1723. Craven, A.P.; Dyke, H.J.; Thomas, E.J. *Tetrahedron* **1989**, *45*, 2417–2429.

1724. Kobayashi, S.; Kobayashi, K.; Hirai, K. *Synlett* **1999**, 909–912.

1725. Bongini, A.; Cardillo, G.; Orena, M.; Porzi, G.; Sandri, S. *Tetrahedron* **1987**, *43*, 4377–4383.

1726. Mazzini, C.; Lebreton, J.; Alphand, V.; Furstoss, R. *Tetrahedron Lett.* **1997**, *38*, 1195–1196.

1727. Fray, M.J.; Bull, D.J.; James, K. *Synlett* **1992**, 709–710.

1728. Kabat, M.M.; Daniewski, A.R.; Burger, W. *Tetrahedron: Asymmetry* **1997**, *8*, 2663–2665.

1729. Chisholm, M.H.; Corning, J.F.; Huffman, J.C. *J. Am. Chem. Soc.* **1983**, *105*, 5925–5926.

1730. Kitamura, M.; Ohkuma, T.; Takaya, H.; Noyori, R. *Tetrahedron Lett.* **1988**, *29*, 1555–1556.

1731. Gopalan, A.S.; Sih, C.J. *Tetrahedron Lett.* **1984**, *25*, 5235–5238.

1732. Deardorff, D.R.; Taniguchi, C.M.; Nelson, A.C.; Pace, A.P.; Kim, A.J.; Pace, A.K.; Jones, R.A.; Tafti, S.A.; Nguyen, C.; O'Connor, C.; Tang, J.; Chen, J. *Tetrahedron: Asymmetry* **2005**, *16*, 1655–1661.

1733. Mulzer, J.; Funk, G. *Synthesis*, **1995**, 101–112.

1734. Puertas, S.; Rebolledo, F.; Gotor, V. *J. Org. Chem.* **1996**, *61*, 6024–6027.

1735. Andruszkiewicz, R.; Barrett, A.G.M.; Silverman, R.B. *Synth. Commun.* **1990**, *20*, 159–166.

1736. Chênevert, R.; Desjardins, M. *Tetrahedron Lett.* **1991**, *32*, 4249–4250.

1737. Kanazawa, A.; Gillet, S.; Delair, P.; Greene, A.E. *J. Org. Chem.* **1998**, *63*, 4660–4663.

1738. Nebois, P.; Greene, A.E. *J. Org. Chem.* **1996**, *61*, 5210–5211.

1739. Bringmann, G.; Künkel, G.; Geuder, T. *Synlett* **1990**, 253–255.

1740. Catasús, M.; Myano, A.; Pericàs, M.A.; Riera, A. *Tetrahedron Lett.* **1999**, *40*, 9309–9312.

1741. Marzi, M.; Minetti, P.; Moretti, G.; Tinti, M.O.; De Angelis, F. *J. Org. Chem.* **2000**, *65*, 6766–6769.

1742. Takano, S.; Yanase, M.; Sekiguchi, Y.; Ogasawara, K. *Tetrahedron Lett.* **1987**, *28*, 1783–1784.

1743. Castejón, P.; Moyano, A.; Pericàs, M.A.; Riera, A. *Tetrahedron* **1996**, *52*, 7063–7086.

1744. Bertelli, L.; Fiaschi, R.; Napolitano, E. *Gazz. Chim. Ital.* **1993**, *123*, 521–524.

1745. Bessodes, M.; Saïah, M.; Antonakis, K. *J. Org. Chem.* **1992**, *57*, 4441–4444.

1746. Saïah, M.; Bessodes, M.; Antonakis, K. *Tetrahedron: Asymmetry* **1991**, *2*, 111–112.

1747. Kolb, H.C.; Bennani, Y.L.; Sharpless, K.B. *Tetrahedron: Asymmetry* **1993**, *4*, 133–141.

1748. Wang, M.-X.; Zhao, S.-M. *Tetrahedron Lett.* **2002**, *43*, 6617–6620.

1749. Azam, S.; D'Souza, A.A.; Wyatt, P.B. *J. Chem. Soc., Perkin Trans. 1* **1996**, 621–627.

1750. Brenner, M.; Seebach, D. *Helv. Chim. Acta* **1999**, *82*, 2365–2379.

1751. Calderari, G.; Seebach, D. *Helv. Chim. Acta* **1985**, *68*, 1592–1604.

1752. Meisterhans, C.; Linden, A.; Hesse, M. *Helv. Chim. Acta* **2003**, *86*, 644–656.

1753. Watanabe, M.; Ikagawa, A.; Wang, H.; Murata, K.; Ikariya, T. *J. Am. Chem. Soc.* **2004**, *126*, 11148–11149.

1754. Licandro, E.; Maiorana, S.; Baldoli, C.; Capella, L.; Perdicchia, D. *Tetrahedron: Asymmetry* **2000**, *11*, 975–980.

1755. Enders, D.; Reinhold, U. *Angew. Chem., Int. Ed. Engl.* **1995**, *34*, 1219–1222.

1756. Enders, D.; Reinhold, U. *Liebigs Ann.* **1996**, 11–26.

1757. Martin, C.J.; Rawson, D.J.; Williams, J.M.J. *Tetrahedron: Asymmetry* **1998**, *9*, 3723–3730.

1758. Wee, A.G.H.; Liu, B.; McLeod, D.D. *J. Org. Chem.* **1998**, *63*, 4218–4227.

1759. Anada, M.; Hashimoto, S. *Tetrahedron Lett.* **1998**, *39*, 79–82.

1760. Morita, T.; Matsunaga, H.; Sugiyama, E.; Ishizuka, T.; Kunieda, T. *Tetrahedron Lett.* **1998**, *39*, 7131–7134.

1761. Yamamoto, T.; Ishibuchi, S.; Ishizuka, T.; Haratake, M.; Kunieda, T. *J. Org. Chem.* **1993**, *58*, 1997–1998.

1762. Itoh, K.; Kanemasa, S. *J. Am. Chem. Soc.* **2002**, *124*, 13394–13395.

1763. Ooi, T.; Fujioka, S.; Maruoka, K. *J. Am. Chem. Soc.* **2004**, *126*, 11790–11791.

1764. Moglioni, A.G.; Brousse, B.N.; Álvarez-Larena, A.; Moltrasio, G.Y.; Ortuño, R.M. *Tetrahedron: Asymmetry* **2002**, *13*, 451–454.

1765. Masson, G.; Zegida, W.; Cividino, P.; Py, S.; Vellée, Y. *Synlett* **2003**, 1527–1529.

1766. Johannesen, S.A.; Albu, S.; Hazell, R.G.; Skrydstrup, T. *Chem. Commun.* **2004**, 1962–1963.

1767. Bravo, P.; Corradi, E.; Pesenti, C.; Vergani, B.; Viani, F.; Volonterio, A.; Zanda, M. *Tetrahedron: Asymmetry* **1998**, *9*, 3731–3735.

1768. Pesenti, C.; Bravo, P.; Corradi, E.; Frigerio, M.; Meille, S.V.; Panzeri, W.; Viani, F.; Zanda, M. *J. Org. Chem.* **2001**, *66*, 5637–5640.

1769. Okamoto, S.; Teng, T.; Fujii, S.; Takayama, Y.; Sato, F. *J. Am. Chem. Soc.* **2001**, *123*, 3462–3471.

1770. Okamoto, S.; Fukuhara, K.; Sato, F. *Tetrahedron Lett.* **2000**, *41*, 5561–5565.

1771. Fernandes, R.A.; Yamamoto, Y. *J. Org. Chem.* **2004**, *69*, 3562–3564.

1772. Gennari, C.; Moresca, D.; Vulpetti, A.; Pain, G. *Tetrahedron* **1997**, *53*, 5593–5608.

1773. Liu, W.-S.; Glover, G.I. *J. Org. Chem.* **1978**, *43*, 754–755.

1774. Rittle, K.E.; Homnick, C.F.; Ponticello, G.S.; Evans, B.E. *J. Org. Chem.* **1982**, *47*, 3016–3018.

1775. Rich, D.H.; Sun, E.T.; Boparai, A.S. *J. Org. Chem.* **1978**, *43*, 3624–3626.

1776. Steulmann, R.; Klostermeyer, H. *Liebigs Ann. Chem.* **1975**, 2245–2250.

1777. Rich, D.E.; Sun, E.T.O. *J. Med. Chem.* **1980**, *23*, 27–33.

1778. Czajgucki, Z.; Sowinski, P.; Andruszkiewicz, R. *Amino Acids* **2003**, *24*, 289–291.

1779. Piveteau, N.; Audin, P.; Pari, J. *Synlett* **1997**, 1269–1270.

1780. Takemoto, Y.; Matsumoto, T.; Ito, Y.; Terashima, S. *Tetrahedron Lett.* **1990** *31*, 217–218.

1781. Mikami, K.; Kaneko, M.; Loh, T.-P.; Terada, M.; Nakai, T. *Tetrahedron Lett.* **1990**, *31*, 3909–3912.

1782. Reetz, M.T.; Fox, D.N.A. *Tetrahedron Lett.* **1993**, *34*, 1119–1122.

1783. Prasad, J.V.N.V.; Rich, D.H. *Tetrahedron Lett.* **1990**, *31*, 1803–1806.

1784. Kano, S.; Yokomatsu, T.; Iwasawa, H.; Shibuya, S. *Tetrahedron Lett.* **1987**, *28*, 6331–6334.

1785. Cook, G.R.; Shanker, P.S. *Tetrahedron Lett.* **1998**, *39*, 3405–3408.

1786. Wuts, P.G.M.; Putt, S.R. *Synthesis* **1989**, 951–953.

1787. Braun, M.; Waldmüller, D. *Synthesis* **1989**, 856–858.

1788. Devant, R.M.; Radunz, H.-E. *Tetrahedron Lett.* **1988**, *29*, 2307–2310.

1789. Woo, P.W.K. *Tetrahedron Lett.* **1985**, *26*, 2973–2976.

1790. Cooke, J.W.B.; Davies, S.G.; Naylor, A. *Tetrahedron* **1993**, *49*, 7955–7966.

1791. Viacrio, J.L.; Rodriguez, M.; Badía, D.; Carrillo, L.; Reyes, E. *Org. Lett.* **2004**, *6*, 3171–3174.

1792. Boger, D.L.; Menezes, R.F. *J. Org. Chem.* **1992**, *57*, 4331–4333.

1793. Gennari, C.; Pain, G.; Moresca, D. *J. Org. Chem.* **1995**, *60*, 6248–6249.

1794. Reetz, M.T.; Rivadeneira, E.; Niemeyer, C. *Tetrahedron Lett.* **1990**, *31*, 3863–3866.

1795. Narita, M.; Otsuka, M.; Kobayashi, S.; Ohno, M.; Umezawa, Y.; Morishima, H.; Saito, S.-i.; Takita, T.; Umezawa, H. *Tetrahedron Lett.* **1982**, *23*, 525–528.

1796. Barrett, A.G.M.; Edmunds, J.J.; Hendrix, J.A.; Malecha, J.W.; Parkinson, C.J. *J. Chem. Soc., Chem. Commun.* **1992**, 1240–1242.

1797. Williams, R.M.; Colson, P.-J.; Zhai, W. *Tetrahedron Lett.* **1994**, *35*, 9371–9374.

1798. Aoyagi, Y.; Williams, R.M. *Tetrahedron* **1998**, *54*, 10419–10433.

1799. Chakraborty, T.K.; Ghosh, A.; Kumar, S.K.; Kunwar, A.C. *J. Org. Chem.* **2003**, *68*, 6459–6462.

1800. Barco, A.; Benetti, S.; Casolari, A.; Pollini, G.P.; Spalluto, G. *Tetrahedron Lett.* **1990**, *31*, 4917–4920.

1801. Bryans, J.S.; Large, J.M.; Parsons, A.F. *Tetrahedron Lett.* **1999**, *40*, 3487–3490.

1802. Rein, T.; Kreuder, R.; von Zezschwitz, P.; Wulff, C.; Reiser, O. *Angew. Chem., Int. Ed. Engl.* **1995**, *34*, 1023–1025.

1803. Misiti, D.; Zappia, G.; Monache, G.D. *Synthesis* **1999**, 873–877.

1804. Wei, Z.-Y.; Knaus, E.E. *Tetrahedron* **1994**, *50*, 5569–5578.

1805. Wei, Z.-Y.; Knaus, E.E. *Synlett* **1993**, 295–296.

1806. Yokokawa, F.; Hamada, Y.; Shioiri, T. *Synlett* **1992**, 703–705.

1807. Reetz, M.T.; Röhrig, D. *Angew. Chem., Int. Ed. Engl.* **1989**, *28*, 1706–1709.

1808. Misiti, D.; Zappia, G. *Tetrahedron Lett.* **1990**, *31*, 7359–7362.

1809. Sakaitani, M.; Ohfune, Y. *J. Am. Chem. Soc.* **1990**, *112*, 1150–1158.

1810. Sakaitani, M.; Ohfune, Y. *Tetrahedron Lett.* **1987**, *28*, 3987–3990.

1811. Reetz, M.T.; Griebenow, N.; Goddard, R. *J. Chem. Soc., Chem. Commun.* **1995**, 1605–1606.

1812. Palomo, C.; Miranda, J.I.; Linden, A. *J. Org. Chem.* **1996**, *61*, 9196–9201.

1813. Palomo, C.; Miranda, J.I.; Cuevas, C.; Odriolzola, J.M. *J. Chem. Soc., Chem. Commun.* **1995**, 1735–1736.

1814. Midland, M.M.; Afonso, M.M. *J. Am. Chem. Soc.* **1989**, *111*, 4368–4371.

1815. Reetz, M.T.; Strack, T.J.; Kanand, J.; Goddard, R. *Chem. Commun.*, **1996**, 733–734.

1816. Reginato, G.; Mordini, A.; Capperucci, A.; Degl'Innocenti, A.; Manganiello, S. *Tetrahedron* **1998**, *54*, 10217–10226.

1817. Trost, B.M.; Roth, G.J. *Org. Lett.* **1999**, *1*, 67–70.

1818. Tseng, C.C.; Terashima, S.; Yamada, S.-i. *Chem. Pharm. Bull.* **1977**, *25*, 29–40.

1819. Trost, B.M.; Bunt, R.C.; Lemoine, R.C.; Calkins, T.L. *J. Am. Chem. Soc.* **2000**, *122*, 5968–5976.

1820. Sutherland, A.; Willis, C.L. *J. Org. Chem.* **1998**, *63*, 7764–7769.

1821. Nishi, T.; Kitamura, M.; Ohkuma, T.; Noyori, R. *Tetrahedron Lett.* **1988**, *29*, 6327–6330.

1822. Lloyd-Williams, P.; Monerris, P.; Gonzalez, I.; Jou, G.; Giralt, E. *J. Chem. Soc., Perkin Trans. 1* **1994**, 1974.

1823. Harris, B.D.; Joullié, M.M. *Tetrahedron* **1988**, *44*, 3489–3500.

1824. Harris, B.D.; Bhat, K.L.; Joullie, M.M. *Tetrahedron Lett.* **1987**, *28*, 2837–2840.

1825. Kazmaier, U.; Krebs, A. *Tetrahedron Lett.* **1999**, *40*, 479–482.

1826. Bösche, U.; Nubbemeyer, U. *Tetrahedron* **1999**, *55*, 6883–6904.

1827. Hoffman, R.V.; Tao, J. *J. Org. Chem.* **1997**, *62*, 2292–2297.

1828. Maibaum, J.; Rich, D.H. *J. Org. Chem.* **1988**, *53*, 869–873.

1829. Reetz, M.T.; Drewes, M.W.; Matthews, B.R.; Lennick, K. *J. Chem. Soc., Chem. Commun.* **1989**, 1474–1475.

1830. Moyer, M.P.; Feldman, P.L.; Rapoport, H. *J. Org. Chem.* **1985**, *50*, 5223–5230.

1831. Nudelman, A.; Marcovici-Mizrahi, D.; Nudelman, A.; Flint, D.; Wittenbach, V. *Tetrahedron* **2004**, *60*, 1731–1748.

1832. Shaw, K.J.; Luly, J.R.; Rapoport, H. *J. Org. Chem.* **1985**, *50*, 4515–4523.

1833. Schuda, P.F.; Greenlee, W.J.; Chakravarty, P.K.; Eskola, P. *J. Org. Chem.* **1988**, *53*, 873–875.

1834. Bonini, B.F.; Comes-Franchini, M.; Fochi, M.; Laboroi, F.; Mazzanti, G.; Ricci, A.; Varchi, G. *J. Org. Chem.* **1999**, *64*, 8008–8013.

1835. Schmidt, U.; Kroner, M.; Griesser, H. *Synthesis* **1989**, 832–835.

1836. Kessler, H.; Schudok, M. *Synthesis* **1990**, 457–458.

1837. Paris, M.; Fehrentz, J.-A.; Heitz, A.; Loffet, A.; Martinez, J. *Tetrahedron Lett.* **1996**, *37*, 8489–8492.

1838. Paris, M.; Fehrentz, J-A.; Heitz, A.; Martinez, J. *Tetrahedron Lett.* **1998**, *39*, 1569–1572.

1839. Joiun, P.; Poncet, J.; Dufour, M.-N.; Maugras, I.; Pantaloni, A.; Castro, B. *Tetrahedron Lett.* **1988**, *29*, 2661–2664.

1840. Reddy, G.V.; Rao, G.V.; Iyengar, D.S. *Tetrahedron Lett.* **1999**, *40*, 775–776.

1841. Smreina, M.; Majer, P.; Majerová, E.; Guerassina, T.A.; Eissenstat, M.A. *Tetrahedron*, **1997**, *53*, 12867–12874.

1842. Jouin, P.; Castro, B.; Nisato, D. *J. Chem. Soc., Perkin Trans I* **1987**, 1177–1182.

1843. Ma, D.; Ma, J.; Ding, W.; Dai, L. *Tetrahedron: Asymmetry* **1996**, *7*, 2365–2370.

1844. Klutchko, S.; O'Brien, P.; Hodges, J.C. *Synth. Commun.* **1989**, *19*, 2573–2583.

1845. Aitken, R.A.; Karodia, N. *Chem. Commun.*, **1996**, 2079–2080.

1846. Aitken, R.A.; Karodia, N.; Massil, T.; Young, R.J. *J. Chem. Soc., Perkin Trans. 1* **2002**, 533–541.

1847. Barlos, K.; Mamos, P.; Papaioannou, D.; Patrianakou, S. *J. Chem. Soc., Chem. Commun.* **1987**, 1583–1584.

1848. Shimamoto, K.; Ishida, M.; Shinozaki, H.; Ohfune, Y. *J. Org. Chem.* **1991**, *56*, 4167–4176.

1849. Wei, Z.-Y.; Knaus, E.E. *J. Org. Chem.* **1993**, *58*, 1586–1588.

1850. Kokotos, G.; Mrkidis, T.; Constantinou-Kokotou, V. *Synthesis* **1996**, 1223–1226.

1851. Silverman, R.B.; Levy, M.A. *J. Org. Chem.* **1980**, *45*, 815–818.

1852. Yee, N.K.; Dong, Y.; Kapadia, S.R.; Song, J.J. *J. Org. Chem.* **2002**, *67*, 8688–8691.

1853. Somfai, P.; He, H.M.; Tanner, D. *Tetrahedron Lett.* **1991**, *32*, 283–286.

1854. Ikota, N. *Heterocycles* **1989**, *29*, 1469–1472.

1855. Wei, Z.Y.; Knaus, E.E. *Synlett* **1994**, 345–346.

1856. Kwon, T.W.; Keusenkothen, P.F.; Smith, M.B. *J. Org. Chem.* **1992**, *57*, 6169–6173.

1857. McAlonan, H.; Stevenson, P.J. *Tetrahedron: Asymmetry* **1995**, *6*, 239–244.

1858. Yoneta, T.; Shibahara, S.; Fukatsu, S.; Seki, S. *Bull. Chem. Soc. Jpn.* **1978**, *51*, 3296–3297.

1859. Renaud, P.; Seebach, D. *Synthesis* **1986**, 424–426.

1860. Yoshifuji, S.; Kaname, M. *Chem. Pharm. Bull.* **1995**, *43*, 1302–1306.

1861. Ha, D.-C.; Kang, S.; Chung, C.-M.; Lim, H.-K. *Tetrahedron Lett.* **1998**, *39*, 7541–7544.

1862. Iriuchijima, S.; Ogawa, M. *Synthesis* **1982**, 41–42.

1863. Hanessian, S.; Schaum, R. *Tetrahedron Lett.* **1997**, 38, 163–166.

1864. Rottmann, A.; Liebscher, J. *Tetrahedron: Asymmetry* **1997**, *8*, 2433–2448.

1865. Silverman, R.B.; Invergo, B.J.; Mathew, J. *J. Med. Chem.* **1986**, *29*, 1840–1846.

1866. Rassu, G.; Pinna, L.; Spanu, P.; Ulgheri, F.; Cornia, M.; Zanardi, F.; Casiraghi, G. *Tetrahedron* **1993**, *49*, 6489–6496.

1867. Pichon, M.; Jullian, J.-C.; Figadère, B.; Cavé, A. *Tetrahedron Lett.* **1998**, *39*, 1755–1758.

1868. Le Coz, S.; Mann, A. *Synth. Commun.* **1993**, *23*, 165–171.

1869. Reetz, M.T.; Strack, T.J.; Mutulis, F.; Goddard, R. *Tetrahedron Lett.* **1996**, *37*, 9293–9296.

1870. Huang, S.-B.; Nelson, J.S.; Weller, D.D. *J. Org. Chem.* **1991**, *56*, 6007–6018.

1871. Doi, T.; Kokubo, M.; Yamamoto, K.; Takahashi, T. *J. Org. Chem.* **1998**, *63*, 428–429.

1872. Camps, P.; Pérez, F.; Soldevilla, N. *Tetrahedron Lett.* **1999**, *40*, 6853–6856.

1873. Fuganti, C.; Grasselli, P. *Tetrahedron Lett.* **1985**, *26*, 101–104.

1874. Yue, T.-Y.; Nugent, W.A. *J. Am. Chem. Soc.* **2002**, *124*, 13692–13693.

1875. Misiti, D.; Zappia, G.; Delle Monache, G. *Gazz. Chim. Ital.* **1995**, *125*, 219–222.

1876. Rajashekhar, B.; Kaiser, E.T. *J. Org. Chem.* **1985**, *50*, 5480–5484.

1877. Bellamy, F.D.; Bondoux, M.; Dodey, P. *Tetrahedron Lett.* **1990**, *31*, 7323–7326.

1878. Bock, K.; Lundt, I.; Pedersen, C. *Acta Chem. Scand. B* **1983**, *37*, 341–344.

1879. Jung, M.E.; Shaw, T.J. *J. Am. Chem. Soc.* **1980**, *102*, 6304–6311.

1880. Lohray, B.B.; Reddy, A.S.; Bhushan, V. *Tetrahedron: Asymmetry* **1996**, *7*, 2411–2416.

1881. Bols, M.; Lundt, I.; Pedersen, C. *Tetrahedron* **1992**, *48*, 319–324.

1882. Wang, G.; Hollingsworth, R.I. *Tetrahedron: Asymmetry* **1999**, *10*, 1895–1901.

1883. Bernardi, A.; Micheli, F.; Potenza, D.; Scolastico, C.; Villa, R. *Tetrahedron Lett.* **1990**, *31*, 4949–4952.

1884. Koot, W.-J.; van Ginkel, R.; Kranenburg, M.; Hiemstra, H.; Louwrier, S.; Moolenaar, M.J.; Speckamp, W.N. *Tetrahedron Lett.* **1991**, *32*, 401–404.

1885. Ohta, T.; Shiokawa, S.; Sakamoto, R.; Nozoe, S. *Tetrahedron Lett.* **1990**, *31*, 7329–7332.

1886. Huang, P.Q.; Wang, S.L.; Ye, J.L.; Ruan, Y.P.; Huang, Y.Q.; Zheng, H.; Gao, J.X. *Tetrahedron* **1998**, *54*, 12547–12560.

1887. Kano, S.; Yuasa, Y.; Yokomatsu, T.; Shibuya, S. *J. Org. Chem.* **1988**, *53*, 3865–3868.

1888. Yanagisawa, H.; Kanazaki, T.; Nishi, T. *Chem. Lett.* **1989**, 687–690.

1889. Mulzer, J.; Büttelmann, B.; Münch, W. *Liebigs Ann. Chem.* **1988**, 445–448.

1890. Shinozaki, K.; Mizuno, K.; Oda, H.; Masaki, Y. *Bull. Chem. Soc. Jpn.* **1996**, *69*, 1737–1745.

1891. Shinozaki, K.; Mizuno, K.; Oda, H.; Masaki, Y. *Chem. Lett.* **1992**, 2265–2268.

1892. Tulshian, D.B.; Gundes, A.F.; Czarniecki, M. *Bioorg. Med. Chem. Lett.* **1992**, *2*, 515–518.

1893. Friestad, G.K.; Marié, J.-C.; Deveau, A.M. *Org. Lett.* **2004**, *6*, 3249–3252.

1894. Ishiuka, T.; Ishibuchi, S.; Kunieda, T. *Tetrahedron* **1993**, *49*, 1841–1852.

1895. Ishibuchi, S.; Ikematsu, Y.; Ishizuka, T.; Kunieda, T. *Tetrahedron Lett.* **1991**, *32*, 3523–3526.

1896. Takahata, H.; Yamazaki, K.; Takamatsu, T.; Yamazaki, T.; Momose, T. *J. Org. Chem.* **1990**, *55*, 3947–3950.

1897. Aubé, J.; Wang, Y.; Ghosh, S.; Langhans, K.L. *Synth. Commun.* **1991**, *21*, 693–701.

1898. Sakagami, H.; Kamikubo, T.; Ogasawara, K. *Synlett* **1997**, 221–222.

1899. Pellegata, R.; Dosi, I.; Villa, M.; Lesma, G.; Palmisano, G. *Tetrahedron* **1985**, *41*, 5607–5613.

1900. Ohfune, Y.; Nishio, H. *Tetrahedron Lett.* **1984**, *25*, 4133–4136.

1901. Li, W.-R.; Han, S.-Y.; Joullié, M.M. *Tetrahedron* **1993**, *49*, 785–802.

1902. Ewing, W.R.; Harris, B.D.; Bhat, K.L.; Joullie, M.M. *Tetrahedron* **1986**, *42*, 2421–2428.

1903. Ewing, W.R.; Joullié, M.M. *Heterocycles* **1988**, *27*, 2843–2850.

1904. Kakinuma, K.; Otake, N.; Yonehara, H. *Tetrahedron Lett.* **1980**, *21*, 167–168.

1905. Kogen, H.; Kadokawa, H.; Kurabayashi, M. *J. Chem. Soc., Chem. Commun.* **1990**, 1240–1241.

1906. Mulzer, J.; Meier, A.; Buschmann, J.; Luger, P. *J. Org. Chem.* **1996**, *61*, 566–572.

1907. Flögel, O.; Amombo, M.G.O.; Reibig, H.-U.; Zahn, G.; Brüdgam, I.; Hartl, H. *Chem.—Eur. J.* **2003**, *9*, 1405–1415.

1908. Otake, N.; Furihata, K.; Kakinuma, K.; Yonehara, H. *J. Antibiotics* **1974**, *27*, 484–486.

1909. Adlington, R.M.; Baldwin, J.E.; Jones, R.H.; Murphy, J.A.; Parisi, M.F. *J. Chem. Soc., Chem. Commun.* **1983**, 1479–1481.

1910. Pruess, D.L.; Scannell, J.P.; Blount, J.F.; Ax, H.A.; Kellett, M.; Williams, T.H.; Stempel, A. *J. Antibiotics* **1974**, *27*, 754–759.

1911. Chênevert, R.; Lavoie, M.; Courchesne, G.; Martin, R. *Chem. Lett.* **1994**, 93–96.

1912. Kobayashi, K.; Miyazawa, S.; Terahara, A.; Mishima, H.; Kurihara, H. *Tetrahedron Lett.* **1976**, 537–540.

1913. Rando, R.R.; Bangerter, F.W. *J. Am. Chem. Soc.* **1976**, *98*, 6762–6764.

1914. Rando, R.R.; Bangerter, F.W. *J. Am. Chem. Soc.* **1977**, *99*, 5141–5145.

1915. Rando, R.R. *Biochemistry* **1977**, *16*, 4604–4610.

1916. Fu, M.; Nikolic, D.; Van Breeman, R.B.; Silverman, R.B. *J. Am. Chem. Soc.* **1999**, *121*, 7751–7759.

1917. Grehn, L.; Ragnarsson, U. *J. Org. Chem.* **1981**, *46*, 3492–3497.

1918. Vázquez, E.; Caamaño, A.M.; Castedo, L.; Mascareñas, J.L. *Tetrahedron Lett.* **1999**, *40*, 3621–3624.

1919. Nishiwaki, E.; Tanaka, S.; Lee, H.; Shibuya, M. *Heterocycles* **1988**, *27*, 1945–1952.

1920. Dervan, P.B. *Science* **1986**, *232*, 464–471.

1921. Taylor, J.S.; Schultz, P.G.; Dervan, P.B. *Tetrahedron* **1984**, *40*, 457–465.

1922. Schultz, P.G.; Taylor, J.S.; Dervan, P.B. *J. Am. Chem. Soc.* **1982**, *104*, 6861–6863.

1923. Yeung, B.K.S.; Boger, D.L. *J. Org. Chem.* **2003**, *68*, 5249–5253.

1924. Grehn, L.; Ding, L.; Ragnarsson, U. *Acta Chem. Scand. B* **1990**, *44*, 67–74.

1925. Krowicki, K.; Lown, J.W. *J. Org. Chem.* **1987**, *52*, 3493–3501.

1926. Grehn, L.; Ding, L.; Ragnarsson, U. *Acta Chem. Scand. B* **1990**, *44*, 75–81.

1927. Baraldi, P.G.; Cozzi, P.; Geroni, C.; Mongelli, N.; Romagnoli, R.; Spalluto, G. *Biorg. Med. Chem.* **1999**, *7*, 251–262.

1928. Sauter, G.; Leumann, C. *Helv. Chim. Acta* **1998**, *81*, 916–931.

1929. Sauter, G.; Stulz, E.; Leumann, C. *Helv. Chim. Acta* **1998**, *81*, 14–34.

1930. Kazuta, Y.; Shuto, S.; Matsuda, A. *Tetrahedron Lett.* **2000**, *41*, 5373–5377.

1931. Shuto, S.; Ono, S.; Imoto, H.; Yoshii, K.; Matsuda, A. *J. Med. Chem.* **1998**, *41*, 3507–3514.

1932. Dinh, T.Q.; Smith, C.D.; Armstrong, R.W. *J. Org. Chem.* **1997**, *62*, 790–791.

1933. Combs, A.P.; Kapoor, T.M.; Feng, S.; Chen, J.K.; Daudé-Snow, L.F.; Schreiber, S.L. *J. Am. Chem. Soc.* **1996**, *118*, 287–288.

1934. Williams, M.A.; Lew, W.; Mendel, D.B.; Tai, C.Y.; Escarpe, P.A.; Laver, W.G.; Stevens, R.C.; Kim, C.U. *Bioorg. Med. Chem. Lett.* **1997**, *7*, 1837–1842.

1935. Lew, W.; Williams, M.A.; Mendel, D.B.; Escarpe, P.A.; Kim, C.U. *Bioorg. Med. Chem. Lett.* **1997**, *7*, 1843–1846.

1936. Kim, C.U.; Lew, W.; Williams, M.A.; Wu, H.; Zhang, L.; Chen, X.; Escarpe, P.A.; Mendel, D.B.; Laver, W.G.; Stevens, R.C. *J. Med. Chem.* **1998**, *41*, 2451–2460.

1937. Storici, P.; Qiu, J.; Schirmer, T.; Silverman, R.B. *Biochemistry* **2004**, *43*, 14057–14063.

1938. Duke, R.K.; Allan, R.D.; Chebib, M.; Greenwood, J.R.; Johnston, G.A.R. *Tetrahedron: Asymmetry* **1998**, *9*, 2533–2548.

1939. Allan, R.D.; Twitchin, B. *Aust. J. Chem.* **1980**, *33*, 599–604.

1940. Allan, R.D.; Fong, J. *Aust. J. Chem.* **1986**, *39*, 855–864.

1941. Nielsen, L.; Brehm, L.; Krogsgaard-Larsen, P. *J. Med. Chem.* **1990**, *33*, 71–77.

1942. Aravinda, S.; Ananda, K.; Shamala, N.; Balaram, P. *Chem.—Eur. J.* **2003**, *9*, 4789–4795.

1943. Woll, M.G.; Lai, J.R.; Guzei, I.A.; Taylor, S.J.C.; Smith, M.E.B.; Gellman, S.H. *J. Am. Chem. Soc.* **2001**, *123*, 11077–11078.

1944. Tsai, J.H.; Waldman, A.S.; Nowick, J.S. *Bioorg. Med. Chem.* **1999**, *7*, 29–38.

1945. Amorín, M.; Castedo, L.; Granja, J.R. *J. Am. Chem. Soc.* **2003**, *125*, 2844–2845.

1946. Amorín, M.; Castedo, L.; Granja, J.R. *Chem.—Eur. J.* **2005**, *11*, 6543–6551.

1947. Ishida, H.; Qi, Z.; Sokabe, M.; Donowaki, K.; Inoue, Y. *J. Org. Chem.* **2001**, *66*, 2978–2989.

1948. Pätzel, M.; Sanktjohanser, M.; Doss, A.; Henklein, P.; Szeimies, G. *Eur. J. Org. Chem.* **2004**, 493–498.

1949. Vázquez, E.; Caamaño, A.M.; Castedo, L.; Gramberg, D.; Mascareñas, J.L. *Tetrahedron Lett.* **1999**, *40*, 3625–36248.

1950. Avram, M.A.; Nenitzescu, C.D.; Maxim, M. *Ber.* **1957**, *90*, 1424–1432.

1951. Kennewell, P.D.; Matharu, S.S.; Taylor, J.B.; Westwood, R.; Sammes, P.G. *J. Chem. Soc. Perkin Trans. 1* **1982**, 2553–2562.

1952. Merchant, R.; Marvel, C.S. *J. Am. Chem. Soc.* **1928**, *50*, 1197–1201.

1953. Mineno, T.; Miller, M.J. *J. Org. Chem.* **2003**, *68*, 6591–6596.

1954. Coldham, I.; Hufton, R. *Tetrahedron* **1996**, *52*, 12541–12552.

1955. Baird, E.E.; Dervan, P.B. *J. Am. Chem. Soc.* **1996**, *118*, 6141–6146.

1956. Griffiths, G.; Mettler, H.; Mills, L.S.; Previdoli, F. *Helv. Chim. Acta* **1991**, *74*, 309–314.

1957. Carruthers, N.I.; Wong, S.-C.; Chan, T.-M. *J. Chem. Res. S* **1996**, 430–431.

1958. Paulini, K.; Reißig, H-U. *Liebigs Ann. Chem.* **1991**, 455–461.

1959. Brighty, K.E.; Castaldi, M.J. *Synlett* **1996**, 1097–1099.

1960. Häusler, J. *Liebigs Ann. Chem.* **1983**, 982–992.

1961. Blythin, D.J.; Kuo, S.-C.; Shue, H.-J.; McPhail, A.T.; Chapman, R.W.; Kreutner, W.; Rizzo, C.; She, H.S.; West, R. *Bioorg. Med. Chem. Lett.* **1996**, *6*, 1529–1534.

1962. Macdonald, S.J.F.; Belton, D.J.; Buckley, D.M.; Spooner, J.E.; Anson, M.S.; Harrison, L.A.; Mills, K.; Upton, R.J.; Dowle, M.D.; Smith, R.A.; Molloy, C.R.; Risley, C. *J. Med. Chem.* **1998**, *41*, 3919–3922.

1963. Kennewell, P.D.; Matharu, S.S.; Taylor, J.B.; Sammes, P.G. *J. Chem. Soc. Perkin Trans. 1* **1980**, 2542–2548.

1964. Bates, D.K.; Li, X.; Jog, P.V. *J. Org. Chem.* **2004**, *69*, 2750–2754.

1965. Kam, B.L.; Oppenheimer, N.J. *J. Org. Chem.* **1981**, *46*, 3268–3272.

1966. Daluge, S.; Vince, R. *Tetrahedron Lett.* **1976**, 3005–3008.

1967. Forti, L.; Ghelfi, F.; Levizzani, S.; Pagnoni, U.M. *Tetrahedron Lett.* **1999**, *40*, 3233–3234.

1968. Bustos, F.; Gorgojo, J.M.; Suero, R.; Aurrecoechea, J.M. *Tetrahedron* **2002**, *58*, 6837–6842.

1969. Harvey, D.F.; Sigano, D.M. *J. Org. Chem.* **1996**, *61*, 2268–2272.

1970. Taylor, S.J.C.; McCague, R., Wisdom, R.; Lee, C.; Dickson, K.; Ruecroft, G.; O'Brien, F.; Littlechild, J.; Bevan, J.; Roberts, S.M.; Evans, C.T. *Tetrahedron: Asymmetry* **1993**, *4*, 1117–1128.

1971. Csuk, R.; Dörr, P. *Tetrahedron: Asymmetry* **1994**, *5*, 269–276.

1972. Annis, D.A.; Helluin, O.; Jacobsen, E.N. *Angew. Chem., Int. Ed. Engl.* **1998**, *37*, 1907–1909.

1973. Müller, M.; Müller, R.; Yu, T.-W.; Floss, H.G. *J. Org. Chem.* **1998**, *63*, 9753–9755.

1974. Xiao, D.; Carroll, P.J.; Mayer, S.C.; Pfizenmayer, A.J.; Joullié, M.M. *Tetrahedron: Asymmetry* **1997**, *8*, 3043–3046.

1975. Adams, H.; Bailey, N.A.; Brettle, R.; Cross, R.; Frederickson, M.; Haslam, E.; MacBeath, F.S.; Davies, G.M. *Tetrahedron* **1996**, *52*, 8565–8580.

1976. Brettle, R.; Cross, R.; Frederickson, M.; Haslam, E.; Davies, G.M. *Bioorg. Med. Chem. Lett.* **1996**, *6*, 291–294.

1977. Baxendale, I.R.; Ernst, M.; Krahnert, W.-R.; Ley, S.V. *Synlett* **2002**, 1641–1644.

1978. Mayer, S.C.; Joullié, M.M. *Synth. Commun.* **1994**, *24*, 2351–2365.

1979. Noyes, W.A.; Nickell, L.F. *J. Am. Chem. Soc.* **1914**, *36*, 118–127.

1980. Noyes, W.A.; Skinner, G.S. *J. Am. Chem. Soc.* **1917**, *39*, 2692–2718.

1981. Weir, J. *J. Chem. Soc.* **1911**, *99*, 1270–1277.

1982. Chênevert, R.; Martin, R. *Tetrahedron: Asymmetry* **1992**, *3*, 199–200.

1983. Trost, B.M.; Stenkamp, D.; Pulley, S.R. *Chem. —Eur. J.* **1995**, *1*, 568–572.

1984. Müller, P.; Imogaï, H. *Tetrahedron: Asymmetry* **1998**, *9*, 4419–4428.

1985. Galeazzi, R.; Mobbili, G.; Orena, M. *Tetrahedron* **1999**, *55*, 261–270.

1986. Langlois, N.; Radom, M.-O. *Tetrahedron Lett.* **1998**, *39*, 857–860.

1987. Gurjar, M.K.; Ghosh, L.; Syamala, M.; Jayasree, V. *Tetrahedron Lett.* **1994**, *35*, 8871–8872.

1988. Hanson, G.J.; Baran, J.S.; Lindberg, T. *Tetrahedron Lett.* **1986**, *27*, 3577–3580.

1989. Beckett, R.P.; Davies, S.G.; Mortlock, A.A. *Tetrahedron: Asymmetry* **1992**, *3*, 123–136.

1990. Niel, G.; Roux, F.; Maisonnasse, Y.; Maugras, I.; Poncet, J.; Jouin, P. *J. Chem. Soc., Perkin Trans. 1* **1994**, 1275–1280.

1991. Bergmeier, S.C.; Lee, W.K.; Rapoport, H. *J. Org. Chem.* **1993**, *58*, 5019–5022.

1992. Podlech, J.; Seebach, D. *Helv. Chim. Acta* **1995**, *78*, 1238–1246.

1993. Galeazzi, R.; Mobbili, G.; Orena, M. *Tetrahedron: Asymmetry* **1997**, *8*, 133–137.

1994. Hanessian, S.; Reinhold, U.; Ninkovic, S. *Tetrahedron Lett.* **1996**, *37*, 8967–8970.

1995. Kim, C.U.; Lew, W.; Williams, M.A.; Liu, H.; Zhang, L.; Swaminathan, S.; Bischofberger, N.; Chen, M.S.; Mendel, D.B.; Tai, C.Y.; Laver, W.G.; Stevens, R.C. *J. Am. Chem. Soc.* **1997**, *119*, 681–690.

1996. Zhang, L.; Williams, M.A.; Mendel, D.B.; Escarpe, P.A.; Kim, C.U. *Bioorg. Med. Chem. Lett.* **1997**, *7*, 1847–1850.

1997. Lew, W.; Wu, H.; Mendel, D.B.; Escarpe, P.A.; Chen, X.; Laver, W.G.; Graves, B.J.; Kim, C.U. *Biorg. Med. Chem. Lett.* **1998**, *8*, 3321–3324.

1998. Burgess, K.; Li, S.; Rebenspies, J. *Tetrahedron Lett.* **1997**, *38*, 1681–1684.

1999. Duwenhorst, J.; Montforts, F.-P. *Synlett* **1999**, 994–996.

2000. Ohno, H.; Mimura, N.; Otaka, A.; Tamamura, H.; Fujii, N.; Ibuka, T.; Shimizu, I.; Satake, A.; Yamamoto, Y. *Tetrahedron*, **1997**, *53*, 12933–12946.

2001. Galeazzi, R.; Mobbili, G.; Orena, M. *Tetrahedron* **1996**, *52*, 1069–1084.

2002. Cardillo, B.; Galeazzi, R.; Mobbili, G.; Orena, M. *Synlett*, **1995**, 1159–1160.

2003. Evans, C.; McCague, R.; Roberts, S.M.; Sutherland, A.G. *J. Chem. Soc., Perkin Trans. 1* **1991**, 656–657.

2004. Lombart, H-G.; Lubell, W.D. *J. Org. Chem.* **1996**, *61*, 9437–9446.

2005. Slomczynska, U.; Chalmers, D.K.; Cornille, F.; Smythe, M.L.; Beusen, D.D.; Moeller, K.D.; Marshall, G.R. *J. Org. Chem.* **1996**, *61*, 1198–1204.

2006. Hanessian, S.; McNaughton-Smith, G.; Lombart, H.-G.; Lubell, W.D. *Tetrahedron* **1997**, *53*, 12789–12854.

2007. von Roedern, E.G.; Lohof, E.; Hessler, G.; Hoffman, M.; Kessler, H. *J. Am. Chem. Soc.* **1996**, *118*, 10156–10167.

2008. Shankaramma, S.C.; Singh, S.K.; Sathyamurthy, A.; Balaram, P. *J. Am. Chem. Soc.* **1999**, *121*, 5360–5363.

2009. Banerjee, A.; Pramanik, A.; Bhattacharjya, S.; Balaram, P. *Biopolymers* **1996**, *39*, 769–777.

2010. Banerjee, A.; Das, A.K.; Drew, M.G.B.; Banerjee, A. *Tetrahedron* **2005**, *61*, 5906–5914.

2011. Gardner, R.R.; Liang, G.-B.; Gellman, S.H. *J. Am. Chem. Soc.* **1995**, *117*, 3280–3281.

2012. Gardner, R.R.; Liang, G.-B.; Gellman, S.H. *J. Am. Chem. Soc.* **1999**, *121*, 1806–1816.

2013. Baldauf, C.; Günther, R.; Hofmann, H.-J. *J. Org. Chem.* **2004**, *69*, 6214–6220.

2014. Wang, J.; Scott, A.I. *Tetrahedron Lett.* **1997**, *38*, 739–740.

2015. Shemin, D.; Russell, C.S.; Abramsky, T. *J. Biol. Chem.* **1955**, *215*, 613–626.

2016. Spencer, J.B.; Stolowich, N.J.; Santander, P.J.; Pichon, C.; Kajiwara, M.; Tokiwa, S.; Takatori, K.; Scott, A.I. *J. Am. Chem. Soc.* **1994**, *116*, 4991–4992.

2017. Vishwakarma, R.A.; Balachandran, S.; Alanine, A.I.D.; Stamford, P.J.; Kiuchi, F.; Leeper, F.J.; Battersby, A.R. *J. Chem. Soc., Perkin Trans 1* **1993**, 2893–2899.

2018. Battersby, A.R.; Hunt, E.; McDonald, E.; Moron, J. *J. Chem. Soc., Perkin Trans. I* **1973**, 2917–2922.

2019. Emery, V.C.; Akhtar, M. *J. Chem. Soc., Chem. Commun.* **1985**, 600–601.

2020. Okazaki, T.; Kurumaya, K.; Kajiwara, M. *Chem. Pharm. Bull.* **1990**, *38*, 1727–1730.

2021. Kurumaya, K.; Okazaki, T.; Seido, N.; Akasaka, Y.; Kawajiri, Y.; Kajiwara, M. ; Kondo, M. *J. Lab. Compds. Radiopharm.* **1989**, *27*, 217–235.

2022. Kajiwara, M.; Hara, K.-i.; Mizutani, M.; Kondo, M. *Chem. Pharm. Bull.* **1992**, *40*, 3321–3323.

2023. Appleton, D.; Duguid, A.B.; Lee, S.-K.; Ha, Y.-J.; Ha, H.-J.; Leeper, F.J. *J. Chem. Soc., Perkin Trans. 1* **1998**, 89–101.

2024. Rogers, L.M.-A.; McGiven, P.G.; Butler, A.R.; MacRobert, A.J.; Eggleston, I.M. *Tetrahedron* **2005**, *61*, 6951–6958.

2025. Berger, Y.; Ingrassia, L.; Neier, R.; Juillerat-Jeanneret, L. *Bioorg. Med. Chem.* **2003**, *11*, 1343–1351.

2026. Kim, J.-N.; Yun, J.-S.; Ryu, H.-W. *J. Chromatogr., A* **2001**, *938*, 137–143.

2027. Bunke, A.; Schmid, H.; Burmeister, G.; Merkle, H.P.; Gander, B. *J. Chromatogr., A* **2000**, *883*, 285–290.

2028. Bergmeier, S.C.; Fundy, S.L. *Biorg. Med. Chem. Lett.* **1997**, *7*, 3135–3138.

2029. Laurent, A.; Jacquault, P.; Di Martino, J.-L.; Hamelin, J. *J. Chem. Soc., Chem. Commun.* **1995**, 1101; corrigenda *Chem. Commun.* **1996**, 885.

2030. Wallach, O. *Ann.* **1900**, *312*, 171–210.

2031. Schniepp, L.E.; Marvel, C.S. *J. Am. Chem. Soc.* **1935**, *57*, 1557–1558.

2032. Fox, S.W.; Dunn, M.S.; Stoddard, M.P. *J. Org. Chem.* **1941**, *6*, 410–416.

2033. Gaudry, R.; Berlinguet, L. *Can. J. Res. B* **1950**, *28*, 245–255.

2034. Ha, H.-J.; Lee, S.-K.; Ha, Y.-J.; Park, J.-W. *Synth. Commun.* **1994**, *24*, 2557–2562.

2035. Neuberger, A.; Scott, J.J. *J. Chem. Soc.* **1954**, 1820–1825.

2036. Iida, K.; Takao, Y.; Ogai, T.; Kajiwara, M. *J. Lab. Cmpds. Radiopharm.* **1997**, *39*, 797–802.

2037. Neuberger, A.; Scott, J.J.; Shuster, L. *Biochem. J.* **1956**, *64*, 137–145.

2038. Shrestha-Dawadi, P.; Lugtenburg, J. *Eur. J. Org. Chem.* **2003**, 4654–4663.

2039. Nudelman, A.; Nudelman, A. *Synthesis* **1999**, 568–570.

2040. Weintraub, L.; Wilson, A.; Goldhamer, D.L.; Hollis, D.P. *J. Am. Chem. Soc.* **1964**, *86*, 4880–4885.

2041. Aschan, W. *Ber.* **1890**, *23*, 3692–3701.

2042. Aschan, W. *Ber.* **1891**, *24*, 2443–2450.

2043. Hansen, T.K.; Ankersen, M.; Hansen, B.S.; Raun, K.; Nielsen, K.K.; Lau, J.; Peschke, B.; Lundt, B.F.; Thøgersen, H.; Johansen, N.L.; Madsen, K.; Andersen, P.H. *J. Med. Chem.* **1998**, *41*, 3705–3714.

2044. Bunce, R.A.; Schilling, C.L., III; Rivera, M. *J. Lab. Cmpds. Radiopharm.* **1997**, *39*, 669–675.

2045. Hunter, C.; Jackson, R.F.W.; Rami, H.K. *J. Chem. Soc., Perkin Trans. 1.* **2001**, 1349–1352.

2046. Schrecker, A.W.; Trail, M.M. *J. Am. Chem. Soc.* **1958**, *80*, 6077–6080.

2047. Pfaltz, A.; Anwar, S. *Tetrahedron Lett.* **1984**, *25*, 2977–2980.

2048. Aurrecocchea, J.M.; Fernández, A.; Gorgojo, J.M.; Sucro, R. *Synth. Commun.* **2003**, *33*, 693–702.

2049. Fischer, E.; Zemplén, G. *Ber.* **1909**, *42*, 4878–4892.

2050. Sparatore, F.; Cumming, W.; North, B.; Shemin, D. *Biochem. Prep.* **1963**, *10*, 6–9.

2051. Marei, A.A.; Raphael, R.A. *J. Chem. Soc.* **1958**, 2624–2626.

2052. Kawakami, H.; Ebata, T.; Matsushita, H. *Agric. Biol. Chem.* **1991**, *55*, 1687–1688.

2053. Dalby, J.S.; Kenner, G.W.; Sheppard, R.C. *J. Chem. Soc.* **1962**, 4387–4396.

2054. Hodgkinson, T.J.; Shipman, M. *Synthesis* **1998**, 1141–1144.

2055. Mavunkel, B.J.; Lu, Z.; Kyle, D.J. *Tetrahedron Lett.* **1993**, *34*, 2255–2258.

2056. Fabiano, E.; Golding, B.T.; Sadeghi, M.M. *Synthesis* **1987**, 190–192.

2057. Trost, B.M.; Lee, C.B. *J. Am. Chem. Soc.* **2001**, *123*, 3687–3696.

2058. Benedetti, F.; Berti, F.; Norbedo, S. *Tetrahedron Lett.* **1999**, *40*, 1041–1044.

2059. Davies, S.B.; McKervey, M.A. *Tetrahedron Lett.* **1999**, *40*, 1229–1232.

2060. Rodriguez, M.; Heitz, A.; Martinez, J. *Tetrahedron Lett.* **1990**, *31*, 7319–7322.

2061. Lygo, B. *Synlett* **1992**, 793–795.

2062. Li, Y.-L.; Luthman, K.; Hacksell, U. *Tetrahedron Lett.* **1992**, *33*, 4487–4490.

2063. Fukuzawa, S.-i.; Miura, M.; Saitoh, T. *J. Org. Chem.* **2003**, *68*, 2042–2044.

2064. Blakskjaer, P.; Høj, B.; Riber, D.; Skryddstrup, T. *J. Am. Chem. Soc.* **2003**, *125*, 4030–4031.

2065. Altmann, K.-H.; Chiesi, C.S.; García-Echeverría, C. *Bioorg. Med. Chem. Lett.* **1997**, *7*, 1119–1122.

2066. Herdeis, C. *Synthesis* **1986**, 232–233.

2067. Zamora, F.; Bueno, M.; Molina, I.; Orgueira, H.A.; Varela, O.; Galbis, J.A. *Tetrahedron: Asymmetry* **1996**, *7*, 1811–1818.

2068. Crosby, S.R.; Hately, M.J.; Willis, C.L. *Tetrahedron Lett.* **2000**, *41*, 397–401.

2069. Davey, A.E.; Horwell, D.C. *Biorg. Med. Chem.* **1993**, *1*, 45–58.

2070. Steurer, S.; Podlech, J. *Eur. J. Org. Chem.* **2002**, 899–916.

2071. Klumpe, M.; Dötz, K.H. *Tetrahedron Lett.* **1998**, *39*, 3683–3684.

2072. Satake, A.; Shimizu, I.; Yamamoto, A. *Synlett* **1995**, 64–68.

2073. Ibuka, T.; Mimura, N.; Ohno, H.; Nakai, K.; Akaji, M.; Habashita, H.; Tamamura, H.; Miwa, Y.; Taga, T.; Fujii, N.; Yamamoto, Y. *J. Org. Chem.* **1997**, *62*, 2982–2991.

2074. Ibuka, T.; Mimura, N.; Aoyama, H.; Akaji, M.; Ohno, H.; Miwa, Y.; Taga, T.; Nakai, K.; Tamamura, H.; Fujii, N.; Yamamoto, Y. *J. Org. Chem.* **1997**, *62*, 999–1015.

2075. Chaperon, A.R.; Engeloch, T.M.; Neier, R. *Angew. Chem., Int. Ed. Engl.* **1998**, *37*, 358–3360.

2076. Demopoulos, B.J.; Anderson, H.J.; Loader, C.E.; Faber, K. *Can. J. Chem.* **1983**, *61*, 2415–2422.

2077. Faber, K.; Anderson, H.J.; Loader, C.E.; Daley, A.S. *Can. J. Chem.* **1984**, *62*, 1046–1050.

2078. Jackson, A.H.; McDonald, D.M.; McDonald, S.F. *J. Am. Chem. Soc.* **1956**, *78*, 505–506.

2079. Frydman, B.; Reil, S.; Despuy, M.E.; Rapoport, H. *J. Am. Chem. Soc.* **1969**, *91*, 2338–2342.

2080. Kenner, G.W.; Rimmer, J.; Smith, K.M.; Unsworth, J.F. *J. Chem. Soc. Perkin Trans. 1* **1977**, 332–338.

2081. Adamczyk, M.; Reddy, R.E. *Tetrahedron Lett.* **1995**, *36*, 9121–9124.

2082. Kenner, G.W.; Smith, K.M.; Unsworth, J.F. *J. Chem. Soc., Chem. Commun.* **1973**, 43–44.

2083. Jones, M.I.; Froussios, C.; Evans, D.A. *J. Chem. Soc., Chem. Commun.* **1976**, 472–473.

2084. Jackson, A.H.; MacDonald, S.F. *Can. J. Chem.* **1957**, *35*, 715–722.

2085. Arsenault, G.P.; MacDonald, S.F. *Can. J. Chem.* **1961**, *39*, 2043–2055.

2086. Ufer, G.; Tjoa, S.S.; MacDonald, S.F. *Can. J. Chem.* **1978**, *56*, 2437–2441.

2087. Leeper, F.J.; Rock, M.; Appleton, D. *J. Chem. Soc., Perkin Trans. 1* **1996**, 2633–2642.

2088. Battersby, A.R.; Moron, J.; McDonald, E.; Feeney, J. *J. Chem. Soc., Chem. Commun.* **1972**, 920–921.

2089. Battersby, A.R.; McDonald, E.; Wurziger, H.K.W.; James, K.J. *J. Chem. Soc., Chem. Commun.* **1975**, 493–494.

2090. Adamczyk, M.; Reddy, R.E. *Tetrahedron* **1996**, *52*, 14689–14700.

2091. de Leon, C.; Ganem, B. *Tetrahedron* **1997**, *53*, 7731–7752.

2092. Chen, Q.; Huggins, M.T.; Lightner, D.A.; Norona, W.; McDonagh, A.F. *J. Am. Chem. Soc.* **1999**, *121*, 9253–9264.

2093. Ehrlich, F. *Ber.* **1901**, *34*, 3366–3377.

2094. Peschke, B.; Madsen, K.; Hansen, B.S.; Johansen, N.L. *Bioorg. Med. Chem. Lett.* **1997**, *7*, 1969–1972.

2095. Qian, Y.; Blaskovich, M.A.; Seong, C.-M.; Vogt, A.; Hamilton, A.D.; Sebti, S.M. *Biorg. Med. Chem. Lett.* **1994**, *4*, 2579–2584.

2096. Jackson, S.; DeGrado, W.; Dwivedi, A.; Parthasarathy, A.; Higley, A.; Krywko, J.; Rockwell, A.; Markwalder, J.; Wells, G.; Wexler, R.; Mousa, S.; Harlow, R. *J. Am. Chem. Soc.* **1994**, *116*, 3220–3230.

2097. Groves, D.R.; von Itzstein, M. *J. Chem. Soc., Perkin Trans. 1* **1996**, 2817–2821.

2098. Toogood, P.L.; Galliker, P.K.; Glick, G.D.; Knowles, J.R. *J. Med. Chem.* **1991**, *34*, 3138–3140.

2099. Ilin, S.; Schlönvogt, I.; Ebert, M.-O.; Jaun, B.; Schwalbe, H. *ChemBioChem* **2002**, *3*, 93–99.

2100. Kühn, C.; Lindeberg, G.; Gogoll, A.; Hallberg, A.; Schmidt, B. *Tetrahedron*, **1997**, *53*, 12497–12504.

2101. Schmidt, B.; Kühn, C. *Synlett* **1998**, 1240–1242.

2102. Smith, P.W.; Trivedi, N.; Howes, P.D.; Sollis, S.L.; Rahim, G.; Bethell, R.C.; Lynn, S. *Biorg. Med. Chem. Lett.* **1999**, *9*, 611–614.

2103. Jones, P.S.; Smith, P.W.; Hardy, G.W.; Howes, P.D.; Upton, R.J.; Bethell, R.C. *Biorg. Med. Chem. Lett.* **1999**, *9*, 605–610.

2104. Shimano, M.; Meyers, A.I. *J. Org. Chem.* **1996**, *61*, 5714–5715.

2105. Shimano, M.; Matsuo, A. *Tetrahedron* **1998**, *54*, 4787–4810.

2106. Marco-Contelles, J.; Bernabé, M. *Tetrahedron Lett.* **1994**, *35*, 6361–6364.

2107. Arndt, H.-D.; Knoll, A.; Koert, U. *Angew. Chem., Int. Ed.* **2001**, *40*, 2076–2078.

2108. Jiang, H.; Léger, J.-M.; Huc, I. *J. Am. Chem. Soc.* **2003**, *125*, 3448–3449.

2109. Parkin, B.A.; Hedrick, G.W. *J. Am. Chem. Soc.* **1958**, *80*, 2899–2902.

2110. Eshghi, H.; Gordi, Z. *Synth. Commun.* **2003**, *33*, 2971–2978.

2111. Kano, S.; Yokomatsu, T.; Shibuya, S. *Heterocycles* **1990**, *31*, 13–16.

2112. Ikota, N. *Heterocycles* **1991**, *32*, 521–528.

2113. Sakai, N.; Ohfune, Y. *Tetrahedron Lett.* **1990**, *31*, 4151–4154.

2114. Sakai, N.; Ohfune, Y. *J. Am. Chem. Soc.* **1992**, *114*, 998–1010.

2115. Kiyooka, S.-i.; Goh, K.; Nakamura, Y.; Takesue, H.; Hena, M.A. *Tetrahedron Lett.* **2000**, *41*, 6599–6603.

2116. Wakamiya, T.; Terashima, S.-i.; Kawata, M.; Teshima, T.; Shiba, T. *Bull. Chem. Soc. Jpn.* **1988**, *61*, 1422–1424.

2117. Shoji, J.; Sakazaki, R.; Wakisaka, Y.; Koizumi, K.; Mayama, M.; Matsuura, S. *J. Antibiotics* **1975**, *28*, 122–125.

2118. Sakai, N.; Ohfune, Y. *Tetrahedron Lett.* **1990**, *31*, 3183–3186.

2119. Wakamiya, T.; Ando, T.; Teshima, T.; Shiba, T. *Bull. Chem. Soc. Jpn.* **1984**, *57*, 142–144.

2120. Hori, K.; Ohfune, Y. *J. Org. Chem.* **1988**, *53*, 3886–3888.

2121. Kondo, S.; Iinuma, K.; Naganawa, H.; Shimura, M.; Sekizawa, Y. *J. Antibiotics* **1975**, *28*, 79–82.

2122. Neuss, N.; Koch, K.F.; Molloy, B.B.; Day, W.; Huckstep, L.L.; Dorman, D.E.; Roberts, J.D. *Helv. Chim. Acta* **1970**, *53*, 2314–2319.

2123. Kiguchi, T.; Shirakawa, M.; Ninomiya, I.; Naito, T. *Chem. Pharm. Bull.* **1996**, *44*, 1282–1284.

2124. Kigichi, T.; Shirakawa, M.; Honda, R.; Ninomiya, I.; Naito, T. *Tetrahedron* **1998**, *54*, 15589–15606.

2125. Randl, S.; Blechert, S. *Tetrahedron Lett.* **2004**, *45*, 1167–1169.

2126. Lucet, D.; Le Gall, T.; Mioskowski, C.; Ploux, O.; Marquet, A. *Tetrahedron: Asymmetry* **1996**, *7*, 985–988.

2127. Boumrah, D.; Campbell, M.M.; Fenner, S.; Kinsman, R.G. *Tetrahedron* **1997**, *53*, 6977–6992.

2128. Harrison, B.A.; Gierasch, T.M.; Neilan, C.; Paternak, G.W.; Verdine, G.L. *J. Am. Chem. Soc.* **2002**, *124*, 13352–13353.

2129. Michielin, O.; Zoete, V.; Gierasch, T.M.; Eckstein, J.; Napper, A.; Verdine, G.; Karplus, M. *J. Am. Chem. Soc.* **2002**, *124*, 11131–11141.

2130. Klein, S.I.; Molino, B.F.; Czekaj, M.; Gardner, C.J.; Chu, V.; Brown, K.; Sabatino, R.D.; Bostwick, J.S.; Kasiewski, C.; Bentley, R.; Windisch, V.; Perrone, M.; Dunwiddie, C.T.; Leadley, R.J. *J. Med. Chem.* **1998**, *41*, 2492–2502.

2131. Frølund, B.; Kristiansen, U.; Brehm, L.; Hansen, A.B.; Krogsgaard-Larsen, P.; Falch, E. *J. Med. Chem.* **1995**, *38*, 3287–3296.

2132. Kitagawa, O.; Vander Velde, D.; Dutta, D.; Morton, M.; Takusagawa, F.; Aubé, J. *J. Am. Chem. Soc.* **1995**, *117*, 5169.

2133. MacDonald, M.; Vander Velde, D.; Aubé, J. *J. Org. Chem.* **2001**, *66*, 2636–2642.

2134. Brandmeier, V.; Sauer, W.H.B.; Feigel, M. *Helv. Chim. Acta* **1994**, *77*, 70–85.

2135. Nesloney, C.L.; Kelly, J.W. *J. Am. Chem. Soc.* **1996**, *118*, 5836–5845.

2136. Basak, A.; Rudra, K.R.; Bag, S.S.; Basak, A. *J. Chem. Soc., Perkin Trans. 1* **2002**, 1805–1809.

2137. Horne, W.S.; Stout, C.D.; Ghadiri, M.R. *J. Am. Chem. Soc.* **2003**, *125*, 9372–9376.

2138. Mieden-Gundert, G.; Klein, L.; Fischer, M.; Vögtle, F.; Heuzé, K.; Pozzo, J.-L.; Vallier, M.; Fages, F. *Angew. Chem., Int. Ed.* **2001**, *40*, 3164–3166.

2139. D'Aléo, A.; Pozzo, J.-L.; Fages, F.; Schmutz, M.; Mieden-Gundert, G.; Vögtle, F.; Caplar, V.; Zinic, M. *Chem. Commun.* **2004**, 190–191.

2140. Caplar, V.; Zinic, M.; Pozzo, J.-L.; Fages, F.; Mieden-Gundert, G.; Vögtle, F. *Eur. J. Org. Chem.* **2004**, 4048–4059.

2141. Mayes, B.A.; Cowley, A.R.; Ansell, C.W.G.; Fleet, G.W.J. *Tetrahedron Lett.* **2004**, *45*, 163–166.

2142. Hunter, D.F.A.; Fleet, G.W.J. *Tetrahedron Asymmetry* **2003**, *14*, 3831–3839.

2143. Ren, R.X.; Zueva, L.D.; Ou, W. *Tetrahedron Lett.* **2001**, *42*, 8441–8443.

2144. Ikushima, Y.; Hatakeda, K.; Sato, M.; Sato, O.; Arai, M. *Chem. Commun.* **2002**, 2208–2209.

2145. Raja, R.; Sankar, G.; Thomas, J.M. *J. Am. Chem. Soc.* **2001**, *123*, 8153–8154.

2146. Sahasrabudhe, K.; Gracias, V.; Furness, K.; Smith, B.T.; Katz, C.E.; Reddy, D.S.; Aubé, J. *J. Am. Chem. Soc.* **2003**, *125*, 7914–7922.

2147. Baruch, J. *Ber.* **1894**, *27*, 172–176.

2148. Goldsobel, A.G. *Ber.* **1894**, *27*, 3121–3129.

2149. Ho, K.-K.; O'Toole, D.C.; Achan, D.M.; Lim, K.T.; Press, J.B.; Leone-Bay, A. *Synth. Commun.* **1996**, *26*, 2641–2649.

2150. Hall, H.K. Jr., *J. Org. Chem.* **1963**, *28*, 3213–3214.

2151. Nesloney, C.L.; Kelly, J.W. *J. Org. Chem.* **1996**, *61*, 3127–3137.

2152. Albert, D.; Feigel, M. *Helv. Chim. Acta* **1997**, *80*, 2168–2181.

2153. Albert, D.; Feigel, M. *Tetrahedron Lett.* **1994**, *35*, 565–568.

2154. von Braun, J. *Ber.* **1907**, *40*, 1834–1846.

2155. Marvel, C.S.; MacCorquodale, D.W.; Kendall, F.E.; Lazier, W.A. *J. Am. Chem. Soc.* **1924**, *46*, 2838–2842.

2156. Musa, A.; Sridharan, B.; Lee, H.; Mattern, D.L. *J. Org. Chem.* **1996**, *61*, 5481–5484.

2157. Lukes, R.; Vesely, Z. *Collect. Czech. Chem. Commun.* **1959**, *24*, 2318–2322.

2158. Ulysse, L.; Chmielewski, J. *Bioorg. Med. Chem. Lett.* **1994**, *4*, 2145–2146.

2159. Torrado, A.; Imperiali, B. *J. Org. Chem.* **1996**, *61*, 8940–8948.

2160. Hagen, T.J. *Synlett* **1990**, 63–66.

2161. Cotarca, L.; Delogu, P.; Maggioni, P.; Nardelli, A.; Bianchini, R.; Sguassero, S. *Synthesis* **1997**, 328–332.

2162. Lee, K.J.; Joo, K.C.; Kim, E.-J.; Lee, M.; Kim, D.H. *Bioorg. Med. Chem.* **1997**, *5*, 1989–1998.

2163. Raghavan, S.; Reddy, S.R. *J. Org. Chem.* **2003**, *68*, 5754–5757.

2164. Pelter, A.; Crump, R.A.N.C.; Kidwell, H. *Tetrahedron: Asymmetry* **1997**, *8*, 3873–3880.

2165. Campos, K.R.; Journet, M.; Cai, D.; Kowal, J.J.; Lee, S.; Larsen, R.D.; Reider, P.J. *J. Org. Chem.* **2003**, *68*, 2338–2342.

2166. Takahata, H.; Banba, Y.; Tajima, M.; Momose, T. *J. Org. Chem.* **1991**, *56*, 240–245.

2167. Panek, J.S.; Liu, P. *Tetrahedron Lett.* **1997**, *38*, 5127–5130.

2168. Thompson, W.J.; Tucker, T.J.; Schwering, J.E.; Barnes, J.L. *Tetrahedron Lett.* **1990**, *31*, 6819–6822.

2169. Kohli, R.M.; Burke, M.D.; Tao, J.; Walsh, C.T. *J. Am. Chem. Soc.* **2003**, *125*, 7160–7161.

2170. Kumar, K.K.; Datta, A. *Tetrahedron* **1999**, *55*, 13899–13906.

2171. Wagner, H.; Harms, K.; Koert, U.; Meder, S.; Boheim, G. *Angew. Chem., Int. Ed. Engl.* **1996**, *35*, 2643–2646.

2172. Kumar, J.S.R.; Datta, A. *Tetrahedron Lett.* **1999**, *40*, 1381–1384.

2173. Golebiowski, A.; Kozak, J.; Jurczak, J. *J. Org. Chem.* **1991**, *56*, 7344–7347.

2174. Golebiowski, A.; Jurczak, J. *J. Chem. Soc., Chem. Commun.* **1989**, 263–264.

2175. Ohfune, Y.; Kurokawa, N. *Tetrahedron Lett.* **1984**, *25*, 1587–1590.

2176. Weber, I.; Potier, P.; Thierry, J. *Tetrahedron Lett.* **1999**, *40*, 7083–7086.

2177. Dondoni, A.; Franco, S.; Junquera, F.; Merchán, F.L.; Merino, P.; Tejero, T.; Bertolasi, V. *Chem.— Eur. J.* **1995**, *1*, 505–520.

2178. Dondoni, A.; Franco, S.; Merchan, F.; Merino, P.; Tejero, T. *Synlett* **1993**, 78–80.

2179. Long, D.D.; Stetz, R.J.E.; Nash, R.J.; Marquess, D.G.; Lloyd, J.D.; Winters, A.L.; Asano, N.; Fleet, G.W.J. *J. Chem. Soc., Perkin Trans. 1* **1999**, 901–908.

2180. Tichy, M.; Holanová, J.; Závada, J. *Tetrahedron: Asymmetry* **1998**, *9*, 3497–3504.

2181. Duursma, A.; Minnaard, A.J.; Feringa, B.L. *J. Am. Chem. Soc.* **2003**, *125*, 3700–3701.

2182. Wenzel, A.G.; Jacobsen, E.N. *J. Am. Chem. Soc.* **2002**, *124*, 12964–12965.

2183. Palomo, C.; Oiarbide, M.; Landa, A.; González-Rego, M.C.; Garcia, J.M.; González, A.; Odriozola, J.M.; Martín-Pastor, M.; Linden, A. *J. Am. Chem. Soc.* **2002**, *124*, 8637–8643.

2184. Wasserman, H.H.; Petersen, A.K. *Tetrahedron Lett.* **1997**, *38*, 953–956.

2185. Windeisen, E.; Pires, R.; Heistracher, E.; Burger, K. *Amino Acids* **1995**, *8*, 397–400.

2186. Schweizer, F.; Otter, A.; Hindsgaul, O. *Synlett* **2001**, 1743–1746.

2187. Edelman M.J.; Gandara D.R.; Hausner P.; Israel V.; Thornton D.; DeSanto J.; Doyle L.A. *Lung Cancer,* **2003**, *39*, 197–199.

2188. Bischoff, L.; David, C.; Martin, L.; Meudal, H.; Roques, B.-P.; Fournié-Zaluski, M.-C. *J. Org. Chem.* **1997**, *62*, 4848–4850.

2189. Taggi, A.E.; Hafez, A.M.; Wack, H.; Young, B.; Drury, W.J., III; Lectka, T. *J. Am. Chem. Soc.* **2000**, *122*, 7831–7832.

2190. Schwalbe, H.; Wermuth, J.; Richter, C.; Szalma, S.; Eschenmoser, A.; Quinkert, G. *Helv. Chim. Acta* **2000**, *83*, 1079–1107.

2191. Chong, J.M.; Sharpless, K.B. *J. Org. Chem.* **1985**, *50*, 1560–1563.

2192. Larchevêque, M.; Henrot, S. *Tetrahedron* **1990**, *46*, 4277–4282.

12 | Modifications of Amino Acids: Resolution, *N*-Alkylation, *N*-Protection, Amidation, and Coupling

12.1 Resolutions of α-Amino Acids

12.1.1 Introduction

The preparation of optically active amino acids via resolution of a racemic reaction product remains an attractive route for some syntheses of α-amino acids, despite the great advances that have been made in asymmetric methodologies. The majority of asymmetric synthetic methods are only suitable for preparing relatively small quantities of product due to lengthy synthetic routes, difficult chromatographic purifications, and/or expensive chiral auxiliaries. In contrast, if significant quantities of a homochiral amino acid are required, a racemic synthesis protocol such as the Strecker synthesis or aminomalonate alkylation can readily provide a substrate for resolution in only a few steps and on a large scale. A 2004 review of industrial methods for the production of optically active intermediates includes a significant section on the industrial production of amino acids, where the continued importance of amino acid resolutions is evident (*1*).

A number of resolution methods are available. Traditional methods, such as crystallization of diastereomeric salts or formation of diastereomeric derivatives, are still employed. However, the enantiomeric purities of many commercial resolving reagents have recently been determined, with a number possessing less than sterling optical purity (*2, 3*). Enzymatic resolutions have also been used for many years; acylases are commonly employed, but a variety of other enzymes can be applied. More recently, some new resolution techniques have been added to the chemist's repertoire. Chromatographic resolution on chiral supports has become more commonplace in recent years (although generally for analytical purposes, rather than preparative, as discussed in

Sections 1.6–1.13), while enantioselective deprotonation/reprotonation is a promising method of rapidly deracemizing amino acids. Reviews of resolution of amino acids by the more traditional methods are contained in the monographs of Greenstein and Winitz (*4*) and Barrett (*5*). An overview of optical resolution methods, with a number of amino acid examples, was published in 2006 (*6*).

One limitation of traditional resolutions is that a maximum of half of the racemic material is obtained as the desired enantiomer. Recycling of the undesired enantiomer via iterative racemization and re-resolution is possible, but time consuming. Conditions for amino acid racemization are included in Section 1.4.5. Efforts have been made to combine resolution with racemization of the remaining enantiomer, allowing for recoveries of >50% in a single step. This area, known as dynamic resolution or dynamic kinetic resolution, was reviewed in 1997 (*7*) and 2003 (*8*), and is also included in a comprehensive review of racemization of organic compounds (*9*) and a more abbreviated review of enzyme-catalyzed deracemization and dynamic kinetic resolution reactions (*10*). The concepts of biocatalytic transformation of racemates into chiral derivatives via dynamic resolution, desymmetrization or deracemization are discussed in a 1999 paper (*11*).

12.1.2 Resolutions via Crystallizations of Salts (see Table 12.1)

The principle of enantiomeric resolution via crystallization originated with the mechanical separation of crystals of sodium ammonium tartrate by Louis Pasteur. Several similar selective crystallizations of racemic amino acids have been reported (*4*), such as the separation of a sulfonate derivative of D-4'-hydroxyphenylglycine (*12*). The enantiomers of α-methyl-4-carboxyphenylglycine were recently resolved by preferential crystallization of a hydantoin intermediate from a racemic solution (*13*). Partially enantiomerically enriched amino acids, products of incomplete stereoselectivity during asymmetric syntheses, can sometimes be recrystallized to give optically pure product. *N*-Fmoc *tert*-butyl ester derivatives of amino acids have been identified as consistently giving higher enantiomeric purity upon recrystallization; a series of 10 compounds with initial 63–92% ee (average 76% ee) gained an average of 13.6% ee upon the first recrystallization (*14*).

However, the usual resolution-by-crystallization procedure employs a chiral acid or base to form a diastereomeric salt with the amino acid, relying on the differential solubilities of the two components of the mixture to effect a separation. Some practical guidelines for successful resolutions via crystallization are presented in the monograph of Greenstein and Winitz, accompanied by a table of amino acids and their resolving agent (*4*). Further examples are given by Barrett (*5*). Most successful separations have been achieved with *N*-acyl amino acids and chiral bases such as brucine, strychnine, ephedrine or quinine. Separations of amino acid esters by crystallization with optically active acids are less common, due to possible side reactions of the ester group. In general, only amino dicarboxylic acids or diaminocarboxylic acids have been separated without *N*-derivatization (*5*), although (−)-1-phenylethanesulfonic acid was recently used to resolve a variety of unprotected amino acids (*15*). A potentially general method of resolution has been described for Ala or 3-fluoro-Ala, using temporary *N*-protection. Reaction of either amino acid with 2,4-pentanedione generated an acid-labile *N*-(1-methyl-2-acetylvinyl) derivative, which was then resolved by salt formation with quinine. The quinine could be removed from the separated diastereomeric salts by extraction of a basic solution, with mild acid hydrolysis then deprotecting the amino group (*16*).

The greatest advantage of amino acid resolution via crystallization of diastereomeric salts is that the method is readily applied to large-scale separations. This advantage is offset by the trial-and-error approach required in order to find a successful resolving agent. A number of examples of resolutions via diastereomeric crystallization are presented in Table 12.1; further examples are contained in the monographs of Greenstein and Winitz (*4*) or Barrett (*5*). Several procedures that appear to have more widespread application are worth further discussion. In 1970, (+)- and (−)-ephedrine were employed to resolve *N*-Cbz-protected Val, Leu, Ile, *erythro*-β-methyl-Leu, *N*,β-dimethyl-Leu, aspartic acid β-methyl ester, Phe, Met, and *O*-benzyloxycarbonyl Ser (*17*). A 1994 systematic examination of the resolution of underivatized amino acids found that (−)-1-phenylethanesulfonic acid was capable of resolving 10 of the 20 amino acids examined, including neutral and basic amino acids and an imino acid (*15*). Several *N*-thiobenzoyl amino acid derivatives have been successfully resolved with various

bases (*18*). Phosphonic acids have been used in recent years: a binaphthol-based phosphonic acid, originally reported in 1978 (*19*), was used for the resolution of an ornithine derivative (*20*), while another phosphonic acid derivative resolved homomethionine (*21*). Crystal structures of the salts formed between (+)-(1*S*)-1,1'-binaphthalene-2,2'-diyl phosphate (bnppa) and L-Ala, L-Val, L-Nva or L-Nle were obtained in order to examine the mechanism of chiral recognition by this ligand; a chiral space between bnppa molecules at the interface between hydrophobic and hydrophilic layers was identified (*22*).

A new method for resolving racemates that avoids the lengthy identification of an optimum resolving reagent was described in 1998. A "family" of resolving reagents, composed of similar reagents such as various chiral phosphoric acids, various dibenzoyl tartaric acids, or various phenylethylamines, is simultaneously mixed with the racemate. The method was successfully applied to over 200 compounds, with yields and enantiomeric excess superior to the products obtained by classical 1:1 mixtures. Surprisingly, the precipitates were found to contain more than one component of the resolving mixture, in non-stoichiometric ratios that were maintained through several recrystallizations. Random combinations of the resolving reagents, instead of families of similar types, generally gave moderate results. Several types of amino acids, including α-amino acids and esters, α,α-dialkyl-α-amino acids and amides, and β-amino acids, were included among the successful examples (*23*). The method has been briefly reviewed (*24*). A somewhat similar effect was employed during resolutions of 4'-fluoro-Phg and 4'-hydroxy-Phg with 10-camphorsulfonic acid, which were only successful if DL- or D-Phg was also added (*25*).

β-Amino acids can also be resolved by salt formation; for example, *threo*-methylphenidate (methyl piperidine-2-phenylacetate, Ritalin; used to treat attention deficit hyperactivity disorder (ADHD)) was enantiomerically purified on a large scale by diastereomeric salt formation with *O*,*O*'-dibenzoyl-D-(+)-tartaric acid in the presence of 4-methylmorpholine (*26*). Tartaric acid was also used to resolve ethyl nipecotate (β²-homoproline, piperidine-3-carboxylic acid) (*27*).

Crystallizations have been combined with a separate or simultaneous step to racemize the remaining material on a number of occasions. One of the earlier uses of this method was in the synthesis of L-α-methyl-DOPA via a Strecker synthesis (*28*). The racemization-sensitive α-aminonitrile intermediate was resolved with *l*-10-camphorsulfonic acid, with the undesired enantiomer then racemized by heating. A resolution of 3,4-dehydroproline by tartaric acid employed racemization of the mother liquor by heating (*29*), as did a more recent resolution of 2-piperazine carboxylic acid with (*S*)-camphor-10-sulfonic acid (*30*). Over 100 patents have described Schiff base-promoted racemization combined with a resolution by selective crystallization (either iterative or simultaneous); the general characteristics of these reactions have been summarized (*9*). The increase in α-carbon acidity of Gly-OMe in the presence of acetone was found to be 7 pK units by an NMR study (*31*). All common amino acids have been converted, generally using aromatic aldehydes as the racemization catalyst. For example, racemic arylglycine esters were converted to the homochiral products using tartaric acid for resolution and benzaldehyde for racemization (*32*). Addition of catalytic benzaldehyde during dibenzoyl tartaric acid resolution of 4-fluorophenylglycine methyl ester allowed for mother liquor racemization at only 40 °C (*33*). In another report, benzaldehyde catalyzed the racemization of the amides of Ala, Leu, Met, Phe,

Phg, and Hfe (homophenylalanine), with crystallization achieved by diastereomeric salt formation with mandelic acid, 2-pyrrolidone-5-carboxylic acid, or L-Asp (*34*). 4′-Chloro-Phe-OMe was resolved by salt formation with (*S,S*)-tartaric acid, with in situ racemization catalyzed by 0.1 equiv of salicylaldehyde to provide a 68% yield of 98% ee purity (*35*). Homocysteine has been resolved via the tartaric acid diastereomeric salt of its cyclic thiazine derivative; the resolution was accompanied by salicaldehyde-catalyzed racemization of the more soluble salt in solution to give resolved yields of >50% (*36*).

Table 12.1 Diastereomeric Salt Formation

P¹	R¹	R²	P²	Resolving Agent	Reference
H	Ph	H	H	10-camphorsulfonic acid	1920 (*1571*) 1922 (*1572*)
H	CH(OH)CO₂H	H	H	strychnine	1922 (*1573*)
CHO	CH₂SBn	H	H	brucine	1939 (*1574*)
Bz	2,5-(MeO)₂-Bn	H	H	(−)-1-Ph-ethylamine	1948 (*1575*)
CHO	CH₂CH₂OPh	H	H	strychnine	1948 (*1576*)
CHO	3-HO-Bn	H	H	brucine	1951 (*1577*)
4-NO₂-Bz	CH₂CH₂OH	H	H	brucine	1951 (*1578*)
CHO	C(Me)₂SBn	H	H	brucine	1953 (*1579*)
Phth	3-CO₂H-4-MeO-Bn	H	H	brucine	1965 (*1580*)
Ac	4-NO₂-Bn	H	H	brucine	1955 (*1581*)
CHO	CH₂CH₂SBn	H	H	brucine	1957 (*1582*)
Ac	CH₂COCH₂OH	H	H	brucine	1960 (*1583*)
CHO		H	H	α-Me-phenethylamine	1962 (*1584*)
H	3-CO₂H-Ph	H	H	L-Arg	1963 (*1585*)
H	*t*Bu	H	Me	dibenzoyl tartaric acid	1964 (*1586*)
H	3-CF₃-Bn	H	H	L-*threo-p*-NO₂-Ph-2-amino-1,3-propanediol	1966 (*1587*)
H		H	H	brucine	1967 (*1588*)
H		H	H	L-Lys	1967 (*1588*)
Ac		H	H	strychnine	1967 (*1588*)
Bn	CH(OH)CO₂H	H	H	ephedrine	1967 (*1589*)
Cbz	(CH₂)₅CO₂H	H	H	D-Tyr-NHNH₂	1968 (*1590*)
Ac	3-MeO-4-HO-Bn	Me	CO₂P² = CN	10-camphorsulfonic acid	1968 (*28*)
H	C(OMe)Me₂	H	H	10-camphorsulfonic acid	1968 (*1591*)
H	CH(OH)CH₂OBn	H	H	L-Tyr-NHNH₂	1969 (*1592*)
Ac	3-Br-4-HO-5-MeO-Bn	H	H	ephedrine	1969 (*1593*)

(Continued)

Table 12.1 Diastereomeric Salt Formation (continued)

P^1	R^1	R^2	P^2	Resolving Agent	Reference
CF_3CO	(isoxazolidin-3-one structure, N–O ring)	H	H	quinine	1969 (*1594*)
Cbz	*i*Pr *i*Bu *s*Bu CH(Me)*i*Pr Bn CH_2OCbz CH_2OMe CH_2CH_2SMe	H	H	ephedrine	1970 (*17*)
Ac	CH(OH)Ph	H	H	quinine	1970 (*1595*)
Fm	$R^1 = R^2$ = 2,2-norbornane	R^1	H	brucine	1972 (*1596*)
Cbz	$CH_2CH(OH)CO_2H$	H	H	brucine	1973 (*1597*)
Bz	$CH_2CH(OH)CO_2H$	H	H	strychnine	1973 (*1597*)
Ac	2,3,4,5,6-Me_5-Bn	H	H	brucine	1973 (*1598*)
CHO	CH(Me)Ph	H	H	brucine	1974 (*1599*)
H	CH_2CH_2(imidazol-4-yl)	H	H	tartaric acid	1975 (*1600*)
PhCS	Et	H	H	strychnine	1975 (*18*)
PhCS	*t*Bu	H	H	brucine	1975 (*18*)
PhCS	$(CH_2)_3$-R^2	R^1	H	morphine	1975 (*18*)
Cbz	CH(OH)[3,4-$(BnO)_2$-Ph]	H	H	ephedrine	1975 (*1601*)
CHO	CH_2(thymin-1-yl)	H	H	α-Me-phenethylamine	1975 (*1602*)
Cbz	CH(Me)Ph	H	H	quinine, quinidine	1976 (*1603*)
CHO	CH(Et)CH_2-R^2	R^1	H	quinine	1977 (*1604*)
C(Me)=CHCOMe	Me	H	H	quinine	1977 (*16*)
C(Me)=CHCOMe	CH_2F	H	H	quinine	1977 (*16*)
H	CF_2CO_2H	H	H	brucine	1977 (*1605*)
Phth	CD(OH)Me	H	H	brucine	1977 (*1606*)
Cbz	$CH_2CH(CO_2tBu)_2$	H	H	quinine, ephedrine	1977 (*1607*)
H	$(CH_2)_3$NH-αCO	Me	CO_2P^2 = CO-R^1	binaphthylphosphoric acid	1978 (*19*)
H	4-HO-Ph	H	H	3-Br-camphor-8-sulfonic acid	1979 (*1608*)
CHO	CH_2(ferrocenyl)	H	H	brucine	1980 (*1609*)
Boc	CH_2(2-thienyl)	H	H	(−)-1-Ph-ethylamine	1980 (*1610*)
H	3-HO-Bn	H	H	brucine	1980 (*1611*)
Bz	CH_2(1-pyrenyl)	H	H	(−)-1-Ph-ethylamine	1983 (*1612*)
Cbz	CH(Ph)CH_2-R^2	R^1	H	brucine	1983 (*1613*)
H	CH(OH)(imidazol-2-yl)	H	H	tartaric acid	1983 (*1614*)
Cbz	CH_2SMe	H	H	ephedrine	1985 (*1615*)
Bz	CH_2(1-pyrenyl)	H	H	(−)-1-Ph-ethylamine	1985 (*1616*)
NH_2	CH_2CH_2SMe	H	H	O,O'-phosphoryl-1-(2-Cl-Ph)-2,3-Me_2-propan-1,3-diol	1986 (*21*)
Ac	CH_2(9-anthryl)	H	H	ephedrine	1986 (*1617*)
Ac	CH_2(9-anthryl)	H	H	ephedrine	1989 (*1618*)
H	*cis* CH(Ph)CH_2-R^2	R^1	Me	dibenzoyl tartaric acid	1988 (*1619*)
Cbz	*trans* CH(Ph)CH_2-R^2	R^1	H	brucine	1988 (*1619*)
Bz	CH_2(1-pyrenyl)	H	H	(+)-1-Ph-ethylamine	1987 (*1620*)
CHO	$C(Me)_2CH_2$-R^2	R^1	H	quinine	1990 (*1621*)
H	Ph 4-HO-Ph CH_2CH_2SMe	H	CO_2P^2 = $CONH_2$	mandelic acid	1992 (*34*)
H	$(CH_2)_3$NH-αCO	Me	CO_2P^2 = CO-R^1	1,1'-binaphthyl-2,2'-diylhydrogen phosphate	1992 (*20*)
H	(quinoxaline ring with $CH_2PO_3H_2$ substituent)	H	Me	dibenzoyl tartaric acid	1993 (*1622*)

Table 12.1 Diastereomeric Salt Formation (continued)

P^1	R^1	R^2	P^2	Resolving Agent	Reference
H	Me Et *i*Pr *i*Bu *n*Bu Ph 4-HO-Ph CH$_2$OH (CH$_2$)$_4$NH$_2$	H	H	(*S*)-1-Ph-ethanesulfonic acid	1994 (*15*)
Boc	*n*Bu, OH (isoxazole structure)	H	H	1-Ph-ethylamine	1998 (*1623*)
H	4-F-Ph, OH (isoxazole structure)	H	H	1-Ph-ethylamine	1998 (*1624*)
H	4-HO-Bn 4-F-Bn Bn	H	H	family of phosphonic acids	1998 (*23*)
H	CH$_2$CH$_2$C(Me)$_2$-(α-N)	H	H	d-tartaric acid	1999 (*1513*) 2003 (*1625*)
CONH-αCO	CO$_2$H CO$_2$H (cyclopropane-oxirane and thiirane structures)	R^1	CO$_2$P^2 = CO-αNH	phenylglycinol	1999 (*1626*)
H H H H H	nipecotic acid (piperidine-3- carboxylate) 4-OH-Ph 4-F-Ph (CH$_2$)$_4$NH-(α-CO)	— H H H	Et H H H CO$_2$P^2 = CO-ϵNH	tartaric acid camphorsulfonic acid camphorsulfonic acid with DL-Phg Ts-L-Phe	2000 (*1627*) 2000 (*25*) 2003 (*1628*)
H	CH(Me)Et	H	H	dibenzoyl tartaric acid	2002 (*1629*)

12.1.3 Resolutions via Diastereomeric Derivatives

12.1.3a Introduction

Diastereomeric derivatives of amino acids can be obtained via esterification, amidation, acylation, or imine formation with homochiral alcohols, amines, acids, or ketones, respectively. The diastereomers are then separated by techniques such as crystallization, chromatography or distillation, with the chiral auxiliary removed after the separation. This method has a greater probability of success than does the crystallization of diastereomeric salts, as several methods of separation are possible, but chromatography (if required) on a large scale can be difficult. Other limitations include the requirement that the resolving reagent has high optical purity, the possibility of chiral auxiliary racemization during diastereomer formation, and the potential for amino acid racemization during removal of the auxiliary. Improvements in chromatography in the past 30 years have resulted in increasing use of diastereomeric

derivatives: Greenstein and Winitz (*4*) reported only two publications (both using fractional crystallization), while Barrett (*5*) listed seven (three using chromatography). Asymmetric syntheses that employ chiral auxiliaries usually incorporate an innate diastereomeric purification step during product isolation, as the asymmetric induction is rarely absolute.

12.1.3b Derivatives of the Carboxylic Acid Group (see Table 12.2)

A variety of amide derivatives of the carboxylic acid group have been formed in order to generate diastereomers suitable for resolution (see Table 12.2). Obrecht and co-workers developed the use of (*S*)-Phe cyclohexylamide as a chiral amine auxiliary for resolution via amide formation (*37*). This was the best of several Phe-based auxiliaries tested (*37, 38*). A range of α,α-disubstituted amino acids with varying side-chain functionality were chromatographically resolved as the dipeptides (*37–41*). An axially asymmetric α,α-disubstituted amino acid

(Bin) was also resolved with this auxiliary (*42*), as were 1-amino-2-phenylcyclohexane-1-carboxylic acid (*43*) and the α-monosubstituted amino acids Val and Phe (*38*). The auxiliary was removed with TFMSA in MeOH at 80 °C (*37*).

Other amines that have been used for amide formation include 1-phenylethylamine (*44–46*) and (*S*)-2-hydroxymethylpyrrolidine (*47*). The γ-carboxyl groups of 2-, 3-, and 4-substituted Glu derivatives (with the α-functional groups complexed as a boroxazolidone) were amidated with (*R*)-1-phenylethylamine or (*R*)-phenylglycinol (*48*). This method was unsuccessful for α-methyl-4-carboxyphenylglycine, and so the α-amino group was *N*-acylated with Boc-Leu instead (*49*).

Ester diastereomeric derivatives are also possible, with the advantage of less harsh deprotection conditions (see Table 12.2).

12.1.3c Derivatives of the α-Amino Group (see Table 12.3)

The α-amino group has also been derivatized to generate diastereomers (see Table 12.3). A range of amino esters were converted into Schiff base derivatives of 2-hydroxypinan-3-one, and then separated by column chromatography on silica (*50, 51*). Lipidic amino acids were also resolved with this auxiliary on a 1 g scale, using silica-based thin-layer chromatography (*52*). The 2-hydroxypinan-3-one Schiff base auxiliary has the advantage of removal under comparatively mild acidic hydrolysis conditions.

Other chiral auxiliaries for the amino group include Boc-Leu, which was used to acylate both 1-amino-2-cyclohexene-1,3-dicarboxylic acid (*53*) and α-methyl-4′-carboxyphenylglycine (*49*). *N*-9-Phenylfluorenyl allylglycine provided resolution of Val-OMe or Phe-OMe; removal of the auxiliary was accomplished by treatment with iodine (*54*). A phthaloyl protecting group has been used as a resolution reagent, with one of the amide carbonyl groups reduced with NaBH$_4$ to give the chiral auxiliary (*55*).

An interesting concept of resolution that mimics an enzymatic acetylation has been developed. A homochiral acetylating reagent, 2-acetylamino-2′-diacetylamino-1,1′-binaphthyl, stereoselectively reacted with racemic Phe-OBn or Leu-OBn to give the acetylated amino acid in 21–24% yield, with 48% ee or 11% ee, respectively (*56*).

12.1.4 Enzymatic Resolutions

12.1.4a Introduction

Enzymatic resolution of amino acids dates from the turn of the century, when Ehrlich isolated the D-isomers of Ala, Leu, Val, Iva, Ser, Phe, Glu, and His by treatment of the racemic mixture with fermenting yeast (*4*). Most resolutions rely on acylases to hydrolyze *N*-acyl amino acids or hydrolytic enzymes (such as proteases,

esterases, and lipases) to hydrolyze amino acid esters. The substrate enantiomer that is deacylated or hydrolyzed generally possesses the natural L-configuration. Enzymatic catalysts can also be used in the reverse direction to prepare acylated, esterified or amidated amino acid derivatives, normally producing the derivatized L-amino acid (*57*). However, this approach is less common, as the unprotected L-amino acid is usually the desired product. One disadvantage of enzymatic resolutions is that some substrates require lengthy reaction times (days). Microwave irradiation has been reported to accelerate subtilisin-catalyzed esterifications and α-chymotrypsin-catalyzed esterifications (*58*); while not applied to enzymatic resolutions in this study, the technique may find use in the future.

The development of immobilized enzymes or cells (e.g., see *59–61*) and of enzymatic reactions in organic solvents (*62*) has helped to increase the popularity of lipases and proteases by making them more amenable to use by synthetic organic chemists. Commercially available stable and active cross-linked enzyme crystals (CLECs) (*63*) will potentially increase the use of enzymatic reactions for amino acid resolutions even more. A review of the properties and synthetic applications of enzymes in organic solvents was published in 2000, though amino acid resolutions were not included as examples (*64*). High enzyme activities in polar organic solvents have been achieved by a simple one-pot method of immobilizing the enzyme on silica. Subtilisin Carlsberg and α-chymotrypsin were used as examples (*65*). Modified peptides bearing two crown ether functionalities have been found to enhance the activity of α-chymotrypsin in organic solvents (*66*). A new approach toward enhancing enzymatic enantioselectivity in organic solvents employed salts of the substrates instead of the free substrates, based on the presumption that a bulky counterion will increase the difference in steric hindrances experienced by the preferred and disfavored substrate enantiomers (*67*).

A number of reviews of enzymatic resolutions have been published, with Greenstein and Winitz providing a comprehensive account of the development of the enzymatic resolution of amino acids up to the late 1950s, including practical experimental procedures (*68*) and a table of amino acids and the enzymes used to resolve them (*4, 69*). Barrett outlined an overview of developments through to the early 1980s (*5*), while Williams' monograph covers most of the 1980s (*70*). A thorough review of modern methods of resolving amino acid racemates was published in 1991 (*57*). Some general reviews of enantioselective syntheses through enzymatic reactions include examples of amino acid preparations: a 1991 review of esterolytic and lipolytic enzymes in organic synthesis (*60*); a 1992 summary covering all types of enzymatic transformations from 1988 to 1991 (*71*); and a 1996 review of enzymatic asymmetrizations of prochiral or *meso* compounds (*72*). Enzymatic resolutions of amino acids via ester hydrolysis were reviewed in 1999 (*73*), while a 2004 review of serine hydrolase acylation of chiral amines included a number

Table 12.2 Resolution of Amino Acids via Diastereomeric Derivatives of Acid Group

Amino Acid	Derivatizing Auxiliary	Resolution Method	Auxiliary Removal	Reference
AMIDATION				
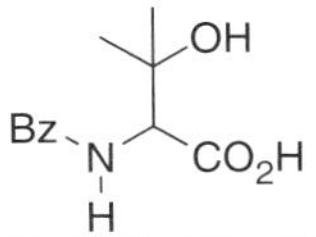 Bz–N(H)– , OH, CO$_2$H	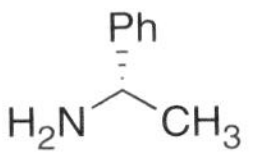 Ph–*CH(CH$_3$)–NH$_2$	crystallization	2N NaOH	1966 (44)
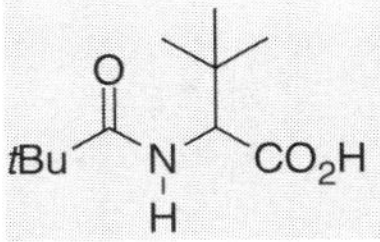 tBu–C(O)–N(H)– , CO$_2$H	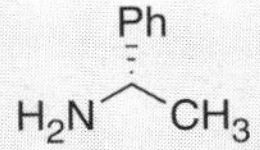 Ph–CH(CH$_3$)–NH$_2$	chromatography	6N HCl	1997 (45)
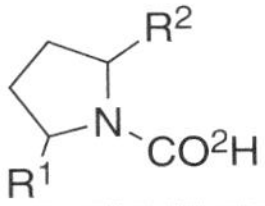 R^2 ; R^1–N–CO$_2$H (pyrrolidine); R^1 = Ac; R^2 = Ph ; R^1 = Boc; R^2 = nPr	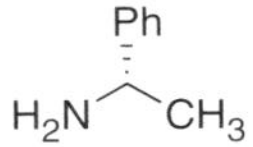 Ph–CH(CH$_3$)–NH$_2$	chromatography	8N HCl	1990 (46)
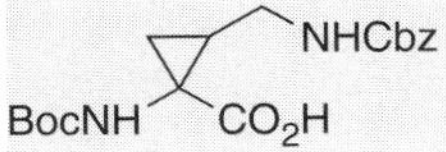 NHCbz ; BocNH , CO$_2$H (cyclopropane)	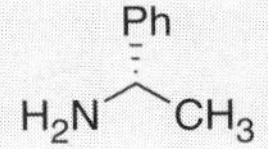 Ph–CH(CH$_3$)–NH$_2$	chromatography	6N HCl	1986 (1630)
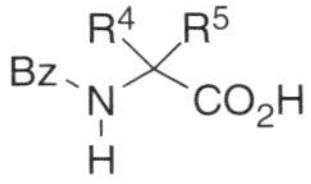 R^4, R^5 ; Bz–N(H)–CO$_2$H ; R^4 = Me; R^5 = iPr, cPr, tBu, Ph, Bn, 4-MeO-Bn ; R^4 = allyl; R^5 = Me ; R^4 = H; R^5 = iPr, Bn ; R^4 = R^5 = -[1,2-(CH$_2$)$_2$-Ph]-, -[1-Ph-2-(CH$_2$)$_3$-Ph]-, -[1-Ph-2-(CH$_2$)$_2$-Ph]-	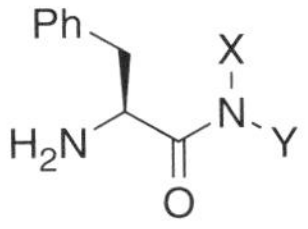 X–N–Y amide, Ph, O ; H$_2$N ; X = Y = Me, -(CH$_2$)$_4$- ; X = H; Y = CH(Bn)CONMe$_2$	flash chromatography SiO$_2$	TFMSA MeOH, 80 °C	1992 (38)
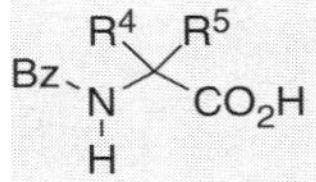 R^4, R^5 ; Bz–N(H)–CO$_2$H ; R^4 = Me; R^5 = iPr, Bn, Ph	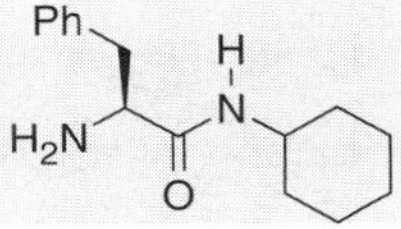 cyclohexyl–N(H)– amide, Ph, O ; H$_2$N	flash chromatography SiO$_2$	TFMSA MeOH, 80 °C	1995 (37)
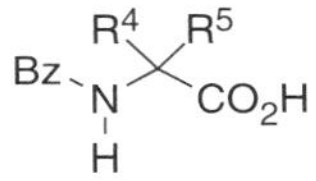 R^4, R^5 ; Bz–N(H)–CO$_2$H ; R^4 = CH$_2$NH$_2$; R^5 = Me, tBu	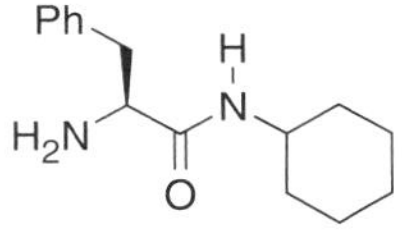 cyclohexyl–N(H)– amide, Ph, O ; H$_2$N	flash chromatography SiO$_2$	TFMSA toluene, 50 °C	1995 (40)

(Continued)

1125

Table 12.2 Resolution of Amino Acids via Diastereomeric Derivatives of Acid Group (continued)

Amino Acid	Derivatizing Auxiliary	Resolution Method	Auxiliary Removal	Reference
$R^4 = CH_2OH$; $R^5 =$ Me, iPr, tBu, Bn, CH_2CO_2tBu, $(CH_2)_2CO_2tBu$ $R^4 = CH_2CO_2tBu$; $R^5 =$ Me, iPr, $(CH_2)_2CO_2tBu$		flash chromatography SiO_2	33% HBr AcOH	1996 (39)
$R^4 = R^5 = -CH(Ph)(CH_2)_4-$		flash chromatography SiO_2	TFMSA MeOH, 80 °C	1998 (43)
$R^4 = R^5 =$		flash chromatography SiO_2	10N HCl	1998 (42)
		flash chromatography SiO_2	—	1999 (1631)
R^1 or R^2 or $R^3 =$ Me or $R^3 = CH_2=$, remainder = H $R^1 = R^3 = -(CH_2)_3-$, $-(CH_2)_2-$		flash chromatography SiO_2	2N HCl, 80 °C	1994 (48)
		flash chromatography SiO_2	6N HCl, reflux	1996 (47)

ESTERIFICATION

Substrate	Product	Method	Reagent	Year (Ref)
cyclopentenyl-CH(NHAc)CO₂H ($Ac-N$H, CO_2H)	Ph-CH(NHAc)-CH₂OH	chromatography	3.6N HCl	1992 (1632)
cyclohexenyl-CH(NHAc)CO₂H	Ph-CH(NHAc)-CH₂OH	chromatography	3.6N HCl	1992 (1633)
tetrahydrocarbazole-CO₂H (NH)	menthol-type (HO, isopropyl cyclohexane)	crystallization	—	1973 (1634)
N,N-dibenzyl piperidine-CO₂H (Bn–N, N–Bn)	menthol-type (HO, isopropyl cyclohexane)	crystallization	BCl₃ in ClCH₂CH₂Cl	1989 (1635)
tetrahydroisoquinoline-CO₂H (N–H)	(HO, Ph, cyclohexane)	chromatography	—	1992 (1636)
azetidine-2,4-dicarboxylic acid (HO₂C, CO₂H, N–H)	menthol-type (HO, isopropyl cyclohexane)	chromatography of diester	NaOH saponification	1993 (395)

Table 12.3 Resolution of Amino Acids via Diastereomeric Derivatives of Amino Group

Amino Acid	Derivatizing Auxiliary	Resolution Method	Auxiliary Removal	Reference
[hydroxyproline structure: ring with OH, N-H, CO_2H]	[Cbz–N(H)–CH–CO_2H structure]	chromatography	6N HCl, 100 °C	1966 (1637)
[cyclohexene with CO_2H and H_2N, CO_2H]	[Boc–N(H)–CH(–CH$_2$CH(CH$_3$)$_2$)–CO_2H (leucine)]	ion exchange chromatography	6N HCl reflux	1990 (53)
[CO_2H-substituted benzene, H_2N, CO_2H]	[Boc–N(H)–CH(–CH$_2$CH(CH$_3$)$_2$)–CO_2H (leucine)]	ion exchange chromatography	6N HCl reflux	1996 (49)
[F$_3$C–CF$_3$, Cbz–N(H)–CO_2H]	[O–Bn substituted benzene, Cbz–N(H)–CH–CO_2H]	chromatography	—	1981 (1638)

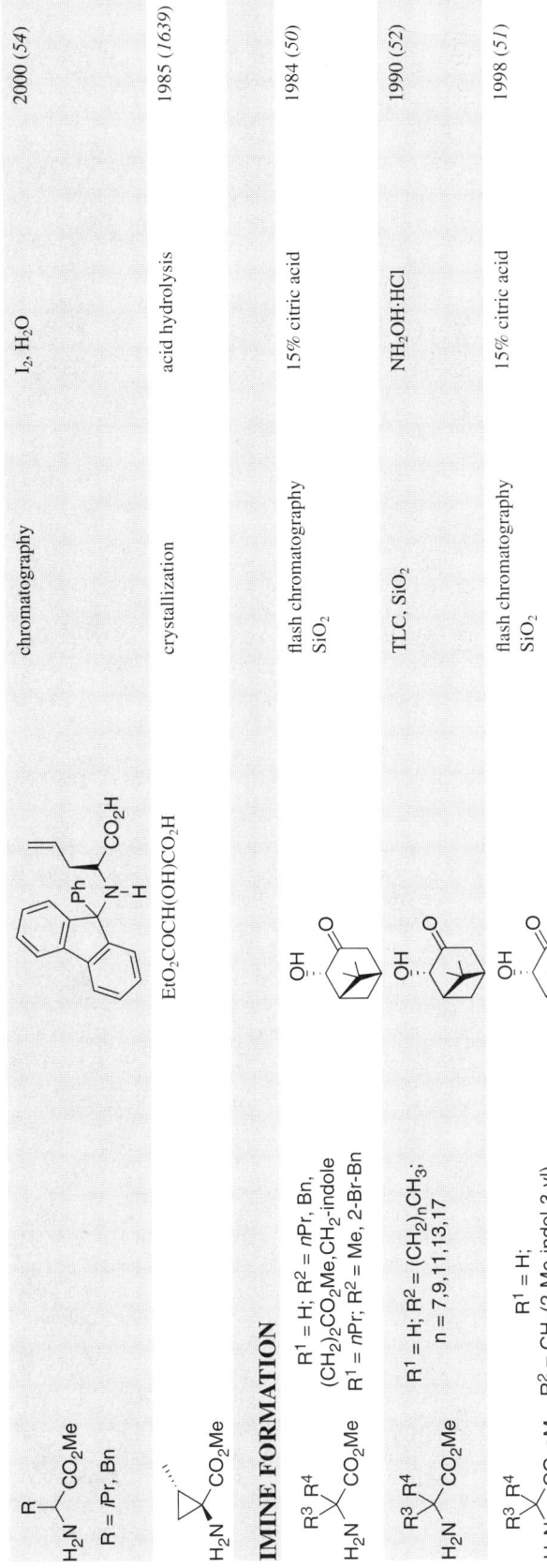

R, H₂N—CO₂Me, R = iPr, Bn

chromatography · I₂, H₂O · 2000 (54)

EtO₂COCH(OH)CO₂H · crystallization · acid hydrolysis · 1985 (1639)

H₂N$\triangleleft$CO₂Me

IMINE FORMATION

R^3 R^4 H₂N—CO₂Me

R^1 = H; R^2 = nPr, Bn, (CH₂)₂CO₂Me, CH₂-indole
R^1 = nPr; R^2 = Me, 2-Br-Bn

flash chromatography SiO₂ · 15% citric acid · 1984 (50)

R^3 R^4 H₂N—CO₂Me

R^1 = H; R^2 = (CH₂)ₙCH₃; n = 7,9,11,13,17

TLC, SiO₂ · NH₂OH·HCl · 1990 (52)

R^3 R^4 H₂N—CO₂eM

R^1 = H; R^2 = CH₂(2-Me-indol-3-yl)

flash chromatography SiO₂ · 15% citric acid · 1998 (51)

of examples of resolutions of β-amino acids via N-acylation by *Candida antarctica* lipase B and penicillin acylase, along with a few α-amino acid resolutions (*74*).

The tables in this section will define the yields of enzymatic resolutions based on the total amount of racemic amino acid substrate, rather than the 50% of enantiomer that can theoretically be converted. Thus, a complete separation of one enantiomer would have a yield of 50%. This definition helps to highlight the loss of compound during an enzymatic resolution and, more importantly, makes it easier to distinguish methods such as enzymatic dynamic resolutions in which >50% of the racemic amino acid is converted into enantiomerically pure product.

12.1.4b Resolutions by Acylases (see Table 12.4)

The discovery in 1948 that aqueous kidney extracts could rapidly hydrolyze one of the enantiomers of N-acetyl DL-Ala was soon developed into a general procedure for the resolution of many amino acids (*4*). The most common L-acylases are from hog kidney or from *Aspergillus oryzae*, as these maintain high enantiospecificity with a wide range of substrates, including non-natural amino acids (*7, 57*). D-Acylases are rarer, and generally have lower enantiospecificities than their L-analogs (*7, 57, 75*), although a D-aminoacylase with high selectivity was indentified in 1992 (*76*). A review of various acylases is contained in a 1991 report (*57*). Some acylases also hydrolyze carbamate groups, but specific carbamatases have also been isolated (*57*).

The substrate specificity of acylase I from both porcine kidney and *Aspergillus* was examined in 1989 with over 50 N-acyl amino acids and analogs (*77*), an extension of earlier studies (*78*) such as an initial study in 1952 where 33 different amino acids were tested, and 17 resolved (*79*). In the 1989 report, 11 of the substrates, including examples with low reactivity such as α-methyl-α-amino acids, were then resolved on a 2–29 g scale (*77*). Acylase from *Aspergillus oryzae* resolved 12 amino acids in 1997 with >99.6% ee and 32–50% yield, including Ala, Abu, Nva, Nle, Val, and several Cys and Met derivatives (*80*). Acylase I has been found to discriminate between mixtures of *ortho*- and *para*-substituted N-acetyl phenylalanines, deacetylating only the *para*-substituted derivatives (*81*). The acylase Amano 30000 (an *Aspergillus* acylase) was used to hydrolyze the N-trifluoroacetyl group from 3,5-dihydroxy-4-methoxyphenylglycine (*82*) and the N-acetyl group from 3,5-dichloro-4-hydroxyphenylglycine (*83*). Aminoacylases have been adsorbed on DEAE-Sephadex and used for laboratory-scale resolutions (6–14 g) of Ala, Met, Phe, Trp, and Val (*61*). Purified acylases are not needed; in 2004, N-acetyl homophenylalanine was resolved using a crude beef kidney acetone powder, giving a 41% yield of Hfe with >99% ee on a 2 g scale (*84*).

N-Alkyl amino acids are generally not substrates for acylases. However, an N-acyl-L-proline acylase has been isolated which allows for N-alkyl,N-acyl amino acids to be resolved, although the substrate spectrum of the enzyme was limited (*85*). Acylase 1 can also catalyze enantioselective ester hydrolysis or transesterification; trimethyl N-carboxymethyl Asp was regio- and enantioselectively converted into the α-butyl ester (*86*). Racemic N-Boc 2-(hydroxymethyl)piperidine was resolved via esterification of the alcohol with vinyl butyrate using *Aspergillus* acylase I. The non-esterified (S)-alcohol was then oxidized to generate pipecolic acid (*87*).

N-Phenylacetyl β-amino acids were resolved using penicillin acylase for the hydrolysis (*88–95*). The cross-linked penicillin G acylase ChiroCLEC-EC was used to acylate racemic β-substituted-β-amino esters with methyl phenylacetate. Five amino esters were studied; four were resolved with 94 to >95% ee (R = Me, Ph, 4-MeO-Ph, 3,4-(MeO)₂-Ph) (*96*). Penicillin acylase has also been used to resolve aminophosphonic and aminophosphonous analogs of Ala (*97*), and for acylation of the L-enantiomers of Phg-OMe and 4′-HO-Phg-OMe, leaving the free D-enantiomer (*98*). A preliminary examination of industrial glutaryl-7-aminocephalosporanic acid acylase found that N-glutaryl amino acid methyl esters were substrates for amide bond cleavage (*99*).

12.1.4c Resolutions by Lipases, Proteases, and Amidases (see Table 12.5)

During the initial investigations of resolutions of α-amino acids by acylases it soon became apparent that some classes of amino acids were not suitable substrates for these enzymes. An alternative route, amide hydrolysis by renal amidases, was found to be effective for many of these compounds (*68*). Proteases and lipases have been used for amino acid resolution since the early 1900's (references in *57*), but until relatively recently they have not been as popular as acylases. One difficulty is that esterases such as pig liver esterase (PLE) can catalyze ester hydrolysis without any enantioselectivity (*100, 101*). Resolutions of amino acids are included in a 1991 review of lipases, esterases, and proteases (*60*).

A detailed study of the enantiopreference of lipase from *Aspergillus niger* (ANL) for hydrolysis of amino acid methyl esters was reported in 1997. Partial purification of the enzyme removed an amidase impurity and improved enantioselectivity. A protonated α-amino substituent was identified as a requirement for high enantio-2selectivity. Small to medium-sized α-amino acids were resolved with high selectivity, giving the L-enantiomer preferentially. β-Amino acids did not react or had poor enantioselectivity (*102*). ANL is reportedly the only lipase showing high enantioselectivity towards a range of amino acid esters. However, in 1989 porcine pancreatic lipase was applied to the hydrolysis of 2,2,2-trifluoroethyl ester groups of N-Cbz amino acids; a range of 15 alkyl and aryl amino acids were resolved with 87 to >99% ee. Very small (Ala), long unbranched alkyl (2-aminooctanoic acid) or alkylaromatic side chains (homoPhe) showed reduced enantioselectivity

Table 12.4 Resolution via Acylases

$$P^1\text{-}NH\text{-}\underset{}{C}(R^1)(R^2)\text{-}CO_2P^2 \xrightarrow{\text{acylase}} H_2N\text{-}\underset{}{C}(R^1)(R^2)\text{-}CO_2P^2 \; + \; P^1\text{-}NH\text{-}\underset{}{C}(R^1)(R^2)\text{-}CO_2P^2$$

P^1	R^1	R^2	P^2	Enzyme	pH	Temp. (°C)	Time	Yield (%)	ee (%)	Reference
$ClCH_2CO$	$(CH_2)_3NH_2$	H	H	hog kidney acylase	7.6	37		28		1951 (1640)
Ac	sBu	H	H	hog kidney acylase						1951 (1641)
$ClCH_2CO$	Et	Me	H	hog kidney acylase	7.5	38	24h	50		1952 (1642)
$ClCH_2CO$	cHex	H	H	acylase 1	7.0	38		30–40		1952 (1643)
	CH_2cHex									
	Ph									
	Bn									
Ac	Me	H	H	hog kidney acylase	7.0	38	16–20h	20–40		1952 (79)
	iPr									
	tBu									
	sBu (threo)									
	sBu (erythro)									
	CH_2CH_2SMe									
	CH_2CH_2SEt									
$ClCH_2CO$	Et	H	H	hog kidney acylase	7.0	38	16–20h	20–40		1952 (79)
	nPr									
	CH_2OH									
	$CH(OH)Me$ (threo)									
	$CH(OH)Me$ (erythro)									
	CH_2CH_2SMe									
	$(CH_2)_3NHCOCH_2Cl$									
	$(CH_2)_4NHCOCH_2Cl$									
	Bn									
Cbz	CH_2CO_2H	H	H	hog kidney acylase	7.0	38	16–20h	20–40		1952 (79)
$ClCH_2CO$	$(CH_2)_2CH(O\text{-}P_2)CH_2NHCbz$	H	side chain	hog kidney acylase	7.2			12–20		1953 (1644)
$ClCH_2CO$	Me	Me	H	acylase 1	7.0	38				1953 (78)
	Et									
	nPr									
	iPr									
	tBu									
Ac	sBu	H	H	hog kidney acylase						1953 (1645)
$ClCH_2CO$	$(CH_2)_3CO_2H$	H	H	acylase 1	7.0	38		46.5		1953 (1646)

(Continued)

Table 12.4 Resolution via Acylases (continued)

P¹	R¹	R²	P²	Enzyme	pH	Temp. (°C)	Time	Yield (%)	ee (%)	Reference
Ac	[imidazole structure]	H	H	acylase 1	7.2	37				1956 (1647)
Ac	(CH₂)₄CO₂H	H	H	acylase 1	7.0	38		38		1957 (1648)
Ac	CH₂OH	H	H	hog kidney acylase	7.5	37				1959 (1649)
Ac	CH₂(4-pyrazolyl)	H	H	Aspergillus acylase, CoCl₂	7.2	37		17		1960 (1650)
ClCH₂CO	CH(O-P₂)CH₂NHCbz	H	side chain	acylase 1	7.2	37	50h	12–20		1961 (1651)
Ac	CH(Me)CO₂H	H	H	hog acylase II	8	38	64h	35		1961 (1652)
Ac	CH(OH)iPr	H	H	Takadiastase	7.2	37		24		1962 (1653)
Ac	cPent	H	H	hog kidney acylase	7.6	37		41		1965 (1654)
Ac	CH₂(pyrazol-1-yl) CH₂(pyrazol-3-yl)	H	H	acylase 1	7.2	rt		47		1966 (1655)
Ac	CH(Et)₂ cHex cPent	H	H	hog kidney acylase, CoCl₂	7.5	37	72h	32–48		1966 (1656)
ClCH₂CO	cPent	H	H	hog kidney acylase, CoCl₂	7.5	37	72h	32–48		1966 (1656)
ClCH₂CO	[pyranone structure]	H	H	aminoacylase	6.8	37	21d	7		1967 (1588)
Ac	4-(CH₂OH)-Bn	H	H	hog kidney acylase	7.5	30	48h	49		1967 (1657)
Ac	CH₂NHMe	H	H	hog kidney acylase	7.6	37	30h	25		1968 (1658)
ClCH₂CO	(CH₂)₅CO₂H	H	H	Takadiastase	7.2	37	8d	42		1968 (1590)
Ac	(CH₂)₃NO₂	H	H	acylase I	7.8	45		38		1969 (1659)
Ac	CH(OH)CH₂OBn	H	H	Takadiastase	6.8	37	4d	33–41		1969 (1592)
Ac	CH(Me)CH₂CH₂F	H	H	hog kidney acylase, CoCl₂	7.5	37	24h	27		1970 (1660)
Ac	CH₂CCH	H	H	hog kidney acylase, Co(OAc)₂	7.5	37	19h	40		1971 (1661)
Ac	2,3,4,5,6-F₅-Bn	H	H	acylase I	7.0	20		26		1971 (1662)
Ac	(CH₂)₃NHAc	H	H	acylase		38		38		1971 (1663)
ClCH₂CO	CHDPh	H	H	acylase I	7.3	38	49h	43		1972 (1664)
Ac	[imidazole (F) structure]	H	H	acylase 1	7.0	rt		37		1973 (1665)

R^1CO	R	(label)		Enzyme	pH	T (°C)	Time	Yield (%)	Year (Ref.)
ClCH$_2$CO	(CH$_2$)$_4$OH	H	H	acylase				10	1974 (1666)
Ac	CH$_2$SBn	^{3}H	H	hog kidney acylase			19h	40	1975 (1667)
Ac	CH$_2$CCH	H	H	hog kidney acylase, Co(OAc)$_2$	7.5	37	4d	32–45	1976 (1668)
Ac	CH$_2$SBn, CD$_2$SBn	^{1}H, ^{2}H	H	hog kidney acylase	7.5	38	70h	26	1976 (1669)
Ac	CH$_2$(2-naphthyl)	H	H	acylase I	7.2	38	64h	35–43	1976 (1670)
Ac	Me; iPr; Bn; CH$_2$CH$_2$SMe; CH$_2$(indol-2-yl)	H	H	aminoacylase adsorbed on Sephadex	7.5	38			1976 (61)
ClCH$_2$CO	(CH$_2$)$_2$O(CH$_2$)$_2$NHAc	H	H	hog kidney acylase	7.2	37	21h	32	1976 (1671)
Ac	CH$_2$(4-Ph-Ph); CH$_2$(1-naphthyl); CH$_2$(2-naphthyl); CH$_2$(3-benzo[b]thienyl); CH$_2$(7-azaindol-3-yl); CH$_2$(3-benzo[b]thiophene-1-oxide)	H	H	*Aspergillus* acylase	7.5	39	42h	20–48	1976 (1672)
Ac	(CH$_2$)$_2$Ph; (CH$_2$)$_3$(4-MeO-Ph)	H	H	Taka-acylase, CoCl$_2$	7.3	38	10d	42	1976 (1673)
ClCH$_2$CO	C(Me)=CH$_2$	H	H	hog kidney acylase	7.2	38		32	1977 (1674)
Ac	Bn (with [2-^{2}H]	H	H	hog kidney acylase	7.1	38	48h	40	1977 (1675)
Ac	CH(Et)CH$_2$-R$_2$	R^1	H	acylase	7	38	o/n	17	1977 (1604)
Ac	CH$_2$CH(Me)Et	H	H	hog kidney acylase	7.5	37	48h	28	1978 (1676)
Ac	(CH$_2$)$_4$OH	H	H	hog kidney acylase	7.5	37	48h	28	1978 (1677)
Ac	(CH$_2$)$_4$OH	H	H	acylase I	7	37	72h	23	1978 (1678)
Ac	2,3-epoxy-cHex	H	H	hog kidney acylase					1978 (1679)
Ac	(4-Cl-pyridin-2-yl structure)	H	H	hog kidney acylase	7.0	38	24h	43	1979 (1680)
Ac	CH$_2$CCH	H	H	hog kidney acylase	7.4	37	36h	17	1979 (1681)
Ac	CH$_2$(2-pyridyl)	H	H	hog kidney acylase					1979 (1682)
ClCH$_2$CO	CH$_2$CO(4-HO-Ph)	H	H	acylase	7.3	38		14	1980 (1683)
Ac	CH$_2$(5-Br-indol-3-yl); CH$_2$(7-Br-indol-3-yl)	H	H	fungal α-amylase				31–33	1980 (1684)
Ac	CH$_2$CCCH$_2$NHBoc	H	H	hog kidney acylase, CoCl$_2$	7.5	37	15h	41	1980 (1685)

(Continued)

Table 12.4 Resolution via Acylases (continued)

P¹	R¹	R²	P²	Enzyme	pH	Temp. (°C)	Time	Yield (%)	ee (%)	Reference
ClCH₂CO	[structure: Cl, O–N ring]	H	H	hog kidney acylase				28		1981 (1686)
Ac	CDMe₂ CTMe₂	H	H	hog kidney acylase	7.0	37		16–26		1981 (1687)
ClCH₂CO	[structure: Cl, O–N ring]	H	H	hog kidney acylase	7.0		1d	29–42		1981 (1688)
ClCH₂CO	Et	Me	H	hog kidney acylase	7.5	38	24h	50		1982 (1689)
Ac	CHDOH	H	H	hog kidney acylase				43–48		1982 (1690)
Ac	CH₂(6-Br-indol-3-yl)	H	H	acylase						1982 (1691)
Ac	C(Me)=CDH	H	H	hog kidney acylase	7.2	37	23h	35		1983 (1692)
Ac	4-HO-Ph	H	H	hog kidney acylase	7.8	37	48h	34		1983 (1693)
CF₃CO	CH₂cPr	H	H	hog kidney acylase	7.0	37	2h	40–45		1983 (1694)
Ac	[structure: MeO isoxazole, N–O]	H	H	immobilized Enzygel aminoacylase	7.1	30	186h	41		1983 (1695)
Ac	C³H(Me)C³HMe ¹³C(³H)₂Me	H	H	acylase I	7.0	38				1983 (1696)
Ac	CH₂CCH	H	H	hog kidney acylase	7.4	37	21h			1983 (1697)
Ac	Bn with [2-²H]	H	H	acylase I	7.5	37	1d	26		1984 (1698)
Ac	CH₂(3-pyridyl)	H	H	hog kidney acylase	7.0–7.2	25	24h			1984 (1699)
ClCH₂CO	CH(Me)CH₂-R²	R¹	H	hog kidney acylase						1985 (1639)
ClCH₂CO	[structure: Cl, O–N ring]	H	H	hog kidney acylase	7.0		1d	29–42		1985 (1700)
Ac	4-(COPh)-Bn	H	H	Aspergillus acylase	7.5	37	18h	40		1986 (1701)
Ac	(CH₂)₄CO₂H	H	H	hog kidney acylase						1986 (1702)
Ac	nBu CH₂COPh	H	Me, tBu	hog kidney acylase	7.2	37	16h	41–43		1988 (1703)
Ac	CH=CHCH₂NH₂	H	H	hog kidney acylase	7.2	37		23–35		1988 (1704)
Ac	CH₂CF₃ (CH₂)₂CF₃	H	H	hog kidney acylase	7.1	36	3.5h	46–49		1988 (1705)
BnCO	CH(OH)(CH₂)₂NHCbz	H	H	immobilized E. coli acylase	7.5	37		31	>95	1988 (1706)

Acyl	R			Enzyme			
Ac, ClCH₂CO, EtCO, sBuCO, Cbz, CHO, NH₂CH₂CO	Me	H	H	hog kidney or *Aspergillus* acylase 1	17–50	91–>99.5	1989 (77)
	Et						
	nPr						
	iPr						
	iBu						
	tBu						
	cPr						
	CH₂-cPr						
	CH₂(adamantyl)						
	(CH₂)ₙCH=CH₂; n = 1,3,6						
	CH₂CH=CHMe						
	CH₂C(Me)=CH₂						
	CH₂CCH						
	(CH₂)₃CCH						
	CH₂CH=CHPh						
	CH₂SPh						
	(CH₂)₃SPh						
	Bn						
	(CH₂)₂Ph						
	CH₂Cl						
	(CH₂)ₙOH: n = 1,3,5,6,8						
	(CH₂)₃CH(Me)OH						
	CH₂C(Me)₂OH						
	(CH₂)₃CN						
	(CH₂)₄CN						
	(CH₂)₃CO₂H						
	CH₂CH=CHCO₂H						
	2-(CO₂H)-cPr						
	CH₂CONH₂						
	(CH₂)₃CH(CO₂H)NHAc						
	CH₂COMe						
	CH₂(2-furyl)						
Ac, ClCH₂CO, EtCO, sBuCO, Cbz, CHO, NH₂CH₂CO	Et	Me	H	hog kidney or *Aspergillus* acylase 1	17–50	91–>99.5	1989 (77)
	CH₂OH						
	(CH₂)₂CO₂H						
	(CH₂)₂SMe						
	Bn						
	4-HO-Bn						
	3,4-(HO)₂-Bn						
	CH₂(indole)						

(Continued)

Table 12.4 Resolution via Acylases (continued)

P¹	R¹	R²	P²	Enzyme	pH	Temp. (°C)	Time	Yield (%)	ee (%)	Reference
Ac, ClCH₂CO, EtCO, sBuCO, Cbz, CHO, NH₂CH₂CO	H, Me, nPr	H	CO₂P² = CH₂CO₂H, CH(Me)CO₂H, PO₃H	hog kidney or Aspergillus acylase 1				17–50	91–>99.5	1989 (77)
Ac	(CH₂)₂CF₃ (CH₂)₃CF₃	H	H	hog kidney acylase	7.0	25	14h	30–40	>99	1989 (1707)
Ac	(Br, OMe isoxazole structure)	H	H	immobilized aminoacylase	7.1	rt	32h	43		1989 (1708)
Ac	CH₂(1-naphthyl)	H	H	Aspergillus acylase	7.2	37	48h	35	>99	1989 (1709)
Ac	CH=CHOMe	Me		hog kidney acylase	7.4	40	16h			1990 (1710)
Ac	(CH₂)₃-αN	H		N-acyl-L-Pro acylase	7	30				1990 (1711)
Ac	CH₂(2-thienyl) CH₂(2-furyl)	H		acylase I, Co(OAc)₂	7–8	38	22h	31–32		1990 (1712)
Bz	CH(OH)(CH₂)₂NHCbz	H		immobilized E. coli acylase acylase I	7.5	37	3h	34		1990 (512)
ClCH₂CO	nOct	H		acylase I						1990 (52)
Ac	CH₂CCH	H		hog kidney acylase	7.5			50		1991 (1713)
Ac, Bz, Cbz	nPr nBu iBu Bn CH₂CH₂SMe	H		Alcaligenes faecalis D-amino acylase	7.8	37	2–10h	10–50	53–100	1992 (76)
ClCH₂CO with N-R	Me Et nPr nBu	H		N-acyl-L-Pro acylase	7			19–50		1992 (85)
Ac	(CH₂)₂CH=CH₂	H		hog kidney acylase	7.5	38		45		1992 (1714)
Ac	CH₂(imidazol-4-yl)	H		hog kidney acylase	7.2	37	72h	37–40	>99.9	1992 (1715)
Ac	CH(Me)CCH	H		hog kidney acylase	7.6	37	5d		>99	1993 (1716)
BnCO	Me Ph CF₃ C₂F₅ C₃F₇ 2-F-Ph 4-F-Ph	H	CO₂P² = CH₂CO₂H	penicillin acylase	6.5–7.5			50	>95	1993 (88)

P1	P2	P3	R	Enzyme	pH	T (°C)	Time	Yield (%)	ee (%)	Year (Ref.)
Ac	H	H	CH2(cyclooctatetraene)	Aspergillus acylase, CoCl2	7.5	40	24h	43	>99	1993 (1717)
Ac	H	H	CH2-(N-Et-carbazol-3-yl)	acylase	—	—	—	—	—	1994 (1718)
BnCO	CO2P2 = CH(Me)CO2H	H	CF3	penicillin acylase	7.5		5h	50	>96	1994 (89)
Ac	H	H	CH(Me)Et	hog kidney acylase	8	38	4d	45	>99	1994 (1719)
Ac	H	H	CH2(6-Me-2-naphthyl)	acylase Amano 15000	7.5	37	o/n	39–50		1994 (1720)
			CH2(6-Cl-2-naphthyl)							
			CH2(5,6,7,8-tetrahydro-2-naphthyl)							
			CH2(3-benzo[b]thienyl)							
			CH2(2,3-dihydro-1,4-benzodioxin-6-yl)							
CICH2CO	H	H	3,5-(HO)2-Ph	aminoacylase	7		3h	45	>99	1995 (1721)
BnCO	CO2P2 = CH2CO2H	H	Ph	penicillin acylase	7.5		5–12h	50	>96	1995 (90)
			4-F-Ph							
			2-F-Ph							
			4-Cl-Ph							
			4-MeO-Ph							
			3,4,5-(MeO)3-Ph							
Ac	H	H	4-CN-Bn	hog kidney acylase		37	4d	34		1995 (1722)
Ac	H	H	CH2(4-F-pyridyl)	hog kidney acylase	7.5	35	24h	45		1995 (1723)
Ac	H	H	(CH2)5Ph	acylase, CoCl2	7.6	37	3d	45–48		1995 (1724)
			(CH2)18Ph							
BnCO	CO2P2 = CH(Me)CO2H, CH(Et)CO2H, CH(Bn)CO2H, CH(allyl)CO2H	H	Me	penicillin G acylase	7	30–35	5h	47–53	94–100	1996 (91)
Ac	H	H	CH2cPr	porcine pancreatic acylase	7.5	37	4h	46		1996 (1725)
Ac	H	Me, Bn	(CH2)4CO2Me/Bn, (CH2)5CO2Me/Bn, (CH2)6CO2Me/Bn	Aspergillus acylase	8	37	20h	32–38		1996 (1726)
Ac	H	H	4-CN-Bn	hog kidney acylase	7.2–7.5, CoCl2	40	3d	35		1996 (1727)
CICH2CO	H	H	CH2CHFCO2Me	hog kidney acylase	6		3h	43	>95	1996 (1728)

(Continued)

Table 12.4 Resolution via Acylases (continued)

P¹	R¹	R²	P²	Enzyme	pH	Temp. (°C)	Time	Yield (%)	ee (%)	Reference
Ac	CH(Me)OBz	H	H	hog kidney acylase	8	38	5d	37		1997 (216)
Ac	Me	H	H	Aspergillus acylase	5.2	37		33–50	>99.6	1997 (80)
	Et									
	nPr									
	nBu									
	iPr									
	CH$_2$SH									
	CH$_2$SMe									
	CH$_2$CH$_2$SMe									
	CH$_2$CH$_2$SSMe									
	CH$_2$CH$_2$SEt									
	CH$_2$CH$_2$SeMe									
	CH$_2$CH$_2$SO$_2$Me									
ClCH$_2$CO	CH$_2$(ferrocene)	H	H	Aspergillus acylase	7	37	48h	35	>95	1997 (1729)
ClCH$_2$CO	nOct	H	H	Aspergillus acylase	7	rt	7d	33–42		1997 (595)
	n-C$_{14}$H$_{29}$									
Bz	CH(OH)iPr	H	H	immobilized penicillin G acylase	7.8	35		36		1997 (1730)
Ac	CH$_2$CH$_2$SCF$_3$	H	H	acylase I	7.4	25	8h	42		1997 (1731)
Ac	CH$_2$(2-anthraquinolyl)	H	H	Aspergillus acylase, CoCl$_2$	7	37	3d	45		1997 (1732)
Ac	2-OH-3,5-(NO$_2$)$_2$-Bn	H	H	Aspergillus acylase	7.5–8.0	40		41		1998 (1733)
Ac	(CH$_2$)$_3$[4-C(-N$_2$-)CF$_3$-Ph]	H	H	acylase, CoCl$_2$	7			26		1998 (1734)
Ac	iBu	H	H	acylase, CoCl$_2$				32–46	96->99.5	1998 (1735)
	cHex									
	CH$_2$CH$_2$SMe									
	4-F-Ph									
	4-Cl-Ph									
	4-MeO-Ph									
	4-Cl-Bn									
Ac	CH$_2$(7-azaindol-3-yl)	H	H	Aspergillus acylase	7.2	40		45		1998 (275)
Ac	CH$_2$(5-CN-2-thiophene)	H	H	acylase	6.5	36				1998 (1736)

		X	Y	Enzyme, conditions	pH	T (°C)	time	ee	Yield (%)	Year (Ref.)
CF$_3$CO	3,5-(OH)$_2$-4-MeO-Ph	H	H	acylase Amano 30000 CoCl$_2$, NaN$_3$	7				42	1998 (82)
Ac	3,5-Cl$_2$-4-OH-Ph	H	H	*Aspergillus* acylase	7.9	37–40		>99.5	26	1998 (83)
Ac	4-*t*Bu-Bn	H	H	hog kidney acylase	7.2–7.5	40	48h	99.9	46	1998 (1737)
Ac	4-HO-Bn 4-Cl-Bn 4-Me-Bn 4-NO$_2$-Bn 4-F-Bn	H	H	hog kidney acylase	7.0	25	2h		36–42	1998 (81)
Ac	4-(CH$_2$OH)-Bn	H	H	acylase 1	8				40	1999 (1738)
Ac	[fused-ring (pyrene / anthracene) CH$_2$ structure]	H	H	acylase		37	2d		46–49	1999 (1481)
Ac	CH$_2$(4-Br-indol-3-yl)	H	H	*Aspergillus* acylase	7.5	37	48h		49	1999 (1739)
Ac	CH$_2$(1-MeO-indol-3-yl)	H	H	*Aspergillus* acylase, CoCl$_2$	7.5	22	48h	—	43	1999 (374)
H	Ph	H	Me	penicillin G acylase: use to acylate with (4-HO-Ph)CH$_2$CO$_2$Me				>98% ee		2000 (98)
	4-HO-Ph									
Ac	2,3-, 2,4-, 2,5-, 2,6-, 3,4-, or 3,5-F$_2$-Bn	H	H	*Aspergillus* acylase	8	38	48h		46	2000 (1740)
Ac	(3S,4S)-CHDCD(CD$_3$)(^{13}CH$_3$)	H	H	*Aspergillus* acylase, CoCl$_2$	8	37			29	2001 (1741)
Ac	(CH$_2$)$_3$Ph	H	H	*Aspergillus* acylase	—	—		>99% ee	—	2001 (1742)
Ac	*n*Hex	H	H	*Aspergillus* acylase, CoCl$_2$	7–8	35–40	39h	>99% ee	35	2002 (373)
Ac	CH(Me)CF$_3$ CH$_2$CH(Me)CF$_3$	H	H	hog kidney acylase I	7.5	25	48h		48	2002 (1743)
Ac	(CH$_2$)$_2$Ph	H	H	crude beef kidney acetone powder	7.5	37	24h	>99%	41 (2g scale)	2004 (84)
Ac	(CH$_2$)$_3$Br (CH$_2$)$_6$Br (CH$_2$)$_7$Br (CH$_2$)$_{10}$Br	H	H	aminoacylase	7	38	24h		33–40	2004 (1744)
Ac	(CH$_2$)$_4$Br	H	H	L- or D-aminoacylase, CoCl$_2$	7	38	24h		35–40% cyclizes to pipecolic acid	2005 (1745)

(21–71% ee) (*103*). *Carica papaya* lipase was applied to transesterification of *N*-Cbz aliphatic amino acid 2,2,2-trifluoroethyl esters with methanol, producing the methyl esters with >99.8% ee (*104*). α-Azido esters can also be resolved using lipases (*105*), while lipases were employed for a synthesis of (*R*)-proline derivatives by asymmetrization of a *meso* precursor (*106*). A new lipase, rice bran lipase, was examined for its ability to hydrolyze esters of *N*-acetyl amino acids. *n*-Butyl esters gave the best results, with >99% ee for Val, Phg, Phe, and Tyr, but poor results with Ala and Ile (*107*). Lipases have been mutated in a combinatorial fashion by simultaneous mutation at two amino acid sites in an attempt to expand substrate specificity; amino esters were not tested as substrates, but the same principle should be useful (*108*).

"Alcalase," an industrial proteolytic enzyme consisting primarily of subtilisin Carlsberg, has been successfully used in supercritical CO_2 to resolve five Cbz-protected amino acid methyl esters possessing unnatural alkyl side chains (*109*). Alcalase was also used to resolve several tetrahydrofuran, tetrahydrothiophene, tetrahydropyran, and tetrahydrothiopyran-based cyclic amino acids (*110*). The resolution of *N*-protected Asn esters by various proteases on a preparative scale (up to 20 g) was studied in 2004, with Alcalase 2.4L giving >99% ee for several protecting group combinations, such as Cbz-D-Asn-OMe (*111*). Subtilisin Carlsberg resolved α-hydroxyglycine-containing dipeptides (*112*) and substituted *N*-trifluoroacetyl phenylalanine methyl esters (*113*), while immobilized subtilisins in two-phase systems were used to prepare optically active D-arylglycines (*114*). The role of conformational flexibility for discrimination of enantiomeric substrates during subtilisin-catalyzed reactions in organic solvents has been examined; enhancement of flexibility by the addition of a denaturing additive such as DMSO improved enantioselectivity (*115*). The racemic *N*-acetyl amino acids produced by alkylation/decarboxylation of diethyl acetamidomalonate are generally resolved using acylases. However, the acyl dialkyl aminomalonate precursors can also be selectively converted into monoesters and decarboxylated to give *N*-protected α-amino esters, with resolution by ester hydrolysis with chymotrypsin or subtilisin. This method allows for resolution of *N*-acyl aminomalonates such as *N*-Cbz aminomalonate to be alkylated, decarboxylated, and resolved into an *N*-protected amino acid ready for peptide synthesis (*116*).

The α-benzyl ester group of dibenzyl esters of Glu, Asp, 2-aminoadipic acid, and 2-aminosuberic acid could be selectively removed by hydrolysis with pronase (a mixture of several *Streptomyces grisues* proteolytic enzymes) (*117*). Thermitase, an alkaline serine protease from *Thermoactinomyces*, was used to hydrolyze the methyl esters of *N*-Boc 4′-chloro-Phe, 2′,4′-dichloro-Phe, 3,4-dichloro-Phe, β-(2-naphthylalanine), and S-benzyl cysteine (*118*), or the benzyl esters of Cbz-Gly-3-thia-Lys, Cbz-Gly-3-thia-Phe, Cbz-Gly-3-thia-*p*-nitro-Phe, and Cbz-Gly-3-thia-Leu (*119*). In 1997 (*120*) and 2005 (*121*) it was reported that *Aspergillus oryzae*

protease (Amano protease A) had greatly enhanced ester hydrolysis rates for *N*-unprotected amino acids over the usual *N*-protected substrates. Enantioselectivity decreased with methyl ester substrates, but could be maintained if longer alkyl esters, such as isobutyl esters, were employed, with further enhancements at low temperatures. A number of aliphatic and arylalanine amino acids were resolved. Apomyoglobin from horse heart myoglobin quickly hydrolyzed the 4-nitrophenol ester of Boc-Phe-ONp, with the remaining ester possessing 98% ee after 60% conversion at 8 min (*122*).

L-Specific aminopeptidase from *Pseudomonas putida* was applied to the resolution of a number of allyl-glycine carboxamides (*123*), while whole bacterial cells containing stereospecific amidases were able to resolve carboxamides of piperazine-2-carboxylic acid and piperidine-2-carboxylic acid (*124*). A peptide amidase from orange peel was capable of selectively hydrolyzing a range of L-amino acid amides, although it was not used for resolution (*125*). The amidase in *Rhodococcus* sp. cells enantioselectively hydrolyzed amides of Phe and a number of arylglycines with 60 to >99% ee (*126*). Papain has been employed to hydrolyze a number of amides, including Boc-Agl(Fmoc)-NH₂ (Agl = α-amino-Gly) (*127*). *N*-Carbamoyl amino acid amides, H₂NCO-Xaa-NH₂, are undesired byproducts during enzymatic amino acid synthesis from hydantoins using L-hydantoinase. Chirazymes P1 and P2 successfully hydrolyzed the amide of Phe, Trp, and Met to give the *N*-carbamoyl amino acids with >98% ee; Pro and Phg did not react (*128*). Aminoacylase I from *Aspergillus melleus*, normally used to resolved *N*-acetyl amino acids, has been found to effectively hydrolyze amino acid amides as well. The amides of Abu, Leu, Phg, 4′-OH-Phg, Tyr, and Hfe were resolved with 92 to >99 ee (*129*).

α,α-Disubstituted amino acids present a particularly challenging substrate for enzymatic resolution. PLE can give moderate enantioselectivity, but has a narrow substrate specificity (*130*). It was used to resolve α-allyl-Phg-OEt, with 96% ee and 33% yield (*131*). An amino acid amidase from *Mycobacterium neoaurum* has a much broader substrate tolerance and gave high enantioselectivities for a range of α-methyl amino amides, although substituents larger than a methyl group reduced the stereoselectivity (*130, 132–134*). This enzyme also resolved Cᵅ-fluoroalkyl-substituted amino acids (*135*), and was used to prepare [¹⁵N]-D-isovaline (*136*). In contrast, amidases from *Ochrobactrum anthropi* were able to hydrolyze bulkier α,α-disubstituted amino acids such as α-propyl-phenylglycine, although the limit was reached with α-benzyl-phenylglycine (*137*). The amidase from *Klebsiella oxytoca* resolved α-trifluoromethyl-Ala amide to give a 50% yield of (*S*)-amide with >99% ee and a 42% yield of (*R*)-acid with 95% ee (*138*). A microbial whole-cell catalyst, amidase-containing *Rhodococcus* sp. AJ270, hydrolyzed primary amides of α-alkyl- and α-aryl-phenylglycines with variable (5–95%) enantioselectivity (*139*).

Enzymes other than PLE have been used for ester hydrolysis during α,α-disubstituted amino acid resolution. A potentially very useful lipase from *Humicola langinosa* was reported in 1995 which uniquely resolved amino esters with two large α-alkyl groups, and discriminated between side chains differing by as little as a single carbon atom. Ethyl esters of amino acids containing alkyl, aryl, and α,α-cyclic side chains were resolved with 53–99% ee (*100*). An esterase purified from commercial *Candida lipolytica* lipase preparations was found to tolerate both α-methyl groups and different types of α-amino groups, including amines, hydrazines, and *N*-acyl-amines (*140*). Several other reports of resolutions of α,α-disubstituted amino acids have appeared (*141*). One method used to overcome the steric hindrance of α,α-disubstituted amino acids is to reduce the carboxyl group to an alcohol and then use a lipase to acylate the alcohol. The amino acid is regenerated by oxidation of the alcohol. This method was applied to the resolution of α-vinyl amino acids (*142*). Another alternative approach used chymotrypsin to enantioselectively decarboxylate α-nitro-α-methyl carboxylic acid (*143*). The remaining enantiomer was hydrogenated to generate the amino group.

There are few examples of the use of lipases or proteases for resolution through ester formation, although Boc-vinylglycine was resolved by this method (*144*). Lipases and proteases formed amino acid amides from the amino acid ester using *tert*-butanol saturated with ammonia (*145*), while *N*-substituted azetidine-2-carboxylate methyl esters were resolved by amidation with *Candida antarctica* lipase (*146*) and several fluoro-substituted phenylalanine derivatives were resolved via hydrazide formation using papain and hydrazine (*147*). Papain was also used to resolve *N*-benzoyl 3′-fluorotyrosine via anilide formation (*148*); this method was employed for a number of other resolutions, such as with *N*α-Ac,*N*δ-tosyl,*O*-benzyl *N*δ-hydroxy-ornithine (*149*). A two-step resolution/acylation of Asp(OMe)-OMe employed lipase SP 526 to *N*-butanoylate the dimethyl ester (65% ee), followed by transesterification with the lipase Novozym 435 to give *n*BuCO-Asp(O*n*Bu)-OMe with >99% ee (*150*). *C. antarctica* lipase A acylated methyl pipecolate with trifluoroethyl butanoate to give a 49% yield of (*S*)-acylated product with 99% ee (*151*).

Lipases have been applied to resolutions of β-amino esters (*152*, *153*), as well as the cyanohydrin precursors of β- and γ-amino acids (*154*). Pig liver esterase hydrolyzed the methyl ester of *N*-benzoyl-3-aminobutanoate (*155*), the ethyl ester of α-ethyl-norleucine (*156*), and desymmetrized dimethyl 3-aminoglutarate (*157*). *threo*-Methylphenidate (Ritalin, used to treat ADHD), was resolved by ester hydrolysis with α-chymotrypsin or subtilisin Carlsberg (*158*). Several *N*-unprotected β-aryl-β-amino acid ethyl esters were resolved by ester hydrolysis with Amano PS, with pH playing a significant role in enantioselectivity. At pH 8, five examples (Ar = Ph, 2-Br-Ph, 4-Br-Ph, 4-F-Ph, 2-naphtyl) were hydrolyzed with 91–99% ee (*159*). Methyl and ethyl esters of the cyclic β-amino acids *N*-Boc homoproline

(pyrrolidine-2-acetic acid), homopipecolic acid (piperidine-2-acetic acid), and 3-carboxymethylmorpholine were resolved by hydrolysis with *Burkholderia cepacia* lipase to give the acids and residual esters with 98 to >99% ee (*160*).

N-Acylation is also employed for β-amino acid resolution. Racemic ethyl 3-aminobutyrate was reacted with *C. antarctica* lipase B in butyl butyrate to produce a mixture of ethyl (*S*)-3-aminobutyrate and butyl (*S*)-3-aminobutyrate, along with *N*-butanoyl butyl (*R*)-3-aminobutyrate (*161*). Ten different cyclic β-amino esters were resolved by acylation with 2,2,2-trifluoroethyl esters using lipase SP 526 from *C. antarctica* or lipase PS from *Pseudomonas cepacia* (*162*). The β³-amino acid analog of Trp was resolved via *N*-acylation with butyl butyrate and *C. antarctica* lipase B (*163*), as were the β³-amino acid analogs of Ala and Phg (*164*), while gram-scale resolutions of four β-heteroaryl-β-amino esters (Ar = 2- or 3-thienyl or furyl) were achieved by acylation with ethyl butanoate (*165*).

12.1.4d Resolutions by Amino Acid Oxidases (see Table 12.5)

Unlike proteases, lipases, or acylases, amino acid oxidases "resolve" DL-α-amino acids by selectively destroying one of the enantiomers via deamination/oxidation to its α-keto acid analog. The D-amino acid oxidase from mammalian kidney or L-amino acid oxidase from snake venom are most commonly employed, resolving a variety of amino acids, including Ala, Arg, Glu, His, Hyp, Leu, Lys, Met, Phe, Ser, Trp, Tyr, and Val (*4*, *5*, *166*). L-Amino acid oxidase was used for a resolution of aryl-substituted Phe analogs (3-CF₃-, 3-Me-, 3-MeO-, 2-Me-, 2-Cl-, 3-Cl-, and 2-F-substituted), giving products with >97% ee in 45–48% yield (*167*). An L-amino acid oxidase has also been used to deracemize amino acids in a dynamic resolution procedure, with >50% yields of the D-amino acids (*168*). Pipecolic acid can be resolved using D-amino acid oxidase, as described in the monograph of Greenstein and Winitz (*169*). Unprotected 4-hydroxy-pipecolic acid was also separated with D-amino oxidase (*170*), while racemic 4,5-didehydropipecolic acid was resolved by a combination of enantiospecific oxidation with D-amino acid oxidase, followed by reduction of the byproduct imine with sodium borohydride to regenerate the racemic substrate; four cycles of this procedure produced 4,5-dehydro-L-Pip with >95% ee (*171*). L-[3-¹¹C]-Ala and L-[3-¹¹C]-Phe were resolved with D-amino acid oxidase immobilized on glass beads (*172*).

12.1.4e Resolutions by Other Enzymes (see Table 12.5)

Carbonic anhydrase can be applied to amino acid resolutions via ester hydrolysis (*173*). The nitrile group of amino nitriles has been hydrolyzed to an amide or acid moiety by various microorganisms, but this method has not been widely used (*57*, *174*). Cyclic derivatives of amino acids such as hydantoins also undergo stereoselective hydrolysis by microorganisms (*57*). A number of aldolase enzymes with varying specificity have been used to degrade one of the enantiomers of racemic *threo*-β-hydroxy amino acid esters, producing the unreacted product enantiomer along with glycine and an aldehyde (*175–177*). D-Lys was obtained by reaction of DL-Lys with lysine decarboxylase (*178*), while DL-Arg was converted into L-Orn by hydrolysis of the guanidino group with L-arginase (*179, 180*). Glutamic decarboxylase, contained in lyophilized *E. coli*, was used to decarboxylate DL-γ-methylene-Glu to provide D-γ-methylene-Glu (*181*).

A post-translational enzyme-catalyzed isomerization of amino acid residues in peptide chains has been observed in the spider *Agelenopsis aperta*, with Ser, Cys, *O*-methyl Ser, and Ala residues in the middle of peptide chains isomerized to the D-amino acid. Both D- and L-amino acid residues can be isomerized, with a common Leu-Xaa-Phe-Ala recognition site (*182*). A dehydroalanine-containing inhibitor of the epimerase has been reported (*183*).

12.1.5 Resolutions by Other Methods

Enantioselective micellar catalysis has been used as a model for enzyme reactions, with hydrolysis of *p*-nitrophenyl esters of D-Phe catalyzed by sugar-derived homochiral surfactant micelles (*184*). Alternatively, metal complexes formed from a chiral macrocyclic or lipophilic ligand were used to form metallomicelles, which hydrolyzed *N*-dodecanoyl amino acid *p*-nitrophenyl esters with enantioselectivities up to $k_S/k_R = 7.8$ (*185, 186*). An active tripeptide (Cbz-L-Phe-L-His-L-Leu) in a coaggregate system of vesicular and micellar surfactants showed an apparent complete stereoselectivity for the hydrolysis of *N*-decanoyl-DL-Phe *p*-nitrophenyl ester (*187*).

An imprinted polymer enantioselectively hydrolyzed Boc-Phe *p*-nitrophenyl esters with nearly twofold faster rate preference for the imprinted enantiomer. The polymer was generated from methacrylic acid, ethylene glycol dimethyacrylate, and an imidazole-containing vinyl monomers, designed to mimic the key catalytic elements of chymotrypsin. The monomers were polymerized around a Phe phosphonate analog designed to mimic an ester hydrolysis tetrahedral transition state (*188*).

A potentially very useful resolution reaction involves enantiodifferentiating coupling reagents, novel derivatives of 2-chloro-4,6-dimethoxy-1,3,5-triazine formed by reaction with chiral amines, such as strychnine, brucine or sparteine. This asymmetric reagent reacted diastereoselectively with 2 equiv of racemic *N*-protected amino acids to preferentially form the activated ester of predominantly one enantiomer (see Scheme 12.1). The active ester then coupled with amino esters to give dipeptides with up to 96% de. However, the diastereoselectivity varied, depending on the amino acid being activated (*189*).

A c3-symmetrical chiral trisoxazoline zinc complex has been developed as an artificial zinc hydrolase enzyme. Kinetic resolution of *N*-benzoyl amino ester phenyl esters via transesterification to the methyl ester proceeded with selectivity factors of up to 4.5; Bz-Phg-OPh was converted into Bz-Phg-OMe with 20% ee at 51% conversion (= selectivity factor of 1.8) (*190*).

12.1.6 Biocatalytic Dynamic Resolutions

Enzymatic resolutions have been combined with several techniques in order to convert a racemic mixture into a single enantiomer, rather than the 50% yield obtained by conventional resolutions (for reviews, see *7, 10*). This method is limited to substrates that are susceptible to racemization—meaning they must possess an α-hydrogen. Traditionally, the racemization has been accomplished by heating the remaining undesired amino acid ester enantiomer to 150–170 °C, in an iterative process. A more advanced version of this method instead employs in situ racemization catalyzed by pyridoxal 5-phosphate (proceeding through a racemization-prone Schiff base intermediate), in combination with protease-catalyzed hydrolysis of the L-amino acid ester. Phe, Tyr, Leu, Nva, and Nle were resolved in one report (*191*). Dynamic kinetic resolution of phenylglycine esters was achieved via lipase-catalyzed ammonolysis using Novozym 435 (*C. antarctica* lipase B) in combination with pyridoxal-catalyzed in situ racemization of the unconverted ester at −20 °C. The amide racemizes much more slowly than the ester at this temperature. D-Phg-NH₂ was produced with 88% ee at 85% conversion (*192*). Racemization of amino acid Schiff base substrates (rather than catalytic Schiff base intermediates) has been induced by DABCO, with ester hydrolysis by chymotrypsin or lipases in mixtures of aqueous and organic solvents. The substrate for resolution is apparently a small amount of Schiff-base-hydrolyzed substrate present in solution, as the resolved L-amino acid conveniently precipitated from the reaction solution as the completely deprotected product (*193*). A dynamic kinetic resolution of Pro-OMe and methyl pipecolate was achieved by acylation with vinyl butanoate by *C. antarctica* lipase A. Acetaldehyde released in situ from the vinyl butanoate in the presence of triethylamine racemized the non-acylated Pro-OMe and Pip-OMe. This allowed for 70–90% of the imino ester to be converted into *N*-butanoyl L-Pro-OMe or L-Pip-OMe with 97% ee (*194*).

Other optically labile substrates include α-bromo esters, potential precursors of amino acids via nucleophilic amination. They were resolved by using lipases in the presence of bromide, conditions that racemize the ester more quickly than the acid (*195*).

Enzymes that racemize amino acids (amino acid racemases) are available. Unfortunately, enzymes that racemize the *N*-acyl amino acid or amino acid ester/amide substrates commonly used for enzymatic resolution are much rarer (*7*). Enzymatic racemizations of amino acids and amino acid derivatives were included in a 1997 review of racemization (*9*). An *N*-acylamino acid

Table 12.5 Resolutions by Lipases, Proteases, Amidases, Oxidases, and Other Enzymes

$$P^1{-}\underset{H}{N}{-}\overset{R^1\ \ R^2}{\diagdown\!\diagup}{-}COP^2 \xrightarrow{\text{lipase}} P^1{-}\underset{H}{N}{-}\overset{R^1\ \ R^2}{\diagdown\!\diagup}{-}CO_2H \ + \ P^1{-}\underset{H}{N}{-}\overset{R^1\ \ R^2}{\diagdown\!\diagup}{-}COP^2$$

P¹	R¹	R²	P²	Enzyme	pH	Temp. (°C)	Time	Yield (%)	ee (%)	Reference
H	2-, 3- or 4-OH-Bn 3,4-(OH)₂-Bn 4-Cl-Bn 4-F-Bn	H	OEt	α-chymotrypsin	5.0		0.5–1.5h	25–40	>99.5	1971 (*1746*) 1980 (*1611*)
Ac	2-Me-Bn CH₃(2-naphthyl)	H	OEt	α-chymotrypsin	7–7.5	37		39		1973 (*116*)
Cbz	Bn	H	OMe	α-chymotrypsin	7–7.5	37		36		1973 (*116*)
Ac	CH₂(2-naphthyl)	H	OMe	α-chymotrypsin	7.6	36	90h	26		1976 (*1670*)
H	4-B(OH)₂-Bn	H	OEt	α-chymotrypsin				48		1980 (*1747*)
Ac	2-F-Bn 4-F-Bn CH₂(1-naphthyl) CH₂(2-naphthyl) CH₂(5-R-indol-3-yl, R = Br, F, Cl, Me, OMe)	H	OMe	α-chymotrypsin	7.0	25				1987 (*1748*)
Bz	CH₂(3-pyridyl)	H	OMe	α-chymotrypsin	7.0			48		1987 (*1749*)
COCF₃	CH=CHCH₂PO₃H₂	H	OEt	α-chymotrypsin	7.8	25		48		1988 (*1750*)
H	CH=CHPO₃Et₂	H	OEt	α-chymotrypsin	7.8	rt				1988 (*1751*)
NHP¹ = NO₂	CH₂CH=CH₂ CH₂C(Me)=CH₂ (CH₂)₂CO₂H Ph CH₂(indol-2-yl)	Me	OMe, OBu	α-chymotrypsin	7.1, DMSO			60	0–>95	1988 (*143*)
Ac	CH₂(4-thiazoyl)	H	OEt	α-chymotrypsin	7.0–7.1		1	47		1990 (*1752*)
Ac, Bz	CH(OH)Me CH(OH)Ph CH(OH)(4-NO₂-Ph)	H	OMe	α-chymotrypsin, subtilisin or bromelain	6.0–7.7		1–70h	33–48	>98	1990 (*1753*)
Ac, Bz, Boc, Cbz, COCH₂Cl	CH₂CH=CH₂ CH₂C(Me)=CH₂ CH₂CH=C(Me)₂	H	OEt	α-chymotrypsin	8.0	37		11–49	86–96	1992 (*1754*)

(Continued)

Table 12.5 Resolutions by Lipases, Proteases, Amidases, Oxidases, and Other Enzymes (continued)

P¹	R¹	R²	P²	Enzyme	pH	Temp. (°C)	Time	Yield (%)	ee (%)	Reference
Boc	CH₂CH=CH₂	H	OEt	α-chymotrypsin	8	37	24h	48		1993 (1755)
H	CH(CF₃)CH₂CO-αN	H	OEt	α-chymotrypsin	7.5	37	3h	33	>96	1996 (1756)
CH(4-Cl-Ph)	Me, iPr, Ph, Bn, 2-F-Bn, 4-F-Bn	H	OMe, OEt	α-chymotrypsin or PPL	DABCO			14-88	85-99.5	1996 (193)
Ac	(CH₂)₄CO₂Me/Bn, (CH₂)₅CO₂Me/Bn, (CH₂)₆CO₂Me/Bn	H	OMe, OBn	α-chymotrypsin or subtilisin (with Aspergillus acylase I)	8	37	20h	32-38		1996 (1726)
Ac	3-OH-6-pyridyl	H	OEt	α-chymotrypsin	6.9		2h	47		1996 (1757)
Ac	CO₂Et	H	OEt	α-chymotrypsin	6.5-7.5		24h			1997 (1758)
Boc	CH₂CH(CN)₂, CH₂CH(CN)CO₂Me	H	OMe	α-chymotrypsin	7.6	37		45-47	95-96	1998 (1759)
H	CH₂CH₂(1,2-Ph)-αN	H	OMe	α-chymotrypsin	7.5			46	97	1998 (1760)
H	4-B(OH)₂-Bn	H	OEt	α-chymotrypsin						1999 (1761)
H	Me, Bn, 4-HO-Bn	H	OMe	alcalase	8.0	30	20min	47-49	86-90	1986 (1762)
Cbz	Et, nPr, nBu, nHex, (CH₂)₃CO₂Me	H	OMe	alcalase	supercritical CO₂		18-48	24-53	82->99	1994 (109)
H	nPr, nBu, iBu, Bn, 4-OH-Bn	H	OBn	alcalase, pyridoxal-5-phosphate	8.5, isopropanol	40	4h	87-92	90-98	1994 (191)
Boc	CH₂CCH	H	OMe	alcalase	0.2% NaHCO₃			48	86	1998 (1763)

		R¹								
Ac	$CH_2OCH_2CH_2\text{-}R^2$, $CH_3SCH_2CH_2\text{-}R^2$, $CH_2CH_2SCH_2CH_2\text{-}R^2$	Me		alcalase	7.0			38	84–92	1998 (*10*)
Cbz, Bz	CH_2CONH_2	H	OMe, OEt, OnBu, OBn	alcalase	6.5 aq.-THF	rt	3–22h	49	>99	2004 (*11*)
Ac	CH_2(6-quinolyl)	H	OEt	subtilisin Carlsberg	7.6	rt		43		1973 (*116*)
Cbz	4-Me-Bn, 3-F-Bn, 4-F-Bn, 2,3,4,5,6-F$_5$-Bn	H	OMe	subtilisin Carlsberg	8	25	24h	21–39		1973 (*1764*)
Ac	$CHPh_2$, 3-CF_3-Bn, 4-CF_3-Bn, CH_2(2-fluorenyl)	H	OMe	subtilisin Carlsberg	7					1982 (*1765*)
Ac	$CH(^3H)$(2-naphthyl)	^3H	OMe	subtilisin Carlsberg	7	rt		46		1984 (*1766*)
Bz	CH_2(3-pyridyl)	H	OMe	subtilisin Carlsberg	7.6			48		1984 (*1767*)
Ac	Ph, 2-, 3-, or 4-Cl-Ph, 4-SMe-Ph, 4-OH-Ph, 2-furyl, 2-thienyl, $(CH_2)_2Ph$, $(CH_2)_2(F_2$-pyridyl$)$	H	OMe	immobilized subtilisin	aq.-org. emulsion			26–50		1985 (*114*)
Ac	CH_2(3-pyridyl), $CH_2$3-benzo[b]thienyl)	H	OEt	subtilisin Carlsberg	6.5–7.0	rt–39		44		1987 (*1768*)
Ac	CH_2(4-thiazoyl)	H	OEt	subtilisin Carlsberg	7.0–7.1	36	1	47		1990 (*1752*)
$COCH_2CH_2$-αN	$CH(OH)CH_2CH_2N_3$	H	OBn	subtilisin Carlsberg	6.5	36	5.5	39		1990 (*513*)
Ac	CH_2(1-naphthyl), CH_2(2-naphthyl)	H	OEt	subtilisin Carlsberg	6.9	39	6h	48		1991 (*1769*)
Boc	4-$(CH_2PO_3Et_2)$-Bn, 4-(CH_2SO_3Na)-Bn, 4-(CH_2CO_2H)-Bn, 4-$(CH_2CONHOH)$-Bn	H	OEt	subtilisin Carlsberg	7.0, dioxane		1–24h	37–48	88	1992 (*1770*)

(Continued)

Table 12.5 Resolutions by Lipases, Proteases, Amidases, Oxidases, and Other Enzymes (continued)

P^1	R^1	R^2	P^2	Enzyme	pH	Temp. (°C)	Time	Yield (%)	ee (%)	Reference
Ac	$CH_2(3\text{-}CH_2CH_2PO_3Et_2\text{-}cHex)$	H	OMe	subtilisin Carlsberg	7.4		12h	19		1993 (204)
Ac	$CH_2(3\text{-}CH_2CH_2PO_3Et_2\text{-}cHex)$	H	OMe	subtilisin Carlsberg	7.4		24h			1993 (205)
H	4-PhO-Bn	H	OMe	subtilisin Carlsberg	7.0		75min	44		1993 (1771)
Ac	$CH_2C(Br)=CH_2$	H	OEt	subtilisin Carlsberg	7.0	37	1h	48.5	97	1993 (1772)
COnPr	Ph	H	OMe	subtilisin	7.8 in 3-methyl-3-3-pentanol	rt	22h	35	65	1995 (1773)
H	$4\text{-}CH_2PO_3Me_2\text{-}Bn$	H	OEt	subtilisin Carlsberg	7.4	rt	12h	44	>95	1996 (1774)
Boc-Val	OBn	H	OMe	subtilisin Carlsberg	7.2, DMF				>99	1997 (112)
COCF$_3$	$3\text{-}F\text{-}4\text{-}NO_2\text{-}Bn$ $3\text{-}NO_2\text{-}4\text{-}F\text{-}Bn$	H	OMe	subtilisin Carlsberg	7.5, CH_2Cl_2				>99	1997 (113)
Ac	CH_2(5-benzofuran)	H	OMe	subtilisin Carlsberg	7.5, 20% DMSO	rt	on	41	93.3	2000 (1775)
Boc	$(CH_2)_4Br$	H	OEt	subtilisin Carlsberg	DMF-H_2O					2004 (1744)
Boc	$(CH_2)_4Br$ $(CH_2)_3CH(Me)Br$	H	OMe	subtilisin Carlsberg	8, 25% DMF	37	3h	40–47		2005 (1745)
H	CH_2[2-(2-pyridyl)-pyrid-4-yl] CH_2[2-(2-pyridyl)-pyrid-5-yl]	H	OMe	alkaline protease		rt	2h	43	93–96	1993 (1776)
=C(Ph)S-P^2	Me iPr nBu iBu $(CH_2)_2$SMe Bn CH_2CONH_2	H	SC(Ph)=-P^1	protease	7.5		12–240h	30–94	57–99	1993 (210)
H	[quinoxaline structure]	H	OMe	alkaline protease	0.2N NaHCO$_3$		1.5h	51	94	1995 (1777)
H	[phenanthroline structure, R = H, Me]	H	OMe	alkaline protease	10% NaHCO$_3$		40min	46	>98	1996 (1778)
H	$3,4\text{-}(OMe)_2\text{-}Bn$	H	OMe	protease type VIII	0.2N NaHCO$_3$		5h	50	88	1996 (1779)

H	Et nPr nBu nPent iBu iPent Bn 2-,3-, or 4-F-Bn 2- or 4-Cl-Bn 4-Br-Bn	H	iBu	*Aspergillus oryzae* protease	7	5–30	5 min to 23 h	24-46	77–>99	1997 (265) 2005 (121)
H	CH_2[2-(4-Me-pyrid-2-yl)-pyrid-4-yl]	H	OMe	alkaline protease					95	1998 (1780)
H	$(CH_2)_3CH(NH_2)CONH_2$	H	NH_2	amidase	8.0	38		44		1957 (1648)
H	iPr iBu Ph Bn 4-OMe-Bn $(CH_2)_2Ph$ Bn	Me	NH_2	amidase from whole-cell *Mycobacterium neoaurum*		37	72h	48		1988 (132)
H	Bn	Et	NH_2	amidase from whole-cell *Mycobacterium neoaurum*		37	72h	48		1988 (132)
H	Me	Et	NH_2	amidase from whole-cell *Mycobacterium neoaurum*	8.5	38	19h	31	>99.5	2000 (136)
H	CF_3 CF_2Cl CF_2Br CF_2H	Me Me Me Bn	NH_2	amidase from *Mycobacterium neoaurum* or *Ochrobactrum anthropi*	8.6	25–37	21–23h	47–58%	amide 98–99.5% ee acid 67.9–96% ee	2004 (135)
H	iPr tBu Ph	H	NH_2	amidase from whole cell *Ochrobactrum anthropi*	7.0					1993 (137)
H	iPr iBu CH=CHPh Ph	Me	NH_2	amidase from whole cell *Ochrobactrum anthropi*	7.0					1993 (137)

(Continued)

Table 12.5 Resolutions by Lipases, Proteases, Amidases, Oxidases, and Other Enzymes (continued)

P¹	R¹	R²	P²	Enzyme	pH	Temp. (°C)	Time	Yield (%)	ee (%)	Reference
H	Et, nPr, CH₂CH=CH₂	Ph	NH₂	amidase from whole cell *Ochrobactrum anthropi*	7.0					1993 (*137*)
H	Me, Et, nPr, iPr, nNon, CH₂CH=CH₂, CH₂CH=CHPh, Bn, Ph	Me	NH₂	*Mycobacterium neoaurum* amidase	8–8.5	37	24–40	13–50	85–>99	1993 (*130*)
H	Et	Bn, Ph	NH₂	*Mycobacterium neoaurum* amidase	8–8.5	37	24–40	13–50	85–>99	1993 (*130*)
H	CH₂CH=CH₂	Bn	NH₂	*Mycobacterium neoaurum* amidase	8–8.5	37	24–40	13–50	85–>99	1993 (*130*)
H	(CH₂)₄-αN, (CH₂)₂NH(CH₂)₂-αN	H	NH₂	whole bacterial cell amidase	7.0	30	36–72h	22g scale	>99	1997 (*124*)
H	Ph, 4-F-Ph, 4-Cl-Ph, 3-Cl-Ph, 2-Cl-Ph, 4-Br-Ph, 4-Me-Ph, 3-Me-Ph, 2-Me-Ph, 4-MeO-Ph, Bn, iPr, cHex	Ph	NH₂	whole bacterial cell amidase from *Rhodococcus*	7.0	30	1–7d	17–36	5–93	2005 (*139*)
H	Me	CF₃	NH₂	amidase from *Klebsiella oxytoca*	—	20		42–50 1g scale	95–>99	2004 (*138*)

	R	X		Enzyme	pH	T (°C)	t	Conv. (%)	ee (%)	Year (Ref)
H	Et iBu Ph 4-HO-Ph CH_2CH_2Ph	NH_2	H	aminoacylase I from *Aspergillus melleus*	7.5	25	0.5–2.5h	48–50	92–>99	2004 (*129*)
H_2NCO	Bn CH_2(indol-3-yl) CH_2CH_2SMe	NH_2	H	Chirazyme P_1 or P_2	7.5	22	24h	40–47	>98	2003 (*128*)
Cbz	Me Et nPr nBu nPent nHex iPr $CH_2CH=CH_2$ Bn 4-thiazolyl	OH	H	*Aspergillus niger* lipase	7.0		12–115h	14–44	85–96	1988 (*1781*)
Cbz	Me Et nPr nBu nPent nHex nHept iBu iPent $CH_2CH=CH_2$ $(CH_2)_2SMe$ $(CH_2)_2SEt$ Bn 2-F-Bn 3-F-Bn 4-F-Bn 4-Cl-Bn $(CH_2)_2Ph$ $(CH_2)_3Ph$ CH_2(4-thiazoyl)	OCH_2CF_3	H	porcine pancreatic lipase	7.0	25	18–47	17–44	21–>99	1989 (*103*)

(Continued)

Table 12.5 Resolutions by Lipases, Proteases, Amidases, Oxidases, and Other Enzymes (continued)

P¹	R¹	R²	P²	Enzyme	pH	Temp. (°C)	Time	Yield (%)	ee (%)	Reference
COnPr, NHP₁ = N₃	CH(OCOnPr)Ph	H	OMe, OEt, CO₂P² = CH(OAc)CO₂M	*Candida cylindrica* lipase	6.5–7		3–96h	0–39	>98	1990 (*105*)
Ac	CH(OAc)CF₃	H	OEt	lipase MY	7.3	40	96h	37	86	1991 (*1782*)
Bz	cHex cHexCH₂ Ph Bn	H	OMe	porcine pancreatic lipase	7.7	30	6–55h	50	>99	1992 (*1783*)
=C(Ph)O–P²	iBu (CH₂)₂SMe CH₂SBn Bn 4-OH-Bn Ph 4-OH-Ph CH₂(2-indolyl)	H	OC(Ph)=P¹	lipases (porcine pancreatic, *Aspergillus niger*)	7.6		5–168	20–>99		1992 (*211*)
=C(Ph)O–P²	Me iPr nBu iBu CH₂SBn (CH₂)₂SMe Ph 4-OH-Ph 4-OMe-Ph 2-furyl 3-furyl 1-naphthyl 2-naphthyl 4-Me-Bn 4-OH-Bn (CH₂)₂Ph (CH₂)₃Ph	H	OC(Ph)=P¹	lipase for dynamic methanolysis, protease for hydrolysis	6.8	50		31–93	85–100	1993 (*210*)
H	(CH₂)₄-αN	H	OnPent	*Aspergillus niger* lipase	5	rt	24h	19	93	1994 (*1784*)

COEt	CH₂-αN	H	OMe	*Candida cylindracea* lipase	7.5	37	30min	60	90	1995 (*1785*)

R	R'	R''	R'''	Enzyme/conditions	pH	T	time	yield	ee	Year (ref)
COEt	CH₂-αN	H	OMe	*Candida cylindracea* lipase	7.5	37	30min	60	90	1995 (*1785*)
H, Ac	3-F-Bn	Me	OEt	lipase L	6.5–7.4		6d	25–40	97.7	1996 (*141*)
Cbz	Me; CH₂CH₂SMe; Ph; Bn	H	OMe	*Aspergillus niger* lipase					93–>99	1997 (*102*)
Ac	Me; iPr; iBu; Ph; Bn; 4-OH-Bn; 4-OMe-Bn	H	OnBu	rice barn lipase	7.5		24h		47–>99	1997 (*107*)
Cbz	Me	H	CO₂P² = CH₂CO₂Et	*Candida antarctica* lipase, 1,4-dioxane			8h	41	98	1997 (*152*)
Bn, CHPh₂, allyl	CH₂CH₂-αN	H	OMe	*Candida antarctica* lipase amidation with NH₃		35			80–97	1998 (*146*)
=C(Ph)O-P²	Me; iPr; iBu; tBu; Bn; CH₂(indol-3-yl); CH₂CH₂SMe	H	OC(Ph)=P¹	*Candida antarctica* lipase dynamic resolution, Et₃N, toluene, MeOH or MeCN, MeOH		37		40–96	14–98	2000 (*213*)
Ph; 4-F-Ph; 4-Br-Ph; 4-Me-Ph; 4-MeO-Ph; 4-NO₂-Ph; 3-Br-4-Me-Ph	CH₂-αN	H	OMe	*Candida rugosaa* lipase	7.5	25	3–5.5h	44–50	7–99	2004 (*1786*)

(Continued)

Table 12.5 Resolutions by Lipases, Proteases, Amidases, Oxidases, and Other Enzymes (continued)

P¹	R¹	R²	P²	Enzyme	pH	Temp. (°C)	Time	Yield (%)	ee (%)	Reference
H	(CH₂)₃-αN (CH₂)₄-αN	H	OMe	*Candida antarctica* lipase. Et₃N dynamic resolution via acylation with nPrCOCH=CH₂		48–56	3–24h	70–90	97	2004 (*194*)
Cbz	Et nPr nBu nPent iBu iPent CH₂cHex (CH₂)₂SMe (CH₂)₂SEt	H	OCH₂CF₃	*Carica papaya* lipase transesterification with MeOH in cyclohexane		45	24h	26–50	95.6–>99.8	2005 (*104*)
Cbz-Gly	CH₂(2-thienyl)	H	OH	carboxypeptidase	7.5	rt		25		1956 (*1787*)
COCH₂Cl	4-Me-Bn 4-Et-Bn	H	OH	carboxypeptidase	7.1		72h			1964 (*1788*)
COCF₃	CHDPh CHTPh CHD(4-OH-Ph) CHT(4-OH-Ph)	H	OH	carboxypeptidase	7.5	37	24h	45		1973 (*1789*)
COCF₃	CH₂(5-F-indol-3-yl)	H	OH	carboxypeptidase	7.0–7.2	38	48h	34		1974 (*1790*)
COCF₃	iPr Bn	Me	OH	carboxypeptidase	7.2	37	16h	35–37		1975 (*1791*)
COCF₃	4-OH-Bn with [2-²H]	H	OH	carboxypeptidase A	8.5	37	24h	43		1977 (*1675*)
COCF₃	4-OH-Bn with [2-¹³C]	H	OH	carboxypeptidase	7.5	37	24h	46		1978 (*1792*)
Cbz- or Boc-Gly	CH₂{4,5,6,7-F₄-indol-3-yl} CH₂(benzo[b]fur-3-yl) CH₂(benzo[b]thien-3-yl)	H	OH	carboxypeptidase A	8.0		2h			1979 (*660*)
COCF₃	[3',5'-¹³C₂]-4-OH-Bn	H	OH	carboxypeptidase A				43		1979 (*1793*)
COCH₂Cl	CH₂(7-Cl-indol-3-yl)	H	OH	carboxypeptidase	7.2	37		20		1981 (*1794*)
COCF₃	CH(Me)Ph	H	OH	carboxypeptidase A	8.0		6d	17–27	93–99.8	1989 (*1795*)
Bz	CH₂(1-naphthyl)	H	OH	carboxypeptidase A	7.2	37	48h	40	>99	1989 (*1709*)
COCF₃	CH(Me)Ph	H	OH	carboxypeptidase A	8.0	37	6d	17–27	93–99.8	1994 (*1796*)

COCF$_3$	2,6-Br$_2$-4-OH-Bn	H	OH	carboxypeptidase	7.4	37		38		1998 (1797)
H	Me	H	OH	snake venom L-oxidase or hog renal D-oxidase	7.2–8.2	37	24–48h	20–46		1958 (1798)
	iPr									
	iBu									
	sBu									
	Bn									
	4-HO-Bn									
	CH$_2$OH									
	CH$_2$(indol-2-yl)									
	CH$_2$(imidazol-2-yl)									
	(CH$_2$)$_4$NHCbz									
	(CH$_2$)$_4$-αN									
	CH$_2$CH(OH)CH$_2$-αN									
	CH$_2$CO$_2$H									
	(CH$_2$)$_2$CO$_2$H									
	(CH$_2$)$_3$NHC(=NH)NH$_2$									
H	CH$_2$CH(Me)CH$_2$-αN	H	OH	D-amino oxidase	8.2	38				1962 (166)
H	CH(OH)CH$_2$CH$_2$-αN	H	OH	D-amino oxidase	8.3	38		50		1963 (627)
H	CH(OMe)CH$_2$CH$_2$-αN	H	OH	D-amino oxidase	8.0	rt		38–42		1963 (1799)
H	CH$_2$CH(OH)CH$_2$CH$_2$-αN	H	OH	D-amino oxidase						1965 (170)
H	iBu with [1-^{11}C]	H	OH	immobilized D-amino acid oxidase	8.3					1983 (1800)
H	Me with [1-^{11}C]	H	OH	immobilized D-amino acid oxidase	8.3					1984 (1801)
H	3-CF$_3$-Bn	H	OH	L-amino acid oxidase				45–48	>97	1993 (167)
	3-Me-Bn									
	3-MeO-Bn									
	2-Me-Bn									
	2-Cl-Bn									
	3-Cl-Bn									
	2-F-Bn									
H	2-OH-Bn	H	OH	L-amino acid oxidase		38				1998 (1733)
	2-OH-3,5-(NO$_2$)$_2$-Bn									

(Continued)

Table 12.5 Resolutions by Lipases, Proteases, Amidases, Oxidases, and Other Enzymes (continued)

P¹	R¹	R²	P²	Enzyme	pH	Temp. (°C)	Time	Yield (%)	ee (%)	Reference
Bn	$CH_2CH_2CH(CO_2Me)$-αN	H	OMe	pig liver esterase, DMSO	7.5			39	100	1987 (106)
H, Ac, NH₂, NHCbz	iBu Bn 3-OMe-4-OH-Bn 2-indolyl	Me	OEt, OnBu	Candida lipolytica esterase	7.4		6h	50	95->99	1992 (140)
H	Me Et nPr nBu $CH_2CH=CH_2$ Bn	Ph	OEt	pig liver esterase	8.0	28		14-66	13-93	1993 (130)
H	Et nPr Ph	H	OEt	*Humicola* amino esterase	8.0			37-74	26-99	1995 (100)
H	Bn	Me	OEt	*Humicola* amino esterase	8.0			37-74	26-99	1995 (100)
H	Me nPr nBu nHex nNon Bn CH=CHMe	Et	OEt	*Humicola* amino esterase	8.0			37-74	26-99	1995 (100)
H	nHex	nPr	OEt	*Humicola* amino esterase	8.0			37-74	26-99	1995 (100)
H	$CH_2CH=CH_2$	Ph	OEt	pig liver esterase	8.0	30	48h	33	96	1997 (131)
H	Et	nBu	OEt	pig liver esterase	8	30	7.5h	31	>99	2000 (156)
Leu	CH_2CHFCO_2Me $CH_2CH(Me)CO_2Me$	H	OH	Leu aminopeptidase	8	36	2–4h			1984 (1802)
H	3-CONH₂-Bn	H	NH₂	Leu aminopeptidase	8.0	38	48h			1961 (1803)
H	$CH(OH)CH_2CH_2$-αN	H	NH₂	Leu aminopeptidase	8.0	37		49		1963 (1804)
H	CH(CH(Me)Et	H	NH₂	Leu aminopeptidase	8	37	40h	48		1981 (1805)
H	$(CH_2)_3SMe$	H	NH₂	*Pseudomonas putida* aminopeptidase		30	20h	46		1986 (21)

	R		Enzyme	pH	T (°C)	Time	ee (%)	ee (%)	Year (Ref.)
H	[structure: purine ring with NH–CH₂CH=C(Me)CH₂OH side chain]	NH₂	*Pseudomonas putida* aminopeptidase	10	40	20h		>95	1990 (1806)
H	iPr, cPent, CH₂CH=CH₂	NH₂	*Pseudomonas putida* aminopeptidase	8.3	37	22h	32–48	>95	1992 (123)
H	Me, Et, iPr, iBu, sBu, Ph, 2-Cl-Ph, 4-OH-Ph, Bn, 4-OH-Bn, (CH₂)₂Ph, 2-naphthyl, 2-thienyl, 2-indolyl, CH₂OH, (CH₂)₂SMe, (CH₂)₃SMe, CH₂CH=CH₂, CH₂C(Me)=CH₂, CH₂C(nHex)=CH₂, (CH₂)₂CO₂H	NH₂	*Pseudomonas putida* L-aminopeptidase	8–9					1992 (34)
H	CH₂N(Boc)CH₂CH₂CH₂-αN	NH₂	Leu aminopeptidase	8.6	rt	7d	36–38	67–88	1995 (1807)
H	Bn, Ph, 2-Me-Ph, 4-MeO-Ph, 3-MeO-Ph	NH₂	amidase in *Rhodococcus* sp.	7.0	30	2–9h	47–56	60–95	2002 (126)

(Continued)

Table 12.5 Resolutions by Lipases, Proteases, Amidases, Oxidases, and Other Enzymes (continued)

P¹	R¹	R²	P²	Enzyme	pH	Temp. (°C)	Time	Yield (%)	ee (%)	Reference
Bz	3-F-4-OH-Bn	H	OH	papain, + H₂NPh	5.0			46		1946 (148)
H	iBu with [1-¹⁴C] (CH₂)₄NH₂ with [1-¹⁴C]	H	OH	papain, + H₂NPh						1950 (1808)
Cbz	(CH₂)₃CONH₂ CH₂CH(Me)CONH₂	H	NH₂	papain	5.0	37	24-46	28-35		1954 (1809)
Ac	CH₂(2-thienyl)	H	OH	papain, + H₂NPh	4.5	40	72h	45		1963 (1810)
Ac	(CH₂)₃CON(Ts)Bn	H	OH	papain, + H₂NPh	6.6-6.20	38	36h	46		1972 (149)
Bz	CH₂NHOBn	H	NHPh	papain	6.15	38	30h	42		1973 (1811)
CO₂Et	2-furyl	H	OMe	papain, 20% DMF	7.0			45	>97	1988 (1812)
Boc	CH=CH₂	H	OH	papain, EtOH, CH₂Cl₂				40		1994 (144)
Cbz	CH₂CH(CO₂tBu)₂	H	OMe	papain		25	20h	45	>99.5	1996 (1813)
Cbz	(CH₂)₄CO₂Me	H	OMe	papain, DMF, 0.2 M NaOAc	7.5	rt	o/n	46		1998 (1814)
Boc	CH₂(4-PO₃Et₂-fur-2-yl)	H	OEt	papain				40	75-99	1999 (1815)
Boc	NHFmoc	H	NH₂	papain	6.2	37	17-20h	42-48	92	2000 (127)
Ac	iBu Bn CH₂CO₂Me (CH₂)₂CO₂H	H	OMe	carbonic anhydrase	7.5			35-40	68->95	1993 (173)
CONHCONH-P₂	Me iPr nBu iBu sBu Ph 4-OH-Ph Bn (CH₂)₂Ph CH₂OH CH(Me)OH (CH₂)₂SMe 2-thienyl	H	NHCONHCO-P¹	D-hydantoinase	8.5	50	0.3-45h	50-95	94->99	1995 (207)

CONHCONH-P^2	Ph	H	NHCONHCO-P^1	D-hydantoinase	9.0	30	—	82	>98	2000 (203)
CONHCONH-P^2	Ph, 4-F-Ph, 4-CF$_3$-Ph	H	NHCONHCO-P^1	immobilized D-hydantoinase	9.0	30	100 min	75–90	>98	2004 (206)
H	CH$_2$C(=CH$_2$)CO$_2$H	H	OH	glutamic decarboxylase	5	37	9h			1963 (181)
=C(Ph)O-P^2	tBu	H	OC(Ph)=-P^1	lipozyme, Et$_3$N, toulene, nBuOH				94	99.5	1995 (212)
Ac	2-F-Bn, 3-F-Bn, 4-F-Bn	H	OEt	lyopholized Saccharomyces cerevisae	7.5	rt		38–43	>96	1988 (1816)
H	(CH$_2$)$_4$NH$_2$	H	OH	Lys decarboxylase	6.0	37	90min	45		1944 (178)
H	CH$_2$CO$_2$Bn, (CH$_2$)$_2$CO$_2$Bn, (CH$_2$)$_3$CO$_2$Bn, (CH$_2$)$_4$CO$_2$Bn	H	OBn	pronase	7.2	25	3h	35–36		1992 (117)
Boc	4-Cl-Bn, 2,4-Cl$_2$-Bn, 3,4-Cl$_2$-Bn, CH$_2$SBn, CH$_2$(2-naphthyl)	H	OMe	thermidase, CaCl$_2$	8	55	16h	35–43		1989 (118)
Cbz-Gly	SPh, S(4-NO$_2$-Ph), S(CH$_2$)$_3$NH$_2$, SiPr	H	OBn	thermidase, CaCl$_2$	7.2–8	45–50		24–47		1992 (119)
Boc	Bn	H	O(4-NO$_2$-Ph)	horse heart apomyoglobin	7.0	4	8 min	40	98	2001 (122)

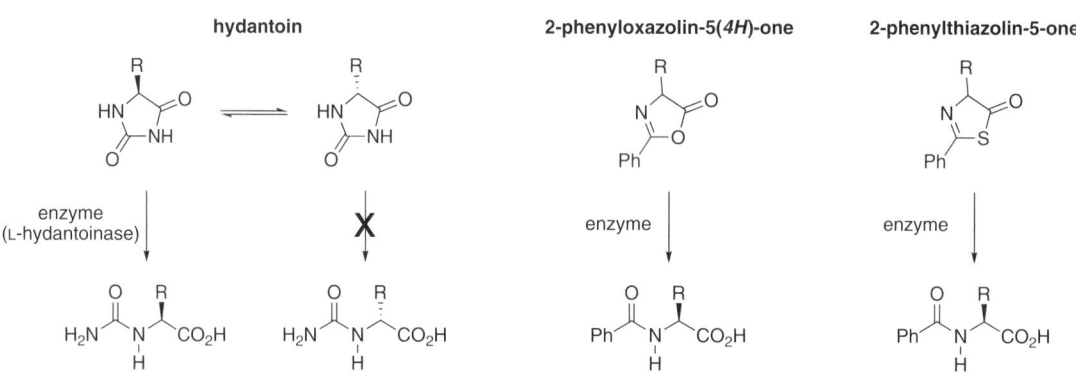

Scheme 12.1 Enantiodifferentiating Coupling Reagents (*189*).

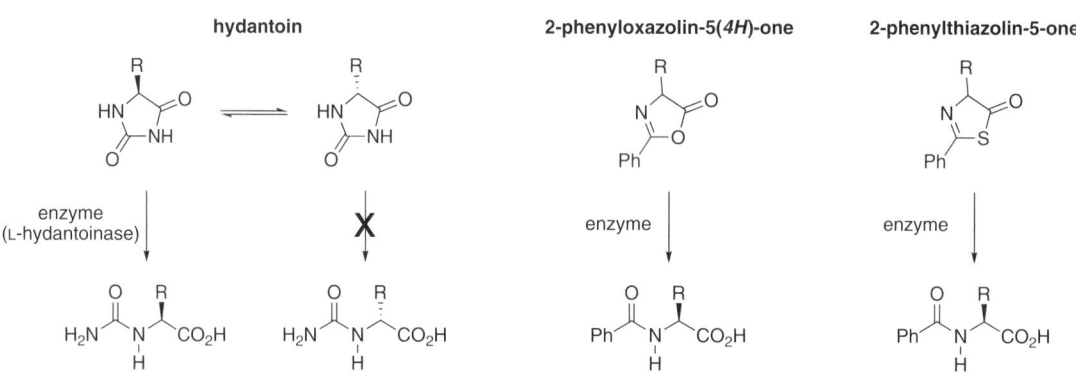

Scheme 12.2 Derivatives Used for Enzymatic Dynamic Resolutions.

racemase, potentially very useful for dynamic resolutions, was isolated from *Streptomyces atratus* in 1994 from a screening of 49,000 strains. The enzyme has a broad substrate tolerance of side chains, but does not act on the deacylated amino acids (*196, 197*). Preliminary reports have also appeared of an amino acid amide racemase (*7, 34*). Site-directed mutagenesis of an alanine racemase from *Geobacillus stearothermophilus* reduced the high Ala specificity and allowed for Ser racemase activity, suggesting that custom tailoring of enzymes might be possible (*198*).

To overcome the lack of racemases for conventional amino acid substrates, other amino acid derivatives have been used (see Scheme 12.2). Racemic hydantoins produced by the Bücherer–Bergs synthesis (see Section 2.2.2) are sensitive to racemization; with aryl substituents they racemize spontaneously under alkaline conditions, while for alkyl substituents hydantoin racemase enzymes are available to epimerize the α-center (*7, 57, 199*). The actual resolution is then carried out by hydantoinases. Both L- and D-hydantoinases are available to stereospecifically hydrolyze the hydantoins to *N*-carbamoyl amino acids (*200*). Degussa employs an L-hydantoinase in conjunction with a racemase and a L-carbamoylase, all within *E. coli* cells, to prepare L-Met, L-Nle, L-Abu, and L-β-(3-pyridyl)-Ala on a scale of several hundreds of kilograms (*201*). A number

of arylglycines were resolved on a scale of up to 18 g using cells of *Agrobacterium radiobacter*, with unoptimized overall yields of 60% (*202*). Phg hydantoin was resolved in 82% with >98% ee (*203*). This method was used to resolve amino acids with phosphate-substituted cyclohexyl side chains, giving optically pure products with >50% yields (*204, 205*). The D-hydantoinase from *Vigna angularis* was immobilized on aminopropyl glass beads, and then used to prepare *N*-carbamoyl D-Phg, D-*p*-F-Phg, and D-*p*-CF$_3$-Phg in 75–90% yield with >98% ee. The immobilized enzyme was reused for up to eight cycles with no loss in activity (*206*). Two new commercially available and thermally highly stable D-hydantoinases from thermophilic microorganisms were applied to the preparation of a variety of D-amino acids, in 1995 with racemization carried out at 50 °C and pH 8.5. The *N*-carbamoyl amino acids were obtained with 50–95% yield and 94 to >99% ee (*207*). The use of hydantoinases and related enzymes as biocatalysts for the synthesis of unnatural chiral amino acids was reviewed in 2001 (*208*), while the use of hydantoinases for the enantioselective production of amino acids was reviewed in 2004 (*209*).

Similarly, oxazol-5(4*H*)-one and 2-phenylthiazolin-5-one derivatives of amino acids spontaneously racemize in situ during lipase-catalyzed hydrolytic or alcoholytic ring opening (*7, 210, 211*). Thiazolinones were directly hydrolyzed with good enantioselectivity using proteases (*210*), while *Aspergillus niger* and porcine pancreatic lipases were applied to dynamic resolutions of oxazolones (*211*). Homochiral *tert*-Leu was prepared in 94% yield with 99.5% ee via a lipase-catalyzed dynamic resolution of its oxazolone derivative (*212*). Oxazolones of Val, Leu, Phe, and Trp were resolved with 90–98% ee in 82–96% yields with *C. antarctica* lipase B, but Met was only produced with 80% ee, and Ala and *tert*-Leu with marginal (10–35%) enantioselectivity (*213*). The enantioselectivity of a dynamic kinetic resolution of phenylglycine oxazolone using Novozym 435 (*C. antarctica* lipase B) was enhanced by the addition of triethylamine (*214*). Another study found that the solid-state buffer pair CAPSO/CAPSO·Na also increased the reaction enantioselectivity (*215*). *Pseudomonas* lipase was used in combination with a protease for the resolution of a range of amino acids, with the lipase used for an oxazolone dynamic resolution to provide amino acid methyl esters with 66–95% ee, and the protease employed to furnish optically pure amino acids (*210*). L- and D-*allo*-threonine have been prepared using an iterative process of acylase resolution and oxazolone epimerization (*216*). Dynamic kinetic resolution conditions were established for *N*-acetyl phenylglycinonitrile and *N*-formyl 4-fluorophenylglycinonitrile, with spontaneous racemization at pH 8. The (*R*)-enantiomer was preferentially hydrolyzed by Nitrilase 5086 to give the phenylglycines in 87–95% yield and 91–99% ee on up to 1 g scales (*217*).

Another method that has been developed to overcome the lack of an amino-racemase enzyme is to employ D-amino acid oxidase to convert half of a racemic amino acid to its α-keto acid. Branched-chain amino acid aminotransferases or amino acid dehydrogenases then enantioselectively reaminate the α-keto acid (*218–220*). This is discussed more thoroughly in Section 5.3.9f. Alternatively, an L-amino acid oxidase from *Proteus myxofaciens* has been used to oxidize the L-amino acid enantiomers from a racemic mixture into an intermediate α-imino acid, with BH₃·NH₃ used to reduce the imine and regenerate the racemic amino acid before further oxidation to the α-keto acid could take place (see Scheme 12.3). Racemic Nva, Nle, Phe, Tyr, Trp, Met, cyclopentylglycine, and allylglycine were converted into the D-amino acids with >99% ee and

79–90% yield; Abu, Val, His, and Ser(Bn) had reduced enantiopurity (28–96% ee) (*168*). A similar strategy employed porcine D-amino acid oxidase and NaCNBH₃ to convert racemic Abu, Leu, α-cyclopentyl-Gly, Phe, Trp, Cys, Pro, and piperazine-2-carboxylic acid to the L-amino acids with 99% ee in 75–90% yield (*221*). By using deuterated reducing agents, isotopically labeled [2-²H]-L-Pro was produced (*221*).

12.1.7 Chemical Dynamic Kinetic Resolutions

There are a number of examples of chemical transformations of racemic amino acid substrates which attempt to make use of rapid epimerization of the α-center to transform a racemic mixture into a single enantiomer within a diastereomeric product. An early version of this chemical dynamic resolution reacted a 2-(trifluoromethyl)-3-oxazolinone derivative of *tert*-leucine with dimethyl L-glutamate to give primarily an *N*-trifluoroacetyl L-L-dipeptide ester, with >60% conversion of DL-Tle to L-Tle (*222*). More recently, ketenes derived from *N*-phthaloyl amino acids were reacted with (*R*)-pantolactone to give diastereomeric amino acid pantolactone esters (*223, 224*). For arylglycines, very high diastereoselectivities were observed (97–98% de), but for alkyl and branched-alkyl side chains the stereoselectivity was reduced (33–94% de). High diastereoselectivity (94% de) was also obtained during reaction of the ketene derived from the β-amino acid 2-phenyl-3-aminopropanoic acid (*225*), with the same reaction subsequently applied to a range of other 2-aryl-3-aminopropanoic acids (*1830*). Alternatively, esterification of *N*-phthaloyl amino acids with (*S*)-α-methylpantolactone in the presence of DCC/DMAP produced predominantly the (*S,S*)-esters in 76–98% yield with 62–80% de (Ala, Abu, Nva, Val, Leu, Phe) (*226*).

Amine displacement of the halide from (*R*)-pantolactone esters of racemic α-bromo acids proceeds preferentially with one of the bromide diastereomers, with concurrent rapid epimerization of the remaining isomer (see Scheme 5.32) (*227–229*). *N*-Benzyl-Pro (61% yield, 75% de), *N* hexyl-Pro (51% yield, 75% de), *N*-benzyl-pipecolic acid (66% yield, 82% de), *N*-benzyl-Phg (72% yield, 82% de), *N,N*-dibenzyl-Phg and 4′-bromo-Phg (70–76% yield, >98% de), *N,N*-dibenzyl-homophenylalanine (56% yield, 82% de),

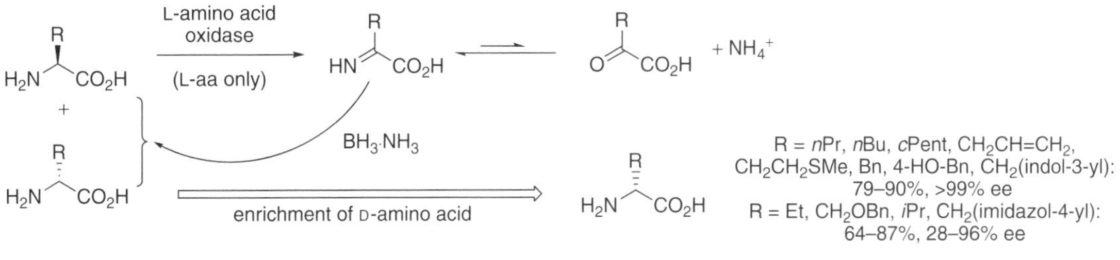

Scheme 12.3 Dynamic Kinetic Resolution of α-Amino Acids via L-Amino Acid Oxidase (*168*).

N,N-dibenzyl-γ-cyclohexyl-Nva (82% yield, 88% de), *N,N*-dibenzyl-Abu (66% yield, 86% de), and *N*-benzyl-Abu (70% yield, 75% de) were synthesized by this method. Resin-immobilized amine nucleophiles, including Gly residues and the *N*-terminal amino acid of tripeptides, have also been employed for the dynamic kinetic resolutions of α-bromo-(*R*)-pantolactone esters. Much higher diastereoselectivity was obtained with resin-bound Gly (87–90% de) than with free Gly amide (50–60% de) (*230*). Diacetone-D-glucose was also employed as an ester chiral auxiliary for dynamic kinetic resolutions of α-chloroarylacetic acids; nucleophilic displacement with a variety of primary and secondary amines produced arylglycine derivatives with 86–94% de (*231*).

The same bromide displacement/racemization procedure has been used with 2-oxo-imidazolidine-4-carboxylate (*232, 233*), 3,4-dimethyl-5-phenylimidazolidin-2-one (*234, 235*), or 3-hydroxy-4,4-dimethyl-1-phenyl-2-pyrrolidinone (*236*) auxiliaries. The latter auxiliary was also used to form ester derivatives of *N*-phthalimido amino acid chlorides (Ala, Val, Phe, and Phg), with in situ epimerization of the amino acid chloride substrate leading to predominantly one diastereomer with up to 96% de (*236*). An improved synthesis of the chiral auxiliary has been reported (*237*). The 3,4-dimethyl-5-phenylimidazolidin-2-one auxiliary has been used to generate both enantiomers of amino acids, with one diastereomer produced by nucleophilic displacement of the bromide with benzylamine under epimerizing conditions, and the other diastereomer obtained by bromide displacement with azide under non-epimerizing conditions (see Scheme 5.33) (*238*). A theoretical study on the inducement of enantioselectivity by imidazolidinone auxiliaries led to the development of a 3-methyl-4-methoxymethyl-5-carboxyimidazolidin-2-one pyrrolidine amide auxiliary, which induced 99% de during amine displacement of 2-bromopropionate, or 100% de if catalytic *n*Bu₄I was added (*239*). Amino acid esters have also been employed as the auxiliary: α-bromophenylacetic acid-(*S*)-Leu-OBn reacted with dibenzylamine and tetrabutylammonium iodide to give (*R*)-Phg-(*S*)-Leu-OBn with 90% de (*240, 241*).

Another approach selectively inverted (*S*)-α-bromo-3-phenylpropionic acid (derived from L-Phe), forming the (*R*)-enantiomer by simple crystallization with (*R*)-bornylamine in the presence of a bromide source (*242*). A rapid screening via parallel experimentation identified conditions suitable for crystallization-induced dynamic resolution of α-bromo-carboxylic acids using various chiral amines in the presence of a catalytic amount of tetrabutylammonium bromide (*243*).

Homocysteine (Hcy) has been resolved via formation of a tartaric acid salt with a cyclic 1,3-thiazane-4-carboxylic acid (THA) derivative of Hcy (*36*). Under the crystallization conditions the remaining THA racemized, allowing for isolation of optically pure THA in up to 81% yield. The same procedure was applied to the 2,2-dimethyl-4-thiazolidinecarboxylic acid formed from Cys and acetone, via formation of the tartaric acid salt in the presence of catalytic salicaldehyde. L- or D-Cys was obtained in 80% yield with 98–100% optical purity from DL-Cys (*244*). A dynamic resolution has also been carried out by crystallizing DL-Pro or DL-Pip with (2*R*,3*R*)-tartaric acid in the presence of catalytic salicaldehyde, producing optically pure L-Pro or (*R*)-Pip in over 70–80% yield (*245*). The same conditions, using (2*S*,3*S*)-tartaric acid, allowed L-Pro to be converted into D-Pro in 85% yield (*245*).

A 1998 report employed crystallization for asymmetric transformations of racemic aminonitriles derived from aldehydes, benzylamine, and HCN or TMSCN. A crystalline 1:1 diastereomeric mix formed with (*R*)-mandelic acid was stirred in ethanol for 12 h to 15 d, during which time the optically labile aminonitriles epimerized to give a single crystalline diastereomer, either the (*R,R*)- or (*S,R*)-product. The pure aminonitrile was recovered after decomposition of the salt with cold aqueous NaHCO₃, and hydrolyzed to the amino acid. The method was suitable for large-scale preparations (50 g), and the auxiliary was readily recovered. Not all derivatives worked, but 8 of 12 different alkyl and aryl side chains were successful (*246*).

Amino acid azlactones are readily racemized (see Scheme 12.4). A chiral ferrocenyl-like analog of

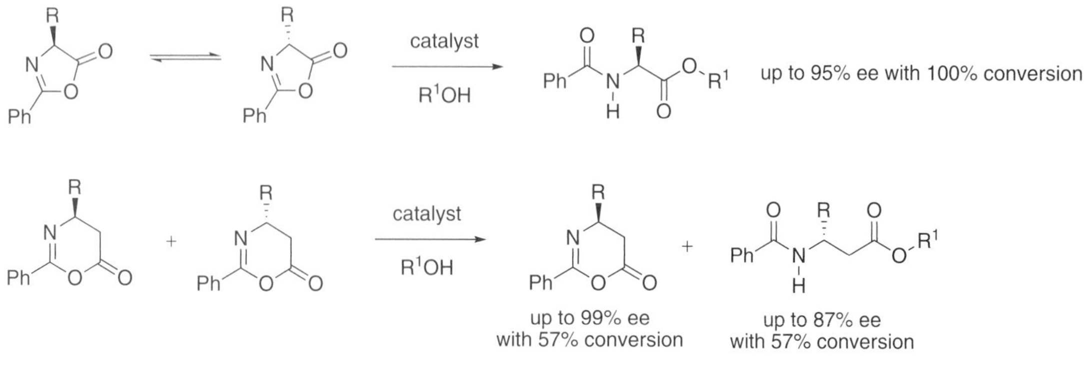

Scheme 12.4 Azlactone Dynamic Kinetic Resolution of α-Amino Acids and Oxazinone Kinetic Resolution of α-Amino Acids.

4-dimethylaminopyridine was employed to catalyze the ring opening of azlactones by methanol, producing *N*-Bz amino esters in 93–98% yield with 44–61% ee (*247*). Seven different derivatives (side chain = Me, Et, $CH_2CH=CH_2$, *i*Bu, CH_2cHex, Bn, CH_2CH_2SMe) were tested. Urea-based bifunctional organocatalysts provided for products with up to 85% ee with 76% conversion during asymmetric ring opening of azlactones with allyl alcohol (*248*). "Second-generation" thiourea-*tert*-amine catalysts gave up to 91% ee with 77% conversion after 48 h (for Leu); Tle azlactone reacted much more slowly with only 28% conversion after 48 h, but with 95% ee (*249*). Urethane-protected *N*-carboxyanhydride (UNCA) derivatives were employed for a cinchona alkaloid-catalyzed kinetic resolution, via asymmetric alcoholysis with methanol (see Scheme 12.5). The enantiomerically enriched products were readily separated by hydrolysis of the remaining anhydride, followed by extractive separation of the neutral ester, acidic amino acid, and basic amine catalyst. The UNCA of Cbz-Phe was resolved on a 4 mmol scale by 10% (DHQD)$_2$AQN catalyst to give Cbz-L-Phe-OH with 93% ee (48% yield), and Cbz-D-Phe-OMe with 97% ee (48% yield). Ten other *N*-Cbz amino acids, plus Fmoc-Phe, Boc-Phe, and Alloc-Phe, were resolved with 84–98% ee for the L-amino acid and 67–97% ee for the D-amino ester (*250*). This procedure was further optimized into a dynamic kinetic resolution of α-aryl amino acids, with a range of α-aryl and α-heteroaryl amino acids produced from racemic UNCAs in 86–95% yield with 89–92% ee (*251*). Both methods were reviewed in 2004 (*252*).

In a similar manner, kinetic resolution of oxazinones using a thiourea-based catalyst and allyl alcohol produced enantiomerically pure β-substituted-β-amino acids (see Scheme 12.4). However, since the oxazinones are configurationally stable, dynamic resolution (with >50% yield of one enantiomer) does not occur (*253*).

Syn-selective hydrogenation of the ketone group of 2-amino-3-oxo carboxylic esters with the chiral catalyst (*R*)-BINAP-Ru was combined with epimerization of the substrate α-center under the reaction conditions to give L-Thr and L-β-hydroxy-DOPA with up to 98% ee and 100% conversion (*254, 255*). A similar hydrogenation was used to prepare *p*-chloro-3-hydroxy-Tyr, a component of the antibiotic vancomycin (*256*). By employing methyl 2-phthalimido-3-ketobutyrate with Ru-((*S*)- or

(*R*)-C$_3$-Tunephos) as the catalyst, D-*allo*-Thr or L-*allo*-Thr were obtained with >99% ee and >94% de *anti* selectivity (*257*). With Ru-SYNPHOS catalyst, α-amino-β-keto ester hydrochlorides gave (2*S*,3*S*) *anti* products in 83–96% yield with 86–99% de and 91–97% ee (side-chain β-substituent = *n*Pr, *i*Pr, $nC_{15}H_{31}$, *c*Pent, *n*BuOBn), while α-benzamido-β-keto esters produced the (2*R*,3*S*) *syn* products in 53–94% yield with 71–99% de and 75 to >99% ee (*258*). A Ru(η^6-arene)-*N*-perfluorosulfonyl-1,2-diamine catalyst produced *threo*-β-hydroxy-3′,4′-dimethoxy-Phe with 90% de and 99% ee in 100% yield from the *N*-Me,*N*-Cbz methyl ester (*259*). In 2004 a dynamic kinetic resolution that was highly selective for the *anti* diastereomers was reported. Hydrogenation of racemic *N*-unprotected α-amino-β-keto benzyl esters using Ru-(*S*)-BINAP catalyst produced a variety of (2*S*,3*S*)-β-hydroxy-α-amino esters with 60–97% ee and over 98% ds for the *anti* isomer. The amino group is proposed to form a five-membered transition state with the Ru catalyst, rather than the six-membered system formed by the ester group of *N*-protected substrates that give *syn* products with the same catalyst (*260*).

One of the more intriguing methods of converting racemic amino acids into homochiral products is deracemization via enantioselective protonation. This area was reviewed in 1996 (*261*) and 2004 (*262*). Duhamel and co-workers have studied this approach (*263–267*), and found that aldimine methyl ester derivatives of amino acids can be deprotonated with LDA or LHMDS and reprotonated with the chiral proton source *O,O′*-dipivaloyl tartaric acid, giving products with 44–70% ee (side chain R = Me, Et, *i*Pr, *n*Bu, *t*Bu, Ph, CH_2CH_2SMe). The enantiomeric selectivity depended on the type of lithium amide and acyl protection of the tartaric acid. In a similar manner, the benzophenone Schiff base methyl esters of Ala, Abu, Leu, Met, Phg, and Phe were deprotonated with mesityllithium in ether at −78 °C, and then reprotonated with a sterically hindered *tert*-L-Leu-based amide at −20 °C to give the amino esters with 23–87% ee (*268*). An amide derivative of piperidine-2-carboxylic acid was enantioselectively protonated with >99% ee using a chiral diamine as the proton source (*269*). Chiral anilines were examined for protonation of enolates of acyl derivatives of amino acid methyl esters; Ala, Abu, and Phe were produced with 65–85% ee (*270*). A promising technology for converting racemic *N*-trifluoroacetyl α-bromo-Tle benzyl ester into Tle with >99% ee via

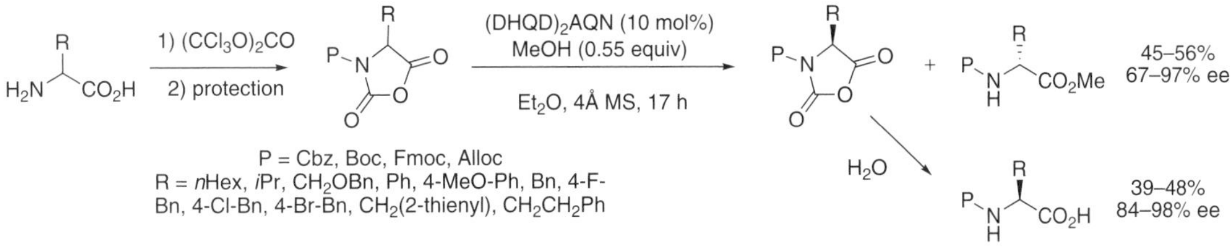

Scheme 12.5 Cinchona Alkaloid-Catalyzed Kinetic Resolution of UNCAs (*250*).

an asymmetric radical protonation process has been developed by an Australian company, Chirogen (*271*). The substrate was treated with a chiral menthyl stannane at –78 °C in toluene in the presence of MgBr$_2$; either L- or D-*tert*-Leu were produced.

Diastereoselective reprotonation has been carried out on chiral metal complexes of Schiff bases of racemic amino acids, originally by Belekon and co-workers (*272*), but more recently, and with greater success, by De and Thomas (*273*). One diastereomer is formed preferentially in the presence of a mild base (such as sodium methoxide) that promotes epimerization of the remaining isomer, with the free amino acid obtained with up to 99% ee and 83% yield. In a similar manner, racemic amino acid methyl esters were converted into Schiff bases of 2-hydroxypinan-3-one. Deprotonation with KO*t*Bu and quenching with a saturated solution of ammonium chloride gave the products with 79 to >98% de (92–96% yield). The diastereomers were chromatographed before imine hydrolysis, giving methyl esters of Val, Leu, Phe, norvaline, and β-(2-naphthyl)-Ala in 72–94% yield with >98% ee (*274*). The method was applied to an attempted deracemization of 7′-azatryptophan, but product with only 66% de was produced (*275*).

A closely related process is enantioselective deprotonation, using chiral bases to selectively deprotonate one enantiomer of a racemic substrate, with the deprotonated species then trapped using a reactive electrophile. This approach was used with Seebach's imidazolidinone glycine template (see Section 7.9). Methylation of the racemic substrate after deprotonation with homochiral lithium bis[(1-phenyl)ethyl]amide gave an Ala derivative and allowed for recovery of the imidazolidinone template with up to 80% ee (*276*).

A chiral cobalt complex formed from β-(6-dimethylaminomethyl-2-pyridyl)-Ala binds amino acids with high stereospecificity, complexing one enantiomer of Ala, Phe or Trp more strongly. Reaction of the diastereomeric mixture with NaOD in D$_2$O epimerized the less-strongly bound complex to give (nearly exclusively) the D-amino acid (*277*). A theoretical study of the reasons behind the stereospecificity of the reaction has been published (*278*).

12.1.8 Other Methods of Resolution

Amino acids can be resolved by chromatography using chiral supports, a method that has found increasing use for the analysis of enantiopurity (see Section 1.8.3). This technique is generally not applicable on a preparative scale due to the expense of chiral columns of sufficient size, but is likely to become increasingly important. For example, a 2004 synthesis of all diastereomers of β-methyl-Phe used an HPLC polysaccharide chiral stationary phase (CSP) to resolve each set of *erythro* or *threo* enantiomers, but required 4–6 h and 33–38 injections to separate 327–760 mg of racemic material (*279*). One of the drawbacks of polysaccharide-derived CSPs is their incompatibility with mobile phases other

than hydrocarbons or alcohols, as the polysaccharide dissolves or swells. Mixed polysaccharide derivatives covalently bonded to an allylsilica gel matrix provide increased stability, and have been used for a semipreparative resolution of *cis*- and *trans*-1-amino-2-phenylcyclobutane-1-carboxylic acid, though (again) multiple (over 20) injections were required to purify 700 mg (*280*). An electrolyte containing hydroxypropyl β-cyclodextrin was used in a preparative-scale continuous free-flow isoelectric focusing separation of dansyl-Trp enantiomers (*281*).

A potentially useful method has been described for the enantioselective extraction of chiral carboxylates from aqueous to organic media, using steroidal guanidinium receptors for the enantioselective recognition of *N*-acyl amino acids. Enantioselectivities of up to 80% ee were achieved for the extraction of seven *N*-acetyl amino acids from aqueous buffer into chloroform (*282, 283*). Diaza-18-crown-6 ethers with arene sidearms demonstrated enantioselective transport of Phe, Phg, and Trp Na or K salts through a liquid membrane (*284, 285*). A ternary complex formed from a binaphthyl-xanthone-based macrocycle, 18-crown-6 ether, and unprotected amino acids showed chiral selectivity for extracting Phe, Phg, Ser, Trp, Val, and Ala from water into chloroform (*286*). Receptors based on a *cis*- or *trans*-tetrahydrobenzoxanthane skeleton with benzoxazole and amidopyridine substituents could complex *N*-triflate amino acids with enantiodifferentiation, and preferentially extract one enantiomer from water into chloroform (*1825, 1826*). This cleft-type tetrahydrobenzoxanthene receptor provided strong chiral discrimination between enantiomers of *N*-dinitrobenzoyl amino acids. Partition of the more strongly-bound complex into organic solvent provides the possibility of large-scale enantioseparations (*1827, 1828*). The enzymes histidine ammonia lyase and phenylalanine ammonia lyase have been mutated to abolish their catalytic activity and act as artificial receptors. They were then immobilized in a membrane to facilitate enantioselective transport across the membrane. Preliminary experiments demonstrated a time-dependent selectivity coefficient of up to 2.5 for Phe and 13.3 for His (*287*).

Countercurrent chromatography (CCC) combines aspects of HPLC and enantioselective extraction. Cinchona-derived anion-exchange-type chiral selectors were selected with the appropriate properties to maintain chiral recognition but also partition correctly in the stationary organic phase (methyl isobutyl ketone) during the CCC run. The amino acid to be resolved was injected in an aqeuous solution. Up to 900 mg of *N*-(3,5-dinitrobenzoyl)-Leu could be resolved in a single 100-minute run, using only 170 mL of stationary phase and 300 mL of mobile phase (*288*). A number of Pro derivatives were evaluated as chiral selectors for countercurrent separations of *N*-(3,5-dinitrobenzoyl)-Leu in a 2005 report (*289*). Enantioseparations in CCC were reviewed in 2001 (*290*), while optimization of high-speed CCC was reviewed in 2005 (*291*).

12.2 N-Alkylation of Amino Acids

12.2.1 Introduction

The N-alkylation of amino acids first attracted significant attention during the late 1960s and early 1970s due to the desire to permethylate peptides in order to increase their volatility for mass-spectral analysis. Many of the conditions developed for peptide methylation were then transferred to the alkylation of individual amino acids. A review of early syntheses of N-alkylated amino acids is included in Greenstein and Winitz (292), while a chapter on mono- and di-alkylation of the amino group is contained in a 1999 monograph on amino acid derivatives (293). There has been renewed interest in the N-alkylation of amino acids due to the importance of N-methyl amino acids in biologically active peptides isolated from bacteria. For example, the immunosuppressant drug cyclosporine, a cyclic undecapeptide, contains N-methyl leucine (MeLeu), MeVal, and MeBmt residues (see Scheme 12.6). Sarcosine (Sar, N-methyl glycine, MeGly) and N-methyl β-hydroxy-Val are found in BBM-928 A, B, and C (luzopeptins), antitumor antibiotics from *Actinomadura luzonensis* (294, 295). Two Sar and two N-methyl L-Val residues are contained in sandramycin, a potent antitumor antibiotic decadepsipeptide that also contains two heteroaromatic chromophores that are responsible for sequence-selective DNA-bis intercalation (296).

N-Methyl D-Ala is found in the cyclodepsipeptide azinothricin, a hexadepsipeptide antibiotic isolated from *Streptomyces* sp. X-14950 (297), while the microcystins, cyclic heptapeptide toxic components from a blue-green algae, contain N-methyl dehydro-Ala (298). N-Methyl Ala, N-methyl Val and N-methyl 5-hydroxy-Leu are contained in BZR-cotoxin II, produced by *Bipolaris zeicola* race 3, the cause of leaf spot disease in corn (299). The actinomycins, from *Streptomyces*, contain N-methyl Val (300). PF1022A is a cyclooctadepsipeptide with strong nematocidal activity. It contains four N-methyl L-Leu residues, with the N-methyl groups critical for activity (301). The nematoicidal cyclodecapeptide omphalotin A, isolated from the basidiomycete *Omphalotus olearius*, has 9 of the 12 amino acid residues N-methylated (2 Ile, 4 Val, 3 Sar) (302). The residues of koshikamide A, a cytotoxic linear peptide isolated from the marine sponge *Theonella* sp., include

N-methyl L-*allo*-Ile, N-methyl Leu, N-methyl Val, and N-methyl Asn (303). The cyclic undecapeptide barangamide A from the sponge *Theonella swinhoei* contains N-methyl D-*allo*-Ile, N-methyl L-Ile, and N-methyl L-Val (304). Lyngbyapeptin A is a tetrapeptide from the cyanobacterium *Lyngbya bouillonii* in which all four residues are N-methylated (305). Another tetrapeptide from a marine cyanobacterium, dragonamide 1, is also permethylated (306). Dolastatins 10–15, cytostatic depsipeptides isolated from a marine mollusk (6–43 mg each from 1600 kg of *Dolabella auricularia*!), contain N-methyl Val, N-methyl Leu, and N,N-dimethyl Val (307–309). N,N-Dimethyl Ile is a component of the depsipeptide zizyphin from *Ziziphus oenoplia* (310), while the cytotoxin apratoxin, from the marine cyanobacterium *Lyngbya majuscula*, has N-methyl Ala and N-methyl Ile (311). The dictyonamides A and B, isolated from a fungus separated from red algae, are undecapeptides containing Sar, five N-methyl Val, and an N-methyl Thr residue (312). The fish attractant strombine, isolated from the conch *Strombus gigas* (8 g from 593 g of conch flesh) is an iminodiacetic acid, N-carboxymethyl Ala (313).

The didemnins, cyclodepsipeptides with antitumor, antiviral, and immunosuppressive activity, isolated from marine tunicates, contain N-methyl D-Leu and N,O-dimethyl Tyr (314–317). N,O-Dimethyl-D-Tyr-N-methyl-L-Val is the dipeptide moiety of the lipodipeptides majusculamides A and B, non-toxic components of the blue-green algae *Lyngbya majuscula* responsible for a contact dermatitis in Hawaii known as "swimmers' itch" (318). Majusculamide C is a cyclic depsipeptide containing N,O-dimethyl-L-Tyr, N-methyl Ile, and N-methyl Val (319). N-Methyl L-Phe is found in the peptide antibiotic staphylomycin, produced by *Streptomyces* (320), N-methyl L-Phg and N,3-dimethyl-L-Leu in the antibiotic etamycin (321) and N-methyl 2'-chloro-3'-hydroxy-4'-methoxy-Phe in the Ras farnesyltransferase inhibitor pepticinnamin E, isolated from *Streptomyces* species (322). An N,N-dimethyl-L-Trp residue is found in waltherine-C, a 14-membered cyclopeptide alkaloid from the *Waltheria douradinha* tree of Brazil (20 mg from 2.8 kg of dried powdered bark). This plant is used in traditional folk medicine to wash wounds, combat laryngitis and as a bronchial anti-inflammatory agent (323). N-methyl 6'-chloro-5'-hydroxy-Trp is a component of keramamide A, isolated from a marine

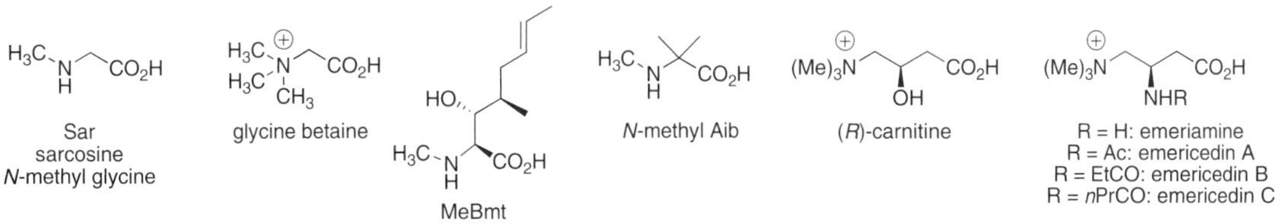

Scheme 12.6 *N-Methyl Amino Acids.*

sponge (*324*). *N*-Methyl 2′-bromo-D-Trp (2-bromo-abrine) is found in jaspamide (also called jasplakinolide), a cyclodepsipeptide with antifungal, antihelminthic, insecticidal, nematocidal, and ichthyotoxic activity (*309, 325–328*). *N,N,N*-Trimethyl 6′-bromo-L-Trp (L-6-bromo-hypaphorine) has been isolated from the sponge *Pachymatisma johnstoni*, with 61 mg obtained from 4.5 kg of damp sponge (*329*).

Glycine betaine (*N,N,N*-trimethylglycine, see Scheme 12.6), is an osmolyte synthesized by marine algae and halophilic plants; it is also produced by bacteria as an osmoprotectant which allows bacteria to grow in hyperosmotic environments such as seawater or other solutions with high concentrations of salts or organic components. Other *N,N,N*-trialkyl glycine derivatives have been examined for toxicity or osmoprotective activity towards the soil bacterium *Sinorhizobium meliloti* (*330*). Carnitine, the *N*-trimethylated derivative of γ-amino-β-hydroxybutyric acid, plays a role in the transportation of fatty acids through mitochondrial membranes (see references within *331, 332*). It was first isolated from a meat extract in 1905, and its structure reported in 1927. (–)-Carnitine can be isolated by alcoholic extraction of beef extract, which contains 2–3% of the amino acid (*333*). Emeriamine, (*R*)-3-amino-4-trimethylaminobutyric acid (an analog of carnitine with an amino group replacing the hydroxyl), is an inhibitor of fatty acid oxidation (*334*).

N-Alkylation has found increasing importance in structure–activity relationship studies, with the *N*-substituent restricting conformational freedom and inducing other effects, such as destabilizing helical peptides (*335*). A systematic conformational analysis of model peptides containing *N*-substituted amino acids was carried out in 1998. *N*-Methylation restricted the conformational flexibility, making β-sheet formation more difficult and increasing the tendency to form periodically helical conformations. Certain types of β-turns were also preferred (*336*). *N*-Methylation of the amide bond between residues *i*+1 and *i*+2 in tetrapeptides was found to induce nucleation of reverse-turn structures (*337*). A theoretical study of tetrapeptides found that Pro-D-NMe-amino acid and D-Pro-NMe-amino acid sequences at the *i*+1 and *i*+2 positions effectively stabilized reverse-turn conformations (*338*). The solution and X-ray structures of peptides containing an *N*-methyl amide have been compared to a standard peptide bond and other isosteres (*339, 340*).

N-Alkylation is a useful method to improve the stability of bioactive peptides, as it greatly reduces proteolysis (*341, 342*). For example, *N*-methyl analogs of endomorphin-2, an opioid tetrapeptide, were highly resistant to peptidases, carboxypeptidase Y, aminopeptidase M, and a rat brain homogenate (e.g., 80–90% intact after 2–20 h vs 0–10% for the parent peptide) (*343*). However, in contrast to their increased resistance to proteolysis, *N*-alkylated peptides have increased lability towards TFA cleavage, and can be degraded under conditions commonly used during solid-phase peptide synthesis (*344–346*). Cleavage occurs between the

N-alkyl derivative and the following (*C*-terminal) residue, and is proposed to proceed via an oxazolinium ion intermediate. *N*-Alkylation of Ser greatly increases the rate of elimination of acetylated (glycosylated) Ser derivatives, as formation of a protective aza-enolate from the amide bond is prevented (*347*). Interchain association is sometimes encountered during peptide synthesis, causing reduced coupling yields. It can be prevented by employing *N*-(2-hydroxy-4-methoxybenzyl) amino acids to form reversibly protected tertiary peptide bonds. The *o*-phenolic group aids the acylation of the secondary amine (*348*).

The peptoids are peptide analogs consisting of *N*-substituted amino acids, generally *N*-alkyl glycine. β-Peptoids, oligomers of *N*-substituted β-aminopropionic acid, have also been reported (*349*). Other *N*-alkylated amino acids include *N*-phosphorylated amino acids, with the monograph of Greenstein and Winitz containing a summary of early developments in their synthesis (*350*). A sulfamic acid (*N*-SO₃H) substituent is contained on an *N*-alkyl glycine residue found in minalemines D–F, isolated from the marine tunicate *Didemnum rodriguesi* (*351*). *N*-Hydroxy and *N*-amino (hydrazino) amino acids have also been synthesized, but are not discussed here.

Acylation of the hindered secondary amine of *N*-alkylated amino acids during peptide synthesis can be difficult, although many reports of poor yields may in fact be due to the acid lability of *N*-alkylated peptide products rather than incomplete coupling (*344–346*). Conditions have been developed which generally give good yields; these are discussed in Sections 12.2.6 and 12.5, and were reviewed in 1997 (*352*).

N-Alkylation can also create problems during determinations of enantiomeric purity. An HPLC method employing a stationary phase of silica-bonded amino acids (Pro, hydroxy-Pro, or Val) in combination with a mobile phase of aqueous/acetonitrile solutions of copper sulfate/acetate successfully resolved a number of *N*-alkyl α-amino acids (*353*). Other methods are described in Sections 1.6–1.13. Methods for modifying amino acid amino groups, acid groups, and side-chain groups, are described in a 1999 monograph (*293*).

12.2.2 N-Methylation

12.2.2a Introduction

N-Methylation of amino acids is the most common modification of the amino group due to the presence of *N*-methyl amino acids in many natural products. A number of highly reactive methylation reagents are available; the difficulty lies in finding a combination of reagents and protecting groups that allow for monomethylation of the amino group. Temporary *N*-protection is generally required during the methylation step in order to prevent polymethylation. A review of the synthetic preparation of *N*-methyl amino acids was published in 2004 (*354*).

12.2.2b Methylation with MeI/NaH
(see Table 12.6)

Peptides were some of the first substrates for N-methylation, in order to increase their volatility for mass-spectral analysis. Permethylation has been achieved with NaH/MeI in DMF (355), DMSO (356–359), or dimethylacetamide (360). The NaH/MeI reagent system was first applied to the N-methylation of individual amino acids by Coggins and Benoiton in 1971. N-Formyl, N-acetyl, N-benzoyl, N-tosyl or N-Cbz aliphatic amino acids were methylated in THF/DMF at 80 °C in 61–96% yield, accompanied by methyl ester formation (361). The same reagents were used to methylate the N^{α}- and/or $N^{hydroxamate}$-nitrogens of O-benzyl-amino acid hydroxamates (362). McDermott and Benoiton subsequently found that N-Cbz amino acids could be selectively N-methylated with NaH and MeI in THF at room temperature, without concomitant ester formation (363). The same conditions were used to methylate N-Boc amino acids (364); reaction at room temperature gave 70–90% yields of the N-Boc,N-methyl amino acids (nine examples), and the products were shown to contain <0.1% of contaminating enantiomer (364). O-Benzyl Thr and O-benzyl Tyr were among those methylated by these conditions; O-benzyl Ser required reaction at 5 °C to avoid elimination to dehydroalanine (364). Other chemists have applied these conditions to syntheses of Boc-(R)-MeLeu-OH (316), Boc-(S)-MeLeu-OH (365), Boc-MeTrp-OMe (366), Boc-MeThr(Bn)-OH (367), Boc-MeCys(Trt)-OH (368), Boc-N,α-Me$_2$-Ser (369), Boc-N,α-Me$_2$-β-fluoro-Ala (369), Boc-Me-β-Ala-OH (365), Boc-Me(4′-BO$_2$C$_2$Me$_4$-Phe)-OMe (370), Cbz-MePhe-OH (371), Cbz-MeTyr(tBu)-OMe (372), Boc-Me[3-(3-benzothienyl)-Ala]-OMe (373), Boc-Me[3-(4-thiazolyl)-Ala]-OMe (373), Boc-Me[3-(2-naphthyl)-Ala]-OMe (373), Boc-Me(1′-Bn-Trp)-OMe (373), Boc-Me(1′-MeO-Trp)-OMe (374), Boc-MeOrn(OMe)-OMe (373), Boc-MeHfe-OMe (373), Boc-MeTyr(Me)-OMe (373), Boc-2-aminoheptanoic acid dialkyl amides (375), Cbz-(S)-MeTyr(Me)-OH (316), and N-methyl,N-Boc 2′-chloro-3′-hydroxy-4′-methoxy-Phe (322). Boc-Gln-OH was dehydrated to Boc-γ-cyano-Abu and then N-methylated with NaH/MeI. Nitrile hydrogenation provided N^{α}-methyl-Orn, with guanylation giving N^{α}-Arg (376).

MeI/NaH has also been employed in conjunction with temporary N-phosphinamide protection to prevent dialkylation. In some cases the methyl esters were also formed (377). Cyclic N-carboxyanhydride (NCA) derivatives of α-CF$_3$-Phe, α-CF$_3$-Phg, and α-CF$_3$-Leu were alkylated with MeI in 42–56% yield using NaH in DMF (378). Isotopic labeling of Boc-Aib-OEt (379) or Boc-L-DOPA (380) was achieved by N-methylation with NaH/^{11}CH$_3$I.

A comparison of N-methylation procedures in 1977 found that N-Boc, N-Cbz, and N-Tos amino acids methylated with NaH/MeI were optically pure (<1% D-isomer), but those prepared via Ag$_2$O/MeI alkylation (see below) had 1–14% of the D-epimer, and those synthesized via reductive alkylations (NaBH$_4$ or H$_2$ reduction) had 0–14% D-isomer (depending on type of amino acid and reaction conditions) (381).

12.2.2c Methylation with MeI/Ag$_2$O
(see Table 12.6)

The Ag$_2$O/MeI reagent combination was introduced in 1967 for the permethylation of peptides (382), although side reactions of some residues have been noted (383). Olsen employed these reagents for the methylation of N-Cbz and N-Boc amino acids in 1970. However, only Ala, Val, Ile, and Phe were derivatized, with Ser and Cys giving mixtures of products (384). In another report, Boc-Abu-OMe, Boc-methallylglycine-OMe, and Boc-homoleucine-OMe were N-methylated in 71–98% yield (385). Dmb-Arg(Mts)-OMe has been methylated with MeI or C^{3}H$_3$I in 87% yield using Ag$_2$O as base (386), as has a Leu analog (387). A 1977 study demonstrated that this method can cause considerable racemization (1–14% D-isomer for N-Boc or N-Cbz Ala, Val, Leu, Ile or Phe) compared to other procedures (381).

12.2.2d Methylation with MeI/Other Bases (see Table 12.6)

Peptides have been permethylated by MeI and KOH in DMSO, with the same conditions applied to the N-methylation of Ac-Gly-OMe and Ac-Glu(OMe)-OMe (388). MeI/KOH in alcohol methylated the dehydroamino acid hydantoin of Tyr in 1912 in 70% yield (389), while an NaOH/MeI combination N^{α}-methylated an N-Ts diaminopropionic acid derivative (390). Didehydroamino acid esters were methylated in 79–91% yield with K$_2$CO$_3$/MeI in DMF (391–393), including residues within peptides (392, 393). The same reagents N-methylated an aminonitrile intermediate in a Strecker-type synthesis (394) or a diester of trans-azetidine-2,4-dicarboxylic acid (395), with MeCN as solvent for the latter reaction. N-Fmoc-dehydo-Phe, attached to a resin, was methylated with MeI/K$_2$CO$_3$ in the presence of 18-crown-6 in DMF (396).

Stronger bases have also been applied to methylation. Peptides that have been multiply deprotonated with LDA can be N-alkylated with MeI or (MeO)$_2$SO$_2$ under certain conditions, although this is usually a temperature-dependent side reaction of the desired C-alkylation (397, 398). Several N-Boc amino acid methyl esters were N-methylated using NaHMDS/MeI, with yields greatly superior to those obtained with NaH (68–72% vs 30–40%) and reportedly no racemization, although no evidence of optical purity was provided (399). Boc-Tyr(TBS)-OH was methylated in 71% yield with tBuLi and MeI in THF (400).

Quaternary amino acid derivatives have been synthesized by reaction of N,N-dimethylamino acids (prepared by reductive methylation) with methyl iodide in nitromethane, with no base. The trimethylammonium iodide derivatives of Ala, Val, and Phe were prepared. Access to the same trimethylamino derivatives via direct permethylation has been associated with racemization (*401*). Several carnitine analogs, *N,N,N*-trimethyl-γ-amino acids, were prepared by trimethylation using Ba(OH)₂/MeI (*402*). *N,N,N*-Trimethyl 5′-bromo-L-Trp (L-5-bromohypaphorine), and the 5′,7′-dibromo-L-Trp analog were obtained by methylation of 5′-bromo-Trp or 5′,7′-dibromo-Trp with MeI/NaOH (*329*). The α-amino groups of Ser, Leu, Thr, Phe-OEt, Gly-Gly-OH, Ala-Gly-OMe, and Gly-Gly-Leu-OMe were quaternized by treatment with methyl iodide and potassium bicarbonate in methanol (*403*).

12.2.2e Methylation with (MeO)₂SO₂ (see Table 12.6)

Cbz-D-Leu was methylated in 99% yield using dimethyl sulfate with KOH and Bu₄N⁺HSO₄⁻ in THF (*404*), while NaH and excess dimethyl sulfate was used for a synthesis of *N*-methyl 2-(1-naphthyl)-Gly (*405*). Schiff base or amidine derivatives of amino acid esters have been methylated with dimethyl sulfate or methyl triflate and then hydrolyzed to yield the *N*-methyl amino acids in 48–75% yield with no racemization (*406*). Dimethyl sulfate was used to form the trimethylamino group of emeriamine, (*R*)-3-amino-4-(trimethylamino)butyric acid (*407*). In 2003 several *N*-Boc amino acids were reacted with dimethyl sulfate in THF in the presence of NaH and catalytic water to give the *N*-Boc,*N*-methyl amino acids in 85–92% yield. The water reacted to generate dry sodium hydroxide, resulting in much faster reaction times than with powdered sodium hydroxide. The reaction could also be applied to residues within a Boc-Pro-Xaa-N(Me)Bn dipeptide (*408*). Methylation of *N*-(*o*-nitrobenzenesulfonyl) amino acid methyl esters has been carried out using dimethyl sulfate and DBU. The methyl ester was cleaved with LiI (avoiding potential racemization with LiOH), with the resulting *N*-methyl,*N*-(*o*-nitrobenzenesulfonyl)-protected amino acid compatible for Fmoc-based SPPS. Yields of amino acids by this route were compared with direct *N*-methylation or 5-oxazolidinone procedures (see below), with the best procedure varying with the substrate (*409*).

12.2.2f Methylation with Other Electrophiles/Bases (see Table 12.6)

Amino acid residues have been *N*-methylated during solid-phase peptide synthesis before coupling of the next residue, via temporary protection of the free amine of a support bound peptide with an *o*-nitrobenzenesulfonyl group. Methylation was accomplished with methyl *p*-nitrobenzenesulfonate and MTBD (a hindered guanidinium base) in DMF, followed by protecting group removal with β-mercaptoethanol/DBU. The temporary *N*-protection was needed to prevent dimethylation (*410*). A range of *N*-protecting groups (trifluoroacetyl, tosyl, alkylsulfonyl) and methylation conditions (Mitsunobu, MTDB/dimethylsulfate) were examined before the optimum conditions were found (*411*). This procedure was applied to a synthesis of cyclosporine analogs (*412*).

Solid-phase methylation has also been accomplished using MeI/K₂CO₃ in DMF in combination with the same temporary *o*-nitrobenzenesulfonyl protecting group (*413*), with diazomethane and *p*-nitrobenzenesulfonyl protection (*414*), or with MeI/DBU using 4-methoxybenzenesulfonyl protection (*415*). The Bts (benzothiazol-2-sulfonyl) group also promotes acidity of the Bts-NH proton. A solid-phase synthesis of a tetrapeptide employed Bts-protected amino acid chlorides for coupling, followed by on-resin methylation with MeI/K₂CO₃, and then Bts removal with PhSH/K₂CO₃. The cycle was repeated three times (*416*). However, one must be aware that coupling sulfonyl-protected amino acids may lead to racemization due to increased acidity of the α-proton (*417*).

The same strategy has been applied to solution-phase syntheses. Temporary *o*-nitrobenzenesulfonyl protection was used on the amino groups of Val-OtBu and Phe-OtBu, allowing for alkylation with K₂CO₃/MeI in DMF in 66–77% yield. The nosyl group was removed with LiOH/mercaptoacetic acid in DMF (*306*). Both 2- and 4-nitrophenylsulfonyl protection was employed during monomethylations of amino acid methyl esters with methyl iodide under solid–liquid PTC conditions (K₂CO₃ and catalytic TEBA in MeCN), with Val, Phe, and Phg methylated in 86–91% yield. Deprotection was achieved with PhSH and K₂CO₃ in MeCN or 0.5 M PhSK (*418*). The Bts group was used to prepare *N*-methyl β,β-dimethyl-Trp by alkylation with MeI in DMF. The Bts group was then removed using thiophenol/K₂CO₃ (*419*). The nosyl protecting/activating strategy was reviewed in 2004 (*420*), and is discussed in greater detail in Sections 12.2.3 and 12.3.3b.

The free base of Aib-OMe·HCl, generated by treatment with PMP in MeOH/CH₃CN, has been methylated with ¹¹CH₃OTf in 60–70% radiochemical yield; with ¹¹CH₃I, the yield was only 10–15% (*421*).

12.2.2g Methylation via Reductive Alkylation with Formaldehyde (see Table 12.7)

Reductive methylation is a popular route for preparing *N*-methyl amino acids, but temporary *N*-protection must be employed to prevent dimethylation of the amino group. *N,N*-Dimethylamino acids were reported in 1950 via reaction of the unprotected amino acid with aqueous formaldehyde under an atmosphere of hydrogen with Pd-C as catalyst (*422*). Similarly, *N,N*-dimethyl Thr was prepared in 82% yield from H-Thr(*t*Bu)-OH using HCHO and H₂/10% Pd/C (*423*); syntheses of *N,N*-dimethyl Ala, Val, and Phe were reported in 1973 (*401*) and *N*ᵅ,*N*ᵅ-dimethyl L-His in 1968 (*424*), with *N*ᵅ,*N*ᵅ,*N*ᵅ-trimethyl L-His (His betaine, or hercynine) then obtained by alkylating the dimethyl derivative with MeI at pH 9 (*424*). These reductive alkylation conditions were also applied to the *N*-terminal end of di- and tripeptides (*425*). *N*ᵅ,*N*ᵅ-Dimethyl Lys and *N*ᵋ,*N*ᵋ-dimethyl Lys were synthesized in 1959 by dimethylation of *N*ᵅ- or *N*ᵋ-monobenzoyl derivatives. The chromatographic

Table 12.6 *N*-Methylations of Amino Acids via Basic Alkylation

$$P^1\diagdown\underset{\underset{H}{|}}{N}-\underset{R^1}{\overset{R^2}{C}}\diagdown\underset{O}{\overset{P^2}{C}} \quad\xrightarrow[\text{MeX}]{\text{base}}\quad P^1\diagdown\underset{\underset{Me}{|}}{N}-\underset{R^1}{\overset{R^2}{C}}\diagdown\underset{O}{\overset{P^2}{C}}$$

P¹	R¹	R²	P²	Base	Methyl Source	Yield (%)	Reference
	peptides			NaH, DMSO	MeI		1968 (359)
	peptides			NaH, DMSO	MeI		1968 (358)
	peptides			NaH, DMA	MeI		1969 (360)
Boc, Cbz, Ac, Bz, CHO, Ts	*i*Bu	H	OH	NaH, THF/DMF	MaI		1971 (361)
	peptides			NaH, DMSO	MeI		1972 (357)
	peptides			NaH, DMF	MeI		1972 (355)
	peptides			NaH, DMSO	MeI		1973 (356)
Cbz	*i*Pr	H	OH	NaH, THF	MeI	43–90	1973 (363)
	*i*Bu						
	*s*Bu						
	CH₂CO₂*t*Bu						
	(CH₂)₂CO₂*t*Bu						
	CH(O*t*Bu)Me						
	(CH₂)₂SMe						
	Bn						
Boc, Cbz, Ts	Me	H	OH	NaH, THF	MeI	<0.1% D-isomer	1977 (381)
	*i*Pr						
	*i*Bu						
	*s*Bu						
	Bn						
Boc	Me	H	OH	NaH	MeI	70–90	1977 (364)
	*i*Pr						
	*i*Bu						
	*s*Bu						
	Bn						
	4-BnO-Bn						
	CH₂OBn						
	CH(OBn)Me						

(Continued)

Table 12.6 *N*-Methylations of Amino Acids via Basic Alkylation (continued)

P¹	R¹	R²	P²	Base	Methyl Source	Yield (%)	Reference
POPh₂	H Me *i*Pr *i*Bu *s*Bu Bn (CH₂)₂SMe	H	OH	NaH, THF	MeI	52–90	1976 (377)
Boc	Me *i*Pr CH₂CO₂Bn (CH₂)₂CO₂Bn CH₂OBn	H	NHOBn	NaH, THF	MeI	67–91	1981 (362)
Bz	2-naphthyl	CO₂Et	OEt	NaH, THF	(MeO)₂SO₂		1982 (405)
Boc	*N*-TBS-indol-3-yl	H	OH	NaH, THF/DMF	MeI	80	1988 (366)
Boc	*i*Bu	H	OH	NaH	MeI	80–90	1989 (316)
Cbz	4-OH-Bn	H	OH	NaH	MeI	80–90	1989 (316)
Boc	*i*Bu	H	OH	NaH	MeI	88	1990 (365)
Cbz	4-O*t*Bu-Bn	H	OMe	NaH	MeI	78	1991 (372)
Boc	3,4-(OMe)₂-Bn	H	OEt	NaH, THF	¹¹CH₃I		1992 (380)
CO₂-P²	*i*Bu Ph Bn	CF₃	OCO-αN	NaH, DMF	MeI	42–56	1994 (378)
Boc	CH₂CH₂CN	H	OH	NaH	MeI	83	1995 (376)
Boc	Me	Me	OEt	NaH, THF	¹¹CH₃I	40	1995 (379)
Boc	CH₂(1-TBS-indol-2-yl)	H	OH	NaH	MeI	73	1995 (1817)
Boc	2-Cl-3-OBn-4-OMe-Bn	H	OH	NaH	MeI	96	1998 (322)
Boc	CH₂(1-MeO-indol-3-yl)	H	OH	NaH, THF	MeI	73	1999 (374)
Boc	CH₂F CH₂OBn	Me	O*t*Bu	NaH	MeI	86–99	2002 (369)

Protecting group	Side chain			Base/Solvent	Reagent	Yield	Year (Ref.)
Boc	CH(OBn)Me	H	OH	NaH	MeI	85	2002 (367)
Boc	CH₂(benzothien-3-yl)	H	OH	NaH, THF	MeI	63–76	2002 (373)
	CH₂(1-Bn-indol-3-yl)						
	CH₂(thiazol-4-yl)						
	CH₂(2-naphthyl)						
	(CH₂)₃N(OMe)Cbz						
	(CH₂)₂Ph						
	4-MeO-Bn						
	4-BnO-Bn						
Boc	CH₂STrt	H	OH	NaH	MeI	84	2003 (368)
Boc or Boc-Pro	iPr	H	OH or N(Me)Bn	NaH + cat. H₂O	(MeO)₂SO₂	80–92	2003 (408)
	iBu	H					
	Bn	H					
	CH₂(2-naphthyl)	H					
		(CH₂)₄-(α-C)					
Cbz	Bn	H	OH	NaH	MeI	97.5	2004 (371)
Boc	4-B(-OCMe₂CMe₂O-)-Bn	H	OMe	NaH	MeI	92	2003 (370)
Boc	nPent	H	NMe₂, NEt₂, N[-(CH₂)₄-]	NaH	MeI	82–90	2004 (375)
	peptides			Ag₂O, DMF	MeI		1967 (382)
	peptides			Ag₂O, DMF	MeI		1968 (383)
Boc, Cbz	Me	H	OH	Ag₂O, DMF	MeI	93–98	1970 (384)
	iPr						
	iBu						
	Bn						
Boc, Cbz, Ts	Me	H	OH	Ag₂O, DMF	MeI	1–14% D-isomer	1977 (381)
	iPr						
	iBu						
	sBu						
	Bn						

(Continued)

Table 12.6 N-Methylations of Amino Acids via Basic Alkylation (continued)

P¹	R¹	R²	P²	Base	Methyl Source	Yield (%)	Reference
2,4-(OMe)₂-Bn	(CH₂)₃NHC(=NH)NHMts	H	OMe	Ag₂O, DMF	MeI, ¹¹CH₃I	87	1993 (386)
Boc	CH₂CH(Me)Et	H	OMe	Ag₂O, DMF	MeI	83	1994 (387)
Boc	Et iPent CH₂C(Me)=CH₂	H	OH, OMe	Ag₂O, DMF	MeI	71–98	1994 (385)
CONH-P²	4-MeO-Bn	H	NHCO-P¹	KOH, ROH	MeI	70	1912 (389)
Ac	peptides			KOH, DMSO	MeI	35–57	1979 (388)
Ac	H (CH₂)₂CO₂Me			KOH, DMSO	MeI	35–57	1979 (388)
Cbz	iBu	H	OH	KOH, nBu₄⁺HSO₄⁻	(MeO)₂SO₂	99	1994 (404)
Boc, Boc-Ala, Boc-MeAla-Leu	=CH₂ =C(Me)₂ =CHPh =CHEt	=R¹	OMe, Gly-OMe	K₂CO₃, DMF	MeI	71–89	1975 (393) 1978 (392)
Ac, Bz, Ts, Cbz	=CHMe =CHEt =CHiPr =CHPh	=R¹	OH	K₂CO₃, DMF	MeI	79–91	1981 (391)
Fmoc	=CHPh	=R¹	Gly-resin	K₂CO₃, DMF, 18-crown-6	MeI	—	2003 (396)
H	CH(CO₂H)CH₂-αN	H	O(8-Ph-menthol)	K₂CO₃, MeCN	MeI		1993 (395)
SO₂(2-NO₂-Ph)	Me iPr Bn 3-indolyl (CH₂)₄NHBoc, CH₂OtBu CH₂CO₂tBu	H	resin	K₂CO₃, DMF	MeI	86–100	1997 (413)

SO$_2$(CH$_2$)$_2$SiMe$_3$	iPr	H	COP^2 = CN	K$_2$CO$_3$, DMF	MeI	85	1997 (394)
SO$_2$(4-NO$_2$-Ph)	Me iPr iBu sBu Bn	H	OMe	—	CH$_2$N$_2$, Et$_2$O	81–97	2003 (414)
SO$_2$(2-NO$_2$-Ph)	iPr Bn	H	OtBu	K$_2$CO$_3$, DMF	MeI		2005 (306)
SO$_2$(4-MeOPh)	H Me iPr iBu sBu Bn 4-HO-Bn (CH$_2$)$_4$NH$_2$ CH$_2$OH CH(OH)Me (CH$_2$)$_2$CO$_2$H (CH$_2$)$_2$SMe	H	resin	DBU, DMSO/NMP	MeI	84–99	1997 (415)
SO$_2$(2- or 4-NO$_2$-Ph)	iPr Bn Ph	H	OMe	K$_2$CO$_3$, and TEBA in MeCN	MeI	86–91	2000 (418)
SO$_2$(benzo-thiazol-2-yl)	Me iPr iBu	H	NH$_2$	DMF	MeI	90	2000 (416)
SO$_2$(benzo-thiazol-2-yl)	C(Me)$_2$(indole-3-yl)	H	NH$_2$	DMF	MeI	90	2001 (419)
SO$_2$(2-NO$_2$-Ph)	Me iBu Bn CH$_2$OtBu, (CH$_2$)$_3$NHC(=NH)NH	H	peptide-resin	MTBD, DMF	(4-NO$_2$-Ph)SO$_2$Me	33–67	1997 (411)

(Continued)

Table 12.6 *N*-Methylations of Amino Acids via Basic Alkylation (continued)

P¹	R¹	R²	P²	Base	Methyl Source	Yield (%)	Reference
SO₂(2-NO₂-Ph)	Me	H	OMe	DMF	(MeO)₂SO₂	73–99	2005 (*409*)
	sBu						
	Ph						
	Bn						
	(CH₂)₄NHCbz						
	(CH₂)₄NHBoc						
	CH₂OtBu						
	CH(Me)OtBu						
	CH₂CO₂tBu						
	CH₂CH₂CO₂tBu						
	(CH₂)₃NHC(=NH)NHPbf						
	CH₂CONHTrt						
	CH₂CH₂NHTrt						
	CH₂(indol-3-yl)						
	4-tBuO-Bn						
	CH₂CH₂SMe						
	CH₂STrt						
	CH₂(1-Trt-imidazol-4-yl)						
Boc	Me	H	OMe	NaHMDS	MeI	68–72	1995 (*399*)
	iBu						
	sBu						
	4-OH-Bn						
Ts	CH₂NMeBoc	H	OH	NaOH	MeI	85	1995 (*390*)
H	Me	Me	OH	PMP	¹¹CH₃OTf	60–70	1995 (*421*)
Boc	4-TBSO-Bn	H	OMe	tBuLi, THF	MeI	71	1988 (*400*)
=CH(4-Cl-Ph), =CHNMe₂	iPr	H	OMe	toluene, Δ	(MeO)₂SO₂ or MeOTf	41–75	1984 (*406*)
	cOct						
	Ph						
	Bn						
	4-Cl-Bn						
=CH(4-Cl-Ph), =CHNMe₂	4-Cl-Bn	Me	OMe	toluene, Δ	(MeO)₂SO₂ or MeOTf	41–75	1984 (*406*)

properties of the products were compared to those of *N*$^\alpha$-methyl-Lys or *N*$^\varepsilon$-methyl-Lys, prepared by a different route (*426*). *N*-Alkyl amino acids can be *N*-methylated under these conditions to give *N*-methyl,*N*-alkyl amino acids (*427*). A modified version of this reaction generates the formaldehyde in situ by air oxidation of methanol over Pd-C, producing much purer products than reactions employing aqueous formaldehyde due to the lack of contaminating paraformaldehyde. The α-amino groups of Leu and Gly and the ε-amino group of Lys were *N*,*N*-dimethylated (*428*).

Amino acids were first mono-*N*-methylated by reductive alkylation with formaldehyde in 1963 via a three-step procedure with temporary *N*-benzyl protection. The amino group was first *N*-benzylated by reductive alkylation with benzaldehyde, and then *N*-methylated using HCHO/HCO$_2$H (no catalyst). Finally, the benzyl group was removed by hydrogenation (*429*). This method was originally used to prepare *N*$^\alpha$-methyl Ala, Val, Leu, Phe, Ser, Lys, and Arg (*429*), with other reports describing syntheses of *N*-methyl L-phenylglycine (*321*) and *N*$^\alpha$-methyl L-His (*424*). Improved conditions were described in 1994, employing paraformaldehyde in anhydrous formic acid for the reductive methylation, and ammonium formate as the hydrogen donor during removal of the benzyl group (*430*). However, a 1977 study demonstrated considerable racemization of *N*-methyl amino acids prepared by imine reduction using HCHO/HCO$_2$H, H$_2$, or NaBH$_4$ (up to 14% D-isomer), while NaBH$_3$CN gave variable results depending on reaction pH (*381*). [^{11}C]-Formaldehyde has been synthesized and used to monomethylate Aib in the presence of formic acid; no temporary *N*-protection was used (*431*).

Fmoc amino acids have been *N*-methylated by a two-step procedure, with an oxazolidinone intermediate prepared from the amino acid and paraformaldehyde subsequently reduced using triethylsilane and TFA (19–94% overall yields) (see Scheme 12.7) (*432*). A similar procedure was reported for *N*-Boc, *N*-Cbz, or *N*-tosyl amino acids. The oxazolidinones were formed from paraformaldehyde and PTSA in refluxing benzene, and then reductively opened with NaBH$_3$CN and TMSCl in acetonitrile. The reduction proceeded in 91–98% yield (19 examples, with Ala, Val, Leu, Ile, Phe, Met, and Tyr) (*434*). The same strategy, with *N*-Cbz protection and TFA/Et$_3$SiH reduction, was applied to suitably side-chain protected or unprotected Ser, Thr, Phg, Tyr, Glu, Gln, Asn, Arg, Cys, Met, Trp, Lys, His, and 2-amino-hex-5-enoic acid (*435–438, 1831*). For *N*-Fmoc or -Cbz

Ser and Thr, protection of the side-chain hydroxy group as TBDMS ethers allowed for oxazolidinone formation and reduction in good yields (72–86% and 78–96%, respectively) (*439*). Alternatively, reductive opening of the oxazolidinones of *N*-Cbz- or *N*-Boc-protected amino acids by catalytic hydrogenation over Pd/C gave the *N*-methyl or *N*-methyl,*N*-Boc amino acids, respectively, in quantitative yields, with no racemization (*440*). Amino acids ring-opened the oxazolidinones formed from *N*-Cbz amino acids to produce *N*-hydroxymethyl,*N*-Cbz dipeptides. These were converted into *N*-methyl dipeptides by treatment with TFAA/triethylsilane (*441*).

The oxazolidine reaction was modified to a one-pot treatment that worked well for δ- and ε-amino acids (51–100% yields), but gave low yields for α- and β-amino acids (27–43%) (*433*). Another report also attempted *N*-methylation of *N*-Cbz β-amino acids via oxazinanone formation and then Et$_3$SiH/TFA reduction, with 28–67% overall yields (*1831*). Microwave heating provided improved yields of the oxazolidinones from *N*-Fmoc α-amino acids, paraformaldehyde, and PTSA in acetonitrile, with reaction times reduced to 3 min at 130 °C. Oxazinanones were prepared from *N*-Fmoc β-amino acids under the same conditions. The reduction with triethylsilane was also microwave heated (1 min at 100 °C), giving 10–96% overall yields (*442*). An earlier report also described an efficient synthesis of oxazolidinones from *N*-Fmoc, -Cbz or -Boc amino acids using paraformaldehyde and PTSA in toluene with microwave heating, giving the cyclized products in 81–98% yield after 3 min (*443*).

Amino acid hexafluoroacetone oxazolidinones undergo a three-component condensation with paraformaldehyde and phosphorus tribromide or thionyl chloride to generate *N*-halomethyl amino acid oxazolidinones in 41–98% yield (see Scheme 12.8). The halide can then be used for a Michaelis–Arbusov reaction with phosphites, phosphinites, and methoxyphosphines to give, after oxazolidinone opening, *N*-phosphinoylmethyl amino acids (*444*). Alternatively, TFA/Et$_3$SiH reduction of the *N*-chloromethyl amino acid hexafluorooxazolidinones produces *N*-methyl amino acid oxazolidinones, which can be opened with *i*PrOH/HCl to form *N*-methyl amino acids, with MeOH/HCl to form *N*-methyl amino methyl esters, or with NH$_2$OH/*i*PrOH to form *N*-methyl α-amino hydroxamic acids (*445*). Glu and Asp, protected as the hexafluoroacetones, were converted into the *N*-methyl amino acids by this procedure, with elaboration of the side chains generating other amino acids (*446*).

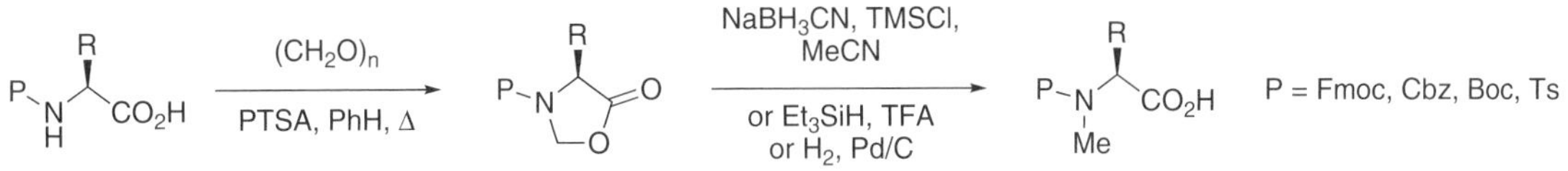

Scheme 12.7 *N*-Methylation via Oxazolidinone Intermediate.

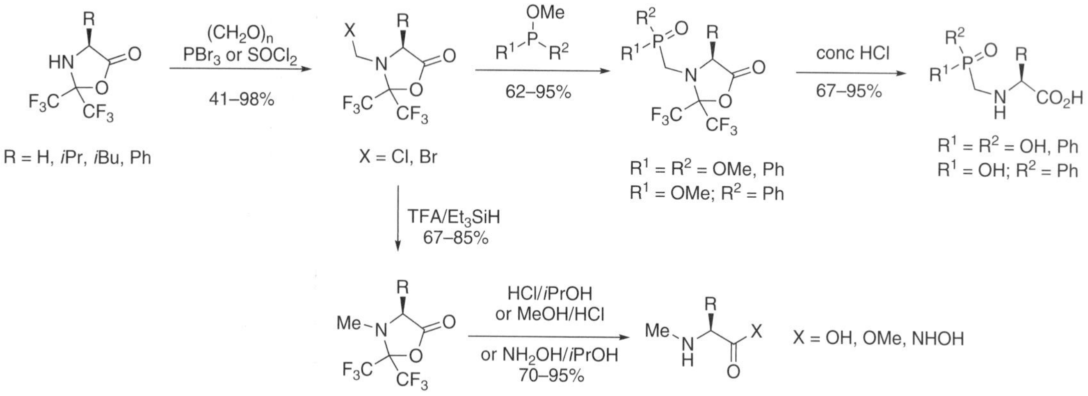

Scheme 12.8 Synthesis of *N*-Methyl and *N*-Phosphinoyl Amino Acids (*444, 445*).

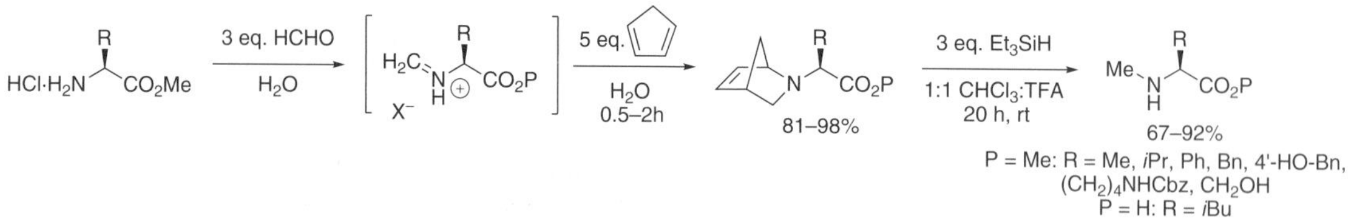

Scheme 12.9 *N*-Methylation of Amino Acids via 2-Azanorbornene (*450*).

N-Hydroxy amino acids (Ala, Val, Leu, Phe; produced by amination of Oppolzer's sultam enolate derivative, see Section 5.3.6c) have been reductively methylated with CH$_2$O/NaBH$_3$CN in 80–100% yield, with the *N*-hydroxyl group subsequently removed with Zn/HCl (*447*). Good yields of monomethylated amino acids were obtained by treatment of amino acid benzophenone Schiff bases with NaBH$_3$CN/aqueous formaldehyde, followed by hydrogenolysis to remove the remaining diphenylmethyl group. This procedure was suitable for difficult substrates such as Trp (*448*). Reductive methylation has been carried out on the *N*-terminal amino groups of resin-bound peptides. The amino group was temporarily protected with a 4,4′-dimethoxydityl (Dod, 4,4′-dimethoxydiphenylmethyl) group to prevent dimethylation, reductively methylated with CH$_2$O/NaBH$_3$CN, and then deprotected with TFA. It was found that 17 of the 19 eligible common amino acids (excluding Pro) were methylated in 86–99% yield; His gave a lower 71% yield, while Gly was not tested (*449*).

A novel method of reductive *N*-methylation traps the initial formaldehyde-derived imine with cyclopentadiene to give a 2-azanorbornene derivative in 81–94% yield (see Scheme 12.9). A retro Diels–Alder reaction in chloroform/TFA in the presence of triethylsilane then gives the *N*-methyl amino acid ester

in 81–94% yield. This reaction is notable for its mild conditions; it proceeds without detectable racemization, works on both methyl ester and free acid substrates, and is effective even with sterically bulky amino acids such as Phg (*450*). It has been applied to the *N*-methylation of β–cyclopropyl-Ala and 4-fluoro-Leu (*385*), as well as N^β-methyl L-α,β-diaminopropionic acid (*451*).

Zinc-modified cyanoborohydride has been reported to be a superior reagent for reductive methylations under mild reaction conditions (*452*), while zinc borohydride was described as an excellent alternative to toxic sodium cyanoborohydride for *N*-methylation of amines when used with paraformaldehyde and zinc chloride (*453*). The latter conditions were used to dimethylate Gly-OEt in 70% yield; monomethylation was not possible (*453*). Sodium triacetoxyborohydride has been demonstrated to be a much more effective reducing reagent than sodium cyanoborohydride or other reagents for a wide variety of reductive aminations (*454*). Several amino acids have been alkylated with this reagent (*454, 455*), but none were methylated.

12.2.2h Methylation via Other Methods

Optically active α-azido acids, which are intermediates in many amino acid synthetic routes (e.g, Evans, and

Table 12.7 *N*-Methylations of Amino Acids via Reductive Alkylation

$$\underset{\text{H}}{\text{P}^1-\text{N}}-\underset{\text{R}}{\overset{\text{R}}{\text{C}}}-\overset{\text{O}}{\underset{}{\text{C}}}-\text{P}^2 \quad \xrightarrow[\text{reducing agent}]{\text{HCHO}} \quad \underset{\text{Me}}{\text{P}^1-\text{N}}-\underset{\text{R}}{\overset{\text{R}}{\text{C}}}-\overset{\text{O}}{\underset{}{\text{C}}}-\text{P}^2$$

P^1	R	P^2	Temporary Protection	Methylation Conditions	Deprotection of Temp. Group	Yield (%)	Reference
H	H Me iPr iBu Bn 4-OH-Bn CH$_2$SMe (CH$_2$)$_4$NH$_2$ CH$_2$CO$_2$H (CH$_2$)$_2$CO$_2$H	OH or amino acid		aq. HCHO, H$_2$, Pd/C: dimethylation			1950 (422) 1950 (425)
Et, nPr, nBu, iBu, nHept	iPr	OH		aq. HCHO, H$_2$, Pd/C			1950 (427)
H	Me iPr iBu Bn CH$_2$OH (CH$_2$)$_4$NHTs (CH$_2$)$_3$NHC(=NH)NHNO$_2$	OH	NaOH, PhCHO then NaBH$_4$ or Pd/H$_2$	HCHO, HCO$_2$H	H$_2$, Pd	74–94	1963 (429)
Bn	CH$_2$(imidazol-4-yl)	OH		HCHO, H$_2$, Pd/C		40	1968 (424)
H	Ph	OH	NaOH, PhCHO then NaBH$_4$	HCHO, HCO$_2$H	H$_2$, Pd	50	1973 (321)
H	Me iPr iBu sBu Bn	OH		HCHO, NaBH$_4$		0.3–14% D-amino acid	1977 (381)
H	H iBu	OH, OMe, Gly-OH		HCHO from MeOH in situ, H$_2$, Pd/C: dimethylation		>90	1978 (428)

(*Continued*)

Table 12.7 *N*-Methylations of Amino Acids via Reductive Alkylation (continued)

P¹	R	P²	Temporary Protection	Methylation Conditions	Deprotection of Temp. Group	Yield (%)	Reference
Fmoc	Me; iPr; Bn; CH₂OBn; (CH₂)₂SMe; (CH₂)₄NPhth; CH₂(3-DNP-imidazol-2-yl)	OH		1) HCHO, TsOH 2) Et₃SiH, TFA		1) 37–96 2) 22–98	1983 (432)
H	Me; iPr; iBu; Bn; 4-OH-Bn; Ph; (CH₂)₄NHCbz; CH₂OH	OMe, Phe-OMe, Ala-Ala-OMe		1) HCHO, H₂O, cyclopentadiene 2) Et₃SiH, TFA		1) 81–98 2) 67–92	1987 (450)
OH	Me; iPr; iBu; Bn; CH₂CO₂Me	bornane-10,2-sultam		HCHO, NaBH₃CN		80–100	1993 (447)
H	Me; iPr; iBu; sBu; Bn; 4-OBrCbz-Bn; CH(Me)Bn, CH₂OBn; CH₂(imidazole); CH₂S(4-Me-Bn); (CH₂)₂SMe; (CH₂)₄NH(2-Cl-Cbz); (CH₂)₃NHC(=NH)NHTs; CH₂CO₂cHex; CH₂CONH₂; (CH₂)₂CO₂cHex; (CH₂)₂CONH₂	peptide-resin	Dod-Cl	HCHO, NaBH₃CN	TFA	56–99	1993 (449)

H	H	OEt	HCHO, Zn(BH$_4$)$_2$, ZnCl$_2$: dimethylation	70	1994 (*453*)
Bn	Bn	OH	(CH$_2$O)$_m$, HCO$_2$H	60	1994 (*430*)
H	CH$_2$cPr CH$_2$CFMe$_2$	OEt	1) HCHO, H$_2$O, cyclopentadiene 2) Et$_3$SiH, TFA	20–31	1994 (*385*)
Fmoc	H CO$_2$H CH$_2$CO$_2$H CH$_2$CONH$_2$ (CH$_2$)$_3$CO$_2$H (CH$_2$)$_4$CO$_2$H, 4-CO$_2$H-Ph 4-CO$_2$H-Bn 4-CO$_2$H-cHex	OH	1) HCHO, TFA 2) Et$_3$SiH	27–100	1996 (*433*)
H	Me with α-Me	OH	H[11]CHO, HCO$_2$H	14 (radiochem.)	1997 (*431*)
=CPh$_2$	Me iBu CH$_2$OH CH(OH)Me CH$_2$(indol-3-yl)	OMe, OEt, OBn, OCHPh$_2$	HCHO, NaBH$_3$CN	H$_2$, Pd	1) 55–90 2) 71–91	1997 (*448*)
Boc, Cbz	Me iPr iBu sBu Bn 4-OH-Bn CH$_2$OTBDPS	OH	1) HCHO, TsOH; 2) H$_2$, Pd/C		quant	1998 (*440*)

(Continued)

Table 12.7 *N*-Methylations of Amino Acids via Reductive Alkylation (continued)

P¹	R	P²	Temporary Protection	Methylation Conditions	Deprotection of Temp. Group	Yield (%)	Reference
$C(CF_3)_2O\text{-}P^2$	Me iPr iBu Ph	$OC(CF_3)_2O\text{-}\alpha N$		1) $(CH_2O)_n$, $SOCl_2$ 2) Et_3SiH, TFA		55–77	1998 (445)
$C(CF_3)_2O\text{-}P^2$	CH_2CO_2H $(CH_2)_2CO_2H$	$OC(CF_3)_2O\text{-}\alpha N$		1) $(CH_2O)_n$, $SOCl_2$ 2) Et_3SiH, TFA	1) H_2O, THF 2) conc. HCl, 80 °C	65–82 methyl-ation 50–63 deprotect-ion	2000 (446)
Cbz	H Me iPr iBu Ph Bn	OH	HCHO, PTSA (form oxazolidinone)	1) ring open with amino acid 2) reduction of N-CH_2OH dipeptide with Et_3SiH/TFAA		29–96	1999 (441)
Cbz	Me iPr iBu sBu Bn 4-BnO-Bn CH_2CH_2SMe	OH	HCHO, PTSA (form oxazolidinone)	reduction of oxazolidinone with $NaBH_3CN$/TMSCl		91–98	1999 (434)
Cbz	CH_2OH Ph 4-HO-Bn $CH_2CH_2CO_2H$ $CH_2CH_2CONH_2$ $(CH_2)_4NH_2$ CH_2CO_2H	OH	HCHO, CSA (form oxazolidinone)	reduction of oxazolidinone with Et_3SiH, TFA		58–91 oxazolidinone 60–79 reduction	2000 (437)

	R		Conditions	Reduction	Yield	Year (Ref.)
Cbz	CH₂CH₂CO₂H	OH	HCHO, CSA (form oxazolidinone)	reduction of oxazolidinone with Et₃SiH, TFA	63 reduction	2000 (438)
Cbz	CH₂OH, CH(OH)Me, 4-HO-Ph, CH₂SH, CH₂CH₂SMe, CH₂CONH₂, CH₂CH₂C=CH₂, CH₂CH₂CO₂H, CH₂(indol-3-yl), CH₂(imidazol-4-yl)	OH	HCHO, PTSA (form oxazolidinone)	reduction of oxazolidinone with Et₃SiH, TFA	70–83	2002 (436), 2003 (435)
Cbz, Fmoc, Cbz, Fmoc	CH₂OH, CH(OH)Me, CH₂OH, CH(OH)Me	OH	HCHO, PTSA (form oxazolidinone) on TBDMS protected	reduction of oxazolidinone with Et₃SiH, TFA	72–96 oxazo-lidinone, 78–96 reduction	2001 (439)
Cbz	(CH₂)₂CONBn₂	OH	HCHO, PTSA (form oxazolidinone)	reduction of oxazolidinone with Et₃SiH, TFA	90 reduction	2005 (1831)
Fmoc	Me, Bn, CH₂OBn, (CH₂)₂NHCbz; COP² = CH₂CO₂H: Me, iPr, Bn	OH	HCHO, PTSA, microwave heating, 3 min (form oxazolidinone from α-amino acid, oxazinanone from β-AA)	reduction of oxazolidinone with Et₃SiH, TFA. microwave heating (1 min)	10–96	2006 (442)
H	CH₂CH₂CO₂H	OH	aq. HCHO, MeOH	H₂, Pd/C	98	2003 (435)

Oppolzer's methods, see Sections 5.3.5, 5.3.6) were reductively methylated with dimethylbromoborane in excellent yield (63–99%) (*456*). The reduction is aided by complexation to the acid group; thus while α-azido acids and β-azido acids react quickly, γ-azido acids are reduced slowly. The amide carbonyl of *N*-formyl amino acids has been reduced by a two-step procedure via treatment with diborane (BH$_3$) followed by mild oxidative cleavage of the borane–amine adduct using I$_2$ and DIEA in AcOH or MeOH. Both solution- and solid-phase *N*-methylations were carried out (*457*). Boc- and Cbz-methyl esters of Gly and Ala were *N*-methylated in 47–57% yield without racemization by treatment with *tert*-butyl perbenzoate in the presence of Cu(II) octanoate (*458*). Unfortunately, bulkier amino acids such as Val were unreactive.

The Mitsunobu reaction can be used to alkylate *N*-arylsulfonamide amino acid esters with methanol in 78–96% yield (*459, 460*), producing optically pure products if TMSI is used for methyl ester deprotection and Na/NH$_3$ for *N*-tosyl removal (*460*). The comparatively acid-labile *N*-Pmc arylsulfonamide group was removed under much milder conditions than the *N*-tosyl group (33% HBr in AcOH with 2% H$_2$O, for 2–5 h) (*459*). Amino acid residues have been *N*-methylated during SPPS before coupling of the next residue. The free amine of a support-bound peptide was temporarily protected with an *o*-nitrobenzenesulfonyl group, methylated with DEAD/PPh$_3$/MeOH, and deprotected with NaSPh/DMF (*413, 461, 462*). Other temporary *N*-protecting arylsulfonyl groups are described in Sections 12.2.3a, 12.2.3b, and 12.3.3b below, as are other methods of removing the temporary *N*-sulfonyl protecting groups

A new method to monoalkylate primary amines without the need for temporary protection relies on a Matteson rearrangement of α-aminoalkylboronic esters (see Scheme 12.10). Resin-supported amino acids were alkylated with pinacol chloromethylboronic ester in the presence of diisopropylethylamine (DIEA). Rearrangement to the *N*-methyl product was followed by treatment with hydrogen peroxide to cleave the boronate group. The rearrangement prevented most dialkylation, as any doubly boronomethylated side product could not undergo a second rearrangement.

Treatment with excess hydrogen peroxide then cleaved the second group (if present) to a give clean monomethylated product (*463*).

12.2.2i Methylation During Syntheses of Amino Acids

Many of the common methods of amino acid synthesis can be adapted to produce *N*-methyl amino acids. The Strecker, Bucherer (see Sections 2.2.1, 2.2.2, 5.2.1), and Ugi four-component condensation (see Sections 2.2.3, 5.2.2) syntheses can be carried out with methylamine or dimethylamine instead of ammonia, leading to products such as Sar (*464*), *N,N*-Me$_2$-Gly (*464*), *N*-Me-Phg (*465, 466*), and others (*85*). Ultrasound and alumina have been used to catalyze the Strecker reaction with KCN and methylamine, giving the *N*-methyl aminonitrile intermediates in high (94–100%) yield (*467*). A Pd-catalyzed aminocarboxylation reaction of aldehydes using CO and acylamines was carried out with *N*-Me,*N*-acyl amines, producing *N*-Me-Leu and *N*-Me-Gly in 86–88% yield (*468*).

Amination reactions include the synthesis of *N,N,N*-trimethyl amino acids by displacement of homochiral α-bromo acids with anhydrous trimethylamine (*401*). Dimethylamine was used to produce the dimethyl derivatives (e.g., *401, 469*). Methylamine can be used for reductive amination of α-keto acids (e.g., *470*) (see Sections 2.3.6, 5.3.9).

Sarcosine was prepared via carboxylation of *N*-Boc, *N*-methyl-(tributylstannyl)methylamine (see Section 2.4.1), as an attempted synthesis via *N*-methylation of Gly resulted in poor yields (*471*). Good yields of *N*-Boc,*N*-methyl phenylglycine were obtained via enantioselective deprotonation of *N*-Boc,*N*-methyl-benzylamine with an *s*BuLi/(−)-sparteine complex followed by carboxylation, producing product with up to 78% ee (*472*).

Side chains can be introduced to sarcosine via alkylation reactions, resulting in new *N*-methyl amino acids. For example, the Myers-pseudoephedrine glycinamide alkylation protocol (see Section 6.2.1b) is effective with pseudoephedrine sarcosinamide as the alkylation substrate, producing *N*-methyl Phe and Abu as examples (*473*). Electrophilic alkylations are also possible, with

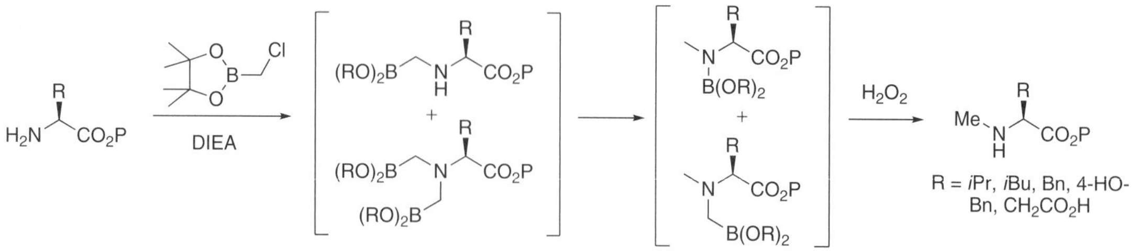

Scheme 12.10 *N*-Methylation of Amino Acids via Matteson Rearrangement (*463*).

γ,δ-unsaturated amino acids synthesized by a Pd-catalyzed allylation of α-tosylsarcosine (*474*).

12.2.3 N-Alkylation
(Other than N-Methylation)

12.2.3a Alkylation via Deprotonation and Reaction with Electrophiles (see Table 12.8)

A variety of combinations of bases and reactive electrophiles have been employed for the N-alkylation of amino acids with substituents other than methyl groups (see Table 12.8). In 2002 a series of alkali or alkali earth metal bases (e.g., $M(OH)_n$ and M_mCO_3; M = Li, Na, K, Rb, Cs, Mg, Ca, Sr, Ba) were tested for their ability to induce monoalkylation of Phe-OMe with activated *p*-nitrobenzyl bromide in DMF. The best, LiOH.H_2O, gave a 95% yield of monoalkyl product (compared to, for example, 55% for NaOH, 60% for KOH, and 64% for Cs_2CO_3). The same conditions gave good yields (73–95%) for monoalkylations of a variety of amino esters with activated alkyl halides, including benzyl bromide, allyl bromide, propargyl bromide, and methyl bromoacetate. Isobutyl bromide did not react, and iodomethane gave a 22% yield of monoalkyl product with 30% of the dialkyl derivative (*475*). Gly-OMe was monoalkylated with allyl bromide, 1-bromo-3-butene, or 1-bromo-4-pentene in acetonitrile and triethylamine in 70–75% yield (*476*), while N-farnesylated Val, Phe, and Met were prepared in good yield by alkylation with farnesyl bromide in the presence of triethylamine. Dialkylation was avoided by using equimolar amounts of reagent (*477*).

N-Boc,N-alkyl glycines, used for peptoid synthesis, have been prepared by several methods, including alkylations of Boc-Gly with allyl, propargyl, and benzyl halides (*478*). N-Boc,N-allyl allylglycine was prepared using NaH/allyl bromide (*479*). N-Ethylated Phe, Tyr, and Met were prepared by deprotonation of the N-Boc derivatives with two equivalents of *t*BuLi, followed by ethylation with triethyloxonium tetrafluoroborate. Less than 1% racemization was observed (*480*). Carbamates derived from amines other than amino acids can be alkylated using alkyl halides (e.g., benzyl chloride, *n*-butyl bromide), cesium carbonate and tetrabutylammonium iodide in DMF; it is not clear whether this procedure would also work with urethane-protected amino esters (*481*).

Temporary N-protection with *o*- or *p*-nitrobenzenesulfonyl (nosyl) groups has been used to prevent dialkylation when using reactive alkyl halides (benzyl or allyl bromide) with cesium carbonate as base. The temporary protection was removed without racemization by treatment with phenylthiolate (*482, 483*). Both 2- and 4-nitrophenylsulfonyl protection was employed during monoalkylations of amino acid methyl esters (Val, Ser, Met, Thr, Gly, Phe, Phg) with alkyl halides (MeI, BnBr, *n*BuBr, allylBr, propargylBr, methallylBr) under

solid–liquid PTC conditions with K_2CO_3 and catalytic TEBA in MeCN, with 76–91% yields. Deprotection was achieved with PhSH and K_2CO_3 in MeCN or 0.5 M PhSK (*418*). However, it has been observed that extended treatment with base for unreactive electrophiles can result in a rearrangement, with the nitrophenyl ring transferred from the sulfonamide to the α-carbon to produce quaternary amino acid derivatives (*484*). Alkylation by Mitsunobu reaction is also possible (see Section 12.2.3b). The Bts (benzothiazol-2-sulfonyl) group has also been employed as temporary protection during alkylation, with removal using thiophenol/K_2CO_3 (*416, 419*). N-Tosyl-allyglycine and -propargylglycine were allylated and propargylated in good yield in a similar fashion (*485*), while the side-chain amino groups of Dap, Dab, Orn, and Lys were protected with an *o*-nosyl group and then alkylated with benzyl bromide/K_2CO_3 while attached to a resin support (*486*). Solid-phase N-alkylation of a peptide N-terminal residue has been accomplished using alkyl halides and DBU as base, with temporary N-(4-methoxybenzene)sulfonyl protection (*415*).

A number of alkylations of specific amino acids have been described. Diaminopropionic acids derived from aziridines were regioselectively $N^α$-alkylated using Cs_2CO_3 and alkyl iodides (*390*). The iodonium salt $(CF_3SO_2)_2NI(Ph)CH_2CF_3$ reacted with Cbz-L-Lys-OH to give $N^ε,N^ε$-bis(2,2,2-trifluoroethyl) L-Lys (*487*). A piperidine-2,3-dicarboxylic acid derivative was synthesized from Asp via N-alkylation of Asp(OMe)-O*t*Bu with $Br(CH_2)_3Cl$, using $NaHCO_3$ as base. The chloro group was subsequently intramolecularly displaced by an Asp β-anion to form the six-membered ring system (*488*). Pro was N-benzylated by KOH/benzyl chloride in isopropanol (*489*), while the imino nitrogen of 4-amino-4-carboxy-Pro was alkylated using RBr/DIEA, RCl/DIEA or RI/K_2CO_3 (*490*). A cyclic N-carboxyanhydride (NCA) derivative of α-CF_3-Phe was alkylated with ethyl iodoacetate in 49% yield using NaH in DMF (*378*). An amidine derivative of Phe-OMe was alkylated with diethyl sulphate and then hydrolyzed to yield N-ethyl Phe 64% yield (*406*).

The iminodiacetic acid strombine (N-carboxymethyl Ala; a fish attractant isolated from the conch *Strombus gigas*) was prepared by alkylation of Ala with monochloroacetic acid (*313*). The N-terminus of Ala-Pro was also alkylated with chloroacetic acid (*491*), while Gly-OBn was monoalkylated with 0.5 equiv of *tert*-butyl bromoacetate in 60% yield (*492*). Other imino diacids have been synthesized by reacting Gly, Ala, or Val with chloroacetic acid or α-bromopropionic acid (*493*), or by reaction of iodoacetic acid esters with N-tosyl amino acid esters in DMF in the presence of Ag_2O (*494*). A series of α,α′-imino- dicarboxylic acids were prepared via N-alkylation of Ala-OEt, Phe-OMe, Tyr-OEt, Lys(Cbz)-OMe, or Pro-OMe with triflate derivatives of α-hydroxy acids (*495*), while the symmetrical imino diacid derivative of Val was synthesized in 96% yield by the displacement of the α-triflate of α-hydroxyisovaleric acid with the amino group of Val (*496*). Ala-OMe was alkylated with homochiral ethyl 2-bromopropionic acid (*497*). The (*R*)-pantolactone ester of 2-bromophenyl-acetic acid was condensed with amino esters (Gly, Phe, Pro) to give primarily the (*S,S*)-iminodiacetic acid with >80% de (*498*). The *N*-carboxymethyl substituent of *N*-carboxymethyl Gly-O*t*Bu was derivatized via reduction to an aldehyde, followed by coupling with a glycosylstannane derivative and then hydroxyl removal, providing an *N*-ethyl-linked sugar moiety (*499*). Iminodiacetic acid has been protected and activated with hexafluoroacetone for coupling with amino acids (*500*).

Analogs of iminodiacetic acids with one carboxy group replaced with a phosphate have been prepared. 4′-Phenyl-Phe-OBn was alkylated with the triflates of hydroxymethyl dialkylphosphonates to give the *N*-phos-phonomethyl amino esters, which were converted into dipeptide inhibitors of neutral endopeptidase (*501*). In a similar manner, dipeptides were *N*-phosphonomethylated using hydroxymethylphosphonate triflates to give inhibitors of human collagenase (*502*). Alternatively, several *N*-phosphorylmethylated amino acids were synthesized by displacement of the heteroatom from *N*-(chloromethyl) or *N*-(alkoxymethyl) amino acid derivatives by phosphoranes (*503*) or phosphorus esters (*504, 505*).

Two amino acids were linked by an *N,N′*-ethylene bridge by treatment with 1,2-dibromoethane (*506*), while two amino acid residues already joined by a bis(amidated) diaminoalkane linker were cyclized by refluxing with *meta*- or *para*-bis(bromoethyl)benzene in acetonitrile with K$_2$CO$_3$/Bu$_4$NBr (*507*). Amino acid imines or amino acids can be reacted with thiiranium ions derived from 2,3-epoxy sulfides, giving highly functionalized chiral *N*-alkyl derivatives of amino acids (see Scheme 12.11) (*508*).

Dipeptide amides with alkylated *C*-terminal and central amide nitrogens have been synthesized by a combinatorial approach on a solid-phase support (*509*). The first *N*-Trt protected amino acid was coupled to an amide-resin, with the amide linker alkylated

by deprotonation with lithium *tert*-butoxide in THF followed by the alkyl halide in DMSO. The bulky trityl group prevented alkylation on the *N*-terminal amino group. The second residue was then coupled, and the amide alkylation repeated. A library of 57,500 compounds was prepared. Peptides have been *N*-perbenzylated or *N*-perallylated by treatment with benzyl or allyl bromide after deprotonation with a P$_4$-phosphazene base in THF, with up to 58% yield for a tetrapeptide (*510*).

Unprotected amino acids undergo Michael addition with acrylonitrile in the presence of 1 N NaOH to give *N,N*-bis(2-cyanoethyl) amino acids. Hydrogenolysis of the cyano groups produced an *N,N*-bis(3-aminopropyl) derivative (*511*). Michael addition of the amino groups of Thr-OBn, Lys(Cbz)-OBn or 3-hydroxy-Lys(Cbz)-OBn to acrylic acid gave *N*-carboxyethyl derivatives. Treatment with di-2-pyridyl disulfide and triphenylphos-phine induced cyclization to enclose the amino group within an azetidin-2-one moiety (*512*). The amino groups of Gly-OEt and Gly-OBn were also incorporated within an azetidin-2-one via acylation with bromopropi-onic acid, followed by intramolecular *N*-alkylation induced with KOH (*513*). The amino group of amino acids has been incorporated into a 2-methylpyrrole ring by reaction with 5-chloro-3-penten-2-one and trieth-ylamine, with 75–81% yield (*514*). *N*-(3-Pyridazinyl)-α-amino esters were synthesized by displacement of 3-chloro-4-cyano-6-phenyl-pyridazine in MeOH with TMEDA as base (*515*).

Alkylations with electrophiles under neutral condi-tions are also possible. *N*-Tosylated or *N*-trifluoro-acetylated amino esters were allylated using allylic carbonate in the presence of Pd(0), with 97–99% yields for the *N*-Ts substrates (*516, 517*). Racemic allylic car-bonates undergo a Pd-catalyzed reaction with amino esters to give *N*-allylated products (see Scheme 12.12). The diastereoselectivity of the newly created stereogenic center was controlled by using chiral ligands, being dictated by the catalyst and not the amino acid (*518*).

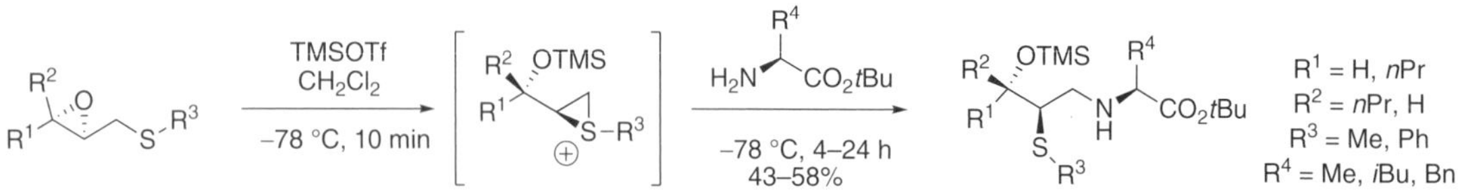

Scheme 12.11 *N*-Alkylation Using Thiiranium Ion (*508*).

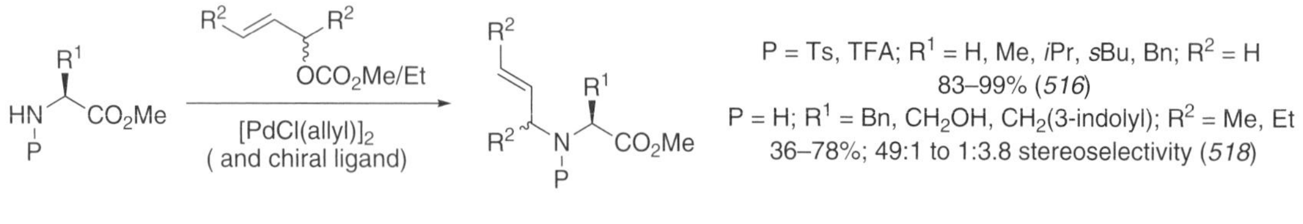

Scheme 12.12 *N*-Allylation of Amino Acids.

Amino esters were also *N*-allylated by Pd-catalyzed reaction with 1,3-diphenylallyl acetate, with diastereoselectivity of up to 70% de observed (depending on the amino ester) (*519*).

12.2.3b Alkylation via Mitsunobu Reaction (see Scheme 12.13)

N-Ts amino acids are sufficiently acidic that they can be used as the acid component of Mitsunobu reactions, reacting with various alcohols to introduce *N*-alkyl groups (*459, 460*). However, the *N*-Ts group requires Na/NH₃ for removal. A somewhat better alternative is the acid-labile Pmc group, which was successfully used for *N*-protection of amino acids undergoing the Mitsunobu alkylation, and can be removed with HBr/AcOH (*459*). *N*-Alkylation of Gly and Phe has also been accomplished on a resin support via Mitsunobu reaction using *N*-4-methoxybenzenesulfonyl or *N*-Ts protection, with a number of alcohols successfully used (BnOH, PhCH₂CH₂OH, PhCH₂CH₂CH₂OH, *c*HexCH₂OH, 26–96%) (*415*).

More usefully, the *p*-nitrobenzenesulfonyl or *o*-nitrobenzenesulfonyl groups were introduced for temporary amine protection during solution-phase Mitsunobu alkylations of amines (*483*). The groups can be readily removed using thiol/base. This approach has also been applied to amino acids attached to a solid-phase resin support (*462, 520*), and to the side chain of 2,3-diaminopropionic acid residues on a solid phase (*521*). The *o*-nitrobenzenesulfonamide group was found to give better results than the *p*-isomer, especially during deprotection. For removal after Mitsunobu alkylation, K₂CO₃/thiophenol resulted in partial saponification of the peptide from the resin, but DBU/2-mercaptoethanol gave good results (*462*). An *N*-ethyl scan of Leu-enkephalin was carried out using this method (*462*). DBU/thioglycerol is another effective reagent combination for removal of the *o*-nitrobenzenesulfonamide group, with the lower volatility of thioglycerol reducing the stench associated with 2-mercaptoethanol (*522*). Another solid-phase synthesis employed *o*-nitrobenzenesulfonyl protection for the amino acid monomers instead of Boc or Fmoc groups, avoiding the need for Fmoc/Boc deprotection and temporary sulfonylation. Each amino acid residue could then be readily *N*-alkylated after coupling, if desired (*461*). However, as mentioned earlier, *N*-nosyl amino acids are prone to racemization during coupling.

A 2005 study examined different Mitsunobu conditions for alkylation of resin-linked Ns-Gly with secondary alcohols. Cyclopentanol reacted in 78% yield using DEAD/PPPh₃/DIEA (*523*). A total synthesis of lipogrammistin-A included *N*-alkylation of a β-amino acid via Mitsunobu reaction of the *N*-nosylated amine with *N*-nosyl,*N*-bromopentyl,*N*-hydroxypropylamine (*524*). The nosyl protecting/activating strategy was reviewed in 2004 (*420*), and is discussed in greater detail in Section 12.3.3b.

Mitsunobu reaction conditions have also been applied to alkylations of resin-bound secondary amino acid hydroiodide salts (e.g., Pro, *N*-benzyl Gly) with benzylic alcohols, giving 66–97% yield (*525*).

12.2.3c Alkylation via Reductive Alkylation (see Table 12.9)

N-Alkyl groups can be introduced to amino acids by reductive alkylation with aldehydes and ketones. An initial report in 1950 examined the effect of substituents on the aldehyde or amino acid components during reductive alkylations, with hydrogenation over a Pd-C catalyst as the reducing agent. When straight-chain aldehydes were combined with unprotected Gly or Ala at room temperature, *N,N*-dialkylamino acids were produced, but other amino acids (Val, Leu or Phg) produced primarily mono-*N*-alkyl amino acids due to increasing side-chain steric hindrance. Similarly, by using aldehydes branched at the α-position, only mono-alkylated amino acids were obtained. Mono-*N*-alkyl amino acids could then be *N*-methylated or *N*-alkylated by a second reductive alkylation with formaldehyde or other aldehydes (*427*). In 2000, Pfizer used these conditions to prepare *N*-alkyl and *N*-alkyl,*N*-methyl amino acid on scales up to 200 g. Unprotected Val, Leu, and His were reacted with ketones in ethanol under H₂ with 20% Pd(OH)₂/C at room temperature to introduce the first alkyl group in 51–93% yield. The tertiary amine was then generated using formaldehyde under the same conditions, though at 50 °C, with 58–92% yield (*526*). *N*-Benzyl Gly or Ser were obtained by reductive alkylation with benzaldehyde using H₂/Raney nickel (*527*). The imino nitrogen of 4-amino-4-carboxy-Pro was alkylated by reductive alkylation with aldehydes and Pd/C-catalyzed hydrogenation (*490*). Unprotected Pro, Ser, and Thr were *N*-monobenzylated by reductive alkylation using benzaldehyde and homogeneous

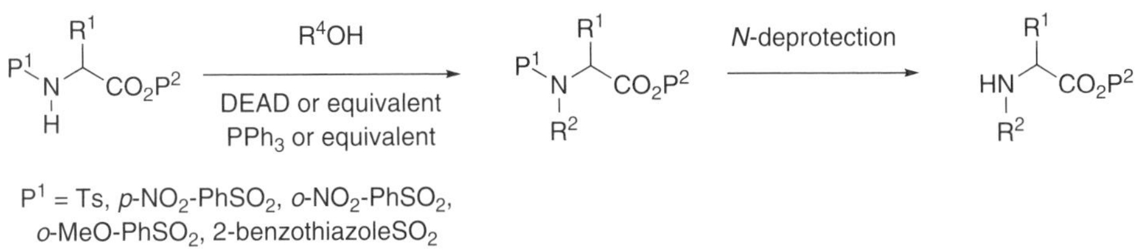

P¹ = Ts, *p*-NO₂-PhSO₂, *o*-NO₂-PhSO₂,
o-MeO-PhSO₂, 2-benzothiazoleSO₂

Scheme 12.13 *N*-Alkylation of Amino Acids – Mitsunobu Reaction.

Table 12.8 *N*-Alkylations of Amino Acids via Basic Alkylation

$$P^1\text{-}N(H)\text{-}C(R^1)(R^2)\text{-}C(=O)\text{-}P^2 \xrightarrow[\text{R}^3\text{X}]{\text{base}} P^1\text{-}N(R^3)\text{-}C(R^1)(R^2)\text{-}C(=O)\text{-}P^2$$

P¹	R¹	R²	P²	Base	Electrophile R³X	Yield (%)	Reference
H	H, Me, iPr	H	OH	NaOH	ClCH₂CO₂H, BrCH(Me)CO₂H		1931 (493)
H	Me	H	OH	Na₂CO₃	ClCH₂CO₂H	52	1975 (313)
H	Me	H	Pro	pH 8–9	ClCH₂CO₂H		1980 (491)
=CHNMe₂	Bn	H	OMe	PhH, reflux	Et₂SO₄	64	1984 (406)
Ts	H, Me, iPr, iBu, sBu, Bn, CH₂CO₂Me, (CH₂)₂CO₂Me	H	OMe, OBn	Ag₂O, DMF	ICH₂CO₂Me	63–83	1984 (494)
Boc	Bn, 4-OBn-Bn, (CH₂)₂SMe	H	OH	tBuLi (2 eq.)	Et₃OBF₄ (1 eq.)		1985 (480l)
H	Me	H	OEt	NaHCO₃	BrCH(Me)CO₂Et	63	1985 (497)
H	Me, iPr, iBu, Ph, Bn, CH₂OH (CH₂)₂SMe	H	OH	NaHSO₃, Na₂SO₃	2-HO-naphthyl, δ	8–33	1986 (572)
Boc	CH₂CO₂Me	H	OtBu	NaHCO₃	Br(CH₂)₃Cl	55	1991 (488)
Boc, Boc-Xaa	peptide	H		P⁴-phosphazene base	BnBr, BrCH₂CH=CH₂	14–80	1992 (510)
H	iBu, Bn, (CH₂)₂SMe	H	OH		Br(CH₂)₂Br (0.5 eq.)		1993 (506)

Boc	H	H	OH	NaH or KOrBu	BnBr / BrCH$_2$CH=CH$_2$ / ClCH$_2$CCH / BrCH$_2$CH=CHCl / BrCH$_2$CH=CHMe	45–87	1994 (478)
CO-O-P^2	CF$_3$	Bn	OCO-αN	NaH, DMF	ICH$_2$CO$_2$Et	49	1994 (378)
Ts	CH$_2$NHBoc	H	OBn	Cs$_2$CO$_3$, THF	EtI / nPrI	57–60	1995 (390)
H	iBu	H	OMe	Et$_3$N	ClP(tBu)R, R = iBu, Ph, convert to P(=S)tBuR, P(=O)tBuR	66–70	1995 (640)
H	iPr, Bn, (CH$_2$)$_2$SMe	H	OMe, OBn	TEA, DMF	BrCH$_2$[C=C(Me)(CH$_2$)$_2$]$_2$C=C(Me)$_2$	90	1996 (477)
H	iPr, Bn, (CH$_2$)$_3$-αN	H	OH	Pd/CuI/TEBA/K$_2$CO$_3$/ DMF/Et$_3$N	PhX	67–95	1996 (568)
Trt-Xaa	30 different side chains	H	NH-resin	LiOrBu	MeI / EtI / BnBr / BrCH$_2$(naphthyl) / BrCH$_2$CH=CH$_2$		1996 (509)
Boc	CH$_2$CH=CH$_2$	H	OMe	NaH, DMF	BrCH$_2$CH=CH$_2$	55	1996 (479)
H	iPr	H	OMe	lutidine	TfOCH(iPr)CO$_2$Bn	96	1997 (496)
SO$_2$(2- or 4-NO$_2$-Ph)	iPr	H	OMe	Cs$_2$CO$_3$	BnBr / BrCH$_2$CH=CH$_2$	72–87	1997 (482)
SO$_2$(4-MeO-Ph)	H	H	Oresin	DBU	RX, R = Bn, 4-Cl-Bn, 2,6-Cl$_2$-Bn, CH$_2$OBz. (CH$_2$)$_2$OBz, CH$_2$COPh, CH$_2$cHex, (CH$_2$)$_2$cHex, (CH$_2$)$_4$NHPhth, CH$_2$CONH(4-Br-Ph), (CH$_2$)$_2$OPh, (CH$_2$)$_3$(4-MeO-Ph), CH$_2$SO$_2$Ph, (CH$_2$)$_2$NHBz, (CH$_2$)$_2$NHPhth, (CH$_2$)$_2$CONH(4-Br-Ph), (CH$_2$)$_2$Ph	24–98	1997 (415)
Ts	CH$_2$CCH, CH$_2$CH=CH$_2$	H	OMe, Oresin	Cs$_2$CO$_3$	BrCH$_2$CH=CH$_2$ / BrCH$_2$CCH / BrCH$_2$CH=CHPh / ClCH$_2$C(Meo=CH$_2$)	60–93	1997 (485)

(Continued)

Table 12.8 *N*-Alkylations of Amino Acids via Basic Alkylation (continued)

P¹	R¹	R²	P²	Base	Electrophile R³X	Yield (%)	Reference
H	(CH₂)₃-αN	H	OH	KOH	BnCl	89	1998 (489)
H	Bn (CH₂)₃-αN	H	OMe, OEt, OtBu	TMEDA	3-Cl-4-CN-6-Ph-1,2-pyridazine	20–77	1998 (515)
H	Me iPr iBu Ph Bn	H	OMe, OEt, OtBu	Pd catalyst	PhCH(OAc)CH=CHPh	26–79, 12–70% de	1998 (519)
H	CH₂C(NHBoc)- (CO₂Et)-CH₂-αN	H	OEt	K₂CO₃ or DIEA	MeI EtI I(CH₂)₂CHPh₂ BrCHPh₂ BrCH₂(2-naphthyl) ClCH₂(1-naphthyl) 2-Ph-BnBr 2-, 3- or 4-Cl-BnCl 3,4-, 2,6-, or 2,4-Cl₂-BnCl 4-Ph-BnCl	60–95	1998 (490)
Ts, COCF₃	H	H	OBn	TEA	BrCH₂CO₂tBu	60	1999 (492)
	H Me iPr sBu Bn =CH₂ CH₂OH	H	OMe, OtBu, OBn	[allylPdCl]₂, PPh₃	H₂C=CHCH₂OCO₂Et	70–99	1999 (517)
SO₂(2- or 4-NO₂-Ph)	H iPr CH₂OH CH₂CH₂SMe CH(OH)Me Bn Ph	H	OMe	K₂CO₃ and TEBA in MeCN	MeI BnBr BrCH₂CH=CH₂ BrCH₂CHCH BrCH₂C(Me)=CH₂ BrCH₂CCH	76–91	2000 (418)

H	H	OMe	LiOH, DMF	BnBr	52–95	2002 (475)
	Me	$OtBu$		4-NO_2-BnBr		
	iPr			2-NO_2-BnBr		
	iBu			$BrCH_2CH=CH_2$		
	Bn			$BrCH_2CCH$		
	4-tBuO-Bn			$BrCH_2CO_2Me$		
	CH_2OtBu			$BrCH_2CH_2CH=CH_2$		
	CH_2CO_2tBu					
	CH_2CONH_2					
	$(CH_2)_3NHC(=NCbz)NHCbz$					
	$(CH_2)_3NHBoc$					
H	H	NH-resin	$Cu(OAc)_2$, TEA	$PhB(OH)_2$	21–75	2002 (571)
	Me			4-Me-$PhB(OH)_2$		
	$(CH_2)_3$-(α-N)			4-MeO-$PhB(OH)_2$		
				4-CF3-$PhB(OH)_2$		
				2-Me-$PhB(OH)_2$		
H	H	OMe	$Cu(OAc)_2$, TEA	4-Me-$PhB(OH)_2$	17–67	2003 (570)
	Me					
	iPr					
	sBu					
	Bn					
	CH_2OtBu					
	$CH(OtBu)Me$					
	4-BnO-Bn					
	CH_2SBn					
	CH_2CH_2SMe					
	$CH_2CH_2CO_2tBu$					
	CH_2CO_2tBu					
	CH_2(indole-3-yl)					
	$(CH_2)_3$-(α-N)					
H	iPr, Bn	$NH(CH_2)_nNHCOCH(R^1)NH_2$	K_2CO_3, Bu_4NBr	meta- or para-$(BrCH_2)_2$-Ph	49–67	2003 (507)

Rh-complexes for hydrogenation (*528*). Hydrogenolysis reductive alkylation conditions have also been applied to the *N*-terminal end of di- and tripeptides (*425*).

Borohydride derivatives are now usually employed as the reducing reagent. Sodium borohydride was the initial reagent of choice, with *N*-isopropyl Ala-OMe prepared from Ala-OMe and acetone (*529*). *N*-Boc, *N*-alkyl glycines (peptoid synthesis monomers) have been prepared by several methods, including NaBH$_4$ reductive alkylation of Gly with benzaldehyde followed by *N*-Boc protection (*478*). Hmb (2-hydroxy-4-methoxy-benzyl) protected amino acids, which are used to improve solid-phase peptide synthesis (SPPS) yields, were prepared by reductive alkylation of the amino acid with 4-methoxy salicaldehyde/NaBH$_4$ (*530*). The reductive alkylation can also be carried out on a solid phase, and has been used to introduce Hmb protection to amino acid residues after they have been coupled by SPPS (*531*). The related 2-hydroxy-6-nitrobenzyl (Hnb) group is also introduced by reductive alkylation (*532*), as is the 3-methylsulfinyl-4-methoxy-6-hydroxybenzyl (SiMB) group (*533*).

Sodium cyanoborohydride soon replaced sodium borohydride as the preferred hydride reduction reagent. A number of amino acids were treated with various aldehydes and ketones in the presence of NaBH$_3$CN/MeOH, giving the alkylated products in 58–96% yield (*534*), while 13 different amino esters were reductively alkylated with Boc-glycinal in 1998 (*535*). This method was used to prepare *N*-(2-Boc-aminoethyl) Gly esters (*536*), as well as *N*-(ω-aminoalkyl), *N*-(ω-thioalkyl) or *N*-(ω-carboxylalkyl) amino acids (*537*). Ser-OMe was alkylated with ω-(2-iodobenzene)alkylaldehydes enroute to constrained Phe analogs via a Heck cyclization (*538*). Iminodiacetic acids were obtained by reductive alkylation of L-Val, L-Leu, and L-Ile with 2-ketoglutaric acid, or of L-Glu with 2-keto analogs of Val or Leu, again using NaBH$_3$CN for the imine reduction (*539*). Similarly, the *N*-terminus of Ala-Pro was reductively alkylated with a range of α-keto acids using NaBH$_3$CN/ MeOH, H$_2$/Pd-C or H$_2$/Raney Ni for the reduction (*491*). NaBH$_3$CN/AcOH/DMF was employed for reductive alkylation of Leu-Val-Phe-Phe (attached to a solid support resin) with an α-keto acid equivalent of Lys (*540*). N^α-(2-Naphthylmethyl), *N*ε-alkyl Lys amides were prepared via attachment of Fmoc-Lys(Aloc) to a resin, Aloc deprotection and reductive N^ε-alkylation with aldehydes and NaBH$_3$CN, and then Fmoc removal and reductive N^α-alkylation with 2-naphthaldehyde and NaBH$_3$CN (*541*). A series of other resin-linked amino acids or amino acid *tert*-butyl esters were also N^α-alkylated with 2-naphthylaldehyde or 3-formylindole (*542*). The amino acid imine substrates for sodium cyanoborohydride reduction have also been generated on a solid support by transamination with ketimines previously prepared from ammonia and various ketones in the presence of titanium tetrachloride (*543*). Microwave irradiation was combined with NaBH$_3$CN reduction for nine different aryl aldehydes and Gly-OMe, with 27–70% yields (*544*).

N-Hydroxy-Phe (prepared via amination of Oppolzer's sultam enolate derivative, see Section 5.3.6c) was also reductively alkylated with NaBH$_3$CN as reagent, with the *N*-hydroxyl group subsequently removed with Zn/HCl (*447*). Similarly, the benzophenone ketimine of Ala was reacted with heptanal/NaBH$_3$CN, followed by hydrogenation to remove the diphenylmethyl group (*448*). A sugar–amino acid hybrid, with the α-amino group of the amino acid contained within a pyranose ring, was synthesized via reductive dialkylation using a dialdehyde and NaBH$_3$CN (*545*).

Sodium triacetoxyborohydride has begun to supplant sodium cyanoborohydride as the reagent of choice for reductive alkylations, and has been demonstrated to be a much more effective reagent than other reducing reagents (including sodium cyanoborohydride) for many reductive aminations, including alkylation of Gly-OEt with 2-butanone or alkylation of ethyl pipecolate with aryl aldehydes (*454*). Another study in 1997 compared NaBH(OAc)$_3$, ZnCl$_2$-mediated NaBH(OAc)$_3$, and NaBH$_3$CN for reductive alkylations of amino esters with amino aldehydes or 3-oxoesters, with the NaBH(OAc)$_3$/ ZnCl$_2$ combination generally giving the best yields (*547*). Sodium triacetoxyborohydride has been applied to several other amino acid N^α-alkylations (*455, 548*), and to the introduction of glycosyl derivatives to the ε-amino group of Lys and the α-amino group of Gly (*549*). A library of *N*-alkyl,*O*-aryl hydroxyproline derivatives was constructed on solid phase, with 34 *N*-alkyl substituents introduced by reductive amination using sodium triacetoxyborohydride in 2.5% acetic acid in CH$_2$Cl$_2$ (*550*). A His-Leu-type imino dicarboxylic acid was synthesized via reductive amination of His(Bn)-OMe with benzyl 2-keto-4-methylpentanoate, producing a mixture of diastereomers. The products were potent inhibitors of the zinc metalloprotease ACE2 (*551*). Phe-OBn and Asp(O*c*Hex)-OAllyl have been *N*-alkylated by a one-pot procedure using aldehydes generated in situ from *S*-ethyl thioesters, with aldehyde formation/ reductive alkylation employing a combination of NaBH(OAc)$_3$, Et$_3$SiH and 10% Pd-C in DMF. The thioesters included some derived from amino acid α- and ω-carboxy groups (*546*).

Other reducing reagents have been employed. Tetrabutylammonium borohydride was applied to the reduction of aryl imines of amino esters, with 95% yields (*552*). A polymer-supported cyanoborohydride reagent was used during a combinatorial alkylation of prolines with aryl aldehydes and ketones (*553*). Peptide isosteres with the amide carbonyl group replaced by a methylene group have been constructed on a solid-phase support by reductive alkylation of amino acids with protected amino acid aldehydes, using LiAlH$_4$ or NaBH$_3$CN as the reducing reagent. Significant racemization of the amino aldehyde was observed with NaBH$_3$CN, but not LiAlH$_4$ (*554*).

The initial imine formed from amino acids and aldehydes can undergo reactions other than reduction. Indium-mediated allylation introduced an allyl group to the RCHO-derived imine, giving *N*-CH(R)CH$_2$CH=CH$_2$

amino substituents with a chiral center (555). The imines formed from amino acid esters and 2-aminocyclo-alkanones were oxidized to generate spirocyclic oxaziri-dines, with photolytic rearrangement producing a conformationally restricted dipeptide in which the α-side chain of the N-terminal residue was cyclized to the amide nitrogen (see Scheme 12.14) (556). The initial iminium ion can also be used for Diels–Alder reactions with cyclopentadiene, giving azanorbornene derivatives in high yields (81–98%) (see Scheme 12.9). The bicyclic compounds were decomposed to form N-methyl amino acids (450).

The intermediate imine is also susceptible to Michael addition reactions of heteroatoms, via three-component condensations of amino acid esters or amides, formalde-hyde, and heteroatoms (see Scheme 12.15). Amides of aromatic acids reacted as the nucleophile component to produce N-(amidomethyl) amino acid esters or amides in 21–90% yield (557–559), while imides of dicarboxy-lic acids added to give N-(imidomethyl) amino acid esters or amides in 41–77% yield (558–560). These N-(amidomethyl)- and N-(imidomethyl)- products were further N-derivatized to generate N-acyl, N-sulfonyl, or N-nitroso products (559, 561), while the ester hydrolysis produced amino acids suitable for coupling to other residues (562). N-(Phosphinoylmethyl) sarcosine was synthesized in good yield by combining sarcosine, paraformaldehyde, and diethyl phosphite (503), while N,N-bis(phosphorylmethyl) amino acids were obtained

from Gly, β-Ala or GABA using paraformaldehyde and phosphorous acid (563). Glyphosate (N-phosphonomethyl Gly) was prepared by addition of phosphonic acid to the imine formed from formaldehyde and N-benzyl Gly, followed by hydrogenolytic removal of the benzyl group (564).

A similar strategy was employed to prepare a variety of functionalized N-substituents via a common interme-diate, N-benzotriazolylmethyl α-amino esters. The intermediates were prepared in 70–90% yield by reaction of amino esters, benzotriazole, and aqueous formaldehyde. The benzotriazole was then displaced by a variety of nucleophiles (HSPh/NaH, P(OEt)$_3$, NaCN), with some nucleophiles requiring the presence of a Lewis acid catalyst (allyl-TMS, H$_2$C=C(Ph)OTMS or Me$_2$C=C(OMe)OTMS with BF$_3$·Et$_2$O)(565). The N-CH$_2$Nu amino ester products (Nu = SPh, PO$_3$Et$_2$, CN, CH$_2$CH=CH$_2$, CH$_2$COPh, CMe$_2$CO$_2$Me) were obtained in 40–98% yield, with no loss of optical purity (565).

Amino acid hexafluoroacetone oxazolidinones have been used as mono-protected substrates. A three-component condensation with paraformaldehyde and phosphorus tribromide or thionyl chloride generated N-halomethyl amino acid oxazolidinones in 41–98% yield (see Scheme 12.8). The halide could then be used for a Michaelis–Arbusov reaction with phosphites, phosphinites, and methoxyphosphines to give, after oxazolidinone opening, N-phosphinoylmethyl amino acids (444).

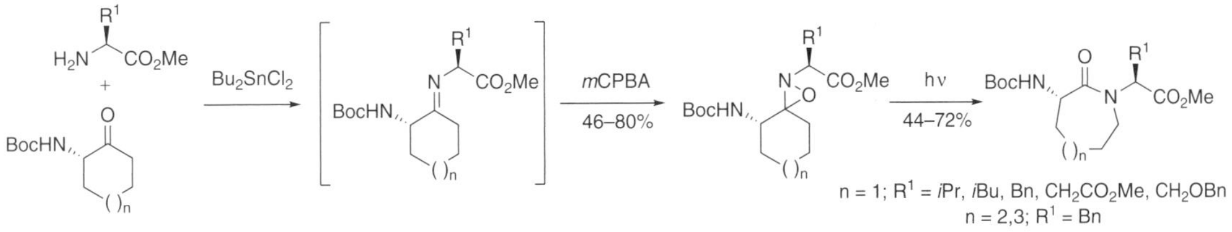

Scheme 12.14 Reductive Amination of Amino Acids Leading to Freidinger Lactams (556).

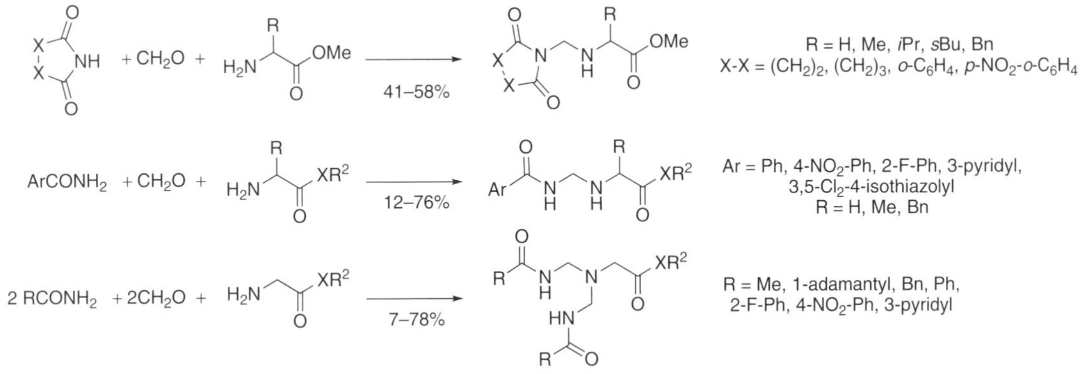

Scheme 12.15 Derivatization of Amino Group via Three-Component Condensations (557, 558, 559, 560, 561, 562).

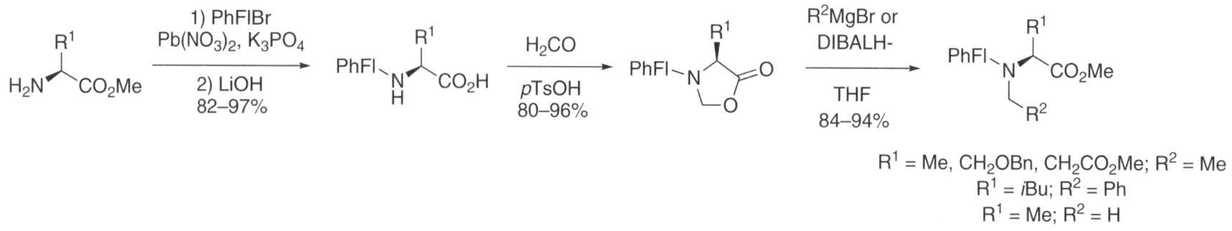

Scheme 12.16 Grignard Opening of Amino Acid Oxazolidinone (*566*).

Fmoc amino acids can be alkylated by a two-step procedure, via acid-catalyzed oxazolidinone formation with aldehydes/ketones, followed by reductive opening of the oxazolidinone with triethylsilane and TFA (see Scheme 12.7) (*432*). Alternatively, N-(9-phenylfluoren-9-yl) oxazolidinones of amino acids were opened with Grignard reagents or reducing reagents to give N-alkyl amino acids in high yield (84–94%) (see Scheme 12.16) (*566*). In a similar manner, cyclization of the side chain and amino groups of Cbz-Ser with pivaldehyde formed an oxazolidine. Hydrogenation under standard conditions removed the Cbz-group but also cleaved the C–O bond, leaving an N-neopentyl group (*567*).

12.2.3d Alkylation via Other Methods

The amide carbonyl of N-acyl amino acids can be reduced by a two-step procedure, using diborane (BH$_3$) followed by mild oxidative cleavage of the borane–amine adduct with I$_2$ and DIEA in AcOH or MeOH. Both solution- and solid-phase N-alkylations were carried out, producing N-methyl, N-ethyl, or N-propyl amino acids from the N-formyl, N-acetyl, or N-propionyl precursors in 66–86% yield (*457*).

N-Aryl α-amino acids were prepared in 67–95% yield by a Pd-Cu catalyzed coupling of the unprotected amino acid (Val, Phe, Pro) with aryl halides (PhI, PhBr). Ser and Glu were not successful substrates (*568*). CuI alone was found to catalyze the coupling of aryl bromides and aryl iodides with a range of unprotected amino acids at 90 °C, giving variable yields of 5–92%. The structure of the amino acid played a critical role in the coupling: Gly, Ser, and Glu produced no product, while Ala, Val, Pro, Phe, Met, Tyr, and Trp were successfully arylated (*569*). Cu(II) acetate promoted the N-arylation of α-amino esters with p-tolylboronic acid at room temperature; a variety of amino esters were arylated in 17–65% yield with miminal racemization (94–99.8% ee) (*570, 571*). Combining 2-naphthol with unprotected amino acids (Ala, Val, Leu, Ser, Met, Phe, Phg) in the presence of Na$_2$SO$_3$/NaHSO$_3$ led to N-(2-naphthyl) α-amino acids in 8–33% yields (*572*). Diazotization of Tyr and reaction with anilines gave a number of (racemic) N-aryl Tyr derivatives via Rh-carbenoid N–H insertion (one example gave 86% yield) (*573, 574*). Refluxing Tyr-OMe with 2-benzoyl-cyclohexanone in anisole in the presence of 10% Pd/C produced an N-(2-benzoyl-phenyl) substituent (*573*).

N-(2,4-Dinitrophenyl) amino acids were obtained via nucleophilic substitution of dinitroaryl halides by unprotected amino acids in H$_2$O/NaHCO$_3$ under microwave irradiation (*575*). Developments in N-arylation chemistry were reviewed in 1998 (*576*).

Nitrogen-centered radicals have been generated on amino acid derivatives and used for intramolecular cyclizations with alkene substituents on either the amino group or α-center (*577*). An N-methanofullerene substituent was introduced to sarcosine by a photochemical radical reaction (*578*). Alternatively, reaction of amino acids with fullerene C$_{60}$ in solution at 50–100 °C gave N-substituted monoadducts (*579*).

12.2.3e Alkylation During Syntheses of Amino Acids

Several aminocarboxylation reactions have been used to prepare N-alkylated compounds (see Sections 2.2, 5.2). The ammonia or ammonium chloride normally used for the Strecker reaction can be replaced with other amine sources such as alkylamines or anilines, leading to N-alkyl and N-aryl amino acids. Amines successfully employed include ethyl (*85*), propyl (*85, 580*), n-hexyl (*580*), cyclohexyl (*581*), phenyl (*580*), benzyl (*582*), and dimethyl (*582*) amines, morpholine (*582*), and pyrrolidine (*582*). Symmetrical or asymmetric amino acid dimers sharing a common amino group (iminodiacetic acids) can be prepared by using an amino acid as the amine component (*583–585*). The Ugi four-component condensation (see Sections 2.2.3, 5.2.2) can also be carried out with a variety of amine components. For example, N-decyl Phg and N-cyclohexyl Val derivatives have been prepared (*586*), as have N-phenethyl dipeptide derivatives (*587*). Unsymmetrical iminodiacetic acids have also been synthesized by Ugi four-component condensations (*588–590*). N-Substituted glycines (peptoid monomers) were prepared in 68–78% yield via a Pd/C-catalyzed amidocarbonylation reaction of N-acyl amines with paraformaldehyde and carbon monoxide (*591*).

Most methods of amino acid synthesis via nucleophilic amination of electrophilic α-substituted acids or esters allow for the introduction of substituted amine components (see Scheme 12.17) (see also Sections 2.3.1, 2.3.2, 5.3.1–5.3.4). Piperidine was used to displace 16 different α- and ω-bromo esters in 1934 (*592*), while five N-alkyl glycines were prepared by displacements of ethyl or benzyl bromoacetates in 1996 (*593*). Monomers

Table 12.9 *N*-Alkylations of Amino Acids via Reductive Alkylation

P^1	R^1	P^2	R^2	R^3	Alkylation Conditions	Yield (%)	Reference
H	H Me *i*Pr *i*Bu Ph	OH, Xaa	Me Et *n*Pr *n*Hex 2-furyl	H	H$_2$, Pd-C (mono or dialkylation)		1950 (427) 1950 (425)
H H	Me H CH$_2$OH	OMe OMe, OEt	Me Ph	Me H	NaBH$_4$ H$_2$, Ra-Ni	23 72–78	1963 (529) 1968 (527)
H	Me	Pro	Me Et *i*Pent Bn (CH$_2$)$_2$Ph (CH$_2$)$_5$NH$_2$	CO$_2$H	NaBH$_3$CN, MeOH or H$_2$, Pd-C, or H$_2$, Ra-Ni		1980 (491)
Fmoc	Me	OH	H Me Bn *c*Hex	H	1) RCHO, TsOH 2) Et$_3$SiH, TFA	1) 30–96 2) 74–88	1983 (432)
H	*i*Pr Bn (CH$_2$)$_2$SMe CH$_2$OH (CH$_2$)$_2$CO$_2$H	OH	Me Et *i*Bu Ph	H	NaBH$_3$CN, MeOH	58–96	1984 (534)
H	*i*Pr Bn (CH$_2$)$_2$SMe CH$_2$OH (CH$_2$)$_2$CO$_2$H	OH	Me	Me	NaBH$_3$CN, MeOH	58–96	1984 (534)

(Continued)

Table 12.9 *N*-Alkylations of Amino Acids via Reductive Alkylation (continued)

P¹	R¹	P²	R²	R³	Alkylation Conditions	Yield (%)	Reference
H	Me, iPr, iBu, Bn, 4-OH-Bn, Ph, $(CH_2)_4NHCbz$, CH_2OH	Me, Phe-OMe, Ala-Ala-OMe	$\longrightarrow$ (structure: R¹–CO–N–P²)		CH_2OH, H_2O, cyclopentadiene	81–98	1987 (450)
H	Me, iPr, iBu, Bn, 4-OBn-Bn, $(CH_2)_4NHFmoc$, CH_2SH, $CH(OBn)Me$, $CH_2(indol-3-yl)$	peptide-resin	$CH(R)NHBoc$ R = Me, iPr, CH_2SH, Bn, 4-OBn-Bn, $(CH_2)_4NHFmoc$, $CH_2(Indol-3-yl)$	H	$LiAlH_4$ or $NaBH_3CN$		1988 (554)
OH	Bn	(2S)-bornane-10,2-sultam	Me, iPr	H	$NaBH_3CN$	71–81	1993 (447)
H	H	OMe, OEt	CH_2NHBoc	H	H_2, Pd/C	64–71	1993 (536)
H	H	OH	Ph	H	1) Et_3N, MeOH 2) $NaBH_4$	59	1994 (478)
H	H	OMe, OEt	H: convert to $CH_2NHCOAr$: Ar = Ph, 3-NO_2-Ph, 4-NO_2-Ph	H	H_2NCOAr	52–90	1994 (557)
H	H, Me, iPr, sBu, Bn	OMe, OEt	H: convert to $CH_2NHCOAr$: Ar = Ph, 3- or 4-NO_2-Ph, 2-F-Ph, 4-Br-Ph, 2-OH-Ph, pyrid-3-yl, 3,5-Cl_2-isothiazolyl	H	$ArCONH_2$	21–90	1994 (557) 1999 (558)
Me	H	OH	H: convert to PO_3Et_2	H	$(EtO)_2P(O)H$	81–86	1995 (503)
H	H	OH	OH: convert to PO_3H_2	OH: convert to PO_3H_2	H_3PO_4	59–83	1995 (563)
H	H, Me, iPr, sBu, Bn	OMe, OEt	H: convert to CH_2R^4: R^4 = NPhth, N[—CO(CH₂)₂CO—], N[CO(CH₂)₃CO]	H	HR^4	41–77	1995 (560) 1996 (558)

R¹	R²	OR / resin	R³	R⁴	Conditions	Yield (%)	Year (Ref.)
H	Me	resin	2-OH-Bn, 4-OMe-Bn	H	1) AcOH, DMF 2) NaBH₃CN	>95	1996 (531)
H	iBu, CH(OH)Me, Bn	OMe	nPent, (CH₂)₃N₃, (CH₂)₃OBn	H	NaBH(OAc)₃, DCE	69–87	1996 (455) 1996 (548)
H	(CH₂)₃-αN	OEt	Et	Me	NaBH(OAc)₃, DCE	88–96	1996 (454)
H	(CH₂)₄-αN	OEt	pyrid-4-yl, 3-NO₂-Ph	H	NaBH(OAc)₃, DCE	88–96	1996 (454)
H	H, iPr, iBu, sBu, (CH₂)₂CO₂H	OH	iPr, iBu, Bn, (CH₂)₂CO₂H	CO₂H	NaBH₃CN		1996 (539)
R⁴		OH	CH₂CH(OnOct)O-CH(CH₂OH)CH₂-αN	H	NaBH₃CN	34–55	1997 (545)
H	Me, CH₂CO₂H, Bn, (indol-3-yl), iPr	OMe	Ph, pyrid-3-yl, cHex, CCTMS, CO₂H	H	1) Na₂SO₄, CH₂Cl₂ 2) In/BrCH₂CH=CH₂	52–80	1997 (555)
H	Me, CH₂CO₂tBu, CH(OtBu)Me	OH	4-OMe-2-OH-Ph	H	NaBH₄	60–80	1997 (530)
=CPh₂	Me, CH₂OH	OMe	nHex	Me	1) NaBH₃CN; 2) H₂, Pd-C	82	1997 (448)
H	iBu, Bn	OMe	(CH₂)ₙ(2-I-Ph) n = 0,1,2,3	H	NaBH₃CN or NaBH₄	53–67	1997 (1818)
H	iBu, Bn	OMe, OtBu	CH₂CO₂Me	Me	NaBH₃CN/ZnCl₂ or NaBH(OAc)₃	33–80	1997 (547)
H		OMe, OtBu	C(Me)₂NHFmoc, C(Me)₂NHBoc, CH(Me)NHAloc, 2-(NHBoc)-cPr, C(Me)₂CO₂Bn, C(Me)₂CO₂Me, CH(Me)CO₂Me, CO₂Et	H	NaBH₃CN/ZnCl₂ or NaBH(OAc)₃	33–80	1997 (547)
H	H	OBn	(peracetylated pyranose: OAc, OAc, AcO, AcO, O)	H	NaBH(OAc)₃	58–85	1997 (549)

(Continued)

Table 12.9 *N*-Alkylations of Amino Acids via Reductive Alkylation (continued)

P¹	R¹	P²	R²	R³	Alkylation Conditions	Yield (%)	Reference
Fmoc	$(CH_2)_4NH_2$	Gly-OBn		H	$NaBH(OAc)_3$	58–85	1997 (549)
H	H *i*Pr *i*Bu *s*Bu Bn 4-O*t*Bu-Bn $(CH_2)_4NHBoc$ $(CH_2)_2CO_2Bn$ CH_2OBn $CH_2)_2SMe$ CH_2(indol-3-yl)	OH	CH_2NHBoc $(CH_2)_2NHBoc$ $(CH_2)_3CO_2tBu$ $(CH_2)_nSBn;\ n = 1,2,3,5$	H	$NaBH_3CN$	19–69	1997 (537)
H	CH_2OH	OMe	$(CH_2)_n$(2-I-Ph) n = 1,2,3	H	$NaBH_3CN$	53–61	1997 (538)
H	H Me *i*Pr *i*Bu Bn 4-OH-Bn $(CH_2)_4NH_2$ $(CH_2)_3NHC(=NH)NH_2$ $(CH_2)_2CONH_2$ CH_2(imidazol-2-yl) CH_2(Indol-3-yl)	OBn, Oallyl	CH_2NHBoc	H	$NaBH_3CN$, MeOH	49	1998 (535)
H	$(CH_2)_3$-αN	OMe, OFm	Ph 4-F-Ph 3,4-Me₂-Ph 4-MeO-Ph 4-NO₂-Ph 4-CF₃-Ph 3,4-(MeO)₂-Ph 3,4-Cl₂-Ph pyrid-2-yl 2-furyl	H	polymer-supported $NaBH_3CN$		1998 (553)

H	nPr iPr iBu Bn	OMe	Ph 2-OH-Ph	H	nBu_4BH_4	95	1998 (552)
H	$CH_2C(NHBoc)-(CO_2Et)$ CH_2-αN	OEt	Bn cHex $CH(Ph)_2$ $(CH_2)_2Ph$	H	H_2, Pd-C	60–95	1998 (490)
H	iBu	NH-Val-Phe-Phe-resin	$CH(R)(CH_2)_4NHBoc$ R = H, NHCbz, NHFmoc	CO_2H	$NaBH_3CN$	60–100	1998 (540)
H	H Me Bn	resin	Ph 4-Cl-Ph 4-MeO-Ph	Me	$NaBH_3CN$	—	1998 (543)
H	H Me Bn	resin	$R^2 = R^3 =$ Me, iPr, Ph, $-CH_2CH_2CH_2(1,2$-Ph$)-$	R^2	$NaBH_3CN$	—	1998 (543)
H	Bn CH_2CO_2cHex	OBn, Oallyl	$(CH_2)_3CH(NHFmoc)CO_2tBu$ $CH(NHBoc)CH_2CO_2Fm$ $CH(NHBoc)(CH_2)_2CO_2tBu$ $(CH_2)_2CH(NHFmoc)CO_2tBu$	H	$NaBH(OAc)_3$, Et_3SiH, 10% Pd-C	76–94	1999 (546)
H	various on solid phase	resin	$6-NO_2-2$-OH-Ph	H	$NaBH_4$	—	2000 (532)
H	iPr iBu CH_2(imidazol-4-yl)	OH	Me Me $-(CH_2)_5-$ $-(CH_2)_2O(CH_2)_2-$	Me Et $-(CH_2)_5-$ $-(CH_2)_2O(CH_2)_2-$	H_2, Pd(OH)$_2$/C, rt	51–92	2000 (526)
iPr, sBu, cHex	Me	OH	Me	H	H_2, Pd(OH)$_2$/C, 50°C	58–92	
H	various	OH	2-HO-4-MeO-5-MeS-Ph-Ph	H	$NaBH_4$	—	2000 (533)
H	$(CH_2)_4NHCH_2$(indol-3-yl)	linkage to solid phase	2-naphthyl	H	$NaBH_3CN$	—	2002 (541)

(Continued)

Table 12.9 *N*-Alkylations of Amino Acids via Reductive Alkylation (continued)

P¹	R¹	P²	R²	R³	Alkylation Conditions	Yield (%)	Reference
H	(CH₂)₃NH₂ (CH₂)₃NHC(=NH)NH₂ CH₂(indol-3-yl) (CH₂)₃-(α-N) CH₂CH(OH)CH₂-(α-N)	tBu, linkage to solid phase	2-naphthyl indol-3-yl 3,4-(MeO)₂-Ph	H	NaBH₃CN	—	2002 (542)
H	CH₂(3-Bn-imidazol-4-yl)	Me	iBu	CO₂Bn	NaBH(OAc)₃, DCE	—	2002 (551)
H	CH₂OH CH(OH)Me (CH₂)₃-(α-N)	H	Ph	H	H₂, Rh-catalyst	34–93%	2004 (528)
H	CH₂CH(OH)CH₂-(α-N)	linkage to solid phase	4-MeS-Ph, 4-Cl-Ph, 3-Br-Ph, 3-F-Ph, 3-Cl-Ph, 4-Me-Ph, 2,4-Me₂-Ph, 3-tBu-Ph, 2-MeO-Ph, 4-EtO-Ph, 2,5-(MeO)₂-Ph, 2,3-(MeO)₂-Ph, 3-CN-Ph, 4-CN-Ph, 3,4-(–CH₂OCH₂–)-5-MeO-Ph, 4-CF₃O-Ph, 4-BnO-Ph, 4-PhO-Ph, 4-(4-Me-PhO)-Ph, 3-CF₃-Ph, 4-CF₃-Ph, 2-thienyl, 5-Me-2-thienyl, 3-thienyl, 3-Me-2-thienyl, 3-benzothienyl, 2-quinolinyl, 4-NO₂-Ph, 4-SO₂Me-Ph, 4-Ph-Ph, 2-naphthyl, 6-MeO-2-naphthyl, (CH₂)₃Ph, 4-CO₂Me-Ph	H	NaBH(OAc)₃, AcOH, CH₂Cl₂	—	2004 (550)
H	H	Me	4-Cl-Ph, 3-NO₂-Ph, 4-NMe₂-Ph, 2,6-Cl₂-Ph, 2,6-Me₂-Ph, 3,5-Me₂-Ph, 3,5-Me₂-4-HO-Ph, 2-Br-4-OMe-5-HO-Ph, 3-HO-4-OMe-Ph, indole-3-yl	H	NaBH₃CN, TEA, MeOH microwave	27–70%	2005 (544)

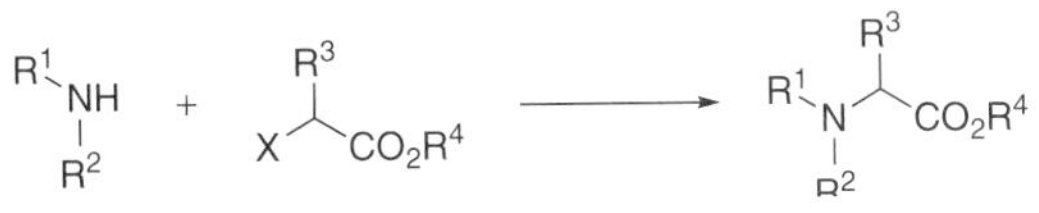

Scheme 12.17 *N*-Alkyl Amino Acids via Amination of Electrophilic Acid.

for peptoid synthesis were prepared by alkylation of the appropriate amines with ethyl bromoacetate, giving analogs of Phe, Leu, Met, Lys, Orn, Arg, Gln, and Tyr (*594*). Peptoid analogs of Lys and Dab (2, 4-diaminobutyric acid) were synthesized by alkylation of BocNH(CH$_2$)$_n$NH$_2$ with benzyl bromoacetate (*492*), while *N*-hexadecyl Gly was synthesized in 41% yield by displacement of ethyl bromoacetate with hexadecylamine (*595*, *596*). Benzylamine and (*R*)- or (*S*)-α-methylbenzylamine have also been reacted with bromoacetates (*597*). The nucleophilic amination method was employed during submonomer solid-phase peptide synthesis of peptoids (including glycopeptoids), with alkyl amines displacing the halide of a resin-linked bromoacetate (*598–603*). Racemic *N*-aryl analogs of L-thyronine were prepared by amination of 2-bromoalkanoic acids (*604*), while amino acids (primarily Gly) with an *N*-phosphono substituent separated by an alkyl, alkenyl, or aryl spacer were prepared by displacing an α-bromo ester with the phosphono-spacer-amine (*605*). *N*,*N*,*N*-trialkyl glycine derivatives, analogs of glycine betaine, have been prepared by amination of chloroacetic acid, esterified to a solid support resin, with a number of tertiary amines (*330*). 3-Hydroxypyridine displaced optically active α-bromo acids with varying degrees of racemization, producing unusual amino acid analogs in which the amino group is a quaternary nitrogen within a pyridine nucleus (*606*).

Alternatively, a dynamic kinetic resolution employs primary and secondary amine displacement of the halide from (*R*)-pantolactone esters of racemic α-bromo acids, proceeding preferentially with one of the bromide diastereomers with concurrent rapid epimerization of the remaining isomer (*227–229*). *N*-Benzyl-Pro (61% yield, 75% de), *N*-hexyl-Pro (51% yield, 75% de), *N*-benzyl-pipecolic acid (66% yield, 82% de), *N*-benzyl-Phg (72% yield, 82% de), *N*,*N*-dibenzyl-Phg and 4′-bromo-Phg (70–76% yield, >98% de), *N*,*N*-dibenzyl-homophenylalanine (56% yield, 82% de), *N*,*N*-dibenzyl-γ-cyclohexyl-Nva (82% yield, 88% de), *N*,*N*-dibenzyl-Abu (66% yield, 86% de), and *N*-benzyl-Abu (70% yield, 75% de) were prepared.

Optically pure *N*-(ω-thioalkyl), *N*-(ω-aminoalkyl), and *N*-(ω-carboxyalkyl) amino acids were synthesized by condensing activated α-hydroxy acid derivatives with ω-thioalkylamines (*607*), α,ω-diaminoalkanes (*608*, *609*), or ω-carboxyalkylamines (*608*), respectively. In a similar manner, various *N*-substituted D-Ala, D-Phe, and D-Asp derivatives were synthesized by displacement of the triflate of optically active α-hydroxyalkanoic acids (*610*). *N*-Methyl,*N*-(triethoxysilylmethyl)-substituted Gly, Ala, β-Ala, and Aib were prepared using

N-methylaminomethyltriethoxysilane in the presence of triethylamine (*611*).

Substituted fumaric acids undergo Michael addition of alkylamines catalyzed by 3-methylaspartase to give optically active *N*-alkyl Asp products (*612*). Racemic Asp derivatives are produced in the absence of enzyme (*613*). Ethyl (*S*)-lactate was aminated by Mitsunobu reaction with *N*-Boc, *N*-Cbz or *N*-Alloc 2-nitrobenzenesulfonamide. The *N*-nosyl group was removed using mercaptoacetic acid to give *N*-protected Ala (*614*). Mitsunobu amination of ethyl (*S*)-lactate with *N*-Troc 2-aminothiazole led to *N*-(2-thiazolyl) Ala (*615*).

α-Substituted amino acids have also been prepared by reductive amination of α-keto acids and esters (see Scheme 12.18) (see Sections 2.3.6, 5.3.9). *N*-Boc,*N*-alkyl glycines (peptoid monomers) were synthesized by reductive amination of glyoxylic acid (*342*, *478*); this was the preferred route for peptoid monomer synthesis in one report (*616*). *N*-(Phosphonomethyl) Gly (glyphosate, a broad-spectrum herbicide) was obtained by reductive amination of glyoxalic acid with (aminomethyl)phosphonic acid, using hydrogenation for the imine reduction step (*617*). Both *N*-(ω-aminoalkyl) and *N*-(ω-thioalkyl) Gly derivatives were prepared using sodium cyanoborohydride as the reducing reagent (*537*). Sodium cyanoborohydride was also employed for reductive amination of *N*-pyruvoyl L-Pro with the α-amino group of Lys(Cbz)-OEt or the imino group of Pro-OBn (*491*). Cobalt complexes of unsubstituted imines of α-keto acids can be *N*-alkylated with a number of halides (RX, R = Me, allyl, Bn), with borohydride reduction of the alkylated imine providing *N*-alkyl amino acids (*618*).

Nucleophilic and electrophilic α-alkylations of *N*-alkyl glycine equivalents can lead to *N*-alkyl amino acids. Symmetrically disubstituted iminodiacetic acid derivatives were prepared by dialkylation of iminodiacetic acid (see Scheme 12.19) (*619*). *N*-(Carboxymethyl) glycinamides, unsymmetrically derivatized iminodiacetic acids, were synthesized via a mono-oxazolidinone derivative prepared with hexafluoroacetone (*620*). *N*-Substituted arylglycines have been synthesized by the addition of arylboronic acids to imines formed from glyoxalic acid and substituted amines. The amino substituents included diphenylmethane, 4-methoxybenzene, and 1-methyladamantyl (*621*). Substituted anilines (or nitrobenzene derivatives reduced in situ) were used to prepare racemic *N*-aryl Ala and Phe derivatives (*622*).

Other methods of *N*-alkyl amino acid synthesis include ring opening of azirines (see Section 8.3) with

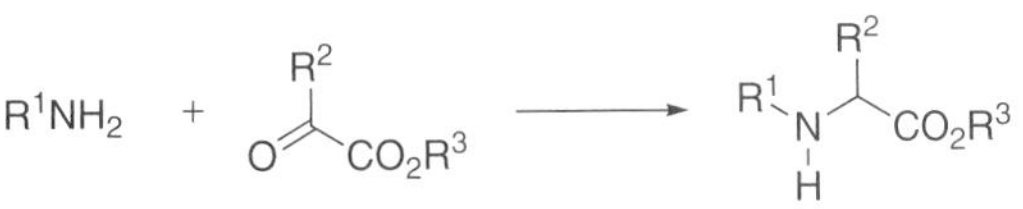

Scheme 12.18 *N*-Alkyl Amino Acids via Reductive Amination of α-Keto Acids and Esters.

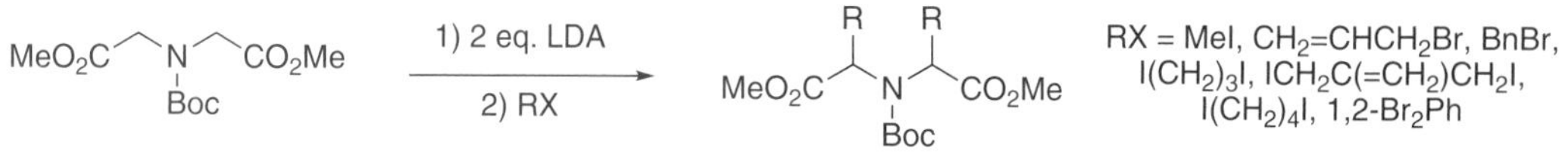

Scheme 12.19 Alkylation of Iminodiacetic Acid (*619*).

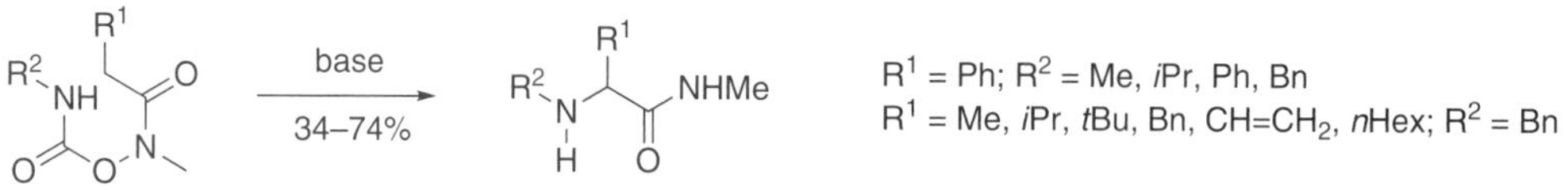

R[1] = Ph; R[2] = Me, *i*Pr, Ph, Bn
R[1] = Me, *i*Pr, *t*Bu, Bn, CH=CH$_2$, *n*Hex; R[2] = Bn

Scheme 12.20 *N*-Alkyl Amino Acids via Intramolecular Rearrangement (*625*).

activated phenols to give *N*-aryl-substituted derivatives (*623*) or with 1,3-diketones to give *N*-alkenyl-substituted derivatives (*624*). An anionic [3,3] sigmatropic rearrangement of *N*-acyl,*N*-methyl hydroxylamine-*O*-carbamates led to racemic *N*-substituted α-amino acid methylamides (see Scheme 12.20) (*625*). Racemic *N,N*-disubstituted amino acids were prepared by a carbenoid-type insertion of ethyl diazoacetate into a C–N bond of a tertiary amine (*626*).

12.2.4 Other *N*-Substituents

Amino acids have been prepared with several heteroatom substituents on the amino group, such as *N*-hydroxy and *N*-amino (hydrazino) amino acids, which are not discussed here. A sulfamic acid (*N*-SO$_3$H) substituent is contained on an *N*-alkyl glycine residue found in minalemines D–F, isolated from the marine tunicate *Didemnum rodriguesi* (*627*). The second residue of Ac-Pro-Xaa-OR dipeptides or Phth-Xaa-Yaa-OR dipeptides has been converted into an *N*-nitroso residue by aprotic nitrosation with N$_2$O$_4$ (*628*).

N-Phosphorylated amino acids (phosphoramidates, (RO)$_2$P(=O)-Xaa), and *N*-phosphonylated amino acids (phosphonamidates, (RO)RP(=O)-Xaa), are found in nature and can be synthesized; the monograph of Greenstein and Winitz contains a summary of early developments (*350*). Syntheses of amino acids with *N*-phosphonomethyl substituents are discussed in Section 12.2.3. Gly and Thr were *N*-phosphorylated with diisopropylchlorophosphate in 1951 (*629*) while *N*-phosphoryl leucinamide was prepared from leucinamide by phosphorylation with dibenzylchlorophosphate (*630*). L-Ala-L-Pro-OMe and other dipeptides were *N*-phosphorylated with dibenzyl chlorophosphite, and then hydrogenolyzed. The product phosphorylated dipeptides were reversible competitive inhibitors of angiotensin-converting enzyme, with K_i of 1.4 nM (*631*). [PhCH$_2$CH$_2$P(=O)(OH)]-L-Ala-L-Pro-OH was even more active, with K_i of 0.5 nM (*632*). Glu has been *N*-phosphorylated or *N*-phosphonylated to give

N-[Ph(MeO)P(=O)]-Glu, *N*-[(MeO)$_2$P)(=O)] Glu, and related derivatives, which were evaluated as inhibitors of γ-glutamyl hydrolase (*633*). *N*-(*O,O′*-dialkyl)phosphoryl-L-Asp undergoes reaction with amino esters without any coupling reagents, giving exclusively the dipeptide formed by reaction with the α-carboxyl group (*634*). Bis(9-fluorenylmethyl)phosphite has been identified as an effective reagent for the *N*-phosphorylation of amino acid methyl esters, generally giving a crystalline product with 80–98% yield (seven examples). Deprotection of both the methyl esters and 9-fluorenylmethyl groups by LiOH gave the *N*-phosphoryl amino acids (PO$_3$$^{2-}$-Xaa) in 78–93% yield (*635*).

Phenyl, methyl or ethylphosphonyl dichlorides were reacted sequentially with alcohols in the presence of catalytic tetrazole, and then with the amino group of Glu(OBn)-OBn in the presence of diethylamine to form phosphonamidate Glu derivatives (*636*). *N,N*-Dialkylphosphoramidite reagents can be employed to phosphonylate amino acids, with *m*CPBA oxidation providing the *N*-phosphoryl amino acids (*637*). Pro-OH was *N*-phosphorylated by diethyl phosphite in carbon tetrachloride (*638*). *N*-(Diisopropylphosphoryl) amino acids ([(*i*PrO)$_2$P(=O)]-Xaa) were prepared in 35–82% yields by slow addition of sodium hypochlorite to a solution of the amino acid and diisopropylphosphite in water. The *N*-phosphoryl moiety was used as a protecting group, with deprotection possible under acidic conditions similar to those used for Boc removal (*639*). One method for preparing *N*-phosphinolated amino acids is by reacting the amino acid ester with phosphinyl chlorides (R$_1$R$_2$PCl, where R$_1$ and R$_2$ = are alkyl groups)), giving an *N*-phosphino intermediate that can be oxidized to either *N*-phosphinoyl (*N*-P(=O)R$_1$R$_2$) or *N*-phosphinothioyl (*N*-P(=S)R$_1$R$_2$) products with good diastereoselectivity at the phosphorus center (*640*).

N-Thiophosphoryl amino acids, N-P(=O)(SH)(OH), were prepared by reacting amino esters with bis(2-cyanoethyl)-*N′,N″*-diisopropylphosphoramidite in the presence of 1*H*-tetrazole, followed by the addition of sulfur (*641*). The *N*-phosphorothioic analog of Leu was synthesized using dimethyl chlorothiophosphate (642). N-Phosphonamidothionate Glu derivatives, with N-P(=S)(R)(OH) substituents (R = Me, Et, nBu, Ph) were prepared via treatment of Glu with NC(CH₂)₂OP(=S)(R), followed by LiOH hydrolysis (643). N-(2-Thiono-1,3,2-oxathiaphospholanyl) amino acid methyl esters were prepared in high yield from amino acid methyl esters and 2-chloro-1,3,2-oxathiaphospholane in pyridine in the presence of elemental sulfur. The products were then treated with methanol, benzyl alcohol, or 3-hydroxypropionitrile to give methyl, benzyl, or 2-cyanoethyl-phosphorothioamidates, -NHP(=O)(S⁻)(OR). The latter products could be treated with concentrated ammonium hydroxide to remove the cyanoethyl group and give N-phosphorothioate amino acids, -NHP(=O)(S⁻)(O⁻). Alternatively, the thio group could be removed by treatment with SeO₂ (644).

N-(2,2-Diphenylvinylidene) amino acids (N=C=CPh₂) have been prepared from benzophenone Schiff bases of Gly, Ala, and Phe by dichlorocarbene addition to give an aziridine intermediate, which was then ring-opened and dechlorinated (645). N-Dithiomethylcarbamate derivatives of amino acids N-C(=S)SMe, can be prepared by reaction with CS₂ in the presence of NaOH, followed by alkylation with methyl iodide. Trp, Pip, and Tic were converted (646). The methylthiocarbonyl analog N-C(=O)SMe was obtained from carbon oxysulfide (COS) followed by methylation, while the methoxythiocarbonyl derivative N-C(=S)OMe was produced by reaction with O-methyl chlorothioformate (646).

12.2.5 N-Alkylation of Side-Chain Amino Groups

Basic amino acids such as Lys, Arg, and His possess another amino group that can also potentially undergo alkylation. A review of N$^\omega$-alkyl diamino acids was published in 1978, focusing primarily on the series of N$^\varepsilon$-alkyl Lys: N$^\varepsilon$-methyl Lys, N$^\varepsilon$,N$^\varepsilon$-dimethyl Lys, and N$^\varepsilon$,N$^\varepsilon$,N$^\varepsilon$-trimethyl Lys (647). The side-chain primary amino groups of Lys and Orn can be functionalized by the same type of reactions as used for the α-amino group, such as reductive alkylation (549). For example, N$^\varepsilon$-methyl L-Lys was prepared from Cbz-Lys via reductive alkylation of the ε-amino group with benzaldehyde/NaBH₄, followed by methylation with formaldehyde/formic acid and removal of the Cbz/benzyl groups by hydrogenolysis (648, 649). The side chain of Cbz-Lys-OMe was reductively alkylated

with sodium borohydride and acetone to give N$^\varepsilon$-isopropyl Lys (529,), or alkylated with isopropyl iodide to give the same product (650). Unprotected Orn was converted into N$^\delta$-benzyl Orn via regioselective imine formation with benzaldehyde followed by hydrogenation (651). N$^\alpha$-(2-Naphthylmethyl), N$^\varepsilon$-alkyl Lys amides were prepared via attachment of Fmoc-Lys(Aloc) to a resin, Aloc deprotection, reductive alkylation with aldehydes and NaBH₃CN, and then Fmoc removal and reductive alkylation with 2-naphthaldehyde and NaBH₃CN (541).

Temporary protection was employed for alkylation of the side-chain amino group of N$^\alpha$-benzoyl, N$^\varepsilon$-p-toluenesulfonyl Lys with methyl iodide, ethyl iodide, or benzyl iodide; the adduct was treated with HBr/HOAc to give the N$^\varepsilon$-alkyl Lys product (649). The same procedure was employed to prepare N$^\delta$-methyl Orn (649). The side-chain amino groups of Bz-Orn(Ts)-OH and Bz-Lys(Ts)-OH were monomethylated by alkylation with dimethylsulfate (652) or by reductive alkylation with formaldehyde/hydrogenation (653), with HBr used for protecting group removal (652). Bz-Orn(Ts)-OH was also methylated using ¹³MeI and NaOH, with both protecting groups removed with 48% aqueous HBr (654).

N$^\varepsilon$,N$^\varepsilon$-dimethyl Lys was synthesized by reductive dimethylation of Ac-Lys or Bz-Lys by hydrogenation with formaldehyde (649), while the side-chain amino groups of poly-L-Lys were dimethylated by repeated reflux with formaldehyde and formic acid, producing polymer with 98% Lys(Me)₂, 1.1% Lys(Me), and 0.5% Lys (655). Attempted polymethylation of the polymer under formaldehyde/hydrogenation conditions gave only 6% conversion to the N$^\varepsilon$,N$^\varepsilon$-dimethyl Lys residues after 7 days, with 37% N$^\varepsilon$-monomethyl Lys and 56% unreacted Lys content (655). The side chains of Cbz-Lys, Boc-Lys, and Boc-Orn were quaternized by treatment with methyl iodide and potassium bicarbonate in methanol (403).

His has been regioselectively alkylated on the τ-nitrogen of the imidazole ring via protection of the π- and α-nitrogens through an intramolecular cyclic urea prepared by reaction with carbonyldiimidazole (656–658) (see Scheme 12.21). Heating with alkyl halides in acetonitrile or diethoxyethane gave the alkylated products in high yield, with acid hydrolysis regenerating the deprotected N$^\tau$-alkyl His derivatives. Regioselective allylation of either the τ- or π-nitrogens has also been reported, using diallylation and selective deprotection or a temporary N$^\tau$-Trt blocking group (659).

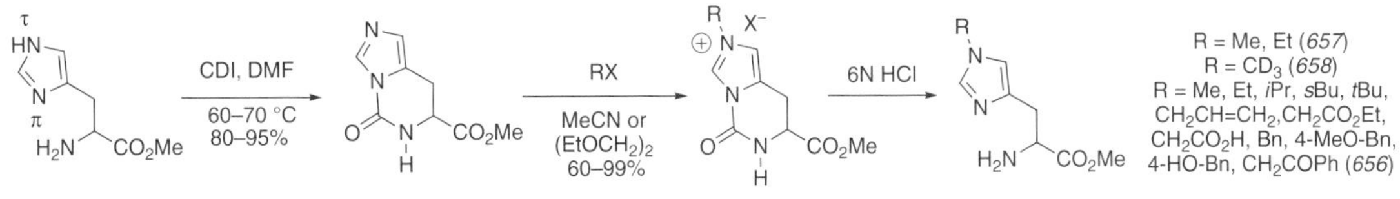

Scheme 12.21 Side Chain N-Alkylation of His.

N^{indole}-Me-L-Trp was prepared by treatment of a solution of L-Trp in liquid ammonia with Na and MeI, giving a 77% yield of product (660). The N-hydroxamate nitrogens of O-benzyl-amino acid hydroxamates were methylated with NaH/MeI (362).

12.2.6 Resolution, Analysis, and Coupling of N-Alkyl Amino Acids

Racemic N-acetyl-N-alkyl amino acids can be difficult to resolve enzymatically as they are unreactive with most common acylases. However, N-acyl-L-proline acylase has been found to resolve a number of N-alkyl amino acids, with reasonable yields as long as the side chain was not longer than two C atoms (85).

Gas chromatographic (6c) analysis is a sensitive tool for determining the optical purity of amino acids that have been converted into volatile derivatives. Unfortunately, conventional derivatization by trifluoro-acetylation is unsuccessful for N-methyl amino acids (661, 662). Instead, N-methyl α-amino acids that had been enantiomerically resolved were analyzed on a chiral GC column after derivatization as oxazoli-dine-2,5-diones by treatment with phosgene (663), or when derivatized as N-alkylureido/N-alkylamides by reaction with isocyanates (661, 664). Alternatively, diastereomeric esters were prepared using (+)-3-methyl-2-butanol (662). Cheung and Benoiton determined the enantiomeric purity of N-methyl amino acids by ion-exchange chromatography of dipeptide derivatives formed by amidation with Lys; the method was sensitive to 0.1% of the epimer (665). N-Alkyl amino acids have been enantioresolved by thin-layer chromatogra-phy (TLC) separation on a reversed-phase silica gel stationary phase impregnated with Cu(II) acetate and a proline derivative, (2S,4R,2'RS)-N-(2'-hydroxydodecyl)-4-hydroxy-Pro (666).

The acylation of N-alkylated amino acids can be problematic during peptide synthesis, with many of the common coupling systems failing to give reasonable yields (although some apparently incomplete couplings may actually be due to cleavage of the N-alkyl amide bond during TFA deprotection/resin cleavage (344–346)). Several reagents have been developed specifically to improve couplings to N-alkyl amino acids, such as Dpp-Cl (667), BOP-Cl (560, 668–670), PyBroP (671), PyAOP (345), and acid chlorides/KOBt (672). The triphosgene reagent BTC was employed for a synthesis of the cyclodecapeptide omphalotin A, in which 9 of the 12 amino acid residues are N-methylated. The BTC method was far superior to TFFH, DCC, or DIC/HOAt (302). BMTB (2-bromo-3,4-dimethylthiazolium bro-mide) was greatly more effective at coupling N-Boc-MeIle with MeIle-OBn (27% conversion after 2 h with 1.5 equiv reagent) compared to PyBOP, EDC/HOBt or BEMT (<5% conversion), and significantly more effective than HATU (16% conversion) (673). These, and other, reagents are discussed in greater detail in Section 12.5.

12.3 N-Acylation and N-Protection of Amino Acids

12.3.1 Introduction

The N-acylation of amino acids is generally used for amide formation during peptide synthesis (see Section 12.5) or to introduce temporary protecting groups for use in peptide synthesis, such as acetyl (Ac), benzoyl (Bz), benzyloxycarbonyl (Cbz or Z), tert-butyloxycarb-onyl (Boc), 9-fluorenylmethoxycarbonyl (Fmoc), or allyloxycarbonyl (Aloc or Alloc) groups. Procedures for acylations with protecting groups are described in monographs such as those of Greene and Wuts (674), Kocienski (675), or Bodansky (676), with comprehen-sive coverage in Houben–Weyl's 1974 German volumes (677) or 2002 updated English versions (678). A chapter on acylation and alkoxycarbonylation of the amino group is contained in a 1999 monograph on amino acid derivatives (293), while amine protecting groups are included in Volume 3 of The Peptides series (679). A 2002 review in the Journal of Combinatorial Chemistry covers protecting groups in solid-phase organic synthesis, including a nice summary of amine protecting groups (680), while a 2000 review covered orthogonal protecting groups for α-amino and carboxy groups in solid-phase peptide synthesis (681). A primer designed for introduction to the field of amino acid and peptide synthesis is part of the Oxford Chemistry Primer series, and includes sections on amino and acid group protection (682). Amino protecting groups were summarized in a 1973 review (683), and have been included in several annual reviews of protecting groups (e.g., 684), while protecting group strategies were discussed in 1996 (685).

Other acylations are used to modify the properties of the amino acid residue. A review of the synthesis, properties, and applications of N-acyl amino acids was published in 1995, with an emphasis on amino acids acylated with long chain (C_{10}–C_{18}) alkyl acids (686). A summary of acylation procedures (acid chloride with Schotten–Baumann method, anhydrides, activated esters, and acids with activating reagents) was also included.

12.3.2 Acylation

Several new acylation conditions have been described in recent years. Modified Schotten–Baumann conditions were described in 2001 for the acylation of amino acids with acyl chlorides, combining amino acids, potassium carbonate, acyl chloride, and a catalytic amount of cationic surfactant (e.g., dodecyltrimethylammonium bromide) in refluxing THF, giving 88–93% yields for nine examples (687). Acylation of amino acids with benzoyl chloride in aqueous alkali can lead to formation of benzoyl dipeptide impurities. Generation of the symmetrical anhydride that leads to the impurity can be circumvented by using diisopropylethylamine as base instead of sodium hydroxide (688). The modified conditions also work well for reactions with ethyl chloroformate, methyl chloroformate, diethyl dicarbonate, and 9-fluorenylmethyl chloroformate, but not with benzyl chloroformate (688). Basic conditions are generally employed for carbamate formation, but a 1998 report found that amino esters could be acylated with alkyl chloroformates using only Zn in benzene, giving 95–98% yields of N-Cbz, N-ethoxycarbonyl, or N-isopropyloxycarbonyl homophenylalanine in under 20 min (689). Lewis acid-mediated acylation with chloroformates in dry acetonitrile is possible in 10–20 min with a yttria–zirconia catalyst, again avoiding the need for bases (690).

An anhydrous solvent system for acylating amino acids is normally not possible due to the insolubility of the zwitterionic amino acids. Acylation of amino acids in DMF with di-tert-butyl dicarbonate (Boc$_2$O), benzyloxycarbonyl O-succinimide (Cbz-OSu), or 9-fluorenylmethoxy O-succinimide (Fmoc-OSu) was made possible by adding strong acids such as TFA, HBF$_4$ or TosOH in combination with an excess of a tertiary base with pK <6, such as pyridine. Boc-Tyr, Cbz-Ala, Fmoc-Trp, and Fmoc-Gln(Mbh) were prepared with 63–92% yields (691). A 19:1 pyridine:chlorotrimethylsilane solution was used to solubilize S-(4-methoxybenzyl) 6-mercapto-2-aminohexanoic acid during acylation with Fmoc-Cl (692). Sterically hindered α,α-disubstituted amino acids and α-alkylated prolines were N-Boc protected in good yield (88–100%, vs 60–85% by previous methods) by using tert-Boc$_2$O in acetonitrile, with the lipophilic base tetramethylammonium hydroxide employed to solubilize the amino acid substrate (693). A one-pot conversion of N-Cbz amino esters into N-acetyl or N-benzoyl derivatives employed SnBr$_2$ in CH$_2$Cl$_2$ with AcBr or BzCl and Et$_3$N, with the products formed in 68–86% yield (694).

A general route for carbamate synthesis was developed with a three-component coupling of the amine, CO$_2$, and alkyl halides in the presence of Cs$_2$CO$_3$ and Bu$_4$NI (695). Alcohols have been reacted with CDI (1,1′-carbonyldiimidazole), followed by an amine to form carbamates, though only benzylic alcohols were tested (696). A modification of the Boc$_2$O procedure normally employed to introduce the N-Boc group, by addition of equiv of DMAP, resulted in enantiopure isocyanates. These were then reacted with alcohols to prepare a number of carbamates; Cbz and Alloc groups were generated as examples (697). Alternatively, an N-trichloroacetamide group was converted into an isocyanate by treatment with Na$_2$CO$_3$ in refluxing DMF; trapping with alcohols and CuCl/nBu$_4$NCl produced carbamates (698). Another general route activated an o-nitro-phenolic resin with the phosgene equivalent BTC (bis-trichloromethyl carbonate) to give a polymer-supported chloroformate equivalent. This was reacted sequentially with an alcohol, then an amine, in a one-pot reaction, though no amino acids were employed (699).

A chemoselective acylating reagent, RCON(Ms) (2-F-Ph) (N-acyl,N-mesyl 2-fluoroaniline), readily transfers the acyl group to primary or secondary amines in THF, with good regioselectivity for primary amines over secondary amines, and unhindered amines over hindered amines. The acyl groups included Cbz, (R = BnO), Bz (R = Ph), and ethoxycarbonyl (R = EtO) (700). Amino acids protected with Cbz, Boc, Fmoc, Adoc, and Bpoc urethane protecting groups have been prepared by acylation with 5-norbornene-2,3-dicarboximido derivatives (701), while Cbz, Alloc, Fmoc, and Troc were introduced using 6-(trifluoromethyl)benzotriazol-1-yl as the activated reagent (702). Similar 1-(tert-butoxycarbonyl)benzotriazole and 1-(p-methoxybenzyl-oxycarbonyl)benzotriazole reagents were used to introduce Boc and Moz protection (703). Carboimido-dithioates, prepared by reaction of the amino group with CS$_2$ followed by methylation, have been converted into Cbz or methoxycarbonyl (Moc) protecting groups by treatment with sodium benzyl alcoholate followed by water, or with MeOH and ZnCl$_2$, respectively (704). A polymeric N-hydroxy succinimide derivative has been used as an acyl transfer reagent, reacting with acyl chlorides to form an activated ester that then acylated amines in good yield, with a simple filtration work-up (705).

N-Formyl amino acids were readily prepared from hydrochlorides of amino acid methyl, ethyl, benzyl or tert-butyl esters by using cyanomethyl formate as the formylating reagent, with 62–97% yields (706). Another procedure reacted amino acid ester hydrochlorides with triethyl orthoformate or trimethyl orthoformate. The amino acid was simply heated in neat reagent, with the unreacted orthoformate removed by distillation followed by distillation of the product. Yields of 50–100% were reported (707). A three-step method with no formylating agent proceeded via imine formation with pivalaldehyde/triethylamine (89–99%), oxidation to an oxaziridine with mCPBA (90–99%), and free-radical-induced ring opening using Fe(II) ions from Möhr's salt (91–99%). No racemization was observed (708). 2,2,2-Trifluoromethyl formate formylated Ser-OBn, Phe-OEt or Pro-OBn in 92–95% yield (709). An environmentally benign process reacted amino ester hydrochlorides with ammonium formate in refluxing dry acetonitrile, with seven N-formyl amino esters prepared in 63–91% yield (710).

Urea formation can be achieved by a one-pot reaction of unprotected amino acids in which the acid group is temporarily protected with TMSCl, followed by urea formation with triphosgene and an amine, and then addition of methanol to effect ester deprotection (711).

N-Acylation can be achieved by methods other than the reaction of amino acids with acylating reagents. The Ugi four-component condensation (e.g., see 586, 588) can provide N-acyl,N-alkyl amino acids, which can otherwise be difficult to synthesize. N-Acetyl and N-lauroyl amino acids were obtained from the Wakamatsu cobalt-catalyzed aminocarboxylation reaction by using acetamide or lauramide, respectively, as the amine source (712). N,N-Dicarbamate-protected amino acids have been synthesized by imidodicarbonate displacement of α-bromo acids or of triflate derivatives of α-hydroxy acids (*713, 714*). Ketenes, generated by photolysis of water-soluble substituted cyclohexa-2,4-dien-1-ones, were trapped by the amino group of amino acids to give amino acids *N*-acylated with substituted hexa-3,5-dienoic acids in 68–90% yield (*715*). Ring opening of 2-dimethylamino-3,3-dimethyl-1-azirine with acids resulted in *N*-acyl Aib dimethyl amides (*624*) (see Section 8.3 for similar reactions).

Removal of acyl groups, unless specifically designed for deprotection, tends to require harsh conditions such as treatment with acid or base at high temperatures. A (PhO)₃P·Cl₂ reagent has been found to efficiently deacylate *N*-monosubstituted amides at low temperatures, including a deprotection of Ac-L-Phe-OMe that proceeded in 95% yield with no racemization (*716*).

Unusual properties can be added to amino acids via acylation with unconventional acids. Acylation of α-amino acids with long-chain acids gives lipophilic amino acids, which often have significantly altered properties. The first fullerene-peptide was prepared by acylation of the *N*-terminal residue of a peptide with a benzoic acid-substituted methanofullerene (*717*). Subsequently, diazoamide derivatives (COCHN₂) of the α-amino group of Phe-OBn and the side-chain amino group of Lys-OEt were reacted with fullerene to form amino acids acylated with carboxy-substituted methanofullerene (*718*). An anthracenyl urethane substituent provides very high UV absorbance and is suitable for both UV absorbance or laser-induced fluorescence detection (*719*), while an Fmoc analog with four additional fused benzene rings was developed as an aid for peptide purification (*720, 721*). An analog of the Cbz group with a boronic acid substituent on the aryl ring was introduced as a method to manipulate solubilizing and affinity properties (*722*).

12.3.3 N-Protecting Groups

12.3.3a Carbamates
(see Schemes 12.22, 12.23)

The standard carbamate protecting groups are Cbz (or Z, benzyloxycarbonyl, commonly removed by hydrogenolysis or strong acid), Boc (*tert*-butyloxycarbonyl, removed by acid cleavage, generally with TFA) or Fmoc (9-fluorenylmethoxycarbonyl, normally removed by treatment with a secondary amine such as piperidine). Some general methods for introducing urethane protecting groups are included in the previous section, while more specific procedures are discussed below. Developments in methods to introduce carbamate reagents were included in a 2002 article (*723*).

The Cbz group was introduced into peptide chemistry by Bergmann and Zervas in 1932 (using Cbz-Cl to form the carbamate) as a removable protecting group that can suppress racemization during coupling (*724*). The *N*-Cbz group can be formed in good yield by a three-component coupling of the amine, CO₂, and benzyl chloride in the presence of Cs₂CO₃ and Bu₄NI (*695*). Cbz analogs can be prepared from substituted benzylic alcohols by

reaction with CDI (1,1′-carbonyldiimidazole), followed by an amine, generating substituted benzyl carbamates (*696*). Protection of amines under neutral conditions is possible by using *N*-Cbz 5,7-dinitroindoline, which reacts by a photolytic mechanism (*725*), or by Lewis acid-mediated acylation with Cbz-Cl in dry acetonitrile for 10–20 min with a yttria–zirconia catalyst (*690*). *N,N*-Cbz₂ protection can be introduced without racemization in 75–91% yield by treatment of *N*-Cbz amino acid esters with NaHMDS and then Cbz-Cl; mixed dicarbamoyl compounds can also be obtained by reacting an *N*-Boc amino ester under the same conditions (*726*). For bis(carbamoyl) protected amines, one group can be selectively removed using LiBr in acetonitrile (*727*).

The Cbz group can be cleaved by hydrogenolysis or strongly acidic conditions (such as TFA or HF with thiol scavengers, or HBr) (*674, 675*). Selective hydrogenation of acetylene, alkene, azide, nitro, and benzyl esters in the presence of a Cbz group is possible by employing 5% Pd/C-ethylenediamine and THF (*728*). A hydroxyapatite-bound Pd catalyst was effective at removing Cbz groups from sterically hindered compounds where other catalysts gave poor yields. High yields of deprotected amino acids could be obtained with much less catalyst (e.g., 92% of Ala from Cbz-Ala with 0.004 equiv of Pd catalyst after 1 h at at 40 °C, compared to 68% with 5% Pd/C catalyst, or 60% with Pd/Al₂O₃) (*729*). A one-pot reaction converted *N*-Cbz amino esters to *N*-Fmoc derivatives in 79–90% by hydrogenation with 10% Pd-C poisoned with 2,2′-dipyridyl, in the presence of Fmoc-OSu (*730*).

Catalytic transfer hydrogenation with cyclohexadiene as the hydrogen donor efficiently removed Cbz groups in the presence of benzyl esters (*731*), while one example of the same selective cleavage has been reported with catechol boron bromide (*732*). In contrast, benzyl ethers were selectively cleaved in the presence of Cbz groups by employing NaBrO₃/Na₂S₂O₄ in EtOAc (*733*). Microwave irradiation greatly enhances the rate of catalytic transfer hydrogenation in isopropanol with Pd/C as catalyst and ammonium formate as the hydrogen donor, resulting in rapid Cbz removal within a few minutes from both solution- and solid-supported substrates (*734*). A combination of magnesium and hydrazinium monoformate in methanol, Mg/H₂NNH₂·HCO₂H, has been identified as a Pd-free transfer hydrogenation system that removed Cbz, Cl-Cbz, and Br-Cbz protecting groups at 25 °C without touching Boc or Fmoc groups (*735*).

Another chemoselective reductive removal method was reported in 1999, with triethylsilane and palladium chloride allowing for the removal of the Cbz group in the presence of an alkene (*736*). Deprotection with boron trifluoride etherate–mercaptoethanol is also possible, avoiding reduction of olefins, acetylenes, imines, halides or nitro groups, or ester hydrolysis (*737*). BCl₃ in dichloromethane generated free amino acids from the *N*-Cbz analogs, though yields were variable (45–89%) (*738*). A Cbz group on a heteroaromatic nitrogen (e.g., indole, imidazole, pyrrole) can be removed in the presence of a

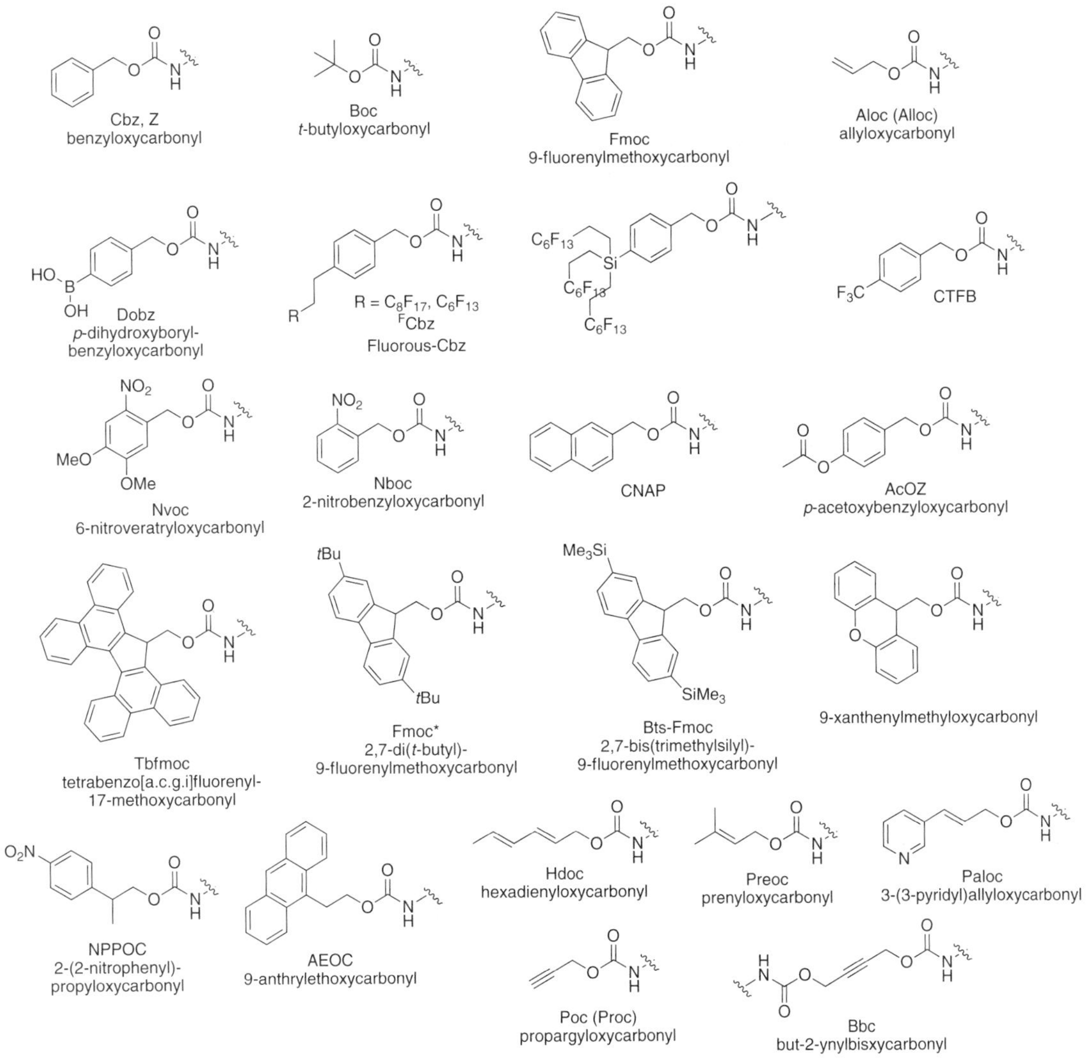

Scheme 12.22 Urethane *N*-Protecting Groups.

Cbz group on a primary or secondary amine using a Ni(0) catalyst with $Me_2NH.BH_3/K_2CO_3$; Cbz-His(Cbz)-OMe was converted into Cbz-His-OMe in 87% yield (*739*). A one-pot conversion of *N*-Cbz amino esters into *N*-acetyl or *N*-benzoyl derivatives employed $SnBr_2$ in CH_2Cl_2 with AcBr or BzCl and Et_3N, with the products formed in 68–86% yield (*694*).

A wide variety of substituted Cbz analogs have been described over the years in order to fine-tune the groups's stability/ease of cleavage or to alter the physical properties of the protected amino acids; some of these are discussed in many of the reviews referenced at the beginning of this section. The *p*-nitrobenzyloxycarbonyl (pNZ) group is removed with 6 M $SnCl_2$ and 1.6 mM HCl/dioxane in DMF, leaving the deprotected amine as a neutral ammonium salt (*740, 741*). The deprotection

conditions leave Boc, Fmoc, and Aloc groups untouched, while TFA, 20% piperidine, or $Pd(PPh_3)_3$/phenylsilane do not remove the pNZ group. The pNZ group was employed as the temporary N^α-protection during SPPS, and prevented both diketopiperazine formation at the dipeptide-resin stage and aspartimide formation, common side reactions during Fmoc-based syntheses (*741*). A 2-naphthylmethyl carbamate group (CNAP) can be hydrogenolyzed in the presence of a 4-trifluoromethylbenzyl carbamate (CTFB) (*742*). An analog of the Cbz group with a boronic acid substituent on the aryl ring was developed as a method of manipulating solubility and affinity properties. The *p*-dihydroxyborylbenzyloxycarbonyl (Dobz) group is introduced with the boronic acid masked as a catechol complex, and is removed under standard hydrogenolysis conditions (*722*).

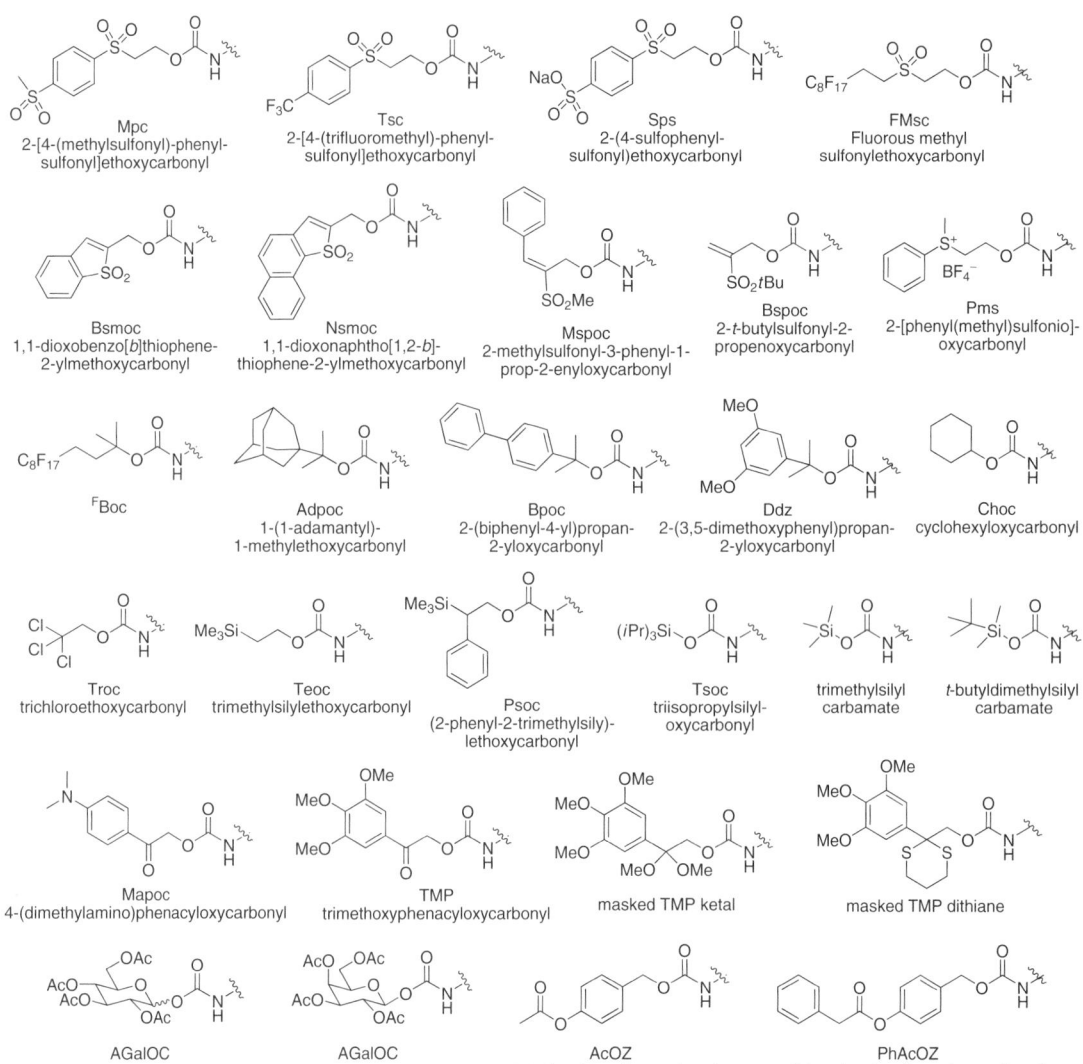

Scheme 12.23 Urethane *N*-Protecting Groups.

Fluorous-Cbz groups (^FCbz, or FZ, FMZ, FEZ)), with *p*-C$_8$F$_{17}$C$_2$H$_4$-, *p*-C$_6$F$_{13}$C$_2$H$_4$-, *p*-C$_6$F$_{13}$C$_2$H$_2$- or *o*-methyl-*p*-C$_8$F$_{17}$C$_2$H$_4$- substituents, were introduced to allow for fluorous purifications via fluorous solid-phase extraction or fluorous chromatography. ^FCbz derivatives of 18 common amino acids (L-enantiomers with the C$_8$F$_{17}$ tag and D-amino acids with the C$_6$F$_{13}$ tag) were reported (*743, 744*). Earlier reports used a *p*-tri(perfluoroalkyl) silyl substituent on the Cbz group for fluorous extraction of amines (*745*).

The *tert*-butyloxycarbonyl (Boc) group was introduced in 1957, when it was compared against urethanes derived from cyclopentanol, cyclohexanol, diisopropyl-carbinol, and *p*-methoxybenzyl alcohol (*746*). The urethane is usually prepared using di-*tert*-butyl dicarbonate (Boc$_2$O), which was reported in 1976 (*747*), although earlier syntheses employed Boc-N$_3$ for formation (*748*). Modified reaction conditions have been developed; the Boc$_2$O reagent rapidly acylated amines in the presence of catalytic (10 mol %) ZrCl$_4$ in acetonitrile at room temperature, with eight different amino esters protected in 80–93% yield within 10–25 min (*749*). Aqueous reaction conditions were reported in 2006: amino esters were simply combined with Boc$_2$O in water at room temperature for 10–60 min with no catalyst and a simple extractive work-up. Phg-OMe, Phe-OMe, and Tyr-OMe were protected in 91–96% yield (*750*). A 1-*tert*-butoxy-2-Boc-1,2-dihydroisoquinoline reagent was described in 2002, chemoselectively acylating 10 different amino acid methyl ester hydrochloride salts (side-chain unprotected Ala, Leu, Val, Phe, Pro, Met, Glu, Ser, Cys, Tyr) in the absence of base in 76–98% yield (*751*).

A second Boc substituent has been added to the amide bond of Boc-Pro-Gly-OMe or the carbamate amine of Boc-Gly-OBn by using Boc$_2$O/DMAP in acetonitrile (752). The same conditions were subsequently applied to the preparation of N,N-Boc$_2$ amino acids from N-Boc amino acid allyl esters via acylation followed by ester deprotection. The bis(Boc) amino acids are stable crystalline compounds that can be used in peptide synthesis (753). The diacylation conditions were also used to develop a procedure for gentle removal of N-acetyl groups. Acylation of the N-acetyl amino acid with Boc$_2$O and DMAP was followed by cleavage of the acetamide by hydrazine as base. A one-pot variation of this procedure converted a number of N-acetyl amino acids to N-Boc amino esters in 80–93% yield with no racemization (754). A one-pot conversion of N-Cbz to N-Boc protection has also been described, using catalytic transfer hydrogenation in the presence of Boc$_2$O (755). The reactivity of the N-Boc group towards various nucleophiles, bases, and electrophiles used during organic synthesis was reviewed in 2002 (756).

The Boc group is removed under acidic conditions with reagents such as 50% trifluoroacetic acid (TFA) in dichloromethane, a 95:5 mixture of TFA:H$_2$O, or saturated HCl solution in solvents such as dichloromethane (e.g., see the monographs of Greene or Kocienski, 674, 675). In 2005 the use of H$_2$SO$_4$ in CH$_2$Cl$_2$ (approximately 1.5 mol H$_2$SO$_4$ per mol substrate) was introduced as a simple, cheap, and eco-friendly alternative to TFA that is suitable for large-scale syntheses (757). Concentrated H$_2$SO$_4$ in tBuOAc or MeSO$_3$H in tBuOAc:CH$_2$Cl$_2$ was able to cleave Boc groups in the presence of tBu esters (758), while a 10% solution of concentrated H$_2$SO$_4$ in 1,4-dioxane cleaved Boc groups on high-load Wang resins with minimal substrate cleavage from the resin (759). It is also possible to remove a Boc group from an amine attached to the acid-sensitive Wang resin with <10% loss of resin loading by using 0.2 M TMSOTf in dichloroethane with 2 equiv of triethylamine for 10 min (760). N-Boc groups are selectively removed in the presence of tert-butyl esters by HNO$_3$ in CH$_2$Cl$_2$ (761). Aqueous phosphoric acid removed Boc groups in the presence of benzyl and methyl esters, TBDMS ethers, and isopropylidene groups, though tert-butyl esters were cleaved (762). A rapid removal of Boc groups from amino acids has been achieved using a toluene solution of p-PTSA and 30sec of microwave irradiation, with 92–98% yields of tosyl salt products (18 examples). Fmoc and Cbz groups, methyl and benzyl esters, and benzyl protecting groups were not affected (763).

More esoteric deprotection methods include electrochemical generation of acid from 1,2-diphenylhydrazine, which allowed for Boc group removal in addressable electrode arrays (764). The acid for Boc deprotection has also been photogenerated from SSb (triphenylsulfonium hexafluoroantimonite), with UV light exposure producing H$^+$SbF$_6^-$ and PPhSPh (765). A solution of chlorotrimethylsilane and phenol in dichloromethane removed the Boc group with high chemoselectivity over benzyl esters, ethers, and urethanes (766), while Me$_3$SiClO$_4$ in benzene induced Boc deprotection in the presence of tert-butyl ethers and N-Cbz groups (767). Treament of N-Boc amino acids with TMSCl/MeOH resulted in a one-pot conversion into amino acid methyl esters in 90–97% yield with minimal (<2%) racemization (768).

Other mild cleavage conditions include boron trifluoride etherate and molecular sieves in dichloromethane at room temperature (769), silica-supported Yb(OTf)$_3$ under solvent-free conditions (770), or Amberlyst-15 acidic ion-exchange resin, which deprotects the Boc group from N-Boc amino acids (5 h at 25 °C in CH$_2$Cl$_2$) and then captures the resulting free amine, which can then be released by treatment with excess ammonia in methanol (771). Montmorillonite K10 clay removes the Boc group from anilines while leaving aliphatic N-Boc amines intact (772), while Boc-Trp(Boc)-OtBu was adsorbed to Si gel and heated to 50 °C under reduced pressure to selectively remove the indole protecting group while leaving the α-amino N-Boc and tBu ester groups intact (773). Boc groups on primary amines were deprotected under basic conditions by treatment with excess sodium tert-butoxide in slightly wet THF, probably via an isocyanate intermediate. N-Boc secondary amines remained intact (774). Methyl, ethyl, benzyl (Cbz), and tert-butyl (Boc) carbamates have been removed using L-selectride. The reagent allows for the selective removal of a methoxycarbonyl group in the presence of a Boc group (775). Some N-Boc protected amines (indoles, anilines) were deprotected using refluxing TBAF in THF, though alkylamines and amino acids were only partly deprotected (776, 777).

Cleavage by SnCl$_4$ is possible in organic solvents such as ethyl acetate, with good yields after <1 h at room temperature (778). Sn(OTf)$_2$ in CH$_2$Cl$_2$, or solvent-free, removed the Boc group in 2–4 h, but did not affect tert-butyl esters or Cbz groups (779). Aluminum chloride removed the N-Boc group in the presence of other acid-sensitive groups such as O-TBS and O-Ac, as well as ester and ether linkages and the N-Cbz group (780, 781). Bismuth(III) trichloride in 50:1 acetonitrile:water at 55 °C selectively removed Boc groups in the presence of acid-labile groups such as Pmc sulfonamides and tert-butyl esters, with no alkylation of Trp, Cys or Met residues (782). Both N-Boc and tert-butyl ester groups were removed in the presence of other acid-sensitive groups (such as N-phenylfluorenyl) with ZnBr$_2$ in dichloromethane (783). Alternatively, tert-butyl esters were removed in the presence of N-Boc groups with CeCl$_3$.7H$_2$O-NaI in acetonitrile (784), with montmorillonite clay in refluxing acetonitrile (leaving tert-butyl ethers, Aloc, Cbz, and benzyl/methyl ester protection intact) (785), or with molecular iodine in refluxing acetonitrile (786). CeCl$_3$.7H$_2$O-NaI also removed one Boc group from Boc$_2$-Ser(THP)-OMe and other N,N-Boc$_2$- or N-Boc,N-Cbz-protected amino acids (787). For bis(Boc)-protected amines, one group can also be selectively removed using LiBr in acetonitrile (727) or by indium or zinc metal in refluxing methanol (788).

Ceric ammonium nitrate in alcohol at room temperature esterified N-Boc amino acids, but when carried out at reflux the esterification was accompanied by simultaneous removal of the Boc group (789).

Fmoc-protected amino acids, introduced in 1970 (790, 791) were originally prepared using Fmoc-chloride as the acylating reagent (791, 792). However, significant quantities (3–7%) of dipeptide byproducts were subsequently observed (793–795). These could be avoided by using Fmoc-O-succinimide for acylation (793, 794). Improved procedures for the use of Fmoc-OSu to prepare a range of Fmoc-amino acids were subsequently reported (796, 797). 9-Fluorenylmethyl pentafluorophenyl carbonate has also been employed as the acylating reagent (798), as has a polymeric N-hydroxy succinimide derivative (705, 799). Protection of amines under neutral conditions is possible by using an N-Fmoc-5,7-dinitroindoline, which reacts by a photolytic mechanism (726). A one-pot conversion of N-Cbz amino esters to N-Fmoc derivatives was accomplished with 79–90% yields by hydrogenation with 10% Pd-C poisoned with 2,2'-dipyridyl, in the presence of Fmoc-OSu (730).

The Fmoc group is generally removed with a solution of 20–25% piperidine in DMF; piperazine, morpholine, and DBU have also been used as the secondary amine (e.g., see 674, 675, 800). For solution-phase deprotections, dimethylamine or diethylamine are employed due to their ease of removal following deprotection, though the dibenzofulvene byproducts can present purification difficulties. Piperidine in dichloromethane is not stable and decomposes with precipitation of piperidine hydrochloride over several days (801). A 5% piperidine in DMF solution was found to be just as efficient as 20% piperidine for solid-phase synthesis; rate studies indicated that piperidine diffusion into the resin is the rate-limiting step for residues that deprotect slowly (802). Since piperidine is a controlled substance in some areas, alternatives can be useful. 4-Methylpiperidine was found to possess identical efficiency as piperidine in Fmoc removal ($t_{1/2}$ = 2.0 min for deprotection of Fmoc-Leu-resin with a 25% solution in DMF), and has similar cost (803). A mild solution of 2% DBU in DMF is another alternative (804). Catalytic DBU in the presence of an aliphatic or polymer-supported thiol scavenger provided deprotected amines in high yields and purity on multigram scales (805). Premature aminolysis of peptides linked to a solid support by a thioester linkage during Fmoc deprotection cycles was avoided by using a DBU/HOBt combination, with the ratio (80 mM DBU/74 mM HOBt in DMF) critical for chemoselectivity (806).

Other Fmoc deprotection reagents include tetrabutylammonium fluoride in DMF (807) and a resin-based piperazino/piperidino-functionalized polystyrene reagent, which was used for deprotection/scavenging (808). An aluminum trichloride/toluene system was identified as a novel reagent for Fmoc removal, which suppressed racemization and generated a highly lipophilic dibenzofulvenetoluene adduct that was readily separated from the reaction. Unfortunately, extended reaction times (3 h)

were required (809). A one-pot reaction for the removal of Fmoc and reprotection with Boc has been reported, using KF, Et$_3$N, and either S-Boc 2-mercapto-4,6-dimethylpyrimidine or di-tert-butyl dicarbonate (810). Resin-bound Fmoc-amino acids can be directly converted into other O-alkyl carbamates via reaction with aliphatic alcohols in the presence of tributylphosphine, DIEA and azodicarboxylic acid dipiperidide (ADDP) (811). Alternatively, a one-pot conversion of an N-Fmoc to N-Cbz group in 90% yield was also described (for Fmoc-Val-OMe), using DMAP to remove the Fmoc group as well as catalyze isocyanate formation in the presence of Boc$_2$O, followed by addition of benzyl alcohol (697).

The Fmoc group is prone to removal during basic hydrolysis of esters elsewhere in the molecule. However, the Fmoc group can be preserved by the addition of CaCl$_2$ during saponification with NaOH/iPrOH/H$_2$O (812). The unprotected basic side chain of Fmoc-Lys, Fmoc-Orn or Fmoc-Dab (but not Fmoc-Dap) can cause premature Fmoc removal during solid-phase synthesis (813). The Fmoc group is not completely stable to the standard Pd/C hydrogenation conditions employed to remove benzyl-type protecting groups, but ammonium formate catalytic transfer hydrogenation in methanol allowed for deprotection of a benzyl ester while retaining most of the N-Fmoc protection (823).

An Fmoc analog with four additional fused benzene rings, tetrabenzo[a.c.g.i]fluorenyl-17-methoxycarbonyl (Tbfmoc) was developed as an aid for peptide purification (720, 721), while a 2-(9-anthryl)ethoxycarbonyl (AEOC) analog of the Fmoc protecting group has very high UV absorbance and is suitable for both UV absorbance or laser-induced fluorescence detection (719). A 2,7-di(tert-butyl)-Fmoc analog (Fmoc*) significantly improved the solubility of some Fmoc-derivatives in organic solvents (814), as did a 2,7-bis(trimethylsilyl)-Fmoc derivative (815). The 2,7-di(tert-butyl)-Fmoc group has been introduced by using a polymer-supported succinimidyl carbonate derivative (816). Fmoc-protected amino acids have been identified as anti-inflammatory agents with a different mechanism of action from other compounds (817). The Fmoc group was reviewed in 1987 (800).

Several other types of urethane protecting groups have been developed (see Schemes 12.22, 12.23). The allyloxycarbonyl (Aloc or Alloc) group is an acid- and base-stable group that is orthogonal to Boc, Fmoc, and Cbz protection. Allyl protecting groups, including Aloc, were extensively review in a 1998 article (818). A polymer-supported N-hydroxysuccinimide derivative has been employed to introduce the Aloc group (819). Aloc urethanes are removed by a palladium catalyst in the presence of a nucleophilic scavenger such as tributyltin hydride, morpholine, phenylsilane, or trimethylsilyldimethylamine (820–822). Combinations of amine–borane complexes and a soluble palladium catalyst were found to be suitable for Aloc deprotection under near-neutral conditions and were applied to a heptapeptide solid-phase peptide synthesis using N^α-Aloc temporary protection (instead of Boc or Fmoc) (824). Aloc groups can be rapidly removed with a Ni(0) catalyst and Me$_2$NH.BH$_3$/Cs$_2$CO$_3$ (739), by a mild Ni(II)-catalyzed electrochemical procedure (825), or by iodine in wet acetonitrile (826). Aloc deprotection in the presence of dimethylallyl esters has been achieved with a water-soluble Pd(OAc)$_2$/TPPTS catalyst system (TPPTS = triphenylphosphinotrisulfonate sodium salt), using diethylamine as scavenger and acetonitrile/water as solvent (827). Aloc-protected amino acids have been deprotected with Pd catalyst/PhSiH$_3$ in the presence of carboxy-activated amino acids, leading to dipeptides in a one-pot reaction (828). In a similar fashion, N-Aloc amines were quickly deprotected with DABCO/Pd(PPh$_3$)$_4$ in the presence of EDC/HOBt and N-Fmoc or N-Boc amino acids (829).

An analog of the Aloc group, 3-(3-pyridyl)allyloxycarbonyl (Paloc), is stable to both TFA and the conditions employed for Rh(I)-catalyzed isomerization and cleavage of allyl esters. It is removed by Pd(0)-catalyzed transfer to N-methylaniline (830). Propargyl carbamates (Poc or Proc) can also be prepared; an N,O, bis(propargyloxycarbonyl) Ser-OMe dervative was obtained in 88% yield from Ser-OMe/HCl and PocCl. Both groups were removed using benzyltrimethylammonium tetrathiomolybdate, but by using only 1.1 equiv it was possible to selectively remove only the O-substituted propargyl carbonate in 84% yield (831). Propargyl pentafluorophenyl carbonate is a stable, crystalline reagent that introduced the N-Poc group in good yield (76–93% for nine examples) (832). A polymer-supported N-hydroxysuccinimide derivative has also been employed to introduce the Poc group (819). N-Poc amino acid chlorides were found to be effective activated amino acids for preparing peptides without racemization; the Poc group was removed with a polymer-supported tetrathiomolybdate (833). An unusual protecting group reagent is the C$_2$-symmetrical but-2-ynylbisoxycarbonyl (Bbc) group, essentially a bis(Poc) protecting group with 1 equiv of bis chloride reagent protecting 2 equiv of amine. The Bbc group is also removed with tetrathiomolybdate under neutral conditions; it is stable to 6 N HCl, 4 N NaOH, TMSI, formic acid, and 20% piperidine, allowing for orthogonal removal of Fmoc, Boc, and Cbz groups (834).

The 2-[4-(methylsulfonyl)-phenylsulfonyl]ethoxycarbonyl (Mpc) group has been used as an alternative to the Fmoc group in peptide synthesis. Like the Fmoc group, it can be introduced as the N-hydroxysuccinimide derivative, with no trace of oligomerization. Deprotection was achieved with 20% piperidine in DMF (835). A fluorous version of the Mpc group, with a perfluoroalkyl substituent, provided a base-labile fluorous protecting group compatible with Fmoc-based SPPS. Deprotection was accomplished with 2% aqueous NH$_3$ solution for 5 min (836). A water-soluble version, the 2-[(4-sulfophenyl)sulfonyl]ethoxycarbonyl (Sps) group, has also been developed. Mild bases, such as aqueous 5% Na$_2$CO$_3$, can be used for deprotection (837). The 2-[phenyl(methyl)sulfonio]ethoxycarbonyl (Pms) group

was also designed as a water-soluble group for use in aqueous solid-phase synthesis. Four different acylating reagents were prepared, and 25 different amino acid derivatives synthesized. Deprotection was achieved under mildly basic conditions, such as 5% NaHCO$_3$ (838). A 4-nitrophenyl carbonate reagent was required in order to introduce this protecting group onto the sulfur-containing amino acids Cys and Met (839). Another base-labile group is the 2-[(4-trifluoromethylphenyl)sulfonyl]ethoxycarbonyl (Tsc) group. This group can be removed under the same conditions as the Fmoc group (20% piperidine in DMF), but has a sufficiently different base susceptibility profile that it can be used as an orthogonal protecting group. The Fmoc group was deprotected in the presence of the Tsc group using 50% 1-methylpyrrolidine in DMF for 1 h at 25 °C, while the Tsc group was selectively removed with 0.1 N aqueous LiOH at 0 °C for 10 min (840). The Tsc group was employed in a solid-phase synthesis of pyrrole-imidazole polyamides (841).

A 1,1-dioxobenzo[b]thiophene-2-ylmethyloxycarbonyl (Bsmoc) protecting group can be cleaved by a Michael-like addition of piperidine, with efficient deprotection by a 2–5% piperidine solution in DMF (compared to the 20% solution normally used for Fmoc removal). Selective deprotection in the presence of the Fmoc group is possible (842, 843). With substrates where even this low percentage of secondary amine causes side reactions, cleavage of the Bsmoc group with 1 equiv of a solid anhydrous base, 4-piperidinopiperidine, was possible (844). Silica-bound piperazine or the polyamine tris(2-aminoethyl)amine have also been used for deblocking, with both reagents useful for solution-phase deprotections (843, 845). The Bsmoc group resulted in less aspartimide formation during peptide syntheses containing Asp sequences than the Fmoc group, due to the reduced basic conditions required for deprotection (843). Premature deblocking during peptide coupling was not observed, unlike the related Bspoc analog (843, 845). A related analog, the 2-methylsulfonyl-3-phenyl-1-prop-2-enyloxycarbonyl (Mspoc) moiety, has similar properties (846), while a naphthyl-based analog (Nsmoc) gave crystalline derivatives for amino acids which formed oils with the Bsmoc group (842).

An acid-sensitive Boc analog, 1-(1-adamantyl)-1-methylethoxycarbonyl (Adpoc), is cleaved by treatment with 3% TFA in dichloromethane for 30 min. It was used as N-protection for a solid phase synthesis of a decapeptide amide with an acid-sensitive handle (847). The 2-(biphenyl-4-yl)propan-2-yloxycarbonyl (Bpoc) and 2-(3,5-dimethoxyphenyl)propan-2-yloxycarbonyl (Ddz) groups are additional highly acid-sensitive Boc analogs, which can be removed with 3% trichloroacetic acid within 2 min (848), or with Mg(ClO$_4$)$_2$ in organic solvents (50°C for 3 h) (789). The cyclohexyloxycarbonyl (Choc) group was originally reported in 1957 (746), but under the deprotection conditions described (HBr) it showed no selectivity over Boc or Cbz groups. However, in 1998 N-Choc was employed as an orthogonal amine-protecting group for the Boc group, being stable to conditions that remove *N*-Boc groups (1 M TMSOTf-thioanisole in THF), but cleaved cleanly with anhydrous HF (*849*). The hexadienyloxycarbonyl (Hdoc) group was cleaved by 1% TFA in CH_2Cl_2 in 10 min, yet was stable to 2 M HCl, Pd(0) conditions, and base (20% piperidine, hydrazine, 2 M NaOH) (*850*). Fluorous analogs of the Boc group, ^FBoc, have been prepared for use in fluorous synthesis and tested on Pro, β-Ala, and isonipecotic acid (*851*).

The 2,2,2-trichloroethoxycarbonyl (Troc) protecting group, which can be introduced using the *O*-succinimide derivative (*793*), is moderately acid stable, with chemoselective removal by reduction with zinc dust (*852*). Removal is also possible with $(Bu_3Sn)_2$ in DMF under microwave heating, useful homogeneous conditions for solid-phase synthesis (*853*). Activated zinc dust in the presence of *N*-methyl imidazole in ethyl acetate or acetone removed the Troc group in the presence of reducible or acid-sensitive functionalities such as *tert*-butyl esters or azido, nitro, chloro, or phenacyl groups (*854*). Indium and NH_4Cl in $EtOH-H_2O$ also removed the Troc group from aliphatic amines in 3–6 h (*855*). Prenyloxycarbonyl (Preoc) protection was removed by a mild two-step one-pot procedure, using iodine in methanol to give 2-iodo-3-methoxy-3-methylbutyl carbamate, which was then cleaved with Zn in methanol. Boc, Fmoc, and methyl ester protecting groups were stable to the deprotection conditions (*856, 857*).

The β-(trimethylsilyl)ethoxycarbonyl (Teoc) (*858*) and triisopropylsilyloxycarbonyl (Tsoc) (*859*) groups are both removed by fluoride ion from sources such as tetra *n*-butyl ammonium fluoride (TBAF), conditions orthogonal to Cbz and Boc protecting groups. They are stable to conditions employed for Fmoc removal. The Tsoc group can also be removed selectively with TBAF in the presence of a TBS silyl ether (*859*). The Tsoc group has been used in combination with an *N*-Fmoc amino acid fluoride and catalytic TBAF for peptide bond formation, allowing for deprotection/coupling under almost neutral conditions in order to minimize diketopiperazine byproduct formation (*860*). The (2-phenyl-2-trimethylsilyl)ethoxycarbonyl (Psoc) group is cleaved under mild conditions (2 equiv of TBAF in CH_2Cl_2) much more rapidly than the Teoc group and orthogonally to the Fmoc group (completely deprotected after 15 min at room temperature; Teoc only 30–40% deprotected after 45 min; Fmoc group not cleaved), and is stable to Cbz, Fmoc and Aloc deprotection conditions (*861*). *tert*-Butyldimethylsilyl carbamates can be prepared from Boc- or Cbz/Alloc-protected amines by treatment with *tert*-butyldimethylsilyl trifluoromethane-sulfonate/2,6-lutidine or *tert*-butyldimethylsilane/Pd(OAc)$_2$, respectively. Deprotection to the free amine was achieved with TBAF. Alternatively, the silyl carbamate was converted into other urethane-type protecting groups via reaction with fluoride ion in the presence of an electrophile, such as methyl iodide, allyl bromide, ethyl iodide or benzyl bromide, to give *N*-methoxycarbonyl, *N*-Alloc, *N*-ethoxycarbonyl, or *N*-Cbz protection (*862*).

Photolabile protecting groups include the 2-nitrobenzyloxycarbonyl (Nboc) and 6-nitroveratryloxycarbonyl (Nvoc) urethanes, (see Scheme 12.22) which are stable to base, relatively stable to acid, and cleavable by photolysis at wavelengths longer than 320 nm (*863*). Removal of the Nvoc group can give poor yields, so a 2-(2-nitrophenyl)propyloxycarbonyl (NPPOC) group was developed which is cleaved twice as quickly using 365 nm radiation (*864*). A 4-(dimethylamino)phenacyloxycarbonyl (Mapoc) group (see Scheme 12.23) was removed by 20 min of photolysis using a high-pressure Hg lamp, and was stable to TFA or 20% piperidine conditions, and had a half-life in normal daylight of 7 days (*865*). A 9-xanthenyl group (see Scheme 12.22) gave good yields during protection but deprotection (photolysis at 300 nm) yields were only 52–65% due to side reactions (*866*). The Nvoc amine protecting group has been used in combination with a photolabile linker or ester protecting group that is cleaved by irradiation at a different wavelength (305 nm) (*867*). The same strategy could potentially be applied to differential amine protection, resulting in a novel orthogonal protection scheme avoiding the use of harsh deprotection reagents. A variation of this orthogonal protection approach employed a 3,4,5-trimethoxyphenacyl (TMP) carbamate group (see Scheme 12.23), which is cleavable by irradiation at 350 nm. Masking the ketone as a dimethyl ketal or 1,3-dithiane produced protecting groups that were inert to irradiation. They were converted back to the photolabile ketone by treatment with aqueous acid or periodic acid, respectively, giving three independent protecting groups that could be sequentially deprotected (*868*).

Enzyme-cleavable urethane protecting groups have been derived from glucose and galactose. The AGlOC (*O*-acetyl-D-glucopyranosyloxycarbonyl) and AGalOC (tetra-*O*-acetyl-β-D-galactopyranosyloxycarbonyl) groups (see Scheme 12.23) were removed in two-step, one-pot reactions using lipase followed by α/β-glucosidase or β-galactosidase (*869*). An earlier enzyme-cleavable group is the PhAcOZ, *p*-(phenylacetoxy)benzyloxycarbonyl, urethane. Penicillin G acylase mediates hydrolysis of the phenylacetate moiety, liberating a phenolate that spontaneously fragments to unmask the amino group (*870–872*). A *p*-acetoxybenzyloxycarbonyl (AcOZ) protecting group is also enzyme labile, with acetyl esterase or lipase cleaving the acetate moiety, leaving a quinone methide that then spontaneously fragments to liberate the amino group (see Scheme 12.24) (*873*).

12.3.3b Other *N*-Protecting Groups (see Scheme 12.25)

The Dde group, 1-(4, 4-dimethyl-2,6-dioxocyclohexylidene)ethyl, was introduced as a quasi-orthogonal group to Fmoc protection. It is stable to piperidine and TFA, allowing for removal of Fmoc and Boc groups, respectively. However, the original conditions for removal (2% hydrazine in DMF) also cleave the Fmoc group. Modified conditions for the orthogonal removal of the Dde group

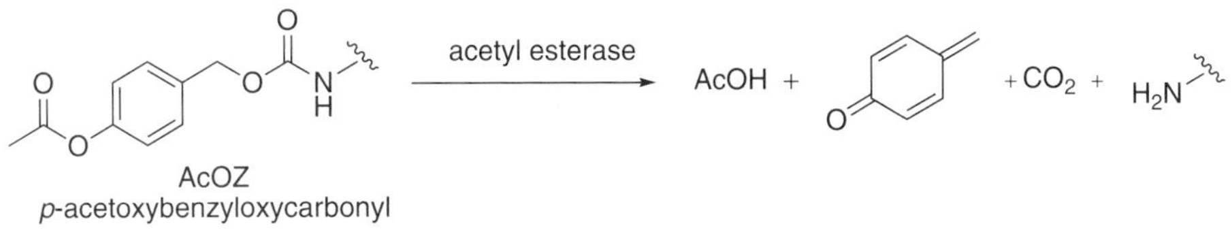

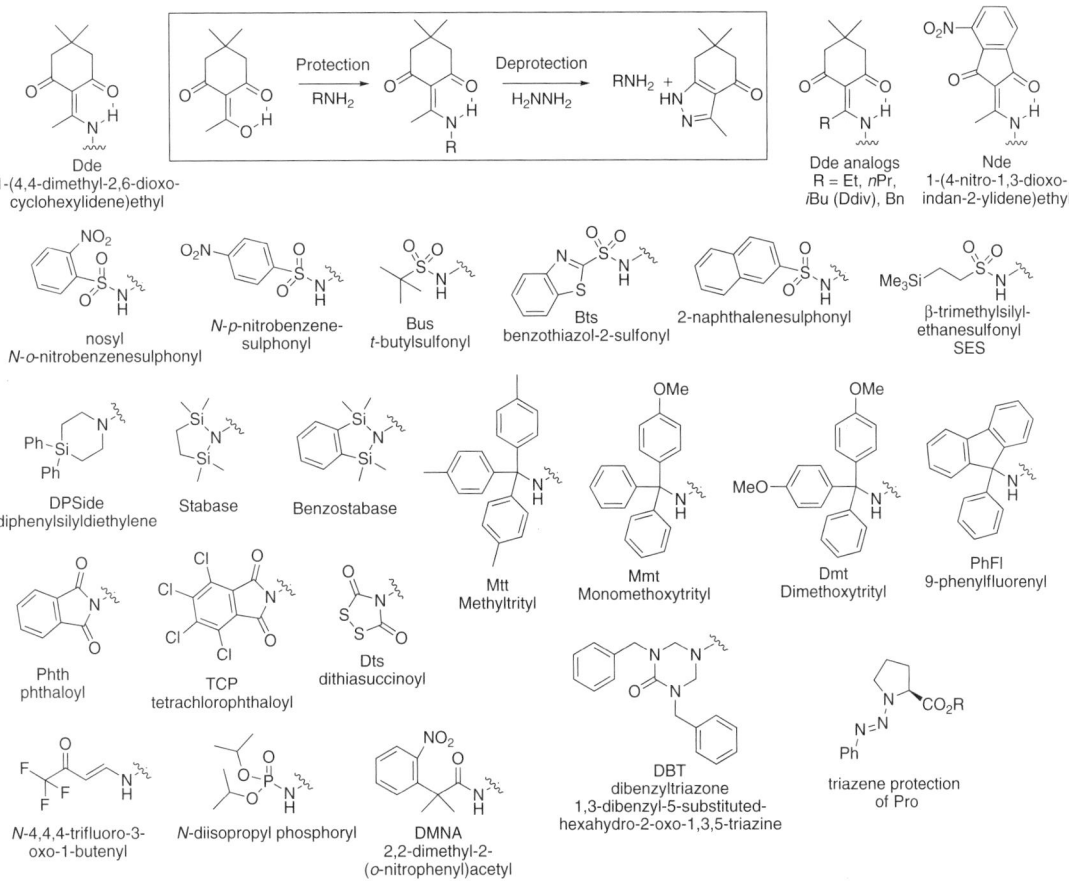

Scheme 12.24 Enzymatic Deprotection of AcOZ *N*-Protecting Group (*873*).

Scheme 12.25 Other *N*-Protecting Groups.

in the presence of an Fmoc group were reported in 2004; a mixture of NH$_2$OH.HCl/imidazole in NMP/CH$_2$Cl$_2$ gave clean deprotection in 3 h (*874*). Although Dde and Aloc are theoretically orthogonal, partial reduction of the Aloc double bond has been observed during removal of the Dde group with 2% hydrazine in DMF. This side reaction can be avoided by addition of allyl alcohol as a scavenger to the cleavage solution, with treatment of an Aloc-containing peptide providing no side product, even with extended reaction times (24 h) and high (10%) hydrazine concentration (*875*). One problem with

Dde protection is a propensity for migration to nearby unprotected amines (*876*).

A Dde analog, Nde, 1-(4-nitro-1,3-dioxoindan-2-ylidene)ethyl was reported in 1998 (*877*). The presence of the benzene ring provides a chromophore allowing for easy visual detection (red color forms upon treatment with 2% hydrazine in DMF), while the nitro group provides electron-withdrawing properties to increase susceptibility to nucleophilic attack. Stability to acid and base, and cleavage with hydrazine were similar to Dde. Four other Dde analogs, with bulkier substituents replacing a methyl group, were also evaluated in 1998. None of the analogs showed premature deblocking with 20% piperidine, conditions in which 3–6% of the Dde group was lost, but all were readily cleaved with 2% hydrazine. They also demonstrated no intermolecular N to N′ migration, as has been observed both inter- and intramolecularly with Dde (878). A mechanistically related Dde analog, Dmab (4{N-[1-(4,4-dimethyl-2,6-dioxocyclohexylidene)-3-methylbutyl]amino}-benzyl alcohol), was used as an orthogonal ester protecting group for Fmoc-Lys(Boc) during cyclopeptide synthesis (879).

N-Arylsulfonyl protection such as the tosyl moiety is required for a number of amino acid procedures, but can be difficult to remove. One must be aware that couplings of sulfonyl-protected amino acids may lead to racemization due to increased acidity of the α-proton (417). Deprotection of several different arylsulfonyl derivatives of Thr by Na/NH₃, HBr/HOAc or electrochemical reduction were compared in 1988, with good yields obtained by electrochemical reduction or HBr hydrolysis of the (2,4-dimethylphenyl)sulfonyl group (880). Other arylsulfonamides were examined in 1999. The 2-naphthalenesulfonyl group was cleanly removed with Mg/MeOH, conditions that allow for selective removal in the presence of a tosyl group (881). The p-nitrobenzenesulfonyl (nosyl) and o-nitrobenzenesulfonyl groups were introduced for temporary amine protection during solution-phase or solid-phase Mitsunobu alkylations of amines. The groups can be removed using K₂CO₃/thiophenol, Cs₂CO₃/thiophenol, DBU/2-mercaptoethanol, LiOH/mercaptoacetic acid, or 5% thiophenol in DMF (414, 461, 462, 483, 486, 520, 521, 882). DBU/thioglycerol is another effective reagent combination for removal of the o-nitrobenzenesulfonamide group, with the lower volatility of thioglycerol reducing the stench associated with 2-mercaptoethanol (522). A fluorous thiol, C₈F₁₇(CH₂)₂SH, has been used for deprotection in combination with either K₂CO₃ or a solid-supported base, allowing for easy reaction purification (883), while a solid-suppported thiol was combined with Cs₂O₃ in THF (24 h at room temperature or 6 min at 80°C with microwave irradiation) (884). The nosyl protecting/activating strategy was reviewed in 2004 (420). The 2,4-dinitrobenzenesulfonamide group has also been employed, with deprotection possible using thiol without base (mercaptoethanol or thiophenol), allowing for selective deprotection of 2,4-dinitrobenzenesulfonamide in the presence of 2-nitrobenzenesulfonamide (882).

In a similar manner, the o-anisylsulfonyl group can be employed for temporary protection during alkylation reactions. Removal was accomplished under Ni⁰-catalyzed conditions, and employed 4.5 equiv of iPrMgCl and 5 mol % [Ni(acac)₂], though the ester group of α-amino esters was reduced under these conditions (885). The Bts (benzothiazol-2-sulfonyl) group, which promotes acidity of the Bts-NH proton, can be removed using Zn/HOAc–EtOH (886), Al–Hg/ether–H₂O (886), NaBH₄/EtOH (416), H₃PO₂ in refluxing THF (416, 886) or thiophenol/K₂CO₃ (416, 419). It allows for coupling of acid chlorides with minimal racemization (416, 886). β-Trimethylsilylethanesulfonyl (SES) is

another sulfonamide-based amine protecting group. It was introduced using β-trimethylsilylethanesulfonyl chloride and removed by treatment with fluoride ion, inducing decomposition to TMS-F, ethylene, SO₂, and the free amine (887). The use of this group for amine protection and activation was reviewed in 2006 (888). The tert-butylsulfonyl (Bus) group can be introduced by reaction of amines with tert-butylsulfinyl chloride followed by oxidation with mCPBA. Deprotection was possible with triflic acid (889).

Silyl protecting groups have also been used on the amino group. N-Activation by silylation has been shown to significantly improve peptide synthesis yields of hindered residues (890, 891). The acidic carbamate proton of Boc-amino acids was removed by silylation with triethylamine and trimethylsilyl triflate reagent. Hydrolysis with 1 N HCl removed the silyl group but left the Boc group intact (892). A procedure for the N-silylation of amino esters under neutral conditions employed TMSCl/zinc dust (893). The "stabase" protecting group removes both amine hydrogens by formation of a five membered ring with 1,2-bis(dimethylchlorosilyl)ethane (894). The "benzostabase" group, formed with 1,2-bis(dimethylsilyl) benzene, is more stable to acidic conditions (895). A new protecting group for primary amines, diphenylsilyldiethylene (DPSide), was introduced in 1999. The amine was reacted with the ditosyl derivative of diphenyldi(2-hydroxyethyl)silane. The group is resistant to acidic, basic or hydrogenolytic conditions, and so is orthogonal to Boc, Cbz, and Phth protection. It was removed with a combination of TBAF/CsF in DMF or THF (896).

An N-phosphoryl group has been used as a protecting group, with deprotection possible under acidic conditions similar to those used for the Boc group. The N-diisopropyl phosphoryl protected amino acids were prepared by slow addition of sodium hypochlorite to a solution of the amino acid and diisopropylphosphite in water, giving 35–82% yields (639).

The methyltrityl (Mtt) group has been employed for amine protection as an orthogonal group to the Fmoc moiety, with removal by weak acids in dichloromethane. Even more acid-labile monomethoxytrityl (Mmt) and dimethoxytrityl (Dmt) groups have been reported, with cleavage using a mixture of acetic acid in trifluoroethanol and dichloromethane (897). Free amino acids and imino acids have been alkylated with 4-ethoxy-1,1,1-trifluoro-3-buten-2-one in aqueous sodium hydroxide, giving stable crystalline N-4,4,4-trifluoro-3-oxobutenyl amino acids. Deprotection is possible with dioxane/hydrochloric acid (898). Lubell and Rapoport developed the 9-(9-phenylfluorenyl) protecting group as a bulky substituent that helps prevent racemization at the α-center by obstructing α-proton removal. The group is introduced by alkylation with 9-bromo-9-phenylfluorene, and cleaved by hydrogenolysis or acid cleavage (899).

The trifluoroacetyl (Tfa) group can be removed under conditions such as mild base hydrolysis. This N-protecting group was recently demonstrated to be effective at suppressing racemization during couplings of acid chlorides, and useful on a process scale (900).

Amino acids have been simultaneously *N*-protected with a Tfa group and activated as *N*-hydroxysuccinimide esters by reaction with *N*-trifluoroacetoxysuccinimide in dichloromethane/pyridine (*901*). A 2,2-dimethyl-2-(*o*-nitrophenyl)acetyl (DMNA) group, added to amines the acyl chloride, is stable under most conditions. Reduction of the nitro group by hydrogenation activates as the protecting group for removal by intramolecular cleavage, with the cyclization promoted by acidic conditions (*902*). Phenylacetic acid can be employed for *N*-acyl protection of phosphopeptides, as it is removable by enzymatic hydrolysis with penicillin G acylase under mild conditions without attack on the peptide bonds, ester groups, or the phosphates (*903*). Picolinic acid (pyridine-2-carboxylic acid) was used to form pyridyl carboxamides of amino acids, with deprotection effected by electrochemical cleavage (*904*). Amino groups have been protected within a dibenzyltriazone ring formed by reaction of the amine with formaldehyde followed by *N*,*N*′-dibenzyl urea. Deprotection proceeds by aqueous hydrolysis in the presence of a secondary amine scavenger. The protecting group is resistant to a range of conditions (reduction, oxidation, C–C bond formation), as demonstrated by several syntheses (*905*).

N-Phthaloyl protection is a traditional method to mask both amino group protons. It is generally introduced by reaction with phthalic anhydride, though for some amino acids the conditions required can induce racemization, and some side chains can interfere with yields. A new reagent was described in 2002 which efficiently *N*-phthaloylates amino acids. The mono-succinimido, monomethyl ester of phthalic acid, methyl 2-[(succinimidooxy)carbonyl]benzoate (MSB), reacted with the sodium salts of amino acids in water–acetonitrile in 81–100% yield (*906*). An efficient, environmentally friendly process for *N*-phthaloylation fused the amino acid with phthalic anhydride at 130–135 °C under reduced pressure (40 mmHg) to remove water. Nine amino acids, including Trp and Tyr (problematic under traditional conditions), were protected in 75–99% yield (*907*). The phthaloyl group is removed by hydrazine under fairly harsh conditions, such as 60% hydrazine in DMF (*674, 675*). Tetrachlorophthaloyl (TCP) has been introduced to provide a phthaloyl derivative that can be removed under milder conditions, with 15% hydrazine in DMF solution at 40 °C removing the group within an hour (*908*). A solid-phase peptide synthesis with this group successfully prepared di- to nonapeptides, though coupling conditions were carefully chosen to avoid epimerization as the phthaloyl group, unlike the urethane protecting groups, does not impede racemization (*908, 909*).

The *N*-carbamoyl group has been used as temporary N^α-protection during peptide synthesis, with removal achieved by nitrosation with gaseous NO/O_2. It was introduced by reaction with aqueous KOCN at 50 °C, followed by acidification. However, depending on the coupling conditions, significant racemization was observed (*910*). Ureas formed with *p*-hydroxyaniline or *m*,*p*-dihydroxyphenethylamine can be cleaved by mushroom tyrosinase (*911*). A dithiasuccinoyl (Dts) amino protecting group is stable to strong acids and photolysis, but can be cleaved under mild conditions by thiolysis with thiols in the presence of catalytic tertiary amines. The group was constructed by reaction of the amine with bis(ethoxythiocarbonyl) sulfide, followed by treatment with (chlorocarbonyl)sulfenyl chloride. A study in 1999 examined various thiol combinations with amines or acids for efficient Dts removal during solid-phase synthesis of PNA oligomers; dithiothreitol with HOAc gave good results (*912*).

α-Azido acids, which are sometimes generated during the synthesis of amino acids, represent another potential α-amino protection strategy. A number of methods have been described for azide reduction, outlined in Section 2.3.1. Amino acids can be efficiently converted into α-azido acids by reaction with Tf_2O/NaN_3, followed by $CuSO_4/K_2CO_3/H_2O$. Twenty-three different amino acids were converted in 41–89% yield with retention of configuration. The azido acids were coupled to a growing peptide chain using DCC/HOBt, and then either "deprotected" to the amine with $Me_3P/$aqueous dioxane or reacted with an Fmoc-amino acid succinimide ester in the presence of Ph_3P to form a peptide bond (*913*). The azido group can be directly converted into an *N*-Boc protected amine by reduction with H_2/Lindlar catalyst (*914*) or decaborane (*915*) in the presence of Boc_2O, conditions that are compatible with reduction-sensitive groups such as alkenes, Cbz, and Fmoc groups. Secondary amines, such as the imino group of Pro, can be protected with a phenyldiazenyl group. The triazene was formed by treatment of Pro-OBn with PhN_2BF_4 in CH_2Cl_2/pyridine, and can be cleaved with ethanol/TFA or ethanol/aq. HCl. The triazine was found to be orthogonal to a Cbz group and was compatible with strong bases, alkylating reagents, and oxidative/reductive conditions (*916, 917*).

Both the amino and acid groups of an amino acid can be temporarily protected by formation of a five-membered ring complex with 9-borabicyclononane (9-BBN). The complex is soluble in organic solvents such as THF (an advantage over more traditional water-soluble copper complexes), and allows for side-chain functionalization with reagents such as acid halides, alkyl halides, ammonia, Arbusov conditions, $POCl_3$, and *m*CPBA. Deprotection is achieved using aqueous HCl or exchange with ethylenediamine in methanol (*918*).

The use of orthogonal amine protection is exemplified in a paper describing a synthesis of a tricyclic undecapeptide somatostatin analog. The cyclization precursor required five-dimensional orthogonal amine protection, with Fmoc, Cbz, Boc, Trt, and Aloc protecting groups employed. The *N*-Trt group was cleaved first (0.75% TFA in CH_2Cl_2), followed by the *N*-Aloc group (Pd-catalysis), the *N*-Boc group (1:1 TFA:CH_2Cl_2), the *N*-Fmoc group (20% piperidine), and finally the *N*-Cbz group (transfer hydrogenolysis with 4% formic acid in methanol with Pd black) (*919*). Cyclic decapeptides containing four Lys residues, with the side chains orthogonally protected with Boc, Fmoc, Dde, and Aloc groups, have been used as templates for synthetic proteins, with the Boc group removed using 50%

TFA/CH$_2$Cl$_2$, the Fmoc with 10% piperidine/DMF, the Aloc with Pd(PPh$_3$)$_4$/PhSiH$_3$, and the Dde with 2% NH$_2$NH$_2$/DMF (920, 921). A similar synthesis employed p-nitro-Cbz, Fmoc, Dde, and Aloc groups, with the Fmoc group removed using 2:98 DBU:DMF, Dde with hydrazine/allyl alcohol, Aloc with Pd(Ph$_3$)$_4$, and the pNZ with SnCl$_2$ (804). Examples of three-dimensional strategies include α,α-bis(aminomethyl)-β-alanine (922) or a triamino analog of cholic acid (923) with N-Aloc, N-Boc, and N-Fmoc protection, and cyclic peptide arrays prepared using a combination of Fmoc, ivDDe, and photolabile NPPOC (924).

A completely novel approach to orthogonal protection potentially allows for an unlimited number of differentially protected amino substituents. The strategy uses poly(N-sec-butylglycyl) protection, with one N-sec-butylglycyl residue on the first amino group to be unmasked, increasing stepwise to a (N-sec-butylglycyl)$_n$ residue on the last (nth) amine to be deprotected. The deprotection cycles are achieved by an Edman degradation, with phenylisothiocyanate (PITC) reacting with the terminal secondary amine of the N-sec-butylglycyl, and then cyclizing and eliminating a phenylthiohydantoin to reveal either the desired amine, or an (N-sec-butylglycyl)$_{n-1}$ protecting group, ready for the next cycle (see Scheme 12.26). The effectiveness of this "unichemo protection" was illustrated by protection of a penta-Lys peptide with mono-to-penta(N-sec-butylglycyl) residues, followed by sequential deprotection and acylation of the Lys side chain with five different acyl groups (925).

12.4 Amidation and Amide Hydrolysis

12.4.1 Amidation

In many cases it is desirable to obtain an amide derivative of an amino acid. Peptidyl amides are found in biologically active molecules such as oxytocin and gonadotropin-releasing hormone. The traditional method of accomplishing this transformation has been to react an alkyl ester of the amino acid with a saturated solution of ethanolic ammonia (926 and references therein). Unfortunately, racemization can occur during this process, with, for example, 5% racemization for Cys(Bn) and 88% racemization for Leu (926 and references therein). For aromatic amino esters, aqueous ammonium hydroxide in toluene gave amides of Phe, Tyr, and Phg with 71–94% yields, although optical purity was not examined (927). Combining 25% ammonium hydroxide with HOBt gave a number of N-protected amino acid amides in 75–90% yield, with no racemization based on optical rotation (928). Similarly, less than 2% racemization was observed during amidation using crystalline salts formed from N-hydroxysuccinimide (HOSu) or 3-hydroxy-1,2,3-benzotriazin-4(3H)-one (HODhbt) and ammonium hydroxide, methylamine or ethylamine. The salts were combined with the amino acid and DCC in DMF and then reacted for 3 h, giving amides of Boc-Leu, Boc-Cys(Bn), or Boc-His(Dnp) in 64–94% yield (926). This amidation procedure was also applied to C-terminal amino acids in peptides with similar success, although some racemization was observed during amidation with benzylamine (929).

Diaminomethane dihydrochloride can be used as a source of ammonia for forming primary amides from activated amino acids such as acyl chorides or 4-nitrophenyl esters. Good yields (80–95%) were obtained for amino acids (Cbz- and Boc-Ala, Val, Phe, Trp, and Gly) and dipeptides, with no racemization detected (930). Another procedure combined ammonium chloride with common coupling reagents such as PyBOP/HOBt, HBTU/HOBt, EDAC/HOBt, or DCC/HOBt. Amidation of Cbz-Tyr(Bz)-OH was complete in 30 min with PyBOP or HBTU, 2 h with EDAC, and 5 h with DCC. Optical integrity was not assessed (931). TOTT and HOTT, 2-mercaptopyridone-1-oxide-based uronium salts, have also been used with ammonium chloride and

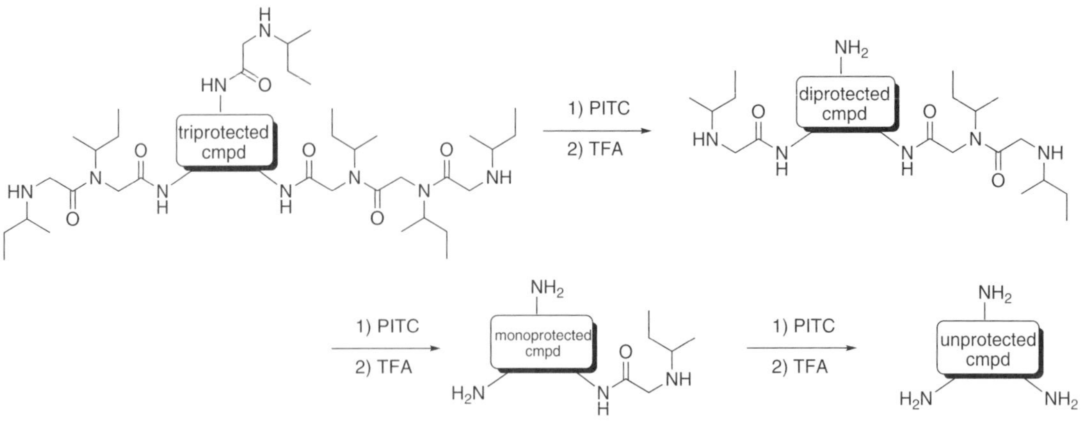

Scheme 12.26 'Unichemo' Orthogonal Protection Strategy.

diisopropylethylamine to form primary amides. Fmoc-Leu-OH, Cbz-Tyr(Bn)-OH, and Boc-Aib-OH, were amidated with 99% yield. The optical rotation for Fmoc-Leu-NH$_2$ indicated significant racemization had occurred (75% ee) (*932*).

Solvent-free conditions have been developed, directly reacting carboxylic acids and silica-supported ammonium chloride in the presence of triethylamine and tosyl chloride. The reaction was rapid (1 min at room temperature) and high yielding (75–90% for 16 examples), but only one amino acid substrate was tested. Primary and secondary amines also reacted (*933*). Another solvent-free microwave-based method prepared primary amides from the acid and urea in the presence of imidazole; Boc-Gly was the only amino acid amidated (79% yield) and so racemization was not investigated (*1823*). A method developed specifically to avoid racemization employed azidotrimethylsilane as the amine source, with Ph$_2$Se$_2$-Bu$_3$P acting as both a carboxyl activation reagent and an in situ reducing reagent to convert the azide to an amine. Cbz-Phe and Cbz-Ala were amidated in 88–89% yield (*934*). Xaa-Gly dipeptides have been converted into amino acid carboxamides by regioselective cleavage of the Gly residue N–C bond using nickel peroxide, although yields were poor to moderate (21–51%) (*935*).

Another method of synthesizing primary amides is to couple an amine such as allyl amine or benzylamine, and then remove the alkyl substituent. One convenient method coupled benzylamine, 4-methoxybenzylamine, or 2,4-dimethoxybenzylamine. The benzyl substituents were deprotected with 4 equiv of *p*-TsOH in refluxing toluene, with no detectable epimerization or Pro, Lys; or Asn. The dimethoxybenzylamide was cleanly dealkylated in the presence of Fmoc, Boc, and Trt protecting groups (*936*).

Substituted amides can be generated by coupling amino acids with amines under the same conditions that are used for peptide synthesis (see Section 12.5). For example, racemic amino acids were amidated via activation with alkyl chloroformates in the presence of a tertiary amine to give a mixed ester anhydride, followed by reaction with the desired amine (*937*). A 2003 process chemistry study of amidation using isobutyl chloroformate determined that the normal sequence of reaction (addition of isobutyl chloroformate to a solution of *N*-protected amino acid and base, followed by addition of amine) often produced significant byproducts, including apparent unreacted amino acid and a urethane corresponding to that formed from the amine component and isobutyl chloroformate. The study found that the byproducts arose from the formation of around 20% of a symmetrical anhydride of the amino acid during the initial activation. Addition of isobutyl chloroformate to a solution of amino acid, base, and amine also led to byproduct formation. However, addition of a solution of amino acid and base to a solution of isobutyl chloroformate, followed by addition of amine, produced <2% of the symmetrical anhydride (as measured by CO$_2$ evolution). Multiple additions of isobutyl chloroformate

were not required to drive the reaction to completion, quantitative yields of crude product were obtained, and the reaction could be scaled up to over 50 kg quantities (*938*). The mixed anhydride coupling method with isobutyl chloroformate and *N*-methyl morpholine was employed for amidation of various *N*-protected amino acids in 2004 (*939*).

The bulky *tert*-leucine residue is extensively racemized (29% D-epimer) when amidated using EDC/HOBt and DIEA in CH$_2$Cl$_2$. Switching to PyAOP with collidine as base in CH$_2$Cl$_2$ gave the desired amide in good yield with minimal (<1%) epimerization; the same reagents in DMF formed 8% of the D-isomer (*940*). A statistical experimental design was employed to optimize amide synthesis with polystyrene-supported *N*-hydroxybenzotriazole resin; the order of addition was found to be important (add acid before DIC), with lower DMF:DCM ratio preferred and 4.4 equiv of DIC optimum (with more or less giving reduced yields) (*941*). *N*-Protected amino esters were converted into amides in a one-pot reaction via hydrolysis followed by polymer-bound carbodiimide activation with HOBt and reaction with amines (*942*). Weakly nucleophilic heteroaromatic amines such as 5-aminotetrazoles or 2-aminothiazoles gave little or no amide when coupled to Boc-Ala with BOP, DCC/HOBt or HBTU. However, POCl$_3$/pyridine gave reasonable yields. No racemization was detected, though amino acids more prone to epimerization were not tested (*943*). *N,N'*-Carbonyldiimidazole (CDI) can be employed to activate acids for amidation reactions; a 2004 report found significant rate enhancements if carbon dioxide was bubbled through the mixture (*944*). Alternatively, tertiary amides have been generated by reaction of secondary amines with CDI, followed by reaction of the resulting carbamoylimidazolium salt with Cbz-Gly or Cbz-Phe, with yields of 80–96% (*945*). A synthesis of amides from amino acid chlorides, amines, and activated zinc has been reported. Yields of 89–96% were obtained (*946*).

Microwave irradiation has been used for solvent-free preparation of amides by simply heating neat primary amines with carboxylic acids, though no amino acids were tested in the original report (*947*). A subsequent optimization of conditions included examples of amidation of Cbz-Pro with benzylamine and α-methylbenzylamine, which proceeded in 75–91% yield and gave product with >98% ee (*948*). Unprotected amino acids can be simultaneously *N*-protected and activated for amidation by treatment with BF$_3$.Et$_2$O, forming a five-membered 2,2-difluoro-1,3,2-oxazoborolidin-5-one ring system. Addition of an amine led to the amidated amino acid, though yields were poor for sterically hindered branched amines (*949*). Unprotected amino acids were converted into a variety of amides in 81–98% yield by reaction with dichlorodimethylsilane and primary amines, presumably reacting through a cyclic silyl intermediate that provides temporary protection of the α-amino group and activation of the carboxyl group. While tested mainly on Phe, other amino acids such as *N*-methyl Gly, β-Ala, and Asp also produced α-amides (*950*).

A potentially useful method for reductive mono-*N*-alkylation of amides was reported in 1999, using TFA/triethylsilane and aldehydes. By this procedure a simple carboxamide can be transformed into a variety of secondary amides (*951*). Azido groups can be reduced to amines and used to amidate acids via in situ reduction by tributylphosphine in the presence of the coupling reagents EDC and HOBt (*952*), or simply by reaction with amino acid thio acids, proceeding via a concerted reduction/amidation process (*953*).

Many methods have been developed for preparing amidated peptides during solid-phase peptide synthesis (SPPS) via variations of the anchoring peptide–resin linkage, with linkers such as acid-labile alkoxy-benzhydrylamine (found in Rink resins) (*954, 955*), alkoxybenzylamine (found in PAL resins) (*956, 957*) or 9-amino-xanthen-3-yloxymethyl (*958, 959*) moieties resulting in *C*-terminal primary amides after peptide cleavage from the resin. Modifications to these resins allow for the synthesis of *N*-alkyl amides, via reductive alkylation of the 9-amino-xanthen-3-yloxy-methyl (*960*) or Rink-substituted benzhydrylamine (*961*) linkers before acylation of the first peptide amino acid residue. More recent modified resins include an acetophenone-based linker that is reductively aminated to introduce the substituted amine, and then acylated for standard peptide synthesis (*962*), and a benzaldehyde-based linker (BAL resin = backbone amide linker resin), which is derivatized by the same procedure (*963, 964*).

N-alkyl amides can also be produced by nucleophilic cleavage of an *o*-nitrobenzyl ester peptide–resin bond by primary and secondary amines (*965*), or by amine cleavage of peptides esterified to the Kaiser oxime resin (*966*). Ammonia can also be used for cleavage of amino acids linked to the Kaiser oxime resin (*967*). Similarly, peptide alkyl and aryl amides have been generated by using a peptide thioester linker to anchor the *C*-terminal amino acid to the resin. The thioester linkage is cleaved by primary and secondary amines in the presence of AgNO$_3$ in less than 1 h, with 50–97% yields of the amide, even with weakly nucleophilic aryl amines (*968*). An oxazolidinone linker has been designed to provide Asn, Asp or Gln *N*-alkyl amides (*969*).

Lipases and proteases can generate amino acid amides from amino acid esters using *tert*-butanol saturated with ammonia (*145*). Racemization is not a problem, as the same system is used to resolve amino acids. Several proteases catalyzed the ammonolysis of *N*-Boc-protected amino acid methyl esters with nearly absolute enantioselectivity, though conversion yields were low (*970*). Preliminary results have been reported using a peptide amidase from orange flavedo (extracts of orange peel) to form dipeptide amides from dipeptides and ammonium hydrogen carbonate (*971*). *Candida antarctica* lipase (CAL) catalyzed the regioselective amidation of Cbz-L-Glu diesters in organic solvents to give the α-monoamide, while Cbz-D-Glu resulted in the γ-monoamide. A variety of alkylamines were coupled (*972, 973*). The same regioselectivity pattern was

observed with L-Asp and D-Asp (*974*). In contrast, both D- and L- *N*-protected 2-aminoadipic acid and 2-aminopimelic acid diethyl esters gave the ω-monoamide under the same conditions (*975*).

Amidation can be achieved during amino acid synthesis, with the Ugi 4CC multicomponent condensation process providing amino acid amides as the product (see Sections 2.2.3, 5.2.2). The isocyanide component provides the amide substituent. 1-Isocyanocyclohexene is a "convertible" isocyanide that produces cyclohexene carboxamides, which can then be transformed into esters, thioesters or acids by reaction with the appropriate nucleophile (*586*).

Amino acid hydroxamic acids (*N*-hydroxy amides), ureas (*N*-acyl amides), and other derivatives have also been synthesized. Treatment of peptide hydrazides with *t*BuONO in the presence of HOAt or HOCt generated active esters that coupled with nucleophiles such as hydroxylamine, *O,N*-dimethylhydroxylamine or hydrazine urea (*976*). α-Amino hydroxamic acids were generated from *N*-Cbz amino acids by coupling the amino acid to ArgoGel-OH resin, and then cleaving the ester linkage with 50% aqueous hydroxylamine for 2 days to form the hydroxamic acid. Twenty-two different amino acid derivatives were prepared in 21–100% yield (*977*). Amino acids linked to an oxime resin can be converted into enantiopure hydroxamic acids by cleavage from the resin with hydroxylamine in a MeOH–CHCl$_3$ solution, with 48–100% yields (*978*). A solid-phase linker that gives hydroxamic acids upon acidic cleavage has also been reported (*979*). α-Amino acyl ureas, Xaa-NHCONHR, have been prepared from amino acid esters by reaction with a urea and NaOMe. If the amino acid substrate was *N*^{α}-unprotected, the product was optically active, but *N*^{α}-Boc amino esters gave racemic products (*980*). Amino acid acyl sulfonamides have been prepared by coupling the sulfonamide with the amino acid using EDC and DMAP, although the optical purity of the products was not established (*981*).

12.4.2 Amide Hydrolysis

Amide cleavage generally requires strongly hydrolytic conditions. A pH rate profile analysis for the hydrolysis of *N*-(phenylacetyl)-Gly-D-Val at 37 °C determined that the half–life of the Gly-Val amide hydrolysis was 267 years at pH 7, while the PhAc-Gly hydrolysis half-life was 243 years, corresponding to rate constants of $k_1 = 2.69 \times 10^{-11}$ and 4.65×10^{-11} s^{-1}, respectively. The rate constants at pH 1 were $k_1 = 3.33 \times 10^{-7}$ and 9.30×10^{-8} s^{-1}, and at pH 13 were 6.23×10^{-8} and 2.32×10^{-8} s^{-1} (*982*). The half-life of a Gly-Gly amide bond has been estimated to be 400 years at 25 °C at neutral pH. However, when the amino group of a polypeptide is exposed (non-acylated), diketopiperazine formation is so rapid that a 100-mer peptide could completely degrade before a single peptide bond was hydrolyzed. Thus, proteins that must survive for extended periods of time are usually *N*-acetylated, while those from microorganisms with short generation times often are not (*983*).

Peptide amidases can be used to cleave primary amides under mild conditions (*68, 123–125, 130, 132–134, 137*). Secondary amides have been hydrolyzed under mild conditions via *N*-nitrosation, treatment with lithium hydroperoxide, and reduction with sodium sulfite. The procedure did not work with Boc-Phg-NHMe, but was effective when *N*-trifluoroacetyl protection was employed (55% yield) (*984*). Alkaline hydrogen peroxide cleaves unactivated amines and peptides, even though ethyl esters are untouched; the kinetics of the cleavage have been examined (*985*). Methyl, *n*-butyl, *iso*butyl, and benzyl amides of *N*-phthaloyl amino acids were converted into amino esters in 91 to >98% yield by treatment with $NaNO_2$ in acetic anhydride, with the reaction proceeding via rearrangement of a nitrosamide intermediate. No racemization was observed (*939*). A procedure for the conversion of secondary and tertiary amides into esters employed triflic anhydride and pyridine in dichloromethane, followed by treatment with an alcohol. Ethyl esters were formed in 56–95% yield (23 examples), and the transformation was successfully carried out without affecting acetate, benzyl, TBDPS or acetal protection of alcohols. However, α-amino acid amides were not tested as substrates (*986*). Under these conditions, primary amides were converted into nitriles (*986*). Dimethylamides have been converted to acids, esters, and thioesters in the presence of *N*-acyl or peptide bonds by treatment with acidic aqueous, alcoholic or thiolic solutions, via a 2-oxazolin-5-one intermediate (*987*).

Pd(II) metal complexes have been employed to catalyze the hydrolysis of peptide bonds, coordinating and cleaving adjacent to Met, His or Trp residues in aqueous solutions. However, in acetone solution, hydrolysis occurred selectively at the *C*-terminal amide bond of Trp (*988*). Further study found that Pd(II) complexes could promote hydrolysis of the amide bond one residue *N*-terminal to Met and His residues with extremely high regioselectivity if the pH was kept at 1.5 or higher (*989*). Some efforts at developing artificial peptidases based on macromolecular catalytic systems were reviewed in 2003 (*990*). A $[Pd(H_2O)_4]^{2+}$ complex selectively cleaved peptides at an Xaa-Pro motif; further selectivity for a Xaa-Pro-Phe sequence was obtained by conjugating the Pd complex to β-cyclodextrin (*991*). Oligopeptides can be hydrolyzed to amino acids by Ce^{IV} under mild conditions, with 10 mM $[Ce(NH_4)_2(NO_3)_6]$ hydrolyzing H-Gly-Phe-OH and H-Gly-Gly-OH at 50 °C and pH 7.0 with half-lives of 2–2.5 h. The urethane and amide groups of Cbz-Gly-Phe-NH_2, Gly-Phe-NH_2, and Cbz-Gly-Phe were not hydrolyzed (*992*). γ-Azido-Abu has been found to be useful as a mild and chemoselective method for activating peptide bonds for cleavage. Treatment of a peptide or protein containing a γ-azido-Abu residue with phosphines or dithiols (e.g., tris(carboxyethyl)phosphine or dithiothreitol) resulted in cleavage at the *C*-terminal side of the residue, with conversion of the γ-azido-Abu into a homoserine lactone (*993*).

Amino acid phenylhydrazides can be converted into an activated intermediate, an acyl diazene, by oxidation with mushroom tyrosinase. The oxidized moiety is then readily hydrolyzed by water (*994*). Phenylhydrazide was subsequently employed as a tyrosinase-labile amino acid protecting group in peptide synthesis (*995*).

12.5 Coupling of Amino Acids

12.5.1 Introduction

The coupling of amino acids is an area of great interest. Much of this research has been driven by the need for reagents that provide extremely high efficiencies during polypeptide synthesis, as a small decrease in yield for each coupling step will substantially reduce the overall yield of polypeptide product. For example, a 99% coupling efficiency for a 20-residue peptide would give an overall yield of 82%, but a slight decrease in efficiency to 97% reduces the overall yield to 54%, along with the accompanying difficulty of purifying the product from single-residue deletion peptides. Couplings of the standard amino acids are generally well-served by established methods, but many of the non-proteinaceous amino acids present difficulties. In particular, poor yields are often observed with acylations of *N*-alkylated amino acids or acylations/amidations of sterically crowded α,α-disubstituted amino acids under standard conditions. *N*-Activation by silylation has been shown to significantly reduce racemization (*996*) and improve coupling yields of hindered residues with several reagent systems (*890, 891*). It should be noted that *N*-alkylated peptides are susceptible towards TFA cleavage between the *N*-alkyl derivative and the following residue, so caution must be exercised during peptide syntheses that use TFA for deprotection (*344–346*).

One concern during coupling is the possibility of racemization (briefly reviewed in *997*, more extensively in *998, 999*), which a number of studies have addressed. A comparison of a number of reagents and coupling conditions found that solvent and the tertiary base employed can have a significant effect in reducing racemization rates, with THF and morpholine bases suppressing racemization compared to DCM and triethylamine (*1000*). Collidine was found to result in less epimerization than *N*-methyl morpholine or DIEA for HATU-induced coupling of peptide segments (*1001*). DMF is often employed as the solvent for coupling reactions in order to solubilize all the components, but unfortunately this solvent generally induces significantly more racemization than non-polar solvents such as DCM (*1001*) or trifluoroethanol–trichloromethane (*1002*) (e.g., an amidation with PyAOP/collidine in DCM caused <1% racemization of an acylated *tert*-Leu residue, compared to formation of 8% of the D-epimer with the same reagents in 2:1 DCM:DMF (*940*)). Claims for the ability of new coupling reagents to suppress racemization must be examined carefully, as the solvents selected for demonstration reactions are often not those of common practical use. Racemization is generally caused by formation of a chirally unstable

2-alkoxy-5-(4*H*)-oxazolone intermediate from activated *N*-alkoxycarbonyl amino acids. The role of oxazolones in racemization during coupling was discussed in 1996 (*1003*). *N*-Acyl amino acids are much more susceptible to racemization than urethane-protected derivatives (see Section 1.4.5 for further details), as illustrated by the almost complete racemization of an *N*-acyl Orn derivative during standard HBTU/DIEA coupling conditions (*1004*).

Another difficulty encountered during peptide synthesis is reduced yields due to interchain association within the resin matrix. Techniques such as temporary *N*-substitution of the peptide bond via a 2-hydroxy-4-methoxybenzyl (Hmb) substituent (*348*) or the use of polar solvents such as DMSO during coupling (*1005*) have been employed to prevent interchain hydrogen bonding and aid in acylation. An improved 2-hydroxy-6-nitrobenzyl (HnB) temporary substituent has also been described (*532*), as has a 3-methylsulfinyl-4-methoxy-6-hydroxybenzyl (SiMB) group (*533*). Several of these techniques were compared in 1994, with the use of DMSO significantly improving synthesis results (*1005*). A TOAC (2,2,6,6-tetramethylpiperidine-*N*-oxyl-4-amino-4-carboxylic acid) residue was employed to optimize the synthesis of aggregating peptide sequences, via analysis of the resin-bound spin-labeled peptides by EPR spectroscopy (*1006*).

The completion of coupling during solid phase peptide synthesis is usually monitored by the Kaiser test, a sensitive reaction that detects free primary amines by treatment with solutions of ninhydrin, phenol and potassium cyanide, producing an intense blue color (*1007*). Secondary amines can be detected with good sensitivity using a chloranil/acetaldehyde solution (*1008*).

Four different approaches have been used for the formation of peptide bonds: (1) coupling of an isolated activated carboxyl derivative (such as an active ester, acid fluoride, or anhydride) to any amino acid with a free amino group; (2) coupling of a free carboxyl group to any amino acid with a free amino group in the presence of an activating reagent (which generally activates the carboxyl group, and often proceeds via a non-isolated active ester generated in situ); (3) enzyme-mediated couplings; and (4) coupling as part of a chemical synthesis of one of the residues involved.

The application of the first two methods to peptide synthesis has been extensively discussed elsewhere, with a comprehensive review of earlier coupling methods (including detailed experimental procedures) by Greenstein and Wnitz (*1009*), a short practical description of coupling using DCC, symmetrical anhydrides, and active esters by Stewart and Young (*1010*), a historical review of peptide bond formation by Wieland and Bodanszky (*1011, 1012*), and definitive accounts of coupling research up to 1984 in *Principles of Peptide Synthesis* (*1013*) and *The Peptides* series (*1014–1016*). A primer designed for introduction to the field of amino acid and peptide synthesis is part of the Oxford Chemistry Primer series, and includes sections on peptide synthesis (*682*), while a 2000 monograph on

Fmoc solid-phase peptide synthesis includes practical coupling methods (*1017*). A review of the synthesis of peptides containing secondary amino acids was published in 1991; this paper also contains a good overview of the more traditional coupling systems (*1018*). Modern coupling methods for hindered peptides containing *N*-methyl Ala, *N*-methyl Aib, or *N*-methyl αAc^5c were compared in 1992 (*1019*). More recent developments are included in the 1997 monograph of Lloyd–Williams, Albericio, and Giralt (*1020*); the same year an extensive review of coupling methods for the incorporation of non-coded amino acids into peptides was published, including sections on *N*-methyl amino acids, α,α-disubstituted amino acids, and dehydroamino acids (*352*). Coupling reagents were also reviewed in a 2002 *Synlett* article (*723*) and in 2004 (*1021*) and 2005 (*1022*) *Tetrahedron* reports. Finally, a new (English) version of *Houben–Weyl*, published in 2003, has a most comprehensive description of coupling methods (*1023*).

Until the 1950s, peptide bond formation between an acylamino acid and amino acid (or ester derivative) proceeded almost exclusively through conversion of the carboxyl group of the acylamino acid to its corresponding acyl azide or acid chloride, followed by condensation with an amino acid in alkaline aqueous solution, or with an amino acid ester in an unreactive organic solvent (*1024*). Variations of these activated amino acids are still employed. The standard coupling conditions employed for many years for solid-phase peptide synthesis (SPPS) simply involved addition of an *N*-protected amino acid with dicyclohexylcarbodiimide (and sometimes an additive) to the growing peptide chain (*1025*). However in the past three decades a broad range of new reagents have been introduced, with the goal of simple reaction conditions, short reaction times, high yields, and minimal racemization. These reagents are generally referred to by their (often similar) acronyms. Some of the traditional methods still compare surprisingly well with the newer (much more expensive!) reagents, even with troublesome residues, and so they should not be ignored when encountering a difficult coupling.

This section presents a summary of coupling methods, emphasizing comparative studies and papers describing demanding couplings. Coupling reagents convert an amino acid into an activated species that can often be isolated, such as an anhydride, active ester, acid halide or acid azide, and thus many coupling methods overlap. Active esters and anhydrides are often prepared using carbodiimide reagents, while symmetrical anhydrides are present as intermediates in carbodiimide-mediated reactions, and active esters are generated by combining coupling reagents with additives. Thus, the separation of methods and reagents in the following sections is somewhat arbitrary. Care must be taken when examining studies that compare different types of coupling reagents, as authors sometimes employ conditions optimized for their reagent. For example, CH_2Cl_2 can suppress racemization during couplings, but this solvent is not as useful as DMF for standard couplings due to amino acid solubility issues. Selection of a coupling method must also consider the final use of the reaction: a large-scale industrial synthesis requires thoroughly optimized reaction conditions and consideration of reagent costs, while a one-time small-scale research application is more focussed on quickly obtaining a reasonable quantity of product. Many reagent combinations possess similar effectiveness for the majority of coupling applications, with differences only becoming apparent when more demanding components are combined. The ideal set of coupling conditions must optimize the reactivity of the activated amino acid, balancing sufficient activation to acylate the amine component with the need to moderate reactivity to prevent side reactions. These requirements will vary with the residues to be coupled, so an ideal reagent combination applicable to all situations is unlikely to emerge.

An interesting related issue is how peptides were first formed under primordial conditions. A 1998 study found that dipeptides were created in experimental settings modeling volcanic or hydrothermal circumstances, using Phe, Tyr or Gly under anaerobic, aqueous, conditions at 100 °C and pH 7–10 in the presence of coprecipitated (Ni, Fe)S and CO in conjuction with H_2S as a catalyst. Unlike previous experiments, dipeptides were formed preferentially over diketopiperazines (*1026*).

12.5.2 Coupling Methods

12.5.2a Amino Acid Halides

A review of peptide synthesis using amino acid halides was published in 1996 (*1027*). Amino acid chlorides were not widely used for peptide synthesis after the 1930s due to their instability and the high reactivity of the reagents required for their preparation (*1009*). Decomposition of *N*-tosyl amino acid chlorides has been observed in the presence of base (*1028*). *N*-Trifluoroacetyl-protected amino acids are very susceptible to racemization during coupling; with acid chlorides under standard conditions, 38–72% epimerization was observed (*996*). However, in 1986 Carpino and co-workers reported that Fmoc amino acid chlorides were stable crystalline derivatives (*1029*), and could be used for both solution- and solid-phase syntheses (*1029–1031*). Unfortunately, the acid-sensitive side-chain protecting groups generally used in combination with *N*-Fmoc protection were not compatible with the preparation of the acid chloride (*1032, 1033*). A new method for acid chloride formation using triphenylphosphine and trichloroacetonitrile under mild non-acidic conditions was reported in 1999, which could potentially overcome this liability (*1034*). In situ generation of *N*-Fmoc acid chlorides is possible by using triphosgene, bis(trichloromethyl)carbonate (BTC), with collidine or diisopropylethylamine. The reagent is particularly suitable for couplings of *N*-methyl amino acids, and no racemization was observed (*1035, 1036*). This reagent was employed for the synthesis of the cyclodecapeptide omphalotin A, in which 9 of the 12 amino acid residues are *N*-methylated. The BTC method was far superior to TFFH, DCC, or

DIC/HOAt (*302*). Cyclosporin O was also synthesized (*1036*). Triphenylphosphine/trichloroisocyanuric acid is another reagent combination for in situ generation of acid chlorides. Boc-Leu was activated and then treated with Pro-OBn to form the dipeptide in 85% yield with no racemization (*1037*).

Fmoc-amino acid chlorides have been used in combination with KOBt (the potassium salt of hydroxybenzotriazole) for the acylation of *N*-methyl amino acid esters; cyclosporin fragments with consecutive *N*-methyl amino acids were prepared in high yield and excellent purity (*672*). Fmoc-Aib-Cl in the presence of KOBt was used to construct difficult sequences, such as Aib-Aib-Aib-Aib and Aib-Pro-Aib-Ala, with 81–85% yield for Aib-Aib dipeptide formation (*1038*). Coupling of Fmoc-amino acid chlorides in organic solvents (THF or toluene) in the presence of commercial zinc dust has been found to proceed without the need for organic or inorganic bases. Dipeptides were formed in 86–90% isolated yield after 10–15 min, with no racemization (*1039*). The coupling rate of Fmoc acid chlorides was greatly improved by using the acid chloride with AgCN in toluene, which proceeds via formation of an oxazolone intermediate. Difficult sequences of up to four sequential secondary amino acids were prepared (*1040*). The nucleophilicity of the amino component being coupled to the acid chloride can be increased by using an *N*-TMS substituent (*893*).

Poc (propargyloxycarbonyl) amino acid chlorides were found to be effective coupling reagents for preparing peptides without racemization; the Poc group was removed with polymer-supported tetrathiomolybdate (*833*). Acid chlorides of α-azido acid equivalents of Aib and Dpg (diphenylglycine) were used to couple up to four consecutive residues, with DTT reduction of the azido group following coupling (*1041*). The acid fluorides (see below) could also be prepared, but were less reactive. Sterically hindered *N*-(benzotriazol-2-sulfonyl) *tert*-Leu was activated as the acid chloride for coupling to a γ-amino acid, with a 70–74% yield (*419*). A synthesis of a tetrapeptide of *N*-methylated amino acids employed *N*-Bts protection to couple the non-methylated amino acid chlorides, followed by on-resin methylation (*416*). In contrast to the earlier (1996) report on the susceptibility of *N*-Tfa amino acid chlorides to racemization, a 2003 paper described conditions which allowed the *N*-trifluoroacetyl amino acid chlorides to be generated with high purity and minimal racemization (using Vilsmeier reagent at −10 °C), and then coupled to amino acid esters under carefully controlled Schotten–Baumann conditions. This procedure was employed for process chemistry on a >100 kg scale (*900*).

Amino acid fluorides (see Scheme 12.27) represent a substantial improvement over acid chlorides. They can be readily prepared by treatment of protected *N*-Fmoc (*1033, 1042*), *N*-Boc (*1043*), *N*-Cbz (*1042, 1043*), or *N,N*-bisprotected (*1044, 1045*) amino acids with cyanuric fluoride. The protected derivatives are mostly crystalline, stable to storage (*1042*), do not form

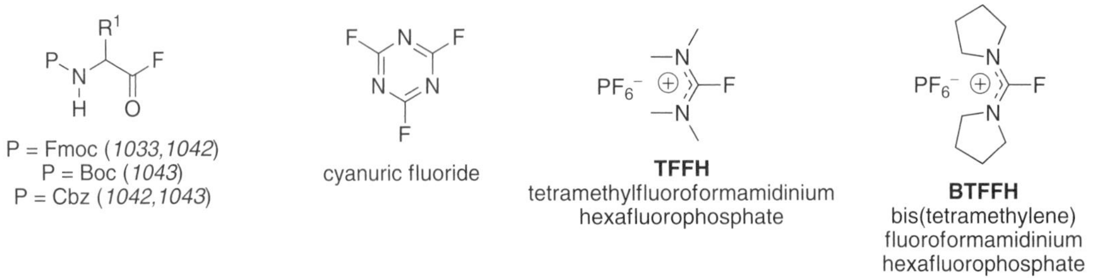

P = Fmoc (*1033,1042*)
P = Boc (*1043*)
P = Cbz (*1042,1043*)

cyanuric fluoride

TFFH
tetramethylfluoroformamidinium
hexafluorophosphate

BTFFH
bis(tetramethylene)
fluoroformamidinium
hexafluorophosphate

Scheme 12.27 Amino Acid Fluorides.

less-reactive oxazolones in the presence of tertiary organic bases (*1033*), and rapidly couple with minimal racemization (*1033, 1043*). The coupling reaction proceeds quickly even in the absence of a tertiary base despite the equivalent of HF that is generated, avoiding a possible source of racemization, and, for Fmoc-protected amino acids, premature deblocking (*1046*).

A new reagent, tetramethylfluoroformamidinium hexafluorophosphate (TFFH), was developed to generate amino acid fluorides as transient intermediates during coupling reactions. If desired, the fluoride could still be isolated. The reagent was used to prepare a pentapeptide containing an Aib-Aib sequence in 88% crude yield with 92% purity (*1047*). A subsequent study found that at least two mechanisms were generating the acid fluorides of *N*-Cbz amino acids: direct conversion of an activated intermediate, or intramolecular cyclization to an oxazolone followed by fluoride nucleophilic ring opening (*1048*). Closely related bis(tetramethylene)fluoroformamidinium hexafluorophosphate (BTFFH) is also effective at generating acid fluorides in situ, and was employed in the synthesis of a number of common peptides, including magainin 1 amide (21-mer) and bradykinin amide (9-mer) (*1049*). Fluoroformamidinium hexafluorophosphate has also been incorporated into a solid-supported polystyrene resin, where it was used for purification-free couplings to give peptides. Unfortunately, and unexpectedly, complete racemization of the amino component was observed (*1050*). Acid fluorides can be generated in the absence of base via carbodiimide activation in the presence of a fluoride source, PTF (Ph$_3$P$^+$Bn H$_2$F$_3^-$). The additive enhances reactivity and reduces racemization (*1051*). A thermally stable DAST analog, [bis(2-methoxyethyl)amino]sulfur trifluoride (Deoxo-Fluor), generated acid fluorides in situ in a one-pot transformation of acids to amides (*1052*).

Amino acid fluorides have been found to be particularly useful for coupling to hindered residues such as Aib (*1053–1055*), and were applied to a synthesis of the alamethicin F30, a 20-mer peptide containing eight Aib residues, including difficult Aib-Pro and Aib-Aib sequences. The syntheses were achieved both with (*1056*) and without (*1046*) added base. A comparative

synthesis of closely related alamethicin acid (Phe-Gln-Glu-Aib-Aib-Val-Pro-Aib-Leu-Gly-Aib-Val-Aib-Gln-Ala-Aib-Ala-Aib-Pro) using amino acid fluorides, UNCAs, or PyBroP activation clearly demonstrated the superiority of the fluorides, as no significant amounts of product were obtained in the UNCA or PyBroP syntheses (*1053, 1054*). Acid fluorides were also used to construct the Aib sequence of an alamethicin-Leu zipper hybrid (*1057*), an Iva-Val-Iva-(α-Me-Val)-(α-Me-Phe)-(α-Me-Val)-Iva sequence (*1058*), and an Aib-(α-methyl-Val)-Aib tripeptide, with two tripeptides then combined via oxazolone activation (*1059*). In 1998 amino acid chlorides were compared to amino acid fluorides for the difficult couplings of Fmoc-Ala-X or Fmoc-Aib-X with H-Aib-OMe or H-MeAib-OMe. The fluoride was far superior for preparation of the Fmoc-Aib-Aib-OMe derivative, but the chloride gave nearly quantitative yields of the more difficult Fmoc-Ala-MeAib-OMe product within 30 min, compared to only a 24% yield after 24 h with the fluoride. No coupling could be achieved between Fmoc-Aib-X and H-MeAib-OMe, but a Pbf-Aib-Cl derivative was successful. The sulfonyl protecting group prevented deactivating oxazolone formation, which is the predominant side reaction observed with urethane-protected amino acid chlorides (*1060*).

A one-pot deprotection/acylation of *N*-Aloc amino acids relied on activated amino acids as the acylating reagent. An Ala-*N*-Me-Aib dipeptide was formed in 76% yield using Fmoc-Ala-F as the activated reagent, while Cbz-Ala-NCA coupled in only 45% yield. Fmoc-Ala-OPfp was even less effective, giving only 29% yield when coupled to less-hindered Aib (*828*). Another one-pot deprotection/coupling sequence employed the Tsoc silyl carbamate group in combination with an *N*-Fmoc amino acid fluoride and catalytic TBAF, allowing for deprotection/coupling under almost neutral conditions in order to minimize diketopiperazine byproduct formation (*860*). For very hindered residues such as *N*-methyl Aib, prior treatment of the amino component by silylation with *N,O*-bis(trimethylsilyl) acetamide (BSA) gave significantly increased rates of reaction for acylations by amino acid fluorides (35% yield after 2 h vs 5%) (*891*). Preformed acid fluorides were also found to be effective for acylating resin-bound secondary amines, using 1.1 equiv of DIEA. Other commonly used reagents (BOP, HATU or HBTU, with or without additives such as HOBt or HOAt) were inefficient (*963*).

In contrast to the above results, other comparative studies have found alternative coupling methods to be more effective than amino acid fluorides. A coupling of Fmoc-Aib-F or Fmoc-Aib-Cl to Pro-O*t*Bu found the chloride to react much more rapidly (10 min vs 6 h (*1061*)). Another study demonstrated that HATU or DIC/HOAt gave superior results when compared to amino acid fluorides for the coupling of Fmoc-Val to MeLeu; the fluoride derivative of Fmoc-MeLeu could not be prepared (*1062*).

Amino acid bromides have recently been applied to peptide synthesis. Up to seven consecutive α-methyl-Val residues were coupled in high yields by using their α-azido acid bromides. The azido acids were obtained from amino acids by treatment with CF₃SO₂N₃, and then converted to the acid bromide in situ using 1-bromo-*N*,*N*-2-trimethyl-1-propenylamine (*1063*). Amino acid bromides were also applied to a solid-phase synthesis of homopeptides containing up to eight successive α-methyl-Val or Aib residues, using the α-azido acid bromide in dichloromethane (*1064*). *N*-2-Nitrobenzenesulfonyl-protected amino acid bromides have also been employed (*1064*) as have *N*-phthalimido amino acid bromides (*1065*) and urethane-protected Pro bromides (*1065*).

12.5.2b Amino Acid Azides

The azide method of peptide coupling is reviewed in *The Peptides* series (*1066*). Formation of peptide bonds through azides was first reported by Curtius in 1902, and has been widely used over the years due to the limited racemization of azide-activated amino acids (*1009, 1013*). Originally prepared via acid hydrazides, the reactive acid azides can be generated in situ from amino acids using DPPA (diphenylphosphoryl azide) (see Scheme 12.28) with good yields and minimal racemization (*1067*). This reagent provided excellent yields for the synthesis of macrocyclic lactams (*1068*). A thio analog of DPPA, dimethylthiophosphinic azide (MPTA) has been reported; it is more stable and causes even less racemization than DPPA (*1069*).

One disadvantage of the azide method is that the azides are unstable. However, peptide azides have recently been used for segment condensation via in situ generation of the azide from peptide hydrazides by treatment with *t*BuONO in the presence of HOAt or

HOCt, generating active esters that efficiently couple with the amino terminus of another peptide fragment (*1070*). The synthesis, isolation, stability, and use of *N*-Fmoc amino acid azides were reported in 2000. They were prepared via acid chlorides or mixed anhydrides, and found to be crystalline solids that were stable to extended room temperature storage. Dipeptides were prepared by solution couplings at room temperature in 77–88% yield without racemization, though reaction times of 16–18 h were required (*1071*).

12.5.2c Amino Acid Active Esters

The use of active esters for peptide synthesis is based on the enhancement of the electrophilic nature of the amino acid carbonyl group by forming an ester with an electron-withdrawing alcohol, with the concept first developed in the 1950s and early 1960s (*1009, 1012, 1013, 1072*). Active esters are reviewed in *The Peptides* series (*1073*). A wide range of reactive esters have been employed for peptide bond formation, with a table of over 40 derivatives in Bodanszky's monograph (*1013*). Even more have been described in the past decade, but most are now generated in situ, in combination with other coupling reagents rather than isolated as distinct entities (e.g., HOBt and HOAt esters). Some of the more successful traditional active ester derivatives (see Scheme 12.29) have been 4-nitrophenyl (e.g., see *1072, 1074*), 2,4-dinitrophenyl (2,4-DNP; e.g., see *1072*), 2,4,5-trichlorophenyl (TCP; e.g., see *1074*), pentafluorophenyl (Pfp; e.g., see *1074, 1075*) or hydroxy succinimide (e.g., see *1074*) derivatives. The active esters are usually prepared by esterification using carbodiimide reagents (*1012*). *p*-Nitrophenyl esters were formed in situ by activation with *p*-nitrophenyl chloroformate in the presence of triethylamine and DMAP; di- and tripeptides were synthesized with negligible racemization detected (*1076*). Active esters formed from 4-nitrophenol or 2,4,5-trichlorophenol were aminolyzed in the presence of additives such as HOBt and HODhbt. Reagents for forming the esters without using DCC were also described (*1077*). Racemization during aminolysis of activated esters in aqueous DMF can be minimized by using Na₂CO₃ as base (*1078*).

Active esters of HODhbt, pentafluorophenol, and 2-mercaptopyridine were prepared from phosphonium reagents (PyDOP, PyPOP, and PyTOP) (*1079*). HODhbt has been used to form isolated active esters (*1080*), with HODhbt active esters of *N*-protected amino acids synthesized from HODhbt, Boc₂O, and DMAP (*1081*).

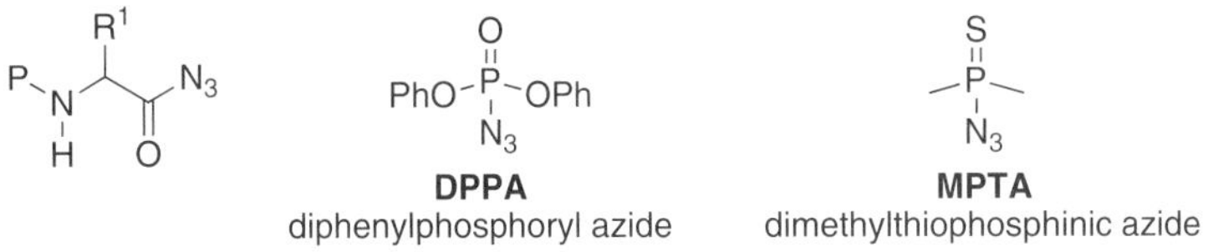

DPPA
diphenylphosphoryl azide

MPTA
dimethylthiophosphinic azide

Scheme 12.28 Amino Acid Azides.

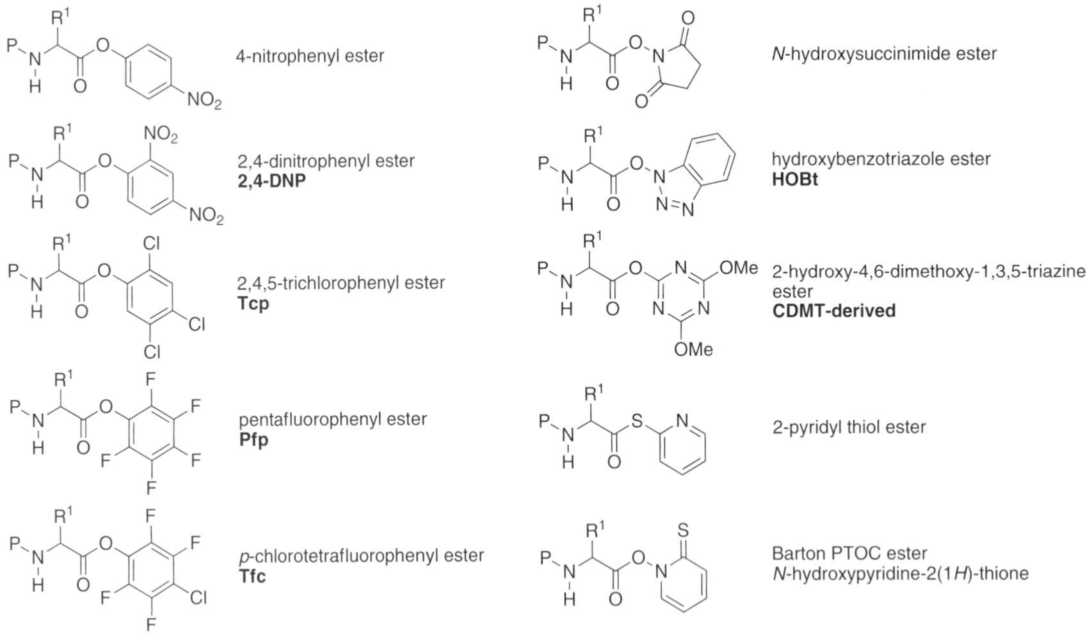

4-nitrophenyl ester

N-hydroxysuccinimide ester

2,4-dinitrophenyl ester
2,4-DNP

hydroxybenzotriazole ester
HOBt

2,4,5-trichlorophenyl ester
Tcp

2-hydroxy-4,6-dimethoxy-1,3,5-triazine ester
CDMT-derived

pentafluorophenyl ester
Pfp

2-pyridyl thiol ester

p-chlorotetrafluorophenyl ester
Tfc

Barton PTOC ester
N-hydroxypyridine-2(1*H*)-thione

Scheme 12.29 Active Esters.

Aryl esters of Boc- and Fmoc- amino acids were recently prepared in good yield by using aryl 4-nitrobenzenesulfonate derivatives in the presence of base and catalytic HOBt; 4-nitrophenyl, pentafluorophenyl, 2,4,5-trichlorophenyl, and pentachlorophenyl esters were initially prepared (*1082*), with further studies on the generation of pentafluorophenyl esters a few years later (*1083*). Pentafluorophenyl-activated esters were employed for solution-phase syntheses of di- or tripeptides of β-amino acids using a microreactor (*1084–1086*). The relative advantages of the pentachlorophenol (increased electron-withdrawing effect) and pentafluorophenol (reduced steric hindrance of fluorine atoms) esters were combined in a more reactive *p*-chlorotetrafluorophenyl ester group (Tfc) (*1087*). Tentagel resin has been linked to 4-carboxy-2,3,5,6-tetrafluorophenol and the phenol used to form an active ester of *N*-ethoxycarbonyl Phe. This activated Phe reagent coupled with unprotected amino acids in aqueous solution at pH 9.2 to give approximately 60% yields of dipeptide, along with 3–8% of hydrolyzed *N*-ethoxycarbonyl Phe (*1088*).

Esters of *N*-hydroxysuccinimide have been popular, in part due to the solubility of the reaction byproduct (*1012*). Yields for couplings of *N*-hydroxysuccinimide amino acid esters with *N*-substituted amino acids were greatly improved by carrying out the couplings at high presure (10 kbar), although for some substrates yields were still <10% (*1089*). Amino acids have been simultaneously *N*-protected with a Tfa group and activated as *N*-hydroxysuccinimide esters by reaction with *N*-trifluoroacetoxysuccinimide in dichloromethane/pyridine (*901*). Polymer-bound HOBt was employed

with DCC to form *N*-hydroxysuccinimide active esters. The polymer could be recycled (*1090*).

Esters of 2-hydroxy-4,6-dimethoxy-1,3,5-triazine, formed from 2-chloro-4,6-dimethoxy-1,3,5-triazine (CDMT) (*1091, 1092*) have been described as "superactive esters," and used to prepare an Aib-Aib-Aib sequence in 81–97% yield (*1093*). However, CDMT was found to generate an azlactone intermediate rather than an active ester in one study; optimized conditions found a one-pot, one-step procedure gave the best yields (*1094*). A potentially very useful application of CDMT involves novel derivatives formed by reaction with chiral amines, such as strychnine, brucine or sparteine. This asymmetric reagent reacts diastereoselectively with 2 equiv of racemic *N*-protected amino acids to preferentially form the activated ester of predominantly one enantiomer (see Scheme 12.1). The active ester then couples with amino esters to give dipeptides with up to 96% de. However, the diastereoselectivity varies depending on the amino acid being activated (*189*). Triazine-based condensing reagents were reviewed in 2000 (*1095*).

The Barton PTOC ester, normally used for radical reactions, is also effective at activating carboxyl groups for reaction with free α-amino acid esters (see Scheme 12.30). Aib residues were coupled to *N*-methyl amino acids in 91–96% yield. Even better results were obtained by employing the ester in combination with *N*-activation of the amine component, by using an *N*-benzenesulfenamide derivative of the amino acid. The couplings were more rapid and proceeded under essentially neutral conditions. The thiocarbonyl moiety

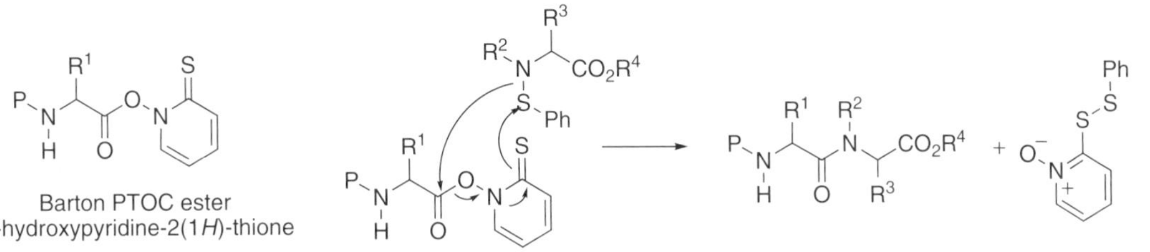

Scheme 12.30 Barton PTOC Ester with Sulfenamide *N*-activation (*1096*).

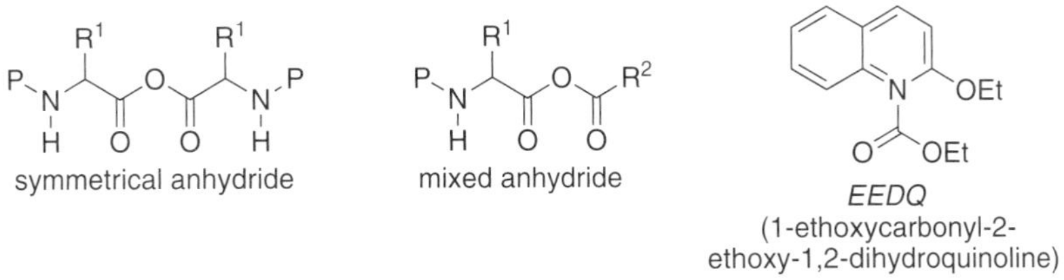

Scheme 12.31 Symmetrical and Mixed Anhydrides Used for Coupling.

of the PTOC ester coordinates to the *S*-atom of the sulfenamide, increasing the amine nucleophilicity and bringing the reactive groups into close proximity (*1096*).

A 2-pyridyl thiol ester was used for a coupling in which the amine component had an unprotected carboxyl group, directly giving the peptide acid. In order to solubilize the zwitterionic unprotected amino acid in organic solvents and increase the amine reactivity, it was protected in situ as an *N*-TMS amino acid TMS ester with 2 equiv of 1-(trimethylsilyl)imidazole (TMSIm). The amino acid thiopyridyl ester was then added. When imino acids were employed as the amine component, coupling in the presence of tertiary amines gave better yields than pretreatment with TMSIm (*1097*).

Hexafluoroacetone reacts with the α-amino and carboxylic acid groups of multifunctional amino acids such as aspartic acid or iminodiacetic acid to form 2,2-bis(trifluoromethyl)-1,3-oxazolidin-5-ones. This cyclic intermediate not only simultaneously protects both groups but also activates the α-acid group as an active ester for coupling with amino acids (*500, 1098*).

12.5.2d Mixed Anhydrides and Symmetrical Anhydrides

Symmetrical or mixed anhydrides (see Scheme 12.31) can be specifically generated as activated derivatives before coupling, but are more often employed as non-isolated active species produced during reactions induced by coupling reagents such as the carbodiimide DCC. Some mixed anhydrides can be very stable, with

crystalline isolated derivatives obtained from α-alkyl imino acids (*1099*). A review of the mixed carbonic anhydride method of peptide synthesis is included in *The Peptides* series (*1100*). Symmetrical anhydrides (*1009, 1012, 1013*) have the disadvantage that only half of the protected amino acid is coupled, which can be a significant factor if the substrate is valuable. Symmetrical anhydrides have been used for coupling the α,α-disubstituted amino acids Aib and Iva into poly-α,α-disubstituted peptides, with 77–98% yields (*1101*), and were shown to give better yields than other methods (PyAOP, PyAOP-HOAt, HATU, BOP-Cl or amino acid fluoride) for coupling Fmoc-Lys(Boc)-OH to Aib, dibenzylglycine or diisobutylglycine residues (*1102*). Symmetrical anhydrides were also found to be effective for acylating resin-bound secondary amines, using CH_2CH_2:DMF (9:1) as solvent. Other commonly used reagents (BOP, HATU or HBTU, with or without additives such as HOBt or HOAt) were inefficient (*963*).

Mixed anhydrides, first applied to amide bond formation in the 1950s (*1009, 1012, 1013*), avoid wasting half of the substrate but instead have the potential to couple the undesired half of the anhydride. For this reason, the reagents employed to form the anhydride employ electronic and steric factors to prevent the side reaction, with isobutyl chloroformate being the most common agent of choice (*1012, 1013*). Despite the age of the mixed anhydride activation method, it is still very competitive with other techniques. A 2003 process chemistry study of amidation using isobutyl chloroformate determined that reaction byproducts (unreacted *N*-protected amino acid and a urethane corresponding to that formed from the amine component and isobutyl chloroformate) stemmed from the formation of around 20% of a symmetrical anhydride during the initial activation step. By adding a solution of *N*-protected amino acid and base to a solution of isobutyl chloroformate, followed by addition of the amine component, symmetrical anhydride formation was avoided. The modified procedure also negated the requirement for multiple additions of isobutyl chloroformate to drive the reaction to completion, generated quantitative yields of crude product, and could be scaled up to over 50 kg quantities (*938*).

A different modification was described in 1999, with the mixed anhydride formed in the presence of the amine component rather than addition of the amine to the pre-formed anhydride. The carboxylate ion reacts more quickly with the isobutyl chloroformate than does the primary amino group, avoiding formation of undesired carbamate. Any symmetrical anhydride formed can acylate the amino group, with the unreacted amino acid subsequently recycled back to a mixed anhydride, allowing for complete consumption, and resulting in improved yields and greatly reduced racemization (*1103*). Mixed anhydrides prepared with isobutyl chloroformate gave dipeptides in good yield when *N*-farnesylated amino acids were coupled, although reaction times of up to 12 h were necessary for complete coupling. HBTU, BOP, and PyBrop were found to be ineffective (*477*). Isobutyl chloroformate activation has also been employed for a reverse-direction solid-phase peptide synthesis. An amino acid, tethered to the resin by an *N*-linked silyl carbamate, was activated and then reacted with an amino acid allyl ester (*1104*). An interesting difference in regioselectivity between coupling methods was observed during acylations of Lys-OMe with 1 equiv of *N*-Boc amino acids. Mixed anhydrides generated by isobutyl chloroformate selectively acylated the N^{ε}-amino group, but BOP-Cl activated acids reacted at the N^{α}-amino group (*1105*).

The coupling reagent EEDQ (1-ethoxycarbonyl-2-ethoxy-1,2-dihydroquinoline) (*1106*) activates the carboxyl group to an ethylcarbonic acid mixed anhydride (*1013*). Addition of copper(II) chloride suppressed racemization during couplings proceeding through mixed anhydrides generated from isobutyl chloroformate or EEDQ (*1107*). EEDQ was the only coupling reagent that minimized racemization (<4%) during couplings of *N*-3,5-dinitrobenzoyl Leu to aminopropyl-functionalized Si gel; DIC-HOBt, DIC-HOAt, DPPA, HATU, and PyBOP all resulted in >30% D-content of the coupled product. The silica appeared to be responsible for the high racemization rate (*1108*).

Mixed pivalic anhydrides were used for a synthesis of a fragment of cyclosporine containing three consecutive *N*-methyl amino acids (*1109*). Activation of Boc-amino acids with di-*tert*-butyl carbonate and coupling with amino methyl esters in the presence of catalytic DMAP was found to produce a number of dipeptides in 67–91% yield with high diastereomeric purity (*1110*). 3,5-Dinitrobenzoyl mixed anhydride activation of the carboxyl component has been employed in combination with in situ generation of the amine component via azide reduction of an α-azido ester. Trialkylphosphines were added as the reducing reagent, with the reaction apparently proceeding via a phosphatriazene intermediate. Little epimerization was observed with substrates known to have a high tendency to racemize (*1111*).

12.5.2e Intramolecular Anhydrides

N-Carboxy anhydrides (NCAs) (see Scheme 12.32) were discovered by Leuchs in 1906 (*1112*), and have been widely used for peptide synthesis (see *1009, 1113* and references cited therein). The use of NCAs in stepwise peptide synthesis and polypeptide synthesis was comprehensively reviewed in 2006 (*1114*). The NCAs have often been employed for polymerization reactions (*1011*), with cobalt and nickel initiators recently reported to induce oligomerization to give block copolypeptides of defined structure (*1115, 1116*). For example, diblock copolymers of varying sizes were formed from poly(L-Glu Na salt) or poly(L-Lys HBr salt) and poly(L-Leu); the products formed strong hydrogels in aqueous solutions (*1117*). A nickelacycle derived from NCAs and zerovalent Ni acted as an efficient polymerization initiator (*1118*). NCAs have been postulated to be intermediates for primordial polypeptide formation, and have been shown to activate inorganic phosphate in aqueous solution to form amino acid phosphate mixed anhydrides (*1119*). A kinetic study of the polymerization of Val NCA in aqueous solution was carried out in 2002 (*1120*).

NCAs are traditionally prepared by reaction of the amino acids with phosgene (*1121, 1122*). A new route to NCAs proceeded via nitrosation of *N*-carbamoyl (*N*-CONH$_2$) amino acids with NO/O$_2$, giving the cyclized NCAs of Gly, Ala, Val, Leu, Phe, and Met in quantitative yield (*910, 1123*). NCAs have also been prepared by a Baeyer–Villiger rearrangement of α-keto-β-lactams (*1124*) or α-hydroxy-β-lactams (see Section 8.5) (*1125–1129*). Very hindered *N*,α,α-trialkyl amino acids were synthesized by *N*-alkylation of the α,α-dialkyl α-amino acid NCA derivative, with the alkylated NCA reacted with amino acids to give dipeptides with 49–95% yields (*378*). For peptide synthesis, addition of KCN to the acylation step can be very beneficial. In one report, simply adding KCN improved coupling yields of α,α-disubstituted NCAs to amino acids from 0–10% to 87–95%, with a 79% yield for coupling to Aib (*1130*).

Unfortunately, NCAs present several difficulties, such as poor stability and the potential for multiple additions during each coupling cycle (*1131*). Urethane-protected amino acid *N*-carboxy anhydrides (UNCAs) (see Scheme 12.32) are a comparatively recent advancement, with an effective synthesis first reported in 1990 by condensing NCAs with urethene acylating reagents in aprotic solvents in the presence of *N*-methylmorpholine (*1131*). UNCAs have also been prepared from *N,N*-bis-urethane-protected amino acids by treatment with the Vilsmeier reagent, SOCl$_2$/DMF (*1044, 1045*). For peptide

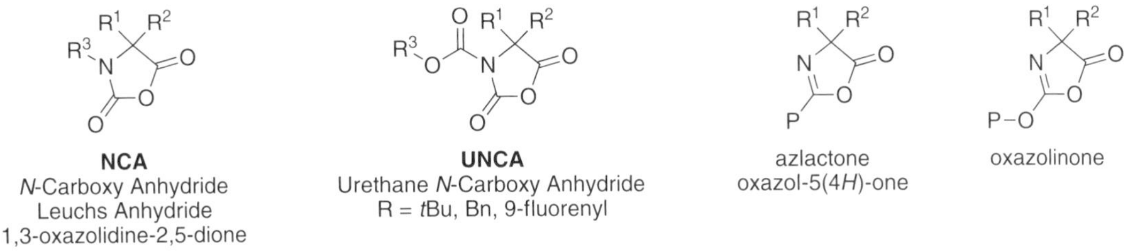

NCA
N-Carboxy Anhydride
Leuchs Anhydride
1,3-oxazolidine-2,5-dione

UNCA
Urethane *N*-Carboxy Anhydride
R = *t*Bu, Bn, 9-fluorenyl

azlactone
oxazol-5(4*H*)-one

oxazolinone

Scheme 12.32 Intramolecular Anhydrides Used for Coupling.

synthesis, UNCAs react rapidly and liberate CO_2 as the only coproduct. Coupling is favored in DMF as solvent, proceeds at a comparable rate to BOP or HBTU couplings (with similar levels of racemization), and can be carried out without the presence of tertiary amines (although yields are reduced) (*1132*). UNCAs can dimerize in the presence of some bases, and so the use of NMM is recommended for peptide synthesis (*1133*). UNCAs have been used to study the intrinsic racemization tendencies of amino acid residues during peptide synthesis, as the UNCA allows epimerization only by proton abstraction and enolization at the α-carbon (*1000*). A cinchona alkaloid-catalyzed resolution of UNCAs was reported in 2001 (*250*).

A comparative study of the coupling of Boc-protected amino acids with hindered *N*-terminal residues such as *N*-Me-Aib found that UNCA-activated derivatives provided nearly complete reaction at 50 °C. In contrast, EDC/HOBt, preactivated Pfp esters, pivaloyl mixed anhydrides or acid fluorides gave <10% yields, while BOP, HBTU, and PyBroP gave 0–63% yields (*1062*). A similar comparison for syntheses of peptides containing MeAla, MeAib, or MeαAc⁵c in 1992 also found UNCAs gave the best results, as long as they were used for prolonged reaction times or at higher temperatures. Pivaloyl mixed anhydrides, Pfp esters or acid fluorides were ineffective, while HBTU or PyBroP gave reasonable yields (*1019*). However, in another study acid fluorides were clearly superior for the synthesis of Aib-rich alamethicin acid (*1053*). None of these comparisons included HOAt-based reagents. Tripeptides have been prepared in a rapid one-pot solution reaction by sequentially adding *N*-Cbz UNCAs for coupling, followed by hydrogenation for Cbz-protecting group removal (*1135*).

UNCAs are generally prepared with *N*-Boc, *N*-Fmoc, or *N*-Cbz protection. Analogs of UNCAs with alternative *N*-protecting groups were evaluated for their use in dipeptide synthesis, and their ability to prevent epimerization. Coupling of *N*-trityl or *N*-phenylfluorenyl *N*-carboxyanhydrides with amino esters in the presence of KCN or NaN₃ catalysts resulted in extensive racemization. However, by heating the reactants in THF without any additive, enantiomerically pure dipeptides were formed in 72–94% yield (*1136*). UNCAs of

β-amino acids have also been reported and used to form amides, though peptide formation was not reported (*1821*).

Another type of intramolecularly activated derivative comprises oxazolones (azlactones), which can be produced from *N*-urethane-protected amino acids under a variety of activating conditions (*1137*). Amino acid oxazolones are normally chemically unstable, although crystal structures of the derivatives of Fmoc-Toac and Cbz-Toac have been determined (*1137*). The oxazolones of normal α-monosubstituted amino acids are also not chirally stable in the presence of base (*1003*). Azlactones have been employed for coupling amino acids for many years (*1009*) and found to be especially useful for activating sterically hindered α,α-dialkylated amino acids, which cannot racemize (*1138–1140*).

More recently, di- and tripeptides containing multiple α,α-dialkylated residues (including Aib and residues with two nucleobase side chains) were successfully coupled by using an oxazolone derivative as the activated carboxyl component. The oxazolone was prepared by dehydration of the Fmoc-amino acid using CIP (2-chloro-1,3-dimethylimidazolidinium hexafluorophosphate) and DIEA (53–77%). Coupling in 1,1,2,2-tetrachloroethane required heating at 100 °C for 10 h, but gave the products in 73–93% yield (*1141*). A hexapeptide of diethylglycine was prepared from oxazolone derivatives in 1999, with EDC in refluxing MeCN employed to form the oxazolone and force the coupling (*1142*). Poly (α-methyl-Phe) homopeptides were synthesized in a similar manner (*1143*), while α-ethyl-Phe was coupled to other α,α-disubstituted residues via its oxazolone in 1995 (*1144*). Peptides containing a *C*-terminal Aib residue were converted into peptide oxazolones by heating in acetic anhydride, followed by reaction with the *N*-terminus of other peptides (*1145*). Other α,α-dialkylglycine peptide syntheses via oxazolones have been reported (*1146–1149*). An α,α-disubstituted amino acid synthesis employed oxazolones, with α-alkylation of the azlactones resulting from DCC dehydration of *N*-Bz protected amino acids. The products were then directly coupled to form dipeptides (*1150*). A review of the preparation, use, and chiral stability of 2-alkoxy-5-(4*H*)-oxazolone was published in 1996 (*1003*).

Carbodiimides

Additives

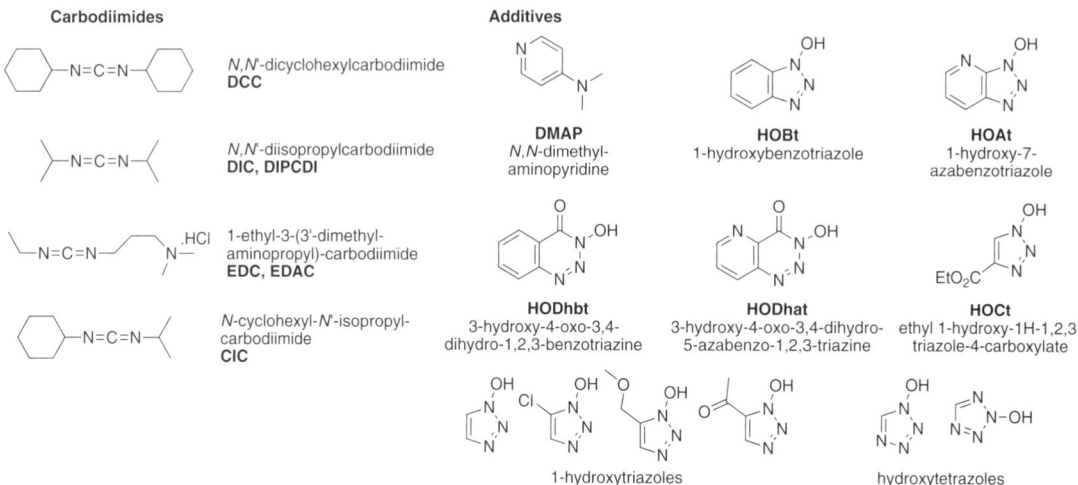

Scheme 12.33 Carbodiimides and Additives for Peptide Synthesis.

12.5.2f Carbodiimides and Additives (see Scheme 12.33)

The carbodiimide method is reviewed in *The Peptide* series (*1151*). The use of dicyclohexylcarbodiimide (DCC) for coupling amino acids was first proposed by Sheehan and Hess in 1955 (*1152*). Disadvantages of DCC for SPPS include formation of a sparingly soluble dicyclohexylurea (DCU) reaction product, which is difficult to separate from the resin (although this is potentially an advantage for solution-phase couplings), and the fact that DCC is a strong allergen to which many people have severe reactions. Other carbodiimides have been prepared (*1009*), with diisopropylcarbodiimide (DIC, DICPDI) (*1153*) and the water-soluble 1-ethyl-3-(3′-dimethylaminopropyl)-carbodiimide (EDC or EDAC) finding increasing favor in recent years, due to reduced toxicity of the reagents, greater ease of handling, and increased solubility of the urea products. EDAC has been used in conjunction with a water-solubilizing N-protecting group, the 2-[phenyl(methyl)sulfonio] ethoxycarbonyl (Pms) group, for aqueous solid-phase synthesis (*838*), and was also used for a solution-phase dipeptide synthesis in a microreactor (*1085*). A hybrid of DCC and DIC, N-cyclohexyl-N′-isopropylcarbodiimide (CIC), was described in 1994. It gave couplings with comparable or better efficiency than DCC without insoluble byproducts (*1154*). A polymer-supported carbodiimide based on EDAC was developed in 1993 (*1155*) and used in a number of amidation syntheses (*942, 1156*). A CH₂Cl₂-soluble oligomeric alkyl cyclohexyl carbodiimide has been reported, which can be precipitated from the reaction via addition of Et₂O or EtOAc (*1157*).

Carbodiimides were initially employed by themselves, generally resulting in the formation of symmetrical anhydrides (as non-isolated intermediates) during couplings (*1013*). They are capable of giving good results: in

1992, DCC was used to prepare a dipeptide containing sterically hindered 4-amino-4-carboxy-2,2,6,6-tetramethylpiperidine oxide in 65% yield. Coupling attempts with active esters, acid chlorides or mixed anhydrides were not successful (*1158*). Symmetrical anhydrides formed by pretreatment of Fmoc-amino acids with DIPCDI for 5 min gave good yields during acylations of secondary amine resins in 1998 (*963*).

In general, however, more effective couplings are achieved by combining carbodiimides with additives that generate active esters in situ. N,N-Dimethylaminopyridine (DMAP) catalyzes both ester and amide formation, but can lead to extensive racemization (e.g., see *1159, 1160*). The most effective additive for peptide synthesis for many years has been 1-hydroxybenzotriazole (HOBt) (see Scheme 12.33), which was introduced by König and Geiger in 1970 along with 20 other related additives (*1161*). The combination of DCC/HOBt suppresses racemization during couplings in non-polar solvents such as dichloromethane (in which racemization is already low) and prevents the carbodiimide side reaction leading to N-acylureas (*1161*). Epimerization can still occur during couplings in polar solvents such as DMF, but is significantly reduced compared to reaction with no additive (*1162*). However, it has been reported that HOBt can react with dichloromethane at 20 °C in the presence of triethylamine to give O-chloromethyl and O,O′-methylbridged dimers in significant quantities after extended times (17% monomer and 28% dimer after 48 h) (*1163*). Racemization during either carbodiimide or carbodiimide/HOBt coupling has been suppressed by addition of anhydrous Cu(II) salts (*1164, 1165*). A polymer-bound HOBt reagent was developed for amide bond formation in combination with DCC (*1166*), while a polymer-bound carbodiimide was employed with HOBt for microwave-catalyzed amidations (*1156*), or for one-pot amidation reactions in combination with a polymer-supported carbonate base to remove HOBt after the reaction (*1167*), or with HOBt for one-pot amidation reactions which included an ester hydrolysis (*942*).

HOBt has often been added to other coupling reagents, such as PyBOP or HBTU, in order to suppress racemization. Unfortunately, in 2005, HOBt monohydrate was reclassified by the UN as a desensitized explosive, resulting in severe restrictions on its transportation and the end of its supply by most manufacturers (*1168*). However, Novabiochem carried out a study comparing coupling using PyBOP with and without added HOBt, and found no difference in coupling efficiency, and minimal difference for a racemization-prone sequence (*1168*). This confirmed the results of Carpino and Albericio, who determined that added HOXt has no effect on coupling efficiencies of HXTU-mediated couplings (*1169*), but was in contrast to the original results of Hudson that suggested adding HOBt to BOP helped to increase reaction speed (*1170*). Ethyl-2-cyano-2-(hydroxyimino)acetate has been developed as a non-explosive replacement for HOBt and HOAt (*1832*), sold by Novabiochem as Oxyma Pure (*1833*).

DIC/HOBt was used in a study comparing high-speed peptide synthesis at elevated temperatures (60 °C for 30 min in DMSO/toluene) with conventional synthesis (room temperature, 90 min, DMF). The higher temperature did not result in any greater racemization, though more side products might be formed (*1171*). A wide range of coupling reagents and conditions were examined for the solid-phase segment condensation of a protected Asp-Phe dipeptide with a growing peptide chain (DCC/Dhbt, DCC/Pfp, BOP/HOBt, HATU, Dpp-Cl/NMM, cyanuric fluoride/pyridine, DIC/HOAt, DIC/HOBt). While a number of these methods gave high yields (>95%), they were often accompanied by significant racemization of the Phe residue, with up to 40% of the undesired epimer. The optimal combination of a good coupling rate (97%) with minimal racemization (0.5%) was achieved with DIC and HOBt in dichloromethane, although Dpp-Cl-mediated reactions also gave good results. The HATU and BOP conditions caused significant racemization (*1172*). In another report EDC/HOBt was used to prepare an isovaline (α-methyl-Abu) homopeptide, although with low (13–63%) coupling yields (*1173*). Similarly, standard DCC/HOBt coupling conditions were found to be ineffective for coupling α-methyl amino acids to a growing peptide chain. However, by silylating the peptide N-terminal amino function with TMSCl before coupling, the same system gave 79–89% yields (*890*). EDC and HOBt were found to be quite effective for peptide couplings in aqueous media (*1174*), and were used for segment condensations of 10-mer peptide segments using a Boc(Bn)-OPac (Pac = phenacyl) protection strategy (*1175*).

1-Hydroxy-7-azabenzotriazole (HOAt), introduced in 1993, was found to be an even more effective additive than HOBt (*1176*). EDC/HOAt gave much better yields than EDC/HOBt for coupling a hindered secondary amine and an Aib-Aib dipeptide, and caused less racemization (*1176*). Similar improvements were observed by replacing HOBt with HOAt during a synthesis of mycobactin analogs using a minimal protection strategy (*1177*). DIC/HOAt gave greatly superior yields (>99%) for the coupling of Fmoc-Val or Fmoc-MeLeu to MeLeu-Ala and Fmoc-MeLeu to MeVal-Ala, compared to yields of 4–88% obtained with PyBroP, BOP, HBTU, BOP or Fmoc-Val-F (*1062*). The use of DIC/HOAt was examined in a 1999 report. It was found that the hindered base collidine enhanced the activation step, with subsequent addition of DIEA improving coupling (*1178*). HOAt was more effective than HOBt or HOBt derivatives at preventing racemization in both DMF and dichloromethane solvents (*1178*), confirming an earlier report (*1001*). The aza regioisomers of HOAt, 4-,5-, and 6-HOAt, were compared for model peptide coupling. The original 7-HOAt ester was 3-times more reactive than the 6-isomer, and more than 10-fold more reactive than the 4- and 5-isomers. It also resulted in the least epimerization (*1179*).

3-Hydroxy-4-oxo-3,4-dihydrobenzo-1,2,3-triazine (HODhbt) is another effective additive (*1002, 1180, 1181*), and both HOBt and HODhbt can be added as catalysts for the aminolysis of other active esters (*1077*). HODhbt active esters release a yellow indicator upon coupling, which is useful for monitoring completion of amide formation (*1034, 1182, 1183*). HODhbt active esters of N-protected amino acids have been synthesized from HODhbt, Boc_2O, and DMAP (*1081*). A 1995 comparison of EDC/HODhbt, EDC/HOAt or EDC/HOBt for coupling Cbz-Phg to Pro-NH_2 found equivalent yields for all three reagents, but racemization for coupling in DMF varied from 7% for HOAt to 25% for HODhbt. With trifluoroethanol/trichloromethane, racemization was reduced to 1% or less (*1002*). A closely related analog, 3-hydroxy-4-oxo-3,4-dihydro-5-azabenzo-1,2,3-triazene (HODhat), has also been described (*1184*). In 1998, six derivatives of 1-hydroxy-1,2,3-triazole or N-hydroxytetrazole were compared against HOBt, HOAt, and HODhbt for coupling efficiency and racemization suppression when used in combination with DIC to couple Fmoc-Ile-OH to Val-Gly-PEGA resin, Z-Phe-Val to Pro-NH-resin, or Fmoc-Aib to Aib-peptide-resin. 5-Chloro-1-hydroxytriazole was very efficient at acylation (better than HOAt for Aib coupling), but was not effective at suppressing racemization. 2-Hydroxytetrazole was a more efficient catalyst than HOAt, and suppressed racemization as well as HOBt, but was explosive in a hammer test (*1185*).

A new additive, HOCt (ethyl 1-hydroxy-1H-1,2,3-triazole-4-carboxylate), was developed to avoid absorption near 300 nm, where other additives such as HOBt absorb strongly. This is the wavelength at which the Fmoc group absorbs; thus, if the coupling reagents are transparent at this frequency, real-time monitoring of each coupling cycle during Fmoc-based continuous flow SPPS is possible by following the disappearance of the Fmoc-amino acid from the coupling solution. Coupling reactions with DIC/HOCt proceeded with little racemization, and gave overall yields nearly 30% higher than coupling with DIC/HOBt for a difficult peptide sequence containing four Aib residues (*1186*). A systematic study on racemization during activation found no evidence for racemization of any Fmoc-amino acid except for Fmoc-His(Trt), which could be minimized by using 3 equiv of HOCt (*1187*).

12.5.2g CDI

Carbonyldiimidazole (CDI), first reported in 1957 (*1188*), forms acylimidazoles as reactive intermediates that are more stable than acid chlorides. However, CDI never achieved the popularity of the carbodiimides due to its expense and moisture sensitivity (*1012*). A 2004 report found significant rate enhancements in amidation reactions employing CDI if carbon dioxide was bubbled through the mixture (*944*).

12.5.2h Phosphonium and Uronium Salts (see Schemes 12.34–12.37)

The concept of activating the carboxyl group with phosphorous derivatives has been investigated for many years (*1012*), but only relatively recently have they become one of the methods of choice for peptide synthesis. The phosphonium and uronium coupling reagents can be used by themselves but are more often employed in conjunction with one of the additives described above (e.g., HOBt, HOAt), usually the one on which the coupling reagent is based. However, as mentioned in Section 12.5.2f, despite earlier reports finding that the addition of HOBt improved coupling reactions with reagents such as BOP (*1170*), Novabiochem (*1168*) and others (*1169*) reported that the addition of HOBt had no effect on the coupling efficiencies of HOBt-based uronium or phosphonium coupling reagents. The first phosphonium coupling reagent, (benzotriazol-1-yloxy) tris(dimethylamino)phosphonium hexafluorophosphate (BOP, not to be confused with BOP-Cl), was introduced by Castro in 1975 (see Scheme 12.34) (*1189–1191*). However, it did not become widely used until after 1987–1988, when several studies demonstrated its effectiveness for SPPS (*1192, 1193*) and compared its advantage over conventional methods (DCC, DCC/HOBt, symmetrical anhydride) for difficult couplings (*1194*), including side-chain-to-side-chain cyclizations (*1195*). BOP was found to suppress diketopiperazine formation during introduction of the third residue to dipeptides (*1196*). BOP alone was quite ineffective at the difficult coupling of Fmoc-Aib to Pro-OtBu, but when used in combination with DMAP, the reaction was complete in 30 min (with racemization of the Aib residue obviously not a concern) (*1061*). Another study demonstrated that BOP caused considerably less racemization of Boc-Ser(OBn) during coupling than PyBroP, TFFH, or HATU, and was also slightly better than HBTU (*1000*).

A number of BOP analogs have been developed (see Scheme 12.34). A more effective reagent was obtained by replacing the tris(dimethylamino)phosphonium group of BOP with a tris(pyrrolidino)phosphonium group, giving the PyBOP (not to be confused with PyBroP!) series of reagents (*1197*). PyBOP gives results at least as good as BOP and avoids the formation of carcinogenic HMPA as a byproduct (*1197, 1198*). It has been used in the synthesis of peptoids (polyN-alkyl glycine) (*342, 616*), to couple a very hindered N-neopentyl residue (*567*), and for coupling Aib residues (*1198*). Solution couplings of Aib using PyBOP/HOBt were enhanced by microwave irradiation (*1199*). PyBOP has also been employed for diethylamide formation (*1200*) and for forming thioamide isostere bonds from amino monothioacids (Boc-AA-SH) and amino acids (*1201*). Several other N-hydroxy and hydroxy derivatives of tris(pyrrolidino)phosphonium salts have been reported, derived from HODhbt (PyDOP), pentafluorophenol (PyPOP) and 2-mercaptopyridine (PyTOP) (*1079*).

An alternative to carbodiimide or phosphonium salt activation of the acid group is to use uronium salts in combination with HOBt (see Scheme 12.34). The first of these, O-(benzotriazol-1-yl)-1,1,3,3-tetramethyluronium hexafluorophosphate (HBTU), was reported in 1978 (*1202*) when it was compared to azides, DCC, active esters, mixed anhydrides, or BOP for couplings. HBTU gave better yields with less racemization in most cases. Further synthetic data was reported in 1984 (*1203*), but HBTU did not become widely used until a third report in 1989, when a range of uronium derivatives were described (*1204*). HBTU and its tetrafluoroborate salt analog, TBTU, gave identical results, and were recommended for SPPS synthesis (*1204*). HBTU has been shown to give much faster couplings than DCC/HOBt (*1205*), and has been used to couple a hindered N-neopentyl residue (*567*). A 1997 study demonstrated that HBTU caused considerably less racemization during couplings of Boc-Ser(OBn) than PyBroP, TFFH, or HATU, although it was less effective than BOP (*1000*). A comparative study of coupling to MeAla, MeAib, and MeαAc⁵c found that HBTU was superior to pivaloyl mixed anhydrides, Pfp esters, or acyl fluorides (*1019*). HBTU/HOBt has also been employed in combination with microwave irradiation to greatly improve yields of β-peptides for longer sequences, when compared to conventional heating (*1206*). Solution couplings of Aib using HBTU/HOBt were also enhanced by microwave irradiation (*1199*).

X-ray analysis of HBTU indicates that it crystallizes as the N-substituted guanidinium isomer, rather than the O-alkylated structure that is traditionally depicted (*1002*). HAPyU (see below) also crystallizes in the guanidinium form, with ¹⁵N NMR spectra indicating the same structure is maintained in solution (*1207*). The structures of most other compounds have not been definitively assigned, and thus have been shown in their traditional representations in Schemes 12.34–12.37. However, a 2002 report disclosed that the uronium O-isomers could in fact be isolated under standard synthesis conditions by simply excluding tertiary amines, with isomerization to the guanidinium form induced by the addition of these organic bases. The O-series possessed significantly enhanced reactivity and reduced

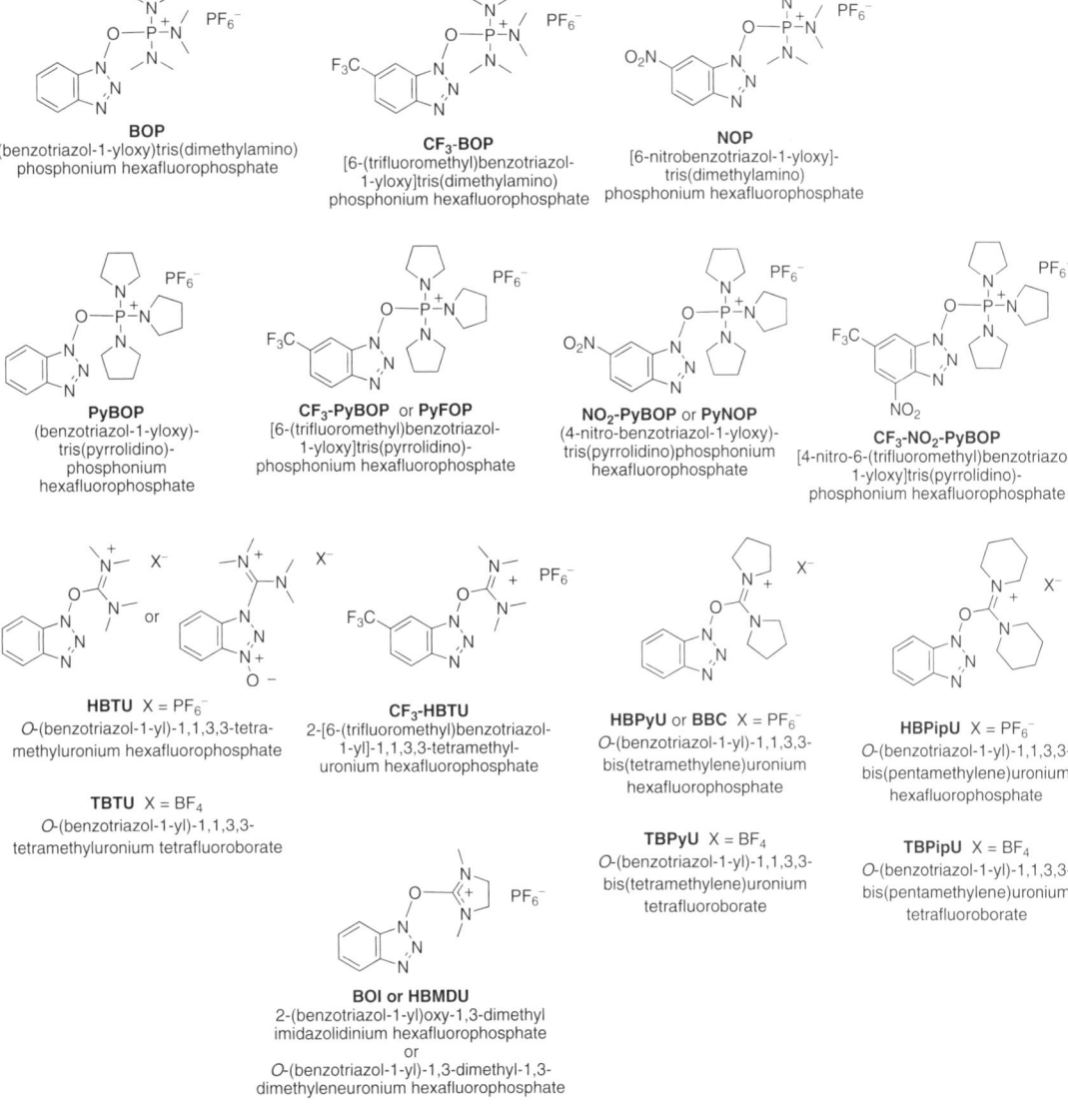

Scheme 12.34 HOBt-Derived Uronium and Phosphonium Coupling Reagents for Peptide Synthesis.

epimerization during couplings, further complicating comparisons of different reagents whose exact structures are unknown (*1208*).

The tetrafluoroborate salt TBTU/HOBt was tested for Boc-based SPPS, and found to give better results than DCC/HOBt at every step of a 22-mer peptide (99.53–99.99% yield at each step, vs 99.36–99.99%) (*1209*). TBTU was employed for a T-Bag synthesis of a series of 10- to 18-mer peptides (*1210*), and TBTU/HOBt for coupling sterically hindered α,α-dialkylamino acids (*1211*). TBTU was found to give better results than CDI, DCC, DIC or HBTU when applied to the coupling of an *N*-alkylated 5-aminopentanoic acid (67% yield), with a similar result provided by the more expensive

PyBroP reagent (*1212*). TBTU was used in a study comparing high-speed peptide synthesis at elevated temperatures (60 °C for 30 min in DMSO/toluene) with conventional synthesis (room temperature, 90 min, DMF). The higher temperature did not result in any greater racemization, though more side products might be formed (*1171*).

Pyrrolidine (HBPyU or BBC) and piperidine (HBPipU) uronium analogs of HOBt-based reagents have also been prepared (*1213, 1214*), with the pyrrolidine salt the more active of the two (*1214*). HBPyU was superior to HBTU, BOP, PyBOP, and DCC for standard couplings (*1213*), but not very effective for coupling residues to *N*-methylated residues (*671*), as has been noted for other

HOBt-containing reagents (671). Other HOBt-based reagents include a dimethylimidazolidinium BOP analog, 2-(benzotriazol-1-yl)oxy-1,3-dimethylimidazo-lidinium hexafluorophosphate (BOI), which was described in 1992 and gave similar coupling results to BOP (1215). Further BOP, PyBOP, and HBTU analogs were derived by substitution on the benzotriazole moiety with nitro (PyNOP) (1079) or trifluoromethyl (CF₃-PyBOP (1217), PyFOP or CF₃-PyBOP (1079), CF₃-HBTU (1217)) groups. The CF₃-derivatives gave better yields than BOP for coupling Aib residues (1217), while both were useful for the synthesis of thioamides from thiocarbonyls and amino acids (1079). A disubstituted CF₃-NO₂-PyBOP analog, with increased electron withdrawal, was much more effective at coupling N-methyl amino acids than PyBroP or HATU (1218).

Another set of coupling reagents replace the HOBt moiety with HOAt (see Scheme 12.35). The phosphonium- and uronium-based derivatives (HATU, TATU, HAPyU, HAPipU, TAPipU, HAMDU, HAMTU, AOP, and PyAOP) were first described in 1993. They showed

enhanced reactivity, with decreased racemization, when compared to the HOBt-derived analogs (1001, 1160, 1176, 1219). HATU is often used without added HOAt in the coupling mixture. A brief summary of HATU/HOAt was published in 2001 (1220). HATU/HOAt has been used to prepare cyclic hexamers, heptamers, and octamers (1221), and was shown to be superior to HBTU in the synthesis of Aib-Aib-containing peptides (1219). HATU or DIC/HOAt gave greatly superior yields (>99%) for the coupling of Fmoc-Val or Fmoc-MeLeu to MeLeu-Ala and Fmoc-MeLeu to MeVal-Ala, compared to yields of 4–88% obtained with PyBroP, BOP, HBTU, BOP or Fmoc-Val-F (1062). HATU was used as the coupling agent in a 1999 description of an accelerated chemical synthesis of small proteins, in which Boc-protected residues were coupled at the rate of 10–15 residues per hour. For SPPS, suitably activated amino acids could acylate a growing peptide chain to >99% completion within 10–100 s on cross-linked divinylbenzene polystyrene resin. DMSO was used as the coupling solvent to improve efficiency and reliability (1222).

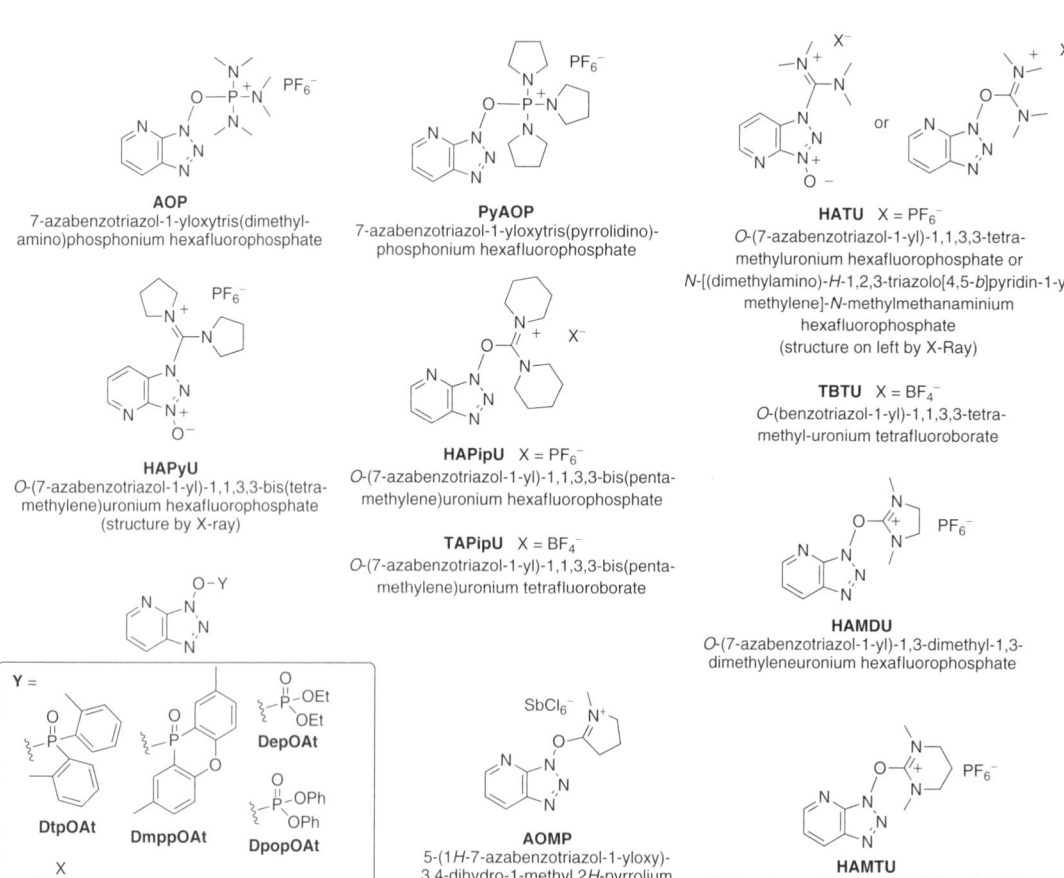

AOP
7-azabenzotriazol-1-yloxytris(dimethyl-amino)phosphonium hexafluorophosphate

PyAOP
7-azabenzotriazol-1-yloxytris(pyrrolidino)-phosphonium hexafluorophosphate

HATU X = PF₆⁻
O-(7-azabenzotriazol-1-yl)-1,1,3,3-tetra-methyluronium hexafluorophosphate or
N-[(dimethylamino)-H-1,2,3-triazolo[4,5-b]pyridin-1-yl-methylene]-N-methylmethanaminium hexafluorophosphate
(structure on left by X-Ray)

TBTU X = BF₄⁻
O-(benzotriazol-1-yl)-1,1,3,3-tetra-methyl-uronium tetrafluoroborate

HAPyU
O-(7-azabenzotriazol-1-yl)-1,1,3,3-bis(tetra-methylene)uronium hexafluorophosphate
(structure by X-ray)

HAPipU X = PF₆⁻
O-(7-azabenzotriazol-1-yl)-1,1,3,3-bis(penta-methylene)uronium hexafluorophosphate

TAPipU X = BF₄⁻
O-(7-azabenzotriazol-1-yl)-1,1,3,3-bis(penta-methylene)uronium tetrafluoroborate

HAMDU
O-(7-azabenzotriazol-1-yl)-1,3-dimethyl-1,3-dimethyleneuronium hexafluorophosphate

Y =

DtpOAt   **DmppOAt**

DepOAt

DpopOAt

4-NBs-4OAt: X = H, Y = NO₂
2-NBs-4-OAt: X = NO₂, Y = H
2,4-DNBs-4-OAt: X = Y = NO₂

AOMP
5-(1H-7-azabenzotriazol-1-yloxy)-3,4-dihydro-1-methyl 2H-pyrrolium hexachloroantimonate

HAMTU
O-(7-azabenzotriazol-1-yl)-1,3-dimethyl-1,3-trimethyleneuronium hexafluorophosphate

Scheme 12.35 HOAt-Derived Uronium and Phosphonium Coupling Reagents for Peptide Synthesis.

Extensive racemization of Fmoc-Ser(*t*Bu)-OH was recently observed during a standard synthesis employing HATU and *N*-methylmorpholine (NMM) as base. A more thorough investigation found that both HATU and HBTU (with or without HOBt/HOAt additives) caused 6–22% D-Ser formation when NMM was used as base. Simply by switching to the hindered base TMP, racemization was reduced to only 0.3–2%. The lowest level of racemization with this base was provided by HATU/HOAt, and the highest by HBTU alone, a reversal of the results with NMM as base! (*1223*). HATU, with stepwise addition of TMP (2,4,6-trimethylpyridine) as base, formed <1% of the D-enantiomer when coupling *N*-acyl Orn derivatives, compared to 4% with a single addition of base, 17% with DIEA as base, and 22% with HBTU/DIEA, increasing to 48% D-enantiomer if the base was added during preactivation (*1004*). Addition of CuCl$_2$ to HATU/HOAt was found to suppress racemization during segment couplings, particularly for protocols involving preactivation (e.g., for coupling Boc-Phe-Val to Val-OBn, reduced from 6.9% epimerization to 0.2%) (*1824*). HATU has been used in room temperature ionic liquid ([bmim]PF$_6$) in a preliminary report, giving good yields of crude peptides with higher purities than provided by conventional solvents (*1224*).

The effectiveness of azabenzotriazole-derived coupling reagents (HAPyU, PyAOP, HATU) were compared against more conventional reagents (DCC/DMAP, BOP, PyBOP, TBTU, DPPA) for the cyclization of peptides, a difficult coupling situation capable of revealing substantial differences between coupling reagents. The HOAt-derived reagents (especially HAPyU) were found to be more effective, with greater yields and less epimerization (<10% compared to 100% epimerization with DCC/DMAP) (*1159, 1160*). The pyrrolidine salt HAPyU was more active than the piperidine analog TAPipU (*1160*). The combination of HAPyU/HOAt with TMP (2,4,6-trimethylpyridine, collidine) as base gave the least racemization of a number of reagent systems tested (EDC/HOBt, EDC/HOAt, EDC/HODhbt, HATU, HDTU, HBTU, BOP, HAPyU) for several couplings (*1002*). PyAOP was found to be better than PyBOP for coupling Aib-Aib sequences and for cyclizations, and was used for both linear synthesis and cyclization of a cyclosporine analog containing seven *N*-methyl amino acids (*345*). The system of TATU/HOAt coupled amino acids to α,α-disubstituted residues with 74–81% yields (*1150*).

The 4-,5-, and 6-aza regioisomers of HATU and HAPyU reagents (the 7-aza isomer is the standard reagent) were compared for model peptide coupling. The 7-isomer was the most reactive for both HAPyU and HATU, and tended to cause the least epimerization (*1179*). Other HOAt reagents have been prepared with 2- or 4-nitro- or 2,4-dinitrophenylsulfonyl esters, with phosphinyl esters, or with phosphate esters. The derivatives were better at minimizing racemization during segment couplings than their uronium/guanidinium counterparts (*1225*).

The HODhbt-derived uronium reagents HDTU (*1002*) and TDBTU (*1204*) (see Scheme 12.36) have also been tested, and found to give similar yields but slightly less racemization than HATU (*1002*). Tetramethyluronium derivatives of other *N*-hydroxy compounds include 2-succinimido (TSTU), 2-2(oxo-1(2H)-pyridyl (TPTU) and 2-(5-norbornene-2,3-dicarboximido) (TNTU) compounds (see Scheme 12.36) (*1204*). The TBTU/HBTU reagent combination was recommended as the best for SPPS, TPTU for segment condensation, and TSTU/TNTU for in situ formation of -OSu and ONB active esters; TDBTU gave the least racemization during test couplings but side reactions were possible (*1204*). A set of reagents derived from 3-hydroxy-4-oxo-3,4-dihydro-5-azabenzo-1,2,3-triazene (HODhat) has also been described, including the uronium HDATU, the pyrrolidine HDAPyU, and the phosphonium PyDAOP. HDATU was more effective than HDTU for synthesis of a model decapeptide (*1184*).

Another uronium-based reagent, PfPyU, was derived from pentafluorophenol instead of a hydroxybenzotriazole. It showed much greater reactivity than TBTU for coupling Fmoc-Ala to Val-O*t*Bu, and similar reactivity to HATU. The reagent was used to prepare a glycopeptide (*1226*). However, another study found that the coupling effectiveness of PfPyU was significantly lower than that of HOAt-derived reagents, and extensive racemization was noted (*1227*). Addition of HOAt to reactions with PfPyU as activating agent greatly improved the coupling results, both with regard to reaction rate and extent of epimerization (*1227*). Active esters of pentafluorophenol or 2-mercaptopyridine were also prepared from the corresponding phosphonium reagents, PyPOP or PyTOP (*1079*).

Uronium reagents derived from 2-mercaptopyridine-1-oxide were synthesized in an attempt to develop less expensive coupling reagents; the hexafluorophosphate salt HOTT and tetrafluoroborate salt TOTT were both tested. TOTT tended to give higher yields that HOTT for couplings of Aib, with similar results as HATU. TOTT also caused less racemization than BOP, PyBOP, HBTU or HATU in coupling Bz-Leu-OH to Gly-OEt (*1228*). In 2001 Albericio et al. provided further examples of coupling with HOTT and TOTT, along with the corresponding 1,3-dimethyl-1,3-trimethylenethiouronium salts HODT and TODT. The 2-mercaptopyridine-1-oxide could also be employed as an additive with other coupling reagents, such as DCC or TBTU, to reduce racemization (*1229*).

Another recently reported family of reagents are HOBt- and HOAt-derived non-symmetrical immonium hexachloroantimonate salts, BOMI, BDMP, BPMP, SOMP, and AOMP (see Scheme 12.36), based on the design principle that replacing one of the two substituted amino groups of the uronium salts would reduce resonance and improve reactivity. Racemization studies during peptide synthesis indicated significantly reduced racemization compared to HBPyU or HAPyU, and much faster couplings (*1230, 1231*). Couplings with BOMI (benzotriazol-1-yloxy-*N*,*N*-dimethylmethaniminium hexachloroantimonate) proceeded with similar effectiveness as other HOBt-based reagents, but racemization

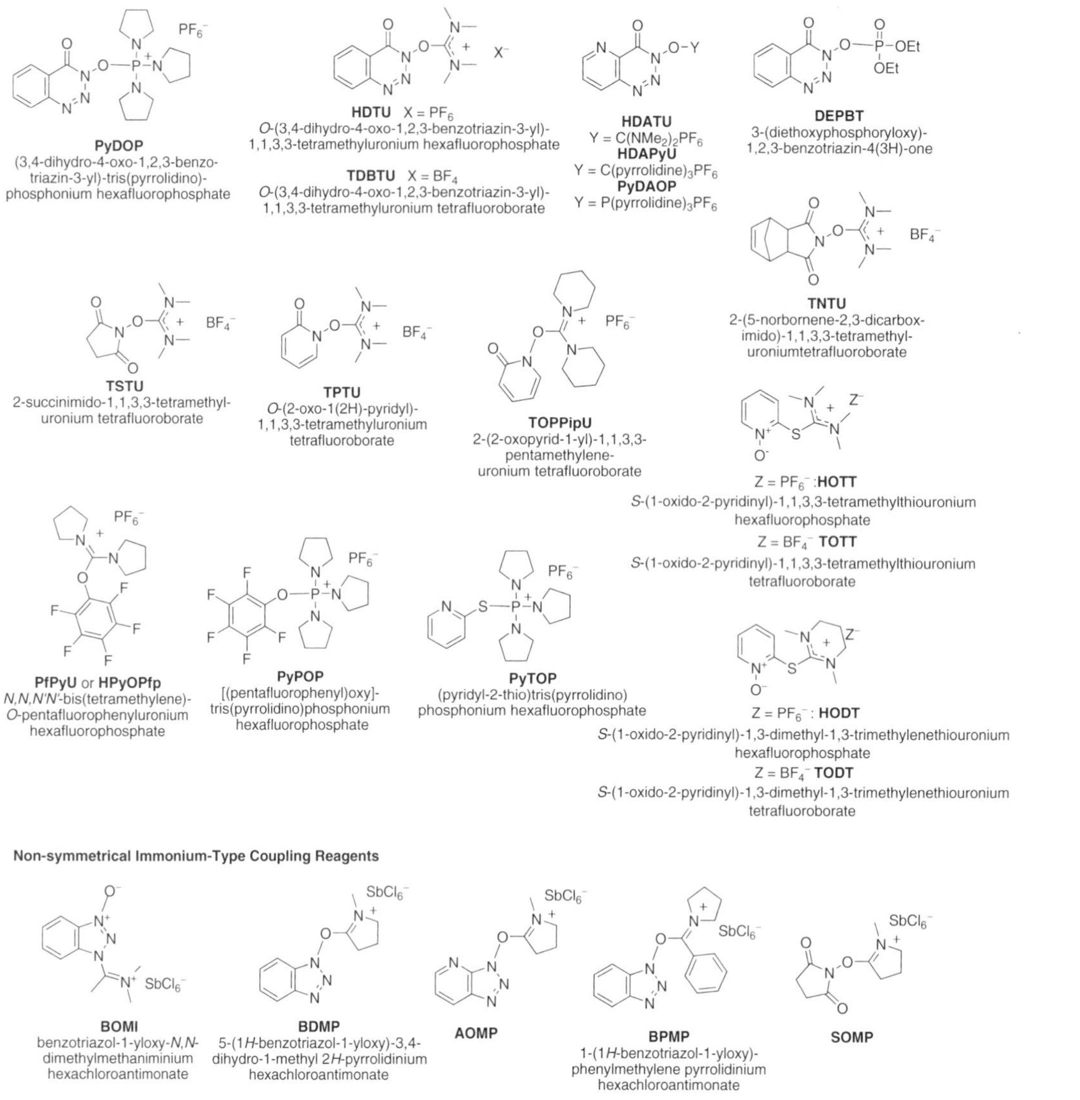

PyDOP
(3,4-dihydro-4-oxo-1,2,3-benzo-
triazin-3-yl)-tris(pyrrolidino)-
phosphonium hexafluorophosphate

HDTU   X = PF₆
O-(3,4-dihydro-4-oxo-1,2,3-benzotriazin-3-yl)-
1,1,3,3-tetramethyluronium hexafluorophosphate

TDBTU   X = BF₄
O-(3,4-dihydro-4-oxo-1,2,3-benzotriazin-3-yl)-
1,1,3,3-tetramethyluronium tetrafluoroborate

HDATU
Y = C(NMe₂)₂PF₆
HDAPyU
Y = C(pyrrolidine)₃PF₆
PyDAOP
Y = P(pyrrolidine)₃PF₆

DEPBT
3-(diethoxyphosphoryloxy)-
1,2,3-benzotriazin-4(3H)-one

TNTU
2-(5-norbornene-2,3-dicarbox-
imido)-1,1,3,3-tetramethyl-
uroniumtetrafluoroborate

TSTU
2-succinimido-1,1,3,3-tetramethyl-
uronium tetrafluoroborate

TPTU
O-(2-oxo-1(2H)-pyridyl)-
1,1,3,3-tetramethyluronium
tetrafluoroborate

TOPPipU
2-(2-oxopyrid-1-yl)-1,1,3,3-
pentamethylene-
uronium tetrafluoroborate

Z = PF₆⁻ :**HOTT**
S-(1-oxido-2-pyridinyl)-1,1,3,3-tetramethylthiouronium
hexafluorophosphate
Z = BF₄⁻ **TOTT**
S-(1-oxido-2-pyridinyl)-1,1,3,3-tetramethylthiouronium
tetrafluoroborate

PfPyU or HPyOPfp
N,N,N'N'-bis(tetramethylene)-
O-pentafluorophenyluronium
hexafluorophosphate

PyPOP
[(pentafluorophenyl)oxy]-
tris(pyrrolidino)phosphonium
hexafluorophosphate

PyTOP
(pyridyl-2-thio)tris(pyrrolidino)
phosphonium hexafluorophosphate

Z = PF₆⁻ : **HODT**
S-(1-oxido-2-pyridinyl)-1,3-dimethyl-1,3-trimethylenethiouronium
hexafluorophosphate
Z = BF₄⁻ **TODT**
S-(1-oxido-2-pyridinyl)-1,3-dimethyl-1,3-trimethylenethiouronium
tetrafluoroborate

Non-symmetrical Immonium-Type Coupling Reagents

BOMI
benzotriazol-1-yloxy-N,N-
dimethylmethaniminium
hexachloroantimonate

BDMP
5-(1H-benzotriazol-1-yloxy)-3,4-
dihydro-1-methyl 2H-pyrrolidinium
hexachloroantimonate

AOMP

BPMP
1-(1H-benzotriazol-1-yloxy)-
phenylmethylene pyrrolidinium
hexachloroantimonate

SOMP

Scheme 12.36 Other Uronium and Phosphonium Coupling Reagents for Peptide Synthesis.

was reduced compared to BOP, HBTU, HBPyU, and HBPipU (1216). BDMP was employed for a synthesis of the N-methylated immunosuppressive undecapeptide cyclosporin O (1232).

The HOBt moiety was replaced with a halide in a set of reagents (see Scheme 12.37) designed for the express purpose of improving coupling yields of N-methylated amino acids (1233). The presence of HOBt had been observed to reduce reaction yields for these substrates (671), and HOBt was also noted to increase racemization during a BOP-Cl-mediated cyclization of a hexapeptide between two N-alkyl residues (670). Indeed, the bromide analog BroP (1234) was found to be superior to BOP for coupling to N-substituted residues in dichloromethane with DIEA as base (1233). Similarly, the phosphonium-based PyBroP (671, 1235) and PyCloP

(671, 1235) halide reagents, and the uronium-based PyClU (671) and TPyClU (1235) analogs, were also much more effective than PyBOP for the coupling of amino acids to N-methylated residues (87–100% yields with <0.3% epimerization vs 58–76% yields and up to 18% epimerization) (671). Further evidence supporting the deleterious effect of HOBt when coupling to N-methyl amino acids were the poor yields given by PyBOP, HBPyU, TBPyU, and DCC/HOBt compared to PyBroP, PyClU, TPyClU or DCC (671, 1235), although the latter series of reagents are not very effective as general reagents in peptide synthesis (1235). Both PyBroP and PyBOP have been used for peptoid (poly (N-alkyl glycine)) synthesis (342, 616). However, PyBroP was ineffective for coupling N-farnesylated residues, with the mixed anhydride method giving good yields (477).

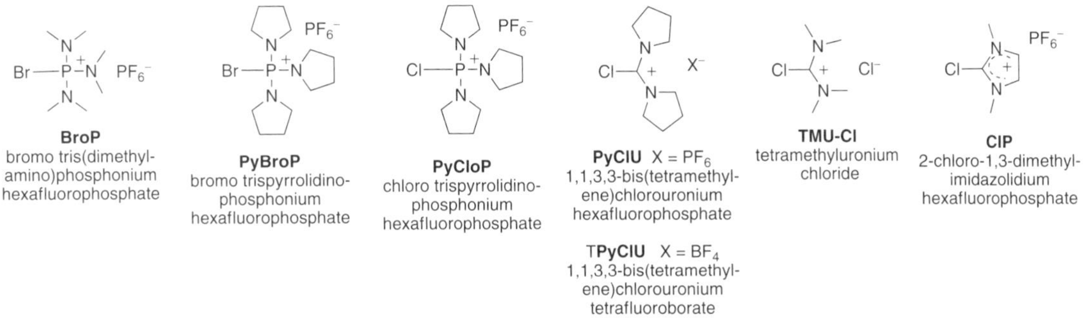

BroP
bromo tris(dimethyl-amino)phosphonium hexafluorophosphate

PyBroP
bromo trispyrrolidino-phosphonium hexafluorophosphate

PyCloP
chloro trispyrrolidino-phosphonium hexafluorophosphate

PyClU X = PF$_6$
1,1,3,3-bis(tetramethyl-ene)chlorouronium hexafluorophosphate

TPyClU X = BF$_4$
1,1,3,3-bis(tetramethyl-ene)chlorouronium tetrafluoroborate

TMU-Cl
tetramethyluronium chloride

CIP
2-chloro-1,3-dimethyl-imidazolidium hexafluorophosphate

Scheme 12.37 Halide-Derived Uronium and Phosphonium Coupling Reagents for Peptide Synthesis.

A mechanism of reaction for the halo reagents has been described (*1235*).

Over the years numerous attempts have been made to compare the effectiveness of the plethora of coupling reagents. A comparison of modern coupling methods for hindered peptides containing MeAla, MeAib, or MeαAc^{5}c was published in 1992. Pivaloyl mixed anhydrides, Pfp esters, acyl fluorides or EDC/HOBt were found to be ineffective for coupling of Boc-protected amino acids to *N*-methyl Aib and other bulky residues (<10% yields), while HBTU and PyBroP gave significant yields (35–50%), and UNCAs were the best, if used for prolonged reaction times or at higher temperatures (50 °C) (*1019*). Several of the phosphonium-based family of compounds (BOP, PyBOP, BroP, and PyBroP) were examined for their ability to couple amino acids to Aib, and to form an Aib-Aib dipeptide. In contrast to the results with *N*-methyl residues, all reagents gave similarly good yields for the first reactions, and much better results for the Aib-Aib formation than provided by DCC, active esters, BOP-Cl, CDI, or mixed anhydride methods. This report also discussed the proposed mechanism of activation (*1198*).

A 1998 study compared the effectiveness of many reagent combinations (DCC/HOAt, HBTU, HATU, HAPyU, HAMDU, AOP, BOP, PyAOP, PyBOP or HDTU) at coupling the hindered amino acid Fmoc-diethylglycine with Phe-OFm (*1236*). The azabenzotriazole derivatives were more effective than the benzotriazole analogs for both activation and coupling steps, with the pyrrolidino reagents (HAPyU, PyAOP, PyBOP) slightly better than the dimethylamino analogs (HATU, BOP). HDTU activated the amino acid quickly, but the resulting active ester was less reactive than the HOAt or HOBt derivatives. All reagents required the presence of a base for efficient activation, with DIEA giving better results for hindered amino acids than weaker bases such as TMP (*1236*). Theoretical calculations have been carried out on HATU, HBTU, HBMDU, HADMU, and HAPyU to attempt to explain the different behaviors of pyrrolidino-, trialkylamine-, and dihydroimidazole-based coupling reagents. Strain within the carbon skeleton and π-electron delocalization were

found to account for the varying reactivities (*1237*). A 1999 report observed that the commercial pyrrolidine-containing phosphonium salts PyAOP, PyBOP, and PyBroP were sometimes contaminated with small amounts (0.5% w/w) of pyrrolidine. This can result in pyrrolidide formation during slow reactions of activated carboxylates, but can be avoided by crystallizing the reagents before use (*1238*).

12.5.2i Other Reagents
(see Scheme 12.38)

CIP (2-chloro-1,3-dimethylimidazolidium hexafluorophosphate) was first used as an esterification agent for anchoring the *C*-terminal amino acid to the Wang resin (*1239*). It was subsequently discovered to be an effective coupling reagent for α,α-dialkyl amino acids (*1240–1243*). Aib-Val dipeptides were prepared with 80–93% yields using CIP/HOAt, in comparison to PyBroP yields of 4–60%. The CIP/HOAt reagent system was then used to couple three consecutive α-methyl-Cys residues in a synthesis of mirabazole C (*1242*), and applied to a synthesis of the Aib-rich peptide lamethicin F-30 (*1243*). CIP/HOAt was also found to be effective for coupling *N*-methyl amino acids, and used to synthesize the depsipentapeptide dolastatin-15 (*307*). The CIP/HOAt combination was also employed for a peptide cyclization, followed by a CIP/DMAP-mediated side-chain cyclization (no racemization potential) to form a bicyclic peptide tachykinin antagonist (*1244*). The CIP coupling is believed to proceed via initial formation of an oxazolone of the Fmoc-amino acid, which is then transformed to a highly reactive ester by the additive (*1241, 1242*). The isolated oxazolone has also been used for coupling (see Section 12.5.2e) (*1141*). A CIP analog, 2-chloro-1,3-dimethyl-1*H*-benzimidazolium hexafluorophosphate (CMBI), was also efficient at coupling hindered *N*-methyl amino acid residues (*1245*).

BEMT, 2-bromo-3-ethyl-4-methylthiazolium tetrafluoroborate, is a thiazolium-based peptide coupling reagent. It was effective for both *N*-methyl and α,α-disubstituted amino acid couplings, with rapid coupling and low racemization. A Cbz-Aib-Aib-OMe dipeptide

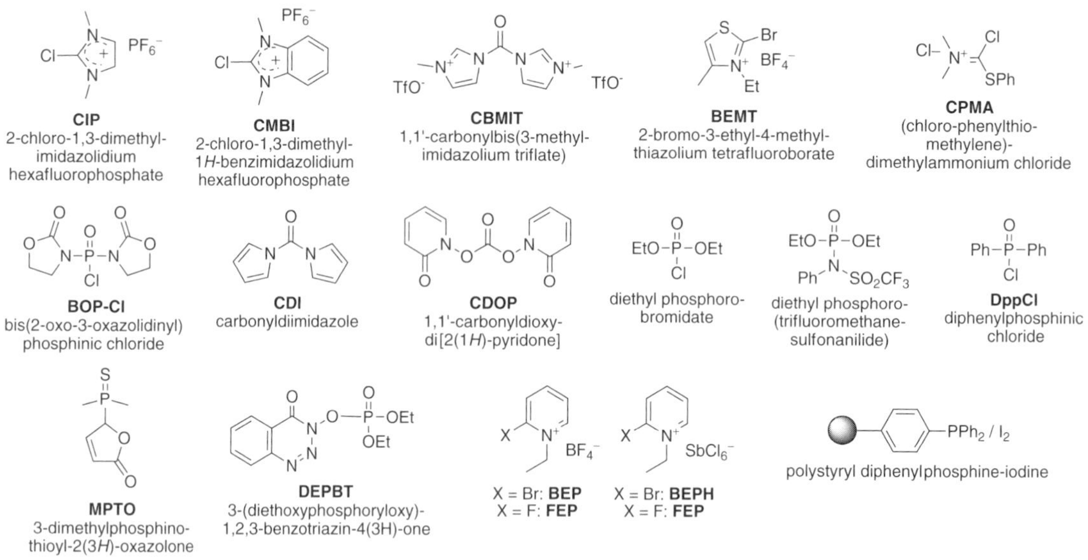

CIP
2-chloro-1,3-dimethyl-imidazolidium hexafluorophosphate

CMBI
2-chloro-1,3-dimethyl-1H-benzimidazolidium hexafluorophosphate

CBMIT
1,1'-carbonylbis(3-methyl-imidazolium triflate)

BEMT
2-bromo-3-ethyl-4-methyl-thiazolium tetrafluoroborate

CPMA
(chloro-phenylthio-methylene)-dimethylammonium chloride

BOP-Cl
bis(2-oxo-3-oxazolidinyl) phosphinic chloride

CDI
carbonyldiimidazole

CDOP
1,1'-carbonyldioxy-di[2(1H)-pyridone]

diethyl phosphoro-bromidate

diethyl phosphoro-(trifluoromethane-sulfonanilide)

DppCl
diphenylphosphinic chloride

MPTO
3-dimethylphosphino-thioyl-2(3H)-oxazolone

DEPBT
3-(diethoxyphosphoryloxy)-1,2,3-benzotriazin-4(3H)-one

X = Br: **BEP**
X = F: **FEP**

X = Br: **BEPH**
X = F: **FEPH**

polystyryl diphenylphosphine-iodine

Scheme 12.38 Other Coupling Reagents.

was obtained in 95% yield; a heptamer containing three N-methyl amino acids was produced in 92% yield. The mechanism of action appears to be a combination of pathways, with formation of the amino acid bromide or direct amidation of an unstable acyloxythiazolium salt predominating (1246). BEMT has also been employed for a synthesis of the N-methylated immunosuppressive undecapeptide cyclosporin O (1232). Both BEMT and a 3-methyl bromide salt analog of BEMT, BMTB (2-bromo-3,4-dimethylthiazolium bromide), were synthesized on a large (0.5 kg scale) in 2003 (673). BMTB was greatly more effective at coupling N-Boc-MeIle with MeIle-OBn (27% conversion after 2 h with 1.5 equiv reagent) compared to PyBOP, EDC/HOBt or BEMT (<5% conversion), and significantly more effective than HATU (16% conversion) (673).

1,1'-Carbonylbis(3-methylimidazolium triflate) (CBMIT) was introduced as an efficient coupling reagent for unhindered residues, although Cu(II) salts were required to suppress racemization (1247). Couplings with Aib or N-methyl-Leu gave reduced yields (36–54%) and some racemization (3–6%). CDOP, 1'-carbonyl-dioxydi[2(1H)-pyridone], generates a 2(1H)-oxo-1-pyridyl ester intermediate. No racemization was observed during segment couplings of dipeptides with Val-OMe, and no additional tertiary amine base was required other than to neutralize the amine component (1248). 2-Chloro-4,6-dimethoxy-1,3,5-triazine (CDMT) forms 2-hydroxy-4,6-dimethoxy-1,3,5-triazine active esters. However, preactivation with CDMT was found to generate an azlactone intermediate rather than an active ester in one report; optimized conditions found a one-pot, one-step procedure gave the best yields (1094). Triazine-based condensing reagents were reviewed in 2000 (1095). A preliminary report of

(chlorophenyl-methylene)dimethylammonium chloride (CPMA) as a new coupling reagent for amide bond formation was published in 2002, though no amino acids were coupled (901).

A polymer-supported Mukaiyama reagent, N-alkyl-2-chloropyridinium triflate, gave 88–99% yield for couplings of Boc-Trp with Val-OMe or Pro-OMe, with <1% racemization (1249). 2-Bromo-N-methylpyridin-ium tetrafluoroborate has also been incorporated into a solid-supported polystyrene resin, where it was used for purification-free couplings to give peptides. Unfortunately, and unexpectedly, complete racemiza-tion of the amino component was observed (1050). The Mukaiyama 2-halopyridinium iodide salts themselves are poorly soluble in conventional solvents; 1-ethyl-2-bromo- or -2-fluoropyridinium tetrafluoroborate or hexachloroantimonate salts (BEP, FEP, BEPH or FEPH) were soluble and highly effective coupling reagents. Highly hindered residues could be coupled with minimal racemization (1250). BEP was employed for a synthesis of the N-methylated immunosuppressive undecapeptide cyclosporin O (1232).

Ethyl propiolate has been identified as a simple, cheap, and convenient peptide coupling reagent. It reacts with N-protected amino acids to form a moderately activated ester of ethyl 3-hydroxyacrylate. Dipeptides were prepared in 65–90% yield, with levels of racemiza-tion similar to that obtained with PyBOP (1251).

Several phosphorus-based reagents have been devel-oped (see Scheme 12.38). Rich and co-workers (668) found that bis(2-oxo-3-oxazolidinyl)phosphinic chloride (BOP-Cl) (1252) was an effective reagent for nearly racemization-free condensation of N-alkylated amino acids and other imino acids (1253), and used the reagent in a synthesis of cyclosporine (1254). It has also been used to cyclize a hexapeptide between two N-alkyl amino acid residues (670), for couplings of residues to sterically hindered 5-tert-butyl-Pro (1255), and applied to the formation of thioamide isostere bonds from amino monothioacids (Boc-AA-SH) and imino acids (1201). The latter reaction proceeded in poor yields with primary amino acids, and PyBOP was the preferred reagent (1201). Unfortunately, BOP-Cl is not generally useful as considerable racemization has been reported (670, 1253), the reagent is ineffective at coupling β-branched amino acids having bulky N-protection (such as Boc-Val), and primary amines are acylated slowly (669). Racemization during couplings with BOP-Cl has been suppressed by the addition of copper(II) chloride (1107).

Diethyl phosphorobromidate, reported in 1978, was able to couple consecutive Aib residues in 84–85% yield (1256). Diphenylphosphinic chloride (DppCl) (1257–1259) has been noted as a useful reagent for solution coupling of N-methyl amino acids (667), while o-phenylene phosphochloridate oligomerized N,O-bis(trimethylsilyl)-α-amino acids (dimer to octamer) (1260). Diethyl phosphate was activated by attaching trifluoromethanesulfonanilide as a leaving group; Bz-Leu was coupled with Gly-OEt in the presence of proton sponge in 85% yield with 2% racemization, while Cbz-Aib coupled to Aib-OMe with HOAt/DIEA in 70% yield (1261). A combination of tributylphosphine and carbon tetrachloride generated tributyltrichloromethylphosphonium chloride, which was used to generate several dipeptides in 70–80% yield without the need for a tertiary base (1262). Phosphonium salt-based reagents such as BOP and BroP have been replaced by a polymer-based polystyryl diphenylphosphine–iodine complex. For solution couplings, this system has the advantage that the phosphine oxide byproduct can be removed by a simple filtration. A number of dipeptides were prepared in 90–99% yield with no apparent racemization (1263). 3-Dimethylphosphinothioyl-2-(3H)-oxazolone (MPTO) is a stable crystalline product that was found to be more effective than DPPA (diphenylphosphoryl azide, see Section 12.5.2b) for standard couplings, and caused substantially less racemization (1069). Another new reagent, 3-(diethoxyphosphoryloxy)-1,2,3-benzotriazin-4(3H)-one

(DEPBT), was employed as a coupling reagent without side-chain protection for Ser, His, and Tyr. Racemization was minimal compared to other reagents such as BOP, and yields of cyclic pentapeptides were significantly higher with DEPBT (52–54%) than with BOP, EDC, HBTU, TBTU or DCC (5–45%) (1264, 1265). The activation is proposed to proceed via an HOOBt ester, similar to the HOBt esters obtained with BOP (1265).

A Ph_2Se_2–Bu_3P system has been employed as the carboxyl activating reagent, and can be combined with in situ generation of the amine component via reduction of α-azido esters to an α-amino acid using the benzeneselenol generated during carboxyl activation (see Scheme 12.39). This method has the advantage that α-azido acids (intermediates in many amino acid synthesis routes proceeding via electrophilic or nucleophilic amination, see Chapters 2 and 5) can be directly coupled into peptides without needing a reduction step (934). Similar in situ reductions employed trialkylphosphines in combination with 3,5-dinitrobenzoyl mixed anhydride activation of the carboxyl component (1111), or reduction by tributylphosphine in the presence of coupling reagents EDC and HOBt (952). Stepwise coupling of azido acids followed by azide reduction has also been employed: highly hindered Aib and Dpg (diphenylglycine) were coupled as their azido acid chlorides and then reduced with DTT; up to four consecutive Aib or Dpg residues were incorporated (1041). Amino acids can be efficiently converted into α-azido acids by reaction with Tf_2O/NaN_3, followed by $CuSO_4/K_2CO_3/H_2O$. Twenty-three different amino acids were converted in 41–89% yield with retention of configuration. The azido acids were coupled to a growing peptide chain using DCC/HOBt, and then either "deprotected" to the amine with Me_3P/aqueous dioxane or reacted with an Fmoc-amino acid succinimide ester in the presence of Ph_3P to form a peptide bond (913).

Alternatively, amide bonds have been formed by reacting amino acid thio acids (generated from a trimethoxybenzyl thioester) with alkyl azides. The coupling appears to proceed via a concerted reduction/amidation process, and no racemization of the thioester component was observed. However, only simple alkyl

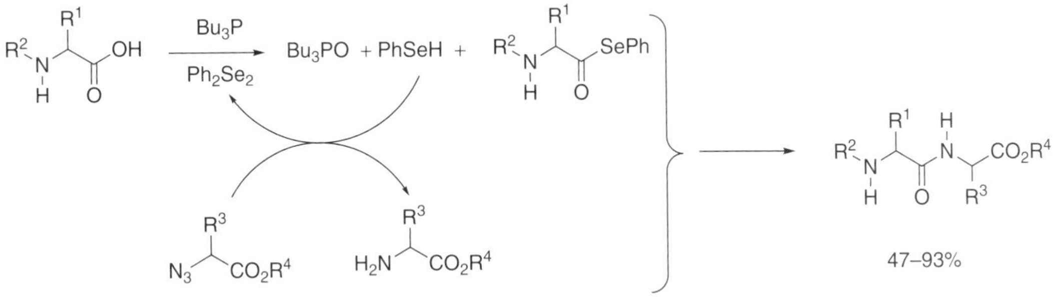

Scheme 12.39 Coupling of α-Azido Acids (934).

azide partners were examined (no azido esters) (*953*). A similar reaction coupled a peptide thioester with an azido ester residue in the presence of 2-diphenylphosphinethiophenol (*1266*). An amino acid or peptide with a *C*-terminal phosphinothioester (Xaa-SCH$_2$PPh$_3$) reacted with *N*-terminal azido-peptides to directly form an amide linkage, ligating the two peptide segments in a Staudinger reaction (*1267, 1268*). This synthesis has been extended to solid-phase-type segment condensation; RNase A(110–111) dipeptide was prepared on a "safety-catch" resin, and then cleaved with diphenylphosphinomethanethiol to give the *C*-terminal phosphinothioester. This was condensed with *N*-terminal azido RNase A(112–124), attached to a PEGA resin. An expressed protein ligation (see following section) was then employed to attach biosynthesized RNase A(1–109) thioester to the *N*-terminal Cys peptide fragment (*1269*). A similar strategy coupled di- or tripeptide *C*-terminal *o*-(diphenylphosphine)phenyl esters, Xaa-O(*o*-PPh$_2$-Ph), with azido di- or tripeptides, though yields were only 6–36% (*1270*).

12.5.2j Other Coupling Methods

An *N*-terminal, *N*-protected α-amino-β-hydroxy acid (e.g. Ser) on a peptide was *O*-acylated with the next residue or peptide of interest. *N*-Deprotection of the hydroxy amino residue resulted in efficient *O*–*N* intramolecular acyl migration. This method potentially overcomes solubility issues during difficult peptide syntheses (*1271*). Solution–phase dipeptide synthesis has been carried out with moderate yields (31–50%) by combining *N*-Boc amino acid methyl esters with unprotected amino acids in the presence of AlMe$_3$. No epimerization was observed as long as *N*-urethane protection was employed, but *N*-acyl derivatives underwent significant racemization (*1272*). *N*-Protected amino acid thioesters of 2-pyridylthiol or *p*-tolylthiol reacted with amino esters in the presence of Me$_3$Al at 0 °C or AgOCOCF$_3$ at 60 °C to form dipeptides (*1273*). The same reaction has been carried out with PEG-supported pyridyl thioesters (*1274*). Copper salts have been found to accelerate amide bond formation between amino acid or peptide thioesters (ethyl 3-thiopropionate derivatives) and the *N*-terminal amino group of a second peptide segment (*1275*). Amino acids can be converted into amino monothio acids and then reacted with amines to form amide bonds (*1276*), a reaction initially developed for segment condensations (*1277, 1278*).

Purified peptide segments have been ligated via a thioester segment condensation approach, which employs a *C*-terminal thioester on the *N*-terminal segment. This is chemoselectively converted to an active ester by reaction with an additive such as HOBt or *N*-hydroxysuccinimide in the presence of silver ion (conditions under which an unprotected carboxy group does not react) (*1279, 1280*). In 2006, two such condensations were employed to join three segments to prepare the 58-residue CCK-58 isoform of cholecystokinin, containing a labile Tyr(SO$_3$H) residue (*442*). Three peptide

segments were also sequentially coupled in a 2005 report, with the thioester/AgCl/HOOBt method used to couple the middle segment to the *C*-terminal fragment and an extended chemical ligation method employed for the second coupling (*1281*). Polypeptide synthesis by the thioester method was employed for syntheses of 90-mer HU-type DNA-binding protein, 110-mer barnase, a phosphorylated cAMP response element binding polypeptide 1, and a glycosylated calcitonin derivative in 1999 (*1282*).

A range of strategies have been developed for peptide ligation, in which coupling reagents or protecting schemes are not employed. Instead, a chemoselective capture step is followed by an intramolecular acyl transfer reaction. Reviews in 1999 (*1283*), 2001 (*1284*), and 2004 (*1285*) examined many of these methods. A brief summary of techniques for chemical ligation of unprotected peptides in aqueous solution has been published (*1286*), while a more comprehensive review in 2000 included native chemical ligation (NCL) strategies (*1287*).

The NCL strategy has found widespread use. This approach employs *C*-terminal thioester peptides and peptides with an *N*-terminal Cys residue (see Scheme 12.40). The side-chain thiol of the *N*-terminal Cys attacks the thioester to join the two segments as a thioester; "conformational assistance" then brings the *N*- and *C*-termini in close proximity for a thioester-to-amide isomerization (*1288*). This method was applied to a synthesis of the antibacterial glycopeptide diptericin, coupling 24- and 58-residue segments (*1289*), for preparation of a keto-containing 68-mer RANTES peptide (*1290*), for extension of a glycopeptide to prepare analogs of prostate-specific antigen (PSA) (*1291*), and for the synthesis of a 90-mer three zinc finger protein, Zif268 (*1829*). A synthesis of the 46-mer protein crambin (homologous with membrane-active plant toxins) coupled three segments (15+16+15 residues) via two native chemical ligations, giving more than a 10-fold improvement in overall yield compared to attempted couplings of 15+31 or 31+15 presynthesized segments (*1292*). A "one-pot" total synthesis was subsequently reported, in which the middle segment *N*-terminal Cys residue was temporarily protected in a thiazolidine ring until after the *C*-terminal segment had been coupled (*1293*). A number of large proteins (C5a, 74 amino acids; MIF, 115 amino acids; GV-PLA$_2$, 118 amino acids) were prepared by consecutive solid-phase ligations of three or four peptide segments (*1294*). Consecutive ligations in one study were aided by the use of photolabile protecting groups on the *N*-terminal Cys residue of the *N*-terminal fragment of the first coupling (*865*).

The unprotected Cys side chain remaining after a ligation synthesis can be protected as a thiosulfonate by reaction with Na$_2$S$_2$O$_3$, allowing for further elaboration of the peptide chain (*1295*), or it can be converted into a dehydroalanine residue via thiol methylation/oxidation or cyanation, followed by elimination (*1296*), or into an Ala residue by desulfurization (*1297*). The latter method was applied to syntheses of microcin J25, a 56-residue

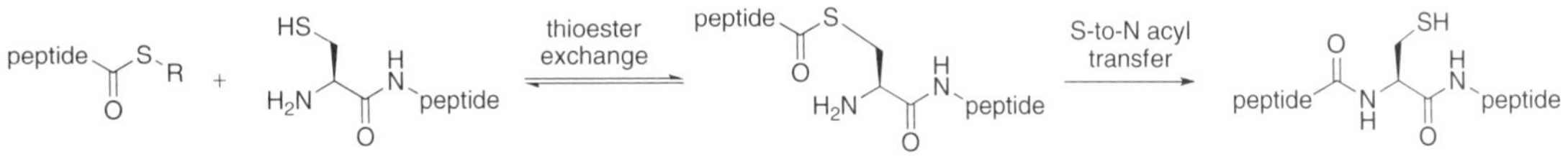

Scheme 12.40 Native Chemical Ligation.

streptococcal protein G B1 domain, and a 110-mer ribonuclease, barnase (*1297*). Homocysteine has also been employed as the *N*-terminal residue, and then methylated after the acylation transfer to form a Met residue (*1298*). Native chemical ligations with *C*-terminal Asp or Glu residues can lead to a side reaction, resulting in amidation of the Asp/Glu side chain rather than the α-carboxyl group. Protecting the Glu side chain as a (phenylsulfonyl)ethyl ester, and the Asp side chain as a 1-methyl-2-oxo-2-phenylethyl ester, prevented the side reaction (*1299*).

The same NCL strategy has been successfully applied to prepare proteins with an internal selenocysteine (Sec) (*1300–1302*) or selenohomocysteine (*1303*) residue. The selenohomocysteine residue was then methylated to give selenomethionine (*1303*). The use of selenocysteine derivatives for native chemical and expressed protein ligations was reviewed in 2002 (*1304*). The ability to convert Sec into dehydroAla (via H_2O_2 oxidation) or Ala (via reduction with H_2/Raney Ni) provides added versatility. These modifications were applied to an unprotected 16-mer peptide with *C*-terminal thioester and *N*-terminal Sec residue that was cyclized upon treatment with thiophenol; the Sec residue was then converted (*1305*).

NCL was used to couple β-peptide thioesters to α-peptides with an *N*-terminal Cys residue, or to β-peptides with an *N*-terminal β³-HCys residue (*1306*). Membrane-embedded peptides were prepared by a chemical ligation using a lipid bilayer to assist the ligation step of two transmembrane domain peptides, with the ligation site forming the extramembrane loop between the two transmembrane segments (*1307*). Two peptide nucleic acid (PNA) segments were linked by a NCL, using a complementary DNA template to increase ligation rates (*1308*). A 33-residue de novo designed peptide ligase catalyzed the template-directed ligation of two fragments, greatly accelerating (10^5) the initial thioester formation, but greatly retarding the subsequent rearrangement to the amide due to the rigidity of the templated complex (*1309*). An oligonucleotide was conjugated to the *C*-terminus of a peptide via a NCL to a Cys residue with an oligonucleotide attached to the carboxy group (*1310*). The ligation strategy can be employed to cyclize peptides attached to a solid-phase resin. A side-chain-anchored peptide was acylated on the *N*-terminal end with Trt-Cys(Xan)-OH, and amidated on the *C*-terminal end with H-Phe-SBn. On-resin removal of the *S*-(9*H*-xanthen-9-yl) and *N*-trityl protecting groups revealed the amino thiol groups,

allowing for cyclic peptide formation with the Phe thioester under aqueous conditions (*1311*).

Several methods have been used to generate the *C*-terminal peptide thioesters required for fragment condensations. A three-segment synthesis of the 71-mer chemokine vMIP I assembled each segment on a solid-phase resin using a safety-catch linker (*1312*). An orthogonal double-linker resin allowed for LCMS monitoring as an *N*-terminal glycopeptide segment was constructed; the resin was activated by alkylation and then cleaved with 3-sulfanylpropionate to generate the thioester fragment. Coupling with RNase B(41–68) produced glycosylated RNase B(30–68) (*1313*). Fmoc-based solid-phase synthesis of peptide thioesters can present difficulties, as the Fmoc deprotection conditions can cause cleavage from the resins normally used. Solid-phase linkers that are stable under conditions compatible with Fmoc/*t*Bu-based chemistry have been developed for producing peptide *C*-terminal thioesters (*1314, 1315*). Premature aminolysis during Fmoc-based SPPS can also be avoided by using a DBU/HOBt combination for Fmoc deprotection (*806*). *C*-terminal peptide thioesters were produced during Fmoc-based solid-phase peptide synthesis by cleavage from conventional PAM or HMBA resins with $AlMe_3$ and thiol (EtSH) (*1316*). Alternatively, peptides were constructed by Fmoc-SPPS on a 2-chloro-trityl resin, and then cleaved from the resin and thioesterified using a thiol (*p*-acetamidothiophenol, thiophenol, 3-mercaptopropionic acid ethyl ester) with PyBOP/DIEA. The *p*-acetamidothiophenol ester was then employed for a NCL (*1317*).

Several variations of the NCL strategy have been reported. An amino acid ester formed with 2-hydroxy-3-mercaptopropionamide undergoes an O-to-S acyl shift under ligation conditions, with the resulting thioester coupling with an *N*-terminal Cys as in the normal NCL approach. The use of a normal ester as starting material avoids incompatibilities of thioesters with the nucleophilic reagents required to cleave Fmoc protecting groups (*1318*). Peptide thioesters have been prepared via an N–S acyl shift from the amide of an *N*-4,5-dimethoxy-2-mercaptobenzyl Ala residue (*1319*). Another version of NCL was developed for complex glycopeptide synthesis, for which it was anticipated that a thioester glycopeptide partner would present synthetic difficulties. Instead, the *N*-terminal glycopeptide coupling partner was derivatized as a less-reactive *C*-terminal ester of *o*-SSEt-phenol. Exposure to neutral aqueous reducing conditions in the presence of the *C*-terminal coupling partner (with an *N*-terminal Cys residue) resulted in

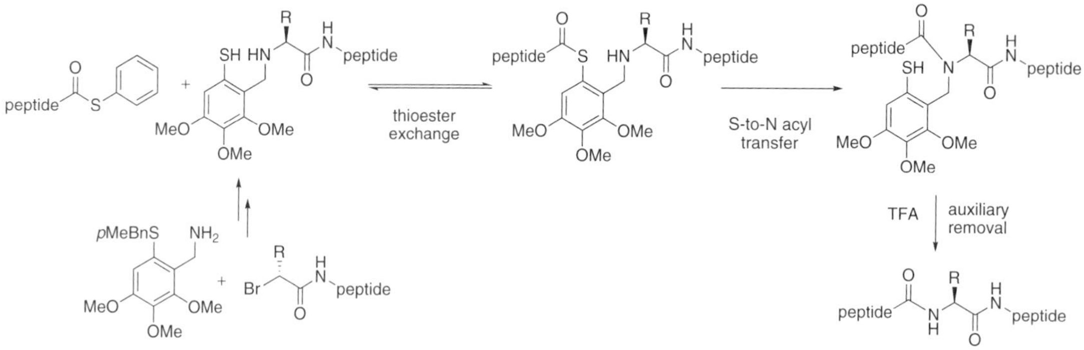

Scheme 12.41 Auxiliary-Mediated Chemical Ligation (*1321*).

peptide formation, possibly via a thioester intermediate generated by intramolecular acyl transfer between the aryl phenol and thiophenol substituents. The methodology was successfully applied to the coupling of two glycopeptide segments, each containing an *N*-linked complex biantennary pentasaccharide (*1320*).

A modified NCL strategy has been developed which eliminates the need for an *N*-terminal Cys residue on the aminocoupling partner. Instead, an *N*-(2-mercaptobenzyl) substituent was employed as the thiol to induce thioester exchange with a peptide thiophenol ester (see Scheme 12.41). The thioester-to-amide transfer occurred as with the Cys residue, leaving an *N*-benzyl-substituted amide bond. The mercaptobenzyl auxiliary was then removed by treatment with TFA, as long as a sufficient number of activating methoxy substituents were present on the aryl ring. The strategy worked for Gly-Gly, Lys-Gly, and Gly-Ala junctions, but gave no product for an Ala-Ala ligation (*1321*). An *N*-(1-phenyl-2-mercaptoethyl) substituent, with one or two methoxy substituents on the phenyl ring, has also been employed. Six different peptides were coupled to an *N*-terminal Gly residue (*1322*). Another analog is an *N*-(1-*o*-nitrophenyl-2-mercaptoethyl) substituent, with the advantage that it can be removed photolytically after the segment condensation (*1281, 1323*). *N*-(3-Amino-3-thiolpropane), *N*-(2-amino-3-thiolpropane) or *N*-(3-amino-3-thiolpropanoic acid) groups were employed as Cys analogs that included the amino group for amide bond formation (*1324, 1325*). Another possible Cys replacement, *N*-[2-*tert*-butylsulfanyl-1-(4-methoxyphenyl)-1-ethane], was described in 2003, but its application to NCL was not reported (*1326*).

The requirement for a segment with an *N*-terminal Cys residue (or equivalent) can also be avoided by using a thioester formed from 2-formyl-4-nitrothiophenol for activation/ligation. The formyl substituent captures the α-amino group of the coupling partner via hemiaminal formation; the peptide acyl group is then transferred from the aryl thiol to the aminal amine with regeneration of the formyl substituent and release of 2-formyl-4-nitrothiophenol. Addition of *N*-methyl maleinimide

improved the efficiency of the ligation via capture of the released thiophenol. Fmoc-Val-SAr coupled to H-Asn(*N*-acetylglucosamine)-O*t*Bu in aqueous DMF in 98% yield (*1327*).

Expressed protein ligation (EPL) is a modification of the NCL technique. Protein splicing is normally an in vivo post-translational event that excises an internal sequence (the intein) from a polypeptide and then ligates the resulting two flanking polypeptides (the exteins) to form the new peptide bond. A split intein can be employed to ligate two fragments (*1328*). The EPL strategy employs in vitro ligation of a chemically synthesized small *C*-terminal protein segment (with an *N*-terminal Cys residue) with a much larger expressed recombinant *N*-terminal protein segment (fused through its *C*-terminus via a thioester linkage to an intein protein splicing element). A mutant version of the splicing protein is used to generate the thioester intermediate (*1329, 1330*). The use of inteins to create these fragments for protein ligation was reviewed in 1999 (*1331*), while the development and application of EPL was reviewed in 2001 (*1332*) and new developmens in EPL in 2004 (*1333*). A study of expressed protein ligation conditions found that moderate concentrations of DMSO and the denaturant guanidine hydrochloride were tolerated during the condensation, overcoming problems due to poor peptide solubility (*1334*). Cyanogen bromide has been employed to cleave proteins at a Cys residue to generate fragments with an *N*-terminal Cys residue suitable for EPL (*1335*).

The EPL technique was first applied to σ^{70} RNA polymerase (*1329*) and Src kinase (*1330, 1334*) analogs, and has since been used to prepare analogs of azurin, a blue copper protein from *Pseudomonas aeruginosa*. The active site Cys residue was used as the site of ligation between a 111-mer protein and 17-mer peptide, and replaced with Sec (*1336, 1337*). Another study replaced the axial coordinating Met residue in azurin (also conveniently located in the *C*-terminal 17 residues) with Leu, norleucine or selenomethionine (*1338*). 5,5-Dimethylproline was incorporated into ribonuclease A as a replacement for Pro114, using EPL to couple RNase A(1–94) with the *C*-terminal peptide containing the modified amino acid (*1339*). Another group applied EPL to couple the Src homology 2 (SH2) domain of the c-Crk-I adaptor protein with an SH2 domain analog in which the two Trp residues were replaced with 7-aza-Trp (*1340*). Both phosphotyrosine (pTyr) sites in the SH2 domain of the protein tyrosine phosphatase SHP-2 were substituted with the pTyr mimetics 4′-phosphonomethylene-Phe and 4′-phosphonodifluoromethylene-Phe (*1341*), while an Asn-Pro reverse-turn sequence within RNase A was replaced with a β-amino acid reverse-turn mimic, (*R*)-nipecotic acid–(*S*)-nipecotic acid. The modified RNase A retained catalytic activity and enhanced the enzyme's conformational stability, whereas a (*R*)-nipecotic acid–(*R*)-nipecotic acid sequence gave inactive enzyme (*1342*).

A 4,4-difluoro-4-phosphono-2-aminobutyric acid residue was substituted for Ser-205 in serotonin *N*-acetyltransferase, giving a semisynthetic protein with enhanced affinity for binding to 14-3-3 protein (a phosphorylation-dependent event) and greater cellular stability (*1343*). Citrulline was incorporated into the *C*-terminal domain of a zinc finger protein by coupling the *C*-terminal 23-mer peptide fragment to an expressed 1–65 region (*1344*). The EPL method was applied to a peptide fragment prepared by an orthogonal ligation approach, based on a Staudinger reaction between an *N*-terminal azide and *C*-terminal phosphinothioester (see previous section) (*1269*). The 304-residue eukaryotic adaptor protein Crk-III was assembled from three polypeptide segments to give a biologically active product, opening the possibility of selective isotopic labeling of internal segments of proteins (*1345*).

The expressed enzymatic ligation method (EEL) was developed to ligate large protein fragments, without requiring an *N*-terminal Cys residue, using an enzyme for the coupling step. A sulfanylacetic acid thioester on the the *N*-terminal 40-mer fragment was coupled to a 28-mer *C*-terminal fragment with an *N*-teminal Ser residue, using protease V8. *N*-terminal Met, Tyr, and Ile residues were also tolerated (*1346*). Another enzymatic ligation strategy employed the enzyme sortase, a transpeptidase found in Gram-positive bacteria that cleaves between Thr and Gly in a LPXTG recognition sequence, and then ligates the Thr carboxyl group with a pentaglycine sequence on cell wall peptidoglycan. Sortase was successfully applied to couple peptides or proteins with a *C*-terminal LPETG sequence to peptides with an *N*-terminal Gly or Gly-Gly residue (*1347*).

Several chemical-based ligations rely upon unnatural linkages between the segments. A pseudoproline-based ligation of peptide segments connects two large unprotected peptide fragments with a 4-thiaproline residue via condensation of a reactive *C*-terminal hydroxyacetaldehyde aldehyde ester with an *N*-terminal Cys residue. The intermediate thiazolidine undergoes a spontaneous *O,N*-acyl transfer from an ester of the hydroxymethyl substituent to acylate the imino nitrogen, aided by entropic activation due to their close proximity (see Scheme 12.42) (*1348, 1349*). This pseudoproline-based ligation has also been applied to peptide segments with *N*-terminal Ser or Thr residues. These residues react with a peptide *C*-terminal hydroxyacetaldehyde ester to form 5-hydroxymethyl-4-oxa-Pro residues (*1349*). Three peptide segments were sequentially ligated via reaction of an unprotected *N*-terminal Cys with a *C*-terminal hydroxyacetaldehyde ester, followed by coupling of a less-reactive Ser/Thr *N*-terminal residue. Alternatively, the glycoaldehyde ester was masked as a glycerol ester during one ligation, and then oxidized to form the aldehyde for the next segment coupling (*1350*). A different strategy for a tandem ligation employed an initial formation of a thiaproline residue, followed by the second ligation of a *C*-terminal thioester with an *N*-terminal Cys residue (*1351*). This strategy has been extended to the preparation of multipartite peptides containing three or four different unprotected peptide segments, including a cell-permeating Arg/Pro-rich sequence, connected in a variety of linear or branched arrangements by three different ligation strategies. One ligation was achieved with a *C*-terminal thioester plus *N*-terminal Cys (forming Xaa-Cys linkage), another by a *C*-terminal glycoaldehyde ester plus *N*-terminal Ser/Thr (forming Xaa-oxaPro), and the third by alkylation of the side chain of the Xaa-Cys thiol with an *N*-terminal chloroacetylated peptide (*1352*).

Ligation with a non-natural thioester-Gly linkage (-COSCH$_2$CO-) is possible by alkylating a *C*-terminal thiol ester with an *N*-terminal bromoacetyl group; this strategy was employed to prepare an HIV protease analog in which one segment also included an ester linkage (*1353*). The Pictet–Spengler reaction of Trp has been employed for a peptide segment ligation reaction, coupling an *N*-terminal Trp residue with a *C*-terminal aminoaldehyde (*1354*). Assisted ligation has been used for difficult peptide cyclizations. An *N*-(6-nitro-2-hydroxybenzyl)

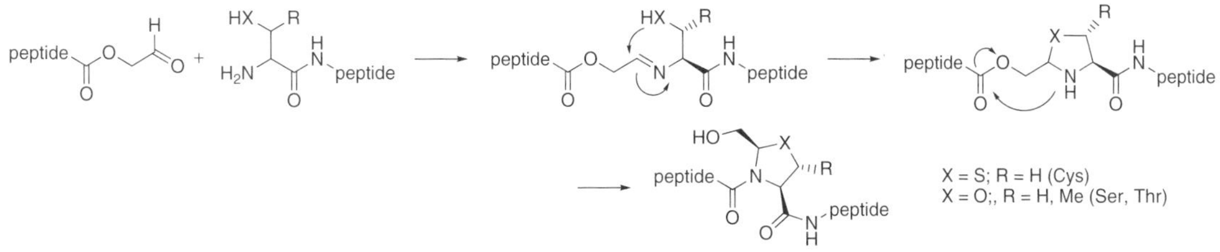

Scheme 12.42 Peptide Ligation via Formation of Pseudoprolines (*1348, 1349*).

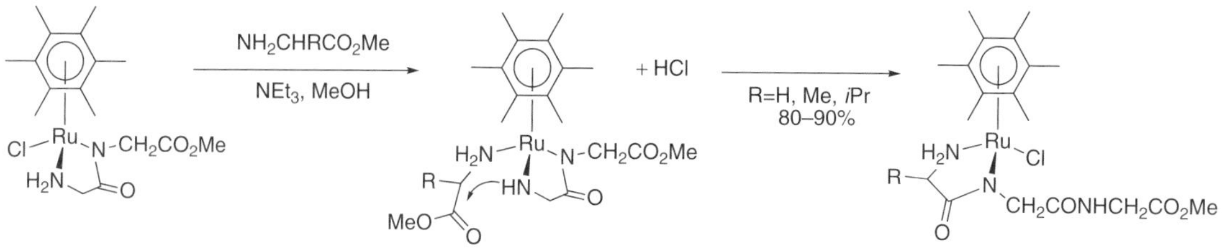

Scheme 12.43 Ruthenium-Catalyzed Peptide Synthesis (*1356–1359*).

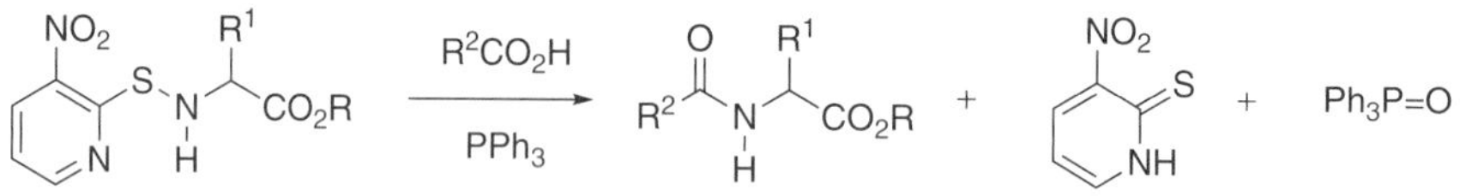

Scheme 12.44 Activation of Amino Group with NPys (*1362*).

group formed a cyclic nitrophenyl ester with a peptide C-terminal acid group that was activated for coupling. An intramolecular O–N acyl transfer step then formed the desired N-substituted amide linkage, with the auxiliary finally removed by photolysis (*1355*).

An enzymatic-like peptide synthesis employed a ruthenium or iridium complex of a dipeptide ester to couple with an amino ester, probably proceeding via nucleophilic attack of a metal-bound amino anion onto the carbonyl of the coordinated amino acid ester (see Scheme 12.43) (*1356, 1357*). The initial complex was prepared by reaction of dipeptide esters with [η^6-arene) MCl$_2$]$_2$ (*1358*). Although only tripeptides (*1357, 1359*) and a tetrapeptide (*1359*) were prepared, the method is intriguing.

DNA templates have been employed to "program" a tripeptide synthesis. The C-terminal residue was attached to a 30-base DNA template, with the next two residues linked to 10-base oligonucleotides complementary to the template DNA bases 11–20 or 1–10, respectively. The amino acid to be coupled was activated with EDC, but product was not formed if reagents with sequence mismatches were employed (*1360*). A similar strategy formed a tripeptide in a single solution by employing amino acids activated via an N-hydroxysuccinimide active ester linkage to an oligonucleotide fragment (*1361*).

Almost all peptide bond formation proceeds via activation of the carboxyl group. However, a 3-nitro-2-pyridinesulfenyl (Npys) group has been used to activate amino groups (see Scheme 12.44). Amide bonds were formed in the presence of an acid by the addition of tertiary phosphine via an oxidation–reduction condensation. The Npys group can be used as a protecting group that is removed during formation of the peptide bond (*1362*). The azirine synthon is another N-activated derivative (see Section 12.5.4).

Peptide synthesis in the unconventional N-to-C direction has been accomplished with no coupling reagents by physically pushing together an N-protected amino acid in close proximity to an amino group, making use of the "local concentraton" principle that, in part, enzymes rely on. The amine component was immobilized on a gold surface in a self-assembled monolayer (SAM), with the Boc- or Fmoc-amino acid dried onto an elastomeric stamp surface. The stamp was placed on the SAM for a period of time, then removed, and the substrate washed vigorously with solvent. The SAM was then treated with reagent to remove the N-protecting group, and the process repeated. An RGD tripeptide was assembled by this method, as was a 20-mer peptide nucleic acid. However, no direct evidence was provided as to the identity of the final products, or the extent of any epimerization (*1363*).

12.5.3 Enzymatic Peptide Syntheses

12.5.3a Introduction

Enzymatic peptide synthesis is very attractive, as it theoretically allows for amide bond formation under mild conditions with no racemization, using substrates with minimal protection (often no side-chain protection) and activation. The lack of racemization is significant, as a study has demonstrated 0.3–0.4% racemization at every step of the SPPS synthesis of a model tripeptide (*1134*). Enzymatic syntheses are particularly suitable for large-scale industrial syntheses where extensive optimization of a fixed reaction is possible, and where minimization of protecting groups (and the accompanying protection/deprotection reaction steps) can result in substantial cost savings. A historical view of enzymatic peptide synthesis is contained in the monograph of Wieland and Bodanszky (*1364*), while an introduction to the use of lipases and proteases in organic solvents was published in 1989 (*62*); more recent efforts are described in a 1991 review of esterolytic and lipolytic enzymes (*60*) and a comprehensive 2002 *Chemical Reviews* article on proteases in organic synthesis (*1365*).

12.5.3b Enzyme-Catalyzed Couplings

Initially, most efforts at peptide synthesis were directed at reversing the action of proteases, but these enzymes have a narrow substrate specificity, an unfavorable thermodynamic equilibrium, and can cause proteolysis of the newly synthesized peptide chain (*1366*). For example, Glu/Asp-specific endopeptidase is highly specific for Glu and can be used to couple peptide segments with a *C*-terminal Glu or Asp; however if Glu is present in the remaining peptide, proteolysis can occur at that site (*1367*). Despite these disadvantages, proteases have found widespread use. The basic principles of protease-catalyzed peptide bond formation were presented in 1985 (*1368*); both thermodynamically and kinetically controlled syntheses are possible. Modifications such as biphasic systems or immobilized enzymes (e.g., see *60*, *1369*) can be used to improve convenience and yields. Freezing was shown to reduce undesired proteolysis while allowing for coupling of specific residues by trypsin, chymotrypsin, or a Glu-specific endopeptidase (*1370*). The ability of freezing conditions to suppress competitive reactions in enzyme-catalyzed peptide synthesis was reviewed in 1996 (*1371*).

One advantage of enzymatic peptide syntheses is that, unlike the majority of SPPS methods, it is possible to emulate ribosomal polypeptide synthesis, which proceeds from the *N*- to the *C*-terminus. A tetrapeptide was assembled in this manner in 1997 from simple esters of unprotected trifunctional amino acids (Ser-OBzl, Tyr-OEt, Arg-OPr, and Ser-OBzl) using the cysteine protease clostripain and/or the serine protease chymotrypsin, with 83–100% coupling yields (*1372*). Another benefit of enzymatic coupling is regioselectivity. *Candida antarctica* lipase (CAL) catalyzed the regioselective α-amidation of Cbz-L-Glu diesters in organic solvents to give the α-monoamide (*972*). Although no amino acid residues were used as the amine component, they should be potential substrates as bulky α-methylbenzylamine was tolerated. When α-chymotrypsin catalyzed formation of a peptide bond between Boc-L-Tyr-OEt and Lys-O*t*Bu, a 7:3 ratio of α-dipeptide to ε-dipeptide was obtained. However, by using subtilisin Carlsberg, amidation took place almost exclusively (>99%) with the ε-amino group (*1373*).

A disadvantage of enzymatic syntheses is the substrate specificity of each enzyme, which varies greatly. A method developed to profile the specificity of new proteases makes use of peptide synthesis on a solid support, as the protease will have specificity for those amino acids that are coupled the most. This method could also be useful to screen for useful proteases for peptide coupling (*1374*). A 1999 report employed *Pseudomonas aeruginosa* elastase to form dipeptides

from *N*-protected amino acids and amino acid amides. Twenty-eight different combinations of Fmoc-, Cbz- or Boc-amino acid and amides of Phe, Leu, Ile, Ala, Trp, Tyr or Val were tested, giving dipeptides in yields of 16–90%. Specificity for Phe or Leu in the P$'_1$ site was observed, with small hydrophobic amino acids generally preferred in the P$_1$ site. *N*-Methyl or α,α-disubstituted amino acids were not accepted (*1375*). The aspartyl protease pepsin has also been used for peptide synthesis (*1376*), while papain catalyzed the condensation of *N*-protected α-amino thioacids (Xaa-SH) with α-amino amides. Good yields were obtained with Cbz-Ala-SH, but bulkier side chains gave low yields (*1377*). A protease-type enzyme isolated from the thermophilic *Clostridium thermohydrosulfuricim* SV 12 showed broad substrate specificity and enantioselectivity towards L-amino acids for dipeptide synthesis from *N*-Cbz amino acids or esters with amino esters (*1378*).

Most enzymatic peptide syntheses have relied on the coded amino acids as substrates, due to the restrictions imposed by protease substrate specificity. One of the few early examples of non-coded amino acids being incorporated into peptides was the sulfoamino acid taurine, which was coupled using substilisin and proteinase K (*1379*). Papain was used to couple dehydroamino acid-containing peptides (*1380*, *1381*). α-Trifluoromethyl amino acids are useful for altering the properties of peptides. Peptide fragments containing these residues were ligated by a number of proteases, as long as the α,α-dialkyl residue was not in the P$_1'$ or P$_1$ positions directly involved in coupling (*1382*). Carboxypeptidase Y was found to catalyze the coupling of *N*-Cbz α-fluoroalkyl-Ala-OMe derivatives with amino amide nucleophiles, with yields of 20–75% (*1383*).

Several methods have been developed to overcome the substrate specificity of proteases. D-Amino acids were incorporated into peptides by using subtilisin in anhydrous organic solvents (*1384*), while D-Phg-L-Xaa dipeptides were constructed from D-Phg-NH$_2$ and amino acids using penicillin acylase (*1385*). Alternatively, subtilisin was engineered to broaden specificity at the P$_1'$ site and to improve stability to non-aqueous conditions (*1386*). The modified subtilisin BPN' catalyzed peptide bond formation in DMF/H$_2$O between peptide fragments containing non-coded amino acids and peptide mimetics (*1387*), and was used to ligate six peptide fragments to prepare fully active ribonuclease A (*1388*). However, close attention to the subsite preferences was still required. Chemically modified mutant enzymes of subtilisin *Bacillus lentus* allowed for greatly broadened stereospecificity, such as couplings with D-amino acid acyl donors and α-branched amino acid acyl acceptors (*1389–1392*), or with β-amino acid acyl donors (*1390*).

A significant difficulty in protease-catalyzed peptide syntheses is that the optimum donor ester substituent varies with each protease, making it difficult to optimize the protease/substrate combinations for different *C*-terminal/*N*-terminal amino acid side-chain combinations. One approach for overcoming substrate specificity utilizes *p*-guanidinophenyl esters. *N*-Boc-α,α-dialkyl amino acid p-guanidinophenyl esters (1393) were cou- pled with amino acid p-nitroanilides in buffer/DMF by *Streptomyces griseus* trypsin, giving the dipeptides in 35–96% yield in 0.5–48 h (1394). The amino acid p-guanidinophenyl esters act as "inverse substrates" (1395), with the p-guanidinophenyl ester leaving group occupying the trypsin-preferred guanidino side-chain specificity pocket in the active site, allowing for a wide substrate tolerance of acyl donor side chains (1396). The set of p- and m-guanidinophenyl and p- and m-guanidi- nomethylphenyl esters were initially tested with both bovine trypsin and *Streptomyces griseus* trypsin to find the best combination (1397). Another paper tested nine inverse substrates (p- and m-guanidinophenyl, p- and m-guanidinomethylphenyl, p-amidinophenyl, and four positional isomers of guanidinonaphthyl esters) of Boc-D- and L-Ala with chum salmon trypsin; the D-amino acids were coupled preferentially (1398).

Similarly, 4-guanidinophenyl esters were used in combination with the cysteine protease clostripain, giving >90% yields for couplings of Boc-amino acids (1399). The same system was applied to enzymatically ligate all-D-peptide fragments to form longer homo-D- peptides (up to 34-mer) (1822), and for coupling Bz-β-Ala with a number of amino esters and amines (1400). Sterically hindered α-methyl-, α-difluoromethyl-, and α-trifluoromethyl-Ala, -Leu, and -Phe residues were coupled using their p-guanidinophenyl esters in combination with trypsin. Freezing reaction conditions (−15 °C) significantly improved yields, while couplings of the α-difluoromethyl residues proceeded more efficiently than the α-methyl- or α-trifluoromethyl derivatives (1401). One problem of using proteases to catalyze couplings at non-specific ligation sites of peptide fragments is undesired proteolysis at other sites containing the preferred substrate sequence of the protease. A trypsin Asp189Glu point mutant shifted the trypsin substrate preference from Lys and Arg to the 4-guanidinophenyl ester substrate mimetics of the frag- ment to be coupled, reducing the undesired proteolysis (1402). Other trypsin mutants were studied in a similar attempt to reduce the proteolytic side reactions, though most modifications also greatly reduced the rate of peptide bond formation (1403).

Another closely related approach to broaden speci- ficity employs a common parent methyl thioester and spontaneous transthioesterifications reactions with other thiols, such as 4-guanidinomercaptophenol (preferred thioester substrate for trypsin), benzyl mercaptan (preferred for chymotrypsin), and 3-mercaptopropionic acid (preferred for V8 protease). The transthioesterifica- tion was carried out either prior to, or in conjunction with, addition of protease, but significant time was saved with the parallel approach (1404). Another reaction scheme converted N-Boc-protected amino acids (or pep- tides) into esters of 1,3-propanediol or 1,4-butanediol using subtilisin; the ester was then a substrate for papain or subtiligase-mediated amidation with an amino amide or peptide (1405). The substrate specificity of proteases can also be circumvented by the use of Pronase, a

commercially available mixture of several *Streptomyces griseus* proteolytic enzymes, including both endopepti- dases and exopeptidases. The mixture was capable of forming a range of dipeptides in yields of 16–95%. The mixture also demonstrated enantioselectivity, as tested by hydrolysis of amino esters (1406).

A significant advancement in enzymatic peptide synthesis was reported by Margolin and Klibanov in 1987, when they disclosed that lipases can act as catalysts in anhydrous organic solvents (1366). Porcine pancreatic lipase (PPL) was found to be quite flexible, catalyzing the synthesis of a range of dipeptides from N-acetyl amino acid 2-chloroethyl esters and amino acid amides or esters in 48–86% yield, although the reactions were quite slow (reaction half-times 1.4–8.3 days) (1366). Yields in aqueous solution using PPL or *Candida cylindracea* lipase (CCL) were much poorer (1407). Enzymatic syntheses of N-protected Leu-enkephalin and other oligopeptides by α-chymotrypsin and thermo- lysin in dichloromethane and tert-amyl alcohol were described in 1998. Coupling racemic amino acids gave optically pure products (1408). The rate of α-chymotrypsin-catalyzed amidation of amino acids with low reactivity, such as Cbz-Ala-OMe, was greatly increased by changing the methyl ester to a 2,2,2-triflu- oromethyl or carbamoylmethyl ester, and by using organic solvents such as acetonitrile with low water con- tent. The carbamoylmethyl ester acyl donor reacted over 100 times more quickly than the methyl ester during acylation of Leu-NH$_2$ (1409, 1410). A range of other esters were also tested (1411). In 2002 a pentapeptide was constructed from amino acid ethyl or trifluoroethyl esters and papain, α-chymotrypsin, and/or thermolysin in dichloromethane, cyclohexane, or tert-amyl alcohol (1412). The superiority of carbamoylmethyl esters over methyl and trifluoroethyl esters was demonstrated in 2002, via α-chymotrypsin-catalyzed couplings of Cbz- Ala-OR with Gly, Ala, Val, Leu, Met or Phe (88–97% yield after 48 h, vs 20–82% for trifluoroethyl esters and 3–7% for methyl esters), or of various Cbz arylalanines with Leu-NH$_2$ (1413). The carbamoylmethyl esters were then used for a segment condensation synthesis of Leu-enkephalin (1414).

Another advancement in enzymatic peptide synthe- sis has been the development of cross-linked enzyme crystals (CLECs), which have greatly increased stability yet are still highly active in either aqueous or neat organic solvents (63, 1415). The CLEC of subtilisin protease was applied to the synthesis of peptides and peptidomimetics in acetonitrile, forming dipeptides in 32–98% yield (generally >90%) after 6–120 h. Both L- or D-amino acid amides or esters could be used as the amine nucleophile, and simple N-Cbz amino acid methyl or benzyl esters as the electrophile. Non- proteinaceous amino acids such as 3-(2-naphthyl)-Ala were also tolerated (1415). The amide formation activity of cross-linked subtilisin Carlsberg crystals in organic solvents could be significantly increased by pretreat- ment of the enzyme crystals with a solution of crown ether in acetonitrile (1416). As an alternative strategy, enzyme activities approaching those of CLEC enzymes in polar organic solvents were achieved by a simple one-pot method of immobilizing the enzyme on silica. Subtilisin Carlsberg and α-chymotrypsin were used as examples (*65*). Zeolite-immobilized enzymes were applied to peptide synthesis in organic solvents, coupling Cbz-Tyr-OEt with Gly-Gly-OEt (up to 69% yield after 1 day with immobilized α-chymotrypsin, vs 53% for the free enzyme), or Cbz-Asp with Phe-OMe (similar yields and reaction rates for free or immobilized thermolysin) (*1417*). Trypsin and chymotrypsin have been immobilized by copolymerization of acrylate derivatives of the enzymes with PEG. Peptide esters were amidated with amino acid or peptide amides with moderate to good yields (*1418*). Thermolysin was adsorbed onto Celite R-640, and then used for peptide synthesis in toluene, with dipeptides formed in 76–93% yield (*1419*).

A potentially significant development in enzyme catalysis was reported in 2002, in which PEGA1900 resin-linked Phe with a free amino terminus was amidated by *N*-Fmoc amino acids using thermolysin in aqueous buffer. Dipeptides were generated with 99% conversion for acylations with Leu, Nle, or Phe; side-chain unprotected Ser, Asp, His, and Gln gave reduced (10–84%) yields. Acylations with racemic Phe produced only the L,L-product. If yields can be optimized, this method has the potential for peptide synthesis without the need for side-chain protection, with negligible racemization, and potentially from racemic acyl donors (*1420*). In a subsequent report, the same enzyme was used to prepare two dipeptide diastereomers, with L-Xaa-L-Phe constructed by acylation of L-Phe-resin with racemic amino acids, and D-Xaa-L-Phe by chemical coupling of the racemic mixture, followed by thermolysin-catalyzed hydrolysis of the L-L dipeptides (*1421*). Other dipeptides have been prepared using this system (*1422*).

12.5.3c Other "Enzyme" Catalysts

There have been some interesting efforts at catalyzing peptide bond formation with RNA catalysts, which has implications for the presence of a prebiotic "RNA world" before the development of proteins. Some of these developments were highlighted in 1998 (*1423*). The mechanism of ribosomal catalysis of peptide bond formation within the all-RNA catalytic site has yet to be elucidated with certainty, but it appears that a hydroxyl group on the peptidyl tRNA substrate plays a pivotal role in the reaction (*1424*).

"Artificial enzymes" have been created by using catalytic antibodies generated against a hapten that resembles the tetrahedral transition state of amide hydrolysis. A Chg-Trp dipeptide phosphonate isostere was prepared as a hapten to generate catalytic antibodies for peptide bond formation (*1425, 1426*). *P*-Nitrophenyl esters of *N*-acetyl Val, Leu, and Phe were coupled with tryptophan amide to give dipeptides with yields of 44–94% (*1425*).

A highly acidic coiled-coil peptide was employed as a pH-tunable peptide ligase, which templated two basic peptide fragments, and then catalyzed their condensation on an analytical (μM) scale (*1427*).

12.5.3d Biosynthetic Coupling Using tRNA

The ultimate extension of using enzymes for peptide synthesis is to hijack the protein biosynthesis machinery of living organisms. This method has the advantage that large proteins can be prepared, but until recently incorporating non-natural amino acids was difficult. In 1989 Schultz and co-workers (*1428*) and Bain et al. (*1429*) independently reported that one of the normal stop codons of a DNA sequence (UAG) could be used to incorporate an unnatural amino acid that had been chemically acylated onto a "suppressor tRNA" with an AUC sequence which recognizes the amber UAG codon. The UAG stop codon is not normally read by tRNAs, but is instead bound by a release factor protein; in some mutants suppressor tRNA can recognize a misplaced stop codon, allowing for continued production of functional protein. In this fashion the mutated gene encodes a protein containing the unnatural amino acid at the desired position. Several reviews of this and related techniques have since appeared (*1430–1438*), including a thorough discussion on methods to amplify the amino acid repertoire of protein biosynthesis in 2004 (*1439*), a comprehensive summary of methods to expand the genetic code by Wang and Schultz in 2005 (*1440*), a chapter in a 2005 monograph (*1441*), and a monograph by Budisa in 2006 (*1442*).

This method has some precedent in nature, in that a UGA codon normally codes for termination of protein synthesis but also codes for selenocysteine (Sec). A number of other factors, such as RNA secondary structure insertion elements, also govern whether Sec becomes incorporated at UGA codon sites (*1443–1445*). The selenocysteine tRNA has unique features that distinguish it from all other tRNAs, with the Sec residue biosynthesized on its own Sec tRNA[Ser]Sec] via a phosphoserine intermediate. A serine attached to the tRNA is phosphorylated, and then converted to the selenocysteine residue. The mammalian kinase responsible for phosphorylating the seryl moiety on the seryl-tRNA[Ser]Sec] was identified by a comparative genomics approach in 2004 (*1446, 1447*). The selenocysteine codon was employed to incorporate selenocysteine into a phage display library of peptides in 2000 (*1448*). In 2002 it was reported that pyrrolysine is also apparently encoded for by a UGA stop codon in *Methanosarcina barkeri* (*1449, 1450*). Synthetic L-pyrrolysine was attached to tRNA_CUA by PylS, an archaeal class II aminoacyl-tRNA synthetase that activated pyrrolysine with ATP and ligated it to the tRNA (*1451, 1452*). When PylS and the corresponding tRNA_CUA were expressed in *E. coli*, pyrrolysine was incorporated into *Methanosarcina barkeri* methyltransferase at the normal position in response to the UAG stop codon, now acting as a sense codon (*1451*). The fact that pyrrolysine was not incorporated into other proteins at the UAG stop codon (which would be expected to cause toxicity) indicates that only "special" UAG sequences are recognized, perhaps by a RNA hairpin secondary structure similar to the selenocysteine insertion element (*1453*).

A wide range of unusual amino acids have been incorporated in proteins by variations of these tRNA-based stop codon methods, including bulky α,α-disubstituted and *N*-methyl amino acids, although D-amino acids are not usually accommodated (*1430, 1454, 1455*). However, the UAG codon method was modified to allow for incorporation of D-amino acids in a cell-free system, via mutagenesis of the peptidyltransferase and helix 89 regions of *E. coli* 23S ribosome (*1456*). A review of unnatural amino acid mutagenesis in 2004 included a chart of structures of 126 unnatural residues incorporated into proteins using nonsense codon suppression (*1434*). 2,3-Methano-Pro, pipecolic acid (Pip), *trans*-3-methyl-Pro, and *cis*- and *trans*-4-hydroxy-Pro were used to replace Pro-34 in human Ras p21 protein, *allo*-Thr for Thr-35, and Val, *tert*-Leu, Leu, Nva, or *O*-methyl Thr for Ile-36 (*1460*). Mutations at Gly^{12}, Gly^{13}, Ala^{59}, Gly^{60}, and Gln^{61} in Ras protein were also examined (*1461*). Ala-82 in *E. coli* T4 lysozyme was successfully replaced with *N*-methyl Ala, 1-aminocyclopropylcarboxylic acid, Aib, Pro, Pip, lactic acid, and thiollactic acid, while azetidine-2-carboxylic acid, *N*-ethyl Ala, and β-alanine were incorporated at very low levels, and D-Ala, was not incorporated at all (*1462*).

Other examples of the use of this procedure include altering Tyr-95 and Pro-17 in adenylate kinase to probe the effects of aromaticity and ring size on activity (*1463*), substituting ring-substituted Tyr and Phe residues into the binding site of the nicotinic acetylcholine receptor (*1464*), replacing Ser-286 in firefly luciferase with a range of other amino acids in order to examine the effects on thermostability, pH dependence, and color of light emitted (*1465*), and including biocytin (a Lys residue side-chain-acylated with biotin) in mRNA display peptide libraries (*1466*). Fluorinated Tyr analogs replaced the active site Tyr in human glutathione transferase A1-1 in order to evaluate the role of the residue in activating the Cys thiol group for nucleophilic addition to substrates (*1467*), while a series of aromatic residues (2'-, 3'- or 4'-fluoro-Phe, 3',4'-difluoro-Phe, 3',4',5'-trifluoro-Phe, 5'-methyl-Trp, 4'-, 5'- or 6'-fluoro-Trp, 7'-aza-Trp, β-(3-benzothienyl)-Ala) were incoporated into thymidylate synthase as replacements for Trp-82 in order to investigate its catalytic role in hydride transfer (*1468*). A photocleavable β-*o*-nitrobenzyl ester of Asp replaced an Asp residue in p21^{ras} (ras) protein, with the ester group blocking interaction with the ras protein downstream effector p120-GAP until it was removed by photolysis (*1469*). Amino acids with ketone side chains have been incorporated into T4 lysozyme in order to investigate if the ketone group could then be used as a handle for site-specific modification/labeling of proteins (*1470*). The amber stop codon system was used to test incorporation levels of spin-labeled or fluorescent amino acids at position 14 in firefly luciferase expressed in *Xenopus* oocytes; incorporation efficiencies were low (0.7–1.7%) but better than those achieved with a quadruplet codon method (see below) (*1471*).

α-Hydroxy acids can be incorporated into protein structures by using nonsense suppression methods coupled with acylated suppressor transfer RNAs. *O*-Unprotected or *O*-TBS-protected α-hydroxy acids were used for acylating the tRNA (*1472*), an improvement over earlier procedures using *O*-nitroveratryl or *O*-nitrobenzyl protection (*1473*). Ester replacements of amide bonds were used to investigate the importance of backbone hydrogen bonds in T4 lysozyme (*1473*), staphylococcal nuclease (*1474*), and Ras protein (*1461*). In a similar manner, hydrazinophenylalanine was incorporated into peptide and protein analogs via misacylated tRNA, forming a 1:1 mixture of the two possible regioisomeric acylation products (primary or secondary nitrogen of the hydrazino moiety) (*1475*).

A significant drawback with using a stop codon as the signal for incorporating acylated suppressor tRNA is that the desired amino acid addition competes with protein synthesis termination; the increased suppressor tRNA concentration also inhibits other protein production (*1476*). However, efficient suppression with low suppressor tRNA concentration can be achieved via in situ deactivation of release factor by specific antibodies (*1476*). Another disadvantage of the stop codon methodology is that no more than two types of amino acids can be incorporated since there are only three stop codons (UAA, UAG, UGA) in the genetic code. One method developed to avoid these problems employs an AGG codon during *E. coli* biosynthesis. This codon is rarely used in *E. coli*., and so chemically misacylated tRNA can compete effectively with the endogenous arginyl tRNA$_{CCU}$ (*1477*).

Alternatively, researchers have employed a frame-shift suppressor tRNA that reads a four-base (quadruplet) codon AGGN and competes effectively against the inefficient native tRNA coding for AGG (*1478*). If the frame-shifted incorporation of the unnatural amino acid does not occur, a termination codon appears, and therefore protein synthesis is truncated. Similarly, frame-shift suppressor tRNA that reads a CGGG four-base codon at the mutation site was used to introduce a *p*-nitro-Phe residue at 22 different sites within streptavidin (*1479*). Another report compared six different four-base codons (AGGU, CGGU, CCCU, CUCU, CUAU, and CGGG) for incorporation of *p*-nitro-Phe into streptavidin, using acylated tRNA containing the complementary four-base anticodons in an *E. coli* in vitro translation system. Both AGG and CGG are minor codons in *E. coli*, and so there is little competition from endogenous tRNAs. The best efficiency, forming 86% of modified vs wild-type protein, was observed with the GGGU codon. Incorporation was as low as 14% with CGCU and 7% with CGAU, but this was still comparable to the 8% given by an amber UAG codon method (*1480*). However, in another study GGGU was used to test incorporation levels of spin-labeled or fluorescent amino acids at position 14 in firefly luciferase expressed in *Xenopus* oocytes; incorporation efficiencies were significantly lower than those achieved with the amber stop codon system (*1471*). The frame-shift suppressor tRNA method was used to examine the relative efficiency of incorporation of different β-arylalanines into strepavidin. Nineteen different β-arylalanines, such as β-(1-naphthyl)-Ala, β-(2-naphthyl)-Ala, β-(2-anthryl)-Ala, β-(9-anthryl)-Ala, β-(9-phenanthryl)-Ala, β-(1-pyrenyl)-Ala, β-(2-pyrenyl)-Ala, β-(ferrocenyl)-Ala, β-(2-anthraquinonyl)-Ala, β-(9-ethylcarbazolyl)-Ala, and β-(9-carbazolyl)-Ala, in addition to 4′-phenyl-Phe, 4′-nitro-Phe, 3′,5′-dinitro-Phe, 4′-dimethylamino-Phe, 4′-phenylazo-Phe, 4′-benzoyl-Phe, and 7′-azatryptophan, were tested in order to define a model for the side chain size allowed into the *E.coli* ribosome (*1481*). In another report, the side-chains of Lys and 4′-amino-Phe were derivatized with 1,5- or 2,6-dansyl chloride and then incorporated into streptavidin using a four-base codon (*1482*).

An important advantage of the four-base codon system over amber suppression is that more than one different type of non-nnatural amino acid can be independently incorporated into a single protein by employing different four-base codons. This was demonstrated by using AGGU and CGGG codons to introduce 2-naphthylalanine and 4′-nitro-Phe as replacements for streptavidin Tyr54 and Tyr83, respectively. The full-length protein was obtained with 64% efficiency (*1480*). An earlier experiment employed CGGG and AGGU to introduce *N*ε-(7-nitrobenz-2-oxa-1,3-diazol-4-yl)-Lys and 2-naphthylalanine into streptavidin at Try54 and Thr57, with only 9% yield with respect to wild-type streptavidin (*1483*). Another experiment introduced both β-anthraniloyl 2,3-diaminopropionic acid and 4′-nitro-Phe into double mutants of streptavidin as a fluorophore-quencher pair, using CGGG and GGGC four-base codons (*1484*). There is some evidence that suggests longer codons (four- or five-base recognition) may have played a role during evolution, with the three-base codon winning out due to an optimum combination of sufficient information (encoding a range of amino acids with some degeneracy) combined with energy conservation (compared to longer systems) (*1485*).

The four-base codon strategy has been combined with other approaches in order to introduce two different amino acids. A four-base CGGG codon was employed to incorporate 7-aza-Trp into dihydrofolate reductase, and a UAG codon to encode *N*β-dabcyl-2,3-diaminopropionic acid, generating a donor–acceptor pair of amino acids for fluorescent resonance energy transfer (FRET) experiments (*1486*). The four-base codon AGGA was combined with a synthetase/tRNA pair derived from archaeal tRNA^Lys to express myoglobin in *E. coli* with a homoglutamine residue at position Gly-24. A double mutation was also carried out, simultaneously using a second orthogonal tRNA/tyrosyl-tRNA synthetase pair from methanocaldococcus *jannaschii* to replace Ala-75 with an *O*-methyl-Tyr residue in response to an amber stop codon (*1487*).

The acylated tRNA required for mutagenesis insertion can be produced by reaction with cyanomethyl active esters of the *N*-protected α-amino acids (*1457*), while a reportedly more effective chemical protection/deprotection strategy was described in 1997 (*1458*). A resin-immobilized ribozyme has been employed to aminoacylate tRNAs with the cyanomethyl esters of L-Phe, α-*N*-biotinyl-L-Phe, and 4'-AcNH-Phe (*1459*).

Further developments to the tRNA protein synthesis technique include non-chemical methods to acylate the tRNA. A hepatitis delta virus ribozyme has been used to improve the synthesis of aminoacyl-tRNAs loaded with non-natural amino acids, giving large amounts of homogeneous product suitable for larger scale-protein synthesis (*1488*). Ordinarily, the suppressor tRNA must not be recognized by the aminoacyl-tRNA synthetases present in the in vitro protein synthesis extract, as it could then be deacylated and/or acylated with a common amino acid. In 1997 Schultz and co-workers reported an engineered tRNA/aminoacyl-tRNA synthetase pair, creating an "orthogonal" suppressor tRNA which is uniquely acylated in *E. coli* by an engineered aminoacyl-tRNA synthetase. The tRNA was prepared by introducing eight mutations in tRNA$_2$^Gln, producing a tRNA that is not acylated by any endogenous *E. coli* tRNA synthetase, yet still is effective at protein translation. Mutagenesis of glutaminyl-tRNA aminoacyl synthetase then produced a mutant GlnRS enzyme that could efficiently acylate the engineered tRNA in vivo (*1489*).

An alternative approach is to combine tRNA acylation proteins from different organisms. A suppressor tRNA from the yeast *Saccharomyces cerevisiae* is "orthogonal" to *Escherichia coli* aminoacyl-tRNA synthetase, yet functions within the *E. coli* translational machinery. Thus, the yeast glutaminyl-tRNA synthetase aminoacylates the yeast suppressor tRNA in *E. coli*, but does not acylate any of the *E. coli* tRNAs, providing an orthogonal tRNA/synthetase pair that can introduce unusual amino acids during protein synthesis in *E. coli* (*1490*). In association with this work, a method was developed to quickly assess which unnatural amino acids can be taken up by bacterial cells. In a similar manner, an orthogonal amber suppression tRNA/aminoacyl-tRNA synthetase pair was derived from *S. cerevisiae* tRNA^Asp and aspartyl-tRNA synthetase, and then used in *E. coli* (*1491*). Wild-type synthetases will accommodate unusual amino acids with conservative alterations, such as substituted phenylalanines in Phe-tRNA synthetase. A virtual screening method correctly predicted which aryl residues could be accommodated by *Thermus thermophilus* PheRS (*1492*).

Further flexibility is introduced by mutating the synthetase to alter substrate selectivity. *E. coli* tyrosyl-tRNA synthetase was mutated so that it aminoacylates *E. coli* tRNA$_{CUA}$ with an unusual aromatic amino acid, but none of the endogenous amino acids. This *E. coli* enzyme–substrate pair is orthogonal to *S. cerevisiae* yeast translational machinery in that the *E. coli* tRNA synthetase does not aminoacylate *S. cerevisiae* cytoplasmic tRNAs, while the *E. coli* tRNA$_{CUA}$ is a poor substrate for

S. cerevisiae aminoacyl tRNA synthetases, yet functions in protein translation in *S. cerevisiae*. The mutations were carried out by genetic selection of random mutations of five residues in the active site region of tyrosyl-tRNA synthetase close to the *para* position of bound Tyr. The derivatizable amino acids 4′-acetyl-L-Phe, 4′-benzoyl-L-Phe, and 4′-azido-L-Phe, along with 4′-iodo-L-Phe (with a heavy atom potentially useful for phasing X-ray structure data), and *O*-methyl L-Tyr (with a methyl group that can be isotopically labeled) were then incorporated into human superoxide dismutase via use of the nonsense codon TAG and misacylated tRNA in the *S. cerevisiae* system. A minimum incorporation purity of 99.8% was obtained (*1493*). In another paper, either 4′-azido-L-Phe or *O*-propargyl-L-Tyr replaced Trp33 in human superoxide dismutase, and then were covalently modified by a [3+2] cycloaddition reaction with an alkyne or azido moiety (respectively) linked to a dye molecule (*1494*). In 2004, *E. coli* leucyl-tRNA synthetase was mutated to accommodate *O*-methyl-L-Tyr, 2-aminooctanoic acid, or *S*-(*o*-nitrobenzyl)-L-Cys. The orthogonal pairing with *E. coli* leucyl suppressor tRNA$_{CUA}$ was then used in a *S. cerevisiae* system to incorporate the amino acids at position 33 in human superoxide dismutase with high efficiency and fidelity. The *S*-(*o*-nitrobenzyl)-L-Cys was subsequently used to replace the active site Cys residue in caspase 3, giving a photocaged, inactive enzyme. Photolysis produced active proteolytic enzyme (*1495*).

Another orthogonal system employed a mutated Trp tRNA synthetase from *Bacillus subtilis* which recognized only 5′-hydroxy-Trp instead of Trp, and a mutated opal suppressor tRNA^{Trp}, in a mammalian expression system (human 293 T cells). The *B. subtilis* TrpRS charged only the cognate mutRNA$^{Trp}_{UCA}$ and no endogenous mammalian tRNAs, while the mutRNA$^{Trp}_{UCA}$ was not charged by any mammalian synthetases. 5′-Hydroxy-Trp was selectively incorporated into the protein foldon as a replacement for Trp-68 by use of a TGA codon, with >97% fidelity. The mutant protein had significantly stronger fluorescence and underwent electrochemical cross-linking (*1496*). Extensive mutagenic modification of leucyl-tRNA synthetase from *Methanobacterium thermoautotrophicum* gave it the ability to recognize and suppress TAG amber codons, TGA opal codons or AGGA four-base codons. This formed an orthogonal pair with leucyl tRNA from *Halobacterium* sp. NRC-1 for reaction in *E. coli* (*1497*).

In 2002 an engineered *E. coli* tyrosyl-tRNA synthetase was modified to recognize 3′-iodo-Tyr (the wild type doesn't recognize the amino acid due to the bulky iodo substituent). When employed in a wheat germ cell-free system, an *E. coli* suppressor tRNA^{Tyr} was not aminoacylated by the wheat germ enzymes, but reacted with the mutated synthetase to give 3′-iodotyrosyl-tRNA. The eukaryotic in vitro translation system then delivered the iodotyrosine to >95% of an amber stop codon position in Ras protein (*1498*). Crystal structures of the mutated *E. coli* tyrosyl-tRNA synthetase, engineered to recognize 3′-iodo-L-Tyr rather than L-Tyr,

have been obtained with both 3-iodo-L-Tyr and L-Tyr bound, showing how the mutant binding pocket is optimized for 3-iodo-L-Tyr recognition (*1499*). Alternatively, *E. coli* amber (UAG) and ochre (UAA) tRNA suppressors have been chemically aminoacylated with Tyr. When transfected into COS1 cells, the Tyr residue was inserted at a stop codon position. Transfection with the same non-aminoacylated suppressor tRNAs did not result in any incorporated amino acids, meaning that the *E. coli* tRNAs were not a substrate for mammalian aminoacyl-tRNA synthetase. This opens up the possibility of being able to simultaneously incorporate two different amino acids at two different positions within a protein, by making use of the two different stop codons (*1500*).

The most widely used orthogonal pairing employs an amber suppressor tRNA/aminoacyl-tRNA synthetase pair derived from the tyrosyl-tRNA synthetase of *Methanocaldococcus jannaschii*, recognizing the TAG amber nonsense codon (*1501*). A library of tyrosyl-tRNA synthetase mutants was generated to alter the amino acid specificity, allowing for incorporation of *O*-methyl Tyr into dihydrofolate reductase with a minimum incorporation purity of >95% (*1502*). Subsequently, [^{15}N]-*O*-Methyl Tyr replaced Ser-4 in whale myoglobin (*1503*). A theoretical study employed structure-based design principles to propose a mutant *Methanococus jannaschii* tyrosyl-tRNA synthetase suitable for incorporation of *O*-methyl Tyr, but experimental proof of the predicted mutations was not carried out (*1504*). In a similar manner, *Methanococcus jannaschii* TyrRS and a mutated tyrosine amber suppressor tRNA (mutRNA$^{Tyr}_{CUA}$) were designed to accept only 4′-acetyl-L-Phe. The amino acid was incorporated into the Z domain of staphyloccocal protein A expressed in *E. coli* cells, with a labeling specificity of 80–90%. The residue was then chemically modified by reaction with fluorescein hydrazide or biotin hydrazide (*1505*). Alternatively, the modified Phe residue was used to form glycoprotein mimetics by reaction with saccharides derivatized with an aminoxy group (*1506*). The same method allowed for incorporation of *O*-allyl Tyr into mutant Z-domain protein, providing an allyl handle for site-specific chemical modifications (*1507*).

Methanocaldococcus jannaschii tyrosyl-tRNA synthetase/mutRNA$^{Tyr}_{CUA}$ pairs were also created to recognize β-(2-naphthyl)-L-Ala (incorporated into mouse dihydrofolate reductase) (*1508*), *O*-(β-*N*-acetylglucosamine)-Ser (expressed in myoglobin, allowing for synthesis of glycoproteins) (*1509*), DOPA (incorporated into sperm whale myoglobin as a redox-active amino acid, with the mutated myoglobin showing a modified redox profile) (*1510*), and *p*-azido-L-Phe (used for site-specific modification of both sperm whale myoglobin and glutathione-*S*-transferase) (*1511*). An orthogonal *Methanocaldococcus jannaschii* tRNA synthetase/mutant tRNA pair was also developed for 4′-amino-Phe, and then inserted into *E. coli* along with *Streptomyces* genes encoding for a synthesis of 4′-amino-Phe. This effectively created a bacterium with a 21-amino acid genetic code encoding 4′-amino-Phe. The residue was incorporated into sperm whale myoglobin (*1512, 1513*). Some of this research was summarized in a 2003 essay (*1432*), while other advances were reviewed in 2002 (*1438*), 2004 (*1433*), and 2005 (*1440*) articles. In 2006 an improved system for efficient incorporation of unnatural amino acids in *E. coli* was reported, using a single-plasmid system to incorporate multiple copies of a gene encoding *Methanocaldococcus jannaschii* tyrosyl-tRNA amber suppressor along with a gene encoding *Methanocaldococcus jannaschii* tyrosyl-tRNA synthetase. The system was tested with 4′-iodo-Phe, 4′-azido-Phe, 4′-acetyl-Phe, and 4′-benzoyl-Phe (*1514*).

Other approaches to incorporating unnatural amino acids have been developed. The fidelity of amino acid pairing with the appropriate tRNA by aminoacyl tRNA synthetases is ensured by a hydrolytic "editing" function at a separate active site. Random mutagenesis generated mutants of *E. coli* that incorrectly aminoacylated tRNA^{Val} with Cys; more than 20% of Val in cellular proteins from such a mutant organism could be replaced with 2-aminobutyric acid (Abu), a residue very similar in structure to Cys (*1515*). Adding steric bulk to the amino acid binding pocket of the editing active site of LeuRS allowed the mutant to be misacylated with Ile (*1516*).

Rather than using nonsense suppressor tRNAs or four-base-anticodon tRNAs to deliver amino acids, it is possible to use an unnatural base pair in a codon in combination with a tRNA that recognizes a codon containing this new base pair. This method would potentially allow for the simultaneous addition of several different unusual amino acids. An initial report in 1992 employed an isoG:isoC pair, allowing for incorporation of 3′-iodo-Tyr by use of chemically synthesized mRNA with an (isoC)Ag codon and an aminoacylated tRNA with a CU(isoG) anticodon (*1517*). Significant improvements were described in 2002 with the development of a 2-amino-6-(2-thienyl)purine (**s**) and pyridin-2-one (**y**) unnatural base pair. A **y**AG codon was inserted in the mRNA encoding position 32 of the human Ras protein, and *S. cerevisiae* tRNA containing the anticodon CU**s** was chemically prepared, and then enzymatically ligated with 3′-chloro-Tyr. The Ras protein was then synthesized in an *E. coli* cell-free system, giving protein with >90% 3′-chloro-Tyr content at position 32 (*1820*). A modified unnatural base-pair system with improved characteristics was subsequently developed (*1518*).

A purified translation system free of amino acid-tRNA synthetases was combined with chemoenzymatically synthesized non-suppressor amino acid-tRNA substrates, allowing for the use of natural codons to encode multiple unnatural amino acids. In one experiment five adjacent allylglycine residues were incorporated into a heptapeptide using an ACC codon, with 30% overall yield compared to the same peptide prepared with five naturally encoded Thr residues. Next, three adjacent different codons (AAC, ACC, and GUU) were employed in a five-codon template, which was translated to give a Met-propargylglycine-allylglycine-*O*-methoxy-Ser-Glu peptide, with approximately 55% yield compared to a translation with the natural amino

acid substrates. Finally, incorporation of a single biotinyllysine residue proceeded with only 20% efficiency, indicating the importance of establishing how well individual amino acids are tolerated (*1519*). This method was also employed to test the ability of the translation machinery to process other non-standard amino acids and analogs: α-hydroxy-acids, *N*-alkyl amino acids, and α,α-disubstituted amino acids were incorporated in decreasing efficiency, while β-amino acids and D-amino acids gave no product. Variants of both Ala and Phe were tested (*1520*).

A potentially versatile approach allows blocks of different amino acids to be incorporated at different sites by changing the response to degenerate codons. Phe is encoded by UUC and UUU, with both codons read by the same tRNA with a GAA anticodon. The UUU codon, read by a G-U wobble base pair, thus forms a less-stable pairing. A mutant yeast tRNA^{Phe} was engineered with an AAA anticodon to read the UUU sequence faster than the wild-type tRNA$^{Phe}_{GAA}$ in *E. coli*. *E. coli* Phe aminoacyl-tRNA synthetase does not effectively charge the yeast tRNA$^{Phe}_{AAA}$, so by adding a mutant yeast PheRS that selectively charged yeast tRNA$^{Phe}_{AAA}$ with 3-(2-naphthyl)-Ala (Nal), an orthogonal system was obtained. Murine dihydrofolate reductase was then expressed in a Phe auxotrophic *E. coli* system in the presence of Nal, producing a protein with predominantly Phe at the five positions encoded by UUU and predominantly Nal at the four positions encoded by UUC. However, the fidelity of the degenerate codons was clearly not complete (*1521*). This approach, and the historical developments leading to it, were summarized in an overview of methods being employed to expand nature's protein repertoire (*1522*).

The capacity of eukaryotes to accommodate unusual amino acids has been assessed with a series of puromycin analogs containing various L- and D-amino acids and L-β-amino acids. Puromycin is a small molecule mimic of aminoacyl tRNA which enters the ribosomal A site and participates in forming a peptide bond with the growing peptide chain, resulting in inhibition of translation. The degree of inhibition caused by the modified puromycins was proposed to reflect the ability of the ribosome to tolerate the unusual amino acids (*1523*). Future research may allow for the puromycins to incorporate unusual amino acids during protein synthesis.

12.5.3e Protein Synthesis

Unnatural amino acids can be incorporated into proteins in vivo by employing auxotrophic expression hosts. This area was included in a 2004 review on methods to amplify the amino acid repertoire of protein biosynthesis (*1439*). Auxotrophic incorporation is commonly employed to incorporate isotopically labeled amino acids into proteins, and differs from the tRNA method in that all residues within the protein corresponding to the auxotrophic strain are replaced. Conservative replacements must be employed. For example, human recombinant annexin V containing 4-thiaproline was prepared by using a proline-auxotrophic *E. coli* strain in minimal media in the presence of thiaproline (*1524*). A Tyr auxotrophic *E. coli* host strain was employed to replace the native Tyr residues in recombinant annexin V (12 residues) or azurin (two residues) with 3′-fluoro-Tyr or 3′-chloro-Tyr (*1525*), while an *E. coli* strain auxotrophic for His introduced 2′- and 4′-fluoro-His into the chaperone protein PapD (*1526*). The two Trp residues in the Src homology 3 domain of the c-Crk-I adaptor protein were replaced with 7-aza-Trp by expression in *E. coli* Trp auxotrophs, giving a domain-specific labeled analog that was then coupled to the non-labeled SH2 domain by expressed protein ligation (*1340*). In contrast, instead of employing a Trp auxotrophic strain, 3-β-indoleacrylic acid was added to inhibit the normal Trp biosynthesis, with 5′-fluoro-Trp incorporated into BIR3 protein in liu of the missing Trp (*1527*).

An *E. coli* strain auxotrophic for Met was used to determine the ability of Met analogs to be incorporated into dihydrofolate reductase; norleucine, 2-amino-5-hexenoic acid, and 2-amino-5-hexynoic acid successfully supported protein synthesis, while norvaline, 2-aminoheptanoic acid, *cis-* or *trans*-2-amino-4-hexenoic acid, or 6,6,6-trifluoro-2-aminohexanoic acid did not (*1528, 1529*). 2-Amino-3-trifluoromethylpentanoic acid or 2-amino-5,5,5-trifluoro-3-methylpentanoic acid replaced >90% of the encoded Ile positions in murine dihydrofolate reductase or cytokine murine interleukin-2 in an Ile auxotrophic *E. coli* strain, producing functional proteins (*1530*). Attempts to incorporate trifluoro-Leu or trifluoro-Met into larger proteins (>10 Da) to create "Teflon" proteins resulted in only partial substitution (5–7% at most), due to high toxicity of the amino acids and an inability to accommodate the analogs into the compact cores of proteins without affecting the protein structural integrity (*1531*). Auxotrophs have also been employed to introduce unusual amino acids into antibiotics. A mutant strain of *Amicolatopsis mediterranei* deficient in β-hydroxy-Tyr biosynthesis was supplemented with 2-fluoro-β-hydroxy-Tyr, 3′-fluoro-β-hydroxy-Tyr or 3′,5′-difluoro-β-hydroxy-Tyr, resulting in the formation of the vancomycin-type antibiotic balhimycin containing fluorine-substituted β-hydroxy-Tyr residues (*1532*).

The auxotroph method can be combined with mutations to allow for greater flexibility in the unusual amino acids that can be tolerated. Multiple residues of 4′-acetyl-Phe were incorporated into dihydrofolate reductase using an auxotrophic *E. coli* strain with a mutated PheRS designed to accommodate the more bulky 4′-acetyl-Phe substituent. The mutant was then treated with biotin hydrazide to derivatize the new amino acids (*1533*). The 4′-acetyl-L-Phe residue has also been incorporated into human growth hormone to allow for specific attachment of PEG polymer, improving the biological half-life of the protein upon in vivo administration (*1534*). An *E. coli* strain auxotrophic for Phe, with a mutant Phe tRNA synthetase with enlarged substrate-binding pocket, was employed to prepare dihydrofolate reductase with

the Phe residues replaced by 4′-bromo-, -iodo-, -cyano-, -ethynyl-, or -azido-Phe, 2′,3′,4′,5′,6′-pentafluoro-Phe, or β-(2-,3-, or 4-pyridyl)-Ala. The extent of Phe replacement varied from 45 to 90%, with only pentafluoro-Phe not incorporated. The pyridylalanines could be incorporated without need of the mutated tRNA synthetase (*1535*). A similar strategy was used for global replacement of Pro within elastin-1, incorporating *cis-* and *trans*-4-fluoro- and 4-hydroxy-Pro, 4,4-difluoro-Pro, 3,4-dehydro-Pro, 4-thia-Pro, pipecolic acid, and azetidine-2-carboxylic acid (*1536*).

An alternative method of (non-specifically) incorporating unusual amino acids is simply to add them during protein overproduction. The protein biosynthesis proofreading mechanism works accurately against naturally occurring amino acids, but is more flexible against nonnatural amino acids with structures similar to a natural one. For example, 3-(1-pyrenyl)-L-Ala was incorporated into β-D-galactosidases overproduced by *E. coli* (*1537*). γ-Azido-Abu (azidohomoalanine) replaced Met in proteins expressed in methionine-depleted *E. coli* bacterial cultures, with murine dihydrofolate reductase containing a >95% replacement of all eight Met residues. A Staudinger ligation of the modified protein produced with a range of adducts produced (*1538*). β-Azido-Ala (azidoalanine) was found to be insufficiently similar to Met to be efficiently activated by the methionyl-tRNA synthetase. However, in another report, β-azido-Ala, γ-azido-Abu, δ-azido-Nva, and ε-azido-Nle were all incorporated into the *E. coli* outer membrane protein OmpC via expression in media depleted of Met and supplemented with the non-canonical amino acid. The surface-exposed azido residues were then employed for a Cu-catalyzed triazole formation with an alkyne linked to biotin, and then visualized by staining with fluorescent avidin. With this detection method, incorporation of β-azido-Ala was observed (*1539*). Val and Ile could be replaced by L-α-(1-cyclobutenyl)glycine in green fluorescent protein overexpressed in *E. coli*, though Leu could not (*1540*).

Attempts to incorporate (2*S*,3*R*)-4,4,4-trifluorovaline into recombinant proteins using *E. coli* cultures depleted in either Ile or Val were unsuccessful. However, by overexpressing the aminoacyl-tRNA synthetase for Ile or Val, the trifluorovaline replaced approximately 92% of the encoded Ile residues in murine dihydrofolate reductase, or 86% of the Val residues (*1541*). Similarly, hexafluoroleucine was incorporated into a coiled-coil leucine zipper protein A1 by expression in *E. coli* in cultures depleted of Leu and supplemented with hexafluoroleucine, with effective incorporation only achieved when the *E. coli* LeuRS was overexpressed (*1542*).

Peptide synthesis can also be accomplished by nonribosomal (NRPSs) peptide synthetases, which activate amino acids by formation of enzyme-bound thioester, with amino acids loaded on a series of modules arranged in the order corresponding to the primary sequence of the final product. The activated amino acids are sequentially transferred to the *N*-terminus of the adjacent residue, with the final thioester linkage cleaved by hydrolysis or cyclization (*1543*). The activity of module components is still being elucidated (*1544*). The NRPSs assemble complex natural products containing many unusual amino acids, with the production of antibiotics particularly well researched (*1545, 1546*). Genetic engineering has been employed to delete a module coding for a Leu residue in the cyclic lipodepsipeptide antibiotic surfactin from *Bacillus subtilis*, resulting in production of an analog with a decreased ring size (*1547*). The apparatus seems ripe for exploitation to incorporate unusual amino acids.

12.5.4 Coupling During Amino Acid Synthesis

A number of amino acid synthesis methods combine amino acid synthesis with peptide bond formation. The most common of these is the Ugi four-component condensation (see Sections 2.2.3, 5.2.2), which can produce tripeptides (reviewed in 1982 (*1548*)). The Ugi reaction has been applied to the synthesis of peptides containing very bulky α,α-disubstituted-α-amino acids (*1549*), such as Aib-Dph-Aib or Dph-Dph-Dph sequences (Dph = α,α-diphenylglycine) (*1550*). α-Trifluoromethyl-α-substituted-α-amino acids are particularly challenging residues for coupling, as in addition to the steric bulk, the amino group has reduced nucleophilicity (pKa for Ala-NH$_2$ = 9.87; for α-trifluoromethyl-Ala-NH$_2$ = 5.91 (*1551*)). Ugi 4CC reactions coupled these residues by using them as the acid or isocyanide components (*1551*).

The Hegedus synthesis of amino acids via photolysis of a chromium carbene complex (see Section 5.4.1a) is also amenable to direct incorporation of the synthesized residue onto a growing peptide chain, as the photolyzed product can be trapped by the peptide *N*-terminal amino group (*1552, 1553*). This method is effective even with very hindered *N*-methyl-α,α-dialkyl-α-amino acids (*1819*). However, it proved to be impractical for multiple couplings on a Merrifield resin (*1552*); somewhat better results were obtained on a polyethylene glycol support (*1554*).

A synthesis of α,α-disubstituted -α-amino acids via alkylation of an amino aldehyde imine, in which the amino group was contained within an oxazolidinone auxiliary, was used to prepare dipeptides by employing an Aib residue to form the imine (*1555*).

Amino acid synthesis can also generate intermediates that are activated for incorporation into peptides via ring-opening reactions. α,α-Dialkyl-α-amino acids can be coupled through 3-amino-2*H*-azirines (see Section 8.3). In contrast to conventional peptide synthesis, the azirine synthon acts as an activated amino component, and reacts quantitatively with the carboxyl group of *N*-protected amino acids-or peptides (*1556–1558*). Aib oligopeptides (*1145, 1559*) and other α,α-dialkyl-α-amino acid-containing peptides (*1557, 1558*) were prepared by this route. Azirines were employed to synthesize cyclic hexapeptides containing up to three Aib and one α-methyl-Phe residues (*1560*), and cyclic hexadepsipeptides containing up to five Aib residues

(*1561*). A solid-phase *N*- to *C*-terminus peptide synthesis procedure employing azirines has been reported (*1562*). Amino acids ready for coupling as their NCAs (*N*-carboxy anhydrides) have been synthesized from α-hydroxy-β-lactams by a Baeyer–Villiger rearrangement of the corresponding α-keto-β-lactams or α-hydroxy-β-lactams (see Section 8.5) (*1124–1127, 1563*). α,α-Disubstituted amino acids were synthesized and coupled by this route, with an α-MeNva-Aib dipeptide formed in 70% yield (*1563*). γ-Amino-β-hydroxy acids have been coupled with α-amino acid esters via ring opening of an α,α-dichloro-β-lactone intermediate, which was produced by a cycloaddition of dichloroketene with *N*-Boc α-amino aldehydes (*1564*). *N*-Boc β-lactams were opened and coupled to α-amino esters by combining the two in DMF in the presence of 1 equiv of sodium azide, allowing for dipeptide synthesis under mild conditions (*1565*).

12.5.5 Peptide Synthesis in the Reverse Direction

Almost all chemical peptide synthesis is conducted with the chain growing in the C-to-N direction. One of the main reasons is that synthesis in the reverse (N→C) direction relies on activation of an *N*-acyl amino acid, which is susceptible to racemization. However, several reverse direction solid-phase syntheses have been described in recent years. In one report an amino acid was tethered to the resin via an *N*-linked silyl carbamate, and then activated with isobutyl chloroformate and reacted with an amino acid allyl ester. Tripeptides were prepared, with no apparent epimerization, even with Phg (*1104, 1566*). Alternatively, the amine of an amino acid allyl ester was attached to an acid-sensitive chlorotrityl resin, deprotected with Pd^0-PhSiH$_3$, activated with HATU/DIEA or DIPCDI/Cu(OBt)$_2$, and reacted with an amino acid allyl ester. The best results were obtained with DIPCDI/Cu(OBt)$_2$, which gave a tetrapeptide that was 56% pure and contained only 0.6% of epimerized diastereomers. A synthesis of the pentapeptide Leu-enkephalin showed no racemization and 57% purity with DIPCDI/Cu(OBt)$_2$, but 5% racemization and 21% purity with HATU/DIEA (*1567*). This compared to an earlier fluorenylmethyl ester approach that gave higher purities, but 40.5% epimerization with TBTU/NMM or 8.2% with HOBt/DIPCDI (*1568*). Another solid-phase *N*- to *C*-terminus peptide synthesis procedure employed azirine to introduce Aib residues, using a PAM-carbamate linker for the *N*-terminal amino acid amine group. A Phe-O*t*Bu residue was incorporated using PyBOP as coupling reagent; with 2.8% racemization (*1562*).

A general solid-phase strategy was reported in 2004, with an amino acid *tert*-butyl ester attached to hydroxymethyl polystyrene resin via a urethane linkage prepared by activation of the resin with phosgene. The *tert*-butyl ester of the first residue was deprotected with 50% TFA/DCM, and the next amino ester coupled with HATU/TMP. Resin deprotection was achieved with 10% TFMSA/TFA. Seven tripeptides and several *C*-terminally modified tripeptides were synthesized; analysis for amino acid racemization found from 1.0 to 1.4% D-amino acid for each residue (Ala, Val, Leu, Orn, Asn, Phe, Tyr) (*1569*). An earlier report coupled an unprotected amino acid to a *p*-nitrophenolchloroformate-activated TentaGel-P resin. The next residue was introduced as an amino acid tri-*tert*-butoxysilyl ester, using HATU/TMP. The ester was deprotected with 5% TFA in CH_2Cl_2, and the cycle repeated. Final cleavage from the resin employed photolytic conditions. Racemization of around 5% per residue was observed, except for 20% for Ser (*t*Bu) (*1570*).

References

1. Breuer, M.; Ditrich, K.; Habicher, T.; Hauer, B.; Keßeler, M.; Stürmer, R.; Zelinski, T. *Angew. Chem., Int. Ed.* **2004**, *43*, 788–824.
2. Armstrong, D.W.; He, L.; Yu, T.; Lee, J.T.; Liu, Y.-S. *Tetrahedron: Asymmetry* **1999**, *10*, 37–60.
3. Armstrong, D.W.; Lee, J.T.; Chang, L.W. *Tetrahedron: Asymmetry* **1998**, *9*, 2043–2064.
4. Greenstein, J.P.; Winitz, N. *Chemistry of the Amino Acids, Vol. 1*; Wiley: New York, 1961, pp 715–760.
5. Barrett, G.C. *Chemistry and Biochemistry of the Amino Acids*; Chapman and Hall: New York, 1985, pp 338–353.
6. Fogassy, E.; Nógrádi, M.; Kozma, D.; Egri, G.; Pálovics, E.; Kiss, V. *Org. Biomol. Chem.* **2006**, *4*, 3011–3030.
7. Stecher, H.; Faber, K. *Synthesis* **1997**, 1–16.
8. Pellissier, H. *Tetrahedron* **2003**, *59*, 8291–8327.
9. Ebbers, E.J.; Ariaans, G.J.A.; Houbiers, J.P.M.; Bruggink, A.; Zwanenburg, B. *Tetrahedron* **1997**, *53*, 9417–9476.
10. Turner, N.J. *Curr. Opin. Chem. Biol.* **2004**, *8*, 114–119.
11. Strauss, U.T.; Felfer, U.; Faber, K. *Tetrahedron: Asymmetry* **1999**, *10*, 107–117.
12. Yamada, S.; Hongo, C.; Chibata, I. *Agric. Biol. Chem.* **1978**, *42*, 1521–1526.
13. Ndzié, E.; Cardinael, P.; Schoofs, A.-R.; Coquerel, G. *Tetrahedron: Asymmetry* **1997**, *8*, 2913–2920.
14. O'Donnell, M.J.; Delgado, F. *Tetrahedron* **2001**, *57*, 6641–6650.
15. Yoshioka, R.; Ohtsuki, O.; Da-Te, T.; Okamura, K.; Senuma, M. *Bull. Chem. Soc. Jpn.* **1994**, *67*, 3012–3020.
16. Gal, G.; Chemerda, J.M.; Reinhold, D.F.; Purick, R.M. *J. Org. Chem.* **1977**, *42*, 142–143.
17. Oki, K.; Suzuki, K.; Tuchida, S.; Saito, T.; Kotake, H. *Bull. Chem. Soc. Jpn.* **1970**, *43*, 2554–2558.
18. Barrett, G.C.; Cousins, P.R. *J. Chem. Soc., Perkin Trans. 1* **1975**, 2313–2315.
19. Bey, P.; Danzin, C.; Van Dorsselaer, V.; Mamont, P.; Jung, M.; Tardif, C. *J. Med. Chem.* **1978**, *21*, 50–55.
20. Tia, Z.; Edwards, P.; Roeske, R.W. *Int. J. Pept. Prot. Res.* **1992**, *40*, 119–126.
21. Vriesema, B.K.; ten Hoeve, W.; Wynberg, H.; Kellogg, R.M.; Boesten, W.H.J.; Meijer, E.M.; Schoemaker, H.E. *Tetrahedron Lett.* **1986**, *26*, 2045–2048.
22. Fujii, I.; Hirayama, N. *Helv. Chim. Acta* **2002**, *85*, 2946–2960.
23. Vries, T.; Wynberg, H.; van Echten, E.; Koek, J.; ten Hoeve, W.; Kellogg, R.M.; Broxterman, Q.B.; Minnaard, A.; Kaptein, B.; van de Sluis, S.; Hulshof, L.; Kooistra, J. *Angew. Chem., Int. Ed. Engl.* **1998**, *37*, 2349–2354.
24. Collet, A. *Angew. Chem., Int. Ed. Engl.* **1998**, *37*, 3239–3241.
25. Kaptein, B.; Elsenberg, H.; Grimbergen, R.F.P.; Broxterman, Q.B.; Hulshof, L.A.; Pouwer, K.L.; Vries, T.R. *Tetrahedron: Asymmetry* **2000**, *11*, 1343–1351.
26. Prashad, M.; Har, D.; Repic, O.; Blacklock, T.J.; Giannousis, P. *Tetrahedron: Asymmetry* **1999**, *10*, 3111–3116.
27. Abele, S.; Vögtli, K.; Seebach, D. *Helv. Chim. Acta* **1999**, *82*, 1539–1558.
28. Reinhold, D.F.; Firestone, D.F.; Gaines, W.A.; Chemerda, J.M.; Sletzinger, M. *J. Org. Chem.* **1968**, *33*, 1209–1213.
29. Scott, J.W.; Focella, A.; Hengartner, U.O.; Parrish, D.R.; Valentine D., Jr., *Synth. Commun.* **1980**, *10*, 529–540.
30. Stingl, K.; Kottenhahn, M.; Drauz, K. *Tetrahedron: Asymmetry* **1997**, *8*, 979–982.
31. Rios, A.; Crugeiras, J.; Amyes, T.L.; Richard, J.P. *J. Am. Chem. Soc.* **2001**, *123*, 7949–7950.
32. Clark, J.C.; Phillips, G.H.; Steer, M.R. *J. Chem. Soc., Perkin Trans. 1* **1976**, 475–481.
33. Moseley, J.D.; Williams, B.J.; Owen, S.N.; Verrier, H.M. *Tetrahedron: Asymmetry* **1996**, *7*, 3351–3352.
34. Kamphuis, J.; Meijer, E.M.; Boesten, W.H.J.; Sonke, T.; van den Tweel, W.J.J.; Schoemaker, H.E. *Ann. N.Y. Acad. Sci.* **1992**, *672*, 510–527.
35. Maryanoff, C.A.; Scott, L.; Shah, R.D.; Villani, F.J. Jr., *Tetrahedron: Asymmetry* **1998**, *9*, 3247–3250.
36. Miyazaki, H.; Ohta, A.; Kawakatsu, N.; Waki, Y.; Gogun, Y.; Shiraiwa, T.; Kurokawa, H. *Bull. Chem. Soc. Jpn.* **1993**, *66*, 536–540.
37. Obrecht, D.; Bohdal, U.; Broger, C.; Bur, D.; Lehmann, C.; Ruffieux, R.; Schönholzer, P.; Spiegler, C.; Müller, K. *Helv. Chim. Acta* **1995**, *78*, 563–580.
38. Obrecht, D.; Spiegler, C.; Schönholzer, P.; Müller, K. Heimgartner, H.; Stierli, F. *Helv. Chim. Acta* **1992**, *75*, 1666–1696.
39. Obrecht, D.; Abrecht, C.; Altorfer, M.; Bohdal, U.; Grieder, A.; Kleber, M.; Pfyffer, P.; Müller, K. *Helv. Chim. Acta* **1996**, *79*, 1315–1337.
40. Obrecht, D.; Karajiannis, H.; Lehmann, C.; Schönholzer, P.; Spiegler, C.; Müler, K. *Helv. Chim. Acta* **1995**, *78*, 703–714.

41. Obrecht, D.; Bohdal, U.; Daly, J.; Lehmann, C.; Schönholzer, P.; Müller, K. *Tetrahedron* **1995**, *51*, 10883–10900.

42. Mazaleyrat, J.-P.; Boutboul, A.; Lebars, Y.; Gaucher, A.; Wakselman, M. *Tetrahedron: Asymmetry* **1998**, *9*, 2701–2713.

43. Avenoza, A.; Busto, J.H.; Cativiela, C.; Peregrina, J.M.; Rodríguez, F. *Tetrahedron* **1998**, *54*, 11659–11674.

44. Oh-Hashi, J.; Harada, K. *Bull. Chem. Soc. Jpn.* **1966**, *39*, 2287–2289.

45. Adger, B.M.; Dyer, U.C.; Lennon, I.C.; Tiffin, P.D.; Ward, S.E. *Tetrahedron Lett.* **1997**, *38*, 2153–2154.

46. Chung, J.Y.L.; Wasicak, J.T.; Arnold, W.A.; May, C.S.; Nadan, A.M.; Holladay, M.W. *J. Org. Chem.* **1990**, *55*, 270–275.

47. Bigge, C.F.; Wu, J.-P.; Malone, T.C.; Taylor, C.P.; Vartanian, M.G. *Bioorg. Med. Chem. Lett.* **1993**, 3, 39–42.

48. Acher, F.; Azerad, R. *Tetrahedron: Asymmetry* **1994**, *5*, 731–744.

49. Coudert, E.; Acher, F.; Azerad, R. *Tetrahedron: Asymmetry* **1996**, *7*, 2963–2970.

50. Bajgrowicz, J.A.; Cossec, B.; Pigière, Ch.; Jacquier, R.; Viallefont, Ph. *Tetrahedron Lett.* **1984**, *25*, 1789–1792.

51. Solladié-Cavallo, A.; Schwarz, J.; Mouza, C. *Tetrahedron Lett.* **1998**, *39*, 3861–3864.

52. Gibbons, W.A.; Hughes, R.A.; Charalambous, M.; Christodoulou, M.; Szeto, A.; Aulabaugh, A.E.; Mascagni, P.; Toth, I. *Liebigs Ann. Chem.* **1990**, 1175–1183.

53. Trigalo, F.; Acher, F.; Azerad, R. *Tetrahedron* **1990**, *46*, 5203–5212.

54. Lodder, M.; Wang, B.; Hecht, S.M. *Tetrahedron* **2000**, *56*, 9421–9429.

55. Easton, C.J. *Pure Appl. Chem.* **1997**, *69*, 489–494.

56. Kondo, K.; Kurosaki, T.; Murakami, Y. *Synlett* **1998**, 725–7726.

57. Verkhovskaya, M.A.; Yamskov, I.A. *Russ. Chem. Rev.* **1991**, *60*, 1163–1179.

58. Roy, I.; Gupta, M.N. *Tetrahedron* **2003**, *59*, 5431–5436.

59. Chibata, I.; Tosa, T. *Ann. Rev. Biophys. Bioeng.* **1981**, *10*, 197–216.

60. Boland, W.; FröBl, C.; Lorenz, M. *Ang. Chem., Int. Ed. Engl.* **1991**, 1049–1072.

61. Chibata, I.; Tosa, T.; Sato, T.; Mori, T. *Methods Enzymol.* **1976**, *45*, 746–759.

62. Chen, C.-S.; Sih, C.J. *Angew. Chem., Int. Ed. Engl.* **1989**, *28*, 695–707.

63. Zelinski, T.; Waldmann, H. *Ang. Chem., Int. Ed. Engl.* **1997**, *36*, 722–724.

64. Carrea, G. Riva, S. *Angew. Chem., Int. Ed.* **2000,** *39*, 2226–2254.

65. Partridge, J.; Halling, P.J.; Moore, B.D. *Chem. Commun.* **1998**, 841–842.

66. Tremblay, M.; Côte, S.; Voyer, N. *Tetrahedron* **2005**, *61*, 6824–6828.

67. Ke, T.; Klibanov, A.M. *J. Am. Chem. Soc.* **1999**, *121*, 3334–3340.

68. Barrett, G.C. *Chemistry and Biochemistry of the Amino Acids*; Chapman and Hall: New York, 1985, pp 1753–1816.

69. Greenstein, J.P.; Winitz, N. *Chemistry of the Amino Acids, Vol. 2*; Wiley: New York, 1961, pp 1753–1816.

70. Williams, R.M. *Synthesis of Optically Active α-Amino Acids*, Organic Chemistry Series, Vol. 7; Pergamon Press: Toronto, 1989.

71. Santaniello, E.; Ferraboschi, P.; Grisenti, P.; Manzocchi, A. *Chem. Rev.* **1992**, *92*, 1071–1140.

72. Schoffers, E.; Golebiowski, A.; Johnson, C.R. *Tetrahedron* **1996**, 3769–3826.

73. Miyazawa, T. *Amino Acids* **1999**, *16*, 191–213.

74. van Rantwijk, F.; Sheldon, R.A. *Tetrahedron* **2004**, *60*, 501–519.

75. Kubo, K.; Ishikura, T.; Fukagawa, Y. *J. Antibiotics* **1980**, *33*, 556–565.

76. Chen, H.-P.; Wu, S.-H.; Tsai, Y.-C.; Yang, Y.-B.; Wang, K.-T. *Bioorg. Med. Chem. Lett.* **1992**, 2, 697–700.

77. Chenault, H.K.; Dahmer, J.; Whitesides, G.M. *J. Am. Chem. Soc.* **1989**, *111*, 6354–6364.

78. Fu, S.-C.J.; Birnbaum, S.M. *J. Am. Chem. Soc.* **1953**, *75*, 918–920.

79. Birnbaum, S.M.; Levintow, L.; Kingsley, R.B.; Greenstein, J.P. *J. Biol. Chem.* **1952**, *194*, 455–470.

80. Bommarius, A.S.; Drauz, K.; Günther, K.; Knaup, G.; Schwarm, M. *Tetrahedron: Asymmetry* **1997**, *8*, 3197–3200.

81. Easton, C.J.; Harper, J.B. *Tetrahedron Lett.* **1998**, *39*, 5269–5272.

82. Bois-Choussy, M.; Zhu, J. *J. Org. Chem.* **1998**, *63*, 5662–5665.

83. Beller, M.; Eckert, M.; Holla, E.W. *J. Org. Chem.* **1998**, *63*, 5658–5661.

84. Regla, I.; Luna, H.; Pérez, H.I.; Demare, P.; Bustoos-Jaimes, I.; Zaldívar, V.; Calcagno, M.L. *Tetrahedron Asymmetry* **2004**, *15*, 1285–1288.

85. Groeger, U.; Drauz, K.; Klenk, H. *Angew. Chem., Int. Ed. Engl.* **1992**, *31*, 195–197.

86. Watanabe, K.; Liljeblad, A.; Aksela, R.; Kanerva, L.T. *Tetrahedron: Asymmetry* **2001**, *12*, 2059–2066.

87. Sánchez-Sancho, F.; Herradón, B. *Tetrahedron: Asymmetry* **1998**, *9*, 1951–1965.

88. Soloshonok, V.A.; Svedas, V.K.; Kukhar, V.P.; Kirilenko, A.G.; Rybakova, A.V.; Solodenko, V.A.; Fokina, N.A.; Kogut, O.V.; Galaev, I.Y.; Kozlova, E.V.; Shishkina, I.P.; Galushko, S.V. *Synlett* **1993**, 339–341.

89. Soloshonok, V.A.; Kirilenko, A.G.; Fokina, N.A.; Kukhar, V.P.; Galushko, S.V.; Svedas, V.K.; Resnati, G. *Tetrahedron: Asymmetry* **1994**, *5*, 1225–1228.

90. Soloshonok, V.A.; Fokina, N.A.; Rybakova, A.V.; Shishkina, I.P.; Galushko, S.V.; Sorochinsky, A.E.; Kukhar, V.P.; Savchenko, M.V.; Svedas, V.K. *Tetrahedron: Asymmetry* **1995**, *6*, 1601–1610.

91. Cardillo, G.; Tolomelli, A.; Tomasini, C. *J. Org. Chem.* **1996**, *61*, 8651–8654.

92. Soloshonok, V.A.; Kirilenko, A.G.; Fokina, N.A.; Shishkina, I.P.; Galushko, S.V.; Kukhar, V.P.; Svedas, V.K.; Kozlova, E.V. *Tetrahedron: Asymmetry* **1994**, *5*, 1119–1126.

93. See Reference *88*.

94. Cardillo, G.; Gentilucci, L.; Tolomelli, A.; Tomasini, C. *J. Org. Chem.* **1998**, *63*, 2351–2353.

95. von Nussbaum, F.; Spiteller, P.; Rüth, M.; Steglich, W.; Wanner, G.; Gamblin, B.; Stievano, L.; Wagner, F.E. *Angew. Chem., Int. Ed. Engl.* **1998**, *37*, 3292–3295.

96. Roche, D.; Prasad, K.; Repic, O. *Tetrahedron Lett.* **1999**, *40*, 3665–3668.

97. Solodenko, V.A.; Belik, M.Y.; Galushko, S.V.; Kukhar, V.P.; Kozlova, E.V.; Mironenko, D.A.; Svedas, V.K. *Tetrahedron: Asymmetry* **1993**, *4*, 1965–1968.

98. Basso, A.; Braiuca, P.; De Martin, L.; Ebert, C.; Gardossi, L.; Linda, P. *Tetrahedron: Asymmetry* **2000**, *11*, 1789–1796.

99. Raimondi, S.; Forti, L.; Monti, D.; Riva, S. *Tetrahedron Asymmetry* **2003**, *14*, 1091–1094.

100. Liu, W.; Ray, P.; Benezra, S.A. *J. Chem. Soc., Perkin Trans.1* **1995**, 553–559.

101. Lam, L.K.P.; Brown, C.M.; De Jeso, B.; Lym, L.; Toone, E.J.; Jones, J.B. *J. Am. Chem. Soc.* **1988**, *110*, 4409–4411.

102. Janes, L.E.; Kazlauskas, R.J. *Tetrahedron: Asymmetry* **1997**, *8*, 3719–3733.

103. Mityazawa, T.; Iwanaga, H.; Ueji, S.; Yamada, T.; Kuwata, S. *Chem. Lett.* **1989**, 2219–2222.

104. Miyazawa, T.; Onishi, K.; Murashima, T.; Yamada, T.; Tsai, S. –W. *Tetrahedron: Asymmetry* **2005**, *16*, 2569–2573.

105. Hönig, H.; Seufer-Wasserthal, P.; Weber, H. *Tetrahedron* **1990**, *46*, 3841–3850.

106. Björkling, F.; Boutelje, J.; Hjalmarsson, M.; Hult, K.; Norin, T. *J. Chem. Soc., Chem. Commun.* **1987**, 1041–1042.

107. Fadnavis, N.W.; Jadhav, V. *Tetrahedron: Asymmetry* **1997**, *8*, 2361–2366.

108. Reetz, M.T.; Bocola, M.; Carballeira, J.D.; Zha, D.; Vogel, A. *Angew. Chem., Int. Ed.* **2005**, *44*, 4192–4196.

109. Chen, S.-T.; Tsai, C.-F.; Wang, K.-T. *Bioorg. Med. Chem. Lett.* **1994**, *4*, 625–630.

110. Lavrador, K.; Guillerm, D.; Guillerm, G. *Biorg. Med. Chem. Lett.* **1998**, *8*, 1629–1634.

111. Iding, H.; Wirz, B.; Rogers-Evans, M. *Tetrahedron* **2004**, *60*, 647–653.

112. Bogenstätter, M.; Steglich, W. *Tetrahedron* **1997**, *53*, 7267–7274.

113. Vergne, C.; Bois-Choussey, M.; Ouazzani, J.; Beugelmans, R.; Zhu, J. *Tetrahedron: Asymmetry* **1997**, *8*, 391–398.

114. Schutt, H.; Schmidt-Kastner, G.; Arens, A.; Preiss, M. *Biopolymers* **1985**, *27*, 420–433.

115. Watanabe, K.; Yoshida, T.; Ueji, S.-I. *Bioorg. Chem.* **2004**, *32*, 504–515.

116. Berger, A.; Smolarsky, M.; Kurn, N.; Bosshard, H.R. *J. Org. Chem.* **1973**, *38*, 457–460.

117. Puniere, M.; Castro, B.; Domergue, N.; Previero, A. *Tetrahedron: Asymmetry* **1992**, *3*, 1015–1018.

118. Lankiewicz, L.; Kasprzykowski, F.; Grzonka, Z.; Kettmann, U.; Hermann, P. *Bioorg. Chem.* **1989**, *17*, 275–280.

119. Hermann, P.; Baumann, H.; Herrnstadt, Ch.; Glanz, D. *Amino Acids* **1992**, *3*, 105–118.

120. Miyazawa, T.; Minowa, H.; Miyamoto, T.; Imagawa, K.; Yanagihara, R.; Yamada, Y. *Tetrahedron: Asymmetry* **1997**, *8*, 367–370.

121. Miyazawa, T.; Imagawa, K.; Minowa, H.; Miyamoto, T.; Yamada, T. *Tetrahedron* **2005**, *61*, 10254–10261.

122. Tomisaka, K.; Ishida, Y.; Konishi, K.; Aida, T. *Chem. Commun.* **2001**, 133–134.

123. Roos, E.C.; Mooiweer, H.H.; Hiemstra, H.; Speckamp, W.N.; Kaptein, B.; Boesten, W.H.J.; Kamphuis, J.; *J. Org. Chem.* **1992**, *57*, 6769–6778.

124. Eichhorn, E.; Roduit, J.-R.; Shaw, N.; Heinzmann, K.; Kiener, A. *Tetrahedron: Asymmetry* **1997**, *8*, 2533–2536.

125. Steinke, D.; Kula, M.-R. *Angew. Chem., Int. Ed. Engl.* **1990**, 29, 1139–1140.

126. Wang, M.-X.; Lin, S.-J. *J. Org. Chem.* **2002**, *67*, 6542–6545.

127. Sypniewski, M.; Penke, B.; Simon, L.; Rivier, J. *J. Org. Chem.* **2000**, *65*, 6595–6600.

128. Trauthwein, H.; May, O.; Dingerdissen, U.; Buchholz, S.; Drauz, K. *Tetrahedron Lett.* **2003**, *44*, 3737–3739.

129. Youshko, M.I.; van Langen, L.M.; Sheldon, R.A.; Svedas, V.K. *Tetrahedron Asymmetry* **2004**, *15*, 1933–1936.

130. Kaptein, B.; Boesten, W.H.J.; Broxterman, Q.B.; Peters, P.J.H.; Schoemaker, H.E.; Kamphuis, J. *Tetrahedron: Asymmetry* **1993**, *4*, 1113–1116.

131. Van Betsbrugge, J.; Tourwé, D.; Kaptein, B.; Kierkels, H.; Broxterman, R. *Tetrahedron* **1997**, *53*, 9233–9240.

132. Kruizinga, W.H.; Bolster, J.; Kellogg, R.M.; Kamphuis, J.; Boesten, W.H.J.; Meijer, E.M.; Schoemaker, H.E. *J. Org. Chem.* **1988**, *53*, 1826–1827.

133. Schoemaker, H.E.; Boesten, W.H.J.; Kaptein, B.; Hermes, H.F.M.; Sonke, T.; Broxterman, Q.B.; van der Tweel, W.J.J.; Kamphuis, J. *Pure Appl. Chem.* **1992**, *64*, 1171–1175.

134. Schoemaker, H.E.; Boesten, W.H.J.; Kaptein, B.; Roos, E.C.; Broxterman, Q.B.; van der Tweel, W.J.J.; Kamphuis, J. *Acta Chem. Scand.* **1996**, *50*, 225–233.

135. Koksch, B.; Quaedfliegg, P.J.L.M.; Michel, T.; Burger, K.; Broxterman, Q.B.; Schoemaker, H.E. *Tetrahedron Asymmetry* **2004**, *15*, 1401–1407.

136. Ogrel, A.; Shvets, V.I.; Kaptein, B.; Broxterman, Q.B.; Raap, J. *Eur. J. Org. Chem.* **2000**, 857–859.

137. van den Tweel, W.J.J.; van Dooren, T.J.G.M.; de Jonge, P.H.; Kaptein, B.; Duchateau, A.L.L.; Kamphuis, J. *Appl. Microbiol. Biotechnol.* **1993**, *39*, 296–300.

138. Shaw, N.M.; Naughton, A.B. *Tetrahedron* **2004**, *60*, 747–752.

139. Wang, M.-X.; Liu, J.; Wang, D.-X.; Zheng, Q.-Y. *Tetrahedron: Asymmetry* **2005**, *16*, 2409–2416.

140. Yee, C.; Blythe, T.A.; McNabb, T.J.; Waits, A.E. *J. Org. Chem.* **1992**, *57*, 3525–3527.

141. Spero, D.M.; Kapadia, S.R. *J. Org. Chem.* **1996**, *61*, 7398–7401.

142. Berkowitz, D.B.; Pumphrey, J.A.; Shen, Q. *Tetrahedron Lett.* **1994**, *35*, 8743–8746.

143. Lalonde, J.J.; Bergbreiter, D.E.; Wong, C.-H. *J. Org. Chem.* **1988**, *53*, 2323–2327.

144. Hallinan, K.O.; Crout, D.H.G.; Errington, W. *J. Chem. Soc., Perkin Trans. 1* **1994**, 3537–3543.

145. de Zoete, M.C.; Ouwehand, A.A.; van Rantwijk, F.; Sheldon, R.A. *Rec. Trav. Chim. Pay-Bas* **1995**, *114*, 171–174.

146. Starmans, W.A.J.; Doppen, R.G.; Thijs, L.; Zwanenburg, B. *Tetrahedron: Asymmetry* **1998**, *9*, 429–435.

147. Bennett, E.L.; Nieman, C. *J. Am. Chem. Soc.* **1950**, *72*, 1800–1803.

148. Niemann, C.; Rapport, M.M. *J. Am. Chem. Soc.* **1946**, *68*, 1671–1672.

149. Isowa, Y.; Takashima, T.; Ohmori, M.; Kurita, H.; Sato, M.; Mori, K. *Bull. Chem. Soc. Jpn.* **1972**, *45*, 1461–1464.

150. Liljeblad, A.; Kanerva, L.T. *Tetrahedron: Asymmetry* **1999**, *10*, 4405–4415.

151. Liljeblad, A.; Lindborg, J.; Kanerva, A.; Katajisto, J.; Kanerva, L.T. *Tetrahedron Lett.* **2002**, *43*, 2471–2474.

152. Sánchez, V.M.; Rebolledo, F.; Gotor, V. *Tetrahedron: Asymmetry* **1997**, *8*, 37–40.

153. Lu, Y.; Miet, C.; Kunesch, N.; Poisson, J.E. *Tetrahedron: Asymmetry* **1991**, *2*, 871–872.

154. Lu, Y.; Miet, C.; Kunesch, N.; Poisson, J.E. *Tetrahedron: Asymmetry* **1993**, 893–902.

155. Estermann, H.; Seebach, D. *Helv. Chim. Acta* **1988**, *71*, 1824–1839.

156. Imawaka, N.; Tanaka, M.; Suemune, H. *Helv. Chim. Acta* **2000**, *83*, 2823–2835.

157. Crossley, M.J.; Fisher, M.L.; Potter, J.J. Kuchel, P.W.; York, M.J. *J. Chem. Soc., Perkin Trans 1* **1990**, 2363–2369.

158. Prashad, M.; Har, D.; Repic, O.; Blacklock, T.J.; Giannousis, P. *Tetrahedron: Asymmetry* **1998**, *9*, 2133–2136.

159. Faulconbridge, S.J.; Holt, K.E.; Sevillano, L.G.; Lock, C.J.; Tiffin, P.D.; Tremayne, N.; Winter, S. *Tetrahedron Lett.* **2000**, *41*, 2679–2681.

160. Pousset, C.; Callens, R.; Haddad, M.; Larchevêque, M. *Tetrahedron Asymmetry* **2004**, *15*, 3407–3412.

161. Gedey, S.; Liljeblad, A.; Fülöp, F.; Kanerva, L.T. *Tetrahedron: Asymmetry* **1999**, *10*, 2573–2581.

162. Kanerva, L.T.; Csomós, P.; Sundholm, O.; Bernáth, G.; Fülöp, F. *Tetrahedron: Asymmetry* **1996**, *7*, 1705–1716.

163. Li, X.-G.; Kanerva, L.T. *Tetrahedron: Asymmetry* **2005**, *16*, 1709–1714.

164. Flores-Sánchez, P.; Escalante, J.; Castillo, E. *Tetrahedron: Asymmetry* **2005**, *16*, 629–634.

165. Solymár, M.; Fülöp, F.; Kanerva, L.T. *Tetrahedron Asymmetry* **2002**, *135*, 2383–2388.

166. Dalby, J.S.; Kenner, G.W.; Sheppard, R.C. *J. Chem. Soc.* **1962**, 4387–4396.

167. Pirrung, M.C.; Krishnamurthy, N. *J. Org. Chem.* **1993**, *58*, 957–958.

168. Alexandre, F.-R.; Pantaleone, D.P.; Taylor, P.P.; Fotheringham.G.; Ager, D.J.; Turner, N.J. *Tetrahedron Lett.* **2002**, *43*, 707–710.

169. Greenstein, J.P.; Winitz, N. *Chemistry of the Amino Acids, Vol. 3*; Wiley: New York, 1961.

170. Jollès, G.; Poiget, G.; Robert, J.; Terlain, B.; Thomas, J.-P. *Bull. Soc. Chim. Fr.* **1965**, 2252–2259.

171. Zabriskie, T.M. *J. Med. Chem.* **1996**, *39*, 3046–3048.

172. Antoni, G.; Långström, B. *J. Lab. Cmpds. Radiopharm.* **1987**, *24*, 125–143.

173. Chênevert, R.; Rhlid, R.B.; Létourneau, M.; Gagnon, R.; D'Astous, L. *Tetrahedron: Asymmetry* **1993**, *4*, 1137–1140.

174. Arnaud, A.; Galzy, P.; Jallageas, J.-C. *Bull. Soc. Chim. Fr.* **1980**, *2*, 87–90.

175. Herbert, R.B.; Wilkinson, B. Ellames, G.J.; Kunec, E.K. *J. Chem. Soc., Chem. Commun.* **1993**, 205–206.

176. Herbert, R.B.; Wilkinson, B.; Ellames, G.J. *Can. J. Chem.* **1994**, *72*, 114–117.

177. Bycroft, M.; Herbert, R.B.; Ellames, G.J. *J. Chem. Soc., Perkin Trans. 1* **1996**, 2439–2442.

178. Neuberger, A.; Sanger, F. *Biochem. J.* **1944**, *38*, 125–129.

179. Bommarius, A.S.; Drauz, K. *Bioorg. Med. Chem.* **1994**, *2*, 617–626.

180. Maehr, H.; Yarmchuk, L.; Leach, M. *J. Antibiotics* **1976**, *29*, 221–226.

181. Marcus, A.; Feeley, J.; Shannon, L.M. *Arch. Biochem. Biophys.* **1963**, *100*, 80–85.

182. Heck, S.D.; Faraci, W.S.; Kelbaugh, P.R.; Saccomano, N.A.; Thadeio, P.F.; Volkmann, R.A. *Proc. Natl. Acad. Sci. U.S.A.* **1996**, *93*, 4036–4039.

183. Murkin, A.S.; Tanner, M.E. *J. Org. Chem.* **2002**, *67*, 8389–8394.

184. Kida, T.; Isogawa, K.; Zhang, W.; Nakatsuji, Y.; Ikeda, I. *Tetrahedron Lett.* **1998**, *39*, 4339–4342.

185. You, J.; Yu, X.; Li, X.; Yan, Q.; Xie, R. *Tetrahedron: Asymmetry* **1998**, *9*, 1197–1203.

186. You, J.; Yu, X.; Liu, K.; Tao, L.; Xiang, Q.; Xie, R. *Tetrahedron: Asymmetry* **1999**, *10*, 243–254.

187. Tanoue, O.; Baba, M.; Tokunaga, Y.; Goto, K.; Matsumoto, Y.; Ueoka, R. *Tetrahedron Lett.* **1999**, *40*, 2129–2132.

188. Sellergren, B.; Karmalkar, R.N.; Shea, K.J. *J. Org. Chem.* **2000**, *65*, 4009–4027.

189. Kaminski, Z.J.; Kolesinska, B.; Kaminska, J.E.; Góra, J. *J. Org. Chem.* **2001**, *66*, 6276–6281.

190. Dro, C.; Bellemin-Laponnaz, S.; Welter, R.; Gade, L.H. *Angew. Chem., Int. Ed.* **2004**, *43*, 4479–4482.

191. Chen, S.-T.; Huang, W.-H.; Wang, K.-T. *J. Org. Chem.* **1994**, *59*, 7580–7581.

192. Wegman, M.A.; Hacking, A.P.J.; Rops, J.; Pereira, P.; van Rantwijk, F.; Sheldon, R.A. *Tetrahedron: Asymmetry* **1999**, *10*, 1739–1750.

193. Parmar, V.S.; Singh, A.; Bisht, K.S.; Kumar, N.; Belekon, Y.N.; Kochetkov, K.A.; Ikonnikov, N.S.; Orlova, S.A.; Tararov, V.I.; Saveleva, T.F. *J. Org. Chem.* **1996**, *61*, 1223–1227.

194. Liljeblad, A.; Kiviniemi, A.; Kanerva, L.T. *Tetrahedron* **2004**, *60*, 671–677.

195. Jones, M.M.; Williams, J.M.J. *Chem. Commun.* **1998**, 2519–2520.

196. Tokuyama, S.; Miya, H.; Hatano, K.; Takahashi, T. *Appl. Microbiol. Biotechnol.* **1994**, *40*, 835–840.

197. Tokuyama, S.; Hatano, K.; Takahashi, T. *Biosci., Biotechnol., Biochem.* **1994**, *58*, 24–27.

198. Patrick, W.M.; Weisner, J.; Blackburn, J.M. *ChemBioChem* **2002**, *3*, 789–792.

199. Lickefett, H.; Krohn, K.; König, W.A.; Gehrcke, B.; Syldatk, C. *Tetrahedron: Asymmetry* **1993**, *4*, 1129–1135.

200. Guivarch, M.; Gillonnier, C.; Brunie, J.-C. *Bull. Soc. Chim. Fr.* **1980**, 1–2, 91–95.

201. Rouhi, A.M. *Chem. Eng. News* **2002**, *July 10*, 43–57.

202. Olivieri, R.; Fascetti, E.; Angelini, L.; Degen, L. *Biotechnol. Bioeng.* **1981**, *23*, 2173–2183.

203. Arcuri, M.B.; Antunes, O.A.C.; Sabino, S.J.; Pinto, G.F.; Oestreicher, E.G. *Amino Acids* **2000**, *19*, 477–482.

204. Hamilton, G.S.; Huang, Z.; Yang, X.-J.; Patch, R.J.; Narayanan, B.A.; Ferkany, J.W. *J. Org. Chem.* **1993**, *58*, 7263–7270.

205. Hamilton, G.S.; Huang, Z.; Patch, R.J.; Narayanan, B.A.; Ferkany, J.W. *Bioorg. Med. Chem. Lett.* **1993**, *3*, 27–32.

206. Arcuri, M.B.; Antunes, O.A.C.; Sabino, S.J.; Pinto, G.F.; Oestreicher, E.G. *Amino Acids* **2004**, *27*, 69–74.

207. Keil, O.; Schneider, M.P.; Rasor, J.P. *Tetrahedron: Asymmetry* **1995**, *6*, 1257–1260.

208. Altenbuchner, J.; Siemann-Herzberg, M.; Syldatk, C. *Curr. Opin. Biotechnol.* **2001**, *12*, 559–563.

209. Burton, S.G.; Dorrington, R.A. *Tetrahedron: Asymmetry* **2004**, *15*, 2737–2741.

210. Crich, J.Z.; Brieva, R.; Marquart, P.; Gu, R.-L.; Flemming, S.; Sih, C.J. *J. Org. Chem.* **1993**, *58*, 3252–3258.

211. Gu, R.-L.; Lee, I.-S.; Sih, C.J. *Tetrahedron Lett.* **1992**, *33*, 1953–1956.

212. Turner, N.J.; Winterman, J.R.; McCague, R.; Parratt, J.S.; Taylor, J.C. *Tetrahedron Lett.* **1995**, *36*, 1113–1116.

213. Brown, S.A.; Parker, M.-C.; Turner, N.J. *Tetrahedron: Asymmetry* **2000**, *11*, 1687–1690.

214. Parker, M.-C.; Brown, S.A.; Robertson, L.; Turner, N.J. *Chem. Commun.* **1998**, 2247–2248.

215. Quirós, M.; Parker, M.-C.; Turner, N.J. *J. Org. Chem.* **2001**, *66*, 5074–5079.

216. Lloyd-Williams, P.; Carulla, N.; Giralt, E. *Tetrahedron Lett.* **1997**, *38*, 299–302.

217. Chaplin, J.A.; Levin, M.D.; Morgan, B.; Farid, N.; Li, J.; Zhu, Z.; McQuaid, J.; Nicholson, L.W.; Rand, C.A.; Burk, M.J. *Tetrahedron: Asymmetry* **2004**, *15*, 2793–2796.

218. Shah, S.A.; Schafer, P.H.; Recchia, P.A.; Ploach, K.J.; LeMaster, D.M. *Tetrahedron Lett.* **1994**, *35*, 29–32.

219. Polach, K.J.; Shah, S.A.; LaIuppa, J.C.; LeMaster, D.M. *J. Lab. Cmpds. Radiopharm.* **1993**, *33*, 809–815.

220. Nakajima, N.; Esaki, N.; Soda, K. *J. Chem. Soc., Chem. Commun.* **1990**, 947–948.

221. Beard, T.M.; Turner, N.J. *Chem. Commun.* **2002**, 246–247.

222. Steglich, W.; Frauendorfer, E.; Weygand, F. *Chem. Ber.* **1971**, *104*, 687–690.

223. Calmes, M.; Daunis, J.; Mai, N. *Tetrahedron: Asymmetry* **1997**, *8*, 1641–1648.

224. Calmes, M.; Daunis, J.; Mai, N.; Natt, F. *Tetrahedron Lett.* **1996**, *37*, 379–380.

225. Calmes, M.; Escale, F. *Tetrahedron: Asymmetry* **1998**, *9*, 2845–2850.

226. Calmes, M.; Glot, C.; Michel, T.; Rolland, M.; Martinez, J. *Tetrahedron: Asymmetry* **2000**, *11*, 737–741.

227. Koh, K.; Ben, R.N.; Durst, T. *Tetrahedron Lett.* **1993**, *34*, 4473–4476.

228. Ben, R.N.; Durst, T. *J. Org. Chem.* **1999**, *64*, 7700–7706.

229. O'Meara, J.A.; Gardee, N.; Jung, M.; Ben, R.N.; Durst, T. *J. Org. Chem.* **1998**, *63*, 3117–3119.

230. Valenrod, Y.; Myung, J.; Ben, R.N. *Tetrahedron Lett.* **2004**, *45*, 2545–2549.

231. Kim, H.J.; Shin, E.-k.; Chang, J.-y.; Kim, Y.; Park, Y.S. *Tetrahedron Lett.* **2005**, *46*, 4115–4117.

232. Nunami, K.-I.; Kubota, H.; Kubo, A. *Tetrahedron Lett.* **1994**, *35*, 8639–8642.

233. Caddick, S.; Jenkins, K.; Treweeke, N.; Candeias, S.X.; Afonso, C.A.M. *Tetrahedron Lett.* **1998**, *39*, 2203–2206.

234. Santos, A.G.; Candeias, S.X.; Afonso, C.A.M.; Jenkins, K.; Caddick, S.; Treweeke, N.R.; Pardoe, D. *Tetrahedron* **2001**, *57*, 6607–6614.

235. Caddick, S.; Afonso, C.A.M.; Candeias, S.X.; Hitchcock, P.B.; Jenkins, K.; Murtagh, L.; Pardoe, D.;

235. Santos, A.G.; Treweeke, N.R.; Weaving, R. *Tetrahedron* **2001**, *57*, 6589–6605.

236. Camps, P.; Pérez, F.; Soldevilla, N.; Borrego, M.A. *Tetrahedron: Asymmetry* **1999**, *10*, 493–509.

237. Camps, P.; Pérez, F.; Soldevilla, N. *Tetrahedron Lett.* **1999**, *40*, 6853–6856.

238. Treweeke, N.R.; Hitchcock, P.B.; Pardoe, D.A.; Caddick, S. *Chem. Commun.* **2005**, 1868–1870.

239. Santos, A.G.; Pereira, J.; Afonso, C.A.M.; Frenking, G. *Chem. —Eur. J.* **2005**, *11*, 330–343.

240. Nam, J.; Chang, J.-y.; Shin, E.-k.; Kim, H.J.; Kim, Y.; Jang, S.; Park, Y.S. *Tetrahedron* **2004**, *60*, 6311–6318.

241. Nam, J.; Chang, J.-Y.; Hahm, K.-S.; Park, Y.S. *Tetrahedron Lett.* **2003**, *44*, 7727–7730.

242. Chen, J.G.; Zhu, J.; Skonezny, P.M.; Rosso, V.; Venit, J.J. *Org. Lett.* **2004**, *6*, 3233–3235.

243. Kiau, S.; Discordia, R.P.; Madding, G.; Okuniewicz, F.J.; Rosso, V.; Venit, J.J. *J. Org. Chem.* **2004**, *69*, 4256–4261.

244. Shiraiwa, T.; Kataoka, K.; Sakata, S.; Kurokawa, H. *Bull. Chem. Soc. Jpn.* **1989**, *62*, 109–113.

245. Shiraiwa, T.; Shinjo, K.; Kurokawa, H. *Bull. Chem. Soc. Jpn.* **1991**, *64*, 3251–3255.

246. Hassan, N.A.; Bayer, E.; Jochims, J.C *J. Chem. Soc., Perkin Trans. 1* **1998**, 3747–3757.

247. Liang, J.; Ruble, J.C.; Fu, G.C. *J. Org. Chem.* **1998**, *63*, 3154–3155.

248. Berkessel, A.; Cleeman, F.; Mukherjee, S.; Müller, T.N.; Lex, J. *Angew. Chem., Int. Ed.* **2005**, *44*, 807–811.

249. Berkessel, A.; Mukherjee, S.; Cleeman, F.; Müller, T.N.; Lex, J. *Chem. Commun.* **2005**, 1898–1900.

250. Hang, J.; Tian, S.-K.; Tang, L.; Deng, L. *J. Am. Chem. Soc.* **2001**, *123*, 12696–12697.

251. Hang, J.; Li, H.; Deng, L. *Org. Lett.* **2002**, *4*, 3321–3324.

252. Tian, S.-K.; Chen, Y.; Hang, J.; Tang, L.; McDaid, P.; Deng, L. *Acc. Chem. Res.* **2004**, *37*, 621–631.

253. Berkessel, A.; Cleemann, F.; Mukherjee, S. *Angew. Chem., Int. Ed.* **2005**, *44*, 7466–7469.

254. Noyori, R.; Ikeda, T.; Ohkuma, T.; Widham, M.; Kitamura, M.; Takaya, H.; Akutagawa, S.; Sayo, N.; Saito, T.; Taketomi, T.; Kumobayashi, H. *J. Am. Chem. Soc.* **1989**, 111, 9134–9135.

255. Genet, J.P.; Pinel, C.; Mallart, S.; Juge, S.; Thorimbert, S.; Laffitte, J.A. *Tetrahedron: Asymmetry* **1991**, *2*, 555–567.

256. Girard, A.; Greck, C.; Ferroud, D.; Genêt, J.P. *Tetrahedron Lett.* **1996**, *37*, 7967–7970.

257. Lei, A.; Wu, S.; He, M.; Zhang, X. *J. Am. Chem. Soc.* **2004**, *126*, 1626–1627.

258. Mordant, C.; Dünkelmann, P.; Ratovelomanana-Vidal, V.; Genet, J.-P. *Eur. J. Org. Chem.* **2004**, 3017–3026.

259. Mohar, B.; Valleix, A.; Desmurs, J.-R.; Felemez, M.; Wagner, A.; Mioskowski, C. *Chem. Commun.* **2001**, 2572–2573.

260. Makino, K.; Goto, T.; Hiroki, Y.; Hamada, Y. *Angew. Chem., Int. Ed.* **2004**, *43*, 882–884.

261. Fehr, C. *Angew. Chem., Int. Ed. Engl.* **1996**, 35, 2566–2587.

262. Duhamel, L.; Duhamel, P.; Plaqueven, J.-C. *Tetrahedron: Asymmetry* **2004**, *15*, 3653–3691.

263. Duhamel, L.; Duhamel, P.; Fourquay, S.; Eddine, J.J.; Peschard, O.; Plaquevent, J.-C.; Ravard, A.; Solliard, R.; Valnot, J.-Y.; Vincens, H. *Tetrahedron* **1988**, *44*, 5495–5506.

264. Duhamel, L.; Fourquay, S.; Plaquevent, J.-C. *Tetrahedron Lett.* **1986**, *27*, 4975–4978.

265. Duhamel, L.; Plaquevent, J.-C. *Tetrahedron Lett.* **1980**, *21*, 2521–2524.

266. Duhamel, L.; Plaquevent, J.-C. *J. Am. Chem. Soc.* **1978**, *100*, 7415–7416.

267. Duhamel, L.; Plaquevent, J.-C. *Bull. Soc. Chim. Fr.* **1982**, *2*, 75–83.

268. Futatsugi, K.; Yanagisawa, A.; Yamamoto, H. *Chem. Commun.* **2003**, 566–567.

269. Martin, J.; Lasne, M.-C.; Plaquevent, J.-C.; Duhamel, L. *Tetrahedron Lett.* **1997**, *38*, 7181–7182.

270. Vedejs, E.; Kruger, A.W.; Suna, E. *J. Org. Chem.* **1999**, *64*, 7863–7870.

271. Rouhi, A.M. *Chem. Eng. News* **2003**, *July 14*, 34–35.

272. Belekon, Y.N.; Zel'tzer, I.E.; Bakhmutov, V.I.; Saporovskaya, M.B.; Ryzhov, M.G.; Yanovsky, A.I.; Struchkov, Y.T.; Belikov, V.M. *J. Am. Chem. Soc.* **1983**, *105*, 2010–2017.

273. De, B.B.; Thomas, N.R. *Tetrahedron: Asymmetry* **1997**, *8*, 2687–2691.

274. Tabcheh, M.; Guibourdenche, C.; Pappalardo, L.; Roumestant, M.-L.; Viallefont, P. *Tetrahedron: Asymmetry* **1998**, *9*, 1493–1495.

275. Lecointe, L.; Rolland-Fulcrand, V.; Roumestant, M.L.; Viallefont, P.; Martinez, J. *Tetrahedron: Asymmetry* **1998**, *9*, 1753–1758.

276. Coggins, P.; Simpkins, N.S. *Synlett* **1991**, 515–516.

277. Chin, J.; Lee, C.S.; Lee, K.J.; Park, S.; Kim, D.H. *Nature* **1999**, *401*, 254–257.

278. Bandyopadhyay, I.; Lee, H.M.; Tarakeshwar, P.; Cui, C.; Oh, K.S.; Chin, J.; Kim, K.S. *J. Org. Chem.* **2003**, *68*, 6571–6575.

279. Alías, M.; López, M.P.; Cativiela, C. *Tetrahedron* **2004**, *60*, 885–891.

280. Lasa, M.; López, P.; Cativiela, C. *Tetrahedron: Asymmetry* **2005**, *16*, 4022–4033.

281. Spanik, I.; Vigh, G. *J. Chromatogr., A* **2002**, *979*, 123–129.

282. Davis, A.P.; Lawless, L.J. *Chem. Commun.* **1999**, 9–10.

283. Lawless, L.J.; Blackburn, A.G.; Ayling, A.J.; Pérez-Payán, M.N.; Davis, A.P. *J. Chem. Soc., Perkin Trans. 1* **2001**, 1329–1341.

284. Demirel, N.; Bulut, Y.; Hosgören, H. *Tetrahedron: Asymmetry* **2004**, *15*, 2045–2049.

285. Demirel, N.; Bulut, Y. *Tetrahedron: Asymmetry* **2003**, *14* 2633–2637.

286. Hernández, J.V.; Oliva, A.I.; Simón, L.; Muñiz, F.M.; Grande, M.; Morán, J.R. *Tetrahedron Lett.* **2004**, *45*, 4831–4833.

287. Skolau, A.; Rétey, J. *Angew. Chem., Int. Ed.* **2002,** *41*, 2960–2962.

288. Franco, P.; Blanc, J.; Oberleitner, W.R.; Maier, N.M.; Lindner, W.; Minguillón, C. *Anal. Chem.* **2002**, *74*, 4175–4183.

289. Delgado, B.; Pérez, E.; Santano, M.C.; Minguillón, C. *J. Chromatogr., A* **2005**, *1092*, 36–42.

290. Foucault, A.P. *J. Chromatogr., A* **2001**, *906*, 365–378.

291. Ito, Y. *J. Chromatogr., A* **2005**, *1065*, 145–168.

292. Greenstein, J.P.; Winitz, N. *Chemistry of the Amino Acids, Vol. 3*; Wiley: New York, 1961, pp 2750–2770.

293. *Amino Acid Derivatives: A Practical Approach;* Barrett, G.C., Ed.; Oxford University Press: Toronto, 1999.

294. Konishi, M.; Ohkuma, H.; Sakai, F.; Tsuno, T.; Koshiyama, H.; Naito, T.; Kawaguchi, H. *J. Am. Chem. Soc.* **1981**, *103*, 1241–1243.

295. Arnold, E.; Clardy, J. *J. Am. Chem. Soc.* **1981**, *103*, 1243–1244.

296. Boger, D.L.; Chen, J.-H.; Saionz, K.W.; Jin, Q. *Biorg. Med. Chem.* **1998**, *6*, 85–102.

297. Maehr, H.; Liu, C.-M.; Palleroni, N.J.; Smallheer, J.; Todaro, L.; Williams, T.H.; Blount, J.F. *J. Antibiotics* **1986**, *39*, 17–25.

298. Namikoshi, M.; Rinehart, K.L.; Sakai, R.; Stotts, R.R.; Dahlem, A.M.; Beasley, V.R.; Carmichael, W.W.; Evans, W.R. *J. Org. Chem.* **1992**, *57*, 866–872.

299. Ueda, K.; Xiao, J.-Z.; Doke, N.; Nakatsuka, S.-I. *Tetrahedron Lett.* **1992**, *33*, 5377–5380.

300. Brockmann, H.; Manegold, J.H. *Chem. Ber.* **1962**, *95*, 1081–1093.

301. Scherkenbeck, J.; Harder, A.; Plant, A.; Dyker, H. *Biorg. Med. Chem. Lett.* **1998**, *8*, 1035–1040.

302. Thern, B.; Rudolph, J.; Jung, G. *Angew. Chem., Int. Ed.* **2002**, *41*, 2307–2309.

303. Fusetani, N.; Warabi, K; Nogata, Y.; Nakao, Y.; Matsunaga, S.; van Soest, R.R.M. *Tetrahedron Lett.* **1999**, *40*, 4687–4690.

304. Roy, M.C.; Ohtani, I.I.; Tanaka, J.; Higa, T.; Satari, R. *Tetrahedron Lett.* **1999**, *40*, 5373–5376.

305. Klein, D.; Braekman, J.-C.; Daloze, D.; Hofman, L.; Castillo, G.; Demoulin, V. *Tetrahedron Lett.* **1999**, *40*, 695–696.

306. Chen, H.; Feng, Y.; Xu, Z.; Ye, T. *Tetrahedron* **2005**, *61*, 11132–11140.

307. Akaji, K.; Hayashi, Y.; Kiso, Y.; Kuriyama, N. *J. Org. Chem.* **1999**, *64*, 405–411.

308. Pettit, G.R.; Kamano, Y.; Herald, C.L.; Fujii, Y.; Kizu, H.; Boyd, M.R.; Boettner, F.E.; Doubek, D.L.; Schmidt, J.M.; Chapuis, J.-C.; Michel, C. *Tetrahedron* **1993**, *49*, 9151–9170.

309. Pettit, G.R.; Kamano, Y.; Kizu, H.; Dufresne, C.; Herald, C.L.; Bontems, R.J.; Schmidt, J.M.;

310. Boettner, F.E.; Nieman, R.A. *Heterocycles* **1989**, *28*, 553–558.

310. Zbiral, E.; Ménard, E.L.; Müller, J.M. *Helv. Chim. Acta* **1965**, *48*, 404–431.

311. Luesch, H.; Yoshida, W.Y.; Moore, R.E.; Paul, V.J.; Corbett, T.H. *J. Am. Chem. Soc.* **2001**, *123*, 5418–5423.

312. Komatsu, K.; Shigemori, H.; Kobayashi, J. *J. Org. Chem.* **2001**, *66*, 6189–6192.

313. Sangster, A.W.; Thomas, S.E.; Tingling, N.L. *Tetrahedron* **1975**, *31*, 1135–1137.

314. Rinehart, K.L.; Kishore, V.; Nagarajan, S.; Lake, R.J.; Gloer, J.B.; Bozich, F.A.; Li, K.-M.; Maleczka, R.E., Jr., Todsen, W.L.; Munro, M.H.G.; Sullins, D.W.; Sakai, R. *J. Am. Chem. Soc.* **1987**, *109*, 6846–6848.

315. Rinehart, K.L., Jr.; Gloer, J.B.; Cook, Jr., J.C. *J. Am. Chem. Soc.* **1981**, *103*, 1857–1859.

316. Hamada, Y.; Kondo, Y.; Shibata, M.; Shiori, T. *J. Am. Chem. Soc.* **1989**, *111*, 669–673.

317. Rinehart, K.L., Jr., Gloer, J.B.; Hughes, R.G., Jr., Renis, H.E.; McGovren, J.P.; Swynenberg, E.B.; Stringfellow, D.A.; Kuentzel, S.L.; Li, L.H. *Science* **1981**, *212*, 933–935.

318. Marner, F.-J.; Moore, R.E.; Hirotsu, K.; Clardy, J. *J. Org. Chem.* **1977**, *42*, 2815–2819.

319. Carter, D.C.; Moore, R.E.; Mynderse, J.S.; Niemczura, W.P.; Todd, J.S. *J. Org. Chem.* **1984**, *49*, 236–241.

320. Vanderhaeghe, H.; Parmentier, G. *J. Am. Chem. Soc.* **1960**, *82*, 4414–4422.

321. Sheehan, J.C.; Ledis, S.L. *J. Am. Chem. Soc.* **1973**, *95*, 875–879.

322. Hinterding, K.; Hagenbuch, P.; Rétey, J.; Waldmann, H. *Angew. Chem., Int. Ed. Engl.* **1998**, *37*, 1236–1239.

323. Morel, A.F.; Flach, A.; Zanatta, N.; Ethur, E.M.; Mostardeiro, M.A.; Gehrke, I.T.S. *Tetrahedron Lett.* **1999**, *40*, 9205–9209.

324. Kobayashi, J.; Sato, M.; Ishibashi, M.; Shigemori, H.; Nakamura, T.; Ohizumi, Y. *J. Chem. Soc., Perkin Trans 1* **1991**, 2609–2611.

325. Zhang, P.; Liu, R.; Cook, J.M. *Tetrahedron Lett.* **1995**, *36*, 9133–9136.

326. Fusetani, N; Matsunaga, S. *Chem. Rev.* **1993**, *93*, 1793–1806.

327. Chu, K.S.; Negrete, G.R.; Konopelski, J.P. *J. Org. Chem.* **1991**, *56*, 5196–5202.

328. Rama Rao, A.V.; Gurjar, M.K.; Nallaganchu, B.R.; Bhandari, A. *Tetrahedron Lett.* **1993**, *34*, 7085–7088.

329. Raverty, W.D.; Thomson, R.H.; King, T.J. *J. Chem., Soc., Perkin Trans. 1* **1977**, 1204–1210.

330. Cosquer, A.; Pichereau, V.; Le Mée, D.; Le Roch, M.; Renault, J.; Carboni, B.; Uriac, P.; Bernard, T. *Biorg. Med. Chem. Lett.* **1999**, *9*, 49–54.

331. Misiti, D.; Zappia, G.; Monache, G. *Synth. Commun.* **1995**, *25*, 2285–2294.

332. Voeffray, R.; Perlberger, J.-C.; Tenud, L.; Gosteli, J. *Helv. Chim. Acta* **1987**, *70*, 2058–2064.

333. Friedman, S.; McFarlane, J.E.; Bhattacharyya, P.K.; Fraenkel, G.; Berger, C.R.A. *Biochem. Prep.* **1960**, *7*, 26–30.

334. Misiti, D.; Santaniello, M.; Zappia, G. *Bioorg. Med. Chem. Lett.* **1992**, *2*, 1029–1032.

335. Chang, C.-F.; Zehfus, M.H. *Biopolymers* **1997**, *41*, 609–616.

336. Möhle, K.; Hofmann, H.-J. *J. Peptide Res.* **1998**, *51*, 19–28.

337. Takeuchi, Y.; Marshall, G.R. *J. Am. Chem. Soc.* **1998**, *120*, 5363–5372.

338. Chalmers, D.K.; Marshall, G.R. *J. Am. Chem. Soc.* **1995**, *117*, 5927–5937.

339. Aubry, A.; Marraud, M. *Biopolymers* **1989**, *28*, 109–122.

340. Marraud, M.; Dupont, V.; Grand, V.; Zerkout, S.; Lecoq, A.; Boussard, G.; Vidal, J.; Collet, A.; Aubry, A. *Biopolymers* **1993**, *33*, 1135–1148.

341. Miller, S.M.; Simon, R.J.; Ng, S.; Zuckermann, R.N.; Kerr, J.M.; Moos, W.H. *Bioorg. Med. Chem. Lett.* **1994**, *4*, 2657–2662.

342. Simon, R.J.; Kania, R.S.; Zuckermann, R.N.; Huebner, V.D.; Jewell, D.A.; Banville, S.; Ng, S.; Wang, L.; Rosenberg, S.; Marlowe, C.K.; Spellmeyer, D.C.; Tan, R.; Frankel, A.D.; Santi, D.V.; Cohen, F.E.; Bartlett, P.A. *Proc. Natl. Acad. Sci. U.S.A.* **1992**, *89*, 9367–9371.

343. Janecka, A.; Kruszynski, R.; Fichna, J.; Kosson, P.; Janecki, T. *Peptides* **2006**, *27*, 131–135.

344. Urban, J.; Vaisar, T.; Shen, R.; Lee, M.S. *Int. J. Pept. Prot. Chem.* **1996**, *47*, 182–189.

345. Albericio, F.; Cases, M.; Alsina, J.; Triolo, S.A.; Carpino, L.A.; Kates, S.A. *Tetrahedron Lett.* **1997**, *38*, 4853–4856.

346. Creighton, C.J.; Romoff, T.T.; Bu, J.H.; Goodman, M. *J. Am. Chem. Soc.* **1999**, *121*, 6786–6791.

347. Sjölin, P.; Kihlberg, J. *Tetrahedron Lett.* **2000**, *41*, 4435–4439.

348. Johnson, T.; Quibell, M.; Owen, D.; Sheppard, R.C. *J. Chem. Soc., Chem. Commun.* **1993**, 369–372.

349. Hamper, B.C.; Kolodziej, S.A.; Scates, A.M.; Smith, R.G.; Cortez, E. *J. Org. Chem.* **1998**, *63*, 708–718.

350. Greenstein, J.P.; Winitz, N. *Chemistry of the Amino Acids, Vol. 2*; Wiley: New York, 1961, pp 1260–1266.

351. Expósito, M.A.; Lopez, B.; Fernández, R.; Vázquez, M.; Debitus, C.; Iglesias, T.; Jiménez, C.; Quiñoá, E.; Riguera, R. *Tetrahedron* **1998**, *54*, 7539–7550.

352. Humphrey, J.M.; Chamberlin, A.R. *Chem. Rev.* **1997**, *97*, 2243–2266.

353. Brückner, H. *Chromatographia* **1989**, *27*, 725–738.

354. Aurelio, L.; Brownlee, R.T.C.; Hughes, A.B. *Chem. Rev.* **2004**, *104*, 5823–5846.

355. Marino, G.; Valente, L.; Johnstone, R.A.W.; Mohammedi-Tabrizi, F.; Sodini, G.C. *J. Chem. Soc., Chem. Commun.* **1972**, 357–358.

356. Morris, H.R.; Dickinson, R.J.; Williams, D.H. *Biochem. Biophys. Res. Commun.* **1973**, *51*, 247–255.

357. Morris, H.R. *FEBS Lett.* **1972**, *22*, 257–260.

358. Thomas, D.W. *Biochem. Biophys. Res. Commun.* **1968**, *33*, 483–486.

359. Vilkas, E.; Lederer, E.; Massot, J.C.; *Tetrahedron Lett.* **1968**, *26*, 3089–3092.

360. Agarwal, K.L.; Kenner, G.W.; Sheppard, R.C. *J. Am. Chem. Soc.* **1969**, *91*, 3096–3097.

361. Coggins, J.R.; Benoiton, N.L. *Can. J. Chem.* **1971**, *49*, 1968–1971.

362. Ramasamy, K.; Olsen, R.K.; Emery, T. *J. Org. Chem.* **1981**, *46*, 5438–5441.

363. McDermott, J.R.; Benoiton, N.L. *Can. J. Chem.* **1973**, *51*, 1915–1919.

364. Cheung, S.T.; Benoiton, N.L. *Can. J. Chem.* **1977**, *55*, 906–910.

365. Yasui, A.; Douglas, A.J.; Walker, B.; Magee, D.F.; Murphy, R.F. *Int. J. Pept. Prot. Res.* **1990**, *35*, 301–305.

366. Grieco, P.A.; Hon, Y.-S.; Perez-Medrano, A. *J. Am. Chem. Soc.* **1988**, *110*, 1630–1631.

367. Hu, T.; Panek, J.S. *J. Am. Chem. Soc.* **2002**, *124*, 11368–11378.

368. Patel, H.M.; Tao, J.; Walsh, C.T. *Biochemistry* **2003**, *42*, 10514–10527.

369. McConathy, J.; Martarello, L.; Malveaux, E.J.; Camp, V.M.; Simpson, N.E.; Simpson, C.P.; Bowers, G.D.; Olson, J.J.; Goodman, M.M. *J. Med. Chem.* **2002**, *45*, 2240–2249.

370. Deng, H.; Jung, J.-K.; Liu, T.; Kuntz, K.W.; Snapper, M.L.; Hoveyda, A.H. *J. Am. Chem. Soc.* **2003**, *125*, 9032–9034.

371. Pettit, G.R.; Hogan, F.; Herald, D.L. *J. Org. Chem.* **2004**, *69*, 4019–4022.

372. Hinds, M.G.; Welsh, J.H.; Brennand, D.M.; Fisher, J.; Glennie, M.J.; Richards, N.G.J.; Turner, D.L.; Robinson, J.A. *J. Med. Chem.* **1991**, *34*, 1777–1789.

373. Chen, Y.; Bilban, M.; Foster, C.A.; Boger, D.L. *J. Am. Chem. Soc.* **2002**, *124*, 5431–5440.

374. Boger, D.L.; Keim, H.; Oberhauser, B.; Schreiner, E.P.; Foster, C.A. *J. Am. Chem. Soc.* **1999**, *121*, 6197–6205.

375. Kells, K.W.; Ncube, A.; Chong, J.M. *Tetrahedron* **2004**, *60*, 2247–2257.

376. Xue, C.-B.; DeGrado, W.F. *Tetrahedron Lett.* **1995**, *36*, 55–58.

377. Coulton, S.; Moore, G.A.; Ramage, R. *Tetrahedron Lett.* **1976**, *44*, 4005–4008.

378. Burger, K.; Hollweck, W. *Synlett* **1994**, 751–753.

379. Schmall, B.; Conti, P.S.; Kiesewetter, D.O.; Alauddin, M.M. *J. Lab. Cmpds. Radiopharm.* **1995**, *37*, 150–152.

380. Horti, A.; Ravert, H.T.; Dannals, R.F.; Wagner, H.N., Jr.; *J. Lab. Cmpds. Radiopharm.* **1992**, *31*, 1029–1036.

381. Cheung, S.T.; Benoiton, N.L. *Can. J. Chem.* **1977**, *55*, 916–921.

382. Das, B.C.; Gero, S.D.; Lederer, E. *Biochem. Biophys. Res. Commun.* **1967**, *29*, 211–215.

383. Agarwal, K.L.; Johnstone, R.A.W.; Kenner, G.W.; Millington, D.S.; Sheppard, R.C. *Nature* **1968**, *219*, 498–499.

384. Olsen, R.K. *J. Org. Chem.* **1970**, *35*, 1912–1915.

385. Papageorgiou, C.; Borer, X.; French, R.R. *Bioorg. Med. Chem. Lett.* **1994**, *4*, 267–272.

386. Landvatter, S.W. *J. Lab. Cmpds. Radiopharm.* **1993**, *33*, 863–868.

387. Papageorgiou, C.; Florineth, A.; Mikol, V. *J. Med. Chem.* **1994**, *37*, 3674–3676.

388. Johnstone, R.A.W.; Rose, M.E. *Tetrahedron* **1979**, *35*, 2169–2173.

389. Johnson, T.B.; Nicolet, B.H. *Am. Chem. J.* **1912**, *47*, 459–475.

390. Solomon, M.E.; Lynch, C.L.; Rich, D.H. *Tetrahedron Lett.* **1995**, *36*, 4955–4958.

391. Shin, C.; Sato, Y.; Hayakawa, M.; Kondo, M. *Heterocycles* **1981**, *16*, 1573–1578.

392. Rich, D.H.; Bhatnagar, P.; Mathiaparanam, P.; Grant, J.A.; Tam, J.P. *J. Org. Chem.* **1978**, *43*, 296–302.

393. Rich, D.H.; Tam, J.; Mathiaparanam, P.; Grant, J. *Synthesis* **1975**, 402–404.

394. Decicco, C.P.; Grover, P. *Synlett* **1997**, 529–530.

395. Kozikowski, A.P.; Tückmantel, W.; Liao, Y.; Manev, H.; Ikonomovic, S.; Wroblewski, J.T. *J. Med. Chem.* **1993**, *36*, 2706–2708.

396. Jiménez, J.C.; Chavarría, B.; López-Macià, À.; Royo, M.; Giralt, E.; Albericio, F. *Org. Lett.* **2003**, *5*, 2115–2118.

397. Miller, S.A.; Griffiths, S.L.; Seebach, D. *Helv. Chim. Acta* **1993**, *76*, 563–595.

398. Seebach, D.; Beck, A.K.; Bossler, H.G.; Gerber, C.; Ko, S.Y.; Murtiashaw, C.W.; Naef, R.; Shoda, S.-I.; Thaler, A.; Krieger, M.; Wenger, R. *Helv. Chim. Acta* **1993**, *76*, 1564–1590.

399. Belagali, S.; Mathew, T.; Himaja, M.; Kocienski, P. *Ind. J. Chem.* **1995**, *34B*, 45–47.

400. Grieco, P.A.; Perez-Medrano, A. *Tetrahedron Lett.* **1988**, *29*, 4225–4228.

401. Gacek, M.; Undheim, K. *Tetrahedron* **1973**, *29*, 863–866.

402. Petter, R.C.; Banerjee, S.; Englard, S. *J. Org. Chem.* **1990**, *55*, 3088–3097.

403. Chen, F.C.M.; Benoiton, N.L. *Can. J. Chem.* **1976**, *54*, 3310–3311.

404. Mayer, S.C.; Ramanjulu, J.; Vera, M.D.; Pfizenmayer, A.J.; Joullié, M.M. *J. Org. Chem.* **1994**, *59*, 5192–5205.

405. Kober, R.; Hammes, W.; Steglich, W. *Angew. Chem., Int. Ed. Engl.* **1982**, *21*, 203–204.

406. O'Donnell, M.J.; Bruder, W.A.; Daugherty, B.W.; Liu, D.; Wojciechowski, K. *Tetrahedron Lett.* **1984**, *25*, 3651–3654.

407. Shinagawa, S.; Kanamaru, T.; Harada, S.; Asai, M.; Okazaki, H. *J. Med. Chem.* **1987**, *30*, 1458–1463.

408. Prashad, M.; Har, D.; Hu, B.; Kim, H.-Y.; Repic, O.; Blacklock, T.J. *Org. Lett.* **2003**, *5*, 125–128.

409. Biron, E.; Kessler, H. *J. Org. Chem.* **2005**, *70*, 5183–5189.

410. Miller, S.C.; Scanlan, T.S. *J. Am. Chem. Soc.* **1997**, *119*, 2301–2302.

411. See Reference *410*.

412. Raman, P.; Stokes, S.S.; Angell, Y.M.; Flentke, G.R.; Rich, D.H. *J. Org. Chem.* **1998**, *63*, 5734–5735.

413. Yang, L.; Chiu, K. *Tetrahedron Lett.* **1997**, *38*, 7307–7310.

414. Di Gioia, M.L.; Leggio, A.; Le Pera, A.; Liguori, A.; Napoli, A.; Siciliano, C.; Sindona, G. *J. Org. Chem.* **2003**, *68*, 7416–7421.

415. Dankwardt, S.M.; Smith, D.B. Porco, J.A., Jr.; Nguyen, C.H. *Synlett* **1997**, 854–855.

416. Vedejs, E.; Kongkittingam, C. *J. Org. Chem.* **2000**, *65*, 2309–2318.

417. Vallee, E.; Loemba, F.; Etheve-Quelquejeu, M.; Valéry, J.-M. *Tetrahedron Lett.* **2006**, *47*, 2191–2195.

418. Albanese, D.; Landini, D.; Lupi, V.; Penso, M. *Eur. J. Org. Chem.* **2000**, 1443–1449.

419. Vedejs, E.; Kongkittingam, C. *J. Org. Chem.* **2001**, *66*, 7355–7364.

420. Kan, T.; Fukuyama, T. *Chem Commun.* **2004**, 353–359.

421. Någren, K.; Lehikoinen, P.; Leskinen, S. *J. Lab. Cmpds. Radiopharm.* **1995** *37*, 156–157.

422. Bowman, R.E.; Stroud, H.H. *J. Chem. Soc.* **1950**, 1342–1345.

423. Poncet, J.; Hortala, L.; Busquet, M.; Guéritte-Voegelein, F.; Thoret, S.; Pierré, A.; Atassi, G.; Jouin, P. *Bioorg. Med. Chem. Lett.* **1998**, *8*, 2849–2854.

424. Reinhold, V.N.; Ishikawa, Y.; Melville, D.B. *J. Med. Chem.* **1968**, *11*, 258–2263.

425. Bowman, R.E. *J. Chem. Soc.* **1950**, 1349–1351.

426. Poduska, K. *Collect. Czech. Chem. Commun.* **1959**, *24*, 1025–1028.

427. Bowman, R.E. *J. Chem. Soc.* **1950**, 1346–1349.

428. Chen, F.M.F.; Benoiton, N.L. *Can. J. Biochem.* **1978**, *56*, 150–152.

429. Quitt, P.; Hellerbach J.; Vogler, K. *Helv. Chim. Acta* **1963**, *46*, 327–333.

430. Kulikov, S.V.; Kepin, O.V.; Samaetsev, M.A. *Russ. J. Gen. Chem.* **1994**, *64*, 312–313.

431. Nader, M.; Theobald, A.; Oberdorfer, F. *J. Lab. Cmpds. Radiopharm.* **1997**, *40*, 730–731.

432. Freidinger, R.M.; Hincle, J.S.; Perlow, D.S.; Arison, B.H. *J. Org. Chem.* **1983**, *48*, 77–81.

433. Luke, R.W.A.; Boyce, P.G.T.; Dorling, E.K. *Tetrahedron Lett.* **1996**, *37*, 263–266.

434. Reddy, G.V.; Iyengar, D.S. *Chem. Lett.* **1999**, 299–300.

435. Aurelio, L.; Box, J.S.; Brownlee, R.T.C.; Hughes, A.B.; Sleebs, M.M. *J. Org. Chem.* **2003**, *68*, 2652–2667.

436. Aurelio, L.; Brownlee, R.T.C.; Hughes, A.B. *Org. Lett.* **2002**, *4*, 3767–3769.

437. Aurelio, L.; Brownlee, R.T.C.; Hughes, A.B.; Sleebs, B.E. *Aust. J. Chem.* **2000**, *53*, 425–433.

438. Hughes, A.B.; Mackay, M..F.; Aurelio, L. *Aust. J. Chem.* **2000**, *53*, 237–240.

439. Luo, Y. Evindar, G.; Fishlock, D.; Lajoie, G.A. *Tetrahedron Lett.* **2001**, *42*, 3807–3809.

440. Reddy, G.V.; Rao, G.V.; Iyengar, D.S. *Tetrahedron Lett.* **1998**, *39*, 1985–1986.

441. Dorow, R.L.; Gingrich, D.E. *Tetrahedron Lett.* **1999**, *40*, 467–470.

442. Govender, T.; Arvidsson, P.I. *Tetrahedron Lett.* **2006**, *47*, 1691–1694.

443. Tantry, S.J.; Babu, K.V.S.; Babu, V.V.S. *Tetrahedron Lett.* **2002**, *43*, 9461–9462.

444. Spengler, J.; Burger, K. *J. Chem. Soc., Perkin Trans. 1* **1998**, 2091–2095.

445. Spengler, J.; Burger, K. *Synthesis* **1998**, 67–70.

446. Burger, K.; Spengler, J. *Eur. J. Org. Chem.* **2000**, 199–204.

447. Oppolzer, W.; Cintas-Moreno, P.; Tamura, O.; Cardinaux, F. *Helv. Chim. Acta* **1993**, *76*, 187–196.

448. Chruma, J.J.; Sames, D.; Polt, R. *Tetrahedron Lett.* **1997**, *38*, 5085–5086.

449. Kaljuste, K.; Undén, A. *Int. J. Pept. Prot. Res.* **1993**, *42*, 118–124.

450. Grieco, P.A.; Bahsas, A. *J. Org. Chem.* **1987**, *52*, 5746–5749.

451. Sakai, N.; Ohfune, Y. *J. Am. Chem. Soc.* **1992**, *114*, 998–1010.

452. Kim, S.; Oh, C.H.; Ko, J.S.; Ahn, K.H.; Kim, Y.J. *J. Org. Chem.* **1985**, *50*, 1927–1932.

453. Bhattacharyya, S.; Chatterjee, A.; Duttachowdhury, S.K. *J. Chem. Soc., Perkin Trans. 1* **1994**, 1–2.

454. Abdel-Magid, A.; Carson, K.G.; Harris, B.D.; Maryanoff, C.A.; Shah, R.D. *J. Org. Chem.* **1996**, *61*, 3849–3862.

455. Ramanjulu, J.M.; Joullié, M.M. *Synth. Commun.* **1996**, *26*, 1379–1384.

456. Dorow, R.L.; Gingrich, D.E. *J. Org. Chem.* **1995**, *60*, 4986–4987.

457. Hall, D.G.; Laplante, C.; Manku, S.; Nagendran, J. *J. Org. Chem.* **1999**, *64*, 698–699.

458. Easton, C.J.; Kociuba, K.; Peters, S.C. *J. Chem. Soc., Chem. Commun.* **1991**, 1475–1476.

459. Wisniewski, K.; Kolodziejczyk, A.S. *Tetrahedron Lett.* **1997**, *38*, 483–486.

460. Papaioannou, D.; Athanassopoulos, C.; Magafa, V.; Karamanos, N.; Stavropoulos, G.; Napoli, A.; Sindona, G.; Aksnes, D.W.; Francis, G.W. *Acta Chem. Scand.* **1994**, *48*, 324–333.

461. Miller, S.C.; Scanlan, T.S. *J. Am. Chem. Soc.* **1998**, *120*, 2690–2691.

462. Reichwein, J.F.; Liskamp, R.M.J. *Tetrahedron Lett.* **1998**, *39*, 1243–1246.

463. Laplante, C.; Hall, D.G. *Org. Lett.* **2001**, *3*, 1487–1490.

464. Eschweiler, W. *Ann.* **1894**, *279*, 39–44.

465. Tiemann, F.; Piest, R. *Ber.* **1881**, *14*, 1982–1984.

466. Dvonch, W.; Fletcher, H., III Alburn, H.E. *J. Org. Chem.* **1964**, *29*, 2764–2766.

467. Hanafua, T.; Ichihara, J.; Ashida, T. *Chem. Lett.* **1987**, 687–690.

468. Beller, M.; Eckert, M.; Vollmüller, F.; Bogdanovic, S.; Geissler, H. *Angew. Chem., Int. Ed. Engl.* **1997**, *36*, 1494–1496.

469. Simmonds, D.H. *Biochem. J.* **1954**, *58*, 520–523.

470. Bycroft, B.W.; Lee, G.R. *J. Chem. Soc., Chem. Commun.* **1975**, 988–989.

471. Ekhato, I.V.; Huang, C.C. *J. Lab. Cmpds. Radiopharm.* **1994**, *34*, 107–115.

472. Voyer, N.; Roby, J. *Tetrahedron Lett.* **1995**, *36*, 6627–6630.

473. Myers, A.G.; Gleason, J.L.; Yoon, T.; Kung, D.W. *J. Am. Chem. Soc.* **1997**, *119*, 656–673.

474. Alonsa, D.A.; Costa, A.; Nájera, C. *Tetrahedron Lett.* **1997**, *38*, 7943–7946.

475. Cho, J.H.; Kim, B.M. *Tetrahedron Lett.* **2002**, *43*, 1273–1276.

476. Hu, Y.-J.; Roy, R. *Tetrahedron Lett.* **1999**, *40*, 3305–3308.

477. Byk, G.; Scherman, D. *Int. J. Pept. Prot. Res.* **1996**, *47*, 333–339.

478. Mouna, A.M.; Nguyen, C.; Rage, I.; Xie, J.; Née, G.; Mazaleyrat, J.P.; Wakselman, M. *Synth. Commun.* **1994**, *24*, 2429–2435.

479. Miller, S.J.; Blackwell, H.E.; Grubbs, R.H. *J. Am. Chem. Soc.* **1996**, *118*, 9606–9614.

480. Hansen, D.W., Jr., Pilipauskas, D. *J. Org Chem.* **1985**, *50*, 945–950.

481. Salvatore, R.N.; Shin, S.I.; Flanders, V.L.; Jung, K.W. *Tetrahedron Lett.* **2001**, *42*, 1799–1801.

482. Bowman, W.R.; Coghlan, D.R. *Tetrahedron* **1997**, *53*, 15787–15798.

483. Fukuyama, T.; Jow, C.-K.; Cheung, M. *Tetrahedron Lett.* **1995**, *36*, 6373–6374.

484. Wilson, M.W.; Ault-Justus, S.E.; Hodges, J.C.; Rubin, J.R. *Tetrahedron* **1999**, *55*, 1647–1656.

485. Bolton, G.L.; Hodges, J.C.; Rubin, J.R. *Tetrahedron* **1997**, *53*, 6611–6634.

486. De Luca, S.; Della Moglie, R.; De Capua, A.; Morelli, G. *Tetrahedron Lett.* **2005**, *46*, 6637–6640.

487. DesMarteau, D.D.; Montanari, V. *Chem. Commun.* **1998**, 2241–2242.

488. Whitten, J.P.; Muench, D.; Cube, R.V.; Nyce, P.L.; Baron, B.M.; McDonald, I.A. *Bioorg. Med. Chem. Lett.* **1991**, *1*, 441–444.

489. Belokon, Y.N.; Tararov, V.I.; Maleev, V.I.; Savel'eva, T.F.; Ryzhov, M.G. *Tetrahedron: Asymmetry* **1998**, *9*, 4249–4252.

490. Valli, M.J.; Schoepp, D.D.; Wright, R.A.; Johnson, B.G.; Kingston, A.E.; Tomlinson, R.; Monn, J.A. *Biorg. Med. Chem. Lett.* **1998**, *8*, 1985–1990.

491. Patchett, A.A.; Harris, E.; Tristram, E.W.; Wyvratt, M.J.; Wu, M.T.; Taub, D.; Peterson, E.R.; Ikeler, T.J.; ten Broeke, J.; Payne, L.G.; Ondeyka, D.L.; Thorsett, E.D.; Greenlee, W.J.; Lohr, N.S.; Hoffsommer, R.D.; Joshua, H.; Ruyle, W.V.;

Rothrock, J.W.; Aster, S.D.; Maycock, A.L.; Robinson, F.M.; Hirschmann, R.; Sweet, C.S.; Ulm, E.H.; Gross, D.M.; Vassil, T..C.; Stone, C.A. *Nature* **1980**, *288*, 280–283.

492. Tran, T.-A.; Mattern, R.-H.; Morgan, B.A.; Taylor, J.E.; Goodman, M. *J. Peptide Res.* **1999**, *53*, 134–145.

493. Abderhalden, E.; Haase, E. *Z. Phys. Chem.* **1931**, *202*, 49–55.

494. Miyazawa, T.; Yamada, T.; Kuwata, S. *Bull. Chem. Soc. Jpn.* **1984**, *57*, 3605–3606.

495. Effenberger, F.; Burkard, U. *Liebigs Ann. Chem.* **1986**, 334–358.

496. Walker, M.A. *Tetrahedron* **1997**, *53*, 14591–14598.

497. Harfenist, M.; Hoerr, D.C.; Crouch, R. *J. Org. Chem.* **1985**, *50*, 1356–1359.

498. Koh, K.; Ben, R.N.; Durst, T. *Tetrahedron Lett.* **1994**, *35*, 375–378.

499. Dechantsreiter, M.A.; Burkhart, F.; Kessler, H. *Tetrahedron Lett.* **1998**, *39*, 253–254.

500. Rühl, T.; Böttcher, C.; Hennig, L.; Burger, K. *Amino Acids* **2004**, *27*, 285–290.

501. De Lombaert, S.; Erion, M.D.; Tan, J.; Blanchard, L.; El-Chehabi, L.; Ghai, R.D.; Sakane, Y.; Berry, C.; Trapani, A.J. *J. Med. Chem.* **1994**, *37*, 498–511.

502. Bird, J.; de Mello, R.C.; Harper, G.P.; Hunter, D.J.; Karran, E.H.; Markwell, R.E.; Miles-Williams, A.J.; Rahman, S.S.; Ward, R.W. *J. Med. Chem.* **1994**, *37*, 158–169.

503. Prishchenko, A.A.; Novikova, O.P.; Livantsov, M.V.; Grigor'ev, E.V. *Russ. J. Gen. Chem.* **1995**, *65*, 1604–1605.

504. Prishchenko, A.A.; Livantsov, M.V.; Novikova, O.P.; Livantsova, L.I.; Luzikov, Y.N. *Russ. J. Gen. Chem.* **1995**, *65*, 1606–1607.

505. Prishchenko, A.A.; Livantsov, M.V.; Livantsova, L.I.; Grigor'ev, E.V.; Pol'shchikov, D.G. *Russ. J. Gen. Chem.* **1995**, *65*, 1604–1605.

506. Takenaka, H.; Miyake, H.; Kojima, Y.; Yasuda, M.; Gemba, M.; Yamashita, T. *J. Chem. Soc., Perkin Trans. 1* **1993**, 933–937.

507. Becerril, J.; Bolte, M.; Burguete, M.E.; Galindo, F.; Garcîa-España, E.; Luis, S.V.; Miravet, J.F. *J. Am. Chem. Soc.* **2003**, *125*, 6677–6686.

508. Gill, D.M.; Pegg, N.A.; Rayner, C.M. *Tetrahedron* **1997**, *53*, 3383–3394.

509. Dörner, B.; Husar, G.M.; Ostresh, J.M.; Houghten, R.A. *Bioorg. Med. Chem.* **1996**, *4*, 709–715.

510. Pietzonka, T.; Seebach, D. *Angew. Chem., Int. Ed. Engl.* **1992**, *31*, 1481–1482.

511. Kim, Y.; Zeng, F.; Zimmerman, S.C. *Chem. —Eur. J.* **1999**, *5*, 2133–2138.

512. Baggaley, K.H.; Elson, S.W.; Nicholson, N.H.; Sime, J.T. *J. Chem. Soc., Perkin Trans 1* **1990**, 1521–1533.

513. Baggaley, K.H.; Elson, S.W.; Nicholson, N.H.; Sime, J.T. *J. Chem. Soc., Perkin Trans 1* **1990**, 1513–1520.

514. Demir, A.S.; Akhmedov, I.M.; Tanyeli, C.; Gercek, Z.; Gadzhili, R.A. *Tetrahedron: Asymmetry* **1997**, *8*, 753–757.

515. Chayer, S.; Essassi, E.M.; Bourguignon, J.-J. *Tetrahedron Lett.* **1998**, *39*, 841–844.

516. Zumpe, F.L.; Kazmaier, U. *Synlett* **1998**, 1199–1200.

517. Zumpe, F.L.; Kazmaier, U. *Synthesis* **1999**, 1785–1791.

518. Trost, B.M.; Clakins, T.L.; Oertelt, C.; Zambrano, J. *Tetrahedron Lett.* **1998**, *39*, 1713–1716.

519. Humphries, M.E.; Clark, B.P.; Williams, J.M.J. *Tetrahedron: Asymmetry* **1998**, *9*, 749–751.

520. Lin, X.; Dorr, H.; Nuss, J.M. *Tetrahedron Lett.* **2000**, *412*, 3309–3313.

521. Rew, Y.; Goodman, M. *J. Org. Chem.* **2002**, *67*, 8820–8826.

522. Urban, J. unpublished data from Molecumetics, Ltd.

523. Olsen, C.A.; Witt, M.; Hansen, S.H.; Jaroszewski, J.W.; Franzyk, H. *Tetrahedron* **2005**, *61*, 6046–6055.

524. Fujiwara, A.; Kan, T.; Fukuyama, T. *Synlett* **2000**, 1667–1669.

525. Zaragoza, F.; Stephensen, H. *Tetrahedron Lett.* **2000**, *41*, 1841–1844.

526. Song, Y.; Sercel, A.D.; Johnson, D.R.; Colbry, N.L.; Sun, K.-L.; Roth, B.D. *Tetrahedron Lett.* **2000**, *41*, 8225–8230.

527. Hardegger, E.; Szabo, F.; Liechti, P.; Rostetter, C.; Zankowska-Jasinska, W. *Helv. Chim. Acta* **1968**, *51*, 78–85.

528. Tararov, V.I.; Kadyrov, R.; Fischer, C.; Börner, A. *Synlett* **2004**, 1961–1962.

529. Schellenberg, K.A. *J. Org. Chem.* **1963**, *28*, 3259–3261.

530. Nicolás, E.; Pujades, M.; Bacardit, J.; Giralt, E.; Albericio, F. *Tetrahedron Lett.* **1997**, *38*, 2317–2320.

531. Ede, N.J.; Ang, K.H.; James, I.E.; Bray, A.M. *Tetrahedron Lett.* **1996**, *37*, 9097–9100.

532. Miranda, L.P.; Meutermans, W.D.F.; Smythe, M.L.; Alewood, P.F. *J. Org. Chem.* **2000**, *65*, 5460–5468.

533. Howe, J.; Quibell, M.; Johnson, T. *Tetrahedron Lett.* **2000**, *41*, 3997–4001.

534. Ohfune, Y.; Kurokawa, N.; Higuchi, N.; Saito, M.; Hashimoto, M.; Tanaka, T. *Chem. Lett.* **1984**, 441–444.

535. Püschl, A.; Sforza, S.; Haaima, G.; Dahl, O.; Nielsen, P.E. *Tetrahedron Lett.* **1998**, *39*, 4707–4710.

536. Dueholm, K.L.; Egholm, M.; Burchardt, O. *Org. Prep. Proc. Int.* **1993**, *25*, 457–461.

537. Bitan, G.; Muller, D.; Kasher, R.; Gluhov, E.V.; Gilon, C. *J. Chem. Soc., Perkin Trans. 1* **1997**, 1501–1510.

538. Gibson, S.E.; Guillo, N.; Middleton, R.J.; Thuilliez, A.; Tozer, M.J. *J. Chem. Soc., Perkin Trans. 1* **1997**, 447–455.

539. Fushiya, S.; Matsuda, M.; Yamada, S.; Nozoe, S. *Tetrahedron* **1996**, *52*, 877–886.

540. Fernández-García, C.; Prager, K.; McKervey, M.A.; Walker, B.; Williams, C.H. *Biorg. Med. Chem. Lett.* **1998**, *8*, 433–436.

541. Mutulis, F.; Mutule, I.; Lapins, M.; Wikberg, J.E.S. *Bioorg. Med. Chem. Lett.* **2002**, *12*, 1035–1038.

542. Mutulis, F.; Mutule, I.; Wikberg, J.E.S. *Bioorg. Med. Chem. Lett.* **2002**, *12*, 1039–1042.

543. Lee, S.-H.; Chung, S.-H.; Lee, Y.-S. *Tetrahedron Lett.* **1998**, *39*, 9469–9472.

544. Santagada, V.; Frecentese, F.; Perissutti, E.; Fiorino, F.; Severino, B.; Cirillo, D.; Terracciano, S.; Caliendo, G. *J. Comb. Chem.* **2005**, *7*, 618–621.

545. Du, M.; Hindsgaul, O. *Synlett* **1997**, 395–397.

546. Han, Y.; Chorev, M. *J. Org. Chem.* **1999**, *64*, 1972–1978.

547. Kim, H.-O.; Carroll, B.; Lee, M.S. *Synth. Commun.* **1997**, *27*, 2505–2515.

548. Ramanjulu, J.M.; Joullié, M.M. *Synth. Commun.* **1996**, *26*, 1379–1384.

549. Arya, P.; Dion, S.; Shimizu, G.K.H. *Bioorg. Med. Chem. Lett.* **1997**, *7*, 1537–1542.

550. Vergnon, A.L.; Pottorf, R.S.; Player, M.R. *J. Comb. Chem.* **2004**, *6*, 91–98.

551. Dales, N.A.; Gould, A.E.; Brown, J.A.; Calderwood, E.F.; Guan, B.; Minor, C.A.; Gavin, J.M.; Hales, P.; Kaushik, V.K.; Stewart, M.; Tummino, P.J.; Vickers, C.S.; Ocain, T.D.; Patane, M.A. *J. Am. Chem. Soc.* **2002**, *124*, 11852–11853.

552. Narasimhan, S.; Swarnalakshmi, S.; Balakumar, R.; Velmathi, S. *Synlett* **1998**, 1321–1322.

553. Ley, S.V.; Bolli, M.H.; Hinzen, B.; Gervois, A.-G.; Hall, B.J. *J. Chem. Soc., Perkin Trans. 1* **1998**, 2239–2241.

554. Coy, D.H.; Hocart, S.J.; Sasaki, Y. *Tetrahedron* **1988**, *44*, 835–841.

555. Loh, T.-P.; Ho, D.S.-C.; Xu, K.-C.; Sim, K.-Y. Tetrahedron Lett. 1997, 38, 865–868.

556. Wolfe, M.S.; Dutta, D.; Aubé, J. *J. Org. Chem.* **1997**, *62*, 654–663.

557. Zlotin, S.G.; Sharova, I.V.; Luk'yanov, O.A. *Russ. Chem. Bull.* **1994**, *43*, 1015–1017.

558. Zlotin, S.G.; Sharova, I.V.; Luk'yanov, O.A. *Russ. Chem. Bull.* **1996**, *45*, 1670–1679.

559. Zlotin, S.G.; Sharova, I.V.; Luk'yanov, O.A. *Russ. Chem. Bull.* **1996**, *45*, 1410–1418.

560. Zlotin, S.G.; Sharova, I.V.; Luk'yanov, O.A. *Russ. Chem. Bull.* **1995**, *44*, 1260–1261.

561. Zlotin, S.G.; Sharova, I.V.; Luk'yanov, O.A. *Russ. Chem. Bull.* **1995**, *44*, 1252–1259.

562. Zlotin, S.G.; Sharova, I.V.; Luk'yanov, O.A. *Russ. Chem. Bull.* **1996**, *45*, 1680–1687.

563. Prishchenko, A.A.; Novikova, O.P.; Livantsov, M.V.; Goncharova, Z.Y.; Merkulova, O.A.; Grigor'ev, E.V. *Russ. J. Gen. Chem.* **1995**, *65*, 1612–1613.

564. Maier, L. *Phosphorus, Sulfur Silicon Relat. Elem.* **1991**, *61*, 65–67.

565. Katritzky, A.R.; Kirichenko, N.; Rogovoy, B.V.; He, H.-Y. *J. Org. Chem.* **2003**, *68*, 9088–9092.

566. Paleo, M.R.; Calaza, I.; Sardina, F.J. *J. Org. Chem.* **1997**, *62*, 6862–6869.

567. Seebach, D.; Sommerfeld, T.L.; Jiang, Q.; Venanzi, L.M. *Helv. Chim. Acta* **1994**, *77*, 1313–1330.

568. Ma, D.; Yao, J. *Tetrahedron: Asymmetry* **1996**, *7*, 3075–3078.

569. Ma, D.; Zhang, Y.; Yao, J.; Wu, S.; Tao, F. *J. Am. Chem. Soc.* **1998**, *120*, 12459–12467.

570. Lam, P.Y.S.; Bonne, D.; Vincent, G.; Clark, C.G.; Combs, A.P. *Tetrahedron Lett.* **2003**, *44*, 1691–1694.

571. Combs, A.P.; Tadesse, S.; Rafalski, M.; Haque, T.S.; Lam, P.Y.S. *J. Comb. Chem.* **2002**, *4*, 179–182.

572. Pirkle, W.H.; Pochapsky, T.C. *J. Org. Chem.* **1986**, *51*, 102–105.

573. Henke, B.R.; Blanchard, S.G.; Brackeen, M.F.; Brown, K.K.; Cobb, J.E.; Collins, J.L.; Harrington, W.W., Jr., Hashim, M.A.; Hull-Ryde, E.A.; Kaldor, I.; Kliewer, S.A.; Lake, D.H.; Leesnitzer, L.M.; Lahmann, J.M.; Lenhard, J.M.; Orband-Miller, L.A.; Miller, J.F.; Mook Jr., R.A.; Noble, S.A.; Oliver, W., Jr.; Parks, D.J.; Plunket, K.D.; Szewczyk, J.R.; Willson, T.M. *J. Med. Chem.* **1998**, *41*, 5020–5036.

574. Cobb, J.E.; Blanchard, S.G.; Boswell, E.G.; Brown, K.K.; Charifson, P.S.; Cooper, J.P.; Collins, J.L.; Dezube, M.; Henke, B.R.; Hull-Ryde, E.A.; Lake, D.H.; Lenhard, J.M.; Oliver, W., Jr.; Oplinger, J.; Pentti, M.; Parks, D.J.; Plunket, K.D.; Tong, W.-Q. *J. Med. Chem.* **1998**, *41*, 5055–5069.

575. Cherng, Y.-J. *Tetrahedron* **2000**, *56*, 8287–8289.

576. Frost, C.G.; Mendonca, P. *J. Chem. Soc., Perkin Trans. 1* **1998**, 2615–2623.

577. Bowman, W.R.; Broadhurst, M.J.; Coghlan, D.R.; Lewis, K.A. *Tetrahedron Lett.* **1997**, *38*, 6301–6304.

578. Zhou, D.; Tan, H.; Luo, C.; Gan, L.; Huang, C.; Pan, J.; Lü, M.; Wu, Y. *Tetrahedron Lett.* **1995**, *36*, 9169–9172.

579. Romanova, V.S.; Tsyryapkin, V.A.; Lyakhovetsky, Y.I.; Parnes, Z.N.; Vol'pin, M.E. *Russ. Chem. Bull.* **1994**, *43*, 1090–1091.

580. Leblanc, J.-P.; Gibson, H.W. *Tetrahedron Lett.* **1992**, *33*, 6295–6298.

581. Bucherer, H. Th.; Fischbeck, H. *J. Prakt. Chem.* **1934**, *140*, 69–89.

582. Mai, K.; Patil, G. *Tetrahedron Lett.* **1984**, *25*, 4583–4586.

583. Stadnikoff, G. *Ber.* **1907**, *40*, 4350–4353.

584. Stadnikoff, G. *Ber.* **1907**, *40*, 4353–4356.

585. Stadnikoff, G. *Ber.* **1907**, *40*, 1014–1019.

586. Keating, T.A.; Armstrong, R.W. *J. Am. Chem. Soc.* **1995**, *117*, 7842–7842.

587. Shibata, N.; Das, B.K.; Takeuchi, Y. *J. Chem. Soc., Perkin Trans. 1* **2000**, 4234–4236.

588. Yanada, T.; Motoyama, N.; Taniguchi, T.; Kazuta, Y.; Miyazawa, T.; Kuwata, S.; Matsumoto, K.; Sugiura, M. *Chem. Lett.* **1987**, 723–726.

589. Demharter, A.; Hörl, W.; Herdtweck, E.; Ugi, I. *Angew. Chem., Int. Ed. Engl.* **1996**, *35*, 173–175.

590. Ugi, I.; Demharter, A.; Hörl, W.; Schmid, T. *Tetrahedron* **1996**, *52*, 11657–11664.

591. Beller, M.; Moradi, W.A.; Eckert, M.; Neumann, H. *Tetrahedron Lett.* **1999**, *40*, 4523–4526.

592. Drake, W.V.; McElvain, S.M. *J. Am. Chem. Soc.* **1934**, *56*, 697–700.

593. Goodfellow, V.S.; Marathe, M.V.; Kuhlman, K.G.; Fitzpatrick, T.D.; Cuadrado, D.; Hanson, W.; Zuzack, J.S.; Ross, S.E.; Wieczorek, M.; Burkard, M.; Whalley, E.T. *J. Med. Chem.* **1996**, *39*, 1472–1484.

594. Kruijtzer, J.A.W.; Hofmeyer, L.J.F.; Heerma, W.; Versluis, C.; Liskamp, R.M.J. *Chem.—Eur. J.* **1998**, *4*, 1570–1580.

595. Koppitz, M.; Huengs, M.; Gratias, R.; Kessler, H.; Goodman, S.L.; Jonczyk, A. *Helv. Chim. Acta.* **1997**, *80*, 1280–1300.

596. Stewart, F.H.C. *Aust. J. Chem.* **1961**, *14*, 654–656.

597. Tran, T.-A.; Mattern, R.-H.; Afargan, M.; Amitay, O.; Ziv, O.; Morgan, B.A.; Taylor, J.E.; Hoyer, D.; Goodman, M. *J. Med. Chem.* **1998**, *41*, 2679–2685.

598. Zuckermann, R.N.; Kerr, J.M.; Kent, S.B.H.; Moos, W.H. *J. Am. Chem. Soc.* **1992**, *114*, 10646–10647.

599. Saha, U.K.; Roy, R. *J. Chem. Soc., Chem. Commun.* **1995**, 2571–2573.

600. Kruijtzer, J.A.W.; Liskamp, R.M.J. *Tetrahedron Lett.* **1995**, *36*, 6969–6972.

601. Saha, U.K.; Roy, R. *Tetrahedron Lett.* **1995**, *36*, 3635–3638.

602. Saha, U.K.; Roy, R. *Tetrahedron Lett.* **1997**, *38*, 7697–7700.

603. Zuckermann, R.N.; Martin, E.J.; Spellmeyer, D.C.; Stauber, G.B.; Shoemaker, K.R.; Kerr, J.M.; Figliozzi, G.M.; Goff, D.A.; Siani, M.A.; Simon, R.J.; Banville, S.C.; Brown, E.G.; Wang, L.; Richter, L.S.; Moos, W.H. *J. Med. Chem.* **1994**, *37*, 2678–2685.

604. Yokoyama, N.; Walker, G.N.; Main, A. J.; Stanton, J.L.; Morrissey, M.M.; Boehm, C.; Engle, A.; Neubert, A.D.; Wasvary, J.M.; Stephan, Z.F.; Steele, R.E. *J. Med. Chem.* **1995**, *38*, 695–707.

605. Bigge, C.F.; Johnson, G.; Ortwine, D.F.; Drummond, J.T.; Retz, D.M.; Brahce, L.J.; Coughenour, L.L.; Marcoux, F.W.; Probert, A.W., Jr. *J. Med. Chem.* **1992**, *35*, 1371–1384.

606. Undheim, K.; Grønneberg, T. *Acta Chem. Scand.* **1971**, *25*, 18–26.

607. Bitan, G.; Gilon, C. *Tetrahedron* **1995**, *51*, 10513–10522.

608. Muller, D.; Zeltser, I.; Bitan, G.; Gilon, C. *J. Org. Chem.* **1997**, *62*, 411–416.

609. Byk, G.; Gilon, C. *J. Org. Chem.* **1992**, *57*, 5687–5692.

610. Effenberger, F.; Burkard, U.; Willfahrt, J. *Liebigs Ann. Chem.* **1986**, 314–333.

611. Lazareva, N.F.; Baryshok, V.P.; Voronkov, M.G. *Russ. Chem. Bull.* **1995**, *44*, 333–335.

612. Gulzar, M.S.; Akhtar, M.; Gani, D. J. *J. Chem. Soc., Chem. Commun.* **1994**, 1601–1602.

613. Zaderenko, P.; López, M.C.; Ballesteros, P. *J. Org. Chem.* **1996**, *61*, 6825–6828.

614. Fukuyama, T.; Cheung, M.; Kan, T. *Synlett* **1999**, 1301–1303.

615. Abarghaz, M.; Kerbal, A.; Bourguignon, J.-J. *Tetrahedron Lett.* **1995**, *36*, 6463–6466.

616. Simon, R.J.; Kania, R.S.; Zuckermann, R.N.; Huebner, V.D.; Jewell, D.A.; Banville, S.; Ng, S.; Wang, L.; Rosenberg, S.; Marlowe, C.K.; Spellmeyer, D.C.; Tan, R.; Frankel, A.D.; Santi, D.V.; Cohen, F.E.; Bartlett, P.A. *Proc. Natl. Acad. Sci. U.S.A.* **1992**, *89*, 9367–9371.

617. Gavagan, J.E.; Fager, S.K.; Seip, J.E.; Clark, D.S.; Payne, M.S.; Anton, D.L.; DiCosimo, R. *J. Org. Chem.* **1997**, *62*, 5419–5427.

618. Drok, K.J.; Harrowfield, J.M.; McNiven, S.J.; Sargeson, A.M.; Skelton, B.W.; White, A.H. *Aust. J. Chem.* **1993**, *46*, 1557–1593.

619. Einhorn, J.; Einhorn, C.; Pierre, J.-L. *Synlett*, **1994**, 1023–1024.

620. Burger, K.; Neuhauser, H.; Worku, A. *Z. Naturforsch.* **1993**, 107–120.

621. Petasis, N.A.; Goodman, A.; Zavialov, I.A. *Tetrahedron* **1997**, *53*, 16463–16470.

622. Fache, F.; Valot, F.; Milenkovic, A.; Lemaire, M. *Tetrahedron* **1996**, *52*, 9772–9784.

623. Chandrasekhar, B.P.; Heimgartner, H.; Schmid, H. *Helv. Chim. Acta* **1977**, *60*, 2270–2287.

624. Vittorelli, P.; Heimgartner, H.; Schmid, H.; Hoet, P.; Ghosez, L. *Tetrahedron* **1974**, *30*, 3737–3740.

625. Endo, Y.; Hizatate, S.; Shudo, K. *Synlett* **1991**, 649–650.

626. West, F.G.; Glaeske, K.W.; Naidu, B.N. *Synthesis* **1993**, 977–980.

627. Irreverre, F.; Morita, K.; Robertson, A.V.; Witkop, B. *J. Am. Chem. Soc.* **1963**, *85*, 2824–2831.

628. Challis, B.C.; Miligan, J.R.; Mitchell, R.C. *J. Chem. Soc., Perkin Trans. 1* **1990**, 3103–3106.

629. Wagner-Jauregg, T.; O'Neill, J.J.; Summerson, W.H. *J. Am. Chem. Soc.* **1951**, *73*, 5202–5206.

630. Kam, C.-M.; Nishino, N.; Powers, J.C. *Biochemistry*, **1979**, *18*, 3032–3038.

631. Galardy, R.E. *Biochemistry* **1982**, *21*, 5777–5781.

632. Galardy, R.E.; Kontoyiannidou-Ostrem, V.; Kortylewicz, Z.P. *Biochemistry* **1983**, *22*, 1990–1995.

633. Rodriguez, C.E.; Holmes, H.M.; Mlodnosky, K.L.; Lam, V.Q.; Berkman, C.E. *Biorg. Med. Chem. Lett.* **1998**, *8*, 1521–1524.

634. Chen, Z.-Z.; Tan, B.; Li, Y.-M.; Zhao, Y.-F. *J. Org. Chem.* **2003**, *68*, 4052–4058.

635. Wu, L.Y.; Berkman C.E. *Tetrahedron Lett.* **2005**, *46*, 5301–5303.

636. Mlodnosky, K.L.; Holmes, H.M.; Lam, V.Q.; Berkman, C.E. *Tetrahedron Lett.* **1997**, *38*, 8803–8806.

637. Chow, C.P.; Berkman, C.E. *Tetrahedron Lett.* **1998**, *39*, 7471–7474.

638. Ma, X.-b.; Zhao, Y.-f. *J. Org. Chem.* **1989**, *54*, 4005–4008.

639. Brands, K.M.J.; Wiedbrauk, K.; Williams, J.M.; Dolling, U.-H.; Reider, P.J. *Tetrahedron Lett.* **1998**, *39*, 9583–9586.

640. Kolodyazhnyi, O.I.; Grishkin, E.V.; Golovatyi, O.R. *Russ. J. Gen. Chem.* **1995**, *65*, 1106–1107.

641. Maung, J.; Campbell, T.Y.; Shieh, C.C.; Berkman, C.E. *Synth. Commun.* **2004**, *34*, 571–577.

642. Thompson, C.M.; Lin, J. *Tetrahedron Lett.* **1996**, 37, 8979–8982.

643. Lu, H.; Mlodnosky, K.L.; Dinh, T.T.; Dastgah, A.; Lam, V.Q.; Berkman, C.E. *J. Org. Chem.* **1999**, *64*, 8698–8701.

644. Baraniak, J.; Kaczmarek, R.; Korczynski, D.; Wasilewska, E. *J. Org. Chem.* **2002**, *67*, 7267–7274.

645. Khlebnikov, A.F.; Novikov, M.S.; Kostikov, R.R. *Synlett* **1997**, 929–930.

646. Saiga, Y.; Iijima, I.; Ishida, A.; Miyagishima, T.; Oh-Ishi, T.; Matsumoto, M.; Matsuoka, Y. *Chem. Pharm. Bull.* **1987**, *35*, 2840–2843.

647. Benoiton, N.L. In *Chemistry and Biochemistry of Amino Acids, Peptides and Proteins. A Survey of Recent Developments, Volume 5;* Weinstein, B., Ed.; Marcel Dekker: New York, 1978, pp 163–211.

648. Benoiton, L.; Berlinguet, L. *Biochem. Prep.* **1966**, *11*, 80–83.

649. Benoiton, L. *Can. J. Chem.* **1964**, *42*, 2043–2047.

650. Warren, S.; Zerner, B.; Westheimer, F.H. *Biochemistry* **1966**, *5*, 817–822.

651. Yamamoto, H.; Hayakawa, T.; Yang, J.T. *Biopolymers* **1974**, *13*, 1117–1125.

652. Yamamoto, H.; Yang, J.T. *Biopolymers* **1974**, *13*, 1093–1107.

653. Skinner, W.A.; Johansson, J.G. *J. Med. Chem.* **1972**, *15*, 427–428.

654. Prabhakaran, P.C.; Woo, N.-T.; Yorgey, P.S.; Gould, S.J. *J. Am. Chem. Soc.* **1988**, *110*, 5785–5791.

655. Seely, J.H.; Benoiton, N.L. *Biochem. Biophys. Res. Commun.* **1969**, *37*, 771–776.

656. Jain, R.; Cohen, L.A. *Tetrahedron* **1996**, *52*, 5363–5370.

657. Noordam, A.; Maat, L.; Beyerman, H.C. *Recl. Trav. Chim. Pays-Bas.* **1978**, *97*, 293–295.

658. Yuan, S.-S.; Ajami, A.M. *J. Lab. Cmpds. Radiopharm.* **1984**, *21*, 97–100.

659. Kimbonguila, A.M.; Boucida, S.; Guibé, F.; Loffet, A. *Tetrahedron* **1997**, *53*, 12525–12538.

660. Rajh, H.M.; Uitzetter, J.H.; Westerhuis, L.W.; van den Dries, C.L.; Tesser, G.I. *Int. J. Pep. Prot. Res.* **1979**, *14*, 68–79.

661. König, W.A.; Benecke, I.; Lucht, N.; Schmidt, E.; Schulze, J.; Sievers, S. *J. Chromatogr.* **1983**, *279*, 555–564.

662. König, W.A.; Benecke, I.; Schulze, J. *J. Chromatogr.* **1982**, *238*, 237–240.

663. König, W.A.; Steinbach, E.; Ernst, K. *Angew. Chem., Int. Ed. Engl.* **1984**, *23*, 527–528.

664. König, W.A.; Steinbach, E.; Ernst, K. *J. Chromatogr.* **1984**, *301*, 129–135.

665. Cheung, S.T.; Benoiton, N.L. *Can. J. Chem.* **1977**, *55*, 911–915.

666. Günther, K. *J. Chromatogr.* **1988**, *448*, 11–30.

667. Galpin, I.J.; Mohammed, A.K.; Patel, A. *Tetrahedron* **1988**, *44*, 1685–1690.

668. Tung, R.D.; Rich, D.H. *J. Am. Chem. Soc.* **1985**, *107*, 4342–4343.

669. Colucci, W.J.; Tung, R.D.; Petri, J.A.; Rich, D.H. *J. Org. Chem.* **1990**, *55*, 2895–2903.

670. Anteunis, M.J.O.; Sharma, N.K. *Bull. Soc. Chim. Belg.* **1988**, *97*, 281–292.

671. Coste, J.; Frérot, E.; Jouin, P.; Castro, B. *Tetrahedron Lett.* **1991**, *32*, 1967–1970.

672. Sivanandaiah, K.M.; Babu, V.V.S.; Shankaramma, S.C. *Int. J. Pept. Prot. Res.* **1994**, *44*, 24–30.

673. Wischnat, R.; Rudolph, J.; Hanke, R.; Kaese, R.; May, A.; Theis, H.; Zuther, U. *Tetrahedron Lett.* **2003**, *44*, 4393–4394.

674. Greene, T.W.; Wuts, P.G.M. *Protective Groups in Organic Synthesis,* 3rd edn; John Wiley and Sons, New York, 1999.

675. Kocienski, P.J. *Protecting Groups;* Thieme Verlag: New York, 2000.

676. Bodansky, M.; Bodansky, A. *The Practice of Peptide Synthesis,* 2nd edn; Springer Verlag: Berlin, 1995. Bodansky, M. *Peptide Chemistry. A Practical Textbook;* Springer Verlag: Berlin, 1988.

677. Wünsch, E. Synthese von Peptiden. In *Houben-Weyl: Methoden der organischen Chemie*, Vols 1/2, Müller, E., Ed.; Georg Thieme: Stuttgart, 1974.

678. Goodman, M.; Felix, A.; Moroder, L.; Toniolo, Eds.; *Houben–Weyl: Methods of Organic Chemistry. Additional and Supplementary Volumes to the 4th Edition. Synthesis of Peptides and Peptidomimetics*; Vols E22a and E22b; Georg Thieme: Stuttgart, 2002.

679. *The Peptides. Analysis, Synthesis, Biology. Volume 3, Protection of Functional Groups in Peptide Synthesis;* Gross, E.; Meienhofer, J, Eds; Academic Press: New York, 1981.

680. Orain, D.; Ellard, J.; Bradley, M. *J. Comb. Chem.* **2002**, *4*, 1–16.

681. Albericio, F. *Biopolymers (Peptide Sci.)* **2000**, *55*, 123–139.

682. Jones, J. *Amino Acid and Peptide Synthesis*; Oxford University Press: New York, 2000.

683. Carpino, L.A. *Acc. Chem. Res.* **1973**, *6*, 191–198.
684. Jarowicki, K.; Kocienski, P. *J. Chem. Soc., Perkin Trans 1* **1998**, 4005–4037.
685. Schelhaas, M.; Waldmann, H. *Angew. Chem., Int. Ed. Engl.* **1996**, *35*, 2056–2083.
686. Mikhalkin, A.P. *Russ. Chem. Rev.* **1995**, *64*, 259–275.
687. Jursic, B.S.; Neumann, D. *Synth. Commun.* **2001**, *31*, 555–564.
688. Chen, F.M.F.; Benoiton, N.L. *Can. J. Chem.* **1987**, *65*, 1224–1227.
689. Yadav, J.S.; Reddy, G.S.; Reddy, M.M.; Meshram, H.M. *Tetrahedron Lett.* **1998**, *39*, 3259–3262.
690. Pandey, R.K.; Dagade, S.P.; Dongare, M.K.; Kumar, P. *Synth. Commun.* **2003**, *33*, 4019–4027.
691. Mitin, Y.V. *Int. J. Pept. Prot. Chem.* **1996**, *48*, 374–376.
692. Heeb, N.V.; Aberle, A.M.; Manbiar, K.P. *Tetrahedron Lett.* **1994**, *35*, 2287–2290.
693. Khalil, E.M.; Subasinghe, N.L.; Johnson, R.L. *Tetrahedron Lett.* **1996**, *37*, 3441–3444.
694. Li, W.-R.; Yo, Y.-C.; Lin, Y.-S. *Tetrahedron* **2000**, *56*, 8867–8875.
695. Salvatore, R.N.; Shin, S.I.; Nagle, A.S.; Jung, K.W. *J. Org. Chem.* **2001**, *66*, 1035–1037.
696. D'Addona, D.; Bochet, C.G. *Tetrahedron Lett.* **2001**, *42*, 5227–5229.
697. Knölker, H.-J.; Braxmeier, T. *Synlett* **1997**, 925–928.
698. Nishikawa, T.; Urabe, D.; Tomita, M.; Tsujimoto, T.; Iwabuchi, T.; Isobe, M. *Org. Lett.* **2006**, *8*, 3263–3265.
699. Mormeneo, D.; Llebaria, A.; Delgado, A. *Tetrahedron Lett.* **2004**, *45*, 6831–6834.
700. Kondo, K.; Sekimoto, E.; Miki, K.; Murakami, Y. *J. Chem. Soc., Perkin Trans. 1* **1998**, 2973–2974.
701. Henklein, P.; Heyne, H.-U.; Halatsch, W.-R.; Niedrich, H. *Synthesis* **1987**, 166–167.
702. Takeda, K.; Tsuboyama, K.; Hoshino, M.; Kishino, M.; Ogura, H. *Synthesis* **1987**, 557–560.
703. Katritzky, A.R.; Fali, C.N.; Li, J.; Ager, D.J.; Prakash, I. *Synth. Commun.* **1997**, *27*, 1623–1630.
704. Anbazhagan, M.; Reddy, T.I.; Rajappa, S. *J. Chem. Soc., Perkin Trans. 1* **1997**, 1623–1627.
705. Barrett, A.G.M.; Cramp, S.M.; Roberts, S.; Zecri, F.J. *Org. Lett.* **2000**, *2*, 261–264.
706. Duczek, W.; Deutsch, J.; Vieth, S.; Niclas, H.-J. *Synthesis* **1996**, 37–38.
707. Chancellor, T.; Morton, C. *Synthesis* **1994**, 1023–1025.
708. Giard, T.; Bénard, D.; Plaquevent, J.-C. *Synthesis* **1998**, 297–300.
709. Hill, D.R.; Hsiao, C.-N.; Kurukulasuriya, R.; Wittenberger, S.J. *Org. Lett.* **2002**, *4*, 111–113.
710. Kotha, S.; Behera, M.; Khedkar, P. *Tetrahedron Lett.* **2004**, *45*, 7589–7509.
711. Weiberth, F.J. *Tetrahedron Lett.* **1999**, *40*, 2895–2898.
712. Wakamatsu, H.; Uda, J.; Yamakami, N. *J. Chem. Soc., Chem. Commun.* **1971**, 1540.
713. Degerbeck, F.; Fransson, B.; Grehn, L.; Ragnarsson, U. *J. Chem. Soc., Perkin Trans. 1* **1992**, 245–253.
714. Clarke, C.T.; Elliott, J.D.; Jones, J.H. *J. Chem. Soc., Perkin Trans. 1* **1978**, 1088–1090.
715. Barton, D.H.R.; Kwon, T.; Taylor, D.K.; Tajbakhsh, M. *Bioorg. Med. Chem.* **1995**, *3*, 79–84.
716. Spaggiari, A.; Blaszczak, L.C.; Pati, F. *Org. Lett.* **2004**, *6*, 3885–3888.
717. Prato, M.; Bianco, A.; Maggini, M.; Scorrano, G.; Toniolo, C.; Wudl, F. *J. Org. Chem.* **1993**, *58*, 5578–5580.
718. Skiebe, A.; Hirsch, A. *J. Chem. Soc., Chem. Commun.* **1994**, 335–336.
719. Engström, A.; Andersson, P.E.; Josefsson, B.; Pfeffer, W.D. *Anal. Chem.* **1995**, *67*, 3018–3022.
720. Ramage, R.; Swenson, H.R.; Shaw, K.T. *Tetrahedron Lett.* **1998**, *39*, 8715–8718.
721. Ramage, R.; Raphy, G. *Tetrahedron Lett.* **1992**, *33*, 385–388.
722. Kemp, D.S.; Roberts, D.C. *Tetrahedron Lett.* **1975**, *52*, 4629–4632.
723. Nájera, C. *Synlett* **2002**, 1388–1403.
724. Bergmann, M.; Zervas, L. *Ber.* **1932**, *65*, 1192–1201.
725. Helgen, C.; Bochet, C.G. *J. Org. Chem.* **2003**, *68*, 2483–2486.
726. Hernández, J.N.; Martin, V.S. *J. Org. Chem.* **2004**, *69*, 3590–3592.
727. Hernández, J.N.; Ramírez, M.A.; Martin, V.S. *J. Org. Chem.* **2003**, *68*, 743–746.
728. Hattori, K.; Sajiki, H.; Hirota, K. *Tetrahedron* **2000**, *56*, 8433–8441.
729. Murata, M.; Hara, T.; Mori, K.; Ooe, M.; Mizugaki, T.; Ebitani, K.; Kaneda, K. *Tetrahedron Lett.* **2003**, *44*, 4981–4984.
730. Dzubeck, V.; Schneider, J.P. *Tetrahedron Lett.* **2000**, *41*, 9953–9956.
731. Bajwa, J.S. *Tetrahedron Lett.* **1992**, *33*, 2299–2302.
732. Boeckman, R.K., Jr.; Potenza, J.C. *Tetrahedron Lett.* **1985**, *26*, 1411–1414.
733. Adinolfi, M.; Guariniello, L.; Iadonisi, A.; Mangoni, L. *Synlett* **2000**, 1277–1278.
734. Daga, M.C.; Taddei, M.; Varchi, G. *Tetrahedron Lett.* **2001**, *42*, 5192–5194.
735. Gowda, D.C. *Tetrahedron Lett.* **2002**, *43*, 311–313.
736. Coleman, R.S.; Shah, J.A. *Synthesis* **1999**, 1399–1400.
737. Bose, D.S.; Thurston, D.E. *Tetrahedron Lett.* **1990**, *31*, 6903–6906.
738. Allevi, P.; Cribiù, R.; Anastasia, M. *Tetrahedron Lett.* **2004**, *45*, 5841–5843.
739. Lipshutz, B.H.; Pfeiffer, S.S.; Reed, A.B. *Org. Lett.* **2001**, *3*, 4145–4148.

740. Isidro-Llobet, A.; Álvarez, M.; Albericio, F. *Tetrahedron Lett.* **2005**, *46*, 7733–7736.

741. Isidro-Llobet, A.; Guasch-Camell, J.; Álvarez, M.; Albericio, F. *Eur. J. Org. Chem.* **2005**, 3031–3039.

742. Papageorgiou, E.A.; Gaunt, M.J.; Yu, J.-q.; Spencer, J.B. *Org. Lett.* **2000**, *2*, 1049–1051.

743. Curran, D.P.; Amatore, M.; Guthrie, D.; Campbell, M.; Go, E.; Luo, Z. *J. Org. Chem.* **2003**, *68*, 4643–4647.

744. Filippov, D.V.; van Zoelen, D.J.; Oldfield, S.P.; van der Marel, G.A.; Overkleeft, H.S.; Drijfhout, J.W.; van Boom, J.H. *Tetrahedron Lett.* **2002**, *43*, 7809–7812.

745. Schwinn, D.; Bannwarth, W. *Helv. Chim. Acta* **2002**, *85*, 255–264.

746. McKay, F.C.; Albertson, N.F. *J. Am. Chem. Soc.* **1957**, *79*, 4686–4690.

747. Moroder, L.; Hallett, A.; Wünsch, E.; Keller, O.; Wersin, G. *Z. Phys. Chem.* **1976**, *357*, 1651–1653.

748. Schnabel, E. *Liebigs Ann. Chem.* **1967**, *702*, 188–196.

749. Sharma, G.V.M.; Reddy, J.J.; Lakshmi, P.S.; Krishna, P.R. *Tetrahedron Lett.* **2004**, *45*, 6963–6965.

750. Chankeshwara, S.V.; Chakraborti, A.K. *Org. Lett.* **2006**, *8*, 3259–3262.

751. Ouchi, H.; Saito, Y.; Yamamoto, Y.; Takahata, H. *Org. Lett.* **2002**, *4*, 585–587.

752. Grehn, L.; Ragnarsson, U. *Angew. Chem., Int. Ed. Engl.* **1985**, *24*, 510–511.

753. Gunnarsson, K.; Ragnarsson, U. *Acta. Chem. Scand.* **1990**, *44*, 944–951.

754. Burk, M.J.; Allen, J.G. *J. Org. Chem.* **1997**, *62*, 7054–7057.

755. Bajwa, J.S. *Tetrahedron Lett.* **1992**, *33*, 2955–2956.

756. Agami, C.; Couty, F. *Tetrahedron* **2002**, *58*, 2701–2724.

757. Strazzolini, P.; Misuri, N.; Polese, P. *Tetrahedron Lett.* **2005**, *46*, 2075–2078.

758. Lin, L.S.; Lanza, T.L., Jr.; de Laszlo, S.E.; Truong, Q.; Kamenecka, T.; Hagmann, W.K. *Tetrahedron Lett.* **2000**, *41*, 7013–7016.

759. Trivedi, H.S.; Anson, M.; Steel, P.G.; Worley, J. *Synlett* **2001**, 1932–1934.

760. Lejeune, V.; Martinez, J.; Cavelier, F. *Tetrahedron Lett.* **2003**, *44*, 4757–4759.

761. Strazzolini, P.; Melloni, T.; Giumanini, A.G. *Tetrahedron* **2001**, *57*, 9033–9043.

762. Li, B.; Bemish, R.; Buzon, R.A.; Chiu, C.K.-F.; Colgan, S.T.; Kissel, W.; Le, T.; Leeman, K.R.; Newell, L.; Roth, J. *Tetrahedron Lett.* **2003**, *44*, 8113–8115.

763. Babu, V.V.S.; Patil, B.S.; Vasanthakumar, G.-R. *Synth. Commun.* **2005**, *35*, 1795–1802.

764. Maurer, K.; McShea, A.; Strathmann, M.; Dill, K. *J. Comb. Chem.* **2005**, *7*, 637–640.

765. Pellois, J.P.; Wang, W.; Gao, X. *J. Comb. Chem.* **2000**, *2*, 355–360.

766. Kaiser, E., Sr.; Picart, F.; Kubiak, T.; Tam, J.P.; Merifield, R.B. *J. Org. Chem.* **1993**, *58*, 5167–5175.

767. Vorbrüggen, H.; Krolikiewicz, K. *Angew. Chem., Int. Ed. Engl.* **1975**, *14*, 818.

768. Chen, B.-C.; Skoumbourdis, A.P.; Guo, P.; Bednarz, M.S.; Kocy, O.R.; Sundeen, J.E.; Vite, G.D. *J. Org. Chem.* **1999**, *64*, 9294–9296.

769. Evans, E.F.; Lewis, N.J.; Kapfer, I.; Macdonald, G.; Taylor, R.J.K. *Synth. Commun.* **1997**, *27*, 1819–1825.

770. Kotsuki, H.; Ohishi, T.; Araki, T.; Arimura, K. *Tetrahedron Lett.* **1998**, *39*, 4869–4870.

771. Liu, Y.-S.; Zhao, C.; Bergbreiter, D.E.; Romo, D. *J. Org. Chem.* **1998**, *63*, 3471–3473.

772. Shaikh, N.S.; Gajare, A.S.; Deshpande, V.H.; Bedekar, A.V. *Tetrahedron Lett.* **2000**, *41*, 385–387.

773. Apelqvist, T.; Wensbo, D. *Tetrahedron Lett.* **1996**, *37*, 1471–1472.

774. Tom, N.J.; Simon, W.M.; Frost, H.N.; Ewing, M. *Tetrahedron Lett.* **2004**, *45*, 905–906.

775. Coop, A.; Rice, K.C. *Tetrahedron Lett.* **1998**, *39*, 8933–8934.

776. Jacquemard, U.; Bénéteau, V.; Lefoix, M.; Routier, S.; Mérour, J.-Y.; Coudert, G. *Tetrahedron* **2004**, *60*, 10039–10047.

777. Routier, S.; Saugé, L.; Ayerbe, N.; Coudert, G.; Mérour, J.-Y. *Tetrahedron Lett.* **2002**, *43*, 589–591.

778. Frank, R.; Schutkowski, M. *Chem. Commun.* **1996**, 2509–2510.

779. Bose, D.S.; Kumar, K.K.; Reddy, A.V.N. *Synth. Commun.* **2003**, *33*, 445–450.

780. Bose, D.S.; Lakshminarayana, V. *Synthesis* **1999**, 66–68.

781. James, K.D.; Ellington, A.D. *Tetrahedron Lett.* **1998**, *39*, 175–178.

782. Navath, R.S.; Pabbisetty, K.B.; Hu, L. *Tetrahedron Lett.* **2006**, *47*, 389–393.

783. Kaul, R.; Brouillette, Y.; Sajjadi, Z.; Handford, K.A.; Lubell, W.D. *J. Org. Chem.* **2004**, *69*, 6131–6133.

784. Marcantoni, E.; Massaccesi, M.; Torregiani, E.; Bartoli, G.; Bosco, M.; Sambri, L. *J. Org. Chem.* **2001**, *66*, 4430–4432.

785. Yadav, J.S.; Reddy, B.V.S.; Rao, K.S.; Harikishan, K. *Synlett* **2002**, 826–828.

786. Yadav, J.S.; Balanarsaiah, E.; Raghavendra, S.; Satyanarayana, M. *Tetrahedron Lett.* **2006**, *47*, 4921–4924.

787. Yadav, J.S.; Reddy, B.V.S.; Reddy, K.S. *Synlett* **2002**, 468–470.

788. Yadav, J.S.; Reddy, B.V.S.; Reddy, K.S.; Reddy, K.B. *Tetrahedron Lett.* **2002**, *43*, 1549–1551.

789. Wildemann, D.; Drewello, M.; Fischer, G.; Schutkowski, M. *Chem. Commun.* **1999**, 1809–1810.

790. Carpino, L.A.; Han, G.Y. *J. Am. Chem. Soc.* **1970**, *92*, 5748–5749.

791. Carpino, L.A.; Han, G.Y. *J. Org. Chem.* **1972**, *37*, 3404–3409.

792. Chang, C.-D.; Waki, M.; Ahmad, M.; Meienhofer, J.; Lundell, E.O.; Haug, J.D. *Int. J. Pept. Prot. Res.* **1980**, *15*, 59–66.

793. Lapatsanis, L.; Milias, G.; Froussios, K.; Kolovos, M. *Synthesis* **1983**, 671–673.

794. Sigler, G.F.; Fuller, W.D.; Chaturvedi, N.C.; Goodman, M.; Verlander, M. *Biopolymers* **1983**, *22*, 2157–2162.

795. Tessier, M.; Albericio, F.; Pedroso, E.; Grandas, A.; Eritja, R.; Giralt, E.; Granier, C.; Van Rietschoten, J. *Int. J. Pept. Prot. Res.* **1983**, *22*, 125–128.

796. de L. Milton, R.C.; Becker, E.; Milton, S.C.F.; Baxter, J.E.J.; Elsworth, J.F. *Int. J. Pept. Prot. Res.* **1987**, *30*, 431–432.

797. Ten Kortenaar, P.B.W.; Van Dijk, B.G.; Peeters, J.M.; Raaben, B.J.; Adams, P.J.H.M.; Tesser, G.I. *Int. J. Pept. Prot. Res.* **1986**, *27*, 398–400.

798. Schön, I.; Kisfaludy, L. *Synthesis* **1986**, 303–305.

799. Chinchilla, R.; Dodsworth, D.J.; Nájera, C.; Soriano, J.M. *Tetrahedron Lett.* **2001**, *42*, 7579–7581.

800. Carpino, L.A. *Acc. Chem. Res.* **1987**, *20*, 201–407.

801. Meienhofer, J.; Waki, M.; Heimer, E.P.; Lambros, T.J.; Makofske, R.C.; Chang, C.-D. *Int. J. Pept. Prot. Res.* **1979**, *13*, 35–42.

802. Zinieris, N.; Leondiadis, L.; Ferderigos, N. *J. Comb. Chem.* **2005**, *7*, 4–6.

803. Hachmann, J.; Lebl, M. *J. Comb. Chem.* **2006**, *8*, 149.

804. Peluso, S.; Dumy, P.; Nkubana, C.; Yokokawa, Y; Mutter, M. *J. Org. Chem.* **1999**, *64*, 7114–7120.

805. Sheppeck, J.E., II; Kar, H.; Hong, H. *Tetrahedron Lett.* **2000**, *41*, 5329–5333.

806. Bu, X.; Xie, G.; Law, C.W.; Guo, Z. *Tetrahedron Lett.* **2002**, *43*, 2419–2422.

807. Ueki, M.; Amemiya, M. *Tetrahedron Lett.* **1987**, *28*, 6617–6620.

808. Carpino, L.A.; Mansour, E.M.E.; Cheng, C.H.; Williams, J.R.; MacDonald, R.; Knapczyk, J.; Carmen, M.; Lopusinski, A. *J. Org. Chem.* **1983**, *48*, 661–665.

809. Leggio, A.; Liguori, A.; Napoli, A.; Siciliano, C.; Sindona, G. *Eur. J. Org.Chem.* **2000**, 573–575.

810. Furlán, R.L.E.; Mata, E.G. *Tetrahedron Lett.* **1998**, *39*, 6421–6422.

811. Zaragoza, F.; Stephensen, H. *Tetrahedron Lett.* **2000**, *41*, 2015–2017.

812. Pascal, R.; Sola, R. *Tetrahedron Lett.* **1998**, *39*, 5031–5034.

813. Farrera-Sinfreu, J.; Royo, M.; Albericio, F. *Tetrahedron Lett.* **2002**, *43*, 7813–7815.

814. Stigers, K.D.; Koutroulis, M.R.; Chung, D.M.; Nowick, J.S. *J. Org. Chem.* **2000**, *65*, 3858–3860.

815. Carpino, L.A.; Wu, A.-C. *J. Org. Chem.* **2000**, *65*, 9238–9240.

816. Chinchilla, R.; Dodsworth, D.J.; Nájera, C.; Soriano, J.M. *Tetrahedron Lett.* **2002**, *43*, 1817–1820.

817. Burch, R.M.; Wietzberg, M.; Blok, N.; Muhlhauser, R.; Martin, D.; Farmer, S.G.; Bator, J.M.; Connor, J.R.; Ko, C.; Kuhn, W.; McMillan, B.A.; Raynor, M.; Shearer, B.G.; Tiffany, C.; Wilkins, D.E. *Proc. Natl. Acad. Sci. U.S.A.* **1991**, *88*, 355–359.

818. Guibé, F. *Tetrahedron* **1998**, *54*, 2967–3042.

819. Chinchilla, R.; Dodsworth, D.J.; Nájera, C.; Soriano, J.M. *Synlett* **2003**, 809–812.

820. Kunz, H.; Unverzagt, C. *Angew. Chem., Int. Ed. Engl.* **1984**, *23*, 436–437.

821. Dangles, O.; Guibé, F.; Balavoine, G.; Lavielle, S.; Marquet, A. *J. Org. Chem.* **1987**, *52*, 4984–4993.

822. Merzouk, A.; Guibé, F. *Tetrahedron Lett.* **1992**, *33*, 477–480.

823. Malkar, N.B.; Lauer-Fields, J.L.; Borgia, J.A.; Fields, G.B. *Biochemistry* **2002**, *41*, 6054–6064.

824. Gomez-Martinez, P.; Dessolin, M.; Guibé, F.; Albericio, F. *J. Chem. Soc., Perkin Trans. 1* **2000**, 2871–2874.

825. Franco, D.; Duñach, E. *Tetrahedron Lett.* **2000**, *41*, 7333–7336.

826. Szumigala, R.H., Jr.; Onofiok, E.; Karady, S.; Armstrong, J.D., III; Miller, R.A. *Tetrahedron Lett.* **2005**, *46*, 4403–4405.

827. Lemaire-Audoire, S.; Savignac, M.; Blart, E.; Bernard, J.-M.; Genêt, J.P. *Tetrahedron Lett.* **1997**, *38*, 2955–2958.

828. Thieriet, N.; Gomez-Martinez, P.; Guibé, F. *Tetrahedron Lett.* **1999**, *40*, 2505–2508.

829. Zorn, C.; Gnad, F.; Salmen, S.; Herpin, T.; Reiser, O. *Tetrahedron Lett.* **2001**, *42*, 7049–7053.

830. von dem Bruch, K.; Kunz, H. *Angew. Chem., Int. Ed. Engl.* **1990**, *29*, 1457–1459.

831. Ramesh, R.; Bhat, R.G.; Chandrasekaran, S. *J. Org. Chem.* **2005**, *70*, 837–840.

832. Bhat, R.G.; Kérourédan, E.; Porhiel, E.; Chandrasekaran, S. *Tetrahedron Lett.* **2002**, *43*, 2467–2469.

833. Bhat, R.G.; Sinha, S.; Chandrasekaran, S. *Chem. Commun.* **2002**, 812–813.

834. Ramesh, R.; Chandrasekaran, S. *Org. Lett.* **2005**, *7*, 4947–4950.

835. Schielen, W.J.G.; Adams, H.P.H.M.; Nieuwenhuizen, W.; Tesser, G.I. *Int. J. Pept. Prot. Res.* **1991**, *37*, 341–346.

836. de Visser, P.C.; van Helden, M.; Filippov, D.V.; van der Marel, G.A.; Drijfhout, J.W.; van Boom, J.H.; Noort, D.; Overkleeft, H.S. *Tetrahedron Lett.* **2003**, *44*, 9013–9016.

837. Hojo, K.; Maeda, M.; Kawasaki, K. *Tetrahedron Lett.* **2004**, *45*, 9095–9097.

838. Hojo, K.; Maeda, M.; Kawasaki, K. *Tetrahedron* **2004**, *60*, 1875–1886.

839. Hojo, K.; Maeda, M.; Takahara, Y.; Yamamoto, S.; Kawasaki, K. *Tetrahedron Lett.* **2003**, *44*, 2849–2851.

840. Choi, J.S.; Kang, H.; Jeong, N.; Han, H. *Tetrahedron* **2005**, *61*, 2493–2503.

841. Choi, J.S.; Lee, Y.; Kim, E.; Jeong, N.; Yu, H.; Han, H. *Tetrahedron Lett.* **2003**, *44*, 1607–1610.

842. Carpino, L.A.; Philbin, M.; Ismail, M.; Truran, G.A.; Mansour, E.M.E.; Iguchi, S.; Ionescu, D.; El-Faham, A.; Riemer, C.; Warrass, R.; Weiss, M.S. *J. Am. Chem. Soc.,* **1997**, *119*, 9915–9916.

843. Carpino, L.A.; Ismail, M.; Truran, G.A.; Mansour, E.M.E.; Iguchi, S.; Ionescu, D.; El-Faham, A.; Riemer, C.; Warrass, R. *J. Org. Chem.* **1999**, *64*, 4324–4338.

844. Greenwald, R.B.; Zhao, H.; Reddy, P. *J. Org. Chem.* **2003**, *68*, 4894–4896.

845. Carpino, L.A.; Philbin, M. *J. Org. Chem.* **1999**, *64*, 4315–4323.

846. Carpino, L.A.; Mansour, E.M.E. *J. Org. Chem.* **1999**, *64*, 8399–8401.

847. Shao, J.; Shekhani, M.S.; Krauss, S.; Grübler, G.; Voelter, W. *Tetrahedron Lett.* **1991**, *32*, 345–346.

848. Zaramella, S.; Yeheskiely, E.; Strömberg, R. *J. Am. Chem. Soc.* **2004**, *126*, 14029–14035.

849. Mezo, G.; Mihala, N.; Kóczán, G.; Hudecz, F. *Tetrahedron* **1998**, *54*, 6757–6766.

850. Lingard, I.; Bhalay, G.; Bradley, M. *Synlett* **2003**, 1791–1792.

851. Luo, Z.; Williams, J.; Read, R.W; Curran, D.P. *J. Org. Chem.* **2001**, *66*, 4261–4266.

852. Carson, J.F. *Synthesis* **1981**, 268–270.

853. Tokimoto, H.; Fukase, K. *Tetrahedron Lett.* **2005**, *46*, 6831–6832.

854. Somsák, L.; Czifrák, K.; Veres, E. *Tetrahedron Lett.* **2004**, *45*, 9095–9097.

855. Mineno, T.; Choi, S.-R.; Avery, M.A. *Synlett* **2002**, 883–886.

856. Vatèle, J.-M. *Tetrahedron* **2004**, *60*, 4251–4260.

857. Vatèle, J.-M. *Tetrahedron Lett.* **2003**, *44*, 9127–9129.

858. Carpino, L.A.; Tsao, J.-H.; Ringsdorf, H.; Fell, E.; Hettrich, G. *J. Chem. Soc., Chem. Commun.* **1978**, 358–359.

859. Lipshutz, B.H.; Papa, P.; Keith, J.M. *J. Org. Chem.* **1999**, *64*, 3792–3793.

860. Sakamoto, K.; Nakahara, Y.; Ito, Y. *Tetrahedron Lett.* **2002**, *43*, 1515–1518.

861. Wagner, M.; Heiner, S.; Kunz, H. *Synlett* **2000**, 1753–1756.

862. Sakaitani, M.; Ohfune, Y. *J. Org. Chem.* **1990**, *55*, 870–876.

863. Amit, B.; Zehavi, U.; Patchornik, A. *J. Org. Chem.* **1974**, *39*, 192–196.

864. Bhushan, K.R.; DeLisi, C.; Laursen, R.A. *Tetrahedron Lett.* **2003**, *44*, 8585–8588.

865. Ueda, S.; Fujita, M.; Tamamura, H.; Fuji, N.; Otaka, A. *ChemBioChem* **2005**, *6*, 1983–1986.

866. Du, H.; Body, M.K. *Tetrahedron Lett.* **2001**, *42*, 6645–6647.

867. Kessler, M.; Glatthar, R.; Giese, B.; Bochet, C.G. *Org. Lett.* **2003**, *5*, 1179–1181.

868. Shaginian, A.; Patel, M.; Li, M.-H.; Flickinger, S.T.; Kim, C.; Cerrina, F.; Belshaw, P.J. *J. Am. Chem. Soc.* **2004**, *126*, 16704–16705.

869. Gum, A.G.; Kappes-Roth, T.; Waldmann, H. *Chem.—Eur. J.* **2000**, *6*, 3714–3721.

870. Sander, J.; Waldmann, H. *Chem. —Eur. J.* **2000**, *6*, 1564–1577.

871. Jeyaraj, D.A.; Prinz, H.; Waldmann, H. *Tetrahedron* **2002**, *58*, 1879–1887.

872. Jeyaraj, D.A.; Waldmann, H. *Tetrahedron Lett.* **2001**, *42*, 835–837.

873. Nägele, E.; Schelhaas, M.; Kuder, N.; Waldmann, H. *J. Am. Chem. Soc.* **1998**, *120*, 6889–6902.

874. Díaz-Mochón, J.J.; Bialy, L.; Bradley, M. *Org. Lett.* **2004**, *6*, 1127–1129.

875. Rohwedder, B.; Mutti, Y.; Dumy, P.; Mutter, M. *Tetrahedron Lett.* **1998**, *39*, 1175–1178.

876. Augustyns, K.; Kraas, W.; Jung, G. *J. Peptide Res.* **1998**, *51*, 127–133.

877. Kellam, B.; Bycroft, B.W.; Chan, W.C.; Chhabra, S.R. *Tetrahedron* **1998**, *54*, 6817–6832.

878. Chhabra, S.R.; Hothi, B.; Evans, D.J.; White, P.D.; Bycroft, B.W.; Chan, W.C. *Tetrahedron Lett.* **1998**, *39*, 1603–1606.

879. Berthelot, T.; Goncales, M.; Laïn, G.; Estieu-Gionnet, K.; Déléris, G. *Tetrahedron* **2006**, *62*, 1124–1130.

880. Roemmele, R.C.; Rapoport, H. *J. Org. Chem.* **1988**, *53*, 2367–2371.

881. Nyasse, B.; Grehn, L.; Maia, H.L.S.; Monteiro, L.S.; Ragnarsson, U. *J. Org. Chem.* **1999**, *64*, 7135–7139.

882. Nihei, K.-i.; Kato, M.J.; Yamane, T.; Palma, M.S.; Konno, K. *Synlett* **2001**, 1167–1169.

883. Christensen, C.; Clausen, R.P.; Begtrup, M.; Kristensen, J.L. *Tetrahedron Lett.* **2004**, *45*, 7991–7993.

884. Cardullo, F.; Donati, D.; Merlo, G.; Paio, A.; Salaris, M.; Taddei, M. *Synlett* **2005**, 2996–2998.

885. Milburn, R.R.; Snieckus, V. *Angew. Chem., Int. Ed.* **2004**, *43*, 892–894.

886. Vedejs, E.; Lin, S.; Klapars, A.; Wang, J. *J. Am. Chem. Soc.* **1996**, *118*, 9796–9797.

887. Weinreb, S.M.; Demko, D.M.; Lessen, T.A.; Demers, J.P. *Tetrahedron Lett.* **1986**, *27*, 2099–2102.

888. Ribière, P.; Declerck, V.; Martinez, J.; Lamaty, F. *Chem. Rev.* **2006**, *106*, 2249–2269.

889. Sun, P.; Weinreb, S.M.; Shang, M. *J. Org. Chem.* **1997**, *62*, 8604–8608.

890. Brunissen, A.; Ayoub, M.; Lavielle, S. *Tetrahedron Lett.* **1996**, *37*, 6713–6716.

891. Wenschuh, H.; Beyermann, M.; Winter, R.; Bienert, M.; Ionescu, D.; Carpino, L.A. *Tetrahedron Lett.* **1996**, 37, 5483–5486.

892. Roby, J.; Voyer, N. *Tetrahedron Lett.* **1997**, *38*, 191–194.

893. Babu, V.V.S.; Vasanthakumar, G.-R.; Tantry, S.J. *Tetrahedron Lett.* **2005**, *46*, 4099–4102.

894. Djuric, S.; Venit, J.; Magnus, P. *Tetrahedron Lett.* **1981**, *22*, 1787–1790.

895. Bonar-Law, R.P.; Davis, A.P.; Dorgan, B.J.; Reetz, M.T.; Wehrsig, A. *Tetrahedron Lett.* **1990**, 31, 6725–6728.

896. Kim, B.M.; Cho, J.H. *Tetrahedron Lett.* **1999**, *40*, 5333–5336.

897. Matysiak, S.; Böldicke, T.; Tegge, W.; Frank, R. *Tetrahedron Lett.* **1998**, *39*, 1733–1734.

898. Gorbunova, M.G.; Gerus, I.I.; Galushko, S.V.; Kukhar, V.P. *Synthesis* **1991**, 207–209.

899. Lubell, W.D.; Rapoport, H. *J. Am. Chem. Soc.* **1987**, *109*, 236–239.

900. Jass, P.A.; Rosso, V.W.; Racha, S.; Soundararajan, N.; Venit, J.J.; Rusowicz, A.; Swaminathan, S.; Livshitz, J.; Delaney, E.J. *Tetrahedron* **2003**, *59*, 9019–9029.

901. Rao, T.S.; Nampalli, S.; Sekher, P.; Kumar, S. *Tetrahedron Lett.* **2002**, *43*, 7793–7795.

902. Jiang, Y.; Zhao, J.; Hu, L. *Tetrahedron Lett.* **2002**, *43*, 4589–4592.

903. Waldmann, H.; Heuser, A.; Schulze, S. *Tetrahedron Lett.* **1996**, *37*, 8725–8728.

904. Auzeil, N.; Dutruc-Rosset, G.; Largeron, M. *Tetrahedron Lett.* **1997**, *38*, 2283–2286.

905. Knapp, S.; Hale, J.J.; Bastos, M.; Molina, A.; Chen, K.Y. *J. Org. Chem.* **1992**, *57*, 6239–6256.

906. Casimir, J.R.; Guichard, G.; Briand, J.-P. *J. Org. Chem.* **2002**, *67*, 3764–3768.

907. Zeng, Q.; Liu, Z.; Li, B.; Wang, F. *Amino Acids* **2004**, *27*, 183–186.

908. Cros, E.; Planas, M.; Mejías, X.; Bardaji, E. *Tetrahedron Lett.* **2001**, *42*, 6105–6107.

909. Cros, E.; Planas, M.; Barany, G.; Bardají, E. *Eur. J. Org. Chem.* **2004**, 3633–3642.

910. Lagrille, O.; Taillades, J.; Boiteau, L.; Commeyras, A. *Eur. J. Org. Chem.* **2002**, 1026–1032.

911. Osborn, H.M.I.; Williams, N.A.O. *Org. Lett.* **2004**, *6*, 3111–3113.

912. Planas, M.; Bardajf, E.; Jensen, K.J.; Barany, G. *J. Org. Chem.* **1999**, *64*, 7281–7289.

913. Lundquist, J.T., IV; Pelletier, J.C. *Org. Lett.* **2001**, *3*, 781–783.

914. Reddy, P.G.; Pratap, T.V.; Kumar, G.D.K.; Mohanty, S.K.; Baskaran, S. *Eur. J. Org. Chem.* **2002**, 3740–3743.

915. Jung, Y.J.; Chang, Y.M.; Lee, J.H.; Yoon, C.M. *Tetrahedron Lett.* **2002**, *43*, 8735–8739.

916. Lazny, R.; Poplawski, J.; Köbberling, J.; Enders, D.; Bräse, S. *Synlett* **1999**, 1304–1306.

917. Lazny, R.; Sienkiewicz, M.; Bräse, S. *Tetrahedron* **2001**, *57*, 5825–5832.

918. Dent, W.H., III; Erickson, R.; Fields, S.C.; Parker, M.; Tromiczak, E.G. *Org. Lett.* **2002**, *4*, 1249–1251.

919. Hirschmann, R.; Yao, W.; Arison, B.; Maechler, L.; Rosegay, A.; Sprengeler, P.A.; Smith, A.B., III. *Tetrahedron* **1998**, *54*, 7179–7202.

920. Dumy, P.; Eggleston, I.M.; Cervigni, S.; Sila, U.; Sun, X.; Mutter, M. *Tetrahedron Lett.* **1995**, *36*, 1255–1258.

921. Xu, Q.; Borremans, F.; Devreese, B. *Tetrahedron Lett.* **2001**, *42*, 7261–7263.

922. Katajisto, J.; Karskela, T.; Heinonen, P.; Lönnberg, H. *J. Org. Chem.* **2002**, *67*, 7995–8001.

923. Zhou, X.-T.; Rehman, A.; Li, C.; Savage, P.B. *Org. Lett.* **2000**, *2*, 3015–3018.

924. Li, S.; Marthandan, N.; Bowerman, D.; Garner, H.R.; Kodadek, T. *Chem. Commun.* **2005**, 581–582.

925. Miranda, L.P.; Meldal, M. *Angew. Chem., Int. Ed.* **2001**, *40*, 3655–3657.

926. Somlai, C.; Szókán, G.; Baláspiri, L. *Synthesis* **1992**, 285–287.

927. Ager, D.J.; Prakash, I. *Synth. Commun.* **1996**, *26*, 3865–3868.

928. Chen, S.-T.; Wu, S.-H.; Wang, K.-T. *Synthesis* **1989**, 37–38.

929. Somlai, C.; Szókán, G.; Penke, B. *Synthesis* **1995**, 683–686.

930. Galaverna, G.; Corradini, R.; Dosena, A.; Marchelli, R. *Int. J. Pept. Prot. Res.* **1993**, *42*, 53–57.

931. Wang, W.; McMurray, J.S. *Tetrahedron Lett.* **1999**, *40*, 2501–2504.

932. Bailén, M.A.; Chinchilla, R.; Dodsworth, D.J.; Nájera, C. *Tetrahedron Lett.* **2000**, *41*, 9809–9813.

933. Khalafi-Nezhad, A.; Parhami, A.; Soltani Rad, M.N.; Zarea, A. *Tetrahedron Lett.* **2005**, *46*, 6879–6882.

934. Ghosh, S.K.; Verma, R.; Ghosh, U.; Mamdapur, V.R. *Bull. Chem. Soc. Jpn.* **1996**, *69*, 1705–1711.

935. Easton, C.J.; Eichinger, S.K.; Pitt, M.J. *Tetrahedron* **1997**, *53*, 5609–5616.

936. Chern, C.-Y.; Huang, Y.-P.; Kan, W.M. *Tetrahedron Lett.* **2003**, *44*, 1039–1041.

937. Kohn, H.; Sawhney, K.N.; LeGall, P.; Conley, J.D.; Robertson, D.W.; Leander, J.D. *J. Med. Chem.* **1990**, *33*, 919–926.

938. Chaudhary, A.; Girgis, M.; Prashad, M.; Hu, B.; Har, D.; Repic, O.; Blacklock, T.J. *Tetrahedron Lett.* **2003**, *44*, 5543–5546.

939. Shendage, D.M.; Fröhlich, R.; Haufe, G. *Org. Lett.* **2004**, *6*, 3675–3678.

940. Fray, M.J.; Ellis, D. *Tetrahedron* **1998**, *54*, 13825–13832.

941. Gooding, O.W.; Vo, L.; Bhattacharyya, S.; Labadie, J.W. *J. Comb. Chem.* **2002**, *4*, 576–583.

942. Kawahata, N.H.; Brookes, J.; Makara, G.M. *Tetrahedron Lett.* **2002**, *43*, 7221–7223.

943. Quéléver, G.; Burlet, S.; Garino, C.; Pietrancosta, N.; Laras, Y.; Kraus, J.-L. *J. Comb. Chem.* **2004**, *6*, 695–698.

944. Vaidyanathan, R.; Kalthod, V.G.; Ngo, D.P.; Manley, J.M.; Lapekas, S.P. *J. Org. Chem.* **2004**, *69*, 2565–2568.

945. Grzyb, J.A.; Batey, R.A. *Tetrahedron Lett.* **2003**, *44*, 7485–7488.

946. Meshram, H.M.; Reddy, G.S.; Reddy, M.M.; Yadav, J.S. *Tetrahedron Lett.* **1998**, *39*, 4103–4106.

947. Perreux, L.; Loupy, A.; Volatron, F. *Tetrahedron* **2002**, *58*, 2155–2162.

948. Gelens, E.; Smeets, L.; Sliedregt, L.A.J.M.; van Steen, B.J.; Kruse, C.G.; Leurs, R.; Orru, R.V.A. *Tetrahedron Lett.* **2005**, *46*, 3751–3754.

949. van Leeuwen, S.H.; Quaedflieg, P.J.L.M.; Broxterman, Q.B.; Milhajlovic, Y.; Liskamp, R.M.J. *Tetrahedron Lett.* **2005**, *46*, 653–656.

950. van Leeuwen, S.H.; Quaedflieg, P.J.L.M.; Broxterman, Q.B.; Liskamp, R.M.J. *Tetrahedron Lett.* **2002**, *43*, 9203–9207.

951. Dubé, D.; Scholte, A.A. *Tetrahedron Lett.* **1999**, *40*, 2295–2298.

952. Tang, Z.; Pelletier, J.C. *Tetrahedron Lett.* **1998**, *39*, 4773–4776.

953. Shangguan, N.; Katukojvala, S.; Greenberg, R.; Williams, L.J. *J. Am. Chem. Soc.* **2003**, *125*, 7754–7755.

954. Penke, B.; Rivier, J. *J. Org. Chem.* **1987**, *52*, 1197–1200.

955. Rink, H. *Tetrahedron Lett.* **1987**, *28*, 3787–3790.

956. Albericio, F.; Barany, G. *Int. J. Pept. Prot. Chem.* **1987**, *30*, 206–216.

957. Albericio, F.; Kneib-Cordonier, N.; Biancalana, S.; Gera, L.; Masada, R.I.; Hudson, D.; Barany, G. *J. Org. Chem.* **1990**, *55*, 3730–3743.

958. Sieber, P. *Tetrahedron Lett.* **1987**, *28*, 2107–2110.

959. Han, Y.; Bontems, S.L.; Hegyes, P.; Munson, M.C.; Minor, C.A.; Kates, S.A.; Albericio, F.; Barany, G. *J. Org. Chem.* **1996**, *61*, 6326–6339.

960. Chan, W.C.; Mellor, S.L. *J. Chem. Soc., Chem. Commun.* **1995**, 1475–1476.

961. Brown, E.G.; Nuss, J.M. *Tetrahedron Lett.* **1997**, *38*, 8457–8460.

962. Bui, C.T.; Bray, A.M.; Ercole, F.; Pham, Y.; Rasoul, F.A.; Maeji, N.J. *Tetrahedron Lett.* **1999**, *40*, 3471–3474.

963. Jensen, K.J.; Alsina, J.; Songster, M.F.; Vágner, J.; Albericio, F.; Barany, G. *J. Am. Chem. Soc.* **1998**, *120*, 5441–5452.

964. del Fresno, M.; Alsina, J.; Royo, M.; Barany, G.; Albericio, F. *Tetrahedron Lett.* **1998**, *39*, 2639–2642.

965. Nicolás, E.; Clemente, J.; Ferrer, T.; Albericio, F.; Giralt, E. *Tetrahedron* **1997**, *53*, 3179–3194.

966. DeGrado, W.F.; Kaiser, E.T. *J. Org. Chem.* **1980** *45*, 1295–1300.

967. Mohan, R.; Chou, Y.-L.; Morrissey, M.M. *Tetrahedron Lett.* **1996**, *37*, 3963–3966.

968. Kaljuste, K.; Tam, J.P. *Tetrahedron Lett.* **1998**, *39*, 9327–9330.

969. Marti, R.E.; Yan, B.; Jarosinski, M.A. *J. Org. Chem.* **1997**, *62*, 5615–5618.

970. López-Serrano, P.; Wegman, M.A.; van Rantwijk, F.; Sheldon, R.A. *Tetrahedron: Asymmetry* **2001**, *12*, 235–240.

971. Cerovsky, V.; Kula, M.-R. *Angew. Chem., Int. Ed. Engl.* **1998**, *37*, 1885–1887.

972. Chamorro, C.; González-Muñiz, R.; Conde, S. *Tetrahedron: Asymmetry* **1995**, *6*, 2343–2352.

973. Conde, S.; López-Serrano, P.; Fierros, M.; Biezma, M.I.; Martínez, A.; Rodríguez-Franco, M.I. *Tetrahedron* **1997**, *53*, 11745–11752.

974. Conde, S.; López-Serrano, P. *Eur. J. Org. Chem.* **2002**, 922–929.

975. Conde, S.; López-Serrano, P.; Castro, A.; Martínez, A. *Eur. J. Org. Chem.* **1999**, 2835–2839.

976. Wang, P.; Shaw, K.T.; Whigham, B.; Ramage, R. *Tetrahedron Lett.* **1998**, *39*, 8719–8720.

977. Dankwardt, S.M. *Synlett* **1998**, 761.

978. Thouin, E.; Lubell, W.D. *Tetrahedron Lett.* **2000**, *41*, 457–460.

979. Floyd, C.D.; Lewis, C.N.; Patel, S.; Whittaker, M. *Tetrahedron Lett.* **1996**, *37*, 8045–8048.

980. Weisz, I.; Roboz, J.; Wolf, J.; Szabo, J.; Bekesi, J.G. *Biorg. Med. Chem. Lett.* **1998**, *8*, 3241–3244.

981. Pelletier, J.C.; Hesson, D.P. *Synlett* **1995**, 1141–1142.

982. Smith, R.M.; Hansen, D.E. *J. Am. Chem. Soc.* **1998**, *120*, 8910–8913.

983. Wolfenden, R.; Snider, M.J. *Acc. Chem. Res.* **2001**, *34*, 938–945.

984. Evans, D.E.; Carter, P.H.; Dinsmore, C.J.; Barrow, J.C.; Katz, J.L.; Kung, D.W. *Tetrahedron Lett.* **1997**, *38*, 4535–4538.

985. Gómez-Reyes, B.; Yatsimirsky, A.K. *Org. Lett.* **2003**, *5*, 4831–4834.

986. Charette, A.B.; Chua, P. *Synlett* **1998**, 163–165.

987. Obrecht, D.; Heimgartner, H. *Helv. Chim. Acta* **1981**, *64*, 482–487.

988. Kaminskaia, N.V.; Johnson, T.W.; Kostic, N.M. *J. Am. Chem. Soc.* **1999**, *121*, 8663–8664.

989. Milovic, N.M.; Kostic, N.M. *J. Am. Chem. Soc.* **2002**, *124*, 4759–4769.

990. Suh, J. *Acc. Chem. Res.* **2003**, *36*, 562–570.

991. Milovic, N.M.; Badjic, J.D.; Kostic, N.M. *J. Am. Chem. Soc.* **2004**, *126*, 696–697.

992. Takarada, T.; Yashiro, M.; Komiyama, M. *Chem.—Eur. J.* **2000**, *6*, 3906–3913.

993. Back, J.W.; David, O.; Kramer, G.; Masson, G.; Kasper, P.T.; de Koning, L.J.; de Jong, L.; van Maarseveen, J.H.; de Koster, C.G. *Angew. Chem., Int. Ed.* **2005**, *44*, 7946–7950.

994. Müller, G.H.; Waldmann, H. *Tetrahedron Lett.* **1999**, *40*, 3549–3552.

995. Völkert, M.; Koul, S.; Müller, G.H.; Lehnig, M.; Waldmann, H. *J. Org. Chem.* **2002**, *67*, 6902–6910.

996. Benouargha, A.; Verducci, J.; Jacquier, R. *Bull. Soc. Chim. Fr.* **1995**, *132*, 824–828.

997. Lloyd-Williams, P.; Albericio, F.; Giralt, E. *Chemical Approaches to the Synthesis of Peptides*

and Proteins; CRC Press: Boca Raton, 1997, pp 114–121.

998. Kemp, D.S. In *The Peptides. Analysis, Synthesis, Biology. Volume 1: Major Methods of Peptide Bond Formation;* Gross, E., Meienhofer, J., Eds; Academic Press: New York, 1979, pp 316–386.

999. Benoiton, N.L. In *The Peptides. Analysis, Synthesis, Biology. Volume 5: Special Methods in Peptide Synthesis, Part B;* Gross, E., Meienhofer, J., Eds; Academic Press: New York, 1983; pp 217–285.

1000. Romoff, T.T.; Goodman, M. *J. Peptide Res.* **1997**, *49*, 281–292.

1001. Carpino, L.A.; El-Faham, A.; Albericio, F. *Tetrahedron Lett.* **1994**, *35*, 2279–2282.

1002. Carpino, L.A.; El-Faham, A.; Albericio, F. *J. Org. Chem.* **1995**, *60*, 3561–3564.

1003. Benoiton, N.L. *Biopolymers* **1996**, *40*, 245–254.

1004. Corradini, R.; Sforza, S.; Dossena, A.; Palla, G.; Rocchi, R.; Filira, F.; Nastri, F.; Marchelli, R. *J. Chem. Soc., Perkin Trans. 1* **2001**, 2690–2696.

1005. Hyde, C.; Johnson, T.; Owen, D.; Quibell, M.; Sheppard, R.C. *Int. J. Pept. Prot. Res.* **1994**, *43*, 431–440.

1006. Cilli, E.M.; Marchetto, R.; Schreier, S.; Nakaie, C.R. *J. Org. Chem.* **1999**, *64*, 9118–9123.

1007. Kaiser, E.; Colescott, R.L.; Bossinger, C.D.; Cook, P.I. *Anal. Biochem.* **1970, 34**, 595–598.

1008. Vojkovsky, T. *Pept. Res.* **1995**, *8*, 236–237.

1009. Greenstein, J.P.; Winitz, N. *Chemistry of the Amino Acids, Vol. 2;* Wiley: New York, **1961**, pp 763–1298.

1010. Stewart, J.M.; Young, J.D. *Solid Phase Peptide Synthesis,* 2nd edn.; Pierce Chemical Company: Rockford, Illinois, **1984**, pp 31–34.

1011. Wieland, T.; Bodanszky, M. *The World of Peptides: A Brief History of Peptide Chemistry;* Springer-Verlag: New York, 1991, pp 36–42.

1012. Wieland, T.; Bodanszky, M. *The World of Peptides: A Brief History of Peptide Chemistry;* Springer-Verlag: New York, 1991, pp 77–102.

1013. Bodanszky, M. *Principles of Peptide Synthesis;* Springer-Verlag: New York, 1984, pp 9–58.

1014. *The Peptides. Analysis, Synthesis, Biology. Vol. 1 Major Methods of Peptide Bond Formation;* Gross, E.; Meienhofer, J., Eds; Academic Press: New York, 1979.

1015. *The Peptides. Analysis, Synthesis, Biology. Vol. 5 Special Method in Peptide Synthesis, Part B;* Gross, E.; Meienhofer, J., Eds; Academic Press: New York, 1983.

1016. Barany, G.; Merrifield, R.B. In *The Peptides. Analysis, Synthesis, Biology. (Vol. 2);* Gross, E., Meienhofer, J., Eds; Academic Press, : New York, 1980.

1017. *Fmoc Solid Phase Peptide Synthesis. A Practical Approach;* Chan, W.C.; White, P.D., Eds; Oxford University Press: New York, 2000.

1018. Ryakhovskii, V.V.; Agafonov, S.V.; Kosyrev, Y.M. *Russ. Chem. Rev.* **1991**, *60*, 924–933.

1019. Spencer, J.R.; Antonenko, V.V.; Delaet, N.G.J.; Goodman, M. *Int. J. Pept. Prot. Chem.* **1992**, 40, 282–293.

1020. Lloyd-Williams, P.; Albericio, F.; Giralt, E. *Chemical Approaches to the Synthesis of Peptides and Proteins;* CRC Press: Boca Raton, Florida, 1997.

1021. Han, S.-Y.; Kim, Y.-A. *Tetrahedron* **2004**, *60*, 2447–2467.

1022. Montalbetti, C.A.G.N.; Falque, V. *Tetrahedron* **2005**, *61*, 10827–20852.

1023. Goodman, M.; Felix, A.; Moroder, L.; Toniolo, Eds; *Houben–Weyl Synthesis of Peptides; Vols E22a, E22b;* Thieme Verlag: New York, 2002.

1024. Greenstein, J.P.; Winitz, N. *Chemistry of the Amino Acids, Vol. 2;* Wiley: New York, 1961, p 943.

1025. Barrett, G.C. *Chemistry and Biochemistry of the Amino Acids;* Chapman and Hall: New York, 1985, pp 326–327.

1026. Huber, C.; Wächtershäuser, G. *Science* **1998**, *281*, 670–672.

1027. Carpino, L.A.; Beyermann, M.; Wenschuh, H.; Bienert, M. *Acc. Chem. Res.* **1996**, *39*, 268–274.

1028. Beecham, A.F. *Chem. Ind.* **1955**, 1120–1121.

1029. Carpino, L.A.; Cohen, B.J.; Stephens, K.E., Jr.; Sadat-Aalaee, S.Y.; Tien, J.-H.; Langridge, D.C. *J. Org. Chem.* **1986**, *51*, 3732–3734.

1030. Beyermann, M.; Bienert, M.; Niedrich, H.; Carpino, L.A.; Sadat-Aalaee, D. *J. Org. Chem.* **1990**, *55*, 721–728.

1031. Sivanandaiah, K.M.; Babu, V.V.S.; Renukeshwar, C. *Int. J. Pept. Prot. Res.* **1992**, *39*, 201–206.

1032. Carpino, L.A.; Sadat-Aalaee, D.; Beyermann, M. *J. Org. Chem.* **1990**, *55*, 1673–1675.

1033. Carpino, L.A.; Sadat-Aalaee, D.; Chao, H.G.; DeSelms, R.H. *J. Am. Chem. Soc.* **1990**, *112*, 9651–9652.

1034. Jang, D.O.; Park, D.J.; Kim, J. *Tetrahedron Lett.* **1999**, *40*, 5323–5326.

1035. Sewald, N. *Angew. Chem., Int. Ed.* **2003**, *41*, 4661–4663.

1036. Thern, B.; Rudolph, J.; Jung, G. *Tetrahedron Lett.* **2002**, *43*, 5013–5016.

1037. da C. Rodrigues, R.; Barros, I.M.A.; Lima, E.L.S. *Tetrahedron Lett.* **2005**, *46*, 5945–5947.

1038. Babu, V.V.S.; Gopi, H.N. *Tetrahedron Lett.* **1998**, *39*, 1049–1050.

1039. Gopi, H.N.; Babu, V.V.S. *Tetrahedron Lett.* **1998**, *39*, 9769–9772.

1040. Perlow, D.S.; Erb, J.M.; Gould, N.P.; Tung, R.D.; Freidinger, R.M.; Williams, P.D.; Veber, D.F. *J. Org. Chem.* **1992**, *57*, 4394–4400.

1041. Meldal, M.; Juliano, M.A.; Jansson, A.M. *Tetrahedron Lett.* **1997**, *38*, 2531–2534.

1042. Bertho, J.-N.; Loffet, A.; Pinel, C.; Reuther, F.; Sennyey, G. *Tetrahedron Lett.* **1991**, *32*, 1303–1306.

1043. Carpino, L.A.; Mansour, E.-S.M.E.; Sadat-Aalaee, D. *J. Org. Chem.* **1991**, *56*, 2611–2614.

1044. Savrda, J.; Wakselman, M. *J. Chem. Soc., Chem. Commun.* **1992**, 812–813.

1045. Wakselman, M.; Mazaleyrat, J.-P.; Savrda, J. *Amino Acids* **1994**, *7*, 67–77.

1046. Wenschuh, H.; Beyermann, M.; El-Faham, A.; Ghassemi, S.; Carpino, L.A.; Bienert, M. *J. Chem. Soc., Chem. Commun.* **1995**, 669–670.

1047. Carpino, L.A.; El-Faham, A. *J. Am. Chem. Soc.* **1995**, *117*, 5401–5402.

1048. Fiammengo, R.; Licini, G.; Nicotra, A.; Modena, G.; Pasquato, L.; Scrimin, P.; Broxterman, Q.B.; Kaptein, B. *J. Org. Chem.* **2001**, *66*, 5905–5910.

1049. El-Faham, A. *Chem. Lett.* **1998**, 671–672.

1050. Barrett, A.G.M.; Bibal, B.; Hopkins, B.T.; Köbberling, J.; Love, A.C.; Tedeschi, L. *Tetrahedron* **2005**, *61*, 12033–121041.

1051. Carpino, L.A.; Ionescu, D.; El-Faham, A.; Beyermann, M.; Henklein, P.; Hanay, C.; Wenschuh, H.; Bienert, M. *Org. Lett.* **2003**, *5*, 975–977.

1052. White, J.M.; Tunoori, A.R.; Tururen, B.J.; Georg, G.I. *J. Org. Chem.* **2004**, *69*, 2573–2576.

1053. Wenschuh, H.; Beyermann, M.; Krause, E.; Brudel, M.; Winter, R.; Schümann, M.; Carpino, L.A.; Bienert, M. *J. Org. Chem.* **1994**, *59*, 3275–3280.

1054. Wenschuh, H.; Beyermann, M.; Krause, E.; Carpino, L.A.; Bienert, M. *Tetrahedron Lett.* **1993**, *34*, 3733–3736.

1055. Bosch, I.; Urpí, F.; Vilarrasa, J. *J. Chem. Soc., Chem. Commun.* **1995**, 91–92.

1056. Wenschuh, H.; Beyermann, M.; Haber, H.; Seydel, J.K.; Krause, E.; Bienert, M.; Carpino, L.A.; El-Faham, A.; Albericio, F. *J. Org. Chem.* **1995**, *60*, 405–410.

1057. Futaki, S.; Fukuda, M.; Omote, M.; Yamauchi, K.; Yagami, T.; Niwa, M.; Sugiura, Y. *J. Am. Chem. Soc.* **2001**, *123*, 12127–12134.

1058. Dehner, A.; Planker, E.; Gemmecker, G.; Broxterman, Q.B.; Bisson, W.; Formaggio, F.; Crisma, M.; Toniolo, C.; Kessler, H. *J. Am. Chem. Soc.* **2001**, *123*, 6678–6686.

1059. Pengo, P.; Pasquato, L.; Moro, S.; Brigo, A.; Fogolari, F.; Broxterman, Q.B.; Kaptein, B.; Scrimin, P. *Angew. Chem., Int. Ed.* **2003**, *42*, 3388–3392.

1060. Carpino, L.A.; Ionescu, D.; El-Faham, A.; Henklein, P.; Wenschuh, H.; Bienert, M.; Beyermann, M. *Tetrahedron Lett.* **1998**, *39*, 241–244.

1061. Ogrel, A.; Bloemhoff, W.; Lugtenburg, J.; Raap, J. *Liebigs Ann./Recueil* **1997**, 41–47.

1062. Angell, Y.M.; García-Echeverría, C.; Rich, D.H. *Tetrahedron Lett.* **1994**, *35*, 5981–5984.

1063. DalPozzo, A.; Ni, M.; Muzi, L.; Caporale, A.; De Castiglione, R.; Kaptein, B.; Broxterman, Q.B.; Formaggio, F. *J. Org. Chem.* **2002**, *67*, 6372–6375.

1064. Ni, M.; Esposito, E.; Kaptein, B.; Broxterman, Q.B.; Dal Pozzo, A.D. *Tetrahedron Lett.* **2005**, *46*, 6369–6371.

1065. DalPozzo, A. Bergonzi, R.; Ni, M. *Tetrahedron Lett.* **2001**, *42*, 3925–3927.

1066. Meienhofer, J. In *The Peptides. Analysis, Synthesis, Biology. Vol. 1: Major Methods of Peptide Bond Formation;* Gross, E.; Meienhofer, J., Eds; Academic Press: New York, 1979, pp 197–240.

1067. Shioiri, T.; Ninomiya, K.; Yamada, S.-I. *J. Am. Chem. Soc.* **1972**, *94*, 6203–6205.

1068. Quin, L.; Sun, Z.; Deffo, T.; Mertes, K.B. *Tetrahedron Lett.* **1990**, *31*, 6469–6472.

1069. Katoh, T.; Ueki, M. *Int. J. Pept. Prot. Res.* **1993**, *42*, 264–269.

1070. Wang, P.; Layfield, R.; Landon, M.; Mayer, R.J.; Ramage, R. *Tetrahedron Lett.* **1998**, *39*, 8711–8714.

1071. Babu, V.V.S.; Ananda, K.; Vasanthakumar, G.-R. *J. Chem. Soc., Perkin Trans. 1* **2000**, 4328–4331.

1072. Glatthard, R.; Matter, M. *Helv. Chim. Acta* **1963**, *46*, 795–804.

1073. Bodanszky, M. In *The Peptides. Analysis, Synthesis, Biology. Vol. 1: Major Methods of Peptide Bond Formation;* Gross, E.; Meienhofer, J, Eds; Academic Press: New York, 1979, pp 105–196.

1074. Morley, J.S. *J. Chem. Soc. C* **1967**, 2410–2421.

1075. Kisfaludy, L.; Löw, M.; Nyéki, O.; Szirtes, T.; Schon, I. *Liebigs Ann. Chem.* **1973**, 1421–1429.

1076. Gagnon, P.; Huang, X.; Therrien, E.; Keillor, J.W. *Tetrahedron Lett.* **2002**, *43*, 7717–7719.

1077. König, W.; Geiger, R. *Chem. Ber.* **1973**, *106*, 3626–3635.

1078. Benoiton, N.L.; Lee, Y.C.; Chen, F.M.F. *Int. J. Pept. Prot. Res.* **1993**, *41*, 512–516.

1079. Høeg-Jensen, T.; Olsen, C.E.; Holm, A. *J. Org. Chem.* **1994**, *59*, 1257–1263.

1080. Atherton, E.; Cameron, L.; Meldal, M.; Sheppard, R.C. *J. Chem. Soc., Chem. Commun.* **1986**, 1763–1765.

1081. Basel, Y.; Hassner, A. *Tetrahedron Lett.* **2002**, *43*, 2529–2533.

1082. Pudhom, K.; Vilaivan, T. *Tetrahedron Lett.* **1999**, *40*, 5939–5942.

1083. Pudhom, K.; Vilaivan, T. *Synth. Commun.* **2001**, *31*, 61–70.

1084. Watts, P.; Wiles, C.; Haswell, S.J.; Pombo-Villar, E. *Tetrahedron* **2002**, *58*, 5427–5439.

1085. Watts, P.; Wiles, C.; Haswell, S.J.; Pombo-Villar, E.; Styring, P. *Chem. Commun.* **2001**, 990–991.

1086. George, V.; Watts, P.; Haswell, S.J.; Pombo-Villar, E. *Chem. Commun.* **2003**, 2886–2887.

1087. Medvedkin, V.N.; Klimenko, L.N.; Mitin, Y.V.; Kretsinger, R.H.; Shabanowitz, J.; Zabolotskikh, V.F.; Podgornova, N.N. *Int. J. Pept. Prot. Res.* **1994**, *44*, 477–484.

1088. Corbett, A.D.; Gleason, J.L. *Tetrahedron Lett.* **2002**, *43*, 1369–1372.

1089. Yamada, T.; Manabe, Y.; Miyazawa, T.; Kuwata, S.; Sera, A. *J. Chem. Soc., Chem. Commun.* **1984**, 1500–1501.

1090. Dendrinos, K.G.; Kalivretenos, A.G. *Tetrahedron Lett.* **1998**, *39*, 1321–1324.

1091. Kaminski, Z.J. *Tetrahedron Lett.* **1985**, *26*, 2901–2904.

1092. Kaminski, Z.J. *Synthesis* **1987**, 917–920.

1093. Kaminski, Z.; *Int. J. Pept. Prot. Res.* **1994**, *43*, 312–319.

1094. Garrett, C.E.; Jiang, X.; Prasad, K.; Repic, O. *Tetrahedron Lett.* **2002**, *43*, 4161–4165.

1095. Kaminski, Z.J. *Biopolymers (Peptide Sci.)* **2000**, *55*, 140–164.

1096. Barton, D.H.R.; Ferreira, J.A. *Tetrahedron* **1996**, *52*, 9367–9386.

1097. Kurokawa, N.; Ohfune, Y. *Tetrahedron* **1993**, *49*, 6195–6222.

1098. Burger, K.; Rudolph, M.; Fehn, S.; Worku, A.; Golubev, A. *Amino Acids* **1995**, *8*, 195–199.

1099. Chan, C.-O.; Cooksey, C.J.; Crich, D. *J. Chem. Soc. Perkin Trans 1* **1992**, 777–780.

1100. Meienhofer, J. In *The Peptides. Analysis, Synthesis, Biology. Vol. 1: Major Methods of Peptide Bond Formation*; Gross, E., Meienhofer, J, Eds; Academic Press: New York, 1979, pp 263–314.

1101. Crsma, M.; Valle, G.; Pantano, M.; Formaggio, F.; Bonora, G.M.; Toniolo, C.; Kamphius, J. *Rec. Trav. Chim. Pay-Bas* **1995**, *114*, 325–331.

1102. Fu, Y.; Hammer, R.P. *Org. Lett.* **2002**, *4*, 237–240.

1103. Shieh, W.-C.; Carlson, J.A.; Shore, M.E. *Tetrahedron Lett.* **1999**, *40*, 7167–7170.

1104. Lipshutz, B.H.; Shin, Y.-J. *Tetrahedron Lett.* **2001**, *42*, 5629–5633.

1105. Shatzmiller, S.; Confalone, P.N.; Abiri, A. *Synlett* **1999**, 963–965.

1106. Belleau, B.; Malek, G. *J. Am. Chem. Soc.* **1968**, *90*, 1651–1652.

1107. Miyazawa, T.; Donkai, T.; Yamada, T.; Kuwata, S. *Int. J. Pept. Prot. Res.* **1992**, *40*, 49–53.

1108. Yang, A.; Gehring, A.P.; Li, T. *J. Chromatogr., A* **2000**, *878*, 165–170.

1109. Wenger, R.M. *Helv. Chim. Acta* **1983**, *66*, 2672–2702.

1110. Mohapatra, D.K.; Datta, A. *J. Org. Chem.* **1999**, *64*, 6879–6880.

1111. Yokum, T.S.; Elzer, P.H.; McLaughlin, M.L. *J. Med. Chem.* **1996**, *39*, 3603–3605.

1112. Leuchs, H. *Ber. Dtsch. Chem. Ges.* **1906**, *39*, 857–859.

1113. Katakai, R. *J. Org. Chem.* **1975**, *40*, 2697–2702.

1114. Kricheldorf, H.R. *Angew. Chem., Int. Ed.* **2006**, *45*, 5752–5784.

1115. Deming, T.J. *Nature* **1997**, *390*, 386–389.

1116. Deming, T.J.; Curtin, S.A. *J. Am. Chem. Soc.* **2000**, *122*, 5710–5717.

1117. Nowak, A.P.; Breedveld, V.; Pine, D.J.; Deming, T.J. *J. Am. Chem. Soc.* **2003**, *125*, 15666–15670.

1118. Deming, T.J. *J. Am. Chem. Soc.* **1998**, *120*, 4240–4241.

1119. Biron, J.-P.; Pascal, R. *J. Am. Chem. Soc.* **2004**, *126*, 9198–9199.

1120. Plasson, R.; Biron, J.Ph.; Cotet, H.; Commeyras, A.; Taillades, J. *J. Chromatogr., A* **2002**, *952*, 239–248.

1121. Katakai, R.; Oya, M.; Uno, K.; Iwakura, Y. *J. Org. Chem.* **1972**, *37*, 327–329.

1122. Katakai, R.; Oya, M.; Toda, F.; Uno, K.; Iwakura, Y. *J. Org. Chem.* **1974**, *39*, 180–182.

1123. Collet, H.; Bied, C.; Mion, L.; Taillades, J.; Commeyras, A. *Tetrahedron Lett.* **1996**, *37*, 9043–9046.

1124. Palomo, C.; Aizpurua, J.M.; Ganboa, I.; Carreaux, F.; Cuevas, C.; Maneiro, E.; Ontoria, J.M. *J. Org. Chem.* **1994**, *59*, 3123–3130.

1125. Palomo, C.; Ganboa, I.; Cuevas, C.; Boschetti, C.; Linden, A. *Tetrahedron Lett.* **1997**, *38*, 4643–4646.

1126. Palomo, C.; Ganboa, I.; Odriozola, B.; Linden, A. *Tetrahedron Lett.* **1997**, *38*, 3093–3096.

1127. Palomo, C.; Aizpurua, J.M.; Ganboa, I. *Russ. Chem. Bull.* **1996**, *45*, 2463–2483.

1128. Palomo, C.; Oiarbide, M.; Esnal, A. *Chem. Commun.* **1997**, 691–692.

1129. Palomo, C.; Oiarbide, M.; Esnal, A.; Landa, A.; Miranda, J.I.; Linden, A. *J. Org. Chem.* **1998**, *63*, 5836–5846.

1130. Palomo, C.; Aizpurua, J.M.; Urchegui, R.; Garcia, J.M. *J. Chem. Soc., Chem. Commun.* **1995**, 2327–2328.

1131. Fuller, W.D.; Cohen, M.P.; Shabankareh, M.; Blair, R.K.; Goodman, M.; Naidar, F.R. *J. Am. Chem. Soc.* **1990**, *112*, 7414–7416.

1132. Xue, C.-B.; Naider, F. *J. Org. Chem.* **1993**, *58*, 350–355.

1133. Pothion, C.; Fehrentz, J.-A.; Aumelas, A.; Loffet, A.; Martinez, J. *Tetrahedron Lett.* **1996**, *37*, 1027–1030.

1134. Riester, D.; Wiesmüller, K.-H.; Stoll, D.; Kuhn, R. *Anal. Chem.* **1996**, *68*, 2361–2365.

1135. Zhu, Y.-F.; Fuller, W.D. *Tetrahedron Lett.* **1995**, *36*, 807–810.

1136. Sim, T.B.; Rapoport, H. *J. Org. Chem.* **1999**, *64*, 2532–2536.

1137. Crisma, M.; Valle, G.; Formaggio, F.; Toniolo, C.; Bago, A. *J. Am. Chem. Soc.* **1997**, *119*, 4136–4142.

1138. Kenner, G.W.; Preston, J.; Sheppard, R.C. *J. Chem. Soc.* **1965**, 6239–6244.

1139. Jones, D.S.; Kenner, G.W.; Preston, J.; Sheppard, R.C. *J. Chem. Soc.* **1965**, 6227–6239.

1140. Leplawy, M.T.; Jones, D.S.; Kenner, G.W.; Sheppard, R.C. *Tetrahedron* **1960**, *11*, 39–51.

1141. Azumaya, I.; Aebi, R.; Kubik, S.; Rebek, J., Jr.; *Proc. Natl. Acad. Sci. U.S.A.* **1995**, *92*, 12013–12016.

1142. Tanaka, M.; Imawaka, N.; Kurihara, M.; Suemune, H. *Helv. Chim. Acta* **1999**, *82*, 494–510.

1143. Pantano, M.; Formaggio, F.; Crisma, M.; Bonora, G.M.; Mammi, S.; Peggion, E.; Toniolo, C.;

Boesten, W.H.J.; Broxterman, Q.B.; Schoemaker, H.E.; Kamphuis, J. *Macromolecules* **1993**, *26*, 1980–1984.

1144. Formaggio, F.; Pantano, M.; Crisma, M.; Bonora, G.M.; Toniolo, C.; Kamphuis, J. *J. Chem. Soc., Perkin Trans 2* **1995**, 1097–1101.

1145. Basu, G.; Bagchi, K.; Kuki, A. *Biopolymers* **1991**, *31*, 1763–1774.

1146. Bonora, G.M.; Toniolo, C.; Di Blasio, B.; Pavone, V.; Pedone, C.; Benedetti, E.; Lingham, I.; Hardy, P. *J. Am. Chem. Soc.* **1984**, *106*, 8152–8156.

1147. Obrecht, D.; Heimgartner, H. *Helv. Chim. Acta* **1981**, *64*, 482–487.

1148. Hardy, P.M.; Lingham, I.N. *Int. J. Pept. Prot. Res.* **1983**, *21*, 392–405.

1149. Maia, H.L.; Ridge, B.; Rydon, H.N. *J. Chem. Soc., Perkin Trans. 1* **1973**, 98–101.

1150. Obrecht, D.; Altorfer, M.; Lehmann, C.; Schönholzer, P.; Müller, K. *J. Org. Chem.* **1996**, *61*, 4080–4086.

1151. Rich, D.H.; Singh, J. In *The Peptides. Analysis, Synthesis, Biology. Vol. 1: Major Methods of Peptide Bond Formation;* Gross, E., Meienhofer, J., Eds; Academic Press, : New York, 1979, pp 241–261.

1152. Sheehan, J.C.; Hess, G.P. *J. Am. Chem. Soc.* **1955**, *77*, 1067–1068.

1153. Sarantakis, D.; Teichman, J.; Lien, E.L.; Fenichel, R.L. *Biochem. Biophys. Res. Commun.* **1976**, *73*, 336–342.

1154. Izdebski, J.; Pachulska, M.; Orlowska, A. *Int. J. Pept. Prot. Res.* **1994**, *44*, 414–419.

1155. Desai, M.J.; Stramiello, L.M.S. *Tetrahedron Lett.* **1993**, *34*, 7685–7688.

1156. Sauer, D.R.; Kalvin, D.; Phelan, K.M. *Org. Lett.* **2003**, *5*, 4721–4724.

1157. Zhang, M.; Vedantham, P.; Flynn, D.L.; Hanson, P.R. *J. Org. Chem.* **2004**, *69*, 8340–8344.

1158. Dulog, L.; Wang, W. *Liebigs Ann. Chem.* **1992**, 301–303.

1159. Ehrlich, A.; Heyne, H.-U.; Winter, R.; Beyermann, M.; Haber, H.; Carpino, L.A.; Bienert, M. *J. Org. Chem.* **1996**, *61*, 8831–8838.

1160. Ehrlich, A.; Rothemund, S.; Brudel, M.; Beyermann, M.; Carpino, L.A.; Bienert, M. *Tetrahedron Lett.* **1993**, *34*, 4781–4784.

1161. König, W.; Geiger, R. *Chem. Ber.* **1970**, *103*, 788–798.

1162. Benoiton, N.L.; Kuroda, K.; Chen, F.M.F. *Int. J. Pept. Prot. Chem.* **1979**, *13*, 403–408.

1163. Ji, J-g.; Zhang, D.-y.; Ye, Y.-h.; Xing, Q.-y. *Tetrahedron Lett.* **1998**, *39*, 6515–6516.

1164. Miyazawa, T.; Otomatsu, T.; Fukui, Y.; Yamada, T.; Kuwata, S. *Int. J. Pept. Prot. Res.* **1992**, *39*, 237–244.

1165. Miyazawa, T.; Otomatsu, T.; Fukui, Y.; Yamada, T.; Kuwata, S. *Int. J. Pept. Prot. Res.* **1992**, *40*, 49–53.

1166. Dendrinos, K.; Jeong, J.; Huang, W.; Kalivretenos, A.G. *Chem. Commun.* **1998**, 499–500.

1167. Lannuzel, M.; Lamothe, M.; Perez, M. *Tetrahedron Lett.* **2001**, *42*, 6703–6705.

1168. *Novabiochem Newsletter*, Merck Biosciences, 2/06.

1169. Carpino et al. *Innovations and Perspectives in Solid Phase Peptide Synthesis, 3rd International Symposium*, Epton, R., Ed.; Mayflower Worldwide: Kingwinford, UK, 1994, p 95.

1170. Hudson, D. *J. Org. Chem.* **1988**, *53*, 617–624.

1171. Souza, M.P.; Tavares, M.F.M.; Miranda, M.T.M. *Tetrahedron* **2004**, *60*, 4671–4681.

1172. Quibell, M.; Packman, L.C.; Johnson, T. *J. Chem. Soc., Perkin Trans. 1* **1996**, 1219–1225.

1173. Jaun, B.; Tanaka, M.; Seiler, P.; Kühnle, F.N.M.; Braun, C.; Seebach, D. *Liebigs Ann./Recueil* **1997**, 1697–1710.

1174. Nozaki, S. *Chem. Lett.* **1997**, 1–2.

1175. Sakakibara, S. *Biopolymers (Peptide Sci.)* **1999**, *51*, 279–296.

1176. Carpino, L. *J. Am. Chem. Soc.* **1993**, *115*, 4397–4398.

1177. Xu, Y.; Miller, M.J. *J. Org. Chem.* **1998**, *63*, 4314–4322.

1178. Carpino, L.A.; El-Faham, A. *Tetrahedron* **1999**, *55*, 6813–6830.

1179. Carpino, L.A.; Imazumi, H.; Foxman, B.M.; Vela, M.J.; Henklein, P.; El-Faham, A.; Klose, J.; Bienert, M. *Org. Lett.* **2000**, *2*, 2253–2256.

1180. König, W.; Geiger, R. *Chem. Ber.* **1970**, *103*, 2024–2033.

1181. König, W.; Geiger, R. *Chem. Ber.* **1970,** *103*, 2034–2040.

1182. Atherton, E.; Holder, J.L.; Meldal, M.; Sheppard, R.C.; Valerio, R.M. *J. Chem. Soc., Perkin Trans. 1* **1988**, 2887–2894.

1183. Atherton, E.; Cameron, L.; Meldal, M.; Sheppard, R.C. *J. Chem. Soc., Chem. Commun.* **1986**, 1763–1765.

1184. Carpino, L.A.; Xia, J.; El-Faham, A. *J. Org. Chem.* **2004**, *69*, 54–61.

1185. Spetzler, J.C.; Meldal, M.; Felding, J.; Vedsø, P.; Begtrup, M. *J. Chem. Soc., Perkin Trans. 1* **1998**, 1727–1732.

1186. Jiang, L.; Davison, A.; Tennant, G.; Ramage, R. *Tetrahedron* **1998**, *54*, 14233–14254.

1187. Robertson, N.; Jiang, L.; Ramage, R. *Tetrahedron* **1999**, *55*, 2713–2720.

1188. Staab, H.A. *Liebigs Ann. Chem.* **1957**, *609*, 75–83.

1189. Castro, B.; Dormoy, J.R.; Evin, G.; Selve, C. *Tetrahedron Lett.* **1975**, *14*, 1219–1222.

1190. Castro, B.; Dormoy, J.-R.; Evin, G.; Selve, C. *J. Chem. Res. S* **1977**, 182.

1191. Castro, B.; Dormoy, J.-R.; Dourtoglou, B.; Evin, G.; Selve, C.; Ziegler, J.-C. *Synthesis*, **1976**, 751–752.

1192. Le-Nguyen, D.; Heitz, A.; Castro, B. *J. Chem. Soc., Perkin Trans. 1* **1987**, 1915–1919.

1193. Seyer, R.; Aumelas, A.; Caraty, A.; Rivaille, P.; Castro, B. *Int. J. Pept. Prot. Chem.* **1990**, *35*, 465–472.

1194. Fournier, A.; Wang, C.-T.; Felix, A.M. *Int. J. Pept. Prot. Chem.* **1988**, 31, 86–97.

1195. Felix, A.M.; Wang, C.-T.; Heimer, E.P.; Fournier, A. *Int. J. Pept. Prot. Chem.*, **1988**, 231.

1196. Gairí, M.; Lloyd-Williams, P.; Albericio, F.; Giralt, E. *Tetrahedron Lett.* **1990**, 31, 7363–7366.

1197. Coste, J.; Le-Nguyen, D.; Castro, B. *Tetrahedron Lett.* **1990**, 31, 205–208.

1198. Frérot, E.; Coste, J.; Pantaloni, A.; Dufour, M-N.; Jouin, P. *Tetrahedron* **1991**, 47, 259–270.

1199. Santagada, V, ; Fiorino, F.; Perissutti, E.; Severino, B.; De Filippis, V.; Vivenzio, B.; Caliendo, G. *Tetrahedron Lett.* **2001**, 42, 5171–5173.

1200. Jones, W.D., Jr.; Ciske, F.L. *J. Org. Chem.* **1996**, 61, 3920–3922.

1201. Le, H.-T.; Gallard, J.-F.; Mayer, M.; Guittet, E.; Michelot, R. *Bioorg. Med. Chem.* **1996**, 4, 2209.

1202. Dourtoglou, V.; Ziegler, J.-C.; Gross, B. *Tetrahedron Lett.* **1978**, 15, 1269–1272.

1203. Dourtoglou, V.; Gross, B.; Lambropoulou, V.; Zioudrou, C. *Synthesis* **1984**, 572–574.

1204. Knorr, R.; Trzeciak, A.; Bannwarth, W.; Gillessen, D. *Tetrahedron Lett.* **1989**, 30, 1927–1930.

1205. Fields, C.G.; Lloyd, D.H.; Macdonald, R.L.; Otteson, K.M.; Noble, R.L. *Peptide Res.* **1991**, 4, 95–101.

1206. Murray, J.K.; Gellman, S.H. *Org. Lett.* **2005**, 7, 1517–1520.

1207. Carpino, L.A.; Henklein, P.; Foxman, B.M.; Abdelmoty, I.; Costisella, B.; Wray, V.; Domke, T.; El-Faham, A.; Mügge, C. *J. Org. Chem.* **2001**, 66, 5245–5247.

1208. Carpino, L.A.; Imazumi, H.; El-Faham, A.; Ferrer, F.J.; Zhang, C.; Lee, Y.; Foxman, B.M.; Henklein, P.; Hanay, C.; Mügge, C.; Wenschuh, H.; Klose, J.; Beyermann, M.; Bienert, M. *Angew. Chem., Int. Ed.* **2002**, 41, 441–445.

1209. Reid, G.E.; Simpson, R.J. *Anal. Biochem.* **1992**, 200, 301–309.

1210. Beck-Sickinger, A.G.; Dürr, H.; Jung, G. *Peptide Res.* **1991**, 4, 88–94.

1211. Bindra, V.A.; Kuki, A. *Int. J. Pept. Prot. Res.* **1994**, 44, 539–548.

1212. Bergmeier, S.C.; Fundy, S.L. *Biorg. Med. Chem. Lett.* **1997**, 7, 3135–3138.

1213. Chen, S.; Xu, J. *Tetrahedron Lett.* **1992**, 33, 647–650.

1214. Coste, J.; Dufour, M.-N.; Le-Nguyen, D.; Castro, B. In *Peptides. Chemistry, Structure and Biology;* Rivier, J.E.; Marshall, G.R., Eds; ESCOM: Leiden, 1990, pp 885–888.

1215. Kiso, Y.; Fujiwara, Y.; Kimura, T.; Nishitani, A.; Akaji, K. *Int. J. Pept. Prot. Res.* **1992**, 40, 308–314.

1216. Li, P.; Xu, J.C. *Tetrahedron Lett.* **1999**, 40, 3605–3608.

1217. Wijkmans, J.C.H.M.; Kruijtzer, J.A.W.; van der Marel, G.; van Boom, J.H.; Bloemhoff, W. *Recl. Trav. Chim. Pays-Bas* **1994**, 113, 394–397.

1218. Wijkmans, J.C.H.M.; Blok, F.A.A.; van der Marel, G.A.; van Boom, J.H.; Bloemhoff, W. *Tetrahedron Lett.* **1995**, 36, 4643–4646.

1219. Carpino, L.A.; El-Faham, A.; Minor, C.A.; Albericio, F. *J. Chem. Soc., Chem. Commun.* **1994**, 201–203.

1220. Kienhöfer, A. *Synlett* **2001**, 1811–1812.

1221. Akamatsu, M.; Roller, P.P.; Chen, L.; Zhang, Z.-Y.; Ye, B.; Burke, T.R., Jr. *Bioorg. Med. Chem.* **1997**, 5, 157–163.

1222. Miranda, L.P.; Alewood, P.F. *Proc. Natl. Acad. Sci., U.S.A.* **1999**, 96, 1181–1186.

1223. Di Fenza, A.; Tancredi, M.; Galoppini, C.; Rovero, P. *Tetrahedron Lett.* **1998**, 39, 8529–8532.

1224. Vallette, H.; Ferron, L.; Coquerel, G.; Gaumont, A.-C.; Plaquevent, J.-C. *Tetrahedron Lett.* **2004**, 4, 1617–1619.

1225. Carpino, L.A.; Xia, J.; Zhang, C.; El-Faham, A. *J. Org. Chem.* **2004**, 69, 62–71.

1226. Habermann, J.; Kunz, H. *Tetrahedron Lett.* **1998**, 39, 265–268.

1227. Klose, J.; El-Faham, A.; Henklein, P.; Carpino, L.A.; Bienert, M. *Tetrahedron Lett.* **1999**, 40, 2045–2048.

1228. Bailén, M.A.; Chinchilla, R.; Dodsworth, D.J.; Nájera, C. *J. Org. Chem.* **1999**, 64, 8936–8939.

1229. Albericio, F.; Bailén, M.A.; Chinchilla, R.; Dodsworth, D.J.; Nájera, C. *Tetrahedron* **2001**, 57, 9607–9613.

1230. Li, P.; Xu, J.-C. *Tetrahedron Lett.* **2000**, 41, 721–724.

1231. Li, P.; Xu, J.-C. *Tetrahedron* **2000**, 56, 4437–4445.

1232. Li, P.; Xu, J.C. *J. Org. Chem.* **2000**, 65, 2591–2958.

1233. Coste, J.; Dufour, M.-N.; Pantaloni, A.; Castro, B. *Tetrahedron Lett.* **1990**, 31, 669–672.

1234. Caastro, B.; Dormoy, J.R. *Tetrahedron Lett.* **1973**, 35, 3243–3246.

1235. Coste, J.; Frérot, E.; Jouin, P. *J. Org. Chem.* **1994**, 59, 2437–2446.

1236. Albericio, F.; Bofill, J.M.; El-Faham, A.; Kates, S.A. *J. Org. Chem.* **1998**, 63, 9678–9683.

1237. Bofill, J.M.; Albericio, F. *Tetrahedron Lett.* **1999**, 40, 2641–2644.

1238. Alsina, J.; Barany, G.; Albericio, F.; Kates, S.A. *Peptide Sci.* **1999**, 6, 243–245.

1239. Akaji, K.; Kuriyama, N.; Kimura, T.; Fujiwara, Y.; Kiso, Y. *Tetrahedron Lett.* **1992**, 33, 3177–3180.

1240. Akaji, K.; Kuriyama, N.; Kiso, Y. *J. Org. Chem.* **1996**, 61, 3350–3357.

1241. Akaji, K.; Kuriyama, N.; Kiso, Y. *Tetrahedron Lett.* **1994**, 35, 3315–3318.

1242. Kuriyama, N.; Akaji, K.; Kiso, Y. *Tetrahedron* **1997**, 53, 8323–8334.

1243. Akaji, K.; Tamai, Y.; Kiso, Y. *Tetrahedron Lett.* **1995**, 36, 9341–9344.

1244. Akaji, K.; Aimoto, S. *Tetrahedron* **2001**, 57, 1749–1755.

1245. Li, P.; Xu, J.C. *Tetrahedron* **2000**, *56*, 9949–9955.
1246. Li, P.; Xu, J.C. *Tetrahedron Lett.* **1999**, *40*, 8301–8304.
1247. Gibson, F.S.; Rapoport, H. *J. Org. Chem.* **1995**, *60*, 2615–2617.
1248. Shiina, I.; Kawakita, Y.-I. *Tetrahedron Lett.* **2003**, *44*, 1951–1955.
1249. Crosignani, S.; Gonzalez, J.; Swinnen, D. *Org. Lett.* **2004**, *6*, 4579–4582.
1250. Li, P.; Xu, J.-C. *Tetrahedron* **2000**, *56*, 8119–8131.
1251. Ioorga, B.; Campagne, J.-M. *Synlett* **2004**, 1826–1828.
1252. Diago-Meseguer, J.; Palomo-Coll, A.L. *Synthesis.* **1980**, 547–551.
1253. van der Auwera, C.; Anteunis, M.J.O. *Int. J. Pept. Prot. Res.* **1987**, *29*, 574–588.
1254. Tung, R.D.; Dhaon, M.K.; Rich, D.H. *J. Org. Chem.* **1986**, *51*, 3350–3354.
1255. Halab, L.; Lubell, W.D. *J. Am. Chem. Soc.* **2002**, *124*, 2474–2484.
1256. Gorecka, A.; Leplawy, M.; Zabrocki, J.; Zwierzak, A. *Synthesis* **1978**, 474–476.
1257. Ramage, R.; Hopton, D.; Parrott, M.J.; Richardson, R.S.; Kenner, G.W.; Moore, G.A. *J. Chem. Soc., Perk in Trans. 1* **1985**, 461–470.
1258. Jackson, A.G.; Kenner, G.W.; Moore, G.A.; Ramage, R.; Thorpe, W.D. *Tetrahedron Lett.* **1976**, *40*, 3627–3630.
1259. Galpin, I.J.; Robinson, A.E. *Tetrahedron* **1984**, *40*, 627–634.
1260. Fu, H.; Li, Z.-L.; Zhao, Y.-F.; Tu, G.-Z. *J. Am. Chem. Soc.* **1999**, *121*, 291–295.
1261. Yasuhara, T.; Nagaoka, Y.; Tomioka, K. *J. Chem. Soc., Perkin Trans. 1* **2000**, 2901–2902.
1262. Lorca, M.; Kurosu, M. *Synth. Commun.* **2001**, *31*, 469–473.
1263. Caputo, R.; Cassano, E.; Longobardo, L.; Mastroianni, D.; Palumbo, G. *Synthesis*, **1995**, 141–143.
1264. Fan, C.-X.; Hao, X.-L.; Ye, Y.-H. *Synth. Commun.* **1996**, *26*, 1455–1460.
1265. Ye, Y.-h.; Li, H.; Jiang, X. *Biopolymers (Peptide Sci.)* **2005**, *80*, 172–178.
1266. Nilsson, B.L.; Kiessling, L.L.; Raines, R.T. *Org. Lett.* **2000**, *2*, 1939–1941.
1267. Nilsson, B.L.; Kiessling, L.L.; Raines, R.T. *Org. Lett.* **2001**, *3*, 9–12.
1268. Soellner, M.B.; Nilsson, B.L.; Raines, R.T. *J. Org. Chem.* **2002**, *67*, 4993–4996.
1269. Nilsson, B.L.; Hondal, R.J.; Soellner, M.B.; Raines, R.T. *J. Am. Chem. Soc.* **2003**, *125*, 5268–5269.
1270. Merkx, R.; Rijkers, D.T.S.; Kemmink, J.; Liskamp, R.M.J. *Tetrahedron Lett.* **2003**, *44*, 4515–4518.
1271. Sohma, Y.; Sasaki, M.; Hayashi, Y.; Kimura, T.; Kiso, Y. *Chem. Commun.* **2004**, 124–125.
1272. Martin, S.F.; Dwyer, M.P.; Lynch, C.L. *Tetrahedron Lett.* **1998**, *39*, 1517–1520.
1273. Kurosu, M. *Tetrahedron Lett.* **2000**, *41*, 591–594.
1274. Benaglia, M.; Guizzetti, S.; Rigamonti, C.; Puglisi, A. *Tetrahedron* **2005**, *61*, 12100–12106.
1275. Ingenito, R.; Wenschuh, H. *Org. Lett.* **2003**, *5*, 4587–4590.
1276. Yamashiro, D.; Blake, J. *Int. J. Pep. Prot. Res.* **1981**, *18*, 383–392.
1277. Yamashiro, D.; Li, C.H. *Int. J. Pep. Prot. Res.* **1988**, *31*, 322–334.
1278. Blake, J. *Int. J. Pep. Prot. Res.* **1981**, *17*, 273–274.
1279. Aimoto, S. *Biopolymers (Peptide. Sci.)* **1999**, *51*, 247–265.
1280. Blake, J.; Li, C.H. *Proc. Natl. Acad. Sci. U.S.A.* **1981**, *78*, 4055–4058.
1281. Kawakami, T.; Tsuchiya, M.; Nakamura, K.; Aimoto, S. *Tetrahedron Lett.* **2005**, *46*, 5533–5536.
1282. See Reference *1279*.
1283. Tam, J.P.; Yu, J.; Miao, Z. *Biopolymers (Peptide Sci.)* **1999**, *51*, 311–332.
1284. Tam, J.P.; Xu, J.; Eom, K.D. *Biopolymers (Peptide Sci.)* **2001**, *60*, 194–205.
1285. Yeo, D.S.Y.; Srinivasan, R.; Chen, G.Y.J.; Yao, S.Q. *Chem. —Eur. J.* **2004**, *10*, 4464–4672.
1286. Walker, M.A. *Angew. Chem., Int. Ed. Engl.* **1997**, *36*, 1069–1071.
1287. Coltart, D.M. *Tetrahedron* **2000**, *56*, 3449–3491.
1288. Beligere, G.S.; Dawson, P.E. *J. Am. Chem. Soc.* **1999**, *121*, 6332–6333.
1289. Shin, Y.; Winans, K.A.; Backes, B.J.; Kent, S.B.H.; Ellman, J.A.; Bertozzi, C.R. *J. Am. Chem. Soc.* **1999**, *121*, 11684–11689.
1290. Tumelty, D.; Carnevali, M.; Miranda, L.P. *J. Am. Chem. Soc.* **2003**, *125*, 14238–14239.
1291. Dudkin, V.Y.; Miller, J.S.; Danishefsky, S.J. *J. Am. Chem. Soc.* **2004**, *126*, 736–738.
1292. Bang, D.; Chopra, N.; Kent, S.B. *J. Am. Chem. Soc.* **2004**, *126*, 1377–1383.
1293. Bang, D.; Kent, S.B.H. *Angew. Chem., Int. Ed.* **2004**, *43*, 243–2538.
1294. Canne, L.E.; Botti, P.; Simon, R.J.; Chen, Y.; Dennis, E.A.; Kent, S.B.H. *J. Am. Chem. Soc.* **1999**, *121*, 8720–8727.
1295. Sato, T.; Aimoto, S. *Tetrahedron Lett.* **2003**, *44*, 8085–8087.
1296. Miao, Z.; Tam, J.P. *Org. Lett.* **2000**, *2*, 3711–3713.
1297. Yan, L.Z.; Dawson, P.E. *J. Am. Chem. Soc.* **2001**, *123*, 526–533.
1298. Pachamuthu, K.; Schmidt, R.R. *Synlett* **2003**, 659–662.
1299. Villain, M.; Gaertner, H.; Botti, P. *Eur. J. Org. Chem.* **2003**, 3267–3272.
1300. Hondal, R.J.; Nilsson, B.L.; Raines, R.T. *J. Am. Chem. Soc.* **2001**, *123*, 5140–5141.
1301. Gieselman, M.D.; Xie, L.; van der Donk, W.A. *Org. Lett.* **2001**, *3*, 1331–1334.

1302. Quaderer, R.; Sewing, A.; Hilvert, D. *Helv. Chim. Acta* **2001**, *84*, 1197–1206.

1303. Roelfes, G.; Hilvert, D. *Angew. Chem., Int. Ed.* **2003**, *42*, 2275–2277.

1304. Gieselman, M.D.; Zhu, Y.; Zhou, H.; Galonic, D.; van der Donk, W.A. *ChemBioChem* **2002**, *3*, 709–716.

1305. Quaderer, R.; Hilvert, D. *Chem. Commun.* **2002**, 2620–2621.

1306. Kimmerlin, T.; Seebach, D.; Hilvert, D. *Helv. Chim. Acta* **2002**, *85*, 1812–1826.

1307. Otaka, A.; Ueda, S.; Tomita, K.; Yano, Y.; Tamamura, H.; Matsuzaki, K.; Fujii, N. *Chem. Commun.* **2004**, 1722–1723.

1308. Ficht, S.; Mattes, A.; Seitz, O. *J. Am. Chem. Soc.* **2004**, *126*, 9970–9981.

1309. Kennan, A.J.; Haridas, V.; Severin, K.; Lee, D.H.; Ghadiri, M.R. *J. Am. Chem. Soc.* **2001**, *123*, 1797–1803.

1310. Stetsenko, D.A.; Gait, M.J. *J. Org. Chem.* **2000**, *65*, 4900–4908.

1311. Tulla-Puche, J.; Barany, G. *J. Org. Chem.* **2004**, *69*, 4101–4107.

1312. Brik, A.; Keinan, E.; Dawson, P.E. *J. Org. Chem.* **2000**, *65*, 3829–3835.

1313. Mezzato, S.; Schaffrath, M.; Unverzagt, C. *Angew. Chem., Int. Ed.* **2005**, *44*, 1650–1654.

1314. Ingenito, R.; Bianchi, E.; Fattori, D.; Pressi, A. *J. Am. Chem. Soc.* **1999**, *121*, 11369–11374.

1315. Camarero, J.A.; Hackel, B.J.; de Yoreo, J.J.; Mitchell, A.R. *J. Org. Chem.* **2004**, *69*, 4145–4151.

1316. Sewing, A.; Hilvert, D. *Angew. Chem., Int. Ed.* **2001**, *40*, 3395–3396.

1317. von Eggelkraut-Gottanka, R.; Klose, A.; Beck-Sickinger, A.G.; Beyermann, M. *Tetrahedron Lett.* **2003**, *44*, 3551–3554.

1318. Botti, P.; Villain, M.; Manganiello, S.; Gaertner, H. *Org. Lett.* **2004**, *6*, 4861–4864.

1319. Kawakami, T.; Sumida, M.; Nakamura, K.; Vorherr, T.; Aimoto, S. *Tetrahedron Lett.* **2005**, *46*, 8805–8807.

1320. Warren, J.D.; Miller, J.S.; Keding, S.J.; Danishefsky, S.J. *J. Am. Chem. Soc.* **2004**, *126*, 6576–6578.

1321. Offer, J.; Boddy, C.N.C.; Dawson, P.E. *J. Am. Chem. Soc.* **2002**, *124*, 4642–4646.

1322. Botti, P.; Carrasco, M.R.; Kent, S.B.H. *Tetrahedron Lett.* **2001**, *42*, 1831–1833.

1323. Kawakami, T.; Aimoto, S. *Tetrahedron Lett.* **2003**, *44*, 6059–6061.

1324. Dose, C.; Seitz, O. *Org. Lett.* **2005**, *7*, 4365–4368.

1325. Ficht, S.; Dose, C.; Seitz, O. *ChemBioChem* **2005**, *6*, 2098–2103.

1326. Clive, D.L.J.; Hisaindee, S.; Coltart, D.M. *J. Org. Chem.* **2003**, *68*, 9247–9254.

1327. Ishiwata, A.; Ichianagi, T.; Takatani, M.; Ito, Y. *Tetrahedron Lett.* **2003**, *44*, 3187–3190.

1328. Lew, B.M.; Mills, K.V.; Paulus, H. *Biopolymers (Peptide Sci.)* **1999**, *51*, 355–362.

1329. Severinov, K.; Muir, T.W. *J. Biol. Chem.* **1998**, *273*, 16205–16209.

1330. Muir, T.W.; Sondh, D.; Cole, P.A. *Proc. Natl. Acad. Sci. U.S.A.* **1998**, *95*, 6705–6710.

1331. Evans, T.C., Jr.; Xu, M.-Q. *Biopolymers (Peptide Sci.)* **1999**, *51*, 333–342.

1332. Muir, T.W. *Synlett* **2001**, 733–740.

1333. Budisa, N. *ChemBioChem* **2004**, *5*, 1176–1179.

1334. Ayers, B.; Blaschke, U.K.; Camarero, J.A.; Cotton, G.J.; Holford, M.; Muir, T.W. *Biopolymers (Peptide Sci.)* **1999**, *51*, 343–354.

1335. Macmillan, D.; Arham, L. *J. Am. Chem. Soc.* **2004**, *126*, 9530–9531.

1336. Berry, S.M.; Gieselman, M.D.; Nilges, M.J.; van der Donk, W.A.; Lu, Y. *J. Am. Chem. Soc.* **2002**, *124*, 2084–2085.

1337. Ralle, M.; Berry, S.M.; Nilges, M.J.; Gieselman, M.D.; van der Donk, W.A.; Lu, Y.; Blackburn, N.J. *J. Am. Chem. Soc.* **2004**, *126*, 7244–7256.

1338. Berry, S.M.; Ralle, M.; Low, D.W.; Blackburn, N.J.; Lu, Y. *J. Am. Chem. Soc.* **2003**, *125*, 8760–8768.

1339. Arnold, U.; Hinderaker, M.P.; Köditz, J.; Golbik, R.; Ulbrich-Hofmann, R.; Raines, R.T. *J. Am. Chem. Soc.* **2003**, *125*, 7500–7501.

1340. Muralidharan, V.; Cho, J.; Trester-Zedlitz, M.; Kowalik, L.; Chait, B.T.; Raleigh, D.P.; Muir, T.W. *J. Am. Chem. Soc.* **2004**, *126*, 14004–14012.

1341. Lu, W.; Shen, K.; Cole, P.A. *Biochemistry* **2003**, *42*, 5461–5468.

1342. Arnold, U.; Hinderaker, M.P.; Nilsson, B.L.; Huck, B.R.; Gellman, S.H.; Raines, R.T. *J. Am. Chem. Soc.* **2002**, *124*, 8522–8523.

1343. Zheng, W.; Schwarzer, D.; Lebeau, A.; Weller, J.L.; Klein, D.C.; Cole, P.A. *J. Biol. Chem.* **2005**, *280*, 10462–10467.

1344. Jantz, D.; Berg, J.M. *J. Am. Chem. Soc.* **2003**, *125*, 4960–4961.

1345. Blaschke, U.K.; Cotton, G.J.; Muir, T.W. *Tetrahedron* **2000**, *56*, 9461–9470.

1346. Machova, Z.; von Eggelkraut-Gottanka, R.; Wehofsky, N.; Bordusa, F.; Beck-Sickinger, A.G. *Angew. Chem., Int. Ed.* **2003**, *42*, 4916–4918.

1347. Mao, H.; Hart, S.A.; Schink, A.; Pollok, B.A. *J. Am. Chem. Soc.* **2004**, *126*, 2670–2671.

1348. Liu, C.F.; Rao, C.; Tam, J.P. *J. Am. Chem. Soc.* **1996**, *118*, 307–312.

1349. Tam, J.P.; Miao, Z. *J. Am. Chem. Soc.* **1999**, *121*, 9013–9022.

1350. Miao, Z.; Tam, J.P. *J. Am. Chem. Soc.* **2000**, *122*, 4253–4260.

1351. Tam, J.P.; Yu, Q.; Yang, J.-L. *J. Am. Chem. Soc.* **2001**, *123*, 2487–2494.

1352. Eom, K.D.; Miao, Z.; Yang, J.-L.; Tam, J.P. *J. Am. Chem. Soc.* **2003**, *125*, 73–82.

1353. Baca, M.; Kent, S.B.H. *Tetrahedron* **2000**, *56*, 9503–9513.

1354. Li, X.; Zhang, L.; Hall, S.E.; Tam, J.P. *Tetrahedron Lett.* **2000**, *41*, 4069–4073.

1355. Meutermans, W.D.F.; Golding, S.W.; Bourne, G.T.; Miranda, L.P.; Dooley, M.J.; Alewood, P.F.; Smythe, M.L. *J. Am. Chem. Soc.* **1999**, *121*, 9790–9796.

1356. Krämer, R. *Angew. Chem., Int. Ed. Engl.* **1996**, *35*, 1197–1199.

1357. Krämer, R.; Maurus, M.; Polborn, K.; Sünkel, K.; Robl, C.; Beck, W. *Chem. —Eur. J.* **1996**, *2*, 1518–1526.

1358. Krämer, R.; Maurus, M.; Bergs, R.; Polborn, K.; Sünkel, K.; Wagner, B.; Beck, W. *Chem. Ber.* **1993**, *126*, 1969–1980.

1359. Beck, W.; Krämer, R. *Angew. Chem., Int. Ed. Engl.* **1991**, *30*, 1467–1468.

1360. Gartner, Z.J.; Kana, M.W.; Liu, D.R. *J. Am. Chem., Soc.* **2002**, *124*, 10304–10306.

1361. Snyder, T.M.; Liu, D.R. *Angew. Chem., Int. Ed.* **2005**, *44*, 7379–7382.

1362. Matsueda, R.; Walter, R. *Int. J. Pept. Prot. Res.* **1980**, *16*, 392–401.

1363. Sullivan, T.P.; van Poll, M.L.; Dankers, P.Y.W.; Huck, W.T.S. *Angew. Chem., Int. Ed.* **2004**, *43*, 4190–4193.

1364. Wieland, T.; Bodanszky, M. *The World of Peptides: A Brief History of Peptide Chemistry*; Springer-Verlag: New York, 1991, pp 57–62.

1365. Bordusa, F. *Chem. Rev.* **2002**, *102*, 4817–4867.

1366. Margolin, A.L.; Klibanov, A.M. *J. Am. Chem. Soc.* **1987**, *109*, 3802–3804.

1367. Bongers, J.; Liu, W.; Lambros, T.; Breddam, K.; Campbell, R.M.; Felix, A.M.; Heimer, E.P. *Int. J. Pept. Prot. Res.* **1994**, *44*, 123–129.

1368. Jakubke, H.-D.; Kuhl, P.; Könnecke, A. *Angew. Chem., Int. Ed. Engl.* **1985**, *24*, 85–93.

1369. Khmel'nitski, Y.L.; Dien, F.K.; Semenov, A.N.; Martinek, K.; Veruovic, B.; Kub´nek, V. *Tetrahedron*, **1984**, *40*, 4425–4432.

1370. Wehofsky, N.; Haensler, M.; Kirbach, S.W.; Wissmann, J.-D.; Bordusa, F. *Tetrahedron: Asymmetry* **2000**, *11*, 2421–2428.

1371. Hänsler, M.; Jakubke, H.-D. *Amino Acids* **1996**, *11*, 379–395.

1372. Bordusa, F.; Ullman, D.; Jakubke, H.-D. *Angew. Chem., Int. Ed. Engl.* **1997**, *36*, 1099–1101.

1373. Kitaguchi, H.; Tai, D.-F.; Klibanov, A.M. *Tetrahedron Lett.* **1988**, *29*, 5487–5488.

1374. Doezé, R.H.P.; Maltman, B.A.; Egan, C.L.; Ulijn, R.V.; Flitsch, S.L. *Angew. Chem., Int. Ed.* **2004**, *43*, 3138–3141.

1375. Rival, S.; Besson, C.; Saulnier, J.; Wallach, J. *J. Peptide Res.* **1999**, *53*, 170–176.

1376. Bemquerer, M.P.; Adlercreutz, P.; Tominaga, M. *Int. J. Pept. Prot. Res.* **1994**, *44*, 448–456.

1377. Mitin, Y.V.; Zapevalova, N.P. *Int. J. Pept. Prot. Res.* **1990**, *35*, 352–356.

1378. Yadav, J.S.; Meshram, H.M.; Prasad, A.R.; Ganesh, Y.S.S.; Rao, A.B.; Seenayya, G.; Swamy,

M.V.; Reddy, M.G. *Tetrahedron: Asymmetry* **2001**, *12*, 2505–2508.

1379. Cerovsky, V.; Jakubke, H.-D. *Int. J. Pept. Prot. Res.* **1994**, *44*, 466–471.

1380. Shin, C.-g.; Kakusho, T.; Arai, K.; Seki, M. *Bull. Chem. Soc. Jpn.* **1995**, *68*, 3549–3555.

1381. Shin, C.-g.; Arai, K.; Hotta, K.; Kakusho, T. *Bull. Chem. Soc. Jpn.* **1997**, *70*, 1427–1434.

1382. Bordusa, F.; Dahl, C.; Jakubke, H.-D.; Burger, K.; Koksch, B. *Tetrahedron: Asymmetry* **1999**, *10*, 307–313.

1383. Thust, S.; Koksch, B. *Tetrahedron Lett.* **2004**, *45*, 1163–1165.

1384. Margolin, A.L.; Tai, D.-F.; Klibanov, A.M. *J. Am. Chem. Soc.* **1987**, *109*, 7885–7887.

1385. Khimiuk, A.Y.; Korennykh, A.V.; van Langen, L.M.; van Rantwijk, F.; Sheldon, R.A.; Svedas, V.K. *Tetrahedron: Asymmetry* **2003**, *14*, 3123–3128.

1386. Sears, P.; Schuster, M.; Wang, P.; Witte, K.; Wong, C.-H. *J. Am. Chem. Soc.* **1994**, *116*, 6521–6530.

1387. Moree, W.J.; Sears, P.; Kawashiro, K.; Witte, K.; Wong, C.-H. *J. Am. Chem. Soc.* **1997**, *119*, 3942–3947.

1388. Jackson, D.Y.; Burnier, J.; Quan, C.; Stanley, M.; Tom, J.; Wells, J. A. *Science* **1994**, *266*, 243–247.

1389. Khumtaveeporn, K.; DeSantis, G.; Jones, J.B. *Tetrahedron: Asymmetry* **1999**, *10*, 2563–2572.

1390. Khumtaveeporn, K.; Ullman, A.; Matsumoto, K.; Davis, B.G.; Jones, J.B. *Tetrahedron: Asymmetry* **2001**, *12*, 249–261.

1391. Matsumoto, K. Davis, B.G.; Jones, J.B. *Chem. Commun.* **2001**, 903–904.

1392. Matsumoto, K.; Davis, B.G.; Jones, J.B. *Chem.— Eur. J.* **2002**, *8*, 4129–4137.

1393. Itoh, K.; Sekizaki, H.; Toyota, E.; Tanizawa, K. *Chem. Pharm. Bull.* **1995**, *43*, 2082–2087.

1394. Sekizaki, H.; Itoh, K.; Toyota, E.; Tanizawa, K. *Tetrahedron Lett.* **1997**, *38*, 1777–1780.

1395. Tanizawa, K.; Kasaba, Y.; Kanaoka, Y. *J. Am. Chem. Soc.* **1977**, *99*, 4485–4488.

1396. Sekizaki, H.; Itoh, K.; Toyota, E.; Tanizawa, K. *Chem. Pharm. Bull.* **1996**, *44*, 1585–1587.

1397. Sckizaki, H.; Itoh, K.; Toyota, E.; Tanizawa, K. *Amino Acids* **1999**, *17*, 285–291.

1398. Sekizaki, H.; Itoh, K.; Toyota, E.; Tanizawa, K. *Amino Acids* **2001**, *21*, 175–184.

1399. Bordusa, F.; Ullman, D.; Elsner, C.; Jakubke, H.-D. *Angew. Chem., Int. Ed. Engl.* **1997**, *36*, 2473–2475.

1400. Günther, R.; Bordusa, F. *Chem. —Eur. J.* **2000**, *6*, 463–467.

1401. Thust, S.; Koksch, B. *J. Org. Chem.* **2003**, *68*, 2290–2296.

1402. Xu, S.; Rall, K.; Bordusa, F. *J. Org. Chem.* **2001**, *66*, 1627–1632.

1403. Rall, K.; Bordusa, F. *J. Org. Chem.* **2002**, *67*, 9103–9106.

1404. Wehofsky, N.; Koglin, N.; Thust, S.; Bordusa, F. *J. Am. Chem. Soc.* **2003**, *125*, 6126–6133.

1405. Liu, C.-F.; Tam, J.P. *Org. Lett.* **2001**, *3* , 4157–4159.

1406. Lobell, M.; Schneider, M.P. *J. Chem. Soc., Perkin Trans. 1* **1998**, 319–312.

1407. West, J.B.; Wong, C.-H. *Tetrahedron Lett.* **1987**, *28*, 1629–1632.

1408. Ye, Y.-H.; Tian, G.-L.; Xing, G.-W.; Dai, D.-C.; Chen, G.; Li, C.-X. *Tetrahedron* **1998**, *54*, 12585–12596.

1409. Miyazawa, T.; Tanaka, K.; Ensatsu, E.; Yanagihara, R.; Yamada, T. *Tetrahedron Lett.* **1998**, *39*, 997–1000.

1410. Miyazawa, T.; Nakajo, S.; Nishikawa, M.; Hamahara, K.; Imagawa, K.; Ensatsu, E.; Yanagihara, R.; Yamada, T. *J. Chem. Soc., Perkin Trans. 1* **2001**, 82–86.

1411. Miyazawa, T.; Tanaka, K.; Ensatsu, E.; Yanagihara, R.; Yamada, T. *J. Chem. Soc., Perkin Trans. 1* **2001**, 87–93.

1412. Liu, P.; Tian, G.-I.; Lee, K.-S.; Wong, M.-S.; Ye, Y.-H. *Tetrahedron Lett.* **2002**, *43*, 2423–2425.

1413. Miyazawa, T.; Ensatsu, E.; Yabuuchi, N.; Yanagihara, R.; Yamada, T. *J. Chem. Soc., Perkin Trans. 1* **2002**, 390–395.

1414. Miyazawa, T.; Ensatsu, E.; Hiramatsu, M.; Yanagihara, R.; Yamada, T. *J. Chem. Soc., Perkin Trans. 1* **2002**, 396–401.

1415. Wang, Y.-F.; Yakovlesky, K.; Zhang, B.; Margolin, A.L. *J. Org. Chem.* **1997**, *62*, 3488–3495.

1416. van Unen, D-J.; Sakodinskaya, I.K.; Engbersen, J.F.J.; Reinhoudt, D.N. *J. Chem. Soc., Perkin Trans. 1* **1998**, 3341–3343.

1417. Xing, G.-W.; Li, X.-W.; Tian, G.-L.; Ye, Y.-H. *Tetrahedron* **2000**, *56*, 3517–3522.

1418. Rolland-Fulcrand, V.; May, N.; Viallefont, P.; Lazaro, R. *Amino Acids* **1994**, *6*, 311–314.

1419. Basso, A.; De Martin, L.; Ebert, C.; Gardossi, L.; Linda, P. *Chem. Commun.* **2000**, 467–468.

1420. Ulijn, R.V.; Baragaña, B.; Halling, P.J.; Flitsch, S.L. *J. Am. Chem. Soc.* **2002**, *124*, 10988–10989.

1421. Ulijn, R.V.; Bisek, N.; Flitsch, S.L. *Org. Biomol. Chem.* **2003**, *1*, 621–622.

1422. Ulijn, R.V.; Bisek, N.; Halling, P.J.; Flitsch, S.L. *Org. Biomol. Chem.* **2003**, *1*, 1277–1281.

1423. Frauendorf, C.; Jäschke, A. *Angew. Chem., Int. Ed. Engl.* **1998**, *37*, 1378–1381.

1424. Weinger, J.S.; Parnell, K.M.; Dorner, S.; Green, R.; Strobel, S.A. *Nat. Struct. Mol. Biol.* **2004**, *11*, 1101–1106.

1425. Hirschmann, R.; Smith, A.B., III; Taylor, C.M.; Benkovic, P.A.; Taylor, S.D.; Yager, K.M.; Sprengeler, P.A.; Benkovic, S.J. *Science* **1994**, *265*, 234–237.

1426. Smith, A.B., III; Taylor, C.M.; Benkovic, S.J.; Hirschmann, R. *Tetrahedron Lett.* **1994**, *35*, 6853–6856.

1427. Yao, S.; Chmielewski, J. *Biopolymers (Peptide Sci.)* **1999**, *51*, 370–375.

1428. Noren, C.J.; Anthony-Cahill, S.J.; Griffith, M.C.; Schultz, P.G. *Science* **1989**, *244*, 182–188.

1429. Bain, J.D.; Glabe, C.G.; Dix, T.A.; Chamberlin, A.R.; Diala, E.S. *J. Am. Chem. Soc.* **1989**, *111*, 8013–8014.

1430. Cornish, V.W.; Mendel, D.; Schultz, P.G. *Angew. Chem., Int. Ed. Engl.* **1995**, *34*, 621–633.

1431. Brunner, J. *Chem. Soc. Rev.* **1993**, 183–189.

1432. Wang, L. *Science* **2003**, *302*, 584–585.

1433. Borman, S. *Chem. Eng. News* **2004**, *January 19*, 64–68.

1434. England, P.M. *Biochemistry* **2004**, *43*, 11623–11629.

1435. van Maarseveen, J.H.; Back, J.W. *Angew. Chem., Int. Ed.* **2003**, *42*, 5926–5928.

1436. Hahn, U.; Palm, G.J.; Hinrichs, W. *Angew. Chem., Int. Ed.* **2004**, *43*, 1190–1193.

1437. Strømgaard, A.; Jensen, A.A.; Strømgaard, K. *ChemBioChem* **2004**, *5*, 909–916.

1438. Wang, L.; Schultz, P.G. *Chem. Commun.* **2002**, 1–11.

1439. Budisa, N. *Angew. Chem., Int. Ed.* **2004**, *43*, 6426–6463.

1440. Wang, L.; Schultz, P.G. *Angew. Chem., Int. Ed.* **2005**, *44*, 34–66.

1441. Walsh, C.T. *Post-translational Modifications of Proteins: Expanding Nature's Inventory;* Roberts and Co.: Englewood, Colorado, 2005.

1442. Budisa, N. *Engineering the Genetic Code;* Wiley-VCH: Weinheim, 2006.

1443. Hatfield, D.L.; Gladyshev, V.N. *Mol. Cell Biol.* **2002**, *22*, 3565–3576.

1444. Diamond, A.M. *Proc. Natl. Acad. Sci. U.S.A.* **2004**, *101*, 13395–13396.

1445. Mehta, A.; Rebsch, C.M.; Kinzy, S.A.; Fletcher, J.E.; Copeland, P.R. *J. Biol. Chem.* **2004**, *279*, 37852–37859.

1446. Carlson, B.A.; Xu, X.-M.; Kryukov, G.V.; Rao, M.; Berry, M.J.; Gladyshev, V.N.; Hatfield, D.L. *Proc. Natl. Acad. Sci. U.S.A.* **2004**, *101*, 12848–12853.

1447. See Reference *1444*.

1448. Sandman, K.E.; Benner, J.S.; Noren, C.J. *J. Am. Chem. Soc.* **2000**, *122*, 960–961.

1449. Srinivasan, G.; James, C.M.; Krzycki, J.A. *Science* **2002**, *296*, 1459–1462.

1450. Hao, B.; Gong, W.; Ferguson, T.K.; James, .M.; Krzycki, J.A.; Chan, M.K. *Science* **2002**, *296*, 1462–1466.

1451. Green-Church, K.B.; Chan, M.K.; Krzycki, J.A. *Nature* **2004**, *431*, 333–335.

1452. Polycarpo, C.; Ambrogelly, A.; Bérubé, A.; Winbush, S.M.; McCloskey, J.A.; Crain, P.F.; Wood, J.L.; Söll, D. *Proc. Natl. Acad. Sci. U.S.A.* **2004**, *101*, 12450–12454.

1453. Schimmel, P.; Beebe, K. *Nature* **2004**, *431*, 257–258.

1454. Mendel, D.; Ellman, J.; Schultz, P.G. *J. Am. Chem. Soc.* **1993**, *115*, 4359–4360.

1455. Bain, J.D.; Wacker, D.A.; Kuo, E.E.; Chamberlin, A.R. *Tetrahedron* **1991**, *47*, 2389–2400.

1456. Dedkova, L.M.; Fahmi, N.E.; Golovine, S.Y.; Hecht, S.M. *J. Am. Chem. Soc.* **2003**, *125*, 6616–6617.

1457. Robertson, S.A.; Ellman, J.A.; Schultz, P.G. *J. Am. Chem. Soc.* **1991**, *113*, 2722–2729.

1458. Lodder, M.; Golovine, S.; Hecht, S.M. *J. Org. Chem.* **1997**, *62*, 778–779.

1459. Murakami, H.; Bonzagni, N.J.; Suga, H. *J. Am. Chem. Soc.* **2002**, *124*, 6834–6835.

1460. Chung, H.-H.; Benson, D.R.; Cornish, V.W.; Schultz, P.G. *Proc. Natl. Acad. Sci. U.S.A.* **1993**, *90*, 10145–10149.

1461. Chung, H.H.; Benson, D.R.; Schultz, P.G. *Science* **1993**, *259*, 806–809.

1462. Ellman, J.A.; Mendel, D.; Schultz, P.G. *Science* **1992**, *255*, 197–200.

1463. Zhao, Z.; Liu, X.; Shi, Z.; Danley, L.; Huang, B.; Jiang, R.-T.; Tsai, M.-D. *J. Am. Chem. Soc.* **1996**, *118*, 3535–3536.

1464. Nowak, M.W.; Kearney, P.C.; Sampson, J.R.; Saks, M.E.; Labarca, C.G.; Silverman, S.K.; Zhong, W.; Thorson, J.; Abelson, J.N.; Davidson, N.; Schultz, P.G.; Dougherty, D.A.; Lester, H.A. *Science* **1995**, *268*, 439–442.

1465. Arslan, T.; Mamaev, S.V.; Mamaeva, N.V.; Hecht, S.M. *J. Am. Chem. Soc.* **1997**, *119*, 10877–10887.

1466. Li, S.; Millward, S.; Roberts, R. *J. Am. Chem. Soc.* **2002**, *124*, 9972–9973.

1467. Thorson, J.S.; Shin, I.; Chapman, E.; Stenberg, G.; Mannervik, B.; Schultz, P.G. *J. Am. Chem. Soc.* **1998**, *120*, 451–452.

1468. Barrett, J.E.; Lucero, C.M.; Schultz, P.G. *J. Am. Chem. Soc.* **1999**, *121*, 7965–7966.

1469. Pollitt, S.K.; Schultz, P.G. *Angew. Chem., Int. Ed. Engl.* **1998**, *37*, 2104–2107.

1470. Cornish, V.W.; Hahn, K.M.; Schultz, P.G. *J. Am. Chem. Soc.* **1996**, *118*, 8150–8151.

1471. Shafer, A.M.; Kálai, T.; Bin Liu, S.Q.; Hideg, K.; Voss, J.C. *Biochem.* **2004**, *43*, 8470–8482.

1472. England, P.M.; Lester, H.A.; Dougherty, D.A. *Tetrahedron Lett.* **1999**, *40*, 6189–6192.

1473. Koh, J.T.; Cornish, V.W.; Schultz, P.G. *Biochem.* **1997**, *36*, 11314–11322.

1474. Chapman, E.; Thorson, J.S.; Schultz, P.G. *J. Am. Chem. Soc.* **1997**, *119*, 7151–7152.

1475. Killian, J.A.; Van Cleve, M.D.; Shayo, Y.F.; Hecht, S.M. *J. Am. Chem. Soc.* **1998**, *120*, 3032–3042.

1476. Agafonov, D.E.; Huang, Y.; Grote, M.; Sprinzl, M. *FEBS Lett.* **2005**, *579*, 2156–2160.

1477. Hohsaka, T.; Sato, K.; Sisido, M.; Takai, K.; Yokoyama, S. *FEBS Lett.* **1994**, *344*, 171–174.

1478. Hohsaka, T.; Ashizuka, Y.; Murakami, H.; Sisido, M. *J. Am. Chem. Soc.* **1996**, *118*, 9778–9779.

1479. Murakami, H.; Hohsaka, T.; Ashizuka, Y.; Sisido, M. *J. Am. Chem. Soc.* **1998**, *120*, 7520–7529.

1480. Hohsaka, T.; Ashizuka, Y.; Taira, H.; Murakami, H.; Sisido, M. *Biochemistry* **2001**, *40*, 11060–11064.

1481. Hohsaka, T.; Kajihara, D.; Ashizuka, Y.; Murakami, H.; Sisido, M. *J. Am. Chem. Soc.* **1999**, *121*, 34–40.

1482. Hohsaka, T.; Muranaka, N.; Komiyama, C.; Matsui, K.; Takaura, S.; Abe, R.; Murakami, H.; Sisido, M. *FEBS Lett.* **2004**, *560*, 173–177.

1483. Hohsaka, T.; Ashizuka, Y.; Sasaki, H.; Murakami, H.; Sisido, M. *J. Am. Chem. Soc.* **1999**, *121*, 12194–12195.

1484. Taki, M.; Hohsaka, T.; Murakami, H.; Taira, K.; Sisido, M. *J. Am. Chem. Soc.* **2002**, *124*, 14586–14590.

1485. Landweber, L.F. *Chem. Biol.* **2002**, *9*, 143.

1486. Anderson, R.D., III; Zhou, J.; Hecht, S.M. *J. Am. Chem. Soc.* **2002**, *124*, 9674–9675.

1487. Anderson, J.C.; Wu, N.; Santoro, S.W.; Lakshman, V.; King, D.S.; Schultz, P.G. *Proc. Natl. Acad. Sci. U.S.A.* **2004**, *101*, 7566–7571.

1488. Röhrig, C.H.; Retz, O.A.; Meergans, T.; Schmidt, R.R. *Biochem. Biophys. Res. Commun.* **2004**, *325*, 731–738.

1489. Liu, D.R.; Magliery, T.J.; Pastrnak, M.; Schultz, P.G. *Proc. Natl. Acad. Sci. U.S.A.* **1997**, *94*, 10092–10097.

1490. Liu, D.R.; Schultz, P.G. *Proc. Natl. Acad. Sci., U.S.A.* **1999**, *96*, 4780–4785.

1491. Pasternak, M.; Magliery, T.J.; Schultz, P.G. *Helv. Chim. Acta* **2000**, *83*, 2277–2286.

1492. Wang, P.; Vaidehi, N.; Tirrell, D.A.; Goddard, W.A., III; *J. Am. Chem. Soc.* **2002**, *124*, 14442–14449.

1493. Chin, J.W.; Cropp, T.A.; Anderson, J.C.; Mukherji, M.; Zhang, Z.; Schultz, P.G. *Science* **2003**, *301*, 964–966.

1494. Deiters, A.; Cropp, T.A.; Mukherji, M.; Chin, J.W.; Anderson, J.C.; Schultz, P.G. *J. Am. Chem. Soc.* **2003**, *125*, 11782–11783.

1495. Wu, N.; Deiters, A.; Cropp, T.A.; King, D.; Schultz, P.G. *J. Am. Chem. Soc.* **2004**, *126*, 14306–14307.

1496. Zhang, Z.; Alfonta, L.; Tian, F.; Bursulaya, B.; Uryu, S.; King, D.S.; Schultz, P.G. *PNAS* **2004**, *101*, 8882–8887.

1497. Anderson, J.C.; Schultz, P.G. *Biochem.* **2003**, *42*, 9598–9608.

1498. Kiga, D.; Sakamoto, K.; Kodama, K.; Kigawa, T.; Matsuda, T.; Yabuki, T.; Shirouzu, M.; Harada, Y.; Nakayama, H.; Takio, K.; Hasegawa, Y.; Endo, Y.; Hirao, I.; Yokoyama, S. *Proc. Nat. Acad. Sci.* **2002**, *99*, 9715–9723.

1499. Kobayashi, T.; Sakamoto, K.; Takimura, T.; Sekine, R.; Vincent, K.; Kamata, K.; Nishimura, S.; Yokoyama, S. *PNAS*, **2005**, *102*, 1366–1371.

1500. Köhrer, C.; Xie, L.; Kellerer, S.; Varshney, U.; RajBhandary, U.L. *Proc. Nat. Acad. Sci.* **2001**, *98*, 14310–14315.

1501. Wang, L.; Magliery, T.J.; Liu, D.R.; Schultz, P.G. *J. Am. Chem. Soc.* **2000**, *122*, 5010–5011.

1502. Wang, L.; Brock, A.; Herberich, B.; Schultz, P.G. *Science* **2001**, *292*, 498–500.

1503. Deiters, A.; Geierstanger, B.H.; Schultz, P.G. *ChemBiochem* **2005**, *6*, 55–58.

1504. Zhang, D.; Vaidehi, N.; Goddard W.A., III; Danzer, J.F.; Debe, D. *Proc. Nat. Acad. Sci.* **2002**, *99*, 6579–6584.

1505. Wang, L.; Zhang, Z.; Brock, A.; Schultz, P.G. *Proc. Nat. Acad. Sci.* **2003**, *100*, 56–61.

1506. Liu, H.; Wang, L.; Brock, A.; Wong, C.-H.; Schultz, P.G. *J. Am. Chem. Soc.* **2003**, *125*, 1702–1703.

1507. Zhang, Z.; Wang, L.; Brock, A.; Schultz, P.G. *Chem. Int. Ed.* **2002**, *41*, 2840–2842.

1508. Wang, L.; Brock, A.; Schultz, P.G. *J. Am. Chem. Soc.* **2002**, *124*, 1836–1837.

1509. Zhang, Z.; Gildersleeve, J.; Yang, Y.-Y.; Xu, R.; Loo, J.A.; Uryu, S.; Wong, C.-H.; Schultz, P.G. *Science* **2004**, *303*, 371–373.

1510. Alfonta, L.; Zhang, Z.; Uryu, S.; Loo, J.A.; Schhultz, P.G. *J. Am. Chem. Soc.* **2003**, *125*, 14662–14663.

1511. Chin, J.W.; Santoro, S.W.; Martin, A.B.; King, D.S.; Wang, L.; Schultz, P.G. *J. Am. Chem. Soc.* **2002**, *124*, 9026–9027.

1512. Mehl, R.A.; Andersen, J.C.; Santoro, S.W.; Wang, L.; Martin, A.B.; King, D.S.; Horn, D.M.; Schultz, P.G. *J. Am. Chem. Soc.* **2003**, *125*, 935–939.

1513. An, S.S.A.; Lester, C.C.; Peng, J.-L.; Li, Y.-J.; Rothwarf, D.M.; Welker, E.; Thannhauser, T.W.; Zhang, L.S.; Tam, J.P.; Scheraga, H.A. *J. Am. Chem. Soc.* **1999**, *121*, 11558–11566.

1514. Ryu, Y.; Schultz, P.G. *Nature Methods* **2006**, *3*, 263–265.

1515. Döring, V.; Mootz, H.D.; Nangle, L.A.; Hendrickson, T.L.; de Crécy-Lagard, V.; Schimmel, P.; Marlière, P. *Science* **2001**, *292*, 501–504.

1516. Mursinna, R.S.; Martinis, S.A. *J. Am. Chem. Soc.* **2002**, *124*, 7286–7287.

1517. Bain, J.D.; Switzer, C.; Chamberlin, R.; Bennert, S.A. *Nature* **1992**, *356*, 537–539.

1518. Hirao, I.; Harada, Y.; Kimoto, M.; Mitsui, T.; Fujiwara, T.; Yokoyama, S. *J. Am. Chem. Soc.* **2004**, *126*, 13298–13305.

1519. Forster, A.C.; Tan, Z.; Nalam, M.N.; Lin, H.; Qu, H.; Cornish, V.W.; Blacklow, S.C. *Proc. Nat. Acad. Sci.* **2003**, *100*, 6353–6357.

1520. Tan, Z.; Forster, A.C.; Blacklow, S.C.; Cornish, V.W. *J. Am. Chem. Soc.* **2004**, *126*, 12752–12753.

1521. Kwon, I.; Kirshenbaum, K.; Tirrell, D.A. *J. Am. Chem. Soc.* **2003**, *125*, 7512–7513.

1522. Dagani, R. *Chem. & Eng. News* **2003**, *June 23*, 40–44.

1523. Starck, S.R.; Qi, X.; Olsen, B.N.; Roberts, R.W. *J. Am. Chem. Soc.* **2003**, *125*, 8090–8091.

1524. Budisa, N.; Minks, C.; Medrano, F.J.; Lutz, J.; Huber, R.; Moroder, L. *Proc. Natl. Acad. Sci. USA* **1998**, *95*, 455–459.

1525. Minks, C.; Alefelder, S.; Moroder, L.; Huber, R.; Budisa, N. *Tetrahedron* **2000**, *56*, 9431–9442.

1526. Eichler, J.F.; Cramer, J.C.; Kirk, K.L.; Bann, J.G. *ChemBioChem* **2005**, *6*, 2170–2173.

1527. Leone, M.; Rodriguez-Mias, R.A.; Pellecchia, M. *ChemBioChem* **2003**, *4*, 649–662.

1528. van Hest, J.C.M.; Kiick, K.L.; Tirrell, D.A. *J. Am. Chem. Soc.* **2000**, *122*, 1282–1288.

1529. Kiick, K.L.; Tirrell, D.A. *Tetrahedron* **2000**, *56*, 9487–9493.

1530. Wang, P.; Tang, Y.; Tirrell, D.A. *J. Am. Chem. Soc.* **2003**, *125*, 6900–6906.

1531. Budisa, N.; Pipitone, O.; Siwanowicz, I.; Rubini, M.; Pal, P.P.; Holak, T.A.; Gelmi, M.L. *Chemistry and Bioiversity* **2004**, *1*, 1465–1475.

1532. Weist, S.; Bister, B.; Puk, O.; Bischoff, D.; Pelzer, S.; Nicholson, G.J.; Wohlleben, W.; Jung, G.; Süssmuth, R.D. *Angew. Chem. Int. Ed.* **2002**, *41*, 3383–3385.

1533. Datta, D.; Wang, P.; Carrico, I.S.; Mayo, S.L.; Tirrell, D.A. *J. Am. Chem. Soc.* **2002**, *124*, 5652–5653.

1534. Service, R.F. *Science* **2005**, *308*, 44.

1535. Kirshenbaum, K.; Carrico, I.S.; Tirrell, D.A. *ChemBioChem* **2002**, *3*, 235–237.

1536. Kim, W.; George, A.; Evans, M.; Conticello, V.P. *ChemBioChem* **2004**, *5*, 928–936.

1537. Udea, T.; Udea, M.; Tanaka, A.; Sisido, M.; Imanishi, Y. *Bull. Chem. Soc. Jpn.* **1991**, *64*, 1576–1581.

1538. Kiick, K.L.; Saxon, E.; Tirrell, D.A.; Bertozzi, C.R. *Proc. Nat. Acad. Sci.* **2002**, *99*, 19–24.

1539. Link, A.J.; Vink, M.K.S.; Tirrell, D.A. *J. Am. Chem. Soc.* **2004**, *126*, 10598–10602.

1540. Jayathilaka, L.P.; Deb. M.; Standaert, R.F. *Org. Lett.* **2004**, *6*, 3659–3662.

1541. Wang, P.; Fichera, A.; Kumar, K.; Tirrel, D.A. *Angew. Chem. Int. Ed.* **2004**, *43*, 3664–3666.

1542. Tang, Y.; Tirrell, D.A. *J. Am. Chem. Soc.* **2001**, *123*, 11089–11090.

1543. Bordusa, F. *ChemBioChem* **2001**, *2*, 405–409.

1544. Yeh, E.; Kohli, R.M.; Bruner, S.D.; Walsh, C.T. *ChemBioChem* **2004**, *5*, 1290–1293.

1545. Walsh, C.T. *ChemBioChem* **2002**, *3*, 124–134.

1546. Mootz, H.D.; Schwarzer, D.; Maahiel, M.A. *ChemBioChem* **2002**, *3*, 490–504.

1547. Mootz, H.D.; Kessler, N.; Linne, U.; Eppelmann, K.; Schwarzer, D.; Marahiel, M.A. *J. Am. Chem. Soc.* **2002**, *124*, 10980–10981.

1548. Ugi, I.; Marquarding, D.; Urban, R. In *Chemistry and Biochemistry of Amino Acids, Peptides and Proteins. A Survey of Recent Developments,*

Volume 6; Weinstein, B. Ed.; Marcel Dekker: New York, 1982, pp 245–289.

1549. Yamada, T.; Yanagi, T.; Omote, Y.; Miyazawa, T.; Kuwata, S.; Sugiura, M.; Matsumoto, K. *J. Chem. Soc., Chem. Commun.* **1990**, 1640–1641.

1550. Yamada, T.; Omote, Y.; Yamanaka, Y.; Miyazawa, T.; Kuwata, S. *Synthesis* **1998**, 991–998.

1551. Burger, K.; Mütze, K.; Hollweck, W.; Koksch, B. *Tetrahedron* **1998**, *54*, 5915–5928.

1552. Pulley, S.R.; Hegedus, L.S. *J. Am. Chem. Soc.* **1993**, *115*, 9037–9047.

1553. Miller, J.R.; Pulley, S.R.; Hegedus, L.S.; DeLombaert, S. *J. Am. Chem. Soc.* **1992**, *114*, 5602–5607.

1554. Zhu, J.; Hegedus, L.S. *J. Org. Chem.* **1995**, *60*, 5831–5837.

1555. Wenglowsky, S.; Hegedus, L.S. *J. Am. Chem. Soc.* **1998**, *120*, 12468–12473.

1556. Heimgartner, H. *Angew. Chem., Int. Ed. Engl.* **1991**, *30*, 238–264.

1557. Heimgartner, H. *Israel J. Chem.* **1986**, *27*, 3–15.

1558. Sahebi, M.; Wipf, P.; Heimgartner, H. *Tetrahedron* **1989**, *45*, 2999–3000.

1559. Obrecht, D.; Heimgartner, H. *Helv. Chim. Acta* **1987**, *70*, 102–115.

1560. Jeremic, T.; Linden, A.; Heimgartner, H. *Chem. Biodiversity* **2004**, *1*, 1730–1761.

1561. Koch, K.N.; Heimgartner, H. *Helv. Chim. Acta* **2000**, *83*, 1881–1900.

1562. Stamm, S.; Heimgartner, H. *Eur. J. Org. Chem.* **2004**, 3820–3827.

1563. Palomo, C.; Aizpurua, J.M.; Ganboa, I.; Odriozola, B.; Urchegui, R.; Görls, H. *Chem. Commun.*, **1996**, 1269–1270.

1564. Palomo, C.; Miranda, J.I.; Linden, A. *J. Org. Chem.* **1996**, *61*, 9196–9201.

1565. Palomo, C.; Aizpurua, J.M.; Cuevas, C. *J. Chem. Soc., Chem. Commun.* **1994**, 1957–1958.

1566. Lipshutz, B.H.; Shin, Y.-J. *Tetrahedron Lett.* **2001**, *42*, 5629–5633.

1567. Thieriet, N.; Guibé, F.; Albericio, F. *Org. Lett.* **2000**, *2*, 1815–1817.

1568. Henkel, B.; Zhang, L.; Bayer, E. *Liebigs Ann. Recueil* **1997**, 2161.

1569. Sasubilli, R.; Gutheil, W.G. *J. Comb. Chem.* **2004**, *6*, 911–915.

1570. Johansson, A.; Akerblom, E.; Ersmark, K.; Lindeberg, G.; Hallberg, A. *J. Comb. Chem.* **2000**, *2*, 496–507.

1571. Marvel, C.S.; Noyes, W.A. *J. Am. Chem. Soc.* **1920**, *42*, 2259–2278.

1572. Ingersoll, A.W.; Adams, R. *J. Am.Chem. Soc.* **1922**, *44*, 2930–2937.

1573. Dakin, H.D. *J. Biol. Chem.* **1922**, *50*, 403–411.

1574. Wood, J.L.; Du Vigneaud, V. *J. Biol. Chem.* **1939**, *130*, 109–114.

1575. Neuberger, A. *Biochem. J.* **1948**, *43*, 599–605.

1576. Armstrong, M.D. *J. Am. Chem. Soc.* **1948**, *70*, 1756–1759.

1577. Sealock, R.R.; Speeter, M.E.; Schweet, R.S. *J. Am. Chem. Soc.* **1951**, *73*, 5386–5388.

1578. Weiss, S.; Stekol, J.A. *J. Am. Chem. Soc.* **1951**, *73*, 2497–2499.

1579. Leach, B.E.; Hunter, J.H.; West, C.A.; Carter, H.E. *Biochem. Prep.* **1953**, *3*, 111–118.

1580. Leonard, F.; Wajngurt, A.; Tschannen, W.; Block, F.B. *J. Med. Chem.* **1965**, *8*, 812–815.

1581. Bergel, F.; Burnop, V.C.E.; Stock, J.A. *J. Chem. Soc.* **1955**, 1223–1230.

1582. Du Vigneaud, V.; Brown, G.B.; Sealock, R.R.; Blaney, D.J.; Law, J.H.; Carter, H.E. *Biochem. Prep.* **1957**, *5*, 84–91.

1583. Miyake, A. *Chem. Pharm. Bull.* **1960**, *8*, 1074–1078.

1584. Dewar, J.H.; Shaw, G. *J. Chem. Soc.* **1962**, 583–585.

1585. Friis, P.; Kjaer, A. *Acta Chem. Scand.* **1963**, *17*, 2391–2396.

1586. Pracejus, H.; Winter, S. *Chem. Ber.* **1964**, *97*, 3173–3182.

1587. Nicolaides, E.; Lipnik, M. *J. Med. Chem.* **1966**, *9*, 958–960.

1588. Senoh, S.; Maeno, Y.; Imamoto, S.; Komamine, A.; Hattori, S.; Yamashita, K.; Matsui, M. *Bull. Chem. Soc. Jpn.* **1967**, *40*, 379–384.

1589. Liwschitz, Y.; Edlitz-Pfeffermann, Y.; Singerman, A. *J. Chem. Soc. C* **1967**, 2104–2105.

1590. Hase, S.; Kiyoi, R.; Sakakibara, S. *Bull. Chem. Soc. Jpn.* **1968**, *41*, 1266–1267.

1591. Edwards, G.W.; Minthorn, M.L., Jr.; *Can. J. Biochem.* **1968**, *46*, 1227–1230.

1592. Okawa, K.; Hori, K.; Hirose, K.; Nakagawa, Y. *Bull. Chem. Soc. Jpn.* **1969**, *42*, 2720–2722.

1593. Crooij, P.; Eliaers, J. *J. Chem. Soc. C* **1969**, 559–563.

1594. Iwasaki, H.; Kamiya, T.; Oka, O.; Ueyanagi, J. *Chem. Pharm. Bull.* **1969**, *17*, 866–872.

1595. Arold, H.; Reissmann, S. *J. Prakt. Chem.* **1970**, *312*, 1130–1144.

1596. Tager, H.S.; Christensen, H.N. *J. Am. Chem. Soc.* **1972**, *94*, 968–972.

1597. Lee, Y.K.; Kaneko, T. *Bull. Chem. Soc. Jpn.* **1973**, *46*, 3494–3498.

1598. Tesser, G.I.; Slits, H.G.A.; van Nispen, J.W. *Int. J. Pepti. Prot. Res.* **1973**, *5*, 119–122.

1599. Arold, H.; Reissmann, S.; Eule, M. *J. Prakt. Chem.* **1974**, *316*, 93–102.

1600. Bloemhoff, W.; Kerling, K.E.T. *Recueil* **1975**, *94*, 182–185.

1601. Hegedüs, B.; Krassó, A.F.; Noack, K.; Zeller, P. *Helv. Chim. Acta* **1975**, *58*, 147–162.

1602. Buttrey, J.D.; Jones, A.S.; Walker, R.T. *Tetrahedron* **1975**, *31*, 73–75.

1603. Kataoka, Y.; Seto, Y.; Yamamoto, M.; Yamada, T.; Kuwata, S.; Watanabe, H. *Bull. Chem. Soc. Jpn.* **1976**, *49*, 1081–1084.

1604. Ichihara, A.; Shiraishi, K.; Sakamura, S. *Tetrahedron Let.* **1977**, *3*, 269–272.

1605. Hageman, J.J.M.; Wanner, M.J.; Koomen, G.-J.;
Pandit, U.K. *J. Med. Chem.* **1977**, *20*, 1677–
1678.

1606. Komatsubara, S.; Kisumi, M.; Chibata, I.;
Gregorio, M.M.V.; Müller, U.S.; Crout, D.H.G.
J. Chem. Soc., Chem. Commun. **1977**, 839–841.

1607. Märki, W.; Oppliger, M.; Thanei, P.; Schwyzer,
R. *Helv. Chim. Acta* **1977**, *60*, 798–806.

1608. Yamada, S.; Hongo, C.; Yoshioka, R.; Chibata, I.
Agric. Biol. Chem. **1979**, *43*, 395–396.

1609. Pospísek, J.; Toma, S.; Fric, I.; Bláha, K. *Collect.
Czech. Chem. Commun.* **1980**, *45*, 435–441.

1610. Lipkowski, A.W.; Flouret, G. *Polish J. Chem.*
1980, *54*, 2225–2228.

1611. Kawai, M.; Chorev, M.; Marin-Rose, J.;
Goodman, M. *J. Med. Chem.* **1980**, *23*, 420–424.

1612. Egusa, S.; Sisido, M.; Imanishi, Y. *Chem. Lett.*
1983, 1307–1310.

1613. Kimura, H.; Stammer, C.H.; Shimohigashi, Y.;
Ren-Lin, C.; Stewart, J. *Biochem. Biophys. Res.
Commun.* **1983**, *115*, 112–115.

1614. Saito, S.-I.; Umezawa, Y.; Yoshioka, T.; Takita,
T.; Umezawa, H. *J. Antibiotics* **1983**, *36*, 92–95.

1615. Hwang, D.-R.; Helquist, P.; Shekhani, M.S.
J. Org. Chem. **1985**, *50*, 1264–1271.

1616. Egusa, S.; Sisido, M.; Imanishi, Y.
Macromolecules **1985**, *18*, 882–889.

1617. Egusa, S.; Sisido, M.; Imanishi, Y. *Bull. Chem.
Soc. Jpn.* **1986**, *59*, 3175–3178.

1618. Sisido, M. *Macromolecules* **1989**, *22*, 4367–
4372.

1619. Mapelli, C.; Stammer, C.H.; Lok, S.; Mierke,
D.F.; Goodman, M. *Int. J. Pept. Prot. Res.* **1988**,
32, 484–495.

1620. López-Arbeloa, F.; Goedeweeck, R.; Ruttens, F.;
De Schryver, F.C.; Sisido, M. *J. Am. Chem. Soc.*
1987, *109*, 3068–3076.

1621. Kirihata, M.; Sakamoto, A.; Ichimoto, I.; Udea,
H.; Honma, M. *Agric. Biol. Chem.* **1990**, *54*,
1845–1846.

1622. Baudy, R.B.; Greenblatt, L.P.; Jirkovsky, I.L.;
Conklin, M.; Russo, R.J.; Bramlett, D.R.;
Emrey, T.A.; Simmonds, J.T.; Kowal, D.M.;
Stein, R.P.; Tasse, R.P. *J. Med. Chem.* **1993**, *36*,
331–342.

1623. Johansen, T.N.; Ebert, B.; Bräuner-Osborne, H.;
Didriksen, M.; Christensen, I.T.; Søby, K.K.;
Madsen, U.; Krogsgaard-Larsen, P.; Brehm, L.
J. Med. Chem. **1998**, *41*, 930–939.

1624. Skjærbæk, N.; Brehm, L.; Johansen, T.N.;
Hansen, L.M.; Nielsen, B.; Ebert, B.; Søby,
K.K.; Stensbøl, T.B.; Falch, E.; Krogsgaard-
Larsen, P. *Biorg. Med. Chem.* **1998**, *6*, 119–131.

1625. Arnold, U.; Hinderaker, M.P.; Köditz, J.; Golbik,
R.; Ulbrich-Hofmann, R.; Raines, R.T. *J. Am.
Chem. Soc.* **2003**, *125*, 7500–7501.

1626. Monn, J.A.; Valli, M.J.; Massey, S.M.; Hansen,
M.M.; Kress, T.J.; Wepsiec, J.P.; Harkness, A.R.;
Grutsch, J.L., Jr.; Wright, R.A.; Johnson, B.G.;
Andis, S.L.; Kingston, A.; Tomlinson, R.; Lewis, R.;
Griffey, K.R.; Tizzano, J.P.; Schoepp, D.D.
J. Med. Chem. **1999**, *42*, 1027–1040.

1627. Chung, Y.J.; Huck, B.R.; Christianson, L.A.;
Stanger, H.E.; Krauthäuser, S.; Powell, D.R.;
Gellman, S.H. *J. Am. Chem. Soc.* **2000**, *122*,
3995–4004.

1628. Sakai, K.; Sakurai, R.; Yuzawa, A.; Hirayama, N.
Tetrahedron: Asymmetry **2003**, *14*, 3713–3718.

1629. Noda, H.; Sakai, K.; Murakami, H. *Tetrahedron:
Asymmetry* **2002**, *13*, 2649–2652.

1630. Wakamiya, T.; Oda, Y.; Fujita, H.; Shiba, T.
Tetrahedron Lett. **1986**, *27*, 2143–2144.

1631. Cativiela, C.; Díaz-de-Villegas, M.D.; Gálvez,
J.A. *Tetrahedron Lett.* **1999**, *40*, 1027–1030.

1632. Santoso, S.; Kemmer, T.; Trowitzsch, W. *Liebigs
Ann. Chem.* **1981**, 658–667.

1633. Santoso, S.; Kemmer, T.; Trowitzsch, W. *Liebigs
Ann. Chem.* **1981**, 642–657.

1634. Maki, Y.; Masugi, T.; Hiramitsu, T.; Ogiso, T.
Chem. Pharm. Bull. **1973**, *21*, 2460–2465.

1635. Aebsicher, B.; Frey, P.; Haerter, H.-P.; Herrling,
P.L.; Mueller, W.; Olverman, H.J.; Watkins, J.C.
Helv. Chim. Acta **1989**, *72*, 1043–1051.

1636. Kammermeier, B.O.T.; Lerch, U.; Sommer, C.
Synthesis **1992**, 1157–1160.

1637. Wolff, J.S.; Ogle, J.D.; Logan, M.A. *J. Biol.
Chem.* **1966**, *241*, 1300–1307.

1638. Vine, W.H.; Hsieh, K.-H.; Marshall, G.R. *J. Med.
Chem.* **1981**, *24*, 1043–1047.

1639. Baldwin, J.E.; Adlington, R.M.; Rawlings, B.J.;
Jones, R.H. *Tetrahedron Lett.* **1985**, *26*, 485–488.

1640. Levintow, L.; Greenstein, J.P. *J. Biol. Chem.*
1951, *188*, 643–646.

1641. Greenstein, J.P.; Levintow, L.; Baker, C.G.;
White, J. *J. Biol. Chem.* **1951**, *188*, 647–663.

1642. Baker, C.G.; Fu, S.-C.J.; Birnbaum, S.M.; Sober,
H.A.; Greenstein, J.P. *J. Am. Chem. Soc.* **1952**,
74, 4701–4702.

1643. Rudman, D.; Meister, A.; Greenstein, J.P. *J. Am.
Chem. Soc.* **1952**, *74*, 551.

1644. Fones, W.S. *J. Am. Chem. Soc.* **1953**, *75*, 4865–
4866.

1645. Greenstein, J.P.; Birnbaum, S.M.; Levintow, L.;
Salzman, N.; Carter, H.E. *Biochem. Prep.* **1953**,
3, 84–95.

1646. Greenstein, J.P.; Birnbaum, S.M.; Otey, M.C.
J. Am. Chem. Soc. **1953**, *75*, 1994–1995.

1647. Ono, T.; Hirohata, R. *Z. Physiol. Chem.* **1956**,
304, 77–81.

1648. Wade, R.; Birnbaum, S.M.; Winitz, M.; Koegel,
R.J.; Greenstein, J.P. *J. Am. Chem. Soc.* **1957**,
79, 648–652.

1649. Akabori, S.; Otani, T.T.; Marshall, R.; Winitz,
M.; Grenstein, J.P. *Arch. Biochem. Biophys.*
1959, *83*, 1–9.

1650. Sugimoto, N.; Watanabe, H.; Ide, A. *Tetrahedron*
1960, *11*, 231–233.

1651. Fones, W.S.; Greenstein, J.P.; Izumiya, N.;
Birnbaum, S.M.; Winitz, M. *Biochem. Prep.*
1961, *8*, 62–69.

1652. Winitz, M.; Birnbaum, S.M.; Greenstein, J.P.; Meister, A.; Scott, S.J. *Biochem. Prep.* **1961**, *8*, 96–99.

1653. Sheehan, J.C.; Maeda, K.; Sen, A.K.; Stock, J.A. *J. Am. Chem. Soc.* **1962**, *84*, 1303–1305.

1654. Hill, J.T.; Dunn, F.W. *J. Org. Chem.* **1965**, *30*, 1321–1322.

1655. Hofmann, K.; Bohn, H. *J. Am. Chem. Soc.* **1966**, *88*, 5914–5919.

1656. Eisler, K.; Rudinger, J.; Sorm, F. *Collect. Czech. Chem. Commun.* **1966**, *31*, 4563–4580.

1657. Smith, S.C.; Sloane, N.H. *Biochim. Biophys. Acta* **1967**, *148*, 414–422.

1658. Vega, A.; Bell, E.A.; Nunn, P.B. *Phytochemistry* **1968**, *7*, 1885–1887.

1659. Maurer, B.; Schierlein, W.K. *Helv. Chim. Acta* **1969**, *52*, 388–396.

1660. Hudlicky, M.; Jelínek, V.; EIsler, K.; Rudinger, J. *Collect. Czech. Chem. Commun.* **1970**, *35*, 498–503.

1661. Scannell, J.P.; Pruess, D.L.; Demny, T.C.; Weiss, F.; Williams, T.; Stempel, A. *J. Antibiotics* **1971**, *24*, 239–244.

1662. Fauchère, J.-L.; Schwyzer, R. *Helv. Chim. Acta* **1971**, *54*, 2078–2080.

1663. Bodanszky, M.; Lindeberg, G. *J. Med. Chem.* **1971**, *14*, 1197–1199.

1664. Wightman, R.H.; Stauton, J.; Battersby, A.R.; Hanson, K.R. *J. Chem. Soc. Perkin Trans. 1* **1972**, 2355–2364.

1665. Kirk, K.L.; Cohen, L.A. *J. Am. Chem. Soc.* **1973**, *95*, 4619–4624.

1666. Dreyfuss, P. *J. Med. Chem.* **1974**, *17*, 252–255.

1667. Bycroft, B.W.; Wels, C.M.; Corbett, K.; Lowe, D.A. *J. Chem. Soc., Chem. Commun.* **1975**, 123.

1668. Leukart, O.; Caviezel, M.; Eberle, A.; Escher, E.; Tun-Kyi, A.; Schwyzer, R. *Helv. Chim. Acta* **1976**, 59, 2181–2183.

1669. Upson, D.A.; Hruby, V.J. *J. Org. Chem.* **1976**, *41*, 1353–1358.

1670. Prasad, K.U.; Roeske, R.W.; Weitl, F.L.; Vilcehz-Martinez, J.A.; Schally, A.V. *J. Med. Chem.* **1976**, *19*, 492–495.

1671. Scannell, J.P.; Pruess, D.L.; Ax, H.A.; Jacoby, A.; Kellett, M.; Stempel, A. *J. Antibiotics* **1976**, *29*, 38–43.

1672. Yabe, Y.; Miura, C.; Horikoshi, H.; Baba, Y. *Chem. Pharm. Bull.* **1976**, *24*, 3149–3157.

1673. Shimohigashi, Y.; Lee, S.; Izumiya, N. *Bull. Chem. Soc. Jpn.* **1976**, *49*, 3280–3284.

1674. Baldwin, J.E.; Haber, S.B.; Hoskins, C.; Kruse, L.I. *J. Org. Chem.* **1977**, *42*, 1239–1241.

1675. Yamamoto, D.M.; Upson, D.A.; Linn, D.K.; Hruby, V.J. *J. Am. Chem. Soc.* **1977**, *99*, 1564–1570.

1676. Gellert, E.; Halpern, B.; Rudzats, R. *Phytochemistry* **1978**, *17*, 802.

1677. Bodanszky, M.; Martinez, J.; Priestley, G.P.; Gardner, J.D.; Mutt, V. *J. Med. Chem.* **1978**, *21*, 1030–1035.

1678. See Reference *1677*.

1679. Dzieduszycka, M.; Smulkowski, M.; Czarnomska, T.; Borowski, E. *Polish J. Chem.* **1978**, *52*, 933–939.

1680. Edgar, M.T.; Pettit, G.R.; Krupa, T.S. *J. Org. Chem.* **1979**, *44*, 396–400.

1681. Fauchère, J.-L.; Leukart, O.; Eberle, A.; Schwyzer, R. *Helv. Chim. Acta* **1979**, *62*, 1385–1395.

1682. Hsieh, K.-H.; Jorgensen, E.C. *J. Med. Chem.* **1979**, *22*, 1199–1206.

1683. Keller-Schierlein, W.; Joos, B. *Helv. Chim. Acta* **1980**, *63*, 250–254.

1684. Allen, M.C.; Brundish, D.E.; Wade, R. *J. Chem. Soc. Perkin Trans. 1* **1980**, 1928–1932.

1685. Sasaki, A.N.; Bricas, E. *Tetrahedron Lett.* **1980**, *21*, 4263–4264.

1686. Baldwin, J.E.; Kruse, L.I.; Cha, J.-K. *J. Am. Chem. Soc.* **1981**, *103*, 942–943.

1687. Baldwin, J.E.; Wan, T.S. *Tetrahedron* **1981**, *37*, 1589–1595.

1688. See Reference *1686*.

1689. Bosch, R; Brückner, H.; Jung, G.; Winter, W. *Tetrahedron* **1982**, *38*, 3579–3583.

1690. Slieker, L.; Benkovic, S.J. *J. Lab. Cmpds. Radiopharm.* **1982**, *19*, 647–657.

1691. Schmidt, U.; Lieberknecht, A.; Grieeser, H.; Bökens, H. *Tetrahedron Lett.* **1982**, *23*, 4911–4914.

1692. Crout, D.H.G.; Lutstorf, M.; Morgan, P.J. *Tetrahedron* **1983**, *39*, 3457–3469.

1693. Townsend, C.A.; Brown, A.M. *J. Am. Chem. Soc.* **1983**, *105*, 913–918.

1694. Muthukumaraswamy, N.; Day, A.R.; Pinon, D.; Liao, C.S.; Freer, R.J. *Int. J. Pept. Prot. Res.* **1983**, *22*, 305–312.

1695. Hansen, J.J.; Lauridsen, J.; Nielsen, E.; Krogsgaard-Larsen, P. *J. Med. Chem.* **1983**, *26*, 901–903.

1696. Cahill, R.; Crout, D.H.G.; Gregorio, M.V.M.; Mitchell, M.B.; Muller, U.S. *J. Chem. Soc., Perkin Trans. 1* **1983**, 173–180.

1697. Cheung, K.-S.; Wasserman, S.A.; Dudek, E.; Lerner, S.A.; Johnston, M. *J. Med. Chem.* **1983**, *26*, 1733–1741.

1698. Fujihara, H.; Schowen, R.L. *J. Org. Chem.* **1984**, *49*, 2819–2820.

1699. Simeno, H.; Fukumoto, Y.; Fuji, M.; Nagamatsu, A. *Chem. Pharm. Bull.* **1984**, *32*, 3620–3625.

1700. Baldwin, J.E.; Cha, J.K.; Kruse, L.I. *Tetrahedron* **1985**, *41*, 5241–5260.

1701. Kauer, J.C.; Erickson-Viitanen, S.; Wolfe, H.R., Jr.; DeGrado, W.F. *J. Biol. Chem.* **1986**, *261*, 10695–10700.

1702. Berges, D.A.; DeWolf, Jr., W.E.; Dunn, G.L.; Grappel, S.F.; Newman, D.J.; Taggart, J.J.; Gilvarg, C. *J. Med. Chem.* **1986**, *29*, 89–95.

1703. Bretschneider, T.; Miltz, W.; Münster, P.; Steglich, W. *Tetrahedron* **1988**, *44*, 5403–5414.

1704. Tolman, V.; Sedmera, P. *Tetrahedron Lett.* **1988**, *29*, 6183–6184.

1705. Tsushima, T.; Kawada, K.; Ishihara, S.; Uchida, N.; Shiratori, O.; Higaki, J.; Hirata, M. *Tetrahedron* **1988**, *44*, 5375–5387.

1706. Baggaley, K.H.; Nicholson, N.H.; Sime, J.T. *J. Chem. Soc., Chem. Commun.* **1988**, 567–568.

1707. Ojima, I.; Kato, K.; Nakahashi, K.; Fuchikami, T.; Fujita, M. *J. Org. Chem.* **1989**, *54*, 4511–4522.

1708. Hansen, J.J.; Nielsen, B.; Krogsgaard-Laresen, P.; Brehm, L.; Nielsen, E.Ø.; Curtis, D.R. *J. Med. Chem.* **1989**, *32*, 2254–2260.

1709. Ranjalahy-Rasoloarijao, L.; Lazaro, R.; Daumas, P.; Heitz, F. *Int. J. Pept. Prot. Res.* **1989**, *33*, 273–280.

1710. Alks, V.; Sufrin, J.R. *Tetrahedron Lett.* **1990**, *31*, 5257–5260.

1711. Groeger, U.; Drauz, K.; Klenk, H. *Angew. Chem., Int. Ed. Engl.* **1990**, *29*, 417–419.

1712. Bladon, C.M. *J. Chem. Soc., Perkin Trans. 1* **1990**, 1151–1158.

1713. Baldwin, J.E., Bradley, M.; Abbott, S.D.; Adlington, R.M. *Tetrahedron* **1991**, 47, 5309–5328.

1714. Baldwin, J.E.; Hulme, C.; Schofield, C.J. *J. Chem. Res S* **1992**, 173.

1715. Furuta, T.; Katayama, M.; Shibasaki, H.; Kasuya, Y. *J. Chem. Soc., Perkin Trans. 1* **1992**, 1643–1648.

1716. Hasegawa, H.; Arai, S.; Shinohara, Y.; Baba, S. *J. Chem. Soc., Perkin Trans 1* **1993**, 489–494.

1717. Pirrung, M.C.; Krishnamurthy, N. *J. Org. Chem.* **1993**, *58*, 954–956.

1718. Taku, K.; Sasaki, H.; Kimura, S.; Imanishi, Y. *Amino Acids* **1994**, *7*, 311–316.

1719. Lloyd-Williams, P.; Monerris, P.; Gonzalez, I.; Jou, G.; Giralt, E. *J. Chem. Soc., Perkin Trans. 1* **1994**, 1974.

1720. Hagiwara, D.; Miyake, H.; Igari, N.; Karino, M.; Maeda, Y.; Fujii, T.; Matsuo, M. *J. Med. Chem.* **1994**, *37*, 2090–2099.

1721. Baker, S.R.; Goldsworthy, J.; Harden, R.C.; Salhoff, C.R.; Schoepp, D.D. *Bioorg. Med. Chem. Lett.* **1995**, *5*, 223–228.

1722. Stüber, W.; Koschinsky, R.; Reers, M.; Hoffman, D.; Czech, J.; Dickneite, G. *Peptide Res.* **1995**, *8*, 78–85.

1723. Andrews, D.M.; Gregoriou, M.; Page, T.C.M.; Peach, J.M.; Pratt, A.J. *J. Chem. Soc., Perkin Trans. 1* **1995**, 1335–1340.

1724. Plummer, M.S.; Shahripour, A.; Kaltenbronn, J.S.; Lunney, E.A.; Steinbaugh, B.A.; Hamby, J.M.; Hamilton, H.W.; Sawyer, T.K.; Humblet, C.; Doherty, A.M.; Taylor, M.D.; Hingorani, G.; Batley, B.L.; Rapundalo, S.T. *J. Med. Chem.* **1995**, *38*, 2893–2905.

1725. Hamon, C.; Rawlings, B.J. *Synth. Commun.* **1996**, *26*, 1109–1115.

1726. Nishino, N.; Arai, T.; Ueno, Y.; Ohba, M. *Chem. Pharm. Bull.* **1996**, *44*, 212–214.

1727. Pearson, D.A.; Lister-James, J.; McBride, W.J.; Wilson, D.M.; Martel, L.J.; Civitello, E.R.; Dean, R.T. *J. Med. Chem.* **1996**, *39*, 1372–1382.

1728. Kokuryo, Y.; Nakatani, T.; Kobayashi, K.; Tamura, Y.; Kawada, K.; Ohtani, M. *Tetrahedron: Asymmetry* **1996**, *7*, 3545–3551.

1729. Kira, M.; Matsubara, T.; Shinohara, H.; Sisido, M. *Chem. Lett.* **1997**, 89–90.

1730. Fadnavis, N.W.; Sharfuddin, M.; Vadivel, S.K.; Bhalerao, U.T. *J. Chem. Soc., Perkin Trans. 1* **1997**, 3577–3578.

1731. Duewel, H.; Daub, E.; Robinson, V.; Honek, J.F. *Biochemistry* **1997**, *36*, 3404–3416.

1732. Sisido, M. *Macromolecules* **1997**, *30*, 2651–2656.

1733. Sun, G.; Slavica, M.; Uretsky, N.J.; Wallace, L.J.; Shams, G.; Weinstein, D.M.; Miller, J.C.; Miller, D.D. *J. Med. Chem.* **1998**, *41*, 1034–1041.

1734. Hashimoto, K.; Yoshioka, T.; Morita, C.; Sakai, M.; Okuno, T.; Shirahama, H. *Chem. Lett.* **1998**, 203–204.

1735. Beller, M.; Eckert, M.; Geissler, H.; Napierski, B.; Rebenstock, H.-P.; Holla, E.W. *Chem.—Eur. J.* **1998**, *4*, 935–941.

1736. Lee, K.; Hwang, S.Y.; Yun, M.; Kim, D.S. *Biorg. Med. Chem. Lett.* **1998**, *8*, 1683–1686.

1737. Jiang, J.; Miller, R.B.; Tolle, J.C. *Synth. Commun.* **1998**, *28*, 3015–3019.

1738. Tamiaki, H.; Onishi, M. *Tetrahedron: Asymmetry* **1999**, *10*, 1029–1032.

1739. Yokoyama, Y.; Hikawa, H.; Mitsuhashi, M.; Uyama, A.; Murakami, Y. *Tetrahedron Lett.* **1999**, *40*, 7803–7806.

1740. Fujita, T.; Nose, T.; Matsushima, A.; Okada, K.; Asai, D.; Yamauchi, Y.; Shirasu, N.; Honda, T.; Shigehiro, D.; Shimohigashi, Y. *Tetrahedron Lett.* **2000**, *41*, 923–927.

1741. Oba, M.; Kobayashi, M.; Oikawa, F.; Nishiyama, K.; Kainosho, M. *J. Org. Chem.* **2001**, *66*, 5919–5922.

1742. Horikawa, E.; Kodaka, M.; Nakahara, Y.; Okuno, H.; Nakamura, K. *Tetrahedron Lett.* **2001**, *42*, 8337–8339.

1743. Xing, X.; Fichera, A.; Kumar, K. *J. Org. Chem.* **2002**, *67*, 1722–1712.

1744. Watanabe, L.A.; Jose, B.; Kato, T.; Nishino, N.; Yoshida, M. *Tetrahedron Lett.* **2004**, *45*, 491–494.

1745. Watanabe, L.A.; Haranaka, S.; Jose, B.; Yoshida, M.; Kato, T.; Moriguchi, M.; Soda, K.; Nishino, N. *Tetrahedron: Asymmetry* **2005**, *16*, 903–908.

1746. Tong, J.H.; Petitclerc, C.; D'Iorio, A.; Benoiton, N.L. *Can. J. Biochem.* **1971**, *49*, 877–881.

1747. Roberts, D.C.; Suda, K.; Samanen, J.; Kemp, D.S. *Tetrahedron Lett.* **1980**, *21*, 3435–3438.

1748. Porter, J.; Dykert, J.; Rivieer, J. *Int. J. Pept. Prot. Res.* **1987**, *30*, 13–21.

1749. Voskuyl-Holtkamp, I.; Schattenkerk, C. *Int. J. Pept. Prot. Res.* **1979**, *13*, 185–194.

1750. Natchev, I.A. *Bull. Chem. Soc. Jpn.* **1988**, *61*, 3711–3715.

1751. Natchev, I.A. *Tetrahedron* **1988**, *44*, 1511–1522.

1752. Hsiao, C.-N.; Leanna, M.R.; Bhagavatula, L.; De Lara, E.; Zydowsky, T.M.; Horrom, B.W.; Morton, H.E. *Synth. Commun.* **1990**, *20*, 3507–3517.

1753. Chênevert, R.; Létourneau, M.; Thiboutot, S. *Can. J. Chem.* **1990**, *68*, 960–963.

1754. Schricker, B.; Thirring, K.; Berner, H. *Bioorg. Med. Chem. Lett.* **1992**, *2*, 387–390.

1755. Schneider, H.; Sigmund, G.; Schricker, B.; Thirring, K.; Berner, H. *J. Org. Chem.* **1993**, *58*, 683–689.

1756. Antolini, L.; Forni, A.; Moretti, I.; Prati, F.; Laurent, E.; Gestmann, D. *Tetrahedron: Asymmetry* **1996**, *7*, 3309–3314.

1757. Cooper, M.S.; Seton, A.W.; Stevens, M.F.G.; Westwell, A.D. *Bioorg. Med. Chem. Lett.* **1996**, *6*, 2613–2616.

1758. Tararov, V.I.; Belekon, Y.N.; Singh, A.; Parmar, V.S. *Tetrahedron: Asymmetry* **1997**, *8*, 33–36.

1759. Dugave, C.; Cluzeau, J.; Ménez, A.; Gaudry, M.; Marquet, A. *Tetrahedron Lett.* **1998**, *39*, 5775–5778.

1760. Katayama, S.; Ae, N.; Nagata, R. *Tetrahedron: Asymmetry* **1998**, *9*, 4295–4299.

1761. Firooznia, F.; Gude, C.; Chan, K.; Marcopulos, N.; Satoh, Y. *Tetrahedron Lett.* **1999**, *40*, 213–216.

1762. Chen, S.-T.; Wang, K.-T.; Wong, C.-H. *J. Chem. Soc., Chem. Commun.* **1986**, 1514–1516.

1763. Wallace, E.M.; Moliterni, J.A.; Moskal, M.A.; Neubert, A.D.; Marcopulos, N.; Stamford, L.B.; Trapani, A.J.; Savage, P.; Chou, M.; Jeng, A.Y. *J. Med. Chem.* **1998**, *41*, 1513–1523.

1764. Bosshard, H.R.; Berger, A. *Helv. Chim. Acta* **1973**, *56*, 1838–1845.

1765. Nestor, J.J., Jr.; Ho, T.L.; Simpson, R.A.; Horner, B.L.; Jones, G.H.; McRae, G.I.; Vickery, B.H. *J. Med. Chem.* **1982**, *25*, 795–801.

1766. Parnes, H.; Shelton, E.J. *J. Lab. Compds. Radiopharm.* **1984**, *21*, 263–284.

1767. Folkers, K.; Kubiak, T.; Stepinski, J. *Int. J. Pept. Prot. Res.* **1984**, *24*, 197–200.

1768. Rao, P.N.; Burdett, J.E., Jr.; Cessac, J.W.; DiNunno, C.M.; Peterson, D.M.; Kim, H.K. *Int. J. Pept. Prot. Res.* **1987**, *29*, 118–125.

1769. Rodriguez, M.; Bernad, N.; Galas, M.C.; Lignon, M.F.; Laur, J.; Aumelas, A.; Martinez, J. *Eur. J. Med. Chem.* **1991**, *26*, 245–253.

1770. Garbay-Jaureguiberry, C.; McCort-Tranchepain, I.; Barbe, B.; Ficheux, D.; Roques, B.P. *Tetrahedron: Asymmetry* **1992**, *3*, 637–650.

1771. Ljungqvist, A.; Bowers, C.Y.; Folkers, K. *Int. J. Pept. Prot. Res.* **1993**, *41*, 427–432.

1772. Leanna, M.R.; Morton, H.E. *Tetrahedron Lett.* **1993**, *34*, 4485–4488.

1773. Orsini, F.; Pelizzoni, F.; Ghioni, C. *Amino Acids* **1995**, *9*, 135–140.

1774. Baczko, K.; Liu, W.-Q.; Roques, B.P.; Garbay-Jaureguiberry, C. *Tetrahedron* **1996**, *52*, 2021–2030.

1775. Behrens, C.; Nielsen, J.N.; Fan, X.-J.; Doisy, X.; Kim, K.-H.; Praetorius-Ibba, M.; Nielsen, P.E.; Ibba, M. *Tetrahedron* **2000**, *56*, 9443–9449.

1776. Imperiali, B.; Prins, T.J.; Fisher, S.L. *J. Org. Chem.* **1993**, *58*, 1613–1616.

1777. Huang, X.; Long, E.C. *Bioorg. Med. Chem. Lett.* **1995**, *5*, 1937–1940.

1778. Cheng, R.P.; Fisher, S.L.; Imperiali, B. *J. Am. Chem. Soc.* **1996**, *118*, 11349–11356.

1779. Torrado, A.; Imperiali, B. *J. Org. Chem.* **1996**, *61*, 8940–8948.

1780. Kise K.J., Jr.; Bowler, B.E. *Tetrahedron: Asymmetry* **1998**, *9*, 3319–3324.

1781. Miyazawa, T.; Takitani, T.; Ueji, S.; Yamada, T.; Kuwata, S. *J. Chem. Soc., Chem. Commun.* **1988**, 1214–1215.

1782. Kitazume, T.; Lin, J.T.; Yamazaki, T. *Tetrahedron: Asymmetry* **1991**, *2*, 235–238.

1783. Bautista, F.M.; Campelo, J.M.; García, A.; Luna, D.; Marinase, J.M. *Amino Acids* **1992**, *2*, 87–95.

1784. Ng-Youn-Chen, M.C.; Serreqi, A.N.; Huang, Q.; Kazlauskas, R.J. *J. Org. Chem.* **1994**, *59*, 2075–2081.

1785. Davoli, P.; Forni, A.; Moretti, I.; Prati, F. *Tetrahedron: Asymmetry* **1995**, *6*, 2011–2016.

1786. Kumar, H.M.; Rao, M.S.; Chakravarthy, P.P.; Yadav, J.S. *Tetrahedron: Asymmetry* **2004**, *15*, 127–130.

1787. Dunn, F.W. *J. Org. Chem.* **1956**, *21*, 1525–1526.

1788. Zhuze, A.L.; Jost, K.; Kasafírek, E.; Rudinger, J. *Collect. Czech. Chem. Commun.* **1964**, *29*, 2648–2662.

1789. Kirby, G.W.; Michael, J. *J. Chem. Soc., Perkin Trans. 1* **1973**, 115–120.

1790. Coy, D.H.; Coy, E.J.; Hirotsu, Y.; Vilcehz-Martinez, J.A.; Schally, A.V.; van Nispen, J.W.; Tesser, G.I. *Biochemistry* **1974**, *13*, 3550–3553.

1791. Turk, J.T.; Panse, G.T.; Marshall, G.R. *J. Org. Chem.* **1975**, *40*, 953–955.

1792. Blumenstein, M.; Hruby, V.J.; Yamamoto, D.M. *Biochemistry* **1978**, *17*, 4971–4977.

1793. Viswanatha, V.; Hruby, V.J. *J. Org. Chem.* **1979**, *44*, 2892–2896.

1794. van Pée, K.-H.; Salcher, O.; Lingens, F. *Liebigs Ann. Chem.* **1981**, 233–239.

1795. Samanen, J.; Narindray, D.; Cash, T.; Brandeis, E.; Adams, W., Jr.; Yellin, T.; Eggleston, D.; DeBrosse, C.; Regoli, D. *J. Med. Chem.* **1989**, *32*, 466–472.

1796. Mosberg, H.I.; Omnaas, J.R.; Lomize, A.; Heyl, D.L.; Nordan, I.; Mousigian, C.; Davis, P.; Porreca, F. *J. Med. Chem.* **1994**, *37*, 4384–4391.

1797. Hasegawa, H.; Shinohara, Y. *J. Chem. Soc., Perkin Trans. 1* **1998**, 243–247.

1798. Parikh, J.R.; Greenstein, J.P.; Winitz, M.; Birnbaum, S.M. *J. Am. Chem. Soc.* **1958**, *80*, 953–958.

1799. Sheehan, J.C.; Whitney, J.C. *J. Am. Chem. Soc.* **1963**, *85*, 3863–3865.

1800. Barrio, J.R.; Keen, R.E.; Ropchan, J.R.; MacDonald, N.S.; Baumgartner, F.J.; Padgett, H.C.; Phelps, M.E. *J. Nucl. Med.* **1983**, *24*, 515–521.

1801. Ropchan, J.R.; Barrio, J.R. *J. Nucl. Med.* **1984**, *25*, 887–892.

1802. Bory, S.; Dubois, J.; Gaudry, M.; Marquet, A.; Lacombe, L.; Weinstein, S. *J. Chem. Soc., Perkin Trans. 1* **1984**, 475–480.

1803. Thompson, J.F.; Morris, C.J.; Asen, S.; Irreverre, F. *J. Biol. Chem.* **1961**, *236*, 1183–1185.

1804. Morita, K.; Irreverre, F.; Sakiyama, F.; Witkop, B. *J. Am. Chem. Soc.* **1963**, *85*, 2832–2834.

1805. Fauchère, J.-L.; Petermann, C. *Int. J. Pept. Prot. Res.* **1981**, *18*, 249–255.

1806. Shadid, B.; van der Plas, H.C.; Boesten, W.H.J.; Kamphuis, J.; Meijer, E.M.; Schoemaker, H.E. *Tetrahedron* **1990**, *46*, 913–920.

1807. Bruce, M.A.; St. Laurent, D.R.; Poindexter, G.S.; Monkovic, I.; Huang, S.; Balasubramanian, N. *Synth. Commun.* **1995**, *25*, 2673–2684.

1808. Borsook, H.; Deasy, C.L.; Haagen-Smit, A.J.; Keighley, G.; Lowy, P.H. *J. Biol. Chem.* **1950**, *184*, 529–543.

1809. Meister, A. *J. Biol. Chem.* **1954**, *210*, 17–35.

1810. Dunn, F.W.; Campaigne, E.; Neiss, E.S. *Biochem. Prep.* **1963**, *10*, 159–165.

1811. Isowa, Y.; Kurita, H.; Ohmori, M.; Sato, M.; Mori, K. *Bull. Chem. Soc. Jpn.* **1973**, *46*, 1847–1850.

1812. Drueckhammer, D.G.; Barbas, C.F., III; Noaki, K.; Wong, C.-H.; Wood, C.Y.; Ciufolini, M.A. *J. Org. Chem.* **1988**, *53*, 1607–1611.

1813. Clapés, P.; Valverde, I.; Jaime, C.; Torres, J.L. *Tetrahedron Lett.* **1996**, *37*, 417–418.

1814. Rivard, M.; Malon, P.; Cerovsky, V. *Amino Acids* **1998**, *15*, 389–392.

1815. Schenkels, C.; Erni, B.; Reymond, J.-L. *Biorg. Med. Chem. Lett.* **1999**, *9*, 1443–1446.

1816. Csuk, R.; Glänzer, B.I. *J. Fluorine Chem.* **1988**, *39*, 99–106.

1817. Ashworth, P.; Broadbelt, B.; Jankowski, P.; Kocienski, P.; Pimm, A.; Bell, R. *Synthesis* **1995**, 199–206.

1818. Gibson, S.E.; Guillo, N.; Tozer, M.J. *Chem. Commun.* **1997**, 637–638.

1819. Dubuisson, C.; Fukumoto, Y.; Hegedus, L.S. *J. Am. Chem. Soc.* **1995**, *117*, 3697–3704.

1820. Hirao, I.; Ohtsuki, T.; Fujiwara, T.; Mitsui, T.; Yokogawa, T.; Okuni, T.; Nakayama, H.; Takio, K.; Yabuki, T.; Kigawa, T.; Kodama, K.; Yokogawa, T.; Nishikawa, K.; Yokoyama, S. *Nat. Biotechnol.* **2002**, *20*, 177–182.

1821. McKiernan, M.; Huck, J.; Fehrentz, J.-A.; Roumestant, M.-L.; Viallefont, P.; Martinez, J. *J. Org. Chem.* **2001**, *66*, 6541–6544.

1822. Wehofsky, N.; Thust, S.; Burmeister, J.; Klussmann, S.; Bordusa, F. *Angew. Chem., Int. Ed.* **2003**, *42*, 677–679.

1823. Khalafi-Nezhad, A.; Mokhtari, B.; Soltani Rad, M.N. *Tetrahedron Lett.* **2003**, *44*, 7325–7328.

1824. Nishiyama, Y.; Ishizuka, S.; Kurita, K. *Tetrahedron Lett.* **2001**, *42*, 8789–8791.

1825. Oliva, A.I.; Simón, L.; Muñiz, F.M.; Sanz, F.; Morán, J.R. *Chem. Commun.* **2004**, 426–427.

1826. Oliva, A.I.; Simón, L.; Muñiz, F.M.; Sanz, F.; Morán, J.R. *Org. Lett.* **2004**, *6*, 1155–1157.

1827. Oliva, A.I.; Simón, L.; Muñiz, F.M.; Sanz, F.; Ruiz-Valero, C.; Morán, J.R. *J. Org. Chem.* **2004**, *69*, 6883–6885.

1828. Pérez, E.M.; Oliva, A.I.; Hernández, J.V.; Simón, L.; Morán, J.R.; Sanz, F. *Tetrahedron Lett.* **2001**, *42*, 5853–5856.

1829. Beligere, G.S.; Dawson, P.E. *Biopolymers (Peptide Sci.)* **1999**, *51*, 363–369.

1830. Calmès, M.; Escale, F.; Glot, C.; Rolland, M.; Martinez, J. *Eur. J. Org. Chem.* **2000**, 2459–2466.

1831. Hughes, A.B.; Sleebs, B.E. *Aust. J. Chem.* **2005**, *58*, 778–784.

1832. Subirós-Funosas, R.; Prohens, R.; Barbas, R.; El-Faham, A.; Albericio, F. *Chem. Eur. J.* **2009**, 15, 9394–9403.

1833. Novabiochem Newsletter, Merck Biosciences, 1/09.

Index

Note: Page numbers followed by *t* denote tables.

TODD ROCKWAY, Ph.D.
ABBOTT LABS
D-47V AP/10
ABBOTT PARK, IL 60064